CHEMISTRY

FOR
DEGREE STUDENTS

B.Sc. Second Year

All Indian Universities

As per UGC Model Curriculum

CHEMISTRY FOR DEGREE STUDENTS

B.Sc. Second Year

All Indian Universities

As per UGC Model Curriculum

Part – I : Inorganic Chemistry
Part – II : Organic Chemistry
Part – III : Physical Chemistry

Dr. R.L. MADAN

M.Sc., Ph.D.

Former, Head of Chemistry Department
Government Postgraduate College, Faridabad
and Principal, Government College, Panchkula

S. CHAND PUBLISHING

S Chand And Company Limited

(ISO 9001 Certified Company)

S Chand And Company Limited

(ISO 9001 Certified Company)

Head Office: Block B-1, House No. D-1, Ground Floor, Mohan Co-operative Industrial Estate, New Delhi – 110 044 | Phone: 011-66672000

Registered Office: A-27, 2nd Floor, Mohan Co-operative Industrial Estate, New Delhi – 110 044
Phone: 011-49731800

www.**schandpublishing.com**; e-mail: **info@schandpublishing.com**

Branches

Chennai	:	Ph: 23632120; chennai@schandpublishing.com
Guwahati	:	Ph: 2738811, 2735640; guwahati@schandpublishing.com
Hyderabad	:	Ph: 40186018; hyderabad@schandpublishing.com
Jalandhar	:	Ph: 4645630; jalandhar@schandpublishing.com
Kolkata	:	Ph: 23357458, 23353914; kolkata@schandpublishing.com
Lucknow	:	Ph: 4003633; lucknow@schandpublishing.com
Mumbai	:	Ph: 25000297; mumbai@schandpublishing.com
Patna	:	Ph: 2260011; patna@schandpublishing.com

First Edition 2011
Reprints 2012, 2013 (Thrice), 2014 (Twice), 2015 (Twice), 2016, 2018, 2019
LPSPE 2019, 2020, 2021 (Thrice)

LPSPE 2022

ISBN: 978-93-528-3725-0 **Product Code:** H6CDY68CHEM10ENAA19L

PRINTED IN INDIA

By Vikas Publishing House Private Limited, Plot 20/4, Site-IV, Industrial Area Sahibabad, Ghaziabad – 201 010 and Published by S Chand And Company Limited, A-27, 2nd Floor, Mohan Co-operative Industrial Estate, New Delhi – 110 044.

PREFACE

I feel pleasure in presenting this book to the teachers and students of B.Sc.-II year of all the Indian Universities. The book has been prepared keeping in view the syllabi prepared by different universities on the basis of Model UGC Curriculum. The present book has evolved out of my experience of more than three decades of classroom teaching at B.Sc. level.

A long span of interaction with students has guided me to give special attention to dark areas which students find difficult to understand. Mathematics background required is simple and the necessary mathematical techniques have been developed within the text. Some salient features of the book are :

- The book has been written in a simple and comprehensible language.
- A large number of illustrations, pictures and interesting examples have been provided to make the reading interesting and understandable.
- The important laws, concepts and principles have been given either in italics or in bold type to help the students to focus on them.
- A large number of solved examples which have appeared in the examination papers of various universities have been given in each chapter, for the benefit of students.
- The questions that have been provided in the Exercise are in tune with the latest pattern of examination.

I believe that the book contains all that is needed to understand the subject in a systematic manner. However, no work is perfect. I would welcome suggestions from the teaching fraternity and students for further improvement of the book.

I express my gratitude to the staff of S.Chand & Company Pvt. Ltd., New Delhi for providing all assistance and bringing out the book in the present shape.

Dr. R.L. Madan
Mobile : 9971775666

UGC MODEL SYLLABUS

CHEMISTRY B.Sc. SECOND YEAR

CH-201 Inorganic Chemistry

I Chemistry of Elements of First Transition Series

Characteristic properties of d-block elements.

Properties of the elements of the first transition series, their binary compounds and complexes illustrating relative stability of their oxidation states, coordination number and geometry.

II Chemistry of Elements of Second and Third Transition Series

General characteristics, comparative treatment with their 3d-analogues in respect of ionic radii, oxidation states, magnetic behaviour, spectral properties and stereochemistry.

III Oxidation and Reduction

Use of redox potential data – analysis of redox cycle, redox stability in water – Frost, Latimer and Pourbaix diagrams. Principles involved in the extraction of the elements.

IV Coordination Compounds

Werner's coordination theory and its experimental verification, effective atomic number concept, chelates, nomenclature of coordination compounds, isomerism in coordination compounds, valence bond theory of transition metal complexes.

V Chemistry of Lanthanide Elements

Electronic structure, oxidation states and ionic radii and lanthanide contraction, complex formation, occurrence and isolation, lanthanide compounds.

VI Chemistry of Actinides

General features and chemistry of actinides, chemistry of separation of Np, Pu and Am from U, similarities between the later actinides and the later lanthanides.

VII Acids and Bases

Arrhenius, Bronsted-Lowry, the Lux-Flood, solvent system and Lewis concepts of acids and bases.

VIII Non-aqueous Solvents

Physical properties of a solvent, types of solvents and their general characteristics, reactions in non-aqueous solvents with reference to liquid NH_3 and liquid SO_2.

CH-202 Organic Chemistry

I Electromagnetic Spectrum : Absorption Spectra

Ultraviolet (UV) absorption spectroscopy–absorption laws (Beer-Lambert Law), molar absorptivity, presentation and analysis of UV spectra, types of electronic transitions, effect of conjugation. Concept of chromophore and auxochrome. Bathochromic, hypsochromic, hyperchromic and hypochromic shifts. UV spectra of conjugated enes and enones.

Infrared (IR) absorption spectroscopy—molecular vibrations, Hooke's law, selection rules, intensity and position of IR bands, measurement of IR spectrum, fingerprint region, characteristic absorptions of various functional groups and interpretation of IR spectra of simple organic compounds.

II Alcohols

Classification and nomenclature.

Monohydric alcohols—nomenclature, methods of formation by reduction of aldehydes, ketones, carboxylic acids and esters. Hydrogen bonding. Acidic nature. Reactions of alcohols.

Dihydric alcohols—nomenclature, methods of formation, chemical reactions of vicinal glycols, oxidative cleavage [$Pb(OAc)_4$ and HIO_4] and pinacol-pinacolone rearrangement.

Trihydric alcohols—nomenclature and methods of formation, chemical reactions of glycerol.

III Phenols

Nomenclature, structure and bonding. Preparation of phenols, physical properties and acidic character. Comparative acidic strengths of alcohols and phenols, resonance stabilization of phenoxide ion. Reactions of phenols—electrophilic aromatic substitution, acylation and carboxylation. Mechanisms of Fries rearrangement, Claisen rearrangement, Gatterman synthesis, Hauben-Hoesch reaction, Lederer-Manasse reaction and Reimer Tiemann reaction.

IV Ethers and Epoxides

Nomenclature of ethers and methods of their formation, physical properties. Chemical reactions—cleavage and autoxidation, Ziesel's method.

Synthesis of epoxides. Acid and base-catalyzed ring opening of epoxides, orientation of epoxide ring opening, reactions of Grignard and organolithium reagents with expoxides.

V Aldehydes and Ketones

Nomenclature and structure of the carbonyl group. Synthesis of aldehydes and ketones with particular reference to the synthesis of aldehydes from acid chlorides, synthesis of aldehydes and ketones using 1, 3-dithianes, synthesis of ketones from nitriles and from carboxylic acids. Physical properties.

Mechanism of nucleophilic additions to carbonyl group with particular emphasis on benzoin, aldol, Perkin and Knoevenagel condensations. Condensation with ammonia and its derivatives. Wittig reaction. Mannich reaction.

Use of acetals as protecting group. Oxidation of aldehydes, Baeyer-Villiger oxidation of ketones, Cannizzaro reaction. MPV, Clemmensen, Wolff-Kishner, $LiAlH_4$ and $NaBH_4$ reductions. Halogenation of enolizable ketones.

An introduction to α, β unsaturated aldehydes and ketones.

VI Carboxylic Acids

Nomenclature, structure and bonding, physical properties, acidity of carboxylic acids, effects of substituents on acid strength. Preparation of carboxylic acids. Reactions of carboxylic acids. Hell-Volhard-Zelinsky reaction. Synthesis of acid chlorides, esters and amides. Reduction of carboxylic acids. Mechanism of decarboxylation.

Methods of formation and chemical reactions of halo acids. Hydroxy acids : malic, tartaric and citric acids.

Methods of formation and chemical reactions of unsaturated monocarboxylic acids.

Dicarboxylic acids : methods of formaton and effect of heat and dehydrating agents.

VII Carboxylic Acid Derivatives

Structure and nomenclature of acid chlorides, esters, amides (urea) and acid anhydrides. Relative stability of acyl derivatives. Physical properties, interconversion of acid derivatives by nucleophilic acyl substitution.

Preparation of carboxylic acid derivatives, chemical reactions. Mechanisms of esterification and hydrolysis (acidic and basic).

VIII Organic Compounds of Nitrogen

Preparation of nitroalkanes and nitroarenes. Chemical reactions of nitroalkanes. Mechanisms of nucleophilic substitution in nitroarenes and their reductions in acidic, neutral and alkaline media. Picric acid.

Halonitroarenes : reactivity. Structure and nomenclature of amines, physical properties. Stereochemistry of amines. Separation of a mixture of primary, secondary and tertiary amines. Structural features effecting basicity of amines. Amine salts as phase-transfer catalysts. Preparation of alkyl and aryl amines (reduction of nitro compounds nitriles), reductive amination of aldehydic and ketonic compounds. Gabriel-phthalimide reaction, Hofmann bromamide reaction.

Reactions of amines, electrophilic aromatic substitution in aryl amines, reactions of amines with nitrous acid. Synthetic transformations of aryl diazonium salts, azo coupling.

CH-203 Physical Chemistry

I Thermodynamics–I

Definition of thermodynamic terms : system, surroundings etc. Types of systems, intensive and extensive properties. State and path functions and their differentials. Thermodynamic process. Concept of heat and work.

First Law of Thermodynamics : statement, definition of internal energy and enthalpy. Heat capacity, heat capacities at constant volume and pressure and their relationship. Joule's law-Joule-Thomson coefficient and inversion temperature. Calculation of w,q, dU & dH for the expansion of ideal gases under isothermal and adiabatic conditions for reversible process.

Thermochemistry : standard state, standard enthalpy of formation. Hess's Law of heat summation and its applications. Heat of reaction at constant pressure and at constant volume. Enthalpy of neutralization. Bond dissociation energy and its calculation from thermo-chemical data, temperature dependence of enthalpy. Kirchhoff's equation.

II Thermodynamics – II

Second law of thermodynamics : need for the law, different statements of the law. Carnot cycle and its efficiency, Carnot theorem. Thermodynamic scale of temperature.

Concept of entropy : entropy as a state function, entropy as a function of V & T, entropy as a function of P & T, entropy change in physical change, Clausius inequality, entropy as a criteria of spontaneity and equilibrium. Entropy change in ideal gases and mixing of gases.

Third law of thermodynamics : Nernst heat theorem, statement and concept of residual entropy, evaluation of absolute entropy from heat capacity data, Gibbs and Helmholtz functions; Gibbs function (G) and Helmholtz function (A) as thermodynamic quantities, A & G as criteria for thermodynamic equilibrium and spontaneity, their advantage over entropy change. Variation of G and A with P, V and T.

III Chemical Equilibrium

Equilibrium constant and free energy. Thermodynamic derivation of law of mass action. Le Chatelier's principle.

Reaction isotherm and reaction isochore–Clapeyron equation and Clausius-Clapeyron equation, applications.

IV Phase Equilibrium

Statement and meaning of the terms-phase, component and degree of freedom, derivation of Gibbs phase rule, phase equilibria of one component system-water, CO_2 and S systems. Phase equilibria of two component system-solid-liquid equilibria, simple eutectic-Bi-Cd, Pb-Ag systems, desilverisation of lead.

Solid solutions-compound formation with congruent melting point (Mg-Zn) and incongruent melting point, ($NaCl$-H_2O), ($FeCl_3$-H_2O) and $CuSO_4$-H_2O) system. Freezing mixtures, acetone -dry ice.

Liquid-liquid mixtures-Ideal liquid mixtures, Raoult's and Henry's law, Non-ideal system-azeotropes-HCl-H_2O and ethanol-water systems.

Partially miscible liquids-Phenol-water, trimethylamine-water, nicotine-water systems. Lower and upper consolute temperature. Effect of impurity on consolute temperature. Immiscible liquids, steam distillation.

Nernst distribution law-thermodynamic derivation, applications.

V Electrochemistry-I

Electrical transport-conduction in metals and in electrolyte solutions, specific conductance and equivalent conductance, measurement of equivalent conductance, variation of equivalent and specific conductance with dilution.

Migration of ions and Kohlrausch law, Arrhenius theory of electrolyte dissociation and its limitations, weak and strong electrolytes, Ostwald's dilution law its uses and limitations. Debye-Huckel-Onsager's equation for strong electrolytes (elementary treatment only). Transport number, definition and determination by Hittorf method and moving boundary method.

Applications of conductivity measurements : determination of degree of dissociation, determination of K_a of acids, determination of solubility product of a sparingly soluble salt, conductometric titrations.

VI Electrochemistry-II

Types of reversible electrodes-gas-metal ion, metal-metal ion, metal-insoluble salt-anion and redox electrodes. Electrode reactions, Nernst equation, derivation of cell E.M.F. and single electrode potential, standard hydrogen electrode-reference electrodes-standard electrode potential, sign conventions, electrochemical series and its significance.

Electrolytic and Galvanic cells-reversible and irreversible cells, conventional representation of electrochemical cells.

EMF of a cell and its measurements. Computation of cell EMF. Calculation of thermodynamic quantities of cell reactions (ΔG, ΔH and K), polarization, over potential and hydrogen overvoltage.

Concentration cell with and without transport, liquid junction potential, application of concentration cells, valency of ions, solubility product and activity conefficient, potentiometric titrations.

Definition of pH and pK_a determination of pH using hydrogen, quinhydrone and glass electrodes, by potentiometric methods.

Buffers-mechanism of buffer action, Henderson-Hazel equation. Hydrolysis of salts. Corrosion-types, theories and methods of combating it.

CONTENTS OF B.Sc. I AND B.Sc. III BOOKS

B.Sc. I

PART–I: INORGANIC CHEMISTRY

PART–II: ORGANIC CHEMISTRY

PART–III: PHYSICAL CHEMISTRY

B.Sc. III

PART–I: INORGANIC CHEMISTRY

PART–II: ORGANIC CHEMISTRY

PART–III: PHYSICAL CHEMISTRY

CONTENTS

PART–I: INORGANIC CHEMISTRY

PART–II: ORGANIC CHEMISTRY

PART–III: PHYSICAL CHEMISTRY

INORGANIC CHEMISTRY

1

Chemistry of Elements of First Transition Series

1.1. INTRODUCTION

The elements which involve the filling of *d*-orbitals in the penultimate shell are called *d*-block elements. In the periodic table, they are placed in between the s- and the *p*-block elements and their properties represent a change from the electropositive *s*-block elements to the electronegative *p*-block elements. That is why they are commonly called **transition** elements.

Transition Elements of *d*-Block Elements

III B	IV B	V B	VI B	VII B	← VIII B		→	IB	II B	Old numbering
3	**4**	**5**	**6**	**7**	**8**	**9**	**10**	**11**	**12**	New numbering
21 Sc	22 Ti	23 V	24 Cr	25 Mn	26 Fe	27 Co	28 Ni	29 Cu	30 Zn	← 3d series
39 Y	40 Zr	41 Nb	42 Mo	43 Tc	44 Ru	45 Rh	46 Pd	47 Ag	48 Cd	← 4d series
57 La	72 Hf	73 Ta	74 W	75 Re	76 Os	77 Ir	78 Pt	79 Au	80 Hg	← 5d series
89 Ac	104 Ku	105 Ha								← 6d series

Transition elements or *d*-block elements may be defined as those elements which have partly filled *d*-orbitals in their elementary form or in their commonly occurring oxidation states.

The *d*-block of the periodic table (shown above) consists of four series, called as first, second, third and the fourth (3*d*, 4*d*, 5*d*, and 6*d*) transition series. The first three series are complete containing 10 elements each and the fourth one is an incomplete series. All these elements are placed in eight sub-groups, viz., III B, IV B, V B, VI B, VII B, VIII B, I B and II B as shown in the periodic table along with the atomic numbers. Lately, these groups are called 3rd, 4th, 5th.......12th groups. Elements of sub-group II B, *viz.* Zn, Cd and Hg, are not included in transition elements as they do not have a partly filled *d*- sub-shell in their elementary state. They do not show the properties of transition elements to any appreciable extent except for their ability to form complexes with ligands such as ammonia, amines, and halide ions, etc. However, they are generally studied with the *d*-block elements in order to maintain a simple classification of elements.

1.2. THE ELECTRONIC CONFIGURATION OF *d*-BLOCK ELEMENTS

Electronic configurations of the first transition series of *d*-block elements are given in Table 1.1. Progressive filling of electrons in 3*d* orbitals takes place.

Table 1.1

Element	*At. No.*	*Electronic Configurations*
Sc	21	$1s^2, 2s^2\,2p^6, 3s^2\,3p^6\,3d^1, 4s^2$ Or [Ar] $3d^1, 4s^2$
Ti	22	$1s^2, 2s^2\,2p^6, 3s^2 3p^6\,3d^2, 4s^2$
V	23	$1s^2, 2s^2\,2p^6, 3s^2\,3p^6\,3d^3, 4s^2$
Cr	24	$1s^2, 2s^2\,2p^6, 3s^2\,3p^6\,3d^5, 4s^1$
Mn	25	$1s^2, 2s^2\,2p^6, 3s^2 3p^6\,3d^5, 4s^2$
Fe	26	$1s^2, 2s^2\,2p^6, 3s^2 3p^6\,3d^6, 4s^2$
Co	27	$1s^2, 2s^2\,2p^6, 3s^2 3p^6\,3d^7, 4s^2$
Ni	28	$1s^2, 2s^2\,2p^6, 3s^2 3p^6\,3d^8, 4s^2$
Cu	29	$1s^2, 2s^2 2p^6, 3s^2\,3p^6\,3d^{10}, 4s^1$
Zn	30	$1s^2, 2s^2\,2p^6, 3s^2 3p^6 3d^{10}, 4s^2$

1.2.1. Electronic configuration of Cr amd Cu

The unexpected electronic configuration of Cr (Ar) $3d^5\,4,s^1$ (in place of (Ar) $3d^4\,4s^2$) and Cu (Ar) $3d^{10}\,4s^1$ (in place of (Ar) $3d^9\,4s^2$) can be explained on the basis of half-filled and completely filled sub-shells which are found to be more stable arrangements. The extra stability of half-filled sub-shells results from the *occupation of each orbital by one electron resulting in spreading and equal distribution of charge around an atom.* In case of copper, it appears that Cu (Ar) $3d^{10}\,4s^1$ with filled $3d$ sub-shell and half-filled $4s$ sub-shell is more stable than (Ar) $3d^9\,4s^2$. Another important factor which explains the stability of these configurations is their large exchange energy.

The electrons in various orbitals of the same sub-shell or in different sub-shell (having similar energy) can exchange their respective positions. During this exchange process a small amount of energy is released. The released energy is called **exchange energy**. It has been found that exchange energy is maximum when all the orbitals of the same sub-shell are either completely filled or exactly half-filled. So the configurations of Cr (Ar) $3d^5 4s^1$ and Cu (Ar) $3d^{10}\,4s^1$ have extra stability.

1.2.2. Electronic configurations of ion of transition metals

Ions of the transition elements are formed by the loss of electrons from the atoms. The configuration obtained after the loss of electrons is the electronic configuration of the ion. Whether the electrons will be lost from the $(n-1)$ d-orbital or ns -orbital depends upon the stability of the electronic configuration of the ion. For example, consider the first transition series. As $4s$ orbitals are occupied first and $3d$-orbitals later, we may expect that $4s$ orbitals are more stable than $3d$ and hence the elements of first transition series would be expected to form ions by the loss of $3d$ electrons. In fact ions are formed by the loss of $4s$ electrons first. This is because prior to occupation of electrons, the $4s$ orbital is energetically more stable than the $3d$ orbitals. But, once the $3d$ orbitals are occupied by electrons, they *repel the 4s electrons to the higher energy level.* Thus, when transition metal atoms form ions, they lose electrons from the $4s$ orbitals before the $3d$ orbitals (Table 1.2).

Table 1.2. Electronic Configuration of the Ions of the Elements of First Transition Series

Atom	Ion	Electronic configuration
Sc (Z = 21)	Sc^{3+}	[Ar] $3d^0$
Ti (Z = 22)	Ti^{2+}	[Ar] $3d^2$
	Ti^{3+}	[Ar] $3d^1$
	Ti^{4+}	[Ar] $3d^0$
V (Z = 23)	V^{2+}	[Ar] $3d^3$
	V^{3+}	[Ar] $3d^2$
	V^{4+}	[Ar] $3d^1$
Cr (Z = 24)	Cr^{2+}	[Ar] $3d^4$
	Cr^{3+}	[Ar] $3d^3$
Mn (Z = 25)	Mn^{2+}	[Ar] $3d^5$
Fe (Z = 26)	Fe^{2+}	[Ar] $3d^6$
	Fe^{3+}	[Ar] $3d^5$
Co (Z = 27)	Co^{2+}	[Ar] $3d^7$
Ni (Z = 28)	Ni^{2+}	[Ar] $3d^8$
Cu (Z = 29)	Cu^{2+}	[Ar] $3d^9$
Zn (Z = 30)	Zn^{2+}	[Ar] $3d^{10}$

1.3. GENERAL CHARACTERISTICS OF *d*-BLOCK ELEMENTS

Some important properties of *d*-block elements are given below:

(*i*) They are practically all hard, strong, high melting, high boiling metals and conduct heat and electricity well.

(*ii*) They form alloys with one another and with other metallic elements.

(*iii*) They are all ductile metals.

(*iv*) They form interstitial compounds with non-metals like C, B, N, H, etc.

(*v*) Because of unpaired electrons they form paramagnetic compounds.

(*vi*) Most of the elements have great tendency to form stable complex ions.

(*vii*) Many of them are electropositive enough to dissolve in mineral acids.

(*viii*) They exhibit catalytic properties and are wonderful catalysts for hydrogenation, oxidation, dehydration etc.

(*ix*) The transition elements form compounds with Lewis acid ligands like CO, NO etc. in which the elements exist in low, zero or even formally negative oxidation states.

(*x*) They exhibit variable oxidation state and their ions and compounds are coloured.

1.4. VARIATION IN ATOMIC RADII IN A PERIOD AND DOWN THE GROUP IN CASE OF TRANSITION ELEMENTS

Along a period. The atomic radius of an element decreases while moving from left to right in a period. This can be understood in terms of increases in nuclear charge with consequent contraction in size.

However, the decrease is very small within some range in the period. For example, in the first transition series, the decrease in atomic radius is very small from chromium to nickel. The same thing

Table 1.3. Important Physical Properties of the Elements of the First Transition Series

Property	Sc	Ti	V	Cr	Mn	Fe	Co	Ni	Cu	Zn
Atomic number	21	22	23	24	25	26	27	28	29	30
Outer electronic configuration	$3d^1\ 4s^2$	$3d^2\ 4s^2$	$3d^3\ 4s^2$	$3d^5\ 4s^1$	$3d^5\ 4s^2$	$3d^6\ 4s^2$	$3d^7\ 4s^2$	$3d^8\ 4s^2$	$3d^{10}\ 4s^1$	$3d^{10}\ 4s^2$
Atomic mass (a.m.u.)	44.96	47.90	50.95	52.01	54.94	55.85	58.94	58.71	63.54	65.38
Atomic radius (Å)	1.44	1.32	1.22	1.17	1.17	1.16	1.16	1.15	1.17	1.25
Stable oxidation states	+3	+4	+3, +4, +5	+2, +3, +6	+2, +3, +4, +7	+2, +3	+2, +3	+2	+1, +2	+2
Atomic volume (c.c.)	15.02	10.60	8.35	7.23	7.30	7.10	6.70	6.60	7.10	9.20
Density (g/c.c.)	3.01	4.51	6.10	7.19	7.43	7.86	8.90	8.90	8.96	7.14
Melting point (°C)	1539	1668	1900	1875	1245	1536	1495	1453	1083	419.5
Boiling point (°C)	2730	3260	3450	2665	2150	3000	2900	2730	2595	906
First ionization energy (kJ mol^{-1})	632.0	656.0	650.0	652.0	717.0	762.0	758.0	736.0	745.0	906.0
Ionic radius (Å): Divalent ion	–	0.91	0.88	0.84	0.80	0.76	0.74	0.72	0.72	0.74
Trivalent ion	0.81	0.76	0.74	0.69	0.66	0.64	0.63	0.63	–	–
Heat of fusion (kJ mol^{-1})	15.9	15.5	17.6	13.8	14.6	15.3	15.2	17.6	13.0	7.4
Heat of vaporisation (kJ mol^{-1})	338.9	445.6	443.5	305.4	224.7	353.9	389.1	380.7	338.9	114.6

happens in other transition series. Another remarkable observation is that the atomic radius starts increasing after that. For example, the value increases from nickel through copper to zinc (See Table 1.3).

As a matter of fact, two opposing forces are simultaneously taking place. In moving from left to right, nuclear charge increases leading to decrease in atomic radius. At the same time, repulsion takes place between the electron population resulting in increase in the atomic radius. *Net value of atomic radius would be determined by the combined effect of these opposing phenomena.* It is found that the radius decreases but very slightly from Cr to Ni. After that repulsive forces outweigh and an increase in radius is observed.

Along a group. While there is some increase in atomic radius in moving from first to second member, the value remains unchanged while moving from second to third member. This can be explained in terms of *lanthanide contraction* which is described in detail in the chapter on *f*-block elements.

1.4.1. Variation in atomic volume and density of transition elements in a period and a group

Atomic volume shows the same trend in the period and the group as the atomic radius. Thus atomic radius volume decreases in the period up to a certain point and then increases thereafter up to the end of the period (See Table 1.3). Reasons are the same as explained in previous section.

In a group the volume increases first and then becomes constant.

Density is the ratio of mass and volume. We observe that the density increases (Density = Mass/volume) in the period as the atomic mass increases and volume shrinks. Towards the end of the period, the trend is reversed. Because of interelectronic repulsion, the atomic volume increases, thereby reducing density (See Table 1.3).

Down the group, the density increases because there is not much variation in atomic volume whereas atomic mass increases.

1.4.2. Variation in melting and boiling points of transition elements

Transition metals have high melting and boiling points. This is due to *strong overlapping between (n – 1)d orbitals and formation of covalent bonds by the d-electrons*. Greater the number of unpaired *d*-electrons, greater the possibility of covalent bond formation, and hence higher the melting and boiling point (See Table 1.3).

Zn, Cd and Hg (melting points 693, 593, and 234.6 K respectively) are notable exceptions because in these elements the $(n-1)d$ orbitals are complete and covalent bond formation in *d*-orbitals is not expected.

The melting and boiling points of these elements increase down the group due to increasing atomic mass. And in a period melting and boiling points reach a maximum near group VB, and VIB. In moving across a transition series from left to right, the number of unpaired *d*-electrons rises to maximum of five in group VI beyond which electron pairing begins. The number of unpaired *d*-electrons in each atom determines the magnitude of binding forces by forming covalent bonds with unpaired electrons of neighbouring atoms.

1.5. TRENDS IN IONISATION ENERGY OF TRANSITION ELEMENTS

The ease with which an electron may be removed from a transition metal atom is called its ionization energy (I.E.). In the transition series the ionization energies increase along the period but the increase is not in a regular manner. However, the increase in I.E. is not as pronounced as in the *s*- and *p*-block elements in the same period. Table 1.4 shows the ionization energies of the 3*d* series. No regular increase in I.E. is observed.

Table 1.4. Ionization Energies of the Elements of 3*d* Series

Elements	*Ionization Energies* ($kJ\ mol^{-1}$)		
(3d series)	*First*	*Second*	*Third*
Sc	632	1245	2450
Ti	656	1320	2721
V	650	1376	2873
Cr	652	1635	2994
Mn	717	1513	3258
Fe	762	1564	2963
Co	758	1647	3237
Ni	736	1756	3400
Cu	745	1961	3560
Zn	906	1736	3838

This may be explained in terms of screening effect of the *d*-electron. The increase in the I.E. expected on account of increase in nuclear charge is almost offset by the extra screening of the nucleus provided by the inner *d*-electrons. The ionization energies of the transition elements are intermediate between those of *s*- and *p*-block elements. This suggests that the transition metals are less electropositive than *s*-block elements.

The *d*-block elements form ionic as well as covalent compounds. Whenever a transition metal gives two series of salts, the lower one is an ionic compound and the higher one is a covalent compound.

From Table 1.4 we find that second I.E. increases almost regularly from Sc to Zn. Values for Cr and Cu are, however, much higher than those of their neighbours on the left as well as on the right. This is due to the stable configuration of Cr^+ ($3d^5$-exactly half filled) and Cu^+ ($3d^{10}$-completely filled). Removal of one electron means they change from a more stable to less stable configuration.

1.6. MAGNETIC PROPERTIES OF TRANSITION ELEMENTS

Transition elements and their compounds have been classified into the following three categories on the basis of their behaviour in the magnetic field.

(*i*) *Paramagnetic substances* are those which are weakly attracted by the magnetic field and this property is associated with the unpaired electrons in an atom or a compound.

(*ii*) *Diamagnetic substances* are those which are slightly repelled by the magnetic field and this property is associated with the paired electrons in an atom or a compound.

(*iii*) *Ferromagnetic substances* are those which are strongly attracted in the magnetic field, *i.e.*, they are more paramagnetic in character and this property is associated with the substances which retain their magnetism even after the removal of magnetic fields. (Due to this property ferromagnetic substances differ from paramagnetic substances which lose their paramagnetism after removing

Magnetic properties of Iron

the magnetic field.) The most familiar example of ferromagnetic substances is iron. Cobalt and nickel are also ferromagnetic.

The paramagnetic or diamagnetic nature of a substance can be measured by weighing the substances (on a Gouy magnetic balance) in the absence of magnetic field and in the presence of magnetic field as shown in Fig. 1.1.

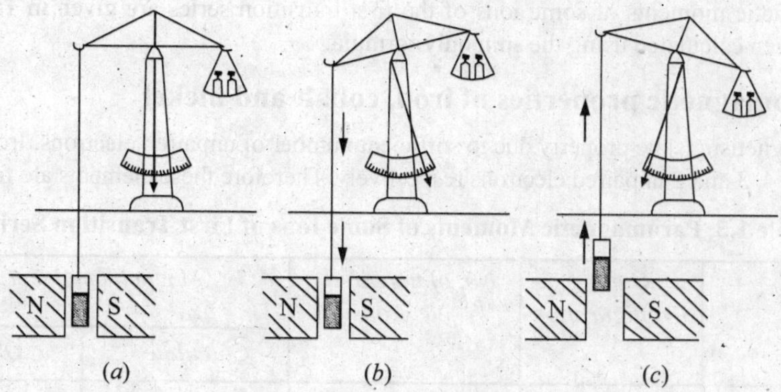

Fig. 1.1. (*a*) Gouy Balance weighing in the absence of magnetic field.
(*b*) Substance attracted by the magnetic field — paramagnetic.
(*c*) Substance repelled by the magnetic field — diamagnetic.

First the substance is weighed in air and then after applying magnetic field. If the weight of the substance increases in a magnetic field, it means that it is attracted by the magnetic field and hence it is paramagnetic. If the weight decreases, it means that it is repelled by the magnetic field and hence it is diamagnetic.

The magnetic moment, which is associated with the number of unpaired electrons and arises due to spin and orbital motion of the electron, can be calculated from the difference in weight of the substance in the Gouy balance. The magnetic moment gives important information about the presence of unpaired electrons in atoms, ions or molecules. In some cases even the structures of transition metal complexes are indicated by it.

Most of the transition elements and their compounds are paramagnetic, *i.e.*, they have a number of unpaired electrons. The relationship between the number of unpaired electrons. *n,* in an atom, ion or molecule and its magnetic moment, μ, in Bohr magnetons is given by the *spin only* formula.

$$\mu_s \text{ or } \mu = \sqrt{n(n+2)} \text{ BM}$$

where n = number of unpaired electrons.

$$1\text{BM} = \frac{eh}{4\mu mc}$$

where e is the electronic charge, h is the Planck's constant, m is the mass of the electron and c is the velocity of light.

The magnetic field produced by unpaired electrons is partly due to their spin and partly to their orbital motion. The magnetic moment μ is given by

$$\mu_S + \mu_L = \sqrt{4S(S+1) + L(L+1)} \quad ...(1.1)$$

(where S = sum of the spin quantum numbers, and L = sum of the orbital angular momentum quantum numbers)

But it has been found that in a number of compounds including first row of the transition elements, the orbital contribution (μ_L) is quenched or suppressed by the electric fields of the surrounding atoms (*i.e.*, $L = 0$) so that the magnetic moment of the species is practically equal to the value due to spin alone, *i.e.*, $\mu_s = \sqrt{4S(S+1)}$, or in terms of unpaired electrons per atom *n*;

$$\mu = \sqrt{n(n+2)} \qquad \text{... (1.2)}$$

The magnetic moments of some ions of the first transition series are given in Table 1.5. The values have been calculated using the spin only formula.

1.6.1. Ferromagnetic properties of iron, cobalt and nickel

Ferromagnetism is the property due to sufficient number of unpaired electrons. Iron, cobalt and nickel possess 4, 3 and 2 unpaired electrons respectively. Therefore these elements are ferromagnetic.

Table 1.5. Paramagnetic Moments of Some Ions of First Transition Series.

Ions	*Outer configuration*	*No. of unpaired electrons*	*Magnetic moment BM*	
			Calculated	*Observed*
Sc^{3+}, Ti^{4+}	$3d^0$	0	0	0
Ti^{3+}, V^{4+}	$3d^1$	1	1.73	1.75
Ti^{2+}, V^{3+}	$3d^2$	2	2.84	2.76
V^{2+}, Cr^{3+} Mn^{4+}	$3d^3$	3	3.87	3.86
Cr^{2+}, Mn^{3+}	$3d^4$	4	4.90	4.80
Mn^{2+}, Fe^{3+}	$3d^5$	5	5.92	5.96
Fe^{2+}, Co^{3+}	$3d^6$	4	4.90	5.0 –5.5
Co^{2+}	$3d^7$	3	3.87	4.4 –5.2
Ni^{2+}	$3d^8$	2	2.84	2.9 – 3.4
Cu^{2+}	$3d^9$	1	1.73	1.4 – 2.2
Cu^{+}, Zn^{2+}	$3d^{10}$	0	0	0

From Table 1.5 we find that the observed magnetic moments agree fairly well with the calculated values. In the later half of the series, the observed values slightly exceed the spin values only. These differences can be explained in terms of some contribution from the orbital magnetic moments also. For the second and third row elements, the orbital motion is not prevented or quenched and therefore a more complex formula (eq. 1.1) is required.

Cause of Magnetic Behaviour

An electron revolves round the nucleus and it also spins around its own axis to produce magnetic field like the magnetic field produced by electric current flowing through a loop of wire. Each electron acts as a tiny magnet having a definite value of magnetic moment. Diamagnetic substances have no unpaired electrons so they have zero magnetic moment. The external magnetic field, however, induces in them a small magnetic moment which is opposed to the external magnetic field. Thus, diamagnetic substances repel lines of force and show a slight decrease in weight. Paramagnetic substances, however, attract lines of force in the applied magnetic field.

1.7. OXIDATION STATES OF TRANSITION ELEMENTS

One of the most striking features of the *d*-block elements is that they exhibit a variety of oxidation states in their compounds. Some common oxidation states of *d*-block elements are given in Table 1.6.

Table 1.6. Oxidation States of *d*-Block Elements 3*d* or First Transition Series

Element	*Sc*	*Ti*	*V*	*Cr*	*Mn*	*Fe*	*Co*	*Ni*	*Cu*	*Zn*
Oxidation States				+1					+1	
	(+2)	+2	+2	+2	+2	+2	+2	+2	+2	+2
	+3	+3	+3	+3	(+3)	+3	+3	(+3)	(+3)	
		+4	+4	(+4)	+4	(+4)	(+4)			
			+5							
				+6	(+6)	(+6)				
					+7					

There is very small difference between the energies of electron in *ns* and $(n-1)d$ orbitals. Thus both *ns* and $(n-1)$ *d* orbitals participate in compound formation showing variable oxidation states. Along with the *s* electrons, 1, 2, or more *d* electrons may be removed or shared. The variable oxidation states shown by the first transition elements may be related to their actual electronic configuration.

The outer electronic configuration of Sc is $3d^1 4s^2$. It shows an oxidation state equal to +2 if both 4*s* electrons are used for bonding and +3 when two *s* and one *d*-electrons are involved.

Similarly, Ti ($3d^2 4s^2$) exhibits oxidation state +2 when it uses only its two *s*-electrons and +3 when one *d*–electron is used in addition to two *s*–electrons and +4 when it uses 2 *s*-electrons and 2*d*-electrons for bonding. On the same ground we can account for the different oxidation states of V, Cr and Mn. The lower oxidation states are generally obtained by loss of *s* electrons, for example, in case of Mn($3d^5 4s^2$), it readily loses the two 4*s* electrons to form Mn^{2+}. But the higher oxidation states are obtained by sharing 4*s* and all or some of the 3*d* electrons. Manganese can exhibit a maximum oxidation state of +7 or almost any positive state between +2 and +7. Thus the correlation between electronic structure and minimum and maximum oxidation state for first five transition elements can be studied.

In the remaining five elements, *i.e.*, Fe ($3d^6 4s^2$), Co ($3d^7 4s^2$), Ni ($3d^8 4s^2$), Cu ($3d^{10} 4s^1$) and Zn ($3d^{10} 4s^2$), the number of 3*d* electrons is more than 5. The minimum oxidation state in each case in equal to the number of 4*s*- electrons, but the maximum oxidation state is not at all related to the total of *ns* and $(n-1)d$ electrons. Their maximum oxidation state is largely confined to +2 or +3 as in the case of iron. The + 4, + 5, and + 6 oxidation states are only rare.

Low oxidation state. In order to save the low oxidation states from oxidation by atmospheric oxygen, syntheses are often carried out in inert nitrogen atmosphere and special solvents (completely deaerated, *i.e.*, air is completely removed) like tetrahydrofuran, dichloromethane and anhydrous ethanol are used. Organic ligands such as CO, C_2H_4, NO, PR_3 etc., are used to form complexes. It is the characteristic feature of these ligands that they form complexes with *d*-block elements only when they are in zero, –1 or +1 oxidation states.

This is due to the fact that the donor atom of these ligands possesses vacant orbital in addition to the lone pair, *e.g.*, CO (*i.e.*,: C : : Ö:) has only six electrons on carbon hence it has vacant orbitals in addition to the lone pair. The ligands first form a normal σ covalent bond by donating its lone pair to the metal atom. The vacant orbitals of the ligand then accept electrons from the filled *d*-orbitals of the metal to form a *d*π–*p*π bond. This is called metal to ligand π bonding in addition to σ bonding. The ability of these ligands to accept back metal electrons is responsible for the stability of these complexes in their lower oxidation states.

High oxidation states. When a metal ion occurs in an oxidation state higher than the one of common occurrence, then it is called the higher oxidation state.

As the oxidation state increases, the metal ions become more acidic, *i.e.*, electron greedy. Hence these elements with high oxidation states can form compounds with the ligands which are highly electronegative and basic, *e.g.*, F^-, Cl^-, NH_3 etc. This is because the small size, highly charged ligands can form strong σ bonds by electron donation which are stable enough for independent existence.

A suitable organic ligand can also stabilise the higher oxidation states by forming stable complexes. For example, Ag (II) has been obtained as cationic complex with pyridine $[Ag\ (Py)_4]^{2+}$ and with dipyridyl $[Ag\ (dipy)_2]^{2+}$, etc.

1.7.1. Relative stability of oxidation states of transition metals

When an element exists in more than one oxidation states, the relative stabilities of the different oxidation states can be predicted on the basis of their ionisation energy and standard electrode potential values.

The factors which determine the relative stability of different oxidation states of transition metals are described as under:

***(a)* Ionisation energy.** The value of ionisation energies gives information regarding the thermodynamic stability of the transition metal compounds in different oxidation states. It has been established that smaller the ionisation energy, greater is the thermodynamic stability of the compound in that particular oxidation state. For example, the first four ionisation energies of nickel and platinum are composed below:

	$(IE_1 + IE_2)$	$(IE_3 + IE_4)$	Total
Ni	2.49×10^3 kJ mol^{-1}	8.80×10^3 kJ mol^{-1}	11.29×10^3 kJ mol^{-1}
Pt	2.66×10^3 kJ mol^{-1}	6.70×10^3 kJ mol^{-1}	9.36×10^3 kJ mol^{-1}

It can be seen from the above table that the sum of first two ionisation energies is less for nickel than for platinum.

$$Ni \longrightarrow Ni^{2+} + 2e^- \qquad I.E. = 2.49 \times 10^3 \text{ kJ mol}^{-1}$$
$$Pt \longrightarrow Pt^{2+} + 2e^- \qquad I.E. = 2.66 \times 10^3 \text{ kJ mol}^{-1}$$

Therefore, ionisation of nickel to Ni^{2+} is favourable as compared to that of platinum. Thus, *the nickel (II) compounds are thermodynamically more stable than platinum (II) compounds.*

On the other hand, the sum of first four ionisation energies is less for platinum than nickel as shown below.

$$Ni \longrightarrow Ni^{4+} + 4e^- \qquad I.E. = 11.29 \times 10^3 \text{ kJ mol}^{-1}$$
$$Pt \longrightarrow Pt^{4+} + 4e^- \qquad I.E. = 9.36 \times 10^3 \text{ kJ mol}^{-1}$$

Thus, *the platinum (IV) compounds are relatively more stable than nickel (IV) compounds.* Therefore, K_2PtCl_6 [having Pt(IV)] is a well-known compound, whereas the corresponding nickel compound does not exist.

(b) Electrode potential: The electrode potential values also guide us to know the stability of different oxidation states. In general, the lower the $E°_{red}$ value, the greater is the stability of that oxidation state. For example, the electrode potential values for copper +1 and copper +2 oxidation states in aqueous solution are given as under:

$$Cu^{+}\,(aq) + 1e^{-} \longrightarrow Cu\,(s)\,;\, E°_{red} = 0.52\text{ V}$$

$$Cu^{2+}\,(aq) + 2e^{-} \longrightarrow Cu\,(s)\,;\, E°_{red} = 0.34\text{ V}$$

It is evident from the reduction potential value that copper in +2 oxidation state is more stable in aqueous solution than copper in +1 oxidation state. However, it may be noted that the electrode potential depends upon many factors such as sublimation energy, hydration energy, etc. The reduction potential involves the following process:

$$M(s) \longrightarrow M^{+}\,(aq) + e^{-}$$

This process may be considred to occur in the following cycle:

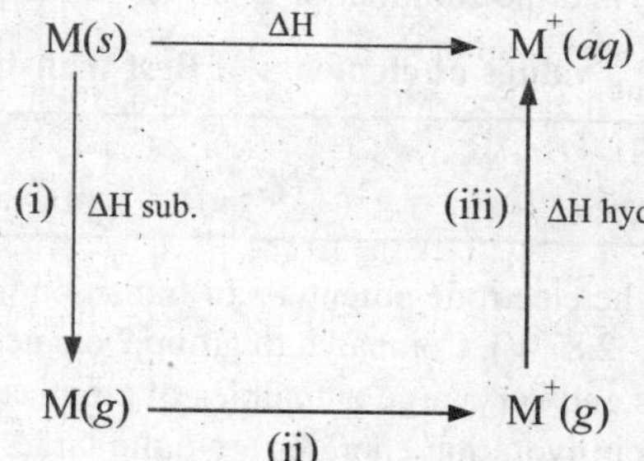

This process actually takes place in the following three steps:

(*i*) In the first step, the atoms get separated from one another and become independent in the gaseous state. The energy needed for this step is known as *enthalpy of sublimation* or *enthalpy of atomisation.*

$$M(s) \longrightarrow M(g) \qquad \textbf{Enthalpy of sublimation, } \Delta H_{sub}$$

(*ii*) In the second step, the outer electron is removed from the isolated atom. The energy required for this change is called *ionisation energy.*

$$M(g) \longrightarrow M^{+}\,(g) + e^{-} \qquad \textbf{Ionisation energy, IE}$$

(*iii*) In the third step, the gaseous ion gets hydrated. In this process energy, known as *hydration energy,* is liberated.

$$M^{+}\,(g) + nH_2O \longrightarrow M^{+}\,(aq) \qquad \textbf{Enthalpy of hydration, } \Delta H_{hyd}$$

The electrode potential, which gives the tendency of the overall change to occur, depends upon the net effect of these steps. The overall energy change that takes place is

$$\Delta H = \Delta H_{sub} + IE + \Delta H_{hyd}$$

Thus, ΔH gives the enthalpy change required to bring the solid metal, M_1 to the monovalent ion in aqueous medium, $M^{+}(aq)$. *Smaller the value of total energy change for a particular oxidation state in aqueous solution, greater will be the stability of that oxidation state.*

It may be noted that energy is required in the steps of sublimation and ionisation while energy is released in the hydration step. Thus, ΔH_{sub} and IE have positive value while ΔH_{hyd} has a negative value.

The electrode potentials are a measure of total energy change. The electrode potentials of different metals can be calculated by forming the cell with standard hydrogen electrode. The following reaction occurs in the cell:

$$2H^+(aq) + M(s) \rightleftharpoons M^{2+}(aq) + H_2(g)$$

Knowing the potential of $2H^+(aq) \mid H_2(g)$, it is possible to determine the potential of $M^{2+}(aq) \mid M$. For the first transition series, the E°, values of $M^{2+}(aq) \mid M$ are given below:

Electrode potentials of elements of first transition series

	Ti	V	Cr	Mn	Fe	Co	Ni	Cu	Zn
E° ($M^{2+}(aq) \mid M$ (in V)	–1.63	–1.20	–0.91	–1.18	–0.44	–0.28	–0.25	0.34	–0.76

We conclude the following from the observed values of E°.

(*i*) There is no regular trend in these values. This is attributed to the irregular variation of ionisation energies ($IE_1 + IE_2$) and the sublimation energies in the period as shown below:

ΔH_{sub} values of elements of first transition series

	Ti	V	Cr	Mn	Fe	Co	Ni	Cu
ΔH_{sub} (kJ mol^{-1})	458	461	318	240	370	404	397	318

(*ii*) It may be noted that the electrode potentials of transition metals are low in comparison to elements of group 2 (e.g. Ca = –2.87 V). Compared to group 2 elements, the transition elements have fairly large ionisation energies and very large enthalpies of atomisation. These factors reduce their electrode potentials though their hydration energies are quite large.

(*iii*) Zinc has low enthalpy of atomisation and fairly large hydration energy. But it has also low electrode potential (–0.76 V) because of its very high ionisation energy ($IE_1 + IE_2$).

(*iv*) The above table shows that copper has positive reduction potential, E°(0.34 V) and this shows that copper is least reactive metal out of the first transition series. This unique behaviour (+ve E value of copper) explains its inability to liberate H_2 from acids. It has been observed that only oxidizing acids (such as nitric acid and hot concentrated sulphuric acid) react with copper in which the acids are reduced. This can be accounted for by the low hydration enthalpy of Cu^{2+} ions.

(*v*) In general, the E° value becomes less negative across the series. This is related to the general increase in the sum of first and second ionisation energies. It is interesting to note that the values of E° of Mn, Ni and Zn are more negative than expected from the general trend. The relatively more negative values of E° for Mn and Zn are due to stability of half filled *d*-subshell in Mn^{2+} ($3d^5$) and the completely filled ($3d^{10}$) configuration in Zn^{2+}. The exceptionally high E° value of Ni from regular trend is related to the highest negative energy of hydration of Ni^{2+} ion.

1.7.2. Ability of transition elements to evolve hydrogen with acids

Standard reduction potentials of some transition metals in acid solution at 298 K are given in increasing order (less negative) in the table below:

Element	*E° (volts)*	Element	*E° (volts)*
$Sc^{3+} + 3e^- \longrightarrow Sc$	– 2.08	$Fe^{2+} + 2e^- \longrightarrow Fe$	– 0.44
$Mn^{2+} + 2e^- \longrightarrow Mn$	– 1.18	$Co^{2+} + 2e^- \longrightarrow Co$	– 0.28
$Zn^{+2} + 2e^- \longrightarrow Zn$	– 0.76	$Ni^{2+} + 2e^- \longrightarrow Ni$	– 0.25
$Cr^{3+} + 3e^- \longrightarrow Cr$	– 0.74	$Cu^{2+} + 2e^- \longrightarrow Cu$	+ 0.34

It is seen that standard reduction potential of a transition element is generally negative, *i.e.*, lower than that of hydrogen (taken as zero, $2H^+ + 2e^- \rightarrow H_2$). Copper is an exception in above cases. All these elements with negative reduction potential liberate hydrogen from dilute acids.

$$M + 2H^+ \longrightarrow M^{2+} + H_2 (g)$$

The rates of evolution of hydrogen are, however, very slow in some cases. Some of these metals (*e.g.*, Cr) are unreactive due to formation of the protective fihn of oxide on their surface.

1.7.3. Though transition metals possess high electrode potentials, they are not good reducing agents

The reducing capacity of a metal depends on its tendency to be oxidised easily to change into ions.

$$M \longrightarrow M^{2+} + 2e^-$$

The transition elements possess this property due to their sufficiently negative standard reduction potentials and should be good reducing agents. But they are not good reducing agents as compared to the elements of groups I, II and III. This is due to their less tendency to form ions because of (*i*) their high ionization energies, (*ii*) low heat of hydration of their ions, and (*iii*) their high heats of sublimation.

1.8. CATALYTIC PROPERTIES OF TRANSITION ELEMENTS

Many transition elements and their compounds are used as a catalyst in the laboratory and in industry. Some of the important catalysts are given below:

(*i*) Finely divided iron and molybdenum are used in the manufacture of ammonia (Haber's process).

(*ii*) Finely divided nickel and palladium are used in the manufacture of vanaspati as well as in hydrogenation reaction in the laboratory.

(*iii*) Zinc chromite is used in the synthesis of methyl alcohol.

(*iv*) Platinum and vanadium pentaoxide are used in the manufacture of ammonia (Haber's process).

(*v*) Fenten's reagent ($FeSO_4 + H_2O_2$) is used for the oxidation of alcohols to aldehydes.

(*vi*) Ziegler-Natta (catalyst Ti Cl_4) is used for the manufacture of polythene.

The different factors contributing to catalytic action are given below:

(*i*) The catalytic properties of the transition elements are probably due to the variable oxidation states, due to which they form unstable intermediate products with various reactants in some cases. These intermediate products decompose to give the final product, regenerating the catalyst (intermediate compound formation theory of catalysis).

A + B + Catalyst	$\longrightarrow$	A.B. Catalyst	$\longrightarrow$	AB + Catalyst
Reactant		Unstable intermediate compound		Product

(*ii*) In some cases the finely divided metals or their compounds provide a large surface area for adsorption of the reactants. The adsorbed reactants react faster due to their closer contact (adsorption theory) on the surface of the catalyst giving rise to products.

1.9. COMPLEX FORMATION IN THE CASE OF TRANSITION ELEMENTS

The transition metal ions have a great tendency to form complex ions with certain molecules (*e.g.*, NH_3 CO, H_2O and 'en' etc.) or ions (*e.g.*, F^-, Cl^-, N^-, etc.). The molecules or ions attaching themselves with the cation are called ligands. The ligands contain lone pairs of electrons which are easily donated to the transition metal ions. This is in contrast to the *s*- and *p*-block elements due to the following reasons:

(*i*) Transition metal cations due to their small size possess high charge density. These can, therefore, accept lone pairs of electrons from the ligands.

(*ii*) They have vacant *d*-orbitals of suitable energy to accept lone pairs of electrons to form complex compounds.

s and *p*-block elements do not meet the above requirements. The number of ligands which are attached to a metal ion by coordinate bonds to form a complex ion is called *coordination number* of the metal ion. The complexes are also called *coordination compounds*. The complex ions have in general linear, square planar, tetrahedral or octahedral structure depending upon the nature of hybridization of the metal ion orbitals.

All transition metal cations, however, do not possess the same capacity for complex formation. There are wide variations between different series but within a series certain trends towards increased stability of complex can be given.

(*i*) In each transition series, the stability of the complexes increases with increasing atomic number of the element and decreasing size of its atoms.

(*ii*) When a transition metal atom exhibits more than one oxidation state, the complex with higher oxidation state has greater stability.

These points can be explained in terms of greater attraction of metal ion for ligand electrons. For example, Co^{2+} does not form a stable complex with ammonia, whereas the corresponding Co^{3+} complex, $[Co(NH_3)_6]^{3+}$, is highly stable.

The formation of a complex, *e.g.*, $[Cr(NH_3)_6]^{3+}$, can be represented as:

$$Cr^{3+} + 6NH_3 \longrightarrow [Cr(NH_3)_6]^{3+}$$

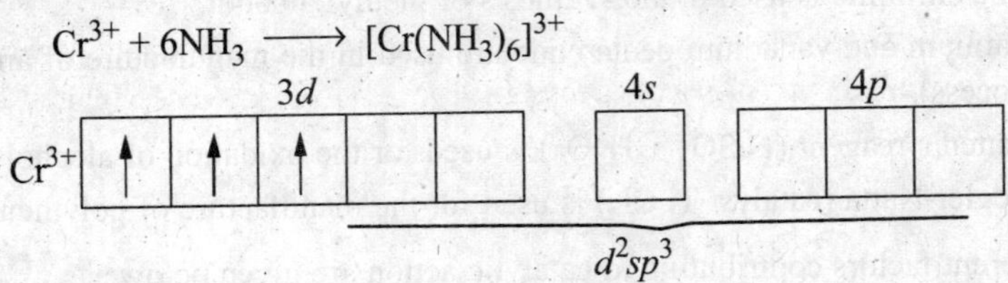

The electronic configuration of Cr is (Ar) $3d^5 4s^1$ and that of Cr^{3+} is (Ar) $3d^3 4s^0$.

The six empty orbitals of Cr^{3+} (*i.e.*, two 3*d*, one 4*s* and three 4*p*) exhibit d^2sp^3 hybridization. The six hybridized orbitals have same energy to accept six pairs of electrons from six molecules of ammonia to form $[Cr(NH_3)_6]^{3+}$. Formation of complexes such as $[Ni(CO)_1]$, $[Ni(CN)_4)^{2-}$, $[Fe(CN_6)]^{4-}$ and $[Fe(CN)_6]^{3-}$ etc., has been explained in detail in the chapter on coordination compounds.

Iron in hemoglobin (a red colouring matter in blood) and cobalt in vitamin B_{12} are the examples of typical complexes of biological importance.

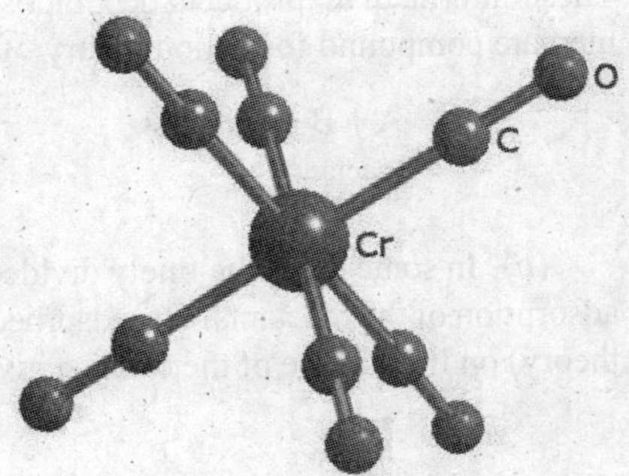

Chromium hexacarbonyl complex

1.10. COLOUR IN TRANSITION METAL COMPOUNDS

Compounds of *s*- and *p*-block elements are almost white whereas ionic and covalent compounds of transition elements are usually coloured. Colour, in general, is associated with the incomplete electron sub-shell and the ability of the compound to promote an electron from one energy level to another by absorption of light of a particular wavelength.

Colour in transition metals is associated with incomplete $(n-1)d$ sub-shell, *i.e.*, it is a property of unpaired *d*-electrons. In transition elements, *d*-electrons are promoted to a higher energy level within the *d*- sub-shell by absorption of a small amount of energy. This energy is obtained by absorption of light in the visible region. The rest of the light is no longer white, *i.e.*, it appears coloured. For example, if red light is absorbed then the transmitted light contains an excess of the light of blue colour. The compound appears blue. For example, Cu^{2+} ion ($3d^9$) absorbs red light from the visible range, for promotion of 3*d* electrons, the ions reflect blue light and appear deep blue.

Anhydrous cobalt compounds also absorb red light for excitation of 3*d*-electrons. When red light is eliminated from white light, the colour appears blue (*i.e.*, blue is the combination of all the remaining colours except red). If all the incident light falling on a substance is absorbed, the substance appears black. If all the incident light is reflected, then the substance appears white. If only a particular wavelength is absorbed from the incident light, then the substance has a characteristic colour.

1.10.1. Colour of Transition Metal Complexes on the basis of crystals field theory

The cause of the colour of the transition metal compounds can be explained on the basis of crystal field theory. In this theory, the splitting of the *d*-orbitals takes place under the influence of surrounding ligands. The *d*-orbitals are split by the surrounding ligands into the upper e_g and lower t_{2g} levels. The energy gap between these two levels is denoted by Δ and is comparatively small between the non-degenerate *d*-orbitals (e_g and t_{2g}) in a transition metal complex. Consequently the excitation of an electron from a lower to a higher level can be achieved by absorption of visible light. This causes the complex to appear coloured.

Fig. 1.2 shows that the *d*-orbitals in Ti^{3+} are split into e_g and t_{2g} orbitals of differing energies when the ligands H_2O molecules approach the cation. The single electron in t_{2g} orbital absorbs green yellow light and is excited to e_g orbital and thus appears purple. Colours exhibited by hydrated ions of the first transition series are given in Table 1.7

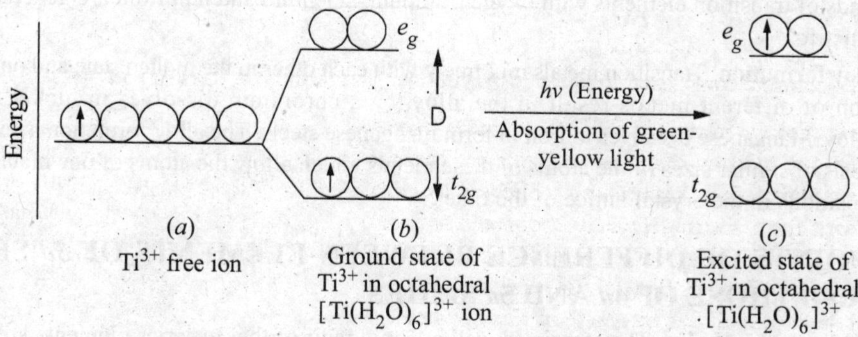

Fig. 1.2. Relative energy levels for the five 3*d*-orbitals of the free and hydrated Ti^{3+} ion.

Table 1.7. Colours of Some Hydrated Ions

Ions	*Outer electronic configuration*	*Colour*	*No. of unpaired electrons*
Sc^{3+}	$3d^0$	colourless	0
Ti^+	$3d^1$	purple	1
V^{3+}	$3d^2$	green	2
Cr^{3+}	$3d^3$	violet	3
Mn^{3+}	$3d^4$	violet	4
Fe^{3+}	$3d^5$	yellow	5
Mn^{2+}	$3d^5$	pink	5
Fe^{2+}	$3d^6$	green	4
Co^2	$3d^7$	pink	3
Ni^{2+}	$3d^8$	green	2
Cu^{2+}	$3d^9$	blue	1
Cu^+	$3d^{10}$	colourless	0
Zn^{2+}	$3d^{10}$	colourless	0

1.11. FORMATION OF INTERSTITIAL COMPOUNDS, NON-STOICHIOMETRIC COMPOUNDS AND ALLOYS

Interstitial compound formation. The transition metals appear to be unique in their ability to react with small atoms of non-metals to give interstitial compounds. The important compounds are hydrides, nitrides, carbides and borides of these elements. Interstitial compounds are those in which the non-metal atoms like nitrogen, carbon etc., occupy holes (interstices) in metal structure.

Formation of non-stoichiometric compounds. A feature of the transition or *d*-block elements is the existence of non-stoichiometric compounds, *i.e.*, compounds of indefinite composi-tion. For example, in FeO, the ratio is not 1:1 but the composition varies between $Fe_{0.94}O$ and $Fe_{0.84}O$. The compounds of transition elements with oxygen, sulphur, selenium and tellurium are generally non-stoichiometric.

Alloy formation. Transition metals mix freely with each other in the molten state and on cooling a solution of different metals result in the alloy, *e.g.*, chromium dissolves in nickel to form Cr–Ni alloy. Manganese dissolves in iron to form manganese steels. The alloy formation is explained on the basis of similar sizes of the atoms of these metals which allow the atoms of one metal to take up the positions in the crystal lattice of the other.

1.12. POINTS OF DIFFERENCE BETWEEN ELEMENTS OF *3d* SERIES AND THOSE OF 4*d* AND 5*d* SERIES

(*i*) Atomic/Ionic size. The atomic as well as ionic radii of the 3*d* series elements are always smaller than the 4*d* and 5*d* series elements but due to lanthanide contraction, the atomic as well as ionic radii of the 4*d* and 5*d* series elements almost have identical size.

(*ii*) Coordination number. The elements of the first series (3*d*) rarely show coordination number greater than six. Coordination numbers 1, 8 or higher are quite common with elements of second (4*d*) and third (5*d*) series. This may be due to their bigger size, so that compounds with higher coordination number may be formed.

***(iii)* Reactivity of elements.** The reactivity of the transition elements generally decreases in going from first transition series to the third transition series. The elements of the first transition series are much more reactive than those of the second and third series. The less reactivity may be due to their higher ionization energies, lower hydration energies, higher sublimation energies, etc.

***(iv)* Lower oxidation state.** The elements of the first transition series have at least one stable lower oxidation state which furnishes stable hydrated cations and oxy salts. On the other hand, the second and third series elements form compounds in their higher oxidation states. The hydrated cations and oxy salts for 4*d* and 5*d* elements generally do not exist.

***(v)* Maximum oxidation state.** The maximum oxidation state in each series is exhibited by iron-group elements. Iron shows a maximum of oxidation state of +6 in FeO_4^{2-} and Ru and Os exhibit maximum oxidation state of +8 in their tetraoxide, MO_4.

***(vi)* Nature of complexes.** The four-coordinate complexes of the 3*d* series elements, with few exceptions, are tetrahedral; for example, Ni^{2+}, Co^{2+}, Cu^{2+} form tetrahedral as well as square planar complexes. The four-coordinate complexes of 4*d* and 5*d* series elements are square planar.

The elements of 3*d* series form both low and high spin complexes but due to high crystals field splitting energies elements of 4*d* and 5*d* series form only low-spin complexes.

***(vii)* Stability of complexes.** Elements of the 3*d* series form more stable complexes with N, O and F donors, whereas the elements of 4*d* and 5*d* series form more stable complexes with P, S and heavier halogen donors.

***(viii)* Application of spin-only formula.** The magnetic data for the 3*d* series elements can be interpreted by spin-only formula. On the other hand, the magnetic data for 4*d* and 5*d* series elements cannot be interpreted by this formula only.

SOME SOLVED PROBLEMS
(From question papers of different universities)

Problem 1. Justify the following statement:

Transition metals are less reactive than alkali and alkaline earth metals.

Solution. Transition metals are less reactive than the alkali and alkaline earth metals because:

(*i*) transition metals are smaller in size compared to alkali and alkaline earth metals; and

(*ii*) transition metals have higher ionisation energies.

Problem 2. Cu^{2+} ($3d^9$) is more stable than Cu^+($3d^{10}$).

Solution. An ion carrying a higher charge and smaller size is associated with greater lattice and hydration energy. That is why Cu^{2+} is more stable than Cu^+. It is expected on the basis of electronic configuration that Cu^+ should be more stable because it has completely filled orbitals. But the lattice and hydration energy factors dominate and as a result Cu^{2+} is more stable.

Problem 3. What justification is there for classifying Zn, Cd and Hg as transition elements?

Solution. Although there are no incomplete *d*-orbitals in Zn, Cd and Hg, these elements have been retained as transition elements for the sake of continuity.

Problem 4. Elements of transition series show common physical and chemical properties. Explain.

Solution. In these elements, the number of electrons in the outermost shell remains the same. Only the penultimate shell is being expanded from 8 to 18 by the addition of *d*-electrons. Their electronic configurations are not very regular. This is due to the reason that energy difference of the electrons in the $(n - 1)$ d and ns levels are not significant and electrons between these levels may change positions mutually. Since chemical properties depend, to a greater extent, on the number of electrons in the outermost shell, which remains the same, the variation in chemical properties amongst these elements of a given period are very small. As is expected, they have physical and chemical properties in common.

Problem 5. "4*s* subshell is filled prior to 3*d* but on ionisation, 4*s* electrons are removed first." Justify.

Solution. 4*s* orbital is spherical and hence has a greater penetrating power as compared to 3*d* orbital which is having a double dumb-bell shape. Therefore 4*s* orbital is associated with lower energy compared to 3*d* and hence the filling of electron takes place first in 4*s* and then in 3*d*.

However, when it comes to removing the electrons, removal first takes place in 4*s*. This is because once the 3*d* orbital has been filled, its energy falls because of greater effective nuclear charge, making it more stable than 4*s* orbital. Hence electrons are first removed from 4*s* orbital which acquires greater energy.

Problem 6. First ionisation energy of copper is higher than those of alkali metals while second and third energies are lower than those of alkali metals. Explain.

Solution. Copper has a smaller size compared to the alkali metals in their period. Due to smaller size and due to poor shielding effect of 3*d* orbital, Cu has a higher first ionisation energy. Second and third I.E. are lower compared to those of alkali metals due to loss of electrons from diffused *d*-orbitals, *d*-electrons have smaller screening effect and give rise to higher effective nuclear charge compared to *s* and *p* electrons.

Problem 7. Why the second ionisation energy for Cu and Cr are anomalously high?

Solution. As a general rule, second ionisation energy is normally higher than the first ionisation energy because after the removal of one electron, the size of the ion shrinks. The outer electrons are pulled with a greater force towards the nucleus and therefore a greater energy is needed to take out the second electron. In the case of Cu and Cr, this effect is much more pronounced. Consider the configuration of Cu and Cr in the outermost orbital.

Cr : .... $3s^2\ 3p^6\ 3d^5\ 4s^1$

Cu : .... $3s^2\ 3p^6\ 3d^{10}\ 4s^1$

whereas it is the 4*s* electron which is involved in first ionisation energy, it is 3*d* electron which is involved in second ionisation energy. It is much more difficult to remove a *d*-electron because of weaker screening effect of remaining *d*-electrons. Moreover, the stable half-filled and fully-filled *d*-orbital configuration of Cr and Cu respectively is distributed if the second electron is to be removed. This requires much more energy and hence the anomaly.

Problem 8. In moving from Ar to K, the electron enters 4*s* orbital instead of 3*d*. Why?

Solution. In argon (Ar), the filling of 3*p* orbitals is complete. As 4*s* orbital is associated with smaller energy, compared to 3*d* orbital, the next electron goes to 4*s* orbital.

Problem 9. Which of the two among Nb, V and Ta are chemically closer and why?

Solution. Vanadium (V) and Tantalum (Ta) are closer to each other because of similar configuration in the outermost orbit.

V .... $3d^3\ 4s^2$

Ta .... $5d^3\ 6s^2$

Problem 10. Explain: K_{19} and Ca_{20} involve filling of 4*s* while the elements Sc_{21} to Zn_{30} involve the filling of 3*d* orbitals.

Solution. From Ca_{20} onwards, the electrons do not go to 4*p* orbitals but to 3*d* orbitals. This is because the next in order of energy are 3*d* orbitals and not 4*p* orbitals. This continues up to Zn_{30} after which the filling takes place in 4*p* orbitals.

Problem 11. Mercury is a liquid metal. Explain.

Solution. As there are no vacant *d*-orbitals in Hg, there is no possibility of mutual overlapping of unpaired *d*-orbitals. Hence the melting point is considerably low and it is a liquid at room temperature.

Problem 12. Transition metals are less electropositive than *s*-block elements. Explain.

Solution. This is so because transition metals have smaller size and higher ionisation energy compared to *s*-block elements. High ionisation energy implies less electropositive nature.

Problem 13. Cu, Ag and Au are typical transition elements while Zn, Cd and Hg are not. Comment.

Solution. The electronic configurations of the elements and their ions are as under:

Cu ....$3d^{10}4s^1$	Cu^{2+} ....$3d^9$
Ag ....$4d^{10}\,5s^1$	Ag^+ ....$4d^{10}$
Au....................$5d^{10}6s^1$	Au^{3+} .... $5d^9$
Zn....................$3d^{10}4s^2$	Zn^{2+} ....$3d^{10}$
Cd....................$4d^{10}5s^2$	Cd^{2+} ....$4d^{10}$
Hg....................$5d^{10}6s^2$	Hg^{2+}....................$5d^{10}$

As zinc, cadmium, mercury salts (ions) have completely filled *d*-orbitals, they do not reflect typical transition behaviour.

In the case of Cu^{2+} and Au^{3+}, there are incomplete *d*-orbitals, hence typical transition characteristics.

Problem 14. Melting points of *d*-block elements in a series reach a maximum upon the middle, thereafter show a decrease. Explain.

Solution. Melting point of a substance depends upon the atomic size and lattice energy. Smaller the size of particles, closer the packing and greater the lattice energy and higher the m.p. In a period, the atomic size decreases to some point and it increases thereafter. Therefore the lattice energy and m.p. increase up to this point and decline after that.

Problem 15. Calculate the number of unpaired electrons in the following:

$$[Fe(H_2O)_6]^{3+},\ [Co(NH_3)_6]^{3+}\ [Fe(CN)_6]^{3-}$$

Which of the above ions do you expect to be coloured?

Solution. Considering the charge on H_2O, NH_3 and CN^- as 0, 0 and –1 the charge on the metals can be calculated as +3, +3 and +3. Thus the ions in question are

$$Fe^{3+} : [Ar]\,3d^5$$

$$Co^{3+} : [Ar]\,3d^6$$

$$Fe^{3+} : [Ar]\,3d^5$$

Thus Fe^{3+} and Co^{3+} contain 5 and 4 unpaired electrons; all of them are coloured.

Problem 16. Why Ti^{3+} (aq) ions are paramagnetic and reddish violet coloured?

Solution. The electronic configurations of Ti and Ti^{3+} are as follows:

Ti	$3d^24s^2$
Ti^{3+}	$3d^1$

Ti^{3+} is paramagnetic due to the presence of one unpaired electron in the presence of ligand water, *d*-orbitals are split into two sets having slightly different energies. The energy difference corresponds to radiation of lower energy. Thus the single electron can absorb radiation of lower energy (red colour) to get excited to higher *d*-level. The transmitted light will be of higher energy having violet colour.

Problem 17. Transition metals are less reactive than alkali and alkaline earth metals. Explain.

Solution. The ionization energies of transition elements are fairly high on account of their small size of the atoms. They have high heat of sublimation due to the presence of covalent bonding. In other words, they require a large amount of energy for a change from the solid to vapour state. Further the metal ions do not get hydrolysed easily, *i.e.*, they have low heat of solvation. All these factors retard the tendency to react. That is why, the *d*-block elements have rather a tendency to remain unreactive or 'noble'. This tendency is quite pronounced in gold and platinum in the third transition series.

On the other hand, alkali and alkaline earth metals possess small ionisation energies. Chemical reactivity is dependent upon the ionisation energy. Hence alkali and alkaline earth metals are more reactive.

Problem 18. Discuss the oxidation states of Cr.

Solution. Chromium exhibits oxidation states of +1, +2, +3 and +6. It also exhibits less common oxidation state of +4. Electronic configurations of chromium in various oxidation state are:

Cr (0)	$3d^54s^1$
Cr (I)	$3d^5$
Cr (II)	$3d^4$
Cr (III)	$3d^3$
Cr (IV)	$3d^2$
Cr (VI)	$3p^6$

Problem 19. Sc^{3+} is more stable than Sc^{2+}. Explain.

Solution. Sc^{3+} possesses greater hydration energy and lattice energy because of greater nuclear charge and smaller size compared to Sc^{2+}.

Problem 20. Zn forms only Zn^{2+} and not Zn^{3+}. Why?

Solution. In the formation of Zn^{3+}, an electron from the *d*-orbitals has to be removed which requires abnormally higher amount of energy because the completely filled ($3d^{10}$) stable configuration is disturbed.

Problem 21. Why are salts of zinc, cadmium and mercury white?

Solution. In simple compounds of zinc, cadmium and mercury (having $Zn^{2+} - 3d^{10}$; $Cd^{2+} - 4d^{10}$ and $Hg^{2+} - 5d^{10}$) there is no possibility of promotion of electrons within the *d* level due to their completely filled sub-shells and hence their compounds are white.

Problem 22. Why are compounds of *s*- and *p*- block elements not coloured?

Solution. In *s*- and *p*-block elements the *d*- orbitals are either missing or completely filled. There cannot be any *d–d* transitions, and the energy to promote an *s*- or *p*- electron to a higher energy level is much greater and corresponds to ultraviolet regions. These compounds absorb no radiations from the visible region and hence the compounds of *s*- and *p*-block elements do not appear coloured in the visible region.

Problem 23. Cu^{2+} ions are coloured and paramagnetic while Zn^{2+} ions are colourless and diamagnetic. Explain.

Solution. Consider the electronic configurations of Cu^{2+} and Zn^{2+} in the outermost orbit.

Cu^{2+} : $3d^9$

Zn^{2+} : $3d^{10}$

Paramagnetism is the property due to the presence of unpaired electrons. Cu($3d^9$) has an unpaired electron whereas Zn^{2+}($3d^{10}$) has no unpaired electron. Therefore the former is paramagnetic whereas the latter is diamagnetic in nature.

Substances which have the possibility of exciting and shifting the electrons from lower levels to vacant *d*-orbitals show colour. In the case of Cu^{2+}, there is one vacancy in 3*d*-orbital to which an electron can be excited. Hence colour is shown by Cu^{2+} compound. In Zn^{2+}, there is no such possibility as 3*d*-orbitals are completely filled ($3d^{10}$).

Problem 24. Cupric salts are blue in colour while cuprous salts are colourless. Explain.

Solution. Electronic configurations of Cu^{2+} and Cu^+ in the outermost orbit are:

Cu^{2+} : $3d^9$

Cu^+ : $3d^{10}$

There is a possibility of excitation of electron from one *d*-level to another *d*-level in Cu^{2+} with the absorption of light from visible range. The transmitted light is blue in colour. Hence Cu^{2+} salts are blue in colour. In Cu^+ salts, there is no possibility of excitation of the electron, hence no colour is shown by such salts.

Problem 25. Cu (I) is diamagnetic whereas Cu (II) is paramagnetic. Explain.

Solution. Magnetic properties of substance are determined by the number of unpaired electrons in the ion. In Cu (II) ion the outermost orbital configuration is $3d^9$. Thus there is one unpaired electron. Hence it is paramagnetic. In Cu (I), the outermost orbital configuration is $3d^{10}$. There is no unpaired electron and therefore Cu (I) is diamagnetic.

Problem 26. In moving down a transition series, the atomic sizes increase from 3*d* to 4*d* but there is hardly any increase from 4*d* to 5*d*. Explain.

Solution. Atomic size decreases while moving from 3*d* to 4*d* due to increased nuclear charge. In moving from 4*d* to 5*d*, increase in nuclear charge is balanced by screening effect of inner orbitals.

Problem 27. Which should be a better oxidising agent: Co^{2+} or Co^{3+} in water?

Solution. Higher the oxidation state of an element, better it is an oxidising agent. Therefore, Co^{3+} shold be the better oxidising agent in water. Standard reduction potentials for Co^{2+} (– 0.28 V) and Co^{3+} (1.80 V) in acid solution also show that Co^{3+} is more easily reduced and better oxidising agent than Co^{2+}.

Problem 28. Which ion should exhibit a higher magnetic moment : Mn^{2+} or V^{2+}?

Solution. Mn^{2+}: [Ar]$3d^5$

V^{2+} : [Ar]$3d^3$

Mn^{2+} has five unpaired electrons whereas V^{2+} has three unpaird electrons. Therefore, Mn^{2+} has higher magnetic moment.

Problem 29. Which of the following pairs of atoms have the most similar atomic radi:

(*i*) Ti, Zr **(*ii*) Cr, Mo** **(*iii*) Mo, W** **(*iv*) Ni, Pb**

Solution. (*iii*) Mo, W

Problem 30. Will Cu^+ undergo disproportionation in an aqueous medium? Standard reduction potentials are 0.158 V for Cu^{2+} to Cu^+ and 0.522 V for Cu^+ to Cu.

Solution. The electrode potentials for Cu^+ and Cu^{2+} can be written as under:

$$Cu^+ + e^- \longrightarrow Cu \qquad E° = 0.522\text{ V} \qquad ...(i)$$

$$Cu^+ \longrightarrow Cu^{2+} + e^- \qquad E° = -0.158\text{ V} \qquad ...(ii)$$

Adding (*i*) and (*ii*), we get

$$2Cu^+ \longrightarrow Cu^{2+} + Cu \qquad E° = 0.364\text{ V}$$

Since E° for disproportionation is positive, Cu^+ will disproportionate.

Problem 31. Calculate in Bohr magneton the expected magnetic moment for the following ions (spin magnetic moment):

(*i*) Fe^{3+} (*ii*) Ni^{2+} (*iii*) Cu^+

Solution. Spin magnetic moment is given as:

$$\mu = \sqrt{n(n+2)}\text{ B.M.}$$

where n is number of unpaired electrons.

(*i*) For Fe^{3+} : [Ar] $3d^5$ No. of unpaired electrons = 5

$$\mu = \sqrt{5(5+2)} = \sqrt{35} = 5.96\text{ B.M.}$$

(*ii*) For Ni^{2+} : [Ar] $3d^8$ No. of unpaired electrons = 2

$$\mu = \sqrt{2(2+2)} = 2\sqrt{2} = 2.828\text{ B.M.}$$

(*iii*) For Cu^+ : [Ar] $3d^{10}$ No. of unpaired electrons = 0

$$\mu = 0$$

Problem 32. Why do transition elements in zero and lower oxidation state form complexes with weak ligands like CO, NO or PR_3?

Solution. Ligands like CO or NO are weak bases and cannot form L → M, σ bonds easily. Because of the weak σ bonds, their complexes would not be stable. However, these ligands possess vacant π-orbitals of the right symmetry and energy so that it can overlap with the filled *d*-orbitals of the transition metals. If the transition metal is in zero or low oxidation state, it possesses electrons in the low energy t_{2g} orbitals which can be easily donated into vacant π* orbitals forming *d*π–*p*π bond as shown below in Fig. 1.3

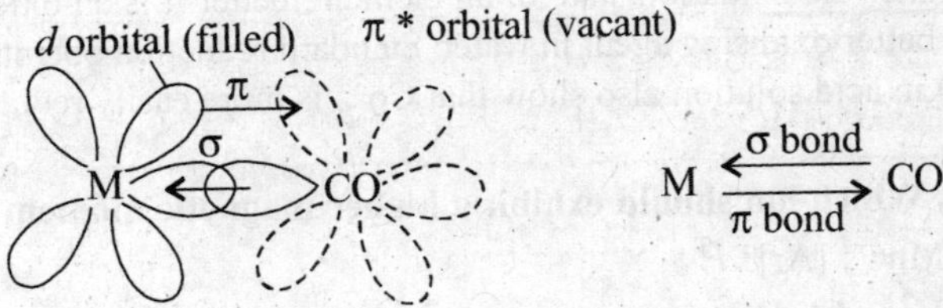

Fig. 1.3. Bonding between transition elements and weak bases (ligands).

Because of the formation of back π bond, these complexes become stable.

Problem 33. (*i*) Calculate the magnetic moment (expected from spin only) for the following ions: Cr^{3+}, Fe^{3+}, Cu^+

(*ii*) Which of the above ions do you expect to be coloured and why?

Solution. (*i*) Cr^{3+} : [Ar] $3d^3$

No. of unpaired electrons = 3, substituting values in the spin-only formula,

$$\mu = \sqrt{3(3+2)}$$
$$= \sqrt{15} = 3.86 \text{ B.M.}$$

(*ii*) Fe^{3+} : [Ar] $3d^5$

No. of unpaired electrons = 5, substituting the values in the spin-only formula,

$$\mu = \sqrt{5(5+2)} = \sqrt{35}$$
$$= 5.96 \text{ B.M.}$$

(*iii*) Cu^+ : [Ar] $3d^{10}$

No. of unpaired electrons = 0

$$\mu = 0$$

(*ii*) Cr^{3+} and Fe^{3+} are coloured because they have unfilled *d*-subshell.

Problem 34. Iron is a transition element but sodium is not. Explain.

Solution. Iron has electronic configuration, $[Ar]3d^64s^2$ and sodium has electronic configuration, $[Ne]3s^1$. In iron the last electron enters a *d*-orbital so it is a *d*-block element or transition element. In sodium, the last electron enters a *s*-orbital, so it is an *s*-block element and not a transition element.

Problem 35. Calculate in Bohr magneton the expected magnetic moment for Mn^{2+} ion (μ spin only).

Solution. Spin magnetic moment is given by the relation:

$$\mu = \sqrt{n(n+2)} \text{ B.M.}$$

where *n* is the number of unpaired electrons.

Mn^{2+} : $3d^5$

No. of unpaired electrons = 5

$$\mu = \sqrt{5(5+2)} = \sqrt{35}$$
$$= 5.96 \text{ B.M.}$$

Problem 36. Why does Mn (II) show maximum paramagnetic character amongst the bivalent ions of first transition series?

Solution. The electronic configuration of Mn (II) is $3d^5$. It contains maximum number of unpaired 3*d* electrons (*i.e.* 5) among the transition elements of first series. Therefore, it has maximum paramagnetic character.

PROPERTIES OF ELEMENTS OF FIRST TRANSITION SERIES AND THEIR BINARY COMPOUNDS AND COMPLEXES

1.13. SCANDIUM

Properties

1. Scandium has a silvery-white lustre but gets tarnished in air due to formation of oxide, Sc_2O_3.
2. It has a melting point of 1539°C and a boiling point of 2730°C.
3. The outer electronic configuration of scandium is $3d^1\ 4s^2$. If forms trivalent ion, Sc^{3+}, in solution. This occurs through the loss of two *s*- and one *d*-electron. Compounds with higher oxidation states are unknown.
4. All the scandium compounds in +3 state are colourless and diamagnetic because the 3*d* subshell in these compounds is empty.

COMPOUNDS OF SCANDIUM

Scandium forms compounds *only* in the +3 oxidation state. Thus, the oxides, halides and oxohalides have the formulae Sc_2O_3, ScX_3 and ScOX, respectively.

Scandium oxide, Sc_2O_3. It is formed readily by heating the salts like scandium carbonate and scandium nitrate.

$$Sc_2(CO_3)_3 \xrightarrow{\text{Heat}} Xc_2O_3 + 3CO_2$$

$$4Sc(NO_3)_3 \xrightarrow{\text{Heat}} 2Sc_2O_3 + 12NO_2 + 3O_2$$

It can also be obtained by the reaction of scandium with water.

$$2Sc + 3H_2O \longrightarrow Sc_2O_3 + 3H_2$$

Scandium halides. When fluoride ions are added to the solution of a Sc(III) compound, the precipitate of ScF_3 is formed first which, in the presence of excess fluoride ions, redissolves to give the complex ion $[ScF_6]^{3-}$

Scandium halides are best prepared by heating the element in a current of the halogen.

Scandium hydroxide. This compound does not have a definite formula $Sc(OH)_3$. It is considered to be a *hydrous oxide,* $Sc_2O_3.nH_3O$. It is less basic than $Ca(OH)_2$ but more basic than $Al(OH)_3$.

Scandium hydride. Scandium forms a hydride which is not quite stoichiometric, the number of hydrogen atoms per atom of scandium being nearer to 2. Its formula may be written as ScH_2. This compound contains hydrogen in the form of hydride ion, H^-. The compound probably contains one Sc^{3+} ion per two H^- ions. The third electron of Sc is present in the conduction band. That is why solid ScH_2 is a conductor of electricity. It reacts with water liberating hydrogen.

Scandium carbide, ScC_2. It is obtained by heating the oxide with carbon at a high temperature of about 1270 K

$$Sc_2O_3 + 7C \longrightarrow 2ScC_2 + 3CO$$

It is an *acetylide* as it reacts with water liberating acetylene. It has been shown by conductivity and magnetic measurements that scandium carbide is an ionic compound having Sc^+ and C_2^- ions.

Complexes. The tendency of an ion to form complexes depends upon its size. Smaller the size of an ion, greater is its tendency to form complexes.

The size of Sc^{3+} ion is smaller than that of other ions of this group. It, therefore, readily forms complex ions, as for example, $[ScF_6]^{3-}$, $[Sc(bipy)_3]^+$, $[Sc(DMSO)_3]^{3+}$, etc., where bipy is bipyridyl and DMSO is dimethylsulphoxide. Some bidentate chelating agents such as $C_2O_4^{2-}$ and acac (acetyl acetonate anion) are also known to form chelates with Sc^{3+} such as, $[Sc(C_2O_4)_2]^-$, $[Sc(acac)_3]$, etc.

1.14. TITANIUM

Properties.

1. Titanium has a high melting point (1668° C) and a high boiling point (3260° C). It is a shining white metal, having rather a low density of 4.5 g per cm^3.

2. Action of air. On exposure to air at ordinary temperatures, titanium reacts with oxygen and nitrogen to form a thin *protective surface coating* of oxide and nitride. This coating protects the metal from corrosion. It remains stable in air even at 400° C.

Titanite a mineral of Ti

3. Action of acids. Titanium is passive and resists attack by acids. The passivity is due to formation of a protective surface coating of oxide and nitride.

The metal dissolves slowly in hot concentrated hydrochloric and nitric acids. It explodes with fuming nitric acid. It dissolves in hydrofluoric acid giving hexafluoro complex compound.

$$Ti + 4HF \longrightarrow TiF_4 + 2H_2$$

$$TiF_4 + 2HF \longrightarrow H_2[TiF_6]$$

4. Action of alkalis. The metal is not attacked by alkalis *even at elevated temperatures.*

5. Combination with non-metals. At high temperatures the metal combines with non-metals like oxygen, nitrogen, halogens, carbon, boron and silicon. The compounds like TiN, TiC, TiB are **interstitial compounds.** These are *hard* and have *refractory properties.*

Titanium is used in building spacecrafts

Electronic Donfiguration and Oxidation States. The outer electronic configuration of titanium is $3d_2\ 4s_2$. It shows oxidation states of +2, +3 and +4. The +4 state is *the most common, the most stable* and also *the most important* oxidation state of the element. It also exercises, though rarely, 0 and +1 oxidation states in the formation of some of its complexes.

COMPOUNDS OF TITANIUM

Compounds of Titanium (IV)

Titanium (IV) oxide, TiO_2 is the most important oxide of titanium. It occurs in nature as *rutile* which is an important ore of titanium.

It is a white powder, insoluble in water.

It is *amphoteric* in character. It dissolves in alkalis yielding titanates.

$$TiO_2 + 2NaOH \longrightarrow Na_2TiO_3 + H_2O$$

When heated with carbon in a stream of chlorine, it gives titanium tetrachloride

$$TiO_2 + 2C + 2Cl_2 \longrightarrow TiCl_4 + 2CO$$

Titanates. When Na_2O is fused with TiO_2, Na_2TiO_3 is formed. When Na_2TiO_3 is heated in an atmosphere of hydrogen, non-stoichiometric compounds of the formulae approximating to $Na_{0.2–0.25}$ TiO_2 are formed. These compounds are known as 'titanium bronzes' which have high electrical conductivity and a metallic bluish lustre.

Calcium titanate, $CaTiO_3$, occurs naturally as *perovskite*. **Iron titanate, $FeTiO_3$,** also occurs naturally as *ilmenite*.

The structure of perovskite comprises of *ccp* arrangement of O^{2-} ions and Ca^{2+} ions in which Ti^{4+} ions occupy 1/4th of octahedral sites formed by O^{2-} ions *only*.

The crystal structure of ilmenite comprises of *hcp* arrangement of O^{2-} ions with 1/3rd of octahedral sites occupied by Ti^{4+} ions and another 1/3rd of octahedral sites occupied by Fe^{3+} ions.

Some titanates such as $MgTi_2O_4$ have a *spinel* structure in which O^{2-} ions form *ccp* arrangement, Mg^{2+} ions occupy the octahedral sites and Ti^{4+} ions occupy the tetrahedral sites.

It has a high refractive index and, therefore, its crystals show great briliance, even greater than diamond. However, being soft, it is not well suited for use in jewellery.

Titanium (IV) chloride, $TiCl_4$. It can be prepared in the laboratory by passing chlorine over a hot mixture of titanium dioxide and carbon.

$$TiO_2 + C + Cl_2 \longrightarrow TiCl_4 + CO_2$$

Properties

1. It is a colourless, fuming liquid at ordinary temperatures and boils at 137° C.

2. It gets readily hydrolysed by water yielding titanium dioxide.

$$TiCl_4 + 2H_2O \longrightarrow TiO_2 + 4HCl$$

If water is not present in excess, hydrolysis takes place only to a limited extent yiélding titanium oxochloride, $TiOCl_2$.

3. It is reduced to metalic titanium on treating with Mg or Na.

4. It dissolves in concentrated hydrochloric acid yielding chloro complexes having formulae $[TiCl_5(H_2O)]^-$ and $[TiCl_6]^{2-}$.

5. Titanium chloride is a strong *Lewis acid* and accepts electrons from strong oxygen-donors like ethers and alcohols yielding addition compounds of the type $(R_2O)TiCl_4$ and $(R_2O)_2TiCl_4$.

Uses. $TiCl_4$ is used for producing dense white smoke screens and for making writings in the sky in defence operations. This is because the spraying of liquid $TiCl_4$ into moist air results in its hydrolysis forming dense white smoke of TiO_2.

It is also used as a catalyst in the manufacture of polyethylene.

Titanium(IV) in solution. Ti^{4+} ion does not exist in aqueous solution but titanyl ion, TiO^{2+}, does exist and forms a number of salts such as $TiOSO_4$ which can be crystalised out from the aqueous solution containing TiO^{2+} and SO_4^{2-} ions.

Complexes of Ti(IV). The colorimetric estimation of TiO^{2+} or $[Ti(OH)_2]^{2+}$ in aqueous solution is carried out by the addition of H_2O_2 to the solution when an orange colour is developed due to the formation of a peroxo complex of Ti^{4+}.

Ti(IV) halides, $TiCl_4$ and TiF_4 react with HCl and HF, respectively, to give hexa coordinate complex ions:

$$TiCl_4 + 2HCl \longrightarrow H_2[TiCl_6]$$

$$TiF_4 + 2HF \longrightarrow H_2[TiF_6]$$

The tetrahalides also react with amines, phosphines and arsines to give hexa coordinate complexes, as, for example, $TiF_4.2PPh_3$.

Tetraalkyl ammonium chloride reacts with $TiCl_4$ to give $[Et_4N]^+ [TiCl_5]^-$ which contains five-coordinated Ti^{4+} in its anion.

Organometallic Compounds of Titanium. When $TiCl_4$ is added to $Al(C_2H_5)_3$ in hexane, it yields a brown solid known as **Ziegler–Natta catalyst** which is used in the conversion of ethene to straight chain polymer polyethene. The probable mechanism is given on the next page.

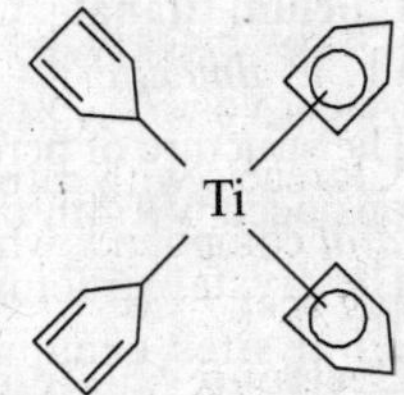

Fig. 1.4. Cyclopentadienyls of titanium.

Several stable cyclopentadienyls, such as $[Ti(\eta^5 - (C_5H_5)_2 (CO)_2]$, $Ti(\eta^5 - C_5H_5)_2]$, etc., are known. More interesting are the compounds in which Ti is attached to four C_5H_5 rings, as shown in Fig. 1.4.

Some alkyls and aryls of Ti are also known but these are quite unstable.

Mechanism of polymerisation of ethene

$$Cl-\underset{Cl}{\overset{Cl}{\underset{|}{\overset{|}{Ti}}}}Cl + Al(C_2H_5)_3 \longrightarrow Cl-\underset{Cl}{\overset{CH_2-CH_3}{\underset{|}{\overset{|}{Ti}}}}-Cl \xrightarrow{C_2H_4}$$

(Ziegler-Natta catalyst)

$$Cl-\underset{Cl}{\overset{CH_2-CH_3}{\underset{|}{\overset{|}{Ti}}}}\begin{matrix} Cl \\ \cdots CH_2 \\ \| \\ CH_2 \end{matrix} \longrightarrow Cl-\underset{Cl}{\overset{CH_3 - CH_2}{\underset{|}{\overset{|}{Ti}}}}\begin{matrix} Cl \\ CH_2 \\ CH_2 \end{matrix} \longrightarrow \overset{CH_2-CH_2-CH_2-CH_3}{\underset{Cl \quad Cl}{Ti}}\,Cl$$

and so on.

1.15. VANADIUM

Properties

1. The pure metal is silvery-grey in colour. It melts at 2173K. The melting point rises appreciably on the addition of carbon. It is *hard* and *corrosion-resistant.*

2. Action of acids. Vandium is unreactive towards mineral acids at ordinary temperatures. It, however, dissolves in hot nitric acid, concentrated sulphuric acid and aqua regia to form meta vanadic acid, HVO_3. The reaction with HNO_3 is

$$V + 5HNO_3 \longrightarrow 2H_2O + 5NO_2 + HVO_3$$

It is also attacked by hydrofluoric acid.

3. Action of alkalis. When fused with alkalis, it gives the corresponding vanadate evolving hydrogen.

$$2V + 6NaOH + 2H_2O \longrightarrow 2Na_3VO_4 + 5H_2$$

4. Combination with non-metals. When heated, it combines with oxygen forming vanadium pentoxide, mixed with lower oxides.

$$4V + 5O_2 \longrightarrow 2V_2O_5$$

5. It combines with nitrogen at high temperature yielding vanadium nitride, VN, which is an interstitial compound. It also forms a sulphide V_2S_3 by passing CS_2 over V_2O_5.

6. It combines with halogens on heating giving compounds like VF_5, VCl_4, VBr_3 and VI_3. These are covalent volatile compounds and get easily hydrolysed by water.

7. It reacts with carbon at high temperature to form carbides, VC and VC_2. The first one is an *interstitial* compound, white and hard. The second one is an *ionic compound* and reacts with water liberating acetylene.

Electronic Configuration and Oxidation States. The outer electronic configuration of vanadium is $3d^3\ 4s^2$. It shows oxidation states of +2, +3, +4 and +5 in its compounds. The compounds in the lower oxidation states are good reducing agents. They are also ionic in character and on account of incomplete electron shells, they are also coloured.

In +5 oxidation state, the *d*-subshell is completely empty. The compounds of vanadium corresponding to this state, therefore, are colourless.

When a colourless compound in which vanadium is in +5 state is reduced with zinc, it first turns *blue*, then *green* and finally *violet*, corresponding to progressive reduction to +4, +3 and +2 oxidation states. Thus, oxidation state of vanadium can be determined by noting the colour of its compounds.

Compounds of Vanadium(IV)

The +4 is the most stable oxidation state of vanadium. It forms a number of compounds in which its oxidation state is +4. Important amongst these are *vanadium dioxide*, VO_2, *halides* like VCl_4 and *oxohalides* like $VOCl_2$. The oxide and the halides are *covalent* while oxohalides are *ionic* in character containing the blue VO^{2+} ion.

Vanadium(IV) oxide, VO_2. It is formed by the reduction of vanadium pentoxide by hydrogen sulphide, sulphur dioxide or oxalic acid.

$$V_2O_5 + H_2S \longrightarrow 2VO_2 + H_2O + S$$

It is a blue amphoteric solid with *rutile structure*. It reacts with non-oxidising acids to form vanadyl and vanadium salts.

$$VO_2 + 2HCl \longrightarrow \underset{\text{Vanadyl chloride}}{VOCl_2} + H_2O$$

$$VO_2 + 4HCl \longrightarrow \underset{\text{Vanadium chloride}}{VCl_4} + 2H_2O$$

It reacts with alkalis to form vanadites.

$$VO_2 + 2NaOH \longrightarrow \underset{\text{Sodium vanadite}}{Na_2VO_3} + H_2O$$

Vanadium(IV) halides. Vanadium tetrachloride can be readily prepared by passing chlorine over powdered vanadium metal at 453 K

$$V + 2Cl_2 \longrightarrow VCl_4$$

It is a dark, reddish-brown oily *liquid* boiling at 427 K. It is unstable and slowly decomposes into trichloride and chlorine.

$$2VCl_4 \longrightarrow 2VCl_3 + Cl_2$$

Vanadium tetrachloride fumes in moist air and is hydrolysed by water giving *vanadyl chloride.*

$$VCl_4 + H_2O \rightleftharpoons VOCl_2 + 2HCl$$

VCl_4 is tetrahedral in gaseous phase and dimeric in the liquid phase.

Vanadium(IV) oxodichloride (Vanadyl chloride), $VOCl_2$, is obtained by the hydrolysis of VCl_4 as shown by the equation:

$$VCl_4 + H_2O \longrightarrow VOCl_2 + 2HCl$$

It can also be prepared by the action of HCl on V_2O_5.

$$V_2O_5 + 6HCl \longrightarrow 2VOCl_2 + 3H_2O + Cl_2$$

Pure $VOCl_2$ is prepared by heating $VOCl_3$ with metallic zinc in a sealed tube.

$$2VOCl_3 + Zn \longrightarrow 2VOCl_2 + ZnCl_2$$

The compound is obtained as grassy green crystals.

Vanadyl chloride is used commercially in dyeing. It is a strong reducing agent.

Vanadium(IV) oxosulphate (Vanadyl sulphate), $VOSO_4$. This is prepared by the reduction of a solution of vanadium pentoxide in concentrated sulphuric acid by sulphur dioxide. The resulting blue solution on evaporation gives bright blue crystals of the hydrated acid vanadyl sulphate, $2VOSO_4.H_2SO_4.xH_2O$, the exact composition depending upon the temperature.

When acid vanadyl sulphate is heated with concentrated sulphuric acid at 533K, a grey-green crystalline powder of *anhydrous vanadyl sulphate*, $VOSO_4$, separates out.

Vanadyl sulphate is used for potentiometric titrations of weak oxidising agents. Its acidic solution is used for removing traces of oxygen from gases.

Vanadium(IV) Complexes. Vanadium tetrachloride and tetrafluoride form complexes of the types VCl_4L and VCl_4L_2, where L stands for a unidentate ligand. These complexes are readily hydrolysed by water.

Several other complexes of V(IV) are also known. Some of these complexes are of the type $[VOX_4]^{2-}$, $[VO(acac)_2]$ and $[VO(bipy)_2Cl]^+$ where X is the halogen anion, bipy is bipyridyl and acac is acetyl acetonate anion. These complexes have square pyramidal structures.

Compounds of Vanadium(V). Vanadium forms only a few stable compounds in which the oxidation state is +5.

Vanadium(V) oxide, V_2O_5. This is the most important compound of vanadium(V). This is obtained as a red or orange solid on thermal decomposition of ammonium metavanadate, NH_4VO_3 as given below:

$$2NH_4VO_3 \longrightarrow V_2O_5 + 2NH_3 + H_2O$$

Vanadium pentoxide is also formed as a brick red precipitate when ammonium vanadate solutions are treated with dilute sulphuric acid.

It is also prepared by heating the metal in oxygen.

It melts at about 923 K and is insoluble in water.

Vanadium(V) oxide has all its outer electrons fully paired in covalent bonds. There is no incomplete electron shell. The compound, therefore, should be colourless. But actually it is orange-red in colour. The reason behind this lies in *defects in the solid state* due to which charge transfer transitions in the visible region are possible.

It is *amphoteric* in character. It dissolves in acid solutions to give salts such as **$VOCl_2$, vanadyl chloride or VCl_5, vanadium pentachloride.**

$$V_2O_5 + 6HCl \longrightarrow 2VOCl_2 + 3H_2O + Cl_2$$

$$V_2O_5 + 10HCl \longrightarrow 2VCl_5 + 5H_2O$$

Chlorine is liberated on dissolving V_2O_5 in hydrochloric acid. This shows that V_2O_5 can act as a moderately strong oxidising agent.

It dissolves in alkalis giving vanadates.

$$V_2O_5 + 6NaOH \longrightarrow \underset{\text{Sod. vanadate}}{2Na_3VO_4} + 3H_2O$$

The vanadate ion, VO_4^{3-}, is stable only in strongly alkaline solutions. On lowering the *p*H, the VO_4^{3-} ion polymerises. The products formed are known as *isopolyvanadates*.

$$VO_4^{3-} \xrightarrow{pH\ 12} [VO_3.OH]^{2-} \xrightarrow{pH\ 10} [V_2O_6.OH]^{3-} \xrightarrow{pH\ 7} [V_3O_9]^{3-}$$

$$[V_3O_9]^{3-} \xrightarrow{pH\ 7} [V_5O_{14}]^{3-}$$

$$[VO_2]^+ \xleftarrow{pH < 1} [V_{10}O_{28}]^{6-} \xleftarrow{pH < 2} [V_2O_5(H_2O)_9] \xleftarrow{pH\ 6.5} [V_5O_{14}]^{3-}$$

Vanadate ion also polymerises with the anions of other acids such as molybdate and tungstate anions.

The pentoxide on reduction yields lower oxides:

$$VO_2 \xleftarrow[Heat]{SO_2} V_2O_5 \xrightarrow[Heat]{H_2} V_2O_3 \xrightarrow[\text{at low pressure}]{\text{Heat with V}} VO$$

Vanadium pentoxide is used as a *catalyst* in various oxidation reactions involving the use of oxygen as the oxidising agent. The well known example is the conversion of sulphur dioxide to sulphur trioxide, in the manufacture of sulphuric acid by Contact process.

$$2SO_2 + O_2 \xrightarrow{V_2O_5} 2SO_3$$

The other catalytic reactions are oxidation of alcohol and hydrogenation of olefins.

Vanadium(V) fluoride. Vanadium combines with fluorine at high temperature to yield VF_5.

$$2V + 5F_2 \xrightarrow{573\ K} 2VF_5$$

It may also be prepared by heating VF_4 at 873 K in an atmosphere of nitrogen.

$$2VF_4 \xrightarrow[Nitrogen]{873\ K} VF_5 + VF_3$$

Vanadium(V) fluoride is a volatile white solid melting at 292.5 K. When heated in air it changes into oxyfluoride.

$$2VF_5 \xrightarrow[(air)]{O_2} 2VOF_3 + 2F_2$$

VF_5 is polymeric in solid and liquid phases consisting of VF_6 octahedral units joined together through V–F–V bridges, as shown in the Fig. 1.5.

In gaseous phase, VF_5 is monomeric and probably has trigonal bipyramidal structure.

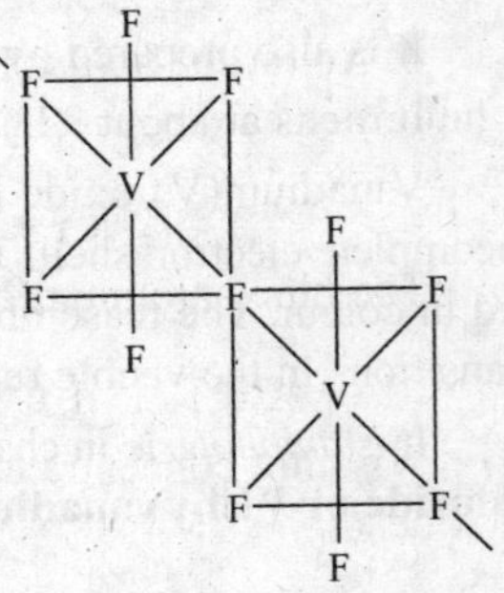

Fig. 1.5. Polymeric Vanadium fluoride

Vanadium(V) oxohalides. Vanadium oxotrichloride, $VOCl_3$, as well as oxotribromide, $VOBr_3$, can be prepared by halogenating V_2O_3 as under:

$$V_2O_3 + 3X_2 \longrightarrow 2VOX_3 + O_2$$

Vanadium oxotrichloride can be prepared more readily by passing chlorine over a mixture of vanadium pentoxide and carbon at red heat.

$$V_2O_3 + 3C + 3Cl_2 \longrightarrow 2VOCl_3 + 3CO$$

Vanadium oxotrifluoride, VOF_3, is also known. It can be prepared by treating oxotrichloride with hydrofluoric acid in the cold.

$$VOCl_3 + 3HF \longrightarrow VOF_3 + 3HCl$$

The oxohalides are simple covalent compounds and are readily hydrolysed as shown by the following equation:

$$VOCl_3 + 2H_2O \longrightarrow HVO_3 + 3HCl$$

Vanadium(V) Complexes. The halides and oxohalides form complexes with other metallic fluorides.

$$VF_5 + 2KF \longrightarrow K_2[VF_7]$$

$$VOF_3 + 3NaF \longrightarrow Na_3[VOF_6]$$

Anionic complexes containing fluorovanadate complex ion are formed by interaction of V_2O_5 with alkali metal fluorides in hydrofluoric acid solution. These complexes are of the type $M[VOF_4]$, $M_2[VOF_5]$ and $M_2[VO_2F_3]$, where M stands for an alkali metal such as Na or K.

1.16. CHROMIUM

Properties

1. Chromium is a bluish-white lustrous metal. It takes a good polish which lasts long because of the formation of an invisible, self-protective oxide coating. It is resistant to corrosive agents. The compounds of chromium are all coloured.

2. Action of air. Chromium is not affected by air at ordinary temperatures and does not tarnish in air. However, on heating in air or oxygen, it is oxidised superficially to green chromic oxide, Cr_2O_3. It burns with brilliance in oxygen at about 2270 K forming chromic oxide.

$$4Cr + 3O_2 \longrightarrow 2Cr_2O_3$$

It reacts with nitrogen also, at elevated temperatures, giving chromium nitride.

$$2Cr + N_2 \longrightarrow 2CrN$$

3. Action of water. Chromium is not at all attacked by water under ordinary conditions. It, however, decomposes steam at red heat to give chromic oxide and hydrogen.

$$2Cr + 3H_2O \longrightarrow Cr_2O_3 + 3H_2$$

4. Action of acids. The metal dissolves in dilute acids evolving hydrogen and giving blue solutions of chromium (II) salts. For example,

$$Cr + 2HCl \longrightarrow CrCl_2 + H_2$$

The blue chromium(II) salts change into green chromium(III) salts due to oxidation in air.

$$4CrCl_2 + 4HCl + O_2 \longrightarrow 4CrCl_3 + 2H_2O$$

With hot concentrated sulphuric acid, chromium gives chromium(III) sulphate and sulphur dioxide.

$$2Cr + 6H_2SO_4 \longrightarrow Cr_2(SO_4)_3 + 6H_2O + 3SO_2$$

Concentrated nitric acid renders the metal 'passive' due to the formation of an insoluble surface layer of chromium(III) oxide as a result of surface oxidation.

The metal is not attacked by alkalis under ordinary conditions.

5. Reducing properties. In the finely divided state, chromium acts as good reducing agent. This is due to the fact that in this form chromium readily reacts with oxygen.

In the undivided state, however, it gets protected by a thin coating of the oxide Cr_2O_3 and is, therefore, extremely resistant to corrosion.

6. Action of halogens. Heated chromium combines directly with fluorine or dry chlorine giving chromium(III) fluoride or chloride.

$$2Cr + 3F_2 \longrightarrow 2CrF_3$$

$$2Cr + 3Cl_2 \longrightarrow 2CrCl_3$$

Electronic Configuration and Oxidation States

The outer electronic configuration of chromium is $3d^5 4s^1$. It exhibits a large number of oxidation states, *viz.*, from 0 to 6. The zero oxidation state occurs in the carbonyl compounds, *e.g.*, $Cr(CO)_6$. These are coordination compounds in which the bonding electrons (lone pairs) are donated by the CO group. The metal electrons themselves remain unchanged and hence oxidation state is said to be zero.

The *stable* and *important oxidation states* of Cr are +2, +3 and +6. These arise from the transference or sharing of one or more of the *d*-electrons as well.

Compounds of Chromium(III). The +3 oxidation state is the most important and the most stable state of chromium. This state also forms a large number of complexes.

The +3 oxidation state is highly stable in acid solutions but it gets oxidised readily to +6 oxidation state in alkali solutions. Thus, Cr_2O_3 is oxidised to Na_2CrO_4 in alkaline medium.

$$2Cr_2O_3 + 8NaOH + 3O_2 \longrightarrow 4Na_2CrO_4 + 4H_2O$$

Compounds in +3 state are prepared by the reduction of Cr(VI) compounds such as chromates and dichromates.

Chromium(III) salts dissolve in water to give violet solution. The violet colour is due to the hydrated chromium(III) ion, $[Cr(H_2O)_6]^{3+}$. This is a highly stable ion and can be exposed to air without undergoing any oxidation or reduction. It has been shown that six water molecules surrounding the Cr(III) ion are located at the apices of a regular octahedron.

The solution of Cr(III) ion is *slightly acidic* on account of hydrolysis.

$$[Cr(H_2O)_6]^{3+} \rightleftharpoons [Cr(H_2O)_5OH]^{2+} + H^+$$

When a base is added to an aqueous solution containing chromium(III) ions, a dirty green precipitate of hydrated chromium(III) hydroxide, $Cr(OH)_3.xH_2O$, is obtained.

$$Cr^{3+} + 3OH^- \xrightarrow[\text{medium}]{\text{Aqueous}} Cr(OH)_3 \text{ (hydrated)}$$

$$\downarrow$$

$$Cr_2O_3.xH_2O$$

The percipitate dissolves when excess of the base is added giving a deep green solution which is characteristic of chromite ion.

$$Cr(OH)_3 + 3OH^- \longrightarrow [Cr(OH)_6]^{3-}$$

Chromium(III) chloride, $CrCl_3$. It is obtained in anhydrous state by passing chlorine either over heated chromium or over a mixture of chromium (III) oxide and carbon heated strongly.

$$Cr_2O_3 + 3C + 3Cl_2 \longrightarrow 2CrCl_3 + 3CO$$

It may also be obtained by passing sulphur chloride vapour over hot chromium (III) oxide.

$$2Cr_2O_3 + 6S_2Cl_2 \longrightarrow 4CrCl_3 + 3SO_2 + 9S$$

It is a pinkish violet solid almost insoluble in water but if even a trace of a chromium (II) salt is present, it dissolves readily giving a green solution.

$$2CrCl_3 \rightleftharpoons 2CrCl_2 + Cl_2$$

It forms three crystalline hexahydrates all having the molecular formula $CrCl_3.6H_2O$. Two of these are green (light green and dark green) in colour and the third is greyish blue. Their structural formulae are:

Dark green (α) *variety*	$[Cr(H_2O)_4Cl_2]Cl.2H_2O$
Light green (γ) *variety*	$[Cr(H_2O)_5Cl]Cl_2.H_2O$
Greyish blue (β) *variety*	$[Cr(H_2O)_6]Cl_3$

The Cl atoms inside the square bracket are attached to the chromium atom by covalent or coordinate bonds and Cl atoms outside the bracket are associated by ionic bonds. Chlorine ions outside the bracket are ionisable and can be precipitated by the addition of silver nitrate.

The coordination number of Cr(III) ion is evidently *six* in all these compounds.

The hexaaqua ion $[Cr(H_2O)_6]^{3+}$ is also present in a number of Cr(III) compounds other than $[Cr(H_2O)_6]Cl_3$. A well known example is chrome alum $(K_2SO_4.Cr_2(SO_4)_3.24H_2O)$ which should better be written as $[K(H_2O)_6]_2[Cr(H_2O)_6]_2(SO_4)_4$ or as $[K(H_2O)_6][Cr(H_2O)_6](SO_4)_2$.

Complexes of Cr(III). Apart from the abovementioned compounds, many complexes of Cr(III) are also known. Some of the common examples are $[Cr(NH_3)_6]^{3+}$, $[Cr(NH_3)_5Cl]^{2+}$, $[Cr(NH_3)Cl_3]$, $[Cr(oxalate)_3]^{3-}$ etc. The last complex shows optical isomerism.

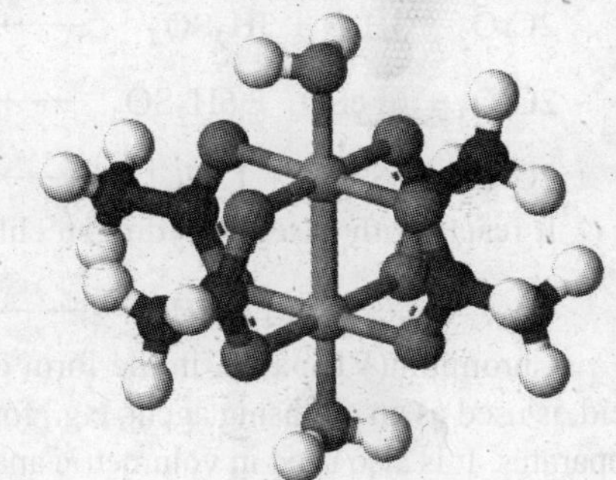

Chromium(II) acetate dimer

Chromium(III) Basic Acetate, $[Cr_3O(CH_3COO)_6(H_2O)_3]^+$. The structure of this complex cation comprises three Cr atoms in a *triangular plane.* These are bonded to an O atom in the middle of the plane. There are two acetato bridges across each edge of the triangular plane, i.e., between any two Cr atoms. The three H_2O molecules occupy the remaining three octahedral sites on the Cr atoms giving the following shape to the complex (Fig. 1.6).

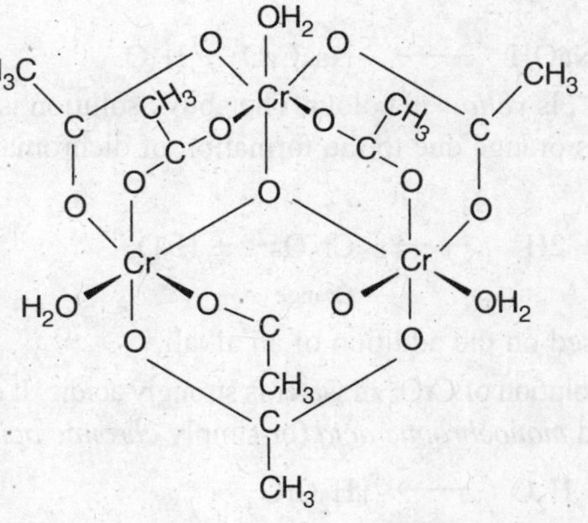

Fig. 1.6. Structure of chromium(III) basic acetate

Chromium(III) Oxide. It is a green solid which is used as a pigment and also as a catalyst in a number of reactions. It is conveniently prepared by heating ammonium dichromate.

$$(NH_4)_2Cr_2O_7 \xrightarrow{\Delta} Cr_2O_3 + N_2 + 4H_2O$$

Compounds of Chromium(VI)

There are only a few compounds of chromium in which its oxidation state is +6. The important ones are the oxide, CrO_3; oxochloride, CrO_2Cl_2; chromates, CrO_4^{2-} and dichromates, $Cr_2O_7^{2-}$.

Chromium(VI) oxide, CrO_3. It is the anhydride of the unknown chromic acid, H_2CrO_4, and is, therefore, called chromic anhydride or 'chromic acid'. It is obtained by the action of concentrated sulphuric acid on a concentrated solution of potassium dichromate.

$$K_2Cr_2O_7 + 2H_2SO_4 \longrightarrow 2KHSO_4 + H_2O + 2CrO_3$$

The mixture is allowed to cool when scarlet needle-like crystals separate out.

It is a red crystalline solid which volatilises at 383K and melts with some decomposition at 463K. On heating at 523K, it decomposes readily evolving oxygen.

$$4CrO_3 \longrightarrow 2Cr_2O_3 + 3O_2$$

It is readily soluble in water giving a red solution which is highly acidic and contains a number of chromic acids.

It is a powerful oxidising agent. It oxidises alcohols, glycerine, ether, oxalic acid, sugar, paper, etc. In acidic medium, it oxidises sulphur dioxide, hydrogen sulphide, stannous salts, ferrous salts, iodine, etc. Chromium(VI) oxide itself is reduced to a chromium(III) salt in every case. Some of the oxidising reactions are given below.

$$2CrO_3 + 3H_2S + 3H_2SO_4 \longrightarrow Cr_2(SO_4)_3 + 3S + 6H_2O$$

$$2CrO_3 + 6FeSO_4 + 6H_2SO_4 \longrightarrow Cr_2(SO_4)_3 + 3Fe_2(SO_4)_3 + 6H_2O$$

$$2CrO_3 + 3SnCl_2 + 12HCl \longrightarrow 2CrCl_3 + 3SnCl_4 + 6H_2O$$

It reacts with gaseous hydrogen chloride to give **chromyl chloride.**

$$CrO_3 + 2HCl \longrightarrow CrO_2Cl_2 + H_2O$$

Chromium(VI) oxide, in the form of a mixture of potassium dichromate and conc. sulphuric acid, is used as an oxidising agent, e.g., for converting ethanol to acetaldehyde and for cleansing glass apparatus. It is also used in volumetric analysis. The solution of chromium(VI) oxide in glacial acetic acid is used as an oxidising agent and as a disinfectant.

It dissolves in alkalis to give chromates. For example,

$$CrO_3 + 2NaOH \longrightarrow Na_2CrO_4 + H_2O$$

Chromate ion, CrO_4^{2-}, is *yellow* in colour. The above solution is, therefore, yellow. But when acidified, the solution turns orange due to the formation of dichromate ion, $Cr_2O_7^{2-}$, as shown by the following equation:

$$\underset{\text{Yellow}}{2CrO_4^{2-}} + 2H^+ \longrightarrow \underset{\text{Orange}}{Cr_2O_7^{2-}} + H_2O$$

The reaction is reversed on the addition of an alkali.

Chromic Acids. The solution of CrO_3 in water is strongly acidic. It contains several chromic acids, the simplest of which called *monochromic acid* (or simply *chromic acid*) is known only in solution.

$$CrO_3 + H_2O \longrightarrow H_2CrO_4$$

The higher chromic acids are supposed to be formed through condensation of 2, 3 or more molecules of chromic acid with the elimination of water.

$$2H_2CrO_4 \longrightarrow \underset{\text{Dichromic acid}}{H_2Cr_2O_7} + H_2O$$

$$3H_2CrO_4 \longrightarrow \underset{\text{Trichromic acid}}{H_2Cr_3O_{10}} + 2H_2O$$

The salts of H_2CrO_4 and $H_2Cr_2O_7$ are known as chromates and dichromates respectively.

Structures of chromate and dichromate ions

Chromate ion has a **tetrahedral structure** with Cr atom lying at the centre. Dichromate ion consists of *two such tetrahedra*, as shown below.

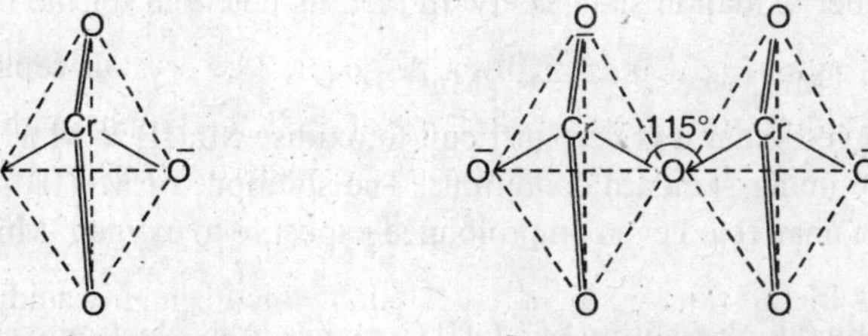

The oxygen atom commonly linked to the two chromium atoms subtends Cr—O—Cr angle equal to 115°.

1.17. MANGANESE

Properties

1. Manganese is a grey metal and exists in three alotropic forms which are stable over various temperature ranges. It is hard and brittle with a specific gravity of 7.43. It melts at 1518 K and boils at 2423 K. It is paramagnetic in nature.

2. Reaction with acids. The reduction potential of the reaction

$$Mn^{2+} + 2e^- \longrightarrow Mn$$

is –1.05 volt. Manganese is, therefore, fairly electropositive. It dissolves readily in dilute acids evolving hydrogen.

$$Mn + 2H^+ \longrightarrow Mn^{2+} + H_2$$

3. Action of water. Manganese decomposes water even in cold, though slowly, with the evolution of hydrogen. The reaction becomes quicker on heating.

$$Mn + 2H_2O \longrightarrow Mn(OH)_2 + H_2$$

Pure manganese, however, does not decompose water at ordinary temperatures.

4. Reactions with non-metals. Manganese is not so reactive towards non-metals at ordinary temperatures. However, when heated, it reacts vigorously. Thus, it combines with fluorine yielding the fluorides MnF_2 and MnF_3. It burns in chlorine to give the chloride $MnCl_2$ and combines with nitrogen to give the nitride Mn_3N_2. It combines with oxygen to give various oxides; with sulphur, it forms manganese sulphide, MnS.

Various Oxidation States of Mangancese

Oxidation State	Comon Compounds
0	$Mn(CO)_5$
+1	$[Mn(CO)_5]Cl$
+2	MnO, $Mn(OH)_2$, MnS, $MnSO_4$, $MnCl_2$
+3	MnF_3, Mn_2O_3, $MnPO_4$, $[Mn(CN)_6]^{3-}$
+4	MnO_2, MnF_4
+6	K_2MnO_4, MnO_3
+7	$KMnO_4$, Mn_2O_7

Aqueous Chemistry of Manganese in Different Oxidation States

Chemistry of Mn(II). Manganese changes readily into Mn^{2+} ion in acid solution on account of its fairly high reduction potential of –1.05 V. Unlike Cr^{2+} ion, Mn^{2+} ion is fairly stable and does not change into higher oxidation state easily. In fact, its potential for the reaction

$$Mn^{3+} + e^- \longrightarrow Mn^{2+}$$

is positive. This suggests that it is very difficult to oxidise Mn(II) to Mn(III). The reverse of it, on the contrary, is easy under standard conditions. The solutions of Mn(II) ions, in neutral or acidic medium, can remain unaffected even on prolonged exposure to oxygen. Mn(II) ion, therefore, is not a reducing agent.

The reason behind high stability of Mn(II) ions lies in its electronic configuration. Mn(II) ion is formed by the loss of both the 4*s*-electrons. Its configuration is, therefore, $3d^5$. This corresponds to exactly half-filled *d*-subshell which is considered a stable structure.

The aqueous solutions of Mn(II) ion, contain complex cation $[Mn(H_2O_6]^{2+}$ and are more or less colourless. However, manganese(II) salts like $MnCl_2$ or $MnSO_4$ have light pink coloration.

Complexes of Mn(II). Complexes of Mn(II) are well known. Some of the hydrated salts, such as $Mn(ClO_4)_2.6H_2O$ contain the complex ion, $[Mn(H_2O)_6]^{2+}$ in which manganese is in +2 oxidation state. The formation of hexaammine ion $[Mn(NH_3)_6]^{2+}$ is also reported by the direct action of ammonia on anhydrous salts of Mn(II). These complexes are *octahedral* and have pink colour. They have five unpaired electrons in 3*d*-subshell. Their mangetic moments are around 6 BM. They are highly stable and resist attack by oxidising as well as reducing agents.

A few octahedral complexes with low magnetic moment (*i.e., low spin complexes)* are also known. The examples are $[Mn(CN)_6]^{4-}$ and $[Mn(CN)_5NO]^{3-}$. They have only one unpaired electron and their magnetic moment is around 2 BM. These complexes, however, are not so stable and get osidised readily as, for example,

$$[Mn(CN)_6]^{4-} \longrightarrow [Mn(CN)_6]^{3-} + e^-$$

Chemistry of Mn(IV). Manganese in +4 state exists primarily as its oxide, MnO_2 which is commonly prepared by heating Mn in oxygen. The oxide is insoluble in water and reacts with hot concentreated hydrochloric acid and concentrated sulphuric acid to evolve Cl_2 and O_2, respecxtively.

$$MnO_2 + 4HCl \longrightarrow MnCl_2 + 2H_2O + Cl_2$$

$$2MnO_2 + 2H_2SO_4 \longrightarrow 2MnSO_4 + 2H_2O + O_2$$

When fused with NaOH, it gives Na_2MnO_4 which is a green solid.

MnO_2 is a strong oxidising agent and is commonly used in the oxidation of organic compounds. A large quantity of pure MnO_2 is used in dry batteries. It is also used in glass industry to produce purple glass. One of its major uses in industry is in the manufacture of potassium permanganate, as shown below:

$$MnO_2 \xrightarrow[\text{NaOH and KNO}_2]{\text{fuse with}} K_2MnO_4$$

$$2K_2MnO_4 + 2H_2O \xrightarrow[\text{oxidation}]{\text{Electrolytic}} 2KMnO_4 + 2KOH + 2H_2$$

MnO_2 is also used as a catalyst in the laboratory preparation of oxygen from $KClO_3$.

However, some complexes of Mn (IV) are well known. These include $K_2[MnF_6]$, $K_2[MnCl_6]$, $K_2[Mn(CN)_6]$, etc.

Chemistry of Manganese(VI). Manganese in +6 state exists in manganate ion, MnO_4^{2-}. The ion, however, is stable only in highly alkaline solutions. In acidic solutions, it changes readily into permanganate ion in which manganese exists in the highest (+7) state.

$$3MnO_4^{2-} + 4H^+ \longrightarrow MnO_2(s) + 2MnO_4^- + 2H_2O$$

Manganate ion changes into permanganate ion even by passing carbon dioxide in its solution, as shown by the following reaction:

$$3MnO_4^- + 2H_2O + 4CO_2 \longrightarrow MnO_2 + 2Mn_4^- + 4HCO_3^-$$

Chemistry of Manganese (VII). Manganese (VII) is best known as permanganate, MnO_4^-, ion. It is one of the *strongest oxidising agents.* The aqueous chemistry of Mn(VII) is, in fact, the chemistry of MnO_4^- ion.

1.18. COBALT

Properties

1. Cobalt is a hard metal, white in colour and is malleable and ductile. It is *ferromagnetic* but much less so than iron. It has a density of 8.9 g per cm^3 and melts at 1768 K. It is not attacked by air but is oxidised on heating.

2. Reaction with acids. Cobalt is slightly electropositive, the standard reduction potential of the reaction,

$$Co^{2+} + 2e^- \longrightarrow Co$$

being – 0.28 volt. It is, therefore, slowly attacked by dilute hydrochloric acid and sulphuric acid evolving hydrogen and forming cobalt(II) salts. Thus,

$$Co + H_2SO_4 \longrightarrow CoSO_4 + H_2$$

It is attacked rapidly by dilute nitric acid.

$$3Co + 8HNO_3 \longrightarrow 3Co(NO_3)_2 + 2NO + 4H_2O$$

It is rendered passive on treatment with concentrated nitric acid.

Electronic Configuration and Oxidation States. The outer electronic configuration of cobalt is $3d^7\ 4s^2$. Its two important oxidation states are +2 and +3. Cobalt also shows oxidation state of zero, like iron, in the formation of carbonyls.

Aqueous Chemistry of Cobalt in Different Oxidation States. Cobalt has two important oxidation states of +2 and +3.

Chemistry of Cobalt(II). Cobalt(II) ion, Co^{2+}, is more stable than Fe^{2+} ion as it cannot be oxidised easily to cobalt(III) state which is much less stable. All cobalt(II) salts are, therefore, stable, However, in basic solutions, oxidation of Co^{2+} to Co^{3+} takes place rather easily. Co^{2+} ion, in water, exists in the hydrated state as $[Co(H_2O)_6]^{2+}$ and is pink in colour. It turns blue when heated due to dehydration. Thus, the *anhydrous* Co(II) salts are blue in colour. Cobalt chloride, $CoCl_2$, is blue when anhydrous. It turns pink when hydrated. Anhydrous $CoCl_2$ is, therefore, used for *testing moisture.* Most of the cobalt(II) compounds are soluble in water.

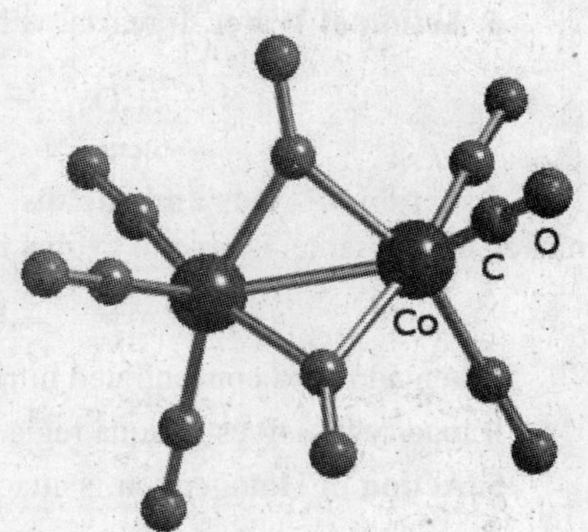

Cobalt complex

When cobalt(II) hydroxide is heated in the *presence of air*, **cobalt(II)-cobalt(III) oxide,** Co_3O_4, is obtained. This oxide has a normal spinel structure with Co^{2+} ions in tetrahedral and Co^{3+} ions in octahedral sites created due to *ccp* arrangement of oxide ions in structure.

Complexes

Co(II) forms a number of complexes which are not as stable as the corresponding complexe of Co(III). It forms tetrahedral complex ions $[CoX_4]^{2-}$ where X is Cl^-, Br^-, I^-, OH^-, etc. When a solution of NH_4SCN is added to an aqueous solution containing Hg^{2+} and Co^{2+} ions, the tetrahedral complex $Hg[Co(NCS)_4]$ is obtained. This blue coloured complex is used for calibrating magnetic balance.

Co(II) also forms a square planar complex with dimethylglyoxime (DMG) whose structure is similar to that of the corresponding Ni(II) complex Ni–DMG.

Chemistry of Cobalt(III). The cobalt(III) ion, Co^{3+}, is a strong oxidising agent and, therefore, gets itself reduced to cobalt(II) state easily. The *reduction potential* of the half reaction,

$$Co^{3+} + e^- \longrightarrow Co^{2+}$$

is quite high, being about 1.84 volts. Such a high value shows that Co^{3+} *can oxidise water to oxygen rapidly.* Consequently, *there is no aqueous chemistry of* Co(III) as it gets reduced readily to Co(II) state.

$$4Co^{3+} + 2H_2O \longrightarrow 4Co^{2+} + O_2(g) + 4H^+$$

Only a few simple cobalt(III) salts are known. The important ones are cobalt(III) fluoride, CoF_3 and cobalt(III) sulphate, $Co_2(SO_4)_3.18H_2O$. These, too, contain the complex ion $[Co(H_2O)_6]^{3+}$ with the solvent or water of crystallisation and decompose in aqueous solution evolving oxygen and producing cobalt(II) ion.

The Co^{3+} ion can exist in stable state only if it is complexed. It is then not easily reduced.

1.19. NICKEL

Properties

1. Nickel is a greyish-white metal with a density of 8.9 g per c.c. It is hard and malleable and melts at 1726 K. It is ferromagnetic though much less than iron. It has high electrical and thermal conductivities.

2. Action of Air and Water. Nickel is quite resistant to attack by air or water at ordinary temperatures. This is due to the formation of a thin protective film of oxide on the surface of the metal. Nickel is particularly resistant to attack in alkaline media. In powder state, however, it is fairly reactive towards air.

3. Action of Water. It reacts with steam at a high temperature giving hydrogen.

$$Ni + \underset{\text{Steam}}{H_2O} \longrightarrow NiO + H_2$$

4. Action of Acids and Alkalis. Due to neagative value (– 0.25 volt) of reduction potential, nickel, all the same, is able to evolve hydrogen from dilute mineral acids. Thus,

$$Ni + 2H^+ \longrightarrow Ni^{2+} + H_2$$

When added to concentrated nitric acid, the metal becomes *passive* like iron and cobalt.

It readily dissolves in aqua regia. It resists the attack of alkalis.

5. Action of Halogens. It is attacked by halogens forming metal halides. For example,

$$Ni + Cl_2 \longrightarrow NiCl_2$$

6. Oxidation. Nickel is more inert to chemical oxidation than iron and cobalt. The standard reduction potential of the change.

$$Ni^{2+}\,(aq) + 2e^- \longrightarrow Ni\,(s)$$

is – 0.25 volt.

Electronic Configuration and Oxidation States. The outer electronic configuration of Ni is $3d^8\,4s^2$. It shows oxidation states of 0, –1, +1, +2, +3 and +4. The anion $[Ni_2(CO)_6]^{2-}$ contains nickel in –1 oxidation state whereas $[Ni(CO)_4]$, $[Ni(PF_3)_2(CO)_2]$ and $K_4[Ni(CN)_4]$ contain nickel in zero oxidation state. Nickel in + 1 state is present in $K_4[Ni_2(CN)_6]$.

The +2 state ($3d^8$) is the only stable state. A few compounds of trivalent nickel ($3d^7$) which are known, are very strong oxidising agents since nickel(III) has a strong tendency to change into nickel(II). Nickel(IV) occurs in certain oxide systems and in a few complexes.

Aqueous Chemistry of Nickel(II). With the general tendency of *decreased stability of higher oxidation states with increase in atomic number*, nickel shows stable oxidation state of +2 only. It may be noted that in the case of iron, both +2 and +3 states are stable. In the case of cobalt, both states are known but +3 state is much less stable. When we come to nickel, the +3 state becomes still less stable.

Nickel(II) ion, Ni^{2+}, is generally hydrated and exists as $[Ni(H_2O)_6]^{2+}$. It is green in colour. The salts of Ni^{2+} are green or blue.

Nickel(II) ion gets precipitated from aqueous solutions as a green gel on the addition of sodium or potassium hydroxide.

$$Ni^{2+} + 2OH^- \longrightarrow Ni(OH)_2$$

The precipitate dissolves in aqueous ammonia due to formation of the complex ion $[Ni(NH_3)_6]^{2+}$. When treated with a mild oxidising agent like bromine in an alkaline medium, a black impure oxide with composition close to $Ni_2O_3.2H_2O$ is precipitated. When treated with strong oxidising agents like chlorine, an impure oxide with composition close to NiO_2 is precipitated. Thus, it is possible to prepare oxides in which nickel is in +3 and +4 states.

Ni^{2+} forms an interesting complex with dimethyl-glyoxime known as *bis(dimethyl glyoxime) nickel(II)*. It is electrically neutral and is red in colour. It is quantitatively precipitated from ammoniacal solution of nickel(II) salt on the addition of dimethylglyoxime. This reaction is used as a *confirmatory test of nickel(II) in qualitative analysis and also in the estimation of nickel.* The formation of this neutral complex may be represented as follows:

```
                 OH                       HO
                /                           \
CH3 — C = N                                  N = C — CH3
      |               + Ni2+ +                   |
CH3 — C = N                                  N = C — CH3
                \                           /
                 OH                       HO
Dimethylglyoxime

                         OH······O
                        /         ↖
                      N             N
                    //  \         /  \\
         H3C — C          ↘     /        C — CH3
  ⟶ 2H+ +      |            Ni           |
         H3C — C          ↗     \        C — CH3
                    \\  /         \  //
                      N             N
                       ↘           /
                         O······HO
```

Ni-DMG (Scarlet red complex)

The complex, bis(dimethylglyoximato)nickel (II) is square planar and diamagnetic. The complex ion $[Ni(CN)_4]^{2-}$ is also square planar and diamagnetic.

Besides $[Ni(H_2O)_6]^{2+}$, some other octahedral complexes of Ni(II) are $[Ni(NH_3)_6]^{2+}$, $[Ni(NH_3)_4(H_2O)_2]^{2+}$, $[Ni(ethylenediamine)_3]^{2+}$, etc. The octahedral Ni(II) complexes are generally paramagnetic.

1.20. COPPER

Properties

1. Copper is a reddish brown metal which melts at 1356 K. It is malleable and ductile and can be drawn into wires. It is an *excellent conductor of heat and electricity*, the best conductor next to silver. The presence of other metals even in small quantities lowers the conductivity appreciably.

2. Action with Air or Oxygen. It is not attacked by dry air at ordinary temperatures. But when heated to redness in air or oxygen, it first forms copper(I) oxide, Cu_2O and finally copper(II) oxide, CuO. If the temperature is too high, copper (II) oxide changes back to copper(I) oxide, Cu_2O.

$$CuO + Cu \longrightarrow Cu_2O$$

In the presence of atmospheric carbon dioxide and moisture it is covered with a green layer of basic carbonate, $CuCO_3.Cu(OH)_2$. This layer protects the rest of the metal from further action.

Copper Metal

3. Action with Water. Copper does not decompose water or steam at ordinary temperatures. However, at very high temperature, steam acts upon it to give the oxide.

$$Cu + H_2O \xrightarrow{\text{White heat}} CuO + H_2$$

4. Action with Acids. Non-oxidizing acids (e.g., hydrochloric acid and dilute sulphuric acid) have no effect in the absence of air or oxygen.

In the presence of air, these acids dissolve the metal forming the corresponding salt and water.

$$Cu + 2HCl + \frac{1}{2}O_2 \text{ (air)} \longrightarrow CuCl_2 + H_2O$$

$$Cu + H_2SO_4 + \frac{1}{2}O_2 \text{ (air)} \longrightarrow CuSO_4 + H_2O$$

Oxidizing acids e.g., nitric acid and concentrated (H_2SO_4) readily attack copper, yielding in each case the copper (II) salts, water and a reduction product of the acid.

$$Cu + 2H_2SO_4 \text{ (conc.)} \longrightarrow CuSO_4 + 2H_2O + SO_2$$

$$3Cu + 8HNO_3 \text{ (dilute)} \longrightarrow 3Cu(NO_3)_2 + 4H_2O + 2NO$$

$$Cu + 4HNO_3 \text{ (conc.)} \longrightarrow Cu(NO_3)_2 + 2H_2O + 2NO_2$$

5. Action with Aqueous Ammonia. Copper is dissolved by aqueous ammonia in the presence of air or oxygen, due to formation of the complex ion, $[Cu(NH_3)_4]^{2+}$.

$$2Cu + 2H_2O + O_2 + 8NH_3 \longrightarrow 2[Cu(NH_3)_4]^{2+} + 4OH^-$$

6. Action with Non-metals. Copper combines with chlorine and other halogens to form halides.

$$Cu + Cl_2 \longrightarrow CuCl_2$$

It combines with sulphur when heated giving a non-stoichiometric compound of the composition $Cu_{1.87}S$.

7. Displacement of Silver and other less Electropositive Metals. Copper displaces silver and mercury from solutions of their salts.

$$Cu + 2Ag^+ \longrightarrow Cu^{2+} + 2Ag$$

8. Action with Cyanides. It also dissolves in aqueous solution of potassium cyanide yielding the complex, $[Cu(CN)_4]^{2-}$.

Electronic Configuration and Oxidation States. The outer electronic configuration of copper is $3d^{10}\,4s^1$. It shows oxidation states of +1 ($3d^{10}$) and +2 ($3d^9$) in its compounds, known as *cuprous* and *cupric* compounds. The +1 state is not stable. It undergoes almost complete disproportionation as

$$2Cu^+ \rightleftharpoons Cu^{2+} + Cu$$

The +2 state, therefore, is much more stable. The +1 state is stable only in insoluble and complex compounds.

Aqueous Chemistry of Copper in Different Oxidation States

Chemistry of Copper(I). Copper(I) ion, Cu^+, has the electronic structure $3d^{10}$. All the *d*-orbitals are completely filled and the electrons are fully paired. The copper(I) compounds are, therefore, **diamagnetic.** Copper(I) ion in solution changes readily to copper (II) ion by disproportionation as shown below

$$2Cu^+\,(aq) \longrightarrow Cu^{2+}\,(aq) + Cu(s)$$

The reduction potentials for the half-reactions involved are given below:

$$Cu^+ + e^- \longrightarrow Cu(s) \qquad E° = +0.52 \text{ volt} \quad ...\,(i)$$

$$Cu^{2+} + e^- \longrightarrow Cu^+ \qquad E° = +0.18 \text{ volt} \quad ...\,(ii)$$

Subtracting eq. (*ii*) from eq. (*i*), we get

$$2Cu^{2+}(aq) \longrightarrow Cu^{2+}(aq) + Cu(s) \qquad E° = +0.34 \text{ volt}$$

The positive value of $E°$ explains why the disproportionation takes place.

This also explains the fact that when the insoluble copper(I) oxide, Cu_2O, is dissolved in dilute sulphuric acid, the product is copper (II) sulphate and *not* copper(I) sulphate.

$$Cu_2O\,(s) + 2H^+ \longrightarrow Cu^{2+} + Cu(s) + H_2O$$

This is again a case of disproportionation. The above reaction is considered to take place in two steps.

$$Cu_2O\,(s) + 2H^+ \longrightarrow 2Cu^+ + H_2O$$

$$2Cu^+ \longrightarrow Cu^{2+} + Cu\,(s)$$

The Cu^+ ion can be stabilised by formation of insoluble substances or complex ions. For example, copper (I) ion can form the insoluble **copper (I) chloride** when brought in contact with chloride ions.

$$2Cu^+ + 2Cl^- \longrightarrow 2CuCl\,(s)$$

The solubility product of CuCl is quite low ($K_{sp} = 3.2 \times 10^{-7}$) and, therefore, there can only be traces of copper (I) ions in solution.

The precipitated copper (I) chloride dissolves in excess of chloride ions yielding the complex ion $[CuCl_2]^-$.

$$CuCl + Cl^- \longrightarrow [CuCl_2]^-$$

If chloride ion is added as potassium chloride, the reaction shall be represented as

$$CuCl + KCl \longrightarrow K[CuCl_2]$$

When CuCl is heated in a stream of ethylene, it forms the stable compound $Cu(C_2H_4)Cl$.

The insoluble copper(I) salts can also be prepared by the reduction of copper (II) salts by iodide ions.

$$2Cu^{2+} + 4I^- \longrightarrow \underset{\text{(Insoluble)}}{2CuI} + I_2$$

Thus, when potassium iodide is added to a solution of copper sulphate, a precipitate of **copper (I) Iodide** mixed with free iodine is obtained.

This reaction involves two steps:

Iodide ion first reduces copper(II) to copper(I) state.

$$2Cu^{2+} + 2I^- \longrightarrow 2Cu^+ + I_2$$

This is followed by the precipitation of copper (I) iodide.

$$2Cu^+ + 2I^- \longrightarrow \underset{\text{(Ppt.)}}{2CuI}$$

This reaction is used for *estimating copper* volumetrically. Molecular iodine formed in the process is estimated by titrating against sodium thiosulphate.

It may be noted that chloride ion cannot bring about this reaction since it is not a reducing agent.

Aqueous Chemistry of Copper(II). The aqueous chemistry of copper in +2 oxidation state is similar to that of +2 ions of the other transition elements. Although most of the *anhydrous* copper(II) salts are colourless, they are blue when hydrated. This is due to the presence of the complex ion $[Cu(H_2O)_4]^{2+}$ which is *blue*. Solutions of copper(II) salts are acidic due to salt hydrolysis.

$$Cu^{2+} + H_2O \rightleftharpoons [Cu(OH)]^+ + H^+$$

The outer electronic configuration of copper(II) is $3d^9$. It has thus an unpaired electron. *The copper(II) compounds are, therefore, paramagnetic*. They are blue or green in colour.

When a base is added to a solution containing Cu^{2+} ions, copper(II) hydroxide, a light blue precipitate is formed.

$$Cu^{2+} + 2OH^- \longrightarrow Cu(OH)_2$$

If some aqueous ammonia is added, copper(II) hydroxide dissolves giving an intense blue solution due to formation of a complex.

$$Cu(OH)_2\ (s) + 4NH_3 \longrightarrow [Cu(NH_3)_4]^{2+} + 2OH^-$$

The copper-ammonia complex decomposes on the addition of acids.

$$[Cu(NH_3)_4]^{2+} + 4H^+ \longrightarrow Cu^{2+} + 4NH_4^+$$

1.21. ZINC

Properties

1. Action of Air. When heated in air to about 770 K, it burns to form a white cloud of zinc oxide which settles to form a woolly flock generally called 'philosophers wool'.

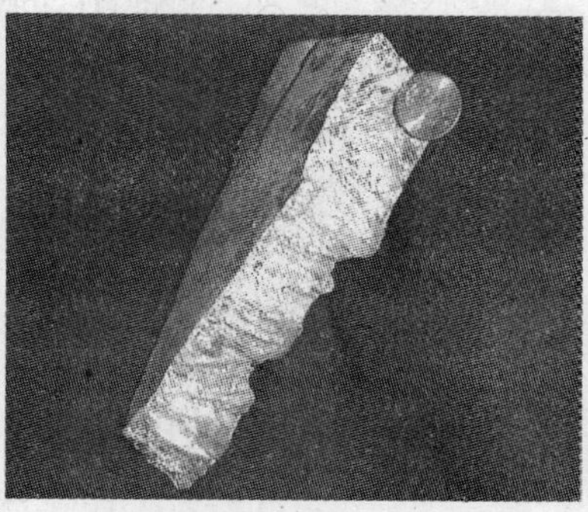

Zinc Metal

$$2Zn + O_2 \longrightarrow 2ZnO$$

It forms basic carbonate in moist air which decreases its lustre and forms a bluish grey film on its surface.

$$4Zn + 3H_2O + 2O_2 + CO_2 \longrightarrow ZnCO_3.3Zn(OH)_2$$

2. Action of Water. Pure zinc does not react with water but impure zinc decomposes steam quite readily with the evolution of hydrogen.

$$Zn + H_2O \longrightarrow ZnO + H_2$$

3. Action of Mineral Acids. Pure zinc does not react with dilute mineral acids but impure metal does react with different acids. The inertness of pure zinc is due to the very large overvoltage of hydrogen at the zinc surface. But when mixed with metals like copper, silver and platinum, on the surface of which hydrogen gas has a low over-voltage, zinc reacts readily evolving hydrogen on the surface of the other metal, with the simultaneous dissolution of zinc.

$$Zn + 2H^+ \longrightarrow Zn^{2+} + H_2$$

If pure zinc is coated with copper, it reacts readily with dilute acids evolving hydrogen. *Zinc-copper couple* is, therefore, used as a reducing agent in a number of organic reactions.

The reaction with sulphuric acid depends upon the concentration of the acid and the temperature of the reaction. If the acid is dilute and cold, zinc sulphate and hydrogen are formed.

$$Zn + H_2SO_4 \longrightarrow ZnSO_4 + H_2$$

If the acid is concentrated and hot, zinc gives sulphur dioxide.

$$Zn + 2H_2SO_4 \longrightarrow ZnSO_4 + 2H_2O + SO_2$$

The reaction with nitric acid is varied. If the acid is very dilute, ammonium nitrate results.

$$Zn + 10HNO_3 \longrightarrow 4Zn(NO_3)_2 + NH_4NO_3 + 3H_2O$$

If the acid is moderately strong, nitric oxide is formed.

$$3Zn + 8HNO_3 \longrightarrow 3Zn(NO_3)_2 + 4H_2O + 2NO$$

If the acid is concentrated and hot, nitrogen dioxide is formed.

$$Zn + 4HNO_3 \longrightarrow Zn(NO_3)_2 + 2H_2O + 2NO_2$$

4. Action with Alkyl Halides. Zinc, like magnesium, forms addition compounds with alkyl halides.

$$Zn + C_2H_5I \longrightarrow ZnC_2H_5I$$

The compound decomposes on heating to form zinc alkyl.

$$2ZnC_2H_5I \longrightarrow \underset{\text{Zinc diethyl}}{Zn(C_2H_5)_2} + ZnI_2$$

5. Action of Alkalis. It reacts with moderately strong alkalis on heating and forms soluble zincates.

$$Zn + 2NaOH \longrightarrow Na_2ZnO_2 + H_2$$

Aqueous Chemistry of Zn(II). Zinc shows +2 oxidation state in all its compounds.

Zinc ion ($3d^{10}$) is colourless because of completely filled *d*-orbitals. It undergoes slight hydrolysis in water giving acidic solution.

$$Zn^{2+} + H_2O \rightleftharpoons Zn(OH)^+ + H^+$$

The degree of hydrolysis is very small, even smaller than that of Cu^{2+} ion. Thus, a solution of Zn^{2+} ions is much less acidic than that of Cu^{2+} ions of the same concentration.

When a base is added to a solution of Zn^{2+} ions, a precipitate of zinc hydroxide is formed.

$$Zn^{2+} + 2OH^- \longrightarrow Zn(OH)_2$$

Zinc hydroxide is amphoteric. It dissolves in dilute acids forming salts.

$$Zn(OH)_2 + 2HCl \longrightarrow ZnCl_2 + 2H_2O$$

It dissolves also in alkalies forming the complex ion, $[Zn(OH)_4]^{2-}$

$$Zn(OH)_2 + 2NaOH \longrightarrow Na_2[Zn(OH)_4]$$

The formation of the complex ions from zinc in alkaline solutions may be represented by the following reaction:

$$Zn(s) + 4OH^- \longrightarrow [Zn(OH)_4]^{2-} + 2e^-$$

The reduction potential for this half-reaction is –1.22 volts, while that in acid solution it is only – 0.76 volt. This shows that zinc metal acts as a stronger reducing agent in alkaline solution than in acid solution.

Complexes of Zn(II). Zn^{2+} ion forms a number of complexes, like ions of other transition elements.

Zn^{2+} ion is hydrated in solution, yielding the complex $[Zn(H_2O)_4]^{2+}$. Thus, a solution of zinc chloride in water may be written as $[Zn(H_2O)_4]Cl_2$.

A precipitate of zinc hydroxide dissolves in ammonia due to formation of the complex ion $[Zn(NH_3)_4]^{2+}$.

$$Zn(OH)_2 + 4NH_3 \longrightarrow [Zn(NH_3)_4](OH)_2$$

Zinc hydroxide also dissolves in cyanide solution giving the complex $[Zn(CN)_4]^{2-}$.

$$Zn(OH)_2 + 4KCN \longrightarrow K_2[Zn(CN)_4] + 2KOH$$

The tetracyano complex is more stable than the tetraammine complex. This is supported by the values of the equilibrium constants of the decomposition reactions of these complexes.

$$[Zn(CN)_4]^{2-} \rightleftharpoons Zn^{2+} + 4CN^-; \qquad K = 1.2 \times 10^{-18}$$

$$[Zn(NH_3)_4]^{2+} \rightleftharpoons Zn^{2+} + 4NH_3; \qquad K = 3.4 \times 10^{-10}$$

All the above complexes have *tetrahedral* shapes.

The complex ion $[Zn(NH_3)_6]^{2+}$ is known in crystals only.

1.22. SOME IMPORTANT COMPOUNDS OF TRANSITION ELEMENTS

TITANIUM NITRATE, $Ti(NO_3)_4$

Preparation. Titanium (IV) nitrate can be prepared by the action of N_2O_5 on titanium (IV) chloride.

$$TiCl_4 + 2N_2O_5 + 2H_2O \longrightarrow Ti(NO_3)_4 + 4HCl$$

Properties

1. It is a white volatile compound.
2. It melts at 58°C.
3. It is highly reactive towards organic substances. While reacting with organic substances, it causes inflammation or explosion. The highly reactive nature of titanium nitrate is probably due to the presence of very reactive nitrate (NO_3) radical in its molecule.

Structure. The structure of $Ti(NO_3)_4$ is shown in Fig. 1.7. In its structure two oxygen atoms from each nitrate group are bonded to titanium atom exhibiting its bidentate nature. The bidentate nitrate groups form a tetrahedron around titanium bonded through two oxygen atoms giving a dodecahedral structure. In this compound titanium attains a coordination number of 8.

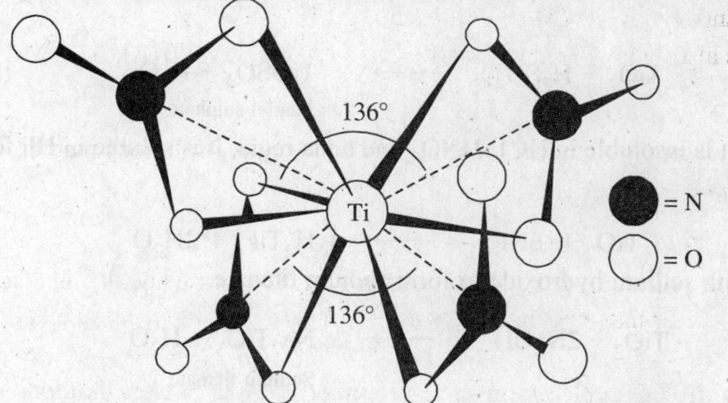

Fig. 1.7. Structure of titanium (IV) nitrate.

TITANIUM DIOXIDE, TiO_2.

Preparation. It is prepared by one of the following processes:

1. From ilmenite ($FeTiO_3$). The powdered ilmenite is digested with excess of conc. H_2SO_4 when $FeSO_4$, $Fe_2(SO_4)_3$ and $TiOSO_4$ (titanyl sulphate) are formed. The resulting product is extracted with water and the solution is then treated with scrap iron to reduce ferric sulphate to ferrous sulphate. The solution is filtered and transferred to vacuum evaporator. During this process crystals of $FeSO_4 \cdot 7H_2O$ are formed which are separated and the solution containing $TiOSO_4$ is boiled with excess water to get hydrated titanium dioxide. The precipitate is filtered, washed and calcined in a rotary kiln to get white pigment grade titanium dioxide. The overall reaction is given below.

$$TiOSO_4 + H_2O \longrightarrow TiO_2 + H_2SO_4$$

This process is also called **sulphate process.**

2. Chloride process. In this process, rutile is heated with chlorine and coke at 900°C forming $TiCl_4$. It is volatile and can be separated from impurities. $TiCl_4$ is heated with oxygen forming pure TiO_2 and Cl_2.

$$\underset{\text{Rutile}}{2TiO_2} + 3C + 4Cl_2 \longrightarrow 2TiCl_4 + 2CO + CO_2$$

$$TiCl_4 + 2O_2 \xrightarrow{\text{Heat}} TiO_2 + 2Cl_2$$

The chlorine is reused. Alternatively, $TiCl_4$ can be hydrolysed completely to form TiO_2.

$$TiCl_4 + 4H_2O \longrightarrow TiO_2 + 2H_2O$$

Properties

1. It is a white solid which turns yellow on heating.
2. It is insoluble in water.
3. It is reduced with hydrogen to form Ti_2O_3.

$$2TiO_2 + H_2 \longrightarrow Ti_2O_3 + H_2O$$

However, on reduction with Ca, Mg or Al, it forms titanium metal.

$$TiO_2 + 2Ca \longrightarrow Ti + 2CaO$$

$$TiO_2 + 2Mg \longrightarrow Ti + 2MgO$$

4. Amphoteric character. It is an amphoteric oxide and dissolves both in acids and bases forming titanates and titanyl compounds. For example, it dissolves in conc. H_2SO_4 forming titanyl sulphate.

$$TiO_2 + H_2SO_4 \longrightarrow \underset{\text{Titanyl sulphate}}{TiOSO_4} + H_2O$$

However, it is insoluble in HCl, HNO_3 and aqua regia. It is soluble in HF forming haxafluoro complex, H_2TiF_6.

$$TiO_2 + 6HF \longrightarrow H_2TiF_6 + 2H_2O$$

It reacts with sodium hydroxide to form sodium titanate.

$$TiO_2 + 2NaOH \longrightarrow \underset{\text{Sodium titanate}}{Na_2TiO_3} + H_2O$$

5. Action with halogens. On passing chlorine or bromine through a mixture of TiO_2 and coke at 950°C, it forms titanium (IV) tetrabromide.

$$2TiO_2 + 3C + 4Cl_2 \xrightarrow{950°C} 2TiCl_4 + CO_2 + 2CO$$

$$2TiO_2 + 3C + 8Br_2 \xrightarrow{950°C} 2TiBr_4 + CO_2 + 2CO$$

6. Action with titanium. On fusion with titanium metal, it produces TiO.

$$TiO_2 + Ti \longrightarrow 2TiO$$

7. Action with carbon. On heating with coke at 1100°C, it forms titanium carbide.

$$TiO_2 + 3C \xrightarrow{1100°C} \underset{\text{Titanium carbide}}{TiC} + 2CO$$

Structure. Its structure is known as rutile structure. It has coordination number 6:3 in which each titanium is surrounded octahedrally by six oxygen atoms while each oxygen atom is surrounded by three titanium atoms disposed towards the corners of an equilateral triangle as shown in Fig. 1.8.

Uses. 1. It is used in large quantities as a white pigment in paints and cosmetic industry. Its properties like high opacity, inertness, good covering power, non-poisonous nature and resistance to temperature make it a very useful compound.

TiO_2 has replaced the earlier pigments such as white lead $2PbCO_3.Pb(OH_2)$ and $BaSO_4$ for this purpose.

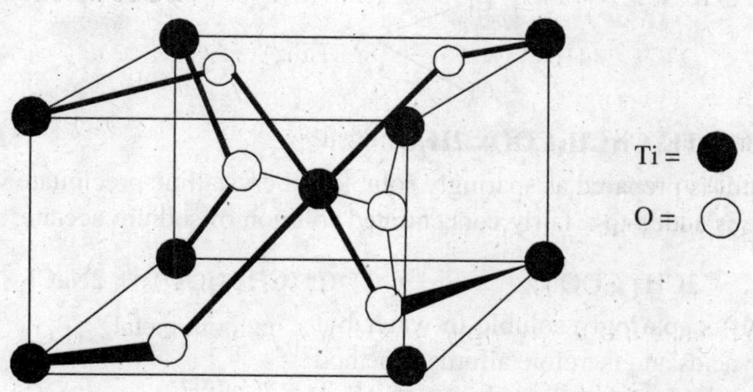

Fig. 1.8. Rutile (TiO_2) structure.

2. It is used as an additive in the manufacture of fibre glass.

3. It is used as a filler in plastics and rubber.

4. It has a high refractive index and, therefore, its crystals have brilliance, even more than diamond. However, because of its softness it is not well-suited for use in jewellery.

5. It is used in enamelling of porcelain goods.

VANADYL CHLORIDE, $VOCl_2$.

Preparation

1. It can be prepared by the reduction of $VOCl_3$ with hydrogen at 300°C.

$$2VOCl_3 + H_2 \xrightarrow{300°C} 2VOCl_2 + 2HCl$$

2. Pure compound is prepared by heating $VOCl_3$ with metallic zinc in a sealed tube.

$$2VOCl_3 + Zn \longrightarrow 2VOCl_2 + ZnCl_2$$

3. It can be prepared by the action of HCl on V_2O_5.

$$V_2O_5 + 6HCl \longrightarrow 2VOCl_2 + 3H_2O + Cl_2$$

4. It can be prepared by the hydrolysis of VCl_4.

$$VCl_4 + H_2O \longrightarrow VOCl_2 + 2HCl$$

Properties

1. The pure compound exists as grassy green crystals.

2. Hydrolysis. It is hydrolysed to yellow brown $VO(OH)_2$ with dilute sodium carbonate solution:

$$VOCl_2 + 2H_2O(aq) \xrightarrow{Na_2CO_3(aq)} VO\,(OH)_2(aq) + 2HCl$$

3. Action with alkalis. It reacts with strong alkalis to form vanadates.

$$VOCl_2 + 4NaOH \longrightarrow \underset{\text{Sodium vanadate}}{Na_2VO_3} + 2NaCl + 2H_2O$$

Uses. Vanadyl chloride is used commercially in dyeing under the name *Divanadyl tetrachloride.* It is a strong reducing agent.

Structure. $VOCl_2$ has been found to contain double (V = O) bond formed by the overlap of π electrons of oxygen into the vacant *d*-orbitals of vanadium. It has trigonal planar structure:

Cl
V═O
Cl

CHROMOUS ACETATE, $Cr(CH_3COO)_2.2H_2O$.

Preparation. It is prepared as sparingly soluble red crystalline precipitate when a solution of chromous chloride is added to a fairly concentrated solution of sodium acetate.

$$2Cr\,Cl_2 + 2CH_3\,COONa \longrightarrow Cr\,(CH_3\,COO)_2 + 2NaCl$$

Properties. It is sparingly soluble in water but readily dissolves in acids and therefore affords a method to prepare various soluble chromous salts.

Uses. A solution of this salt in hydrochloric acid is employed to determine the proportion of oxygen in a gas mixture.

Structure. Chromous acetate exists as a dimeric molecule and has the formula

$$[(CH_3COO)_2\,Cr]_2 \cdot 2H_2O$$

Its structure is similar to the acetate of Cu (II) involving a very short [2.46 Å] metal-metal bond and has the H_2O molecule as end groups [Fig. 1.9].

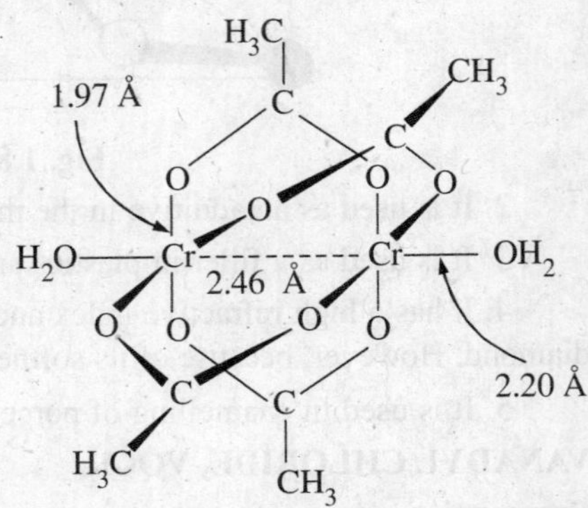

Fig. 1.9. Dimeric structure of chromous acetate.

Chromium (II) acetate monohydrate is diamagnetic.

This compound exists as a dimeric molecule and has the formula

$$[(CH_3\,COO)_2\,Cr]_2 \cdot 2H_2O$$

Interaction between two metal ions is so strong that four unpaired electrons of each Cr^{2+} become fully paired and therefore the compound becomes diamagnetic.

POTASSIUM CHROMATE, K_2CrO_4.

Preparation. A mixture of powdered chromite ore, excess of potassium carbonate and a little lime is heated in the presence of air.

$$\underset{\text{Chromite ore}}{8FeCr_2O_4} + 8K_2CO_3 + 7O_2 \xrightarrow{\text{Lime}} \underset{\text{Potassium chromate}}{8K_2CrO_4} + 2Fe_2O_3 + 8CO_2$$

The fused mass is extracted with water to get potassium chromate while ferric oxide, Fe_2O_3, is left behind.

2. It can be prepared by treating aqueous solution of potassium dichromate with potassium carbonate or potassium hydroxide.

$$K_2Cr_2O_7 + K_2CO_3 \longrightarrow 2K_2CrO_4 + CO_2$$

$$K_2Cr_2O_7 + 2KOH \longrightarrow 2K_2CrO_4 + H_2O$$

3. It can also be prepared by dissolving chromium (III) oxide, Cr_2O_3, or hydroxide, $Cr(OH)_3$, in excess of alkali in the presence of an oxidising agent.

$$2Cr(OH)_3 + 4KOH + 3O \longrightarrow 2K_2CrO_4 + 5H_2O$$

$$Cr_2O_3 + 4KOH + 3O \longrightarrow 2K_2CrO_4 + 2H_2O$$

Properties. 1. It occurs as lemon yellow crystalline solid. Its crystals do not contain water of crystallisation.

2. Its crystals are isomorphous with potassium sulphate, K_2SO_4.

3. It is freely soluble in water.

4. Action with acids. When sulphuric acid is added to yellow solution of potassium chromate, its colour changes to orange due to the formation of potassium dichromate.

$$\underset{\text{Yellow}}{2K_2CrO_4} + H_2SO_4 \longrightarrow \underset{\text{Orange}}{K_2Cr_2O_7} + K_2SO_4 + H_2O$$

The reaction is reversed on the addition of an alkali.

$$\underset{\text{Orange}}{K_2Cr_2O_7} + 2KOH \longrightarrow \underset{\text{Yellow}}{2K_2CrO_4} + H_2O$$

In alkaline solution, chromate ions are present while in acidic solution dichromate ions are present.

5. Action with barium, lead and silver salts. (*i*) With barium chloride solution, it gives yellow precipitate of barium chromate.

$$BaCl_2 + K_2CrO_4 \longrightarrow \underset{\text{Yellow ppt}}{BaCrO_4} + 2KCl$$

(*ii*) With lead nitrate solution, it gives yellow precipitate of lead chromate.

$$Pb(NO_3)_2 + K_2CrO_4 \longrightarrow \underset{\text{Yellow ppt}}{PbCrO_4} + 2KNO_3$$

(*iii*) With silver nitrate solution, a brick red precipitate of Ag_2CrO_4 is obtained.

$$2AgNO_3 + K_2CrO_4 \longrightarrow \underset{\text{Silver chromate}}{Ag_2CrO_4} + 2KNO_3$$

6. Oxidising properties. The acidic solution of potassium chromate acts as an oxidising agent. It oxidises

(*i*) *Hydrogen sulphide to sulphur*

$$2K_2CrO_4 + 5H_2SO_4 \longrightarrow 2K_2SO_4 + Cr_2(SO_4)_3 + 5H_2O + 3O$$

$$H_2S + O \longrightarrow H_2O + S] \times 3$$

$$2K_2CrO_4 + 5H_2SO_4 + 3H_2S \longrightarrow 2K_2SO_4 + Cr_2(SO_4)_3 + 8H_2O + 3S$$

(*ii*) *Sulphur dioxide to sulphuric acid*

$$2K_2CrO_4 + 5H_2SO_4 \longrightarrow 2K_2SO_4 + Cr_2(SO_4)_3 + 5H_2O + 3O$$

$$SO_2 + H_2O + O \longrightarrow H_2SO_4] \times 3$$

$$2K_2CrO_4 + 3SO_2 + 2H_2SO_4 \longrightarrow 2K_2SO_4 + Cr_2(SO_4)_3 + 2H_2O$$

(iii) Hydrochloric acid to chlorine

$$2K_2CrO_4 + 5H_2SO_4 \longrightarrow 2K_2SO_4 + Cr_2(SO_4)_3 + 5H_2O + 3O$$
$$[2HCl + O \longrightarrow H_2O + Cl_2] \times 3$$
$$2K_2CrO_4 + 5H_2SO_4 + 6HCl \longrightarrow 2K_2SO_4 + Cr_2(SO_4)_3 + 8H_2O + 3Cl_2$$

(iv) Liberates iodine from potassium iodide

$$2K_2CrO_4 + 5H_2SO_4 \longrightarrow 2K_2SO_4 + Cr_2(SO_4)_3 + 5H_2O + 3O$$
$$[2KI + H_2SO_4 + O \longrightarrow K_2SO_4 + H_2O + I_2] \times 3$$
$$2K_2SO_4 + 8H_2SO_4 + 6KI \longrightarrow 5K_2SO_4 + Cr_2(SO_4)_3 + 8H_2O + 3I_2$$

Uses. (*i*) As an indicator in silver nitrate titrations.

(*ii*) In the manufacture of pigments, basic lead chromate $PbCrO_4 \cdot PbO$ and basic zinc chromate $ZnCrO_4 \cdot ZnO$.

(*iii*) As a reagent for testing silver, lead and barium ions.

Structure of Chromate Ion CrO_4^{2-}

Four oxygens around chromium are arranged in a tetrahedral manner as shown in Fig. 1.10.

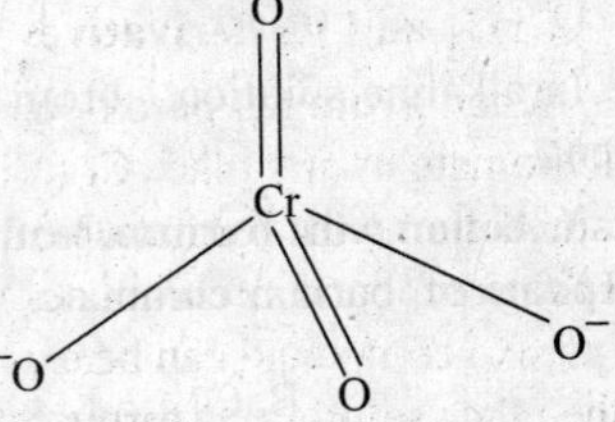

Fig. 1.10. Structure of chromate ion.

CHROMYL CHLORIDE, CrO_2Cl_2.

Preparation. 1. It is prepared by heating a mixture of sulphuric acid, potassium dichromate and potassium chloride.

$$K_2Cr_2O_2 + 4KCl + 3H_2SO_4 \xrightarrow{\Delta} 2CrO_2Cl_2 + 3K_2SO_4 + 3H_2O$$

2. It can also be obtained by treating chromium (VI) oxide with gaseous hydrogen chloride.

$$CrO_3 + 2HCl \longrightarrow CrO_2Cl_2 + H_2O$$

Properties

1. It is a deep red liquid boiling at 116°C under atmospheric pressure.

2. It is soluble in carbon disulphide and nitrobenzene.

3. Action of heat. It decomposes on passing through red hot tube.

$$4CrO_2Cl_2 \longrightarrow 2Cr_2O_3 + 4Cl_2 + O_2$$

4. Action with sodium chloride. When vapours of chromyl chloride are passed into concentrated solution of a sodium chloride, a chlorochromate is formed.

$$CrO_2Cl_2 + NaCl + H_2O \longrightarrow \underset{\text{Sodium chloro-chromate}}{NaCrO_3Cl} + 2HCl$$

5. Action with alkalis. With alkalis, it forms chromates.

$$CrO_2Cl_2 + 4NaOH \longrightarrow Na_2CrO_4 + 2NaCl + 2H_2O$$

6. Action with water. It is hydrolysed by water giving hydrochloric acid and chromic acid.

$$CrO_2Cl_2 + 2H_2O \longrightarrow 2HCl + H_2CrO_4$$

7. As an oxidising agent. Chromyl chloride is a powerful oxidising agent. It oxidises toluene to benzaldehyde.

$$\underset{\text{Toluene}}{C_6H_5CH_3} \xrightarrow{CrO_2Cl_2} \underset{\text{Benzaldehyde}}{C_6H_5CHO}$$

This reaction is known as Etard's reaction.

Uses. (*i*) It is used in quantitative analysis for detection and confirmation of chloride radical.

(*ii*) It is used in Etard's reaction.

Structure. It has a tetrahedral structure as shown in Fig. 1.11.

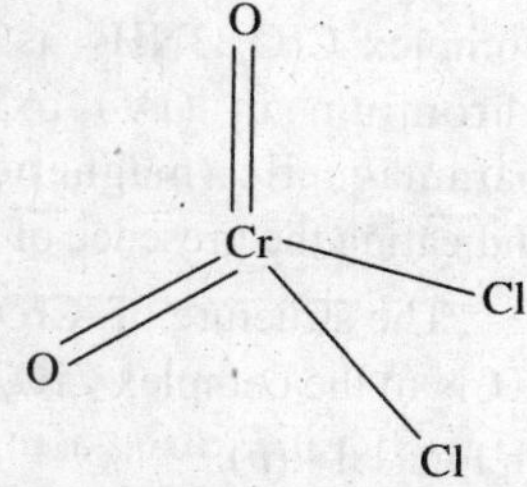

Fig. 1.11 Structure of chromyl chloride.

CrO_5 and its derivatives

When hydrogen peroxide is added to an acidified solution of a dichromate or any other Cr (VI) compound, a deep blue colour is obtained due to the formation of $CrO(O_2)_2$. This colour fades away and the compound decomposes rapidly in aqueous solution to Cr (III) ions and oxygen. However, the peroxo compound can be extracted into ether, where it remains stable. On the addition of pyridine to the ether solution, it forms the adduct Py.CrO_5. Blue colour in the above reaction is due to the formation of chromium peroxide, CrO_5 and the following reactions occur:

$$\underset{\text{Orange}}{Cr_2O_7^{2-}} + 2H^+ + 4H_2O_2 \longrightarrow \underset{\text{Blue}}{2CrO_3} + 5H_2O$$

$$\underset{\text{Yellow}}{CrO_4^{2-}} + 2H^+ + 2H_2O_2 \longrightarrow \underset{\text{Blue}}{2CrO_5} + 3H_2O$$

The CrO_5 is essentially $CrO(O_2)$ and has the structure shown in Fig. 1.12 (*a*). This structure is pentagonal in which chromium is at the centre of pentagon. In ether, it forms hexacoordinated complex with ether and gets stabilized [Fig. 1.12 (*b*)]. On adding pyridine to the etheral solution, it forms solid pyridine adduct, Py.CrO_5.

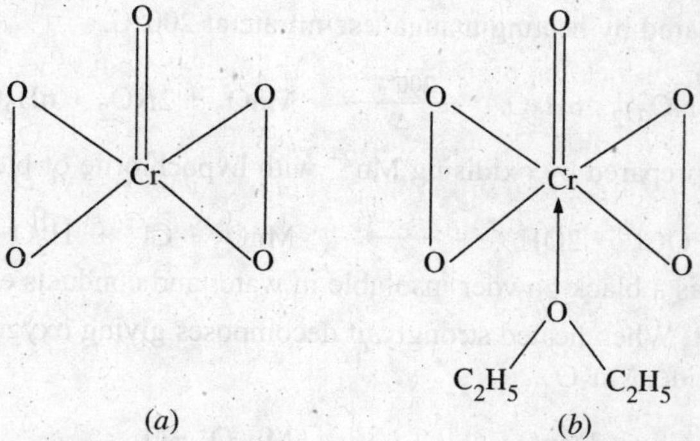

Fig. 1.12. Structures of (*a*) CrO_5 (*b*) CrO_5 –ether complex.

Compounds containing $[CrO_8]^{3-}$ ion.

On adding 30% H_2O_2 solution to alkaline ammonium dichromate or potassium dichromate, a red brown compound $(NH_4)_3CrO_8$ ammonium peroxochromate or K_3CrO_8, is formed. This compound contains the tetraperoxo species $[Cr(O_2)_4]^{3-}$ This contains chromium in (V) oxidation state and is paramagnetic (magnetic moment = 1.8 BM) indicating the presence of one unpaired electron.

The structure of this ion is *dodecahedral* in which the eight oxygen atoms form a regular dodecahedron around chromium (V) ion as shown in Fig. 1.13.

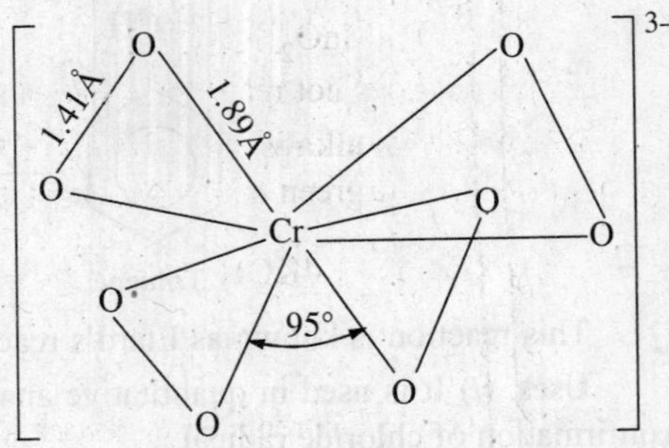

Fig. 1.13. Structure of $[Cr_3O_8]^{3-}$ ion.

Compounds containing CrO_4.

When ammonium peroxochromate $(NH_4)_3CrO_5$ is heated to 50°C, a reddish brown complex $CrO_4.3NH_3$ is formed. This contains chromium in (IV) oxidation state and is paramagnetic (magnetic moment = 2.8 BM) indicating the presence of two unpaired electrons.

The structure of $CrO_4.3NH_3$ is pentagonal bipyramidal as shown in Fig. 1.14 (*a*). On adding KCN to the complex $CrO_4.3NH_3$ another complex $CrO_4(CN)_3^{2-}$ is obtained. Its structure is shown in Fig. 1.14 (*b*).

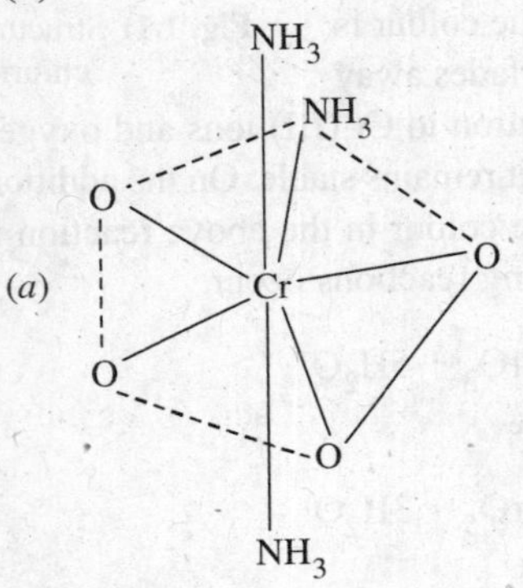

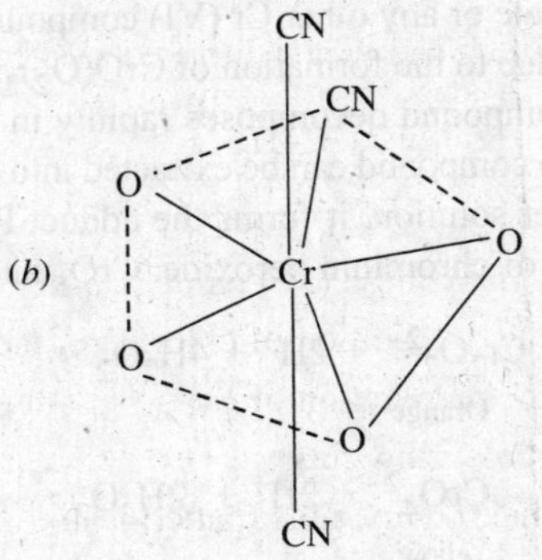

Fig. 1.14. (*a*) Structure of $CrO_4.3NH_3$ (*b*) Structure of $CrO_4(CN)_3^{3-}$.

MANGANESE DIOXIDE, MnO_2

Preparation. 1. It occurs as pyrolusite in nature.

2. It may be prepared by heating manganese nitrate at 200°C.

$$Mn(NO_3)_2 \cdot 6H_2O \xrightarrow[\Delta]{200°C} MnO_2 + 2NO_2 + 6H_2O$$

3. It can also be prepared by oxidising Mn^{2+} with hypochlorite or bleaching powder.

$$Mn^{2+} + OCl^- + 2OH^- \longrightarrow MnO_2 + Cl^- + H_2O$$

Properties. 1. It is a black powder insoluble in water and conducts electricity.

2. **Action of heat.** When heated strongly, it decomposes giving oxygen and leaving a residue of trimanganese tetroxide, Mn_3O_4.

$$3MnO_2 \xrightarrow{\Delta} Mn_3O_4 + O_2$$

3. **Action with acids.** When heated with cold concentrated hydrochloric acid, it gives Mn (II) chloride and chlorine.

$$MnO_2 + 4HCl \longrightarrow MnCl_2 + 2H_2O + Cl_2$$

When heated with concentrated sulphuric acid, manganese (II) sulphate and oxygen are produced.

$$2MnO_2 + 2H_2SO_4 \longrightarrow 2MnSO_4 + 2H_2O + O_2$$

Nitric acid does not react with manganese dioxide.

4. **Action with alkalis.** On fusion with alkalis in the presence of air or oxidising agents such as $NaNO_3$ or $NaClO_3$, green mass of K_2MnO_4 is formed.

$$2MnO_2 + 4KOH + O_2 \longrightarrow \underset{\substack{\text{Potassium manganate} \\ \text{Green mass}}}{2K_2MnO_4} + 2H_2O$$

5. **Reduction.** It is reduced to manganese by aluminium.

$$2MnO_2 + 4Al \longrightarrow 3Mn + 2Al_2O_3$$

6. **As an oxidising agent.** Manganese dioxide in acid solution acts as a strong *oxidising agent* when it is reduced to Mn (II) ion.

$$MnO_2 + 4H^+ + 2e^- \longrightarrow Mn^{2+} + 2H_2O$$

It oxidises chloride ion to free chlorine and oxalate ion to carbon dioxide.

$$MnO_2 + 4H^+ + 2Cl^- \longrightarrow Mn^{2+} + 2H_2O + Cl_2$$

$$MnO_2 + 4H^+ + C_2O_4^{2-} \longrightarrow 2CO_2 + Mn^{2+} + 2H_2O$$

Uses. Manganese (IV) oxide is used

1. in the preparation of chlorine in the laboratory.
2. in the preparation of oxygen from potassium chlorate where it acts as a catalyst.
3. for colouring glass.
4. in making dry cells (batteries).
5. in the production of ferromanganese.

POTASSIUM PERMANGANATE, ($KMnO_4$)

Preparation. 1. Potassium permanganate is prepared by fusing MnO_2 or pyrolusite with KOH or K_2CO_3 in the presence of atmospheric oxygen or any other oxidizing agent, *e.g.*, KNO_3. The fused mass is green due to the formation of potassium manganate.

$$2MnO_2 + 4KOH + O_2 \xrightarrow{\text{heat}} 2K_2MnO_4 + 2H_2O$$

$$2MnO_2 + 2K_2CO_3 + O_2 \longrightarrow \underset{\text{Potassium manganate}}{2K_2MnO_4} + 2CO_2$$

The fused mass is extracted with water and treated with a current of ozone, chlorine or CO_2 when K_2MnO_4 is oxidized to potassium permanganate. The purple solution thus obtained is concentrated when it deposits dark purple, needle-like crystals of $KMnO_4$.

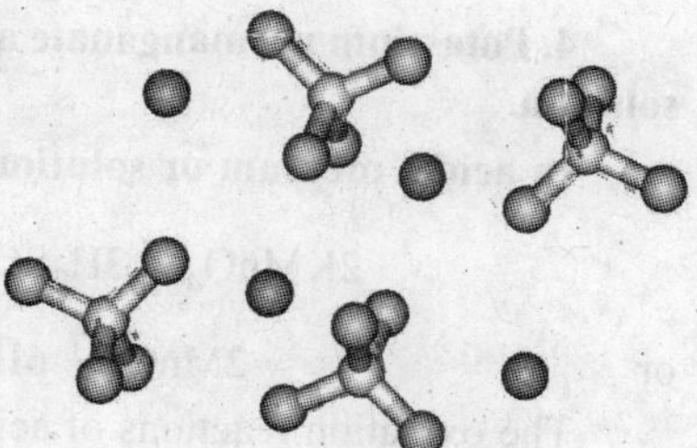
Potassiun permanganate unit cell

$$2K_2MnO_4 + O_3 + H_2O \longrightarrow 2KMnO_4 + 2KOH + O_2$$

$$2K_2MnO_4 + Cl_2 \longrightarrow 2KMnO_4 + 2KCl$$

CO_2 neutralizes the alkali formed and facilitates the reaction.

$$3K_2MnO_4 + 2CO_2 \longrightarrow 2KMnO_4 + 2K_2CO_3 + MnO_2$$

2. It is manufactured by the electrolytic oxidation of a basic solution of potassium manganate. The manganate solution as placed in a cell provided with a rotating nickel wire-netting anode and an inlet pipe made of iron for the liquor which acts as cathode (Fig. 1.15). The oxygen liberated at the anode converts manganate into permanganate.

$$MnO_4^{2-} - e^- \longrightarrow MnO_4^-$$

Hydrogen is evolved in the cathodic compartment.

$$2H^+ + 2e^- \longrightarrow 2H \rightarrow H_2$$

Fresh water ionisation

The purple solution containing $KMnO_4$ is evaporated under controlled condition when $KMnO_4$ crystallizes out.

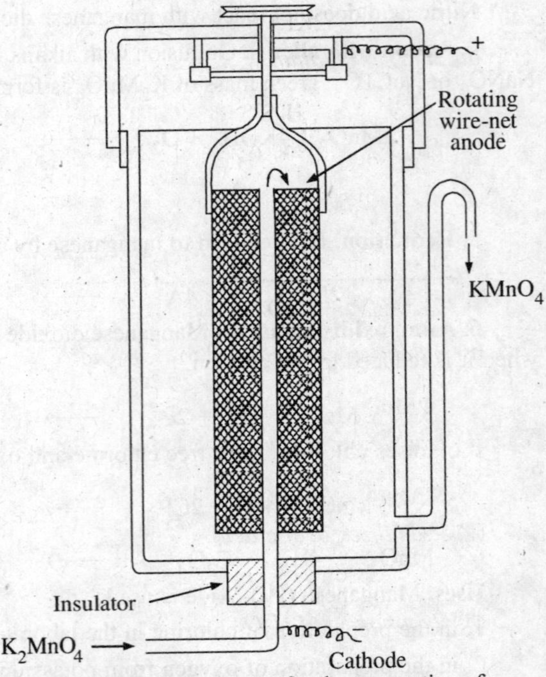

Fig. 1.15. Electrolytic process for the preparation of $KMnO_4$.

Properties. 1. Potassium permanganate is obtained as dark purple needle-like crystals with a metallic lustre. It is moderately soluble in water giving a purple solution. Acidic solution of permanganate is unstable and decomposes slowly to MnO_2.

$$4MnO_4^- + 4H^+ \longrightarrow 3O_2 + 2H_2O + 4MnO_2\downarrow$$

2. Action of Heat.

$$\underset{}{2KMnO_4} \xrightarrow{\text{Heat}} \underset{\text{Pot. manganate}}{K_2MnO_4} + \underset{\text{Manganese dioxide}}{MnO_2} + O_2$$

3. With conc. H_2SO_4, it gives [Mn (VII)] oxide Mn_2O_7, a dark brown highly explosive liquid which further decomposes into MnO_2 and oxygen.

$$2KMnO_4 + H_2SO_4 \longrightarrow Mn_2O_7 + K_2SO_4 + H_2O$$

$$2Mn_2O_7 \longrightarrow 4MnO_2 + 3O_2$$

4. Potassium permanganate acts as a powerful oxidising agent in neutral, alkaline or acidic solution.

In acidic medium or solution. The main reaction is

$$2KMnO_4 + 3H_2SO_4 \longrightarrow K_2SO_4 + 2MnO_4 + 3H_2O + 5\,[O]$$

or

$$2MnO_4 + 6H^+ \longrightarrow 2Mn^{2+} + 3H_2O + 5\,(O)$$

The oxidation reactions of acidified $KMnO_4$ are catalysed by the Mn (II) ion.

(a) It oxidizes ferrous salts to ferric salts.

$$2FeSO_4 + H_2SO_4 + [O] \longrightarrow Fe_2(SO_4)_3 + H_2O$$

(*b*) *It liberates halogens from halides.*

$$2KI + [O] + H_2O \longrightarrow 2KOH + I_2$$

(*c*) *H_2S to sulphur.*

$$H_2S + [O] \longrightarrow H_2O + S$$

(*d*) *Sulphur dioxide is oxidized to sulphuric acid.*

$$2KMnO_4 + 3H_2SO_4 \longrightarrow K_2SO_4 + 2MnSO_4 + 3H_2O + 5\,[O]$$

$$SO_2 + H_2O + [O] \longrightarrow H_2SO_4\,] \times 5$$

$$2KMnO_4 + 5\,SO_2 + 2H_2O \longrightarrow K_2\,SO_4 + 2MnSO_4 + 2H_2SO_4$$

(*e*) *Sulphites are oxidized to sulphates.*

$$2MnO_4^- + 6H^+ + 5SO_3^{2-} \longrightarrow 2Mn^{2+} + 5SO_4^{2-} + 3H_2O$$

(*f*) *Halogen acids are oxidized to halogens.*

$$KMnO_4 + 3H_2SO_4 + 10HX \longrightarrow K_2SO_4 + 2MnSO_4 + 8H_2O + 5X_2$$

(*g*) *It oxidizes oxalic acid to CO_2.*

$$H_2C_2O_4 + [O] \longrightarrow 2CO_2 + H_2O$$

(*h*) *Reaction with H_2O_2*

$$2KMnO_4 + 3H_2SO_4 \longrightarrow K_2SO_4 + 2MnSO_4 + 3H_2O + 5\,[O]$$

$$H_2\,O_2 + [O] \longrightarrow H_2O + O_2] \times 5$$

$$2KMnO_4 + 3H_2\,SO_4 + 5H_2O_2 \longrightarrow K_2SO_4 + 2MnSO_4 + 8H_2O + 5O_2$$

(*i*) *Ethyl alcohol is oxidized to acetaldehyde*

$$CH_3CH_2OH + [O] \xrightarrow{KMnO_4} CH_3CHO + H_2O$$

In alkaline solution. The Main reaction is

$$2KMnO_4 + 2KOH \longrightarrow 2K_2MnO_4 + H_2O + [O]$$

(*a*) *It oxidizes ethylene to ethylene glycol.*

$$C_2H_4 + [O] + H_2O \longrightarrow HOCH_2 \cdot CH_2OH$$

(*b*) *It oxidizes toluene to benzoic acid*

$$C_6H_5CH_3 + 3\,[O] \longrightarrow C_6H_5COOH + H_2O$$

(*c*) *It oxidizes KI to KIO_3.*

$$KI + 3\,[O] \longrightarrow KIO_3$$

In neutral solution. The main reaction is

$$2KMnO_4 + H_2O \longrightarrow 2KOH + 2MnO_2 + 3\,[O]$$

(*a*) *It oxidizes H_2S to sulphur.*

$$H_2S + [O] \longrightarrow H_2O + S$$

(b) Sodium thiosulphate is oxidized to sodium sulphate.

$$8KMnO_4 + 3Na_2S_2O_3 + H_2O \longrightarrow 8MnO_2 + 3Na_2SO_4 + 3K_2SO_4 + 2KOH$$

(c) It oxidizes $MnSO_4$ to MnO_2,

$$MnSO_4 + H_2O + [O] \longrightarrow MnO_2 + H_2SO_4$$

Uses. 1. Potassium permanganate is used as oxidizing agent.

2. It is also used in volumetric estimation of ferrous salts, oxalic acid or oxalates, hydrogen peroxide etc.

3. It finds application as a disinfectant.

4. It is used in qualitative analysis for detecting halides, sulphites, oxalates etc.

5. It is employed in the manufacture of saccharin, benzoic acid, and acetaldehyde.

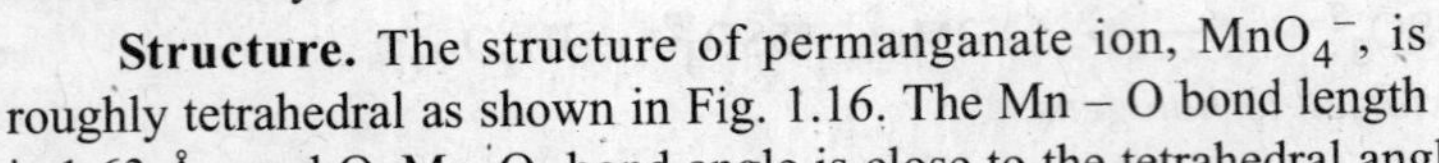

Structure. The structure of permanganate ion, MnO_4^-, is roughly tetrahedral as shown in Fig. 1.16. The Mn – O bond length is 1.63 Å and O–Mn–O bond angle is close to the tetrahedral angle of 109°.

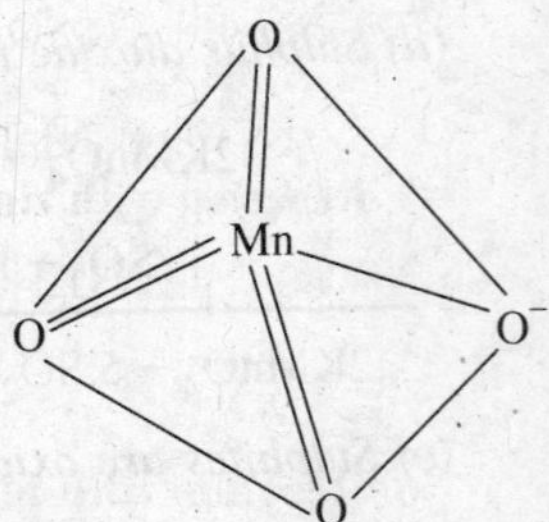

Fig. 1.16. Structure of MnO_4^- ion.

POTASSIUM FERROCYANIDE, $K_4[Fe(CN)_6]$

Preparation. 1. It is obtained during the purification of coal gas. Coal gas contains some cyanogon $(CN)_2$ as an impurity. The gas is passed over hydrated iron oxide.

$$(CN)_2 + H_2O \longrightarrow 2HCN + \frac{1}{2}O_2$$

$$Fe(OH)_2 + 2HCN \longrightarrow Fe(CN)_2 + 2H_2O$$

$$Fe(OH)_3 + 3HCN \longrightarrow Fe(CN)_3 + 3H_2O$$

$$3Fe(CN)_2 + 4Fe(CN)_3 \longrightarrow \underset{\text{Ferric ferrocyanide (Prussian blue)}}{Fe_4[Fe(CN)_6]_3}$$

Ferric ferrocyanide formed is treated with $Ca(OH)_2$ when calcium ferrocyanide is formed.

$$Fe_4[Fe(CN)_6]_3 + 6Ca(OH)_2 \longrightarrow \underset{\text{Soluble}}{3Ca_2[Fe(CN)_6]} + \underset{\text{ppt.}}{4Fe(OH)_3}$$

It is extracted with water and then treated with K_2CO_3 when potassium ferrocyanide is obtained on crystallisation.

$$Ca_2[Fe(CN)_6] + 2K_2CO_3 \longrightarrow K_4[Fe(CN)_6] + 2CaCO_3$$

2. It is manufactured by heating fused KCNS with iron in the absence of air. Potassium ferrocyanide is obtained by concentrating the solution.

$$6KCNS + 6Fe \longrightarrow 6KCN + 6FeS$$

$$6KCN + FeS \longrightarrow K_2S + K_4[Fe(CN)_6]$$

Properties. 1. Potassium ferrocyanide is a yellowish crystalline substance and it is soluble in water.

2. **Action of heat.** On heating above 373 K, it decomposes as

$$K_4[Fe(CN)_6] \longrightarrow 4KCN + FeC_2 + N_2$$

3. **Reducing action.** It is a reducing agent and is oxidized by Cl_2, H_2O_2 or acidified $KMnO_4$.

$$2K_4[Fe(CN)_6] + Cl_2 \longrightarrow 3K_3[Fe(CN)_6] + 2KCl$$

4. **Action with sulphuric acid.** (*i*) With conc. H_2SO_4, it produces carbon monoxide

$$K_4[Fe(CN)_6] + 6H_2SO_4 + 6H_2O \longrightarrow 2K_2SO_4 + 3(NH_4)_2SO_4 + FeSO_4 + 6CO$$

(*ii*) With dil. H_2SO_4, it gives hydrogen cyanide.

$$K_4[Fe(CN)_6] + 3H_2SO_4 \longrightarrow 2K_2SO_4 + FeSO_4 + 6HCN$$

5. **Reaction with zinc salts.**

$$K_4[Fe(CN)_6] + 2ZnSO_4 \longrightarrow \underset{\text{Zinc ferrocyanide white ppt.}}{Zn_2[Fe(CN)_6]} + 2K_2SO_4$$

6. **Reaction with ferric salts.** With ferric salts, it produces Prussian blue colour.

$$Fe^{3+} + K_4[Fe(CN)_6] \longrightarrow \underset{\text{Soluble Prussian blue}}{K\{\overset{III}{Fe}[\overset{II}{Fe}(CN)_6]\}} + 3K^+$$

7. **Reaction with copper salts.**

$$K_4[Fe(CN)_6] + 2CuSO_4 \longrightarrow \underset{\text{Cupric ferrocyanide [Reddish brown ppt.]}}{Cu_2[Fe(CN)_6]} + 2K_2SO_4$$

Uses. 1. It is used in making Prussian blue pigment.

2. It is used for making blueprints.

3. It is used as a reagent in the laboratory for testing Fe^{3+}, Cu^{2+} and Zn^{2+} ions.

Structure. Structure of ferrocyanide ion $[Fe(CN)_6]^{4-}$ is octahedral, shown in Fig. 1.17. This structure is explained below by valence bond theory of complexes. The compound is diamagnetic as there is no unpaired electron present.

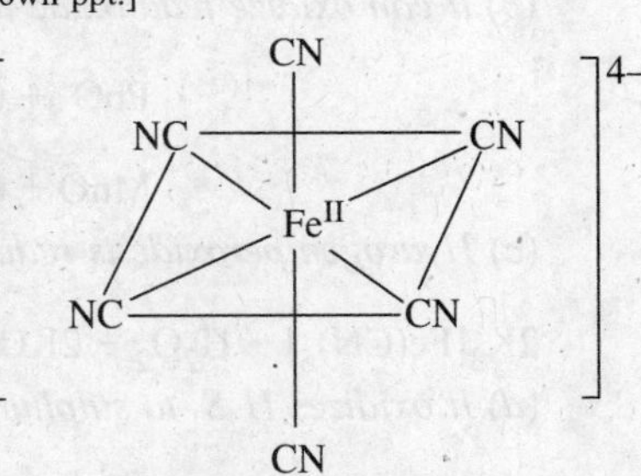

Fig. 1.17. Octahedral structure of $[Fe(CN)_6]^{4-}$ ion.

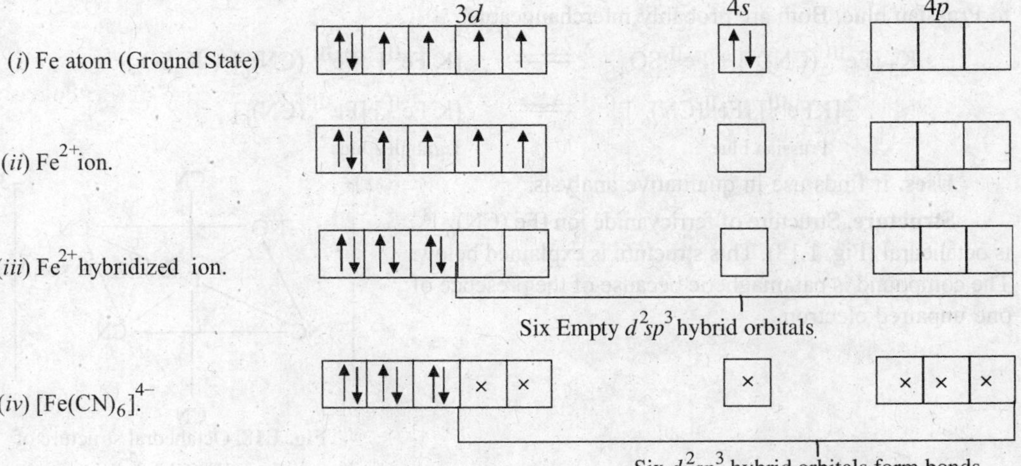

POTASSIUM FERRICYANIDE

Preparation. 1.On passing chlorine gas through a solution of potassium ferrocyanide, potassium ferricyanide is obtained.

$$2K_4 [Fe (CN)_6] + Cl_2 \longrightarrow 2K_3[Fe (CN)_6] + 2KCl$$

The ferricyanide is extracted from the mixture by fractional crystallisation.

Properties. 1. It forms dark-red anhydrous crystals, giving brown yellow solution with water.

2. **Oxidizing properties.** It is a powerful oxidizing agent both in acid and alkaline media.

Oxidation in Acidic Medium

(*a*) It *oxidizes H_2S to sulphur*

$$3H_2S + 6K_3 [Fe (CN)_6] \longrightarrow 4K_4 [Fe^{II} (CN)_6] + [K_2 Fe^{II}] [Fe^{II} (CN)_6] + 6HCN + 3S$$

(*b*) *It evolves chlorine from hydrochloric acid.*

$$K_3 [Fe (CN)_6] + 3HCl \longrightarrow 3KCl + H_3 [Fe (CN)_6]$$

$$2H_3 [Fe (CN)_6] + 2HCl \longrightarrow 2H_4 [Fe (CN)_6] + Cl_2\uparrow$$

In alkaline medium. (*a*) *It can oxidize chromic oxide to potassium chromate.*

$$2K_3 [Fe (CN)_6] + 2KOH \longrightarrow 2K_4 [Fe^{II}(CN)_6] + H_2O + O$$

$$Cr_2O_3 + 4KOH + 3O \longrightarrow 2K_2CrO_4 + 2H_2O$$

(*b*) *It can oxidize lead oxide to dioxide and manganous oxide to dioxide.*

$$PbO + O \longrightarrow PbO_2$$

$$MnO + O \longrightarrow MnO_2$$

(*c*) *Hydrogen peroxide is reduced to water and oxygen.*

$$2K_3[Fe(CN)_6] + H_2O_2 + 2KOH \longrightarrow 2K_4 [Fe (CN)_6] + H_2O + O_2$$

(*d*) *It oxidizes H_2S to sulphur and SO_2 to sulphuric acid.*

$$2K_3 [Fe (CN)_6] + 2KOH + H_2S \longrightarrow 2K_4 [Fe (CN)_6] + 2H_2O + S$$

$$2K_3 [Fe (CN)_6] + 2KOH + SO_2 \longrightarrow 2K_4 [Fe (CN)_6] + H_2 SO_4$$

3. Potassium ferrocyanide reacts with ferrous salt to give Turnbull's blue which is very similar to Prussian blue. Both are probably interchangeable.

$$K_3 [Fe^{III} (CN)_6] + Fe^{II} SO_4 \rightleftharpoons [K Fe^{II}] [Fe^{III} (CN)_6] + K_2SO_4$$

$$\underset{\text{Prussian blue}}{[KFe^{III}] [Fe^{II}(CN)_6]} \rightleftharpoons \underset{\text{Turnbull's blue}}{[K Fe^{II}] [Fe^{III} (CN)_6]}$$

Uses. It finds use in qualitative analysis.

Structure. Structure of ferricyanide ion $[Fe (CN)_6]^{3-}$ is octahedral (Fig. 1.18). This structure is explained below. The compound is paramagnetic because of the presence of one unpaired electron.

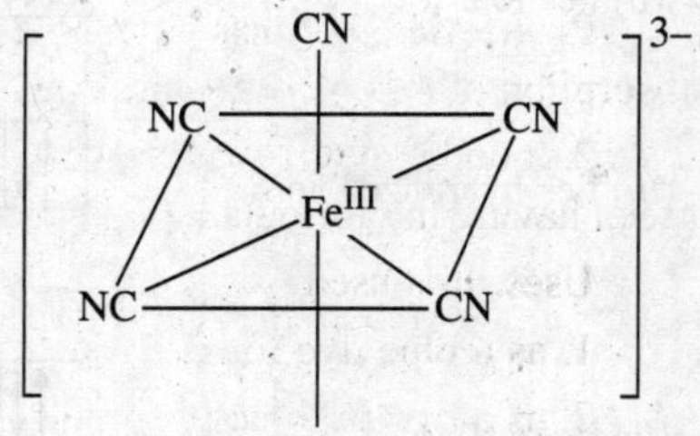

Fig. 1.18. Octahedral structure of $[Fe(CN)_6]^{3-}$ ion.

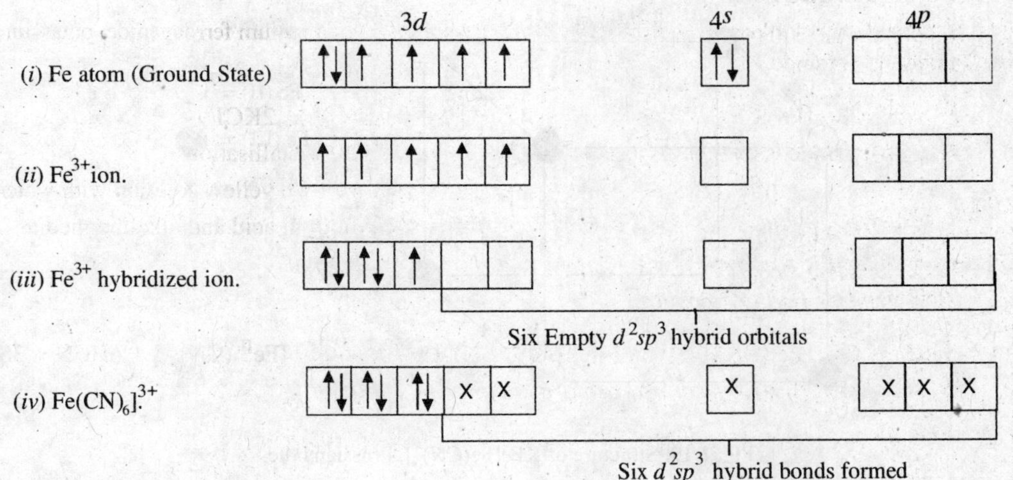

Problem. Assign reasons for the following:

(*i*) Chromium (II) acetate monohydrate is diamagnetic.

(*ii*) Potassium ferrocyanide is used for testing copper salts.

(*iii*) Sodium cobaltinitrite is used in the detection of K^+ ions.

Solution. (*i*) Chromium (II) acetate monohydrate is diamagnetic because of coupling of spins of unpaired electrons on two chromium atoms in the dimeric structure of chromium acetate.

(*ii*) $K_4Fe(CN)_6 + 2\,CuSO_4 \longrightarrow Cu_2[Fe(CN)_6] + 2K_2SO_4$

Cupric ferrocyanide (Reddish brown ppt.)

(*iii*) $Na_3[Co(NO_2)_6] + K^+ \longrightarrow K[Co(NO_2)_6] + Na^+$

Yellow ppt.

Prussian Blue, Turnbull's Blue and Berlin Green

Preparation **Prussian blue.** It is prepared by the action of Fe^{3+} ion ($FeCl_3$) with potassium ferrocyanide solution.

$$FeCl_3 + K_4[Fe(CN)_6] \longrightarrow KFe^{III}[Fe^{II}(CN)_6] + 3KCl$$

Prussian blue

The reaction may also be written as

$$4FeCl_3 + 3K_4[Fe(CN)_6] \longrightarrow Fe_4^{III}[Fe^{II}(CN)_6] + 12\,KCl$$

Properties. 1. It has deep blue colour and is soluble in water. The deep blue colour is due to absorption of visible light causing electron transfer between Fe (II) and Fe (III) states.

2. It undergoes oxidation with nitric acid or chlorine and forms green pigment known as Berlin green having the formula $Fe^{III}[Fe(CN)_6]$.

Uses. It is used

1. as a blue dye for fabrics.
2. as a dye for blue print paper and for ink.

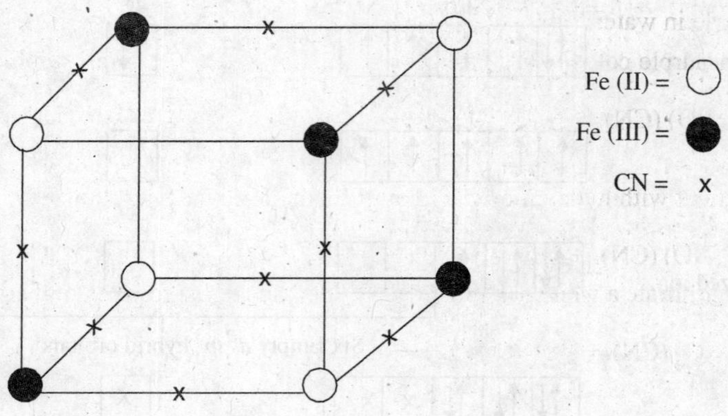

Fig. 1.19. Structure of $KFe[Fe(CN)_6]$, Prussian blue.

Berlin's green

When $KFe^{III}[Fe^{II}(CN)_6]$ is oxidised by means of Cl_2 or HNO_3, it gives a green pigment called Berlin's green, $Fe^{III}[Fe^{III}(CN)_6]$, ferriferricyanide.

$$2KFe^{III}[Fe^{II}(CN)_6] + Cl_2 \longrightarrow 2KCl + \underset{\text{Berlin's green}}{Fe^{III}[Fe^{III}(CN)_6]}$$

Structure. It has simple cubic structure in which iron atoms occupy each corner of the cube and CN^- ions occupy the middle of each side of the cube (Fig. 1.19). In this arrangement, half of the iron atoms have six carbon atoms as the nearest neighbours while the other half have six nitrogen atoms as the nearest neighbours. Potassium atoms occupy only half the cubic centres.

Turnbull's blue

Another compound chemically identical to Prussian blue is Turnbull's blue. It is obtained by the action of potassium ferricyanide with aqueous ferrous sulphate.

$$4K_3[Fe^{III}(CN)_6] + 3Fe^{II}SO_4 \longrightarrow \underset{\text{Turnbull's blue}}{Fe_4^{III}[Fe^{II}(CN)_6]_3} + 6KCN + 3K_2SO_4$$

Recent studies accept Prussian blue and Turnbull's blue as same compound.

SODIUM NITROPRUSSIDE, $Na_2[Fe(NO)(CN)_5] \cdot 2H_2O$.

Preparation 1. It is prepared by warming a solution of potassium ferrocyanide with 50% HNO_3 till a brown solution is obtained. Potassium nitrate crystallises out on cooling. Mother liquor after neutralization with soda ash is filtered and concentrated. Sodium nitroprusside is obtained on cooling the filtrate.

$$K_4Fe(CN)_6 + 4HNO_3 \longrightarrow H_4[Fe(CN)_6] + 4KNO_3$$

$$3H_4[Fe(CN)_6] + HNO_3 \longrightarrow 3H_3[Fe(CN)_6] + NO + 2H_2O$$

$$H_3[Fe(CN)_6] + NO \longrightarrow H_2[Fe(NO)(CN)_5] + HCN$$

$$H_2[Fe(NO)(CN)_5] + Na_2CO_3 \longrightarrow Na_2[Fe(NO)(CN)_5] + H_2O + CO_2\uparrow$$

2. It can be obtained by passing nitric oxide into an acidified solution of sodium ferricyanide.

$$Na_3Fe(CN)_6 + NO \longrightarrow Na_2[Fe(NO)(CN)_5] + NaCN$$

Properties. 1. Sodium nitroprusside exists as red rhombic shaped crystals.

2. It is soluble in water.

3. It gives a purple colour with freshly prepared solution of an alkali sulphide.

$$Na_2Fe(NO)(CN)_5 + Na_2S \longrightarrow \underset{\text{Purple colour}}{Na_4Fe(NOS)(CN)_5}$$

4. On treatment with hydrochloric acid, it gives the following reaction:

$$Na_2Fe(NO)(CN)_5 + 2HCl \longrightarrow H_2Fe(NO)(CN)_5 + 2NaCl$$

5. With silver nitrate a white precipitate is obtained due to the formation of silver nitroprusside.

$$Na_2Fe(NO)(CN)_5 + 2AgNO_3 \longrightarrow \underset{\text{(White ppt.)}}{Ag_2Fe(NO)(CN)_5} + 2NaNO_3$$

6. **Action with alkalis:**

$$6Na_2[Fe(CN)_5NO] + 14NaOH \longrightarrow 5Na_4[Fe(CN)_6] + Fe(OH)_2 + 6NaNO_3 + 6H_2O$$

7. **Action with sulphite ions:**

$$Na_2[Fe(CN)_5NO] + Na_2SO_3 \longrightarrow \underset{\text{Red colour}}{Na_4[Fe(CN)_5NO \,.\, SO_3]}$$

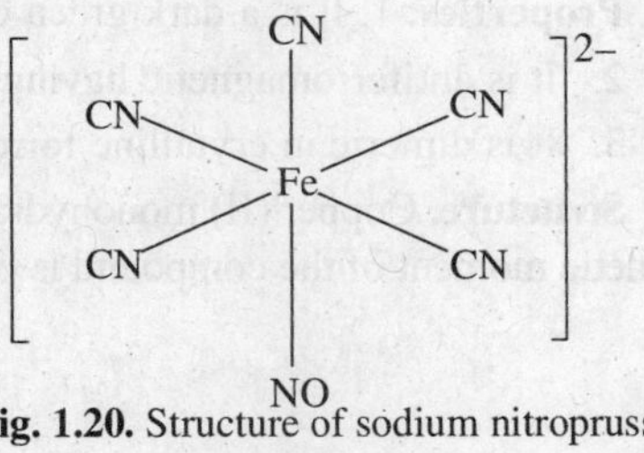

$$\left[\begin{array}{c} \text{CN} \\ \text{CN} \quad\quad \text{CN} \\ \text{Fe} \\ \text{CN} \quad\quad \text{CN} \\ \text{NO} \end{array}\right]^{2-}$$

Fig. 1.20. Structure of sodium nitroprusside.

Uses. It finds use in

(*i*) detection of sulphide ion in inorganic analysis.

(*ii*) detection of ketones as it gives a violet colour with ketones.

Structure. The six groups are attached to iron atom in an octahedral manner as shown in Fig. 1.20.

SODIUM COBALTINITRITE, $Na_3[Co(NO_2)_6]$.

Preparation. 1. It can be prepared by mixing cobalt nitrate with sodium nitrite solution. A 5% solution of acetic acid is added followed by a current of air.

$$Co(NO_3)_2 + 5NaNO_2 + 2HNO_2 \longrightarrow Na_3[Co(NO_2)_6] + 2NaNO_3 + NO + H_2O$$

The solution is filtered and alcohol is added to it. Sodium cobaltinitrite is precipitated.

Properties. 1. It is a yellow powder readily soluble in water.

2. It is used as an analytical reagent for the analysis of potassium salts.

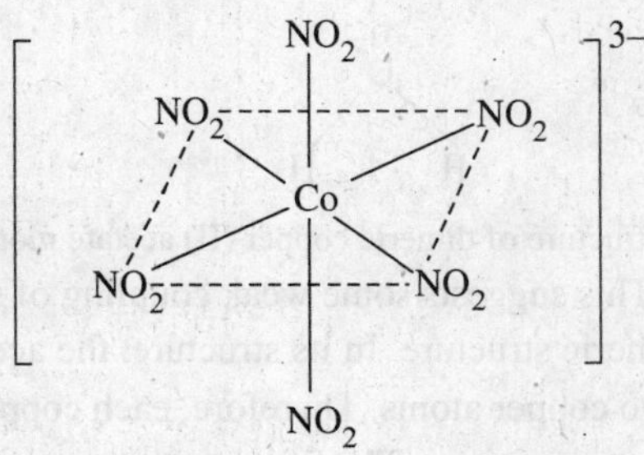

Fig. 1.21. Structure of cobaltinitrite ion

$$Na_3[Co(NO_2)_6] + K^+ \longrightarrow \underset{\text{Yellow ppt.}}{K[Co(NO_2)_6]} + Na^+$$

3. $2NH_4Cl + Na_3[Co(NO_2)_6] \longrightarrow (NH_4)_3[Co(NO_2)_6] + 3NaCl$

Structure. Cobaltinitrite ion (Hexanitrito cobalate (III) ion) has octahedral structure as illustated in Fig. 1.21.

COPPER ACETATE MONOHYDRATE, $Cu(CH_3COO)_2.H_2O$.

Preparation. It is prepared by dissolving copper carbonate or copper oxide in acetic acid and evaporating the solution to crystallisation.

$$2CuCO_3 + 4CH_3COOH \longrightarrow [Cu(CH_3COO)_2 . H_2O]_2 + 2CO_2$$

$$CuO + CH_3COOH \longrightarrow Cu(CH_3COO)_2 . H_2O$$

Properties. 1. It is a dark green crystalline solid soluble in alcohol and water.

2. It is antiferromagnetic having magnetic moment of 1.43 BM at 25°C.

3. It is dimeric in crytalline form and undergoes dissociation in solution.

Structure. Copper (II) monohydrate exists as dimer and its structure is shown in Fig. 1.22. The magnetic moment of the compound is only 1.43 BM which is less than the normal value of 1.75–2.2 BM for one unpaired electron. This suggests some weak coupling of spins of unpaired electrons on the two copper atoms in the dimeric structure. In its structure, the acetate groups act as a bidentate bridging groups between the two copper atoms. Therefore, each copper atom is surrounded by four oxygen atoms in a square planar arrangement. The fifth position around each Cu atom is occupied by the oxygen atom from a water molecule. The other copper atom occupies the sixth of the octahedral positions. The structure is similar to that of chromous acetate.

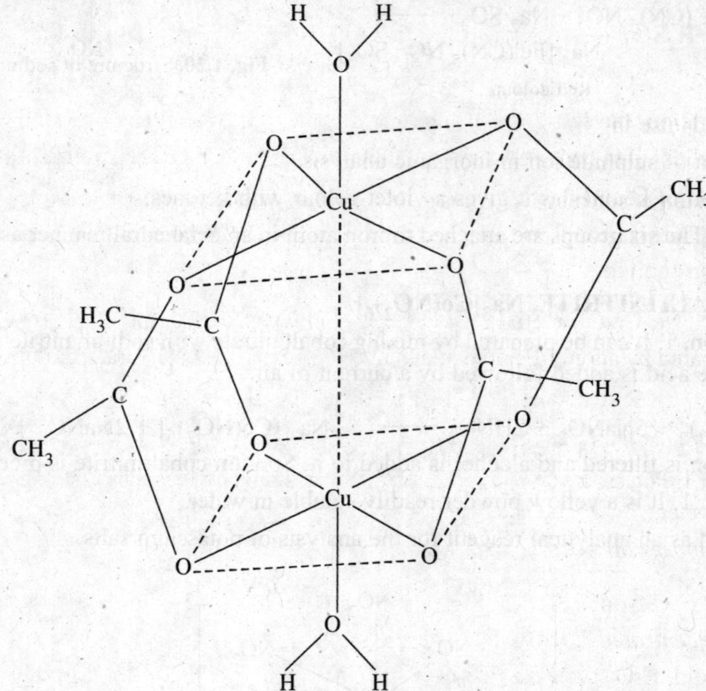

Fig. 1.22. Structure of dimeric copper (II) acetate monohydrate.

POTASSIUM DICHROMATE ($K_2Cr_2O_7$)

Preparation. 1. It is manufactured from chromite ore through the following steps:

***(i)* Preparation of sodium chromate.** Finely powdered chromite is mixed with soda-ash and quick lime and then roasted in a furnace in presence of excess of air.

$$2Fe(CrO_2)_2 + O \longrightarrow Fe_2O_3 + 2Cr_2O_3$$

$$Cr_2O_3 + 2Na_2CO_3 + 3O \longrightarrow 2NaCrO_4 + 4CO_2] \times 2$$

$$2Fe(CrO_2)_2 + 4Na_2CO_3 + 7O \longrightarrow 4Na_2CrO_4 + Fe_2O_3 + 4CO_2$$

or

$$4Fe(CrO_2)_2 + 8Na_2CO_3 + 7O_2 \longrightarrow 8Na_2CrO_4 + 2Fe_2O_3 + 8CO_2$$

Function of quick lime is to keep the mass porous and thus help oxidation by air. The yellow mass is extracted with water when ferric oxide is left behind.

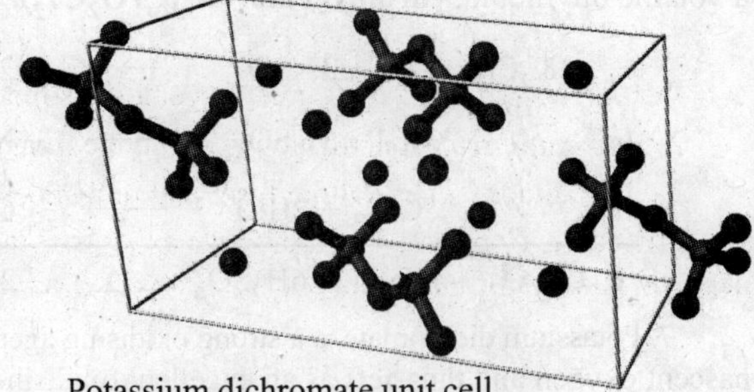

Potassium dichromate unit cell

***(ii)* Conversion of chromate into dichromate.** Sodium chromatic solution obtained in *(i)* is treated with conc. H_2SO_4 when it is converted into sodium dichromate.

$$2Na_2CrO_4 + H_2SO_4 \longrightarrow Na_2Cr_2O_7 + Na_2SO_4 + H_2O$$

On concentration, the less soluble sodium sulphate, $Na_2SO_4 \cdot 10H_2O$ crystallizes out. This is filtered hot and allowed to cool when sodium dichromate, $Na_2Cr_2O_7$, separates on standing.

***(iii)* Conversion of sodium dichromate into potassium dichromate.** Hot concentrated solution of sodium dichromate is treated with a calculated quantity of KC1 when $K_2Cr_2O_7$, being less soluble, crystallizes out on cooling.

$$Na_2Cr_2O_7 + 2KCl \longrightarrow K_2Cr_2O_7 + 2NaC1$$

2. It is prepared by adding a calculated quantity of H_2SO_4 to a saturated solution of potassium chromate.

$$2K_2CrO_4 + H_2SO_4 \longrightarrow K_2Cr_2O_7 + K_2SO_4 + H_2O$$

Properties. 1. Potassium dichromate is obtained as orange red crystals with melting point 671K. It is soluble in water.

2. *It decomposes on heating as follows:*

$$4K_2Cr_2O_7 \longrightarrow 4K_2CrO_4 + 2Cr_2O_3 + 3O_2$$

3. *With concentrated sulphuric acid.*

(i) in the cold, CrO_3 is formed.

$$K_2Cr_2O_7 + 8H_2SO_4 \longrightarrow 2CrO_7 + 2KHSO_4 + H_2O$$

(ii) On heating, oxygen is evolved.

$$2K_2Cr_2O_7 + 8H_2SO_4 \longrightarrow 2K_2SO_4 + 2Cr_2(SO_4)_3 + 8H_2O + 3O_2$$

4. With an alkali, it gives a chromate. Chromates and dichromates are interconvertible in solution. Even weak acids cause the change from yellow chromate to orange dichromate, while this is reversed by addition of alkali or a carbonate solution.

$$2CrO_4^{2-} + 2H^+ \longrightarrow 2HCrO_4^- \rightleftharpoons H_2O + Cr_2O_7^{2-}$$

$$HCrO_4^- + OH^- \longrightarrow H_2O + CrO_4^{2-}$$

$$2HCrO_4^- + CO_3^{2-} \longrightarrow 2CrO_4^{2-} + H_2O + CO_2$$

5. Potassium dichromate solution gives insoluble chromates with soluble salts of lead and barium.

$$2Pb(NO_3)_2 + K_2Cr_2O_7 + H_2O \longrightarrow \underset{\text{Yellow ppt.}}{2PbCrO_4} + 2KNO_3 + 2HNO_3$$

6. Solid $K_2Cr_2O_7$ when heated with conc. H_2SO_4 and a chloride gives orange-red vapour of a volatile oily liquid, **chromyl chloride (CrO_2Cl_2).**

$$K_2Cr_2O_7 + 2H_2SO_4 \longrightarrow 2KHSO_4 + 2CrO_3 + H_2O$$

$$NaCl + H_2SO_4 \longrightarrow NaHSO_4 + HCl\ ] \times 4$$

$$CrO_3 + 2HCl \longrightarrow CrO_2Cl_2 + H_2O\ ] \times 2$$

$$K_2Cr_2O_7 + 4NaCl + 6H_2SO_4 \longrightarrow 2CrO_2Cl_2 + 2KHSO_4 + 4NaHSO_4 + 3H_2O$$

7. Potassium dichromate is a strong oxidising agent. In the neutral or acidic medium, it furnishes nascent oxygen and thus acts as an excellent oxidising agent.

$$K_2Cr_2O_7 + 4H_2O \longrightarrow 2KOH + 2Cr(OH)_3 + 3O$$

$$K_2Cr_2O_7 + 4H_2SO_4 \longrightarrow K_2SO_4 + Cr_2(SO_4)_3 + 4H_2O + 3O$$

(i) It liberates iodine from KI solution.

$$K_2Cr_2O_7 + 4H_2SO_4 \longrightarrow K_2SO_4 + Cr_2(SO_4)_3 + 4H_2O + 3O$$

$$2KI + H_2SO_4 \longrightarrow K_2SO_4 + 2HI\ ] \times 3$$

$$2HI + O \longrightarrow H_2O + I_2\ ] \times 3$$

$$K_2Cr_2O_7 + 7H_2SO_4 + 6KI \longrightarrow K_2SO_4 + Cr_2(SO_4)_3 + 7H_2O + 3I_2$$

or $$Cr_2O_7^{2-} + 6I^- + 14H^+ \longrightarrow 2Cr^{3+} + 7H_2O + 3I_2$$

(ii) It oxidises ferrous sulphate to ferric sulphate.

$$K_2Cr_2O_7 + 4H_2SO_7 \longrightarrow K_2SO_4 + Cr_2(SO_4)_3 + 3Fe_2(SO_4)_3 + 4H_2O + 3(O)$$

$$2FeSO_4 + H_2SO_4 + O \longrightarrow Fe_2(SO_4)_3 + H_2O\ ] \times 3$$

$$K_2Cr_2O_7 + 7H_2SO_4 + 6FeSO_4 \longrightarrow K_2SO_4 + Cr_2(SO_4)_3 + 3Fe_2(SO_4)_3 + 7H_2O$$

or $$Cr_2O_7^{2-} + 14H^+ + 6Fe^{2+} \longrightarrow 2Cr^{3+} + 7H_2O + 6Fe^{3+}$$

(iii) It oxidizes SO_2 to H_2SO_4.

$$Cr_2O_7^{2-} + 2H^+ + 3SO_2 \longrightarrow 2Cr^{3+} + 3SO_4^{2-} + H_2O$$

(iv) It oxidizes H_2S to S.

$$Cr_2O_7^{2-} + 8H^+ + 3H_2S \longrightarrow 2Cr^{3+} + 3S + 7H_2O$$

(*v*) *It oxidizes sulphite ion* $(SO_3)^{2-}$ *to sulphate ion* (SO_4^{2-})

$$Cr_2O_7^{2-} + 3SO_3^{2-} + 8H^+ \longrightarrow 2Cr^{3+} + 3SO_4^{2-} + 4H_2O$$

(*vi*) *It oxidizes alcohol first to aldehydes and then to acids.*

$$\underset{\text{Ethyl alcohol}}{CH_3CH_2OH} \xrightarrow{[O]} \underset{\text{Acetaldehyde}}{CH_3CHO} \xrightarrow{[O]} \underset{\text{Acetic acid}}{CH_3COOH}$$

Uses. It is used:

(*i*) For preparation of other chromium compounds like chrome-alum, chrome red, zinc chromate and lead chromate.

(*ii*) As an oxidizing agent in organic chemistry.

(*iii*) In chrome tanning.

(*iv*) In dyeing and calico printing.

(*v*) As a volumetric reagent for esstimation of ferrous ions, iodine etc.

(*vi*) Mixture of $K_2Cr_2O_7$ and conc. H_2SO_4 is used as a cleaning agent for glasswares.

(*vii*) In photography.

Structure: The dichromate ion $(Cr_2O_7)^{2-}$ consists of two CrO_4 tetrahedra linked through one common oxygen atom at a common corner. Each of the Cr atoms is at the centre of a tetrahedron. The Cr–O–Cr bond angle is 115° and the common O–Cr bond distance is 1.9 Å. The other O–Cr bonds are shorter being 1.6 Å each as shown in Fig. 1.23.

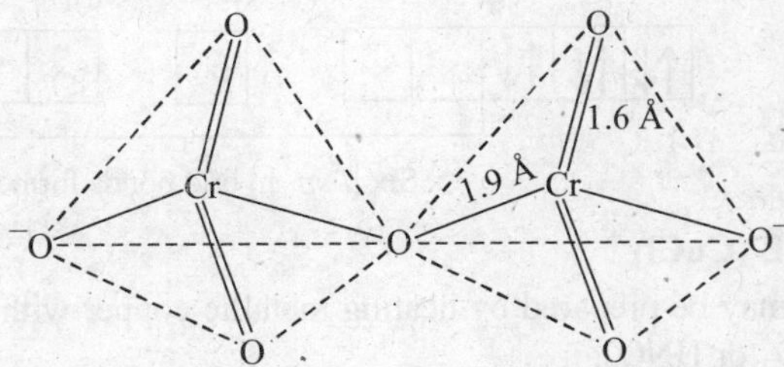

Fig. 1.23. Structure of dichromate ion $(Cr_2O_7)^{2-}$.

HEXAAMINE COBALT (III) CHLORIDE $[Co(NH_3)_6]Cl_3$

Preparation. 1. It is prepared by passing a current of air through an aqueous solution of cobaltous chloride containing ammonium chloride and ammonia for several hours. The reaction is catalysed by activated charcoal.

$$4CoCl_2 + 4NH_4Cl + 20NH_3 + O_2 \longrightarrow 4\,[Co(NH_3)_6]\,Cl_3 + 2H_2O$$

2. It is obtained by adding 20 volume hydrogen peroxide to a solution containing ammonium chloride, cobaltous chloride, ammonia and activated charcoal cooled to 283 K. The mixture solution is heated to 323-333 K and filtered when crystals separate out on cooling.

$$2CoCl_2 + 2NH_4Cl + 10NH_3 + H_2O_2 \longrightarrow 2\,[Co(NH_3)_6]Cl_3 + 2H_2O$$

Properties. 1. It is a golden brown crystalline compound and is soluble in water.

2. Hexaammine cobaltic chloride is very sensitive to hydroxyl ion and is decomposed to cobaltic oxide, Co_2O_3.

$$2\,[Co\,(NH_3)_6]\,Cl_3 + 6\,NaOH \longrightarrow Co_2O_3 + 6NH_3 + 6NaCl + 3H_2O$$

3. It is extremely stable and is not decomposed even by conc. H_2SO_4 which converts chloride to sulphate.

$$2[Co(NH_3)_6]Cl_3 + 3H_2SO_4 \longrightarrow [Co(NH_3)_6]_2(SO_4)_3 + 6HCl$$

Structure. $[Co(NH_3)_6]^{3+}$ ion is an octahedral complex formed by d^2sp^3 hybridisation (Fig. 1.24). The complex has no unpaired electron and therefore, it is diamagnetic.

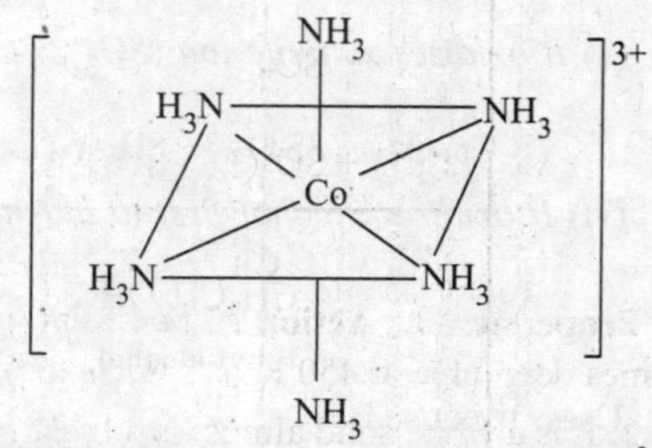

Fig. 1.24. Octahedral structure of $[Co(NH_3)_6]^{3+}$ ion.

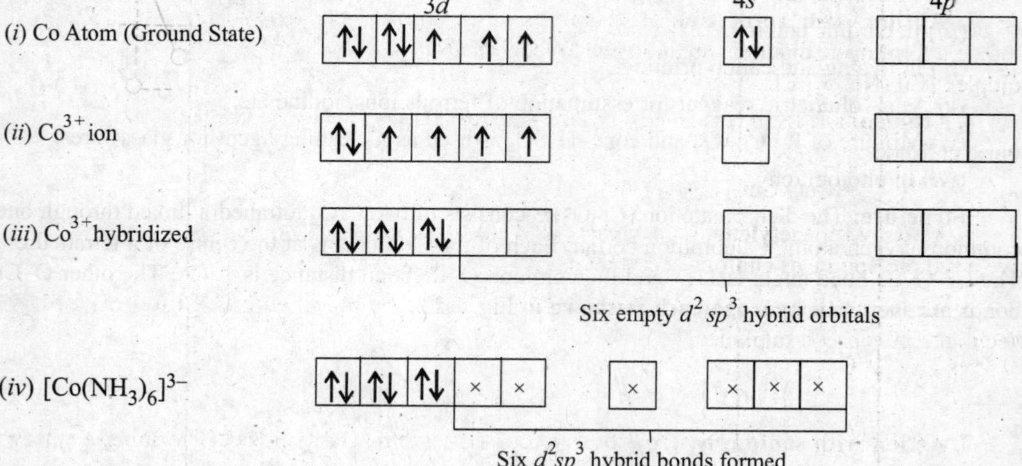

CUPROUS CHLORIDE (CuCl)

Preparation. 1. It may be prepared by heating metallic copper with excess of conc. HCl in presence of a little $KClO_3$ or HNO_3.

$$Cu + 2HCl + O \longrightarrow CuCl_2 + H_2O$$

$$CuCl_2 + Cu \xrightarrow{KClO_3} 2CuCl$$

The reaction mixture is poured in water when cuprous chloride is precipitated.

2. It is obtained by heating cupric oxide, cupric chloride or copper sulphate with excess of copper turnings and conc. hydrochloric acid.

$$CuO + Cu + 2HCl \longrightarrow 2CuCl + H_2O$$

$$CuCl_2 + Cu \longrightarrow 2CuCl$$

$$CuSO_4 + Cu + 2HCl \longrightarrow 2CuCl + H_2SO_4$$

3. It is also formed when $CuCl_2$ is strongly heated (710 K).

$$2CuCl_2 \longrightarrow 2CuCl + Cl_2$$

4. It is also prepared by passing SO_2 through a solution of $CuSO_4$ containing NaCl when CuCl is precipitated.

$$Cu^{2+} + e^- \longrightarrow Cu^+] \times 2$$

$$SO_2 + 2H_2O \longrightarrow SO_4^{2-} + 4H^+ + 2e^-$$

$$2Cu^{2+} + SO_2 + 2H_2O \longrightarrow 2Cu^+ + SO_4^{2-} + 4H^+$$

$$Cu^+ + Cl^- \longrightarrow CuCl$$

Properties. 1. Action of heat. On heating white CuCl becomes deep blue at 450 K and melts to a deep green liquid.

2. It is a white solid almost insoluble in water but soluble in conc. HCl with which it gives the complex ions.

$$CuCl + 3Cl^- \rightleftharpoons [CuCl_4]^{3-}$$

3. Action with ammonia. It dissolves in aqueous ammonia forming a colourless solution due to the formation of complex $[Cu(NH_3)_4]$ Cl.

4. Action of air. In air it is slowly oxidized to green basic cupric chlroide.

$$3CuO.CuCl_2.3H_2O$$

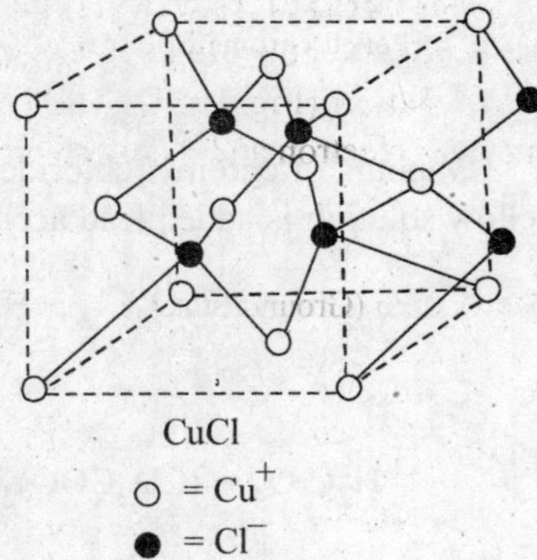

Fig. 1.25. Structure of CuCl.

5. Action with acetylene. It also reacts with acetylene to give a red precipitate of cuprous acetylide, Cu_2C_2.

6. Action with hydrogen sulphide. Hydrogen sulphide reacts with CuCl to give a black precipitate of cuprous sulphide.

$$2CuCl + H_2S \longrightarrow Cu_2S + 2HCl$$

7. Action with sodium hydroxide. Cuprous chloride reacts with NaOH forming a yellow precipitate of cuprous oxide which turns red on boiling.

$$2CuCl + 2NaOH \longrightarrow Cu_2O + 2NaCl + H_2O$$

8. Action of oxidizing agent. When treated with oxidizing agents, it is converted into cupric chloride

$$2CuCl + 2HCl + (O) \longrightarrow 2CuCl_2 + H_2O$$

Uses. Cuprous Chloride is used in the gas analysis for absorbing carbon monoxide and acetylene.

Structure. CuCl has a covalent, tetrahedral zinc blende [ZnS] structure with the metal atom tetrahedrally surrounded by halogen atoms. Cu^+ ions occupy half of the suitable tetrahedral sites in a face centred cubic lattice formed by Cl^- ions. Thus each Cl^- ion is also surrounded tetrahedrally by four Cu^+ ions. (Fig. 1.25).

In vapour state, CuCl is polymeric and contains a cyclic ring of alternating Cu and Cl atoms.

Problem 1. Write the reactions involved in chromyl chloride test.

Solution. Chromyl chloride test is performed to identify chloride in a sample. The sample is heated along with solid potassium dichromate and concentrated sulphuric acid when red vapours of chromyl chloride are evolved. The reactions may be represented as:

$$K_2Cr_2O_7 + 2H_2SO_4 \longrightarrow 2KHSO_4 + 2CrO_3 + H_2O$$

$$NaCl + H_2SO_4 \longrightarrow NaHSO_4 + HCl\ ] \times 4$$

$$CrO_3 + 2HCl \longrightarrow CrO_2Cl_2 + H_2O\ ] \times 2$$

$$\underset{\text{Pot. dichromate}}{K_2Cr_2O_7} + 4NaCl + 6H_2SO_4 \longrightarrow \underset{\text{Chromyl chloride}}{2CrO_2Cl_2} + 2KHSO_4 + 4NaHSO_4 + 3H_2O$$

Vapours of chromyl chloride are passed through water. A yellow solution is obtained. To the yellow solution is added lead acetate. A yellow precipitate of lead chromate is formed.

$$CrO_2Cl_2 + H_2O \longrightarrow \underset{\substack{\text{Chromic acid} \\ \text{(yellow)}}}{H_2CrO_4} + 2HCl$$

$$H_2CrO_4 + (CH_3COO)_2Pb \longrightarrow \underset{\substack{\text{Lead chromate} \\ \text{(yellow ppt.)}}}{PbCrO_4} + 2CH_3COOH$$

Problem 2. Potassium dichromate solution oxidises KI into I_2 in presence of dilute HCl. Balance the equation for the above reaction by ion electron method. Find the equivalent weight of $K_2Cr_2O_7$.

Solution. The balanced equation for the oxidation of KI into I_2 in presence of dil HCl is :

$$Cr_2O_7^{2-} + 6I^- + 14H^+ \longrightarrow 2Cr^{3+} + 3I_2 + 7H_2O$$

As six moles of iodide ions are converted into iodine with one mole of dichromate ion, the equivalent weight of potassium dichromate will be one-sixth of the molecular weight.

$$\text{Equivalent weight of } K_2Cr_2O_7 = \frac{\text{Molecular wt. of } K_2Cr_2O_7}{6}$$

$$= \frac{(2\times39)+(2\times52)+(7\times16)}{6}$$

$$= \frac{294}{6} = 49$$

EXERCISES
(Based on Question Papers of Different Universities)

Multiple Choice Questions

Choose the correct option.

1. The number of unpaired electrons in Fe^{3+} (Z = 26) is :

(*a*) 4 (*b*) 5

(*c*) 6 (*d*) 2.

2. In first transition series, the divalent compound having maximum magnetic moment is of:
   (*a*) Mn (*b*) Fe
   (*c*) Cu (*d*) Cr.
3. Which of the following does not belong to first transition series?
   (*a*) Silver (*b*) Cobalt
   (*c*) Chromium (*d*) Vanadium.
4. *d*-block elements show all the following properties except:
   (*a*) variable oxidation states (*b*) catalytic properties
   (*c*) natural radioactivity (*d*) colour of the compounds.
5. In the first transition series, the highest oxidation state is exhibited by :
   (*a*) Cr (*b*) Mn
   (*c*) Co (*d*) Cu.
6. Which of the following ions is not expected to be coloured ?
   (*a*) Mn^{2+} (*b*) Fe^{3+}
   (*c*) Ti^{3+} (*d*) Cu^{+}
7. Which of the following is used as calibrant in magnetic measurements ?
   (*a*) $CoCl_4^{2-}$ (*b*) $KCu(CN)_2$
   (*c*) $Co Hg(CNS)_4$ (*d*) Fe_3O_4
8. The expected magnetic moment of Ti^{3+} ion is :
   (*a*) 4.90 (*b*) 2.84
   (*c*) 1.73 (*d*) 0.
9. Which of the following is expected to have the highest magnetic moment ?
   (*a*) Co^{2+} (*b*) Cu^{2+}
   (*c*) Ni^{2+} (*d*) Mn^{2+}
10. Which of the following is expected to be diamagnetic ?
    (*a*) $CrCl_3$ (*b*) $CuCl_2$
    (*c*) $ZnCl_2$ (*d*) $CuSO_4$.
11. The first transition series comprises of elements from
    (*a*) scandium to zinc (*b*) lanthanium to hafnium
    (*c*) yttrium to cadmium (*d*) None of these

ANSWERS

1. (*b*)	**2.** (*a*)	**3.** (*b*)	**4.** (*c*)	**5.** (***b***)
6. (*d*)	**7.** (*c*)	**8.** (*c*)	**9.** (*d*)	**10.** (*c*)
11. (*a*)				

Short Answer Questions

1. What is the justification for classifying Zn, Cd and Hg as transition elements ?
2. Cu^{2+} is more stable than Cu^{+}. Explain.
3. Give an example each of complex of manganese showing different oxidation states.
4. How do you explain the anomalous electronic configurations of Cr and Cu ?
5. Which of the *d*-block elements are not regarded as transition elements ?

6. Why do transition metals have high m.p. and b.p. than representative elements ?
7. Give one example each in which cobalt shows +2 and +3 oxidation states in their complexes. Also discuss their magnetic behaviour.
8. Calculate the spin magnetic moment of V^{3+} ion.
9. Both potassium and copper have $4s^1$ electron in their valence shell but copper is far less reactive than potassium. Comment.
10. Why are transition metal compounds often coloured ?
11. What will happen when acidified $K_2Cr_2O_7$ is treated with H_2O_2?
12. Compounds of *s* and *p*-block elements are generally colourless but those of *d*-block are coloured. Explain.
13. Calculate the magnetic moment (spin only) for the complex $K_4[Mn(NCS)_6]$.
14. How is $VOCl_2$ obtained from $VOCl_3$?
15. What is the formula of Prussian blue ?
16. Why is the electronic configuration of Cr(Z = 24) [Ar]$3d^5 4s^1$ rather than [Ar] $3d^4 4s^2$?
17. Give two examples of Fe(0) compounds.
18. 4*s* subshell is filled prior to 3*d* but on ionization 4s electrons are removed first. Explain. why ?

General Questions

1. (*a*) What are transition elements ? Which of the *d*-block elements are not regarded as transition elements ?

   (*b*) In what way the electronic configurations of transition elements are different from those of representative elements ?

2. (*a*) What are *d*-block elements ? Why are they called transition elements ?

   (b) Discuss the position of transition elements

3. What are transition elements ? Why do transition metals

   (*a*) show variable oxidation states ?

   (*b*) form a large number of complexes ?

   (*c*) give coloured and paramagnetic ions ?

   (*d*) exhibit good catalytic properties ?

4. (*a*) In the transition series starting from lanthanum (At. No. 57), the next element hafnium has atomic number 72; why do we observe this jump in atomic number ?

   (*b*) Write down the electronic configurations of the following atoms :

   (*i*) Mn (Z = 25) (*ii*) Co (Z = 27)
   (*iii*) W (Z = 74) (*iv*) Mo (Z = 42)
   (*v*) Ag (Z = 47) (*vi*) Hg (Z = 80)
   (*vii*) Zr (Z = 40) (*viii*) Cd (Z = 48)

   (*c*) How do you explain the anomalous electronic configurations of Cr($4s^1$ $3d^5$) and Cu($4s^1$ $3d^{10}$)?

5. (*a*) Give a brief account of peroxo compounds of chromium.

   (*b*) Why does Mn(II) ion show maximum paramagnetic character amongst the bivalent ions of first transition series ?

6. Explain the following :

(*a*) Transition metals are less reactive than *s*-block elements.

(*b*) 4*s* orbital is filled prior to the 3*d* orbital but on ionization the 4*s* electrons are lost first.

(*c*) Zinc, cadmium and mercury are not included in *d*-block elements.

(*d*) The compounds of transition elements are generally coloured while those of *s* and *p*-block are generally colourless.

(*e*) The ionisation energies of 3*d* elements do not vary much with increasing atomic number.

7. What are the various oxidation states exhibited by manganese ? Discuss the binary compounds and compounds of stablest oxidation state of manganese.

8. (*a*) What are transition elements and why are they named as such ?

(*b*) What justification is there for calssifying Zn, Cd and Hg as transition elements ?

(*c*) Explain the trends of the variations of the melting and boiling points of the *d*-block elements.

(*d*) Discuss the electronic configurations of *d*-block elements.

(*e*) What are the various oxidation states of Ti and V ?

(*f*) Mercury is a liquid metal. Explain.

9. (*a*) What are the important oxidation states of copper ? Discuss the chemistry of +1 and +2 oxidation states.

(*b*) Give an account of complexes of Fe (II) and Fe (III). Discuss their structures.

(*c*) Explain the structure of copper (II) acetate.

10. Explain:

(*a*) Transition elements show variable oxidation states.

(*b*) Chromium and copper have irregular electronic configurations.

(*c*) Cu^{2+} ions are coloured and paramagnetic while Zn^{2+} ions are colourless and diamagnetic.

(*d*) The first ionisation energy of Hg is higher than that of Cd.

(*e*) The transition elements exhibit good catalytic properties.

11. (*a*) The radii and densities of *d*-block elements decrease up to group VIII but show increase towards the end. Explain.

(*b*) Calculate in Bohr magnetons the expected magnetic moment for the following ions (spin magnetic moment only):

(*i*) Fe^{3+} (*ii*) Mn^{2+} (*iii*) Ni^{2+} (*iv*) Cu^{+}

12. Assign reasons for the following :

(*a*) Cu^{2+} is coloured and paramagnetic while Zn^{2+} ions are colourless and diamagnetic.

(*b*) Transition elements show variable oxidation states.

13. Discuss binary compounds and complexes of Cr (II) and Cr (III).

14. (*a*) Explain the structures of

(*i*) CrO_5 (*ii*) CrO_5 py (*iii*) $(CrO_8)^{3-}$

(*b*) Explain the structure of chromium (III) acetate and copper (II) acetate. Discuss the main differences between their structures.

15. (*a*) Compare the stability of Cu^{+} and Cu^{2+}.

(*b*) List important complexes of Cu (II).

16. (*a*) How would you account for +2, +3 and +4 oxidation states of titanium ? Which of these states is most stable and why ?

(*b*) Discuss the aqueous chemistry of Ti (IV).

17. (*a*) Discuss the compounds and complexes of V (V).

(*b*) Explain the halides of titanium.

18. Assign reasons for the following :

(*i*) The melting points of *d*-block elements in a series reach a maximum up to the middle and thereafter show a decrease.

(*ii*) Compounds of *s* and *p*-block elements are generally colourless but those of *d*-block elements coloured.

(*iii*) 4*s*-subshell is filled prior to 3*d* but on ionisation, 4*s*-electrons are removed first.

(*iv*) $Cu^{2+}(3d^9)$ is more stable than Cu^+ $(3d^{10})$.

(*v*) Transition elements in zero or low oxidation states only form complexes with weak ligands (like CO, NO, etc.)

19. (*a*) Assign the reason that Cu^{2+} ions are coloured and paramagnetic while Zn^{2+} ions are colourless and diamagnetic in nature.

(*b*) What justification is there for classifying Zn, Cd and Hg as transition elements.

20. (*a*) How would you define a transition element ?

(*b*) What is characteristic about the electronic configuration of the transition elements?

(*c*) Account for the fact that transition elements and many of their compounds show paramagnetic character.

21. (*a*) Why is first ionisation energy of copper higher than those of alkali metals while second and third energies are lower ? Explain.

(*b*) What are *d*-block elements ? Why are they called transition elements?

(*c*) Why does Mn(II) show maximum paramagnetic character amongst the bivalent ions of first transition series ?

22. Give reasons for the following :

(*a*) Transition metals generally form a large number of complexes.

(*b*) K_{19} and Ca_{20} involve the filling of 4*s* while the elements Sc_{21}, Zn_{30} involve the filling of 3*d* orbitals.

(*c*) The compounds of *d*-block elements are coloured while those of *s* and *p*-blocks are generally colourless.

23. (*a*) Explain: "Transition metals are good catalyst."

(*b*) What is meant by ionisation energy ? Explain its variation among transition elements of first series.

(*c*) Why are transition metal salts coloured ?

(*d*) Why do transition metals form more stable complexes ?

(*e*) Explain :

Cu^{2+} is more stable than Cu^+.

24. (*a*) What are *d*-block elements ?

(*b*) Why are the transition metal salts paramagnetic ?

(*c*) Explain ferromagnetism.

(*d*) Differentiate between first and second transition series.

(*e*) The transition metals are good catalysts. Explain.

25. (*a*) Why do transition metals form large number of complexes ?

(*b*) Why do the transition elements in zero or lower oxidation states only form complexes with weak ligands like CO and NO ?

(*c*) Why do most of the transition metal complexes show paramagnetism ?

26. (*a*) What is ionisation energy ? How do ionisation energies vary through the first transition series ? Give reasons.

(*b*) Discuss the position of *d*-block elements in the long form of periodic table. Why are they called transition elements ?

(*c*) Zn, Cd and Hg are classified as transition elements. Explain.

27. (*a*) (*i*) The transition metals are much less reactive. Explain it.

(*ii*) Why is Cu^{2+} more common than Cu^{+} ?

(*b*) Define ionisation energy. How does it vary through the first transition series ? Explain.

(*c*) Explain the magnetic properties of the transition metal salts.

28. Discuss the structure of

(*a*) VF_5 (*b*) VO $(acac)_2$ (*c*) $(NH_4)_3\ CrO_4$

(*d*) Py.CrO $(O_2)_2$ (*e*) $Cr_2(CH_3COO)_4.\ 2H_2O$

29. (*a*) Explain the following in case of transition elements :

(*i*) Variable oxidation states (*ii*) Catalytic nature

(*iii*) Formation of complex ions (*iv*) Formation of interstitial compounds.

(*b*) Why do transition elements in zero or lower oxidation states form complexes with ligands like CO and NO ?

30. (*a*) What are various oxidation states of Ti and V ?

(*b*) Cu^{2+} is more stable than Cu^{+}. Explain.

(*c*) Explain "Transition metals are good catalysts."

31. (*a*) Give an account of the chemistry of iron showing particularly its transition character.

(*b*) How do you explain the existence of oxy cations such as TiO^{2+} and VO^{2+} in aqueous solution rather than hydrated Ti^{4+} and V^{4+} cations ?

(*c*) Write down the number of unpaired electrons in each of the following ions : Ti^{2+}, V^{3+}, Cr^{3+}, Mn^{2+}, Fe^{2+}, Fe^{3+} and Co^{2+}.

Indicate how would you expect the five 3*d* orbitals to be occupied for these hydrated ions (octahedral).

32. Explain:

(*i*) Transition elements exhibit good catalytic properties.

(*ii*) The ionisation energy of 3*d* elements do not vary much with increasing atomic number.

(*iii*) Cu^{2+} is coloured and paramagnetic while Zn^{2+} is colourless and diamagnetic.

(*iv*) Co^{2+} does not form stable complexes with ammonia whereas Co^{3+} forms a stable complex $[Co(NH_3)_6]^{3+}$

(*v*) The radii and densities of *d*-block elements decrease up to group VIII but show increase towards the end.

33. (*a*) Explain the general characteristics of *d*-block elements w.r.t.
(*i*) size (*ii*) variable oxidation states (*iii*) ionisation energies.

34. Discuss the chemistry of (special reference to complexes):
(*a*) Mn (II) (*b*) Cr (III) (*c*) Fe (II)
(*d*) Cu (II)

35. (*a*) Why do transition elements form coloured compounds ?
(*b*) Why Cu^{2+} ($3d^9$) is more stable than Cu^+ ($3d^{10}$) ?
(*c*) How do the size of atoms and ions of transition elements vary ?

36. (*a*) Explain the structure of copper (II) acetate monohydrate and account for its low magnetic moment.
(*b*) Discuss the behaviour of Ti (IV) halides.
(*c*) How are Ti (II) halides prepared ?

37. (*a*) Discuss the lower and zero oxidation states of iron, copper and chromium.
(*b*) Account for Cu (III) state.

38. (*a*) List five characteristic properties of transition elements.
(*b*) Why *d*-block elements have greater tendency to form complexes ?

39. (*a*) Describe the preparation of potassium permanganate from pyrolusite. Describe one of its oxidising reactions in each of the acidic, alkaline and basic medium.
(*b*) How is Prussian blue prepared ? Give its two properties and discuss its structure in brief.
(*c*) Give the chemistry of chromyl chloride test for the detection of chloride ions.

40. (*a*) Why the ionisation potential of 3*d* elements do not vary much with increasing atomic number ?
(*b*) What is the cause of origin of paramagnetism in transition metal complexes ?
(*c*) Calculate the magnetic moment (expected from spin only) for the following ions :
Cr^{3+}, Fe^{3+}, Cu^+
(*d*) Which of the above ions do you expect to be coloured and why ?

41. (*a*) Transition elements exhibit good catalytic properties. Explain.
(*b*) "Melting points of *d*-block elements in a series reach a maximum up to middle and thereafter show a decrease." Assign reason to above statement.
(*c*) 4*s*-subshell are filled prior to 3*d* but on ionization 4*s* electrons are removed first.

42. Explain why:
(*i*) Iron is a transition element but potassium is representative one.
(*ii*) Iron forms two chlorides $FeCl_2$ and $FeCl_3$ while potassium only one *i.e.* KCl.
(*iii*) Both Fe^{2+} and Fe^{3+} are coloured in water solution, while K^+ is not coloured.
(*iv*) In the process of filling of electrons in $_{26}Fe$, 4*s* subshell is filled prior to 3*d* subshell while in the process of removal again 4*s* electrons are removed prior to 3*d* electrons.
(*v*) Large number of *d*-block elements and their compounds are used as catalysts.

43. (*a*) Describe the essential features transition metals and their compounds, using, examples from the first transition series.
(*b*) State, with reason, whether you would consider copper and tin to be transition elements or not.

44. (*a*) Justify reducing nature of Fe^{2+} (write equations only).

(*b*) Why transition elements show variable oxidation states ? Explain.

(*c*) Why Zn, Cd, Hg are not typical transition elements ? Explain.

45. (*a*) Draw the structure of $[Cr(CH_3COO)_2H_2O]_2$. What is the bond multiplicity between two atoms in it ?

(*b*) What is the equivalent weight of potassium permanganate in acidic, neutral and basic solution? Explain.

(*c*) Explain the following :

(*i*) Zn forms Zn^{2+} and not Zn^{3+} ion.

(*ii*) Salts of Zn, Cd and Hg are colourless.

(*iii*) Transition elements show the catalytic character.

46. (*a*) Why do metals of first transition series mainly form high spin complexes, while those of second and third transition series form low spin complexes ?

(*b*) Why do transition metals in zero or low oxidation states form complexes with ligands like CO, NO, PR_3 etc. ?

47. (*a*) What are *d*-block elements ? Give their general electronic configuration, position in the periodic table and number of series of these elements.

(*b*) Explain why :

(*i*) Coinage metals and II B group elements are included in the *d*-block of elements.

(*ii*) All elements of *d*-block are metals.

(*iii*) *d*-block elements form complex ions.

48. (*a*) What is *d–d* transition ? How does it cause the colour in the compounds of transition elements ?

(*b*) Explain the chemistry of titanium in oxidation state (III).

(*c*) Discuss the structure of $[Cr\ (OOCCH_3)_2]_2 .\ H_2O$.

49. (*a*) Explain that transition metals are good catalysts.

(*b*) Calculate in Bohr magneton the expected magnetic moment for Mn^{2+} ion (spin only).

(*c*) On the basis of relative stability of different oxidation states of Cr, explain the oxidizing ability of $K_2Cr_2O_7$ in sulphuric acid (dilute).

(*d*) Draw and explain the structure of copper (II) acetate monohydrate. Account for its low magnetic moment.

50. (*a*) Explain :

(*i*) Why copper having $3d^{10}$ configuration is less stable than copper with $3d^9$ configuration ?

(*ii*) Why is hydrated form of copper sulphate blue and the anhydrous one white?

(*b*) Give any four properties of chromyl chloride.

(*c*) Why do transition elements form complexes ? Explain.

(*d*) Why do transition elements exhibit good catalytic properties ? Discuss.

51. (*a*) Describe the oxidising character of $KMnO_4$ in acidic, basic and neutral solutions. What is the equivalent weight of $KMnO_4$ in each of these media ?

(*b*) Explain : Anhydrous $CuSO_4$ is white while hydrated $CuSO_4$ is blue in colour.

(*c*) Hg is liquid at room temperature. Explain.

52. (*a*) Cu^{2+} ($3d^9$) is more stable than Cu^+ ($3d^{10}$). Explain.

(*b*) Which should be better oxidising agent Co^{2+} or Co^{3+} in water ?

53. (*a*) Define a transition metal and list some of the important properties associated with transition elements.

(*b*) Cu^{2+} ions are generally coloured whereas Zn^{2+} ions are colourless and diamagnetic. Comment.

54. (*a*) What justification is there for classifying Zn, Cd, Hg as transition elements.

(*b*) Draw structures of chromium (II) acetate hydrated and copper (II) acetate hydrated and discuss main differences between their structures.

(*c*) Why are transition metals less reactive than *s*-block elements?

55. (*a*) Explain the characteristics of *d*-block elements w.r.t.

(*i*) variable oxidation states (*ii*) ionisation energy.

(*b*) First ionisation energy of zinc, cadmium and mercury are high. Explain.

56. What are the various oxidation states of titanium ? Discuss their stability.

57. Give the ionic equations, showing oxidising character of $KMnO_4$ in acidic and neutral media.

58. (*a*) Why is $[Ti(H_2O)_6]^{3+}$ ion violet ?

(*b*) 4*s* orbital is filled prior to 3*d* but upon ionization 4*s* electrons get removed prior to 3*d*. Explain.

(*c*) Calculate the magnetic moment (spin only) for the complex $K_4[Mn(NCS)_6]$.

59. Explain :

(*a*) All transition elements are *d*-block elements, but all *d*-block elements are not transition elements.

(*b*) Transition elements are not good reducing agents.

(*c*) Decrease in atomic radii in first transition series is not regular.

(*d*) Melting and boiling points of transition elements are very high.

(*e*) Magnetic moment of transition elements is calculated on the basis of spin formula only.

60. (*a*) In transition elements, 4*s*-subshell is first filled prior to 3*d*, but upon ionization the 4*s* electrons are removed first. Explain.

(*b*) Why is Cu^{2+} coloured and paramagnetic while Zn^{2+} ions are colourless and diamagnetic ?

(*c*) Discuss the trends of melting points and boiling points of transition elements.

(*d*) Why do transition elements show large number of oxidation states ?

61. (*a*) 4*s* subshell is filled prior to 3*d* but on ionisation 4*s* electrons are removed first. Explain why ?

(*b*) Most of transition metal complexes are coloured but those of zinc, cadmium and mercury are colourless. Why?

(*c*) Iron is a transition element but sodium is not. Give reasons.

62. What do you understand by the term magnetic properties of transition metals? Define the terms ferromagnetic and anti-ferromagnetic substances.

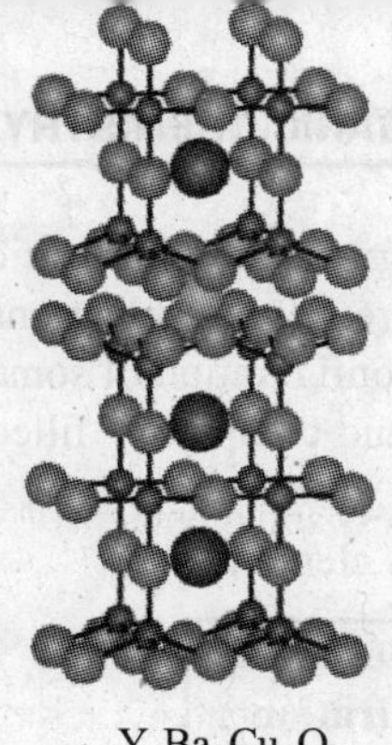

Y $Ba_2Cu_3O_7$ superconductor

2

Chemistry of Elements of Second and Third Transition Series

2.1. INTRODUCTION

Chemistry of elements of first transition (3*d*) series has been studied in the previous chapter. In this chapter we shall study the general characteristics of the elements of second (4*d*) and third (5*d*) transition series. Also comparison of second and third transition series elements and their compounds will be drawn against their analogues in first transition series. There are ten elements in the second series. These are ytterium, zirconium, niobium, molybdenum, technetium, ruthenium, rhodium, palladium, silver and cadmium. The first member of third transition series is lanthanum. Because of its resemblance with lanthanide (4*f*) series, it has been discussed in chapter 5 separately. Other 9 members of the third transition series are: hafnium, tantalum, tungsten, rhenium, osmium, iridium, platinum, gold and mercury.

2.2. ELECTRONIC CONFIGURATION OF ATOMS AND IONS

The general electronic configuration of atoms of the transition series is:

$(n-1)\ d^{1-10}\ n\ s^{1-2}$

Just as the first transition series involves filling of 3*d* orbitals, the second transition series involves filling of 4*d* and the third transition series involve filling of 5*d* orbitals. Table 2.1 gives the electronic configuration of second transition series.

Table 2.1. Electronic configurations of second transition series elements

Element	Symbol	Atomic number	Outer electronic configuration
Yttrium	Y	39	– $4d^1\ 5s^2$
Zirconium	Zr	40	– $4d^2\ 5s^2$
Niobium	Nb	41	– $4d^4\ 5s^1$
Molybdenum	Mo	42	– $4d^5\ 5s^1$
Technetium	Tc	43	– $4d^5\ 5s^2$
Ruthenium	Ru	44	– $4d^7\ 5s^1$
Rhodium	Rh	45	– $4d^8\ 5s^1$
Palladium	Pd	46	– $4d^{10}\ 5s^1$
Silver	Ag	47	– $4d^{10}\ 5s^1$
Cadmium	Cd	48	– $4d^{10}\ 5s^2$

Anomalous configuration of Mo, Tc, Pd and Ag can be explained in terms of greater stability of half-filled and completely filled *d*-orbitals.

Third transition series involves filling of 5*d* orbitals. Table 2.2 gives the electronic configuration of third transition series elements. Between La and Hf, there are fourteen electrons called lanthanoids or lanthanides). Again we find there are anomalies in the configuration of some elements which can be explained in terms of greater stability of half-filled and completely filled d-orbitals and factors like nucleus-electron and electron-electron forces.

Table 2.2. Electronic configuration of third transition series elements

Element	Symbol	Atomic number	Outer electronic configuration
Lanthanum	La	57	$- 5d^1\ 6s^2$
Hafnium	Hf	72	$- 5d^2\ 6s^2$
Tantalum	Ta	73	$- 5d^3\ 6s^2$
Tungsten	W	74	$- 5d^4\ 6s^2$
Rhenium	Re	75	$- 5d^5\ 6s^2$
Osmium	Os	76	$- 5d^6\ 6s^2$
Iridium	Ir	77	$- 5d^7\ 6s^2$
Platinum	Pt	78	$- 5d^9\ 6s^1$
Gold	Au	79	$- 5d^{10}\ 6s^1$
Mercury	Hg	80	$- 5d^{10}\ 6s^2$

Fourth transition series involves filling of 6d orbitals. Fourteen elements between Ac and Rf are called actinides, which possess similar characteristics. These actinides involve progressive filling of 5f orbital. Fourth transition series is an incomplete one and contains nine elements. Table 2.3 gives the configuration of fourth transition series elements.

Table 2.3. Electronic configuration of fourth transition series

Element	Symbol	Atomic number	Outer electronic configuration
Actinium	Ac	89	$- 6d^1 7s^2$
Rutherfordium	Rf	104	$- 5f^{14}\ 6d^2 7s^2$
Hahnium	Ha	105	$- 5f^{14}\ 6d^3 7s^2$
Seaborgium	Sg	106	$- 5f^{14}\ 6d^4 7s^2$
Bohrium	Bh	107	$- 5f^{14}\ 6d^5 7s^2$
Hassium	Hs	108	$- 5f^{14}\ 6d^6 7s^2$
Meitnerium	Mt	109	$- 5f^{14}\ 6d^7 7s^2$
Darmstadium	Ds	110	$- 5f^{14}\ 6d^8 7s^2$
Rontgenium	Rg	111	$- 5f^{14}\ 6d^{10} 7s^1$

2.3. GENERAL PROPERTIES OF TRANSITION ELEMENTS

Some general properties of elements of second and third transition series are given in Tables 2.4 and 2.5 respectively.

We shall discuss the general trends in some of the important properties.

1. Atomic radii

The atomic radii in a particular series decrease with increase in atomic number. This is due to increase in nuclear charge. However, it has been observed that the decrease in atomic radius becomes smaller after midway. For example, for the elements of second transition series, the atomic radii decrease gradually from yttrium to cadmium but from ruthenium onwards it remains practically

Table 2.4. Important properties of Elements of the Second Transition Series

Elements and their atomic numbers	Outer electronic configuration	Atomic mass (a.m.u)	Atomic volume (ml)	Density (g ml^{-1})	Melting point (°C)	Boiling point (°C)	Atomic radius (Å)	Stable oxidation state	Ionic radius (Å) for stable oxid. state	First ionization energy (kJ mol^{-1})	Heat of fusion (kJ mol^{-1})	Heat of vaporisation (kJ mol^{-1})
Y (39)	$4d^1 5s^2$	88· 90	19·8	4·47	1509	2927	1·80	+3	0·93	635·9	11·3	389·1
Zr (40)	$4d^2 5s^2$	91·22	14·1	6·49	1852	3580	1·58	+4	0·80	669·4	16·7	502·0
Nb (41)	$4d^4 5s^1$	92·91	10·8	8·40	2415	3300	1·46	+5	0·70	652·7	26·7	—
Mo (42)	$4d^5 5s^1$	95·94	9·4	10·21	2610	5560	1·39	+6	0·62	694·5	27·6	535·5
Te (43)	$4d^5 5s^2$	(99)	—	11·50	2200	—	1·36	+4	—	698·7	23·0	502·5
Rn (44)	$4d^7 5s^1$	101·70	8·3	12·20	2500	4900	1·34	+3	0·69	723·8	25·5	602·6
Rh (45)	$4d^8 5s^1$	102· 90	8·3	12·40	1966	4500	1·34	+3	0·86	744·7	21·7	577·4
Pd (46)	$4d^{10} 5s^0$	106·40	8·9	12·00	1552	3980	1·37	+2	0·50	803·3	16·7	380·7
Ag (47)	$4d^{10} 5s^1$	107·87	10·3	10·51	960.8	2210	1·44	+1	1·26	732·2	11·3	286·2
Cd (48)	$4d^{10} 5s^2$	200·57	13·1	8·65	320.9	765	1·54	+2	0·97	866·1	6·1	99·9

Table 2.5. Important properties of Elements of the Third Transition Series

Elements and their atomic numbers	Outer electronic configuration	Atomic mass (a.m.u)	Atomic volume (ml)	Density (g ml^{-1})	Melting point (°C)	Boiling point (°C)	Atomic radius (Å)	Stable oxidation states	Ionic radius	First ionization energy (kJ mol^{-1})	Heat of fusion for stable oxid. state (kJ mol^{-1})	Heat of vaporisation (kJ mol^{-1})
Hf (72)	$5d^2\ 6s^2$	178·49	13·60	13·1	2222	5400	1.58	+4	0·81	531.4	21·7	648·5
Ta (73)	$5d^3\ 6s^2$	180·95	10·90	16·6	2996	5425	1.46	+5	0·73	577.4	28·4	753·1
W (74)	$5d^5\ 6s^1$	183·85	9·53	19·3	3410	5930	1.39	+6	0·68	769.8	33·7	774·0
Re (75)	$5d^5\ 6s^2$	186·21	8·85	21·0	3180	5900	1·37	+2 to +7	—	761.5	33·0	636·0
Os (76)	$5d^6\ 6s^2$	190.21	8·43	22·6	2700	5500	1·26	+4, + 6	0.67 (+4)	841.0	2628	690·0
Ir (77)	$5d^7\ 6s^2$	192·30	8.54	22·5	2454	5300	1·26	+3, +4	0·60 (+ 4)	887.1	27·6	648·5
Pt (78)	$5d^8\ 6s^2$	195· 10	9·10	21·4	1769	4530	1·29	+2, + 4	0·52 (+2)	866.1	21·7	564·8
Au (79)	$5d^{10}\ 6s^1$	196·97	10·20	19·3	1063	2970	1·44	+1, + 3	1·37 (+1)	891.2	12·7	365·3
Hg (80)	$5d^{10}\ 6s^2$	200·59	14·81	13·6	–38.4	357	1·57	+2	1·10	1008.3	2·3	58·2

constant or increases only slightly towards the end. This is because in the atoms of *d*-block elements, the addition of electrons takes place in inner $(n-1)d$ subshell which supplements the primary screening effect. The additional electrons screen the outer *ns* electrons quite *effectively* from the added nuclear charge. As a result the effective nuclear charge increases only slightly. Thus, there is only a little change in the atomic radii.

The increase in atomic radii of elements towards the end of the series can also be explained in terms of increased forces of repulsion between the added electrons which exceed the attractive forces due to increased nuclear charge. Therefore, the electron cloud expands and size increases. (Refer to Tables 2.4 and 2.5)

The atomic radii in general, increase considerably while going down the group. This is due to the fact that the electrons occupy energy levels farther from the nucleus. However, the transition metals of third series have nearly the same radii as metals of second transition series above them. This is because of a phenomenon known as *lanthanide contraction* which involves filling of *f*-orbitals. Because of the shape, screening effect of *f*-orbitals is minimal.

2. Ionic radii

The ionic radii also follow the same trend as the atomic radii. Since the metals exhibit different oxidation states, the radii of ions also differ. In general, the ionic radii decrease with increase in oxidation states. For the same oxidation state, the ionic radii generally decrease with increase in nuclear charge. (Refer to Table 2.4 and 2.5)

3. Density

The densities of transition metals, in general are very high. The densities in second and third series reach a maximum value at Rh and Ir respectively and then decrease. This is due to their small radii and closed packed structure in the elements of group 10 (VIII B). After group 10, there are increased forces of repulsion between added electrons and these more than counterbalance attractive force due to increased nuclear charge. Thus, atomic volume increases after group 10 and consequently, density decreases. The densities of second row are high and those of third row are even higher. This is because the atomic sizes of the elements of second and third transition series remain almost the same but their atomic masses become two fold. As a result of this, the densities of the elements of the third transition series are generally twice those of the corresponding elements of the second transition series. (Refer to Table 2.4 and 2.5)

4. Melting and boiling points

The transition elements have very high melting and boiling points. It is observed that the melting and boiling points of these elements reach a maximum value near group 7 or 8 and then decrease with increase in atomic number. However, technitium has abnormally low melting point. The high melting points of the transition metals reveal that the atoms in these elements are held together by strong metallic bonds. This is also evident from the fact that these metals have high enthalpies of atomization (sublimation)

5. Ionisation energies

The ionisation energies increase as we move across second and third transition series, though not quite regularly. In any particular series, the difference in ionisation energies between any two successive *d*-block elements is much less as compared to the difference in case of successive s-block or *p*-block elements. The increase in ionisation energy along a given transition series is due to the effect of increasing nuclear charges. Consequently, ionisation energy is expected to increase. However, in case of transition elements, the addition of electron takes place in the penultimate shell i.e., $(n-1)$ *d* subshell and this increases the screening effect, the outer electrons are shielded more and more. Thus, the effect of increasing nuclear charge is opposed by the additional screening effect of the nucleus and consequently, ionisation energy increases but quite slowly among a period of *d*-block elements. (See Tables 2.4 and 2.5)

In general the first ionisation energies of third transition series are higher than those of second transition series. This is due to the presence of *f*-orbitals in the atoms of third transition series. The 4*f*-orbitals have very poor shielding effect. As a result, the outer electrons have greater effective nuclear charge acting on the outer valence electrons. Therefore, their ionisation energies are higher.

6. Oxidation states

The transition elements exhibit a large number of oxidation states. This is one of the remarkable features of transition elements. Like first series, the elements of second and third transition series also exhibit variable oxidation states as given in Table 2.6.

Table 2.6 Different oxidation states of elements of second and third transition series.

Second transition series

Y	Zr	Nb	Mo	Tc	Ru	Rh	Pd	Ag	Cd
+3	(+3)	(+2)	+2	+2	+2	+2	+2	+1	+2
	+4	(+3)	+3	(+4)	+3	+3	(+3)	(+2)	
		+4	+4		+4	+4			
		+5	+5		(+5)	(+6)		+3	
					(+6)				
					(+7)				
					(+8)				

Third transition series

La	Hf	Ta	W	Re	Os	Ir	Pt	Au	Hg
+3	(+3)	(+2)	+2	(−1)	+2	+2	+2	+1	+1
	+4	(+3)	(+3)	(+1)	+3	+3	(+3)	+3	+2
		(+4)	+4	(+2)	+4	+4	+4		
		+5	+5	+3	+6	(+6)	(+5)		
			+6	+4	+8		(+6)		
				+5					
				(+6)					
				+7					

The variable oxidation states are due to the participation of inner $(n–1)$ *d* and outer *ns* electrons.

In group 8 the second and third row elements show a maximum oxidation state of +8 compared with (+6) for iron. This is because the electronic configurations of the atoms in the second and third transition series do not always follow the pattern of their analogues of first series. Therefore the stability of oxidation states exhibited by second and third transition series are different from those of elements of first transition series. For example, the ground state electronic configurations of Pd ($4d^{10}$) and coinage metals, Cu, Ag and Au $[(n-1)\, d^{10}\, ns^1]$, have filled *d*-subshells but they behave as typical transition metals. This is because in their most common oxidation states, Cu (II) Pd (II), and Au (III) have partly filled *d*-sublevel. However, in Zn, Cd and Hg, the ions Zn^{2+}, Cd^{2+} and Hg^{2+} have d^{10} configuration. Because of this, these elements do not show characteristics of transition elements.

In general, in the elements of second and third transition series the higher oxidation states are more stable than those of corresponding first transition series elements. For example, the heavier elements such as Mo, W, Tc, Re, Ru and Os from oxoanions of the formula $[MO_4]^{n-}$ in higher valence states which are less easily reduced than those of analogus elements of first transition series. The elements in groups 6 to 8 prefer their highest oxidation states.

7. Magnetic properties

The transition elements involve the partial filling of *d*-subshells, hence most of their compounds have unpaired electrons in *d*-orbitals and therefore, give rise to paramagnetic character. The magnetic character is expressed in terms of magnetic moment. The larger the number of unpaired electrons in a substance, the greater is the paramagnetic character and larger is the magnetic moment. The magnetic moment is expressed in Bohr magneton abbreviated as B.M.

In the case of elements of first transition series, the magnetic moments arise only from the spin of the electrons. This can be calculated from the relation:

$$\mu = \sqrt{n(n+2)} \quad \text{B.M.}$$

Where n is the number of unpaired electrons and B.M. represents Bohr magneton.

However, in the case of elements of second and third transitions, it has been seen that the magnetic moment is higher than calculated by the above formula. This suggests that there is a contribution of orbital motion of the electron also. Therefore, the magnetic moments of elements of second and third transition series are due to both electron spin and orbital motions.

8. Catalytic properties

Many transition metals and their compounds act as good catalysts for various reactions. For example,

(*i*) Pd is used for hydrogenation reactions in organic chemistry.

(*ii*) Pt/PtO is used as Adam's catalysts in reduction reactions.

(*iii*) Pt/Rh is used in Ostwald process in the manufacture of HNO_3.

(*iv*) $PdCl_2$ is used in tracker in proces for converting C_2H_4 to CH_3CHO.

GENERAL COMPARISON OF SECOND AND THIRD TRANSITION SERIES WITH FIRST TRANSITION SERIES

The elements of second and third transition series have very similar chemical propereties but in many cases they show large difference from the elements of first transition series. The important difference and their causes are given below:

1. Anomalous electronic configuration

In the first transition series only two elements Cr and Cu have anomalous electronic configurations. These are due to extra stabilities associated with half filled and completely filled (n–1) d subshells. However, in the case of second and third transition series in addition to anomalous configurations due to half filled and completely filled *d*-orbitals [For example, Mo($4d^5$), Ag($4d^{10}$) and Au($5d^{10}$)], many other elements have irregular configurations [e.g. Nb, Ru, Rh, Pd; second transition series; Pt, Ir; third transition series). These irregularities in electronic configurations can be explanned in terms of nucleus-electron and electron-electron forces.

2. Atomic and ionic radii

The atomic and ionic radii of elements of only first transition series are well known. On the other hand, the radii of elements of second and third transition series are not known with certainty except in a few cases.

The atomic and ionic radii of elements of second and third transition series are very similar in a group due to lanthanide contraction, whereas the corresponding radii of first transition series smaller.

The elements in group 3 of *d*-block show increase in size from Sc $\longrightarrow$ Y $\longrightarrow$ La. However, in the next groups (4-12) there is increase in radii between the first and the second member, but there is hardly any change between the second and the third element.

This is due to the fact that there are 14 intervening elements between lanthanum and hafnium (called lanthanides) in which the antipenultimate 4*f* shell of electrons is filled. There is gradual decrease in size of the 14 elements from Ce to Lu and this is called **lanthanide contraction** (discussed later). This lanthanide contraction counterbalances the normal size increase on descending a group of transition elements. Therefore, the radii of next members of third transition series become almost same as that of corresponding elements of second transition series.

3. Magnetic properties

The magnetic behaviour of second and third transition series elements is complex and cannot be easily explained on the basis of spin only contributions. It has been observed that the transition elements of first series form mainly **high spin complexes** while those of second and third series form low spin complexes. This is due to the fact that the filling of electrons in second and third transition series involves 4*d* and 5*d* subshells respectively. The 4*d* and 5*d*-orbitals are large more spread out in space than 3*d*-orbitals. There are smaller interelectronic repulsions in 4*d* and 5*d*-orbitals compared to 3*d*-orbitals. With less interelectronic repulsions, ligands can approach the metal orbitals in complexes. Due to this close approach, they produce very large crystal field splitting energies. Therefore, these elements from low spin (spin paired) complexes.

The spin only formula gives satisfactory agreement relating to the observed magnetic moments of first row transition metal complexes to the number of unpaired electrons. However, in the case of second and third row elements, the orbital contribution also plays its part. In addition to spin, orbit coupling may also occur. Therefore, the interpretation of magnetic moments in terms of the number of unpaired electrons cannot be extended to the second and third row elements.

4. Oxidation states

For the first transition series, the lower oxidation states are more stable than for the elements of second and third transition series. The elements of first transition series have at least one stable lower oxidation state which form stable oxysalts and hydrated cations. This does not happen for the elements of second and third transition series. The chemistry of complexes and aqua ions of the lower valence states especially +2 and +3 which are significant in the first transition series are of no importance for second the third transition series except for Ru, Pd and Pt. For example, the first row forms a large number of extremely stable complexes such as $[Cu^{III} Cl_6]^{3-}$ and $[Co^{III} (NH_3)_6]^{3+}$ but corresponding examples of Mo or W or Rh or Ir are not known. While +2 state is known for all elements of the first transition series, it is relatively unimportant for the heavier elements. Cadmium is nearly restricted to +2 oxidation state and Hg (II), Pd (II) and Pt (II) are the only other important dipositive ions. Chromium (III) is most stable oxidation state of chromium but both molybdenum and tungsten are strongly reducing in that oxidation state with +6 oxidation state being much more important.

5. Geometry of complexes

The four coordinated complexes of first transition series are mainly tetrahedral with a few exceptions. For example, the four coordinated complexes of Ni^{2+}, Co^{2+}, Cu^{2+} are tetrahedral as well as square planar. On the other hand, the four coordinated complexes of elements of second and third transition series are generally square planar. In a given oxidation state, they show multiple stereochemistries. For example, in $Na_3[ZnF_7]$, the $[ZnF_7]^{3-}$ is pentagonal bipyramial, and in $(NH_4)_3[ZrF_7]$, it is capped trigonal prism. In $Cu_2[ZrF_8]$, the Zr is at the centre of square anti prism.

COMPARATIVE STUDY OF SECOND AND THIRD TRANSITION SERIES

Yttrium and Lanthanum (Group 3 or III B)

First member of this group scandium forms compounds in the + 3 oxidation state. The oxide, halides and oxohalides have the formulae Sc_2O_3, ScX_3 and ScOX respectively.

COMPOUNDS OF YTTRIUM

1. Yttrium forms the oxide Y_2O_3. $Y(OH)_3$ is obtained as gelatinous precipitate when alkali is added to an aqueous solution of Y^{3+} salt. $Y(OH)_3$ is basic in nature.

2. With the exception of YF_3, all halides are soluble in water. These are prepared by heating the metal with the halogens.

3. Yttrium forms a non-stoichiometric hydride with the formula approximately equal to YH_3.

4. Yttrium forms a number of complexes with EDTA, acac, NCS^- with formulae, [Y(EDTA)], $[Y(acac)_3 . H_2O]$, $[Y(NCS)_6]^{3-}$, respectively. Complexes of Y^{3+} with coordination numbers 8 and 10 are also known.

5. **Organometallic Compounds.** Yttrium forms $Y(C_5H_5)_3$ when YCl_3 reacts with NaC_5H_5. The bonding in $Y(C_5H_5)_3$ is predominately ionic.

Lanthanum is a part of lanthanide series. Its properties have been given separately in chapter 5.

Zirconium and Hafnium (Group 4 or IV B)

The first member of this group is titanium. It forms oxide, tetanates chloride, complexes and organometallic compounds in oxidation state +4.

Compounds of Zirconium

1. Oxides and Hydroxides. $ZrO_2 . nH_2O$ is obtained as a gelatinous precipitate or adding an alkali to zirconium (IV) solutions. On strong heating, the hydrous oxide gives ZrO_2.The hydrous oxide is also obtained by vapour phase oxidation of $ZrCl_4$. This is done by passing the vapour of $ZrCl_4$ mixed with air through a special flame produced by burning a hydrocarbon in excess of oxygen.

2. Halides. $ZrCl_4$ is made by passing chlorine over the dioxide heated with carbon.

It gets easily hydrolysed yielding oxochloride.

$$ZrCl_4 + H_2O \longrightarrow ZrOCl_2 + 2HCl$$

The oxochloride crystallises from the solution as the octahydrate, $ZrOCl_2 . 8H_2O$. $ZnCl_4$ resembles $TiCl_4$ in most of its properties.

3. $ZrBr_4$ and **ZrI_4** are also known. These are made by direct combination of the elements and are similar in properties to $ZrCl_4$. $ZrBr_4$ is yellow while ZrI_4 is red-brown in colour.

4. Zr (IV) in Aqueous Solutions. The zirconyl ion, ZrO^{2+} exists in aqueous solutions alongwith the usual anions such as NO_3^-, SO_4^{2-}, etc. The compounds of ZrO^{2+} can be easily isolated from its aqueous solutions. $ZrO(NO_3)_2$ in solid state contains an oxygen-bridged chain, as shown:

```
 Zr           Zr           Zr           Zr
/  \        /    \        /  \        /    \
    O      O      O      O    O      O
     \    /        \    /      \    /
      Zr            Zr          Zr
```

Zirconyl nitrate solution is used for the removal of phosphate ions in qualitative analysis.

5. Complexes of Zirconium. Apart from the complex ions $[ZrX_6]^{2-}$ (X = halogen), some 7-coordinated complex ions of Zr are also known. Well known example is $[ZrF_7]^{3-}$

The atomic and ionic radii of Zr and Hf are identical and therefore their chemistry is remarkably similar. Hf forms similar compounds as Zr.

Niobium and tantalum (Group 5 or VB)

The first member of the group vanadium forms oxide, halides, oxochloride, oxosulphate and complexes in +4 oxidation state and oxides, fluoride, oxohalides and complexes in +5 oxidation state. Atomic radii of Nb and Ta are identical. Therefore, they show similar chemical reactivity with some difference here and there .

Niobium and Tantalum show oxidation states of +5, +4, +3 and +2 in their various compounds. The most stable oxidation state, however, is + 5.

Compounds with +5 Oxidation State. 1. Oxides. Nb_2O_5 and Ta_2O_5.

Both the compounds are inert and can be brought into solution only on treating them with hydrofluoric acid or by fusing them with sodium carbonate or sodium hydroxide.

$$Nb_2O_5 + 3Na_2CO_3 \longrightarrow 2Na_3NbO_4 + 3CO_2$$

Tantalum pentoxide reacts in the same manner but slowly.

Both the pentoxides are more stable than V_2O_5. Consequently, it is more difficult to reduce them to the lower oxides.

2. Halides. The pentafluorides are obtained by heating the metals, the pentoxides or pentachlorides in fluorine. They are white, volatile solids.

The other pentahalaides are yellow to brown or purple-red solids. These are obtained by direct combination of the elements as well as by heating the pentoxides with carbon and the halogen concerned.

$$2\,Ta + 5I_2 \longrightarrow 2TaI_5$$

$$Nb_2O_5 + 5C + 5Cl_2 \longrightarrow 2NbCl_5 + 5CO$$

All the pentahalides can be sublimed in atmosphere or the corresponding halogen without decomposition. These halides are soluble in organic liquids including carbon tetrachloride and ether. These are readily hydrolysed by water to yield hydrous pentoxides and hydrohalic acids.

$$2TaCl_5 + 5H_2O \longrightarrow Ta_2O_5 + 10HCl$$

The chlorides act as Friedel-Craft catalysts.

Both NbF_5 and NbF_5 have cyclic tetrameric structures, as shown below:

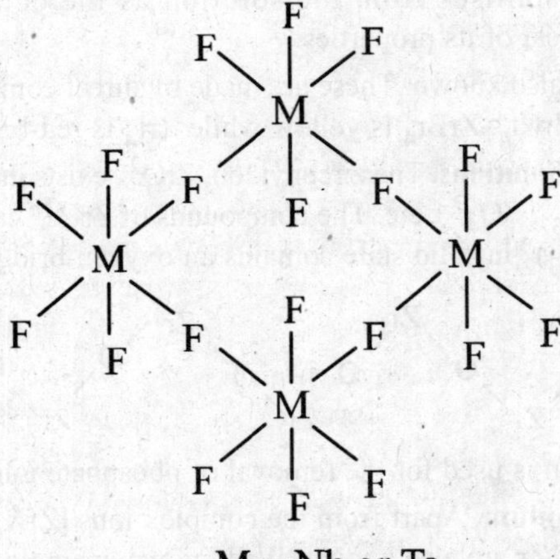

NbF_5 and TaF_5 react with $F^-(aq)$ to give $[MF_6]^-$ and $[MF_7]^{2-}$ (M is Nb or Ta). With very high concentration of F^- ions in aqueous solutions, TaF_5 forms $[TaF_8]^{3-}$ and NbF_5 forms $[NbOF_6]^{3-}$.

3. Oxohalides. The oxohalides can be prepared by the oxidation of the corresponding pentahalides. Thus,

$$2NbCl_5 + O_2 \longrightarrow 2NbOCl_3 + 2Cl_2$$

$$2TaCl_5 + O_2 \longrightarrow 2TaOCl_3 + 2Cl_2$$

The oxofluorides can also be prepared by heating the pentafluorides with silica. Thus,

$$2TaF_5 + SiO_2 \longrightarrow 2TaOF_3 + SiF_4$$

4. Compounds of Lower Oxidation States. Niobium (IV) and Tentalum (IV) Oxides. NbO_2 is obtained by the reduction of Nb_2O_5 in a current of hydrogen at elevated temperature.

$$Nb_2O_5 + H_2 \longrightarrow 2NbO_2 + H_2O$$

TaO_2 is obtained by reduction of Ta_2O_5 by carbon at a high temperature:

$$Ta_2O_5 + C \longrightarrow 2TaO_2 + CO$$

5. The Lower Halides. The lower halides of Nb and Ta are prepared by reduction of the pentahalides on heating either with the metal itself or with aluminium. The oxidation state of the final product depends upon the intensity of the reduction process. Higher the temperature and larger the time of reduction, lower is the oxidation state of the product obtained.

It was previously thought that the lower halides of Nb and Ta are non-stoichiometric in nature. It has now been confirmed that most of them are **cluster compounds** whose structures are based on $[M_6X_{12}]^{2+}$ cluster units.

The tetrahalides are all dark solids and are diamagnetic. This is possible only if there is extensive M-M bonding in these compounds. They are also volatile above 300°C with the exception of NbF_4.

6. Organometallic Compounds of Niobium and Tantalum. Nb and Ta form a number of organometallic compounds with cyclopentadienyl group. Some of these compounds are $[Nb(\eta^5\text{–}C_5H_5)_2(\eta^1\text{–}C_5H_5)_2]$, $[Nb(\eta^5\text{–}C_5H_5)_2Cl_2]$, $[Nb(\eta^5\text{–}C_5H_5)_2Cl_3]$ and $[Ta(\eta^5\text{–}C_5H_5)_2Cl_2]$.

Molybdenum and Tungsten (Group 6 or VI B)

The first member of the group, chromium, forms oxide, chloride and complexes in +3 oxidation state; CrO_3 and chromic acid in +6 oxidation state.

Compounds of Molybdenum

Compounds of Molybdenum (VI).

1. Molybdenum trioxide, MoO_3. It is obtained by roasting molybdenite ore in air or boiling it with conc. nitric acid. It is an acidic oxide and dissolves in a strong alkali solution to give molybdate which contains MoO_4^{2-} ion. Lowering of *p*H of the molybdate solution yields a polymolybdate. MoO_3 is thermally and kinetically more stable than CrO_3. It may be pointed out here the whereas CrO_3 is a strong oxidising agent, MoO_3 has very mild oxidising character. This indicates higher stability of +6 oxidation of Mo compared to Cr.

It dissolves in ammonia and when the solution in excess of ammonia is allowed to cool, crystals of **ammonium molybdate** separate out. Ammonium molybdate has *not* been assigned the simple formula $(NH_4)_2MoO_4$ but is represented as $(NH_4)_2Mo_7O_{24}.4H_2O$. Ammonium molybdate is used as a laboratory reagent for the detection as well as elimination of phosphates.

When added to a solution of a phosphate it produces a yellow perrcipitate of ammonium phosphomolybdate, usually reprepsented as $(NH_4)_3PO_4.12MoO_3$.

2. Polymolybdates. The salts of molybdic acid, *i.e.*, molybdates, give polymolybdates on acidification. The composition of polymolybdate formed depends upon the *p*H of the solution, as shown below:

$$MoO_4^{2-} \xrightarrow{pH6} [Mo_7O_{24}]^{6-} \xrightarrow{pH2} [Mo_8O_{26}]^{4-} \xrightarrow{pH1} MoO_3.2H_2O$$

The polymolybdates obtained by the condensation of MoO_4^{2-} units alone are called **isopolymolybdates.** $[Mo_7O_{24}]^{6-}$ is an example.

3. Heteropolymolybdic Acids and Their salts. Heteropolymolybdates are formed when molybdate solutions containing other oxo anions, usually phosphate, silicate and arsenate, are allowed to crystallise together. The ratio of MoO_3 to PO_4^{3-} or SiO_4^{4-}, etc., condensed together is usually 12:1 or 6:1.

The structure of heteropolymolybdates consists of tetrahedral PO_4^{3-}, SiO_4^{3-} or BO_4^{3-} ions surrounded by MoO_6 octahedral units. A well known example of heteropolymolybdate is the product obtained by mixing solutions of PO_4^{3-} and ammonium polymolybdate, *i.e.*, $(NH_4)_3$ $[PO_4Mo_{12}O_{36}]$.

4. Mixed Oxides of Mo (VI). When MoO_3 is fused with the oxides of alkali and alkaline earth metals, mixed oxides are produced. The structures of these mixed oxides are quite complicated.

5. Molybdenum hexafluoride, MoF_6. This is another important compound of Mo (VI). It is formed by direct combination of the elements. It forms colourless crystals melting at 17°C.

It reacts slowly with NaF to give a crystalline solid of composition $MoF_6.2NaF$. It is more stable than CrF_6.

6. Other Halides and Oxohalides of Mo (VI). The dark coloured $MoCl_6$ and $MoBr_6$ are well known. These solids hydrolyse easily to produce molybdic acids.

7. Compounds of Molybdenum (V). Molybdenum pentoxide, Mo_2O_5. It is obtained by heating molybdenum powder with molybdenum trioxide in nitrogen at 750°C.

$$Mo + 5MoO_3 \longrightarrow 3Mo_2O_5$$

8. Molybdenum pentafluroide, MoF_5. It is prepared by the action of fluorine diluted with nitrogen on the metal at 400°C. It is also formed by the reduction of hexafluoride by the carbonyl of the metal.

$$5MoF_6 + Mo(CO)_6 \xrightarrow{25°C} 6MoF_5 + 6CO$$

9. Molybdenum pentachloride, $MoCl_5$. It is prepared by heating the metal in chlorine. It is highly deliquescent and is hydrolysed in water to give the oxochloride $MoOCl_3$.

$$MoCl_5 + H_2O \longrightarrow MoOCl_3 + 2HCl$$

10. Compounds of Molybdenum (IV). Molybdenum dioxide, MoO_2. It is obtained by heating the trioxide at 450°C in hydrogen.

$$2MoO_3 + 2H_2 \longrightarrow 2MoO_2 + 2H_2O$$

11. Molybdenum tetrafluoride, MoF_4, is obtained by treating MoF_5 with molybdenum carbonyl at –75°C when a compound of composition Mo_2F_9 is obtained which on heating to 150°C gives MoF_4.

12. Molybdenum tetrachloride, $MoCl_4$, is obtained by refluxing the solution of $MoCl_5$ in benzene, in a reducing atmosphere. It is also obtained on treating MoO_2 with carbon tetrachloride at 250°C.

$$MoO_2 + CCl_4 \longrightarrow MoCl_4 + CO_2$$

13. Compounds of Molybdenum (III). The existence of **molybdenum sesquioxide**. Mo_2O_3, has not been firmly established. However, the **trihydroxide, $Mo(OH)_3$**, appears as a black precipitate by adding alkali to a solution of molybdate reduced by zinc.

14. Compounds of Molybdenum (II). The dichloride, $MoCl_2$, is obtained on heating the trichloride in an atmosphere of carbon dioxide.

$$2MoCl_3 \longrightarrow MoCl_2 + MoCl_4$$

The tetrachloride, being more volatile, distils off. It can also be prepared by passing chlorine over molybdenum heated at 120°C. It is an amorphous yellow solid, insoluble in water. The **dibromide** has the molecular formula, Mo_6Br_{12}. It is similar to chloride in most of its properties.

Blude Oxides. Blue oxides of molybdenum, known as *molybdenum blues*, are prepared by the reduction of acidified solutions of molybdates by H_2S or SO_2. They contain oxides and hydroxides. Their compositions vary although the colour in each case remains blue. Thus, they have the compositions $MoO_{2.33}.H_2O$, $Mo_{2.5}(OH)_{2.5}$, $MoO_2(OH)$, etc.

Compounds containing Molybdenum in Unusual Oxidation States. The organometallic compound $(C_6H_6)_2Mo^+$ has Mo in + 1 oxidation state whereas metal carbonyls $Mo(CO)_6$, $[Mo_2(CO)_{10}]^{2-}$ and $[M(CO)_5]^{2-}$contain the metal in zero, –1 and –2 oxidation states. This is in agreement with the general statement that pi-acceptor ligands stabilise lower oxidation states of the metal.

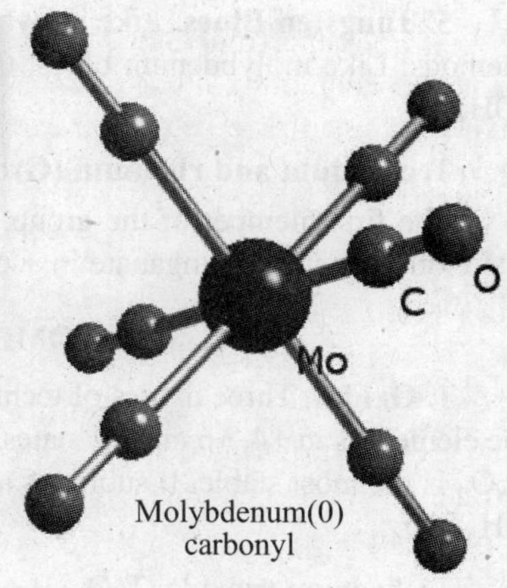

Molybdenum(0) carbonyl

COMPOUNDS OF TUNGSTEN

Tungsten shows oxidation states of 2, 3, 4, 5 and 6. The compounds in which tungsten is in + 6 state are more important.

1. Oxides. Tungsten Trioxide, WO_3. This is the most important oxide of tungsten. It is obtained from *wolframite*. It is a yellow powder, insoluble in water and has amorphous nature. It dissolves in alkalis forming *tungstates*. When heated mildly in a current of hydrogen, it changes into **tungsten dioxide, WO_2** which is a brown powder. It is soluble in water. It has a complicated structure containing W-W bonds.

2. Tungstic Acid and Tungstates. The acid H_2WO_4 is obtained as a precipitate when a solution of any tungstate is treated with hydrochloric acid.

When WO_3 dissolves in sodium hydroxide solution, sodium tungstate, Na_2WO_4, is formed. While CrO_4^{2-} is is strongly oxidising, WO_4^{2-} is a very mild oxidising agent.

On lowering the *p*H of tungstate solution, **isopolytungstates** are formed. These compounds have complicated structures containing the WO_6 octahedral units joined in a variety of ways by sharing the corners or edges of the octahedral units.

When WO_4^{2-}, cocondenses with PO_4^{3-}, SiO_4^{4-} and BO_4^{3-}, **heteropolytungstates** are formed. A well known example of a hetero-polytungstic acid is phosphotungstic acid, $H_3PO_4.12WO_3$.

3. Halides. Tungsten also forms a number of halides which are covalent. The examples are: WF_6, WCl_6, WCl_5, WCl_4, Their structures and properties are similar to those of the corresponding halides of molybdenum.

4. Tugsten Bronzes. On reducing alkali metal tungstates (*e.g.*, Na_2WO_4 or K_2WO_4) with tungsten, *deeply coloured* substances are obtained. These substances, called tungsten bronzes, are unreactive but are good conductors of electricity. They have the general formula M_xWO_3 (where M = Li, Na or K) The value of $x<1$. These are, evidently, non-stoichiometric substances in which M^+ ion is always present in interstitial sites.

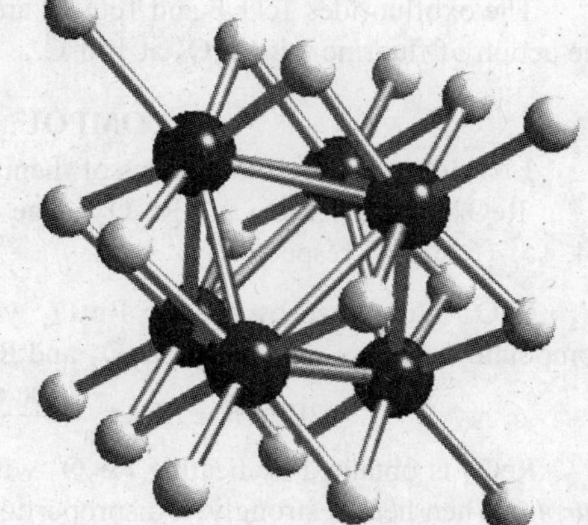
Tungsten(III) chloride

5. Tungsten Blues. Like molybdenum blues, tungsten also forms **tungsten blues** by similar methods. Like molybdenum blues, they have varied compositions as $WO_{2.67}(OH)_{0.33}$ and $WO_{2.7}(OH)_{0.1}$.

Technitium and rhenium (Group 7 or VII B)

The first member of the group, manganese forms complexes in +2 oxidation state, oxide in + 4 oxidation state, manganate in + 6 oxidation state and permanganate in + 7 oxidation state.

COMPOUNDS OF TECHNETIUM

1. Oxides. Three oxides of technetium having compositions TcO_2, TcO_3 and Tc_2O_7, in which the element is in +4, +6 and +7 states, respectively, are known. Amongst these, technetium dioxide, TcO_2 is the most stable. It sublimes at 1000°C. It is obtained by heating ammonium pertechnetate, NH_4TcO_4.

Technetium trioxide, TcO_3, is obtained by heating TcO_3Br.

Technetium heptoxide, Tc_2O_7, is made by heating the metal or other oxides to 200°C in air. It is a pale-yellow hygroscopic solid and melts at 300°C.

2. Pertechnic Acid and Pertechnetates. Pertechnic acid is obtained by dissolving Tc_2O_7 in water.

$$Tc_2O_7 + H_2O \longrightarrow 2HTcO_4$$

The solution, on evaporation, yields dark red crystals of the anhydrous acid. Thus, unlike $HMnO_4$, which is known only in solution, $HTcO_4$ is known in solid state as well.

Pertechnetates of alkali metals like $KTcO_4$ are also known. These are obtained by dissolving Tc_2O_7 in alkalis. These are more stable than the permanganates and do not have oxidising properties. Pertechnetates react with potassium in ethanol to give $K_2[TcH_2]$.

3. Halides. The known halides of technetium are $TcCl_4$, $TcCl_6$ and TcF_6. The hexafluoride, TcF_6, has been studied in greater details. It exists as golden yellow crystals melting at 33°C. It is obtained by heating the metal in fluorine.

The oxofluorides TcO_3F and TcO_3Cl are also known. The oxofluoride, $TcOF_3$, is obtained by the action of fluorine with TcO_2 at 150°C.

COMPOUNDS OF RHENIUM

1. Oxides. Amongst the oxides of rhenium are the dioxide, trioxide, pentoxide and heptoxide (*i.e.*, ReO_2, ReO_3, Re_2O_5 and Re_2O_7). The oxidation states of the metal in these compounds are +4, +5, +6 and +7 respectively.

ReO_2 is obtained by heating Re_2O_7 with the metal at 600°C. At higher temperatures, this compound decomposes back to Re_2O_7 and Re.

$$2Re_2O_7 + 3Re \xrightarrow{600°C} 7\,ReO_2$$

ReO_3 is obtained by heating Re_2O_7 with rhenium at 300°C. It exists in two forms: *red and purple.* When heated strongly, it disproportionates as under:

$$3ReO_3 \longrightarrow Re_2O_7 + ReO_2$$

2. Perrhenic Acid and Perrhenates. Perrhenic acid is obtained as colourless solution on dissolving Re_2O_7 in water:

$$Re_2O_7 + H_2O \longrightarrow 2HReO_4$$

The acid is also obtained by oxidising the metal on heating in nitric acid or by passing chlorine in a suspension of ReO_2.

$$2ReO_2 + 3Cl_2 + 4H_2O \longrightarrow 2HReO_4 + 6HCl$$

It has not been possible to obtain anhydrous perrhenic acid. The solution when heated gives back Re_2O_7.

Perrhenic acid is a strong monobasic acid. It reacts with magnesium, iron and zinc, evolving hydrogen. It is not a strong oxidising agent. However, It can oxidise HBr to bromine and HI to iodine.

Alkali metal perrhenates (*e.g.*, $KReO_4$ and $NaReO_4$) are colourless and stable compounds. Sodium perrhenate and ammonium perrhenate are very soluble in water. Potassium perrhenate, however, is much less soluble. It appears as a precipitate when KCl and $HReO_4$ solutions are mixed.

$$KCl + HReO_4 \longrightarrow KReO_4 + HCl$$

3. Halides and Oxohalides. A number of halides of rhenium in which the element is in +3, +4, +5, +6 and +7 oxidation states, are known, A number of **oxohalides** are also known.

$ReBr_3$ is obtained by direct combination of the elements at 300°C. It is a dark green solid. It has a trimeric structure as shown. The lines between the atoms in this structure do not show the exact bond order.

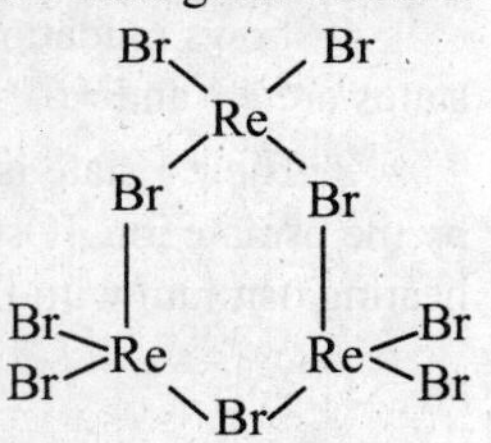

$ReCl_5$ is also made by heating the metal in chlorine. When heated in an inert atmosphere, it decomposes into **$ReCl_3$** which is a violet crystalline solid. It reacts with water to give perrhenic acid

$$3ReCl_5 + 8H_2O \longrightarrow HReO_4 + 2ReO_2\text{ (hydrated)} + 15HCl$$

$ReBr_4$ and **ReI_4** are obtained by reducing perrhenic acid with HBr and HI, respectively.

Unusual Oxidation States of Rhenium. Re has zero oxidation state in $[Re_2(CO)_{10}]$ and +1 oxidation state in $[ReCl(CO)_5]$.

Ruthenium and osmium (Group 8)

The chemistry of Ru and Os shows little resemblance to that of Fe, the first number of group 8, except in sulphides or phosphoric and in some complexes with π bonding ligands like CO, PR_3 or $\eta^5-C_5H_5$. The higher oxides are more readily obtained than for Fe.

COMPOUNDS OF RUTHENIUM

1. Ruthenium (II) compounds, $RuCl_2$ and RuO are known to exist. However, ruthenium (II) complexes such as $[Ru\ (NH_3)_6]^{2+}$ and $[Ru\ (CN)_6]^{4-}$ are more stable than simple ruthenium (II) compounds.

Ruthenium shows special affinity towards π acid ligands such as CO, PR_3, etc. with which it forms tetrahedral complexes. π acid ligand NO also forms a large number of mixed complexes with Ru (II). Examples are $[Ru(NO)Cl_5]^{2-}$ and Ru $[NO(CN)_5]^{2-}$. The Ru – NO linkage in the mixed nitroso complex is either straight or bent, *i.e.*, Ru – N = O or Ru – N=O (bent)

2. Ruthenium (III) complexes are also quite stable. Examples of such examples are $[Ru(NH_3)_6]^{3+}$ and $[Ru\ (NH_3)_5H_2O]^{3+}$.

Ruthenium shows special affinity towards π acid ligands such as CO and even N_2. The affinity for N_2 is shown by the following reaction:

$$[Ru(NH_3)_5H_2O]^{3+} \xrightarrow{NaN_3} [Ru\ (NH_3)_5\ N_2]^{3+}$$

RuO_2 is also knwon. It is a bluish black solid which contains ruthenium in +4 oxidation state.

3. RuO_4 containing Ru (VIII) is a yellow solid which dissolves in alkalis to give perruthenate and ruthenate, as shown.

$$4RuO_4 + 4OH^- \longrightarrow \underset{\substack{\text{Perruthenate} \\ \text{containing Ru (VIII)}}}{4RuO_4^-} + H_2O + O_2$$

$$4RuO_4 + 4OH^- \longrightarrow \underset{\substack{\text{Ruthenate} \\ \text{containing Ru (VI)}}}{4RuO_4^{2-}} + 2H_2O + O_2$$

Unusual Oxidation States of Ruthenium. Ruthenium exists in –2 oxidation state in $[Ru(CO)_4]^{2-}$ and in the zero oxidation state in Ru $(CO)_5$, $RU_3(CO)_{12}$, etc.

COMPOUNDS OF OSMIUM

1. Osmium reacts with alkaline oxidising agents to give compounds like Na_2OsO_4 and K_2OsO_4. When heated it reacts with sulphur giving OsS_4, with fluorine giving OsF_6 and with chlorine giving $OsCl_4$.

It shows oxidation states of +2, +3, +4, +6 and +8 in its compounds. The important oxidation states are +4 and +6.

2. The +4 state is exhibited in the compounds, OsO_2 and $OsCl_4$. The + 6 state is represented by the osmate ion, $[OsO_4]^{2-}$ in sodium or potassium osmate and also in OsF_6 which is obtained by heating osmium with flourine.

$$Os + 3F_2 \longrightarrow Os\,F_6$$

3. The +8 state is exhibited in the oxide OsO_4. This is the most important compound of osmium and is also known as **osmic acid**.

OsO_4. Osmium tetroxide is a solid melting at 40°C and boiling at 100°C. It is highly volatile and its vapours are corrosive to animal tissue and hence very dangerous, particularly for the eyes.

It is a strong oxidising agent and gets reduced easily even by dust or grease to OsO_2.

4. Osmium forms a number of complex compounds with ammonia and cyanide ions as the ligands.

Unusual Oxidation States. Like ruthenium, osmium also forms a number of complexes with π acid ligands such as CO in which osmium is present in –2 or zero oxidation state. Examples are $[Os(CO)_4]^{2-}$, $[Os(CO)_5]$, $[Os_3(CO)_{12}]$, etc.

Rhodium and Iridium (Group 9)

The first member of group 9 is cobalt. It forms complexes in + 2 and +3 oxidation states.

COMPOUNDS OF RHODIUM

1. The highest oxidation state of rhodium is +6 which the metal exhibits in RhF_6. This compound is unstable and highly reactive.

2. The compounds of Rh (V) are very rare and are unstable. RhF_5 is actually a cyclic tetramer in which from RhF_6 octahedra form a ring through Rh–F–Rh bonds.

3. The compound of Rh (IV) is RhO_2 which is formed by the oxidation of an aqueous solution of Rh (III) compounds by a strong oxidising agent such a $NaBiO_3$ in the presence of an alkali.

The halocomplexes of Rh (IV) are slightly more stable than the simple Rh (IV) compounds. Thus, $K_2\,[RhCl_6]$ and $K_2[RhF_6]$ are slighlty more stable than RhF_4 and RhO_2.

4. The + 3 oxidation state of Rh is more stable than any other oxidation state of the metal. All the trihalides are known and are found to be quite stable. These halides, however, cannot be represented by simple structures.

The complexes of Rh(III) are also very stable. Most of these are diamagnetic in character. The complex cation $[Rh(H_2O)_6]^{3+}$ is quite stable in aqueous solution, whereas $[Co(H_2O)_6)^{3+}$ is oxidising. This indicates that Rh(III) salt solutions in water are more stable than Co(III) salt solutions. The anion $[RhCl_6]^{3-}$ is also quite stable.

5. The +2 oxidation state of Rh in not very stable. However, the pi acceptor ligands stabilise this oxidation state. Rhodocene. $Rh(II)(\eta^5 - C_5H_5)_2$ is similar to ferrocene and cobaltocene in structure but it is much less stable.

6. The +1 oxidation state of Rh is quite common in complexes containing pi acid ligands such as PR_3 and CO.

Wilkinson's Catalyst. The complex $[RhCl\ (PPh_3)_3]$ is known as Wilkinson's catalyst. It has a square planar structure and is red-violet in colour. It is obtained by refluxing $RhCl_3\ 3H_2O$ with PPh_3. It is used as a catalyst in the **OXO** process in which CO and H_2 are added to the **end double bond** of alkenes forming aldehydes which on oxidation give alcohols.

COMPOUNDS OF IRIDIUM

1. The highest oxidation state of iridium is + 6. The compound IrF_6 is known but is quite unstable.

2. Iridium pentafluoride IrF_5 has a structure similar to that of RhF_5 and MoF_5 and is better represented as $(IrF_5)_4$. The oxidation state of Ir in IrF_5 is evidently +5.

3. The +4 oxidation state of Ir is quite stable. The tetrahalides of Ir are known. When Ir is heated in a current of oxygen, its gets converted to IrO_2.

$$Ir + O_2 \longrightarrow IrO_2$$

A number of complexes of Ir (IV) are also known. Some of these are $[IrCl_6]^{2-}$, $[IrCl_3\ (H_2O)_3]^+$, $[IrCl_4(H_2O)_2]^+$, $[IrCl_5(H_2O]^-$ and $[Ir(C_2O_4)_3]^{2-}$ The last complex is optically active.

4. The +2 oxidation state of Ir is not very common although $IrCl_2$ (having a polymeric structure with Ir-Cl-Ir bridges) is known. Some phosphine complexes of this state are known.

Vsaka's compound, trans $[IrCl(CO)\ (PPh_3)_2]$, is yellow in colour and has the capability of coordinating aerial O_2. The yellow coloured Vsaka's compound turns orange when it takes up oxygen. When flushed with nitrogen, it releases oxygen and again turns yellow.

$[Ir(PF_3)_4]^-$ and $[Ir_4\ (CO)_{12}]$ contain Ir in –1 and zero oxidation state, respectively.

Palladium and Platinum (Group 10)

The first member of this group, mickel forms compounds in +2 and +3 oxidation states. The anion $[Ni_2(CO)_6]^{2-}$ contain nickel in –1 oxidation state. $[Ni\ (CO)_4]$, $K_2[Ni\ (CN)_4]$ contain Ni in zero oxidation state. Nickel in +1 state is present in $K_4[Ni_2(CN)_6]$.

COMPOUNDS OF PALLADIUM

1. Palladium (0) Compounds. Palladium shows its zero oxidation state in its complexes formed by σ donor and π acceptor ligands. One such complex is $[Pd(CO)\ (Ph_3)_3]$.

2. Palladium (II) Compounds. Palladium forms a large number of compounds in which it has the oxidation state of +2. Some of the important compounds of Pd (II) are described beloow.

(a) Palladous oxide, PdO. It is obtained by heating spongy palladium at 800°C in oxygen. It is a black powder. It is insoluble in acids and even in aqua regia. It is a strong oxidising agent as on heating it decomposes into the metal and oxygen.

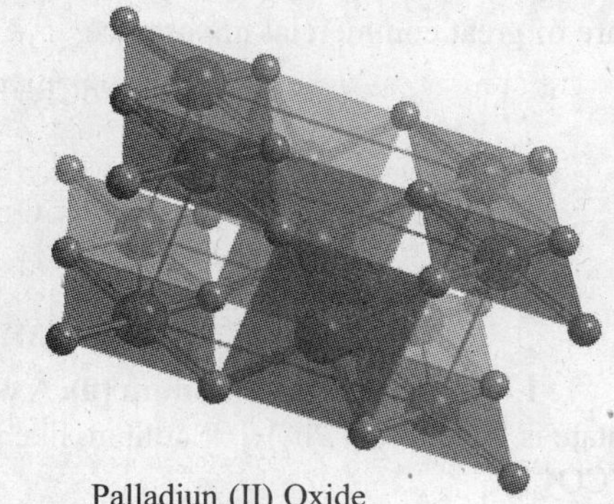

Palladiun (II) Oxide

(b) Palladous chloride, $PdCl_2$. It is obtained by heating palladium in chlorine. It is a dark red solid which is highly soluble in water. The solution on evaporation yields crystals of $PdCl_2.2H_2O$. $PdCl_2$ exists as a chain polymer in the solid state, having the following structure:

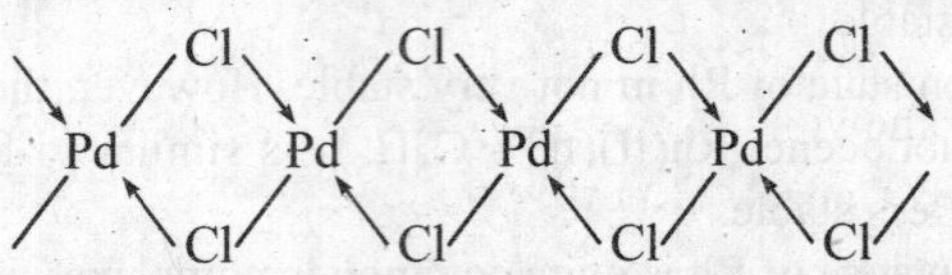

Cl Cl Cl Cl
Pd Pd Pd Pd
Cl Cl Cl Cl

(c) Palladous nitrate, $Pd(NO_3)_2$. It is obtained by dissolving palladium metal or palladous hydroxide in dilute nitric acid. The solution on evaporation yields brown deliquescent prism-like crystals.

(d) Palladous sulphate, $PdSO_4.2H_2O$. This is obtained by dissolving palladium hydroxide in dilute sulphuric acid. It is a brown deliquescent solid.

Complex Compounds of Palladium (II). A number of complex compounds of divalent palladium are known. Palladium ion, Pd^{2+}, in aqueous solution is hydrated and exists as the complex $[Pd(H_2O)_4]^{2+}$. It also forms a complex with ammonia $[Pd(NH_3)_4]^{2+}$. A solution of palladous chloride in hydrochloric acid contains the divalent complex anion, $[PdCl_4]^{2-}$.

Ammine complex tetraamminepalladium (II) chloride, $[Pd (NH_3)_4]Cl_2$, is a pink solid and **dichlorodiamminepalladium (II)**, $[Pd(NH_3)_2Cl_2]$, is a yellow solid. The latter exists in two forms:

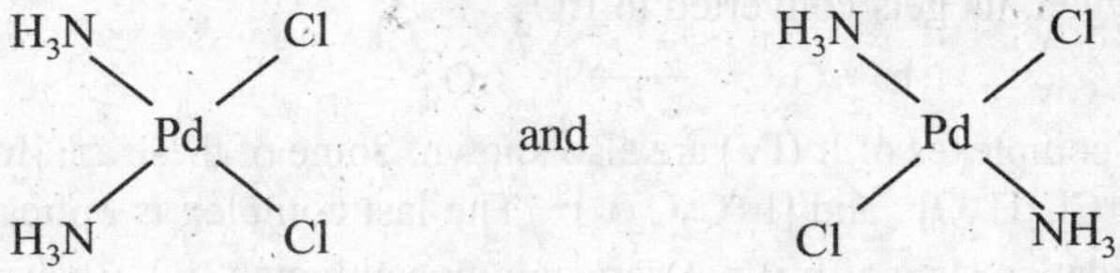

H_3N Cl / Pd / H_3N Cl — and — H_3N Cl / Pd / Cl NH_3

cis-form *trans*-form

3. Palladium (III) Compounds. Palladium (III) compounds are rather uncommon. Amongst the few compounds known, $PdF_3$is important. This is obtained by heating the metal in a current of fluorine at 500°C.

4. Palladium (IV) Compounds. Very few compounds of Pd (IV) are known. The complexes of Pd (IV) are more stable than the Pd (IV) compounds. Well known complexes of Pd (IV) are $Na_2[PdX_6]$ and Pd $[X_4(NH_3)_2]$ (X= F, Cl, Br).

Organometallic Compounds of Palladium. $PdCl_2$ and K_2PdCl_4 react with olefins to produce complexes of the type $[PdCl_3(C_2H_4)]^-$, $Pd[Cl_2\ (C_2H_4)_2]$, etc. The Pd-olefin bonding in these complexes is unique in the sense that olefin acts both as σ donor through its filled pi bonding orbitals and π acceptor through its vacant pi antibonding orbital. The reactions involving $PdCl_2$ and olefins are of great commercial importance.

$$C_2H_4 + PdCl_2 + H_2O \longrightarrow \underset{\text{Ethanal}}{CH_3CHO} + Pd + 2HCl$$

$$C_2H_4 + CO + H_2 \xrightarrow{PdCl_2} \underset{\text{Propanoic acid}}{CH_3CH_2COOH}$$

COMPOUNDS OF PLATINUM

1. Compounds of Platinum (0). A well known compound in which Pt occurs in zero oxidation state is $Pt[(CO)_2\ (PPh_3)_2]$. Platinum, like palladium, does not form a pure binary metal carbonyl, Pt $(CO)_4$.

2. Compounds of Platinum (I)

Platinum monochloride, PtCl. Univalent platinum is represented by platinum monochloride, PtCl. It is obtained by heating platinum (II) chloride with platinum at 581–583°C.

$$PtCl_2 + Pt \longrightarrow 2PtCl$$

However, it is not known in pure state.

3. Compounds and Complexes of Platinum (II)

Platinum (II) chloride, $PtCl_2$. The typical bivalent platinum is represented by platinum (II) chloride. This is obtained by heating platinum (IV) chloride at 580°C.

$$PtCl_4 \longrightarrow PtCl_2 + Cl_2$$

With hydrochloric acid and potassium chloride, the following complex compounds are found.

$$PtCl_2 + 2HCl \longrightarrow H_2[PtCl_4]$$

$$PtCl_2 + 2KCl \longrightarrow K_2[PtCl_4]$$

Complex platinum (II) chlorides are also obtained by the reduction of complex platinum (IV) chlorides by sulphur dioxide, etc.

$$H_2[PtCl_6] + SO_2 + 2H_2O \longrightarrow H_2SO_4 + 2HCl + H_2[PtCl_4]$$

The complex anion $[PtCl_4]^{2-}$ is diamagnetic. It has a square planar geometry.

Potassium tetracyanoplatinate (II), $K_2[Pt(CN)_4]$. $3H_2O$. It is obtained by mixing a solution of potassium cyanide with platinum (II) chloride in proper proportions and evaporating to crystallisation.

$$PtCl_2 + 2KCN \longrightarrow Pt(CN)_2 + 2KCl$$

$$Pt(CN)_2 + 2KCN \longrightarrow K_2[Pt(CN)_4]$$

Potassium tetracyanoplatinate (II) is used in photography.

Diamminedichloroplatinum (II). This complex exists in two isomeric forms, *viz*, *cis* diamminedichloroplatinum (II) and *trans* diamminedichloroplatinum (II), whose structures are given below.

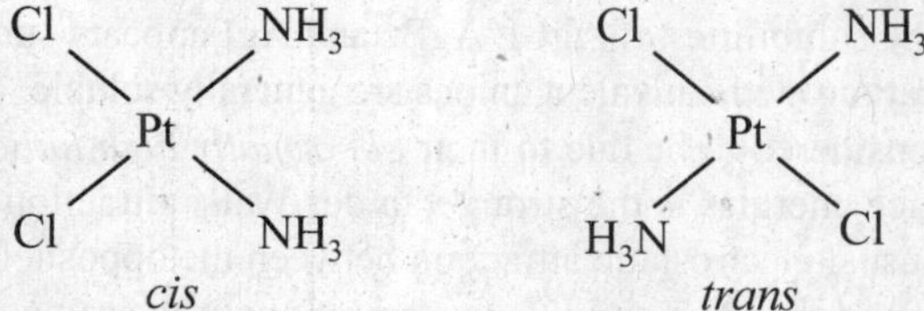

The *cis* isomer, commonly known as *cis* **platin**, is used in the treatment of cancer.

3. Compounds and Complexes of Platinum (IV).

Platinum (IV) chloride, $PtCl_4$. This is one of the best known compounds of platinum. It is obtained by heating chloroplatinic acid in a current of chlorine at 370°C.

$$H_2[PtCl_6] \longrightarrow PtCl_4 + 2HCl$$

It is a reddish brown, crystalline solid, readily soluble in water, the solution having acidic reaction.

Chloroplatinic acid, $H_2[PtCl_6]$. It is the best known complex of platinum (IV). It is prepared by the action of aqua regia on platinum.

$$2HNO_3 + 8HCl + Pt \longrightarrow H_2[PtCl_6] + 4H_2O + 2NOCl$$

The solution is evaporated to drive off nitric acid before crystallising chloroplatinic acid.

Chloroplatinic acid is a reddish brown crystalline substance associated with six molecules of water of crystallisation, $H_2[PtCl_6]$. $6H_2O$.

On heating in chlorine at 360° – 370°C, it decomposes into platinum (IV) chloride.

$$H_2[PtCl_6] \longrightarrow PtCl_4 + 2HCl$$

Chloroplatinic acid is a strong dibasic acid, ionising as

$$H_2[PtCl_6] \longrightarrow 2H^+ + [PtCl_6]^{2-}$$

Some other well known complexes of Pt (IV) are $[Pt(NH_3)_6]^{4+}$, $[Pt(NH_3)_5Cl]^{3+}$, etc.

4. Organometallic Compounds of Platinum. Platinum forms a number of organometallic compounds. The first organometallic compound was reported way back in 1825. It is known as **Zeis's salt** and has the formula $K[Pt(C_2H_4)Cl_3].H_2O$.

It may be pointed out that C_2H_4 in this compound acts as σ donor and pi acceptor to Pt (II) with the result that C_2H_4 ligates, perpendicularly to the plane containing Pt^{2+} and three Cl^- ions. The Zeis's salt is commercially employed as a catalyst in the synthesis of a number of organic compounds.

Silver and gold (Group 11)

Copper is the first member of group 11. It exhibits oxidation states of +1 and +2 in the formation of oxides and complex compounds.

COMPOUNDS OF SILVER

The normal oxidation state of silver is +1 in its compounds. A few compounds in which silver is in oxidation state of +2 have been reported but these are highly unstable. Silver is reduced readily in these compounds to +1 state. These compounds, therefore, act as strong oxidising agents. For example, the oxide, AgO, in which silver is in +2 oxidation state, is formed by passing ozone over silver. This decomposes readily to Ag_2O, giving out oxygen and thus acts as a strong oxidising agent,

$$2AgO \longrightarrow Ag_2O + O$$

However, some complexes of silver in +2 oxidation state are more stable than its salts. Thus, the complexes, $[Ag(py)_4]^{2+}$, $[Ag(1, 10\ \text{phenanthrolin})_2]^{2+}$ are known to be reasonably stable and contain Ag^{2+} ion in a square planar environment.

Solubilities of Silver Halides. The solubility products of AgCl, AgBr and AgI at the room temperature are 2.8×10^{-10}, 5.0×10^{-13} and 8.5×10^{-17}, respectively. Thus, AgI is the least soluble silver halide.

The extremely low solubilities of AgCl, AgBr and AgI appears surprising in view of the fact that salts of univalent cations with univalent anions are generally soluble. The abnormal insolubilities of silver halides are considered to be due to their *abnormally high lattice energies* and the reason for unusually high lattice energies is the strong van der Walls attraction between silver and halide ions in addition to the usual electrostatic attraction between the oppositely charged ions Comparing AgCl with KCl, it is found that Ag^+ and K^+ ions are of about the same size, the ionic radii of Ag^+ and K^+ being 1.26 and 1.33 Å. However, Ag^+ ion has 46 electrons while K^+ ion has only 18 electrons. As a general rule, **the larger the number of electrons in an ion, the stronger is its van der Walls attraction for neighbouring ions**. Consequently, the lattice energy of AgCl is much higher than that of KCl. The lattice energy of AgCl is 895·4 kJ mol^{-1} while that of KCl is 702·9 $kJmol^{-1}$. Thus, more energy is needed to break the lattice of AgCl than that of KCl. That is why silver chloride is much less soluble than potassion chloride.

Silver(I) Fluoride

The lattice energies of AgBr and AgI are even higher because of greater number of electrons in their anions. Consequently, they are even less soluble than AgCl.

The lattice energy of AgF is the lowest because of the smallest number of electrons in its anion. This salt, therefore, is soluble in water.

Silver bromide is soluble in solutions containing high concentration of thiosulphate ion due to complex formation.

$$AgBr(s) + 2S_2O_3^{2-} \longrightarrow [Ag(S_2O_3)_2]^{3-} + Br^-$$

This reaction is made use of an fixing the image on the photographic plate. Sodium thiosulphate dissolves away the whole of the unaffected silver bromide from the exposed film.

COMPOUNDS OF GOLD

Gold commonly exhibits oxidation states of +1 and +3. The +3 state is more stable. Gold (I) ion is unstable as it undergoes disproportionation in solution.

$$3Au^+ \rightleftharpoons Au^{3+} + 2Au$$

The equilibrium constant of the reaction, given by

$$K = [Au^{3+}]/[Au^+]^3 = 10^{10},$$

shows that there can be only traces of gold (I) ion in aqueous solution.

Gold (I) is, however stable in complex compounds.

The most important compound of gold, in which it exhibits a stable +3 oxidation state, is $AuCl_3$. In the solid state, it has a dimeric structure in which Cl atoms form the bridges. $AuCl_3$ when dissolved in organic solvents is called **liquid gold** which is used in paints.

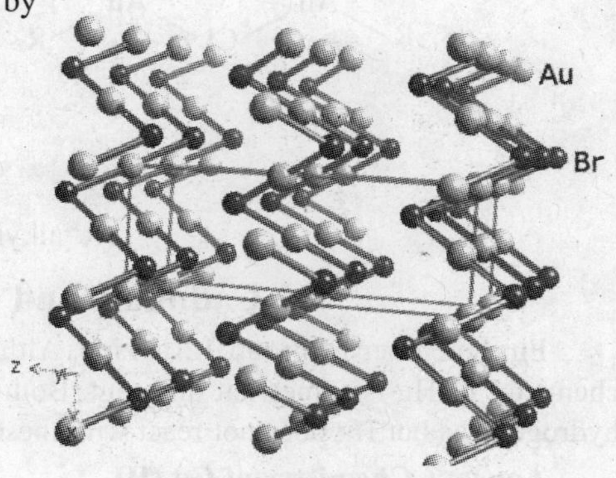

Gold (I) bromide

The only known compound of gold in its +5 oxidation state in AuF_5. This is a dark red polymeric solid which is diamagnetic in character.

Complex Compounds of Gold. Gold, like copper and silver, forms complex compounds both in +1 and +3 states. Some of the important complex compounds are given below.

Potassium dicyanoaurate (I), K [Au (CN)$_2$]. It is obtained by the following methods:

1. By dissolving gold (I) oxide in excess of potassium cyanide solution.

$$Au_2O + 4KCN + H_2O \longrightarrow 2K[Au(CN)_2] + 2KOH$$

2. By treating gold with a dilute solution of KCN in the presence of atmospheric oxygen.

$$4Au + 8KCN + O_2 + 2H_2O \longrightarrow 4K[Au(CN)_2] + 2KOH$$

The $[Au(CN)_2]^-$ complex ion is linear in geometry.

Sodium dithiosulphatoaurate (I), $Na_3[Au(S_2O_3)_2]$. This is obtained by adding gold (III) chloride to *hypo* solution.

$$4Na_2S_2O_3 + AuCl_3 \longrightarrow Na_3[Au(S_2O_3)_2] + Na_2S_2O_3 + 3NaCl$$

Hypo is oxidized to sodium tetrathionate while trivalent gold is reduced to monovalent state.

Potassium tetracyanoaurate (III) , K[Au(CN)$_4$]. It is obtained by treating a hot saturated solution of gold (III) chloride with potassium cyanide. The solution on crystallization yields crystals of potassium tetracyanoaurate (III).

$$AuCl_3 + 4KCN \longrightarrow K[Au(CN)_4] + 3KCl$$

Chloroauric acid, H[AuCl$_4$]. When gold is dissolved in aqua regia, a yellow solution of chloroauric acid is obtained. The sodium and potassium salts of this acid can be prepared by treating calculated amounts of alkali chlorides with gold trichloride.

$$AuCl_3 + KCl \longrightarrow K[AuCl_4]$$

Fulminating Gold, AuN_2H_3. When concentrated ammonia is added to gold (III) hydroxide or oxide, a yellowish or greenish substance is formed which is named **fulminating gold.**

Fulminating gold has a variable composition, formula approximating to NH=Au–NH_2. It is a highly explosive substance. It explodes violently when heated or struck with a hammer.

Organometallic Compounds of Gold. A number of organometallic compounds of gold are also known. These contain strong Au–C bonds in their structures. The structures of some of these compounds are given below:

```
R        Cl        R
   Au        Au
R        Cl        R

     R            R
     |            |
R — Au — CN — Au — R
     |            |
     C            C
     N            N
     |            |
R — Au — CN — Au — R
     |            |
     R            R
```

(R=alkyl or aryl group)

Cadmium and mercury (Group 12)

First member of group 12 is zinc. Although the chemistries of Zn and Cd are similar but chemistry of Hg is somewhat different. Both Zn and Cd react with non-oxidising acids evolving hydrogen gas but Hg does not react with these acids.

Aqueous Chemistry of Cd (II)

The usual oxidation state of cadmium is +2, The Cd^{2+} ion is hydrated but to a smaller extent than Zn^{2+} ion. When hydrogen sulphide is passed through a solution containing Cd^{2+} ions, yellow precipitate of CdS results. Its solubility product is 1.0×10^{-28}. It is used as a pigment under the name *cadmium yellow*.

Cadmium forms a number of complexes like $[Cd(NH_3)_4]^{2+}$, $[Cd(CN)_4]^{2-}$, $[CdI_4]^{2-}$. A precipitate of cadmium hydroxide dissolves in ammonia due to the formation of the complex $[Cd(NH_3)_4]^{2+}$

A number of octahedral complex of Cd(II) are also known. Examples are $[Cd(NH_3)_6]^{2+}$, $[Cd(en)_3]^{2+}$, etc.

Chemistry of Cd (I). The Cd-Cd bond is much weaker than Hg-Hg bond. That is why Hg_2^{2+} ion is very unstable. Cd_2^{2+} is formed only in the fused melts of the type Cd-$CdCl_2$.

Oxidation States of Mercury

1. Mercury displays oxidation states of +1 and +2 in its compounds. In this respect it differs from zinc and cadmium which show +2 oxidation state only. Mercurous ion (*i.e.*, mercury in +1 state) is unique in the sense that it consists of two atoms linked by a covalent bond formed by 6*s*-electrons of the two Hg^+ ions. Hg_2^{2+} ion is diamagnetic in nature.

The Hg_2^{2+} ion in aqueous solution is stable towards disproportionation to Hg and Hg^{2+} ion, *i.e*, the reaction

$$Hg_2^{2+}\ (aq) \rightleftharpoons Hg^{2+}\ (aq) + Hg$$

proceeds forward to a very small extent only. But, if we add to the above aqueous solution a reagent which *removes* Hg^{2+} in the form of a precipitate, the equilibrium shifts towards the right meaning therby that more of Hg_2^{2+} disproportionates into Hg and Hg^{2+}.

Thus, when hydrogen sulphide is passed through an aqueous solution of mercury (I) nitrate, the products are mercury (II) sulphide and free mercury.

$$Hg_2^{2+} + H_2S \longrightarrow HgS + Hg + 2H^+$$

Similarly, when sodium hydroxide is added to an aqueous solution of mercury (I) ions, mercury (II) oxide and mercury are obtained.

$$Hg_2^{2+} + 2OH^- \longrightarrow HgO + Hg + H_2O$$

Thus, it is not possible to prepare mercury (I) sulphide or mercury (I) oxide.

When aqueous ammonia is added to mercury (I) chloride, the solid turns black due to the formation of finely divided mercury.

$$Hg_2Cl_2(s) + 2NH_3 \longrightarrow Hg(NH_2)Cl + Hg + NH_4^+ + Cl^-$$

The compound Hg (NH_2) Cl is white but it is masked by the dark colour of mercury.

This serves as a test for distinguishing between AgCl and Hg_2Cl_2 in the analysis of cations of first group. The former dissolves while the latter turns black on the addition of aqueous ammonia.

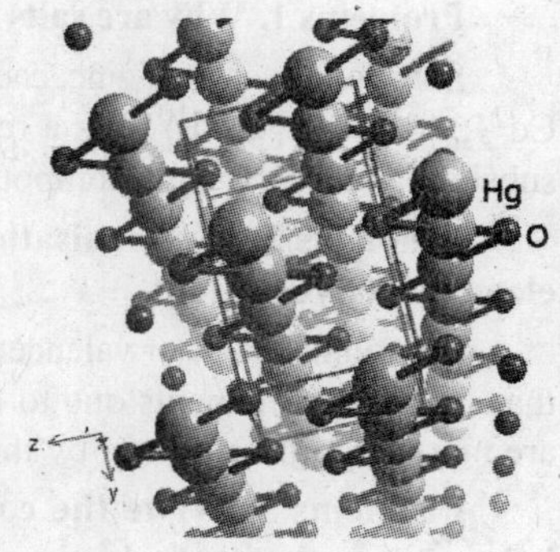

Mercury(II) Oxide

Complexes of Hg (II). Mercury in +2 state is found in soluble or insoluble salts and in complex ions. A solution of $HgCl_2$ (mercury (II) chloride) contains only a small concentration of Hg^{2+} or Cl^- ions in comparison to the concentration of undissociated $HgCl_2$ molecules. This is because $HgCl_2$ is a covalent compound. If excess of chloride ion is added to the solution of the salt, the complex $[HgCl_4]^{2-}$ is formed.

When dissolved in ammonia, complexes containing one, two, three and four ammonia molecules are known to be formed. With four molecules, the ammonia complex has the formula $[Hg(NH_3)_4]^{2+}$.

Organometallic Compounds of Mercury. A large number of organometallic compounds of mercury are known. The compounds RHgX and R_2Hg (R = alkyl or aryl group) can be easily prepared from Grignard reagents, as shown below:

$$RHgCl_2 + RMgCl \longrightarrow RHgCl + MgCl_2$$

$$RHgCl + RMgCl \longrightarrow R_2Hg + MgCl_2$$

RHgX compounds are extremely toxic and hence are used for treating seeds and also as pesticides and fungicides.

Mercuric salts react with olefins and accetylenes as follows:

$$HgCl_2 + \;>C=C< \;\longrightarrow\; Cl-\overset{|}{\underset{|}{C}}-\overset{|}{\underset{|}{C}}-Hg-Cl$$

These organometallic compounds can be easily hydrolysed and, therefore, they act as intermediates in the synthesis of aldehydes and alcohols from acetylenes and olefins.

$$HC \equiv CH \xrightarrow{HgCl_2} \left[\begin{matrix} C & & HgCl \\ & C=C & \\ H & & H \end{matrix}\right] \xrightarrow[-Hg(OH)Cl]{H_2O} ClHC = CHOH \xrightarrow{HgCl_2}$$

$$\left[\begin{array}{c} Cl \quad\quad\quad H \\ H - C - C - HgCl \\ Cl \quad\quad\quad OH \end{array}\right] \xrightarrow[-Hg(OH)Cl]{H_2O} [Cl_2HCCH(OH)_2] \xrightarrow{-H_2O} Cl_2HCCHO$$

SOME SOLVED PROBLEMS
(From Question Papers of Different Universities)

Problems 1. Why are salts of zinc, cadmium and mercury white?

Solution. In salts of zinc, cadmium and mercury, the *d*-subshell are completely filled (Zn^{2+}; $3d^{10}$; Cd^{2+}; $4d^{10}$; Hg^{2+}; $5d^{10}$). As a result, there is no possibility of excitation of *d*-electrons within *d*-subshell and hence their compounds are white.

Problems 2. First ionisation energies of 5*d* elements are higher than those of 3*d* and 4*d* elements. Give reasons.

Solution. The outer valence electrons of 5*d* elements experience greater effective nuclear charge than 3*d* and 4*d* elements due to the poor shielding of the nucleus by 4*f*-electrons. Therefore, they are more strongly attracted by the nucleus and their first ionization energies are higher.

Problems 3. Write the corresponding members of second and third transition series belonging to the group having the following first member:

(*i*) Mn **(*ii*) Ni** **(*iii*) V** **(*iv*) Zn**

Solution. (*i*) Technetium (Tc) and Rhenium (Re)

(*ii*) Palladium (Pd) and Platinum (Pt)

(*iii*) Niobium (Nb) and Tantalum (Ta)

(*iv*) Cadmium (Cd) and Mercury (Hg)

Problem 4. Write the disproportionation reactions of Cu+ and Au+.

Solution. Cu+ disproportionates as :

$2Cu^+ = Cu + Cu^{2+}$

Au^+ disproportionates as :

$3Au^+ = 2Au + Au^{3+}$

Problem 5. Why are the physical properties of second and third series of transition elements more similar to each other than they are to the first transition series ?

Solution. This is due to lanthanide contraction which results in almost similar atomic radii of the second and third series of transition elements. But, there is a significant difference in atomic radii of first and second series of transition elements and therefore, their physical properties are different.

Problem 6. The first ionization energy of Hg is higher than that of Cd. Explain.

Solution. In mercury, the presence of 4*f* orbitals cause poor shielding effect and therefore the effective nuclear charge on the outermost electrons becomes higher. As a result, the outermost electrons are strongly attracted by the nucleus, and require higher energy to remove the electron. Thus, the first ionization energy of Hg is higher than that of Cd.

Problem 7. Which of the following pairs of atoms have the most similar atomic radii ?

(*i*) Ti, Zr (*ii*) Cr, Mo (*iii*) Mo, W (*iv*) Ni, Pb

Solution. (*iii*) Mo, W. Elements of second and third transition series in a group have similar characteristics because of lanthanide contraction.

Problem 8. Write the stereochemistry of the following complexes :

(*i*) $[Ag(CN)_2]^-$ (*ii*) $Rh(CO)_4^-$ (*iii*) $KRuF_6$

Solution. (*i*) $[Ag(CN)_2]^-$: Linear (*ii*) Rh $(CO)_4^-$; Tetrahedral (*iii*) $KRuF_6$: Octahedral

EXERCISES
(Based on Question Papers of Different Universities)

Multiple Choice Questions

Choose the correct option.

1. Re (IV) has the electronic configuration:
   (*a*) d^2 (*b*) d^3
   (*c*) d^4 (*d*) d^5
2. Au^+ disproportionates as
   (*a*) $3\,Au^+ \rightleftharpoons 2Au^{3+} + Au$ (*b*) $2Au^+ \rightleftharpoons Au^{2+} + Au$
   (*c*) $3\,Au^+ \rightleftharpoons Au^{3+} + 2Au$ (*d*) does not disproportionate
3. The third member of the group of Mn and Re is
   (*a*) Mo (*b*) Tc
   (*c*) Nb (*d*) Pd
4. Which of the following do not belong to the same group?
   (*a*) Cr, Mo, W (*b*) Ni, Pd, Pt
   (*c*) Mn, Ru, Re (*d*) Cu, Ag, Au
5. The highest oxidation state shown by rhenium is
   (*a*) +5 (*b*) +6
   (*c*) +7 (*d*) +8
6. Mo and W belong to group of
   (*a*) Cu (*b*) Mn
   (*c*) Fe (*d*) Cr
7. M-M bond is present in
   (*a*) $[Mo(CN)_7]^{5-}$ (*b*) $WF_8{}^{2-}$
   (*c*) Fe_2Cl_6 (*d*) $Mo_2(OR)_6$
8. The stablest oxidation state of Au is
   (*a*) I (*b*) III
   (*c*) V (*d*) –I
9. Stable oxidation states of Zr and Nb are respectively
   (*a*) +3 and +5 (*b*) +4 and +5
   (*c*) +5 and +6 (*d*) +3 and +3
10. Atomic radii of third transition series, along the period
    (*a*) increases regularly (*b*) decreases regularly
    (*c*) first increases and then decreases (*d*) first decreases and then increases
11. The metal which is used as Adam's catalyst, for hydrogenation reactions and as a catalyst in the contact process for the manufacture of sulphuric acid is
    (*a*) Ti (*b*) V
    (*c*) Pt (*d*) Cr

ANSWERS

1. (*b*)	2. (*c*)	3. (*b*)	4. (*c*)	5. (*c*)	6. (*d*)
7. (*d*)	8. (*b*)	9. (*b*)	10. (*d*)	11. (*c*)	

Short Answer Questions

1. Write the electronic configurations of Pd (Z = 46) and Rh (Z = 45).

2. Give electronic configurations of W (Z = 74) and Mo (Z = 42).

3. Which are the two common oxidation states of gold? Which of these is unstable to disproportionation?

4. How do the stability constants sequence for halogen complexes vary for Ag (I)?

5. Name the first and last element of second transition series.

6. Name two elements belonging to group of copper.

7. Name two elements belonging to second or third transition series which are used as catalysts.

8. Give two examples of complexes of Pt (IV).

9. What are the maximum oxidation states exhibited by Re and Os?

10. What is the basic difference in the magnetic properties of elements of first and second transition series?

11. How many elements are present in second transition series?

12. The densities of third transition elements are double than those of second transition series. Explain.

13. Why do Zr and Hf have similar properties?

General Questions

1. Give a general comparison of second and third transition series with first transition series.

2. Give a brief account of oxidation states and sterochemistry of palladium and platinum.

3. Compare the important characteristics of first transition series with those of elements of second and third transition series.

4. Compare the oxidation states and metal-metal bonding tendency of elements of first transition series and their analogues heavier elements.

5. Why do Zr and Hf have similar properties?

6. Discuss the variation of

(*i*) ionic radii and

(*ii*) oxidation states among the elements of second and third transition states.

7. Discuss a comparative treatment of elements of second and third transition series with their analogues of first transition series with respect to the following properties:

(*a*) ionic radii (*b*) oxidation states (*c*) stereochemistry

8. Explain

(*i*) Complexes of first transition series are mainly high spin while those of second and third transition series are of low spin.

(*ii*) Elements of second and third transition series exhibit catalytic properties.

(*iii*) The heavier transition elements are more prone to metal-metal bonding.

9. Give a comparative account of complexes of Rh and Ir with Co.

10. (*a*) First ionization energies of 5*d* elements are higher than those of 3*d* and 4*d* elements. Give reasons.

(*b*) The salts of zinc, cadmium and mercury are white. Explain.

(c) Out of Au (I) and Au (III) which state disproportionates and how?

11. (*a*) Discuss the variation of radii of atoms and ions of second and third transition series in comparison with first transition series.

(*b*) Compare the general behaviour of oxidation states of second and third transition series with first transition series.

12. (*a*) List the main differences between Zr and Hf with their analogue Ti.

(*b*) Give a brief account of aqueous chemistry of Zr and Hf.

13. Give main differences of technetium and rhenium from manganese.

14. What is metal-metal bond? Discuss with examples.

15. (*a*) Why do Zr and Hf have similar properties? Discuss.

(*b*) Out of Au (I) and Au (III) which state disproportionates and how?

(*c*) Write the corresponding members of second and third transition series belonging to the groups having the first member :

(*i*) Mn (*ii*) Ni

16. (*a*) Elements of second and third transition series have similar chemical properties but show large differences from the elements of first transitions series. Comment and explain.

(*b*) Why is mercury a liquid metal? Explain.

(*c*) First ionization energy of Hg is higher than that of Cd. Explain.

17. (*a*) Why do Zr and Hf have similar properties?

(*b*) What is metal-metal bond? The heavier elements are more prone to metal-metal bonding comment.

(*c*) Why are 1st transition series complexes mainly high spin while that of 3rd and 2nd series of low spin?

18. What are the differences between the first row and the other two rows of the transition metals?

19. First ionisation energies of 5*d* elements are higher than those of 3*d* and 4*d* elements. Explain.

20. (*a*) Write the general electronic configuration of second and third transition series

(*b*) Draw the structures of

$[TaF_7]^{2-}$ and $[TaF_8]^{3-}$

(*c*) How does the reactivity of Hg differ from that of Zn and Cd? Explain

21. (*a*) Which transition metal is liquid at room temperature?

(*b*) Which transition elements has highest melting point? Give its melting point.

(*c*) Explain the stereochemistry of the following:

(*i*) $[Au(CN)_2]^+$ (*ii*) $[HgI_3]^-$ (*iii*) $[PtCl_3]^{2-}$

(*iv*) $[NbF_7]^{2-}$ (*v*) $[ReF_8]^{2-}$

22. (*a*) First I.E. of Hg is higher than that of Cd. Why?

(*b*) The densities of third transition series metals are double than those of second transition series metals Why?

(*c*) The melting and boiling points of transition series are high. Why?

(*d*) Mercury is a liquid metal. Comment.

23. Explain the following:

(*i*) The first ionisation energy of Hg is higher than that of cadmium.

(*ii*) Zirconium and Hafnium exhibit similar properties.

24. (*a*) Compare second and third transition elements series with those of first series in the following properties:

(*i*) Aqueous chemistry (*ii*) Magnetic properties (*iii*) Metal-metal bonding.

(*b*) First ionisation energy of 5*d* elements is higher than those of 3*d* and 4*d* elements. Why?

(*c*) What is highest oxidation state shown by transition elements? Name the elements.

(*d*) What are platinum metals? Why are these called noble metals?

25. Compare and contrast the first transition series elements with those of second and third transition series elements.

26. (*a*) The first ionisation energies of 5*d* elements are higher than that of 3*d* and 4*d* elements. Explain.

(*b*) Densities of the third transition series metals are double than those of second transition series metals. Explain.

(*c*) Write the electronic configuration of

Pd (Z = 46) and Nb (Z = 41)

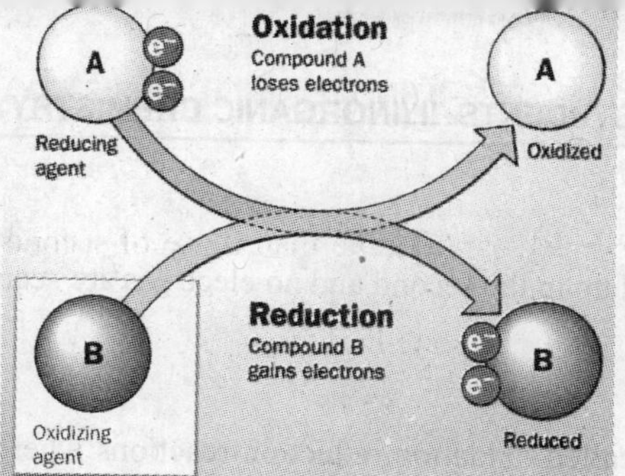

3

Oxidation and Reduction

3.1. INTRODUCTION

We come across oxidation and reduction reactions very frequently in our daily life. We find that iron gets rusted in humid atmosphere. The lead acid batteries in the inverters and in our cars produce electricity. The food that we consume generates heat which keeps us moving. Metals are extracted from their ores. All there are examples of oxidation reduction reactions. The phenomenon of oxidation and reduction has been explained in different ways. The treatise changes with the advancement in the subject.

3.1.1. Older concept of oxidation and reduction

According to this concept, addition of oxygen or removed of hydrogen is called **oxidation** which addition of hydrogen or removal of oxygen is called **reduction**. This is illustrated with the help of the following examples:

Examples of oxidation

$$H_2S + Cl_2 \longrightarrow 2HCl + 5 \qquad \textbf{Removal of hydrogen}$$

$$2\,KMg + O_2 \longrightarrow 2\,Mg \qquad \textbf{Addition of oxygen}$$

Examples of reduction

$$Fe_2O_3 + 3CO \longrightarrow 2Fe + 3CO_2 \qquad \textbf{Removal of oxygen}$$

$$Br_2 + H_2 \longrightarrow 2HBr \qquad \textbf{Addition of hydrogen}$$

There are reactions which are examples of oxidation reduction but cannot be explained with the above definitions. There was therefore a need to give a wider definitions of the above terms. This is how electronic concept came into being.

3.1.2. Oxidation reductions based on electronic concept

Consider the reaction between sodium and chlorine to give sodium chloride. Sodium atom loses an electron to become sodium ion and chloride atom gains an electron to become chloride ion.

$$Na \longrightarrow Na^+ + \overline{e}$$

$$Cl + \overline{e} \longrightarrow Cl^-$$

The overall reaction is:

$$Na + Cl \longrightarrow Na^+\,Cl^-$$

This reaction is an electron transfer rection, *i.e.*, electron is transferred from sodium to chlorine atom.

Consider another reaction where aluminium powder is burnt in oxygen to form aluminium oxide. Here, aluminium atom loses electrons to form aluminium ions and oxygen gains electrons to form oxide as follows:

$$Al \longrightarrow Al^{3+} + 3e] \times 4$$

$$O + 2e \longrightarrow O^{2-}] \times 6$$

The electron produced in the first reaction are used up in the second and no electrons are seen in the net reaction:

$$4Al + 3O_2 \longrightarrow 2Al_2O_3$$

Such electrons transfer reactions are commonly called oxidation reduction reactions where oxidation and reduction reactions occur side by side in terms of change of electrons. This is also commonly stated as: *oxidation and reduction reactions go hand in hand in every electron transfer reaction.* There can be no oxidation without reduction and *vice-versa.*

3.1.3. Oxidation process in terms of electronic concept

A substance is said to have been oxidised if it loses electrons. Oxidation is a process which involves the loss of electrons. According to this definition of oxidation, substances like oxygen, halogens and other non-metals act as oxidizing agents. Any substance that accepts electron is called an *oxidising agent.* If the chlorine gas is passed through an iodide solution, the solution becomes brown due to the formation of iodine.

$$Cl_2 + 2I^- \longrightarrow 2Cl^- + I_2$$

This means, chlorine oxidizes iodide ion to iodine. The loss of electrons during oxidation may result in:

(*i*) Increase of positive charge on atom or ion undergoing oxidation.

For example,

$$Na - e \longrightarrow Na^+$$

$$Al - 3e \longrightarrow Al^{+3}$$

$$Fe^{2+} - e \longrightarrow Fe^{3+}$$

$$Sn^{2+} - 2e \longrightarrow Sn^{4+}$$

(*ii*) Decrease of negative charge on atom or ion undergoing oxidation.

For example,

$$2I^- - 2e \longrightarrow I_2$$

$$S^{2-} - 2e \longrightarrow S$$

$$MnO_4^{2-} - e \longrightarrow MnO_4^-$$

$$[Fe(CN_6)]^{4-} - e \longrightarrow [Fe(CN)_6]^{3-}$$

3.1.4. Reduction process in terms of electronic concept

Reduction is a process which involves the gain of electrons. Any substance from which electrons can be readily removed acts as a reducing agent. A **reducing agent** is a substance that can give electrons. The gain of electrons during reduction may lead to the following situation:

(*i*) Decrease of positive charge on atom or ion undergoing reduction.

For example,

$$Na^+ + \overline{e} \longrightarrow Na$$

$$Fe^{3+} + e \longrightarrow Fe^{2+}$$

$$Sn^{4+} + 2e \longrightarrow Sn^{2+}$$

$$O_2 + 2e \longrightarrow 2O^{2-}$$

***(ii)* Increase of negative charge on atom or ion undergoing reduction.**

For example,

$$MnO_4^- + e \longrightarrow MnO_4^{2-}$$

$$[Fe(CN_6]^{3-} + e \longrightarrow [Fe(CN)_6]^{4-}$$

Definition of oxidation and reduction and of oxidising agents can be summarised by an equilibrium:

$$\text{Reducing agent} \underset{\text{Reduction}}{\overset{\text{Oxidation}}{\rightleftharpoons}} \text{Oxidising agent + Electrons}$$

Reduction oxidation reactions are also abbreviated as **redox** reactions.

3.2. OXIDATION NUMBER

We can explain the redox changes in terms of electron transfer in the case of ionic reaction. But this is not so in the case of covalent compounds. To explain the redox reactions uniformly in the case of all type of substance and reactions, the concept of oxidation number was mooted.

The oxidation number is a measure of the extent of oxidation of an element in its compounds.

Alternatively: *It is the charge which an atom appear to have when all other atoms are removed from it as ions.*

It is determined by using a set of simple rules:

(*i*) An element in any of its allotropic forms is assigned an oxidation number of zero.

(*ii*) The oxidation number of a monoatomic ion is equal to its charge.

Thus Na^+ is assigned an oxidation number +1 and magnesium ion Mg^{+2} is assigned a number equal to +2. Singly charged monoatomic, negative ions such as Cl^- are assigned an oxidation number of –1 and double charged negative ions such as O^{2-} have an oxidation number of –2.

(*iii*) In any neutral molecule the sum of the oxidation numbers is zero; in an ion sum of the oxidation numbers is equal to the charge on the ion.

(*iv*) Fluorine, the most electronegative element, has an oxidation number of –1 in all its compounds.

(*v*) Oxygen has an oxidation number –2 in its compounds. However, there are three exceptions to this generalization:

(*a*) When oxygen is combined with fluorine as in the compounds F_2O, its oxidation number is +2.

(*b*) In compounds containing oxygen-oxygen bonds, such as H_2O_2, the oxidation number of oxygen is –1.

(*c*) In super oxides like KO_2, the oxidation number of oxygen atom is –1/2.

(*vi*) The halogens other than fluorine (*i.e.*, chlorine, bromine, iodine) have an oxidation number of –1 except when they are combined with a more electronegative element, that is, with oxygen or with a more electronegative halogen.

(*vii*) Hydrogen in its compounds exhibits an oxidation number of +1. The only exceptions occur in metal hydrides in which hydrogen has an oxidation number of –1, for example, in sodium hydride, $Na^{+1} H^{-1}$.

(*viii*) When an atom is bonded to another atom of same electronegativity (for example, a molecule consisting of two identical atoms (H_2 or O_2), then the bond electrons are divided equally between the two atoms. Each atom may thus have an oxidation number equal to zero. Similarly, phosphorus in P_4 or sulphur in S_8 have oxidation number zero.

The concept of oxidation number gives us another definition of oxidation and reduction: *Oxidation may be defined as any change in which the oxidation number of an atom increases. Reduction may be defined as any change in which the oxidation number of an atom decreases.*

(*ix*) For complex ions the sum of oxidation numbers of all the atoms is equal to the charge on the ion. For example, in NO_3^- ion, the sum of the oxidation numbers of nitrogen atom and 3 oxygen atoms must be equal to –1.

3.2.1. Calculation of oxidation number

Keeping in view the above rules, we can calculate the oxidation number of any atom in a molecule/ion.

Step 1. Write down the formula of the given molecule/ion leaving some space between the atoms.

Step 2. Write oxidation number of each atom on its top. Write *x* for the oxidation number of the atom whose oxidation number has to be found out.

Step 3. Determine the total oxidation number of each element by multiplying oxidation no. of each atom with the number of atoms of that kind in the molecule/ion.

Step 4. Add up the oxidation number of each element and equate it equal to charge on the molecule/ion.

Step 5. Solve for the value of *x*.

Example 1. Determine oxidation number of

(*i*) S in H_2SO_3

(*ii*) P in PO_4^{3-}

(*iii*) Cr in $Cr_2O_7^{2-}$ and Cr O_4^{2-}

(*iv*) Fe in Fe_2O_3

Solution. (*i*) H_2SO_3. Since the oxidation number of hydrogen is +1 and that of oxygen is –2, we have

$$\overset{+1}{H}_2 \quad \overset{x}{S} \quad \overset{-2}{O}_3$$

∴ Total oxidation numbers of all the atoms in H_2SO_3

$$= (+1 \times 2) + (x \times 1) + (-2 \times 3)$$

$$= 2 + x - 6$$

$$= x - 4$$

Since the sum of oxidation numbers of all the atoms of a neutral molecule like H_2SO_3 is zero, therefore,

$$x - 4 = 0$$

or

$$x = +4$$

Hence, the oxidation number of sulphur is H_2SO_3 is +4.

(*ii*) *PO_4^{3-} ion.* Let the oxidation number of P be *x* and that of oxygen is –2. We have

$$\overset{x}{P} \quad \overset{-2}{O}_4$$

∴ Total oxidation numbers of all the atoms in anion like PO_4^{3-} is equal to –3; therefore,

$$x - 8 = -3$$
$$x = -3 + 8$$
$$x = +5$$

Hence the oxidation number of phosphorus is PO_4^{3-} ion is +5.

(iii) $\boldsymbol{Cr_2O_7^{2-}}$ *ion.* Let the oxidation number of Cr be x. Oxygen has oxidation number –2. Thus we have

$$\overset{x}{Cr_2}\qquad\overset{-2}{O_7}$$

∴ Total oxidation number of all the atoms in $Cr_2O_7^{2-}$ ion

$$= (x \times 2) + (-2 \times 7)$$
$$= 2x - 14$$

Since the sum of the oxidation numbers of all the atoms in an ion $Cr_2O_7^{2-}$ is equal to (–2), therefore,

$$2x - 14 = -2$$

or
$$2x = -2 + 14$$
$$2x = +12$$
$$x = \frac{12}{2} = 6$$
$$x = +6$$

Hence, oxidation number of chromium in $Cr_2O_7^{2-}$ is + 6. Similarly, oxidation number of Cr in CrO_4^{2-} is also + 6.

(iv) $\boldsymbol{Fe_2O_3}$. Let the oxidation number of Fe be x, that of oxygen is –2. We have

$$\overset{x}{Fe}_{2}\qquad\overset{-2}{O}_{3}$$

∴ Total oxidation numbers of all the atoms in Fe_2O_3

$$= (x \times 2) + (-2 \times 3)$$
$$= 2x - 6$$

Since the sum of the oxidation number of all the atoms in a neutral molecule like Fe_2O_3 is zero,

Therefore $\quad 2x - 6 = 0$

or $\quad x = +3$

Example 2. Assign oxidation number (O.N.) to each of the elements in the following compounds or ionic species.

(*a*) BaF_2 (*b*) CO_2 (*c*) SCl_2 (*d*) S_2Cl_2

Solution. (*a*) BaF_2 is an ionic compound: Ba^{+2} $(F^-)_2$. Therefore O.N. of Ba is +2 and O.N. of F is –1.

(*b*) Each oxygen has O.N. = –2 and therefore, C has O.N. = +4.

(*c*) Each chlorine has O.N. = –1 and therefore S has an O.N. +2.

(*d*) Chlorine has an oxidation number = –1, so

$$2(-1) + 2(x) = 0$$

or
$$x = 1$$

Hence oxidation number of S = 1.

Example 3. Find out the oxidation number of carbon in each of its compounds.

(*i*) CH_4 (*ii*) CH_3Cl (*iii*) CH_2Cl_2 (*iv*) $CHCl_3$ (*v*) CCl_4

Let the oxidation number of carbon be represented by *x*

Solution. (*i*) $\overset{+1}{C}H_4 : x + 4 = 0$ or $x = -4$

(*ii*) $\overset{+1}{C}H_3\overset{-1}{Cl} : x + 3 - 1 = 0$ or $x = -2$

(*iii*) $\overset{+1}{C}H_2\overset{-1}{Cl_2}$: 2H = + 2, 2Cl = –2 ∴ $x + 2 - 2 = 0$ or $x = 0$

(*iv*) $\overset{+1}{C}H\overset{-1}{Cl_4} : x + 1 - 3 = 0$ or $x = 2$

(*v*) $C\overset{-1}{Cl_4} : x - 4 + 0$ or $x = 4$

Example 4. Give the oxidation number of sulphur in each of the following:

(*i*) $H_2S_2O_2$ (*ii*) $H_2S_2O_4$ (*iii*) $H_2S_2O_6$ (*iv*) $S_2O_7^{2-}$

Let the oxidation number of sulphur be represented by *x*.

Solution. (*i*) $\overset{+1}{H_2}S_2\overset{-2}{O_2} : 2 + 2x - 4 = 0$ or $x = 1$

(*ii*) $\overset{+1}{H_2}S_2\overset{-2}{O_4} : 2 + 2x - 8 = 0$ or $x = 3$

(*iii*) $\overset{+1}{H_2}S_2\overset{-2}{O_6} : 2 + 2x - 12 = 0$ or $x = 5$

(*iv*) $S_2O_7^{2-} : 2x - 14 = -2$ or $x = 6$

Example 5. Find out the oxidation of the central atom in each of the following fluro derivatives:

(*i*) ClF_3 (*ii*) OF_2 (*iii*) $XeOF_2$ (*iv*) HSO_3F (*v*) IF_7

Let the oxidation number of the central atom be *x*.

Solution . (*i*) $Cl\overset{-1}{F_3} : x - 3 = 0$ or $x = 3$

(*ii*) $O\overset{-1}{F_2} : x - 2 = 0$ or $x = 2$

(*iii*) $Xe\overset{-2}{O}\overset{-1}{F_2} : x - 2 - 2 = 0$ or $x = 4$

(*iv*) $\overset{+1}{H}S\overset{-2}{O_3}\overset{-1}{F} : 1 + x - 6 - 1 = 0$ or $x = 6$

In certain compounds, the knowledge of structure is important to arrive at correct oxidation number.

1. Oxidation of Cr in CrO_5

Ordinarily, we would calculate the oxidation number of Cr as

$x + 5(-2) = 0$ or $x = +10$

But this is not possible because maximum number of Cr can be + 6 ($3d^5\ 4s^1$). This anomaly arises because of the structure of CrO_5 which is

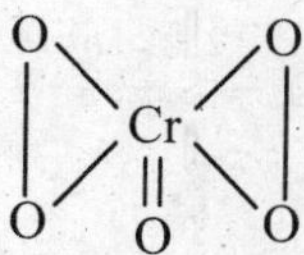

There are two peroxo bonds (— O — O —) in which the oxidation number of each O is –1. There are two negative charges for two oxygen atoms O_2^{-2}. We can now calculate the oxidation number of Cr in the light of the above.

$x + 1 \times (-2) + 4\,(-1) = 0 \qquad \text{or} \qquad x = +8$

2. Oxidation number of S in Caro's acid H_2SO_5 (Peroxomonosulphuric acid)

Normal calculation of oxidation number of S will give the value as + 8 , as follows:

$2 \times (+1) + x + 5\,(-2) = 0 \qquad \text{or} \qquad x = +8$

But this is not correct because S cannot have oxidation number more than + 6 (it belongs to oxygen group). Let us consider its structure before finding its oxidation number. The structure of the compound is

```
          O
          ||
H—O —S—O —O—H
          ||
          O
```

Thus, there is one peroxo (— O — O —) group in which each O has –1 oxidation number. We can recalculate the oxidation number

$2\,(+1) + x + 2\,(-1) + 3\,(-2) = 0 \qquad \text{or} \qquad x = +6$

3. Oxidation number of S in $S_2O_8^{2-}$ ion

Normal calculation of oxidation number of S in $S_2O_8^{2-}$ would give its value as + 7. But this is not a correct result. We have to take into account the structure of $S_2O_8^{2-}$ ion.

```
     O                  O
     ||                 ||
⁻O— S —O— O—S — O⁻
     ||                 ||
     O                  O
```

The ion contains one peroxide bond, in which the oxidation number of each oxygen is –1. Now calculate the oxidation number.

$2x + 2\,(-1) + 6\,(-2) = -2 \qquad \text{or} \qquad x = +6$

4. Oxidation number of Cl in bleaching powder $CaOCl_2$

$CaOCl_2$ has the structure $Ca^{2+}\ (OCl^-)\ Cl^-$. Here, the oxidation number of Cl in OCl^- is + 1 and in Cl^-, it is –1. The average of the two oxidation numbers become zero.

5. Oxidation number of 5 in $S_4O_6^{2-}$ (tethathionate ion)

If x is the O.N. of S, then

$4x + 6\,(-2) = -2 \qquad \text{or } x = +\frac{5}{2}$

This is in fact the average oxidation number. The O.N. of different 5 atoms are shown below:

$$\mathrm{O-\overset{\overset{\displaystyle O}{\|}}{\underset{\underset{\displaystyle O}{\|}}{S}}{}^{+5}-\overset{0}{S}-\overset{0}{S}-\overset{\overset{\displaystyle O}{\|}}{\underset{\underset{\displaystyle O}{\|}}{S}}-O^-}$$

6. Oxidation number of N in HNO_4 (peroxynitric acid)

In the structure of HNO_4, there is one peroxy group, where each O has oxidation number –1. The O.N. of N can be calculated as under

$+1 + x + 2\,(-1) + 2\,(-2) = 0$ **or** $x = +5$

PRACTICE PROBLEMS

Q.1. What is the oxidation number of

(*i*) N in HNO_3, NO, NO_2

(*ii*) P in PH_3, P_2O_5, H_3PO_4

(*iii*) S in $Na_2S_2O_3$, H_2SO_4

Ans. (*i*) +5, +2 and +4

(*ii*) –3, +5 and +5

(*iii*) +2, +6

Q. 2. Calculate the oxidation number in the following:

(*i*) Si in $Be_3Al_2Si_6O_{18}$

(*ii*) P in $HP_2O_7^-$

(*iii*) U in UO_2SO_4

(*iv*) Cr in CrO_2Cl_2 and K_2CrO_4.

Ans. (*i*) +4 (charge on Si_6O_{18} is –12)

(*ii*) +5 (charge on P_2O_7 is – 4)

(*iii*) +6

(*iv*) +6, +6

Example 6. In the given reactions, indicate (*i*) oxidation, (*ii*) reduction, (*iii*) oxidising agent, and (*iv*) reducing agent:

(*a*) $2\,FeCl_3 + SnCl_2 \longrightarrow 2\,FeCl_2 + SnCl_4$

(*b*) $3\,Fe + 4H_2O \longrightarrow Fe_3O_4 + 4H_2$

Can we classify these as electron-transfer reactions?

Solution. (*a*) (*i*) Oxidation of $SnCl_2$ occurs to form $SnCl_4$.

(*ii*) Reduction of $FeCl_3$ occurs to form $FeCl_2$.

(*iii*) $FeCl_3$ acts as an oxidizing agent.

(*iv*) $SnCl_2$ acts as a reducing agent.

(*b*) (*i*) Oxidation of Fe occurs to form Fe_3O_4.

(*ii*) Reduction of H_2O occurs to form H_2.

(*iii*) H_2O acts as an oxidising agent.

(*iv*) Fe acts as a reducing agent. Yes we can classify these as electron-transfer reactions.

3.2.2. Points of Distinction Between Valency and Oxidation Number

Valency	*Oxidation number*
(1) It refers to combining capacity of an element. It refers to number of hydrogen atoms which combine with one atom of an element.	It refers to the residual charge which an atom appears to have when all other atoms are removed from it as ions.
(2) It is always a whole number.	It can be even in fractions, *e.g.*, oxidation number of iron in Fe_3O_4 is 2.60 and S in $S_4O_6^{2-}$ is 2.5.
(3) In general, elements like C, N and S have fixed valency. For example, the valency of nitrogen in all its compounds is always three.	Even elements like C, N, S have variable oxidation states depending upon the compounds formed by them. For example, oxidation no. of nitrogen in various compounds are : *Compound* — *Oxidation no.* N_2O — +1 NO — +2 N_2O_3 — +3 NO_2 — +4 N_2O_5 — +5 N_2 — zero NH_3 — –3

3.2.3. Oxidation and reduction in terms of oxidation number

In terms of oxidation number, oxidation and reduction may be defined as follows:

Oxidation. It is a process in which the oxidation number of an atom shows an increase. For example, in the oxidation of ferrous Fe^{+2} to ferric Fe^{+3} the oxidation number of iron increases from +2 to +3. Similarly, conversion of iodide I^- to I_2 iodine is oxidation because the oxidation number of I increases from –1 to 0.

Reduction. It is reverse of oxidation and refers to a process in which the oxidation number shows decrease. For example, in the reduction of stannic Sn^{4+} to stannous Sn^{2+}, the oxidation no. of tin decreases from +4 to +2. Similarly, conversion of chlorine Cl_2 to chloride Cl^- is reduction because the oxidation number of Cl decreases from zero to –1.

Since oxidation and reduction go hand in hand, therefore, in a chemical reaction, whenever there is an increase in oxidation number of any species, there is a corresponding decrease in oxidation number of some other species. This is illustrated below in the chemical reaction between chlorine and potassium iodide.

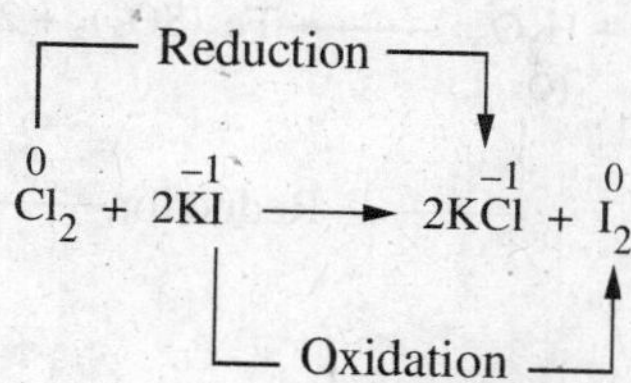

Here, the oxidation number of Cl decreases from zero in Cl_2 to –1 in Cl^-. This refers to reduction of Cl_2 to Cl^-. There is a corresponding increase in oxidation number of I from –1 in I^- to zero in I_2. This indicates oxidation of iodide to iodine.

An oxidizing agent is sometimes abbreviated as **oxidant** and a reducing agent as **reductant.** Changes in oxidation number in oxidation, reduction, oxidising agent or reducing agent are summarised as under:

Oxidation: Increase of oxidation number or loss of electrons.
Reduction: Decrease of oxidation number or gain of electrons.
Oxidising agent: Shows decrease in oxidation number or is an electron acceptor.
Reducing agent: Shows increase in oxidation number or is an electron donor.

This is illustrated further with the help of following reactions:

(*a*) $4NH_3 + 5O_2 \longrightarrow 4NO + 6H_2O$

(*b*) $4Sn + 10HNO_3 \longrightarrow 4Sn(NO_3)_2 + 3H_2O + NH_4NO_3$

(*c*) $2FeSO_4 + H_2SO_4 + H_2O_2 \longrightarrow Fe_2(SO_4)_3 + 2H_2O$

(*a*)

$$\underset{\text{(Red.)}}{4NH_3} + \underset{\text{(Oxid.)}}{5\overset{0}{O}_2} \longrightarrow 4N\overset{+2-2}{O} + 6H_2\overset{-2}{O}$$

Reduction (O: 0 → –2); Oxidation (N in NH_3 → +2 in NO)

Reductant and oxidant are abbreviated as Red. and Oxid. respectively.

(*b*)

$$\underset{\text{(Red.)}}{4\overset{0}{Sn}} + \underset{\text{(Oxid.)}}{10H\overset{+5}{N}O_3} \longrightarrow 4\overset{+2}{Sn}(NO_3)_2 + \overset{-3}{N}H_4NO_3 + 3H_2O$$

Oxidation (Sn: 0 → +2); Reduction (N: +5 → –3)

In this reaction oxygen is reduced to oxide (O^{-2}) present both in NO as well as H_2O.

Nitric acid HNO_3 (O.N of N = + 5) behaves as an oxidant. All N atoms on the right hand side remain the same. (O.N. = +5) except one N of ammonium ion in NH_4NO_3 which shows an oxidation state = –3.

(*c*)

$$\underset{\text{(Red.)}}{2\overset{+2}{Fe}SO_4} + H_2SO_4 + \underset{\text{(Oxid.)}}{H_2\overset{-1}{O}_2} \longrightarrow \overset{+3}{Fe}_2(SO_4)_3 + 2H_2\overset{-2}{O}$$

Oxidation (Fe: +2 → +3); Reduction (O: –1 → –2)

3.2.4. Balancing of redox equations by oxidation number method

Various steps for balancing a redox equation by this method are as follows:

Step I. Assign oxidation number of atoms undergoing oxidation reduction and write separate equations for such atoms only.

Step II. Add electrons on the right side for oxidation and on the left side for reduction. The number of electrons added in each case should be equal to the change in oxidation number.

Step III. Balance the number of electrons in the two equations and add the two equations to cancel out electrons.

Step IV. Then write the complete equation in its original form taking the coefficients of the oxidants and reductants in step II. Balance other atoms.

Step V. In acid medium, if oxygen remains unbalanced add water on the side deficient in O. If hydrogen still remains unbalanced then add H^+ on the required side. Be sure that all atoms and charges are balanced in the fmal equation.

Step VI. In basic medium, if oxygen remains unbalanced then add same number of H_2O as the unbalanced atoms on the side with excess and add double the number of OH^- on the other side. This will balance oxygen.

In basic medium, if H remains unbalanced, then add same number of OH^- as the difference on the side with the excess H and add same number of H_2O on the other side.

Be sure that all the participants and charges are balanced in the final equation.

Consider the reaction:

$$\mathbf{Zn + HNO_3 \longrightarrow N_2O + Zn(NO_3)_2 + ...}$$

This is balanced by oxidation number method as follows:

Step I.

$$\overset{0}{Zn}+H\overset{+5}{N}O_3 \longrightarrow \overset{+1}{N_2}O+\overset{+2}{Zn}(NO_3)_2$$

$$\overset{0}{Zn} \longrightarrow \overset{2+}{Zn} \text{ (oxidation)}$$

$$2\overset{+5}{N} \longrightarrow 2\overset{+1}{N} \text{ (reduction)}$$

Step II.

$$\overset{0}{Zn} \longrightarrow \overset{+2}{Zn}+2e \quad ...(i)$$

$$2\overset{+5}{N}+8e \longrightarrow 2\overset{+1}{N} \quad ...(ii)$$

Step III. Multiply (*i*) by 4 and add

$$4\overset{0}{Zn} \longrightarrow 4\overset{2+}{Zn}+8\overline{e}$$

$$2\overset{+5}{N}+8\overline{e} \longrightarrow 2\overset{+1}{N}$$

$$4\overset{0}{Zn}+2\overset{+5}{N} \longrightarrow 4\overset{+2}{Zn}+2\overset{+1}{N}$$

Step IV. $4Zn + 2HNO_3 \longrightarrow 4Zn(NO_3)_2 + N_2O + ...$

H_2O would be formed in the reaction and balancing H on the two sides,

$$4Zn + 10HNO_3 \longrightarrow 4Zn(NO_3)_2 + N_2O + 5H_2O$$

Consider another reaction:

$$CrO_4^{2-} + SO_3^{2-} \longrightarrow Cr(OH)_4^- + SO_4^{2-}$$

This is balanced by oxidation number method as follows:

Step I. $\overset{+6}{Cr}O_4^{2-} + \overset{+4}{S}O_3^{2-} \longrightarrow \overset{+3}{Cr}(OH)_4^- + \overset{+6}{S}O_4^{2-}$

$$\overset{+6}{Cr} \longrightarrow \overset{+3}{Cr} \text{ (reduction)}$$

$$\overset{+4}{S} \longrightarrow \overset{+6}{S} \text{ (oxidation)}$$

Step II. $\overset{+6}{Cr} + 3e \longrightarrow \overset{+3}{Cr}$ ... (*i*)

$\overset{+4}{S} \longrightarrow \overset{+6}{S} + 2\bar{e}$ ...(*ii*)

Step III. Multiply (*i*) by 2 and (*ii*) by 3 to cancel electrons and add

$$2\overset{+2}{Cr} + 6e \longrightarrow 2\overset{+3}{Cr}$$

$$3\overset{+4}{S} \longrightarrow 3\overset{+6}{S} + 6e$$

$$2\overset{+6}{Cr} + 3\overset{+4}{S} \longrightarrow 2\overset{+3}{Cr} + 3\overset{+6}{S}$$

Step IV. $2CrO_4^{2-} + 3SO_3^{2-} \longrightarrow 2Cr(OH)_4^- + 3SO_4^{2-}$

Step V. In basic medium as per rule VI add $3H_2O$ on R.H.S. and $6OH^-$ on L.H.S. to balance oxygen atoms.

$$2CrO_4^{2-} + 3SO_3^{2-} + 6OH \longrightarrow 2Cr(OH)_4^- + 3SO_4^{2-} + 3H_2O$$

In order to balance H as per rule VI, there is an excess of 8H on R.H.S. and $8OH^-$ has to be added on R.H.S. Add $8H_2O$ on L.H.S. to balance total hydrogen atoms.

$$2CrO_4^{2-} + 3SO_3^{2-} + 6OH^- + 8H_2O \longrightarrow 2Cr(OH)_4^- + 3SO_4^{2-} + 3H_2O + 8OH^-$$

Rearranging we get

$$2CrO_4^{2-} + 3SO_3^{2-} + 5H_2O \longrightarrow 2Cr(OH)_4^- + 3SO_4^{2-} + 2OH^-$$

Example 1. Balance the following equation taking place in acidic medium:

$$Cr_2O_7^{2-} + H_2SO_3 \longrightarrow Cr^{+3} + HSO_4^-$$

Solution.

Step I. $Cr_2O_7^{2-} + H_2\overset{+4}{S}O_3 \longrightarrow \overset{+3}{Cr} + H\overset{+6}{S}O_4^-$

$$\overset{+6}{2\text{Cr}} \longrightarrow \overset{+3}{2\text{Cr}} \text{ (reduction)}$$

$$\overset{+4}{\text{S}} \longrightarrow \overset{+6}{\text{S}} \text{ (oxidation)}$$

Step II. $2Cr^{+6} + 6e \longrightarrow 2Cr^{+3}$ ....(*i*)

$S^{4+} \longrightarrow S^{+6} + 2e$ ...(*ii*)

Step III. To cancel electrons multiply (*ii*) by 3

$$2Cr^{+6} + 6e \longrightarrow 2Cr^{+3}$$

$$3S^{+4} \longrightarrow 3S^{+6} + 6e$$

$$2Cr^{+6} + 3S^{+4} \longrightarrow 2Cr^{3+} + 3S^{+6}$$

Step IV. $Cr_2O_7^{2-} + 3H_2SO_3 \longrightarrow 2Cr^{3+} + 3HSO_4^-$

Step V. Except oxygen and hydrogen atoms, all other atoms are balanced. To balance oxygen, add water on the side deficient in oxygen [*i.e.*, R.H.S.]. We get

$$Cr_2O_7^{2-} + 3H_2SO_3 \longrightarrow 2Cr^{3+} + 3HSO_4^- + 4\,H_2O$$

Since this reaction occurs in acidic medium, therefore, add H^+ on L.H.S. to balance hydrogen atom.

$$Cr_2O_7^{2-} + 3H_2SO_3 + 5H^+ \longrightarrow 2Cr^{3+} + 3HSO_4^- + 4\,H_2O$$

Example 2. Balance the equation:

$$\mathbf{HCl + KMnO_4 \longrightarrow Cl_2 + MnCl_2 + KCl + H_2O}$$

Solution.

Step I. $\text{H}\overset{-1}{\text{Cl}} + \text{K}\overset{+7}{\text{Mn}}\text{O}_4 \longrightarrow \overset{0}{\text{Cl}_2} + \overset{+2}{\text{Mn}}\text{Cl}_2 + \text{KCl} + \text{H}_2\text{O}$

$\overset{-1}{\text{Cl}}$ is oxidised to $\overset{0}{\text{Cl}}$ and $\overset{+7}{\text{Mn}}$ is reduced to $\overset{+2}{\text{Mn}}$

$$\overset{-1}{\text{Cl}} \longrightarrow \overset{0}{\text{Cl}_2} \text{ (oxidation)}$$

$$\overset{+7}{\text{Mn}} \longrightarrow \overset{2+}{\text{Mn}} \text{ (reduction)}$$

Step II. $2Cl^- \longrightarrow Cl_2 + 2e$ ...(*i*)

$\overset{+7}{\text{Mn}} + 5e \longrightarrow \overset{2+}{\text{Mn}}$ ...(*ii*)

Step III. To cancel electrons, multiply (*i*) by 5 and (*ii*) by 2 and then add

$$10\,Cl^- \longrightarrow 5Cl_2 + 10e$$

$$2\,\overset{+7}{\text{Mn}} + 10e \longrightarrow 2\,\overset{+2}{\text{Mn}}$$

$$2\,\overset{+7}{\text{Mn}} + 10\text{Cl}^- \longrightarrow 2\,\overset{+2}{\text{Mn}} + 5\text{Cl}_2$$

Step IV. $2KMnO_4 + 10HCl \longrightarrow 2MnCl_2 + 5Cl_2 + 2KCl + H_2O$

Balance K by multiplying KCl by 2.

$$2KMnO_4 + 10HCl \longrightarrow 2MnCl_2 + 5Cl_2 + 2KCl + H_2O$$

Then balance 16Cl on R.H.S. by making 16 HCl on the L.H.S. and we get

$$2KMnO_4 + 16HCl \longrightarrow 2MnCl_2 + 5Cl_2 + 2KCl + H_2O$$

Both oxygen and hydrogen get balanced by multiplying H_2O with 8. We get the final balanced equation

$$2KMnO_4 + 16\ HCl \longrightarrow 2MnCl_2 + 2KCl + 5Cl_2 + 8H_2O$$

Example 3. Balance the following equation: $KMnO_4 + NH_3 \longrightarrow KNO_3 + MnO_2 + KOH + H_2O$

Solution.

Step I. $K\overset{+7}{Mn}O_4 + \overset{-3}{N}H_3 \longrightarrow K\overset{+5}{N}O_3 + \overset{+4}{Mn}O_2 + OH^- + H_2O$

$$\overset{+7}{Mn} \longrightarrow \overset{+4}{Mn}\text{(reduction)}$$

$$\overset{-3}{N} \longrightarrow \overset{+5}{N}\text{ (oxidation)}$$

Step II. $\overset{+7}{Mn} + 3e \longrightarrow \overset{+4}{Mn}$ ... (*i*)

$\overset{-3}{N} \longrightarrow \overset{+5}{N} + 8\bar{e}$ ...(*ii*)

Step III. Multiply (*i*) by 8 and (*ii*) by 3 and add

$$8\overset{+7}{Mn} + 3\overset{-3}{N} \longrightarrow 8\overset{+4}{Mn} + 3\overset{+5}{N}$$

Step IV. $8KMnO_4 + 3NH_3 \longrightarrow 8MnO_2 + 3KNO_3 + KOH + H_2O$

To balance K we have 5KOH on R.H.S.

$$8KMnO_4 + 3NH_3 \longrightarrow 8MnO_2 + 3KNO_3 + 5KOH + H_2O$$

In order to balance hydrogen and oxygen atoms, we add one H_2O on R.H.S.

$$8KMnO_4 + 3NH_3 \longrightarrow 8MnO_2 + 3KNO_3 + 5KOH + 2H_2O$$

Example 4. Balance the following equation:

$$\mathbf{KClO_3 + H_2SO_4 \longrightarrow KHSO_4 + HClO_4 + ClO_2 + H_2O}$$

Solution.

Step I. $K\overset{+5}{Cl}O_3 + H_2SO_4 \longrightarrow KHSO_4 + H\overset{+7}{Cl}O_4 + \overset{+4}{Cl}O_2 + H_2O$

$$\overset{+5}{Cl} \longrightarrow \overset{+7}{Cl}\text{ (oxidation)}$$

$$\overset{+5}{Cl} \longrightarrow \overset{+4}{Cl}\text{ (reduction)}$$

Step II. $\overset{+5}{Cl} \longrightarrow \overset{+7}{Cl} + 2e$ ... (*i*)

$\overset{+5}{Cl} + e \longrightarrow \overset{+4}{Cl}$ ...(*ii*)

Step III. Multiply (*ii*) by 2 and, add, we get

$$3\overset{+5}{Cl} \longrightarrow \overset{+7}{Cl} + 2\overset{+4}{Cl}$$

Step IV. One $\overset{+7}{Cl}$ comes from 1 $HClO_4$ and $2\overset{+4}{Cl}$ from $2ClO_2$.

$\therefore$ $3KClO_3 + H_2SO_4 \longrightarrow KHSO_4 + HClO_4 + 2ClO_2 + H_2O$

This is balanced.

Example 5. Balance and complete the following equations:

(*i*) $Mn^{+2} + PbO_2 \longrightarrow MnO_4^- + H_2O + Pb^{+2}$ **(acidic medium)**

(*ii*) $ClO_2 + SbO_2^- \longrightarrow ClO_2 + Sb(OH)_6^-$ **(basic medium)**

Solution.

(*i*) Step I. $\overset{+2}{Mn} + \overset{+4}{Pb}O_2 \longrightarrow \overset{+7}{Mn}O_4^- + H_2O + \overset{+2}{Pb}$

$\overset{+2}{Mn} \longrightarrow \overset{+7}{Mn}$(oxidation)

$\overset{+4}{Pb} \longrightarrow \overset{+7}{Pb}$ (reduction)

Step II. $\overset{+2}{Mn} \longrightarrow \overset{+7}{Mn} + 5e$ ...(*i*)

$\overset{+4}{Pb} + 2e \longrightarrow \overset{+2}{Pb}$ ...(*ii*)

Step III. Multiply (*i*) by 2 and (*ii*) by 5, then add:

$$2\overset{+2}{Mn} + 5\overset{+4}{Pb} \longrightarrow 2\overset{+7}{Mn} + 5\overset{+2}{Pb}$$

Step IV. $2Mn^{+2} + 5PbO_2 \longrightarrow 2MnO_4^- + 5Pb^{2+} + H_2O$

Step V. Add H_2O on the side deficient in O atoms. To balance H we add $4H^+$ on L.H.S. and the final equation becomes:

$$2Mn^{2+} + 5PbO_2 + 4H^+ \longrightarrow 2MnO_4^- + 5Pb^{2+} + 2H_2O$$

This balances all the charges and all the atoms.

(*ii*) Step I. $\overset{+4}{Cl}O_2 + \overset{+3}{Sb}O_2^- \longrightarrow \overset{+3}{Cl}O_2^- + \overset{+5}{Sb}(OH)_6^-$

$\overset{+4}{Cl} \longrightarrow \overset{+3}{Cl}$ (reduction)

$\overset{+3}{Sb} \longrightarrow \overset{+5}{Sb}$(oxidation)

Step II. $\overset{+4}{Cl} + e \longrightarrow \overset{+3}{Cl}$ ... (*i*)

$\overset{+3}{Sb} \longrightarrow \overset{+5}{Sb} + 2e$ ...(*ii*)

Step III. Multiply (*i*) by 2 and add

$$2\overset{+4}{Cl} + \overset{+3}{Sb} \longrightarrow 2\overset{+3}{Cl} + \overset{+5}{Sb}$$

Step IV. $2ClO_2 + SbO_2^- \longrightarrow 2ClO_2^- + Sb(OH)_6^-$

Step V. As per rule V we add $4H_2O$ on R.H.S. and $8OH^-$ on L.H.S.

$$2ClO_2 + SbO_2^- + 8OH^- \longrightarrow 2ClO_2^- + Sb(OH)_6^- + 4H_2O$$

In order to balance H atoms we add $6OH^-$ on R.H.S. As it has six excess H, to balance this, we add $4H_2O$ on L.H.S.

$$2ClO_2 + SbO_2^- + 8OH^- + 6H_2O \longrightarrow 2ClO_2^- + Sb(OH)_6^- + 4H_2O + 6OH^-$$

Rearranging we get

$$2ClO_2^- + SbO_2^- + 2OH^- + 2H_2O \longrightarrow 2ClO_2^- + Sb(OH)_6^-$$

Example 6. Balance the following equation:

$$\mathbf{KMnO_4 + MnSO_4 + H_2O \longrightarrow MnO_2 + K_2SO_4 + H_2SO_4}$$

Solution.

Step I. $K\overset{+7}{Mn}O_4 + \overset{+2}{Mn}SO_4 + H_2O \longrightarrow \overset{+4}{Mn}O_2 + K_2SO_4 + H_2SO_4$

$\overset{+7}{Mn} \longrightarrow \overset{+4}{Mn}$ (reduction)

$\overset{+2}{Mn} \longrightarrow \overset{+4}{Mn}$ (oxidation)

Step II. $\overset{+7}{Mn} + 3e \longrightarrow \overset{+4}{Mn}$ ...(*i*)

$\overset{+2}{Mn} \longrightarrow \overset{+4}{Mn} + 2e$ ...(*ii*)

Step III. Multipling (*i*) by 2 and (*ii*) by 3 and add, we get

$$2\overset{+7}{Mn} + 3\overset{+2}{Mn} \longrightarrow 5\overset{+4}{Mn}$$

Step IV. $2KMnO_4 + 3MnSO_4 + H_2O \longrightarrow 5MnO_2 + K_2SO_4 + H_2SO_4$

To balance SO_4 on the two sides we have $2H_2SO_4$ on R.H.S. That is,

$$2KMnO_4 + 3MnSO_4 + H_2O \longrightarrow 5MnO_2 + K_2SO_4 + 2H_2SO_4$$

Both H and O get balanced if we add one H_2O on L.H.S.

$$2KMnO_4 + 3MnSO_4 + 2H_2O \longrightarrow 5MnO_2 + K_2SO_4 + 2H_2SO_4$$

Example 7. Balance the following equations:

(*i*) $Mn^{2+} + S_2O_8^{2-} \longrightarrow MnO_4^- + HSO_4^-$ **(in acidic medium)**

(*ii*) $AsO_3^{3-} + MnO_4^- \longrightarrow AsO_4^{3-} + MnO_2$ **(in alkaline medium)**

Solution.

(*i*) **Step I.** $\overset{+2}{Mn} + \overset{+7}{S}_2O_8^{2-} \longrightarrow \overset{+7}{Mn}O_4^- + H\overset{+6}{S}O_4^-$

$\overset{+2}{Mn} \longrightarrow \overset{+7}{Mn}$ (oxidation)

$2\overset{+7}{S} \longrightarrow 2\overset{+6}{S}$ (reduction)

Step II. $\overset{+2}{Mn} \longrightarrow \overset{+7}{Mn} + 5e$ ...(*i*)

$2\overset{+7}{S} + 2e \longrightarrow 2\overset{+6}{S}$ ...(*ii*)

Step III. Multiply (*i*) by 2 and (*ii*) by 5 and add

$2\overset{+2}{Mn} + 10\overset{+7}{S} \longrightarrow 2\overset{+7}{Mn} + 10\overset{+6}{S}$

Step IV. $2Mn^{+2} + 5S_2O_8^{2-} \longrightarrow 2MnO_4^- + 10HSO_4^-$

Step V. Add $8H_2O$ on L.H.S. to balance deficient O

$2Mn^{2+} + 5S_2O_8^{2-} + 8H_2O \longrightarrow 2MnO_4^- + 10HSO_4^-$

Balance H by adding $6H^+$ on R.H.S.

$2Mn^{+2} + 5S_2O_8^{2-} + 8H_2O \longrightarrow 2MnO_4^- + 10HSO_4^- + 6H^+$

(*ii*) **Step I.** $\overset{+3}{As}O_3^{3-} + \overset{+7}{Mn}O_4^{3-} \longrightarrow \overset{+5}{As}O_4^{3-} + \overset{+4}{Mn}O_2$

$\overset{+3}{As} \longrightarrow \overset{+5}{As}$ (oxidation)

$\overset{+7}{Mn} \longrightarrow \overset{+4}{Mn}$ (reduction)

Step II. $\overset{+3}{As} \longrightarrow \overset{+5}{As} + 2e$ ...(*i*)

$\overset{+7}{Mn} + 3e \longrightarrow \overset{+4}{Mn}$ ... (*ii*)

Step III. Multiply (*i*) by 3 and (*ii*) by 2 and add

$3\overset{+3}{As} + 2\overset{+7}{Mn} \longrightarrow 3\overset{+5}{As} + 3\overset{+4}{Mn}$

Step IV. $3AsO_3^{3-} + 2MnO_4^- \longrightarrow 3AsO_4^{3-} + 2MnO_2$

Step V. One oxygen atom on L.H.S. is in excess, therefore one H_2O is added on L.H.S. and $2OH^-$ on R.H.S.

$$3AsO_3^{3-} + 2MnO_4^- + H_2O \longrightarrow 3AsO_4^{3-} + 2MnO_2 + 2OH^-$$

Example 8. Balance the following equations:

(i) $\mathbf{MnO_4^- + H^+ + H_2O \longrightarrow Mn^{+2} + O_2 + H_2O}$

(ii) $\mathbf{Zn + NaNO_3 + NaOH \longrightarrow Na_2ZnO_2 + NH_3 + H_2O}$

Solution.

(i) Step I. $\overset{+7}{Mn}O_4^- + H^+ + H_2\overset{-1}{O_2} \longrightarrow \overset{+2}{Mn} + \overset{0}{O_2} + H_2O$

$$Mn^{+7} \longrightarrow Mn^{+2} \text{ (reduction)}$$

$$2\overset{-1}{O} \longrightarrow \overset{0}{O_2} \text{ (oxidation)}$$

Step II. $Mn^{+7} + 5e \longrightarrow \overset{+2}{Mn}$ ...(i)

$$2\overset{-1}{O} \longrightarrow O_2 + 2\overline{e} \quad ...(ii)$$

Step III. Multiply (*i*) by 2 and (*ii*) by 5 and add

$$2\overset{+7}{Mn} + 10O^- \longrightarrow 2\overset{+2}{Mn} + 5O_2$$

Step IV. $2MnO_4^- + 5H_2O_2 + H^+ \longrightarrow 2\overset{+2}{Mn} + 5O_2 + H_2O$

Step V. Add water on the side deficient in oxygen. Consequently, we add $7H_2O$ on R.H.S. to get a total of $8H_2O$.

$$2MnO_4^- + 5H_2O_2 + H^+ \longrightarrow 2Mn^{2+} + 5O_2 + 8H_2O$$

Add H^+ on L.H.S. to balance hydrogens.

$$2MnO_4^- + 5H_2O_2 + 6H^+ \longrightarrow 2Mn^{2+} + 5O_2 + 8H_2O$$

(ii) Step I. $\overset{0}{Zn} + Na\overset{+5}{N}O_3 + NaOH \longrightarrow Na_2\overset{+2}{Zn}O_2 + \overset{-3}{N}H_3 + H_2O$

$$\overset{0}{Zn} \longrightarrow \overset{+2}{Zn} \text{ (oxidation)} \quad ...(i)$$

$$N^{+5} \longrightarrow N^{-3} \text{ (reduction)} \quad ...(ii)$$

Step II. $\overset{0}{Zn} \longrightarrow \overset{+2}{Zn} + 2e$ ...(i)

$$N^{+5} + 8e \longrightarrow N^{-3} \quad ...(ii)$$

Step III. Multiply (*i*) by 4 and add (*ii*) to get

$$4\overset{0}{Zn} + \overset{+5}{N} \longrightarrow 4\overset{+2}{Zn} + \overset{-3}{N}$$

Step IV. $4Zn + NaNO_3 + NaOH \longrightarrow 4Na_2ZnO_2 + NH_3 + H_2O$

Balancing sodium atoms,

$$4Zn + NaNO_3 + 7NaOH \longrightarrow 4Na_2ZnO_2 + NH_3 + H_2O$$

Add one H_2O on R.H.S. to balance all the atoms.

$$4Zn + NaNO_3 + 7NaOH \longrightarrow 4Na_2ZnO_2 + NH_3 + 2H_2O$$

Example 9. Complete and balance the following equations:

(*i*) $K_2Cr_2O_7 + KI + HCl \longrightarrow KCl + H_2O + ... + ...$

(*ii*) $MnO_4^- + Fe^{2+} \longrightarrow Mn^{2+} + H_2O + ...$(in acidic medium)

(*iii*) $(NH_4)_2SO_4 + NO + NO_2 \longrightarrow H_2SO_4 + H_2O + ...$

(*iv*) $Zn + HNO_3 \longrightarrow ......+ H_2O + N_2O$

(*v*) $CrO_4^{2-} + CN^- \longrightarrow Cr(OH)_3 + ...$ (in alkaline medium)

Solution. (*i*) $K_2\overset{+6}{Cr_2}O_7 + K\overset{-1}{I} + HCl \longrightarrow KCl + H_2O + \overset{0}{I_2} + \overset{+3}{Cr}Cl_3$

Balancing oxidation number, we get

$$K_2Cr_2O_7 + 6KI + HCl \longrightarrow KCl + H_2O + 3I_2 + 2CrCl_3$$

Balancing K and Cl, we get

$$K_2Cr_2O_7 + 6KI + 14HCl \longrightarrow 8KCl + H_2O + 3I_2 + 2CrCl_3$$

Hydrogen and oxygen are balanced by adding H_2O on R.H.S.

$$K_2Cr_2O_7 + 6KI + 14HCl \longrightarrow 8KCl + 7H_2O + 3I_2 + 2\,CrCl_3$$

(*ii*) $MnO_4^- + 5Fe^{2+} \longrightarrow Mn^{2+} + H_2O + Fe^{3+}$

Balancing oxidation numbers, we get

$$MnO_4^- + 5Fe^{2+} \longrightarrow Mn^{2+} + 5Fe^{3+} + H_2O$$

Balancing oxygen by adding H_2O on R.H.S. and the H by adding H^+ on L.H.S. (acidic medium)

$$MnO_4^- + 5Fe^{2+} + 8H^+ \longrightarrow Mn^{2+} + 5Fe^{3+} + 4H_2O$$

(*iii*) $(\overset{-3}{N}H_4)_2SO_4 + \overset{+2}{N}O + \overset{+4}{N}O_2 \longrightarrow H_2SO_4 + H_2O + ...$

The oxidation numbers on nitrogen add up to give total $(-3 \times 2 + 2 + 4 = 0)$ of zero. Therefore, nitrogen on R.H.S. is in the form of N_2.

$$\therefore \quad (NH_4)_2SO_4 + NO + NO_2 \longrightarrow H_2SO_4 + 3H_2O + 2N_2$$

(*iv*) $Zn + HNO_3 \longrightarrow ..... + H_2O + N_2O$

Product of zinc and nitric acid would be zinc nitrate

$$\overset{0}{Zn} + H\overset{+5}{N}O_3 \longrightarrow \overset{+2}{Zn}(\overset{+5}{N}O_3)_2 + H_2O + \overset{+1}{N_2}O$$

Balancing oxidation numbers, we get

$$4Zn + 2HNO_3 \longrightarrow 4Zn(NO_3)_2 + N_2O + H_2O$$

Balancing nitrogen atoms, we get

$$4Zn + 10HNO_3 \longrightarrow 4Zn(NO_3)_2 + N_2O + H_2O$$

Balancing (O) by adding H_2O, we get

$$4Zn + 10HNO_3 \longrightarrow 4Zn(NO_3)_2 + N_2O + H_2O$$

(*v*) $$Cr\,O_4^{2-} + 3CN^- \longrightarrow 2Cr(OH)_3 + CNO^-$$

In this equation Cr^{+6} is reduced to Cr^{+3} and CN^- is oxidised to CNO^-.

Balancing these, we get

$$2CrO_4^{2-} + 3CN^- \longrightarrow 2Cr(OH)_3 + 3CNO^-$$

Balancing oxygen in alkaline medium, we get

$$2CrO_4^{2-} + 3CN^- + 2OH^- \longrightarrow 2Cr(OH)_3 + 3CNO^- + H_2O$$

Balancing H by adding $6H_2O$ on L.H.S. and $6OH^-$ on R.H.S., we get

$$2CrO_4^{2-} + 3CN^- + 2OH^- + 6H_2O \longrightarrow 2Cr(OH)_3 + 3CNO^- + H_2O + 6OH^-$$

$$Or_2CrO_4^{2-} + 3CN^- + 5H_2O \longrightarrow 2Cr(OH)_3 + 3CNO^- + 4OH^-$$

3.2.5. Ion electron method of balancing chemical equations

Various steps used in the balancing of equation by this method are given below.

Step I. Divide equation into oxidation half and reduction half.

Step II. Balance atoms in each half equation. Then balance oxygen by adding H_2O and hydrogen by adding H^+. In case of alkaline medium use OH^- and H_2O as in oxidation number method.

Step III. Balance charges in each equation by adding electrons.

Step IV. Use appropriate integers to make loss of electrons equal to gain of electrons and then add, so that electrons get cancelled.

Step V. Check balancing of all the atoms and charges.

This is explained with the help of the following reactions:

(*i*) $$Cl_2 + KIO_3 \longrightarrow Cl^- + KIO_4$$

(*ii*) $$Zn + NO_3^- \longrightarrow ZnO_2^{2-} + NH_3$$

(*i*) Step I. $$Cl_2 \longrightarrow 2Cl^-$$

$$IO_3^- \longrightarrow IO_4^-$$

Step II. $$Cl_2 \longrightarrow 2Cl^-$$

$$IO_3^- + 2OH \longrightarrow IO_4^- + H_2O \text{ (being alkaline)}$$

Step III. $Cl_2 + 2e \longrightarrow 2Cl^-$

$IO_3^- + 2OH^- \longrightarrow IO_4^- + H_2O + 2e$

Step IV. Adding, we get

$$Cl_2 + IO_3^- + 2OH^- \longrightarrow 2Cl^- + IO_4^- + H_2O$$

$$Cl_2 + KIO_3^- + 2OH^- \longrightarrow 2Cl^- + KIO_4 + H_2O$$

(*ii*) Step I. $Zn \longrightarrow ZnO_2^{2-}$

$NO_3^- \longrightarrow NH_3$

Step II. $Zn + 4OH^- \longrightarrow ZnO_2^{2-} + 2H_2O$

$NO_3^- + 3H_2O \longrightarrow NH_3 + 6OH^-$

To balance H atom, we add hydroxide on R.H.S. and water on L.H.S.

$$NO_3^- + 3H_2O + 3H_2O \longrightarrow NH_3 + 6OH^- + 3OH^-$$

Step III. $Zn + 4OH^- \longrightarrow ZnO_2^{2-} + 2H_2O + 2e$ ...(*i*)

$NO_3^- + 6H_2O + 8e \longrightarrow NH_3 + 9OH^-$ ...(*ii*)

Step IV. Multiplying (*i*) by 4 and adding to (*ii*), we get

$$4Zn + 16OH^- + NO_3^- + 6H_2O \longrightarrow 4ZnO_2^{2-} + 8H_2O + NH_3 + 9OH^-$$

or $$4Zn + 7OH^- + NO_3^- \longrightarrow ZnO_2^{2-} + NH_3 + 2H_2O$$

Again consider the following reactions. These are balanced by ion-electron method as follows:

(*i*) $K_2Cr_2O_7 + H_2SO_4 + (COOH)_2 \longrightarrow K_2SO_4 + H_2O + Cr_2(SO_4)_3 + CO_2$

(*ii*) $Zn + H_2SO_4 + H_2MoO_4 \longrightarrow ZnSO_4 + H_2O + Mo_2(SO_4)_3$

(*i*) Step I. $Cr_2O_7^{2-} \longrightarrow 2Cr^{3+}$

$C_2O_4^{2-} \longrightarrow 2CO_2$

Step II. $Cr_2O_7^{2-} + 14H^+ \longrightarrow 2Cr^{3+} + 7H^2O$

$C_2O_4^{2-} \longrightarrow 2CO_2$

Step III. $Cr_2O_7^{2-} + 14H^+ + 6e \longrightarrow 2Cr^{+3} + 7H_2O$

$C_2O_4^{2-} \longrightarrow 2CO_2 + 2e$

Step IV. $Cr_2O_7^{2-} + 14H^+ + 6e \longrightarrow 2Cr^{3+} + 7H_2O$

$3C_2O_4^{2-} \longrightarrow 6CO_2 + 6e$

$$Cr_2O_7^{2-} + 14H^+ + 3C_2O_4^{2-} \longrightarrow 2Cr^{3+} + 7H_2O + 6CO_2$$

Step V. While writing in the form given in the question, we take care not to write H_2SO_4.

$$K_2Cr_2O_7 + 14H^+ + 3(COOH)_2 \longrightarrow Cr_2(SO_4)_3 + 7H_2O + 6CO_2$$

Accounting for $6H^+$ added for oxalic acid we are left with $8H^+$ which mean $4H_2SO_4$.

$$K_2Cr_2O_7 + 4H_2SO_4 + 3(COOH)_2 \longrightarrow Cr_2(SO_4)_3 + 7H_2O + 6CO_2$$

(*ii*) $$H_2MoO_4 + Zn + H_2SO_4 \longrightarrow Mo_2(SO_4)_3 + ZnSO_4 + H_2O$$

Step I.
$$Zn \longrightarrow Zn^{2+}$$
$$MoO_4^{2-} \longrightarrow Mo^{3+}$$

Step II.
$$Zn \longrightarrow Zn^{2+}$$
$$MoO_4^{2-} + 8H^+ \longrightarrow Mo^{3+} + 4H_2O$$

Step III.
$$Zn \longrightarrow Zn^{2+} + 2e$$
$$MoO_4^{2-} + 8H^+ + 3e \longrightarrow Mo^{3+} + 4H_2O$$

Step IV.
$$3\ Zn \longrightarrow 3Zn^{2+} + 6e$$
$$2MoO_4^{2-} + 16H^+ + 6e \longrightarrow 2Mo^{3+} + 8H_2O$$
$$\overline{3Zn + 2MoO_4^{2-} + 16H^+ \longrightarrow 3Zn^{2+} + 2Mo^{3+} + 8H_2O}$$

StepV. $3Zn + 2H_2MoO_4 + 16\ H^+ \longrightarrow 3ZnSO_4 + Mo_2(SO_4)_3 + 8H_2O$

Balancing H and SO_4, we get 6 H_2SO_4 on L.H.S. because $4H^+$ are taken care of by $2H_2MoO_4$

$$3Zn + 2H_2\ MoO_4 + 6H_2SO_4 \longrightarrow 3ZnSO_4 + Mo_2(SO_4)_3 + 8H_2O$$

PROBLEMS FOR PRACTICE

Q. 1. Balance the following equations by oxidation number method:

(*i*) $CH_4 + O_2 \longrightarrow CO_2 + H_2O$

(*ii*) $FeCl_3 + SnCl_2 \longrightarrow SnCl_4 + FeC_2$

(*iii*) $P_4 + HNO_3 \longrightarrow H_3PO_4 + NO_2 + H_2O$

(*iv*) $Zn + HNO_3 \longrightarrow Zn(NO_3)_2 + NO + H_2O$

(*v*) $NH_3 + O_2 \longrightarrow NO + H_2O$

(*vi*) $Mn^{2+} + BiO_3^- + H^+ \longrightarrow HMnO_4 + Bi^{3+} + H_2O$

(*vii*) $Fe^{2+} + H_2O_2 + H^+ \longrightarrow Fe^{3+} + H_2O$

Ans. (*i*) $CH_4 + 2O_2 \longrightarrow CO_2 + 2H_2O$

(*ii*) $2FeC1_3 + SnC1_2 \longrightarrow SnC1_4 + 2FeC1_2$

(*iii*) $P_4 + 20HNO_3 \longrightarrow 4H_3PO_4 + 20NO_2 + 4H_2O$

(*iv*) $3Zn + 8HNO_3 \longrightarrow 3Zn(NO_3)_2 + 2NO + 4H_2O$

(*v*) $4NH_3 + 5O_2 \longrightarrow 4NO + 6H_2O$

(*vi*) $2Mn^{2+} + 5BiO_3^- + 16H^+ \longrightarrow 2HMnO_4 + 5Bi^{3+} + 7H_2O$

(*vii*) $2Fe^{2+} + H_2O_2 + 2H^+ \longrightarrow 2Fe^{3+} + 2H_2O$

Q. 2. Balance the following equations by oxidation number method:

(*i*) $H_2S + MnO_4^- + H^+ \longrightarrow S + Mn^{2+} + H_2O$

(*ii*) $Cr_2O_7^{2} + H^+ + Fe^{2+} \longrightarrow 2Cr^{3+} + Fe^{3+} + H_2O$

(*iii*) $AsO_3^{3-} + IO_3^- \longrightarrow AsO_4^{3-} + I^-$

(*iv*) $HNO_3 + I_2 \longrightarrow HIO_3 + NO_2 + H_2O$

(*v*) $MnO_4^- + Cl^- + H^+ \longrightarrow Mn^{2+} + Cl_2 + H_2O$

Ans. (*i*) $5H_2S + 2MnO_4^- + 6H^+ \longrightarrow 5S + 2Mn^{2+} + 8H_2O$

(*ii*) $Cr_2O_7^{2-} + 6Fe^{2+} + 14H^+ \longrightarrow 2Cr^{3+} + 6Fe^{3+} + 7H_2O$

(*iii*) $3AsO_3^{2-} + IO_3^- \longrightarrow 3AsO_4^{3-} + I^-$

(*iv*) $I_2 + 10HNO_3 \longrightarrow 2HIO_3 + 10NO_2 + 2H_2O$

(*v*) $2MnO_4^- + 10Cl^- + 16H^+ \longrightarrow 5Cl_2 + 2Mn^{2+} + 8H_2O$

Q. 3. Balance the following by oxid. No. and Ion electron methods:

(*i*) $IO_3^- + Cl^- \longrightarrow Cl_2 + I_2$

(*ii*) $Zn + NO_3^- \longrightarrow Zn^{2+} + NH_4^+$

(*iii*) $MnO_2 + BiO_3^- \longrightarrow MnO_4^- + Bi^{3+}$ (acidic medium)

(*iv*) $Ni(CN)_4^{2-} + Br_2 \longrightarrow Ni_2O_3 + Br^- + CN^-$

(*v*) $Cl_2 + OH^- \longrightarrow Cl^- + ClO_3^-$

(*vi*) $C_6H_{12}O_6 + H_2SO_4 \longrightarrow SO_2 + CO_2$

(*vii*) $NO_2^- \longrightarrow NO + HNO_3$

(*viii*) $H_3AsO_4 + KI + HCl \longrightarrow H_3AsO_3 + I_2 + KCl + H_2O$

(*ix*) $K_2MnO_4 + H_2O \longrightarrow MnO_2 + KMnO_4 + KOH$

Ans. (*i*) $2IO_3^- + 10Cl^- + 12H^+ \longrightarrow I_2 + 5Cl_2 + 6H_2O$

(*ii*) $4Zn + 10H^+ + NO_3^- \longrightarrow NH_4^+ + 4Zn^{2+} + 3H_2O$

(*iii*) $2MnO_2 + 3BiO_3 + 12H^+ \longrightarrow 2HMnO_4 + 3Bi^{3+} + 5H_2O$

(*iv*) $2Ni(CN)_4^{2-} + 6OH^- + Br_2 \longrightarrow Ni_2O_3 + 3H_2O + 2Br^- + 8CN^-$

(*v*) $3Cl_2 + 6OH^- \longrightarrow 5Cl^- + ClO_3^- + 3H_2O$

(*vi*) $C_6H_{12}O_6 + 12H_2SO_4 \longrightarrow 12SO_2 + 6CO_2 + 18H_2O$

(*vii*) $3NO_2^- + 3H^+ \longrightarrow NO + HNO_3 + H_2O$

(*viii*) $H_3AsO_4 + 2HCl + 2KI \longrightarrow H_3AsO_3 + I_2 + H_2O + 2KCl$

(*ix*) $3K_2MnO_4 + 2H_2O \longrightarrow 2KMnO_4 + MnO_2 + 4KOH$

Example. Complete and balance the following equations:

(*i*) $PCl_5 + NO_2^- \longrightarrow PCl_3 + Cl^- + \ldots$

(*ii*) $P + HNO_3 \longrightarrow H_3PO_4 + H_2O + \ldots$

(*iii*) $S + OH^- \longrightarrow S_2O_3^{2-} + 2S^{2-}$

(*iv*) $MnO_4^{2-} \longrightarrow MnO_4^- + MnO_2$ **(acidic medium)**

Solution. (*i*) Nitrogen dioxide is formed in this redox reaction.

$$PCl_5 + 2NO_2^- \longrightarrow PCl_3 + 2Cl^- + NO_2$$

(*ii*) NO_2 is evolved in this reaction

$$P + HNO_3 \longrightarrow H_3PO_4 + H_2O + NO_2$$

Without going into each step, we have

Oxidation $\quad P \longrightarrow PO_4^{3-}$

$$P + 4H_2O \longrightarrow PO_4^{3-} + 8H^+ + 5e \quad \ldots(i)$$

Reduction $\quad NO_3^- \longrightarrow NO_2$

$$NO_3^- + 2H^+ \longrightarrow NO_2 + H_2O + e^- \quad \ldots(ii)$$

Multiply (*ii*) by 5 and add to (*i*), and rearrange:

$$P + 5HNO_3 \longrightarrow H_3PO_4 + 5NO_2 + H_2O$$

(*iii*) $\quad S + OH^- \longrightarrow S^{2-} + S_2O_3^{2-}$

This reaction takes place in basic medium.

Oxidation $\quad 2S^0 \longrightarrow S_2O_3^{2-}$ (O.N. of S in $S_2O_3^{2-}$ is + 2)

$$2S + 6OH^- \longrightarrow S_2O_3^{2-} + 3H_2O + 4e \quad \ldots(i)$$

Reduction $\quad S \longrightarrow S^{2-}$

$$S + 2e \longrightarrow S^{2-} \quad \ldots(ii)$$

Multiplying (*ii*) by 2 and adding to (*i*), we get

$$4S + 6OH^- \longrightarrow 2S^{2-} + S_2O_3^{2-} + 3H_2O$$

(*iv*) $\quad MnO_4^{2-} \longrightarrow MnO_4^- + MnO_2$

Oxidation $\quad MnO_4^{2-} \longrightarrow MnO_4^- + e \quad \ldots(i)$

Reduction $\quad MnO_4^{2-} + 4H^+ + 2e \longrightarrow MnO_2 + 2H_2O \quad \ldots(ii)$

Myltiplying (*i*) by 2 and adding to (*ii*)

$$3MnO_4^{2-} + 4H^+ \longrightarrow 2MnO_4^- + MnO_2 + 2H_2O$$

3.3. AUTO OXIDATION

It is a process in which a substance undergoes oxidation and reduction by itself. It may be compared with *disproportionation* reaction but in case of auto-oxidation the process occurs slowly without any action of heat, light or electricity. Some simple auto-oxidation reactions are given below:

(*i*) $\quad 3HNO_2 \longrightarrow HNO_3 + 2NO + 2H_2O$

Here nitrous acid undergoes simultaneous oxidation and reduction into HNO_3 and NO respectively.

$$\overset{+3}{N}O_2^- \longrightarrow \overset{+5}{N}O_3^- \text{ (oxidation)}$$

$$\overset{+3}{N}O_2^- \longrightarrow \overset{+2}{N}O \text{ (reduction)}$$

(*ii*) $$H_2\overset{+6}{Mn}O_3 \longrightarrow 2H\overset{+7}{Mn}O_4 + \overset{+4}{Mn}O_2 + 2H_2O$$

Manganic acid undergoes self-oxidation and reduction into peramanganic acid and manganese dioxide.

$$\overset{+6}{Mn}O_4^{2-} \begin{cases} \longrightarrow \overset{+7}{Mn}O_4^- & \text{(oxidation)} \\ \longrightarrow \overset{+4}{Mn}O_2 & \text{(reduction)} \end{cases}$$

(*iii*) Hypochlorous acid undergoes auto-oxidation as follows:

$$3H\overset{+1}{Cl}O \longrightarrow H\overset{+5}{Cl}O_3 + 2H\overset{-1}{Cl}$$

Here $$Cl^{+1} \begin{cases} \longrightarrow Cl^{+5} & \text{(oxidation)} \\ \longrightarrow Cl^{-1} & \text{(reduction)} \end{cases}$$

3.4. WORKING OF GALVANIC CELL BASED ON REDOX REACTIONS

Consider the redox reaction

$$Zn + CuSO_4 \longrightarrow ZnSO_4 + Cu$$

taking place in a galvanic cell. The above reaction in ionic form may be written as

$$Zn + Cu^{+2} \longrightarrow Zn^{+2} + Cu$$

which can be split up into two half reactions as given below:

$$Zn \longrightarrow Zn^{+2} + 2e^- \quad \text{(oxidation half reaction)}$$

$$Cu^{+2} + 2e^- \longrightarrow Cu \quad \text{(reduction half reaction)}$$

On adding the two-half reactions, we get the overall reaction again. In order to generate electricity, we allow this reaction to take place in an indirect manner in a cell. We may proceed as under:

Take two beakers A and B containing dilute solution of zinc sulphate and copper sulphate respectively in them (Fig. 3.1). Put a zinc metal rod in beaker A and a copper metal rod in beaker B.

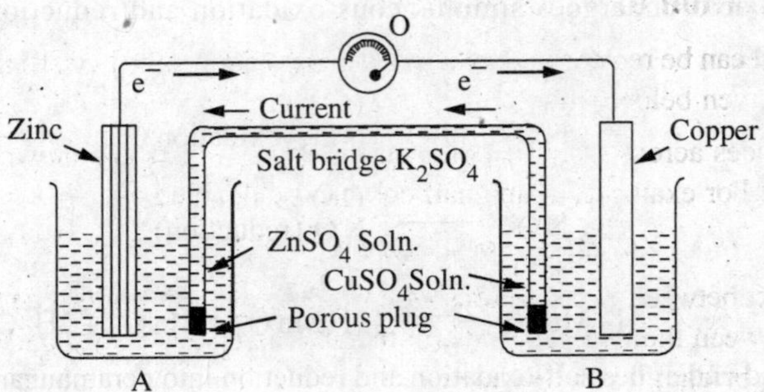

Fig. 3.1. Zn-$CuSO_4$ reaction in a Galvanic cell

The two rods are connected by a wire through an ammeter and a key externally. The two solutions are connected by a salt bridge (a U-shaped tube containing concentrated solution of an electrolyte like KC1 or K_2SO_4 or partially solidified solution of these electrolytes in agar-agar and gelatin). As soon as the connections are made by inserting the key, the current starts flowing which is indicated by the deflection produced in ammeter. Some important features of the device called cell may be summarised as below:

(*i*) The direction of the flow of electrons is from zinc to copper rod. However, the direction of conventional current (positive electricity) is from copper rod to zinc rod as indicated by the direction of deflection in the ammeter.

(*ii*) The zinc rod at which oxidation takes place is called **anode** and the copper rod at which reduction takes place is called **cathode.** Each of the beakers is called a **half cell or electrode.**

(*iii*) Since electrons are produced at the zinc electrode ($Zn \longrightarrow Zn^{+2} + 2e^-$), it is negative pole whereas the electrons are consumed at copper electrode ($Cu^{+2} + 2e^- \longrightarrow Cu$), it acts as the +ve pole of cell.

(*iv*) The chemical reaction $Zn + Cu^{+2} \longrightarrow Zn^{+2} + Cu$ starts taking place as soon as the connections are made. No heat is produced, whereas large amount of heat (≈ 12.5 kJ per mole) is produced when this reaction is done directly in a single vessel. The obvious inference is that an approximately equivalent amount of electrical energy is produced when the reaction is done in a galvanic cell.

(*v*) The two-half reaction taking place separately at the zinc rod and copper rod can be represented as:

$$Zn \longrightarrow Zn^{+2} + 2e^- \quad \text{(oxidation half reaction)}$$

$$Cu^{+2} + 2e^- \longrightarrow Cu \quad \text{(reduction half reaction)}$$

(*vi*) The oxidation of Zn into Zn^{+2} ions produces excess of Zn^{2+} ions and the reduction of Cu^{+2} ions into Cu atoms leave the excess of SO_4^{2-} ions in the two half cells A and B respectively. The salt bridge maintains electrical neutrality in the two half cells by allowing the diffusion of the excess of ions to the other beakers respectively.

(*vii*) The concentration of copper sulphate solution decreases with the operation of the cell due to reduction of Cu^{+2} ions into copper atoms and migration of SO_4^{-2} ions to the other side. Consequently the current strength falls with the passage of time. The cell based on the above reaction is commonly called *Daniel Cell.* The salt bridge is replaced by a porous pot through which SO_4^{2-} ions migrate from copper sulphate solution (taken in porous pot) to zinc sulphate solution. In the similar manner many other galvanic cells based upon different redox reactions can be set up. For example,

$$Cu + 2AgNO_3 \longrightarrow Cu(NO_3)_2 + 2Ag$$

$$Zn + 2HCl \longrightarrow ZnCl_2 + H_2$$

3.4.1. Representation of a galvanic cell

A galvanic cell can be represented by a shorthand notation called **cell diagram**. The conventions adopted for it are given below:

(*i*) The interfaces across which a potential difference exists are shown by a semicolon ; or a single verticle line. For example, a zinc half cell can be denoted as:

$$Zn; Zn^{+2} \quad \text{or} \quad Zn/Zn^{+2}$$

(*ii*) The contact between two solutions with the help of a salt bridge can be denoted by a double vertical line (||) between the two solutions. In the above cell Zn^{+2} ions and Cu^{+2} ions solution are connected by a salt bridge; it can be shown as:

$$Zn^{+2} \quad || \quad Cu^{+2}$$

(*iii*) The half cell in which oxidation takes place is written on left hand side and the half cell in which reduction takes place is written on R.H.S. with the metal electrodes always on the outside extreme.

(*iv*) The concentration of the two solutions, partial pressure of gases involved and their physical states are shown by suitable signs within brackets.

(*v*) The polarity of the two electrodes are indicated by (–) and (+) signs on the metal electrodes.

(*vi*) The e.m.f. of the cell is stated in volts along with the temperature at which it is applicable. Thus the cell diagram for the Daniel cell is.

$$(+)Zn \mid Zn^{+2} \mid\mid Cu^{+2} \mid Cu\ (-) \qquad E_{298} = 1.1 \text{ volts.}$$

3.4.2. Function of salt bridge in a galvanic cell

The salt bridge serves three important functions in a galvanic cell:

(*i*) It prevents any net charge accumulation in other beaker by permitting the excess of ions to diffuse from one half cell to other through the salt bridge. Thus, it maintains electrical neutrality in the two half cells.

(*ii*) It physically separates the two electrodes, but completes the electrical circuit with electrons flowing from one electrode to another externally and the ions flowing between the two compartments through salt bridge.

(*iii*) It minimizes the liquid function potential [voltage produced where two dissimilar solutions are in contact] between the two solutions connected through salt bridge.

3.5. ELECTRODE POTENTIAL

When a metal rod is dipped in water or the aqueous solution of its ions (Fig. 3.2), the chemical potential of the metal ions in the metal rod is higher than in aqueous phase. This leads to a flow of metal ions from the metal rod to the aqueous phase. The electrons are not solubilized in aqueous phase and, therefore, are left on the metal surface. This creates an electric field between the metal rod and the aqueous solution, the metal being negatively charged and the solution is positively charged.

$$M \longrightarrow M^{+n} + ne^{-}$$

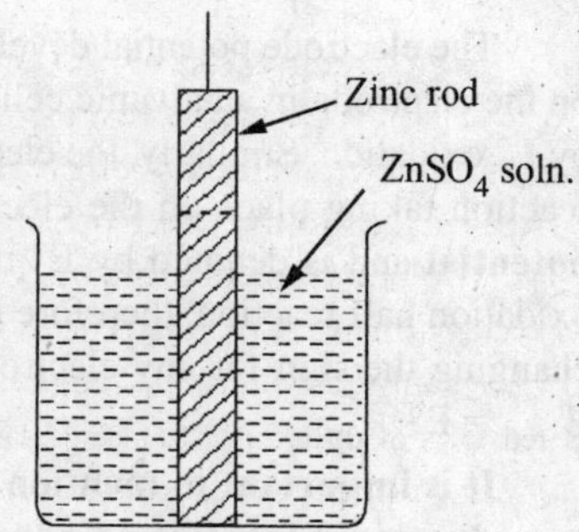

Fig. 3.2. Zn metal electrode.

An equilibrium state is established after a short while due to the reverse reaction initiated by the collisions of metal ions (M^{+n}) with the surface of metal rod and combining with electrons available on it.

$$M^{+n} + ne^- \longrightarrow M$$

At this stage a definite potential difference is developed between the metal rod and the solution. The equilibrium potential difference between the metal electrode and its surrounding solution is called **electrode potential.** Thus it is the tendency of an electrode to lose electrons when kept in contact with the solution of its own ions.

In a galvanic cell two electrodes are combined together so as to oppose their tendencies of losing electrons. Oxidation takes place at one electrode and reduction at the other. In other words, one of the electrodes has greater tendency of losing the electrons than the other. The other electrode may be considered as having greater tendency of gaining the electrons than the former. *Thus, the tendency of an electrode to lose or gain electrons in a cell is called electrode potential.*

The tendency of losing the electrons is called the **oxidation electrode potential** and the tendency of gaining the electrons is called the **reduction potential** of the electrode. They can be denoted by E_{oxid} and E_{red} respectively. The magnitude of electrode potential of an electrode depends upon several factors such as:

(*i*) Nature of the electrode.

(*ii*) Concentration of metal cations.

(*iii*) Temperature.

Under similar condition the two tendencies of an electrode are equal in magnitude but opposite in direction. This is because the equilibrium state remains unaltered from, whatever direction it is approached. Therefore,

$$E_{red} = -E_{oxid}$$

3.5.1. Standard electrode potential

If in an electrode set up, all active constituents or reactants are at unit activity, it is said to be **normal or standard electrode.** This can be achieved, when the concenteration of the electrolyte taken is one molar and the electrode is a pure solid or gas at a pressure of one atmosphere. The electrode is set up at a temperature of 298K. The electrode potential now developed is called **standard electrode potential**. It is denoted by E°, such an electrode is called standard or normal electrode. For example, a standard zinc electrode is set up by dipping a zinc metal rod into one molar zinc sulphate solution kept at 298K and one atmosphere pressure. It can be represented as $Zn(s) \mid Zn^{+2}$ (1M).

The electrode potential developed in a standard electrode due to oxidation reaction taking place on the electrode in a galvanic cell is called **standard oxidation electrode potential** and is denoted by E° oxidation. Similarly, the electrode potential developed in a standard electrode due to reduction reaction taking place on the electrode in a galvanic cell is called **standard reduction electrode potential** and is denoted by E° reduction. Since the reduction half reaction is just the reverse of oxidation half reaction, therefore reduction potential is obtained from oxidation potential by simply changing the sign for any electrode as these are equal in magnitude but opposite in direction. So $E^\circ_{red} = E^\circ_{oxid}$.

It is important to mention here that it is a convention to express all electrode potentials as reduction potential as per the recommendations of IUPAC. Therefore, E_{red} may be expressed as E only and E°_{red} as E°.

3.5.2. E.M.F. of a cell

The electromotive force of a galvanic cell is the potential difference, which causes the flow of current from the electrode at higher potential to the electrode at lower potential. It depends upon the chemical nature of the reactants and products. In other words, it represents the driving force of the chemical reaction taking place in the cell. This is because when the cell is opposed by a greater e.m.f. the spontaneous reaction is reversed and electrolysis takes place. The e.m.f. of a cell is denoted by E_{cell}.

Thus the e.m.f. of galvanic cell is the sum of the tendencies of the two electrodes of transporting the electrons in the same direction in the galvanic cell. The tendency of transporting the electrons by anode can be measured by its reduction electrode potential. Therefore,

$$\text{E.M.F. of cell} = E_{oxid} \text{ of anode } + E_{red} \text{ of cathode}$$

$$E_{cell} = E_{oxid.} \text{ (Anode) } + E_{red} \text{ (cathode)}$$

$$\text{or } E_{cell} = E_{red} \text{ (cathode)} - E_{red} \text{ (anode)}$$

$$E_{cell} = E_{cathode} - E_{anode} \quad [\because E_{oxid} = -E_{red}]$$

In the cell diagram of a galvanic cell the cathode is written are right hand side and the anode is written on left hand side. Therefore, the e.m.f. of a cell can be expressed as:

$$E_{cell} = E_{right} - E_{left} = E_R - E_L$$

If both the electrodes are standard electrodes, the e.m.f. of such a cell is called standard e.m.f. and is denoted by E° cell. Hence

$$E^\circ_{cell} = E^\circ_{oxid\cdot} \text{ (anode)} + E^\circ_{(Red)} \text{ (cathode)}$$

$$E^\circ_{cell} = E^\circ_{cathode} - E^\circ_{Anode}$$

$$= E^\circ_{right} - E^\circ_{left} = E^\circ_R - E^\circ_L$$

3.5.3. Measurement of e.m.f. of a cell

The potential difference between the two electrodes of a galvanic cell can be measured by connecting a voltmeter between them. The value of e.m.f. so obtained is not very correct due to the following reasons:

(*i*) The voltmeter draws some current from the cell, thereby it changes the e.m.f. of the cell.

(*ii*) With appreciable flow of current a part of e.m.f. will be utilized in overcoming the internal resistance of the cell.

Therefore, it is evident that e.m.f. of a cell should be measured by a method which involves little or no flow of current. The method is based upon Poggendorff s compensation principle and is called potentiometric method.

Procedure. In this method the unknown e.m.f. is opposed by another known e.m.f. until the two become equal and there is no net flow of current in the circuit as shown in Fig. 3.3.

(*i*) It consists of a uniform wire *AB* of high resistance and connected with a storage battery *B* of e.m.f. larger than the e.m.f. of the cell to be measured.

(*ii*) The cell *X* whose e.m.f. is to be measured is connected to the end A and the sliding contact *D,* through a galvanometer in such a manner that it sends e.m.f. in a direction opposite to that of storage battery *B.*

(*iii*) The sliding contact D is moved along the wire AB till at a particular point D no current flows through the galvanometer. The e.m.f. of the cell say E_x is proportional to the length AD, *i.e.,*

$$E_x \propto AD$$

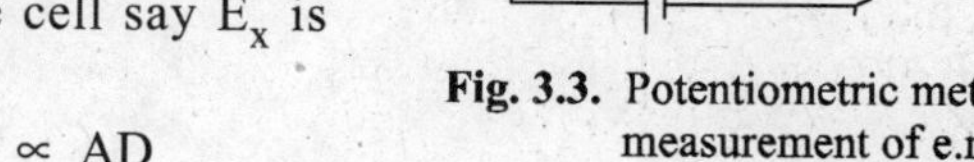

Fig. 3.3. Potentiometric method of measurement of e.m.f.

(*iv*) The cell X is now replaced by a standard cell the e.m.f. of which say E_s is known. The null point is obtained now at D′.

Then $E_s \propto AD'$.

Comparing the two equations

$$E_x = \frac{AD}{AD'} \times E_s$$

Knowing the values of E_S, AD and AD′ the value of E_x can be calculated, which is the e.m.f. of the cell. The standard cell must give a constant and reproducible e.m.f. without any variation with temperature. A Weston cadmium standard cell can be used for this purpose.

Hydrogen electrode

Hydrogen gas at one atmospheric pressure is bubbled into 1 molar hydrochloric acid solution in which a platinized platinum foil remains dipped (Fig. 3.4). The electrode can be represented as:

$$Pt, H_2\ (P = 1\ atm)\ |\ H^+\ (1\ M)$$

and electrode reactions taking place on the electrode as:

$$H_2 \rightleftharpoons 2H^+\ (1M) + 2e^-$$

Oxidation takes place when it acts as anode and reduction when acts as cathode.

When it acts as cathode, following reaction takes place:

$$2H^+ + 2e^- \rightleftharpoons H_2\ (g)$$

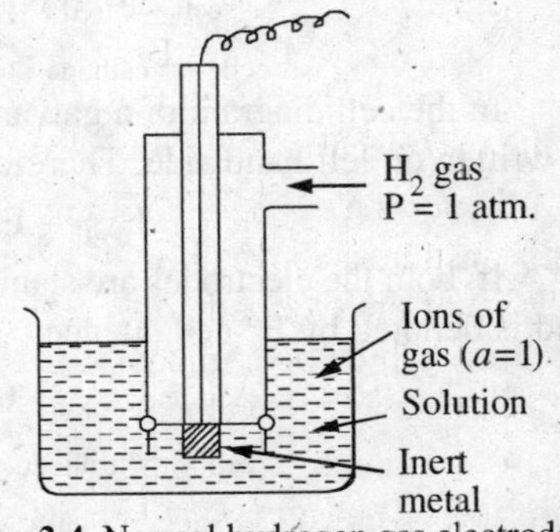

Fig. 3.4. Normal hydrogen gas electrode.

Calomel electrode

Hg (*l*), Hg_2Cl_2 (*s*)/KCl (*g*) combination is known as calomel electrode. The electrode is set up by taking pure mercury along with mercurous chloride paste at the bottom of the tube (Fig. 3.5) which is then filled with KCl solution of desired concenteration (0.1 N, 1 N or saturated) from the side tube (*A*). Another side tube *B* ending in a fine jet is used for connecting to the other half cell in order to set up a cell. A platinum wire is dipped in the tube so that it remains immersed into mercury and acts as inert metal electrode. The electrode can be represented as:

$$Hg\ (l), Hg_2Cl_2(s)\ |\ KCl\ (1\ N)$$

The electrode reactions are

$$2Hg\ (l) + 2Cl^- \longrightarrow Hg_2Cl_2 + 2e^-\ \text{(as anode)}$$

$$Hg_2Cl_2 + 2e^- \longrightarrow 2Hg\ (l) + 2Cl^-\ \text{(as cathode)}$$

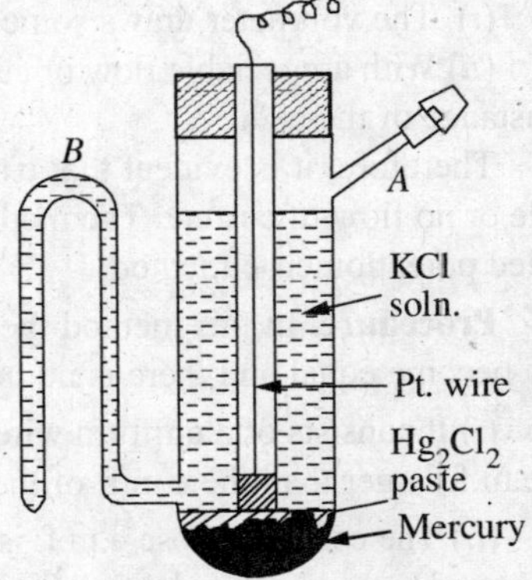

Fig. 3.5. Calomel electrode (metal metal insoluble salt anion electrode).

3.5.4. Measurement of standard electrode potential

The standard electrode potential is an important characteristic of a metal or non-metal. It predicts the relative ease of oxidation or reduction of the various metals and non-metals.

It is not possible to measure the electrode potential of a single electrode, because as soon as metallic conductor is placed in the solution of electrolyte to make the electrical circuit, a second electrode comes into operation. Hence in order to measure the standard electrode potential of an electrode, a standard electrode is set up and is coupled with a normal hydrogen electrode (NHE) to form a galvanic cell. The NHE is a reference electrode and its electrode potential is arbitrarily taken as zero. The e.m.f. of the constituted cell, which is the difference of potential between two electrodes can be experimentally measured. Since the electrode potential of one of them is zero, that of the other is equal to the observed e.m.f.

$$E° = E°_{Cathode} - E°_{Anode}$$
$$= E°_{right} - E°left$$

[where E° is the cell voltage of the cell. $E°_{cathode}$ and $E°_{anode}$ are standard electrode potentials of the two electrodes expressed as reduction potential.]

Hence the standard electrode potential of the electrode is numerically equal to the e.m.f. of the cell, constituted above, the sign of standard electrode potential of the electrode is determined by the polarity of the electrode or nature of the reaction occurring at the electrode. The polarity of the electrode can be determined by the direction of the flow of current, which can be noted from the deflection in ammeter connected in the circuit.

For example, the standard electrode potential of zinc electrode can be found by coupling it with NHE. The cell voltage is found to be 0.763 volt and electrons flow from zinc metal electrode to the hydrogen electrode. Therefore, zinc electrode acts as anode and hydrogen gas electrode as cathode.

$$\xrightarrow{\quad e \quad}$$

$$Zn \mid Zn^{+2}\ (a=1) \parallel H^{+}\ (a=1) \mid H_2\ (g)\ (1\ atm),\ Pt$$

$$E° = E°_{cathode} - E°_{anode}$$
$$0.763 = E°_{hydrogen} - E°_{zinc}$$
$$= 0 - E°_{Zn}$$
$$\therefore \quad E°_{Zn} = -\,0.763 \text{ volt}$$

If the electrode under experiment acts as negative pole when coupled with NHE, it will have –ve electrode potential and if it acts as a +ve pole then it will have +ve electrode potential. The standard electrode potential of copper electrode can be found in a similar manner. The galvanic cell formed when it is coupled with NHE is:

$$\xrightarrow{\quad e \quad}$$

$$Pt,\ H_2\ (g)\ (1\ atm) \mid H^{+}\ (a=1) \parallel Cu^{++}\ (a=1) \mid Cu$$

The cell voltage is measured as 0.34 volt and the electrons flow from hydrogen electrode to copper electrode,

$$E° = E°_{Cathode} - E°_{Anode}$$
$$= E°_{Copper} - E°_{Hydrogen}$$
$$0.34 = E°_{Copper} - 0$$
$$E°_{Copper} = 0.34 \text{ volt}$$

Since the copper electrode acts as positive pole in the constituted cell, it will have positive potential.

3.5.5. Electrochemical series

The table that gives the relative reactivities of metals or relative tendencies of reactions is called electrochemical series. In fact this is the table of standard electrode reduction potential. (See Table 3.1.) Reduction potential of an electrode denotes the tendency of the relevant reaction to take place in the direction of reduction (*i.e.*, to get reduced). It may be remembered that taking up of electrons means reduction.

Table 3.1. Some standard reduction potentials

← Oxidising Power decreases (upwards)

Reducing Power decreases → (downwards)

Half-Reaction				$E°$, *V*
$3/2N_2$	$+H^+ + e$	$\rightleftharpoons$	HN_3 (aq)	– 3.09
Li^+	$+ e$	$\rightleftharpoons$	Li	– 3.045
K^+	$+e$	$\rightleftharpoons$	K	–2.925
Cs^+	$+e$	$\rightleftharpoons$	Cs	–2.923
Ba^{2+}	$+ 2e$	$\rightleftharpoons$	Ba	–2.906
Sr^{2+}	$+2e$	$\rightleftharpoons$	Sr	–2.888
Ca^{2+}	$+2e$	$\rightleftharpoons$	Ca	–2.866
Na^+	$+e$	$\rightleftharpoons$	Na	–2.714
Mg^{2+}	$+2e$	$\rightleftharpoons$	Mg	–2.363
$1/2H_2$	$+e$	$\rightleftharpoons$	H^-	–2.25
Be^{2+}	$+2e$	$\rightleftharpoons$	Be	–1.79
Al^{3+}	$+3e$	$\rightleftharpoons$	Al	–1.662
V^{2+}	$+2e$	$\rightleftharpoons$	V	–1.186
Mn^{2+}	$+ 2e$	$\rightleftharpoons$	Mn	– 1.180
TiO^{2+}	$+ 2H^+ + 4e$	$\rightleftharpoons$	$Ti + H_2O$	–0.88
Zn^{2+}	$+2e$	$\rightleftharpoons$	Zn	–0.763
Cr^{3+}	$+3e$	$\rightleftharpoons$	Cr	–0.744

(Contd.)

← Oxidising Power decreases

Half -Reaction				$E°$, V
As	$+ 3H^+ + 3e$	$\rightleftharpoons$	AsH_3	–0.607
Fe^{2+}	$+ 2e$	$\rightleftharpoons$	Fe	– 0.440
Cr^{3+}	$+ e$	$\rightleftharpoons$	Cr^{2+}	– 0.41
Cd^{2+}	$+2e$	$\rightleftharpoons$	Cd	–0.403
Co^{2+}	$+2e$	$\rightleftharpoons$	Co	–0.277
H_3PO_4	$+ 2H^+ + 2e$	$\rightleftharpoons$	$H_3PO_3 + H_2O$	–0.276
V^{3+}	$+e$	$\rightleftharpoons$	V^{2+}	–0.256
$V(OH)_4^+$	$+ 4H^+ + 5e$	$\rightleftharpoons$	$V + 4H_2O$	–0.254
Ni^{2+}	$+2e$	$\rightleftharpoons$	Ni	–0.25
Sn^{2+}	$+ 2e$	$\rightleftharpoons$	Sn	– 0.136
O_2	$+H^+ + e$	$\rightleftharpoons$	HO_2	–0.13
Pb^{2+}	$+2e$	$\rightleftharpoons$	Pb	–0.126
$2H^+$	$+ 2e$	$\rightleftharpoons$	H_2	0.0000
$S(r)$	$+ 2H^+ + 2e$	$\rightleftharpoons$	H_2S	+0.142
Sn^{4+}	$+ 2e$	$\rightleftharpoons$	Sn^{2+}	+ 0.154
Cu^{2+}	$+ e$	$\rightleftharpoons$	Cu^+	+ 0.153
SO_4^{2-}	$+ 2H^+ + 2e$	$\rightleftharpoons$	$SO_3^{2-} + H_2O$	+ 0.172
Hg_2Cl_2	$+ 2e$	$\rightleftharpoons$	$2Hg + 2Cl^-$	+ 0.2676
Cu^{2+}	$+ 2e$	$\rightleftharpoons$	Cu	+ 0.337
VO^{2+}	$+2H^+ + e$	$\rightleftharpoons$	$V^{3+} + H_2O$	+0.36
$[Fe(CN)_6]^{3+}$	$+ e$	$\rightleftharpoons$	$[Fe(CN)_6]^{4-}$	+ 0.36
Cu^+	$+ e$	$\rightleftharpoons$	Cu	+ 0.52
I_2	$+2e$	$\rightleftharpoons$	$2I^-$	+0.5355
MnO_4^-	$+ e$	$\rightleftharpoons$	MnO_4^{2-}	+ 0.564
$2Hg^{2+}$	+ 2e	$\rightleftharpoons$	Hg^{2+}	+ 0.64
O_2	$+ 2H^+ + 2e$	$\rightleftharpoons$	H_2O_2	+0.6824
Fe^{3+}	$+ e$	$\rightleftharpoons$	Fe^{2+}	+ 0.771

Reducing Power decreases →

(Contd.)

← Oxidising Power decreases

Reducing Power decreases →

Half-Reaction				$E°, V$
Hg_2^{2+}	$+2e$	⇌	Hg	+0.788
Ag^+	$+e$	⇌	Ag	+0.7991
$2NO_3^-$	$+ 4H^+ + 2e$	⇌	$N_2O_4 + 2H_2O$	+ 0.803
NO_3^-	$+4H^+ +3e$	⇌	$NO + 2H_2O$	+0.96
Pd^{2+}	$+ 2e$	⇌	Pd	+ 0.987
N_2O_4	$+4H^+ +4e$	⇌	$2NO + 2H_2O$	+1.03
Br_2 (1)	$+ 2e$	⇌	$2\ Br^-$	+ 1.0652
Br_2 (aq)	$+ 2e$	⇌	$2\ Br^-$	+ 1.087
O_2	$+4H^+ +4e$	⇌	$2H_2O$	+1.229
MnO_2	$+4H^+ +2e$	⇌	$Mn^{2+} + 2H_2O$	+ 1.23
$C1O_4^-$	$+2H^+ +2e$	⇌	$ClO_3^- +H_2O$	+1.230
$Cr_2O_7^{2-}$	$+ 14H^+ + 6e$	⇌	$2Cr^{3+} + 7H_2O$	+1.33
Cl_2	$+ 2e$	⇌	$2C1^-$	+ 1.3595
PbO_2	$+4H^+ +2e$	⇌	$Pb^{2+} + 2H_2O$	+ 1.455
Au^{3+}	$+ 3e$	⇌	Au	+ 1.50
MnO_4^-	$+8H^+ +5e$	⇌	$Mn^{2+} + 4H_2O$	+1.51
$2BrO_3^-$	$+12H^+ + 10e$	⇌	$Br_2 + 6H_2O$	+ 1.51
Ce^{4+}	$+ e$	⇌	Ce^{3+}	+ 1.61
Au^+	$+ e$	⇌	Au	− 1.69
MnO_4^-	$+4H^+ +3e$	⇌	$MnO2 + 2H_2O$	+1.695
H_2O_2	$+ 2H^+ + 2e$	⇌	$2H_2O$	+ 1.776
Co^{3+}	$+e$	⇌	Co^{2+}	+1.81
Ag^{2+}	$+e$	⇌	Ag^+	+ 1.98
$S_2O_8^{2-}$	$+ 2e$	⇌	$2SO_4^{2-}$	+ 2.01
$O_3 + 2H^+$	$+2e$	⇌	$O_2 + H_2O$	+2.07
F_2 (g)	$+ 2e$	⇌	$2F^-$	+2.87

A negative value of reduction potential for an electrode denotes that there is little tendency for the electrode reduction. Rather such electrodes show oxidation tendencies, *i.e.*, to lose the electrons. For example, potassium electrode (see Table 3.1) has a potential of –2.925 volts. Thus it has little tendency for reduction. But fluorine electrode has a potential of +2.87 volts. Thus it has strong tendency for reduction and fluorine takes up electrons to change into fluoride ions.

3.6. USE OF REDOX POTENTIAL DATA

Major applications of redox potential data are given below :

***(i)* Calculation of standard e.m.f. of the cell.** It is possible to determine the e.m.f. of a cell using the electrochemical series. The electrode having the higher reduction potential is taken as cathode and the one having lower value is taken as anode. e.m.f. of the cell E° is then obtained by using the relation.

$$E^\circ = E^\circ_{Cathode} - E^\circ_{Anode}$$

where $E^\circ_{cathode}$ and E°_{anode} are the potentials of the cathode and anode respectively.

***(ii)* Comparison of the reactivity of metals.** The metals which occupy higher position in electrochemical series have lower reduction potentials. This indicates that ions of such metals will not be reduced to metals easily. On the contrary, such metals would be easily oxidised to their ions by losing electrons. Therefore, from the position of a metal in the electrochemical series, it is possible to predict the relative reactivities of metals. A metal having smaller reduction potential (*i.e.*, greater negative reduction potential) can displace metals having greater reduction potential (*i.e.*, smaller negative reduction potential) from the solution of their salts. In other words, a metal occupying higher position in the series can displace the metals lying below it from the solution of their salts.

We conclude that the *metals occupying higher positions in the electrochemical series are more reactive in displacing the other metals from the solutions of their salts.* For example, zinc lies above copper in the series and hence, it can displace the copper from copper sulphate solution. Copper cannot displace zinc from zinc sulphate solution because it lies below zinc in the series and hence is less reactive.

***(iii)* Predicting the feasibility of a redox reaction.** With the help of electromotive series we can predict whether a given redox reaction will take place or not. From the given equation the substances undergoing oxidation and reduction are identified. The substance undergoing oxidation will act as anode and the substance undergoing reduction will act as cathode. The e.m.f. of this hypothetical cell is calculated as under:

$$\text{E.M.F.} = E^\circ_{cathode} - E^\circ_{anode}$$

If e.m.f. comes out to be +ve the given redox reaction will take place and if e.m.f. comes out to be –ve the given redox reaction will not take place.

Let us say, we want to know whether the following reaction is feasible or not:

$$Sn^{+2} + Cu \longrightarrow Sn + Cu^{+2}$$

$E^\circ_{Sn^{+2}/Sn} = -0.136$ **volts and** $E^\circ_{Cu^{+2}/Cu} = 0.34$

The above reaction can be split into two half reactions as given below:

$$Cu \longrightarrow Cu^{+2} + 2e^- \quad \text{(oxidation) Assuming that the above}$$
$$Sn^{+2} + 2e^- \longrightarrow Sn \quad \text{(reduction) reaction takes place}$$

Since at copper electrode oxidation takes place, so it acts as anode and at tin electrode reduction takes place so it acts as cathode.

$$\begin{aligned} \text{EMF} &= E^\circ_{cathode} - E^\circ_{anode} \\ &= E^\circ_{Sn^{+2}/Sn} - E^\circ_{Cu^{+2}/Cu} \\ &= -0.136 - 0.34 \\ &= -0.476 \text{ volts} \end{aligned}$$

Since the e.m.f. comes out to be –ve, the reaction as given above will not take place. However the reverse reaction,

$$Sn + Cu^{+2} \longrightarrow Sn^{2+} + Cu, \text{ does take place.}$$

Example 1.

(*i*) $Cr_2O_7^{2-} + 14H^+ + 6e \rightleftharpoons 2Cr^{3+} + 7H_2O$ $E° = 1.33\ V$

$S_2O_8^{2-} + 2e \rightleftharpoons 2\ SO_4^{2-}$ $E° = 2.01\ V$

(*ii*) $Cl_2 + 2e \rightleftharpoons 2Cl^-$ $E° = 1.3595\ V$

$2I^- \rightleftharpoons I_2 + 2e$ $E° = -0.5355\ V$

(*iii*) $Fe^{3+} + e \rightleftharpoons Fe^{2+}$ $E° + 0.771\ V$

$Sn^{2+} \rightleftharpoons Sn^{4+} + 2e$ $E° + -0.154\ V$

Write balanced equations for the three pairs and also give the appropriate cell potentials.

Solution (*i*) Since $S_2O_8^{2-} / SO_4^{2-}$ has higher reduction potential, therefore, $S_2O_8^{2-}$ acts as a stronger oxidising agent and the cell equation would be

$[S_2O_8^{2-} + 2e \rightleftharpoons 2SO_4^{2-}] \times 3$ $E° = 2.01\ V$

$2Cr^{3+} + 7H_2O \rightleftharpoons Cr_2O_7^{2-} + 14H^+ + 6e$ $E° = -1.33\ V$

On addition we get,

$$2Cr^{3+} + 3S_2O_8^{2-} + 7H_2O \rightleftharpoons Cr_2O_7^{2-} + 6\ SO_4^{2-} + 14H^+$$

$$E°_{cell} = 2.01 - 1.33 = 0.68\ V$$

Note: While multiplying electrochemical equations, E° values are not multiplied and taken as such.

(*ii*) In this case the second equation does not represent standard reduction potential but is written to indicate standard oxidation potential. On addition of the two, we get

$$Cl_2 + 2I^- \rightleftharpoons I_2 + 2Cl^-$$

$E°_{cell} = 1.3595 - 0.5355 = 0.8240\ V$

(*iii*) Multiply eq. (1) by 2 and add to eq. (2)

$$2Fe^{3+} + Sn^{2+} \rightleftharpoons 2Fe^{2+} + Sn^{4+}$$

$$E°_{cell} = 0.771V - 0.154\ V = 0.617\ V$$

Example 2. Given: $Ce^{4+} + e \longrightarrow Ce^{3+}$ $E° = 1.61\ V$ ... (*i*)

$MnO_4^- + 8H^+ + 5e \rightleftharpoons Mn^{2+} + 4H_2O$ $E° = 1.51V$ ... (*ii*)

Comment on the feasibility of the reaction.

Solution. E° values indicate that Ce^{4+} is a stronger oxidant than acidic MnO_4^-. Therefore, acidic MnO_4^- cannot oxidise Ce^{3+} to Ce^{4+} but theoretically Ce^{4+} can oxidise Mn^{2+} to MnO_4^-. That is, theoretically the following reaction is possible:

$$Ce^{4+} + e \rightleftharpoons Ce^{3+}] \times 5$$

$$Mn^{2+} + 4H_2O \rightleftharpoons MnO_4^- + 8H^+ + 5e$$

$$5Ce^{4+} + Mn^{2+} + 4H_2O \rightleftharpoons 5Ce^{3+} + MnO_4^- + 8H^+$$

The cell potential for such a reaction is:

$$E°_{call} = 1.61 - 1.51 = 0.10 \text{ V}$$

Since $E°_{cell}$ is +ve therefore this reaction is theoretically possible. However $E°_{cell} = 0.10$ V has a very small cell potential. Therefore such a reaction would be slow or may not occur under normal conditions.

Example 3. Given: $Co^{3+} + e \rightleftharpoons Co^{2+}$ $E° = 1.81$ V

$Fe^{3+} + e \rightleftharpoons Fe^{2+}$ $E° = 0.77$ V

Write the appropriate redox reaction and the correct cell potential.

Solution. Based on E° values the correct redox reaction would be:

$$Co^{3+} + e \rightleftharpoons Co^{2+} \qquad E° = 0.181 \text{ V}$$

$$Fe^{2+} \rightleftharpoons Fe^{3+} + e \qquad E° = -0.77 \text{ V}$$

Adding to cancel out electrons, we get

$$Co^{3+} + Fe^{2+} \rightleftharpoons Co^{2+} + Fe^{3+}$$

$E° = 1.81 - 0.77 = 1.04$ V

A positive value $E° = 1.04$ V indicates that the cell is properly set up.

3.7. EQUILIBRIUM CONSTANT FROM REDOX POTENTIAL

The oxidation reduction reactions are reversible in nature and therefore equilibrium is established between oxidising agent and reducing agent. The value of equilibrium constant gives the extent to which the reaction proceeds in the forward direction. Consider the oxidation reduction reaction of Zn and Cu. The cell reaction is:

$$Zn\ (s) + Cu^{2+}\ (aq) \rightleftharpoons Zn^{2+}(aq) + Cu\ (s)$$

Applying law of chemical equilibrium, we have

$$K = \frac{[Cu(s)][Zn^{2+}]}{[Zn(s)][Cu^{2+}]}$$

Concentrations of solids Cu (s) and Zn (s) remain constant and may be assumed equal to 1.

$$K = \frac{[Zn^{2+}]}{[Cu^{2+}]} \qquad \text{... } (i)$$

The equilibrium constant for any oxidation reduction reaction is related to $E°_{cell}$ as :

$$E°_{cell} = \frac{2.303\,RT}{nF} \log K_c$$

or

$$E°_{cell} = \frac{0.059}{n} \log K_c \text{ at } 25°C$$

Here $E°_{cell} = E°_{cathode} - E°_{anode}$ and n is the number of electrons involved in the reaction. For the reaction:

$$Zn\ (s) + Cu^{2+}\ (aq) \rightleftharpoons Zn^{2+}\ (aq) + Cu(s)$$

The two-half cell reactions the electrode potentials are:

$$Zn^{2+} + 2e^- \rightleftharpoons Zn\,(s) \qquad E° = -0.763\text{ V}$$

$$Cu^{2+} + 2e^- \rightleftharpoons Cu\,(s) \qquad E° = +0.337\text{ V}$$

$$E°_{cell} = E°_{cathode} - E°_{anode}$$

$$= 0.337 - (-0.763) = 1.10\text{V}$$

Now $$E°_{cell} = \frac{0.059}{n}\log K \qquad \text{... }(ii)$$

Substituting for $E°_{cell}$ and n, we have

$$1.10 = \frac{0.059}{n}\log K$$

or $$\log K = \frac{2\times 1.10}{0.059} = 37.2$$

or $$K = 10^{37..2}$$

Since the value of K is very large, it means that the concentration of Zn^{2+} [Eq (i)] is very large and that of Cu^{2+} is very small. In other words, the reaction virtually goes to completion.

Similarly let us study the oxidation of Fe^{2+} by MnO_4^- ions:

$$MnO_4^- + 8H^+ + 5Fe^{2+} \rightleftharpoons Mn^{2+} + 5Fe^{3+} + 4H_2O$$

$$E°_{cell} = 1.510 - 0.771\text{V} = 0.739\text{ V}.$$

$$\log K = \frac{5\times 0.739}{0.059} = 62.6$$

$$K = 10^{62.6}$$

As the value of K is again very large, Fe^{2+} ions are oxidised to a large extent in the presence of acidified MnO_4^- ions.

3.8. FREE ENERGY CHANGE AND REDOX POTENTIALS

The standard electrode potential E°, of the cell is related to the standard Gibbs free energy change, Δ G° of the reaction by the equation:

$$\Delta G° = -n\,FE° \qquad \text{... }(i)$$

The free energy change Δ G° is related to enthalpy change Δ H° and entropy change of the reaction, Δ S° at temperature T as :

$$\Delta G° = \Delta H° - T\Delta S° \qquad \text{... }(ii)$$

Since the overall radox reaction is the difference of two half reactions, the standard Gibbs free energy change of the overall reaction is the difference of the standard Gibbs free energies of the two half reactions. The overall reaction is spontaneous or favourable (*i.e.* K > 1) in the direction which gives to the negative value of the resulting overall Δ G°. We choose one-half reaction which has Δ G° = 0 and express all other values relative to it. As in case of electrode potential, the half reaction of reduction of hydrogen ions is chosen for which Δ G° = 0 at all temperatures:

$$2H+\,(aq) + 2e^- \longrightarrow H_2(g) \qquad \Delta G°=0$$

This can be used for calculating ΔG° for redox reaction. For example for the redox reaction,

$$Zn^{2+}\,(aq) + H_2(g) \longrightarrow Zn\,(s) + 2H^+\,(aq) \qquad \Delta G° = +\,147\text{ kJ mol}^{-1}$$

The H^+ reduction half reaction has zero value of Gibbs free energy (according to the convention), therefore,

$$Zn^{2+}\ (aq) + 2e^- \longrightarrow Zn\ (s) \qquad \Delta G° = +147\ kJ\ mol^{-1}$$

Standard Gibbs free energies may be measured by setting up an electrochemical cell having one hydrogen electrode. The potential difference between the electrodes is measured. From this, Gibbs free energy is calculated using the relation:

$$\Delta G° = -nFE^a \qquad ...(iii)$$

The potential E° is called standard potential.

Since the standard Gibbs energy is the difference of ΔG° for the half reactions, E° for the overall reaction is also the difference of the two standard potentials of the reduction half reactions contributing to the overall reaction. Thus, from the half reactions:

$$2H^+\ (aq) + 2e^- \longrightarrow H_2(g) \qquad E° = 0.0\ V$$

$$Zn^{2+}\ (aq) + 2e^- \longrightarrow Zn(s) \qquad E° = -0.76V$$

The difference of the two-half cells is

$$2H^+\ (aq) + Zn(s) \longrightarrow H_2(g) + Zn^{2+}\ (aq) \qquad E° = 0-(-0.76) = 0.76\ V$$

Since $\Delta G° = -nFE°$, the reaction is favourable if E° > 0. For the above reaction, E°> 0 therefore, zinc thermodynamically reduces H^+ ions under standard condition (acidic solution, *p*H = 0 and unit activity of Zn^{2+}). All the couples having negative electrode potentials can reduce H^+ ions.

3.9. ANALYSIS OF REDOX CYCLE

The electrode potential of a metal depends upon a number of factors. The conversion of M to M^+ in aqueous solution may be considered to occur in a series of steps:

(1) Sublimation. This converts solid metal to the gaseous state. The energy required for this process is enthalpy of sublimation represented by $\Delta H°_s$.

(2) Ionization. The energy equal to ionization energy is required to remove electron from the gaseous atom. It is represented by I.

(3) Hydration. When gaseous ion is formed, it gets hydrated in aqueous solution. The amount of energy released is enthalpy of hydration represented by ΔH_{aq}.

For the formation of a cation in aqueous solution, we may consider a cycle as shown below:

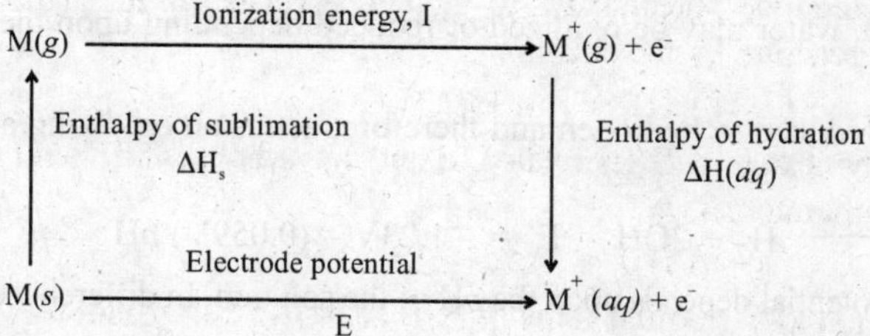

The overall, energy change E which gives electrode potential may be written as:

$$E = \Delta H_s + I + \Delta H\ (aq)$$

The enthalpy of sublimation and the ionization energy are positive because energy required to sublime and ionise while enthalpy of hydration is negative because energy is evolved.

If we are considering formation of an anion in solution the ionization energy may be replaced by electron affinity, EA when the gaseous atoms goes to gaseous anion. Thus, for the anion, the following relation will hold:

$$E = \Delta H_s + EA + \Delta H\,(aq)$$

It may be realised that the electrode potential values as obtained above are from enthalpy changes only. We need entropy values also for calculating Gibbs free energy changes.

When similar systems are being compared (e.g. Fe^{2+} | Fe and Zn^{2+} | Zn) the entropy changes will be very similar in each so that the difference in entropy contribution to the systems are negligible.

Therefore, for approximate treatments, the entropy changes may be neglected especially for comparison purposes. Thus, we may use the approximate relation:

$$-\,nFE^\circ = \Delta H^\circ$$

Variation in electrode potential in a group

Alkali metals have low melting points and hence low enthalpies of sublimation and also the atoms are large and therefore, have low ionization energies. Thus, the electrode potentials are high and the metals are very reactive. The decrease in electrode potential takes place from sodium to cesium due to decrease in ionization energy and sublimation energy with increasing size down the group. Lithium shows anomalous behaviour because of very large hydration energy released for the small cation.

In case of halogens, the sum of the atomization energy (half the bond energy of X_2 molecules) and the electron affinity is approximately constant from F to I. The decrease in electrode potential from fluorine to iodine is due to decrease in hydration energy from F to I as the size of the ion increases.

Most transition metals have high melting points and therefore, their sublimation energies are high. Because of their small size, their ionization energies are also high. Therefore, they have low values of electrode potential and such metals have less tendency to lose electrons.

3.10. REDOX STABILITY IN WATER

The ion or molecule may undergo oxidation or reduction depending upon the type of the solvent. For example, strong oxidizing agents will oxidize water to O_2 while strong reducing agents will reduce water to H_2. Thus, strong oxidizing and reducing agents may not exist in aqueous solutions. As a result, water may be oxidized or reduced depending upon the nature of the ions or molecules present.

Water may be reduced to hydrogen and therefore, act as an oxidizing agent according to the reaction :

$$2H_2O + 2\,e^- \longrightarrow H_2 + 2OH^- \quad E^\circ = -\,1.23V - (0.059V)\,pH$$

The electrode potential depends upon the *p*H of the solution. In different solutions, the following half reactions occur:

Reduction of Water

Neutral solution: $H_2O + e^- \longrightarrow OH^- + \frac{1}{2}H_2$ $\quad E^\circ = -\,0.414V$

1.0 M acid solution: $H_3O^+ + e^- \longrightarrow H_2O + \frac{1}{2}H_2$ $\quad E^\circ = 0.00V$

1.0 M basic solution: $H_2O + e^- \longrightarrow OH^- + \frac{1}{2}H_2$ $\quad E^\circ = -0.828V$

Oxidation of water

Neutral solution: $\frac{1}{2}O_2 + 2H^+ + 2e^- \longrightarrow H_2O$ $E° = +0.185$ V

1.0 M acid solution: $\frac{1}{2}O_2 + 2H^+ + 2e^- \longrightarrow H_2O$ $E° = +1.229$V

1.0 M basic solution : $\frac{1}{2}O_2 + H_2O + 2e^- \longrightarrow 2OH^-$ $E° = +0.401$V

From the above data minimum reduction potentials required to oxidize water to dioxygen should be more than + 0185 V in neutral solution, more than +1.229V in 1.0 M acid solution and more than +0.401V in 1.0 M basic solution.

Similarly, the half reactions with reduction potentials less than 0.00V (*i.e.* negative) in acid solution, less than –0.414 V (E° < – 0.414V) in neutral solution and less than –0.828 V (E° < – 0.828 V) in basic solution, should reduce water to H_2.

Reactions of metals with water or aqueous acids are usually the oxidation of metal by water or hydrogen ions. The overall reaction may be represented as:

$$M(s) + H_2O(l) \longrightarrow M^+(aq) + \frac{1}{2}H_2(g) + OH^-(aq)$$

or

$$M(s) + H^+(aq) \longrightarrow M^+(aq) + \frac{1}{2}H_2(g)$$

In view of the reduction potentials given above, these reactions are favourable for metals of s-block and first transition series from Group 4 to Group 7.

As for as oxidation is concerned, only a few oxidizing agents can oxidize water to give appreciable of O_2 evolution. The standard reduction potentials greater than +1.23 V occur for redox couples such as acidified dichromate couple, $Cr_2O_7^{2-} \mid Cr^{3+}$ (1.38V) and acidified permanganate couple, $MnO_4^- \mid Mn^{2+}$ (+1.51 V).

Representation of potential data

Electrode potentials data help us to understand the relative oxidising or reducing abilities of couples arranged in same diagram. They also provide information about the oxidation states of an element in a convenient way. These are commonly known as diagrams. Three types of diagrams are described below.

3.11. LATIMER DIAGRAMS

These are the simplest diagrams which give the reduction potentials for the half reactions corresponding to different oxidation states that an element exhibits in a single reduction potential diagram. When an element can exist in several oxidation states, it is convenient to display the reduction potentials corresponding to various half reactions in a diagrammatic form. In this diagram,

(i) highest oxidation state is written at the left and lowest oxidation state is written at the right and species such as electrons, H^+ and H_2O are omitted.

(ii) The value of the standard potential (in volts) is written over a horizontal line connecting species with the element in different oxidation states.

Latimer diagram of Iron

Iron exists in different oxidation states like Fe (VI) in FeO_4^{2-}, Fe (III) in Fe^{3+}, Fe (II) in Fe^{2+} and Fe (0) in Fe. The half cell reactions involving these changes are:

$$Fe^{2+} + 2e^- \longrightarrow Fe(s) \qquad E° = -0.47V$$

$$Fe^{3+} + 3e^- \longrightarrow Fe(s) \qquad E° = -0.057V$$

$$Fe^{3+} + e^- \longrightarrow Fe^{2+} \qquad E° = 0.77V$$

$$FeO_4^{2-} + 3e^- + 8H^+ \longrightarrow Fe^{3+} + 4H_2O \qquad E° = +2.20V$$

The Latimer diagram for iron giving all the above information can be represented as :

$$\overset{VI}{FeO_4^{2-}} \xrightarrow{+2.20} \overset{III}{Fe^{3+}} \xrightarrow{0.77} \overset{II}{Fe^{2+}} \xrightarrow{-0.47} \overset{0}{Fe}$$

$$Fe^{3+} \xrightarrow{-0.057} Fe$$

The species such as H_2O, H^+, OH^- and e^- are omitted in the diagram. The oxidation numbers are written under or over the species. Potential for $Fe^{3+} \longrightarrow Fe$, according to the above diagram should be (0.77 – 0.47) *i.e* 0.30 V. Actually, it is shown as – 0.057V. It may be noted that the potentials may not be added for half reactions because the electrons may not balance. In such cases potential are converted into free energies using the relation $\Delta G° = -nFE°$ (where n is the number of electrons involved in the reaction and F is the Faraday). Since Gibbs free energy is a thermodyanmic function, the free energies may be added and the resulting Gibbs free energy is converted back to E° value using the relation: $E° = -\Delta G°/nF$

$$Fe^{3+} + e^- \longrightarrow Fe^{2+} \qquad E° = 0.77V$$

$$\Delta G° = -1\,(+0.77)\,F = -0.77\,F$$

$$Fe^{2+} + 2e^- \longrightarrow Fe \qquad E° = -0.47V$$

$$\Delta G° = -2\,(-0.47)\,F = 0.94\,F$$

Adding

$$Fe^{3+} + 3e^- \longrightarrow Fe \qquad \Delta G° = (-0.77+0.94)\,F \text{ or } +0.17F$$

Now, E° for the above reaction , $Fe^{3+} + 3e^- \longrightarrow Fe$ may be calculated by converting free energy change into potential as follows:

$$E° = \frac{\Delta G°}{-nF} = \frac{+0.17F}{-3F} = -0.057$$

General calculation of standard potential for non-adjacent species (Calculation of E values for skip step couples)

The basis of calculation is that the overall ΔG° for two successive steps is the sum of the individual values that is

$$\Delta G° \text{ (overall)} = \Delta G°_1 + \Delta G°_2$$

Where $\Delta G°_1$ and $\Delta G°_2$ are individual free energy changes for the two steps. We convert individual E° values to ΔG° values by multiplying with relevant factor $-n$F and add them. Then the sum is converted back to E° for the non-adjacent couple by dividing by $-n$F for the overall electron transfer. Thus,

$$-nFE° = -n_1FE_1° - n_2FE_2°$$

where

$$n = n_1 + n_2$$

$$-nFE° = -(n_1E°_1 + n_2E°_2)F \quad \text{or} \quad E° = \frac{n_1E°_1 + n_2E°_2}{n}$$

$$\text{or } E° = \frac{n_1E°_1 + n_2E°_2}{n_1 + n_2}$$

or

$$E° = \frac{\Sigma n_i E_i^\circ}{\Sigma n_i}$$

This general procedure is followed to calculate the electrode potential for any non-adjacent species.

The electrode potential for the reduction of FeO_4^{2-} to Fe^{3+} is + 2.20V.

$$FeO_4^{2-} \xrightarrow{2.20} Fe^{3+}$$

As $\Delta G° = -nFE°$, $\Delta G°$ for this change will be large and negative. Thus, the reaction is thermodynamically possible and **FeO_4^{2-} can be easily reduced.** In order words, **FeO_4^{2-} acts as a strong** oxidizing agent.

The electrode potentials are measured on a scale with

$$H^+ (aq) + e^- \longrightarrow \frac{1}{2} H_2 \qquad E° = 0.0V$$

We know that reactions with negative value for E° are more strongly reducing than hydrogen and we call them as strongly reducing. The E° values for the changes $Fe^{3+} \longrightarrow Fe$ and $Fe^{2+} \longrightarrow Fe$ are both negative so that $\Delta G°$ is positive. Thus, neither Fe^{3+} nor Fe^{2+} has any tendency to get reduced to Fe.

Latimer diagram for chlorine in acidic solution

VII V III I 0 –I

$$ClO_4^- \xrightarrow{+1.20} ClO_3^- \xrightarrow{+1.18} HClO_2 \xrightarrow{+1.65} HClO \xrightarrow{+1.67} Cl_2 \xrightarrow{+1.36} Cl^-$$

Latimer diagram of redox potentials of Cl_2 (volts) at pH = 0

Information conveyed by Latimer diagram of chlorine

(i) $ClO^-_4 + 2H^+ + 2e^- \longrightarrow ClO_3^- + H_2O \qquad E° = + 1.20\ V$

Similarly

(ii) $ClO_3^- \xrightarrow{1.18} HClO_2$ represents the following half reaction:

$$2\,ClO_3^- + 6H^+ + 4e^- \rightarrow 2HClO_2 + 2H_2O \qquad E° = +1.18V$$

(iii) $HClO_2 \xrightarrow{1.65} HClO$ represent the following half reaction:

$$HClO_2 + 2H^+ + 2e^- \longrightarrow HClO + H_2O \qquad E° = +1.65V$$

(iv) $HClO \xrightarrow{1.67} Cl_2$ represents the following half reaction:

$$2HClO + 2H^+ + 2e^- \longrightarrow Cl_2 + 2H_2O \qquad E° = 1.67\ V$$

(v) $Cl_2 \xrightarrow{1.36} Cl^-$ represents the following half reaction:

$$Cl_2 + 2e^- \longrightarrow 2Cl^- \qquad E° = + 1.36V$$

The half reactions have been balanced by H^+ or H_2O as explained in the previous section.

Latimer diagram for chlorine in basic medium (pH = 14)

VII V III I 0 –I

$$ClO_4^- \xrightarrow{+0.36} ClO_3^- \xrightarrow{+0.33} ClO_2^- \xrightarrow{+0.66} OCl^- \xrightarrow{+0.42} \frac{1}{2}Cl_2 \xrightarrow{+1.36} Cl^-$$

The equations in basic medium may be balanced by the appropriate number of H_2O and OH^- as explained in the previous section.

(i) $ClO^-_4 \xrightarrow{+0.36} ClO_3^-$ represents the following half reaction:

$$ClO^-_4 + H_2O + 2e^- \longrightarrow ClO_3^- + 2OH^- \qquad E° = +0.36\ V$$

(ii) $$ClO^-_3 + H_2O + 2e^- \longrightarrow ClO_3^- + 2OH^- \qquad E° = +0.33\ V$$

and so on.

Disproportionation

Disproportionation is a process in which one oxidation state decomposes forming some ions in a higher oxidation state and some in a lower oxidation state. One of the most important information which can be obtained from Latimer diagram is whether any of the oxidation states are unstable with respect to disproportionation. This can be understood in terms of electrode potentials.

Consider the disproportionation reaction:

$$2M^+ (aq) \longrightarrow M (s) + M^{2+} (aq) \qquad ...(i)$$

The reaction will be spontaneous only if E° > 0. Let us express the overall reaction as the difference of two-half reactions.

$$M^+ (aq) + e^- \longrightarrow M (s) \qquad E° (R) \qquad ...(ii)$$

$$M^{2+} (aq) + e^- \longrightarrow M^+ (aq) \qquad E° (L) \qquad ...(iii)$$

Eq (*i*) can be obtained by subtracting eq. (*iii*) from eq (*ii*).

The deignation R and L refer to relative positions right and left of the couple in the Latimer diagram. More highly oxidized species is always placed on the left. The standard potential for the overall reaction is:

$$E° = E°(R) - E° (L)$$

It will be positive only if E° (R) > E° (L).

Thus, a *species has a thermodynamic tendency to disproportionate into its neighbours if the potential on the right of the species in the Latimer diagram is higher than the potential on the left.*

Latimer diagram of iron shows that reduction potential values become progressively less positive (or more negative) on moving left to right and therefore, Fe^{3+} and Fe^{2+} are stable with respect to disproportionation, that is, they do not disproportionate.

Consider reduction potential diagram for copper in acid solution. The Latimer diagram is:

$$\begin{array}{ccccc} \text{II} & & \text{I} & & 0 \\ Cu^{2+} & \xrightarrow{+0.15} & Cu^{+} & \xrightarrow{+0.50} & Cu \end{array}$$

$$Cu^{2+} \xrightarrow{+0.35V} Cu$$

The potential and hence the free energy released when Cu^{2+} is reduced to Cu^+ are both very small and therefore Cu^{2+} is not an oxidizing agent but is stable. On moving from left to right the potentials become more positive. Hence the species in the middle is unstable to disproportionation. In other words, it behaves both as a self oxidizing and self reducing agent because it is energetically favourable. This can also be shown as:

$$Cu^+ + e^- \longrightarrow Cu \qquad E° = +0.50\ V$$

$$Cu^{2+} + e^- \longrightarrow Cu^+ \qquad E° = +0.15\ V$$

$$\text{Overall} \qquad 2Cu^+ \longrightarrow Cu + Cu^{2+} \qquad E° = 0.35V$$

Since $E° > 0$, the reaction takes place i.e, Cu^+ disproportionates to Cu^{2+} and Cu in solution and hence Cu^+ is only found in the solid state.

Consider Latimer diagram for O_2.

$$\overset{0}{O_2} \xrightarrow{+0.682} \overset{-I}{H_2O_2} \xrightarrow{+1.776} \overset{-II}{H_2O}$$

$$O_2 \xrightarrow{+1.229} H_2O$$

In this case, on moving from left to right, the reduction potential increases and therefore H_2O_2 is unstable with respect to disproportionation. This can be verified by considering the two-half reactions:

$$H_2O_2(aq) + 2H^+ + 2e- \longrightarrow 2H_2O(l) \qquad E° = 1.776 \text{ V} \qquad ...(i)$$

$$O_2(g) + 2H^+ + 2e^- \longrightarrow H_2O_2\ (aq) \qquad E° = +\ 0.682 \text{ V} \qquad ...(ii)$$

Subtracting (*ii*) from (*i*), we get

$$2H_2O_2(aq) \longrightarrow 2H_2O(l) + O_2\ (g)\ E° = 1.094\text{V}$$

Since E° is more than 0, *the disproportion takes place.*

Consider the Latimer diagram for Cl_2 in basic medium.

$$\overset{VII}{ClO_4^-} \xrightarrow{+0.36} \overset{V}{ClO_3^-} \xrightarrow{+0.33} \overset{III}{ClO_2} \xrightarrow{+0.66} \overset{I}{OCl^-} \xrightarrow{+0.42} \overset{0}{\tfrac{1}{2}Cl_2} \xrightarrow{+1.36} \overset{-I}{Cl^-}$$

From the above diagram we can infer that both ClO_2 and Cl_2 should disproportionate. ClO_2 should disproportionates into ClO_3^- and OCl^- and Cl_2 disproportionates into OCl^- and Cl^-.

The diagram indicates that OCl^- should be stable with regards to disproportionation. But it is not true. Now, if we calculate electrode potentials for the changes ClO_3^- to OCl^- and OCl^- to Cl^-, by the procedure explained earlier for non-adjacent species, we get the Latimer diagram as:

$$\overset{VII}{ClO_4^-} \xrightarrow{+0.36} \overset{V}{ClO_3^-} \xrightarrow{+0.33} \overset{III}{ClO_2} \xrightarrow{+0.66} \overset{+I}{OCl^-} \xrightarrow{+0.42} \overset{0}{\tfrac{1}{2}Cl_2} \xrightarrow{+1.36} \overset{-I}{Cl_2}$$

$$ClO_3^- \xrightarrow{+0.50} OCl^- \qquad OCl^- \xrightarrow{+0.88} Cl_2$$

From this diagram it is clear that the potentials around OCl^- and ClO_3^- do not decrease from left to right as:

$$ClO_3^- \xrightarrow{+0.50} OCl^- \xrightarrow{+0.88} Cl^-$$

Therefore, OCl^- is unstable to disproportionation and it disproportionates into ClO_3^- and Cl^-

Similarly, for the sequence,

$$ClO_4^- \xrightarrow{+0.36} ClO_3^- \xrightarrow{+0.50} OCl^-$$

We find that ClO_3^- *is unstable with respect to disproportionation into* ClO_4^- *and* OCl^-.

Another utility of Latimar diagrams is in predicting the products of the reactions in which the element have several oxidation states. For example, consider the reaction between acidified solution of $KMnO_4$ and KI. The reduction potential diagrams are:

VII VI IV III II 0

$$MnO_4^- \xrightarrow{+0.56} MnO_4^{2-} \xrightarrow{+2.26} MnO_2 \xrightarrow{+0.95} MnO^{3+} \xrightarrow{+1.51} Mn^{2+} \xrightarrow{-1.19} Mn$$

$MnO_4^- \to MnO_2$: +1.69; $MnO_2 \to Mn^{2+}$: +1.23; $MnO_4^- \to Mn^{2+}$: +1.51

VII V III 0 −1

$$IO_4^{-1} \xrightarrow{+165} IO_3^- \xrightarrow{+1.34} HOI^* \xrightarrow{+1.44} \frac{1}{2}I_2(s) \xrightarrow{+0.54} I^-$$

$$H_5IO_6 \xrightarrow{+160} IO_3^-$$

$IO_3^- \to \frac{1}{2}I_2(s)$: +1.19; $HOI \to I^-$: + 0.99

We can rule out certain species that will not be obtained as products because they have a tendency for disproportionation. We observe that there are three species, HOI, Mn^{3+} and MnO_4^{2-} in this diagram which are unstable towards disproportionation. Therefore, these are not formed. The half reaction, $Mn^{2+} \longrightarrow Mn$ has a large negative value and therefore, $\Delta G°$ will have large positive value for this. Hence this step can be ignored. Thus, the reduction potential diagram may be simplified as:

VII V IV 0 II −I

$$MnO_4^- \xrightarrow{+1.70} MnO_2 \xrightarrow{+1.23} Mn^{2+}$$

$$IO_4^- \xrightarrow{+1.65} IO_3^- \xrightarrow{+1.19} I_2(s) \xrightarrow{+0.54} I^-$$

$$H_5IO_6 \xrightarrow{+1.60} IO_3^-$$

The reduction potentials of some common couples are given in Table 3.2

Table 3.2 Redox potential data for some common couples

In acid solution

1. $$Cr_2O_7^{2-} \xrightarrow{1.33} Cr^{3+} \xrightarrow{-0.41} Cr^{2+} \xrightarrow{-0.91} Cr$$

$Cr^{3+} \to Cr$: − 0.74; $CrCl_2^+ \to Cr$: −0.76; $Cr_2O_7^{2-} \to Cr$: 0.295

2. $$MnO_4^- \xrightarrow{0.56} MnO_4^{2-} \xrightarrow{2.26} MnO_2 \xrightarrow{0.95} Mn^{3+} \xrightarrow{1.51} Mn^{2+} \xrightarrow{-1.19} Mn$$

$MnO_4^{2-} \xrightarrow{(2.0)} MnO_3^-$; $MnO_3^- \xrightarrow{(2.5)} MnO_2$; $MnO_2 \to Mn^{2+}$: 1.23

3. $$TiO^{2+} \xrightarrow{(0.1)} Ti^{3+} \xrightarrow{-0.37} Ti^{2+} \xrightarrow{(-1.63)} Ti$$

$TiO^{2+} \to Ti$: (−0.86)

4. $TcO_4^- \xrightarrow{(0.7)} TcO_2 \xrightarrow{(0.6)} Tc^{2+} \xrightarrow{(0.4)} Tc$

$TcO_4^- \xrightarrow{(0.6)} Tc$; $TcO_2 \xrightarrow{(0.5)} Tc$

5. $V(OH)_4^+ \xrightarrow{1.0} VO^{2+} \xrightarrow{0.36} VO^{3+} \xrightarrow{-0.25} V^{2+} \xrightarrow{-1.2} V$

$V(OH)_4^+ \xrightarrow{-0.253} V$

6. $H_2MoO_4 \xrightarrow{(+0.4)} MoO_2^+ \xrightarrow{(+0.0)} Mo^{3+} \xrightarrow{-0.2} Mo$

$H_2MoO_4 \xrightarrow{(0.0)} Mo$

7. $RuO_4 \xrightarrow{1.00} RuO_4^- \xrightarrow{0.59} RuO_4^{2-} \xrightarrow{0.3} Ru^{2+} \xrightarrow{0.45} Ru$

8. $Ti^{3+} \xrightarrow{1.25} Ti^+ \xrightarrow{0.3363} Ti$

9. $BrO_4^- \xrightarrow{1.82} BrO_3 \xrightarrow{1.49} HOBr \xrightarrow{1.59} Br_2 \xrightarrow{1.07} Br^-$

$BrO_3 \xrightarrow{1.52} Br_2$; $Br_2(aq) \xrightarrow{1.087} Br^-$

10. $ReO_4^- \xrightarrow{(0.4)} ReO_3 \xrightarrow{(0.6)} ReO_2 \xrightarrow{(0.2)} Re^{3+} \xrightarrow{(0.3)} Re \xrightarrow{(-0.4)} Re^-$

$ReO_4^- \xrightarrow{0.51} ReO_2$; $ReO_2 \xrightarrow{0.252} Re$

11. $H_5IO_6 \xrightarrow{1.7} IO_3^- \xrightarrow{1.14} HOI \xrightarrow{1.45} I_2 \xrightarrow{0.535} I^-$

$IO_3^- \xrightarrow{1.20} I_2$

12. $FeO_4^{2-} \xrightarrow{1.9} Fe^{3+} \xrightarrow{0.771} Fe^{2+} \xrightarrow{-0.440} Fe^\circ$

13. $Hg^{2+} \xrightarrow{0.920} Hg_2^{2+} \xrightarrow{0.789} Hg$

$Hg^{2+} \xrightarrow{0.854} Hg$

14. $ClO_4^- \xrightarrow{1.19} ClO_3^- \xrightarrow{1.21} HClO_2 \xrightarrow{1.645} HOCl \xrightarrow{1.63} Cl_2 \xrightarrow{1.3595} Cl^-$

$ClO_3^- \xrightarrow{1.15} ClO_2 \xrightarrow{1.27} HClO_2$; $ClO_3^- \xrightarrow{1.47} Cl_2$

15. $Cd^{2+} \xrightarrow{< -0.6} Cd_2^{2+} \xrightarrow{> -0.2} Cd$

$Cd^{2+} \xrightarrow{-0.403} Cd$

16. $In^{3+} \xrightarrow{-0.45} In_2^{4+} \xrightarrow{-0.35} In^+ \xrightarrow{-0.25} In$

$In^{3+} \xrightarrow{-0.342} In$

17. $Ga^{3+} \xrightarrow{-0.65} In_2^{4+} \xrightarrow{-0.45} Ga$

($Ga^{3+} \rightarrow Ga$: −0.53)

In Basic Solution

1. $BrO_3^- \xrightarrow{0.54} BrO^- \xrightarrow{0.45} Br_2^\circ \xrightarrow{1.07} Br^-$

($BrO_3^- \rightarrow Br^-$: 0.61; $BrO^- \rightarrow Br^-$: 0.76)

2. $FeO_4^{2-} \xrightarrow{0.9} Fe(OH)_3 \xrightarrow{-0.56} Fe(OH)_2 \xrightarrow{-0.887} Fe$

3. $Cu(OH)_2 \xrightarrow{-0.08} Cu_2O \xrightarrow{0.358} Cu$

4. $ClO_4^- \xrightarrow{0.36} ClO_3^- \xrightarrow{0.33} ClO_2^- \xrightarrow{0.66} ClO^- \xrightarrow{0.42} Cl_2 \xrightarrow{1.3595} Cl^-$

($ClO_3^- \xrightarrow{-0.50} ClO_3^- \xrightarrow{1.16} ClO_2^-$; $ClO_3^- \rightarrow ClO^-$: 0.50; $ClO^- \rightarrow Cl^-$: 0.88)

5. $H_3IO_6^{2-} \xrightarrow{(0.7)} IO_3^- \xrightarrow{0.14} IO^- \xrightarrow{0.45} I_2(s) \xrightarrow{0.535} I^-$

($IO^- \rightarrow I^-$: −0.49)

6. $MnO_4^- \xrightarrow{0.564} MnO_4^{2-} \xrightarrow{0.60} MnO_2 \xrightarrow{0.2} Mn(OH)_3 \xrightarrow{0.1} Mn(OH)_2 \xrightarrow{-1.55} Mn$

($MnO_4^{2-} \xrightarrow{0.34} MnO_3^- \xrightarrow{0.84} MnO_2$; $MnO_4^- \rightarrow MnO_2$: 0.588; $MnO_2 \rightarrow Mn(OH)_2$: 0.05)

7. $CrO_4^{2-} \xrightarrow{-0.13} Cr(OH)_3 \xrightarrow{-1.1} Cr(OH)_2 \xrightarrow{-1.4} Cr$

($Cr(OH)_3 \rightarrow Cr$: −1.3; $CrO_2^- \rightarrow Cr$: −1.2)

8. $CoO_2 \xrightarrow{0.7} Co(OH)_3 \xrightarrow{0.14} Co(OH)_2 \xrightarrow{-0.72} Co$

Example 19. The Latimer diagram is

$ClO^- \xrightarrow{+0.42} Cl_2 \xrightarrow{+1.36} Cl^-$.

Calculate the value for the reduction of ClO^- to Cl^- in aqueous basic medium.

Solution. The half reactions may be written as:

$$ClO^- + H_2O + e^- \longrightarrow \frac{1}{2}Cl_2 + 2OH^- \qquad E° = +0.42$$

$$\frac{1}{2}Cl_2 + e^- \longrightarrow Cl^- \qquad E° = +1.36$$

Here $n_1 = 1$ and $n_2 = 1$

$\therefore$ E° for ClO^- to Cl^- is calculated by the formula

$$E° = \frac{n_1 E_1^\circ + n_2 E_2^\circ}{n_1 + n_2}$$

Subustituting the values, we have

$$E° = \frac{(1\times0.42) + (1\times1.36)}{2} = +0.89 \text{ V}$$

Example 20. Latimer Diagram for Iron is:

$$Fe^{3+} \xrightarrow{0.771 \text{ V}} Fe^{2+} \xrightarrow{0.44 \text{ V}} Fe \qquad (Fe^{3+} \to Fe: -0.04 \text{ V})$$

How do we derive the potential for $Fe^{3+} + 3e^- \longrightarrow Fe$

Solution. Convert E° values into ΔG° values = – nFE° or E° = –ΔG°/nF)

$$\Delta G°_{net} = -(1)(F)(+0.771 \text{ V}) + [-2(F)(-0.44 \text{ V})]$$

$$= +0.109 \text{ (F) V}$$

$$E°_{net} = [+0.109 \text{ (F)}/3F$$

$$= -0.0363\text{V}$$

Example 21. The Latimer diagram for manganese system is:

$$\mathbf{MnO_4^- \xrightarrow{0.56} MnO_4^{2-} \xrightarrow{2.26} MnO_2 \xrightarrow{0.95} Mn^{3+} \xrightarrow{0.51} Mn^{2+}}$$

Calculate the E° value for $MnO_4^{2-} | Mn^{2+}$ couple.

Solution. The process ocurs as:

$$MnO_4^{2-} \xrightarrow[n=2]{2.26} MnO_2 \xrightarrow[n=1]{0.95} Mn^{3+} \xrightarrow[n=1]{0.51} Mn^{2+} \qquad (MnO_4^{2-} \to Mn^{2+}: E° = ?,\ n = 4)$$

As we know

$$\Delta G° \text{ (Overall)} = \Delta G_1° + \Delta G_2° + \Delta G_3°$$

$$E° = \frac{n_1 E°_1 + n_2 E°_2 + n_3 E°_3}{n}$$

$$= \frac{2\times2.26 + 1\times0.95 + 1\times0.51}{4}$$

$$= 1.75 \text{ V}$$

Example 22. Show that disproportionation of H_2O_2 into O_2 and H_2O is spontaneous under acidic conditions. Given that

$$O_2 \xrightarrow{+0.70} H_2O_2 \xrightarrow{+1.76} H_2O$$

Solution. The disproportionation of H_2O_2 can be written as:

$$2H_2O_2 \longrightarrow 2H_2O + O_2$$

The reaction may be split into two half reactions as:

$$H_2O_2 + 2H^+ + 2e^- \longrightarrow 2H_2O \qquad E° = +1.76\ V$$

$$O_2 + 2H^+ + 2e^- \longrightarrow H_2O_2 \qquad E° = 0.70\ V$$

Subtracting the two equations, we have

$$2H_2O_2 \longrightarrow 2H_2O + O_2 \qquad E° = 1.06\ V$$

Since E° is positive, the reaction is spontaneous.

Thus, H_2O_2 disproportionates into H_2O and O_2

3.12. FROST DIAGRAM

A **Frost diagram** of an element is a **plot of product of oxidation state (N) and reduction potential (E°) against oxidation number of the element (N).** For example, for an element X, it is plot of NE° for the couple X(N) | X (0) against the oxidation number, N.

$$X(N) + Ne^- \longrightarrow X(0) \qquad E°$$

We know that NE° proportional to the standard reaction Gibbs energy for the conversion of the species X(N) to the element (NE° = – Δ G°/F). A Frost diagram can also be regarded as *a plot of standard reaction Gibbs energy against oxidation number.*

In general, the following observations may be made from Frost diagram:

1. Frost diagram is regarded as a plot of Gibbs free energy against oxidation number. Therefore, **the most stable oxidation state of an element corresponds to the species that lies lowest in the Frost diagram.**

For example, the most stable oxidation state in a Frost diagram shown below corresponds to point B.

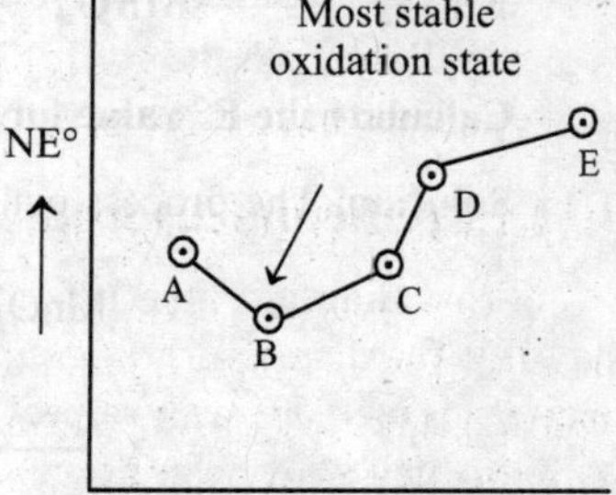

2. The slope of the line joining any two points in the Frost diagram is equal to the standard potential of the couple formed by the two species represented by the points. We can easily predict the spontaneity of the reaction between any two couples by comparing the slopes of the corresponding lines.

- More positive the slope, the greater is the oxidizing power of the couple.
- More negative the slope, the greater is the reducing power of the couple.

These diagram are plotted with reference to the $H^+ | H_2$ or $OH^- | H_2$ couple having slope of zero.

The Frost diagram for nitrogen under basic conditions is shown in Fig. 3.6

(*i*) The reducing agent in the couple with less positive slope (i.e. less positive E°) for example NO_2^- is likely to undergo **oxidation**.

(*ii*) The oxidising agent in the couple with more positive slope (i.e. more positive E°) for example NO_3^- is likely to undergo **reduction**.

(*iii*) Steeper the line joining two points in a Frost diagram, higher is the potential of the corresponding couple.

(*iv*) Predicts disproportionation and comproportionation.

For example, a steep slope connecting HNO_3 (NO_3^-) to lower oxidation state (NO_2^-) in Fig 3.6 indicates that HNO_3 is good oxidizing agent.

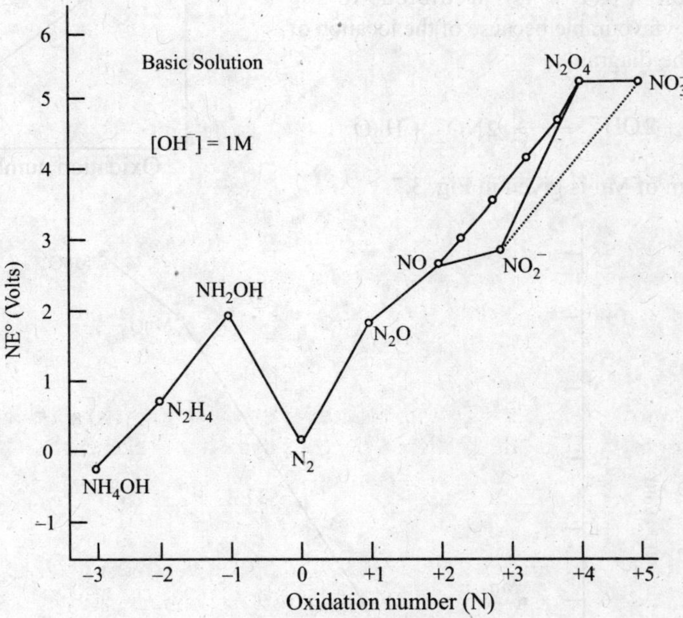

Fig. 3.6. Frost diagram for nitrogen

The oxidizing and reducing ability of any couple can be easily predicted from the Frost diagram on the basis of sign and magnitude of the slope. The couples $N_2O_4 \mid NO_2^-$, NO| N_2O and $NO_3^- \mid NO_2^-$ have large positive slope and therefore are **strong oxidising agents.**

Similarly, both the $N_2H_4 \mid N_2$ and $NH_2OH \mid N_2$ couples are strong reducing couples because their tie lines have large negative slopes.

3.12.1 Predicting disproportionation with the help of Frost diagram

Frost diagram also helps to predict the relative stability for disproportionation. **A species in a Frost diagram is unstable with respect to disproportionation if its point lies above the line connecting two adjacent species.** In other words, **a species will undergo disproportionation only if its point lies above the line connecting two adjacent species.**

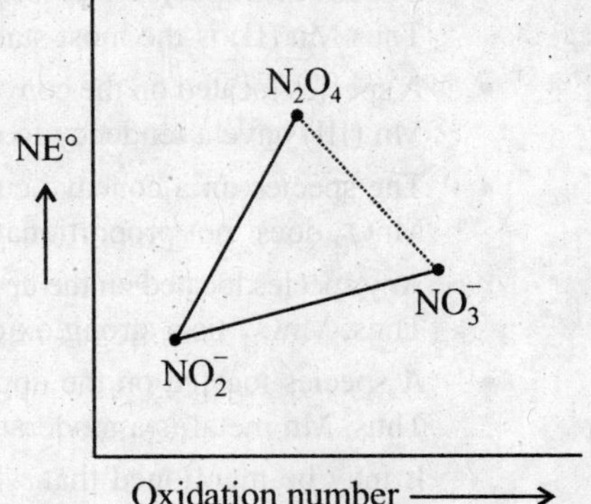

For example, as shown in the figure, N_2O_4 lies above the joining NO_3^- and NO_2^-. Therefore, it is unstable with respect to disproportionation. Disproportionation takes place as under :

$$N_2O_4 + 2OH \longrightarrow NO_2^2 + 2OH$$

Reverse of disproportionation is called comproportionation. It may also be predicted from the Frost diagram. Two species will tend to comproportionate into an intermediate species that lies below the straight line joining the terminal species. For example, the reaction between nitrogen (II) oxide and dinitrogen tetraoxide to form nitrite ion (NO_2^-) is predicted to be thermodynamically favourable because of the location of these species in the diagram.

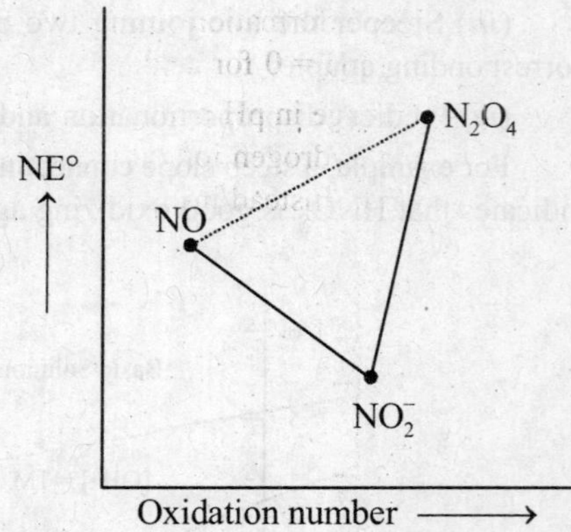

$$NO + \frac{1}{2} N_2O_4 + 2OH^- \longrightarrow 2NO_2^- + H_2O$$

Frost diagram of Mn is given in Fig. 3.7.

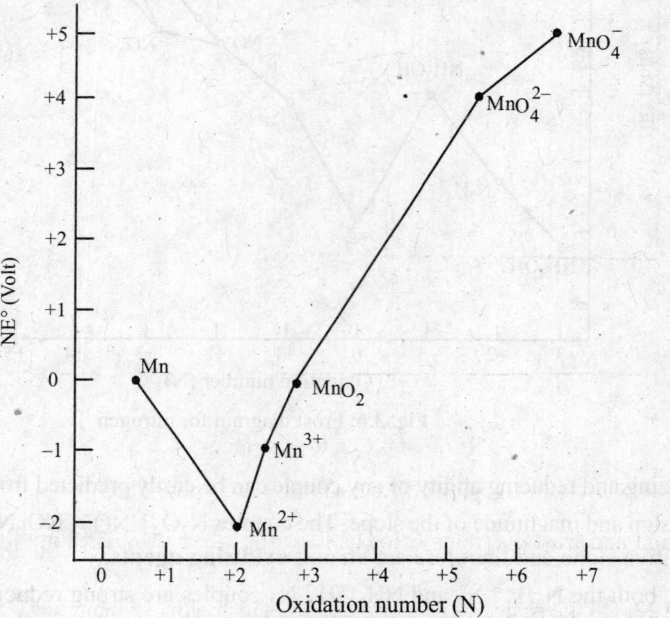

Fig. 3.7 Frost diagram for manganese.

The following inferences can be drawn from the Frost diagram of Mn (Fig. 3.7)

- The lower a species is present on the diagram, the more thermodynamically stable it is. Thus Mn (II) is the most stable species.
- A species located on the convex curve can undergo disproportionation. Thus, MnO_4^{2-} and Mn (III) have a tendency to disproportionate.
- The species on a concave curve shows no tendency towards disproportionation. Thus, MnO_2 does not proportionate.

  Any species located on the upper right side of the diagram will be a strong oxidising agent. Thus, MnO_4^- is a strong oxidising agent.
- A species located on the upper left side of the diagram will work as a reducing agent. Thus, Mn metal is a moderate reducing agent.

  It may be mentioned that while a given species may be thermodynamically unstable towards reduction, the kinetics of the reaction may be slow.

- The information obtained from a Frost diagram is for species under standard conditions i.e. pH = 0 for acidic solution and pH = 14 for basic solution.
- Change in pH may change the relative stability. The potential of any process involving the hydrogen ion will change with pH. Under basic conditions aqueous Mn^{2+} does not exist. Instead insoluble Mn $(OH)_2$ is formed. (Fig. 3.8).

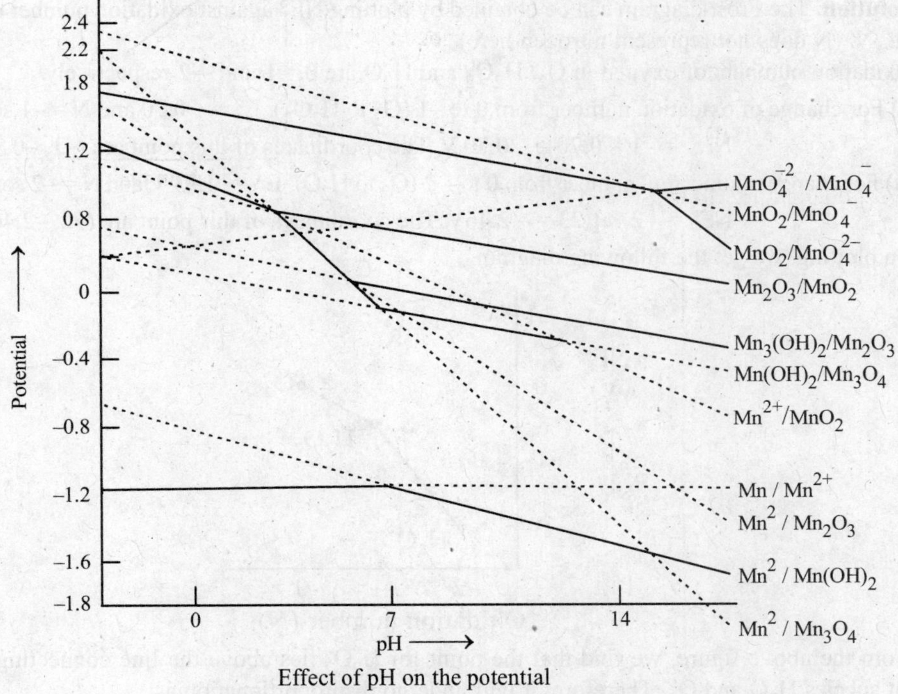

Fig. 3.8 Effect of pH on the potential.

Latimer and and frost diagrams helps us to arrive at the principles involved in the extraction of elements.

Example 22. On the basis of Latimer diagram given below, explain why Cu^+ disproportionates in aqueous solution.

$$Cu^{2+} \xrightarrow{+0.15} Cu^+ \xrightarrow{+0.50} Cu$$

Solution. We know that a species has a thermodynamic tendency to disproportionate into its two neighbours if the potential on the right of species is higher than potential on the left. In this case, the potential on the right is higher than on the left of Cu^+, therefore Cu^+ has a tendency to disproportionate.

This can also be shown as:

	$Cu^+ + e^-$	$\longrightarrow$ Cu	E° = + 0.50 V
	$Cu^{2+} + e^-$	$\longrightarrow$ Cu^+	E° = + 0.15 V
Subtracting	$2\,Cu^+$	$\longrightarrow$ $Cu + Cu^{2+}$	E° = + 0.35 V

Since E° > 0 in solution, Cu^+ disproportionates into Cu^{2+} and Cu.

Example 24. Construct a Frost diagram from the Latimer diagram given below:

$$O_2 \xrightarrow{+0.70} H_2O_2 \xrightarrow{1.76} H_2O \quad (O_2 \to H_2O: +1.23)$$

According to the diagram, which species is thermodynamically least stable and hence will undergo disproportionation.

Solution. The Frost diagram can be obtained by plotting NE° against oxidation number of the element, N. (N does not represent nitrogen here)

Oxidation numbers of oxygen in O_2, H_2O_2 and H_2O are 0, –1 and – 2 respectively.

(i) For change of oxidation number from 0 to –1 (O_2 to H_2O_2), E° = + 0.70 and N = –1 so that

NE° = –1 × 0.70 = – 0.70 V. The coordinates of this point are (–1, –0.7)

(ii) For change of oxidation number from 0 to – 2 (O_2 to H_2O), E° = +1.23 V and N = –2 so that

NE° = –2 × 1.23 = – 2.46V. The coordinates of this point are (–2, –2.46)

On plotting, we get the following diagram

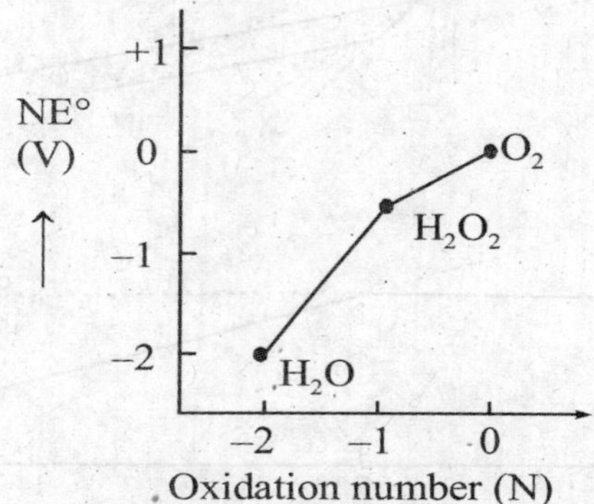

From the above figure, we find that the point for H_2O_2 lies above the line connecting two adjacent species H_2O and O_2. Therefore, it will undergo disproportionation as:

$$2H_2O_2 \longrightarrow 2H_2O + O_2$$

Example 25. Consider the Latimer diagram for thallium:

$$Tl^{3+} \xrightarrow{+1.26\ V} Tl^{+} \xrightarrow{-0.34V} Tl \quad (Tl^{3+} \to Tl: +0.73)$$

Construct a Frost diagram.

(*i*) Predict the stability or unstability of Tl^+

(*ii*) Which of these species is strong oxidant?

Solution. The oxidation number of thallium in Tl^{3+}, Tl^+ and Tl is + 3, +1 and 0 respectively.

For change of Tl (0) to Tl^+ (I), E ° = – 0.34V and N = 1 so that NE° = 1 × (–0.34) = –0.34V

For change of Tl (0) to Tl^{3+} (III), E° = + 0.73 and N = 3 so that

NE° = 3 × 0.73 = 2.19 V.

Coordinate of points corresponding to Tl, Tl^+ and Tl^{3+} are (0, 0), (1, – 0.34) and (3, 2.19) respectively.

Plotting the graph, we get the Frost diagram:

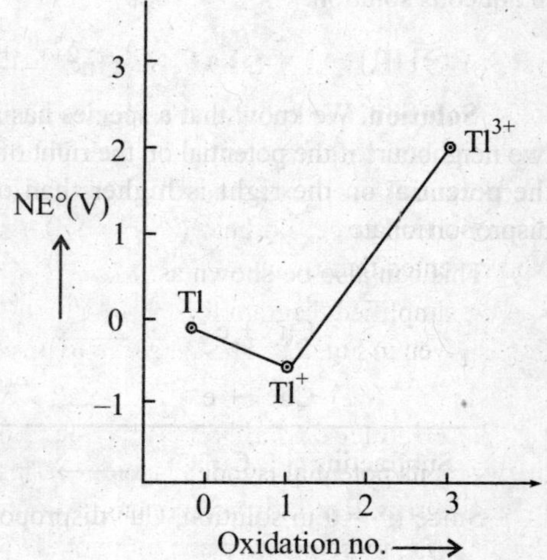

(*i*) Tl^+ is stable with respect to disproportionation because it does not lie above the line joining two adjacent species (Tl and Tl^{3+}). Since it lies lowest in the diagram, it is most stable oxidation state.

(*ii*) Tl^{3+} is strong oxidant because the slope of the line connecting it with either lower oxidation state is large and positive.

Example 23. Consider the Frost diagram for manganese in acidic medium.

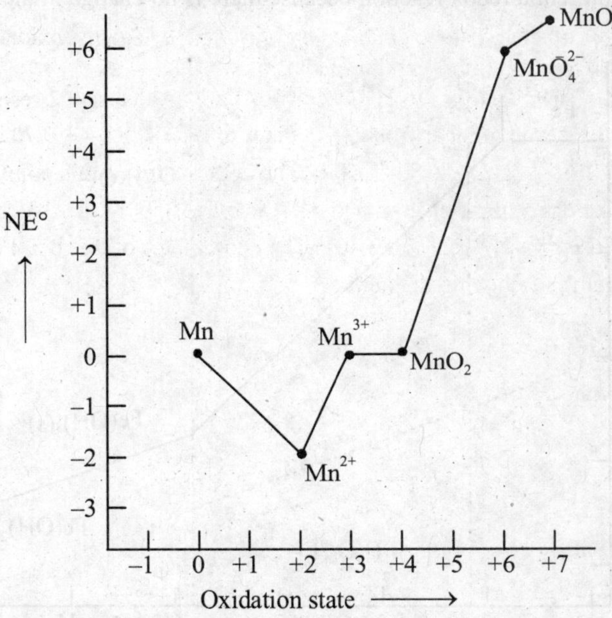

Comment on the stability of manganese (III) and manganate ion with respect to disproportionation. Write the reactions.

Solution. Mn^{3+} lies above the line joining Mn^{2+} and MnO_2. Therefore, it will disproportionate as per the following equation:

$$2Mn^{3+} + 2H_2O \longrightarrow Mn^{2+} + MnO_2 + 4H^+$$

Manganate ion (MnO_4^{2-}) lies above the line joining MnO_2 and Mn (MnO_4^-). Therefore, it will disproportionate as per the following reaction:

$$3\,MnO_4^{2-} + 4H^+ \longrightarrow MnO_2 + 2\,MnO_4^- + 2H_2O$$

3.13. POURBAIX DIAGRAMS

Mercel Pourbaix

Pourbix diagrams, also known as potential/pH diagrams map out possible stable phases of an aqueous electrochemical system. Predominant ion boundaries are represented by lines. The diagrams are named after Marcel Porbaix (1904 - 98), the Russian born chemist, who invented them.

A simplified diagram for some naturally occurring compounds of iron is given in Fig. 3.9. This diagram involves the following reactions:

$$Fe^{3+}\,(aq) + e^- \longrightarrow Fe^{2+}\,(aq) \quad E° = +\,0.77V$$

The reduction half reaction does not involve H^+ ions and therefore, its potential is independent of *p*H. This gives horizontal line in the diagram. If the solution contains a couple having potential above this line (*i.e.* more positive and more oxidizing couple) then oxidized species, Fe^{3+} will be the major product. Hence the horizontal line at the top left of the diagram represents a boundary that seperates the regions of Fe^{3+} and Fe^{2+}.

Consider the reaction

$$Fe^{3+}\ (aq)\ 3H_2O(l) \longrightarrow Fe(OH)_3\ (s) + 3H^+\ (aq).$$

This does not represent a redox reaction because there is no change in the oxidation number of any element.

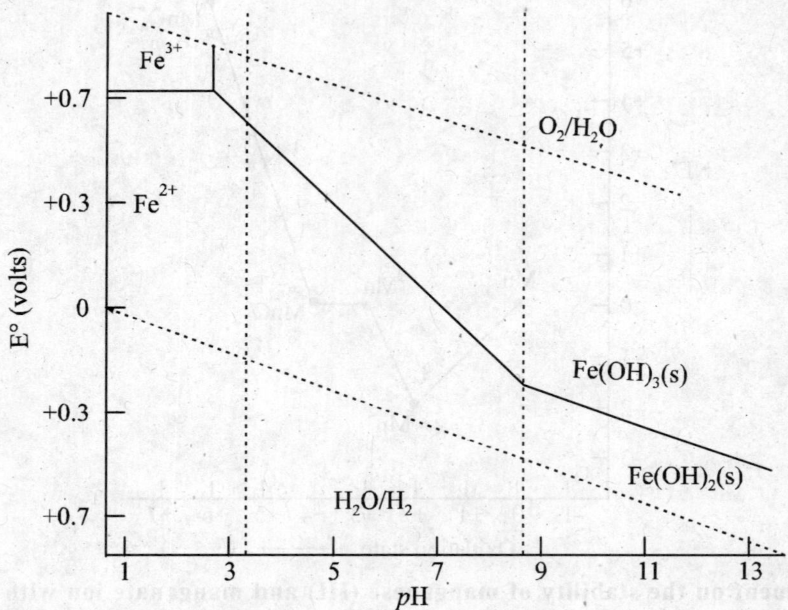

Fig. 3.9. Pourbix diagram for some important naturally occurring compounds of iron.

The boundary between the region where Fe^{3+} is soluble and the region where Fe $(OH)_3$ precipitates is independent of any redox couples. However, this boundary depends upon *p*H. Fe^{3+} (*aq*) is favoured by low *p*H while the Fe $(OH)_3$ is favoured by high *p*H. The equilibrium concentration varies with *p*H and the vertical boundary represents the *p*H at which Fe^{3+} becomes dominant according to this definition.

Now consider the reaction.

$$Fe\ (OH)_3\ (s) + 3H^+\ (aq) + e^- \longrightarrow Fe^{2+}\ (aq) + 3H_2O\ (l)$$

This reaction depends upon the *p*H because H^+ ion concentration appears in the half reaction as shown by Nernst equation:

$$E = E° - 0.059\ V \log \frac{\left(Fe^{2+}\right)}{\left(H^+\right)^3} = E° - 0.059\ V\ [\log (Fe^{2+} - 3 \log H^+]$$

$$= E° - 0.059\ V \log [Fe^{2+}] + 0.177\ V \log H^+$$

$$= E° - 0.059\ V \log (Fe^{2+}) - (0.177\ V\ pH)$$

Fig. 3.9 shows that the potential falls lineraly as the *p*H increases. The regions of potential and *p*H that lie above this line correspond to conditions in which the oxidized species Fe $(OH)_3$ is stable. The regions below this line correspond to conditions under which the reduced species Fe^{2+} is stable. We observe that a strong oxidizing agent is required to oxidize and precipitate Fe^{2+} ions in acidic media than in basic media.

The vertical line at $pH = 9$ divides the regions in which either the reactants or products are stable in the reactions:

$$Fe^{2+}\ (aq) + 2H_2O\ (l) \longrightarrow Fe\ (OH)_2\ (s) + 2H^+\ (aq)$$

Similarly, the line separates the region in which $Fe\ (OH)_2$ and $Fe\ (OH)_3$ are stable.

$$Fe\ (OH)_3\ (s) + H^+\ (aq) + e^- \longrightarrow Fe\ (OH)_2\ (s) + H_2O(l)$$

Thus Pourbaix diagram is graph of conditions of potential and pH under which species are stable in water.

3.15. THERMODYNAMIC PRINCIPLES OF OF METALLURGY

For any process, Gibbs free energy change (Δ G) is given by the equation,

$$\Delta G = \Delta H - T\Delta S \quad ...(i)$$

where ΔH is the enthalpy change and ΔS is the entropy change and T is the absolute temperature. The free energy change is also related to the equilibrium constant K of the reactant-product system at temperature T by the equation.

$$\Delta G = -RT\ In\ K \quad ...(ii)$$

If ΔG is –ve, then K will be positive. This means that the reaction will proceed in the forward direction. We draw the following conclusions:

1. The criterion of feasibility of a reaction at any temperature is that the ΔG of the reaction must be negative. Thus, if ΔS is +ve in equation (i), then on increasing the temperature (T), the value of T ΔS will increase and when $T\Delta S > \Delta H$, ΔG will be –ve and the reaction will proceed in the forward direction.

2. A reaction with ΔG positive can be still made to occur by coupling it with another reaction having large negative ΔG so that the net ΔG of the two reactions in negative.

Such coupling reactins can be explained in terms of Ellingham diagram, i.e., ΔG° vs T plots for formation of oxides.

3.16. ELLINGHAM DIAGRAM

Ellingham diagram consists of posts of $\Delta_f G°$ vs T for formation of oxides of element (Fig 3.10). Similar diagrams can also be constructed for sulphides and halides. Such diagrams help us in predicting the feasibility of thermal reduction of an ore.

Let us consider the formation of a metal oxide from the metal

$$2xM\ (s) + O_2\ (g) \longrightarrow 2M_xO\ (s) \quad ...(i)$$

In this reaction, dioxygen – a gas has been used up and metal oxide – a solid is formed. Gases have higher entropy than liquids and solids. Therefore, during this reaction, ΔS becomes negative. Thus, if temperature is increased, TΔS becomes more and more negative. Therefore, ΔG becomes less and less negative according to equation (*i*) in Art. 3.15.

Thus, ΔG increases or becomes less – ve with increase in temperature (normally, ΔG decreases with increase in temperature). In other words, $\Delta_f G°$ vs T lines have +ve slopes for most of the reactions involving the formation of metal oxides, $M_xO(s)$ as shown in Ellingham diagram. Each plot is a straight line except when some change in phase (solid → liquid or liquid → gas) takes places. The temperature at which such a change occurs, is indicated by an increase in the slope on the +ve side. For example in the Zn – ZnO plot, at the boiling point of zinc (1180 K), there is an abrupt increase in the +ve slope of the curve. Similarly, the +ve slope of the Mg – MgO curve increases

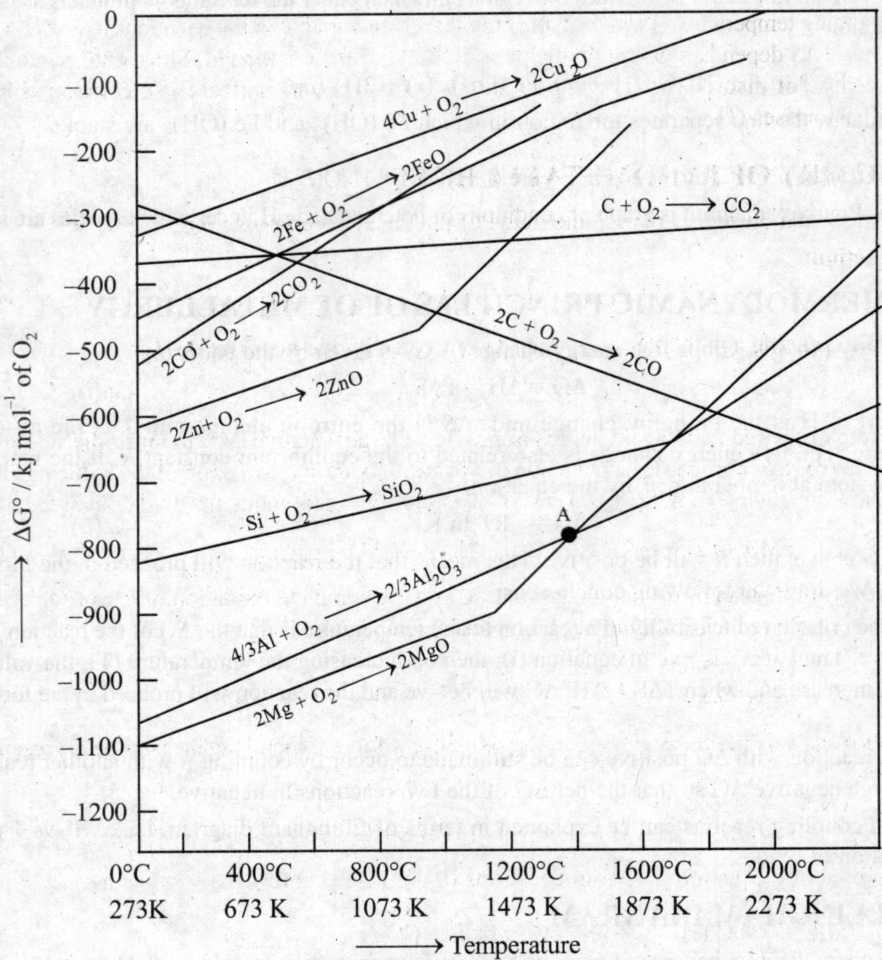

Fig 3.10. Gibbs energy (ΔG°) vs T plots (schematic) for formation of some oxides (Ellingham diagram).

abruptly at the melting point of Mg at 1363 K and that of Si–SiO_2 curve increases abruptly at the melting point of Si at 1966 K. The curves in such cases move upwards suddenly.

3.17. LIMITATIONS OF ELINGHAM DIAGRAM

Ellingham diagram although quite useful have some limitations as well:

(i) Kinetics of reduction: Ellingham diagrams are based on thermodynamic concepts, therefore, these diagrams simply suggest whether the reduction of a given metal oxide with a particular reducing agent is possible or not. It, however, does not tell anything about the rate of the reduction process.

(ii) Reactant/product equilibrium: The interpretation of $\Delta G°$ is based upon K (i.e., $\Delta G° = -RT \text{ in } K$). Thus, it is assumed that reactants and products are in equilibrium.

$$M_xO + A_{red} \rightleftharpoons xM + AO_{ox}$$

But, this is not always true because the reactant/product may be in solid phase.

However, it nicely explains why the reactions are slow when every species is in the solid state and smooth when the oxide one melts.

It is interesting to note that ΔH and ΔS values for any chemical reaction remain nearly constant even on changing temperature. Therefore, the only dominant variable in the equation $\Delta G = \Delta H - T\Delta S$ is T. However, ΔS depends much upon the physical state of the compound. Since entropy depends upon the degree of disorder or randomness in the system, it will increase if a compound melts ($s \rightarrow l$) or vapourises ($l - g$).

3.18. THEORY OF PYROMETALLURGY

We know that during reduction, the oxide of the metal decomposes, *i.e.*,

Reduction:

$$M_xO\,(s) \longrightarrow xM\,(s \text{ or } l) + \tfrac{1}{2}O_2\,(g) \quad ...(i)$$

and the reducing agent takes away the oxygen given by the metal oxide, *i.e.*,

$$C(s) + \tfrac{1}{2}O_2\,(g) \longrightarrow CO\,(g) \quad ...(ii)$$

The role of the reducing agent here is to provide a large negative $\Delta G°$ to make the sum of the $\Delta G°$ of the above two reactions (*i.e.*, reduction of the metal oxide and oxidation of the reducing agent) negative. Now equation (*i*) is actually reverse of equation (*iii*) representing the oxidation of the metal.

$$xM(s \text{ or } l) + \frac{1}{2}O_2\,(g) \longrightarrow M_xO\,(s);\ \Delta G°_{(M,M_xO)} \quad ...(iii)$$

However, if, instead of partial oxidation of C to CO, complete oxidation of C to CO_2 occurs, the oxidation of the reducing agent may be represented as follows:

$$\frac{1}{2}C\,(s) + \frac{1}{2}O_2\,(g) \longrightarrow \frac{1}{2}CO_2\,(g);\ \frac{1}{2}\Delta G°_{(C,CO_2)} \quad ...(iv)$$

If CO is used as the reducing agent, the oxidation of the reducing agent may be represented as follows:

$$CO\,(g) + \frac{1}{2}O_2\,(g) \longrightarrow CO_2\,(g);\ \Delta G°_{(CO,CO_2)} \quad ...(v)$$

On subtracting equation (*iii*) from equations (*ii*), (*iv*) and (*v*) respectively, we have

$$M_xO\,(s) + C\,(s) \longrightarrow xM\,(s \text{ or } l) + CO\,(g) \quad ...(vi)$$

$$M_xO\,(s) + \frac{1}{2}C\,(s) \longrightarrow xM(s \text{ or } l) + \frac{1}{2}CO_2\,(g) \quad ...(vii)$$

$$M_xO\,(s) + CO\,(g) \longrightarrow xM(s \text{ or } l) + CO_2\,(g) \quad ...(viii)$$

These three equations (*vi*), (*vii*) and (*viii*) describe the actual reduction of the metal oxide, M_xO to the metal, M. The $\Delta G°$ values for these reactions, in general, can be obtained by subtraction of the corresponding $\Delta G°$ values obtained from the Ellingham diagram. If these values are –ve, the reduction is feasible otherwise not. Without actual calculations for $\Delta_r G°$, we can predict the feasibility of a reduction process simply by looking at the Ellingham diagram. Metals for which free energy of formation ($\Delta_f G°$) of their oxides is more negative can reduce those metals for which the free energy of formation ($\Delta_f G°$) of their respective oxides is less negative.

For example, at any given temperature, any metal will reduce the oxide of other metals which lie above it in the Ellingham diagram because the free energy change ($\Delta_f G°$) for the overall redox reaction comes out to be negative.

Greater the difference (Net ΔG) easier is the reduction. For example, Al reduces ZnO, FeO and CuO more readily than Zn reduces FeO and Cu_2O.

3.19. EFFECT OF TEMPERATURE ON THE FREE ENERGY CHANGE ($\Delta_r G°$) OF THE OVERALL REDUCTION PROCESS

For any process,

$$\Delta G = \Delta H - T\Delta G$$

On increasing the temperature, the value of ΔH and ΔS nearly remain constant, therefore, the value of $\Delta_r G°$ becomes more negative. This means that if a particular reduction process does not occur at a lower temperature, it may occur at a higher temperature. We need to select the temperature in such a way that the $\Delta_r G°$ of the overall redox reaction becomes –ve.

In the Ellingham diagram, this temperature is indicated by the intersection of the two curves *i.e.*, curve for formation of M_xO and the curve for the oxidation of the reducing agent. At the point of intersection of the two curves, $\Delta G° = 0$. Below this temperature, the $\Delta G°$ is –ve compared so that of the reducing agent and hence the oxide is stable. Above this temperature, $\Delta G°$ is +ve compared to that of the reducing agent and hence the oxide is unstable, *i.e.*, it would give metal and oxygen. For example, the temperature at the intersection point 'A' of the Al $\longrightarrow$ Al_2O_3 and Mg $\longrightarrow$ MgO curves is approx. 1623 K. Therefore, below 1623 K, MgO is more stable and hence Mg can reduce Al_2O_3 to Al but above 1623 K, Al_2O_3 is more stable and hence Al can reduce MgO to Mg.

3.20. EXTRACTION OF NON-METALS BY OXIDATION

Non-metals occur in the combined state in the reduced form. Therefore, they are generally isolated by oxidation of their compounds.

An example is discussed below:

Isolation of chlorine: A very common example of extraction based on oxidation is the isolation of chlorine from brine.

$$2\,Cl^-(aq) + 2\,H_2O(l) \longrightarrow 2\,OH^-(aq) + H_2(g) + Cl_2(g)$$

Since oxidation cannot be carried out by ordinary chemical methods, it is accomplished by electrolysis. The $\Delta G°$ for this reaction is + 422 kJ. The minimum potential difference required for oxidation is calculated as under:

$$\Delta G° = -nFE°$$

$$\text{or } 422 \times 1000\text{J} = -2 \times 96500 \times E°$$

$$\text{or } E° = -2.18\ \text{V} - \simeq -2.2\text{V}.$$

However, due to overvoltage, actual electrolysis requires an external emf which is greater than 2.2 V.

During electrolysis, Cl_2 is liberated at the cathode and H_2 at the anode while NaOH is obtained in the solution. In other words, H_2 and NaOH are the by products of this electrolysis.

Electrolysis of moltent NaCl can also be carried out but in this case Na metal is liberated at the cathode.

SOME SOLVED EXAMPLES

Example 1. Given $\quad Sn^{4+} + 2e^- \rightleftharpoons Sn^{2+} \quad$ **E° = 0.154 V**

$\quad Fe^{3+} + e^- \rightleftharpoons Fe^{2+} \quad$ **E° = 0.771 V**

Determine the cell voltage.

Solution. Since E° of Fe^{3+} is greater it will act as an oxidising agent and will oxidise Sn^{2+} to Sn^{4+}. The cell potential can be determined as:

$$E°_{cell} = E°_{cathode} - E°_{anode}$$

$$= 0.771 - 0.154 = 0.617\ \text{V}$$

Example 2. Given : $Ce^{4+} + e^- \rightleftharpoons Ce^{3+}$ E°=1.61 V E° = 1.61 V

$$MnO_4^- + 8H^+ + 5e^- \rightleftharpoons Mn^{2+} + 4H_2O$$ E° = 1.51 V

Comment on the feasibility of the reaction.

Solution. E° values indicate that Ce^{4+} is a stronger agent than acidic MnO_4^-. Therefore, acidic MnO_4^- cannot oxidise Ce^{3+} to Ce^{4+} but Ce^{4+} can oxidise Mn^{2+} to MnO_4^-, theoretically.

For this reaction,

$$[Ce^{4+} + e^- \rightleftharpoons Ce^{3+}] \times 5$$

$$Mn^{2+} + 4H_2O \rightleftharpoons MnO_4^- + 8H^+ + 5e^-$$

$$5Ce^{4+}\ Mn^{2+} + 4H_2O \rightleftharpoons 5Ce^{3+} + MnO_4^- + 8H^+$$

The cell potential is:

$$E°_{cell} = 1.61 - 1.51 = 0.10 \text{ V}$$

Since $E°_{cell}$ is positive, the reaction is theoretically feasible. However, as the value of $E°_{cell}$ is very small, such reaction would be very slow.

Example 3. Will it be possible to oxidise Cl^- and Co^{2+} with acidic $Cr_2O_7^{2-}$?

E° ($Cl_2|2Cl^-$) = 1.3595 V, E° ($Co^{3+}|Co^{2+}$) = 1.81 V and E° ($Cr_2O_7^{-2}|2Cr^{3+}$) = 1.33 V?

Solution. As reduction potentials of both Cl^- and Co^{2+} are greater than that of $Cr_2O_7^{-2}|Cr^{3+}$ acidified dichromate cannot oxidise Cl^- to Cl_2 or Co^{2+} to Co^{3+}.

Example 4. Will permangate ion oxidise H_2O to evolve O_2 in acidic solution given that:

$$MnO_4^- + e^- \rightleftharpoons MnO_4^{2-} \qquad E° = +0.564 \text{ V}$$

$$\frac{1}{2}O_2 + 2H^+ + 2e^- \rightleftharpoons H_2O(l) \qquad E° = 1.23 \text{ V}$$

Solution. We want to check the possibility of the following reaction:

$$2MnO_4^- + H_2O \longrightarrow 2MnO_4^{2-} + \frac{1}{2}O_2 + 2H^+$$

Given that

$$2MnO_4^{-2} + 2e^- \rightleftharpoons 2MnO_4^{2-} \qquad E° = 0.564 \text{ V} \qquad ...(i)$$

$$\frac{1}{2}O_2 + 2H^+ + 2e^- \rightleftharpoons H_2O \qquad E° = 1.23 \text{ V} \qquad ...(ii)$$

Subtracting eq. (*ii*) from (*i*), we have

$$2MnO_4^- + H_2O \rightleftharpoons 2MnO_4^- + \frac{1}{2}O_2 + 2H^+ \qquad E° = -0.666 \text{ V}$$

Since E° is negative, the above reaction will not occur.

Example 5. The Latimer diagram for amercium is:

$$AmO_2^{2+} \xrightarrow{+1.70} AmO_2^+ \xrightarrow{+0.86} Am^{4+} \xrightarrow{+2.62} Am^{3+} \xrightarrow{-.207} Am$$

($AmO_2^+ \rightarrow Am^{3+}$: 1.724; $AmO_2^{2+} \rightarrow Am^{3+}$: 1.726)

Predict the species which can disproportionate and into which species.

Solution. AmO_2^+ will disproportionate to AmO_2^{2+} and Am^{3+} because the potential the potential for $AmO_2^{2+} \longrightarrow AmO_2^+$ is less and for $AmO_2^+ \longrightarrow Am^{3+}$ is more.

Similarly, Am^{4+} will disproportionate into AmO_2^+ and Am^{3+}

Example 6. The reduction potentials are:

$$Cu^{2+} + 2e \longrightarrow Cu^{+} \qquad E° = 0.15\ V$$

$$Cu^{+} + e \longrightarrow Cu \qquad E° = +0.50\ V$$

Calculate the value for $Cu^{2+} \longrightarrow Cu$ and draw Latimer diagram.

Solution. Given

$$Cu^{+} + e \longrightarrow Cu \qquad E° = 0.50 \qquad ...(i)$$

$$Cu^{2+} + e \longrightarrow Cu^{+} \qquad E° = 0.15\ V \qquad ...(ii)$$

Subtracting (*ii*) from (*i*)

$$2Cu^{+} \longrightarrow Cu^{2+} + Cu \quad E° = 0.35V$$

Latimer diagram is

$$Cu^{2+} \xrightarrow{+0.15} Cu^{+} \xrightarrow{+0.50} Cu$$

$$Cu^{2+} \xrightarrow{+0.35} Cu$$

EXERCISE
(Based on Question Papers of Different Universities)

Multiple Choice Questions (Choose the Correct Options)

1. $E°_{cell}$ and K_C at 25°C are related as

(*a*) $E°_{cell} = -\dfrac{0.059}{n} \log K_C$ (*b*) $E°_{cell} = \dfrac{0.059}{n} \log K_C$

(*c*) $E°_{cell} = 0.059.\ n \log \dfrac{1}{K_C}$ (*d*) $E°cell = \dfrac{n}{0.059} \log K_C$

2. Consider the Latimer diagram:

$$O_2 \xrightarrow{-0.33} O_2^{-} \xrightarrow{-1.69} H_2O_2 \xrightarrow{1.77} H_2O$$

The electrode potential for O_2/H_2O half reaction is

(*a*) 0.08 (*b*) 1.23

(*c*) 1.39 (*d*) 1.44

3. In the diagram:

$$MnO_4^{-} \xrightarrow{+1.69} MnO_2 \xrightarrow{+0.95} Mn^{3+} \xrightarrow{1.51} Mn^{2+}$$

$$MnO_2 \xrightarrow{+1.23} Mn^{2+}$$

Which species disproportionates?

(*a*) MnO_2 (*b*) Mn^{3+}

(*c*) Both MnO_2 and Mn^{3+} (*d*) MnO_4^{-}

4. Cu^{+} disproportionates into:

(*a*) Cu only (*b*) Cu^{2+} and Cu^{3+}

(*c*) Cu^{2+} and Cu (*d*) Cu and Cu^{-}

5. In which of the following compounds, Mn has the highest oxidation number?
(*a*) $MnCl_2$ (*b*) Mn_3O_4
(*c*) Mn (*d*) K_2MnO_4

6. According to reduction potential diagram:

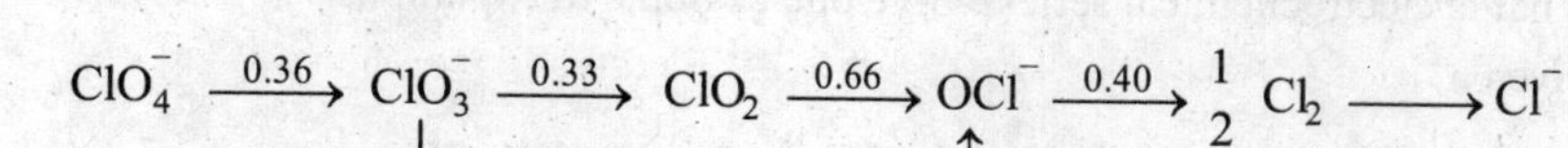

ClO_3^- should disproportionate into
(*a*) ClO_4^- and OCl^- (*b*) ClO_4^- and ClO_2
(c) ClO_2 and OCl^- (*d*) ClO_2 and Cl^-

7. In the diagram:

$$FeO_4^{2-} \xrightarrow{2.20} Fe^{3+} \xrightarrow{0.77} Fe^{2+} \xrightarrow{-0.47} Fe$$

The electrode potential for Fe^{3+} to Fe is
(*a*) 0.30 (*b*) 1.24
(*c*) –0.057 (*d*) +0.057

8. The value of x in the diagram:

$$Tl^{3+} \xrightarrow{+1.25} Tl^{+} \xrightarrow{-0.34} Tl \quad (Tl^{3+} \xrightarrow{x} Tl)$$

(*a*) +1.59 (*b*) +1.23
(*c*) 0.91 (*d*) –0.91

9. Frost diagram is the plot of electrode potentials of couples X (N) | X (0) as:
(*a*) NE° vs. E° (*b*) NE° vs. N
(*c*) E° vs. N (*d*) NE° vs. *p*H

10. In a Latimer diagram, the highest and lowest oxidation states are written respectively on
(*a*) left and right sides (*b*) right and left sides
(*c*) depends upon the value of oxidation states (*d*) none of these

ANSWERS

1. (*b*)	**2.** (*b*)	**3.** (*b*)	**4.** (*c*)	**5.** (*d*)	**6.** (*a*)
7. (*c*)	**8.** (*b*)	**9.** (*b*)	**10.** (*a*)		

Short Answer Questions

1. How is electrode potential of a couple related to Gibbs free energy change?
2. Define oxidation and reduction on the basis of electronic concept.
3. What is meant by disproportionation?
4. In the reaction:
$$3MnO_2 + 4Al \longrightarrow 3Mn + 2Al_2O_3$$
Predict the oxidant and reductant.
5. What is the oxidation number of Cr in CrO_2Cl_2 and K_2CrO_4?
6. Why is it possible to oxidise CO^{2+} and Cl^- by acidic $Cr_2O_7^{2-}$?

7. Construct a Frost diagram from given Latimer diagram for thallium

$$Tl \xrightarrow{+1.26\,V} Tl^{+1} \xrightarrow{-0.34V} Tl$$
(Tl to Tl: +0.73V)

8. What is electrochemical series? Give one example to explain it.

9. Given

$$Ce^{4+} + e^- \rightleftharpoons Ce^{3+} \qquad E° = 1.61\ V$$

$$MnO_4^- + 8H^+ + 5e^- \rightleftharpoons Mn^{2+} + 4H_2O \qquad E° = 1.51V$$

Comment on the feasibility of their reactions.

10. Define cell voltage. How does it explain feasibility of a reaction?

General Questions

1. Discuss the redox stability in water with examples.
2. What is meant by oxidation and reduction? Show that oxidation and reduction go hand in hand.
3. What is meant by redox cycle? Explain.
4. Define oxidation and reduction in terms of oxidation number. How is electrode potential related to Gibbs free energy and equilibrium constant?
5. What is Frost diagram? Discuss Frost diagram for nitrogen under basic conditions.
6. The Latimer diagram for the copper system is:

$$Cu^{2+} \xrightarrow{0.15\,V} Cu^+ \xrightarrow{0.52} Cu$$

Calculate the electrode potential and equilibrium constant for the reaction:

$$2\ Cu^+ \rightleftharpoons Cu^{2+} + Cu$$

7. From the following Latimer diagram:

$$O_2 \xrightarrow{0.125} HO_2 \xrightarrow{1.510} H_2O_2$$

Calculate the net E° value for the following reaction:

$$2HO_2 \longrightarrow O_2 + H_2O_2$$

8. What is meant by Latimer diagram? How does it predict whether a given species can disproportionate or not? In the following diagram, predict which species can disproportionate and how?

$$ClO_4^- \xrightarrow{+0.37} ClO_3^- \xrightarrow{+0.30} ClO_2^- \xrightarrow{+0.68} ClO^- \xrightarrow{+0.42} Cl_2 \xrightarrow{+1.36} Cl^-$$
(ClO^- to Cl^-: +0.89)

9. Draw a Frost diagram for mercury in acid solution from the following Latimer diagram:

$$Hg^{2+} \xrightarrow{0.911} Hg_2^{2+} \xrightarrow{0.796} Hg$$

Comment on the tendency of any of the species to act as an oxidising agent, a reducing agent or to undergo disproportionation.

10. Discuss the following Latimer diagram:

$$O_2 \xrightarrow{+0.682} H_2O_2 \xrightarrow{+1.776} H_2O$$
(O_2 to H_2O: +1.229)

11. Discuss two important uses of redox potential data.

12. Calculate the potential for the oxidation of UO_2 to UO_2^{2+} in acid solution from the following information.

$$UO^{2+}\ e^- \longrightarrow UO_2 \qquad E° = 0.66\ V$$

$$UO_2^{2+} + e^- \longrightarrow 4O_2^+ \qquad E° = 0.16\ V$$

13. (*a*) Write a comprehensive note on analysis of redox cycle.

(*b*) Construct Frost diagram from the Latimer diagram given below:

$$O_2 \xrightarrow{+0.70} H_2O_2 \xrightarrow{+1.76} H_2O$$
$$O_2 \xrightarrow{+1.23} H_2O$$

(*c*) Explain Pourbaix diagrams for some naturally occurring iron compounds.

14. The Latimer diagram is:

$$BrO_3^- \xrightarrow{0.54} BrO^- \xrightarrow{0.45} \frac{1}{2}Br_2 \xrightarrow{+1.07} Br^-$$

Calculate E° for the half cell $BrO_3^- \longrightarrow Br^-$ and $BrO^- \longrightarrow Br^-$.

15. (*a*) Explain why Cu^+ disproportionates in solution?

(*b*) Explain why the standard reduction potentials for $Cu^{2+} \longrightarrow Cu^+$ and $Cu^+ \longrightarrow Cu$ are + 0.15 and + 0.50 V, respectively, yet that for Cu^{2+}/Cu is + 0.34 V?

16. (*a*) Given $Sn^{4+} \longrightarrow Sn^{2+} + 2e$, E° =0.154 V. Explain what does it signify.

(*b*)What are Latimer diagrams? How can you use Latimer diagrams to

(*i*) Predict disproportionation of a metal ion.

(*ii*) Find redox potential of a non-adjacent couple.

(*c*)Though fluorine has lower electron affinity than chlorine, yet it has highest potential. Why?

17. (*a*) Explain the term disproportionation and comproportionation using Frost diagram.

(*b*) Discuss redox stability of water with examples.

18. (*a*) What is electrochemical series? Give one example to explain it.

(*b*) Construct a Frost diagram for oxygen from Latimer diagram.

$$O_2 \xrightarrow{+0.70} H_2O_2 \xrightarrow{+1.76} H_2O$$
$$O_2 \xrightarrow{+1.23} H_2O$$

(c) Given;

$$Ce^{4+} + e^- \rightleftharpoons Ce^{3+} \qquad E° = 1.61\ V$$

$$MnO_4^- + 8H^+ + 5e^- \rightleftharpoons Mn^{2+} + 4H_2O \qquad E° = 1.51\ V$$

Comment on the feasibility of their reaction.

19. (*a*) What is Redox Cycle? Discuss its implications.

(*b*) Given : $Fe^{3+} + e^- \rightleftharpoons Fe^{2+} \qquad E° = 0.771V$

$$MnO_4^- + 8H^+ + 5e^- \rightleftharpoons Mn^{2+} + 4H_2O \quad E° = 1.51\ V$$

Comment on the feasibility of their reaction.

(*c*) Discuss:

(*i*) Reduction by carbon

(*ii*) Self-reduction method involved in extraction of elements.

20. (*a*) Explain what happens when ?

(*i*) I_2 is added to KBr (1M)

(*ii*) Zn metal is added to 1M $AgNO_3$ Solution.

Given: $E^\circ_{I_2/I^-} = 0.54V$, $E^\circ_{Br_2/Br^-} = 1.09V$, $E^\circ_{zn^{2+}/zn} = -0.76V$ and

$E^\circ_{Ag/Ag^+} = -0.80\,V$.

(*b*) What information is conveyed by Pourbaix diagrams?

21. (*a*) Given;

$$Ce^{4+} + e^- \rightleftharpoons Ce^{3+} \qquad E^\circ = 1.61\ V$$

$$MnO_4^- + 8H^+ + 5e^- \rightleftharpoons Mn^{2+} + 4H_2O \qquad E^\circ = 1.51\ V$$

Comment on the feasibility of their reaction.

(*b*) Construct a Frost diagram for oxygen from Latimer diagram.

(*c*) What is meant by Latimer diagram? How does it predict whether a given species can disproportionate or not?

22. On the basis of Latimer diagram given below explain why Cu^+ disproportionates in aqueous solution.

$$Cu^{2+} \xrightarrow{+0.15} Cu^+ \xrightarrow{+0.50} Cu.$$

23. What is Frost diagram? Construct a Frost diagram from the Latimer diagram given below:

$$O_2 \xrightarrow{+0.34} H_2O_2 \xrightarrow{+1.76} H_2O$$

$$O_2 \xrightarrow{+1.23} H_2O$$

According to the diagram, which species is thermodynamically least stable and hence will undergo disproportionation?

24. (*a*) What is disproportionation? Discuss with the help of an example.

(*b*) Give equation for the disproportionation of hypochlorous acid in aqueous solution.

25. (*a*) The Latimer diagram is:

$$BrO_3^- \xrightarrow{0.55\,V} BrO^- \xrightarrow{+0.44\,V} \frac{1}{2}Br_2 \xrightarrow{1.08\,V} Br^-$$

Calculate the electrode potential for half cell reaction: $BrO_3^- \longrightarrow Br^-$

(*b*) Consider the Latimer-diagram for thalium:

$$Tl \xrightarrow{1.26\,V} Tl^{+1} \xrightarrow{-0.34\,V} Tl$$

Construct the Frost diagram. Which species is the strongest oxidising agent? Predict the stability of Tl^{+1}.

26. (*a*) What is Frost diagram? Construct a Frost diagram from given Latimer diagram for thallium.

$$Tl^{3+} \xrightarrow{1.26\ V} Tl^{+1} \xrightarrow{0.34V} Tl$$

(overall, Tl^{3+} to Tl: + 0.73 V)

(*b*) From the following Latimer diagram predict whether Cu^+ is unstable w.r.t. disproportionation in aqueous solution or not.

II I 0

$$Cu^{2+} \xrightarrow{+0.15} Cu^{+} \xrightarrow{+0.50} Cu$$

(overall, Cu^{2+} to Cu: +0.35V)

27. (*a*) Making use of the following Latimer diagram :

$$ClO^- \xrightarrow{+0.42\ V} Cl_2 \xrightarrow{+1.36\ V} Cl^-$$

Calculate E° value for the reduction of ClO^- to Cl^- in aqueous basic medium.

(*b*) Construct a Frost diagram

$$O_2 \xrightarrow{+0.70} H_2O_2 \xrightarrow{+1.76} H_2O$$

(overall, O_2 to H_2O: + 1.23)

(*c*) Discuss redox stability of metal ions in water.

28. The choice of reducing agent (in the metallogy of an element) in a particular case depends on thermodynamic factor. How far do you agree with the statement ? Support your answer with examples.

29. Describe the principle of extraction of each of the following:

(*i*) Sn from SnO_2 (*ii*) Zn from ZnO (*iii*) Cr from Cr_2O_3

30. Which is a better reducing agent, C or CO at 710°C (983 K)?

31. What are Pourbaix diagrams ? What is their utility?

4 Coordination Compounds

4.1. INTRODUCTION

When we add a solution of ammonia to a solution of copper salt, a dark blue colour due to the formation of $[Cu (NH_3)_4]^{2+}$ appears. We know that when we add ammonia solution to a precipitate of silver chloride, the precipitate dissolves due to the formation of $[Ag (NH_3)_2] Cl$. Again, when we add ammonia to a solution of nickel chloride, a violet colour due to the formation of $[Ni (NH_3)_6]^{2+}$ is obtained. These are all examples of coordination compound formation. Thus

$$[Cu (NH_3)_4]^{2+}, [Ag (NH_3)_2]^- \text{ and } [Ni (NH_3)_6]^{2+}$$

are examples of **coordination** compounds. These are also known as **complex** compounds. Transition elements, in particular, have a tendency to form coordination compounds. The species in the square bracket remain as a single entity. They do not ionise to give the components of the square bracket separately.

Such compounds are widely present in minerals, plants and animals and play important functions. Haemoglobin and chlorophyll are coordination compounds of iron and magnesium respectively as central metal ions. Coordination compounds find wide applications in electrophating, textile dyening and analysis of trace elements, etc.

4.1.1. Difference between double salts and coordination compounds

Difference between double salts and coordination can be visualised as under:

(*i*) Those compounds which lose their identity in solution are called *double salts.*

(*ii*) Those compounds which do not lose their identity in solution are called *coordination* or *complex* compounds.

The double salts when dissolved in water give tests for each constituent ion *e.g.*, a solution of carnallite ($KCl.MgCl_2.6H_2O$) gives tests for Mg^{2+} and Cl^- ions. A solution of Mohr's salt similarly gives tests for Fe^{2+}, NH_4^+ and SO_4^{2-} ions. They only exist in crystals. The coordinate compounds or complex compounds when dissolved in water do not give tests of each constituent ion, *e.g.*, a solution of potassium ferrocyanide $K_4Fe(CN)_6$ in water does not give the usual test of Fe^{2+} and CN^- ions, rather a new ferrocyanide ion is $[Fe(CN)_6]^{4-}$ formed and similarly the solution of potassium argentocyanide $K[Ag(CN)_2]$ in water does not give test of Ag^+ and CN^- ions but instead a new argentocyanide ion is formed. Ferrocyanide and argentocyanide ions are called complex ions and compounds containing such ions are called *complex or coordination compounds.*

4.1.2. Applications of coordination compounds

Some important applications of coordination compounds are given as under:

1. A well known Ziegler-Natta catalyst, a complex of aluminium and titanium, is used for low pressure polymerisation of ethylene and propene to propare polyalkenes and rubber.

2. Complex formation is used extensively in analytical chemistry for detection, estimation and separation of metal ions.

3. Coordination compounds are also used in soil treatment, water softening, corrosion control, extraction of metal, electroplating etc.

4. Chlorophyll which is a vital pigment in photosynthesis is a magnesium complex and haemoglobin which carries O_2 to animal cell is a complex of iron.

$Fe(NH_4)_2(SO_4)_2$ (Mohr salt) is a salt but $K_4Fe(CN)_6$ is a complex

$Fe(NH_4)_2(SO_4)_2$ is a salt because its solution gives reactions of Fe^{2+}, NH_4^+ and SO_4^{2-}. $K_4Fe(CN)_6$ is not a salt but it is a complex because the solution of this compound furnishes only K^+ and $[Fe(CN)_6]^{4-}$. It does not give tests of Fe^{2+} and CN^-.

4.1.3. Distinction between the following

(*i*) $[Cr(H_2O)_6]Cl_3$ (I) and $[Cr(H_2O)_5Cl]Cl_2 \cdot H_2O$ (II)

(*ii*) $[Co(NH_3)_5Br]SO_4$ (I) and $[Co(NH_3)_5SO_4]Br$ (II) ?

(*i*) Compound I will give precipitate of AgCl with $AgNO_3$ that will correspond to the presence of three chloride ions whereas compound II will give precipitate corresponding to only two chloride ions.

(*ii*) Compound I gives the test for sulphate ions and not for bromide ion whereas compound II gives the test for bromide ion not for sulphate ions.

4.1.4. (*i*) Ligands (*ii*) Coordination number (*iii*) Coordination sphere

***(i)* Ligands:** The ion or neutral molecule attached to central atom in a coordination compound is called Ligand. Ligand may be neutral molecule such as ammonia, water, ethylenediamine. It may also be an ion such as Cl^-, CN^-, NO_2^-, SCN^- etc.

***(ii)* Coordination number.** Total number of ligands which are directly attached to the central metal atom is known as coordination number of that metal ion. Coordination number varies from 1-12 but more common are four and six. The coordination number of silver in $[Ag(NH_3)_2]^+$ is two and that of platinum in $[Pt(Cl)_6]^{-2}$ is 6. A metal atom may exhibit more than two coordination numbers in different complexes. For example, in complexes $[FeCl_4]^-$ and $[Fe(CN)_6]^{3-}$ iron atom shows co-ordination numbers 4 and 6 respectively. Similarly copper with ethylenediamine (abbreviated as en) shows coordination number 6 in $[Cu(en)_3]^{2+}$ but with ammonia in $[Cu(NH)_4]^{2+}$ it shows coordination number 4.

***(iii)* Coordination sphere.** The central metal atom and ligands that are directly attached to it enclosed in square brackets are collectively known as coordination sphere. The ligands and metal atom inside the square brackets behave as single constituent unit.

4.1.5. Calculation of charge on the complex ion

The charge carried by a complex ion is the algebraic sum of the charges carried by central metal ion and the ligands coordinated to the central metal ion. For example, in complex ion $[Cu(NH_3)_4]^{2+}$ Cu ion carries +2 charge and NH_3 molecule is neutral, therefore the net charge on the complex ion is +2. Similarly in the complex ion $[Ag(CN)_2]^{-1}$, Ag ion carries a charge of +1 and each CN carries a charge of −1. Therefore the net charge on the complex ion $[Ag(CN)_2]$ is $+1-2=-1$.

Some **basic rules for the determination of oxidation number of central atom** in the complex compound are given below:

(*i*) In a complex compound, group I elements (Li, Na, K, etc.) have an oxidation state of +1 and group II elements (Be, Mg, Ca, etc.) have an oxidation state of + 2.

(*ii*) In a compound oxidation number of hydrogen is generally +1 except in hydride where it is −1.

(*iii*) In a compound oxidation number of halogen is generally taken as –1.

(*iv*) The sum of the charges carried by atoms, ions or molecules in a complex compound is equated to the actual charge of the complex ion.

4.1.6. Types of ligands

If a ligand is attached to the central metal atom only by one coordination bond then the ligand is called as *monodentate or unidentate ligand.* Example of unidentate ligands are Cl^-, NO_2^- Br^-, CN^-, OH^-, H_2O, NH_3, CO, etc.

Ligands having two or more than two donor atoms are called *polydentate ligands.* Depending upon the number of donor atoms or coordinating atoms, the ligand may be classified as bidentate (two), tridentate (three donor atoms) and those with four, five and six atoms are called tetra, penta and hexa dentate ligands respectively.

For example, in tetraamine copper (II) complex ion, *i.e.*, $[Cu(NH_3)_4]^{2+}$, I, NH_3 is a monodentate ligand and ethylene diamine is a bidentate ligand in ethylene diamine copper (II) complex ion II.

$$\left[\begin{array}{ccc} NH_3 & & NH_3 \\ & \searrow Cu \swarrow & \\ NH_3 & \nearrow \quad \nwarrow & NH_3 \end{array}\right]^{2+} \qquad \left[\begin{array}{ccc} CH_2-H_2N & & NH_2-CH_2 \\ | & \searrow Cu \swarrow & | \\ CH_2-H_2N & \nearrow \quad \nwarrow & NH_2-CH_2 \end{array}\right]^{2+}$$

I II

ethylenediamine ligand $\left(\begin{array}{l} CH_2 - NH_2 \\ | \\ CH_2 - NH_2 \end{array}\right)$ is a typical bidentate ligand.

4.2. CHELATES

Chelates are the special types of coordination compounds containing one or more rings in their structure. The ring structures are formed linking metal ion with polydentate ligand. The ring structure compounds are called *chelates*. The polydentate ligands are called *chelating agents* and the process of formation is called *chelation*.

$$Cu^{2+} + 2 \begin{array}{l} CH_2-NH_2 \\ | \\ CH_2-NH_2 \end{array} \longrightarrow \left[\begin{array}{ccc} CH_2-NH_2 & & NH_2-CH_2 \\ | & \searrow Cu \swarrow & | \\ CH_2-NH_2 & \nearrow \quad \nwarrow & NH_2-CH_2 \end{array}\right]^{2+}$$

For example, Ethylenediamine, a bidentate ligand, forms a complex with Cu ions.

This complex contains a pair of five-membered ring and this is called a chelate.

Another familiar example of chelation is nickel-DMG complex which is formed when nickel ion reacts with DMG in an ammoniacal medium

$$2 \begin{array}{l} CH_3-C=NOH \\ \quad\;\; | \\ CH_3-C=NOH \end{array} + Ni^{2+} \longrightarrow \begin{array}{c} H \\ O \cdots \quad O \\ \uparrow \qquad | \\ CH_3-C=N \qquad N=C-CH_3 \\ | \quad \searrow Ni \swarrow \quad | \\ CH_3-C=N \qquad N=C-CH_3 \\ | \qquad \downarrow \\ O \quad \cdots O \\ H \end{array} + 2H^+$$

Dimethyl glyoxime (DMG) Chelate

Iron form complex with oxalate ion as given below

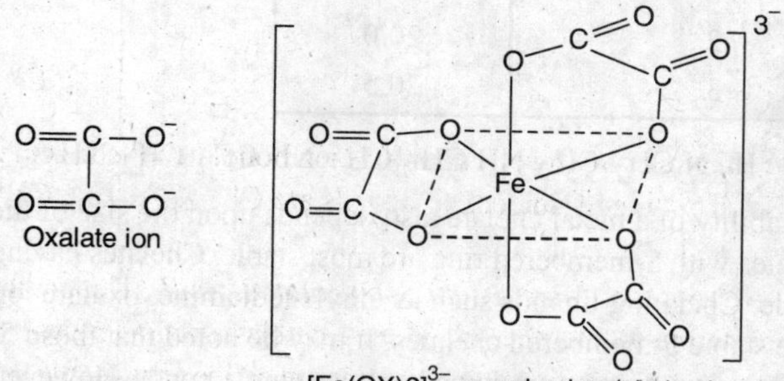

EDTA (hexadentate ligand) is widely used for chelating metal ions

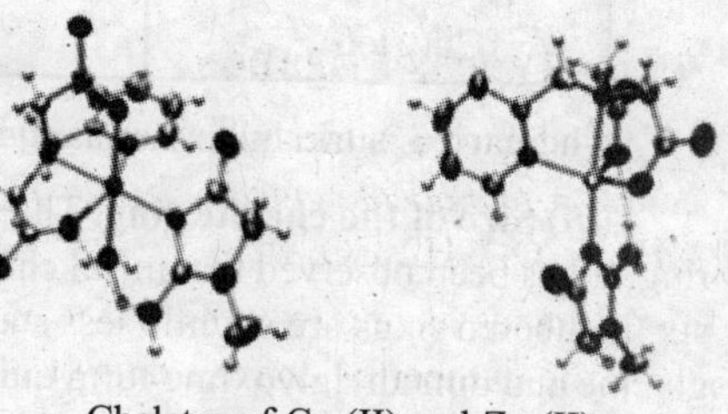

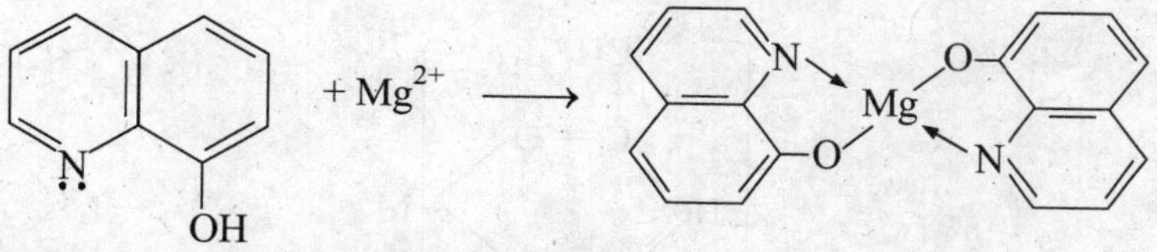

Chelates have been classified as bidentate, tridentate, quadridentate and hexadentate depending upon whether the chelating agent is bidentate, tridentate, quadridentate or hexadentate respectively.

4.2.1. Factors affecting stability of chelates

The stability of chelates depends upon the following factors :

***(i)* Chelate effect.** The complexes formed by chelating ligands (polydentate ligands) such as ethylene diamine (en), EDTA, etc are more stable than those formed by monodentate ligands such as H_2O or NH_3. *This enhanced stability of complexes containing chelating ligands* is called **chelate effect.** For example, the nickel complex with chelating ligand, en is more stable than corresponding complex with NH_3 ligand. This is expressed in terms of stability constant β or log β

$$Ni^{2+}\,(aq) + 6NH_3\,(aq) \rightleftharpoons [Ni(NH_3)_6]^{2+}\,(aq) \qquad \log\beta = 8.61$$

$$Ni^{2+}\,(aq) + 3\text{ en}\,(aq) \rightleftharpoons [Ni(en)_3]^{2+}\,(aq) \qquad \log\beta = 18.28$$

log β is a measure of stability of the complex. Larger the value of log β, greater is the stability of complex.

The chelate effect can be easily understood on the basis of favourable entropy change during chelation. During chelation, there is entropy increase because several molecules of the solvent in the solvated cation are replaced by one or more number of chelating ligand molecules. For example, consider the example of Ni^{2+} complexes with NH_3 and en. During complexation with NH_3, six NH_3 molecules displace six H_2O molecules making no change in the number of independent molecules. However, in reaction with ethylene diamine, it takes only 3 en molecules to displace six H_2O molecules. As a result, there is increase in entropy which favours chelate effect.

***(ii)* Number of chelate rings.** Greater the number of chelate rings in a complex, greater is the stability of a complex. This is clear from the following example.

Complex	No. of chelate rings	log β
$[Cu(NH_3)_4]^{2+}$	0	12.6
$[Cu(en)_2]^{2+}$	2	20.0
$[Cu(trien)]^{2+}$	3	20.5

where trien is triethylenetetraamine : $H_2\ \ddot{N}\ CH_2\ CH_2\ \ddot{N}H\ CH_2\ CH_2\ \ddot{N}H\ CH_2\ CH_2\ \ddot{N}H_2$

***(iii)* Size of the chelate ring.** The stability of a metal chelate also depends upon the size of the ring. It has been observed that metal chelates with 5-membered ring are most stable. Chelates having six membered rings are slightly less stable. Chelating ligands such as ethylenediamine, oxalate ion glycine and dimethylglyoxime form quite stable 5-membered chelates. It may be noted that those 5-membered chelates are more stable when atoms in the ring are joined only by single bonds. However, 6-membered chelates having double bond or unsaturation in the ring are more stable than corresponding 5-membered chelates. For example, acetylacetonate complexes containing 6-membered rings and double bonds in the ring form stable complexes.

$$\left[\begin{array}{c} CH_3 \\ | \\ C{=}O \searrow \\ HC \quad\quad\quad M \\ \| \quad\quad \nearrow \\ C{-}O \\ | \\ CH_3 \end{array}\right]_n$$

Acetylacetonato complex

4.2.2. Applications of chelate compounds

(1) A number of chelates are of biological importance. For example,

(*i*) Vitamin B_{12}, essential for human body, is a cobalt complex.

(*ii*) Haemoglobin in red blood cells contains certain porphyrin complex.

(*iii*) Chlorophyll is a green plant pigment containing a magnesium porphyrin complex.

(*iv*) Metal poisoning by lead, copper, chromium, nickel, zinc, etc. in the body can be removed by treating with EDTA. It forms metal–EDTA chelate which is easily excreted in the urine.

(2) In analytical chemistry chelates are used in both qualitative and quantitative analysis. For example, Ni^{2+}, Mg^{2+}, Cu^{2+} etc. are quantitatively precipitated by chelating agent as insoluble *compounds.*

(3) Standard EDTA, a bidendate ligand solution, is widely used in volumetric analysis to determine hardness of water.

Alfred Werner

4.3. WERNER'S THEORY OF COORDINATION COMPOUNDS

Important postulates of this theory are:

(1) Metals possess two types of valencies, *primary or ionizable* valency and *secondary or non-ionizable* valency.

(*a*) Primary valencies of the central metal ion are satisfied by negative ions. For example, in simple salts such as $CoCl_3$, $CuSO_4$ and AgCl, the primary valencies of Co, Cu and Ag are 3, 2 and 1 respectively. In complex compounds its attachment to metal ion is shown by dotted lines. In modem terminology it corresponds to oxidation state of metal atom or ion and is *ionizable.*

(*b*) The secondary valencies are satisfied by either –ve ions or neutral molecules. Its attachment to metal is shown by thick lines and is non-ionizable. It corresponds to coordination number of metal ion or atom.

(2) Every metal ion has a tendency to satisfy both its primary and secondary valencies. In order to meet this requirement, a negative ion must satisfy both types of valencies.

(3) Every metal has fixed number of secondary valencies.

(4) Secondary valencies (or ligands satisfying the secondary valencies) are always directed towards fixed positions in space around the central metal ion.

The geometry of a complex is determined by the number and arrangement of such ligands in space. For example, if a metal ion has secondary valency of four, the geometry of complex can be tetrahedral or square planar. On the other hand, if secondary valency is six, the complex is expected to be *octahedral.*

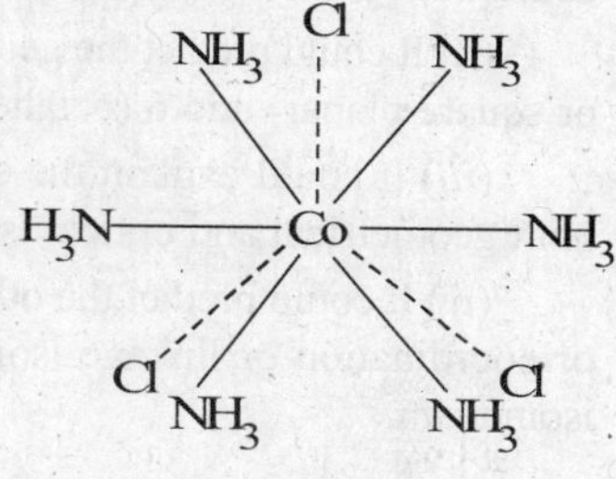

$CoCl_3.6NH_3$

The primary valencies are non-directional and play no important role in determining the geometry of a complex compound. Thus on the basis of Werner's theory the complex $CoCl_3 \cdot 6NH_3$ may be represented as shown in the figure.

Cobalt has three primary or ionizable valencies (shown by dotted lines) satisfied by chlorine. It has six secondary valencies (shown by dark lines) and are satisfied by ammonia molecules.

In modern notation the complex ion is shown in square brackets with ionable ligands outside the coordination sphere. The primary valencies are ionic bonds, whereas secondary valencies are coordinate bonds. Thus the complexes can be rewritten as shown in Table 4.1.

Table 4.1.

S. No.	*Werner formula*	*Modern formula*	*Cation*	*Anion*	*Total no. of ions*
1.	$CoCl_3 . 6NH_3$	$[Co(NH_3)_6]Cl_3$	$[Co(NH_3)_6]^{3+}$	$3Cl^-$	4
2.	$CoCl_3 . 5NH_3$	$[Co(NH_3)_5Cl]Cl_2$	$[Co(NH_3)_5 Cl]^{2+}$	$2Cl^-$	3
3.	$CoCl_3 . 4NH_3$	$[Co(NH_3)_4Cl_2]Cl$	$[Co(NH_3)_4Cl_2]^+$	Cl^-	2
4.	$CoCl_3 . 3NH_3$	$[Co(NH_3)_3Cl_3]$	non-ionizable	—	—

4.3.1. Experimental Verification of Werner's Theory

The following evidence supports Werner's theory:

1. Validity of Werner's Co(III)-Ammonia complexes (Structure 1 to 4) given in Table 4.1 is supported by chemical reactivity or by precipitation data, *i.e.*, by precipitating the number of chlorine as chloride ion by adding silver nitrate solution to freshly prepared solution of the complexes. The number of chloride ions precipitated are 3, 2, 1 and zero respectively providing primary or ionizable valencies of complexes as given by Werner.

2. Measurement of molar conductivities of solutions of coordination compounds also helps to estimate the number of ions in solution. For example, a value of 500 ohm^{-1} or more indicates the presence of 4 ions, between 200-250 ohm^{-1} indicates the presence of 3 ions and about 100 ohm^{-1} indicates 2 ions. It may also be noted that when the number of coordinate anions is same as the oxidation number of central cation, the coordination compound is a non-electrolyte.

The validity of formulating the Werner's compounds $CoCl_3$. $6NH_3$ and $CoCl_3$ $5NH_3$ as $[Co(NH_3)_6]$ Cl_3 and $[Co(NH_3)_5$ $Cl]$ Cl_2 respectively is supported by conductance measurement. Experimentally the conductivities of $CoCl_3.6NH_3$ and $CoCl_3$ $5NH_3$ fall in the range expected for 4 and 3 ions respectively and hence Werner's compounds are represented as $[Co(NH_3)_6Cl_3]$ (4 ions) and $[Co(NH_3)_5Cl]Cl_2$ (3 ions).

4.3.2. Applications of Werner's theory of coordination compound

Important applications of Werner's theory of coordination compounds are listed below:

(*i*) It is successful in explaining the formation and behaviour of coordination compounds which could not be otherwise explained on the basis of electronic theory of valency.

(*ii*) It could predict the geometry of different complexes with coordination number 4 (tetrahedral or square planar) and 6 (octahedral).

(*iii*) It could explain the existence of isomerism in coordination compounds. It could predict both geometrical and optical isomerism in coordination compound.

(*iv*) It could predict the other forms of isomerism in coordinational compounds, *e.g.*, ionisation or coordination or linkage isomerism etc. which are different from both geometrical and optical isomerism.

4.4. IUPAC RULES FOR NOMENCLATURE OF COORDINATION COMPOUND

Coordination compounds are named according to following rules:

1. The order of naming ions. (*i*) *The positive ions (whether in a complex ion or in a simple salt) is named first followed by negative ions*. For example.

HCl-hydrogen chloride, $FeCl_2$ -iron (II) chloride, KCl-potassium chloride, $FeCl_3$-Iron (III) chloride $[Cu(NH_3)_4$ $SO_4]$ – tetrammine copper (II) sulphate, K_4 $[Fe(CN)_6]$–Potassium hexacyanoferrate (II).

(*ii*) *Non-ionic or molecular complexes are given a one-word name*

$[Co(NH_3)_3Cl_3]$ Triamminetrichlorocobalt (III)

$[Pt(NH_3)_2Cl_2]$ Diamminedichloroplatinum (II)

$[Pt(NH_3)_2Cl_4]$ Diammnetetrachloroplatinum (IV)

2. Naming the ligands. The names of negative ligands end in –o and of positive ligands (which are rare) end in –ium. Neutral ligands are named as such like molecules. This rule is illustrated below:

(*i*) *Negative ligands end in –o* :

F^-	fluorido*	NO_3^-	nitrato
Cl^-	chlorido	NO_2^-	nitrito–N
Br^-	bromido	ONO^-	nitrito–O
OH^-	hydroxo	SO_4^{2-}	sulphato
CN^-	cyano	COO^-	
CH_3COO^-	acetato	|	oxalato
SCN^-	thiocyanato	COO^-	
NCS^-	isothiocyanato	O^{2-}	oxo
CO_3^{2-}	carbonato	O_2^{2-}	peroxo

* Ligands like F^-, Cl^-, Br^-, etc. were named as fluoro, chloro, bromo, etc. Latest IUPAC nomenclature recommends names as fluorido, chlorido, bromido, etc. However, both nomenclatures for such ligands are in vogue and acceptable. – NO_2^- which was earlier named nitro is now called nitrito – N and – ONO which was earlier called nitrito is now called nitrito–O

NH^{2-}	imido	N_3^-	azido
NH_2^-	amido	CH_3O^-	methoxo
H^-	hydrido	$C_6H_5^-$	phenyl
S^{2-}	thio	$C_5H_5^-$	cyclopentadienyl
HS^-	mercapto (hydrogen sulphido)		

(*ii*) *Positive ligands end in –ium*

NO^+	nitrosonium	NO_2^+	nitronium
$NH_2NH_3^+$	hydrazinium		

(*iii*) *Neutral ligands are named as such*

$NH_2CH_2CH_2NH_2$	ethylenediamine
C_5H_5N	pyridine
$(C_6H_5)_3P$	triphenylphosphine

However, there are a few exceptions in naming neutral ligands. For example,

H_2O	aqua	NH_3	ammine
NO	nitrosyl	CO	carbonyl

Note the spelling of ammine with two m's. This is used for NH_3 while all other amines are spelled with usual one m.

(*iv*) *Prefixes di–, tri–, tetra– etc. are used to indicate the no. of ligands of one type.*

For complex ligands which already have di–, tri–, etc. on their names, the prefixes bis, tris, tetrakis etc. are used and the group is placed within brackets. For example $[Cu(en)_2]Cl_2$ is named as bis (ethylenediamine) copper (II) chloride

(v) *Abbreviations for complicated molecules.* For complicated names of ligands, abbreviations are generally used. For example, for ethylenediamine the abbreviation *en* is commonly used. Some common abbreviations are given below :

Name of molecule	Formula	Abbreviation
Ethylenediamine	$H_2NCH_2CH_2NH_2$	en
pyridine	C_5H_5N	py
2, 3-bipyridine		bipy
acetylacetonate		acac
oxalate		ox
8-quinoline		oxine
diethylenetriamine		dien
triethylenetetraamine		tren
ethylenediaminetetraacetate		EDTA
1, 10-phenathroline		phen
urea		ur
thiourea		tu
glycine ion		gly
dimethyl glyoxime		dmg
dimethyl sulphoxide		dmso

3. Naming of a complex ion. (*a*) Ligands are named first followed by central metal atom at the end. *The ligands are listed in alphabetical order regardless of charge and number of each.* The name of ligand is treated as a unit and the prefixes di- tri- etc. are not considered for alphabetial order. For example, "diammine" is listed under *'a'* and not *'d'*.

According to old rules, negative ligands were written first followed by neutral and positive ligand.

(*b*) When a coordination sphere or complex is a cation or a neutral molecule, the name of central metal atom is written as such, followed immediately by a Roman numeral in parenthesis to indicate its oxidation state, *e.g.*, copper (II), cobalt (II) etc. When the complex is anionic, the suffix **ate** is added to the name of metal, along with the oxidation state, *e.g.*, ferrate (II), argentate (I), manganate (II), nickelate (I) etc.

Names of some typical coordination compounds are given below:

(*a*) $K[Ag(CN)_2]$	Potassium dicyanoargentate (I)
$[Pt(NH_3)_4(NO_2\backslash)Cl]SO_4$	Tetraamminechloridonitrito–N platinum(IV) sulphate
$K_2[HgCl_4]$	Potassium tetrachloridomercurate(II)
$K_4[Cu(CN)_6]$	Potassium hexacyanocuprate(II)
$[Co(NH_3)_3(NO_2)Cl_2]$	Triamminedichloridonitrooito–N cobalt(III)
$Na[Au(CN)_2]$	Sodium dicyanoaurate(I)
$[(C_6H_5)_3P_3)Rh]Cl$	Tris (triphenyl phosphine) rhodium(I) chloride
$[Cu(H_2O)_2(NH_3)_4]SO_4$	Tetraamminediaquacopper(II) sulphate
$Na_3[Co(NO_2)_6]$	Sodium hexanitrito–N cobaltate(III)
$Na[B(NO_3)_4]$	Sodium tetranitratoborate(III)
$[Cr(H_2O)_6]Cl_3$	Hexaaquachromium(III) chloride
$K_2[Fe(CN)_5NO]$	Potassium pentacyanonitrosylferrate(III)
$K_3[Fe(C_2O_4)_3]$	Potassium trioxalatoferrate(III)
$[Pt(NH_3)_6Cl_4]$	Hexaammineplatinum(IV) chloride
$K_4[Ni(CN)_4]$	Potassium tetracyanonickelate(0) In this case oxidation number of Ni is 0.
$K_2[PdCl_4]$	Potassium tetrachloridopalladate(II)
$[PtCl(NH_2CH_3)(NH_3)_2Cl]$	Diamminechlorido(methylamine) platinum(II) chloride
$[Ru(HSO_3)_2\ (NH_3)_4]$	Tetraamminedihydrogensulphitoruthenium(II)
$[Co(NH_2CH_2CH_2NH_2)_2Cl_2]_2SO_4$	Dichloridobis (ethylenediamine) cobalt(III) sulphate
$[Cr(NH_3)_6]^{3+}$	Hexaamminechromium(III) ion
$[Pt(NH_3)_4]\ [PtCl_4]$	Tetraammineplatinum (II) tetrachloridoplatinate(II) In this case the cation as well as anion are complex ions.
$[Pt(NH_3)_4Cl_2]\ [PtCl_4]$	Tetraamminedichloridoplatinum(IV) tetrachloridoplatinate (II)
$K_2[Zn(OH)_4]$	Potassium tetrahydroxozincate(II)
$Na_2[SiF_6]$	Sodium hexafluorosilicate(IV)
$Hg[Co(CNS)_4]$	Mercury tetrathiocyanatocobaltate(II)
$[Co(en)_2Br_2]Cl$	Dibromobis (ethylenediamine) cobalt(III) chloride
$K[BF_4]$	Potassium tetrafluoridoborate(III)
$Na_3[Fe(C_2O_4)_3]$	Sodium trioxalatoferrate(III)
$K_4[Mo(CN)_8]$	Potassium octacyanomolybdate(IV)
$Mn_3(CO)_{12}$	Dodecacarbonyltrimanganese(0)

$[VO(acac)_2]$	Bis (acetylacetonato) oxovanadium(IV) The ligands are oxo (O^{2-}) and $acac^{-1}$
$Fe(C_5H_5)_2$	Bis (cyclopentadienyl)iron(II)
$Na_3[Ag(S_2O_3)_2]$	Sodium bis (thiosulphato) argentate(I)
$[Ru(NH_3)_5N_2]Cl_2$	Pentaammine (dinitrogen) ruthenium(II) chloride
$[Pt(py)_4][PtCl_4]$	Tetrapyridineplatinum(II) tetrachloridoplatinate(II)
$K_2[OsCl_5N]$	Potassium pentachloridonitridoosmate (VI)
$K[PtCl_3(C_2H_4)]$	Potassium trichlorido (ethylene) platinate(II)
$[CoCl_3(NH_3)_2\{(CH_3)_2NH\}]$	Diamminetrichlorido (dimethylamine) cobalt(III)
$[CuCl_2(CH_3NH_2)_2]$	Dichloridobis(methylamine) copper(II)
$[PtCl_2(C_5H_5N)(NH_3)]$	Amminedichlorido (pyridine) platinum(II)
$[PtClBr(NH_3)(py)]$	Amminebromidochloridopyridineplatinum(II)
$[Cr(PPh_3)(CO)_5]$	Pentacarbonyltriphenylphosphinechromium (0)
$K[CrF_4O]$	Potassium tetrafluoridoxochromate (V)
$K_2[Cr(CN)_2O_2(O_2)NH_3]$	Potassium amminedicyanodioxoperoxochromate (VI)
$[CuCl_2(CH_3NH_2)_2]$	Dichloridobis (methylamine) cobalt (II) chloride
$[Cr(C_6H_6)_2]$	Bis(benzene) chromium (0)
$K[PtCl_3(C_2H_4)]$	Potassium trichloridoethyleneplatinate (II)
$Fe[Fe(CN)_6]$	Iron hexacyanoferrate (III)

(*b*) If the complex contains two or more metal ions, it is called as **polynuclear** complex. The ligands which link the two metal atoms are called bridge groups and are separated from rest of the complex by hyphens and denoted by Greek letter 'μ'.

$$\left[(H_2O)_4\,Fe \begin{matrix} OH \\ OH \end{matrix} Fe\,(H_2O)_4\right](SO_4)_2$$

Octaaqua-μ-dihydroxodiiron (III) sulphate

$$\left[(NH_3)_4\,Co \begin{matrix} OH \\ OH \end{matrix} Co\,(NH_3)_4\right](NO_3)_4$$

Octaammine-μ-amido-μ-nitrodicobalt (III) nitrate

(*c*) Geometrical isomers are named by the terms **cis** or **trans** depending upon whether similar ligands are on the same side or on opposite sides of the metal ion, *e.g.*, $[Pt(NH_3)_2Cl_2]$ exists as

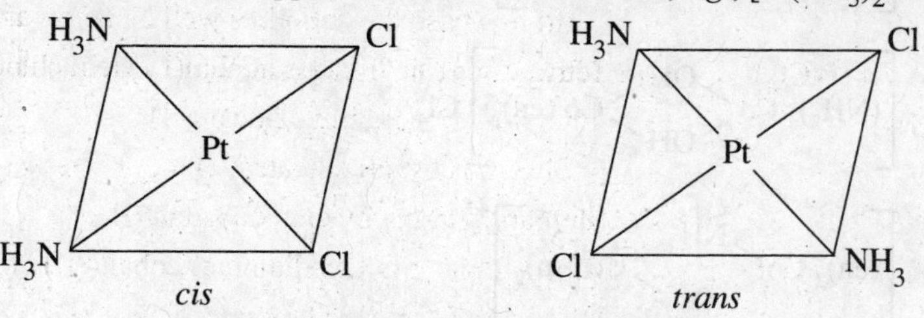

Diamminedichloridoplatinum (II)

(*d*) Dextro and laevorotatory optically active compounds are designated by (+) or *d*- and (–) or 1- respectively, *e.g.*, *d*-$K_3[Cr(C_2O_4)_3]$–potassium trioxalatochromate (III).

(*e*) Complicated ligands are generally denoted by abbreviations such as 'en' for ethylenediamine, 'py' for pyridine, 'phen' for 1, 10 phenanthroline etc.

PROBLEMS FOR PRACTICE

Q. 1. Give the names of following coordination compounds:

(1) $[Co(ONO)(NH_3)_5]SO_4$, (2) $K(BF_4)$, (3) $[Cr(NH_3)_6]Cl$, (4) $Na[Co(CO)_4]$, (5) $[(CH_3)_4N]_2[Co(NCS)_4]$, (6) $Na_3[Ag(S_2O_3)_2]$, (7) $K_3[Al(C_2O_4)_3]$, (8) $[Co(en)_3]_2(SO_4)_3$, (9) Cis-$[Cr(H_2O)_4Cl_2]Cl$, (10) $(NH_4)_2TiCl_6$, (11) $[Pt(NH_3)_4][PtCl_4]$, (12) $H[AuCl_4]$, (13) $Na_2[Fe(CN)_5NO]$, (14) $K[Ag(CN)_2]$, (15) $[Co(NH_3)_6][CdCl_5]$, (16) $[Co(NH_3)_4.H_2O.Br](NO_3)_2$.

Ans. 1. Pentamminenitrito cobalt (III) sulphate.

2. Potassium tetrafluoridoborate (III)
3. Hexamminechromium (III) chloride
4. Sodium tetracarbonylcobaltate (I)
5. Tetramethylammonium tetrathiocyanato–N–cobaltate (II)
6. Sodium dithiosulphato argentate (I)
7. Potassium trioxalatoaluminate (III)
8. Tris (ethylenediamine) cobalt (III) sulphate
9. Cis-tetraaquadichloridochrominium (III) chloride
10. Ammonium hexachloridotitanate (IV)
11. Tetrammine platinum (II) tetrachloridoplatinum (II)
12. Hydrogen tetrachloridoaurate (III)
13. Sodium pentacyanonitrosylferrate (III)
14. Potassium dicyanoargenate (I)
15. Hexaamminecobalt (III) pentachloridocadmate (II)
16. Tetraammineaquabromidocobalt (III) nitrate.

Q. 2. Write down the structures of following polynuclear complexes:

1. **Decaammine-μ-hydroxodichromium (III) bromide**
2. **Tetrabis (ethylenediammine)-μ-amido-μ-hydroxodicobalt (III) sulphate**
3. **Tetrammine-μ-dihydroxo bis (ethylenediammine) dicobalt (III) chloride**
4. **Bis (ethylenediammine) cobalt (III)-μ-amido-μ-peroxo-bis (ethylenediammine)**
5. **Triammine cobalt (III)-μ-trihydroxo-triammine cobalt (III) ion.**

Ans. 1. $[(NH_3)_5Cr-OH-Cr(NH_3)_5]Br_5$

2. $\left[(en)_2Co \begin{smallmatrix} NH_2 \\ OH \end{smallmatrix} Co(en)_2\right](SO_4)_2$

3. $\left[(NH_3)_4Co \begin{smallmatrix} OH \\ OH \end{smallmatrix} Co(en)_2\right]Cl_4$

4. $\left[(en)_2Co \begin{smallmatrix} NH_2 \\ O_2 \end{smallmatrix} Co(en)_2\right]^{4+}$

5. $\left[(NH_3)_3Co \begin{smallmatrix} OH \\ OH \\ OH \end{smallmatrix} Co(NH_3)_3\right]^{3+}$

Q. 3. Write down the chemical formulae of the following compounds:

(*i*) Chloridonitrito–N– bis (ethylenediammine) cobalt (III) ion.

(*ii*) Cis-chloridonitrito–N–diammineplatinum (II)

(*iii*) Sodium trioxalatoaluminate (III)

Ans. (*i*) $[CoCl(NO_2)(en)2]^+$

(*ii*) $\left[\begin{array}{ccc} Cl & & NO_2 \\ & \searrow Pt \swarrow & \\ & \nearrow \quad \nwarrow & \\ NH_3 & & NH_3 \end{array}\right]$

(*iii*) $Na_3[Al(C_2O_4)_3]$

Q. 4. Write down the IUPAC nomenclature of the following compounds:

(*i*) $[Co(NH_3)_6]Cl_3$, (*ii*) $K_4Fe(CN)_6$ and (*iii*) $[Cr(NH_3)_6]Cl_3$

Ans. (*i*) Hexaamminecobalt (III) Chloride

(*ii*) Potassium hexacyanoferrate (II)

(*iii*) Hexaamminechromium (III) Chloride

4.5. ISOMERISM IN COORDINATION COMPOUNDS

Compounds having the same moleculer formula but different structures and properties are called isomers and this phenomenon is called isomerism. Coordination compounds frequently exhibit isomerism. Isomerism is of two types

1. Structural isomerism

2. Stereoisomerism

These are being discussed separately

STRUCTURAL ISOMERISM

Isomers which have the same moleculer formula but different structural arrangement of atoms are called structural isomer. These are of the following types.

1. Ionisation isomerism. This type of isomerism occurs due to exchange of groups between the complex ion and ion outside it. Such compounds behave differently when dissolved in water. For example $[Co(NH_3)_5Br]SO_4$ is red-violet in solution and gives a precipitate with $BaCl_2$ confirming the presence of SO_4^{2-} ion.

On the other hand, $[Co(NH_3)_5SO_4]Br$ is red and does not give tests for sulphate ion in solution but instead gives precipitate of AgBr with $AgNO_3$. These two compounds are ionisation isomers.

2. Hydrate isomerism. These isomers differ in number of water molecules attached to metal ion as ligands. Thus three isomers of $CrCl_3\ 6H_2O$ are known as given below. From the conductivity measurement and quantitative precipitation of ionized chlorine, they have the following formulae:

(*i*) $[Cr(H_2O)_6]Cl_3$ Violet (three ionic chlorine)

(*ii*) $[Cr(H_2O)_5Cl]Cl_2 \cdot H_2O$ Green (two ionic chlorine)

(*iii*) $[Cr(H_2O)_4Cl_2]Cl \cdot 2H_2O$ Dark green (one ionic chlorine)

3. Linkage isomerism. This type of isomerism occurs when more than a single atom in a coordinated group may act as a donor. For example, in NO_3^- ion, either oxygen or nitrogen can act as donor giving two different isomers. Two isomers of cobalt having ligands ammonia and NO_2^- showing linkage isomerism are $[Co(NH_3)_5NO_2]^{2+}$ pentamminenitrito–Ncobalt (III) ion and $[Co(NH_3)_5 ONO]^{2+}$ pentammine nitrito–O–cobalt (II) ion. They are linkage isomers of each other.

Another ligand which can exhibit linkage isomerism is thiocyanate ion SCN^-, *e.g.*,
It can form M–SCN and M–NCS.

4. Coordination isomerism. When both positive and negative ions are complex, isomerism is possible due to the interchange of ligands between the two complex ions. This type of isomerism is called coordination isomerism and is illustrated by the following examples:

$[Co(NH_3)_6]\,[Cr(CN)_6]$ and $[Cr(NH_3)_6]\,[Co(CN)_6]$

$[Co(NH_3)_6]\,[Cr(NO_2)_6]$ and $[Cr(NH_3)_6]\,[Co(NO_2)_6]$

In $[Co(NH_3)_6]\,[Cr(CN)_6]$, $[Co(NH_3)_6]^{3+}$ is a cation and $[Cr(CN)_6]^{2-}$ is an anion. Ligands interchange their positions to form $[Cr(NH_3)_6]\,[Co(CN)]_6$. Cation is written first.

5. Coordinate position isomerism. In polynuclear complexes, an interchange of ligands between two metal atoms gives rise to another type of isomerism called coordination position isomerism, *e.g.*,

$$\left[(NH_3)_4\,Co \begin{array}{c} NH_2 \\ O_2 \end{array} Co\,(NH_3)_2Cl_2\right] Cl_2$$

Tetrammine cobalt (III) μ-amido-μ superoxodiamminidichloridocobalt (III) chloride and

$$\left[Cl(NH_3)_3\,Co \begin{array}{c} NH_2 \\ O_2 \end{array} Co\,(NH_3)_3Cl\right] Cl_2$$

Triamminechloridocobalt (III)-μ-amido-μ-superoxotriamminechloridocobalt (III) are coordination position isomers. Here O_2, superoxo, is one bridging group and the other is amido (NH_2).

6. Polymerization isomerism. This is not a true isomerism since it occurs between compounds having the same empirical formula but different molecular formula. Thus $[Pt(NH_3)_2Cl_2]$, $[Pt(NH_3)_4]$ $[PtCl_4]$, are polymerization isomers and have same empirical formula but have molecular weight in the ratio of 1:2.

STEREOISOMERISM

Compounds which have the same position of atoms or groups but differ in the spatial arrangements around the central atom are called stereoisomers and this phenomenon is called stereoisomerism. It is of two types:

Geometrical isomerism

Optical isomerism.

These are discussed separately as under:

4.5.1. Geometrical isomerism in coordination compounds

This type of isomerism is exhibited by complexes in which same two ligands occupy positions adjacent to each other (cis-isomer) or opposite to each other (trans-isomer). This type of isomerism is not possible in complexes having coordination number 2, 3 and tetrahedral complexes (C.N. 4). This is due to the fact that in these arrangements all the positions are adjacent to each other.

This type of isomerism occurs in four coordinated square planar complexes and six coordinated octahedral complexes.

(A) Cis-trans isomerism in square planar complexes

(*i*) MA_2B_2 *Type*. It was first shown by Werner that Pt (II) complex $[Pt(NH_3)_2Cl_2]$ existed in two geometrical forms and it must have a planar structure. The cis- and trans- forms of diammine dichloro platinum (II) complex are shown below.

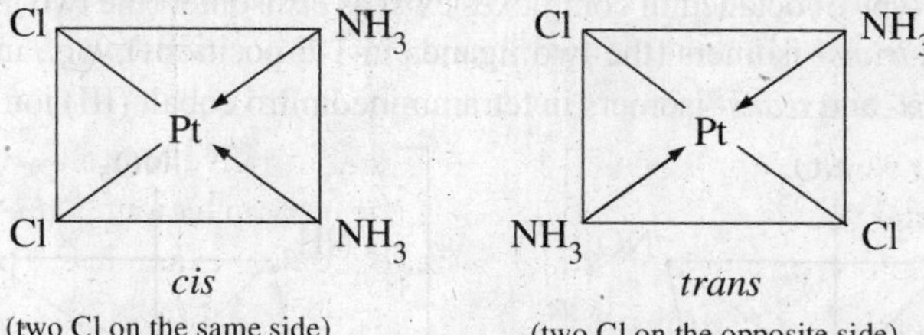

(*ii*) *MA_2BC type of complexes* also exist in cis- and trans- forms *e.g.*, $[Pt(NH_3)_2ClNO_2]$ can exist in the following two geometrical forms.

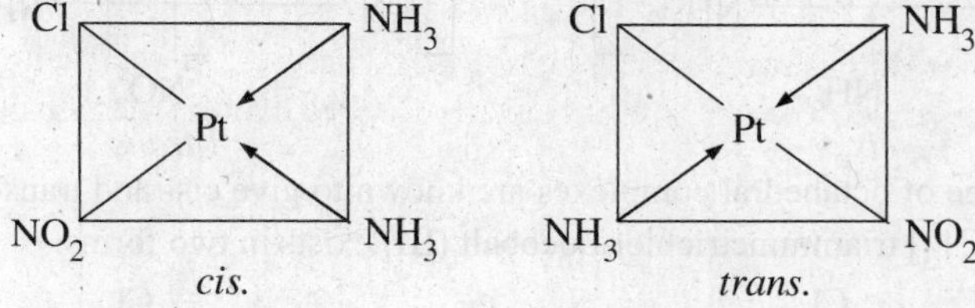

In *cis* form the two ammine groups are on the same side of Pt whereas in *trans* form, the two ammine groups are on the opposite side of Pt.

(*iii*) *Complexes of the type MABCD* exhibit three isomers, *e.g.*, $[Pt(Py)(NH_3)BrCl]$ can exist as:

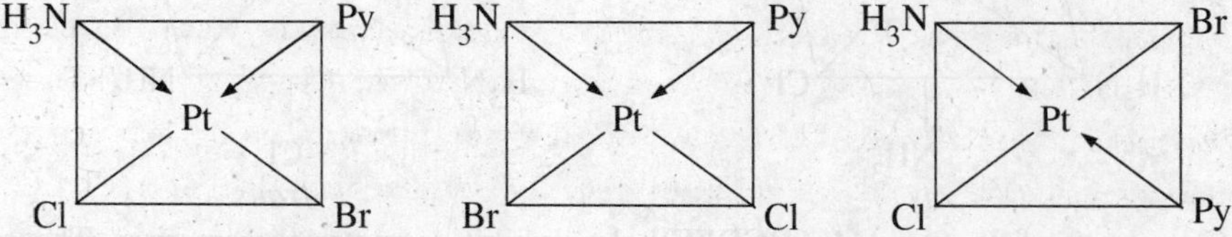

These can be written by fixing any one ligand to one position (say NH_3) and then placing other ligands *trans* to it. Three isomers of the complex $[Pt(Py)(NH_3)(NH_2OH)NO_2]$ have actually been isolated and identified.

(*iv*) *Geometrical isomers are also found in square planar complexes* which have unsymmetrical bidentate ligands, *e.g.*, the Pt (II)-glycinate complex exists in *cis*- and *trans*-forms as shown below:

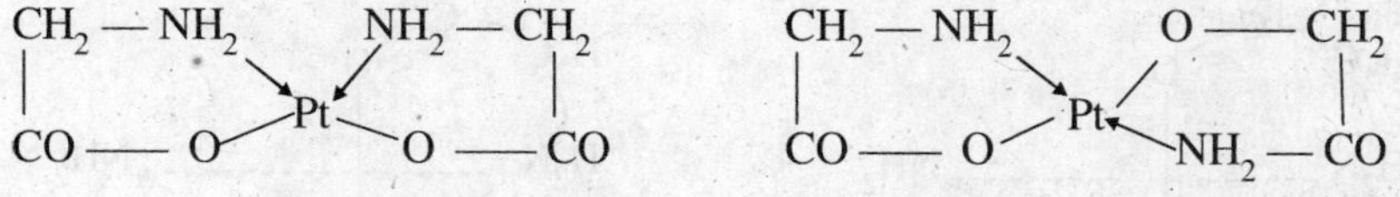

CH_2 — NH_2 NH_2 — CH_2
Pt
CO — O O — CO

CH_2 — NH_2 O — CH_2
Pt
CO — O NH_2 — CO

(*v*) *Geometrical isomers is also shown by bridged binuclear planar complexes* of the type $M_2A_2X_4$, *e.g.*, $[PtCl_2PEt_3]_2$ where PEt_3 is triethylenephosphine.

Cl Cl Cl
Pt Pt
Et_3P Cl PEt_3
cis

Et_3P Cl Cl
Pt Pt
Cl Cl PEt_3
trans

(B) Geometrical Isomerism in Octahedral Complexes

(*i*) For monodentate ligands A and B, no isomers are possible for $[MA_5B]$ or $[MAB_5]$ type of complexes. $[MA_4B_2]$ type of octahedral complexes exist as *cis* isomer (the two monodentate ligands in 1, 2-positions) and *trans*- isomers (the two ligands in 1-6-positions), *e.g.*, in tetramminedinitro cobalt (III) ion. The *cis*- and *trans*- isomers in tetramminedinitro cobalt (III) ion are given below.

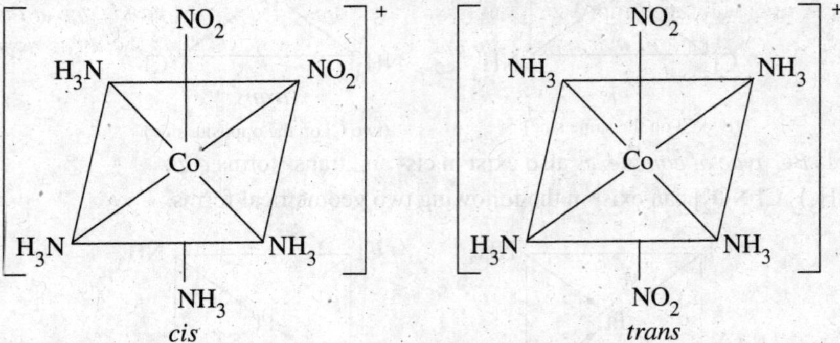

(*ii*) $[MA_3B_3]$ type of octahedral complexes are known to give cis- and trans- isomers, *e.g.*, the complex $[Co(NH_3)_3Cl_3]$ triamminetrichloridocobalt (III) exists in two forms.

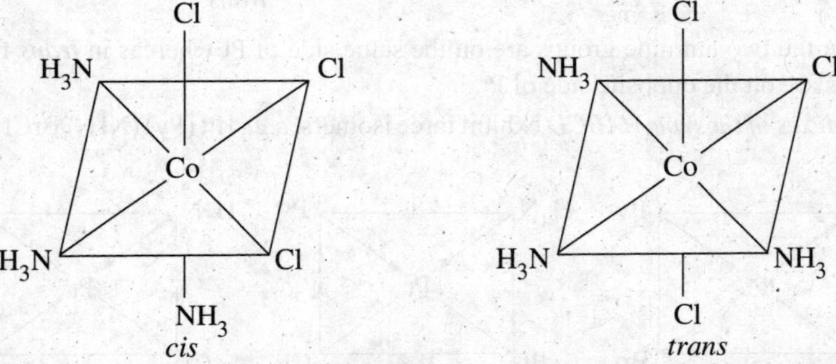

(*iii*) Complexes of the type [M(ABCDEF)] also exhibit geometrical isomerism. Theoretically fifteen different isomers are possible but in actual practice for a complex, [Pt (Py) (NH_3) (NO_2) Cl, Br, I] only three isomers have been isolated.

(*iv*) Unsymmetrical bidentate ligands in octahedral complexes also exhibit geometrical isomerism (similar to square planar complexes). For example, the chromium-triglycinato complex is found to exist in *cis*- and *trans*- forms.

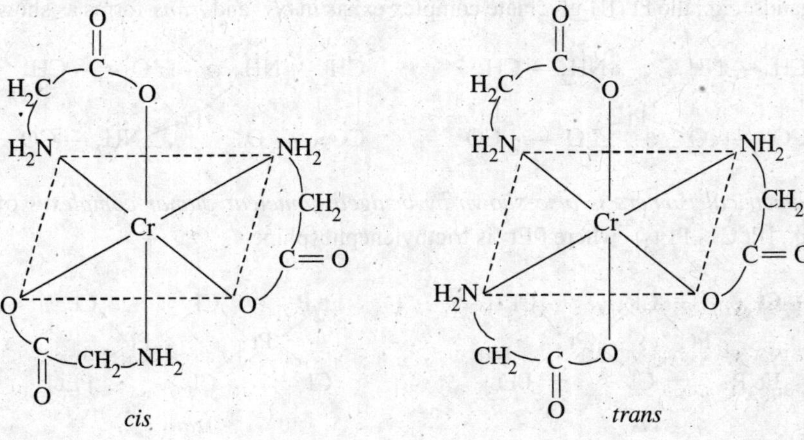

fac and *mer* notation of Geometrical Isomers

Octahedral complexes of the type $[MA_3B_3]$ like $[Co(NO_2)_3(NH_3)_3]$ also exist in two geometrical isomers. When the three ligands (with same donor atoms) are on the same triangular face of the octahedron, the compound is called **facial** or **fac isomer.** When the three ligands are on the same equatorial plane of the octahedron i.e. around the meridian of the octahedron, the compound is called **meridional** or **mer isomer.** Simply speaking, *in facial isomer, the three ligands are at the corners of a triangular face while in meridional isomer, the three ligands are at the three corners of a square plane* as shown below:

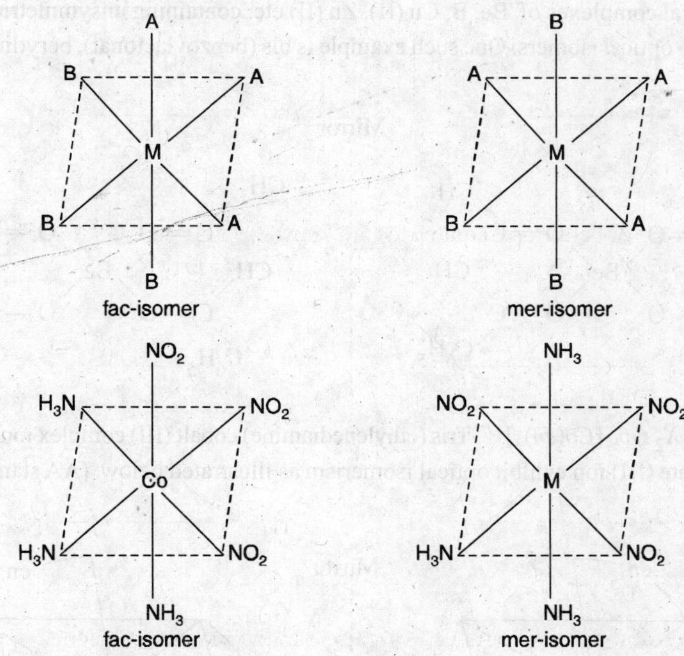

fac- and *mer-* isomers of $[RhCl_3(Py)_3]$ are given below:

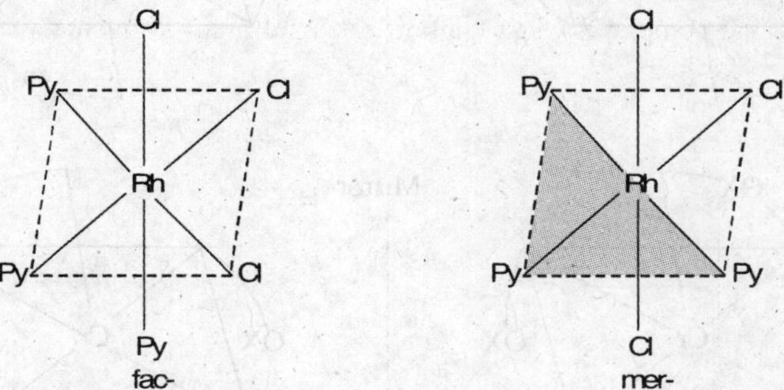

4.5.2. Optical isomerism in Coordination Compounds

Optical isomerism arises when the structure of one form is the mirror image of the other and the two are not superimposable. Such compounds are called optical isomers. For optical activity a compound must not have a plane of symmetry. An optically active compound has a property of rotating the plane of polarised light either to its right (called dextro-rotatory, *d or* (+)) or its left (called laevorotatory *l* or (–)). The two isomers *"d"* and *"l"* of a compound are called *enantiomers.*

(*i*) Square planar compounds do not exhibit optical activity as they possess a plane of symmetry. However, the property is exhibited by tetrahedral *complexes.*

In tetrahedral complexes of type (Mabcd) (analogous to tetrahedral carbon atom) it is expected that a metal complex with four different groups attached to metal ion should exhibit optical activity. However it has not been possible to isolate optically active compound of this type.

(*ii*) Tetrahedral complexes of Be, B, Cu (II), Zn (II) etc. containing unsymmetrical ligands have been resolved into optical isomers. One such example is bis (benzoylactonat), berytlium (II) complex as shown below.

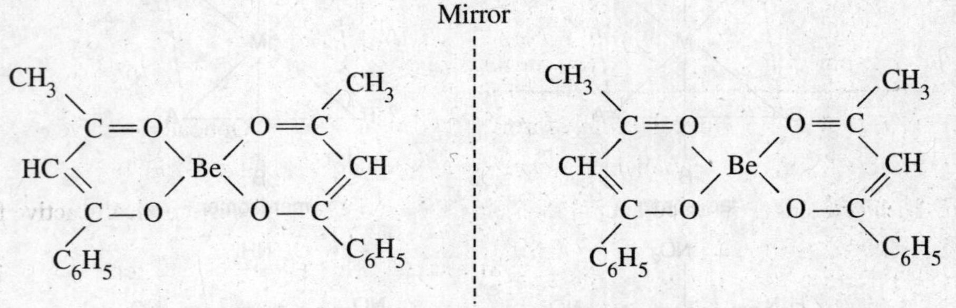

Mirror

CH_3 C=O HC C–O C_6H_5 Be O=C CH_3 CH O–C C_6H_5 | CH_3 C=O CH C–O C_6H_5 Be O=C CH_3 CH O–C C_6H_5

(*iii*) *M(AA)*$_2$, *X*$_2$ *type* $[Co(en)_3]^{3+}$. Tris (ethylenediamine) cobalt (III) complex ion and $[Cr(OX_3)I^-$ trioxalato-chromate (III) ion exhibit optical isomerism as illustrated below. (AA stands for bidentate ligand).

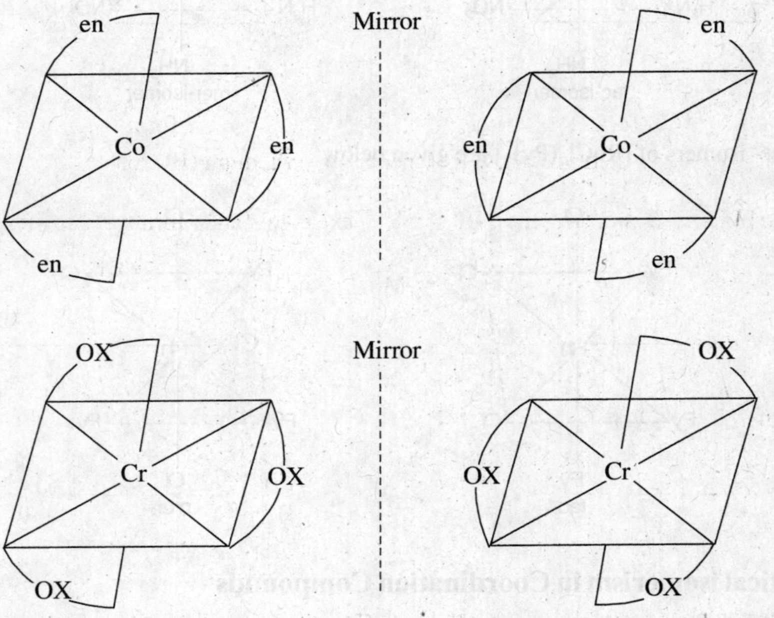

(*iv*) *Type* $[M(AA)_2 X_2]$. Such complexes show both geometrical and optical isomerism. The *trans-*

isomers will be optically inactive due to the presence of a plane of symmetry, only *cis*- form is unsymmetrical and can form two optical isomers. For example, in the complex ion *cis*-chloridobis (ethylenediammine) cobalt (III) ion $[Co(en)_2Cl_2]^+$ three isomers are formed.

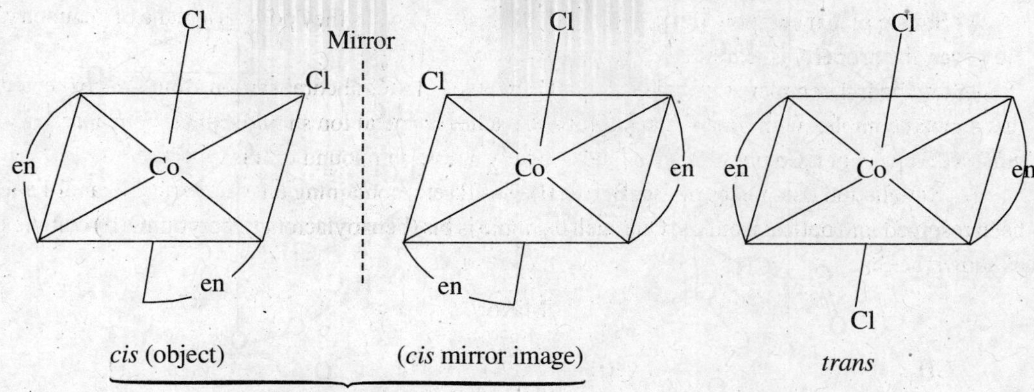

cis dichloridobis (ethylenediamine) cobalt (III) ion.

cis-Dichloridobis (ethylenediamine) rhodium (III) ion also exists in two optically active forms.

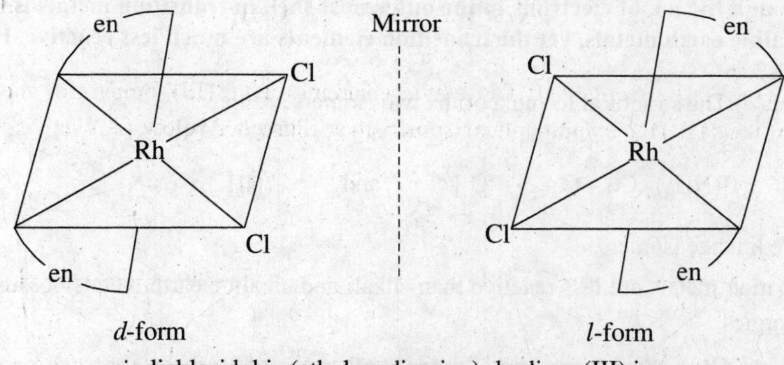

cis dichloridobis (ethylenediamine) rhodium (III) ion

(*v*) *Type* $[M(AA), B_2X_2]$. $[Co(en)(NH_3)_2Cl_2]^+$ exists in *d* and *l* forms as shown below

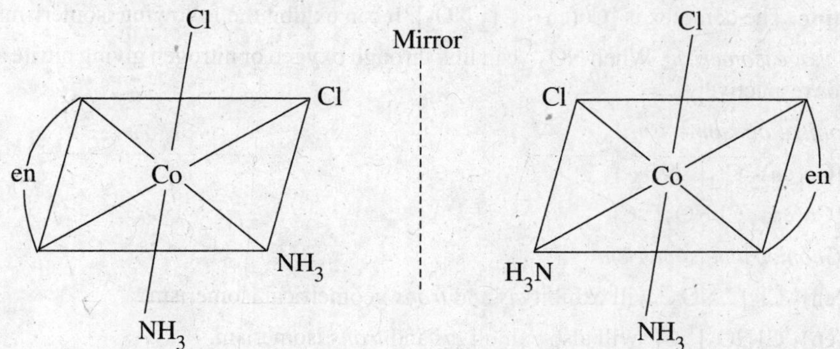

(*vi*) The compound containing hexadentate ligand have also been resolved into optical isomers. For example, the ethylenediaminetetra acetic acid (EDTA), a hexadentate ligand, forms $[Co(EDTA)]^-$, a complex amion which exists in two isomeric forms.

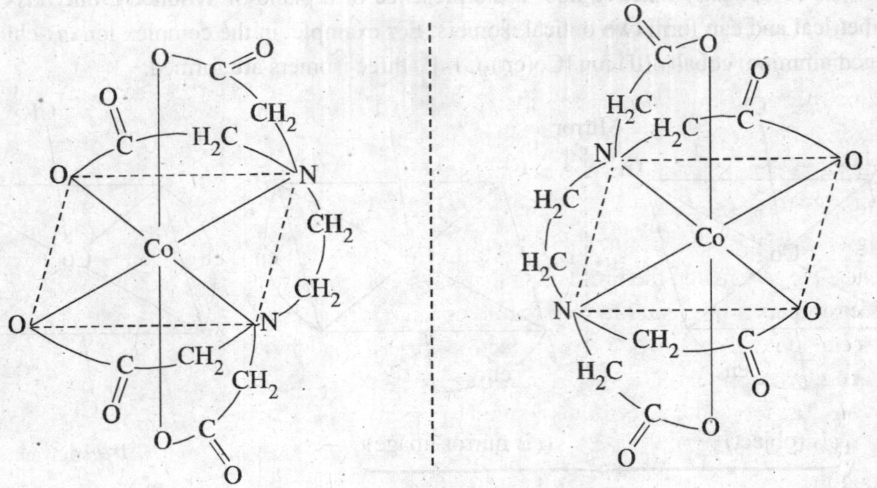

Mirror

Example 5. (*a*) Draw the structural formula of two isomers of the complex ion $[Co(NH_3)_5NO_2]^{2+}$. What sort of isomers are these?

(*b*) Although the no. of electrons in the outermost shell in transition metals is the same as alkali or alkaline earth metals, yet the transition elements are much less reactive. How do you account for it?

Solution. (*a*) The structural formula of the two isomers are

$$[(NH_3)_5\,Co-O-N=O\,]^{2+} \quad \text{and} \quad \left[(NH_3)_5\,Co-N\!\begin{smallmatrix}\nearrow O\\ \searrow O\end{smallmatrix}\right]^{2+}$$

These are linkage isomers.

(*b*) Transition metals are less reactive than alkali and alkaline earth metals because of higher ionisation energies.

Example 6. Give the names and illustrate all types of isomers that are possible in an octahedral complex compound of one cobalt (III) ion, two en molecules, two chloride ions and one nitrate ion.

Solution. The complex is $[Co(en)_2\,C1_2\,NO_2]$. It can exhibit the following isomerisms:

(*i*) *Linkage isomerism.* When NO_2^- can link through oxygen or nitrogen giving nitrite and nitrito compounds respectively.

(*ii*) *Ionisation isomerism*

$[Co(en)_2\,Cl_2]^+\,NO_2^-$

$[Co(en)_2\,Cl\,NO_2]^+\,Cl^-$

(*iii*) *Geometrical isomerism*

$[Co\,(en)_2\,Cl_2]^+\,NO_2^-$ will exhibit *cis* and *trans* geometrical isomerism.

$[Co\,(en)_2\,C1NO_2]^+\,Cl^-$ will also exhibit *cis* and *trans* isomerism.

Both the *cis* as well as *trans* isomers will then show optical *isomerism.*

4.6. SIDGWICK THEORY (EFFECTIVE ATOMIC NO: EAN CONCEPT) OF COORDINATION COMPOUNDS

Nevil Sidgwick

Although Werner's theory could, to a large extent, explain the formation of coordination compounds, it lacked theoretical justification. Attempts were made to find the basis of bond formation in coordination compounds. The first attempt in this direction was made by Sidgwick.

Sidgwick in 1927 explained the bonding in coordination compounds. He considered the formation of complexes by donation of an electron pair from ligand to metal atom or ion. He suggested that the metal ion will continue to accept electron pair till it achieves the next inert gas configuration. This is known as the *effective atomic number rule.* The total no. of electrons acquired by the metal is called effective atomic number (EAN) of metal. This can be illustrated by calculating the EAN of cobalt (III) in $[Co(NH_3)_6]_2^{3+}$ -hexammine cobalt (III) ion. Atomic number of cobalt = 27, it has 27 electrons. Number of electrons in Co (III) or $Co^{3+} = 27 - 3 = 24$. Number of electrons contributed by $6NH_3$ molecules $= 6 \times 2 = 12$.

$\therefore$ EAN of Co^{3+} in the complex $[Co(NH_3)_6]^+ = 24 + 12 = 36$ which is the electronic configuration of next inert gas, *i.e.*, krypton.

Thus according to Sidgwick, this will be a stable complex. Though in many cases the EAN is the same as atomic number of next inert gas (Table 4.2), yet it is not always so. In few cases it may be either one or two units less or more than the inert gas.

Thus EAN rule permits coordination of large number of ligands to the metal ion in a coordination compound. It has, only qualitative significance. However this theory does not predict the geometry of the molecule and fails to explain stability after the accumulation of negative charge from ligands.

Table 4.2. Complexes obeying effective atomic number rule.

Complex	Central metal ion	At. No.	No. of electrons in metal ion	Electrons gained from ligands	EAN
$[Fe(CN)_6]^{4-}$	Fe^{2+}	26	24	12	36 (Kr)
$[Fe(CO)_5]$	Fe	26	26	10	36 (Kr)
$[Cr(CO)_6]$	Cr	24	24	12	36 (Kr)
$[Ni(CO)_4]$	Ni	28	28	8	36 (Kr)
$[Cu(CN)_4]^{3-}$	Cu^+	29	28	8	36 (Kr)
$[Mo(CO)_6]$	Mo	42	42	12	54 (Xe)
$[Pd(NH_3)_6]^{4+}$	Pd^{4+}	46	42	12	54 (Xe)
$[PtCl_6]^{2-}$	Pt^{4+}	78	74	12	86 (Rn)
$[Pd(NH_3)_6]^{4+}$	Pd^{4+}	46	42	12	54 (Xe)

However, there are many exceptions to the EAN rule. Many stable complexes are known which the EAN rule is not obeyed. These are given in Table 4.3.

Table 4.3. Complexes obeying effective atomic number rule.

Complex	Central metal ion	At. No.	No. of electrons in metal ion	No. of electrons from ligands	EAN
$[V(CO)_6]$	V	23	23	12	35
$[Mn(CN)_6]^{4-}$	Mn^{2+}	25	23	12	35
$[Ni(CN)_4]^{2-}$	Ni^{2+}	28	26	8	34
$[Ni(NH_3)_6]^{2+}$	Ni^{2+}	28	26	12	38
$[Pt(NH_3)_4]^{2+}$	Pt^{2+}	78	76	8	84
$[Fe(CN)_6]^{3-}$	Fe^{3+}	26	23	12	35
$[Cr(NH_3)_6]^{3+}$	Cr^{3+}	24	21	12	33
$[CoCl_4]^{2-}$	Co^{2+}	27	25	8	35
$[Cu(NH_3)_4]^{2+}$	Cu^{2+}	29	27	8	35
$[Ag(NH_3)_2]^{+}$	Ag^{+}	47	46	4	50
$[PdCl_4]^{2-}$	Pd^{2+}	46	44	8	52

Example 7. Calculate the effective atomic number of the central metal ion in the following complexes and write their IUPAC names.

(*i*) $[Pt(NH_3)_4Cl_2]Cl_2$ **(*ii*) $Na_2[Fe(CN)_5NO]$**

(*iii*) $[Co(en)_2Cl_2]Cl$ **(*iv*) $K[Ag(CN)_2]$**

Solution. (*i*) **$[Pt^{IV}(NH_3)_4Cl_2]Cl_2$**

E.A.N of Pt^{4+} $= (78 - 4) + (2 \times 4) + (2 \times 2)$

$= 74 + 8 + 4 = 86$

IUPAC name : Tetraamminedichloridoplatinum (IV) chloride

(*ii*) $Na_2[Fe^{III}(CN)_5NO]$

E.A.N. of $Fe^{3+} = (26 - 3) + (2 \times 5) + 2$ There are a total of six monodentate ligands.

$= 23 + 10 + 2 = 35$ Each ligand contributes two electrons

IUPAC name : Sodium pentacyanonitrosylferrate (III)

(*iii*) $[Co(en)_2Cl_2]Cl$

E.A.N. of $Co^{3+} = (27 - 3) + (4 \times 2) + (2 \times 2)$ Each (en) contributes 4 electrons

$= 24 + 8 + 4 = 36$ as it is a bidentate ligand.

IUPAC name : Dichloridobis ethylenediammine cobalt (III) chloride

(*iv*) $K(Ag(CN_2))$

E.A.N of Ag+ $= (47 - 1) + 2 \times 2 = 50$

IUPAC name : Potassium dicyanoargentate (I)

4.7. VALENCE-BOND THEORY (VBT) OF COORDINATION COMPOUNDS

The main postulates of this theory are as under:

(1) The metal atom or ion in the complex has a number of empty orbitals which can accommodate electrons donated by the ligands to form coordinate bonds, *e.g.*, the Cr^{3+} ion has $3d$, $4s$, and $4p$ vacant orbitals.

(2) Since the energy of the $(n-1)$ *d*-orbitals are near the energies of *s* and *p* orbitals of the nth orbit, these orbitals are together hybridised to give new equivalent orbitals called hybridized orbitals which take part in bond formation with ligand orbitals.

(3) The hybridized metal orbitals overlap with ligand orbitals to form strong bonds. Greater the extent of overlapping, greater is the stability of the complex.

(4) Each ligand donates a pair of electrons to the central metal ion.

(5) If the complex contains unpaired electrons, the complex is paramagnetic in nature whereas, if it has paired electrons, the complex is diamagnetic.

(6) The non-bonding metal electrons present in the inner orbitals do not take part in chemical bonding.

(7) Under the influence of a strong ligand, the electrons can be forced to pair up against the Hund's rule of maximum multiplicity. Either $(n-1)\,d$ inner orbitals or outer *nd* orbitals may be used for hybridization.

This is illustrated by taking the examples of formation of the following complex ions.

(*a*) Formation of $[Cr(NH_3)_6]^{3+}$ ion. The electronic configurations of only 3*d*, 4*s* and 4*p* outer orbitals ($3d^5 4s^1$) are taken into account.

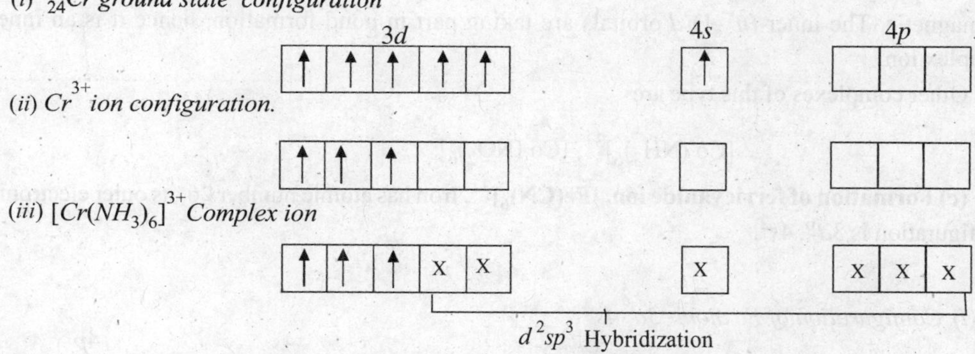

The donated electron pairs from ligands are shown as *crosses.* (×)

The two 3*d*-orbitals, one 4*s* orbital and three 4*p* orbitals are available to form six hybrid orbitals d^2sp^3, which can accept six electron pairs (one electron pair from each nitrogen atom of NH_3) from the ligands. The complex has a octahedral geometry and should be paramagnetic due to the presence of 3-unpaired electrons and actually it is so. Other complexes of Cr with similar inner structure are $[Cr(CN)_6]^{3-}$ and $[Cr(H_2O)_6]^{3+}$. These are inner orbital complexes since inner *d*-orbitals (3*d*) are used in bond formation.

(*b*) Formation of $[Co(CN)_6]^{3-}$ ion. Cobalt (atomic number is 27) has outer electronic configuration as

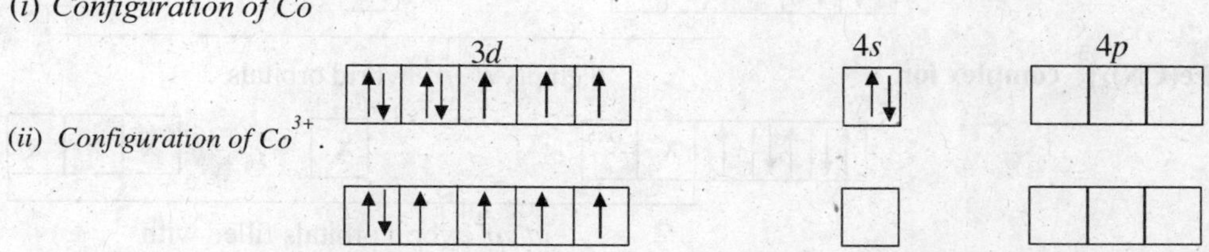

Co^{3+} does not contain any empty 3*d*-orbitals. Now according to VBT, in presence of CN^- (a strong ligand) the electrons in 3*d*-orbital are forced to pair up against the Hund's rule in order to make room for electrons donated by the ligands. Thus after pairing up against Hund's rule, the electronic configuration of Co^{3+} ion becomes as follows:

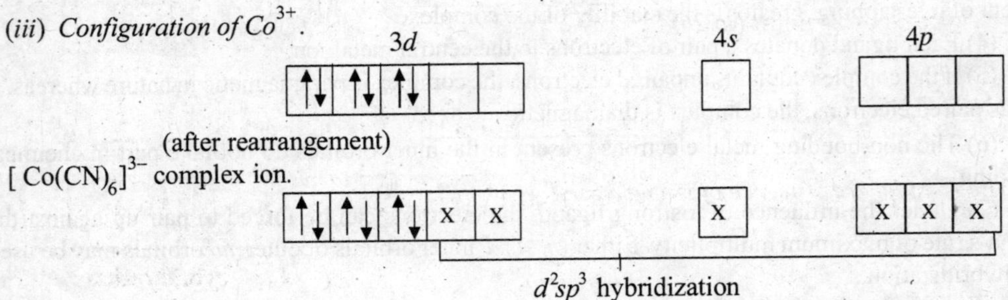

The two $3d$-orbitals, one $4s$ orbital and three $4p$ orbitals are now available to form six hybrid orbitals d^2sp^3, which can accept six electron pairs (one pair from each ligand) from the ligands.

d^2sp^3 hybridization gives octahedral geometry and also it has no unpaired electrons, so it is diamagnetic. The inner $(n-1)$ d orbitals are taking part in bond formation, hence it is an inner complex ion.

Other complexes of this type are

$$[Co(NH_3)_6]^{3+}, [Co(NO_2)_6]^{3-}$$

(*c*) Formation of ferricyanide ion, $[Fe(CN)_6]^{3-}$. Iron has atomic number 26, its outer electronic configuration is $3d^6, 4s^2$.

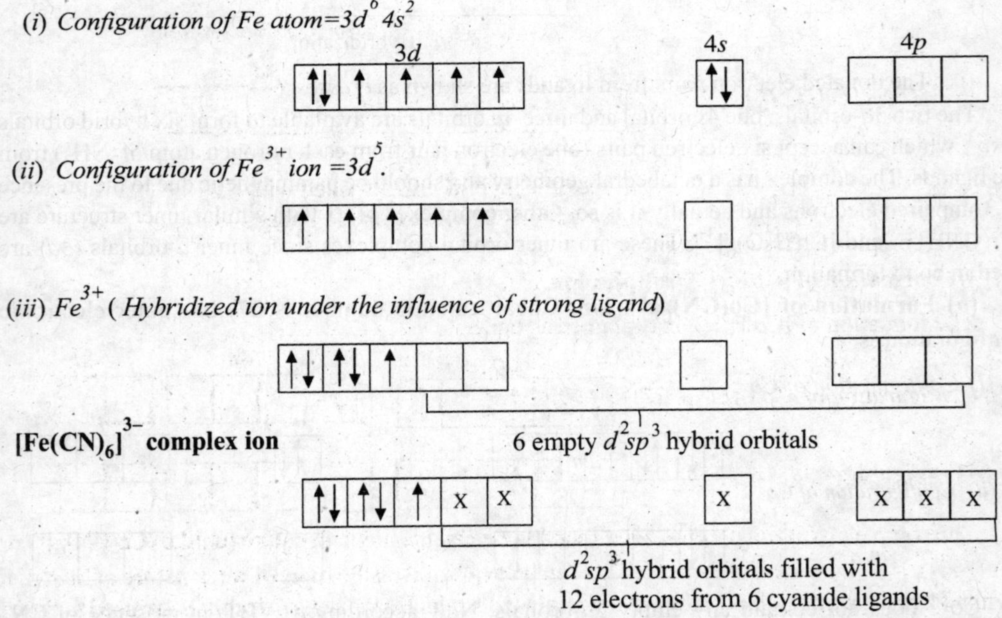

The resulting complex is inner octahedral complex but due to one unpaired electron, it is a little paramagnetic. Experimentally, it has been found to be so.

(*d*) Formation of ferrocyanide ion $[Fe(CN)_6]^{4-}$. This is explained as under:

The resulting complex is inner octahedral complex and it should be diamagnetic due to absence of unpaired electrons and actually it is so.

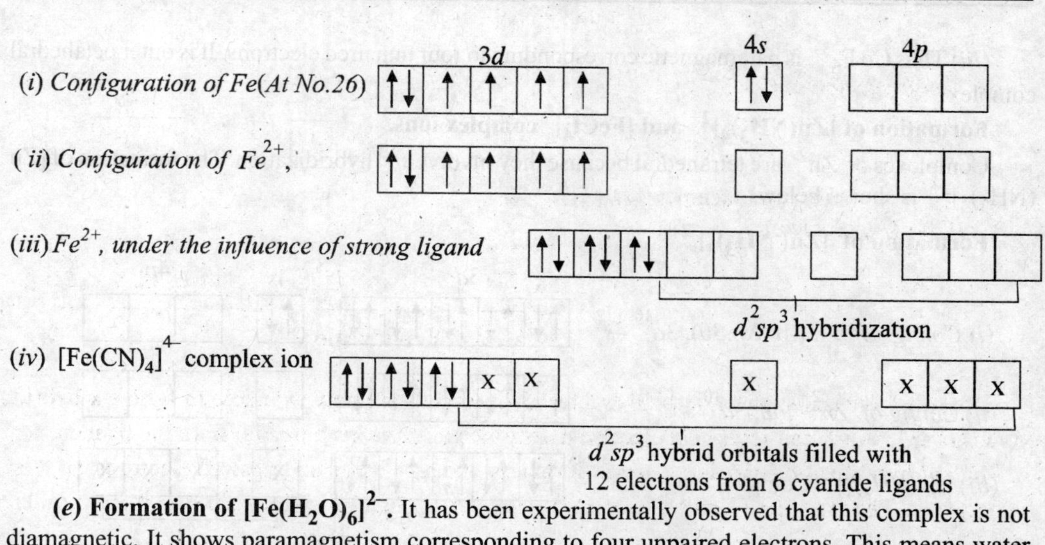

(e) Formation of $[Fe(H_2O)_6]^{2-}$. It has been experimentally observed that this complex is not diamagnetic. It shows paramagnetism corresponding to four unpaired electrons. This means water molecule is too weak a ligand to effect pairing of electrons in 3*d* orbitals. In such cases, outer *d* orbitals, 4*d* in this case, are used for bond formation with ligands. Thus configuration will be as under in this case:

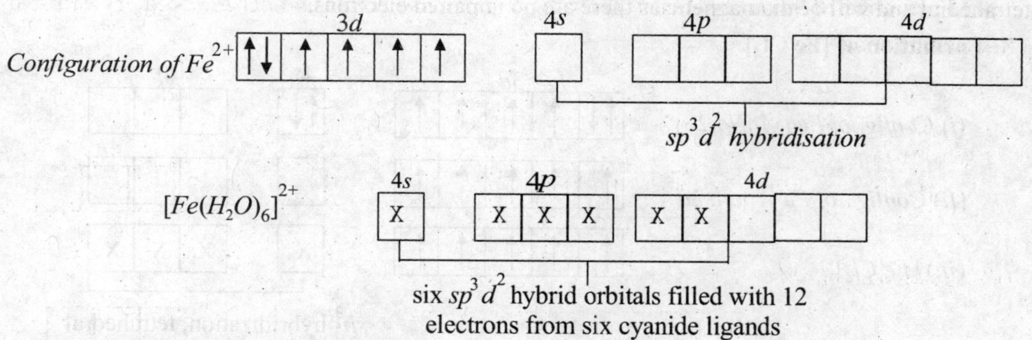

This complex is thus outer complex.

(f) Formation of $[CoF_6]^{3-}$ complex ion.

The formation of $[Co\ F_6]^{3-}$ is explained as under:

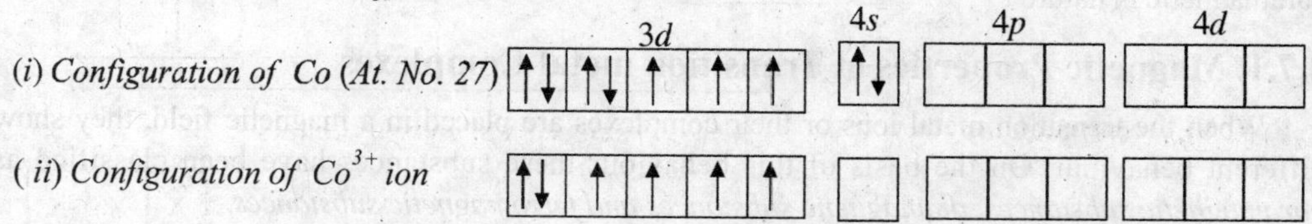

The complex $[(Co\ (Fe_6)]^{3-}$ has been found to be paramagnetic in nature (unlike $[Co(CN)_6]^{3-}$ or $[Co(NH_3)_6]^{3+}$ which are diamagnetic. This can be explained on the basis of weak nature of ligand. It cannot force the pairing up of electrons against Hund's rule. Hence outer 4*d* orbitals are used for bond formation with ligands.

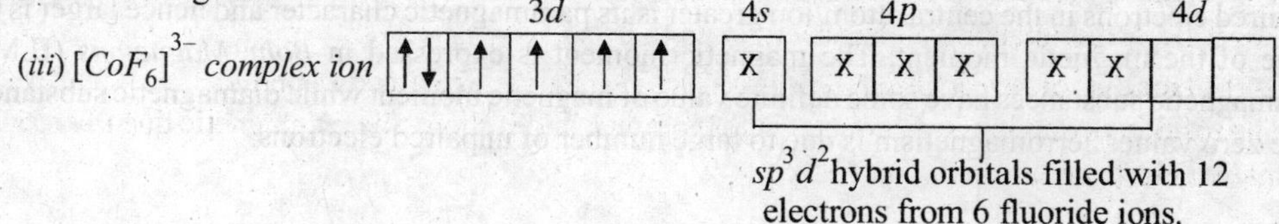

(*iii*) Thus CoF_6^{3-} is paramagnetic corresponding to four unpaired electrons. It is outer octahedral complex.

Formation of $[Zn(NH_3)_4]^{2+}$ and $[FeCl_4]^-$ complex ions.

Complexes of Zn^{2+} are tetrahedral because they involve sp^3 hybridization. The formation of $[Zn(NH_3)_4]^{2+}$ is shown below.

Formation of $[Zn(NH_3)_4]^{2+}$

3d 4s 4p

(*i*) *Config. Of Zn (At.No.30), $3d^{10}4s^2$*

(*ii*) *Config.of Zn^{2+} ion, $3d^{10}$*

(*iii*) *$[Zn(NH_3)_4]^{2+}$* Complex ion

sp^3 hybrid orbitals filled with 8 electrons from 4 NH_3 ligands.

In (*ii*) above the 3*d* orbitals are fully occupied, therefore they cannot participate in hybridization. Four sp^3 hybrid orbitals can be formed as indicated. The resulting $[Zn(NH_3)_4]^{2+}$ complex ion will be tetrahedral and will be diamagnetic as there are no unpaired electrons.

Formation of $[FeCl_4]^-$

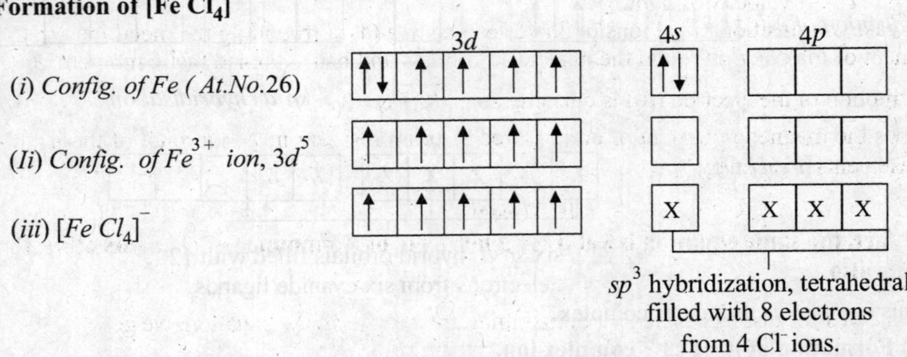

Cl^- is a weak ligand. It cannot force the pairing up of electrons against Hund's rule. Hence it is paramagnetic in nature.

4.7.1. Magnetic Properties of Transition metal Complexes

When the transition metal ions or their complexes are placed in a magnetic field, they show different behaviour. On the basis of this behaviour these substances have been classified as *paramagnetic substances, diamagnetic substances and ferromagnetic substances.*

The transition metal complexes whose central atom/ion contains one or more unpaired electrons are paramagnetic while those with no unpaired electrons behave as diamagnetic substances. The paramagnetic character increases with the increase in the number of unpaired electrons. The paramagnetism of a substance is expressed in terms of its magnetic moment (μ). Larger the number of unpaired electrons in the central atom/ion greater is its paramagnetic character and hence larger is the value of the magnetic moment. The magnetic moment is expressed in *Bohr Magnetons* (B.M.). Paramagnetic substances have some definite value of magnetic moment while diamagnetic substances have zero value. Ferromagnetism is due to large number of unpaired electrons.

Formula for calculating the magnetic moment of transition metal complexes

Consider a transition metal complex of 3*d*-elements whose central metal ion contains only unpaired electrons in its 3*d* orbitals. The magnetic moment (μ) of an unpaired electron is equal to the sum of its magnetic moment due to its orbital motion, called *orbital magnetic moment* (μ_l) and that due to its spin motion, called *spin magnetic moment* (μ_s).

We may write

$$\mu = \mu_l + \mu_s$$

But $$\mu_l = \sqrt{l(l+1)}\frac{eh}{4\pi mc}$$

and $$\mu_s = 2\sqrt{s(s+1)}\frac{eh}{4\pi mc}$$

$\therefore$ $$\mu = [\sqrt{l(l+1)} + 2\sqrt{s(s+1)}]\frac{eh}{4\pi mc}$$

$$= [\sqrt{l(l+1)} + 2\sqrt{s(s+1)}]\text{B.M}$$

where B.M. is called Bohr Magneton which is equal to $\frac{eh}{4\pi mc}$. B.M. is the unit of magnetic moment. One B.M. is equal to 9.27×10^{-21} ergs/gauss.

Here e = charge on the electron,
h = Planck's quantum number,
m = mass of the electron and
c = velocity of light.

In case of transition metal ions of 3*d*-series, the ligands surrounding the metal ions restrict the movement of the electron round the nucleus in orbitals and hence the magnetic moment due to the orbital motion of the electron (μ_l) is quenched or $\sqrt{l(l+1)} = 0$.

Thus the magnetic moment of an unpaired electron residing in a 3*d*-orbital of the ions of 3*d*-series elements is given by:

$$\mu = 2\sqrt{s(s+1)}\ \text{B.M.}$$

In fact, the same equation is valid for a large no. of compounds of elements other than 3*d* elements also.

Value of *s* for one electron = $\frac{1}{2}$. Substituting for *s* in the above equation, we get,

$$\mu = 2\sqrt{\frac{1}{2}\left(\frac{1}{2}+1\right)}\ \text{B.M.}$$

With *n* unpaired electrons, $$\mu = 2\sqrt{\frac{n}{2}\left(\frac{n}{2}+1\right)}\ \text{B.M.}$$

$$= \sqrt{n(n+2)}\ \text{B.M.}$$

Since the value of μ is only due to the spin motion of *n* unpaired electrons, it is also called *spin-only magnetic moment* and is represented as μ_s. Thus:

$$\mu = \mu_s = \sqrt{n(n+2)}\ \text{B.M.} \qquad .....(i)$$

Equation (*i*) shows that the magnitude of μ depends on the number of unpaired electrons (n) present in a given species. The value of μ for different number of unpaired electrons can be calculated with the help of equation (*i*) as shown below:

For $n = 0$ $\quad \mu_s = \sqrt{0(0+2)} = 0.0$ B. M.

For $n = 1$ $\quad \mu_s = \sqrt{1(1+2)} = \sqrt{3} = 1.73$ B.M.

For $n = 2$ $\quad \mu_s = \sqrt{2(2+2)} = \sqrt{8} = $ B.M
For $n = 3$ $\quad \mu_s = 3.87$ B.M.
For $n = 4$ $\quad \mu_s = 4.90$ B.M.
For $n = 5$ $\quad \mu_s = 5.92$ B.M.

High spin and low spin complexes

Complexes which are formed by using inner *d* orbitals such as $[Co(CN)_6]^{3-}$ and $[Fe(CN)_6]^{3-}$ have smaller number of unpaired electrons and are low spin complexes. Complexes such as $[Fe(H_2O)_6]^{2+}$ which are formed by using outer *d* orbitals have a large number of unpaired electrons and have high spin. Thus outer complexes are high spin complexes.

Example 8. Discuss the formation of $[Ni(CN)_4]^{2-}$ complex ion.

Solution. Hybridization of one '*d*' one '*s*' and two '*p*' orbitals gives four equivalent hybrid orbitals dsp^2 which are planar and mutually at right angles pointing towards the corners of a square.

The nickelocyanide ion or tetracyanonickelate (II) ion $[Ni(CN)_4]^{2-}$ formation can be represented as follows:

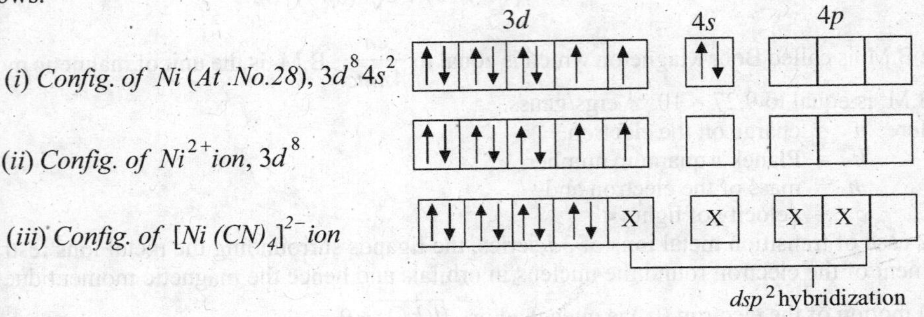

In this complex, the ligand is strong enough to force the two unpaired electrons to pair up. The result is that one 3*d* orbital becomes available for hybridization with 4*s* and two 4*p* orbitals to give four dsp^2 hybrid orbitals. These orbitals accept an electron pair from each of the four CN^- ligands. The shape of complex will be square planar and it will be diamagnetic as there is no unpaired electron in the complex.

4.7.2. Limitations of the valence bond theory of coordination compounds

(1) The basic defect in the theory is that attention is focussed only on the orbitals of central atom and the importance of the ligands are not properly stressed.

(2) It does not explain the relative energies of different structures.

(3) It offers no explanation of the colour observed for complex ion.

(4) It cannot predict whether a 4-coordinate complex will be tetrahedral or square planar. Consider, for example, a Cu (II) complex $[Cu(NH_3)_4]^{2+}$. The VBT predicts the complex to be tetrahedral.

In order to explain the experimentally determined square planar geometrically of Cu (II) complex it is assumed that one electron from the 3*d* orbital is promoted to the 4*p* orbital to vacate a *d*–orbital for accommodating a pair of electrons from the ligand.

(5) It does not take into account the splitting of *d*-energy levels.

(6) For the same metal in some complexes electrons pair up against the Hund's rule while in others the electronic configuration is not disturbed. VBT does not offer any strong explanation for their anomaly.

(7) It does not explain the spectra of complexes.

(8) The VBT does not provide any satisfactory explanation for the existence of inner orbital and outer orbital complexes.

(9) It cannot explain why certain complexes are more labile than others. Labile complexes are those in which one ligand can be easily displaced by another ligand. On the other hand, inert complexes are those in which displacement of ligand is slow.

4.7.3. Shape of *d* orbitals

There are five *d*-orbitals viz., d_{xy}, d_{yz}, d_{zx}, $d_{x^2-y^2}$ and d_{z^2}. Spatial orientation of *d*-orbitals are shown in Fig. 21.1. All the five *d*-orbials are not identical and there are in fact two sets t_{2g} and e_g. The t_{2g} orbitals point in between the axes *x*, *y* and *z* and e_g orbitals point along the axes (Fig. 4.1). Alternate names for t_{2g} and e_g are d_ε (called as *d*-epsilon, a Greek alphabet) and d_γ (called as *d*-gamma, a Greek alphabet) respectively.

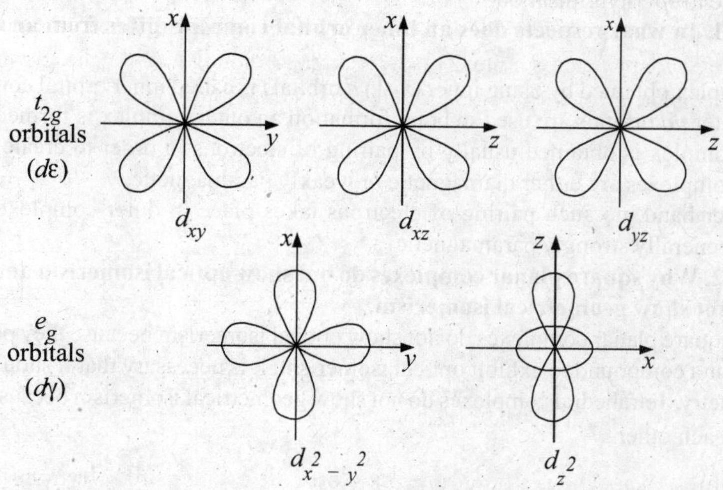

Fig. 4.1. Spatial orientation of *d*-orbitals.

Example 9. A square planar complex of divalent metal outer electronic configuration $ns^1(n-1)\,d^{10}$ is not oxidised. Can you explain its formation of the basis on V.B. theory?

Solution. $ns^1(n-1)\,d^{10}$ refers to Cu. The divalent metal ion will have configuration d^9. X-ray studies of the complex $[Cu(NH_3)_4]^{2+}$ show that it has square planar geometry. Thus it involves dsp^2 (or sp^2d) hybridisation. We could shift one election from 3*d* orbital to 4*p* orbital and then explain the square planar geometry by hybridisation of one 3*d*, one 4*s* and two 4*p* orbitals. One electron left in 4*p* orbital should be removed on oxidation. But that does not happen.

$[Cu(NH_3)_4]^{2+}$

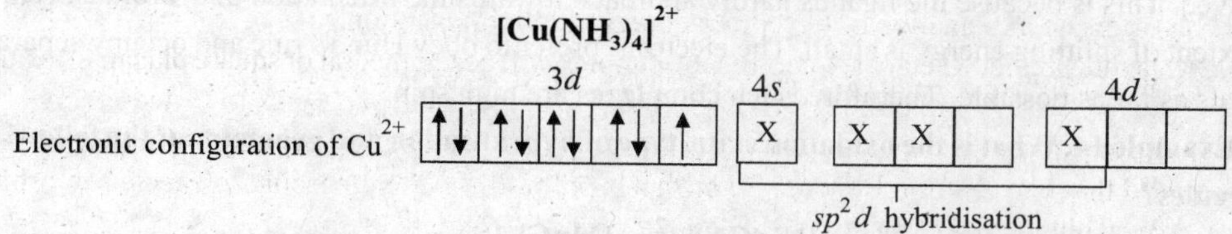

Formation of $[Cu(NH_3)_6]^{2+}$ is explained by assuming that one unpaired electron remains in 3*d*. Hybridisation of one 4*s*, two 4*p* and one 4*d* orbitals takes place (sp^2d) to give square planar shape. As 3*d* electron possesses higher energy, oxidation of $[Cu(NH_3)_4]^{2+}$ or removal of 3*d* electron does not take place.

This is how the shape of the molecule is explained.

Example 10. Explain why $Ni(CO)_4$ is diamagnetic but tetrahedral.

Solution. The oxidation state of Ni in $Ni(CO)_4$ is zero. The electronic configuration of Ni atom is

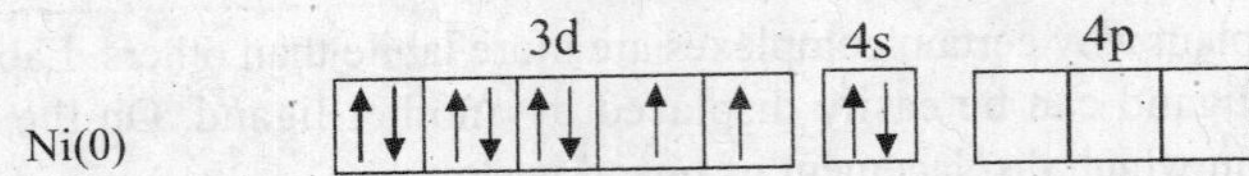

However 4*s* electrons are shifted 3*d* and sp^3 hybridisation takes place as follows.

3*d* 4*s* 4*p*

$Ni(CO)_4$ ↑↓ ↑↓ ↑↓ ↑↓ ↑↓ X X X X

sp^2d hybridisation

Thus there are no unpaired electrons and the complex is diamagnetic. The shape of the complex is tetrahedral due to sp^3 hybridisation.

Example 11. In what respects does an inner orbital complex differ from an outer orbital complex?

Ans. A complex obtained by using inner (*n*–1) *d* orbitals is called inner orbital complex. On the other hand if outer *nd* orbitals are used in bond formation an outer complex is formed.

An inner complex is obtained usually by pairing of electrons in order to create some (*n*–1)*d* orbitals. Such complexes are either diamagnetic or weakly paramagnetic.

On the other hand, no such pairing of electrons takes place in outer complexes. Thus such complexes are generally strongly paramagnetic.

Example 12. Why square planar complexes do not show optical isomerism and tetrahedral complexes do not show geometrical isomerism.

Solution. Square planar complexes do not show optical isomerism because they possess a plane of symmetry. For a compound to exhibit optical isomerism, it is necessary that it should not possess a plane of symmetry. Tetrahedral complexes do not show geometrical isomerism because all positions are adjacent to each other.

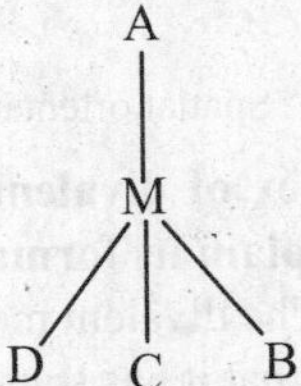

Example 13. Tetrahedral complexes are generally high spin. Why?

Solution. Tetrahedral complexes have generally high spin even when strong field ligands are involved. This is because the ligands hardly approach towards the orientation of *d*-orbitals. Hence the extent of splitting energy is small. The electrons prefer to obey Hund's rule and occupy separate orbitals as far as possible. Therefore, such complexes are high spin.

Example 14. What is the oxidation state, type of hybridisation and geometry of the following molecules?

$$[Ni\ (CN)_4]^{2-},\ [MnCl_4]^{2-}$$

Solution. $[Ni\ (CN)_4]^{2-}$

Oxidation state: $x - 4 = -2$ or $x = 2$

Hybridisation: sp^2d

Geometry: Square planar.

$[MnCl_4]^{-2}$

Oxidation state $x - 4 = 2$ or $x = 2$

Hybridisation: sp^3

Geometry: Tetrahedral.

Example 15. Explain the given magnetic moment of the following compound

Compound	*Magnetic moment (B M.)*
$K_3 [Fe F_6]$	5.9
$[Co (NH_3)_6] Cl_3$	0.0
$[Ni (H_2O)_6] Cl_2$	2.9
$K_4 [Mn (CN)_6]$	1.9
$K_4 [Mn F_6]$	5.9

Solution. K_3 [Fe F_6]. Fluoride ion is a weak field high spin ligand. Iron has oxidation state of +3. Fe^{3+} has five unpaired electrons in 3*d* orbitals.

$\therefore$ Magnetic moment = $\sqrt{n \times (n+2)}$ or $\sqrt{35} = 5.9$

[Co (NH_3)$_6$] Cl_3. It is a strong field, low spin complex. Cobalt has oxidation state +3. Co^{3+} has electrons in *d* orbitals which get paired due to strong ligand leaving no unpaired electron. Therefore, magnetic moment = $\sqrt{0 \times 0} = 0$

[Ni (H_2O)$_6$] Cl_2. Nickel is in +2 oxidation state. Ni^{2+} has 8 electrons in 3*d* orbitals distributed as follows:

↑↓	↑↓	↑↓	↑	↑

Water being a weak ligand cannot force pairing of electrons. Hence there are two unpaired electrons.

Magnetic moment = $\sqrt{2 \times 4} = 2.9$ **B.M.**

K_4 [Mn (CN)$_6$]. Mn is in +2 oxidation state but cyanide is a strong ligand. Mn^{2+} has 5 electrons in 3*d* orbitals which get paired leaving only one unpaired electron.

Hence, magnetic moment = $\sqrt{1 \times 3} = 1.9$ **B.M.**

K_4 [Mn F_6]. Mn^{2+} has 5 electrons in 3*d* orbital, but fluoride being weak ligand, pairing of electrons does not take place.

Therefore, magnetic moment = $\sqrt{5 \times 7} = 5.9$ **B.M.**

4.8. STABILITY CONSTANT OF A COMPLEX ION

Complex ions dissociate, in aqueous solutions to establish an equilibrium with the central metal ion and the ligands from which they are derived. Thus, in solution containing silver-ammonia complex ion, $[Ag(NH_3)_2]^+$, the following equilibrium exists :

$$[Ag(NH_3)_2]^+ \rightleftharpoons Ag^+ + 2NH_3$$

The equilibrium constant, *K,* given by the following expression,

$$K = \frac{[Ag^+][NH_3]^2}{[Ag(NH_3)_2]^+}$$

is 6×10^{-8}. This shows that the degree of dissociation of the complex ion is extremely small. In other words, the complex ion is highly stable.

The above equilibrium constant is called the *dissociation constant* of the complex ion. Smaller the value of the dissociation constant, greater will be the stability of the complex. The *reciprocal of dissociation constant* is **stability constant** of the complex ion. Greater the value of the stability constant, greater will be the stability of the complex ion.

The stability constant of the complex ion $[Ag(NH_3)_2]^+$ is the equilibrium constant of the reverse reaction, $Ag^+ + 2NH_3 \rightleftharpoons [Ag(NH_3)_2]^+$.

Thus,

$$K' = \frac{1}{K} = \frac{[Ag(NH_3)_2]^+}{[Ag^+][NH_3]^2}$$

The stability constant K' of the complex ion $[Ag(NH_3)_2]^+$ is $1/(6 \times 10^{-8}) = 1.67 \times 10^7$.

4.8.1. Factors which effect the stability of a complex ion

Main factors that affect the stability of a complex are discussed as under:

1. The nature of the central ion. In general, *greater the charge density on the central ion, greater is the stability of its complexes.* In other words, *greater the charge and the smaller the size of an ion, i.e.,* the larger the charge/radius ratio of an ion, *greater is the stability of its complexes.* For instance, Fe^{3+} ion carries a higher charge and has a smaller size than Fe^{2+} ion. Hence, charge density is higher on Fe^{3+} than on Fe^{2+} ion. The complexes of Fe^{3+} ion thus show greater stability. This is illustrated by the fact that while stability constant of $[Fe(CN)_6]^{3-}$ is 10^{31}, that of $[Fe(CN)_6]^{4-}$ is only about 10^6.

2. Electronegativity. Electronegativity of the central ion also influences the stability of its complexes. *Greater the electronegativity of the central ion, greater is the stability of its complexes.* This is because the bonding between a central ion and a ligand takes place due to the donation of electron pairs by the ligands. Hence, *a strong electron-attracting central ion will give stable complexes.*

3. The nature of the ligand. In general, *more basic a ligand, greater is the ease with which it can donate its lone pairs of electrons and, therefore, the greater is the stability of the bonds formed by it.* Thus, CN^- and F^- ions and NH_3 molecules, which are strong bases, are good ligands to form stable complexes.

For anionic ligands, *higher the charge and smaller the size, more stable is the complex formed.* Thus, F^- ion gives more stable complexes than Cl^- ion.

4. The chelate effect. The complexes containing chelate rings are found to be *more stable* than those containing similar but unidentate ligands. This effect is known as **chelate effect.**

Consider the formation of complexes from the hydrated cadmium ion $[Cd(H_2O)_4]^{2+}$ with methylamine and ethylenediamine (en) as the ligands, in aqueous medium. The reactions can be represented as under:

$$[Cd(H_2O)_4]^{2+} + 4CH_3NH_2 \rightleftharpoons [Cd(CH_2NH_2)_4]^{2+} + 4H_2O \quad .....(1)$$

$$[Cd(H_2O)_4]^{2+} + 2(en) \rightleftharpoons [Cd(en)_2]^{2+} + 4H_2O \quad .....(2)$$

The complex formed by ethylenediamine is a chelate, as shown below :

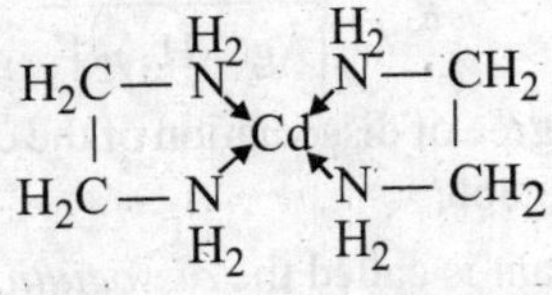

```
      H2    H2
  H2C— N    N— CH2
   |    ↘  ↙   |
   |     Cd    |
   |    ↗  ↖   |
  H2C— N    N— CH2
      H2    H2
```

The stability constants of the complexes with CH_3NH_2 and ethylene diamine are known to be

3.10×10^6 and 3.2×10^{10}, respectively. Evidently, the chelate is 1000 times more stable than the other complex. This is explained by increase in entropy in the case of complex formation with ethylenediamine.

The value of $\Delta H°$ for both the above reactions are known to be practically the same, *viz.*, about – 57 kJ. Thermodynamically, the stability of a specie depends upon the value of $\Delta G°$ which should be large and negative. $\Delta G°$, in turn, is equal to $\Delta H° - T\Delta S°$. The greater stability of $[Cd(en)_2]^{2+}$ complex than that of $[Cd(CH_3NH_2)_4]^{2+}$ complex should be due to its greater positive value of $\Delta S°$ since $\Delta H°$ is the same for both the reactions.

In the first reaction, 4 molecules of CH_3NH_2 displace 4 molecules of H_2O coordinated to Cd^{2+} ion to form the complex ion $[Cd(CH_3NH_2)_4]^{2+}$ so that the number of molecules on the products side is the *same* as the number of molecules on the reactants side. As a consequence, *there is very little change in the extent of randomness in going from reactants to products* ; that is, $\Delta S°$ *for the reaction is nearly zero.* Therefore, for this reaction, the contribution of $\Delta S°$ to $\Delta G°$ is nearly zero so that $\Delta G° \simeq \Delta H° \simeq -57$ kJ. The driving force for reaction (*i*) thus comes mainly from the decrease in the enthalpy of the reaction.

In the second reaction, two molecules of ehtylenediamine displace four molecules of H_2O. There is thus release of *two extra molecules resulting in an increase in the number of molecules on the products side and hence an increase in the disorder or entropy of the system.* $\Delta S°$ is thus positive for this reaction. Consequently, $\Delta G°$ is equal to – 57 kJ – $T\Delta S°$. Evidently, $\Delta G°$ *has a larger negative value. This accounts for the greater stability of the* $[Cd(en)_2]^{2+}$ *complex.*

Note : A summary of shape and magnetic nature of the complexes of Cr, Mn, Fe, Co, Ni and Cu has been provided on the following pages. This will be found as a quick and useful ready reckoner on complexes, by the reader. Electronic configuration of the metal ions, oxidation state of the metal, type of hybridisation and number of unpaired electrons have also been given alongside.

Summary of shape and magnetic nature of complexes of Cr, Mn, Fe, Co, Ni and Cu.

Atom/ion/ complex	Electronic configuration	Oxidation state of metal involved	Type of hypridisation	Shape of the molecule	No.of un-paired, electrons	Magnetic nature
$Cr^{3+}(d^3)$	3d: ↑ ↑ ↑ _ _ ; 4s: _ ; 4p: _ _ _	+3			3,	Paramagnetic
$[Cr(NH_3)_6]^{3+}$	3d: ↑ ↑ ↑ : : ; 4s: : ; 4p: : : : (d^2sp^3)	+3	d^2sp^3 (Inner)	Octahedral	3,	Paramagnetic
$[Cr(H_2O)_6]^{3+}$	3d: ↑ ↑ ↑ _ _ ; 4s: : ; 4p: : : : ; 4d: : : _ _ _ (sp^3d^2)	+3	sp^3d^2 (Outer)	Octahedral	3,	Paramagnetic
$Mn^{2+}(d^5)$	3d: ↑ ↑ ↑ ↑ ↑ ; 4s: _ ; 4p: _ _ _	+2			5,	Paramagnetic
$[Mn(CN)_6]^{4-}$	3d: ↑↓ ↑↓ ↑↓ : : ; 4s: : ; 4p: : : : (d^2sp^3)	+2	d^2sp^3 (Inner)	Octahedral	1,	Paramagnetic
$[MnCl_4]^{2-}$	Rearrangement; 3d: ↑ ↑ ↑ ↑ ↑ ; 4s: : ; 4p: _ : : (sp^3)	+2	sp^3	Tetrahedral	5,	Paramagnetic
$Fe^{2+}(d_6)$	↑↓ ↑ ↑ ↑ ↑ ; _ ; _ _ _	+2			4,	Paramagnetic

$[Fe(CN)_6]^{4-}$	↑↓ ↑↓ ↑↓ : : | : | : : : (Rearrangement; d^2sp^3)	+2	d^2sp^3(Inner)	Octahedral	0	Diamagnetic
$[Fe(H_2O)_6]^{2+}$	↑↓ ↑ ↑ ↑ ↑ | : | : : : | : : (empty empty empty) (sp^3d^2)	+2	sp^3d^2(Outer)	Octahedral	4	Paramagnetic
$[Fe(NH_3)_6]^{2+}$	same as above	+2	sp^3d^2 (Inner)	Octahedral	4	Paramagnetic
$Fe^{3+}(d^5)$	↑ ↑ ↑ ↑ ↑ | (empty) | (empty empty empty)	+3			5	Paramagnetic
$[Fe(CN)_6]^{3-}$	↑↓ ↑↓ ↑ : : | : | : : : (d^2sp^3)	+3	d^2sp^3 (Inner)	Octahedral	1	Paramagnetic
Fe	↑↓ ↑ ↑ ↑ ↑ | ↑↓ | (empty empty empty)	0			4	Paramagnetic
$Fe(CO)_5$	↑ ↑ ↑ ↑ : | : | : : : (dsp^3)	0	dsp^3(Inner)	Trigonal bipyramidal	0	Diamagnetic
$Co^{3+}(d^6)$	↑↓ ↑ ↑ ↑ ↑ | (empty) | (empty empty empty)	+3			4	Paramagnetic

$[CoF_6]^{3-}$	↑↓ ↑ ↑ ↑ ↑	:	: : :	: : _ _ _ (sp^3d^2)	+3	sp^3d^2(Outer)	Octahedral	4	Paramagnetic
$[Co(NH_3)_6]^{3+}$	↑↓ ↑↓ ↑↓ : : Rearrangement	:	: : : (d^2sp^3)		+3	d^2sp^3 (Inner)	Octahedral	0	Diamagnetic
$Co^{2+}(d^7)$	↑↓ ↑↓ ↑ ↑ ↑	_	_ _ _		+2			3	Paramagnetic
$[Co(H_2O)_6]^{2+}$	↑↓ ↑↓ ↑ ↑ ↑	:	: : :	: : _ _ _ (sp^3d^2)	+2	sp^3d^2 (Outer)	Octahedral	3	Paramagnetic
$Ni^{2+}(d^8)$	3*d*: ↑↓ ↑↓ ↑↓ ↑ ↑	3*s*: _	3*p*: _ _ _		+2			2	Paramagnetic
$[NiCl_4]^{2-}$	↑↓ ↑↓ ↑↓ ↑ ↑	:	: : : (sp^3)		+2	sp^3	Tetrahedral	2	Paramagnetic
$[Ni(CN)_4]^{2+}$	↑ ↑ ↑ ↑ : Rearrangement	:	: : _ (dsp^2)		+2	dsp^2	Square Planer	0	Diamagnetic
Ni	↑↓ ↑↓ ↑↓ ↑ ↑	↑↓	_ _ _		0			2	Paramagnetic

$Ni(CO)_4$	↑↓ ↑↓ ↑↓ ↑↓ ↑↓	:	: : : (Rearrangement; sp^3)	0	sp^3	Tetrahedral	0	Diamagnetic
$[Ni(NH_3)_6]^{2+}$	3d: ↑↓ ↑↓ ↑↓ ↑ ↑	3s: :	3p: : : :	4d: : : _ _ _ (sp^3d^2)	+2	sp^3d^2 (Outer)	Octahedral	2	Paramagnetic
$Cu^{2+}(d^9)$	↑↓ ↑↓ ↑↓ ↑↓ ↑	_	_ _ _	+2			1	Paramagnetic
$[CuCl_4]^{2-}$	↑↓ ↑↓ ↑↓ ↑↓ ↑	:	: : : (sp^3)	+2	sp^3	Tetrahedral	1	Paramagnetic
$[Cu(NH_3)_4]^{2+}$	↑↓ ↑↓ ↑↓ ↑ :	:	: : ↑ (dsp^2) one electron is shifted from 3*d*- to 4*p*- orbtial.	+2	dsp^2	Square planer	1	Paramagnetic

SOLVED PROBLEMS
(Drawn from Different University Papers)

Q.1. The molar conductance of $PtCl_4.4NH_3$ in water corresponds to a total of three ions. Write its structure on the basis of Werner's theory.

Ans. Platinum exhibits a primary valency of four and secondary valency of six. In $PtCl_4.4NH_3$, the four secondary valencies are satisfied by four NH_3 molecules and two Cl^- ions. The remaining two Cl^- ions satisfy only primary valency. Therefore, the complex may be represented as:

$[Pt(NH_3)_4Cl_2]Cl_2$

It will ionise in water as:

$$[Pt(NH_3)_4Cl_2]Cl_2 \rightleftharpoons [Pt\,(NH_3)_4Cl_2]^{2+} + 2Cl^-$$

Thus, the complex ionises in solution giving three ions.

Q.2. (*a*) How many chloride ions will be precipitated in the following complexes:

(*i*) $CoCl_3 . 5NH_3$ **(*ii*) $PtCl_4 . 3NH_3$**
(*iii*) $PtCl_4 . 6NH_3$ **(*iv*) $CoCl_3 . 3NH_3$**
(*v*) $PtCl_4 . 2NH_3$

(*b*) Write the structural formula of the following complexes and predict the number of ions per mole of the complex :

(*i*) $CoCl_3 . 4NH_3$ **(*ii*) $PtCl_4 . 5NH_3$**

Ans. (*a*) The number of Cl^- ions which will be precipitated :

(*i*) $CoCl_3 . 5NH_3$ or $[Co\,(NH_3)_5\,Cl]Cl_2$: 2 Cl^- ions
(*ii*) $PtCl_4 . 3NH_3$ or $[Pt\,(NH_3)_3\,Cl_3]Cl$: 1 Cl^- ion
(*iii*) $PtCl_4 . 6NH_3$ or $[Pt\,(NH_3)_6]\,Cl_4$: 4 Cl^- ions
(*iv*) $CoCl_3 . 3NH_3$ or $[Co\,(NH_3)_3\,Cl_3]$: no Cl^- ion
(*v*) $PtCl_4 . 2NH_3$ or $[Pt\,(NH_3)_2\,Cl_4]$: no Cl^- ion

(*b*) Number of ions per mole of complex :

(*i*) $CoCl_3 . 4NH_3$. It may be formulated as :

$$[Co\,(NH_3)_4\,Cl_2]\,Cl \rightleftharpoons [Co\,(NH_3)_4\,Cl_2]^+ + Cl^-$$

Two ions (one complex ion and one simple ion)

(*ii*) $PtCl_4 . 5NH_3$. It may be formulated as :

$$[Pt(NH_3)_5\,Cl]\,Cl_3 \rightleftharpoons [Pt(NH_3)_5\,Cl]^{3+} + 3Cl^-$$

Four ions (one complex ion and three simple ions)

Q.3. Name the type of isomerism exhibited by the following isomers :

(*a*) $[Co(NH_3)_5(ONO)]Cl_2$ — $[Co(NH_3)_5NO_2]Cl_2$
(*b*) $[Pt(NH_3)_4][PtCl_6]$ — $[Pt(NH_3)_4Cl_2][PtCl_4]$
(*c*) $[Co(en)_2Cl_2]Cl.H_2O$ — $[Co(en)_2(H_2O)Cl]Cl_2$
(*d*) $[Co(NH_3)_5SO_4]Br$ — $[Co(NH_3)_5Br]SO_4$
(*e*) $[Co(NH_3)_6][Cr(CN)_6]$ — $[Cr(NH_3)_6][Co(CN)_6]$

Ans. The types of isomers exhibited by the pairs of compounds are :

(*a*) Linkage isomerism
(*b*) Coordination isomerism

(*c*) Hydrate isomerism (*d*) Ionisation isomerism

(*e*) Coordination isomerism

Q.4. The complex $Co(NH_3)_5\ CO_3.Cl$ has two ionisation isomers. Write their structural formulae and give their IUPAC names.

Ans. The two ionisation isomers are :

$[Co(NH_3)_5CO_3]Cl$ Pentaamminecarbonatocobalt(III) chloride

$[Co(NH_3)_5Cl]CO_3$ Pentaamminechloridocobalt(III) carbonate

Q.5. Write the formula for the complex dichlorobis (ethylenediamine) cobalt(III) ion. Draw its geometrical and optical isomers.

Ans. The formula for the complex: dichlorobis (ethylenediamine) cobalt (III) ion is $[Co(en)_2Cl_2]^+$.

It exists in two geometrical isomers as :

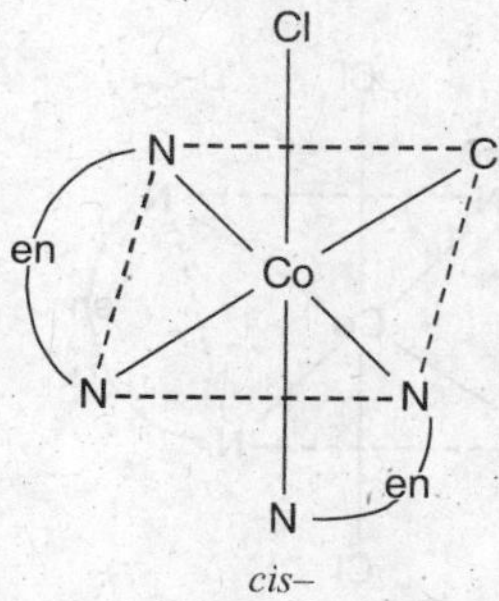

cis–

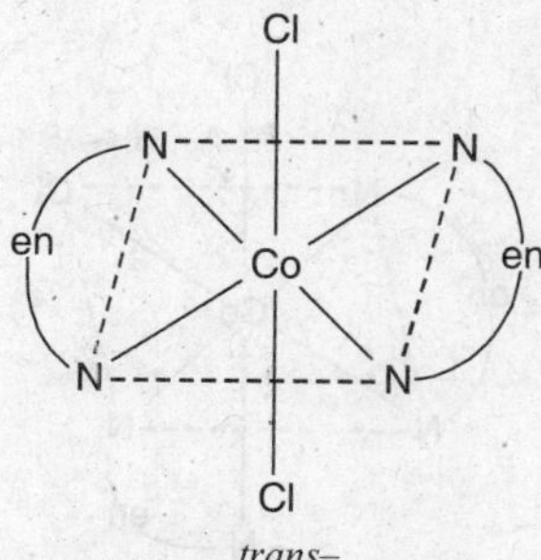

trans–

Out of *cis-* and *trans-* forms, *cis*-form is optically active while *trans*-form is optically inactive. The optical isomers of *cis* form are shown below:

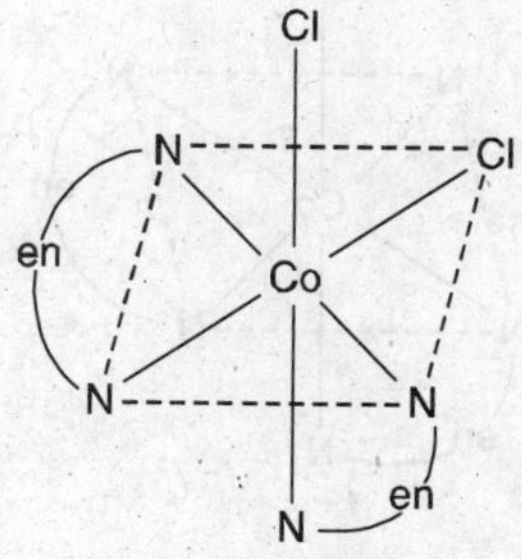

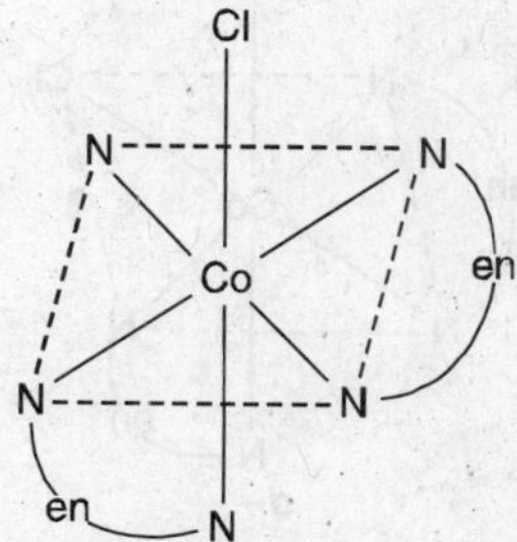

Thus three isomers exist for this complex ion.

Q.6. A coordination compound has the formula $CoCl_3.4NH_3$. It precipitates silver ions as AgCl and its molar conductance corresponds to a total of two ions.

(*a*) Deduce its structural formula and name the complex.

(*b*) Discuss geometrical isomerism exhibited by the complex.

(*c*) Replace ammonia molecules by two molecules of ethylenediamine and discuss isomerism in the resulting complex ion.

Ans. (*a*) On the basis of molar conductance and its reaction with Ag^+ ion, the compound has the molecular formula :

$[Co(NH_3)_4Cl_2]Cl$.

Tetraamminedichloridocobalt (III) chloride

(*b*) It will exhibit geometrical isomerism. The *cis* and *trans* isomers are :

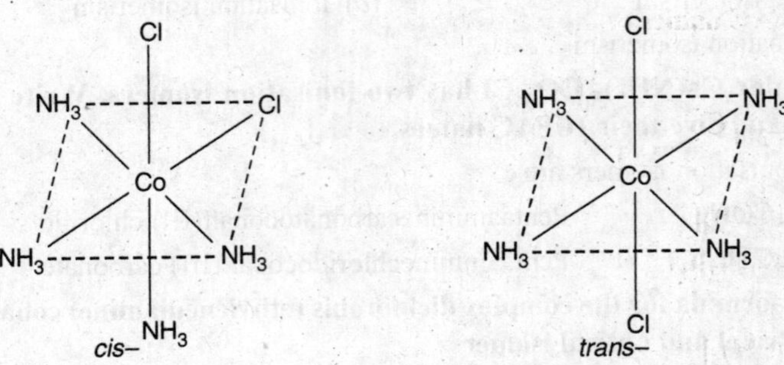

(*c*) Replacing ammonia by ethylene diamine (en) gives $[Co(en)_2\,Cl_2]^+\,Cl^-$

It shows geometrical isomers *cis* and *trans*. Out of these *cis* isomer shows optical isomerism as well.

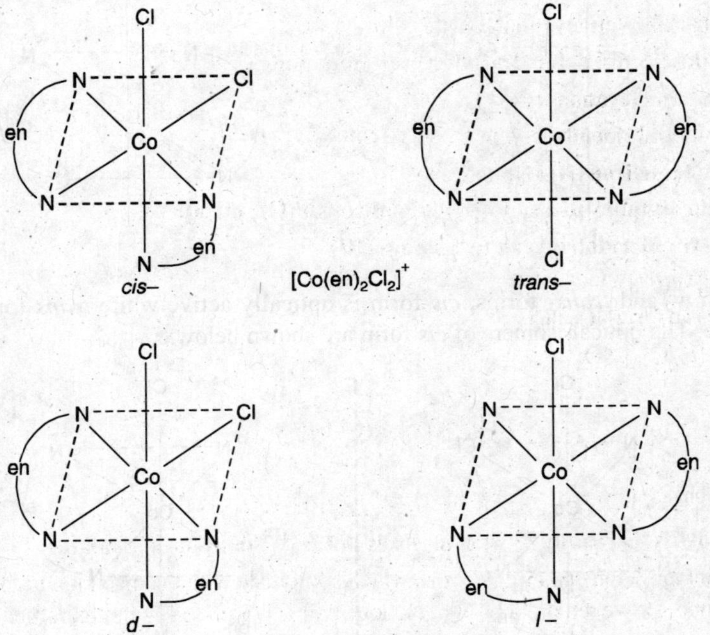

Optical isomers of the *cis* form

Q.7. Predict the number of different types of isomers for the folowing complexes :

(*a*) $Co[EDTA]^-$ **(*b*) $[Pt(NH_3)_3\,Br]NO_2$**

(*c*) $[Cr(en)_2Cl_2]Cl$ **(*d*) $[Cr(ox)_3]^{3-}$**

(*e*) $[Pt(NH_3)_4][PtCl_4]$

Ans. Different types of isomers that exist for the complexes are :

(*a*) $Co[EDTA]^-$: Optical isomers

(*b*) $[Pt(NH_3)_3Br]NO_2$: Ionisation isomers

(*c*) $[Cr(en)_2Cl_2]Cl$:

(*i*) *cis* and *trans.* geometrical isomers (*ii*) *cis* isomer will be optically active.

(*d*) $[Cr(ox)_3]^{3-}$: Optical isomers

(*e*) $[Pt(NH_3)_4][PtCl_4]$: Coordination isomers

Q.8. Write IUPAC names of the following :

(*a*) $\left[(NH_3)_4Co\begin{smallmatrix}NH_2\\ \\ NO_2\end{smallmatrix}Co(NH_3)_4\right](NO_3)_4$

(*b*) $Na[Co(CO)_4]$

(*c*) $[Cr(en)_2Cl_2]Cl$

(*d*) $K[BF_4]$

(*e*) $K_4[Ni(CN)_4]$

Ans. (*a*) Tetraammine cobalt (III) – μ –amido–μ–nitrotetraammine cobalt (III) nitrate

(*b*) Sodium tetra carbonyl cobaltate (–I)

(*c*) Dichloridobis(ethylenediamine) cobalt (III) chloride

(*d*) Potassium tetrafluoridoborate (III)

(*e*) Potassium tetracyanonickelate (0)

Q.9. Write the formulae for the following compounds :

(*a*) Sodium dicyanoaurate (I)

(*b*) Pentaamminenitrocobalt (III) chloride

(*c*) Hexaaquairon (II) sulphate

(*d*) Octa ammine–μ–amido–μ–nitritodicobalt (III) nitrate

(*e*) cis–trichloridotripyridinerhodium (III)

Ans. (*a*) $Na[Au(CN)_2]$

(*b*) $[Co(NH_3)_5(NO_2)]Cl_2$

(*c*) $[Fe(H_2O)_6]SO_4$

(*d*) $\left[(NH_3)_4Co\begin{smallmatrix}NH_2\\ \\ ONO\end{smallmatrix}Co(NH_3)_4\right](NO_3)_4$

(*e*) cis–$[Rh\,Cl_3(Py)_3]$

Q.10. Why does NH_3 readily form complexes but NH_4^+ does not ? Explain.

Ans. NH_3 contains a lone pair of electrons which coordinate with the metal ion to form the complex compound. However, in NH_4^+ ion, the lone pair is bound to H^+ and therefore, is not available for bonding to metal ion. Therefore, NH_4^+ does not form complexes.

Q.11. Draw the structural formula of two isomers of the complex ion $[Co(NH_3)_5NO_2]^{2+}$. Name the type of isomerism and give their IUPAC names.

Ans. Two isomers are :

$[Co(NH_3)_5-(O—N=O)]^{2+}$ and $[Co(NH_3)_5(NO_2)]^{2+}$

Pentaamminenitritocobalt (III) ion   Pentaamminenitrocobalt (III) ion

These are called linkage isomers.

Q.12. What are the oxidation states and coordination numbers of the metal ions in the following complexes :

(*a*) $K_4[Ni(CN)_4]$ (*b*) $[CO(en)_3]_2(SO_4)_3$ (*c*) $K_3[Al(C_2O_4)_3]$

Ans. (*a*) Oxidation state = 0, coordination number = 4

(*b*) Oxidation state = +3, coordination number = 6

(*c*) Oxidation state = +3, coordination number = 6.

Q.13. A compound with the empirical formula $Co(NH_3)_5BrSO_4$ exists in two forms : red and violet. Solution of red gives a precipitate of AgBr on additioon of $AgNO_3$. The violet form gives no precipitate on the addition of $AgNO_3$ but gives a white precipitate on addition of $BaCl_2$ solution. From these observations draw the structure of each compound and explain the observations.

Ans. Two ionization isomers are possible with the given empirical formula

1. Red compound is $[Co(NH_3)_5SO_4]Br$

   It ionises to give Br^- ions in solution and therefore, forms yellow ppt. of AgBr with $AgNO_3$ solution.

$$[Co(NH_3)_5SO_4]\,Br \rightleftharpoons [Co(NH_3)_5SO_4]^+ + Br^-$$

$$Br^- + AgNO_3 \longrightarrow \underset{\text{Yellow ppt.}}{AgBr} + NO_3^-$$

2. Violet compound is $[Co(NH_3)_5Br]SO_4$. It ionises to give SO_4^{2-} ions which give white ppt. with $BaCl_2$ solution.

$$[Co(NH_3)_5\,Br]SO_4 \rightleftharpoons [Co(NH_3)_5Br]^{2+} + SO_4^{2-}$$

$$SO_4^{2-} + BaCl_2 \longrightarrow \underset{\text{White ppt.}}{BaSO_4} + 2Cl^-$$

Q.14. Explain how $[Pt(NH_3)_2Cl_2]$ and $[Pt(NH_3)_6]Cl_4$ differ in their molar conductivities ?

Ans. $[Pt(NH_3)_2Cl_2]$ is non conductor because it is a molecular complex while $[Pt(NH_3)_6]Cl_4$ is a conductor and its conductance is expected to correspond to a total of 5 ions.

Q.15. The complex $CoBr_3.4NH_3.2H_2O$ has molar conductivity corresponding to 3 : 1 electrolyte. Write the molecular formula and its IUPAC name.

Ans. $[Co(NH_3)_4(H_2O)_2]\,Br_3$

IUPAC name : Tetraamminediaquacobalt (III) bromide.

Q.16. Sketch the structures of the following ions :

(*a*) trans $[Co(NH_3)_4Cl_2]^+$ **(*b*) cis $[Pt(en)_2(CN)_2]^{2+}$**

Ans. (*a*)

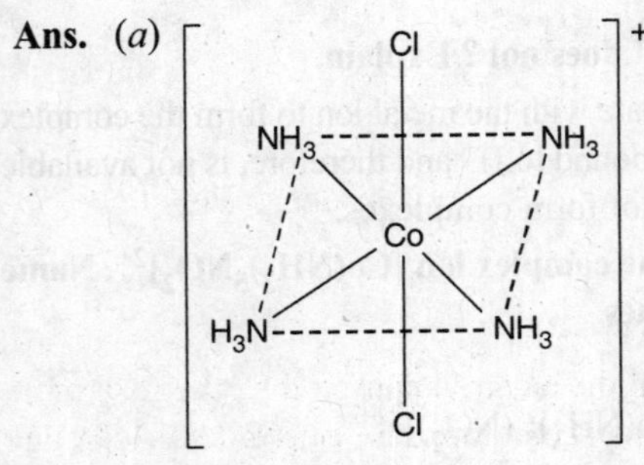

(*b*)

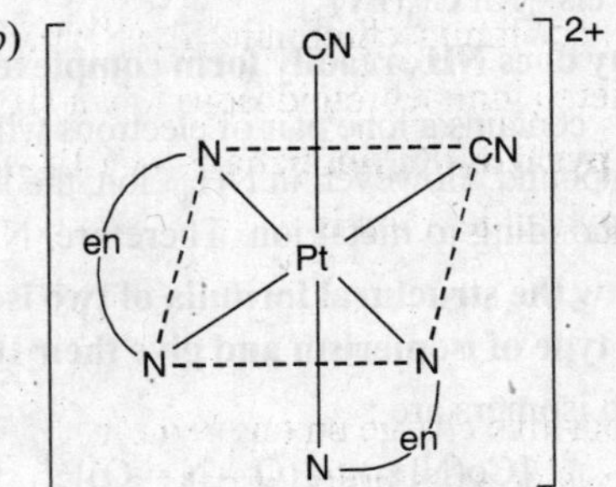

Q.17. Write the following :

(*a*) Linkage isomer of $[Co(NH_3)_5ONO]Cl_2$

(*b*) Polymerisation isomer of $[Co(NH_3)_3(NO_2)_3]$

(*c*) Ionisation isomer of $[Co(NH_3)_5Br]SO_4$

(*d*) Coordination isomer of $[Cu(NH_3)_4][PtCl_4]$

Ans. (*a*) $[Co(NH_3)_5NO_2]Cl_2$ (*b*) $[Co(NH_3)_6][Co(NO_2)_6]$

(*c*) $[Co(NH_3)_5SO_4]Br$ (*d*) $[Pt(NH_3)_4][CuCl_4]$

Q.18. Why is $K_4[Fe(CN)_6]$ not toxic whereas KCN is highly toxic ?

Ans. The compound $K_4[Fe(CN)_6]$ ionises in water as :

$$K_4[Fe(CN)_6] \longrightarrow 4K^+ + [Fe(CN)_6]^{4-}$$

There are no free CN^- in $[Fe(CN)_6]^{4-}$ whereas KCN ionizes as :

$$KCN \longrightarrow K^+ + CN^-$$

Thus free CN^- ions are produced . It is the free CN^- ions which are toxic. As $K_4[Fe(CN)_6]$ does not produce free CN^- ions, it is not toxic.

Q.19. Write IUPAC names of

(*a*) $Hg[Co(CNS)_4]$

(*b*) $[Pt(NH_3)_4(NO_2)Cl]SO_4$

(*c*) $[Pt(py)_4][PtCl_4]$

(*d*) $K[B(NO_3)_4]$

(*e*) $[Co(NH_3)_4(NCS)Cl]NO_3$

(*f*) $K_2[Zn(OH)_4]$

(*g*) $[Cr(NH_3)_6][Co(CN)_6]$

(*h*) $[Co(NH_3)_4(H_2O)Br](NO_3)_2$

Ans. (*a*) Mercury tetrathiocyanatocobaltate (II)

(*b*) Tetraamminechloridonitroplatinum (IV) sulphate

(*c*) Tetrapyridineplatinum (II) tetrachloridoplatinate (II).

(*d*) Potassium tetranitratoborate (III)

(*e*) Tetraamminechloridoisothiocyanatocobalt (III) nitrate

(*f*) Potassium tetrahydroxozincate (II)

(*g*) Hexaamminechromium (III) hexacyanocobaltate (III)

(*h*) Tetraamminebromidoaquacobalt (III) nitrate

Q.20. Can pyrazinium ion behave as a ligand ?

Ans. The pyrazinium ion is :

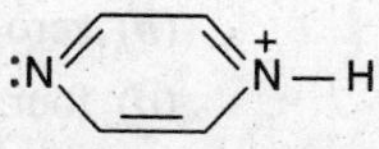

:N(ring)N⁺—H

The positive charge on one N atom is far away from the electron pair on the other N atom. There is no possibility of delocalisation of electrons. Therefore, the electron pair can be easily donated and hence pyrazinium ion can act as a ligand.

Q.21. Write the IUPAC names of the following :

(*a*) $[Pt(NH_3)_4][CoCl_4]$

(*b*) $[Pd(H_2O)_2(ONO)_2I_2]$

(*c*) $VO(acac)_2$

(*d*) $Na_2[Fe(CN)_5NO]$

(*e*) $[Cr(CO)_5\{P(C_6H_5)_3\}]$

Ans. (*a*) Tetraammineplatinum (II) tetrachloridocobaltate (II)
(*b*) Diaquadiiododinitritopalladium (IV)
(*c*) Bis (acetylacetonato) oxovanadium (IV)
(*d*) Sodium pentacyanonitrosylferrate (III)
(*e*) Pentacarbonyltriphenylphosphinechromium (0)

Q.22. Oxalic acid is commonly used to remove rust stains. Justify.

Ans. Rust is hydrated iron (III) oxide, $Fe_2O_3.xH_2O$. Iron (III) forms soluble complex with chelating oxalate ion furnished by oxalic acid.

$$Fe_2O_3 + 6C_2O_4^{2-} + 6H_3O^+ \longrightarrow \underset{\text{Soluble complex}}{[2Fe(C_2O_4)_3]^{3-}} + 9H_2O$$

The soluble complex containing iron (rust) is washed away.

EXERCISES
(Based on Question Papers of Different Universities)

Multiple Choice Questions

Choose the correct option

1. Which of the following does not obey EAN rule ?
   (*a*) $[Cu(CN)_4]^{3-}$ (*b*) $[Pt(NH_3)_4]^{2+}$
   (*c*) $[Pd(NH_3)_6]^{4+}$ (*d*) $[Cr(CO)_6]$.
2. Which of the following exhibits geometrical isomerism ?
   (*a*) $[Cr(en)_2Cl_2]^+$ (*b*) $[Cr(NH_3)_6][Co(CN)_6]$
   (*c*) $[Co(NH_3)_4(NO_2)Cl]^+$ (*d*) $[Co(NH_3)_6]Cl_3$.
3. Which of the following has lowest oxidation state of central atom ?
   (*a*) $Fe_3[Fe(CN)_6]$ (*b*) $Na[Co(CO)_4]$
   (*c*) $Fe(CO)_5$ (*d*) $[Co(en)_3]Cl_3$.
4. The complexes $[Co(NH_3)_5\ NO_2]\ Cl_2$ and $[Co(NH_3)_5(ONO)]Cl_2$ are examples of
   (*a*) geometrical isomers (*b*) coordination isomers
   (*c*) linkage isomers (*d*) position isomers.
5. The number of unpaired electrons in $NiCl_4^{2-}$ (tetrahedral) are
   (*a*) two (*b*) zero
   (*c*) one (*d*) four.
6. The formula of the complex tris (ethylenediamine) cobalt (III) sulphate is :
   (*a*) $[Co(en)_3]SO_4$ (*b*) $[Co(en)_3]_2SO_4$
   (*c*) $[Co(en)_3SO_4]$ (*d*) $[Co(en)_3]_2\ (SO_4)_3$.
7. Which of the following does not show optical isomerism ?
   (*a*) $[Cr(ox)_3]^{3-}$ (*b*) $[Co(en)(NH_3)_2\ Cl_2]^+$
   (*c*) *trans* $[Pt(NH_3)_2Cl_2]$ (*d*) $[Co(EDTA)]^-$
8. The IUPAC name of $Na_3\ [Fe(C_2O_4)_3]$ is :
   (*a*) Sodium trioxalatoferrate (III) (*b*) Sodium trioxalatoiron (III)
   (*c*) Sodium tris(oxalato) ferrate (III) (*d*) Sodium tris(oxalato) iron (III).

9. Which of the following ligands is bidentate ?
   (*a*) EDTA (*b*) Ethylenediamine
   (*c*) Acetato (*d*) Pyridine.

10. The IUPAC name of $K_4[Ni(CN)_4]$ is :
   (*a*) Potassium tetracyanonickel (III) (*b*) Potassium tetracyanonickel (II)
   (*c*) Potassium tetracyanonickelate (0) (*d*) Potassium tetracyanonickelate (IV).

11. The IUPAC name of $K_4[Fe(CN)_6]$ is :
   (a) Potassium hexacyanoferrate (III) (b) Potassium ferrocyanide
   (c) Potassium hexacyanoiron (II) (d) Potassium hexacyanoferrate (II).

12. The number of ions per mole of the complex $CoCl_3 . 5NH_3$ in aqueous solution will be
   (*a*) three (*b*) four
   (*c*) two (*d*) nine.

ANSWERS

1. (*b*)	**2.** (*b*)	**3.** (*b*)	**4.** (*c*)	**5.** (*a*)	**6.** (*d*)
7. (*c*)	**8.** (*a*)	**9.** (*b*)	**10.** (*c*)	**11.** (*d*)	**12.** (*a*)

Short Answer Questions

1. What is Effective Atomic Number rule ?
2. Give one example of an outer orbital complex.
3. What are linkage isomers ? Give one example.
4. Give IUPAC names for the following complexes:
   (*i*) $K[Ag(CN)_2]$ (*ii*) $[Cu(NH_3)_4]SO_4$
5. Define coordination sphere.
6. Differentiate between a ligand and a chelate.
7. Draw the geometrical isomers of triglycinate chromium (III).
8. Sketch possible isomers of $[Co(en)_2Cl_2)]^+$
9. How many chloride ions will be precipitated in the following complexes ?
   (*i*) $CoCl_3 . 5NH_3$ (*ii*) $PtCl_4 . 3NH_3$
   (*iii*) $CoCl_3 . 3NH_3$ (*iv*) $PtCl_4 . 6NH_3$
10. How many geometrical and optical isomers are possible for the complex $[Co(en)_2Cl_2]^+$ ion ?
11. Which of the two : $[PdCl_4]^{2-}$ or $[Pd(NH_3)_6]^{4+}$ obey EAN rule ?
12. Write the IUPAC name of $Na[Co(CO)_4]$.
13. What are ambidentate ligands ? Give one example.
14. Why does NH_3 readily form complexes but NH_4^+ does not ?
15. What are chelates ? Give one example.
16. Name the type of isomerism exhibited by the following isomers :
   (*i*) $[Co(NH_3)_5(ONO)]Cl_2$, $[Co(NH_3)_5(NO_2)]Cl_2$
   (*ii*) $[Pt(NH_3)_4][PtCl_6]$, $[Pt(NH_3)_4Cl_2][PtCl_4]$

General Questions

1. Give IUPAC names of the following complexes :

(*a*) $K_2[PtCl_6]$ (*b*) $Ca_2[Fe(CN)_6]$

(*c*) $NH_4[Cr(NH_3)_2 (NCS)_4]$ (*d*) $K_3[Al (C_2O_4)_3]$

(*e*) $[Co (Cn)_2 Cl_2]_2 SO_4$

2. (*a*) How does Valence Bond Theory explain the following complexes :

(*i*) $[Ni(CN)_4]^{2-}$ is diamagnetic and square planar.

(*ii*) $[NiCl_4]^{2-}$ is paramagnetic and tetrahedral and

(*iii*) $[Ni(CO)_4]$ is diamagnetic and tetrahedral ?

(Z for Ni = 28)

(*b*) Draw the geometrical isomers for the complex ion : dichlorobis (ethylene diamine) rhodium (III). Indicate which of these exhibits optical isomerism.

3. What are basic postulates of Werner's coordination theory ? How does it account for non-ionic nature of $CoCl_3 . 3NH_3$ complex ?

4. (*a*) What are cationic and anionic complexes ? Give two examples of each.

(*b*) Name the following complexes :

(*i*) $K_2Ni\left[\left(\begin{matrix} S-C=O \\ | \\ S-C=O \end{matrix}\right)_2\right]$ (*ii*) $K_4[Mo(CN)_8]$

(*iii*) $[VO(acac)_2]$ (*iv*) $Fe(C_5H_5)_2$.

5. (*a*) Name the different kinds of structural isomers in coordination compounds. Give one example of each kind of isomer.

(*b*) Why do square planar complexes not exhibit optical isomerism ?

(*c*) Draw all possible isomers of

(*i*) $[Cr(en)_2Cl_2]^+$ (*ii*) $[Co(NH_3)_4Cl_2]$ (*iii*) $[Co(EDTA)]^-$

6. Write the chemical formula for the following complexes :

(*a*) Tetraaquadichlorochromium (III) nitrate.

(*b*) Hexaaquanickel (II) perchlorate

(*c*) *trans*-diamminechloronitroplatinum (II)

(*d*) Lithium tetrahydridoaluminate (III).

7. (*a*) What do you understand by chelate and a chelating ligand ? Explain with examples.

(*b*) Name the following complexes :

(*i*) $Na_2[CrOF_4]$ (*ii*) $K_3[Fe(CN)_5CO]$

(*iii*) $[Pd(H_2O_2) (ONO)_2]I_2$ (*iv*) $[Pt(phen)_2] [Pt(CO_3)_2]$.

8. (*a*) How does Valence Bond theory account for the fact that $[CoF_6]^{3-}$ is paramagnetic while $[Co(NH_3)_6]^{3+}$ is diamagnetic though both are octahedral ?

(*b*) Discuss geometrical isomerism in octahedral complexes and optical isomerism in $[Co(en)_3]^{3+}$, $[Cr(ox)_3]^{3-}$, dichlorobis (ethylene diamine) rhodium (III) and $[Co(en)(NH_3)_2Cl_2]^+$.

9. (*a*) What are isomers ? Give one example for each of the following types of isomerism in metal complexes :

(*i*) linkage (*ii*) ionization

(*iii*) coordination position

(*b*) Draw the optical isomer representations of $[Rh(en)_2Cl_2)]^+$

10. (*a*) Explain the rule for naming a bridging group in a compound.

(*b*) Give IUPAC names of the following bridged complexes :

(*i*) Ph_3P, Cl on Pd; Pd bridged by two Cl to Pd; Pd with Cl, PPh_3 — where $PPh_3 = P(C_6H_5)_3$

$$\begin{array}{c} Ph_3P \quad\;\; Cl \quad\;\; Cl \\ \quad Pd \qquad\quad Pd \\ Cl \quad\;\; Cl \quad\;\; PPh_3 \end{array}$$

(*ii*) $\left[(en)_4Co \begin{array}{c} NH \\ \\ OH \end{array} Co(en)_2\right]$

11. (*a*) What are the important postulates of valence bond theory of coordination compounds?

(*b*) Using Valence Bond theory, predict the geometry of the following :

(*i*) $FeCl_4^-$ (five unpaired electrons)

(*ii*) $[FeF_6]^{3-}$ (five unpaired electrons)

(*iii*) $[Ni(CO)_4]$ (diamagnetic)

12. (*a*) Describe the basic postulates of Werner's theory of coordination.

(*b*) How are $PtCl_4.6NH_3$, $PtCl_4.5NH_3$, $PtCl_4.4NH_3$, $PtCl_4.3NH_3$ and $PtCl_4.2NH_3$ formulated on the basis of Werner's theory if four, three, two, one and zero chloride ions are precipitated immediately as AgCl from their water solutions when $AgNO_3$ solution is added to them ?

(*c*) Explain hydrate isomerism taking the example of hydrated chromium (II) chloride.

13. Write the formulae of the following complex entities :

(*i*) Tetraammineaquabromocobalt (III) nitrate

(*ii*) Sodium hexanitrocobaltate (II)

(*iii*) Potassium bis(thiosulphato) argenate (I)

(*iv*) Bis (acetylacetonato) copper (II)

(*v*) Tris (ethylenediamine) cobalt (III) sulphate.

14. (*a*) Discuss important postulates of valence bond theory of coordination compounds.

(*b*) Discuss the geometry of the following on the basis of Valence Bond Theory :

(*i*) $[Ni(CN)_4]^{2-}$ (diamagnetic) (*ii*) $[Ni(CO)_4]$ diamagnetic

15. (*a*) A coordination compound $CrCl_3.4H_2O$ precipitates silver chloride with $AgNO_3$. The molar conductance of the solution corresponds to a total of two ions. Write the structural formula of the compound and the complex ion.

(*b*) Give one example for each type of 6-coordinate compounds exhibiting optical isomerism.

16. (*a*) What are the postulates of Werner's coordination theory ? How does it account for non-ionic nature of $CoCl_3.3NH_3$ complex ?

(*b*) How does valence bond theory explain the following :

(*i*) $[NiCl_4]^{2-}$ is paramagnetic and tetrahedral.

(*ii*) $[Ni(CN)_4]^{2-}$ is diamagnetic and square planar.

17. (*a*) Write the formulae of the following compounds:

(*i*) Hexaaquairon (II) sulphate

(*ii*) Potassium tetracyanonickelate (0)

(*iii*) *trans*-chloronitrodiammine platinum (II).

(*b*) Why do square planar complexes not exhibit optical isomerism ?

18. (*a*) Write the formulae of the following compounds.

(*i*) Tris(acetylacetonato) manganate (II)

(*ii*) Sodium tetrachloroplatinate (II)

(*iii*) Chloronitritotetraammineplatinum (IV) sulphate

(*iv*) Potassium hexacyanoferrate (II).

(*b*) Write the IUPAC names of the following compounds :

(*i*) $[Fe(C_2O_4)_3]^{3-}$ (*ii*) $[(NH_3)_5Co—NH_2—Co(NH_3)_5]Cl_5$

(*iii*) $[Pt(NH_3)_4][CoCl_4]$

19. (*a*) Describe the bonding in $[Fe(H_2O)_6]^{3+}$ and $[Fe(CN)_6]^{3-}$ on the basis of Valence Bond Theory.

(*b*) The molar conductance of $PtCl_4.4NH_3$ in water corresponds to a total of three ions. Write its structure on the basis of Werner's theory.

20. (*a*) What is the difference between inner orbital and outer orbital complexes ? Explain with examples.

(*b*) Write the formula of the following :

(*i*) Octaammine-μ-amido-μ-nitrito dicobalt (III) nitrate

(*ii*) Sodium dicyanoaurate (I)

(*iii*) Hexaaquairon (II) sulphate.

21. (*a*) Write the IUPAC name of the following complexes :

(*i*) $K_3[Ir(C_2O_4)_3]$ (*ii*) $[Cr(en)_2Cl_2]Cl$

(*iii*) $\left[(NH_3)_4Co\begin{matrix} NH_2 \\ NO_2 \end{matrix}Co(NH_3)_4\right](NO_3)_4$

(*b*) Define the terms :

(*i*) Ligand (*ii*) Co-ordination sphere

(*iii*) Complex ion.

22. (*a*) What is the difference between inner orbital and outer orbital complexes ? Explain with examples.

(*b*) Do square planar complexes exhibit geometrical isomerism ? Explain.

(*c*) Write the formulae of the following compounds :

(*i*) Sodium dicyanoaurate (I)

(*ii*) Pentaamminecobalt (III) chloride

(*iii*) cis-trichlorotripyridinerhodium (III).

23. Write the IUPAC names of the following :

(i) $H[AuCl_4]$ (*ii*) $[Pt(NH_3)_4Cl_2][PtCl_4]$

(*iii*) $Na_2[Fe(CN)_5NO]$ (*iv*) $[Cr(H_2O)_4Cl_2]NO_3$

(*v*) $[Cr(NH_3)_4Cl_2]^+$

24. (*a*) Following the IUPAC system, name the following compounds :

(*i*) $K_3[Fe(ox)_3]$ (*ii*) $K_2[PtCl_6]$

(*iii*) *trans*-$[Pt(NH_3)_2Cl.NO_2]$

(*iv*) $\left[(H_2O)_4Co\begin{matrix} \diagup OH \diagdown \\ \diagdown OH \diagup \end{matrix}Co(H_2O)_4\right]Cl_4$

(*b*) Define the following terms :

ligand and chelate.

25. (*a*) Write the IUPAC name of the following compounds :

(*i*) $Hg\,[Co(NCS)_4]$ (*ii*) $[Co(NO_2)_3\,(NH_3)_3]$

(*iii*) $[Pt(NH_3)_4\,Cl_2]\,[Pt\,Cl_4]$ (*iv*) $[Cr\,(PPh_3)\,(CO)_5]$.

(*b*) Write the formulae of the following compounds :

(*i*) Lithium tetrahydridoaluminate (III)

(*ii*) hexaaquanickel (II) perchlorate

(*iii*) Amminechloroopyridineplatinum (II).

26. (*a*) Write the IUPAC name of the folowing :

(*i*) $K_4[Ni(CN)_4]$ (*ii*) $[Ru(NH_3)_5\,N_2]^{2+}$

(*iii*) $[Ag\,(S_2O_3)_2]^{3-}$ (*iv*) $K[Ag\,(CN)_2]$

(*b*) Write the formulae of the following :

(*i*) Potassium tetrafluorooxochromate (V)

(*ii*) Triamminetrinitrocobalt (III)

(*iii*) Sodium hexanitrocobaltate (II)

(*iv*) Lithium tetrahydridoaluminate (III).

27. Write the IUPAC names of the following compounds :

(*i*) $[Ru\,(NH_3)_5\,N_2]\,Cl_2$ (*ii*) $[Co(ONO)\,(NH_3)_5]\,SO_4$

(*iii*) $[Pt(Py)_4]\,[Pt\,Cl_4]$ (*iv*) $K_3[Fe(CN)_5NO]$

28. (*a*) Write the IUPAC names of the following compounds :

(*i*) $Ca_2[Fe(CN)_6]$ (*ii*) $[Co(NH_3)_5Cl]^{2+}$

(*iii*) $[Co(NH_3)_3\,(NO_2)\,Cl_2]$

(*b*) Write the formulae of the following compounds :

(*i*) Bis (acetylacetonate) copper (II)

(*ii*) Tetraaquadichlorochromium (III) nitrate

(*iii*) Sodium pentacyanonitrosyl ferrate (III)

(*d*) What is the oxidation number of Fe in $K_3[Fe(CN)_6]$ and $Fe(CO)_5$?

29. (*a*) List the basic postulates of Werner's coordination theory. How will you explain the bonding in $CoCl_3.\ 3NH_3$ and $CoCl_3.\ 5NH_3$?

(*b*) Write IUPAC names for the following :

(*i*) $Fe[Fe(CN)_6]$

(*ii*) $K_3[Al(C_2O_4)_3]$

(*iii*) $[Co(en)_3]Cl_3$

30. (*a*) Write the IUPAC name of the following compounds :

1. $K_3\,[Fe(C_2O_4)_3]$

2. $[Ru (NH_3)_5N_2] Cl_2$
3. $[Co (en)_2 Br_2]Cl$
4. $[Pt (NH_3)_4 Cl_2] [PtCl_4]$

(*b*) Write the formulae of the following compounds :
1. Lithium tetrahydridoaluminate (III)
2. Bis (cyclopentadienyl)iron (II)
3. Dichloro bis (ethylenediamine) copper (II)

(*c*) 1. What are ambidenatate ligands ? Give two examples.
2. Calculate the charge on the central metal ion in the following compounds : $Fe(CO)_5$, $K[PtCl_3 (C_2H_4)]$, $K_4 [Fe (CN)_6]$

31. (*a*) Why don't we come across geometrical isomers in tetrahedral complexes ?
(*b*) Illustrate with suitable examples the following types of isomerism in co-ordination complexes :
(*i*) ionization (*ii*) linkage
(*c*) Give IUPAC names of the following compounds :
(*i*) $K_3 [Fe(CN)_6]$ (*ii*) $[Pd(H_2O)_2(ONO)_2I_2]$
(*iii*) $[Co (en)_3] Cl_3$

32. (*a*) How does the VBT account for the fact that $[CoF_6]^{3-}$ is paramagnetic while $[Co(NH_3)_6]^{3+}$ is diamagnetic ?
(*b*) Write the structures of two ionisation isomers of the complex $[Co(NH_3)_5CO_3]Cl$ and give their IUPAC names.
(*c*) Write IUPAC names for the following :
(*i*) $[Co (NH_3)_3 NO_2 Cl.CN]$ (*ii*) $[Cr(en)_3]Cl_3$
(*iii*) $AlCl_3.(EtO_4)_4$

33. (*a*) Write the IUPAC names of the following compounds :
(*i*) $[Co(en)_3]_2 (SO_4)_3$ (*ii*) $K_4[Ni(CN)_4]$
(*iii*) $[Cr(NH_3)_6] [CoF_6]$
(*iv*) $\left[(H_2O)_4Co \begin{smallmatrix} OH \\ OH \end{smallmatrix} Co(H_2O)_4\right] Cl_4$ (two OH bridges)
(*b*) Write the formulae of the following compounds :
(*i*) Iron hexacyanoferrate (III)
(*ii*) Triamminetrinitrocobalt (III)
(*iii*) Hexaaquanickel (II) perchlorate.
(*c*) What do you understand by a chelating ligand and a chelate ? Explain with examples.

34. (*a*) Write the names of the following complexes :
(*i*) $[Co (NH_3)_4 (NCS) Cl]NO_3$
(*ii*) $[Pt (NH_3)_4 (NO_2) Cl] SO_4$
(*b*) What is valence bond theory of transition metal complexes ? Discuss in brief.

35. (*a*) What is the difference between inner and outer d-orbitals complexes ? Explain by taking suitable examples.
(*b*) Square planar complexes with coordination number four exhibit geometrical isomerism, whereas tetrahedral complexes donot. Explain.

(*c*) The complex dichlorobis (ethylenediamine) cobalt (III) ion is optically active. Draw its optical isomers.

(*d*) Give IUPAC names of

(*i*) $Li[AlH_4]$

(*ii*) $K_3[Al(C_2O_4)_3]$

(*iii*) $Fe(CO)_5$

(*iv*) $\left[(H_2O)_4Fe \begin{smallmatrix} OH \\ OH \end{smallmatrix} Fe(H_2O)_4\right] SO_4$

36. (*a*) Draw the isomers of $PtCl_2 (NH_3)_2$. How can they be distinguished by dipole moment measurements ?

(*b*) Square planar complexes donot show optical isomerism. Explain.

(*c*) Write the IUPAC names of the following :

(*i*) $\left[(NH_3)_4Co \begin{smallmatrix} NH \\ NH \end{smallmatrix} Co(NH_3)_4\right] SO_4$

(*ii*) $[Pt (NH_3)_4] [CoCl_4]$

(*iii*) $Fe(CO)_5$

(*d*) According to V.B. theory $[Ni (CN)_4]^{2-}$ is square planar while $[NiCl_4]^{2-}$ and $[Ni(CO)_4]$ are tetrahedral. Calculate number of unpaired electrons in them.

37. What are the postulates of Werner's coordination theory and its experimental verification?

38. (*a*) Why $Co(NH_3)_6{}^{3+}$ is inner complex while $CoF_6{}^{3-}$ is outer complex ?

(*b*) Draw the optical isomers of $[Co (EDTA)]^-$.

(*c*) $Fe (CN)_6{}^{4-}$ is octahedral and diamagnetic while $Cr (CN)_6{}^{3-}$ is octahedral and paramagnetic. Explain.

(*d*) Write the IUPAC names of

(*i*) $K [B (NO_3)_4]$ (*ii*) $Hg [Co(CNS)_4]$

(*iii*) $[Cr(NH_3)_6].[Co(CN)_6]$

39. (*a*) What is isomerism ? Discuss various types of isomerism in coordination compounds with one example in each case.

(*b*) Give an example of square planar complex of nickel and discuss the anomalous properties of nickel (II) complexes in brief.

(*c*) What are the limitations of valence bond theory of transition metal complexes.

40. What is isomerism in coordination compounds ? Discuss any three types of isomerism with two examples in each case.

41. (*a*) $[Co (NH_3)_6]^{3+}$ and $[CoF_6]^{3-}$ are both octahedral but shows marked differences in their magnetic properties. Explain.

(*b*) Molar conductance of $PtCl_4.4NH_3$ in water corresponds to total of three ions. Write the structures on the basis of Werner's theory.

(*c*) Explain with examples linkage and coordination isomerism.

(*d*) Write the names of the following coordination compounds :

$Fe_4 [Fe (CN)_6]$, $[Cu (HSO_3)_2 (NH_3)_4]$, $K_2 [Zn(OH)_4]$, $[Pt(py)_4] [PtCl_4]$, $Li [AlH_4]$, $[Co (NH_3)_4 (H_2O) Br] (NO_3)_2$.

42. (*a*) What are chelates ? Discuss the factors which effect the stability of chelates.

(*b*) Discuss hybridisation, structure and magnetic properties (on the basis of VBT) in $[Cr(NH_3)_6]^{3+}$ ion.

(*c*) Which of the following do and which do not obey EAN rule ?

(*i*) $Cr(CO)_6$

(*ii*) $Cr(C_6H_6)(CO)_4$

(*iii*) $[Mn(CN)_6]^{4-}$

(*iv*) $[Fe(H_2O)_6]^{2+}$

(*d*) Give IUPAC name of the following :

(*i*) $[PtCl_2(C_5H_5N)(NH_3)]$

(*ii*) $Fe[Fe(CN)_6]$

(*iii*) $[Pt(py)_4][PtCl_4]$

43. (*a*) Write IUPAC names of the following coordination compounds.

(*i*) $Hg[Co(CNS)_4]$

(*ii*) $[Cr(NH_3)_6][Co(CN)_6]$

(*iii*) $[Pt(NH_3)_4(NO_2)Cl)]SO_4$

(*b*) What are the postulates of Werner's coordination theory ? How does it account for non-ionic nature of $CoCl_3.3NH_3$ complex ?

(*c*) What is effective atomic number ?.

44. (*a*) What are the basic postulates of Werner's theory of coordination.

(*b*) $[Co(NH_3)_6]^{3+}$ is diamagnetic but $[CoF_6]^{3-}$ is paramagnetic. Discuss.

(*c*) Define and explain :

(*i*) Linkage isomerism.

(*ii*) Co-ordination position isomerism.

45. (*a*) Which of the following complexes follow EAN rule ? (record electron count in each case).

(*i*) $[Cr(NH_3)_6]^{3+}$

(*ii*) $[PtCl_6]^{4-}$

(*iii*) $[Ni(NH_3)_6]^{2+}$

(*iv*) $[Co(NO_2)_6]^{3-}$.

(*b*) What is the coordination number of Fe in $[Fe(edta)]^-$?

(*c*) Explain with the help of VBT, why $[Fe(CN)_6]^{4-}$ is diamagnetic while $[Fe(CN)_6]^{3-}$ is paramagnetic in nature.

46. (*a*) Square planar complexes do not exhibit optical isomerism and tetrahedral complexes the geometrical isomerism. Explain.

(*b*) With the help of valence bond theory explain the structure of $[Ni(CO)_4]$

(*c*) Draw all possible isomers of complex ion $[Co(en)_2Cl_2]^+$.

47. (*a*) Discuss the naming of coordination compounds according to Werner's coordination theory with suitable examples.

(*b*) What informations do we get from valence bond theory of transition metal complexes ? Explain in brief.

(*c*) What are chelates ? Give suitable examples of chelates and discuss how do chelate ring size and number of chelate rings effect the stability of a complex.

48. (*a*) Using valence bond theory, explain the geometry and magnetic character of CoF_6^{-3} and $Co(CN)_6^{-3}$.

(*b*) Write IUPAC names of

(*i*) $[Pt(NH_3)_4Cl_2]\ [PtCl_4]$

(*ii*) $\left[(H_2O)_4Fe\begin{smallmatrix} OH \\ OH \end{smallmatrix}Fe\,(e_n)_2\right]^{4+}$

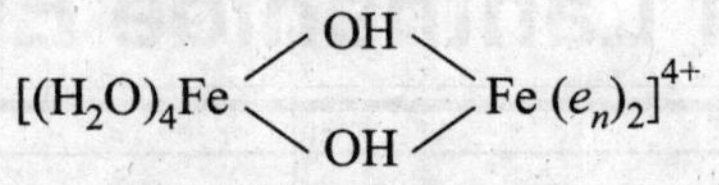

(*iii*) $Hg\ [Co(CNS)_4]$

(*c*) A co-ordination compound $CrCl_3.6H_2O$ gives precipitates of AgCl with $AgNO_3$ solution. The molar conductance of the solution of the compound to total three ions. Write the structure of the compound. What is its IUPAC name ?

49. (*a*) What are chelates ? Discuss the factors which affect stability of chelates.

(*b*) How does V.B.T. explain the following complexes :

(*i*) $[NiCl_4]^{2-}$ is paramagnetic and tetrahedral.

(*ii*) $[Ni(CN)_4]^{2-}$ is diamagnetic and square planer.

(*c*) What is effective atomic number rule ?

50. (*a*) How many chloride ions will be precipitated in the following complexes ? Explain :

(*i*) $CoCl_3.4NH_3$

(*ii*) $CoCl_3.5NH_3$

(*iii*) $CoCl_3.6NH_3$

(*b*) Discuss geometrical isomerism in compounds with coordination number 3 and 4 ?

(*c*) Discuss giving one example of each kind of the following isomerism :

(*i*) Ionization isomerism

(*ii*) Coordination isomerism

Structure of Ce (*iv*) oxide

Chemistry of Lanthanide Elements

1																	18
H	2											13	14	15	16	17	He
Li	Be	Transition Elements										B	C	N	O	F	Ne
Na	Mg	3	4	5	6	7	8	9	10	11	12	Al	Si	P	S	Cl	Ar
K	Ca	21 Sc	22 Ti	23 V	24 Cr	25 Mn	26 Fe	27 Co	28 Ni	29 Cu	30 Zn	Ga	Ge	As	Se	Br	Kr
Rb	Sr	39 Y	40 Zr	41 Nb	42 Mo	43 Tc	44 Ru	45 Rh	46 Pd	47 Ag	48 Cd	In	Sn	Sb	Te	I	Xe
Cs	Ba	57 La*	72 Hf	73 Ta	74 W	75 Re	76 Os	77 Ir	78 Pt	79 Au	80 Hg	Tl	Pb	Bi	Po	At	Rn
Fr	Ra	89 Ac*	104 Ku														

f- block elements

Lanthanide series (4*f*)	58 Ce	59 Pr	60 Nd	61 Pm	62 Sm	63 Eu	64 Gd	65 Tb	66 Dy	67 Ho	68 Er	69 Tm	70 Yb	71 Lu
Actinide series (5*f*)	90 Th	91 Pa	92 U	93 Np	94 Pu	95 Am	96 Cm	97 Bk	98 Cf	99 Es	100 Fm	101 Md	102 No	103 Lr

5.1. INTRODUCTION

The fifteen elements from lanthanum to lutetium (at numbers 57 to 71) are called **lanthanides** or **rare earths** (see periodic table). These elements involve the filling of 4*f* orbitals. Strictly speaking, the first member lanthanum (configuration $6s^2\ 5d^1 4f^0$) and the last member, lutetium (configuration $6s^2\ 5d^1 4f^{14}$) have no partially filled 4*f* orbital in their ground state yet they are considered to belong to lanthanide series, because their chemical properties are very closely similar to the other thirteen elements of the series.

Lanthanum Metal

These elements are also called the *inner transition elements* because the filling of electrons takes place in the 4*f* subshell, *i.e.*, two levels below the outer 5*p* and 6*s* subshells which are already filled in these elements. The 4*f* electrons are not exposed to their surroundings as they do not extend out to the surface of the atoms or ions. As a result, the properties of lanthanides are very similar to one another and do not show much variation. Decrease in their radii as we move across from La to Lu is responsible for small variation in their properties.

5.1.1. Electronic configuration of lanthanides

In lanthanum (57) the incoming electron enters the $5d$ subshell and therefore it has the configuration [Xe] $6s^2\ 5d^1\ 4f^0$. However, as we move one step further, the $4f$ subshell becomes more stable than the $5d$ subshell and the next electron in cerium enters the $4f$ subshell. Cerium has the configuration [Xe] $6s^2\ 4f^1 5d^1$. The filling of $4f$ subshell continues till we reach ytterbium which has the configuration [Xe] $6s^2\ 4f^{14}\ 5d^0$. Now after $6s$ and $4f$ subshells have been filled, the $5d$ subshell is the next choice for filling electrons. At lutetium the electron enters the $5d$ subshell and it has the configuration [Xe] $6s^2\ 4f^{14}\ 5d^1$. The electronic configuraiton and oxidation states of lanthanides are listed in Table 5.1. In general, their electronic configuraiton may be written as:

$$[Xe]\ 6s^2\ 4f^{0-14}\ 5d^{0-1}$$

Since the lanthanides have similar electronic configuration, they all exhibit +3 oxidation state and have similar chemical properties. Therefore they are all grouped together at one position of the periodic table.

Table 5.1. Electronic configuration of lanthanides

Atomic number	*Name*	*Symbol*	*Electronic configuration*	*Oxidation states*
57	Lanthanum	La	[Xe] $4f^0\ 5d^1\ 6s^2$	+3
58	Cerium	Ce	[Xe] $4f^1\ 5d^1\ 6s^2$	+3, +4
59	Praseodymium	Pr	[Xe] $4f^3\ 6s^2$	+3, +4
60	Neodymium	Nd	[Xe] $4f^4\ 6s^2$	+2, +3
61	Promethium	Pm	[Xe] $4f^5\ 6s^2$	+3
62	Samarium	Sm	[Xe] $4f^6\ 6s^2$	+2, +3
63	Europium	Eu	[Xe] $4f^7\ 6s^2$	+2, +3
64	Gadolinium	Gd	[Xe] $4f^7\ 5d^1\ 6s^2$	+3
65	Terbium	Tb	[Xe] $4f^9\ 6s^2$	+3, +4
66.	Dysprosium	Dy	[Xe] $4f^{10}\ 6s^2$	+3, +4
67	Holmium	Ho	[Xe] $4f^{11}\ 6s^2$	+3
68	Erbium	Er	[Xe] $4f^{12}\ 6s^2$	+3
69	Thulium	Tm	[Xe] $4f^{13}\ 6s^2$	+2, +3
70	Ytterbium	Yb	[Xe] $4f^{14}\ 6s^2$	+2, +3
71	Lutetium	Lu	[Xe] $4f^{14}\ 5d^1\ 6s^2$	+3

5.1.2. Oxidation states of lanthanide elements

The common oxidation state shown by all the lanthanides is +3. Some elements display +2 and +4 states, but these are always less stable than the group valency +3. The unusual oxidation states such as +2 or +4 exhibited by the lanthanides can be explained on the basis of extra stability associated with empty, half-filled or fully filled $4f$ subshell. Some examples are as follows:

***(i)* Eu and Yb exhibit +2 oxidation states.** This is because in +2 oxidation state europium and ytterbium acquire f^7 and f^{14} electronic configuration as:

$$Eu^{2+} : [Xe]\ 4f^7,\ Yb^{+2} : [Xe]\ 4f^{14}$$

(*ii*) Ce and Tb exhibit +4 oxidation states. This is because in +4 oxidation state cerium and terbium acquire f^0 and f^7 electronic configuration which are supposed to be stable configurations

$$Ce^{4+} : [Xe]\ 4f^0,\ Tb^{4+} : [Xe]\ 4f^7$$

(*iii*) La, Gd and Lu exhibit +3 oxidation states. This is because in +3 oxidation states La^{3+} has [Xe] $4f^0$, Gd^{3+} has [Xe] $4f^7$ and Lu^{3+} has [Xe] $4f^{14}$ electronic configurations which are stable due to empty, half-filled and fully filled 4*f* subshell respectively.

Exceptions. There are a few elements which show +2 and +4 oxidation states but their existence can't be easily justified. For example, Sm and Tm show +2 state having f^6 and f^{13} electrons, when they are only one electron short of the necessary half-filled or completely filled configuration. We have other examples where Pr and Nd show +4 state even though they are left with f^1 and f^2 electrons. Oxidation states of lanthanides are shown in Table 5.1.

SOLVED EXAMPLES

Example 1. Give the important oxidation states of lanthanides. How would you account for them? Why is +3 most stable?

Solutions. In reactions of lanthanides, it is the two electrons from 6*s* orbital and one electron from either the 5*d* or 4*f* that are removed. That is why these elements exhibit +3 oxidation state. In some cases +2 and +4 oxidation states are also experienced. It is on energy consideration of orbitals that +3 oxidation state is most stable.

Example 2. In transition series starting from La (57) the next element is Hf (72). Why is this jump observed?

Solution. In the transition series from La (57) to Hg (80) the filling of electrons takes place in 5*d* orbitals, La containing one electron and Hg containing 10 electrons in 5*d* orbital. After La, filling starts taking place in 4*f* orbitals on energy consideration and build up of elements from cerium to lutetium takes place. It is only after Lu (71) that the electrons are again filled up in 5*d* orbital and the next element is Hf (72).

Example 3. Ytterium is taken as a practical member of lanthanide series though it doesn't belong to this series. Explain.

Solution. Y and La belong to group 3 of the periodic table. They have similar outer electronic configuration, that of Y being $4d^1\ 5s^2$ and that of La being $5d^1\ 6s^2$. Therefore they show identical properties. Moreover La along with other lanthanides show identical properties. Therefore *Y* is taken as a practical member of lanthanide series.

5.1.3. Lanthanide contraction

From Table 5.2 and Fig. 5.1 we find that as we move along the lanthanide series from lanthanum to lutetium there is a continuous decrease in the size of lanthanide ion with increase in atomic number. This decrease is known as *lanthanide contraction. Lanthanide contraction may be defined as the small steady decrease in size of lanthanide ions with increase in atomic number.* Thus in the lanthanide series, lanthanum has the largest radius while lutetium has the smallest radius.

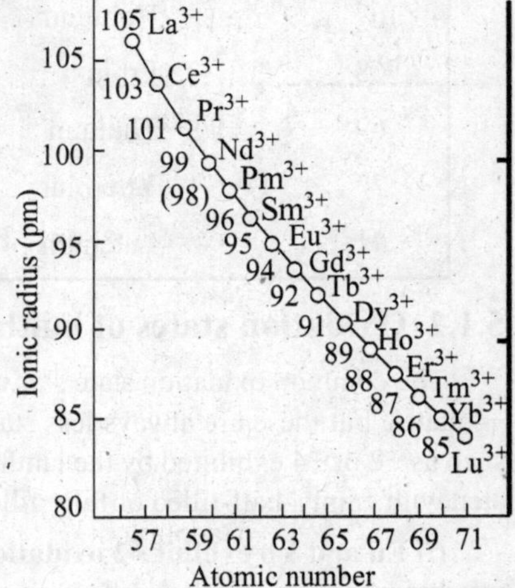

Fig. 5.1. Ionic radii of trivalent lanthanide ions.

In moving from Ce^{3+} to Lu^{3+}, 14 electrons are added to $4f$ orbital, while the decrease in ionic radius is from 103 pm to 85 pm *i.e.* 18 pm only. This is a very small decrease in comparison to elements of other periods.

Cause of Lanthanide contraction. The cause of lanthanide contraction is generally attributed to imperfect shielding of one $4f$ electron by another in the same shell. Thus as we move along the lanthanide series, the nuclear charge and the number of $4f$ electrons increase by one unit at each step. However due to imperfect shielding (because the shape of f orbitals is very much diffused) the net effective nuclear charge increases which causes contraction although small in the size of the electron charge cloud and thus each ion shrinks in comparison with its predecessor. For example, in going from La^{3+} to Lu^{3+} the ionic radii decrease from 106 pm to 85 pm.

Consequences of Lanthanide Contraction

(i) **Electronegativity.** There is a slight increase in electronegativity of trivalent ions.

(ii) **Ionic radius.** There is a steady decrease in the ionic size.

(iii) **Basicity difference.** There is a decrease in basic strength of the oxides and hydroxide of lanthanides with increase in atomic numbers. Thus La $(OH)_3$ is the most basic while Lu $(OH)_3$ is the least basic. Due to lanthanide contraction the decrease in the size of the cation increases the covalent character between the lanthanide ion and the hydroxide ion (according to Fajan's rules) thereby reducing the basic character of the lanthanide hydroxides.

(iv) **E° value.** There is a small but smooth increase in standard electrode potential values, E° for the process.

$$M^{3+} + 3e \longrightarrow M\,(g)$$

ranging from –2.52 V for lanthanum to –2.25 V for lutetium.

(v) **Resemblance of second and third transition series.** We find that there is almost no change in ionic radius when we move from Zr to Hf, Nb to Ta and Mo – W, etc. As the filling takes place in inner f orbitals, the ionic radii of the above couples remains almost constant.

(vi) **Similarity among lanthanides.** Because of very small change in the ionic radius, the properties of lanthanide elements show similar properties. This creates problems in the separation of lanthanides from one another.

Table 5.2. Ionic radii of lanthanides.

Atomic number	*Name*	*Symbol*	*Electronic configuration*	M^{3+} *radii (pm)*
57	Lanthanum	La	[Xe] $4f^0\ 5d^1\ 6s^2$	106
58	Cerium	Ce	[Xe] $4f^1$ 5d^1 $6s^2$	103
59	Praseodymium	Pr	[Xe] $4f^3\ 6s^2$	101
60	Neodymium	Nd	[Xe] $4f^4\ 6s^2$	100
61	Promethium	Pm	[Xe] $4f^5\ 6s^2$	98
62	Samarium	Sm	[Xe] $4f^6\ 6s^2$	96
63	Europium	Eu	[Xe] $4f^3\ 6s^2$	95
64	Gadolinium	Gd	[Xe] $4f^7\ 5d^1\ 6s^2$	94
65	Terbium	Tb	[Xe] $4f^9\ 6s^2$	92

66	Dysprosium	Dy	[Xe] $4f^{10}$ $6s^2$	91
67	Holmium	Ho	[Xe] $4f^{11}$ $6s^2$	89
68	Erbium	Er	[Xe] $4f^{12}$ $6s^2$	88
69	Thulium	Tm	[Xe] $4f^{13}$ $6s^2$	87
70	Ytterbium	Yb	[Xe] $4f^{14}$ $6s^2$	86
71	Lutetium	Lu	[Xe] $4f^{14}5d^1 6s^2$	85

5.1.4. Colour of lanthanide compounds

The lanthanide metals are silvery white but the trivalent lanthanide ions show different colours. Colour of the ions depends on the number of unpaired electrons because the elements with xf electrons often have a similar colour to those of $(14 - x)\,f$ electrons. This is evident from Table 5.3. The colour of lanthanide ions is due to the presence of partly filled f-orbitals. As a result it is possible to absorb certain wavelength from the visible region of the spectrum. This results in transition from one $4f$ orbital to another $4f$ orbital known as f-f transition. The colour transmitted is the colour shown by the substance.

Table 5.3.

No. of unpaired electrons ions		*Colour*
0	La^{3+}, Lu^{3+}	Colourless
1	Ce^{3+}, Yb^{3+}	Colourless
2	Pr^{3+}, Tm^{3+}	Green
3	Nd^{3+}, Er^{3+}	Red
4	Pm^{3+}, Ho^{3+}	Pink and yellow respectively
5	Sm^{3+} Dy^{3+}	Yellow
6	Eu^{3+}, Tb^{3+}	Pink
7	Gd^{3+}	Colourless

5.1.5. Variation in melting and boiling points of lanthanides

Lanthanides have fairly high melting and boiling points. However, their melting and boiling points do not show definite trend as we move from La to Lu. Both Eu and Yb form +2 ions and so it results in slightly weaker metallic bonding as there are fewer than three electrons per atom in the metallic bonding. As a consequence, melting and boiling points have lower values for these two elements as compared to the values for other lanthanide elements. Also abnormally greater radii for these two elements resulting in weaker packing account for lower melting and boiling points. Melting and boiling points of lanthanides are given in Table 5.4.

Table 5.4.

Atomic number	*Name*	*Density (g/ml)*	*Melting point (K)*	*Boiling point (K)*
57	Lanthanum	6.17	1193	3742
58	Cerium	6.77	1068	3741
59	Praseodymium	6.78	908	3400
60	Neodymium	7.00	1297	3300
61	Promethium	7.20	–	–
62	Samarium	6.93	1345	2173
63	Europium	5.24	1099	1712
64	Gadolinium	7.95	1585	3273
65	Terbium	8.33	1629	3073
66	Dysprosium	8.56	1680	2873
67	Holmium	8.76	1734	2873
68	Erbium	9.16	1770	3173
69	Thulium	9.35	1818	2000
70	Ytterbium	7.01	1097	1700
71	Lutetium	9.74	1925	3600

5.1.6. Magnetic properties of lanthanides

Magnetism is a property associated with unpaired electrons. All lanthanide ions with the exception of La^{3+}, Lu^{3+}, Yb^{3+} and Ce^{4+} are paramagnetic because they contain unpaired electrons in the 4*f* orbital. Their magnetic moments do not obey the simple "spin only" formula (except for f^6, f^7 and f^{14} cases) for the magnetic moment.

$\mu_{\text{spin only}} = \sqrt{n(n+2)}$ where *n* is equal to the number of unpaired electrons.

In case of lanthanides, the magnetic effect arising from the orbital motion of the electron, as well as that arising from the electrons spinning on its axis, contributes to the total magnetic moment. Therefore magnetic moments of lanthanides are calculated by taking into consideration spin as well as orbital contributions and a more complex formula.

$$\mu = \sqrt{4S(S+1) + L(L+1)}$$

which involves the orbital quantum number, L and spin quantum number, *S* is used.

Magnetic moment values of lanthanides vary depending upon the atomic number. La^{3+} is diamagnetic (due to f^0) and the value of magnetic moment increases attaining maximum value at neodymium. It then suddenly drops to 1.47 for samarium. It increases again reaching maximum value for dyprosium and holmium. Thereafter, it again falls touching zero at lutetium which is diamagnetic because it contains 14 electrons in its *f* orbitals. Change in magnetic moment with atomic number is shown in Fig. 5.2. Values of magnetic moments (Theoretical and experimental) for tripositive ions are given in Table 5.5.

The unusual shape of the curve arises because the spin and orbital magnetic moments work in opposite directions when the *f*-sub shell is less than half full. But they supplement each other when *f*-sub shell is more than half full

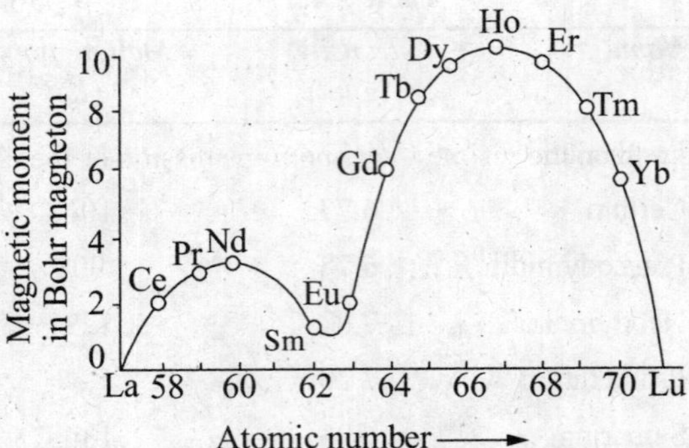

Fig. 5.2. Experimental values of the magnetic moments of the lanthanide ions.

Table 5.5. Magnetic moments (theoretical and experimental) for tripositive lanthanide ions.

Ion	Magnetic moment (BM)		Ion	Magnetic moment (BM)	
	Experimental	Theoretical		Experimental	Theoretical
La^{3+}	0	0	Gd^{3+}	7.8 — 8.0	7.94
Ce^{3+}	2.3 — 2.5	2.54	Tb^{3+}	9.4 — 9.6	9.72
Pr^{3+}	3.4 — 3.6	3.58	Dy^{3+}	10.4 — 10.5	10.65
Nd^{3+}	3.5 — 3.6	3.62	Ho^{3+}	10.3 — 10.5	10.60
Pm^{3+}	2.7	2.68	Er^{3+}	9.4 — 9.6	9.57
Sm^{3+}	1.4 — 1.7	0.84	Tm^{3+}	7.1 — 7.4	7.63
Eu^{3+}	3.4 — 3.6	0	Yb^{3+}	4.4 — 4.9	4.50
			Lu^{3+}	0	0

SOLVED EXAMPLES

Example 4. Explain why

(*i*) Ce^{3+} is more basic than Ce^{4+}

(*ii*) La^{3+} is colourless but Lu^{4+} is orange red.

Solution. (*i*) Configuration of Ce^{3+} and Ce^{4+} are:

Ce^{3+} : [Xe] $4f^1$

Ce^{4+} : [Xe]

Ce^{3+} has a tendency to lose $4f^1$ electron to attain the inert gas structure, whereas Ce^{4+} wouldn't lose the electron because that would disrupt inert gas structure. Therefore Ce^{3+} is more basic than Ce^{4+}.

(*ii*) Configuration of La^{3+} and Lu^{4+} are:

La^{3+} : [Xe] $4f^0$

Lu^{4+} : [Xe] $4f^{13}$

Transition within the *f* orbitals is not possible in La^{3+} as there are no electrons in *f*-orbitals. Hence no absorption of visible light is possible; it is therefore colourless.

In the case of Lu^{4+}, there is one vacancy in 4*f* orbitals. Hence *f*- *f* transition takes place absorbing a radiation from visible light and transmitting orange-red light.

Example 5. What are *d–d* and *f–f* transitions? How do they differ?

Solution. Transitions within the *d*-orbitals of an element are called *d–d* transition, while transitions within *f* orbitals of an element are called *f– f* transitions. These transitions can be brought about by the absorption of radiation from the visible light and transmission of the complementary colour. The compound thus looks coloured.

5.1.7. Variation in properties of Lanthanides

***(i)* Metallic radii.** The metallic radii (see Table 5.6) decrease with increase in atomic number. The metallic radii of lanthanides fall in the range 174-208 pm which are quite comparable with those of *s*-block elements, particularly calcium having metallic radius of 174 pm. This indicates fairly large size of the atoms of lanthanide elements. Eu and Yb show very surprisingly irregular sizes. The reason is that these elements suffer repulsions between greater number of *f*-electrons. Oxidation state of +2 for these elements is also responsible for anomalies.

***(ii)* Density.** Lanthanides have low densities which range between 6.77 and 9.74 g cm^{-3}. However, densities of lanthanides don't show definite trends with rise in atomic number. Eu and Yb have low values of density than expected (see Table 5.4).

***(iii)* Ionisation energies.** Lanthanides have fairly low ionisation energy values (IE_I is about 585 kJ/mol and IE_{II} is about 1150 kJ mol^{-1}) due to their fairly large size which are quite comparable with those of alkaline earth metals particularly calcium (IE_I = 589 kJ/mol, IE_{II} = 1145 kJ mol^{-1}). First ionisation energies of lanthanides are given in Table 5.6.

***(iv)* Electronegativity values.** Their electronegativity values fall in the range 1.0 –1.15 (Allred and Rochow scale) and are quite comparable with the electronegativity values of the *s*-block elements, particularly calcium. Therefore, lanthanide ions are expected to form ionic compounds.

Table 5.6. Properties of lanthanides.

Name	*Symbol*	*Electronic configuration*	*Metallic radii (pm)*	*First Ionisation energy (kJ/mol^{-1})*	*E° (V) M^{3+}/M*
Lanthanum	La	[Xe] $4f^0$ $5d^1$ $6s^2$	188	541.3	–2.52
Cerium	Ce	[Xe] $4f^1$ $5d^1$ $6s^2$	182	540.3	–2.48
Praseodymium	Pr	[Xe] $4f^3$ $6s^2$	183	528.7	–2.47
Neodymium	Nd	[Xe] $4f^4$ $6s^2$	182	530.7	–2.44
Promethium	Pm	[Xe] $4f^5$ $6s^2$	183	536.0	–2.42
Samarium	Sm	[Xe] $4f^6$ $6s^2$	180	540.3	–2.41
Europium	Eu	[Xe] $4f^7$ $6s^2$	208	547.1	–2.41
Gadolinium	Gd	[Xe] $4f^7$ $5d^1$ $6s^2$	180	594.4	–2.40
Terbium	Tb	[Xe] $4f^9$ $6s^2$	178	577.0	–2.39
Dysprosium	Dy	[Xe] $4f^{10}$ $6s^2$	177	567.0	–2.35
Holmium	Ho	[Xe] $4f^{11}$ $6s^2$	177	574.0	–2.32
Erbium	Er	[Xe] $4f^{12}$ $6s^2$	176	580.6	–2.30
Thulium	Tm	[Xe] $4f^{13}$ $6s^2$	175	560.6	–2.28
Ytterbium	Yb	[Xe] $4f^{14}$ $6s^2$	193	600.1	–2.27
Lutetium	Lu	[Xe] $4f^{14}$ $5d^1$ $6s^2$	173	593.4	–2.25

Chemistry of all lanthanides is so identical

All the lanthanides have similar outer electronic configuration and display mainly +3 oxidation state in their compounds, therefore lanthanides have exceedingly similar chemical properties. Their similarity is much closer than that of ordinary transition elements because lanthanides differ mainly in the number of 4*f* electrons which are buried deep in the atoms of lanthanides and thus do not influence their properties. Moreover, due to lanthanide contraction there is a very small difference in the size of all the fifteen trivalent lanthanide ions. Thus, for all practical purposes, the size of these ions is almost identical which results in similar chemical properties of these elements.

5.1.8. Chemical properties of lanthanides

Some chemical properties of Lanthanides are given below:

***(i)* Action with air.** They are easily tarnished in moist air. If ignited in air, they burn to give +3 or +4 oxides like Ln_2O_3 or CeO_2.

***(ii)* Action with nitrogen.** They react with nitrogen to form metal nitrides of the general formula MN.

***(iii)* Action with halogens.** They combine directly with halogens to give trihalides MX_3.

***(iv)* Action with water.** Their standard electrode potential values, E°, indicate that these metals should react vigorously with water. In fact they do liberate hydrogen slowly from cold water and rapidly from hot water.

***(v)* Action with hydrogen.** They react with hydrogen especially at elevated temperatures and form non-stoichiometric hydrides having composition MH_2 and MH_3.

5.1.9. Tendency to form Complexes

Although the lanthanide ions have a high charge (+3), yet the size of their ions is very large (yielding small charge to size ratio). As a consequence, they have poor tendency to form complexes. They form complexes mainly with chelating agents such as β-diketones, EDTA, L-L′ dipyridyl and β-hydroxyquinoline. Complex formation tendency and stability increases with increasing atomic number. This fact is utilized in the separation of lanthanides.

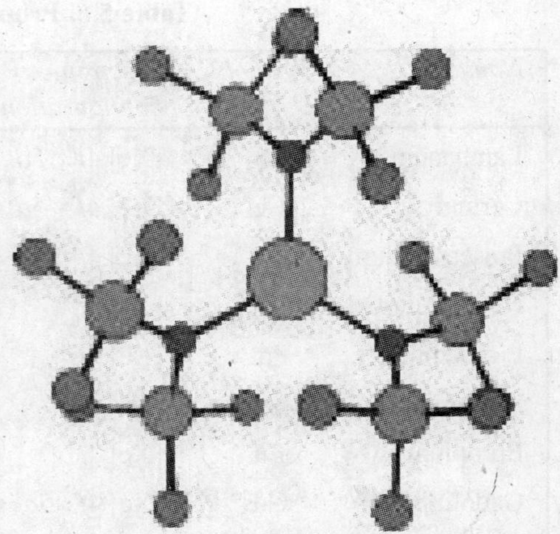

Crystal structure of Lanthanide silylamide

Coordination number 6 is not very common. Coordination number below 6 occurs only with bulky ligands such as (2,6-dimethyl phenyl)$^-$ and $[N(SiMe_3)_2]$ ligands. However, coordination numbers 7, 8 and 9 are common and these give a variety of stereochemistries. For example, the coordination numbers for $[Ln(OH_2)_n]^{3+}$ in aqueous solution are thought to be 9 for early lanthanides and 8 for the latter smaller members of the series. The smaller ytterbium ion, Yb^{3+} forms seven coordinated complex $[Yb(acac)_3(OH_2)]$ and the larger La^{3+} is eight coordinated in $[La(acac)_3(OH_2)_2]$ where acac is acetylacetonate. The structures of these complexes are shown in Fig. 5.3. Coordination numbers 10 and 12 occur with the larger (lighter) lanthanides and small chelating ligands such as NO_3^- and SO_4^{2-}.

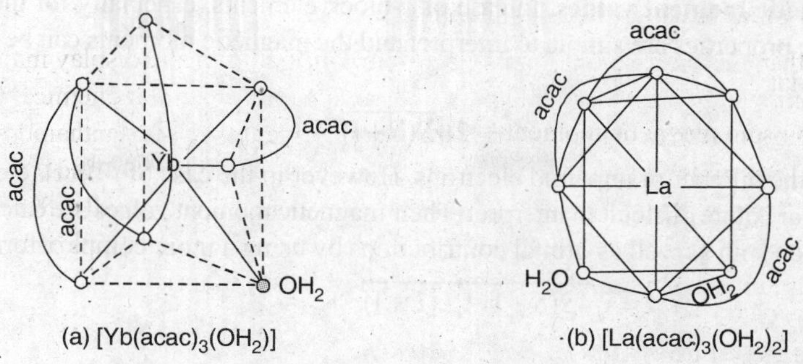

Fig. 5.3. Structures of (*a*) $[Yb(acac)_3(OH_2)]$ and (*b*) $[La(acac)_3(OH_2)_2]$

Ce^{4+} is smaller and highly charged and forms $[Ce(NO_3)_6]^{2-}$ in the non-aqueous solvent N_2O_4. This is 12 coordinated Ce in which each NO_3^- ion uses two oxygen atoms to coordinate to the metal (Fig. 5.4.).

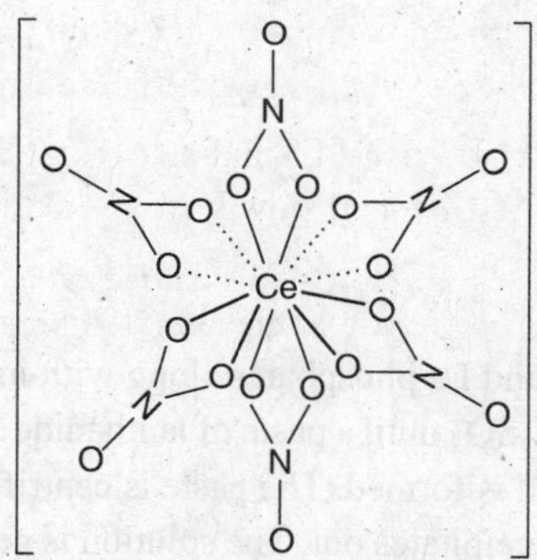

Fig. 5.4. 12-coordinated nitrato complex, $[Ce(NO_3)_6]^{2-}$

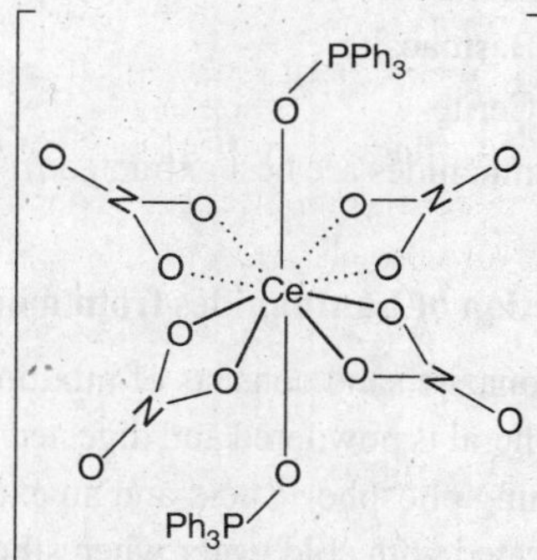

Fig. 5.5. 10-coordinated $[Ce(NO_3)_4(OPPh_3)_2]$

It gives a 10-coordinated orange complex with the formula $Ce(NO_3)_4(OPPh_3)_2]$ when two *trans* nitrate groups are replaced by Ph_3PO groups (Fig. 5.5.)

5.1.10. Comparison of characteristics of *d*- and *f*- block elements

(*i*) Size. The sizes of the atoms and ions of *d*-block elements are relatively smaller than those of the *f*-block elements.

(*ii*) Basic character. The oxides and hydroxides of *d*-block elements are less basic than those of *f*-block elements.

(*iii*) Ionisation energies. The first ionization values of *d*-block elements are much higher as compared to those of *f*-block elements.

(*iv*) Complex formation. The elements of *d*-block form complexes easily as compared to *f*-block elements. The latter mainly form metal chelates with strong chelating agents.

(*v*) Reactions with unsaturated hydrocarbons. Both *d*- and *f*-block elements react with unsaturated hydrocarbons such as olefins and cyclopentadienes. The former form π-type of complexes while the latter form salts such as Lu $(C_5H_5)_3$ known as lanthanide cyclopentadenides.

(*vi*) Electronic spectra. Electronic spectra of salts and complexes of *d*-block elements show broad bands while those of *f*-block elements exhibit sharp line-like bands.

***(vii)* Magnetic moment values.** In case of *d*-block elements, especially for the first transition series, magnetic properties are simple to interpret and the magnetic moments can be calculated from spin-only formula:

$$\text{spin magnetic moment} = 2\sqrt{S(S+1)}$$

where S is half the number of unpaired electrons. However in the case of *f*-block elements the magnetic properties are quite difficult to interpret. Their magnetic moment values are calculated by taking into consideration spin as well as orbital contributions by using a more complex formula:

$$\mu = \sqrt{4S(S+1) + L(L+1)}$$

5.2. OCCURRENCE AND EXTRACTION OF LANTHANIDES FROM MONAZITE SANDS

Major minerals containing Lanthanides are :

1. Monazite
2. Bastnaesite
3. Cerite

Lanthanides are best extracted from monazite

Extraction of Lanthanides from monazite

Monazite sand consists of mixture of Th, Ce, Nd, Pr and La phosphates along with impurities. The mineral is powdered and digested with concentrated H_2SO_4 until a paste of lanthanide sulphates containing phosphoric acid and an excess of sulphuric acid is formed. The paste is centrifuged and then treated with cold water when silica being insoluble precipitates out. The solution is neutralised with a previously prepared mixture of lanthanide oxides when thorium, zirconium and titanium precipitate out as pyrophosphates. Sodium sulphate is added to the clear mother liquor so that the light lanthanides (La to Sm) may precipitate out as double salts while the heavy lanthanides (Gd to Lu) still remain in solution.

Hot sodium hydroxide is added to the precipitated double sulphates of light lanthanides to form a mixture of hydrated oxides. This is washed (till free form Na_2SO_4) and dried in air at 373 K whereby cerium is completely oxidized to CeO_2 while other lanthanides change into their oxides. At this stage, the approximate composition of residual solids is CeO_2 (50%), Nd_2O_3 (20%), La_2O_3 (17%), Pr_2O_3 (8%), Sm_2O_3 (5%). This mixture is reacted with dilute nitric acid when all the lanthanide oxides (except CeO_2) get dissolved and filtered out. The crude CeO_2 is dissolved in 85% HNO_3 and cerium is precipitated from it as the red basic nitrate, $Ce(OH)(NO_3)_3 \cdot 2H_2O$ by the addition of excess of dilute H_2SO_4. After the removal of cerium, the filtrate containing lanthanides are separated into individual components.

The method has been summed up in the following flow sheet.

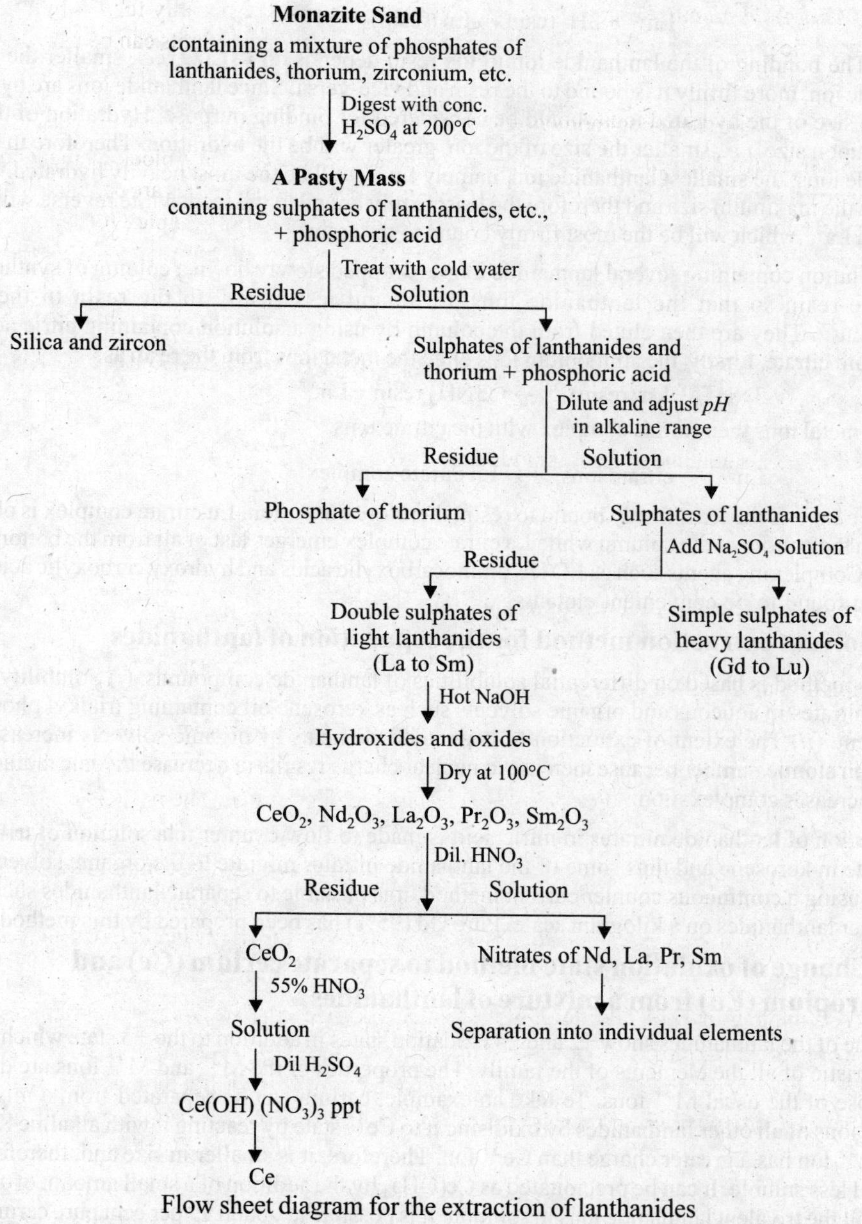

Monazite Sand
containing a mixture of phosphates of
lanthanides, thorium, zirconium, etc.

↓ Digest with conc. H_2SO_4 at 200°C

A Pasty Mass
containing sulphates of lanthanides, etc.,
+ phosphoric acid

↓ Treat with cold water

Residue → Silica and zircon

Solution → Sulphates of lanthanides and thorium + phosphoric acid

↓ Dilute and adjust *pH* in alkaline range

Residue → Phosphate of thorium

Solution → Sulphates of lanthanides

↓ Add Na_2SO_4 Solution

Residue → Double sulphates of light lanthanides (La to Sm)

Solution → Simple sulphates of heavy lanthanides (Gd to Lu)

Double sulphates of light lanthanides ↓ Hot NaOH

Hydroxides and oxides

↓ Dry at 100°C

CeO_2, Nd_2O_3, La_2O_3, Pr_2O_3, Sm_2O_3

↓ Dil. HNO_3

Residue → CeO_2

Solution → Nitrates of Nd, La, Pr, Sm → Separation into individual elements.

CeO_2 ↓ 55% HNO_3

Solution

↓ Dil H_2SO_4

$Ce(OH)(NO_3)_3$ ppt

↓

Ce

Flow sheet diagram for the extraction of lanthanides

As the properties of Lanthanide elements are quite similar, separation of these elements from one another poses a challenge. However, the following methods have been employed with a fair degree of success.

5.2.1. Separation of lanthanides by ion exchange method

This is the most rapid and effective method for the separation and purification of the lanthanides. This method is based on the following principles:

(*i*) Lanthanide ions on contact with synthetic resins containing –COOH or $–SO_3H$ group undergo proton exchange.

$$Ln^{3+} + 3H\text{–resin} \rightarrow Ln\,(resin)_3 + 3H^+$$

(*ii*) The bonding of the lanthanide ion to the resin depends on its size, *i.e.*, smaller the size of lanthanide ion, more firmly it is bound to the resin and vice-versa. Since lanthanide ions are hydrated, therefore size of the hydrated ions should be considered for binding purpose. Hydration of the ions depends upon size, *i.e.*, smaller the size of the ion, greater will be the hydration. Therefore in case of lanthanide ions, the smallest lanthanide ion, namely Lu^{3+}, will be the most heavily hydrated. Thus it will have the maximum size and therefore the least firmly bound to the resin while reverse will be the case with La^{3+} which will be the most firmly bound to the resin.

A solution containing several lanthanide ions is dropped slowly down a column of synthetic ion exchange resin so that the lanthanide ions are bound less firmly to the resin in the order La^{3+} to Lu^{3+}. They are then eluted from the column by using a solution containing citric acid and ammonium citrate. Firstly, the ammonium ions elute the metal ions from the resin as:

$$3NH^+{}_4 + Ln\,(resin)_3 \;\rightarrow\; 3NH_4\,resin + Ln^{3+}$$

The metal ions then form a complex with the citrate ions.

$$Ln^{3+} \;+\; citrate\ ions \;\rightarrow\; Ln\ citrate\ complex$$

Since Lu^{3+} is the least firmly bound to resin, therefore on elution, Lu citrate complex is obtained first from the bottom of the column while La citrate complex emerges last of all from the bottom of the column. Complexing agents such as EDTA, aminocarboxylic acids and hydroxy carboxylic acids have also been found to be convenient elutants.

5.2.2. Solvent extraction method for the separation of lanthanides

This method is based on differential solubilities of lanthanide compounds. (*i*) Solubility of lanthanide nitrates in aqueous and organic solvents such as kerosene oil containing trialkyl phosphates is different. (*ii*) The extent of extraction of lanthanide elements by organic solvents increases with increase in atomic number because increase in nuclear charge results in decrease in ionic radius which in turn increases complexation.

Solution of lanthanide nitrates in nitric acid is made to flow counter to a solution of tri-n-butyl phosphate in kerosene and thus some of the lanthanide nitrates migrate to the organic solvent layer. Thus by using a continuous counter current method, it is possible to separate lanthanides such as Gd from other lanthanides on a kilogram scale. Pure Gd (95%) has been prepared by this method.

5.2.3. Change of oxidation state method to separate cerium (Ce) and europium (Eu) from a mixture of lanthanides

Some of the lanthanides show +2 and +4 oxidation states in addition to the +3 state which is most characteristic of all the elements of the family. The properties of the M^{2+} and M^{4+} ions are different from those of the usual M^{3+} ions. To take an example, cerium can be separated from a mixture of trivalent ions of all other lanthanides by oxidising it to Ce^{4+} state by reacting it with alkaline $KMnO_4$. Now, Ce^{4+} ion has a greater charge than Ce^{3+} ion. Therefore, it is smaller in size and, therefore, less basic and less soluble. It can be precipitated as $Ce(OH)_4$ by the addition of a small amount of an alkali leaving all the trivalent lanthanide ions in solution. It is possible to obtain 99 per cent pure cerium from a mixture containing only 40 per cent cerium by this method.

Europium can be separated from a mixture of trivalent lanthanide ions by reducing it first to the +2 state by means of zinc amalgam. Eu^{2+} can then be precipitated as $EuSO_4$ which is insoluble in water. The sulphates of all the trivalent lanthanides, on the other hand, are soluble in water.

5.2.4. Extraction of Lanthanide metal from its salt

The following steps are involved in the process of extraction of the metal from its salts

(*i*) Reduction of anhydrous chlorides with sodium. Lighter lanthanides such as La, Ce and Gd can be easily prepared by the reduction of anhydrous lanthanide chlorides with sodium at 1273 K.

$$La\,Cl_3 + 3Na \rightarrow La + 3NaCl$$

***(ii)* Reduction of anhydrous fluorides with calcium or magnesium.** Heavier lanthanides such as Lu are generally prepared by the reduction of anhydride fluorides with calcium or magnesium metal at 1273 K.

$$2LuF + 3Mg \rightarrow 3MgF + 2Lu$$

***(iii)* Electrolysis of fused chloride.** Electrolysis of fused lanthanide chlorides is carried out. This method is similar to that of preparation of calcium by the electrolysis of calcium chloride.

5.2.5. Lanthanide Compounds

OXIDES

Only Ce, Pr, Tb and Nb form oxides in +4 oxidation state. CeO_2 is pale yellow inert material when prepared by ignition. However, in the hydrous freshly precipitated form it redissolves quite readily in acids. The analogous dark coloured PrO_2 and TbO_2 are obtained from their stable oxides Pr_6O_{11} and Tb_4O_7 under extreme conditions. For example, PrO_2 is obtained by O_2 at 282 atm. pressure and 400°C while TbO_2 requires atomic oxygen at 450°C.

Lanthanides burn with oxygen forming the oxides Ln_2O_3. **(Ln for Lanthanide in general)**. These are also obtained by the ignition of hydroxides, carbonates, nitrates etc.

$$2Ln + 3O_2 \longrightarrow Ln_2O_3$$

$$Ln_2(CO_3)_2 \longrightarrow Ln_2O_3 + 3CO_2$$

$$4Ln(NO_3)_3 \longrightarrow 2Ln_2O_3 + 2NO_2 + 3O_2$$

The elements Ce, Pr and Tb are exceptions which form the most stable products CeO_2, Pr_6O_{11} and Tb_4O_7 respectively.

Ln_2O_3 are strongly basic and their basic character decreases as atomic number increases. for example, La_2O_3 is strongly basic while Lu_2O_3 is least basic. The lighter more basic oxides resemble the oxides of alkaline earth metals. All are soluble in water but form hydroxides. They dissolve readily in aqueous acids to form solution which contain the $[Ln(H_2O)_x]^{3+}$. Hydrous hydroxides can be precipitated from these solutions by the addition of ammonia or aqueous alkali.

Crystalline $Ln(OH)_3$ has a 9 coordinated tricapped, trigonal prismatic structure and may be prepared by prolonged treatment of Ln_2O_3 with conc. NaOH at high temperature and pressure.

Chalcogenides (sulphide, selenide, talurides etc.)

These are known for many lanthanides. In general, these are stable in dry air but are hydrolysed in the presence of moisture. If heated in air they oxidise (especially sulphides) to basic salts of the corresponding oxo-anion and they are attacked by acids with the evolution of H_2Z.

Monochalcogenides LnZ (Z = S, Se, Te) have been prepared for all the lanthanides except for Pm, mostly by direct combination. They are almost black and with the exceptions of SmZ, EuZ, YbZ, TmSe and TmTe, they have high metallic conductivity and consist of $Ln^{III} + Z^{2-}$ ions.

Trivalent chalcogenides, Ln_2Z_3 have been obtained by direct combination or action of H_2S on the chloride or oxide. When Ln_2Z_3 are heated with an excess of chalcogen in a sealed tube at 600°C, products with composition upto or nearing LnZ_2 are obtained. These are polychalcogenides with the metal in +3 oxidation states.

HALIDES

Lanthanides form halides of the types LnX_4, LnX_3 and LnX_2.

Fluorides of Ce(IV), Pr(IV) and Tb(IV) have been prepared. CeF_4 is white and is comparatively stable and can be prepared either directly from the elements or by the action of F^- on aqueous solutions of Ce^{IV} when it crystallizes as the monohydrate. It can also be prepared by treating cerium (III) fluoride with fluorine at room temperature.

$$2CeF_3 + F_2 \longrightarrow 2CeF_4$$

Anhydrous Lanthanide (III) fluorides are prepared by adding HF or NaF solution to a salt solution containing Ln^{3+} ions :

$$Ln^{3+} + 3F \longrightarrow LnF_3$$

Lanthanide fluorides are sparingly soluble in water.

Anhydrous lanthanide chlorides are prepared by heating the metal in the atmosphere of dry chlorine or by heating lanthanide oxides with carbonyl chloride as

$$2Ln + 3Cl_2 \longrightarrow 2\,LnCl_3$$

$$Ln_2O_3 + 3COCl_2 \longrightarrow 2\,LnCl_3 + 3CO_2$$

Simple anhydrous lanthanide halides cannot be prepared simply by heating the hydrated salts as they form lanthanum oxychlorides (LnOCl).

5.2.6 Uses of lanthanides

(*i*) Cerium glass cuts off heat and ultraviolet light and so is used in glare-reducing spectacles.

(*ii*) Cerium-magnesium alloys are used in flashlight powders.

(*iii*) Neodymium oxide dissolved in selenium oxychloride is one of the most powerful liquid lasers known so far.

(*iv*) Ceria and thoria are used in gas lamp mantles.

(*v*) Cerium salts are used in analysis, dyeing, lead accumulators, medicines and as catalysts.

(*vi*) Lanthanide elements and their compounds are being used in nuclear control, shielding and fluxing devices.

(*vii*) Since lanthanides improve the workability of steel when heated, so alloys of lanthanides with iron are well known.

(*viii*) Gadolinium sulphate heptahydrate has been used to produce very low temperature.

SOLVED EXAMPLES

Q.1. Can lanthanum ion (Z = 57) exist in +4 oxidation state ? Jusitfy your answer.

Ans. Lanthanum (Z = 57) has the electronic configuration [Xe] $5d^1 6s^2$ and forms La^{3+} by the removal of all the three electrons of outermost shell resulting in noble gas configuration of xenon. To form La^{4+} ion, the electron has to be removed from stable xenon core, which is energetically difficult. Thus, La^{4+} cannot exist.

Q.2. Why is europium (II) more stable than cerium (II) ?

Ans. Europium (II) has electronic configuration [Xe] $4f^7 5d^0$ while cerium (II) has electronic configuration [Xe] $4f^1\ 5d^1$. In Eu^{2+}, $4f$ subshell is half filled and $5d$-subshell is empty. Since half filled and completely filled electronic configurations are more stable, Eu^{2+} ion is more stable than Ce^{2+} in which neither $4f$ subshell nor $5d$ subshell is half filled or completely filled.

Q.3. Lanthanides have poor tendency to form complexes. Explain.

Ans. Although lanthanides have high charge (+3), but the size of their ions is also very large resulting in small charge density. Therefore, they have poor tendency to form complexes. However, they form complexes with a few chelating ligands such as β–diketones, oximes, EDTA, α, α′–dipyridyl etc.

Q.4. Write the following in order of decreasing reducing strength : Eu^{2+}, Sm^{2+}, Yb^{2+}.

Ans. The decreasing order of reducing strength is :

$Sm^{2+} > Eu^{2+} > Yb^{2+}$

Q.5. Among La (57), Sm (62), Gd (64) and Yb (70) :

(*i*) Which element will give coloured ion ?

(*ii*) Which element will give paramagnetic ion ?

(*iii*) What are the possible oxidation states ?

Ans. (*i*) Sm^{3+} gives coloured ion

(*ii*) Sm^{3+} and Gd^{3+} are paramagnetic because of the presence of unpaired electrons in 4*f* orbitals.

(*iii*) La and Gd show oxidation state of + 3 only ; $La^{3+} = [Xe]4f^0$; $Gd^{3+} = [Xe]4f^7$.

Yb also shows oxidation state of + 2 besides + 3

$$Yb^{2+} = [Xe]4f^{14}$$

Sm also show oxidation state of + 2 besides + 3, although Sm^{2+} has neither half filled nor completely filled 4*f* orbitals.

$$Sm^{2+} = [Xe]4f^6$$

Q.6. What are the problems in separation of lanthanides from one another ?

Ans. All lanthanides have almost similar properties because they have similar +3 charge and almost similar size. Therefore, it becomes difficult to separate the lanthanides from one another. However, recently, methods have been developed based on the properties such as solubility complex formation, hydration of ions etc., which can help in the separation process.

Q.7. What is the consequence of lanthanide contraction on the basic strength of lanthanide hydroxides ?

Ans. Due to lanthanide contraction, the size of the lanthanide ions decrease regularly with increase in atomic number. According to Fajan's rules, the smaller positive ion can cause greater polarisation of the anion and therefore, covalent character between La^{3+} and OH^- ions increases. Consequently the basic character of the hydroxides decreases with increase in atomic number. Thus $La(OH)_3$ is most basic while $Lu(OH)_3$ is the least basic.

Q.8. Explain the following :

(*i*) Which is more basic Gd_2O_3 or YbO ?

(*ii*) Which is thermally more stable $La(NO_3)_3$ or $Tm(NO_3)_3$?

Ans. (*i*) Gd_2O_3 is more basic than YbO because basic character decreases with increase in atomic number. Due to lanthanide contraction, the decrease in size with increase in atomic number increases the covalent character. Basic character therefore decreases.

(*ii*) $La(NO_3)_3$ is thermally more stable than $Tm(NO_3)_3$. Thermal stability of nitrates decreases along the series due to increasing polarising power of the cations with decrease in size along the series.

Q.9. What happens when cerium (III) nitrate is treated with alkaline $KMnO_4$?

Ans. Ce(III) gets oxidised to Ce(IV)

$$3Ce^{3+} + MnO_4^- + 2H_2O \longrightarrow 3Ce^{4+} + MnO_2 + 4OH^-$$

Q.10. Lanthanides do not form oxocations. Explain.

Ans. Lanthanides do not form oxocations because they do not show high oxidation states such as +6. Therefore, the lanthanide ions do not have high charge density. Hence they do not form oxocations.

Q.11. La^{3+}, Lu^{3+} and Ce^{4+} ions are diamagnetic while Sm^{3+} shows low paramagnetism. Explain.

Ans. The electronic configuration of the ions are :

La^{3+} : [Xe] $4f^0$

Lu^{3+} : [Xe] $4f^{14}$

Ce^{4+} : [Xe] $4f^0$

Sm^{3+} : [Xe] $4f^5$

The above electronic configurations show that La^{3+} and Ce^{4+} have no electron in 4*f*-subshell while Lu^{3+} has fully filled *f*-orbitals. Therefore, they are diamagnetic. On the other hand, Sm^{3+} has five unpaired electrons in 4*f*-subshell which is responsible for its paramagnetism.

EXERCISES
(Based on Question Papers of Different Universities)

Multiple Choice Questions

Choose the correct option

1. Which of the following does not belong to lanthanides ?
   (*a*) Nd (*b*) Tm (*c*) Cm (*d*) Ce
2. The separation of lanthanides in ion-exchange method is based on :
   (*a*) basicity of lanthanides (*b*) oxidation state of the ion
   (*c*) solubility of their nitrates (*d*) size of the ions
3. Which of the following pair has same number of unpaired electrons ?
   (*a*) Eu^{3+} — Sm^{3+}
   (*b*) Lu^{3+} — Yb^{2+}
   (*c*) La3+ — Eu^{2+}
   (*d*) Gd^{3+} — Sm^{2+}
4. Which of the following trivalent lanthanide ions is not coloured ?
   (*a*) Er^{3+} (*b*) Yb^{3+} (*c*) Pm^{3+} (*d*) Sm^{3+}
5. Which of the following trivalent lanthanide ions is diamagnetic ?
   (*a*) Lu^{+3} (*b*) Nd^{+3} (*c*) Pr^{+3} (*d*) Tm^{+3}
6. Which lanthanide has the configuration $4f^7 5d^1 6s^2$?
   (*a*) Sm (*b*) Gd (*c*) Eu (*d*) Tb
7. Which of the following has smallest ionic size ?
   (*a*) La^{3+} (*b*) Cd^{3+} (*c*) Dy^{3+} (*d*) Lu^{3+}
8. Which of the following forms stable +4 oxidation state ?
   (*a*) La (Z = 57) (*b*) Ce (Z = 58) (*c*) Eu (Z = 63) (*d*) Cd (Z = 64)
9. The principal oxidation state of lanthanides is :
   (*a*) +2 (*b*) +3 (*c*) +4 (*d*) zero
10. The coordination number of Ce in $[Ce(NO_3)_6]^{2-}$ is :
    (*a*) 6 (*b*) 8 (*c*) 12 (*d*) 4
11. Decreasing order of ionic radii of Sm^{3+}, Pr^{3+} and La^{3+} is
    (*a*) $La^{3+} > Pr^{3+} > Sm^{3+}$ (*b*) $La^{3+} > Sm^{3+} > Pr^{3+}$
    (*c*) $Pr^{3+} > La^{3+} > Sm^{3+}$ (*d*) $Sm^{3+} > La^{3+} > Pr^{3+}$
12. Out of La^{3+}, Pr^{3+} and Nd^{3+} which is colourless ?
    (*a*) Pr^{3+} (*b*) La^{3+} (*c*) Nd^{3+} (*d*) None of these

ANSWERS

1. (*c*)	2. (*d*)	3. (*b*)	4. (*b*)	5. (*a*)	6. (*b*)
7. (*d*)	8. (*b*)	9. (*b*)	10. (*c*)	11. (*a*)	12. (*b*)

Short Answer Questions

1. Why do Ce and Tb exhibit +4 oxidation states ?
2. Landanides have poor tendency to form complexes. Why ?
3. Which of the two : Ce^{3+} or Gd^{3+} has smaller size ?
4. Why actinides have greater tendency to form complexes than lanthanides ?
5. Write the general electronic configuration of lanthanides.
6. Compare basic character of $La(OH)_3$ and $Lu(OH)_3$.
7. Why do Zr and Hf have similar properties ?
8. Can lanthanum (Z = 57) exist in the +4 oxidation state ? Explain.
9. Why $La(OH)_3$ is more basic than $Lu(OH)_3$?
10. Why is europium (II) more stable than cerium (II) ?
11. Define lanthanide contraction.
12. Why do Eu and Yb exhibit +2 oxidation states ?
13. Which of the two La^{3+} or Ce^{3+} is paramagnetic ?

General Questions

1. What are lanthanides ? Discuss their electronic structures and the similarity of their properties.
2. Discuss the following properties of lanthanides :
   (*a*) Tendency to form complexes
   (*b*) Oxidation states.
3. Give the important oxidation states of lanthanides. How would you account for these ? Why is +3 most stable ?
4. What are the problems in the separation of lanthanides from one another ? Discuss the ion exchange and solvent extraction methods for the separation of lanthanides.
5. Why does lanthanum exhibit only +3 oxidation state whereas certain other elements of *f*-block show +2 and +4 also ?
6. Give the general electronic configuration of lanthanides. Explain.
   (*i*) the filling of 4f-subshell in the lanthanide series.
   (*ii*) the anomalous oxidation states of +2 and +4 shown by some elements in the series.
7. What is lanthanide contraction ? How would you account for it ? What are its important consequences ?
8. Explain the following :
   (*i*) Filling of 4*f*-sub-level is not regular in the lanthanide series.
   (*ii*) +3 oxidation state is the characteristic oxidation satate of lanthanides, though their atoms contain only $2(6s^2)$ outermost electrons.
9. Explain why :
   (*i*) Zr and Hf have almost similar properties.
   (*ii*) Cerium (IV) sulphate is used in redox reactions.
   (*iii*) La, Gd and Lu show only +3 oxidation states whereas other lanthanides can exhibit +2 and +4 oxidation states.

10. How is lanthanide contraction explained ? What is its effect on the following :

(*i*) properties of second and third transition series.

(*ii*) extraction of lanthanides by ion exchange method.

(*iii*) basic strength of lanthanide hydroxides.

11. (*a*) Why do the lanthanides form a closely knit group with similar chemical and physical properties ?

(*b*) Can lanthanum (Z = 57) ion exist in +4 oxidation state ? Justify your answer.

12. (*a*) Give reasons for the existence of lanthanide contraction and its consequence. Compare the following properties of lanthanides with those of transition elements :

(*i*) magnetic properties

(*ii*) colour

(*iii*) complex formation.

13. (*a*) Discuss separation of lanthanides by ion-exchange method.

(*b*) What are lanthanides ? They are also called rare earth elements. Explain.

14. Explain why :

(*i*) Lanthanides form fewer complexes as compared to actinides ?

(*ii*) Values of atomic radii for Eu and Yb is more than expected ?

15. Write short notes on :

(*a*) Lanthanide contraction.

(*b*) Ion-exchange method for separation of lanthanides.

16. (*a*) Why are lanthanides called inner transition elements ?

(*b*) Discuss the electronic configuration of the lanthanides and their position in the periodic table.

(*c*) What is lanthanide contraction ? Discuss the cause of lanthanide contraction and its effects.

17. Why is chemistry of all lanthanides identical ?

18. (*a*) Describe oxidation states of lanthanides. Why is +3 state most stable ?

(*b*) What are double salts of lanthanides ?

19. Explain 'lanthanide contraction', its cause and consequences. Is there similar actinide contraction ? If so, explain why ?

20. (*a*) What are lanthanides ? Give the important oxidation states of lanthanides. How would you account for them ?

(*b*) Discuss briefly ion exchange method for the separation of lanthanides.

21. What is lanthanide contraction ? Discuss its effect on post lanthanide elements.

22. Write notes on :

(*i*) Stability of oxidation states of lanthanides.

(*ii*) Variation of ionic radii among lanthanides.

23. (*a*) What are lanthanides ? Give the important oxidation states of lanthanides. How would you account for these ?

(*b*) Discuss briefly the ion-exchange method for the separation of lanthanides.

24. (*a*) Why is separation of lanthanides difficult ?

(*b*) Which is more basic in nature and why $La(OH)_3$ or $Lu(OH)_3$?

25. Suggest reasons as to why :
(*a*) The chemistry of lanthanides is so identical, and
(*b*) Basic strength of lanthanide oxides (M_2O_3) decreases with increasing atomic number.

26. (*a*) Discuss ion exchange method for separation of lanthanides.
(*b*) Compare basic character of $La(OH)_3$ and $Lu(OH)_3$.

27. Explain the reasons why most of the lanthanide ions are coloured and paramagnetic.

28. (*a*) What is lanthanide contraction ? What are its consequences ?
(*b*) How are lanthanides separated from each other by ion-exchange method ?
(*c*) Calculate the magnetic moment (μ) of Dy^{3+} ion having outer electronic configuration $4f^9$.

29. (*a*) Give two examples of lanthanide ions which have the same number of unpaired *f*-electrons as well as same colour.
(*b*) Give the suitable reasons for the following :
(*i*) Yb, Ho, Er occur together in natural minerals.
(*ii*) Eu and Yb have lower m.p. and b.p. as compared to other lanthanide elements.
(*c*) Describe the separation of lanthanides by ion exchage method.

30. (*a*) What are the applications of lanthanides ?
(*b*) Discuss separation of lanthanides by ion-exchange method.
(*c*) Lanthanides have poor tendency to form complexes. Explain.

31. Discuss three methods including ion exchange methods to separate various lanthanides from mixture.

32. (*a*) Explain :
(*i*) Which is more basic and why : Gd_2O_3 or YbO ?
(*ii*) Chemistry of all lanthanides is identical.
(*b*) What is lanthanide contraction ? What are its consequences ? Discuss.
(*c*) Name the Lanthanides in correct order.

33. (*a*) Describe briefly ion-exchange method for the separation of lanthanides.
(*b*) Discuss and compare the electronic structure and position of lanthanides in the periodic table.

34. Why is the chemistry of lanthanides so similar ?
Describe ion exchange method for their separation.

35. (*a*) What are lanthanides? Give their applications.
(*b*) $La(OH)_3$ is more basic than $Lu(OH)_3$. Explain.

36. (*a*) Give the important oxidation states of lanthanides. How would you account for them? Why is +3 state more stable.
(*b*) Why separation of lanthanides is difficult ? Discuss the ion exchange method for the separation of lanthanides.

37. (*a*) Can lanthanum ion (57) exist in +4 oxidation state ? Justify your answer.
(*b*) Why is Eu (II) more stable than Ce (II) ?

38. Work out the number of unpaired electrons in the ground state of the following ions :
La^{3+}, Ce^{4+} and Yb^{2+}

39. What are the characteristic precipitation reactions of lanthanides in +2, +3 and +4 ions? How are the lanthanide ions separated from each other ?

40. What is lanthanide contraction and what is its effect on the extraction of lanthanides by ion-exchange method ?

41. (*a*) What are the lanthanide ions which are used as n m r shift reagents and how it helps in assigning and interpreting the peaks in lanthanide complexes ?

(*b*) What is lanthanide contraction ? How could you account for it ? What are its consequeces and effects ?

(*c*) Explain why all the oxidation states of lanthanides cannot be explained on the basis of extrastability of f^0, f^1 and f^{14} configurations.

42. (*a*) Describe ion-exchange method for the separation of lanthanides.

(*b*) What is lanthanide contraction ? Give its consequences.

43. What is lanthanide contraction ? What bearing does it have on the separation of their elements and on the atomic radii of the atoms of elements which succeed this group of metals in the periodic table.

44. (*a*) Chemistry of Lanthanides is similar. comment.

(*b*) Lanthanides have poor tendency to form complexes. Why?

(*c*) Compare basic character of $La(OH)_3$ and $Lu(OH)_3$.

45. (*a*) (*i*) Why is europium (II) more stable than cerium (II) ?

(*ii*) Write the following in order of decreasing reducing strength :

Eu^{2+}, Sm^{2+}, Yb^{2+}

(*b*) (*i*) Which is thermally more stable and why ?

$La(NO_3)_3$ or $Tm\ (NO_3)_3$

(*ii*) Which is more basic and why ?

Gd_2O_3 or YbO

46. (*a*) Give points of similarities between lanthanides and actinides.

(*b*) Describe the extraction of lanthanides from monazite.

47. (*a*) Discuss ion-exchange method for the separation of lanthanides.

(*b*) Give reasons for the existance of lanthanide contraction. Also discuss its consequences.

(*c*) Can lanthanum ion exist in +4 oxidation state ? Justify your answer.

48. Explain the following :

(*i*) Eu and Yb are lower m.pt. and b.pt. lanthanides.

(*ii*) The calculated magnetic moment Eu^{+3} is zero, while observed value is not zero.

(*iii*) The trans-uranium elements are generally radioactive in nature.

49. (*a*) Which is more basic and why. $La(OH)_3$ or $Lu(OH)_3$?

(*b*) Why is Eu(II) more stable than Ce(II)?

(*c*) Discuss ion-exchange method for separation of lanthanides.

(*d*) What are the applications of lanthanides?

50. (*a*) Do you think lanthanides form double salts ? If yes, cite an example.

(*b*) What is lanthanide contraction? What are the consequences of lanthanide contraction?

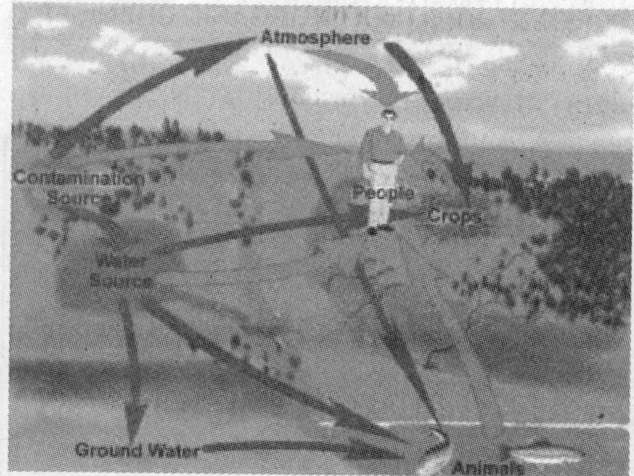

Disaster of nuclear waste

6

Chemistry of Actinides

6.1. INTRODUCTION

The fifteen elements from actinium (Z = 89) to lawrencium (Z = 103) in which the 5*f* orbitals are being filled up are called actinide series. The name actinide is derived from actinium, the very first member of the series. These are also known as **actinons**. Only the first four elements of this series, namely, actinium, thorium, protoactinium and uranium, occur in nature, the other elements are made artificially by nuclear bombardment. All the actinides are toxic to humans.

6.1.1. Electronic Configuration of Actinides

After mercury we follow six elements from thalium to radon in which electrons enter the 6*p* subshell. Radon has the configuration [Xe] $6s^2\ 4f^{14}\ 5d^{10}\ 6p^6$. At this stage the 7*s* and 6*d* subshells have smaller energy relative to the 5*f* subshell and the next two electrons are added to the 7*s* subshell and we get francium and radium which are *s*-block elements. We find that the 5*f* and 6*d* subshells penetrate the 7*s* orbital quite significantly and are therefore stabilized relative to the 7*p* subshell. The next few electrons enter either of these two subshells, *i.e.*, 5*f* and 6*d*. It soon becomes obvious that as we fill electrons from actinium onwards (*i.e.*, actinide series), the energies of 6*d* and 5*f* subshell do not vary significantly and the allocation of exact configuration becomes difficult. It is because of this very reason that different authors have tried to suggest different electronic configurations for actinides. However, the electronic configurations for actinides set out in Table 6.1 have been generally accepted. An important point is that around actinium both 6*d* and 5*f* subshells are of very similar energies, but as we keep filling the electrons the 5*f* subshell definitely acquires greater stability. From uranium onwards there is no doubt that the 5*f* subshell is more stable. The general electronic configuration of actinides is

(L-R) Glennseaborg and Edwin McMilan discover Np and Pu

$$[Rn]\ 5f^{0-14}\ 6d^{0-2}\ 7s^2$$

Table 6.1. Properties of Actinides

Element	*Symbol*	*Atomic number*	*Electronic configuration*	*Oxidation states*	*Ionic radii (pm)* M^{3+}	M^{4+}
Actinium	Ac	89	[Rn] $5f^0\ 6d^1\ 7s^2$	3	111	–
Thorium	Th	90	[Rn] $5f^0\ 6d^2\ 7s^2$	4	108	99
Protactinium	Pa	91	[Rn] $5f^2\ 6d^1\ 7s^2$	3, 4, 5	105	96
Uranium	U	92	[Rn] $5f^3\ 6d^1\ 7s^2$	3, 4, 5, 6	103	93
Neptunium	Np	93	[Rn] $5f^5\ 6d^0\ 7s^2$	3, 4, 5, 6, 7	101	92
Plutonium	Pu	94	[Rn] $5f^6\ 6d^0\ 7s^2$	3, 4, 5, 6, 7	100	90
Americium	Am	95	[Rn] $5f^7\ 6d^0\ 7s^2$	3, 4, 5, 6	99	89
Curium	Cm	96	[Rn] $5f^7\ 6d^1\ 7s^2$	3, 4	98	88
Berkelium	Bk	97	[Rn] $5f^9\ 6d^0\ 7s^2$	3, 4	96	–
Californium	Cf	98	[Rn] $5f^{10}\ 6d^0\ 7s^2$	3, 4	95	–
Einsteinium	Es	99	[Rn] $5f^{11}\ 6d^0\ 7s^2$	2, 3	–	–
Fermium	Fm	100	[Rn] $5f^{12}\ 6d^0\ 7s^2$	2, 3	–	–
Mendelevium	Md	101	[Rn] $5f^{13}\ 6d^0\ 7s^2$	2, 3	–	–
Nobelium	No	102	[Rn] $5f^{14}\ 6d^0\ 7s^2$	2, 3	–	–
Lawrencium	Lr	103	[Rn] $5f^{14}\ 6d^1\ 7s^2$	3	–	–

As there is not much difference between 5*f* and 6*d,* it becomes difficult to know whether the electron has entered 5*f* or 6*d*. This makes predicting electronic configuration difficult.

6.1.2. Oxidation states

It is found that most stable oxidation states of the elements up to uranium is the one involving all the valency electrons. Neptunium forms the +7 state using all the valency electrons but this is oxidising and the most stable state is +5.

Plutonium also shows states up to +7 and americium up to +6 but the most stable state drops to Pu (+4) and Am (+3) and latter elements tend to be stable in +3 oxidation state.

Berkelium in +4 state is strongly oxidising but is more stable than curium and americium in +4 state due to f^7 configuration. Similarly, nobelium is markedly stable in +2 state due to its f^{14} configuration. Oxidation states of actinides are given in Table 6.1.

When the oxidation state of an actinide increases to +6, the ion is no longer a simple ion. **There is a high charge density on the ion which leads to the formation of oxygenated ion, *e.g.*, UO_2^{2+}**. Such oxycations are stable in acid and aqueous solutions.

6.1.3. Ionic radius and actinide contraction

It is seen from Table 6.1 that there is a regular decrease in ionic radii with increase in atomic number from actinium to lawrencium. This is called actinide contraction. This contraction is similar to the lanthanide contraction. The decrease in ionic radii of actinides (M^{3+} and M^{4+} ions) is shown in Fig. 6.1.

The actinide contraction is caused due to imperfect shielding of one 5*f* electron by another in the same shell. Thus as we move along the actinide series, the nuclear charge and the number of 5*f* electrons increase by one unit at each step. However, due to imperfect shielding (because the shape of *f* orbitals is very much diffused) the net effective nuclear charge increases which causes contraction in the size of the electron cloud.

However in actinide contraction there are bigger jumps in contraction between the consecutive elements as compared to lanthanides. This is due to lesser shielding of 5*f* electrons which results in stronger pull by the nucleus.

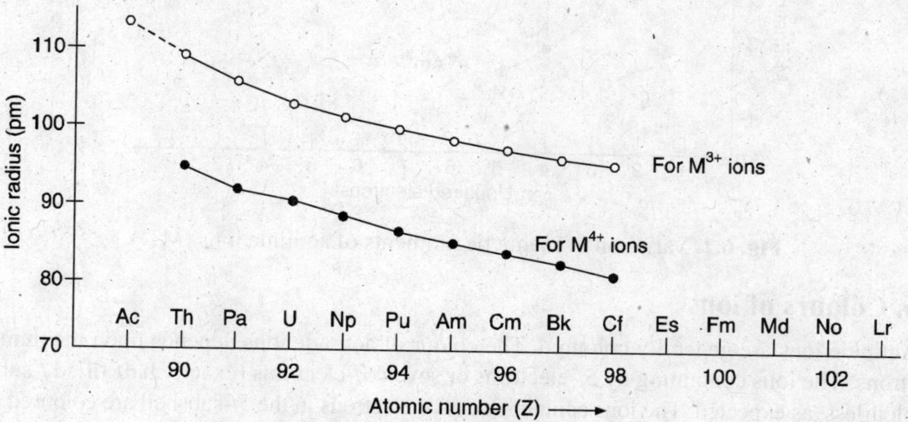

Fig. 6.1. Variation of ionic radii of actinides.

6.1.4. Variation in m.p., b.p. and density

(*i*) Melting and boiling points. There is no regular trend in the melting and boiling points of actinides as seen in Table 6.2.

Table 6.2. Melting points, boiling points and densities of some actinides.

Elements	*Ac*	*Th*	*Pa*	*U*	*Np*	*Pu*	*Am*	*Cm*
m.p. (K)	1323	2033	1503	1405	910	913	1273	1613
b.p. (K)	—	4123	—	4086	—	3508	2880	—
Density (g/cm^3)	—	11.72	15.37	19.07	19.5	19.82	11.7	7.0

(*ii*) Density. Densities of actinides increase from left to right as shown in Table 6.2. This is due to progressive increase in atomic mass and decrease in atomic size.

6.1.5. Magnetic Properties

The magnetic properties of actinides are smaller than the theoretical values. This is due to the fact that 5*f* electrons of actinides are more exposed to crystal field and therefore, are less effectively shielded which results in quenching of the orbital contributions.

The magnetic moment of tri positive actinides decreases from U^{3+} to give a minimum at Pu^{3+} beyond which it increases and shows a maximum at Cf^{3+}. Thereafter, it starts decreasing again as illustrated in Fig. 6.2.

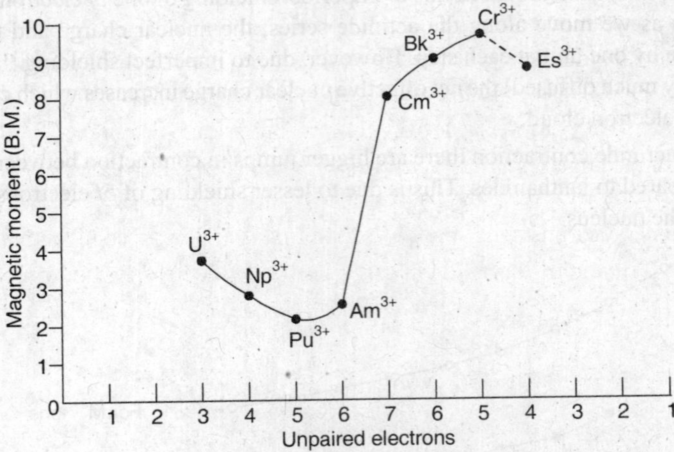

Fig. 6.2. Variation of magnetic moments of actinide ions (M^{3+}).

6.1.6. Colours of ions

Actinide ions are generally coloured. The colour of actinide ions depends upon the number of *f*-electrons. The ions containing no 5*f*-electrons or seven 5*f*-electrons (exactly half filled *f*-subshell) are colourless, as expected. The ions containing 2 to 6 electrons in the 5*f*-subshell are coloured (Table 6.3). The colours arise due to electronic transitions between various energy states *within the f-subshell* (f-f transitions). Table 6.3 below gives the colour of actinide ions in aqueous solution.

Table 6.3. Colours of actinide ions in aqueous solutions.

Ion	Number of electrons	Colour
Ac^{3+}	f^0	colourless
U^{3+}	f^3	red
Np^{3+}	f^4	blue or purple
Pu^{3+}	f^5	violet
Am^{3+}	f^6	pink
Cm^{3+}	f^7	colourless
Th^{4+}	f^0	colourless
Pa^{4+}	f^1	colourless
U^{4+}	f^2	green
Np^{4+}	f^3	yellow-green
Pu^{4+}	f^4	orange
Am^{4+}	f^5	pink
Cm^{4+}	f^6	pale-yellow

6.1.7. Formation of complexes

Actinides have a greater tendency to form complexes because of higher nuclear charge and small size of their atoms. Complexes formed by actinides are described below.

1. Small and highly charged M^{4+} ions exhibit greatest tendency towards complex formation, *e.g.*, Pu^{4+} forms very strong anion complexes.

2. Most actinide halides form adducts with alkali metal halides. For example, $ThCl_4$ forms the complexes $MThCl_5$, M_2ThCl_6 and M_3ThCl_7 where M is the metal alkali.

3. With pyridine, $ThCl_4$ as well as $ThBr_4$ form monopyridine complexes.

4. They also form complexes with acetylacetone, oxine, EDTA, such as tertrakis– (acetyl acetonato) thorium, Th $(acac)_4$ and $UO_2(C_9H_6NO)_2 \cdot C_9H_7ON$.

5. $UO^2{}_2^+$ forms rather unstable complexes with EDTA as compared to lanthanides.

6. The degree of complex formation in case of actinides decreases in the order

$$M^{4+} > MO_2^{2+} > M^{3+} > MO_2^+$$

where M is the element of actinide series.

6.1.8. Comparison in properties of Lanthanides and Actinides

Similarities.

(*i*) Oxidation state of +3 dominates in both the series.

(*ii*) In both the series *f* orbitals are being progressively filled.

(*iii*) Just like lanthanide contraction we have actinide contraction.

(*iv*) The nitrates, perchlorates and sulphates of trivalent actinides as well as lanthanides are soluble while the hydroxides, fluorides and carbonates are insoluble.

(*v*) In the absorption spectra of the elements of both the series sharp line-like bands appear due to *f*–*f* transitions.

Differences

Lanthanides	*Actinides*
(*i*) Binding energies of 4*f* are higher.	Binding energies of 5*f* are lower.
(*ii*) 4*f* electrons have comparatively greater shielding effect.	5 *f* electrons have poor shielding effect.
(*iii*) Maximum oxidation state exhibited by lanthanides is +4, *e.g.*, Ce^{4+}.	Due to lower binding energies they show higher oxidation states such as +4, +5 and +6.
(*iv*) They do not form oxocations.	They form oxocations such as UO_2^{2+}, UO_2^+ NpO_2^+, PuO_2^+.
(*v*) Except prometheium they are non-radioactive.	All of them are radioactive.
(*vi*) They do not form complexes easily. Their complexes with π-bonding ligands are not known.	They form complexes even with π-bonding ligands such as alkyl phosphines, thioethers etc.
(*vii*) They are paramagnetic but magnetic properties can be easily explained.	They are also paramagnetic but it is difficult to explain magnetic properties.

6.2. EXTRACTION OF THORIUM FROM MONAZITE SANDS

Various steps involved in the extraction of thorium from monazite sands are as follows:

(*a*) Conversion into thoria. Monazite is powdered and digested with conc. H_2SO_4 until pasty mass of sulphates become soluble in water. Since thorium is more basic than other rare earths, addition of sodium hydroxide or magnesium oxide to the acidic solution of the salts results in the precipitation of thorium phosphate. The washed thorium phosphate is heated with sodium carbonate when it changes into crude thoria ThO_2. The crude thoria contains lanthanides as impurities and is purified by conversion into soluble oxalate, recrystallization as nitrate and, finally, calcination into thoria.

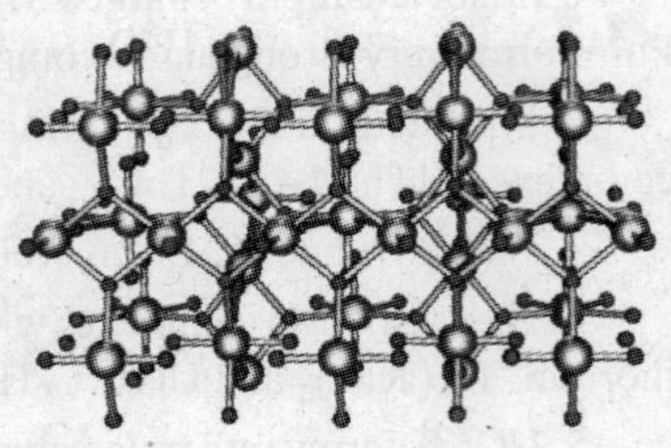
Structure of ThH_2

***(b)* Conversion to $ThCl_4$ and then into thorium.** Pure thoria is heated with phosgene when thorium tetrachloride is formed which on reduction with calcium forms thorium.

$$ThO_2 + 2COCl_2 \longrightarrow ThCl_4 + 2CO_2$$

$$ThCl_4 + 2Ca \longrightarrow 2CaCl_2 + Th$$

***(c)* Refining of crude thorium by Van Arkel method.** The metal thus prepared is slightly impure and contains thoria as impurity. It is purified by heating in iodine vapours when it forms volatile ThI_4. The vapours of ThI_4 are dissociated into the metal on a hot filament (1400-1500 K).

$$Th + 2I_2 \longrightarrow ThI_4$$

$$ThI_4 \xrightarrow{1400-1500K} \underset{\text{Pure}}{Th} + 2I_2$$

6.3 EXTRACTION OF URANIUM FROM ITS ORE PITCHBLENDE

Pitchblende contains uranium as U_3O_8 or UO_2. The various steps involved in the extraction of uranium are as under:

***(a)* Concentration.** The ore is crushed to a fine powder and then concentrated by gravity separation in order to remove lighter impurities.

***(b)* Roasting.** The concentrated ore is then roasted in a reverberatory furnace in the presence of excess of air in order to remove volatile impurities such as sulphur, arsenic and antimony.

$$S + O_2 \longrightarrow SO_2\uparrow$$

$$4As + 3O_2 \longrightarrow 2As_2O_3\uparrow$$

$$4Sb + 3O_2 \longrightarrow 2Sb_2O_3\uparrow$$

The roasted ore is then mixed with sodium carbonate and sodium nitrate. The whole mass is then fused on the hearth of a reverberatory furnace resulting in the formation of sodium uranate.

$$U_3O_8 + 3Na_2CO_3 + [O] \longrightarrow \underset{\text{Sodium uranate}}{3Na_2UO_4} + 3CO_2\uparrow$$

***(c)* Leaching with sulphuric acid.** The fused mass is leached with dilute sulphuric acid whereby sodium uranate changes into soluble uranyl sulphate while impurities such as $PbSO_4$ and $RaSO_4$ remain insoluble and are filtered off.

$$Na_2UO_4 + 2H_2SO_4 \longrightarrow \underset{\text{Uranyl sulphate}}{UO_2SO_4} + Na_2SO_4 + 2H_2O$$

***(d)* Treatment with sodium carbonate.** The solution containing uranyl sulphate is treated with excess of sodium carbonate to precipitate iron, aluminium, nickel, cobalt and manganese as their insoluble carbonates while uranyl sulphate reacts with sodium carbonate to form soluble sodium tricarbonatouranate $Na_4 [UO_2 (CO_3)_3]$.

$$UO_2SO_4 + 3Na_2CO_3 \longrightarrow Na_4 [UO_2 (CO_3)_3] + Na_2SO_4$$

***(e)* Treatment with sulphuric acid.** The insoluble carbonates are removed by filtration and the filtrate is treated with dilute sulphuric acid, boiled and evaporated, when heavy yellow precipitate of sodium diuranate $Na_2U_2O_7 \cdot 6H_2O$ is obtained.

$$2Na_4 [UO_2 (CO_3)_3] + 3H_2SO_4 \rightarrow Na_2U_2O_7 \cdot 6H_2O + 3CO_2 + 3Na_2SO_4$$

***(f)* Treatment with ammonium carbonate.** The yellow precipitate of sodium diuranate is digested with ammonium carbonate resulting in the formation of soluble ammonium uranyl carbonate $(NH_4)_4 [UO_2 (CO_3)_3]$ while any insoluble impurity is removed by filtration.

The filtrate is boiled when lemon yellow crystals of ammonium diuranate $(NH_4)_2U_2O_7$ are obtained.

***(g)* Conversion into U_3O_8 and then UO_2.** Ammonium diuranate on calcination yields black oxide of uranium, U_3O_8 which on heating with charcoal or hydrogen is reduced to uranium dioxide UO_2.

***(h)* Conversion of UO_2 into U.** It can be achieved by either of the two processes:

(*i*) UO_2 is heated with chlorine to form yellow crystals of uranyl chloride UO_2Cl_2 This on heating with carbon in presence of chlorine yields volatile uranium tetrachloride UCl_4. This on heating with sodium metal gives uranium metal.

$$UO_2 + Cl_2 \longrightarrow UO_2Cl_2$$

Flow sheet diagram for the extraction of uranium from pitch blende is given below:

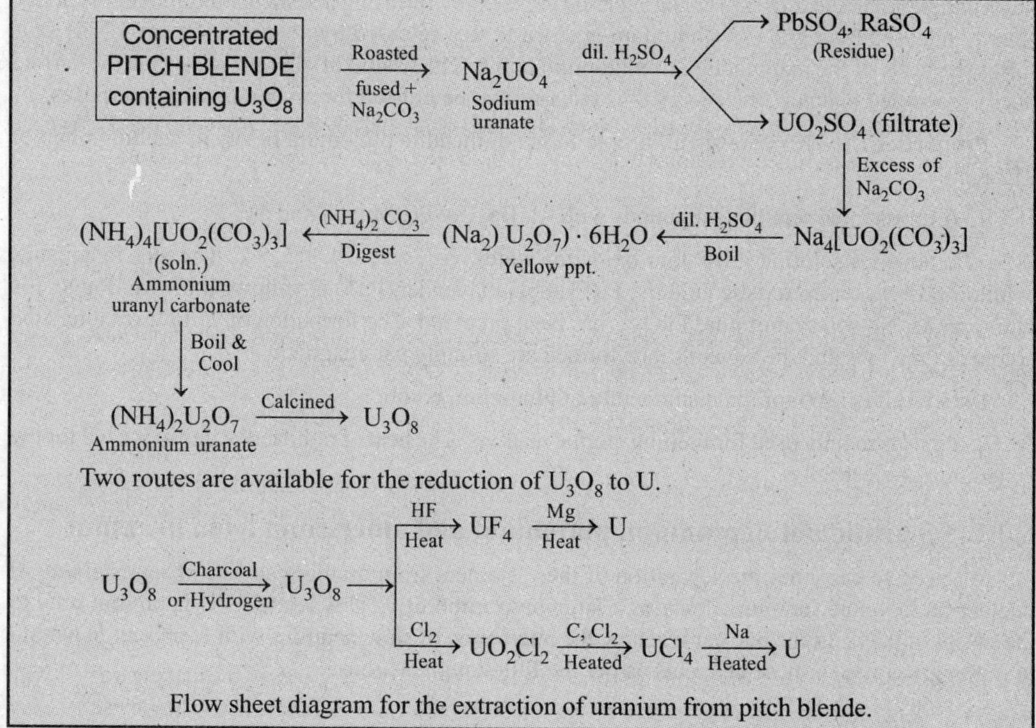

Flow sheet diagram for the extraction of uranium from pitch blende.

$$UO_2O_2 + Cl_2 \xrightarrow{\text{Carbon}} UCl_4 + O_2$$

$$UCl_4 + 4Na \longrightarrow U + 4NaCl$$

(*ii*) UO_2 is heated in hydrofluoric acid vapours to form uranium tetrafluoride UF_4 which on heating with magnesium in pressure vessel yields uranium metal.

$$UO_2 + 4HF \longrightarrow UF_4 + 2H_2O$$

$$UF_4 + 2Mg \longrightarrow U + 2MgF_2 \text{ (slag)}$$

6.4. OCCURRENCE AND PRODUCTION OF PLUTONIUM

Occurrence. It occurs in minute concentration in uranium-bearing ores like Canadian Pitchblende. The isotope involved is Pu^{239}. No other isotope has been found to occur in nature. The important ores of plutonium are *Colorado Pitchblende* and *Camolite.*

Production of Plutonium. It is produced mainly from uranium in nuclear reactors. Natural uranium contains a large proportion of U^{238} whereas U^{235} is present only to the extent of 0.7%. The interaction of slow neutrons with U^{238} present in natural uranium gives rise to plutonium in a series of following nuclear reactions:

$${}_{92}U^{238} + {}_0n^1 \longrightarrow {}_{92}U^{239} + \gamma$$

$${}_{92}U^{239} \longrightarrow {}_{93}Np^{239} + \beta$$

$${}_{93}Np^{239} \xrightarrow{\text{2-3 days}} {}_{94}Pu^{239} + \beta$$

Plutonium, being chemically different from uranium, can be separated relatively easily. Plutonium undergoes nuclear fission changing into U^{235} thereby releasing a tremendous amount of energy. Energy released by a gram of plutonium is equal to that released by about 20 tons of T.N.T. This forms the basis of the manufacture of plutonium bomb. The fission of plutonium can also be carried out in controlled manner. The energy thus released can be utilised for several peaceful purposes.

Properties : 1. There are six allotropic forms of metallic plutonium below its melting point of 912 K.

2. It forms intermetallic compounds with Al, Be, Co, Fe, Mn, Ni and Ag.

3. In aqueous solution it has four oxidation states, *viz.*, + 3, + 4, + 5, + 6. The ions in aqueous solution possess characteristic colours, Pu^{3+} is blue lavender, Pu^{4+} is yellowish brown, PuO_2^{2+} is pink-orange. No solution of pure PuO_2^{2-} has been prepared. The formation of complexes alters the colours of the ions and the aqueous ions formed are strongly paramagnetic.

Uses : 1. It is used for the manufacture of plutonium bomb.

2. Pu^{239} is mainly used for fuelling atomic reactors. The heat of radioactive decay is used for the generation of electricity.

6.4.1. Separation of neptunium, plutonium and americium from uranium

We need to carry out the separation of these element from uranium as part of reprocessing of nuclear fuel. Natural uranium contains a large proportion of ^{238}U whereas ^{235}U is present only to the extent of 0.7%. In the nuclear reactors, the interaction of slow neutrons with U present in natural uranium gives rise to these elements as per the following reactions.

$$^{235}_{92}U \xrightarrow{^{1}_{0}n} {}^{236}_{92}U \xrightarrow{^{1}_{0}n} {}^{237}_{92}U \xrightarrow{-\beta} {}^{237}_{93}Np$$

$$^{238}_{92}U \xrightarrow{^{1}_{0}n} {}^{239}_{92}U \xrightarrow{-\beta} {}^{239}_{93}Np \xrightarrow{-\beta} {}^{239}_{94}Pu$$

Americium is formed by succession of (n, γ) reactions starting with Pu in a nuclear reactor as follows:

$$^{239}_{92}Pu \xrightarrow{(n,\gamma)} {}^{240}_{94}Pu \xrightarrow{(n,\gamma)} {}^{241}_{94}Pu \xrightarrow{-\beta} {}^{241}_{95}Am$$

These elements are extracted from spent fuel rods which have been used in a nuclear reactor. Their separation is very difficult and hazardous. It is because they are not only mixed with highly radioactive fission products, but the material itself is also fissile. Moreover, Pu is extremely toxic and corcinogenic. The reprocessing of nuclear fuels is an important new technology. In addition to Pu, Np and Am, the irradiated nuclear fuel contains isotopes of over 30 fission product elements. The extraction of Np, Pu and Am involves the following steps:

Fuel rods removed from a nuclear reactor are immersed in cooling ponds for about 100 days. This keep them cool and the highly radioactive isotopes with short half lives such as $^{131}_{53}I$ lose most of their activity.

Flow sheet diagram for the processing of the nuclear fuel

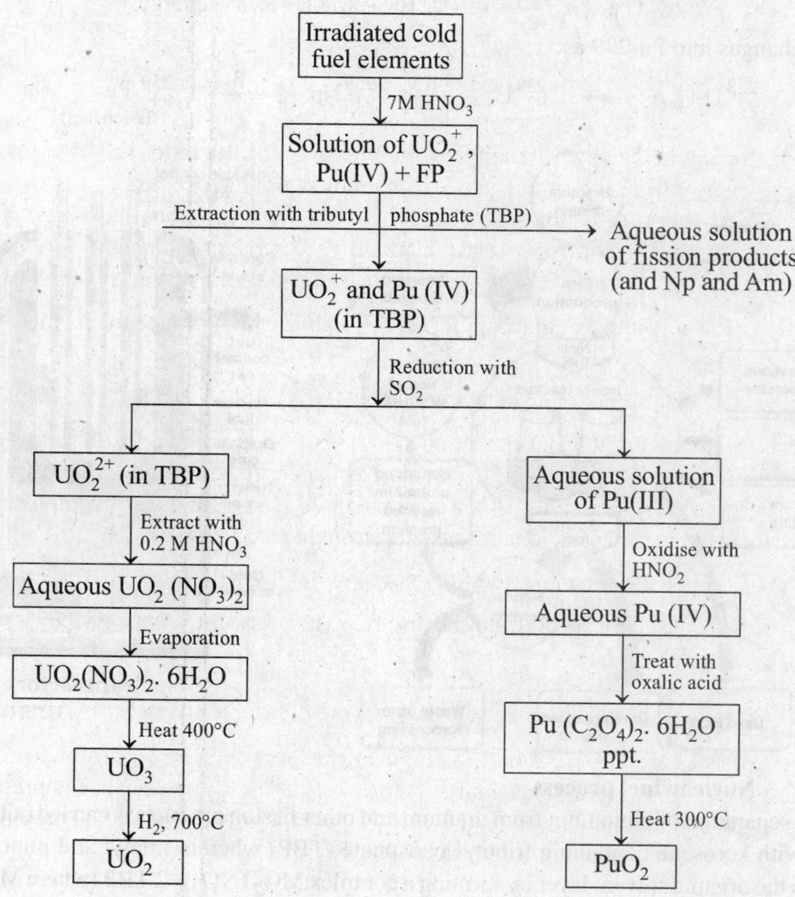

The extraction of Np, Pu and Am is done by Purex method as described below.

1. The fuel rods are dissolved in 7M HNO_3 which give solution containing UO_2^{2+} and Pu(IV). These elements are then extracted by 20% tributyl phosphate (TBP) in kerosene [as UO_2^{2+} and Pu (IV)] leaving other trans-uranium elements (Np, Am and Cm) together with other fission products (FP) in the aqueous phase.
2. Then the solution containing UO_2^{+} and Pu (IV) is carefully reduced with SO_2, or NH_2OH which gives Pu (III). This is easily separated from UO_2^{2+} and U^{4+} by solvent extraction.
3. From this Pu is easily precipitated as fluoride or oxalate which on heating gives PuO_2.
4. U is precipitated as uranyl nitrate which on heating and subsequent treatment with H_2 at 700°C give UO_2.

6.4.2. Nuclear fuels

Uranium (U-233, U-235) and plutonium (Pu-239) are fusion fuels that can be directly used in atomic reactors for fission while uranium (U-238) and thorium (Th-232) can not be used as fission fuels directly. However, by nuclear reactions they give fission fuels. U-238 gives Pu-239 while Th-232 gives U-233.

In all naturally occuring uranium the isotopic abundance of fission fuel U-235 is only 0.7%; the balance being U-238. Enrichment of U-235 is generally carried out by gaseous diffusion using UF_6. Since India has no uranium enrichment facilities, therefore natural uranium as fuel for reactor is used with moderators such as heavy water, graphite, whereby U-235 undergoes fission as:

$$^{235}_{92}U + ^{1}_{0}n \longrightarrow ^{141}_{56}Ba + ^{91}_{36}Kr + 3^{1}_{0}n + \text{energy}$$

U-238 changes into Pu-239 as:

$$^{238}_{92}U + ^{1}_{0}n \longrightarrow ^{239}_{92}U \xrightarrow{-\beta} ^{239}_{93}Np \xrightarrow{-\beta} ^{239}_{94}Pu \text{ (fission fuel)}$$

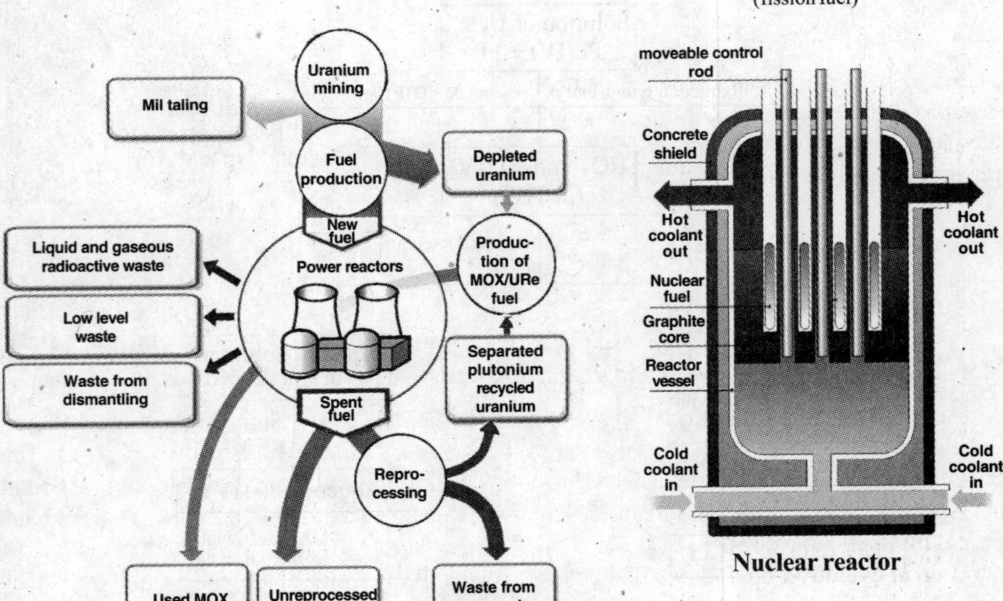

Nuclear fuel process **Nuclear reactor**

Large-scale separation of plutonium from uranium and other fission products is carried out by solvent extraction with kerosene containing tributyl phosphate (TBP) wherein uranyl and plutonyl nitrate migrate into the organic solvent layer by forming a complex $MO_2(NO_3)_2\,2\,TBP$ (where M = U or Pu) while the fission products remain in the aqueous phase.

The organic solvent layer containing uranium and plutonium is separated, washed with 0.1 M HNO_3 containing a reducing agent whereby Pu^{VI} is reduced to Pu^{III} and migrates to the aqueous phase from which it is recovered while U^{VI} remains in the organic solvent layer and is obtained by washing it with water.

Another fission fuel, U-233, which is not available in nature but can be produced by neutron irradiation of Th-232 in breeder reactors as:

$$^{232}_{90}\text{Th} + ^{1}_{0}n \longrightarrow ^{233}_{91}\text{Pa} \xrightarrow{-\beta} ^{233}_{92}\text{U}$$

Thorium is abundantly available in India and is likely to be the major fuel for future atomic reactors.

6.4.3. Transuranic elements

Elements with atomic numbers higher than 92 (that of uranium) are called transuranic elements. In minerals only Np and Pu occur and that too in small amounts. Their extraction is tedious and expensive. Elements from neptunium to lawrencium have been synthesised by artificial means. There are two main nuclear reactions which are used in the synthesis of transuranium elements:

(i) By capture of neutron by the nucleus of an element which increases its neutron-proton ratio. This results in instability of "host" nucleus. Then a neutron is converted to a proton with emission of a β-particle and the increase of atomic number by one unit. For example,

$$^{238}_{92}\text{U} + ^{1}_{0}n \longrightarrow ^{239}_{92}\text{U} \xrightarrow{-\beta} ^{239}_{93}\text{Np} \xrightarrow{-\beta} ^{239}_{94}\text{Pu}$$

(ii) By the capture of unstable nuclei of light elements ranging from helium to neon which increases atomic number by several units in one step. Bombardment with alpha particles has been used quite often to synthesise trans-uranium elements. For example,

$$^{242}_{95}\text{Am} + ^{4}_{2}\text{He} \longrightarrow ^{243}_{97}\text{Bk} + 3^{0}_{1}n$$

To synthesise very heavy elements, bombarding nuclei with heavier than alpha particles is required. Stripped boron and carbon atoms accelerated in the cyclotron have recently been used to synthesise elements with atomic numbers 102 and 103.

$$^{246}_{96}\text{Cm} + ^{12}_{6}\text{C} \longrightarrow ^{254}_{102}\text{No} + 4^{1}_{0}n$$

$$^{250}_{98}\text{Cf} + ^{11}_{5}\text{B} \longrightarrow ^{257}_{103}\text{Lr} + 4^{1}_{0}n$$

Earnest Lawrence discovered cyclotron

6.4.4. Later actinides and their resemblance with later lanthanides

The elements from curium onwards i.e. curium, berkelium, californium, einsteinium, fermium, mendelevium, nobelium and lawrencium are called later actinides. The isotope ^{242}Cm was first isolated among the products of α-bambardment of ^{239}Pu followed by the discovery of americium. This is available in macroscopic quantities while all others are available only in tracer amounts. Isotopes of other elements were first identified in products from the first hydrogen bomb explosion (1952) or in cyclotron bombardments. Not much information is available about these elements because of their limited availability and their instability.

The chemical interest in these later actinide elements is mainly because of their resemblance with the corresponding lanthanides quite closely. However, inspite of their similarities, these actinides can be easily separated from the lanthanides because actinides form complexes more readily. When both groups of ions are absorbed on cation exchange column, the actinides can be eluted with conc. HCl because of complex formation. The actinide ions are separated from each other by the ion exchange using nitrate solutions to elute them. The order in which the actinide ions are eluted shows a close similarity to the order in which the lanthanides are eluted.

As in lanthanides, the +III oxidation state is the most stable for all the elements from atomic numbers 96-103 *i.e.* Cm, Bk, Cf, Es, Md and Lw except No. The most stable state of No is +II because it has a favourable f^{14} electronic configuration. The elements from 98-102 *i.e.* californium to nobelium, show +II oxidation state also. Except for nobelium, this state is strongly reducing in nature. Higher oxidation state of +IV are also found in Cm and other elements. Lawrencium exists only in +III state and resists both oxidation and reduction because of stable fully filled 5*f* orbitals. The oxidation states of later actinides are similar to lanthanides. This can be seen by comparing Cm and Gd, Bk and Tb, No and Yb and Lr and Lu.

The spectral and magnetic properties of some compounds of isoelectronic pairs of ions such as CmF_3 and GdF_3 are also similar due to almost similar amount of energy required for the promotion of $5f \rightarrow 6d$ in later actinides as in the case of lanthanides ($4f \rightarrow 5d$).

Later actinides resemble later lanthanides because the respective electronic transitions $5f - 6d$ and $4f - 5d$ involve similar amount of energy. While earlier actinides do not resemble earlier lanthanides because of different energies associated with their electronic transitions. Less energy is needed for $5f - 6d$ transition in case of actinides than for $4f - 5d$ tranistion in case of lanthanides.

SOLVED EXAMPLES

Example 1. What are transuranic elements? Name the elements having atomic number 95, 97 and 99.

Solution. Elements having atomic number higher than 92 are called transuranic elements.

At. No.	Element
95	Americium, Am
97	Berkelium, Bk
99	Einsteinium, Es

Example 2. Actinides have greater tendency to form complexes than lanthanides. Explain.

Solution. Due to small size, higher nuclear charge and higher charge density, actinides have greater tendency to form complexes than lanthanides.

Example 3. Complete the following reactions :

(*a*) $^{254}Es \longrightarrow {}^{254}Fm + ...$

(*b*) $^{239}Pu + {}^{4}_{2}He \longrightarrow .... + {}^{1}_{0}n$

(*c*) $... + {}^{14}_{7}N \longrightarrow {}^{247}Es + 5\,{}^{1}_{0}n$

(*d*) $^{253}Es + {}^{4}_{2}He \longrightarrow .... + {}^{1}_{0}n$

(*e*) $^{239}_{92}U \longrightarrow .... + {}^{0}_{-1}e$

Solution. (*a*) ${}^{0}_{-1}e$ (*b*) ^{242}Cm (*c*) ^{238}U (*d*) ^{256}Md (*e*) ^{239}Np

Example 4. Write synthesis reaction for the preparation of following transuranic elements :

(*i*) Curium

(*ii*) Nobelium

(*iii*) Einsteinium

Solution. (*i*) $^{239}_{94}Pu + ^{4}_{2}He \longrightarrow ^{242}_{96}Cm + ^{1}_{0}n$

(*ii*) $^{246}_{96}Cm + ^{12}_{6}C \longrightarrow ^{254}_{102}No + 4\,^{1}_{0}n$

(*iii*) $^{238}_{92}U + ^{14}_{7}N \longrightarrow ^{247}_{99}Es + 5\,^{1}_{0}n$

Example 5. Actinides form oxocations but lanthanides do not. Why?

Solution. When the oxidation state of actinide ions increase to +6, then they do not form simple ions. The higher oxidation state leads to high charge density on actinide ion which favours the formation of oxocation such as UO_2^{2+} in solid state and in solution. These oxocations are stable in acids as well as aqueous solutions.

On the other hand, lanthanides do not show high oxidation states such as +6. Therefore, lanthanide ions do not have high charge density and hence do not form oxocations.

Example 6. Oxocations MO^{2+} are formed by U, Pu, Np and Am only while heavier members of the actinide series do not form such cations. Explain.

Solution. The tendency towards disproportionation of these oxocations increases with increase in atomic number. Therefore, oxocations of heavier members do not exist.

Example 7. (*i*) Which element of actinide series has the highest melting and boiling points?

(*ii*) Which of the following are radioactive?

Ac, U, Pu, Cm

(*iii*) Which element is produced by β-decay of neptunium-239?

Solution.(*i*) Thorium

(*ii*) All are radioactive

(*iii*) Plutonium

Example 8. It is difficult to interpret the paramagnetic behaviour of actinides. Explain.

Solution. 5*f* electrons of actinides are less effectively shielded which results in quenching of orbital contribution. Therefore, magnetic moments of actinide ions are less than the theoretically predicted values.

Example 9. Write a brief account of conversion of ammonium diuranate to uranium.

Solution. (*i*) Ammonium diuranate $(NH_4)_2U_2O_7$ is calcinated to get U_3O_8.

$$3(NH_4)_2U_2O_7 \longrightarrow 2U_3O_8 + 6NH_3 + 3H_2O + O_2$$

(*ii*) U_3O_8 is then reduced by heating with charcoal or hydrogen to uranium dioxide UO_2 as:

$$U_3O_8 + 2H_2 \longrightarrow 3UO_2 + 2H_2O$$

(*iii*) UO_2 is then heated with HF at 670-770 K to form uranium tetrafluoride UF_4 as

$$UO_2 + 4HF \longrightarrow UF_4 + 2H_2O$$

(iv) UF_4 is then heated with calcium or magnesium to form uranium as:

$$UF_4 + 2Ca \longrightarrow 2CaF_2 + U$$

Example 10. What happens when

(*i*) Uranyl sulphate is treated with excess of Na_2CO_3.

(*ii*) Ammonium diuranate is calcined.

(*iii*) Sodium diuranate is digested with ammonium carbonate.

Solution. (*i*) Soluble sodium tricarbonatournate is formed

$$UO_2SO_4 + 3Na_2CO_3 \longrightarrow Na_4[UO_2(CO_3)_3] + Na_2SO_4$$

(*ii*) On calcination, it gives U_3O_8

$$(NH_4)_2U_2O_7 \longrightarrow 2U_3O_8 + 6NH_3 + 3H_2O + O_2$$

(*iii*) $Na_2U_2O_7 \xrightarrow{(NH_4)_2CO_3} (NH_4)_4[UO_2(CO_3)_3] \longrightarrow (NH_4)_2U_2O_7$

EXERCISES
(Based on Question Papers of Different Universities)

Multiple Choice Questions

Choose the correct answer

1. Neptunium is radioactive and emits a β-particle giving

(*a*) plutonium (*b*) samarium

(*b*) curium (*d*) californium

2. Which element of actinide series has the highest melting point?

(*a*) Th (*b*) Pu

(*c*) U (*d*) Np

3. The element between Pu and Cm is

(*a*) Am (*b*) Np

(*c*) U (*d*) Pa

4. Which of the following statements is not correct for actinides and lanthanides?

(*a*) Oxidation state of +3 is predominant in both the series.

(*b*) Both shows contraction in their ionic radii.

(*c*) The elements of both series are radioactive.

(*d*) Both involve the filling of *f*-orbitals.

5. β-decay of neptunium gives

(*a*) Americium (*b*) Curium

(*c*) Fermium (*d*) Plutonium

6. Which of the following is a naturally occurring actinide?

(*a*) Cf (*b*) Pu

(*c*) Th (*d*) No

7. Which of the following is not radioactive?

(*a*) Np (*b*) Bk

(*c*) Cm (*d*) Eu

8. Which of the following ion is colourless in its aqueous solution?

(*a*) NP^{3+} (*b*) Th^{4+}

(*c*) Am^{3+} (*d*) Cm^{3+}

9. The oxidation state of U in UO_2^{2+} is

(*a*) 4 (*b*) 2

(*c*) 6 (*d*) 3

10. The most stable state of nobelium is
(a) +3 (b) +2
(c) +5 (d) +4

11. The elements before and after U are respectively
(a) Th and Np (b) Pa and Pu
(c) Pa and Np (d) Pu and Am

12. Which trivalent actinide ion is red in colour
(a) Pu^{3+} (b) Am^{3+} (c) Ac^{3+} (d) U^{3+}

ANSWERS

1. (a)	2. (a)	3. (a)	4. (c)	5. (d)	6. (c)
7. (d)	8. (b)	9. (c)	10. (b)	11. (c)	12. (d)

Short Answer Questions

1. Give one use of UF_6.
2. Why do actinides have a greater tendency to form complexes than lanthanides?
3. Why does uranium form oxygenated ions such as UO_2^+?
4. How is curium-96 prepared from plutonium-94? Give its reactions.
5. Name the element which is formed by β-decay of neptunium.
6. What are transuranic elements?
7. Draw the structure of $UO_2(NO_3)_2 \cdot (H_2O)_2$
8. Name two minerals of uranium.
9. "Ions of actinides are coloured." Explain.
10. Why do actinides show +4, +5 and +6 oxidation states?

General Questions

1. Write notes on :
(i) Transuranic elements (ii) Uranium as nuclear fuel.
2. What are transuranic elements? Give preparation of one such element.
3. (a) What are actinides? Why are they so called?
(b) Complete the following reactions:

(i) ${}^{241}_{95}Am + ? \longrightarrow ? + \gamma$ (ii) ${}^{239}_{94}Pu + ? \longrightarrow ? + {}^{1}_{0}n$

(iii) ${}^{242}_{95}Am + ? \longrightarrow ? + \gamma$ (iv) ${}^{235}_{92}U + ? \longrightarrow ? + {}^{1}_{0}n$

4. Compare the oxidation states of actinides with lanthanides.
5. Discuss the electronic structure and position of actinides in periodic table. Compare their important properties with those of lanthanides. Why is the chemistry of actinides difficult to study? Give reasons.
6. (a) In what respects do actinides differ from lanthanides?
(b) Actinides exhibit oxidation states higher than +3. Explain.
(c) It is difficult to interpret paramagnetic character of actinides. Explain.

7. Discuss briefly:
(*i*) Nuclear fuels. (*ii*) Complex formation by actinides.

8. Compare and contrast the properties of lanthanides and actinides.

9. Give the preparation of transuranic elements.

10. Discuss
(*i*) Oxidation states
(*ii*) Paramagnetism and
(*iii*) Complex ion formation in case of actinides.

11. Discuss
(*i*) oxidation states (*ii*) paramagnetism

12. Give a brief account of the following:
(*i*) Complex formation by actinides.
(*ii*) Conversion of ammonium diuranate to uranium.
(*iii*) Nuclear fuels. (*iv*) Transuranic elements.

13. Discuss the chemistry of various steps involved in the extraction of uranium from pitch blende.

14. (*a*) Actinides have greater tendency than lanthanides to form complexes. Why is it so?
(*b*) Compare the properties of actinides with those of lanthanides.

15. (*a*) What are transuranic elements? List them in order.
(*b*) What is the maximum oxidation state shown by actinides?

16. (*a*) Actinides have greater tendency to form complexes than lanthanides. Why?
(*b*) What are transuranic elements?

17. (*a*) Discuss the chemistry involved in separation of Plutonium from Uranium by
(*i*) Hexone method (*ii*) Purex method.
(*b*) What are later actinides? Discuss.
(*c*) Explain
(*i*) Actinides form oxocations but lanthanides do not.
(*ii*) Actinides have uncertain configurations.

18. Compare the properties of lanthanides with actinides with respect to
(*i*) oxidation states (*ii*) basic character
(*iii*) complex formation.

19. Discuss briefly the method of separation of Np, Pu and Am from U.

20. Starting from monozite ore, how is Th extracted?

21. Explain the following in case of actinides:
(*i*) Oxidation states (*ii*) Magnetic properties
(*iii*) Actinide contraction.

22. Discuss complex formation by actinides.

23. (*a*) Actinides form oxocations but lanthanides do not. Explain.
(*b*) What is nuclear fusion? Give two reactions to explain.
(*c*) Actinides have a great tendency to form complexes than lanthanides. Explain.

24. (*a*) Explain why heavier members of the actinide series do not form oxocations (MO^{n+}).

(*b*) Giving two examples, describe method of preparation of transuranium elements through transmutation with high energy particles.

25. Discuss the extraction of thorium from monazite sand.

26. Discuss briefly the following compounds of actinides:

(*i*) Oxides and chalcogenides

(*ii*) Halides

27. (*a*) What are actinides? Why are they so called? Give their electronic configurations.

(*b*) Discuss the way in which actinides resemble their lanthanide congeners.

(*c*) Actinides form oxocations but lanthanides do not. Why?

28. Discuss the ways in which the actinides resemble their lanthanide congeners.

29. Discuss extraction of uranium from its ores.

30. (*a*) Discuss the following.

(*i*) Actinides have uncertain configurations.

(*ii*) Actinides exhibit oxidation states higher than +3.

(*iii*) Actinides have greater tendency for complex formation than lanthanides.

(*b*) Discus similarities between later lanthanides and later actinides.

31. What are later actinides? How do they closely resemble lanthanides?

32. Give the differences between actinides and lanthanides.

33. Give two examples of transuranic elements.

34. Give a brief account of: (*a*) Nuclear fuels; (*b*) Transuranic elements.

35. (*a*) Actinides have stronger tendency to form complexes than lanthanides. Explain

(*b*) What are nuclear fuels? Give the preparation of plutonium.

(*c*) Give points of similarities between actinides and lanthanides.

36. Explain the reasons why actinides are capable of forming complexes with π-bonding ligands.

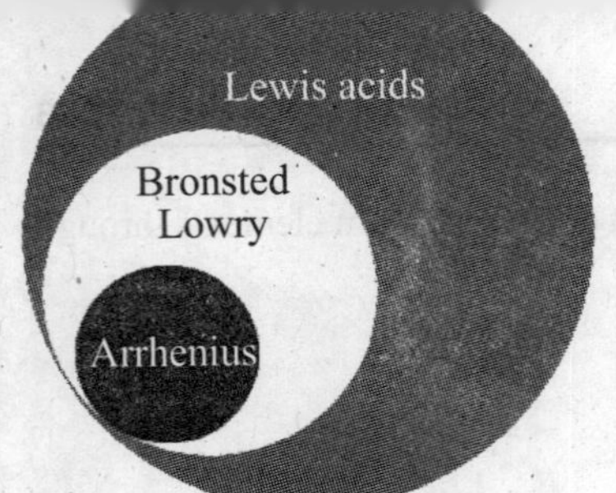

7

Acids and Bases

7.1. INTRODUCTION

In earlier times, acids and bases were defined according to their observed properties. For example, acids were those substance which had sour taste and corroded the metals. Similarly, bases were substances, which had bitter taste and neutralised the acids. A little later, it became possible to identify acids and bases with the help of litmus paper. An acid turned blue litmus into red and the base turned red litmus into blue. Thus, different definitions were proposed for acids, and bases as time passed by. Every theory or definition of acid and base had some limitations or drawbacks. To overcome the limitation of a theory, a new theory was proposed which was broader than the earlier one. In this chapter, we shall review some theories of acid and base.

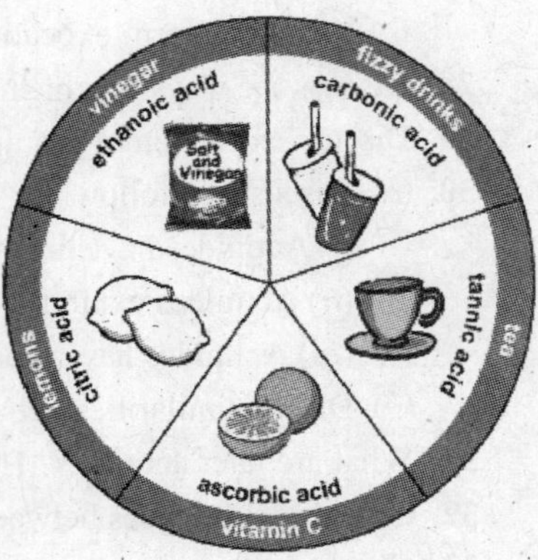

Acids in the home

7.2. ARRHENIUS CONCEPT OF ACIDS AND BASES

Svante Arrhenius, as a part of his theory of ionisation, proposed a theory of acids and bases. According to Arrhenius theory:

An acid is a substance that dissociates in aqueous solution to give hydrogen ions. Thus, HC1, HNO_3, H_2SO_4, CH_3COOH and HCN are acids because they liberate H^+ ions when dissolved in water.

Svante Arrhenius

$$HCl \rightleftharpoons H^+ + Cl^-$$

$$HNO_3 \rightleftharpoons H^+ + NO_3^-$$

$$H_2SO_4 \rightleftharpoons 2H^+ + SO_4^{2-}$$

$$HCN \rightleftharpoons H^+ + CN^-$$

A base is a substance that dissociates in *aqueous solution to give hydroxyl ions.* Thus, NaOH, KOH, $Ba(OH)_2$ and NH_3 are bases because they liberate (OH^-) ions when dissolved in water.

$$NaOH \rightleftharpoons Na^+ + OH^-$$

$$KOH \rightleftharpoons K^+ + OH^-$$

$$Ba(OH)_2 \rightleftharpoons Ba^{2+} + 2OH^-$$

$$NH_3 \overset{H_2O}{\rightleftharpoons} NH_4^+ + OH^-$$

Hydronium Ion (H_3O^+)

According to Arrhenius, an acid releases H^+ ions in aqueous medium. However, it has been found that H^+ ion cannot exist independently.

In fact, H^+ ion is unique among ions. It has no electron. Its radius is equal to that of nucleus (10^{-13} cm). The concentration of charge in such a small volume makes the attraction between water *and H^+ ions extremely large.* H^+ ion accepts a pair of electrons from water molecules [H – O – H] to form *hydrated proton* or **hydronium ion.**

$$H^+ + H-O-H \longrightarrow H_3O^+ \text{ or } \left\{ \begin{array}{c} H \quad H \\ \diagdown \; \diagup \\ O \\ | \\ H \end{array} \right\}^{\oplus} \text{ (Hydronium ion)}$$

Dissociation of hydrocloric acid in aqueous solution may be represented as:

$$HCl + H_2O \rightleftharpoons H_3O^+ + Cl^-$$

Limitations of Arrhenius Theory to explain acid-base concept

1. Limited scope. Arrhenius concept is limited to *aqueous medium* only. It fails to explain the behaviour of acids and bases in *non-aqueous* solvents like ammonia, sulphur dioxide, alcohol, etc.

2. Proton in aqueous solutions. According to Arrhenius concept acids are those which furnish H ions in water. H^+ -ion is nothing but a proton, which is not capable of independent existence. As already discussed it may exist as hydronium ion, *i.e.* H_3O^+.

3. Acids and bases without H^+ and OH^- ions. There are a number of substances (like CO_2, SO_2, NO_2, SO_3 etc.) which do not contain H^+ion, but behave as acids. Similarly, compounds like CaO, MgO, anhydrous NH_3, Na_2CO_3 etc. which do not contain OH^- ions, behave as bases.

4. Neutralisation without H^+ and OH^- ions. Neutralization between HCl(g) and NH_3 (g) does not involve H^+ ions and OH^- ions *e.g.*

$$NH_3(g) + HCl(g) \rightarrow NH_4\,Cl(s)$$

$$CaO\,(s) + SO_2(g) \rightarrow CaSO_3(s)$$

7.3. BRONSTED-LOWRY CONCEPT OF ACID AND BASE

J.N. Bronsted and J.M. Lowry independently and simultaneously (1923) put forward a more general concept for acids and bases. According to this concept :

(*i*) *Acid is a substance (molecule or ion) which has a tendency to donate a proton to any substance, and*

(*ii*) *A base is a substance (molecule or ion) which has a tendency to accept a proton from any other substance.*

In other words, an *acid is a proton donor and base is a proton acceptor.* In fact, acid-base reaction involves transference of proton. That is why, this concept is also called *proton transfer theory* of acids and bases. In accordance with this broader definition, an acid or base may be molecular or ionic, *e.g.,*

Molecular acids : HF, HC1, HNO_3, H_2SO_4, CH_3COOH, H_2S, H_3PO_4, H_2O, etc.

Ionic acids : H_3O^+, NH_4^+, HSO_4^-, HSO_3^-, HPO_4^{2-} etc.

Molecular bases :

RNH_2, NH_3, H_2O, R_2NH etc.

Ionic bases :

OH^-, S^{2-}, CO_3^{2-}, Cl^-, Br^-, NO_3^-, etc.

J.N. Bronsted

J.M. Lowry

It is important to note that no single substance is an acid or a base, A single substance cannot donate a proton unless and until some other substance which accepts the proton is also present. This proton transfer may or may not involve any medium, *e.g.,* the reaction does not involve any medium. Here, HC1 has denoted a proton to NH_3. Hence, HC1 is an acid with respect to NH_3, which is a base.

$$NH_3\,(g) + HC1\,(g) \rightarrow NH_4^+\,C\bar{1}\,(s)$$

Consider some more acid-base reactions.

$$\underset{\text{Base}}{NH_3} + \underset{\text{Acid}}{H_2O} \rightleftharpoons NH_4^- + OH^-$$

$$\underset{\text{Acid}}{CH_3COOH} + \underset{\text{Base}}{NH_3} \rightleftharpoons CH_3COO^- + NH_4^+$$

$$\underset{\text{Acid}}{HCl} + \underset{\text{Base}}{H_2O} \rightleftharpoons H_3O^+ + Cl^-$$

$$\underset{\text{Acid}}{H_2SO_4} + \underset{\text{Base}}{H_2O} \rightleftharpoons H_3O^+ + HSO_4^-$$

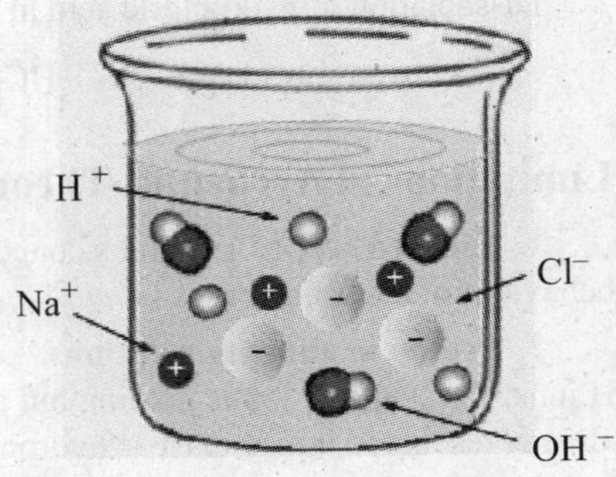

Bronsted Lowry Theory

A careful study of the above equations reveals that water is an acid w.r.t. NH_3 whereas it is a base w.r.t. HC1 or H_2SO_4. Such substances like H_2O, HCO_3^-, HSO_4^- etc. which can act as acids (proton donor) as well as bases (proton acceptor) are called **amphoteric** or **amphiprotic** substances.

Conjugate Acid-Base pairs. When an acid loses a proton, the residual part of it will have a tendency to accept a proton. Thus, it will behave as a base. *Such pairs of substances which differ from one another by a proton are known as conjugate acid-base pairs.* Consider a general example of an acid :

$$\underset{\text{Acid}}{HA} \rightleftharpoons \underset{\text{Proton}}{H^+} + \underset{\text{Conjugate base}}{A^-}$$

Consider the reaction between HC1 and H_2O.

$$\underset{\text{Acid}}{HCl} + \underset{\text{Base}}{H_2O} \rightleftharpoons \underset{\text{Acid}}{H_3O^+} + \underset{\text{Base}}{Cl^-}$$

HC1 donates a proton to water in the forward reaction.

$\therefore$ HC1 is an acid and H_2O is a base.

In the backward reaction, H_3O^+ ion donates its proton to Cl^- ion.

$\therefore$ H_3O^+ is an Acid and, Cl^- is a base.

It is clear that HC1 loses a proton and forms Cl^- ion (which is a base). HC1 and Cl^- ions are called *conjugate acid-base pair.*

$$\underset{\text{Acid}}{HCl} \rightleftharpoons H^+ + \underset{\text{Conjugate base}}{Cl^-}$$

Similarly, H_2O (base) accepts a proton and forms H_3O^+ (acid). Here, H_2O and H_3O^+ are also conjugate base-acid pair.

$$\underset{\text{Base}}{H_2O + H^+} \rightleftharpoons \underset{\text{Conjugate acid}}{H_3O^+}$$

In the above example, consider HC1 as acid-1 and Cl^- as base-1. Similarly, consider H_2O as base-2 and H_3O^+ as acid-2. The overall reaction between HC1 and H_2O, can, hence, be written as :

$$\underset{\text{Acid–1}}{HCl} + \underset{\text{Base–2}}{H_2O} \rightleftharpoons \underset{\text{Acid–2}}{H_3O^+} + \underset{\text{Base–1}}{Cl^-}$$

Some other examples of acid base reactions and conjugate acid- base pairs are given in the following table 7.1.

Table: 7.1. Conjugate acid-base pairs.

Acid-1		Base-2		Acid-2		Base-1
H_2SO_4	+	H_2O	$\rightleftharpoons$	H_3O^+	+	HSO_4^-
H_2O	+	NH_3	$\rightleftharpoons$	NH_4^+	+	OH^-
CH_3COOH	+	H_2O	$\rightleftharpoons$	H_3O^+	+	CH_3COO^-
H_2S	+	H_2O	$\rightleftharpoons$	H_3O^+	+	HS^-
H_3O^+	+	HCO_3^-	$\rightleftharpoons$	H_2CO_3	+	H_2O
H_3O^+	+	ClO_4^-	$\rightleftharpoons$	$HC1O_4$	+	H_2O
H_3PO_4	+	H_2O	$\rightleftharpoons$	H_3O^+	+	$H_2PO_4^-$

All Bronsted bases may not be Arrhenius bases.

According to Arrhenius, a base must furnish OH^- ions in aqueous solution. According to *Bronsted theory,* a base must accept a proton. Consider for example carbonate ion, CO_3^{2-}.

According to Bronsted theory, CO_3^{2-} ***ion is a base as it accepts a proton to form*** HCO_3^- ***ion.***

$$CO_3^{2-} + H^+ \rightarrow HCO_3^-$$

According to Arrhenius concept, CO_3^{2-} is not a base as it does not furnish OH^- ions in aqueous solution. Hence, all Arrhenius acids are Bronsted acids but all Bronsted bases are not Arrhenius bases.

7.4. LUX-FLOOD THEORY OF ACIDS AND BASES

This concept was proposed by Lux and later extended by Flood. It visualises the acid – base reaction in terms of the oxide ion and is mainly used for molten oxides at high temperatures. According to this concept, an acid *is a substance which accepts the oxide ion and a* **base** *is a substance which donates the oxide ion.* In other words, *an* **acid** *is an oxide ion-acceptor and a* **base** *is an oxide ion-donor.* For example, in the reaction

$$\underset{\text{Base}}{CaO} + \underset{\text{Acid}}{SiO_2} \longrightarrow CaSiO_3,$$

calcium oxide is a base and silicon dioxide is an acid. Similarly, in the reaction

$$PbO + SO_3 \longrightarrow PbSO_4,$$

PbO is a base while SO_3 is an acid.

This concept, has only a limited scope.

7.5. LEWIS CONCEPT OF ACID AND BASE

G. N. Lewis

According to Lewis, **an acid** is defined as *any molecule, ion or radical that can accept one or more electron pairs, that is, an acid is an electron pair acceptor.* In other words, a substance which has vacant orbitals to accommodate one or more pairs of electrons can act as a Lewis acid. A **base** is defined as *a species that can donate one or more electron pairs, i.e., a base is an electron pair donor*. Consequently any substance with unshared pair of electrons can behave as a Lewis base. Acid base or neutralisation reactions may, therefore, be termed as donor-acceptor reactions or coordinate covalent bond formation reactions as shown below:

S. No.	*Acceptor*	*Donor (Acid)*	*Donor-acceptor compound (Base) Neutralization product*
1.	BF_3	NH_3	$H_3N \longrightarrow BF_3$
2.	$SnCl_4$	$2R_2O$	$(R_2O)_2\ SnCl_4$
3.	H^+	NH_3	NH_4^+
4.	$FeCl_3$	Cl^-	$(FeCl_4)^-$
5.	SO_3	O^{2-}	SO_4^{2-}

Different kinds of Lewis acids

These can be classified into the following two categories:

(*i*) **All the cations are Lewis acids** because they have tendency to accept electron pairs. Larger cations of alkali and alkaline earth metals behave as relatively weaker Lewis acid due to their lesser tendency to accept electrons. Smaller cations like H^+, Li^+, Be^{++} etc., are likely to act as stronger Lewis acids due to their higher tendency to accept electrons and form stronger bonds. Also higher charged cations have a higher tendency to accept electrons relative to the lower charged cations of a given metal. For example, Fe^{3+} behaves as a stronger Lewis acid as compared to Fe^{2+}. Similarly, Sn^{4+} is a stronger Lewis acid than Sn^{2+}. Examples of some cations acting as Lewis acids are given below:

S. No.	*Lewis Acid*	*Lewis Base*	*Complex*
1	Ag^+	$2\ NH_3$	$[Ag\ (NH_3)_2]^+$
2	Cu^{2+}	$4\ Cl^-$	$[CuCl_4]^{2-}$
3	Mn^{2+}	$6\ H_2O$	$[Mn\ (H_2O)_6]^{2+}$
4	Ni^{2+}	$4\ CN^-$	$[Ni\ (CN)_4]^{2-}$

(*ii*) **Electron-deficient** compounds in which the central atom can expand its valence shell can behave as Lewis acid. BF_3, $AlCl_3$, $GaBr_3$, $BeCl_2$ are electron-deficient compounds and have tendency to accept electron pairs to complete their octet.

$$BF_3 + F^- \longrightarrow [BF_4]^-$$

Compounds like SiF_4, $SnCl_4$, $SbCl_3$, $SbCl_5$, $TeBr_4$, SO_3 etc also represent Lewis acids where central atom can expand its valence shell.

$$SnCl_4 + 2Cl^- \longrightarrow [SnCl_6]^{2-}$$

$$SO_3 + R_2O \longrightarrow [R_2O\ (SO_3)]$$

$$SiF_4 + 2F^- \longrightarrow [\ SiF_6]^{2-}$$

Different kinds of Lewis bases

Following two types of compounds constitute Lewis bases:

(*i*) **All the anions can act as Lewis bases** because they have a tendency to donate electron pairs. For example, oxides (O^{2-}), hydroxides (OH^-), halides (X^-), amides (NH_2^-) etc., act as Lewis bases. The greater the charge density on the anion, the stronger the base.

(*ii*) Substances having **unshared or lone pairs of electrons** act as Lewis bases. For example, water (H_2O :), ammonia (NH_3), ether (R_2O :), Ketones ($R_2\ C = O$:) etc., act as electron donors and are thus Lewis bases.

Extended Lewis definition of acid and base

The extended Lewis acids include unsaturated species like CO_2 and CS_2, where the central atom is bonded to atoms of different electronegativities by one or more multiple bonds. Although both the molecules lack acceptor character of the central atom because carbon atom cannot extend its octet, yet these have been indirectly treated as Lewis acids. This is due to the polarization of the p-electron cloud on approach of a Lewis base towards carbon atom of CO_2 or CS_2. The polarization leads to a partial +ve charge on carbon and a partial – ve charge on the more electronegative atom like O. The +vely charged carbon then acts as an electron acceptor as shown in the reaction of CO_2 with Lewis bases.

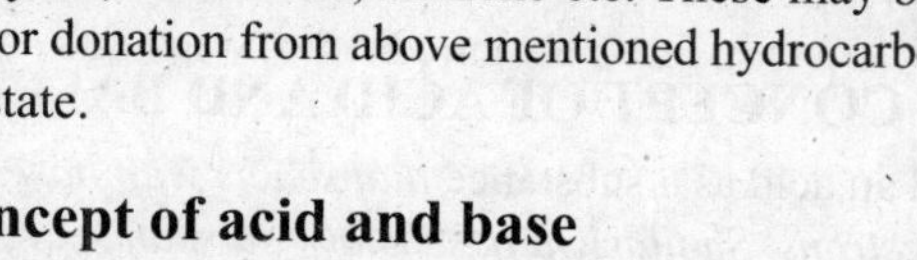

$$O{=}C{=}O \;+\; :\ddot{\underset{\cdot\cdot}{O}}H^- \longrightarrow \;:O{=}C(O^-)(O{-}H)$$

Bicarbonate

The extended Lewis bases include substance like carbon monoxide (CO) and unsaturated hydrocarbons like alkenes, alkynes, butadienes, benzene etc. These may be treated as Lewis bases because of a lone pair on : CO or donation from above mentioned hydrocarbons to form π-complexes with metals in low oxidation state.

Limitations of Lewis concept of acid and base

Some limitations of Lewis concepts of acid are base and given below:

(*i*) It does not provide us any uniform scale of acid or base strength.

(*ii*) It fails to tell us whether salt formation is related to acid - base behaviour or not.

(*iii*) Normal acid - base reactions are fast but complex formation in terms of donor - acceptor bond formation is a slow process.

(*iv*) It is very general and includes formation of $Ni(CO)_4$ and $Cr(C_6H_6)_2$ as Lewis - acid base reactions. However, bonding in these compounds is more complex and beyond the scope of Lewis concept.

7.6. CADY-ELSEY THEORY (SOLVENT SYSTEM CONCEPT) OF ACID AND BASE.

An acid is defined as a solute which directly or indirectly increases the concentration of cations characteristic of the solvent and base as a solute which increases the concentration of the anions characteristic of the solvent.

$$2H_2O \rightleftharpoons \underset{\text{(acid}}{H_3O^+} + \underset{\text{(base)}}{OH^-}$$

$$2NH_3 \rightleftharpoons \underset{\text{(acid)}}{NH^+_4} + \underset{\text{(base)}}{NH^-_2}$$

$$2SO_2 \rightleftharpoons \underset{\text{(acid)}}{SO^{2+}} + \underset{\text{(base)}}{SO_3^{2-}}$$

Thus NH_4Cl behaves as an acid in ammonia because it furnishes NH^+_4 which are cations characteristic of solvent and KNH_2 is a base in NH_3 because it gives NH_2^- which are anions characteristic of the solvent. Thus on addition of NH_4Cl in NH_3, the concentration of cations characteristic of solvent *i.e.*, NH_4^+ is increased and on addition of KNH_2 in NH_3, conc. of anions characteristic of solvent *i.e.*, NH_2^- is increased.

Consider the auto-ionization of liquid nitrogen tetroxide :

$$N_2O_4 \rightleftharpoons \underset{\text{Acid ion}}{NO^+} + \underset{\text{Base ion}}{NO_3^-}$$

According to solvent– system concept, in liquid nitrogen tetroxide, the substances which furnish nitrosyl ions (NO^+) behave as acids and those which furnish nitrate ions (NO_3^-) behave as bases. Accordingly, NOC1 behaves as an acid and $NaNO_3$ behaves as a base, in this solvent.

$$\underset{\text{Acid}}{NOC1} \xrightleftharpoons{\text{liq.}N_2O_4} \underset{\text{Solvent cation}}{NO^+} + Cl^-$$

$$NaNO_3 \xrightleftharpoons{\text{liq.}N_2O_4} NO^+ + \underset{\text{Solvant anion}}{NO_3^-}$$

Autoionization of sulphuric acid is represented as

$$2H_2SO_4 \rightleftharpoons H_3SO_4^+ + HSO_4^-$$

Thus, any species giving $H_3SO_4^+$ ion acts as an acid and any species giving HSO_4^- ion acts as a base.

7.7. USANOVICH CONCEPT OF ACID AND BASE

Usanovich defined an acid as a substance *that reacts with bases to form salts, gives up cations or takes up anions or electrons. Similarly a base is any substance or species that neutralizes an acid, gives up anions or electrons or takes up cations.*

Usanovich's definition includes all substances that can be defined as acids or bases in terms of other acid base concepts. Over and above, his definition includes all redox-reactions (electron transfer reactions) as acid-base reactions.

An example of each kind of Usanovich acid or base along with plausible explanation of the concept is given in Table 7.2.

Table 7.2. Usanovich acids and bases and their reactions.

	Acid	*Base*	*Salt*	*Justifications*
1.	HCl	NaOH	$Na^+ Cl^- + H_2O$	Salt formation
2.	HCl	NH_3	$NH^+{}_4Cl^-$	Acid gives up cation H^+
3.	$2NH_4Cl$	$Ba(OH)_2$	$Ba^{2+}\ Cl_2^{2-} + 2NH_3 + H_2O$	Acid gives up cation NH_4^+
4.	SO_3	CaO	$Ca^{2+}\ SO^2{}_4{}^-$	Acid SO_3 accepts anion O^{2-}
5.	Cl_2	Cu	$Cu^{2+}\ Cl_2{}^{2-}$	Acid gains an electron
6.	$Fe(CN)_2$	4KCN	$K_4^{4+}\ [Fe(CN_6)]^{4-}$	Base KCN gives up anionCN$^-$
7.	F_2	Mg	$Mg^{2+}\ F^2{}_2{}^-$	Base Mg loses electrons
8.	H_2SO_4	NH_3	$(NH_4)^2{}_2{}^+\ SO_4{}^{2-}$	BaseNH$_3$ takes up cations H^+

Although Usanovich concept lays more stress on the cationic and anionic aspects of acids and bases, yet it does not ignore unsaturation in cation acid - base reaction. The acidic function is determined by the presence of coordinately unsaturated species and the basic function by the presence of coordinately saturated species as shown below:

$$\underset{\text{(acid)}}{O{=}C{=}O} + \underset{\text{(base)}}{OH^-} \longrightarrow \underset{\text{(bicarbonate)}}{HO{-}C({=}O){-}O^-}$$

$$\underset{\text{(acid)}}{S{=}C{=}S} + \underset{\text{(base)}}{R_2NH} \longrightarrow \underset{\text{(dithiocarbonic acid)}}{R_2N{-}C({=}S){-}SH}$$

Hard and soft acids and bases

According to Usanovich concept, cations come under the category of acids. Hard acids (cations) are those which fulfil the following conditions:

(*i*) *Cations of smaller size or lighter elements,*

(*ii*) *Cations of higher charge,*

(*iii*) *Not easily polarizable, and*

(*iv*) *Molecules or ions with lesser number of valence electrons.*

Examples of hard acids are alkali metal ions, alkaline earth metal ions, B, Al, Si ions, lighter transition metal ions such as Ti^{4+}, Cr^{3+}, Fe^{3+}, Co^{2+}, etc.

Similarly soft acids are those which satisfy the following features:

(*i*) *Cations of larger size or heavier elements*

(*ii*) *Cations of lower charge*

(*iii*) *Easily polarizable*

(*iv*) *Molecules or ions with larger number of valence electrons.*

Examples of soft acids are heavy transition metal ions (2nd and 3rd row) such as Hg^{2+}, Pd^{2+}, Cd^{2+} etc. and those in lower oxidation state such as Cu^+, Ag^+, Hg^+ etc.

According to Usanovich concept ligands come in the category of bases. Hence hard bases (ligands) are those which have the following features:

(*i*) *donor atoms of higher electronegativity;*

(*ii*) *donor atoms with low polarizability.*

Examples of hard bases are molecules or ions that have donor atoms N, O, F, Cl etc. like H_2O, NH_3, F^-, ROH.

Similarly soft bases usually have:

(*i*) *donor atoms of lower electronegativity*

(*ii*) *donor atoms of higher polarizability.*

Molecules or ions where the donor atom is P, As, S, Se etc. like R_3P, R_2S, I^- are examples of soft bases. It may be noted that hard acids have the tendency to combine with hard bases. Likewise soft acids combine with soft bases.

It may be noted that there is no sharp division between hard and soft acids or bases.

HSAB (Pearson's) principle

R. G. Pearson

R.G. Pearson suggested a simple rule for predicting the stability of co-ordinate compound formed between acids and bases (metal ions and ligands) and is often termed as the HSAB principle. It states that *hard acids prefer to bond to hard bases and soft acids prefer to combine with soft bases.*

Thus, hard acids will show the following order of preference for bonding with donor atoms or ions within a group:

$$F^- \gg Cl^- > Br^- > I^-$$
$$R_3O \gg R_2S$$
$$R_3N \gg R_3P$$

Soft acids will show the opposite trend and the following order of preference for bonding the donor atoms or ions within a group is observed:

$$I^- \gg Br^- > Cl^- > F^-$$
$$R_2S \gg R_2O$$
$$R_3P \gg R_3N$$

Hard acids and bases are not easily polarizable and their interactions are predominantly *ionic.* On the other hand, soft acids and soft bases are usually polarizable and hence their interactions are predominantly *covalent*.

Basis of hard-hard and soft-soft interactions

According to the most accepted view, the cause of *hard– hard interaction* is an electrostatic interaction. Bonding between hard acids and hard bases is predominantly ionic. The electrostatic energy between a positive and negative ion pair is inversely proportional to the internuclear distance. Therefore, smaller the ions, the lesser would be the internuclear distance and the greater would be the electrostatic attraction between the two ions. Consequently, the resulting compound would be highly stable.

Electrostatic interaction cannot explain the *soft– soft interaction* because the sizes of soft species are comparatively very large. Polarisation of the species, however, plays an important role in explaining their interactions. Most of the soft acids have 6 to 10 *d* electrons in their electronic configurations. These electrons get easily polarised favouring covalent bonding between them and the soft bases which are also easily polarisable. Thus, bonding between soft acids and soft bases is assumed to be *largely covalent.*

Considering hard – hard interactions as ionic and soft – soft interactions as covalent, Misons and coworkers (1967) proposed the following relation which can tell whether a given species is hard or soft:

$$pK = -\log K = aX + bY + c$$

where K is the equilibrium constant for the dissociation of acid-base; X, Y are parameters for the acids ; a, b are parameters of the bases and c is a constant required to adjust the pK values in such a way that all of them lie on the same scale.

The values of parameter Y for some of the cations are given below :

Hard acids	Li^+	Al^{3+}	Mg^{2+}	Na^+	Ca^{2+}	Fe^{3+}	Co^{2+}	Cs^+
	0.36	0.70	0.87	0.93	1.62	2.37	2.56	2.73

Soft acids	Sn^{2+}	Tl^{3+}	Cu^+	Pb^{2+}	Tl^+	Hg^{2+}	Au^+
	3.17	3.2	3.45	3.58	3.78	4.25	5.95

If the value of Y is less than 2.8, the acid is hard and if the value is more than 3.2, the acid is soft.

The value of parameter b for some of the bases are given below :

	Hard bases			Soft bases	
OH^-	NH_3	Cl^-	Br^-	I^-	$S_2O_3^{2-}$
0.40	1.08	2.48	5.58	7.17	12.4

The value of b, evidently, increases with the softness of the base. If the value is less than 3, the base is hard and if the value is more than 5, the base is soft.

In soft-soft interactions, the contribution of π bonding is also important.

Applications of HSAB principle.

HSAB principle has the following applications.

***(a)* Stability of complexes:** This principle explains the special stability of complexes formed by hard-hard and soft-soft interactions. This has nothing to do with the actual acidic or basic strengths of acids or bases. For example, both OH^- and F^- are hard bases and although OH^- is a far stronger base than F^-, yet these are treated equally by the HSAB principle. We can predict the stability of certain complexes on the basis of hard-hard and soft-soft preferences. For example, the complex between a soft acid Ag^+ and soft base I^- is quite stable in contrast to the one between Ag^+ and hard base F^-

$$\underset{\text{Soft acid}}{Ag^+} + \underset{\text{Soft base}}{2I^-} \longrightarrow \underset{\text{Stable}}{AgI_2^-}$$

$$\underset{\text{Soft acid}}{Ag^+} + \underset{\text{Soft base}}{2F^-} \longrightarrow \underset{\text{Unstable}}{AgF_2^-}$$

***(b)* Course of reaction:** We can also predict the course of a reaction by the use of this principle. For example, reaction between LiI and CsF always proceeds to give LiF and CsI

$$\underset{\text{hard-soft}}{LiI} + \underset{\text{soft-hard}}{CsF} \longrightarrow \underset{\text{hard-hard}}{LiF} + \underset{\text{soft-soft}}{CsI}$$

There is preferential stability due to hard-hard and soft-soft combinations. Similarly reaction between CaS and H_2O proceeds towards the right to give CaO.

$$\underset{\text{hard-soft}}{CaS} + \underset{\text{hard}}{H_2O} \longrightarrow \underset{\text{hard-hard}}{CaO} + \underset{\text{soft}}{H_2S}$$

Formation of cuprous iodide by reaction between cupric iodide and cuprous fluoride is another example of hard-hard preference.

$$\underset{\text{hard-soft}}{CuI_2} + \underset{\text{soft-hard}}{2CuF} \longrightarrow \underset{\text{hard-hard}}{CuF_2} + \underset{\text{soft-soft}}{2CuI}$$

Symbiosis

Soft ligands have a tendency to combine wilh a centre already associated with soft ligands and hard ligands have a tendency to combine with a centre already associated with hard ligands. This tendency has been termed as **symbiosis.** Thus, F^- ion, a hard ligand, readily combines with BF_3 to form a stable complex BF_4^-. Similarly, H^- ion, a soft ligand, readily combines with BH_3 to form a stable complex BH_4^- :

$$BF_3 + F^- \longrightarrow BF_4^-$$
$$BH_3 + H^- \longrightarrow BH_4^-$$

Mixed substituents are generally not preferred. Thus, compounds such as BF_3H^- and BH_3F^-, having mixed substituents, interact to yield BH_4^- and BF_4^-, *i.e.*, compounds with the same substituents:

$$BH_3F^- + BF_3H^- \longrightarrow BH_4^- + BF_4^-$$

Likewise, the following reaction occurs spontaneously:

$$CH_3F + CHF_3 \longrightarrow CH_4 + CF_4$$

Example 1. Which of the following reactions will proceed to right hand side?

(*i*) $\mathbf{CuCl_2 + 2KI \longrightarrow 2KCl + CuI + \frac{1}{2} I_2}$

(*ii*) $\mathbf{BeI_2 + HgF_2 \longrightarrow BeF_2 + HgI_2}$

(*iii*) $\mathbf{CdCl_2 + H_2S \longrightarrow 2HCl + CdS}$

Solution. On the basis of hard-hard and soft-soft preference we can see that all the above reactions proceed towards R.H. S.

7.8. RELATIVE STRENGTH OF ACIDS AND BASES

Strength of acids. The strengths of acids like HCl, HNO_3, H_2SO_4, HCN, HF can be compared by their tendencies to give protons. Acids like HCl, HNO_3, or H_2SO_4 are strong acids while CH_3COOH, HF, HCN are weak acids. Ionisation of an acid in water measures the relative strength of the acid.

For instance, ionisation of an acid HA may be represented as

$$HA \rightleftharpoons H^+ + A^-$$

The equilibrium constant for the reaction is written as

$$K_a = \frac{[H^+][A^-]}{[HA]}$$

where K_a is known as **dissociation constant** or **ionisation constant** of an acid and measures the relative strength of an acid. The stronger the acid, the larger will be its K_a.

The term pK_a, that is $-\log K_a$, is commonly used to express the dissociation of an acid.

$$pK_a = -\log K_a = \frac{1}{\log K_a}$$

Thus the higher the value of pK_a, the weaker the acid.

PK_a values of some acids in water is given in Table 7.3.

Table 7.3. pK_a values of some acides in water.

Acid	Dissociation reaction in water			pK_a
CH_3COOH	CH_3COOH	$\rightleftharpoons$	$CH_3COO^- + H^+$	4.76
HF	HF	$\rightleftharpoons$	$H^+ + F^-$	3.30
HCOOH	HCOOH	$\rightleftharpoons$	$H^+ + HCOO^-$	3.70
H_2CO_3	H_2CO_3	$\rightleftharpoons$	$H^+ + HCO_3^-$	6.36
H_3BO_3	H_3BO_3	$\rightleftharpoons$	$H^+ + H_2BO_3^-$	9.2
HCN	HCN	$\rightleftharpoons$	$H^+ + CN^-$	6.4

* pK_a is analogous to pH.

Just as $pH = -\log [H^+]$, $pK_a = -\log k_a$

Strength of bases. Consider the ionisation of base MOH:

$$MOH \rightleftharpoons M^+ + OH^-$$

The equilibrium constant may be expressed as:

$$K_b = \frac{[M^+][OH^-]}{[MOH]}$$

where K_b is known as the **ionisation** or **dissociation constant** of the base. The higher the value of K_b, the stronger is the base.

K_b may be expressed as

$$pK_b = -\log K_b \frac{1}{\log K_b}$$

Larger the value of pK_b, lesser is the dissociation and weaker is the base.

PK_b values of some bases in water is given in Table 7.4.

Table 7.4 pK_b values of some bases in water.

Base	Dissociation reaction in water			pK_b
$Ca(OH)_2$	$Ca(OH)_2$	$\rightleftharpoons$	$Ca^{2+} + 2OH^-$	96.3
NH_4OH	NH_4OH	$\rightleftharpoons$	$NH_4^+ + OH^-$	3.25
$C_6H_5NH_2$	$C_6H_5NH_2 + H_2O$	$\rightleftharpoons$	$C_6H_5NH_3^+ + OH^-$	8.82

C_5H_5N	$C_5H_5N + H_2O$	$\rightleftharpoons$	$C_5H_5NH^+ + OH^-$	9.60
$C_2H_5NH_2$	$C_2H_5NH_2 + H_2O$	$\rightleftharpoons$	$C_2H_5N_3^+ + OH^-$	7.75
AgOH	AgOH	$\rightleftharpoons$	$Ag^+ OH^-$	2.43

7.8.1. Trends in acid strength of hydro acids

Substances like HCl, H_2S, PH_3, and H_2 Te are called hydro acids. *There is a general increase in hydro acid strength in going from left to right along a period.* For example, the following trend of acid strength is observed along the second period:

$$HF > H_2O > NH_3 > CH_4$$

This trend can be explained on the basis of increase in electronegativity as we move from left to right along a period.

$$C < N < O < F$$

With the increase in electronegativity of the central atom, the tendency to pull the electron pair from H increases, and dissociation of H becomes easier and acidic character increases.

The above trend based on the electronegativity holds true along a period. It does not hold good as we move down a group. In fact, acidity in a group increases with decrease in electronegativity as we go down the group. For example, in the hydro acids of group VII, HI is the strongest acid and HF is the weakest acid.

$$HI > HBr > HCl > HF$$

which is opposite to the trend along a period.

Similarly in group VI the acid strength increases as we move down the group with H_2O as the weakest acid.

$$H_2O < H_2S < H_2Se < H_2Te$$

and H_2Te is the strongest acid in spite of H_2O being the most polar on the basis of electronegativity.

Example 2. Giving reasons arrange the following in order of increasing acid strength.

HF, HCI, HBr, HI

Solution. The order of increasing acid strength is

$$HF < HCl < HBr < HI$$

This is because bond dissociation energy of the halogen acids decreases while going from HF to HI.

Example 3. Arrange the following in decreasing basic strength:

NH_3, PH_3 and AsH_3

Solution. On moving down a group, the hydrides show increasing acid strength or decreasing basic strength. Hence the order of decreasing basic strength will be

$$NH_3 > PH_3 > AsH_3$$

7.8.2. Trend in acid strength of oxo-acids

Substances like $HClO_4$, H_3PO_4, HSO_4^-, containing oxygen are called oxo-acids. The strength of oxo-acids depends upon the following factors:

(i) Size of the central atom

(ii) Electronegativity of the central atom in the molecule.

(iii) Inductive effects of substituents on central atom.

Consider an oxo-acid X – O – H.

where *X* is the central atom. A small and highly electronegative atom *X* will tend to attract the shared pair of electrons more towards itself. This will create a high polarity in O – H bond thereby allowing the proton to dissociate and make the compound acidic. Thus higher the electronegativity of *X* more easily is the proton lost and greater is the acidity of oxo-acid.

As we go down a group, the electronegativity decreases as the size increases and acidic character decreases. For example, the acidic character of oxo-acids of halogens decreases in the given order as is clear from pK_a values.

$$\begin{array}{llll} & HClO > & HBrO > & HIO \\ pK_a & 7.4 & 8.7 & 11 \end{array}$$

It has been observed that the compound XOH behaves as an acid only if X has electronegativity value greater than 1.8. For example, ClOH is an acid while LiOH is a base because electronegativity of Cl is 3.0 while that of Li is 1.0.

Presence of additional electronegative group like oxygen increases the acidity. The relative acidity increases with increase in number of oxygen atoms. For example, in the series of oxo-acids of chlorine, the acid strength increases in the order

	$HClO <$	$HClO_2 <$	$HClO_3 <$	HC1O
Oxidation no. of Cl	1	3	5	7
pK_a value	7.52	1.94	–3	–10

In these compounds, the acidity increases with increasing oxidation numbers of the atom (X). All these four acids have only one protonated *oxygen i.e.*, OH group. The rest of the oxygen atoms are surrounding the central atom. With increase in the number of oxygen atoms on the central atom, the extent of electron withdrawal from the O – H bond increases. As a result, the polarity of the O – H bond or tendency for the proton formation increases.

$$Cl - OH < OCl - OH < O_2Cl - OH < O_3Cl - OH$$

The increase in acidic strength can also be explained in terms of the stability of conjugate base formed by the loss of H^+ ion from the acid.

The anions ClO^-, ClO_2^-, ClO^-_3 and ClO^-_4 formed by the loss of proton are stabilized by the charge distribution among O atoms. The charge stabilization is minimum in ClO^- and maximum in ClO^-_4. This means that ClO^- will have minimum stability and would trend to combine with H^+ thus making it the weakest acid and ClO^-_4 will have maximum stability and minimum attraction for H^+ therefore making it the strongest acid amongst the four.

$$[O \overset{....}{=} Cl]^- < [O \overset{....}{=} Cl \overset{....}{=} O]^- < [O \overset{....}{=} \overset{\overset{O}{\|}}{Cl} \overset{....}{=} O]^- < [O \overset{....}{=} \underset{\underset{O}{\|}}{\overset{\overset{O}{\|}}{Cl}} \overset{....}{=} O]^-$$

$$\xrightarrow{\text{increase in stability}}$$

Likewise, ON(OH) nitrous acid (pK_a = 3.3) is a weak acid but O_2N (OH) (nitric acid) is a strong (pK_a = – 1.4) acid.

Among organic acids, the inductive effect of substituents is important. Acetic acid H_3CCOOH is a weak acid but replacement of methyl hydrogens by a more electronegative chloro group increases the acidity of the acid as shown by their pK_a values. As the number of electronegative substituents increases the acidity of the acid increases further.

Acid	pK_a
CH_3COOH	4.8
$ClCH_2COOH$	2.9
$Cl_2CHCOOH$	1.3
Cl_3CCOOH	0.7

7.8.3. Effect of polarity and dielectric constant on relative strengths of acids and bases.

The polarity and dielectric constant of a solvent influence considerably the acidic and basic strength of ionic species dissolved in that solvent. According to Coulomb's law, the force of attraction between two oppositely charged ions is given by

$$F = \frac{q_1 q_2}{Dr^2} \quad \text{.... (1)}$$

where q_1 and q_2 are the charges on the ions, r is the distance between the two ions and D is the dielectric constant of the solvent. Higher the dielectric constant of the solvent, lower is the magnitude of F. In solvents with low dielectric constants, the force of attraction between the ions of the solute will be relatively large and consequently the dissociation of the solute would be small. In solvents with high dielectric constants, the force of attraction between the ions would be small and the extent of dissociation would be large. In other words, the acidic or basic strength of a species would increase with an increase in the dielectric constant of the solvent. The dielectric constant and the dipole moment are known to be directly related. Hence, the greater the polarity of a solvent, the greater would be the acidic or basic strength of the species dissolved in that solvent.

7. 8.4. Effect of substituents on relative strengths of acids and bases.

The substituents present in acids and bases have a marked effect on their relative strengths. The increase or decrease in acid or base strength depends upon the nature of the substituents, their number and the type of bonding involved in them. This is illustrated below.

(*i*) Electron-releasing or Electron-withdrawing Nature of Substituents. An electron-withdrawing substituent increases and an electron-releasing substituent decreases the acid strength of a species. Thus, chloroacetic acid is stronger than acetic acid while propionic acid is weaker than acetic acid. This is because Cl is an electron-withdrawing and CH_3 is an electron-releasing group.

Similarly, an electron-releasing substituent would increase and an electron-withdrawing substituent would decrease the basic strength of a species. Thus, while NH_3 is basic, NF_3 is almost neutral. Likewise, $(CH_3)_2NH$ is more basic than CH_3NH_2 and NH_3. The CH_3 group, being electron-releasing, increases the basic character of ammonia.

The inductive effect decreases as the substituent moves away from the COOH group. This is why 1-chloro propionic acid ($CH_3CHClCOOH$) is a stronger acid than 2-chloro propionic acid ($ClCH_2CH_2COOH$).

(*ii*) Steric Effect of Substituents. Strangely enough, $(CH_3)_3N$ is less basic than $(CH_3)_2NH$ although due to the presence of an additional electron-releasing CH_3 group in trimethyl amine, it is expected to be more basic than dimethyl amine. This anomaly is due to the *steric effect.* Approach of an electron-acceptor towards the N atom to accept the electron pair is hindered because of the presence of as many as three CH_3 groups around the N atom. This restricts the donation of the electron pair of N atom in $(CH_3)_3N$ and hence reduces the basicity of $(CH_3)_3N$.

(iii) Type of Bonding Involved in Substituents. The presence of a pi bond in a substituent alters the acidic or basic character of the species. For instance, although fluorine is more electron-withdrawing than chlorine or bromine yet BF_3 is *less acidic* than BCl_3 or BBr_3. This may be explained as follows : In a B – F bond in BF_3, one of the filled *p* orbitals of fluorine of proper symmetry forms a pi bond by overlapping with a vacant *p* orbital of boron, causing a shift of electronic charge from fluorine to boron. This decreases the electron-accepting power of boron in BF_3. The *p* orbitals of Cl or Br possess much higher energy and are not suited for pi bonding with boron. Therefore, the electron-accepting power of boron in BCl_3 or BBr_3 is more than that in BF_3.

The following factors also influence the relative strengths of acids and bases.

(a) Charge on the Species. All other factors remaining constant, *an increase in the positive charge would increase the acidic strength of the species and an increase in the negative charge would increase the basic strength of the species*. This is because an increase in positive charge increases the electron-accepting capacity and an increase in negative charge increases the electron-donating capacity of the species.

(b) Electronegativity. In the species HX, the higher the electronegativity (electron-pulling capacity) of X, the easier would be the release of proton (H) from HX. In other words, the acid strength (proton-releasing capacity) of HX would increase with increase in the electronegativity of X. For example, the electronegativity of non-metals of the second period are in the order C < N < O < F. The acid strength of their hydrides also varies in the same order, *i.e.*, $CH_4 < NH_3 < H_2O < HF$. Thus, within a period of the Periodic table, higher the electronegativity of a non-metal X, greater would be the acid strength of HX.

7.8.5. Effect of hydration on the strength of acid.

On the basis of electronegativities of the halogens, the strength of the halogen acids should change in the order HI < HBr < HCl < HF. But, actual order is just the reverse, *i.e.*, HF < HCl < HBr < HI. This anomaly may be explained as follows :

The ionization of HX (X = halogen) in an aqueous solution, in fact, involves a number of energy factors. We can write the following cycle similar to Born-Haber cycle, for the ionization of HX into H^+ (*aq*) and X^- (*aq*) ions.

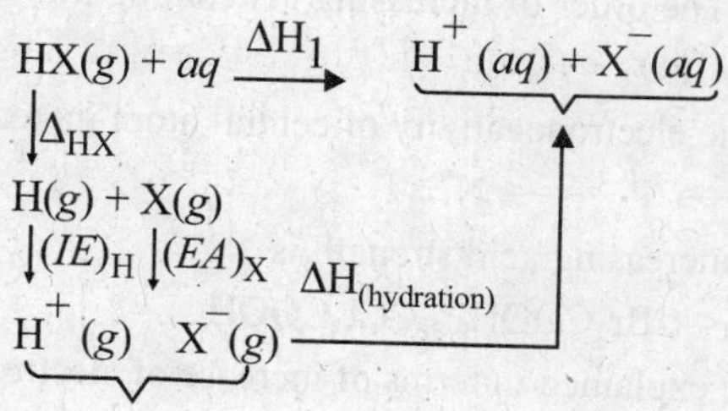

$$\begin{array}{lcl} HX(g) + aq & \xrightarrow{\Delta H_1} & H^+(aq) + X^-(aq) \\ \downarrow \Delta_{HX} & & \uparrow \\ H(g) + X(g) & & \\ \downarrow (IE)_H \quad \downarrow (EA)_X & & \Delta H_{(hydration)} \\ H^+(g) \quad X^-(g) & \longrightarrow & \end{array}$$

From Hess's law,

$$\Delta H_1 = \Delta_{HX} + (IE)_H - (EA)_X - \Delta H_{hydration}$$

where ΔH_1 = Enthalphy of ionization of HX in water

Δ_{HX} = Dissociation energy of HX(g)

$(IE)_H$ = Ionization energy of H atom.

$(EA)_X$ = Electron affinity of X atom.

$\Delta H_{hydration}$ = Enthalpy of hydration of H^+ and X^- ions.

Both $(EA)_X$ and $\Delta H_{hydration}$ are exothermic processes and are written with negative signs in the Hess's law expression for ΔH_1.

Evidently, the ionization of HX, *i.e.*, the acid strength of HX would be favoured if ΔH_1 has a large negative value. To this end, while Δ_{HX} should be small, $(EA)_X$ and $\Delta H_{hydration}$ should be large since $(IE)_H$ is a constant factor. The enthalpies of ionization of the various halogen acids determined as above, fully support the variation of their strengths in the order HF < HC1 < HBr < HI.

Example 4. How does the strength of acids depend upon:

(i) Oxidation no. of central atom.

(ii) Inductive effect of substituents.

Solution (*i*) Acid strength of an acid increases with the increase in oxidation number of central atom. For example, the acid strength increases in the order:

$$HClO < HClO_2 < HClO_3 < HClO_4$$

because the oxidation number of chlorine increases from 1 for HClO to 3 for $HClO_2$, 5 for $HClO_3$ and 7 for $HClO_4$.

(*ii*) An electron releasing group decreases the acid strength because it hinders the release of protons. For example, the acid strength for the following acids shows the order:

$$HCOOH > CH_3COOH > CH_3CH_2COOH$$

Thus replacements by CH_3 and CH_3CH_2 – for H – in HCOOH results in decrease of acid strength. An electron withdrawing groups increases the acid strength because it helps in release of protons. The acid strength of the following acids exhibits the order:

$$CCl_3COOH > CHCl_2COOH > CH_2ClCOOH > CH_3COOH$$

Thus replacements of hydrogen by electron withdrawing group (Cl –) increases the acid strength. Greater the number of chlorine groups, greater the acid strength.

Example 5. (*a*) Arrange the following in order of increasing acid strength:

(*i*) HNO_3, HPO_3, and $HAsO_3$,

(*ii*) CCl_3COOH, CBr_3COOH, CI_3COOH

(*b*) Write the conjugate base for the following:

$HClO_4$, HNO_3, H_2CO_3, $H_2PO^-{}_4$

Solution. (*a*) (*i*) The order of increasing acid strength is

$$HAsO_3 < HPO_3 < HNO_3$$

This is because the electronegativity of central atom increases in the order:

$$As \longrightarrow P \longrightarrow N$$

(*ii*) The order of increasing acid strength is

$$CI_3COOH < CBr_3COOH < CCl_3COOH$$

This can again be explained in terms of increase of electronegativity from I through Br to Cl.

(*b*) The conjugate base for the acids are

$$ClO^-{}_4,\ NO^-{}_3,\ HCO^-{}_3\ HPO^2{}_4{}^-$$

Example 6. Which of the following bases is the weakest and why?

Cl^-, I^-, Br^-, F^-

Solution. Cl^-, I^-,Br^- and F^- are conjugate bases of HC1, HI, HBr and HF respectively. Since conjugate base of strong acid is weak, I^- is expected to be weakest because it is the conjugate base of the strongest acid HI.

$$\underset{\text{Strong acid}}{HI} \longrightarrow H^+ + \underset{\substack{\text{Weak base}\\\text{(conjugate base)}}}{I^-}$$

Example 7. Arrange the following in order of decreasing acid strength $HClO$, $HClO_2$, $HClO_3$, and $HClO_4$

Solution. The order is $HClO_4 > HClO_3 > HClO_2 > HClO$

This can be explained in terms of oxidation number of the central atom. Greater the oxidation number, greater the acid strength. The oxidation numbers of Cl in $HClO_4$, $HClO_3$, $HClO_2$ and HClO are 7, 5, 3 and 1 respectively. The above order can also be explained in terms of the stability of the anions (base) released by the acids. ClO^-_4 released from $HClO_4$ has the maximum stability because the negative charge can be dissipated with the greatest ease because there are 3 (maximum) other oxygens, as shown below.

$$O{=}\overset{\overset{O}{\|}}{\underset{\underset{O}{\|}}{Cl}}{-}O^- \longleftrightarrow O{=}\overset{\overset{O^-}{|}}{\underset{\underset{O}{\|}}{Cl}}{=}O \longleftrightarrow {}^-O{-}\overset{\overset{O}{\|}}{\underset{\underset{O}{\|}}{Cl}}{=}O \longleftrightarrow O{=}\overset{\overset{O}{\|}}{\underset{\underset{O^-}{|}}{Cl}}{=}O$$

There are four resonating structures of the anion released. In the case of $HClO_3$, $HClO_2$ and HClO, there will be 3, 2 and 1 resonating structures respectively of the anions released by the acids. Thus the stability of the anion will go on decreasing. A weak anion has always the tendency to combine with proton to give back the original acid in the undissociated form, thus reducing acid strength.

$$\text{Acid} \rightleftharpoons H^+ + \text{Anion}$$

Example 8. Give reasons for the following order of acid strength:

(*i*) $HF > H_2O > NH_3$

(*ii*) $H_2Fe > H_2Si > H_2S > H_2O$

(*iii*) $HNO_3 > HNO_2$

Solution. (*i*) In the case of hydro acids involving elements in a period, the acid strength increases due to increase of electronegativity. HF, H_2O and NH_3 are hydro acids of F, O, and N. As electronegativity decreases in the order F > O > N, the acid strength follows the order: HF > H_2O > NH_3.

(*ii*) This can be explained on the basis of increasing bond dissociation energy.

(*iii*) $HNO_3 > HNO_2$. Nitrogen has greater oxidation number in HNO_3 than HNO_2. Therefore HNO_3 is stronger than HNO_2.

7.9. LINUS PAULING'S AND RICCI'S OBSERVATIONS ON THE STRENGTH OF ACIDS

Linus Pauling made the following observations regarding strength of acids.

(*i*) For a given polybasic acid, the successive $pK_{a_1} pK_{a_2} pK_{a_3}$ values approximate to a ratio 1 : 4 : 8 as shown below:

Acid	PK_{a1}	PK_{a2}	pK_{a3}
H_3PO_4	2.1	7.2	12.0
H_3AsO_4	2.3	7.0	13.0
$H_4P_2O_7$	0.8	2.2	6.7

(*ii*) The magnitude of pK_a is given by the formulation: $O_m\,E\,(OH)_n$.

For

$m = 0$; $pK_a \geq 7$;	very weak acids; *e.g.*, $B(OH)_3$
$m = 1$; $pK_a \cong 2$;	weak acid; *e.g.*, $OCl(OH)$
$m = 2$; $pK_a \cong -3$;	strong acid; *e.g.*, $O_2Cl(OH)$
$m = 3$; $pK_a \cong -8$;	very strong acid; *e.g.*, $O_3Cl(OH)$

Observations of Ricci are given as under in the form of generalisation:

$$pK_a = 8.0 - 9.0\,m + 4.0\,n \qquad \text{... }(i)$$

where m = formal charge on the central atom E

n = the no. of non-hydroxyl oxygen atoms in the acid. Formal charge is calculated as follows:

m = no. of valence shell electrons on the central atom $- \frac{1}{2}$(no. of electron shared with other atoms) – (no. of unshared valence shell electrons)

Thus, formal charge on chlorine in different acids can be obtained as under:

$:\ddot{\underset{..}{Cl}} - OH$	$7 - \frac{1}{2}(2) - 6 = 0$
$O - \ddot{\underset{..}{Cl}} - OH$	$7 - \frac{1}{2}(4) - 4 = +1$
$O_2 > Cl - OH$	$7 - \frac{1}{2}(6) - 2 = +2$
$O_3 > -Cl - OH$	$7 - \frac{1}{2}(8) - 0 = +3$

For example in $HClO_4$; $m = +3$; $n = 3$

Substituting these values in the empirical equation (*i*) given above we can calculate pK_a

$$pK_a = 8.0 - 9(+3) + 4.0(3) = -7$$

Similarly for $HClO_3$, $m = +2$; $n = 2$

$$pK_a = 8.0 - 18 + 8 = -2$$

For $H_3PO_4 = OP(OH)_3$; $m = 5 - 4 - 0 = 1$; $n = 1$

$$pK_a = 8 - 9 + 4 = -3$$

The empirical values calculated according to above equation are quite close to the reported pK_a values. Thus, knowing the values of formal charge on the central atom in the oxo-acid and the number of non-hydroxyl oxygen atoms on the acid, it is possible to get an approximate of acid strength.

7.10. TREND IN ACID STRENGTH OF HYDRATED CATIONS

Hydration energy is the heat change accompany the formation of 1 mole of a hydrated substance. Hydration energies of $[Ba(H_2O)_2]^{2+}$, $[Cu(H_2O)_6]^{2+}$ are approximately 30.75 and 320 kJ/mol respectively. This large variation in hydration energy is associated with acidity, that is, charge-to-size ratio of the cation. Larger value of charge-to-size ratio for the cation ($Al^{3+} > Cu^{2+} > Ba^{2+}$) is responsible for high hydration energy. There is the phenomenon of hydrolysis which takes place along with hydration as illustrated below.

$$Fe^{3} + 6H_2O \xrightarrow{\text{Hydration}} [Fe(H_2O)_6]^{3+} \xrightarrow{\text{Hydrolysis}} [Fe(H_2O)_5(OH)]^{2+} + H^+$$

Small and highly charged cations with larger charge-to-size ratio can lead to the breaking of O – H bonds in the hydrated cation to yield hydrolysed species and the hydrated proton.

The larger the charge-to-size ratio (acidity) of the cation, the more polar the O – H bond in the coordinated H_2O, and the more easily a proton is released as H_3O^+. On the basis of charge-to-size ratio, the relative acid strength decreases as:

$[Fe(H_2O)_6]^{3+} > [V(H_2O)_6]^{3+} > [Cr(H_2O)_6]^{3+} > [Sc(H_2O)_6]^{3+} > [Al(H_2O)_6]^{3-} > [La(H_2O)_6]^{3+}$

Among divalent hydrated cations, the acid strength decreases in the order:

$[Be(H_2O)_n]^{2+} > [Cu(H_2O)_n]^{2+} > [Mg(H_2O)_n]^{2+} > [Ca(H_2O)_n]^{2+} > [Ba(H_2O)_n]^{2+}$

7.11. RELATIVE STRENGTH OF LEWIS ACIDS

Relative strength of Lewis acids depends mainly on following factors:

(*i*) Inductive effect: The electron withdrawing inductive effect makes a Lewis acid stronger and an electron releasing effect will decrease acid strength. For example, if H in BH_3 is replaced by electron withdrawing *F,* its acidic strength increases. On the other hand, if H in BH_3 is replaced by electron releasing CH_3, then its acidic strength decreases.

$$Me_3B < BH_3 < BF_3$$

(*ii*) Resonance: The inductive effect rule fails in explaining the relative acidity of boron halides BF_3, BCl_3, BBr_3. Based on electron withdrawing effect of substituents, BF_3 should have been the strongest Lewis acid and BBr_3 the weakest Lewis acid but a reverse order of acid strength is observed.

$$BBr_3 > BCl_3 > BF_3$$

This anomaly is attributed to resonance effect occurring due to the tendency of halogen atoms to contribute electron density toward boron atom. This is known as **back donation** from halogen to boron as shown in Fig. 7.1.

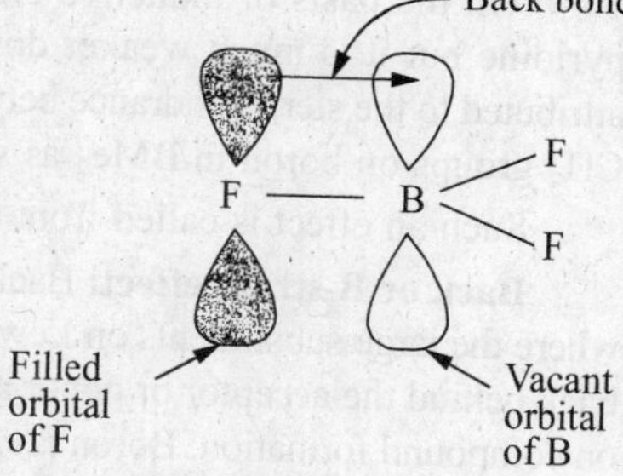

Fig. 7.1. Back bonding.

Back donation is more effective in case of BF_3 due to the easier overlap of filled 2*p* orbitals of fluorine with an empty 2*p* orbitals of boron. Since both the overlapping orbitals belong to the same energy level, therefore the overlap is more effective and leads to an additional bond known as *p*π-*p*π back bond. This type of back bonding can oscillate among the three B – F bonds in BF_3 as shown below:

$$\overset{(+)}{F}{=}\overset{(-)}{B}(F)_2 \longleftrightarrow F{-}\overset{(-)}{B}({=}\overset{(+)}{F})(F) \longleftrightarrow F{-}\overset{(-)}{B}(F)({=}\overset{(+)}{F}) \longleftrightarrow F{\overset{..}{=}}B(F)(F)$$

The resonance stabilized molecule BF_3 due to back donation from F to B, is no longer so electron deficient with the result that the Lewis acid character is reduced. This tendency of back donation from filled *p*-orbitals of halogen to unfilled 2*p*-orbitals of B is less pronounced in case of BCl_3 and BBr_3 due to higher energy donor 3*p*-orbitals of Cl and 4*p*-orbitals of Br relative to 2*p*-orbital of B. Consequently, resonance stabilisation of BCl_3 is small and that of BBr_3 is smaller. Hence the acidity order is:

$$BBr_3 > BCl_3 > BF_3$$

Similarly, B $(OR)_3$ is weaker than BR_3 due to resonance stabilization of the former.

Example 9. Explain the order of acid strength of the following:

(*i*) $BF_3 < BCl_3 < BBr_3 < BI_3$

(*ii*) $Me_3B < BH_3 < BF_3$

Soution. (*i*) Boron forms back bonding with F effectively

The electrons are thus donated to B making it less electron deficient and thus weaker acid. This back bonding decreases as we move from BF_3 towards BI_3. Thus BI_3 is the strongest of these acids.

(*ii*) Methyl group (Me) is electron donating whereas F^- is electron withdrawing. Therefore BF_3 is the strongest and Me_3B is the weakest acid.

7.12. F-STRAIN AND B-STRAIN EFFECTS AND STABILITY OF ACID-BASE COMPLEXES

Front or F-strain effect: In reactions between Lewis acid and bases, such as pyridine, 2- picoline and 4-picoline with the same acid trimethylborane BMe_3, the adduct bond energies are 71, 42 and 76 kJ per mol, respectively.

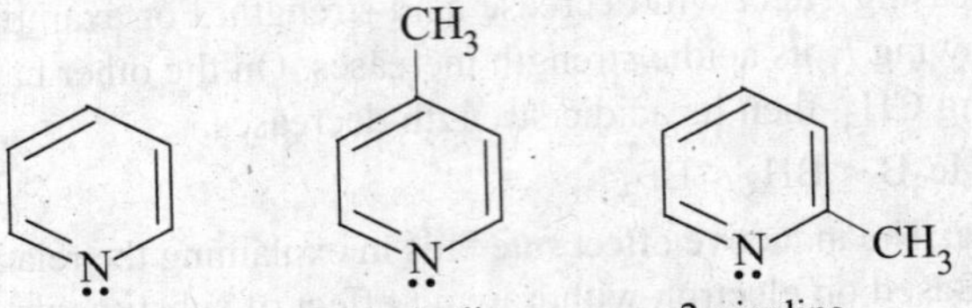

Pyridine 4-picoline 2-picoline

On the basis of inductive effect of methyl group 2-picoline should be stronger donor than pyridine but it is much weaker donor towards BMe_3. This may be attributed to the steric hindrance between CH_3 group of 2-picoline and CH_3 groups on boron in BMe_3 as shown:

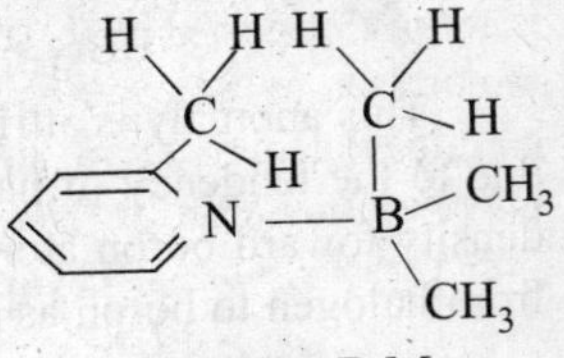

2-picoline-B Me_3

Such an effect is called front or "F-strain" effect.

Back or B-strain effect: Back or "B-strain" is observed in case where the large substituents on Lewis acids or bases are forced to fold back behind the acceptor or donor atom due to changed hybridization on compound formation. Boron is sp^2 hybridised in BR_3 (R = alkyl) but changes to sp^3 hybridization in its complexes with amines $R_3B: NR_3$. However if alkyl groups in BR_3 are larger, they may branch off to reduce its acid strength. Consequently, a large molecule like trimesityl boron is almost inert towards trialkylamines because its plane triangular (sp^2 hybridization, bond angle = 120°) geometry is forced to take tetrahedral configuration (sp^3 hybridization bond angle 109.5°) on complexation. This causes too much crowding or a considerable steric congestion due to forced fold back of mesityl groups behind boron atom in going from planar (120°) to tetrahedral (109.5°) configuration as shown in Fig. 7.2.

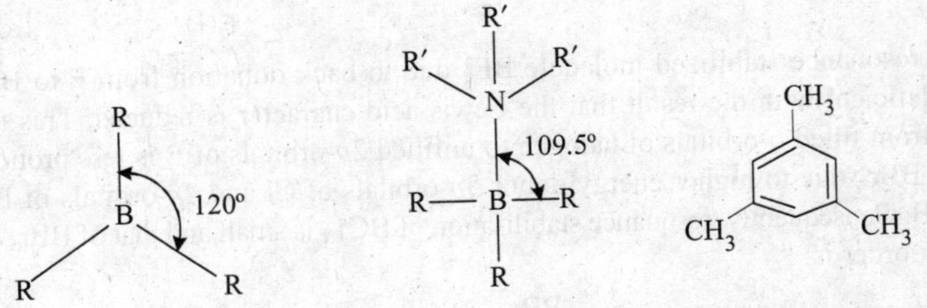

sp^2 hybridisation sp^3 hybridisation Mesitylene

Fig. 7.2. Forced foldback of mesityl groups.

Example 10. Which one of these bases, pyridine, α-picoline and γ-picoline, forms strongest complex with BMe_3? Give reasons for your answer.

Solution. γ-picoline forms the strongest complex followed by pyridine followed by α -picoline. In α-picoline, complex formation is hindered because of *F*-strain.

γ-picoline has a greater basicity due to electron donating effect of methyl group. Hence it has a greater tendency for complex formation. Pyridine lies in between.

7.13. ORDER OF BASIC STRENGTH OF PRIMARY, SECONDARY AND TERTIARY AMINES IN WATER

Methyl group has electron releasing inductive effect and expected order of basic strength is:

$$R_3N > R_2NH > RNH_2$$

The observed order however is $R_2NH > RNH_2 > R_3N$

This trend could be explained on the basis of steric effect but in gaseous state the basicity of methyl amines shows a regular gradation.

$$R_3N > R_2NH > RNH$$

showing that steric effect has no role to play. This means that the reasons of lowered basicity of Me_3N lies only in the effect of solvent. When amines are dissolved in water ammonium ion is formed which undergoes hydration through hydrogen bonding (Fig. 7.3). The greater the number of hydrogen bonds formed by the ammonium ion of the base amine, the larger would be the heat of hydration and greater be the stability.

The energy of hydration of ammonium ions decreases in the order $RNH_3^+ > R_2NH_2^+ > R_3NH^+$ obtained from RNH_2, R_2NH and R_3N as shown in Fig. 7.3. This order of basicity due to solution is totally opposite of the basicity expected from inductive effect. Thus actual basicity of an amine in solution is the combined effect of inductive effect of substituents and extent of solvation of the conjugate acid of the base. As a consequence of the combined effect, dialkyl amines are the strongest bases among amines followed by primary amines and then followed by tertiary amines.

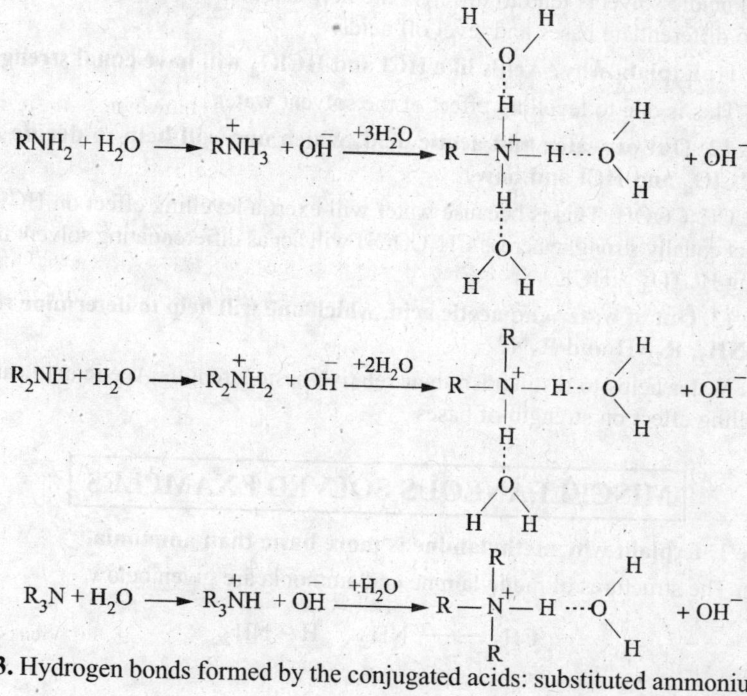

$$RNH_2 + H_2O \longrightarrow R\overset{+}{N}H_3 + \overset{-}{O}H \xrightarrow{+3H_2O} R-\overset{+}{N}(H\cdots OH_2)_3 + OH^-$$

$$R_2NH + H_2O \longrightarrow R_2\overset{+}{N}H_2 + \overset{-}{O}H \xrightarrow{+2H_2O} R_2\overset{+}{N}(H\cdots OH_2)_2 + OH^-$$

$$R_3N + H_2O \longrightarrow R_3\overset{+}{N}H + \overset{-}{O}H \xrightarrow{+H_2O} R_3\overset{+}{N}-H\cdots OH_2 + OH^-$$

Fig. 7.3. Hydrogen bonds formed by the conjugated acids: substituted ammonium ions.

7.14. LEVELLING EFFECT AND DIFFERENTIATING EFFECT OF SOLVENT

Solvents play an important role in altering the strength of an acid or a base. Non-aqueous solvents like ammonia, amides ($RCONH_2$), hydrazine etc., are *good proton* acceptor and hence will help acids to ionise in them. As a result of this all acids tend to become strong in a basic solvent. Such solvents are said to exert a *levelling effect* on the strength of acids. The reverse is true of acidic solvents like CH_3COOH, HCOOH, hydrogen halides etc., in which all bases tend to become strong. Such acidic solvents however tend to decrease acid strength of the acids and tend to exert differentiating effect on the strength of acids. Thus water behaves as levelling solvent for acids such as $HClO_4$, HI, HBr, HNO_3 and HCl and bases such as NaH, $NaNH_2$ and $NaOC_2H_5$. So in aqueous medium the acid strength is similar due to levelling effect of water. There will not be much difference in the acid strength due to levelling effect of the solvent. Therefore to compare their relative strengths of strong acids and bases, such a solvent should be selected that ionizes the strong acids and bases to different extent. Such a solvent is called *differentiating solvent.*

For example, acetic acid acts as differentiating solvent towards halogen acid and their ionization increases with increase in atomic number of halogen.

That is, acid strength decreases as

$$HI > HBr > HCl$$

This way, acetic acid has been used extensively to determine the relative strengths of strong acids like $HClO_4$, H_2SO_4, HNO_3 halogen acids, etc., and the following trend has been observed:

$$HClO_4 > HI > HBr > HCl > H_2SO_4 > HNO_3$$

However acetic acid exerts a levelling effect (*i.e.*, it brings basic strength on the same level) on the strength of bases. Consequently, the relative strength of ammonia, primary amines, secondary amines, tertiary amines etc., cannot be distinguished and their basic strength gets enhanced and levelled.

In general acidic solvents tend to differentiate between acids and level off bases while basic solvents tend to differentiate bases and level off acids.

Example 11. Explain why? Acids like HCl and $HClO_4$ will have equal strength in water.

Solution. This is due to levelling effect of the solvent water.

Example 12. Out of water and acetic acid, which one will help to decide the relative strengths of $HClO_4$ and HCl and how?

Solution. CH_3COOH. This is because water will exert a levelling effect on HCl and $HClO_4$ and both will act equally strong, whereas CH_3COOH will act as differentiating solvent for acids. The acid trend being $HClO_4 > HCl$.

Example 13. Out of water and acetic acid, which one will help to determine relative basic strength of RNH_2, R_2NH and R_3N?

Solution. Water helps to distinguish their relative basic strength. Acetic acid cannot be used due to its levelling effect on strength of bases.

MISCELLANEOUS SOLVED EXAMPLES

Example 1. Explain why methylamine is more basic than ammonia.

Solution. The structures of methylamine and ammonia are given below

$$CH_3 \rightarrow \ddot{N}H_2 \qquad H - \ddot{N}H_2$$

Due to electron repelling effect of CH_3 – group, There is greater electron density on N in methylamine.

Example 2. $(SiH_3)_3$ N is weaker base than $(CH_3)_3$ N. Explain.

Solution. $(CH_3)_3$ group is more electron repelling than $(SiH_3)_3$ group.Therefore $(CH_3)_3$ N is a stronger base

Example 3. Classify, giving reasons, the following either as Lewis acid or Lewis base: $AlCl_3$, S^{2-}

Solution. $AlCl_3$ is electron acceptor hence Lewis acid. S^{2-} is electron donor, hence Lewis base.

Example 4. Which is stronger acid HCN or HOCN?

Solution. HOCN is expected to be a stronger acid than HCN because H atom in HOCN is bonded to a highly electronegative O atom. This induces a positive charge on the H atom and therefore, it can be easily ionized as H^+ ion in water.

Example 5. Why does benzene sulphonic acid act as Bronsted acid?

Solution. Benzene sulphonic acid is $H-\ddot{\underset{..}{O}}-\underset{\underset{\ddot{O}:}{\|}}{\overset{\overset{\ddot{O}:}{\|}}{S}}-C_6H_5$

Here, S atom is surrounded by three highly electronegative oxygen atoms. Due to the inductive effect of these atoms, a positive charges is induced on H atom. This facilitates the release of H atom as H^+ ion. Therefore, benzene sulphonic acid acts as a Bronsted acid.

Example 6. Which one is stronger acid BBr_3 or BCl_3 and why?

Solution. BBr_3 is stronger acid than BCl_3. This is because of resonance stabilization of BCl_3 due to back bonding from Cl to B. This increases electron density of B which then no longer remain electron deficients. Hence its Lewis strength decreases. In BBr_3 back bonding is not possible.

Example 7. H_2Se is stronger acid than H_2S. Explain.

Solution. When we move from S to Se, atomic size increases and as a result dissociation of hydride increases. Therefore, H_2Se can more easily dissociate than H_2S and hence is stronger acid.

Example 8. Explain why Cl— OH is an acid whereas NaOH is a base.

Solution. In general acidic or basic character of Z– O – H depends upon electronegativity of Z.

(*i*) If Z is a highly electronegative atom (electronegativity greater than 1.8), it attracts the shared pair of electron more towards itself. As a result O – H bond becomes polar and allows the proton to dissociate as H^+ and therefore, behaves as an acid. In the case of Cl – O – H, the electronegativity of Cl is 3.0 and therefore, it behaves as an acid.

(*ii*) If electronegativity of Z is less than 1.8, it does not attract the shared electron pair towards itself. As a result O – H bond does not break. Rather Z – O bond breaks forming Z^+ and OH^- ions. Therefore, Z – O – H behaves as a base. In the case of NaOH, sodium has electronegative of 0.9 and therefore, NaOH is a base.

Example 9. Which is stronger acid in each of the following pairs?

(*a*) HF OF HI **(*b*) HBrO or HClO**

(*c*) HNO_2 or HNO_3 **(*d*) BH_3 or $B(CH_3)_3$**

(*e*) B_2O_3 or Al_2O_3 **(*f*) $HClO_2$ or $HClO_3$**

(*g*) H_2S or H_2O **(*h*) H_3PO_3 or H_3PO_4**

(*i*) FCH_2COOH or $ClCH_2COOH$ **(*j*) PH_3 or H_2S**

Solution.

(*a*) HI (*b*) HClO

(*c*) HNO_3 (*d*) BH_3

(*e*) B_2O_3 (*f*) $HClO_3$

(*g*) H_2S (*h*) H_3PO_3

(*i*) FCH_2COOH (*j*) H_2S

Example 10. Why is ammonia termed as a base though it does not contain OH^- ions?

Solution. The basic nature of ammonia is due to its tendency to donate electron pair. Therefore, it is a Lewis base.

Example 11. Which one is stronger acid BF_3 or BCl_3?

Solution. BCl_3

Example 12. Arrange the following in the increasing order of Lewis acid strength: $SnCl_4$, $GeCl_4$, $SiCl_4$, CCl_4

Solution. Lewis acid strength increases as :

$$CCl_4 < SiCl_4 < GeCl_4 < SnCl_4$$

Example 13. Which one among the following is more acidic?

(*a*) H_2S or H_2Se **(*b*) HF or HCl**

(*c*) BF_3 or BBr_3 **(*d*) H_2S or PH_3**

(*e*) Fluoroacetic acid or chloroacetic acid

(*f*) $O_3Mn(OH)$ or $O_2Mn(OH)_2$ **(*g*) H_3PO_4 or H_3AsO_4**

(*h*) H_2SO_4 or H_2SO_3 **(*i*) H_2SO_4 or $H_2S_2O_7$**

Solution. (*a*) H_2Se (*b*) HCl

(*c*) BBr_3 (*d*) H_2S

(*e*) Fluoroacetic acid (*f*) $O_3Mn(OH)$

(*g*) H_3PO_4 (*h*) H_2SO_4

(*i*) $H_2S_2O_7$

Example 14. Write the formula of conjugate base of each of the following:

HSO_4^-, CH_3COOH, H_3PO_4, $CH_3NH_3^+$

Solution. Conjugate bases:

HSO_4^-	:	SO_4^{2-}
CH_3COOH	:	CH_3COO^-
H_3PO_4	:	$H_2PO_4^-$
$CH_3NH_3^-$	:	CH_3NH_2

Example 15. What is the conjugate base of $[Al(H_2O)_6]^{3+}$ ion?

Solution. $[Al(H_2O)_5OH]^{2+}$

Example 16. Arrange the members in the following groups in order of decreasing basicity: (*i*) OH^-, SH^-, SeH^- (*ii*) F^-, Cl^-, Br^-, I^- (*iii*) CH_3OH, CH_3NH_2, CH_3F.

Solution. Order of decreasing basicity.

(*i*) $OH^- > SH^- > SeH^-$ (*ii*) $F^- > Cl^- > Br^- > I^-$ (*iii*) $CH_3NH_2 > CH_3OH > CH_3F$

Example 17. Explain the following:

Trimethylamine is more basic than NH_3.

Solution. Methyl (CH_3) group is electron releasing group. When it is attached to the N atom, it increases the electron density on N atom. As a result, electron pair becomes readily available and basic strength increases. Thus, $(CH_3)_3N$ is more basic than NH_3.

$$H_3C \rightarrow \ddot{N}(CH_3)(CH_3) \qquad H-\ddot{N}(H)-H$$

Example 18. Classify the following as:

(*i*) Lewis acids

(*ii*) Border line Lewis acids.

[CS_2, $TeCl_4$, SO_3, CO_2, Mg^{++}, $GaCl_3$]

Solution. Lewis acids: $TeCl_4$, $GaCl_3$, Mg^{++}, SO_3

Border line Lewis acids: CS_2, CO_2.

Example 19. Which of the following can be classified as:

(*i*) Lewis bases and

(*ii*) Border line Lewis bases?

{C_2H_4, Pyridine, CO, water, benzene, diethyl ether, acetone, 1, 3–butadiene.}

Solution. Lewis bases: Pyridine, water, diethyl ether, acetone.

Border line Lewis bases: CO, C_2H_4, benzene, 1,3-butadiene.

Example 20. Categories following species into:

(*i*) Lewis acid-base cases and

(*ii*) Border line cases (Extended Lewis concept):

$Cr(CO)_6$, $Fe(C_5H_5)_2$, $[Al(H_2O)_6]Cl_3$, $SbCl^-_6$, R_3NH^+, $[Co(NH_3)_6]^{3-}$, $PdCl_2(C_2H_4)$

Solution. Lewis acid-base: R_3NH^+, $SbCl^-_6$, $[Al(H_2O)_6]Cl_3$, $[Co(NH_3)_6]^{3-}$.

Border line cases $Cr(CO)_6$, $Fe(C_5H_5)_2$, $PdCl_2(C_2H_4)$.

Example 21. Classify the following as Lewis acids or bases giving reasons:

(*i*) BF_3 (*ii*) NH_3 (*iii*) Cl^- (*iv*) Zn^{2+}

Solution. BF_3 and Zn^{2+} are Lewis acids. Electron-deficient and positively-charged species act as Lewis acids.

NH_3 and Cl are Lewis bases. Molecules with lone pair of electrons and negatively charged species act as Lewis bases.

Example 22. Explain the expected trends in the relative acid strengths of $HClO_4$, $HBrO_4$ and HIO_4.

Solution. The number of oxygen atoms on the central atom remains the same (= 3) but the electronegativity of central atom decreases as the size increases from Cl to I. Consequently, the polarity of O – H bond or the acidic character decreases in going from $HClO_4$ to HIO_4. Thus, the order of acid strength is $HClO_4 > HBrO_4 > HIO_4$.

Example 23. Justify the relative order of acid strength

$H_3PO_2 > H_3PO_3 > H_3PO_4$

Solution. The structures of the acids are:

```
      O              O               O
      ‖              ‖               ‖
  H ／P＼ OH     H ／P＼ OH     HO ／P＼ OH
      |              |               |
      H              OH              OH
```

Number of oxygen substituents on the central P atom is the same (*i. e.,* one) in all the cases but the number of OH bonds increases from one in H_3PO_2 to two in $H_3\ PO_3$ to three in $H_3\ PO_4$. Consequently, a higher polarity will be created in a single OH bond than in two OH bonds which in turn would be more than in three OH bonds. Hence the acidity changes in the order $H_3\ PO_2 > H_3\ PO_3 > H_3\ PO_4$. The P – H does not ionise and has no influence on the acidity of phosphorus oxy-acids.

Example 24. Arrange the following phosphoric acids in the decreasing order of relative acid strength: H_3PO_4, $H_4P_2O_7$ and $H_5P_3O_{10}$. Give reasons for your answer.

Solution. The structures of $H_3\ PO_4$, $H_4P_2O_7$ and $H_5P_3O_{10}$ are given below:

```
      O             O     O              O     O     O
      ‖             ‖     ‖              ‖     ‖     ‖
  HO ／P＼ OH   HO ／P＼ O ／P＼ OH   HO ／P＼ O ／P＼ O ／P＼ OH
      |             |     |              |     |     |
      OH            OH    OH             OH    OH    OH
```

Acidic strength depends upon the number of unprotonated oxygen atom on P atom (P = O) which polarise a given number of OH bonds. We have to count the number of OH groups per P = O for the three acids. In this series, the number of OH groups per P = O decreases in the order $H_3PO_4 > H_4P_2O_7 > H_5\ P_3\ O_{10}$. Therefore acidity increases in the order $H_5\ P_3\ O_{10} > H_4\ P_2\ O_7 > H_3\ PO_4$.

Example 25. Explain the order of relative acidity

$O_3Mn(OH) > O_2Mn(OH)_2 > OMn(OH)_3$

Solution. The structures of acids are

```
      O               O                   O
      ‖               ‖                   ‖
  O = Mn — O — H   O = Mn — O — H   H—O — Mn — O — H
      ‖               |                   |
      O               O                   O
                      |                   |
                      H                   H
```

The oxygen linked to Mn by double bond attracts the electrons towards itself facilitating removal of protons. Number of protons released will depend upon the number of hydroxyl groups per Mn = O group. Smaller the number of hydroxyl group per Mn = O group, greater will be proton liberation and thus greater the acid strength. The above order of acid strength is therefore justified.

Example 26. Explain why Cl – OH is an acid whereas Na – OH is a base.

Solution. This is due to electronegativity value greater than 1.8 for Cl in HClO and less than 1.8 for Na in NaOH.

Example. 27. Explain acidity trends

(*i*) $H_3PO_4 > H_3AsO_4$ **(*ii*) $H_2SO_4 > H_2SO_3$**

(*iii*) $H_2S_2O_7 > H_2SO_4$ **(*iv*) $[Fe(H_2O)_6]^{3+} > [Fe(H_2O)_6]^{2+}$**

Solution. (*i*) Electronegativity of central As is less than that of P, therefore $H_3\ PO_4$ is a stronger acid than $H_3\ AsO_4$.

(*ii*) The number of oxygen substituents on S in H_2SO_4 is two, whereas it is one in H_2SO_3. Therefore sulphuric acid is stronger than sulphurous acid.

(*iii*) Acidity trend is explained by the structures:

O O O O
S S S
O O O OH OH
OH OH

Disulphuric acid $H_2S_2O_7$ has more number of oxygen substituents on S per OH group than sulphuric acid. Therefore $H_2S_2O_7$ is stronger than H_2SO_4.

(*iv*) Larger charge-to-radius ratio of Fe^{3+} relative to Fe^{2+} determines the acid strength. $[Fe(H_2O)_6]^{3+}$ is stronger than $[Fe\ (H_2O)_6]^{2+}$.

Example 28. In each pair of acids, state which is stronger and why?

(*i*) H_2SO_4 and H_2SO_3

(*ii*) HF and HCl

(*iii*) CH_3COOH and HCOOH

(*iv*) C_6H_5COOH and CH_3COOH.

Solution. (*i*) H_2SO_4 is stronger than H_2SO_3 because there are more non-hydroxyl oxygen on H_2SO_4. Non-hydroxyl oxygen atoms withdraw electrons towards themselves and help in the removal of protons.

(*ii*) HCl is stronger than HF. The bond between H and fluorine becomes weaker with increase in the size of halogen. Weaker the bond, stronger the acid.

(*iii*) HCOOH is stronger than CH_3COOH. In acetic acid CH_3 – group has electron donating effect and thus would prevent release of protons.

(*iv*) C_6H_5COOH is stronger than CH_3COOH because in C_6H_5COOH, the phenyl group is electron withdrawing whereas CH_3 – group in CH_3COOH is electron donating.

EXERCISES
(Based on Question Papers of Different Universities)

Multiple Choice Questions

Choose the correct option

1. Which of the following is not a Lewis base?

(*a*) CN^- (*b*) $AlCl_3$

(*c*) ROH (*d*) NH_3

2. The pK_a value according to Ricchi formula for H_3PO_4 is:

(*a*) 1 (*b*) 3

(*c*) 5 (*d*) –7

3. The basic character of NMe_3, NH_3 and NF_3 follows the order.

(*a*) $NF_3 > NH_3 > NMe_3$ (*b*) $NMe_3 > NH_3 > NF_3$

(*c*) $NMe_3 > NF_3 > NH_3$ (*d*) $NH_3 > NF_3 > NMe_3$

4. Which of the following statement is not correct regarding Lewis acids and bases?

(*a*) NH_3 and H_2O both behave as Lewis bases.

(*b*) Substances which can donate a pair of electrons are called Lewis bases.

(*c*) All Lewis bases are also called Bronsted bases.

(*d*) Lewis bases must contain an atom having less than an octet of electrons.

5. The correct order of acidic character of hydro acids of group VII is:

(*a*) HI > HBr > HCl > HF (*b*) HI > HBr > HF > HCl

(*c*) HF > HCl > HBr > HI (*d*) HI > HF > HBr > HCl.

6. Which of the following has largest value of pK_a?

(*a*) Me_3CCH_2COOH (*b*) Me_3SiCH_2COOH

(*c*) $Cl_2CHCOOH$ (*d*) HCOOH

7. Which of the following statments is false?

(*a*) The higher the value of pK_b the stronger is base.

(*b*) pK_a is defined as – log K_a

(*c*) Out of HCN (pK_a = 6.4) and HCOOH (pK_a = 3.70), HCOOH is a stronger acid.

(*d*) The larger the value of K_a, the stronger is the acid.

8. The acidic character of BF_3, BMe_3 and BH_3 follows the order

(*a*) $BF_3 > BH_3 > BMe_3$ (*b*) $BF_3 > BMe_3 > BH_3$

(*c*) $BMe_3 > BH_3 > BF_3$ (*d*) $BH_3 > BF_3 > BMe_3$.

9. The acidic character of oxo acids of phosphorus follows the trend

(*a*) $H_3PO_2 > H_3PO_3 > H_3PO_4$ (*b*) $H_3PO_4 > H_3PO_3 > H_3PO_2$

(*c*) $H_3PO_4 > H_3PO_2 > H_3PO_3$ (*d*) $H_3PO_2 > H_3PO_4 > H_3PO_3$.

10. Which of the following is the strongest acid?

(*a*) CH_3COOH (*b*) $ClCH_2COOH$

(*c*) Cl_2CHOOH (*d*) Cl_3CCOOH

ANSWERS

1. (*b*)	**2.** (*b*)	**3.** (*b*)	**4.** (*d*)	**5.** (*a*)	**6.** (*b*)
7. (*a*)	**8.** (*a*)	**9.** (*b*)	**10.** (*d*)		

Short Answer Questions

1. Define Lewis acid and a Lewis base.

2. HCl is a strong acid but Lewis concept is unable to explain this. Explain.

3. Which one is stronger out of H_2S and H_2Se and why?

4. Out of BCl_3 and $(CH_3)_3B$ which one will you expect to be a better Lewis acid?

5. Out of RNH_2 or R_3N which one will you expect to be a better base?

6. Arrange the following is order of increasing basicity: NF_3, NMe_3 and NH_3

7. Show Lewis acid-base reaction between CaO and SO_3.

8. Which of the two NH_3 or NF_3 is weaker and why?

9. Differentiate between a Lewis acid and Lewis base with the help of an example.

10. Cl– OH is an acid whereas Na – OH is a base. Explain

11. How does relative strength of an acid vary with the oxidation number of the central atom?

12. Why is pyridine a border line base while ammonia is a hard base?

General Questions

1. (*a*) What is Bronsted-Lowry concept of acids and bases? How does this concept differ from Arrhenius concept of acids and bases
   (*b*) Giving suitable examples explain the terms conjugate acid and conjugate base.
   (*c*) Define Lewis and Lewis base.
2. (a) What are Lewis acids and Lewis bases? Give one example in each case.
   (*b*) Which would you expect to be a better Lewis acid: BCl_3 or BF_3? Explain.
   (*c*) List some important drawbacks of Lewis acid-base concept.
3. (a) Arrange the following in increasing acidic strength and justify your answer:
   $HClO_2$, $HClO$, $HClO_3$, $HClO_4$
   (*b*) Arrange the following in increasing basic strength in aqueous solution and give reasons:
   RNH_2, R_2NH, R_3N
   (*c*) What is Lux-Flood concept of acids and bases? Explain with examples.
4. Classify giving reasons the following compounds as Lewis acids and Lewis bases:
   (*i*) $AlCl_3$ (*ii*) H_2O (*iii*) Zn^{2+} (*iv*) I^-
5. (a) Arrange the following acids in the order of their increasing strengths:
   (*i*) $HClO_4$, $HClO_3$, $HClO_2$, $HClO$ (*ii*) H_2O, H_2S, H_2Se, H_2Te.
   (b) Which is a stronger acids in each of the following pairs? Give reasons:
   (*i*) FCH_2COOH or $ClCH_2COOH$ (*ii*) HF or HCl
   (*iii*) BF_3 or BBr_3 (*iv*) H_3PO_3 or H_3PO_4
6. (a) Name the factors that govern the relative strength of Lewis acids and bases.
   (*b*) Arrange the following in order of increasing Lewis acid strength.
   (*i*) $SnCl_4$, $GeCl_4$, $SiCl_4$, CCl_4
   (*ii*) CH_4, NH_3, H_2O, HF
   (*iii*) BF_3, BCl_3, BBr_3
   (iv) BF_3 $B(CH_3)_3$, BH_3
   (*b*) Show the Lewis acid-base reaction between :
   (*i*) CaO and SO_3
   (*ii*) NH_3 and BF_3
   (*iii*) NH_3 and H^+
   (*d*) Justify by Lewis concept that CO_2 is an acid.
7. (a) Arrange in order of increasing acidic strength:
   (*i*) Sulphuric, selenic and telluric acids.
   (*ii*) Nitrous, phosphorous and arseneous acids.
   (*iii*) Seleneous, sulphurous and tellurous acids.
   (*b*) Which would you expect to be better Lewis acid: BCl_3 or $B(CH_3)_3$? Explain.
8. (a) 2-Methyl pyridine is weaker base towards $B(CH_3)_3$ than pyridine.
   (*b*) $(CH_3)_3N$ is weaker base in aqueous solution than $(CH_3)_2NH$.
9. (a) Define pK_a. How its value can give the idea of the relative strength of acids?

(*b*) the acidic strength of HOI, HOBr and HOCl is found to decrease as:
HOI < HOBr < HOCl.
Is this order correct? Discuss.

(*c*) Estimate the pK_a as values for H_3PO_4 and H_2SeO_4 according to Ricci empirical relationship.

(*d*) Why is NF_3 much weaker base than NH_3?

10. (a) Discuss the significance of inductive effect and resonance effect in the acidic strength of organic carboxylic acids.

(*b*) Arrange the following acids in the increasing order of their acidic strength:
$HClO_4$, H_2CO_3, H_2SO_4, H_3PO_4

11. (a) What are levelling and differentiating solvents? Discuss one example in each case.

(*b*) Strengths of acids can be ranked on the basis of their ability to co-ordinate. Comment

(*c*) Arrange the following in the decreasing order basicity

(*i*) OH^-, SH^-, SeH^-

(*ii*) F^-, Cl^-, Br^-, I^-

(*iii*) CH_3H, CH_3NH_2, CH_3F

12. (a) Describe Pauling's empirical rules for rationalising acid strengths of HXO acids.

(*b*) Arrange the following in the decreasing base strength:

(*i*) NH_3, PH_3, AsH_3

(*ii*) Pyridine, 2-methyl pyridine, 4-methyl pyridine (towards B $(CH_3)_3$ as Lewis acids.)

(*c*) Estimate pK_a values for $HBrO_4$, HSO_3F, $HClO_3$.

13. (a) Explain the following effects in determining the relative strengths of Lewis acids and bases:

(*i*) inductive effect (*ii*) resonance (*iii*) steric effect.

(*b*) Account for the basic character:
$NH_3 < R_3N < RNH_2 < R_2NH$

14. (*a*) How does solvent affect the strength of acids and bases?

(*b*) Differentiate between a Lewis acid and a Bronsted acid.

(*c*) Which of the following can act as Lewis acids?

(*i*) SO_3 (*ii*) OH (*iii*) H_2O (*iv*) $CaCl_2$

(*v*) CO_2 (*vi*) Ag^+.

15. (a) Why is $HClO_4$ is stronger acid than $HClO_3$?

(*b*) Why is BF_3 a weaker Lewis acid than $HClO_3$?

(*c*) Which one out of H_2S or H_2Se is a stronger acid and why?

16. (a) How does the relative strength of acids depend upon:

(*i*) oxidation number of central atom

(*ii*) inductive effect of the substituents.

(*b*) Why is 2-methyl pyridine weaker base than pyridine.

17. (a) Discuss acid-base reactions according to Lewis theory. What are the limitations of Lewis concept?

(*b*) Account for the decreasing acidic strength in the order:
HClO > HBrO > HIO

(*c*) Account for the increasing acidic strength in the order:

$HClO < HClO_2 < HClO_3 < HClO_4$

18. (a) Discuss the effects of F– and B– strains on the stabilities of Lewis acid-base complexes giving definite examples.

(*b*) Discuss Lewis acids and Lewis bases with suitable examples in each case.

(*c*) What are the properties of hard Lewis bases and soft Lewis bases?

(*d*) Explain the resonance effect of substituent on the strengths of Lewis acids and Lewis bases.

(*e*) Arrange the following in the decreasing basic strength:

(*i*) NH_3, PH_3, AsH_3

(*ii*) Pyridine, 2-methylpyridine, 4-methylpyridine (towards $(B(CH_3)_3$ as Lewis acid).

19. (a) Compare the basic strength of 2–methyl pyridine and 4–methyl pyridine towards trimethylborane.

(*b*) Arrange the following acids in the order of increasing acidic strength and give reasons for your answer.

(*c*) Arrange the following in the decreasing order of their basic strength:

F^-, I^-, Cl^-, Br^-.

20. (a) What do you understand by conjugate acids and bases? Show that a strong acid has a weak conjugate base and *vice versa*.

(*b*) Discuss the relative strength of protonic acid of the type HY.

21. (a) Discuss the weaker basicity of NMe_3 as compared with that of Me_2NH in aqueous solution.

(*b*) Classify, giving reasons, the following molecules/ions as Lewis acids and Lewis bases:

(*i*) $AlCl_3$ (*ii*) H_2O (*iii*) Zn^{2+} (*iv*) I^-

(*c*) How does the strength of an acid and a base change with the effect of substituents and solvent?

22. (a) Describe Bronsted-Lowery concept of acids and bases. Give two examples.

(*b*) Suggest a reason why the order of acidic strength in the oxo acids of chlorine is:

$HClO_4 > HClO_3 > HClO_2 > HClO$

(*c*) What do you understand by levelling effect of solvent?

(*d*) Arrange the following in the order of increasing acidic strength:

$CHCl_2COOH$, $CH_2ClCOOH$, CCl_3COOH, CH_3COOH

23. (a) What do you understand by Lewis definition of acids and bases? Discuss applications and drawbacks of Lewis acid-base concept.

(*b*) Explain why HI is a stronger acid than HCl.

(*c*) Which of these: Pyridine, α-picoline and γ-picoline form strongest complex with BMe_3? Give reasons for your answer.

(*d*) Explain with examples, the effect of electron releasing substituents on the relative strengths of acids.

24. (a) What is the effect of electron withdrawing substituents on the strength of acids?

(*b*) Explain the following:

(*i*) BF_3 is weaker acid than BBr_3.

(*ii*) $HClO_4$ is stronger acid than $HClO_3$.

25. (a) How do solvents affect the strength of acids and bases?

(*b*) Arrange the following in the order of decreasing basicity:

(*i*) OH^-, SH^- an SeH^-

(*ii*) I^-, Cl^-, Br^- and F^-

26. (*a*) How will you account for the increasing acidic strength:

$HClO < HClO_2 < HClO_3 < HClO_4$.

(*b*) Discuss the effect of solvent on the strength of acids and bases.

27. (*a*) Account for the increasing acidic strength in the order of $HClO < HClO_2 < HClO_3 < HClO_4$.

(*b*) How does solvent affect the strength of bases?

(*c*) Explain:

(*i*) 2-picoline does not form stable complex with trimethyl borane.

(*ii*) H_2Se is stronger acid than H_2S.

28. (a) Which out of H_2S and H_2Se is stronger acid and why?

(*b*) Account for the increasing acidic strength in order

$HClO < HClO_2 < HClO_3 < HClO_4$

(*c*) Explain the following:

(*i*) Why is NF_3 much weaker than NH_3?

(*ii*) How does solvent effect strength of bases?

(*d*) Arrange the following in increasing order of their acidic strength and explain in detail.

$HClO_4$, $HClO_3$, $HClO_2$, $HClO$

29. (*a*) Acidic character of halides of boron is of the following order:

$BBr_3 > BCl_3 > BF_3$. Explain

(*b*) Arrange the following in increasing order of their acidic strength and justify it.

(*i*) HF, H_2O, NH_3

(*ii*) Me_3B, BH_3, BF_3

30. (*a*) Which out of BBr_3 and BCl_3 is stronger acid and why?

(*b*) The strength of acids is as given below:

(*i*) $FCH_2COOH > ClCH_2COOH > BrCH_2COOH$

(*ii*) $HCOOH > CH_3COOH$.

Explain.

(*c*) What is the effect of substituents and solvents on the strength of acids and bases?

31. (*a*) $HClO_3$ is a weaker acid than $HClO_4$. Explain.

(*b*) How does relative strength of an acid depends upon oxidation number of central atom?

32. (*a*) Describe Lewis concept of acids and bases with suitable examples. What are its drawbacks?

(*b*) Giving reasons arrange the following in order of increasing acidic strength:

HF, HCl, HBr, HI.

(*c*) In each pair of acids, state which is stronger and why?

(*i*) H_2SO_4 and H_2SO_3

(*ii*) HF and HCl

(*iii*) CH_3COOH and HCOOH

(*iv*) C_6H_5COOH and CH_3COOH

33. (*a*) What is pK_a? Giving reasons arrange the following in order of increasing pK_a values: $HClO_2$, $HClO_3$, $HClO$, $HClO_4$

(*b*) (i) Which one is stronger base and why NH_3 or $(CH_3)_3N$?

(*ii*) Which one is stronger acid and why [B $(OH)_3$ or $Al(OH)_3$]?

(*c*) Define Lewis acid and Lewis base. Give one example of each acid and of the neutralisation reaction.

34. (*a*) Explain solvent system concept of acids and bases taking an example of ammonia as a solvent.

(*b*) Which one is stronger acid BBr_3 or BCl_3? Explain.

(*c*) Arrange the following in decreasing order of basic strength and explain:

PH_3, NH_3, AsH_3

35. (*a*) What is pK_a? Giving reasons arrange the following in order of their increasing pK_a values

$HClO_2$, $HClO_3$, $HClO$, $HClO_4$,

(*b*) Discuss solvent system concept of acids and bases.

(*c*) Arrange the following in the decreasing order of basic strength:

PH_3, NH_3 and AsH_3

36. (*a*) Give the electron concept of acids and bases.

(*b*) Give the types of Lewis bases.

(*c*) Explain why H_2S is stronger acid than H_2O.

(*d*) Explain why RNH_2 is stronger base than NH_3.

(*e*) Explain why HNO_3 is stronger acid than HNO_2.

37. Discuss in brief the Lewis concept of acids and bases with suitable examples.

38. (*a*) Methyl pyridine is weaker base than pyridine. Explain.

(*b*) Define acids and base in term of Lux Flood concept.

(*c*) Justify by Lewis concept that CO_2 is an acid.

39. (*a*) Why is $HClO_4$ stronger acid than $HClO$? Explain

(*b*) Explain: BF_3 behaves as weaker Lewis acid than BI_3.

(*c*) Why is R_2NH stronger base than R_3N ($R = CH_3$)? Explain.

(*d*) Explain reasons of increasing acidic character in

H_3PO_4, H_2SO_4, $HClO_4$

40. (*a*) Giving reasons arrange the following in order of increasing acidic strength:

BCl_3, BBr_3, BF_3, BI_3

(*b*) Arrange the following in order of increasing acid strength. Give reasons:

$HClO$, HIO, $HBrO$

(*c*) Explain Lewis concept of acids and bases.

41. (*a*) Which one is stronger acid, BBr_3 or BCl_3? Explain.

(*b*) Methyl pyridine is a weaker base than pyridine. Explain

(*c*) Describe the effect of substituents on the strength of acids and bases.

42. (*a*) Explain acid and base in terms of Lux Flood concept.

(*b*) What are the limitations of Lewis concept of acids and bases?

(*c*) Explain why trisilylamine is weak base than trimethylamine?

43. Discuss acids and bases according to solvent system concept with suitable examples.

44. (*a*) Arrange the following in the increasing order of their acid strength:

CH_4, NH_3, H_2O and HF

Justify your answer.

(*b*) 2-Methyl pyridine is weaker base than pyridine. Why?

(*c*) Classify giving reasons the following into Lewis acid or Lewis base:

(*i*) H^+ (*i*) $SnCl_4$ (*iii*) CH_3NH_2 (*iv*) $CH_3 - O - CH_3$

45. (*a*) Write the formula of conjugate base of each of following:

HSO_4^-, CH_3COOH, H_3PO_4, $CH_3NH_3^+$

(*b*) (*i*) Arrange the following in order of increasing basicity:

OH^-, SH^-, SeH^-

(*ii*) Which one is stronger acid out of H_2S and H_2Se and why?

(*c*) Why $(CH_3)_3$ N is stronger base than NH_3

46. (*a*) Give Bronsted-Lowry definition for acids and bases.

(*b*) Discuss solvent system concept to express acids and bases with suitable examples?

47. (*a*) Explain the following:

(*i*) Br^- is Lewis as well as Bronsted base.

(*ii*) BBr_3 is a Lewis acid but not a Bronsted acid.

(*iii*) CO_2 is acidic in nature according to Lewis concept.

(*b*) Explain with example, Lux-Flood concept of acids and bases.

(*c*) What are polyprotic bases? Give example.

48. (*a*) Account for the increasing acidic strength in $HClO < HClO_2 < HClO_4$.

(*b*) What do you understand by labelling and differentiating solvents? Discuss with one example in each case.

(*c*) Arrange the following bases in the decreasing order of basic strength and explain by giving reasons for their decreasing order of basic strength.

NH_3, PH_3, AsH_3

49. (*a*) Explain the following:

(*i*) Every Arrhenius base also is Bronsted base but every Bronsted base is not Arrhenius base.

(*ii*) HNO_3 is stronger acid than HNO_2.

(*iii*) Explain the order of acidic strength: $H_3PO_4 < H_2SO_4 < HClO_4$

(*b*) Explain with the examples: Each strong Bronsted acid has weak conjugate base and vice-versa.

(*c*) What is the effect of solvent on strength of acids and bases.

50. (*a*) Explain the following:

(*i*) Trimethylamine is more basic than NH_3.

(*ii*) NF_3 is much weaker base than NH_3.

(*iii*) Cl – OH is an acid whereas Na – OH is a base.

(*b*) Describe with example Lux-Flood concept of acids and bases.

(*c*) What are the limitations of Lewis concepts of acids and bases ?

(*d*) Write the conjugate acid of the following :

(*i*) NH_3 (*ii*) H_2O (*iii*) S^{2-} (*iv*) CN^-

(*v*) CH_3COO^-

51. (*a*) Who proposed the Electron-Donor-Acceptor system concept for acids and bases in solution? Discuss in brief with suitable examples

(*b*) What are the main points of criticism of the Arrhenius theory of acids and bases in solution?

(*c*) Who proposed donor acceptor concept system of an acids and bases? Discuss in brief with suitable examples

52. (*a*) Give one example of each, a coordinating and a non-coordinating non-aqueous solvent.

(*b*) Out of NH_3 and NF_3, which is more base and why ?

(*c*) Justify the following order of acidic nature of halides of boron: $BBr_3 > BCl_3 > BF_3$.

53. (*a*) Discuss acids and bases according to Bronsted-Lowry definition with suitable examples.

(*b*) Discuss Lux-Flood theory for acids and bases in brief with suitable examples.

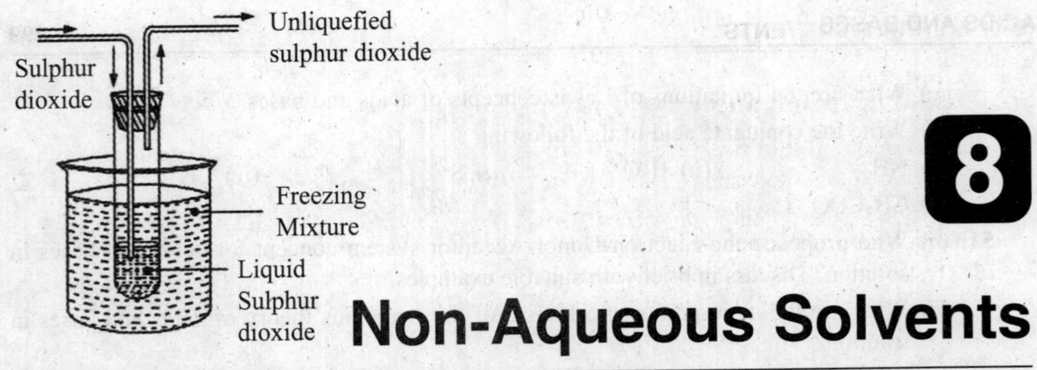

8

Non-Aqueous Solvents

8.1. INTRODUCTION

Since early times, chemical reactions have been carried out in water medium. This was because of easy availability of water in pure form. This is almost a free resource and then there are other advantages like favourable physical properties, with a long range (0–100°C) for its liquid state and high dielectric constant, which helps in the dissolution of other substances in it. Being neutral, non-toxic, odourless and non-poisonous, enhances its utility as a solvent. Lately, attempts have been made to find some other substances which could be used as suitable solvents for inorganic substances. We have met success in the development of substances such as liquid ammonia, liquid sulphur dioxide, liquid hydrofluoric acid and liquid nitrogen tetroxide. In the present chapter, we shall discuss liquid ammonia and liquid sulphur dioxide as solvents.

8.2. PHYSICAL PROPERTIES OF SOLVENTS AND THEIR ROLE IN CHEMICAL REACTIONS

The utility of a solvent is determined from its properties that may help to carry out the chemical reactions and separate the products conveniently. Some of the important physical properties of the solvents are discussed below:

8.2.1. Melting point and boiling point (Liquid range)

Most of the reactions in a solvent system are carried out in liquid phase, therefore, the melting point and boiling point give the temperature range within which the solvent is useful. The melting and boiling points for some common solvents are given in Table 8.1.

Table 8.1 Melting points and boiling points of some common solvents.

Solvent	m.p. (°C)	b.p. (°C)
H_2O	0.0	100
NH_3	–77.7	–33.4
HF	–89.3	19.5
SO_2	–75.5	–10.2
CH_3COOH	16.6	118.1

It is clear from the above table that *water has excellent liquid range (0 to 100°C). Acetic acid can also act as solvent at ordinary temperatures while NH_3 and SO_2 work as solvents only at low temperatures*.

8.2.2. Heat of fusion and heat of vaporisation

The values of heat of fusion and heat of vaporisation help to know the nature and strength of forces which hold the molecules of the solvent together in the solid or the liquid state. Larger the value of heat of fusion or heat of vaporisation, the stronger are the intermolecular binding forces in the solid or the liquid respectively. Thus the magnitudes of molar heat of fusion and molar heat of vaporisation give us an idea of the nature and strength of the intermolecular forces. This helps to determine the suitability of a substance as a solvent.

Trouton

A better idea of the intermolecular forces in a solvent is less obtained by the ratio of molar heat of vaporisation to the boiling point. This ratio is called *Trouton constant, i.e.*,

$$\frac{\Delta H_{evap}}{T_b} \approx \text{Trouton constant}$$

For normal liquids, if molar heats of vaporisations are expressed in calories and boiling point in absolute scale (0 K), then the Trouton constant has the value of about 90 for normal liquids. Such liquids exist as independent molecules without any bonds between them. *A higher value of Trouton constant indicates association in molecules*. Common solvents like water, ammonia, hydrogen fluoride have Trouton constant value greater than 90 and are, therefore, associated liquids. The molecules of these liquids are polar. The association of polar molecules is due to their tendency to orient themselves so that the oppositely charged ends of the molecules come near to each other. Table 8.2 gives the molar heats of fusion and vaporisation of some solvents.

Table 8.2. Molar heats of fusion and vaporisation of some solvents

Solvent	**Molar heat of fusion ($kJ\ mol^{-1}$)**	**Molar heat of vaporisation ($kJ\ mol^{-1}$)**
Water	6.0	40.65
Ammonia	5.65	23.56
Sulphur dioxide	7.40	24.94
Hydrogen fluoride	4.58	30.26

The above table shows that the values of molar heats of fusion for water and ammonia are almost same. This means that the forces which hold molecules together in water and ammonia are of same magnitude. Thus, these substances will serve as similar solvents.

8.2.3. Dielectric constant

This property helps to estimate the solubility of polar or non-polar substances in the solvent. According to coulombic law, the force of attraction between two charged ions is given by

$$F = \frac{q^+q^-}{Dr^2}$$

where q+ and q^- are the charges of the cation and anion respectively, r is the distance between them D is a constant called the *dielectric constant* of the solvent. If D is large, the force F will be small. Thus, only a small amount of energy will be required to separate the ions and hence it will be easy to dissolve an ionic substance. With the decrease in dielectric constant, the solubility of ionic compounds will decrease. The dielectric constants of some common solvents are provided in Table 8.3.

Table 8.3 Dielectric constants of some comon solvents.

Solvent	Dielectric constant (D)
H_2O	81.7 (at 18°C)
NH_3	22.0 (at –33.5°C)
SO_3	17.3 (at –16°C)
CH_3COOH	9.7 (at 18°C)
HF	83.6 (at 0°C)
HCN	123 (at 16°C)
H_2SO_4	100 (at 25°C)

From the above Table we find that solvents like HF and H_2O are good solvents for polar compounds because of their high dielectric constant values. On the other hand, solvents like NH_3 and SO_2 with smaller values of dielectric constant show smaller tendency to dissolve ionic compounds.

8.2.4. Viscosity

Viscosity gives a measure of the fluidity of the solvent. Solvents like water, ethyl alcohol and carbon tetrachloride have low viscoscities and flow rapidly under ordinary temperatures. In solvents of low viscosities, the operations such as precipitation, crystallization, filtration etc. can be easily carried out without any difficulty. With increasing viscotiy of a liquid, the difficulty of such operations increases. Solvents like anhydrous sulphuric acid have higher viscosities and this reduces their usefulness as solvent.

8.3. TYPES OF SOLVENTS

Ionizing and Non-Ionizing Solvents

***(i)* Ionizing solvents.** These are polar in nature and dissolve ionic compounds. In these solvents, ionic reactions occur. They also show self-ionization. Ammonia, sulphur dioxide and hydrofluoric acid are examples of ionizing solvents.

$$NH_3 + NH_3 \rightleftharpoons NH_4^+ + NH_2^-$$

$$SO_2 + SO_2 \rightleftharpoons SO^{2+} + SO_3^{2-}$$

$$3HF \rightleftharpoons H_2F^+ + HF_2^-$$

$$N_2O_4 \rightleftharpoons NO^+ + NO_3^-$$

***(ii)* Non-ionizing solvents.** These solvents are non-polar and dissolve only non-polar or neutral compounds. No ionic reactions occur in these solvents. They do not undergo self-ionization. Benzene and carbon tetrachloride are examples of non-ionizing solvents.

Protic and aprotic solvents

Solvents such as NH_3 or HF which have hydrogen as their constituent are called *protic solvents.* Solvents such as SO_2 which have no hydrogen in their molecule are *aprotic solvents.*

Auto-ionization (solvent system) theory of solvent

This theory is based on the mode of self-ionization of the solvent. Each solvent is considered to consist of an acidic part and a basic part and on auto-ionization is split into these two parts, *i.e.*, acidic part and basic part. A substance that produces or increases the concentration of acidic part of the solvent behaves as an acid. Similarly a substance that increases the concentration of basic part of the solvent behaves as a base. The behaviour of any solute is thus determined in terms of the release of a characteristic cation or anion of that solvent. For example, ammonium chloride (NH_4Cl) is defined as an acid and sodium amide ($NaNH_2$) as a base in liquid ammonia.

$$2NH_3 \rightleftharpoons NH_4^+ + NH_2^- \text{ (Self–ionization)}$$

$$NH_4Cl \rightleftharpoons NH_4^+ + Cl^- \text{ (Acid)}$$

$$NaNH_2 \rightleftharpoons Na^+ + NH_2^- \text{ (Base)}$$

Thus ammonium chloride acts as acid and sodomide as base.

Similarly, the behaviour of acidic and basic solutes in phosphorus oxy-chloride ($POCl_3$) solvent may be studied as:

$$2POCl_3 \rightleftharpoons POCl_2^+ + POCl_4^- \text{ (Self -ionization)}$$

$$FeCl_3 + POCl_3 \rightleftharpoons POCl_2^+ + FeCl_4^- \text{ (Acid)}$$

Thus $FeCl_3$ acts as an acid in $POCl_3$ solvent

$$(CH_3)_4N^+Cl^- + POCl_3 \rightleftharpoons (CH_3)_4N^+ + POCl_4^- \text{ (base)}$$

Thus tetramethyl ammonium chloride acts as a base in $POCl_3$ solvent.

The reaction between tetramethyl ammonium chloride and ferric chloride may be investigated as a titration in $POCl_3$.

$$(CH_3)_4N^+Cl^- + FeCl_3 \xrightarrow{POCl_3} (CH_3)_4N^+ + FeCl_4^-$$

Behaviour of acidic and basic solutes is determined by the self-ionization species of the solvents. Therefore, self-ionization of the solvent to give ionic species is very essential in solvent system concept.

8.4. COORDINATION MODEL OF SOLVENT

Doubtful existence of ionic species in some solvents led to the development of coordination concept of solvent. Coordination of solutes by solvent molecules is quite important. When a solute dissolves in a solvent, dissolution is always accompanied by solvent coordination. However, behaviour of solutes in solvents has generally ignored the importance of coordination by the solvent and more emphasis has been placed on the self-ionization of the solvent. Lately a number of chemists have criticized the overemphasis on ionic reactions in solutions as laid down by the proponents of solvent system concept. Meek and Drago found that the reaction between tetramethyl ammonium chloride and ferric chloride could be as well carried out in triethyl phosphate PO $(OC_2H_5)_3$ as in phosphorus oxychloride $POCl_3$. This is very surprising since triethyl phosphate did not contain any chloride ion and yet $FeCl_4^-$ ion is formed.

$$FeCl_3 + POCl_3 \xrightarrow{POCl_3} POCl_2^+ + FeCl_4^-$$

Meek and Drag suggested the 'Coordination Model' as an alternative way to 'solvent system' to account for these observations. They suggested that in triethyl phosphate the chloride ion transfer takes place from one $FeCl_3$ molecule to another with the formation of a cationic iron (III) species.

$$2FeCl_3 \xrightarrow{PO(OC_2H_5)_3} [FeCl_2[OP(OC_2H_5)_3]_4]^+ + FeCl_4^-$$

Similarly, ionisation of $FeCl_3$ in $POCl_3$ solvent may be represented as:

$$2\,FeCl_3 \xrightarrow{POCl_3} [Fe\,Cl_2\,(OPCl_3)_n]^+ + FeCl_4^-$$

Thus, they could easily account for the presence of $FeCl_4^-$ species both in $POCl_3$ and $PO(OC_2H_5)_3$ solutions simply by suggesting that the solvent molecules coordinate with interacting solutes. These authors emphasized that similarities in physical properties of the two solvents, mainly their coordinating ability and dielectric constants, are more important than the presence or absence of self-ionization species.

8.5 TYPES OF REACTIONS IN NON-AQUEOUS SOLVENTS

The following types of reactions ocur in non-aqueous solvents:

8.5.1 Acid-base reactions

The acid-base reactions in non-aqueous solvents can be explained on the basis of solvent concept. According to **solvent system concept:**

an acid is a substance that by direct dissociation or reaction with the solvent gives the cation characteristic of the solvent. And, a base is a substance that gives the anion characteristic of the solvent.

In the case of protonic solvents, cation is the solvated proton and the Bronsted protonic concept is same as the solvent system concept. For example, in liquid ammonia solvent, NH_4^+ ions act as acid and NH_2^- ion acts as base. The neutralisation reaction can be represented as

$$\underset{\text{Acid}}{NH_4Cl} + \underset{\text{Base}}{NaNH_2} \xrightarrow{\text{liq. }NH_3} \underset{\text{Salt}}{NaCl} + \underset{\text{Solvent}}{2NH_3}$$

In non-protonic solvents such as benzene, carbon tetrachloride, Lewis acid-base type of reactions can occur because of non-ionisation of the solvents.

The acid-base behaviour of a solute in a solvent depends upon the acidity or basicity of the solvent. For example, acetic acid behaves as strong acid in ammonia, weak acid in water but a base in sulphuric acid, as shown by the following chemical equations:

$$\underset{\text{(Strong acid)}}{CH_3COOH} + NH_3 \rightleftharpoons CH_3COO^- + NH_4^+$$

$$\underset{\text{(Weak acid)}}{CH_3COOH} + H_2O \rightleftharpoons CH_3COO^- + H_3O^+$$

$$\underset{\text{(Weak acid)}}{CH_3COOH} + H_2SO_4 \rightleftharpoons CH_3COOH_2^+ + HSO^-_4$$

8.5.2. Metathetical or precipitation reactions

The reactions in which precipitation occurs on mixing two solutions are called metathetical or precipitation reactions. These types of reactions can be predicted if the values of the solubility products of the compounds formed in the solvent are known. We know silver nitrate and barium chloride to give a precipitate of silver chloride leaving barium nitrate in the solution.

$$2AgNO_3 + BaCl_2 \xrightarrow{\text{Water}} \underset{\text{ppt.}}{2AgCl\downarrow} + Ba(NO_3)_2$$

On the other hand, a precipitate of silver chloride combines with barium nitrate in liquid ammonia to give precipitate of barium chloride leaving silver nitrate in solution.

$$2AgCl + Ba\,(NO_3)_2 \xrightarrow{\text{liq. } NH_3} \underset{\text{ppt.}}{BaCl_2\downarrow} + 2AgNO_3$$

The reaction is thus reversed on changing the solvent.

8.5.3. Salt formation

The reactions between appropriate acidic and basic substances to form salts are called salt formation reactions. In such cases, the solvent used must be one in which the appropriate base can exist. For example, sodium ureide cannot be obtained by the reaction of urea on hydroxide in water. It is because the strong base available in water (the OH^- ion) will not take a proton from urea molecule i.e. the sodium salt of urea is unstable in water and ureide ion will take a proton from water to form hydroxide ion.

$$\underset{\text{Sodium ureide}}{Na+\,(NHCONH_2)^-} + H_2O \longrightarrow Na^+ + OH^- + \underset{\text{Urea}}{NH_2CONH_2}$$

However, this salt formation can be carried out in liquid ammonia solution by the reaction of strong base sodamide.

$$\underset{\text{Urea}}{NH_2CONH_2} + \underset{\text{Sodamide}}{Na^+NH_2^-} \xrightarrow{\text{liq. } NH_3} \underset{\text{Sodium ureide}}{Na^+\,(NHCONH_2)^-} + NH_3$$

8.5.4. Solvation

The reaction in which the molecules of the solvent get attached to a solute species (cation, anion or molecule) are called solvation reactions. The species formed is called solvate. For example, solvation reactions in water, ammonia and alcohol give hydrate, ammoniate and alcoholate respectively

$$CuSO_4 + 4NH_3 \longrightarrow \underset{\text{Ammoniate}}{CuSO_4.4NH_3}$$

$$CuCl_2 + 4H_2O \longrightarrow \underset{\text{Hydrate}}{[Cu\,(H_2O)_4]^{2+}\,2Cl^-}$$

$$BaSO_4 + 3H_2SO_4 \longrightarrow \underset{\text{Solvate of sulphuric acid}}{BaSO_4.3H_2SO_4}$$

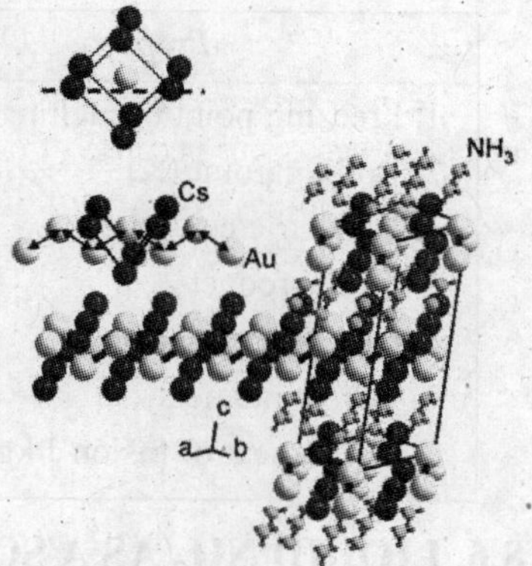

Structure of cesium auride ammoniate

Solvation of a cation occurs through ion-dipole interactins between the positive cation and the negative end of the dipole or through coordinate bond. On the other hand, solvation of anion occurs through ion-diple interactions between the negative ion and the positive end of the solvent dipole or through hydrogen bonding in case of solvents such as water, ammonia or hydrogen fluoride.

8.5.5. Solvolytic reactions or solvolysis

The reactions in which the solvent molecules react with the solute in such a way that the solvent molecules split up into two parts, one or both of which get attached to the solute molecule or ion are called solvolytic reactions. These reactions occur through self ionisation of the solvent therefore the concentrations of either the cation or the anion is changed. Solvolytic reactions in water, ammonia and ethanol are given below:

(*i*) Water as solvent:

$$2H_2O \rightleftharpoons H_3O^+ + OH^- \quad \text{(auto ionization)}$$

$$SO_2Cl_2 + 4H_2O \rightleftharpoons SO_2(OH)_2 + 2H_3O^+ + 2Cl^- \quad \text{(Conc. of } H_3O^+ \text{ is increased)}$$

$$F^- + H_2O \rightleftharpoons HF + OH^- \quad \text{(Conc. of } OH^- \text{ is increased)}$$

(*ii*) Alcohol as solvent:

$$2C_2H_5OH \rightleftharpoons C_2H_5OH_2^+ + OC_2H_5^- \quad \text{(auto ionization)}$$

$$SO_2Cl_2 + 4C_2H_5OH \rightleftharpoons SO_2(OC_2H_5)_2 + 2C_2H_5OH_2^+ + 2Cl^- \quad \text{(Conc. of } C_2H_5OH_2^+ \text{ is increased)}$$

$$CH_3^- + C_2H_5OH \rightleftharpoons OC_2H_5^- + CH_4 \quad \text{(Conc. of } OC_2H_5^- \text{ in increased)}$$

(*iii*) Ammonia as solvent:

$$2NH_3 \rightleftharpoons NH_4^+ + NH_2^- \quad \text{(auto ionization)}$$

$$SO_2Cl_2 + 4NH_3 \rightleftharpoons SO_2(NH_2)_2 + 2NH_4^+ + 2Cl^- \quad \text{(Conc. of } NH_4^+ \text{ is increased)}$$

$$H^- + NH_3 \rightleftharpoons NH_2^- + H_2 \quad \text{(Conc. of } NH_2^- \text{ in increased)}$$

Comparison of physical properties of ammonia and water

Some physical properties of water and ammonia as solvent are given in Table 8.4

Table 8.4 Physical Properties of ammonia and water.

Property	*Ammonia*	*Water*
1. Freezing point or melting point (K)	195.3	273
2. Boiling point (K)	239.6	373
3. Dielectric constant	22 at 239.6 K	81 at 291 K
4. Ionic product	1.9×10^{-3} at 223 K	1×10^{-14} at 298 K
5. Specific conductance	9.0×10^{-11} at 195 K	6×10^{-6} at 298 K
6. Dipole moment	1.148 at 233 K	1.84 at 291 K
7. Latent heat of fusion J/kg	0.34	0.33

8.6. LIQUID NH_3 AS A SOLVENT

Important characteristics of ammonia as a solvent.

Some important points about liquid ammonia are:

(*i*) Like water, ammonia molecules are associated through hydrogen bonding in liquid ammonia as it contains a strongly electronegative element-nitrogen, which is responsible for hydrogen bonding in it.

(*ii*) The lower value of its dielectric constant as compared to that of water shows that liquid ammonia is a poor electrolytic solvent than water.

(*iii*) Lower viscosity of liquid ammonia tends to promote greater ionic mobilities and thereby overcome, to some extent, the dielectric constant effect. Most salts actually conduct electric current better when dissolved in liquid ammonia than when dissolved in water. This is not due to high ionizing power of liquid ammonia but due to low viscosity which increases the mobility of ions.

(*iv*) The melting point and boiling point of ammonia are abnormal with respect to PH_3, AsH_3 and SbH_3 due to intermolecular association occurring in it as a result of H-bonding. The freezing and boiling points of liquid ammonia are lower than those of water. Therefore, more precautions are to be taken while handling liquid ammonia.

(*v*) Among the soluble compounds in liquid ammonia are nitrates, nitrites, thiocyanates and most cyanides. The majority of fluorides are insoluble, as are oxides, hydroxides, sulphates, carbonates, phosphates, sulphites and most sulphides. Among the halides, other than the fluorides, solubility decreases from the iodides to the chlorides. Most iodides (even that of silver) are soluble, the bromides are less soluble and among the chlorides only the ammonium, beryllium and sodium compounds are soluble.

(*vi*) Because of its ability to donate lone pair of electrons to form a covalent coordinate link as well as by ion-dipole attractions, ammonia forms many solvates which are analogues of salt-hydrates.

(*vii*) In general, solubilities of covalent organic compounds in liquid ammonia are considerably higher than in water probably due to greater dispersion energy interactions.

(*viii*) The specific heat of ammonia is higher than that of water.

(*ix*) Liquid ammonia possesses unusual high thermal coefficient of expansion, *i.e.*, its density decreases with rise in temperature.

(*x*) It has high critical temperature and pressure.

(*xi*) Liquid ammonia is a very poor conductor of electricity.

(*xii*) Liquid ammonia is an ionizing solvent, *i.e.*, undergoes ionization though to a lesser extent as compared to water.

8.6.1 Auto-ionization of liquid ammonia

The ionization of ammonia is supposed to take place just like that of water,

$$2NH_3 \rightleftharpoons \underset{\text{Ammonium ion}}{NH_4^+} + \underset{\text{Amide ion}}{NH_2^-}$$

$$2H_2O \rightleftharpoons \underset{\text{Hydronium ion}}{H_3O^+} + \underset{\text{Hydroxide ion}}{OH^-}$$

But the ionization of ammonia is very small as is evident from its low conductance and also from its ionic product, *i.e.*,

$$K_{H_2O} = [H_3O^+] \times [OH^-] = 1 \times 10^{-14} \text{ at } 291 \text{ K}$$

$$K_{NH_3} = \left[NH_4^+\right] \times \left[NH_2^-\right] = 1.9 \times 10^{-33} \text{ at } 223 \text{ K}$$

Ammonium ion in liquid ammonia is analogous to hydronium ion in aqueous solution and amide ion to hydroxyl ion. Hence any substance which gives ammonium ions in liquid ammonia behaves as acid and is known as ammonia acid, *e.g.*, NH_4Cl. Any substance which yields amide ion acts as base and is known as ammonia base, *e.g.*, sodamide $NaNH_2$.

On the basis of Bronsted-Lowery concept, ammonium ions would behave as acids, and amides as bases in liquid ammonia.

$$NH_3 + NH_3 \rightleftharpoons NH_4^+ + NH_2^-$$

$$\underset{\text{Acid I}}{H_2O} + \underset{\text{Base II}}{H_2O} \rightleftharpoons \underset{\text{Acid II}}{H_3O^+} + \underset{\text{Base I}}{OH^-}$$

8.6.2. Types of reactions in liquid ammonia as a solvent

Various reactions are described as under:

(1) Acid-base reactions. As substances producing H_3O^+ ions in water are acids, in the same way ammonium salts giving NH_4^+ ions in liquid ammonia are acids. Similarly, substances producing OH^- ions in water and NH_2^- (amide) ions in liquid ammonia are bases in that solvent. Thus acid base neutralization can occur in both solvents.

$$\underset{\text{Acid}}{HCl} + \underset{\text{Base}}{NaOH} \rightleftharpoons NaCl + H_2O \quad \text{(in water)}$$

$$\underset{\text{Acid}}{NH_4Cl} + \underset{\text{Base}}{NaNH_2} \longrightarrow \underset{\text{Salt}}{NaCl} + 2NH_3 \quad \text{(in ammonia)}$$

The course of neutralization reactions in liquid ammonia may be followed by means of conductometric titrations, and in certain cases through the indicators. For example, phenolphthalein is colourless in liquid ammonia but develops an intense red colour in the presence of alkali amides.

(2) Redox Reactions. Redox reactions take place in liquid ammonia just as they do in water as illustrated hereunder:

(*i*) Solutions of alkali and alkaline earth metals in liquid ammonia may act as electron donors by yielding their conduction electrons to electron acceptors. Hence the addition of metallic sodium in liquid ammonia to NH_4Br is a redox reaction.

$$2Na + 2NH_4Br \xrightleftharpoons{\text{Liq. } NH_3} 2NaBr + 2NH_3 + H_2$$

(*ii*) Nitrous oxide oxidizes metallic potassium in liquid ammonia to potassium ion and is itself reduced to nitrogen in the process.

$$N_2O + 2K + NH_3\,(l) \rightleftharpoons KNH_2 + KOH + N_2$$

(3) Precipitation Reactions or Metathesis Reactions. Because of difference in solubilities of various substances in liquid ammonia and water, a large number of reactions, which do not occur in water, have been reported to be carried out in ammonia. The courses of these reactions are determined by the solubilities of products. For example,

(*i*) Lithium chloride may be precipitated by the reaction of ammonium chloride and lithium nitrate in liquid ammonia.

$$NH_4Cl + LiNO_3 \xrightleftharpoons{NH_3(l)} NH_4NO_3 + LiCl \downarrow$$

(*ii*) When solutions of potassium iodide and ammonium chloride in liquid ammonia are brought together, a white ppt. of KC1 is produced.

$$KI + NH_4\,Cl \xrightleftharpoons{\text{Liq. NH}_3} KC1 \downarrow + NH_4\,I$$

(*iii*) The precipitation of barium sulphide occurs from ammonia solution of barum nitrate and ammonium sulphide.

$$Ba\,(NO_3)_2 + (NH_4)_2\,S \xrightarrow{\text{Liq.NH}_3} BaS \downarrow + 2NH_4\,NO_3$$

(*iv*) When solutions of metal nitrates and ammonium halides in liquid ammonia are brought together, iodides and bromides get precipitated.

$$2NH_4\,I + Zn\,(NO_3)_2 \xrightleftharpoons{NH_3(l)} Zn\,I_2 \downarrow + 2NH_4\,NO_3$$

$$2NH_4\,Br + Sr\,(NO_3)_2 \xrightleftharpoons{NH_3(l)} Sr\,Br_2 \downarrow + 2NH_4\,NO_3$$

(4) Complex Formations. Some reactions which have been studied are quite similar to complex formation occurring in the water system, *e.g.*, the zinc ion is precipitated as insoluble zinc amide which dissolves in excess of amide to form the ammonozincate.

$$Zn\,(NO_3)_2 + 2KNH_2 \xrightleftharpoons{NH_3(l)} \underset{\text{Insoluble}}{Zn(NH_2)_2} + 2KNO_3$$

$$Zn(NH_3)_2 + \underset{\text{Excess}}{2KNH_2} \xrightleftharpoons{NH_3(l)} \underset{\text{Soluble}}{K_2[Zn(NH_2)_4]}$$

(5) Solvolysis Reactions. These reactions are usually referred to as *ammonolysis or ammonolytic* reactions. Reactions of this type are analogous to the familiar hydrolysis reactions of water. In other words, an ammonolytic reaction is one of double decomposition in which ammonia is a reactant.

(*i*) Alkali metal and alkaline earth metal hydrides are ammonolysed to the corresponding metal amide and hydrogen.

$$M^+H^- + NH_3\,(l) \longrightarrow M^+\,NH_2^- + H_2$$

(*ii*) Many covalent halides undergo ammonolysis in liquid ammonia. In some cases elevated temperature and pressure are maintained to carry out the reactions. Some examples are given below:

$$HgC1 \xrightarrow{\text{Liq.NH}_3} Hg\,NH_2\,Cl$$

An ammonolytic reaction, familiar to most chemists, is the action of ammonia on mercurous chloride.

$$Hg_2\,C1_2 + 2NH_3\,(l) \longrightarrow Hg + HgNH_2\,Cl + NH_4^+ + Cl^-$$

$$SiC1_4 \xrightarrow{\text{Liq.NH}_3} \underset{\text{Tetramide silane}}{Si(NH_2)_4}$$

$$Si\,H_3\,Cl \xrightarrow{\text{Liq.NH}_3} (Si\,H_3)_3\,N$$

$$CCl_4 \xrightarrow[\text{elevated temp and pressure}]{\text{Liq.}NH_3} C(NH)(NH_2)_2$$

$$PCl_3 \xrightarrow{\text{Liq.}NH_3} P(NH_2)_3 \xrightarrow{273\text{ K}} P_2(NH)_2$$

(*iii*) When organic halides are ammonolysed at the boiling point of liquid ammonia, primary, secondary and tertiary amines are formed.

$$RX + 2NH_3\,(l) \longrightarrow RNH_2 + NH_4X$$

$$RX + RNH_2 + NH_3\,(l) \longrightarrow R_2NH + NH_4X$$

$$RX + R_2NH + NH_3\,(l) \longrightarrow R_3N + NH_4X$$

(6) Reaction with Metals. An important property of liquid ammonia is its ability to dissolve the alkali and alkaline earth metals (except Be and Mg) to give homogeneous solution. Dilute metal ammonia solutions are blue but when more metal is added, a bronze phase (copper-coloured) separates and floats on the blue solution. The metal can be recovered unchanged by evaporating the solvent.

$$2M + 2NH_3\,(l) \longrightarrow 2MNH_2 + H_2 \quad (M = \text{an alkali metal}) \quad ...(1)$$

$$M' + 2NH_3(l) \longrightarrow M'(NH_2)_2 + H_2 \quad (M' = \text{Alkaline earth metal}) \quad ...(2)$$

Studies have led to the theory that they contain solvated metal cations and solvated electrons.

$$M \rightleftharpoons M^+ + e^- \quad \text{or}$$

$$M + x\,NH_3 \longrightarrow M(NH_3)_x^+ + e^-$$

$$e^- + y\,NH_3 \longrightarrow e\,(NH_3)_y^-$$

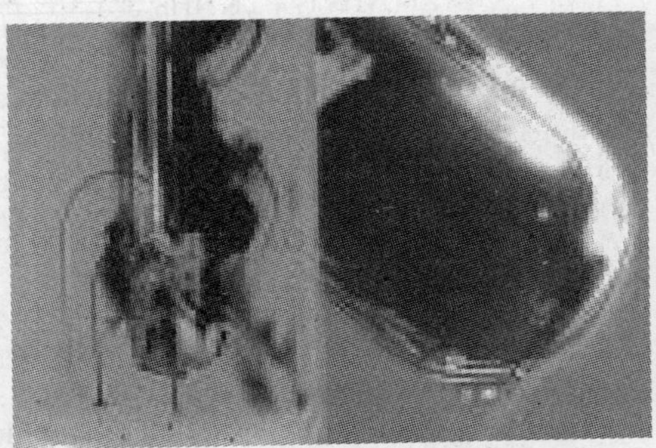

Solution of sodium in liquid ammonia

The blue colour in these metal solutions can be attributed to the presence of free electrons. The fact that the metals may be reclaimed by evaporation of the solvent shows that these electrons are returned to their parent ions. The blue solutions of metals in liquid ammonia are unique as they conduct an electric current and this bears evidence to the presence of free electrons. It is also true that the same free electrons render their solutions as most powerful reducing agents.

(7) Formation of Ammoniates. Ammonia forms ammoniates with many inorganic salts, such as $MgI_2 .6NH_3$, $NiI_2 \cdot 6NH_3$, $NaI \cdot 4NH_3$ in a similar way as water forms hydrates. Ammonia molecules are coordinated to the cation in these ammoniates.

(8) Extensive Dissociation of Weak Acids. Because liquid ammonia accepts protons readily, it enhances the ionization of so called weak acids such as acetic acid.

$$NH_3\,(l) + CH_3COOH \rightleftharpoons CH_3COO^- + NH_4^+$$

Removal of H^+ by NH_3 causes the reaction to proceed in the forward direction. Thus acetic acid has a pH value of 5 in water but is almost completely ionized in liquid ammonia. Ammonia thus reduces the difference between strength of acids and is called a *levelling solvent*.

8.6.3. Applications of liquid ammonia as a solvent

Application of liquid ammonia as a solvent are given as under:

(*i*) *Metal ammonia solutions are powerful reducing agents* because they add electrons or produce nascent hydrogen. Thus it is possible to synthesize unusual compounds which are otherwise difficult to prepare. Metal ammonia solutions react with substances that are acidic in liquid ammonia to form metal salts and hydrogen gas. Many covalent hydrides and their derivatives react with metal ammonia solutions to form metal salts that are stable in liquid ammonia.

$$AsH_3 + Na \xrightarrow{NH_3(l)} NaAsH_2 + \frac{1}{2}H_2$$

$$PH_3 + Na \xrightarrow{NH_3(l)} NaPH_2 + \frac{1}{2}H_2$$

$$PH_3 + 2Na \xrightarrow{NH_3(l)} Na_2PH + H_2$$

(*ii*) The most useful property of liquid ammonia as a reaction medium in organic chemistry comes from its ability to form stable solutions of alkali metals that are highly reducing and from its unusual ability to dissolve a variety of organic compounds. These properties allow reduction reactions to be carried out in a homogeneous phase than by catalytic hydrogenation. In organic synthesis, metal ammonia solutions react with conventional transition metal complexes to yield compounds in which the metal atom has a formal valency of zero, assuming K = + 1 and CN = – 1.

$$K_3[Cr(CN)_6] + 3K \xrightarrow{NH_3(l)} K_6[Cr(CN)_6]$$

$$K_2[Ni(CN)_4] + 2K \xrightarrow{NH_3(l)} K_4[Ni(CN)_4]$$

$$K_2[Pt(CN)_4] + 2K \xrightarrow{NH_3(l)} K_4[Pt(CN)_4]$$

Oxidation state of Cr, Ni and Pt in these products comes out to be zero.

8.6.4. Advantages of using liquid Ammonia as solvent

(*i*) One of the greatest advantages of using liquid ammonia as solvent is that liquid ammonia dissolves alkali metals. The dissolved alkali metals can be recovered simply by evaporating the alkali metal-ammonia solution.

(*ii*) If a reducing agent stronger than hydrogen is to be used in an aqueous medium, it will react with H_2O to liberate H_2 and hence cannot be used in aqueous solution. In such cases, many alkali-ammonia solutions which contain mobile ammoniated electrons $e^-(NH_3)$ can be used for reducing the materials soluble in liquid ammonia.

8.6.5. Disadvantages of using Liquid Ammonia

(*i*) The use of liquid ammonia as a solvent and a reaction medium requires elaborate equipment and special techniques are to be used.

(*ii*) Due to the hygroscopic nature of liquid ammonia, reactions are done in sealed tubes so as not to allow ammonia to come in contact with moisture.

(*iii*) The boiling point of liquid ammonia is 239.5 K and hence while working with this solvent, a lower temperature is to be maintained, if high pressures are to be avoided.

8.7. LIQUID SO_2 AS A SOLVENT

Characteristics of liquid sulphur dioxide as a solvent.

Some of the properties by liquid SO_2 as a solvent are given in Table 8.5.

Table 8.5 Some important physical properties of liquid SO_2.

Property	Value
Freezing point (K)	196.5
Boiling point (K)	2.63
Density (g cm^{-3})	1.46 (at – 10°C)
Viscosity (cP)	0.428 (at –10°C)
Dielectric constant	17.4 (at –10°C)
	15.6 (at 0°C)
Dipole moment	1.61
Specific conductance (ohm^{-1} cm^{-1})	4 × 10–8 (at – 10°C)

(*i*) It is clear from the table that its liquid range is quite high, therefore, it can work as a good solvent.

(*ii*) The bond angle of 119.5° in O – S – O indicates that SO_2 molecule is in triangular valence state, *i.e.*, sp^2 hybridization with two hybrid orbitals to form bonds with oxygen and the third orbital occupied by a lone pair of electrons. The S – O bond length of 1.46 Å indicates some double bond character in the molecule. The SO_2 molecule may be shown to exist in the following two resonating structures.

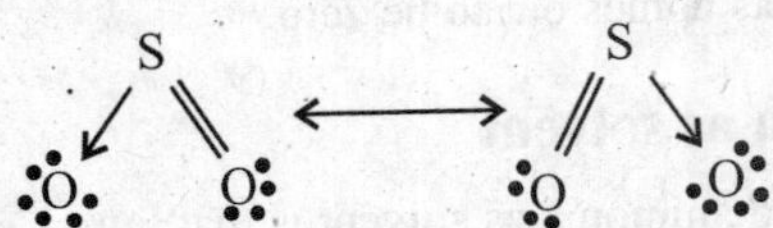

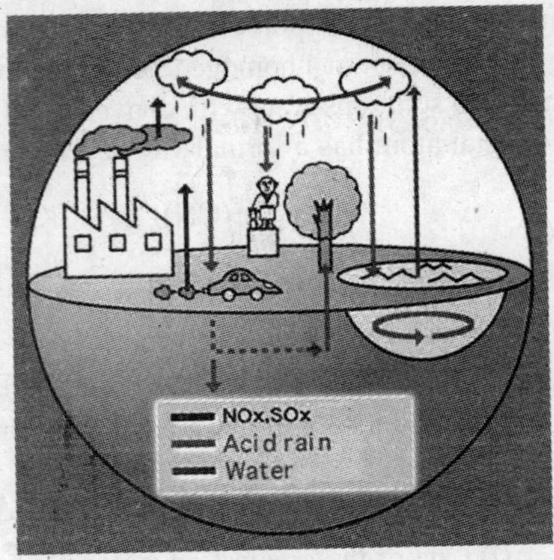

Acid rain

As S-atom contains a vacant bonding orbital, it can act as an electron acceptor and form molecular complexes.

(*iii*) The auto-ionization of liquid SO_2 takes place as follows:

$$SO_2 + SO_2 \rightleftharpoons \underset{\text{Thionyl ion}}{SO^{2+}} + \underset{\text{Sulphite ion}}{SO_2^{2-}}$$

(*iv*) Solubility of inorganic substances in liquid SO_2. Iodides are found to be most soluble in liquid SO_2. In case of alkali metal halides the order of solubility is found to be MI > MBr > MC1 > MF. Alkali metal sulphites, sulphates, acetates, cyanides and thiocyanates are also found to be soluble. Metal sulphates, sulphites, oxides and hydroxides are sparingly soluble. Ammonium, mercury and thalium salts are also soluble.

Many substances like halogens, interhalogens (ICI, IBr etc.), phosphorus halides ($PC1_3$, PBr_3 etc.), oxyhalides of phosphorus and sulphur ($POCl_3$, $SOCl_2$, etc.), their halides like $CC1_4$, $SiC1_4$, $GeCl_4$, $SnCl_4$, $AsCl_3$ and CS_3 are found to be soluble in liquid SO_2.

(v) Conductivity of salt solutions in SO_2. Because of its high dipole moment value, it dissolves many inorganic compounds in it. Therefore, it behaves as a moderately good conductor for many alkali metals, ammonium and tetra alkyl ammonium chlorides, bromides, iodides and thiocyanates while di-, tri-alkyl ammonium salts are very poor conductors.

It is found that the conductivity of electrolyte solutions in liquid SO_2 increases with the size of the cation. Thus the following order of conductivity is found:

$$Na^+ < NH_4^+ < K^+ < (CH_3)_3 S^+ < (CH_3)_4 N^+ \text{ etc.}$$

The conductivity of the anions is found to increase in the order:

$$SCN^- < ClO_4^- < Cl^- < I^- \text{ etc.}$$

(vi) Solubility of organic compounds in liquid SO_2. SO_2 is found to be a better medium for the dissolving of covalent compounds. Aromatic hydrocarbons and alkenes are more soluble than aliphatic hydrocarbons. SO_2 has also been used as a solvent for carrying out many organic reactions such as Friedel-Craft's reactions, sulphonation, bromination etc.

8.7.1 Chemical reactions in liquid SO_2

These reactions are discussed under the following headings.

1. Acid-Base Reactions. Liquid SO_2 undergoes auto-ionization as follows:

$$\underset{Acid_1}{SO_2} + \underset{Base_2}{SO_2} \rightleftharpoons \underset{Acid_2}{SO^{2+}} + \underset{Base_1}{SO_3^{2-}}$$

Thus according to solvent system concept, any substance which makes available SO^{2+} ions would behave as an acid, whereas the one making available SO_3^{2-} would behave as base. The compounds such as $SOCl_2$ and $SOBr_2$ behave as acids in liquid SO_2 and Na_2SO_3 would behave as a base. The titration of potassium sulphite (K_2SO_3) against thionyl chloride or acetate can be followed conductometrically.

$$\underset{Acid}{SOCl_2} + \underset{Base}{K_2SO_3} \longrightarrow \underset{Salt}{2KCl} + \underset{Solvent}{2SO_2}$$

Some other acid-base neutralization reactions are:

$$\underset{Acid}{SOCl_2} + \underset{Base}{Cs_2SO_3} \longrightarrow \underset{Salts}{2CsCl} + \underset{Solvent}{2SO_2}$$

$$SOBr_2 + [N(CH_3)_4]_2SO_3 \longrightarrow 2[N(CH_3)_4]Br + 2SO_2$$

$$SO(SCN)_2 + K_2SO_3 \longrightarrow 2K(SCN) + 2SO_2$$

Amphoteric behaviour is shown by the zinc (II) or aluminium (III) salts in liquid SO_2.

$$ZnCl_2 + K_2SO_3 \xrightarrow{SO_2(l)} ZnSO_3 + 2KCl \text{ (Acidic character)}$$

$$ZnSO_3 + K_2SO_3 \xrightarrow{SO_2(l)} K_2[Zn(SO_3)_2] \text{ (Basic character)}$$

$$Al_2(SO_3)_2 + 3K_2SO_3 \xrightarrow{SO_2(l)} 2K_3[Al(SO_3)_3] \text{ (Basic character)}$$

$$2AlCl_3 + 3K_2SO_3 \xrightarrow{SO_2(l)} Al_2(SO_3)_3 + 6KCl \text{ (Acidic character)}$$

2. Redox Reactions. Many reactions of gaseous sulphur dioxide are known where it behaves both as oxidizing and as reducing agent. But sulphur dioxide, in liquid state, does not have any marked reducing or oxidising tendency and acts as a medium for many redox reactions. Iodine is liberated from KI in the presence of $SbCl_5$.

$$6KI + 3SbCl_5 \xrightleftharpoons{SO_2(l)} 3I_2 + SbCl_3 + 2K_3[SbCl_6]$$

3. Solvolysis Reactions. Certain covalent halides are found to undergo solvolysis reactions in liquid SO_2. For example,

$$PCl_5 + SO_2(l) \longrightarrow POCl_3 + SOCl_2$$

$$AsCl_5 + SO_2(l) \longrightarrow AsOCl_3 + SOCl_2$$

4. Metathetical reactions. The precipitation of a substance depends upon its solubility in the solvent. There are many chemical reactions which have been carried out in this solvent.

$$BaI_2 + Zn(CNS)_2 \xrightarrow{SO_2(l)} Ba(CNS)_2\downarrow + ZnI_2$$

$$SbCl_3 + 3LiI \xrightarrow{SO_2(l)} SbI_3\downarrow + 3LiCl$$

$$PbF_2 + Li_2SO_4 \xrightarrow{SO_2(l)} PbSO_4\downarrow + 2LiF$$

Thionyl chloride in liquid SO_2 has helped to prepare many compounds, *e.g.*,

$$2CH_3COOAg + SOCl_2 \xrightarrow{SO_2(l)} 2AgCl\downarrow + \underset{\text{Thionyl acetate}}{SO(OCOCH_3)_2}$$

$$2KBr + SOCl_2 \xrightarrow{SO_2(l)} 2KCl\downarrow + SOBr_2$$

$$2NH_4(SCN) + SOCl_2 \xrightarrow{SO_2(l)} 2NH_4Cl\downarrow + SO(SCN)_2$$

5. Sulphur dioxide of crystallization. These reactions are also called solvation reactions. Salts having sulphur dioxide of crystallization have been obtained.

$$NaI + 4SO_2(l) \longrightarrow NaI \cdot 4SO_2$$

$$KI + 4SO_2(l) \longrightarrow KI \cdot 4SO_2$$

6. Complex formation. Complex formation in the presence of liquid SO_2 has been reported by Walden and coworkers. Sulphur dioxide itself does not take part in complexation, but provides medium for the purpose.

$$SbCl_3 + 3KCl \xrightarrow{SO_2(l)} K_3^+[SbCl_6]^{3-}$$

$$SbCl_5 + KCl \xrightarrow{SO_2(l)} K^+[SbCl_6]^-$$

$$SbCl_5 + [N(CH_3)_4]^+Cl^- \xrightarrow{SO_2(l)} [N(CH_3)_4]^+[SbCl_6]^-$$

$$SbCl_5 + CH_3COCl \xrightarrow{SO_2(l)} [CH_3CO]^+[SbCl_6]^-$$

7. Organic reactions in liquid sulphur dioxide. Organic reactions where relatively inert and non-inflammable solvents are required, liquid SO_2 serves the best medium. Some of the important reactions are:

(*a*) *Sulphonation.* Good yields have been reported for the sulphonation of aromatic compounds in liquid SO_2.

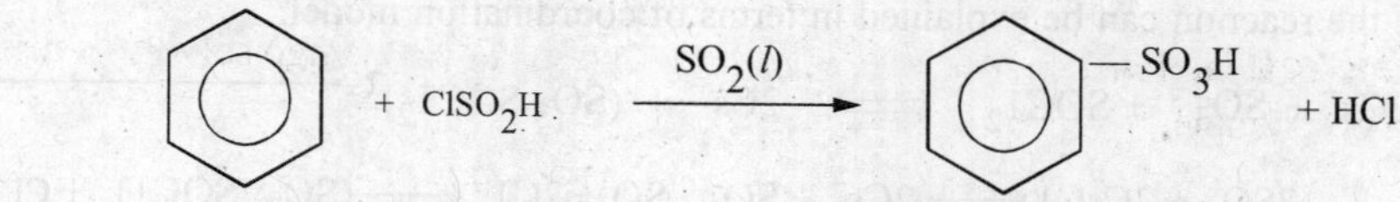

$$C_6H_6 + ClSO_2H \xrightarrow{SO_2(l)} C_6H_5\text{—}SO_3H + HCl$$

Chloro sulphonic acid — Benzene sulphonic acid

(*b*) *Bromination.* Both for addition and substitution reactions with bromine, liquid SO_2 has been used.

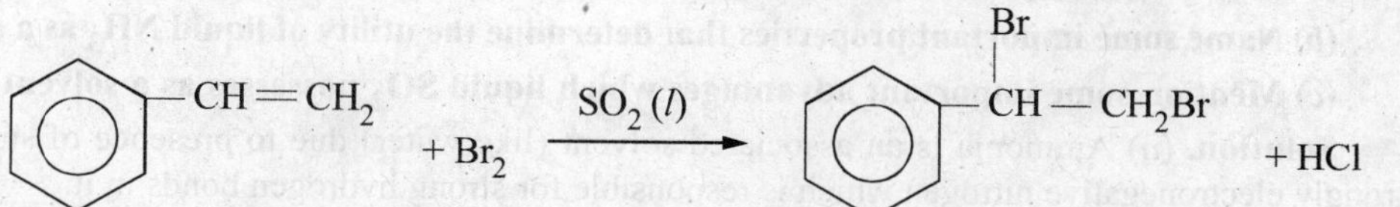

$$C_6H_5\text{—}CH{=}CH_2 + Br_2 \xrightarrow{SO_2\,(l)} C_6H_5\text{—}CH(Br)\text{—}CH_2Br + HCl$$

Styrene

$$C_6H_5OH + Br_2 \xrightarrow{SO_2\,(l)} p\text{-}BrC_6H_4OH + HBr$$

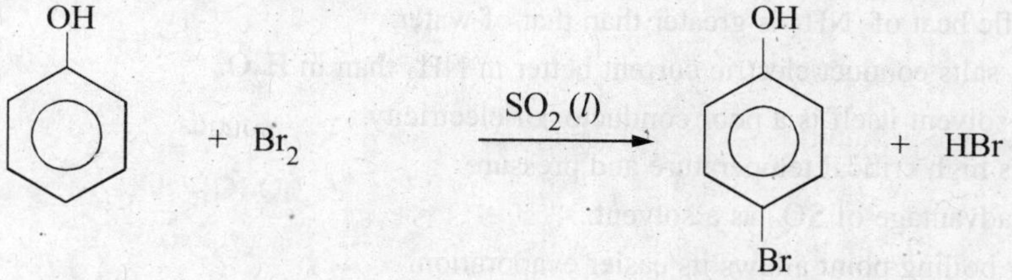

Phenol — *p*-Bromophenol

(*c*) *Friedel-Crafts reaction.* $AlCl_3$ is highly soluble in liquid SO_2 and is, therefore, a preferred solvent over ether which is highly inflammable.

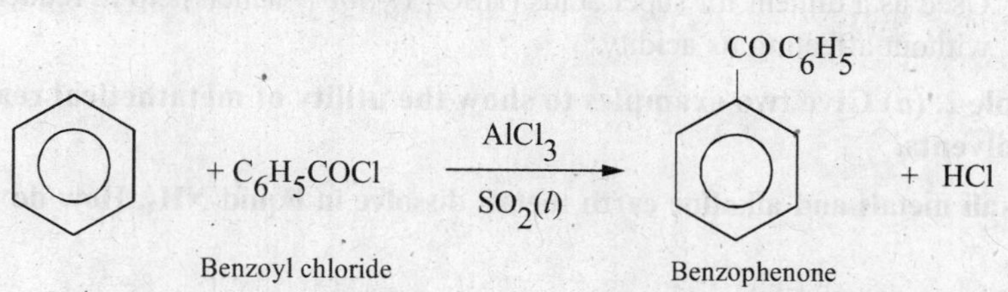

$$C_6H_6 + C_6H_5COCl \xrightarrow[SO_2(l)]{AlCl_3} C_6H_5\text{—}CO\,C_6H_5 + HCl$$

Benzoyl chloride — Benzophenone

8.7.2. Circumstances under which the solvent system model fails

The main difficulty in the application of solvent system arises when there is no proof for the existence of self -ionization species. This concept has been applied even in almost non-conducting solvents like SO_2 where the existence of species SO^{2+} is doubtful. Reaction of $SOC1_2$ with $Cs_2\,SO_3$ in terms of solvent system is explained as:

$$2SO_2 \rightleftharpoons SO^{2+} + SO_3^{2-}$$

$$Cs_2\,SO_3 \rightleftharpoons 2Cs^{+} + SO_3^{2-} \text{ (base)}$$

$$SOCl_2 \rightleftharpoons SO^{2+} + 2Cl \quad \text{(acid)}$$

Neutralisation occurs as:

$$2Cs^+ + SO_3^{2-} + SO^{2+} + 2Cl^- \longrightarrow 2CsCl + 2SO_2$$

However, there is no experimental support for the existence of SO^{2+}. Probably self-ionization is not required and the reaction can be explained in terms of coordination model.

$$2Cs^+ + SO_3^{2-} + SOCl_2 \rightleftharpoons 2Cs^+ + (SO_3\,SOCl_2)^{2-}$$

$$2SO_2 + 2CsCl \longleftarrow 2Cs^+ + SO_3 \cdot SO + 2Cl^- \longleftarrow (SO_3 \cdot SOCl)^- + Cl^- + 2Cs^+$$

SOLVED EXAMPLES

Example 1. (*a*) Explain that liquid NH_3 is an associated liquid.

(*b*) Name some important properties that determine the utility of liquid NH_3 as a solvent.

(*c*) Mention some important advantages which liquid SO_2 possesses as a solvent.

Solution. (*a*) Ammonia is an associated solvent (like water) due to presence of small and strongly electronegative nitrogen which is responsible for strong hydrogen bonds in it.

(*b*) Important properties determining utility of ammonia as a solvent better than water are:

(*i*) Specific heat of NH_3 is greater than that of water.

(*ii*) Most salts conduct electric current better in NH_3 than in H_2O.

(*iii*) The solvent itself is a poor conductor of electricity.

(*iv*) It has high critical temperature and pressure.

(*c*) Special advantage of SO_2 as a solvent.

(*i*) Low boiling point allows its easier evaporation.

(*ii*) Low melting point allows reaction at low temperatures.

(*iii*) Inert solvent.

(*iv*) Liquid SO_2 has unusual solvating power for both covalent and ionic compounds.

(*v*) Used as a diluent for super acids (HSO_3F–SbF_3) which help in reducing its viscosity without affecting its acidity.

Example 2. (*a*) Give two examples to show the utility of metathetical reactions in non-aqueous solvents.

(*b*) Alkali metals and alkaline earth metals dissolve in liquid NH_3. How do these processes differ?

(*c*) Classify the following as acids and bases in suitable non-aqueous solvents:

$(NH_4)_2SO_4$, $NaNH_2$, SOF_2, $Cr_2(SO_3)_3$, $[(CH_3)_3N]_2SO_3$.

Give one reaction in each case to justify your answer.

Solution. (*a*) Water soluble $BaCl_2$ can be precipitated from the solution of silver chloride and $Ba(NO_3)_2$ in liquid ammonia.

$$Ba(NO_3)_2 + 2AgCl \xrightarrow{NH_3} BaCl_2\downarrow + 2AgNO_3$$

Similarly, LiCl can be precipitated out of the reaction of NH_4Cl and $LiNO_3$ in liquid NH_3.

$$NH_4Cl + LiNO_3 \xrightarrow{NH_3} NH_4NO_3 + LiCl\downarrow$$

Metathetical reactions in SO_2 are:

$$SOCl_2 + 2CH_3COOAg \xrightarrow{SO_2} 2AgCl \downarrow + SO(OOCCH_3)_2$$

$$2NH_4SCN + SOCl_2 \xrightarrow{SO_2} 2NH_4Cl \downarrow + SO(SCN)_2$$

(*b*) Alkaline earth metals are soluble in ammonia giving blue solutions but on evaporation of the solvent they are found to be present as ammoniates, $Ca(NH_3)_6$, $Sr(NH_3)_6$, $Ba(NH_3)_6$. On the other hand, alkali metals are obtained as such on evaporation of ammonia.

(*c*) Ammonium salts $(NH_4)_2SO_4$ act as acids and amides as bases in liquid ammonia ($2NH_3 \rightleftharpoons NH_4^+ + NH_2^-$).

$$(NH_4)_2SO_4 + 2NaNH_2 \xrightarrow{NH_3} \underset{\text{Salt}}{Na_2SO_4} + 4NH_3$$

In SO_2 which ionises as $2SO_2 \rightleftharpoons SO^{2+} + SO_3^{2-}$, SOF_2 acts as acid and $Cr_2(SO_3)_2$ or $[(CH_3)_4N]_2SO_3$ as a base.

$$3SOF_2 + Cr_2(SO_3)_3 \xrightarrow{SO_2} 2CrF_3 + 6SO_2$$

$$SOF_2 + [(CH_3)_4N]_2SO_3 \xrightarrow{SO_2} 2(CH_3)_4NF + 2SO_2$$

Example 3. Complete the following reactions:

(*i*) $\mathbf{SO_2 + SO_3 \xrightarrow{liq.SO_2}}$

(*ii*) $\mathbf{SOCl_2 + Cs_2SO_3 \xrightarrow{liq.SO_2}}$

(*iii*) $\mathbf{AgNO_3 + KNH_2 \xrightarrow{liq.\ NH_3}}$

(*iv*) $\mathbf{Na_2O + NH_4Cl \xrightarrow{liq.\ NH_3}}$

(*v*) $\mathbf{NbC1_5 + SO_2 \xrightarrow{liq.SO_2}}$

(*vi*) $\mathbf{CH_3COOAg + SOC1_2 \xrightarrow{liq.SO_2}}$

(*vii*) $\mathbf{AlC1_3 + [(CH_3)_4N]_2SO_3 \xrightarrow{liq.SO_2}}$

(*viii*) $\mathbf{SnI_4 + KNH_2 \xrightarrow{liq.SO_2}}$

(*ix*) $\mathbf{C_5H_5N + SOC1_2 \xrightarrow{liq.SO_2}}$

Solution. (*i*) SO_3 dissolves in SO_2.

(*ii*) $SOCl_2 + Cs_2SO_2 \xrightarrow{SO_2} 2CsCl + 2SO_2$

(*iii*) $AgNO_3 + KNH_2 \xrightarrow{NH_3} KNO_3 + AgNH_2$

(*iv*) $Na_2O + 2NH_4Cl \xrightarrow{NH_3} 2NaCl + 2NH_3 + H_2O$

(*v*) $NbC1_5 + SO_2 \xrightarrow{SO_2} NbOC1_3 + SOCl_2$

(*vi*) $2CH_3COOAg + SOCl_2 \xrightarrow{SO_2} 2AgCl + SO(OOCCH_3)_2$

(*vii*) $2AlCl_3 + 3[(CH_3)_4N]_2SO_3 \xrightarrow{SO_2} 6(CH_3)_4NCl + Al_2(SO_3)_3$

Example 4. (*a*) Account for the following:

(*i*) K_2SO_3 is soluble in SO_2.

(*ii*) SO_2 acts both as a Lewis base and acid.

(*b*) Complete the following reactions:

(*i*) $PbI_2 + KNH_2 \xrightarrow{liq.NH_3}$

(*ii*) $SO_2Cl_2 + NH_3 \xrightarrow{liq.NH_3}$

(*iii*) $C_6H_6 + SO_3 \xrightarrow{liq.SO_2}$

(*iv*) $NH_4Cl + NaNH_2 \xrightarrow{liq.NH_3}$

(*v*) $Na + NH_4Cl \xrightarrow{liq.NH_3}$

Solution. (*a*) (*i*) It dissolves because of solvate formation.

(*ii*) SO_2 can behave both as Lewis acid and Lewis base. It acts as a donor (Lewis base) in the following solvates: $CaI_2.4SO_3$,$LiI.4SO_2$: where it donates electrons through an oxygen atom in SO_2. It acts as an acceptor (Lewis acid) in the following: $C_5H_5N \cdot SO_2$, $C_6H_7N \cdot SO_2$ where it accepts electrons through sulphur atom in the molecule SO_2.

(*b*) (*i*) $PbI_2 + KNH_2 \xrightarrow{NH_3} PbNH + HI + KI$

(*ii*) $SO_2Cl_2 + NH_3 \xrightarrow{NH_3} SO_2(NH_2)_2$

(*iii*) $C_6H_6 + SO_3 \xrightarrow{SO_2} C_6H_5SO_3H$

(*iv*) $NH_4Cl + NaNH_2 \xrightarrow{NH_3} NaCl + 2NH_3$

(*v*) $NH_4Cl + Na \xrightarrow{NH_3} NaCl + NH_3 + 1/2\ H_2$

Example 5. (*a*) The electrical conductivity of liquid ammonia is increased when ammonium chloride is dissolved in it.

(*b*) A liquid ammonia solution of phenolphthalein is colourless but becomes red coloured on addition of KNH_2.

(*c*) A number of metals dissolve in liquid ammonia to give strongly reducing solutions but react with water.

Solution (*a*) Liquid ammonia ionises as follows:

$$2NH_3 \rightleftharpoons NH_4^+ + NH_2^-$$

Substances yielding ammonium ions in liquid ammonia will act as ammono acids. Addition of NH_4Cl increases NH_4^+ concentration in liquid ammonia with the result that the solution shows increased conductance.

(*b*) Addition of KNH_2 to ammonia increases amide ion NH_2^- concentration and behaves as ammono base. As soon as concentration of NH_2^- ions increases in solution, the solution behaves as a base and imparts red colour to phenolphthalein just as an aqueous basic solution imparts red (pink) colour.

(*c*) Metals with low ionization energy and higher solvation energy, that is, with higher electrode potential, dissolve in liquid ammonia (without chemical reaction) to give blue solutions which show strong reducing properties and high electrical conductance. Ammoniated electron is responsible for strong reducing action and high conductance.

$$Na \rightleftharpoons Na^+ + e^-$$

$$Na^+ + (NH_3)_x \rightleftharpoons [Na\,(NH_3)_x\,]^+$$

$$e^- + x\,(NH_3) \rightleftharpoons [e\,(NH_3)_x]^-$$

Such metals react with water because water is much stronger oxidising agent than ammonia.

$$Na + H_2O \longrightarrow NaOH + 1/2\,H_2$$

Example 6. (*a*) Suggest reasons for reverse order of solubility for alkali halides in liquid SO_2 and liquid NH_3.

(*b*) With the help of examples, point out the difference between solvation and solvolysis reactions.

(*c*) Liquid NH_3 is a poor electrical conductor, yet salt solutions in this solvent conduct electricity better than their aqueous solution.

Solution. (*a*) Alkali halides are soluble in liquid ammonia due to complexation. For example, NaI dissolves in ammonia due to the formation of complex $NaI.4NH_3$. However, alkali halides are not soluble in liquid SO_2 due to the inability of SO_2 to form complexes with alkali metal halides.

(*b*) Solvation involves coordination of the solvent molecules to the metal ion. Solvation in water is usually stated as hydration. Take the case of $MgI_2.6NH_3$. Here ammonia molecules are bound to MgI_2. This complex is called a solvate or an ammoniate. Similarly $MgC1_2 . 6H_2O$ is called a solvate or a hydrate. Therefore, solvation refers to simple coordination on complexation. On the other hand, solvalytic reaction in liquid ammonia means attachment of ammonium NH_4^+ or amide NH_2^- ions to the reactant. This is analogous to hydrolysis in water. The following are solvolytic reactions in NH_3.

$$SiCl_4 \xrightarrow{+NH_3} Si\,(NH_2)_4$$

$$SO_2\,Cl_2 \xrightarrow{+NH_3} SO_2\,(NH_2)_2$$

(*c*)Most salts conduct electric current better when dissolved in liquid NH_3 than when dissolved in H_2O. This is not due to higher ionizing power of liquid NH_3. (It is actually 1/3 of H_2O) but due to its low viscosity which greatly increases the mobility of ions.

Example 7. Liquid ammomic exhibits association whereas liquid SO_2 does not. Explain.

Solution. Liquid NH_3 associates itself like water through hydrogen bonding. Nitrogen is sufficiently electronegative to cause hydrogen bonding. This passbility does not exist with sulphur. Hence SO_2 does not associate.

Example 8. Acetamide behaves as a weak base in aqueous solution but shows acidic properties in liquid NH_3. Why?

Solution. In aqueous solution, acetamide gives OH^- ions. Therefore it behaves as basic:

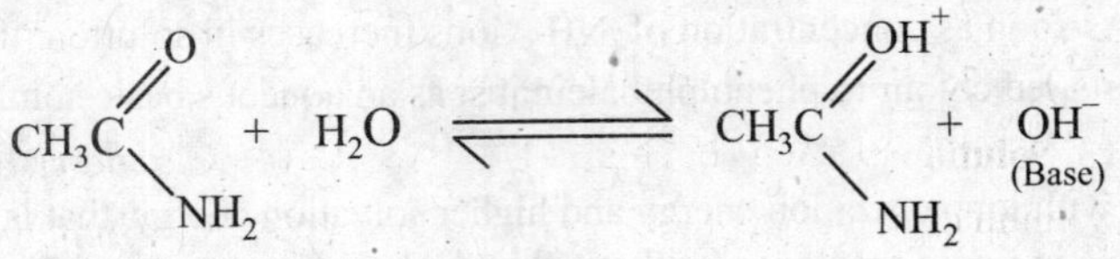

$$CH_3C(=O)NH_2 + H_2O \rightleftharpoons CH_3C(=OH^+)NH_2 + \underset{\text{(Base)}}{OH^-}$$

In liquid NH_3, it gives NH_4^+, which is acidic.

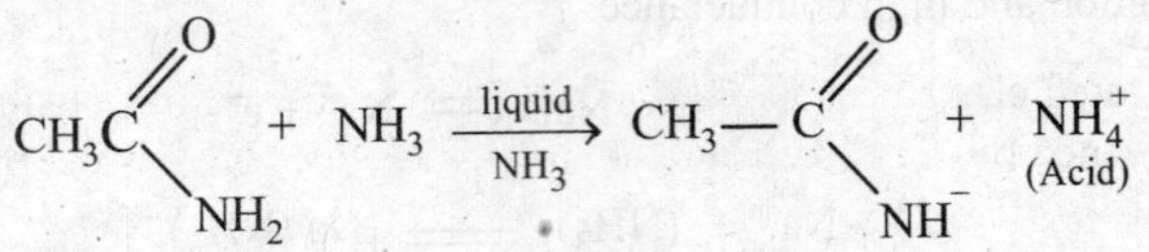

$$CH_3C(=O)NH_2 + NH_3 \xrightarrow[NH_3]{\text{liquid}} CH_3—C(=O)NH^- + \underset{\text{(Acid)}}{NH_4^+}$$

Example 9. Complete the following reactions:

(*i*) $PbI_2 + KNH_2 \xrightarrow{\text{liq.}NH_3}$

(*ii*) $SiCl_4 + NH_3 \xrightarrow{\text{liq.}NH_3}$

(*iii*) $Zn(NO_3)_2 + KNH_2 \xrightarrow{\text{liq.}NH_3}$

Solution. (i) $PbI_2 + KNH_2 \xrightarrow{\text{liq. }NH_3} PbNH \downarrow + HI + KI$

(*ii*) $SiCl_4 + NH_3 \xrightarrow{\text{liq. }NH_3} Si(NH_2)_4 \xrightarrow{0°C} Si(NH)(NH_2)_2 \xrightarrow{1200°C} Si_3N_4$

(*iii*) $Zn(NO_3)_2 + 2KNH_2 \xrightarrow{\text{liq. }NH_3} Zn(NH_2)_2 \downarrow + 2KNO_3$

Example 10. Acount for solubility of AgI in liquid NH_3.

Solution. AgI dissolves in liquid NH_3 due to the formation of a soluble complex, $[Ag(NH_3)_2]$ I.

Example 11. In liquid NH_3, what are the species which are characateristic of acids and bases?

Solution. NH_3 ionises as:

$$2NH_3 \rightleftharpoons \underset{\text{(Acidic)}}{NH_4^-} + \underset{\text{(Basic)}}{NH_2^-}$$

Acidic species : NH_4^-

Basic species : NH_2^-

Example 12. Complete the following reactions in liquid SO_2:

(*i*) $HgI_2 + KI \longrightarrow$

(*ii*) $SbCl_5 + NOCl \longrightarrow$

(*iii*) $Cr_2(SO_3)_3 + SOCl_2 \longrightarrow$

Solution. (i) $HgI_2 + 2KI \xrightarrow{\text{liq. }SO_2} \underset{\text{(Complex)}}{K_2[HgI_4]}$

Solubility of HgI_2 increases due to the formation of complex compound $K_2[HgI_4]$.

(ii) $SbCl_5 + NOCl \xrightarrow{\text{liq. } SO_2} NO[SbCl_6]$

(iii) $Cr_2(SO_3)_3 + 3SOCl_2 \xrightarrow{\text{liq. } SO_2} 2CrCl_3 + 6SO_2$

Example 13. Solutions of sodium in liquid NH_3 are blue and good reducing agents. Explain.

Solution. Sodium dissolves in liquid NH_3 and ionises to give sodium ion and valence electron which get solvated by ammonia molecules.

$$Na \xrightarrow[NH_3]{\text{liquid}} [Na(NH_3)_x]^+ + [e(NH_3)_y]^-$$

The ammoniated electrons absorb at about 5000 Å and are responsible for blue colour of the solution. The presence of radily available ammoniated electrons also explains its strong reducing character.

Example 14. The blue colour of alkali metals in liquid ammonia is due to .... Complete the statement.

Solution. Ammoniated electrons.

Example 15. Predict the behaviour of following solutes in liquid SO_2:

(*i*) Cs_2SO_3 **(*ii*) $(CH_3)_4NBr$** **(*iii*) $SOCl_2$**

Solution. (*i*) basic (*ii*) salt (*iii*) acidic

Example 16. Give two properties showoing ammonia to be a better solvent than water.

Solution. (*i*) Liquid ammonia is less viscous than water. Therefore, salts when dissolved in liquid ammonia conduct electric current better than in water.

(*ii*) Liquid ammonia is less associated than water. Therefore it is a better solvent for organic compounds.

Example 17. Complete the reactions:

(*i*) $HgI_2 + KI \xrightarrow{\text{liq. } SO_2}$

(*ii*) $2KI + SOCl_2 \xrightarrow{\text{liq. } SO_2}$

Solution. (*i*) $HgI_2 + 2KI \xrightarrow{\text{liq. } SO_2} K_2HgI_4$

(*ii*) $2KI + SOCl_2 \longrightarrow 2KCl + SOI_2$

Example 18. Give the solvolytic reaction of ammonium acetate in liquid SO_2.

Solution. $2CH_3\overset{\overset{O}{\|}}{C}ONH_4 + 2SO_2 \longrightarrow (NH_4)_2SO_3 + (CH_3\overset{\overset{O}{\|}}{C}O)_2SO$

$$(CH_3\overset{\overset{O}{\|}}{C}O_2)SO \rightleftharpoons SO_2 + (CH_3\overset{\overset{O}{\|}}{C}-)_2O$$

Example 19. Explain : metal-liquid NH_3 solutions are good reducing agents.

Solution. Metal-liquid NH_3 are good reducing agents due to the presence of ammoniated electrons

$$Na \xrightarrow[NH_3]{\text{liquid}} [Na(NH_3)_x]^+ + [e(NH_3)_y]^-$$

Example 20. Give autoionization of SO_2.

Solution. $2SO_2 \rightleftharpoons SO^{2+} + SO_3$

Example 21. In liquid ammonia, what are the species characteristic of acids and bases?

Solution. $2NH_3 \rightleftharpoons NH_4^+ + NH_2^-$

NH_4^+ are acidic species and NH_2^- are basic species.

Example 22. Complete the following:

(i) $PbI_2 + KNH_2 \xrightarrow{\text{liq. } NH_3}$

(ii) $SO_2Cl_2 + NH_3 \xrightarrow{\text{liq. } NH_3}$

(iii) $AlCl_3 + [(CH_3)_4N]_2\ SO_3 \xrightarrow{\text{liq. } SO_2}$

(iv) $C_6H_6 + SO_3 \xrightarrow{\text{liq. } SO_2}$

(v) $SnI_4 + KNH_2 \xrightarrow{\text{liq. } NH_3}$

Solution. (*i*) $PbI_2 + KNH_2 \xrightarrow{\text{liq. } NH_3} \underset{\text{Lead amide}}{PbNH} + KI + HI$

(*ii*) $SO_2Cl_2 + 4NH_3 \xrightarrow{\text{liq. } NH_3} \underset{\text{Sulphamide}}{SO_2(NH_2)_2} + 2NH_4Cl$

(*iii*) $2AlCl_3 + 3[(CH_3)_4N]_2SO_3 \xrightarrow{\text{liq. } SO_2} Al_2SO_3 + 6[(CH_3)_4N]\ Cl$

$Al_2(SO_3)_3 + 3[(CH_3)_4N]\ SO_3 \xrightarrow{\text{liq. } SO_2} \underset{\text{Soluble complex}}{2[(CH_3)_4\ N]_3\ [Al(SO)_3]}$

(*iv*) $C_6H_6 + SO_3 \xrightarrow{\text{liq. } SO_2} C_6H_5SO_3H$

(*v*) $SnI_4 + 4KNH_2 \xrightarrow{\text{liq. } NH_3} 4KI + Sn\ (NH_2)_4$

Example 23. Complete the following reactions:

(i) $NbCl_5 \xrightarrow{\text{liq. } SO_2}$

(ii) $Ag(CH_3COO) + SOCl_2 \xrightarrow{\text{liq. } SO_2}$

(iii) $AlCl_3 + [(CH_3)_4N]_2\ SO_3 \xrightarrow{\text{liq. } SO_2}$

(iv) $SnI_4 + KNH_2(NH_3) \xrightarrow{\text{liq. } SO_2}$

Solution. (*i*) $NbCl_5 \xrightarrow{\text{liq. } SO_2} NbOCl_3 + SOCl_2$

(*ii*) $2Ag\ (CH_3COO) + SOCl_2 \xrightarrow{\text{liq. } SO_2} 2AgCl + \underset{\text{Thionyl acetate}}{SO(OOCH_3)_2}$

(*iii*) $AlCl_3 + [(CH_3)_4\ N]_2SO_3 \xrightarrow{\text{liq. } SO_2} Al_2(SO_3)_2 + 6[(CH_3)_4N]^+\ Cl^-$

(*iv*) $SnI_4 + 4KNH_2\ (NH_3) \xrightarrow{\text{liq. } SO_2} Sn(NH_2)_4 + 4KI$

Example 24. Complete the following reactions:

(*i*) $SOBr_2 + [(CH_3)_4N]\ SO_3 \xrightarrow{\text{liq. } SO_2}$

(*ii*) $C_6H_6 + SO_3 \xrightarrow{\text{liq. } SO_2}$

(*iii*) $PbI_2 + KNH_2 \xrightarrow{\text{liq. } NH_3}$

Solution. (*i*) $SOBr_2 + [(CH_3)_4N]_2\ SO_3 \xrightarrow{\text{liq. } SO_2} 2[N(CH_3)_4]\ Br + 2SO_2$

(*ii*) $C_6H_6 + SO_3^- \xrightarrow{\text{liq. } SO_2} C_6H_6SO_3H$

(*iii*) $PbI_2 + KNH_2 \xrightarrow{\text{liq. } NH_3} \underset{\text{Lead imide}}{PbNH} + KI + HI$

Example 25. Complete the following reactions:

(*i*) $(NH_2)_2CO + NH_3 \xrightarrow{\text{liq. } NH_3}$

(*ii*) $SOCl_2 + Cs_2SO_3 \xrightarrow{\text{liq. } SO_2}$

(*iii*) $NbCl_5 \xrightarrow{\text{liq. } SO_2}$

Solution. (*i*) $(NH_2)_2C{=}O + NH_3 \xrightarrow{\text{liq. } NH_3} (NH^-)(NH_2)C{=}O + NH_4^+$

(*ii*) $SOCl_2 + Cs_2SO_3 \xrightarrow{\text{liq. } SO_2} 2CsCl + 2SO_2$

(*iii*) $NbCl_5 \xrightarrow{\text{liq. } SO_2} NbOCl_3 + SOCl_2$

Example 26. Complete the following reactions:

(*i*) $PbI_2 + KNH_2 \xrightarrow{\text{liq. } NH_3}$

(*ii*) $(CH_3)(NH_2)C{=}O + NH_3 \longrightarrow$

(*iii*) $HgI_2 + 2KI \xrightarrow{\text{liq. } SO_2}$

(*iv*) $Cr_2(SO_3)_3 + SOCl_2 \xrightarrow{\text{liq. } SO_2}$

Solution. (*i*) $PbI_2 + KNH_2 \xrightarrow{\text{liq. } NH_3} \underset{\text{Lead imide}}{PbNH} \downarrow + KI + HI$

(*ii*) $\begin{matrix} CH_3 \\ NH_2 \end{matrix} \rangle C=O + NH_3 \longrightarrow CH_3-C\begin{matrix} \nearrow O \\ \searrow NH^- \end{matrix} + NH_4^+$

(*iii*) $HgI_2 + 2KI \xrightarrow{\text{liq. } SO_2} K_2HgI_4$

(*iv*) $Cr_2(SO_3)_3 + 3SOCl_2 \xrightarrow{\text{liq. } SO_2} 2CrCl_3 + 6SO_2$

EXERCISES
(Based on Question Papers of Different Universities)

Multiple Choice Questions

Choose the correct option

1. The reactions in which the molecules of the solvent get attached to the solute species are called:

(*a*) solvation reactions (*b*) solvolytic reactions
(*c*) metathetical reactions (*d*) redox reactions.

2. AgCl is soluble in ammonia due to the formation of:

(*a*) $AgNH_2$ (*b*) $AgCl.NH_3$
(*c*) $[Ag(NH_3)_2]Cl$ (*d*) $NH_4[AgCl_2]$

3. Alkali metals dissolve in liquid ammonia to form blue coloured solutions. The blue colour is due to

(*a*) alkali metal (*b*) alkali metal ion
(*c*) ammoniated electron (*d*) ammoniated alkali metal ion

4. Trouton constant is defined as:

(*a*) $\Delta H_{evap}/T_b$ (*b*) $\Delta H_{evap}/H_{fusion}$
(*c*) $\Delta H_{evap}.T_b$ (*d*) T_b/T_f

5. Which of the following show amphoteric behaviour in liquid ammonia?

(*a*) $Zn(NH_2)_2$ (*b*) NH_4Cl
(*c*) KNH_2 (*d*) $(NH_2)_2SO$

6. Which of the following is not a protonic solvent?

(*a*) HF (*b*) H_2O
(*c*) $CHCl_3$ (*d*) H_2SO_4

7. The reaction:

$NH_2CONH_2 + NH_3 \longrightarrow {}^-NHCONH_2 + NH_4^+$ is called

(*a*) protolysis reaction (*b*) solvation reaction
(*c*) ammonolysis reaction (*d*) metathetical reaction

8. Which of the following compounds does not act as acid in liquid ammonia?

(*a*) $(NH_2)_2SO$ (*b*) $(NH_2)_2CO$
(*c*) NH_4Cl (*d*) KNH_2

9. Which of the following is not true for alkali metal-ammonia solutions?

(*a*) These are blue in colour and the colour is independent of the metal dissolved.

(*b*) These are paramagnetic and paramagnetic character increases with increasing concentration.

(*c*) The solutions are conducting.

(*d*) They show reducing properties.

10. In the reaction:

$SOCl_2 + NH_3 \longrightarrow$

The main product is

(*a*) SO_2 (*b*) NH_2SO_2Cl

(*c*) $NH_2SO_2NH_2$ (*d*) $(NH_4)_2\ SO_3$

ANSWERS

1. (*a*)	**2.** (*c*)	**3.** (*c*)	**4.** (*a*)	**5.** (*a*)	**6.** (*c*)
7. (*a*)	**8.** (*d*)	**9.** (*b*)	**10.** (*c*)		

Short Answer Questions

1. Complete the reaction:

$SO_2Cl_2 + NH_3 \xrightarrow{\text{liq. } NH_3}$

2. Give solvolytic reaction of PCl_5 in liquid SO_2.

3. What are ionising and non-ionising solvents? Give one example of each.

4. K_2SO_3 is soluble in liquid SO_2. Explain.

5. Show amphoteric behaviour of zinc amide in liquid ammonia.

6. Account for the solubility of AgI in liquid NH_3.

7. Why does sulphamic acid behave as dibasic in liquid ammonia?

8. Why are solutions of alkali metals in liquid ammonia blue in colour?

9. What is the principal effect of dielectic constant on the solubility of an ionic compound?

10. Why solvents with high dipole moment dissolve polar substances easily?

General Questions

1. (*i*) What are non-aqueous solvents?

(*ii*) What do you understand by ionising and non-ionising solvents? Give examples.

2. Describe the following reactions in liquid ammonia giving suitable examples:

(*i*) Acid-base reactions (*ii*) Precipitation reactions

(*iii*) Complex formation reactions (*iv*) Ammonolysis

3. Discuss examples of solvolytic reactions in:

(*i*) liquid NH_3 (*ii*) liquid SO_2

4. (*a*) List some important advantages of liquid SO_2 as a non-aqueous solvent inspite of its toxic nature.

(*b*) Which of the following act as an acid or a base in liquid SO_2 and why?

(*i*) Thionyl bromide (*ii*) Cesium sulphite

(*iii*) Trimethylammonium sulphite (*iv*) $SO(SCN)_2$.

5. Discuss the classification of solvents as :

(*i*) Protonic and non-protonic solvents.

(*ii*) Acidic, basic or amphoteric solvents.

(*iii*) Ionising and non-ionising solvents.

6. (*a*) Give equations for the auto-ionization of liquid ammonia.

(*b*) Discuss acid-base reactions in liquid ammonia.

7. (a) Solutions of sodium in liquid ammonia are blue coloured and paramagnetic. The magnetic character decreases with increasing metal concentration. Explain.

(*b*) Complete and balance the following equations:

(*i*) $C_6H_6 + SO_3 \xrightarrow{\text{liq. } SO_2}$

(*ii*) $SO(SCN)_2 + K_2SO_3 \xrightarrow{\text{liq. } SO_2}$

(*iii*) $SiCl_4 + NH_3 \xrightarrow{\text{liq. } NH_3}$

8. How do the following physical properties characterize a non-aqueous solvent?

(*i*) Dielectric constant. (*ii*) Heat of fusion and heat of vaporisation.

(*ii*) Melting point and boiling point.

9. Discuss the chemistry of liquid SO_2 to act as a non-aqueous solvent.

10. (*i*) What is the principal effect of dielectric constant?

(*ii*) Discuss the acid-base behaviour in liquid NH_3 in terms of solvent system concept.

11. (*i*) Discuss the chemistry of liquid ammonia as a solvent.

(*ii*) What are the advantages and disadvantages of liquid ammonia as a solvent?

12. Complete the following equations (balance if necessary):

(*a*) $HgI_2 + KNH_2 \xrightarrow{NH_3}$

(*b*) $SbCl_5 + NOCl \xrightarrow{SO_2}$

(*c*) $NbCl_5 + SO_2 \xrightarrow{SO_2}$

(*d*) $AlCl_3 + [(CH_3)_4N]_2SO_3 \text{ (excess)} \xrightarrow{SO_2}$

(*e*) $SO_2Cl_2 \xrightarrow{NH_3}$

13. Explain Why:

(*a*) Sulphite salts behave as weak acids in aqueous solvents but exhibit strong acidic properties in liquid ammonia.

(*b*) Liquid NH_3 behaves as a differentiating solvent for NaH_2.

(*c*) Metal ammonia solutions are good electrolytic conductors.

(*d*) SO_2 is a better solvent for organic compounds.

(*e*) In SO_2, there is no appreciable association as exhibited by liquid ammonia.

14. (*a*) Define an aprotic and a coordinating solvent. Give one example in each case. Describe their mode of auto ionization. Describe a reaction to show amphoterism of the aprotic solvent.

(*b*) Define the following:

(*i*) Ammonolysis of $SiCl_4$ at variable temperatures.

(*ii*) Redox reactions in liquid NH_3.

(*c*) Account for the following:

Cesium sulphate and thionyl chloride undergo neutralisation reaction in liquid SO_2.

15. (*a*) Liquid NH_3 exhibits association whereas liquid SO_2 does not. Explain.

(*b*) Complete the following reactions in liquid SO_2:

(*i*) $HgI_2 + KI \longrightarrow$

(*ii*) $SbCl_5 + NaCl \longrightarrow$

(*iii*) $Cs_2SO_3 + SOCl_2 \longrightarrow$

(*c*) Account for the following:

(*i*) AgI is highly soluble in liquid NH_3.

(*ii*) Thionyl bromide behaves as an acid in liquid SO_2.

(*d*) Describe detailed accounts of coordination model for reactions in non-aqueous solvents. How does it differ from solvent system concept?

16. (*a*) Explain the following briefly:

(*i*) K_2SO_3 is soluble in liquid SO_2

(*ii*) Solvolytic reactions in liquid NH_3.

(*b*) Describe the role of the following physical properties on the role of a solvent in chemical reactions:

(*i*) Dielectric constant.

(*ii*) Heat of fusion and heat of vaporisation.

17. Complete the following reactions:

(*i*) $NbCl_5 \xrightarrow{\text{liq. } SO_2}$

(*ii*) $Ag\,(CH_3COO) + SOCl_2 \xrightarrow{\text{liq. } SO_2}$

(*iii*) $AlCl_3 + [N(NH_3)_4]_2\,SO_2 \xrightarrow{\text{liq. } SO_2}$

(*iv*) $SnI_4 + KNH_2\ (\text{liq. } NH_3) \xrightarrow{\text{liq. } SO_2}$

18. (*a*) Discuss acid base reactions in liquid ammonia.

(*b*) What are the advantages and disadvantages of liquid ammonia?

(*c*) When alkali metal is dissolved in liquid ammonia, conducting blue coloured solution is obtained? Explain chemistry of it.

(*d*) Why is ammonia a better solvent than water?

19. (*a*) Explain the structure of SO_2.

(*b*) Discuss with examples advantages and disadvantages of liquid NH_3 as solvent.

(*c*) What are non-aqueous solvents? Explain classification of solvents.

(*d*) Explain acid-base reactions in liquid NH_3.

20. (*a*) Discuss acid-base reactions in sulphur dioxide.

(*b*) How do the following reactions proceed:

$HgI_2 + KNH_2 \xrightarrow{NH_3}$

$Cs_2SO_3 + SOCl_2 \xrightarrow{SO_2}$

21. (*a*) Name some important properties that determine the utility of liquid NH_3 as a solvent.

(*b*) Complete the following reactions:

(*i*) $PbI_2 + KNH_2 \xrightarrow{\text{liq. } NH_3}$

(*ii*) $C_6H_6 + SO_2 \xrightarrow{\text{liq. } SO_2}$

(*iii*) $NH_4Cl + NaNH_2 \xrightarrow{\text{liq. } NH_3}$

(*c*) Discuss solvolytic reactions in

(*i*) liquid NH_3 (*ii*) liquid SO_2

22. (*a*) Describe the role of the following physical properties on the role of the solvent in chemical reactions:

(*i*) Melting and boiling point

(*ii*) Heat of fusion and heat of vaporisation

(*b*) Describe the following processes in liquid amonia with suitable examples:

(*i*) Precipitation reaction

(*ii*) Ammonolysis

(*c*) Complete the following reactions:

(*i*) $KCl + HF \xrightarrow{\text{Liquid HF}}$

(*ii*) $PBr_5 + SO_2 \xrightarrow{\text{Liquid } SO_2}$

(*iii*) $H_2O + H_2SO_4 \xrightarrow{100\%\ H_2SO_4}$

23. (*a*) Account for the following:

(*i*) K_2SO_3 is soluble liquid SO_2.

(*ii*) SO_2 acts both as a Lewis base and acid.

(*b*) Complete the following reactions:

(*i*) $PbI_2 + KNH_2 \xrightarrow{\text{liq. } NH_3}$

(*ii*) $SO_2Cl_2 + NH_2 \xrightarrow{\text{liq. } NH_3}$

(*iii*) $C_6H_6 + SO_3 \xrightarrow{\text{liq. } SO_2}$

(*iv*) $NH_4Cl + NaNH_2 \xrightarrow{\text{liq. } NH_3}$

(*v*) $Na + NH_4Cl \xrightarrow{\text{liq. } NH_3}$

(*c*) Explain

(*i*) Liquid NH_3 is better solvent than H_2O.

(*ii*) Liquid SO_2 possesses special advantages over other solvents.

24. (*a*) Complete the following reactions:

(*i*) $PbI_2 + KNH_2 \xrightarrow{NH_3(l)}$

(*ii*) $AlCl_3 + [(CH_3)_4N]_2 SO_3 \xrightarrow{SO_2(l)}$

(*iii*) $C_6H_6 + SO_3 \xrightarrow{\text{liq. } SO_2}$

(*b*) Account for the following:

(*i*) SO_2 can act both as Lewis acid and base.

(*ii*) In aqueous solution, alkali metals liberate H_2 gas but are stable in NH_3 solution.

(*iii*) The electrical conductivity of liquid NH_3 increases by adding NH_4Cl.

25. Explain the following:

(*a*) The electrical conductance of ammonia increases on addition of NH_4 Cl in it.

(*b*) The solubility of I_2 in SO_2 increases by addition of KI in it.

(*c*) AgCl is soluble in liquid ammonia.

26. (*a*) Define aprotic and coordinating solvents. Give one example of each.

(*b*) Alkali metal ammonia solutions are good electrolytic conductors. Explain.

(*c*) K_2SO_3 is soluble in liquid SO_2. Explain.

(*d*) Complete the following

(*i*) $PbI_2 + KNH_2 \xrightarrow{\text{liq. } NH_3}$

(*ii*) $(CH_3)(NH_2)C{=}O + NH_3 \longrightarrow$

(*iii*) $HgI_2 + KI \xrightarrow{\text{liq. } SO_2}$

(*iv*) $Cr_2(SO_4)_3 + SOCl_2 \xrightarrow{\text{liq. } SO_2}$

27. (*a*) Compare the properties of liquid ammonia and water.

(*b*) Discuss acid base reactions in liquid SO_2.

(*c*) Acetic acid acts as strong acid in liquid ammonia. Explain.

28. Discuss with suitable examples the acid-base reactions and solvolysis reactions in liquid ammonia as nonaqueous solvent.

29. (*a*) Discuss solvolytic reactions in liquid ammonia as solvent.

(*b*) Strong oxidising agents apparently do not exist in liquid ammonia. Give reasons underlying this statement.

(*c*) Alkali metals react with water but dissolve without chemical action in liquid ammonia. Give reasons for this difference in behaviour.

(*d*) Discuss liquid SO_2 as a solvent and what are the most effective bases of liquid SO_2. Give their reactions in this solvent.

30. (*a*) Discuss the following reactions of SO_2:

(*i*) Auto-ionisation

(*ii*) Precipitation reaction

31. Discuss the importance of four physical properties which are crucial in determining the usefulness of any solvent in chemical process.

32. (*a*) What are the species characteristics of acids and bases in liquid ammonia?

(*b*) Compare the properties of non-aqueous solvent liquid ammonia with water.

(*c*) Complete the following reactions:

(*i*) $(NH_2)_2CO + NH_3 \xrightarrow{\text{liq. } NH_3}$

(*ii*) $SOCl_2 + Cs_2SO_3 \xrightarrow{\text{liq. } SO_2}$

33. (*a*) Liquid SO_2 is a better solvent for organic reactions. Why?

(*b*) Complete the following reaction:

(*i*) $SOBr_2 + [(CH_3)_4N]\,SO_3 \xrightarrow{\text{liq. } SO_2}$

(*ii*) $CH_6 + SO_3 \xrightarrow{\text{liq. } SO_2}$

(*iii*) $PI_2 + KNH_2 \xrightarrow{\text{liq. } NH_3}$

34. (*a*) Define a protonic and coordinating solvent. Give one example in each case.

(*b*) Solutions of alkali metal in liquid NH_3 are reducing in nature. Why?

(*c*) Complete the following:

(*i*) $PbI_2 + KNH_2 \xrightarrow{\text{Liq. } NH_3}$

(*ii*) $HgI_2 + KI \xrightarrow{\text{Liq. } NH_3}$

35. (*a*) How: (*i*) heat of fusion, and (*ii*) dipole moment, explain the solubilities of substances in solvents?

(*b*) Describe: (*i*) Complex Formation and (*ii*) Ammonolysis reactions in liquid ammonia giving two examples in each case.

(*c*) Complete the following reactions in lig. SO_2.

(*i*) $HgI_2 + KI \longrightarrow$

(*ii*) $SbCl_5 + NOCl \longrightarrow$

(*iii*) $Cs_2SO_3 + SOCl_2 \longrightarrow$

36. (*a*) Give some important properties that determine the utility of liquid NH_3 as a solvent.

(*b*) Complete the following reactions:

(*i*) $PbI_2 + KNH_2 \xrightarrow{\text{liq. } NH_3}$

(*ii*) $SO_2Cl_2 + 4NH_3 \xrightarrow{\text{liq. } NH_3}$

(*iii*) $CdI_2 + RbI \xrightarrow{\text{liq. } SO_2}$

(*iv*) $Cs_2SO_3 + RbI \xrightarrow{\text{liq. } NH_3}$

(*v*) C_6H_5OH (Phenol) $+ Br_2 \xrightarrow{\text{liq. } SO_2}$

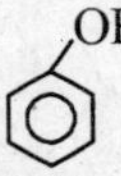

37. (*a*) Complete the following reactions:

(*i*) $HgI_2 + KNH_2 \xrightarrow{NH_3}$

(*ii*) $SbCl_5 + NOCl \xrightarrow{SO_2}$

(*iii*) $SO_2Cl_2 + 4NH_3 \longrightarrow$

38. (*a*) Explain the acid-base reactions in liquid SO_2 as solvent.

(*b*) Complete the following reaction:

(*i*) $BCl_3 \xrightarrow{\text{Liq. } NH_3}$

(*ii*) $PCl_3 \xrightarrow{\text{Liq. } NH_3}$

(*iii*) $SbCl_5 + NaCl \xrightarrow{\text{Liq.} SO_2}$

(*iv*) $C_6H_6 + C_6H_5COCl \xrightarrow{\text{Liq.} SO_2}$

(*c*) Alkali metal solution in liquid ammonia is blue in colour, paramagnetic in character and conduct electricity. Explain.

39. (*i*) Discuss the self auto-ionization of the liquid SO_2 as non-aqueous solvent. What are the most effective acids and bases of this solvent system? Give reactions.

(*ii*) Complete the following reactions:

(*i*) $HgI_2 + KNH_2 \xrightarrow{\text{liq. } NH_3}$

(*ii*) $CdI_2 + KI \xrightarrow{\text{liq. } SO_2}$

(*iii*) $C_6H_6 + t - C_5H_{11}Cl \xrightarrow{\text{liq. } SO_2}$

ORGANIC CHEMISTRY

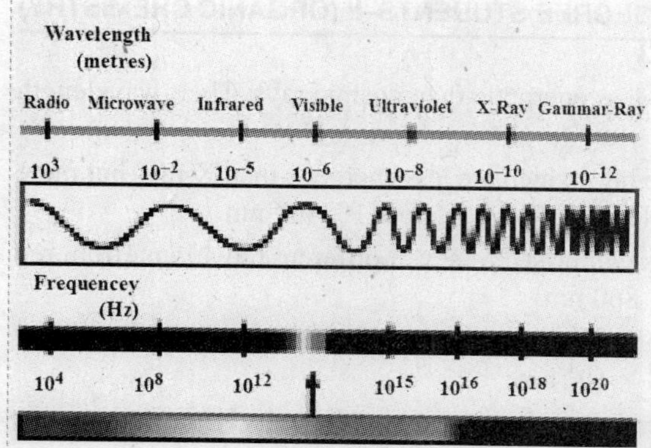

1 Electromagnetic Spectrum: Absorption Spectra

1.1. INTRODUCTION

We always need to characterise an organic compound whether it has been synthesised in the laboratory or it has been extracted from a natural source. The sequence of characterisation is as follows: First of all, the elements present in the compound are detected by usual analytic methods. Then we find out the molecular mass and molecular formula of the compound. This is followed by the determination of functional groups in the compound by usual chemical tests. Melting point, boiling point and formation of derivatives are carried out to arrive at the structure of the compound. During the last few decades, we have been taking recourse to spectroscopic methods for the determination of structure in a more clean and convenient manner. Spectroscopic methods give almost foolproof insight into the detailed structure and stereochemistry of the compound. Spectroscopic methods of analysis include ultraviolet (UV), infrared (IR), nuclear magnetic resonance (NMR) and mass spectroscopies. In this book, we shall restrict ourselves to the first two spectroscopies *viz*. UV spectroscopy and IR spectroscopy.

1.2. ELECTROMAGNETIC SPECTRUM

When a beam of sunlight is passed through a prism, it splits into seven different colours. This set of colours obtained by splitting the white light is called *spectrum*. (Fig 1.1) Different colours are associated with different energies and wavelengths. Red light has the smallest energy and the longest wavelength. Energy goes on increasing and wavelength goes on decreasing as we move from red colour to violet colour, thus the violet light has the maximum energy and minimum wavelength.

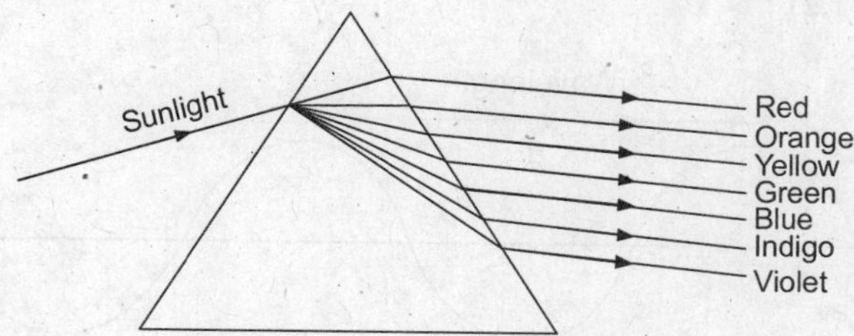

Fig. 1.1. Visible portion of electromagnetic spectrum

Electromagnetic spectrum. In addition to the white light with its seven constituent radiations, there are many more different types of radiations. Some of them are more energetic and some less energetic than the visible white light. The electromagnetic spectrum comprises the following radiations.

Cosmic rays. These rays come from sun and are known to be the radiations of highest energy. They have wavelength less than 10^{-3} nm (nanometre).

γ-rays and X-rays. γ-rays and X-rays are less energetic than cosmic rays. Their wavelength ranges between $10^{-3} - 10^{-1}$ nm.

Utraviolet light (U.V. light). These are the rays which are less energetic than X-rays but more energetic than visible light. The range of wavelength of U.V. light is 10–400 nm.

Visible light. It is ordinary sunlight and is comprised of seven different radiations (from red to violet). It has the wavelength range 400 nm–860 nm.

Infrared light are radiations weaker in energy but larger in wavelength than visible light. They have their wavelength in the range $8 \times 10^2 - 3 \times 10^5$ nm.

Microwaves are radiations of still smaller energy but larger wavelengths lying in the range of $3 \times 10^5 - 1 \times 10^7$ nm.

Radiowaves have the longest wavelength, more than 10^7 nm.

All the above radiations are part of the electromagnetic spectrum. Their wavelengths are shown as follows:

Cosmic Rays	γ-rays *X*-rays	UV	Visible	IR	Microwaves	Radiowaves
$<10^3$	$10^{-3} - 10^{-1}$	10 – 400	400 – 860	$8 \times 10^2 - 3 \times 10^5$	$3 \times 10^5 - 1 \times 10^7$	$>10^7$

Wavelength in nanometre (nm)

The wavelengths of different constituents of visible light are given below:

Colour	*Wavelength, (nm)*
Violet, indigo	400–435
Blue	435–480
Green-blue	480–490
Blue-green	490–500
Green	500–560
Yellow-green	560–580
Yellow	580–595
Orange	595–610
Red	610–750

1.3. CHARACTERISTICS OF ELECTROMAGNETIC RADIATIONS

***(a)* Frequency.** This is one of the parameters of a radiation which has the form of a wave as shown below (Fig 1.2).

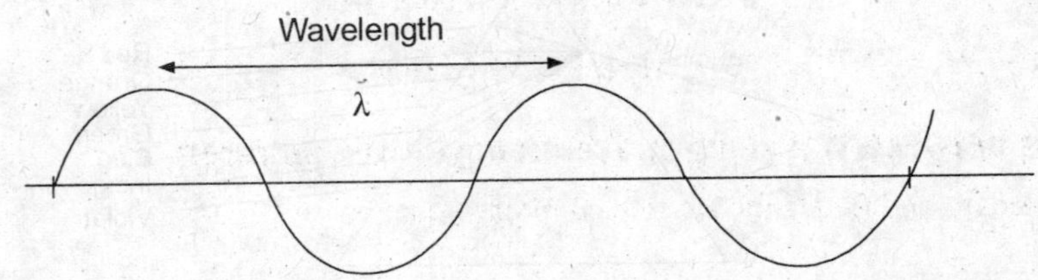

Fig. 1.2. Frequency and wavelength

Frequency is the no. of times a wave crosses a particular point in one second. It is denoted by the Greek letter ν (nu). It is expressed as cycles per second or Hertz.

$$1 \text{ Hz} = 1 \text{ cycle sec}^{-1}$$

Bigger units of frequency are kilocycle and megacycle per second.

$$1 \text{ kilocycle} = 10^3 \text{ Hz}$$

$$1 \text{ megacycle} = 10^6 \text{ Hz}$$

(b) Wavelength. This is another parameter of a wave. It is the distance between two successive crests of the wave. It is denoted by the Greek letter λ (lambda). Different regions of the electromagnetic spectrum possess different wavelengths. Different units for expressing wavelength have been used in the past. National Bureau of Standards 1963 has recommended nano metre (nm) as the unit of wavelength.

$$1 \text{ nm} = 10^{-9} \text{ m}$$

Sometimes, one comes across other units of wavelength, too. These are given here for the benefit of readers.

The other units that are used are, centimetre (cm), angstrom (Å), micron (μ), millimicron (mμ). Their relations between one another are reproduced below:

(i) $1 \text{ cm} = 10^{-2} \text{ m}$

(ii) $1 \text{ Å} = 10^{-8} \text{ cm} = 10^{-10} \text{ m}$

(iii) $1 \ \mu = 10^{-4} \text{ cm} = 10^{-6} \text{ m}$

(iv) $1 \text{ m}\mu = 10^{-7} \text{ cm} = 10^{-9} \text{ m}$

(v) $1 \text{ nm} = 10 \text{ Å} = 1 \text{ m}\mu$

(c) **Wave number $\overline{\nu}$ (nu bar).** Frequency of a wave is more conveniently used in the form of wave number. (Fig. 1.3) It is defined on the number of waves per centimetre. Its relation with wavelength is

$$\overline{\nu} = \frac{1}{\lambda}$$

$$\text{No. of waves in 1 cm} = \overline{\nu}$$

$$\text{No. of waves in 1 cm} = \frac{1}{\lambda}$$

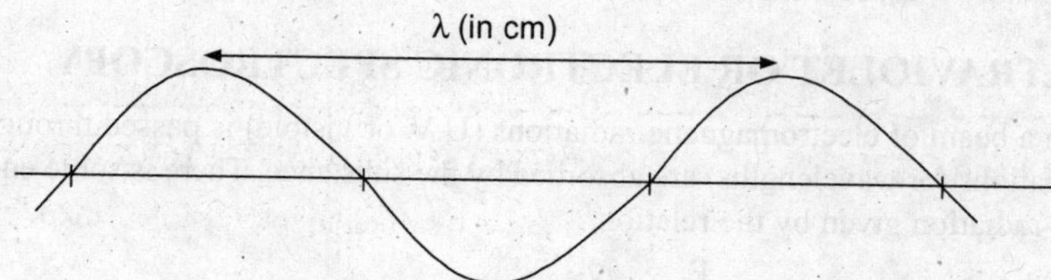

Fig. 1.3. Wave number

Relations between wavelength, frequency, wave number

Wavelength and frequency are related by the equation

$$\lambda \times \nu = c \qquad \text{... (1)}$$

where λ and ν stand for wavelength and frequency of the wave, c is the velocity of light.

$$\overline{\nu} = \frac{1}{\lambda} \qquad \text{... (2)}$$

But $$\lambda = \frac{c}{\nu} \text{ or } \frac{1}{\lambda} = \frac{\nu}{c} \qquad \text{... (3)}$$

From (2) and (3)

$$\bar{\nu} = \frac{\nu}{c}$$

A wave having frequency ν has energy given by the relation

$$E = h\nu \text{ where } h \text{ is Planck's constant}$$

1.4. ADVANTAGES OF SPECTROSCOPIC METHODS OF ANALYSIS OVER THE CONVENTIONAL METHODS

Spectroscopic methods have the following advantages over the conventional methods:

1. These methods require microquantities of the sample for investigation compared to a few grams required in classical methods.

2. These methods are quick and very sensitive.

3. These methods generally give reliable and reproducible results and provide a permanent record of observations in the form of spectra.

4. The material can be retrieved unchanged after the investigation and can be reused for other studies, except in case of mass spectrometry, where the sample is destroyed.

5. Continuous monitoring of changes taking place is possible in spectroscopic methods which can give information on kinetics of changes.

1.5. KINDS OF SPECTRA

There are two types of spectra:

***(a)* Emission spectrum.** When light emitted by a substance is passed through a prism, we obtain certain lines, which are called emission spectrum. Emission spectrum is studied in cases of atoms.

***(b)* Absorption spectrum.** When white light is passed through a substance, certain portion of the light is absorbed by it. Missing wavelengths leave dark lines or bands at their places. The spectrum obtained is called absorption spectra. This is used to study the structure of organic compounds.

1.6. ULTRAVIOLET OR ELECTRONIC SPECTROSCOPY

When a beam of electromagnetic radiations (U.V. or visible) is passed through a compound, certain radiations (or wavelengths) are absorbed by the substance. There is some energy associated with every radiation given by the relation

$$E = h\nu$$

where h is the Planck's constant and ν is the frequency of radiation.

This energy absorbed by the substance produces some changes within the molecule. If the energy absorbed is high (in UV or visible range), it causes electronic excitation, *i.e.*, electrons are excited to higher levels.

It is important to note the molecule does not absorb just any radiation. It absorbs only that radiation which posses the appropriate energy to bring about permitted transition to higher electronic, vibrational or rotational levels. This all depends upon the structure of the molecule. The radiation absorbed is something which is characteristic of a substance. The absorption of radiation is observed with the help of a spectrophotometer by passing radiations of continuously changing wavelength through a substance and analysing the intensity of transmitted radiations. The graph between the amount of radiation absorbed by the sample and the wavelength of radiation is called *absorption spectrum.* The absorption spectrum consists of bands with peaks of maximum intensity. Visible region extends from 400–800 nm while UV region extends from 100 – 400 nm.

The range of electronic spectroscopy is 200–800 nm.

1.7. WORKING OF A UV SPECTROPHOTOMETER

UV Spectrophotometers record the spectra of compounds in the range 200–800 nm. This range consists of two parts, 200–400 nm is the range of ultraviolet radiations whereas 400–800 nm is the range of visible light; spectra cannot be taken below 200 nm because oxygen present in the spectrophotometer absorbs this radiation. To record spectra of substances below 200 nm, special vacuum techniques are needed.

Fig 1.4 gives the schematic representation of spectrophotometer.

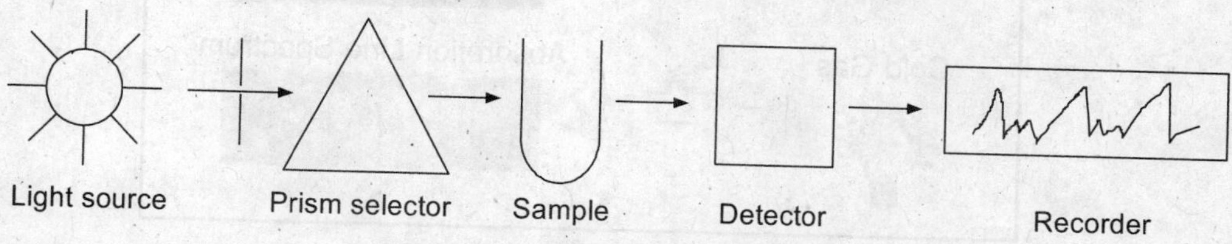

Fig. 1.4. UV spectrophotometer

For recording the UV spectrum, the given sample is dissolved in a suitable solvent which itself does not absorb light in that range. Commonly employed solvents are 95% ethanol, hexane and water. The positions of the absorption peaks are slightly shifted with the change of solvent. A quartz cell of 1 cm path length is used as a container for the sample solution. The solution is exposed to UV/visible light by the prism selector. Prism selector is rotating continuously to emit lights of varying wavelengths. Hydrogen discharge lamps and tungsten lamps are used for emitting UV and visible light respectively.

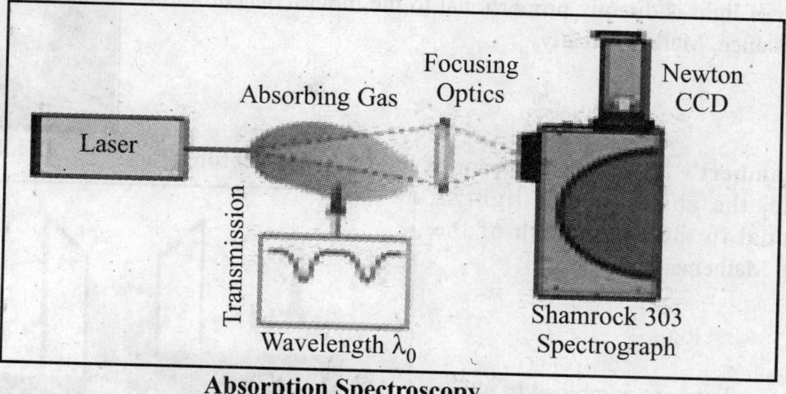

Absorption Spectroscopy

The instruments provide a running graph between wavelength of radiation absorbed and the intensity of absorption. (Fig. 1.5) There are a number of peaks or humps in the spectrum. The wavelengths corresponding to tops of the humps correspond to the absorption wavelengths and are denoted as λ_{max}. Intensity of absorption corresponding to that wavelength is called molar extinction coefficient expressed as $\in$ (epsilon).

A specimen UV spectrum is given in the graph above. It indicates λ_{max} at 290 nm.

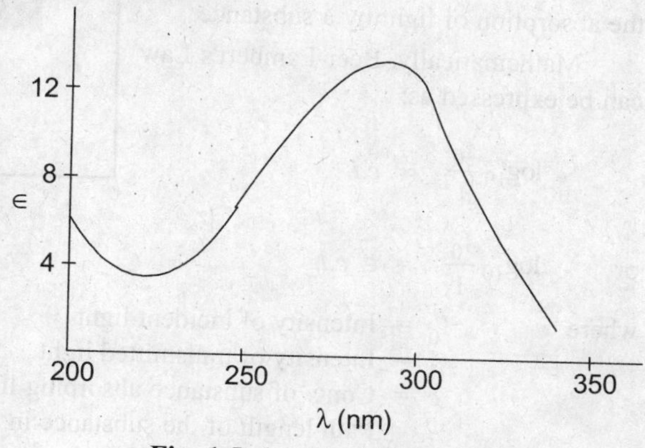

Fig. 1.5. A typical UV spectrum

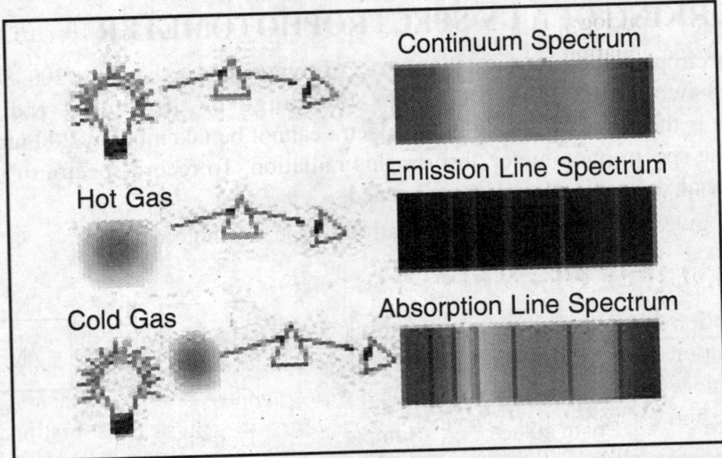

Absorption Spectrum

1.8. ABSORPTION LAWS

J.H. Lambert

Spectroscopy works on the laws governing absorption of light by a substance. There are two laws which govern the absorption of light.

1. Beer's Law. When a beam of monochromatic light is passed through a substance dissolved in a non-absorbing medium, the absorption of light is directly proportional to the molar concentration of the substance. Mathematically,

$$\log_{10}\frac{I_0}{I} \propto c$$

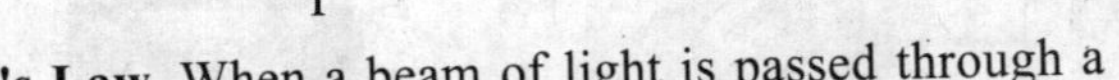

2. Lambert's Law. When a beam of light is passed through a substance, the absorption of light is proportional to the path length of the substance. Mathematically,

$$\log_{10}\frac{I_0}{I} \propto l$$

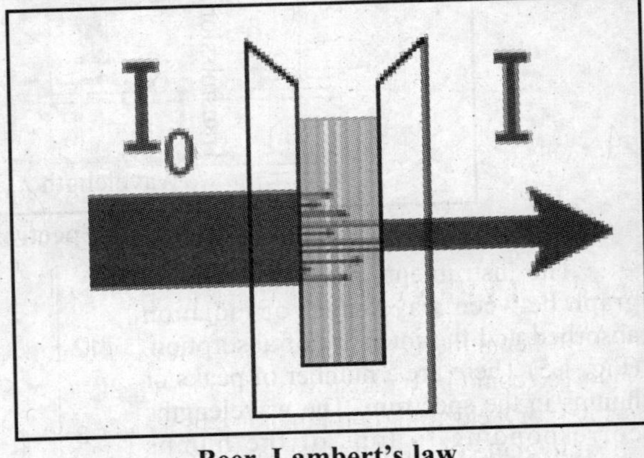

Beer–Lambert's law

The two laws are combined to obtain the absorption of light by a substance.

Mathematically, Beer-Lambert's Law can be expressed as:

$$\log_{10}\frac{I_0}{I} \propto c.l.$$

or $$\log_{10}\frac{I_0}{I} = \in c.l. \qquad \text{... (1)}$$

where I_0 = Intensity of incident light
I = Intensity of transmitted light
c = Conc. of substance absorbing light in *moles per litre.*
l = Path length of the substance in *cm*
$\in$ = Proportionality constant known as molar extinction constant.

The expression $\log_{10} I_0/I$ is termed as optical density or **absorbance** of the substance and denoted as A. So equation (1) above can be written as:

$$A = \log_{10} \frac{I_0}{I} = \in.c.l.$$

$$\text{Percent transmission of sample} = 100 \times \frac{I}{I_0}$$

1.9. PRINCIPLE OF UV SPECTROSCOPY

Absorption of UV and visible light causes a change in the electronic states of the molecules due to excitation of an electron from a lower to a higher energy level. It may be pointed out that each electronic level in a molecule is associated with a number of vibrational sub-levels with smaller energy gaps, and each vibrational level is associated with a number of rotational sub levels with still smaller energy gaps. Therefore, it can be said that while we are considering electronic excitation from a lower to a higher level, the electron can move from any vibrational or rotational sub-level in the excited state.

If vibrational and rotational levels were not associated with the electronic levels, then an electronic excitation of the molecule would have given sharp lines in the spectrum. In actual practice, we get bands rather than lines. These bands consist of closely spaced lines corresponding to excitations involving small difference in energies (Fig. 1.6)

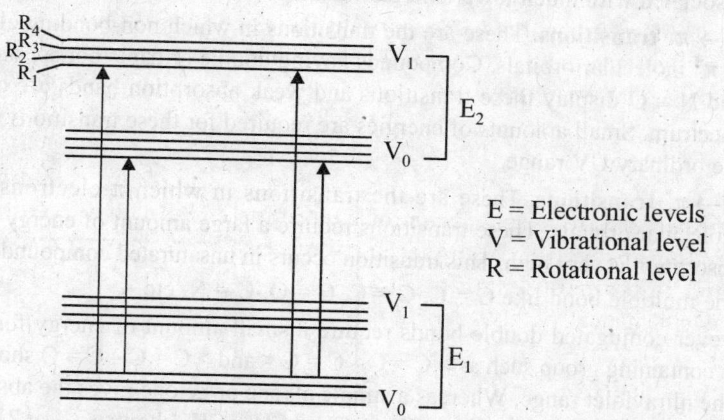

Fig. 1.6. Electronic transition from ground state to the excited state.

1.10. TYPES OF ELECTRONIC TRANSITIONS

When a molecule absorbs radiations, the electrons are excited to higher levels. The electron involved could be σ electron (occuping σ molecular orbital) or π electron (occupying a π molecular orbital) or *n* electron (non-bonding). In the diagram below, σ, π and *n* electrons have been indicated, in a molecule of aldehyde RCHO.

R σ, H σ — C =(π, σ)= Ö: — *n* (non-bonding)

The following electronic transitions are possible:

(*a*) $\sigma \longrightarrow \sigma^*$ (σ antibonding)

(*b*) $n \longrightarrow \sigma^*$ (σ antibonding)

(*c*) $n \longrightarrow \pi^*$ (π antibonding)

(*d*) $\pi \longrightarrow \pi^*$ (π antibonding)

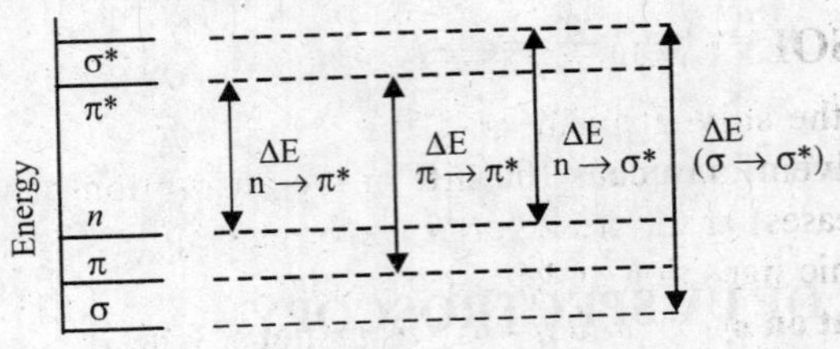

$\Delta E\,(n \rightarrow \pi^*) < \Delta E\,(\pi \rightarrow \pi^*) < \Delta E\,(n \rightarrow \sigma^*) < \Delta E\,(\sigma \rightarrow \sigma^*)$

A brief description of each is given below.

(*a*) $\sigma \rightarrow \sigma^*$ transition. Transition in which a σ electron is excited to σ^* orbital is called $\sigma \rightarrow \sigma^*$ transition. A high amount of energy is required for this transition, this energy is supplied by UV radiation of wavelength less than 200 nm. Such a transition does not take place in ordinary UV range > 200 nm. Saturated hydrocarbons in which all the electrons are involved in the formation of σ bonds come in this category.

(*b*) $n \rightarrow \sigma^*$ transitions. This transition corresponds to the excitation of a non-bonding electron to σ^* molecular orbital. Compounds having a hetero atom like O, S or N and having lone-pairs of electrons, not participating in bonding, show this transition. These transitions are associated with much lower energies than in $\sigma \rightarrow \sigma^*$ transitions.

(*c*) $n \rightarrow \pi^*$ transitions. These are the transitions in which non-bonding electrons are excited to π^* molecular orbitals. Compounds having double or triple bonds, such as C = O, C = S and N = O display these transitions and weak absorption bands are obtained in the UV spectrum. Small amounts of energies are required for these transitions which take place in the ordinary UV range.

(*d*) $\pi \rightarrow \pi^*$ transitions. These are the transitions in which π electrons are excited to π^* molecular orbitals. These transitions require a large amount of energy but the intensity of absorption is very high. This transition occurs in unsaturated compounds containing at least one multiple bond like C = C, C ≡ C, C = O, C = N, etc.

However conjugated double bonds require a small amount of energy for $\pi - \pi^*$ transition. Molecules containing group such as >C = C – C = C< and >C = C – C = O show $\pi - \pi^*$ absorption bands in the ultraviolet range. Whereas a simple alkene or ketone gives the absorption peak at 170 nm, a conjugated diene like butadiene $CH_2 = CH - CH = CH_2$ shows λ_{max} at 217 nm ($\in$ = 20,900). As the no. of conjugated bonds increases, the absorption takes place at longer wavelength. If there are enough conjugated bonds in the molecule, λ_{max} will move into visible region and the compound will look coloured. β-carotene, which contains 11 double bonds in conjugation, is coloured because it absorbs radiation of wavelengths 451 nm.

In the study of organic compounds, we are particularly interested in

$n \rightarrow \pi^*$ transitions

and $\pi \rightarrow \pi^*$ transitions

Probability of electronic transition

Electronic transitons have been divided into two types.

(*a*) Allowed transitions. Molecules don't absorb just any radiation. They absorb only such radiations which have appropriate energy to excite the electrons to allowed higher levels. These are the transitions which have $\varepsilon = 10^4$ or more. $\pi - \pi^*$ transitions conform to this requirement. 1–3 butadiene which shows absorption at 217 nm and has ε = 21000 undergoes an **allowed transition.**

(b) Forbidden transitions. These are the transitions having ε value less than 10^4, $n - \pi^*$ transitions generally belong to this category. A saturated aldehyde or ketone shows an absorption band at 290 nm at $\varepsilon < 100$. It involves a $n \rightarrow \pi^*$ forbidden transition.

1.11. ROLE OF SOLVENT IN THE ELECTRONIC TRANSITIONS

The polarity of the solvent influences the extent of electronic transitions in the UV spectroscopy. The solvent may cause a shift in the absorption band either towards longer wavelength (λ_{max} increases) or towards shorter wavelength (λ_{max} decreases). This depends upon the type of the electronic transition under consideration.

1. Effect of solvent on $n - \pi^*$ and $n - \sigma^*$ transitions. When a particular group is more polar in the ground state than in the excited state, the *n*- electrons in the ground state are more stabilised in comparison with the excited state because of hydrogen bonding or dipole-dipole interactions with molecules of the polar solvent.

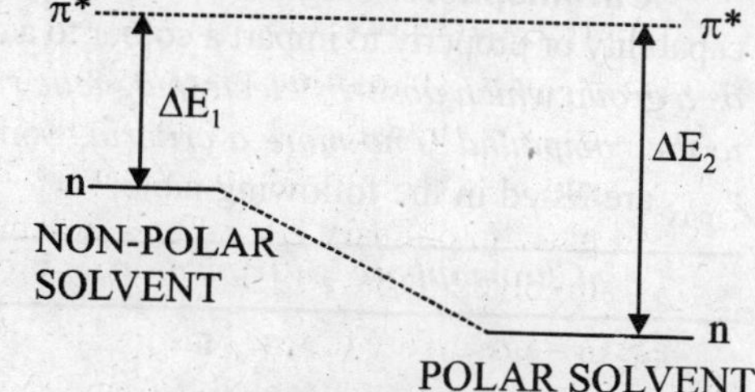

Fig. 1.7 Role of solvent in $n \rightarrow \pi^*$

Consequently the energy difference between the ground state and the excited state increases (Fig 1.7). *Thus, the absorption shifts towards the region of shorter wavelength.* Take, for example, the case of carbonyl group.

A carbonyl group is more polar in the ground state than in the excited state as illustrated below.

$$\overset{\delta^+}{>C}=\overset{\delta^-}{\ddot{O}}: \qquad\qquad >\overset{\delta^+}{\dot{C}}=\overset{\delta^-}{\dot{O}}:$$

Ground state (More polar) — Excited state (Less polar)

As a result, the extent of hydrogen bonding or dipole-dipole interactions in the ground state will be more as compared to that in the excited state. Therefore, the ground state energy of the compound containing this group will tend to decrease as the energy is released during the interactions. Thus, in such groups absorption shifts to shorter wavelength. For example, λ_{max} for acetone is 279 nm in hexane solvent (non-polar in nature) but it drops to 264 nm in water (polar solvent).

2. Effect of solvent on $\pi \rightarrow \pi^*$ transition. When a particular group is more polar in the excited state than in the ground state, the absorption shifts to a region of longer wavelength. Actually, the extent of hydrogen bonding or the dipolar interactions in the excited state will be more, compared to the ground state. In other words, the excited state will be relatively more stabilised as compared to the ground state or the energy of the excited state will decrease (Fig 1.8). This will cause a shift in the absorption band to longer wavelength. Take the example of ethylene.

π* — ΔE1 — π — NON-POLAR SOLVENT; π* — ΔE2 — π — POLAR SOLVENT

Fig. 1.8 Effect of solvent in $\pi \rightarrow \pi^*$ transition

The molecule of ethylene is more polar in the excited state than in the ground state due to $\pi \rightarrow \pi^*$ electronic transition.

$$>C=C< \qquad\qquad >\overset{\delta^+}{C}-\overset{\delta^-}{C}<$$

Ground state (Less polar) — Excited state (More polar)

As a result, in the polar solvent, π* M.O. will be relatively more stabilised than in the ground state of the molecule. Hence, the absorption will shift to a region of longer wavelength.

A Suitable solvent for ultraviolet spectroscopy should meet the following requirements:

(*i*) It should not itself absorb radiations in the region under the solute molecules.

(*ii*) It should be less polar so that it has minimum interaction with the solute molecules.

The most commonly employed solvent is 95% ethanol. It is cheap, has good dissolving power and does not absorb radiations above 210 nm. Thus, it is transparent above 210 nm. Commercial ethanol should not be used as it contains some benzene which undergoes absorption in the UV range at about 280 nm.

1.12. CHROMOPHORE AND AUXOCHROME

Chromophore. Originally, chromophore was considered as a functional group which has a capability or property to impart a colour to a compound. In a broader sense, *a chromophore is defined as a group which absorbs electromagnetic radiation in the visible or UV range. Imparting a colour to the compound is no more a criteria.* Some prominent chromophores along with their λ_{max} and ε_{max} are listed in the following table:

Chromophore	*Transition*	λ_{max} *(nm)*	ε *max*	*Solvent*
$>C=C<$	$\pi-\pi^*$	175	15000	Vapour
$-C\equiv C-$	$\pi-\pi^*$	175	10000	Hexane
$>C=O$	$\pi-\pi^*$	180	10000	Hexane
	$n-\pi^*$	160	18000	
	$n-\sigma^*$	285	15	
$-\overset{OH}{\overset{\vert}{C}}=O$	$n-\pi^*$	205	60	Methanol
$-N=N-$	$n-\pi^*$	338	5	Ethanol
$-NO_2$	$n-\pi^*$	274	15	Methanol

Auxochrome. *An auxochrome is a group which when attached to chromophore shifts the absorption maximum towards longer wavelengths along with an increase in the intensity of absorption.* By itself, it does not show any absorption above 200 nm. We may compare auxochromes with promoters of catalysts. Promoters by themselves have no catalytic ability but they improve upon the activity of the catalyst. Some prominent auxochromic groups are –OH, – NH_2, –OR, –NHR and –NR_2. When auxochrome –NH_2 is attached to benzene ring, the absorption changes from 255 nm (ε_{max} = 203) to 280 nm (ε_{max} = 1430).

The secret behind the capability of auxochromes, to shift absorption maximum towards longer wavelength, is that it helps in extending conjugation by providing lone pair of electrons as shown below:

$$\underset{\text{Chromophore}}{CH_2=CH} — \underset{\text{Auxochrome}}{\ddot{N}R_2} \longleftrightarrow :\bar{C}H_2-CH=\overset{+}{N}R_2$$

1.13. TYPES OF SHIFTS

Substituent groups can produce four types of effects on adsorption.

(*a*) Bathochromic or red shift. Auxochromes bring about an increase in the value of λ_{max}. The absorption is shifted towards higher wavelength or towards red portion of the spectrum. This shift of absorption of light towards higher wavelength is known as **bathochromic or red shift.** This effect may be produced by change of solvent. It is also produced if two or more chromophores are present in conjugation. Ethylene, for example, shows $\pi - \pi^*$ transition at 170 nm whereas 1, 3 butadiene shows absorption at 217 nm.

***(b)* Hypsochromic or blue shift.** If we remove conjugation from a system, the absorption maximum is shifted towards lower wavelength or towards blue portion of the spectrum. This shifting of absorption maximum towards shorter wavelength by removing conjugation from a system is called **hypsochromic or blue shift.** For example, protonation of aniline causes a blue shift from 280 nm to 203 nm because the anilinium ion has no electrons to participate in conjugation with benzene ring.

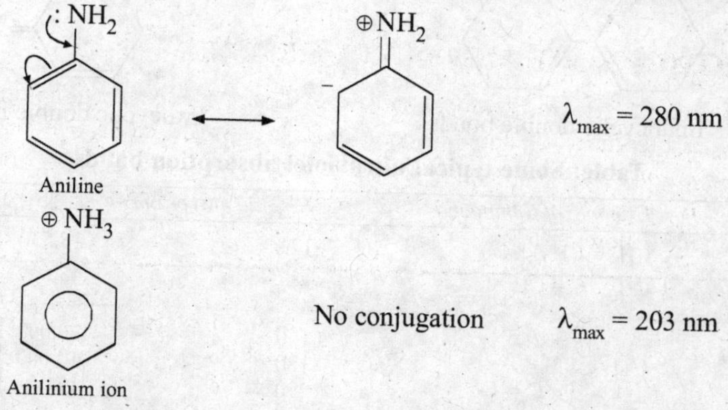

$: NH_2$ ⟷ $\oplus NH_2$ $\lambda_{max} = 280$ nm

Aniline

$\oplus NH_3$

No conjugation $\lambda_{max} = 203$ nm

Anilinium ion

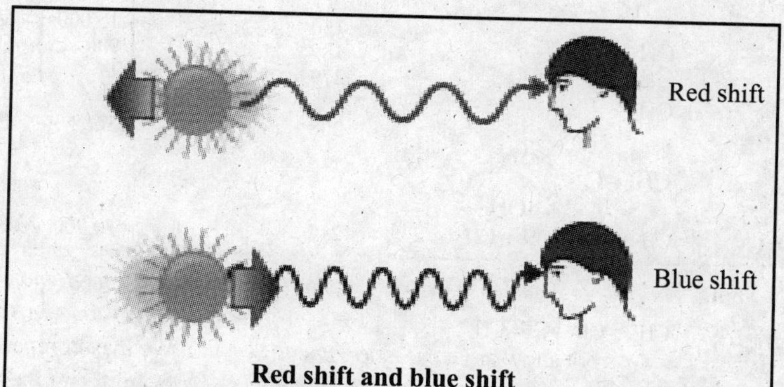

Red shift

Blue shift

Red shift and blue shift

***(c)* Hyperchromic effect.** Introduction of an auxochrome into a system brings about an increase in the intensity of absorption. This effect is known as **hyperchromic effect.** A methyl group at no. 2 position in pyridine increases $\in_{max}$ from 2750 to 3560 for $\pi - \pi^*$ transition.

***(d)* Hypochromic effect.** Groups which distort the geometry of molecule bring about a decrease in intensity of absorption ($\in_{max}$). If we introduce a methyl group at position no. 2 in biphenyl, distortion of the geometry of the molecule causes the hydrochromic effect.

1.14. UV ABSORPTION BANDS DISPLAYED BY SOME PROMINENT COMPOUNDS, INDICATING THE CHROMOPHORES, λ_{max} AND $\in_{max}$

Absorption bands displayed by some prominent compounds are given in the table on next page.

1.15. UV ABSORPTION IN CYCLIC DIENES

Types of cyclic dienes. There are two types of cyclic dienes *i.e.* homoannular and heteroanular. A homoannular diene is the one in which the two double bonds are present in the same ring while the heteroannular diene is one in which the two double bonds are present in the adjacent rings.

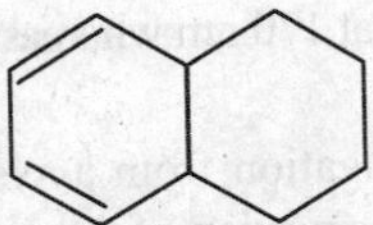
Homoannular diene

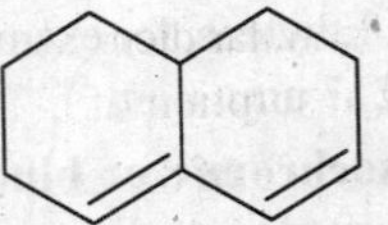
Heteroannular diene

Nature of double bonds in the cyclic dienes. A double bond is said to be endocylic if it is present within a particular ring and is known as exocylic if it lies outside a particular ring.

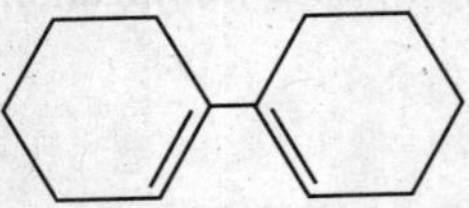
Endocyclic double bonds

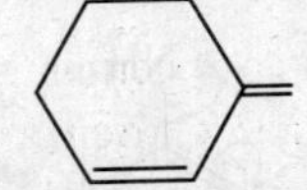
Exocyclic double bond

Table: Some typical ultraviolet absorption bands

Group	*Example of Compound*	λ_{max} *(in mm or nm)*	ε_{max}	*Solvent*
$>C=C<$	$CH_2=CH_2$	171	15,530	Vapour
$-C\equiv C-$	$CH_3-C\equiv CH$	187	450	Cyclohexane
$>C=O$	$CH_3(H)C=O$	160 180 290	20,000 10,000 17	Vapour Hexane
	$(CH_3)_2C=O$	166 189 279	16,000 900 17	Vapour Hexane
$-COOH$	CH_3-COOH	208	32	Ethanol
$>C=C-C=C<$	$CH_2=CH-CH=CH_2$	217	20,900	Hexane
$>C=C-C\equiv C-$	$CH_2=CH-C\equiv CH$	219 228	7600 7800	Ethanol
$>C=C-C=O$	$CH_3-CH=CH-CH=O$	218 320	18,000 30	Hexane
	$CH_3-CH=CH-C(CH_3)=O$	224 314	9750 38	Ethanol
Benzenoid ring	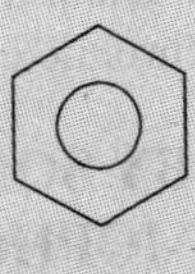 C_6H_6 (benzene)	203.5 254	7400 204	Water
	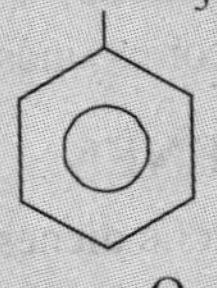 $C_6H_5-OCH_3$	217 269	600	Water
	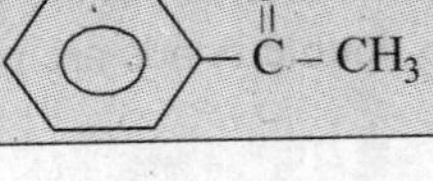 $C_6H_5-CO-CH_3$	245.5	9800	Water

It has been observed that

(1) a homoannular diene absorbs at a greater wavelength than the heteroannular diene. However, the intensity of absorption of homoannular diene is smaller.

(2) the presence of the exocylic double bond generally increases the wavelength of absorption of a cyclic conjugated diene.

1.16. WOODWARD RULES FOR CALCULATING λ_{max} IN CONJUGATED DIENES, TRIENES AND POLYENES

R.B. Woodward

Woodward gave certain empirical rules for calculating the λ_{max} in conjugated dienes, trienes and polyenes. These rules were later modified by Fieser. According to these rules, each diene has a certain fixed basic value and its actual λ_{max} depends upon the following factors:

(*i*) The number of alkyl substituents or ring residues present on the diene system.

(*ii*) The number of double bonds extending conjugation beyond the diene system.

(*iii*) The presence of auxochromes *i.e.* the polar groups such as – Cl, Br, – OR, – SR; – NH_2, – $OCOCH_3$ etc.

The contribution by the different substituents or auxochromes, if present, is added to the λ_{max} value or the basic value of the diene to obtain the net value of λ_{max}. These values are given in the following table.

Table. Values of λ_{max} for different substituents

Parent open-chain diene	217 nm
Parent heteroannular diene	214 nm
Parent homoannular diene	253 nm
For each double bond extending conjugation add	30 nm
For each alkyl substituent (R) or ring residue add	5 nm
For each exocyclic double bond add	5 nm
Auxochromes	
For an OAc add	0 nm
For an OR add	6 nm
For an SR add	30 nm
For Cl or a Br atom add	5 nm

This is illustrated with the help of the following examples:

Example 1. Calculate the λ_{max} in the UV spectrum of 2, 4–hexadiene.

Solution. $CH_3 – CH = CH – CH = CH – CH_3$

The parent unit of 2, 4-hexadiene is a butadiene. It has two methyl substituents attached to either sides. Its λ_{max} can be calculated using Woodward rules as under:

$$\begin{aligned} \text{Basic value} &= 217 \text{ nm} \\ \text{Two alkyl substituents } (2 \times 5) &= 10 \text{ nm} \\ \text{Total calculated } \lambda_{max} &= 227 \text{ nm} \end{aligned}$$

Example 2. What λ_{max} value do you expect for the following compound?

(Structure: cyclohexane ring with exocyclic double bond =C(CH₃)–C(CH₃)=CH₂; labels: CH_3, $C – C = CH_2$, CH_3)

Solution. It is a conjugated diene system which has two alkyl substituents, two ring residues and one exocyclic double bond.

Basic value	=	217 nm
Two alkyl substituents (2 × 5)	=	10 nm
Two ring residues (2 × 5)	=	10 nm
One exocyclic double bond	=	5 nm
Total calculated λ_{max}	=	242 nm

1.17. APPLICATIONS OF ULTRAVIOLET SPECTROSCOPY

The important applications of the study of UV spectra of compounds are as given below:

***(i)* Detection of conjugation.** With the help of UV spectrum, we can establish the presence of conjugation in a compound. Conjugation can be

$$-C=C-C= \text{ or } -C\equiv C-C\equiv \text{ or } C=C-C=O \text{ or } C_6H_5-C=C$$

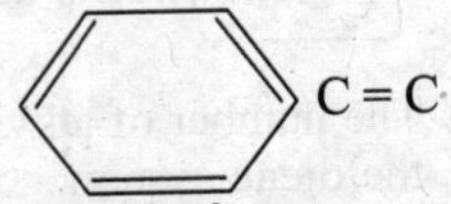

By observing the λ_{max} values, we can also predict the location of substituents.

***(ii)* Detection of functional groups.** It is possible to detect certain functional groups with the help of UV spectrum. The negative test is also of value. Absence of absorption above 200 nm is a sure indication of the absence of conjugation, carbonyl group and benzene rings in the compound.

***(iii)* Detection of geometrical isomers.** When a compound exhibits geometrical isomerism, the trans isomer shows absorption at higher wavelength with larger values of extinction coefficients ($\in_{max}$) compared to cis isomers. Out of two stilbenes ($C_6H_5CH = CH\ C_6H_5$), the trans isomer shows absorption at 294 nm, ($\in_{max}$ = 24000) while the cis isomer absorbs at 278 nm ($\in_{max}$ = 9350).

SOME SOLVED STRUCTURAL PROBLEMS

Example 1. The following dienes have λ_{max} at 176 nm, 211 nm and 215 nm. Find out which is which?

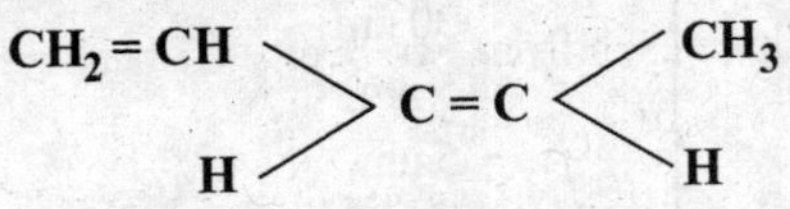

(*a*) $(CH_2=CH)(H)C=C(CH_3)(H)$ — $CH_2=CH$ and CH_3 on the same side

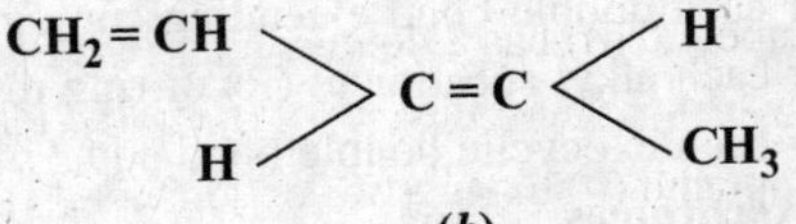

(*b*) $(CH_2=CH)(H)C=C(H)(CH_3)$ — $CH_2=CH$ and CH_3 on opposite sides

(*c*) $CH_2=CH-CH_2-CH=CH_2$

Solution. (*i*) λ_{max} = 176 nm indicates that there is no conjugation in the molecule. Compound (*c*) has no conjugation. Hence the compound showing λ_{max} at 176 nm is (*c*) *i.e.*, $CH_2=CH-CH_2-CH=CH_2$ or 1, 4 Pentadiene.

(*ii*) Structures (*a*) and (*b*) are geometrical isomers of the compound 1, 3, pentadiene, (*a*) is *cis*-isomer and (*b*) is *trans*-isomer. As a rule the trans isomer shows absorption at higher wavelength. hence the compound showing λ_{max} at 215 is (*b*) or *trans*-1, 3 pentadiene.

(*iii*) Compound showing λ_{max} at 211 nm is (*a*) ,*i.e.*, *cis*-1, 3 pentadiene.

Example 2. Which of the following compounds, if any, can give a UV spectrum showing λ_{max} at (*i*) 218 (ε = 18000) (*ii*) 245.5 (ε = 9800) (*iii*) 186.5 (ε = 450) and 415 (ε = 63000).

(*a*) $CH_3CH_2CH_2CH_3$ (*b*) $CH_3-C\equiv CH$

(*c*) $CH_3-CH=CH-CHO$ (*d*) 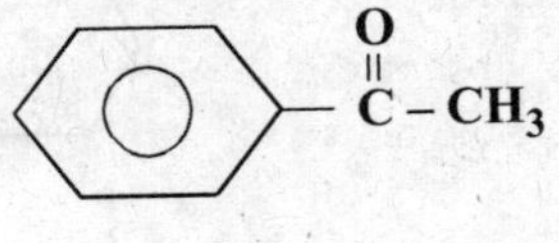 $C_6H_5-\overset{O}{\overset{\|}{C}}-CH_3$

Solution. (*a*) $CH_3CH_2CH_2CH_3$ does not have any conjugation. It does not have even one double or triple bond. Therefore none of the λ_{max} values corresponds to this compound.

(*b*) $CH_3 - C \equiv CH$ has a triple bond but the electrons are not much mobile. This compound absorbs at $\lambda_{max} = 186.5$ nm.

(*c*) $CH_3 - CH = CH - \underset{\underset{O}{\|}}{CH}$

This compound has conjugation with possibility of resonance. The compound absorbs at 218 nm (λ_{max} = 218).

(*d*) $C_6H_5 - \overset{\overset{O}{\|}}{C} - CH_3$ Methyl phenyl ketone

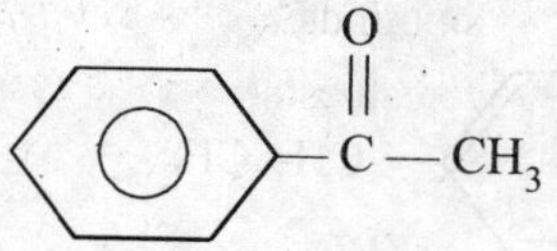

In this compound, three double bonds of the ring and double bond in C = O are in conjugation. The no. of double bonds in conjugation is much more than in compound (*c*). Thus compound (*d*) absorbs at 245.5 nm.

Example 3. Indicate the increasing order of wavelength of λ_{max} of the following compounds in the UV region of the spectrum.

(*a*) C_6H_5-**CH = CH – CH = CH**$-C_6H_5$

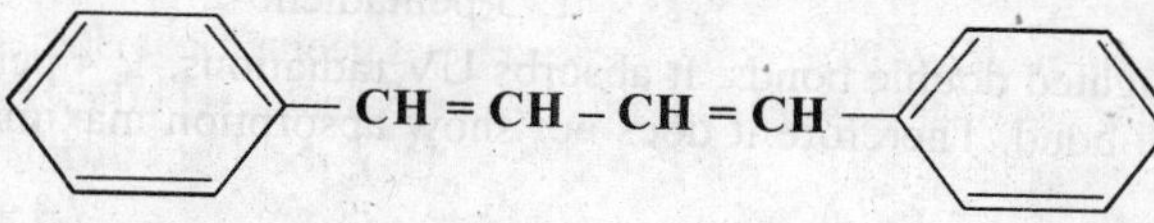

(*b*) $CH_2 = CH - CH = CH_2$

(*c*) $CH_2 = CH - CH_2 - CH = CH_2$

Solution. Compound (*c*) has no conjugation in the molecule. Therefore it will have the smallest value of λ_{max}.

Compound (*a*) has 8 double bonds in conjugation (three each from two benzene rings and two outside the ring). Therefore it will have the maximum value of λ_{max}).

Compound (*b*) having two double bonds in conjugation will have intermediate value of λ_{max} between the compounds (*a*) and (*c*).

Hence increasing order of wavelength of λ_{max} for the compounds (*a*), (*b*) and (*c*) is (*c*) < (*b*) < (*a*).

Example 4. What type of compounds absorb UV radiations? Select the compounds in support of your answer from the following list:

(*i*) 1, 3-butadiene **(*ii*) 1-hexene**
(*iii*) cyclobutane **(*iv*) 1, 3-cyclohexadiene**
(*v*) chlorobenzene **(*vi*) nitrobenzene**
(*vii*) chlorocyclohexane

Solution. Compounds having conjugated double bonds absorb U.V. radiations. Out of the list given, the following compounds will absorb U.V. radiations:

1, 3-butadiene, 1, 3-cyclohexadiene, chlorobenzene and nitrobenzene.

Example 5. Arrange the following in increasing order of U.V. absorption maxima.

(*i*) anthracene **(*ii*) ethylene**
(*iii*) naphthalene **(*iv*) butadiene**

Solution. UV absorption maxima depends upon the no. of conjugated double bonds in the compounds. Greater the no. of such double bonds, greater will be absorption maxima. Out of the list given above, anthracene has 7 conjugated double bonds, naphthalene has 5 conjugated double bonds, butadiene has two conjugated double bonds and ethylene has no conjugated double bond. Therefore increasing order of U.V. absorption maximum of the compounds will be as under:

Ethylene < butadiene < naphthalene < anthracene

Example 6. Can UV spectral data be useful to distinguish between the following pairs? Give reasons.

(*i*) Ethylbenzene and Styrene

(*ii*) $CH_2 = CH - CH_2 - CH = CH_2$ and $CH_3 - CH = CH - CH = CH_2$

Soution. (*i*)

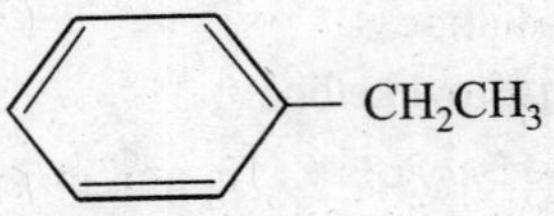

C_6H_5–CH_2CH_3 (Ethylbenzene) C_6H_5–$CH{=}CH_2$ (Styrene)

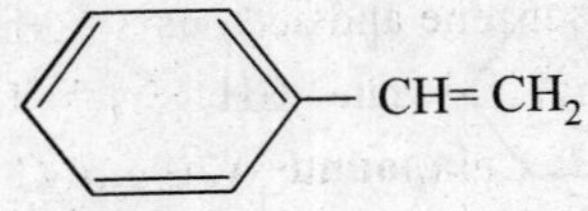

Ethylbenzene has 3 conjugated double bonds and styrene has four conjugated double bonds. Hence these two compounds can be differentiated by observing the λ_{max} value. Styrene will absorb at higher wavelength compared to ethylbenzene.

(*ii*) $CH_2 = CH - CH - CH = CH_2$ (1, 4 pentadiene) $CH_3 - CH = CH - CH = CH_2$ (1, 3 pentadiene)

1, 3 pentadiene has two conjugated double bonds. It absorbs UV radiations. 1, 4 pentadiene does not have a conjugated double bond. Therefore it does not show absorption maxima in UV range.

Example 7. Which will have greater λ_{max}?

(*i*) C_6H_5–NH_2 (*ii*) C_6H_5–$\overset{+}{N}H_3Cl^-$

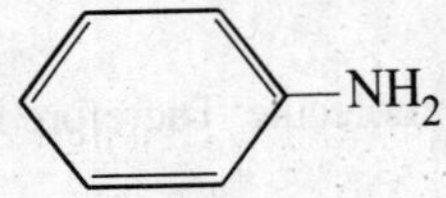

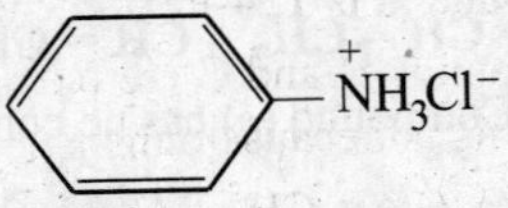

Solution. Lone pair of electrons on nitrogen is available for conjugation with the double bonds of benzene in structure (*i*). In structure (*ii*), there are no electrons on nitrogen for conjugation with the benzene ring. Hence structure (*i*) shows greater λ_{max} compared to structure (*ii*).

Example 8. Mesityl oxide, a condensation product of acetone, is known to consist of two isomers (I) and (II) shown below. One isomer exhibits a maximum at 235 nm with ε = 12000 while the other shows no high intensity absorption above 220 nm. Which of the two isomers absorbs at 235 nm?

$CH_2 = C(CH_3) - CH_2 - COCH_3$ **(I)** $CH_3 - CH = C(CH_3) - COCH_3$ **(II)**

Solution. Isomer (II) absorbs at 235 nm because its double bond system is conjugated while (I) has unconjugated double bonds and would be expected to absorb at shorter wavelengths.

Example 9. How will you differentiate between the following compounds using UV spectroscopy?

CH_3COCH_3 and $CH_3COC_6H_5$

Solution. In $CH_3COC_6H_5$, there is a sort of conjugation within the ring as well as outside the ring.

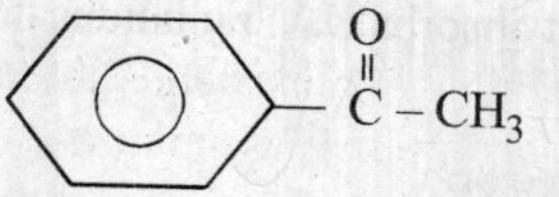

$C_6H_5-\overset{O}{\overset{\|}{C}}-CH_3$

whereas CH_3COCH_3 gives absorption maximum of 166 nm ($\in = 16000$), methylphenyl ketone will show absorption maximum at a much higher wavelength due to conjugated double bonds in the ring. Also C = O bond is in conjugation with the double bonds of benzene ring.

Example 10. Which of the following absorb and which don't absorb U.V. radiations?

(*i*) Ethyl alcohol **(*ii*) Benzene**

(*iii*) Water **(*iv*) Acetone**

Solution. Benzene and acetone absorb U.V. radiations.

Ethyl alcohol and water don't absorb U.V. radiations.

Example 11. Compounds A, B and C have the formula C_5H_8 and on hydrogenation yield *n*-pentane. Their ultraviolet spectra show the following values of λ_{max}:

A : 176 mμ **B : 211 mμ**

C : 215 mμ

1-pentene has λ_{max} = 178 mμ

What are the likely structures for A, B and C?

Solution. The formula C_5H_8 suggests that the molecule has two double bonds (C_nH_{2n-2}) in all the three compounds. If the two double bonds are conjugated, the absorption takes place around 220 mμ.

Accordingly

(*i*) Compound A is 1, 4-Pentadiene

(*ii*) Compounds B and C are both 1, 3-Pentadiene. Out of the two, B is *cis* isomer and C is *trans* isomer. This is because trans isomer absorbs at a higher wavelength. Thus

A $CH_2 = CH - CH_2 - CH = CH_2$

B $(CH_2=CH)(H)C=C(CH_3)(H)$

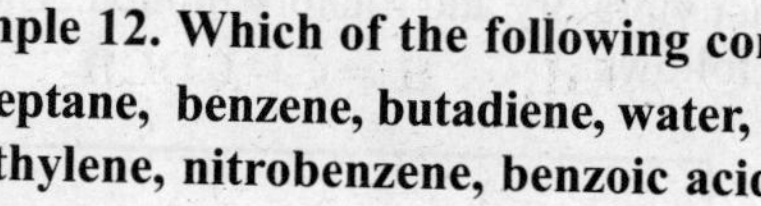

C $(CH_2=CH)(H)C=C(H)(CH_3)$

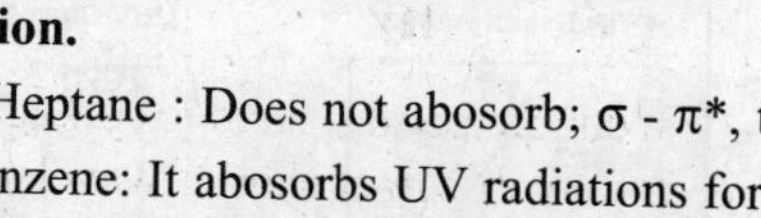

Example 12. Which of the following compounds absorb UV radiations:

n - Heptane, benzene, butadiene, water, 1-heptene, chlorohexane, ethanol, n-butyl amine, acetone, ethylene, nitrobenzene, benzoic acid?

Solution.

1. *n*-Heptane : Does not abosorb; σ - π^*, transition requires very high energy
2. Benzene: It abosorbs UV radiations for the $\pi \rightarrow \pi^*$ transition.
3. Butadiene: It absorbs strongly due to $\pi \rightarrow \pi^*$ transition.
4. Water, H_2O : Does not abosorb.
5. 1- Heptene, $CH_3(CH_2)_4CH = CH_2$: It absorbs due to $\pi \rightarrow \pi^*$ transition.
6. Chlorohexane, $CH_3(CH_2)_5Cl$: It does not abosorb.
7. Ethanol, CH_3CH_2OH: It does not absorb.
8. n-Butyl amine, $CH_3(CH_2)_3NH_2$: It does not absorb.

9. Acetone, $CH_3-\underset{\underset{O}{\parallel}}{C}-CH_3$: It absorbs U.V. radiations due to $\pi \rightarrow \pi^*$ and $n \rightarrow \pi^*$ transitons.

10. Ethylene, $CH_2 = CH_2$: It absorbs due to $\pi \rightarrow \pi^*$ transitions.

11. Nitrobenzene, $C_6H_5-\overset{\overset{O}{\uparrow}}{N}=O$: It absorbs due to $\pi \rightarrow \pi^*$ and $n \rightarrow \pi^*$ transitions.

12. Benzoic acid, $C_6H_5-\underset{|}{\overset{OH}{C}}=O$: It absorbs due to $\pi \rightarrow \pi^*$ and $n \rightarrow \pi^*$ transitions.

Example 13. The UV spectrum of acetone shows two important peaks at λ_{max} = 279 nm (ε_{max} 15) and λ_{max} = 189 nm (ε_{max} 900). Identify the electronic transition for each peak.

Solution. The peak at λ_{max} = 279 nm (ε_{max} 15) is due to $n \rightarrow \pi^*$ transition. That is:

$$(CH_3)_2C=\ddot{O}: \longrightarrow (CH_3)_2C\dot{=}\dot{O}:$$

The peak λ_{max} = 189 nm (ε_{max} = 900) is due to $\pi \rightarrow \pi^*$ transition. That is :

$$(CH_3)_2C=\ddot{O}: \longrightarrow (CH_3)_2\dot{C}-\dot{\ddot{O}}:$$

INFRARED SPECTROSCOPY

1.18. PRINCIPLES OF INFRARED SPECTROSCOPY

The atoms in a molecule are in a state of constant vibratory and rotatory motion. The different ways in which the vibrations can take place are as follows:

For diatomic molecules

Let us say we have a molecule XY. It can stretch the X-Y bond inwards or outwards in the following ways:

(X)–ʘʘʘʘʘʘʘʘʘʘ–(Y) Extended

(X)–ʘʘʘʘʘʘʘʘ–(Y) Normal

(X)–ʘʘʘ–(Y) Compressed

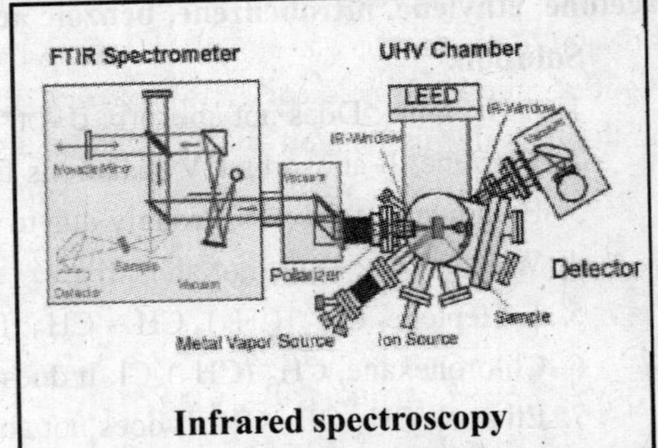

Infrared spectroscopy

Thus the compression and extension of the bond takes place along the X–Y bond (see Fig. above).

For triatomic molecule

Let there be a molecule XYZ with a bond angle ∠XYZ. Here we have two types of movements:

(1) Stretching. As in the case of XY diatomic molecule, there will be stretching in the XYZ molecule, too. Here two bonds are involved viz, XY and YZ bonds. Let us assume that the molecule is linear. The different stretching forces that are manifested are as follows:

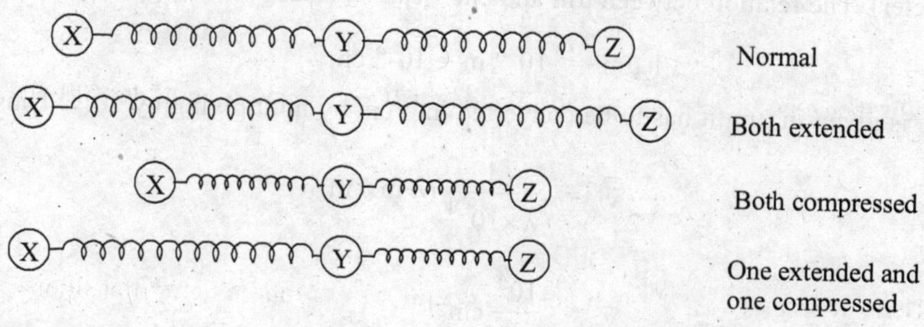

(2) Bending. If the molecule is angular, there will still be stretching of XY and YZ bonds. Besides, there will be some bending forces between XY and YZ bonds, as a result of which the ∠ XYZ will change, it will increase and decrease continuously as shown below:

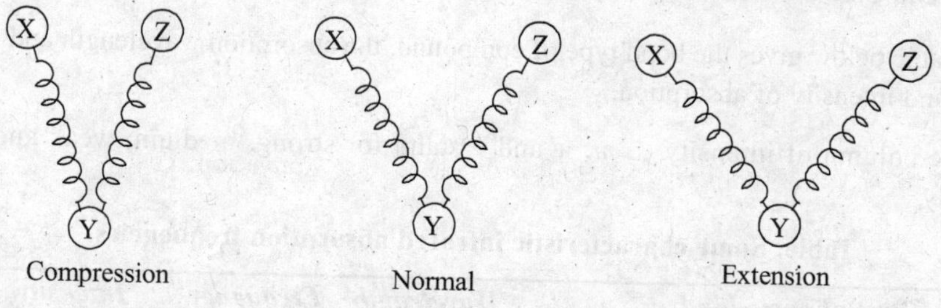

Thus there are a number of vibrational and rotational changes taking place in the molecule. On absorption of light, the molecules are excited from lower vibrational and rotational levels to higher ones. For effecting such transitions, frequencies in the infrared range are good enough. Molecules will absorb such frequencies as are needed to excite molecules from lower vibrational levels to permitted higher levels. Absorption of energy by a bond increases the amplitude of vibrations. Every bond and every functional group has a specific absorption frequency which is calculated to excite the molecule to higher vibrational or rotational level. A graph between the amount of absorption (or emission) of infrared light against the frequency (or wavelength) of this light is called **Infrared (or IR) spectrum.** The following diagram illustrates various frequencies in the IR range associated with various bonds in an alcohol molecule.

C — O — H
(6.7 – 8.3 μm)

Rest of the molecule — C ⟷ O — H

O — H stretching
(2.7 – 3.3 μm)

C — O
stretching
(8.3 – 10 μm)

1.19. RANGE AND UNITS IN IR SPECTROSCOPY

IR region has the range of 0.8 to 20 μm or 800 to 20000 nm. The region between 0.8 and 2.5 μm is called the near infrared region and the one between 15 and 20 μm is called far infrared region. The region between 2.5 and 15 μm is the proper infrared region. Generally wave number ($\bar{\nu}$), which is the no. of vibration of the radiation per centimetre, is used for describing IR spectra. The unit of wave number is cm^{-1}. The other unit which is used in connection with IR spectra is μm (micrometer). The relation between μm and cm^{-1} can be established as given below:

$$1\ \mu m = 10^{-6}\ m = 10^{-4}\ cm$$

If λ is given in μm, it has to be converted into cm by multiplying by 10^{-4}. Thus

$$\bar{\nu} = \frac{1}{\lambda \times 10^{-4}} (\lambda \text{ in } \mu m)$$

or

$$\bar{\nu} = \frac{10^4}{\lambda}\ cm^{-1}$$

1.20. IR ABSORPTION BANDS OF SOME IMPORTANT BONDS AND FUNCTIONAL GROUPS

The table below gives the bond type of compound, the absorption wavelength and frequency (in cm^{-1}) and intensity of absorption.

In the column of intensity, *s*, *m*, *w* and *v* stand for strong, medium, weak and variable respectively.

Table: Some characteristic infrared absorption frequencies.

Bond	*Type of compound*	*Wavelength (in μm)*	*Frequency (in cm^{-1})*	*Intensity*
$-\overset{\vert}{\underset{\vert}{C}}-H$	Alkanes	3.28–3.51	2850–2960	*m–s*
$=\overset{\vert}{C}-H$	Alkenes (terminal)	3.22–3.32	3010–3100	*m*
$\equiv C-H$	Alkynes (terminal)	3.03	3300	*s*
>–C – H	Aromatic rings	3.22–3.33	3000–3100	*m*
$-\overset{\Vert}{\underset{\vert}{C}}-H$	Aldehydes	3.47–3.77	2650–2880	*w* (often two bands)
$-\overset{\vert}{\underset{\vert}{C}}-\overset{\vert}{\underset{\vert}{C}}-$	Alkanes	6.66–16.66	600–1500	*w* (seldom useful for identification)
$>C=C<$	Alkenes	5.95 –6.2	1620–1680	*v*
$-C\equiv C-$	Alkynes	4.42–4.76	2100–2600	*v*

Bond	*Type of compound*	*Wavelength (in μm)*	*Frequency (in cm^{-1})*	*Intensity*
C = C	Aromatic rings	6.25–6.66	1500–1600	*v*
– C – O –	Alcohols, ethers, carboxylic acids, esters	7.7–10	1000–1300	*s*
>C = O	Aldehydes, ketones, carboxylic acids, esters	5.68–5.95	1680–1760	*s*
– O – H	Monomeric alcohols and phenols	2.74–2.78	3600–3650	*v*
– O – H	Hydrogen bonded alcohols and phenols Hydrogen bonded acids	2.94–3.11 3.33–4	3200–3400 2500–3000	*s* (*broad*) *v* (*broad*)
– N – H	Amines	2.86–3.03	3300–3500	*m*
C – N(C)(C)	Amines	7.36–8.49	1180–1360	*v*
– C ≡ N	Nitriles	4.42–4.52	2210–2260	*v*
– NO_2	Nitro compounds	6.5–7.5	1330–1540	Pair of strong bands

1.21. FUNCTIONAL GROUP REGION AND FINGERPRINT REGION

Functional group region. The region of 2.5 to 7.4 μm (4000 to 1430 cm^{-1}) is related with vibrational changes in the functional groups in the molecule. By observing absorption in this range, we can draw some conclusions regarding the functional groups present in the molecule. Hence this region is called the functional group region.

Fingerprint region. It is found that the region 7–11 μm (1430–910 cm^{-1}) gives a good deal of information regarding structure of the compound and various other groups besides the functional groups in the molecule and their relative positions. This region is called the fingerprint region. Frequencies in the fingerprint region can lead us to identify a given compound.

Intensity of absorption bands

It has been observed that greater the polarity of the bond, greater will be the intensity of absorption. For example, the carbonyl bond (C = O) is highly polar and therefore gives rise to intense absorption bond.

1.22. ABSORPTION FREQUENCY CHANGES FROM COMPOUND TO COMPOUND IN A GROUP

In the table of IR absorption frequencies, it is the range and not a particular frequency, which is given. This is because the absorption frequency depends not only upon the group but also upon the environments of that group, *i.e.* the other bonds and groups which are associated with that particular group in the molecule. We find that there is a slight variation in the absorption frequency if the environments are changed. This is illustrated in the table below which gives the absorption frequencies due to the carbonyl group in different compounds.

Table. Absorption frequencies due to carbonyl groups

Compound	*Frequency*	*Compound*	*Frequency*
$CH_3-\overset{O}{\overset{\parallel}{C}}-CH_3$ Acetone	1700 cm^{-1}	Cyclobutanone	1780 cm^{-1}
$CH_3-\overset{O}{\overset{\parallel}{C}}-CH_2CH_3$ Methyl ethyl ketone	1710 cm^{-1}	Cyclopentanone	1750 cm^{-1}
$CH_3-\overset{O}{\overset{\parallel}{C}}-CH=CH_2$ Methyl vinyl ketone	1680 cm^{-1}	Cyclohexanone	1710 cm^{-1}

Important

Negative evidence is very reliable in interpreting the IR spectrum of a compound. If an absorption frequency corresponding to a particular group is missing from the spectrum, it is almost sure that the group is absent in the molecule.

1.23. FACTORS INFLUENCING VIBRATIONAL FREQUENCY

It has been observed that absorption fequency of a bond in the IR region as calculated by Hook's law is never equal to the experimental value. This is because the vibration of each group is influenced by the structure of the molecule present in the nearest neighbourhood. The following factors are responsible for a shift in the vibration frequency (or wave number) from its theoretical or calculted value:

1. Electronic effects. The electronic effects such as inductive and resonance effects influence the stretching vibrational frequencies of both C = O group and C = C bonds. Normally, the two effects work together. In some cases, the inductive effect has a greater magnitude while in the other, the resonance effect dominates. As a result of these effects, there may be an increase or decrease in the value of force constant and consequently wave constants. Let us consider the examples involving inductive and resonance effects.

(*i*) Inductive effect. The inductive effect of the group present in the vicinity of the carbonyl group or double bond may be either + I effect (electron releasing) or – I effect (electron withdrawing). The alkyl groups have +1 effect and their presence tends to reduce the value of force constant. Therefore, the absorption shifts to lower wave number (as per Hook's law) when such groups are present. For example, wave number decreases when we move from formaldehyde to acetaldehyde to acetone.

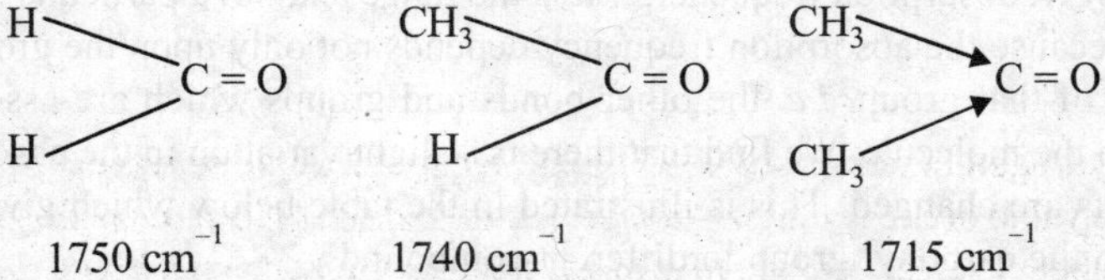

Thus, greater the + I effect, smaller will be the magnitude of wave number.

Similarly, the atoms such as Cl, Br, etc. have – I effect. Their presence in the molecule increases the value of force constant. Therefore, the absorption shifts to higher wave numbers. For example,

$$\begin{matrix} CH_3 \\ CH_3 \end{matrix} \!\!> C=O \qquad \begin{matrix} ClCH_2 \\ CH_3 \end{matrix} \!\!> C=O \qquad \begin{matrix} ClCH_2 \\ ClCH_2 \end{matrix} \!\!> C=O$$

$$1715\ cm^{-1} \qquad 1725\ cm^{-1} \qquad 1740\ cm^{-1}$$

Thus, greater the – I effect, higher will be the wave number for the absorption.

(*ii*) **Resonance effect**. The conjugation in the unsaturated hydrocarobons or in the aromatic ring decreases the magnitude of the force constant. Their absorption therefore shifts to lower wave number. For example,

$$CH_3-\overset{\overset{\displaystyle O}{\|}}{C}-CH=CH_2$$

Methyl vinyl ketone — Absorption shifts from 1715 to 1690 cm^{-1}

$$C_6H_5-\overset{\overset{\displaystyle O}{\|}}{C}-H$$

Benzaldehyde — Absorption shifts from 1715 to 1700 cm^{-1}

(2) Hydrogen Bonding. In phenols, carboxylic acids or amines which show hydrogen bonding, the value of the force constant decreases because of attraction and this shifts absorption towards lower wave numbers. The following points may be noted in this connection:

(a) Intermolecular hydrogen bonding gives broader bands while intramolecular hydrogen bonding gives sharp and well defined bands.

(b) The absorption bands in the intermolecular hydrogen bonding are concentration dependent while those in case of intramolecular hydrogen bonding are independent of concentration, since it takes place in the molecule itself.

We shall now consider the effect of the hydrogen bonding on the value of the absorption frequencies or wave numbers in some of the molecules.

(*i*) **Alcohols and phenols**. In case of both alcohols and phenols, the free O – H groups absorb strongly in the region of 3650 – 3584 cm^{-1}. In case the O – H group is involved in the H-bonding, a broad absorption band appears around 3350 cm^{-1}.

$$\cdots\cdots \overset{\delta^+}{H}-\underset{\underset{\displaystyle R}{|}}{\overset{\delta^-}{O}} \cdots\cdots \overset{\delta^+}{H}-\underset{\underset{\displaystyle R}{|}}{\overset{\delta^-}{O}} \cdots\cdots \overset{\delta^+}{H}-\underset{\underset{\displaystyle R}{|}}{\overset{\delta^-}{O}} \cdots\cdots$$

Because of hydrogen bonding, the O – H bond length tends to increase or the bond becomes weak, therefore, the value of the force constant decreases and so is the value of absorption frequency to lower wave number. The appearance of a broad band in the hydrogen bonded molecules is due to the reason that the molecules exist as dimers, trimers, tetramers etc. because of association. Therefore, the spectrum of closely spaced O – H absorption frequencies results in a broad band.

(*ii*) **Carboxylic acids.** Carboxylic acids are known to exist as dimers in solutions as illustrated below:

```
         O..........H – O
       //                \
R – C                     C – R
       \                 //
         O – H..........O
```

The H-bonds in the dimers are quite strong. Therefore, O – H stretching frequencies appear at much lower wave numbers (2700 – 2500 cm^{-1}) which are smaller than the stretching frequencies of C – H vibrations. Again, because of hydrogen bonding C = C stretching vibrations appear at lower wave numbers (1720 cm^{-1} as compared to free C = O stretching vibrations which appear around 1760 cm^{-1}.

(*iii*) **Amines.** The hydrogen bonding in amines is less as compared to the alcohols or the carboxylic acids because of lower polarity of N – H bond. Thus, shifts in the strecthing frequencies of the amines are lower than those observed in alcohols or carboxylic acids. For example, amines show N – H stretching at 3500 cm^{-1} in dilute solution and 3300 cm^{-1} in condensed phase.

1.24. RECORDING IR SPECTRA OF A COMPOUND

IR spectrophotometer consists of the following components:

(*a*) IR radiation source. A Nernst glower, which is a rod made of an alloy of zirconium, yttrium and erbium oxides, is used as IR radiation source. It is electrically heated to 1750 K.

(b) Rock-salt cells and prisms. Glass and quartz absorb IR radiations, hence these materials cannot be used for making cells for the sample and for prism which are used in the optics of the IR spectrophotometers. Rock-salt or some similar material is used for this purpose.

Sample preparation. Generally the sample in the form of solid is avoided because it scatters the radiations too much. For obtaining better spectra the sample is taken in the form of potassium bromide disc (a mixture of the sample and KBr). Second technique is to grind the sample with nujol which is a highly purified form of petroleum to obtain nujol mull which is subjected to investigation. Third technique is to dissolve the sample in pure $CHCl_3$ or CCl_4 . Solvents like water and alcohol are avoided. In special cases the substance, if gaseous, may be taken as such.

Recording of spectra. The sample to be investigated is placed in the path of IR rays. The intensity of transmitted light is measured automatically and a graph between the intensity of transmitted light and the wavelength is obtained.

1.25. HOOKE'S LAW

The vibrations of a diatomic molecule may be compared to those of a simple harmonic oscillator in which the force tending to restore an atom to its original position is proportional to the displacement of the atom from that position. This is called *Hooke's Law*. In other words, a covalent bond may be considered as a weightless spring. The restoring force F acting upon it when it is stretching to a distance (x) is given by Hooke's Law:

$$F = -kx$$

where k is the force constant of the spring (or bond).

A molecule may, therefore, be considered as a collection of balls and springs, where the balls are atoms and the springs are the chemical bonds. As a direct consequence, a vibrating diatomic molecule can

Robert Hooke

function as a simple harmonic oscillator.

The **frequency of its vibration** (ν) is given by

$$\bar{\nu} = \frac{1}{2\pi c}\sqrt{\frac{k}{R}}$$

where R is termed the **reduced mass** of the diatomic system. It is calculated from the expression, $R = m_1 m_2/(m_1 + m_2)$; where m_1 and m_2 are the masses of the two balls, c is the velocity of light.

Hooke's law

Such a simple mechanical analogy explains the position (frequency or wavelength) of various absorptions in the infrared spectrum from the nature of the atoms and the strengths of the bonds between them as illustrated below.

(*i*) *The* **smaller** *the constituent atoms of a diatomic molecule or group, smaller is the reduced mass (R) and, therefore,* **higher** *its vibrational frequency.*

The bonds between hydrogen and other atoms have the highest known fundamental frequencies, but covalent compounds of halogens of most metals show infrared absorptions of very low frequencies which are too low to be recorded on a routine spectrophotometer.

(*ii*) The force constant (k) is a measure of the strength of the bond and, therefore, it increases with the increase in bond order. As such, vibrational frequency also increases with the increase in bond strength or bond order. For example, the stretching frequencies increase as we move from C–C to C = C to C ≡ C (see Table under sec. 1.20).

1.26. NUMBER OF FUNDAMENTAL VIBRATIONS IN CASE OF DIATOMIC AND POLYATOMIC MOLECULES

The number of fundamental absorption bands exhibited by a molecule is related to the fundamental ways of vibrating or **vibrational modes** available to a molecule. In general:

$$\frac{\textit{Number of fundamental}}{\textit{absorption band}} = \frac{\textit{Number of fundamental}}{\textit{vibrational modes}}$$

Since a diatomic molecule has only one fundamental vibrational mode (*i.e.* stretching vibrations), it gives rise to one fundamental absorption band. However, polyatomic molecules can have more than one kind of vibrational movement; they exhibit more than one fundamental absorption bands as given below.

(*a*) Non-linear polyatomic molecules. It has been found that a non-linear polyatomic molecule can have $3n - 6$ modes of fundamental vibrations where n is the number of atoms present in the molecule. For example, water ($n = 3$) has three vibrational modes (*i.e.* $3 \times 3 - 6$), each of which can give an absorption band.

(*b*) Linear polyatomic molecules. A linear polyatomic molecule having n atoms can have $3n - 5$ fundamental vibrations. For example, carbon dioxide ($n = 3$) has four vibrational modes (*i.e.*, $3 \times 3 - 5$).

In case of larger molecules, the number of fundamental vibrations increases very rapidly with the increase in the value of n. But it has been observed that *the number of fundamental bands actually obtained is usually less than that expected from the theoretical number of fundamental vibrations*. This may be due to the following reasons:

(*i*) Some of the fundamental vibrations may be very weak and, therefore, may not be recorded as bands.

(*ii*) Some of the fundamental vibrations may be very close and as a result overlapping of bands may take place.

1.27. SELECTION RULES

Most of the molecules possess dipole moment due to unequal sharing of electrons within the bonds linking the constituent atoms. When a polar bond undergoes stretching vibration along the internuclear axis, the electron distribution changes. As a result dipole moment also undergoes a change. In other words, a vibration produces a fluctuating dipole moment. Interaction between the oscillating dipole moment and the infrared radiation leads to the absorption of energy from the radiation. This interaction can occur only if the dipole moment at one extreme of the vibration is different from the dipole moment at the other extreme of the vibration in a molecule.

There are certain rules, known a **selection rules,** which determine whether a vibration would be effective for a particular type of spectrum or not. For infrared spectrum, the rule is that *only those vibrations are effective in causing absorption which are not* **centro-symmetric** (*i.e.* the vibrations are not symmetrical about the centre of the molecule). Since most of the functional groups in organic chemistry are not centro-symmetric, they respond very well to infrared spectroscopy.

1.28. I.R SPECTRUM OF SOME TYPICAL ORGANIC MOLECULES

(1) I.R. spectrum of acetic acid $(CH_3-\overset{\overset{\Large O}{||}}{C}-OH)$

IR spectrum of acetic acid can be interpreted on the basics of group frequencies as under:

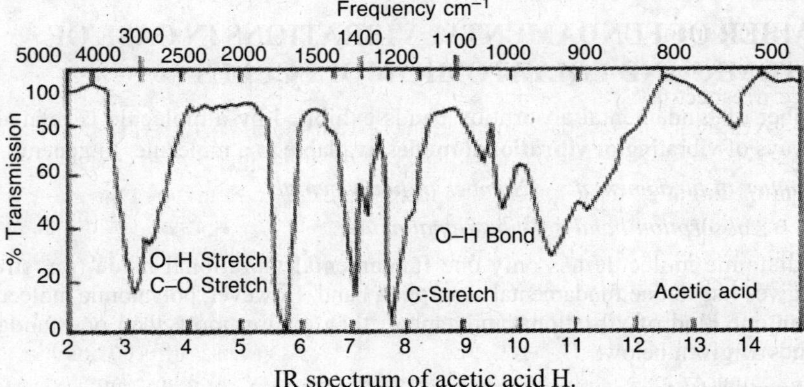

IR spectrum of acetic acid H.

The important bands in the above spectrum are as follows:

3030 cm^{-1} (3.3 μm); a broad band of hydrogen bonded O – H stretching of a carboxylic acid.

1710 cm^{-1} (5.85 μm); C = O stretching of a saturated chain carboxylic acid.

1280 cm^{-1} (7.8 μm); C – O stretching of a carboxylic acid.

945 cm^{-1}(10.6 mm); O — H bending of a carboxylic acid.

(2) IR spectrum of benzaldehyde, $C_6H_5-\overset{\overset{\Large O}{||}}{C}-H$

IR spectrum of benzaldehyde is reproduced as under

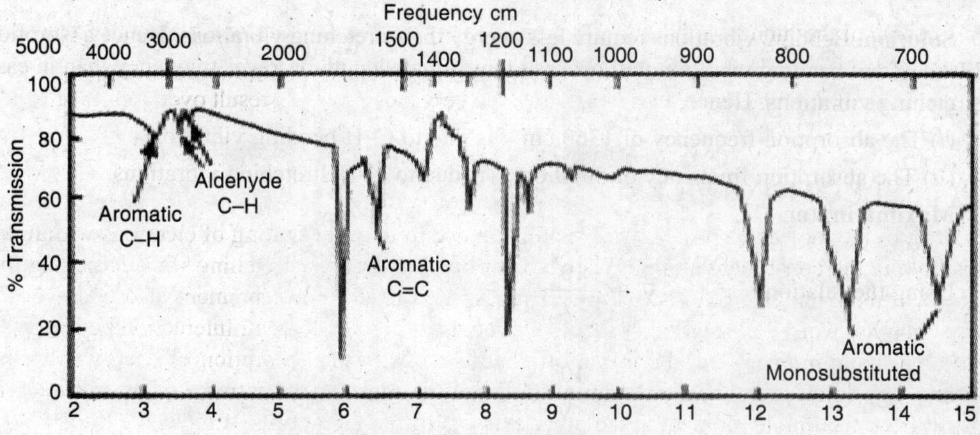

Fig. IR spectrum of benzaldehyde.

Based on the data for group frequencies, the spectrum can be interpreted as under:

3080 cm^{-1} (3.25 μm)	C–H stretching of aromatic ring.
2860 cm^{-1} (3.50 μm) 2775 cm^{-1} (3.60 μm)	C – H stretching of aldehyde
1700 cm^{-1} (5.88 μm)	C = O stretching of aromatic aldehyde
1595 cm^{-1} (6.27 μm)	C = C stretching of aromatic ring
745 cm^{-1} (13.42 μm)	Characteristic bands of aromatic substitution
685 cm^{-1} (14.55 μm)	Substitution (C – H deforming)

1.29. APPLICATIONS OF INFRARED SPECTROSCOPY

Some of the important applications of IR spectroscopy are given below:

(*i*) **Identification of the functional group.** Every functional group absorbs IR radiation in a particular range irrespective of the compound in which such group is present. For example, a carbonyl group gives IR absorption band in the range of 1650–1750 cm^{-1}. Hence by observing the absorption frequency, we can tell something about the functional group in the compound.

(*ii*) **Determination of structure.** Investigation of the IR spectra in the range 2.5–7 μm gives us information regarding functional groups present in the molecule. Further, the IR spectra in the fingerprint region gives us an insight into various other groups and their relative positions. The two studies together help us to arrive at the exact compound.

(*iii*) **Testing purity of a compound.** Generally, a pure compound has well defined and sharp absorption frequencies whereas the spectra of an impure compound has diffused and blurred lines due to additional impurity materials present in the sample. IR spectra of many pure compounds are available in literature. We can compare the IR spectra of the unknown sample with that of the pure sample. If the two tally exactly, the sample is pure, otherwise not.

(*iv*) **Following progress of a reaction.** During the course of a reaction, many intermediate compounds are formed which change into the products. These intermediate compounds can be identified by IR spectroscopy. IR spectrophotometer records the spectra continuously and by interpreting them, we can follow the reaction. The mechanism of the reaction can thus be proposed.

SOLVED STRUCTURAL PROBLEMS

Example 1. IR spectrum of acetone gives two maxima due to C–H vibrations at 1360 cm^{-1} and 3000 cm^{-1}. Identify the stretching and bending mode. Express the maxima in μm.

Solution. Bending vibrations require less energy than stretching vibrations. Hence absorption of IR in case of bending vibration will occur at longer wavelength or lower frequency than in case of stretching vibrations. Hence,

(*i*) The absorption frequency of 1360 cm^{-1} is due to C–H bending vibration.

(*ii*) The absorption frequency of 3000 cm^{-1} is due to C–H stretching vibrations.

Maxima in μm

Using the relation $\bar{\nu} = \dfrac{10^4}{\lambda}$

or $l = \dfrac{10^4}{\nu}$

(*i*) Substituting $\bar{\nu} = 1360$, $\lambda = \dfrac{10^4}{1360} = 7.35\ \mu m$

(*ii*) Substituting $\bar{\nu} = 3000$, $\lambda = \dfrac{10^4}{3000} = 3.33\ \mu m$

Example 2. Indicate the expected absorption regions in the IR spectra of the following compounds: isopropyl alcohol, dimethyl ether and toluene.

Solution. Isopropyl alcohol

The absorption bands anticipated in the IR spectra of isopropyl alcohol $(CH_3)_2$ CHOH are as follows:

(*i*)	A strong and broad band due to hydrogen bonded O–H stretching	3200–3400 cm^{-1}
(*ii*)	Due to C–H stretching	2850–2950 cm^{-1}
(*iii*)	Due to C–O stretching	1000–1300 cm^{-1}

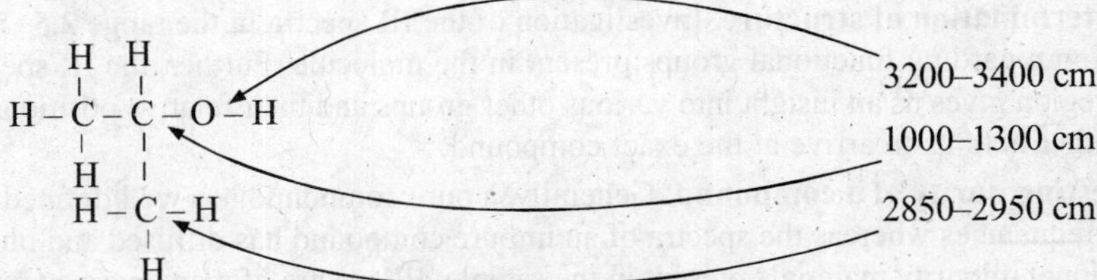

Dimethyl ether

(*i*)	Due to C–O stretching	1060–1150 cm^{-1}
(*ii*)	Due to C–H stretching of methyl groups	2850–2950 cm^{-1}

Toluene C_6H_5–CH_3

(*i*)	Due to C–H stretching of methyl groups	2850–2950 cm^{-1}
(*ii*)	Due to C–H stretching of aromatic ring	3000–3100 cm^{-1}
(*iii*)	Due to C–C stretching of aromatic ring	1500–1600 cm^{-1}

Example 3. Give the approximate IR bands in the following compounds:

(*a*) CH_3CH_2OH (*b*) $CH_3-\underset{\underset{O}{\|}}{C}-CH_3$ (*c*) $C_6H_5-NO_2$

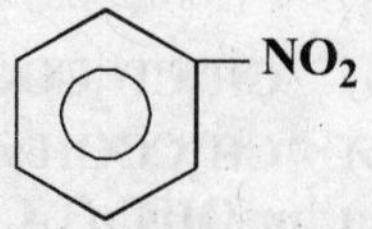

Solution. IR absorption bands in the compounds are obtained as given below:

CH_3CH_2OH

(*i*)	Due to hydrogen bonded O–H stretching	3200–3400 cm+$^{-1}$
(*ii*)	Due to C–H stretching	2850–2950 cm^{-1}
(*iii*)	Due to C–O stretching	1000–1300 cm^{-1}

$CH_3-\underset{\underset{O}{\|}}{C}-CH_3$

(*i*)	Due to C–H stretching	2850–2950 cm^{-1}
(*ii*)	Due to C = O stretching	1700 cm^{-1}

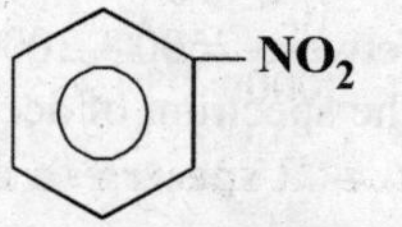

$C_6H_5-NO_2$

(*i*)	Due to C – H stretching of aromatic ring	3000–3100 cm^{-1}
(*ii*)	Due to C – C stretching of aromatic ring	1500–1600 cm^{-1}
(*iii*)	Due to $-NO_2$ group stretching	1330–1540 cm^{-1}

Example 4. Give the approximate position of characteristic absorption bands of carbonyl group in the IR spectra of the following compounds:

(*i*) CH_3CHO **(*ii*)** C_6H_5CHO

(*iii*) $C_6H_5COCH_3$ **(*iv*)** $HO-C_6H_4-COCH_3$ (para)

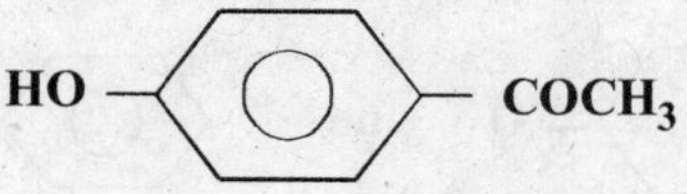

(*v*) $C_6H_4(OH)COCH_3$ (OH ortho to $COCH_3$)

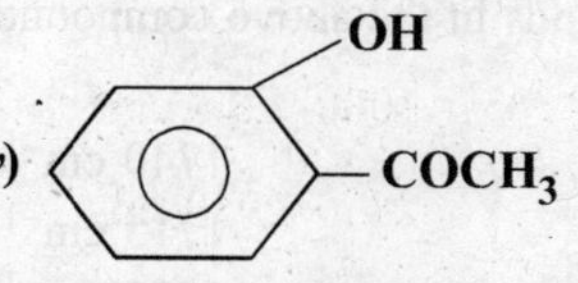

Solution. The absorption bands of the compounds are as given below:

(*i*)	CH_3CHO	~ 1740 cm^{-1}
(*ii*)	C_6H_5CHO	~ 1700 cm^{-1}
(*iii*)	$C_6H_5COCH_3$	~ 1690 cm^{-1}
(*iv*)	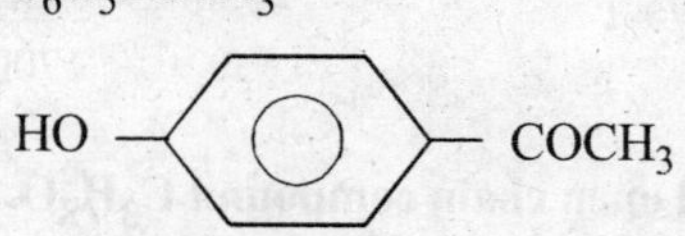 $HO-C_6H_4-COCH_3$	~ 1690 cm^{-1}
(*v*)	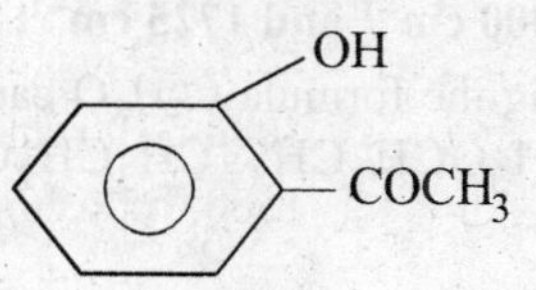 $C_6H_4(OH)COCH_3$ (ortho)	~ 1690 cm^{-1} (Due to chelation)

Example 5. How will you distinguish between the following pairs of compounds using IR spectra?

(*a*) $CH_3CH_2COCH_3$ and $CH_3OCH=CH_2$

(*b*) CH_3COOH and CH_3COCH_3

(*c*) *o*-OHC_6H_4COOH and *m*-$OH.C_6H_4COOH$.

Solution. It is possible to distinguish between the compounds as given below :

(*a*) $CH_3CH_2COCH_3$ and $CH_3OCH = CH_2$ (methyl ethyl ketone and methyl vinyl ether)

(*i*) A strong absorption band at about 1715 cm^{-1} due to C–O stretching in methyl ethyl ketone will be obtained. No such band will be seen in the spectrum of methyl vinyl ether.

(*ii*) An absorption band at about 3050 cm^{-1} due to C–H stretching of $CH_2 = CH$ – groups and another band at about 1100 cm^{-1} due to C–O stretching will be observed in the IR spectrum of methyl vinyl ether. Such bands will not be formed in the spectrum of methyl ethyl ketone.

(*b*) CH_3COOH and CH_3COCH_3 (acetic acid and acetone)

(*i*) Both the compounds will show an absorption band at about 1700 cm^{-1} due to C = O stretching.

(*ii*) Spectrum of acetic acid will show an absorption band between 2500 – 3000 cm^{-1} due to hydrogen bonded carboxy group which will not be observed in the spectrum of acetone.

(*c*) *o*-hydroxy benzoic acid and *m*-hydroxybenzoic acid

In *o*-hydroxy benzoic acid, there is a possibility of hydrogen bonding between the two groups as they are quite near to each other. Hence, we shall observe a band in the range 2500–3000 cm^{-1} while in *m*-hydroxy benzoic acid (no hydrogen bonding) we shall see an absorption band in the range 3200–3400 cm^{-1}.

Example 6 Give the approximate positions of characteristic absorption bands of the carbonyl group in the IR spectrum of the following compounds:

CH_3CHO, $CH_3COCH_2CH_3$, $CH_3COCH = CH_2$

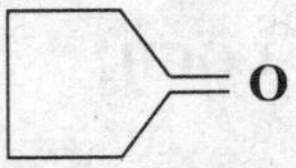

=O and –CHO

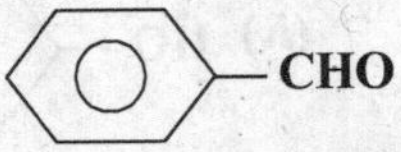

Solution. The approximate positions of the absorption bands in the above compounds are given below:

(*i*)	CH_3CHO	1740 cm^{-1}
(*ii*)	$CH_3COCH_2CH_3$	1715 cm^{-1}
(*iii*)	CH_3 $COCH = CH_2$	1675 cm^{-1}
(*iv*)	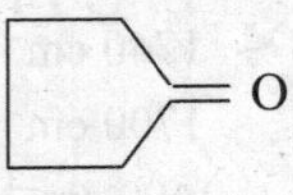 =O	1740 cm^{-1}
(*v*)	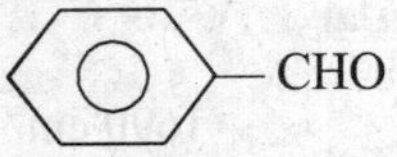 –CHO	1700 cm^{-1}

Example 7. Deduce the structure of saturated open chain compound C_3H_8O, which has an infrared absorption band at 2950 cm^{-1} but none near 3300 cm^{-1} and 1725 cm^{-1}.

Soution. (*i*) A saturated open chain compound having the formula C_3H_8O can be either an alcohol or ether. Thus the possible structures are $CH_3OCH_2CH_3$, $CH_3CH_2CH_2$ OH and $CH_3CHOHCH_3$.

(*ii*) There is no absorption band at 3300 cm^{-1} (due to hydrogen bonded alcohol). Hence the possibility of $CH_3CH_2CH_2OH$ and $CH_3CHOHCH_3$ is ruled out.

(*iii*) Absence of a band at 1725 cm^{-1} rules out the possibility of a carbonyl group (aldehyde or ketone).

(*iv*) The only possibility is $CH_3OCH_2CH_3$. The band at 2950 cm^{-1} is due to C–H stretching of alkyl groups.

Example 8. Which of the following compounds could give rise to infrared spectrum in the figure below?

Acetone, propionic acid, *n*-propyl alcohol or ethyl formate?

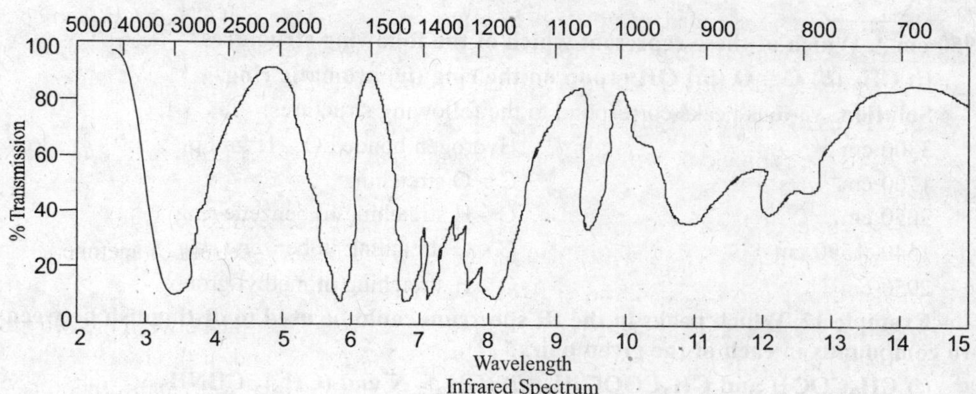

Infrared Spectrum

Solution. (*i*) We observe a band at 3000 cm^{-1}. This can be due to hydrogen bonded alcoholic or carboxylic group.

(*ii*) Since acetone and ethyl formate are not expected to give this band, the possibility of the compound being acetone or ethyl formate is ruled out.

(*iii*) The spectrum shows a band at 1700 cm^{-1} which is due to C = O of the propionic acid. No such group is present in propyl alcohol.

(*iv*) The band near 1300 cm^{-1} confirms the presence of propionic acid. Hence the spectrum relates to propionic acid.

Example 9. IR spectrum of ethyl acetate shows three important bands at 3002 cm^{-1}, 1742 cm^{-1} and 1240 cm^{-1}. Attribute these peaks to the following features of the molecule: CH_3 or CH_2 –CH_3, C = O, C–O.

Solution. The features responsible for each of the bands are as follows:

3002 cm^{-1}	C–H stretching of CH_3 or CH_2–CH_3
1742 cm^{-1}	C = O stretching
1240 cm^{-1}	C– O stretching

Example 10. Two isomers (A) and (B) having the molecular formula C_3H_6O exhibit the following peaks in the IR spectrum. (A) 1710 cm^{-1} (B) 3300 cm^{-1}, 1640 cm^{-1}. Write the structures of A and B on this basis.

Solution. (*i*) The compound with molecular formula C_3H_6O could be an aldehyde, ketone or an unsaturated alcohol, A and B are isomers of each other.

(*ii*) In the compound (A), we do not have an absorption band at 2600–2800 which is due to aldehydic $-\overset{\|}{C}-H$ stretching. Hence the possibility of the compound being an aldehyde is ruled out. A peak at 1710 cm^{-1} shows that it is CH_3COCH_3 *i.e.* acetone.

(*iii*) In the compound (B), the peaks are obtained at 3300 cm^{-1} and 1640 cm^{-1}. There is no peak at 1700 cm^{-1} thus ruling out the possibility of carboxyl group. The band at 3300 cm^{-1} is due to hydrogen bonded alcoholic group and at 1640 cm^{-1} is due to C = C stretching. Thus simultaneous presence of an alcoholic group and a C = C is indicated. The probable structure is $CH_2 = CH - CH_2OH$ *i.e.* allyl alcohol.

Example 11. The infrared spectrum of methyl salicylate (benzene ring bearing **OH** and **$COOCH_3$**) **shows the following peaks: 3300, 1700, 3050, 1540, 1590 and 2950 cm^{-1}. Which of these represent which of the following structures?**

(*i*) CH_3 (*ii*) C = O (*iii*) OH group on the ring (*iv*) aromatic ring.

Solution. Various peaks correspond to the following structures:

3300 cm^{-1}	Hydrogen bonded O – H group
1700 cm^{-1}	C = O stretching
3050 cm^{-1}	C – H stretching in benzene ring
1540, 1590 cm^{-1}	C – C stretching in benzene ring
2950 cm^{-1}	C – H stretching in methyl group

Example 12. Which peaks in the IR spectrum could be used to distinguish between the two compounds in each of the given pairs?

(*i*) CH_3COOH and $CH_3COOC_2H_5$ (*ii*) $(CH_3)_3$ N and $(CH_3)_2$ $CHNH_2$

Solution. CH_3COOH and $CH_3COOC_2H_5$

A peak around 2500–3000 cm^{-1} can distinguish between acetic acid and ethyl acetate because such a peak will be given by acetic acid only. This peak corresponds to hydrogen bonded –OH stretching.

$(CH_3)_3$ N and $(CH_3)_2$ $CHNH_2$

A peak around 3300–3500 cm^{-1} can distinguish between the two compounds. The peak corresponds to N–H stretching and will be given by $(CH_3)_2$ $CHNH_2$ only.

Example 13. A compound having molecular formula C_8H_8O shows strong peaks at 1685 cm^{-1}. Which of the following is the likely structure of the compound (*i*) C_6H_5 $COCH_3$ (*ii*) C_6H_5 CH_2CHO (*iii*) $C_6H_5OCH = CH_2$?

Solution. (*i*) A band at 1685 cm^{-1} shows the presence of a carbonyl group. Hence the structure $C_6H_5OCH = CH_2$ is ruled out.

(*ii*) There are two possibilities, phenyl methyl ketone and phenyl acetaldehyde. Aliphatic aldehydes and phenyl substituted aliphatic aldehydes give a peak at about 1740 cm^{-1}. Hence the possibility of $C_6H_5CH_2$ CHO which is phenyl substituted acetaldehyde is ruled out. The only possibility left out is $C_6H_5COCH_3$.

Example 14. An aromatic compound having molecular formula C_8H_8O shows an IR absorption band at 1690 cm^{-1} but no band at 3300 cm^{-1}. Deduce the structure of the compound.

Solution. An absorption band at 1690 cm^{-1} shows that a carbonyl group (> C = O) is present in the molecule. Absence of a band at 3300 cm^{-1} shows that the molecule does not contain an alcoholic group. Hence the possible structure of the compound is

$$C_6H_5-\overset{O}{\overset{\|}{C}}-CH_3$$

methyl phenyl ketone

Example 15. Draw conclusion about the structure of the compound from the following IR spectrum. The molecular formula of the compound is C_7H_8O.

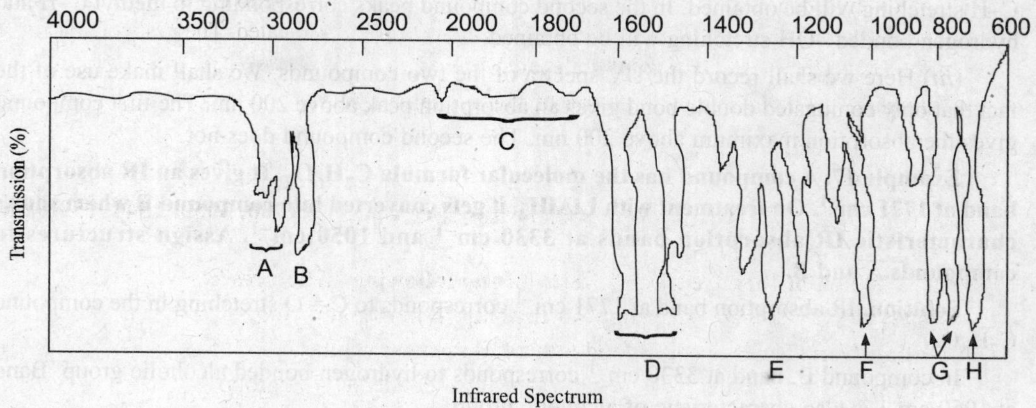

Infrared Spectrum

Solution. The absorption frequencies as observed in the spectrum are related to the groups as follows:

A	Aromatic C–H stretching	3060, 3030 & 3000 cm^{-1}
B	Methyl C–H stretching	2950, 2835 cm^{-1}
C	Overtone	2000–1650 cm^{-1}
D	C ----- C ring stretching	1590, 1480 cm^{-1}
E & F	C–O–C stretching	1245, 1030 cm^{-1}
G	C–H bending	800, 740 cm^{-1}
H	C ----- C bending	680 cm^{-1}

Based on the above observations the structure of the compound is

(benzene ring)–OCH_3 Anisole

Example 16. Indicate whether you will use IR or UV spectroscopy for distinguishing between the following pairs of compounds. Explain on what basis distinction can be made.

(*i*) $CH_2 = CH - CH_2 - OCH_3$ and $CH_3CH_2COCH_3$

(*ii*) $CH_3 - O - CH_3$ and $CH_3 - CH_2 - OH$

(*iii*) $CH_2 = CH\overset{O}{\overset{\|}{C}}(CH_2)_2 - CH_3$ and $CH_2 = CH - (CH_2)_2 - \overset{O}{\overset{\|}{C}} - CH_3$

Solution. (*i*) We shall use IR spectroscopy for the first pair because in the first compound, we shall obtain characteristic absorption peaks corresponding to the C = C, C–O–C and C–H stretching. In the second compound, the absorption peaks will correspond to the C = O and C–H stretching vibrations.

(*ii*) In the second pair, again IR spectrum will be used for distinguishing between the compounds. In the first compound, the peaks corresponding to C–O–C and methyl C–H stretching will be obtained. In the second compound peaks corresponding to methyl C–H and hydrogen bonded –OH stretching will be obtained.

(*iii*) Here we shall record the UV spectra of the two compounds. We shall make use of the fact that only conjugated double bond gives an absorption peak above 200 nm. The first compound gives the absorption maximum above 200 nm. The second compound does not.

Example 17. A compound has the molecular formula $C_7H_6O_2$. It gives an IR absorption band at 1771 cm^{-1}. On treatment with $LiAlH_4$, it gets converted into compound B which shows characteristic IR absorption bands at 3330 cm^{-1} and 1050 cm^{-1}. Assign structures to compounds A and B.

Solution. IR absorption band at 1771 cm^{-1} corresponds to C = O stretching in the compound $C_7H_6O_2$.

In compound B, band at 3330 cm^{-1} corresponds to hydrogen bonded alcoholic group. Band at 1050 cm^{-1} is also characteristic of alcoholic group.

With the formula, $C_7H_6O_2$, there are two possible compounds which conform to the given data.

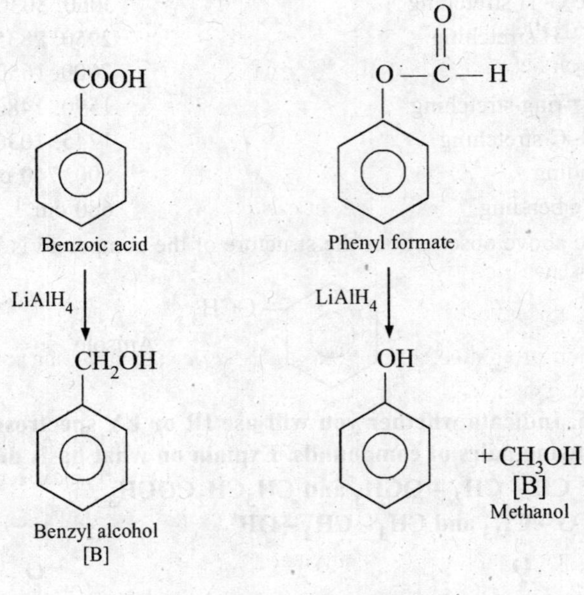

COOH

Benzoic acid

$LiAlH_4$

CH_2OH

Benzyl alcohol
[B]

O—C(=O)—H

Phenyl formate

$LiAlH_4$

OH

+ CH_3OH
[B]
Methanol

Example 18. Distinguish between $CH_2 = CH - CH_2$ OH and CH_3 CH_2 CHO with IR spectroscopy.

Solution. (*i*) The compound that absorbs around 1700 cm^{-1} is CH_3CH_2CHO.

(*ii*) The compound that absorbs around 3400 cm^{-1} (due to O–H stretching) and around 1620 cm^{-1} (due to C = C stretching) is $CH_2 = CH - CH_2$ OH.

Example 19. What absorption in I.R. spectrum could be used to distinguish between acetone and ethanol?

Solution. Acetone gives a band at 1700 cm^{-1} due to C = O stretching whereas ethanol gives a band at 3400 cm^{-1} due to O–H stretching (hydrogen bonded).

Example 20. Oxidation of 2-propanol to propanone is being carried out. How would you study the progress of the reaction with I.R. spectrum?

Or

2-propanal on oxidation gives a ketone and on dehydration gives an olefin. Explain the use of IR spectroscopy in the identification of the starting material and the products.

Solution. 2-propanol gives a strong band due to hydrogen bonded O–H stretching at 3400 cm^{-1}. On being converted into propanone on oxidation, it gives an absorption band at 1700 cm^{-1} due to C = O stretching.

EXERCISES
(Based on Question Papers of Different Universities)

ULTRAVIOLET SPECTROSCOPY

Objective Questions

Multiple Choice Questions (Choose the Correct Option)

1. The range of U.V. light is

(*a*) 500 – 1000 nm (*b*) 10 – 400 nm

(*c*) 1000 – 1500 nm (*d*) 1500 – 2000 nm

2. The relation between λ, ν and c is

(*a*) $\lambda \times \gamma = c$ (*b*) $\nu \times c = \lambda$

(*c*) $\lambda \times c = \nu$ (*d*) $\lambda \times \nu \times c = 1$

3. The equation $\log_{10} \frac{I_0}{I} = \in C \ell$ is an expression of

(*a*) Beer's law (*b*) Lambert's law

(*c*) Beer-Lambert's law (*d*) Hooke's law

4. Introduction of an auxochrome into a system brings about an increase in the intensity of absorption. This is called

(*a*) hypsochromic effect (*b*) bathochromic effect

(*c*) hypochromic effect (*d*) hyperchromic effect

5. Increasing order of UV absorption maxima of ethylene, butadiene and naphthalene is

(*a*) butadiene < naphthalene < ethylene

(*b*) ethylene < naphthalene < butadiene

(*c*) ethylene < butadiene < naphthalene

(*d*) naphthalene < butadiene < ethylene

6. Which of the following do not absorb UV radiations?

(*a*) Ethyl alcohol (*b*) Benzene

(*c*) Water (*d*) Acetone

7. Which of the following cannot be detected with the help of UV absorption spectra?

(a) Conjugation (b) Functional group

(c) Geometrical isomerism (d) Optical isomerism

ANSWERS

1. (*b*)	**2.** (*a*)	**3.** (*c*)	**4.** (*d*)	**5.** (*c*)	**6.** (*a*), (c)
7. (d)					

Short Answer Questions

1. Which is Lambert-Beer Law? Express it mathematically and give significance of each term.
2. Explain how you will distinguish between $CH_3-\overset{\overset{\large O}{||}}{C}-CH_3$ and $CH_2=CH-\overset{\overset{\large O}{||}}{C}-CH_3$ with the help of UV spectroscopy.
3. Why is absorption spectroscopy used to study the spectra of organic molecules?
4. Which of the following absorbs UV radiations?
   (*i*) Benzoic acid (*b*) Ethyl alcohol
   (*ii*) Butadiene (*d*) Benzene
5. Define 'Electronic spectroscopy'. What is its range?
6. Discuss the principle of UV spectroscopy.
7. Following the Woodward and Fieser rules, calculate the value of absorption maxima for the following compound:

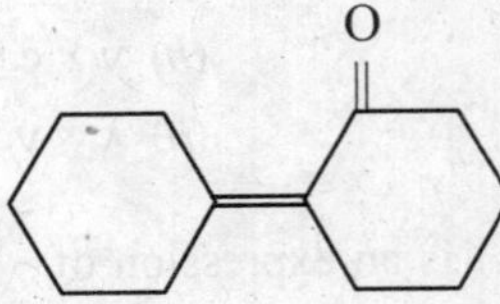

8. Explain the following:
   (*i*) Chromophores (*ii*) Auxochromophores
9. Why does the transition of $\pi \rightarrow \pi^*$ require more energy than required for $n \rightarrow \pi^*$ transition?
10. Which part of organic molecules is excited by UV radiations?
11. Explain (*i*) Chromophores (*ii*) Auxochromophores.
12. The UV spectra of acetone shows two peaks at λ_{max} = 280 nm ($\in_{max}$ = 15) and λ_{max} = 190 nm ($\in_{max}$ = 100). Identify.
13. Discuss the following terms:
    (*i*) Hyperchromic shift (*ii*) Hypochromic shift
14. What type of electronic excitations are possible when acetone absorbs electromagnetic radiations in UV region ? Arrange the excitations in decreasing order of energy.

General Questions

1. Illustrate four important applications of UV spectroscopy.
2. What is meant by $n \rightarrow \pi^*$, $\pi \rightarrow \pi^*$ and $\sigma \rightarrow \sigma^*$ transitions? Arrange the excitations in order of decreasing energy.
3. (*a*) What is the principle of ultraviolet spectroscopy?
   (*b*) What is Lambert-Beer Law? Express it mathematically and give significance of each term.
4. Explain the following:
   (*i*) Beer-Lambert Law (*ii*) Chromophores (*iii*) Woodward rules.

5. (a) Write notes on:

(*i*) Chromophores (*ii*) Auxochromophores

(*b*) Give four advantages of spectroscopic methods over classical methods.

6. (*a*) Discuss the principle of UV spectroscopy.

(*b*) A saturated carbonyl compound showed absorption at 190 nm and 300 nm. What type of transition is associated with each absorption?

(*c*) State and explain Beer-Lambert Law.

7. (*a*) Explain the following terms:

(*i*) Chromophores (*ii*) Auxochromophores

(*b*) Compounds A, B and C have the formula C_5H_8 and upon hydrogen yield n- pentane showing following λ_{max} values:

A = 1760 mμ, B = 211 mμ, C = 251 mμ,

(1-pentane has λ_{max} = 1780 *m*μ). Assign structures to A, B and C.

8. Describe the Woodward and Fieser rules for calculating λ_{max} in dienenes and α, β - unsaturated ketones.

9. (*a*) Explain the principle of UV spectroscopy.

(*b*) Define and explain the term chromophore.

(*c*) Which of the following compounds absorb U.V. raditions? Give reasons and explain the type of electronic transitions taking place.

(*i*) Nitrobenzene (*ii*) Heptane (*iii*) Benzene

10. Explain the various transitions in UV spectroscopy with suitable examples.

INFRARED SPECTROSCOPY

Objective Questions

Multiple Choice Questions (Choose the Correct Option)

1. IR radiations have a range of

(*a*) 800 – 20000 nm (*b*) 10 – 400 nm

(*c*) 400 – 800 nm (*d*) 20000 – 40000 nm

2. Aldehyde group shows IR absorption frequency in the range of

(*a*) 3300 cm^{-1} (*b*) 2650 – 2880 cm^{-1}

(*c*) 3010 – 3100 cm^{-1} (*d*) 2850 – 2960 cm^{-1}

3. Hooke's law for the vibrations of a diatomic molecules may be represented as

(*a*) $F = kx$ (*b*) $F = -kx$

(*c*) $x = kF$ (*d*) $x = -kF$

4. For infrared spectrum, only those vibrations are effective in causing absorption which are not centro-symmetric. This is the statement of

(*a*) Beer's law (*b*) Hooke's law

(*c*) Lambert's law (*d*) Selection rule

5. Which compound will show maximum IR absorption frequency?

(*a*) Acetone (*b*) Methyl ethyl ketone

(*c*) cyclobutanone (*d*) Methyl vinyl ketone

6. What is the effect of intramolecular hydrogen bonding on the IR absorption?

(*a*) Broader bands are obtained

(*b*) Diffused bands are obtained

(*c*) Sharp and well defined bands are obtained

(*d*) None of the above

ANSWERS

1. (*a*)	**2.** (*b*)	**3.** (*b*)	**4.** (*d*)	**5.** (*c*)	**6.** (*c*)

Short Answer Questions

1. Describe the principle of infra-red spectroscopy.
2. What do you understand by stretching and bonding vibrations?
3. Give a brief account of the bending vibrations.
4. How would IR peaks of the following compounds differ?

   (*i*) Acetone and ethanol (*ii*) Acetic acid and methanol
5. The infra-red structure of a compound is often referred to as fingerprint. Explain.
6. How will you differentiate between $CH_3-\overset{\overset{\displaystyle O}{\|}}{C}-CH=CH_2$ and CH_3COCH_3 using IR and UV spectra?
7. Why does hydrogen bonding shift the absorption spectrum to greater wavelength?
8. How does infra-red spectroscopy show whether a particular OH group in the organic compound is associated or not?
9. Give the characteristic I.R. peaks of the following compounds:

   (*i*) $CH_3-C\equiv CH$ (*ii*) CH_3-CH_2OH
10. How many fundamental vibrations each of the following molecules can have?

    (*i*) C_6H_6 (*ii*) $CH\equiv CH$ (*iii*) CH_3OH
11. Give the approximate positions of the characteristic infra-red bands in the following compounds:

    (*i*) CH_3CH_2OH (*ii*) CH_3COCH_3 (*iii*) $CH_2=CHCOCH_3$
12. What is the principal difference between UV and IR spectroscopy?
13. How will you distinguish between the following pairs of compounds with the help of IR spectroscopy?

    (*i*) $CH_3-CH_2-\overset{\overset{\displaystyle O}{\|}}{C}-CH_3$ and $CH_2-CH-O-CH_3$

    (*ii*) CH_3COOH and $CH_3\ COCH_3$
14. Describe the various molecular vibrations in IR spectroscopy.

General Questions

1. (*a*) What is meant by stretching vibrations? How many types of stretching vibrations a molecule of H_2O can have?

(*b*) How do the IR spectrum of the following compounds differ?

(*i*) Acetone and ethanol (*ii*) Acetic acid and methanol

2. (*a*) What is the principle of IR spectroscopy?

(*b*) What absorption in IR spectrum can be used to distinguish between methanol and ethanol?

3. (*a*) What is the principal difference between UV and IR spectroscopy ?

(*b*) How will you distinguish between following pairs of compounds on the basis of IR spectroscopy ?

(*i*) Ethyl alcohol and diethyl ether (*ii*) Acetic acid and ethyl acetate

(*iii*) Acetic acid and acetone.

(*c*) Discuss the various molecular vibrations in IR spectroscopy.

4. (*a*) Discuss the principle of IR spectroscopy.

(*b*) Give the appropriate position of the characteristic absorption bonds of the carbonyl groups in IR spectrum of acetaldehyde.

(*c*) Two isomers A and B of the molecular formula C_3H_6O give IR absorption bands near 1650 cm^{-1} and 1710 cm^{-1}.

Assign structural formula to A and B consistent with their IR absorption spectroscopy.

5. Indicate whether you would use IR and UV spectroscopy for distinguishing between the following pairs of compounds. Explain on what basis the distinction can be made:

(*i*) $CH_2=CH-CH_2-O-CH_3$ and $CH_3-CH_2-\overset{\overset{\displaystyle O}{\|}}{C}-CH_3$

(*ii*) CH_3-O-CH_3 and CH_3-CH_2-OH

(*iii*) $CH_2=CH-\overset{\overset{\displaystyle O}{\|}}{C}-(CH_2)_2-CH_3$ and $CH_2=CH-(CH_2)_2-\overset{\overset{\displaystyle O}{\|}}{C}-CH_3$

6. (*a*) Write in brief the principle of IR spectroscopy.

(*b*) How will you distinguish between the following pairs of compounds by IR spectroscopy?

(*i*) CH_3COCH_3 and CH_3COOH (*ii*) CH_3COCH_3 and CH_3CH_2OH

(*c*) Indicate whether you would prefer IR or UV spectroscopy for distinguishing between:

(*i*) CH_3OCH_3 and CH_3CH_2OH

(*ii*) $CH_2=CHOCH_2CH_3$ and $CH_2=CHCH_2CH_2COCH_3$

7. How can infra-red data be used to distinguish the compounds in the following pairs?

$HO-C_6H_4-O-C_6H_4-OH$ and $C_6H_5-\overset{\overset{\displaystyle O}{\|}}{C}-C_6H_5$

8. Account for the following:

   (*a*) O – H bond absorbs at higher wave number than N – H bond though the mass of oxygen is higher than that of nitrogen.

   (*b*) Triple bond absorbs at higher wave number than double bond.

9. Discuss the factors which influence the positions of absorption frequencies from their normal values.

10. (*a*) Which absorption in IR spectrum would be used to distinguish the following?

    (*i*) CH_3COOH and CH_3COCH_3 (*ii*) $CH_3CH_2NHCH_3$ and $(CH_3)_3N$

    (*b*) What do you understand by stretching and bending vibrations?

    (*c*) Oxidation of 2-propanol to propanone is being carried out. How would you study the progress of reaction with the help of IR spectroscopy?

11. Enlist the various applications of infra-red spectroscopy.

2

Alcohols

2.1. CLASSIFICATION

Alcohols are organic compounds having one or more –OH groups attached to carbon in the molecule.

Alcohols may be classified as under:

(*a*) Aliphatic alcohols.

(*i*) **Monohydric alcohols.** Alcohols having one alcoholic group (–OH) in the molecule are called monohydric alcohols. For example,

$CH_3 - OH$ (Methyl alcohol) C_2H_5OH (Ethyl alcohol)

(*ii*) **Dihydric alcohols.** Alcohols having two alcoholic groups in the molecule are called dihydric alcohols. For example,

$$\begin{array}{l} CH_2OH \\ | \\ CH_2OH \end{array} \quad \text{Ethylene glycol}$$

(*iii*) **Trihydric alcohols.** Alcohols having three alcoholic groups in the molecule are called trihydric alcohols. For example,

$$\begin{array}{l} CH_2OH \\ | \\ CHOH \\ | \\ CH_2OH \end{array} \quad \text{Glycerol}$$

It may be mentioned that more than one alcoholic groups attached to one carbon make the molecule unstable with liberation of a water molecule.

(*b*) Aromatic alcohols. These are aromatic compounds in which the alcohol group is attached to the side-chain of the benzene nucleus. For example,

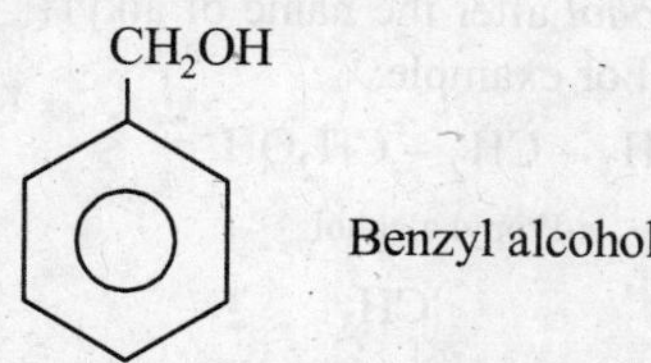

Benzyl alcohol

It may be noted that aromatic compounds containing –OH group directly attached to nucleus don't come in the category of alcohols. Such compounds are phenols and are different from alcohols in many respects.

2.2. PRIMARY, SECONDARY AND TERTIARY ALCOHOLS

Monohydric alcohols are classified as under:

(*a*) **Primary alcohols** *are those alcohols which have just one (or none) alkyl group linked to the carbon attached directly to hydroxyl group.*

They have the general formula $R - CH_2 - OH$

For example,

$$\underset{\text{Methyl alcohol}}{H-\overset{\displaystyle H}{\underset{\displaystyle H}{\overset{|}{\underset{|}{C}}}}-OH} \qquad \underset{\text{Ethyl alcohol}}{CH_3-\overset{\displaystyle H}{\underset{\displaystyle H}{\overset{|}{\underset{|}{C}}}}-OH}$$

Therefore, primary alcoholic group is $- CH_2OH$.

(*b*) **Secondary alcohols** *are those alcohols which have two alkyl groups linked to the carbon attached directly to hydroxyl group*. They have the general formula:

$$\begin{matrix} R \\ R \end{matrix} \!\!> CHOH$$

For example, $\begin{matrix} CH_3 \\ CH_3 \end{matrix} \!\!> CHOH$ *Iso*-propyl alcohol

Therefore, secondary alcoholic group is $> CHOH$

(*c*) **Tertiary alcohols** *are those alcohols which have three alkyl groups linked to the carbon attached directly to hydroxyl group*. They have the general formula:

$$\begin{matrix} R \\ R \end{matrix} \!\!> \underset{\displaystyle R}{\underset{|}{C}} - OH$$

For example, $CH_3-\overset{\displaystyle CH_3}{\underset{\displaystyle CH_3}{\overset{|}{\underset{|}{C}}}}-OH$ *tert*-Butyl alcohol

Therefore, tertiary alcoholic group is $\geqslant C - OH$

2.3. NOMENCLATURE OF ALCOHOLS

Three different systems of nomenclature are in use as described hereunder.

(*i*) **The common system.** According to this system, which is used only for the simpler alcohols, alcohols are named by adding the word *alcohol* after the name of alkyl group present in the molecule. For example:

CH_3OH Methyl alcohol CH_3CH_2OH Ethyl alcohol $CH_3 - CH_2 - CH_2OH$ *n*-Propyl alcohol

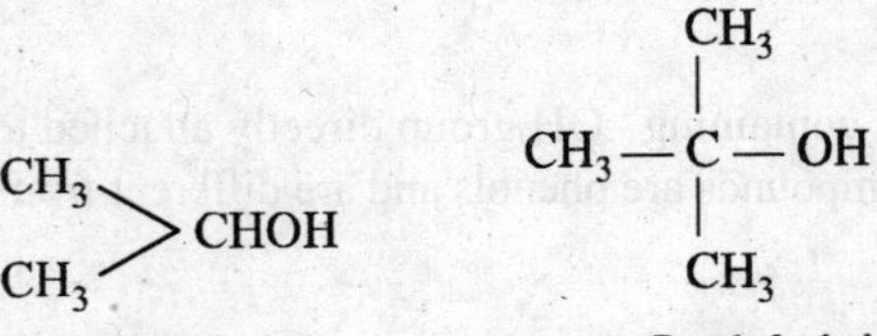

$\begin{matrix} CH_3 \\ CH_3 \end{matrix} \!\!> CHOH$ Isopropyl alcohol $CH_3-\overset{\displaystyle CH_3}{\underset{\displaystyle CH_3}{\overset{|}{\underset{|}{C}}}}-OH$ tert-Butyl alcohol

Alcoholic beverages

(*ii*) **The carbinol system.** According to this system, all alcohols are named as derivatives of methyl alcohol, which itself is known as carbinol.

For example,

$CH_3 — CH_2OH$
Methyl carbinol

$(CH_3)_2CHOH$
Dimethyl carbinol

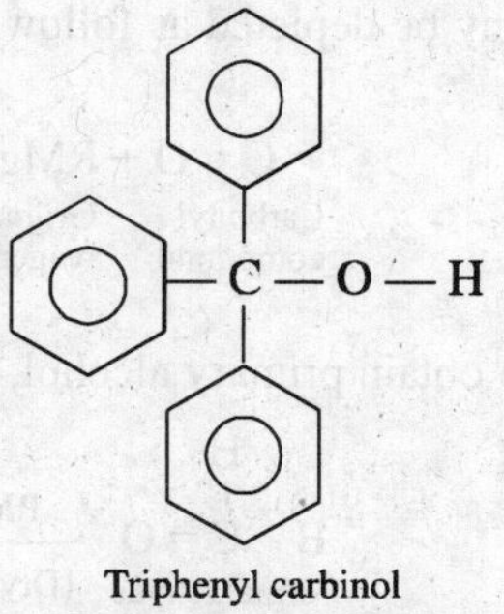

$(C_6H_5)_3C—O—H$
Triphenyl carbinol

However, this system is used only rarely and has been replaced by IUPAC system.

(*iii*) **The IUPAC system.** IUPAC names of alcohols end with the suffix *ol*. Of course, in case of higher alcohols, the longest chain of carbon atoms bearing the –OH groups is selected as the parent chain and the positions of –OH groups are indicated by suitable numbers.

CH_3OH
Methanol

$CH_3CHOHCH_3$
2-Propanol

$CH_3 - C(CH_3)_2 - OH$
2-Methylpropan-2-ol

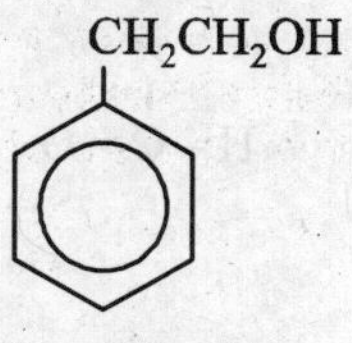

$C_6H_5CH_2CH_2OH$
2-Phenylethanol

The common and IUPAC names of a few alcohols are given in the table below

Formula	*Common Name*	*IUPAC Name*
CH_3OH	Methyl alcohol	Methanol
CH_3CH_2OH	Ethyl alcohol	Ethanol
$CH_3CH_2CH_2OH$	n-Propyl alcohol	Propan-1-ol
$CH_3CHOHCH_3$	Iso-Propyl alcohol	Propan-2-ol
$CH_3 - CH_2 - CH_2 - CH_2OH$	n-Butyl alcohol	Butan-1-ol
$CH_3 - CH_2 - CHOH - CH_3$	Sec-Butyl alcohol	Butan-2-ol
$CH_3 — CH(CH_3) — CH_2OH$	Iso-butyl alcohol	2-Methylpropan-1-ol
$CH_3 — C(CH_3)_2 — OH$	Tert-butyl alcohol	2-Methylpropan-2-ol
$(CH_3)_2CHCH_2CH_2OH$	Isoamyl alcohol	3-Methylbutan-1-ol

2.4. METHODS OF PREPARATION OF ALCOHOLS

Some important methods of preparation of alcohols are described as follows:

1. From Grignard reagent. The most important method for the preparation of primary, secondary and tertiary alcohols is by the reaction between a Grignard reagent and a carbonyl compound such as an aldehyde or a ketone. The addition product initially formed is decomposed with water, or preferably dilute mineral acid such as HCl or H_2SO_4 to get the alcohols. The general reaction may be depicted as follows:

$$\underset{\text{Carbonyl compound}}{>C=O} + \underset{\text{Grignard reagent}}{RMgX} \xrightarrow{\text{Dry ether}} \underset{\text{Addition product}}{-\overset{|}{\underset{R}{\underset{|}{C}}}-OMgX} \xrightarrow[-Mg(OH)X]{H_2O} \underset{\text{Alcohol}}{-\overset{|}{\underset{R}{\underset{|}{C}}}-OH}$$

(*i*) To obtain primary alcohol, formaldehyde is taken as the carbonyl compound.

$$\underset{\text{Formaldehyde}}{H-\overset{H}{\overset{|}{C}}=O} \xrightarrow[\text{(Dry ether)}]{RMgX} H-\overset{H}{\overset{|}{\underset{CH_3}{\underset{|}{C}}}}-OMgX \xrightarrow[-Mg(OH)X]{H_2O} \underset{\text{Primary alcohol}}{H-\overset{H}{\overset{|}{\underset{R}{\underset{|}{C}}}}-OH}$$

For example,

$$H-\overset{H}{\overset{|}{C}}=O \xrightarrow[\text{(Dry ether)}]{CH_3MgI} H-\overset{H}{\overset{|}{\underset{CH_3}{\underset{|}{C}}}}-OMgI \xrightarrow[-Mg(OH)X]{H_2O} \underset{\text{Ethyl alcohol}}{H-\overset{H}{\overset{|}{\underset{CH_3}{\underset{|}{C}}}}-OH}$$

(*ii*) To obtain a secondary alcohol, the Grignard reagent is treated with an aldehyde other than formaldehyde.

$$\underset{\text{Some aldehyde other than Formaldehyde}}{R'-\overset{H}{\overset{|}{C}}=O} \xrightarrow[\text{(Dry ether)}]{RMgX} R'-\overset{H}{\overset{|}{\underset{R}{\underset{|}{C}}}}-OMgX \xrightarrow[-Mg(OH)X]{H_2O} \underset{\text{Secondary alcohol}}{R'-\overset{H}{\overset{|}{\underset{R}{\underset{|}{C}}}}-OH}$$

R and R′ could be same or different.

For example,

$$\underset{\text{Acetaldehyde}}{CH_3-\overset{H}{\overset{|}{C}}=O} \xrightarrow[\text{(Dry ether)}]{CH_3MgI} CH_3-\overset{H}{\overset{|}{\underset{CH_3}{\underset{|}{C}}}}-OMgI \xrightarrow[-Mg(OH)I]{H_2O} \underset{\text{Isopropyl alcohol}}{CH_3-\overset{H}{\overset{|}{\underset{CH_3}{\underset{|}{C}}}}-OH}$$

(*iii*) Use of a ketone to react with Grignard reagent leads to the ultimate formation of a **tertiary alcohol**. For obtaining tertiary alcohol, a suitable Grignard reagent and a ketone are taken together.

$$\underset{\text{Ketone}}{R''-\overset{R'}{\overset{|}{C}}=O} \xrightarrow[\text{(Dry ether)}]{R-MgX} R'-\overset{R''}{\overset{|}{\underset{R}{\underset{|}{C}}}}-OMgX \xrightarrow[-Mg(OH)X]{H_2O} \underset{\text{Tertiary alcohol}}{R'-\overset{R''}{\overset{|}{\underset{R}{\underset{|}{C}}}}-OH}$$

For example,

$$\underset{\text{Acetone}}{CH_3-\overset{CH_3}{\overset{|}{C}}=O} \xrightarrow[\text{(Dry ether)}]{CH_3MgI} CH_3-\overset{CH_3}{\overset{|}{\underset{CH_3}{\underset{|}{C}}}}-OMgI \xrightarrow[-Mg(OH)I]{H_2O} \underset{\textit{Tert}\text{-Butyl alcohol}}{CH_3-\overset{CH_3}{\overset{|}{\underset{CH_3}{\underset{|}{C}}}}-OH}$$

It may be noted that in all these reactions, the number of carbon atoms in the alcohol obtained is equal to the sum of the carbon atoms in the aldehyde or ketone and Grignard reagent.

Mechanism of reaction

The carbon-magnesium bond of the Grignard reagent is a highly polar bond; carbon being negatively charged and magnesium being positively charged. The negatively charged carbon of the alkyl group becomes attached to the positively polarised carbon of the carbonyl group. Simultaneously, the positively charged magnesium end of the remaining part of Grignard reagent gets attached to negatively polarised oxygen. Magnesium salt of the feebly acidic alcohol is formed as shown below.

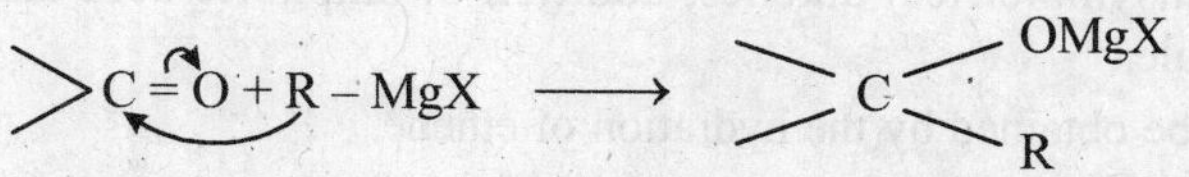

$$>C=O + R-MgX \longrightarrow >C<^{OMgX}_{R}$$

The addition complex thus formed is decomposed by water in an acidic solution to form alcohols.

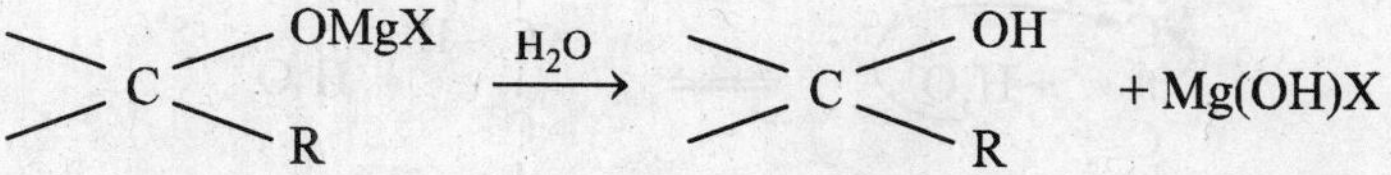

$$>C<^{OMgX}_{R} \xrightarrow{H_2O} >C<^{OH}_{R} + Mg(OH)X$$

(*iv*) If a higher homologue of alcohol is required to be prepared, use is made of ethylene oxide.

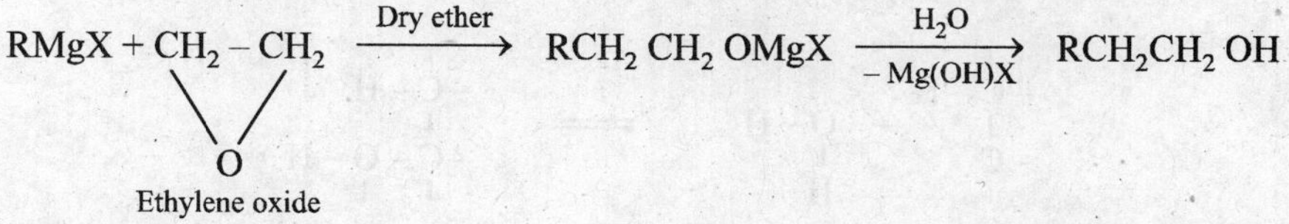

$$RMgX + \underset{\text{Ethylene oxide}}{CH_2-CH_2(-O-)} \xrightarrow{\text{Dry ether}} RCH_2\,CH_2\,OMgX \xrightarrow[-\,Mg(OH)X]{H_2O} RCH_2CH_2\,OH$$

For example,

$$CH_3MgBr + CH_2-CH_2(-O-) \xrightarrow{\text{Dry ether}} CH_3CH_2CH_2\,OMgI \xrightarrow[-\,Mg(OH)X]{H_2O} \underset{n\text{-Propyl alcohol}}{CH_3CH_2CH_2OH}$$

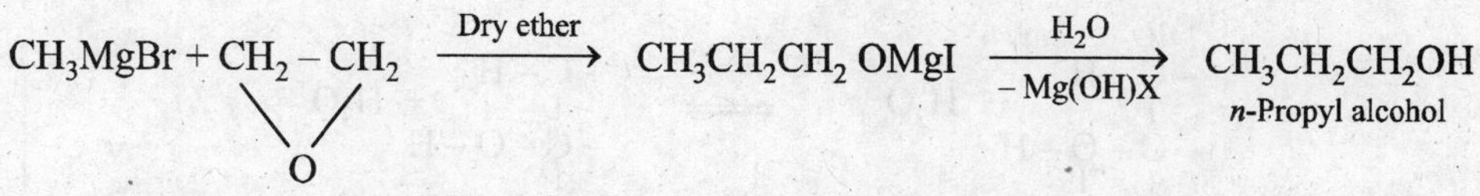

It may be noted that the alcohol obtained contains two carbon atoms more than the alkyl group of Grignard reagent.

Mechanism of reaction

The nucleophilic alkyl group of the Grignard reagent attaches itself to positively polarised carbon while the electrophilic magnesium attaches itself to negatively polarised oxygen of the C–O bond of ethylene oxide; the three-membered ring getting ruptured during the process. The resulting addition product is then decomposed as usual.

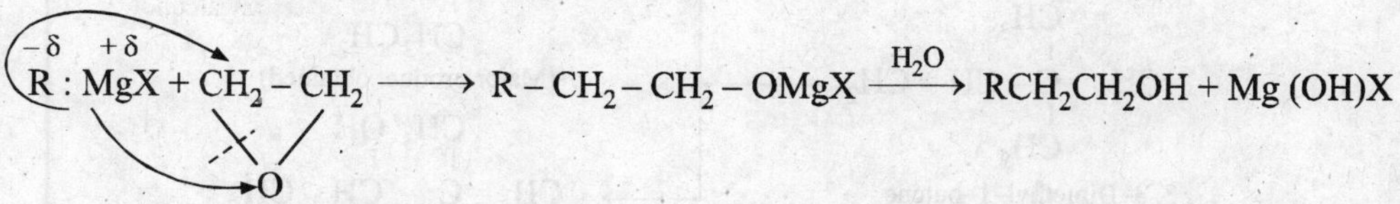

$$\overset{-\delta}{R}:\overset{+\delta}{MgX} + CH_2-CH_2(-O-) \longrightarrow R-CH_2-CH_2-OMgX \xrightarrow{H_2O} RCH_2CH_2OH + Mg\,(OH)X$$

(2) By the hydration of alkenes. An alkene adds a molecule of sulphuric acid to give alkyl hydrogen sulphate which is hydrolysed with water to obtain an alcohol.

$$\underset{\text{Alkene}}{>C=C<} \xrightarrow{H_2SO_4} \underset{\text{Alkyl hydrogen sulphate}}{-\overset{|}{\underset{\underset{H}{|}}{C}}-\overset{|}{\underset{|}{C}}-HSO_4} \xrightarrow{H_2O} \underset{\text{Alcohol}}{-\overset{|}{\underset{\underset{H}{|}}{C}}-\overset{|}{\underset{|}{C}}-OH} + H_2SO_4$$

For example,

$$\underset{\text{Propylene}}{CH_3-CH=CH_2} \xrightarrow{H_2SO_4} \underset{\text{Isopropyl hydrogen sulphate}}{CH_3CH(HSO_4)CH_3} \xrightarrow[\text{Heat}]{H_2O} \underset{\text{Isopropyl alcohol}}{CH_3CHOHCH_3} + H_2SO_4$$

In case of unsymmetrical alkenes, addition of sulphuric acid takes place according to Markownikoff's rule.

Ethanol can be obtained by the hydration of ethene.

Mechanism of Hydration

$$\underset{\text{Alkene}}{>C=C<} + H_3O^+ \rightleftharpoons \underset{\text{Carbocation}}{-\overset{|}{C}-H \atop -\underset{|}{C^+}} + H_2O$$

$$-\overset{|}{C}-H \atop -\underset{|}{C^+} + :\ddot{O}-H \atop \quad\ \ \underset{H}{|} \rightleftharpoons \underset{\text{Protonated Alcohol}}{-\overset{|}{C}-H \atop -\underset{|}{C}-\overset{+}{O}-H \atop \qquad\ \ \underset{H}{|}}$$

$$-\overset{|}{C}-H \atop -\underset{|}{C}-\overset{+}{O}-H \atop \qquad\ \ \underset{H}{|} + H_2O \rightleftharpoons \underset{\text{Alcohol}}{-\overset{|}{C}-H \atop -\underset{|}{C}-O-H} + H_3O^+$$

The first step in the hydration of alkenes involves the formation of a carbocation. The carbocation thus formed may sometimes undergo rearrangement to a more stable carbocation. *This leads the formation of* **unexpected products** *in the reaction*. For example, hydration of 3, 3-dimethyl-1-butene gives rise to 2, 3-dimethyl-2-butanol as the major product while the product expected is 3, 3-dimethyl-2-butanol.

$$\underset{\text{3, 3–Dimethyl–1–butene}}{CH_3-\overset{\overset{CH_3}{|}}{\underset{\underset{CH_3}{|}}{C}}-CH=CH_2} \longrightarrow \underset{\text{(Major product obtained)}}{CH_3-\overset{\overset{OH}{|}}{\underset{\underset{CH_3}{|}}{C}}-\underset{\underset{CH_3}{|}}{CH}-CH_3}$$

$$\longrightarrow \underset{\text{(Expected product)}}{CH_3-\overset{\overset{CH_3}{|}}{\underset{\underset{CH_3}{|}}{C}}-\overset{\overset{OH}{|}}{CH}-CH_3}$$

This is due to the fact that the carbocation formed in the first step of the reaction is a secondary carbocation which rearranges to a more stable tertiary carbocation by alkyl shift as shown below.

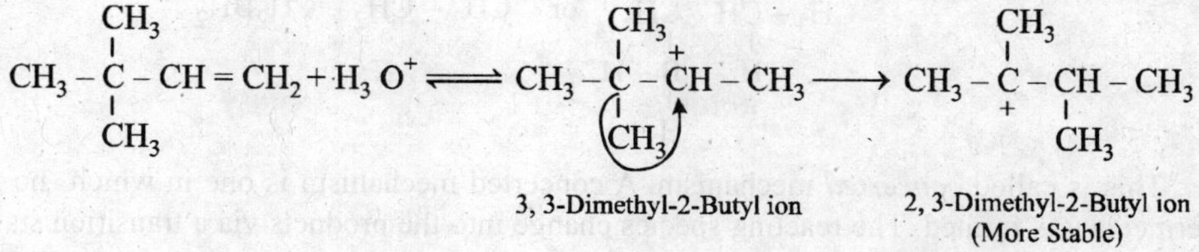

$$CH_3-\underset{CH_3}{\overset{CH_3}{\underset{|}{\overset{|}{C}}}}-CH=CH_2+H_3O^+ \rightleftharpoons CH_3-\underset{CH_3}{\overset{CH_3}{\underset{|}{\overset{|}{C}}}}-\overset{+}{C}H-CH_3 \longrightarrow CH_3-\overset{CH_3}{\overset{|}{\underset{+}{C}}}-\underset{CH_3}{\underset{|}{CH}}-CH_3$$

3, 3-Dimethyl-2-Butyl ion — 2, 3-Dimethyl-2-Butyl ion (More Stable)

The tertiary ion formed by rearrangement then gives rise to 2, 3-dimethyl-2-butanol as the major product.

(3) By the hydroboration-oxidation of alkenes. When treated with diborane $(BH_3)_2$, alkenes are converted into alkyl boranes which on oxidation with alkaline hydrogen peroxide yield alcohols. For example:

$$6CH_2=CH_2+(BH_3)_2 \longrightarrow \underset{\text{Triethylborane}}{2(CH_3-CH_2)_3B}$$

$$(CH_3-CH_2)_3B+3H_2O_2 \xrightarrow{OH} 3CH_3CH_2OH+H_3BO_3$$

Diborane required in the reaction is prepared *in situ* by reaction between sodium borohydride $(NaBH_4)$ and boron trifluoride (BF_3). It is not necessary to isolate the alkyl boranes, rather they are treated in the same reaction vessel with alkaline hydrogen peroxide.

A significant feature of hydroboration-oxidation is that it always leads to **anti-Markownikoff** addition of water to alkenes. As such, this method can be used to prepare alcohols which cannot be prepared by other methods of hydration of alkenes. Thus, we have:

$$6CH_3-CH=CH_2 \xrightarrow{(BH_3)_2} \underset{\text{Tri-n-propylborane}}{2(CH_3-CH_2-CH_2)_3B} \xrightarrow{H_2O_2/OH^-} \underset{n\text{-Propyl alcohol}}{6CH_3CH_2CH_2OH}+2H_3BO_3$$

Due to anti-Markownikoff orientation, only primary alcohols can be obtained by this method.

Mechanism of the reaction : In a normal HX addition, H-would have attached itself to C_1 and X– to C_2 . But in this case, electron-deficient boron attaches itself to C_1 giving rise to an intermediate product as illustrated below:

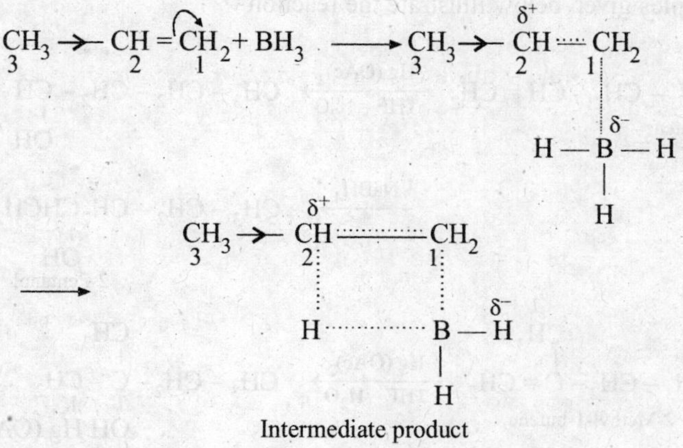

$$\underset{3}{CH_3}\rightarrow\underset{2}{CH}=\underset{1}{CH_2}+BH_3 \longrightarrow \underset{3}{CH_3}\rightarrow\underset{2}{\overset{\delta^+}{CH}}\cdots\underset{1}{CH_2}\cdots\overset{\delta^-}{B}H_3$$

$$\longrightarrow \underset{3}{CH_3}\rightarrow\underset{2}{\overset{\delta^+}{CH}}\cdots\cdots\underset{1}{CH_2}\quad\begin{matrix}\vdots & & \vdots\\ H & \cdots\cdots & \overset{\delta^-}{B}-H\\ & & |\\ & & H\end{matrix}$$

Intermediate product

The transition state ultimately generates the alkyl borane in which B–H bond has been cleaved and the C_2 – H bond has been completely formed.

$$\underset{\displaystyleH}{CH_3-\underset{|}{CH}}-\underset{\underset{\displaystyle H}{|}}{\underset{|}{B-H}}\!\!\!\!\!\!\!\!\!\!\!\!CH_2 \quad \text{or} \quad CH_3-CH_2-CH_2BH_2$$

This is called *concerted* mechanism. A concerted mechanism is one in which no reaction intermediate is formed. The reacting species change into the products via a transition state.

4. Oxymercuration-demercuration of alkenes. It is a convenient two-step method for preparing alcohols from alkenes.

When treated with mercuric acetate in a water-tetrahydrofuran solution, alkenes form hydroxy mercurial compounds. These mercury compounds are reduced by sodium borohydride ($NaBH_4$) to yield alcohols.

$$(i)\ \underset{\text{Alkene}}{>C=C<} + H_2O + \underset{\substack{\text{Mercuric}\\ \text{acetate}}}{Hg(OOCCH_3)_2} \xrightarrow[\text{(Oxymercuration)}]{\text{THF}} \underset{\text{OH HgOOCCH}_3}{-\overset{|}{\underset{|}{C}}-\overset{|}{\underset{|}{C}}-} + CH_3COOH$$

$$(ii)\ \underset{\text{OH HgOOCCH}_3}{-\overset{|}{\underset{|}{C}}-\overset{|}{\underset{|}{C}}-} \xrightarrow[\text{(Demercuration)}]{NaBH_4} \underset{\substack{\text{OH H}\\ \text{Alcohol}}}{-\overset{|}{\underset{|}{C}}-\overset{|}{\underset{|}{C}}-} + Hg + CH_3COO^-$$

In the first step water and mercuric acetate add to be double bond. This is known as **oxymercuration.** In the second step, sodium borohydride reduces the mercurial compound and the mercuri-acetate group gets replaced by hydrogen. This is known as **demercuration.** Both the steps take place very rapidly and conveniently at room temperature. The organo-mercurial compound formed in the first step is not isolated but is reduced *in situ* by sodium borohydride to form alcohol.

The alcohol formed is in *accordance with Markownikoff addition* of water (*i.e.* H– and –OH) to the carbon-carbon double bond.

Thus the overall reaction may be written as follows:

$$R-CH=CH_2 \xrightarrow[(ii)\ NaBH_4]{(i)Hg(Oac_2)\,/\,THF-H_2O} R-\underset{\underset{\displaystyle OH}{|}}{CH}-CH_3$$

The examples given below illustrate the reaction

$$\underset{\text{1-Pentene}}{CH_3-CH_2-CH_2-CH=CH_2} \xrightarrow[THF-H_2O]{Hg(OAc)_2} CH_3-CH_2-CH_2-\underset{\underset{\displaystyle OH}{|}}{CH}-\underset{\underset{\displaystyle Hg(OAc)}{|}}{CH_2}$$

$$\xrightarrow{NaBH_4} \underset{\text{2-Pentanol}}{CH_3-CH_2-CH_2\underset{\underset{\displaystyle OH}{|}}{CH}CH_2}$$

$$\underset{\text{2-Methyl-1-butene}}{CH_3-CH_2-\overset{\overset{\displaystyle CH_3}{|}}{C}=CH_2} \xrightarrow[THF-H_2O]{Hg(OAc)_2} CH_3-CH_2-\overset{\overset{\displaystyle CH_3}{|}}{\underset{\underset{\displaystyle OH}{|}}{C}}-\underset{\underset{\displaystyle Hg(OAc)}{|}}{CH_2}$$

$$\xrightarrow{NaBH_4} CH_3-CH_2-\overset{\overset{\displaystyle CH_3}{|}}{\underset{\underset{\displaystyle OH}{|}}{C}}-CH_3$$

2-Methyl-2-butanol

It may be mentioned that oxymercuration – demercuration is *regioselective* in nature. We get only one alcohol as the product according to the Markownikoff's rule. This is unlike acidic hydration of unsymmetrical alkenes in which there is a possbility of obtaining a mixture of two alcohols because of the formation of more stable carbocation.

Mechanism of Oxymercuration

Electrophilic attack by the mercury species, $\overset{+}{H}gOAc$, at the less substituted carbon of the double bond (*i.e.* the carbon which carries greater number of hydrogens) takes place to form a mercury-bridged carbocation. This cation is immediately attacked by nucleophilic solvent *i.e.* water to yield the hydroxymercurial compound as shown below.

$$\underset{\text{Alkene}}{R-CH=CH_2} + \overset{+}{H}gOAc \longrightarrow R-\overset{\delta+}{CH}-\underset{\underset{\underset{\delta+}{\displaystyle HgOAc}}{|}}{CH_2}$$

Mercury-bridged carbocation

The mechanism of demercuration step is still not clearly understood.

5. By the hydrolysis of alkyl halides. An alcohol may be obtained by the hydrolysis of an alkyl halide with sod. or pot. hydroxide solution or moist silver oxide

$$\underset{\text{Alkyl halide}}{R-X} + \underset{aq}{KOH} \longrightarrow \underset{\text{Alcohol}}{R-OH} + KX$$

For example,

$$\underset{\text{Ethyl bromide}}{C_2H_5Br} + \underset{aq}{KOH} \longrightarrow \underset{\text{Ethyl alcohol}}{C_2H_5OH} + KBr$$

6. By the reduction of aldehydes, ketones, acids and esters. This is another useful method for the preparation of alcohols. The reduction may be brought about with hydrogen in the presence of a catalyst, with nascent hydrogen from some suitable source such as reaction between sodium and ethanol or with complex metal hydrides such as sodium borohydride ($NaBH_4$) or lithium aluminium hydride ($LiAlH_4$).

For example,

$$\underset{\text{Acetaldehyde}}{CH_3-\overset{\overset{\displaystyle H}{|}}{C}=O} + H_2 \xrightarrow{\text{Pt or Ni}} \underset{\text{Ethyl alcohol}}{CH_3-\overset{\overset{\displaystyle H}{|}}{\underset{\underset{\displaystyle H}{|}}{C}}-OH}$$

$$\underset{\text{Acetone}}{CH_3-\overset{\overset{\displaystyle O}{||}}{C}-CH_3} + 2H \xrightarrow{Na/C_2H_5OH} \underset{\text{Isopropyl alcohol}}{CH_3-\overset{\overset{\displaystyle OH}{|}}{\underset{\underset{\displaystyle H}{|}}{C}}-CH_3}$$

$$\underset{\text{Butyric acid}}{CH_3-CH_2-CH_2-\overset{\overset{\displaystyle O}{\|}}{C}-OH} \xrightarrow{LiAlH_4} \underset{n\text{-Butyl alcohol}}{CH_3-CH_2-CH_2-\overset{\overset{\displaystyle H}{|}}{\underset{\underset{\displaystyle H}{|}}{C}}-OH}$$

$$\underset{\text{Ethyl acetate}}{CH_3-\overset{\overset{\displaystyle O}{\|}}{C}-OC_2H_5} \xrightarrow{LiAlH_4} \underset{\text{Ethyl alcohol}}{2C_2H_5OH}$$

$NaBH_4$ is particularly suited for the preparation of unsaturated alcohols by the reduction of unsaturated carbonyl compounds as it does not affect the carbon-carbon double bond.

For example,

$$\underset{\text{Crotonaldehyde}}{CH_3-CH=CH-CHO} \xrightarrow{NaBH_4} \underset{\text{But-2-ene-1-ol}}{CH_3-CH=CH-CH_2OH}$$

7. By the hydrolysis of esters. The hydrolysis of esters with alkali solution is generally employed for the preparation of alcohols from naturally occurring esters.

$$\underset{\text{Ester}}{RCOOR'} + NaOH \longrightarrow \underset{\text{Sodium salt of acid}}{RCOONa} + \underset{\text{Alcohol}}{R'OH}$$

For example,

$$\underset{\text{Cetyl palmitate}}{C_{15}H_{31}-COOC_{16}H_{33}} + \underset{aq}{NaOH} \longrightarrow \underset{\text{Sodium palmitate}}{C_{15}H_{31}COONa} + \underset{\text{Cetyl alcohol}}{C_{16}H_{33}OH}$$

8. By the fermentation of carbohydrates. This is one of the oldest methods of obtaining alcohols, particularly ethyl alcohol. Glucose and fructose are obtained by the enzymatic hydrolysis of canesugar. Then there is further conversion to ethyl alcohol. Similarly starch can be fermented into ethyl alcohol.

The reactions taking place are described below.

From sugar:

$$C_{12}H_{22}O_{11} + H_2O \xrightarrow{\textit{Invertase}} \underset{\text{Glucose}}{C_6H_{12}O_6} + \underset{\text{Fructose}}{C_6H_{12}O_6}$$

$$C_6H_{12}O_6 \xrightarrow{\textit{Zymase}} \underset{\text{Ethyl alcohol}}{2C_2H_5OH} + 2CO_2$$

From starch:

$$2(C_6H_{10}O_5)_n + nH_2O \xrightarrow{\textit{Diastase}} \underset{\text{Maltose}}{nC_{12}H_{22}O_{11}}$$

$$C_{12}H_{22}O_{11} + H_2O \xrightarrow{\textit{Maltase}} 2C_6H_{12}O_6$$

$$C_6H_{12}O_6 \xrightarrow{\textit{Zymase}} \underset{\text{Ethyl alcohol}}{2C_2H_5OH} + 2CO_2$$

Example. Name the following alcohols by Carbinol and by IUPAC methods:

(*a*) $CH_3-\underset{\underset{\displaystyle OH}{|}}{CH}-CH=CH_2$

(b) $C_6H_5-\underset{CH_2CH_3}{\overset{CH_3}{\overset{|}{\underset{|}{C}}}}-OH$

(c) $CH_3-CH_2-\underset{CH_2CH_2CH_3}{\overset{CH_3}{\overset{|}{\underset{|}{C}}}}-OH$

Solution.

	Carbinol System	IUPAC System
(*a*)	Methyl vinyl carbinol	But-3-en-2-ol
(*b*)	Ethyl methyl phenyl carbinol	2-Phenylbutan-2-ol
(*c*)	Methyl-ethyl-n-propyl carbinol	3-Methylhexan-3-ol.

2.5. PHYSICAL PROPERTIES OF ALCOHOLS

(1) Physical state. The lower members are coloureless liquids with characteristic alcoholic smell and a burning taste. The higer members ar colourless and waxy solids.

(2) Boiling points. Alcohols have generally higher boiling points as campared to haloalkanes, ethers and alkanes of camparable molecular masses. This is due to the polar nature of the O — H bond and also due to intermolecular hydrogen bonding present in the molecules of alkanols.

$$\cdots\cdots\underset{\delta^-}{\overset{R}{\overset{|}{O}}}-\overset{\delta^+}{H}\cdots\cdots\underset{\delta^-}{\overset{R}{\overset{|}{O}}}-\overset{\delta^+}{H}\cdots\cdots\underset{\delta^-}{\overset{R}{\overset{|}{O}}}-\overset{\delta^+}{H}\cdots\cdots$$

Hydrogen bonding

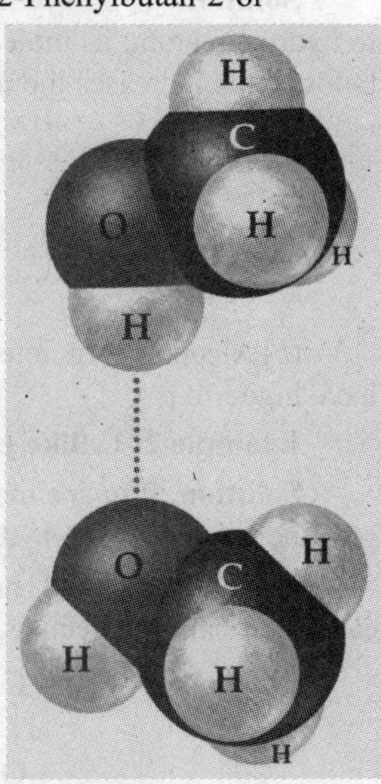

Hydrogen bonding in Alcohols

Energy is needed to overcome the attractive forces resulting from the hydrogen bonding. As a result, the boiling points of alkanols are higher than the members of other families. This is evident from the boiling points of ethanol, methoxymethane, chloromethane and propane which have nearly the same molecular masses.

Compound	C_2H_5OH	CH_3OCH_3	CH_3Cl	C_3H_8
b.p. (K)	351	249	249	231

Among the isomeric alkanols, the boiling points tend to decrease with the increase in branching and as a result the boiling points vary as primary > secondary > tertiary. This is supported by following data:

Compound	$CH_3CH_2CH_2CH_2OH$	$CH_3CH(OH)C_2H_5$	$(CH_3)_3COH$
b.p. (K)	391	373	353

Decrease in the boiling point takes place due to decreasing van der Waal's forces between the molecules on increased branching.

(3) Solubility. Alcohols being organic in nature are expected to be insoluble in water. But these are mostly water soluble due to the presence of the intermolecular hydrogen bonding in the molecule of alcohol and water

$$\cdots\overset{\delta^+}{H}-\underset{\underset{R}{|}}{\overset{\delta^-}{O}}\cdots\cdots\overset{\delta^+}{H}-\underset{\underset{R}{|}}{\overset{\delta^-}{O}}\cdots\cdots\overset{\delta^+}{H}-\underset{\underset{R}{|}}{\overset{\delta^-}{O}}\cdots\cdots\overset{\delta^+}{H}-\underset{\underset{H}{|}}{\overset{\delta^-}{O}}\cdots$$

Energy is released during attraction and as a result, the energy of system decreases. This helps alcohol to dissolve in water. However, the higher homologues of alcohols are insoluble in water. Thus hexanol and higher homologues are insoluble in water. This is because a higher alcohol behaves almost like the corresponding alkane. Thus hexanol behaves like hexane. Moreover, a bulky alkyl group in an alcohol hinders the formation of smooth hydrogen bonds.

Example 1. (*a*) Alcohol has higher boiling point than alkyl halide or alkane of comparable mol. wt. Why?

(*b*) Why alcohols have higher boiling points than ethers of comparable molecular weights.

Solution. (*a*) Since in alcohol (ROH), strongly electronegative oxygen atom is directly linked to hydrogen atom, its molecules get associated through intermolecular hydrogen bonding as shown below. This increases the molecular mass and hence the boiling point of alcohol rises.

$$\cdots\overset{\overset{C_2H_5}{|}}{\overset{\delta^-}{O}}-\overset{\delta^+}{H}\cdots\overset{\overset{C_2H_5}{|}}{\overset{\delta^-}{O}}-\overset{\delta^+}{H}\cdots\overset{\overset{C_2H_5}{|}}{\overset{\delta^-}{O}}-\overset{\delta^+}{H}\cdots$$

(*b*) Alcohols form intermolecular hydrogen bonding whereas ethers do not. That is why alcohols have higher b.p.

Example 2. Unlike propane or butane, propanol is soluble in water. Why?

Solution. Propane molecules cannot join with water molecules through hydrogen bonding as propane has no oxygen-hydrogen bond in its molecule. Consequently, propane is not soluble in water.

On the other hand, propanol is soluble in water as its molecule contains –OH bond, which causes propanol molecules to form hydrogen bonds with water molecules as shown below:

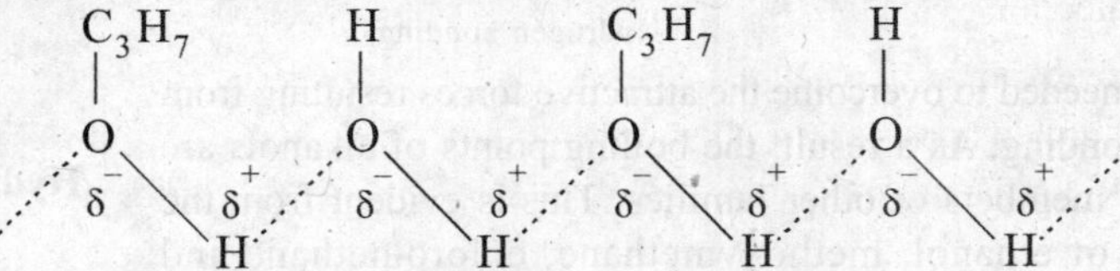

Consequently propanol is soluble in water.

Example 3. Why is *n*-hexanol insoluble in water?

Solution. Alcohols have the general formula ROH. As the R group becomes larger, ROH resembles more closely with the hydrocarbon.

In *n*-hexanol molecule, the carbon content is quite high which causes it to resemble closely with hexane (which is not soluble in water). Moreover, the large hexyl ($C_6H_{13}-$) group obstructs the formation of hydrogen bonds between water and *n*-hexanol molecules. As a result, *n*-hexanol is not soluble in water.

Example 4. Why are lower members of alcohols soluble in water while higher members are not ?

Solution. Lower members of alcohols are soluble in water because they form hydrogen bonds with water. Higher members behave more like hydrocarbons which are insoluble in water. Formation of hydrogen bonds in the case of higher members is hindered because of steric reasons as the alkyl groups in such cases are quite bulky. As the hydrogen bond is not formed, they are insoluble in water.

Example 5. The boiling point of 1-propanol (molecular mass 60) is 370 K while the boiling point of 1, 2-ethanediol (molecular mass 62) is 470 K. Explain the reason for high boiling point of 1, 2-ethanediol.

Solution. Since 1, 2-ethanediol molecule contains two –OH groups, it is capable of forming more intermolecular hydrogen bonds and hence is more highly associated substance. Therefore, its boiling point is very high.

On the other hand, 1-propanol molecule has only one –OH group, its molecules are comparatively less associated through hydrogen bonding. Consequently, its boiling point is lower than that of 1, 2-ethanediol.

2.6. CHEMICAL PROPERTIES

Reactions involving the cleavage of O – H bond

1. Acidic nature. Alcohols have a acidic character as they react with active metals like sodium or potassium liberating hydrogen. For example,

$$C_2H_5OH + Na \longrightarrow \underset{\text{Sodium ethoxide}}{C_2H_5ONa} + 1/2\ H_2$$

However, alcohols are weak acids. This is because alcohols have an electron releasing alkyl group (+I effect) which increases electron density around oxygen, so that the release of proton is rendered difficult. That is why alcohols are weak acids.

$$C_2H_5 \rightarrow \ddot{\underset{\cdot\cdot}{O}} - H$$

Electron releasing group (+ I effect)

Acidic character of the alcohols shows the following order:

Primary alcohol > Secondary alcohol > Tertiary alcohol

The acidic character of alcohols depends upon the release of proton from O–H bond. Electron releasing (+I effect) increases from primary alcohols (having one alkyl radical) to secondary alcohols (having two alkyl radicals) to tertiary alcohols (having three alkyl radicals).

$$CH_3 \rightarrow \overset{H}{\underset{H}{C}} - O - H \qquad CH_3 \rightarrow \overset{CH_3\downarrow}{\underset{H}{C}} - O - H \qquad CH_3 \rightarrow \overset{CH_3\downarrow}{\underset{\uparrow CH_3}{C}} - O - H$$

Primary alcohol (Acidic) | Secondary alcohol (Less acidic) | Tertiary alcohol (Least acidic)

As a result, in tertiary alcohols, the release of proton is hindered making them weakest acids of the three types of alcohols. Therefore tertiary alcohols are the weakest acids.

Example. Giving reasons arrange the following in decreasing order of acidity.

Tert. butyl alcohol, ethyl alcohol, methyl alcohol and isopropyl alcohol.

Solution. In view of the discussion above the acidity decreases in the following order:

$$\underset{\substack{\text{Methyl alcohol}\\(\textit{Most acidic})}}{CH_3OH} > \underset{\substack{\text{Ethyl alcohol}\\(\textit{Less acidic})}}{C_2H_5OH} > \underset{\substack{\text{Isopropyl alcohol}\\(\textit{Still less acidic})}}{CH_3-\overset{\overset{CH_3}{|}}{\underset{\underset{H}{|}}{C}}-OH} > \underset{\substack{\textit{tert.}\text{ butyl alcohol}\\(\textit{Least acidic})}}{CH_3-\overset{\overset{CH_3}{|}}{\underset{\underset{CH_3}{|}}{C}}-OH}$$

It may be noted that ethyl group has greater + I effect compared to methyl group.

2. Reaction with Grignard reagents. Alcohols react with Grignard reagents to produce hydrocarbons as shown below.

$$\underset{\text{Alcohol}}{R'OH} + \underset{\text{Grignard reagent}}{RMgX} \longrightarrow \underset{\text{Hydrocarbon}}{RH} + R'OMgX$$

For example,

$$\underset{\substack{\text{Methyl}\\\text{alcohol}}}{CH_3OH} + \underset{\substack{\text{Ethyl}\\\text{magnesium}\\\text{iodide}}}{C_2H_5MgI} \longrightarrow \underset{\text{Ethane}}{C_2H_6} + CH_3OMgI$$

3. Reaction with organic acids. Alcohols react with organic acids to form esters. This reaction is termed as esterification. It is generally carried out in the presence of concentrated sulphuric acid or dry hydrogen chloride.

$$\underset{\text{Acid}}{RCOOH} + \underset{\text{Alcohol}}{HOR'} \xrightleftharpoons{\text{Conc. } H_2SO_4} \underset{\text{Ester}}{RCOOR'} + H_2O$$

For example,

$$\underset{\text{Acetic acid}}{CH_3COOH} + \underset{\text{Ethyl alcohol}}{HOC_2H_5} \xrightleftharpoons{H^+} \underset{\text{Ethyl acetate}}{CH_3COOC_2H_5} + H_2O$$

Mechanism. The reaction is believed to proceed through the following mechanism:

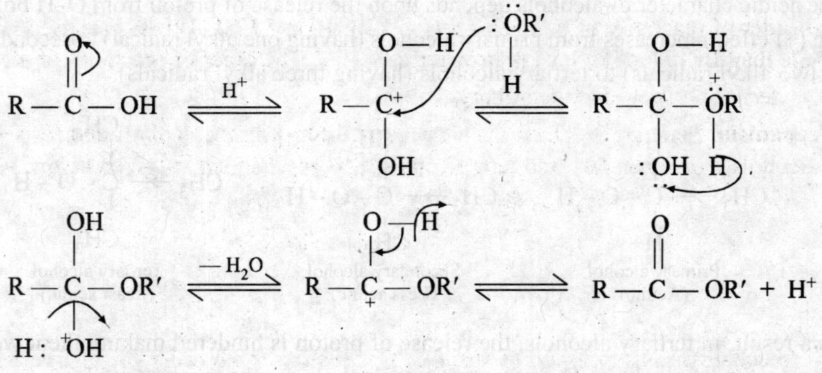

$$R-\overset{\overset{O}{||}}{C}-OH \xrightleftharpoons{H^+} R-\overset{\overset{O-H}{|}}{\underset{\underset{OH}{|}}{C^+}} \xrightleftharpoons{:\ddot{O}R' \atop |\ H} R-\overset{\overset{O-H}{|}}{\underset{\underset{:\ddot{O}H}{|}}{C}}-\overset{+}{\ddot{O}}R\underset{H}{|}$$

$$R-\overset{\overset{OH}{|}}{\underset{\underset{H:\overset{+}{O}H}{|}}{C}}-OR' \xrightleftharpoons{-H_2O} R-\overset{\overset{O-H}{|}}{\underset{+}{C}}-OR' \rightleftharpoons R-\overset{\overset{O}{||}}{C}-OR' + H^+$$

It has been proved beyond doubt by isotopic tracer technique that the reaction involves the cleavage of O–H bond of alcohols and the C–OH bond of organic acids.

4. Reaction with acid halides or acid anhydrides. When alcohols are heated with an acid halide or acid anhydride, the hydrogen of –OH group is replaced by acyl group (RCO–) and an ester is formed. The reaction takes place more rapidly than esterification with carboxylic acids. Thus, we have:

$$\underset{\text{Acetyl chloride}}{CH_3COCl + HOC_2H_5} \longrightarrow \underset{\text{Ethyl acetate}}{CH_3COOC_2H_5} + HCl$$

$$\underset{\text{Acetic anhydride}}{\begin{matrix} CH_3CO \\ CH_3CO \end{matrix}\!\!>\!O} + HOC_2H_5 \longrightarrow \underset{\text{Ethyl acetate}}{CH_3COOC_2H_5} + \underset{\text{Acetic acid}}{CH_3COOH}$$

Mechanism. The reaction mechanism may be depicted as follows:

$$R-\overset{\overset{\ddot{O}:}{\|}}{C}-Cl + :\underset{\underset{H}{|}}{\ddot{O}R'} \longrightarrow R-\overset{:\ddot{O}:^{-}}{\underset{\underset{\underset{H}{|}}{:\overset{+}{O}R'}}{C}}-Cl \xrightarrow[-Cl^-]{} R-\overset{\overset{O}{\|}}{C}-\underset{\underset{H}{|}}{\overset{+}{\ddot{O}}R'} \xrightarrow[-H^+]{} R-\overset{\overset{O}{\|}}{C}-OR'$$

Reactions involving cleavage of C — OH bond

1. Reaction with hydrogen halides. Alcohols react with hydrogen halides to form alkyl halides and water.

$$\underset{\text{Alcohol}}{ROH + HX} \longrightarrow \underset{\text{Alkyl halide}}{RX} + H_2O$$

For example,

$$\underset{\text{Ethyl alcohol}}{C_2H_5OH} + HCl \xrightarrow{ZnCl_2} \underset{\text{Ethyl chloride}}{C_2H_5Cl} + H_2O$$

Hydrogen bromide and hydrogen iodide react in the same way but without the presence of a catalyst. However, if an alcohol is heated with hydrogen iodide in the presence of red phosphorus, it gets reduced to a hydrocarbon. For example,

$$\underset{\text{Methyl alcohol}}{CH_3OH} + 2HI \xrightarrow{P} \underset{\text{Methane}}{CH_4} + H_2O + I_2$$

The order of reactivity of halogen acids is HI > HBr > HCl. This is because I^- is a stronger nucleophile than Br^- which in turn is stronger than Cl^-. The order of reactivity of alcohols is

tertiary > secondary > primary

Mechanism. Separate mechanisms are proposed for primary alcohols and for secondary and tertiary alcohols. For secondary and tert. alcohols, SN^1 mechanism, as given below, is proposed.

$$R-OH \xrightarrow{\text{Slow}} \underset{\text{Carbonium ion}}{R^+} + OH^-$$

$$R^+ + \underset{\text{Halide ion}}{X^-} \xrightarrow{\text{Fast}} R-X$$

This mechanism is supported by the fact that the observed order of relative reactivities of alcohols is the same as the order of stabilities of carbonium ions.

For primary alcohols, SN^2 mechanism involving a transition state is proposed.

$$R-OH + X^- \xrightarrow{\text{Slow}} \underset{\text{Transition state}}{[X......R.....OH]^-} \xrightarrow{\text{Fast}} R-X + OH^-$$

***(ii)* Reaction with phosphorus halides.** Alcohols react with phosphorus halides, *i.e.* PCl_5, PCl_3, PBr_3 or PI_3, to yield corresponding alkyl halides.

$$\underset{\text{Alcohol}}{ROH} + PCl_5 \longrightarrow \underset{\text{Alkyl chloride}}{RCl} + POCl_3 + HCl$$

$$3\underset{\text{Alcohol}}{ROH} + PX_3 \longrightarrow 3\underset{\text{Alkyl halide}}{RX} + \underset{\text{Phosphorus acid}}{H_3PO_3}$$

$$3\underset{\text{Ethyl alcohol}}{C_2H_5OH} + PI_3 \longrightarrow \underset{\text{Ethyl iodide}}{3C_2H_5I} + H_3PO_3$$

$$\underset{\text{Benzyl alcohol}}{C_6H_5-CH_2OH} + PCl_5 \longrightarrow \underset{\text{Benzyl chloride}}{C_6H_5-CH_2Cl} + POCl_3 + HCl$$

Mechanism. The mechanism of reaction with $\mathbf{PX_3}$ may be illustrated by considering the reaction with PCl_3 as given below:

$$R-\ddot{O}(H): + PCl_3 \longrightarrow \underset{\text{Protonated Alkyl Dichloro phosphite}}{R-\overset{+}{O}(H)-PCl_2} + :Cl^-$$

$$\bar{Cl}: + R-\overset{+}{O}(H)-PCl_2 \longrightarrow R-Cl + HOPCl_2$$

In the first step, nucleophilic attack of alcohol on PCl_3 takes place to form a protonated alkyl dichlorophosphite. In the second step, chloride ion acts as a nucleophile and displaces $HOPCl_2$ from the phosphite initially formed.

The mechanism for reaction between alcohol and PCl_5 is as follows:

$$2\,PCl_5 \longrightarrow \overset{+}{P}Cl_4 + \overset{-}{P}Cl_6$$

$$R-\ddot{O}(H): + \overset{+}{P}Cl_4 \longrightarrow R-\overset{+}{O}(H)-PCl_4$$

$$\underset{(\text{From } \overset{-}{P}Cl_5)}{\bar{Cl}:} + R-\overset{+}{O}(H)-PCl_4 \longrightarrow R-Cl + H-O-PCl_4$$

$$H-O-PCl_4 \longrightarrow HCl + O=PCl_3$$

3. Reaction with thionyl chloride. Alcohols react with thionyl chloride to form alkyl chlorides.

$$\underset{\text{Alcohol}}{ROH} + \underset{\text{Thionyl chloride}}{SOCl_2} \longrightarrow \underset{\text{Alkyl chloride}}{RCl} + SO_2 + HCl$$

$$\underset{\text{Ethyl alcohol}}{C_2H_5OH} + SOCl_2 \longrightarrow \underset{\text{Ethyl chloride}}{C_2H_5Cl} + SO_2 + HCl$$

Mechanism. The reaction is supposed to have the following mechanism.

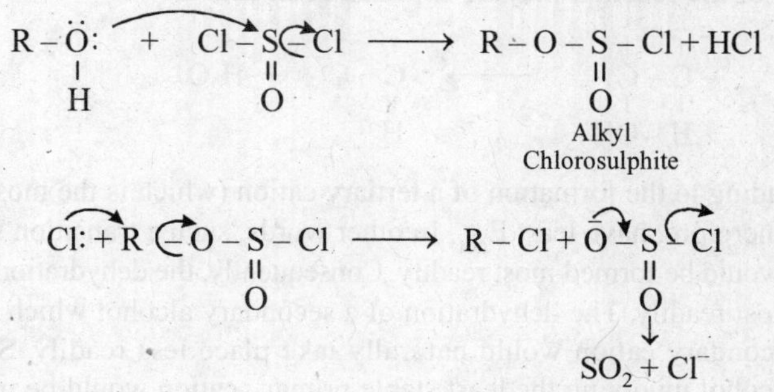

$$R-\ddot{O}(H): + Cl-S(=O)-Cl \longrightarrow \underset{\text{Alkyl Chlorosulphite}}{R-O-S(=O)-Cl} + HCl$$

$$Cl^{-} + R-O-S(=O)-Cl \longrightarrow R-Cl + O-S(=O)-Cl \;\rightarrow\; SO_2 + Cl^-$$

Reactions involving cleavage of alkyl and hydroxyl groups

1. Dehydration. When alcohols are heated with sulphuric acid, they get dehydrated to alkenes. For example,

$$\underset{\text{Ethyl alcohol}}{CH_3CH_2OH} \xrightarrow{H_2SO_4\ (433-443\ K)} \underset{\text{Ethylene}}{CH_2=CH_2} + H_2O$$

Mechanism of dehydration

The mechanism of dehydration of alcohols involves the formation of a carbocation as an intermediate. The reaction is found to take place in three steps.

1. In the first step alcohol combines with a hydrogen ion (from the acid) to form an oxonium ion.
2. The second step involves the dissociation of oxonium ion and formation of the carbocation.
3. Finally, the carbocation loses a hydrogen ion to yield the alkene as illustrated below.

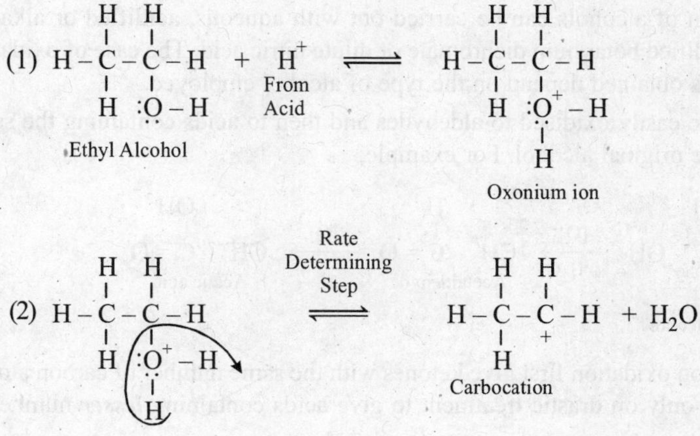

$$(1)\ \underset{\text{Ethyl Alcohol}}{H-C(H)(H)-C(H)(:O-H)-H} + \underset{\text{From Acid}}{H^+} \rightleftharpoons \underset{\text{Oxonium ion}}{H-C(H)(H)-C(H)(:\overset{+}{O}(H)-H)-H}$$

$$(2)\ H-C(H)(H)-C(H)(:\overset{+}{O}(H)-H)-H \xrightleftharpoons{\text{Rate Determining Step}} \underset{\text{Carbocation}}{H-C(H)(H)-\overset{+}{C}(H)-H} + H_2O$$

$$\text{(3)}\quad H-\overset{H}{\underset{H}{\overset{|}{\underset{|}{C}}}}-\overset{H}{\overset{|}{\underset{+}{C}}}-H \longrightarrow \underset{\text{Ethylene}}{\underset{H}{\overset{H}{>}}C=C\underset{H}{\overset{H}{<}}} + \underset{\text{Taken up by a new molecule of alcohol}}{H^+}$$

Relative ease of dehydration of alcohols. It has been observed that the ease of dehydration of different classes of alcohols follows the order: tertiary > secondary > primary.

Rate determining step of the reaction may be depicted as under.

$$-\underset{H}{\underset{|}{\overset{|}{C}}}-\underset{{}^+OH_2}{\underset{|}{\overset{|}{C}}}- \longrightarrow \underset{H}{\underset{|}{\overset{|}{C}}}-\underset{+}{\overset{|}{C}}- + H_2O$$

The transition state leading to the formation of a tertiary cation (which is the most stable ion) would be most stable and, therefore, have least E_{act}. In other words, such a transition state and the carbocation generated by it would be formed most readily. Consequently, the dehydration of a tertiary alcohol would take place most readily. The dehydration of a secondary alcohol which involves the formation of less stable secondary cation would naturally take place less readily. Similarly the dehydration of a primary alcohol involving the least stable primary cation would be most difficult to take place. Thus the relative ease of dehydration of alcohols follows the same order as that of ease of formation of carbocations.

Orientation of dehydration. Many alcohols give rise to mixtures of isomeric alkenes on dehydration. The ratio of the alkenes formed in a particular reaction depends upon the reaction conditions. However in general it may be stated that dehydration of an alcohol favours the formation of more stable alkene. For example:

$$\underset{\text{Sec- Butyl alcohol}}{CH_3-CH_2-CHOH-CH_3} \xrightarrow[-H_2O]{H_3PO_4-Al_2O_3,\ 298\ K} \begin{cases} \underset{\text{1-Butene (24.1\%) (less stable)}}{CH_3-CH_2-CH=CH_2} \\ \underset{\text{2-Butene (75.9\%) (more stable)}}{CH_3-CH=CH-CH_3} \end{cases}$$

2. Oxidation. Oxidation of alcohols can be carried out with aqueous, acidified or alkaline potassium permanganate, acidified potassium dichromate or dilute nitric acid. The ease of oxidation and the nature of the products obtained depend on the type of alcohol employed.

(*a*) *Primary alcohols* are easily oxidised to aldehydes and then to acids containing the same number of carbon atom as the original alcohol. For example,

$$\underset{\text{Ethyl alcohol}}{CH_3-\overset{H}{\underset{H}{\overset{|}{\underset{|}{C}}}}-OH} \xrightarrow[-H_2O]{[O]} \underset{\text{Acetaldehyde}}{CH_3-\overset{H}{\overset{|}{C}}=O} \xrightarrow{[O]} \underset{\text{Acetic acid}}{CH_3-\overset{OH}{\overset{|}{C}}=O}$$

(*b*) *Secondary alcohols* on oxidation first give ketones with the same number of carbon atoms. Ketones are further oxidised only on drastic treatment to give acids containing *lesser* number of carbon atoms. For example:

$$\underset{\text{Isopropyl alcohol}}{\overset{CH_3}{\underset{CH_3}{>}}C\overset{H}{\underset{OH}{<}}} \xrightarrow[-H_2O]{[O]} \underset{\text{Acetone}}{\overset{CH_3}{\underset{CH_3}{>}}C=O} \xrightarrow[-CO_2,\,-H_2O]{4[O]} \underset{\text{Acetic acid}}{CH_3COOH}$$

(*c*) *Tertiary alcohols* fail to undergo oxidation in neutral or alkaline solution. However, with acid oxidising agents they are oxidised to give ketones and acids containing smaller number of carbon atoms than the original alcohols. For example:

$$\underset{\textit{Tert}\text{-Butyl alcohol}}{(CH_3)_3C-OH} \xrightarrow{4[O]} \underset{\text{Acetone}}{\overset{CH_3}{\underset{CH_3}{>}}C=O} + CO_2 + 2H_2O$$

$$\overset{CH_3}{\underset{CH_3}{>}}C=O \xrightarrow{4[O]} \underset{\text{Acetic acid}}{CH_3COOH} + CO_2 + H_2O$$

3. Dehydrogenation. Different classes of alcohols furnish different products when their vapours are passed over copper at high temperature 473–573 K.

(*a*) *Primary alcohols* are dehydrogenated to form aldehydes.

For example:

$$CH_3-\underset{H}{\overset{H}{\underset{|}{\overset{|}{C}}}}-OH \xrightarrow{Cu,\ 473-573\ K} \underset{\text{Acetaldehyde}}{CH_3-\overset{H}{\overset{|}{C}}=O} + H_2$$

(*b*) *Secondary alcohols* are also dehydrogenated to give ketones. For example:

$$\underset{\text{Isopropyl alcohol}}{\overset{CH_3}{\underset{CH_3}{>}}C\overset{H}{\underset{OH}{<}}} \longrightarrow \underset{\text{Acetone}}{\overset{CH_3}{\underset{CH_3}{>}}C=O} + H_2$$

(*c*) *Tertiary alcohols* are dehydrated to give alkenes. For example:

$$\underset{\text{Tert-Butyl alcohol}}{CH_3-\underset{CH_3}{\overset{CH_3}{\underset{|}{\overset{|}{C}}}}-OH} \longrightarrow \underset{\textit{Iso}\text{-butylene}}{CH_3-\overset{CH_3}{\overset{|}{C}}=CH_2} + H_2O$$

Example 1. The ease of dehydration of alcohols is as: tert. alcohol > sec. alcohol > primary alcohol. Explain.

Solution. As per the mechanism of dehydration of alcohols, carbonium ion is produced as intermediate product. The reactivity of the alcohol will depend upon the stability of the carbonium ion. It is found that the stabilities of carbonium ions follows the order :

tert. carbonium ion > sec. carbonium ion > primary carbonium ion

Therefore, the ease of dehydration of alcohols also follows the same order:

tert. alcohol > sec. alcohol > primary alcohol

Example 2. 2-Butene is the major product when *n*-butyl alcohol is heated with conc. H_2SO_4. Explain.

Solution.

$$CH_3-CH_2-\underset{H}{\overset{H}{C}}-\underset{H}{\overset{H}{C}}-OH + H^+$$

$$\downarrow$$

$$\underset{\text{Protonated alcohol}}{CH_3-CH_2-\underset{H}{\overset{H}{C}}-\underset{H}{\overset{H}{C}}-\overset{+}{O}-H} \longrightarrow \underset{n\text{-Butyl carbonium ion}}{CH_3-CH_2-\underset{H}{\overset{H}{C}}-\underset{H}{\overset{H}{C^+}}}$$

n-Butyl carbonium ion is converted into more stable sec. butyl carbonium ion by hydride shift from the neighbouring carbon atom as follows:

$$CH_3-\underset{H}{\overset{H}{C}}-\underset{H}{\overset{H}{C}}-\underset{H}{\overset{H}{C^+}} \longrightarrow CH_3-\underset{H}{\overset{H}{C}}-\overset{H}{\overset{+}{C}}-\underset{H}{\overset{H}{C}}-H$$

$$\longrightarrow \underset{\text{2-Butene}}{CH_3-\overset{H}{C}=\overset{H}{C}-CH_3}$$

This is how 2-butene and not 1-butene is obtained.

2.7. DISTINCTION BETWEEN PRIMARY, SECONDARY AND TERTIARY ALCOHOLS

The following methods enable us to distinguish between primary, secondary and tertiary alcohols.

1. Oxidation. Primary, secondary and tertiary alcohols yield different products on oxidation. Primary alcohol gives an aldehyde and then an acid containing the same number of carbon atoms as the original alcohol. Secondary alcohols form ketones containing the same number of carbon atoms as the original alcohol but further oxidation of ketones occurs only with strong oxidising agents to form acids containing lesser number of carbon atoms. Tertiary alcohols are very difficult to oxidise but if oxidised they yield ketones and then acids, both containing lesser number of carbon atoms than the alcohol. Thus, the identification of oxidation products can reveal the nature of alcohols.

$$R-CH_2OH \xrightarrow{[O]} R-CHO \qquad R_2CHOH \xrightarrow{[O]} R_2C=O$$

$$(CH_3)_3C-OH \xrightarrow{4[O]} CH_3COCH_3 + CO_2 + 2H_2O$$

(2) Reaction with hot reduced copper. When the vapours of alcohols are passed over hot reduced copper at 473–573 K, primary and secondary alcohols lose hydrogen to form aldehydes and ketones respectively while tertiary alcohols get dehydrated to form alkenes as already explained.

(3) Lucas test. This test consists in treating the alcohol with Lucas reagent which is an equimolar mixture of concentrated hydrochloric acid and anhydrous zinc chloride. Appearance of cloudiness in the reaction mixture indicates the conversion of alcohol into alkyl chloride. It has been observed that a tertiary alcohol reacts immediately, a secondary alcohol reacts within five minutes while a primary alcohol does not react appreciably at room temperature.

(4) Victor Meyer's method. This reaction is carried out as follows:

The given alcohol is treated with phosphorous and iodine when an alkyl iodide is obtained. It is distilled with silver nitrite to yield the corresponding nitroalkane. The nitroalkane is finally treated with nitrous acid (*i.e.*, sodium nitrite + dil. sulphuric acid) and the solution made alkaline.

The formation of *blood red* colouration shows the original alcohol to be primary, a *blue* colouration indicates a secondary alcohol while the solution remains *colourless* in case of tertiary alcohol. The reactions are explained in tabular form as under:

Primary alcohol	*Secondary alcochol*	*Tertiary alcohol*
RCH_2OH	R_2CHOH	R_3C-OH
↓ $P + I_2$	↓ $P + I_2$	↓ $P + I_2$
RCH_2I	R_2CHI	R_3CI
↓ $AgNO_2$	↓ $AgNO_2$	↓ $AgNO_2$
RCH_2NO_2	R_2CHNO_2	R_3CNO_2
↓ HONO Nitrous acid	↓ HONO Nitrous acid	↓ HNO_2
$R-C(=NOH)-NO_2$	$R_2C(NO)-NO_2$	No action
Nitrolic acid	Pseudo nitrol	
Produces blood red colour with alkali	*Blue colour*	*Colourless solution*

Example 1. How will you achieve the following conversion?

(*i*) Ethyl alcohol ⟶ Sec. butyl alcohol

(*ii*) *n*-propyl alcohol ⟶ Isopropyl alcohol

(*iii*) Isopropyl alcohol ⟶ Tert. butyl alcohol

(iv) Isopropyl alcohol ⟶ *n*-Propyl alcohol

(*v*) 2-Methyl-2-butanol ⟶ 3-Methyl-2-butanol

(*vi*) Isobutyl alcohol ⟶ Tert. butyl alcohol.

Solution. (*i*) Ethyl alcohol into sec. butyl alcohol

$$\underset{\substack{\text{Ethyl alcohol} \\ (\textit{Primary})}}{CH_3CH_2OH} \xrightarrow{K_2Cr_2O_7} \underset{\text{Acetaldehyde}}{CH_3CHO}$$

$$CH_3CHO + CH_3CH_2MgBr \longrightarrow CH_3-\underset{\underset{\displaystyle CH_3}{|}}{\underset{\displaystyle CH_2}{\underset{|}{\overset{\overset{\displaystyle H}{|}}{C}}}}-OMgBr \xrightarrow{H_2O^+}$$

$$CH_3-CHOH-CH_2-CH_3 + Mg(OH)Br$$

sec. Butyl alcohol (*Secondary*)

(*ii*) *n*-Propyl alcohol into isopropyl alcohol

$$\underset{\substack{n\text{-Propyl alcohol}\\(Primary)}}{CH_3CH_2CH_2OH} \xrightarrow[Al_2O_3,\ 625\ K]{\text{Dehydration}} \underset{\text{Propylene}}{CH_3-CH=CH_2} \xrightarrow{HI}$$

$$\underset{\text{Isopropyl iodide}}{CH_3-CHI-CH_3} \xrightarrow[\substack{\text{(Moist silver}\\\text{oxide)}}]{AgOH} \underset{\substack{\text{Isopropyl alcohol}\\(Secondary)}}{CH_3-CHOH-CH_3} + AgI$$

(*iii*) Isopropyl alcohol into tert. butyl alcohol

$$\underset{\substack{\text{Isopropyl alcohol}\\(Secondary)}}{CH_3-CHOH-CH_3} \xrightarrow{K_2Cr_2O_7} \underset{\text{Acetone}}{CH_3COCH_3} \xrightarrow{CH_3MgI} CH_3-\underset{\underset{\displaystyle CH_3}{|}}{\overset{\overset{\displaystyle CH_3}{|}}{C}}-OMgI$$

$$\xrightarrow{H_2O^+} CH_3-\underset{\underset{\displaystyle CH_3}{|}}{\overset{\overset{\displaystyle CH_3}{|}}{C}}-OH + Mg(OH)I$$

tert-Butyl alcohol (*Tertiary*)

(*iv*) Isopropyl alcohol into n-propyl alcohol

$$\underset{\substack{\text{Isopropyl alcohol}\\(Secondary)}}{CH_3-CHOH-CH_3} \xrightarrow[Al_2O_3,\ \text{Heat}]{\text{Dehydration}} CH_3-CH=CH_2$$

$$\downarrow (BH_3)_2$$

$$(CH_3-CH_2-CH_2)_3B$$

$$\downarrow$$

$$CH_3-CH_2-CH_2OH$$

n-Propyl alcohol (*Primary*)

(*v*) 2-Methyl-2-butanol into 3-methyl-2-butanol

$$\underset{\substack{\text{2-Methyl-2-butanol}\\(Tertiary)}}{CH_3-\underset{\underset{\displaystyle OH}{|}}{\overset{\overset{\displaystyle CH_3}{|}}{C}}-CH_2-CH_3} \xrightarrow[Al_2O_3,\ 625\ K]{\text{Dehydration}} \underset{\text{2-methyl-2-butene}}{CH_3-\overset{\overset{\displaystyle CH_3}{|}}{C}=CH-CH_3} \xrightarrow{(BH_3)_2}$$

$$\left[\begin{array}{c} CH_3 \\ | \\ CH_3-CH-CH- \\ | \\ CH_3 \end{array}\right]_3 B \xrightarrow{H_2O_2/OH^-} \underset{\substack{\text{3-methyl-2-butanol} \\ (Secondary)}}{CH_3-\overset{\overset{CH_3}{|}}{CH}-CHOH-CH_3}$$

(*vi*) Isobutyl alcohol into tert. butyl alcohol

$$\underset{\substack{\text{Isobutyl alcohol} \\ (Primary)}}{CH_3-\overset{\overset{CH_3}{|}}{CH}-CH_2OH} \xrightarrow[Al_2O_3,\ 625\ K]{\text{Dehydration}} \underset{\text{Isobutylene}}{CH_3-\overset{\overset{CH_3}{|}}{C}=CH_2} \xrightarrow{HI}$$

$$\underset{\text{Tert-Butyl iodide}}{CH_3-\overset{\overset{CH_3}{|}}{\underset{\underset{I}{|}}{C}}-CH_3} \xrightarrow{AgOH} \underset{\substack{\text{Tert-Butyl alcohol} \\ (Tertiary)}}{CH_3-\overset{\overset{CH_3}{|}}{\underset{\underset{OH}{|}}{C}}-CH_3} + AgI$$

Example 2. Why does dehydration of 1-phenyl-2-propanol in acid form 1-phenyl-1-propene rather than 1-phenyl-2-propene?

Solution. The structure of 1-phenyl-2-propanol is

$$\underset{\text{I}}{H-\overset{\overset{H}{|}}{\underset{\underset{H}{|}}{C}}-\overset{\overset{H}{|}}{\underset{\underset{OH}{|}}{C}}-\overset{\overset{H}{|}}{\underset{\underset{H}{|}}{C}}-C_6H_5}$$

On reaction with a protonic acid, it first forms an oxonium ion (II) which then loses a water molecule to form a carbocation (III).

$$\underset{\text{II}}{H-\overset{\overset{H}{|}}{\underset{\underset{H}{|}}{C}}-\overset{\overset{H}{|}}{\underset{\underset{\overset{+}{O}H_2}{|}}{C}}-\overset{\overset{H}{|}}{\underset{\underset{H}{|}}{C}}-C_6H_5} \qquad \underset{\text{III}}{H-\overset{\overset{H}{|}}{\underset{\underset{H}{|}}{\overset{3}{C}}}-\overset{\overset{H}{|}}{\overset{2}{\overset{+}{C}}}-\overset{\overset{H}{|}}{\underset{\underset{H}{|}}{\overset{1}{C}}}-C_6H_5}$$

The carbocation having structure III loses a proton to stabilize it. This proton can be ejected either from carbon atom No. 1 or carbon atom No. 3. Accordingly, two different alkenes are possible as:

$$H-\overset{\overset{H}{|}}{\underset{\underset{H}{|}}{\overset{3}{C}}}-\overset{\overset{H}{|}}{\overset{2}{\overset{+}{C}}}-\overset{\overset{H}{|}}{\underset{\underset{H}{|}}{\overset{1}{C}}}-C_6H_5 \xrightarrow[\text{from C}-1]{\text{Loss of }H^+} \underset{\substack{\text{1-Phenyl-1-propene} \\ \text{IV}}}{H-\overset{\overset{H}{|}}{\underset{\underset{H}{|}}{C}}-\overset{\overset{H}{|}}{C}=\overset{\overset{H}{|}}{C}-C_6H_5}$$

Or

$$H-\overset{\overset{H}{|}}{\underset{\underset{H}{|}}{\overset{3}{C}}}-\overset{\overset{H}{|}}{\overset{2}{\overset{+}{C}}}-\overset{\overset{H}{|}}{\underset{\underset{H}{|}}{\overset{1}{C}}}-C_6H_5 \xrightarrow[\text{from C}-3]{\text{Loss of }H^+} \underset{\substack{\text{1-Phenyl-2-propene} \\ \text{V}}}{H-\overset{\overset{H}{|}}{C}=\overset{\overset{H}{|}}{C}-\overset{\overset{H}{|}}{\underset{\underset{H}{|}}{C}}-C_6H_5}$$

Alkene IV is more substituted and therefore more stable than alkene V. Moreover, alkene IV has additional stability as its double bond is conjugated with the benzene ring. Consequently, only alkene IV will be formed.

Example 3. Outline all the steps in the synthesis of following compounds from *n*-butyl alcohol using necessary inorganic reagents.

(*a*) n-Butyl bromide **(*b*) 1-Butene**

(*c*) *n*-Butyl hydrogen sulphate **(*d*) n-Butyraldehyde**

(*e*) *n*-Butyric acid **(*f*) 1-Butyne.**

Solution. (*a*) *n*-Butyl alcohol into *n*-Bromide.

$$\underset{n\text{-Butyl alcohol}}{CH_3\,CH_2\,CH_2\,CH_2OH} \xrightarrow{P + Br_2} \underset{n\text{-Butyl bromide}}{CH_3CH_2CH_2CH_2Br}$$

(*b*) *n*-Butyl alcohol into 1-butene.

$$\underset{n\text{-Butyl alcohol}}{CH_3CH_2CH_2CH_2OH} \xrightarrow{SOCl_2} \underset{n\text{-Butyl chloride}}{CH_3CH_2CH_2CH_2Cl} \xrightarrow[\text{KOH}]{\text{Alcoholic}} \underset{\text{1-Butene}}{CH_3CH_2CHCH = CH_2}$$

(*c*) *n*-Butyl alcohol into *n*-butyl hydrogen sulphate.

$$\underset{n\text{-Butyl alcohol}}{CH_3CH_2CH_2CH_2OH} \xrightarrow[\text{383 K}]{H_2SO_4\text{ at}} \underset{n\text{-Butyl hydrogen sulphate}}{CH_3CH_2CH_2CH_2HSO_4 + H_2O}$$

(*d*) *n*-Butyl alcohol into *n*-butyraldehyde.

$$\underset{n\text{-Butyl alcohol}}{CH_3CH_2CH_2CH_2OH} \xrightarrow[\text{573 K}]{\text{Hot reduced}} \underset{n\text{-Butylaldehyde}}{CH_3CH_2CH_2CHO + H_2}$$

(*e*) *n*-Butyl alcohol into *n*-butyric acid.

$$\underset{n\text{-Butyl alcohol}}{CH_3CH_2CH_2CH_2OH} \xrightarrow[\text{Cu, 573 K}]{\text{Hot reduced}} \underset{n\text{-Butyraldehyde}}{CH_3CH_2CH_2CHO} \xrightarrow[K_2Cr_2O_7 + H_2SO_4]{\text{Oxidation}} \underset{n\text{-Butyric acid}}{CH_3CH_2CH_2COOH}$$

(*f*) *n*-Butyl alcohol into 1-butyne.

$$\underset{n\text{-Butyl alcohol}}{CH_3CH_2CH_2CH_2OH} \xrightarrow{SOCl_2} \underset{n\text{-Butyl chloride}}{CH_3CH_2CH_2CH_2Cl} \xrightarrow[\text{KOH}]{\text{Alcoholic}} \underset{\text{1-Butene}}{CH_3CH_2CHCH = CH_2}$$

$$\xrightarrow{Br_2} \underset{\text{1, 2-Dibromobutane}}{CH_3CH_2\underset{|\atop Br}{CH} - \underset{|\atop Br}{CH_2}} \xrightarrow[\text{liq. }NH_3]{NaNH_2\text{ in}} \underset{\text{Butyne-1}}{CH_3CH_2C \equiv CH}$$

Example 4. How will you convert:

(*a*) Isopropyl alcohol into 2-methyl-2-hexanol.

(*b*) Secondary butyl alcohol into 3-methyl-3-pentanol.

Solution. (*a*) Isopropyl alcohol into 2-methyl-2-hexanol

$$\underset{\text{Isopropyl alcohol}}{CH_3 - \underset{|\atop OH}{CH} - CH_3} \xrightarrow[H_2SO_4]{K_2Cr_2O_7} \underset{\text{Acetone}}{CH_3 - \underset{\|\atop O}{C} - CH_3} \xrightarrow{CH_3CH_2CH_2CH_2MgI}$$

$$CH_3 - \overset{CH_3\atop |}{\underset{|\atop OMgI}{C}} - CH_2CH_2CH_2CH_3 \xrightarrow{H_2O^+} \underset{\text{2-Methyl-2-hexanol}}{CH_3 - \overset{CH_3\atop |}{\underset{|\atop OH}{C}} - CH_2CH_2CH_2CH_3}$$

(b) Secondary butyl alcohol into 2-methyl-3-pentanol.

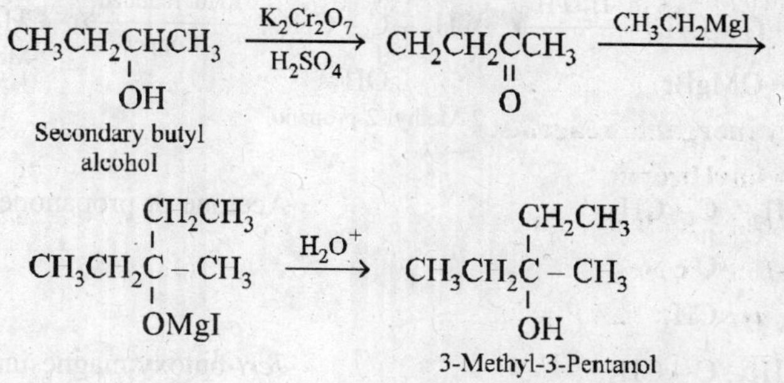

$$\underset{\text{Secondary butyl alcohol}}{CH_3CH_2\underset{OH}{\underset{|}{C}}HCH_3} \xrightarrow[H_2SO_4]{K_2Cr_2O_7} CH_2CH_2\underset{O}{\underset{\|}{C}}CH_3 \xrightarrow{CH_3CH_2MgI}$$

$$CH_3CH_2\overset{CH_2CH_3}{\overset{|}{\underset{OMgI}{\underset{|}{C}}}}-CH_3 \xrightarrow{H_2O^+} \underset{\text{3-Methyl-3-Pentanol}}{CH_3CH_2\overset{CH_2CH_3}{\overset{|}{\underset{OH}{\underset{|}{C}}}}-CH_3}$$

Example 5. Predict the products of the following reactions. Write the mechanism involved:

$$(a)\ CH_3CH_2\overset{CH_3}{\overset{|}{C}}H.CH_2OH \xrightarrow{H^+} \qquad (b)\ CH_3\overset{CH_3}{\overset{|}{C}}H-\overset{CH_3}{\overset{|}{\underset{OH}{\underset{|}{C}}}}-CH_3 \xrightarrow{H^+}$$

Solution. (a)

$$CH_3CH_2\overset{CH_3}{\overset{|}{C}}H.CH_2OH + H^+ \longrightarrow CH_3CH_2\overset{CH_3}{\overset{|}{C}}H-\overset{H}{\overset{|}{\underset{H}{\underset{|}{C}}}}-\overset{H}{\overset{|}{\underset{+}{O}}}H \xrightarrow{-H_2O}$$

$$CH_3-CH_2\overset{CH_3}{\overset{|}{C}}H-\overset{H}{\overset{|}{\underset{H}{\underset{|}{C^+}}}} \xrightarrow[\text{shift}]{\text{1, 2-Methyl}} CH_3CH_2-\overset{+}{\underset{H}{\underset{|}{C}}}-\overset{CH_3}{\overset{|}{\underset{H}{\underset{|}{C}}}}-H \xrightarrow{-H^+} \underset{\text{2-Pentene}}{CH_3CH=CHCH_2CH_3}$$

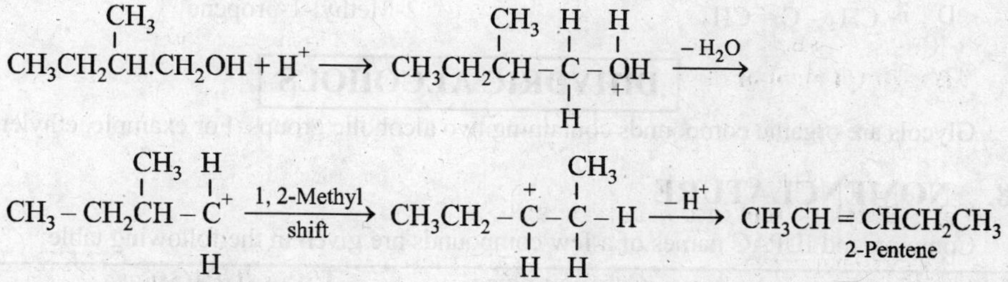

A more substituted alkene is more stable.

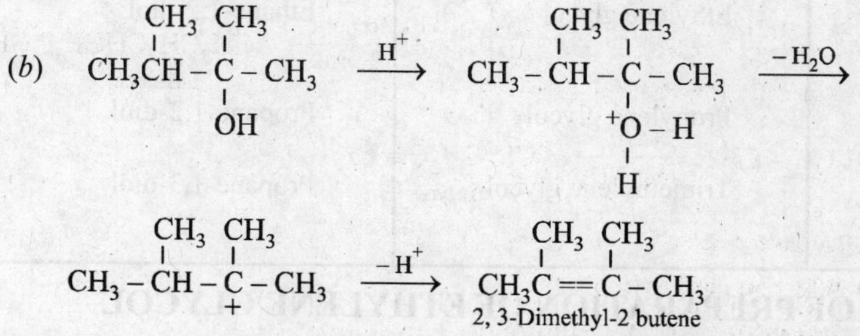

$$(b)\quad CH_3\overset{CH_3}{\overset{|}{C}}H-\overset{CH_3}{\overset{|}{\underset{OH}{\underset{|}{C}}}}-CH_3 \xrightarrow{H^+} CH_3-\overset{CH_3}{\overset{|}{C}}H-\overset{CH_3}{\overset{|}{\underset{^+O-H}{\underset{|}{C}}}}-CH_3 \xrightarrow{-H_2O}$$

(where ^+O-H bears a further H below)

$$CH_3-\overset{CH_3}{\overset{|}{C}}H-\overset{CH_3}{\overset{|}{\underset{+}{C}}}-CH_3 \xrightarrow{-H^+} \underset{\text{2, 3-Dimethyl-2 butene}}{CH_3\overset{CH_3}{\overset{|}{C}}=\overset{CH_3}{\overset{|}{C}}-CH_3}$$

Example 6. Identify the products (A, B, C and D) in the following sequence of reactions.

$$\textbf{Isopropyl alcohol} \xrightarrow[H_2SO_4]{K_2Cr_2O_7} \textbf{A} \xrightarrow{CH_3MgBr} \textbf{B} \xrightarrow[H^+]{H_2O} \textbf{C} \xrightarrow{\text{Hot reduced copper}} \textbf{D}$$

Solution.

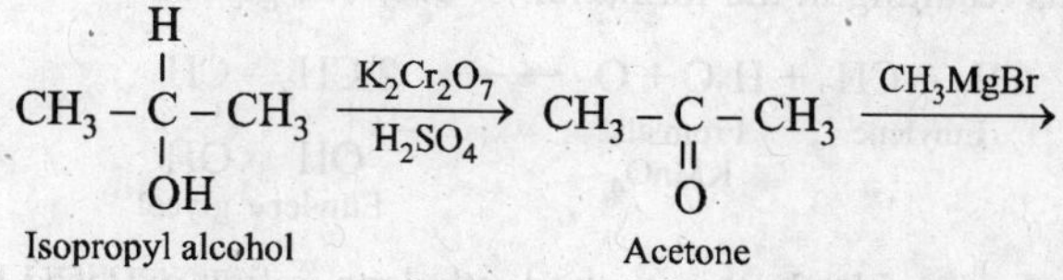

$$\underset{\text{Isopropyl alcohol}}{CH_3-\overset{H}{\overset{|}{\underset{OH}{\underset{|}{C}}}}-CH_3} \xrightarrow[H_2SO_4]{K_2Cr_2O_7} \underset{\text{Acetone}}{CH_3-\underset{O}{\underset{\|}{C}}-CH_3} \xrightarrow{CH_3MgBr}$$

$$\underset{\text{OMgBr}}{\overset{\text{CH}_3}{CH_3-\underset{|}{\overset{|}{C}}-CH_3}} \xrightarrow{H_2O/H^+} \underset{\text{2-Methyl-2-propanol}}{\underset{\text{OH}}{\overset{\text{CH}_3}{CH_3-\underset{|}{\overset{|}{C}}-CH_3}}} \xrightarrow[\text{Copper}]{\text{Hot reduced}} \underset{\text{2-Methyl-1-propene}}{\overset{\text{CH}_3}{CH_3\overset{|}{C}=CH_2}}$$

Therefore,

A is $CH_3-\underset{\underset{O}{\|}}{C}-CH_3$ Acetone or propanone-2

B is $CH_3-\underset{\underset{OMgBr}{|}}{\overset{\overset{CH_3}{|}}{C}}-CH_3$ *Tert*-butoxy magnesium bromide

C is $CH_3-\underset{\underset{OH}{|}}{\overset{\overset{CH_3}{|}}{C}}-CH_3$ *Tert*-butyl alcohol or 2-Methyl-2-propanol

D is $CH_3-\overset{\overset{CH_3}{|}}{C}=CH_2$ 2-Methyl-1-propene

DIHYDRIC ALCOHOLS

Glycols are organic compounds containing two alcoholic groups. For example, ethylene glycol.

2.8. NOMENCLATURE

Common and IUPAC names of a few compounds are given in the following table:

Structure	*Common name*	*IUPAC name*
$\underset{OH}{\underset{\vert}{CH_2}}-\underset{OH}{\underset{\vert}{CH_2}}$	Ethylene glycol	Ethane-1,2-diol
$CH_3-\underset{OH}{\underset{\vert}{CH}}-\underset{OH}{\underset{\vert}{CH_2}}$	Propylene glycol	Propane-1,2-diol
$\underset{OH}{\underset{\vert}{CH_2}}-CH_2-\underset{OH}{\underset{\vert}{CH_2}}$	Trimethylene glycol	Propane-1,3-diol

2.9. METHODS OF PREPARATION OF ETHYLENE GLYCOL

Ethylene glycol may be prepared by one of the following methods:

(1) Hydroxylation of ethylene. When ethylene is treated with a cold alkaline solution of potassium permanganate (known as Baeyer's reagent), a hydroxyl group is added to each of the doubly bonded carbon atoms resulting in the formation of ethylene glycol.

$$\underset{\text{Ethylene}}{CH_2=CH_2} + H_2O + \underset{\substack{\text{From alk.}\\ KMnO_4}}{O} \longrightarrow \underset{\text{Ethylene glycol}}{2\ \underset{OH}{\underset{|}{CH_2}}-\underset{OH}{\underset{|}{CH_2}}}$$

(2) Hydrolysis of ethylene oxide. In this method, ethylene is first oxidised by heating in air at 525 K in the presence of silver catalyst to form ethylene oxide or ethylene epoxide. The epoxide, thus formed, is hydrolysed in the presence of an acid or base to form ethylene glycol.

$$CH_2 = CH_2 + \frac{1}{2}O_2 \xrightarrow{525\text{ K, Ag}} \underset{\text{Ethylene epoxide}}{CH_2 - CH_2 \text{ (bridged by O)}}$$

$$\underset{\text{(epoxide, bridged by O)}}{CH_2 - CH_2} + H_2O \xrightarrow{\text{Acid or base}} \underset{\text{Ethylene glycol}}{\underset{|\quad\ \ |}{CH_2 - CH_2}\atop{OH\quad OH}}$$

Ethylene can also be converted into ethylene epoxide by treatment with peroxyformic acid (HCO_2 OH)

$$CH_2 = CH_2 + HCO_2OH \longrightarrow \underset{\text{(bridged by O)}}{CH_2 - CH_2} + HCOOH \xrightarrow{\text{Acid or base}} CH_2(OH) - CH_2(OH)$$

It may be noted that hydroxylation by $KMnO_4$ gives *cis* compound whereas peroxyformic acid produces *trans* compound.

(3) By the hydrolysis of ethylene dichloride or dibromide or chlorohydrin.

$$\begin{array}{l}CH_2Br\\ |\\ CH_2Br\end{array} + Na_2CO_3 + H_2O \longrightarrow \begin{array}{l}CH_2OH\\ |\\ CH_2OH\end{array} + 2NaBr + CO_2$$

The yield of ethylene glycol is, however, not good as the hydrolysis is also accompanied by the formation of vinyl bromide through an elimination reaction.

$$\begin{array}{l}CH_2Br\\ |\\ CH_2Br\end{array} + Na_2CO_3 \longrightarrow \begin{array}{l}CH_2\\ \|\\ CHBr\end{array} + NaBr + NaHCO_3$$

(4) Bimolecular reduction of carbonyl compound. Symmetrical glycols, other than ethylene glycol, can be obtained by the bimolecular reduction of aldehydes or ketones, with magnesium in the presence of benzene.

$$2\ \begin{array}{c}\diagdown\ \diagup\\ C\\ \|\\ O\end{array} \xrightarrow[\text{Benzene (Bimolecular reduction)}]{\text{Mg.}} \underset{\text{Pinacol}}{\begin{array}{c}|\quad\ \ |\\ -C-C-\\ |\quad\ \ |\\ OH\ \ OH\end{array}}$$

For example,

$$\underset{\text{Acetone (2-molecules)}}{\begin{array}{c}CH_3\\ |\\ CH_3-C=O\\ CH_3-C=O\\ |\\ CH_3\end{array}} + Mg \xrightarrow{\text{Benzene}} \begin{array}{c}CH_3CH_3\\ |\quad\ |\\ H_3C-C-C-CH_3\\ |\quad\ |\\ O\quad O\\ \diagdown\ \diagup\\ Mg\end{array} \xrightarrow{2H_2O} \underset{\text{Pinacol (2, 3-Dimethylbutane-2, 3-diol)}}{\begin{array}{c}CH_3CH_3\\ |\quad\ |\\ CH_3-C-C-CH_3\\ |\quad\ |\\ OH\ OH\end{array}}$$

Physical properties of ethylene glycol. It is a colourless syrupy liquid (b.p. 470 K) having a sweet taste. Its high boiling point is due to the presence of two sites for hydrogen bonding per molecule (*i.e.*, two –OH groups). For the same reason it is also highly soluble in water and ethyl alcohol.

2.10. CHEMICAL PROPERTIES OF ETHYLENE GLYCOL

Like monohydric alcohols, ethylene glycol exhibits the usual reactions of hydroxyl group. However, it is not necessary that both the alcoholic groups should react with a particular reagent, or in other words, it is possible to carry out reactions in which only one of the two alcoholic groups is involved.

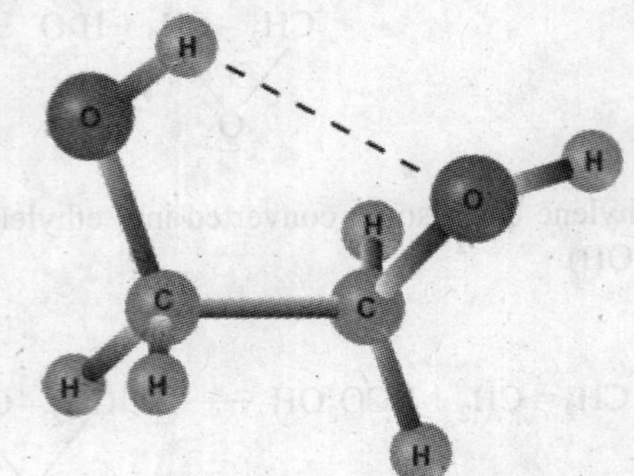

Interaction in glycol molecule

(1) Reaction with metals. Ethylene glycol reacts with metals like sodium to form mono and disodium derivatives.

$$\underset{\text{Ethylene glycol}}{\begin{matrix} CH_2OH \\ | \\ CH_2OH \end{matrix}} \xrightarrow[\frac{1}{2}H_2]{\text{Na, 333 K}} \underset{\text{Monosodium glycolate}}{\begin{matrix} CH_2ONa \\ | \\ CH_2OH \end{matrix}} \xrightarrow[-\frac{1}{2}H_2]{\text{Na, 433 K}} \underset{\text{Disodium glycolate}}{\begin{matrix} CH_2ONa \\ | \\ CH_2ONa \end{matrix}}$$

(2) Reaction with halogen acids. It reacts with halogen acids to form two different products as shown below:

$$\underset{\text{Ethylene glycol}}{\begin{matrix} CH_2OH \\ | \\ CH_2OH \end{matrix}} + HCl \xrightarrow[-H_2O]{433\text{ K}} \underset{\text{Ethylene chlorohydrin}}{\begin{matrix} CH_2Cl \\ | \\ CH_2OH \end{matrix}} \xrightarrow[-H_2O]{\text{HCl, 473 K}} \underset{\text{Ethylene dichloride}}{\begin{matrix} CH_2Cl \\ | \\ CH_2Cl \end{matrix}}$$

With HI, ethylene diodide first formed decomposes to give ethylene.

$$\begin{matrix} CH_2OH \\ | \\ CH_2OH \end{matrix} \xrightarrow[-2H_2O]{2HI} \underset{\text{Unstable}}{\begin{matrix} CH_2I \\ | \\ CH_2I \end{matrix}} \longrightarrow \begin{matrix} CH_2 \\ \| \\ CH_2 \end{matrix} + I_2$$

3. Reaction with phosphorus trihalides. When treated with PCl_3 or PBr_3, ethylene glycol forms ethylene dichloride or dibromide.

$$3\begin{matrix} CH_2OH \\ | \\ CH_2OH \end{matrix} + 2PCl_3 \longrightarrow 3\begin{matrix} CH_2Cl \\ | \\ CH_2Cl \end{matrix} + 2H_3PO_3$$

4. Reaction with PCl_5 or thionyl chloride. It reacts with phosphorus pentachloride or thionyl chloride to form ethylene dichloride:

$$\begin{matrix} CH_2OH \\ | \\ CH_2OH \end{matrix} + 2PCl_5 \longrightarrow \begin{matrix} CH_2Cl \\ | \\ CH_2Cl \end{matrix} + POCl_3 + 2HCl$$

$$\begin{matrix} CH_2OH \\ | \\ CH_2OH \end{matrix} + 2SOCl_2 \longrightarrow \begin{matrix} CH_2Cl \\ | \\ CH_2Cl \end{matrix} + 2SO_2 + 2HCl$$

5. Reaction with organic acids. It reacts with organic acids to form mono- or di-esters depending upon whether one or two moles of the acid are used per mole of glycol.

$$\begin{array}{c} CH_2OH \\ | \\ CH_2OH \end{array} \xrightarrow[-H_2O]{HOOCCH_3} \underset{\text{Glycol monoacetate}}{\begin{array}{c} CH_2O.OCCH_3 \\ | \\ CH_2.OH \end{array}} \xrightarrow[H_2O]{HOOCCH_3} \underset{\text{Glycol diacetate}}{\begin{array}{c} CH_2O.OCCH_3 \\ | \\ CH_2O.OCCH_3 \end{array}}$$

6. Oxidation. Ethylene glycol gives different products on oxidation under different conditions.

(*i*) *When oxidised with nitric acid*, ethylene glycol ultimately gives rise to oxalic acid through the intermediate formation of many products which may also be isolated.

$$\underset{\text{Ethylene glycol}}{\begin{array}{c} CH_2OH \\ | \\ CH_2OH \end{array}} \xrightarrow[(-H_2O)]{[O]} \underset{\text{Glycollic aldehyde}}{\begin{array}{c} CHO \\ | \\ CH_2OH \end{array}} \xrightarrow{[O]} \underset{\text{Glycollic acid}}{\begin{array}{c} COOH \\ | \\ CH_2OH \end{array}}$$

$$\text{Glycollic aldehyde} \xrightarrow[(-H_2O)]{[O]} \underset{\text{Glyoxal}}{\begin{array}{c} CHO \\ | \\ CHO \end{array}} \xrightarrow{[O]} \underset{\text{Glyoxalic acid}}{\begin{array}{c} COOH \\ | \\ CHO \end{array}} \xrightarrow{[O]} \underset{\text{Oxalic acid}}{\begin{array}{c} COOH \\ | \\ COOH \end{array}}$$

$$\text{Glycollic acid} \xrightarrow[(-H_2O)]{[O]} \text{Glyoxalic acid}$$

(*ii*) *When treated with periodic acid* (HIO_4) *or lead tetra acetate* [$Pb(OOCH_3)_4$], compounds containing two (or more) –OH groups attached to adjacent carbon atoms undergo oxidative cleavage. The reaction breaks carbon-carbon bonds and yields carbonyl compounds (aldehydes, ketones, etc) as oxidation products.

$$\begin{array}{c} | \\ -C-OH \\ | \\ -C-OH \\ | \end{array} \xrightarrow[\text{or } Pb(OOCCH_3)_4]{HIO_4} 2-\overset{|}{C}=O$$

For example,

$$\underset{\text{Propane-1, 2-diol}}{\begin{array}{c} H \\ | \\ H-C-OH \\ | \\ H-C-OH \\ | \\ CH_3 \end{array}} + HIO_4 \longrightarrow \begin{array}{c} H \\ | \\ H-C=O \\ + \\ H-C=O \\ | \\ CH_3 \end{array} + HIO_3 + H_2O$$

$$\begin{array}{c} R \\ | \\ R-C-OH \\ | \\ H-C-OH \\ | \\ R' \end{array} + Pb(OOCCH_3)_4 \longrightarrow \begin{array}{c} R \\ | \\ R-C=O \\ + \\ H-C=O \\ | \\ R' \end{array} \begin{array}{l} +2CH_3COOH \\ +Pb(OOCCH_3)_2 \end{array}$$

It may be noted that for *every C–C bond broken, a C–O bond is formed at each carbon.*

It may be added that in these reactions *periodic acid is found to work well in* **aqueous solutions** *while lead tetraacetate gives good results in* **organic solvents.**

Mechanism

(*a*) **Oxidation with periodic acid.** Oxidative cleavage with periodic acid is believed to take place through a cyclic intermediate formed by the attack of periodic ion (IO_4^-) on glycol as shown below:

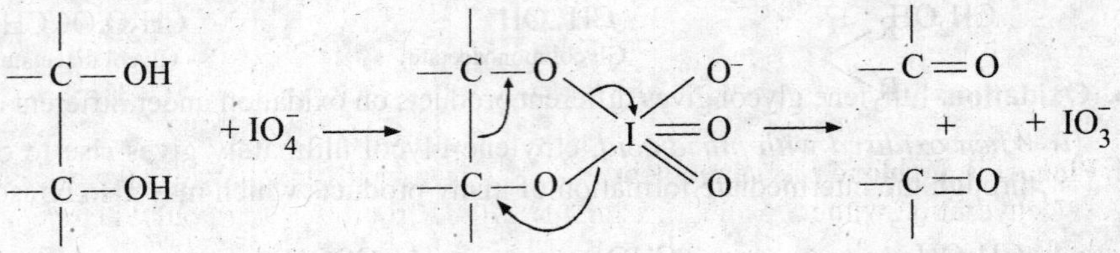

$$\begin{matrix} -\overset{|}{C}-OH \\ -\underset{|}{C}-OH \end{matrix} + IO_4^- \longrightarrow \begin{matrix} -\overset{|}{C}-O \\ -\underset{|}{C}-O \end{matrix}\!\!>\!I\!\begin{matrix} O^- \\ =O \\ O \end{matrix} \longrightarrow \begin{matrix} -\overset{|}{C}=O \\ + \\ -\underset{|}{C}=O \end{matrix} + IO_3^-$$

(*b*) **Oxidation with lead tetraacetate.** This reaction is also believed to occur through a cyclic intermediate which undergoes an oxidation-reduction process leading to the cleavage of carbon-carbon bond.

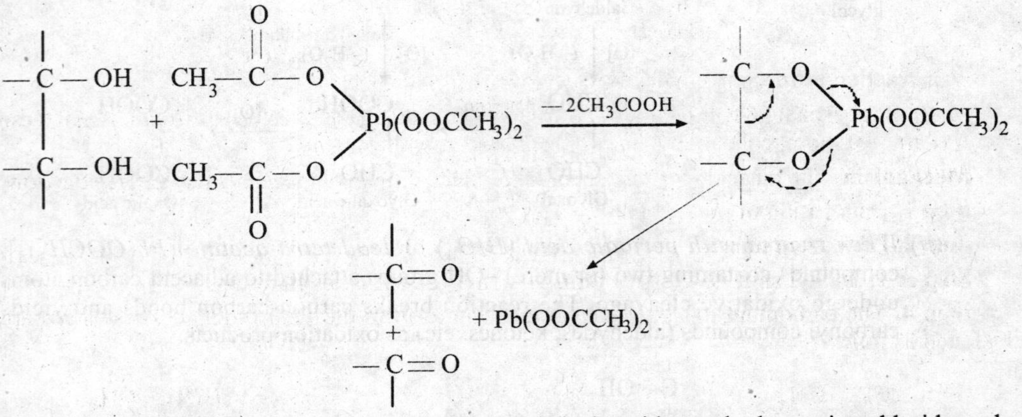

$$\begin{matrix} -\overset{|}{C}-OH \\ -\underset{|}{C}-OH \end{matrix} + \begin{matrix} CH_3-\overset{O}{\overset{\|}{C}}-O \\ CH_3-\underset{O}{\underset{\|}{C}}-O \end{matrix}\!\!>\!Pb(OOCCH_3)_2 \xrightarrow{-2CH_3COOH} \begin{matrix} -\overset{|}{C}-O \\ -\underset{|}{C}-O \end{matrix}\!\!>\!Pb(OOCCH_3)_2$$

$$\longrightarrow \begin{matrix} -\overset{|}{C}=O \\ + \\ -\underset{|}{C}=O \end{matrix} + Pb(OOCCH_3)_2$$

(7) Dehydration. When treated with dilute sulphuric acid or anhydrous zinc chloride under pressure in a sealed tube, ethylene glycol gets dehydrated to form acetaldehyde (formed through arrangement of vinyl alcohol).

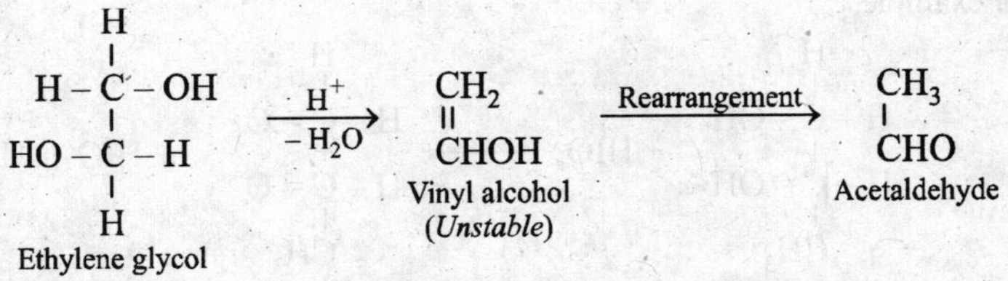

$$\underset{\text{Ethylene glycol}}{\begin{matrix} H \\ | \\ H-C-OH \\ | \\ HO-C-H \\ | \\ H \end{matrix}} \xrightarrow[-H_2O]{H^+} \underset{\substack{\text{Vinyl alcohol} \\ (\textit{Unstable})}}{\begin{matrix} CH_2 \\ \| \\ CHOH \end{matrix}} \xrightarrow{\text{Rearrangement}} \underset{\text{Acetaldehyde}}{\begin{matrix} CH_3 \\ | \\ CHO \end{matrix}}$$

But when distilled with conc. H_2SO_4, glycol forms dioxane as shown below.

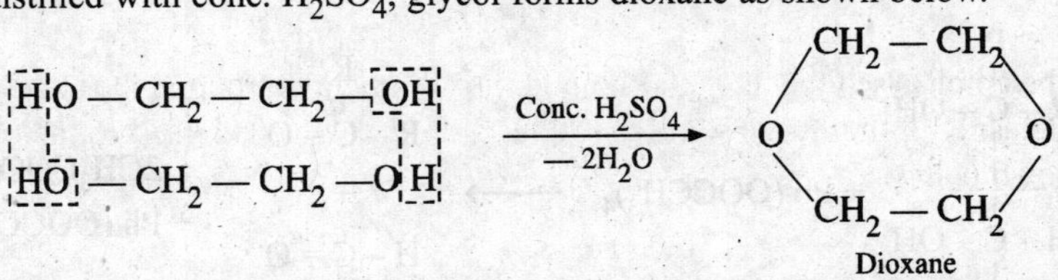

$$\begin{matrix} H\,O-CH_2-CH_2-OH \\ HO-CH_2-CH_2-O\,H \end{matrix} \xrightarrow[-2H_2O]{\text{Conc. } H_2SO_4} \underset{\text{Dioxane}}{\begin{matrix} CH_2-CH_2 \\ O \qquad\quad O \\ CH_2-CH_2 \end{matrix}}$$

8. Reaction with aldehydes and ketones. Glycol undergoes condensation with aldehydes and ketones in the presence of *p*-toluene sulphonic acid (commonly called PTS) to form cyclic acetals and ketals respectively. For example:

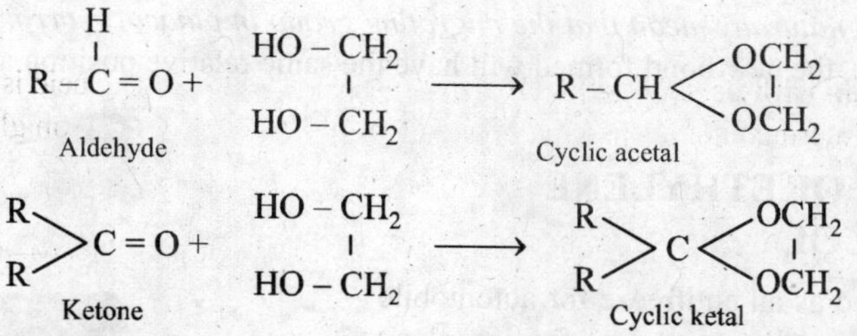

$$R-\overset{H}{\overset{|}{C}}=O + \begin{matrix} HO-CH_2 \\ | \\ HO-CH_2 \end{matrix} \longrightarrow R-CH\left\langle\begin{matrix} OCH_2 \\ | \\ OCH_2 \end{matrix}\right.$$

Aldehyde — Cyclic acetal

$$\begin{matrix} R \\ R \end{matrix}\rangle C=O + \begin{matrix} HO-CH_2 \\ | \\ HO-CH_2 \end{matrix} \longrightarrow \begin{matrix} R \\ R \end{matrix}\rangle C \left\langle\begin{matrix} OCH_2 \\ | \\ OCH_2 \end{matrix}\right.$$

Ketone — Cyclic ketal

9. Pinacol-pinacolone rearrangement. It is found that on treatment with sulphuric acid, pinacol undergoes dehydration with molecular rearrangement to form methyl *tert*-butyl ketone, also known as **pinacolone.**

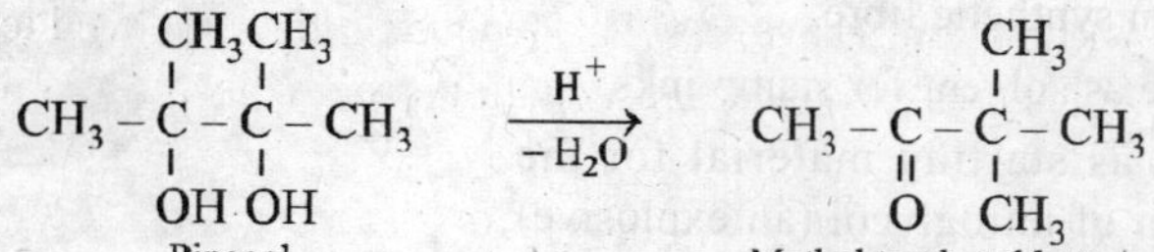

$$CH_3-\underset{OH}{\overset{CH_3}{C}}-\underset{OH}{\overset{CH_3}{C}}-CH_3 \xrightarrow[-H_2O]{H^+} CH_3-\underset{O}{\overset{}{\underset{\|}{C}}}-\underset{CH_3}{\overset{CH_3}{C}}-CH_3$$

Pinacol — Methyl tert butyl ketone or Pinacolone

This reaction is known as pinacol-pinacolone rearrangement.

Many other vicinal glycols are now known to undergo similar acid-catalysed reactions. All these reactions are, therefore, collectively known as pinacol rearrangements.

Mechanism. The pinacol rearrangement takes place through a mechanism as depicted below.

Step 1. Protonation of pinacol takes place.

Step 2. Protonated pinacol loses a water molecule to form carbonium ion.

Step 3. 1, 2 alkyl shift takes place forming a more stable carbonium ion.

Step 4. The carbonium ion loses a hydrogen ion to form the final product. These steps are illustrated as follows:

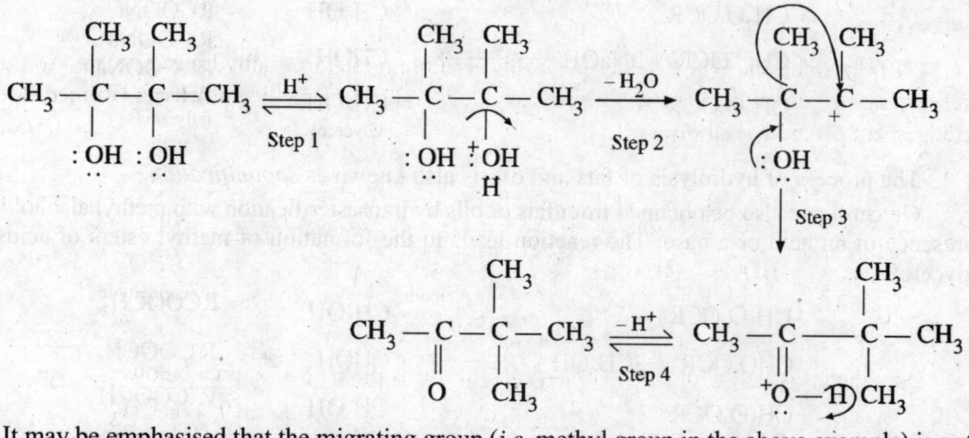

$$CH_3-\underset{:\ddot{O}H}{\overset{CH_3}{C}}-\underset{:\ddot{O}H}{\overset{CH_3}{C}}-CH_3 \underset{\text{Step 1}}{\overset{H^+}{\rightleftharpoons}} CH_3-\underset{:\ddot{O}H}{\overset{CH_3}{C}}-\underset{\overset{+}{:O}H\text{–}H}{\overset{CH_3}{C}}-CH_3 \xrightarrow[\text{Step 2}]{-H_2O} CH_3-\underset{:\ddot{O}H}{\overset{CH_3}{C}}-\overset{CH_3}{\underset{+}{C}}-CH_3$$

Step 3 ↓

$$CH_3-\underset{O}{\underset{\|}{C}}-\underset{CH_3}{\overset{CH_3}{C}}-CH_3 \underset{\text{Step 4}}{\overset{-H^+}{\rightleftharpoons}} CH_3-\underset{^{+}\ddot{O}-H}{\underset{\|}{C}}-\underset{CH_3}{\overset{CH_3}{C}}-CH_3$$

It may be emphasised that the migrating group (*i.e.* methyl group in the above example) is not completely detached from the carbon it is leaving till it has attached itself to the electron-deficient carbon as shown below.

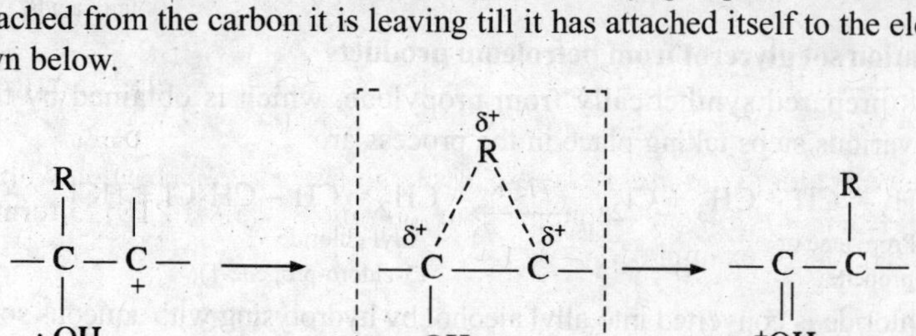

$$-\underset{:\ddot{O}H}{\overset{R}{C}}-\underset{+}{C}- \longrightarrow \left[-\underset{:\ddot{O}H}{\overset{\delta+}{C}}\cdots\overset{\delta+\ R\ \delta+}{}\cdots\overset{\delta+}{C}-\right] \longrightarrow -\underset{^{+}\ddot{O}H}{\underset{\|}{C}}-\overset{R}{C}-$$

This would naturally mean that the migrating group in pinacol rearrangement will retain its configuration i.e. the new bond formed will have the same relative position as that held previously be the old bond.

2.11. USES OF ETHYLENE GLYCOL

(*i*) It is used as an antifreeze for automobile radiators and as a coolant for aeroplane motors under the trade name Prestone.

(*ii*) It is used in the manufacture of Dacron, a well-known synthetic fibre.

(*iii*) It finds use as solvent for stamp inks.

(*iv*) It is used as starting material for the preparation of nitroglycol (an explosive) and solvents such as cellosolve and diglyme.

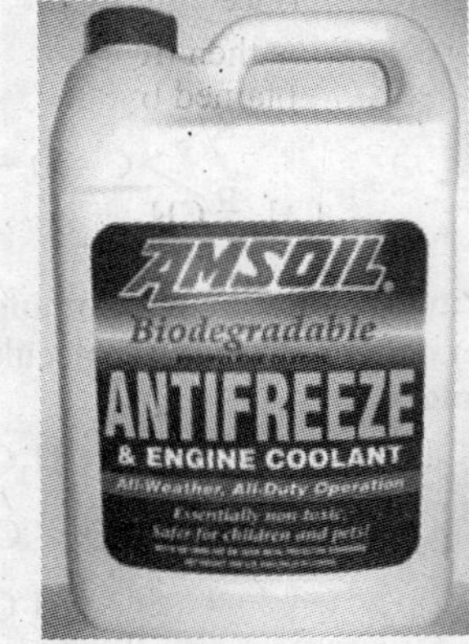

Ethylene glycol is used as antifreeze

TRIHYDRIC ALCOHOLS

2.12. PREPARATION

1. Preparation of glycerol from fats

Glycerol occurs in many natural fats and oils in the form of glyceryl esters of long chain fatty acids. Hydrolysis of these fats or oils with sodium hydroxide gives sodium salts of fatty acids and glycerol.

$$\underset{\text{Fat or oil}}{\begin{array}{l}CH_2O.OCR\\ |\\ CHO.OCR' + 3NaOH\\ |\\ CH_2O.OCR''\end{array}} \longrightarrow \underset{\text{Glycerol}}{\begin{array}{l}CH_2OH\\ |\\ CHOH\\ |\\ CH_2OH\end{array}} + \underset{\text{Sodium salts of fatty acids or soap}}{\begin{array}{l}RCOONa\\ R'COONa\\ R''\,COONa\end{array}}$$

The process of hydrolysis of fats and oils is also known as *saponification.*

Glycerol can also be obtained from fats or oils by transesterification with methyl alcohol in the presence of an acid or a base. The reaction leads to the formation of methyl esters of acids and glycerol.

$$\underset{\text{Fat or oil}}{\begin{array}{l}CH_2O.OCR\\ |\\ CHO.OCR' + 3CH_3OH\\ |\\ CH_2O.OCR''\end{array}} \longrightarrow \underset{\text{Glycerol}}{\begin{array}{l}CH_2OH\\ |\\ CHOH\\ |\\ CH_2OH\end{array}} + \underset{\text{Methyl esters of fatty acids}}{\begin{array}{c}RCOOCH_3\\ +\\ R'COOCH_3\\ +\\ R''COOCH_3\end{array}}$$

2. Preparation sof glycerol from petroleum products

Glycerol is prepared synthetically from propylene, which is obtained by the cracking of petroleum. The various steps taking place in the process are:

$$\underset{\text{Propylene or propene}}{CH_2 = CH - CH_3} + Cl_2 \xrightarrow{773\text{ K}} \underset{\text{Allyl chloride (3-chloro-propene-1)}}{CH_2 = CH - CH_2Cl} + HCl$$

The allyl chloride is converted into allyl alcohol by hydrolysing with aqueous sodium carbonate at 423 K and at a pressure of 12 atmospheres.

$$CH_2=CH-CH_2Cl \xrightarrow[423\text{ K }12\text{ atm}]{aq.\ Na_2CO_3} CH_2=CH-CH_2OH$$

Allyl chloride → Allyl alcohol

Allyl alcohol is then treated with hypochlorous acid (HOCl) to give rise to chlorohydrin from which glycerol is obtained by treatment with sodium hydroxide.

$$CH_2=CH-CH_2OH \xrightarrow{HOCl} \underset{OH}{CH_2}-\underset{Cl}{CH}-\underset{OH}{CH_2}$$

Glycerol β-mono chlorohydrin

$$\xrightarrow{Aq.\ NaOH} \underset{OH}{CH_2}-\underset{OH}{CH}-\underset{OH}{CH_2} + NaCl$$

Glycerol

Complete synthesis of glycerol

Complete synthesis of glycerol starting from the elements is described as under.

Structure of glycerol

$$2C + H_2 \xrightarrow{\text{Electric arc}} C_2H_2 \xrightarrow[H_2SO_4/HgSO_4,\ 333\text{ K}]{H_2O} CH_3CHO$$

Acetylene, Acetaldehyde

$$\xrightarrow{\text{Oxidation}} CH_3COOH \xrightarrow[\text{of calcium salt}]{\text{Distillation}} CH_3COCH_3$$

Acetic acid, Acetone

$$\xrightarrow{\text{Reduction}} CH_3CHOH.CH_3 \xrightarrow[\text{dehydration}]{H_2SO_4} CH_3.CH=CH_2$$

Isopropyl alcohol, Propylene

$$\xrightarrow[773\text{ K}]{Cl_2} Cl.CH_2.CH=CH_2 \xrightarrow{\text{dil. NaOH}} HOCH_2-CH=CH_2$$

Allyl chloride, Allyl alcohol

$$\xrightarrow{Cl_2/H_2O} \underset{OH}{CH_2}-\underset{Cl}{CH}-\underset{OH}{CH_2} \xrightarrow{Aq.\ NaOH} \underset{OH}{CH_2}-\underset{OH}{CH}-\underset{OH}{CH_2}$$

Glycerol β-mono chlorohydrin, Glycerol

2.13. CHEMICAL PROPERTIES OF GLYCEROL

Important reactions of glycerol are described as under:

(1) Reaction with sodium. When treated with metals like sodium at room temperature, one of the two primary alcoholic groups reacts forming monosodium glycerolate. At higher temperature, the second primary alcoholic group also reacts giving disodium glycerolate but the secondary alcoholic group is not affected.

$$\underset{\text{Glycerol}}{\begin{array}{l}CH_2OH\\ |\\ CHOH\\ |\\ CH_2OH\end{array}} \xrightarrow[-H_2]{\text{Na, room temp.}} \underset{\text{Mono sodium glycerolate}}{\begin{array}{l}CH_2ONa\\ |\\ CHOH\\ |\\ CH_2OH\end{array}} \xrightarrow[-H_2]{\text{Na, high temp.}} \underset{\text{Disodium glycerolate}}{\begin{array}{l}CH_2ONa\\ |\\ CHOH\\ |\\ CH_2ONa\end{array}}$$

(2) Reaction with HCl. When treated with equimolar quantities of hydrogen chloride at 383K, glycerol gives a mixture of α- and β-monochlorohydrins. If excess of hydrogen chloride is used, however, then a mixture of dichlorohydrins is formed.

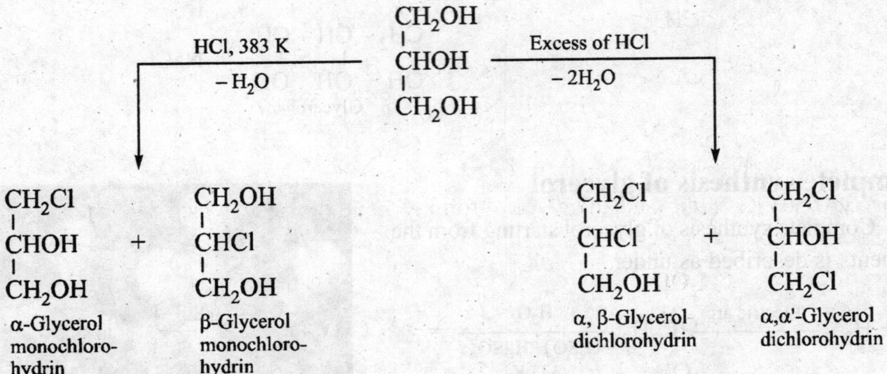

(3) Reaction with HI. Different products are formed under different conditions as depicted below.

(*i*) When heated with a *small* amount of hydrogen iodide, glycerol produces mainly allyl iodide.

$$\underset{\text{Glycerol}}{\begin{array}{l}CH_2OH\\ |\\ CHOH\\ |\\ CH_2OH\end{array}} \xrightarrow[-3H_2O]{+3HI} \underset{\textit{Unstable}}{\begin{array}{l}CH_2I\\ |\\ CHI\\ |\\ CH_2I\end{array}} \xrightarrow{-I_2} \underset{\text{Allyl iodide}}{\begin{array}{l}CH_2\\ \|\\ CH\\ |\\ CH_2I\end{array}}$$

(*ii*) When a large amount of hydrogen iodide is used, the main product is isopropyl iodide, which, in fact, results by the subsequent action of hydrogen iodide on the allyl iodide first formed.

$$\underset{\text{Allyl iodide}}{\begin{array}{l}CH_2\\ \|\\ CH\\ |\\ CH_2I\end{array}} \xrightarrow{HI} \begin{array}{l}CH_3\\ |\\ CHI\\ |\\ CH_2I\end{array} \xrightarrow{-I_2} \begin{array}{l}CH_3\\ |\\ CH\\ \|\\ CH_2\end{array} \xrightarrow{HI} \underset{\text{Isopropyl iodide}}{\begin{array}{l}CH_3\\ |\\ CHI\\ |\\ CH_3\end{array}}$$

(4) Reaction with monocarboxylic organic acids. When heated with organic acids like acetic acid, glycerol forms mono-, di- and tri-esters depending upon the amount of acid and reaction conditions employed.

$$\underset{\text{Glycerol monoacetate}}{\begin{array}{l}CH_2O.OCCH_3\\ |\\ CHOH\\ |\\ CH_2OH\end{array}} \qquad \underset{\text{Glycerol diacetate}}{\begin{array}{l}CH_2O.OCCH_3\\ |\\ CHO.OCCH_3\\ |\\ CH_2OH\end{array}} \qquad \underset{\text{Glycerol triacetate}}{\begin{array}{l}CH_2O.OCCH_3\\ |\\ CHO.OCCH_3\\ |\\ CH_2O.OCCH_3\end{array}}$$

(5) Reaction with oxalic acid

(*i*) When heated with oxalic acid at 383 K, it forms glycerol mono-oxalate. This loses CO_2 to give glycerol monoformate which undergoes hydrolysis to yield formic acid.

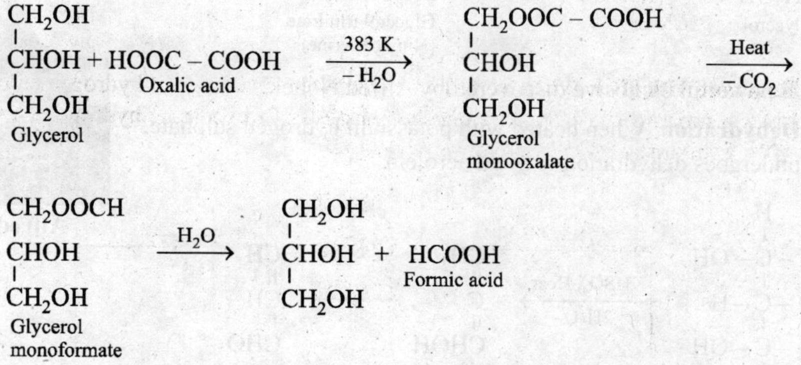

$$\underset{\text{Glycerol}}{\begin{array}{l}CH_2OH\\|\\CHOH\\|\\CH_2OH\end{array}} + \underset{\text{Oxalic acid}}{HOOC-COOH} \xrightarrow[-H_2O]{383\text{ K}} \underset{\text{Glycerol monooxalate}}{\begin{array}{l}CH_2OOC-COOH\\|\\CHOH\\|\\CH_2OH\end{array}} \xrightarrow[-CO_2]{\text{Heat}}$$

$$\underset{\text{Glycerol monoformate}}{\begin{array}{l}CH_2OOCH\\|\\CHOH\\|\\CH_2OH\end{array}} \xrightarrow{H_2O} \begin{array}{l}CH_2OH\\|\\CHOH\\|\\CH_2OH\end{array} + \underset{\text{Formic acid}}{HCOOH}$$

(*ii*) At 503 K, oxalic acid and glycerol form glyceryl dioxalate which decomposes to give allyl alcohol and carbon dioxide.

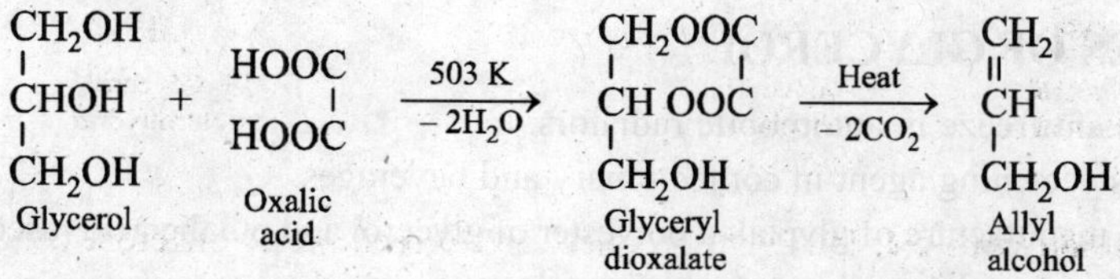

$$\underset{\text{Glycerol}}{\begin{array}{l}CH_2OH\\|\\CHOH\\|\\CH_2OH\end{array}} + \underset{\text{Oxalic acid}}{\begin{array}{l}HOOC\\|\\HOOC\end{array}} \xrightarrow[-2H_2O]{503\text{ K}} \underset{\text{Glyceryl dioxalate}}{\begin{array}{l}CH_2OOC\\|\qquad\;\;|\\CH\,OOC\\|\\CH_2OH\end{array}} \xrightarrow[-2CO_2]{\text{Heat}} \underset{\text{Allyl alcohol}}{\begin{array}{l}CH_2\\\|\\CH\\|\\CH_2OH\end{array}}$$

(6) Oxidation. Glycerol gives a variety of products on oxidation with different oxidising agents. With mild oxidising agents such as hydrogen peroxide in the presence of ferrous sulphate (Fenton's reagent) it gives a mixture of glyceraldehyde and dihydroxy acetone. With stronger oxidising agents it is ultimately converted into carbon dioxide and water as shown below.

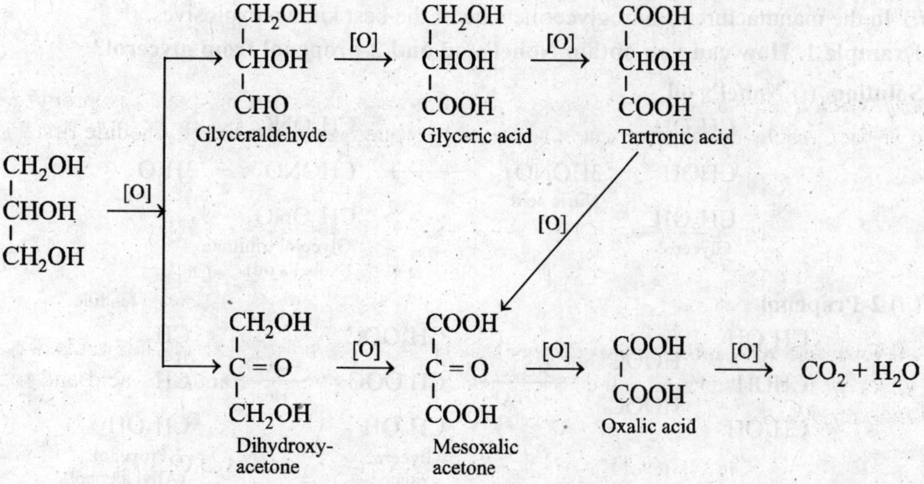

$$\begin{array}{l}CH_2OH\\|\\CHOH\\|\\CH_2OH\end{array} \xrightarrow{[O]} \underset{\text{Glyceraldehyde}}{\begin{array}{l}CH_2OH\\|\\CHOH\\|\\CHO\end{array}} \xrightarrow{[O]} \underset{\text{Glyceric acid}}{\begin{array}{l}CH_2OH\\|\\CHOH\\|\\COOH\end{array}} \xrightarrow{[O]} \underset{\text{Tartronic acid}}{\begin{array}{l}COOH\\|\\CHOH\\|\\COOH\end{array}}$$

$$\begin{array}{l}CH_2OH\\|\\CHOH\\|\\CH_2OH\end{array} \xrightarrow{[O]} \underset{\text{Dihydroxy-acetone}}{\begin{array}{l}CH_2OH\\|\\C=O\\|\\CH_2OH\end{array}} \xrightarrow{[O]} \underset{\text{Mesoxalic acetone}}{\begin{array}{l}COOH\\|\\C=O\\|\\COOH\end{array}} \xrightarrow{[O]} \underset{\text{Oxalic acid}}{\begin{array}{l}COOH\\|\\COOH\end{array}} \xrightarrow{[O]} CO_2 + H_2O$$

(Tartronic acid $\xrightarrow{[O]}$ Mesoxalic acetone)

(7) Reaction with nitric acid. When added slowly to a mixture of concentrated nitric acid and concentrated sulphuric acid maintained at 283–298 K, glycerol forms glyceryl trinitrate which is commonly known as nitroglycerine.

Alfred Nobel

$$\underset{\text{Glycerol}}{\begin{array}{l}CH_2OH\\ |\\ CHOH\\ |\\ CH_2OH\end{array}} + 3HONO_2 \longrightarrow \underset{\substack{\text{Glycerol trinitrate}\\ \text{(Nitroglycerine)}}}{\begin{array}{l}CH_2O.NO_2\\ |\\ CHO.NO_2\\ |\\ CH_2O.NO_2\end{array}} + 3H_2O$$

It is a powerful explosive discovered by Alfred Nobel.

(8) Dehydration. When heated with potassium hydrogen sulphate, glycerol undergoes dehydration to yield acrolein.

$$\underset{\text{Glycerol}}{\begin{array}{c}H\\ |\\ H-C-OH\\ |\\ HO-C-H\\ |\\ H-C-OH\\ |\\ H\end{array}} \xrightarrow[2H_2O]{KHSO_4,\ Heat} \underset{\substack{\textit{Unstable}\\ \text{enol}}}{\begin{array}{c}CH_2\\ \|\\ C\\ \|\\ CHOH\end{array}} \longrightarrow \underset{\text{Acrolein}}{\begin{array}{c}CH_2\\ \|\\ CH\\ |\\ CHO\end{array}}$$

2.14. USES OF GLYCEROL

(*i*) As an antifreeze in automobile radiators.
(*ii*) As a sweetening agent in confectionary and beverages.
(*iii*) In the manufacture of glyptal–a polyester of glycerol and pthalic acid–used as an alkyl resin.
(*iv*) In making medicines like cough syrup, lotions, etc.
(*v*) As a preservative for fruits and other eatables which require to be kept moist. This is because glycerol is sufficiently hygroscopic to absorb moisture from air.
(*vi*) In making non-drying printing inks, stamp colours, etc.
(*vii*) In the preparation of good quality toilet soaps and cosmetics.
(*viii*) In the manufacture of nitroglycerine–one of the best known explosives.

Example 1. How can you obtain Nobel's oil and 2-propenol from glycerol?

Solution. (*i*) Nobel's oil

$$\underset{\text{Glycerol}}{\begin{array}{l}CH_2OH\\ |\\ CHOH\\ |\\ CH_2OH\end{array}} + \underset{\text{Nitric acid}}{3HONO_2} \longrightarrow \underset{\substack{\text{Glycerol trinitrate}\\ \text{(Nobel's oil)}}}{\begin{array}{l}CH_2ONO_2\\ |\\ CHONO_2\\ |\\ CH_2ONO_2\end{array}} + 3H_2O$$

(*ii*) **2-Propenol**

$$\begin{array}{l}CH_2OH\\ |\\ CHOH\\ |\\ CH_2OH\end{array} + \begin{array}{l}HOOC\\ |\\ HOOC\end{array} \xrightarrow[-2H_2O]{503\ K} \underset{\substack{\text{Glycerol}\\ \text{oxalate}}}{\begin{array}{l}CH_2OOC\\ |\qquad |\\ CH\ OOC\\ |\\ CH_2OH\end{array}} \xrightarrow[Heat]{-2CO_2} \underset{\substack{\text{2-Propenol}\\ \text{(Allyl alcohol)}}}{\begin{array}{l}CH_2\\ \|\\ CH\\ |\\ CH_2OH\end{array}}$$

Example 2. A neutral liquid A on treatment with concentrated HCl in the presence of zinc chloride produces a compound B which upon treatment with alcoholic KOH yields a compound C. The compound C decolourises bromine water and upon ozonolysis forms methanal only. Deduce this structure of A, B, and C and explain the reactions.

Solution. The structure of compound C can be obtained by joining together the carbonyl carbon atoms of the methanal molecules formed by ozonolysis.

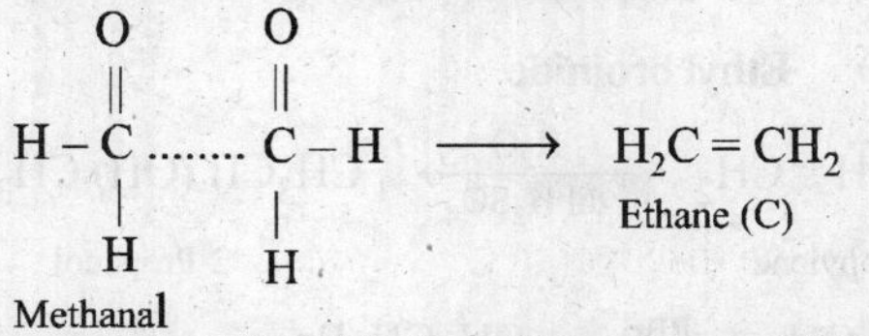

$$\underset{\text{Methanal}}{\mathrm{H-\overset{\overset{O}{\|}}{\underset{\underset{H}{|}}{C}}\ \ldots\ldots\ \overset{\overset{O}{\|}}{\underset{\underset{H}{|}}{C}}-H}} \longrightarrow \underset{\text{Ethane (C)}}{H_2C=CH_2}$$

The compound B is expected to be CH_3CH_2Cl as C has been obtained by the action of alcoholic KOH on B. Since B has been formed by treating A with conc. HCl, the compound A must be CH_3CH_2OH. The reaction can be written as under:

$$\underset{\substack{\text{Ethanol}\\(A)}}{CH_3CH_2OH} \xrightarrow[ZnCl_2]{HCl} \underset{\substack{\text{Chloroethane}\\(B)}}{CH_3CH_2Cl} \xrightarrow{\text{alc KOH}} \underset{\substack{\text{Ethene}\\(C)}}{CH_2=CH_2} \xrightarrow{\text{Ozonolysis}} \underset{\text{Methanal}}{2H-\overset{\overset{O}{\|}}{C}-H}$$

Example 3. The acid catalysed hydration of the compound A produces the compound B and C and not D. Account for this

$$\underset{(A)}{(CH_3)_3CCH=CH_2} \xrightarrow{H_3O^+} \underset{(B)}{(CH_3)_2C(OH)\,CH\,(CH_3)_2} + \underset{(C)}{(CH_3)_3CH(OH)CH_3} + \underset{\substack{(D)\\(\textit{Not possible})}}{(CH_3)_3CHH_2CH_2OH}$$

Solution. In the above reaction, the product (D) is not possible because the primary carbocation which is formed by the attack of proton is highly unstable in nature. However, the secondary and tertiary carbocations are comparatively more stable than the primary carbocation and we obtain the products (C) and (B) respectively.

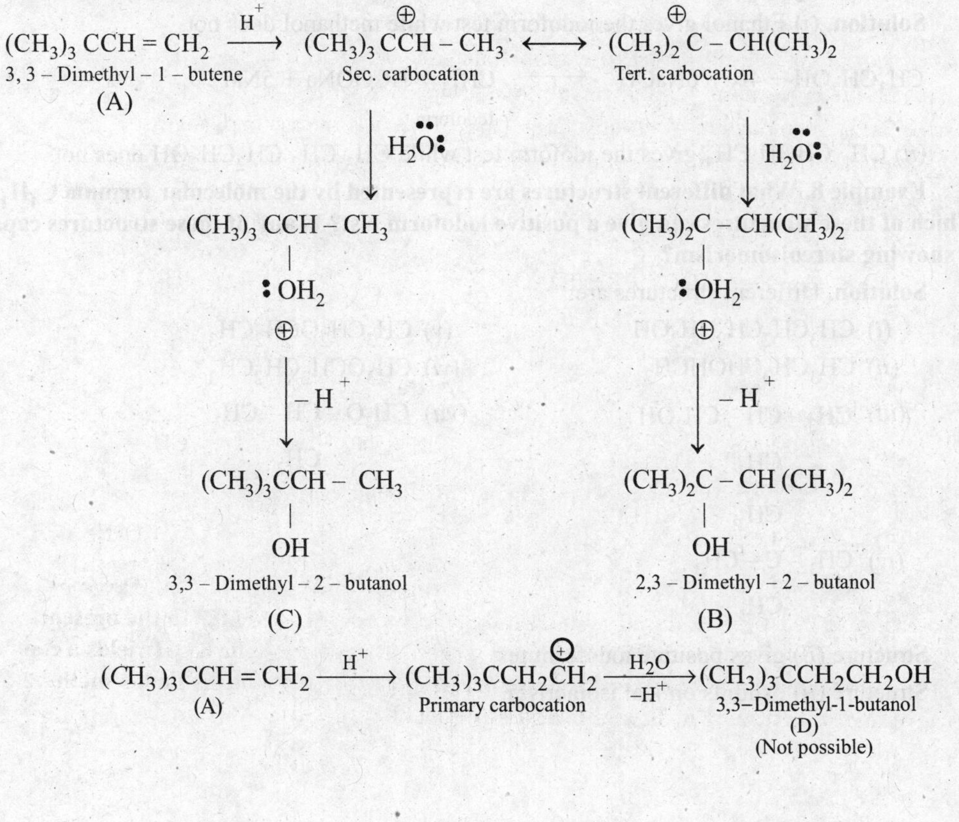

$$\underset{\substack{\text{3, 3 – Dimethyl – 1 – butene}\\(A)}}{(CH_3)_3\,CCH=CH_2} \xrightarrow{H^+} \underset{\text{Sec. carbocation}}{(CH_3)_3\,C\overset{\oplus}{C}H-CH_3} \longleftrightarrow \underset{\text{Tert. carbocation}}{(CH_3)_2\,\overset{\oplus}{C}-CH(CH_3)_2}$$

$$\downarrow H_2\ddot{O}: \qquad\qquad\qquad \downarrow H_2\ddot{O}:$$

$$(CH_3)_3\,CCH(\overset{\oplus}{:OH_2})-CH_3 \qquad\qquad (CH_3)_2C(\overset{\oplus}{:OH_2})-CH(CH_3)_2$$

$$\downarrow -H^+ \qquad\qquad\qquad \downarrow -H^+$$

$$\underset{\substack{\text{3,3 – Dimethyl – 2 – butanol}\\(C)}}{(CH_3)_3CCH(OH)-CH_3} \qquad\qquad \underset{\substack{\text{2,3 – Dimethyl – 2 – butanol}\\(B)}}{(CH_3)_2C(OH)-CH\,(CH_3)_2}$$

$$\underset{(A)}{(CH_3)_3CCH=CH_2} \xrightarrow{H^+} \underset{\text{Primary carbocation}}{(CH_3)_3CCH_2\overset{\oplus}{C}H_2} \xrightarrow[-H^+]{H_2O} \underset{\substack{\text{3,3–Dimethyl-1-butanol}\\(D)\\\text{(Not possible)}}}{(CH_3)_3CCH_2CH_2OH}$$

Example 4. How will you do the following conversions?

(*i*) Propylene $\longrightarrow$ 2 Propanal

(ii) Ethanol $\longrightarrow$ Ethyl bromide

Solution. (*i*)
$$\underset{\text{Propylene}}{CH_3CH=CH_2} \xrightarrow[\text{dil } H_2SO_4]{H_2O} \underset{\text{2-Propanol}}{CH_3CH\,(OH)\,CH_3}$$

(*ii*)
$$\underset{\text{Ethanol}}{CH_3CH_2OH} \xrightarrow{PBr_3} \underset{\text{Ethyl bromide}}{CH_3CH_2Br}$$

Example 5. Why 3° alcohols are easily dehydrated as compared to 2° or 1° alcohols ?

Solution. The intermediate carbocation formed in case of 3° alcohol is more stable than that formed from 2° or 1° alcohol.

Example 6. Which alcohol will form alkene on heating with Cu at 300°C?

(*i*) CH_3CH_2OH (*ii*) $(CH_3)_3COH$ (*iii*) $(CH_3)_2CHOH$ (*iv*) $C_6H_5-\underset{\underset{OH}{|}}{CH}-CH_3$

Solution. (*ii*)
$$CH_3-\overset{\overset{CH_3}{|}}{\underset{\underset{CH_3}{|}}{C}}-OH \xrightarrow[300°C]{Cu} \underset{\text{2-methylbutene}}{CH_3-\overset{\overset{CH_3}{|}}{C}=CH_2} + H_2O$$

Example 7. How will you distinguish between following pairs of the compounds:

(*i*) CH_3OH and C_2H_5OH

(ii) $CH_3CH_2CH_2OH$ and $CH_3CHOHCH_3$

Solution. (*i*) Ethanol gives the iodoform test while methanol does not.

$$CH_3CH_2OH + 4I_2 + 6NaOH \longrightarrow \underset{\text{Iodoform}}{CHI_3} + HCOONa + 5NaI$$

(*ii*) CH_3 CHOH CH_3 gives the idoform test while CH_3 CH_2 CH_2CH_2OH does not.

Example 8. What different structures are represented by the molecular formula $C_4H_{10}O$? Which of these structures can give a positive iodoform test? Is any of these structures capable of showing stereoisomerism?

Solution. Different structures are:

(*i*) $CH_3CH_2CH_2CH_2OH$

(*ii*) $CH_3CH_2CHOHCH_3$

(*iii*) $CH_3-\underset{\underset{CH_3}{|}}{CH}-CH_2OH$

(*iv*) $CH_3-\overset{\overset{CH_3}{|}}{\underset{\underset{CH_3}{|}}{C}}-CH_3$

(*v*) $CH_3CH_2OCH_2CH_3$

(*vi*) $CH_3OCH_2CH_2CH_3$

(*vii*) $CH_3O-\underset{\underset{CH_3}{|}}{CH}-CH_3$

Structure (*ii*) gives positive iodoform test

Structure (*ii*) exhibits optical isomerism.

EXERCISES
(Based on Question Papers of Different Universities)

Multiple Choice Questions (Choose the Correct Option)

1. Which of the following gives positive Iodoform test?

(*a*) 1-Propanol (*b*) 2-Propanol

(*c*) 3-Propanol (*d*) None of these

2. When ethanol is treated with sodium metal:

(*a*) Sodium ethoxide is formed

(*b*) The sodium is oxidised

(*c*) The acidic hydrogen is reduced

(*d*) All these occur

3. The compound which reacts most readily with Lucas reagent is

(*a*) CH_3CH_2Cl (*b*) $(CH_3)_2CHOH$

(*c*) CH_3CH_2OH (*d*) $(CH_3)_3$ COH

4. Which of the following reagents will replace –OH group by a halogen atom?

(*a*) HOCl (*b*) Br_2

(*c*) $SOCl_2$ (*d*) I_2

5. Which of the following compounds is the least soluble in water?

(*a*) $HOCH_2CH_2OH$ (*b*) $CH_3CH_2CH_2OH$

(*c*) CH_3CH_2OH (*d*) $CH_3CH_2CH_2CH_2OH$

6. Ethyl alcohol can react with concentrated H_2SO_4 to give

(*a*) Ethylene (*b*) Diethyl ether

(*c*) Ethyl hydrogen sulphate (*d*) All of these

7. The major product of acid-catalysed dehydration of 2-butanol is:

(*a*) 2-Butene (*b*) 2-Butyne

(*c*) 1-Butene (*d*) 1- Butyne

8. Hydroboration-oxidation of propene gives:

(*a*) Isopropyl alcohol (*b*) n-Propyl alcohol

(*c*) Isobutyl alcohol (*d*) tert-Butyl alcohol

9. Oxidation of a primary alcohol with $KMnO_4/OH^-$ produces

(*a*) a carboxylic acid (*b*) an ether

(*c*) a ketone (*d*) an ester

10. Which is the best reagent for carying out the following conversion?

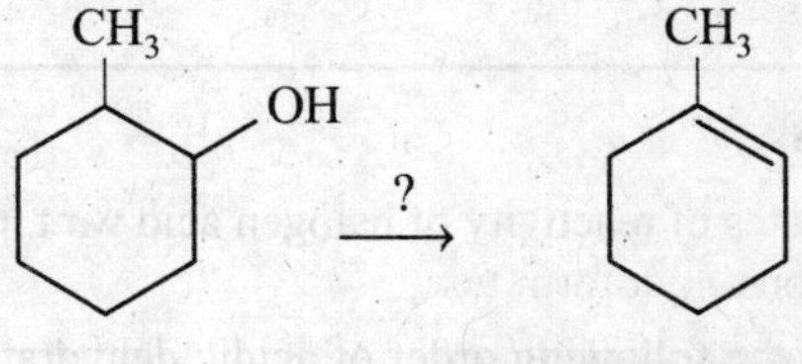

CH_3 OH $\xrightarrow{?}$ CH_3

(*a*) $LiAlH_4$ (*b*) Conc H_2SO_4

(*c*) H_2/Ni (*d*) NaOH

11. Oxidation of a secondary alcohol with $K_2Cr_2O_4/H^+$ produces

(*a*) a carboxylic acid (*b*) a ketone
(*c*) an aldehyde (*d*) an ester

12. Isopropyl alcohol can be converted to acetone by treatment with

(*a*) $HCl/ZnCl_2$ (*b*) $Na_2Cr_2O_7/H_2SO_4$
(*c*) NaOH (*d*) $LiAlH_4$

13. The mechanism of dehydration of an alcohol to give an ether involves formation of:

(*a*) Carbonium ions (*b*) Carbanions
(*c*) Free radicals (*d*) Carbenes

14. In cold countries ethylene glycol is added to water in car radiators. This helps to

(*a*) reduce the viscosity (*b*) make water a better lubricant
(*c*) lower the freezing point (*d*) lower the boiling point

15. Ethylene glycol reacts with excess of PCl_5 to give

(*a*) Chloroethane (*b*) 1,2- Dichloroethane
(*c*) Hexachloroethane (*d*) 1,3-Dichloroethane

16. Ethylene glycol undergoes oxidation with hot acidic $KMnO_4$ to form:

(*a*) Formic acid (*b*) Formaldehyde
(*c*) Acetic acid (*d*) Acetaldehyde

17. When ethylene glycol is heated with concentrated HNO_3, it forms

(*a*) Oxalic acid (*b*) Ethylene oxide
(*c*) Dioxane (*d*) Diethylene glycol

18. Compound (A) reacts with sodium metal to form one mole of H_2. The compound (A) can be

(*a*) $CH_3CH_2CH=CH_2$ (*b*) $HOCH_2CH_2CH_2CH_2OH$
(*c*) $CH_2=CH-CH=CH_2$ (*d*) $CH_3CH_2CH_2CH_2OH$

19. When glycerol is heated with potassium hydrogen sulphate ($KHSO_4$). it forms

(*a*) Acrolein (*b*) Acetic acid
(*c*) Allyl alcohol (*d*) Propionic acid

20. When glycerol is treated with a mixture of concentrated $HNO_3 + H_2SO_4$, it forms

(*a*) Nitroethane (*b*) 1- Nitropropane
(*c*) Nitroglycerine (*d*) 2- Nitropropane

ANSWERS

1. (*b*)	**2.** (*d*)	**3.** (*d*)	**4.** (*c*)	**5.** (*d*)	**6.** (*d*)
7. (*a*)	**8.** (*b*)	**9.** (*a*)	**10.** (*b*)	**11.** (*b*)	**12.** (*b*)
13. (*a*)	**14.** (*c*)	**15.** (*b*)	**16.** (*a*)	**17.** (*a*)	**18.** (*b*)
19. (*a*)	**20.** (*c*)				

Short Answer Questions

1. Discuss the relative rates of reactivity of halogen acid w.r.t. primary, secondary and tertiary alcohols, giving reasons.

2. How will you justify the following order of acidic dehydration of alcohols?
Tertiary alcohols > Secondary alcohols > Primary alcohols

3. Complete the following reaction and discuss the mechanism.

$$CH_3-\overset{\overset{\displaystyle CH_3}{|}}{C}=CH_2+H_2O \xrightarrow{H^+}$$

4. How are alkanols classified? Give one example of each type.
5. What are primary, secondary and tertiary alcohols? How will you distinguish them from one another with the help of Lucas reagent?
6. How do primary, secondary and tertiary alcohols differ in their reactions towards oxidation?
7. How will you distinguish between primary, secondary and tertiary alcohols with the help of Lucas reagent?
8. Arrange the following as stated:
   (*i*) Increasing order of reactivity towards sodium metal:
   1-Butanol, 2-Butanol, 2-Methyl-2- propanol
   (*ii*) Decreasing order of acid strength:
   CH_3OH, H_2O, C_6H_5OH
9. Starting from Grignard reagent, how will you prepare primary and secondary alcohols?
10. Discuss the mechanism of acidic hydration of alkenes.
11. How will you explain that the solubility of alcohols in water decreases with increasing molecular weight?
12. How will you covert ethyl alcohol into secondary butyl alcohol?
13. Explain why ethyl alcohol has higher boiling point than dimethyl ether.
14. Account for the fact that the dehydration of n-butyl alcohol yields mainly 2-butene rather than 1-butene.
15. Outline the mechanism of the acidic dehydration of alcohols.
16. Account for the fact that ethyl alcohol does not react with NaBr but on the addition of conc. H_2SO_4, ethyl bromide is formed.
17. What happens when acetone is treated with Grignard reagent and the product is hydrolysed
18. Give the mechanism of acidic dehydration of ethyl alcohol?
19. Compare the behaviour of n-propyl alcohol and isopropyl alcohol towards oxidation.
20. Both ethyl alcohol and n-hexyl alcohol contain hydroxyl group but the former is soluble in water while the latter is not. Give reason for the difference.

General Questions

1. How will you account for the following?
   (*i*) Alcohols are weakly acidic in nature.
   (*ii*) Ethyl alcohol has a higher boiling point than dimethyl ether.
   (*iii*) Order of reactivity of alcohols towards sodium metal is:
   primary > secondary > tertiary.
2. How do the following react with ethyl alcohol?
   (*i*) Hot reduced copper (*ii*) Phosphorus pentachloride
   (*iii*) Acetyl chloride
3. What are primary, secondary and tertiary alcohols? How will you distinguish them with the help of oxidation and dehydrogenation reactions?

4. How will you convert the following?
   (*i*) Ethyl alcohol to ethane (*ii*) Ethyl alcohol to ethyl acetate
   (*iii*) Isopropyl alcohol to acetone (*iv*) Tertiary butyl alcohol to isobutylene
5. (*a*) How will you account for the weekly acidic character of alcohols? Discuss their important chemical properties.
   (*b*) How does ethyl alcohol react with:
   (*i*) Na (*ii*) PCl_5
   (*iii*) CH_3COOH (*iv*) CH_3MgBr
   (*c*) "Oxidation of alcohols can help in differentiating primary, secondary and tertiary alcohols". Comment.
6. Explain what happens when:
   (*i*) Ethanol reacts with thionyl chloride
   (*ii*) Sec. butyl alcohol is heated with dilute H_2SO_4 at 170°C
   (*iii*) Isopropyl alcohol is heated with acidified $K_2Cr_2O_7$
7. (*a*) Give the mechanism of pinacol-pinacolone rearragement.
   (*b*) Describe the mechanism of acidic dehydration of n-propyl alcohol.
   (*c*) How will you prove that there are three hydroxyl groups in glycerol?
8. (a) Write a short note on Pinacol-Pinacolone rearrangement.
   (*b*) Starting from Grignard reagent, how will you prepare secondary and tertiary alcohols?
   (*c*) Write the different oxidation products of Glycol.
   (*d*) Write a short note on hydroboration oxidation of alkenes.
9. (*a*) n- Butyl alcohol forms 1-butene and 2-butene on acidic hydration. What is the major product? Support your answer giving mechanism.
   (*b*) Why are acohols weaker acids than water?
10. (*a*) Why are the boiling points of alcohols higher than those of corresponding alkanes?
   (*b*) What is the main product of dehydration of 3,3-dimethyl-2 butanol? Illustrate your answer with suitable mechanism.
11. (*a*) Describe the mechanism of hydroboration oxidation.
   (*b*) Discuss the mechanism of the following reaction:

$$\begin{array}{l} CH_2-OH \\ \,| \qquad\qquad + HIO_4 \longrightarrow \\ CH_2-OH \end{array}$$

   (*c*) How will you justify the order of reactivity of primary, secondary and tertiary alcohols with sodium metal?
12. What happens when:
   (*i*) Glycol is heated with hydrogen iodide?
   (*ii*) Glycol is treated with lead tetra- actate?
   (*iii*) Glycol is heated with fuming sulphuric acid?
13. Discuss the preparation of alcohols by the acidic hydration of alkenes and also give the mechanism of the reaction.
14. Discuss the mechanism of the following reactions:
   (*i*) Pinacol-Pinacolone rearrangement.
   (*ii*) Hydroboration-oxidation.

15. How is oxymercuration-demercuration of alkenes done? In what way is it different from the acidic hydration?

16. An alcohol A ($C_4H_{10}O$), on oxidation with acidic sodium dichromate, gives a carboxylic acid B (C_4H_8O). Treatment of A with hot concentrated sulphuric acid brings about dehydration and gives compound C (C_4H_8). Treatment of C with warm aqueous sulphuric acid gives D ($C_4H_{10}O$), a new alcohol isomeric with compound A. Compound D is resistant to oxidation. Identify A, B, C and D. Give equations for all the reactions.

A is 2-Methylpropan -1-ol B is 2-Methylprepanoic acid

C is 2-Methylpropene D is 2-Methylpropane -2- ol

17. A hydrocarbon (A) reacts with hydrogen iodide to form compound (B) which reacts with aqueous potassium hydroxide to give compound (C). Oxidation of (C) gives acetone. What are the structural formulas of (A), (B) and (C)?

(A) Propene (B) Isopropyl iodide

(C) Isopropyl alcohol

18. An organic compound of formula $C_3H_8O_3$ on acetylation with acetyl chloride (or acetic anhydride) gives a compound of formula $C_9H_{14}O_6$. How many hydroxyl groups are present in the compound? What is its possible structure?

(**Ans.** $CH_2OH - CHOH - CH_2OH$, glycerol)

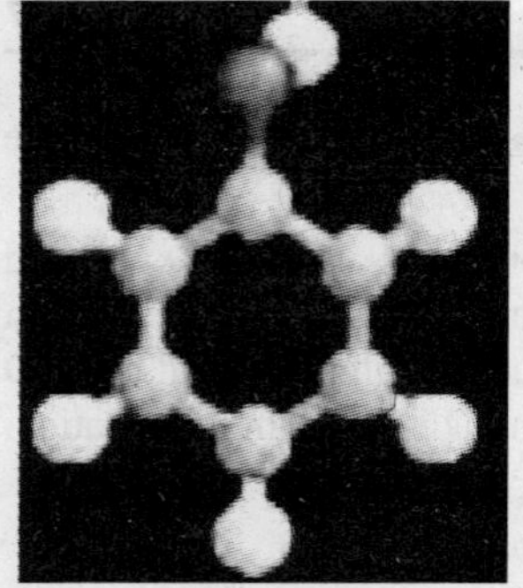

3 Phenols

3.1. CLASSIFICATION AND NOMENCLATURE

Phenols are organic compounds containing at least one –OH group attached directly to benzene ring.

Depending upon the number of hydroxyl groups attached to benzene ring, phenols can be classified as monohydric, dihydric and trihydric phenols.

Monohydric phenols. The simplest member of the series is hydroxybenzene, commonly known as phenol, while others are named as substituted phenols. The three isomeric hydroxy toluenes are known as cresols.

OH OH OH OH

CH_3 CH_3 CH_3

Phenol or Hydroxybenzene or Carbolic acid

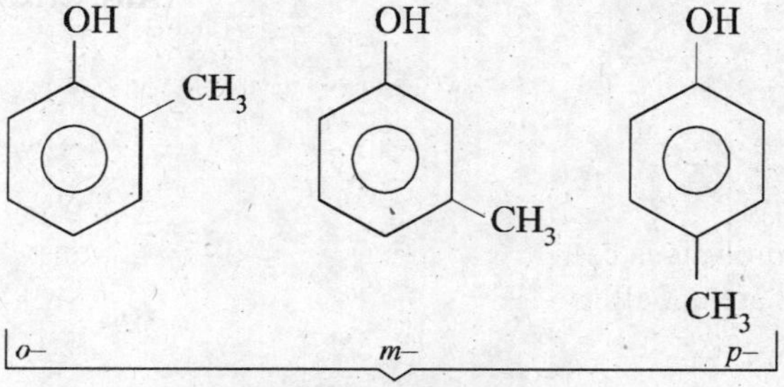

o– *m*– *p*–

Cresols or Hydroxy toluenes

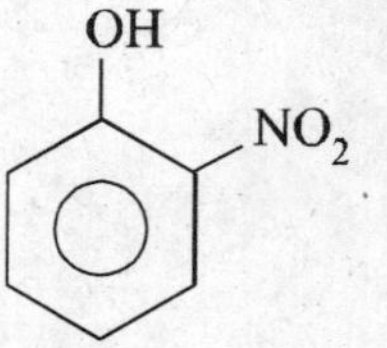

OH OH OH

NO_2 Cl

NH_2 etc.

o-Nitrophenol 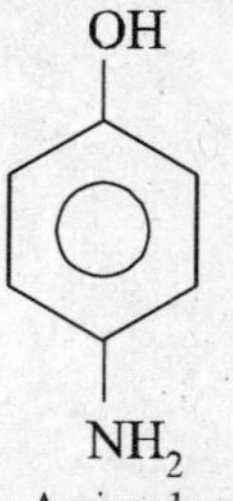 *p*-Aminophenol 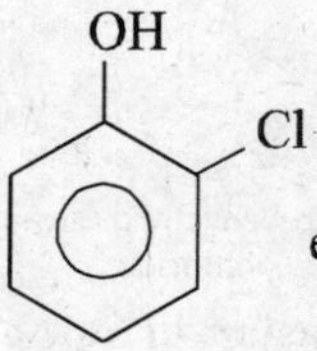 *o*-Chlorophenol

Dihydric phenols. The three isomeric dihydroxy benzenes are better known by their common names as given below :

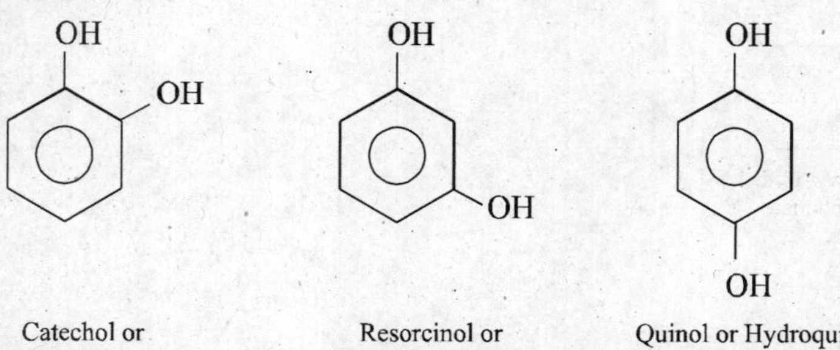

Catechol or *o*-Dihydroxybenzene

Resorcinol or *m*-Dihydroxybenzene

Quinol or Hydroquinone or *p*-Dihydroxybenzene

Trihydric phenols. Trihydroxy benzenes are known by their common names as given below:

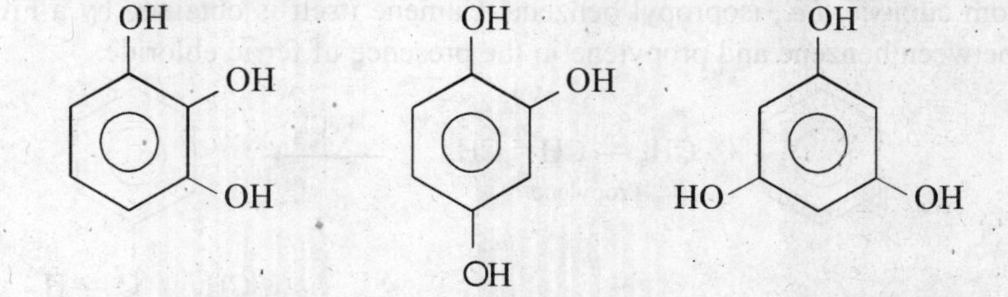

1, 2, 3-Trihydroxybenzene or Pyrogallol

1, 2, 4-Trihydroxybenzene or Hydoxy quinol

1, 3, 5-Trihydroxybenzene or Phloroglucinol

3.2. COMMON METHODS OF PREPARATION OF MONOHYDRIC PHENOLS

The commonly employed methods of preparation of phenols are described as under :

1. Hydrolysis of diazonium salts. In this method, the reaction is usually carried out by slowly adding diazonium salt solution to excess of boiling dilute sulphuric acid.

$$ArN_2{}^+X^- + H_2O \xrightarrow{H^+,\ heat} Ar-OH + HX + N_2$$

For example :

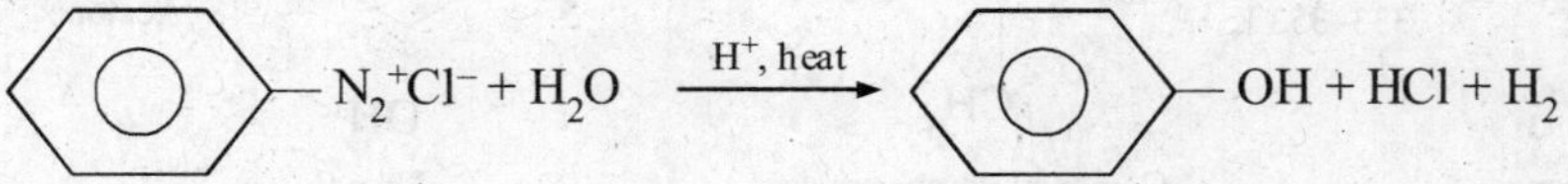

$$C_6H_5-N_2{}^+Cl^- + H_2O \xrightarrow{H^+,\ heat} C_6H_5-OH + HCl + H_2$$

Benzene diazonium chloride

Since hydrolysis is carried out under very mild conditions, other groups in the molecule of diazonium salt are not affected.

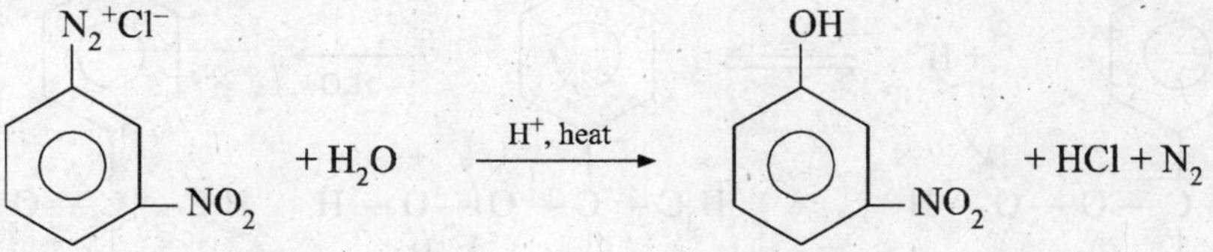

$$m\text{-}NO_2C_6H_4N_2{}^+Cl^- + H_2O \xrightarrow{H^+,\ heat} m\text{-}NO_2C_6H_4OH + HCl + N_2$$

m-Nitrobenzene diazonium chloride → *m*-Nitrophenol

2. Fusion of sulphonates with alkali. In this method, phenols are obtained by fusion of sodium salts of aromatic sulphonic acids with sodium hydroxide, followed by acidification.

$$\underset{\text{Sodium sulphonate}}{ArSO_3Na} + 2NaOH \xrightarrow[Na_2SO_3,\ H_2O]{Fuse} Ar-ONa \xrightarrow{HCl} \underset{\text{Phenol}}{Ar-OH} + NaCl$$

For example :

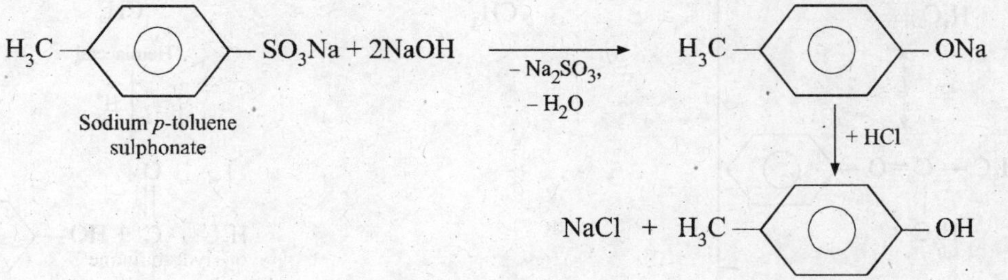

$$H_3C-C_6H_4-SO_3Na + 2NaOH \xrightarrow[-H_2O]{-Na_2SO_3,} H_3C-C_6H_4-ONa$$

Sodium *p*-toluene sulphonate

$$H_3C-C_6H_4-ONa \xrightarrow{+HCl} NaCl + H_3C-C_6H_4-OH$$

p-cresol

3. From cumene hydroperoxide. A recently developed process involves the preparation of phenol from cumene, *i.e.*, isopropyl benzene. Cumene itself is obtained by a Friedel-Craft type reaction between benzene and propylene in the presence of ferric chloride.

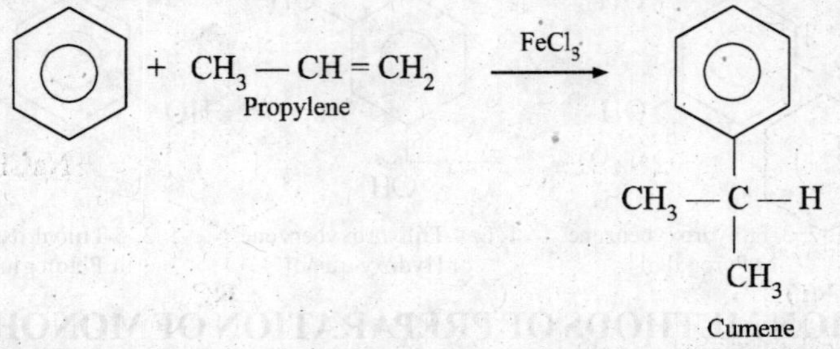

$$C_6H_6 + \underset{\text{Propylene}}{CH_3-CH=CH_2} \xrightarrow{FeCl_3} \underset{\text{Cumene}}{C_6H_5-CH(CH_3)_2}$$

Cumene is readily oxidised by air to form cumene hydroperoxide which is converted into phenol and acetone by treatment with an aqueous acid.

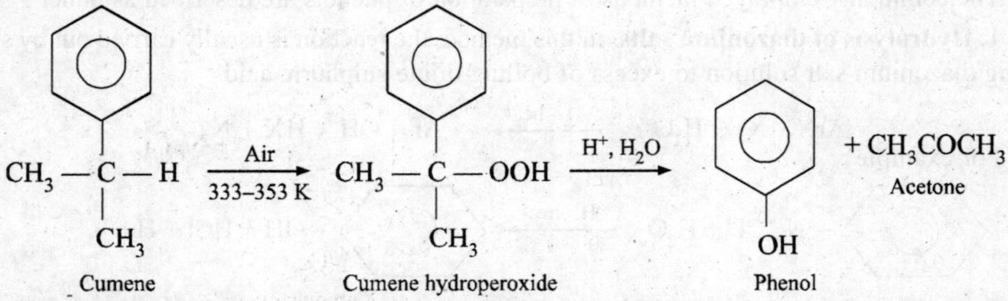

$$\underset{\text{Cumene}}{C_6H_5-C(CH_3)_2-H} \xrightarrow[333\text{–}353\ K]{\text{Air}} \underset{\text{Cumene hydroperoxide}}{C_6H_5-C(CH_3)_2-OOH} \xrightarrow{H^+,\ H_2O} \underset{\text{Phenol}}{C_6H_5OH} + \underset{\text{Acetone}}{CH_3COCH_3}$$

The mechanism of conversion of cumene hydroperoxide into phenol involves a sequence of five steps as shown below :

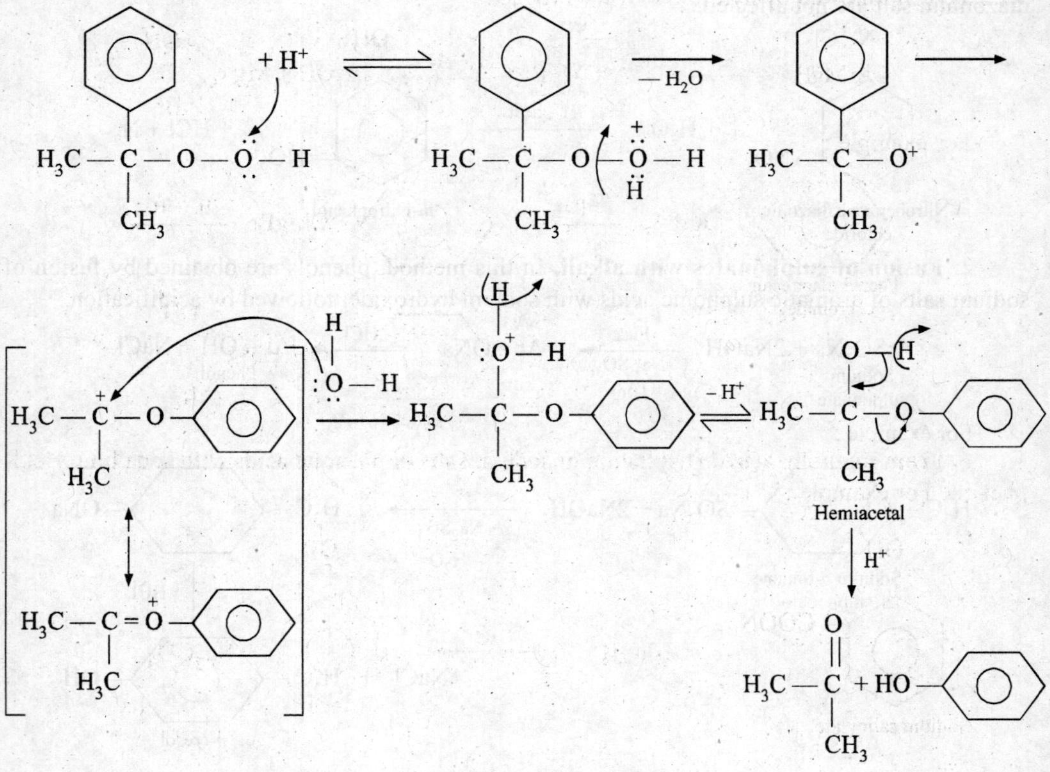

$$C_6H_5-C(CH_3)_2-O-\ddot{O}-H \underset{}{\xrightarrow{+H^+}} C_6H_5-C(CH_3)_2-O-\overset{+}{\ddot{O}}H_2 \xrightarrow{-H_2O} C_6H_5-C(CH_3)_2-O^+ \longrightarrow$$

$$\left[H_3C-\overset{+}{C}(H_3C)-O-C_6H_5 \longleftrightarrow H_3C-C(H_3C)=\overset{+}{O}-C_6H_5\right] \xrightarrow{H_2\ddot{O}:} H_3C-C(CH_3)(\overset{+}{O}H_2)-O-C_6H_5 \underset{}{\overset{-H^+}{\rightleftharpoons}} \underset{\text{Hemiacetal}}{H_3C-C(CH_3)(OH)-O-C_6H_5}$$

$$\xrightarrow{H^+} H_3C-\overset{O}{\overset{\|}{C}}-CH_3 + HO-C_6H_5$$

4. From aryl halides. Substituted phenols having electron withdrawing groups like $-NO_2$ group in *ortho-para* positions to the –OH group can be easily prepared from the corresponding aryl halides by treatment with aqueous sodium hydroxide. For example :

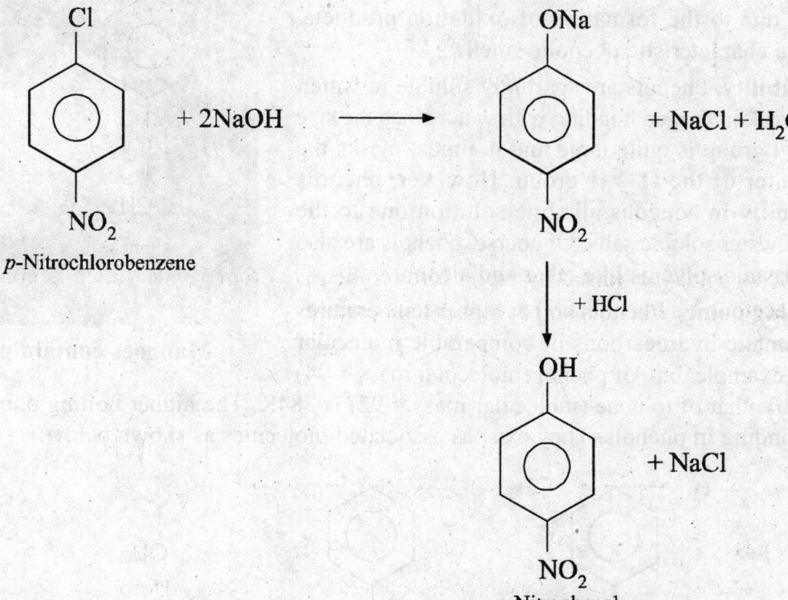

$$p\text{-}O_2N{-}C_6H_4{-}Cl + 2NaOH \longrightarrow p\text{-}O_2N{-}C_6H_4{-}ONa + NaCl + H_2O$$

p-Nitrochlorobenzene

$$p\text{-}O_2N{-}C_6H_4{-}ONa \xrightarrow{+\,HCl} p\text{-}O_2N{-}C_6H_4{-}OH + NaCl$$

p-Nitrophenol

5. From Grignard reagents. Phenols may be prepared by treating aryl magnesium bromides with oxygen followed by hydrolysis of the product.

$$2ArMgBr \xrightarrow{O_2} 2ArOMgBr \xrightarrow{H_2O/H^+} 2ArOH + Mg\begin{matrix}OH\\Br\end{matrix}$$

For example :

$$C_6H_5{-}MgBr \xrightarrow{[O]} C_6H_5{-}OMgBr \xrightarrow{H_2O/H^+} C_6H_5{-}OH + Mg\begin{matrix}OH\\Br\end{matrix}$$

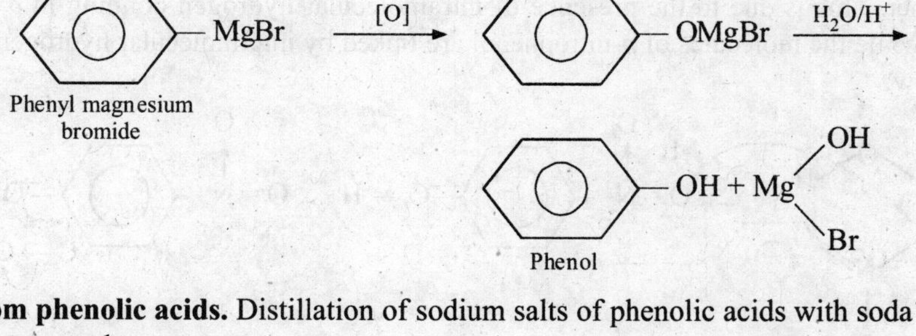

Phenyl magnesium bromide → Phenol

6. From phenolic acids. Distillation of sodium salts of phenolic acids with soda lime yields phenols. For example :

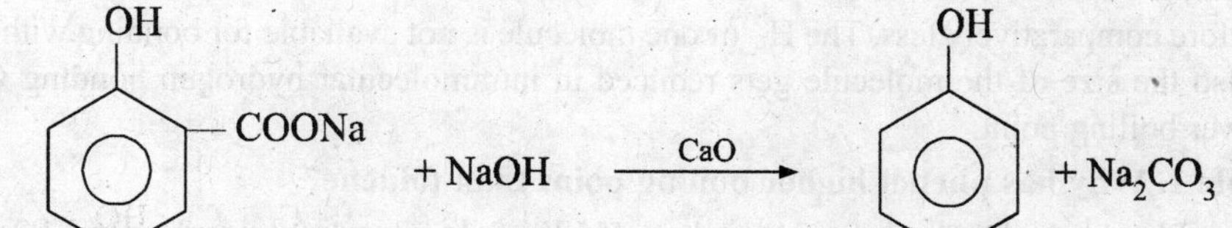

$$o\text{-}HO{-}C_6H_4{-}COONa + NaOH \xrightarrow{CaO} C_6H_5{-}OH + Na_2CO_3$$

Sodium salicylate

3.3. PHYSICAL PROPERTIES OF PHENOLS

1. State and mell. Phenols are colourless crystalline solids or liquids. But they attain colour when exposed to atmosphere due to the formation of oxidation products. Phenols have characteristic phenolic smell.

Mangoes contain phenols

2. Solubility. Phenols are sparingly soluble in water. They form weak hydrogen bonding with water since the size of the phenyl group is quite large and it almost masks the polar character of the O – H group. However, phenols dissolve readily in aqueous alkaline solution due to the formation of water soluble salts. Of course, phenols are also soluble in organic solvents like ether and alcohol.

3. Boiling points. Phenols boil at higher temperatures than the aromatic hydrocarbons of comparable molecular masses. For example, b.p. of phenol (molecular mass = 94) is 455 K while that of toluene (molecular mass = 92) is 384K. The higher boiling point is due to hydrogen bonding in phenols. They exist as associated molecules as shown below:

$$\cdots\cdots \overset{\delta^-}{O}(C_6H_5) - \overset{\delta^+}{H} \cdots \overset{\delta^-}{O}(C_6H_5) - \overset{\delta^+}{H} \cdots\cdots \overset{\delta^-}{O}(C_6H_5) - \overset{\delta^+}{H} \cdots\cdots \overset{\delta^-}{O}(C_6H_5) - \overset{\delta^+}{H} \cdots\cdots$$

The nature and position of the substituent influences the boiling point of phenol. For example, *o*-nitrophenol is steam volatile while *p*-nitrophenol is not. Thus *o*-nitrophenol has smaller b.p. than the *p*- isomer. This is due to the presence of intramolecular hydrogen bonding in *o*-nitrophenol molecules while the molecules of *p*-nitrophenol are linked by intermolecular hydrogen bonding as shown below:

$$\text{(o-nitrophenol, intramolecular H-bond)} \qquad \cdots O{=}\overset{\overset{O}{\uparrow}}{N}-C_6H_4-O-H\cdots O{=}\overset{\overset{O}{\uparrow}}{N}-C_6H_4-O-H\cdots$$

Therefore, the attractive forces in the former are less as compared to the latter and its boiling point is therefore comparatively less. The H– of one molecule is not available for bonding with other molecule. Also the size of the molecule gets reduced in intramolecular hydrogen bonding which results in lower boiling point.

Example 1. Why has phenol higher boiling point than toluene?

Solution. Phenol (molecular mass 94) boils at 455 K while toluene (molecular mass 92) boils at 383 K only. The reason for higher boiling point of phenol as compared to toluene is that phenol forms intermolecular hydrogen bonds leading to association of its molecule. Consequently, its molecular mass increases manifold and hence its boiling point is high. Additional energy is required to break the hydrogen bonds which raises its boiling point. Hydrogen bonding in phenol is depicted below :

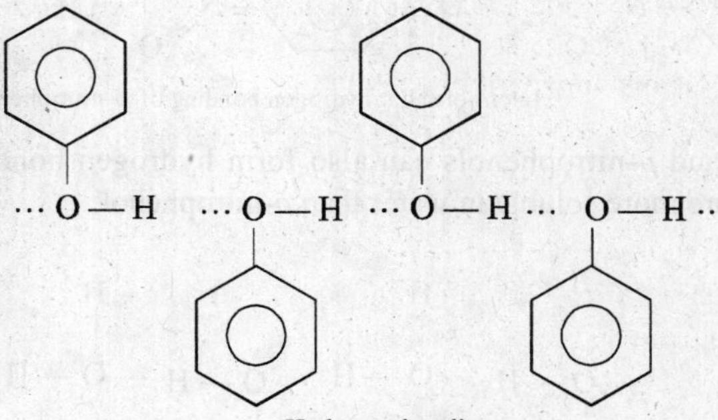

··· O—H ··· O—H ··· O—H ··· O—H ···

Hydrogen bonding

On the other hand, toluene does not contain any strong electronegative atom attached to hydrogen. Therefore, it does not form intermolecular hydrogen bonding. Hence, its boiling point is lower than that of phenol.

Example 2. Why is phenol more soluble in water than toluene?

Solution. Phenol is somewhat soluble in water as phenol molecules can form hydrogen bonds with water molecules as shown below :

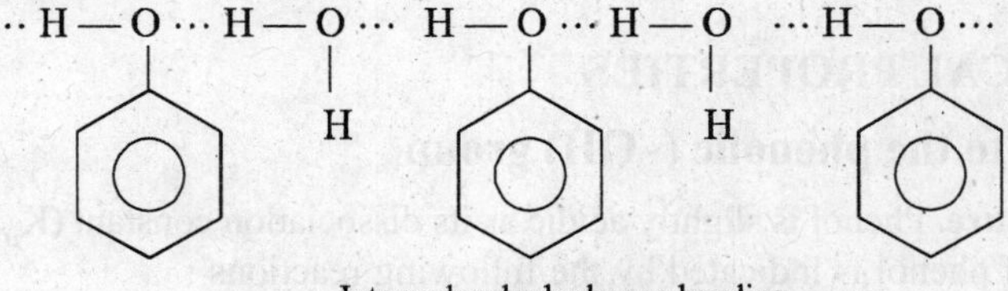

··· H—O ··· H—O ··· H—O ··· H—O ··· H—O ···
H H

. Intermolecular hydrogen bonding

On the other hand, toluene is not capable of forming hydrogen bonds with water molecules, hence toluene is not soluble in water. It can be said, in general, that a compound is soluble in water if it can form hydrogen bond with it.

Example 3. Account for the lower boiling point and decreased water solubility of *o*-nitrophenol as compared with their *m*- and *p*-isomers.

Solution. *o*-Nitrophenol forms *intramolecular* hydrogen bonding resulting in the formation of a six-membered ring.

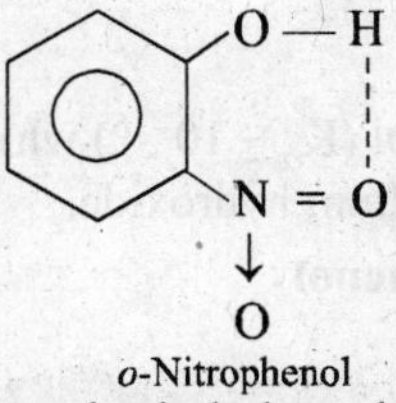

O—H
N = O
↓
O

o-Nitrophenol
Intramolecular hydrogen bonding

Due to intramolecular hydrogen bonding, *o*-nitrophenol is incapable of forming hydrogen bonding with other *o*-nitrophenol molecules, *i.e.*, intermolecular hydrogen bonding. Hence, it does not exist as associated molecule. Moreover, the molecular size shrinks as a consequence of intramolecular bonding. Therefore, it boils at a lower temperature.

Similarly, due to intramolecular hydrogen bonding, *o*-nitrophenol does not form hydrogen bonds with water molecules also. Therefore, it is not very soluble in water.

m-nitrophenol as well as *p*-nitrophenol form intermolecular hydrogen bonding resulting in association of molecules. Consequently, they have high boiling points.

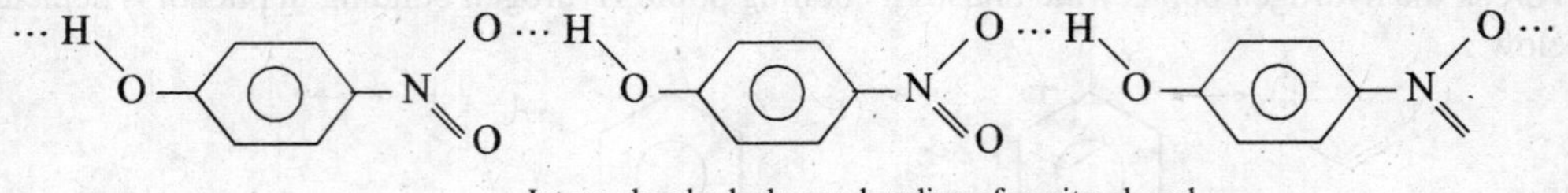

Intermolecular hydrogen bonding of *p*-nitrophenol

Moreover, *m*- and *p*-nitrophenols can also form hydrogen bonds with water molecules. Consequently, they are more soluble in water than *o*-nitrophenol.

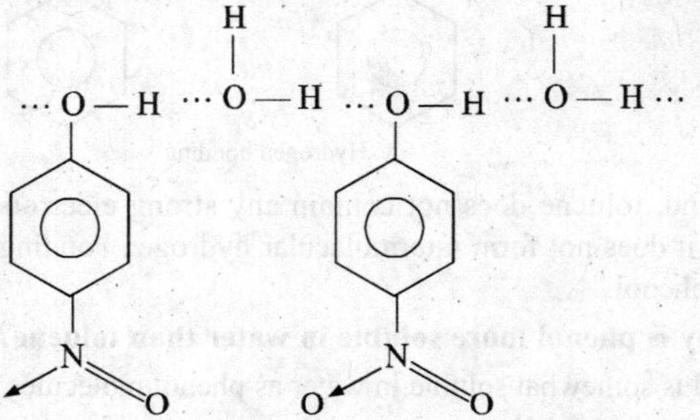

3.4. CHEMICAL PROPERTIES

Reactions due to the phenolic (–OH) group

1. Acidic nature. Phenol is slightly acidic as its dissociation constant (K_a) is 1.3×10^{-10}. The acidic character of phenol is indicated by the following reactions :

(*i*) Phenol turns blue litmus red.

(*ii*) Phenol produces hydrogen with sodium.

(*iii*) Phenol neutralises sodium hydroxide to form sodium phenolate and water as:

$$C_6H_5OH + NaOH \longrightarrow C_6H_5\overset{-}{O}\overset{+}{Na} + H_2O$$

Phenol (*acid*) + NaOH (*Base*) → Sodium phenolate (*Salt*) + H_2O (*Water*)

This is in contrast with ethyl alcohol ($K_a \sim 10^{-16}$) which is almost neutral as it does not turn blue litmus red nor does it neutralise sodium hydroxide.

Reason for acidic character of phenol

In water, phenol ionises as :

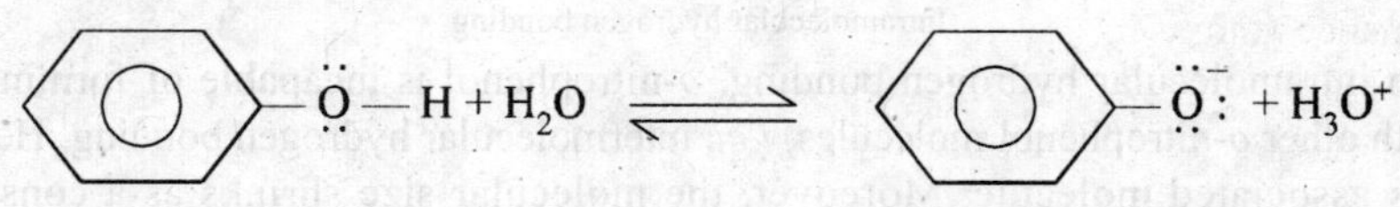

$$C_6H_5\ddot{O}—H + H_2O \rightleftharpoons C_6H_5\ddot{O}:^{-} + H_3O^+$$

Phenol → Phenoxide ion

Now phenol as well as phenoxide ion cannot be represented by a single valence bond formula, rather both are resonance hybrids of various contributing structures as given below :

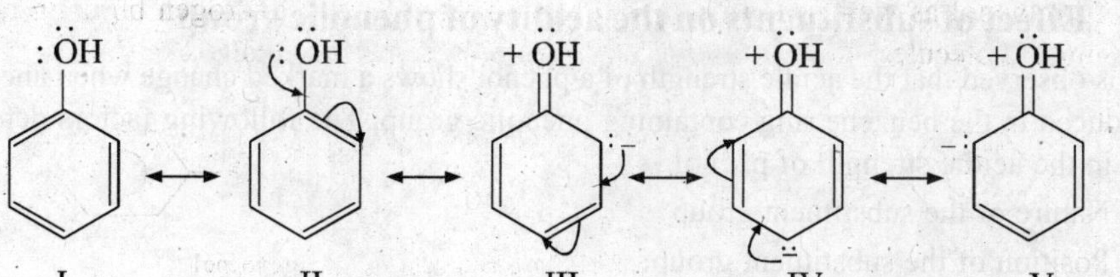

I II III IV V

Resonating structures of phenol

It is evident from the contributing structures III, IV and V that oxygen atom acquires partial positive charge due to resonance. As a result, bond between O–H becomes weak which helps the release of proton and makes phenol acidic.

After the release of proton, the phenoxide ion is formed which is also resonance hybrid of various contributing structures (VI–X).

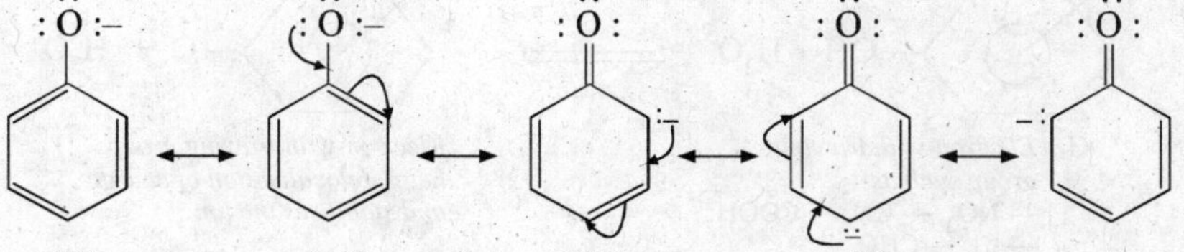

VI VII VIII IX X

Thus, phenol as well as phenoxide ion are stabilised by resonance.

However, phenoxide ion is more stabilized by resonance than phenol. This is because in phenol contributing structures III to V involve separation of charge and are therefore less stable than structures I and II. On the other hand, in phenoxide ions none of the contributing structures VI to X involve charge separation and hence all the contributing structures in phenoxide ion are almost equally stable. In short,

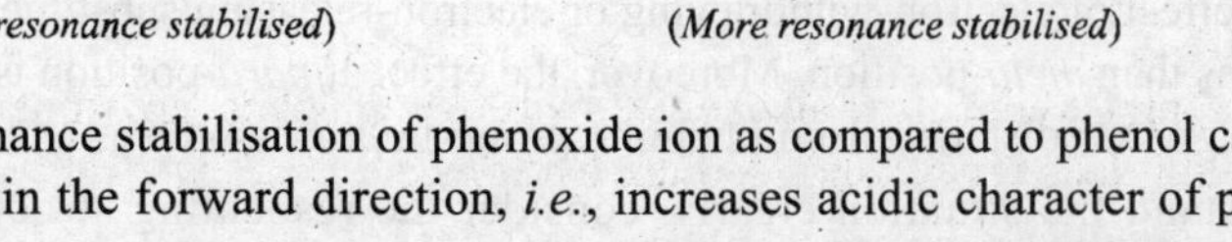

$$\text{Phenol} + H_2O \rightleftharpoons \text{Phenoxide ion} + H_3O^+$$

Phenol (*Less resonance stabilised*) Phenoxide ion (*More resonance stabilised*)

The greater resonance stabilisation of phenoxide ion as compared to phenol causes the above reaction to take place in the forward direction, *i.e.*, increases acidic character of phenol.

On the other hand, ethyl alcohol is neutral as neither ethyl alcohol molecule nor the ethoxide ion produced from it has resonating structures. In other words, neither ethyl alcohol nor ethoxide ions is resonance stabilised.

$$C_2H_5-\ddot{\underset{\cdot\cdot}{O}}-H + H_2O \rightleftharpoons C_2H_5-\ddot{\underset{\cdot\cdot}{O}}:^- + H_3O^+$$

Ethyl alcohol (*No resonance stabilization*) Ethoxide ion (*No resonance stabilization*)

3.4.1. Effect of substituents on the acidity of phenolic group

It is observed that the acidic strength of a phenol shows a marked change when another group is introduced in the benzene ring containing phenolic group. The following factors determine the change in the acidic strength of phenol:

1. Nature of the substituent group
2. Position of the substituent group

These two factors are discussed separately as under.

1. Nature of substituent in the benzene ring. Electron-withdrawing substituents such as halogens, nitro group, carboxylic group increase the acidic character of phenol. This is because electron-withdrawing groups help in the delocalization of the negative charge of phenoxide ion and thus help to stabilize the phenoxide ion and thereby increase the acidic character of phenol. For example, nitrophenols are more acidic than phenol.

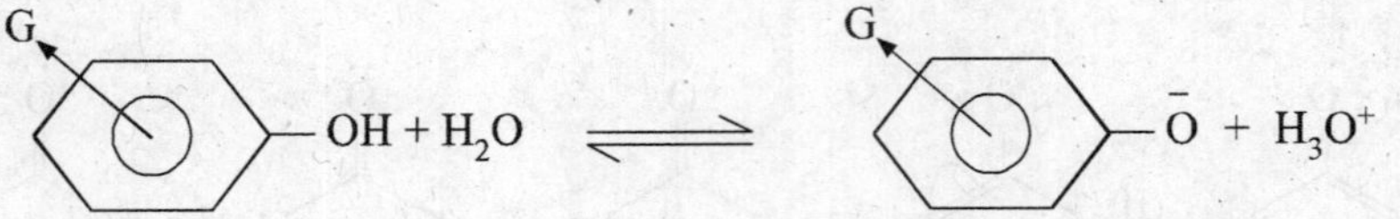

G — OH + H_2O ⇌ G — $\bar{O}$ + H_3O^+

G = *Electron-withdrawing group such as* — NO_2, — CN, — COOH, — Cl, — CHO

Electron-withdrawing group helps delocalisation of charge and stabilises the ion.

On the other hand, electron releasing groups such as $-CH_3$, –OH, –OR tend to destabilise the phenoxide ion by adding to its negative charge and thereby reduce the acidic character of phenol. For example, cresols are less acidic than phenol.

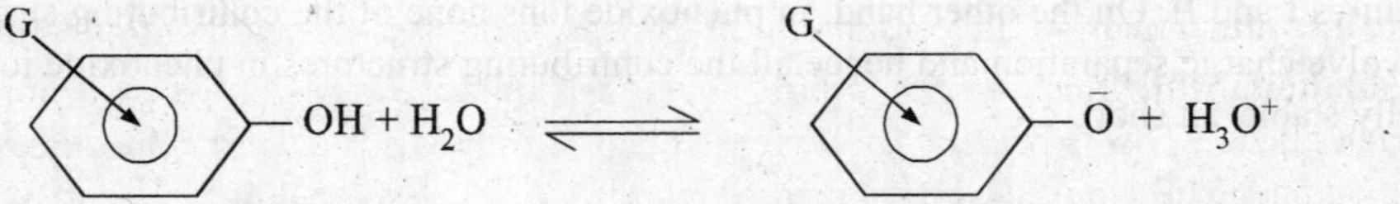

G — OH + H_2O ⇌ G — $\bar{O}$ + H_3O^+

G = *Electron-releasing groups such as* — CH_3, — OH, — OR, — NR_2

Electron-releasing group intensifies the negative charge and destabilizes the ion.

2. Position of substituent relative to hydroxyl group. Not only the nature of substituent, but also the position of the substituent relative to hydroxyl group has influence on the acidic character of phenol. The effect of electron-withdrawing or electron-releasing substituent is felt more at *ortho*- or *para*-positions than *meta*-position. Moreover, the effect at *para*-position is less than that at *ortho*-position due to greater distance from phenolic group in *para* isomer.

To understand the situation clearly, consider the resonating structures of *o*-, *m*- and *p*-nitrophenate ions as shown below:

o-nitrophenate ion

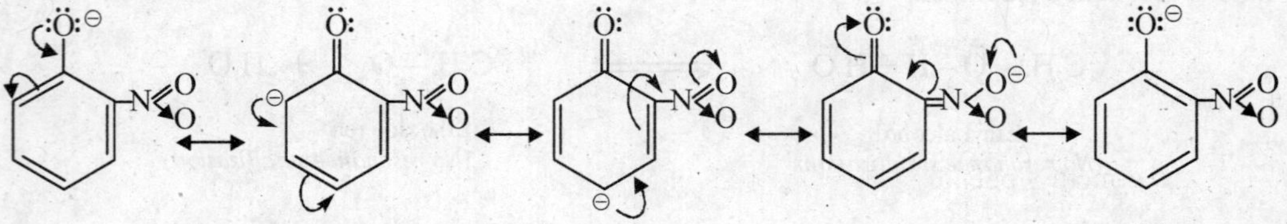

I ⟷ II ⟷ III ⟷ IV ⟷ (V)

p-nitrophenate ion

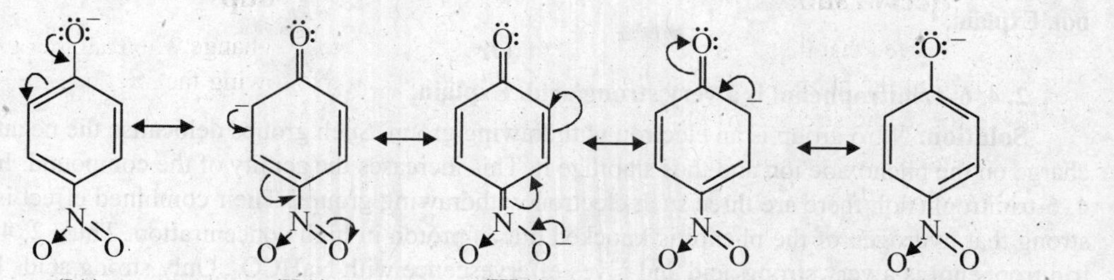

(IA) (IIA) (IIIA) (IVA) (VA)

m-nitrophenate ion

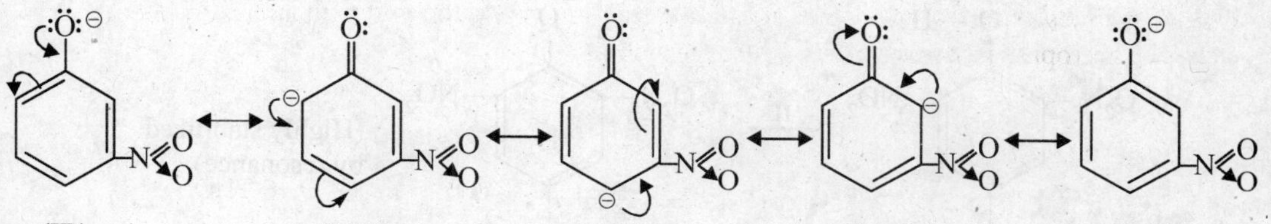

(IB) (IIB) (IIIB) (IVB) (VB)

In all the three cases, the number of the constituting structures for the ion are the same but in case of *o*-and *p*-substituted ion, structures IV and III A have a negative charge on the oxygen atom of the nitro group. This makes the contributing structure relatively more stable than the rest. No such structure happens in the case of *m*- substituted ion. Therefore, the hybrid in this case is less stable than the hybrid for the *o*- and *p*-substituted ions. Consequently, *o*-nitrophenol and *p*-nitrophenol are stronger acids than the *m*- isomer.

Example 1. Introduction of nitro group in aromatic nucleus increases the acidity of phenols but the introduction of –CH_3 group into the nucleus decreases the acidity of phenols. Explain.

Solution. Nitro group increases the acidity of phenol. For example, *o*-nitrophenol is a stronger acid than phenol. This is because of the reason that nitro group is able to delocalise the negative charge on phenoxide ion and thus able to stabilize it as illustrated below :

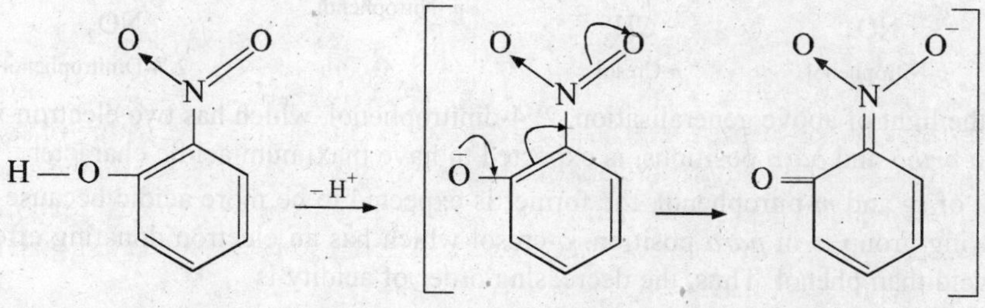

$\xrightarrow{-H^+}$

For this reason, nitro group increases the acidity of phenol. On the other hand, a methyl group, which is electron repelling, further intensifies negative charge on the phenoxide ion, making it unstable.

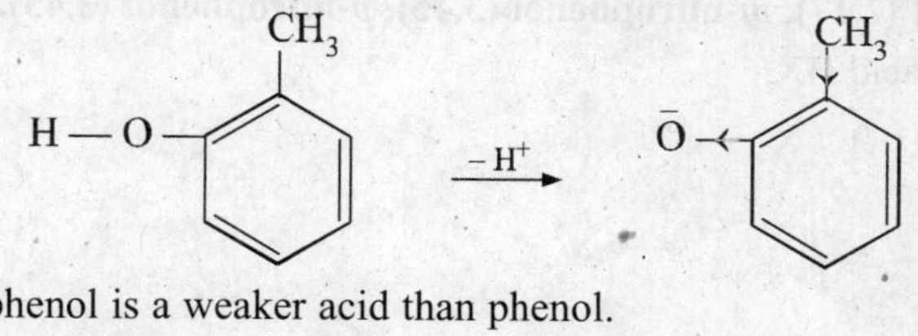

$\xrightarrow{-H^+}$

Thus *o*-methyl phenol is a weaker acid than phenol.

Example 2. 2, 4, 6-Trinitrophenol gives effervescence with sodium bicarbonate but phenol does not. Explain.

Or

2, 4, 6-Trinitrophenol is a very strong acid. Explain.

Solution. Nitro group is an electron withdrawing group. Such groups delocalise the negative charge on the phenoxide ion and thus stabilize it. This increases the acidity of the compound. In 2, 4, 6-trinitrophenol, there are three such electron withdrawing groups. Their combined effect is so strong that hydrogen of the phenol is knocked out as proton in high concentration. Thus, 2, 4, 6-trinitrophenol is a very strong acid and gives effervescence with $NaHCO_3$. Only strong acids like mineral acids give effervescence with sodium bicarbonate. In phenol (C_6H_5OH), the concentration of H^+ is not sufficient to react with sod. bicarbonate.

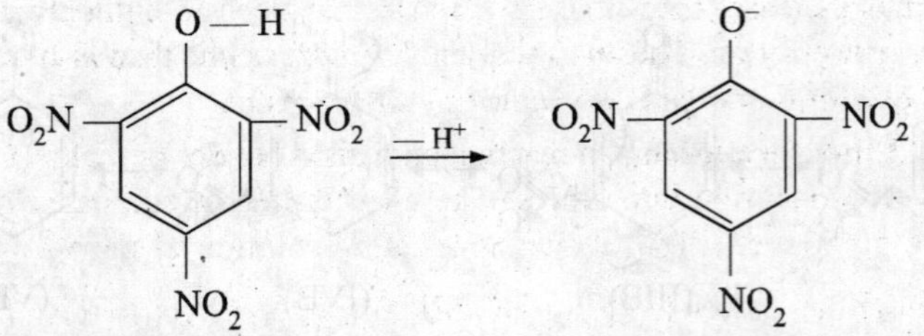

O—H, O_2N, NO_2, NO_2 $\xrightarrow{-H^+}$ O^-, O_2N, NO_2, NO_2 (Highly stabilized by resonance)

Example 3. Arrange the following in decreasing order of acidity :
***p*-nitrophenol, *p*-cresol, *m*-nitrophenol, 2, 4-dinitrophenol.**

Solution. Electron-withdrawing groups increase the acidity of phenols and the effect is more pronounced in *ortho* and *para* positions compared to the *meta* positions.

On the other hand, electron-donating groups tend to decrease the acidity of the phenol.

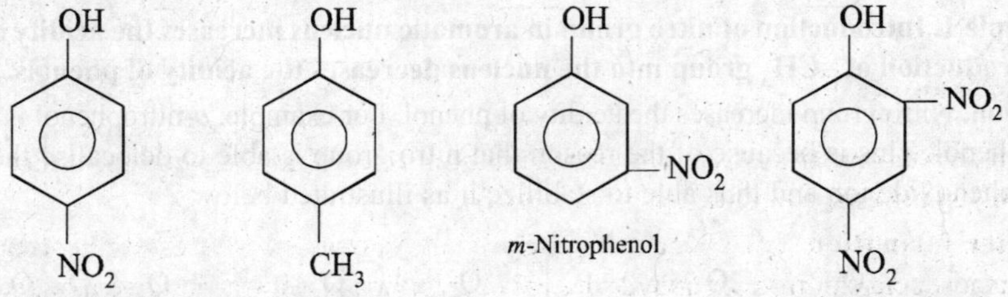

OH, NO_2 — *p*-Nitrophenol; OH, CH_3 — *p*-Cresol; OH, NO_2 — *m*-Nitrophenol; OH, NO_2, NO_2 — 2, 4-Dinitrophenol

In the light of above generalisation, 2, 4-dinitrophenol, which has two electron-withdrawing groups in *ortho* and *para* positions, is expected to have maximum acidic character.

Out of *p*- and *m*-nitrophenol, the former is expected to be more acidic because the electron withdrawing group is in *para* position. *p*-cresol which has an electron donating effect will be a weaker acid than phenol. Thus, the decreasing order of acidity is

2, 4-dinitrophenol > *p*-nitrophenol > *m*-nitrophenol > *p*-cresol.

Example 4. Explain the order of acidic strength of various nitrophenols (pK_a values are given) : *o*-nitrophenol (2.17), *m*-nitrophenol (3.45), *p*-nitrophenol (3.43).

Solution. For an acid HX,

$$HX \rightleftharpoons H^+ + X^-$$

$$K_a = \frac{[H^+][X^-]}{[HX]}$$

$$pK_a = -\log K_a$$

Thus, lower the value of pK_a, greater is the value of K_a and thus greater the $[H^+]$ concentration or greater acidity. The order of acidic strength in the light of pK_a values can be written as:

m-nitrophenol (OH, NO_2) < p-nitrophenol (OH, NO_2) < o-nitrophenol (OH, NO_2)

Nitro group has the effect of withdrawing electrons towards itself through resonance. This facilitates removal of H^+ and thus increases the acidity. It is found that electron-withdrawing effect works more in *p*-position than in *m*-position. Thus *p*-nitrophenol is more acidic than *m*-nitrophenol. Maximum acidity in the case of *o*-nitrophenol is explained by *ortho-effect.*

2. Williamson synthesis. Ether formation. On reaction with alkyl halides or dialkyl sulphates in alkaline medium, phenols are converted into ethers. The alkyl halide or sulphate undergoes nucleophilic substitution by the phenoxide ion present in alkaline solution of phenol. Thus we have :

$$ArOH + NaOH \longrightarrow ArO^- Na^+ + H_2O$$

$$ArONa + RX \longrightarrow Ar - O - R + NaX$$

For example :

$$C_6H_5ONa + C_2H_5I \longrightarrow C_6H_5-O-C_2H_5 + NaI$$

Sod. phenoxide → Ethyl phenyl ether or Phenetole

This reaction constitutes an important method of synthesising ethers, under the name **Williamson's synthesis,** from phenols (or alcohols).

3. Ester formation. Like alcohols, phenols are converted into esters by reaction with carboxylic acids, acid chlorides or anhydrides. Acid chlorides or anhydrides give a better yield than the acids.

$$ArOH + R-\overset{O}{\overset{\|}{C}}-Cl \longrightarrow Ar-O-\overset{O}{\overset{\|}{C}}-R + HCl$$

For example :

$$C_6H_5OH + C_2H_5-\overset{O}{\overset{\|}{C}}-Cl \longrightarrow C_6H_5-O-\overset{O}{\overset{\|}{C}}-C_2H_5 + HCl$$

Phenol → Phenyl propionate

$$C_6H_5OH \text{ (Phenol)} + (CH_3CO)_2O \xrightarrow{\text{Pyridine}} C_6H_5O{-}CO{-}CH_3 \text{ (Phenyl acetate)} + CH_3COOH$$

4. Fries rearrangement. If an ester of a phenol is heated with aluminium chloride, the acyl group migrates from the phenolic oxygen to the *ortho* or *para* position on the ring. This is called **Fries rearrangement.**

For example :

$$C_6H_5O{-}CO{-}CH_2CH_3 \text{ (Phenyl propionate)} \xrightarrow[413-423\text{ K}]{AlCl_3} p\text{-}HOC_6H_4{-}CO{-}CH_2CH_3 + o\text{-}HOC_6H_4{-}CO{-}CH_2{-}CH_3$$

Phenyl propionate → *p*-Hydroxyphenyl ethyl ketone (About 50%) + *o*-Hydroxyphenyl ethyl ketone (About 35%)

At low temperature, *para* product predominates while at high temperature *ortho* isomer is the main product.

Mechanism. Fries rearrangement probably involves the formation of an acylium ion (RCO^+) which then attacks the aromatic ring as in Friedel-Crafts acylation. The complete mechanism may be summed up as follows :

$$C_6H_5\ddot{O}COR \underset{}{\overset{AlCl_3}{\rightleftharpoons}} C_6H_5\overset{+}{O}(\overset{-}{A}lCl_3)COR \rightleftharpoons C_6H_5\ddot{O}\overset{-}{A}lCl_3 + RCO^+ \rightleftharpoons \text{(cyclohexadienone: } \overset{+}{=}OAlCl_3\text{, H, COR)} \xrightarrow{-H^+Cl^-} o\text{-}(\ddot{O}AlCl_2)C_6H_4COR \xrightarrow{H_2O} o\text{-}HOC_6H_4COR$$

5. Reaction with zinc dust. When distilled with zinc dust, phenols form corresponding hydrocarbons. For example:

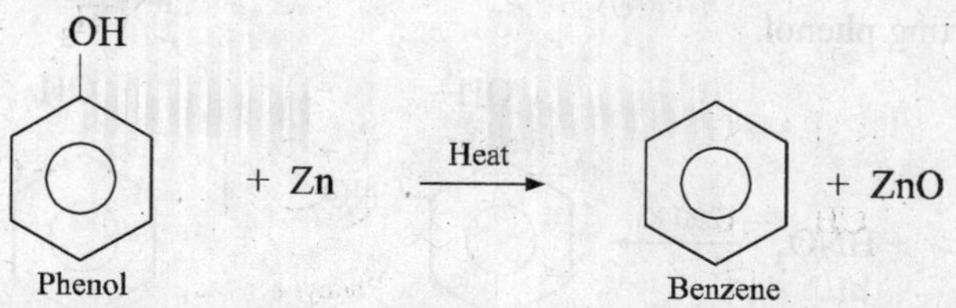

$$\underset{\text{Phenol}}{C_6H_5OH} + Zn \xrightarrow{\text{Heat}} \underset{\text{Benzene}}{C_6H_6} + ZnO$$

6. Reaction with ammonia. When heated with ammonia at high temperature and pressure in the presence of catalysts like anhydrous zinc chloride, –OH group of phenols is replaced by $-NH_2$ group. For example :

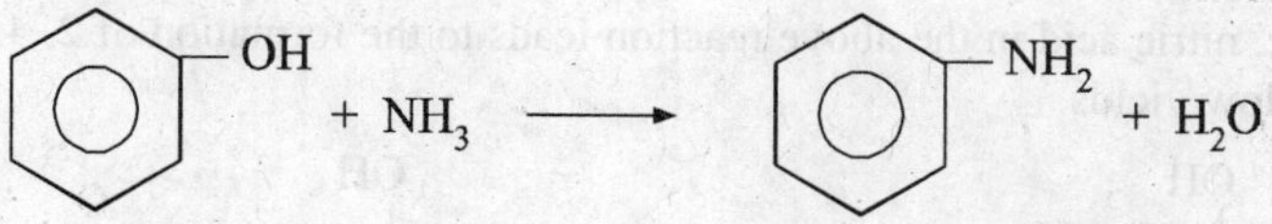

$$C_6H_5OH + NH_3 \longrightarrow C_6H_5NH_2 + H_2O$$

Reactions due to benzene ring

Electrophilic substitution reactions of phenols

The important ring substitutions are described below :

1. Halogenation. On treatment with halogens, phenols are very rapidly halogenated forming polyhalogen derivatives. For example, reaction of phenol with an aqueous solution of bromine immediately produces tribromophenol.

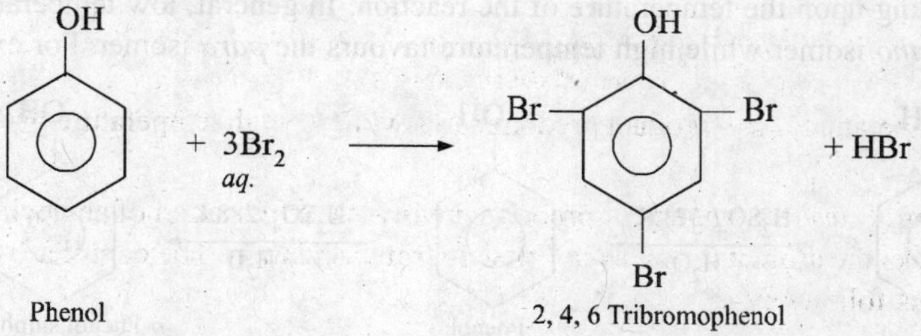

$$\underset{\text{Phenol}}{C_6H_5OH} + \underset{aq.}{3Br_2} \longrightarrow \underset{\text{2, 4, 6 Tribromophenol}}{C_6H_2Br_3OH} + HBr$$

Halogenation can, however, be limited to monohalogenation stage by performing it at low temperature and in non-polar solvents such as carbon tetrachloride or carbon disulphide. For example :

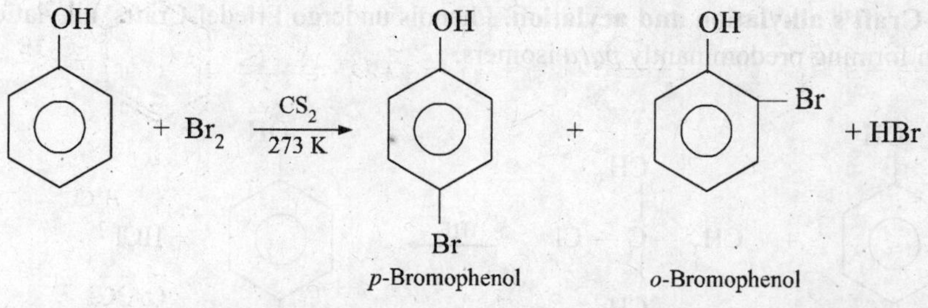

$$C_6H_5OH + Br_2 \xrightarrow[273\text{ K}]{CS_2} \underset{p\text{-Bromophenol}}{p\text{-}BrC_6H_4OH} + \underset{o\text{-Bromophenol}}{o\text{-}BrC_6H_4OH} + HBr$$

It may be noted that phenol ionises in aqueous medium to form phenoxide ion. The negative charge on oxygen in phenoxide ion is donated to the benzene ring. Consequently, the ring gets activated towards substitution. We obtain trisubstitution products. However, *the situation is different in non-polar solvents. Ionisation of phenol does not take place. Electrons of –OH group are donated to a small extent by resonance. Therefore only monosubstitution takes place.*

2. Nitration. Nitration of phenols gives poor yields of the expected nitro compounds. This is because phenols are easily oxidised and nitric acid can bring about not only nitration but also oxidation of the starting phenol.

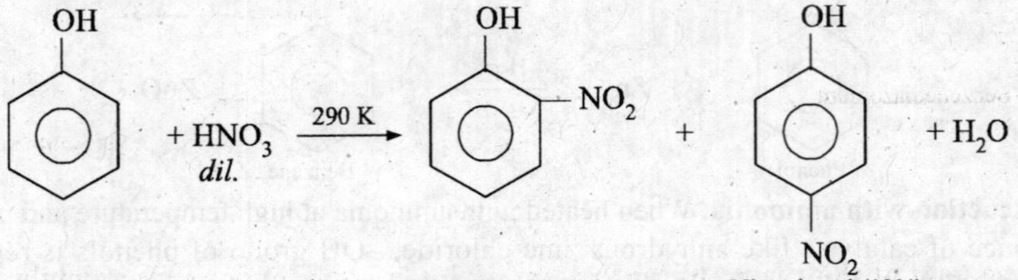

$$C_6H_5OH + HNO_3 \text{ (dil.)} \xrightarrow{290\text{ K}} o\text{-}NO_2C_6H_4OH + p\text{-}NO_2C_6H_4OH + H_2O$$

o-Nitrophenol (40%) *p*-Nitrophenol (13%)

Use of conc. nitric acid in the above reaction leads to the formation of 2, 4, 6-trinitrophenol or picric acid in low yields.

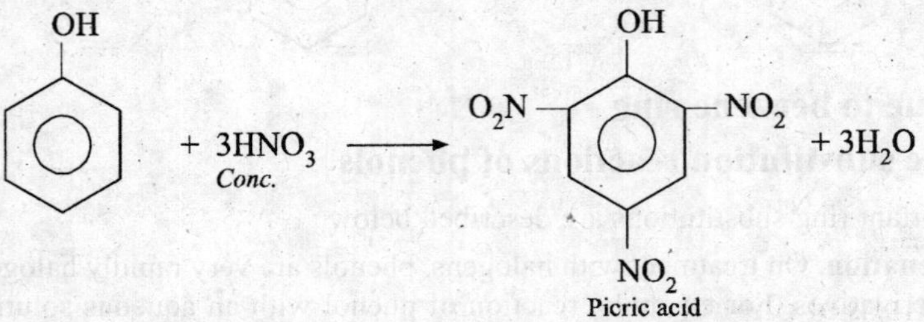

$$C_6H_5OH + 3HNO_3 \text{ (Conc.)} \longrightarrow (O_2N)_3C_6H_2OH + 3H_2O$$

Picric acid

3. Sulphonation. Sulphonation of phenols gives either the *ortho* or the *para* isomer as the main product depending upon the temperature of the reaction. In general, low temperature favours the formation of *ortho* isomer while high temperature favours the *para* isomer. For example :

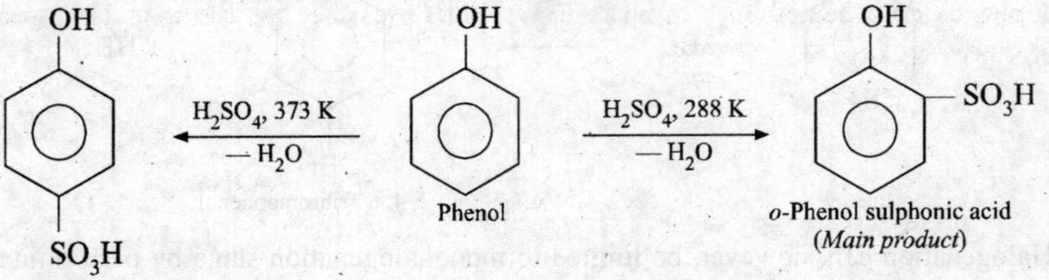

$$p\text{-}HO_3SC_6H_4OH \xleftarrow[-H_2O]{H_2SO_4,\ 373\text{ K}} C_6H_5OH \xrightarrow[-H_2O]{H_2SO_4,\ 288\text{ K}} o\text{-}HO_3SC_6H_4OH$$

p-Phenol sulphonic acid (*Main product*) Phenol *o*-Phenol sulphonic acid (*Main product*)

4. Friedel-Craft's alkylation and acylation. Phenols undergo Friedel-Crafts' alkylation as well as acylation forming predominantly *para* isomers.

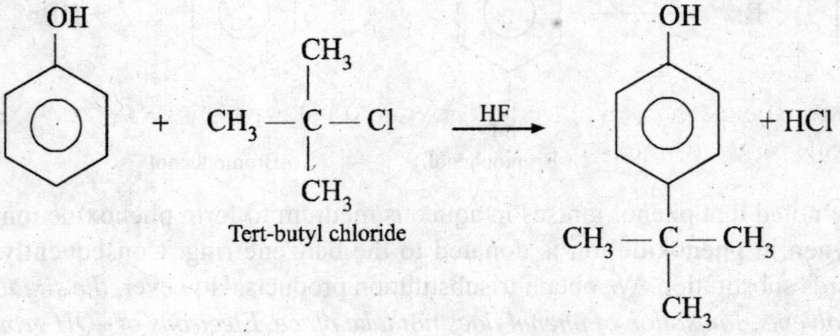

$$C_6H_5OH + CH_3-\underset{CH_3}{\overset{CH_3}{C}}-Cl \xrightarrow{HF} p\text{-}HOC_6H_4-\underset{CH_3}{\overset{}{C}}(CH_3)-CH_3 + HCl$$

Tert-butyl chloride *p*-Tert. butyl phenol

5. Coupling with diazonium salts. When treated with diazonium salts in weakly alkaline solution, phenols undergo coupling to form hydroxyazo compounds. For example :

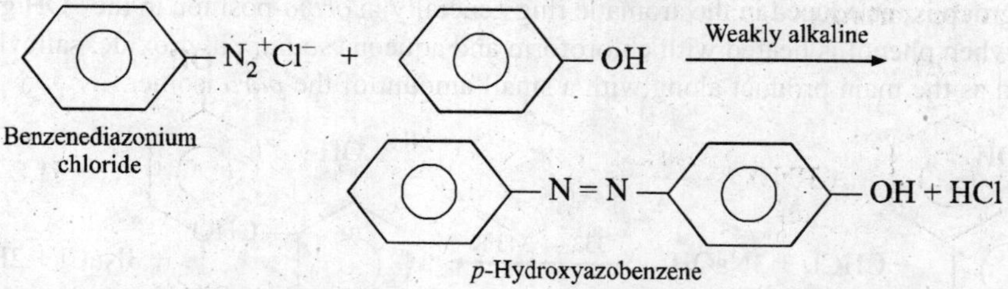

$$C_6H_5-N_2^+Cl^- + C_6H_5OH \xrightarrow{\text{Weakly alkaline}} C_6H_5-N=N-C_6H_4-OH + HCl$$

Benzenediazonium chloride

p-Hydroxyazobenzene

Mechanism of the reaction. Mechanism of the reaction involves electrophilic attack of diazonium salt on the phenate ion.

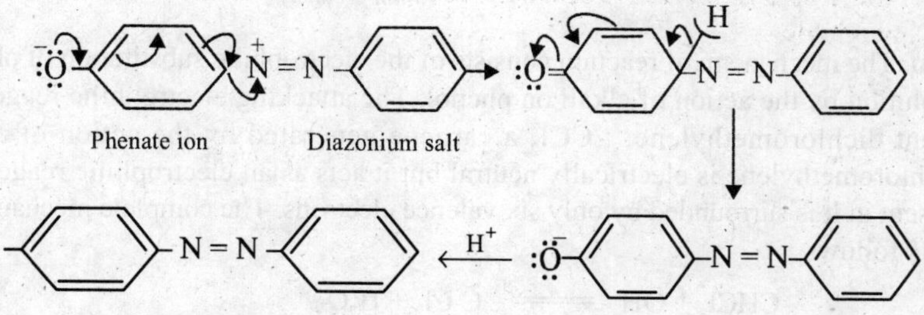

Phenate ion Diazonium salt

$$HO-C_6H_4-N=N-C_6H_5 \xleftarrow{H^+} {}^-:\ddot{O}-C_6H_4-N=N-C_6H_5$$

Special reactions of phenols

1. Kolbe's reaction

When heated with carbon dioxide under pressure, sodium salts of phenols lead to the formation of salts of phenolic acids; the ortho isomers being the main product. For example, when sodium phenoxide is heated with carbon dioxide under pressure, sodium salicylate is obtained as the major product.

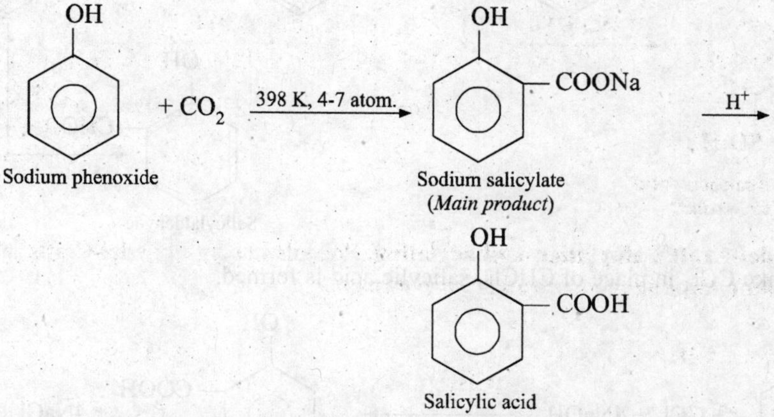

$$C_6H_5ONa + CO_2 \xrightarrow{398\ K,\ 4\text{-}7\ atom.} o\text{-}HOC_6H_4COONa \xrightarrow{H^+} o\text{-}HOC_6H_4COOH$$

Sodium phenoxide

Sodium salicylate (*Main product*)

Salicylic acid

Mechanism. The mechanism of Kolbe's reaction is believed to involve the electrophilic attack by carbon dioxide on the phenoxide ion as shown below :

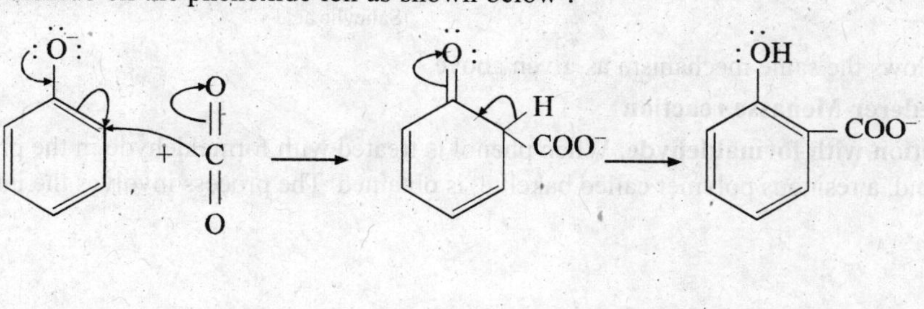

$$C_6H_5\ddot{O}:^- + O=C=O \longrightarrow \text{(H, COO}^-\text{ intermediate)} \longrightarrow o\text{-}HOC_6H_4COO^-$$

2. Reimer-Tiemann reaction

On heating with chloroform and an alkali, phenols are converted into phenolic aldehydes; an aldehyde group is introduced in the aromatic ring generally in *ortho*-position to the –OH group. For example, when phenol is heated with chloroform and aqueous sodium hydroxide, salicylaldehyde is obtained as the main product along with a small amount of the *para* isomer.

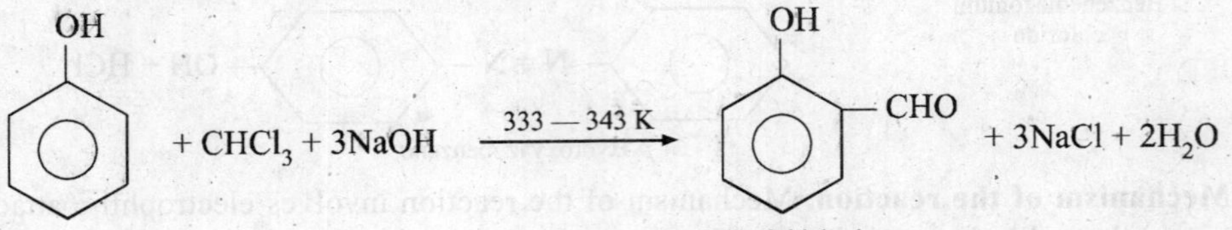

$$C_6H_5OH + CHCl_3 + 3NaOH \xrightarrow{333-343\text{ K}} o\text{-}HOC_6H_4CHO + 3NaCl + 2H_2O$$

Salicylaldehyde
(*Main product*)

Mechanism. The mechanism of reaction consists of the electrophilic substitution of phenoxide ion formed in solution by the action of alkali on phenol. The attacking electrophilic reagent is the electron deficient dichloromethylene; : CCl_2 a carbene generated by the action of alkali on chloroform. Dichloromethylene is electrically neutral but it acts as an electrophilic reagent as the carbon atom present in it is surrounded by only six valence electrons. The complete mechanism may be summed up as follows :

$$CHCl_3 + \bar{O}H \rightleftharpoons \bar{C}Cl_3 + H_2O$$
$$\bar{C}Cl_3 \longrightarrow :CCl_2 + Cl^-$$

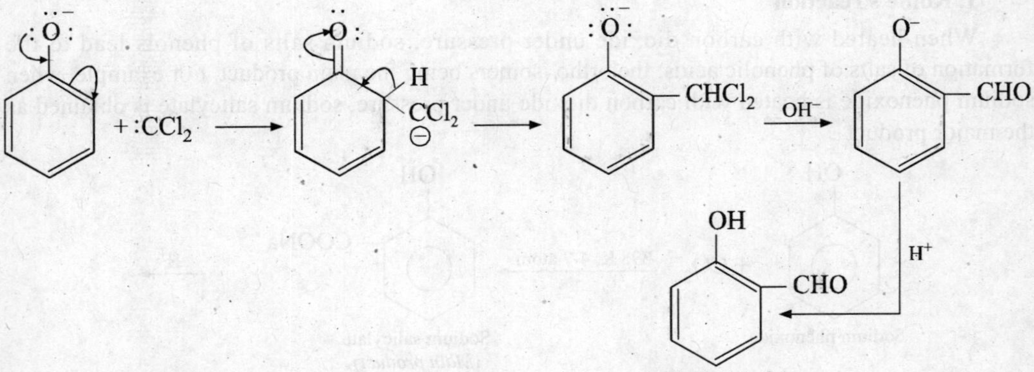

$$C_6H_5\ddot{O}:^- + :CCl_2 \longrightarrow \text{(intermediate: H, } \overset{\ominus}{C}Cl_2\text{)} \longrightarrow o\text{-}CHCl_2C_6H_4\ddot{O}:^- \xrightarrow{OH^-} o\text{-}CHOC_6H_4O^- \xrightarrow{H^+} o\text{-}HOC_6H_4CHO$$

Salicylaldehyde

If we take CCl_4 in place of $CHCl_3$, salicylic acid is formed.

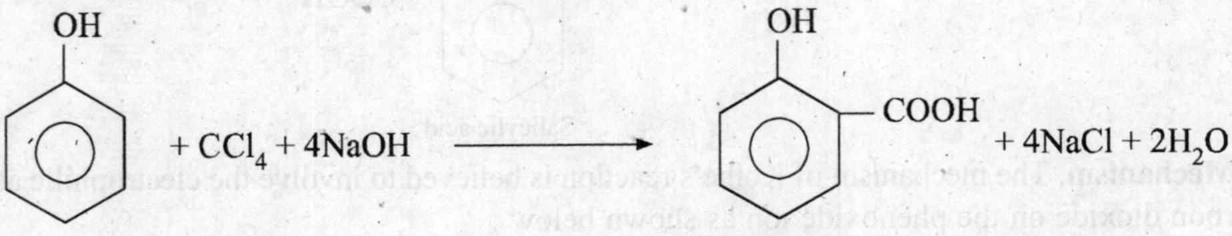

$$C_6H_5OH + CCl_4 + 4NaOH \longrightarrow o\text{-}HOC_6H_4COOH + 4NaCl + 2H_2O$$

Salicylic acid

It follows the same mechanism as given above.

3. Lederer-Menasse reaction

Reaction with formaldehyde. When phenol is treated with formaldehyde in the presence of alkali or acid, a resinous polymer called bakelite, is obtained. The process involves the interlinking of many phenol rings by – CH_2 – groups in *ortho* and *para* positions with respect to the phenolic group as depicted below :

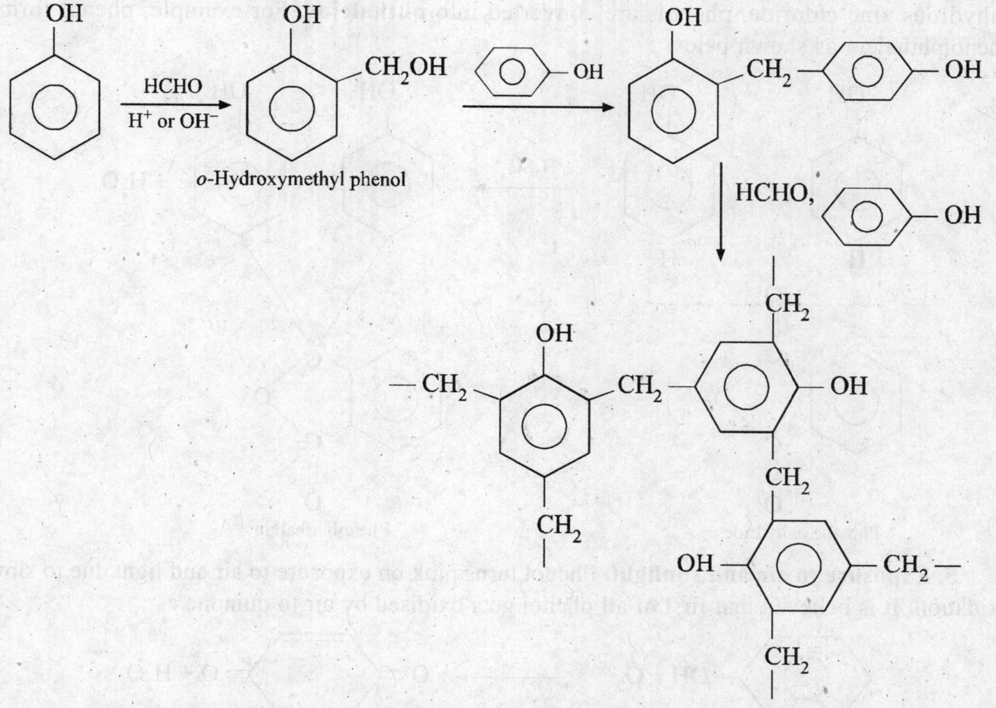

OH OH OH

HCHO / H^+ or OH^- ; CH_2OH ; OH ; CH_2 ; OH

o-Hydroxymethyl phenol

HCHO, OH

CH_2 ; OH ; — CH_2 ; CH_2 ; OH ; CH_2 ; CH_2 ; OH ; CH_2 — ; CH_2

Mechanism. The mechanism of the first stage of reaction, *i.e.,* formation of *o*-hydroxymethyl phenol, involves the electrophilic substitution on the aromatic ring. The reaction may take place between phenol and protonated formaldehyde (*acid catalysis*) or between phenoxide ion and formaldehyde (*base catalysis*) as shown below :

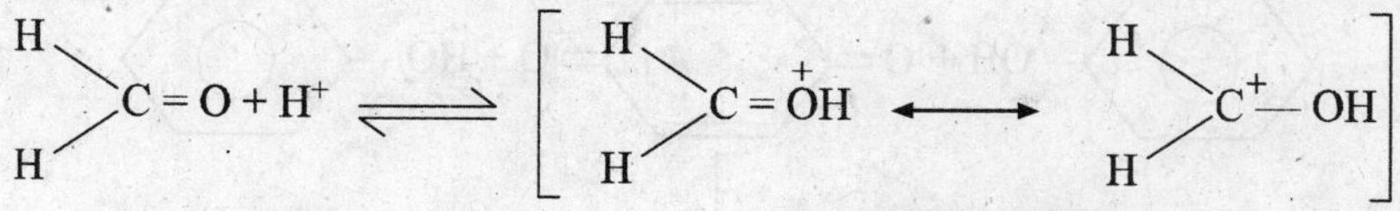

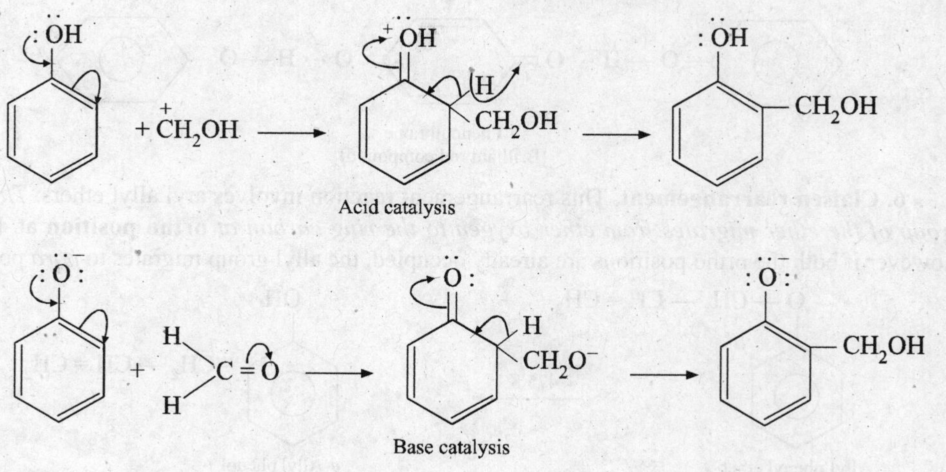

$$H_2C=O + H^+ \rightleftharpoons \left[H_2C=\overset{+}{O}H \longleftrightarrow H_2\overset{+}{C}-OH \right]$$

Protonated formaldehyde

:ÖH + $\overset{+}{C}H_2OH$ ⟶ $\overset{+}{O}H$ (H, CH_2OH) ⟶ :ÖH, CH_2OH

Acid catalysis

$:\ddot{O}:^-$ + $H_2C=O$ ⟶ :Ö: (H, CH_2O^-) ⟶ $:\ddot{O}:^-$, CH_2OH

Base catalysis

4. Phthalein reaction

On heating with phthalic anhydride in the presence of concentrated sulphuric acid or anhydrous zinc chloride, phenols are converted into **phthaleins.** For example, phenol forms phenolphthalein as shown below :

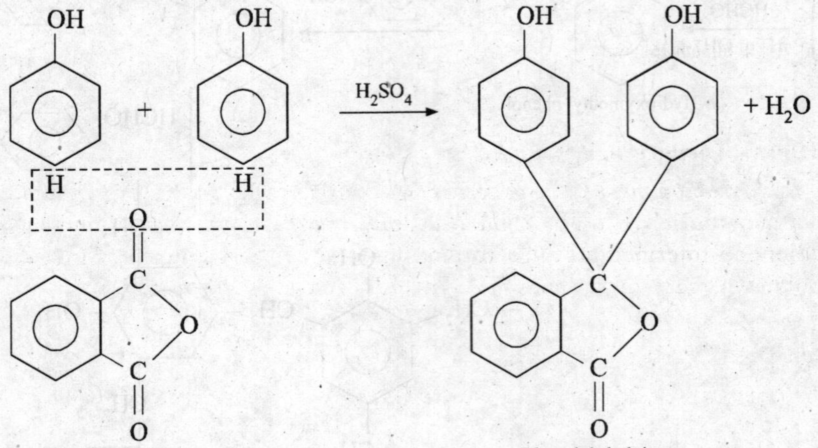

OH OH H H O C O C O $\xrightarrow{H_2SO_4}$ OH OH C O C O $+ H_2O$

Phthalic anhydride Phenolphthalein

5. Exposure to air and sunlight. Phenol turns pink on exposure to air and light due to slow oxidation. It is believed that first of all phenol gets oxidised by air to quinone as :

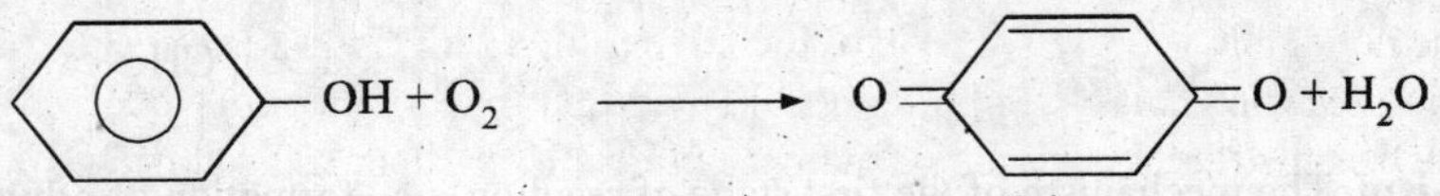

$$\text{OH} + O_2 \longrightarrow O= \;\; =O + H_2O$$

Phenol Quinone

Then quinone combines with phenol through hydrogen bonding to form a brilliant red addition product known as **phenoquinone.**

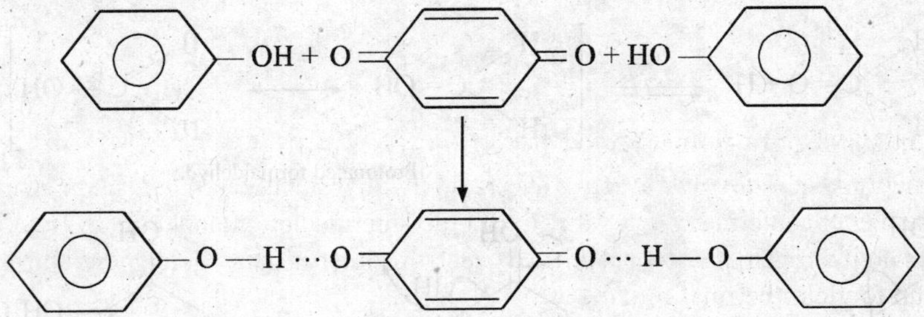

$$\text{OH} + O= \;\; =O + \text{HO}$$

$$\downarrow$$

$$\text{O—H} \cdots O= \;\; =O \cdots \text{H—O}$$

Phenoquinone
(Brilliant red compound)

6. Claisen rearrangement. This rearrangement reaction involves aryl allyl ethers. *The allyl group of the ether migrates from ether oxygen to the ring carbon at* **ortho position** at 475 K. However if both the ortho positions are already occupied, the allyl group migrates to *para* position.

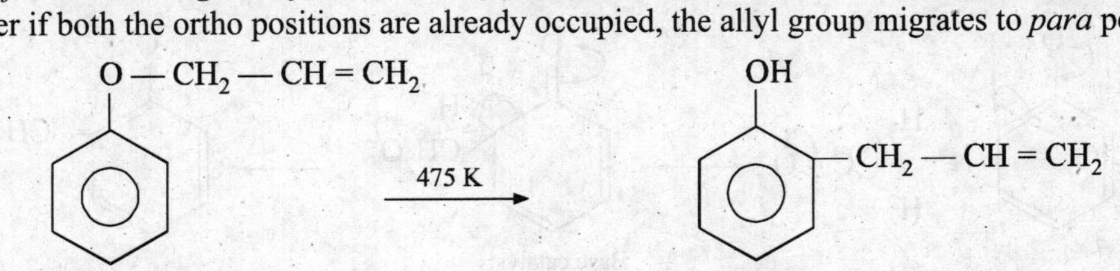

$$O—CH_2—CH=CH_2 \xrightarrow{475\text{ K}} \text{OH}, \; CH_2—CH=CH_2$$

Allyl phenyl ether *o*-Allyl phenol

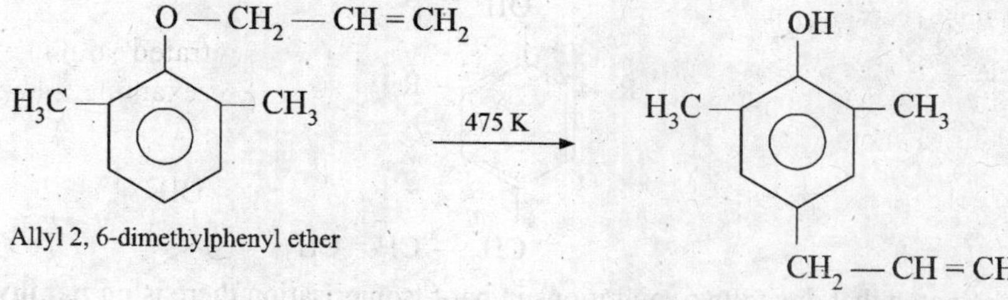

$O-CH_2-CH=CH_2$, H_3C, CH_3 → (475 K) → OH, H_3C, CH_3, $CH_2-CH=CH_2$

Allyl 2, 6-dimethylphenyl ether

4-Allyl-2, 6-dimethylphenol

Mechanism of ortho isomerisation

Ortho isomerisation involves a concerted mechanism in which the cleavage of allyl-oxygen bond and formation of allyl-carbon bond at the *ortho* position take place simultaneously. The cyclohexadienone intermediate thus formed undergoes a tautomeric change to give the o-allylphenol in which the ring regains the aromatic character as shown below:

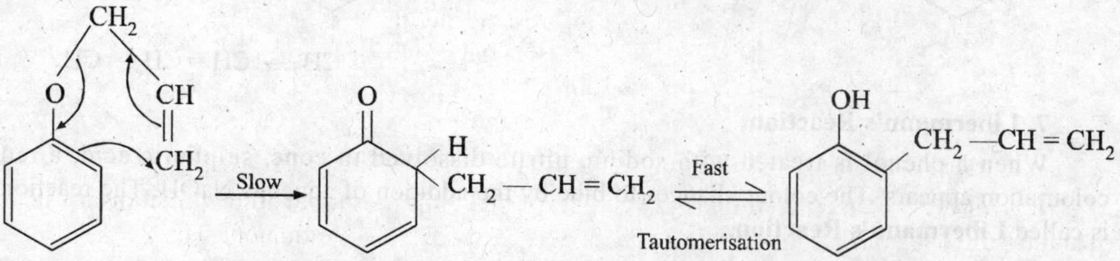

CH_2, O, CH, CH_2 —Slow→ O, H, $CH_2-CH=CH_2$ ⇌ (Fast, Tautomerisation) OH, $CH_2-CH=CH_2$

It may be noted that it is the γ-carbon of the allyl group (with respect to oxygen) which attaches itself to the ring carbon. This is supported by the fact that *ortho migration* involves an **inversion** in the position of substituents in the allyl group with respect to that of the starting compound. The following reaction illustrates this.

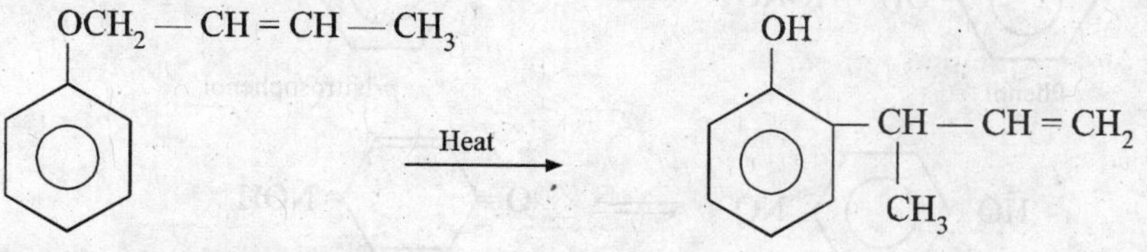

$OCH_2-CH=CH-CH_3$ —Heat→ OH, $CH-CH=CH_2$, CH_3

Mechanism of para isomerisation

The *para isomerisation* involves two stages. In the first stage, *ortho migration* takes place and the cyclohexadienone intermediate is formed. But this intermediate cannot undergo tautomerisation since there is no hydrogen at *ortho* position. As such this intermediate undergoes another migration of allyl group to yield the final product.

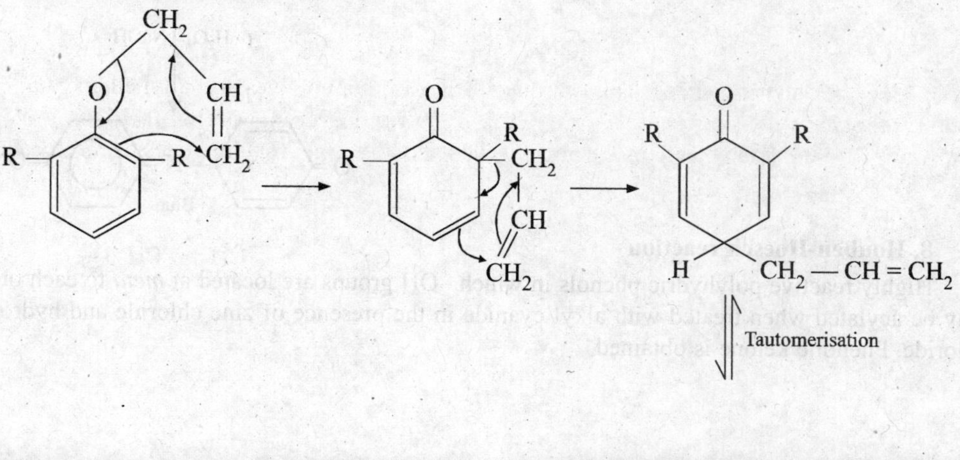

CH_2, O, CH, R, R, CH_2 → O, R, R, CH_2, CH, CH_2 → O, R, R, H, $CH_2-CH=CH_2$

⇓ Tautomerisation

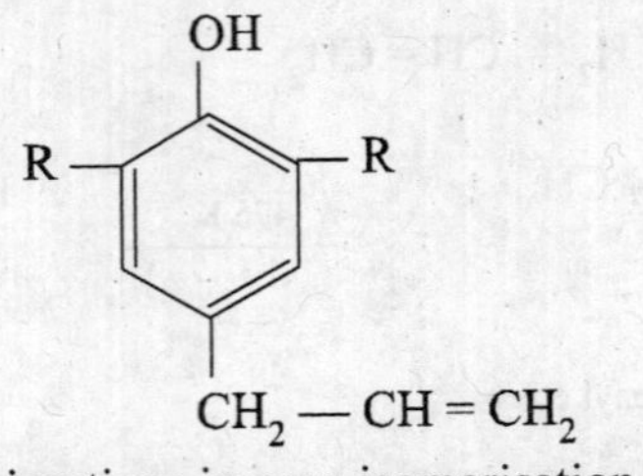

OH

R — — R

$CH_2 - CH = CH_2$

It may be noted that due to two migrations in para-isomerisation there is **no net inversion** in the position of the substituents with respect to the starting compound. For example :

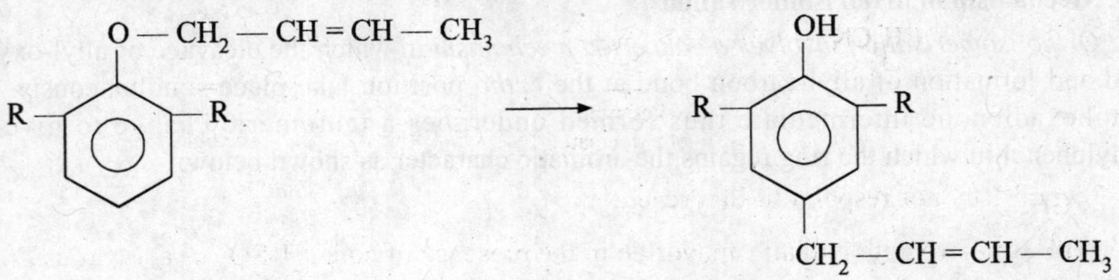

$O - CH_2 - CH = CH - CH_3$ (R, R) ⟶ OH (R, R) $CH_2 - CH = CH - CH_3$

7. Libermann's Reaction

When a phenol is treated with sodium nitrite dissolved in conc. sulphuric acid, a red colouration appears. The colour changes to blue by the addition of aqueous NaOH. The reaction is called **Libermann's Reaction.**

$$2NaNO_2 + H_2SO_4 \longrightarrow Na_2SO_4 + 2HNO_2$$

Nitrous acid

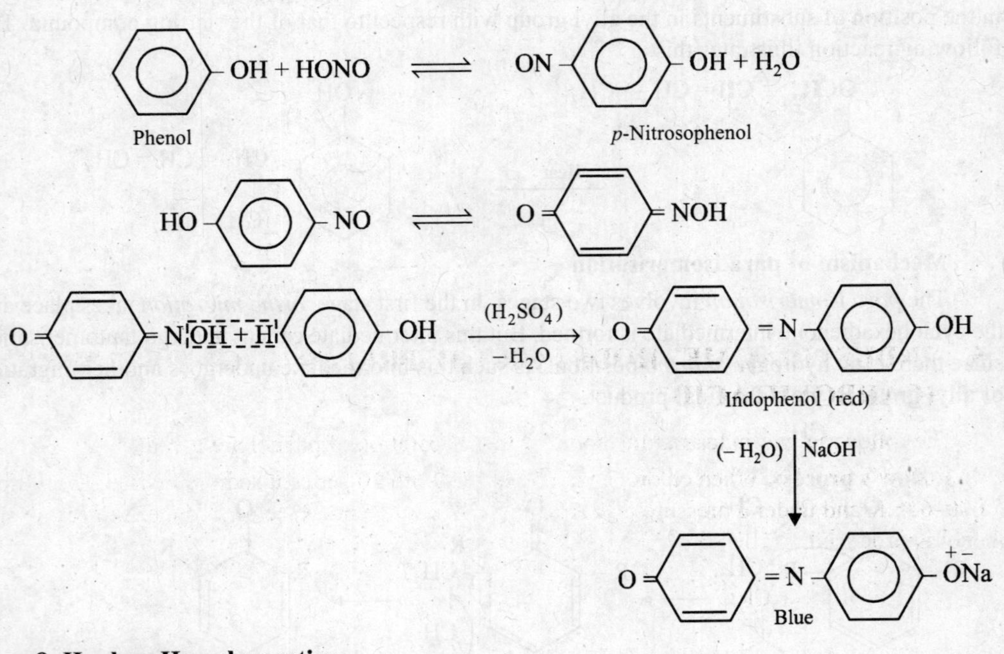

$C_6H_5-OH + HONO \rightleftharpoons ON-C_6H_4-OH + H_2O$

Phenol p-Nitrosophenol

$HO-C_6H_4-NO \rightleftharpoons O=C_6H_4=NOH$

$O=C_6H_4=N\,\boxed{OH + H}\,C_6H_4-OH \xrightarrow[-H_2O]{(H_2SO_4)} O=C_6H_4=N-C_6H_4-OH$

Indophenol (red)

$(-H_2O)$ ↓ NaOH

$O=C_6H_4=N-C_6H_4-\overset{+\,-}{ONa}$

Blue

8. Houben-Hoesch reaction

Highly reactive polyhydric phenols in which –OH groups are located at *meta* to each other, may be acylated when treated with alkyl cyanide in the presence of zinc chloride and hydrogen chloride. Phenolic ketone is obtained.

$$\text{Resorcinol} + RCN + H_2O \xrightarrow[\text{Ether}]{ZnCl_2/HCl} \text{Phenolic ketone} + NH_3$$

For example

$$\text{Resorcinol} + CH_3CN \longrightarrow \text{ketimine (intermediate)} \xrightarrow{H_2O} \text{Resorinol acetophenone} + NH_3$$

Phenol does not respond to this reaction.

(*vi*) Resorcinol with phthalic anhydride in the presence of conc. H_2SO_4 or anhydrous $ZnCl_2$ to form fluorescein.

$$2\,\text{Resorcinol} + \text{Phthalic anhydride} \xrightarrow{\text{Conc. } H_2SO_4} \text{Fluorescein}$$

3.5. INDUSTRIAL METHODS FOR THE PREPARATION OF PHENOL (CARBOLIC ACID)

The following two industrial methods for the preparation of phenol are given.

1. Dow's process. When chlorobenzene is heated with 10 per cent sodium hydroxide solution at 600–625 K and under a pressure of 200 atmospheres in the presence of a copper salt (catalyst), phenol is produced.

$$\text{Chlorobenzene} + NaOH \xrightarrow[\text{200 atm.}]{600-625\text{ K}} \text{Phenol} + NaCl$$

2. Raschig's process. A recent method consists in preparing chlorobenzene by passing a mixture of benzene vapours, hydrogen chloride and air over copper chloride (catalyst) heated to 500K.

$$2\,C_6H_6 + 2HCl + O_2 \xrightarrow[CuCl_2]{500\ K} 2\,C_6H_5Cl + 2H_2O$$

Benzene; O_2 (From air); Chlorobenzene

Chlorobenzene formed is hydrolysed with steam at 700 K in the presence of silica catalyst.

$$C_6H_5Cl + H.OH \xrightarrow[SiO_2]{700\ K} C_6H_5OH + HCl$$

Chlorobenzene; Phenol

3.6. DISTINCTION BETWEEN ALCOHOLS AND PHENOLS

Distinguishing features of alcohols and phenols are given in the tabular form as under.

Distinction between Alcohols and Phenols

Alcohols	Phenols
1. Alcohols have no effect on litmus paper.	Phenols turn blue litmus red.
2. Alcohols do not give any colour with ferric chloride solution.	An aqueous or alcoholic solution of phenol gives a characteristic violet, red or green colour with ferric chloride solution.
3. Alcohols show no reaction with NaOH solution.	Phenols are neutralised by NaOH solution to form salt and water.
4. Alcohols do not form any precipitate with bromine water.	Phenols form a white precipitate with bromine water.
5. Alcohols do not form any dye with benzene diazonium chloride.	Phenols form coloured dyes with icc cold solution of benzene diazonium chloride.
6. Alcohols do not respond to Libermann's test.	Phenols respond to Libermann's test and give characteristic coloration.

SOLVED CONCEPTUAL PROBLEMS

Example 1. *o*-nitrophenol is more acidic than *o*-aminophenol. Explain.

Solution. OH, NO_2 (o-nitrophenol) OH, NH_2 (o-aminophenol)

Nitro group is an electron withdrawing group. It helps in delocalising the negative charge on phenoxide ion. The phenoxide ion is stabilized and can exist independently without combining back with H^+ to give undissociated acid. Thus nitrophenol is a stronger acid.

Amino group is electron donating group. It intensifies the negative charge on the phenoxide ion. The phenoxide ion is thus destabilized. It combines back with H^+ to give undissociated acid. Thus aminophenol is a weaker acid.

Example 2. Give one test to distinguish between phenol and benzyl alcohol.

Solution. Benzyl alcohol reacts with PCl_5 to form benzyl chloride. It is easily hydrolysed with aqueous KOH to form water soluble KCl which gives white precipitate of AgCl on treating with an aqueous solution of silver nitrate thus:

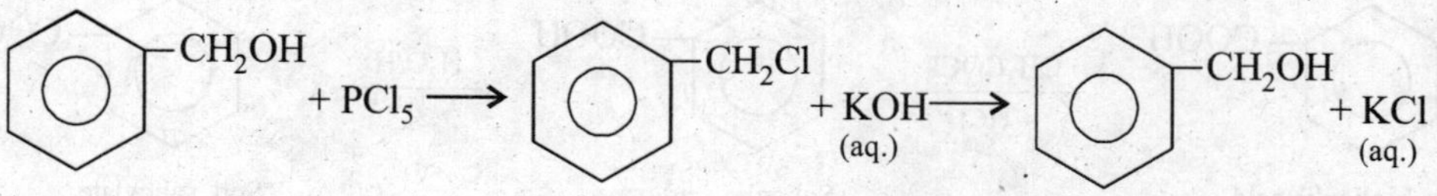

$$C_6H_5CH_2OH + PCl_5 \longrightarrow C_6H_5CH_2Cl \underset{(aq.)}{+ KOH} \longrightarrow C_6H_5CH_2OH + \underset{(aq.)}{KCl}$$

$$\underset{(aq.)}{KCl} + \underset{(aq.)}{AgNO_3} \longrightarrow \underset{(aq.)}{KNO_3} + \underset{(White\ ppt)}{AgCl}$$

Phenol does not give the above test

Example 3. Identify A, B and C in the following reaction.

$$\text{Phenol} \xrightarrow{AC_2O} A \xrightarrow[AlCl_3]{anhy} B + C$$

Solution.

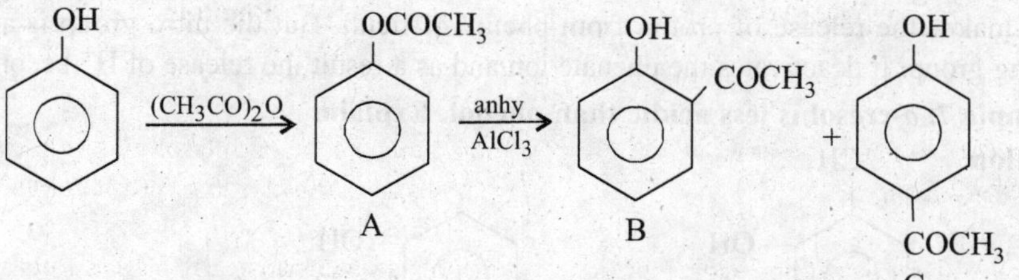

$$C_6H_5OH \xrightarrow{(CH_3CO)_2O} \underset{A}{C_6H_5OCOCH_3} \xrightarrow[AlCl_3]{anhy} \underset{B}{o\text{-}HOC_6H_4COCH_3} + \underset{C}{p\text{-}HOC_6H_4COCH_3}$$

Example 4. *p*-nitrophenol is more acidic than phenol. Explain

Solution. This is because phenoxide ion from *p*-nitrophenol is able to stabilize itself more with the help of delocalisation of negative charge.

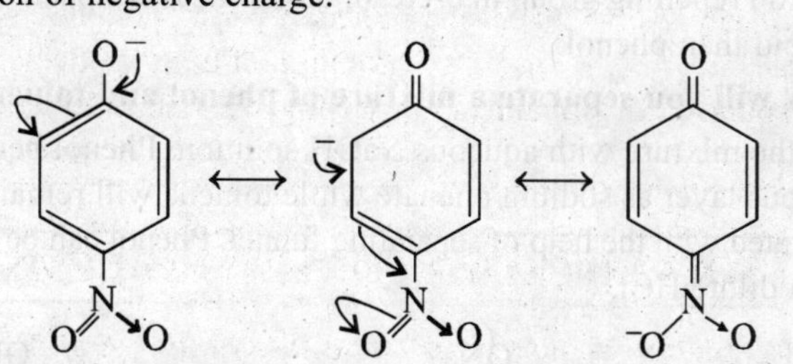

$$p\text{-}O^-C_6H_4NO_2 \longleftrightarrow \longleftrightarrow O{=}C_6H_4{=}N^+(O^-)O^-$$

Example 5. An organic compound '*A*' having molecular formula C_6H_6O gives a characteristic colour with aqueous $FeCl_3$ solution. When '*A*' is treated with CO_2 and NaOH at 400 K under pressure '*B*' is obtained. The compound '*B*' upon acidification gives compound '*C*' which reacts with acetyl chloride to form '*D*' which is a popular pain killer. Deduce the structures of *A*, *B*, *C*, *D*.

Solution. The data suggests that the compound A is phenol because it gives characteristic colour with $FeCl_3$ solution. The reactions may be written as under:

Phenol (A) [OH] $\xrightarrow[-H_2O]{NaOH}$ Sod. phenate [ONa] $\xrightarrow[\text{(Pressure)}]{CO_2/400\ K}$ Phenyl sod. carbonate [OCOONa] (Rearrangement) ⟶ Sod. salicylate (B) [OH, COONa] $\xrightarrow{H_2O/H^+}$ Salicylic acid (C) [OH, COOH] $\xrightarrow[-HCl]{CH_3COCl}$ Acetyl salicylic acid or Aspirin (D) [$OCOCH_3$, COOH]

Aspirin is a popular pain killer.

Example 6. Arrange the following in order of increasing acidic strength and assign suitable reason.

(*i*) Phenol (*ii*) *p*-nitrophenol (*iii*) *p*-cresol

Solution. The increasing order of acidic strength is : *p*-cresol < phenol < *p*-nitrophenol.

The methyl group in *p*-cresol has + I effect and increases the electron density on the phenate ion which makes the release of proton from phenol difficult. But the nitro group is an electron withdrawing group. It deactivates the phenate ion and as a result the release of H^+ becomes easier.

Example 7. *o*-cresol is less acidic than phenol. Explain.

Solution. *o*-Cresol [CH_3, OH] Phenol [OH]

Presence of electron repelling group in *o*-cresol destabilizes the phenoxide ion. Consequently *o*-cresol is a weaker acid than phenol.

Example 8. How will you separate a mixture of phenol and toluene ?

Solution. Shake the mixture with aqueous NaOH solution. Phenol being acidic will dissolve and pass into the aqueous layer as sodium phenate while toluene will remain in the oily layer. The two layers can be separated with the help of separating funnel. Phenol can be regenerated by treating the aqueous layer with dilute HCl.

Phenol (In the mixture) [OH] $\xrightarrow{NaOH}$ Sod. phenate (Aqueous layer) [ONa] $\xrightarrow{HCl}$ Phenol (Phenol regenerated in pure form) [OH] + NaCl

EXERCISES
(Based on Question Papers of Different Universities)

Multiple Choice Questions (Choose the correct option)

1. When phenol is treated with neutral $FeCl_3$ solution, it develops
(*a*) Violet colour (*b*) Yellow colour
(*c*) Green colour (*d*) Nothing happens

2. Phenol reacts with excess bromine water to give
(*a*) *o*-plus *p*-Bromophenol (*b*) Bromobenzene
(*c*) 2, 4, 6-Tribromophenol (*d*) *m*-Bromophenol

3. Phenol is used
(*a*) in alcoholic beverages (*b*) as anesthetic
(*c*) in antiseptics (*d*) as moth repellant

4. Sodium phenoxide reacts with CO_2 at 125°C under 5 atm pressure to give salicylic acid. This reaction is called
(*a*) Kolbe's reaction (*b*) Perkin reaction
(*c*) Wurtz reaction (*d*) HVZ reaction

5. Which of the following compounds is aspirin?
(*a*) Methyl salicylate (*b*) Salicylic acid
(*c*) Phenyl salicylate (*d*) Acetylsalicylic acid

6. Which of the following compounds will react with PCl_5 least vigorously ?
(*a*) Ethanol (*b*) Acetic acid
(*c*) Phenol (*d*) Benzoic acid

7. The compound which is most capable of hydrogen bonding is :
(*a*) $CH_3OCH_2CH_3$ (*b*) $CH_3CH_2CH_2CH_3$
(*c*) C_6H_5—OH (*d*) CH_3CH_2—S—S—CH_2CH_3

8. Resorcinol on distillation with zinc dust gives
(*a*) Benzene (*b*) Cyclohexane
(*c*) Toluene (*d*) *m*-Xylene

9. Phenol is acidic because of
(*a*) Resonance (*b*) Electromeric effect
(*c*) Inductive effect (*d*) Peroxide effect

10. Rank the following molecules in decreasing order of boiling points :

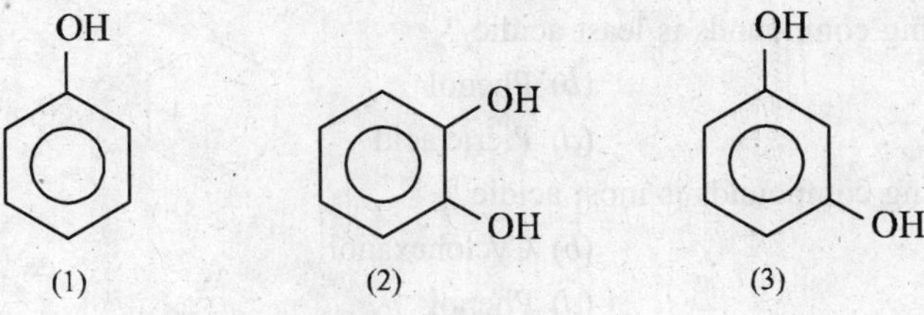

(*a*) (3) > (2) > (1)
(*b*) (2) > (3) > (1)
(*c*) (3) > (1) > (2)
(*d*) (2) > (1) > (3)

[**Hint.** (*a*) Hycroxyl groups can more easily form intramolecular H-bonds in the *ortho* position than the *meta* or *para* positions. Intramolecular H-bond decreases the boiling point. Therefore, compound (3) has a higher boiling point than compound (2). Compound (2), with two hydroxyl groups, has a still higher boiling point than phenol.

11. Bakelite has the following characteristics:
(*a*) A polymer made from phenol and formaldehyde
(*b*) A thermosetting plastic
(*c*) Can be used as an adhesive
(*d*) All of the above

12. Anisole on heating with concentrated HI gives
(*a*) Iodobenzene
(*b*) Phenol + CH_3I
(*c*) Iodobenzene + CH_3OH
(*d*) Phenol + CH_3OH

13. Rank the following compounds in order of increasing boiling point :

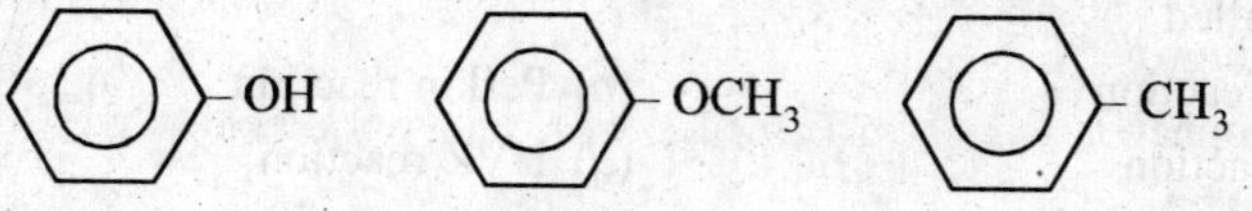

C_6H_5–OH (1) C_6H_5–OCH_3 (2) C_6H_5–CH_3 (3)

(*a*) (1) < (2) < (3)
(*b*) (3) < (1) < (2)
(*c*) (1) < (3) < (2)
(*d*) (3) < (2) < (1)

14. Which group forms the strongest H-bonds to water molecules ?
(*a*) Alcohols
(*b*) Ethers
(*c*) Phenols
(*d*) All equally strong

15. Anisole is formed when phenol is treated with
(*a*) $CH_3I/NaOH$
(*b*) $CH_3CH_2I/NaOH$
(*c*) $CHCl_3/NaOH$
(*d*) Acetic anhydride

16. Phenol is
(*a*) stronger acid than acetic acid
(*b*) weaker acid than acetic acid
(*c*) stronger base than methylamine
(*d*) weaker base than methylamine

17. Which of the following compounds is most acidic ?
(*a*) Formic acid
(*b*) Ethanol
(*c*) Acetic acid
(*d*) Phenol

18. Which of the following compounds is most acidic ?
(*a*) *o*-Cresol
(*b*) *p*-Nitrophenol
(*c*) *p*-Cresol
(*d*) *p*-Chlorophenol

19. Which of the following compounds is least acidic ?
(*a*) HCl
(*b*) Phenol
(*c*) Acetylene
(*d*) Picric acid

20. Which of the following compounds is most acidic ?
(*a*) Water
(*b*) Cyclohexanol
(*c*) Ethanol
(*d*) Phenol

ANSWERS

1. (*a*)	**2.** (*c*)	**3.** (*c*)	**4.** (*a*)	**5.** (*d*)	**6.** (*c*)
7. (*c*)	**8.** (*a*)	**9.** (*a*)	**10.** (*a*)	**11.** (*d*)	**12.** (*b*)
13. (*d*)	**14.** (*c*)	**15.** (*a*)	**16.** (*b*)	**17.** (*a*)	**18.** (*b*)
19. (*c*)	**20.** (*d*)				

Short Answer Questions

1. In Claisen rearrangement, the ortho rearrangement is accompanied by inversion. Explain.

2. Give the mechanism of coupling by phenol with diazonium salt.

3. Why is *p*-nitrophenol a stronger acid than the *m*-isomer ?

4. Why are phenols more acidic than alcohols ?

5. How will you show that *o*-cresol is less acidic than phenol ?

6. Why is *o*-nitrophenol steam volatile while *p*-nitrophenol is not ?

7. Give the mechanism of Reimer Tiemann reaction.

8. How will you prepare phenolphthalein from phenol? Why does it acquire a pink colour in the basic medium ?

9. 2, 4, 6-trinitrophenol gives effervescence with sodium carbonate while phenol does not. Explain.

10. Complete the following reaction :

$$\text{Phthalic anhydride} + \text{Phenol} \xrightarrow[H_2SO_4]{\text{Conc.}}$$

11. Discuss the mechanism of Fries rearrangement.

12. Describe the formation of bakelite from phenol. Give the mechanism of the reaction.

13. Why is *p*-nitrophenol a stronger acid than *m*-nitrophenol ?

14. Why is Anhydrous $AlCl_3$ not suitable for carrying out the alkylation of phenol ?

15. Give two tests to distinguish phenol from ethyl alcohol.

16. Give two reactions of phenol which support its acidic character.

17. Why is *o*-cresol less acidic than phenol ?

18. When phenyl acetate is heated with anhydrous aluminium chloride, name the rearrangement involved.

19. Why does the introduction of NO_2 group in the ring increase acidic character of phenol ?

20. Picric acid has no carboxyl group, but it is still a very strong acid. Why ?

General Questions

1. What happens when :

(*i*) Phenol is heated with zinc dust.

(*ii*) CO_2 gas is passed through sodium phenate under pressure and the product is hydrolysed.

(*iii*) Allyl aryl ether is heated to 473 K.

2. What happens when :

(*a*) Phenol is treated with bromine dissolved in CCl_4.

(*b*) Phenol is heated with phthalic anhydride in the presence of conc. H_2SO_4.

(*c*) Phenyl acetate is heated in the presence of anhydrous $AlCl_3$.

3. Give the order of the decreasing acidic strength of the following compounds. Give reason:

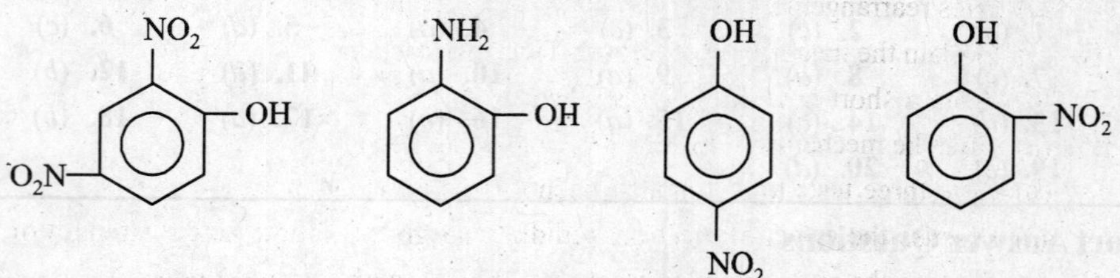

4. Two isomeric compounds (*A*) and (*B*) have molecular formula C_7H_8O. Compound (*A*) readily dissolves in aqueous NaOH, but compound (*B*) does not. Compound (*A*) is insoluble in aqueous $NaHCO_3$ and reacts with bromine water to form (C), $C_7H_5OBr_3$. Identify (*A*), (*B*) and (*C*).

Answer. (A) = *m*-Cresol
(B) = Anisole
(C) = 2, 4, 6-Tribromo-3-methylphenol

5. (*a*) Describe the mechanism of Fries rearrangement.
(*b*) Write a short note on Kolbe's reaction.
(*c*) *o*-cresol is less acidic than phenol.

6. How will you account for the following?
(*i*) Boiling point of ethyl alcohol is more than that of phenol.
(*ii*) Phenol is a stronger acid than ethyl alcohol.
(*iii*) Compared to toluene, phenol is more soluble in water.

7. How will you convert phenol into :
(*i*) Bakelite (*ii*) Phenolphthalein (*iii*) Picric acid

8. Discuss the effect of substituents on the acidic character of phenol.

9. Compound (A), C_6H_6O, is soluble in NaOH. When treated with $CHCl_3$ and NaOH, it forms (B) (Reimer-Tiemann reaction). Compound (B), an oxidation, gives (C) which reacts with acetic anhydride in the presence of a small amount of H_2SO_4 to form (D), $C_9H_8O_4$. Deduce the structural formulas of (A), (B), (C) and (D). Write equations for the reactions involved.

Answer. (A) = Phenol (B) = Salicylaldehyde
(C) = Salicylic acid (D) = Acetylsalicylic acid (Aspirin)

10. How will you prepare the following from phenol ?
(*i*) Salicylic acid (*ii*) Benzene
(*iii*) Picric acid (*iv*) *p*-Hydroxy benzoic acid

11. How will you convert phenol into :
(*i*) Aspirin (*ii*) Salol (*iii*) Benzene

12. Write notes on :
(*i*) Fries rearrangement (*ii*) Reimer-Tiemann reaction.

13. Give the mechanism of :
(*i*) Fries rearrangement (*ii*) Kolbe's schmidt reaction.

14. How will you account for the fact that the bromination of phenol in aqueous medium gives 2, 4, 6-bromo phenol while in CS_2 solvent the product is a mixture of *o*-bromo and *p*-bromophenol ?

15. Write short notes on the following :

(*i*) Fries rearrangement (*ii*) Kolbe's reaction.

16. (*a*) Explain the mechanism of Reimer-Tiemann reaction.

(*b*) Write a short note on Kolbe's reaction.

17. (*a*) Give the mechanism of Reimer-Tiemann reaction

(*b*) Give three tests to distinguish phenol from ethyl alcohol.

(*c*) Discuss the effect of electron withdrawing groups on the acidic strength of phenols.

18. (*a*) Discuss the mechanism of reaction of phenol with formaldehyde.

(*b*) Benzyl alcohol fails to give the characteristic reactions of the phenolic group. Why ?

19. What is coupling reaction ? Give the mechanism of coupling of phenol by diazonium salt.

20. Which is more reactive towards electrophilic substitution–benzene or phenol? Give a suitable explanation.

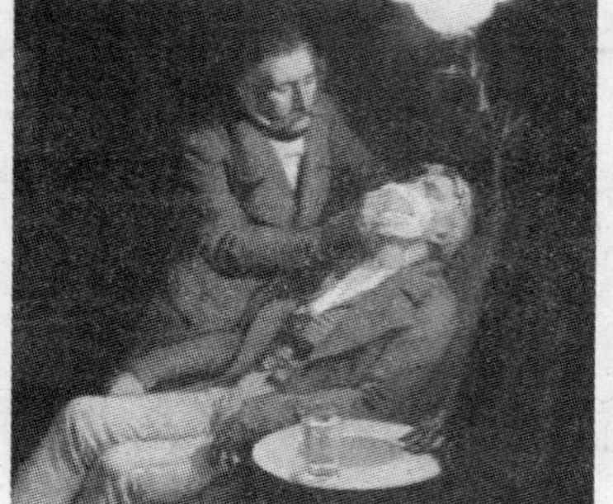

Ethers were among the early anaesthetics

4 Ethers and Epoxides

4.1. CLASSIFICATION

Ethers *may be regarded as derivatives of water in which both the hydrogen atoms have been replaced by alkyl (or aryl) groups*. Ethers have the general formula R–O–R′ where R and R′ may be similar or dissimilar alkyl or aryl radicals. For example,

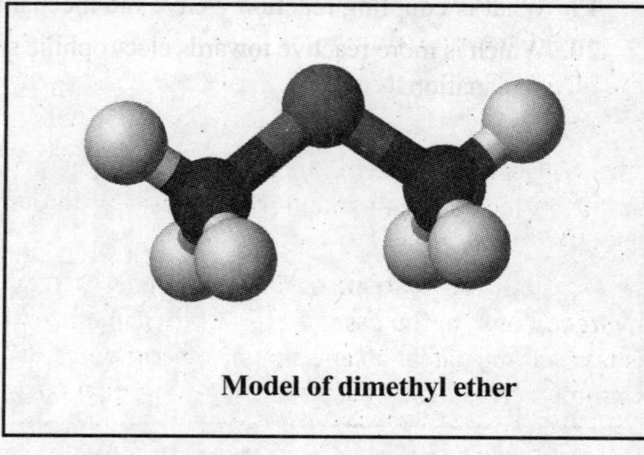

Model of dimethyl ether

$CH_3 — O — CH_3$ Dimethyl ether

$C_6H_5 — O — C_2H_5$ Ethyl phenyl ether

Ethers may be classified into two types :

(*a*) Simple (or Symmetrical) Ethers *are those ethers in which groups R and R′ are the same.* For example,

$CH_3 — O — CH_3$ Dimethyl ether

$C_2H_5 — O — C_2H_5$ Diethyl ether

(*b*) Mixed (or Unsymmetrical) Ethers *are those ethers in which groups R and R′* are different. For example,

$CH_3 – O – C_2H_5$ Ethyl methyl ether

$CH_3 – O – C_6H_5$ Methyl phenyl ether

4.2. STRUCTURE OF DIETHYL ETHER

Ethers are expected to have shape and structure similar to that of water as shown below :

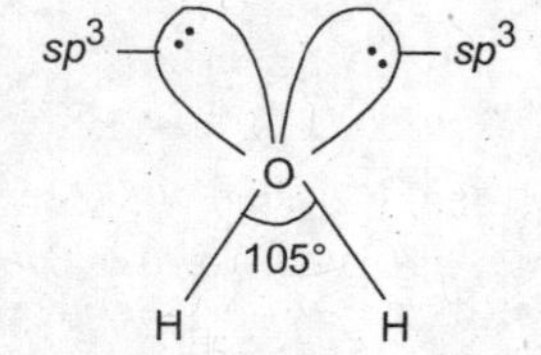

Orbital structure of water molecule

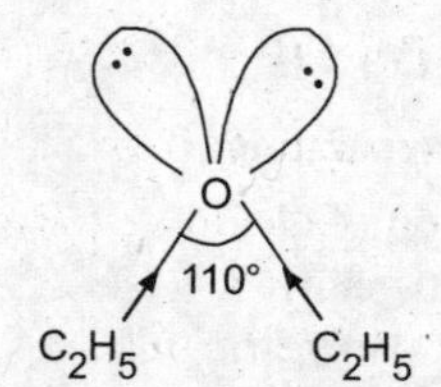

Orbital structure of diethyl ether

The $C_2H_5 - O - C_2H_5$ bond angle has been found to be 110° which is close to the tetrahedral angle of 109°, 28′. However, the $C_2H_5 - O - C_2H_5$ bond angle of 110° is bigger than $H - O - H$ bond angle in water of 105°. This is due to the fact that the steric hindrance between two bulkier alkyl groups tends to increase the bond angle while the lone pair-lone pair repulsions tend to decrease the bond angle. The net effect is that the tetrahedral bond angle is almost preserved.

Diethyl ether has a dipole moment of 1.18 D. The dipole moment value of diethyl ether reveals that ether is not a linear molecule otherwise there would have been no resultant dipole moment. Hence the ether molecule is planar and angular.

4.3. NOMENCLATURE OF ETHERS

There are two systems of naming ethers.

(*i*) **Common system**

(*ii*) **IUPAC system**

These are described separately as under :

(*i*) **Common system.** According to this system, the alkyl or aryl groups on either side of oxygen are identified. The compound is named by writing these groups in alphabetical order followed by the word 'ether'.

(*ii*) **IUPAC system.** According to this system, ethers are considered as alkoxy derivatives of hydrocarbons. In the case of unsymmetrical ethers, the group having greater no. of carbon atoms is taken as the parent alkane and the other group along with the oxygen is taken as the alkoxy group. Common and IUPAC names of some common compounds are tabulated below.

Compound	*Common name*	*IUPAC name*
CH_3OCH_3	Dimethyl ether	Methoxymethane
$C_2H_5OC_2H_5$	Diethyl ether	Ethoxyethane
$CH_3OC_2H_5$	Ethyl methyl ether	Methoxyethane
$C_6H_5-O-C_6H_5$	Diphenyl ether	Phenoxybenzene
$C_6H_5-OCH_3$	Methyl phenyl ether	Methoxybenzene
$C_6H_5-OC_2H_5$	Ethyl phenyl ether	Ethoxybenzene

Sometimes special names are also used for some of the ethers. For example,

$C_6H_5-OCH_3$ is popularly known as anisole while $C_6H_5-OC_2H_5$ is called phenetole.

4.4. METHODS OF PREPARATION

Methods of preparation of alcohols are discussed as under :

1. Dehydration of alcohols. When an excess of alcohol is heated with concentrated sulphuric acid (or glacial phosphoric acid) at suitable temperature, two moles of the alcohol lose one mole of water to form a symmetrical ether. Thus, we have :

$$2\,ROH \xrightarrow[413\,K]{H_2SO_4} R - O - R + H_2O$$

For example,

$$\underset{\text{Ethyl alcohol}}{2C_2H_5OH} \xrightarrow{H_2SO_4,\ 413\ K} \underset{\text{Diethyl ether}}{C_2H_5-O-C_2H_5} + H_2O$$

The formation of alkenes, which is a competing reaction, is suppressed by the use of excess of alcohol and regulation of temperature.

Mechanism. The mechanism of formation of ethers involves nucleophilic substitution with the protonated alcohol as the substrate and a second molecule of alcohol as nucleophile. The reaction can follow either first-order kinetics (SN^1) or second-order kinetics (SN^2) depending upon the nature of alcohol used. In general, secondary and tertiary alcohols react by SN^1 mechanism. In both the mechanisms, the first step involves the formation of protonated alcohol. This is followed by other steps which differ in the two mechanisms as illustrated below :

SN^1 mechanism :

(*i*) $$R:\ddot{O}:H + \underset{\text{(From acid)}}{H^+} \rightleftharpoons \underset{\text{Protonated alcohol}}{R:\overset{H}{\overset{|}{\ddot{O}^+}}:H}$$

(*ii*) $$R:\overset{H}{\overset{|}{\ddot{O}^+}}:H \xrightarrow{\text{Slow}} \underset{\text{Carbonium ion}}{R^+} + H_2O$$

(*iii*) $$R:\overset{H}{\overset{|}{\ddot{O}}} + R^+ \overset{\text{Fast}}{\rightleftharpoons} \underset{\text{Protonated ether}}{R:\overset{H}{\overset{|}{\ddot{O}^+}}:R}$$

(*iv*) $$R:\overset{H}{\overset{|}{\ddot{O}}}:R^+ \overset{\text{Fast}}{\rightleftharpoons} \underset{\text{Ether}}{R:\ddot{O}:R} + H^+$$

SN^2 mechanism

(*i*) $$R:\ddot{O}:H + \underset{\text{(From acid)}}{H^+} \rightleftharpoons \underset{\text{Protonated alcohol}}{R:\overset{H}{\overset{|}{\ddot{O}^+}}:H}$$

(*ii*) $$R-\overset{H}{\overset{|}{\ddot{O}}}: + R\overset{+}{O}H_2 \xrightarrow{\text{Slow}} \underset{\text{Transition state}}{\left[R-\overset{H}{\overset{|}{O}}\ldots\ldots R\ldots\ldots OH_2\right]^+} \xrightarrow{-H_2O} R-\overset{H}{\overset{|}{O^+}}-R$$

(*iii*) $$\underset{\text{Protonated ether}}{R-\overset{H}{\overset{|}{O^+}}-R} \rightleftharpoons \underset{\text{Ether}}{R-O-R} + H^+$$

2. By using diazomethane

Preparation of methyl ethers by the use of diazomethane. In the presence of fluoroboric acid, diazomethane reacts with alcohols or phenols to produce ethers in excellent yields. For example, *n*-hexyl alcohol, $C_6H_{13}OH$, is easily converted into hexyl methyl ether by this method.

Mechanism

(*a*) $$\underset{\text{Diazomethane}}{:\bar{C}H_2-\overset{+}{N}\equiv N} + \underset{\text{Fluoro boric acid}}{\overset{+}{H}BF_4^-} \longrightarrow \left[CH_3-\overset{+}{N}\equiv N:\bar{B}F_4\right] \longrightarrow \overset{+}{C}H_3 + N_2 + BF_4^-$$

$$(b)\quad CH_3CH_2-\ddot{O}H+\overset{\oplus}{C}H_3 \longrightarrow CH_3CH_2-\underset{+}{\overset{CH_3}{O}}-H \xrightarrow{-H^+} CH_3CH_2OCH_3$$

3. Williamson's synthesis

Alexander Williamson

This is the most important and widely used method for the preparation of both symmetrical and unsymmetrical ethers. It is carried out by treating an alkyl or substituted alkyl halide with sodium alkoxide or sodium phenoxide.

$$\underset{\text{Sod.alkoxide}}{R'ONa} + RX \longrightarrow R'-O-R+NaX$$

$$\underset{\text{Sod. phenoxide}}{C_6H_5-ONa} + RX \longrightarrow C_6H_5-O-R+NaX$$

For example,

$$\underset{\text{Sod.ethoxide}}{C_2H_5ONa} + \underset{\text{Ethyl iodide}}{C_2H_5I} \longrightarrow \underset{\text{Diethyl ether}}{C_2H_5-O-C_2H_5} + NaI$$

$$\underset{\text{Sod.ethoxide}}{C_2H_5ONa} + \underset{\text{Methyl bromide}}{CH_3Br} \longrightarrow \underset{\text{Ethyl methyl ether}}{C_2H_5-O-CH_3} + NaBr$$

$$\underset{\text{Sod. phenoxide}}{C_6H_5-ONa} + \underset{\text{Ethyl iodide}}{C_2H_5I} \longrightarrow \underset{\text{Phenetole}}{C_6H_5-OC_2H_5} + NaI$$

The reaction, of course, involves the **nucleophilic substitution** of alkoxide or phenoxide ion for the halide ion of alkyl halide.

$$\overset{\delta+}{R}-\overset{\delta-}{X} + \underset{\text{Alkoxide ion}}{\bar{O}R'} \longrightarrow [R'O\,....R\,.....X]^- \longrightarrow R'-O-R+X^-$$

For the preparation of methyl aryl ethers, methyl sulphate, $(CH_3)_2SO_4$ is generally used in place of methyl halides.

$$2\underset{\text{Sod. phenoxide}}{C_6H_5-ONa} + \underset{\text{Methyl sulphate}}{(CH_3)_2SO_4} \longrightarrow 2\underset{\text{Anisole}}{C_6H_5-OCH_3} + Na_2SO_4$$

Limitations in Williamson's synthesis

Limitations in Williamson's synthesis are explained as under :

(*i*) *Since a halogen attached to an aromatic ring cannot be readily replaced by a nucleophilic reagent, aryl halides are not generally used in the Williamson synthesis.* Therefore in the preparation of alkyl-aryl ethers, the reaction is carried out between an alkyl halide and sodium phenoxide and **not** between aryl halide and sodium alkoxide. For example, as shown above, phenetole is prepared by reaction between ethyl iodide and sodium phenoxide and not between phenyl iodide and sodium ethoxide.

Diaryl ethers are not generally prepared by this reaction.

(*ii*) As the Williamson synthesis involves the attack of strongly nucleophilic alkoxide or phenoxide ion at the alkyl halide, *elimination can also take place as a competing reaction to substitution.* Now, tertiary alkyl halides have a greater tendency to undergo elimination than secondary or primary alkyl halides. Therefore, in the preparation of unsymmetrical ethers where there is choice of alkoxides and alkyl halides, one should select the alkyl halide that has a greater tendency to take part in substitution and a smaller tendency to undergo elimination. Therefore, primary alkyl halides are used in preference to secondary alkyl halides which in turn are preferred to tertiary halides. On the other hand, tertiary alkoxides are preferred to secondary or primary alkoxides.

For example, in the preparation of ethyl *tert*-butyl ether, we employ ethyl halide and sodium tert-butoxide and not *tert*-butyl halide and sodium ethoxide.

$$\underset{\text{Sod.}tert\text{-butoxide}}{CH_3-\underset{CH_3}{\overset{CH_3}{|}}{C}-ONa} + C_2H_5I \longrightarrow \underset{\text{Ethyl } tert\text{-butyl ether}}{CH_3-\underset{CH_3}{\overset{CH_3}{|}}{C}-OC_2H_5} + NaI$$

Example 1. Select the preferred combination of halide and oxide in the preparation of following ethers by Williamson synthesis.

(*a*) Ethyl phenyl ether

(*b*) Isopropyl methyl ether

Solution. (*a*) Preparation of ethyl phenyl ether. There are two combinations of halide and oxide as given below :

(*i*) $\underset{\text{Chlorobenzene}}{C_6H_5-Cl} + \underset{\text{Sod. ethoxide}}{NaOC_2H_5}$

(*ii*) $\underset{\text{Sod. phenate}}{C_6H_5-ONa} + \underset{\text{Chloroethane}}{ClC_2H_5}$

Williamson synthesis involves nucleophilic reaction. In the combination (*i*), the nucleophile $C_2H_5O^-$ attacks the chlorobenzene molecule. But it is difficult to break the C–Cl bond in aromatic compounds. Hence, the reaction does not take place easily.

$$C_6H_5-Cl + \bar{O}C_2H_5 \longrightarrow \left[C_2H_5O \cdots C_6H_5-Cl\right]^-$$

C–Cl bond is difficult to break and hence no reaction.

In combination (*ii*), the nucleophile $C_6H_5O^-$ attacks the chloroethane molecule, knocking out Cl^- easily. The reactions can be depicted as under :

$$C_6H_5O^- + C_2H_5-Cl \longrightarrow [C_6H_5O \ldots C_2H_5 \ldots Cl]^-$$

$$\downarrow$$

$$\underset{\text{Ethyl phenyl ether}}{C_6H_5OC_2H_5} + Cl^-$$

(*b*) Isopropyl methyl ether. There are two combinations of halide and oxide.

(*i*) $(CH_3)_2CHONa$ + $BrCH_3$
Sod.isopropoxide Methyl bromide

(*ii*) $(CH_3)_2CHBr$ + $NaOCH_3$
Isopropyl bromide Sod. methoxide

Out of the two combinations, the first is preferred

$$(CH_3)_2\,CHONa + BrCH_3 \longrightarrow \underset{\text{Methyl isopropyl ether}}{(CH_3)_2\,CHO\,CH_3}$$

With the second combination, elimination instead of substitution will take place resulting in the formation of propene.

$$CH_3 - \underset{\substack{|\\CH_3}}{CH} - Br + NaOCH_3 \longrightarrow CH_3 - \underset{\substack{\|\\CH_2\\\text{Propene}}}{CH} + NaBr + CH_3OH$$

Example 2. Why cannot diphenyl ether be prepared by dehydration of phenol with conc. H_2SO_4?

Solution. Diphenyl ether cannot be prepared by dehydration of phenol with conc. H_2SO_4 due to the following reasons :

(*i*) Phenol is attacked by sulphuric acid to form hydroxy benzene sulphonic acids.

(*ii*) In case of phenols, initial protonation leading to oxonium ion does not take place readily because the oxygen atom in phenol has some positive charge (*i.e.*, less electron density) as the electrons of the oxygen atom tend to enter the benzene ring.

4.5. PHYSICAL PROPERTIES OF ETHERS

1. Physical State. The first two members viz. dimethyl ether and methyl ethyl ether are gases at room temperature while diethyl and higher homologue are liquids with a characteristic ethereal smell.

2. Boiling points. Ethers are low boiling substances because there are no appreciable molecular interactions like hydrogen bonding. Alcohols having comparable molecular masses as ethers have comparatively higher boiling points because of the presence of hydrogen bonding. Hydrogen bonding holds the molecules together and results in higher boiling point.

Example 1. Why do ethers have lower boiling points than isomeric alcohols?

Solution. Ethers have lower boiling points than isomeric alcohols. For example, ethyl alcohol boils at 351 K while diethyl ether boils at 308 K. This is because in ether the oxygen atom is not directly linked with hydrogen atom. Therefore, hydrogen bonding is not possible between the ether molecules. Hence, their boiling points are not high.

On the other hand, alcohols such as ethyl alcohol exhibit hydrogen bonding as they contain oxygen atom which is directly linked with hydrogen atom. As a result, the molecules of alcohol get associated due to hydrogen bonding. Due to strong intermolecular interaction between the molecules of alcohol, they have high boiling points.

Example 2. Which compound of the following pair would have the higher boiling point and why?

$$CH_3 - CH_2 - CH_2 - OH \quad \text{or} \quad CH_3 - CH_2 - O - CH_3$$

Solution. If the molecules of a compound get associated due to hydrogen bonding, it results in rise of boiling point. Greater the extent of hydrogen bonding, greater the boiling point of the liquid.

n-propyl alcohol has a higher boiling point than methyl ethyl ether. This is because of occurrence of hydrogen bonding in the former. Representing $CH_3-CH_2-CH_2-$ by R–, the hydrogen bonding can be depicted as under :

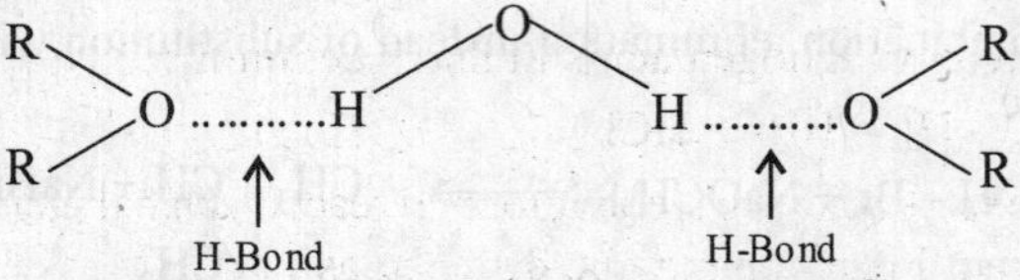

There is no possibility of hydrogen bond in methyl ethyl ether.

Example 3. Why do alcohols and isomeric ether have comparable water solubilities?

Solution. Like alcohols, ethers can also form hydrogen bonds with water. Consequently, ethers have some solubility in water.

Moreover, ethers are weakly polar molecules. This is also the reason for their solubility in water. However, the solubility of ethers in water decreases as the size of alkyl group increases. Ethers are more soluble in alcohol due to hydrogen bonding and similarity of structure.

4.6. CHEMICAL PROPERTIES OF ETHERS

Important chemical properties of ethers are described as under:

1. Formation of oxonium salts. Ethers dissolve in concentrated solutions of strong mineral acids to form oxonium salts, *i.e.*, ethers behave as weak bases. For example :

$$CH_3-\ddot{\underset{\cdot\cdot}{O}}-CH_3+HCl \longrightarrow \left[CH_3-\overset{H}{\underset{\cdot\cdot}{\overset{|+}{O}}}-CH_3\right]Cl^-$$

Dimethyl oxonium chloride

$$C_2H_5-\ddot{O}-C_2H_5+H_2SO_4 \longrightarrow \left[C_2H_5-\overset{H}{\overset{|+}{\ddot{O}}}-C_2H_5\right]HSO_4^-$$

Diethyl oxonium hydrogen sulphate

The salts are stable at low temperature in strongly acidic medium. However, they get decomposed by water to give back ether and the acid.

2. Formation of peroxides. On standing in air or oxygen, particularly in the presence of light, ethers form peroxides. The structure and composition of ether peroxides are not known with certainty. One of the proposed structures is as shown below.

$$CH_3-CH_2-O-CH_2-CH_3+O_2 \longrightarrow CH_3-CH_2-O-\underset{\underset{O-O-H}{|}}{CH}-CH_3$$

Diethyl ether peroxide

Another structure that has been proposed for diethyl ether peroxide is given as under :

$$C_2H_5-O-C_2H_5+O \xrightarrow[\text{air}]{\text{From}} C_2H_5-\overset{\overset{O}{\uparrow}}{O}-C_2H_5$$

Diethyl peroxide

The ether peroxides are unstable compounds and decompose explosively when heated. Their presence in ethers can be detected by the formation of a red colour on shaking with an aqueous solution of ferrous ammonium sulphate and potassium thiocyanate. Ethers must be made free of peroxides before being distilled.

3. Cleavage by acids. On heating with concentrated acids (particularly hydroiodic acid or hydrobromic acid), the ether linkage breaks to form an alcohol or phenol and an alkyl halide.

$$R - O - R' + HX \longrightarrow RX + R'OH$$

For example,

$$\underset{\text{Ethyl ether}}{C_2H_5 - O - C_2H_5} + HI \longrightarrow \underset{\text{Ethyl iodide}}{C_2H_5I} + \underset{\text{Ethyl alcohol}}{C_2H_5OH}$$

The order of reactivity of halogen acids in this reaction is :

$$HI > HBr > HCl$$

In the presence of excess acid, the alcohol formed reacts with halogen acid to form alkyl halide. Thus in that case the overall reaction may be depicted as :

$$R - O - R' + 2HX \longrightarrow RX + R'X + H_2O$$

The mechanism of the reaction is believed to be as follows :

(*a*) Initially the ether reacts with the acid to form a protonated ether as shown below.

$$R - O - R' + HX \longrightarrow \underset{\text{Protonated ether}}{\left[R - \overset{\overset{\large H}{|}}{\ddot{O}^+} - R' \right]} + X^-$$

(*b*) The protonated ether is then attacked by the nucleophilic halide ion by SN[1] or SN[2] mechanism depending upon the nature of R and R′ groups and the reaction conditions. Presence of primary alkyl groups favours SN[2] mechanism whereas secondary or tertiary alkyl groups tend to react by SN[1] mechanism. Thus, we have :

SN[1] mechanism

$$\left[R - \overset{\overset{\large H}{|}}{\underset{..}{O^+}} - R' \right] \xrightarrow{\text{Slow}} R^+ + R'OH$$

$$R^+ + X^- \longrightarrow RX$$

SN[2] mechanism

$$\left[R - \overset{\overset{\large H}{|}}{\underset{..}{O^+}} - R' \right] + X^- \longrightarrow \left[\overset{\delta^-}{X} \ldots R \ldots \overset{\overset{\large H}{|}\,\delta^+}{O} - R' \right] \longrightarrow RX + R'OH$$

Point of cleavage. In case of mixed ethers, the alkyl halide produced depends upon the nature of groups attached to oxygen.

(*i*) If one group is methyl and the other a *primary* or *sec*-alkyl group, it is methyl halide which is produced. For example,

$$CH_3 - O - C_2H_5 + HI \longrightarrow CH_3I + C_2H_5OH$$

It is assumed that the reaction proceeds by SN[2] mechanism in this case and the halide ion attaches itself to a smaller alkyl group on steric consideration.

(*ii*) When the ether is a methyl-*tert*-alkyl ether, we get a *tert*-alkyl halide and methyl alcohol. For example,

$$CH_3-\underset{CH_3}{\overset{CH_3}{\underset{|}{\overset{|}{C}}}}-O-CH_3 + HI \longrightarrow CH_3-\underset{CH_3}{\overset{CH_3}{\underset{|}{\overset{|}{C}}}}-I + CH_3OH$$

Methyl *tert*-butyl ether — *tert*-butyl iodide

This can be explained by SN^1 mechanism, the carbonium ion produced being *tert*-alkyl because it is more stable than primary carbonium ion.

Cleavage of alkyl-aryl and diaryl ethers. In case of an alkyl-aryl ether, the cleavage of *aromatic* carbon-oxygen bond is much more difficult than the cleavage of *alkyl* carbon-oxygen bond. Thus, the cleavage of such ethers by halogen acids yields a phenol and an alkyl halide. For example,

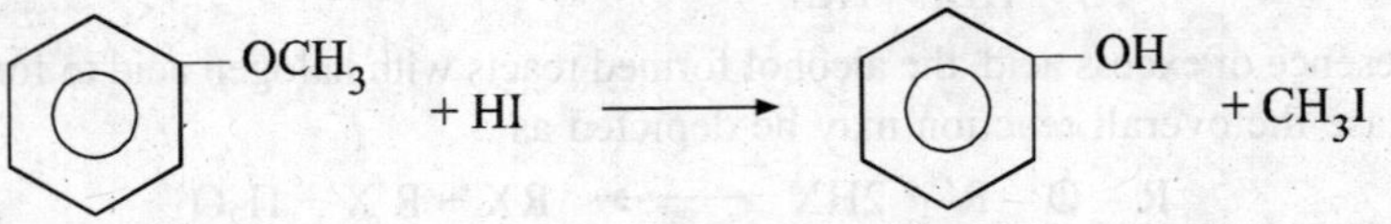

$$C_6H_5-OCH_3 + HI \longrightarrow C_6H_5-OH + CH_3I$$

Anisole — Phenol

Diaryl ethers are very difficult to cleave as both carbons attached to oxygen are part of aromatic rings. For example, diphenyl ether is not cleaved by HI even at 525 K.

4. Halogenation. On treatment with halogens in the presence of sunlight, ethers are readily halogenated to form polyhalogenated ethers. For example :

$$C_2H_5-O-C_2H_5 \xrightarrow{Cl_2/hv} C_2Cl_5-O-C_2Cl_5$$

Ethyl ether — Perchlorodiethyl ether

If the reaction is carried out in the absence of sunlight, halogenation takes place at the α-position only. For example :

$$CH_3-CH_2-O-CH_2-CH_3 \xrightarrow{Cl_2} CH_3-CHCl-O-CHCl-CH_3$$

α-α′-Dichlorodiethyl ether

5. Reaction with PCl_5

$$C_2H_5OC_2H_5 + PCl_5 \longrightarrow 2C_2H_5Cl + POCl_3$$

Diethyl ether — Ethyl chloride

6. Electrophilic substitution. Aromatic ethers such as anisole and phenetole undergo aromatic electrophilic substitution like nitration, halogenation, etc. It may be recalled that the alkoxy group is moderately activating and *ortho-para* directing.

For example,

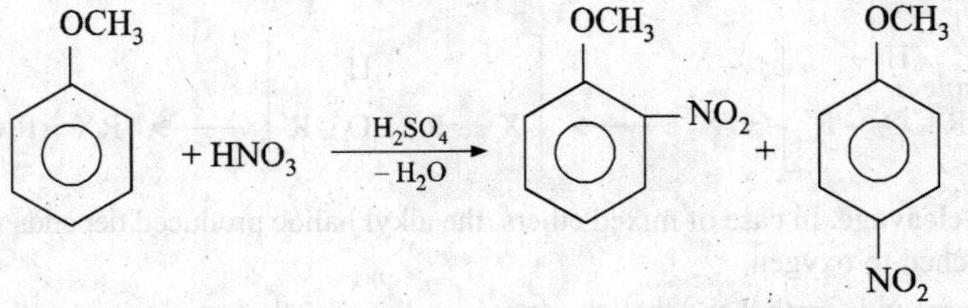

$$C_6H_5OCH_3 + HNO_3 \xrightarrow[-H_2O]{H_2SO_4} o\text{-}NO_2C_6H_4OCH_3 + p\text{-}NO_2C_6H_4OCH_3$$

o-and *p*-Nitroanisols

Alkoxy group has – I inductive effect and + M resonance effect. The latter effect being stronger, there is greater electron density on the ring. This leads to a faster electrophilic substitution.

The *ortho-para* directing effect may be explained in terms of the contributing structures of the carbonium ions formed by *ortho*, *para* and *meta* attacks as shown below, taking Y^+ as the electrophile.

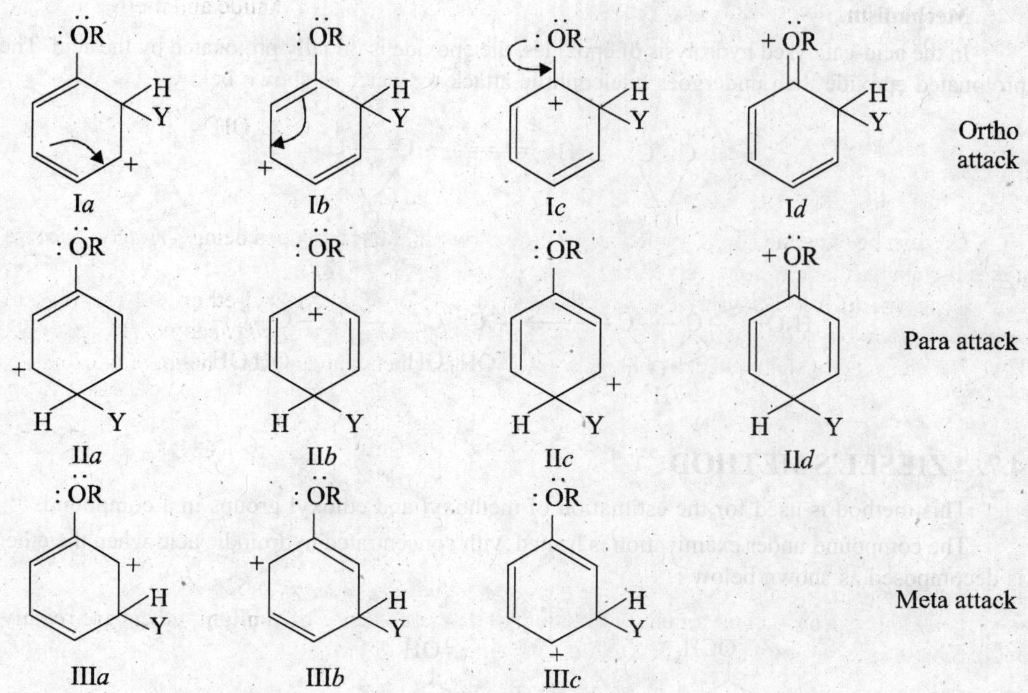

Ia Ib Ic Id — Ortho attack

IIa IIb IIc IId — Para attack

IIIa IIIb IIIc — Meta attack

Carbonium ions formed by *ortho* and *para* attacks are more stable than the meta carbonium ion due to the additional contributions by particularly stable contributing structures I*d* and II*d* respectively. As a result, *ortho* and *para* carbonium ions are formed in preference to meta ion so that substitution occurs mostly at *ortho* and *para* positions.

7. Hydrolysis of epoxides to glycols. Ethers are relatively unreactive compounds which react with only a few reagents. However, the three-membered ring present in epoxides is highly strained and undergoes ring-opening quite easily. These reactions take place through the cleavage of one of the carbon-oxygen bonds and may be catalysed by either acids or bases.

One of the most important cleavage reactions of epoxides is **acid-catalysed hydrolysis** to form glycols.

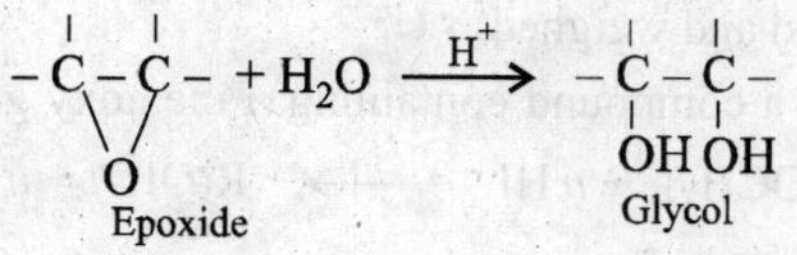

$$\underset{\text{Epoxide}}{-\overset{|}{C}-\overset{|}{C}-}\ (\text{bridged by } O) + H_2O \xrightarrow{H^+} \underset{\text{Glycol}}{-\underset{OH}{\overset{|}{C}}-\underset{OH}{\overset{|}{C}}-}$$

For example :

$$\underset{\text{Ethylene oxide}}{CH_2-CH_2}\ (\text{bridged by } O) + H_2O \xrightarrow{H^+} \underset{\text{Ethane-1, 2-diol (or Ethylene glycol)}}{\underset{OH}{CH_2}-\underset{OH}{CH_2}}$$

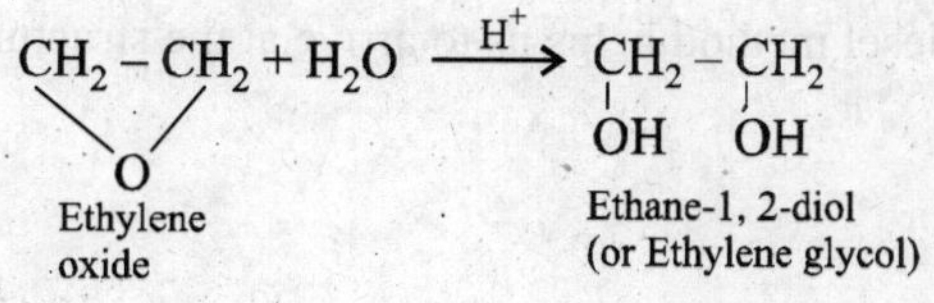

Mechanism

In the acid-catalysed hydrolysis of epoxides, the epoxide is initially protonated by the acid. The protonated epoxide then undergoes nucleophilic attack by water as shown below :

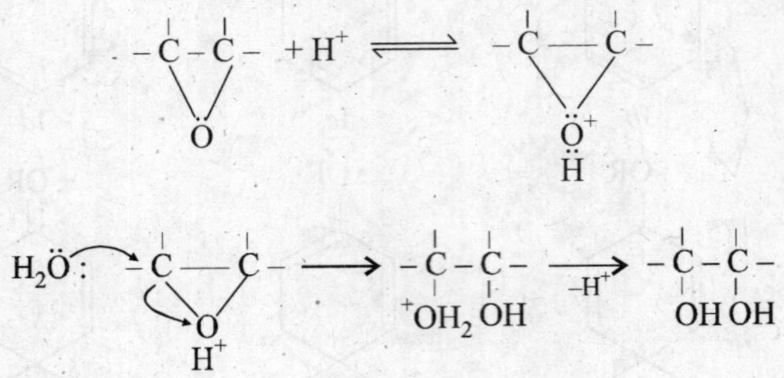

$$-\overset{|}{C}-\overset{|}{C}- \text{(epoxide, bridged by } O\text{)} + H^+ \rightleftharpoons -\overset{|}{C}-\overset{|}{C}- \text{(bridged by } \ddot{O}^+-H\text{)}$$

$$H_2\ddot{O}: \curvearrowright -\overset{|}{C}-\overset{|}{C}- \text{(bridged by } O^+H\text{)} \longrightarrow -\overset{|}{C}-\overset{|}{C}- \text{ with } {}^+OH_2 \text{ and } OH \xrightarrow{-H^+} -\overset{|}{C}-\overset{|}{C}- \text{ with } OH\ OH$$

4.7. ZIESEL'S METHOD

This method is used for the estimation of methoxyl and ethoxyl groups in a compound.

The compound under examination is heated with concentrated hydroiodic acid when the ether is decomposed as shown below :

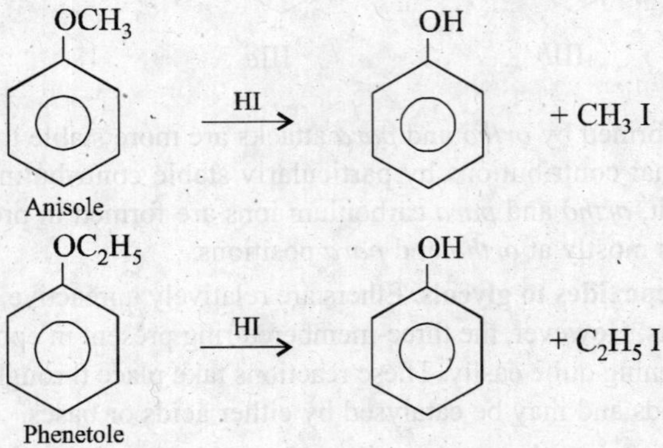

$$\underset{\text{Anisole}}{C_6H_5OCH_3} \xrightarrow{HI} C_6H_5OH + CH_3\,I$$

$$\underset{\text{Phenetole}}{C_6H_5OC_2H_5} \xrightarrow{HI} C_6H_5OH + C_2H_5\,I$$

The alkyl halide (CH_3I or C_2H_5I) formed is absorbed by an ethanolic solution of silver nitrate. Silver iodide formed is filtered and weighed.

A general reaction with a compound containing *n* methoxy groups can be written as :

$$R(OCH_3)_n + n\,HI \longrightarrow R(OH)_n + n\,CH_3I$$

$$n\,CH_3\,I + n\,AgNO_3 \longrightarrow n\,AgI + n\,CH_3\,NO_3$$

From the weight of silver iodide, we can estimate the number of $-OCH_3$ or $-OC_2H_5$ groups present in the molecule. Ziesel method helps us to arrive at the structure and molecular weight of an ethereal compound.

CROWN ETHERS

Cyclic ethers having crown-like shapes are called crown ethers.
For example

[15] Crown-5 [18] Crown-6

The first number within square brackets in the name denotes the size of the ring while the second number represents the number of oxygen atoms.

Crown ethers show the property of forming complexes with cations like Na^+, K^+ and Li^+. Because of this property, ionic compounds can be dissolved in organic solvents containing a certain amount of crown ether. For example, potassium permangamate is soluble in water but insoluble in benzene. If some [18] crown-6 is added to benzene, it is possible to extract the potassium permanganate from water into benzene.

SOLVED EXAMPLES

Example 1. How many isomeric ethers are possible for $C_4H_{10}O$? Give their structures and IUPAC names.

Solution. Three isomeric ethers are possible having the molecular formula $C_4H_{10}O$. Their structures and IUPAC names are given below :

```
       H      H   H   H                              H
       |      |   |   |                              |
1.  H-C-O-C-C-C-H                      H    H-C-H  H
       |      |   |   |                    |      |      |
       H      H   H   H              2. H-C-O-C------C-H
                                            |      |      |
                                            H      H      H
```

1-Methoxypropane 2-Methoxypropane

```
       H   H       H   H
       |   |       |   |
3.  H-C-C-O-C-C-H
       |   |       |   |
       H   H       H   C
```

Ethoxyethane

Example 2. The dipole moment of diethylene ether (1.18 D) is lower than that of water (1.84 D). Explain, why?

Solution. Due to the inductive effect of the ethyl groups, the separation of charges becomes weaker and, therefore, dipole moment of diethylene ether becomes smaller.

Example 3. Predict the products in the following reactions :

(*a*) $C_6H_5-O-CH_2CH_3 \xrightarrow{HI}$

(*b*) $(CH_3)_2\ CHO\ C_3H_7 \xrightarrow{HBr}$ **(*c*)** $CH_3\ O\ C(CH_3)_3 \xrightarrow{HI}$

(*d*) $C_6H_5-O-C_6H_5 \xrightarrow{HBr}$

Solution. (*a*)

$$C_6H_5-O-CH_2CH_3 \xrightarrow{HI} C_6H_5-OH + CH_3CH_2I$$

Phenetole → Phenol

This is because aromatic C–O bond is difficult to break. Instead of this, alkyl carbon-oxygen bond breaks forming ethyl carbonium ion which combines with iodide ion to form ethyl iodide.

(*b*) $(CH_3)_2\,CHO\,C_3H_7 \xrightarrow{HBr} (CH_3)_2\,CHOH + C_3H_7Br$

2-Propoxy propane (Propyl-isopropyl ether) → Propanol-2 + 1-Bromopropane

Here we assume the reaction to take place by SN^2 mechanism. Bromide ion will attach itself to less hindered *n*-propyl ion, so that *n*-propyl bromide will be one of the products.

(*c*) $CH_3\,OC\,(CH_3)_3 \longrightarrow CH_3OH + (CH_3)_3\,C\,I$

Methyl *tert.* butyl ether → Methyl alcohol + *tert.* Butyl iodide

In this reaction, SN^1 reaction is assumed. The protonated ether will decompose giving more stable *tert.* butyl carbonium ion which will combine with iodide ion to give the above product.

(*d*) We don't expect the reaction to take place because of the difficulty in breaking aromatic carbon and oxygen bonds.

$$C_6H_5-O-C_6H_5 \xrightarrow{HBr} \text{No reaction.}$$

Example 4. How would you convert ethanol into ethoxy ethane and vice-versa?

Solution. (*i*) *Ethanol into ethoxyethane.*

$$2C_2H_5OH \xrightarrow[413\text{ K}]{\text{Conc. } H_2SO_4} C_2H_5-O-C_2H_5 + H_2O$$

Ethanol → Ethoxyethane

(*ii*) *Ethyoxyethane into ethanol*

$$C_2H_5-O-C_2H_5 + HI \longrightarrow C_2H_5OH + C_2H_5I$$

Ethoxyethane → Ethanol

Example 5. How do you say that ethers have weak basic character?

Solution. Ethers have the structure $R-\ddot{O}-R'$

There are two lone pairs of electrons on oxygen in the molecule. Ethers can donate these electrons and thus they have basic character.

However, they have weak basic nature as they can dissolve only in strong mineral acids like HCl or H_2SO_4 forming salts.

$$C_2H_5-\ddot{O}-C_2H_5 + HCl \longrightarrow [C_2H_5-\overset{H}{\ddot{O}}-C_2H_5]^+\,Cl^-$$

Diethyl oxonium hydrogen chloride

$$C_2H_5-\ddot{O}-C_2H_5 + H_2SO_4 \longrightarrow [C_2H_5-\overset{H}{\ddot{O}}-C_2H_5]^+\,HSO_4^-$$

Diethyl oxonium hydrogen sulphate

4.8. EPOXIDES

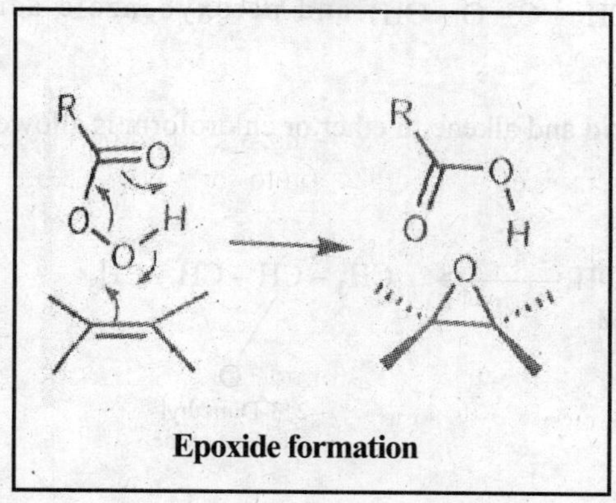

Epoxide formation

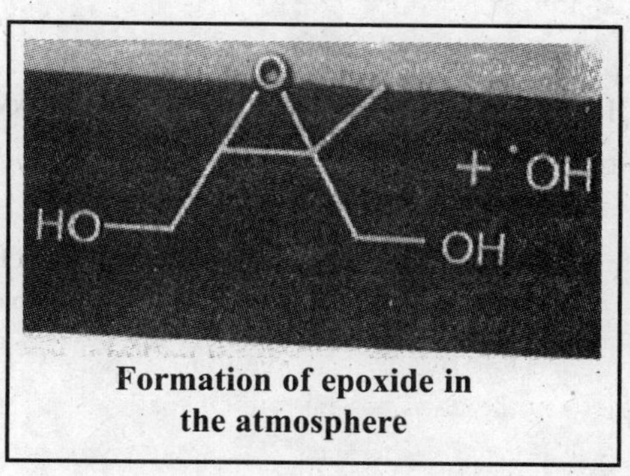

Formation of epoxide in the atmosphere

Epoxides are cyclic ethers with three-membered rings. They are called oxiranes according to IUPAC nomenclature. Some examples of epoxides are given below

$$\underset{\text{(O}^1\text{ bridging C}^2\text{ and C}^3)}{\overset{2}{C}H_2 - \overset{3}{C}H_2}$$

IUPAC name : oxirane
Common name : ethylene oxide

$$\underset{\text{(O}^1\text{ bridging C}^2\text{ and C}^3)}{CH_3 - \overset{2}{C}H - \overset{3}{C}H_2}$$

IUPAC name : 2-methyl oxirane
Common name : Propylene oxide

Preparation of ethylene oxide

1. By distilling ethylene chlorohydrin with potassium hydroxide solution

$$\underset{\text{Ethylene chlorohydrin}}{CH_2OH - CH_2Cl} + KOH \longrightarrow \underset{\diagdown O \diagup}{CH_2 - CH_2} + H_2O + KCl$$

2. It is manufactured by passing ethylene and oxygen under pressure over a silver catalyst at 250°C

$$\underset{\text{Ethylene}}{CH_2 = CH_2} + ½\,O_2 \xrightarrow[250°C]{Ag} \underset{\diagdown O \diagup}{CH_2 - CH_2}$$

3. By epoxidation of alkenes

Epoxides (cyclic ethers) are prepared by reaction between an alkene and an organic *peroxy acid* (also known as **peracids**). The reaction is known as **epoxidation.**

$$\underset{\text{Alkene}}{RCH = CHR} + \underset{\text{Peracid}}{R' - \overset{\overset{O}{\|}}{C} - O - OH} \xrightarrow{\text{Expoxidation}} \underset{\text{Epoxide}}{\underset{\diagdown O \diagup}{RCH - CHR}} + R' - \overset{\overset{O}{\|}}{C} - OH$$

Commonly used peroxy acids in this reaction are **peroxyformic** acid

$(H-\overset{O}{\overset{\|}{C}}-O-OH)$, **peroxyacetic acid** $(CH_3-\overset{O}{\overset{\|}{C}}-O-OH)$ and peroxybenzoic acid $C_6H_5-\overset{O}{\overset{\|}{C}}-O-OH$. When a solution of peroxy acid and alkene in ether or chloroform is allowed to react, epoxide is formed.

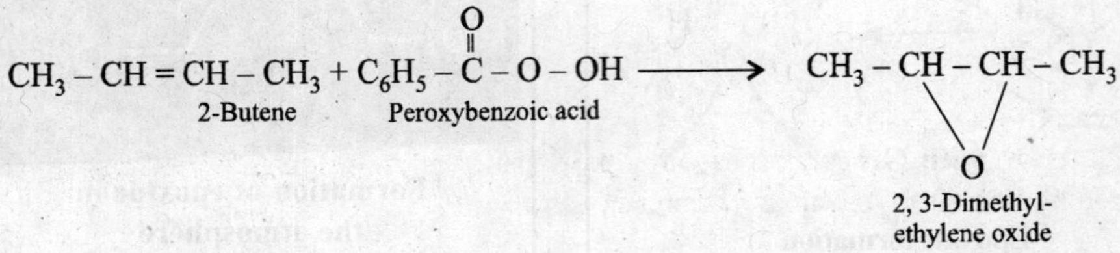

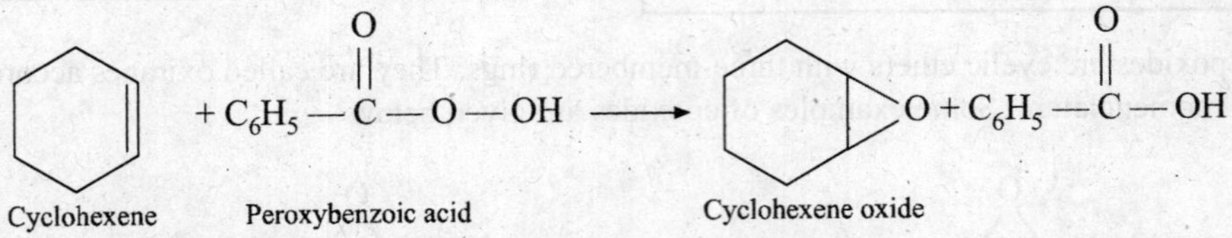

$$\underset{\text{2-Butene}}{CH_3-CH=CH-CH_3} + \underset{\text{Peroxybenzoic acid}}{C_6H_5-\overset{O}{\overset{\|}{C}}-O-OH} \longrightarrow \underset{\text{2, 3-Dimethyl-ethylene oxide}}{CH_3-\overset{\quad O \quad}{CH-CH}-CH_3}$$

$$\underset{\text{Cyclohexene}}{C_6H_{10}} + \underset{\text{Peroxybenzoic acid}}{C_6H_5-\overset{O}{\overset{\|}{C}}-O-OH} \longrightarrow \underset{\text{Cyclohexene oxide}}{C_6H_{10}O} + C_6H_5-\overset{O}{\overset{\|}{C}}-OH$$

Mechanism. The mechanism of epoxidation involves the transfer of an oxygen atom from peroxy acid to alkene as shown below :

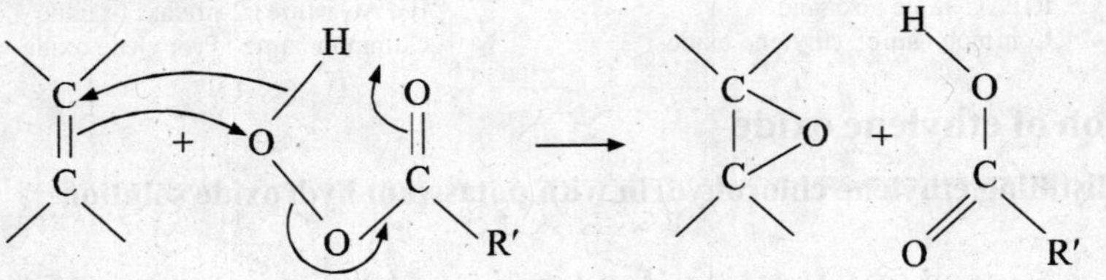

$$>C=C< \; + \; H-O-O-\overset{O}{\overset{\|}{C}}-R' \longrightarrow >C\overset{O}{-}C< \; + \; H-O-\overset{O}{\overset{\|}{C}}-R'$$

Properties of ethylene oxide

1. Ethylene oxide on heating undergoes rearrangement and forms acetaldehyde

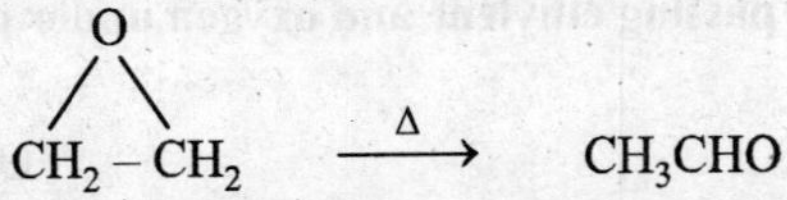

$$\overset{\quad O \quad}{CH_2-CH_2} \xrightarrow{\Delta} CH_3CHO$$

2. Ring opening reactions in acidic medium. Ethylene oxide reacts with H_2O, HCl, HCN etc. in acidic medium to form addition products by cleavage of the oxide ring.

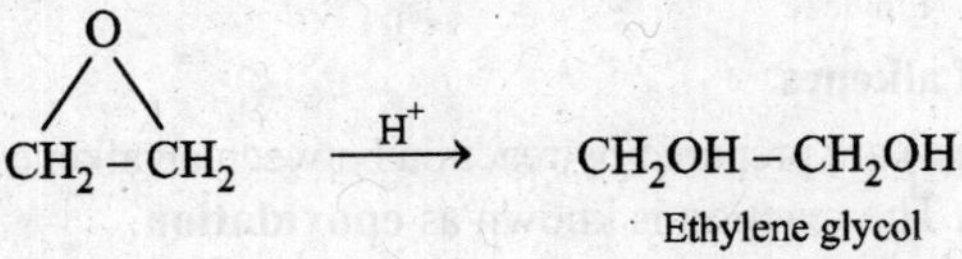

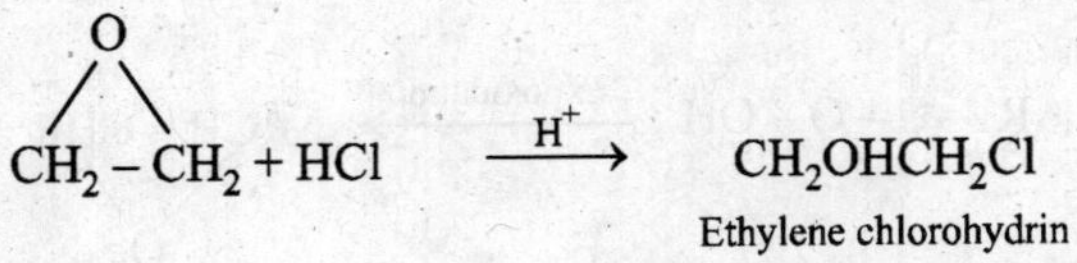

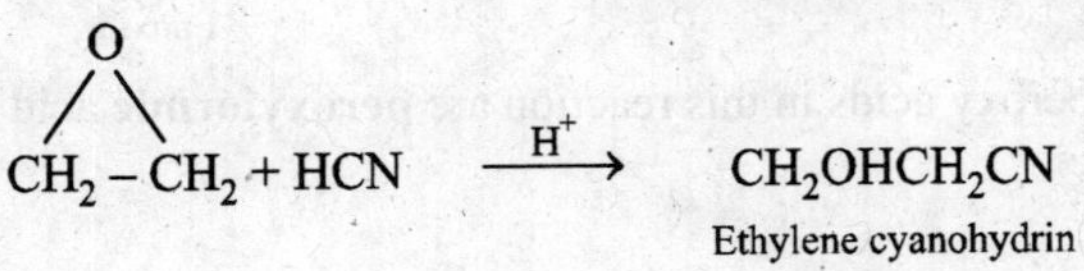

$$\overset{\quad O \quad}{CH_2-CH_2} \xrightarrow{H^+} \underset{\text{Ethylene glycol}}{CH_2OH-CH_2OH}$$

$$\overset{\quad O \quad}{CH_2-CH_2} + HCl \xrightarrow{H^+} \underset{\text{Ethylene chlorohydrin}}{CH_2OHCH_2Cl}$$

$$\overset{\quad O \quad}{CH_2-CH_2} + HCN \xrightarrow{H^+} \underset{\text{Ethylene cyanohydrin}}{CH_2OHCH_2CN}$$

3. Ring opening in basic medium. Ethylene oxide reacts with ethyl alcohol and ammonia in basic medium or in the presence of basic catalysts to form addition products

$$\underset{\text{(epoxide ring, O bridging)}}{CH_2-CH_2} + C_2H_5OH \longrightarrow \underset{\text{Glycol monoethyl ether}}{CH_2OHCH_2OC_2H_5}$$

$$CH_2-CH_2 + NH_3 \longrightarrow \underset{\text{Ethanolamine}}{CH_2NH_2CH_2OH}$$

4. Reaction with Grignard reagents. It forms addition products with Grignard reagents, which upon hydrolysis give primary alcohols.

$$CH_2-CH_2 + CH_3MgI \longrightarrow CH_3CH_2CH_2OMgI \xrightarrow{H_2O} \underset{\text{Propanol}}{CH_3CH_2CH_2OH} + Mg(OH)I$$

5. Reaction with $LiAlH_4$.

$$CH_2-CH_2 + 2[H] \xrightarrow{LiAlH_4} \underset{\text{Ethanol}}{CH_3CH_2OH}$$

EXERCISES

(Based on Question Papers of Different Universities)

Multiple Choice Questions (Choose the correct option)

1. The reaction of a sodium alkoxide with an alkyl halide is called

(*a*) Wurtz-Fittig reaction (*b*) Perkin reaction

(*c*) Williamson's synthesis (*d*) Aldol condensation

2. Ethers are stored in brown bottles. This is because on exposure to air and light ethers are converted to

(*a*) Peroxide (*b*) Oxonium ions

(*c*) Ozonides (*d*) Electrophiles

3. Which of the following reagents readily react with ethyl methyl ether ?

(*a*) NaOH (*b*) Conc HI (*c*) $KMnO_4$ (*d*) H_2O

4. Diethyl ether on heating with excess concentrated HI gives

(*a*) Methyl iodide (*b*) Isopropyl iodide

(*c*) Ethyl iodide (*d*) *n*-Propyl iodide

5. Ethanol is isomeric with

(*a*) Dimethyl ether (*b*) Ethanal

(*c*) Diethyl ether (*d*) Propanone

6. Diethyl ether reacts with sodium metal to give
(*a*) Ethanol (*b*) Sodium ethoxide
(*c*) Ethanal (*d*) Nothing happens

7. Diethyl ether and methyl propyl ether are
(*a*) Conformational isomers (*b*) Metamers
(*c*) Geometrical isomers (*d*) Enantiomers

8. Ethyl alcohol reacts with concentrated H_2SO_4 at 140°C to form
(*a*) Acetone (*b*) Ethylene
(*c*) Diethyl ether (*d*) Acetic acid

9. Diethyl ether reacts with excess of hot concentrated HI to form
(*a*) $CH_3CH_2I + CH_3CH_2OH$ (*b*) Only CH_3CH_2OH
(*c*) $CH_3CH_2OH + CH_2 = CH_2$ (*d*) Only CH_3CH_2I

10. Ethylene oxide reacts with ammonia to give
(*a*) I-Aminoethanol (*b*) Ethylamine
(*c*) 2-Aminoethanol (*d*) Acetamide

11. Ethylene oxide reacts with HBr to give
(*a*) 1-Bromoethanol (*b*) Ethyl bromide
(*c*) 2-Bromoethanol (*d*) Ethylene glycol

12. The following compounds have identical molecular weights. Which would have the lowest boiling point ?
(*a*) 1-Methoxypropane (*b*) 1-Butanol
(*c*) 1, 1-Dimethylethanol (*d*) 2-Butanol

13. The compound with the lowest boiling point is
(*a*) H_2O (*b*) CH_3CH_2OH
(*c*) CH_3OCH_3 (*d*) $CH_3CH_2CH_3$

14. The product of the reaction of ethylene oxide with acidic methanol is
(*a*) $CH_3OCH_2CH_2OH$ (*b*) $CH_3CH_2CH_2CH_2OH$
(*c*) $HOCH_2CH_2OH$ (*d*) $CH_3CH_2OCH_2CH_3$

15. Cyclic ethers with three-membered ring are called
(*a*) Lactones (*b*) Oxiranes
(*c*) Alkoxides (*d*) Epoxy resins

ANSWERS

1. (*c*)	**2.** (*a*)	**3.** (*b*)	**4.** (*c*)	**5.** (*a*)	**6.** (*d*)
7. (*b*)	**8.** (*c*)	**9.** (*d*)	**10.** (*c*)	**11.** (*c*)	**12.** (*a*)
13. (*d*)	**14.** (*a*)	**15.** (*b*)			

Short Answer Questions

1. Write the IUPAC name of di-isopropyl ether.
2. Write all the structural isomers with molecular formula $C_4H_{10}O$ and give their IUPAC names.
3. What are epoxides ? Why are epoxides more reactive than ethers ?

4. Give the preparation of the following :
   (*i*) Methyl tertiary butyl ether (*ii*) Methyl *n*-propyl ether.
5. Though ethers are the derivatives of water, the two differ in bond angles. Explain.
6. Differentiate between symmetrical and unsymmetrical ethers giving one example of each.
7. Why acidic dehydration of alcohols is not the suitable method for the preparation of ethers?
8. Complete the following reaction and discuss its mechanism :

   $$CH_3 - O - C_2H_5 + HI \longrightarrow$$

9. Dimethyl ethers and ethyl alcohol are functional isomers. Which has a higher boiling point and why ?
10. Give the reaction of diethyl ether with H_2SO_4.
11. The alkoxy group in ether is ortho and para directing in nature. Explain.
12. What are the products formed when anisole is heated with HI ?
13. Give the mechanism of Williamson's synthesis.
14. Name the pair of alkyl halide and alkoxide for the preparation of ethyl tertiary butyl ether.

General Questions

1. Outline a scheme for the synthesis of :
   (*i*) Ethyl methyl ether (*ii*) Tertiary butyl methyl ether
   (*iii*) Anisole.
2. Give reasons for the following :
   (*i*) Ethers have a tendency to form salts with cold mineral acids.
   (*ii*) The boiling points of ethers are lower than those of the isomeric alcohols.
   (*iii*) The acidic dehydration method is not suitable for converting a tertiary alcohol into ether.
3. How does anisole react with
   (*i*) Bromine dissolved in CS_2
   (*ii*) Nitrating mixture of conc. HNO_3 and conc. H_2SO_4
   (*iii*) Excess of HI at 373 K ?
4. What are ethers ? Give two examples of simple and mixed ethers.
5. Explain why cleavage of phenyl methyl ether with HI produces phenol and methyl iodide and not iodobenzene and methyl alcohol.
6. Give two methods by which diethyl ether can be prepared from ethyl alcohol.
7. (*a*) Discuss the following reactions of ethers with special reference to the mechanism involved.
   (*i*) Cleavage by acids (*ii*) Electrophilic substitution in aromatic ether.
   (*b*) Give the mechanism of Williamson's synthesis.
8. (*a*) Why alkoxy group is ortho and para directing in nature although the oxygen atom is quite electronegative in nature ?
   (*b*) How does the ether react with :
   (*i*) Concentrated sulphuric acid (*ii*) Water (*iii*) Hydrogen iodide ?
9. Discuss briefly the following reactions of ether :
   (*i*) Action with air (*ii*) Formation of oxonium salt.
   (*iii*) Formation of complexes.

10. (*a*) Give the mechanism of cleavage of ether by HI.

(*b*) Give the mechanism of Williamson's synthesis.

(*c*) Explain why ethyl alcohol has higher boiling point than dimethyl ether ?

11. Give the products of the following :

(*i*) Sodium phenoxide + Ethyl bromide (*ii*) Ethyl bromide + Sodium tert. butoxide

(*iii*) Anisole + hot. conc. HI.

12. Identify X, Y and Z in the following sequence of reactions :

(*i*) $C_2H_5OH \xrightarrow{Na} X \xrightarrow[heat]{C_2H_5I} Y \xrightarrow{Cl_2(hv)} Z$

(*ii*) $C_2H_5ONa \xrightarrow{C_2H_5I} X \xrightarrow[(373\,K)]{\text{Excess of HBr}} Y \xrightarrow{\text{Alc. KOH}} Z$

(*iii*) $C_2H_5OC_2H_5 \xrightarrow[(Pressure)]{Dil.\,H_2SO_4} X \xrightarrow[443\,K]{Conc.\,H_2SO_4} Y \xrightarrow{HBr} Z$

13. (*a*) How does acidic dehydration of alcohols lead to the preparation of ethers ? Give mechanism.

(*b*) Discuss the mechanism of cleavage of an ether with halogen acid.

14. How will you account for the fact that the alkoxy group when present in the ring is ortho and para directing in nature ?

15. Give the mechanism of Williamson's synthesis of ethers.

16. What are epoxides ? How are they formed ? Discuss the mechanism of epoxide formation.

17. Give the synthesis of ethers by Williamson's method. Explain its mechanism.

18. Compound (A), $C_4H_{10}O$ is found to be soluble in sulphuric acid. (A) does not react with sodium metal or potassium permanganate. When (A) is heated with excess of HI, it is converted into a single alkyl iodide. What is the structural formula of (A) ?

Ans. (A) is $CH_3CH_2OCH_2CH_3$ (Diethyl ether)

19. Compound (A) $C_4H_{10}O_3$ shows the properties of both an alcohol and an ether. When heated with iodide, (A) yields only ICH_2CH_2I. Draw a structural formula for (A).

Ans. (A) is $HOCH_2CH_2OCH_2CH_2OH$)

20. An ether (A), $C_6H_{14}O$, when heated with excess of hot concentrated HI produced two alkyl halides which on hydrolysis form compounds (B) and (C). Oxidation of B gave an acid (D), whereas oxidation of (C) gave a ketone (E). Deduce the structural formulas of (A), (B), (C), (D) and (E).

Ans. (A) = $CH_3-CH_2-O-\underset{|}{\overset{CH_3}{CH}}-CH_2CH_3$ (2-Ethoxybutane)

(B) = CH_3CH_2OH (C) = $CH_3-\overset{OH}{\overset{|}{CH}}-CH_2CH_3$ (2-Butanol)

(D) = $CH_3-\overset{O}{\overset{\|}{C}}-OH$ (E) = $CH_3-\overset{O}{\overset{\|}{C}}-CH_3$

5

Aldehydes and Ketones

5.1. INTRODUCTION

Aldehydes and ketones are organic compounds, containing the carbonyl group *i.e.*, >C=O. Aldehydes can be represented as :

$$\begin{matrix} R \\ H \end{matrix} \!\!>\! C = O$$

where R stands for an alkyl or aryl group.

Ketones can be represented as :

$$\begin{matrix} R \\ R' \end{matrix} \!\!>\! C = O$$

where R and R′ are alkyl or aryl groups. R and R′ could be same or different groups. If R and R′ are same, the ketones are *simple ketones* and if they are different, they are called *mixed ketones*.

Thus, we find that in case of aldehydes, one of the groups that is attached to the carbonyl group is an alkyl or aryl group and the second group that is attached is hydrogen (H). In case of ketones, the two groups that are attached to carbonyl group are alkyl or aryl groups.

The presence of the common carbonyl group in the two classes of compounds makes them display similar chemical properties. However aldehydes are more reactive than ketones because of the presence of free hydrogen atom.

5.2. NOMENCLATURE OF ALDEHYDES

There are two systems of naming aldehydes.

(*i*) **Common system.** According to this system, an aldehyde is named after the acid which it produces on oxidation. Thus the aldehyde which produces acetic acid on oxidation is called acetaldehyde, CH_3CHO. The aldehyde which produces propionic acid on oxidation is called propionaldehyde CH_3CH_2CHO. The aldehyde which produces benzoic acid on oxidation is called benzaldehyde C_6H_5CHO and so on.

(*ii*) **IUPAC system.** Rules governing IUPAC system of nomenclature are employed in naming aldehydes in this system. The longest chain containing the —CHO group is identified. Total number of carbon atoms in this chain is counted. The aldehyde is named as the derivative of the corresponding hydrocarbon. For example, if there are 4 carbon atoms, it is a derivative of butane. If there are 5 carbon atoms in the chain, it is a derivative of pentane and so on. The ending *-e* of the hydrocarbon is changed into *-al*. The position of other substituents, if any, is indicated by suitable numbers. Aldehydic carbon is given number 1 for this purpose. For example, the following compound is named as 2-methylbutanal.

$$\overset{4}{CH_3} - \overset{3}{CH_2} - \overset{2}{\underset{\underset{CH_3}{|}}{CH}} - \overset{1}{CHO}$$

Aromatic aldehydes are named by writing carbaldehyde after the aromatic part.

Table 5.1 compares common and IUPAC names of some common aldehydes.

Table 5.1. Common and IUPAC names of some aldehydes

Formula	Common Name	IUPAC Name
$HCHO$	Formaldehyde	Methanal
$CH_3 - CHO$	Acetaldehyde	Ethanal
$CH_3 - \underset{\underset{CH_3}{\vert}}{CH} - CHO$	Isobutyraldehyde	2-Methylpropanal
$CH_3 - CH = CH - CHO$	Crotonaldehyde	2-Butenal
C_6H_5—CHO (benzene ring)—CHO	Benzaldehyde	Benzenecarbaldehyde

5.3. NOMENCLATURE OF KETONES

Like in the case of aldehydes, there are two systems of naming ketones.

1. Common system. Here the two alkyl or aryl groups constituting the compound are written followed by the word 'ketone'. For example, $CH_3COC_2H_5$ is named as ethyl methyl ketone, $CH_3COC_6H_5$ is named as methyl phenyl ketone.

2. IUPAC system. The longest chain containing the —CO— group is identified and total number of carbon atoms in that chain is counted. The ketone is named as the derivative of the corresponding hydrocarbon. The suffix *–e* of the hydrocarbon is converted into **—one.** The position of other substituents, if any, is indicated by suitable numbers. The ketonic carbon is given the lowest possible number as per the rules. The compound

$$\overset{1}{CH_3} - \overset{2}{CO} - \overset{3}{\underset{\underset{CH_3}{|}}{CH}} - \overset{4}{CH_2} - \overset{5}{CH_3}$$

is named as 3-methylpentanone-2.

Common and IUPAC names of some common ketones are listed in Table 5.2.

Table 5.2. Common and IUPAC names of some ketones

Formula	Common Name	IUPAC Name
CH_3COCH_3	Acetone or dimethyl ketone	Propanone
$CH_3COCH_2CH_3$	Methyl ethyl ketone	Butanone
$CH_3CH_2COCH_2CH_3$	Diethyl ketone	Pentan-3-one
$CH_3COCH_2CH_2CH_3$	Methyl propyl ketone	Pentan-2-one
$CH_3 - CO - \underset{\underset{CH_3}{\vert}}{CH} - CH_3$	Methyl isopropyl ketone	3-Methylbutan-2-one
$C_6H_5 - CO - CH_3$	Methyl phenyl ketone	—
$C_6H_5 - CO\ C_6H_5$	Diphenyl ketone	—

5.4. STRUCTURE OF THE CARBONYL GROUP

The bond between carbon and oxygen is a double bond. It comprises a σ and a π bond. Carbonyl carbon atom uses sp^2 hybridisation of orbitals around it, the fourth one remains unhybridised. Three sp^2 hybrid orbitals form three σ bonds, one with oxygen and two with the other groups attached to it.

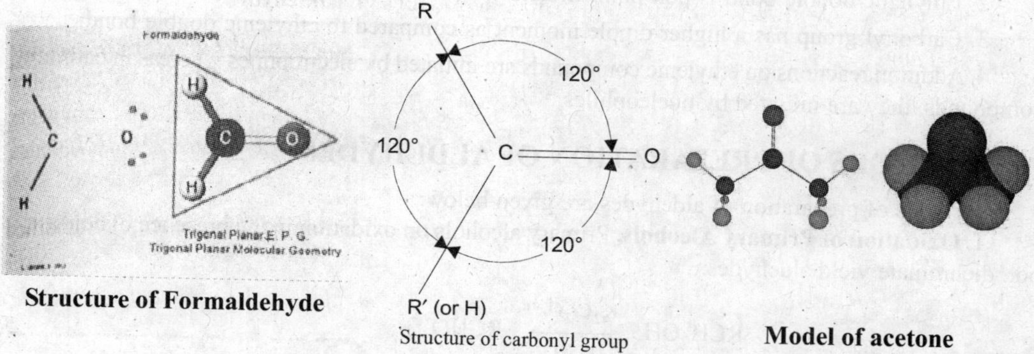

Structure of Formaldehyde Structure of carbonyl group **Model of acetone**

The fourth orbital (unhybridised) of carbon forms a π bond with oxygen. As is well known, three sp^2 hybridised orbitals are inclined at an angle of 120° in the same plane. Hence R, R′ (or H), C and O lie in the same plane.

Another observation that we make is that there is a difference of electronegativity between carbon and oxygen, oxygen being more electronegative than carbon. That makes the shared pair of electrons between C and O to shift towards oxygen, thereby making carbon slightly positive and oxygen slightly negative.

$$\gt \overset{\delta+}{C} = \overset{\delta-}{O}$$

(*i*) As the carbonyl group is planar in shape, there is a possibility for the attacking species to attach themselves to the carbonyl group from above the plane or below the plane of the carbonyl group. That doubles the probability of the attachment of reacting species to the carbonyl group making it a reactive group.

(*ii*) The polar nature of the carbonyl group supplements the rate of reaction. Positive charge on carbon invites the negatively charged groups or ions (nucleophiles) to attack this centre thereby making it highly reactive.

5.5. COMPARISON BETWEEN THE PROPERTIES OF THE CARBONYL GROUP AND ETHYLENIC DOUBLE BOND

Similarities

1. Both the carbonyl group and the ethylenic double bond consist of one σ bond and one π bond.
2. sp^2 hybridisation takes place on the carbon atom in both cases.
3. Both of them are planar in shape giving an angle of 120°.
4. Carbonyl group as well as the ethylenic double bond give addition reactions.

Dissimilarities

1. Ethylenic double bond is longer and weaker than the carbonyl group bond as shown below :

	Bond Length	*Bond Energy*
$>C=O$	1.23 Å	745 kJ mol^{-1}
$>C=C<$	1.44 Å	610 kJ mol^{-1}

2. Ethylenic double bond is non-polar in nature whereas carbonyl group is polar in nature.

3. Carbonyl group has a higher dipole moment as compared to ethylenic double bond.

4. Addition reactions on ethylenic compounds are initiated by electrophiles whereas in carbonyl compounds, they are initiated by nucleophiles.

5.6. METHODS OF PREPARATION OF ALDEHYDES

Methods of preparation of aldehydes are given below:

1. Oxidation of Primary Alcohols. Primary alcohols on oxidation in the presence of acidified pot. dichromate yield aldehydes.

$$RCH_2OH \xrightarrow[H_2SO_4]{K_2Cr_2O_7} RCHO$$

where R stands for an alkyl or aryl group. For example,

$$\underset{n\text{-Propyl alcohol}}{CH_3CH_2CH_2OH} \xrightarrow[H_2SO_4]{K_2Cr_2O_7} \underset{\text{Propionaldehyde}}{CH_3CH_2CHO}$$

$$\underset{\text{Benzyl alcohol}}{C_6H_5-CH_2OH} \xrightarrow[H_2SO_4]{K_2Cr_2O_7} \underset{\text{Benzaldehyde}}{C_6H_5-CHO}$$

A demerit of this method of preparation is that the reaction does not completely stop on the production of aldehyde. Further oxidation of the aldehyde to carboxy acid could take place.

Oxidation is normally carried by potassium dichromate mixed with dilute sulphuric acid (or chromic acid, H_2CrO_4). In case, the alcohol has very poor solubility in water, then a mixture of chromic anhydride (CrO_3), acetone and sulphuric is used. This is known as **Jone's reagent.**

The above oxidation is successful only in case the aldehyde is low boiling and can be readily removed by distillation. However, if the aldehyde has comparatively high boiling point (373 K or above) then special oxidising agents listed below are employed.

(a) Sarett's reagent. It is a complex compound formed by the careful addition of CrO_3 in excess of pyridine.

$$\underset{\text{Chromic oxide}}{CrO_3} + 2\underset{\text{Pyridine}}{C_5H_5N} \text{ or } (2C_5H_5N) \longrightarrow \underset{\text{Chromic oxide Pyridine complex}}{CrO_3 \cdot 2C_5H_5N}$$

The complex is dissolved in methylene dichloride (CH_2Cl_2) and the solution oxidises primary alcohol to aldehydic stage only.

$$\underset{\text{Primary alcohol}}{R-CH_2-CH_2OH} \xrightarrow[(CH_2Cl_2)]{\text{Sarett's reagent}} \underset{\text{Aldehyde}}{R-CH_2-CHO}$$

(b) Corey's reagent. It is another complex of chromic oxide formed by the reaction between chromic acid and pyridinium chloride. It has the formula pyridinium chlorochromate (PCC) $C_5H_5\overset{+}{N}HCrO_3\overset{-}{Cl}$. The solution of the complex in methylene dichloride is used to oxidise primary alcohol to aldehyde.

(c) Pyridinium dichromate (PDC). This is also a complex of pyridine with chromic acid and oxidises primary alcohol to aldehydic stage.

The complexes listed above are used in non-aqueous medium. They check the oxidation of aldehyde to the carboxylic acid. Actually, in the aqueous medium, the aldehyde gets hydrated to form aldehyde hydrate RCH $(OH)_2$. The aldehyde hydrate is more readily oxidised as compared to aldehyde. Therefore it is not so easy to stop the oxidation at the aldehydic stage. However, in non-aqueous solvent, the hydrate is not formed and therefore the oxidation of aldehyde to carboxylic acid does not take place.

2. Dehydrogenation of Primary Alcohols. Alcohol vapours when passed over heated copper at 475–575 K get dehydrogenated and yield aldehydes.

$$RCH_2OH \xrightarrow[-2[H]]{Cu,\ 475-575\ K} RCHO$$

$$C_6H_5-CH_2OH \xrightarrow[-2[H]]{Cu,\ 475-575\ K} C_6H_5-CHO$$

There is no risk of the aldehyde getting oxidised to carboxy acid in this method of preparation.

3. Oxidation of toluene. Aromatic aldehydes like benzaldehyde can be obtained by the oxidation of methylbenzene or toluene using chromium trioxide in acetic anhydride.

$$\underset{\text{Toluene}}{C_6H_5-CH_3} \xrightarrow[(CH_3CO)_2O]{Cr_2O_3} \underset{\text{Benzylidene acetate}}{C_6H_5-CH(OOCCH_3)_2} \xrightarrow{H_2O} \underset{\text{Benzaldehyde}}{C_6H_5-CHO} + 2\,CH_3COOH$$

4. Etard's Reaction. This method of preparation of aldehydes involves treatment of methylbenzene with chromyl chloride (CrO_2Cl_2) which acts as an oxidising agent.

$$\underset{\text{Toluene}}{C_6H_5CH_3} \xrightarrow{CrO_2Cl_2} \underset{\text{Addition complex}}{\left[C_6H_5CH_3\,.\,2\,CrO_2Cl_2\right]} \xrightarrow{H_2O} \underset{\text{Benzaldehyde}}{C_6H_5CHO}$$

5. Halogenation of toluene followed by hydrolysis. The controlled halogenation of toluene in the presence of sunlight gives a gem dihalide which upon hydrolysis gives benzaldehyde as follows:

$$\underset{\text{Toluene}}{C_6H_5CH_3} \xrightarrow[\text{sunlight}]{Cl_2} \underset{\text{Benzyl chloride}}{C_6H_5CH_2Cl} \xrightarrow[\text{sunlight}]{Cl_2} \underset{\text{Benzal chloride}}{C_6H_5CHCl_2}$$

$$\underset{\text{Benzal chloride}}{C_6H_5CHCl_2} \xrightarrow[-2\,HCl]{2\,H_2O} \underset{\text{(Unstable)}}{C_6H_5CH(OH)_2} \xrightarrow{-H_2O} \underset{\text{Benzaldehyde}}{C_6H_5CHO}$$

6. Reduction of Acid Chloride (Rosenmund's Reduction). Hydrogen gas is passed through a hot solution of acid chloride in xylene in the presence of palladium as catalyst based on barium sulphate, when the partial reduction of acid chloride to aldehyde takes place.

$$RCOCl + H_2 \xrightarrow[\text{Xylene solution}]{Pd/BaSO_4} RCHO + HCl$$

$$\underset{\text{Acetyl chloride}}{CH_3COCl} + H_2 \xrightarrow{Pd/BaSO_4} \underset{\text{Acetaldehyde}}{CH_3CHO} + HCl$$

$$\underset{\text{Benzoyl chloride}}{C_6H_5COCl} + H_2 \xrightarrow{Pd/BaSO_4} \underset{\text{Benzaldehyde}}{C_6H_5CHO} + HCl$$

7. By the hydrolysis of Gem dihalides. Gem dihalides on hydrolysis with an alkali solution produce aldehydes.

$$\underset{\text{Gem dihalide}}{RCHCl_2} \xrightarrow{NaOH} \underset{\text{Unstable}}{RCH(OH)_2} \xrightarrow[-H_2O]{} \underset{\text{Aldehyde}}{RCHO}$$

$$\underset{\text{Ethylidene dichloride}}{CH_3CHCl_2} \xrightarrow{NaOH} \underset{\text{Unstable}}{CH_3CH(OH)_2} \longrightarrow \underset{\text{Acetaldehyde}}{CH_3CHO} + H_2O$$

8. Ozonolysis of Alkenes. An alkene on treatment with ozone forms an ozonide which gets decomposed into aldehydes on hydrolysis in the presence of some reducing agents like zinc and acetic acid or H_2 and Pd.

$$CH_3-CH_2-CH=CH_2+O_3 \longrightarrow CH_3-CH_2-\underset{|}{CH}\overset{O}{\frown}CH_2 \ (O-O)$$

Butene-1 — Ozonide

$$\xrightarrow{H_2O} CH_3-CH_2CHO + HCHO$$

Propionaldehyde Formaldehyde

9. Stephen's Reduction. Alkyl cyanides on reduction with stannous chloride and HCl produce aldehydes. The reaction takes place via formation of aldimine intermediate.

$$CH_3-C\equiv N+2[H]+HCl \xrightarrow{SnCl_2/HCl} CH_3CH=NHHCl$$

Acetaldimine hydrochloride

$$\downarrow H_2O$$

$$CH_3CHO + NH_4Cl$$

Acetaldehyde

10. Gattermann-Koch Reaction. Aromatic aldehydes can be prepared by treating benzene with carbon monoxide in the presence of $AlCl_3$ and hydrochloric acid.

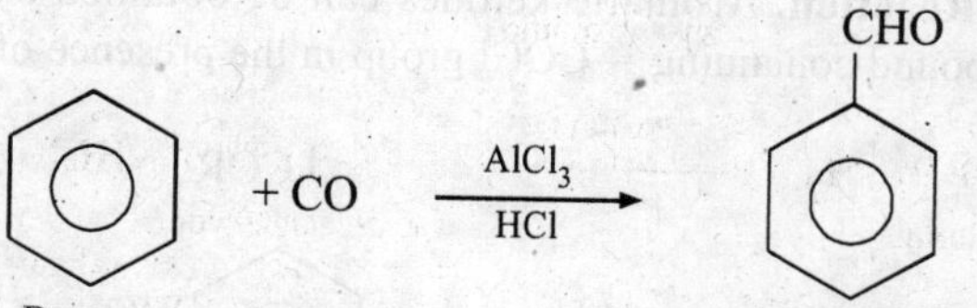

$$C_6H_6 + CO \xrightarrow[HCl]{AlCl_3} C_6H_5CHO$$

Benzene — Benzaldehyde

11. Reimer-Tiemann Reaction. Phenolic aldehydes can be prepared conveniently by this reaction. A phenol is treated with chloroform and aqueous sodium hydroxide solution.

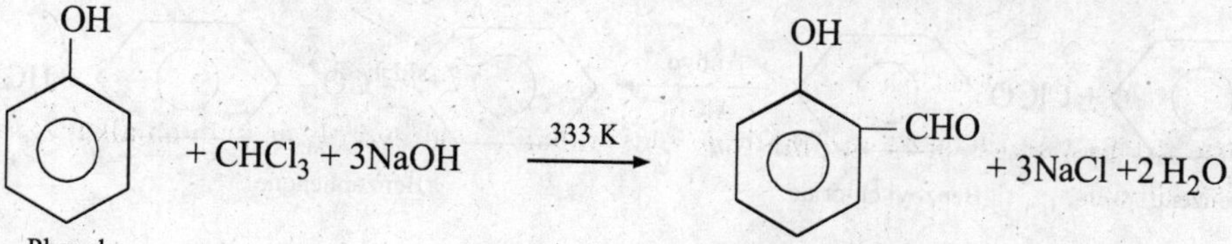

$$C_6H_5OH + CHCl_3 + 3NaOH \xrightarrow{333\ K} C_6H_4(OH)CHO + 3NaCl + 2H_2O$$

Phenol — Salicylaldehyde

5.7. METHODS OF PREPARATION OF KETONES

Methods for the preparation of ketones are given as under:

1. Oxidation of Secondary Alcohols. Secondary alcohols on oxidation in the presence of acidified potassium dichromate or potassium permanganate or chromium trioxide yield ketones.

$$\begin{matrix} R \\ H \end{matrix}\!\!>CHOH \xrightarrow[{[O]}]{K_2Cr_2O_7} \begin{matrix} R \\ H \end{matrix}\!\!>C=O$$

Secondary alcohol — Ketone

$$\underset{\text{Isopropyl alcohol}}{(CH_3)_2CHOH} \xrightarrow[\text{[O]}]{K_2Cr_2O_7} \underset{\text{Acetone}}{(CH_3)_2C=O}$$

2. Oppenauer Oxidation. This method is suitable for the oxidation of alcohols containing a double bond. The double bond is unaffected. Oxidation is carried out by heating the secondary alcohol in presence of acetone as solvent and aluminium isopropoxide $[(CH_3)_2CHO]_3$ Al or Aluminium tert-butoxide $[(CH_3)_3CO]_3$ Al.

$$\underset{\text{Pent-3-en- ol-2}}{CH_3CH=CHCHOHCH_3} \xrightarrow[\text{isopropoxide}]{\text{Aluminium}} \underset{\text{Pent-3-en-one-2}}{CH_3CH=CHCOCH_3}$$

3. Dehydrogenation of Secondary Alcohols. Ketones can be prepared by dehydrogenation of secondary alcohols. Dehydrogenation can be achieved by passing vapours of the alcohols over heated copper at 475-575 K.

$$\underset{\text{Isopropyl alcohol}}{(CH_3)_2CHOH} \xrightarrow[\text{475-575 K}]{\text{Cu}} \underset{\text{Acetone}}{(CH_3)_2C=O} + H_2$$

4. Friedel-Crafts Reaction. Aromatic ketones can be obtained by treating an aromatic hydrocarbon with a compound containing —COCl group in the presence of anhyd. $AlCl_3$.

$$\underset{\text{Benzene}}{C_6H_6} + \underset{\text{Acid chloride}}{RCOCl} \xrightarrow[AlCl_3]{\text{Anhyd.}} \underset{\text{Ketone}}{C_6H_5COR} + HCl$$

$$\underset{\text{Benzene}}{C_6H_6} + \underset{\text{Benzoyl chloride}}{ClCOC_6H_5} \xrightarrow[AlCl_3]{\text{Anhyd.}} \underset{\text{Benzophenone}}{C_6H_5COC_6H_5} + HCl$$

5. Reaction of Acid Chlorides with Organocadmium Compounds.

$$\underset{\text{Acid chloride}}{2RCOCl} + \underset{\text{Dialkyl or Diaryl cadmium}}{R'_2Cd} \longrightarrow \underset{\text{Ketone}}{2RCOR'} + CdCl_2$$

where R and R′ stand for alkyl or aryl groups.

$$\underset{\text{Acetyl chloride}}{2CH_3COCl} + \underset{\text{Diethyl cadmium}}{(C_2H_5)_2Cd} \longrightarrow \underset{\text{Methyl ethyl ketone}}{2CH_3COC_2H_5} + CdCl_2$$

It may be noted that Grignard's reagents also react with acetyl chloride to form ketone but the ketone will further react with Grignard reagent to form a tertiary alcohol as given by the following equation:

$$\underset{\text{Acetyl chloride}}{CH_3-\overset{\overset{O}{\|}}{C}-Cl} + \underset{\text{Methyl mag. chloride}}{CH_3MgCl} \longrightarrow \underset{\text{Acetone}}{CH_3-\overset{\overset{O}{\|}}{C}-CH_3} + MgCl_2$$

$$\underset{\text{Acetone}}{CH_3-\overset{\overset{O}{\|}}{C}-CH_3} + CH_3MgCl \longrightarrow \left[CH_3-\overset{\overset{OMgCl}{|}}{\underset{\underset{CH_3}{|}}{C}}-CH_3\right] \xrightarrow{H_2O} \underset{\text{Tertiary butyl alcohol}}{CH_3-\overset{\overset{OH}{|}}{\underset{\underset{CH_3}{|}}{C}}-CH_3} + Mg\begin{matrix}OH\\Cl\end{matrix}$$

Dialkyl cadmium has no tendency to react with ketone and is therefore a preferred reagent.

Nowadays acid chlorides are converted into ketones by treating with lithium organo cuprates [R_2CuLi] because there is no side reaction and the yield of ketone is also quite good.

$$\underset{\text{Acid chloride}}{2R-\overset{\overset{O}{\|}}{C}-Cl} + R_2'CuLi \longrightarrow \underset{\text{Ketone}}{2R-\overset{\overset{O}{\|}}{C}-R'} + CuCl + LiCl$$

Lithium organo cuprates are obtained by the reaction between cuprous halide and organo lithium compound.

$$2\,RLi + CuX \longrightarrow R_2CuLi + LiX$$

6. Preparation from Carboxylic Acids. Carboxylic acid is treated with alkyl or aryl lithium to form an addition compound which is decomposed by water to form a ketone as given by the following equations:

$$R-\overset{\overset{O}{\|}}{C}-OH + 2R'Li \xrightarrow{-R'H} \underset{\text{Addition compound}}{\left[R-\overset{\overset{OLi}{|}}{\underset{\underset{R'}{|}}{C}}-OLi\right]} \xrightarrow{2H_2O} R-\overset{\overset{OH}{|}}{\underset{\underset{R'}{|}}{C}}-OH + 2LiOH$$

$$\downarrow -H_2O$$

$$\underset{\text{Ketone}}{R-\underset{\underset{R'}{|}}{C}=O}$$

$$\underset{\text{Acetone}}{CH_3-CH_2-\overset{\overset{\displaystyle O}{\|}}{C}}-OH + \underset{\text{Methyl lithium}}{2CH_3Li} \xrightarrow{-CH_4} \underset{\text{Addition compound}}{\left[CH_3-CH_2-\underset{\underset{\displaystyle CH_3}{|}}{\overset{\overset{\displaystyle OLi}{|}}{C}}-OLi\right]}$$

$$\xrightarrow{2H_2O} CH_3-CH_2-\underset{\underset{\displaystyle CH_3}{|}}{\overset{\overset{\displaystyle OH}{|}}{C}}-OH \xrightarrow[-H_2O]{KOH} \underset{\text{Propanone}}{CH_3-CH_2-\underset{\underset{\displaystyle CH_3}{|}}{C}=O}$$

7. By distillation of calcium salts of carboxylic acids. Ketones can also be obtained by distilling calcium salts of carboxylic acids.

$$\underset{\text{Cal. acetate}}{\begin{matrix}CH_3COO \\ CH_3COO\end{matrix}\!\!>\!Ca} \xrightarrow{\text{Distil}} \underset{\text{Acetone}}{CH_3-\overset{\overset{\displaystyle O}{\|}}{C}-CH_3} + CaCO_3$$

$$\underset{\text{Cal. benzoate}}{\begin{matrix}C_6H_5COO \\ C_6H_5COO\end{matrix}\!\!>\!Ca} \xrightarrow{\text{Distil}} \underset{\text{Benzophenone}}{C_6H_5-\overset{\overset{\displaystyle O}{\|}}{C}-C_6H_5} + CaCO_3$$

8. Hydration of Alkynes. Hydration of alkynes containing three or more carbons produces ketones. The alkyne is passed through a solution of dil. H_2SO_4 containing $HgSO_4$. It may be mentioned that acetylene on hydration produces acetaldehyde.

$$\underset{\text{Propyne}}{CH_3-C\equiv CH} + H_2O \xrightarrow[HgSO_4]{H_2SO_4} \underset{\text{Unstable}}{CH_3-\underset{\underset{\displaystyle OH}{|}}{C}=CH_2} \longrightarrow \underset{\text{Acetone}}{CH_3-\underset{\underset{\displaystyle O}{\|}}{C}-CH_3}$$

$$\underset{\text{2-Butyne}}{CH_3-C\equiv C-CH_3} + H_2O \xrightarrow[HgSO_4]{H_2SO_4} \underset{\text{Unstable}}{CH_3-CH=\underset{\underset{\displaystyle OH}{|}}{C}-CH_3}$$

$$\downarrow$$

$$\underset{\text{Butanone}}{CH_3-CH_2-\underset{\underset{\displaystyle O}{\|}}{C}-CH_3}$$

9. Ozonolysis of Alkenes. Ozonolysis of alkenes followed by reductive hydrolysis produces ketones.

$$\underset{\text{Alkene}}{R_1-\overset{\overset{R_2}{|}}{C}=\overset{\overset{R_3}{|}}{C}-R_4} + O_3 \longrightarrow \underset{\text{Ozonide}}{R_1-\overset{\overset{R_2}{|}}{C}\underset{\underset{O\text{---}O}{|\qquad|}}{\overset{O}{\frown}}\overset{\overset{R_3}{|}}{C}-R_4}$$

$$\xrightarrow{Zn/H_2O} \underset{\text{Ketone}}{R_1-\overset{\overset{R_2}{|}}{C}=O} + \underset{\text{Ketone}}{R_3-\overset{\overset{R_4}{|}}{C}=O} + H_2O_2$$

Here R_1, R_2, R_3 and R_4 stand for alkyl groups. It may be mentioned that ozonolysis of alkenes with branching at double bond produces ketones and ozonolysis of alkenes with no branching at the double bond produces aldehydes.

10. From Acetoacetic Ester. Hydrolysis of alkyl derivatives of ethyl acetoacetates and subsequent heating yields methyl ketones.

$$\underset{\text{Alkyl acetoacetic ester}}{CH_3CO\underset{\underset{R}{|}}{C}HCOOC_2H_5} \xrightarrow{H_2O} CH_3CO-\underset{\underset{R}{|}}{C}HCOOH$$

$$\xrightarrow{\text{Heat}} \underset{\text{Ketone}}{CH_3-CO-CH_2R} + CO_2$$

5.8. PHYSICAL PROPERTIES OF ALDEHYDES AND KETONES

The important physical properties of aldehydes and ketones are described as follows:

1. Physical state. Formaldehyde is a gas at room temperature. The other members of the aldehyde family and ketones are liquids or solids.

2. Solubility. The lower members of aldehydes and ketones containing up to four carbon atoms are soluble in water. This is due to the presence of hydrogen bonding between the carbonyl compound and water, both of which are of polar nature. However, the solubility in water decreases with increase in the size of the alkyl group and the compounds with more than four carbon atoms are practically insoluble in water.

$$>\overset{\delta^+}{C}=\overset{\delta^-}{O}\ .....\ \overset{\delta^+}{H}—\overset{\overset{H}{|}}{\overset{\delta^-}{O}}\ .....$$

Aromatic aldehydes and ketones are practically insoluble in water because of very small scope of hydrogen bonding.

3. Boiling points. Both aldehydes and ketones have lower boiling points than alcohols and acids of comparable molecular masses. This is because there is no intermolecular hydrogen bonding in their molecules. Actually, they have C — H bonds which are not involved in the hydrogen bonding because of their low polarity. However, the carbonyl compounds have dipolar interactions because of their polar nature. Therefore, their boiling points are greater than those non-polar compounds such as hydrocarbons. Ketones have slightly higher boiling points than the aldehydes having the same number of carbon atoms.

The melting and boiling points of some aliphatic aldehydes and ketones arranged in increasing order of their molecular masses are given in Table 5.3.

Table 5.3. Boiling points of some aldehydes & ketones

Name	Formula	bp°(K)
Aldehydes:		
Formaldehyde	HCHO	252
Acetaldehyde	CH_3CHO	293
Propionaldehyde	CH_3CH_2CHO	323
n-Butyraldehyde	$CH_3(CH_2)_2CHO$	348
n-Valeraldehyde	$CH_3(CH_2)_3CHO$	376
n-Caproaldehyde	$CH_3(CH_2)_4CHO$	404
Ketones:		
Acetone	CH_3COCH_3	329
Methyl ethyl ketone	$CH_3COCH_2CH_3$	353
2-Pentanone	$CH_3COCH_2CH_2CH_3$	375
3-Pentanone	$CH_3CH_2COCH_2CH_3$	375
2-Hexanone	$CH_3CO(CH_2)_3CH_3$	423
3-Hexanone	$CH_3CH_2COCH_2CH_2CH_3$	

Example 1. The boiling point of acetone is 329 K while that of propanal is 322 K although both have the same molecular mass.

Solution. Boiling point is a property which is related to the polarity in the molecule. Greater the polarity in the molecule, greater will be the intermolecular attraction and hence higher will be the boiling point. Acetone has a higher boiling point compared to propanal because of double inductive effect of two alkyl groups.

$$\begin{matrix} CH_3 \searrow \\ \quad\quad >C=O \\ CH_3 \nearrow \end{matrix} \qquad CH_3 \rightarrow\!\!\!— CH_2 \rightarrow\!\!\!— CHO$$

In propanal, there is inductive effect of only one alkyl group. Hence greater polarity will be induced in acetone.

Example 2. *n*-butyl alcohol boils at 391 K while *n*-butyraldehyde boils at 349 K though their molecular masses are 74 and 72 respectively. Account for the higher b.p. of the alcohol.

Solution. There is hydrogen bonding in the case of *n*-butyl alcohol which results in association of *n*-butyl alcohol molecules.

$$\cdots\cdots H-\underset{\displaystyle C_4H_9}{\underset{|}{O}}\cdots\cdots H-\underset{\displaystyle C_4H_9}{\underset{|}{O}}\cdots\cdots H-\underset{\displaystyle C_4H_9}{\underset{|}{O}}\cdots\cdots$$

No such hydrogen bonding takes place in *n*-butyraldehyde. Hence these molecules are not associated. As such, *n*-butyl alcohol boils at a higher temperature compared to *n*-butyraldehyde. Boiling point is a property which is dependent upon the size of the molecule. Associated molecules which become bigger and heavier require a higher temperature to boil.

Example 3. Why carbonyl compounds have lower boiling points than the alcohols from which they are derived?

Solution. Carbonyl compounds have lower boiling points than the alcohols containing an equal no. of carbon atoms. For example, acetaldehyde has lower boiling point than ethyl alcohol. Acetone and propionaldehyde have lower boiling points than propyl alcohol etc.

This is explained in terms of hydrogen bonding. In aldehydes and ketones, there is no hydrogen bonding as there is not enough polarity in the molecules. But alcohols exhibit hydrogen bonding between hydrogen of one and oxygen of second molecule.

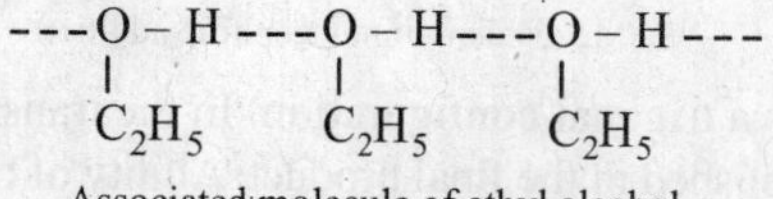

$$\text{---O}-\text{H---O}-\text{H---O}-\text{H---}$$
$$C_2H_5 \qquad C_2H_5 \qquad C_2H_5$$

Associated molecule of ethyl alcohol

As a consequence of hydrogen bonding, the molecular weight of alcohol increases. Hence, it boils at a higher temperature.

As the carbonyl compounds have no such bonding, they boil at temperatures below the boiling point of alcohols.

Example 4. *o*-salicylaldehyde has a lower boiling point compared to *m*- or *p*- isomer. Explain.

Solution.

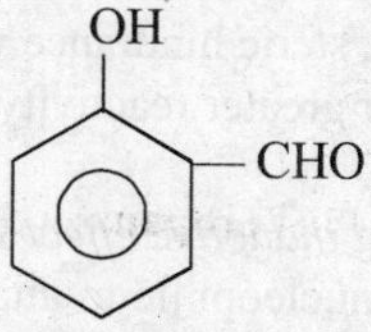

OH, CHO — *o*-salicylaldehyde

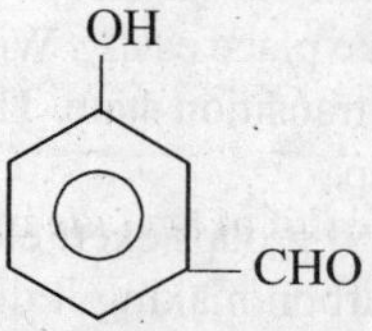

OH, CHO — *m*-salicylaldehyde

OH, CHO — *p*-salicylaldehyde

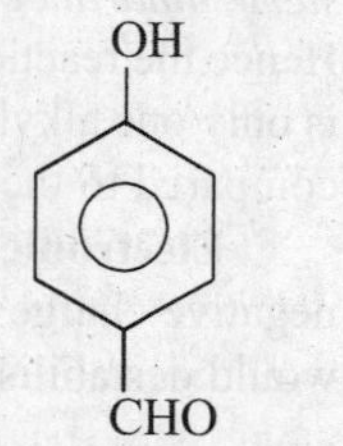

Because of the proximity of – OH and – CHO groups in *o*-salicylaldehyde, there is possibility of hydrogen bonding which reduces the size of the molecule. Consequently, the boiling point is lowered.

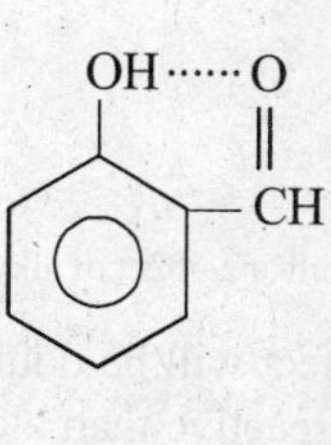

OH······O
‖
CH

5.9. CHEMICAL PROPERTIES

(A) Nucleophilic Addition Reactions

A nucleophilic addition reaction is a reaction in which the addition of a molecule to the carbonyl compound takes place with the initiation of the negative part of the adding molecule. The negative part of the adding molecule (e.g. CN in HCN) adds first to the positive carbon of the carbonyl group, followed by the attachment of the positive part of the adding molecule to the negative oxygen of the carbonyl group. The addition of a nucleophile (: Z) to a carbonyl compound may be represented as shown below.

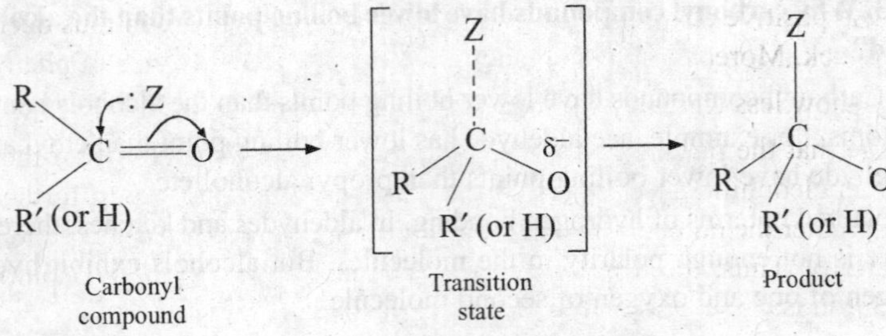

Carbonyl compound → Transition state → Product

Mechanism of nucleophilic addition

Carbonyl compound has a trigonal configuration. In the transition state, it starts acquiring sp^3 configuration which is accomplished in the final product. Ability of oxygen to acquire negative charge is the reason for nucleophilic reactions in carbonyl compounds.

Acetaldehyde is more reactive than acetone towards nucleophilic addition reactions:

Aldehydes are observed to be more reactive than ketones particularly in giving nucleophilic reactions for the following reasons:

Steric Factors. As has been explained earlier, the nucleophile attaches itself to the positive carbon of the carbonyl group. With the ketone, the transition state (see figure above) will experience *steric hindrance* because of bulky alkyl groups. Consequently the transition state will not be stable. Hence the reaction will not take place easily. With an aldehyde, steric hindrance will be less as there is only one alkyl group in the transition state. This accounts for greater reactivity of aldehyde group compared to the ketonic group.

Electronic Factors. Alkyl groups exert electron releasing *inductive effect* which increases the negative charge density on carbon making it difficult for the nucleophile to attach to carbon. This would destabilise the transition state (see figure below).

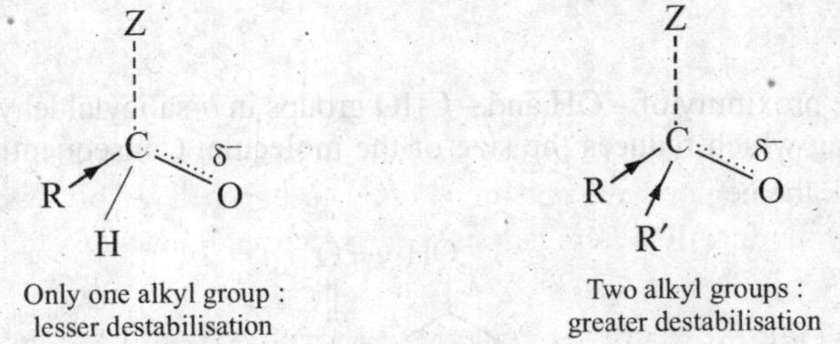

Only one alkyl group : lesser destabilisation

Two alkyl groups : greater destabilisation

Destabilising effect of alkyl groups

With ketones, this destabilising effect will be doubled because of two alkyl (or aryl) groups. This is the reason why ketones are less reactive than aldehydes.

Aromatic aldehydes and ketones are less reactive than aliphatic aldehydes and ketones:

Phenyl group has electron withdrawing inductive effect. Thus we would expect the transition state to get stabilized in the nucleophilic addition reaction of an aromatic aldehyde or ketone. But this is not actually so. As a matter of fact, aromatic aldehydes and ketones are less reactive than their aliphatic counterparts. This is so because aromatic aldehydes and ketones get stabilised due to resonance as shown below :

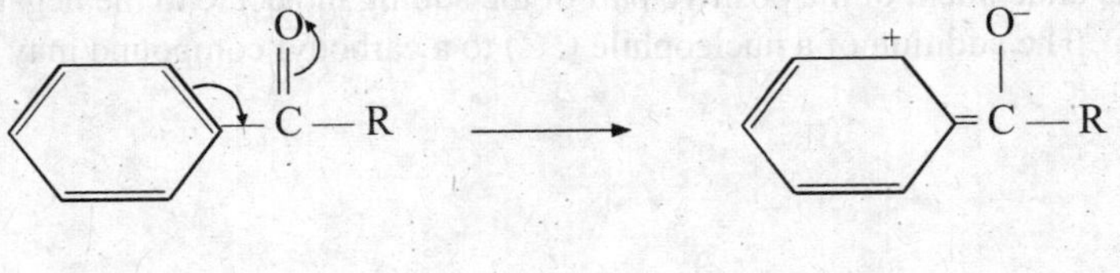

This creates a positive charge on the ring instead of carbonyl carbon thus decreasing chances of nucleophilic attack. Moreover, there are steric factors due to bulky size of phenyl group.

Hence they show less tendency for nucleophilic reactions.

We conclude that the presence of alkyl group decreases the reactivity of the aldehydes and ketones towards nucleophilic addition while the presence of phenyl group further decreases the reactivity. Thus, greater the number of such groups present or bigger the size of the groups, smaller will be the reactivity. Thus, the decreasing order of reactivity of carbonyl compounds towards nucleophilic addition is:

$$(H)(H)C{=}O > (CH_3)(H)C{=}O > (C_6H_5)(H)C{=}O > (CH_3)(H)C{=}O > (C_6H_5)(C_6H_5)C{=}O$$

Comparison of Addition reactions in Alkenes and Carbonyl compounds

Both aldehydes and ketones take part in the nucleophilic addition reactions and the addition is catalysed in the presence of acid. However, alkenes take part in the electrophilic addition reactions. If we compare the nature of the double bond in the members of both the families, we observe that in alkenes, the double bond is non-polar in nature. As there is a high electron density in the double bond, it is expected to undergo an electrophile attack to form a carbocation which then combines with the other part of the attacking reagent to give an addition product finally as shown by the following scheme:

$$>C{=}C< \;+\; \overset{\delta^+}{A}{-}\overset{\delta^-}{B} \xrightarrow{\text{Slow}} >\overset{\oplus}{C}{-}C(A)< + \bar{B}$$

Carbocation

$$\bar{B} + >C(A){-}\overset{\oplus}{C}< \xrightarrow{\text{Fast}} >C(B){-}C(A)<$$

However, elcectrophilic addition is not possible in carbonyl compounds. That will require the electrophile to attack the negative oxygen atom while carbon will be left with a sextet of electrons. This means the nucleophile will preferably attack the electron deficient carbon atom of the carbonyl group.

Addition of certain nucleophilic reagents to aldehydes and ketones is catalysed by the presence of an acid:

It has been observed that a strong nucleophile gets attached to the carbonyl group easily but some difficulty is felt by the weak nucleophile in attaching itself to the carbonyl group. In such cases, the presence of a protonic acid containing (H^+) proves helpful as illustrated below :

$$R(R'\text{ (or H)})C{=}O \underset{}{\overset{H^+}{\rightleftharpoons}} R(R'\text{ (or H)})C{=}OH^+ \xrightarrow{:Z^-} \left[R(R'\text{ (or H)})(Z)C\cdots OH \right] \longrightarrow R(R'\text{ (or H)})(Z)C{-}OH$$

Transition state — Product

Role of the acid catalyst in nucleophilic reaction

The proton from the acid attaches itself to the negative oxygen. This paves the way for the attachment of even the weakest nucleophile. The energy of activation for the nucleophilic attack is decreased thereby facilitating the reaction.

α-hydrogen in aldehydes and ketones is acidic in nature:

Electronegative nature of oxygen atom in the carbonyl group directs the flow of electrons towards oxygen. This helps in the removal of hydrogen as in the presence of a base (: B) as shown below :

$$\underset{|}{\overset{\text{H}}{\underset{\alpha}{\overset{|}{-\text{C}}}}}-\overset{\text{O}}{\overset{\|}{\text{C}}}-+:\underset{\text{Base}}{\text{B}} \rightleftharpoons \text{B}-\text{H}+\left[-\overset{-}{\underset{|}{\text{C}}}-\overset{\text{O}}{\overset{\|}{\text{C}}}- \longleftrightarrow -\underset{|}{\text{C}}=\overset{\text{O}^-}{\overset{|}{\text{C}}}-\right]$$

The carbanion obtained after the removal of hydrogen gets resonance stabilized. This is all due to the ability of oxygen to accommodate the negative charge.

Important nucleophilic addition reactions of aldehydes and ketones are discussed as under:

1. Addition of sodium bisulphite. All aldehydes and ketones react with a saturated solution of sodium bisulphite to form the solid bisulphate compound.

$$\underset{\text{Aldehyde or ketone}}{>\text{C}=\text{O}} + \text{NaHSO}_3 \rightleftharpoons \underset{\text{Bisulphite compound}}{>\text{C}<^{\text{OH}}_{\text{SO}_3\text{Na}}}$$

$$\underset{\text{Acetone}}{(\text{CH}_3)_2\text{C}=\text{O}} + \text{NaHSO}_3 \rightleftharpoons \underset{\text{Acetone bisulphite}}{(\text{CH}_3)_2\text{C}<^{\text{OH}}_{\text{SO}_3\text{Na}}}$$

$$\text{C}_6\text{H}_5-\text{CHO} + \text{NaHSO}_3 \rightleftharpoons \underset{\text{Benzaldehyde bisulphite}}{\text{C}_6\text{H}_5-\text{CH}<^{\text{OH}}_{\text{SO}_3\text{Na}}}$$

Ketones containing bulky groups fail to give the reaction due to steric hindrance.

Mechanism of the Reaction

The reaction is initiated by the attack of the bisulphite ion on the carbonyl carbon followed by the shifting of H to carbonyl oxygen. The complete mechanism is illustrated hereunder:

$$>\text{C}=\text{O} + :\overset{-}{\text{S}}\text{O}_3\text{H} \rightleftharpoons >\text{C}<^{\text{O}^-}_{\text{SO}_3\text{H}} \rightleftharpoons >\text{C}<^{\text{OH}}_{\text{SO}_3^-} \xrightleftharpoons{\text{Na}^+} >\text{C}<^{\text{OH}}_{\text{SO}_3\text{Na}}$$

Utility of the Reaction. The bisulphite compound as obtained above gets decomposed on treatment with dil. HCl giving back the aldehyde or ketone. Thus the reaction can be used to purify or separate the carbonyl compound from impurities.

$$>\text{C}<^{\text{OH}}_{\text{SO}_3\text{Na}} + \text{HCl} \longrightarrow >\text{C}=\text{O} + \text{NaCl} + \text{H}_2\text{O} + \text{SO}_2$$

2. Addition of alcohol to carbonyl compounds

Alcohols add to aldehydic compounds forming acetals. Generally the reaction takes place in two steps. In the first step, one molecule of alcohol reacts with the carbonyl compound to form hemi-acetal which is generally unstable. Therefore, hemi-acetal reacts with the second molecule of alcohol to form acetal. The reaction takes place in the presence of HCl gas. Ketones do not participate in such a reaction.

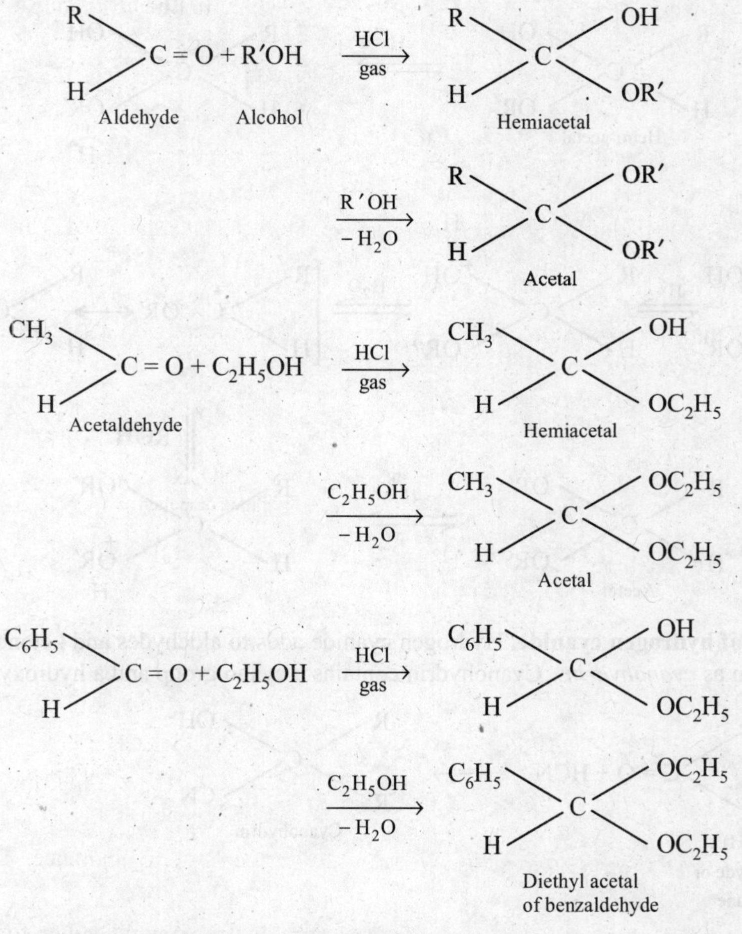

$$\underset{\text{Aldehyde}}{(R)(H)C{=}O} + \underset{\text{Alcohol}}{R'OH} \xrightarrow[\text{gas}]{HCl} \underset{\text{Hemiacetal}}{(R)(H)C(OH)(OR')}$$

$$\xrightarrow[-H_2O]{R'OH} \underset{\text{Acetal}}{(R)(H)C(OR')(OR')}$$

$$\underset{\text{Acetaldehyde}}{(CH_3)(H)C{=}O} + C_2H_5OH \xrightarrow[\text{gas}]{HCl} \underset{\text{Hemiacetal}}{(CH_3)(H)C(OH)(OC_2H_5)}$$

$$\xrightarrow[-H_2O]{C_2H_5OH} \underset{\text{Acetal}}{(CH_3)(H)C(OC_2H_5)(OC_2H_5)}$$

$$(C_6H_5)(H)C{=}O + C_2H_5OH \xrightarrow[\text{gas}]{HCl} (C_6H_5)(H)C(OH)(OC_2H_5)$$

$$\xrightarrow[-H_2O]{C_2H_5OH} \underset{\text{Diethyl acetal of benzaldehyde}}{(C_6H_5)(H)C(OC_2H_5)(OC_2H_5)}$$

R & R′ in the above reaction stand for alkyl groups.

Mechanism of the Reaction

Step 1. A molecule of alcohol adds to the carbonyl group to form a hemiacetal. This is made possible by the presence of H^+ as it produces a positive centre on the carbonyl carbon. Alcohol molecule approaches this positive centre. H^+ is regenerated in this step.

Step 2. The proton now removes the –OH group from the hemi-acetal, thus reproducing positive centre. This enables the second alcohol molecule to attach itself to the hemi-acetal. Finally, the proton is again given out. These steps are illustrated below :

$$\begin{matrix}R\\H\end{matrix}\!\!>C=O \;\overset{H^+}{\rightleftharpoons}\; \left[\begin{matrix}R\\H\end{matrix}\!\!>C=\overset{+}{O}H \longleftrightarrow \begin{matrix}R\\H\end{matrix}\!\!>\overset{+}{C}-OH\right]$$

$$\Big\Downarrow R'OH$$

$$\underset{\text{Hemi-acetal}}{\begin{matrix}R\\H\end{matrix}\!\!>C<\!\!\begin{matrix}OH\\OR'\end{matrix}} \;\overset{-H^+}{\rightleftharpoons}\; \begin{matrix}R\\H\end{matrix}\!\!>C<\!\!\begin{matrix}OH\\\overset{+}{O}R'\\|\\H\end{matrix}$$

$$\begin{matrix}R\\H\end{matrix}\!\!>C<\!\!\begin{matrix}OH\\OR'\end{matrix} \;\overset{H^+}{\rightleftharpoons}\; \begin{matrix}R\\H\end{matrix}\!\!>C<\!\!\begin{matrix}H\\|\\\overset{+}{O}H\\OR?\end{matrix} \;\overset{-H_2O}{\rightleftharpoons}\; \left[\begin{matrix}R\\H\end{matrix}\!\!>\overset{+}{C}-OR' \longleftrightarrow \begin{matrix}R\\H\end{matrix}\!\!>C=\overset{+}{O}R'\right]$$

$$\Big\Downarrow ROH$$

$$\underset{\text{Acetal}}{\begin{matrix}R\\H\end{matrix}\!\!>C<\!\!\begin{matrix}OR'\\OR'\end{matrix}} \;\overset{-H^+}{\rightleftharpoons}\; \begin{matrix}R\\H\end{matrix}\!\!>C<\!\!\begin{matrix}OR'\\\overset{+}{O}R'\\H\end{matrix}$$

3. Addition of hydrogen cyanide. Hydrogen cyanide adds to aldehydes and ketones to form compounds known as *cyanohydrins*. Cyanohydrin contains a cyano group and a hydroxy group.

$$\underset{\substack{\text{Aldehyde or}\\\text{ketone}}}{\begin{matrix}R\\R\\\text{(or H)}\end{matrix}\!\!>C=O} + HCN \longrightarrow \underset{\text{Cyanohydrin}}{\begin{matrix}R\\R\end{matrix}\!\!>C<\!\!\begin{matrix}OH\\CN\end{matrix}}$$

$$\underset{\text{Acetaldehyde}}{\begin{matrix}CH_3\\H\end{matrix}\!\!>C=O} + HCN \longrightarrow \underset{\substack{\text{Acetaldehyde}\\\text{cyanohydrin}}}{\begin{matrix}CH_3\\H\end{matrix}\!\!>C<\!\!\begin{matrix}OH\\CN\end{matrix}}$$

$$\underset{\text{Acetone}}{\begin{matrix}CH_3\\CH_3\end{matrix}\!\!>C=O} + HCN \longrightarrow \underset{\text{Acetone cyanohydrin}}{\begin{matrix}CH_3\\CH_3\end{matrix}\!\!>C<\!\!\begin{matrix}OH\\CN\end{matrix}}$$

Mechanism of the Reaction

Step 1. Cyanide ion attaches itself to the carbonyl carbon by nucleophilic addition process.

Step 2. H^+ then attaches itself to the oxygen, which carries a negative charge, to produce the final product.

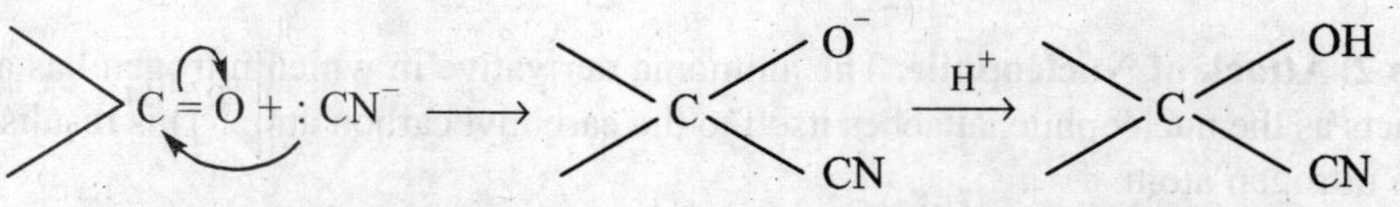

$$>C=O + :CN^- \longrightarrow >C(O^-)(CN) \xrightarrow{H^+} >C(OH)(CN)$$

Cyanohydrins are important compounds as they yield α-hydroxy carboxylic compounds on hydrolysis.

$$\underset{\text{Acetaldehyde cyanohydrin}}{CH_3CH(OH)CN} \xrightarrow{\text{Hydrolysis}} \underset{\text{Lactic acid}}{CH_3CH(OH)COOH}$$

4. Reaction of ammonia derivatives. Derivatives of ammonia *viz.* hydroxylamine, hydrazine, phenyl hydrazine and semicarbazide react with the carbonyl compounds to give oximes, hydrazones, phenyl hydrazones and semi-carbazones respectively. as shown below :

(*i*) $>C=O + \underset{\text{Hydroxyl amine}}{H_2NOH} \longrightarrow \underset{\text{Oxime}}{>C=NOH} + H_2O$

(*ii*) $>C=O + \underset{\text{Hydrazine}}{H_2NNH_2} \longrightarrow \underset{\text{Hydrazone}}{>C=NNH_2} + H_2O$

(*iii*) $>C=O + \underset{\text{Phenylhydrazine}}{H_2NNHC_6H_5} \longrightarrow \underset{\text{Phenylhydrazone}}{>C=N\,NHC_6H_5} + H_2O$

(*iv*) $>C=O + \underset{\text{Semicarbazide}}{H_2N\,NHCONH_2} \longrightarrow \underset{\text{Semicarbazone}}{>C=N\,NHCONH_2} + H_2O$

(*v*) $>C=O + H_2NNH-C_6H_3(NO_2)_2$ (2 : 4 Dinitrophenyl hydrazine)

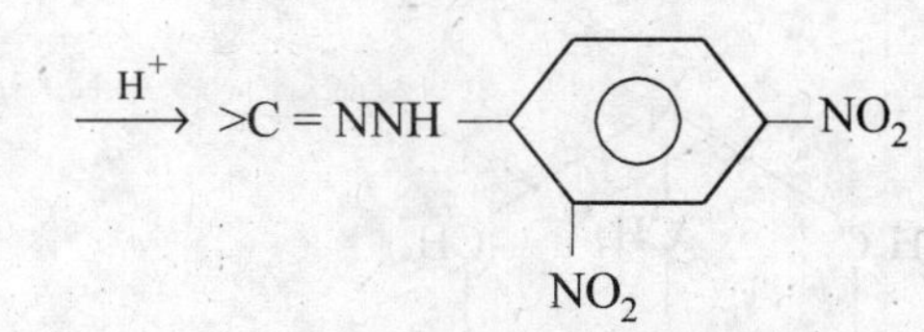

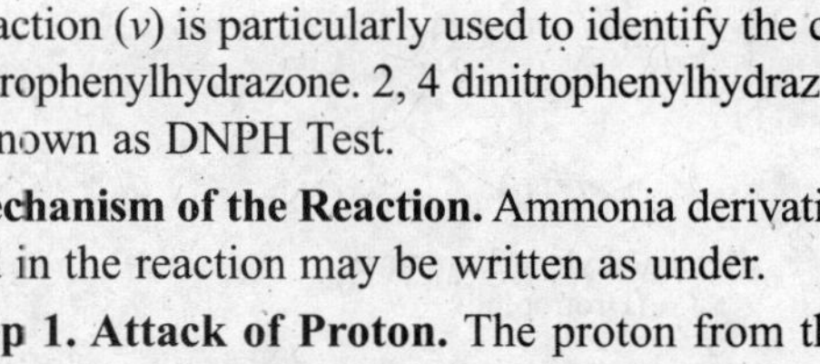

$\xrightarrow{H^+} >C=NNH-C_6H_3(NO_2)_2$ (2, 4, Dinitrophenylhydrazone)

Reaction (*v*) is particularly used to identify the carbonyl group as it gives yellow derivative viz. 2, 4 dinitrophenylhydrazone. 2, 4 dinitrophenylhydrazine is abbreviated as DNPH. Hence this reaction is also known as DNPH Test.

Mechanism of the Reaction. Ammonia derivatives may be represented by H_2NG. Various steps involved in the reaction may be written as under.

Step 1. Attack of Proton. The proton from the acid which is used to catalyse the reaction attaches itself to the electron rich carbonyl oxygen atom as shown below :

$$>C=O \overset{H^+}{\rightleftharpoons} >C=\overset{+}{O}H \xrightarrow{H_2\ddot{N}-G} \left[>C<\begin{matrix}OH\\ \overset{+}{H_2N}-G\end{matrix}\right] \xrightarrow{-H^+} \left[>C<\begin{matrix}OH\\ HN-G\end{matrix}\right] \xrightarrow{-H_2O} >C=N-G$$

Step 2. Attack of Nucleophile. The ammonia derivative in which nitrogen has a lone pair of electron acts as the nucleophile, attaches itself to the carbonyl carbon atom. This results in a positive charge on nitrogen atom.

Step 3. Loss of Proton. The intermediate formed in step 2 is unstable and loses a proton.

Step 4. Loss of Water Molecule. The product formed in step 3 loses a water molecule to form a crystalline compound having a double bond between carbonyl carbon and nitrogen ammonia derivative.

$$\underset{\text{Acetaldehyde}}{CH_3CHO} + \underset{\text{Hydrazine}}{NH_2NH_2} \longrightarrow \underset{\text{Acetaldehyde hydrazone}}{CH_3CH=N-NH_2} + H_2O$$

The mechanism of the reaction is as follows :

$$\begin{matrix}CH_3\\ H\end{matrix}>C=O \overset{H^+}{\rightleftharpoons} \begin{matrix}CH_3\\ H\end{matrix}>C=\overset{+}{O}H \qquad \text{(Attack of proton)}$$

$$\begin{matrix}CH_3\\ H\end{matrix}>C=\overset{+}{O}H \xrightarrow{H_2\ddot{N}-NH_2} \begin{matrix}CH_3\\ H\end{matrix}>C<\begin{matrix}OH\\ H_2\overset{+}{N}-NH_2\end{matrix} \qquad \text{(Attack of nucleophile)}$$

$$\begin{matrix}CH_3\\ H\end{matrix}>C<\begin{matrix}OH\\ H_2\overset{+}{N}-NH_2\end{matrix} \xrightarrow{-H^+} \begin{matrix}CH_3\\ H\end{matrix}>C<\begin{matrix}OH\\ HN-NH_2\end{matrix} \qquad \text{(Removal of } H^+\text{)}$$

$$\begin{matrix}CH_3\\ H\end{matrix}>C<\begin{matrix}OH\\ HNHH_2\end{matrix} \xrightarrow{-H_2O} \begin{matrix}CH_3\\ H\end{matrix}>C=N-NH_2 \qquad \text{(Removal of } H_2O\text{)}$$

5. Reaction of ammonia with formaldehyde, acetaldehyde and benzaldehyde.

(*i*) Ammonia reacts with formaldehyde to give a white crystalline solid known as hexamethylene tetramine.

$$\underset{\text{Formaldehyde}}{6HCHO + 4NH_3} \longrightarrow (CH_2)_6N_4 + 6H_2O$$

Hexamethylene Tetramine (Urotropine)

Hexamethylene tetramine is used as a urinary antiseptic medicine.

(*ii*) Ammonia reacts with acetaldehyde to form a crystalline product called acetaldehyde ammonia. This product is unstable and loses a water molecule to yield aldimine.

$$\underset{H}{\overset{CH_3}{>}}C=O + NH_3 \longrightarrow \underset{\text{Acetaldehyde ammonia}}{\underset{H}{\overset{CH_3}{>}}C\underset{NH_2}{\overset{OH}{<}}} \xrightarrow{-H_2O} \underset{\text{Aldimine}}{\underset{H}{\overset{CH_3}{>}}C=NH}$$

Aldimine polymerises as under :

$$nCH_3CHNH \longrightarrow (CH_3CHNH)_n$$

(*iii*) Ammonia reacts with benzaldehyde to form hydrobenzamide.

$$3\,C_6H_5CHO + 2NH_3 \longrightarrow \underset{\text{Hydrobenzamide}}{C_6H_5CH=N-CH(C_6H_5)-N=CHC_6H_5} + 3H_2O$$

6. Reaction with Grignard reagent

Carbonyl compounds (aldehydes and ketones) react with Grignard reagents forming addition compounds which get decomposed by dil. HCl to yield different alcohols.

Formation of Primary Alcohols. Primary alcohols are obtained by the reaction of formaldehyde with the Grignard reagent.

$$\underset{H}{\overset{H}{>}}C=O + \underset{\text{Ethyl mag. iodide}}{C_2H_5MgI} \longrightarrow \underset{\text{Addition compound}}{\underset{H}{\overset{H}{>}}C\underset{C_2H_5}{\overset{OMgI}{<}}}$$

$$\xrightarrow[H_2O]{\text{dil. HCl}} \underset{\text{Propyl alcohol}}{\underset{H}{\overset{H}{>}}C\underset{C_2H_5}{\overset{OH}{<}}} + \underset{\text{Magnesium Hydroxy iodide}}{Mg(OH)I}$$

A desired primary alcohol can be obtained by treating formaldehyde with a suitable Grignard reagent. For example, ethyl alcohol will be obtained by the action of methyl magnesium iodide on formaldehyde. Butyl alcohol will be obtained if formaldehyde is reacted with propyl magnesium iodide and so on.

Victor Grignard

Formation of Secondary Alcohol. A secondary alcohol is obtained when an aldehyde other than formaldehyde is treated with a Grignard reagent.

$$\underset{\text{Acetaldehyde}}{\underset{H}{\overset{CH_3}{>}}C=O} + CH_3MgBr \longrightarrow \underset{H}{\overset{CH_3}{>}}C\underset{CH_3}{\overset{OMgBr}{<}}$$

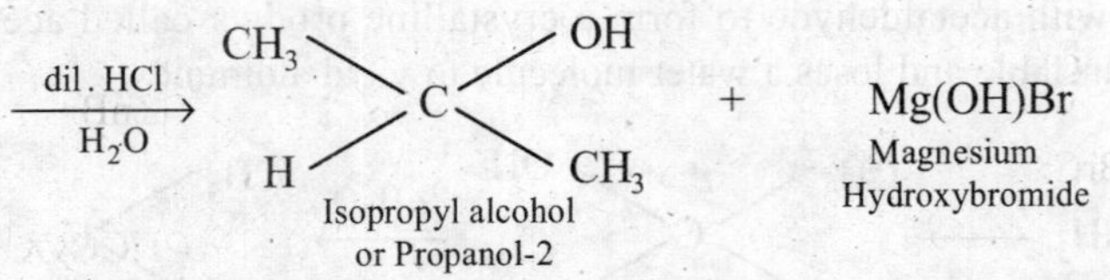

$$\xrightarrow[H_2O]{dil.\,HCl} \begin{matrix} CH_3 \\ H \end{matrix}\!\!>\!C\!<\!\!\begin{matrix} OH \\ CH_3 \end{matrix} + Mg(OH)Br$$

Isopropyl alcohol or Propanol-2 — Magnesium Hydroxybromide

Again a desired secondary alcohol can be produced by combining a suitable aldehyde with Grignard reagent.

Formation of Tertiary Alcohol. A tertiary alcohol is formed by the reaction of ketone with a Grignard reagent followed by hydrolysis.

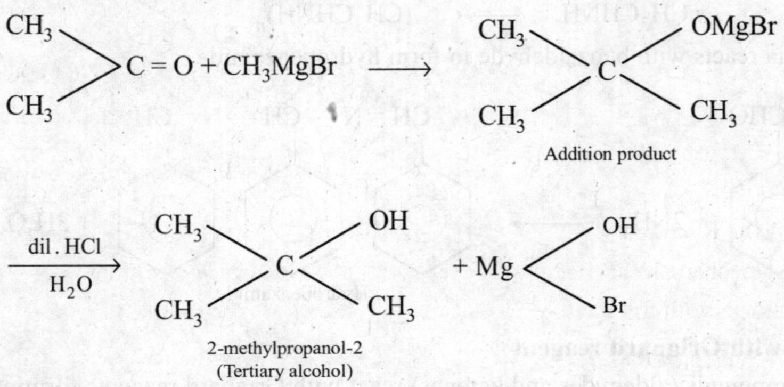

$$\begin{matrix} CH_3 \\ CH_3 \end{matrix}\!\!>\!C=O + CH_3MgBr \longrightarrow \begin{matrix} CH_3 \\ CH_3 \end{matrix}\!\!>\!C\!<\!\!\begin{matrix} OMgBr \\ CH_3 \end{matrix}$$

Addition product

$$\xrightarrow[H_2O]{dil.\,HCl} \begin{matrix} CH_3 \\ CH_3 \end{matrix}\!\!>\!C\!<\!\!\begin{matrix} OH \\ CH_3 \end{matrix} + Mg\!<\!\!\begin{matrix} OH \\ Br \end{matrix}$$

2-methylpropanol-2 (Tertiary alcohol)

Mechanism of the Reaction

The reaction proceeds as a nucleophilic addition reaction. Grignard reagent (R Mg X) consists of two parts R^- and MgX^+. The nucleophile R^- attacks the carbonyl group at carbon followed by the attachment of positive group MgX^+ at carbonyl oxygen, forming the addition product. The product is decomposed by water in the presence of HCl as described above.

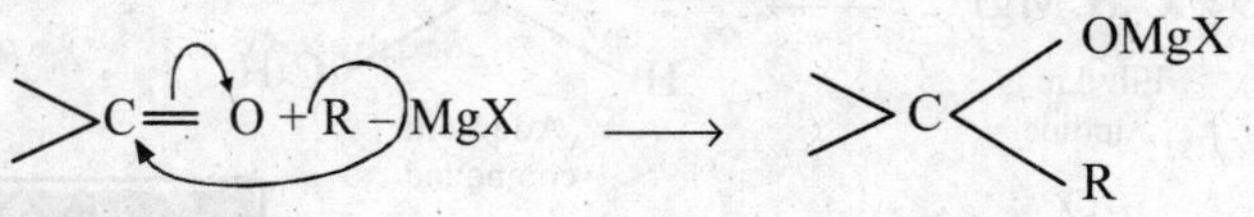

$$>C=O + R-MgX \longrightarrow >C\!<\!\!\begin{matrix} OMgX \\ R \end{matrix}$$

(B) CONDENSATION REACTIONS OF ALDEHYDES AND KETONES

Reformatsky reaction

Reformatsky reaction is performed in order to obtain β-hydroxy esters from aldehydes and ketones using α-bromo ester and zinc in the presence of dry ether. The reactions with an aldehyde and a ketone are depicted below :

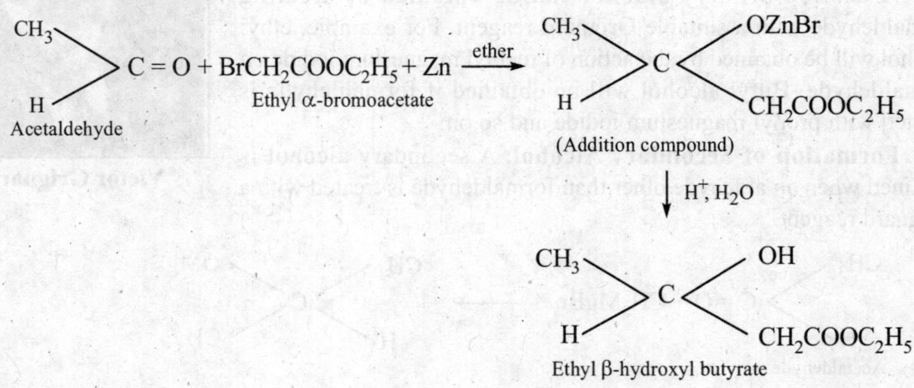

$$\begin{matrix} CH_3 \\ H \end{matrix}\!\!>\!C=O + BrCH_2COOC_2H_5 + Zn \xrightarrow{ether} \begin{matrix} CH_3 \\ H \end{matrix}\!\!>\!C\!<\!\!\begin{matrix} OZnBr \\ CH_2COOC_2H_5 \end{matrix}$$

Acetaldehyde — Ethyl α-bromoacetate — (Addition compound)

$$\downarrow H^+, H_2O$$

$$\begin{matrix} CH_3 \\ H \end{matrix}\!\!>\!C\!<\!\!\begin{matrix} OH \\ CH_2COOC_2H_5 \end{matrix}$$

Ethyl β-hydroxyl butyrate

$$\underset{\text{Acetone}}{(CH_3)_2C=O} + BrCH(CH_3)COOC_2H_5 + Zn \xrightarrow{\text{Ether}} \underset{\text{(Addition product)}}{(CH_3)_2C(OZnBr)CH(CH_3)COOC_2H_5}$$

$$\xrightarrow{H^+, H_2O} \underset{\text{Ethyl 2, 3-dimethyl 3-hydroxy butyrate.}}{(CH_3)_2C(OH)CH(CH_3)COOC_2H_5}$$

Mechanism of the Reaction

The following mechanism has been proposed for the Reformatsky reaction.

Step 1. Zinc reacts with the ester to form an organometallic compound.

$$Zn + BrCH_2COOC_2H_5 \xrightarrow{\text{Ether}} BrZnCH_2COOC_2H_5$$

Step 2. Organometallic compound reacts with the carbonyl compound by nucleophilic addition.

$$>C=O + BrZnCH_2COOC_2H_5 \longrightarrow \underset{\text{Addition Product}}{>C(OZnBr)CH_2COOC_2H_5}$$

Step 3. The addition product formed above gets hydrolysed to give the final product.

$$>C(OZnBr)CH_2COOC_2H_5 \xrightarrow{H^+, H_2O} >C(OH)CH_2COOC_2H_5 + Zn(OH)Br$$

2. Aldol condensation. Two molecules of an aldehyde or a ketone having α hydrogen atom react in the presence of dil. alkali to form β-hydroxyaldehyde or β-hydroxy ketone.

$$\underset{\text{Acetaldehyde}}{CH_3-\overset{O}{\overset{\|}{C}}-H} + \underset{\text{Acetaldehyde}}{HCH_2CHO} \xrightarrow{OH^-} \underset{\text{3-hydroxybutanol (Aldol)}}{CH_3-\overset{OH}{\overset{|}{C}}(H)-CH_2CHO}$$

$$\underset{\text{Acetone}}{CH_3-\overset{O}{\overset{\|}{C}}-CH_3} + \underset{\text{Acetone}}{HCH_2COCH_3} \xrightarrow{OH^-} \underset{\text{Diacetone alcohol}}{CH_3-\overset{OH}{\overset{|}{C}}(CH_3)-CH_2COCH_3}$$

Mechanism of the Reaction

Step 1. Formation of carbanion. The hydroxide ion from a base removes the proton from the α-carbon atom to give a carbanion. This carbanion gets resonance stabilised.

Step 2. Attack by carbanion. The carbanion thus formed attacks the second molecule by nucleophilic addition mechanism to form an alkoxide ion.

Step 3. Attachment of a hydrogen. The alkoxide ion attaches a hydrogen atom removed from water molecule.

These steps as mentioned above are sketched as under :

$$H\bar{O} + H-CH_2-\underset{H}{\underset{|}{C}}=O \rightleftharpoons \left[\bar{C}H_2-\underset{H}{\underset{|}{C}}=O \longleftrightarrow CH_2=\underset{H}{\underset{|}{C}}-\bar{O}\right] + H_2O$$

$$CH_3-\overset{H}{\overset{|}{C}}=O \rightleftharpoons CH_3-\overset{H}{\overset{|}{\underset{O^-}{\underset{|}{C}}}}-CH_2-CHO \overset{H_2O}{\rightleftharpoons} CH_3-\overset{H}{\overset{|}{\underset{OH}{\underset{|}{C}}}}-CH_2-CHO+\bar{O}H$$

It may be noted that aldol condensation is a property of aldehydes and ketones having hydrogen atoms attached to α-carbon atoms. Hence this reaction is not observed with compounds like formaldehyde, benzaldehyde which don't process α-hydrogen atoms.

Another point which may be noted is, that β-hydroxy-aldehydes and β-hydroxy ketones which are obtained in the base-catalysed reaction easily get dehydrated on warming with dilute acids to form α - β unsaturated carbonyl compounds.

$$\underset{\text{Aldol}}{CH_3\underset{OH}{\underset{|}{CH}}CH_2CHO} \xrightarrow[-H_2O]{H^+} \underset{\text{Crotonaldehyde}}{CH_3CH=CHCHO}$$

This is the reason we avoid obtaining aldol by acid-catalysed reaction, although, in principle, it is possible to carry out aldol condensation in the presence of an acid.

Aldol reaction provides an important tool in the hands of a chemist to synthesize α, β unsaturated compounds.

3. Crossed Aldol Condensation. We may come across cases involving reaction between two different aldehydes or ketones or an aldehyde and a ketone. Such reactions are known as crossed aldol condensation. If both the reacting carbonyl compounds have α-hydrogen atoms, we may get four different products which are difficult to separate. But by adjusting experimental condition it is possible to obtain one product in abundance. Out of the two carbonyl compounds, the one having an α-hydrogen atom forms the carbanion and the other having no α-hydrogen atom acts as the site for attack by the nucleophile carbanion.

$$\underset{\text{Formaldehyde}}{H-\overset{H}{\overset{|}{C}}=O} + \underset{\text{Acetaldehye}}{H-CH_2CHO} \xrightarrow{OH^-} \underset{\text{3-Hydroxypropionaldehyde}}{H-\overset{H}{\overset{|}{\underset{CH_2CHO}{\underset{|}{C}}}}-OH}$$

$$\underset{\text{Formaldehyde}}{H-\overset{\displaystyle H}{\overset{|}{C}}=O} + \underset{\text{Acetone}}{HCH_2COCH_3} \xrightarrow{OH^-} \underset{\text{4-Hydroxy-2-butanone}}{H-\overset{\displaystyle H}{\overset{|}{C}}-OH \atop {|\atop CH_2COCH_3}}$$

Claisen

Mechanism of crossed aldol condensation is the same as that of aldol condensation.

4. Claisen-Schmidt reaction. This reaction may be considered as a special case of crossed aldol condensation. It involves condensation of aromatic carbonyl compound with an aliphatic carbonyl compound in the presence of a base. Hydroxy derivative first obtained loses a water molecule to yield α, β- unsaturated aldehyde or ketone. For example

$$\underset{\text{Benzaldehyde}}{C_6H_5-CHO} + \underset{\text{Acetaldehyde}}{CH_3CHO} \xrightarrow{OH^-} \underset{\text{Addition compound}}{C_6H_5-\overset{\displaystyle OH}{\overset{|}{C}H}-CH_2CHO} \xrightarrow{-H_2O} \underset{\text{Cinnamaldehyde}}{C_6H_5-CH=CHCHO}$$

$$C_6H_5-CHO + CH_3COCH_3 \xrightarrow{OH^-} \underset{\text{Addition compound}}{C_6H_5-\overset{\displaystyle OH}{\overset{|}{C}H}-CH_2COCH_3} \xrightarrow{-H_2O} \underset{\text{Benzalacetone}}{C_6H_5-CH=CHCOCH_3}$$

Mechanism. It involves the following steps:

Step 1. Formation of carbanion. α-hydrogen of the aliphatic carbonyl compound is removed by the hydroxide ions, thus creating carbanions.

Step 2. Attack by carbanion on the aromatic carbonyl compound. The carbanion obtained in step 1 attacks the aromatic carbonyl group by nucleophilic addition mechanism to form the alkoxide ion.

Step 3. Attachment of a Proton. The alkoxide ion removes a proton from water and attaches to itself to give β-hydroxy compound.

Step 4. α-hydrogen atom in the hydroxy compound is removed by the base followed by the removal of —OH group, thus removing a water molecule finally. The sequence of the above steps is shown below :

$$HO^- + H - CH_2 - CHO \rightleftharpoons \bar{C}H_2 - CHO + H_2O$$

$$C_6H_5 - \underset{\underset{O}{\|}}{\overset{H}{\overset{|}{C}}} + \bar{C}H_2CHO \rightleftharpoons C_6H_5 - \underset{\underset{O^-}{|}}{\overset{H}{\overset{|}{C}}} - CH_2CHO \xrightleftharpoons{H_2O} C_6H_5 - \underset{\underset{OH}{|}}{\overset{H}{\overset{|}{C}}} - \underset{\underset{H}{|}}{CH} - CHO + OH^-$$

$$C_6H_5 - \underset{\underset{OH}{|}}{\overset{H}{\overset{|}{C}}} - \underset{\underset{H}{|}}{CH} - CHO \xrightleftharpoons{OH^-} C_6H_5 - \underset{\underset{OH}{|}}{CH} - \bar{C}H - CHO + H_2O$$

$$\Big\downarrow\!\!\Big\uparrow - OH^-$$

$$C_6H_5 - CH = CHCHO$$

5. Benzoin Condensation. Aldol condensation is a property of carbonyl compounds containing α-hydrogen atoms. Thus benzaldehyde which does not contain α-hydrogen atom is not capable of giving this reaction. Such aromatic aldehydes condense in the presence of KCN (which acts as a base) to yield *benzoins*. This reaction is known as *Benzoin condensation*.

$$C_6H_5-CHO + OHC-C_6H_5 \xrightarrow{KCN} C_6H_5-\underset{}{\overset{OH}{\overset{|}{CH}}}-\overset{O}{\overset{\|}{C}}-C_6H_5$$

Benzaldehyde Benzaldehyde Benzoin

Mechanism. It involves the following steps :

Step 1. Formation of a carbanion. Direct formation of a carbanion is not possible, as there are no α-hydrogen atoms. Hence this carbanion formation takes place after some rearrangement. It involves attachment of CN^- to the carbonyl carbon with simultaneous shifting of electrons to oxygen. Hydrogen is then shifted from carbon to oxygen. This results in the formation of carbanion.

Step 2. Attack of the carbanion (nucleophile) on the second molecule. Carbanion obtained in step 1 above attacks the carbonyl group of the second molecule forming an addition product, an alkoxide.

Step 3. Loss of cyanide ion. The alkoxide loses cyanide ion with simultaneous loss of hydrogen from the same carbon atom. Hydrogen combines with the oxygen on the neighbouring carbon atom giving the final products. The sequence of steps is illustrated below:

$$C_6H_5 - \overset{O}{\overset{\|}{C}} - H + CN^- \rightleftharpoons C_6H_5 - \underset{\underset{CN}{|}}{\overset{O^-}{\overset{|}{C}}} - H \rightleftharpoons C_6H_5 - \underset{\underset{CN}{|}}{\overset{OH}{\overset{|}{C^-}}}$$

$$C_6H_5-\underset{CN}{\overset{OH}{C^-}} + \underset{H}{\overset{O}{\overset{\|}{C}}}-C_6H_5 \rightleftharpoons C_6H_5-\underset{CN}{\overset{O-H}{C}} - \underset{H}{\overset{O^-}{C}}-C_6H_5$$

$$\rightleftharpoons C_6H_5-\overset{O}{\overset{\|}{C}} - \underset{H}{\overset{OH}{C}}-C_6H_5 + CN^-$$

6. Perkin's reaction. It is a kind of aldol condensation in which an aromatic aldehyde condenses with acetic anhydride in the presence of sodium acetate to yield unsaturated carboxylic acid. Thus benzaldehyde on treatment with acetic anhydride and sod. acetate yields cinnamic acid.

$$\underset{\text{Benzaldehyde}}{C_6H_5CHO} + \underset{\text{Acetic anhydride}}{(CH_3CO)_2O} \xrightarrow[180^\circ C]{\text{Sod. acetate}} C_6H_5-\underset{OH}{CH}-CH_2CO-O-COCH_3$$

$$\xrightarrow{-CH_3COOH} \underset{\text{Cinnamic acid}}{C_6H_5-CH=CHCOOH}$$

Mechanism. It involves the following steps :

Step 1. Formation of a carbanion. Acetic anhydride in the presence of acetate ion loses a hydrogen forming a carbanion.

Step 2. Attack of the carbanion on the carbonyl group. Carbanion obtained in the above step attacks the carbonyl carbon atom (nucleophilic attack) forming on alkoxide.

Step 3. Protonation of the alkoxide ion. Alkoxide gets protonated to form an aldol type compound.

Step 4. Dehydration. The hydroxy group and neighbouring hydrogen are removed as water forming an unsaturated product.

Step 5. Hydration. The alkenic compound obtained above gets hydrolysed giving the final product.

William Henry Perkin

The above steps are illustrated as given below :

$$C_6H_5-\underset{H}{C}=O + \underset{\text{Carbanion}}{\overset{\ominus}{C}H_2-\overset{O}{\overset{\|}{C}}-O-\underset{O}{\underset{\|}{C}}-CH_3} \longrightarrow C_6H_5-\underset{CH_2-\overset{O}{\overset{\|}{C}}-O-\underset{O}{\underset{\|}{C}}-CH_3}{\overset{H}{C}}-O^-$$

$$\downarrow H_2O$$

$$C_6H_5-\underset{\underset{CH_3-C(=O)-O-C(=O)-CH}{|}}{\overset{H}{C}}(OH)\xrightarrow{-H_2O} C_6H_5-CH=C\text{ (cyclic anhydride adduct)}$$

$$C_6H_5-CH=CH-\overset{O}{\overset{\|}{C}}-O-H + CH_3COO^{\ominus}$$

Cinnamic acid

7. Knoevenagal Reaction. An aldehyde or a ketone on condensation with a compound having an active methylene group (such as malonic ester or acetoacetic ester) in the presence of a base such as pyridine gives an unsaturated compound.

$$CH_3-\overset{H}{\overset{|}{C}}=\boxed{O+H_2}C\begin{matrix}COOC_2H_5\\COOC_2H_5\end{matrix}\xrightarrow[-H_2O]{\text{Pyridine-}}CH_3-\overset{H}{\overset{|}{C}}=C\begin{matrix}COOC_2H_5\\COOC_2H_5\end{matrix}$$

Acetaldehyde Malonic ester

$$\xrightarrow{H_2O,\ H^+\text{ and heat}} CH_3CH=CHCOOH$$

Crontonic acid

$$CH_3-\overset{CH_3}{\overset{|}{C}}=\boxed{O+H_2}C\begin{matrix}COCH_3\\COOC_2H_5\end{matrix}\xrightarrow[-H_2O]{\text{Pyridine}}CH_3-\overset{CH_3}{\overset{|}{C}}=C\begin{matrix}COCH_3\\COOC_2H_5\end{matrix}$$

Acetoacetic ester

$$\xrightarrow{H_2O,\ H^+} CH_3-\overset{CH_3}{\overset{|}{C}}=\underset{\underset{COOC_2H_5}{|}}{C}HCOCH_3$$

Mechanism. Mechanism of the reaction is explained by taking the example of reaction between an aldehyde and aceto acetic ester.

Step 1. Formation of a carbanion. The base B : attracts a hydrogen from the acetoacetic ester forming a carbanion.

Step 2. Attack by the carbanion. The carbanion as obtained above attacks the carbonyl carbon by nucleophilic chains forming an alkoxide ion.

Step 3. Protonation. The alkoxide draws a proton to convert itself into a hydroxy compound.

Step 4. Dehydration. Hydroxy and hydrogen from the neighbouring carbon atoms are removed as water, and an unsaturated compound is obtained. The above steps are illustrated as follows :

$$B:+H_2C\begin{matrix}COCH_3\\COOC_2H_5\end{matrix}\longrightarrow H\overline{C}\begin{matrix}COCH_3\\COOC_2H_5\end{matrix}+BH^+$$

$$R-\underset{\underset{O}{\|}}{\overset{H}{\overset{|}{C}}}+H\bar{C}\begin{matrix}COCH_3\\COOC_2H_5\end{matrix} \longrightarrow R-\underset{\underset{O^-}{|}}{\overset{H}{\overset{|}{C}}}-\underset{\underset{COCH_3}{|}}{CH}-COOC_2H_5$$

$$\xrightarrow{H^+} R-\underset{\underset{OH}{|}}{\overset{H}{\overset{|}{C}}}-\underset{\underset{COCH_3}{|}}{CHCOOC_2H_5} \xrightarrow{-H_2O} R-CH=\underset{\underset{COCH_3}{|}}{C}-COOC_2H_5$$

8. Mannich Reaction. Formaldehyde on condensation with an amine and an active methylene compound gives a product which is called *Mannich Base*. This reaction is called Mannich reaction.

$$C_6H_5-\overset{\overset{O}{\|}}{C}-CH_3+HCHO+(CH_3)_2NH \xrightarrow{-H_2O} \underset{\text{Mannich base}}{C_6H_5-\overset{\overset{O}{\|}}{C}-CH_2-CH_2-N(CH_3)_2}$$

Mechanism. It involves the following steps :

Step 1. Formation of Imminium ion. Formaldehyde reacts with the amine to form an imminium salt which gives rise to the imminium ion.

Step 2. Formation of the carbanion. The active methylene compound in the presence of a base gives the carbanion.

Step 3. Attack by the carbanion. Carbanion formed in the above step attacks the imminium ion to give the final product. The above steps are illustrated below.

$$(CH_3)_2\ddot{N}-H+\underset{\underset{H}{|}}{\overset{H}{\overset{|}{C}}}=O \rightleftharpoons \underset{\text{Imminium salt}}{(CH_3)_2\ddot{N}-\underset{\underset{H}{|}}{\overset{H}{\overset{|}{C}}}-OH} \rightleftharpoons \underset{\text{Imminium ion}}{(CH_3)_2\overset{+}{N}=CH_2}+OH^-$$

$$R-\overset{\overset{O}{\|}}{C}-CH_3 \underset{}{\overset{\text{Base}}{\rightleftharpoons}} \left[R-\overset{\overset{O}{\|}}{C}-\bar{C}H_2 \longleftrightarrow R-\overset{\overset{O^-}{|}}{C}=CH_2\right]$$

$$H_2C=\overset{+}{N}(CH_3)_2 \longrightarrow R-\overset{\overset{O}{\|}}{C}-CH_2-CH_2-N(CH_3)_2$$

9. Wittig's Reaction. This reaction takes place between an aldehyde or ketone and phosphorus ylides to form substituted alkenes. The phosphorus ylides required in the reaction are obtained by the action of a base on suitable alkyl triphenyl phosphonium halides (prepared from alkyl halide and triphenyl phosphine). For example,

$$(C_6H_5)_3\,P + CH_3\,I \longrightarrow [(C_6H_5)_3\,\overset{+}{P}CH_3]\,\overset{-}{I} \xrightarrow{C_6H_5Li\ (Base)} (C_6H_5)_3\,P = CH_2 + C_6H_6 + LiI$$

Triphenyl phosphorus; Methyl iodide; Methylene triphenyl phosphorane (ylide)

George Wittig

The ylides obtained as shown as above are not separated but are made to react as such with the aldehydes or ketones in solution.

Some examples of Wittig reaction are given below:

$$>C = \boxed{O + (C_6H_5)_3\,P =}\ CRR' \longrightarrow >C = C<^{R}_{R'} + (C_6H_5)_3\,PO$$

Alkene; Triphenyl phosphine oxide

$$C_6H_5-CHO + (C_6H_5)_3\,P = CH_2 \longrightarrow C_6H_5-CH = CH_2 + (C_6H_5)_3\,PO$$

Methylene triphenyl Phosphorane; Styrene

$$(CH_3)_2C = O + (C_6H_5)_3\,P = CH-CH_3 \longrightarrow (CH_3)_2C = CH-CH_3 + (C_6H_5)_3\,PO$$

Ethylidene triphenyl phosphorane; 2-Methylbutene-2

Mechanism. The following steps are involved in the reaction :

Step 1. Attack by ylides. The phosphorus ylides exist as a resonance hybrid of two structures.

$$\underset{[A]}{(C_6H_5)_3\,P = CRR'} \longleftrightarrow \underset{[B]}{(C_6H_5)_3\,P^+ - \overset{-}{C}RR'}$$

Structure [B] has a negative charge on carbon, thus it acts as a carbanion and initiates the nucleophilic attack on the carbonyl carbon. The resulting addition product is known as *betaine*.

Step 2. Elimination of Triphenyl phosphine oxide. Betaine obtained above undergoes elimination of triphenyl phosphine oxide to yield the alkene. The above steps are illustrated hereunder.

$$>C=O + \overset{-}{C}RR'-\overset{+}{P}(C_6H_5)_3 \longrightarrow \underset{\text{Betaine}}{-\underset{O^-}{\underset{|}{C}}-\underset{^+P(C_6H_5)_3}{\underset{|}{C}RR'}} \longrightarrow >C = CRR' + (C_6H_5)_3\,PO$$

Example 1. Suggest a mechanism for the following conversion. Name the product.

$$C_6H_{10}{=}O \longrightarrow C_6H_{10}{=}CH_2$$

Solution. Cyclohexanone $=O + (C_6H_5)_3P = CH_2 \longrightarrow$ Methylene cyclohexane $= CH_2 + (C_6H_5)_3PO$

Cyclohexanone | Methylene triphenyl phosphorane | Methylene cyclohexane | Triphenyl phosphine oxide

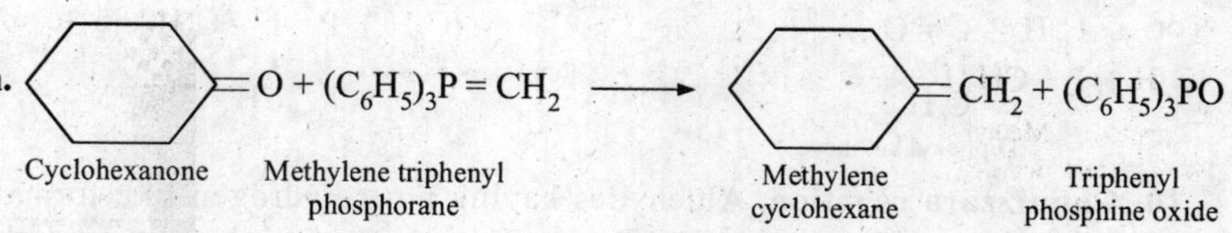

Mechanism of the Reaction

$$(C_6H_5)_3P = CH_2 \longrightarrow (C_6H_5)_3\overset{+}{P} - \overset{-}{C}H_2$$

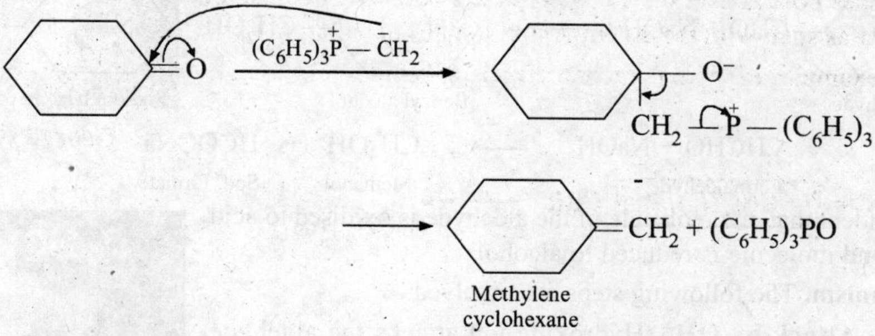

$$\text{Cyclohexanone}{=}O \xrightarrow{(C_6H_5)_3\overset{+}{P} - \overset{-}{C}H_2} \text{cyclohexyl}(-O^-)(-CH_2-\overset{+}{P}-(C_6H_5)_3)$$

$$\longrightarrow \text{cyclohexane}{=}CH_2 + (C_6H_5)_3PO$$

Methylene cyclohexane

Example 2. Which ylides and carbonyl compounds would you use to prepare

(*a*) $C_6H_5 - CH = C(CH_3) - CH_3$

(*b*) $CH_3 - C(CH_3) = C(C_6H_5) - CH_3$

(*c*) $C_2H_5 - C(CH_3) = C(C_6H_5) - C_6H_5$

Solution. The following pairs of carbonyl compounds and ylides are used to prepare the above compounds.

	Carbonyl compound		Ylide
(*a*)	$C_6H_5 - CHO$ (Benzaldehyde)	+	$(C_6H_5)_3P = C(CH_3) - CH_3$
or	$CH_3 - C(CH_3) = O$ (Acetone)	+	$(C_6H_5)_3P = CHC_6H_5$
(*b*)	$CH_3 - C(CH_3) = O$ (Acetone)	+	$(C_6H_5)_3P = C(C_6H_5) - CH_3$
or	$CH_3 - C(C_6H_5) = O$ (Methyl phenyl ketone)	+	$(C_6H_5)_3P = C(CH_3) - CH_3$
(*c*)	$C_2H_5 - C(CH_3) = O$ (Methyl ethyl ketone)	+	$(C_6H_5)_3P = C(C_6H_5) - C_6H_5$

or $C_6H_5-\underset{\underset{C_6H_5}{|}}{C}=O$ + $(C_6H_5)_3P=\underset{\underset{CH_3}{|}}{C}-C_2H_5$

Diphenyl ketone

10. Cannizzaro reaction. Aldehydes having no α-hydrogen like formaldehyde and benzaldehyde undergo self oxidation and reduction. This reaction is given when an aldehyde is treated with a conc. solution of sodium or potassium hydroxide. An alcohol and a salt of carboxylic acid are produced. For example:

$$2\,C_6H_5-CHO + NaOH \longrightarrow C_6H_5-CH_2OH + C_6H_5-COONa$$

Benzaldehyde — Benzyl alcohol — Sodium bezoate

$$2HCHO + NaOH \longrightarrow CH_3OH + HCOONa$$

Formaldehyde — Methanol — Sod. formate

It is evident that one molecule of the aldehyde is oxidised to acid and the second molecule is reduced to alcohol.

Cannizzaro

Mechanism. The following steps are involved:

Step 1. Attack by OH^-. Hydroxide ion attacks the aldehyde molecule to form an anion.

Step 2. Hydride ion transfer. Oxidation-reduction takes place between the anion formed in step 1 and the second molecule of aldehyde through transfer of a hydride ion. These two steps are illustrated hereunder.

$$C_6H_5-\overset{\overset{O}{\|}}{C}-H + \bar{O}H \rightleftharpoons C_6H_5-\overset{\overset{O^-}{|}}{\underset{\underset{OH}{|}}{C}}-H$$

$$C_6H_5-\overset{\overset{O^-}{|}}{\underset{\underset{OH}{|}}{C}}-H \quad C_6H_5-\overset{\overset{O}{\|}}{C}-H \longrightarrow C_6H_5-\overset{\overset{O}{\|}}{\underset{\underset{OH}{|}}{C}} + C_6H_5-\overset{\overset{O^-}{|}}{\underset{\underset{H}{|}}{C}}-H$$

$$\downarrow -H^+ \qquad\qquad \downarrow +H^+$$

$$C_6H_5COO^- \qquad C_6H_5CH_2OH$$

(C) HALOGENATION OF CARBONYL COMPOUNDS

The α- hydrogen atoms in the carbonyl compounds can be replaced by halogens in the presence of acids and bases. These are discussed below separately as base-promoted halogenation and acid-promoted halogenation.

Base-promoted halogenation. One or more α- hydrogen atoms in the carbonyl compound can be replaced by halogens in the presence of a base like OH^- or CH_3COO^- ions.

$$CH_3COCH_3 + Br_2 + OH^- \longrightarrow CH_3COCH_2Br + H_2O + Br^-$$

Acetone — Bromo acetone

Mechanism

Kinetics studies reveal that the rate of above reaction is proportional to the concentration of the carbonyl compound and the base. It is independent of the concentration of bromine.

Rate of halogenation ∝ [Carbonyl compound] × [Base].

This is explained in terms of formation of a carbanion intermediate as the slow and rate determining step followed by fast bromination process as illustrated hereunder.

$$CH_3-\overset{\overset{\large O}{\|}}{C}-CH_3 + :B \xrightleftharpoons[\text{Slow}]{-BH} \left[CH_3-\overset{\overset{\large O}{\|}}{C}-\overset{-}{C}H_2 \longleftrightarrow CH_3-\overset{\overset{\large O^-}{|}}{C}=CH_2 \right] \xrightarrow[\text{Fast}]{Br_2} CH_3-\overset{\overset{\large O}{\|}}{C}-CH_2Br + Br^-$$

As the base is actually consumed in the rate determining step, we call it base-promoted reaction rather than base-catalysed reaction.

Certain observations have been made in respect of base-promoted halogenation reaction.

(1) In base promoted halogenation of unsymmetrical ketones (RCOR′), halogenation occurs preferably at the methyl group rather than the methylene group.

$$CH_3CH_2COCH_3 + Br_2 \xrightarrow{\text{Base}} CH_3CH_2COCH_2Br + HBr$$

This is because the electron releasing effect of R alkyl group (CH_3-CH_2- in the above case) decreases the acidity of the methylene hydrogen atom and prevents the formation of the carbanion. This results in the preferential formation of carbanion (I) which is more stable than the carbanion (II).

$$\underset{\text{(I)}}{CH_3 \rightarrow\!\!- \bar{C}H_2-\overset{\overset{\large O}{\|}}{C}-\bar{C}H_2} \qquad \underset{\text{(II)}}{CH_3-\bar{C}H-\overset{\overset{\large O}{\|}}{C}-CH_3}$$

(2) Base-promoted halogenation of a monohalogenated ketone takes place preferentially at the carbon already halogenated as long as it has a hydrogen atom. This is because the hydrogen atom attached to carbon containing halogen atoms are more acidic and likely to be replaced by halogen as per the above mechanism. The two carbanions resulting from the halogenated ketone are written below. Obviously carbanion III is more easily formed than IV. Hence further halogenation occurs at carbon already halogenated.

$$\underset{\text{(III)}}{CH_3-\overset{\overset{\large O}{\|}}{C}-\bar{C}H \rightarrow\!\!- Br} \qquad \underset{\text{(IV)}}{\bar{C}H_2-\overset{\overset{\large O}{\|}}{C}-CH_2Br}$$

Acid-promoted Halogenation. Here halogenation of the carbonyl compound takes place in the presence of H^+ ions.

$$CH_3-CO-CH_3 + Br_2 \xrightarrow{H^+} CH_3-\overset{\overset{\large O}{\|}}{C}-CH_2Br + HBr$$

Mechanism. Like in case of base-promoted reaction, here, too, the rate of the reaction is proportional to the concentration of the carbonyl compound and the acid. It is independent of the concentration of the halogen.

Rate of halogenation ∝ [Carbonyl compound.] [H^+].

This shows the possibility of formation of intermediate in the rate-determining slow step of the reaction followed by fast attachment of the halogen. The intermediate in acid-promoted halogenation is an *enol*. The steps involved in the reaction are reproduced as follows:

$$\text{(I)}\quad CH_3-\overset{\overset{\displaystyle O}{\|}}{C}-CH_3+H-Z \xrightleftharpoons{\text{Fast}} CH_3-\overset{\overset{\displaystyle \overset{+}{O}H}{\|}}{C}-CH_3+\overset{-}{Z}$$

$$\text{(II)}\quad CH_3-\overset{\overset{\displaystyle \overset{+}{O}H}{\|}}{C}-CH_2-H+\overset{-}{Z} \xrightarrow{\text{Slow}} \underset{\text{Enol}}{CH_3-\overset{\overset{\displaystyle :\ddot{O}H}{|}}{C}=CH_2}+H-Z$$

$$\text{(III)}\quad CH_3-\overset{\overset{\displaystyle :\ddot{O}H}{|}}{C}=CH_2+X-X \xrightarrow{\text{Fast}} \underset{\text{Protonated Haloketone}}{CH_3-\overset{\overset{\displaystyle \overset{+}{O}H}{\|}}{C}-CH_2X}+\overset{-}{X}$$

$$\text{(IV)}\quad CH_3-\overset{\overset{\displaystyle \overset{+}{O}-H}{\|}}{C}-CH_2X+\overset{-}{Z} \xrightarrow{\text{Fast}} CH_3-\overset{\overset{\displaystyle O}{\|}}{C}-CH_2X+H-Z$$

Following observations have been made in respect of acid-promoted reactions.

(1) In the acid-catalysed halogenation of ketones, halogenation takes place preferably at the methylene group rather than the methyl group. Thus in the compound $CH_3CH_2COCH_3$, halogenation would yield $CH_3CHBr\ COCH_3$ and not $CH_3CH_2COCH_2Br$. This is explained by the fact that enols which are more branched are more stable. The enols which would be formed with halogenation at the methylene and methyl groups respectively are given below :

$$\underset{\text{(I)}}{CH_3CH_2-\overset{\overset{\displaystyle OH}{|}}{C}=CH_2} \qquad \underset{\text{(II)}}{CH_3CH=\overset{\overset{\displaystyle OH}{|}}{C}-CH_3}$$

Less branched less stable   More branched more stable

(2) Further halogenation of the monohalogen ketones does not take place readily because the halogen already present tends to withdraw the electrons towards itself. This prevents the attachment of hydrogen ion on the carbonyl oxygen.

Aromatic aldehydes and ketones undergo electrophilic substitution reaction with the halogen in the presence of halogen carriers like FeX_3. The halogen enters at the meta position w.r.t. the carbonyl group.

$$C_6H_5CHO + Cl_2 \xrightarrow{FeCl_3} m\text{-}ClC_6H_4CHO + HCl$$

m-chloro-benzaldehyde

(D) REDUCTION REACTIONS

Reduction to hydrocarbons

(a) Clemmensen reduction. Aldehydes and ketones give hydrocarbons on reduction with zinc amalgam and concentrated hydrochloric acid.

$$\underset{\text{Acetone}}{CH_3-\overset{\overset{O}{\|}}{C}-CH_3} \xrightarrow{Zn-Hg/HCl} \underset{\text{Propane}}{CH_3CH_2CH_3} + H_2O$$

$$C_6H_5-\overset{\overset{O}{\|}}{C}-CH_3 \xrightarrow{Zn-Hg/HCl} \underset{\text{Ethyl benzene}}{C_6H_5-CH_2-CH_3} + H_2O$$

Mechanism. The mechanism involves reduction under acidic conditions. The carbonyl group gets protonated and the protonated species takes up electrons from the metal atom.

$$R_2C{=}O \overset{H^+}{\rightleftharpoons} \left[R_2C{=}\overset{+}{O}-H \longleftrightarrow R_2\overset{+}{C}-OH\right] \xrightarrow{2e^-\ (Zn)} R_2\overset{-}{C}-OH \xrightarrow{2H^+\ (acid)} R_2\underset{H}{\underset{|}{C}}-\overset{+}{O}H_2$$

$$R_2\underset{H}{\underset{|}{C}}-\overset{+}{O}H_2 \xrightarrow{-H_2O} R_2\underset{H}{\underset{|}{C^+}} \xrightarrow{2e^-\ (Zn)} R_2\underset{H}{\underset{|}{C^-}} \xrightarrow{H^+} R_2CH_2$$

(b) Wolff-Kishner reduction. Both aldehydes and ketones can be reduced to hydrocarbons on heating with hydrazine and KOH in the presence of ethylene glycol or dimethyl sulphoxide acting as the solvents.

$$\underset{\text{Acetaldehyde}}{CH_3CH{=}O} \xrightarrow[-H_2O]{NH_2NH_2} \underset{\text{Hydrazone}}{CH_3CH{=}NNH_2} \xrightarrow{KOH} \underset{\text{Ethane}}{CH_3CH_3} + N_2$$

$$\underset{\text{Acetone}}{CH_3-\underset{CH_3}{\underset{|}{C}}{=}O} \xrightarrow[-H_2O]{NH_2NH_2} \underset{\text{Hydrazone}}{CH_3-\underset{CH_3}{\underset{|}{C}}{=}NNH_2} \xrightarrow{KOH} CH_3CH_2CH_3 + N_2$$

Mechanism of the reaction is understood to be as

$$\underset{\text{Hydrazone}}{R_2C{=}N-NH_2} \xrightarrow[-H_2O]{OH^-} R_2C{=}N-\overset{\ominus}{N}H \longleftrightarrow R_2\overset{\ominus}{C}-N{=}N-H \xrightarrow{H^+\ shift} R_2C(H)-N{=}\overset{-}{N}$$

$$R_2C(H)-N{=}\overset{-}{N} \xrightarrow{-N_2} R_2\overset{-}{C}-H \xrightarrow[-OH^-]{H_2O} \underset{\text{Alkane}}{R_2CH_2}$$

(c) Reduction with HI in the presence of red phosphorus. Both aldehydes and ketones are reduced to hydrocarbons

$$\underset{\text{Acetaldehyde}}{CH_3CHO} + 4HI \xrightarrow[\text{Heat}]{\text{Red P}} \underset{\text{Ethane}}{CH_3CH_3} + H_2O + 2I_2$$

$$CH_3COCH_3 + 4HI \xrightarrow[\text{Heat}]{\text{Red P}} CH_3CH_2CH_3 + H_2O + 2I_2$$

Reduction to alcohols

Reduction of aldehydes and ketones into alcohols can be achieved by the following different methods.

1. Hydrogenation. Aldehydes and ketones on hydrogenation with molecular hydrogen in the presence of nickel produce primary and secondary alcohols respectively.

$$\underset{\text{Propanal}}{CH_3CH_2CHO} + H_2 \xrightarrow{\text{Ni}} \underset{\text{Propanol}}{CH_3CH_2CH_2OH}$$

$$\underset{\text{Propanone}}{CH_3COCH_3} + H_2 \xrightarrow{\text{Ni}} \underset{\text{Propanol-2}}{CH_3CHOHCH_3}$$

However, this method has the disadvantage of reducing some other reducible group such as a double bond in the compound also.

$$\underset{\text{Crotonaldehyde}}{CH_3CH = CHCHO} + H_2 \xrightarrow{\text{Ni}} \underset{\text{Butanol-1}}{CH_3CH_2CH_2CH_2OH}$$

2. Reduction with $LiAlH_4$. This reagent reduces the carbonyl group into alcoholic group without touching the double bond.

$$\underset{\text{Crotonaldehyde}}{CH_3CH = CHCHO} \xrightarrow[\text{[H]}]{LiAlH_4} \underset{\text{Crotonyl alcohol}}{CH_3CH = CHCH_2OH}$$

3. Reduction with $NaBH_4$. Sodium borohydride has the same function as $LiAlH_4$ although the former is much milder than the latter.

$$\underset{\text{Cinnamaldehyde}}{C_6H_5{-}CH = CHCHO} \xrightarrow{NaBH_2} \underset{\text{Cinnamyl alcohol}}{C_6H_5{-}CH = CHCH_2OH}$$

Sodium borohydride is uneffective on ethylenic double bond, nitro group and nitrile group.

Mechanism of reduction of acetone with lithium aluminium hydride

It involves the following steps :

Step 1. Formation of alkoxide. Hydride ion from $LiAlH_4$ combines with carbonyl carbon as shown ahead.

$$LiAlH_4 \longrightarrow Li^+ + Al^{3+} + 4H^-$$

$$CH_3{-}\overset{\overset{\displaystyle O}{\|}}{C}{-}CH_3 + :\bar{H} \longrightarrow CH_3{-}\underset{\underset{\displaystyle H}{|}}{\overset{\overset{\displaystyle O^-}{|}}{C}}{-}CH_3$$

Step 2. Lithium ion combines with the alkoxide ion.

$$CH_3-\underset{\underset{H}{|}}{\overset{\overset{O^-}{|}}{C}}-CH_3+Li^+ \longrightarrow CH_3-\underset{\underset{H}{|}}{\overset{\overset{OLi}{|}}{C}}-CH_3$$

Step 3. Lithium alkoxide is changed into alcohol by adding dilute acid.

$$CH_3-\underset{\underset{H}{|}}{\overset{\overset{OLi}{|}}{C}}-CH_3+\overset{+}{H} \longrightarrow \underset{\text{Isopropyl alcohol}}{CH_3-\underset{\underset{H}{|}}{\overset{\overset{OH}{|}}{C}}-CH_3}+\overset{+}{Li}$$

4. Meerwein Pondorof Verely reduction. The reduction is carried by heating a ketone with isopropyl alcohol in the presence of aluminium isopropoxide. Acetone formed is distilled off and the product left is a secondary alcohol.

$$\underset{\text{Ketone}}{\begin{matrix}R\\R\end{matrix}\!\!>\!C=O}+\begin{matrix}CH_3\\CH_3\end{matrix}\!\!>\!CHOH \underset{}{\overset{Al[OCH(CH_3)_2]_3}{\rightleftharpoons}} \underset{\text{Sec. alcohol}}{\begin{matrix}R\\R\end{matrix}\!\!>\!CHOH}+\begin{matrix}CH_3\\CH_3\end{matrix}\!\!>\!C=O$$

Mechanism. The mechanism of the reduction proceeds as follows:

$$\begin{matrix}R\\R\end{matrix}\!\!>\!C=O \longleftrightarrow \begin{matrix}R\\R\end{matrix}\!\!>\!C=O\rightarrow Al[OCH(CH_3)_2]_3 \;(\text{with } H-C(H_3C)(CH_3)-\ddot{O}-H) \longrightarrow \begin{matrix}R\\R\end{matrix}\!\!>\!C\!-\!H\!-\!O^-\!-\!\overline{Al}[OCH(CH_3)_2]_3 + \overset{+}{C}(CH_3)(CH_3)-\ddot{O}-H$$

$$\downarrow$$

$$Al[OCH(CH_3)_2]_3+\underset{\text{Sec. alcohol}}{\begin{matrix}R\\R\end{matrix}\!\!>\!CHOH}+\begin{matrix}CH_3\\CH_3\end{matrix}\!\!>\!C=O$$

5. Bimolecular Reduction. Aldehydes and ketones undergo bimolecular reduction in the presence of Mg amalgam to form symmetrical glycols known as **pinacols.**

$$2\underset{\text{Acetone}}{\begin{matrix}CH_3\\CH_3\end{matrix}\!\!>\!C=O} \xrightarrow[\text{Benzene}]{\text{Mg Amalgam}} \begin{matrix}CH_3-\overset{\overset{CH_3}{|}}{C}-O\\CH_3-\underset{\underset{CH_3}{|}}{C}-O\end{matrix}\!\!>\!Mg \xrightarrow{H_2O} \underset{\text{Pinacol}}{\begin{matrix}CH_3-\overset{\overset{CH_3}{|}}{C}-OH\\CH_3-\underset{\underset{CH_3}{|}}{C}-OH\end{matrix}}$$

Mechanism. It involves formation of a bond between two carbonyl carbon atoms as shown below:

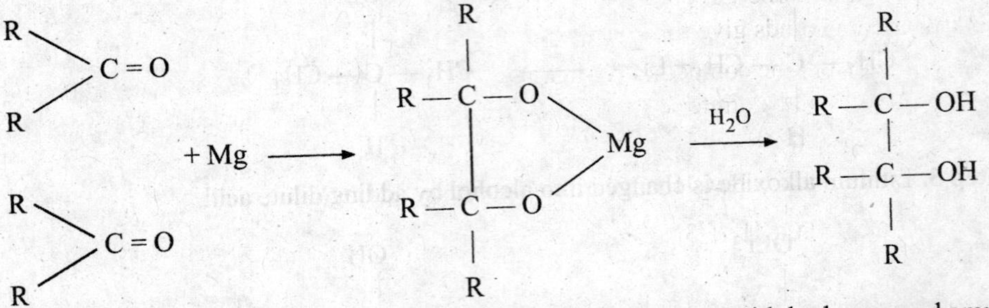

$$2\,R_2C{=}O + Mg \longrightarrow \begin{matrix} R \\ | \\ R-C-O \\ | \\ R-C-O \\ | \\ R \end{matrix}\!\!>Mg \xrightarrow{H_2O} \begin{matrix} R \\ | \\ R-C-OH \\ | \\ R-C-OH \\ | \\ R \end{matrix}$$

(*ii*) **Reductive Amination.** Aldehydes and ketones on treatment with hydrogen and ammonia in the presence of nickel are converted into amines. This reaction is known as reductive amination.

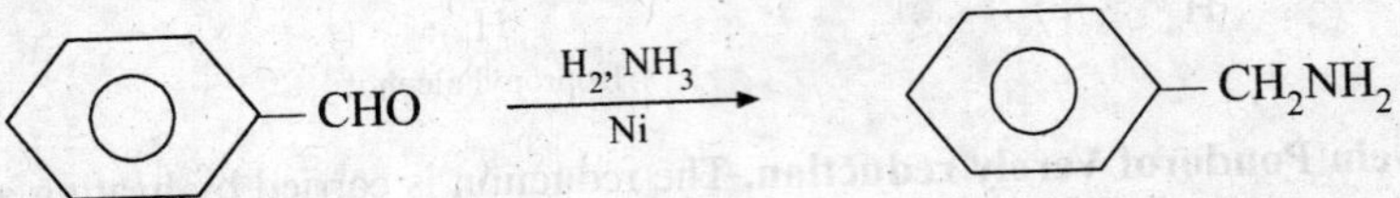

$$C_6H_5CHO \xrightarrow[Ni]{H_2,\ NH_3} C_6H_5CH_2NH_2$$

Benzaldehyde → Benzylamine

$$CH_3CH_2COCH_3 \xrightarrow[Ni]{H_2,\ NH_3} CH_3CH_2CH(NH_2)-CH_3$$

Methyl ethyl ketone → 2-Aminobutane

(E) OXIDATION REACTIONS

1. Oxidation of Aldehydes

(*a*) With Tollen reagent. Tollen's reagent is a mixture of silver nitrate and ammonium hydroxide. Aldehydes reduce the Tollen's reagent on heating, silver is deposited on the inner walls of the test tube. This is used as a test for the aldehydic group and is known as **silver mirror test.**

$$RCHO + 2[Ag(NH_3)_2]^+[OH]^- \longrightarrow RCOONH_4 + \underset{\text{Silver mirror}}{2\,Ag} + 3NH_3 + H_2O$$

(*b*) With Fehling solution. Fehling solution is a mixture of copper sulphate, sodium hydroxide and sod. pot. tartarate. The resulting solution liberates very small amounts of copper ions. Aldehydes on warming with Fehling solution produces a red precipitate of cuprous oxide.

$$RCHO + \underset{\text{Cupric ion}}{2Cu^{2+}} + Na^+ + 5[OH]^- \longrightarrow RCOONa + \underset{\text{Cuprous oxide (Red ppt.)}}{Cu_2O} + 3H_2O$$

2. Oxidation of Ketones

Ketones don't contain an easily oxidisable group. As such they don't react with Tollen's reagent or Fehling solution. However, in the presence of strong oxidising agents and on prologned heating, these are oxidised to carboxy acids containing fewer carbon atoms than the original ketone.

$$CH_3CH_2CH_2COCH_2CH_3 \xrightarrow[K_2Cr_2O_7/H^+]{[O]} \underset{\text{Butanoic acid}}{CH_3CH_2CH_2COOH} + \underset{\text{Ethanoic acid}}{CH_3COOH}$$

3. Haloform Reaction

It is a reaction given by a compound which contains $CH_3CO–$ group or a group oxidisable to this group. Thus acetone CH_3COCH_3, Acetaldehyde CH_3CHO, acetophenone $CH_3COC_6H_5$ and similar other compounds give this reaction.

In this reaction, the compound containing $CH_3CO–$ group is treated with a halogen (chlorine, bromine or iodine) and a dilute solution of sodium hydroxide to form haloform CHX_3 (chloroform, bromoform or iodoform).

$$NaOH + X_2 \longrightarrow NaOX + HX$$

$$CH_3COCH_3 + 3NaOX \longrightarrow \underset{\text{Haloform}}{CHX_3} + CH_3COONa + 2NaOH$$

Compounds like ethanol and ethanal also give this reaction.

$$CH_3CH_2OH + X_2 \longrightarrow CH_3CHO + 2HX$$

$$CH_3CHO + 3X_2 \longrightarrow \underset{\substack{\text{Trihalo}\\\text{acetaldehyde}}}{CX_3CHO} + 3HX$$

$$CX_3CHO + NaOH \longrightarrow \underset{\text{Haloform}}{CHX_3} + \underset{\text{Sod. formate}}{HCOONa}$$

Thus any compound which contains a $CH_3CO–$ group or compounds like CH_3CH_2OH or CH_3CHO or $CH_3CHOHCH_3$ which on oxidation would produce $CH_3CO–$ group, give this test. Generally we use I_2 / NaOH mixture in which case iodoform is produced as result of the reaction. Iodoform being a yellow solid is easily identified.

Mechanism. When methyl ketone is treated with halogen in the presence of alkali, then the hydrogen atoms of the methyl group get replaced by the halogen atoms. The carbonyl carbon atom is then attacked by the base OH^- ion. This results in the formation of carboxylate ion and haloforms as per the following equations:

$$\underset{\text{Acetone}}{R-\overset{\overset{\displaystyle O}{\|}}{C}-CH_3} + 3X_2 \xrightarrow[-3HX]{OH^-} R-\overset{\overset{\displaystyle O}{\|}}{C}-CX_3 \xrightarrow{:OH^-} R-\overset{\overset{\displaystyle O^{\ominus}}{|}}{\underset{\underset{\displaystyle OH}{|}}{C}}-CX_3 \longrightarrow R-\overset{\overset{\displaystyle O}{\|}}{\underset{\underset{\displaystyle OH}{|}}{C}} + \bar{C}X_3$$

$$R-\overset{\overset{\displaystyle O}{\|}}{\underset{\underset{\displaystyle O-H}{|}}{C}} + \bar{C}X_3 \rightleftharpoons \underset{\text{Salt of acid}}{R-\overset{\overset{\displaystyle O}{\|}}{\underset{\underset{\displaystyle O^-}{|}}{C}}} + \underset{\text{Haloform}}{CHX_3}$$

Importance of the Reaction. The reaction assumes analytical importance as it can be used to detect a $CH_3CO–$ group in a compound. If the substance responds to this test, it is certain that there is a $CH_3CO–$ group or a group oxidisable to this group, in the compound. As has been explained above, the substance is treated with a mixture of iodine and dil. NaOH when a yellow precipitate of iodoform is formed.

$$CH_3COCH_3 + 3NaOI \longrightarrow CH_3COCI_3 + 3\ NaOH$$

$$CH_3COCI_3 + NaOH \longrightarrow CH_3COONa + \underset{\substack{\text{Iodoform}\\\text{(yellow ppt.)}}}{CHI_3}$$

Example 1. Identify the compounds which respond to haloform test : CH_3CO_2H, CH_3CH_2OH, CH_3CONH_2, CH_3COCl, $CH_3CHOHCH_3$, CD_3CHO, $C_6H_5COCD_3$ and $CH_3CH_2COCH_2CH_3$

Solution. Compounds which contain $CH_3CO–$ group or a group which can be oxidised to this group give haloform reaction.

In the list given above, the compounds that respond to haloform test are :

CH_3CH_2OH, $CH_3CHOHCH_3$, CD_3CHO, $C_6H_5COCD_3$

D stands for deuterium *i.e.* heavy hydrogen. The compounds containing heavy hydrogen give the reaction but at a slower speed.

Example 2. Complete the following reactions indicating against each the name of the reaction.

(*i*) $C_6H_5OH + CHCl_3 + NaOH \longrightarrow$

(*ii*) $CH_3CHO + HCN \xrightarrow{H_3O^+}$

(*iii*) $C_6H_5CHO + (CH_3CO)_2O \xrightarrow{CH_3COONa}$

(*iv*) (cyclohexanone) $=O + (C_6H_5)_3 P = CH_2 \longrightarrow$

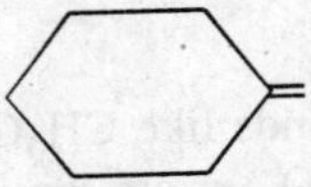

Solution. (*i*) $C_6H_5OH + CHCl_3 + 3NaOH$

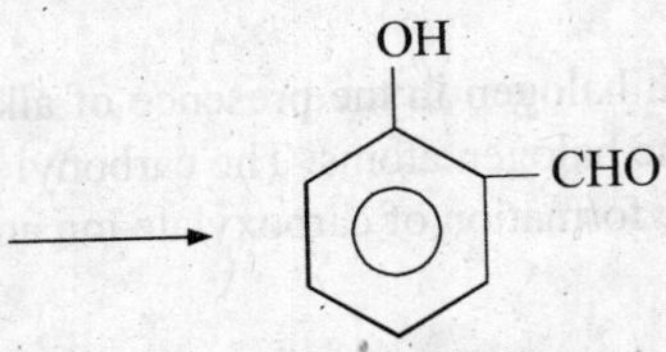

$\longrightarrow$ $C_6H_4(OH)CHO$ (o-) $+ 3\,NaCl + 2H_2O$

Salicyladehyde

This reaction is Reimer-Tiemann Reaction

(*ii*) $CH_3CHO + HCN \longrightarrow CH_3CH(OH)—CN \xrightarrow{H_3O^+} CH_3CH(OH)COOH$

α-hydroxy propionic acid

This is nucleophilic addition of HCN to acetaldehyde followed by hydrolysis.

(*iii*) $C_6H_5CHO + (CH_3CO)_2O \xrightarrow{CH_3COONa} C_6H_5CH = CHCOOH$

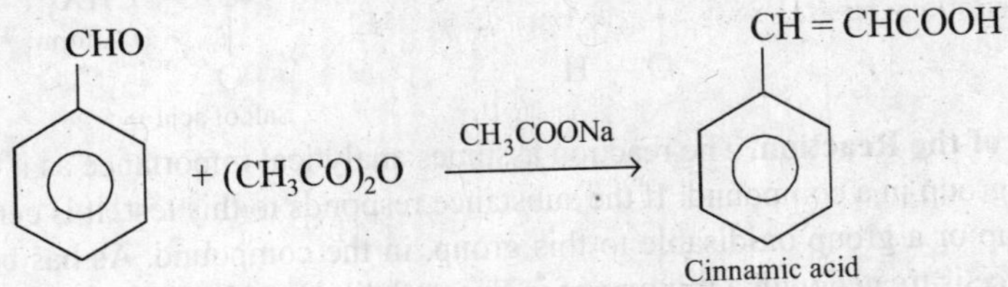

Cinnamic acid

This is a case of Perkin's Reaction.

$$(iv)\quad \text{Cyclohexanone}{=}O + (C_6H_5)_3P{=}CH_2 \longrightarrow \text{Methylene cyclohexane}{=}CH_2 + (C_6H_5)_3PO$$

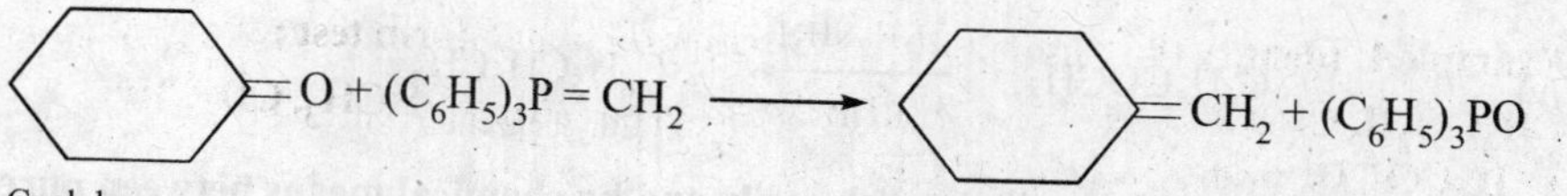

Cyclohexanone — Methylene cyclohexane

This is Wittig's reaction.

Example 3. Complete the following reactions and identify the products:

(*i*) Acetic anhydride + $CH_3COO^- \longrightarrow$ A + ?

(*ii*) A + $C_6H_5CHO \longrightarrow$ B

(*iii*) B is hydrolysed and heated $\longrightarrow$ C + ?

Solution.

$$(i)\quad (CH_3CO)_2O + CH_3COO^- \longrightarrow \underset{\text{Carbanion (A)}}{\bar{C}H_2CO{-}O{-}COCH_3} + CH_3COOH$$

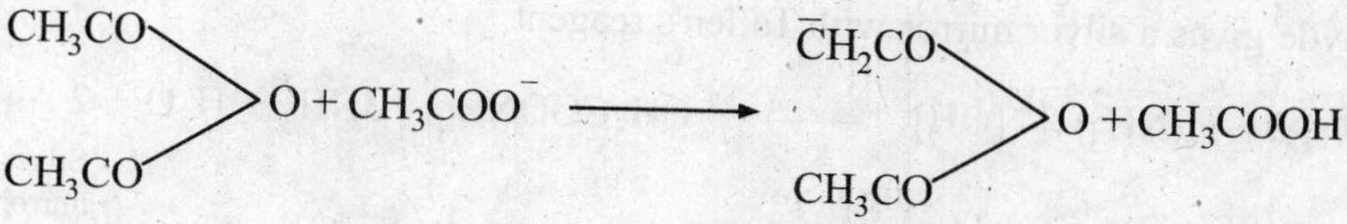

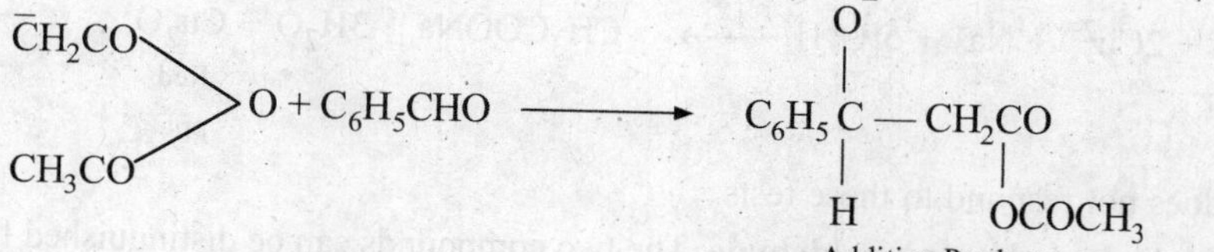

$$(ii)\quad \bar{C}H_2CO{-}O{-}COCH_3 + C_6H_5CHO \longrightarrow \underset{\text{Addition Product (B)}}{C_6H_5\,\overset{O^-}{\underset{H}{C}}-\underset{OCOCH_3}{CH_2CO}}$$

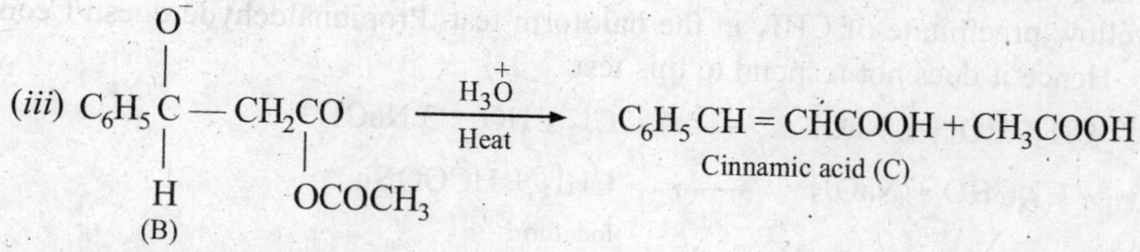

$$(iii)\quad \underset{(B)}{C_6H_5\,\overset{O^-}{\underset{H}{C}}-\underset{OCOCH_3}{CH_2CO}} \xrightarrow[\text{Heat}]{H_3\overset{+}{O}} \underset{\text{Cinnamic acid (C)}}{C_6H_5CH=CHCOOH} + CH_3COOH$$

Example 4. What happens when

(*i*) Butanone is treated with $LiAlH_4$?

(*ii*) $CH_3CH = CHCHO$ is reduced with $NaBH_4$?

(*iii*) Butanone is reduced with amalgamated Zn and HCl?

(*iv*) Acetophenone is treated with hydrazine and KOH?

Solution.

$$(i)\quad \underset{\text{Butanone}}{CH_3CH_2COCH_3} \xrightarrow[\text{[H]}]{LiAlH_4} \underset{\text{Butanol-2}}{CH_3CH_2CHOHCH_3}$$

$$(ii)\quad CH_3CH=CHCHO \xrightarrow[\text{[H]}]{NaBH_4} \underset{\text{But-2-en-1-ol}}{CH_3CH=CH-CH_2OH}$$

$$(iii)\quad \underset{\text{Butanone}}{CH_3CH_2COCH_3} \xrightarrow[\text{[H]}]{Zn/HCl} \underset{\text{Butane}}{CH_3CH_2CH_2CH_3}$$

(*iv*) $$C_6H_5COCH_3 \xrightarrow[KOH]{NH_2NH_2} \underset{\text{Ethylbenzene}}{C_6H_5CH_2CH_3}$$

Example 5. How would you distinguish easily and by chemical means between pure samples of the following pairs of compounds?

(*a*) CH_3CHO and CH_3COCH_3

(*b*) CH_3CHO and CH_3CH_2CHO

(*c*) $CH_3COCH_2CH_2CH_3$ and $CH_3CH_2COCH_2CH_3$

(*d*) C_6H_5 CHO and CH_3CHO

Solution. (*a*) CH_3CHO and CH_3COCH_3. The above two compounds namely acetaldehyde and acetone can be distinguished from each other by performing Tollen's reagent and Fehling solution tests.

Acetaldehyde gives a silver mirror with Tollen's reagent.

$$CH_3CHO + 2\,[Ag(NH_3)_2]^+\,[OH]^- \longrightarrow CH_3COONH_4 + 3NH_3 + H_2O + \underset{\substack{\text{Silver}\\\text{mirror}}}{2\,Ag}$$

Also it gives a red ppt. of Cu_2O with Fehling solution.

$$CH_3CHO + 2Cu^{2+} + Na^+ + 5[OH]^- \longrightarrow CH_3COONa + 3H_2O + \underset{\substack{\text{Red}\\\text{ppt.}}}{Cu_2O}$$

Acetone does not respond to these tests.

(*b*) Acetaldehyde and propionaldehyde. The two compounds can be distinguished from each other by performing the haloform test. Acetaldehyde (CH_3CHO) contains CH_3CO– group. Hence it would give a yellow precipitate of CHI_3 in the haloform test. Propionaldehyde doesn't contain CH_3CO– group. Hence it does not respond to this test.

$$CH_3CHO + 3\,NaOI \longrightarrow CI_3\,CHO + 3\,NaOH$$

$$CI_3CHO + NaOH \longrightarrow \underset{\text{Iodoform}}{CHI_3} + HCOONa$$

(*c*) Pentanone-2 ($CH_3COC_3H_7$) and pentanone-3 ($C_2H_5COC_2H_5$). Pentanone-2 contains CH_3CO– group, hence it would give the iodoform test. Pentanone-3 does not contain CH_3CO– group. Hence it would not respond to this test. This is how the two compounds can be distinguished from each other.

$$CH_3COC_3H_7 + 3NaOI \longrightarrow CI_3\,COC_3H_7$$

$$CI_3COC_3H_7 + NaOH \longrightarrow \underset{\text{Iodoform}}{CHI_3} + \underset{\text{Sod. butyrate}}{C_3H_7COONa}$$

(*d*) Benzaldehyde and acetaldehyde. Benzaldehyde on oxidation with alkaline $KMnO_4$ would produce benzoic acid which is a white solid. Acetaldehyde on oxidation under the same condition produces acetic acid which is not a solid and has the typical vinegar smell.

$$C_6H_5CHO + [O] \longrightarrow \underset{\text{Benzoic acid}}{C_6H_5COOH}$$

$$CH_3CHO + [O] \longrightarrow \underset{\text{Acetic acid}}{CH_3COOH}$$

Example 6. Give equations for the reactions, if any, of propionaldehyde with the following reagents:

(*i*) H_2 in the presence of nickel

(*ii*) Ethanol and HCl gas

(*iii*) Saturated aqueous $NaHSO_3$ solution

(*iv*) $K_2Cr_2O_7 + H_2SO_4$

(*v*) Dil. NaOH solution

(*vi*) NaOH + Br_2 (or sod. hypobromite)

(*vii*) Benzaldehyde with dil. NaOH solution

(*viii*) Phenyl hydrazine

Solution. (*i*) **H_2 in the presence of Ni**

$$\underset{\text{Propionaldehyde}}{CH_3CH_2CHO} + H_2 \xrightarrow{Ni} \underset{\text{Propyl alcohol}}{CH_3CH_2CH_2OH}$$

(*ii*) **Ethanol and HCl gas**

$$CH_3CH_2CHO + 2C_2H_5OH \xrightarrow[-H_2O]{HCl} CH_3CH_2CH\begin{cases} OC_2H_5 \\ OC_2H_5 \end{cases}$$

Acetal

(*iii*) **Saturated aqueous $NaHSO_3$ solution**

$$CH_3CH_2CHO + NaHSO_3 \longrightarrow CH_3CH_2CH\begin{cases} OH \\ SO_3Na \end{cases}$$

Bisulphite compound

(*iv*) **$K_2\ Cr_2O_7 + H_2SO_4$**

$$CH_3CH_2CHO + O \xrightarrow[H_2SO_4]{K_2Cr_2O_7} \underset{\text{Propanoic acid}}{CH_3CH_2COOH}$$

(*v*) **Dil. NaOH solution**

$$CH_3CH_2CHO + CH_3CH_2CHO \xrightarrow{NaOH} \underset{\text{3-hydroxy-2-methyl pentanal}}{CH_3CH_2CH(OH)\overset{\overset{CH_3}{|}}{CH}CHO}$$

(*vi*) **NaOH + Br_2**

There will be no reaction as there is no CH_3CO– group in propionaldehyde.

(*vii*) $C_6H_5CHO + CH_3CH_2CHO \xrightarrow{NaOH} C_6H_5CH(OH)—CH(CH_3)CHO$

3-hydroxy-3-phenyl 2-methylpropanal

$\xrightarrow{-H_2O} C_6H_5CH=C(CH_3)—CHO$

α-Methyl Cinnamaldehyde

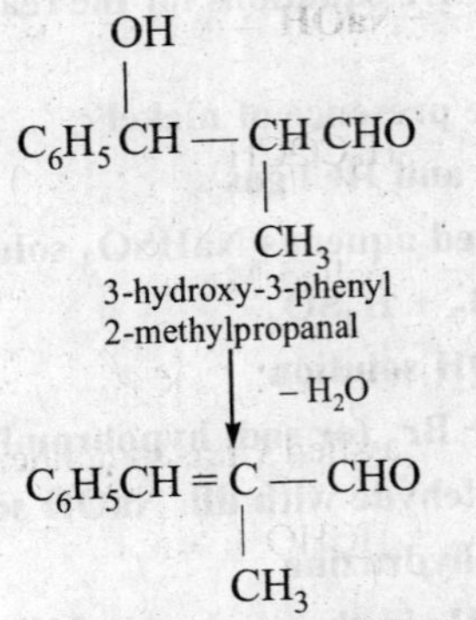

(*viii*) **Phenylhydrazine**

$$CH_3CH_2CHO + C_6H_5NHNH_2 \longrightarrow CH_3CH_2CHNNHC_6H_5$$

Acetaldehyde phenyl hydrazone

Example 7. Identify A, B and C in the following :

(*i*) **Benzaldehyde** $\xrightarrow{NaHSO_3}$ **A**

(*ii*) **Benzaldehyde** $\xrightarrow[2.\ H_3O^+]{1.\ CH_3MgI}$ **A** $\xrightarrow[Oxid.]{Mild}$ **B** $\xrightarrow[OH^-]{Br_2}$ **C**

Solution. (*i*) C_6H_5CHO (Benzaldehyde) $\xrightarrow{NaHSO_3} C_6H_5CH(OH)(SO_3Na)$

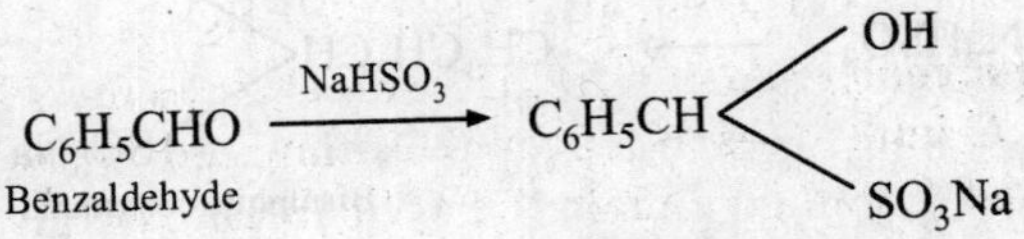

Bisulphite Compound [A]

(*ii*) $C_6H_5CHO \xrightarrow[H_3O^+]{CH_3MgI} C_6H_5CH(OMgI)(CH_3)$

$\xrightarrow{H_3O^+} C_6H_5CH(OH)\,CH_3 + Mg(OH)I$

Phenyl Methyl Carbinol [A]

$\xrightarrow[Oxid.]{Mild} (C_6H_5)(CH_3)CO$ — Acetophenone [B]

$\xrightarrow[KOH]{Br_2} C_6H_5COONa + CHBr_3$

Sod. benzoate [C]; Bromo-form

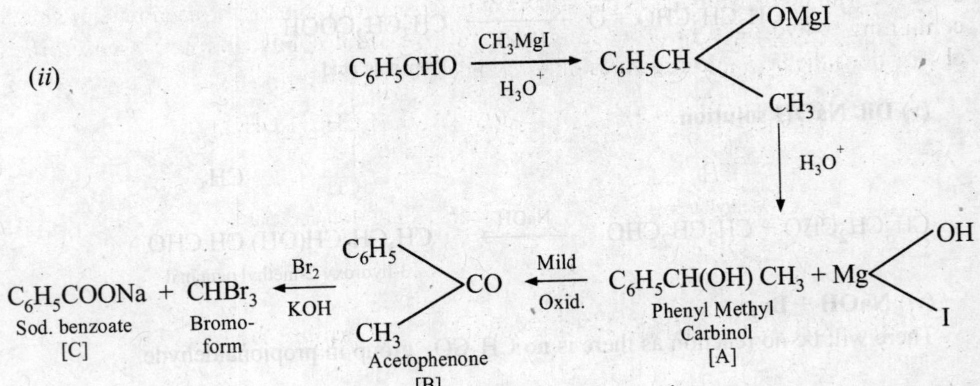

Example 8. Complete the following equations and give the names of the reactions involved in each case:

(*i*) $C_6H_5COCH_3 + HCHO + (CH_3)_2NH \longrightarrow$

(*ii*) $C_6H_5CHO + CH_3COCH_3 \xrightarrow{Base}$

***(iii)* 2HCHO + NaOH** $\longrightarrow$

Solution. *(i)* $C_6H_5COCH_3 + HCHO + (CH_3)_2NH \xrightarrow{-H_2O} C_6H_5COCH_2CH_2-N(CH_3)_2$ (Mannich base)

This reaction is called **Mannich** reaction.

(ii) $C_6H_5CHO + CH_3COCH_3 \xrightarrow{NaOH} C_6H_5CH(OH)CH_2COCH_3$

This reaction is called **Claisen** condensation.

(iii) $HCHO + HCHO + NaOH \longrightarrow \underset{\text{Methanol}}{CH_3OH} + \underset{\text{Sol. formate}}{HCOONa}$

This reaction is called **Cannizzaro** reaction.

Example 9. Arrange the following in increasing order of reactivity: CH_3CHO, HCHO, CH_3CH_2CHO:

Solution. The reactivity in the case of the carbonyl compounds depends upon the size of the alkyl groups attached to carbonyl groups.

$$\begin{matrix} R \\ \text{(or H) } R \end{matrix} \Big> C=O$$

Smaller the size of the alkyl groups, greater is the reactivity. Hence with this criteria, the reactivity in the decreasing order will be as follows :

$$CH_3CH_2CHO < CH_3CHO < HCHO$$

Example 10. An organic compound A, $C_5H_{10}O$ forms chain isomers B and C. B undergoes Cannizzaro reaction and C undergoes aldol condensation and forms 3-hydroxy-2-propyl heptanal. Identify A, B and C and give their IUPAC names. Also give the outlines of the mechanism of Cannizzaro reaction as applicable to B.

Solution. It is obvious from the above that A, B and C are carbonyl compounds. Also B is an aldehyde with no α-hydrogen as only such compounds give Cannizzaro reaction. C is an aldehyde containing α-hydrogen in order to be able to give aldol condensation. Based on the above observations the probable structures of A, B and C are as follows :

[A] $CH_3CH_2\underset{\underset{CH_3}{|}}{CH}-CHO$ (2-Methylbutanal) or $CH_3\underset{\underset{CH_3}{|}}{CH}CH_2CHO$ (3-Methylbutanal)

[B] $CH_3-\overset{\overset{CH_3}{|}}{\underset{\underset{CH_3}{|}}{C}}-CHO$ (2, 2 Dimethypropanal) [C] $CH_3CH_2CH_2CH_2CHO$ (Pentanal)

The compounds as suggested above give the products which tally with those given in the problem

$$2CH_3-\underset{\underset{[B]}{CH_3}}{\overset{CH_3}{\overset{|}{\underset{|}{C}}}}-CHO + NaOH \longrightarrow CH_3-\underset{CH_3}{\overset{CH_3}{\overset{|}{\underset{|}{C}}}}-CH_2OH + CH_3-\underset{CH_3}{\overset{CH_3}{\overset{|}{\underset{|}{C}}}}-COONa$$

Also $\underset{[C]}{CH_3CH_2CH_2CH_2CHO} + \underset{[C]}{CH_3CH_2CH_2CH_2CHO}$

$$\xrightarrow{NaOH} CH_3CH_2CH_2CH_2CH(OH)\underset{CH_2\,CH_2\,CH_3}{\underset{|}{C}HCHO}$$

3-hydroxy-2-propylheptanal

Example 11. Complete the following reactions:

1. $(CH_3CH_2CH_2CH_2)_2\,Cd + 2\,\begin{matrix}CH_3\\CH_3\end{matrix}\!\!>CHCOCl \longrightarrow$

2. $C_6H_5-CH_2COCl \xrightarrow{H_2,\ Pd-BaSO_4}$

3. $NO_2-C_6H_4-COCl \xrightarrow{[C_4H_9(t)O]_3Al}$

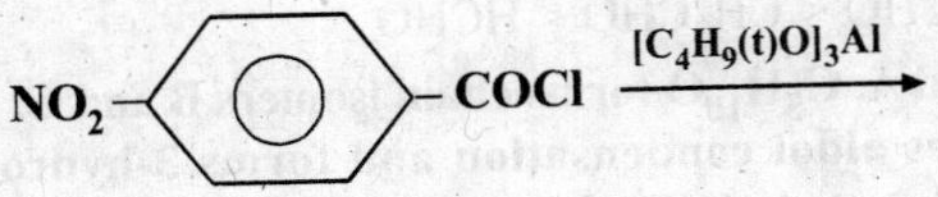

Solution. 1. $\underset{\text{Dibutyl cadmium}}{(CH_3CH_2CH_2CH_2)_2\,Cd} + 2\,\underset{\text{Isobutyl chloride}}{\begin{matrix}CH_3\\CH_3\end{matrix}\!\!>CHCOCl}$

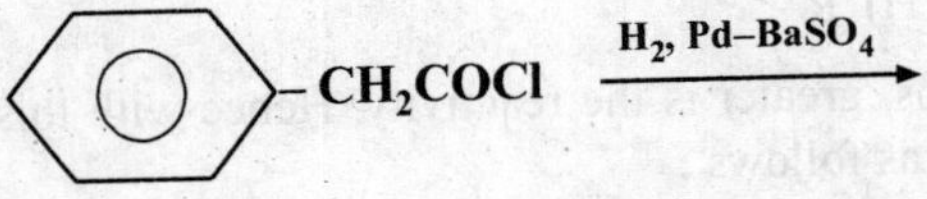

$$\longrightarrow \underset{\text{2-Methylheptan-3-one}}{CH_3CH_2CH_2CH_2\,COCH\!<\!\begin{matrix}CH_3\\CH_3\end{matrix}} + 2\,CdCl_2$$

2. 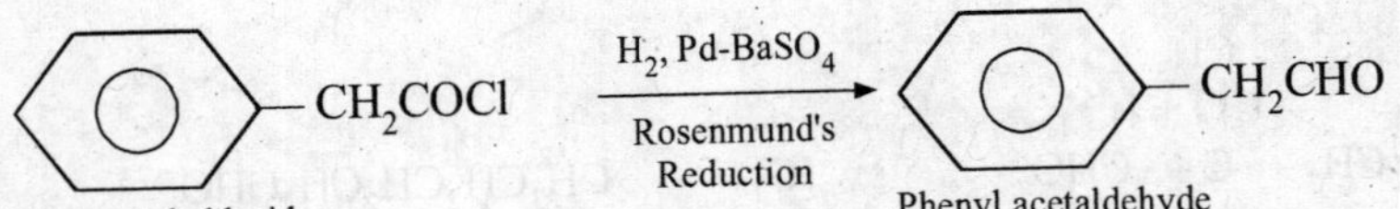 $\underset{\text{Phenyl acetyl chloride}}{C_6H_5-CH_2COCl} \xrightarrow[\text{Rosenmund's Reduction}]{H_2,\ Pd\text{-}BaSO_4} \underset{\text{Phenyl acetaldehyde}}{C_6H_5-CH_2CHO}$

3. 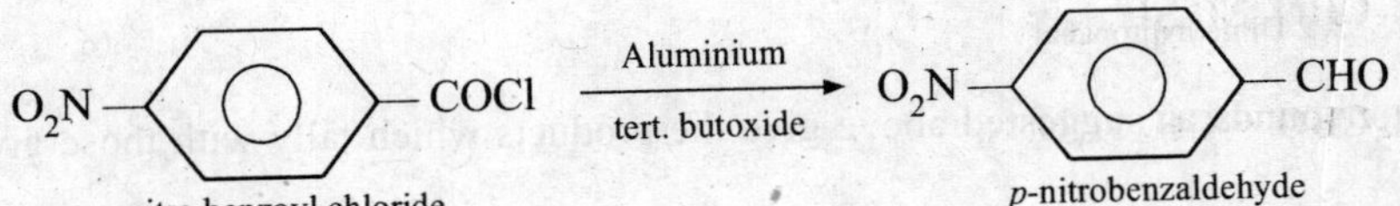 $\underset{p\text{-nitro benzoyl chloride}}{O_2N-C_6H_4-COCl} \xrightarrow[\text{tert. butoxide}]{\text{Aluminium}} \underset{p\text{-nitrobenzaldehyde}}{O_2N-C_6H_4-CHO}$

Example 12. Two isomeric compounds A and B each with molecular formula C_3H_6O react with phenyl hydrazine to form phenyl hydrazones. A reduces Fehling solution and B does not. Upon oxidation A gives an acid of formula $C_3H_6O_2$ and B on similar treatment gives an acid of formula $C_2H_4O_2$. Identify compounds A and B and explain the reactions.

Solution. Reaction with phenyl hydrazine is a property characteristic of carbonyl compounds. Two possible compounds with the formula C_3H_6O are

$$\underset{\text{Propionaldehyde}}{CH_3CH_2CHO} \qquad [A]$$

$$\underset{\text{Acetone}}{(CH_3)_2C=O} \qquad [B]$$

A reduces Fehling solution and B does not, as we know that aldehydes reduce the Fehling solution and the ketones do not give this reaction.

$$\underset{[A]}{CH_3CH_2CHO} + \underset{\text{Cupric ions}}{2Cu^{2+}} + Na^+ + 5[OH]^- \longrightarrow CH_3CH_2COONa + \underset{\text{Cuprous oxide}}{Cu_2O} + 3H_2O$$

$$\underset{[B]}{(CH_3)_2C=O} + 2\,Cu^{2+} + Na^+ + 5\,[OH]^- \longrightarrow \text{No reaction}$$

$$\underset{[A]}{CH_3CH_2CHO} \xrightarrow{[O]} \underset{\text{Propionic acid}}{CH_3CH_2COOH} \qquad (\text{Mol. Formula } C_3H_6O_2)$$

$$\underset{[B]}{(CH_3)_2C=O} \xrightarrow{[O]} \underset{\substack{\text{Acetic acid} \\ (\text{Mol. Formula} \\ C_2H_4O_2)}}{CH_3COOH} + CO_2 + H_2O$$

Hence compound A is CH_3CH_2CHO *i.e.* propionaldehyde and compound B is CH_3COCH_3 *i.e.* acetone

Example 13. Complete and name the following reactions:

(*i*) $(CH_3)_3\,CCHO + (CH_3)_3\,CCHO \xrightarrow{50\%\ NaOH}$

(*ii*) $(C_2H_5)_2CO + (C_6H_5)_3\,P=CH-CH_3 \longrightarrow$

(*iii*) $C_6H_5CHO + (C_6H_5)_3\,P = CH_2 \longrightarrow$

(*iv*) $C_6H_5CHO + CH_2\,(COOC_2H_5)_2 \xrightarrow{\text{Base}}$

(*v*) $C_6H_5CHO + BrCH_2 - COOC_2H_5 \xrightarrow[\text{Ether}]{\text{Zn}}$

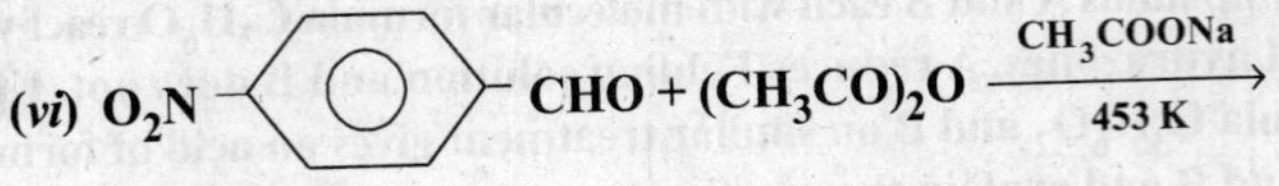

(*vi*) $O_2N-C_6H_4-CHO + (CH_3CO)_2O \xrightarrow[453\ K]{CH_3COONa}$

Solution. (*i*) $(CH_3)_3CCHO + (CH_3)_3\ CCHO$ Cannizzaro Reaction

$$\downarrow 50\%\ NaOH$$

$$(CH_3)_3CCOONa + (CH_3)_3\ CCH_2OH$$

Sod. Trimethylacetate Neopentyl alcohol

(*ii*) $(C_2H_5)_2C=O + (C_6H_5)_3\ P=CH-CH_3$ Wittig Reaction

Diethyl ether

$$\longrightarrow (C_2H_5)_2C=CH-CH_3 + (C_6H_5)_3\ PO$$

3-Ethylpent-2-ene Triphenyl phosphine oxide

(*iii*) $C_6H_5CHO + (C_6H_5)_3\ P=CH_2$ Wittig Reaction

$$\longrightarrow C_6H_5CH=CH_2 + (C_6H_5)_3\ PO$$

Styrene Triphenyl phosphine oxide

(iv) $C_6H_5CHO + CH_2\ (COOC_2H_5)_2$ Knoevenagal Reaction

Malonic ester

$$\xrightarrow{Base} C_6H_5CH=C(COOC_2H_5)_2 + H_2O$$

Benzal malonic ester

(*v*) $C_6H_5CHO + BrCH_2COOC_2H_5 + Zn$ Reformatsky Reaction

$$\xrightarrow{Ether} C_6H_5CH(OH)\ CH_2COOC_2H_5 + Zn\ (OH)\ Br$$

Ethyl 3-hydroxy-3-phenyl propionate

(*vi*) $O_2N-C_6H_4-CHO + (CH_3CO)_2\ O$ Perkin's Reaction

$$\xrightarrow[453]{CH_3COONa} O_2N-C_6H_4-CH=CH-COOH + CH_3COOH$$

p-nitro cinnamic acid

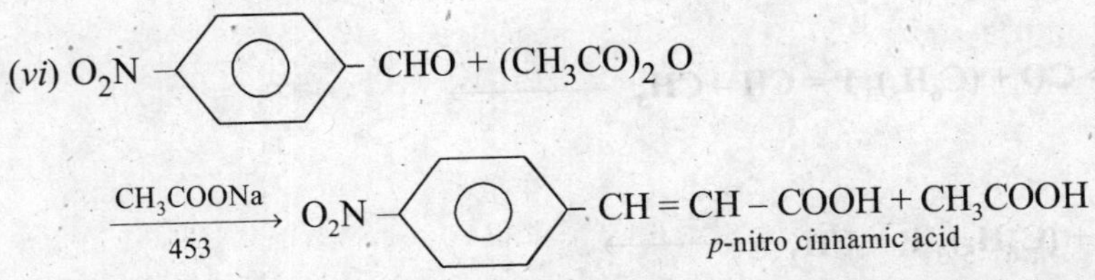

Example 14. How do you distinguish a ketone and an aldehyde by chemical method?

Solution. Both aldehydes and ketones give a yellow precipitate of 2, 4-dinitrophenyl hydrazone on treatment with 2, 4-dinitrophenyl hydrazine.

Aldehydes on boiling with ammoniacal silver nitrate solution give a shining layer of silver metal on the inner wall of the test tube (Tollen's reagent test). Similarly aldehydes on boiling with Fehling solution give a red ppt. of cuprous oxide.

Ketones give neither the Tollen's reagent test nor Fehling solution test.

Example 15. Alkenes undergo electrophilic addition reactions while carbonyl compounds undergo nucleophilic addition reactions. Explain.

Solution. This is due to the electronegative nature of carbonyl oxygen that a positive charge is created on the carbonyl carbon. This attracts nucleophiles towards it giving nucleophilic reactions.

$$>C=O \longrightarrow >\overset{+}{C}-\overset{-}{O}$$

In alkenes, such positive centre is not created. Instead, the π electrons on C = C attract the electrophiles towards it.

Example 16. Why does HCN not add across carbonyl double bond in benzophenone?

Solution. This is because the positive charge created on carbonyl carbon is removed by the resonance in two benzene rings in the compound.

$$C_6H_5-\overset{+}{C}(O^-)-C_6H_5 \longleftrightarrow {}^{+}C_6H_5=C(O^-)-C_6H_5$$

As the positive centre on carbonyl carbon is destroyed, nucleophilic addition of HCN becomes difficult.

Example 17. Nucleophilic addition of ammonia derivatives to carbonyl group is catalysed by acid but in presence of large excess of the acid, the rate of reaction is lowered considerably.

Solution. Presence of a mineral acid promotes the addition of ammonia derivatives to carbonyl group, as this helps to stabilize positive charge on carbonyl carbon.

$$>C=O+\overset{+}{H} \longrightarrow >\overset{+}{C}-OH$$

But in the presence of excess of the acid, the nucleophile also gets protonated and thus loses tendency to attach to the positive centre

$$R-\ddot{N}H_2+H^+ \longrightarrow R-\overset{+}{N}H_3$$

Example 18. Acid catalyses the addition of semi carbozide to acetone. Yet, too much acidity of the medium is harmful in this reaction.

Solution. In the presence of excess acid, the nucleophile gets protonated and loses ability to approach the positive centre

$$\ddot{N}H_2NHCONH_2+\overset{+}{H} \longrightarrow \overset{+}{N}H_3NHCONH_2$$

Example 19. Formaldehyde does not give aldol condensation. Explain.

Solution. Aldol condensation is given by compounds containing α-hydrogen atoms. Since formaldehyde does not contain α-hydrogen atoms, it does not give aldol condensation.

Example 20. *o*-hydroxy benzaldehyde is steam volatile while *p*-hydroxy benzaldehyde is not. Why?

Solution.

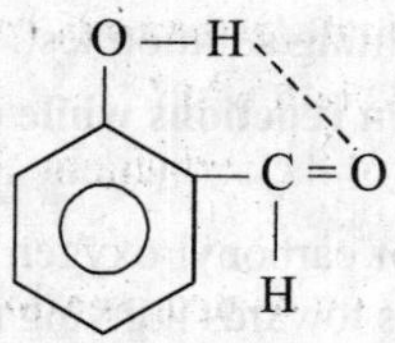

o-hydroxy benzaldehyde

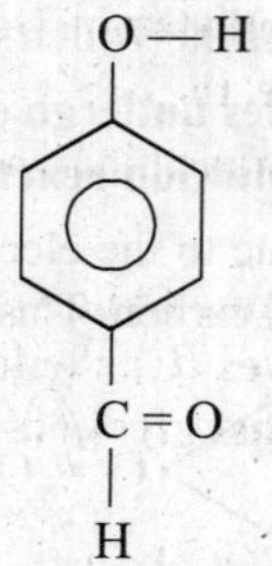

p-hydroxy benzaldehyde

Intramolecular hydrogen bonding between hydrogen of phenolic group and oxygen of aldehydic group takes place in *o*-hydroxy benzaldehyde because of proximity of the two groups, thus lowering the boiling point. Hence *o*-hydroxybenzaldehyde is steam volatile. This type of hydrogen bonding does not take place in the para isomer because the two groups are apart.

Example 21. Write all the possible isomers of the compound with molcular formula C_4H_8O and give their IUPAC names.

Solution. Possible isomers are:

1. $CH_3CH_2CH_2CHO$ — Butanal
2. $CH_3-CH(CH_3)-CHO$ — 2-Methyl propanal
3. $CH_3COCH_2CH_3$ — Butan-2-one

Example 22. How will you distinguish between aldehydes and ketones?

Solution. Both aldehydes and ketones give a yellow precipitate with 2, 4, Dinitrophenylhydrize. Aldehydes respond to Fehling solution test and silver mirror test whereas ketones don't respond to these tests.

Example 23. How will you differentiate between acetophenone and benzophenone?

Solution. Acetophenone reacts with hydrogen cyanide, while benzophenone does not.

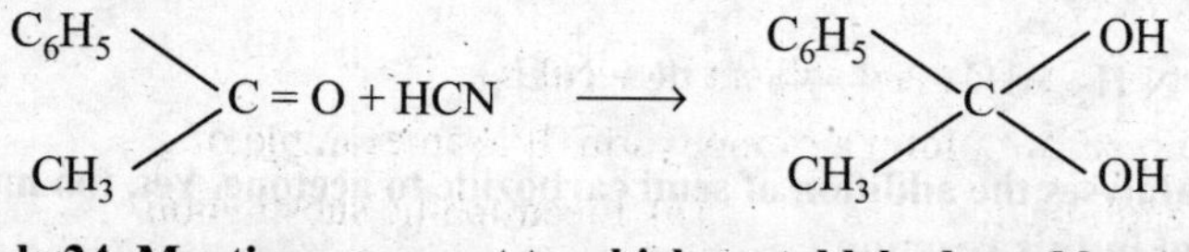

$$(C_6H_5)(CH_3)C=O + HCN \longrightarrow (C_6H_5)(CH_3)C(OH)(OH)$$

Example 24. Mention a reagent to which acetaldehyde and benzaldehyde react similarly and another reagent to which these compounds react differently.

Solution. (*i*) HCN (*ii*) I_2/NaOH

Example 25. Predict product (s) formed in the following reaction.

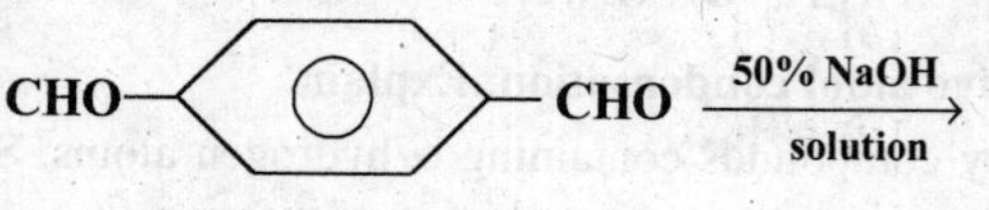

$$CHO-C_6H_4-CHO \xrightarrow[\text{solution}]{50\%\ NaOH}$$

Solution. $CHO-C_6H_4-CHO \xrightarrow[\text{solution}]{50\%\ NaOH}$

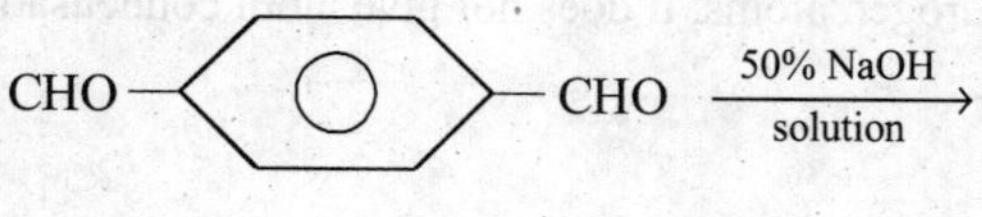

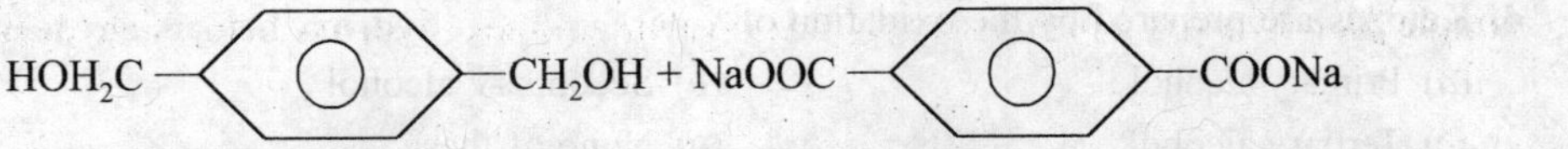

$$HOH_2C-C_6H_4-CH_2OH + NaOOC-C_6H_4-COONa$$

Example 26. ***p*-dimethylaminobenzaldehyde fails to undergo benzoin condensation but when mixed with benzaldehyde, the condensation does occur. Explain why?**

Solution. There is steric hindrance in the attachment of the carbanions formed in the first step to the second molecule.

Example 27. A ketone (A) gives iodoform test. (A) on hydrogenation gives (B) which on heating with H_2SO_4 gives (C). Action of O_3 on (C) gives (D) which when treated with water in the presence of Zn dust gives only acetaldehyde. Identify (A), (B), (C), (D) and explain the reactions involved.

Solution. $A = CH_3COCH_2CH_3$ $\quad B = CH_3CHOHCH_2CH_3$

$C = CH_3CH = CH - CH_3$

$E = CH_3CHO$ $\quad D = CH_3 - \underset{\text{(ozonide ring: CH–O–CH bridged by O–O)}}{CH \quad\quad CH} - CH_3$

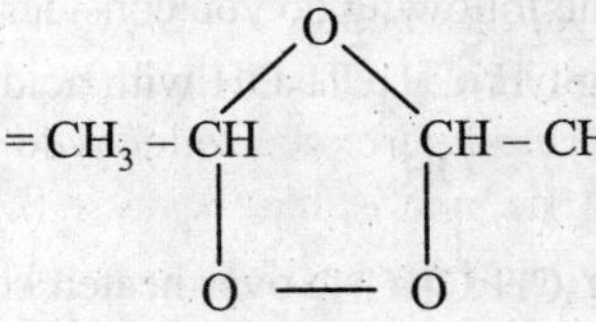

$$CH_3COCH_2CH_3 \xrightarrow{H_2} CH_3CHOHCH_2CH_3 \xrightarrow[-H_2O]{H_2SO_4} CH_3CH = CH - CH_3$$

$$\downarrow O_3,\ Zn$$

$$2\ CH_3CHO$$

EXERCISES
(Based on Question Papers of Different Universities)

Multiple Choice Questions (Choose the correct option)

1. Which action best accounts for the solubility of aldehydes and ketones in water?

(*a*) Polar interactions between solute molecules.

(*b*) H-bonding between solute molecules.

(*c*) van der Waals forces

(*d*) H-bonding between solute and solvent molecules

2. Acetone reacts with HCN to form a cyanohydrin. It is an example of

(*a*) Electrophilic addition (*b*) Electrophilic substitution

(*c*) Nucleophilic addition (*d*) Nucleophilic substitution

3. Boiling acetaldehyde (CH_3CHO) reacts with chlorine gas to give

(*a*) $CH_3-\overset{O}{\overset{\|}{C}}-Cl$ (*b*) $CCl_3-\overset{O}{\overset{\|}{C}}-H$

(*c*) CH_3CHCl_2 (*d*) $CHCl_3$

4. Ketones are prepared by the oxidation of

(*a*) Primary alcohol (*b*) Secondary alcohol

(*c*) Tertiary alcohol (*d*) None of these

5. Which of the following compounds reacts with sodium bisulphite and ammoniacal silver nitrate solution?

(*a*) $CH_3CH_2-\overset{\overset{\displaystyle O}{\|}}{C}-OH$ (*b*) $CH_3CH_2-\overset{\overset{\displaystyle O}{\|}}{C}-H$

(*c*) $CH_3CH_2CH=CH_2$ (*d*) $CH_3C\equiv CCH_3$

6. Which of the following reagents will react readily with both aldehydes and ketones?

(*a*) Grignard reagent (*b*) Fehling's reagent

(*c*) Tollens' reagent (*d*) Schiff's reagent

7. Which of the following do you consider to be suitable for obtaining acetone (CH_3COCH_3)?

(*a*) Heating $CH_3CH_2CH_2OH$ with acidic $Na_2Cr_2O_7$

(*b*) Passing $CH_3\underset{}{\overset{\overset{\displaystyle OH}{|}}{C}}HCH_3$ over heated copper

(*c*) Oxidation of $CH_3CH_2CH_3$ with concentrated HNO_3

(*d*) Heating $CH_3CH=CH_2$ with dilute H_2SO_4

8. Acetone undergoes reduction with hydrazine in the presence of NaOH to form propane. The reaction is known as

(*a*) Clemmensen reduction (*b*) Wolf-Kishner reduction

(*c*) Rosenmund reduction (*d*) Reformatsky reaction

9. Reduction of acetaldehyde with H_2/Ni gives

(*a*) Ethyl alcohol (*b*) Acetic acid

(*c*) Ethylene (*d*) Ethane

10. Aldehydes undergo oxidation with $KMnO_4/H^+$ to give

(*a*) Alcohols (*b*) Acetals

(*c*) Ketones (*d*) Acids

11. The appearance of a silver mirror in Tollens' test indicates the presence of:

(*a*) an aldehyde (*b*) a ketone

(*c*) an alcohol (*d*) an alkene

12. Which of the following compounds will give a positive test with Fehling's solution?

(*a*) Formaldehyde (*b*) Acetone

(*c*) Ethyl acetate (*d*) Acetic acid

13. Cannizzaro reaction is not given by

(*a*) Formaldehyde (*b*) Trimethylacetaldehyde

(*c*) Acetaldehyde (*d*) Benzaldehyde

14. A hydrazone will result from the reaction of hydrazine with

(*a*) a phenol (*b*) an aldehyde

(*c*) an alcohol (*d*) an acid

15. Organic compound (A), C_4H_8O, does not react with sodium or PCl_5. However, it reduces an alkaline solution of a copper (II) salt on heating. Compound (A) could be

(*a*) an aldehyde (*b*) a primary alcohol
(*c*) a ketone (*d*) a secondary alcohol

16. Which of the following compounds will give a positive iodoform test?

(*a*) Benzaldehyde (*b*) 2-Pentanone
(*c*) 3-Hexanone (*d*) 3-Pentanone

17. Which of the following will react with Fehling's solution giving a red precipitate?

(*a*) CH_3CH_2CHO (*b*) $CH_3CH_2OCH_2CH_3$
(*c*) $CH_3CH_2CH_2OH$ (*d*) $CH_3-\overset{O}{\overset{\|}{C}}-CH_3$

18. A compound (A) gave a positive iodoform test, but did not reduce silver nitrate in ammonia solution. Compound (A) could be

(*a*) $CH_3-\overset{O}{\overset{\|}{C}}-H$ (*b*) $CH_3CH_2-\overset{O}{\overset{\|}{C}}-CH_2CH_3$
(*c*) $C_6H_5-\overset{O}{\overset{\|}{C}}-CH_2CH_3$ (*d*) $C_6H_5-\overset{O}{\overset{\|}{C}}-CH_3$

19. An unknown compound gave a positive Tollens' test. Treatment of the unknown with I_2/NaOH gave a solid which was identified as iodoform. The unknown compound was?

(*a*) Ethanal (*b*) Acetone
(*c*) Propanal (*d*) Acetophenone

20. An organic compound (A), C_3H_8O, on oxidation gives (B), C_3H_6O. The compound (A) could be

(*a*) an aldehyde (*b*) a ketone
(*c*) an alcohol (*d*) an ester

ANSWERS

1. (*d*)	**2.** (*c*)	**3.** (*b*)	**4.** (*b*)	**5.** (*b*)	**6.** (*a*)
7. (*b*)	**8.** (*b*)	**9.** (*a*)	**10.** (*d*)	**11.** (*a*)	**12.** (*a*)
13. (*c*)	**14.** (*b*)	**15.** (*a*)	**16.** (*b*)	**17.** (*a*)	**18.** (*d*)
19. (*a*)	**20.** (*a*)				

Short Answer Questions

1. Complete the following equations :

(*i*) $C_6H_5CHO + CH_3COCH_3 \longrightarrow$ (*ii*) $2HCHO + NaOH \longrightarrow$

2. Write a note on Wolf-Kishner reduction.

3. Why are aromatic aldehydes and ketones less reactive than the corresponding aliphatic aldehydes and ketones ?

4. Why is the *p*H of the solution carefully controlled during the addition of ammonia derivatives to carbonyl compounds ?

5. How will you convert

(*i*) Ethanol into acetone (*ii*) Formaldehyde into *n*-butane ?

6. Give the chemical tests to distinguish between acetaldehyde and acetone.

7. What happens when acetone is treated with CH_3MgBr followed by hydrolysis ?

8. Arrange the following in decreasing order of reactivity towards nucleophilic addition :

$HCHO, CH_3CHO, CH_3CH_2CHO$

9. Why do carbonyl compounds take part in nucleophilic addition and not in electrophilic addition reactions ?

10. What is Tollen's reagent ? How does it help in identifying an aldehyde ?

11. What happens when

(*i*) $CH_3CH = CHCHO$ is reduced with $NaBH_4$?

(*ii*) $CH_3COCH_2CH_3$ is reduced with Zn-Hg/HCl ?

12. Out of acetaldehyde and acetone, which has higher boiling point and why ?

13. How will you convert :

(*i*) Acetaldehyde into tertiary butyl alcohol ?

(*ii*) Acetaldehyde into crotonaldehyde ?

14. Discuss the mechanism of aldol condensation.

15. Why is α-hydrogen atom in aldehydes and ketones acidic in nature ?

16. How will you distinguish benzaldehyde from acetaldehyde and acetone ?

17. Formaldehyde gives Cannizzaro's reaction while acetaldehyde does not. Assign a suitable reason.

18. How will you purify an impure sample of benzaldehyde ?

19. Why do carbonyl compounds have lower boiling points than the alcohols from which they are formed ?

20. Complete the following reaction and give mechanism :

$$C_6H_5CHO + BrCH_2COOC_2H_5 \xrightarrow[(ii)\ H_2O]{(i)\ Zn}$$

21. Why is α-hydrogen in aldehydes or ketones of acidic nature ?

22. Sketch the mechanism of the nucleophilic reaction between HCN and acetaldehyde.

23. Addition of certain nucleophilic reagents to aldehydes and ketones is catalysed by presence of an acid. Explain.

24. Write chemical equation to distinguish between CH_3CH_2CHO and CH_3CHO.

25. Give chemical equations for the formation of

(*i*) acetal (*ii*) oxime (*iii*) cyanohydrin.

26. Explain why acetaldehyde is more reactive than acetone towards nucleophilic addition.

27. Explain Wolf-Kishner reduction with mechanism.

28. Arrange the following compounds in order of decreasing reactivity towards nucleophilic addition and assign suitable explanatioin :

$$C_6H_5-\overset{\overset{\displaystyle O}{\|}}{C}-C_6H_5,\ CH_3CHO,\ CH_3-\overset{\overset{\displaystyle O}{\|}}{C}-CH_3$$

29. Give the product of the following and explain the mechanism: $2C_6H_5CHO \xrightarrow{\text{Conc.NaOH}}$

30. Write a short note on keto-enol tautomerism.

31. Why is α-hydrogen in aldehydes and ketones acidic in nature?

General Questions

1. What is reductive amination ? How does it help in forming primary, secondary and tertiary amines ?

2. Suggest a mechanism of the following conversions. Name the product in each case :

(*i*) 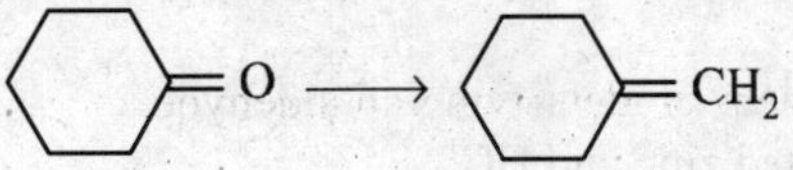 Cyclohexanone $\longrightarrow$ methylenecyclohexane (ring=O $\longrightarrow$ ring=CH_2)

(*ii*) $CH_3-\overset{O}{\overset{\|}{C}}-CH_3 \longrightarrow BrCH_2-\overset{O}{\overset{\|}{C}}-CH_3$

(*iii*) $C_6H_5CHO + CH_3CHO \longrightarrow C_6H_5CH=CHCHO$

3. Discuss the addition of $NaHSO_3$ to carbonyl compounds. What is the utility of the reaction ?

4. Complete the following reactions :

(*i*) $2C_6H_5CHO \xrightarrow{OH^-}$ (*ii*) $CH_3COCH_3 + BrCH_2COOC_2H_5 \xrightarrow{Zn}$

(*iii*) $2C_6H_5CHO + KOH\ (Conc.) \longrightarrow$

5. Give the mechanism of following reactions :

(*i*) Perkin's reaction (*ii*) Cross-aldol condensation

(*iii*) Cannizzaro's reaction

6. How will you account for the reactive nature of carbonyl compounds ?

7. Give mechanism of the following reactions :

(*i*) Aldol condensation (*ii*) Reformatsky reaction.

8. Give the mechanism of the following reactions :

(*i*) Reaction of acetone with Grignard reagent.

(*ii*) Condensation of acetaldehyde with NaOH.

9. Whereas aldehydes act as reducing agents, ketones fail to act. Explain.

10. Give an account of Reformatsky reaction along with mechanism.

11. How will you distinguish between the following pairs with the help of chemical test ?

(*i*) CH_3CHO and CH_3COCH_3 (*ii*) CH_3CHO and CH_3CH_2CHO

(*iii*) $CH_3COCH_2CH_2CH_3$ and $CH_3CH_2COCH_2CH_3$

(*iv*) C_6H_5CHO and CH_3CHO

12. Describe the mechanism of nucleophilic addition reactions of carbonyl compounds and the usefulness these reactions in the synthesis.

13. (*a*) Describe Reformatsky reaction and explain its mechanism.

(*b*) Write notes on :

(*i*) Aldol condensation (*ii*) Perkin reaction

14. (*a*) Write a note on keto-enol tautomerism.

(*b*) Describe the mechanism of the chromic acid oxidation of an aldehyde into acid.

15. Discuss the various methods employed for the reduction of carbonyl compounds.

16. Complete the following reactions indicating the product in each case :

(*i*) $CH_3CHO + HCN \xrightarrow{H_3O^+}$ (*ii*) $C_6H_5CHO + (CH_3CO)_2O \xrightarrow{CH_3COONa}$

(*iii*) (cyclohexanone) $=O + (C_6H_5)_3P = CH_2 \longrightarrow$

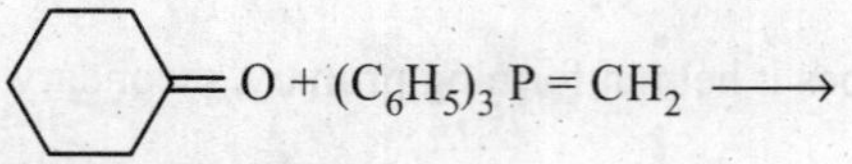

17. What happens when :

(*i*) Butanone is treated with $LiAlH_4$

(*ii*) $CH_3CH = CHO$ is reduced with $LiAlH_4$

(*iii*) Butanone is reduced with amalgamated zinc and HCl

(*iv*) Acetone is treated with hydrazine and KOH ?

18. Write notes on :

(*i*) Reductive amination (*ii*) Aldol condensation

19. Complete the following :

(*i*) $C_6H_5CHO + KOH \longrightarrow$ (*ii*) (benzene ring)$-CH = CH - CHO \xrightarrow{NaBH_4}$

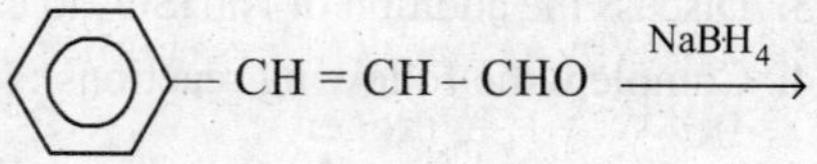

(*iii*) $CH_3CH_2CHO \xrightarrow{\text{Dilute base}}$ (*iv*) $CH_3COCH_3 + 3NaIO \longrightarrow$

20. Write notes on :

(*i*) Reformatsky reaction (*ii*) Reductive amination

(*iii*) Perkin reaction.

21. Give the mechanism of the following :

(*i*) Base-catalysed aldol condensation (*ii*) Perkin reaction

(*iii*) Cannizzaro's reaction

22. Discuss the mechanism of reduction of carbonyl compounds with the help of $LiAlH_4$.

23. How will you convert ?

(*i*) Ethanal into lactic acid (*ii*) Ethanal into 2-hydroxy-2-butenoic acid

(*iii*) Acetylene into acetone (*iv*) Acetophenone into ethylbenzene.

24. (*a*) What is the primary condition for a compound to undergo Cannizzaro's reaction ? Which of the following compounds would undergo Cannizzaro's reaction ? Explain its mechanism.

(*i*) $(CH_3)_2CHCHO$ (*ii*) $CH_3-C(C_2H_5)(CH_3)-CHO$ (*iii*) $H-C(CH_3)(CH_3)-CH(CHO)-C(CH_3)(CH_3)-CH_3$

25. (*a*) Give the mechanism of the following reactions :

(*i*) Perkin reaction (*ii*) Haloform reaction

(*b*) "Alkenes undergo electrophilic addition reactions while carbonyl compounds (>C = O) undergo nucleophilic addition reactions." Explain.

26. (*a*) Discuss the mechanism of Perkin's reaction.

(*b*) Give chemical reaction to distinguish between HCHO and C_6H_5CHO

(*c*) Discuss suitable mechanism for the reaction.

$$C_6H_5CHO \xrightarrow{KCN} C_6H_5CH(OH)COC_6H_5.$$

27. Give the mechanism of the following reactions :

(*i*) Knovenaegal Reaction (*ii*) Perkin Reaction

(*iii*) Wittig Reaction

28. (*a*) Arrange the following compounds in order of decreasing reactivity towards nucleophiles (Most reactive first)

(*i*) $C_6H_5COCH_3$ (*ii*) CH_3CHO

(*iii*) CH_3COCH_3 (*iv*) $CH_3COCH_2CH_3$

(*b*) Give the mechanism of Perkin's reaction.

29. (*a*) Give mechanism of the following reactions :

(*i*) Benzoin condensation (*ii*) Cannizzaro's reaction

(*b*) Explain :

(*i*) Why ketones are less reactive than aldehydes towards nucleophilic addition ?

(*ii*) Why α-hydrogen in aldehydes or ketones is acidic in nature ?

30. An alkene C_6H_{12} after ozonolysis yielded two products. One of these gave a positive iodoform reaction but a negative Tollen's test. The other product gave a positive Tollen's test, but negative iodoform reaction. What is the structure and IUPAC name of the alkene ?

Ans. The alkene is 2-Methylpent-2-ene

31. Compound (A), $C_5H_{11}Br$, forms a Grignard reagent to form *n*-pentane. When (A) is treated with aqueous potassium hydroxide, compound (B), $C_5H_{12}O$, is formed. Compound (B) gives a positive Lucas test. When (B) is treated with potassium dichromate in sulphuric acid, (C) is formed. Compound (C) gives a negative iodoform test. Deduce the structural formulas of (A), (B), (C) and (D). **Ans.** (A) = 3-Bromopentane, (B) = Pentan-3-ol

(C) = Pentan-3-one

32. Compound (A), $C_4H_{10}O$ undergoes oxidation to give (B), $C_4H_8O_2$. (B) forms an oxime but does not give Tollen's test. (B) reacts with iodine and potassium hydroxide to give iodoform. Deduce the structures of (A) and (B) and explain the reactions.

Ans. (A) Butan-2-ol, (B) = Butan-2-one)

6

Carboxylic Acids

6.1. INTRODUCTION

Carboxylic acids are organic compounds containing –COOH or carboxy group. Such compounds can be represented by the general formula R–COOH, where R stands for some aliphatic or aromatic group. Carboxylic compounds can be classified as follows:

Monocarboxylic compounds: Those containing one carboxy group in the molecule such as acetic acid, butyric acid.

Dicarboxylic compounds : Those containing two carboxy groups in the molecule such as oxalic acid, malonic acid.

Tricarboxylic compounds :Those containing three carboxy groups in the molecule such as citric acid.

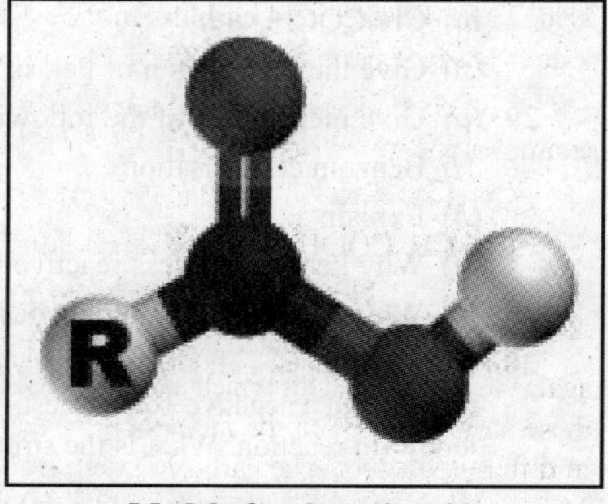

Model of carboxylic acid

The carboxy group $-\overset{\overset{O}{\|}}{C}-OH$ consists of two groups viz. $-\overset{\overset{O}{\|}}{C}-$ (carbonyl group) and –OH (hydroxyl group). The word carboxy is derived from the two groups – carbonyl and hydroxy.

6.2. NOMENCLATURE OF ALIPHATIC AND AROMATIC CARBOXYLIC COMPOUNDS

There are two systems of nomenclature of carboxylic compounds:

1. Common system **2. IUPAC system**

1. Common System. In this system of naming, the carboxylic acid is named after the source from which it is obtained. For example, HCOOH, formic acid, is so named because it is obtained from *formica* i.e. ants. Acetic acid gets its name from *acetum* i.e. vinegar which forms the source of acetic acid. Butyric acid C_3H_7COOH gets its name from *butyrum* i.e. butter.

Substituted carboxylic acids are named by indicating the group and the position where such group is attached. The positions on the carbon chain in a carboxylic acid are determined as under:

$$\overset{\delta}{C}-\overset{\gamma}{C}-\overset{\beta}{C}-\overset{\alpha}{C}-COOH$$

Thus the position next to the carboxy group is α position. Next to that is β position, still next are γ and δ positions.

The compound $CH_3-\underset{}{\overset{\overset{Cl}{|}}{C}H}-COOH$ is named as α-chloropropionic acid. All the carbons in the longest chain containing carboxylic acid are taken into consideration for determination of the parent carboxylic acid. Some examples are given below:

$$\overset{\gamma}{CH_3} - \underset{\underset{CH_3}{|}}{\overset{\beta}{CH}} - \overset{\alpha}{CH_2} - COOH$$ β-methyl butyric acid

$$\overset{\delta}{CH_3} - \underset{\underset{CH_3}{|}}{\overset{\gamma}{CH}} - \underset{\underset{CH_3}{|}}{\overset{\beta}{CH}} - \overset{\alpha}{CH_2} - COOH$$ β, γ-dimethyl valeric acid

2. IUPAC System. In this system, carboxylic acids are known as *alkanoic acids*. Total carbon atoms in the longest chain containing the carboxylic group and including the carboxylic carbon atom are counted. The carboxylic compound is the derivative of the alkane with as many carbons as the acid. Thus if there are 4 carbons in the acid, it is a derivative of butane; if there are 5 carbons, it is a derivative of pentane; and so on.

The **e** of the alkane is changed into **oic acid.** Some examples are given below:

CH_3COOH Ethanoic acid

CH_3CH_2COOH Propanoic acid
(Common name propionic acid)

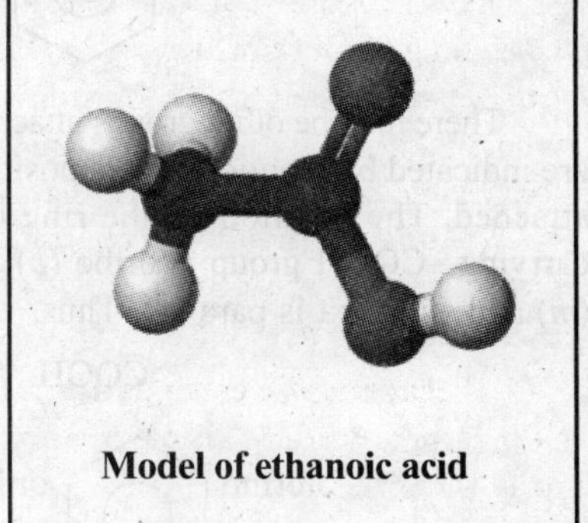
Model of ethanoic acid

For naming substituted acids, according to IUPAC system, again we identify groups which are attached and positions where these are attached. But the numbering of the positions is done in different manner. The carboxy group carbon is given no. 1 position, next one no. 2 position and so on.

$$\overset{5}{C} - \overset{4}{C} - \overset{3}{C} - \overset{2}{C} - \overset{1}{COOH}$$

Thus the following compounds will be named, according to IUPAC system, as shown below against the compounds.

$$\overset{3}{CH_3} - \underset{\underset{Cl}{|}}{\overset{2\quad 1}{CHCOOH}}$$

2-chloropropanoic acid

$$\overset{4}{H_2NCH_2}\overset{3}{CH_2}\overset{2}{CH_2}\overset{1}{COOH}$$

4-aminobutanoic acid

Table 6.1 gives the common and IUPAC names of some common Carboxylic compounds.

Table 6.1: Common and IUPAC names of some monocarboxylic acids

Monocarboxylic acid	Common name	IUPAC name
$HCOOH$	Formic acid	Methanoic acid
CH_3COOH	Acetic acid	Ethanoic acid
CH_3CH_2COOH	Propionic acid	Propanoic acid
$CH_3CH_2CH_2COOH$	*n*-Butyric acid	Butanoic acid
$\underset{\underset{CH_3}{|}}{CH_3CHCOOH}$	Isobutyric acid (α–Methylpropionic acid)	2-Methylpropanoic acid
$CH_3(CH_2)_3COOH$	Valeric acid	Pentanoic acid
$CH_3(CH_2)_4COOH$	Caproic acid	Hexanoic acid
$CH_3(CH_2)_6COOH$	Caprylic acid	Octanoic acid
$CH_3(CH_2)_8COOH$	Capric acid	Decanoic acid

$CH_3(CH_2)_{10}COOH$	Lauric acid	Dodecanoic acid
$CH_3(CH_2)_{12}COOH$	Myristic acid	Tetradecanoic acid
$CH_3(CH_2)_{14}COOH$	Palmitic acid	Hexadecanoic acid
$CH_3(CH_2)_{16}COOH$	Stearic acid	Octadecanoic acid

Nomenclature of Aromatic Carboxylic Acids

The simplest aromatic carboxy acid is benzoic acid.

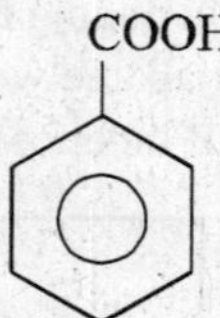

Formic acid in the sting of ants destroys banana crop.

There may be other groups attached to the ring. These are indicated by mentioning the positions where these are attached. The position on the ring next to the position carrying –COOH group is ortho (*o*), next to that is meta (*m*) and still next is para (*p*). Thus,

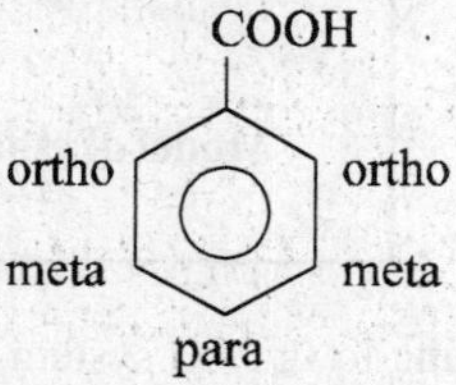

COOH

ortho ortho

meta meta

para

In IUPAC system, carboxy group position is taken as no. 1 and then the positions are counted clockwise.

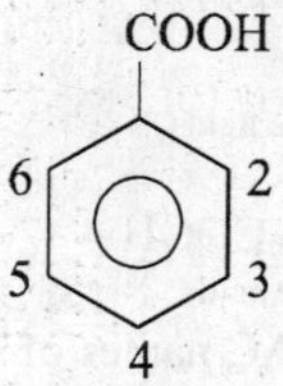

COOH

6 2

5 3

4

The name benzoic acid is accepted in IUPAC nomenclature. It is also named as benzene carboxylic acid in IUPAC nomenclature.

Names of some carboxylic compounds are given below.

(*i*) COOH, Cl

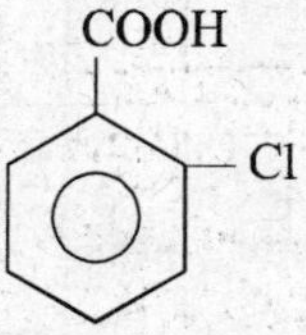

Ortho chlorobenzoic acid

(*ii*) COOH, CH_3

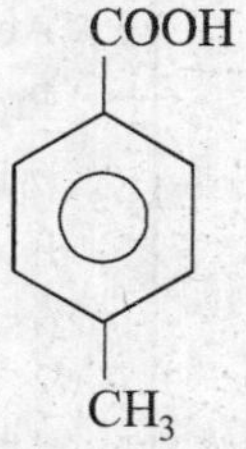

Para methylbenzoic acid

(iii) 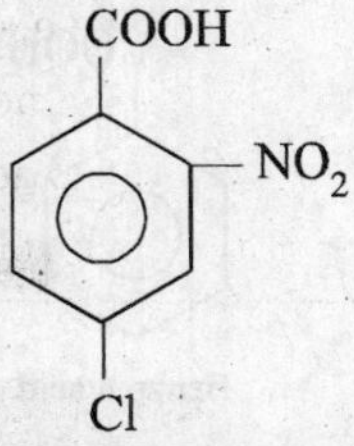 2-nitro-4-chlorobenzoic acid (COOH, NO_2, Cl on benzene ring)

(iv) 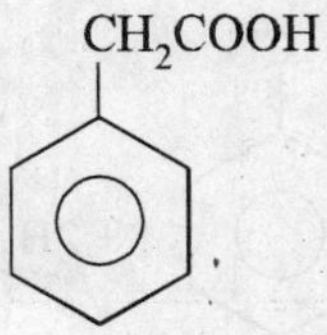 Phenyl acetic acid (CH_2COOH on benzene ring).

6.3. METHODS OF PREPARATION OF CARBOXYLIC ACIDS

Carboxylic acids (aliphatic and aromatic) can be prepared by the following methods:

1. Oxidation of alcohols and aldehydes. Oxidation can be carried out in the presence of acidified $KMnO_4$ or $K_2Cr_2O_7$ or dilute nitric acid.

$$R-CH_2OH \xrightarrow[K_2Cr_2O_7\,/\,H_2SO_4]{[O]} R-CHO \xrightarrow{[O]} R-COOH$$

Examples

$$\underset{\text{Ethanol}}{CH_3CH_2OH} \xrightarrow[K_2Cr_2O_7\,/\,H_2SO_4]{[O]} \underset{\text{Ethanal}}{CH_3CHO} \xrightarrow[K_2Cr_2O_7\,/\,H_2SO_4]{[O]} \underset{\text{Ethanoic acid}}{CH_3COOH}$$

$$C_6H_5CH_2OH \xrightarrow[K_2Cr_2O_7\,/\,H_2SO_4]{[O]} \underset{\text{Benzaldehyde}}{C_6H_5CHO} \xrightarrow[H_2SO_4]{K_2Cr_2O_7} \underset{\text{Benzoic acid}}{C_6H_5COOH}$$

2. By oxidation of alkyl benzenes. Alkyl benzenes on oxidation with potassium permanganate yield carboxylic acid. Whatever the length of the attached alkyl group, it is converted simply into – COOH.

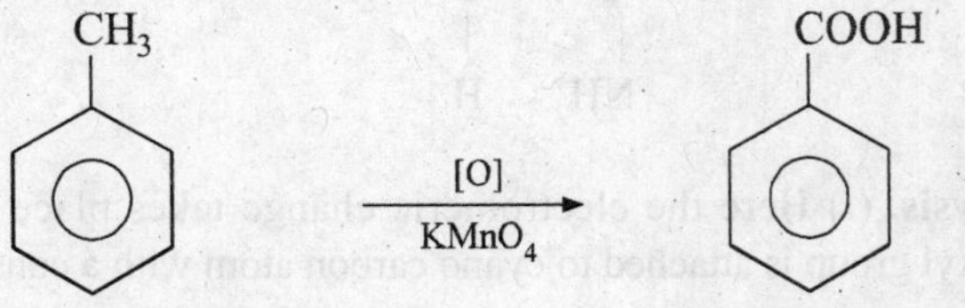

$$\underset{\text{Toluene}}{C_6H_5CH_3} \xrightarrow[KMnO_4]{[O]} \underset{\text{Benzoic acid}}{C_6H_5COOH}$$

$$\underset{\text{Ethylbenzene}}{C_6H_5C_2H_5} + 6\,[O] \xrightarrow{KMnO_4} \underset{\text{Benzoic acid}}{C_6H_5COOH} + CO_2 + 2H_2O$$

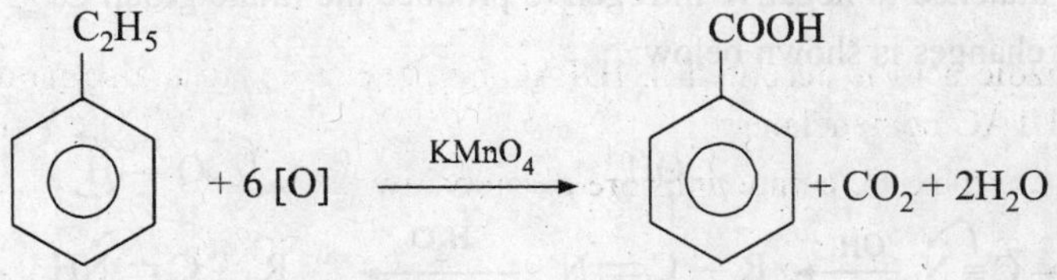

3. By hydrolysis of nitriles or cyanides

$$R-CN + 2H_2O \xrightarrow[\text{Alkali}]{\text{Acid or}} R-COOH + NH_3$$

$$\underset{\text{Methyl cyanide}}{CH_3-CN} + 2H_2O \xrightarrow[\text{Alkali}]{\text{Acid or}} \underset{\text{Acetic acid}}{CH_3COOH} + NH_3$$

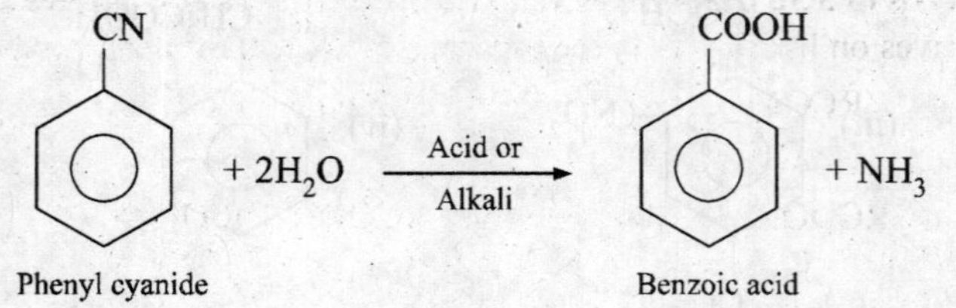

$$C_6H_5CN + 2H_2O \xrightarrow[\text{Alkali}]{\text{Acid or}} C_6H_5COOH + NH_3$$

Phenyl cyanide → Benzoic acid

Mechanism

Acidic hydrolysis. (*i*) Hydrogen ion attaches itself to the cyano nitrogen atom with the shifting of charge to nitrogen and creation of positive charge on carbon.

(*ii*) Water molecule is attached to the carbonium ion.

The sequence of changes that take place subsequently is shown hereunder.

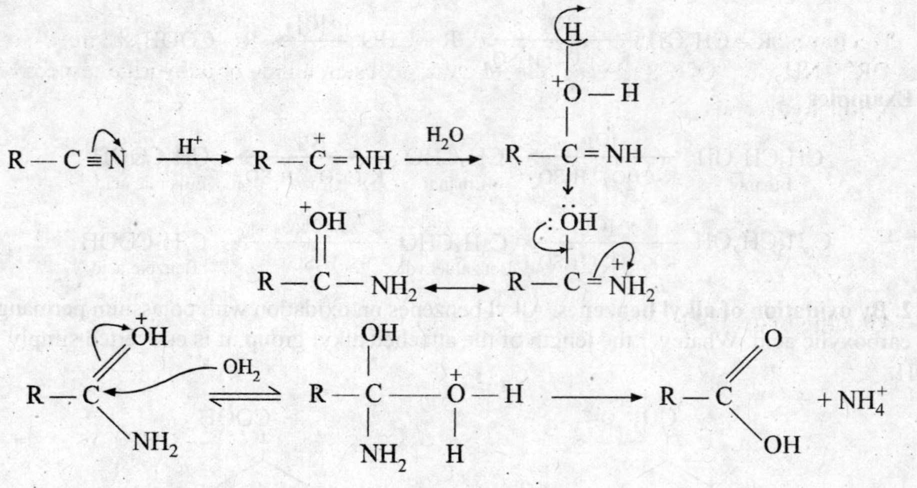

$$R-C\equiv N \xrightarrow{H^+} R-\overset{+}{C}=NH \xrightarrow{H_2O} R-\underset{\overset{+}{O}H_2}{C}=NH$$

$$R-\underset{:\ddot{O}H}{C}=\overset{+}{N}H_2 \longleftrightarrow R-\underset{\overset{+}{O}H}{\overset{\|}{C}}-NH_2$$

$$R-C(=\overset{+}{O}H)NH_2 \xrightarrow{OH_2} R-C(OH)(NH_2)-\overset{+}{O}H_2 \longrightarrow R-C(=O)OH + NH_4^+$$

Alkaline hydrolysis. (*i*) Here the electromeric change takes place in the presence of the hydroxyl group. Hydroxyl group is attached to cyano carbon atom with a consequent negative charge on nitrogen.

(*ii*) A H^+ is attached to negative nitrogen to produce the imine group C = NH.

Sequence of changes is shown below:

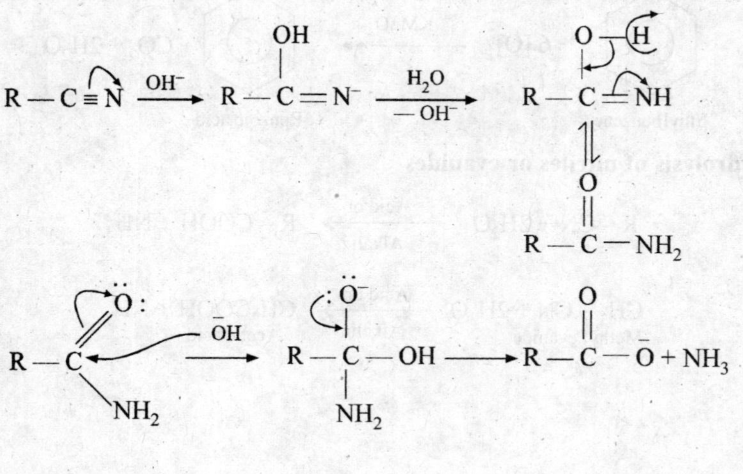

$$R-C\equiv N \xrightarrow{OH^-} R-C(OH)=N^- \xrightarrow[-OH^-]{H_2O} R-C(OH)=NH \rightleftharpoons R-C(=O)-NH_2$$

$$R-C(=\ddot{O}:)NH_2 \xrightarrow{\bar{O}H} R-C(:\ddot{O}^-)(NH_2)-OH \longrightarrow R-C(=O)-\bar{O} + NH_3$$

4. By hydrolysis of acid derivatives viz. acid chlorides, esters, amides and anhydrides. The above acid derivatives on hydrolysis in the presence of an acid or alkali produce carboxylic acids.

$$\underset{\text{Acid chloride}}{RCOCl} + H_2O \longrightarrow RCOOH + HCl$$

$$\underset{\text{Ester}}{RCOOR'} + H_2O \longrightarrow \underset{\text{Acid}}{RCOOH} + R'OH$$

$$RCONH_2 + H_2O \longrightarrow RCOOH + NH_3$$

$$\underset{\text{Acid Anhydride}}{(RCO)_2O} + H_2O \longrightarrow \underset{\text{Acid}}{2\,RCOOH}$$

Mechanism. An acid derivative could be represented as $R-\overset{O}{\overset{\|}{C}}-Y$ where Y stands for Cl^-, $-OR'$, $-NH_2$ or $-OOCR$ in case of acid chloride, ester, amide or anhydride respectively.

Acidic hydrolysis

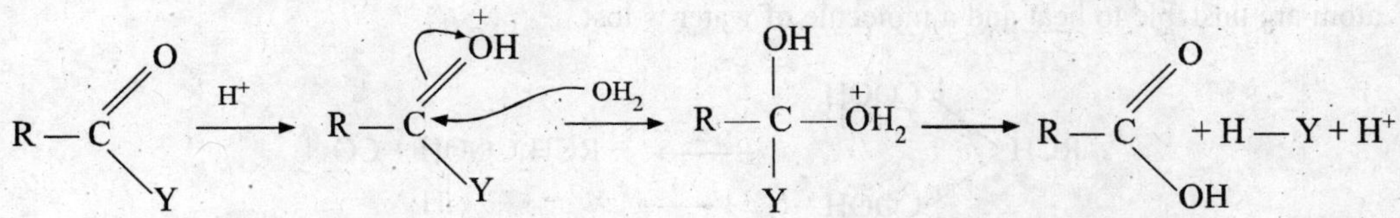

$$R-C(=O)-Y \xrightarrow{H^+} R-C(=\overset{+}{O}H)-Y \xrightarrow{OH_2} R-C(OH)(Y)-\overset{+}{O}H_2 \longrightarrow R-C(=O)-OH + H-Y + H^+$$

Alkaline hydrolysis

$$R-C(=\ddot{O}:)-Y \xrightarrow{^-OH} R-C(:\ddot{O}:^-)(Y)-OH \longrightarrow R-\overset{O}{\overset{\|}{C}}-OH + Y^- \xrightarrow{OH^-} R-\overset{O}{\overset{\|}{C}}-O^- + H_2O$$

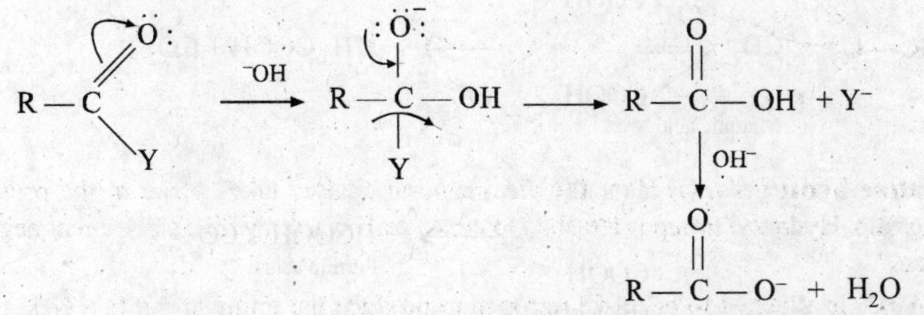

5. By the action of carbon dioxide on alkyl or phenyl magnesium bromide. Grignard reagent reacts with carbon dioxide forming an addition product which yields carboxylic acid on hydrolysis.

$$\overset{O}{\overset{\|}{C}}=O + \underset{\text{Methyl magnesium bromide}}{CH_3\,Mg\,Br} \longrightarrow \underset{\text{Addition Product}}{CH_3-\overset{O}{\overset{\|}{C}}-OMgBr}$$

$$\xrightarrow{H_2O} \underset{\text{Acetic acid}}{CH_3-\overset{O}{\overset{\|}{C}}-OH} + Mg\langle^{OH}_{Br}$$

$$O=C=O + C_6H_5MgBr \longrightarrow C_6H_5-\overset{\overset{O}{\|}}{C}-OMgBr$$

$$\xrightarrow{H_2O} \underset{\text{Benzoic acid}}{C_6H_5-\overset{\overset{O}{\|}}{C}-OH} + Mg\begin{matrix} OH \\ Br \end{matrix}$$

Mechanism. The nucleophile alkyl group is attached to carbon of the carbon dioxide molecule whereas –MgBr group is attached to one of the two oxygen atoms. This is shown schematically as under:

$$R-MgX \quad + \quad O=C=O \longrightarrow R-\overset{\overset{O}{\|}}{C}-OMgX$$

6. By decarboxylation of dicarboxy acids. Two carboxy groups attached to the same carbon atom are unstable to heat and a molecule of water is lost.

$$RCH\begin{matrix} COOH \\ COOH \end{matrix} \xrightarrow{\Delta} RCH_2COOH + CO_2\uparrow$$

$$\underset{\text{Malonic acid}}{CH_2\begin{matrix} COOH \\ COOH \end{matrix}} \longrightarrow CH_3\,COOH + CO_2$$

$$\underset{\text{Oxalic acid}}{\begin{matrix} COOH \\ | \\ COOH \end{matrix}} \xrightarrow{\Delta} \underset{\text{Formic acid}}{HCOOH} + CO_2$$

7. By hydrolysis of trihalogen compounds.

$$RCCl_3 + 3H_2O \xrightarrow[-3HCl]{KOH} RC\begin{matrix} OH \\ OH \\ OH \end{matrix} \xrightarrow{\Delta} RCOOH + H_2O$$

$$\underset{\text{Trichloro-ethane}}{CH_3CCl_3} + 3H_2O \xrightarrow[-3HCl]{KOH} CH_3\,C\begin{matrix} OH \\ OH \\ OH \end{matrix} \xrightarrow{\Delta} \underset{\text{Acetic acid}}{CH_3COOH} + H_2O$$

$$\underset{\text{Benzo-trichloride}}{C_6H_5CCl_3} + 3H_2O \xrightarrow[-3HCl]{KOH} C_6H_5\,C\begin{matrix} OH \\ OH \\ OH \end{matrix} \xrightarrow{\Delta} \underset{\text{Benzoic acid}}{C_6H_5COOH} + H_2O$$

8. From Malonic ester and Acetoacetic ester. Malonic ester synthesis. A desired acid can be prepared from malonic ester having the formula

$$CH_2\begin{cases} COOC_2H_5 \\ COOC_2H_5 \end{cases} \qquad \text{Diethyl malonate}$$

It involves the following steps:

(*i*) Diethyl molecule is treated with sod. ethoxide to produce sodium malonic ester.

$$CH_2(COOC_2H_5) \xrightarrow[-C_2H_5OH]{C_2H_5ONa} \overset{+}{Na}\overset{-}{C}H\,(COOC_2H_5)_2$$

(*ii*) Sodiomalonic ester is then treated with an appropriate alkyl halide to produce alkyl malonic ester.

$$\overset{+}{Na}\overset{-}{C}H\,(COOC_2H_5)_2 + RX \longrightarrow RCH\,(COOC_2H_5)_2$$

(*iii*) Alkyl malonic ester is hydrolysed and heated to produce carboxylic acid.

$$RCH\,(COOC_2H_5)_2 \xrightarrow{\text{Hydrolysis}} RCH\,(COOH)_2 \xrightarrow{\Delta} RCH_2COOH$$

By a proper choice of the alkyl halide, any desired carboxylic acid can be produced.

Acetoacetic ester synthesis. Like malonic ester, acetoacetic ester also possesses an α-hydrogen atom. This α-hydrogen can be replaced by sodium to produce sodium acetoacetic ester, which is then treated with an alkyl halide. The addition product so obtained is subjected to alkaline hydrolysis to produce acids.

$$\underset{\text{Acetoacetic ester}}{CH_3COCH_2COOC_2H_5} \xrightarrow{\text{Na / Acohol}} \underset{\text{Sod. salt of the ester}}{CH_3CO\overset{\overset{+}{Na}\,|\,-}{C}HCOOC_2H_5}$$

$$\xrightarrow{RBr} \underset{\text{Alkyl acetoacetic ester}}{CH_3\,CO\overset{\overset{R}{|}}{C}HCOOC_2H_5} \xrightarrow{\text{Hydrolysis}} CH_3COOH + \underset{\text{Desired acid}}{RCH_2COOH} + C_2H_5OH$$

A proper choice of the alkyl halide is required in order to obtain a particular acid. For example, to obtain propionic acid, the alkyl halide needed is methyl bromide, to obtain valeric acid, the alkyl halide to be used is *n*-butyl bromide and so on.

Example 1. Give probable mechanism of alkaline hydrolysis of an ester.

Or

Give mechanism in support of the fact that during alkaline hydrolysis of an ester, it is the acyl oxygen bond that cleaves.

Solution. Hydrolysis of an ester can be represented by the following equation:

$$\underset{\text{Ethyl acetate}}{CH_3-\overset{\overset{O}{\|}}{C}-OC_2H_5} + NaOH \longrightarrow CH_3-\overset{\overset{O}{\|}}{C}-ONa + C_2H_5OH$$

The mechanism involves the following steps:

***(i)* Nucleophilic attack of hydroxide ion**

$$CH_3-\overset{\overset{\displaystyle O}{\|}}{C}-OC_2H_5 \xrightarrow{OH^-} CH_3-\underset{\underset{\displaystyle OH}{|}}{\overset{\overset{\displaystyle O^-}{|}}{C}}-OC_2H_5$$

(II)

***(ii)* Elimination of ethoxide ion from (II)**

$$CH_3-\underset{\underset{\displaystyle OH}{|}}{\overset{\overset{\displaystyle O^-}{|}}{C}}-OC_2H_5 \longrightarrow CH_3-\underset{\underset{\displaystyle OH}{|}}{\overset{\overset{\displaystyle O}{\|}}{C}} + C_2H_5O^-$$

(III)

***(iii)* Removal of proton from the acid (III)**

$$CH_3-\overset{\overset{\displaystyle O}{\|}}{C}-O-H + C_2H_5O^- \longrightarrow CH_3-\overset{\overset{\displaystyle O}{\|}}{C}-O^- + C_2H_5OH$$

Example 2. Sketch the mechanism of hydrolysis of ethyl acetate in the presence of dil. HCl.

Solution. Mechanism of hydrolysis of ethyl acetate in the presence of dil. HCl is as given below.

$$CH_3-C\begin{matrix}\nearrow O\\ \searrow OC_2H_5\end{matrix} \xrightarrow{H^+} CH_3-\overset{+}{C}\begin{matrix}OH\\ OC_2H_5\end{matrix} \xrightarrow{H_2O} CH_3-\underset{\underset{\displaystyle OC_2H_5}{|}}{\overset{\overset{\displaystyle OH}{|}}{C}}-\overset{+}{O}H_2$$

Ethyl acetate

$$\longrightarrow CH_3-\overset{\overset{\displaystyle O}{\|}}{C}-OH + C_2H_5OH + H^+$$

6.4. STRUCTURE OF CARBOXYLIC GROUP

Carboxy acid is represented as:

$$R-\overset{\overset{\displaystyle O}{\|}}{C}-O-H$$

where R is an alkyl or aryl group. The carboxy group actually exhibits resonance. The two resonating structures of the carboxy group are:

$$\underset{I}{-\overset{\overset{\displaystyle O}{\|}}{C}-\ddot{O}H} \longleftrightarrow \underset{II}{-\overset{\overset{\displaystyle O^-}{|}}{C}=\overset{+}{O}H}$$

Structure II is not so stable energetically because of the separation of charges. It does not contribute much to the actual structure of the carboxylic group. Thus the two structures do not contribute equally to the actual structure. Consequently, resonance energy, which estimates the stability of a compound, is less. On the other hand, the carboxylate ion obtained after removing one hydrogen atom is much more stable.

$$R-\overset{\overset{\displaystyle O}{\|}}{C}-O-H \longrightarrow \underset{\text{Carboxylate ion}}{R-\overset{\overset{\displaystyle O}{\|}}{C}-\bar{O}} + H^+$$

$$\underset{III}{R-\overset{\overset{\displaystyle O}{\|}}{C}-O^-} \longleftrightarrow \underset{IV}{R-\overset{\overset{\displaystyle O^-}{|}}{C}=O}$$

Structures III and IV which are resonances hydrids of carboxylate ion are similar and contribute equally to the actual structure. In such a case as per the rules of resonance phenomenon, the value of resonance energy is quite high which accounts for the stability of carboxylate ion. This would imply that carboxylate ion and the hydrogen ion can exist separately. This explains the highly acidic nature of the carboxylate ion.

Resonance stabilization does not take place to this extent in the case of phenol or alcohol. Hence carboxylic compounds are stronger in acidic nature than phenols and alcohols.

Since the double bond is changing its position, the carboxylate ion could be represented as

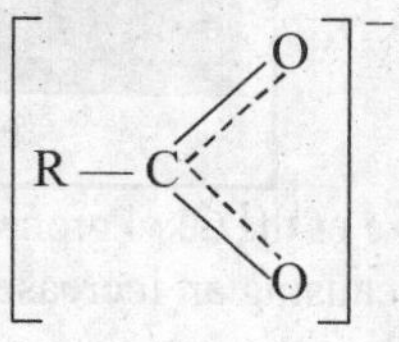

$$\left[R - C \begin{matrix} \overset{\cdot\cdot}{\nearrow} O \\ \searrow O \end{matrix} \right]^-$$

6.5. PHYSICAL PROPERTIES OF CARBOXYLIC ACIDS

1. State. The monocarboxylic acids containing up to ten carbon atoms are colourless liquids. The higher acids are colourless waxy solids. The aromatic acids are colourless crystalline solids.

2. Smell. Lower aliphatic carboxylic acids (up to three carbons) possess a pungent smell. Higher homologues up to eight carbon atoms are slightly unpleasant to smell. Carboxy acids having more than eight carbons possess no smell. Aromatic carboxy acids have no smell.

3. Miscibility. The lower members of carboxylic acids are miscible with water due to the presence of intermolecular hydrogen bonding with H_2O molecules.

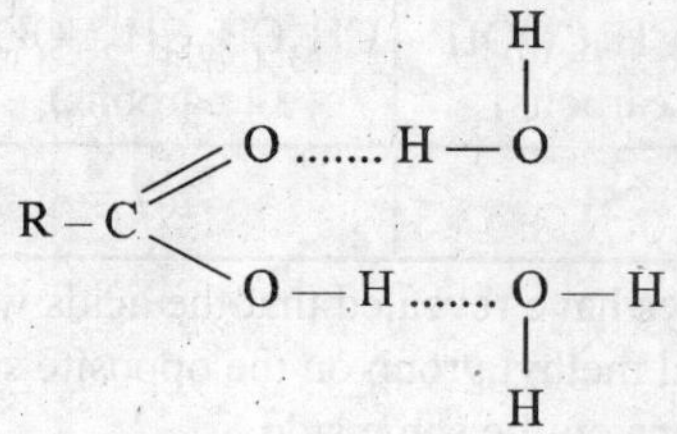

```
                    H
                    |
          O ....... H — O
        //
  R – C
        \
          O — H ....... O — H
                        |
                        H
```

However, the solubility in water decreases gradually with the increase in the size of alkyl group. The bulky alkyl group tends to mask the carboxyl group which takes part in the hydrogen bonding. Aromatic acids have little solubility in water. Both aliphatic and aromatic carboxylic acids are soluble in organic solvents like benzene, alcohol, ether, etc.

4. Boiling points. Monocarboxylic acids have higher boiling points compared to the alcohols of comparable molecular masses. For example, both *n*-propyl alcohol and acetic acid have molecular mass equal to 60. While alcohol boils at 370 K, the b.p. of the acid is about 391 K. The higher boiling point of the acid as compared to that of alcohol is due to the greater extent of intermolecular hydrogen bonding in the former compared to the latter as shown below :

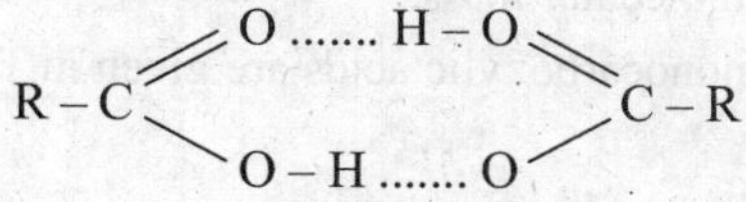

```
        O ....... H – O                          R           R           R
      //               \\                       /           /           /
R – C                    C – R      – – – H — O ....... H — O ..... H – O .....
      \                 /
        O – H ....... O
```

Hydrogen bonding in carboxylic acids

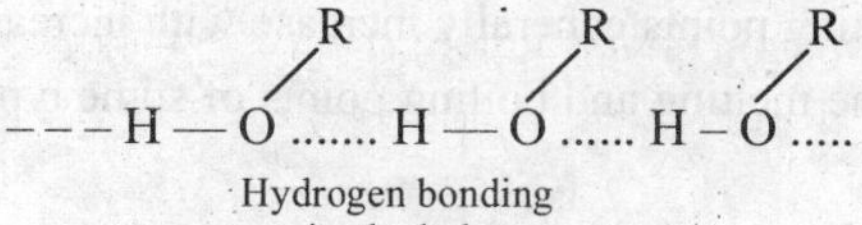

Hydrogen bonding in alcohols

Most of the carboxylic acids exist as dimers, *i.e.*, two molecules of carboxylic acids are linked by strong hydrogen bonds. This has been confirmed by the determination of molecular mass of acetic acid in solvents like benzene by the depression in freezing point method. The molecular mass is found to be 120 whereas in the monomeric state, the molecular of the acid is only 60.

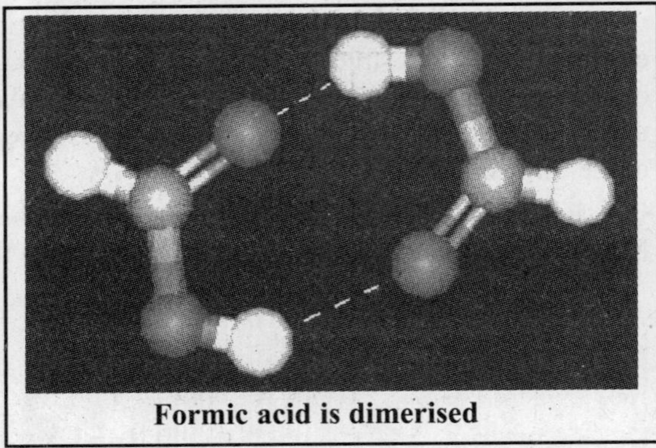

Formic acid is dimerised

The boiling point increases with increase in the molecular mass. As the size of the alkyl group increases, the magnitude of the attractive forces (van der Waals forces) increases causing an increase in boiling point. The boiling points of a few monocarboxylic acids are given below :

Acid	HCOOH	CH_3COOH	C_2H_5COOH	C_3H_7COOH
Boiling point (K)	373	391	424	437

The boiling points of the aromatic acids are higher than those of aliphatic acids due to greater size of the phenyl group as compared to that of the alkyl group.

5. Melting points. X-ray diffraction studies have shown that the acids containing even number of carbon atoms have higher melting points compared to the immediately next lower and higher members in the family which contain odd number of carbon atoms. This is evident from the melting point data.

Acid	CH_3CH_2COOH (3 carbons)	$CH_3CH_2CH_2COOH$ (4 carbons)	$CH_3CH_2CH_2CH_2COOH$ (5 carbons)
Melting point (K)	251	267	239

The X-ray diffraction studies have revealed that the acids with even number of carbon atoms have carboxyl group and terminal methyl group on the opposite sides while those with odd number of carbon atoms have these groups on the same side.

CH_3–CH_2–CH_2–COOH (zig-zag)

(Terminal groups are on the opposite side)

CH_3–CH_2–CH_2–CH_2–COOH (zig-zag)

(Terminal groups are on the same side)

There are greater attractive forces between the molecules in acids containing even no. of carbon atoms because the molecules with even number of carbon atoms fit into the crystal lattice better than the molecules with odd number of carbon atoms. The melting points of the former are, therefore, higher. This trend is, however, limited to members having up to ten carbon atoms. In higher members, the melting points generally increase with increase in molecular mass.

The melting and boiling points of some typical monocarboxylic acids are given in Table 6.2.

Table 6.2. Melting and boiling points of some Monocarboxylic acids

Compound	Formula	M.P. (K)	B.P. (K)
Formic acid	HCOOH	281	373
Acetic acid	CH_3COOH	289	391
Propionic acid	CH_3CH_2COOH	251	424
Butyric acid	$CH_3CH_2CH_2COOH$	267	437
Isobutyric acid	$CH_3-CH(CH_3)-COOH$	226	427
Valeric acid	$CH_3CH_2CH_2CH_2COOH$	239	460
Isovaleric acid	$CH_3-CH(CH_3)-CH_2COOH$	235	449
Caproic acid	$CH_3-(CH_2)_4-COOH$	270	578
Palmitic acid	$CH_3-(CH_2)_{14}-COOH$	336	625
Benzoic acid	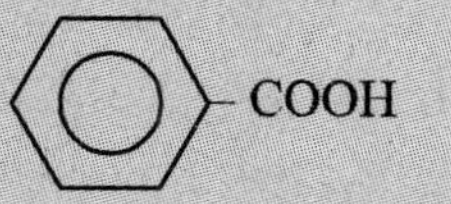 C_6H_5-COOH	395	523
o-Toluic acid	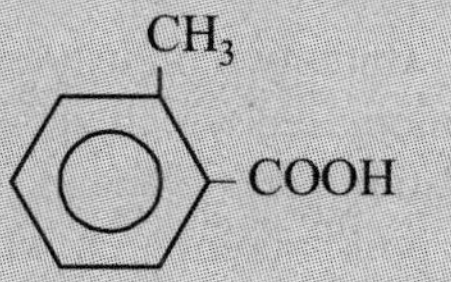 $CH_3C_6H_4-COOH$ (CH₃ ortho)	379	532
m-Toluic acid	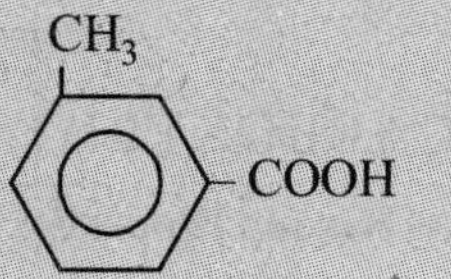 $CH_3C_6H_4-COOH$ (CH₃ meta)	385	536
p-Toluic acid	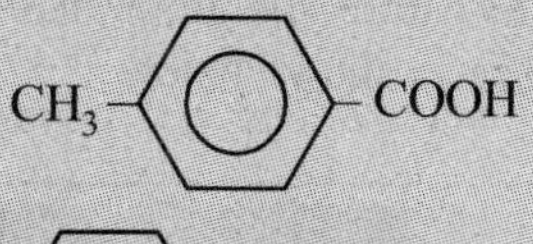 $CH_3-C_6H_4-COOH$	433	548
Phenylacetic acid	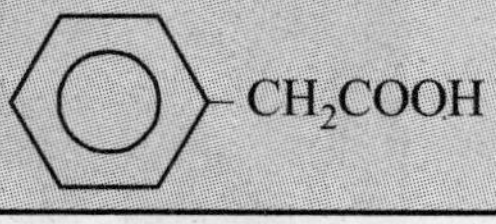 $C_6H_5-CH_2COOH$	350	539

Example 1. Explain why the first few members of carboxylic acids are soluble and higher members insoluble in water.

Solution. Lower members from formic to butyric acid are soluble in water because of the formation of hydrogen bond between water and the acid. Because of electronegativity difference between oxygen and hydrogen, hydrogen bond is formed as shown below:

(*i*)

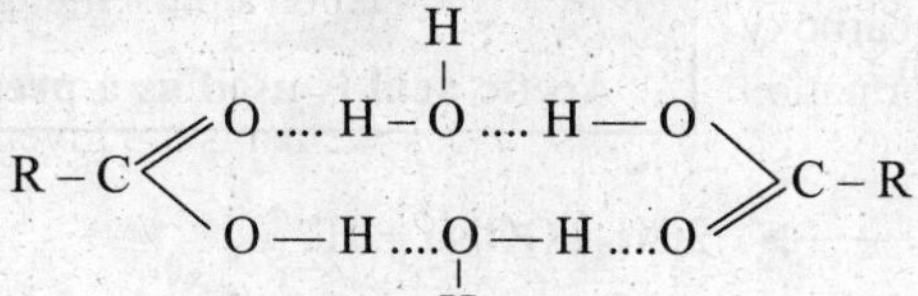

```
                    H
                    |
          O .... H - O .... H — O
R - C  <                          >  C - R
          O — H .... O — H .... O
                     |
                     H
```

$$\begin{array}{ccccccc} & R & & R & & & \\ & | & & | & & & \\ & C=O & H & C=O & H & & \\ & | & | & | & | & & \\ (ii) \quad ...... & O — H ...... & O — H ..... & O — H ...... & O — H..... \end{array}$$

As the size of the alkyl group increases, the solubility in water decreases because the strength of hydrogen bond decreases. This is because a long fatty chain hinders the formation of hydrogen bond. Hence higher fatty acids are insoluble in water. They are, however, soluble in organic solvents like benzene and ether.

Example 2. Explain why carboxylic acids show much higher boiling point compared to alcohols of comparable molecular masses.

OR

Most of the carboxy acids exist as dimers. Explain.

OR

Mol. wt. of acetic acid is double than the calculated value. Explain.

Solution. Carboxylic compounds exhibit hydrogen bonding between two acid molecules as shown below:

$$R—C\begin{matrix} \nearrow O ...... H—O \searrow \\ \searrow O — H .....O \nearrow \end{matrix}C — R$$

Thus the acid molecule exists as a **dimer** in which two molecules have been associated into one molecule. Thus, the size of the molecule is increased which accounts for the higher boiling point of carboxylic compounds. No such dimerisation takes place in between alcohol molecules. Hence carboxy compounds have higher boiling points compared with alcohols of comparable molecular weights.

6.6. CHEMICAL PROPERTIES OF CARBOXYLIC ACIDS

(A) Reactions Due to H-Atom of the Carboxyl (Group Properties due to ionisable hydrogen):

The hydroxyl group, which is a part of the carboxylic group, has ionisable hydrogen. This is because of the electronic change that takes place as shown below:

$$\overset{\displaystyle O}{\overset{\|}{-C}} \leftarrow O \leftarrow H$$

This facilitates in the release of protons (H^+).

Various reactions of the carboxylic group due to ionisable hydrogen are given below:

(*i*) **Acidic Nature.** Carboxy compounds turn blue litmus red. This is indicative of acidic nature of the group.

$$RCOOH + H_2O \rightleftharpoons RCOO^- + H_3O^+$$

Acetic acid is used as a preservative

(*ii*) **Liberation of Hydrogen.** Electropositive elements like Na, K and Ca react with carboxy acids to produce hydrogen gas with the formation of the salts.

$$2CH_3\ COOH + 2Na \longrightarrow 2CH_3COONa + H_2 \uparrow$$

$$2CH_3COOH + Ca \longrightarrow (CH_3COO)_2\ Ca + H_2 \uparrow$$

(*iii*) **Salt Formation.** Carboxy acids react with alkalies and metallic oxides and decompose bicarbonates with the formation of salts. Reaction takes place also with ammonia.

$$CH_3COOH + KOH \longrightarrow \underset{\text{Pot. acetate}}{CH_3COOK} + H_2O$$

$$2CH_3COOH + HgO \longrightarrow \underset{\text{Mercuric acetate}}{(CH_3COO)_2Hg} + H_2O$$

$$CH_3COOH + NaHCO_3 \longrightarrow CH_3COONa + CO_2 + H_2O$$

$$CH_3COOH + NH_3 \longrightarrow \underset{\text{Ammonium acetate}}{CH_3COONH_4}$$

(B) Properties involving replacement of hydroxyl group:

The –OH group of the acid can be replaced by –OR′, –Cl, –NH_2 and OOCR′ groups to form esters, acid chlorides, amides and anhydrides respectively. These reactions are given separately as under:

(*i*) Reaction with alcohols (formation of esters). Carboxylic acids on heating with an alcohol in the presence of a strong mineral acid (such as H_2SO_4) form **esters.** This reaction is known as **esterification** and is reversible in nature.

$$\underset{\text{Acid}}{RCOOH} + \underset{\text{Alcohol}}{R'OH} \underset{}{\overset{H^+}{\rightleftharpoons}} RCOOR' + H_2O$$

$$\underset{\text{Acetic acid}}{CH_3COOH} + \underset{\text{Ethanol}}{C_2H_5OH} \overset{H^+}{\rightleftharpoons} \underset{\text{Ethyl acetate}}{CH_3COOC_2H_5} + H_2O$$

Mechanism. H^+ from the mineral acid protonates carbonyl oxygen. As a result, carbon atom becomes positive and undergoes nucleophilic attack by the alcohol molecule. Finally ester is obtained as depicted below:

$$R-\overset{\overset{O}{\|}}{C}-O-H \xrightarrow{H^+} R-\overset{\overset{OH}{|}}{\underset{+}{C}}-O-H \xrightarrow{\overset{OR'}{H}} R-\overset{\overset{OH}{|}}{\underset{\underset{\underset{(H)}{|}}{+OR'}}{C}}-O-H$$

$$\underset{\text{Ester}}{R-\overset{\overset{O}{\|}}{\underset{\underset{OR'}{|}}{C}}} \xleftarrow{-H_2O} R-\overset{\overset{O-H}{|}}{\underset{\underset{OR'}{|}}{C}}-\overset{+}{O}-\underset{H}{\mathbf{H}}$$

To summarise, esterification involves the cleavage of O–H bond of alcohol and C–OH bond of the carboxy acid. Thus,

$$R-\overset{\overset{O}{\|}}{C}-\boxed{OH + H}\;OR' \longrightarrow \underset{\text{Ester}}{R-\overset{\overset{O}{\|}}{C}-OR'}$$

This has been confirmed by radioactive tracer studies.

For example, when benzoic acid is esterified with methyl alcohol containing O – 18 oxygen (labelled alcohol), the labelled oxygen after the reaction is present in the ester and not water.

$$\underset{\text{Benzoic acid}}{C_6H_5-\overset{\overset{O}{\|}}{C}-OH} + \underset{\text{Labelled alcohols}}{CH_3O^{18}H} \overset{H^+}{\rightleftharpoons} \underset{\text{Labelled ester}}{C_6H_5-\overset{\overset{O}{\|}}{C}-O^{18}CH_3} + H_2O$$

Had water been labelled, then OH part of alcohol and H atom of carboxyl group might have been involved which, of course, is not the case. This technique of studying the mechanism of a reaction is known as isotopic labelling.

***(ii)* Reaction with PCl_5, PCl_3 or $SOCl_2$ (Formation of acid chloride).** An acid chloride is obtained when an acid is treated with PCl_5 or $SOCl_2$ (Thionyl chloride).

$$\underset{}{RCOOH} + PCl_5 \longrightarrow \underset{\text{Acid chloride}}{RCOCl} + POCl_3 + HCl$$

$$3RCOH + \underset{\text{Phosphorus trichloride}}{PCl_3} \longrightarrow RCOCl + \underset{\text{Phosphorus acid}}{H_3PO_3}$$

$$CH_3COOH + \underset{\text{Thionyl chloride}}{SOCl_2} \longrightarrow \underset{\text{Acetyl chloride}}{CH_3COCl} + SO_2\uparrow + HCl\uparrow$$

Out of the above mentioned reagents, thionyl chloride is preferred because the side products SO_2 and HCl are gaseous and escape. So we do not need to carry out fractional distillation to extract the main product.

Mechanism. The mechanism involves the nucleophile attack of the carboxylic acid on the phosphorus atom of phosphorous halide on sulphur atom or thionyl chloride as follows :

$$R-\overset{O}{\overset{\|}{C}}-\ddot{O}-H + PCl_4\,(Cl) \xrightarrow{-HCl} R-\overset{O}{\overset{\|}{C}}-O-\underset{Cl}{\underset{|}{P}}Cl_3 \longrightarrow R-\overset{O}{\overset{\|}{C}}-Cl + O=PCl_3$$

$$R-\overset{O}{\overset{\|}{C}}-\ddot{O}-H + \overset{O}{\overset{\|}{\underset{Cl}{\underset{|}{S}}}}-Cl \xrightarrow{-HCl} R-\overset{O}{\overset{\|}{C}}-O-\overset{O}{\overset{\|}{S}}-Cl \longrightarrow R-\overset{O}{\overset{\|}{C}}-Cl + SO_2$$

***(iii)* Reaction with ammonia (formation of acid amides).** Ammonia reacts with acids to form ammonium salts which on heating liberate a water molecule giving amides.

$$\underset{\text{Acid}}{RCOOH} + NH_3 \longrightarrow \underset{\text{Amm.salt}}{RCOONH_4} \xrightarrow[-H_2O]{\Delta} \underset{\text{Amide}}{RCONH_2}$$

$$\underset{\text{Benzoic acid}}{C_6H_5-COOH} + NH_3 \longrightarrow \underset{\text{Ammonium benzoate}}{C_6H_5-COONH_4} \xrightarrow[-H_2O]{\Delta} \underset{\text{Benzamide}}{C_6H_5-CONH_2}$$

***(iv)* Formation of acid anhydride.** When the vapours of a carboxy acid are passed over P_2O_5, one molecule of water is eliminated from two molecules of the acid resulting in the formation of acid anhydride.

$$\begin{matrix} RCOOH \\ RCOOH \end{matrix} \xrightarrow[-H_2O]{P_2O_5} \underset{\text{Acid anhydride}}{\begin{matrix} RCO \\ RCO \end{matrix}\!\!>O}$$

Mechanism. The reaction is supposed to proceed by the following mechanism :

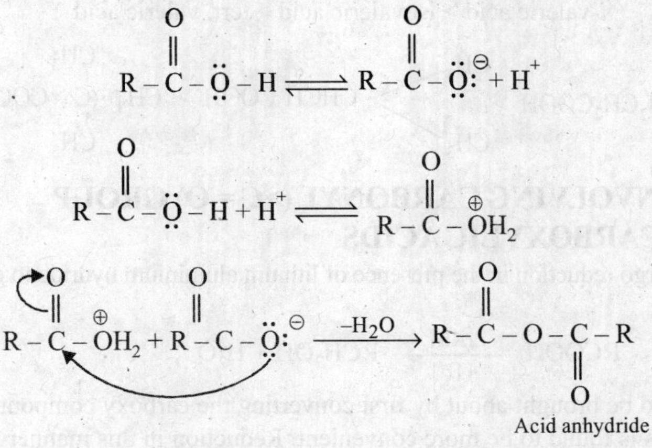

$$R-\overset{\overset{O}{\|}}{C}-\ddot{\underset{\cdot\cdot}{O}}-H \rightleftharpoons R-\overset{\overset{O}{\|}}{C}-\ddot{\underset{\cdot\cdot}{O}}:^{\ominus} + H^{+}$$

$$R-\overset{\overset{O}{\|}}{C}-\ddot{\underset{\cdot\cdot}{O}}-H + H^{+} \rightleftharpoons R-\overset{\overset{O}{\|}}{C}-\overset{\oplus}{O}H_2$$

$$R-\overset{\overset{O}{\|}}{C}-\overset{\oplus}{O}H_2 + R-\overset{\overset{O}{\|}}{C}-\ddot{\underset{\cdot\cdot}{O}}:^{\ominus} \xrightarrow{-H_2O} R-\overset{\overset{O}{\|}}{C}-O-\underset{\underset{O}{\|}}{C}-R$$

Acid anhydride

An anhydride can also be obtained by treating an acid with an acid chloride in the presence of pyridine which removes HCl molecules from the mixture.

$$\underset{\text{Acetic acid}}{CH_3COOH} + \underset{\text{Acetyl chloride}}{CH_3COCl} + \text{Pyridine} \longrightarrow \underset{\text{Acetic anhydride}}{\begin{matrix} CH_3CO \\ CH_3CO \end{matrix}\!\!>\! O} + \text{Pyridine hydrochloride}$$

Mechanism

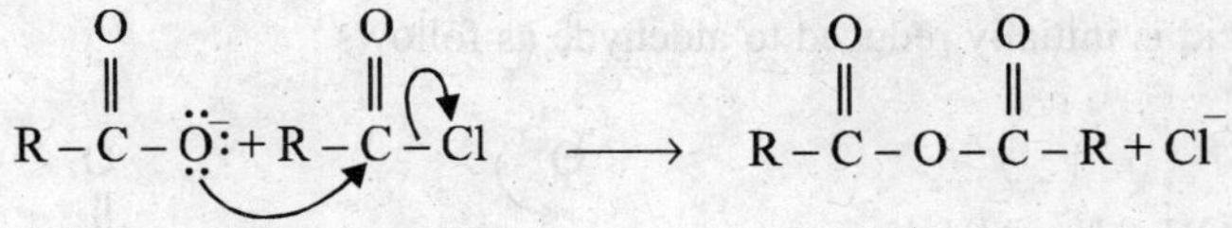

$$R-\overset{\overset{O}{\|}}{C}-\ddot{\underset{\cdot\cdot}{O}}:^{-} + R-\overset{\overset{O}{\|}}{C}-Cl \longrightarrow R-\overset{\overset{O}{\|}}{C}-O-\overset{\overset{O}{\|}}{C}-R + Cl^{-}$$

Example 1. (*a*) Acetic acid reacts with the following alcohols separately. Arrange the alcohols in decreasing order of reactivity of reaction:

(*i*) *n*-butyl alcohol (*ii*) isobutyl alcohol (*iii*) tert. butyl alcohol

(*b*) Ethanol reacts with the following acids separately. Arrange the acids in order of reactivity of reaction.

(*i*) *n*-valeric acid (*ii*) isovaleric acid (*iii*) tert. valeric acid.

Solution. (*a*) As per the mechanism of esterification the alcohol molecule attaches itself to the carboxy acid by nucleophilic addition mechanism. This nucleophilic addition will be easy if the alkyl group of the alcohol is small. Large alkyl groups and branched alkyl groups will pose a problem in the nucleophile addition of the alcohol molecule to the carboxy acid. Hence the order of reactivity will be:

n-butyl alcohol > isobutyl alcohol > tert. butyl alcohol.

or

$$CH_3CH_2CH_2CH_2OH > \begin{matrix} CH_3 \\ CH_3 \end{matrix}\!\!>\! CHCH_2OH > CH_3-\overset{\overset{CH_3}{|}}{\underset{\underset{CH_3}{|}}{C}}-OH$$

(*b*) By the same analogy as given above, the order of reactivity of the acid with ethanol will be:

n-valeric acid > isovaleric acid > tert. valeric acid

or

$$CH_3CH_2CH_2CH_2COOH > \begin{matrix} CH_3 \\ CH_3 \end{matrix}\!\!>\! CHCH_2COOH > CH_3-\underset{CH_3}{\overset{CH_3}{\overset{|}{\underset{|}{C}}}}-COOH$$

(C) REACTIONS INVOLVING CARBONYL (>C = O) GROUP

REDUCTION OF CARBOXYLIC ACIDS

Carboxylic acids undergo reduction in the presence of lithium aluminium hydride to give primary alcohols.

$$RCOOH \xrightarrow[4\,[H]]{LiAlH_4} RCH_2OH + H_2O$$

The reduction can also be brought about by first converting the carboxy compound into ester followed by reduction. This is found to be more convenient. Reduction in this manner is known as **Beauveault Blanc** reaction.

$$RCOOH + R'OH \xrightarrow{H^+} RCOOR' \xrightarrow{Na/Alcohol} RCH_2OH + R'OH$$

Mechanism. The reduction by lithium aluminium hydride proceeds by the transfer of hydride ion (H^-) as follows :

$$LiAlH_4 \rightleftharpoons Li^+ + AlH_4^-$$

The carboxylic acid is initially reduced to aldehyde as follows :

$$R-\overset{\overset{O}{\|}}{C}-OH + H-\overset{\ominus}{Al}H_3 \longrightarrow R-\underset{\underset{H}{|}}{\overset{\overset{O^{\ominus}}{|}}{C}}-OH \longrightarrow \underset{\text{Aldehyde}}{R-\underset{\underset{H}{|}}{\overset{\overset{O}{\|}}{C}}} + OH^-$$

The aldehyde is then further reduced to primary alcohol as follows :

$$H_3\bar{A}l-H + \underset{R\quad H}{\overset{\overset{O}{\|}}{C}} \longrightarrow H_3\bar{A}l-O-CH_2R \xrightarrow[\text{Repeated}]{RCHO} \bar{A}l(OCH_2R)_4$$

$$\bar{A}l(OCH_2R)_4 \xrightarrow{H_2O} \underset{\text{Primary alcohol}}{4RCH_2OH + Al(OH)_3}$$

(D) REACTIONS DUE TO CARBOXYL GROUP

(*i*) Decarboxylation of carboxylic acids. Carboxylic acid gets decomposed if heated in the presence of soda-lime. The reaction takes place in two steps:

$$CH_3COOH + NaOH \longrightarrow CH_3COONa + H_2O$$

$$CH_3COONa + NaOH \xrightarrow{Heat} \underset{\text{Methane}}{CH_4} + Na_2CO_3$$

$$C_6H_5COOH + NaOH \longrightarrow C_6H_5COONa + H_2O$$

$$C_6H_5COONa + NaOH \xrightarrow{\text{Heat}} \underset{\text{Benzene}}{C_6H_6} + Na_2CO_3$$

Carboxylic acid containing electron withdrawing groups such as $-NO_2$ or CH_3CO- get decarboxylated by heating alone. Thus,

$$CH_3-\overset{\overset{O}{\|}}{C}-CH_2COOH \xrightarrow{\Delta} CH_3-\overset{\overset{O}{\|}}{C}-CH_3 + CO_2\uparrow$$

Mechanism. The mechanism involves the loss of a molecule of CO_2 from carboxylate anion as follows :

$$R-CH_2-\overset{\overset{:\ddot{O}}{\|}}{C}-ONa \rightleftharpoons \underset{\text{Carboxylate anion}}{R-CH_2-\overset{\overset{:\ddot{O}}{\|}}{C}-\ddot{\underset{..}{O}}:^{\ominus}} + Na^+$$

$$R-CH_2-\overset{\overset{:\ddot{O}}{\|}}{C}-\ddot{O}:^- \xrightarrow[-CO_2]{} R\bar{\ddot{C}}H_2 \xrightarrow{H_2O} \underset{\text{Alkane}}{RCH_3} + \bar{O}H$$

***(ii)* Kolbe's electrolytic reaction.** If an electric current is passed through an aqueous solution of alkali salt of a carboxy acid, we obtain hydrocarbons.

Adolf Kolbe

$$2RCOONa \longrightarrow 2RCOO^- + 2Na^+$$

At the anode: $2RCOO^- \xrightarrow{-2e^-} \underset{\text{Unstable}}{[2\,RCOO]} \longrightarrow \underset{\text{Hydrocarbon}}{R-R + CO_2}$

At the cathode: $2Na^+ \xrightarrow{+2e^-} 2Na \xrightarrow{2H_2O} 2NaOH + H_2$

Thus, if we take sodium acetate as the salt, the hydrocarbon produced will be ethane.

***(iii)* Hundsdiecker Reaction.** In this reaction, silver salt of the carboxylic acid reacts with bromine in the presence of dry carbon tetrachloride to form alkyl bromide as the product.

$$\underset{\text{Silver acetate}}{CH_3COOAg} + Br_2 \xrightarrow{(CCl_4)} \underset{\text{Methyl bromide}}{CH_3Br} + CO_2 + AgBr$$

The reaction is called Hundsdiecker reaction.

Mechanism of the reaction. The mechanism involves the initial electrophile attack of Br^+ on carboxylate ion of the silver salt.

$$R-\overset{\overset{O}{\|}}{C}-\bar{\ddot{O}}:Ag + Br-Br \xrightarrow{-AgBr} R-\overset{\overset{O}{\|}}{C}-\ddot{\underset{..}{O}}-Br \xrightarrow{-\dot{B}r} \underset{\text{Free radical}}{R-\overset{\overset{O}{\|}}{C}-\ddot{\underset{..}{O}}\cdot}$$

$$\downarrow -CO_2$$

$$R-Br \xleftarrow[-\dot{Br}]{Br-Br} \dot{R} \text{ (Free radical)}$$

***(iv)* Distillation of calcium salt.** Calcium salts of carboxylic acid on distillation form aldehydes and ketones.

$$\underset{\text{Cal.formate}}{(HCOO)_2Ca} \xrightarrow{\text{Distil}} \underset{\text{Formaldehyde}}{HCHO} + CaCO_3$$

$$\underset{\text{Cal.acetate}}{(CH_3COO)_2Ca} + \underset{\text{Cal. formate}}{Ca(OOCH)_2} \xrightarrow{\text{Distil}} \underset{\text{Acetaldehyde}}{2\,CH_3CHO} + 2CaCO_3$$

$$\underset{\text{Cal.acetate}}{(CH_3COO)_2Ca} \xrightarrow{\text{Distil}} \underset{\text{Acetone}}{CH_3COCH_3} + CaCO_3$$

(E) HALOGENATION OF CARBOXYLIC ACIDS

In the presence of a small amount of red phosphorus, α-hydrogen atoms of the alkyl group are replaced by chlorine or bromine to form α-halogen acids. This reaction is known as **Hell-Volhard-Zelinsky** (Abbreviated as HVZ) reaction.

$$\underset{n\text{-Butyric acid}}{CH_3CH_2CH_2COOH} \xrightarrow[\text{Heat}]{Br_2/P} \underset{\alpha\text{-bromo butyric acid}}{CH_3CH_2CHBr\,COOH} \xrightarrow{Br_2/P} \underset{\alpha,\alpha\text{-dibromobutyric acid}}{CH_3CH_2CBr_2COOH}$$

Halogen substitution will take place as long as α-hydrogen atoms are available. It is worth mentioning that bromination cannot be carried out beyond α- position whereas chlorination beyond α-position is possible provided all α-hydrogen atoms have been replaced by chlorine.

Mechanism of halogenation. The reaction is believed to take place by the following steps:

(*i*) $$2P + \underset{\text{Halogen}}{3X_2} \longrightarrow \underset{\text{Phosphorus trihalide}}{2PX_3}$$

(*ii*) $$3RCH_2COOH + PX_3 \longrightarrow \underset{\text{Acid halide}}{3RCH_2COX} + H_3PO_3$$

(*iii*) $$RCH_2\,COX + X_2 \longrightarrow RCHXCOX + HX$$

(*iv*) $$RCHXCOX + RCH_2COOH \longrightarrow \underset{\text{Halogen substituted carboxy acid}}{RCHXCOOH} + RCH_2COX$$

The C – X bond in α-haloacid is highly reactive in nature. It can be cleaved very easily and the halogen atom gets replaced by a number of nucleophiles leading to a variety of products.

Thus, such a reaction is of great synthetic importance. The following examples illustrate.

(*i*) **With aq KOH :** α-hydroxy acids are formed

$$\underset{\alpha\text{–Chloroacetic acid}}{ClH_2C.COOH} + KOH\,(aq) \longrightarrow \underset{\substack{\alpha\text{–Hydroxy acetic acid}\\ \text{[Glycollic acid]}}}{HOH_2CCOOH} + KCl$$

(*ii*) **With alcoholic ammonia :** α-amino acids are formed

$$\underset{\alpha\text{–Chloroacetic acid}}{ClCH_2COOH} + NH_3\,(\text{alcoholic}) \longrightarrow \underset{\substack{\alpha\text{–Amino acetic acid}\\ \text{[Glycine]}}}{NH_2CH_2COOH} + HCl$$

(*iii*) **With alcoholic KCN followed by acid hydrolysis.** Dicarboxylic acids are formed

$$\underset{\alpha\text{–Chloro acetic acid}}{Cl.CH_2COOH} + KCN\,(\text{alcohol}) \xrightarrow{-KCl} \underset{\alpha\text{–Cyano acetic acid}}{CNCH_2-COOH} \xrightarrow{H_2O/H^+} \underset{\text{Malonic acid}}{CH_2(COOH)_2}$$

(F) REACTIONS OF BENZENE RING IN AROMATIC ACIDS

Aromatic acids give the usual replacement reactions of the benzene ring, *i.e.* halogenation, nitration, sulphonation etc. The mechanism is electrophilic in nature. The carboxyl (– COOH) group shows conjugation and as a result the ortho and para positions in the ring get positively charged.

[Resonance structures of benzoic acid showing positive charge at ortho and para positions]

The meta position in the ring is comparatively the centre of higher electron density. Therefore, disubstitution takes place at the meta position in the ring. Since the ring has been deactivated due to the electron withdrawing nature of carboxyl group, the disubstitution reactions in the benzene ring take place with difficulty. Nitration, sulphonation and halogenation reactions of benzoic acid are given below :

$$\underset{\text{Benzoic acid}}{C_6H_5COOH} \xrightarrow[\text{(Nitration)}]{HNO_3/H_2SO_4} \underset{\text{3-Nitrobenzoic acid (meta)}}{m\text{-}NO_2C_6H_4COOH} + H_2O$$

$$\underset{\text{Benzoic acid}}{C_6H_5COOH} \xrightarrow[\text{(Sulphonation)}]{H_2SO_4(SO_3)} \underset{\text{3-Sulphobenzoic acid (meta)}}{m\text{-}HO_3SC_6H_4COOH} + H_2O$$

$$C_6H_5COOH \xrightarrow[\text{(Halogenation)}]{Cl_2/FeCl_3} 3\text{-}ClC_6H_4COOH + H_2O$$

Benzoic acid 3-Chlorobenzoic acid (meta)

Example 1. A carboxylic acid does not form phenyl hydrazone when treated with phenyl hydrazine. Explain.

Solution. In a carboxylic acid, there is a pair of electrons on the oxygen in conjugation with carbonyl oxygen.

$$R-\overset{\overset{O}{\|}}{C}-\ddot{O}-H \longrightarrow R-\overset{\overset{O^-}{|}}{C}=\overset{+}{O}-H$$

This prevents the creation of a positive centre at carboxyl carbon. Therefore nucleophilic addition of phenyl hydrazine to give phenyl hydrazone is not possible.

Example 2. A carboxy acid does not form an oxime. Explain.

Solution. An oxime is obtained in a reaction between a carboxyl compound and hydroxylamine

$$>C=O + NH_2OH \longrightarrow \underset{\text{Oxime}}{>C=N-OH}$$

This is a nucleophilic reaction in which the hydroxylamine molecule first attaches itself to the carbonyl carbon which is positive due to electronegative oxygen attached to it.

In the case of carboxylic compounds, this positive centre is destroyed because of lone pair of electrons present in conjugation with the carbonyl group

$$R-\overset{\overset{O}{\|}}{C}-\ddot{O}-H$$

Hence the reaction does not take place.

6.7. ACID STRENGTH OF CARBOXY ACIDS (EFFECT OF SUBSTITUENTS)

As explained earlier, carboxylic compounds are acidic in nature because they liberate carboxylate and hydrogen ions. Hydrogen ions do not exist as such but they are associated with water molecules as H_3O^+ ions. Carboxylate ion is able to stabilize itself by resonance energy and this is the cause of strongly acidic nature of carboxylic compounds. The acid is dissociated as:

$$RCOOH + H_2O \longrightarrow RCOO^- + H_3O^+$$

This is a reaction in equilibrium and according to the law of mass action, its equilibrium constant, K_a, which may also be referred to as acidity constant, is given by the relation

$$K_a = \frac{[RCOO^-][H_3O^+]}{[RCOOH][H_3O]}$$

The quantities within square brackets represent the molar concentration of these species. Since water exists in large quantities, its concentration almost remains constant and hence may be omitted. The equilibrium constant for acetic acid may be rewritten as:

$$K_a = \frac{[CH_3COO^-][H_3O^+]}{[CH_3COOH]}$$

The above equation shows that greater the value of $[H_3O^+]$, greater will be the value of K_a. It implies that a greater value of K_a signifies a stronger acid. Hence greater the value of K_a, greater is the acid strength of the carboxy compound.

Lately, there has been a practice to represent the strength of an acid in terms of pK_a value which is related to K_a by the following equation:

$$pK_a = -\log K_a$$

Thus an acid having K_a value equal to 1.75×10^{-5} will have its pK_a value as 4.75. It may be observed that pK_a value changes inversely as the K_a value. Thus a strong acid will have a higher value of K_a but a smaller value of pK_a.

Effect of Substituents on the Acid Strength

The strength of a carboxylic compound changes when certain groups are introduced in it at certain positions. As a matter of fact, we expect an increase in the acid strength if some electron-withdrawing group is introduced in the molecule. A reverse effect is observed when the group introduced is electron repelling. If the extra group attached is represented as –G, then the situation in the two cases can be represented as below:

$$G \leftarrow CH_2 \leftarrow C\begin{cases} O \\ O^- \end{cases} \qquad \text{Acid strength increases}$$

$$G \rightarrow CH_2 \rightarrow C\begin{cases} O \\ O^- \end{cases} \qquad \text{Acid strength decreases}$$

When an electron-attracting group is attached, it will stabilise the carboxylate ion to a greater extent by dissipating the negative charge, it will help in the release of more protons. In the case of an electron-repelling group, there will be a greater concentration of negative charge on the carboxylate group which will ultimately reduce the stability of the carboxylate ion. Hence the liberation of protons will be suppressed. The effect of substituents on the acid strength can be studied under two headings:

1. Aliphatic acids. The following observations are made here:

(*i*) Consider the case of acetic acid, it has a pK_a value of 4.75. If we replace one of the hydrogens of the methyl group by a halogen group, an electron-attracting group, the halogenated acid is found to have a lower value of pK_a *i.e.* it is found to be stronger than acetic acid. Further, the acid strength changes with the halogen. As fluorine is more electronegative than chlorine and chlorine is more electronegative than bromine, we find that fluoroacetic acid is more acidic than chloroacetic acid which in turn is more acidic than bromoacetic acid.

$$F-CH_2COOH > Cl-CH_2COOH > Br-CH_2COOH$$

(*ii*) The increase in the acid strength of an acid depends upon the no. of such electron-withdrawing groups. Greater the no. of electron-withdrawing groups, greater is the strength of the acid. Thus dichloroacetic acid is stronger than monochloro acetic acid and trichloroacetic acid is stronger than dichloroacetic acid.

$$Cl \leftarrow \overset{\overset{Cl}{\uparrow}}{\underset{\underset{Cl}{\downarrow}}{C}} \leftarrow COOH > \begin{matrix} Cl \searrow \\ Cl \nearrow \end{matrix} CH \leftarrow COOH > Cl \leftarrow CH_2 \leftarrow COOH > CH_3 - COOH$$

Trichloroacetic acid | Dichloroacetic acid | Monochloroacetic acid | Acetic acid

(*iii*) The strength of the acid depends upon the position of the electron-withdrawing group relative to the carboxy group. Nearer the electron-withdrawing group to the carboxy group, greater is the acid strengths. Thus α-chloropropionic acid is stronger than β-chloropropionic acid.

$$\underset{\alpha\text{-chloropropionic acid}}{CH_3-\underset{|}{\overset{Cl}{CH}}-COOH} \quad > \quad \underset{\beta\text{-chloropropionic acid}}{ClCH_2-CH_2-COOH}$$

(*iv*) The substitution of an electron-repelling group in the carboxy acid decreases the acid strength. Greater the number of such constituents, greater will be the effect.

$$\underset{\text{I}}{CH_3\rightarrow\underset{\uparrow\atop CH_3}{\overset{CH_3\atop\downarrow}{C}}\twoheadrightarrow COOH} < \underset{\text{II}}{\overset{CH_3\searrow}{\underset{CH_3\nearrow}{}}CH-CH_2\twoheadrightarrow COOH} < \underset{\text{III}}{CH_3\rightarrow CH_2\rightarrow CH_2\rightarrow CH_2\rightarrow COOH}$$

Thus, consider the above three compounds. Compound I will experience the maximum electron repelling effect as the three methyl groups are directly attached to carbon two positions next to the carboxy group. In compounds II and III, the number of methyl (or alkyl) groups attached to carbon next to the carboxy group is 2 and 1 respectively. As a consequence, compound III will be the strongest acid and compound I the weakest in the series.

It may be mentioned that a propyl group will have a greater electron-repelling effect than ethyl group which in turn will have greater repelling effect than methyl group. Thus, the first four homologues of carboxylic compounds have the strength in the following order:

Formic acid > Acetic acid > Propionic acid > Butyric acid

2. Aromatic acids. The following observations are made here:

(*i*) **Electron-withdrawing** groups like $-NO_2$, $>C=O$, $-CN$, $-Cl$ when substituted into the benzene ring raise the strength of the acid by inductive effect and resonance effect. Resonance effect is known to be stronger than inductive effect. But the position where such groups are linked is more important. The above groups produce a positive charge at the ortho and para positions but not at the meta position as shown below:

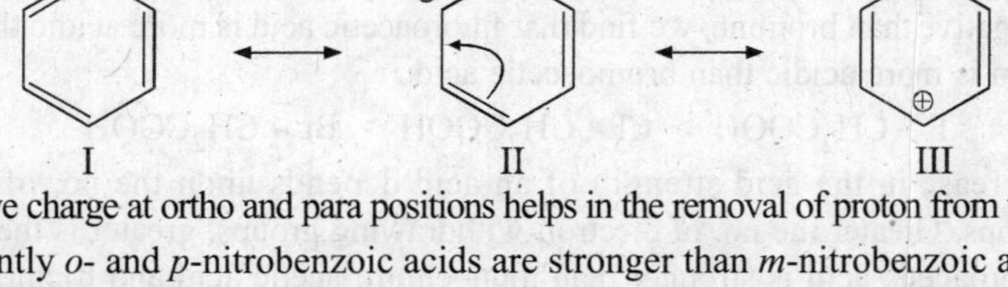

O ← N = O (I) ⟷ O ← N — O$^{\ominus}$ ($\oplus$ at ortho) (II) ⟷ O ← N — O$^{\ominus}$ ($\oplus$ at para) (III)

The positive charge at ortho and para positions helps in the removal of proton from the carboxy group.

Consequently *o*- and *p*-nitrobenzoic acids are stronger than *m*-nitrobenzoic acid. But *m*-nitrobenzoic acid is definitely stronger than benzoic acid as nitro group draws the electron, although not strongly, by inductive effect at the meta position.

(*ii*) Reverse is the case when **electron-donating** group is linked to the benzene ring. Hydroxy group is an electron-donating group in the resonance phenomenon as shown below:

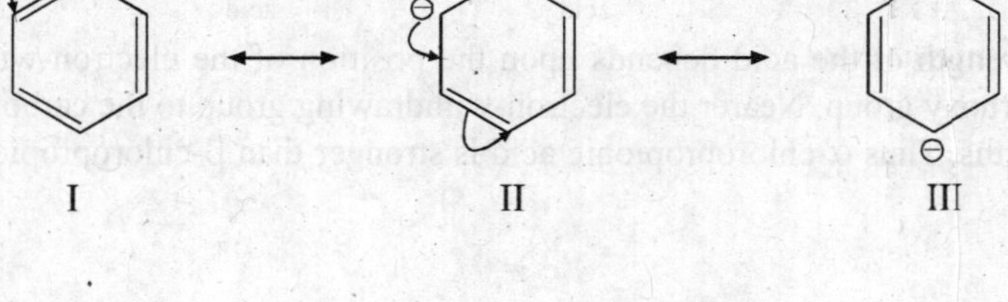

:OH (I) ⟷ $\oplus$OH ($\ominus$ at ortho) (II) ⟷ $\oplus$OH ($\ominus$ at para) (III)

Thus para position, where the carboxy group is attached, become negative. This will decrease the strength of *p*-hydroxy benzoic acid as compared to benzoic acid. But hydroxy group will produce a slight positive charge at the meta position due to inductive effect. Thus *m*-hydroxy benzoic acid is stronger than benzoic acid. Ortho hydroxy benzoic acid, although against expectation, is a stronger acid than benzoic acid. As a general observation, a group in the ortho position always increases the acid strength. This is called *ortho effect*. Thus out of *o*-, *m*- and *p*-substituted benzoic acids, *o*-compound will be strongest acid. However, there is no satisfactory explanation for this effect.

The K_a (dissociation constant) values of some substituted aromatic acids are given in the following table.

Table 6.3. Acidity contains (Ka) of substituted benzoic acids

(K_a of $C_6H_5COOH = 6.3 \times 10^{-5}$)

Substituent (nature)	K_a value at para position	K_a value at meta position	K_a value at ortho position
CH_3 (electron releasing)	4.3×10^{-5}	5.5×10^{-5}	12.5×10^{-5}.
OCH_3 (electron releasing)	3.2×10^{-5}	8.3×10^{-5}	8.2×10^{-5}
OH (electron releasing)	2.5×10^{-5}	8.5×10^{-5}	105×10^{-5}
Cl (electron withdrawing)	10.4×10^{-5}	15.0×10^{-5}	120×10^{-5}.
NO_2 (electron withdrawing)	35.0×10^{-5}	32.0×10^{-5}	600×10^{-5}.

Example 1. Arrange the following in the order of their increasing acidity. Give reasons for your answer.

COOH (benzoic acid), COOH with CH_3 at para position, COOH with NO_2 at para position

Solution. The order of increasing acidity is :

p-$CH_3C_6H_4COOH$ < C_6H_5COOH < p-$NO_2C_6H_4COOH$

Methyl group is electron-donating group. It creates a negative centre on the carbon linking the carboxy group. This creates difficulty in the removal of proton.

Nitro group is electron-withdrawing group. This creates a positive charge on the carbon linking the carboxy group. This helps in the release of proton from the carboxy group.

Example 2. Arrange the following in order of increasing acidity: benzoic acid, *p*-hydroxy benzoic acid, *p*-nitrobenzoic acid.

Solution. The order of increasing acidity is:

p-hydroxy benzoic acid < benzoic acid < *p*-nitrobenzoic acid

An electron-donating group like –OH creates negative charge on the carbon linking the carboxy group. This hinders the removal of proton from the carboxy group.

An electron-attracting group like $-NO_2$ creates a positive centre on the carbon carrying the carboxy group. This helps in the release of protons.

Example 3. Arrange the following compounds in the increasing order of acidity:
(*a*) (*i*) Benzoic acid (*ii*) *p*-nitrobenzoic acid (*iii*) phenol.
(*b*) (*i*) *p*-toluic acid (*ii*) Benzoic acid (*iii*) *p*-chlorobenzoic acid
(*iv*) 2, 4, 6-Trichlorobenzoic acid.
(*c*) (*i*) *p*-Toluic acid (*ii*) *p*-chlorobenzoic acid (*iii*) *m*-chlorobenzoic acid (*iv*) Benzoic acid.

Solution. (*a*) Phenol < benzoic acid < *p*-nitrobenzoic acid

(*b*) *p*-Toluic acid < benzoic acid < *p*-chlorobenzoic acid < 2, 4, 6-Trichlorobenzoic acid.

(c) *p*-Toluic acid < benzoic acid < *p*-clorobenzoic acid < *m*-chlorobenzoic acid.

Example 4. Arrange the following compounds in the increasing order of their acid strength and account for it.

$$\mathbf{ClCH_2COOH,\ BrCH_2COOH,\ CNCH_2COOH,\ CH_3COOH,\ CH_3CH_2CH_2COOH}$$

Solution. The acid strength of an acid is affected by the substituents. An electron-attracting group in the chain increases the acid strength and an electron-releasing group decreases the acid strength due to inductive effect of the group. Out of –Cl, –Br and –CN which are all electron-withdrawing group, –CN has the strongest inductive effect followed by –Cl and –Br groups. In butyric acid $CH_3CH_2CH_2COOH$, it is the $CH_3 - CH_2 -$ group which is present in the chain. It has electron-releasing effect. Hence the compounds are arranged in increasing order of acid strength as under:

$CH_3CH_2 - CH_2COOH$
$H - CH_2COOH$
$Br - CH_2COOH$ ↓ Acid strength increases
$Cl - CH_2COOH$
$CN - CH_2COOH$

Example 5. Complete the following:

(*i*) $\mathbf{CH_2\langle^{COOH}_{COOH} \xrightarrow{\Delta}}$

(*ii*) $\mathbf{CH_3CONH_2 + X \longrightarrow CH_3COOH + Y}$

(*iii*) $\mathbf{(CH_3COO)_2\,Ca \xrightarrow{\Delta}}$

(*iv*) $\mathbf{CH_3COOH + Cl_2 \longrightarrow X + Y}$

(*v*) $\mathbf{CH_3COOH + P_2O_5 \longrightarrow X}$

(*vi*) $\mathbf{C_6H_5COOH + SOCl_2 \longrightarrow X + Y + Z}$

Solution. (*i*) $\underset{\text{Malonic acid}}{CH_2\langle^{COOH}_{COOH}} \xrightarrow{\Delta} CH_3COOH + CO_2\uparrow$

(*ii*) $\underset{\text{Acetamide}}{CH_3CONH_2} + H_2O \longrightarrow \underset{\text{Acetic acid}}{CH_3COOH} + NH_3$

(*iii*) $(CH_3COO)_2Ca \xrightarrow{\Delta} CH_3COCH_3 + CaCO_3$

Cal.acetate — Acetone

(*iv*) $CH_3COOH + Cl_2 \longrightarrow CH_2ClCOOH + HCl$

Chloroacetic acid

(*v*) $2CH_3COOH \xrightarrow{P_2O_5} (CH_3CO)_2O + H_2O$

Acetic anhydride

(*vi*) $C_6H_5COOH + SOCl_2 \longrightarrow C_6H_5COCl + SO_2 + HCl$

Example 6. An amide (A) having molecular formula C_3H_7ON on hydrolysis gives an acid $C_3H_6O_2$ which upon chlorination in the presence of red phosphorus (HVZ reaction) produces chloro acid. The latter on boiling with aqueous NaOH and subsequent acidification forms lactic acid. Trace the reactions and deduce the structural formula of A.

Solution. $\underset{(A)}{RCH_2CONH_2} \xrightarrow{H_2O} \underset{(B)}{RCH_2COOH} \xrightarrow{Cl_2/P} \underset{(C)}{RCHClCOOH} \xrightarrow{H_2O} \underset{(D)}{RCH(OH)COOH}$

Lactic acid has the formula $CH_3CH(OH)COOH$.

It means R in the above compounds A, B and C is methyl group (CH_3 —).

Hence the compounds A, B and C are

(A) $CH_3CH_2CONH_2$ Propanamide (B) CH_3CH_2COOH Propanoic acid (C) $CH_3CHClCOOH$ 2-Chloropropanoic acid

Example 7. Although *p*-hydroxybenzoic acid is less acidic than benzoic acid, *o*-hydroxybenzoic acid (salicylic acid) is 15 times more acidic than benzoic acid. Explain.

Solution. Carboxylate ions of *o*-hydroxybenzoic acid is stabilized by means of effective hydrogen bonding leading to the formation of a ring.

Salicylic acid $\longrightarrow$ Salicylate ion stabilized by intramolecular hydrogen bonding $+ H^+$

Example 8. Arrange the following in decreasing order of their acid strength:

(*i*) C_6H_5COOH, *p*-OH.C_6H_4COOH, *p*-CH_3.C_6H_4COOH, *p*-Cl.C_6H_4COOH, *p*-Br.C_6H_4COOH, *p*-NO_2.C_6H_4.COOH

(*ii*) HCOOH, C_6H_5COOH, C_6H_5OH, HCl

Solution. (*i*) *p*-NO_2.C_6H_4COOH (–I and –R effect of –NO_2) > *p*-ClC_6H_4COOH (–I effect of –Cl) > *p*-BrC_6H_4COOH) (–I effect of –Br) > C_6H_5COOH >

p–CH_3.C_6H_4.COOH (+ I effect of –CH_3) > *p*–OHC_6H_4.COOH (+ R effect of –OH)

Note: – I is electron-withdrawing inductive effect

+ I is electron-donating inductive effect

– R is electron-withdrawing resonance effect

+ R is electron-donating resonance effect

(*ii*) $HCl > HCOOH > C_6H_5COOH > C_6H_5OH$

Benzoic acid is weaker than formic acid due to electron releasing resonance effect of $-C_6H_5$.

Example 9. Explain:

(*a*) 2, 4, 6-Trimethylbenzoic acid fails to esterify under normal conditions of esterification, while 2, 4, 6-trimethylphenylacetic acid is esterified readily under normal conditions.

(*b*) 2, 4, 6-Trihydroxybenzoic acid decarboxylates readily on heating, but benzoic acid does not.

Solution. (*a*) 2, 4, 6-trimethylbenzoic acid fails to esterify under normal conditions of esterification, because of steric hindrance of methyl groups.

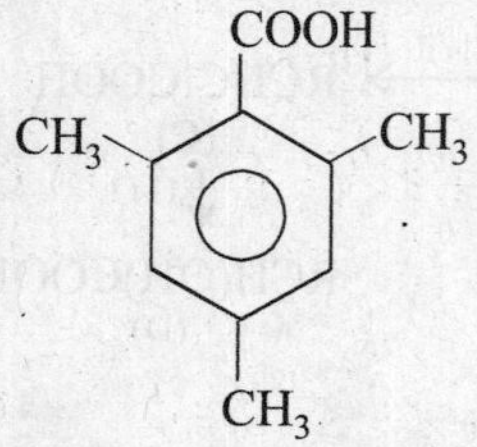

COOH, CH_3, CH_3, CH_3

2, 4, 6-Trimethylbenzoic acid

There is no such hindrance due to methyl groups in trimethylphenylacetic acid as methyl groups are at a distance from the carboxylic group.

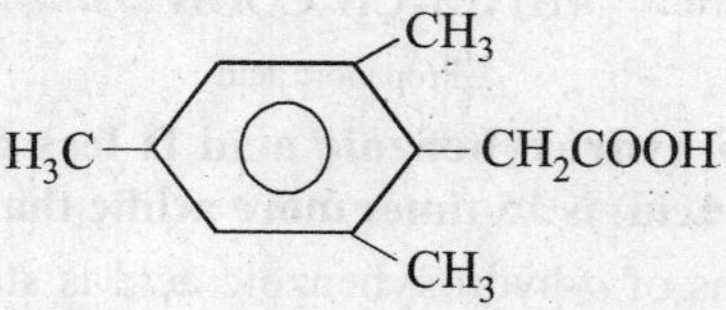

CH_3, H_3C—, CH_2COOH, CH_3

(*b*) 2, 4, 6-trihydroxybenzoic acid decarboxylates easily due to electron withdrawing inductive effect of the –OH group.

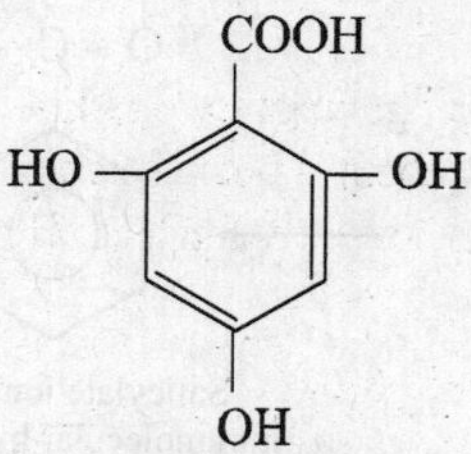

COOH, HO, OH, OH

It can also be explained in terms of **ortho effect.**

Example 10. Arrange the following in descending order of strength and justify your answer:

***p*-toluic acid, phenylacetic acid, *m*-nitrobenzoic acid, benzoic acid.**

Solution. *p*-Toluic acid H_3C—⟨benzene ring⟩—COOH

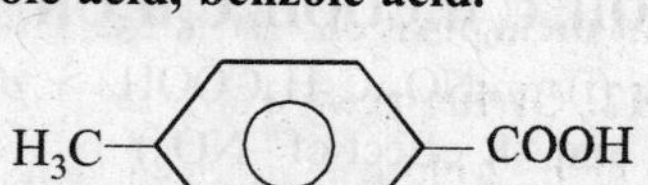

Phenylacetic acid — C_6H_5–CH_2COOH

m-Nitrobenzoic acid — NO_2, COOH (benzene ring)

Benzoic acid — C_6H_5–COOH

The decreasing order of strength is as follows:

NO_2–C_6H_4–COOH > C_6H_5–COOH > C_6H_5–CH_2COOH >
– I Effect

H_3C–C_6H_4–COOH
+ I Effect

Example 11. Arrange the following in order of decreasing acid strength and account for your answer.

(*i*) *p*-Nitrobenzoic acid, *m*-nitrobenzoic acid

(*ii*) *p*-Hydroxybenzoic acid, *m*-hydroxybenzoic acid.

Solution. (*i*) O_2N–C_6H_4–COOH > (NO_2) C_6H_4–COOH

This is because the nitro group brings about a positive charge at para position by inductive and resonance effects, whereas it brings about a positive charge at meta position by inductive effect only. Resonance effect is stronger than inductive effect. Hence *p*-isomer is a stronger acid than *m*-isomer.

(*ii*) (OH) C_6H_4–COOH > HO–C_6H_4–COOH

Hydroxy group brings about a negative charge at para position by resonance effect, thereby decreasing the acid strength. Hence *p*-hydroxybenzoic acid is a weaker acid. On the other hand, hydroxy group at meta position has –I inductive effect, which results in increased acid strength.

Example 12. Giving reasons, arrange the following in increasing order of their acidity: Propionic acid, 2-chloropropionic acid, 2-fluoropropionic acid.

Solution. Fluorine has –I inductive effect, therefore, it tends to draw the electrons towards itself. It helps in the release of protons from the acid. Fluorine being more electronegative than chlorine, has greater inductive effect than chlorine.

The increasing order of acid strength is as follows:

$$CH_3CH_2COOH < Cl-CH_2CH_2COOH < F-CH_2CH_2COOH$$

Example 13. Arrange the following in order of increasing acidity. Give reasons.

Butanoic acid, 2-fluorobutanoic acid, 2-chlorobutanoic acid.

Solution. $CH_3CH_2CH_2COOH < CH_3CH_2\underset{\displaystyle Cl}{\underset{|}{C}}HCOOH < CH_3-CH_2-CH_2-\underset{\displaystyle F}{\underset{|}{CH}}-COOH$

The reason is the same as in the Example 12.

Example 14. Arrange the following in order of increasing acidity. Give reason.

$F-CH_2COOH$, $Cl-CH_2COOH$, CH_3-COOH, $H-COOH$

Solution. (*i*) Out of Cl and F, the latter has greater inductive effect (–I). Hence, fluoroacetic acid is more acidic than chloroacetic acid which, in turn, is more acidic than acetic acid.

(*ii*) Out of acetic and formic acids, the former has a weaker acid strength because it has +I inductive effect.

The increasing order of acid strength is as follows:

$$CH_3COOH < Cl-CH_2COOH < F-CH_2COOH < H-COOH$$

Example 15. Giving reasons, justify that trimethyl acetic acid is a weaker acid whereas trichloroacetic acid is a much stronger acid than acetic acid.

Solution.

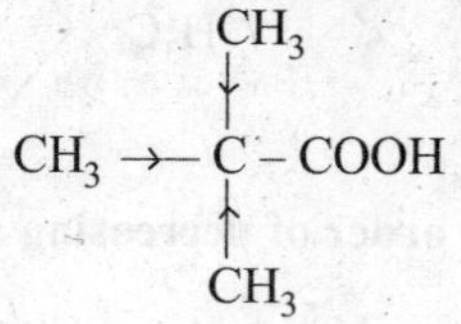

$$CH_3 \rightarrow \overset{\displaystyle CH_3}{\underset{\displaystyle CH_3}{\overset{\downarrow}{\underset{\uparrow}{C}}}} - COOH \qquad\qquad Cl \rightarrow \overset{\displaystyle Cl}{\underset{\displaystyle Cl}{\overset{\uparrow}{\underset{\downarrow}{C}}}} - COOH$$

Trimethylacetic acid 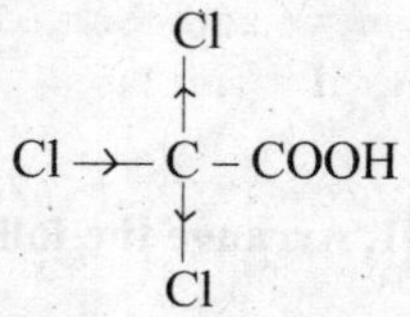   Trichloroacetic acid

As shown above, three methyl groups together exert +I inductive effect, as a result of which, the release of protons from the acid is suppressed. On the other hand, three chloro groups exert –I inductive effect, which helps in the release of protons from the acid. Therefore trichloroacetic acid is much stronger than acetic acid and trimethyl acetic acid is a much weaker acid than acetic acid.

Example 16. Chloroacetic acid is stronger acid than acetic acid while acetic acid is itself weaker than formic acid. Why?

Solution. Chloro group exerts inductive effect and pulls the electrons towards itself, thus helping in the release of protons

$$Cl \leftarrow CH_2 \leftarrow \overset{\displaystyle O}{\overset{\|}{C}} \leftarrow O-H$$

CH_3- group in acetic acid pushes the electrons away from itself thus hindering removal of protons

$$CH_3 \rightarrow \overset{\displaystyle O}{\overset{\|}{C}} \rightarrow O-H$$

In formic acid, there is no electron-pushing alkyl group therefore chloroacetic acid is stronger than acetic acid which in turn is weaker than formic acid.

Example 17. When K_a of an acid is large, pK_a is small. Why?

Solution. This is because $pK_a = -\log K_a$.

Hence increase in the value of K_a leads to decrease in the value of pK_a.

Example 18. Formic acid acts both as an acid as well as an aldehyde. Explain.

Solution. Besides acting as an acid, formic acid also gives the properties of an aldehyde. This is because, like the aldehydes, formic acid also gets oxidised easily.

$$HCOOH \xrightarrow{[O]} CO_2 + H_2O$$

Example 19. Formic acid is stronger than acetic acid. Why?

Solution.

$$\underset{\text{Acetic acid}}{CH_3 \rightarrow\!\!- \underset{\underset{O}{\|}}{C} \rightarrow\!\!- O \rightarrow\!\!- H} \qquad \underset{\text{Formic acid}}{H - \underset{\underset{O}{\|}}{C} - O - H}$$

Methyl group in acetic acid displaces the electrons towards hydrogen in acetic acid as shown by arrow mark. This diminishes the chances of removal of hydrogen as protons. There is no such situation in formic acid. Hence formic acid is stronger than acetic acid.

Example 20 . Esterification of benzoic acid is easier than that of orthodimethyl benzoic acid. Why ?

Solution. Esterification of benzoic acid is easier than orthodimethyl benzoic acid because orthodimethyl benzoic acid experiences steric hindrance due to the methyl groups.

Example 21. Compound (A), C_3H_7Cl reacts with alcoholic KOH to form (B), C_3H_6. Compound (B) decolourises Br_2/CCl_4 solution. Reaction of (A) with Mg in ether and subsequent treatment with CO_2 and dil.acid gives a compound (C) whose molecular formula is $C_4H_8O_2$; When we add compound (C) to aqueous $NaHCO_3$ solution, bubbles are evolved. Give the structural formulas of (A), (B) and (C), and write equations for all the reactions involved.

Solution.

$$(A) = CH_3 - \overset{\overset{Cl}{|}}{CH} - CH_3 \quad \text{(2-Chloropropane)}$$

$$(B) = CH_3 - CH = CH_2 \quad \text{(Propene)}$$

$$(C) = CH_3 - \overset{\overset{CH_3}{|}}{CH} - COOH \quad \text{(2-Methylpropanoic acid)}$$

Example 22. How will you distinguish between benzoic acid and cinnamic acid ?

Solution. As cinnamic acid has a double bond, it will discharge the colour of Br_2 / CCl_4;

$$C_6H_5CH = CHCOOH + \underset{\text{red}}{Br_2} \xrightarrow{CCl_4} \underset{\text{(Colourless)}}{C_6H_5CHBrCHBrCOOH}$$

Benzoic acid will not give this test.

Example 23. How will you distinguish between phenol and salicylic acid ?

Solution. Phenol on heating with phthalic anhydride and conc. H_2SO_4 produces phenolphthalein which gives pink colour when NaOH solution is added to the reaction mixture. Salicylic acid does not give this test.

HALO ACIDS

Halogen substituted or halo acids are compounds in which one or more hydrogen atoms of the hydrocarbon chain of carboxylic acid are replaced by corresponding number of halogen atoms.

6.8. NOMENCLATURE OF HALO ACIDS

Common system

Common names of halogen substituted acids (or halo acids) are derived from the common names of parent carboxylic acids. The position of halogen atoms are indicated by Greek letters α, β, γ, δ etc., where α is the position of carbon adjacent to the carboxylic group, next position being β, still next positions are γ and δ

$$\overset{\delta}{C}-\overset{\gamma}{C}-\overset{\beta}{C}-\overset{\alpha}{C}-COOH$$

IUPAC System

In IUPAC system of nomenclature, the position of halogen is indicated by numerals 1, 2, 3, 4..., number 1 being the position of carboxylic carbon atom

$$\overset{4}{C}-\overset{3}{C}-\overset{2}{C}-\overset{1}{COOH}$$

The common & IUPAC names of some halogen substituted acids are given in Table 6.4.

Table 6.4. Common and IUPAC names of halogen acids

Halo Acid	*Common Name*	*IUPAC Name*
$ClCH_2COOH$	Chloroacetic acid	Chloroethanoic acid
$CH_3CH(Cl)COOH$	α-chloropropionic acid	2-chloropropanoic acid
$CH_3CH(Br)CH_2COOH$	β-bromobutyric acid	3-bromobutanoic acid
$CH_2(I)CH_2CH_2COOH$	γ-iodobutyric acid	4-iodobutanoic acid
$CH_3CH(Br)CH(Br)COOH$	α, β-dibromobutric acid	2, 3-dibromobutanoic acid
$C_6H_4(Cl)COOH$ (Cl and COOH on adjacent ring carbons)	α-Chlorobenzoic acid	2–chlorobenzene carboxylic acid

6.9. METHODS OF PREPARATION OF HALO ACIDS

General methods of preparation of halo acids are given below.

1. Hell-Volhard-Zelinsky Reaction (HVZ Reaction). α-hydrogen atoms of a carboxylic acid can be replaced by halogen on treatment with the halogen in the presence of red phosphorus

$$\underset{\text{Propanoic acid}}{CH_3CH_2COOH} + Br_2 \xrightarrow[-\ HBr]{\text{Red P}} \underset{\text{2-Bromopropanoic acid}}{CH_3CHBrCOOH} \xrightarrow{Br_2/P} \underset{\text{2,2-Dibromopropanoic acid}}{CH_3CBr_2COOH}$$

The substitution takes place at the α-position. And the reaction can be stopped at a particular stage by using appropriate quantities of the halogen and phosphorus. It is observed that chlorination can take place beyond positions as well.

In the reaction between acetic acid and Cl_2, all the three hydrogen atoms can be successively replaced by chlorine.

$$CH_3COOH + Cl_2 \xrightarrow[-HCl]{} \underset{\text{Monochloro-acetic acid}}{CH_2ClCOOH} \xrightarrow[-HCl]{Cl_2} \underset{\text{Dichloroacetic acid}}{CHCl_2COOH}$$

$$\xrightarrow[-HCl]{Cl_2} \underset{\text{Trichloroacetic acid}}{CCl_3COOH}$$

2. Action of sulphuryl chloride on carboxylic acid. Carboxylic acid on treatment with sulphuryl chloride in the presence of iodine yields halo acid.

$$\underset{\text{Propanoic acid}}{CH_3CH_2COOH} + \underset{\text{Sulphuryl chloride}}{2SO_2Cl_2} \xrightarrow{I_2} \underset{\text{2-chloropropanoyl chloride}}{CH_3CHClCOCl} + 2SO_2 + 2HCl$$

$$\xrightarrow[H_2O]{\text{Boil}} \underset{\text{2-Chloropropanoic acid}}{CH_3CHClCOOH} + HCl$$

3. Halogenation of malonic acid and alkyl malonic acids

$$\underset{\text{Malonic acid}}{CH_2\begin{cases} COOH \\ COOH \end{cases}} + Br_2 \longrightarrow \underset{\text{Bromomalonic acid}}{CHBr\begin{cases} COOH \\ COOH \end{cases}} \xrightarrow[-CO_2]{\Delta} \underset{\text{Bromoacetic acid}}{CH_2BrCOOH}$$

$$\underset{\text{Methyl malonic acid}}{CH_3CH\begin{cases} COOH \\ COOH \end{cases}} + Br_2 \longrightarrow \underset{\alpha\text{-bromomethyl-malonic acid}}{CH_3CBr\begin{cases} COOH \\ COOH \end{cases}} \xrightarrow[\Delta]{-CO_2} \underset{\alpha\text{-Bromopropionic acid}}{CH_3CHBrCOOH}$$

4. From α-hydroxy acids. α-hydroxy acids can be converted into α-halo acids on treatment with halogen acids.

$$\underset{\text{Lactic acid}}{CH_3CH(OH)COOH} + HBr \longrightarrow \underset{\alpha\text{-bromopropionic acid}}{CH_3CHBrCOOH} + H_2O$$

5. From α, β unsaturated carboxylic acids. Such acids on addition of halogens and halogen acids produce halo acids as given below.

(*i*) $\underset{\text{Crotonic acid}}{CH_3CH = CHCOOH} + Br_2 \longrightarrow \underset{\alpha,\ \beta\text{-dibromobutyric acid}}{CH_3CHBr\ CHBrCOOH}$

(*ii*) $\underset{\text{Crotonic acid}}{CH_3CH = CHCOOH} + HBr \longrightarrow \underset{\beta\text{-bromobutyric acid}}{CH_3CHBr\ CH_2COOH}$

This affords us a method to prepare β haloacids. It may be mentioned that in the above reaction (*ii*), the addition of HBr to the double bond takes place contrary to the Markownikoff's rule. This is because of the electron withdrawing (– I) inductive effect of the carboxylic group. Consequently β-carbon becomes slightly positive and α-carbon becomes slightly negative. Therefore Br^- and H^+ (of HBr) are attached to α and β positions respectively.

$$CH_3-\overset{\delta^+}{C}H=\overset{\delta^-}{C}H-COOH \xrightarrow{\overset{\delta^+}{H}-\overset{\delta^-}{Br}} CH_3\overset{+}{C}HCH_2COOH \xrightarrow{Br^-} CH_3CHBrCH_2COOH$$

6. From α, β unsaturated aldehydes. Addition of halogen acid to α, β unsaturated aldehydes followed by oxidation yields β-halogen acids

$$\underset{\text{Acrolein}}{\overset{\delta^+}{C}H_2=\overset{\delta^-}{C}H-CHO} \xrightarrow{\overset{\delta^+}{H}-\overset{\delta^-}{Cl}} \underset{\beta\text{-chloropropion aldehyde}}{CH_2ClCH_2CHO} \xrightarrow{[O]} \underset{\beta\text{-chloropropionic acid}}{CH_2ClCH_2COOH}$$

6.10. PROPERTIES OF HALO ACIDS

Halo acids give all the properties of the parent carboxylic acid. In addition, they give the properties due to the halogen group. Important properties of halo acids are as follows.

1. Acidic strength. They show acidic properties due to the presence of carboxylic group in the molecule. In fact, the halo acids are more acidic than the normal acids because of the inductive effect of halogen. Halogen withdraws the electron towards itself, thereby helping in the release of proton.

$$R-\underset{\underset{X}{\downarrow}}{\overset{H}{\overset{|}{C}}}\leftarrow\overset{O}{\overset{\|}{C}}\leftarrow O\leftarrow H \longleftrightarrow R-\underset{\underset{X}{|}}{\overset{H}{\overset{|}{C}}}-\overset{O}{\overset{\|}{C}}-O^- + H^+$$

Following observations are made in this connection.

(*i*) Among various halo acids, chloro acids are stronger in acidic nature than bromo acids which are, in turn, stronger than iodo acids. This is due to greater inductive effect of chlorine compared to bromine and greater effect of bromine compared to iodine.

(*ii*) For halo acids, having the same halogen atom, the acid strength decreases as the position of halogen goes away from the carboxy group. Thus 3-chloropropionic acid will be a weaker acid than 2-chloropropionic acid.

2. Reactions due to the carboxylic group. The halogen does not interfere in the reactions of carboxylic group and these properties are given in the normal manner. Halo acids decompose carbonates and bicarbonates, react with alcohols to form esters and with phosphorus pentachloride to form acids chlorides.

$$\underset{\text{Chloroacetic acid}}{ClCH_2COOH} + NaHCO_3 \longrightarrow \underset{\text{Sod. chloroacetate}}{ClCH_2COONa} + CO_2 + H_2O$$

$$2ClCH_2COOH + Na_2CO_3 \longrightarrow 2ClCH_2COONa + CO_2 + H_2O$$

$$ClCH_2COOH + C_2H_5OH \xrightarrow{H^+} \underset{\text{Ethyl-chloroacetate}}{ClCH_2COOC_2H_5} + H_2O$$

$$\underset{\text{Chloroacetic acid}}{ClCH_2COOH} + PCl_5 \longrightarrow \underset{\text{Chloroacetyl chloride}}{ClCH_2COCl} + POCl_3 + HCl$$

3. Reactions due to the halogen atom. (*i*) **Action of alkalis.** Different products are obtained when halo acids are heated with aqueous solution of alkali, depending upon the position of halogen atom with respect to the carboxylic group.

α-halogen substituted acids give α-hydroxy acids on hydrolysis

$$\underset{\alpha\text{-halo acid}}{\underset{|\atop Cl}{RCH} - COOH} \xrightarrow{NaOH} \underset{\alpha\text{-hydroxy acid}}{R - \underset{|\atop OH}{CH} - COOH}$$

$$\underset{\alpha\text{-chloropropionic}}{CH_3CHClCOOH} \xrightarrow{NaOH} \underset{\text{Lactic acid}}{CH_3CH(OH)COOH}$$

β-haloacids on treatment with alkali yield a mixture of β-hydroxy acid and α, β unsaturated acid. This is because β-hydroxy acid is unstable and eliminates a water molecule to give more stable α, β unsaturated acid.

$$\underset{\beta\text{-chloropropionic acid}}{CH_2ClCH_2COOH} \xrightarrow{NaOH} \underset{\beta\text{-hydroxypropionic acid}}{CH_2(OH)CH_2COOH}$$

$$\downarrow -H_2O$$

$$\underset{\text{Acrylic acid}}{CH_2 = CHCOOH}$$

(*ii*) **Nucleophilic substitution reactions.** Halogen of the halo acid can be substituted by nucleophiles like CN^-, NH_3, $-OC_2H_5$.

$$\underset{\alpha\text{-chloropropionic acid}}{CH_3\underset{|\atop Cl}{CH} - COOH} + KCN \longrightarrow \underset{\alpha\text{-cyanopropionic acid}}{CH_3\underset{|\atop CN}{CH} - COOH} + HCl$$

$$\underset{\alpha\text{-chloropropionic acid}}{CH_3\underset{|\atop Cl}{CH} - COOH} + NH_3 \longrightarrow \underset{\alpha\text{-aminopropionic acid}}{CH_3\underset{|\atop NH_2}{CH} - COOH} + HCl$$

$$CH_3\underset{|\atop Cl}{CH} - COOH + \underset{\text{Sod. ethoxide}}{NaOC_2H_5} \longrightarrow \underset{\alpha\text{-ethoxypropionic acid}}{CH_3 - \underset{|\atop OC_2H_5}{CH} - COOH}$$

Example 1. Explain the following:

(*a*) Chloroacetic acid is stronger than acetic acid.

(*b*) Dichloroacetic acid is stronger than chloroacetic acid.

(*c*) Trichloroacetic acid is strong like a mineral acid.

(*d*) Monochloroacetic acid is stronger than monobromoacetic acid.

Solution. (*a*) Because of – I inductive effect of – Cl, chloroacetic acid is stronger than acetic acid.

$$\underset{\text{Chloroacetic acid}}{Cl \leftarrow CH_2 \leftarrow COOH} > \underset{\text{Acetic acid}}{CH_3 - COOH}$$

(*b*)

$$\underset{\text{Dichloroacetic acid}}{{Cl \atop Cl}\!\!>\!CH - COOH} > \underset{\text{Chloroacetic acid}}{Cl \leftarrow CH_2COOH}$$

There are two chloro groups in dichloroacetic acid and one chloro group in chloroacetic acid. Hence in the former compound, there is twice the electron withdrawing inductive effect and hence greater acid strength.

$$\text{(c)}\quad \begin{array}{c} Cl \\ \uparrow \\ Cl \leftarrow C - COOH \\ \downarrow \\ Cl \end{array} \quad \text{Trichloroacetic acid}$$

There is the electron-withdrawing effect of three chloro groups. The effect is so strong that it produces a great acid strength equivalent to that of a mineral acid.

$$\text{(d)}\quad \underset{\text{Chloroacetic acid}}{Cl \leftarrow CH_2 - COOH} \; > \; \underset{\text{Bromoacetic acid}}{Br \leftarrow CH_2COOH}$$

Both chloro and bromo have – I inductive effect but out of the two, chloro has greater inductive effect than bromo. Therefore, chloroacetic acid is stronger than bromoacetic acid.

6.11. HYDROXY ACIDS

Nomenclature

A carboxylic acid with a hydroxy group in its hydrocarbon chain is called hydroxy acid. It can be obtained by substituting a hydrogen atom in its hydrocarbon chain by hydroxy group. Hydroxy group could be present in α, β, γ, δ positions with respect to the carboxylic group (according to common system) or 1, 2, 3, 4 positions (according to IUPAC system). The nomenclature of hydroxy acids is given in Table 6.5:

Table 6.5. Common and IUPAC names of hydroxy acids

Hydroxy acid	*Common name*	*IUPAC name*
$HOCH_2COOH$	Hydroxyacetic acid (Glycollic acid)	Hydroxyethanoic acid
$HOCH_2CH_2COOH$	β-hydroxypropionic acid	3-hydroxypropanoic acid
$CH_3CHOHCOOH$	α-hydroxypropionic acid (lactic acid)	2-hydroxypropanoic acid

6.12. METHODS OF PREPARATION OF HYDROXY ACIDS

[A] Preparation of α-hydroxy acids

1. Hydrolysis of α-haloacids. Hydrolysis of α-haloacids using alkali solution or moist silver oxide yields α-hydroxy acids.

$$R - \underset{\substack{| \\ Cl}}{CH} - COOH + NaOH \longrightarrow R - \underset{\substack{| \\ OH}}{CH} - COOH + NaCl$$

$$R - \underset{\substack{| \\ Cl}}{CH} - COOH + AgOH \longrightarrow R - \underset{\substack{| \\ OH}}{CH} - COOH + AgOH$$

$$\underset{\text{Chloroacetic acid}}{CH_2ClCOOH} + AgOH \longrightarrow \underset{\text{Glycollic acid}}{CH_2OHCOOH} + AgCl$$

2. Action of Nitrous acids. α-amino acid reacts with nitrous acid to produce α-hydroxy acid. A mixture of sodium nitrite and hydrochloric acid yields the nitrous acid.

$$CH_3-\underset{\underset{NH_2}{|}}{CH}-COOH + HNO_2 \xrightarrow[HCl]{NaNO_2} CH_3-\underset{\underset{OH}{|}}{CH}-COOH + N_2 + H_2O$$

α-aminopropionic acid — Lactic acid

3. Hydrolysis of Cyanohydrins. Cyanohydrins are obtained by the action of HCN on aldehydes or ketones. These are then hydrolysed in the presence of mineral acid to produce hydroxy acids.

$$CH_3\overset{\overset{O}{||}}{C}-H + HCN \longrightarrow CH_3-\underset{\underset{H}{|}}{\overset{\overset{OH}{|}}{C}}-CN \xrightarrow[H^+]{H_2O} CH_3-\underset{\underset{H}{|}}{\overset{\overset{OH}{|}}{C}}-COOH$$

Acetaldehyde — Acetaldehyde cyanohydrin — Lactic acid

$$CH_3-\underset{\underset{CH_3}{|}}{\overset{\overset{O}{||}}{C}} + HCN \longrightarrow CH_3-\underset{\underset{H}{|}}{\overset{\overset{OH}{|}}{C}}-CN \xrightarrow[H^+]{H_2O} CH_3-\underset{\underset{CH_3}{|}}{\overset{\overset{OH}{|}}{C}}-COOH$$

Acetone — α-hydroxy α-methyl-propionic acid

4. Reduction of ketoacids. Reduction of keto acid with sodium amalgam yields hydroxy acid

$$CH_3COCOOH + 2[H] \xrightarrow{Na/Hg} CH_3CHOHCOOH$$

α-ketopropionic acid pyruvic acid — Lactic acid

5. Oxidation of 1, 2 dihydroxy compounds. 1, 2 dihydroxy compounds of the following type upon oxidation with suitable oxidising agents give hydroxy acids.

$$OHCH_2CH_2OH + 2[O] \xrightarrow{dil.\ HNO_3} OHCH_2COOH + H_2O$$

Glycol — Glycollic acid

$$CH_3CHOHCH_2OH + 2[O] \xrightarrow[HNO_3]{dil.} CH_3CHOHCOOH + H_2O$$

1, 2 Propylene glycol — Lactic acid

[B] Preparation of β-hydroxy acids

1. Action of HNO_2. β-amino acid reacts with nitrous acid to produce β-hydroxy acid

$$\underset{\underset{NH_2}{|}}{CH_2}CH_2COOH + HNO_2 \longrightarrow \underset{\underset{OH}{|}}{CH_2}CH_2COOH + N_2 + H_2O$$

β-aminopropionic acid — β-hydroxypropionic acid

2. From chlorohydrin. A chlorohydrin is treated with potassium cyanide and then hydrolysed with water.

$$HOCH_2CH_2Cl + KCN \xrightarrow[-KCl]{} OHCH_2CH_2CN \xrightarrow{H_2O} OHCH_2CH_2COOH$$

Ethylene chlorohydrin — β-hydroxypropionic acid

3. Oxidation of 1, 3 dihydroxy compounds

$$\underset{\text{1,3 Propyleneglycol}}{HOCH_2CH_2CH_2OH} + 2[O] \xrightarrow[HNO_3]{dil.} \underset{\beta\text{-hydroxypropionic acid}}{HOCH_2CH_2COOH}$$

4. Reformatsky reaction. An aldehyde or ketone is treated with an α-bromoester and zinc metal in ethereal solution. The addition product initially formed is hydrolysed by dilute mineral acid to yield β-hydroxy acid.

$$\underset{\text{Acetaldehyde}}{CH_3\overset{O}{\overset{\|}{C}}H} + BrCH_2COOC_2H_5 + Zn$$

$$\big\downarrow \text{Ether}$$

$$CH_3-\overset{OZnBr}{\overset{|}{\underset{H}{\underset{|}{C}}}}-CH_2COOC_2H_5 \xrightarrow[H^+]{H_2O} \underset{\beta\text{-hydroxybutyric acid}}{CH_3CH(OH)CH_2COOH}$$

6.13. PROPERTIES OF HYDROXY ACIDS

Proportion of hydroxy acids are described below :

1. Solubility. Hydroxy acids are soluble in water. This is because of the presence of hydroxy and carboxylic groups both of which form hydrogen bonds with water.

2. Reaction with acetyl chloride. Acetyl derivatives of hydroxy acids are obtained on treatment with acetyl chloride or acetic anhydride.

$$\underset{\text{Lactic acid}}{CH_3\underset{OH}{\underset{|}{C}}HCOOH} + CH_3COCl \longrightarrow \underset{\text{Acetyl deri. of lactic acid}}{CH_3\underset{OCOCH_3}{\underset{|}{C}}H-COOH} + HCl$$

$$\underset{\beta\text{-hydroxy propionic acid}}{HOCH_2CH_2COOH} + CH_3COCl \longrightarrow \underset{\text{Acetyl derivative}}{CH_3COOCH_2CH_2COOH} + HCl$$

3. Oxidation. Hydroxyl group of the hydroxy acid is oxidised to aldehydic or ketonic group (as the case may be)

$$\underset{\text{Glycollic acid}}{HOCH_2COOH} \xrightarrow[\text{dil. }HNO_3]{[O]} \underset{\text{Glyoxalic acid}}{OHCCOOH} \xrightarrow{[O]} \underset{\text{Oxalic acid}}{HOOCCOOH}$$

$$\underset{\text{Lactic acid}}{CH_3CH(OH)COOH} \xrightarrow[\text{dil. }HNO_3]{[O]} \underset{\text{Pyruvic acid}}{CH_3\overset{O}{\overset{\|}{C}}COOH}$$

4. Reduction. On heating with hydroiodic acid, hydroxy acids are reduced to the parent acids.

$$\underset{\text{Lactic acid}}{CH_3\underset{OH}{\underset{|}{C}}HCOOH} + 2HI \xrightarrow{P} \underset{\text{Propionic acid}}{CH_3CH_2COOH} + H_2O + I_2$$

5. Formation of esters. On treatment with alcohols, hydroxy acid forms esters

$$\underset{\substack{|\\ OH}}{CH_2}CH_2COOH + C_2H_5OH \xrightarrow{H^+} \underset{\substack{|\\ OH}}{CH_2}CH_2COOC_2H_5 + H_2O$$

β-hydroxypropionic acid → Ethyl β-hydroxy propionate

$$CH_3\underset{\substack{|\\ OH}}{CH}COOH + C_2H_5OH \xrightarrow{H^+} CH_3\underset{\substack{|\\ OH}}{CH}COOC_2H_5 + H_2O$$

Lactic acid → Ethyl lactate

6. Formation of salts. Salts are obtained when hydroxy acids are treated with alkali.

$$CH_3\underset{\substack{|\\ OH}}{CH}COOH + NaOH \longrightarrow CH_2\underset{\substack{|\\ OH}}{CH}COONa + H_2O$$

Lactic acid → Sod. lactate

7. Reaction with sodium metal. Both the hydroxy and carboxy groups are reacted when treated with sodium metal.

$$CH_3\underset{\substack{|\\ OH}}{CH}COOH + Na \longrightarrow CH_3\underset{\substack{|\\ ONa}}{CH}COONa + H_2$$

Lactic acid → Disodium salt

8. Action of PCl_5. Hydroxy acids react with PCl_5, when both the hydroxy and carboxy groups are chlorinated.

$$CH_3\underset{\substack{|\\ OH}}{CH}COOH + PCl_5 \longrightarrow CH_3\underset{\substack{|\\ Cl}}{CH}COCl + POCl_3 + H_2O$$

Lactic acid → 2-chloropropanoyl chloride

9. Action of heat. Hydroxy acids on heating yield different products depending upon the position of the hydroxy group relative to the carboxylic group.

(*i*) **α-hydroxy acids.** Two molecules of α-hydroxy acid on heating form cyclic diesters known as lactides by the interaction of – OH from one molecule and – COOH from the other molecule

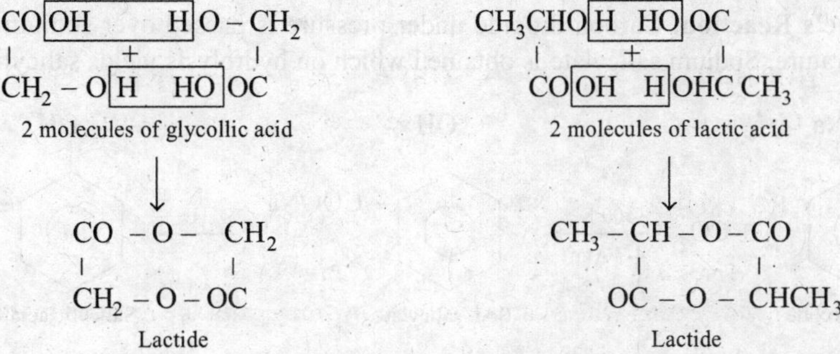

(*ii*) **β-hydroxy acid.** β-hydroxy acids on heating liberate water molecule and give α, β unsaturated carboxylic acid.

$$R-\underset{\substack{|\\ OH}}{CH}-CH_2COOH \xrightarrow[-H_2O]{\Delta} R-CH=CHCOOH$$

β-hydroxy acid → α, β unsaturated acid

$$\underset{\beta\text{-hydroxy propionic acid}}{\underset{|}{\overset{}{CH_2}}\!\!-CH_2COOH}\ \xrightarrow[-H_2O]{\Delta}\ \underset{\text{Acrylic acid}}{CH_2=CHCOOH}$$

(with OH on the first CH_2)

This reaction is catalysed by acids. The mechanism of reaction is as follows:

$$R-\underset{OH}{\underset{|}{CH}}-CH_2-COOH \xrightarrow{H^+} R-CH\overset{H}{\overset{|}{-}}CH-COOH \text{ (with } \overset{+}{O}H_2 \text{ on CH)} \xrightarrow{-H_2O,\ H^+} \underset{\alpha,\ \beta\text{ unsaturated acid}}{R-CH=CH-COOH}$$

Example 1. Of the following pairs of hydroxy acids, which are more acidic? Explain.

(*a*) (*i*) $HOCH_2COOH$ (*ii*) $HOCH_2CH_2COOH$

(*b*) (*i*) $CH_3CHOHCOOH$ (*ii*) $OHCH_2CH_2COOH$

(*c*) (*i*) $OHCH_2COOH$ (*ii*) $CH_3CHOHCOOH$

Solution. (*a*) (*i*) is more acidic than (*ii*)

(*b*) (*i*) is more acidic than (*ii*)

(*c*) (*i*) is more acidic than (*ii*)

Hydroxy group is an electron-withdrawing group. Thus this is expected to increase the acidity of the acid. However, electron-withdrawing effect of the hydroxy group depends upon its position. Nearer its position from the carboxy group, greater the electron-withdrawing effect and hence stronger the acid. Thus we would expect a hydroxy group at α position to cause greater increase in the acid strength than at β position. Based on above principle, compound (*i*) is stronger than (*ii*) in pairs (*a*) and (*b*). In pair (*c*) compound (*i*) is stronger acid than (*ii*) although in both of the compounds, the hydroxy group is at α position. This is so because in compound (*ii*), the methyl group has electron-donating effect which diminishes the electron-withdrawing effect of the hydroxy group.

SALICYLIC ACID

6.14. METHODS OF PREPARATION

Salicylic acid is prepared by the following methods.

1. Kolbe's Reaction. Carbon dioxide under pressure is passed over sodium phenoxide at 400 K temperature. Sodium salicylate is obtained which on hydrolysis yields salicylic acid.

$$\underset{\text{Sod. phenoxide}}{C_6H_5ONa} + CO_2 \xrightarrow[4\text{-}7\text{ Atm}]{400\text{ K}} \underset{\text{Sod. salicylate}}{C_6H_4(OH)COONa} \xrightarrow{H^+} \underset{\text{Salicylic acid}}{C_6H_4(OH)COOH}$$

Mechanism. Mechanism of the reaction is as follows:

1. Carbon dioxide attacks the phenoxide ion by nucleophilic addition mechanism. A proton is removed from the position where CO_2 is attached. Salicylic ion is produced.

2. Hydrolysis leads to the final product viz. salicylic acid. These steps are illustrated as under.

+ C=O (O) ⟶ [cyclohexadienone with H, C(=O)—O⁻] ⟶ [O⁻, C(=O)—O⁻] —$H^{\oplus}$⟶ [OH, C(=O)—OH]

2. Reimer-Tiemann's Reaction. Sodium phenoxide on heating with carbon tetrachloride in alkaline medium gives sodium salicylate which upon hydrolysis gives salicylic acid.

Sodium phenoxide (ONa) $+ CCl_4$ —NaOH⟶ (OH, CCl_3) —NaOH⟶ (ONa, COONa) —H^+⟶ (OH, COOH) Salicylic acid

Mechanism of the reaction is given below

(O⁻) $+ CCl_3$—Cl —$-Cl^-$⟶ (=O, H, CCl_3) ⟶ (OH, C(Cl)(Cl)(Cl), $^{\ominus}OH$) —$-Cl^{\ominus}$⟶ (OH, C(Cl)(Cl)—O—H, $^{\ominus}OH$) —$-Cl^{\ominus}$⟶ (OH, C(Cl)=O, $^{\ominus}OH$) ⟶ (OH, C(OH)(Cl)—O⁻) —$-Cl^{\ominus}$⟶ (OH, C(=O)—Cl) —$^{\ominus}OH$⟶ (O⁻, C(=O)—O⁻) —$H^{\oplus}$⟶ (OH, COOH)

3. Action of nitrous acid on anthranilic acid

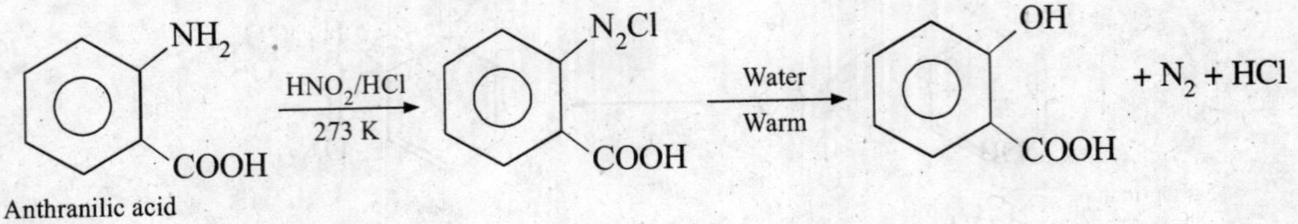

$$\text{C}_6\text{H}_4(\text{NH}_2)(\text{COOH}) \xrightarrow[273\text{ K}]{HNO_2/HCl} \text{C}_6\text{H}_4(\text{N}_2\text{Cl})(\text{COOH}) \xrightarrow[\text{Warm}]{\text{Water}} \text{C}_6\text{H}_4(\text{OH})(\text{COOH}) + N_2 + HCl$$

Anthranilic acid

6.15. PROPERTIES OF SALICYLIC ACID

1. Reaction with alkalis and carbonates

Alkalis react with both the phenolic and carboxylic groups whereas carbonates react only with carboxylic group

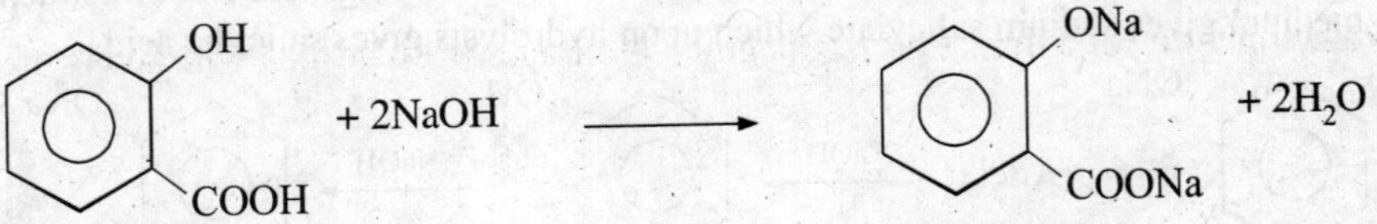

$$\text{C}_6\text{H}_4(\text{OH})(\text{COOH}) + 2NaOH \longrightarrow \text{C}_6\text{H}_4(\text{ONa})(\text{COONa}) + 2H_2O$$

Salicylic acid — Disodium salicylate

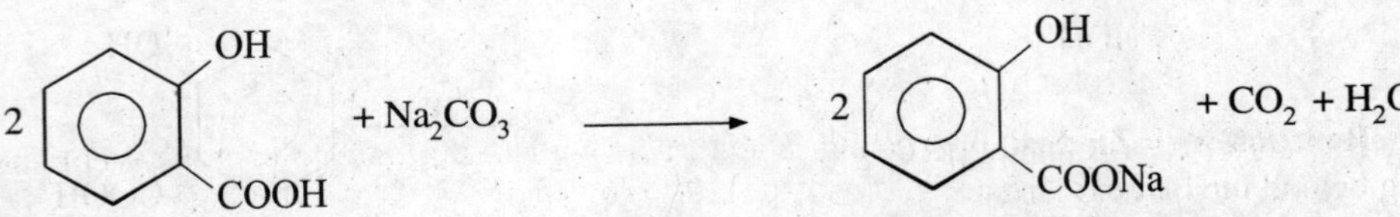

$$2\,\text{C}_6\text{H}_4(\text{OH})(\text{COOH}) + Na_2CO_3 \longrightarrow 2\,\text{C}_6\text{H}_4(\text{OH})(\text{COONa}) + CO_2 + H_2O$$

Salicylic acid — Monosodium salicylate

2. Reaction with alcohols and phenols. Formation of esters takes place with both alcohols and phenols.

$$\text{C}_6\text{H}_4(\text{OH})(\text{COOH}) + C_2H_5OH \xrightarrow{\text{HCl gas}} \text{C}_6\text{H}_4(\text{OH})(\text{COOC}_2\text{H}_5) + H_2O$$

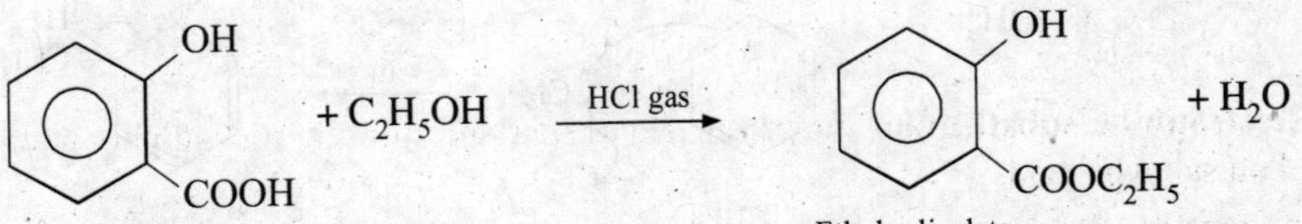

Salicylic acid — Ethanol — Ethyl salicylate

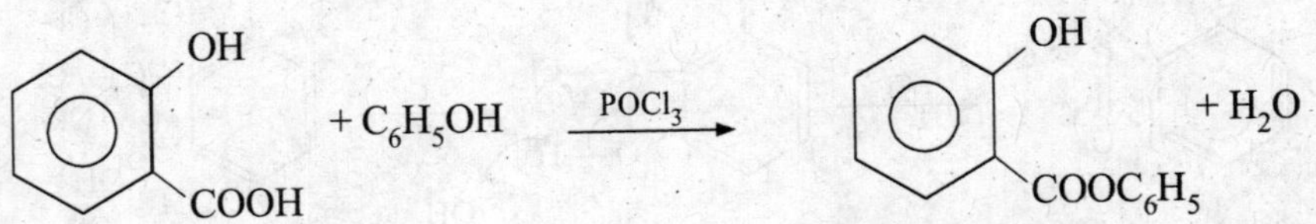

$$\text{C}_6\text{H}_4(\text{OH})(\text{COOH}) + C_6H_5OH \xrightarrow{POCl_3} \text{C}_6\text{H}_4(\text{OH})(\text{COOC}_6\text{H}_5) + H_2O$$

Phenyl salicylate (salol)

3. Reaction with phosphorus pentachloride

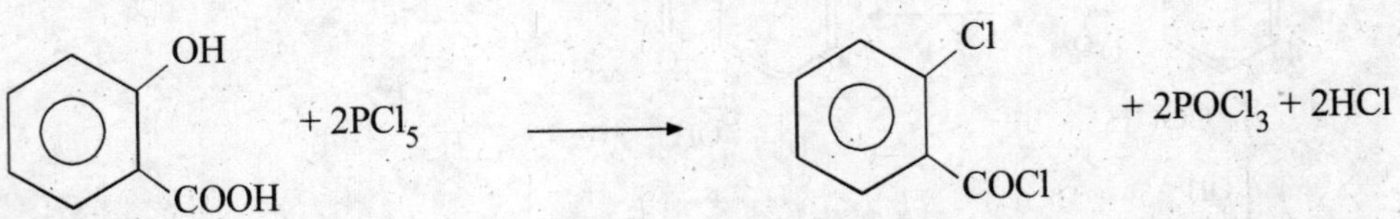

$$\text{C}_6\text{H}_4(\text{OH})(\text{COOH}) + 2PCl_5 \longrightarrow \text{C}_6\text{H}_4(\text{Cl})(\text{COCl}) + 2POCl_3 + 2HCl$$

Salicylic acid — *o*-chloro benzoyl chloride

4. Reaction with Acetyl chloride and Acetic anhydride

Salicylic acid (OH, COOH) + CH_3COCl (Acetyl chloride) ⟶ Acetyl salicylic acid (Aspirin) (O.COCH$_3$, COOH) + HCl

Salicylic acid (OH, COOH) + $(CH_3CO)_2O$ (Acetic anhydride) ⟶ Acetyl salicylic acid (O.COCH$_3$, COOH) + CH_3COOH

5. Reaction with soda lime. Carboxylic group is eliminated on heating with soda lime leaving the phenolic group.

(OH, COOH) + CaO ⟶ Phenol (OH) + $CaCO_3$

6. Reaction with Zn dust. On heating salicylic acid with Zn dust, phenolic group is eliminated leaving behind the carboxy group.

Salicylic acid (OH, COOH) + Zn $\xrightarrow{\text{Distil.}}$ Benzoic acid (COOH) + ZnO

7. Electrophilic substitution reactions. Halogenation, nitration and sulphonation can be performed on salicylic acid.

Sulphonation. Sulphonic group takes the para position with respect to phenolic group.

Salicylic acid (OH, COOH) + H_2SO_4 ⟶ Salicylic acid sulphonate (OH, COOH, HO_3S) + H_2O

Bromination. During bromination, carboxy group is decomposed and a tribromo product is formed.

(OH, COOH) + Br_2 (Aq. solution) ⟶ 2, 4, 6-Tribromophenol (OH, Br, Br, Br) + 3HBr + CO_2

Nitration. As in case of bromination, nitration takes place with the simultaneous decomposition of carboxylic group.

$$\text{Salicylic acid} + 3HNO_3 \xrightarrow{H_2SO_4} \text{2, 4, 6-Trinitrophenol} + 3H_2O + CO_2$$

Salicylic acid

2, 4, 6-Trinitrophenol

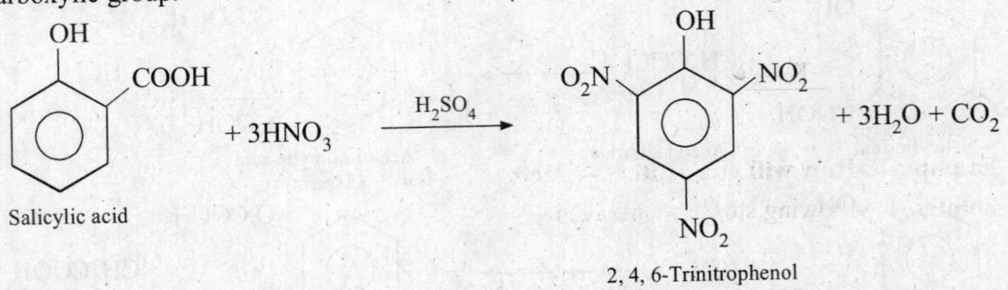

Mechanism of decomposition of carboxylic group during the course of halogenation and nitration is illustrated below.

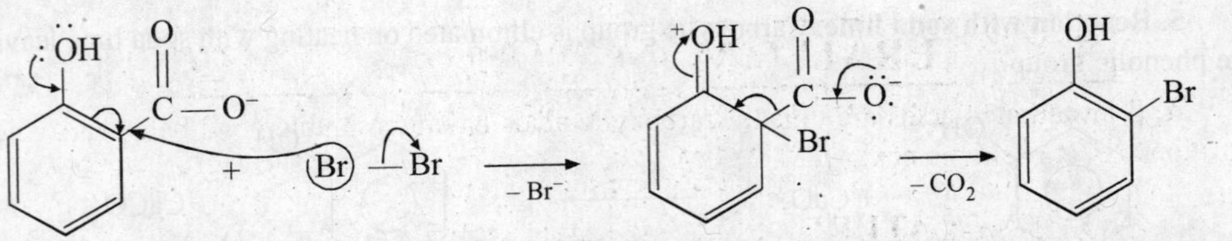

$$\text{Salicylate} + Br-Br \xrightarrow{-Br^-} \text{intermediate} \xrightarrow{-CO_2} \text{o-Bromophenol}$$

Example 1. *o*-hydroxy benzoic acid (Salicylic acid) melts at 432 K, *m*-hydroxy benzoic acid at 474 K and *p*-hydroxy benzoic acid at 487 K. Explain the difference in melting points.

Solution. Intramolecular (within the same molecule) hydrogen bonding takes place in *o*-hydroxy benzoic acid due to the close proximity of the phenolic and carboxylic groups.

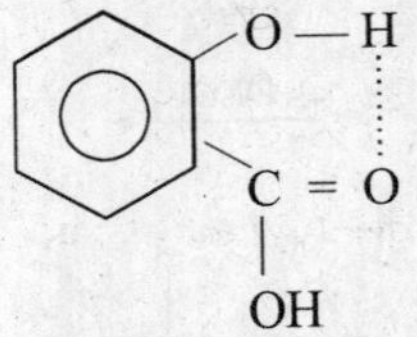

This results in shrinking of size of the molecule. Consequently the melting point is lowered.

On the other hand *m*- & *p*-isomers exhibit intermolecular (between different molecules) hydrogen bonding resulting in increase in the size of molecules.

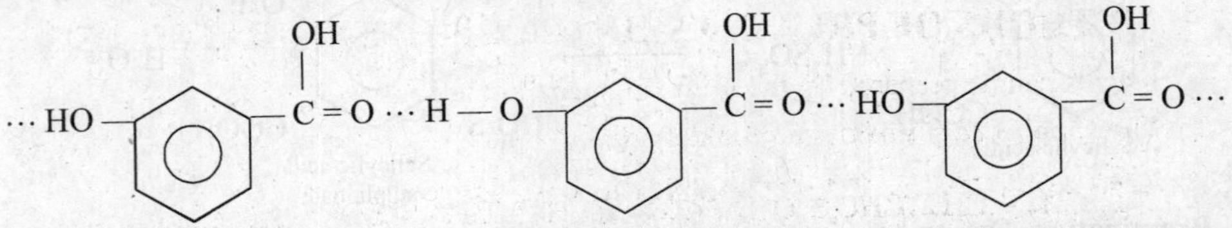

$$\cdots HO-C_6H_4-C(OH)=O\cdots H-O-C_6H_4-C(OH)=O\cdots HO-C_6H_4-C(OH)=O\cdots$$

Consequently there is an increase in the melting point. In the *p*-isomer, there is more effective H-bonding because there is no steric hindrance. *m*-isomer faces some steric hindrance

Example 2. How will you synthesise lactic acid from acetaldehyde ?

Solution. Following steps are involved :

$$\underset{\text{Acetaldehyde}}{CH_3-\overset{O}{\overset{\|}{C}}-H} \xrightarrow{HCN} CH_3-\overset{OH}{\overset{|}{C}H}-CN \xrightarrow{H_2O/H^+} \underset{\text{Lactic acid}}{CH_3-\overset{OH}{\overset{|}{C}H}-COOH}$$

Example 3. How will you synthesise lactic acid from ethyl alcohol ?

Solution. Following steps are involved :

$$\underset{\text{Ethyl alcohol}}{CH_3CH_2OH} \xrightarrow{K_2Cr_2O_7/H^+} CH_3\overset{O}{\overset{||}{C}}H \xrightarrow{HCN} CH_3\overset{OH}{\overset{|}{C}}HCN \xrightarrow{H_2O/H^+} \underset{\text{Lactic acid}}{CH_3\overset{OH}{\overset{|}{C}}HCOOH}$$

Example 4. How will you synthesise lactic acid from acetylene ?

Solution. Following steps are involved:

$$\underset{\text{Acetylene}}{HC \equiv CH} \xrightarrow[HgSO_4]{H_2O/H_2SO_4} CH_3\overset{O}{\overset{||}{C}}H \xrightarrow{HCN} CH_3\overset{OH}{\overset{|}{C}}HCN \xrightarrow{H_2O/H^+} \underset{\text{Lactic acid}}{CH_3\overset{OH}{\overset{|}{C}}HCOOH}$$

UNSATURATED MONOCARBOXYLIC ACIDS

α, β unsaturated acids are organic carboxylic acids having a double bond between α and β positions *i.e.* at positions next and still next to the carboxylic group.

6.16. NOMENCLATURE

Nomenclature of a few important α, β unsaturated acids is given in Table 6.6.

Table 6.6. Common and IUPAC names of unsaturated acids

α, β unsaturated acid	*Common name*	*IUPAC name*
$CH_2 = CH - COOH$	Acrylic acid	Propenoic acid
$CH_3 - CH = CH - COOH$	Crotonic acid	But-2-enoic acid
$CH_2 = \overset{CH_3}{\overset{\mid}{C}} - COOH$	Methyl acrylic acid	2-methylpropenoic acid
$C_6H_5-CH = CHCOOH$	Cinnamic acid	3-phenylpropenoic acid

6.17. METHODS OF PREPARATION OF α, β-UNSATURATED ACIDS

Such acids can be prepared by the following general methods:

1. Oxidation of α, β unsaturated aldehyde.

$$\underset{\text{Acrolein}}{CH_2 = CH - CHO + O} \xrightarrow[\text{reagent}]{\text{Tollen's}} \underset{\text{Acrylic acid}}{CH_2 = CH - COOH}$$

2. Dehydrohalogenation of β-halo acids. α, β unsaturated acids can be obtained by the dehydrohalogenation of β-halo acid using alcoholic potassium hydroxide

$$\underset{\text{3-chlorobutanoic acid}}{CH_3 - \underset{Cl}{\underset{|}{C}}H - CH_2 - COOH} \xrightarrow[KOH]{Alc.} \underset{\text{But-2-enoic acid}}{CH_3 - CH = CH - COOH} + HCl$$

The reaction is supposed to take place by the following mechanism:

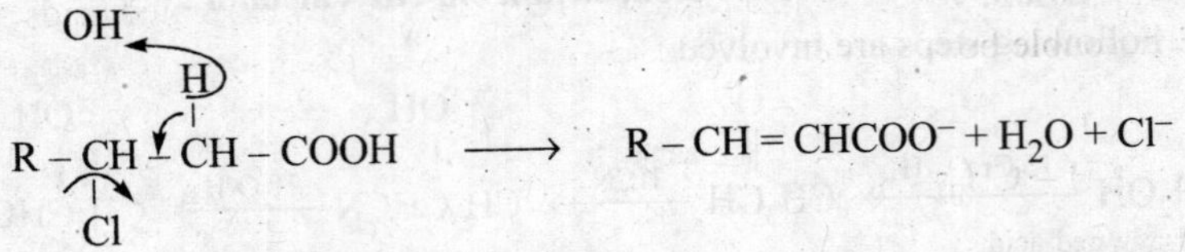

$$R-\underset{\underset{Cl}{|}}{CH}-\overset{\overset{H}{|}}{CH}-COOH \xrightarrow{OH^-} R-CH=CHCOO^- + H_2O + Cl^-$$

3. From alkynes. Acetylene is treated with HCN to give an addition product which is subjected to hydrolysis.

$$\underset{\text{Acetylene}}{CH \equiv CH} + HCN \longrightarrow \underset{\text{Vinyl cyanide}}{CH_2 = CHCN} \xrightarrow{H_2O} \underset{\text{Acrylic acid}}{CH_2 = CHCOOH}$$

4. Dehydration of β-hydroxy acid

$$\underset{\beta\text{-hydroxy acid}}{R-\underset{\underset{OH}{|}}{CH}-CH_2-COOH} \xrightarrow{\Delta} \underset{\alpha,\ \beta\text{-unsaturated acid}}{RCH = CHCOOH}$$

$$\underset{\beta\text{-hydroxypropanoic acid}}{CH_2OH-CH_2-COOH} \xrightarrow{\Delta} \underset{\text{Acrylic acid}}{CH_2 = CH-COOH}$$

5. From Grignard reagent

$$\underset{\text{Vinyl magnesium halide}}{CH_2 = CHMgX} + CO_2 \xrightarrow[\text{furan (THF)}]{\text{Tetrahydro}} \underset{\text{(Addition compound)}}{CH_2 = CH-\overset{\overset{O}{||}}{C}-OMgX}$$

$$\xrightarrow{H_2O/H^+} \underset{\text{Acrylic acid}}{CH_2 = CH-COOH} + Mg\begin{matrix} OH \\ X \end{matrix}$$

6. Perkin's reaction

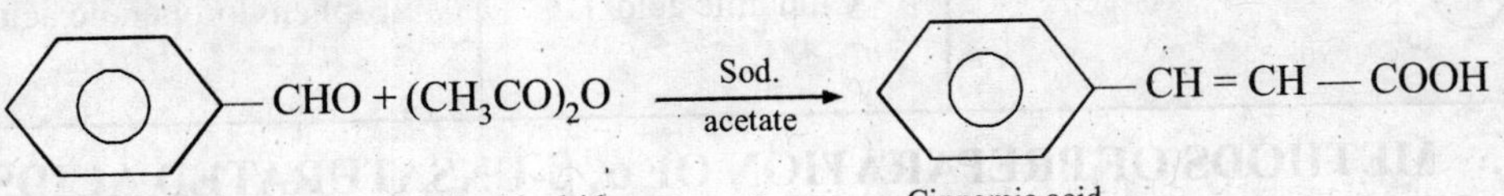

$$\underset{\text{Benzaldehyde}}{C_6H_5-CHO} + \underset{\text{Acetic anhydride}}{(CH_3CO)_2O} \xrightarrow[\text{acetate}]{\text{Sod.}} \underset{\text{Cinnamic acid}}{C_6H_5-CH = CH-COOH}$$

7. Knoevenagel reaction

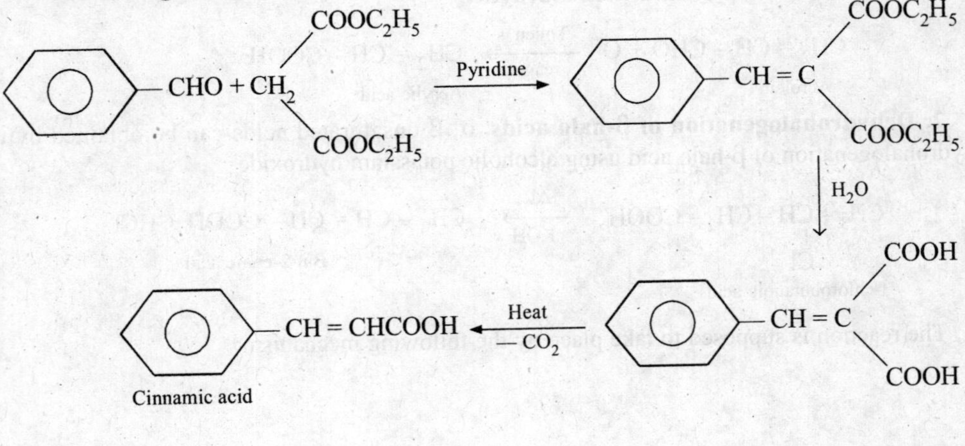

$$C_6H_5-CHO + CH_2\begin{matrix} COOC_2H_5 \\ COOC_2H_5 \end{matrix} \xrightarrow{\text{Pyridine}} C_6H_5-CH = C\begin{matrix} COOC_2H_5 \\ COOC_2H_5 \end{matrix}$$

$$\xrightarrow{H_2O} C_6H_5-CH = C\begin{matrix} COOH \\ COOH \end{matrix} \xrightarrow[-CO_2]{\text{Heat}} \underset{\text{Cinnamic acid}}{C_6H_5-CH = CHCOOH}$$

8. By isomerisation. When unsaturated acids other than α, β unsaturated acids are boiled with strong alkali, the double bond shifts to the α, β position. On acidification, α, β unsaturated acid is obtained.

$$\underset{\beta,\,\gamma \text{ unsaturated acid}}{CH_3CH = CH - CH_2 - COOH} \xrightarrow[\text{Boil}]{\text{NaOH}} \underset{\alpha,\,\beta \text{ unsaturated acid}}{CH_3CH_2CH = CHCOOH}$$

Mechanism

$$CH_3 - CH = CH - \underset{\textcircled{H}}{\underset{|}{CH}} - \overset{O}{\overset{\|}{C}} - O - H + OH^-$$

$$\downarrow$$

$$CH_3 - CH = CH - \overset{\ominus}{CH} - \overset{O}{\overset{\|}{C}} - O - H$$

$$\downarrow$$

$$CH_3 - \overset{\ominus}{CH} - CH = CH - \overset{O}{\overset{\|}{C}} - O^-$$

$$\downarrow H^+$$

$$CH_3 - CH_2 - CH = CH - \overset{O}{\overset{\|}{C}} - O^- \xrightarrow{H^+} CH_3CH_2CH = CHCOOH$$

6.18. PROPERTIES OF α, β–UNSATURATED ACIDS

Important properties of unsaturated acids are given below.

1. Electrophilic addition reactions. α, β unsaturated acids undergo addition reactions characteristics of carbon-carbon double bond. Addition of halogen, acids and water takes place on the ethylenic double bond but the reaction is slow because of the conjugation of the double bond with carboxylic acid.

It may be noted that during hydration and hydrohalogenation, hydroxyl and halogen always add on β-carbon, which is contrary to the Markownikoff's rule.

(*i*) $$\underset{\text{Acrylic acid}}{CH_2 = CH - COOH} + Cl_2 \longrightarrow \underset{\alpha,\,\beta \text{ dichloropropionic acid}}{\underset{Cl}{\underset{|}{CH_2}} - \underset{Cl}{\underset{|}{CH}} - COOH}$$

(*ii*) $$CH_2 = CH - COOH + HBr \longrightarrow \underset{\beta\text{-bromopropionic acid}}{\underset{Br}{\underset{|}{CH_2}} - CH_2 - COOH}$$

(*iii*) $$CH_2 = CH - COOH + H_2O \longrightarrow \underset{\beta\text{-hydroxypropionic acid}}{\underset{OH}{\underset{|}{CH_2}} - CH_2 - COOH}$$

Mechanism of addition

1. A proton gets attached to carbonyl oxygen of the carboxyl group giving resonance stabilised positive ion.

2. The nucleophile (X^- or OH^-) attacks the carbonium ion at β position to give enol form of the acids.

3. The enol form tautomerises to more stable keto form.

These steps are mechanistically shown below.

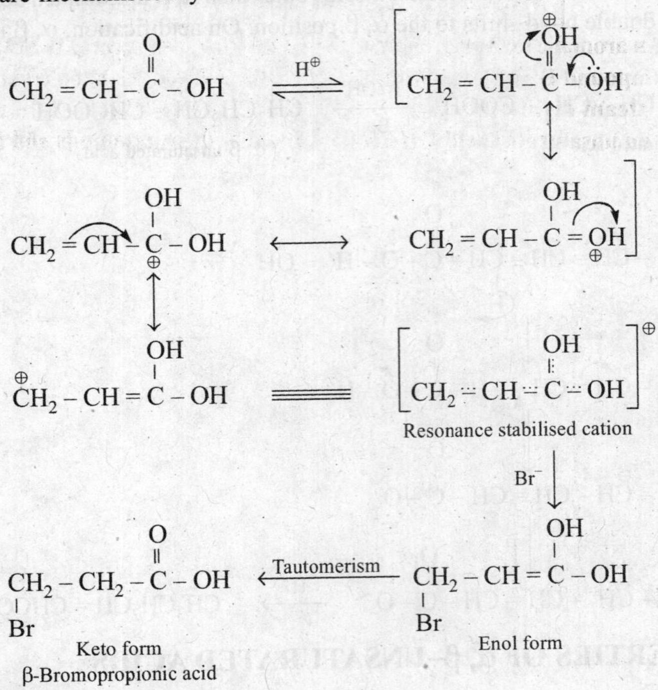

$$CH_2=CH-\overset{\overset{\displaystyle O}{\|}}{C}-OH \underset{}{\overset{H^{\oplus}}{\rightleftharpoons}} \left[CH_2=CH-\overset{\overset{\displaystyle \overset{\oplus}{O}H}{\|}}{C}-\ddot{O}H \longleftrightarrow CH_2=CH-\overset{\overset{\displaystyle OH}{|}}{C}=\overset{\oplus}{O}H\right]$$

$$CH_2=CH-\overset{\overset{\displaystyle OH}{|}}{\underset{\oplus}{C}}-OH \longleftrightarrow$$

$$\overset{\oplus}{C}H_2-CH=\overset{\overset{\displaystyle OH}{|}}{C}-OH \equiv \left[CH_2\overset{...}{=}CH\overset{...}{=}\overset{\overset{\displaystyle OH}{\vdots}}{C}\overset{...}{=}OH\right]^{\oplus}$$

Resonance stabilised cation

$$\downarrow Br^-$$

$$\underset{\underset{\displaystyle Br}{|}}{CH_2}-CH_2-\overset{\overset{\displaystyle O}{\|}}{C}-OH \xleftarrow{\text{Tautomerism}} \underset{\underset{\displaystyle Br}{|}}{CH_2}-CH=\overset{\overset{\displaystyle OH}{|}}{C}-OH$$

Keto form (β-Bromopropionic acid) — Enol form

2. Reaction of the carboxylic group

(*i*) **Formation of salt**

$$\underset{\text{Crotonic acid}}{CH_3-CH=CH-COOH}+NaOH \longrightarrow \underset{\text{Sod. crotonate}}{CH_3-CH=CHCOONa}+H_2O$$

$$\underset{\text{Acrylic acid}}{2CH_2=CHCOOH}+Na_2CO_3 \longrightarrow \underset{\text{Sod. acrylate}}{2CH_2=CHCOONa}+CO_2+H_2O$$

(*ii*) **Esterification**

$$\underset{\text{Acrylic acid}}{CH_2=CHCOOH}+C_2H_5OH \xrightarrow{H_2SO_4} \underset{\text{Ethyl acrylate}}{CH_2=CHCOOC_2H_5}+H_2O$$

(*iii*) **Action of PCl_5**

$$\underset{\text{Crotonic acid}}{CH_3CH=CHCOOH}+PCl_5 \longrightarrow \underset{\text{Crotonoyl chloride}}{CH_2CH=CHCOCl}+POCl_3+HCl$$

3. Reduction. Double bond of the unsaturated acid gets saturated on catalytic hydrogenation or treatment with nascent hydrogen.

$$\underset{\text{Acrylic acid}}{CH_2=CHCOOH}+2[H] \xrightarrow{Na/C_2H_5OH} \underset{\text{Propanoic acid}}{CH_3CH_2COOH}$$

4. Reaction with alkali. Unsaturated acid (α, β or others) on fusion with alkali breaks to give acetic acid and one more acid with two carbons less.

$$\underset{\beta,\gamma \text{ unsaturated acid}}{R-CH=CH-CH_2-COOH} \xrightarrow{NaOH} \underset{\alpha,\beta \text{ unsaturated acid}}{R-CH_2-CH=CHCOOH}$$

$$\downarrow \text{NaOH fuse}$$

$$R-CH_2COOH+CH_3COOH$$

MISCELLANEOUS QUESTIONS

Example 1. An aromatic hydrocarbon A, C_6H_6 on treatment with conc. H_2SO_4 gives a compound B ($C_6H_6SO_3$). Compound B on fusion with NaCN yields C (C_7H_5N). C on treatment with Sn Cl_2/HCl followed by steam distillation gives D (C_7H_6O). D reacts with sodium acetate and acetic anhydride to give an unsaturated acid E ($C_9H_8O_2$). Identify the compounds and name the reactions involved.

Solution.

$$\underset{A}{C_6H_6} \xrightarrow[\text{Sulphonation}]{\text{Conc. } H_2SO_4} \underset{B}{C_6H_5SO_3H} \xrightarrow[\text{NaCN}]{\text{Fusion}} \underset{C}{C_6H_5CN}$$

$$\xrightarrow[\text{HCl, Stephen's reaction}]{SnCl_2} \underset{D}{C_6H_5CHO} \xrightarrow[(CH_3CO)_2O]{CH_3COONa} \underset{\text{E (cinnamic acid)}}{C_6H_5CH=CHCOOH}$$

Example 2. How will you synthesise lactic acid from acetaldehyde?

Solution. Following steps are involved:

$$CH_3-\overset{O}{\overset{\|}{C}}-H \xrightarrow{HCN} CH_3-\overset{OH}{\overset{|}{C}H}-CN \xrightarrow{H^+/H_2O} \underset{\text{Lactic acid}}{CH_3-\overset{OH}{\overset{|}{C}H}-COOH}$$

DICARBOXYLIC ACIDS

6.19. NOMENCLATURE

Dicarboxylic acids are those organic compounds which contain two carboxylic groups in the molecule.

Nomenclature of some prominent dicarboxylic and tricarboxylic acids is given in Table 6.7.

Table 6.7. Nomenclature of dicarboxylic acids

Compound	*Common name*	*IUPAC name*
$COOH-COOH$	Oxalic acid	Ethanedioic acid
$CH_2(COOH)_2$	Malonic acid	Propane-1, 3-dioic acid
$CH_2COOH-CH_2COOH$	Succinic acid	Butane-1, 4-dioic acid
$CH_2COOH-CH(OH)COOH$	Malic acid	Butan-2-ol-1, 4-dioic acid

CH(OH)COOH | CH(OH)COOH	Tartaric acid	Butan-2, 3 diol-1, 4-dioic acid
CHCOOH ‖ CHCOOH	Maleic acid	But-2-ene-1, 4-dioic acid
CH_2COOH | CH(OH)COOH | CH_2COOH	Citric acid	3-hydroxy-3-carboxypentane-1, 5-dioic acid

OXALIC ACID

It occurs as potassium hydrogen oxalate in the wood *sorrel* and tomatoes. The insoluble calcium oxalate is found in some stony deposits in kidneys and bladder in human body.

Model of carboxylic acid

6.20. PREPARATION OF OXALIC ACID

Oxalic acid is prepared industrially by heating sodium formate to 400°C.

$$\underset{\text{Sod formate}}{2\,HCO\overset{-}{O}\overset{+}{Na}} \xrightarrow{400^{o}} \underset{\text{Sod oxalate}}{\overset{+}{Na}\overset{-}{O}OC - CO\overset{-}{O}\overset{+}{Na}} + H_2$$

Sodium oxalate thus formed is dissolved in water and calcium hydroxide added to precipitate calcium oxalate. The solution is filtered and the filtrate treated with calculated quantity of dilute sulphuric acid to liberate the oxalic acid, as shown by the following equations :

$$\underset{\text{Sod oxalate}}{\begin{matrix} CO\overset{-}{O}\overset{+}{Na} \\ | \\ CO\overset{-}{O}\overset{+}{Na} \end{matrix}} + Ca(OH)_2 \longrightarrow \underset{\text{Calcium oxalate}}{\begin{matrix} CO\overset{-}{O} \\ | \\ CO\overset{-}{O} \end{matrix}\; Ca^{2+}\downarrow} + 2NaOH$$

$$\underset{\text{Calcium oxalate}}{\begin{matrix} CO\overset{-}{O} \\ | \\ CO\overset{-}{O} \end{matrix}\; Ca^{2+}} + H_2SO_4 \longrightarrow \underset{\text{Oxalic acid}}{\begin{matrix} COOH \\ | \\ COOH \end{matrix}} + CaSO_4\downarrow$$

Calcium sulphate is removed by filtration and oxalic acid is crystallised from the filtrate as the hydrate, $(COOH)_2.2H_2O$.

(2) **Laboratory Method.** Oxalic acid is prepared by oxidation of sucrose or molasses with concentrated nitric acid in the presence of vanadium pentoxide as catalyst.

$$C_{12}H_{22}O_{11} + 18[O] \xrightarrow[V_2O_5]{HNO_3} 6 \underset{\text{Oxalic acid}}{\begin{matrix} COOH \\ | \\ COOH \end{matrix}} + 5H_2O$$

The –CHOH.CHOH– units present in sucrose molecule are broken and oxidised to oxalic acid.

6.21. PROPERTIES OF OXALIC ACID

Physical. When crystallised from water, colourless prismatic crystals of oxalic acid dihydrate, $(COOH)_2.2H_2O$, are obtained. The hydrated oxalic acid melts at 101.5°C, while the anhydrous acid melts at 189.5°C. The hydrated acid becomes anhydrous when carefully heated to 150°C. Oxalic acid is an active poison. It depresses the central nervous system and causes malfunction of kidneys.

Chemical. It gives all the usual reactions of COOH group twice. Also, the acid gives some peculiar reactions which involve the cleavage of the weakened linkage between the two highly oxidised carbon atoms.

1. Formation of Mono- and Di- derivatives. Oxalic acid is a much stronger acid than acetic acid and readily forms two series of salts, esters, acid halides and amides by reaction with alkali, alcohol, PCl_5 and ammonia respectively.

$$\begin{array}{cccc} COONa & COOC_2H_5 & COCl & CONH_2 \\ | & | & | & | \\ COONa & COOC_2H_5 & COCl & CONH_2 \\ \text{Disodium oxalate} & \text{Diethyl oxalate} & \text{Oxalyl chloride} & \text{Oxamide} \end{array}$$

2. Action with Glycerol. Oxalic acid reacts with glycerol to form formic acid or allyl alcohol, depending upon experimental conditions. (*See under reactions of glycerol*)

3. Oxidation. It is readily oxidised, for example with acidified potassium permanganate.

$$\underset{\text{Oxalic acid}}{HOOC-COOH} + [O] \xrightarrow[H_2SO_4]{KMnO_4} 2CO_2 + H_2O$$

4. Action of Heat. When heated at 150°C, it decarboxylates to give formic acid.

$$\underset{\text{Oxalic acid}}{HOOC-COOH} \xrightarrow{150^{\circ}} \underset{\text{Formic acid}}{HCOOH} + CO_2$$

5. Action with H_2SO_4. When heated with concentrated sulphuric acid, it is decomposed to give carbon dioxide, carbon monoxide and water.

$$\underset{\text{Oxalic acid}}{HOOC-COOH} \xrightarrow[\Delta]{\text{Conc. } H_2SO_4} CO_2 + CO + H_2O$$

6.22. USES OF OXALIC ACID

Oxalic acid is used

(1) For removing ink stains and for bleaching straw

(2) As a mordant in dyeing and calico printing

(3) In manufacture of inks and metal polishes

(4) For preparing allyl alcohol and formic acid in the laboratory and industry

(5) In *Redox Titrations.*

MALONIC ACID

6.23. METHODS OF PREPARATION OF MALONIC ACID

1. **By oxidation of malic acid**

$$\underset{\text{Malic acid}}{\begin{matrix}COOH\\|\\CH(OH)\\|\\CH_2\\|\\COOH\end{matrix}} \xrightarrow[K_2Cr_2O_7]{[O]} \underset{\text{Oxalacetic acid}}{\begin{matrix}COOH\\|\\CO\\|\\CH_2\\|\\COOH\end{matrix}} \xrightarrow[K_2Cr_2O_7]{[O]} \underset{\text{Malonic acid}}{\begin{matrix}COOH\\|\\CH_2\\|\\COOH\end{matrix}}$$

2. **From potassium chloroacetate**

$$\underset{\text{Potassium chloroacetate}}{\begin{matrix}CH_2Cl\\|\\COOK\end{matrix}} + KCN \xrightarrow{-KCl} \underset{\text{Potassium cyanoacetate}}{\begin{matrix}CH_2CN\\|\\COOK\end{matrix}} \xrightarrow[H^+]{H_2O} \underset{\text{Malonic acid}}{\begin{matrix}CH_2COOH\\|\\COOH\end{matrix}}$$

6.24. PROPERTIES OF MALONIC ACID

1. Decarboxylation (Action of heat). On heating to 410–420 K malonic acid decomposes to give acetic acid

$$CH_2\begin{matrix}\diagup COOH\\ \diagdown COOH\end{matrix} \xrightarrow[410–420\ K]{\Delta} CH_3COOH + CO_2$$

Mechanism of decarboxylation

$$\begin{matrix}HO & & CH_2 & & O\\ & C & & C & \\ & \| & & | & \\ & O & & O & \\ & & & H & \end{matrix} \xrightarrow{-CO_2} \underset{\text{Enolic form of acetic acid}}{\begin{matrix}HO & & CH_2\\ & C & \\ & | & \\ & OH & \end{matrix}} \rightleftharpoons \underset{\text{Acetic acid}}{CH_3COOH}$$

2. Dehydration. Malonic acid loses two water molecules on heating in the presence of P_2O_5

$$\begin{matrix} & H & OH\\ & | & |\\ O=C- & C- & C=O\\ | & | & \\ OH & H & \end{matrix} \xrightarrow{-2H_2O} \underset{\text{Carbon suboxide}}{O=C=C=C=O}$$

3. Action of nitrous acid. On treatment with nitrous acid followed by hydrolysis, malonic acid gives mesoxalic acid

$$\underset{\text{Malonic acid}}{CH_2\begin{matrix}\diagup COOH\\ \diagdown COOH\end{matrix}} \xrightarrow[-H_2O]{HNO_3} HON=C\begin{matrix}\diagup COOH\\ \diagdown COOH\end{matrix} \xrightarrow{H_2O} \underset{\text{Mesoxalic acid}}{O=C\begin{matrix}\diagup COOH\\ \diagdown COOH\end{matrix}}$$

4. Action of bromine

$$\underset{\text{Malonic acid}}{CH_2\begin{matrix}\diagup COOH\\ \diagdown COOH\end{matrix}} + Br_2 \longrightarrow \underset{\text{Monobromo-malonic acid}}{BrCH\begin{matrix}\diagup COOH\\ \diagdown COOH\end{matrix}} + HBr$$

5. Action with aldehyde

$$CH_2\langle^{COOH}_{COOH} + RCHO \longrightarrow RCH=C\langle^{COOH}_{COOH}$$

Malonic acid

$$\xrightarrow{\text{Heat}} RCH=CHCOOH + CO_2$$

α, β unsaturated acid

MALEIC ACID

6.25. METHODS OF PREPARATION OF MALEIC ACID

1. From malic acid (Note the spelling difference between malic and maleic acids)

$$\begin{matrix} CH_2COOH \\ | \\ CH(OH)COOH \end{matrix} \xrightarrow[-2H_2O]{523\text{ K}} \begin{matrix} CHCO \\ \| \\ CHCO \end{matrix}\rangle O \xrightarrow[\text{Boil}]{H_2O} \begin{matrix} CHCOOH \\ \| \\ CHCOOH \end{matrix}$$

Malic acid — Maleic anhydride — Maleic acid

2. From benzene (Manufacture)

$$\text{Benzene} + 9O_2 \xrightarrow[-4CO_2,\,-2H_2O]{V_2O_5,\ 673\text{–}700\text{ K}} \begin{matrix} CHCO \\ \| \\ CHCO \end{matrix}\rangle O \xrightarrow{2H_2O} \begin{matrix} CHCOOH \\ \| \\ CHCOOH \end{matrix}$$

Benzene — Maleic anhydride — Maleic acid

6.26. PROPERTIES OF MALEIC ACID

1. Formation of salts. It reacts with alkalis to form salts. Two series of salts viz., mono and dialkali metal maleates, can be obtained.

$$\begin{matrix} CHCOOH \\ \| \\ CHCOOH \end{matrix} + NaOH \xrightarrow{-H_2O} \begin{matrix} CHCOONa \\ \| \\ CHCOOH \end{matrix} \xrightarrow[-H_2O]{NaOH} \begin{matrix} CHCOONa \\ \| \\ CHCOONa \end{matrix}$$

Maleic acid — Mono sodium maleate — Disodium maleate

2. Formation of esters. Two series of esters viz., mono and dialkyl maleates, are formed with alcohols.

$$\begin{matrix} CHCOOH \\ \| \\ CHCOOH \end{matrix} + C_2H_5OH \xrightarrow[-H_2O]{\text{Conc.}H_2SO_4} \begin{matrix} CHCOOC_2H_5 \\ \| \\ CHCOOH \end{matrix} \xrightarrow[-H_2O]{C_2H_5OH,\ H^+} \begin{matrix} CHCOOC_2H_5 \\ \| \\ CHCOOC_2H_5 \end{matrix}$$

Mono ethyl maleate — Diethyl maleate

3. Action of heat. Maleic acid on heating forms anhydride.

$$\begin{matrix} CHCOOH \\ \| \\ CHCOOH \end{matrix} \xrightarrow[\Delta\ (403\text{ K})]{-H_2O} \begin{matrix} CH-CO \\ \| \\ CH-CO \end{matrix}\rangle O$$

Maleic anhydride

The ease of formation of anhydride suggests that the two carboxylic groups are on the same side of the double bond *i.e.* maleic acid is a *cis* compound. Two carboxylic groups in *trans* positions would make it difficult for dehydration to take place.

4. Isomerisation. On heating continuously for sufficient time at 423 K or exposure to UV light, it changes into fumaric acid, its *trans* isomers, which is more stable.

$$\underset{\text{Maleic acid}}{\begin{array}{c} H-C-COOH \\ \| \\ H-C-COOH \end{array}} \xrightarrow[\Delta]{423\text{ K}} \underset{\text{Fumaric acid}}{\begin{array}{c} H-C-COOH \\ \| \\ HOOC-C-H \end{array}}$$

5. Addition of hydrogen. On reduction with sod-amalgam and water or catalytic hydrogenation in presence of Ni, maleic acid converts into succinic acid.

$$\underset{\text{Maleic acid}}{\begin{array}{c} CHCOOH \\ \| \\ CHCOOH \end{array}} + 2[H] \text{ or } H_2 \xrightarrow[\text{or } H_2,\text{ Ni}]{Na/Hg-H_2O} \underset{\text{Succinic acid}}{\begin{array}{c} CH_2COOH \\ | \\ CH_2COOH \end{array}}$$

6. Addition of bromine

$$\underset{\text{Maleic acid}}{\begin{array}{c} CHCOOH \\ \| \\ CHCOOH \end{array}} + Br_2 \longrightarrow \underset{\substack{\text{Dibromo-}\\ \text{succinic acid}\\ (d \text{ and } l)}}{\begin{array}{c} CHBrCOOH \\ | \\ CHBrCOOH \end{array}}$$

7. Addition of HBr

$$\underset{\text{Maleic acid}}{\begin{array}{c} CHCOOH \\ \| \\ CHCOOH \end{array}} + HBr \longrightarrow \underset{\text{Bromosuccinic acid}}{\begin{array}{c} CH_2COOH \\ | \\ CHBrCOOH \end{array}}$$

8. Oxidation. On oxidation with alkaline potassium permanganate, maleic acid changes into mesotartaric acid.

$$\underset{\text{Maleic acid}}{\begin{array}{c} CHCOOH \\ \| \\ CHCOOH \end{array}} + H_2O + [O] \xrightarrow[KMnO_4]{\text{Alk.}} \underset{\text{Mesotartaric acid}}{\begin{array}{c} CH(OH)COOH \\ | \\ CH(OH)COOH \end{array}}$$

FUMARIC ACID

6.27. METHODS OF PREPARATION OF FUMARIC ACID

1. From maleic acid. On heating for a long time at 423 K maleic acid converts into more stable isomer fumaric acid because of less steric hindrance.

$$\underset{\text{Maleic acid}}{\begin{array}{c} H-C-COOH \\ \| \\ H-C-COOH \end{array}} \xrightarrow[423\text{ K}]{\Delta} \underset{\text{Fumaric acid}}{\begin{array}{c} H-C-COOH \\ \| \\ HOOC-C-H \end{array}}$$

2. By the action of alcoholic potash on bromosuccinic acid

$$\underset{\text{Bromosuccinic acid}}{\begin{array}{c} CHBrCOOH \\ | \\ CH_2COOH \end{array}} \xrightarrow[KOH]{\text{Alc.}} \underset{\text{Fumaric acid}}{\begin{array}{c} H-C-COOH \\ \| \\ HOOC-C-H \end{array}} + KBr + H_2O$$

3. From malonic acid

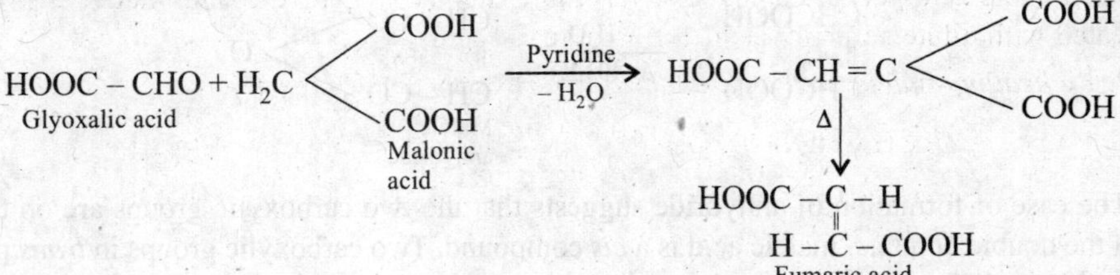

4. Industrial preparation. Fermentation of carbohydrate using *Rhizopus nigricans* produces fumaric acid

$$\text{Carbohydrate} \xrightarrow[\text{nigricans}]{\text{Rhizopus}} \text{Fumaric acid}$$

6.28. PROPERTIES OF FUMARIC ACID

1. Dehydration. Compared to maleic acid, fumaric acid gets dehydrated with greater difficulty at 540 K to give maleic anhydride.

$$\underset{\text{Fumaric acid}}{\begin{array}{c} H-C-COOH \\ \| \\ HOOC-C-H \end{array}} \xrightarrow[-H_2O]{540\ K} \underset{\text{Maleic anhydride}}{\begin{array}{c} CHCO \\ \| \\ CHCO \end{array} > O}$$

2. Reduction

$$\underset{\text{Fumaric acid}}{\begin{array}{c} H-C-COOH \\ \| \\ HOOC-C-H \end{array}} + 2[H] \longrightarrow \underset{\text{Succinic acid}}{\begin{array}{c} CH_2COOH \\ | \\ CH_2COOH \end{array}}$$

3. Addition of Br_2

$$\underset{\text{Fumaric acid}}{\begin{array}{c} H-C-COOH \\ \| \\ HOOC-C-H \end{array}} + Br_2 \longrightarrow \underset{\text{Dibromosuccinic acid}}{\underset{d \text{ and } l}{\begin{array}{c} CHBrCOOH \\ | \\ CHBrCOOH \end{array}}}$$

4. Addition of HBr

$$\underset{\text{Maleic acid}}{\begin{array}{c} CHCOOH \\ \| \\ CHCOOH \end{array}} + HBr \longrightarrow \underset{\text{Bromosuccinic acid}}{\begin{array}{c} CH_2COOH \\ | \\ CH(Br)COOH \end{array}}$$

5. Oxidation. When oxidised with alkaline $KMnO_4$, fumaric acid gives a racemic mixture of *d* and *l* tartaric acids.

$$\underset{\text{Fumaric acid}}{\begin{array}{c} H-C-COOH \\ \| \\ HOOC-C-H \end{array}} + H_2O + O \xrightarrow[KMnO_4]{\text{Alk.}} \underset{\text{Tartaric acid}}{\underset{d \text{ and } l}{\begin{array}{c} CH(OH)COOH \\ | \\ CH(OH)COOH \end{array}}}$$

MALIC ACID, HOOC – CHOH – CH_2 – COOH

6.29. METHODS OF PREPARATION OF MALIC ACID

1. It is best prepared from green mountain-ash berries. The extract from the fruit is treated with lime when malic acid gets precipitated as insoluble calcium malate. Calcium malate is filtered off and treated with dilute sulphuric acid to obtain malic acid.

2. *By heating maleic acid with dilute sulphuric acid under pressure*

$$\underset{\text{maleic acid}}{\begin{array}{c} CH\ COOH \\ \| \\ CH\ COOH \end{array}} + \begin{array}{c} OH \\ | \\ H \end{array} \xrightarrow[\text{Pressure}]{H_2SO_4} \underset{\text{malic acid}}{\begin{array}{c} HO-CH-COOH \\ | \\ CH_2-COOH \end{array}}$$

3. *By partial reduction of tartaric acid with hydriodic acid*

$$\begin{array}{l} HO-CH-COOH \\ \quad\;\; | \\ HO-CH-COOH \end{array} + 2HI \longrightarrow \begin{array}{l} HO-CH-COOH \\ \quad\;\; | \\ \quad\; CH_2-COOH \end{array}$$

tartaric acid — malic acid

4. *Laboratory preparation.* By hydrolysis of bromosuccinic acid with moist silver oxide

$$\begin{array}{l} Br-CH-COOH \\ \quad\;\; | \\ \quad\; CH_2-COOH \end{array} + Ag\,OH \longrightarrow \begin{array}{l} HO-CH-COOH \\ \quad\;\; | \\ \quad\; CH_2-COOH \end{array} + Ag\,Br$$

bromosuccinic acid — malic acid

6.30. PROPERTIES OF MALIC ACID

1. *Reduction with HI.* With HI, succinic acid is obtained

$$\begin{array}{l} HO-CH-COOH \\ \quad\;\; | \\ \quad\; CH_2-COOH \end{array} + 2\,HI \longrightarrow \begin{array}{l} CH_2-COOH \\ | \\ CH_2-COOH \end{array} + H_2O + I_2$$

malic acid — succinic acid

2. *Action of Heat.* On heating it forms fumaric acid, maleic acid and maleic anhydride

$$\begin{array}{l} HO-CH-COOH \\ \quad\;\; | \\ \quad\; CH_2-COOH \end{array} \xrightarrow[-H_2O]{\Delta} \begin{array}{l} HOOC-C-H \\ \qquad\quad \| \\ \quad\;\; H-C-COOH \end{array} \xrightarrow{\Delta} \begin{array}{l} H-C-COOH \\ \quad\;\; \| \\ H-C-COOH \end{array} \xrightarrow{\Delta} \begin{array}{l} H-C-CO \\ \quad\;\; \| \qquad \rangle O \\ H-C-CO \end{array}$$

3. *Oxidation.* On oxidation, it is converted into oxal-acetic acid which exhibits keto-enol tautomerism.

$$\begin{array}{l} HO-CH-COOH \\ \quad\;\; | \\ \quad\; CH_2-COOH \end{array} + [O] \xrightarrow[-H_2O]{} \begin{array}{l} O=C-COOH \\ \qquad | \\ \quad\; CH_2-COOH \end{array} \rightleftharpoons \begin{array}{l} HO-C-COOH \\ \qquad\;\; \| \\ \quad\;\; CH-COOH \end{array}$$

oxal acetic acid (keto form) — (enol form)

TARTARIC ACID

$$\begin{array}{c} COOH \\ | \\ CHOH \\ | \\ CHOH \\ | \\ COOH \end{array}$$

6.31. METHODS OF PREPARATION OF TARTARIC ACID

1. From grape juice. Tartaric acid occurs as potassium hydrogen tartarate in the grape juice. On fermentation, it separates out as a brown mass known as *argol*, which is crystallised to obtain colourless crystals of pure substance, known as *cream of tartar*. This cream of tartar is treated with lime and subsequently with calcium chloride. Calcium tartarate obtained during the process is separated by filtration. This is neutralised with calculated quantities of sulphuric acid. Calcium sulphate produced during the reaction is removed by filtration. Filtrate is concentrated to obtain crystals of dextro –tartaric acid.

$$\underset{\text{Pot.hydrogen tartarate}}{2KHC_4H_4O_6} + \underset{\text{Cal. hydroxide}}{Ca(OH)_2} \longrightarrow \underset{\text{Pot. Tartarate}}{K_2C_4H_4O_6} + \underset{\text{Cal. Tartarate}}{CaC_4H_4O_6} + 2H_2O$$

$$\underset{\text{Pot.tartarate}}{K_2C_4H_4O_6} + CaCl_2 \longrightarrow \underset{\text{Cal. tartarate}}{CaC_4H_4O_6} + 2KCl$$

$$\underset{\text{Calcium tartarate}}{CaC_4H_4O_6} + H_2SO_4 \longrightarrow \underset{\text{(+) Tartaric acid}}{H_2C_4H_4O_4} + CaSO_4\downarrow$$

2. By hydroxylation of maleic and fumaric acids. Maleic acid and fumaric acids on hydroxylation with alkaline pot. permanganate give meso tartaric acid and racemic tartaric acid respectively.

$$\underset{\text{Maleic acid}}{\begin{array}{c} HC-COOH \\ \| \\ HC-COOH \end{array}} + H_2O + O \xrightarrow{KMnO_4} \underset{\text{Mesotartaric acid}}{\begin{array}{c} COOH \\ | \\ H-C-OH \\ | \\ H-C-OH \\ | \\ COOH \end{array}}$$

$$\underset{\text{Fumaric acid}}{\begin{array}{c} H-C-COOH \\ \| \\ HOOC-C-H \end{array}} + H_2O + O \xrightarrow{KMnO_4} \underset{\text{(+) Tartaric acid}}{\begin{array}{c} COOH \\ | \\ H-C-OH \\ | \\ HO-C-OH \\ | \\ COOH \end{array}} + \underset{\text{(–) Tartaric acid}}{\begin{array}{c} COOH \\ | \\ HO-C-H \\ | \\ H-C-OH \\ | \\ COOH \end{array}}$$

3. Synthesis from acetylene

$$\underset{\text{Acetylene}}{\begin{array}{c} CH \\ ||| \\ CH \end{array}} \xrightarrow[BaSO_4]{H_2/pd} \underset{\text{Ethylene}}{\begin{array}{c} CH_2 \\ \| \\ CH_2 \end{array}} + Br_2 \longrightarrow \underset{\text{Dibromo ethane}}{\begin{array}{c} CH_2Br \\ | \\ CH_2Br \end{array}} \xrightarrow[KCN]{Alc.} \underset{\text{Ethylene cyanide}}{\begin{array}{c} CH_2CN \\ | \\ CH_2CN \end{array}}$$

$$\xrightarrow{H_2O/H^+} \begin{array}{c} CH_2COOH \\ | \\ CH_2COOH \end{array} \xrightarrow{Br_2} \underset{\text{Dibromo-succinic acid}}{\begin{array}{c} CHBrCOOH \\ | \\ CHBrCOOH \end{array}} \xrightarrow{AgOH} \underset{\text{Tartaric acid}}{\begin{array}{c} CH(OH)COOH \\ | \\ CH(OH)COOH \end{array}}$$

6.32. PROPERTIES OF TARTARIC ACID

1. Optical isomerism. Due to the presence of two chiral carbon atoms in the molecule of tartaric acid, it exists in two optically active forms and one optically inactive meso form.

2. Formation of salts and esters. As this compound contains two carboxylic groups, treatment with alkalis and alcohols gives salts and esters respectively. Two series of salts and esters are obtained depending upon whether one or two hydrogens of tartaric acid are replaced.

$$\underset{\text{Tartaric acid}}{\begin{array}{c} CH(OH)COOH \\ | \\ CH(OH)COOH \end{array}} + NaOH \xrightarrow[-H_2O]{} \underset{\text{Monosodium tartarate}}{\begin{array}{c} CH(OH)COONa \\ | \\ CH(OH)COOH \end{array}} \xrightarrow[-H_2O]{NaOH} \underset{\text{Disodium tartarate}}{\begin{array}{c} CH(OH)COONa \\ | \\ CH(OH)COONa \end{array}}$$

$$\underset{\text{Tartaric acid}}{\begin{array}{c} CH(OH)COOH \\ | \\ CH(OH)COOH \end{array}} + C_2H_5OH \xrightarrow[-H_2O]{H^+} \underset{\text{Monoethyl tartarate}}{\begin{array}{c} CH(OH)COOC_2H_5 \\ | \\ CH(OH)COOH \end{array}} \xrightarrow[-H_2O]{C_2H_5OH} \underset{\text{Diethyl tartarate}}{\begin{array}{c} CH(OH)COOC_2H_5 \\ | \\ CH(OH)COOC_2H_5 \end{array}}$$

It also forms mixed salts as given below

$$\underset{\text{Sod. pot. tartarate (Rochelle salts)}}{\begin{array}{c} CH(OH)COONa \\ | \\ CH(OH)COOK \end{array}} \qquad \underset{\text{Pot. antimonyl tartarate (Tartar emetic)}}{\begin{array}{c} CH(OH)COO.SbO \\ | \\ CH(OH)COOK \end{array}}$$

3. Dehydrogenation. On treatment with hydrogen peroxide in the presence of ferrous salts, tartaric acid is dehydrogenated to dihydroxy fumaric acid.

$$\underset{\text{Tartaric acid}}{\begin{array}{c} CH(OH)COOH \\ | \\ CH(OH)COOH \end{array}} \xrightarrow[Fe^{2+}]{H_2O_2} \underset{\text{Dihydroxy fumaric acid}}{\begin{array}{c} HO-C-COOH \\ \| \\ HOOC-C-OH \end{array}}$$

4. Action of heat

$$\begin{array}{l} CH(OH)COOH \\ | \\ CH(OH)COOH \end{array} \xrightarrow{\Delta} \underset{\text{Pyruvic acid}}{CH_3-CO-COOH} + CO_2 + H_2O$$

On prolonged heating at 423 K, cyclic anhydride is obtained.

$$\underset{\text{Tartaric acid}}{\begin{array}{l} CH(OH)COOH \\ | \\ CH(OH)COOH \end{array}} \xrightarrow[\text{Prolonged heating}]{\Delta} \begin{array}{l} CH(OH)-C{=}O \\ | \qquad\qquad\quad >O \\ CH(OH)-C{=}O \end{array}$$

5. Reduction. Hydrogen iodide reduces tartaric acid first to malic acid and then to succinic acid

$$\underset{\text{Tartaric acid}}{\begin{array}{l} CH(OH)COOH \\ | \\ CH(OH)COOH \end{array}} \xrightarrow{HI} \underset{\text{Malic acid}}{\begin{array}{l} CH_2COOH \\ | \\ CH(OH)COOH \end{array}} \xrightarrow{HI} \underset{\text{Succinic acid}}{\begin{array}{l} CH_2COOH \\ CH_2COOH \end{array}}$$

6. Oxidation

$$\underset{\text{Tartaric acid}}{\begin{array}{l} CH(OH)COOH \\ | \\ CH(OH)COOH \end{array}} \xrightarrow{[O]} \underset{\text{Tartronic acid}}{\begin{array}{l} CH(OH)COOH \\ | \\ COOH \end{array}} \xrightarrow{[O]} \underset{\text{Oxalic acid}}{\begin{array}{l} COOH \\ | \\ COOH \end{array}}$$

CITRIC ACID

6.33. METHODS OF PREPARATION OF CITRIC ACID

1. From Lemon Juice. Lemon juice is boiled to coagulate proteinous matter which is filtered off. The hot filtrate is neutralised with lime. Citric acid in the filtrate is converted into calcium citrate which is filtered off. Calcium citrate is decomposed to citric acid by treating with a calculated amount of sulphuric acid. Solution containing citric acid is concentrated to crystallise out pure citric acid.

2. From Molasses. Fermentation of molasses with the micro-organism *Aspergillus niger* in the presence of some inorganic salts yields citric acid. The fermentation takes place at a pH of 1-2 and takes about 10 days at a temperature 301-303 K.

3. Synthesis of citric acid from glycerol

$$\underset{\text{Glycerol}}{\begin{array}{l} CH_2OH \\ | \\ CHOH \\ | \\ CH_2OH \end{array}} \xrightarrow{Cl_2} \underset{\text{1, 3-dichloro-propanol-2}}{\begin{array}{l} CH_2Cl \\ | \\ CHOH \\ | \\ CH_2Cl \end{array}} \xrightarrow[HNO_3]{[O]} \underset{\text{1, 3-Dichloro-acetone}}{\begin{array}{l} CH_2Cl \\ | \\ C=O \\ | \\ CH_2Cl \end{array}} \xrightarrow{HCN} \underset{\text{1, 3-Dichloroacetone cyanohydrin}}{\begin{array}{l} CH_2Cl \\ | \\ C{<}^{OH}_{CN} \\ | \\ CH_2Cl \end{array}}$$

$$\xrightarrow{KCN} \begin{array}{l} CH_2CN \\ | \\ C{<}^{OH}_{CN} \\ | \\ CH_2CN \end{array} \xrightarrow{H_2O / H^+} \underset{\text{Citric acid}}{\begin{array}{l} CH_2COOH \\ | \\ C(OH)COOH \\ | \\ CH_2COOH \end{array}}$$

6.34. PROPERTIES OF CITRIC ACID

1. Formation of salts and esters. As it is a tricarboxylic compound, there is a possibility to form three series of salts and esters depending upon whether one, two or three carboxy hydrogens are substituted.

$$\begin{array}{l} CH_2COOH \\ | \\ C(OH)COOH \\ | \\ CH_2COOH \end{array} \xrightarrow{NaOH} \underset{\text{Mono sodium citrate}}{\begin{array}{l} CH_2COONa \\ | \\ C(OH)COOH \\ | \\ CH_2COOH \end{array}} \xrightarrow{NaOH} \underset{\text{Di. sodium citrate}}{\begin{array}{l} CH_2COONa \\ | \\ C(OH)COOH \\ | \\ CH_2COONa \end{array}}$$

$$\xrightarrow{NaOH} \underset{\text{Tri. sodium citrate}}{\begin{array}{l} CH_2COONa \\ | \\ C(OH)COONa \\ | \\ CH_2COONa \end{array}}$$

Similarly, formation of mono, di and tri alkyl citrate takes place on reaction with an alcohol.

2. Acetylation. Acetyl derivatives are obtained with acetic anhydride or acetyl chloride.

$$\underset{\text{Citric acid}}{\begin{array}{l} CH_2COOH \\ | \\ C(OH)COOH \\ | \\ CH_2COOH \end{array}} + CH_3COCl \longrightarrow \underset{\text{Monoacetyl citric acid}}{\begin{array}{l} CH_2COOH \\ | \\ C(OCOCH_3)\ COOH \\ | \\ CH_2COOH \end{array}} + HCl$$

3. Reduction. On reduction with HI, citric acid gives tricarballylic acid.

$$\underset{\text{Citric acid}}{\begin{array}{l} CH_2COOH \\ | \\ C(OH)COOH \\ | \\ CH_2COOH \end{array}} + 2HI \longrightarrow \underset{\text{Tricarballylic acid}}{\begin{array}{l} CH_2COOH \\ | \\ CHCOOH \\ | \\ CH_2COOH \end{array}} + H_2O + I_2$$

4. Action of fuming sulphuric acid

$$\begin{array}{l} CH_2COOH \\ | \\ C(OH)COOH \\ | \\ CH_2COOH \end{array} \xrightarrow[H_2SO_4]{\text{Fuming}} \underset{\text{Acetone dicarboxylic acid}}{\begin{array}{l} CH_2COOH \\ | \\ CO \\ | \\ CH_2COOH \end{array}} + H_2O + CO$$

5. Action of heat. When heated to 423 K, citric acid loses a water molecule forming aconitic acid. This reaction is characteristic of β-hydroxy acids.

$$\begin{array}{l} CH_2COOH \\ | \\ C(OH)COOH \\ | \\ CH_2COOH \end{array} \xrightarrow[-H_2O]{\Delta} \underset{\text{Aconitic acid}}{\begin{array}{l} CHCOOH \\ \| \\ C-COOH \\ | \\ CH_2COOH \end{array}}$$

On heating to higher temperatures, aconitic acid further loses a molecule of carbon dioxide to form a mixture of itaconic acid and citraconic acid and mesaconic acid which change into their anhydrides as shown on page 582.

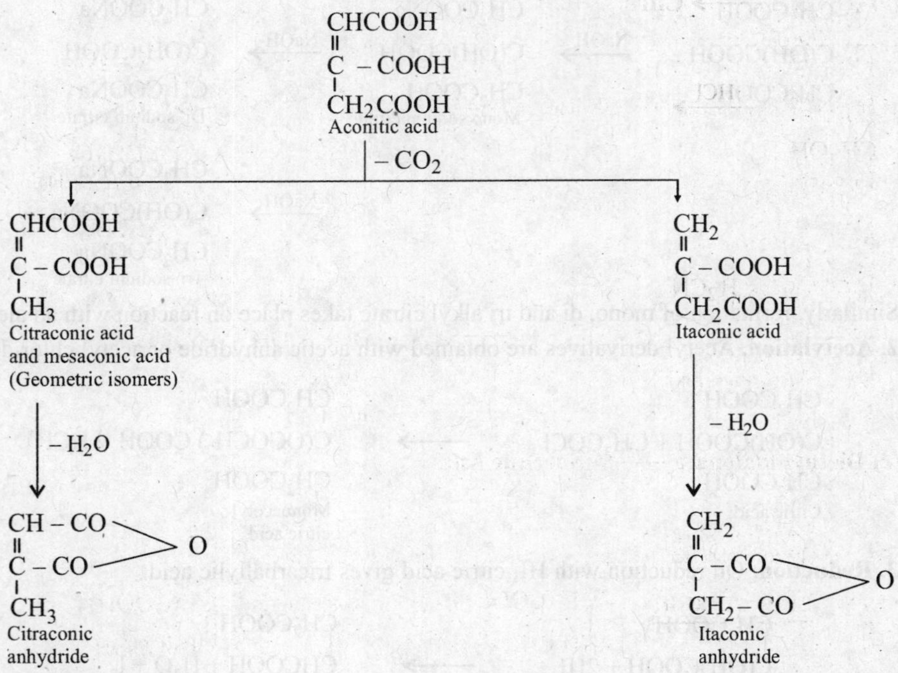

6. Complex Formation. Citric acid prevents the precipitation of heavy metal hydroxides by forming soluble complexes. This property of citric acid is used in the making of *Benedict solution* which contains a mixture of copper sulphate, sodium carbonate and sodium citrate.

SOLVED EXAMPLES

Example 1. How will you convert ?

(*a*) Acetylene ⟶ Tartaric acid

(*b*) Glycerol ⟶ Citric acid

(*c*) Diethyl malonate ⟶ Succinic acid

(*d*) Tartaric acid ⟶ Maleic acid.

Solution. (*a*) Acetylene ⟶ Tartaric acid

$$\underset{\text{Acetylene}}{\overset{CH}{\underset{CH}{|||}}} \xrightarrow[BaSO_4]{H_2/Pd} \underset{\text{Ethylene}}{\overset{CH_2}{\underset{CH_2}{||}}} + Br_2 \longrightarrow \underset{\text{Dibromoethane}}{\overset{CH_2Br}{\underset{CH_2Br}{|}}} \xrightarrow[KCN]{Alc.} \underset{\text{Ethylene cyanide}}{\overset{CH_2CN}{\underset{CH_2CN}{|}}}$$

$$\xrightarrow{HO_2/H^+} \underset{\text{Succinic acid}}{\overset{CH_2COOH}{\underset{CH_2COOH}{|}}} \xrightarrow{Br_2/P} \underset{\text{Dibromo-succinic acid}}{\overset{CH_2BrCOOH}{\underset{CH_2BrCOOH}{|}}} \xrightarrow{AgOH} \underset{\text{Tartaric acid}}{\overset{CH(OH)COOH}{\underset{CH(OH)COOH}{|}}}$$

(b) Glycerol ⟶ Citric acid

$$\underset{\text{Glycerol}}{\begin{array}{l}CH_2OH\\ |\\ CHOH\\ |\\ CH_2OH\end{array}} \xrightarrow{HCl} \underset{\text{1,3 Dichloro-propanol-2}}{\begin{array}{l}CH_2Cl\\ |\\ CHOH\\ |\\ CH_2Cl\end{array}} \xrightarrow[HNO_3]{[O]} \underset{\alpha,\alpha'\text{ dichloro-acetone}}{\begin{array}{l}CH_2Cl\\ |\\ C=O\\ |\\ CH_2Cl\end{array}} \xrightarrow{HCN} \underset{\alpha,\alpha'\text{ Dichloro-acetone cyanohydrin}}{\begin{array}{l}CH_2Cl\\ |\\ C(OH)(CN)\\ |\\ CH_2Cl\end{array}}$$

$$\xrightarrow{KCN} \begin{array}{l}CH_2CN\\ |\\ C(OH)(CN)\\ |\\ CH_2CN\end{array} \xrightarrow{H_2O/H^+} \underset{\text{Citric acid}}{\begin{array}{l}CH_2COOH\\ |\\ C(OH)COOH\\ |\\ CH_2COOH\end{array}}$$

(c) Diethyl malonate ⟶ Succinic acid

$$\underset{\text{Malonic ester}}{CH_2(COOC_2H_5)_2} \xrightarrow{C_2H_5ONa} \underset{\text{Sod. salt}}{[CH(COOC_2H_5)_2]Na}$$

$$\xrightarrow{BrCH_2COOC_2H_5} \begin{array}{l}CH(COOC_2H_5)_2\\ |\\ CH\,COOC_2H_5\end{array} \xrightarrow{\text{Hydrolysis}} \begin{array}{l}CH(COOH)_2\\ |\\ CH_2COOH\end{array}$$

$$\xrightarrow{\Delta} \underset{\text{Succinic acid}}{\begin{array}{l}CH_2COOH\\ |\\ CH_2COOH\end{array}} + CO_2$$

(d) Tartaric acid ⟶ Maleic acid

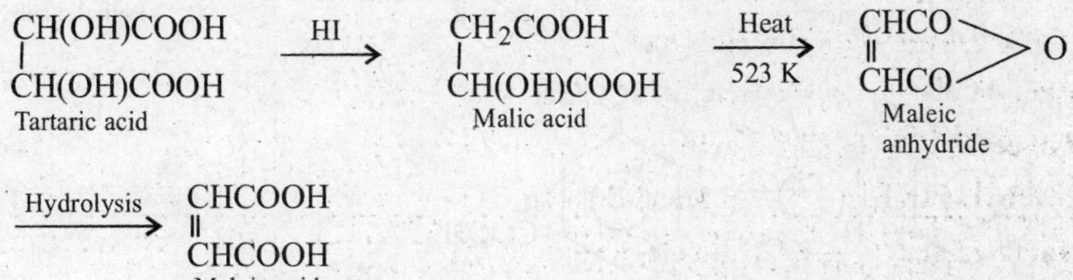

Example 2. Write a note on stereoisomerism in tartaric acid.

Solution. Stereoisomers (optical isomers) of tartaric acid are:

$$\underset{d\text{-tartaric acid}}{\begin{array}{c}COOH\\ |\\ H-C-OH\\ |\\ HO-C-H\\ |\\ COOH\end{array}} \qquad \underset{l\text{-tartaric acid}}{\begin{array}{c}COOH\\ |\\ HO-C-H\\ |\\ H-C-OH\\ |\\ COOH\end{array}} \qquad \underset{meso\text{-tartaric acid}}{\begin{array}{c}COOH\\ |\\ H-C-OH\\ |\\ H-C-OH\\ |\\ COOH\end{array}}$$

Example 3. What is the action of heat on

(*i*) glutaric acid (*ii*) adipic acid ?

Solution. (*i*)

$$\underset{\text{Glutaric acid}}{CH_2\left\langle\begin{array}{l}CH_2COOH\\ CH_2COOH\end{array}\right.} \xrightarrow{\Delta} \underset{\text{Butyric acid}}{CH_3CH_2CH_2COOH} + CO_2\uparrow$$

(*ii*) $\begin{array}{l} CH_2-CH_2COOH \\ | \\ CH_2-CH_2-COOH \end{array} \xrightarrow{\Delta} \underset{\text{Valeric acid}}{CH_3CH_2CH_2CH_2COOH} + CO_2\uparrow$

Example 4. Prove that citric acid is a monohydroxy tribasic acid and has three –COOH groups on different carbon atoms.

Solution. 1. Citric acid forms three series of salts and esters corresponding to successive removal of one, two or three carboxy hydrogens.

2. On treatment with HI, citric acid produces tricarballylic acid which is known to be a tricarboxylic acid with three carboxy groups on different carbon atoms. Action with HI reduces one –OH group to –H.

The above reactions prove that citric acid is a monohydroxy tribasic acid.

Example 5. How will you convert acetic acid into malonic acid ?

Solution.

$$\underset{\text{Acetic acid}}{CH_3COOH} \xrightarrow{Cl_2,\ P} \underset{\text{Chloroacetic acid}}{CH_2\begin{cases} Cl \\ COOH \end{cases}} \xrightarrow{KCN} CH_2\begin{cases} CN \\ COOH \end{cases} \xrightarrow{KMnO_4} \underset{\text{Malonic acid}}{CH_2\begin{cases} COOH \\ COOH \end{cases}}$$

Example 6. Write the names and structures of three isomeric dicarboxylic acids obtained when three isomeric xylenes are oxidized. Indicate, which diacarboxylic acid gives anhydride on dehydration ?

Solution. 1. o-Xylene ($C_6H_4(CH_3)_2$, 1,2-) $\xrightarrow{O}$ Phthalic acid ($C_6H_4(COOH)_2$, 1,2-) $\xrightarrow{\text{dehydration}}$ Phthalic anhydride ($C_6H_4(CO)_2O$)

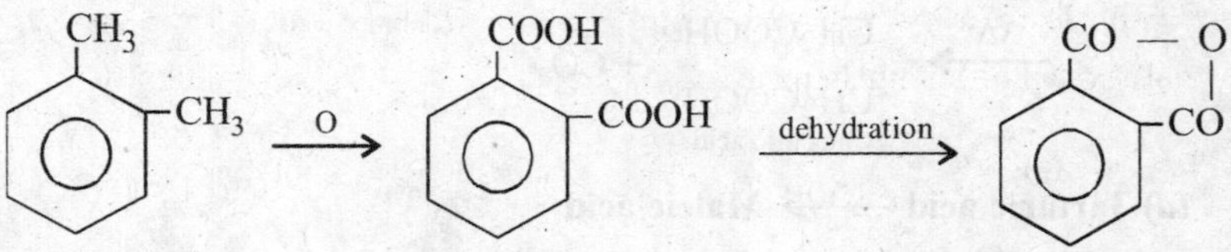

o-Xylene — Phthalic acid — Phthalic anhydride

2. *m*-Xylene (CH_3, CH_3 at 1,3-) $\longrightarrow$ Isophthalic acid ($COOH$, $COOH$ at 1,3-)

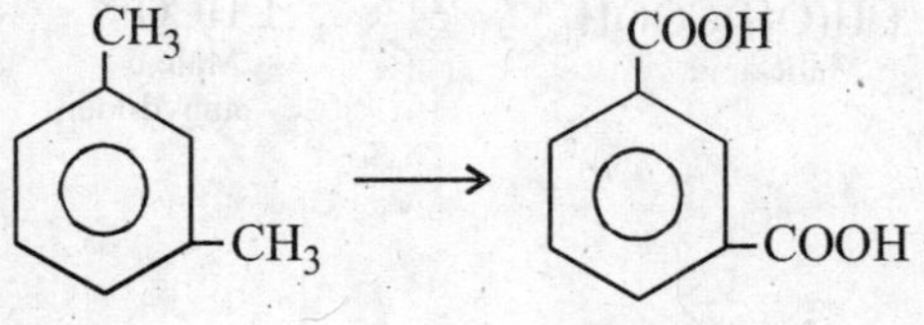

Isophthalic acid

m-Xylene

3. *p*-Xylene (CH_3, CH_3 at 1,4-) $\longrightarrow$ Terephthalic acid ($COOH$, $COOH$ at 1,4-)

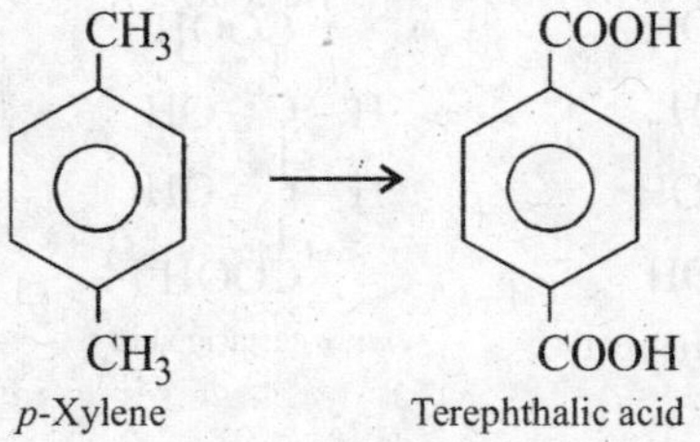

p-Xylene — Terephthalic acid

Example 7. How many optically active stereoisomers of tartaric acid are possible? Draw their configurations.

Solution.

$$\begin{array}{ccc} \text{COOH} & \text{COOH} & \text{COOH} \\ | & | & | \\ \text{HO}-\text{C}-\text{OH} & \text{H}-\text{C}-\text{OH} & \text{H}-\text{C}-\text{OH} \\ | & | & | \\ \text{H}-\text{C}-\text{OH} & \text{HO}-\text{C}-\text{OH} & \text{H}-\text{C}-\text{OH} \\ | & | & | \\ \text{COOH} & \text{COOH} & \text{COOH} \\ \text{I } (l\text{-form}) & \text{II } (d\text{-form}) & \text{III (meso)} \end{array}$$

I and II are optically active forms while III is optically non-active.

Example 8. How will you distinguish between maleic acid and fumaric acid ?

Solution. Maleic acid on heating gives maleic anhydride (m.p. 250°C). No anhydride is formed when furmaric acid is heated.

$$\underset{\text{Maleic acid}}{\begin{array}{c}\text{CHCOOH}\\ \| \\ \text{CHCOOH}\end{array}} \xrightarrow{\text{Heat}} \underset{\text{Maleic anhydride}}{\begin{array}{c}\text{CHCO}\\ \| \\ \text{CHCO}\end{array}\!\!>\!\text{O}} + H_2O$$

Example 9. How will you distinguish between oxalic acid and acetic acid ?

Solution. Oxalic acid gives carbon dioxide on heating which forms white precipitate with lime-water. Acetic acid does not give this test.

$$\underset{\text{Oxalic acid}}{\text{HOOC}-\text{COOH}} \xrightarrow{\Delta} \underset{\text{Formic acid}}{\text{HCOOH}} + CO_2$$

Example 10. How will you synthesise oxalic acid from formic acid ?

Solution. Following steps are involved:

$$\underset{\text{Formic acid}}{\text{H}-\overset{\overset{\text{O}}{\|}}{\text{C}}-\text{OH}} \xrightarrow{\text{NaOH}} \underset{\text{(2 molecules)}}{\text{H}-\overset{\overset{\text{O}}{\|}}{\text{C}}-\text{ONa}} \xrightarrow[300^\circ]{\text{NaOH}} \begin{array}{c}\text{COONa}\\ | \\ \text{COONa}\end{array} \xrightarrow{H_2SO_4} \underset{\text{Oxalic acid}}{\begin{array}{c}\text{COOH}\\ | \\ \text{COOH}\end{array}}$$

Example 11. How will you synthesise malonic acid from acetic acid ?

Solution. Following steps are involved :

$$\underset{\text{Acetic acid}}{CH_3COOH} \xrightarrow[P]{Cl_2} ClCH_2COOH \xrightarrow{\text{NaCN}} CNCH_2COOH$$

$$\xrightarrow{H_2O/H^+} \text{HOOC}-CH_2-\text{COOH (Malonic acid)}$$

Example 12. How will you synthesise maleic acid from succinic acid ?

Solution. Following steps are involved :

$$\underset{\text{Succinic acid}}{\begin{array}{c}CH_2COOH\\ | \\ CH_2COOH\end{array}} \xrightarrow[\text{(HVZ Reaction)}]{Cl_2/P} \begin{array}{c}Cl-CHCOOH\\ | \\ CH_2COOH\end{array} \xrightarrow{\text{alc. KOH}} \underset{\text{Maleic acid}}{\begin{array}{c}\text{CHCOOH}\\ \| \\ \text{CHCOOH}\end{array}}$$

Example 13. How will you distinguish between oxalic acid and succinic acid ?

Solution. Oxalic acid gives carbon dioxide on heating, which forms white precipitate with lime-water. No carbon dioxide is evolved when succinic acid is heated. (When succinic acid is heated, succinic anhydride is formed.)

$$\underset{\text{Oxalic acid}}{HOOC-COOH} \xrightarrow{\Delta} \underset{\text{Formic acid}}{HCOOH + CO_2}$$

$$CO_2 + Ca(OH)_2 \longrightarrow \underset{\text{white ppt.}}{CaCO_3} + H_2O$$

Example 14. How will you synthesise succinic acid from ethylene ?

Solution. Following steps are involved :

$$\underset{\text{Ethylene}}{\begin{matrix} CH_2 \\ \| \\ CH_2 \end{matrix}} \xrightarrow{Br_2} \begin{matrix} CH_2Br \\ | \\ CH_2Br \end{matrix} \xrightarrow{NaCN} \begin{matrix} CH_2CN \\ | \\ CH_2CN \end{matrix} \xrightarrow{H_2O/H^+} \underset{\text{Succinic acid}}{\begin{matrix} CH_2COOH \\ | \\ CH_2COOH \end{matrix}}$$

Example 15. How will you synthesise oxalic acid from ethylene ?

Solution. Following steps are involved:

$$\underset{\text{Ethylene}}{\begin{matrix} CH_2 \\ \| \\ CH_2 \end{matrix}} \xrightarrow[250°]{O_2/Ag} \begin{matrix} H_2C \\ | \\ H_2C \end{matrix}\!\!>\!O \xrightarrow{H_2O/H^+} \begin{matrix} CH_2OH \\ | \\ CH_2OH \end{matrix} \xrightarrow{KMnO_4/H^+} \underset{\text{Oxalic acid}}{\begin{matrix} COOH \\ | \\ COOH \end{matrix}}$$

Example 16. How will you synthesise adipic acid from benzene ?

Solution. Following steps are involved :

$$\underset{\text{Benzene}}{C_6H_6} \xrightarrow[\Delta]{3H_2/Ni} C_6H_{12} \xrightarrow[H_3BO_3/\Delta]{O_2} C_6H_{11}OH \xrightarrow[[O]]{HNO_3} C_6H_{10}O \xrightarrow[[O]]{HNO_3} \underset{\text{Adipic acid}}{\begin{matrix} CH_2CH_2COOH \\ | \\ CH_2CH_2COOH \end{matrix}}$$

Example 17. How will you synthesise tartaric acid from ethylene ?

Solution. Following steps are involved :

$$\underset{\text{Ethylene}}{\begin{matrix} CH_2 \\ \| \\ CH_2 \end{matrix}} \xrightarrow{Br_2} \begin{matrix} CH_2Br \\ | \\ CH_2Br \end{matrix} \xrightarrow{NaCN} \begin{matrix} CH_2CN \\ | \\ CH_2CN \end{matrix} \xrightarrow{H_2/HO^+} \begin{matrix} CH_2COOH \\ | \\ CH_2COOH \end{matrix}$$

$$\xrightarrow[\text{HVZ reaction}]{Br_2/P} \begin{matrix} BrCHCOOH \\ | \\ BrCHCOOH \end{matrix} \xrightarrow[(AgOH)]{\text{moist } Ag_2O} \underset{\text{Tartaric acid}}{\begin{matrix} HO-CH-COOH \\ | \\ HO-CH-COOH \end{matrix}}$$

Example 18. Compound (A), $C_4H_6O_4$, when heated with acetic anhydride gives (B), $C_4H_4O_3$. Compound (B) reacts with ammonia to give (C), $C_4H_{12}O_4N_2$. Compound (C) on heating to 300°C forms (D), $C_4H_5O_2N$. Both (B) and (C) can be hydrolysed to (A). Identify (A), (B), (C), and (D).

Solution. (A) = Succinic acid

(B) = Succinic anhydride

(C) = Ammonium succinate

(D) = Succinimide

Example 19. How will you synthesise N-bromosuccinimide from succinic acid ?

Solution. Following steps are involved :

$$\underset{\text{Succinic acid}}{\begin{matrix} CH_2COOH \\ | \\ CH_2COOH \end{matrix}} \xrightarrow{2NH_3} \begin{matrix} CH_2COONH_4 \\ | \\ CH_2COONH_4 \end{matrix} \xrightarrow{\Delta} \begin{matrix} CH_2-CO \\ | \\ CH_2-CO \end{matrix}\!\!>\!NH \xrightarrow[NaOH]{Br_2} \underset{\substack{N\text{-Bromosuccinimide} \\ \text{(NBS)}}}{\begin{matrix} CH_2-CO \\ | \\ CH_2-CO \end{matrix}\!\!>\!NBr}$$

Example 20. How will you distinguish between citric acid and tartaric acid ?

Solution. When tartaric acid is treated with ammoniacal silver nitrate solution (Tollens' reagent), silver mirror is formed. Citric acid does not give this test.

$$\underset{\text{Tartaric acid}}{\begin{matrix} CH(OH)COOH \\ | \\ CH(OH)COOH \end{matrix}} + 2Ag(NH_3)_2NO_3 \longrightarrow \begin{matrix} CH(OH)CO\bar{O} \\ | \\ CO\bar{O} \end{matrix} + \underset{\substack{\text{silver} \\ \text{mirror}}}{2Ag} + 2NH_4NO_3 + 2NH_3$$

Example 21. Justify the observation that *o*-hydroxybenzoic acid is a stronger acid than *o*-methoxybenzoic acid.

Solution. This can be explained in terms of relative ease of formation of carboxylate ion from *o*-hydroxy benzoic acid due to stabilisation by intra molecular hydrogen bonding as shown below:

Stabilised by hydrogen bonding

No hydrogen bonding possible

In case of *o*-methoxybenzoic acid, no such stabilisation of the carboxylate ion takes place.

Example 22. Peroxy acids are found to be weaker than parent carboxylic acids. Explain.

Solution. Peroxy acids are found to be weaker than parent carboxylic acids because the former do not ionise easily as the anions formed by them are not resonance stabilised. On the other hand, the anions formed by carboxylic acids are stabilised by resonance which favours their formation.

$$CH_3COOH \overset{+H_2O}{\rightleftharpoons} CH_3COO^- + H_3O^+$$

Resonance stabilised; more easily formed.

$$CH_3-CO-O-OH \overset{+H_2O}{\rightleftharpoons} CH_3CO-O-O^- + H_3O^+$$

No resonance stablisation; less easily formed.

Example 23. 2, 4, 6-Trimethylbenzoic acid is very difficult to esterify directly. Explain.

Solution. In 2, 4, 6-trimethylbenzoic acid, the carboxylic group is not easily attacked by alcohol due to strong steric hindrance caused by the presence of three methyl groups attached to benzene ring. Hence its direct esterification does not take place.

EXERCISES
OBJECTIVE QUESTIONS

Multiple Choice Questions (Choose the correct option)

1. In the conversion of wine to vinegar
(*a*) Ethanol is oxidised to acetic acid (*b*) Ethanol is reduced to acetic acid
(*c*) Methanol is oxidised to acetic acid (*d*) Methanol is reduced to acetic acid

2. In succinic acid, $HOOC(CH_2)_nCOOH$, *n* is equal to
(*a*) 1 (*b*) 2 (*c*) 3 (*d*) 4

3. Which compound has the highest boiling point ?
(*a*) CH_3CH_3 (*b*) CH_3OCH_3 (*c*) CH_3CH_2OH (*d*) CH_3COOH

4. The greater acidity of carboxylic acids compared to alcohols arises primarily from :
(*a*) the electron-donating effect of the hydroxyl group
(*b*) the electron-withdrawing effect of the carboxyl oxygen
(*c*) the acidity of α-hydrogens of carboxylic acids
(*d*) the resonance stability associated with the carboxylate ion

5. The compound in which hydrogen bonding is not possible is
(*a*) CH_3OCH_3 (*b*) H_2O (*c*) CH_3CH_2OH (*d*) CH_3COOH

6. Which of the following statements is false about the acid-strength of acetic acid ?
(*a*) Acetic acid is a stronger acid than monochloroacetic acid.
(*b*) Acetic acid is a stronger acid than propionic acid.
(*c*) Acetic acid is a weaker acid than trichloroacetic acid.
(*d*) Acetic acid is a weaker acid than formic acid.

7. Check the incorrect statement :
(*a*) Acetic acid is present in sour milk
(*b*) Formic acid is present in insect bites
(*c*) Tartaric acid is present in grapes
(*d*) Citric acid is a tricarboxylic acid

8. Vinegar is a
(*a*) 5% solution of acetic acid in water (*b*) 25% solution of acetic acid in water
(*c*) 50% solution of acetic acid in water (*d*) 40% solution of formic acid in water

9. When carboxylic acids and dicarboxylic acids have similar molecular weights, how do their melting points compare ?
(*a*) Carboxylic acids have greater melting points.
(*b*) Dicarboxylic acids have greater melting points.
(*c*) Both acids have similar melting points.
(*d*) No consistent trend exists.

10. Which of the following compounds will react with Tollens' reagent to give metallic silver?
(*a*) Formic acid (*b*) Ethyl alcohol (*c*) Acetic acid (*d*) Acetone

11. Acetic acid reacts with methyl alcohol in the presence of an acid catalyst to give
(*a*) Methyl formate (*b*) Ethyl formate (*c*) Methyl acetate (*d*) Ethyl acetate

12. Hunsdiecker reaction is used for the preparation of
(*a*) Alkyl chlorides and bromides (*b*) Alkyl nitrates and nitrites
(*c*) Ketenes (*d*) Alcohols

13. Ammonium acetate on strong heating gives
(*a*) Urea (*b*) Formamide (*c*) Uric acid (*d*) Acetamide

14. Which of the following compounds on treatment with $NaHCO_3$ will liberate CO_2 ?
(*a*) Acetic acid (*b*) Ethylamine (*c*) Acetone (*d*) Ethyl alcohol

15. Butyric acid reacts with PCl_5 to give
(*a*) Benzoyl chloride (*b*) 1-Chlorobutane
(*c*) Butyryl chloride (*d*) 1-Chloropropane

16. Which reaction does not yield an ester as one of the products ?
(*a*) A carboxylic acid is heated with an alcohol
(*b*) A Grignard reagent is added to a carboxylic acid
(*c*) An acid halide is treated with an alcohol
(*d*) An alkyl halide is heated with the salt of a carboxylic acid

17. Acetic acid undergoes reduction with $LiAlH_4$ to give
(*a*) Ethanol (*b*) Ethane (*c*) Ethanal (*d*) Ethyne

18. Which of the following reagents will convert acetic into acetyl chloride ?
(*a*) NaCl (*b*) $HCl/ZnCl_2$ (*c*) $SOCl_2$ (*d*) HCl

19. Calcium acetate on strong heating gives
(*a*) Methane + $CaCO_3$ (*b*) Ethane + $CaCO_3$
(*c*) Acetone + $CaCO_3$ (*d*) Ethane + CaO

20. Silver acetate reacts with Br_2 to form methyl bromide, carbon dioxide, and AgBr. This is an example of
(*a*) Wurtz reaction (*b*) Etard's reaction
(*c*) Hunsdiecker reaction (*d*) Perkin reaction

ANSWERS

1. (*a*)	**2.** (*b*)	**3.** (*d*)	**4.** (*d*)	**5.** (*a*)	**6.** (*a*)
7. (*a*)	**8.** (*a*)	**9.** (*b*)	**10.** (*a*)	**11.** (*c*)	**12.** (*a*)
13. (*d*)	**14.** (*a*)	**15.** (*c*)	**16.** (*b*)	**17.** (*a*)	**18.** (*c*)
19. (*c*)	**20.** (*c*)				

Short Answer Questions

1. Compare the acidic strength of *o*-chlorobenzoic acid and its *p*-isomer.

2. Giving reasons, arrange the following in the increasing order of their acidity : 2-Chloropropionic acid, 2-fluoropropionic acid and propionic acid.

3. Account for the fact that C – O bond length is shorter in RCOOH than in ROH.

4. Benzoic acid reacts with sodium bicarbonate to give CO_2 but phenols do not. Justify the statement.

5. Arrange $CH_3 - CHCl - COOH$, CH_3COOH, $F - CH_2 - COOH$ in order of increasing acidity.

6. Arrange the following in order of their acidic strength :
p-nitrobenzoic acid, *o*-nitrobenzoic acid, *m*-nitrobenzoic acid.

7. Giving reasons, arrange the following in order of increasing acidity :

[Structures: benzoic acid (C_6H_5COOH); p-nitrobenzoic acid (O_2N–C_6H_4–COOH); m-nitrobenzoic acid (COOH, NO_2)]

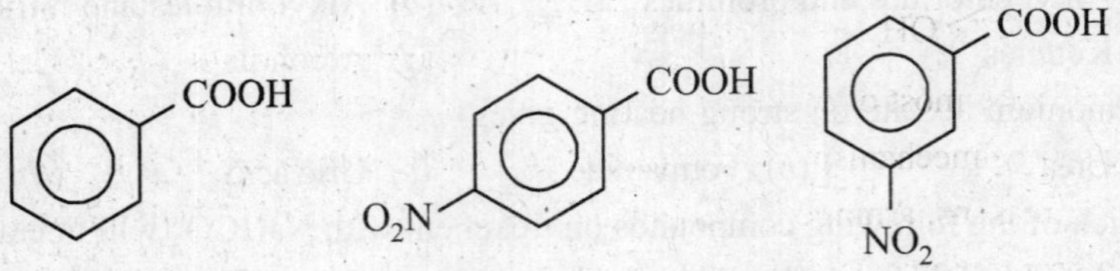

8. $(CH_3)_3C - COOH$ does not get esterified in acidic medium. Why ?

9. Why does a carboxylic acid not form an oxime or phenyl hydrazone ?

10. Complete the following :

(*i*) $C_6H_5COONa + NaOH \xrightarrow{CaO}$

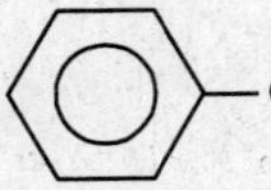

(*ii*) $CH_3CH_2COOH \xrightarrow{Br_2/P}$

11. Arrange the following in order of increasing acidity giving reasons :

(*i*) FCH_2COOH (*ii*) CH_3COOH

(*iii*) $ClCH_2COOH$ (*iv*) $HCOOH$

12. Out of acetic acid and *n*-propyl alcohol, which is expected to have higher boiling point and why ?

13. How are carboxylic acids obtained from CO ?

14. Complete the following :

(*i*) $CH_3CH_2COOH \xrightarrow{Br_2/P}$

(*ii*) $C_6H_5COONa + NaOH \xrightarrow{CaO}$

15. Complete the following reactions, giving the names of main products:

$$C_6H_5COOH + HNO_3 \xrightarrow{H_2SO_4}$$

$$C_6H_5CH_2CH_2COOH + Cl_2 \xrightarrow{Red\ P}$$

16. *o*-Hydroxybenzoic acid is a stronger acid than *o*-methoxy benzoic acid. Explain.

17. Arrange the following acids in increasing order of their acidic characteristics :
CH_3COOH, CCl_3COOH, $CH_2ClCOOH$, $CH_3 - CHBr - COOH$

18. Arrange the following acids in descending order of their acidic strengths : *p*-Toluic acid, *p*-Bromobenzoic acid, *m*-Bromobenzoic acid, Benzoic acid, Phenylacetic acid.

19. Chloroacetic acid is a stronger acid than acetic acid while acetic acid itself is weaker than formic acid. Justify.

20. Why is benzoic acid weaker than formic acid ?

21. Why do alcohols behave as weaker acids than carboxylic acid ?

22. *p*-Nitrobenzoic acid is stronger than benzoic acid. Explain.

23. Out of $CH_2ClCOOH$ and $CHCl_2COOH$, which is stronger and why ?

24. Which is the stronger acid in each of the following pairs ?

(*i*) $H_3C-C(=O)-OH$ or $ClH_2C-C(=O)-OH$

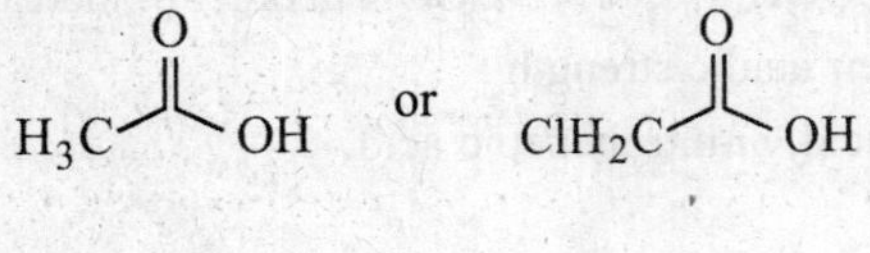

(*ii*) $H_3C-C(=O)-OH$ or $Me_3\overset{+}{N}H_3C-C(=O)-OH$

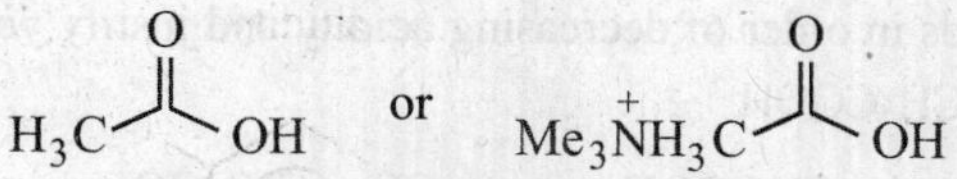

25. Explain why most of carboxylic acids exist as cyclic dimers ?

26. Discuss the mechanism of esterification of acetic acid with ethyl alcohol.

27. Giving reasons, arrange the following in increasing order of their acidity.
2-Chloropropionic acid, 2-fluoropropionic acid and propionic acid.

28. Which is a stronger acid and why ?
(*i*) $CH_2ClCOOH$ and $CH_2BrCOOH$
(*ii*) $CH_3CH_2CHClCOOH$ and $CH_3CHCl\ CH_2COOH$

29. How will you explain that :
(*i*) monocarboxylic acid is stronger than phenol ?
(*ii*) Fluoroacetic acid is stronger than chloroacetic acid ?

30. Complete the following reactions :

(*i*) $CH_3COOH + CH_3COCl \xrightarrow{Py}$

(*ii*) $C_6H_5COOH \xrightarrow[\text{Heat}]{HNO_3/H_2SO_4}$

(*iii*) $CH_3CONH_2 + Br_2 + 4KOH \longrightarrow$

31. Give the mechanism of esterification.

32. Write a short note on acidity of carboxylic acids.

33. Give the mechanism of decarboxylation.

34. Write briefly about the following :
(*i*) Hell-Volhard-Zelinsky reaction.
(*ii*) *p*-Nitrobenzoic acid is stronger than benzoic acid.

35. Sketch the mechanism of the following reaction :

$$CH_3COOH + CH_3CH_2OH \xrightarrow{H^+} CH_3COOC_2H_5 + H_2O$$

General Questions

1. Give one method of preparation (each method should employ different reaction) of each of the following :
(*i*) *n*-Valeric acid (*ii*) Propionic acid
(*iii*) Benzoic acid (*iv*) Phenyl acetic acid

2. Write short notes on
(*i*) HVZ reaction (*ii*) Hunsdiecker reaction
(*iii*) Acidity of carboxylic acids

3. Write the equation for the reaction, if any, of valeric acid $CH_3(CH_2)_3COOH$ with each of the following :
(*i*) KOH (*ii*) $NaHCO_3$
(*iii*) NH_3, then heat (*iv*) $LiAlH_4$, then H_2O, H^+
(*v*) PCl_5 (*vi*) *n*-Propyl alcohol, H^+
(*vii*) $SOCl_2$

4. Arrange the following compounds in order of decreasing acidity and justify your answer :

(*i*) $ClCH_2COOH$, $Cl_2CHCOOH$, CH_3COOH

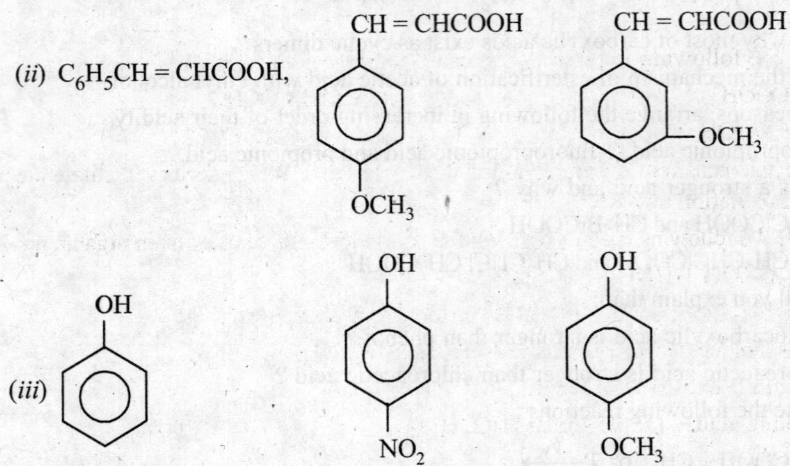

5. Show with the help of equations, how would you prepare each of the following carboxylic acids from organometallic compounds :

(*i*) 2-Methylbutanoic acid (*ii*) 3-Methylbutanoic acid

(*iii*) Phenyl acetic acid (*iv*) Benzoic acid

(*v*) Butyric acid

6. Complete the following :

(*i*) $C_6H_5COONa + NaOH \xrightarrow{CaO}$

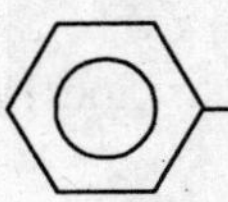

(*ii*) $CH_3CH_2COOH \xrightarrow{Br_2/P}$

(*iii*) $C_6H_5COOH + SOCl_2 \longrightarrow$

7. Why do carboxylic acids behave as acids whereas alcohols don't although both have an – OH as a part of their structures ?

8. Comment upon the structure of carboxyl group and discuss the relative acidities of various carboxylic acids. How will you convert a simple aliphatic carboxylic acid into its acid chloride, anhydride, amide and ethyl ester ?

9. Which acid of each pair shown below would you expect to be stronger ? Why ?

(*i*) CH_2FCOOH or $CH_2ClCOOH$

(*ii*) $HOCH_2COOH$ or NO_2CH_2COOH

(*iii*) $CH_2Cl\ CH_2COOH$ or $CH_3CH_2CHClCOOH$

(*iv*) $(CH_3)_3N^+-C_6H_4-COOH$ or C_6H_5-COOH

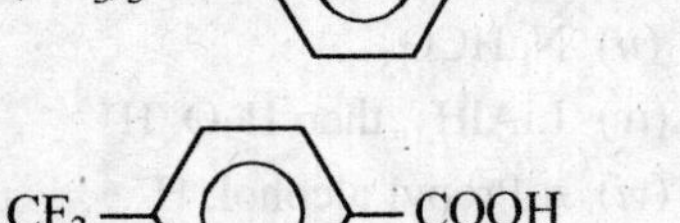

(*v*) $CF_3-C_6H_4-COOH$ or $CH_3-C_6H_4-COOH$

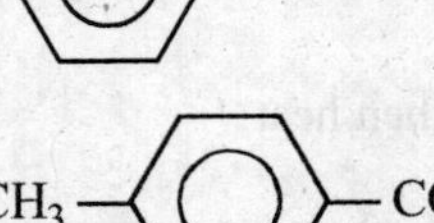

(*vi*) C_6H_5–COOH or $CH_3-CH_2-C_6H_4-COOH$

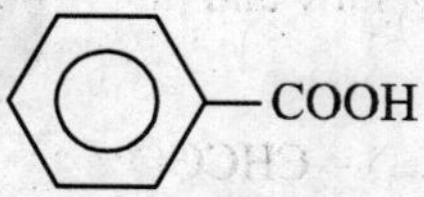

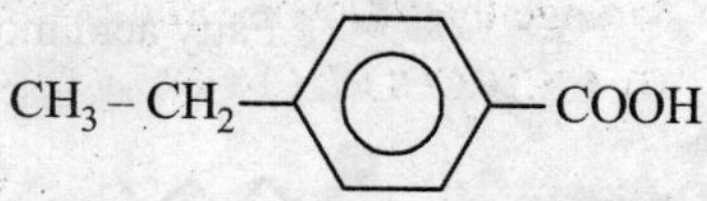

10. Arrange the following in order of increasing acidity giving reasons :

(*i*) CH_3COOH (*ii*) CCl_3COOH

(*iii*) $CH_2ClCOOH$ (*iv*) $CH_3CHBrCOOH$

11. Give the mechanism of decarboxylation of simple acids. Which factors facilitate the process of decarboxylation and why ?

12. Complete the following reactions giving the names and structures of main organic products :

(*i*) $C_6H_5COOH + HNO_3 + H_2SO_4$

(*ii*) $C_6H_5CH_2CH_2COOH + Cl_2$ in the presence of phosphorus

(*iii*) Cyclohexane carboxylic acid + $C_2H_5OH + H_2SO_4$

(*iv*) Sodium benzoate + Soda lime

(*v*) *p*-Toluic acid + $LiAlH_4$, then H_2O, H^+

(*vi*) Butanoic acid + NH_3, then heat.

13. Explain the following :

(*i*) 2, 4, 6-Trimethylbenzoic acid is very difficult to esterify.

(*ii*) Why *p*-nitrobenzoic acid is stronger than *m*-nitrobenzoic acid ?

14. (*a*) What is acidity of carboxylic acid due to ? Why is it stronger acid as compared to alcohols ?

(*b*) Describe the effect of substituents on the acidity of an acid. How can acidity of acids be measured ?

15. What are carboxylic acids ? Draw the resonating structures of carboxylate ion. How are these converted into esters and acid anhydrides ?

16. Discuss the mechanism of acidic and alkaline hydrolysis of acid derivatives.

Glycerin

Fatty acid molecule

Fatty acid molecule

Fatty acid molecule

7

Carboxylic Acid Derivatives

7.1. GENERAL

A carboxylic acid derivative is compound which is derived or obtained from a carboxylic acid by performing a reaction.

There are four different types of derivatives which we commonly come across. These are:

1. Esters
2. Acid Anhydrides
3. Acid Chlorides
4. Acid Amides.

The four types of acid derivatives are obtained replacing the –OH group of the carboxylic acid by – OR′, –OCOR, –Cl and – NH_2 groups, respectively.

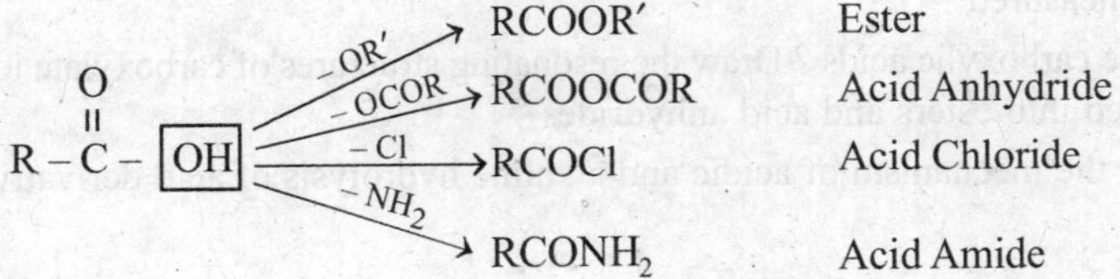

7.2. NOMENCLATURE OF ESTERS

Esters are named as follows by two different systems of nomenclature:

1. Common System. As an ester is obtained by the combination of a carboxylic acid and an alcohol (or phenol), it is named accordingly. The alkyl part of the alcohol is written first followed by the acid part (with the ending – *ate*).

2. IUPAC System. Naming an ester according to this system is not much different from the common system, except that the IUPAC name of the relevant carboxylic acid is used.

Names of a few compounds according to both systems of nomenclature are given in Table 7.1.

Table 7.1. Common and IUPAC names of esters

Compound	*Common Name*	*IUPAC Name*
$CH_3COOC_2H_5$	Ethyl acetate	Ethyl Ethanoate
$HCOOC_2H_5$	Ethyl formate	Ethyl methanoate
$CH_3COOC_6H_5$	Phenyl acetate	Phenyl ethanoate
$C_6H_5COOC_6H_5$	Phenyl benzoate	Phenyl benzoate
$C_6H_5COOCH_3$	Methyl benzoate	Methyl benzoate

7.3. METHODS OF PREPARATION OF ESTER

Esters can be prepared by the following general methods:

1. By the combination of a carboxylic acid and an alcohol (or phenol)

$$\underset{\text{Carboxylic acid}}{RCOOH} + \underset{\text{Alcohol}}{R'OH} \xrightarrow{H^+} RCOOR' + H_2O$$

$$CH_3COOH + C_2H_5OH \xrightarrow{H^+} \underset{\text{Ethyl ethanoate}}{CH_3COOC_2H_5} + H_2O$$

$$\underset{\text{Benzoic acid}}{C_6H_5COOH} + CH_3OH \xrightarrow{H^+} \underset{\text{Methyl benzoate}}{C_6H_5COOCH_3} + H_2O$$

$$CH_3COOH + \underset{\text{Phenol}}{C_6H_5OH} \xrightarrow{H^+} \underset{\text{Phenyl acetate}}{CH_3COOC_6H_5} + H_2O$$

Esters are condensation products of acids and alcohols

This reaction is known as esterification. Its mechanism is described in the chapter on carboxylic acids.

2. By the combination of acid chlorides or anhydrides with alcohols or phenols

$$\underset{\text{Acid chloride}}{RCOCl} + \underset{\text{Alcohol or phenol}}{R'OH} \longrightarrow \underset{\text{Ester}}{RCOOR'} + HCl$$

$$\underset{\text{Acid anhydride}}{(RCO)_2O} + \underset{\text{Alcohol or phenol}}{R'OH} \longrightarrow \underset{\text{Ester}}{RCOOR'} + \underset{\text{Acid}}{RCOOH}$$

$$\underset{\text{Ethanoyl chloride}}{CH_3COCl} + \underset{\text{Ethanol}}{C_2H_5OH} \longrightarrow \underset{\text{Ethyl ethanoate}}{CH_3COOC_2H_5} + HCl$$

$$\underset{\text{Ethanoic anhydride}}{(CH_3CO)_2O} + C_2H_5OH \longrightarrow \underset{\text{Ethyl ethanoate}}{CH_3COOC_2H_5} + CH_3COOH$$

$$\underset{\text{Ethanoyl chloride}}{CH_3COCl} + \underset{\text{Phenol}}{C_6H_5OH} \longrightarrow \underset{\text{Phenyl ethanoate}}{CH_3COOC_6H_5} + HCl$$

Unlike esterification involving a carboxylic acid and an alcohol, these reactions do not involve equilibrium between the reactants and products. Hence we can expect better yields in such reactions.

3. Trans-esterification. Treatment of an ester with an alcohol different from the constituent alcohol part of the ester results in the displacement reaction where the two alcohols are interchanged.

$$\underset{\text{Ethyl acetate}}{CH_3COOC_2H_5} + \underset{\text{Butyl alcohol}}{C_4H_9OH} \overset{H^+}{\rightleftharpoons} \underset{\text{Butyl acetate}}{CH_3COOC_4H_9} + C_2H_5OH$$

4. By heating silver salt of an acid with an alkyl halide

$$\underset{\text{Silver salt of acid}}{RCOOAg} + \underset{\text{Alkyl halide}}{R'X} \longrightarrow \underset{\text{Ester}}{RCOOR'} + \underset{\text{Silver halide}}{AgX}$$

$$\underset{\text{Silver acetate}}{CH_3COOAg} + \underset{\text{Ethyl bromide}}{C_2H_5Br} \longrightarrow \underset{\text{Ethyl acetate}}{CH_3COOC_2H_5} + AgBr$$

5. Reaction of a carboxylic acid with diazomethane

$$\underset{\text{Acetic acid}}{CH_3COOH} + \underset{\text{Diazomethane}}{CH_2N_2} \xrightarrow{\text{Ether}} \underset{\text{Methyl acetate}}{CH_3COOCH_3} + N_2\uparrow$$

$$\underset{\text{Benzoic acid}}{C_6H_5COOH} + CH_2N_2 \xrightarrow{\text{Ether}} \underset{\text{Methyl benzoate}}{C_6H_5COOCH_3} + N_2\uparrow$$

7.4. PHYSICAL PROPERTIES OF ESTERS

1. Sweet smell and fragrance in fruits and flowers is due to the presence of esters. Esters present in various substances are as follows:

Fruit	Ester
Orange	Octyl acetate
Banana	Isoamyl acetate
Apple	Isoamyl valerate
Pineapple	Methyl butyrate
Apricot	Amyl butyrate

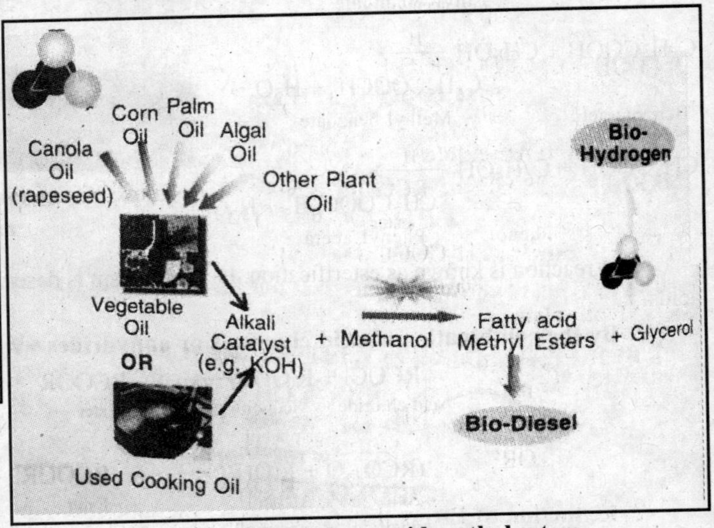

Biodiesels are fatty acid methyl esters

2. Esters are insoluble in water and show lower boiling points than the corresponding acids.

This is due to absence of hydrogen bonding in esters.

7.5. CHEMICAL PROPERTIES OF ESTERS

1. Nucleophilic substitution. Esters undergo nucleophilic substitution that is typical of carboxylic acid derivatives. Attack occurs at electron-deficient acyl carbon and results in the replacement of – OR′ group by – OH, – OR″ or – NH_2.

$$R-\overset{\overset{O}{\|}}{C}-OR' + :Z^- \longrightarrow R-\underset{\underset{OR'}{|}}{\overset{\overset{\bar{O}}{|}}{C}}-Z \longrightarrow R-\overset{\overset{O}{\|}}{C}-Z + :\bar{O}-R'$$

The following reactions are examples of nucleophilic substitution:

Hydrolysis. Esters on hydrolysis in the presence of an alkali give the alcohol (or phenol) and the sodium salt of the carboxylic acid.

$$\underset{\text{Ester}}{RCOOR'} + NaOH \longrightarrow RCOONa + \underset{\text{Alcohol}}{R'OH}$$

$$\underset{\text{Ethyl acetate}}{CH_3COOC_2H_5} + NaOH \longrightarrow \underset{\text{Sod. acetate}}{CH_3COONa} + C_2H_5OH$$

In the presence of an inorganic acid, the hydrolysis takes place as follows:

$$\underset{\text{Ethyl acetate}}{CH_3COOC_2H_5} + H_2O \xrightarrow{H^+} \underset{\text{Acetic acid}}{CH_3COOH} + C_2H_5OH$$

The mechanism of hydrolysis of an ester in the presence of an acid and an alkali is explained in the chapter on carboxylic acids.

Action of PCl_5 or $SOCl_2$. Esters, on treatment with phosphorus pentachloride or thionyl chloride, are converted into acid chloride.

$$\underset{\text{Ester}}{RCOOR'} + PCl_5 \longrightarrow \underset{\text{Acid chloride}}{RCOCl} + POCl_3 + R'Cl$$

$$\underset{\text{Ester}}{R\,COOR'} + \underset{\text{Thionyl chloride}}{SOCl_2} \longrightarrow \underset{\text{Acid chloride}}{RCOCl} + R'Cl + SO_2$$

$$\underset{\text{Ethyl acetate}}{CH_3COOC_2H_5} + PCl_5 \longrightarrow \underset{\text{Acetyl chloride}}{CH_3COCl} + \underset{\text{Ethyl chloride}}{C_2H_5Cl} + POCl_3$$

$$\underset{\text{Ethyl acetate}}{CH_3COOC_2H_5} + SOCl_2 \longrightarrow \underset{\text{Acetyl chloride}}{CH_3COCl} + \underset{\text{Ethyl chloride}}{C_2H_5Cl} + SO_2$$

Action of Ammonia

$$\underset{\text{Ester}}{RCOOR'} + NH_3 \longrightarrow \underset{\text{Amide}}{RCONH_2} + \underset{\text{Alcohol}}{R'OH}$$

$$\underset{\text{Ethyl acetate}}{CH_3COOC_2H_5} + NH_3 \longrightarrow \underset{\text{Acetamide}}{CH_3CONH_2} + \underset{\text{Ethanol}}{C_2H_5OH}$$

Mechanism

$$R-\overset{\overset{\ddot{O}:}{\|}}{\underset{\underset{:\ddot{O}R'}{|}}{C}} + \ddot{N}H_3 \longrightarrow \left[R-\overset{\overset{\ominus}{:\ddot{O}:}}{\underset{\underset{:\ddot{O}R'}{|}}{C}}-\overset{\oplus}{N}H_3\right] \longrightarrow R-\overset{\overset{O}{\|}}{C}-NH_2 + R'OH$$

2. Reduction of Esters. Esters get reduced to alcohols in the presence of Na and alcohol or $LiAlH_4$.

$$RCOOR' \xrightarrow[\text{or Na/Alcohol}]{LiAlH_4} RCH_2OH + R'OH$$

$$\underset{\text{Ethylacetate}}{CH_3COOC_2H_5} \xrightarrow[\text{or Na/Alcohol}]{LiAlH_4} \underset{\text{Ethylalcohol}}{2CH_3CH_2OH}$$

3. Reaction with Grignard reagents. Esters of formic acid give secondary alcohols and those of other acids give tertiary alcohols on reaction with Grignard reagents.

Reaction between ethyl formate and methyl magnesium bromide

$$(i) \quad H-\overset{\overset{O}{\|}}{C}-OC_2H_5 + \overset{-}{C}H_3\overset{+}{Mg}Br \longrightarrow H-\overset{\overset{\bar{O}}{|}}{\underset{\underset{CH_3}{|}}{C}}-OC_2H_5 \xrightarrow{\overset{+}{Mg}Br}$$

$$H-\overset{\overset{OMgBr}{|}}{\underset{\underset{CH_3}{|}}{C}}-OC_2H_5 \xrightarrow{-\,Mg(Br)OC_2H_5} \underset{\text{Acetaldehyde}}{H-\overset{\overset{O}{\|}}{\underset{\underset{CH_3}{|}}{C}}}$$

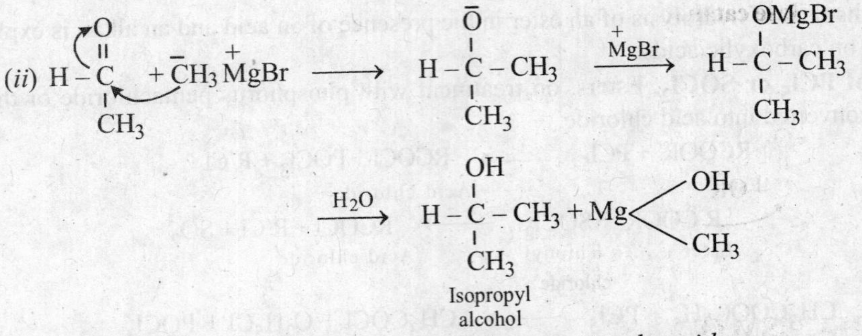

$$(ii)\ H-\underset{CH_3}{\underset{|}{C}}=O + \overset{-}{C}H_3\overset{+}{M}gBr \longrightarrow H-\underset{CH_3}{\underset{|}{C}}(\bar{O})-CH_3 \xrightarrow{\overset{+}{M}gBr} H-\underset{CH_3}{\underset{|}{C}}(OMgBr)-CH_3$$

$$\xrightarrow{H_2O} H-\underset{CH_3}{\underset{|}{C}}(OH)-CH_3 + Mg\langle^{OH}_{CH_3}$$

Isopropyl alcohol

Reaction between ethyl acetate and methyl magnesium bromide

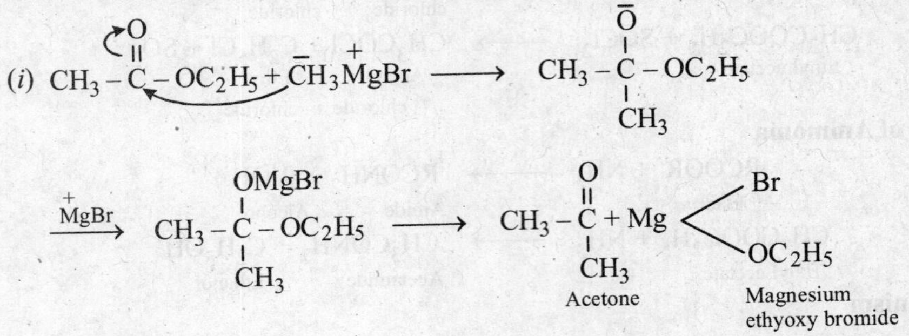

$$(i)\ CH_3-\overset{O}{\overset{||}{C}}-OC_2H_5 + \overset{-}{C}H_3\overset{+}{M}gBr \longrightarrow CH_3-\underset{CH_3}{\underset{|}{C}}(\bar{O})-OC_2H_5$$

$$\xrightarrow{\overset{+}{M}gBr} CH_3-\underset{CH_3}{\underset{|}{C}}(OMgBr)-OC_2H_5 \longrightarrow CH_3-\underset{CH_3}{\underset{|}{C}}=O + Mg\langle^{Br}_{OC_2H_5}$$

Acetone　　Magnesium ethyoxy bromide

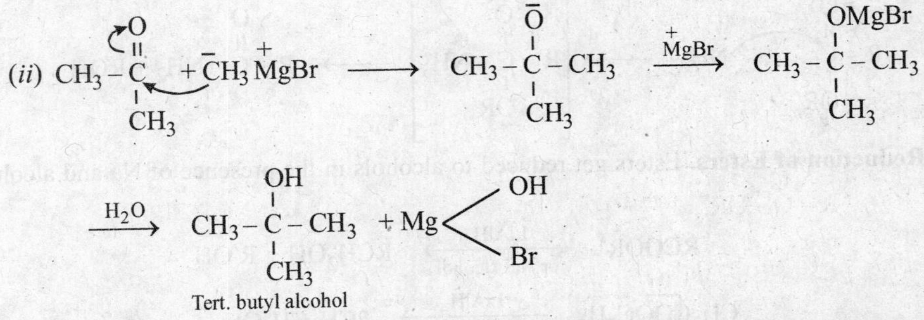

$$(ii)\ CH_3-\underset{CH_3}{\underset{|}{C}}=O + \overset{-}{C}H_3\overset{+}{M}gBr \longrightarrow CH_3-\underset{CH_3}{\underset{|}{C}}(\bar{O})-CH_3 \xrightarrow{\overset{+}{M}gBr} CH_3-\underset{CH_3}{\underset{|}{C}}(OMgBr)-CH_3$$

$$\xrightarrow{H_2O} CH_3-\underset{CH_3}{\underset{|}{C}}(OH)-CH_3 + Mg\langle^{OH}_{Br}$$

Tert. butyl alcohol

4. Trans-esterification. As mentioned in the methods of preparation of esters, trans-esterification is a reaction involving an ester and an alcohol which is different from the alcohol part of the ester.

$$RCOOR' + R''OH \xrightarrow[\text{or alkali}]{\text{Acid}} RCOOR'' + R'OH$$

The reaction takes place in the presence of an acid or alkali.

Mechanism (Acid catalysed)

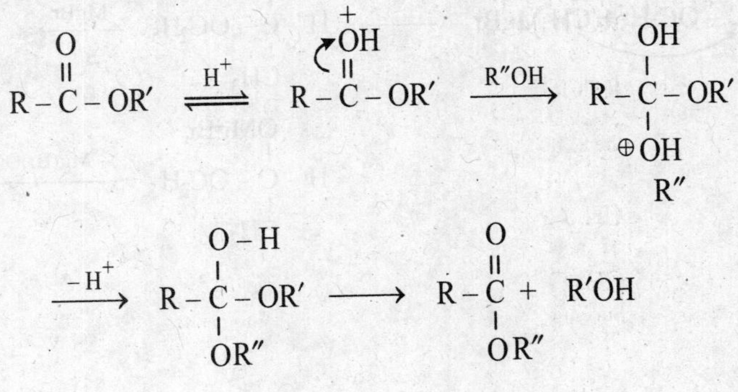

$$R-\overset{O}{\overset{||}{C}}-OR' \underset{}{\overset{H^+}{\rightleftharpoons}} R-\overset{\overset{+}{O}H}{\overset{||}{C}}-OR' \xrightarrow{R''OH} R-\underset{\oplus \underset{R''}{O}H}{\underset{|}{\overset{OH}{\overset{|}{C}}}}-OR'$$

$$\xrightarrow{-H^+} R-\underset{OR''}{\underset{|}{\overset{O-H}{\overset{|}{C}}}}-OR' \longrightarrow R-\underset{OR''}{\underset{|}{\overset{O}{\overset{||}{C}}}} + R'OH$$

Mechanism (Base catalysed)

$$R''OH + NaOH \longrightarrow R''ONa + H_2O$$

$$R''ONa \longrightarrow R''O^- + Na^+$$

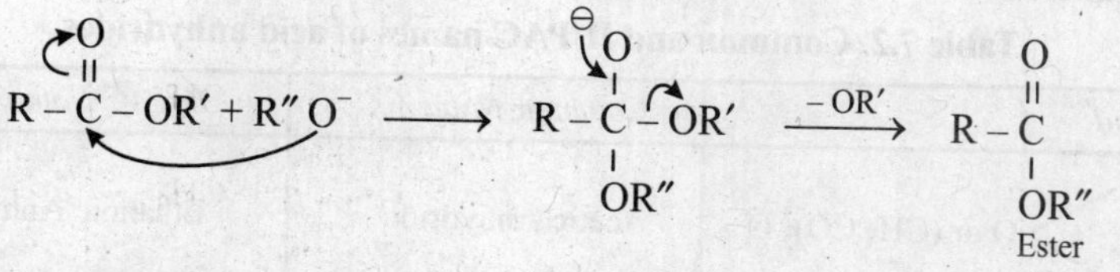

$$R-\overset{O}{\overset{\|}{C}}-OR' + R''O^- \longrightarrow R-\underset{OR''}{\underset{|}{\overset{O^\ominus}{\overset{|}{C}}}}-OR' \xrightarrow{-OR'} \underset{\text{Ester}}{R-\underset{OR''}{\underset{|}{\overset{O}{\overset{\|}{C}}}}}$$

7.6. MECHANISM OF ACID-CATALYSED AND BASE-CATALYSED HYDROLYSIS OF AN ESTER

Both these reactions are acyl nucleophilic substitution reactions.

Acid-catalysed

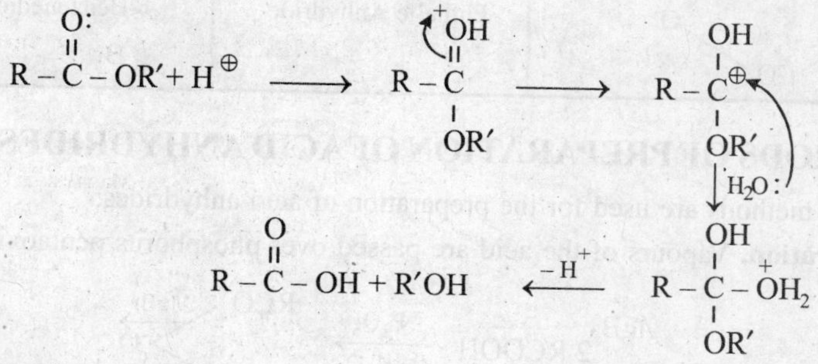

$$R-\overset{O:}{\overset{\|}{C}}-OR' + H^\oplus \longrightarrow R-\underset{OR'}{\underset{|}{\overset{\overset{+}{O}H}{\overset{\|}{C}}}} \longrightarrow R-\underset{OR'}{\underset{|}{\overset{OH}{\overset{|}{C^\oplus}}}} \xrightarrow{H_2O:} R-\underset{OR'}{\underset{|}{\overset{OH}{\overset{|}{C}}}}-\overset{+}{O}H_2 \xrightarrow{-H^+} R-\overset{O}{\overset{\|}{C}}-OH + R'OH$$

Base-catalysed

$$R-\overset{O}{\overset{\|}{C}}-OR' \xrightarrow{OH^-} R-\underset{OR'}{\underset{|}{\overset{O^-}{\overset{|}{C}}}}-OH \longrightarrow R-\overset{O}{\overset{\|}{C}}-OH + RO^-$$

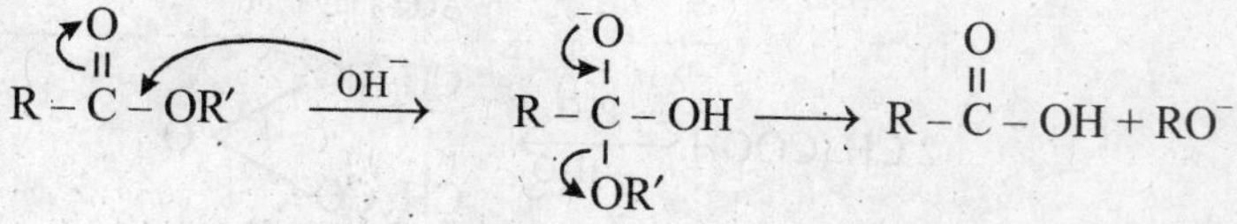

ACID ANHYDRIDES

7.7. GENERAL

Acid anhydrides are organic substances obtained by the dehydration of carboxylic acids. Generally one water molecule is removed from the two molecules of monocarboxylic acid. In the case of dicarboxylic acids, one water molecule is taken out from one molecule of the acid. Thus

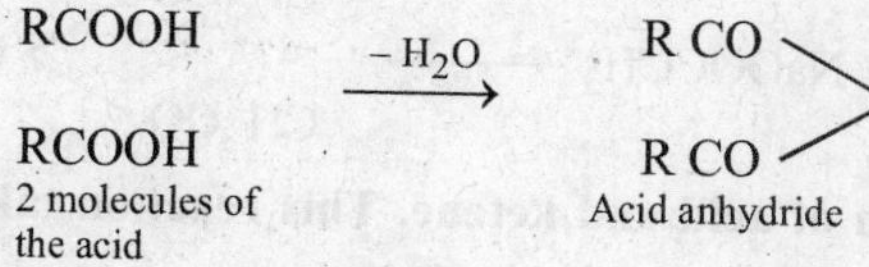

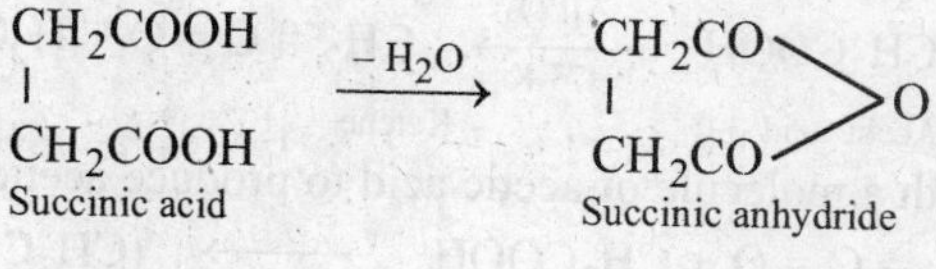

$$\underset{\text{2 molecules of the acid}}{\begin{matrix} RCOOH \\ RCOOH \end{matrix}} \xrightarrow{-H_2O} \underset{\text{Acid anhydride}}{\begin{matrix} RCO \\ RCO \end{matrix}\!\!>O}$$

$$\underset{\text{Succinic acid}}{\begin{matrix} CH_2COOH \\ | \\ CH_2COOH \end{matrix}} \xrightarrow{-H_2O} \underset{\text{Succinic anhydride}}{\begin{matrix} CH_2CO \\ | \\ CH_2CO \end{matrix}\!\!>O}$$

7.8. NOMENCLATURE OF ACID ANHYDRIDES

Such compounds are named by adding the word "anhydride" after the name of the acid as indicated in Table 7.2.

Table 7.2. Common and IUPAC names of acid anhydrides

Compound	*Common Name*	*IUPAC Name*
$\begin{matrix} CH_3CO \\ CH_3CO \end{matrix} \rangle O$ or $(CH_3CO)_2O$	Acetic Anhydride	Ethanoic Anhydride
$\begin{matrix} CH_2CO \\ \vert \\ CH_2CO \end{matrix} \rangle O$	Succinic Anhydride	Butanedioic Anhydride
$C_6H_4 \begin{matrix} CO \\ CO \end{matrix} \rangle O$	Phthalic Anhydride	*o*-Benzenedioic Anhydride

7.9. METHODS OF PREPARATION OF ACID ANHYDRIDES

Following methods are used for the preparation of acid anhydrides:

1. Dehydration. Vapours of the acid are passed over phosphorus pentaoxide.

$$2\,RCOOH \xrightarrow[-H_2O]{P_2O_5} \underset{\text{Acid anhydride}}{\begin{matrix} RCO \\ RCO \end{matrix} \rangle O}$$

$$2\,CH_3COOH \xrightarrow[-H_2O]{P_2O_5} \underset{\text{Acetic anhydride}}{\begin{matrix} CH_3CO \\ CH_3CO \end{matrix} \rangle O}$$

2. By the action of acid chloride on the sod. salt of acid:

$$\underset{\substack{\text{Acid} \\ \text{chloride}}}{RCOCl} + \underset{\substack{\text{Sod. salt} \\ \text{of acid}}}{NaOOCR} \longrightarrow \underset{\text{Acid anhydride}}{\begin{matrix} RCO \\ RCO \end{matrix} \rangle O} + NaCl$$

$$CH_3COCl + NaOOCCH_3 \longrightarrow \begin{matrix} CH_3CO \\ CH_3CO \end{matrix} \rangle O + NaCl$$

3. By the combination of acid and ketene. This reaction takes place in two steps:

(*i*) Dehydration of acid at high temp. gives ketene.

$$\underset{\text{Acetic acid}}{CH_3COOH} \xrightarrow[973\text{ K}]{AlPO_4} \underset{\text{Ketene}}{CH_2 = C = O} + H_2O$$

(*ii*) Ketene reacts with a molecule of acetic acid to produce acetic anhydride.

$$\underset{\text{Ketene}}{CH_2 = C = O} + CH_3COOH \longrightarrow \underset{\text{Acetic anhydride}}{(CH_3CO)_2O}$$

7.10. PROPERTIES OF ACID ANHYDRIDES

1. Hydrolysis. Acid anhydrides get hydrolysed by water to produce acids.

$$\underset{\text{Acetic anhydride}}{(CH_3CO)_2O} + H_2O \longrightarrow \underset{\text{Acetic acid}}{2\,CH_3COOH}$$

2. Alcoholysis. With ethyl alcohol, acetic anhydride reacts to form ethyl acetate.

$$\underset{\text{Acetic anhydride}}{(CH_3CO)_2\,O} + \underset{\text{Ethyl alcohol}}{C_2H_5OH} \longrightarrow \underset{\text{Ethyl acetate}}{CH_3COOC_2H_5} + CH_3COOH$$

Better yield of the ester is obtained by this method.

3. Ammonolysis. Ammonia reacts with acetic anhydride to produce acetamide.

$$\underset{\text{Acetic anhydride}}{(CH_3CO)_2\,O} + NH_3 \longrightarrow \underset{\text{Acetamide}}{CH_3CONH_2} + CH_3COOH$$

4. Action of primary amine

$$(CH_3CO)_2O + \underset{\text{Ethyl amine}}{C_2H_5NH_2} \longrightarrow \underset{\text{N-Ethyl acetamide}}{CH_3CONHC_2H_5} + CH_3COOH$$

5. Action of PCl_5 or HCl. With either of these, acid chlorides are formed.

$$\underset{\text{Acetic anhydride}}{(CH_3CO)_2\,O} + PCl_5 \longrightarrow \underset{\text{Acetyl chloride}}{2CH_3COCl} + \underset{\text{Phosphorus oxytrichloride}}{POCl_3}$$

6. Reduction. Acid anhydrides on reduction produce alcohols. Reducing agents used are sodium-alcohol or $LiAlH_4$.

$$(CH_3CO)_2\,O + 8[H] \xrightarrow[\text{or } LiAlH_4]{\text{Na/Alcohol}} 2\,C_2H_5OH + H_2O$$

7. Friedel-Crafts acylation. Acetic anhydride can be used for acylating aromatic compounds in the presence of anhyd. $AlCl_3$

$$\underset{\text{Benzene}}{C_6H_6} + (CH_3CO)_2O \xrightarrow[AlCl_3]{\text{Anhyd.}} \underset{\text{Acetophenone}}{C_6H_5COCH_3} + CH_3COOH$$

ACID AMIDES

Acid amides are organic compounds having the general formula $RCONH_2$ where R stands for some alkyl or aryl group. These are acid derivatives as these compounds can be obtained from acid.

7.11. NOMENCLATURE OF ACID AMIDES

These compounds are obtained by replacing the – OH group of the carboxy acids RCOOH by NH_2 group. Nomenclature of amides according to common system and IUPAC system are given in Table 7.3.

Table 7.3. Common and IUPAC names of acid amides

Compound	*Corresponding acid*	*Common name*	*IUPAC name*
$HCONH_2$	HCOOH	Formamide	Methanamide
CH_3CONH_2	CH_3COOH	Acetamide	Ethanamide
$CH_3CH_2CONH_2$	CH_3CH_2COOH	Propionamide	Propanamide
$C_6H_5CONH_2$	C_6H_5COOH	Benzamide	Benzenamide

7.12. METHODS OF PREPARATION OF AMIDES

1. By the action of NH_3 on acid derivatives

(*i*) $(CH_3CO)_2O + NH_3 \longrightarrow CH_3CONH_2 + CH_3COOH$
Acetic anhydride → Acetamide

(*ii*) $CH_3COCl + NH_3 \longrightarrow CH_3CONH_2 + HCl$
Acetyl chloride → Acetamide

(*iii*) $CH_3COOC_2H_5 + NH_3 \longrightarrow CH_3CONH_2 + C_2H_5OH$
Ethyl acetate → Acetamide

Mechanism

These are all nucleophilic substitution reactions. Mechanism of the reaction is given below:

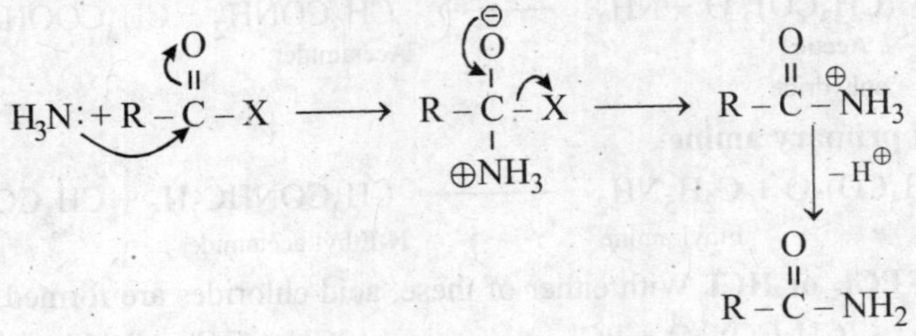

$$H_3N\!: + R-\overset{\overset{\displaystyle O}{\|}}{C}-X \longrightarrow R-\overset{\overset{\displaystyle O^{\ominus}}{|}}{\underset{\underset{\displaystyle \oplus NH_3}{|}}{C}}-X \longrightarrow R-\overset{\overset{\displaystyle O}{\|}}{C}-\overset{\oplus}{N}H_3 \xrightarrow{-H^{\oplus}} R-\overset{\overset{\displaystyle O}{\|}}{C}-NH_2$$

Here X stands for – $OCOCH_3$, – Cl or – OC_2H_5 group.

2. From ammonium salts. Ammonium salts of carboxylic acids on heating give amides.

$$CH_3COOH + NH_3 \longrightarrow CH_3COONH_4$$

$$CH_3COONH_4 \xrightarrow{\Delta} CH_3CONH_2 + H_2O$$

3. Partial hydrolysis of cyanides

$$CH_3CN + H_2O \xrightarrow[H_2O_2]{\text{Alkaline}} CH_3CONH_2$$

7.13. PROPERTIES OF AMIDES

1. Hydrogen bonding. There is intermolecular hydrogen bonding between amide molecules. This leads to higher boiling points of amides. Hydrogen bonding involves H–O bonds as shown below:

$$\ldots H-NH-\underset{\underset{\displaystyle R}{|}}{CO}\ldots H-NH-\underset{\underset{\displaystyle R}{|}}{CO}\ldots H-NH-\underset{\underset{\displaystyle R}{|}}{CO}\ldots$$

2. Reactivity. Acid amides are quite reactive towards nucleophilic substitution. This is because of the positive charge on acyl carbon atom, which is so necessary for nucleophilic substitution and the intermediate compound is stabilised as shown below:

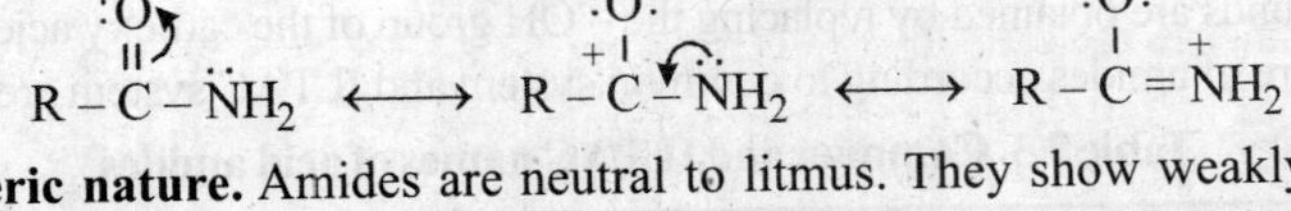

$$R-\overset{\overset{\displaystyle :\ddot{O}}{\|}}{C}-\ddot{N}H_2 \longleftrightarrow R-\overset{\overset{\displaystyle :\ddot{O}:^{-}}{|}}{\overset{+}{C}}-\ddot{N}H_2 \longleftrightarrow R-\overset{\overset{\displaystyle :\ddot{O}:^{-}}{|}}{C}=\overset{+}{N}H_2$$

3. Amphoteric nature. Amides are neutral to litmus. They show weakly acidic as well as basic properties.

$CH_3CONH_2 + HCl \longrightarrow CH_3CONH_2HCl$ (Acetamide hydrochloride) — Basic Property

$CH_3CONH_2 + Na \longrightarrow CH_3CONHNa + 1/2\ H_2$ (Sod. acetamide) — Acidic Property

The low basicity is because of the following transition:

$$R-\overset{\overset{O}{\|}}{C}-\underset{\underset{H}{|}}{N}-H \longrightarrow R-\overset{\overset{O^{\ominus}}{|}}{C}=\overset{\oplus}{N}H_2$$

4. Hydrolysis. They are hydrolysed to parent acids in the presence of a mineral acid or alkali.

$$CH_3CONH_2 + H_2O \xrightarrow[\text{or } OH^-]{H^+} CH_3COOH + NH_3$$

Mechanism (In the presence of an acid)

$$R-\overset{\overset{O}{\|}}{C}-NH_2 + \overset{\oplus}{H} \longrightarrow R-\underset{\underset{NH_2}{|}}{\overset{\overset{OH}{|}}{C^+}} + H\ddot{O}-H \longrightarrow R-\underset{\underset{NH_2}{|}}{\overset{\overset{OH}{|}}{C}}-\overset{\oplus}{O}H_2$$

$$\xrightarrow{-H^{\oplus}} R-\overset{\overset{O}{\|}}{C}-OH + NH_3 \longrightarrow RCOONH_4$$

Mechanism (In the presence of a base)

$$R-\overset{\overset{O}{\|}}{C}-NH_2 \xrightarrow{OH^{\ominus}} R-\underset{\underset{NH_2}{|}}{\overset{\overset{O^{\ominus}}{|}}{C}}-O-H \longrightarrow RCOO^- + NH_3$$

5. Dehydration. On heating with some dehydrating agent such as phosphorus pentoxide, amides get converted into cyanides or nitriles.

$$\underset{\text{Acetamide}}{CH_3CONH_2} \xrightarrow{P_2O_5} \underset{\text{Methyl cyanide}}{CH_3CN + H_2O}$$

6. Reduction. On reduction in the presence of sodium and alcohol, or lithium aluminium hydride, amides get converted into primary amines.

$$\underset{\text{Acetamide}}{CH_3CONH_2} \xrightarrow[\text{Na/Alcohol}]{LiAlH_4} \underset{\text{Ethyl amine}}{CH_3CH_2NH_2}$$

7. Action of nitrous acid

$$CH_3CONH_2 + HNO_2 \longrightarrow CH_3COOH + N_2 + H_2O$$

8. Hoffmann Bromamide or Hoffmann Degradation reaction. Amides on treatment with bromine and potassium hydroxide yield amines with one carbon atom less than the original amides. This reaction is employed to step-down the series.

$$\underset{\text{Acetamide}}{CH_3CONH_2} + Br_2 + KOH\ (\text{alc.}) \longrightarrow \underset{\text{Methyl amine}}{CH_3NH_2} + 2KBr + K_2CO_3 + 2H_2O$$

For mechanism, see the chapter on amines.

ACID CHLORIDES

Acid chlorides are organic compounds having the general formula RCOCl where R stands for some alkyl or aryl group. These are acid derivatives and can be obtained by substituting – OH group of the carboxy acids by –Cl group. These can also be obtained by replacing – NH_2, – OCOR or – OR group by – Cl group.

7.14. NOMENCLATURE OF ACID CHLORIDES

Nomenclature of acid chlorides according to common and IUPAC systems is given in Table 7.4.

Table 7.4. Common and IUPAC names of acid chlorides

Acid chloride	*Corresponding acid*	*Common name*	*IUPAC name*
HCOCl	HCOOH	Formyl chloride	Methanoylchloride
CH_3COCl	CH_3COOH	Acetyl chloride	Ethanoylchloride
CH_3CH_2COCl	CH_3CH_2COOH	Propionyl chloride	Propanoylchloride
C_6H_5COCl	C_6H_5COOH	Benzoyl chloride	–

7.15. METHODS OF PREPARATION OF ACID CHLORIDES

1. Using PCl_5 or PCl_3. Acid chlorides can be prepared by the action of PCl_5 or PCl_3 on a carboxylic acid.

$$CH_3COOH + PCl_5 \longrightarrow \underset{\text{Acetyl chloride}}{CH_3COCl} + \underset{\text{Phosphorus oxytrichloride}}{POCl_3} + HCl$$

$$3CH_3COOH + PCl_3 \longrightarrow 3CH_3COCl + \underset{\text{Phosphorus acid}}{H_3PO_3}$$

2. By the action of thionyl chloride on carboxylic acid

$$CH_3COOH + \underset{\text{Thionyl chloride}}{SOCl_2} \longrightarrow CH_3COCl + SO_2\uparrow + HCl$$

3. By the action of PCl_3 or thionyl chloride on sodium or calcium salts of fatty acids

$$3CH_3COONa + PCl_3 \longrightarrow \underset{\text{Acetyl chloride}}{CH_3COCl} + Na_3PO_3$$

$$\underset{\text{Cal.acetate}}{(CH_3COO)_2\,Ca} + \underset{\text{Thionyl chloride}}{SOCl_2} \longrightarrow \underset{\text{Acetyl chloride}}{2CH_3COCl} + CaSO_3$$

7.16. PROPERTIES OF ACID CHLORIDES

Acid chlorides are the most reactive of carboxylic acid derivatives. They undergo nucleophilic substitution reactions. Important reactions of acid chlorides are given below:

1. Hydrolysis. Acid chlorides are hydrolysed by water to form the parent carboxylic acids.

$$\underset{\text{Acetyl chloride}}{CH_3COCl + H_2O} \longrightarrow \underset{\text{Acetic acid}}{CH_3COOH} + HCl$$

$$\underset{\text{Benzoyl chloride}}{C_6H_5COCl + H_2O} \longrightarrow \underset{\text{Benzoic acid}}{C_6H_5COOH} + HCl$$

Mechanism of reaction is as follows:

$$CH_3-\overset{\overset{O}{\|}}{C}-Cl + H-\ddot{O}-H \longrightarrow CH_3-\overset{\overset{O^{\ominus}}{|}}{\underset{\underset{\oplus OH_2}{|}}{C}}-Cl \xrightarrow{-Cl} CH_3-\overset{\overset{O}{\|}}{C}-\overset{\oplus}{O}H_2 \xrightarrow{-H^{\oplus}} CH_3-\overset{\overset{O}{\|}}{C}-O-H$$

2. Reaction with alcohol. An acid chloride reacts with alcohol to form an ester.

$$CH_3COCl + C_2H_5OH \longrightarrow CH_3COOC_2H_5 + HCl$$

The reaction takes place as a nucleophilic substitution with the following mechanism:

$$CH_3-\underset{Cl}{\overset{O}{\overset{\|}{C}}}+C_2H_5\ddot{O}-H \longrightarrow CH_3-\underset{Cl}{\overset{O^{\ominus}}{\overset{|}{C}}}-\underset{C_2H_5}{\overset{\oplus}{O}H} \xrightarrow{-Cl^-} CH_3-\overset{O}{\overset{\|}{C}}-\underset{C_2H_5}{\overset{\oplus}{O}H} \xrightarrow{-H^{\oplus}} CH_3-\overset{O}{\overset{\|}{C}}-OC_2H_5$$

3. Ammonolysis. An acid chloride reacts with ammonia to form an amide.

$$\underset{\text{Acetyl chloride}}{CH_3COCl} + NH_3 \longrightarrow \underset{\text{Acetamide}}{CH_3CONH_2} + HCl$$

$$\underset{\text{Benzoyl chloride}}{C_6H_5COCl} + NH_3 \longrightarrow \underset{\text{Benzamide}}{C_6H_5CONH_2} + HCl$$

Mechanism of the reaction is given below. It is again a nucleophilic substitution reaction.

$$CH_3-\underset{Cl}{\overset{O}{\overset{\|}{C}}}+\ddot{N}H_3 \longrightarrow CH_3-\underset{Cl}{\overset{O^{\ominus}}{\overset{|}{C}}}-\overset{\oplus}{N}H_3 \xrightarrow{-Cl^{\ominus}} CH_3-\overset{O}{\overset{\|}{C}}-\overset{\oplus}{N}H_3 \xrightarrow{-H^{\oplus}} CH_3-\overset{O}{\overset{\|}{C}}-NH_2$$

4. Reaction with amines. An acid chloride on reaction with an amine gives substituted amide.

$$CH_3COCl + C_2H_5NH_2 \longrightarrow \underset{\text{N-Ethyl acetamide}}{CH_3CONHC_2H_5} + HCl$$

$$CH_3COCl + (C_2H_5)_2NH \longrightarrow \underset{\text{N,N-Diethyl acetamide}}{CH_3CON\,(C_2H_5)_2} + HCl$$

5. Formation of acid anhydride. There is a reaction with sodium salt of fatty acid and acid anhydride is obtained.

$$\underset{\text{Acetyl chloride}}{CH_3COCl} + \underset{\text{Sod. acetate}}{NaOOCCH_3} \longrightarrow \underset{\text{Acetic anhydride}}{(CH_3CO)_2\,O} + NaCl$$

6. Reaction with Grignard reagent

$$\underset{\text{Acetyl chloride}}{CH_3COCl} + \underset{\text{Methyl mag. chloride}}{ClMgCH_3} \longrightarrow \underset{\text{Acetone}}{CH_3COCH_3} + MgCl_2$$

7. Reaction with organocadmium compound

$$\underset{\text{Diethyl cadmium}}{(C_2H_5)_2\,Cd} + \underset{\text{Acetyl chloride}}{2CH_3COCl} \longrightarrow \underset{\text{Methyl ethyl ketone}}{2CH_3COC_2H_5} + CdCl_2$$

8. Rosenmund's reduction. Hydrogenation of an acid chloride in the presence of palladium based on $BaSO_4$ using xylene as a solvent gives aldehyde.

$$\underset{\text{Acetyl chloride}}{CH_3COCl} + H_2 \xrightarrow[\text{Xylene soln.}]{Pd\,/\,BaSO_4} \underset{\text{Acetaldehyde}}{CH_3CHO} + HCl$$

9. Friedel-Crafts reaction

$$C_6H_6 + CH_3COCl \xrightarrow[AlCl_3]{Anhy.} C_6H_5COCH_3 + HCl$$

Acetophenone

10. Action with halogens. Treatment with a halogen brings about substitution at the alkyl group, α-hydrogen atoms undergo substitution with halogen as shown below:

$$\underset{\text{Propionyl chloride}}{CH_3CH_2COCl} + Cl_2 \longrightarrow \underset{\text{Chloropropionyl chloride}}{CH_3CHCl\ COCl} + HCl$$

11. Reaction with carboxy acid. An acid chloride reacts with a carboxylic acid in the presence of pyridine to produce acid anhydride.

$$\underset{\text{Acetyl chloride}}{CH_3COCl} + CH_3COOH \xrightarrow{\text{Pyridine}} \underset{\text{Acetic anhydride}}{(CH_3CO)_2\,O} + HCl$$

Example 1. The following compounds are subjected to hydrolysis:

(*a*) Acetic anhydride (*b*) Acetamide (*c*) Ethyl acetate (*d*) Acetyl chloride

Arrange the reactants in decreasing order of reactivity.

Or

Arrange the acid chlorides, amides, anhydrides and esters in order of increasing reactivity towards nucleophilic attack. Give suitable reasons.

Solution. The order of reactivity of the above reactions in the decreasing order is as follows:

Acetyl chloride > Acetic anhydride > Ethyl acetate > Acetamide

1. All the above reactions are nucleophilic substitution reactions where – OCOR group of the anhydride, – NH_2 of the amide, – OC_2H_5 of the ester and – Cl of the acid chloride are replaced by – OH group.

2. We expect greater reactivity in a nucleophilic substitution when the acyl carbon has the maximum positive charge so that it could attract the nucleophile (water in this case).

3. All the groups which are attached to the acyl group (CH_3CO–) donate the electrons to acyl carbon thereby rendering it less positive. However, it is observed that –Cl being most electronegative supplies the minimum electrons to the acyl carbon atom. Thus carbon remains positive to the maximum extent in the case of acid chloride. Hence reactivity is maximum with an acid chloride. This is illustrated as below:

$$R-\overset{\delta^+}{C}(=\overset{\delta^-}{O})-\ddot{Cl}: \;>\; R-\overset{\delta^+}{C}(=\overset{\delta^-}{O})-\ddot{O}:-\overset{\delta^+}{C}(=\overset{\delta^-}{O})-R \;>\; R-\overset{\delta^+}{C}(=\overset{\delta^-}{O})-\ddot{O}R' \;>\; R-\overset{\delta^+}{C}(=\overset{\delta^-}{O})-\ddot{N}H_2$$

This can also be explained on the basis of ease of removal of the bases Cl^-, $RCOO^-$, $R'O^-$ and NH_2^- respectively.

Example 2. Complete the following equations:

(i) $CH_3CH_2CONH_2 + Br_2 + KOH \longrightarrow$

(ii) $CH_3COCl + H_2 \xrightarrow{Pd / BaSO_4}$

(iii) $CH_3COCl + CH_3COONa \longrightarrow$

Solution. (i) $\underset{\text{Propionamide}}{CH_3CH_2CONH_2} + Br_2 + KOH \longrightarrow \underset{\text{Ethyl amine}}{CH_3CH_2NH_2} + 2KBr + K_2CO_3 + 2H_2O$

(ii) $\underset{\text{Acetyl chloride}}{CH_3COCl} + H_2 \xrightarrow{Pd / BaSO_4} \underset{\text{Acetaldehyde}}{CH_3CHO} + HCl$

(iii) $CH_3COCl + CH_3COONa \xrightarrow{\text{Pyridine}} \underset{\text{Acetic anhydride}}{(CH_3CO)_2\,O} + NaCl$

Example 3. Complete the following reactions:

(i) Propionic acid + $SOCl_2$

(ii) Benzoyl chloride + $NH\,(CH_2\,CH_3)_2$

(iii) Butanoyl chloride + 1-butanol

(iv) Toluene + Acetyl chloride + Anhy. $AlCl_3$

(v) Propionamide + $NaNO_2$ **+ HCl**

(vi) $(CH_3)_2CHCOOH + CH_2N_2$

Solution. (i) $\underset{\text{Propionic acid}}{CH_3CH_2COOH} + \underset{\text{Thionyl chloride}}{SOCl_2} \longrightarrow \underset{\text{Propionoyl chloride}}{CH_3CH_2COCl} + SO_2 + HCl$

(ii) $\underset{\text{Benzyol chloride}}{C_6H_5COCl} + \underset{\text{Diethyl amine}}{NH\,(CH_2CH_3)_2} \longrightarrow \underset{\text{N, N Diethyl benzamide}}{C_6H_5CON\,(CH_2CH_3)_2} + HCl$

(iii) $\underset{\text{Butanoyl chloride}}{CH_3CH_2CH_2COCl} + \underset{\text{1-Butanol}}{C_4H_9OH} \longrightarrow \underset{\text{Butyl butanoate}}{CH_3CH_2CH_2COOC_4H_9} + HCl$

(iv) Toluene ($C_6H_5CH_3$) + $CH_3\,COCl$ (Acetyl chloride) $\xrightarrow[AlCl_3]{\text{Anhyd.}}$ α-methyl acetophenone (CH_3, $COCH_3$ on benzene ring) + HCl

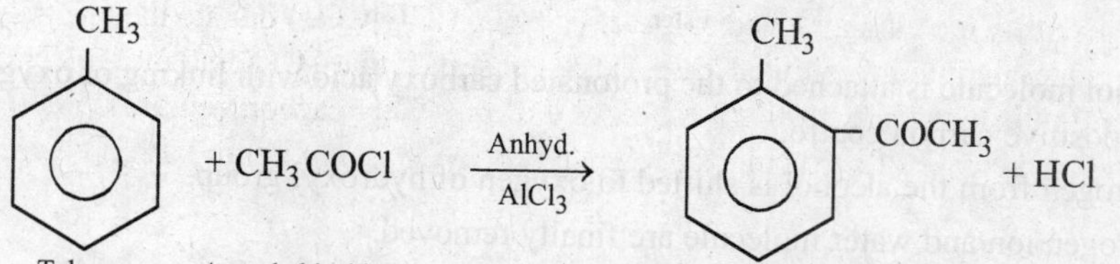

(v) $\underset{\text{Propionamide}}{CH_3CH_2CONH_2} + HNO_2 \xrightarrow[HCl]{NaNO_2} \underset{\text{Propionic acid}}{CH_3CH_2COOH} + N_2 + H_2O$

(vi) $\underset{\text{Isobutyric acid}}{(CH_3)_2\,CHCOOH} + \underset{\text{Diazo methane}}{CH_2N_2} \longrightarrow \underset{\text{Methyl iso-butyrate}}{(CH_3)_2\,CHCOOCH_3} + N_2 \uparrow$

Example 4. Give the mechanism of alkaline hydrolysis of amides.

Solution. $CH_3CONH_2 + NaOH \longrightarrow CH_3COONa + NH_3 \uparrow$

The mechanism of the reaction is as follows:

$$(i)\ CH_3-\overset{O}{\overset{\|}{C}}-NH_2 \xrightarrow{\bar{O}H} CH_3-\underset{OH}{\overset{-O}{C}}-NH_2$$

$$(ii)\ CH_3-\underset{OH}{\overset{-O}{C}}-NH_2 \longrightarrow CH_3-\overset{O}{\overset{\|}{C}}-OH + NH_2^-$$

$$(iii)\ CH_3-\overset{O}{\overset{\|}{C}}-OH + NH_2^- \longrightarrow CH_3-\overset{O}{\overset{\|}{C}}O^- + NH_3$$

Example 5. Discuss the probable mechanism of ester formation from carboxylic acid and alcohol.

Solution. $RCOOH + R'OH \overset{H^+}{\rightleftharpoons} \underset{\text{Ester}}{RCOOR'} + H_2O$

Mechanism of the reaction is as follows:

(*i*) In the first step, proton from the mineral acid is attached to carbonyl oxygen forming protonated carboxy acid with a positive centre on carbon.

$$R-\overset{O}{\overset{\|}{C}}-\ddot{O}-H \xrightarrow{H^+} R-\overset{OH}{\overset{|}{C^+}}-\ddot{O}-H \xrightarrow{:OR'H} R-\underset{\underset{H}{+OR}}{\overset{OH}{C}}-\ddot{O}-H$$

$$R-\underset{\underset{\text{Ester}}{OR'}}{\overset{O}{\overset{\|}{C}}} \xleftarrow{-H_2O} R-\underset{\underset{\text{Ester}}{OR'H}}{\overset{O-H}{C}}-\overset{+}{O}-H$$

(*ii*) Alcohol molecule is attached to the protonated carboxy acid with linking of oxygen of the alcohol to the positive carbon centre.

(*iii*) Hydrogen from the alcohol is shifted to oxygen of hydroxyl group.

(*iv*) Hydrogen ion and water molecule are finally removed.

Example 6. What is nucleophilic acyl substitution? Give its mechanism.

Solution. These are the substitution reactions of acid derivatives of carboxylic acids in which – Cl, – OOCR, – NH_2 or – OR′ of the acid chloride, anhydride, amide and ester is replaced by other nucleophiles. The reaction can be represented in general as:

$$R-C\overset{\nearrow O}{\underset{\searrow Y}{}} + :Z \longrightarrow R-C\overset{\nearrow O}{\underset{\searrow Z}{}} + :Y$$

where Y stands for – Cl, – OOCR, – NH_2 or – OR′ and : Z stands for the nucleophile like OH^-.

Since the nucleophile attaches itself to the carbon of the acyl group, such reactions are called nucleophilic substitution reactions.

The mechanism of the reaction is as follows:

(*i*) $R-C(=\ddot{O}:)(-Y) + :Z \longrightarrow R-C(-:\ddot{O}:^-)(-Y)-Z$

(*ii*) $R-C(-:\ddot{O}:^-)(-Y)-Z \longrightarrow R-C(=\ddot{O}:)-Z + :Y$

Example 7. Giving suitable reasons, arrange acetamide, acetyl chloride, ethyl acetate and acetic anhydride in order of increasing reactivity towards nucleophilic attack.

Solution. Refer to Example 6, Y stands for – Cl of acid chloride or – OOCR of anhydride or – NH_2 of amide or – OR of the ester.

The reactivity of the acid derivative to give the product will depend upon the ease with which – Y can be removed from the intermediate product in step (*ii*) of the reaction as shown earlier.

The ease of removal of Y is governed by two factors:

1. The ease with which Y can be removed depends upon its basicity. Weaker the base, easier it is removed from the intermediate compound. The basicity of Y increases in the order:

$$Cl < RCOO^- < NH_2 < OR$$

Thus – Cl can be removed most easily followed by RCOO followed by NH_2 and finally followed by OR. This can also be interpreted in terms of inductive effects of Cl, O and N. Hence reactivity follows the order:

Ester < Acid amide < Acid anhydride < Acid chloride

Example 8. Compare and contrast acyl nucleophilic substitution reactions with alkyl nucleophilic substitution reactions.

Solution. The two types of reactions can be expressed in a general form as follows:

$$R-Y + :Z \longrightarrow R-Z + :Y$$ (Alkyl nucleophilic substitution reaction)

$$R-C(=O)-Y + :Z \longrightarrow R-C(=O)-Z + :Y$$ (Acyl nucleophilic substitution reaction)

The main points of difference between the two are as under:

1. In alkyl nucleophilic reaction, the nucleophile attaches itself to the alkyl carbon whereas in acyl nucleophilic reaction it attaches itself to the acyl carbon.

$$R-Y + Z \longrightarrow Z \;......\; R \;......\; Y$$

$$R-C(=O)-Y + Z \longrightarrow R-\overset{O}{\overset{\|}{C}}(\cdots Y) \;.........\; Z$$

2. Acyl nucleophilic reaction takes place faster as compared to alkyl nucleophilic reaction. The intermediate product in the former is less stable because the pentavalent carbon becomes more crowded. In contrast, in acyl reaction the intermediate product is less crowded with tetravalent carbon and hence more stable.

Alkyl nucleophilic substitution reaction

$$Z:+ \text{(C)}-Y \xrightarrow{SN^2} Z.......\text{(C)}.......Y \longrightarrow Z-\text{(C)} + :Y$$

Acyl nucleophilic substitution reaction

$$Z:+ \; R(Y)C{=}O \longrightarrow Z(R)(Y)C{-}O \longrightarrow R(Z)C{=}O + :Y$$

Further the attachment of nucleophile to acyl carbon requires the breaking of a π-bond whereas attachment of a nucleophile to alkyl carbon requires partial breaking of a σ-bond which requires greater energy.

Example 9. Explain why acetylating activities of an acid anhydride is lower than that of acid chloride.

Or

Which of the following is more reactive towards nucleophilic attack and why?

Acetic anhydride, acetyl chloride

Solution. Acetic anhydride and acetyl chloride are used to acetylate alcohols, phenols and amines.

(*i*) $CH_3-\overset{O}{\overset{\|}{C}}-Cl + \bar{O}R \longrightarrow CH_3-\overset{O}{\overset{\|}{C}}(OR) + \bar{Cl}$

(*ii*) $(CH_3-\overset{O}{\overset{\|}{C}})_2O + \bar{O}R \longrightarrow CH_3-\overset{O}{\overset{\|}{C}}-OR + CH_3CO\bar{O}$

Reaction (*i*) takes place faster as compared to reaction (*ii*) because $\bar{Cl}$ is a weaker base and can be removed easily compared to $CH_3CO\bar{O}$ which is a strong base.

Conjugate bases of strong acids are weak bases and conjugate bases of weak acids are strong bases. Accordingly $\bar{Cl}$, which is a conjugate base of strong acid HCl, is weak and $CH_3CO\bar{O}$, which is a conjugate base of weak acid acetic acid, is strong.

Example 10. Out of the following, which is more reactive towards nucleophilic attack and why?

(*i*) Acetic anhydride

(*ii*) Acetamide

Solution. Acetic anhydride is more reactive than acetamide because after the nucleophile has been linked to carbonyl carbon, it is easier to reomove CH_3COO^- than NH_2^-

$$(CH_3-\overset{O}{\overset{\|}{C}})_2O + \bar{O}R \longrightarrow CH_3-\overset{O}{\overset{\|}{C}}-OR + CH_3CO\bar{O}$$

$$CH_3-\overset{O}{\overset{\|}{C}}-NH_2 + \bar{O}R \longrightarrow CH_3-\overset{O}{\overset{\|}{C}}-OR + NH_2^-$$

Example 11. Reason out why acetamide is slowly hydrolysed with water but rapidly in the presence of dil H_2SO_4.

Solution. Presence of dil H_2SO_4 helps in creating positive centre at the carbonyl carbon which further enables the nucleophilic attack of water molecule on this carbon.

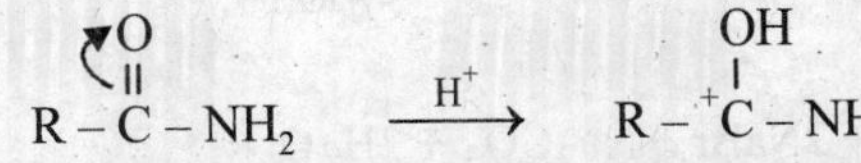

$$R-\overset{\overset{O}{\|}}{C}-NH_2 \xrightarrow{H^+} R-\overset{\overset{OH}{|}}{\overset{+}{C}}-NH_2$$

Example 12. Explain why acetyl chloride reacts vigorously with water at room temperature but benzoyl chloride reacts slowly.

Solution. This is because positive charge developed on carbonyl carbon in benzoyl chloride is destroyed due to resonance in the benzene ring

So the attack of nucleophile H_2O becomes mild.

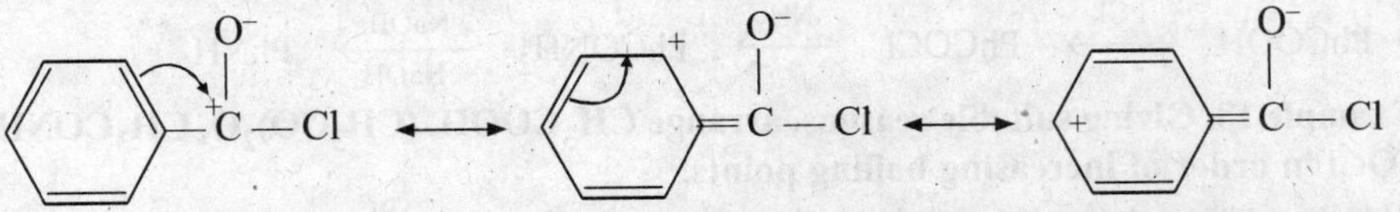

$$C_6H_5-\overset{\overset{O^-}{|}}{\overset{+}{C}}-Cl \longleftrightarrow \overset{+}{C_6H_5}=\overset{\overset{O^-}{|}}{C}-Cl \longleftrightarrow \overset{+}{C_6H_5}=\overset{\overset{O^-}{|}}{C}-Cl$$

Example 13. Explain why does acetyl chloride boil at lower temperature than acetic acid?

Solution. Acetic acid exists as a dimer because of hydrogen bonding as under:

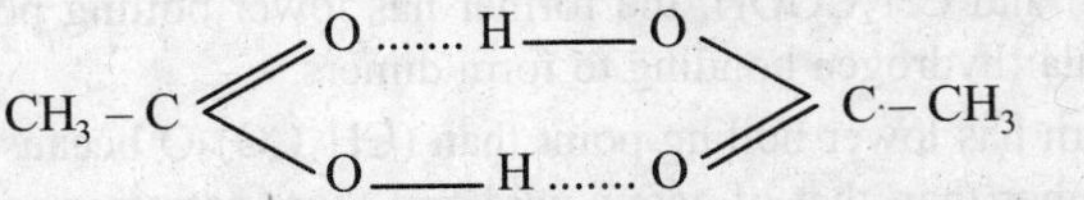

$$CH_3-C\begin{matrix}\nearrow O \cdots\cdots H - O \searrow \\ \searrow O - H \cdots\cdots O \nearrow\end{matrix}C-CH_3$$

This type of hydrogen bonding does not take place in acetyl chloride. Hydrogen bonding results in bigger molecules, boiling at higher temperature.

Example 14. Amides are much weaker bases than ammonia. Explain.

Solution. This is because the lone pair of electrons on nitrogen which is responsible for basic nature of a compound is drawn towards carbon as given below.

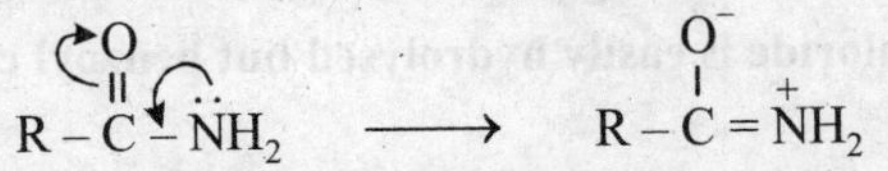

$$R-\overset{\overset{O}{\|}}{C}-\ddot{N}H_2 \longrightarrow R-\overset{\overset{O^-}{|}}{C}=\overset{+}{N}H_2$$

Example 15. The b.p. of acid anhydrides are higher than those of acids from which they are derived. Explain.

Solution. This is because an anhydride is obtained by the combinations of two molecules of the acid. This is supposed to have a bigger molecular size and thus higher b.p.

Example 16. Give suitable reagent for the following conversion :

$$C_6H_5\ CONH_2 \longrightarrow C_6H_5\ NH_2$$

Solution. $C_6H_5\ CONH_2 \xrightarrow{Br_2/KOH(alc)} C_6H_5NH_2$

Example 17. Complete the following reactions:

1. 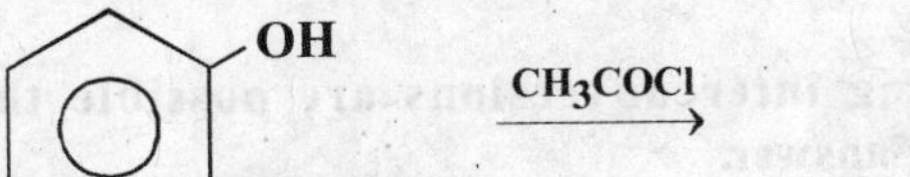 $C_6H_5-CH_2-\overset{\overset{O}{\|}}{C}-NH_2 \xrightarrow[NaOH]{Br_2}$

2. $C_6H_4(OH)(COOH) \xrightarrow{CH_3COCl}$ (o-hydroxybenzoic acid)

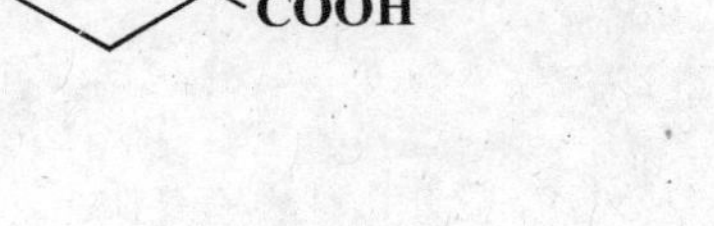

3. PhCOOH $\xrightarrow{PCl_5}$ **A** $\xrightarrow{NH_3}$ **B** $\xrightarrow[NaOH]{NaOBr}$

Solution. 1. $C_6H_5-CH_2-\overset{O}{\overset{||}{C}}-NH_2 \xrightarrow[4NaOH]{Br_2}$

$C_6H_5-CH_2NH_2 + 2NaBr + Na_2CO_3 + 2H_2O$

2. $C_6H_4(OH)(COOH) \xrightarrow{CH_3COCl} C_6H_4(OCOCH_3)(COOH) + HCl$

3. $PhCOOH \xrightarrow{PCl_5} PhCOCl \xrightarrow{NH_3} PhCONH_2 \xrightarrow[NaOH]{NaOBr} PhNH_2$

Example 18. Giving suitable reasons, arrange CH_3COOH, $(CH_3CO)_2O$, CH_3CONH_2 and CH_3COCl in order of increasing boiling points.

Solution. The order of increasing boiling points is : $CH_3COCl < CH_3COOH < (CH_3CO)_2O < CH_3CONH_2$. This can be explained as under :

Out of CH_3COCl and CH_3COOH, the former has lower boiling point because CH_3COOH undergoes intermolecular hydrogen bonding to form dimers.

CH_3COOH in turn has lower boiling point than $(CH_3CO)_2O$ because molecular mass of the anhydride is much higher than that of acetic acid and there are strong intermolecular forces of attraction. Similarly CH_3COOH has lower boiling point than CH_3CONH_2 because CH_3COOH forms only cyclic dimers due to hydrogen bonding. On the other hand, CH_3CONH_2 involves extensive intermolecular hydrogen bonding which leads to the association of a large number of molecules. Thus, the boiling point of CH_3CONH_2 is higher than that of $(CH_3CO)_2O$ also.

Hence the increasing order of boiling points is as given above.

Example 19. Acetyl chloride is easily hydrolysed but benzoyl chloride is not. Explain the reason.

Solution. The positive charge on acyl carbon of benzoyl chloride is less compared to that on acetyl chloride due to the strong electron releasing resonance effect of benzene ring present in benzoyl chloride. Therefore, it does not undergo nucleophilic attack. As such, its hydrolysis does not take place easily. It takes place only with OH^- which is a strong nucleophilic reagent.

Example 20. Why C–N bond in amides is shorter than C – N bond in amines ?

Solution. The carbon of C – N bond in amides is sp^2 hybridised while the carbon of C – N bond in amines is sp^3 hybridised. Being relatively smalller in size sp^2 hybrid carbon forms shorter carbon-nitrogen bond.

Example 21. Acetic anhydride is preferred over acetyl chloride for acetylation reactions. Explain.

Solution. Acetyl chloride is much more reactive and as a result its acetylation reactions are very fast and difficult to control. On the other hand, acetic anhydride reacts at slower speed and can be easily controlled. Hence acetic anhydride is preferred over acetyl chloride for carrying out acetylation reactions.

Example 22. Which of the following interconversions are possible through nucleophilic acyl substitution. Justify your answer.

(*i*) CH_3COCl into CH_3CONH_2

(*ii*) $(CH_3CO)_2O$ into $CH_3COOC_2H_5$

(*iii*) CH_3CONH_2 into $(CH_3CO)_2O$.

Solution. All the compounds involved in the conversions are acid derivatives. A conversion will depend upon the relative reactivity of the reactant and the product. It is generally possible to convert a more reactive acid derivative into a less reactive one. Hence the conversions (*i*) and (*ii*) are possible while the conversion (*iii*) is not possible.

Example 23. An unknown ester $C_5H_{12}O_2$ was hydrolysed with water and acid to give carboxylic acid (A) and alcohol (B). Treatment of (B) with PBr_3 gave an alkyl bromide (C). When (C) was treated with KCN, a product (D) was formed which on hydrolysis with water and acid gave a carboxylic acid (A). Give the structure and name of the original ester. Identify (A) through (D) and write equations for the reactions involved.

Solution. Unknown ester is $C_2H_5COOC_2H_5$: On hydrolysis it gives C_2H_5COOH (A) and C_2H_5OH(B)

$$\underset{(B)}{C_2H_5OH} \xrightarrow{PBr_3} \underset{(C)}{C_2H_5Br} \xrightarrow{KCN} \underset{(D)}{C_2H_5CN} \xrightarrow{H_2O} \underset{(A)}{C_2H_5COOH}$$

UREA

HISTORICAL IMPORTANCE

It is the end product of human metabolism of nitrogen containing foods (proteins). In 1773, Roulle isolated it from urine and named it urea. It has historical significance as it was the first organic compound synthesized in the laboratory (Wohler 1828) from inorganic material.

Wohler produced urea

7.17. METHODS OF PREPARATION OF UREA

Methods of preparation of urea are described as under :

1. From urine. Urine is evaporated to a small bulk and nitric acid is added to it to precipitate sparingly soluble urea nitrate $(NH_2CONH_2.HNO_3)$. Urea nitrate is treated with barium carbonate to remove the acid and then extracted with alcohol, barium nitrate being insoluble in alcohol.

$$\underset{\text{Urea nitrate}}{2NH_2CONH_2.HNO_3} + BaCO_3 \longrightarrow \underset{\text{Urea}}{2NH_2.CO.NH_2} + Ba(NO_3)_2 + H_2O + CO_2$$

2. By heating a solution containing a mixture of potassium cyanate and ammonium sulphate to dryness (Wohler synthesis). When this mixture is heated to dryness, ammonium cyanate is formed, which undergoes molecular rearrangement (isomeric change) to give urea. Urea is extracted from this mixture by dissolving in hot absolute alcohol. Ammonium sulphate being insoluble may be filtered off and the filtrate on cooling deposits urea.

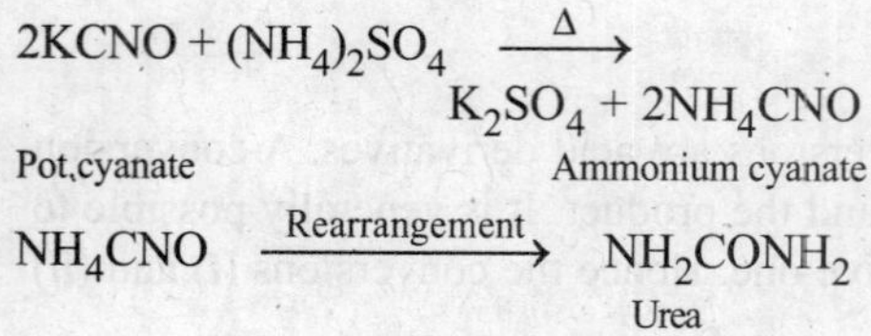

$$2KCNO + (NH_4)_2SO_4 \xrightarrow{\Delta} \underset{\text{Pot. cyanate}}{} K_2SO_4 + \underset{\text{Ammonium cyanate}}{2NH_4CNO}$$

$$NH_4CNO \xrightarrow{\text{Rearrangement}} \underset{\text{Urea}}{NH_2CONH_2}$$

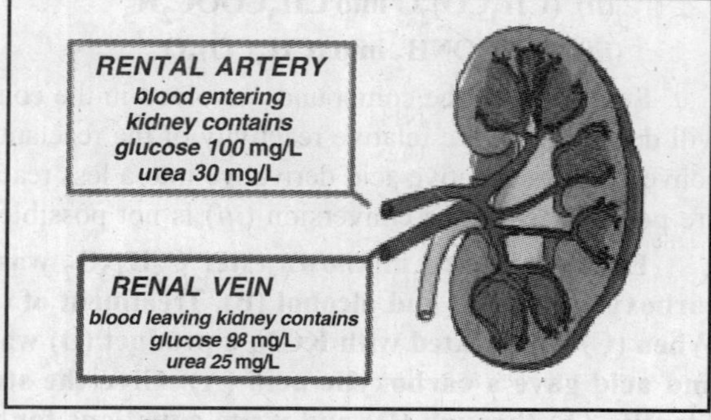

Glucose-urea cycle in the kidney

3. By partial hydrolysis of cyanamide (Manufacture). Cyanamide for this purpose is obtained from calcium cyanamide which is manufactured by passing nitrogen through heated calcium carbide at about 1073 K.

$$\underset{\text{Calcium carbide}}{CaC_2} \xrightarrow[1073\text{ K}]{N_2} \underset{\text{Calcium cyanamide}}{CaN.C\equiv N} \xrightarrow[-CaSO_4]{H_2SO_4} \underset{\text{Cyanamide}}{H_2N.C\equiv N} \xrightarrow{H_2O} \underset{\text{Urea}}{NH_2CONH_2}$$

4. By the action of ammonia on carbonyl chloride (Laboratory preparation).

$$\underset{\text{Carbonyl chloride}}{O=C\begin{matrix}\diagup Cl \\ \diagdown Cl\end{matrix}} + \underset{\text{Ammonia}}{\begin{matrix}H\,NH_2 \\ H\,NH_2\end{matrix}} \longrightarrow \underset{\text{Urea}}{O=C\begin{matrix}\diagup NH_2 \\ \diagdown NH_2\end{matrix}} + 2HCl$$

5. By the action of liquid carbon dioxide on liquid ammonia at 423 K and 35 atmospheric pressure (Manufacture).

$$O=C=O + \begin{matrix}H\,NH_2 \\ H\,NH_2\end{matrix} \xrightarrow[423\text{ K}]{35\text{ atm}} \underset{\text{Urea}}{O=C\begin{matrix}\diagup NH_2 \\ \diagdown NH_2\end{matrix}} + H_2O$$

7.18. PROPERTIES OF UREA

Physical Properties of Urea

Urea is a white crystalline solid. It melts at 405 K. It is soluble in water and ethyl alcohol but insoluble in ether. It is odourless and has a cooling taste.

Chemical Properties of Urea

Urea contains an amide group attached to an amino group and gives reactions of both these groups.

1. Basic nature (Salt formation). An aqueous solution of urea is neutral to litmus but behaves as a weak monoacid base and forms salt with strong acids. Thus, when nitric acid is added to a strong solution of urea, a crystalline precipitate of urea nitrate is formed.

$$\underset{\text{Urea}}{NH_2.CO.NH_2} + HNO_3 \longrightarrow \underset{\text{Urea nitrate}}{[NH_2.CO.NH_3]^+ NO_3^-}$$

2. Hydrolysis. It is rapidly hydrolysed into ammonia and carbon dioxide when boiled with dilute acids or alkalis.

$$O=C\begin{matrix}NH_2\\NH_2\end{matrix} + 2H_2O \xrightarrow{-2NH_3} O=C\begin{matrix}OH\\OH\end{matrix} \longrightarrow CO_2 + H_2O$$

Urea Unstable

Thus, upon hydrolysis with dil. hydrochloric acid urea forms ammonium chloride and with caustic soda it forms sodium carbonate.

$$NH_2.CO.NH_2 + H_2O + 2HCl \longrightarrow CO_2 + 2NH_4Cl$$
$$NH_2.CO.NH_2 + 2NaOH \longrightarrow Na_2CO_3 + 2NH_3$$

Enzyme, urease also brings about the hydrolysis of urea.

$$NH_2.CO.NH_2 \xrightarrow[\text{Urease}]{H_2O} 2NH_3 + CO_2$$

3. Reaction with nitrous acid. Urea reacts with nitrous acid to produce carbon dioxide, nitrogen and water.

$$\begin{matrix}HO\ \ NO\\H_2N-CO-NH_2\\ON-HO\end{matrix} \longrightarrow 2N_2 + 2H_2O + O=C\begin{matrix}OH\\OH\end{matrix} \longrightarrow H_2O + CO_2$$

Unstable

4. Acetylation. Urea undergoes acetylation with acetyl chloride.

$$CH_3CO\,\boxed{Cl + H}\,NHCONH_2 \longrightarrow CH_3CONHCONH_2 + HCl$$

Acetyl chloride Urea Acetyl urea

5. Reaction with sodium hypohalites. With excess of alkaline sodium hypobromite, urea is converted to nitrogen and sodium carbonate.

$$NH_2.CO.NH_2 + 3NaBrO \longrightarrow N_2 + CO_2 + 2NaBr + 2H_2O$$
$$CO_2 + \underset{\text{excess}}{2NaOH} \longrightarrow Na_2CO_3 + H_2O$$

6. Action of heat.

(*a*) On heating gently at about 405 K urea melts and after evolution of ammonia, a solid compound called **biuret** is produced.

Biuret contains – CO – NH – group and hence gives a violet colouration with alkaline dilute copper sulphate solution (1%).

$$NH_2-\overset{O}{\overset{\|}{C}}-\boxed{NH_2 + H}NH-\overset{O}{\overset{\|}{C}}-NH_2 \xrightarrow[405\ K]{heat} NH_2-\overset{O}{\overset{\|}{C}}-NH-\overset{O}{\overset{\|}{C}}-NH_2 + NH_3$$

Urea Urea Biuret

(*b*) When heated rapidly, ammonia is liberated and isocyanic acid is produced which trimerizes to cyanuric acid.

$$NH_2.CO.NH_2 \xrightarrow[-NH_3]{heat} HNCO \xrightarrow{Trimerizes} (HNCO)_3$$

Urea Isocyanic acid Cyanuric acid

7. Reaction with formaldehyde. Urea reacts with formaldehyde to form urea formaldehyde resins or plastics.

$$H-\overset{H}{\overset{|}{C}}=O + H_2N-CO-NH_2 \longrightarrow HO-\underset{H}{\underset{|}{\overset{H}{\overset{|}{C}}}}-NH-CO-NH_2$$

Methyl urea

$$HO-\underset{\underset{H}{|}}{\overset{\overset{H}{|}}{C}}-NH-CO-NH_2 + O=\overset{\overset{H}{|}}{C}-H \longrightarrow \underset{\text{Dimethyl urea}}{HO-\underset{\underset{H}{|}}{\overset{\overset{H}{|}}{C}}-NH-CO-NH-\underset{\underset{H}{|}}{\overset{\overset{H}{|}}{C}}-OH}$$

As the end – CHOH groups further react with – NH_2, the chain grows in length. Again the – NH – groups of different chains react with formaldehyde which gives rise to cross linkages forming valuable polymers.

Uses of Urea: It is used:

1. in the manufacture of formaldehyde-urea plastics or resins.
2. as a stabilizer for explosives.
3. for making barbiturates used as hypnotics and sedatives.
4. as a fertilizer.

7.19. ESTIMATION OF UREA

Two methods are generally employed for the estimation of urea : (*a*) *the Hypobromite method*; and (*b*) *urease method.* These are described as under :

1. The Hypobromite Method. This is a very rapid and sufficiently accurate method for medical requirements. A known volume of urine (or urea solution) is treated with sodium hypobromite solution containing excess of sodium hydroxide.

$$\underset{60\text{ g}}{CO(NH_2)_2} + 3NaOBr \longrightarrow CO_2 + 3NaBr + 2H_2O + \underset{\substack{22.4\text{ litres}\\ \text{at NTP}}}{N_2}$$

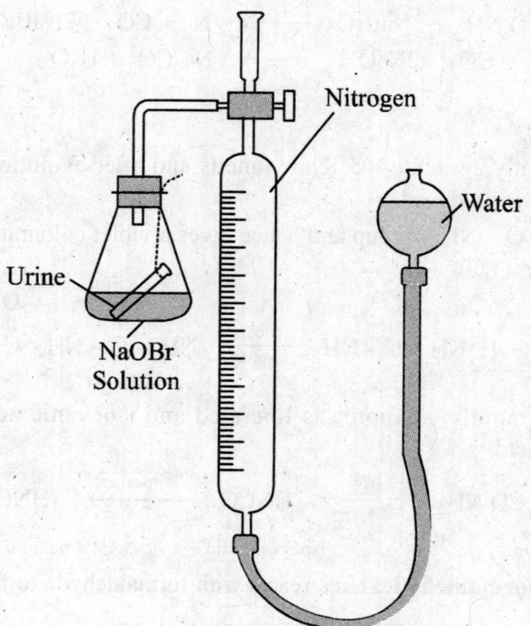

Fig. 7.1. Estimation of Urea by the Hypobromite method.

The carbon dioxide evolved is absorbed by sodium hydroxide and the volume of nitrogen evolved is measured in the nitrometer. As shown by the chemical equation, 22.4 litres of nitrogen at NTP are produced by 60 g of urea. Therefore, the amount of urea in the sample of urine is calculated as

$$= \frac{60}{22,400} \times v \text{ grams}$$

where v is the volume of nitrogen gas (reduced to NTP conditions) collected in the nitrometer.

Procedure. Place about 30 ml of freshly prepared sodium hypobromite solution containing excess of NaOH in the 250 ml conical flask (Fig. 7.1). Measure out exactly 5 ml of the given sample of urine into the small test-tube and fit the rubber stopper securely. Adjust water to the same level in the two limbs of the nitrometer. Tilt the flask so that the urine and hypobromite solution come into contact. The reaction starts and nitrogen collects in the nitrometer while CO_2 is absorbed by excess of sodium hydroxide in the flask. The volume of N_2 is noted again by adjusting the water levels in the nitrometer. It is reduced to NTP.

2. The Urease Method. This method is based on the conversion of urea to ammonium carbonate by hydrolysis with *urease* (soyabean extract). The ammonium carbonate solution so produced is estimated by titration with N/10 HCl.

$$\underset{60\text{ g (1 mole)}}{CO(NH_2)_2} + 2H_2O \xrightarrow[60^\circ C]{urease} \underset{(1\text{ mole})}{(NH_4)_2CO_3}$$

$$(NH_4)_2CO_3 + \underset{2 \times 36.5\ (2\text{ moles})}{2HCl} \longrightarrow 2NH_4Cl + CO_2 + H_2O$$

From the volume of N/10 HCl used by a known weight of the sample, the amount of urea can be calculated as under :

1 g mole of urea ≡ 2000 ml of N HCl (2 × 36.5 g)

i.e., 60 g of urea ≡ 2000 ml of N HCl

or 1 mol of N HCl ≡ 0.003 g urea

Procedure. Weigh accurately about 0.1 g of the crude urea in a 250 ml flask. Add to it 20 ml of warm water to bring the temperature to 60°C (optimum condition for *urease*). Add 3 drops of phenophthalein and 10 ml of *urease* solution, when the solution turns pink. Now add to it N/10 HCl dropwise till the pink colour is just discharged. This gives a pH of about 8. Continue to add acid carefully so as to keep the colour discharged. The reaction (hydrolysis to ammonium carbonate) is complete in about five minutes. The resulting solution is then titrated against N/10 HCl using methyl orange as indicator. The volume of the acid used is noted and calculations are made as shown above.

EXERCISES

(Based on Question Papers of Different Universities)

Multiple Choice Questions (Choose the correct option)

1. Which of the following is not a carboxylic acid derivative ?

(*a*) Ester (*b*) Acid amide

(*c*) Ethers (*d*) Acid chloride

2. The IUPAC name of $HCOOC_6H_5$ is

(*a*) phenyl formate (*b*) phenyl methanoate

(*c*) benzyl methanoate (*d*) none of the above

3. Esters are obtained by the reaction between

(*a*) acid chloride and alcohol (*b*) acid anhydride and phenol
(*c*) acid chloride and phenol (*d*) all the above

4. Isoamyl valerate has the fragrance of

(*a*) apple (*b*) banana
(*c*) orange (*d*) apricot

5. Reaction between ethyl formate and methyl magnesium bromide gives

(*a*) acetic acid (*b*) acetaldehyde
(*c*) ethyl alcohol (*d*) acetone

6. Higher boiling point of acid amide is due to

(*a*) intramolecular hydrogen bond (*b*) intermolecular hydrogen bonding
(*c*) inductive effect (*d*) electromeric effect

7. Arrange the acid derivative ethyl acetate, acetyl chloride, acetamide and acetic anhydride in order of decreasing activity.

(*a*) Acetyl chloride > Acetic anhydride > Ethyl acetate > Acetamide
(*b*) Acetic anhydride > Acetyl chloride > Acetamide > Ethyl acetate
(*c*) Acetamide > Ethyl acetate > Acetic anhydride > Acetyl chloride
(*d*) Ethyl acetate > Acetic anhydride > Acetyl chloride > Acetamide

ANSWERS

1. (*c*) **2.** (*b*) **3.** (*d*) **4.** (*a*) **5.** (*b*) **6.** (*b*) **7.** (*a*)

Short Answer Questions

1. Give the reaction and reagent for the formation of acetyl chloride.

2. Complete the following :

(*i*) $CH_3CONH_2 + P_2O_5 \longrightarrow$

(*ii*) $CH_3CONH_2 + Na\ NO_2 + HCl \longrightarrow$

(*iii*) $CH_3COOH + P + Br_2 \longrightarrow$

3. Write down the product of the reaction:

$$C_2H_5 - \underset{\underset{O}{\|}}{C} - NH_2 \xrightarrow{OBr^-} ?$$

4. Which of the following can be most easily hydrolysed and why ?

(*i*) $(CH_2CO)_2O$ (*ii*) CH_3COCl (*iii*) CH_3CONH_2 (*iv*) $CH_3COOC_2H_5$

5. Complete the following :

$$CH_3CONH_2 + \xrightarrow[\Delta]{P_2O_5}$$

6. Complete the following :

$$(i)\ CH_3 - \overset{\overset{O}{\|}}{C} - OC_2H_5 \xrightarrow{Na/C_2H_5OH} C$$

(ii) $C_6H_5 - COCl + H_2 \xrightarrow[BaSO_4]{Pd} D$

7. Which of the following can be most easily hydrolysed and why ?
CH_3COCl, C_2H_5Cl, $(CH_3CO)_2O$, CH_3CONH_2

8. Discuss the order of ease of hydrolysis of
(i) acid chlorides (ii) esters
(iii) acid amides (iv) acid anhydrides

9. Why acetic anhydride is preferred to acetyl chloride for acetylation reactions ?

10. Complete the following :

$C_6H_5CONH_2 \xrightarrow[\Delta]{P_2O_5}$

11. Write briefly why CH_3CONH_2 is amphoteric in nature.

12. Starting with malonic ester, how will you prepare succinic acid ?

13. Why acetyl chloride is more reactive than acetamide ?

14. How would you prepare ethylamine by Hoffmann degradation of amides ?

15. How will you convert an acid chloride into an amide, an ethyl ester and an amide ?

16. Which of the following can be hydrolysed most easily and why ?
acetic anhydride, acetyl chloride, acetamide and ethyl acetate.

17. Write a short note on Hoffmann Bromamide reaction.

18. CH_3CONH_2 is amphoteric in nature. Explain.

19. Complete the following reactions :

(i) $CH_2 = C = O \xrightarrow{CH_3COOH} A \xrightarrow{C_2H_5OH} B \xrightarrow{NH_3} C$

(ii) $C_6H_5COOC_2H_5 \xrightarrow{NH_3} A \xrightarrow{P_2O_5} B \xrightarrow{\text{Hydrolysis}} C$

(iii) $CH_3CH_2COOH \xrightarrow{PCl_5} A \xrightarrow{CH_3COONa} B$

20. Write a short note on trans-esterification.

21. Why does acetyl chloride get hydrolysed more easily than ethyl acetate ? Explain.

22. Explain the following :
(i) Trans-esterification (ii) Amphoteric nature of amides

23. Give the mechanism of alkaline hydrolysis of esters.

24. Account for : Acyl halides are most reactive of all the acid derivatives.

25. Give the mechanism of esterification.

26. How does the presence of electron-donating group retard the rate of basic hydrolysis ?

27. Give the mechanism of esterification.

28. Discuss Hoffmann's Bromamide reaction of amides. Give its mechanism.

29. Discuss the acid or basic hydrolysis of amides.

30. Give a brief account of trans-esterification reaction. Give its mechanism.

31. Discuss amphoteric behaviour of amides.

32. Give reasons why acetamide is hydrolysed very slowly with water but rapidly in the presence of an acid or alkali.

33. Discuss the mechanism of hydrolysis of ethyl acetate.

34. Outline the mechanism of laboratory preparation of ethyl acetoacetate.

35. Why does acetyl chloride get hydrolysed more easily than ethyl acetate ?

36. What are amides ? Discuss amphoteric nature of amides.

37. What is esterification ? Discuss its mechanism.

General Questions

1. What are acid chlorides ? How can acetyl chloride be converted into :

(*i*) acetamide (*ii*) acetic anhydride

(*iii*) ethyl acetate (*iv*) ethyl methyl ketone ?

2. What are acid amides ? How can they be prepared from other acid derivatives ? Discuss their important physical properties and chemical reactions.

3. Why C – N bond in amides is shorter than C – N bond in amines ? Give how amides can be prepared from :

(*i*) an acid (*ii*) an acid chloride (*iii*) an acid anhydride ?

4. Discuss the relative stability and order of reactivity of acid derivatives towards nucleophilic substitution reactions.

5. What are acid anhydrides ? How can they be prepared ? Discuss their important physical and chemical properties.

6. What is the mechanism of nucleophilic acyl substitution ? How can you justify the observed order of reactivity of acid derivatives in terms of this mechanism ?

7. Describe the methods of preparation and important chemical reactions of esters.

8. Arrange the following acid derivatives in decreasing order of reactivity with reason.

$$RCOOR,\ RCOCl,\ RCONH_2,\ \begin{matrix} RCO \diagdown \\ \quad O \\ RCO \diagup \end{matrix}$$

9. An amide (A) having the molecular formula 'C_3H_7ON' on hydrolysis gives an acid '$C_3H_6O_2$' which upon chlorination in presence of red phosphorus produces chloro acid. The latter on boiling with aq. NaOH and subsequent acidification forms lactic acid. Trace the reaction and deduce the structural formula of (A).

10. What do you understand by nucleophilic acyl substitution ? What is the role of an acid in it ? Give its mechanism. How does it differ from nucleophilic alkyl substitution ?

11. How can acetyl chloride be converted into

(*i*) Acetamide (*ii*) Acetic anhydride

(*iii*) Ethyl acetate (*iv*) Acetaldehyde ?

12. Give the mechanism of acidic and basic hydrolysis of acid derivatives.

13. Arrange these in order of decreasing order of reactivity towards nucleophilic substitution. Is an ester a carboxylic acid derivative or not ?

14. Describe the mechanism of acid-catalysed and base-catalysed hydrolysis of esters.

15. How will you explain the greater reactivity of acyl halides than alkyl halides towards nucleophilic attack ?

16. Describe the synthetic applications of ethylacetoacetic ester.

17. Give the mechanism and reaction of acetyl chloride with the following reagents :

(*i*) Ethanol (*ii*) C_2H_5OH/Na (*iii*) Acetic acid

18. How can you convert a carboxylic acid into:

(*a*) Amide (*b*) Acid chloride

(*c*) Anhydride ?

19. Explain why acetyl chloride gives nucleophilic substitution reactions much easily than ethyl chloride.

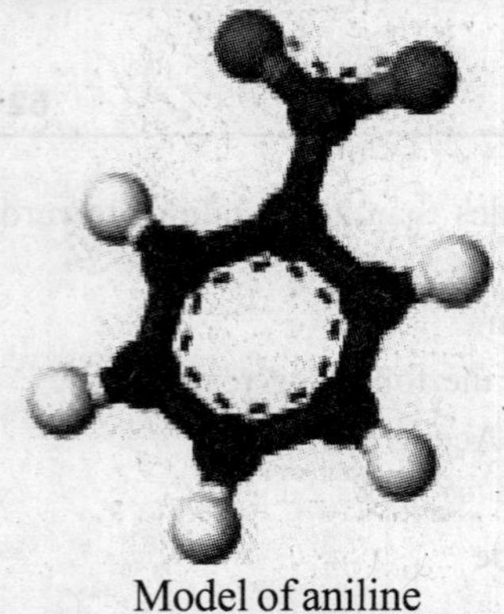

Model of aniline

Organic Compounds of Nitrogen

NITRO COMPOUNDS

8.1. INTRODUCTION

Nitro compounds are the compounds which contain at least one nitro group ($-NO_2$) in the molecule *e.g.,* nitromethane (CH_3NO_2), nitrobenzene ($C_6H_5NO_2$) etc.

Structure of Nitro group. Nitro group has the structure

$$-N\begin{matrix} \nearrow\!\!\!= O \\ \searrow O \end{matrix} \qquad \text{(I)}$$

Thus there is one nitrogen-oxygen double bond and one nitrogen-oxygen co-ordinate bond. It is also represented as

$$-\overset{+}{N}\begin{matrix} = O \\ - O^- \end{matrix} \qquad \text{(II)}$$

Structure II contributes to a greater extent to the actual structure of nitro group, as it is resonance stabilized

$$-\overset{+}{N}\begin{matrix} = O \\ - O^- \end{matrix} \longleftrightarrow -\overset{+}{N}\begin{matrix} - O^- \\ = O \end{matrix}$$

The above two resonating structures have equal energies. This imparts extra stability to the structure of nitro group. Single and double bonds are exchanging their positions, hence single structure of $-NO_2$ group could be written as

$$\left[-\overset{\oplus}{N}\begin{matrix} \cdots O \\ \cdots O \end{matrix}\right]^{\ominus}$$

Orbital diagram of nitro group is represented as

p-orbital of nitrogen

p-orbital of oxygen

R — N

O

O

Difference between nitro and nitrite groups

Consider the following two compounds

$$R-\overset{+}{N}\begin{matrix}\nearrow O \\ \searrow O^-\end{matrix} \qquad\qquad R-O-N=O$$

(*a*) (*b*)

Out of the two, compound (*a*) is a nitro compound whereas compound (*b*) is a nitrite. The difference between the two is obvious.

In nitro compound, the alkyl (or aryl) group is directly linked to nitrogen atom, whereas in the nitrite, alkyl group is linked to oxygen.

Some points of distinction between nitro compounds and nitrites are given in the table below.

Distinction between Nitroalkanes and Alkyl nitrites

No.	*Property*	*Nitroalkanes*	*Alkyl Nitrite*
1.	*Boiling point*	High	Low
2.	*Reduction (Sn/HCl)*	A primary amine is formed. $RNO_2 + 6H \xrightarrow{Sn/HCl} RNH_2 + 2H_2O$	Alcohol is formed. $RONO + 4H \longrightarrow ROH + NH_2OH$
3.	*Action of alkalis*	Salt formation takes places. $RCH_2NO_2 + NaOH \longrightarrow RCHNaNO_2 + H_2O$	Hydrolysis takes place with the formation of alcohol. $RONO + NaOH \longrightarrow ROH + NaNO_2$
4.	*Action of HNO_2*	Nitrolic acid is formed. $RCH_2NO_2 + HNO_2 \longrightarrow R-\underset{\underset{NOH}{\parallel}}{C}-NO_2 + H_2O$	No action.

8.2. NOMENCLATURE OF NITRO COMPOUNDS

Nitro compounds are named by writing the word nitro before the name of parent compound. Nitro compounds may be divided into two categories *viz.*, aliphatic nitro compounds and aromatic nitro compounds. The nomenclature of compounds from both categories is given below.

Aliphatic nitro compounds

Compound	*Name*
CH_3NO_2	Nitromethane
$CH_3CH_2NO_2$	Nitroethane
$CH_3CH_2CH_2NO_2$	1-nitropropane
$CH_3CH_2\underset{\underset{NO_2}{\mid}}{CH}-CH_3$	2-nitrobutane

Aromatic nitro compounds

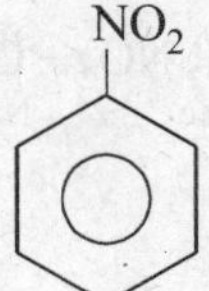

Nitrobenzene

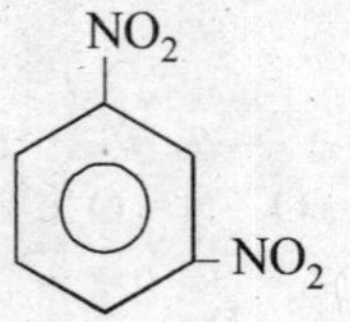

NO_2 / NO_2 (benzene ring structure) — *m*-dinitrobenzene

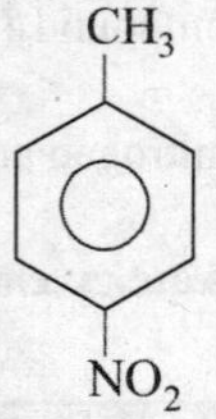

CH_3 / NO_2 (benzene ring structure) — *p*-nitrotolune

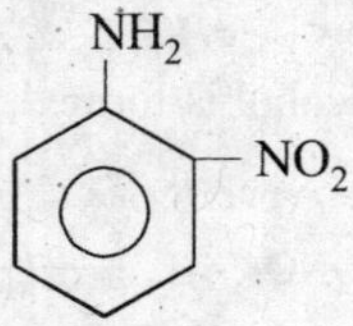

NH_2 / NO_2 (benzene ring structure) — *o*-nitroaniline

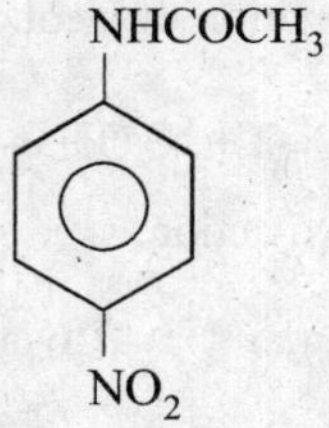

$NHCOCH_3$ / NO_2 (benzene ring structure) — *p*-nitroacetanilide

8.3. METHODS OF PREPARATION OF NITROALKANES OR ALIPHATIC NITRO COMPOUNDS

Methods of preparation of nitroalkanes are given below:

1. Nitration of alkanes. It is carried out by heating the alkane with nitric acid in vapour phase at 700 K. One hydrogen of the alkane is replaced by nitro group.

$$\underset{\text{Alkane}}{R-H} + HNO_3 \longrightarrow \underset{\text{Nitroalkane}}{R\,NO_2} + H_2O$$

However, this reaction does not give one single product in pure state. At the high temp. of 700 K, hydrocarbons are broken into smaller fragments and consequently we obtain a mixture of different nitroalkanes.

$$\underset{\text{Propane}}{CH_3-CH_2-CH_3} \xrightarrow[700\text{ K}]{HNO_3} \underset{\text{1-Nitropropane}}{CH_3-CH_2-CH_2\,NO_2} + \underset{\text{2-Nitropropane}}{CH_3\overset{\overset{NO_2}{|}}{C}H-CH_3}$$

$$+ \underset{\text{Nitroethane}}{CH_3\,CH_2\,NO_2} + \underset{\text{Nitromethane}}{CH_3\,NO_2}$$

2. Action of nitrous acid on α-halogen acid. On boiling an aqueous solution of sodium nitrite with an α-halogen carboxylic acid, nitroalkane is obtained

$$\underset{\text{Chloroacetic acid}}{\underset{|\atop Cl}{CH_2COOH}} + NaNO_2 \longrightarrow \underset{\text{Nitroacetic acid}}{\underset{|\atop NO_2}{CH_2COOH}} \xrightarrow{\Delta} \underset{\text{Nitro-methane}}{CH_3NO_2} + CO_2$$

3. Action of silver nitrite on alkyl halide. Alkyl halides undergo nucleophilic substitution with silver nitrite to yield nitroalkanes.

$$\underset{\text{Ethyl iodide}}{C_2H_5I} + AgNO_2 \longrightarrow \underset{\text{Nitroethane}}{C_2H_5NO_2} + AgI$$

4. Oxidation of amines. An amino group attached to a tertiary carbon atom is oxidised with potassium permanganate to give nitroalkane.

$$\underset{\text{Tert.butylamine}}{H_3C-\overset{CH_3}{\underset{CH_3}{C}}-NH_2} + 3[O] \longrightarrow \underset{\text{Tert. nitrobutane}}{CH_3-\overset{CH_3}{\underset{CH_3}{C}}-NO_2} + H_2O$$

5. Hydrolysis of α-nitroalkenes. An α-nitroalkene on hydrolysis in the presence of an acid or alkali produces nitroalkane.

$$\underset{\text{2-Methyl-1 nitro-propene-1}}{CH_3-\overset{CH_3}{\overset{|}{C}}=CHNO_2} + H_2O \longrightarrow \underset{\text{Acetone}}{CH_3-\overset{CH_3}{\overset{|}{C}}=O} + \underset{\text{Nitromethane}}{CH_3NO_2}$$

6. Oxidation of oximes. Aldoximes and ketoximes on oxidation with trifluoroperoxy acetic acid yield primary and secondary nitroalkanes respectively.

$$\underset{\text{Aldoxime}}{RCH=NOH} \xrightarrow{[O]} RCH=\overset{O^-}{\overset{|}{N^+}}-OH \rightleftharpoons \underset{1^\circ\ \text{Nitroalkane}}{RCH_2NO_2}$$

$$\underset{\text{Ketoxime}}{R_2C=NOH} \xrightarrow{[O]} R_2C=\overset{O^-}{\overset{|}{N^+}}-OH \rightleftharpoons \underset{2^\circ\ \text{Nitroalkane}}{R_2CHNO_2}$$

8.4. PHYSICAL PROPERTIES OF NITRO COMPOUNDS

1. Physical state. Nitroalkanes are colourless oily liquids in the pure form.

2. Smell. They have a pleasant smell.

3. Solubility and density. Nitroalkanes are sparingly soluble in water. These compounds have density more than one *i.e.* they are heavier than water.

4. Boiling points. Nitro compounds have strong dipole-dipole interaction. The dipole moment is of the order of 3–4 D. Consequently, they posses higher boiling points than hydrocarbons of comparable molecular masses. However, most of them are steam volatile

8.5. CHEMICAL PROPERTIES OF NITROALKANES

1. Acidic nature of nitroalkanes. (Tautomerism) Nitroalkanes containing α-hydrogen atom exhibit acidic behaviour and form salts with strong alkalis.

$$RCH_2NO_2 + NaOH \longrightarrow [RCHNO_2]^- Na^+ + H_2O$$

Acidic nature of α-hydrogen containing nitroalkanes is due to the electron-withdrawing inductive effect of $-NO_2$ group. Also the anion obtained after releasing one proton gets resonance stabilized.

$$R-\overset{H}{\underset{H}{C}}-\overset{+}{N}(=O)-O^- \xrightarrow{OH^-} \left[R-\overset{-}{C}H-\overset{+}{N}(=O)-O^- \longleftrightarrow R-CH=\overset{+}{N}(-O^-)-O^-\right] + H_2O$$

Primary and secondary nitroalkanes exhibit tautomerism, the tautomeric forms being nitro form and acinitro form

$$\underset{\text{Nitro form}}{R-CH_2-\overset{+}{N}(=O)-O^-} \rightleftharpoons \underset{\text{Acinitro form}}{R-CH=\overset{+}{N}(-OH)-O^-}$$

2. Reduction. Nitroalkanes on reduction under different conditions give different products

(*a*) *In acidic medium*

$$\underset{\text{Nitroalkane}}{RNO_2} + 6[H] \xrightarrow{\text{Metal / Acid}} \underset{\text{Primary amine}}{RNH_2} + 2H_2O$$

(*b*) *Catalytic reduction*

$$\underset{\text{Nitro alkane}}{RNO_2} + 3H_2 \xrightarrow{Ni} \underset{\text{Primary amine}}{RNH_2} + 2H_2O$$

(*c*) *In neutral reducing medium*

$$RNO_2 + 4[H] \xrightarrow{Zn / NH_4Cl} \underset{\text{N-alkyl hydroxylamine}}{RNHOH} + H_2O$$

(*d*) *With $SnCl_2$ and HCl*

$$RCH_2NO_2 \xrightarrow[HCl]{SnCl_2} \underset{\text{N-alkyl hydroxylamine}}{RCH_2NHOH} + \underset{\text{Aldoxime}}{RCH=NOH}$$

3. Hydrolysis. (*a*) Primary nitro compounds on boiling with HCl are converted into a mixture of carboxylic acid and hydroxylamine

$$\underset{1^\circ\ \text{Nitroalkane}}{RCH_2NO_2} + H_2O \xrightarrow{HCl} \underset{\text{Carboxylic acid}}{RCOOH} + \underset{\text{Hydroxyl amine}}{NH_2OH}$$

(*b*) Secondary nitroalkanes are hydrolysed to ketones on boiling with HCl

$$\underset{2^\circ\ \text{Nitroalkane}}{2R_2CHNO_2} \xrightarrow{HCl} \underset{\text{Ketone}}{2R_2CO} + \underset{\text{Nitrous oxide}}{N_2O} + H_2O$$

Tertiary nitroalkanes are not hydrolysed by acids.

4. Reaction with nitrous acid. (*a*) Primary nitroalkanes react with nitrous acid to form nitrolic acid which dissolves in alkali to give red coloured solution.

$$\underset{\substack{\text{Primary nitro}\\\text{compound}}}{R-\underset{|}{\overset{}{CH_2}}(NO_2)} + O=NOH \longrightarrow \underset{\text{Nitrolic acid}}{R-C(NO_2)=NOH} + H_2O$$

(*b*) Secondary nitroalkanes react with nitrous acid to produce pseudo nitroles which give blue colour with alkali.

$$\underset{2^\circ\text{ Nitro compound}}{R_2C\begin{smallmatrix}H\\NO_2\end{smallmatrix}} + \underset{\text{Nitrous acid}}{HO-NO} \longrightarrow R_2C\begin{smallmatrix}NO\\NO_3\end{smallmatrix} \quad \text{Pseudo nitrole}$$

Tertiary nitroalkanes do not react with nitrous acid since they do not contain any α-hydrogen atoms.

5. Halogenation. Primary and secondary nitro-compounds are halogenated in α position in alkaline solution.

(*a*)

$$\underset{\text{Primary nitro alkane}}{RCH_2NO_2} \xrightarrow{OH^-} [RCH=N^-O_2]Na^+ \xrightarrow[NaOH]{Br_2} \underset{\alpha\text{-bromo nitroalkane}}{RCH(Br)-NO_2} + NaBr$$

$$\xrightarrow{NaOH} [RCBr=NO_2^-]Na^+ \xrightarrow{Br_2/NaOH} \underset{\alpha,\ \alpha\text{-dibromo-nitroalkane}}{RCBr_2NO_2} + NaBr$$

(*b*)

$$\underset{\text{Secondary nitroalkane}}{R_2CHNO_2} \xrightarrow{NaOH} [R_2C=NO_2]Na^+ \xrightarrow[NaOH]{Br_2} \underset{\text{2-bromo product}}{R_2CBrNO_2} + NaBr$$

Nitromethane has three α-hydrogens and gives trihalogen product

$$\underset{\text{Nitro-methane}}{CH_3NO_2} + 3Cl_2 + NaOH \longrightarrow \underset{\text{Chloro-picrin}}{Cl_3CNO_2} + 3NaCl + 3H_2O$$

6. Reaction with Grignard Reagent. The aci-form of nitroalkane reacts with Grignard reagent as follows.

$$\underset{\text{Nitroalkane}}{RCH=N\begin{smallmatrix}OH\\O\end{smallmatrix}} + CH_3MgI \longrightarrow \underset{\text{Methane}}{CH_4} + RCH=N\begin{smallmatrix}OMgI\\O\end{smallmatrix}$$

7. Action of heat. Nitroalkanes on heating to 573 K produce alkenes

$$\underset{\text{Nitroalkane}}{RCH_2CH_2NO_2} \xrightarrow{573\text{ K}} \underset{\text{Alkene}}{RCH=CH_2} + HNO_2$$

8. Condensation with aldehydes. Primary and secondary nitroalkanes undergo aldol type reaction with aldehydes and ketones in the presence of a base to form nitroalcohols

$$\underset{\text{Acetaldehyde}}{CH_3-\underset{H}{C}=O} + \underset{\text{Nitroethane}}{\overset{CH_3}{CH_2}-NO_2} \longrightarrow \underset{\text{3-nitro-2-butanol}}{CH_3-\underset{OH}{CH}-\overset{CH_3}{CH}-NO_2}$$

$$R-\overset{H}{\underset{H}{C}}-\overset{+}{N}(=O)O^- + OH^- \longrightarrow \left[CH_3-\overset{-}{\underset{H}{C}}-\overset{+}{N}(=O)O^- \longrightarrow CH_3-\underset{H}{C}=\overset{+}{N}(O^-)O^-\right] + H_2O$$

$$CH_3-\overset{H}{C}(=O) + \left[\overset{CH_3}{\underset{H}{\overset{|}{C}^-}}-\overset{+}{N}(=O)O^- \longleftrightarrow \overset{CH_3}{\underset{H}{C}}=\overset{+}{N}(O^-)O^-\right] \longrightarrow CH_3-\overset{H}{\underset{O^-}{C}}-\overset{CH_3}{CH}-\overset{+}{N}(=O)O^- \xrightarrow{H-O-H} CH_3-\overset{H}{\underset{OH}{C}}-\overset{CH_3}{CH}-NO_2 + OH^-$$

9. Mannich reaction. Nitro compounds having α-hydrogen atom condense with formaldehyde and ammonia as follows.

$$R_2CHNO_2 + HCHO + NH_4Cl \longrightarrow R_2\overset{NO_2}{C}-CH_2NH_2HCl + H_2O$$

Nitroamine hydrochloride

8.6. METHODS OF PREPARATION OF AROMATIC NITRO COMPOUNDS

Methods of preparation of aromatic nitro compounds are given below:

Aromatic nitro compounds are prepared by the nitration of benzene or its derivatives. Following nitrating agents are generally employed.

(*i*) In the case of aromatic compounds containing electron releasing agents such as $-NH_2$; $-OH$ and $-CH_3$, the reaction takes place with conc. HNO_3 as the nitrating agent. In some cases even dil. HNO_3 can carry out nitration at room temp.

$$\underset{\text{Phenol}}{C_6H_5OH} + HNO_3\text{ (dil.)} \xrightarrow{-H_2O} \underset{o\text{-nitrophenol}}{o\text{-}O_2NC_6H_4OH} + \underset{p\text{-nitrophenol}}{p\text{-}O_2NC_6H_4OH}$$

As the electron releasing groups are activating groups, further substitution in the benzene ring is easier and the nitro group takes either the *ortho* or *para* position relative to the group already present.

(*ii*) For aromatic compounds containing electron-withdrawing group in the ring, or when it is desired to introduce more than one nitro group, a strong nitrating agent is required and the reaction requires a higher temperature. Here a mixture of conc. HNO_3 and conc. H_2SO_4 is used as the nitrating agent

$$\text{Benzene} \xrightarrow[\text{328 K}]{\text{Conc. HNO}_3 / \text{Conc. H}_2\text{SO}_4} \text{Nitrobenzene} \xrightarrow[\text{373 K}]{\text{Conc. HNO}_3 / \text{Conc. H}_2\text{SO}_4} m\text{-dinitrobenzene}$$

Mechanism of Nitration

Nitration of aromatic compounds is an electrophilic substitution reaction, in which the hydrogen of the ring is substituted by $-NO_2$ group. The electrophile involved is NO_2^+ nitronium ion which is released from the mixture of HNO_3 and H_2SO_4.

Different steps involved in the process of nitration are as follows:

(*i*) **Formation of electrophile**

$$HNO_3 + H_2SO_4 \longrightarrow \underset{\text{Nitronium ion}}{\overset{+}{N}O_2} + H_3\overset{+}{O} + 2HSO_4^-$$

(*ii*) **Attachment of NO_2^+ to the ring.**

$$\overset{\oplus}{NO_2} + C_6H_6 \longrightarrow [\text{resonance structures of } C_6H_6NO_2^+] \quad \text{Or} \quad \text{Intermediate product}$$

Intermediate product

(*iii*) **Proton transfer to yield the final product.**

$$C_6H_6NO_2^+ + HSO_4^{\ominus} \longrightarrow \underset{\text{Nitrobenzene}}{C_6H_5NO_2} + H_2SO_4$$

8.7. PHYSICAL PROPERTIES OF AROMATIC NITRO COMPOUNDS

1. Physical state Nitrobenzene is a yellow oily liquids. Other compounds are yellow crystalline solids. Nitrobenzene has the order of bitter almonds. It is also known as *oil of mirbane*.

2. Solubility. Aromatic nitro compounds are insoluble in water but readily soluble in organic solvents such as alcohol and other.

3. Melting and boiling points of nitrobenzenes gradually increase with increase in the number of nitro groups. This is evident from the table below:

Compound	**M.P. (K)**	**B.P. (K)**
Ntrobenzene	287.7	483.0
m-Dinitrobenzene	363.8	576.0
Sym-Trinitrobenzene	395.0	–

8.8. CHEMICAL PROPERTIES

1. Electrophilic Substitution. Nitro group is a deactivating group. So further substitution does not take place easily in nitro compounds. To convert nitro benzene into *m*-dinitrobenzene, fuming HNO_3 is required.

$$C_6H_5NO_2 + HNO_3 \text{ (fuming)} \xrightarrow[373\text{ K}]{H_2SO_4} C_6H_4(NO_2)_2 + H_2O$$

Nitrobenzene → *m*-dinitrobenzene

2. Nucleophilic substitution. Nitro compounds easily undergo nucleophilic substitution reaction because of positive charge on the ring as a result of electron withdrawing effect of nitro group.

$$C_6H_5NO_2 + OH^- \longrightarrow \text{o-}HOC_6H_4NO_2 + \text{p-}HOC_6H_4NO_2$$

Nitrobenzene → *o*-nitrophenol + *p*-nitrophenol

The mechanism of this reaction (taking the case of *p*-nitrophenol) is as follows:

Nitrobenzene + $\bar{O}H$ ⟷ [resonance structures of the intermediate: negative charge on ring positions ortho/para to the H, OH-bearing carbon and finally on O of N–O⁻] → *p*-nitrophenol (loss of H)

3. Substitution of nitro group in polynitro compounds. On or more nitro groups can be replaced by other nucleophilic groups.

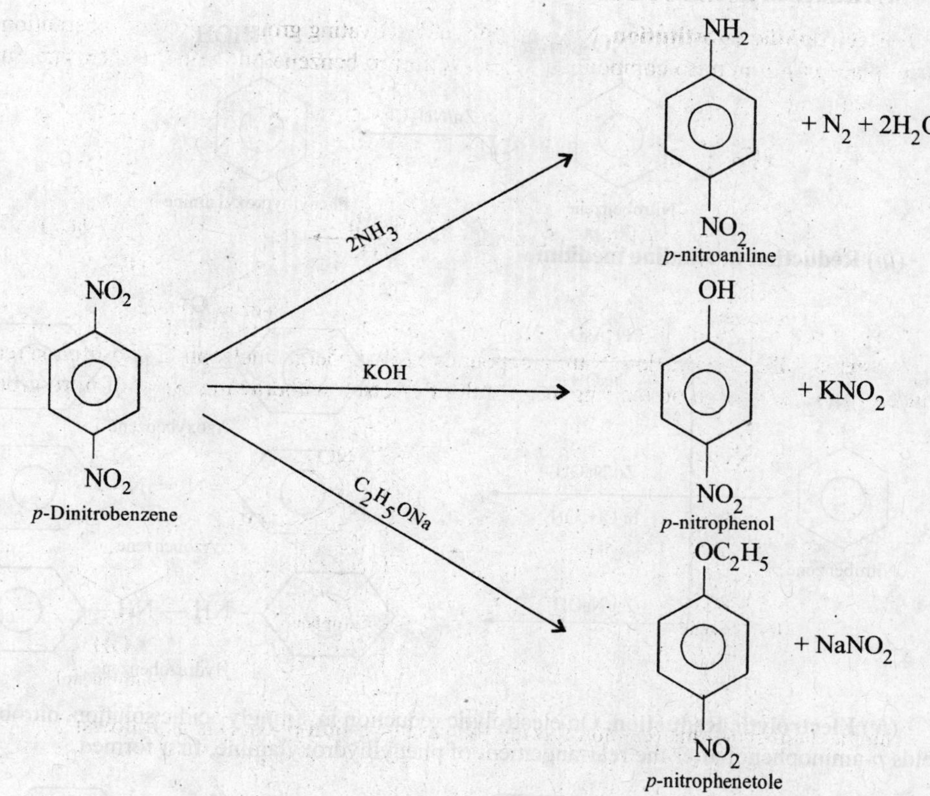

4. Reduction. Nitro compounds undergo reduction to give different products with different reducing agents. The reduction proceeds in stages involving the following intermediate compounds.

$$C_6H_5NO_2 \xrightarrow{[H]} C_6H_5NO \xrightarrow{[H]} C_6H_5NHOH \xrightarrow{[H]} C_6H_5NH_2$$

Nitrobenzene → Nitrosobenzene → Phenyl hydroxylamine → Aniline

Actual product obtained depends upon the reducing agent used

(*i*) Reduction in acidic medium.

$$C_6H_5NO_2 \xrightarrow[\text{or Sn/HCl}]{\text{Zn/HCl}} C_6H_5NH_2 + 2H_2O$$

Nitrobenzene → Aniline

(*ii*) **Reduction in neutral medium**

$$C_6H_5NO_2 \xrightarrow{Zn/NH_4Cl} C_6H_5NHOH$$

Nitrobenzene → Phenylhydroxyl amine

(*iii*) **Reduction in alkaline medium**

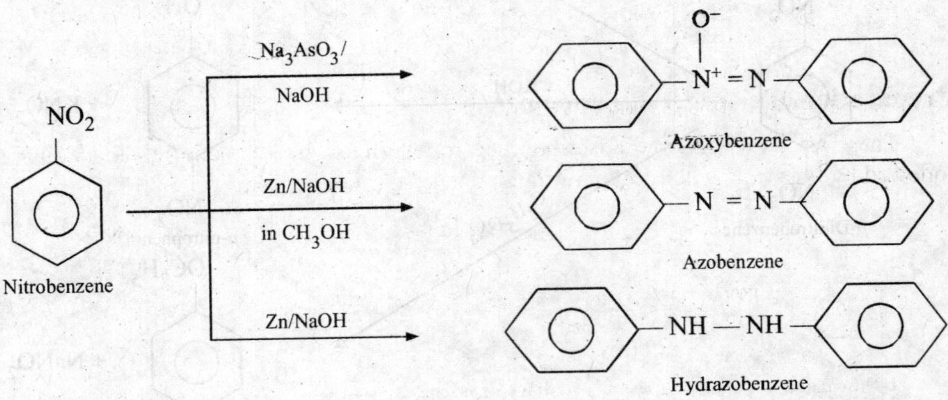

(*iv*) **Electrolytic Reduction.** On electrolytic reduction in strongly acidic solution, nitrobenzene yields *p*-aminophenol after the rearrangement of phenylhydroxylamine, first formed.

$$C_6H_5NO_2 \xrightarrow{\text{Reduction}} C_6H_5NHOH \xrightarrow{\text{Rearrange-ment}} HO{-}C_6H_4{-}NH_2$$

Nitrobenzene → Phenyl-hydroxylamine → *p*-aminophenol

(*v*) **Catalytic Reduction**

$$C_6H_5NO_2 + 3H_2 \xrightarrow{Ni} C_6H_5NH_2 + 2H_2O$$

Nitrobenzene → Aniline

(*vi*) **Selective reduction.** It is possible to reduce one nitro group out of two in a dinitro compound by taking a mild reducing agent such as amm. hydrosulphide NH_4SH or sodium polysulphide Na_2S_x.

$$C_6H_4(NO_2)_2 + 3NH_4SH \longrightarrow NO_2{-}C_6H_4{-}NH_2 + 3S + 2H_2O + 3NH_3$$

m-Dinitrobenzene → *m*-Nitroaniline

8.9. DINITROBENZENES

There are three dinitrobenzenes. They differ in the relative position of nitro groups.

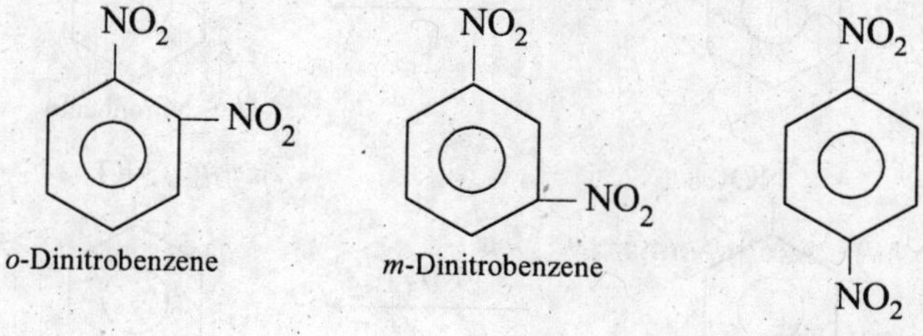

o-Dinitrobenzene *m*-Dinitrobenzene *p*-Dinitrobenzene

Preparation of *o*- and *p*-dinitrobenzene

These are obtained from corresponding nitroanilines, by subjecting them to diazotisation followed by treatment with sodium nitrite in the presence of copper.

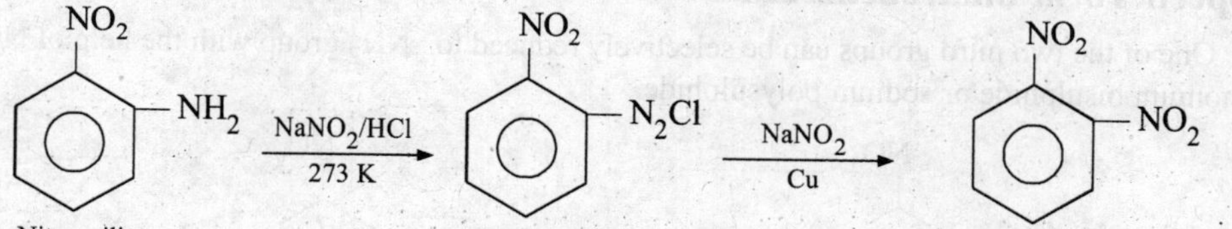

$$o\text{-Nitroaniline} \xrightarrow[273\text{ K}]{NaNO_2/HCl} o\text{-Nitrobenzene diazonium chloride } (N_2Cl) \xrightarrow[Cu]{NaNO_2} o\text{-Dinitrobenzene}$$

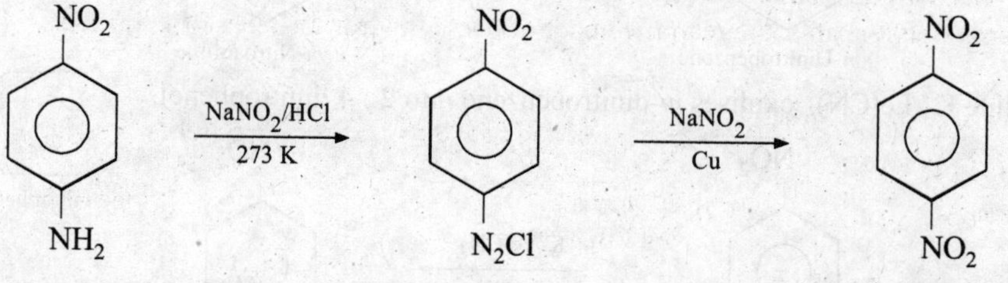

$$p\text{-Nitroaniline} \xrightarrow[273\text{ K}]{NaNO_2/HCl} p\text{-Nitrobenzene diazonium chloride } (N_2Cl) \xrightarrow[Cu]{NaNO_2} p\text{-Dinitrobenzene}$$

Preparation of *m*-dinitrobenzene

m-dinitrobenzene is prepared by heating nitrobenzene with a mixture of conc. HNO_3 and conc. H_2SO_4 at 373 K.

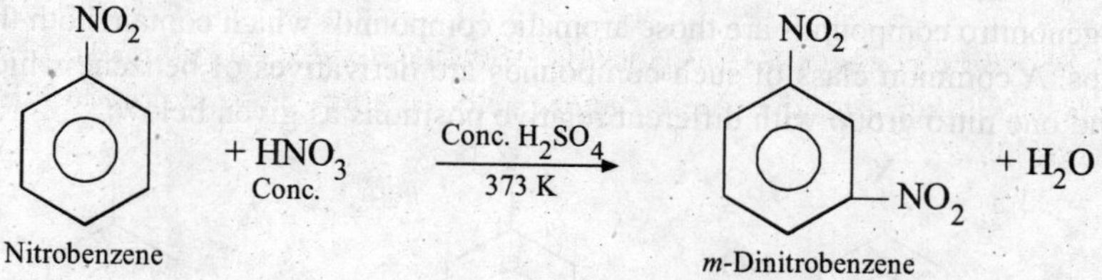

$$\text{Nitrobenzene} + \underset{\text{Conc.}}{HNO_3} \xrightarrow[373\text{ K}]{\text{Conc. } H_2SO_4} m\text{-Dinitrobenzene} + H_2O$$

Properties of *o*- and *p*-dinitrobenzenes

o- and *p*-dinitrobenzenes allow one of the nitro groups to be replaced by groups like $-NH_2$ and $-OH$ in nucleophilic substitution reactions.

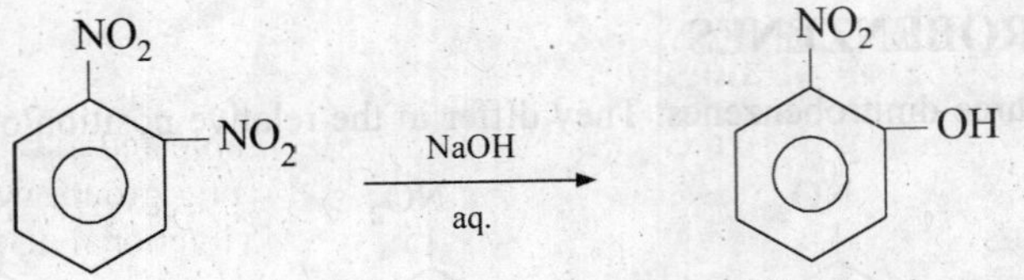

NO_2, NO_2 (o-dinitrobenzene) $\xrightarrow[\text{aq.}]{NaOH}$ NO_2, OH

o-Nitrophenol

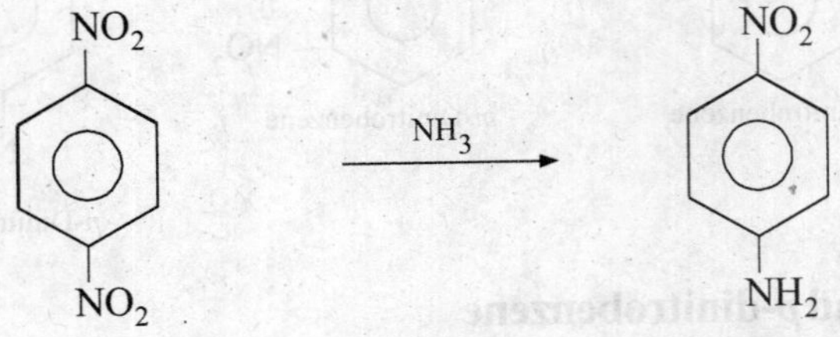

NO_2, NO_2 $\xrightarrow{NH_3}$ NO_2, NH_2

p-Dinitrobenzene *p*-Nitroaniline

Properties of *m*-dinitrobenzenes

One of the two nitro groups can be selectively reduced to $-NH_2$ group with the help of NH_4HS ammonium bisulphide or sodium polysulphide.

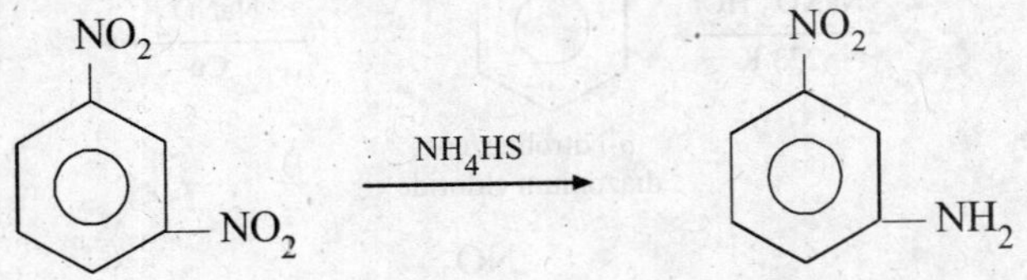

NO_2, NO_2 $\xrightarrow{NH_4HS}$ NO_2, NH_2

m-Dinitrobenzene *m*-Nitroaniline

Alkaline K_3 $Fe(CN)_6$ oxidises *m*-dinitrobenzene into 2 : 4 dinitrophenol.

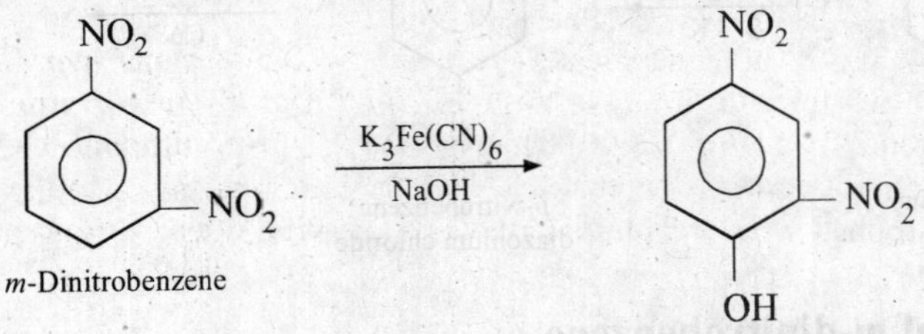

NO_2, NO_2 $\xrightarrow[NaOH]{K_3Fe(CN)_6}$ NO_2, NO_2, OH

m-Dinitrobenzene 2, 4 dinitrophenol

8.10. HALOGENONITRO COMPOUNDS

Halogenonitro compounds are those aromatic compounds which contain both the halogen and nitro groups. A common class of such compounds are derivatives of benzene which contain one halogen and one nitro group with different relative positions as given below:

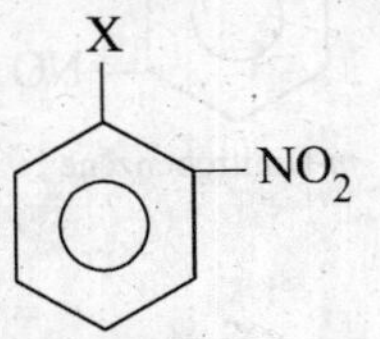

X, NO_2 X, NO_2 X, NO_2

o-halogenoitrobenzene 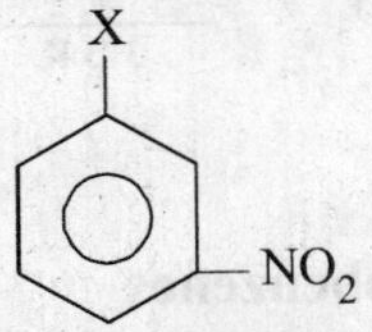 *m*-halogenonitrobenzene  *p*-halogenonitrobenzene

Preparation of *o*- and *p*-halogenonitrobenzenes

Halogen compound will be the starting materials. It is subjected to nitration with a mixture of nitric and sulphuric acids. As the halogen group is *o-p* directing group, we obtain a mixture of *o*- and *p*-halogenonitrobenzene. The mixture can be subjected to fractional distillation to separate them into pure substances.

$$\text{Chlorobenzene} + HNO_3 \xrightarrow[H_2SO_4]{\text{Conc.}} o\text{-chloronitrobenzene} + p\text{-chloronitrobenzene}$$

Preparation of *m*-halogenonitrobenzene

Here the starting material is nitrobenzene. It is subjected to halogenation in the presence of iron. As the nitro group is meta directing, we shall obtain *m*-halogenonitrobenzene.

$$\text{Nitrobenzene} \xrightarrow[Fe]{Cl_2} m\text{-chloronitrobenzene} + HCl$$

Properties of halogenonitrobenzenes

Nucleophilic substitution reactions. *o*- and *p*-halogenonitrobenzenes give nucleophilic substitution reactions easily. Thus the halogen group is easily replaced by –OH, $-NH_2$, –OR groups. There is one more observation that greater the member of nitro groups in *ortho* or *para* position easier is the nucleophilic substitution. We compare the ease of substitution of chloro group in 4-chloronitrobenzene, 2, 4-Dinitrochlorobenzene and 2, 4, 6-trinitrochlorobenzene. In the first case substitution reaction takes place with 15% NaOH solution, in the second case with a much weaker base, sodium carbonate and in the third case, warming with water is enough to perform the reaction.

$$\text{4-Chloronitrobenzene} \xrightarrow[423\ K]{15\%\ \text{Aq. NaOH}} \text{4-Nitrophenol} + NaCl$$

$$\text{2, 4-Dinitrochlorobenzene} \xrightarrow[403\ K]{\text{Aq. } Na_2CO_3} \text{2, 4-Dinitrophenol}$$

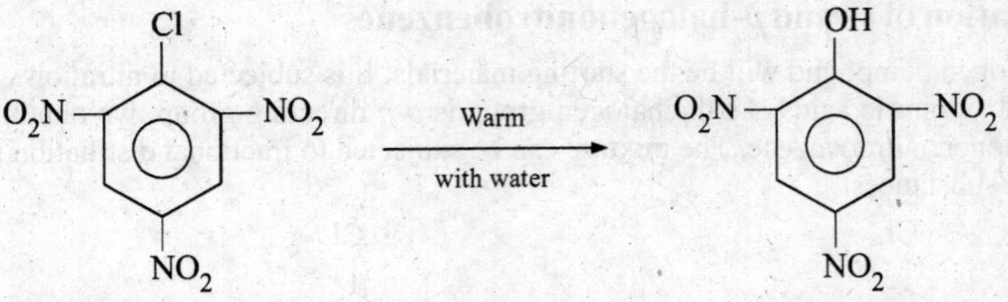

2, 4, 6 Trinitrochlorobenzene $\xrightarrow[\text{with water}]{\text{Warm}}$ 2, 4, 6 Trinitrophenol

Influence of nitro groups on the reactivity of halogen towards nucleophilic substitution.

Compare the reactivity of two compounds *viz.*, chlorobenzene and *o*-nitrochlorobenzene towards nucleophilic substitution reaction.

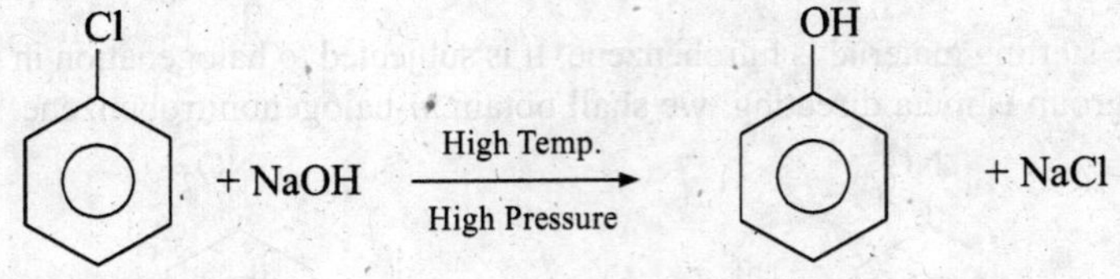

$$\text{Chlorobenzene} + NaOH \xrightarrow[\text{High Pressure}]{\text{High Temp.}} \text{Phenol} + NaCl$$

$$o\text{-Nitrochlorobenzene} + NaOH\ (15\%\ \text{Aq. sol}) \longrightarrow o\text{-Nitrophenol} + NaCl$$

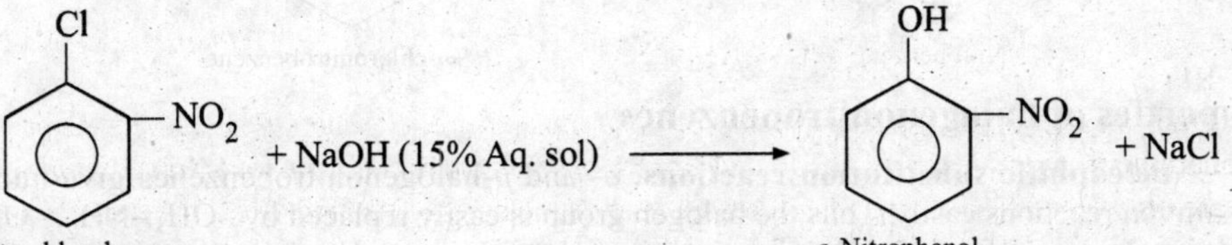

We find that in the second case, much easier conditions are required to perform the reaction. In a nucleophilic substitution reaction, the following steps are involved.

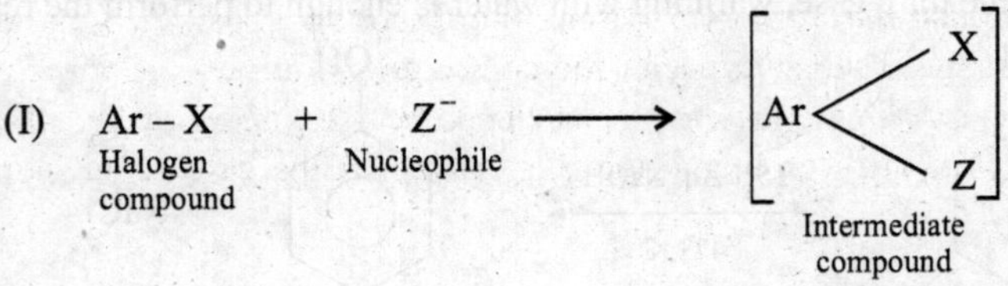

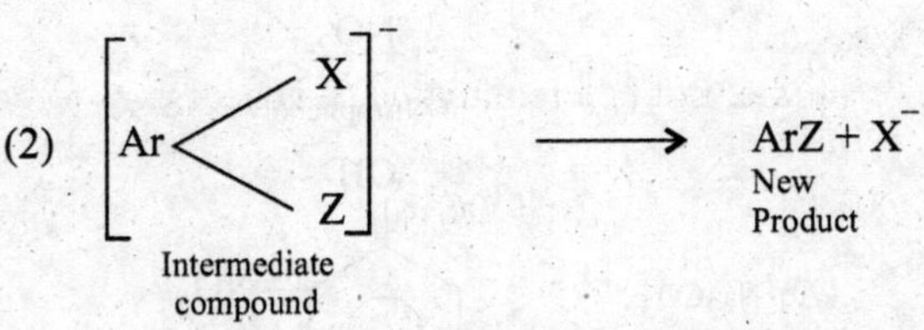

(I) $Ar-X$ (Halogen compound) $+$ Z^- (Nucleophile) $\longrightarrow$ $\left[Ar\langle^{X}_{Z}\right]^-$ (Intermediate compound)

(2) $\left[Ar\langle^{X}_{Z}\right]^-$ (Intermediate compound) $\longrightarrow$ $ArZ + X^-$ (New Product)

Thus for the reaction to take place efficiently, the intermediate compound needs to be stabilized. The intermediate compound, a carbonion, is resonance-stabilized in the presence of nitro group. But stabilisation is more if the nitro group is present is ortho or para position with respect to halogen group. This is shown below.

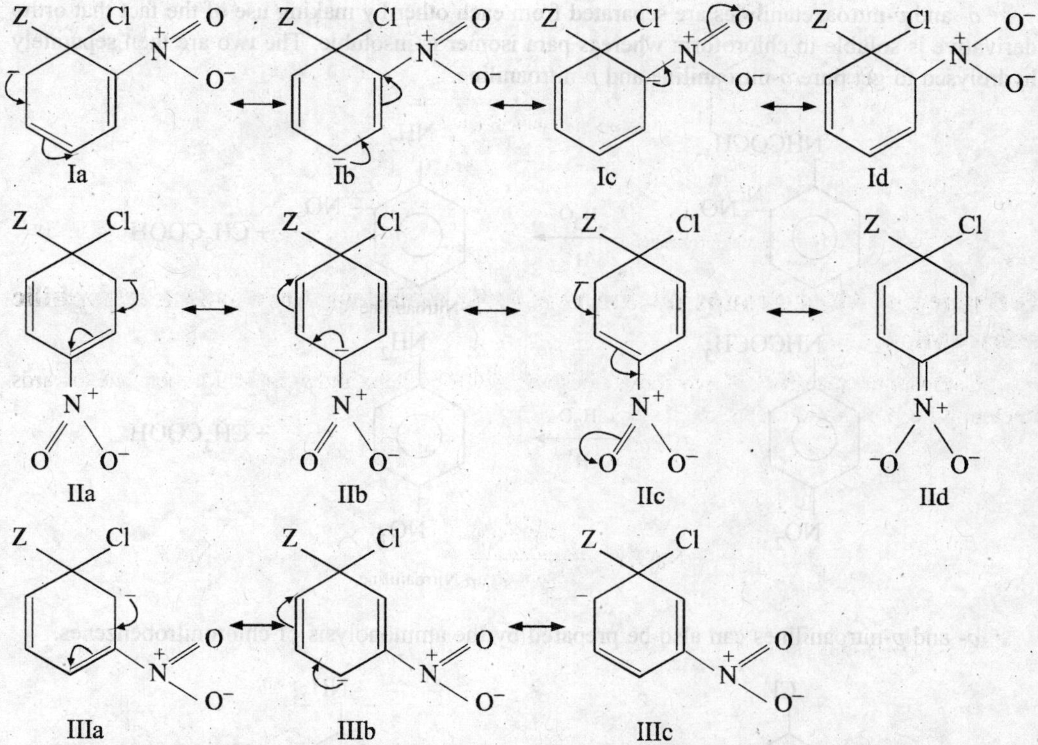

Ia Ib Ic Id

IIa IIb IIc IId

IIIa IIIb IIIc

It may be noted that in case of *ortho* or *para* nitrochlorobenzene, the intermediate product has 4 resonating structures each Ia, Ib, Ic and Id (for *ortho*) and IIa, IIb, IIc and IId (for *para*) whereas there are only three resonating structures in case of meta IIIa, IIIb and IIIC. Greater the number of resonating structures for a compound, greater is the stabilization. Hence intermediate compound is more stabilized in case of *ortho* and *para* nitrochlorobenzenes compared to *m*-nitrochlorobenzene.

Moreover structures Ic and IIb are particularly stable as there is a negative charge on the ring directly to neutralize the +ve charge on nitro group. No such stable structure is present in the case of meta isomer. Thus it is proved that nitro group if present in ortho or para position with respect to chloro group, enhances nucleophilic substitution for the chloro group.

Again, every nitro group has its effect to increase the nucleophilic substitution ability of chloro group. Hence greater the number of nitro groups in the ring, greater is the ease of nucleophilic substitution reaction.

8.11. *o*-, *m*- *p*-NITROANILINES

Preparation of *o*- and *p*-nitroanilines

These are prepared by the nitration of aniline. But before nitration, it is necessary to protect the $-NH_2$ group by acetylation.

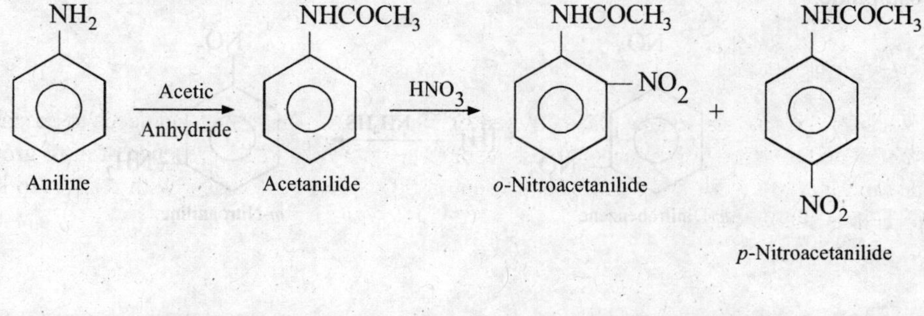

NH_2 (Aniline) $\xrightarrow{\text{Acetic Anhydride}}$ $NHCOCH_3$ (Acetanilide) $\xrightarrow{HNO_3}$ $NHCOCH_3$, NO_2 (*o*-Nitroacetanilide) + $NHCOCH_3$, NO_2 (*p*-Nitroacetanilide)

o- and *p*-nitroacetanilides are separated from each other by making use of the fact that ortho derivative is soluble in chloroform whereas para isomer is insoluble. The two are then separately hydrolysed to get pure *o*-nitroaniline and *p*-nitroaniline.

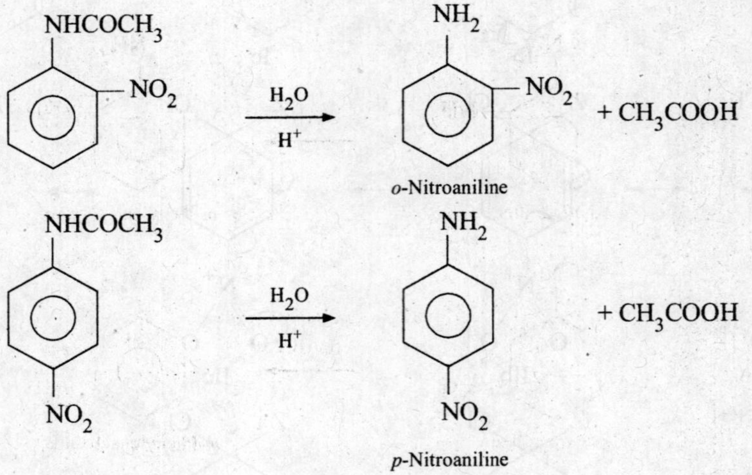

$$\text{(}o\text{-}NHCOCH_3, NO_2\text{ benzene)} \xrightarrow[H^+]{H_2O} \text{(}o\text{-}NH_2, NO_2\text{ benzene)} + CH_3COOH$$

o-Nitroaniline

$$\text{(}p\text{-}NHCOCH_3, NO_2\text{ benzene)} \xrightarrow[H^+]{H_2O} \text{(}p\text{-}NH_2, NO_2\text{ benzene)} + CH_3COOH$$

p-Nitroaniline

o- and *p*-nitroanilines can also be prepared by the ammonolysis of chloronitrobenzenes.

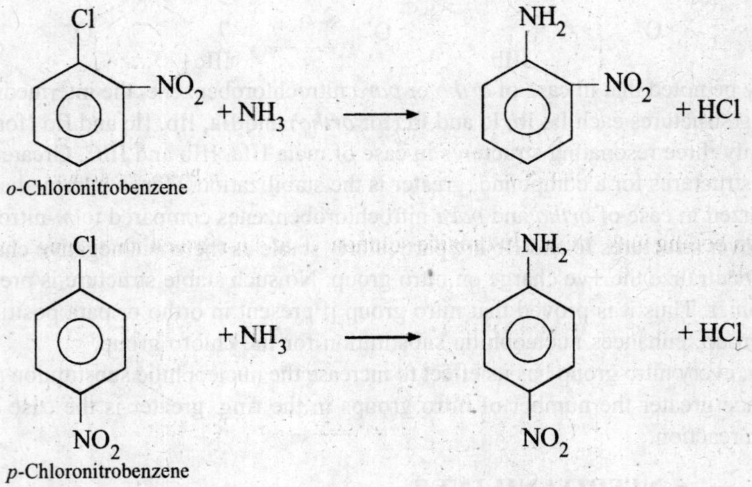

$$\text{(}o\text{-Chloronitrobenzene)} + NH_3 \longrightarrow \text{(}o\text{-}NH_2, NO_2\text{ benzene)} + HCl$$

o-Chloronitrobenzene

$$\text{(}p\text{-Chloronitrobenzene)} + NH_3 \longrightarrow \text{(}p\text{-}NH_2, NO_2\text{ benzene)} + HCl$$

p-Chloronitrobenzene

Preparation of *m*-nitroaniline

m-nitroaniline can be prepared by selective reduction of *m*-dinitrobenzene with ammonium hydrosulphide.

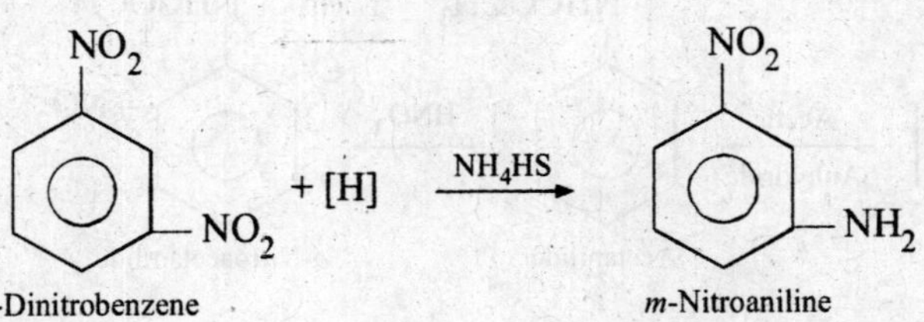

$$\text{(}m\text{-Dinitrobenzene)} + [H] \xrightarrow{NH_4HS} \text{(}m\text{-}NO_2, NH_2\text{ benzene)}$$

m-Dinitrobenzene *m*-Nitroaniline

Properties of nitroanilines

(*i*) *o*-, *m*- and *p*-nitroanilines on reduction with metal and acid, yield the corresponding phenylene diamines.

o-Nitroaniline $\xrightarrow{Sn/HCl}$ *o*-Phenylene diamine

m-Nitroaniline $\xrightarrow{Sn/HCl}$ *m*-Phenylene diamine

p-Nitroaniline $\xrightarrow{Sn/HCl}$ *p*-Phenylene diamine

(*ii*) On boiling with sodium hydroxide solution, *o*- and *p*-nitroanilines change into corresponding nitrophenols.

o-Nitroaniline $\xrightarrow{NaOH}$ *o*-Nitrophenol

p-Nitroaniline $\xrightarrow{NaOH}$ *p*-Nitrophenol

8.12. NITROPHENOLS

There are three nitrophenols as given below:

OH, NO_2 — *o*-Nitrophenol; OH, NO_2 — *m*-Nitrophenol; OH, NO_2 — *p*-Nitrophenol

Methods of preparation of *o*- and *p*-nitrophenols

(*i*) **By nitration of phenols.** Phenolic group is *o-p* directing group. Hence it will direct the nitro group in *o*- and *p*-positions on nitration of phenols

$$\text{Phenol} \xrightarrow[\text{HNO}_3]{\text{Dil.}} \textit{o}\text{-Nitrophenol} + \textit{p}\text{-Nitrophenol} + H_2O$$

The two are separated by steam distillation. *Ortho* isomer distils over leaving behind the *para* isomer.

(*ii*) **By hydrolysis of *o*- and *p*-chloronitrobenzenes**

$$\textit{o}\text{-Chloronitrobenzene} + NaOH \longrightarrow \textit{o}\text{-Nitrophenol} + NaCl$$

$$\textit{p}\text{-Chloronitrobenzene} + NaOH \longrightarrow \textit{p}\text{-Nitrophenol} + NaCl$$

Preparation of *m*-nitrophenol

It is prepared from *m*-dinitrobenzene as follows:

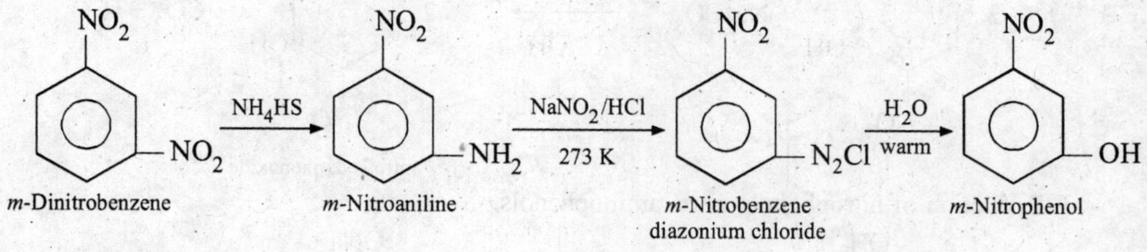

m-Dinitrobenzene $\xrightarrow{NH_4HS}$ *m*-Nitroaniline $\xrightarrow[273\ K]{NaNO_2/HCl}$ *m*-Nitrobenzene diazonium chloride (N_2Cl) $\xrightarrow[warm]{H_2O}$ *m*-Nitrophenol

Properties of nitrophenols

1. *Ortho* nitrophenol behaves differently as regards solubility in water. Out of the three isomers, *ortho* nitrophenol is least soluble in water. It can however be steam distilled out of the three isomers, *ortho* has the lowest m.p. followed by *meta* and *para* isomers. The different physical properties of *ortho* isomer from the other two can be explained in terms of *intramolecular* hydrogen bonding *i.e.* hydrogen bonding within the same molecule. The nitro and phenolic groups are so close to each other that a hydrogen bond, as shown below is formed.

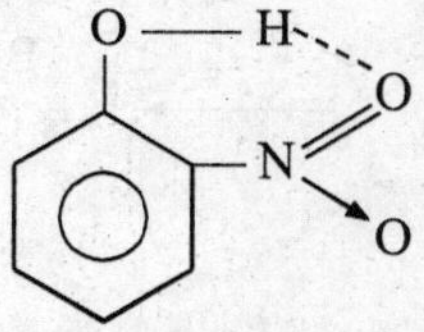

This eliminates the chances of hydrogen bond between the molecules of *o*-nitrophenol and water and hence accounts for its insolubility in water. The other isomers form intermolecular hydrogen bond with water as shown below and are water soluble.

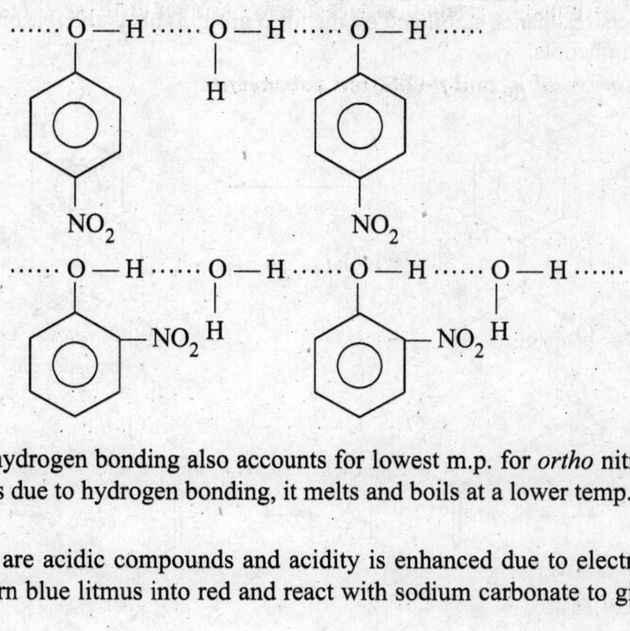

Intramolecular hydrogen bonding also accounts for lowest m.p. for *ortho* nitrophenol. As the molecular size shrinks due to hydrogen bonding, it melts and boils at a lower temp. compared to *m*- and *p*-isomers.

2. Nitrophenols are acidic compounds and acidity is enhanced due to electron withdrawing $-NO_2$ group. They turn blue litmus into red and react with sodium carbonate to give salts.

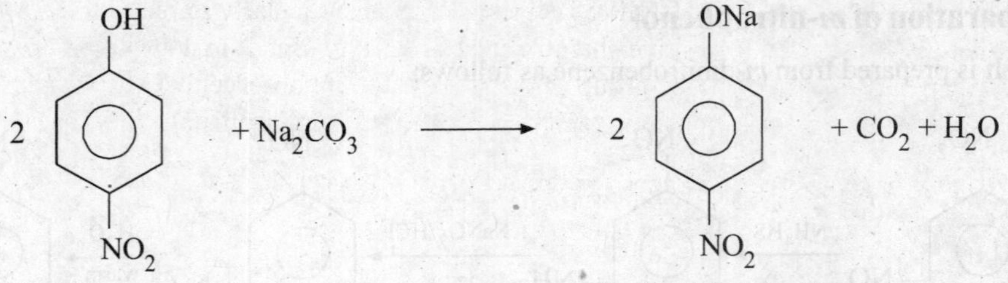

$$2\ p\text{-}NO_2C_6H_4OH + Na_2CO_3 \longrightarrow 2\ p\text{-}NO_2C_6H_4ONa + CO_2 + H_2O$$

p-nitro sod. phenoxide

3. Reduction of nitrophenols gives aminophenols.

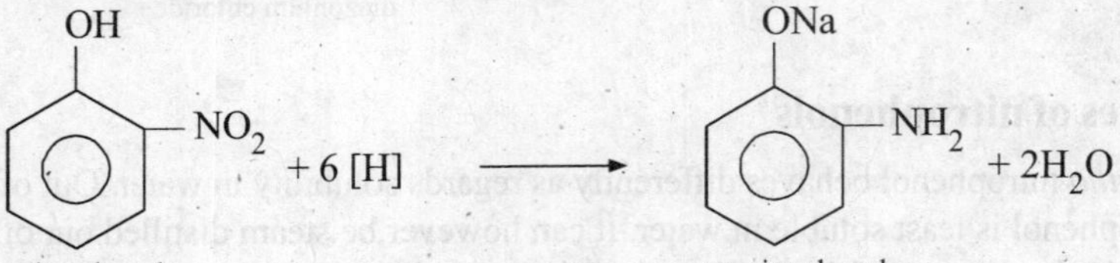

$$o\text{-}NO_2C_6H_4OH + 6\,[H] \longrightarrow o\text{-}NH_2C_6H_4ONa + 2H_2O$$

o-nitrophenol — *o*-aminophenol

4. Bromination yields tribromophenol

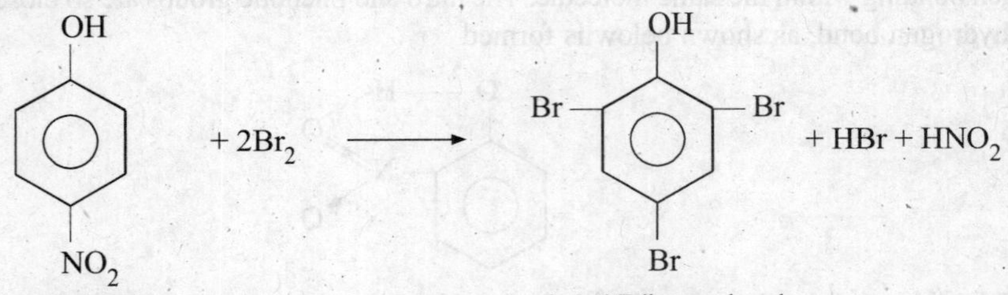

$$p\text{-}NO_2C_6H_4OH + 2Br_2 \longrightarrow C_6H_2Br_3OH + HBr + HNO_2$$

p-nitrophenol — 2, 4, 6-Tribromophenol

Effect of nitro group on the reactivity of phenolic group

Nitro group is an electron-withdrawing group. Hence it helps in the release of protons. The phenoxide ion gets resonance stabilized by the nitro group. This is the reason why nitrophenols are more acidic then phenols.

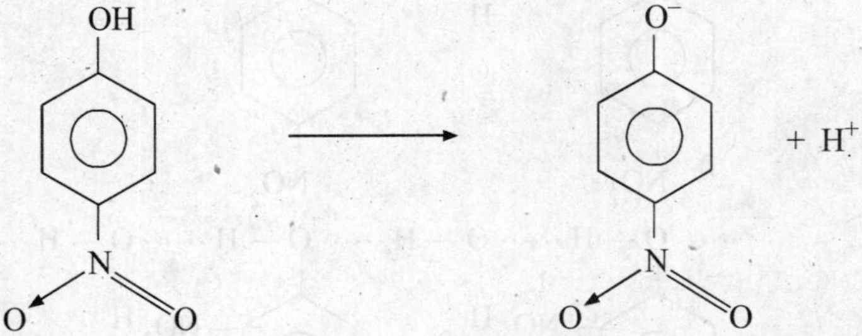

$$p\text{-}NO_2C_6H_4OH \longrightarrow p\text{-}NO_2C_6H_4O^- + H^+$$

Phenoxide ion

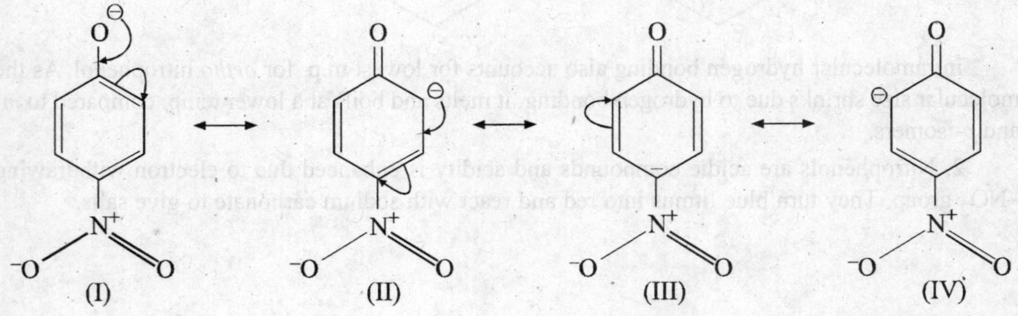

(I) ⟷ (II) ⟷ (III) ⟷ (IV)

Out of resonating structures I, II, III & IV, structure III is particularly stable as it has a negative charge to directly neutralise the positive charge on nitro nitrogen. But, it may be noted that position of nitro group relative to the phenolic group is also important. We observe that nitro group in *ortho* and *para* positions and not *meta* position, enhance the acidity of nitrophenols.

8.13. 2, 4, 6-TRINITROPHENOL (PICRIC ACID)

Methods of preparation

1. From chlorobenzene

Chlorobenzene $\xrightarrow{HNO_3}$ 2, 4 Dinitrochlorobenzene $\xrightarrow{Aq.Na_2CO_3}$ 2, 4 Dinitrophenol $\xrightarrow[H_2SO_4]{HNO_3}$ Picric acid

2. From phenol

Phenol $\xrightarrow{H_2SO_4}$ (phenol-2,4-disulphonic acid, SO_3H groups) $\xrightarrow{HNO_3}$ Picric acid

Properties. 1. Due to the presence of three NO_2 groups, it is much stronger acid than phenol. It can decompose bicarbonates, a property normally given by carboxylic compounds and not phenolic compounds.

Picric acid $+ NaHCO_3 \longrightarrow$ Sod. picrate $+ CO_2 + H_2O$

2. With phosphorus pentachloride, it gives picryl chloride

Picric acid $\xrightarrow{PCl_5}$ Picryl chloride

3. On selective reduction with sodium sulphide, picric acid gives picramic acid.

OH, O_2N, NO_2, NO_2 —(Na_2S, Reduction)→ OH, O_2N, NH_2, NO_2

Picric acid Picramic acid

Example 1. A compound $C_7H_5O_6N_3$ (A) undergoes oxidation with acidified potassium dichromate to give a monocarboxylic acid $C_7H_3O_8N_3$ (B). When (B) is heated in acetic acid solution, $C_6H_3O_6N_3$ (C) is formed. What are (A), (B) and (C)? Give the reactions involved.

Solution. Compound A contains nitrogen, it could be a nitro or amino compound. Upon oxidation, (A) gives a carboxylic acid (B). It is the alkyl groups in benzene rings that are oxidised to –COOH. On heating (B), the compound (C) is formed which contains two oxygens less than (B). This step appears to be decarboxylation reaction. Based on these observations, the compounds, (A), (B) and (C) are:

CH_3, O_2N, NO_2, NO_2 (A) $COOH$, O_2N, NO_2, NO_2 (B) O_2N, NO_2, NO_2 (C)

CH_3, O_2N, NO_2, NO_2 —($K_2Cr_2O_7$)→ $COOH$, O_2N, NO_2, NO_2 —(Heat, CH_3COOH solution)→ O_2N, NO_2, NO_2

Trinitrotoluene (A) 2, 4, 6 Trinitrobenzoic acid (B) 1, 3, 5 Trinitrobenzene (C)

Example 2. Identify the unknown compounds (A), (B) and (C) in the following:

(*i*) Benzene —(HNO_3, H_2SO_4)→ **(A)** —(HNO_3, H_2SO_4)→ **(B)** —($(NH_4)_2S$)→ **(C)**

(*ii*) Benzene —(HNO_3, H_2SO_4)→ **(A)** —(Br_2, $FeBr_3$)→ **(B)** —(Sn/HCl)→ **(C)**

Solution.

(i) Benzene $\xrightarrow[H_2SO_4]{HNO_3}$ Nitrobenzene (A) $\xrightarrow[H_2SO_4]{HNO_3}$ *m*-Dinitrobenzene (B) $\xrightarrow{(NH_4)_2S_x}$ *m*-Nitroaniline (C)

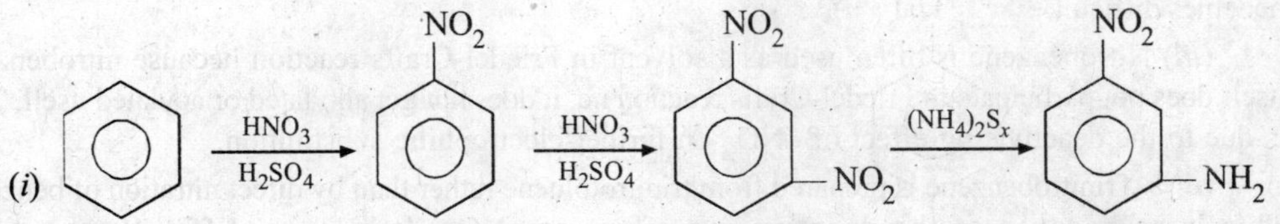

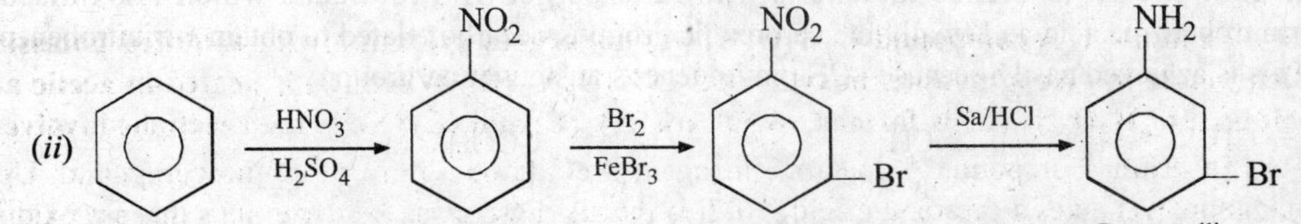

(ii) Benzene $\xrightarrow[H_2SO_4]{HNO_3}$ Nitrobenzene (A) $\xrightarrow[FeBr_3]{Br_2}$ *m*-Bromonitrobenzene (B) $\xrightarrow{Sa/HCl}$ *m*-Bromoaniline

Example 3. How will you synthesise the following:

(*a*) 2, 4, 6 Trinitrotoluene from benzene.

(*b*) Nitrosobenzene from nitrobenzene.

Solution. (*a*) **2, 4, 6 Trinitrotoluene from benzene**

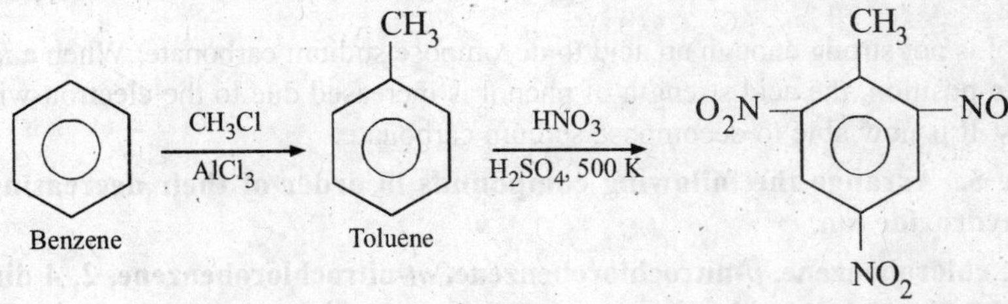

Benzene $\xrightarrow[AlCl_3]{CH_3Cl}$ Toluene $\xrightarrow[H_2SO_4,\ 500\ K]{HNO_3}$ Trinitrotoluene

(*b*) **Nitrosobenzene from nitrobenzene**

Nitrobenzene ($C_6H_5-NO_2$) $\xrightarrow[NH_4Cl]{Zn}$ Phenyl hydroxylamine (C_6H_5-NHOH) $\xrightarrow[H_2SO_4]{Na_2Cr_2O_7}$ Nitrosobenzene (C_6H_5-NO)

Example 4. Explain the following:

(*i*) Nitration of toluene is easier than that of benzene.

(*ii*) Nitration of nitrobenzene is more difficult than that of benzene.

(*iii*) Nitrobenzene is often used as a solvent in Friedel-Crafts reaction.

(*iv*) s-Trinitrobenzene is prepared from TNT rather than by direct nitration of benzene.

(*v*) p-nitrophenol decomposes sodium carbonate whereas phenol does not.

Solution. (*i*) Nitration of aromatic hydrocarbons is an electrophilic substitution reaction. Toluene contains a methyl group which exerts an electron repelling inductive effect. It makes the benzene ring richer in electron cloud thereby facilitating the nitration process.

(*ii*) Nitrobenzene is difficult to nitrate than benzene. Nitro group is an electron withdrawing group. It deactivates the benzene ring for further electrophilic substitution because it creates a positive centre on the ring due to the electron withdrawing effect of nitro group. Hence further nitration becomes difficult.

(*iii*) Nitrobenzene is often used as a solvent in Friedel-Crafts reaction because nitrobenzene itself does not participate in Friedel-Crafts reaction i.e. it does not get alkylated or acylated itself. This is due to the deactivating effect of $-NO_2$ on further electrophilic substitution.

(*iv*) *s*-Trinitrobenzene is prepared from trinitrotoluene rather than by direct nitration of benzene. This is so because of the deactivating effect of $-NO_2$. It is quite difficult to prepare *m*-dinitrobenzene from nitrobenzene and still more difficult to prepare *s*-trinitrobenzene from *m*-dinitrobenzene. Hence toluene is trinitrated to get trinitrotoluene which is oxidised to trinitrobenzoic acid and finally the carboxylic group is decarboxylated to obtain *s*-trinitrobenzene. This is achieved easily because $-CH_3$ in toluene is an activating group.

$$\text{Toluene} \xrightarrow[H_2SO_4]{HNO_3} \text{T.N.T.} \xrightarrow{[O]} \text{2,4,6-trinitrobenzoic acid (COOH)} \xrightarrow[\text{heat}]{\text{Sodalime}} s\text{-Trinitrobenzene}$$

Toluene   T.N.T.   *s*-Trinitrobenzene (*s* stands for symmetrical or 1, 3, 5 substituted)

(*v*) Phenol is not strong enough an acid to decompose sodium carbonate. When a nitro group is inducted in *p*-position, the acid strength of phenol is increased due to the electron-withdrawing effect of $-NO_2$. It is now able to decompose sodium carbonate.

Example 5. Arrange the following compounds in order of their decreasing activity towards the hydroxide ion.

Benzene, chlorobenzene, *p*-nitrochlorobenzene, *m*-nitrochlorobenzene, 2, 4 dinitrochlorobenzene and 2, 4, 6-trinitrochlorobenzene.

Solution. Nitro group in the *ortho* and *para* position has the effect of increasing reactivity towards nucleophilic substitution in aromatic compounds. Greater the number of such nitro groups, more pronounced will be the effect. Hence the order of decreasing activity towards hydroxide ion will be as follows.

2, 4, 6-trinitrochlorobenzene > 2, 4-dinitrochlorobenzene > *p*-nitrochlorobenzene > *m*-nitrochlorobenzene (due to –I effect) > chlorobenzene > benzene.

Example 6. Which reagent is used for the reduction of one of nitro groups of *m*-dinitrobenzene?

Or

How will you synthesise *m*-nitroaniline from nitrobenzene.

Solution. Ammonium hydrogen sulphide is used for partial reduction of *m*-dinitrobenzene.

$$\text{nitrobenzene} \xrightarrow[H_2SO_4]{HNO_3} m\text{-dinitrobenzene} \xrightarrow{NH_4HS} m\text{-nitroaniline}$$

nitrobenzene   *m*-dinitrobenzene   *m*-nitroaniline

Example 7. (*a*) *o*-nitrophenol has lower m.p. and b.p. as compared to *m*- and *p*-isomers. Explain.

(*b*) Nitrophenols are more acidic than phenols. Explain.

Solution. (*a*) *o*-nitrophenol has lower m.p. and b.p. as compared to *m*- and *p*-isomers because hydrogen bonding takes place in *o*-isomer as depicted below.

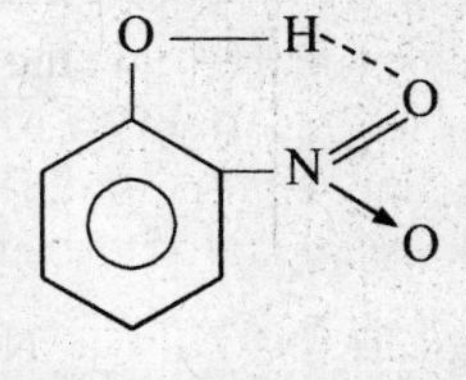

This results in shrinking of size of the molecule resulting in lower m.p. and b.p.

(*b*) Nitrophenols are more acidic than phenols because nitro group is electron attracting. This helps in the release of protons.

Also nitrophenoxide ion obtained after the release of proton stabilizes itself by resonance

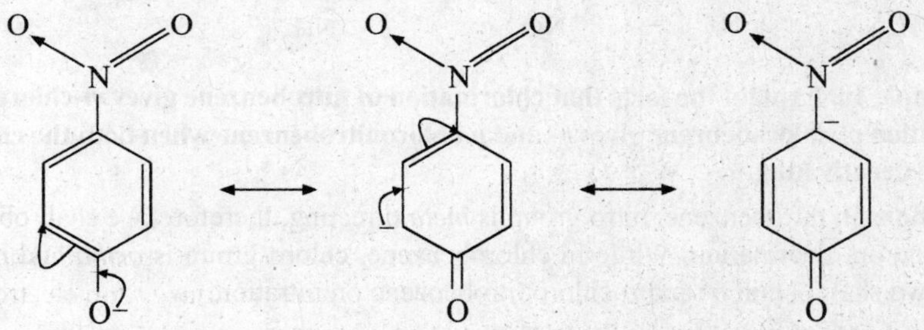

Example 8. *o*-nitrophenol is steam volatile but *p*-nitrophenol is not. Explain.

Solution. *o*-nitrophenol exhibits intramolecular hydrogen bonding which results in lowering of boiling point. Hence it can be steam distilled. *p*-nitrophenol shows no such phenomenon, as such it has a higher boiling point. It is not steam volatile.

Example 9. Account for the following conversions giving only equation with reaction conditions.

(*i*) H_2N–C_6H_4–C_6H_4–NH_2 ⟶ F–C_6H_4–C_6H_4–F

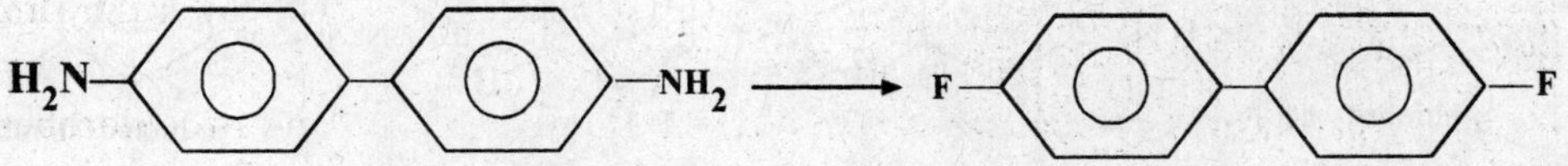

(*ii*) p-$CH_3C_6H_4NH_2$ ⟶ p-$CH_3C_6H_4NO_2$

(*iii*) p-$NO_2C_6H_4NH_2$ ⟶ p-$NO_2C_6H_4NO_2$

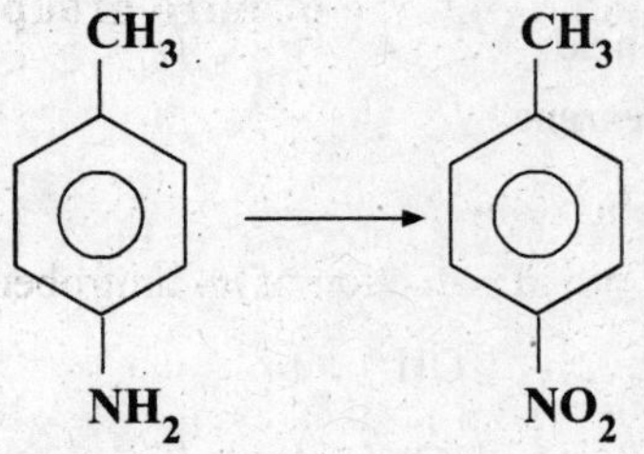

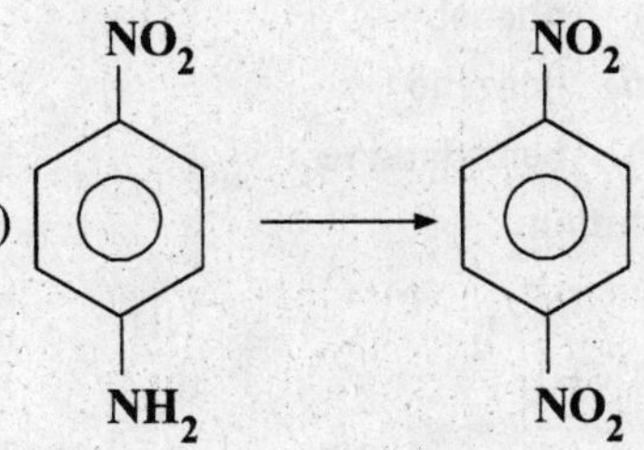

Solution.

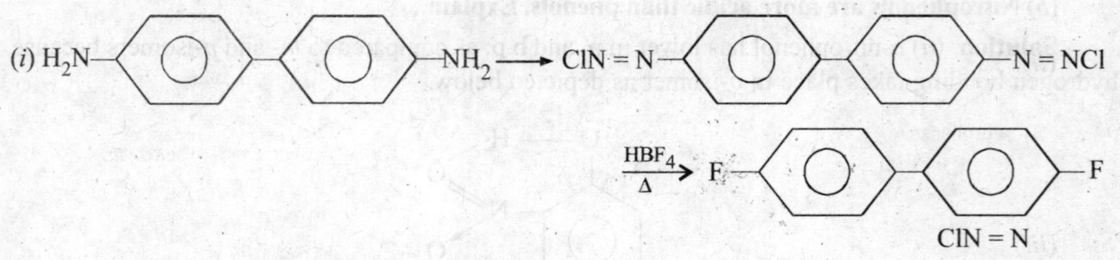

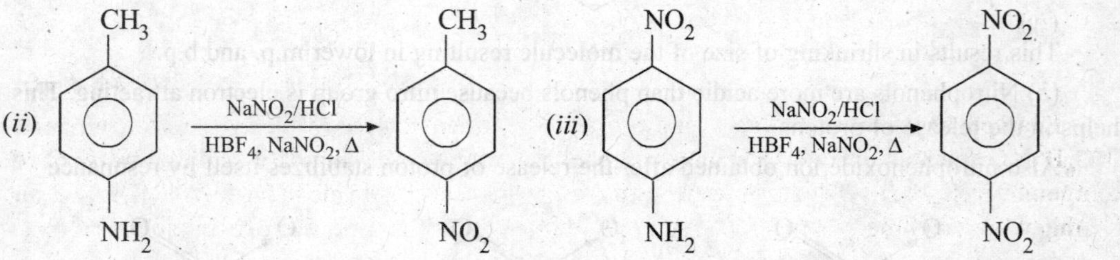

(*i*) H_2N–C$_6$H$_4$–C$_6$H$_4$–NH_2 ⟶ ClN = N–C$_6$H$_4$–C$_6$H$_4$–N = NCl $\xrightarrow[\Delta]{HBF_4}$ F–C$_6$H$_4$–C$_6$H$_4$–F

ClN = N

(*ii*) CH_3, NH_2 $\xrightarrow[HBF_4, NaNO_2, \Delta]{NaNO_2/HCl}$ CH_3, NO_2

(*iii*) NO_2, NH_2 $\xrightarrow[HBF_4, NaNO_2, \Delta]{NaNO_2/HCl}$ NO_2, NO_2

Example 10. Explain the facts that chlorination of nitrobenzene gives *m*-chloronitrobenzene, while nitration of chlorobenzene gives *o*- and *p*-chloronitrobenzene when both the chloro and nitro groups are deactivating.

Solution. In nitrobenzene, nitro group is *meta* directing, therefore, we shall obtain *m*-chloronitrobenzene on chlorination. While in chlorobenzene, chloro group is *ortho* and *para* directing, therefore, we shall obtain *o*- and *p*-chloronitrobenzene on nitration.

Example 11. How will you do the following conversion :

Nitrobenzene → *m*-bromochlorobenzene

Solution.

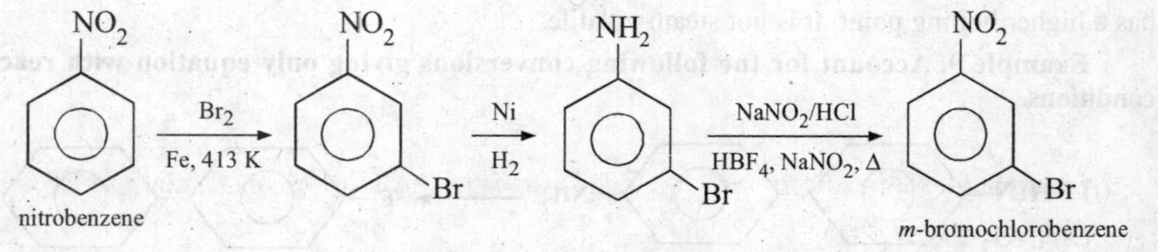

NO_2 (nitrobenzene) $\xrightarrow[Fe, 413 K]{Br_2}$ NO_2, Br $\xrightarrow[H_2]{Ni}$ NH_2, Br $\xrightarrow[HBF_4, NaNO_2, \Delta]{NaNO_2/HCl}$ NO_2, Br (*m*-bromochlorobenzene)

Example 12. Starting from nitrobenzene, how will you get the following compounds:

(*i*) **phenol** (*ii*) **benzoic acid**

(*iii*) **benzene** (*iv*) **bromobenzene**

(*v*) **benzylamine**

Solution.

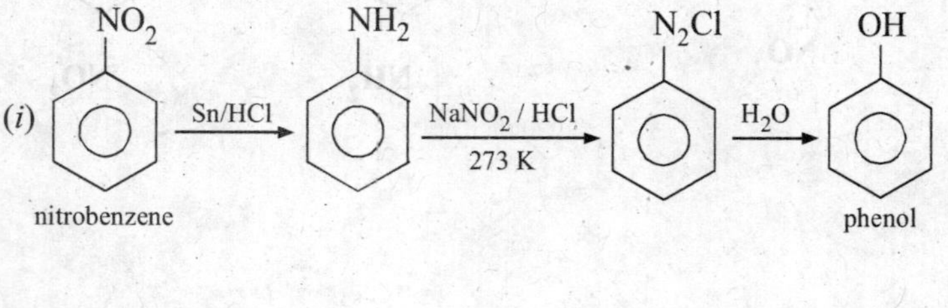

(*i*) NO_2 (nitrobenzene) $\xrightarrow{Sn/HCl}$ NH_2 $\xrightarrow[273 K]{NaNO_2 / HCl}$ N_2Cl $\xrightarrow{H_2O}$ OH (phenol)

(ii) nitrobenzene ($C_6H_5NO_2$) $\xrightarrow{Sn/HCl}$ $C_6H_5NH_2$ $\xrightarrow[273\ K]{NaNO_2\ /\ HCl}$ $C_6H_5N_2Cl$ $\xrightarrow{CuCN}$ C_6H_5CN $\xrightarrow{H_2O}$ C_6H_5COOH (benzoic acid)

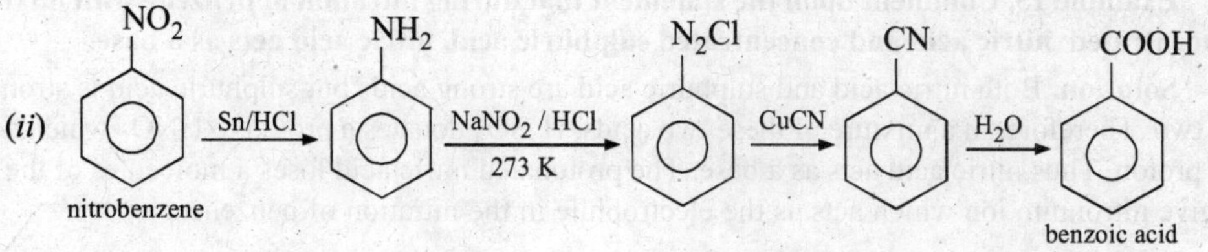

(iii) & (iv) nitrobenzene ($C_6H_5NO_2$) $\xrightarrow{Sn/HCl}$ $C_6H_5NH_2$ $\xrightarrow[273\ K]{NaNO_2\ /\ HCl}$ $C_6H_5N_2Cl$ $\xrightarrow[H_2O]{HPO_2}$ benzene $\xrightarrow[Br_2]{Fe}$ C_6H_5Br (Bromobenzene)

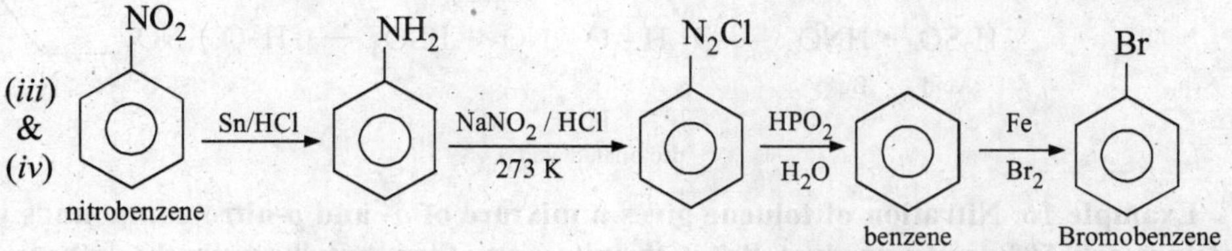

Example 13. An aromatic compound A ($C_6H_5O_2N$) when reduced with Fe/HCl gave compound B(C_6H_7N), which on diazotisation gave compound C which on boiling with acidified water gave compound D (C_6H_6O). D in alkaline medium reacts with carbon dioxide to form E ($C_7H_6O_3$), E on heating with soda lime gives back D. With explanations and equations, identify compounds A to E.

Solution. A = $C_6H_5NO_2$ B = $C_6H_5NH_2$ C = $C_6H_5N_2Cl$

D = C_6H_5OH E = $C_6H_5(OH)\ COOH$

Equations

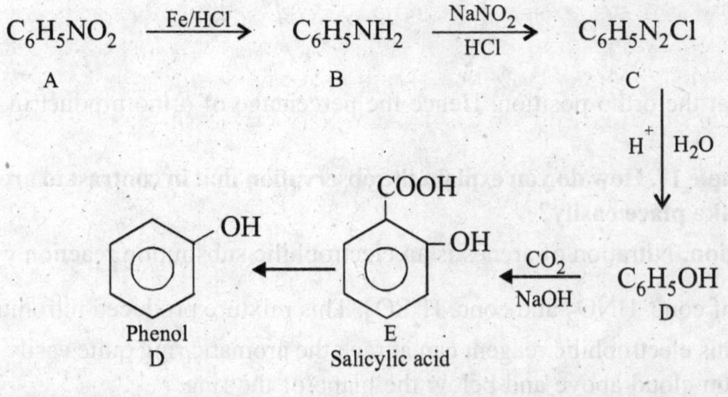

$$\underset{A}{C_6H_5NO_2} \xrightarrow{Fe/HCl} \underset{B}{C_6H_5NH_2} \xrightarrow[HCl]{NaNO_2} \underset{C}{C_6H_5N_2Cl}$$

$$\underset{C}{C_6H_5N_2Cl} \xrightarrow{H^+,\ H_2O} \underset{D}{C_6H_5OH}$$

$$\underset{D}{C_6H_5OH} \xrightarrow[NaOH]{CO_2} \underset{\text{E Salicylic acid}}{C_6H_4(OH)COOH} \longrightarrow \underset{\text{Phenol D}}{C_6H_5OH}$$

Example 14. Name the product obtained in the following reaction and explain its formation.

$CH_3NO_2 + C_6H_5CHO \xrightarrow{KOH\ (alc)}$.

Solution. We obtain β-nitrostyrene, $C_6H_5CH = CHNO_2$.

Due to the presence of strong electron-attracting nitro group, hydrogen atoms of CH_3 – group of nitromethane are weakly acidic in nature. In the presence of a base, it forms a carbanion which undergoes nucleophilic attack at the carbonyl group of benzaldehyde to form an aldol like product which undergoes dehydration to yield to final product as shown below:

$$CH_3-NO_2 \xrightarrow[-H_2O]{OH^-} \underset{\text{Carbanion}}{:\bar{C}H_2-NO_2} \xrightarrow{C_6H_5CHO} \underset{\text{Aldol type product}}{C_6H_5-\overset{OH}{\overset{|}{C}H}-CH_2NO_2} \xrightarrow{-H_2O} \underset{\beta-\text{Nitrostyrene}}{C_6H_5-CH=CHNO_2}$$

Example 15. Comment upon the statement that during nitration of benzene with mixture of concentrated nitric acid and concentrated sulphuric acid, nitric acid acts as a base.

Solution. Both nitric acid and sulphuric acid are strong acids but sulphuric acid is stronger of the two. Therefore, in a mixture of these two acids, H_2SO_4 donates a proton to HNO_3 which accepts the proton. Thus nitric acid acts as a base. The protonated nitric acid loses a molecules of the water to give nitronium ion which acts as the electrophile in the nitration of benzene.

$$\underset{\text{Acid}}{H_2SO_4} + \underset{\text{Base}}{HNO_3} \longrightarrow \underset{\text{protonated acid}}{H-\overset{+}{\underset{|\atop H}{O}}-NO_2} + HSO_4^- \longrightarrow H_2O + NO_2^+$$

Example 16. Nitration of toluene gives a mixture of *o*- and *p*-nitro compounds which contain nearly 50% ortho product. But in the nitration of tert-butylbenzene the ortho product constitutes only about 16% of the nitro compound obtained. Explain.

Solution. In the nitration of toluene, nitronium ion (NO_2^+) can attack both the *ortho* and *para* position quite easily so that ortho product is formed in large quantities.

But in tert.- butyl toluene $\left(C_6H_5-\underset{CH_3}{\overset{CH_3}{\underset{|}{\overset{|}{C}}}}-CH_3\right)$, the bulky *tert.* butyl group offers strong steric hindrance at the ortho position. Hence the percentage of ortho product in the mixtrue obtained is very low.

Example 17. How do you explain the observation that in contrast to arenes, nitration of alkanes does not take place easily?

Solution. Nitration of arenes is an electrophilic substiution reaction which is carried out with a mixture of conc. HNO_3 and conc. H_2SO_4. This mixture produced nitronium ion (NO_2^+) even in the cold. This electrophilic reagent can attrack the aromatic ring quite easily because of the presence of π-electron cloud above and below the plane of the ring.

In case of alkanes, electrophilic attack on the molecule is not possible since alkanes contain only α bonds and no π-electron are available. Therefore nitration of alkanes takes place by different mechanims involving the formation of free radicals. We require a higher temperature to carry out the reaction with nitric acid in the vapour phase when nitric acid undergoes homolytic cleavage to generate nitro free radicals $(N\dot{O}_2)$. The free radical then attacks the alkane molecule to ultimately form the nitro product. Due to high temperature used, cleavage of carbon-carbon bonds of alkanes also takes place to yield a mixture of products.

AMINES

8.14. CLASSIFICATION OF AMINES

Amines are organic compounds containing nitrogen. They may be considered as derivatives of ammonia (NH_3), in which one or more hydrogen atoms have been replaced by alkyl or aryl groups. They may be classified as under:

Primary amines. These are the compounds in which one hydrogen atom in ammonia is replaced by an alkyl or aryl group. For example,

$CH_3 - NH_2$	$CH_3 - CH_2 - NH_2$	$C_6H_5NH_2$
Methyl amine	Ethyl amine	Phenyl amine

Secondary amines. These are the compounds in which two hydrogen atoms are replaced by two, same or different, alkyl or aryl groups.

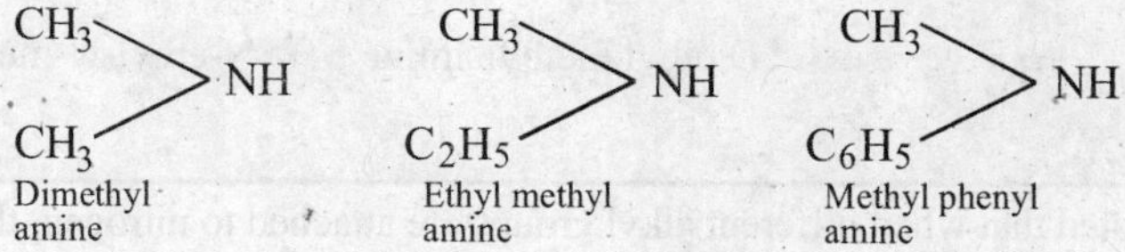

Tertiary amines. These are the compounds in which all the three hydrogens of ammonia are replaced by three alkyl or aryl groups.

$(CH_3)_3N$ — Trimethyl amine

$(CH_3)_2(C_2H_5)N$ — Ethyl dimethyl amine

$(CH_3)(C_2H_5)(C_6H_5)N$ — Ethyl methyl Phenyl amine

There is another classification of amines, too.

Alkyl amines. The amines which contain only alkyl groups in the molecule are known as alkyl amines.

Aryl amines. Amines containing only aryl groups are known as aryl amines.

Alkylaryl amines. Amines containing both alkyl and aryl groups in the molecule are known as alkylaryl amines.

Secondary and tertiary amines are classified as under:

Simple amines. Amines containing same alkyl or aryl groups attached to nitrogen are called single amines.

Mixed amines. If the alkyl or aryl groups attached to nitrogen are different, the amine is called mixed amine.

8.15. NOMENCLATURE OF AMINES

Nomenclature of amines is discussed as under:

Compound	*Common Name*	*Name IUPAC*
Primary Amines		
$CH_3 - NH_2$	Methyl amine	Aminomethane
$CH_3 - CH_2 - NH_2$	Ethyl amine	Aminoethane
$CH_3 - CH_2 - CH_2 - NH_2$	Propyl amine	Aminopropane
$CH_3 - \underset{\mid}{CH} - NH_2$ with CH_3 on CH	Isopropyl amine	2-methylaminoethane
Secondary Amines		
$CH_3 - NH - CH_3$	Dimethyl amine	N-methylaminomethane

Compound	*Common Name*	*Name IUPAC*
$CH_3 - NH - C_2H_5$	Ethyl methyl amine	N-methylaminoethane
$CH_3 - NH - C_3H_7$	Methyl propyl amine	N-methylaminopropane
Tertiary Amines		
$CH_3 - N(CH_3) - CH_3$	Trimethyl amine	N, N-dimethylaminomethane
$CH_3 - N(C_2H_5) - C_2H_5$	Diethyl methyl amine	N-ethyl-N-methylaminoethane

It may be noted that when different alkyl groups are attached to nitrogen, the biggest alkyl group forms part of aminoalkane. The notation N-methyl means that methyl group is attached to nitrogen.

Aromatic amines

Simplest aromatic amine is aminobenzene or aniline.

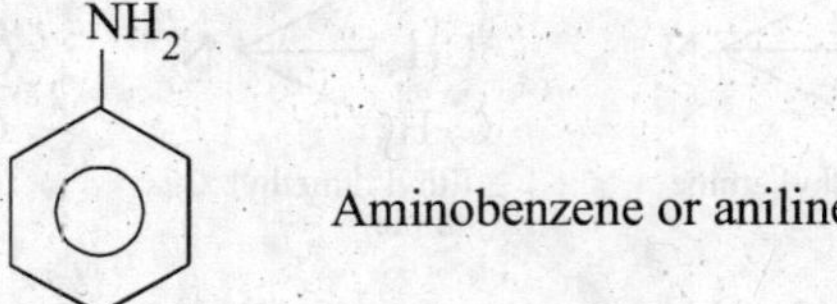

Aminobenzene or aniline

IUPAC name of aniline is benzenamine.

There are amines in which other positions in the benzene ring are occupied by other groups. Their position, relative to – NH_2 group is identified. For this purpose, we give number 1 to the position where – NH_2 group is attached. Then the number increases as we move clockwise.

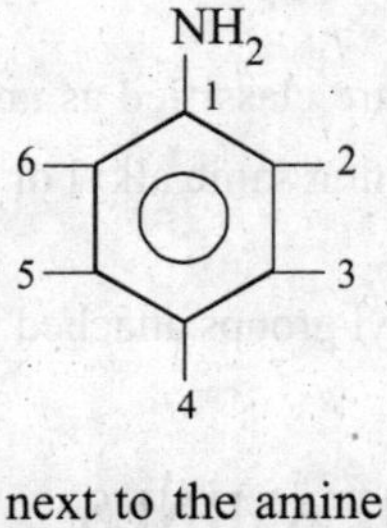

NH_2
1
6 2
5 3
4

By another convention, the position next to the amine group is ortho (*o*); with gap of one position, meta (*m*) and that vertically opposite to – NH_2 group is para (*p*).

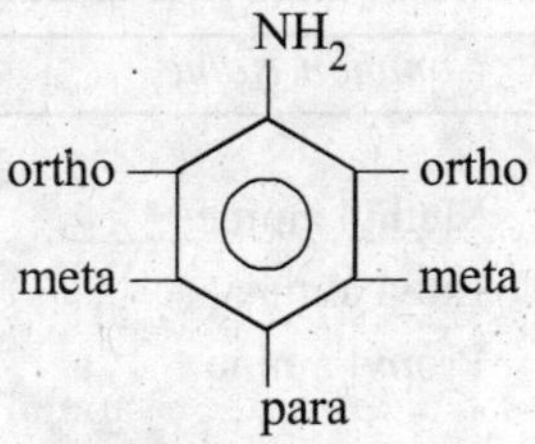

NH_2
ortho ortho
meta meta
para

The compounds are named accordingly. Sometimes there are special names, too.

Compound	Name	Special name
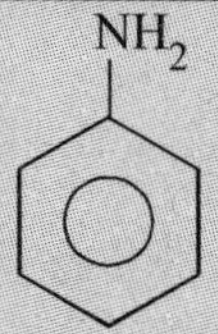 NH_2	Aminobenzene (IUPAC name is benzenamine)	Aniline
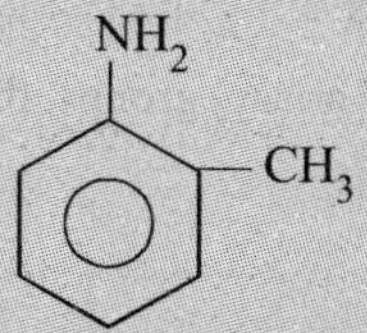 NH_2, CH_3	2-methylaniline (IUPAC name is 2-methylbenzenamine)	*o*-Toluidine
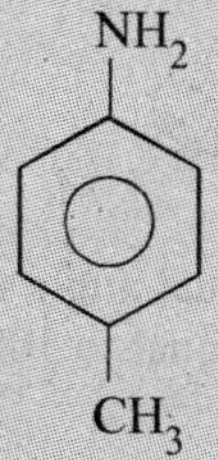 NH_2, CH_3	4-methylaniline	*p*-Toluidine
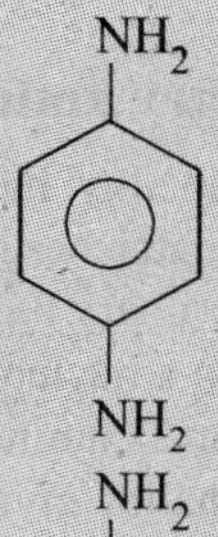 NH_2, NH_2	*p*-diaminobenzene	*p*-Phenylene diamine
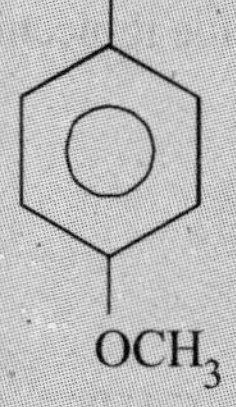 NH_2, OCH_3	*p*-methoxy aniline	Anisidine
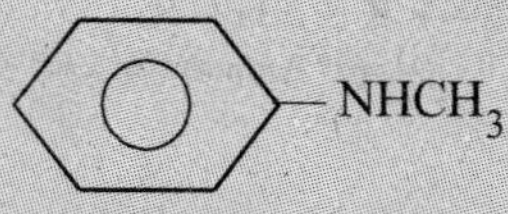 $NHCH_3$	N-methyl aniline	–
 NH	N-phenyl aniline	Diphenyl amine

8.16. STRUCTURE AND SHAPE OF AMINE MOLECULES

As amines are derivatives of ammonia, the shape of the former is basically the same as that of the latter. We know that ammonia uses sp^3 hybridisation. Out of the four, three hybridised orbitals are oriented towards the corners of a tetrahedron. The fourth orbital contains a lone pair of electrons. Thus the shape of NH_3 could be represented as:

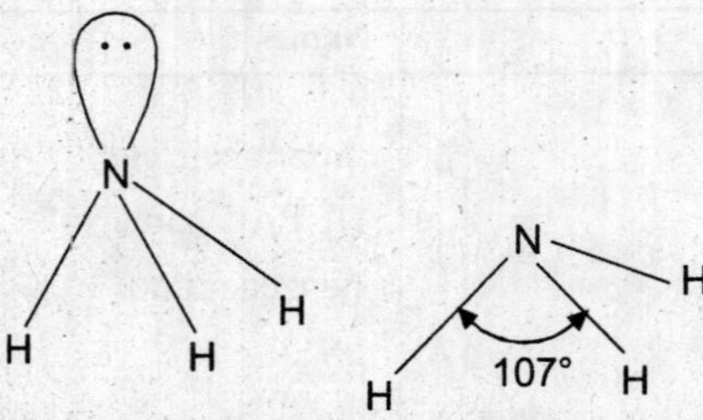

As amines molecules are obtained by replacing hydrogen by alkyl or aryl groups, their shape could be represented as:

Sp^2 hybridised

N

CH_3 CH_3 CH_3

N 108°

CH_3 CH_3 CH_3

As the angle ∠HNH in ammonia is known to be 107°, we expect an equal angle in case of amines also. It has been observed that trimethyl amine has an angle ∠CNC equal to 108°.

8.17. METHODS OF PREPARATION OF AMINES

Common methods for preparing primary, secondary and tertiary amines

1. Ammonolysis (Hoffmann's method). This reaction involves treating an alkyl halide or an aryl halide of the type $C_6H_5CH_2$ X with ammonia. Ammonia is a nucleophile. It attaches itself with the alkyl group and the halogen is removed as halide to give a molecule of primary amine or 1° amine. But the reaction does not stop here. The primary amine formed acts as a nucleophile and attaches itself to the alkyl halide molecule giving rise to secondary amine or 2° amines. Again secondary amine acts as nucleophile and attaches itself to a molecule of alkyl halide giving rise to tertiary amine or 3° amine. Finally tertiary amine attaches itself to the alkyl halide molecule to form the quaternary ammonium salt (4°). These reactions are explained as under.

$$H_3\ddot{N} + R - X \longrightarrow H_3\overset{\oplus}{N} - R + \overset{\ominus}{X} \xrightarrow{-H^{\oplus}} R - \ddot{N}H_2 \quad \text{(1° Amine)}$$

$$R - \ddot{N}H_2 + R - X \longrightarrow R - \overset{\oplus}{N}H_2 - R + \overset{\ominus}{X} \xrightarrow{-H^{\oplus}} R - NH - R \quad \text{(2° Amine)}$$

$$R_2\ddot{N} + R - X \xrightarrow{-X^{\ominus}} R_3\overset{\oplus}{N}H \xrightarrow{-H^{\oplus}} R_3N \quad \text{(3° Amine)}$$

$$R_3\ddot{N} + R - X \longrightarrow R_4\overset{\oplus}{N}\overset{\ominus}{X} \quad \text{Quaternary ammonium salt (4°)}$$

For example, $CH_3 - Cl + NH_3 \xrightarrow{-HCl} \underset{\text{Methyl amine}}{CH_3NH_3} \xrightarrow[-HCl]{CH_3Cl} \underset{\text{Dimethyl amine}}{(CH_3)_2NH}$

$$\xrightarrow[-\,HCl]{CH_3Cl} \underset{\text{Trimethyl amine}}{(CH_3)_3N} \xrightarrow{CH_3Cl} \underset{\text{Tetramethyl ammonium chloride}}{(CH_3)_4N^+Cl^-}$$

Thus, we generally obtain a mixture of the above substance on the ammonolysis of methyl chloride.

However, reaction can be made to stop at the first step *i.e.* at the stage of primary amine by taking large excess of NH_3 so that alkyl halide is not available to the primary amine once it is formed.

For preparing aryl amines, aryl halides are treated with ammonia, but drastic conditions are used.

$$\underset{\text{Chlorobenzene}}{C_6H_5Cl} + NH_3 \xrightarrow[60\ \text{Atm.}]{473\ K} \underset{\text{Aniline}}{C_6H_5NH_2} + HCl$$

But it is easier to ammonolise aromatic compounds with a side chain.

$$\underset{\text{Benzyl chloride}}{C_6H_5CH_2Cl} + NH_3 \longrightarrow \underset{\text{Benzyl amine}}{C_6H_5CH_2NH_2} + HCl$$

2. By the action of ammonia on alcohols or phenols. On passing a mixture of alcohol and ammonia over heated alumina at 633 K, a mixture of primary, secondary and tertiary amines is obtained

$$C_2H_5OH + NH_3 \xrightarrow[633\ K,\ -H_2O]{Al_2O_3} C_2H_5NH_2 \xrightarrow[Al_2O_3\,(633\ K),\ -H_2O]{C_2H_5OH}$$

$$\underset{\text{Diethyl amine}}{(C_2H_5)_2NH} \xrightarrow[Al_2O_3\,(633\ K),\ -H_2O]{C_2H_5OH} \underset{\text{Triethyl amine}}{(C_2H_5)_3N}$$

$$\underset{\text{Phenol}}{C_6H_5OH} + NH_3 \xrightarrow{\text{Anhy. } AlCl_3} \underset{\text{Aniline}}{C_6H_5NH_2} + H_2O$$

3. Reductive amination of aldehydes and ketones. Aldehydes and ketones on reduction with hydrogen and ammonia change into primary amines at 373 K and 150 atm. pressure and in the presence of nickel as catalyst.

$$R-\overset{H}{\overset{|}{C}}=O+NH_3 \xrightarrow{-H_2O} \underset{\text{Aldimine}}{R-\overset{H}{\overset{|}{C}}=NH} \xrightarrow{H_2,\ Ni} \underset{\text{Primary amine}}{RCH_2NH_2}$$

$$R-\overset{R'}{\overset{|}{C}}=O+NH_3 \xrightarrow{-H_2O} \underset{\text{Ketimine}}{R-\overset{R'}{\overset{|}{C}}=NH} \xrightarrow{H_2,\ Ni} \underset{\text{Primary amine}}{R-\overset{R'}{\overset{|}{C}}HNH_2}$$

$$\underset{\text{Propanal}}{CH_3CH_2CHO} \xrightarrow[373\ K,\ 150\ Atm.]{NH_3,\ H_2,\ Ni} \underset{\text{Propyl amine}}{CH_3CH_2CH_2NH_2}$$

$$(CH_3)_2C=O \xrightarrow[373\ K,\ 150\ Atm.]{NH_2,\ H_2,\ Ni} \underset{\text{Isopropyl amine}}{(CH_3)_2CH-NH_2}$$

Mechanism

Reductive amination involves the formation of amine as an intermediate and proceeds through the following mechanism.

$$>C=\ddot{O}+:NH_3 \longrightarrow >C(O^-)(\overset{+}{N}H_3) \longrightarrow \left[>C(OH)-\overset{}{N}H-H\right] \xrightarrow{-H_2O} >C=NH \xrightarrow{H_2,\ Ni} -\overset{|}{C}H-NH_2$$

4. Reduction of nitro compounds. Nitro compounds on reduction or hydrogenation give rise to amines. Because of ready availability of nitrobenzene, this method is used to prepare aniline particularly.

(*i*) **Catalytic hydrogenation.** Nitro compound is treated with hydrogen gas in the presence of finely divided nickel or platinum.

$$\underset{\text{Nitrobenzene}}{C_6H_5-NO_2}+3H_2 \xrightarrow{Ni\ or\ Pt} \underset{\text{Aniline}}{C_6H_5-NH_2}+2H_2O$$

We can't use this method to hydrogenate the nitro group in a compound containing ethylenic double bond, as the double bond would also get saturated.

(*ii*) **Reduction.** Here the nitro compound is dissolved in hydrochloric acid and treated with tin metal.

$$C_6H_5-NO_2 \xrightarrow[6[H]]{Sn/HCl} C_6H_5-NH_2+2H_2O$$

(*iii*) Lithium aluminium hydride could also be used as a reducing agent.

(*iv*) **Partial Reduction.** One of the two nitro groups in a dinitro compound can be reduced keeping the second intact, using ammonium hydrogen sulphide as the reducing agent.

$$C_6H_4(NO_2)_2 \xrightarrow{NH_4HS} C_6H_4(NO_2)(NH_2)$$

m-dinitro benzene → *m*-nitro aniline

5. Reduction of nitriles or cyanides. Aliphatic and aromatic nitriles (of the type $C_6H_5CH_2CN$), can be reduced by catalytic hydrogenation or chemical reduction to produce amines. When the reducing agent is sodium and alcohol, the reduction is known as **Mendius Reaction or Reduction.**

$$\underset{\text{Nitrile}}{R-C\equiv N} + 2H_2 \xrightarrow[433\text{ K}]{Ni} \underset{\text{Primary amine}}{RCH_2NH_2}$$

$$\underset{\text{Ethane nitrile}}{CH_3CN + 2H_2} \xrightarrow[433\text{ K}]{Ni} \underset{\text{Ethylamine}}{CH_3CH_2NH_2}$$

$$\underset{\text{Phenyl methyl cyanide}}{C_6H_5-CH_2CN + 2H_2} \xrightarrow[433\text{ K}]{Ni} \underset{\text{Phenyl ethylamine}}{C_6H_5-CH_2CH_2NH_2}$$

6. Reduction of oximes and amides. Hydrogenation or chemical reduction of oximes (which are obtained from aldehydes and ketones) and amides yields amines.

$$\underset{\text{Acetaldoxime}}{CH_3CH=NOH} + 4[H] \xrightarrow[\text{Alcohol}]{Na} CH_3CH_2NH_2 + H_2O$$

$$\underset{\text{Acetone oxime}}{(CH_3)_2C=NOH} + 4H \xrightarrow[\text{Alcohol}]{Na} \underset{\text{Isopropyl amine}}{(CH_3)_2CHNH_2} + H_2O$$

$$CH_3CONH_2 + 4[H] \longrightarrow CH_3CH_2NH_2 + H_2O$$

7. Hofmann's degradation of amides. On treatment with bromine and potassium hydroxide, amides are converted into amines.

$$\underset{\text{Amide}}{R-\overset{O}{\overset{\|}{C}}NH_2} + Br_2 + 4KOH \longrightarrow \underset{\text{Amine}}{RNH_2} + K_2CO_3 + 2KBr + 2H_2O$$

$$\underset{\text{Acetamide}}{CH_3CONH_2} + Br_2 + 4KOH \longrightarrow \underset{\text{Methyl amine}}{CH_3NH_2} + K_2CO_3 + 2KBr + 2H_2O$$

Mechanism

They hypobromite ion OBr^- produced by alkaline solution of Br_2 attacks the amide molecule. Various steps which are involved are illustrated hereunder.

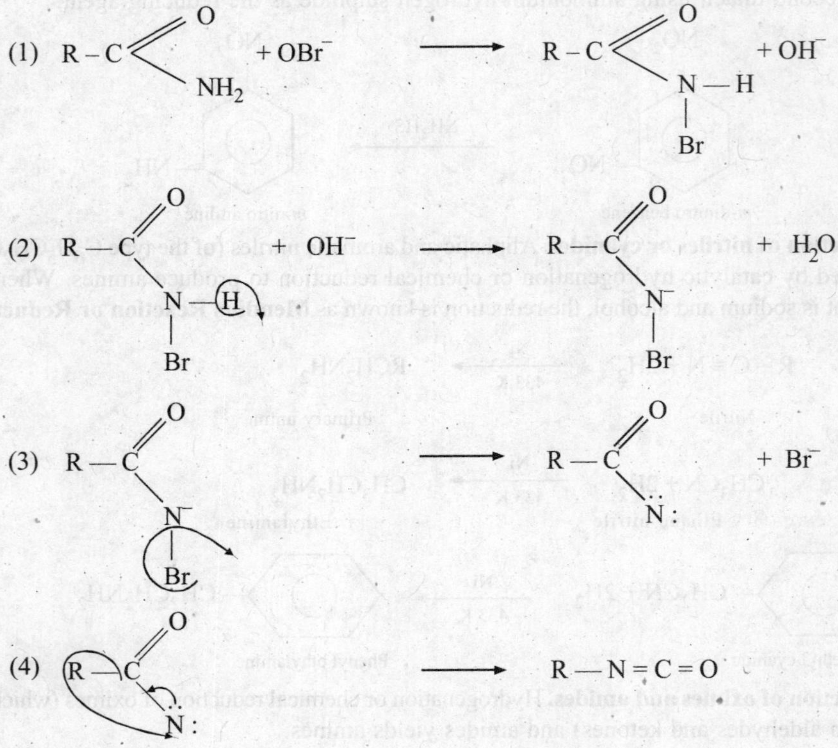

(1) $R-C(=O)NH_2 + OBr^- \longrightarrow R-C(=O)-N(Br)-H + OH^-$

(2) $R-C(=O)-N(Br)-H + OH^- \longrightarrow R-C(=O)-N^-(Br) + H_2O$

(3) $R-C(=O)-N^-(Br) \longrightarrow R-C(=O)-\ddot{N}: + Br^-$

(4) $R-C(=O)-\ddot{N}: \longrightarrow R-N=C=O$

$$R-N=C=O+2OH^- \longrightarrow R-NH_2+CO_3^{2-}$$

There is rearrangement in step 4 where the alkyl group leaves carbon and attaches itself to nitrogen to form isocyanate which is hydrolysed to amine.

8. Gabriel Synthesis. Phthalimide is converted into potassium salt on treatment with KOH. Pot. phthalimide on treatment with an alkyl or arylalkyl halide forms N-substituted phthalimide, which is hydrolysed to produce phthalic acid and a primary amine. Hydrolysis is performed with 20% HCl under pressure.

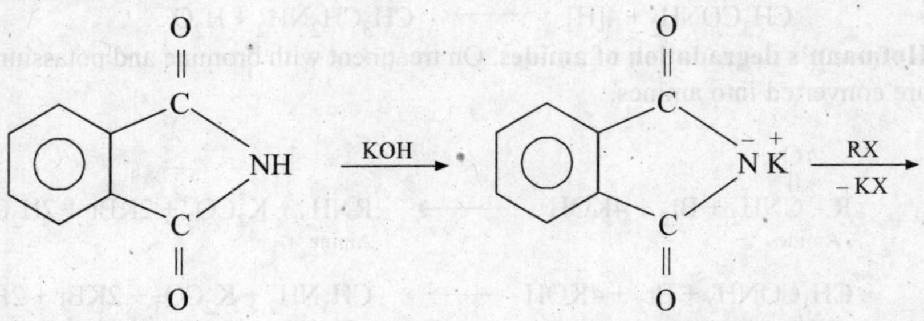

$$C_6H_4(CO)_2NH \xrightarrow{KOH} C_6H_4(CO)_2\overset{-}{N}\overset{+}{K} \xrightarrow[-KX]{RX}$$

Phthalimide Pot. salt of phthalimide

$$\text{N-alkyl Phthalimide} \xrightarrow{H^+, HO_2} \text{Phthalic acid} + \underset{\text{Primary amine}}{RNH_2}$$

9. From Grignard Reagent. Grignard regents react with chloramine in ether to produce amines.

$$CH_3MgBr + ClNH_2 \longrightarrow \underset{\text{Methyl amine}}{CH_3NH_2} + \underset{\text{Magnesium chloro bromide}}{Mg\langle^{Cl}_{Br}}$$

Additional Methods for Preparing Secondary Amines

(*i*) **Reduction of isonitriles or isocyanides**

$$\underset{\text{Alkyl isocyanide}}{R-NC} + 2H_2 \xrightarrow{Ni} \underset{\text{Methyl alkyl amine}}{RNHCH_3}$$

(*ii*) **Reduction of N-alkyl amides**

$$\underset{\text{N-alkyl amide}}{R-\overset{O}{\overset{\|}{C}}-NHR'} \xrightarrow{LiAlH_4} \underset{\text{Sec. amine}}{RCH_2NHR'}$$

$$\underset{\text{N-methyl acetamide}}{CH_3CONHCH_3} + 2H_2 \xrightarrow[-H_2O]{Ni} \underset{\text{Ethyl methyl amine}}{CH_3CH_2NHCH_3}$$

(*iii*) From Grignard reagent. Substituted chloramine reacts with Grignard reagent to produce a secondary amine.

$$\underset{\text{Methyl mag. chloride}}{CH_3MgCl} + \underset{\text{Methyl chloramine}}{ClNHCH_3} \longrightarrow \underset{\text{Dimethyl amine}}{CH_3NHCH_3} + MgCl_2$$

Additional Methods for Preparing Tertiary Amines

(*i*) **Reduction of N, N Disubstituted amides**

$$\underset{\text{N, N Disubstituted amide}}{R\overset{O}{\overset{\|}{C}}-N\langle^{R'}_{R'}} + 2H_2 \longrightarrow \underset{\text{Tertiary amine}}{RCH_2N\langle^{R'}_{R'}} + H_2O$$

(*ii*) **From Grignard reagent.** Disubstituted chloramine reacts with Grignard reagent to produce tertiary amine.

$$RMgCl + ClN\begin{matrix} R' \\ R' \end{matrix} \longrightarrow \underset{\text{Tertiary amine}}{R-N\begin{matrix} R' \\ R' \end{matrix}} + MgCl_2$$

(*iii*) **Decomposition of quaternary ammonium hydroxide.** Quaternary ammonium salts are hydrolysed with moist silver oxide to produce quaternary ammonium hydroxides which are decomposed on heating to give tertiary amine.

$$[R_4N]^+X^- + AgOH \longrightarrow [R_4N^+]OH^- + AgX \xrightarrow{\Delta} \underset{\text{Tert amine}}{R_3N} + ROH$$

Example 1.

(*a*) **How will you prepare ethylamine and benzylamine using Gabriel phthalimide syntheses?**

(*b*) **Can this method be used to prepare aniline? Give reasons for your answer.**

Solution.

(*a*) **Preparation of ethylamine and benzylamine.**

$$\text{Phthalimide } C_6H_4(CO)_2NH \xrightarrow{KOH} C_6H_4(CO)_2N^-K^+$$

$$C_6H_4(CO)_2N^-K^+ \xrightarrow[-KCl]{C_6H_5CH_2Cl} \underset{\text{N-Benzyl phthalimide}}{C_6H_4(CO)_2N-CH_2C_6H_5}$$

$$C_6H_4(CO)_2N^-K^+ \xrightarrow[-KCl]{C_2H_5Cl} \underset{\text{N-Ethyl phthalimide}}{C_6H_4(CO)_2NC_2H_5}$$

$$\text{N-Benzyl phthalimide} \xrightarrow{H^+\ H_2O} \underset{\text{Phthalic acid}}{C_6H_4(COOH)_2} + \underset{\text{Benzylamine}}{C_6H_5CH_2NH_2}$$

$$\text{N-Ethyl phthalimide} \xrightarrow{H^+\ H_2O} \underset{\text{Phthalic acid}}{C_6H_4(COOH)_2} + \underset{\text{Ethylamine}}{C_2H_5NH_2}$$

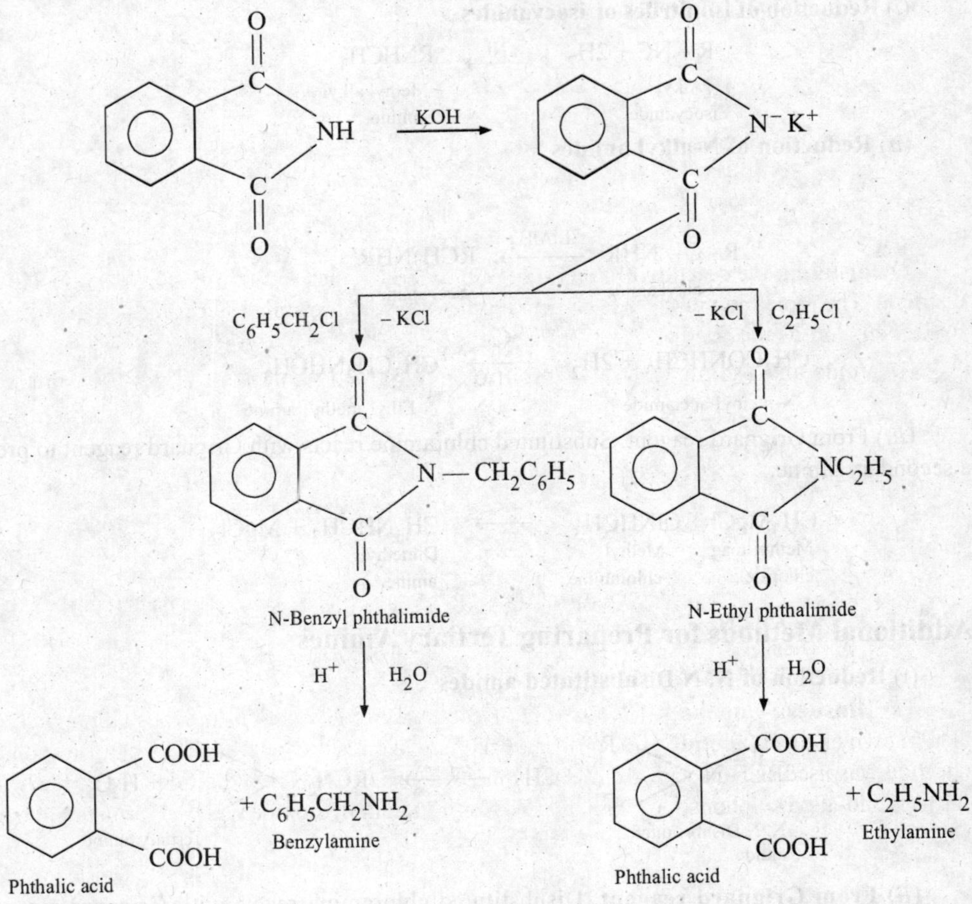

(*b*) Aniline cannot be prepared by this method. This is because chlorobenzene has no tendency to combine with potassium phthalimide to give N-phenyl phthalimide. This step is nucleophilic substitution and nuclear aromatic halogen compounds have no reactivity towards nucleophilic substitution reactions.

8.18. SEPARATING A MIXTURE OF PRIMARY, SECONDARY AND TERTIARY AMINES

Certain methods of preparation of amines lead to the formation of a mixture of primary, secondary and tertiary amines and quaternary ammonium salts. When a mixture of amines (or salts of amines) and quaternary ammonium salts is to be separated, it is first distilled with potassium hydroxide. The quaternary ammonium salt, being non-volatile, is left behind while the mixture of amines distils over. This mixture is then separated by the following methods:

(1) Hofmann's method. The mixture of amines is treated with diethyl oxalate when the primary amine forms a substituted oxamide (a solid), the secondary amine forms dialkyl oxamic ester (**a liquid**) and the tertiary amine does not react.

$$\underset{\text{Diethyl oxalate}}{\begin{matrix} CO\,OC_2H_5 \\ | \\ CO\,OC_2H_5 \end{matrix}} + \underset{\text{Primary amine}}{\begin{matrix} H\,NHR \\ \\ H\,NHR \end{matrix}} \longrightarrow \underset{\substack{\text{Substituted oxamide} \\ (Solid)}}{\begin{matrix} CONHR \\ | \\ CONHR \end{matrix}} + 2C_2H_5OH$$

$$\underset{\text{Diethyl oxalate}}{\begin{matrix} COOC_2H_5 \\ | \\ COOC_2H_5 \end{matrix}} + \underset{\text{Secondary amine}}{HNR_2} \longrightarrow \underset{\substack{\text{Dialkyl oxamic ester} \\ (Liquid)}}{\begin{matrix} CONR_2 \\ | \\ COOC_2H_5 \end{matrix}} + C_2H_5OH$$

On distilling the reaction mixture, the tertiary amine distils over first and is collected as the distillate. The remaining mixture is cooled and filtered when oxamic ester passes as the filtrate and oxamide is left as the residue.

Oxamide and oxamic ester are now separately boiled with alkali to get back the original amines.

$$\underset{\text{Oxamide}}{\begin{matrix} CONHR \\ | \\ CONHR \end{matrix}} + 2KOH \longrightarrow \underset{\text{Pot. oxalate}}{\begin{matrix} COOK \\ | \\ COOK \end{matrix}} + \underset{\text{Primary amine}}{2RNH_2}$$

$$\underset{\text{Oxamic ester}}{\begin{matrix} CONR_2 \\ | \\ COOC_2H_5 \end{matrix}} + 2KOH \longrightarrow \underset{\text{Pot. oxalate}}{\begin{matrix} COOK \\ | \\ COOK \end{matrix}} + C_2H_5OH + \underset{\text{Sec. amine}}{R_2NH}$$

(2) Hinsberg's method. The mixture of amines is treated with *p*-toluenesulphonyl chloride, also known as tosyl chloride (TSCl) or Hinsberg's reagent (in the earlier days benzene sulphonyl chloride was used as Hinsberg's reagent) in the presence of sodium hydroxide. The primary amine forms mono-alkyl sulphonamide which dissolves in sodium hydroxide with the formation of sodium salt.

$$p-CH_3-C_6H_4-SO_2Cl + H_2NR \longrightarrow p\text{-}CH_3C_6H_4-SO_2NHR$$

p-Toluene sulphonyl chloride; Primary amine; Alkyl sulphonamide

$$\downarrow NaOH$$

$$p-CH_3-C_6H_4-SO_2NNaR$$

Sod. salt of alkyl sulphonamide (*Soluble*)

The secondary amine forms dialkyl sulphonamide which is insoluble in alkali.

$$p-CH_3-C_6H_4-SO_2Cl + HNR_2 \longrightarrow p-CH_3-C_6H_4-SO_2NR_2 + HCl$$

Secondary amine; Dialkyl sulphonamide (*Insoluble in alkali*)

The tertiary amine does not react at all.

The alkaline mixture obtained above is extracted with ether when the unreacted tertiary amine and sulphonamide of secondary amine pass into ether layer while the sodium salt of sulphonamide of primary amine remains in aqueous layer. From the ethereal layer, ether is removed by evaporation and the remaining mixture is distilled. As a result tertiary amine (being volatile) distils off while sulphonamide of secondary amide is left behind. This is hydrolysed by hydrochloric acid when the hydrochloride of secondary amine is obtained. This is distilled over sodium hydroxide to get the secondary amine.

$$p-CH_3-C_6H_4-SO_2NR_2 + H_2O + HCl \longrightarrow p-CH_3C_6H_4-SO_2NHR + R_2NH.HCl$$

Dialkyl sulphonamide; *p*-Toluene sulphonic acid; Sec. amine hydrochloride

$$R_2NH.HCl + NaOH \longrightarrow R_2NH + NaCl + H_2O$$

Sec. amine

The aqueous layer containing the sodium salt of sulphonamide of primary amine is acidified with hydrochloric acid when the sodium salt changes to free sulphonamide which gets hydrolysed to yield the hydrochloride of primary amine. This is followed by distillation over sodium hydroxide to get the free amine.

$$p-CH_3-C_6H_4-SO_2NNaR + HCl \longrightarrow p-CH_3-C_6H_4-SO_2NHR + NaCl$$

Sod. salt of sulphonamide

$$\downarrow H.OH + HCl$$

$$p-CH_3-C_6H_4SO_2OH + RNH_2.HCl$$

Primary amine hydrochloride

$$RNH_2.HCl + NaOH \longrightarrow RNH_2 + NaCl + H_2O$$

Primary amine

8.19. PHYSICAL PROPERTIES AND STEREOCHEMISTRY OF AMINES

(*i*) Lower amines are colourless gases, higher ones are liquids and still higher are solids.

(*ii*) Lower aliphatic amines have ammoniacal odour which vanishes as we move to higher one.

(*iii*) Amines are polar compounds and form intermolecular hydrogen bonding as shown below

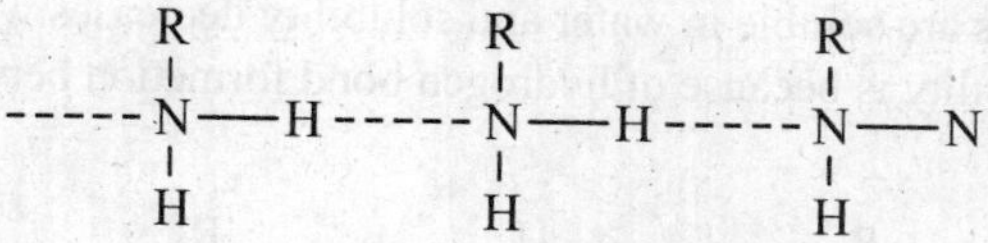

```
      R              R              R
      |              |              |
-----N—H-----N—H-----N—N
      |              |              |
      H              H              H
```

Primary amine

```
      R              R              R
      |              |              |
-----N—H-----N—H-----N—N
      |              |              |
      R              R              R
```

Secondary amine

Tertiary amines do not form hydrogen bonding as there is no hydrogen attached to nitrogen.

As a result of hydrogen bonding, amines have higher boiling points as compared to non-polar compounds of comparable molecular masses. However, the strength of hydrogen bond in amines is lower as compared to that in alcohols and carboxylic acids. Therefore, amines boil at lower temperatures as compared to alcohols and carboxylic acids of comparable molecular masses. Melting and boiling points of some amines are given in Table 8.1.

Table 8.1. Melting and boiling points of some amines.

Compound	*Formula*	*M.P. (K)*	*B.P (K)*
Methylamine	CH_3NH_2	181	265.5
Dimethylamine	$(CH_3)_2NH$	177	280.5
Trimethylamine	$(CH_3)_3N$	156	276
Ethylamine	$CH_3CH_2NH_2$	193	290
Diethylamine	$(CH_3CH_2)NH$	234	328
Triethylamine	$(CH_3CH_2)_3N$	158	362
n-Propylamine	$CH_3CH_2CH_2NH_2$	190	322
Isopropylamine	$CH_3-CH(CH_3)-NH_2$	172	307
Benzylamine	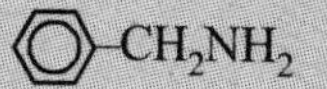 $C_6H_5-CH_2NH_2$	–	458
Aniline	 $C_6H_5-NH_2$	267	457
N-Methylaniline	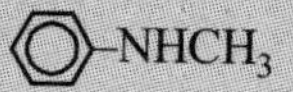 $C_6H_5-NHCH_3$	216	469
N, N-Dimethylaniline	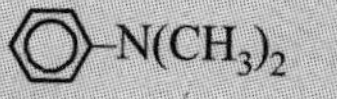 $C_6H_5-N(CH_3)_2$	276	467
o-Toluidine	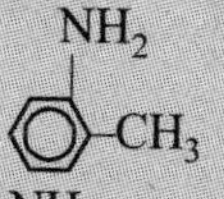 o-$CH_3C_6H_4NH_2$	245	473
m -Toluidine	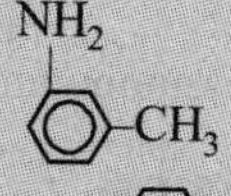 m-$CH_3C_6H_4NH_2$	243	476
p-Toluidine	 $H_2N-C_6H_4-CH_3$	317	473
Diphenyl amine	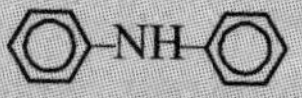 $C_6H_5-NH-C_6H_5$	326	575
Triphenylamine	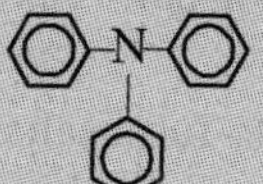 $(C_6H_5)_3N$	400	538

(*iv*) Lower amines are soluble in water and solubility decreases with the increase in size of the alkyl group. This solubility is because of hydrogen bond formation between the molecules of amine and water

$$\cdots\cdots \underset{\underset{H}{|}}{\overset{\overset{R}{|}}{N}}-H\cdots\cdots\overset{\overset{H}{|}}{O}-H\cdots\cdots\underset{\underset{R}{|}}{\overset{\overset{R}{|}}{N}}-H\cdots\cdots\overset{\overset{H}{|}}{O}-H$$

Stereochemistry of amines

Amines in which nitrogen carries three different alkyl groups can exist in two forms, which are non-superimposable mirror images of each other. Such an amine is dissymmetric and should exhibit optical activity. In actual practice, amines have not been reported to exhibit optical activity.

$$\begin{matrix} CH_3 \\ C_2H_5 \\ C_3H_7 \end{matrix}\Big\rangle N \xrightleftharpoons{\text{Rapid inversion}} N\Big\langle \begin{matrix} CH_3 \\ C_2H_5 \\ C_3H_7 \end{matrix}$$

This is because the two forms undergo rapid inversion from one form to the other and hence the two forms become indistinguishable.

It may be noted that quaternary ammonium salts in which four different groups are attached to the central atom exhibit optical activity. This is because quaternary ammonium salts have tetrahedral arrangement of the groups around the central nitrogen atom. This is unlike pyramidal arrangement of amines. Tetrahedral arrangement does not allow flipping from one form to the other to take place. Therefore, the two enantiomeric forms exists independently: This is explained below with the example of methyl allyl phenyl benzyl ammonium chloride.

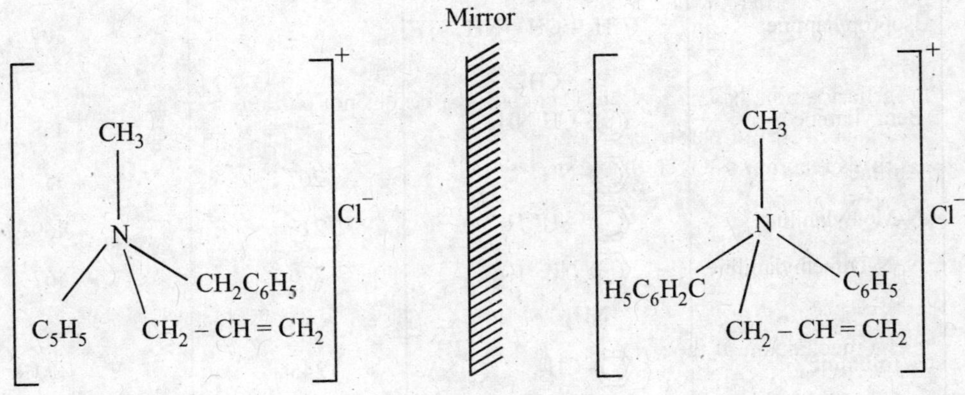

Mirror image forms of methyl allyl phenyl benzyl ammonium chloride (No flipping possible)

8.20. CHEMICAL PROPERTIES OF AMINES

Chemical properties of amines are given below

1. Basic character. Due to the presence of a lone pair of electrons on nitrogen, amines act as bases and give the following reactions:

(*a*) **Formation of salts.** They react with acids to form salts

$$RNH_2 + HCl \longrightarrow R\overset{+}{N}H_3\overset{-}{Cl} \quad \text{Alkyl ammonium chloride}$$

$$R_2NH + HCl \longrightarrow R_2\overset{+}{N}H_2Cl^- \quad \text{Dialkyl ammonium chloride}$$

$$R_3N + HCl \longrightarrow R_3\overset{+}{N}HCl^- \quad \text{Trialkyl ammonium chloride}$$

(*b*) **Formation of hydroxide.** Water soluble amines form substituted ammonium hydroxides which ionise to give hydroxyl ions.

$$RNH_2 + H_2O \rightleftharpoons RNH_3OH \rightleftharpoons RNH_3^+ + OH^-$$

$$R_2NH + H_2O \rightleftharpoons R_2NH_2OH \rightleftharpoons R_2NH_2^+ + OH^-$$

$$R_3N + H_2O \rightleftharpoons R_3NHOH \rightleftharpoons R_3NH^+ + OH^-$$

Thus the aqueous solution of amines are basic in nature.

2. Alkylation. All the three types of amines, (primary, secondary and tertiary) undergo alkylation with alkyl halides. The ultimate compound is quaternary ammonium halide.

$$\underset{\text{Primary amine}}{RNH_2} \xrightarrow[-HX]{RX} \underset{\text{Secondary amine}}{R_2NH} \xrightarrow[-HX]{RX} \underset{\text{Tertiary amine}}{R_3NH} \xrightarrow[-HX]{RX} \underset{\text{Quaternary amm. salt.}}{[R_4N]^+X^-}$$

3. Acylation. Primary and secondary amines on treatment with acetyl chloride or acetic anhydride undergo acylation to give N-substituted amides.

$$RCOCl + R'NH_2 \longrightarrow \underset{\text{N-alkyl amide}}{RCONHR'} + HCl$$

$$RCOCl + R'_2NH \longrightarrow \underset{\text{N, N dialkyl amide}}{RCONR'_2} + HCl$$

Tertiary amine has no hydrogen and as such it does not participate in acylation.

In case the acid chloride or anhydride is aromatic, the reaction is known as **Schotten-Baumann's** reaction as given below:

$$\underset{\text{Benzoyl chloride}}{C_6H_5COCl} + \underset{\text{Aniline}}{H_2NC_6H_5} \longrightarrow \underset{\text{N-Phenyl benzamide}}{C_6H_5CONHC_6H_5}$$

The mechanism of the reaction is as follows:

$$R-\overset{R'}{\underset{H}{N}}: + R-\overset{O}{\overset{\|}{C}}-Cl \longrightarrow R-\overset{R'}{\underset{H}{\overset{\oplus}{N}}}-\overset{O}{\overset{\|}{C}}-R + Cl^{\ominus}$$

(R = Alkyl or Aryl group)
(R' = H or Alkyl or Aryl group)

$$\xrightarrow{R-\ddot{N}H-R'} R-\overset{R'}{N}-\overset{O}{\overset{\|}{C}}-R + R-\overset{\oplus}{N}H_2-R'$$

Since tertiary amines cannot lose the proton after attachment to carbon, they do not react.

4. Reaction with Hinsberg's reagent (sulphonyl chloride). Primary and secondary amines react with sulphonyl chlorides such as benzene or toluene sulphonyl chloride to form substituted sulphonamides.

$$H_3C-C_6H_4-SO_2Cl + CH_3CH_2NH_2 + NaOH \longrightarrow C_6H_5-SO_2NHCH_2CH_3 + NaCl + H_2O$$

p-Toluene sulphonyl chloride; Ethyl amine; N-Ethyl *p*-Toluene sulphonamide

This reaction is used for the separation of three classes of amines.

5. Reaction with nitrous acid. (*i*) **Primary aromatic amines** react with nitrous acid (or a mixture of $NaNO_2$ and HCl) in cold to form diazonium salts.

$$C_6H_5-NH_2 + NaNO_2 + HCl \xrightarrow{273\ K} C_6H_5-N_2Cl + NaCl + 2H_2O$$

Benzene diazonium chloride

(*ii*) **Secondary amines,** both aliphatic and aromatic, react with nitrous acid to give N-nitroso-amines which are yellow oily compounds insoluble in dilute mineral acids.

$$R_2NH + NaNO_2 + HCl \longrightarrow R_2N-N=O + NaCl + H_2O$$

N-nitrosodialkyl. amine

$$C_6H_5-\underset{\displaystyle CH_3}{\underset{|}{N}}H + NaNO_2 + HCl \longrightarrow C_6H_5-\underset{\displaystyle CH_3}{\underset{|}{N}}-N=O + NaCl + H_2O$$

N-nitroso N-methyl aniline

(*iii*) **Tertiary aromatic amines** react with nitrous acid to give substituted product *p*-nitrosoamine which is green in colour.

$$(CH_3)_2N-C_6H_5 + NaNO_2 + HCl \xrightarrow{273\ K} (CH_3)_2N-C_6H_4-N=O + NaCl + H_2O$$

p-Nitroso, *n*-Dimethyl aniline

Tertiary aliphatic amines react with nitrous acid to form water soluble amine salts, trialkyl ammonium nitrite. There is no change in colour.

$$R_3N + NaNO_2 + HCl \longrightarrow [R_3NH]^+ NO_2^- + NaCl$$

Trialkyl ammonium nitrite

6. Carbylamine Reaction. Alkyl isocyanide is formed when a primary amine is heated with chloroform and alcoholic potash. Alkyl isocyanide gives a very disagreeable odour.

$$RNH_2 + CHCl_3 + KOH\ (alc.) \longrightarrow RNC + 3KCl + 3H_2O$$

Primary amine; Carbyl-amine

This reaction is not given by secondary and tertiary amine and hence this reaction can be used as a test to distinguish primary amines from secondary and tertiary amines.

Mechanism of the reaction

The reaction is believed to involve the formation of a **nitrogen ylid** by the addition of dichlorocarbene initially formed. This is followed by proton transfer and elimination of hydrogen chloride to give the product.

$$HO^- + H-CCl_3 \xrightarrow[-H_2O]{} :\bar{C}Cl_3 \longrightarrow :CCl_2 + Cl^-$$

Dichloro carbene

$$R-\underset{H}{\overset{H}{|}}\!\!N: + :CCl_2 \longrightarrow R-\underset{H}{\overset{H}{|}}\!\!\overset{+}{N}-\bar{C}\begin{matrix}Cl\\Cl\end{matrix}$$

$$\xrightarrow{-H^+} R-\ddot{N}-\overset{H}{\underset{Cl}{C}}-Cl \longrightarrow R-\underset{H}{\overset{+}{N}}=C\begin{matrix}H\\Cl\end{matrix}$$

$$\xrightarrow{-H^+} R-\ddot{N}=C\begin{matrix}H\\Cl\end{matrix} \longrightarrow R-\overset{+}{N}\equiv C-H \xrightarrow{OH^-} R-\overset{+}{N}\equiv C^-$$

7. Formation of Schiff's base. Schiff's base is obtained when primary amine reacts with aldehyde

$$\underset{\text{Aldehyde}}{RCHO} + \underset{\text{Primary amine}}{H_2NR'} \longrightarrow \underset{\text{Schiff's base}}{RCH = NR'} + H_2O$$

8. Action of carbon disulphide. On warming with carbon disulphide in the presence of mercuric chloride, primary amines yield isothiocyanate which has the smell of mustard oil.

$$RNH_2 + S = C = S \xrightarrow{HgCl_2} \underset{\text{Isothiocyanate}}{R-N=C=S} + H_2S$$

9. Acidic nature. Although predominantly basic in nature, primary and secondary amines also exhibit some acidic properties when treated with alkali metals or Grignard reagents.

$$\underset{1^\circ\text{ Amine}}{2RNH_2} + 2Na \longrightarrow RNHNa^+ + H_2$$

$$R_2NH + R'Mg\,X \longrightarrow R_2NMgX + R'H$$

10. Ring substitution in aromatic amines. Halogen, nitro or sulphonic acid groups can be introduced in the benzene ring by electrophilic substitution mechanism. Amino group is an activating group, it activates the *ortho* and *para* positions for substitution. This is because, there is a lone pair of electrons on nitrogen of the amines group. This lone pair of electrons is donated to the benzene ring by resonance at *ortho* and *para* positions which become negative and attract the electrophiles like X^+, NO_2^+ and SO_3. The products obtained are therefore, *ortho* and *para* substituted amines.

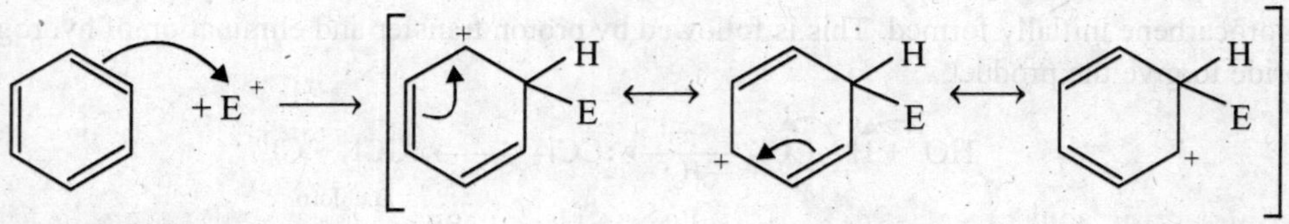

$$C_6H_6 + E^+ \longrightarrow \left[C_6H_6E^+ \longleftrightarrow C_6H_6E^+ \longleftrightarrow C_6H_6E^+ \right]$$

Halogenation

$$\underset{\text{Aniline}}{C_6H_5NH_2} + 3Br_2 \longrightarrow \underset{\text{2, 4, 6 Tribromo aniline}}{C_6H_2Br_3NH_2} + 3HBr$$

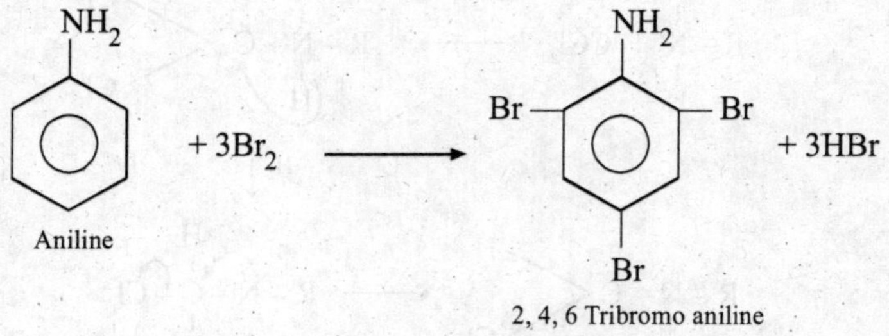

If a monohalogen derivatives is desired, amino group is first acetylated. Acetylation slows down the rate of electrophilic substitution as well as protects the amino group. Thus

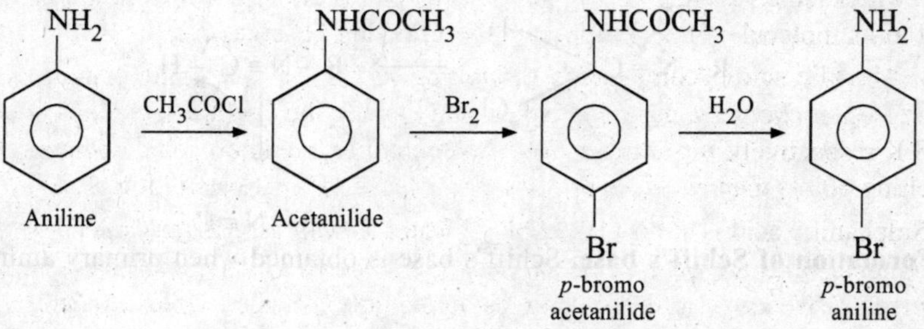

$$\underset{\text{Aniline}}{C_6H_5NH_2} \xrightarrow{CH_3COCl} \underset{\text{Acetanilide}}{C_6H_5NHCOCH_3} \xrightarrow{Br_2} \underset{p\text{-bromo acetanilide}}{Br\text{-}C_6H_4\text{-}NHCOCH_3} \xrightarrow{H_2O} \underset{p\text{-bromo aniline}}{Br\text{-}C_6H_4\text{-}NH_2}$$

Nitration. Nitric acid is a strong oxidising agent. Hence there is a danger that amines will get decomposed by nitric acid. However, if aniline is first protected by acetylation and then nitrated, we obtain *p*-nitro aniline.

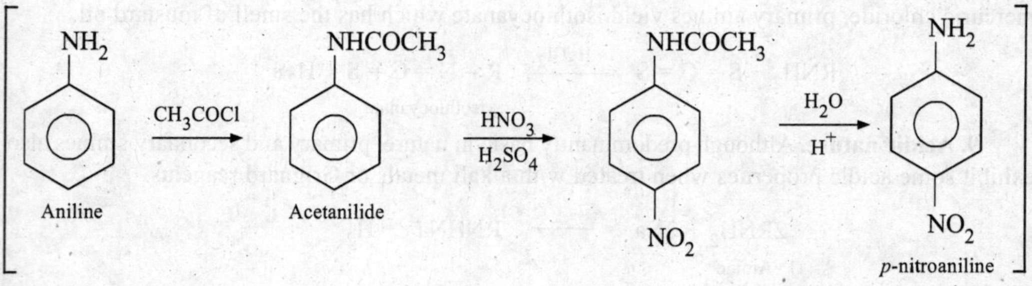

$$\underset{\text{Aniline}}{C_6H_5NH_2} \xrightarrow{CH_3COCl} \underset{\text{Acetanilide}}{C_6H_5NHCOCH_3} \xrightarrow[H_2SO_4]{HNO_3} O_2N\text{-}C_6H_4\text{-}NHCOCH_3 \xrightarrow[H^+]{H_2O} \underset{p\text{-nitroaniline}}{O_2N\text{-}C_6H_4\text{-}NH_2}$$

Sulphonation

Sulphonation of aromatic amines takes place under drastic conditions. The formation of sulphanilic acid takes place through the amine salt, anilinium hydrogen sulphate

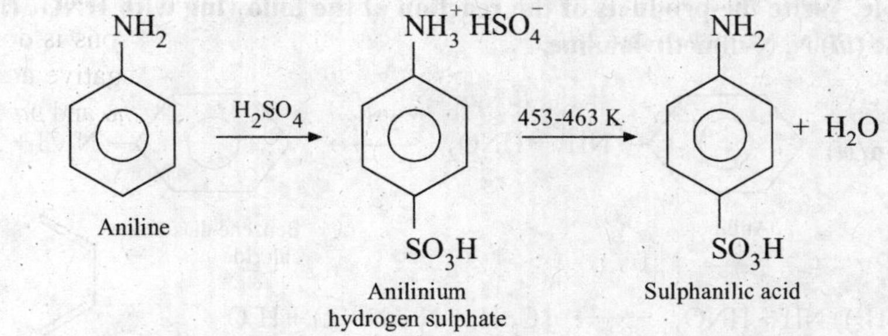

$$C_6H_5NH_2 \text{ (Aniline)} \xrightarrow{H_2SO_4} p\text{-}HSO_3C_6H_4\overset{+}{N}H_3\,HSO_4^- \text{ (Anilinium hydrogen sulphate)} \xrightarrow{453\text{-}463\ K} p\text{-}H_2NC_6H_4SO_3H \text{ (Sulphanilic acid)} + H_2O$$

Sulphanilic acid contains both an acidic and basic group. These groups react to form intramolecular salt. This type of internal salts are known as dipolar or Zwitter ions as shown below:

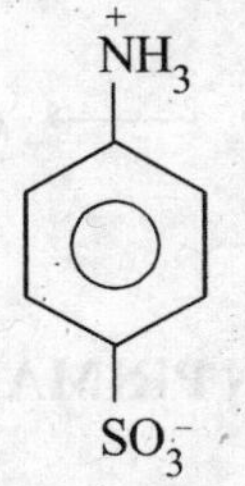

$$p\text{-}\overset{+}{N}H_3C_6H_4SO_3^-$$

Internal salt of sulphanilic acid

Evidence in support of internal salt (zwitter ion) structure

The zwitter ionic structure of sulphanilic acid means the free amino or sulphonic acid groups are absent in its molecule. This is supported by following observations:

(*i*) Sulphanilic acid decomposes with melting at 553 - 573 K while benzene sulphonic acid (having the free sulphonic acid group) and aniline (having the free amino groups) melt at only 339 K and 268 K respectively, *i.e.* at much lower temperatures, we know ionic compounds have much higher melting points than covelent compounds because of electrostatic forces.

(*ii*) Sulphanilic acid is almost insoluble in water as well as most organic solvents. In contrast to this, sulphonic acids are generally highly soluble in water while amines are generally soluble in organic solvents. We know ionic compounds are soluble in water while covalent compounds are insoluble.

(*iii*) Sulphanilic acid is insoluble in aqueous mineral acids whereas amines go into solution with the formation of salts. Sulphanilic acid, however, dissolves in aqueous bases such as NaOH. This is because the strongly basic hydroxide ion can abstract a proton from $-NH_3$ to form the water soluble sodium salt.

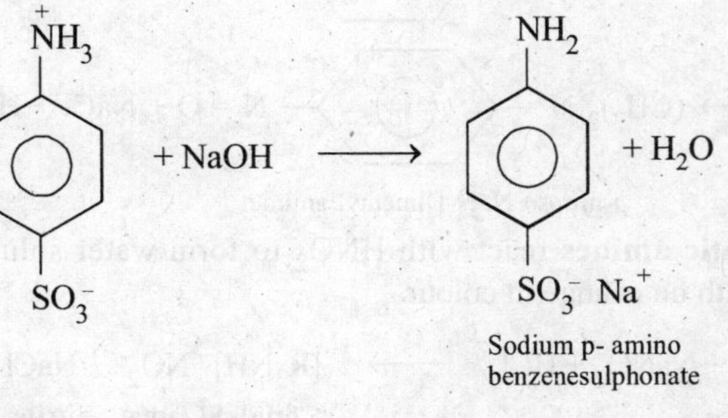

$$p\text{-}\overset{+}{N}H_3C_6H_4SO_3^- + NaOH \longrightarrow p\text{-}H_2NC_6H_4SO_3^-\,Na^+ + H_2O$$

Sodium p- amino benzenesulphonate

Example. Write the products of the reaction of the following with HNO_2 (*i*) Aniline (*ii*) Diethylamine (*iii*) N, N-dimethylaniline.

Solution. (*i*)

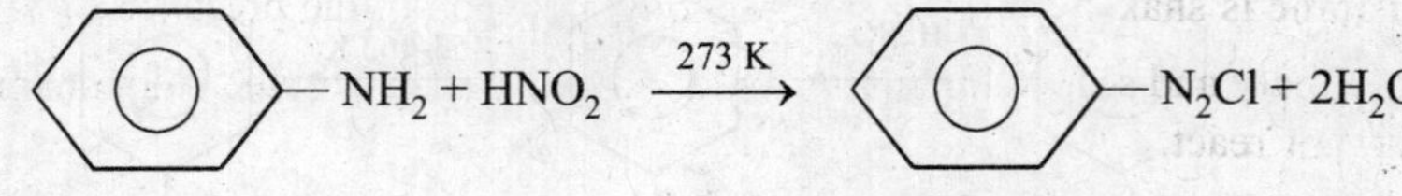

$$C_6H_5-NH_2 + HNO_2 \xrightarrow{273\text{ K}} C_6H_5-N_2Cl + 2H_2O$$

Aniline — Benzene diazonium chloride

(*ii*) $(C_2H_5)_2NH + HNO_2 \longrightarrow C_2H_5-N(C_2H_5)-N=O + H_2O$

N-nitroso diethyl amine

(*iii*) $(CH_3)_2N-C_6H_5 + HNO \longrightarrow (CH_3)_2N-C_6H_4-N=O + H_2O$

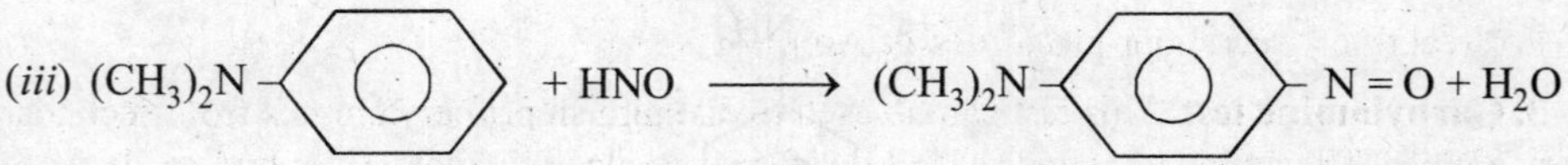

N-nitroso-N, N-dimethyl aniline

8.21. DISTINCTION BETWEEN PRIMARY, SECONDARY AND TERTIARY AMINES

Different methods to distinguish between different amines are discussed as under:

1. Nitrous acid method

Aliphatic primary amines react with HNO_2 (or a mixture of $NaNO_2$ and HCl) to give an effervescence of nitrogen gas.

$$RNH_2 + NaNO_2 + HCl \longrightarrow [RN_2Cl] \xrightarrow{H_2O} ROH + HCl + N_2$$

Aromatic primary amines react with HNO_2 to form diazonium salt.

$$ArNH_2 + NaNO_2 + HCl \xrightarrow{273\text{ K}} ArN_2^+ Cl^- + NaCl + H_2O$$

Secondary amines react with HNO_2 to give yellow oily N-nitrosoamines

$$R_2NH + NaNO_2 + HCl \longrightarrow R_2N-N=O + NaCl + H_2O$$

N-nitroso dialkyl amine

Tertiary aromatic amines react with nitrous acid to yield green-coloured *p*-nitrosoamine.

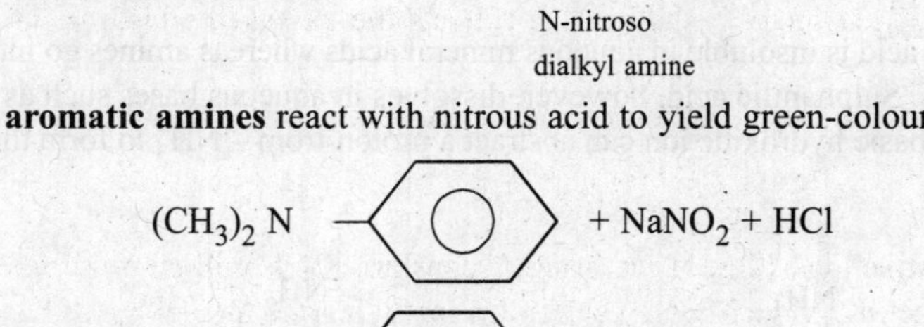

$$(CH_3)_2N-C_6H_5 + NaNO_2 + HCl \xrightarrow{273\text{ K}} (CH_3)_2N-C_6H_4-N=O + NaCl + H_2O$$

p-nitroso N, N-Dimethyl aniline

Tertiary aliphatic amines react with HNO_2 to form water soluble amine salts trialkyl ammonium nitrites with no change in colour.

$$R_3N + NaNO_2 + HCl \longrightarrow [R_3NH]^+ NO_2^- + NaCl$$

Trialkyl amm. nitrite

2. Hinsberg's method. It is a very reliable test for distinguishing between amines and consists of following steps:

(*i*) Amine is shaken with toluene sulphonyl chloride in the presence of an alkali.

(*ii*) Primary and secondary amines form mono and disubstituted sulphonamides while tertiary amines do not react.

(*iii*) The reaction mixture is treated with excess of sodium hydroxide. Monosubstituted sulphonamide forms a soluble sodium salt which on acidification regenerates the insoluble amide.

(*iv*) The disubstituted sulphonamide from secondary amine produces an insoluble compound unaffected by the action of acids.

(*v*) If the unreacted amine remains insoluble in alkali but dissolves on acidification, it indicates a tertiary amines.

For reactions, see under properties of Amines

3. Carbylamine test. This test can be used to distinguish primary amines from secondary and tertiary amines. The amine is heated with chloroform and alcoholic potassium hydroxide. If a strong offensive smell (of RNC) is noticed, it is a primary amine.

$$\underset{\text{Primary amine}}{R\,NH_2} + CHCl_3 + 3KOH \longrightarrow \underset{\text{Carbyl amine}}{R-NC} + 3KCl + 3H_2O$$

8.22. BASICITY OF AMINES

An aqueous solution of an amine exists in the following equilibrium state.

$$RNH_2 + H_2O \rightleftharpoons R\overset{+}{N}H_3 + OH^-$$

For a reaction in equilibrium, we can have the following equation:

$$K_b = \frac{[R\overset{+}{N}H_3][OH^-]}{[RNH_2]}$$

Concentration of water has been omitted from the above equation as it remains constant.

K_b is called the basicity constant of the base. It reflects the extent of basic character of a base. Higher the value of K_b, stronger is the base. It reflects the extent to which amine can accept the proton from water.

pK_b. This is another term which is used to express the basic strength of a base. It is related to K_b as follows:

$$pK_b = -\log K_b$$

Thus, it is negative logarithm of the basicity constant K_b. It will be observed that higher the K_b value, lower is the pK_b value. For example, for methylamine the values of K_b and pK_b are as follows:

$$K_b = 4.4 \times 10^{-4}]$$
$$pK_b = 3.36$$

So for two bases X and Y, if X has a higher value of K_b, it will have a lower value of pK_b. pK_b values of a few important amines are given in the table below:

pK_b values of some amines

(pK_b *of ammonia* = 4.75)

Compound	*Formula*	pK_b
Methylamine	CH_3NH_2	3.36
Dimethylamine	$(CH_3)_2NH$	3.29
Trimethylamine	$(CH_3)_3N$	4.2
Ethylamine	$CH_3CH_2NH_2$	3.34
Diethylamine	$(CH_3CH_2)_2NH$	3.02
Triethylamine	$(CH_3CH_2)_3N$	3.26
n-Propylamine	$CH_3CH_2CH_2NH_2$	3.42
Benzylamine	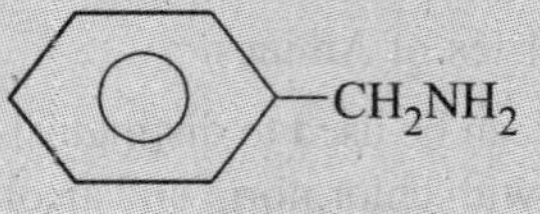 C_6H_5–CH_2NH_2	4.64
Aniline	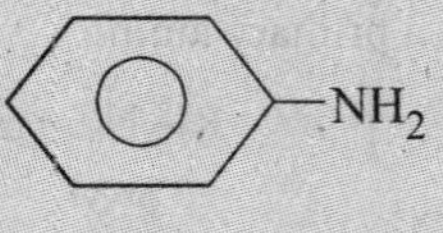 C_6H_5–NH_2	9.38
N-Methylaniline	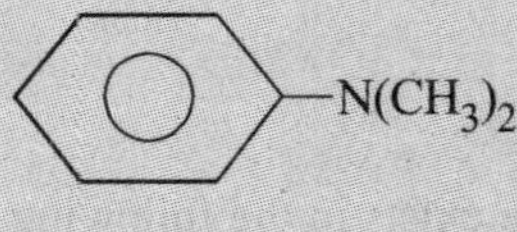 C_6H_5–$N(CH_3)_2$	9.15
N, N-Dimethylaniline	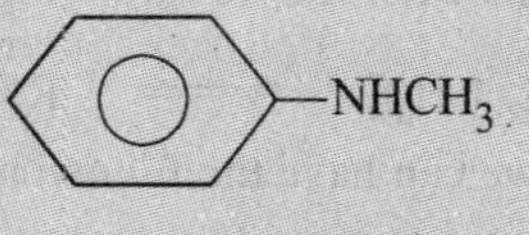 C_6H_5–$NHCH_3$	8.95
Diphenylaniline	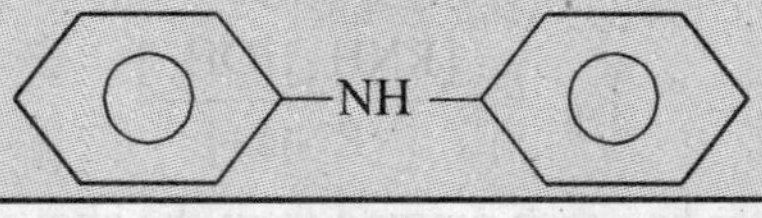 C_6H_5–NH–C_6H_5	13.15

Following observations may be made on perusal of the above table.

(*i*) Aliphatic amines are stronger bases compared to ammonia.

(*ii*) Aromatic amines are weaker than ammonia and aliphatic amines.

(*iii*) Mixed aliphatic-aromatic secondary and tertiary amines are slightly stronger than pure aromatic amines.

8.23. EXPLANATION OF RELATIVE BASICITIES

According to modern acid-base theory the basicity of amine depends upon the availbility of electron pair on nitrogen atom and the stability of the substituted ammonium ion formed. Any factor that increases the availability of electron pair on nitrogen or increases the stability of ammonium ion relative to the parent amine will increase the basicity of amine. Any factor that has the opposite effect will decreases the basicity. Different cases are considered as under:

A. Aliphatic amines: Aliphatic amines have greater basicity as compared to ammonia. This can be explained in terms of inductive effect as explained below:

(1) Inductive effect. *The greater basicity of aliphatic amines as compared to ammonia* can be explained in terms of the electron releasing inductive effect of alkyl group. Aliphatic amines contain one or more alkyl groups in place of hydrogen atoms of ammonia. An alkyl group, being electron releasing, increases the electron density on nitrogen and thus makes the electron pair on nitrogen more available for sharing.

Also the electron releasing inductive effect of alkyl group decreases the positive charge of the substituted ammonium ion formed by the amine and thus increases its stability.

$$R \rightarrow \underset{H}{\overset{H}{N}}: + H^+ \rightleftharpoons R \rightarrow \underset{H}{\overset{H}{\overset{+}{N}}}-H$$

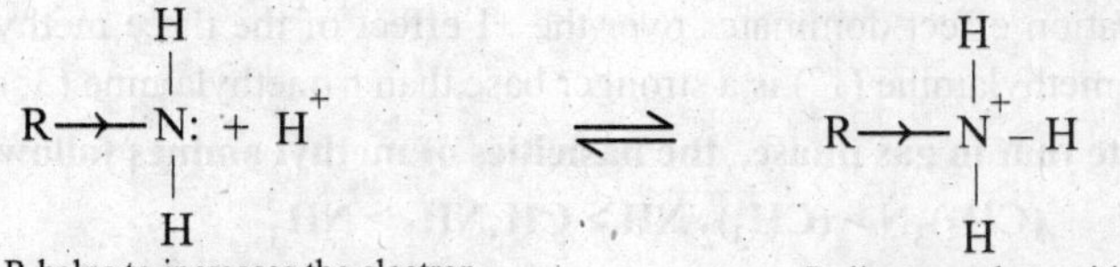

R helps to increases the electron density around nitrogen and thus makes the electron pair more available

R disperses the positive charge of the ion and thus increases its stability

It may be noted that in terms of inductive effect alone, a tertiary amine shoud be more basic than a secondary amine which in turn should be more basic than a primary amine. While it is true that secondary amines are stronger bases than primary amines, the tertiary amines have been found to be weaker than secondary amines and sometimes even weaker than primary amines as shown below:

$$(CH_3)_2\,NH > CH_3\,NH_2 > (CH_3)_3\,N$$
$$(C_2H_5)_2\,NH > (C_2\,H_5)_3N > C_2H_5NH_2$$

This means that the relative basicities of *primary, secondary and tertiary amines cannot be explained in terms of inductive effect alone*. There is another factor i.e. solvation effects, which also plays its role as explained below:

(2) *Solvation effect*. This factor strongly influences the basicity of amines in aqueous solution. The substituted ammonium ions formed by amines by taking up a proton get highly solvated in water through hydrogen bonding and are, therefore, strongly stabilised. But the extent of solvation and consequent stabilisation depends upon the number of hydrogen atoms in substituted ammonium ion which can form hydrogen bonds as shown below:

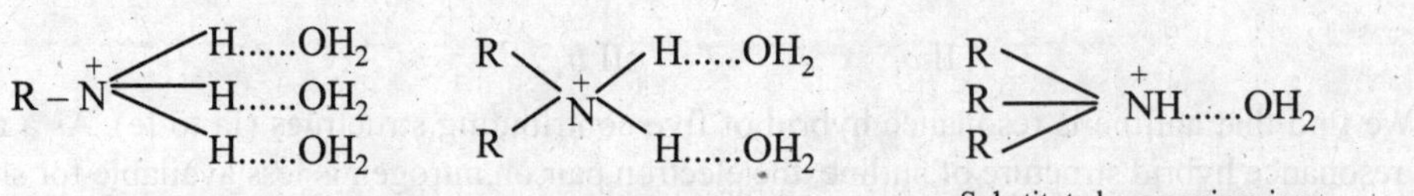

$$R-\overset{+}{N}\begin{cases}H......OH_2\\H......OH_2\\H......OH_2\end{cases}$$

Substituted ammonium ion from primary; maximum stabilisation by solvation

$$\begin{matrix}R & & H......OH_2\\ & \overset{+}{N} & \\ R & & H......OH_2\end{matrix}$$

Substituted ammonium ion from secondary amine

$$\begin{matrix}R\\R\\R\end{matrix} \overset{+}{N}H......OH_2$$

Substituted ammonium ion from temutiary amine; minimum stabilisation by solvation

As the number of hydrogen atoms on substituted ion formed goes on decreasing in going from 1° to 3° amine, the extent of hydrogen bonding and the stabilisation of the ion by solvation also decreases.

As such the basicity of the amines should decreases correspondingly in the order:

$$RNH_2 > R_2NH > R_3N$$

Thus, we observe that *inductive effect and solvation effect have opposing influence on the basicities of aliphatic amines*. Therefore the net basicity of an amine will depends upon the combined effect of inductive effect and solvation effect.

The least stabilisation of ammonium ion obtained from 3° amine is responsible for the lesser basicity of such amines than expected on the baisis of inductive effect alone.

The fact that triethylamine (3°) is a stronger base than ethylamine (1°) while trimethylamine (3°) is a weaker basic methylamine (1°) can be explained as follows: Ethyl group has a greater electron releasing effect than methyl group. Therefore, in ethylamines the +I effect of three ethyl groups in triethylamine becomes the dominating factors as compared with the solvation effect. As a result triethylamine (3°) is a stronger base than ethylamine (1°). But in case of trimethylamine (3°) and methylamine (1°), the solvation effect dominates over the +I effect of the three methyl groups in trimethylamine. Therefore, methylamine (1°) is a stronger base than trimethylamine (3°).

It is interesting to note that in gas phase, the basicities of methyl amines follow the order:

$$(CH_3)_3N > (CH_3)_2NH > CH_3NH_2 > NH_3$$

This is obviously due to the fact that no solvation effect is involved here only the inductive effect plays its part

B. Aromatic amines : Aromatic amines are weaker as compared to ammonia and aliphatic amines. Consider the contributing structures of aniline and anilinium ion shown below:

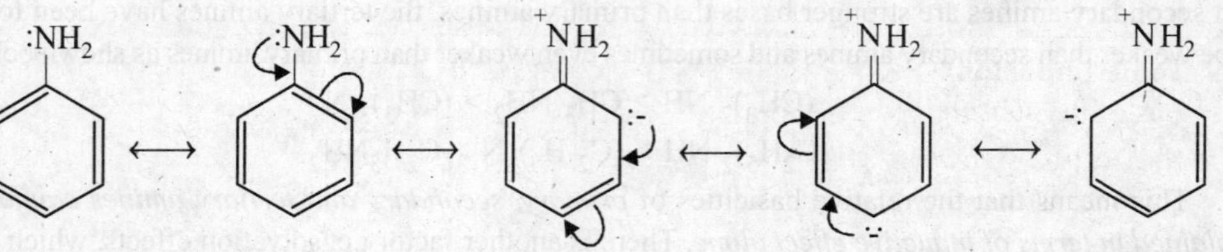

$:NH_2$ ⟷ $:NH_2$ ⟷ $\overset{+}{N}H_2$ ⟷ $\overset{+}{N}H_2$ ⟷ $\overset{+}{N}H_2$

I*a* I*b* I*c* I*d* I*e*

$\overset{+}{N}H_3$ ⟷ $\overset{+}{N}H_3$

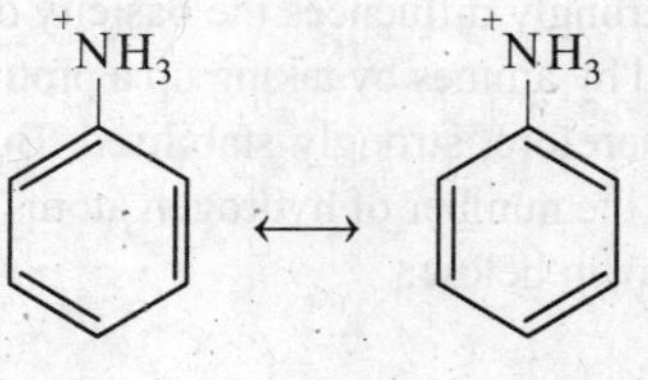

II*a* II*b*

We find that aniline is resonance hybrid of five contributing structrues (Ia to Ie). As a result, in the resonance hybrid structure of aniline, the electron pair on nitrogen is less available for sharing with a proton.

It may be further noted that aniline is **more stabilised** by resonance than the anilinium ion because aniline is a resonance hybrid of five contributing structures while anilinium ion has only two contributing structures (II*a* and II*b*).

Thus aniline is a weaker base than ammonia or aliphatic amines because of the lesser electron availability on nitrogen and lesser stability of the anilinum ion relative to aniline.

Mixed alphatic-aromatic secondary and tertiary amines. These are slightly stronger bases than purely aromatic amines. This is because of the presence of one or more electron releasing alkyl groups in their molecules. Thus, the electron pair on nitrogen of mixed amines is relatively more available for sharing than the electron pair of purely aromatic amines. Also, the substituted ammonium ion obtained from mixed amines gets stabilised due to dispersal of the positive charge by alkyl groups but such stabilisation is not possible in case of purely aromatic amines.

8.24. EFFECT OF SUBSTITUENTS ON THE BASICITY OF AROMATIC AMINES

Aniline has the pK_b value of 9.387. If some group is introduced in the benzene ring of aniline molecule, we find that there is appreciable change in the pK_b value. It means substituents change the basic strength of aniline. The pK_b values of the some substituted anilines are given in Table 8.2:

Table 8.2. pK_b values of ring substituted anilines

(pK_b of aniline = 9.387)

Substituent	pK_b		
	Para isomer	Meta isomer	Ortho isomer
$-CH_3$	8.93	9.31	9.6
$-OCH_3$	8.83	9.7	9.53
$-NH_2$	7.83	9.13	9.5
$-Cl$	9.82	10.53	11.31
$-NO_2$	13	11.5	13.46

Some important observations. Some important observations related to the nature and positions of the substituents on the basicity are made as under.

(*i*) *Electron releasing groups exert greater base strengthening effect when present in para position than when present in meta position.* This is clear from the pK_b values of the following compounds.

NH_2 NH_2 NH_2

> >

CH_3

CH_3

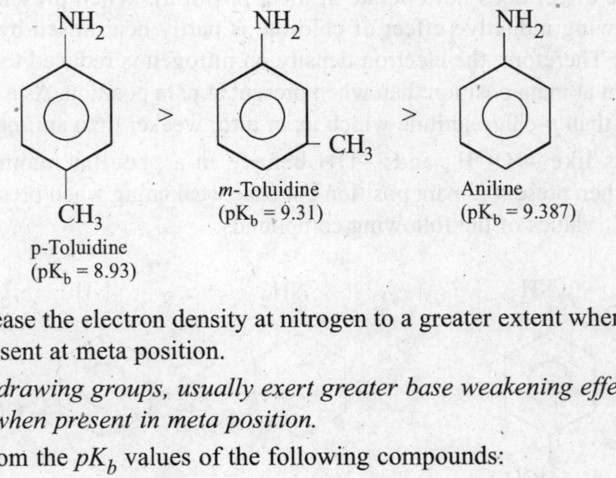

p-Toluidine (pK_b = 8.93)

m-Toluidine (pK_b = 9.31)

Aniline (pK_b = 9.387)

Such groups increase the electron density at nitrogen to a greater extent when present at para position than when present at meta position.

(*ii*) *Electron withdrawing groups, usually exert greater base weakening effect when present in para position than when present in meta position.*

This is evident from the pK_b values of the following compounds:

NH_2 NH_2 NH_2

< <

NO_2

NO_2

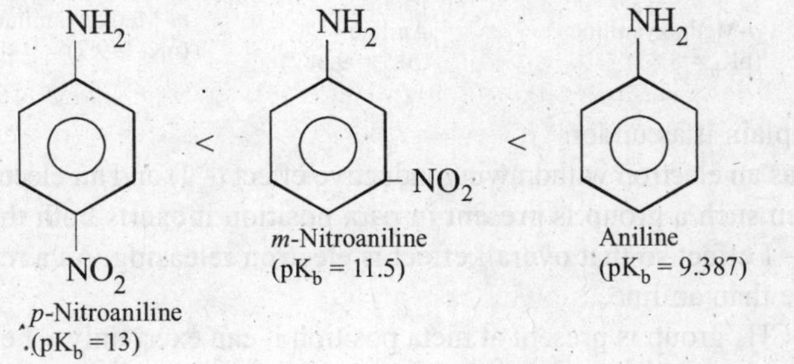

p-Nitroaniline (pK_b =13)

m-Nitroaniline (pK_b = 11.5)

Aniline (pK_b = 9.387)

The above observations are explained as follows. Nitro groups has an electron withdrawing inductive effect as well as electron withdrawing resonance effect. While the inductive effect operates at both meta and para positions, the resonance effect does not operate at meta position. The net result is that electron density on nitrogen is reduced to a greater extent by the group when present at para position than at meta position. Hence *p*-nitroaniline is a weaker base than *m*-nitroaniline which itself is weaker than aniline.

(*iii*) *Halogens such as chlorine and bromine exert greater base weakening effect when present in meta position than when present in para position*. For example, the basic strengths of the *meta* and *para* chloroanilines are expressed as their pK_b values.

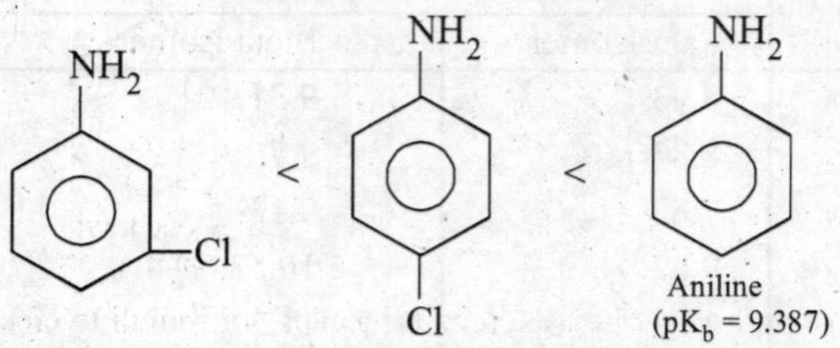

m-Chloroaniline ($pK_b = 10.53$) < *p*-Chloroaniline ($pK_b = 9.98$) < Aniline ($pK_b = 9.387$)

It may be recalled that chlorine has an electron withdrawing inductive effect and electron-releasing resonance effect.

When chlorine is present in meta position, chlorine exert only electron withdrawing inductive effect (resonance effect does not operate at meta position). When present at para position, the electron withdrawing inductive effect of chlorine is partly neutralised by its electron releasing resonance effect. Therefore, the electron density on nitrogen is reduced to a greater extent when chlorine is present at meta position than when present at para position. As a result, *m*-chloroaniline is a weaker base than *p*-chloroaniline which is, in turn, weaker than aniline.

(*iv*) Groups like $—OCH_3$ and —OH behave in a peculiar manner. They act as **base strengthening** when present in para position but base weakening when present in meta position as shown by the pK_b values of the following compounds.

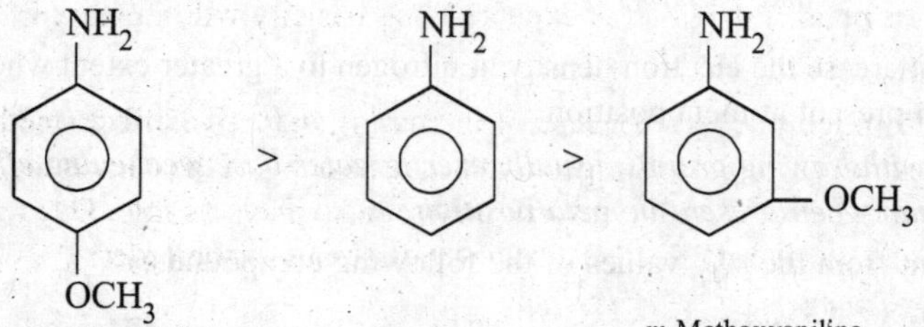

p-Methoxyaniline ($pK_b = 8.83$) > Aniline ($pK_b = 9.387$) > *m*-Methoxyaniline ($pK_b = 9.7$)

We may explain it as under:

$—OCH_3$ has an electron withdrawing inductive effect (– I) and an electron releasing resonance effect (+R). When such a group is present in para position it exerts both the effects but +R effect is stronger than – I effect so that overall effect is electron releasing. As a result, *p*-methoxyaniline is a stronger base than aniline.

When $—OCH_3$ group is present at meta position it can exert only – I effect. Resonance effect does not work at meta position. This decreases the electron density on nitrogen. That is why *m*-methoxyaniline is a weaker base than aniline.

Ortho-effect

Even electron-releasing groups decrease the basic strength when present in ortho position. The phenomenon of decreasing of basic strength with electron-releasing as well as electron withdrawing groups in the ortho position is known as **ortho effect.** There is no satisfactory explanation of this observation.

Example 1. Explain why aniline is less basic than cyclohexylamine?

Solution.

Aniline   Cyclohexyl amine

Smaller basicity of aniline is due to delocalisation of non-bonding electrons on the amino group in aniline

No such delocalisation of electrons takes place in cyclohexyl amine.

Example 2. Arrange the following compounds in increasing order of their basic strength NH_3, CH_3NH_2, $(CH_3)_2NH$, $(CH_3)_3N$ in

(a) Polar solvent like ethanol (or water)

(b) In non-polar solvent or in gas phase

Solution. (*a*) In polar solvent. It is expected that basicity will increase in the order:

$NH_3 < CH_3NH_2 < (CH_3)_2NH < (CH_3)_3N$

As we move from left to right, because of increasing inductive effect (methyl is an electron repelling group), electron density on nitrogen and hence basicity of the compound should increases. But, this is not found to be so. Here the solvation effect also plays its role. Due to combined effect of inductive effect and solvation effect, the basicity increases in the order:

$NH_3 < (CH_3)_3N < CH_3NH_2 < (CH_3)_2NH$

This is explained by saying that the proton of alcohol (or water) finds it difficult to attach to nitrogen in tertiary amine because of steric hindrance as there are already three bulky methyl groups attached. Thus tertiary amine is not easily solvated by alcohol or water.

(*b*) In non-polar solvent or in gas phase

As expected, the basicity will increase in the order

$NH_3 < CH_3NH_2 < (CH_3)_2NH < (CH_3)_3N$

Then is no attachment of protons from the solvent in this case. In the gas phase, no solvent is involved.

8.25. EXHAUSTIVE METHYLATION OF AMINES AND HOFMANN'S ELIMINATION

A substituted quaternary ammonium hydroxide on heating gives an alkene, a tertiary amine and water. This reaction is known as Hofmann elimination. The substituted quaternary ammonium hydroxide is prepared by exhaustive methylation of an amine by a series of reactions given below:

$$\underset{1^\circ \text{ Amine}}{RNH_2N^+} \xrightarrow[-HX]{RX} \underset{2^\circ \text{ Amine}}{R_2NH} \xrightarrow[-HX]{RX} \underset{3^\circ \text{ Amine}}{R_3N} \xrightarrow{RX} \underset{\text{Quaternary amm. halide}}{[R_4N^+]X^-}$$

Conversion of an amine into quaternary ammonium salt by treatment with alkyl halide successively is called **exhaustive alkylation.** If the alkyl halide is a methyl halide, it is known as **exhaustive methylation.**

The solution of quaternary ammonium halide as obtained above is treated with moist silver oxide. Silver halide is precipitated and quaternary ammonium hydroxide is obtained in solution.

$$2R_4N^+\,X^- + Ag_2O + H_2O \longrightarrow \underset{\text{Quaternary hydroxide}}{2R_4N^+\,OH^-} + 2AgX$$

Quaternary ammonium hydroxide on heating strongly at 400 K undergoes an elimination reaction as under.

$$\left[R-\overset{\overset{\large R}{|}}{\underset{\underset{\large R}{|}}{N^+}}-R\right] OH^- \xrightarrow{\text{Heat}} R-\overset{\overset{\large R}{|}}{\underset{\underset{\large R}{|}}{N}} + \text{Alkene} + \text{Water}$$

Taking a particular example

$$\underset{\text{Trimethyl-ethyl ammonium hydroxide}}{\left[CH_3-\overset{\overset{\large CH_3}{|}}{\underset{\underset{\large CH_3}{|}}{N^+}}-CH_2CH_3\right] OH^-} \xrightarrow{\text{Heat}} \underset{\text{Trimethyl amine}}{CH_3-\overset{\overset{\large CH_3}{|}}{\underset{\underset{\large CH_3}{|}}{N}}} + \underset{\text{Ethylene}}{CH_2=CH_2} + H_2O$$

This reaction is known as Hofmann's elimination.

Utility of Hofmann's elimination. This reaction is of great importance in determining the structure of unknown amines. Consider for example that we obtain a mixture of trimethylamine, 1-butene and water in Hofmann elimination *i.e.* after subjecting the amine to exhaustive methylation, treatment with moist silver oxide, followed by heating. By tracing the steps backward in the above reaction, we can see that the amine in the present case is $CH_3CH_2CH_2NH_2$ or *n*-butyl amine.

$$CH_3CH_2CH_2CH_2NH_2 \xrightarrow[\text{Methylation}]{\text{Exhaustive}} CH_3CH_2CH_2CH_2N(CH_3)_3X$$

$$\xrightarrow[H_2O]{Ag_2O} [CH_3CH_2CH_2CH_2N^+(CH_3)_3]OH^-$$

$$\xrightarrow{\text{Heating}} \underset{\text{1-Butene}}{CH_3CH_2CH = CH_2} + \underset{\text{Trimethyl amine}}{(CH_3)_3N} + H_2O$$

Mechanism of Hofmann's elimination

The reaction involves the mechanism given below:

The hydroxide ion removes the hydrogen from the carbon as shown hereunder, creating a double bond and separating the tertiary amine from rest of the molecule.

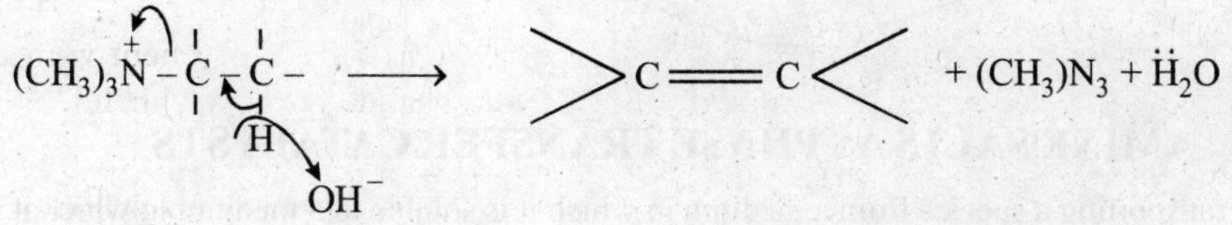

$$(CH_3)_3\overset{+}{N} - \overset{|}{\underset{|}{C}} - \overset{|}{\underset{H}{C}} - \quad (OH^-) \longrightarrow \;>C{=}C< \; + (CH_3)N_3 + \ddot{H}_2O$$

8.26. COPE ELIMINATION

Cope elimination is a reaction which is performed to identify the unknown tertiary amine. It involves treatment of tertiary amine with hydrogen peroxide to obtain amine oxide which on heating gives N, N dialkylhydroxyl amine and an alkene.

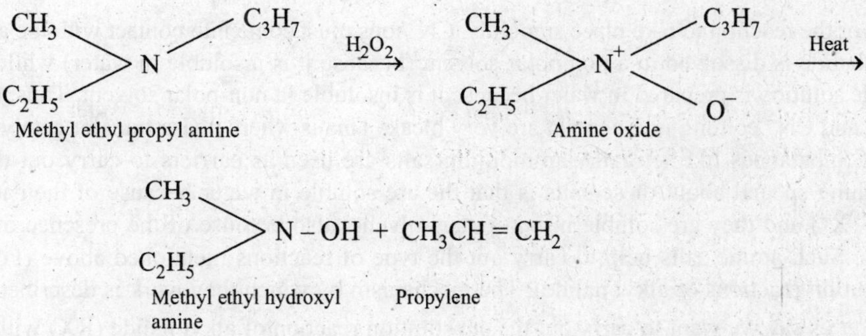

$$\underset{\text{Methyl ethyl propyl amine}}{(CH_3)(C_2H_5)N(C_3H_7)} \xrightarrow{H_2O_2} \underset{\text{Amine oxide}}{(CH_3)(C_2H_5)\overset{+}{N}(C_3H_7)O^-} \xrightarrow{\text{Heat}}$$

$$\underset{\text{Methyl ethyl hydroxyl amine}}{(CH_3)(C_2H_5)N-OH} + \underset{\text{Propylene}}{CH_3CH = CH_2}$$

The dialkyl hydroxylamine can be again converted into tertiary amine on treatment with CH_3I and Ag_2O. Again Cope elimination reaction is performed provided there is still β-hydrogen atom in the molecule.

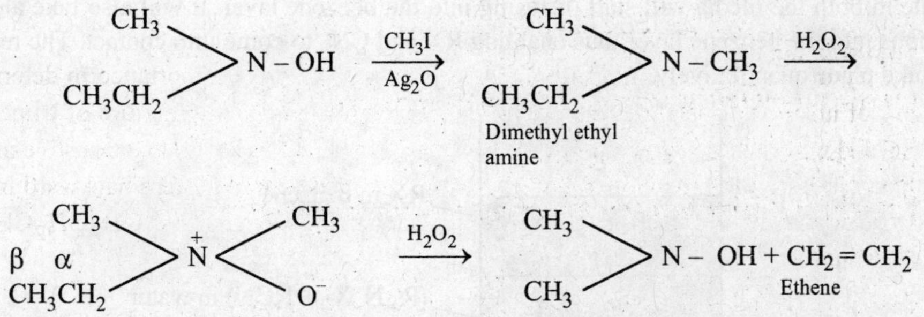

$$(CH_3)(CH_3CH_2)N-OH \xrightarrow[Ag_2O]{CH_3I} \underset{\text{Dimethyl ethyl amine}}{(CH_3)(CH_3CH_2)N-CH_3} \xrightarrow{H_2O_2}$$

$$(CH_3)(\overset{\beta\ \alpha}{CH_3CH_2})\overset{+}{N}(CH_3)O^- \xrightarrow{H_2O_2} (CH_3)(CH_3)N-OH + \underset{\text{Ethene}}{CH_2 = CH_2}$$

It may be noted that in the first Cope elimination process propylene was obtained and in the second Cope elimination process ethene was obtained. This lead us to say that one propyl and one ethyl groups were present in the original tertiary amine. Of course, the third alkyl group was methyl because further repetition of Cope elimination will not give us an alkene. In fact, no β-hydrogen is left in the substituted dimethyl hydroxylamine that we obtain in the second Cope reaction and hence there is no possibility of Cope elimination taking place any more.

The following mechanism has been proposed for Cope elimination reaction:

$$\mathrm{R_2\overset{+}{N}(\text{—}\ddot{\underset{..}{O}}\text{:}^-)\text{—}C\text{—}C\text{—}H} \longrightarrow \mathrm{R_2N\text{—}OH} + \mathrm{>C{=}C<}$$

8.27. AMINE SALTS AS PHASE TRANSFER CATALYSTS

Transporting a species from a medium in which it is soluble to a medium in which it is ordinarily insoluble in called **phase transfer**. The substances which makes such a process feasible is called **phase transfer catalyst**. Phase transfer catalysis is a new technique which enables certain reactions to take place, which might not be possible by conventional methods.

Consider for example, the displacement reaction of an alkyl halide (particularly the higher homologues) with potassium cyanide.

$$\mathrm{R — X + CN^-} \longrightarrow \mathrm{[CN .... R .... X]^-}$$

For the reaction to take place smoothly, CN^- ions must come into contact with the alkyl halide. Alkyl halide is dissolved in a non polar solvent (because it is insoluble in water) while potassium cyanide solution is prepared in water because it is insoluble in non-polar solvent. Thus, the chances of RX and CN^- coming into contact are very bleak. This is where phase transfer catalyst comes in handy. Substances like tetraalkylammonium salts are used as carriers to carry out the process. Something special about these salts is that the are soluble in water because of their ionic nature ($R_4\,N^+\,X^-$) and they are soluble in non-polar solvents also because of the presence of four alkyl groups. Such amine salts help to carry out the type of reactions mentioned above (For example, substitution reactions of alkyl halide). The mechanism by which they work is described as under:

Let us say we want to carry out the substitution reaction of alkyl halide (RX) with CN^- ions. We want the two reactants to come in contact. What is done is we dissolve the alkyl halide in benzene and potassium cyanide in water. We also add a small quantity of the amine salt tetraalkylammonium halide ($R_4\,N^+\,X^-$) to it. This salt being ionic will dissolve in water. We add the benzene solution of alkyl halide to the aqueous solution of potassium cyanide containing a little of $R_4N^+X^-$ (see Fig. 8.3) $R_3N^+X^-$ being soluble in both the media will start diffusing into the benzene layer. It will also take along with it CN^- ions into the benzene layer thus enabling RX and CN^- to come into contact. The reaction will thus take place quantitatively.

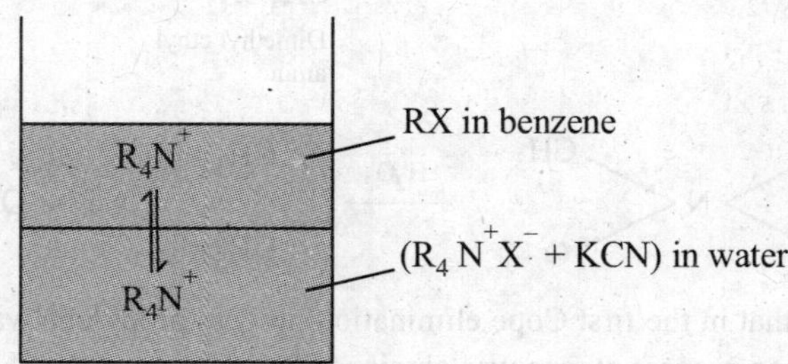

Fig. 8.3. Phase transfer catalysis

Example 1. Arrange the following in decreasing order of basicity:
(*i*) Aniline (*ii*) *p*-Nitroaniline (*iii*) *m*-nitroaniline (*iv*) Diphenylamine

Solution. (*i*) Nitro group has electron-withdrawing effect. Hence nitro anilines are expected to be weaker bases than aniline. Out of *p* and *m* nitroanilines, *p*-nitroaniline is weaker because nitro group has greater electron-withdrawing effect at *p*-position.

(*ii*) Diphenylamine is expected to be weaker than aniline because there are two benzene rings which draw the electrons by resonance. In fact, of the four amines, this is the weakest amine. Hence order of decreasing basicity of the amines is as under:

Aniline > *m*-nitroaniline > *p*-nitroaniline > diphenylamine

Example 2. K_b values of *p*-nitroaniline and *m*-nitroaniline are respectively 0.001×10^{-10} and 0.032×10^{-10}. Which is a weaker base and why?

Solution. Higher the K_b value of a base, greater is the strength of the base. In this problem *m*-nitroaniline has a greater value of K_b, hence it is a stronger base than *p*-nitroaniline. It may be noted that a reverse sequence is followed when we use pK_b values.

Example 3. Arrange the following compounds in increasing order of basicity:
Aniline, *p*-nitroaniline, *p*-methoxyaniline, ammonia, ethylamine, diethylamine, benzylamine and *o*-toluidine.

Solution. Based upon the rules discussed in this chapter, the compounds are arranged in increasing order of basic strength as given below:

p-nitro aniline (NH_2, NO_2) < *o*-Toluidine (NH_2, CH_3) < *p*-methoxy aniline (NH_2, OCH_3) < Aniline (NH_2) < NH_3 Ammonia

< Benzyl amine (CH_2NH_2) < $C_2H_5NH_2$ Ethylamine < $(C_2H_5)_2NH$ Diethylamine

Example 4. Explain the following:

(*a*) Aniline is less basic than N-methyl aniline.

(*b*) Diphenylamine is a much weaker base than aniline.

(*c*) Triphenylamine and N, N-dimethyl aniline are both tertiary amines. Triphenylamine is insoluble in HCl but N, N-dimethylaniline readily dissolves in HCl.

(*d*) 2, 4, 6-trinitroaniline is termed picramide even though it contains no amide linkage.

Solution. (*a*) Aniline is less basic than N-methylaniline

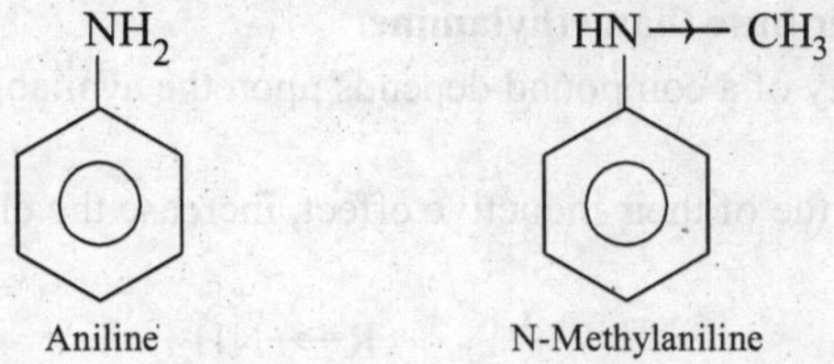

In N-methylamine, there is electron donating methyl group which increases electron density on nitrogen thereby increasing the basicity of the compound. Hence, N-methylaniline is a stronger base than aniline.

(*b*) Diphenylamine is a much weaker base than aniline.

NH_2

Aniline

NH

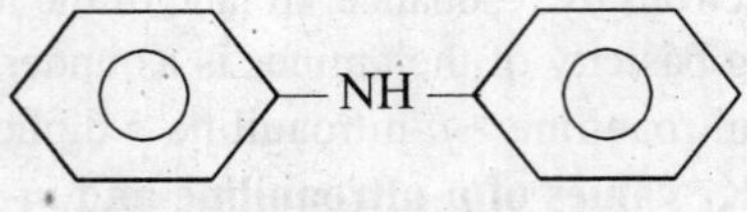

Diphenylamine

Phenyl groups are electron-attracting groups. They attract the electron pair on nitrogen, towards themselves, thereby reducing the basic strengths. Hence diphenylamine is a weaker base than aniline.

(*c*)

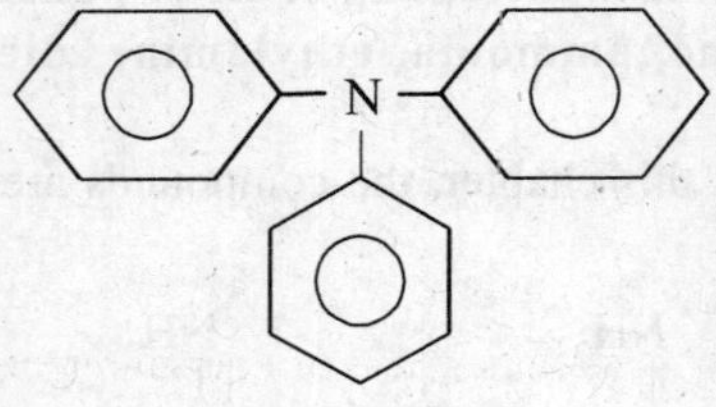

N

Triphenylamine

$N(CH_3)_2$

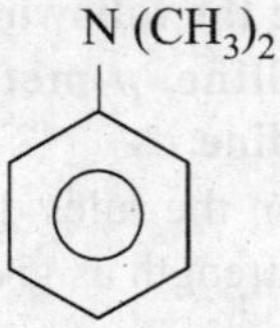

N, N-Dimethyl aniline

Triphenylamine is a very weak base because of three electron-withdrawing phenyl groups. Hence it does not dissolve in HCl.

N, N, Dimethylaniline is comparatively a stronger base because of electron donating effect of two methyl groups. Hence it dissolves in HCl.

(*d*)

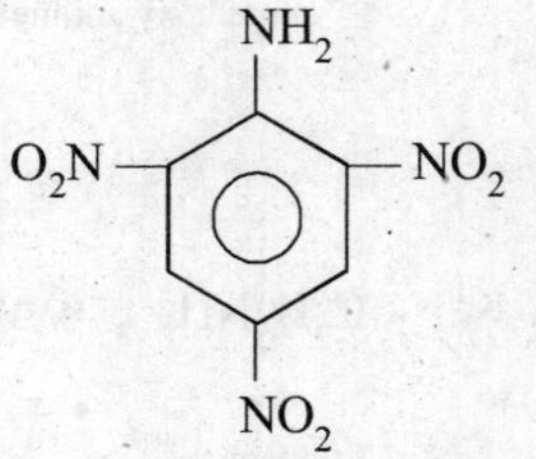

NH_2

O_2N — NO_2

NO_2

2, 4, 6-Trinitro aniline

The combined electron-withdrawing effect of three nitro groups makes it almost a neutral compound comparable to an amide.

Example 5. Explain the following:

(*a*) 3°Amines have greater pK_b values (less basic) than 1° and 2° amines.

(*b*) Aromatic amines are weaker bases than aliphatic amines.

(*c*) Aniline is weaker base than ammonia.

(*d*) Aniline is weaker base than ethylamine.

Solution. (*a*) Basicity of a compound depends upon the availability of electrons (on nitrogen in case of amines).

Alkyl groups, by virtue of their inductive effect, increase the electron density on nitrogen in the amine.

$$R \rightarrow \dot{N}H_2$$

Therefore, primary amine is more basic than ammonia, secondary amine is more basic than primary amine because of the inductive effect of two alkyl groups. But this trend does not extend further. Tert. amine is not more basic than secondary amine. In fact, it is weaker than both primary and secondary (1° and 2°) amines. Therefore, it has a smaller value of K_b and greater value of pK_b. This is because of steric hindrance. Protonated tert. amine is unstable because of steric hindrance.

$$R_3N: + H^+ \longrightarrow R_3N^+-H$$

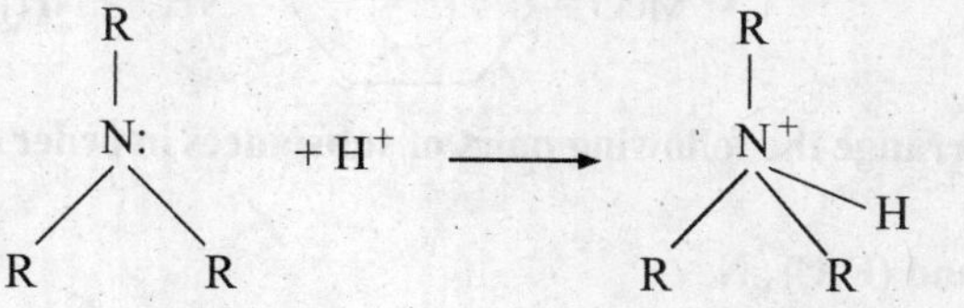

Unstable

(*b*) Aromatic amines are weaker bases than aliphatic amines. This is because the electron pair on nitrogen in aromatic compounds takes part in resonance with benzene ring. This decreases the availability of electrons on nitrogen making it a weaker base.

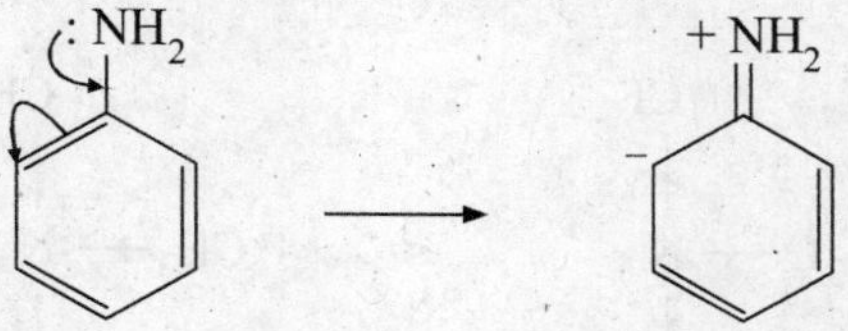

$$C_6H_5-\ddot{N}H_2 \longrightarrow \overset{+}{N}H_2= \text{(ring with } - \text{ charge)}$$

and so on

(*c*) Aniline is weaker base than ammonia. This is again because of the same reasons as in part (*b*) of this question.

(*d*) Aniline is a weaker base than ethylamine.

$$C_6H_5\ddot{N}H_2 \qquad CH_3CH_2\ddot{N}H_2$$

Electron pair on nitrogen in aniline is comparatively less available because of resonance with the benzene ring, thus making it less basic.

Example 6. Given below are pK_a values of a few aromatic amines with substituent

Z in $Z-C_6H_4-NH_2$

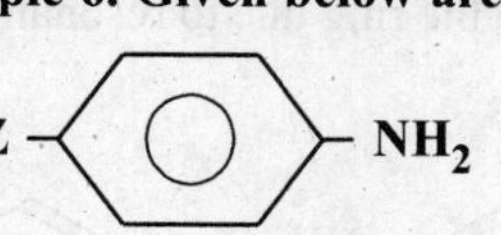

Z =	**pK_a values**
CH_3	**5.12**
NO_2	**1.02**
Br	**3.91**
MeO	**5.29**
NH_2	**6.08**

Arrange them in order of decreasing basic strength.

Solution. Greater the strength of a base, lower the pK_a value.

Note: Here the term pK_a has been used in place of pK_b. The increasing order of pK_a values is

1.02, 3.91, 5.12, 5.29, 6.08

Hence the order of decreasing basic strength will be

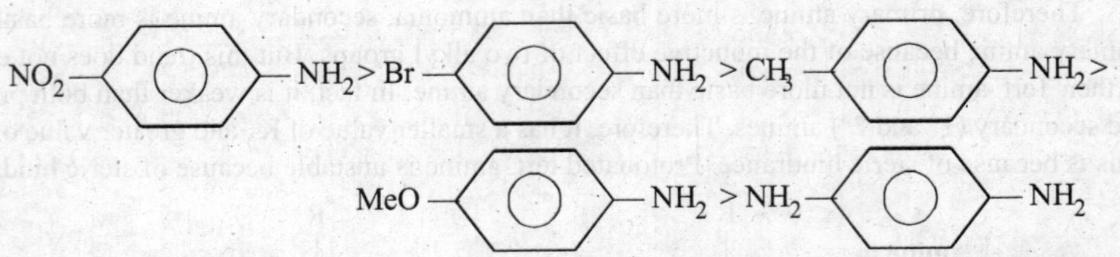

$NO_2-C_6H_4-NH_2 > Br-C_6H_4-NH_2 > CH_3-C_6H_4-NH_2 >$

$MeO-C_6H_4-NH_2 > NH_2-C_6H_4-NH_2$

Example 7. Arrange the following pairs of substances in order of increasing strength. Show your reasoning.

(*i*) $(CH_3)_3N$ and $(F_3C)_3N$

(*ii*) $C_6H_5-NH_2$ and $C_2H_5NH_2$

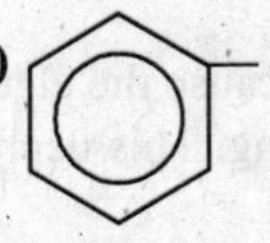

Solution. (*i*)

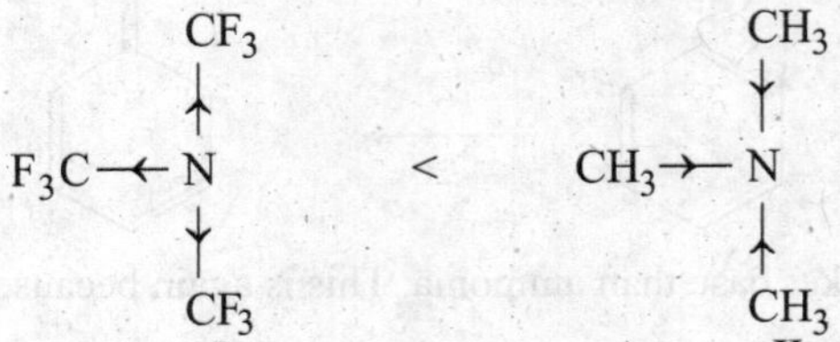

$$N(CF_3)_3 \; (\text{I}) < N(CH_3)_3 \; (\text{II})$$

In compound II, there are three electron-donating methyl groups, attached to nitrogen. They increase electron density on nitrogen. In compound I, there are three electro-withdrawing CF_3^- groups attached to nitrogen (fluorine is highly electronegative). They decrease the electron density on nitrogen. Hence, compound II is more basic than compound I.

(*ii*)

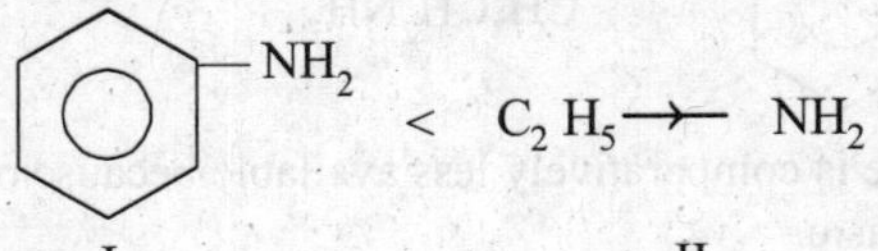

$$C_6H_5-NH_2 \; (\text{I}) < C_2H_5 \rightarrow NH_2 \; (\text{II})$$

In compound I, the electron pair on nitrogen is drawn by the ring due to resonance, as shown below:

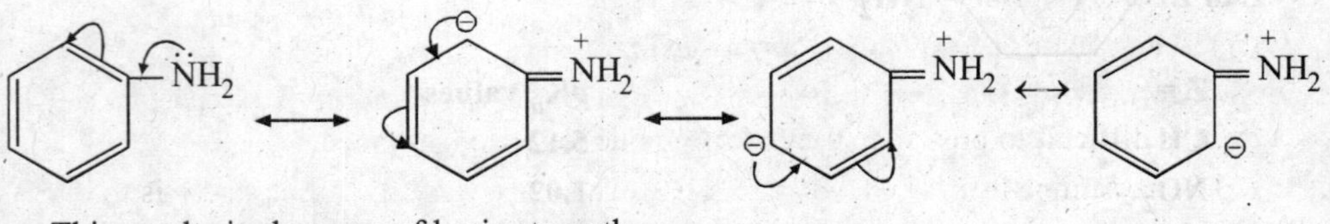

$$C_6H_5-\ddot{N}H_2 \longleftrightarrow \overset{\ominus}{C_6H_5}=\overset{+}{N}H_2 \longleftrightarrow \overset{\ominus}{C_6H_5}=\overset{+}{N}H_2 \longleftrightarrow \overset{\ominus}{C_6H_5}=\overset{+}{N}H_2$$

This results in decrease of basic strength.

In ethylamine II, the electron-releasing effect of ethyl group increases the electron density and basic strength. Hence, ethylamine is more basic than aniline.

Example 8. How will you convert

(*i*) Methylamine to ethylamine

(*ii*) Ethylamine to methylamine

(*iii*) Aniline to benzoic acid?

Solution. (*i*) Methylamine to ethylamine.

$$\underset{\text{Methylamine}}{CH_3NH_2} \xrightarrow{HNO_2} CH_3OH \xrightarrow{PCl_5} CH_3Cl$$

$$\xrightarrow{KCN} CH_3CN \xrightarrow[\text{[H]}]{\text{Na/Alcohol}} \underset{\text{Ethylamine}}{CH_3CH_2NH_2}$$

(*ii*) Ethylamine to methylamine.

$$\underset{\text{Ethylamine}}{CH_3CH_2NH_2} \xrightarrow{HNO_2} CH_3CH_2OH \xrightarrow{[O]} CH_3CHO$$

$$\xrightarrow{[O]} CH_3COOH \xrightarrow[\Delta]{NH_3} CH_3CONH_2 \xrightarrow[KOH]{Br_2} \underset{\text{Methylamine}}{CH_3NH_2}$$

(*iii*) Aniline to benzoic acid.

$$\underset{\text{Aniline}}{C_6H_5NH_2} \xrightarrow[\text{HCl (273 K)}]{NaNO_2} C_6H_5N_2Cl \xrightarrow{CuCN} C_6H_5CN \xrightarrow{H_2O} \underset{\text{Benzoic acid}}{C_6H_5COOH}$$

Example 9. Asymmetric amines do not show optical isomerism. Explain.

Solution. This is because the mirror images of the compound undergo rapid inversion or flipping into each other.

$$(H)(R)(R')\!>\!N \longleftrightarrow N\!<\!(H)(R)(R')$$

Therefore, it is not possible to distinguish between the two different forms.

Example 10. Which is more basic amine in the following pairs and why?

(*i*) $(C_2H_5)_3$ N and $(C_2H_5)_2$ NH

(*ii*) Aniline and benzylamine

(*iii*) Diphenylamine and methylphenylamine

(*iv*) Triethylamine and N, N-dimethylaniline.

Solution. (*i*) $(C_2H_5)_2$ NH is more basic than $(C_2H_5)_3$ N contrary to expectation. This is because it is difficult to protonate triethylamine due to steric hindrance.

(*ii*) Benzylamine is more basic than aniline. This is because in aniline, there is the electron-withdrawing phenyl group attached directly to $-NH_2$ but in benzylamine, no such electron-withdrawing group is present with the result that electrons are not delocalised and therefore it is a stronger base.

(*iii*) Methylphenylamine is a stronger base than diphenylamine because in the former there is only one electron-withdrawing benzene ring whereas in diphenylamine there are two benzene rings.

(*iv*) Triethylamine is a stronger base than N, N-dimethylaniline. This is because the latter contains an electron withdrawing group. Steric factors are common to both.

Example 11. How will you convert ethylamine to acetic acid?

Solution. The following steps are involved in the conversion.

$$C_2H_5NH_2 \xrightarrow{HNO_3} CH_3CH_2OH \xrightarrow{[O]} CH_3CHO \xrightarrow{[O]} CH_3COOH$$

Example 12. Why are amides weaker bases than amines?

Solution. This is because carbonyl group is electron-attracting. It attracts the lone pair of electrons on nitrogen and thus reduces its basicity

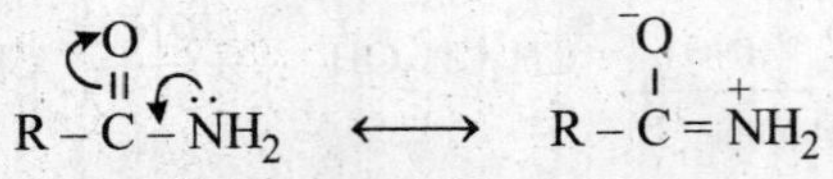

$$R-\overset{O}{\overset{\|}{C}}-\ddot{N}H_2 \longleftrightarrow R-\overset{O^-}{\overset{|}{C}}=\overset{+}{N}H_2$$

Example 13. Arrange the following compounds in order of their increasing basic character.

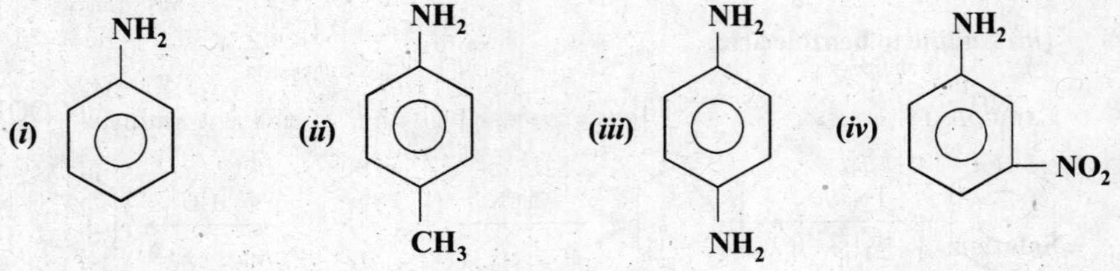

(*i*) Aniline ($C_6H_5NH_2$) (*ii*) p-toluidine (NH_2 with CH_3 para) (*iii*) p-phenylenediamine (NH_2 with NH_2 para) (*iv*) m-nitroaniline (NH_2 with NO_2 meta)

Solution. m-Nitroaniline (NH_2, NO_2) < Aniline (NH_2) < p-Toluidine (NH_2, CH_3) < p-Phenylenediamine (NH_2, NH_2)

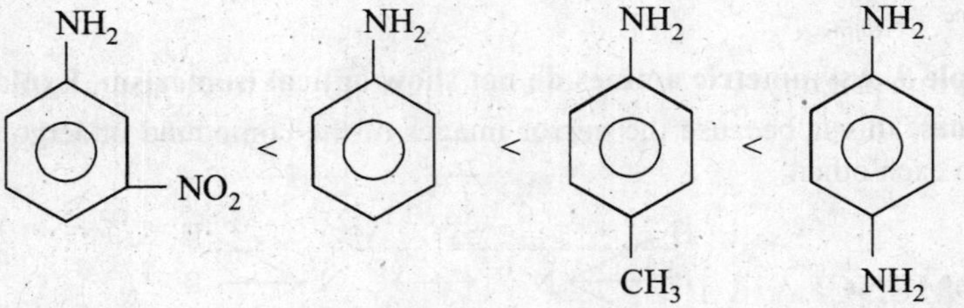

Example 14. How will you do the following conversions :

(*i*) Aniline ⟶ Azobenzene

(*ii*) *o*–Chlorotoluene ⟶ *o*–Toluic acid

Solution. (*i*)

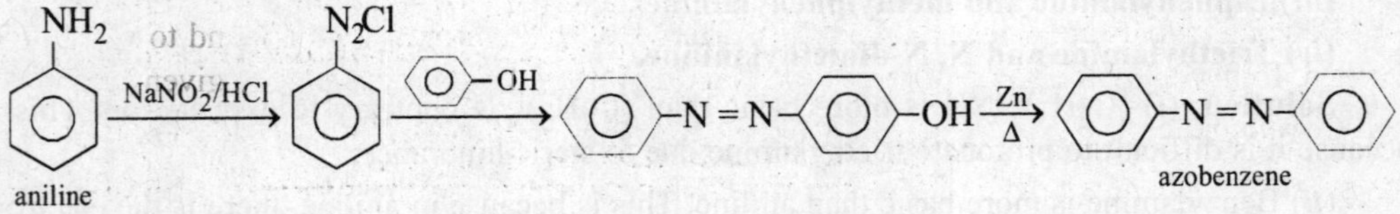

$$\underset{\text{aniline}}{C_6H_5NH_2} \xrightarrow{NaNO_2/HCl} C_6H_5N_2Cl \xrightarrow{C_6H_5OH} C_6H_5-N=N-C_6H_4-OH \xrightarrow[\Delta]{Zn} \underset{\text{azobenzene}}{C_6H_5-N=N-C_6H_5}$$

(*ii*)

$$\underset{o\text{-chlorotoluene}}{CH_3C_6H_4Cl} \xrightarrow{KCN} CH_3C_6H_4CN \xrightarrow{H_2O} \underset{o\text{-toluic acid}}{CH_3C_6H_4COOH}$$

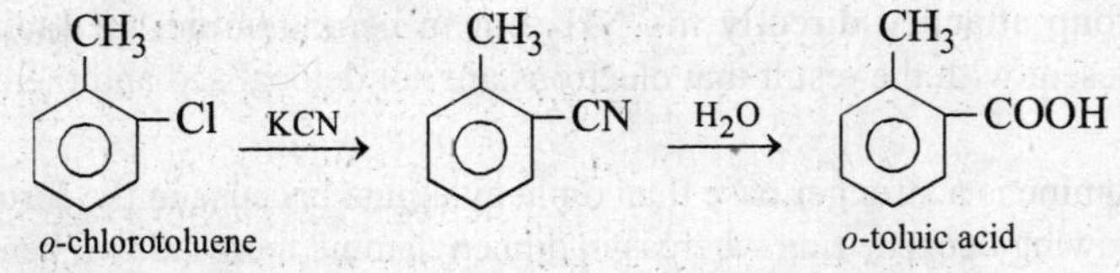

Example 15. Why cannot benzylamine be diazotised?

Solution. Benzylamine cannot be diazotised because – NH_2 group is not directly attached to benzene ring.

Example 16. Complete the following reactions :

aniline + acetic anhydride ⟶

Solution.

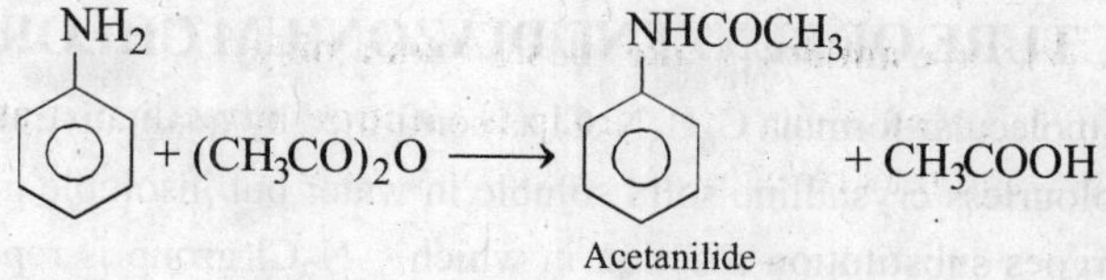

$$C_6H_5NH_2 + (CH_3CO)_2O \longrightarrow C_6H_5NHCOCH_3 + CH_3COOH$$

Acetanilide

Example 17. Suggest a method for conversion of aniline into nitrobenzene.

Solution.

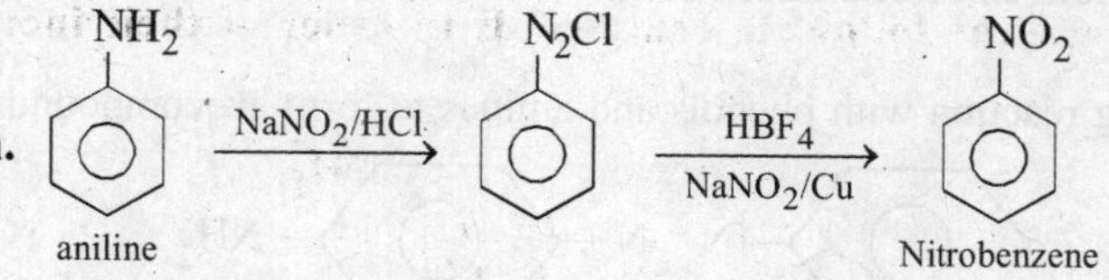

$$C_6H_5NH_2 \xrightarrow{NaNO_2/HCl} C_6H_5N_2Cl \xrightarrow[NaNO_2/Cu]{HBF_4} C_6H_5NO_2$$

aniline Nitrobenzene

Example 18. Explain why aniline is a poorer base than cyclohexyl amine.

Solution. $C_6H_5\ddot{N}H_2$ (Aniline) $C_6H_{11}\ddot{N}H_2$ (Cyclohexylamine)

The electron pair on nitrogen in aniline is delocalised due to resonance with benzene ring. This is not so in case of cyclohexyl amine. Thus aniline is less basic than cyclohexyl amine.

DIAZONIUM SALTS

8.28. NOMENCLATURE

Diazonium salts are aromatic organic compounds containing $-N_2X$ group where X is a halogen. It could also be a $-NO_2$, $-HSO_4$ or $-BF_4$ group. The name **diazo** stands for two nitrogens. The word **onium** is obtained from ammonium as these compounds resemble ammonium salts in some respects.

Nomenclature

Diazonium salts are named by adding **'diazonium'** to the aromatic compound to which it is attached, followed by the name of the anion. Names of some diazonium salts are given below:

Compound	*Name*
$C_6H_5-N_2Cl$	Benzenediazoniumchloride
$C_6H_5-N_2HSO_4$	Benzenediazoniumhydrogensulphate
$CH_3-C_6H_4-N_2Br$	*p*-Toluenediazoniumbromide

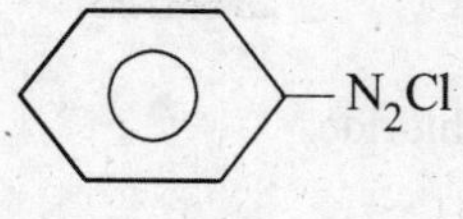

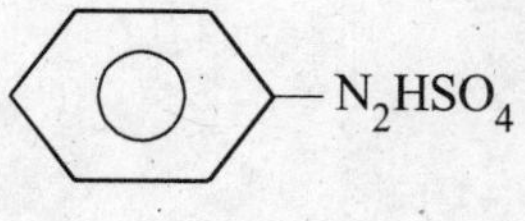

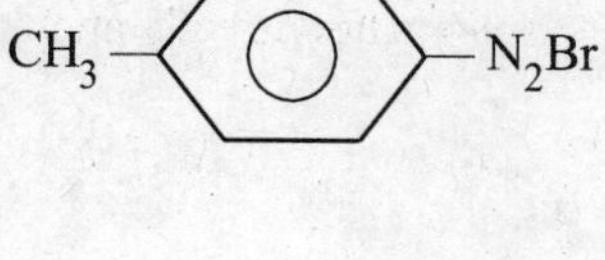

$NO_2-C_6H_4-N_2BF_4$ *p*-Nitrobenzenediazoniumfluoroborate

8.29. STRUCTURE OF BENZENE DIAZONIUM CHLORIDE

(*i*) It has the molecular formula $C_6H_5N_2Cl$ as confirmed by qualitative and quantitative analysis.

(*ii*) It is a colourless crystalline solid soluble in water but insoluble in alcohol and ether.

(*iii*) It undergoes substitution reaction in which $-N_2Cl$ group is replaced by a monovalent group like $-Cl$, $-Br$, $-CN$ or $-NO_2$.

(*iv*) Benzene diazonium chloride exhibits salt-like character. This is indicated by conductivity measurement.

(*v*) It gives coupling reaction with phenols and amines to form azocompounds of the type

$$C_6H_5-N=N-C_6H_4-NH_2$$

(*vi*) On reduction with stannous chloride and hydrochloric acid, it gives phenylhydrazine.

(*vii*) It forms salts called diazotate, in alkaline solution, which exist as geometrical isomers.

Based on the above observations, different scientists gave their structure for benzene diazonium chloride, which are represented below.

Griess Formula. Griess was the discoverer of benzene diazonium chloride. He proposed the structure for this compound as given aside:

However, this formula was not consistent with the properties of the compound. It did not explain well the following properties:

(*a*) Its salt-like character.

(*b*) Replacement of $-N_2Cl$ by other monovalent groups.

(*c*) Reduction to phenylhydrazine in which only one nitrogen is attached to the benzene ring.

(*d*) Formation of azocompounds in coupling reactions.

Kekule Structure

Kekule gave his structure of benzene diazonium chloride as under:

$$C_6H_5-N=N-Cl$$

This formula explained satisfactorily replacement of $-N_2Cl$ by monovalent groups, coupling reactions and reduction to phenylhydrazine. But it didn't account for the salt-like or ionic nature of the compound.

Blomstrand formula

Blomstrand gave the following structure for benzene diazonium chloride.

$$C_6H_5-\underset{\large Cl}{\underset{|}{N}}\equiv N$$

But this has been modified to

$$\left[C_6H_5-\overset{+}{N}\equiv N: \right] Cl^-$$

This modified structure resembles that a ammonium chloride, which can be written as

$$[H-\overset{+}{N}\equiv H_3]\,Cl^-$$

This structure very well explains the following:

(*a*) Salt-like or ionic structure for the compound as indicated by conductivity measurement.

(*b*) Solubility in water and solubility in organic solvents. It is a common observation that polar compounds are soluble in polar solvents.

(*c*) Presence of nitrogen-nitrogen triple bond as indicated by infrared spectrum.

(*d*) Existence of geometrical isomers of diazoates

$$\underset{\text{Benzene diazonium chloride}}{C_6H_5\overset{+}{N}\equiv N-Cl} \xrightleftharpoons{NaOH} \underset{\text{Benzene diazonium hydroxide}}{C_6H_5\overset{+}{N}\equiv NOH^-}$$

$$\underset{\text{Sod. diazotate}}{C_6H_5-N=N\overset{-}{O}\overset{+}{Na}} \longleftarrow \underset{\text{Diazohydroxide}}{C_6H_5N=N-OH}$$

Due to the presence of double bond between N–N, there exists a possibility of occurrence of geometrical isomers as given below:

$$\underset{\text{(Cis) diazotate}}{\begin{array}{c} C_6H_5N \\ \| \\ Na^+O^- - N \end{array}} \qquad \underset{\text{(Trans) diazotate}}{\begin{array}{c} C_6H_5-N \\ \| \\ N-O^-Na^+ \end{array}}$$

8.30. LABORATORY PREPARATION OF BENZENE DIAZONIUM CHLORIDE

Diazonium salts are prepared by reaction between aromatic primary amine and nitrous acid obtained from sodium nitrite and dil. hydrochloric acid *in situ*. For preparing benzene diazonium chloride, the following method is adopted.

Aniline is dissolved in dil. hydrochloric acid and the solution is cooled to 273-278 K. Now a cooled solution of sodium nitrite is added to the first solution taking care that the temperature does not rise beyond 278 K. A few drops of the mixture are added to starch-potassium iodide solution. As soon as a blue colour is obtained, further addition of sodium nitrite is stopped. Thus the excess of nitrous acid is avoided. This is the completion of diazotisation.

$$C_6H_5NH_2 + NaNO_2 + HCl \xrightarrow{273-278\text{ K}} C_6H_5N_2^+Cl^- + NaCl + 2H_2O$$

Benzene diazonium chloride obtained above is used as such in the solution. It is not extracted or crystallised out. This is because dry benzene diazonium chloride is explosive.

Mechanism of diazotisation. It is believed that the amine molecule undergoes nucleophilic reaction by attaching itself to nitrous anhydride obtained from nitrous acid. The complete mechanism is given below:

$$H-\ddot{O}-\ddot{N}=O + \overset{\oplus}{H} \underset{\text{Mineral acid}}{} \rightleftharpoons H-\overset{H}{\underset{\oplus}{\ddot{O}}}-N=O \rightleftharpoons \overset{\oplus}{N}=O + H_2O \text{ (Nitrosoniumion)}$$

$$Ar-\ddot{N}H_2 + \overset{\oplus}{N}=O \longrightarrow Ar-\overset{H}{\underset{H}{\overset{\oplus}{N}}}-N=O \xrightarrow{-H^{\oplus}} Ar-\underset{H}{\ddot{N}}-\ddot{N}=\ddot{O}:$$

$$\xrightarrow{H^+} Ar-\underset{H}{\ddot{N}}-\ddot{N}=\overset{\oplus}{O}H \xrightarrow{-H^{\oplus}} Ar-\ddot{N}=\ddot{N}-\ddot{O}H \xrightarrow{H^{\oplus}} Ar-\ddot{N}=\ddot{N}-\overset{\oplus}{O}H_2$$

$$\xrightarrow{-H_2O} \left[Ar-\overset{\oplus}{N}\equiv N \longleftrightarrow Ar-\ddot{N}=N^{\oplus} \right]$$

Aromatic amines carrying electron withdrawing groups in the benzene ring are difficult to diazotise. This is because mechanistically, it is a nucleophilic reaction. Electron withdrawing groups decrease the electron density on amino nitrogen, thereby reducing the chances of attack on nitrous acid and nitrous anhydride.

$$O_2N-C_6H_4-NH_2 \longleftrightarrow {}^{-}O(O^{-})N=C_6H_4=\overset{\oplus}{N}H_2$$

Example 1. Which of the following gives most stable diazo compound on treatment with HNO_2 and why?

$C_2H_5NH_2$, Aniline, *o*-nitroaniline

Solution. Aniline gives the most stable diazo compound viz. benzene diazonium chloride on treatment with HNO_2.

Aliphatic amines don't give diazo compound, alcohol is formed on treatment with HNO_2. Extent of diazotisation is decreased if an electron-withdrawing group like – NO_2 group is attached to the benzene ring.

8.31. CHEMICAL PROPERTIES OF BENZENE DIAZONIUM CHLORIDE

Properties of benzene diazonium chloride can be discussed under two heads.

(A) Reaction in which N_2 is evolved and the – N_2Cl group is replaced by some other monovalent group.

(B) Coupling reaction of reduction, in which nitrogen atoms are retained.

These reactions are being given separately as under:

Reactions in which $-N_2Cl$ is replaced by some other group.

1. Replacement by – Cl, – Br or – CN. This is a reaction in which group is replaced by one of above mentioned groups. The reaction is carried out by treating a solution of diazonium salt with cuprous chloride, bromide or cyanide. The reactions are given as under:

$$C_6H_5\overset{+}{N}_2Cl^- \xrightarrow{CuCl} C_6H_5N_2Cl + N_2\uparrow$$

$$C_6H_5\overset{+}{N}_2\overset{-}{Cl} + CuBr \longrightarrow C_6H_5Br + N_2\uparrow + CuCl$$

$$C_6H_5\overset{+}{N}_2\overset{-}{Cl} + CuCN \longrightarrow C_6H_5CN + N_2\uparrow + CuCl$$

This reaction is known as **Sandmeyer's reaction.**

In a slighly different reaction, known as **Gattermann reaction,** replacement by – Cl, – Br or – CN is done by taking copper powder and halogen acid.

$$C_6H_5\overset{+}{N}_2Cl^- \xrightarrow{Cu\,/\,HCl} C_6H_5Cl + N_2$$

$$C_6H_5\overset{+}{N}_2Cl^- \xrightarrow{Cu\,/\,HBr} C_6H_5Br + N_2$$

$$C_6H_5\overset{+}{N}_2Cl^- \xrightarrow{Cu\,/\,KCN} C_6H_5CN + N_2$$

2. Replacement by iodine. On warming diazonium salt solution with potassium iodide solution, $-N_2Cl$ group in replaced by – I group.

$$C_6H_5\overset{+}{N}_2\,Cl^- + KI \longrightarrow C_6H_5I + N_2 + KCl$$

3. Replacement by fluorine. Replacement by fluorine can be brought about by the following two-step reaction. It is known as Balz Schiemann reaction.

$$C_6H_5\overset{+}{N}_2\,Cl^- + \underset{\text{Fluoro boric acid}}{HBF_4} \longrightarrow \underset{\text{Diazonium fluoborate (insoluble)}}{C_6H_5\overset{+}{N}_2BF_4^-} + HCl$$

$$C_6H_5\overset{+}{N}_2BF_4^- \xrightarrow{Heat} \underset{\text{Fluorobenzene}}{C_6H_5F} + N_2 + BF_3$$

The reaction proceeds by SN^1 mechanism

$$Ar-\overset{+}{N}\equiv N \longrightarrow \overset{+}{Ar} + N_2$$

$$\overset{+}{Ar} + F-\overset{-}{B}F_3 \longrightarrow Ar-F + BF_3$$

4. Replacement by – OH. On warming the solution of diazonium salt, $-N_2Cl$ group is replaced by – OH group to yield phenol.

$$\underset{\text{Benzene diazonium chloride}}{C_6H_5N_2Cl} + H_2O \longrightarrow \underset{\text{Phenol}}{C_6H_5OH} + N_2 + HCl$$

5. Replacement by – NO_2. Substitution by nitro group is achieved by decomposing diazonium fluoborate with sodium nitrite solution in the presence of copper powder.

$$C_6H_5\overset{+}{N}_2\,Cl^- + HBF_4 \longrightarrow C_6H_5\overset{+}{N}_2\,BF_4^- \xrightarrow[Cu]{NaNO_2} C_6H_5-NO_2 + N_2 + NaBF_4$$

6. Replacement by aryl group. Diazonium salts react with other aromatic compounds in the presence of alkali to give diphenyl. This is known as **Gomberg-Bachmann** reaction.

$$C_6H_5-\overset{+}{N}_2\overset{-}{Cl} + C_6H_6 \xrightarrow{NaOH} C_6H_5-C_6H_5 + N_2 + HCl$$

7. Replacement by hydrogen. If diazonium salt solution is allowed to stand in contact with hypophosphorus acid in the presence of cuprous ions, – N_2Cl group is replaced by – H.

$$C_6H_5N_2Cl + H_3PO_2 + H_2O \longrightarrow C_6H_6 + H_3PO_3 + N_2 + HCl$$

Mechanism of the reaction is as follows:

$$Ar\overset{+}{N} \equiv N + \underset{\text{(Cuprous)}}{Cu^+} \longrightarrow Ar - N \equiv N + \underset{\text{(Cupric)}}{Cu^{2+}}$$

$$\downarrow$$

$$\dot{A}r + N_2$$

$$\dot{A}r + H-\overset{\overset{O}{\|}}{\underset{\underset{H}{|}}{P}}-OH \longrightarrow ArH + H-\overset{\overset{O}{\|}}{\underset{\bullet}{P}}-OH$$

$$ArN_2^+ + H-\overset{\overset{O}{\|}}{\underset{\bullet}{P}}-OH \longrightarrow \dot{A}r + N_2 + H-\overset{\overset{O}{\|}}{\underset{+}{P}}-OH \xrightarrow{H_2O} H-\overset{\overset{O}{\|}}{\underset{\underset{OH}{|}}{P}}-OH + H^+$$

Synthetic Utility. The reaction offers a convenient method of removing an – NH_2 or –NO_2 groups from an aromatic ring. The complete process involving diazotisation of an amine and subsequent reduction of diazonium salt is called **deamination.** This can be a very useful step in a number of organic conversions. For example,

$$\underset{\text{Aniline}}{C_6H_5NH_2} \xrightarrow{\text{Bromine water}} \underset{\text{2, 4, 6- Tri-bromoaniline}}{C_6H_2Br_3NH_2} \xrightarrow{NaNO_2/HCl} \underset{\text{2, 4, 6-Tribromo-benzene diazonium chloride}}{C_6H_2Br_3\overset{+}{N}_2Cl^-} \xrightarrow{H_3PO_2} \underset{\text{Symm, or 1, 2, 3, Tribromobenzene}}{C_6H_3Br_3}$$

Reactions in which nitrogen atoms are retained in the product:

8. Reduction. On treatment with stannous chloride and HCl, diazonium salts are reduced to hydrazines.

$$\underset{\text{Benzene diazonium chloride}}{C_6H_5\overset{+}{N}_2\overset{-}{Cl}} + 4[H] \xrightarrow[\text{or } Na_2SO_3]{SnCl_2/HCl} \underset{\text{Phenyl hydrazine hydrochloride}}{C_6H_5NHNH_2HCl}$$

9. Coupling reaction. It is an important reaction of diazonium salts and involves reaction of diazonium ions which act as electrophilic agents, with aromatic compounds containing strong electron-releasing groups such as – OH, – NHR and – NHR_2. The reaction is known as coupling and leads to the formation of azo compounds having the general formula

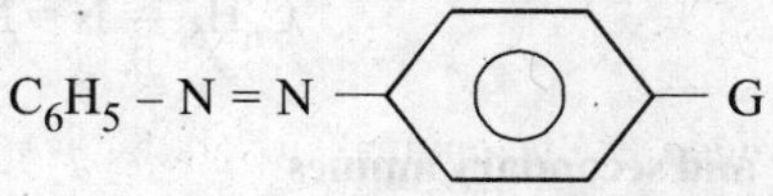

$$C_6H_5 - N = N - C_6H_4 - G$$

where G is one of above mentioned electron releasing groups. The coupling reaction in general may be written as

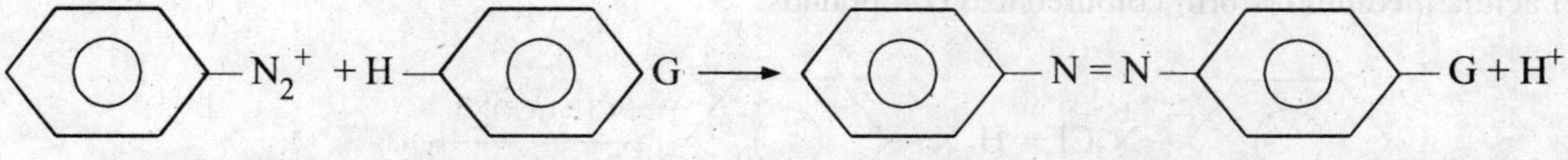

$$C_6H_5-N_2^+ + H-C_6H_4-G \longrightarrow C_6H_5-N=N-C_6H_4-G + H^+$$

***(a)* Coupling with phenol.** Benzene diazonium salt reacts with phenol in weakly alkaline solution to form hydroxy azo compound.

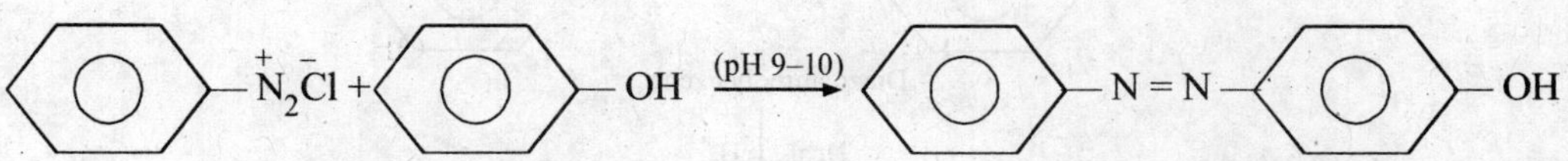

$$C_6H_5-\overset{+}{N}_2\overset{-}{Cl} + C_6H_5-OH \xrightarrow{(\text{pH } 9\text{–}10)} \underset{p\text{-hydroxyazobenzene}}{C_6H_5-N=N-C_6H_4-OH} + HCl$$

Mechanism of reaction

It involves the electrophilic attack of diazonium ion on phenoxide ion as shown below:

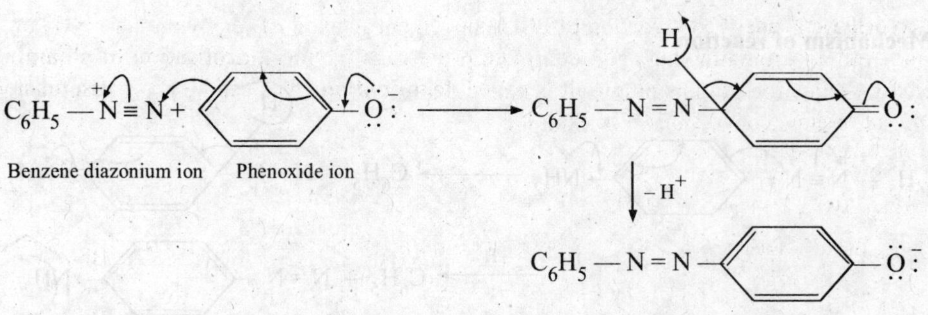

$$\underset{\text{Benzene diazonium ion}}{C_6H_5-\overset{+}{N}\equiv N} + \underset{\text{Phenoxide ion}}{C_6H_5-\ddot{O}:^-} \longrightarrow C_6H_5-N=N-C_6H_4(H)=\ddot{O}: \xrightarrow{-H^+} C_6H_5-N=N-C_6H_4-\ddot{O}:^-$$

***(b)* Coupling with tertiary amine.** This coupling takes place in acidic solution. Diazonium salt reacts with tertiary amine to form dialkylamino azo compound.

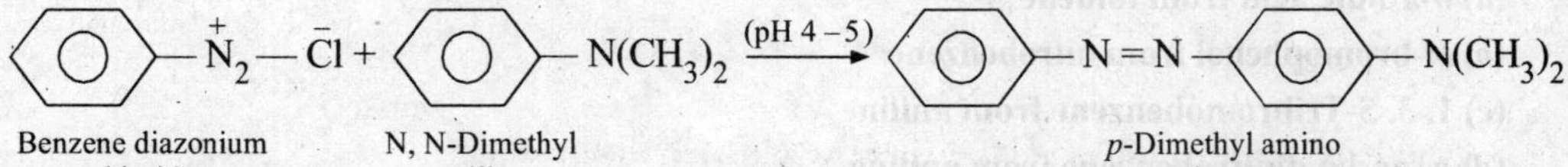

$$\underset{\text{Benzene diazonium chloride}}{C_6H_5-\overset{+}{N}_2-\overset{-}{Cl}} + \underset{\text{N, N-Dimethyl aniline}}{C_6H_5-N(CH_3)_2} \xrightarrow{(\text{pH } 4-5)} \underset{p\text{-Dimethyl amino azobenzene}}{C_6H_5-N=N-C_6H_4-N(CH_3)_2}$$

The mechanism of the reaction is as follows:

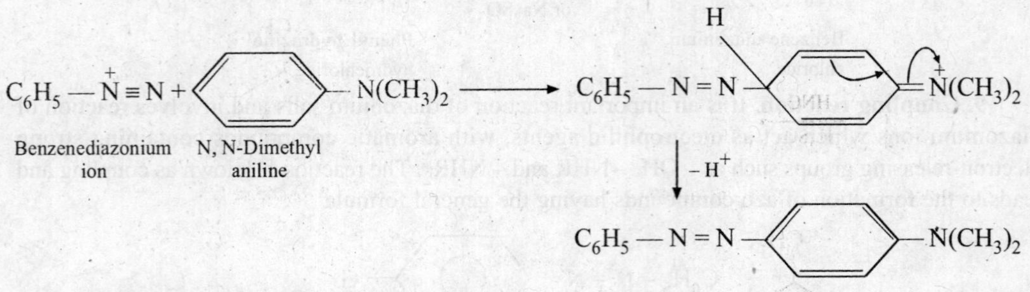

$$C_6H_5-\overset{+}{N}\equiv N + C_6H_5N(CH_3)_2 \longrightarrow C_6H_5-N=N-C_6H_5(H)=\overset{+}{N}(CH_3)_2 \xrightarrow{-H^+} C_6H_5-N=N-C_6H_4-N(CH_3)_2$$

Benzenediazonium ion; N, N-Dimethyl aniline

(c) Coupling with primary and secondary amines

Primary and secondary amines react with diazonium salts in weakly alkaline medium to form colourless diazoamino compounds. However, these compounds undergo rearrangement when heated in acidic medium to form coloured azo compounds.

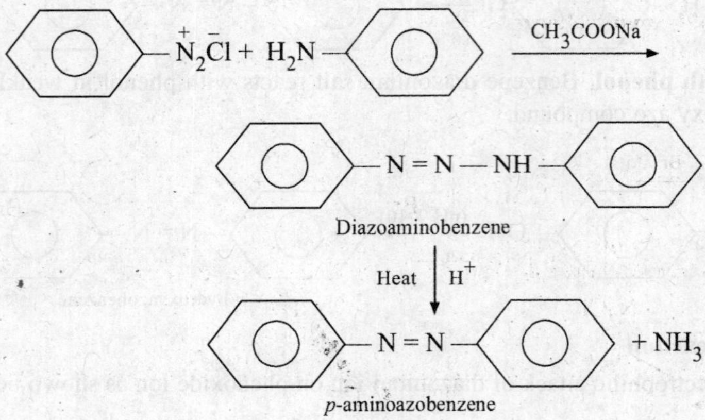

$$C_6H_5\overset{+}{N}_2\overset{-}{Cl} + H_2N-C_6H_5 \xrightarrow{CH_3COONa} C_6H_5-N=N-NH-C_6H_5 + HCl$$

Diazoaminobenzene

$$\xrightarrow[\text{Heat}]{H^+} C_6H_5-N=N-C_6H_4-NH_2 \; (p) + NH_3$$

p-aminoazobenzene

Mechanism of reaction

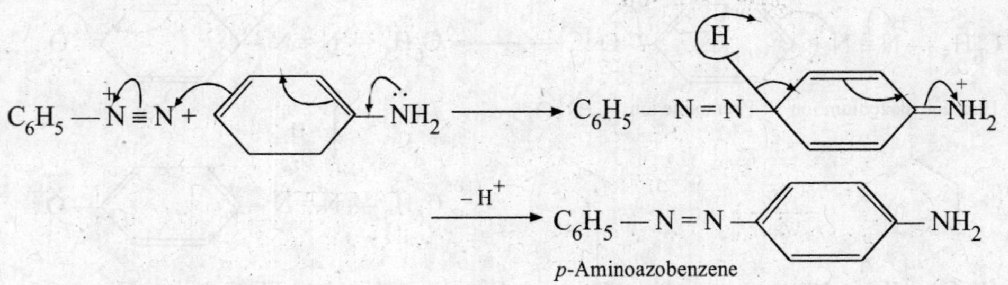

$$C_6H_5-\overset{+}{N}\equiv N + C_6H_5\ddot{N}H_2 \longrightarrow C_6H_5-N=N-C_6H_5(H)=\overset{+}{N}H_2 \xrightarrow{-H^+} C_6H_5-N=N-C_6H_4-NH_2$$

p-Aminoazobenzene

Example 1. How will you synthesise:

(*a*) *o*-Toluic acid from toluene

(*b*) *m*-bromophenol from nitrobenzene

(*c*) 1, 3, 5-Tribromobenzene from aniline

(*d*) *o*- and *p*-dinitrobenzene from aniline

Solution. (*a*) *o*-Toluic acid from toluene

Toluene $\xrightarrow[H_2SO_4]{HNO_3}$ *o*-Nitrotoluene (CH_3, NO_2) $\xrightarrow[SnCl_2/HCl]{[H]}$ *o*-Toluidine (CH_3, NH_2)

$\xrightarrow[HCl]{NaNO_2}$ *o*-Toluene diazonium chloride (CH_3, N_2Cl) $\xrightarrow{CuCN}$ (CH_3, CN) $\xrightarrow{H_2O}$ *o*-Toluic acid (CH_3, $COOH$)

(*b*) *m*-bromophenol from nitrobenzene

Nitrobenzene (NO_2) $\xrightarrow{Br_2/AlCl_3}$ *m*-Bromonitrobenzene (NO_2, Br) $\xrightarrow{[H]}$ *m*-Bromoaniline (NH_2, Br)

$\xrightarrow{NaNO_2/HCl}$ *m*-Bromo benzene diazonium chloride ($\overset{+}{N_2}\overset{-}{Cl}$, Br) $\xrightarrow[Boil]{H_2O}$ Phenol (OH, Br) $+ N_2 + HCl$

(*c*) Aniline (NH_2) $\xrightarrow[Water]{Br_2}$ 2, 4, 6-Tribromoaniline (NO_2, Br, Br, Br) $\xrightarrow[HCl]{NaNO_2}$ 2, 4, 6-Tribromobenzene diazonium chloride ($\overset{+}{N_2}\overset{-}{Cl}$, Br, Br, Br)

$\xrightarrow{H_3PO_2}$ 1, 3, 5-Tribromobenzene (Br, Br, Br)

(*d*) *o*- and *p*-dinitrobenzene from aniline

NH_2 Aniline $\xrightarrow{(CH_3CO)_2O}$ $NHCOCH_3$ Acetanilide $\xrightarrow[H_2SO_4]{HNO_3}$ *o*-Nitroacetanilide + *p*-Nitroacetanilide

$\xrightarrow{H_2O/H^+}$ *o*-Nitroaniline + *p*-Nitroaniline

$\downarrow$ $NaNO_2/HCl$

$\overset{+}{N_2}\overset{-}{Cl}$ (o-NO_2) + $\overset{+}{N_2}\overset{-}{Cl}$ (p-NO_2)

$\downarrow$ HBF_4 | $NaNO_2/Cu$

o-Dinitrobenzene + *p*-Dinitrobenzene

Example 2. How will you bring about the following conversions?

(*a*) *p*-nitroaniline into 3, 4, 5-triiodo nitrobenzene.

(*b*) *p*-toluidine into *m*-bromotoluene.

(*c*) Nitrobenzene into *p*-bromobenzonitrile.

(*d*) *m*-dinitrobenzene into *m*-nitrochlorobenzene.

Solution. (*a*) *p*-nitroaniline into 3, 4, 5-triiodo nitrobenzene:

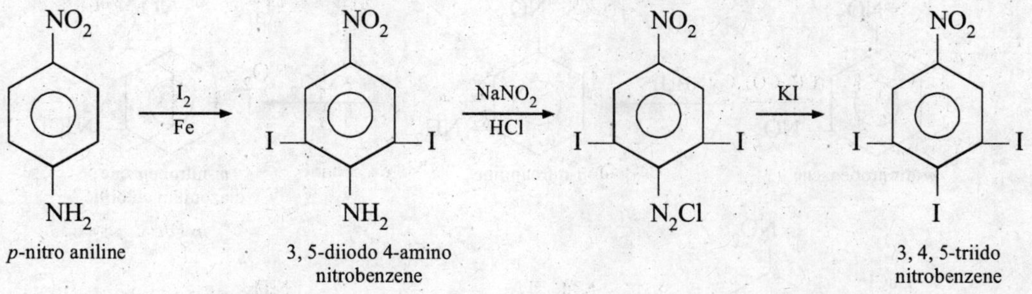

NO_2, NH_2 (*p*-nitro aniline) $\xrightarrow[Fe]{I_2}$ NO_2, I, I, NH_2 (3, 5-diiodo 4-amino nitrobenzene) $\xrightarrow[HCl]{NaNO_2}$ NO_2, I, I, N_2Cl $\xrightarrow{KI}$ NO_2, I, I, I (3, 4, 5-triido nitrobenzene)

(*b*) *p*-toluidine into *m*-bromotoluene

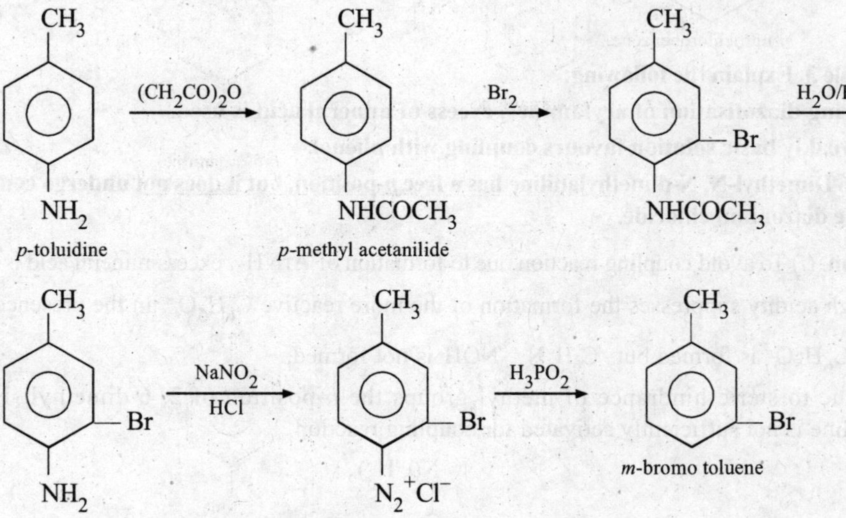

CH_3, NH_2 (*p*-toluidine) $\xrightarrow{(CH_2CO)_2O}$ CH_3, $NHCOCH_3$ (*p*-methyl acetanilide) $\xrightarrow{Br_2}$ CH_3, Br, $NHCOCH_3$ $\xrightarrow{H_2O/H^+}$

CH_3, Br, NH_2 $\xrightarrow[HCl]{NaNO_2}$ CH_3, Br, $N_2^+Cl^-$ $\xrightarrow{H_3PO_2}$ CH_3, Br (*m*-bromo toluene)

(*c*) Nitrobenzene into *p*-bromobenzonitrile

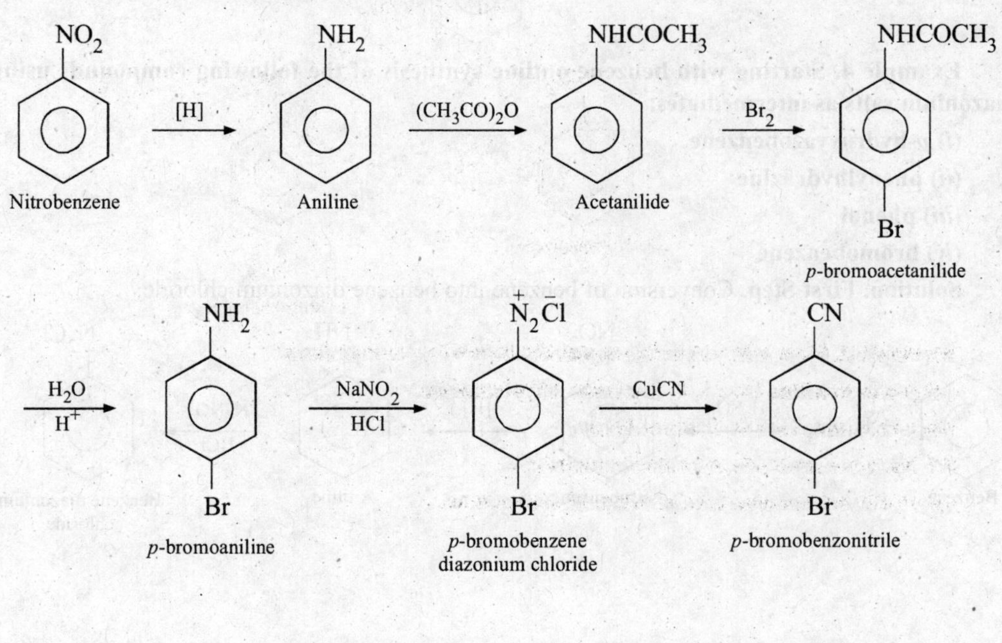

NO_2 (Nitrobenzene) $\xrightarrow{[H]}$ NH_2 (Aniline) $\xrightarrow{(CH_3CO)_2O}$ $NHCOCH_3$ (Acetanilide) $\xrightarrow{Br_2}$ $NHCOCH_3$, Br (*p*-bromoacetanilide)

$\xrightarrow[H^+]{H_2O}$ NH_2, Br (*p*-bromoaniline) $\xrightarrow[HCl]{NaNO_2}$ $\overset{+}{N_2}\overset{-}{Cl}$, Br (*p*-bromobenzene diazonium chloride) $\xrightarrow{CuCN}$ CN, Br (*p*-bromobenzonitrile)

(*d*) *m*-dinitrobenzene into *m*-nitrochlorobenzene

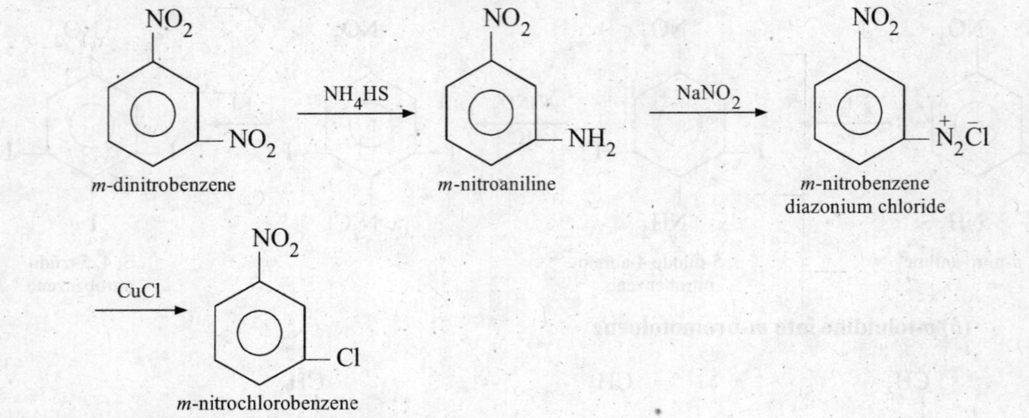

NO_2, NO_2 (m-dinitrobenzene) $\xrightarrow{NH_4HS}$ NO_2, NH_2 (m-nitroaniline) $\xrightarrow{NaNO_2}$ NO_2, $\overset{+}{N}_2\overset{-}{Cl}$ (m-nitrobenzene diazonium chloride)

m-dinitrobenzene

m-nitroaniline

m-nitrobenzene diazonium chloride

$\xrightarrow{CuCl}$ NO_2, Cl

m-nitrochlorobenzene

Example 3. Explain the following:

(*i*) During diazotisation of arylamines, excess of mineral acid is used.

(*ii*) A weakly basic solution favours coupling with phenol.

(*iii*) 2, 6 Dimethyl-N, N-dimethylaniline has a free *p*-position, but it does not undergo coupling with benzene diazonium chloride.

Solution. (*i*) To avoid coupling reaction due to formation of $Ar\overset{+}{N}H_3$, excess mineral acid is used.

(*ii*) High acidity suppresses the formation of the more reactive $C_6H_5O^-$. In the presence of a weak base, $C_6H_5O^-$ is formed but $C_6H_5N = \overset{+}{N}OH$ is not formed.

(*iii*) Due to steric hindrance of methyl groups the *p*-position of 2, 6 dimethyl -N, N-dimethylaniline is not sufficiently activated for coupling reaction.

$N(CH_3)_2$, H_3C, CH_3

Example 4. Starting with benzene outline synthesis of the following compounds using diazonium salts as intermediates:

(*i*) *p*-hydroxyazobenzene

(*ii*) phenylhydrazine

(*iii*) phenol

(*iv*) bromobenzene

Solution. First Step. Conversion of benzene into benzene diazonium chloride.

Benzene $+ HNO_3 \xrightarrow{H_2SO_4}$ Nitrobenzene (NO_2) $\xrightarrow{[H]}$ Aniline (NH_2) $\xrightarrow[HCl]{NaNO_2}$ Benzene diazonium chloride (N_2Cl)

Benzene

Nitrobenzene

Aniline

Benzene diazonium chloride

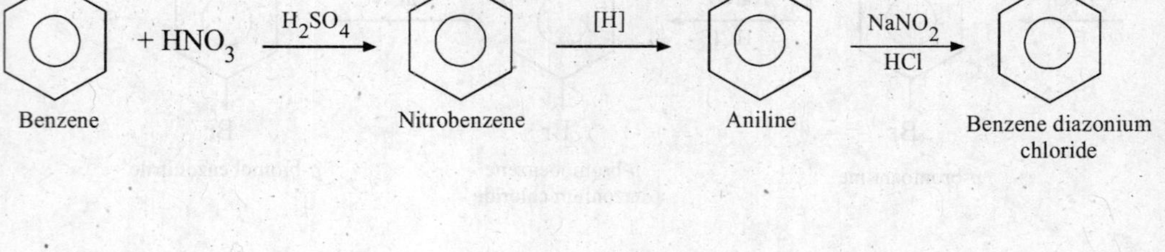

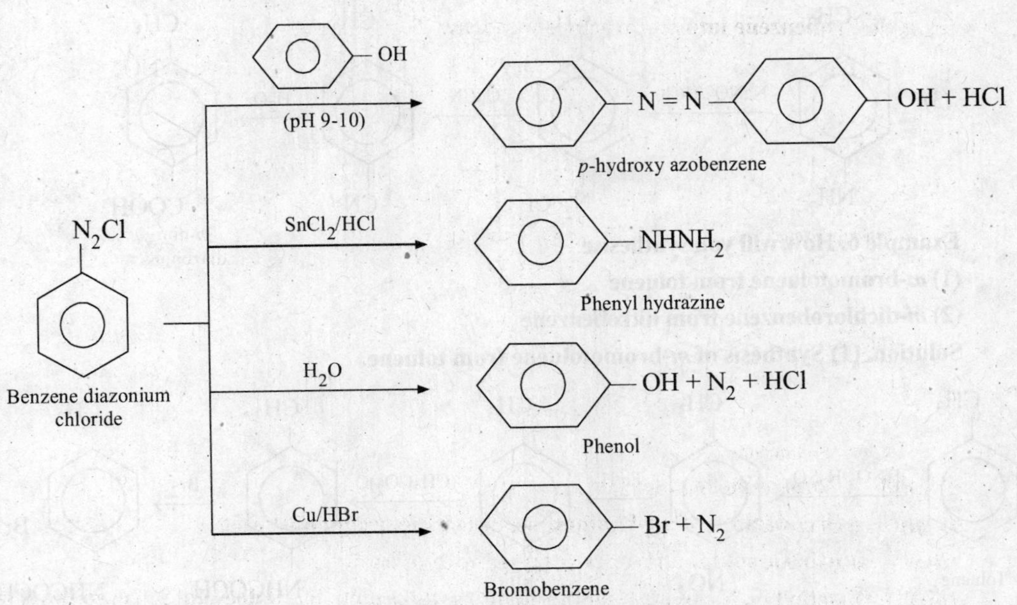

Benzene diazonium chloride ($C_6H_5N_2Cl$)

- $\xrightarrow[\text{(pH 9-10)}]{C_6H_5OH}$ $C_6H_5-N=N-C_6H_4-OH + HCl$ (*p*-hydroxy azobenzene)
- $\xrightarrow{SnCl_2/HCl}$ $C_6H_5-NHNH_2$ (Phenyl hydrazine)
- $\xrightarrow{H_2O}$ $C_6H_5-OH + N_2 + HCl$ (Phenol)
- $\xrightarrow{Cu/HBr}$ $C_6H_5-Br + N_2$ (Bromobenzene)

Example 5. How will you bring about the following conversions:

(1) Nitrobenzene into *m*-nitrophenol

(2) *m*-Nitro aniline into *m*-chloroiodobenzene

(3) *p*-Toluidine into *p*-toluic acid.

Solution.

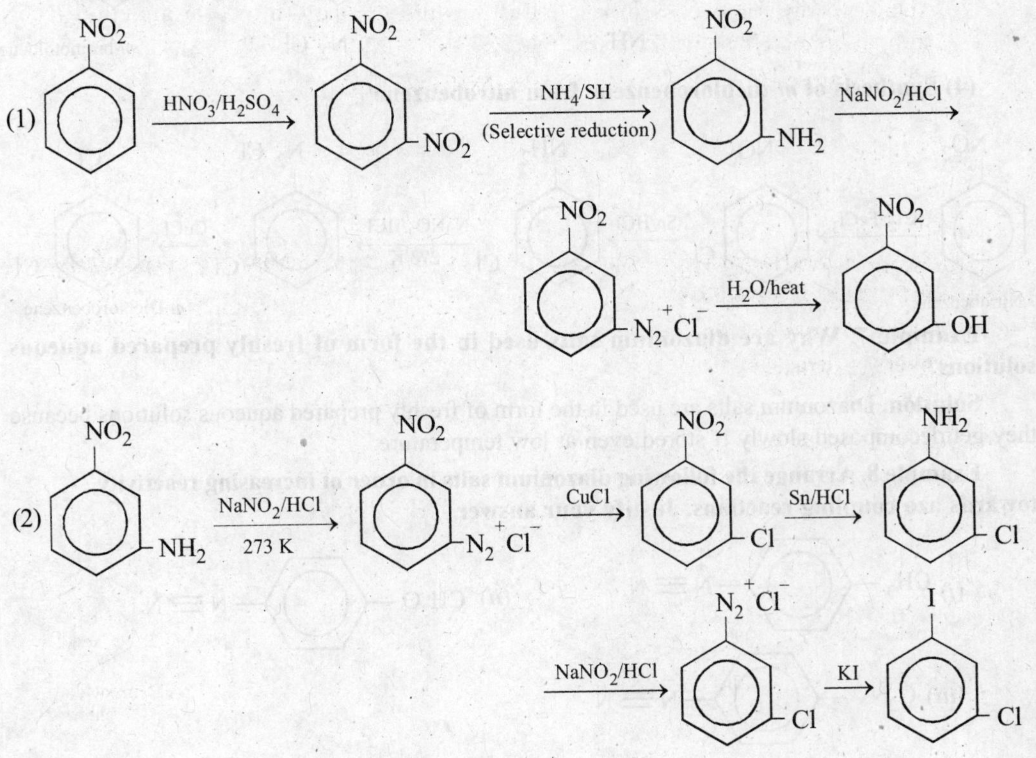

(1) $C_6H_5NO_2 \xrightarrow{HNO_3/H_2SO_4}$ m-$C_6H_4(NO_2)_2 \xrightarrow[\text{(Selective reduction)}]{NH_4/SH}$ m-$NO_2C_6H_4NH_2 \xrightarrow{NaNO_2/HCl}$ m-$NO_2C_6H_4N_2^+Cl^- \xrightarrow{H_2O/heat}$ m-$NO_2C_6H_4OH$

(2) m-$NO_2C_6H_4NH_2 \xrightarrow[273\text{ K}]{NaNO_2/HCl}$ m-$NO_2C_6H_4N_2^+Cl^- \xrightarrow{CuCl}$ m-$NO_2C_6H_4Cl \xrightarrow{Sn/HCl}$ m-$NH_2C_6H_4Cl \xrightarrow{NaNO_2/HCl}$ m-$ClC_6H_4N_2^+Cl^- \xrightarrow{KI}$ m-ClC_6H_4I

(3) p-Toluidine (CH_3, NH_2) $\xrightarrow{NaNO_2/HCl}$ $CH_3C_6H_4\overset{+}{N_2}\overset{-}{Cl}$ $\xrightarrow{CuCN}$ $CH_3C_6H_4CN$ $\xrightarrow{H_2O}$ $CH_3C_6H_4COOH$

Example 6. How will you synthesise

(1) *m*-bromotoluene from toluene

(2) *m*-dichlorobenzene from nitrobenzene

Solution. (1) Synthesis of *m*-bromotoluene from toluene

Toluene $\xrightarrow{HNO_3/H_2SO_4}$ p-nitrotoluene (CH_3, NO_2) $\xrightarrow{Fe/H^+}$ p-toluidine (CH_3, NH_2) $\xrightarrow{(CH_3CO_2)O}$ (CH_3, $NHCOCH_3$) $\xrightarrow{Br_2}$ (CH_3, Br, $NHCOCH_3$)

$\xrightarrow{H_2O/H}$ (CH_3, Br, NH_2) $\xrightarrow{NaNO_2/HCl}$ (CH_3, Br, $\overset{+}{N_2}\overset{-}{Cl}$) $\xrightarrow{H_3PO_2}$ (CH_3, Br) *m*-bromotoluene

(4) Synthesis of *m*-dichlorobenzene from nitrobenzene

Nitrobenzene (NO_2) $\xrightarrow{Cl_2,\ FeCl_3}$ (NO_2, Cl) $\xrightarrow{Sn/HCl}$ (NH_2, Cl) $\xrightarrow[273\ K]{NaNO_2/HCl}$ ($\overset{+}{N_2}\overset{-}{Cl}$, Cl) $\xrightarrow{CuCl}$ (Cl, Cl) *m*-Dichlorobenzene

Example 7. Why are diazonium salts used in the form of freshly prepared aqueous solutions?

Solution. Diazonium salts are used in the form of freshly prepared aqueous solutions because they get decomposed slowly if stored even at low temperature.

Example 8. Arrange the following diazonium salts in order of increasing reactivity towards azo coupling reactions. Justify your answer.

(*i*) $CH_3-C_6H_4-\overset{+}{N}\equiv N$ (*ii*) $CH_3O-C_6H_4-\overset{+}{N}\equiv N$

(*iii*) $O_2N-C_6H_4-\overset{+}{N}\equiv N$

Solution. Presence of electron withdrawing groups in the benzene ring increases the reactivity of diazonium salts towards azo coupling reaction while presence of electron releasing groups decreases the reactivity. Therefore the increasing order of reactivity is:

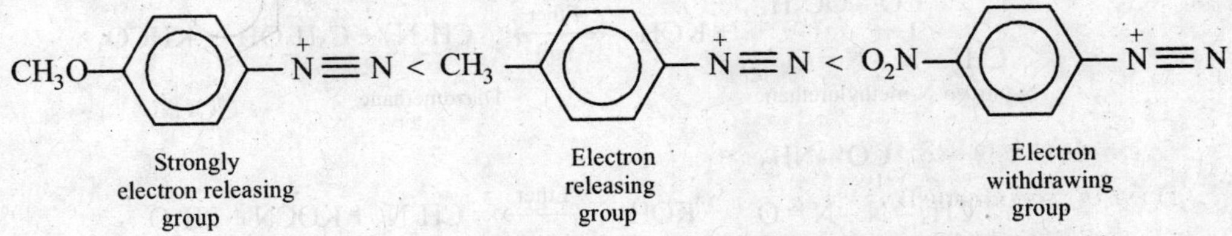

$$CH_3O-C_6H_4-\overset{+}{N}\equiv N < CH_3-C_6H_4-\overset{+}{N}\equiv N < O_2N-C_6H_4-\overset{+}{N}\equiv N$$

Strongly electron releasing group; Electron releasing group; Electron withdrawing group

Example 9. How will you convert aniline into 1, 2, 3, 5-tetrabromobenzene?

Solution.

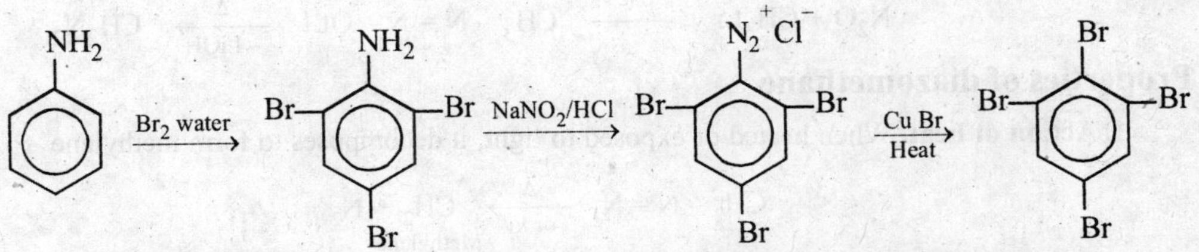

$$C_6H_5NH_2 \xrightarrow{Br_2\ water} \text{2,4,6-tribromoaniline} \xrightarrow{NaNO_2/HCl} \text{2,4,6-tribromobenzenediazonium chloride } (\overset{+}{N_2}\overset{-}{Cl}) \xrightarrow[Heat]{Cu\ Br_2} \text{1,2,3,5-tetrabromobenzene}$$

Example 10. Why do we use excess of mineral acid during preparation of diazonium salts?

Solution. During preparation of diazonium salts by diazotisation of arylamines, two moles of mineral acid are required per mole of arylamine. One mole of mineral acid is used in dissolving the amine while one mole is used in liberating HNO_2 from $NaNO_2$. In actual practice we use three moles of mineral acid. The excess acid is necessary to maintain proper pH of the reaction mixture to prevent the coupling reaction of diazonium salts with the undiazotised amine to give an azo dye.

Example 11. Aromatic diazonium salts are more stable than aliphatic diazonium salts. Explain.

Solution. Aromatic diazonium salts are more stable than aliphatic diazonium salts due to the dispersal of positive charge on nitrogen atom due to resonance interaction with the benzene ring as shown below:

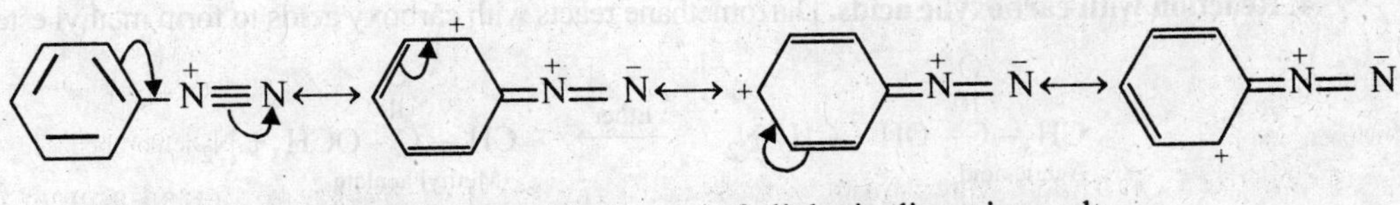

$$C_6H_5-\overset{+}{N}\equiv N \longleftrightarrow \overset{+}{C_6H_5}=\overset{+}{N}=\overset{-}{N} \longleftrightarrow \overset{+}{C_6H_5}=\overset{+}{N}=\overset{-}{N} \longleftrightarrow \overset{+}{C_6H_5}=\overset{+}{N}=\overset{-}{N}$$

There is no such resonance possible in case of aliphatic diazonium salts.

8.32. DIAZOMETHANE

Structure

It exists as a resonance hybrid of two structures

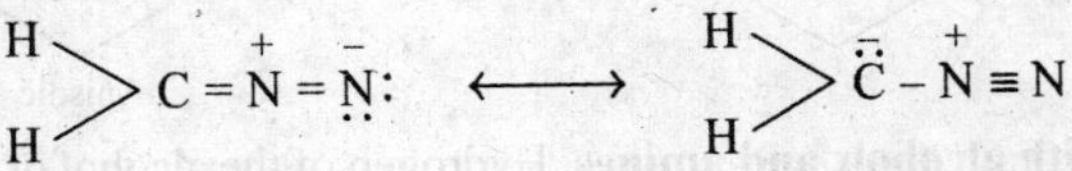

$$H_2C=\overset{+}{N}=\overset{-}{\ddot{N}}: \longleftrightarrow H_2\overset{-}{\ddot{C}}-\overset{+}{N}\equiv N$$

Methods of Preparation

1. It can be obtained from N-nitroso-N-methyl compounds as shown below:

$$\underset{\text{N-nitroso-N-methylurethane}}{CH_3-\overset{\overset{\displaystyle CO-OC_2H_5}{|}}{N}-N=O} + KOH \xrightarrow[\text{boil}]{\text{Ether}} \underset{\text{Diazomethane}}{CH_2N_2} + C_2H_5OH + KHCO_3$$

$$\underset{\text{N-nitroso-N-methyl urea}}{CH_3-\overset{\overset{\displaystyle CO-NH_2}{|}}{N}-N=O} + KOH \xrightarrow{\text{Ether}} \underset{\text{Diazomethane}}{CH_2N_2} + KOCN + 2H_2O$$

2. By passing nitrous oxide (N_2O) through an ethereal solution of methylithium (CH_3Li)

$$N_2O + CH_3Li \longrightarrow CH_3-N=N-OLi \xrightarrow[-LiOH]{\Delta} CH_2N_2$$

Properties of diazomethane

1. Action of heat. When heated or exposed to light, it decomposes to form methylene

$$\bar{C}H_2-\overset{+}{N}\equiv N \xrightarrow[\text{or } h\nu]{\Delta} \underset{\text{Methylene}}{CH_2:} + N_2$$

Methylene is very reactive and adds to alkanes to form higher homologues

$$\underset{\text{Ethane}}{CH_3\,CH_3} + CH_2: \longrightarrow \underset{\text{Propane}}{CH_3CH_2CH_3}$$

2. Reduction. On treatment with Na / Hg, it gives methyl hydrazine

$$CH_2N_2 + 4[H] \xrightarrow{\text{Na / Hg}} \underset{\text{Methyl hydrazine}}{CH_3NHNH_2}$$

3. Reaction with mineral acids.

$$CH_2N_2 + HCl \longrightarrow \underset{\text{Intermediate compound}}{[CH_3N_2Cl]} \longrightarrow \underset{\text{Methyl chloride}}{CH_3Cl} + N_2\uparrow$$

4. Reaction with carboxylic acids. Diazomethane reacts with carboxy acids to form methyl esters

$$\underset{\text{Acetic acid}}{CH_3-\overset{\overset{\displaystyle O}{||}}{C}-OH} + CH_2N_2 \xrightarrow{\text{Ether}} \underset{\text{Methyl acetate}}{CH_3-\overset{\overset{\displaystyle O}{||}}{C}-OCH_3} + N_2$$

5. Reaction with phenols. On treatment with carboxy acids, phenols get methylated

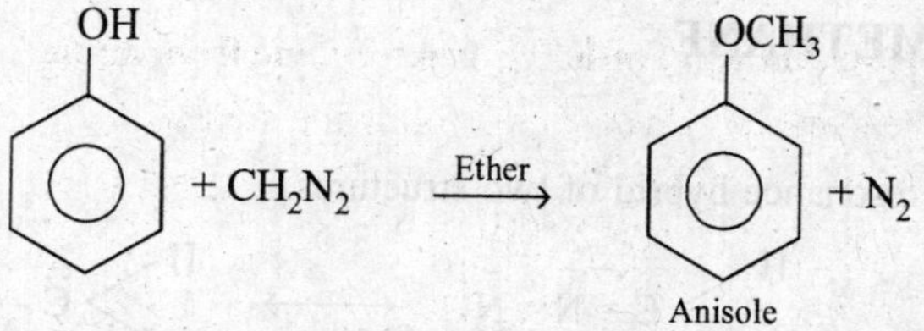

$$C_6H_5OH + CH_2N_2 \xrightarrow{\text{Ether}} \underset{\text{Anisole}}{C_6H_5OCH_3} + N_2$$

6. Reaction with alcohols and amines. Hydrogen of the alcohol or amine is replaced by methyl group in the presence of BF_3.

$$C_2H_5OH + CH_2N_2 \xrightarrow{BF_3} \underset{\text{Ethyl methyl ether}}{C_2H_5OCH_3} + N_2$$

$$C_2H_5NH_2 + CH_2N_2 \xrightarrow{BF_3} \underset{\text{N-methyl ethan-amine}}{C_2H_5NHCH_3} + N_2$$

7. Reaction with carbonyl compounds. Diazomethane converts aldehydes into ketones, while ketones are converted into higher homologues

$$\underset{\text{Acetaldehyde}}{CH_3-\overset{O}{\overset{||}{C}}-H} + CH_2N_2 \longrightarrow \underset{\text{Acetone}}{CH_3-\overset{O}{\overset{||}{C}}-CH_3} + N_2$$

$$-\overset{O}{\overset{||}{C}}-CH_3 + CH_2N_2 \longrightarrow \underset{\text{Methyl ethyl ketone}}{CH_3-\overset{O}{\overset{||}{C}}-CH_2CH_3} + N_2$$

8. Ring expansion. Cyclohexanone reacts with diazomethane to produce cycloheptanone.

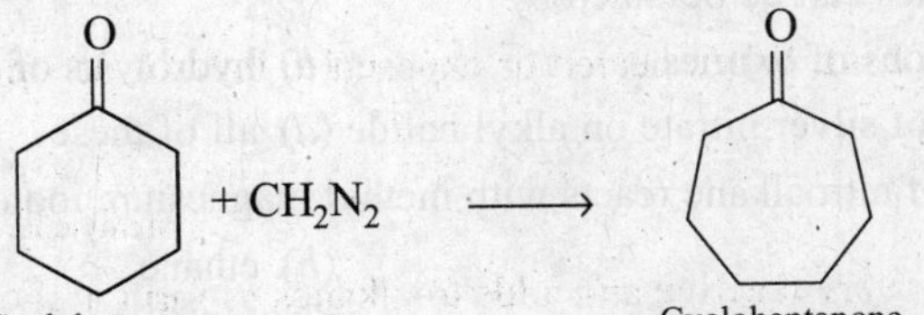

$$\underset{\text{Cyclohexanone}}{C_6H_{10}O} + CH_2N_2 \longrightarrow \underset{\text{Cycloheptanone}}{C_7H_{12}O}$$

9. Arnst-Eistert Synthesis. This method is used to produce the higher homologue of a carboxylic acid by the following sequence of reactions:

$$R-COOH \xrightarrow{SO_2Cl_2} RCOCl \xrightarrow{CH_2N_2} RCOCHNH_2 \xrightarrow{Ag_2O} \underset{\text{Higher acid}}{RCH_2COOH}$$

Example 1. Explain why the following reaction is carried out in weakly alkaline medium

$$C_6H_5N_2Cl + C_6H_5OH \xrightarrow[\text{medium}]{\text{weakly alkaline}}$$

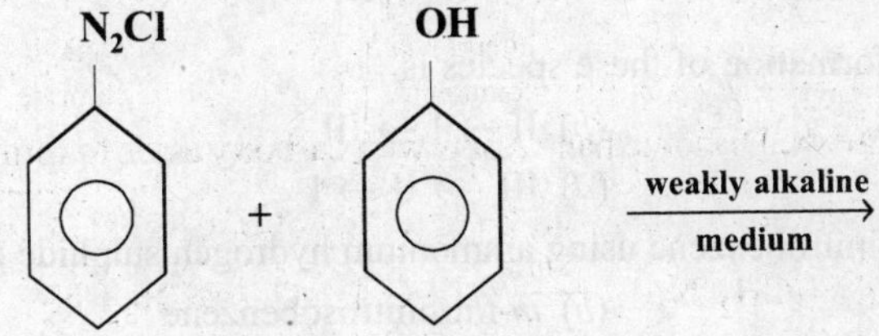

Solution. Weakly alkaline medium helps in the removal of the proton in the intermediate step of the reaction.

Example 2. How will you synthesise benzylamine from aniline.

Solution. The synthesis involves the following steps:

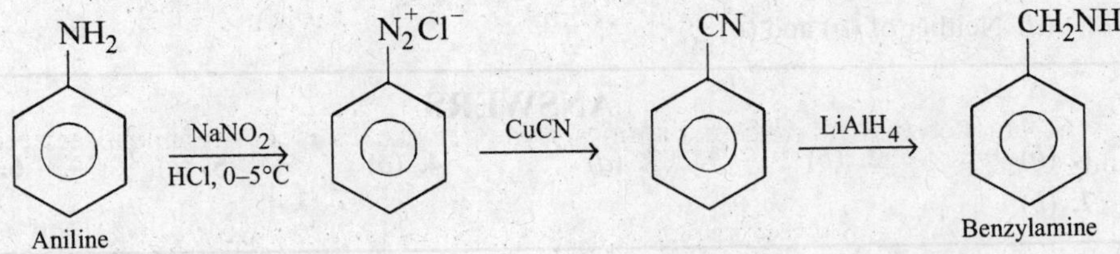

$$\underset{\text{Aniline}}{C_6H_5NH_2} \xrightarrow[\text{HCl, 0–5°C}]{NaNO_2} C_6H_5N_2^+Cl^- \xrightarrow{CuCN} C_6H_5CN \xrightarrow{LiAlH_4} \underset{\text{Benzylamine}}{C_6H_5CH_2NH_2}$$

EXERCISE

Based On Questions Paper of Different Universities

NITRO COMPOUNDS

Multiple Choice Questions (Choice the Correct Option)

1. Nitroalkanes on reaction with Sn/HCl give

(*a*) alcohol (*b*) primary amine

(*c*) secondary amine (*d*) none of these

2. Nitro compounds show ——— boiling point than hydrocarbons of comparable molecular masses.

(*a*) lower (*b*) higher

(*c*) equal (*d*) none of these

3. Nitro alkanes can be obtained by

(*a*) oxidations of oximes (*b*) hydrolysis of α-nitroalkenes

(*c*) action of silver nitrate on alkyl halide (*d*) all of these

4. Aci-form of nitroalkane reacts with methyl magnesium iodide to form

(*a*) methane (*b*) ethane

(*c*) nitric acid (*d*) none of above

5. The reduction of nitrobenzene involves the following species

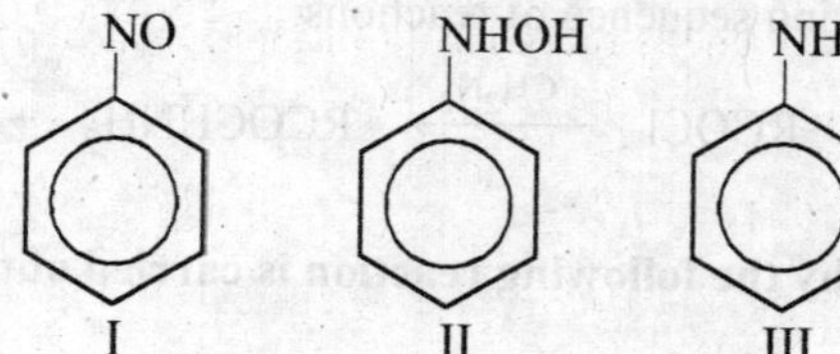

The correct sequences of formation of these species is

(*a*) I → II → II (*b*) II → I → III

(*c*) I → II → III (*d*) III → II → I

6. Selective reduction of m-dinitrobenzene using ammonium hydrogen sulphide gives

(*a*) *m*-nitroaniline (*b*) *m*-nitronitrosobenzene

(*c*) 1,3-Benzenediamine (*d*) *m*-nitrohydroxylamine

7. *o*-Nitrophenol involves

(*a*) Intermolecular hydrogen bonding

(*b*) Intramolecular hydrogen bonding

(*c*) Both the above

(*d*) Neither of (*a*) and (*b*)

ANSWERS

1. (*b*)	**2.** (*b*)	**3.** (*d*)	**4.** (*a*)	**5.** (*c*)	**6.** (*a*)
7. (*b*)					

Short Answer Questions

1. Complete the following reactions:

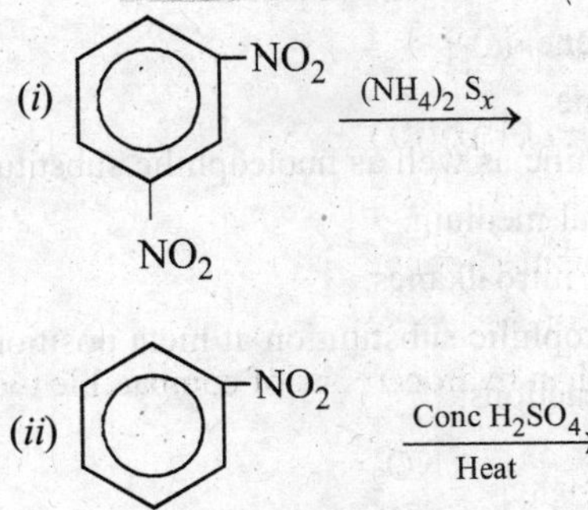

2. What do you understand by vapour phase nitration of alkanes? What are its limitations?
3. How can ortho-nitrotoluene be prepared from toluene?
4. Give a suitable mechanism for the following transformation:

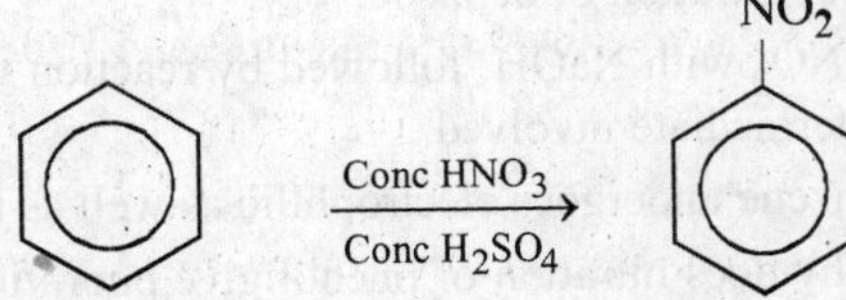

5. How will you explain toluene is more easily nitrated than benzene?
6. Complete the following:

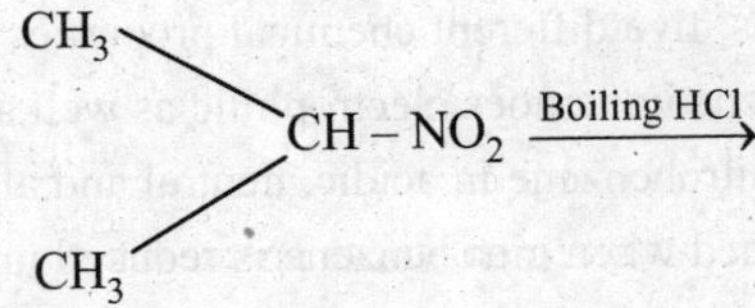

7. Outline the reduction of nitrobenzene in acidic, neutral and alkaline medium.
8. Reduction of nitrobenzene under different conditions yields different products. Give three examples.
9. Nature of reduction product of nitrobenzene depends upon the pH value of the medium of reduction and upon the nature of the reducing agent. Elucidate.
10. What is meant by nitration of aromatic compounds? State the mechanism of the reaction.
11. Write a short note on the tautomerism shown by nitroalkanes.
12. Discuss the reduction of nitrobenzene in acidic, neutral and alkaline medium. (No mechanism).
13. Between chlorobenzene and *o*-nitrochlorobenzene, which one is more reactive towards hydroxide ion and why?
14. Explain the reduction of nitroalkanes and nitroarenes in acidic and neutral media separately.
15. What is meant by a nitrating agent? List at least four such agents and discuss the citeria for the choice of a nitrating agent to be used in a particular case.
16. Give two methods of preparation of picric acid. Why does it act as a strong acid?

17. Give mechanism of nitration of benzene. Also name the electrophile involved in the reaction.
18. (a) Give mechanism of nitration of benzene.
    (b) Convert nitrophenol into picric acid.
19. Give two methods of preparation of nitrobenzene.
20. Discuss selective reduction of m-dinitrobenzene.
21. Explain why nitrobenzene undergoes electrophilic as well as nucleophilic substitution.
22. Discuss the reduction of nitrobenzene in neutral medium.
23. Give one test to distinguish between 1°, 2°, 3° nitroalkanes.
24. Explain why does nitrobenzene undergo electrophilic substitution at meta position.
25. Give a suitable mechanism for the following reactions:

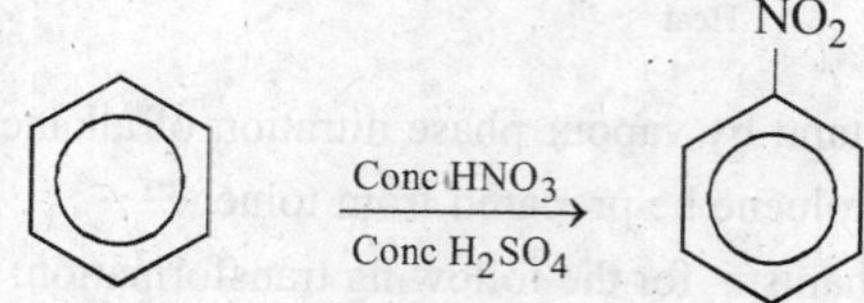

26. Give the mechanism of nitration of benzene.
27. Treatment of $CH_3 - NO_2$ with NaOH, followed by reaction with propanal leads to what? Name the reactive intermediate involved.
28. Explain why nitrobenzene undergoes electrophilic as well as nucleophilic substitution.
29. What is nitration? Why does nitration of phenol give poor yield of picric acid?
30. Outline the reduction of nitrobenzene in acidic and neutral medium.

General Questions

1. Describe with equations, five different chemical properties of aromatic nitro compounds.
2. Explain why nitrobenzene undergoes electrophilic as well as nucleophilic substitution?
3. Give the reduction of nitrobenzene in acidic, neutral and alkaline media.
4. What products are formed when nitrobenzene is reduced under different conditions? Give necessary equations.
5. How is nitrobenzene prepared on a large scale? Give an account of the products obtained by its reduction, under different conditions.
6. What are nitroalkanes? How are they classified? Give their important methods of preparation.
7. State the justify the effect of nito group on the reactions of halogens in aromatic nitro compounds.
8. Disuss the reduction of nitrobenzene under the following conditions:
   (*i*) Acidic medium (*b*) Catalytic reduction
   (*iii*) Selective reduction.
9. (*a*) Describe the mechanism of nucleophilic substitution in nitroarenes.
   (*b*) Write on the reducing products of nitrobenzene obtained by using different reducing agents.
10. Discuss on the basis of mechanism the order of reactivity of the following compounds towards hydroxyl ions: Chlorobenzene; 2,4-Dinitrochlorobenzene.

11. Discuss the behaviour of aromatic nitro compounds on reduction by considering the case of nitrobenzene.

12. Give the main chemical reactions of nitroalkanes.

13. Presence of nitro group at ortho and para positions with respect to chlorine atom in chlorobenzene facilitates nucleophilic substitution; justify.

14. Outline the reduction of nitrobenzene in acidic, neutral and alkaline medium.

15. Describe the methods of preparation and chemical reactions of picric acid.

AMINO COMPOUNDS

Multiple Choice Questions (Choose the Correct Option)

1. Anisidine is the name for the compound

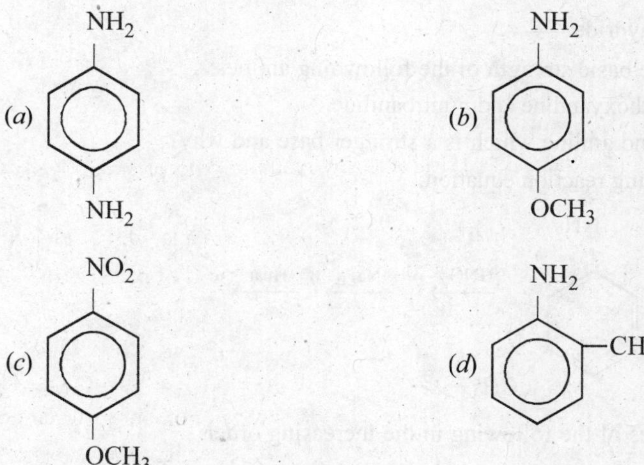

2. Amines have the

(*a*) pyramidal shape (*b*) square planar shape
(*c*) trigonal shape (*d*) octahedral shape

3. When the reduction is carried out using sodium and alcohol, it is called

(*a*) Clammensen reduction (*b*) Mendius reduction
(*c*) Stephen's reduction (*d*) None of the above

4. Gabriel synthesis is carried out for the preparation of

(*a*) aldehydes and ketones (*b*) alcohols
(*c*) primary amines (*d*) carboxylic acids

5. Secondary amines can be obtained by the reduction of

(*a*) isonitriles (*b*) N-alkyl amides
(*c*) either of the two (*d*) neither of the two

6. Hinsberg method is used for the separation of a mixture of

(*a*) primary, secondary and tertiary alcohols
(*b*) primary, secondary and tertiary amines
(*c*) *o*-, *m*- and *p*-nitrophenols
(*d*) All the above

7. Amines with three different alkyl groups exhibit

(*a*) optical isomerism (*b*) geometrical isomerism

(*c*) position isomerism (*d*) none

ANSWERS

1. (*b*)	**2.** (*a*)	**3.** (*b*)	**4.** (*c*)	**5.** (*c*)	**6.** (*b*)
7. (*d*)					

Short Answer Questions

1. Arrange the following in the decreasing order of their basicity:
Dimethylamine, aniline, ammonia, ethylamine, *p*-nitroanilne, *p*-anisidine, *p*-toluidine.

2. Write equations to show the following reactions.

Aniline + Acetic anhydride $\longrightarrow$

3. Compare the relative basic strength of the followoing amines:
m-nitroaniline, *p*-methoxyaniline and *p*-nitroaniline.

4. Out of ethylamine and aniline which is a stronger base and why?

5. Complete the following reaction equation:

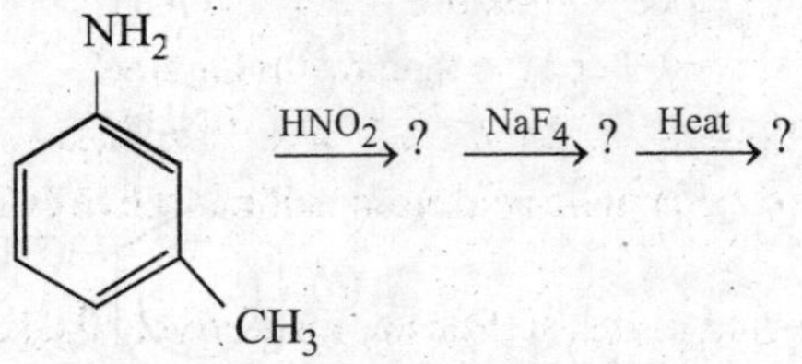

$$\text{(benzene ring with } NH_2 \text{ and } CH_3 \text{ meta)} \xrightarrow{HNO_2} ? \xrightarrow{NaF_4} ? \xrightarrow{Heat} ?$$

6. Arrange the basicities of the following in the increasing order:

(*i*) $C_6H_5NH_2$ (*ii*) *m*-$NO_2—C_6H_4—NH_2$

(*iii*) *p*-$NO_2 — C_6H_4\ NH_2$ (*iv*) *o*-$NO_2 — C_6H_4 — NH_2$

7. Which is more basic in the following pair: Aniline and p-nitroaniline

8. Starting from aniline how will you obtain:

(*i*) Sulphanilic acid (*ii*) p-Nitroaniline

9. Arrange the following in increasing order of basicity. Give reasons for your answer.
Aniline, ammonia, dimethylamine, *p*-nitroaniline, *p*-anisidine, *p*-toluidine

10. Why ammonia is a stonger base than aniline?

11. How will you distinguish between 1° and 2° amines? Give a chemical test.

12. How will you distinguish between ethylamine and aniline?

13. Explain why aniline is a weaker base than methylamine.

14. Why is amino group protected before carrying out the nitration of aniline?

15. Why ethylamine is a stronger base than aniline?

16. Arrange the following pairs of substances in order of expected base strengths:

(i) $(CH_3)_3$ N and $(F_3C)_3$ N (*ii*) Benzylamine and p-aminotoluene.

17. Out of aniline and hexylamine, which is less basic and why?

18. What do you understand by a phase transfer catalyst? Give one example.

19. Explain the following reactions:

(*i*) Gabriel phthalimide reaction (*ii*) Hofmann's Bromamide reaction.

20. Why does acylation of aniline tend to deactivate it towards electrophilic aromatic substitution?

21. Write short notes on

(*i*) Carbylamine reaction (*ii*) Hofmann Bromamide reaction

(*iii*) Gabriel phthalimide reaction.

22. Explain Hinsberg's method of separation of primary and secondary amines.

23. (*a*) Arrange the following in order of decreasing basicities giving reasons:

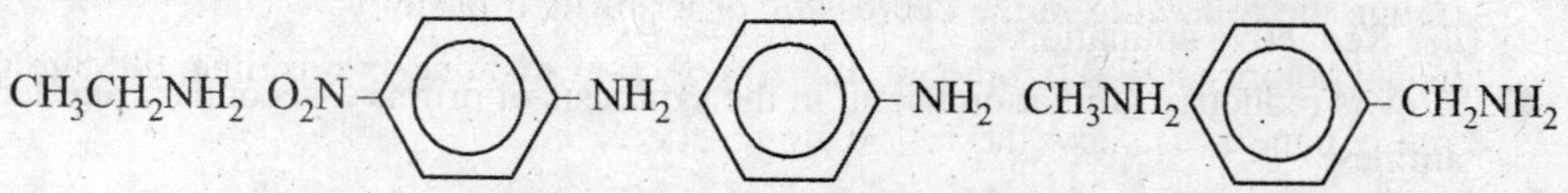

$CH_3CH_2NH_2$ $O_2N-C_6H_4-NH_2$ $C_6H_5-NH_2$ CH_3NH_2 $C_6H_5-CH_2NH_2$

(*b*) Distinguish between 1°, 2° and 3° amines using p-toluene sulphonyl chloride reagent.

24. Explain the carbylamine reaction with its mechanism.

25. Give two methods of preparation of N-methyl ethyl amine.

26. Write the mechanism of carbylamine reaction.

27. What is ortho effect? How does it influence the basicity of aromatic amines?

28. Give the mechanism of Cope elimination.

29. Explain why aniline is a weaker base than methyl amine?

30. Discuss the action of nitrous acid ion 1°, 2° and 3° amines.

31. How will you prepare sulphanilic acid from aniline ? Give evidence in favour of its zwitter ionic structure.

32. K_b for *m*-nitroaniline and *p*-nitroaniline are respectively 0.032×10^{-10} and 0.001×10^{-10}. Which is a weaker base and why?

33. N, N-Dimethylaniline is soluble in HCl whereas triphenylamine is insoluble even though both are tertiary amines. Explain.

34. Bromination of aniline gives mainly 2,4,6- tribromo aniline whereas nitration of aniline with mixed acids (nitric and sulphuric) gives mainly m-nitroaniline. Explain.

35. Name a reaction shown only by primary amines. Also give the mechanism of this reaction.

36. Give the mechanism of Hofmann's Bromamide reaction.

37. Complete the following:

(*i*) $C_6H_5NH_2 \xrightarrow{Aq\ Br_2}$

(*ii*) $CH_3\ NH_2 + Na\ NO_2 + 2\ HCl \xrightarrow{Cold}$

38. Give the mechanism of the following reaction:

$$C_6H_5CONH_2 + Br_2 + 4KOH \longrightarrow C_6H_5NH_2 + 2KBr + K_2CO_3 + H_2O$$

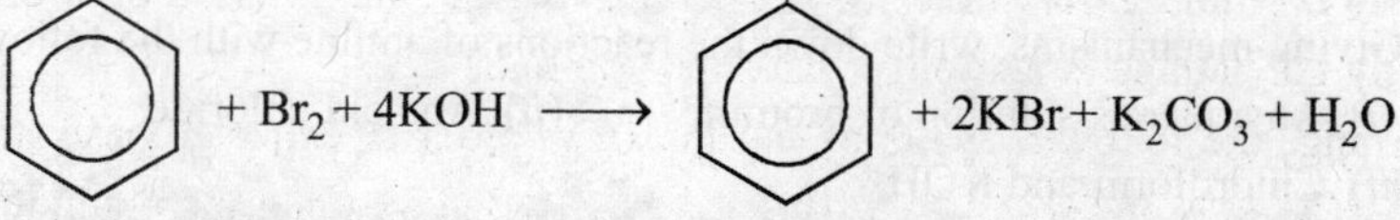

39. What is the ortho effect? How does it influence the basicity of aromatic amines?

40. What is the order of basicity of primary, secondary and tertiary alkyl amines? Justify your answer.

41. Discuss the mechanism of carbylamine reaction.

42. Briefly discuss Gabriels-phthalimide synthesis.

43. How the reductive amination is useful in the synthesis of primary, secondary and tertiary alkyl amines?

44. Starting from aniline, how will you obtain?

(*i*) Sulphanilic acid (*ii*) *p*-Nitroaniline

45. Discuss the mechanism of Hoffmann's Bromamide reaction.

46. How would you prepare ethylamine by:

(*i*) Gabriel phthalimide reaction

(*ii*) Hofmann degradation of amides

(*iii*) Reductive amination.

47. How is reductive amination useful in the synthesis of primary, secondary and tertiary alkyl amines?

48. How will you purify amines from non-basic impurities?

General Questions

1. Draw structures and give the names of various isomeric amines of the formula C_7H_9N which contain a benzene ring. How would you distinguish them from one another?

2. Write equations for each of the following reactions:

(*i*) Aniline + H_2SO_4 + Heat (*ii*) Acetanilide + Bromine

(*iii*) Dimethylamine + $NaNO_2$ + HCl (*iv*) Tetramethyl ammonium chloride + NaOH

(*v*) Benzamide, $LiAlH_4$, then hydrolysis (*vi*) Aniline + Bromine

(*vii*) p-Toluidine +$NaNO_2$+HCl at 278 K (*viii*) Cyclohexylamine + CH_3COCl.

3. What are primary, secondary and tertiary amines? How are these distinguish from each other?

4. (*a*) Write equations for five methods by which you can prepare amines.

(*b*) Which of these methods can be used for preparing aliphatic primary amines and which can be used for aromatic primary amines?

5. Write equations showing the reactions, if any, of each of the following with *n*-butylamine.

(*a*) HCl (*ii*) CH_3I,

(*iii*) CH_3COCl, (*iv*) $C_6H_5SO_2Cl$ + NaOH,

(*v*) $NaNO_2$+ HCl, (*vi*) Final product of (*ii*) + moist Ag_2O.

6. Compare the behaviour of aniline, N-methylaniline and N,N-dimethylaniline towards each of the following reagents.

(*i*) Dil. HCl (*ii*) $NaNO_2$ + HCl (aq)

(*iii*) Benzene sulphonyl chloride +NaOH (*iv*) Acetic anhydride

(*v*) Bromine water and (*vi*) CH_3I

7. What is meant by phase transfer catalyst? Explain the use of tetraalkyl ammonium salts as phase transfer catalysts.

8. Giving mechanisms, write down the reactions of aniline with the following:

(*i*) An aqueous solution of bromine (*ii*) An acid chloride

(*iii*) Chloroform and KOH.

9. Write short notes on:

(*i*) Carbylamine reaction

(*ii*) Hoffmann Bromamide reaction

(*iii*) Gabriel phthalimide reaction.

10. Starting from aniline how will you obtain

(*i*) Sulphanilic acid (*ii*) p-Nitroaniline

11. Arrange the following pairs of substances in order of expected base strengths. Show your reasoning.

(*i*) $(CH_3)_3N$ and $(F_3C)_3N$

(*ii*) $C_6H_5-CH_2NH_2$ and $CH_3-C_6H_4-NH_2$

(*iii*) $ClCH_2CH_2NH_2$ and $CH_3CH_2CH_2NH_2$

(*iv*) 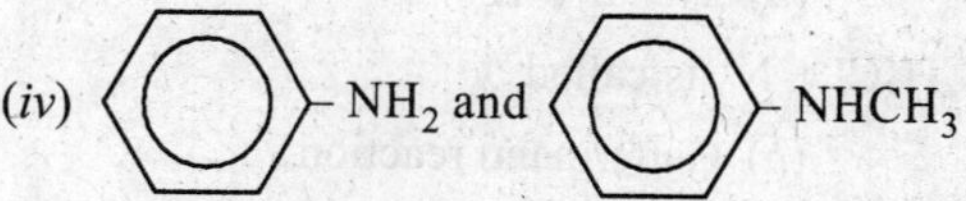

12. Write down the various possible isomers of compound having the molecular formula C_3H_9N and how will you separate these isomers from their mixture?

13. Discuss the orientation of electrophilic substitution of aniline in terms of relative stability of intermediates formed.

14. Write equations to show the reaction of aniline with the following reagents:

(*i*) Bromine water (*ii*) Acetic anhydride

(*iii*) Toluene sulphonyl chloride (*iv*) Conc. sulphuric acid (453 – 463 K)

(*v*) Chloroform and alkali.

15. Explain the following with mechanism:

(*i*) Gabriel phthalimide reaction

(*ii*) Hoffmann's Bromamide reactions

16. What happens when:

(*i*) Aniline is oxidised in the presence of chromic acid

(*ii*) Aniline is treated with Cl_2 water.

(*iii*) Aniline is treated with phenol at 523 K in presence of anhydrous $ZnCl_2$.

(*iv*) Aniline is treated with C_6H_5COCl and NaOH.

17. Write short notes on:

(*a*) Gabriel phthalimide reaction

(*b*) Amine salts as phase transfer catalysts.

18. How can you differentiate between primary, secondary and tertiary amines?

19. Using Hinsberg's reagent, how will you distinguish between aliphatic primary, secondary and tertiary amines? Give reactions.

20. Why are amines basic in character? Inductive effect is not the only criterion for the basic character of different amines. Explain.

21. Write what do you know about the following:

(*i*) Carbylamine reaction (*ii*) Sulphanilic acid

22. Explain Hinsberg's method of separation of 1°, 2° and 3° amines.

DIAZONIUM COMPOUNDS

Multiple Choice Questions (Choose the Correct Option)

1. The correct structure of benzene diazonium chloride which explain all the properties of the compound was given by
   (*a*) Greiss (*b*) Kekule
   (*c*) Bloomstrand (*d*) none
2. Diazotisation of aniline can be carried out at the temperature range
   (*a*) 73–78K (*b*) 173–178 K
   (*c*) 373–378 K (*d*) 273–278 K
3. The reaction $C_6H_5N_2Cl \xrightarrow{Cu/HCl} C_6H_5Cl + N_2$ is called
   (*a*) Sandmeyer's reaction (*b*) Gattermann reaction
   (*c*) Gomberg Bachmann reaction (*d*) Balz Schiemann reaction
4. $- N_2Cl$ group can be replaced by – H by means of
   (*a*) hypophosphorus acid (*b*) phosphoric acid
   (*c*) phosphorus acid (*d*) meta phosphoric acid
5. Coupling reaction involves the reaction between benzene diazonium chloride and
   (*a*) ether (*b*) carboxylic acid
   (*c*) carboxylic acid derivative (*d*) phenol
6. Benzene diazonium chloride can be converted into fluorobenzene using
   (*a*) CuF (*b*) Cu/HF
   (*c*) HBF_4 (*d*) F_2
7. The coupling reaction between benzene diazonium chloride and animine takes place in
   (*a*) strongly alkaline medium (*b*) weakly alkaline medium
   (*c*) strongly acidic medium (*d*) weakly acidic medium

ANSWERS

1. (*c*) 2. (*d*) 3. (*b*) 4. (*a*) 5. (*d*) 6. (*c*)
7. (*b*)

Short Answer Questions

1. Why are diazonium salts used in the form of freshly prepared aqueous solutions?
2. Complete the following equation:

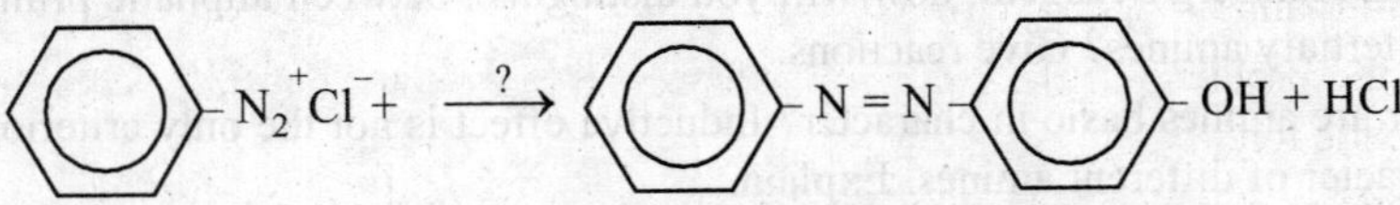

3. Starting from benzene, outline the synthesis of *m*-bromophenol.
4. Why is diazotisation always carried below 5°C?
5. Explain diazotisation. How is phenol obtained from aniline?
6. What is diazotisation? Why is excess mineral acid required for this process of diazotisation?
7. Convert nitrobenzene into *m*-bromophenol.
8. How will you prepare iodobenzene from benzene diazonium chloride?

9. What are diazonium salts? Why are they so named?
10. Give one example of Gomberg-Bachmann reaction.
11. How can you obtain the following from benzene diazonium chloride?
    (*i*) An azodye (*ii*) Phenol
12. Write the product of the following reactions:

$$m\ Br - C_6H_4 - NH_2 \xrightarrow[\text{(ii) } H_2O/\text{heat}]{\text{(i) } NaNO_2/HCl}$$

13. Complete the following reaction

$$C_6H_5N_2Cl \xrightarrow{SnCl_2/HCl}$$

14. Discuss the mechanism of diazotisation of amines.
15. Complete the following reaction:

$$C_6H_5N_2^+Cl^- + C_6H_5OH \xrightarrow[pH = 9-10]{\text{Ice cold}}$$

16. (*i*) How will you convert benzene into benzene diazonium chloride and *vice versa*?
    (*ii*) *p*-Hydroxyazobenzene.
17. How will you prepare benzene diazonium chloride starting from benzene?
18. Why do we use mineral acid in excess during the preparation of diazonium salts?
19. How does benzene diazonium chloride react with N, N-dimethylaniline?
20. How will you prepare 1, 3, 5- tribromobenzene from nitrobenzene?
21. Convert nitrobenzene to *m*-dichlorobenzene.
22. How will you prepare benzene diazonium chloride starting from benzene?
23. How is benzene diazonium chloride prepared? Give its mechanims.
24. Discuss the mechanism of coupling reaction.
25. Write a note on azo-coupling.
26. How will you convert:
    (*i*) p-Nitroaniline into 3, 4, 5-tri-iodonitrobenzene.
    (*ii*) Aniline into 1,3,5-tribromobenzene
27. How will you synthesise:
    (*i*) *m*-Bromophenol from benzene
    (*ii*) 1,3, 5- Tribromobenzene from benzene.
28. How would you prepare:
    (*i*) benzonitrile from aniline
    (*ii*) p-nitroaniline starting from aniline
29. Write the products in the reactions of benzene diazonium chloride with CuCl, Cul and phenoxide ion.
30. How will you obtain the following from benzene (attempt any two):
    (*i*) 1, 2, 3- Tribromobenzene
    (*ii*) *m*-Bromphenol
31. Discuss the role of pH in the coupling reactions of diazonium salts.

General Questions

1. Starting with benzene or toluene, how would you prepare the following compounds by employing the intermediate formation of diazonium salts as one of the steps:
   (*i*) 1, 3, 5, Tribromobenzene (*ii*) *m*-Bromotoluene
   (*iii*) *p*-Hydroxyazobenzene (*iv*) Phenylhydrazine
   (*v*) Diazoaminobenzene (*vi*) Phenol
2. Give the reagent and any special conditions necessary to convert o-toluenediazonium chloride into :
   (*i*) Toluene (*ii*) o-Cresol
   (*iii*) o-Chlorotoluene (*iv*) o-Fluorotoluene
   (*v*) o-Tolunitrile (*vi*) 2-Methyl-4 (N,N-dimethylamino)-azobenzene
3. Write equations for the reactions of *p*-nitrobenzenediazonium hydrogen sulphate with:
   (*i*) HBr + Cu, (*ii*) Hot dil. H_2SO_4
   (*iii*) *p*-Cresol, (*iv*) CuCN,
   (*v*) KI, (*vi*) HBF_4, then heat
   (*vii*) H_3PO_2.
4. Describe five reactions of diazonium salts in which gaseous nitrogen is evolved and two reactions in which the two nitrogen atoms of the diazonium salts are retained in the molecule of the product.
5. Write with equations six important synthetic applications of benzene diazonium chloride and name the products.
6. Discuss the process of diazotisation and the molecular structure of the diazonium salts.
7. How will you convert aniline to
   (*i*) *Sym*-tribromobenzene (*ii*) *m*-bromophenol
   (*iii*) *p*-hydroxyazobenzene (*iv*) *p*-iodobenzoic acid
   (*v*) *p*-nitrofluorobenzene?
8. Discuss the structure of benzene diazonium chloride.
9. Explain the mechanism of diazotisation. Which of the following gives more stable diazo compound on treatment with HNO_2 and why? Aniline, o-Nitroaniline.
10. How will you convert diazonium salts into:
    (*i*) Diphenyl (*ii*) Phenol
    (*iii*) Benzene and (*iv*) Phenyl hydrazine
11. What is diazotisation? How is benzene diazonium chloride prepared in the laboratory? Give its mechanism.
12. Starting from benzene diazonium chloride, how will you prepare:
    (*i*) Iodobenzene (*ii*) Phenol
    (*iii*) Aminoazobenzene (*iv*) Benzene
13. Discuss the mechanism of coupling of diazonium salts with phenol and aniline. Discuss the role of pH in these reactions.
14. What are diazonium salts? Compare the stability of alkyl and aryl diazonium salts. How will you prepare the following from *p*-toluidine:
    (*i*) *p*-bromotoluene (*ii*) toluene

15. What is diazotisation? Discuss the structure of benzene diazonium chloride.

16. What is Sandmeyer reactions. Give its mechanism. Using Sandmeyer reaction, how will you bring about the following conversions:

(*i*) Aniline into 1, 2, 3, 5-tetrabromobenzene

(*ii*) Nitrobenzene into m-dichlorobenzene

(*iii*) Aniline into p-iodonitrobenzene

17. What is diazotisation? Give its mechanism. Discuss the structure of benzene diazonium chloride.

18. Give the synthesis and mechanism of diazotisation. Which of the following gives most stable compound of treatment with HNO_2 and why? $C_6H_5NH_2$, $C_2H_5NH_2$, o-nitroaniline.

19. What is diazotisation? Give its mechanism.

20. Discuss the structure of benzene diazonium chloride.

21. Complete the following conversions:

(*i*) Aniline $\longrightarrow$ 1,3, 5 tribromobenzene

(*ii*) Nitrobenzene $\longrightarrow$ *m*-Bromophenol.

22. Give the synthetic application of diazonium salts.

PHYSICAL CHEMISTRY

Thermodynamics–I

(First Law of Thermodynamics and Thermochemistry)

1.1. THERMODYNAMICS

Thermo means heat and *dynamics* means motion or mechanical work. Hence thermodynamics means that branch of science which deals with the conversion of heat into mechanical work and vice versa. In a wider sense *thermodynamics is a branch of science that deals with quantitative relationships between heat and other forms of energy.* Study of thermodynamics is based on three laws *viz.* first, second and third laws of thermodynamics. There is no theoretical proof of the laws, but nothing contrary to the laws has been reported so far.

Objectives of Thermodynamics

(*i*) **To predict the feasibility of a process:** Whether or not, a process will occur under the given set of conditions, can be predicted by applying the principles of thermodynamics.

(*ii*) **To estimate the yield of the products:** Thermodynamic relations help to predict the yield of the products obtainable in a process or a reaction. It is possible to know the extent upto which the reaction takes place before attaining equilibrium.

(*iii*) **To deduce some important relationships in physical chemistry:** It has been possible to deduce some important and useful results such as Raoult's law of lowering of vapour pressure and expressions for depression in freezing point, elevation in boiling point, distribution law and phase rule from the study of thermodynamics.

1.1.1. Limitations of Thermodynamics

Thermodynamics suffers from certain limitations. These are :

(*i*) It deals with the properties like temperature, pressure *etc.* of the matter in bulk *i.e.* with macroscopic quantities and not with the microscopic quantities. Thus, it deals with large groups of atoms, molecules and ions rather than individual atoms, molecules or ions.

(*ii*) It helps to predict the feasibility of a process but does not tell anything about the time taken for the process to complete or the rate at which the process would proceed.

(*iii*) It concerns itself only with the initial and final states of a system and is not concerned with the path by which the change is brought about *i.e.* it does not reveal the mechanism of a process.

1.2. DEFINITION OF CERTAIN THERMODYNAMICS TERMS

1. System and surroundings: *The part of the universe chosen for thermodynamic consideration (i.e. to study the effect of temperature, pressure etc.) is called a system.*

The remaining portion of the universe, excluding the system, is called *surroundings.*

2. Open, closed and isolated systems

(*a*) **Open system:** *A system is said to be an open system if it can exchange both matter and energy with the surroundings.* For example, if some water is kept in an open vessel (Fig. 1.1), exchange of both matter and energy takes place between the system and the surroundings.

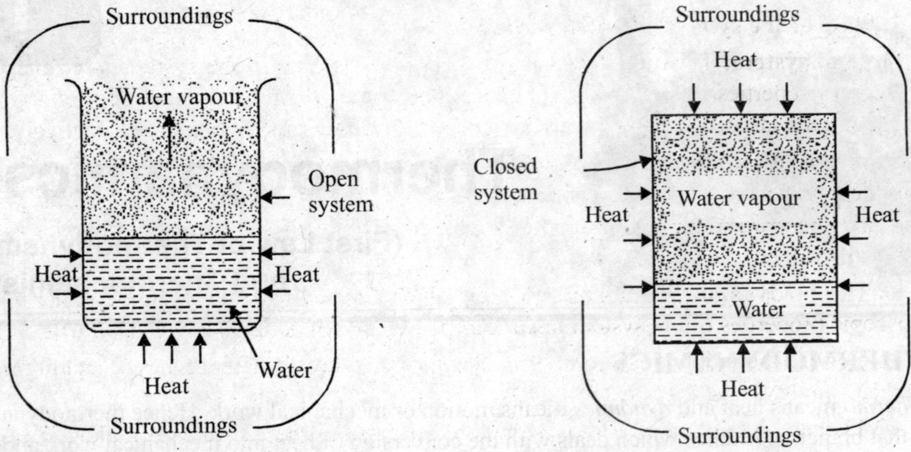

Fig. 1.1. Example of open system: Water kept in an open vessel.

Fig. 1.2. Water placed in an closed vessel.

(*b*) **Closed system:** *If a system can exchange only energy with the surroundings but not matter, it is called a closed system.* For example, if some water is placed in a closed metallic vessel (Fig. 1.2), then as the vessel is closed, no exchange of matter between the system and the surroundings can take place. But, as the vessel has conducting walls, exchange of energy can take place between the system and the surroundings.

(*c*) **Isolated system:** *If a system can exchange neither matter not energy with the surroundings, it is called an isolated system.* For example, if water is placed in a vessel which is closed as well as insulated, no exchange of matter or energy is possible between the system and the surroundings. This constitutes an isolated Fig. 1.3.

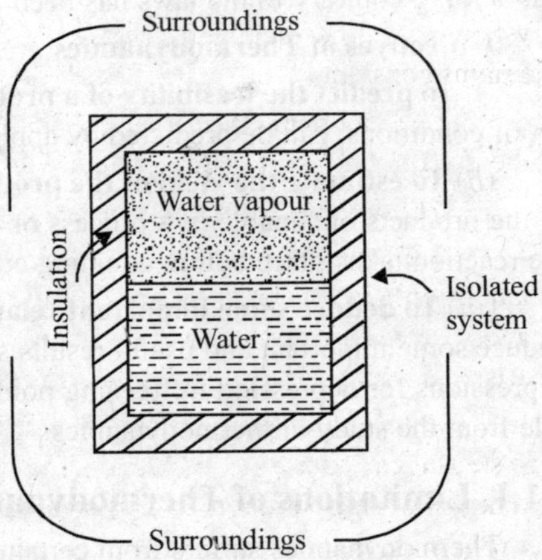

Fig. 1.3. An example of an isolated system.

3. Homogeneous system: *A system is said to be homogeneous if it is uniform throughout.* For example, a gas or a mixture of gases or a pure liquid or a pure solid or a solution of a solid in a liquid are all homogeneous systems. Such a system has one phase only.

4. Heterogeneous system: *If a system is not uniform throughout, it is said to be heterogeneous.* It then consists of two or more phases which are separated from each other by definite bounding surfaces. A few common examples of the heterogeneous systems are (*a*) a system consisting of a liquid and its vapour, (*b*) two or more immiscible liquids, (*c*) a mixture of two or more solids.

5. Macroscopic system: *Macroscopic system means a system containing a large amount of the substance.* In other words, a macroscopic system is one in which there are large number of particles *i.e.* atoms, ions or molecules.

6. Macroscopic property: *A property associated with the collective behaviour of particles in a macroscopic system is called a macroscopic property.* A few examples of these properties are pressure, volume, temperature, surface tension, viscosity, density, refractive index etc.

7. State of the system and state variables

State of system: The state of a system means a condition of the system when the various macroscopic properties like pressure, volume, temperature etc. of the system have definite values. The first and the last state of the system are called *initial state* and *final state* respectively.

State variables: Macroscopic properties are known as state variables. The four most common macroscopic properties which are sufficient to define the state of a system are *composition, pressure, volume* and *temperature.* If these properties are fixed, all other physical properties of the system are automatically fixed.

8. Thermodynamic equilibrium: A system is said to be in thermodynamic equilibrium if the macroscopic properties of the system in various phases do not undergo any change with time.

In fact, the thermodynamic equilibrium denotes that three different types of equilibria which must exist simultaneously. These are :

(*i*) **Thermal equilibrium:** This equilibrium demands that there should be no flow of *heat* from one portion of the system to another. This can happen if the temperature of the system remains constant.

(*ii*) **Mechanical equilibrium:** This equilibrium implies that no work should be done by one part of the system over another *i.e.* no macroscopic movement of matter should occur within the system or of the system with respect to its surroundings. This is possible if the pressure of the system remains constant.

(*iii*) **Chemical equilibrium:** This equilibrium demands that no change in composition should take place in any part of the system with passage of time.

9. Extensive and intensive properties

(*i*) **Extensive properties:** These are the properties which depend upon the quantity of the matter contained in the system. Examples of extensive properties are *mass, volume, energy, heat capacity* etc. The total value of an extensive property is equal to the sum of the values for the separate parts into which the system may be divided for the sake of convenience.

(*ii*) **Intensive properties:** These are those properties which depend only upon the nature of the substance and are independent of the amount of the substance present in the system. The common examples of these properties are *temperature, pressure, refractive index, viscosity, density, surface tension, specific heat* etc. It is because pressure and temperature are intensive properties independent of the quantity of the matter present that they are frequently used as variables to describe the state of a system.

Interestingly an extensive property may become intensive property by specifying unit amount of the substance concerned. Thus mass and volume are extensive properties but density is intensive property, heat capacity is an extensive property but specific heat is intensive.

10. Thermodynamic processes

(*i*) **Isothermal process:** *When a process is carried out in such a manner that the temperature remains constant throughout the process, it is called an isothermal process.* Obviously, when such a process occurs, heat can flow from the system to the surroundings and vice versa in order to keep the temperature of the system constant.

(*ii*) **Adiabatic process:** *When a process is carried out in such a manner that no heat can flow from the system to the surroundings* or *vice versa, i.e., the system is completely insulated from the surroundings, it is called an adiabatic process.*

(*iii*) **Isochoric process:** *It is a process during which the volume of the system is kept constant.*

(*iv*) **Isobaric process:** *It is a process during which the pressure of the system is kept constant.*

11. Reversible and irreversible processes

A reversible process is a process which is carried out infinitesimally slowly so that all changes occurring in the direct process can be exactly reversed and the system remains almost in a state of equilibrium at all times. In other words, a reversible process may be defined as a process which is conducted in such a manner that at every stage, driving force is only infinitesimally greater than the opposing force and which can be reversed by increasing the opposing force by an infinitesimal amount.

Strictly speaking, a reversible process is almost impossible as it would require an infinite time for completion. A process can be made very nearly reversible in some cases. For example, if an opposing E.M.F. is applied to a cell, then no current flows if the opposing E.M.F is exactly equal to that of the cell. If the opposing E.M.F. is very slightly (infinitesimally) less than that of the cell, a very small current is given out by the cell. If the opposing E.M.F. is slightly greater than that of the cell, a very small current flows in the opposite direction.

A reversible process can also be demonstrated with the help of the following experiment as shown in Fig. 1.4.

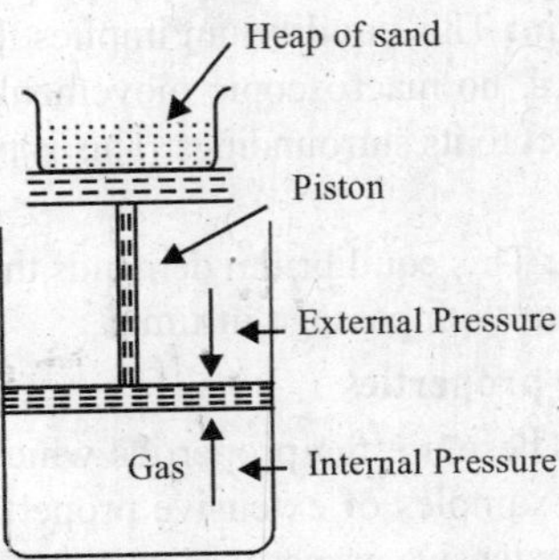

Fig. 1.4 Demonstration of a reversible process

Take a cylinder and fit a frictionless piston into it as shown in Fig. 1.4. The cylinder contains a gas at a moderate pressure, due to which the piston will start rising upwards. Balance it by placing a petri dish containing fine sand over the piston. In the balanced position, the piston will be static.

Now if we add a very small sand grain to the petri dish, the piston will start moving ownwards very slowly. On the other hand, if we take out a very small sand grain from the petri dish, the piston will start moving upwards very slowly. In both cases, it demonstrates reversible process.

Irreversible process. A process which does not meet the above requirements is called an **irreversible process.** In other words, an irreversible process is defined as that process which is not carried out infinitesimally slowly (instead, it is carried out rapidly) so that the successive steps of the direct process cannot be retraced and any change in the external conditions disturbs the equilibrium.

It is interesting to note that all naturally occurring processes like the flow of water down a hill or the flow of heat from a hot end to a cold end of an iron bar etc. are thermodynamically irreversible.

1.3. INTERNAL ENERGY AND CHANGE IN INTERNAL ENERGY

Internal energy: The evolution or absorption of energy in different processes clearly shows that every substance must be associated with some definite amount of energy, the actual value of which depends upon the nature of the substances (*i.e.* arrangement of atoms and electrons within the molecules) and the conditions of temperature, pressure, volume and composition. The energy associated with a substance is called its *internal energy* and is usually denoted by the symbol E or U.

Change of internal energy: It is not possible to find the absolute value of internal energy of a substance because it involves certain quantities (like translational, vibrational and rotational kinetic energies) which cannot be measured.

However, it is not required to know the absolute value of internal energy possessed by any substance (or a system). What is required in different processes is simply the *change of internal energy* when the reactants change into products or when system changes from initial state to the final state. This is easily measurable and is represented by ΔE (or ΔU).

Thus if the internal energy of a system in the initial state is E_1 (or U_1) and in the final state, it is E_2 (or U_2) then the change of internal energy (ΔE or ΔU) may be given by

$$\Delta E \text{ (or } \Delta U) = E_2 - E_1$$

Similarly, in a chemical reaction if E_R is the internal energy of the reactants and E_p is the internal energy of the products, then energy change accompanying the process would be

$$\Delta E = E_P - E_R \text{ or } \Delta U = U_P - U_R$$

Sign conventions in energy: Obviously, if $E_1 > E_2$ (or $E_R > E_P$), the extra energy possessed by the system in the initial state (or the reactants) would be given out, and ΔE will be negative according to the above equations. Similarly if $E_1 < E_2$ (or $E_R < E_P$), energy will be absorbed in the process and ΔE will be positive.

Hence ΔE (or ΔU) *is negative* if energy *is evolved*

and ΔE (or ΔU) *is positive* if energy *is absorbed*

Units of U. The unit of energy is erg or joule, 1 Joule = 10^7 ergs

1.4. EXPRESSION FOR WORK OF EXPANSION AGAINST CONSTANT PRESSURE

The expression for such a work may be derived as follows :

Consider a gas enclosed in a cylinder fitted with a frictionless piston. (Fig.1.5).

Area of cross-section of the cylinder = a sq cm

Pressure on the piston (which is slightly less than internal pressure of the gas so that the gas can expand) = P.

Distance through which gas expands = dl cm

As pressure is force per unit area, force (f) acting on the piston will be

$$f = P \times a$$

$\therefore$ Work done by the gas (*i.e.* the system)

$$= \text{Force} \times \text{Distance} = f \times dl$$
$$= P \times a \times dl$$

But $a \times dl = dV$ is a small increase in the volume of the gas. Hence the small amount of work (δW) done by the gas can be written as

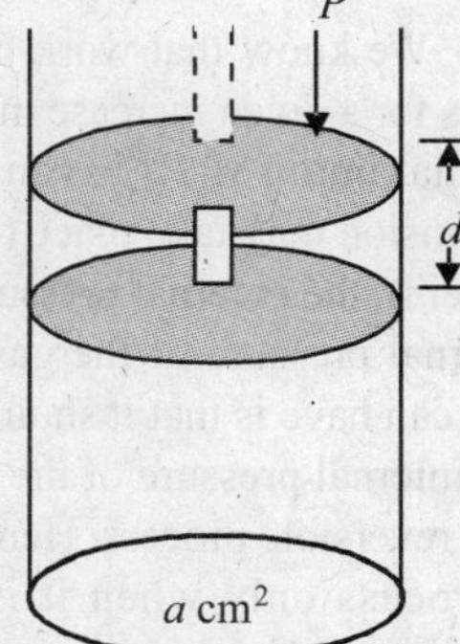

Fig. 1.5. Expansion of a gas.

$$\delta W = P\, dV$$

If the gas expands from initial volume V_1 to the final volume V_2, then the total work done (w) will be given by

$$w = \int_{V_2}^{V_1} P\, dV$$

If the external pressure P against which the gas expands remains almost constant throughout the process, the above result may be written as

$$w = P\int_{V_2}^{V_1} dV = P(V_2 - V_1)$$

$$= P.\Delta V$$

where $\Delta V = (V_2 - V_1)$ is the total change in the volume of the gas.

If the external pressure (P) is slightly greater more than the pressure of the gas, the gas will contract *i.e.* the work will be done by the surroundings on the system. However, the same formulae will apply for the work done.

It may be mentioned here that P is the *external* pressure and hence is sometimes written as P_{ext} so that

$$w = P_{ext} \times \Delta V.$$

Note: According to the latest SI conventions of sign, w is taken as positive if work is done on the system (*i.e.* $w_{contraction}$ is positive) and it is taken as negative if work is done by the system (*i.e.* $w_{expansion}$ is negative). Thus $w_{exp} = -P\Delta V$ and $w_{contraction} = +P\Delta V$.

Electrical work. The electrical work done is expressed as under:

Electrical work done = Electromotive force (emf) × Quantity of electricity

Gravitational work. Gravitational work done when an object of mass (m) moves against the force of gravitation (g) and the displacement of the object is h, is given by:

Work done = $m \times g \times h$

1.4.1. Work done in a reversible process is the maximum work obtainable

The work done by a system is the maximum work if the process is carried out under reversible conditions. This may be explained as follows :

We know that work done by the system = $P_{ext} \times \Delta V$. Thus for a given increase in volume (ΔV), the work done can be maximum if P_{ext} is maximum. But in order that the expansion may take place (*i.e.* the work may be done by the system), the external pressure should always be less than the internal pressure of the gas. Thus the maximum value that P_{ext} can have is that it should be *infinitesimally smaller* than the internal pressure of the gas. But these are the conditions of a reversible process. Hence the maximum work is done in a process only when it is carried out under reversible conditions. Hence in reversible processes, w is sometimes replaced by w_{max}.

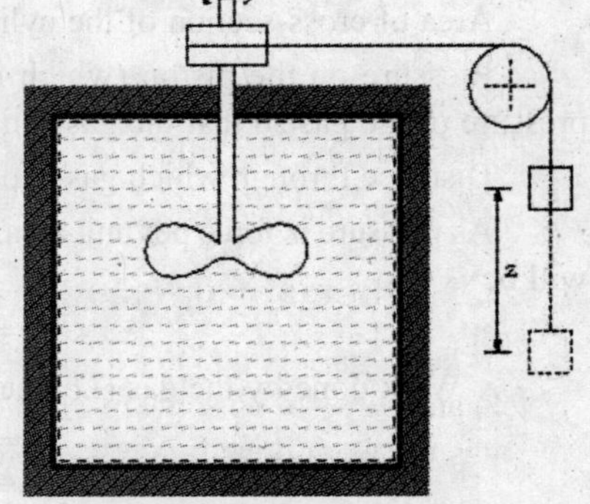

Joule's apparatus

1.5. HEAT

Heat is a mode of energy exchanged between the system and the surroundings as a result of the difference of temperature between them. It is usually represented by the letter q.

Sign of '*q*'. When heat is given by the system to the surroundings, it is **given a negative** sign. When heat is **absorbed by the system** from the surroundings, it is given a **positive** sign.

Units of '*q*'. Heat is usually measured in terms of calories. *A calorie is defined as the quantity of heat required to raise the temperature of one gram of water through 1°C.*

In the S.I. units, heat is expressed in terms of **joules**. The two units are related to each other as under:

1 calorie = 4.184 joules

Heat flows from higher temperature to lower temperature

1.5.1. Zeroth law of thermodynamics

The zeroth law of thermodynamics states: *When two bodies have equality of temperature with a third body, they in turn have equality of temperature with each other.*

We are so familiar with this fact that it does not need any further experiments for its support. However, since this fact is not derivable from other laws and since it logically precedes the first and the second law of thermodynamics, it has been called as the zeroth law of thermodynamics.

1.6. FIRST LAW OF THERMODYNAMICS

First law of thermodynamics can be stated in a number of ways. Some statements of the law are:

(*i*) Energy can neither be created nor destroyed, although it may be converted from one form to another.

(*ii*) The total energy of an isolated system remains constant, although it may undergo transformation from one form to the other.

(*iii*) Whenever certain quantity of some form of energy disappears, an exactly equivalent amount of some other form of energy must be produced.

(*iv*) It is impossible to construct a perpetual motion machine *i.e.* a machine which would produce work continuously without consuming energy.

Einstein in 1905 showed that the energy could be created by the destruction of mass. The energy produced (E) by the destruction of mass m is given by the equation

$$E = mc^2 \text{ (called Einstein equation)}$$

where c is the velocity of light. In the light of this idea, the law of conservation of energy has been modified and now it is called *law of conservation of mass and energy.* It may be stated as follows:

The total mass the energy of an isolated system remains constant in chemical thermodynamics.

1.6.1. Mathematical formulation of the first law of thermodynamics

If heat q is supplied to a system, it may be used up partly to increase the internal energy of the system and partly to do some mechanical work. If ΔU is the increase in internal energy of the system and w is the work done by the system, then we have

$$\Delta U = q + w \quad \text{... } (i)$$

If the work done is work of expansion,

$$w = -P\Delta V$$

or

$$q = \Delta U + P\Delta V$$

If the changes are very small, it may be written as

$$\delta q = dU + PdV \quad \text{... } (ii)$$

1.6.2. Internal energy is a state function but work and heat are not

Internal energy: The internal energy is a state function, that is, when a process occurs, the change in its value depends only upon its value in the initial state and that in the final state and is independent of the path by which the change is brought about. For example, consider a system having a particular pressure, volume and temperature represented by the point A in Fig. 1.6. Suppose the pressure, volume and temperature are changed in such a way that the system is brought to the point B by path I. If the system is returned to the point A by path II, then the energy change involved in path I must be equal in magnitude to that involved in path II. If the energy changes involved in paths I and II are not the same, then suppose that the increase in internal energy by path I is greater than the decrease in internal energy along path II. Thus by carrying out the process $A \rightarrow B$ by path I and the reverse process $B \rightarrow A$ by path II, though the original state has been restored, some energy would have been created. This is contrary to the first law of thermodynamics. Thus it may be inferred that the energy change involved in the process $A \rightarrow B$ must be equal in magnitude to that involved in the reverse process $B \rightarrow A$ by path II. In other words, *the energy change accompanying a process is independent of the path followed and depends only upon the initial and the final states.* Thus if U_A is the internal energy of the system in the initial state and U_B that in the final state, then the change in internal energy accompanying the process would be given by

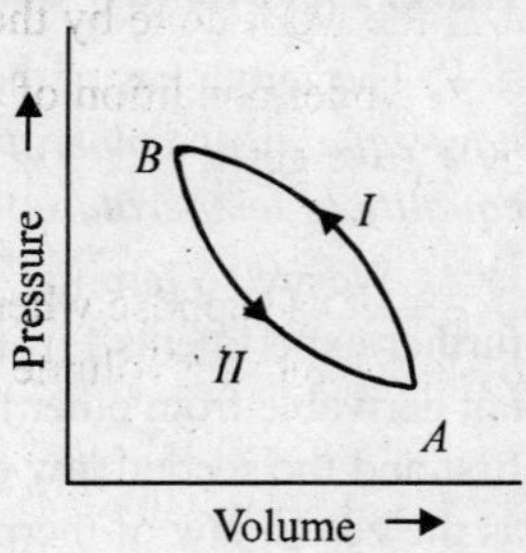

Fig. 1.6 Energy changes in direct and reverse paths.

$$\Delta U = U_B - U_A$$

Work and heat: Unlike internal energy, the work done during a process does not depend merely upon the initial and the final states of the system but it depends upon the path followed. Hence work is not a state function but is a path function. This is explained as under :

Suppose we want to change the system from state A to state B. This may be done by following different paths as shown in the different figures below (Fig. 1.7). Confining ourselves to the work of expansion only, we know that the work done for a small change in volume dV is PdV where P is the external pressure. Thus the total work done is the sum of PdV terms and this is equivalent to the area under the curve in the P–V diagrams. From the figures given below it is obvious that the areas under the curves are different and hence the work done is different when different paths are followed.

That heat is also not a state function follows directly from the mathematical formulation of the first law of thermodynamics *viz.,*

$$q = \Delta U - w$$

Thus the heat q is the sum of two quantities *i.e.* the internal energy change, ΔU and the work done, w. Now ΔU is a state function whereas w is not. Now since w depends upon the path followed, therefore, q will also depend upon the path followed. In other words, q is also not a state function but is a path function.

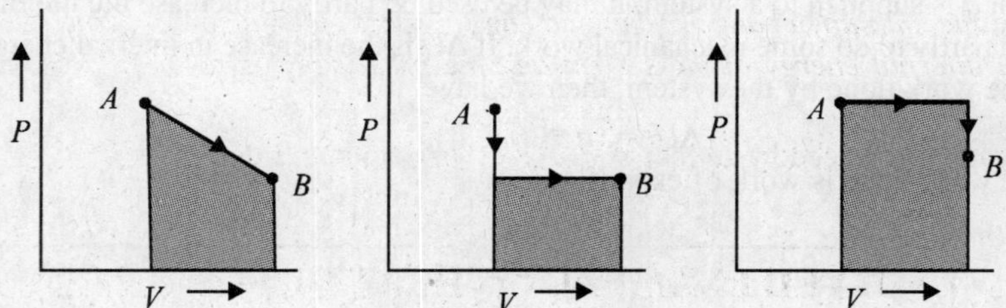

Fig 1.7 Different paths followed in going from state A to state B.

1.7. ENTHALPY OR HEAT CONTENT

If a process is carried out at constant pressure, the work of expansion is given by

$$w = P\Delta V \qquad ... (i)$$

where ΔV is the increase in volume and P is the constant pressure.

According to first law of thermodynamics, we know that

$$q = \Delta U - w \quad \text{... (ii)}$$

where q is the heat absorbed by the system, ΔU is the increase in internal energy of the system and w is the work done by the system.

Under condition of constant pressure, putting $w = -P\Delta V$ and representing the heat absorbed by q_p, we get

$$q_p = \Delta U + P\Delta V \quad \text{... (iii)}$$

Let us suppose when the system absorbs q_p calories of heat, its internal energy increases from E_1 to E_2 and the volume increases from V_1 to V_2. Then we have

$$\Delta U = U_2 - U_1 \quad \text{... (iv)}$$

and $$\Delta V = V_2 - V_1 \quad \text{... (v)}$$

Putting these values in equation (*iii*) above, we get

$$q_p = (U_2 - U_1) + P\,(V_2 - V_1)$$

$$q_p = (U_2 + PV_2) - (U_1 + PV_1) \quad \text{... (vi)}$$

Now as *U, P* and *V* are the functions of state, therefore, the quantity $U + PV$ must also be a state function. The thermodynamic quantity $U + PV$ is called the **heat content or enthalpy** of the system and is represented by the symbol *H i.e.* the enthalpy may be defined mathematically by the equation

$$H = U + PV$$

Thus if H_2 *is* the enthalpy of the system in the final state and H_1 is the value in the initial state, then

$$H_2 = U_2 + P\,V_2$$

and $$H_1 = U_1 + PV_1$$

Putting these values in equation (*vi*), we get

$$q_p = H_2 - H_1$$

or $$q_p = \Delta H \quad \text{... (vii)}$$

where $\Delta H = H_2 - H_1$ is the enthalpy change of the system. Hence **enthalpy change** *of a system is equal to the heat absorbed by the system at constant pressure.*

Further, putting the value of q_p, from equation (*vii*) in equation (*iii*) we get

$$\Delta H = \Delta U + P\Delta V \quad \text{... (viii)}$$

Hence *the enthalpy change accompanying a process may also be defined as the sum of the increase in internal energy of the system and the pressure-volume work done i.e., the work of expansion.*

$P\Delta V$ can be replaced by ΔnRT, so that eq. (*viii*) can be written as

$$\Delta H = \Delta U + \Delta nRT$$

SOLVED EXAMPLE ON HEAT OF FORMATION

Example. The heat of formation of methane at 27°C is –19.3 kcal when the measurements are made at constant pressure. What will be the heat of formation at constant volume?

Solution. Apply the relationship

$$\Delta H = \Delta U + \Delta n\,.\,RT \quad \text{... (i)}$$

ΔH is the heat of reaction at constant pressure whereas ΔU is the heat of reaction at constant volume.

$$\Delta H = -19.3 \text{ kcal}$$

For the reaction

$$C(s) + 2H_2(g) \longrightarrow CH_4(g)$$

No. of moles of the gaseous reactants = 2

No. of moles of the gaseous products = 1

Thus $\Delta n = 1 - 2 = -1$

Substituting the values in eq. (*i*) above

$$-19.3 \times 1000 \text{ cal} = \Delta U + (-1) \times (1.987 \text{ cal}) \times 300$$

or $\Delta U = -19300 - 5961$ cal

or $\Delta U = -13339$ cal or -13.339 kcal.

1.8. HEAT CAPACITY

Heat capacity of a system is defined as the amount of heat required to raise the temperature of the system through I° C.

If q is the amount of heat supplied to a system to raise the temperature from T_1 to T_2, then the heat capacity of the system is given by

$$C = \frac{q}{T_2 - T_1} = \frac{q}{\Delta T} \qquad \text{... (i)}$$

The value of C has to be considered in a narrow range because the heat capacity varies with the temperature.

If δq is the small amount of heat absorbed by a system which raises the temperature of the system by a small amount dT, then the heat capacity of the system is given by

$$C = \frac{\delta q}{\delta T} \qquad \text{... (ii)}$$

There are two types of heat capacities, *viz.*,

(*i*) Heat capacity at constant volume (C_v)

(*ii*) Heat capcity at constant pressure (C_p)

(*i*) Heat capacity at constant volume (C_v)

According to first law of Thermodynamics

$$\delta q = dU + PdV \qquad \text{... (iii)}$$

Substituting for δq from eq. (*iii*) into eq. (*ii*)

$$C = \frac{dU + PdV}{dT} \qquad \text{... (iv)}$$

When the volume is kept $dV = 0$ and the eq. (*iv*) becomes

$$C_v = \left(\frac{\partial U}{\partial T}\right)_v \qquad \text{...(v)}$$

or *for an ideal gas,* this equation may simply be written as

$$C_v = \frac{dU}{dT} \quad \text{... } (vi)$$

C_v is the heat capacity at constant volume.

Thus *the heat capacity at constant volume may be defined as the rate of change of internal energy with temperature at constant volume.*

(*iii*) Heat capacity at constant pressure, C_p

When the pressure is kept constant during the absorption of heat, equation (*iv*) becomes

$$C_p = \left(\frac{\partial U}{\delta T}\right)_p + P\left(\frac{\partial V}{\partial T}\right)_p \quad \text{... } (vii)$$

We know that the heat content or enthalpy of a system is given by

$$H = E + PV$$

Differentiating w.r.t. T at constant P, we get

$$\left(\frac{\partial H}{\partial T}\right)_p = \left(\frac{\partial U}{\partial T}\right)_p + P\left(\frac{\partial V}{\partial T}\right)_p \quad \text{...} (viii)$$

Combining equations (*vii*) and (*viii*), we get

$$C_p = \left(\frac{\partial H}{\partial T}\right)_p \quad \text{... } (ix)$$

or for any ideal gas, this equation may simply be put in the form

$$C_p = \frac{dH}{dT} \quad \text{.. } (x)$$

C_p is the heat capacity at constant pressure.

Thus *the heat capacity at constant pressure may be defined as the rate of change of enthalpy with temperature at constant pressure.*

1.8.1. Relation between C_p and C_v

If the volume of the system is kept constant when the heat is supplied to a system, then no work is done by the system. Thus the heat absorbed by the system is used up completely to increase the internal energy of the system. And if the pressure of the system is kept constant when the heat is supplied to the system, then some work of expansion is also done by the system in addition to the increase in internal energy. Thus if at constant pressure, the temperature of the system is to be raised through the same value as at constant volume, then some extra heat is required for doing the work of expansion. Thus $C_p > C_v$.

We have the following equations for C_v and C_p.

$$C_v = \frac{dU}{dT} \quad \text{... } (i)$$

$$C_p = \frac{dH}{dT} \quad \text{... } (ii)$$

The difference between the heat capacities of an ideal gas can be obtained by subtracting equation (*i*) from equation (*ii*). So we have

$$C_p - C_v = \frac{dH}{dT} - \frac{dU}{dT} \quad ... (iii)$$

But $H = U + PV$ (by definition)

and $PV = RT$ (for 1 mole of an ideal gas)

$\therefore$ $H = U + RT.$

Differentiating this equation w.r.t *T*, we get

$$\frac{dH}{dT} = \frac{dU}{dT} + R \quad ... (iv)$$

or

$$\frac{dH}{dT} - \frac{dU}{dT} = R \quad ... (v)$$

Combining equation (*iii*) and (*v*),

$$\boldsymbol{C_p - C_v = R} \quad \text{(for 1 mole of an ideal gas)}$$

Thus C_p is greater then C_v by the gas constant *R, i.e.,* approximately 2 calories or 8.314 joules. The ratio of specific heats (C_p/C_v) is represented by γ

Thus

$$\gamma = \frac{C_p}{C_v}$$

It is found that the value of γ comes out to be about 1.67 for monoatomic gases, 1.40 for diatomic gases and 1.30 for triatomic gases.

1.8.2. Joule's Law

The law can be stated as: *The change of internal energy of an ideal gas with volume at constant temperature is zero.*

Mathematically, $\left(\frac{\partial U}{\partial V}\right)_T = 0$

This result has been obtained from the Joule Thomson experiment as explained hereunder.

A combination of bulbs A and B connected through a stop cock S is immersed in water taken in a trough (Fig. 1.8 *a*). Bulb A contains as ideal gas at certain pressure with the stop cock closed. There is complete vaccum in the flask B. The bulbs reach a thermal equilibrium after some time. The temperature of water is noted.

The stop cock S is now opened and the gas is allowed to move to bulb B. The bulbs reach a thermal equilbrium after some time. The temperture of water is again noted.

We find that there is no change in temperature. On opening the stop cock, the gas from the bulb A expands into the evacuated bulb B *i.e* the opposing force $P_{\text{opposing}} = 0$. Thus work of expansion

$$dW = -P_{\text{opposing}}\, dV = 0$$

From the first law of thermodynamics, $dv = dq + dw$

As $dW = 0$, we have $dV = dq$

As the temperature of water has not changed, $dq = 0$. Hence $dU = 0$. The mathematical form of Joule's law can be arrived at as follows:

Taking *U* as a function of volume and temperature

$$U = f(V, T)$$

Differentiating

$$dU = \left(\frac{\partial U}{\partial T}\right)_V dT + \left(\frac{\partial U}{\partial V}\right)_T dV \quad \dots (i)$$

We have found earlier that in such a situation $dU = 0$ and $dT = 0$.

Substituting for dU and dT in eq. (i) above, we have

$$\left(\frac{\partial U}{\partial V}\right)_T dV = 0 \quad \text{but } dV \neq 0$$

$$\therefore \quad \left(\frac{\partial U}{\partial V}\right)_T = 0$$

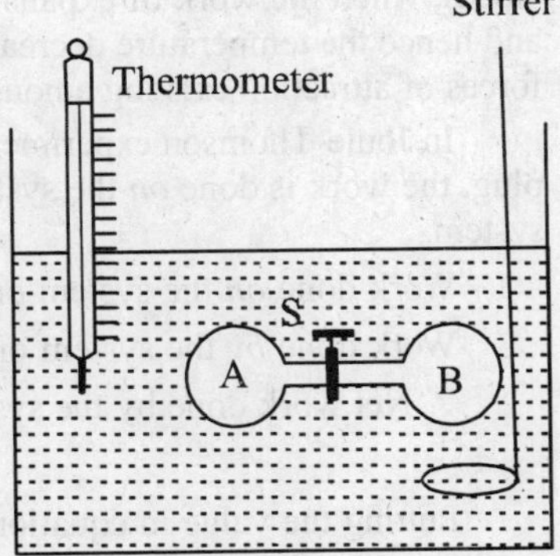

Fig. 1.8a. Joule Thomson effect

1.8.3. Joule-Thomson effect

It was observed by Joule and Thomson that *when a real gas at a certain pressure expands adiabatically through a porous plug or a fine hole into a region of low pressure, it is accompanied by cooling (except for hydrogen and helium which get warmed up)*. This phenomenon is known as **Joule-Thomson effect.**

James P. Joule

Fig. 1.8*b*. shows the experimental set-up of Joule and Thomson experiment.

A tube thoroughly insulated is fitted with a porous plug in the middle and pistons on either side as shown in the figure. The pressures of the gas are kept different on the two sides of the plug, being P_1 on the left side and P_2 on the right side where $P_1 > P_2$. By applying pressure on the piston on the left slowly enough (so as not to change the pressure P_1), a volume of the gas equal to V_1 is forced slowly through the porous plug and then allowed to expand to the pressure P_2 and volume V_2 by moving the piston on the right outward. When the process is being carried out as above, the temperature on the two sides of the porous plug is recorded accurately. The experiment is repeated with a different gases. It is observed that a fall in temperature of the gas takes place in case of all gases except hydrogen and helium when the experiments are carried out at ordinary temperature *i.e.* at room temperature.

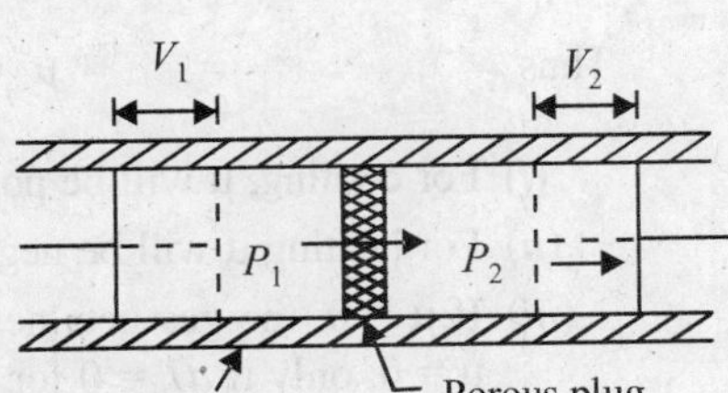

Fig. 1.8b Joule-Thomson experiment (Expansion of real gas through a porous plug).

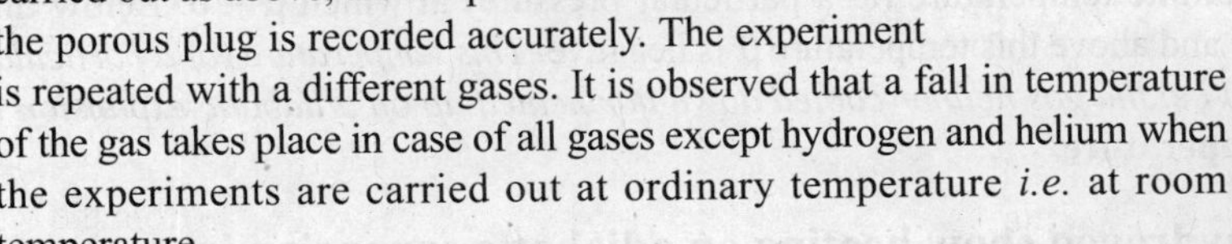

Sir William Thomson

As the process is carried out adiabatically,

$$q = 0$$

$\therefore$ By first law of thermodynamics *viz.* $\Delta U = q + w$

$$\Delta U = w$$

or

$$w = \Delta U \quad \dots (i)$$

It can be said that the work done, during the expansion of the gas under adiabatic conditions, is at the cost of the internal energy. In other words, when the work of expansion is done adiabatically, the internal energy of the gas decreases and hence the temperature decreases. In fact, a part of the internal energy is used up to overcome the forces of attraction existing among the molecules (van der Waals' forces).

In Joule-Thomson experiment, which is an example of adiabatic process, on the left side of the plug, the work is done *on* the system whereas on the right side of the plug, the work is done by the system.

Work done *on* the system on the left side = $P_1 V_1$

Work done *by* the system on the right side = $-P_2 V_2$

∴ Net work done by the system is

$$w = P_1 V_1 - P_2 V_2 \quad \text{... } (ii)$$

Putting the value in equation (*i*), we get

$$\Delta U = P_1 V_1 - P_2 V_2$$

Further putting $\Delta U = U_2 - U_1$

we get $U_2 - U_1 = P_1 V_1 - P_2 V_2$

or $U_2 + P_2 V_2 = U_1 + P_1 V_1$

or $H_2 = H_1$

i.e. $\Delta H = H_2 - H_1 = 0$

Thus it leads to the conclusion that *when the expansion of a gas takes place adiabatically through a porous plug of a fine hole, the enthalpy of the system remains constant.*

1.8.4. Joule-Thomson coefficient

Joule-Thomson coefficient μ may be defined as the temperature change in degrees produced by a drop of one atmospheric pressure when the gas expands under conditions of constant enthalpy. It is expressed as :

Thus
$$\mu = \left(\frac{\partial T}{\partial P}\right)_H$$

(*i*) For cooling, μ will be positive (because dT and dP both will be negative).

(*ii*) For heating μ will be negative (because dT is positive while dP is negative).

(*iii*) If μ = 0, the gas gets neither heated up not cooled on adiabatic expansion (because μ = 0, only if $dT = 0$ for any value *of* dP).

1.8.5. Inversion temperature

Every gas has a definite temperature (at a particular pressure) at which μ = 0. Below this temperature μ is positive and above this temperature μ is negative. *This temperature (at a particular pressure) at which μ = 0 i.e. the gas neither cooled down nor heated up on adiabatic expansion is called the* **inversion temperature.**

1.8.6. Helium and hydrogen show heating on adiabatic expansion in Joule-Thomson experiment

The inversion temperatures of most of the gases are very high (much higher than the room temperature). That is why these gases undergo cooling on adiabatic expansion under ordinary conditions of temperature. On the other hand, the inversion temperatures of hydrogen and helium are very low (being much below 0°C). Hence these gases undergo heating on adiabatic expansion under ordinary conditions of temperature. But, if these gases are first cooled to temperatures below their inversion temperatures and then allowed to expand adiabatically, these gases also show cooling effect.

1.9. RELATIONSHIP OF μ WITH OTHER THERMODYNAMIC QUANTITIES

Enthalpy H is a state function. Taking temperature T and pressure P as the independent variables.

$$H = f(T, P) \qquad \text{... (i)}$$

$$\therefore \quad dH = \left(\frac{\partial H}{\partial P}\right)_T dP + \left(\frac{\partial H}{\partial T}\right) dT \qquad \text{... (ii)}$$

In Joule-Thomson effect, the enthalpy of the system remains constant *i.e.*,

$$dH = 0$$

Putting this value in the above equation, we get

$$\left(\frac{\partial H}{\partial P}\right)_T dP + \left(\frac{\partial H}{\partial T}\right)_p dT = 0 \qquad \text{... (iii)}$$

Dividing this equation throughout by dP and imposing the condition of constant enthalpy, we get

$$\left(\frac{\partial H}{\partial P}\right)_T + \left(\frac{\partial H}{\partial T}\right)_P \left(\frac{\partial T}{\partial P}\right)_H = 0 \qquad \text{... (iv)}$$

But $\left(\frac{\partial H}{\partial T}\right)_P = C_p$, heat capacity at constant pressure,

and $\left(\frac{\partial T}{\partial P}\right)_H = \mu$, Joule-Thomson coefficient ... (v)

Putting these values in equation (*iv*), we get

$$\left(\frac{\partial H}{\partial P}\right)_T + \mu C_p = 0$$

or

$$\mu = -\frac{1}{C_P}\left(\frac{\partial H}{\partial P}\right)_T \qquad \text{...(vi)}$$

Thus μ can be calculated from $\left(\frac{\partial H}{\partial P}\right)_T$ and *vice versa.*

SOLVED EXAMPLE ON JOULE-THOMSON COEFFICIENT

Example . Explain why an ideal gas shows neither heating nor cooling in Joule-Thomson experiment ?

OR

Why is Joule-Thomson coefficient of an ideal gas zero ?

Solution. An ideal gas, undergoes neither cooling nor heating on adiabatic expansion in Joule-Thomson experiment. This is expected as explained below:

The intermolecular forces of attraction (*i.e.* van der Waals forces) in an ideal gas are negligible. Hence no energy is used up in overcoming these forces of attraction when the gas expands adiabatically. Thus the internal energy of the gas does not fall and therefore the temperature also does not fall.

1.10. EXPRESSION FOR WORK DONE IN ISOTHERMAL REVERSIBLE EXPANSION OF A GAS

A general expression for the work of expression in an isothermal reversible process may be derived as follows :

Let P be the pressure of the gas within a system undergoing an isothermal expansion by a reversible process. Then in order that the expansion may take place very slowly (*i.e.* reversibly) the external pressure should be $P - dP$, where dP is very small quantity. The small amount of work (δw) done by the system when its volume increases by an infinitesimal amount, dV, is equal to the product of the external pressure and the volume change *i.e.*,

$$\delta w = -(P - dP)\, dV \qquad \text{... (i)}$$

Neglecting the term $dP \times dV$ which is very small (being the product of two very small quantities), equation (*i*) becomes

$$\delta w = -P.\, dV \qquad \text{... (ii)}$$

Hence the total work (w) done by the system when its volume increases from V_1 to V_2 will be given by

$$w = -\int_{V_1}^{V_2} PdV \qquad \text{... (iii)}$$

Now, for n moles of an ideal gas, the equation of state is

$$PV = nRT$$

so that

$$P = \frac{nRT}{V} \qquad \text{... (iv)}$$

Substituting this value is equation (*iii*), we get

$$w = -\int_{V_1}^{V_2} \frac{nRT}{V}\, dV \qquad \text{... (v)}$$

For an isothermal process, T is constant. Also n and R are constant quantities. Hence equation (*v*) may be written as

$$w = -nRT \int_{V_1}^{V_2} \frac{dV}{V}$$

$$= -nRT\, ln\, \frac{V_2}{V_1} \qquad \text{...(vi)}$$

or

$$w = -2{\cdot}303\, nRT \log \frac{V_2}{V_1} \qquad \text{... (vii)}$$

A relation can also be obtained in terms of pressure.

As the process is isothermal *i.e.* the temperature remains constant, therefore by Boyle's law

$$P_1 V_1 = P_2 V_2 \qquad \text{... (viii)}$$

where P_1, V_1 are the initial pressure and volume and P_2, P_2 are the final pressure and volume of the system.

Equation (*viii*) can be written as

$$\frac{V_2}{V_1} = \frac{P_1}{P_2}$$

Substituting this value in equation (*vii*), we get

$$w = -2{\cdot}303\ nRT \log \frac{P_1}{P_2} \qquad \text{... }(ix)$$

Work done in a *reversible process* is the *maximum* work obtainable from a system. Hence w in equations (*vii*) and (*ix*) may be replaced by w_{max}. Expressions (*vii*) and (*ix*) may thus be written as

$$w_{max} = -2{\cdot}303\ nRT \log \frac{V_2}{V_1}$$

$$= -2{\cdot}303\ nRT \log \frac{P_1}{P_2}$$

Internal energy of a system is a function of temperature. As the temperature remains constant in an isothermal process, internal energy does not change or $\Delta U = 0$.

Expression for q (heat absorbed): According to first law of thermodynamics

$$q = \Delta U - w$$

But $\Delta U = 0$ (*as explained above*)

$\therefore$ $q = -w$

Thus all the heat absorbed is used up completely in doing the work of expression only. Hence the expressions for the heat absorbed in an isothermal reversible expansion of an *ideal gas* are same as for the work done *i.e.*

$$q = -w = 2{\cdot}303\ nRT \log \frac{V_2}{V_1} = 2{\cdot}303\ nRT \log \frac{P_1}{P_2}$$

1.10.1. Values for ΔU: and ΔH in an isothermal expansion process

Expression for ΔU : Since the internal energy of an *ideal gas* is function of temperature only, therefore, in an isothermal process as the temperature remains constant, the internal energy of the system (containing an ideal gas) remains constant

i.e $\Delta U = 0$

Expression for ΔH

$$\Delta H = \Delta (U + PV)$$

$$= \Delta U + \Delta(PV)$$

$$= \Delta U + \Delta(nRT)$$

$$= \Delta U + \Delta nR\ \Delta T$$

$$= 0 + 0 \quad \begin{cases} \Delta U = 0 \text{ as explained above and} \\ \Delta T = 0 \text{ as the process is isothermal} \end{cases}$$

i.e. $\Delta H = 0$

1.10.2. Values of *q, w,* ΔU and ΔH in case of isothermal irreversible expansion when

(*i*) opposing pressure is zero (*ii*) opposing pressure is constant and less than P_2 but greater than zero.

The expressions for *w, q,* ΔU and ΔH for the above expansion may be obtained as follows :

(*i*) **For free expansion:** Here $P_{opp} = 0$

$$\therefore \quad w = -\int P_{opp}\, dV = 0$$

Hence
$$q = w = 0$$
$$\Delta U = \Delta H = 0$$

(as the gas is *ideal* and expansion is *isothermal*)

(*ii*) **For intermediate expansion:** Here P_{opp} = constant and lies in the range $0 < P_{opp} < P_2$. Hence

$$w = -\int_{V_1}^{V_2} P_{opp}\, dV = -P_{opp}\,(V_2 - V_1)$$

$$\therefore \quad q = -w = P_{opp}\,(V_2 - V_1)$$

$$\Delta U = \Delta H = 0$$ (as the gas is *ideal* and expansion is *isothermal*)

It may be noted that the work done in this process will be less than the work done in the reversible process.

1.11. SOLVED EXAMPLES ON ISOTHERMAL REVERSIBLE EXPANSION OF GAS

Example 1. Ten moles of an ideal gas at the initial pressure of one atmosphere at 0°C were expanded reversibly under isothermal conditions to a final pressure of 0·1 atmosphere. Calculate the work done by the gas, the change in internal energy and the heat absorbed by the system ($R = 8{\cdot}314$ J K^{-1} deg^{-1} mol^{-1}).

Solution. Here we are given that

$$n = 10 \text{ moles}$$
$$P_1 = 1 \text{ atm}$$
$$P_2 = 0{\cdot}1 \text{ atm}$$
$$T = 0°\text{C} = 273 \text{ K}$$
$$R = 8.314 \text{ JK}^{-1} \text{ mol}^{-1}$$

$$\therefore \quad w_{max} = -2.303\, nRT \log \frac{P_1}{P_2}$$

$$= -2.303 \times 10 \times 8{\cdot}314 \times 273 \times \log \frac{1}{0.1}$$

$$= \mathbf{-52271.7\ J} \qquad (\because \log 10 = 1.0)$$

For an ideal gas during isothermal process

$$\Delta U = 0$$

$\therefore$ By first law of thermodynamics ($q = \Delta U - w$)

$$q = -w_{max} = 52271.7 \text{ J}$$

Example 2. One mole of an ideal gas is heated at constant pressure from 0°C to 100°C (*i*) Calculate the work involved (*ii*) If the gas were expanded isothermally and reversibly at 0°C from 1 atmosphere to some other pressure *P*, what must be the final pressure if the isothermal work is equal to the work involved in (*i*)?

Solution. Work involved when the gas is heated at constant pressure from 0°C to 100°C is obtained as follows :

(*i*)
$$\begin{aligned} w &= -P\Delta V = -P(V_2 - V_1) \\ &= -(PV_2 - PV_1) \\ &= -(nRT_2 - nRT_1) \\ &= -nR(T_2 - T_1) \\ &= 1 \times 8.314\,(373 - 273) \\ &= \mathbf{831{\cdot}4\ J} \end{aligned}$$

(*ii*) Final pressure if the process is carried out isothermally involving the work as in step (*i*) is obtained as under

$$w = -2.303\, nRT \log \frac{P_1}{P_2}$$

Put $P_1 = 1$ atm, $T = 273$ K, $R = 8.314$ JK^{-1} and $n = 1$

This on solving gives

$$P = \mathbf{0.6933\ atm}$$

Example 3. Calculate the pressure-volume work performed by the system during reversible isothermal expansion of two moles of an ideal gas from 2 litres to 10 litres at 20°C.

Solution. Here we are given that

$$\begin{aligned} n &= 2 \text{ moles} \\ V_1 &= 2 \text{ litres} \\ V_2 &= 10 \text{ litres} \\ T &= (20 + 273)\text{ K} = 293\text{ K} \end{aligned}$$

Taking $R = 8.314$ Joules degree^{-1} mol^{-1}

$$\begin{aligned} w_{max} &= -2{\cdot}303\, nRT \log \frac{V_2}{V_1} \\ &= -2.303 \times 2 \times 8.314 \times 293 \times \log \frac{10}{2} \\ &= -11220.2 \times \log 5 \\ &= -112202 \times 0.6990 \\ &= \mathbf{7842.9\ joules} \end{aligned}$$

PROBLEMS FOR PRACTICE

1. Find the work required to compress reversibly 1 mole of a perfect gas from 1 to 100 atmospheres at 27°C. **[Ans.** 11488 J]
2. Calculate the work done during isothermal expansion of one mole of an ideal gas from 10 atmoshpheres to 1 atmosphere at 27°C. **[Ans.** – 5.74 kJ]
3. Ten moles of an ideal gas are compressed isothermally and reversibly at 100°C from pressure of 1 atmosphere to a pressure of 10 atmosphere. Calculate the following quantities:

   (*i*) W and (*ii*) ΔH **[Ans.** $W = 71419$ J, $\Delta H = 0$]

4. Calculate the work done by 10 litres of an ideal gas at 0°C and 10 atm in an isothermal reversible expansion to a final pressure of 1 atm (R = 8.314 joules).

[**Hint.** (*i*) Reduce the given volume to volume at S.T.P.]

$$\frac{P_1 V_1}{T_1} = \frac{P_2 V_2}{T_2}$$

$$\frac{10 \times 10}{273} = \frac{1 \times V_2}{273}$$

or $$V_2 = 100 \text{ litres at S.T.P.}$$

(*ii*) Calculate the number of moles as follows:

22.4 litres at S.T.P. = 1 mole

100 litres at S.T.P = 4.46 moles

$$w = -2.303\, nRT \log \frac{P_1}{P_2}$$

or $$w = -23.3 \text{ kJ}]$$

5. Calculate the maximum work per mole when 5 moles of an ideal gas expand isothermally and reversibly from 20 litres to 100 litres at 27°C. [R = 8.314 × 10^7 ergs mol^{-1} deg^{-1}]

[**Ans.** – 200.76 × 10^5 ergs]

6. Estimate the work done in joules when the volume of 16.0 g of oxygen at 300 K changes isothermally from 5.0 to 25.0 litres [R = 8.3143 Jk^{-1} mol^{-1}] [**Ans.** – 2007.6 J]

7. One mole of an ideal gas is heated at constant pressure from 0°C to 100°C.

(*i*) Calculate the work involved (*ii*) If the gas were expanded isothermally and reversibly at 0°C from 1 atmosphere to some other pressure P, what must be the final pressure if the isothermal work is equal to the work involved in (*i*)

[**Hint.** (*i*) w = $-P\Delta V = -P(V_2 - V_1) = (PV_2 - PV_1) = -(nRT_2 - nRT_1)$
$= -nR(T_2 - T_1) = -1 \times 8.314\ (373 - 273) = -831.4$ J]

(*ii*) $w = 2.303 - nRT \log \frac{P_1}{P_2}$. Put P_1 = 1 atm, T = 273 K, R = 8.314 JK^{-1} mol^{-1}, n = 1

On solving, we get P_2 = 0.6933 atm]

8. Calculate the value of w, ΔU and ΔH in the isothermal and reversible expansion of 64 grams of oxygen at 410 K from an initial volume of 20 litres to a final volume of 60 litres. Assume oxygen behaves ideally. [**Ans.** w = –7490.8J, $\Delta H = \Delta U$ = zero]

9. Estimate the maximum work which can be obtained by the isothermal expansion of 1 mole of an ideal gas from 2.24 litres to 22.4 litres at 273 K (R = 8.3143 JK^{-1} mol^{-1}).

[**Ans.** – 5227.4 J].

10. Calculate the work done when 5 moles of an ideal gas at 27°C expand from an initial pressure of 5 atm to 1 atm reversibly.

1.12. CALCULATION OF VALUES OF *w*, *q*, ΔU AND ΔH IN ADIABATIC EXPANSION OF AN IDEAL GAS

An adiabatic process is the one in which no heat enters or leaves the sýstem of any stage. For every infinitesimal change of the process,

$$\delta q = 0$$

Putting this value in the equation of the first law *viz.*

$$dU = \delta q + \delta w, \text{ we get}$$

or $$dU = \delta w \qquad \text{... }(i)$$

If the work involved is the work of expansion only, then if dV is the small increase in volume and P is the pressure of the gas, then

$$\delta w = -PdV \qquad \text{... }(ii)$$

Putting this value in equation (*i*), we get

$$dU = -PdV \qquad \text{... }(iii)$$

For an ideal gas, $$C_v = \frac{dE}{dT}$$

or $$dE = C_v\, dT$$

Putting this value in equation (*iii*) we get

$$C_v dT = -PdV \qquad \text{... }(v)$$

Also
$$\begin{aligned} \Delta H &= \Delta\,(U + PV) \\ &= \Delta U + \Delta(PV) \\ &= \Delta U + \Delta(RT) \\ &= \Delta U + R\,\Delta T \\ &= C_v\,\Delta T + R\,\Delta T \\ &= (C_v + R)\,\Delta T \\ &= (C_v + R)\,(T_2 - T_1) \\ &= C_p\,(T_2 - T_1) \qquad \text{... }(vi) \end{aligned}$$

In a nutshall,

For any type of adiabatic process (reversible or irreversible)

$$\begin{aligned} q &= 0 \\ w &= \Delta E \\ \Delta U &= C_v\,\Delta T = C_v\,(T_2 - T_1) \\ \Delta H &= C_p\,\Delta T = C_p\,(T_2 - T_1) \end{aligned}$$

1.12.1. Reversible adiabatic expansion

I. Relationship between temperature and volume in reversible adiabatic expansion

Consider 1 mol of the *ideal gas.* Then according to the gas equation

$$PV = RT$$

or $$P = \frac{RT}{V} \qquad \text{... }(i)$$

Also $$C_v\, dT = -PdV \qquad \text{... }(ii)$$

Substituting for P from eq. (*i*) in (*ii*)

$$C_v dT = -\frac{RT}{V}\,dV$$

or $$C_v \frac{dT}{T} = -R\frac{dV}{V} \quad \text{... } (iii)$$

Now suppose the volume of the gas changes from V_1 to V_2 when the temperature changes from T_1 to T_2. Then assuming C_v to be independent of temperature, and integrating equation (*iii*) between the limits T_1, T_2 and V_1, V_2, we get

$$C_v \int_{T_1}^{T_2} \frac{dT}{T} = -R \int_{V_1}^{V_2} \frac{dV}{V}$$

$$C_v [\ln T]_{T_1}^{T_2} = -R[\ln V]_{V_1}^{V_2}$$

$$C_v \ln\frac{T_2}{T_1} = -R\ln\frac{V_2}{V_1}$$

$$C_v \ln\frac{T_2}{T_1} = R\ln\frac{V_1}{V_2}$$

Now, for 1 mol of an ideal gas we know that

$$C_p - C_v = R$$

Putting this value of R in equation (*iv*), we get

$$C_v \ln\frac{T_2}{T_1} = (C_p - C_v)\ln\frac{V_1}{V_2}$$

Dividing throughout by C_v we get

$$\ln\frac{T_2}{T_1} = \left(\frac{C_p}{C_v} - 1\right)\ln\frac{V_1}{V_2}$$

Putting $\frac{C_p}{C_v} = \gamma$, the ratio of the two heat capacities, we get

$$\ln\frac{T_2}{T_1} = (\gamma - 1)\ln\frac{V_1}{V_2}$$

or $$\ln\frac{T_2}{T_1} = \ln\left(\frac{V_1}{V_2}\right)^{\gamma-1}$$

Taking antilogarithm of both sides, we get

$$\frac{T_2}{T_1} = \left(\frac{V_1}{V_2}\right)^{\gamma-1} \quad \frac{T_2}{T_1} = \left(\frac{V_1}{V_2}\right)^{\gamma-1}$$

or $$T_1 V_1^{\gamma-1} = T_2 V_2^{\gamma-1}$$

or $$\boxed{TV^{\gamma-1} = \text{constant}}$$

II. Relationship between temperature and pressure in adiabatic expansion of an ideal gas

For an ideal gas, $P_1V_1 = RT_1$

and $P_2V_2 = RT_2$

i.e. $V_1 = \frac{RT_1}{P_1}$

and $V_2 = \frac{RT_2}{P_2}$

$$\therefore \quad \frac{V_1}{V_2} = \frac{\frac{RT_2}{P_1}}{\frac{RT_2}{P_2}} = \frac{T_1P_2}{T_2P_1} \quad \text{... (i)}$$

$$\text{Also} \quad \frac{T_2}{T_1} = \left(\frac{V_1}{V_2}\right)^{\gamma-1} \quad \text{... (ii)}$$

(The relation between T and V in case of adiabatic expansion)

From (*i*) and (*ii*)

$$\frac{T_2}{T_1} = \left(\frac{T_1P_2}{T_2P_1}\right)^{\gamma-1}$$

or
$$\frac{T_2}{T_1}\times\left(\frac{T_2}{T_1}\right)^{\gamma-1} = \left(\frac{P_2}{P_1}\right)^{\gamma-1}$$

or
$$\left(\frac{T_2}{T_1}\right)^{\gamma} = \left(\frac{P_2}{P_1}\right)^{\gamma-1}$$

or
$$\left(\frac{T_2}{T_1}\right)^{\gamma} = \left(\frac{P_1}{P_2}\right)^{1-\gamma}$$

Taking the γth root of both sides, we get ... (*iii*)

$$\frac{T_2}{T_1} = \left(\frac{P_1}{P_2}\right)^{(1-\gamma)/\gamma}$$

or
$$T_1P_1^{(1-\gamma)/\gamma} = T_2P_2^{(1-\gamma)/\gamma}$$

or
$$\boxed{TP^{(1-\gamma)/\gamma} = \textbf{constant}}$$

III. Relationship between pressure and volume in adiabatic expansion of an ideal gas

Consider the relations between temperature and volume and temperature and pressure

$$\frac{T_2}{T_1} = \left[\frac{V_1}{V_2}\right]^{\gamma-1} \quad \dots (i)$$

$$\frac{T_2}{T_1} = \left[\frac{P_1}{P_2}\right]^{(1-\gamma)/\gamma} \quad \dots (ii)$$

From (*i*) and (*ii*)

$$\left(\frac{P_1}{P_2}\right)^{(1-\gamma)/\gamma} = \left(\frac{V_1}{V_2}\right)^{\gamma-1}$$

or
$$\left(\frac{P_1}{P_2}\right)^{(1-\gamma)/\gamma} = \left(\frac{V_2}{V_1}\right)^{1-\gamma}$$

Taking (1 – γ)th root of both sides, we have

$$\left(\frac{P_1}{P_2}\right)^{1/\gamma} = \frac{V_2}{V_1}$$

Raising both sides to the power γ, we have

$$\frac{P_1}{P_2} = \left(\frac{V_2}{V_1}\right)^{\gamma}$$

or
$$PV_1^{\gamma} = P_2V_2^{\gamma}$$
$$PV^{\gamma} = \text{constant}$$

1.12.2. Work done in an adiabatic reversible expansion of an ideal gas

The expression for the work done in an adiabatic reversible expansion of an ideal gas may be obtained as under:

For adiabatic process, we have

$$PV^{\gamma} = \text{constant}$$

Differentiating this equation, we get

$$P\,(\gamma V^{\gamma-1})\,dV + V^{\gamma}\,dP = 0$$

or
$$\gamma PV^{\gamma-1}\,dV + V^{\gamma}\,dP = 0$$

Dividing throughout by $V^{\gamma-1}$, we get

$$\gamma PdV + VdP = 0$$

or
$$VdP = -\gamma PdV \quad \dots (i)$$

For *n* moles of an ideal gas, $PV = nRT$

Differentiating this equation completely, we get

$$PdV + VdP = nRdT \quad \dots (ii)$$

Substituting for VdP from equation (*i*) in this equation we get

$$PdV - \gamma PdV = nRdT$$

or
$$(1-\gamma)\,PdV = nRdT$$

or $$PdV = \frac{nRdT}{1-\gamma} \qquad \text{... }(iii)$$

Work done in the adiabatic process is given by general equation

$$w = -\int_{V_1}^{V_2} PdV \qquad \text{... }(iv)$$

Substituting for PdV from equation (*iii*) in this equation, and changing the limits from volumes to temperatures, we get

$$w = -\int_{T_1}^{T_2} \frac{nRdT}{1-\gamma}$$

$$= -\frac{nR}{1-\gamma}\int_{T_1}^{T_2} dT$$

$$= -\frac{nR}{1-\gamma}(T_2 - T_1) \qquad \text{... }(v)$$

Hence $$\boldsymbol{w = -\frac{nR(T_2 - T_1)}{1-\gamma}}$$

Expression for work done may be obtained in a different form.

Putting $R = C_p - C_v$ and $\gamma = C_p / C_v$ equation (*v*) becomes

$$w = -\frac{n(C_p - C_v)(T_2 - T_1)}{1 - \frac{C_p}{C_v}}$$

$$= -\frac{n(C_p - C_v)(T_2 - T_1)}{\frac{(C_v - C_p)}{C_p}}$$

$$= -\frac{n(C_p - C_v)(T_2 - T_1)}{\frac{-(C_p - C_v)}{C_v}}$$

$= nC_v\,(T_2 - T_1)$, (for n moles of the ideal gas)

$= C_v\,(T_2 - T_1)$, (for 1 mol of the ideal gas) ... (*vi*)

For n moles of an ideal gas,

$P_1\,V_1 = nRT_1$ (for the initial state)

and $P_2V_2 = nRT_2$ (for the final state)

Substituting these values in equation (*v*), we get

$$\boldsymbol{w = -\frac{P_2V_2 - P_1V_1}{1-\gamma}} \qquad \text{... }(vii)$$

SOLVED EXAMPLES ON ADIABATIC REVERSIBLE EEXPANSION OF AN IDEAL GAS

Example 1. For a certain gas C_p = 8.58J. Two moles of the gas are expanded adiabatically from an initial temperature of 20°C to a final temperature of – 45.4°C. Calculate the work done in the above case.

Solution. $C_p = 8.58\text{J}, C_v = 8.58 - 2 = 6{\cdot}58\text{J}$

Therefore $r = \frac{C_p}{C_v} = \frac{8{\cdot}58}{6{\cdot}58} = 1.3$

$$n = 2, R = 8.314\text{J K}^{-1}\text{ mol}^{-1}$$

Work done
$$w = -\frac{nR(T_2 - T_1)}{1-\gamma}$$
$$= -\frac{2\times 8.314\times(-65.4)}{(1-1.3)}$$
$$= -\frac{2\times 8.314\times 65.4}{0.3}\text{J}$$
$$= \mathbf{-\ 3624.9\ J}$$

Example 2. Three moles of a perfect gas with C_v = 5.0 cal/mol/degree at 1.0 atmosphere are to be compressed adiabatically and reversibly from a volume of 75 litre at 1·0 atmosphere to a pressure of 100 atmospheres. Predict (*a*) the final volume of the gas (*b*) the final temperature of the gas.

Solution. Here we are given that

$$n = 3 \text{ moles}$$
$$C_v = 5.0 \text{ cal/mol/degree}$$
$$P_1 = 1.0 \text{ atm}$$
$$P_2 = 100 \text{ atm}$$
$$V_1 = 75 \text{ litres}$$

For a perfect gas
$$C_p = C_v + R$$
$$= 5{\cdot}0 + 2.0 = 7.0 \text{ cal } (\text{R} \cong 2 \text{ cal})$$

$\therefore$
$$\gamma = \frac{C_p}{C_v} = \frac{7.0}{5.0} = 1.4$$

(*a*) *To calculate the final volume*

$$P_1V_1^\gamma = P_2V_2^\gamma$$

or
$$\left(\frac{V_1}{V_2}\right)^\gamma = \frac{P_2}{P_1}$$

or
$$\left(\frac{75}{V_2}\right)^{1{\cdot}4} = \frac{100}{1.0} = 100$$

$$\therefore \quad 1.4\,[\log 75 - \log V_2] = \log 100$$

$$\text{or} \quad 1.4\,(1.8751 - \log V_2) = 2$$

$$\text{or} \quad 1.4 \log V_2 = 0.62514$$

$$\text{or} \quad \log V_2 = 0.4465$$

$$V_2 = \text{antilog}\,(0.4465)$$

$$= \mathbf{2{\cdot}796\ litres}$$

(*b*) *To calculate the final temperature*

For a perfect gas,

$$PV = nRT$$

When $P = 100$ atm

$V = 2.796$ litres (calculated above)

$n = 3$ moles (given)

$R = 0{\cdot}0821$ litre atm deg^{-1} mol^{-1}

$$\therefore \quad T = T = \frac{P \times V}{n \times R} = \frac{100 \times 2.796}{3 \times 0.0821} = \mathbf{1135.2\ K}$$

Example 3. Two moles of H_2 are compressed adiabatically from S.T.P. conditions to occupy a volume of 4.48 litres. Calculate the final temperature (γ for H_2= 1.41).

Solution. We are given that $n = 2$

so that

$$\left.\begin{aligned} V_1 &= 2 \times 22.4 = 44.8 \text{ litres} \\ T_1 &= 273 \text{ K} \end{aligned}\right\} \text{at S.T.P.}$$

$$V_2 = 4.48 \text{ litres}$$

$$T_2 = ?$$

$$\gamma = 1.41$$

Substituting the values in the formula

$$T_1 V_1^{\gamma-1} = T_2 V_2^{\gamma-1}$$

$$\text{or} \quad \frac{T_2}{T_1} = \left(\frac{V_1}{V_2}\right)^{\gamma-1}$$

$$\text{We get} \quad \frac{T_2}{273} = \left(\frac{44.8}{4.48}\right)^{1{\cdot}41-1} = (10)^{0{\cdot}41}$$

$$\text{or} \quad T_2 = 273 \times (10)^{0{\cdot}41}$$

$$\therefore \quad \log T_2 = \log 273 + 0.41 \log 10$$

$$= 2{\cdot}4362 + 0.41$$

$$= 2{\cdot}8462$$

$$\text{or} \quad T_2 = \text{antilog}\ 2{\cdot}8462 = \mathbf{701.8\ K}$$

Example 4. A dry gas at NTP is expanded adiabatically from 1 litre to 5 litre. Calculate the final temperature assuming ideal behaviour (C_p / C_v = 1.4)

Solution. Apply the relation $TV^{\gamma - 1}$ = constant

$$T_1 V_1^{\gamma-1} = T_2 V_2^{\gamma-1} \quad \text{or} \quad \frac{T_1}{T_2} = \left(\frac{V_2}{V_1}\right)^{\gamma-1}$$

$$T_1 = 273 \text{ K}, \quad T_2 = ? \quad V_1 = 1 \text{ litre}, \quad V_2 = 5 \text{ litre}$$

Substituting the values in the equation above,

$$\frac{273}{T_2} = 5^{1.4-1} \quad \text{or} \quad \frac{273}{T_2} = 5^{0.4}$$

Taking logarithm of both sides

$$\log 273 - \log T_2 = 0.4 \log 5$$

or $$\log T_2 = \log 273 - 0.4 \log 5$$

or $$\log T_2 = 2.4362 - (0.4 \times 00.6990)$$

$$= 2.1566$$

or $$T_2 = 143.4 \text{ K}$$

PROBLEMS FOR PRACTICE

1. Two moles of hydrogen at S.T.P are compressed adiabatically to a volume of 10 litres. Calculate the final pressure and temperature of the gas. Given that γ for hydrogen is 1.41.

[**Ans.** Final Pressure = 8.30 atm, Final temperature = 505 K]

[**Hint.** $V_1 = 2 \times 22.4 = 44.8$ litres, $V_2 = 10$ litres]

2. 14 grams of N_2 at 290 K are compressed adiabatically from 8 to 5 litres. Calculate the final temperature and the work done on the gas. Assume $C_p = \frac{7}{2}R$.

[**Ans.** $T = 349.9$ K, $w = 622.5$ J]

3. A quantity of air ($\gamma = 1.4$) at 300 K is compressed adiabatically to one third of its volume. Find the change in its temperature. [**Ans.** 465.5 K]

4. Calculate the change in energy for 2 moles of H_2 warmed at constant volume from 25° to 50°C. Given that for the gas near room temperature, C_v is constant and is about 5 cal/deg

[**Ans.** 250 ergs.]

5. For a certain ideal gas $C_p = 8.58$ cal mol^{-1}. What will be the final volume and temperature when 2 moles of the gas at 20°C and 15 atm are allowed to expand adiabatically and reversibly to 5 atm pressure? Also calculate the work done in the above process.

[**Ans.** $V = 7.44$ litres $t = -45.4$°C, $w = -3625$ J]

[**Hint.** First calculate T_2, then find out w using the formular $w = -\frac{nR(T_2 - T_1)}{1-\gamma}$

$T_1\gamma = 293$ K, $T_2 = 227.6$ K

$\gamma = \frac{8.58}{6.58} = 1.30$, $n = 2$, R = 8.314 JK^{-1} mol^{-1}]

1.12.3. Comparison of isothermal and adiabatic expansion of an ideal gas

For a reversible adiabatic process, pressure and volume are related as

$$PV^{\gamma} = \text{constant}$$

While *for a reversible isothermal process,* the relationship is simply given by Boyle's law, *viz.*

$$PV = \text{constant}$$

Since C_p is always greater than C_v,

therefore the ratio $\frac{C_p}{C_v} = \gamma$ is greater than unity.

Hence the increase of volume for a given decrease of pressure will be less in an adiabatic than in an isothermal expansion. Therefore the plot of pressure against volume *i.e.,* pressure volume curve will be steeper for an adiabatic than for an isothermal process, starting at the same point, as shown is Fig. 1.9.

Further, since the area under the pressure-volume curve gives the work of expansion, from Fig. 1.9, it may be observed that for the same increase of volume, the work done in an adiabatic process is less than in an isothermal process.

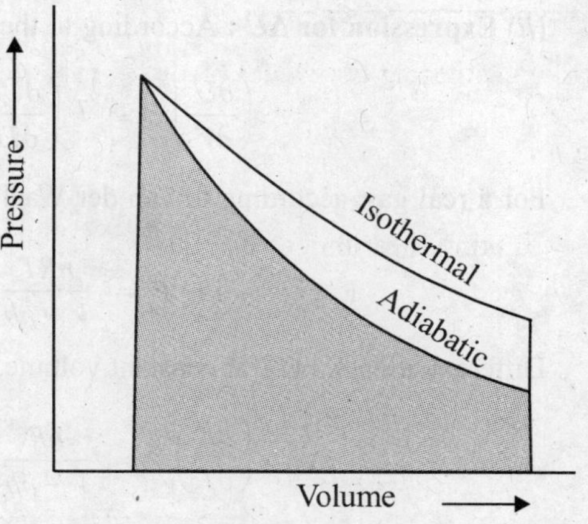

Fig .1.9 Isothermal and adiabatic curves

1.13. CALCULATION OF *w*, *q*, Δ*U* AND Δ*H* IN THE EXPANSION OF A REAL GAS IN ISOTHERMAL REVERSIBLE EXPANSION

(*i*) **Expression for *w*,**

$$dw = -P_{ext}\, dV$$

For the reversible expansion

$$P_{ext} = P_{int} - dP \simeq P_{int}$$

where P_{int} is the pressure (internal pressure) of the gas

Hence $$dw = -P_{int}\, dV \quad \text{... }(i)$$

As the gas is real, P_{int} can be obtained from van der Waal's equation *i.e.*

$$\left(P + \frac{an^2}{V^2}\right)(V - nb) = nRT$$

We get

$$P_{int} = \frac{nRT}{V - nb} - \frac{an^2}{V^2} \quad \text{... }(ii)$$

Putting this value in eqn (*i*), we get

$$-dw = \left(\frac{nRT}{V - nb} - \frac{an^2}{V^2}\right) dV$$

$$\therefore \quad -w = \int_{V_1}^{V_2} \left(\frac{nRT}{V - nb} - \frac{an^2}{V^2}\right) dV$$

$$= nRT\left[\ln(V - nb)\right]_{V_1}^{V_2} - an^2\left[-\frac{1}{V}\right]_{V_1}^{V_2}$$

or $$-w = nRT \ln\frac{V_2 - nb}{V_1 - nb} + an^2\left(\frac{1}{V_2} - \frac{1}{V_1}\right) \quad \text{... }(iii)$$

(*ii*) **Expression for ΔU :** According to thermodynamic equation of state

$$\left(\frac{\partial U}{\partial V}\right)_T = T\left(\frac{\partial U}{\partial T}\right)_V - P \qquad \text{... }(iv)$$

For a real gas, according to van der Waal's equation

$$P = \frac{nRT}{V-nb} - \frac{an^2}{V^2} \qquad \text{... }(v)$$

Differentiating w.r.t. T at constant volume, we get

$$\therefore \quad \left(\frac{\partial P}{\partial T}\right)_V = \frac{nR}{V-nb} \qquad \text{... }(vi)$$

Substituting the values from eqns. (*v*) and (*vi*) in eqn. (*iv*) we get

$$\left(\frac{\partial U}{\partial V}\right)_T = \frac{nRT}{V-nb} - \left(\frac{nRT}{V-nb} - \frac{an^2}{V^2}\right) = \frac{an^2}{V^2} \qquad \text{... }(vii)$$

For dU, we can write

$$dU = \left(\frac{\partial U}{\partial T}\right)_V dT + \left(\frac{\partial U}{\partial V}\right)_T dV \qquad \text{... }(viii)$$

For the isothermal process, $dT = 0$

$$\therefore \quad dU = \left(\frac{\partial U}{\partial V}\right)_T dV = \frac{an^2}{V^2} dV \quad \text{[Substituting value from eqn } (vii)]$$

Hence for a finite process,

$$\Delta U = \int_{V_1}^{V_2} \frac{an^2}{V^2} dV$$

or

$$\boldsymbol{\Delta U = -an^2\left(\frac{1}{V_2} - \frac{1}{V_1}\right)} \qquad \text{... }(ix)$$

(*iii*) **Expression for q :** According to first law of thermodynamics

$$q = \Delta U - w$$

Substituting the values of w and ΔE from eqns. (*iii*) and (*ix*), we get

$$q = -an^2\left(\frac{1}{V_2} - \frac{1}{V_1}\right) + nRT \ \ln \frac{V_2 - nb}{V_1 - nb} + an^2\left(\frac{1}{V_2} - \frac{1}{V_1}\right)$$

or

$$\boldsymbol{q = nRT \ln \frac{V_2 - nb}{V_1 - nb}} \qquad \text{... }(x)$$

(*iv*) **Expression for ΔH :** We have

$$\Delta H = \Delta(U + PV)$$
$$= \Delta U + \Delta(PV)$$

Substituting the value of ΔU and P from eqns (*ix*) and (*ii*), we get

$$\Delta H = -an^2\left(\frac{1}{V_2}-\frac{1}{V_1}\right)+\Delta\left[\left(\frac{nRT}{V-nb}-\frac{an^2}{V^2}\right)V\right]$$

$$= -an^2\left(\frac{1}{V_2}-\frac{1}{V_1}\right)+nRT\left(\frac{V_2}{V_2-nb}-\frac{V_1}{V_1-nb}\right)-an^2\left(\frac{1}{V_2}-\frac{1}{V_1}\right)$$

$$= -2an^2\left(\frac{1}{V_2}-\frac{1}{V_1}\right)+nRT\left[\frac{V_2-nb+nb}{V_2-nb}-\frac{V_1-nb+nb}{V_1-nb}\right]$$

$$= -2an^2\left(\frac{1}{V_2}-\frac{1}{V_1}\right)+nRT\left[\left(1+\frac{nb}{V_2-nb}\right)-\left(1+\frac{nb}{V_1-nb}\right)\right]$$

$$\mathbf{\Delta H = -2an^2\left(\frac{1}{V_2}-\frac{1}{V_1}\right)+n^2bRT\left(\frac{1}{V_2-nb}-\frac{1}{V_1-nb}\right)} \qquad ...(xi)$$

1.13.1. *q, w,* ΔE and ΔH in adiabatic reversible expansion of a real gas

(*i*) Value for *q*: For the adiabatic process, as already explained,

$$q = 0$$

(*ii*) Expressions for *w* and ΔU: By first law of thermodynamics,

$$\Delta U = q + w$$

As $q = 0$, therefore $\Delta U = w$

The expression for ΔU and hence for w can be obtained as follows :

$$dU = \left(\frac{\partial U}{\partial T}\right)_v dT + \left(\frac{\partial U}{\partial V}\right)_T dV$$

Putting $(\partial U/\partial T)_v = nC_v$ (for n moles) and $\left(\frac{\partial U}{\partial V}\right)_T = \frac{an^2}{V^2}$, we get

$$dU = nC_v dT + \frac{an^2}{V^2}dV \qquad ...(i)$$

Hence for a finite process,

$$\Delta U = \int_{T_1}^{T_2} nC_v dT + \int_{V_1}^{V_2}\frac{an^2}{V^2}dV$$

Taking C_v to be independent of temperature, we have

$$\mathbf{\Delta U = nC_v(T_2-T_1) - an^2\left(\frac{1}{V_2}-\frac{1}{V_1}\right)} \qquad ...(ii)$$

This expression can be used for calculation of ΔU if T_1, T_2, V_1 and V_2 are known. However, if T_2 is not given, it is first calculated as follows :

$$dU = dw = -PdV$$

or

$$dU = -PdV \qquad ...(iii)$$

For a real gas $$P = \frac{nRT}{V-nb} - \frac{an^2}{V^2}$$

and $$dU = nC_v dT + \frac{an^2}{V^2} dV$$ Refer Eqn. (*i*) above

Substituting these values in eqn. (*iii*), we get

$$nC_v dT + \frac{an^2}{V^2} dV = -\frac{nRT}{V-nb} dV + \frac{an^2}{V^2} dV$$

$$nC_v dT = -\frac{nRT}{V-nb} dV$$

$$C_v dT = -\frac{RT}{V-nb} dV$$

Separating the variables, it can be rewritten as

$$C_v \frac{dT}{T} = -R \frac{dV}{V-nb}$$

Integrating between suitable limits, we get

$$\int_{T_1}^{T_2} C_v \frac{dT}{T} = -R \int_{V_1}^{V_2} \frac{dV}{V-nb}$$

or $$C_v [\ln T]_{T_1}^{T_2} = -R \left[\ln(V-nb)\right]_{V_1}^{V_2}$$

or $$C_v \ln \frac{T_2}{T_1} = -R \ln \frac{V_2 - nb}{V_1 - nb}$$

or $$\ln\left(\frac{T_2}{T_1}\right)^{C_v} = \ln\left(\frac{V_1 - nb}{V_2 - nb}\right)^R$$

or $$\left(\frac{T_2}{T_1}\right)^{C_v} = \left(\frac{V_1 - nb}{V_2 - nb}\right)^R$$

or $$\frac{T_2}{T_1} = \left(\frac{V_1 - nb}{V_2 - nb}\right)^{R/C_v}$$

or $$\mathbf{T_2 = T_1 \left(\frac{V_1 - nb}{V_2 - nb}\right)^{R/C_v}} \quad \text{... (iv)}$$

Knowing V_1, V_2 and T_1, T_2 (temperature after expansion) can be calculated. This may be substituted in eqn. (*ii*) to get ΔU and hence *w*.

(*iii*) **Expression for ΔH**

$$\Delta H = \Delta(U + PV) = \Delta U + \Delta(PV)$$

$$= \Delta U + \Delta\left[\left(\frac{nRT}{V - nb} - \frac{an^2}{V^2}\right)V\right]$$

$$= \Delta U + nR\left(\frac{V_2T_2}{V_2 - nb} - \frac{V_1T_1}{V_1 - nb}\right) - an^2\left(\frac{1}{V_2} - \frac{1}{V_1}\right)$$

Substituting the value of ΔU from eqn. (*ii*), we get

$$\Delta H = nC_v(T_2 - T_1) - 2an^2\left(\frac{1}{V_2} - \frac{1}{V_1}\right) + nR\left(\frac{V_2T_2}{V_2 - nb} - \frac{V_1T_1}{V_1 - nb}\right) \quad \text{... (v)}$$

1.14. THERMOCHEMISTRY

The branch of chemistry which deals with heat changes accompanying chemical reactions is called **thermochemistry.**

A chemical reaction involves rearrangement of atoms, *i.e.*, some bonds within the reactant molecules are broken while some new bonds in molecules of the products are formed. Energy is consumed in the dissociation of bonds while energy is evolved when the bonds are formed. If the total energy evolved is greater than the total energy consumed, the net result is the *release of energy.* On the other hand, if the energy consumed is greater than the energy evolved, the net result is the *absorption of energy.*

All chemical reactions are usually accompanied by energy changes. These energy changes appear in the form of heat, light, work, electricity, etc. These energy changes are of great practical utility both for domestic and industrial purposes. A few reactions alongwith the form in which the accompanying energy changes appear and the use to which they are put are listed below :

(*i*) Burning of coal in air produces *heat.* That is why it is used as a fuel.

$$C\,(s) + O_2\,(g) \longrightarrow CO_2\,(g) + \text{heat}$$

(*ii*) *Digestion and metabolism of carbohydrates* in biological systems produce *heat* which maintains the body temperature.

(*iii*) *Burning of candle* (*wax*) *in air* produces heat and light. It is, therefore, used as a source of light. Wax is a hydrocarbon.

(*iv*) *Reaction that takes place in a galvanic cell* produces electrical energy which may be used to run an electrical motor or to ring a bell, etc.

$$Zn\,(s) + Cu^{2+}\,(aq) \longrightarrow Zn^{2+}\,(aq) + Cu\,(s)$$

Units of Heat and Work. The heat changes are measured in calories, kilocalories, joules or kilojoules.

1 calorie = 4.184 J　　　　1 kcal = 4.184 kJ

The S.I. unit of heat is joule or kilojoule.

Work is measured in ergs or joules. The S.I. unit of work is joule.　　1 Joule = 10^7 ergs

1.14.1. Enthalpy and enthalpy change

When a process is carried out at constant pressure and temperature, there is a change in volume.

If the volume increases, some mechanical work is done by the system so that the internal energy decreases. Similarly, if the volume decreases, some mechanical work is done on the system so that the internal energy increases. In all such processes occurring at constant temperature when pressure remains constant, the total energy of a system is given by a thermodynamic quantity called **heat content** or **enthalpy** of the system and not by internal energy. It is denoted by H and is defined mathematically by the equation

$$H = E + PV$$

where E is the internal energy, P and V are the pressure and volume of the system respectively.

The **enthalpy** *of a system may be defined as the sum of the internal energy and the product of pressure and volume of the system.*

The *physical significance of enthalpy* may be put as under :

The **enthalpy** *of a substance is the amount of energy stored within the substance, i.e., available for conversion into heat.*

Like internal energy, *the heat content or enthalpy of a substance is also a state function, i.e.,* depends only on the state of the system and is independent of the manner by which the state has been attained.

Enthalpy change. Like internal energy, the absolute value of enthalpy of a substance cannot be determined. However, the change in the enthalpy when a system undergoes a change can be determined experimentally accurately and that serves our purpose.

A relation between enthalpy change and internal energy change exists as follows :

$$\Delta H = \Delta E + P\,\Delta V \qquad \dots (ii)$$

where $\Delta E = E_2 - E_1$ represents the change in internal energy

and $\Delta V = V_2 - V_1$ represents the change in volume of the system.

Thus, **enthalpy change** *of a system may be defined as the sum of internal energy change and the pressure-volume work done.*

If q_p is the heat absorbed at constant pressure, then $q_p = \Delta H$ ... (*iii*)

Thus, enthalpy change *is the heat absorbed or evolved at constant pressure.*

Standard enthalpy change. The enthalpy change of a reaction when all the reactants and the products are in their standard states, *i.e.*, at 298 K and under a pressure of one atmosphere is known as the standard enthalpy change. It is usually represented by $\Delta H°$.

As a chemical reaction involves breaking of bonds of the reactants and the formation of new bonds in the products, some net energy change (called the enthalpy of reaction) takes place.

$$\text{Enthalpy change of a reaction} = \begin{bmatrix}\text{Energy consumed to} \\ \text{break the bonds of} \\ \text{the reactants}\end{bmatrix} - \begin{bmatrix}\text{Energy released in the} \\ \text{formation of the bonds in} \\ \text{the products}\end{bmatrix}$$

If energy consumed > energy released, the net result is the absorption of energy and the reaction is *endothermic*.

If energy released > energy consumed, the net result is the evolution of energy and the reaction is *exothermic*.

LATEST SIGN CONVENTIONS FOR *q* OR ΔH, ΔU AND *w*.

(*i*) When heat is absorbed by the system, *q* is positive.
(*ii*) When heat is given out by the system, *q* is negative.
(*iii*) When energy is absorbed by the system, ΔU is positive.
(*iv*) When energy is given out by the system, ΔU is negative.
(*v*) When work is done on the system, *w* is positive.
(*vi*) Work done by the system, *w* is negative.

1.14.2. Exothermic and endothermic reactions

Exothermic Reactions *are those reactions which are accompanied by the evolution of heat.*

The quantity of heat produced is shown alongwith the products with a 'plus' sign. A few examples of exothermic reactions are given below :

$$C(s) + O_2(g) \longrightarrow CO_2(g) + 393.5 \text{ kJ}$$

$$H_2(g) + \frac{1}{2}O_2(g) \longrightarrow H_2O(g) + 248.8 \text{ kJ}$$

$$CH_4(g) + 2O_2(g) \longrightarrow CO_2(g) + 2H_2O(g) + 890.4 \text{ kJ}$$

Endothermic Reactions *are those reactions which are accompanied by absorption of heat.*

Heat absorbed can be written with the products with a 'minus' sign. A few examples of endothermic reactions are given below :

$$N_2(g) + O_2(g) \longrightarrow 2NO(g) - 180.7 \text{ kJ}$$

$$C(s) + H_2O(g) \longrightarrow CO(g) + H_2(g) - 131.4 \text{ kJ}$$

$$C(s) + 2S(s) \longrightarrow CS_2(l) - 92.0 \text{ kJ}$$

The enthalpy change (ΔH) accompanying a reaction is given by

$$\Delta H = \text{Heat content of products} - \text{Heat content of reactants} = H_P - H_R$$

A reaction is exothermic, if the total heat content of the reactants is more than that of the products, i.e., $H_R > H_P$. Thus, ΔH will be negative for an exothermic process.

Thus, the exothermic reactions given above may be written in terms of ΔH as :

$$C(s) + O_2(g) \longrightarrow CO_2(g), \quad \Delta H = -393.5 \text{ kJ}$$

$$H_2(g) + \frac{1}{2}O_2(g)\, O(l) \longrightarrow H_2O(l), \quad \Delta H = -285.8 \text{ kJ}$$

$$CH_4(g) + 2O_2(g) \longrightarrow CO_2(g) + 2H_2O(l), \quad \Delta H = -890.4 \text{ kJ}$$

A reaction is endothermic if the total heat content of reactants is less than that of the products, i.e.,

$$H_R < H_P$$

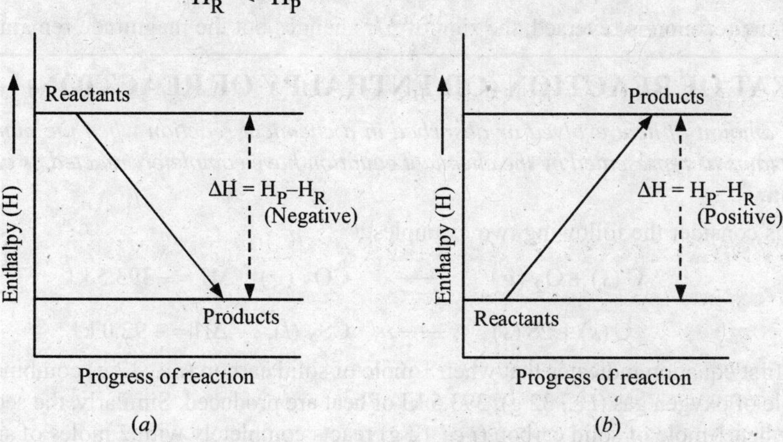

Fig. 1.10 (*a*) Exothermic reaction (*b*) Endothermic reaction.

Thus, ΔH will be positive for an endothermic reaction.

Endothermic reactions given above may be written in terms of ΔH as :

$$N_2(g) + O_2(g) \longrightarrow 2NO(g), \quad \Delta H = +180.7 \text{ kJ}$$

$$C(s) + H_2O(g) \longrightarrow CO(g) + H_2(g), \quad \Delta H = +131.4 \text{ kJ}$$

$$C(s) + 2S(s) \longrightarrow CS_2(l), \quad \Delta H = +92.0 \text{ kJ}$$

Exothermic and endothermic reactions may be represented graphically as shown in Fig. 1.10(*a*) and 1.10(*b*).

1.14.3. Thermochemical equation

A balanced chemical equation not only indicating the quantities of the different reactants and products but also indicating the amount of heat evolved or absorbed is called a **thermochemical equation.**

Fractional coefficients are allowed in writing a thermochemical equation. For example, the formation of water is written as

$$H_2(g) + \frac{1}{2}O_2(g) \longrightarrow H_2O(l), \quad \Delta H = -285.8 \text{ kJ}$$

Thus, 285.8 kJ of heat is produced when 1 mole of hydrogen reacts with 0.5 mole of oxygen. If the quantities of reactants are doubled, the heat produced will also be doubled. For example, in the above case, we may write

$$2H_2(g) + O_2(g) \longrightarrow 2\ H_2O(l), \quad \Delta H = -571.6 \text{ kJ}$$

SOME POINTS TO REMEMBER ABOUT THERMOCHEMICAL EQUATIONS

(*i*) Unless otherwise mentioned, ΔH values are for the standard state of the substances (*i.e.*, 298 K and 1 atm pressure).

(*ii*) The physical state (*s, l, g,* etc.) of the different substances must be mentioned as the heat evolved or absorbed depends upon the physical state of the substances.

(*iii*) If the coefficients of the substances are multiplied or divided by some number, the value of ΔH is also multiplied or divided by the same number.

(*iv*) If the reaction is reversed, the sign of ΔH changes but the magnitude remains the same.

1.15. HEAT OF REACTION (OR ENTHALPY OF REACTION)

The amount of heat evolved or absorbed in a chemical reaction when the number of moles of the reactants as represented by the chemical equation have completely reacted, is called the **heat of reaction.**

Let us consider the following two examples :

$$C(s) + O_2(g) \longrightarrow CO_2(g) \quad \Delta H = -393.5 \text{ kJ}$$

$$C(s) + 2S(s) \longrightarrow CS_2(l), \quad \Delta H = +92.0 \text{ kJ}$$

The first equation indicates that when 1 mole of solid carbon (*i.e.*, 12 g) combines completely with 1 mole of oxygen gas (*i.e.*, 32 g), 393.5 kJ of heat are produced. Similarly, the second equation tells that when 1 mole of solid carbon (*i.e.*, 12 g) reacts completely with 2 moles of sulphur (64 g), 92 kJ of heat are absorbed.

Heat of reaction, ΔH = Heat content of products – Heat content of reactants

If the reaction is reversed, the sign of ΔH is also reversed.

1.15.1. Factors Influencing Heat of Reaction

The heat of a reaction depends upon a number of factors as explained below :

(i) Quantities of the reactants involved. If the quantities of the reactants are changed, the heat of reaction changes proportionally.

(ii) Physical state of the reactants and products. Since latent heat is involved in the change of state, therefore, the physical state of the reactants and products affects the heat of reaction. For example, when hydrogen and oxygen gases combine to form liquid water, the heat of reaction is different from the value when they combine to form water vapours.

$$H_2\,(g) + \frac{1}{2}O_2\,(g) \longrightarrow H_2O\,(l), \quad \Delta H = -\,285.8 \text{ kJ}$$

$$H_2\,(g) + \frac{1}{2}O_2\,(g) \longrightarrow H_2O\,(g), \quad \Delta H = -\,248.8 \text{ kJ}$$

It is, therefore, essential to write the symbols *s, l, g* or *aq* to indicate whether a particular substance is solid, liquid, gaseous or an aqueous solution.

(iii) Allotropic modifications. Heat of reaction changes if a different allotropic form is involved in the reaction. For example,

$$C\,(\text{diamond}) + O_2\,(g) \longrightarrow CO_2\,(g), \quad \Delta H = -\,395.4 \text{ kJ}$$

$$C\,(\text{graphite}) + O_2\,(g) \longrightarrow CO_2\,(g), \quad \Delta H = -\,393.5 \text{ kJ}$$

(iv) Concentration of solutions. If the solutions are involved in a reaction, their concentrations affect the heat of reaction.

(v) Temperature. The heat of reaction depends upon the temperature at which the reactants and products are taken. Generally, the heat of reaction increases with temperature.

(vi) Conditions of constant pressure or constant volume, i.e., whether the reaction takes place at constant pressure or at constant volume.

If q_p and q_v represent the heats of reaction at constant pressure and volume respectively, then the following equation is applicable.

$$q_p = q_v + \Delta n_g\, RT$$

where
$$\Delta n_g = n_p - n_r$$

n_p and n_r represent the number of moles of gaseous reactants and products respectively. R and T have the usual meanings.

1.15.2. Heat (or enthalpy) of combustion

The heat of combustion of a substance is defined as the heat change (usually the heat evolved) when 1 mole of substance is completely burnt or oxidised in oxygen. For example,

$$C\,(s) + O_2\,(g) \longrightarrow CO_2\,(g), \Delta H = -\,393.5 \text{ kJ}$$

When one mole of carbon (12 g) burns completely in oxygen, 393.5 kJ of heat are evolved. Thus, heat of combustion of carbon is 393.5 kJ.

The significance of the words *burns completely* must be clearly understood. For example, carbon may be oxidised to carbon monoxide (CO) or carbon dioxide (CO_2). The heats evolved in the two cases are different, *viz.*,

$$C\,(s) + \frac{1}{2}O_2\,(g) \longrightarrow CO\,(g), \quad \Delta H = -110.5 \text{ kJ}$$

$$C\,(s) + O_2\,(g) \longrightarrow CO_2\,(g), \quad \Delta H = -393.5 \text{ kJ}$$

Complete oxidation means oxidation to CO_2 and not to CO. Hence, the heat of combustion of carbon is 393.5 kJ mol^{-1}.

1.15.3. Calorific value of foods and fuels

For the working of the human machine, we eat carbohydrates, fats, etc., in the form of food. The carbohydrates are first decomposed in our body by the enzymes to form glucose which is then oxidised by the oxygen that we inhale to produce energy as given by the following equation.

$$C_6H_{12}O_6\,(s) + 6O_2\,(g) \longrightarrow 6CO_2\,(g) + 6H_2O\,(g) + 2900 \text{ kJ}$$

This oxidation reaction is usually called '*combustion of food*'.

Different foods have different heats of combustion. These are usually expressed in terms of their calorific values.

The calorific value of a fuel or food is the amount of heat in calories or joules produced from the complete combustion of one gram of the fuel or the food.

Thus, according to the above reaction, 1 mole of glucose, *i.e.*, 180 g of glucose produce 2900 kJ of heat.

$$\text{Hence, calorific value of glucose} = \frac{2900}{180} = 16.1 \text{ kJ g}^{-1}$$

Calorific values of some foods and fuels are given in Tables 1.1 and 12.

Table 1.1: Calorific Values of Foods per Gram

Food	Calorific value kJ / gram	Food	Calorific value kJ / gram
Ghee	37.65	Fish	5.2
Butter	30.50	Meat	12.0
Cheese	14.56	Egg (yolk)	5.4
Milk (buffalo's)	4.90	Egg (white)	14.0
Milk (cow's)	2.20	Sugar	16.0
Curd	2.51	Rice	16.7
Soyabean	18.07	Potatoes	4.0

Table 1.2 : Calorific Values of some Common Fuels in kJ per kg

Wood	16,000
Charcoal	29,000
Natural gas	46,000
Kerosene	38,000

A normal youngman needs 10,000 kJ of energy per day for normal functioning.

SOLVED EXAMPLE ON CALORIFIC VALUE

Example . A cylinder of indane gas contains 11.2 kg of butane. A normal family needs 25000 kJ energy per day for cooking. How long will the cylinder of gas last assumping 100% combustion?

If the efficiency of combustion is 85%, how long will the cylinder last? Heat of combustion of butane is 2658 kJ mol^{-1}.

Solution. Mol. mass of butane is 58 g mol^{-1}.

Thus, 58 g butane on complete combustion produces = 2658 kJ

$\therefore$ 11200 g butane on complete combustion produces $= \dfrac{2658 \times 11200}{58}$ kJ = 513268.96 kJ.

The cylinder will last $= \dfrac{513268.96}{25000} = 20.5$ days

If the efficiency of combustion is only 85%, the cylinder will last = 0.85 × 20.5 ≈ 17.5 days.

1.16. STANDARD STATE

The standard molar enthalpy of every element at 1 atm pressure and 298.15 K in the most stable state is taken as zero. These conditions of 1 atm pressure and 298.15 K are called standard state conditions. It is not possible to determine the absolute value of molar enthalpy, but we can use the above convention to find out the standard molar enthalpy of a substance. The most stable form of hydrogen is gaseous H_2, that of bromine is liquid Br_2 while that of iodine is solid I_2. Their molar standard enthalpies are taken as zero at 1 atmospheric pressure and 298.15 K temperature. In case of elements exhibiting allotrophy, we consider the most stable form under the conditions of standard state.

1.16.1. Enthalpy of formation and standard enthalpy of formation

Heat of Formation (or Enthalpy of Formation) : *The enthalpy of formation of a substance is defined as the heat change that takes place when 1 mole of the substance is formed from its elements under given conditions of temperature and pressure. It is usually represented by* ΔH_f.

Standard enthalpy of formation *of a substance is defined as the heat change accompanying the formation of 1 mole of the substance in the standard state from its elements, also taken in the standard state (i.e., 298 K, and 1 atmospheric pressure). It is usually represented by* $\Delta H^o{}_f$.

For example, in the reaction

$$C\ (s) + O_2\ (g) \longrightarrow CO_2\ (g), \quad \Delta H^o{}_f = -393.5 \text{ kJ}$$

When 1 mole of CO_2 (*g*) is formed from its elements, viz., C (*s*) and O_2 (*g*) (all substances being taken in the standard state), 393.5 kJ of heat is produced. Hence, the standard heat of formation of gaseous CO_2 is 393.5 kJ mol^{-1}.

Importance of standard enthalpies of formation. Knowing the standard enthalpies of formation of the different compounds involved in a chemical reaction, the standard enthalpy change of the given reaction can be obtained using the formula.

$$\Delta H^o{}_{\text{reaction}} = \begin{bmatrix}\text{Sum of the standard heats}\\ \text{of formation of products}\end{bmatrix} - \begin{bmatrix}\text{Sum of the standard heats}\\ \text{of formation of reactants}\end{bmatrix}$$

i.e.,

$$\Delta H^o{}_{\text{reaction}} = \Sigma\, \Delta H^o{}_f \text{ (Products)} - \Sigma\, \Delta H^o{}_f \text{ (Reactants)}$$

Thus, for a general reaction

$$aA + bB \longrightarrow cC + dD$$

$$\Delta H^o_{reaction} = [c\,\Delta H^o{}_f(C) + d\,\Delta H^o{}_f(D)] - [a\,\Delta H^o{}_f(A) + b\,\Delta H^o{}_f(B)]$$

Remember $\Delta H^o{}_f$ is taken as zero for the elements in a reaction. Suppose we are interested in finding the heat of combustion of methane. The reaction is represented as :

$$CH_4\,(g) + 2O_2(g) \longrightarrow CO_2\,(g) + 2H_2O\,(l)$$

$$\Delta H^o_{reaction} = [\Delta H^o{}_f \text{ for } CO_2 + (2\times\Delta H^o{}_f \text{ for } H_2O)] - [\Delta H^o{}_f \text{ for } CH_4 + (2\times\Delta H^o{}_f \text{ for } O_2)]$$

Putting the values of heats of formation in the standard state

$$\Delta H^o_{reaction} = [-393.5 + 2\times(-285.8)] - [-74.9 + 0] = -965.1 + 74.9 = -890.2 \text{ kJ}$$

SOLVED EXAMPLES ON ENTHALPIES OF FORMATION

Example 1. Enthalpies of formation of CO_2 (*g*) and H_2O (*l*) under standard conditions are 394.65 kJ and 285.84 kJ per mole. If the standard enthalpy of combustion of acetaldehyde (CH_3CHO) be 1167.62 kJ per mole, find its enthalpy of formation.

Solution. From the given data

(*a*) $C + O_2 \longrightarrow CO_2\,(g);\quad \Delta H = -394.65$ kJ

(*b*) $H_2 + \frac{1}{2}O_2 \longrightarrow H_2O\,(l);\quad \Delta H = -285.84$ kJ

(*c*) $CH_3CHO + \frac{5}{2}O_2 \longrightarrow 2CO_2 + 2H_2O\,(g);\quad \Delta H = -1167.62$ kJ

Our aim is to find out the ΔH value for the reaction

$$2C + 2H_2 + \frac{1}{2}O_2 \longrightarrow CH_3CHO$$

Using standard enthalpy values

$\Delta H^o{}_f$ for $CO_2 = -394.65$ kJ, $\quad \Delta H^o{}_f$ for $H_2O = -285.84$ kJ and $\Delta H^o{}_f$ for $O_2 = 0$

($\because$ Enthalpy of formation of elements is taken as zero).

Let $\Delta H^o{}_f$ for CH_3CHO be x kJ.

ΔH value for any reaction = Enthalpies of the products – Enthalpies of the reactants.

From the equation

$$CH_3CHO + \frac{5}{2}O_2 \longrightarrow 2CO_2\,(g) + 2H_2O\ ;\quad \Delta H = -1167.62 \text{ kJ}$$

$$\therefore\ -1167.62 = [2\times\Delta H^o{}_f\,(CO_2) + 2\times\Delta H^o{}_f(H_2O)] - [\Delta H^o{}_f(CH_3CHO) + \frac{5}{2}\times\Delta H^o{}_f(O_2)]$$

$$= 2\times(-394.65) + 2\times(-285.84) - x + 0 = -789.30 - 571.68 - x$$

$$-1167.62 = -1360.98 - x \quad \text{or} \quad x = -193.36 \text{ kJ}$$

Hence, enthalpy of formation of CH_3CHO is **–193.36 kJ.**

Example 2. Calculate ΔH^o for the reaction : $CO_2(g) + H_2(g) \longrightarrow CO(g) + H_2O(g)$ given that ΔH^o_f for $CO_2(g)$, $CO(g)$ and $H_2O(g)$ are – 393.5, – 111.3 and – 241.8 kJ mol^{-1} respectively.

Solution. ΔH^o for the reaction $CO_2(g) + H_2(g) \longrightarrow CO(g) + H_2O(g)$ is

$$\Delta H^o = [H^o{}_{f\,(\text{products})} - H^o{}_{f\,(\text{reactants})}]$$

$$= [(-111.3) + (-241.8)] - [(-393.5) + 0] = (-111.3 - 241.8) - (-393.5)$$

$$= -353.1 + 393.5$$

ΔH^o = **+ 40.4 kJ.** [Enthalpy of every element in standard state is assumed as zero].

Example 3. Enthalpies of formation of C_2H_5OH (*l*), CO_2 (*g*) and H_2O (*l*) are –277.0, – 393.5 and – 285.8 kJ mole^{-1} respectively. Calculate enthalpy change for reaction.

$$C_2H_5OH\,(l) + 3O_2\,(g) \longrightarrow 2CO_2(g) + 3H_2O\,(l)$$

Solution. ΔH^o for the reaction $C_2H_5OH\,(l) + 3O_2\,(g) \longrightarrow 2CO_2\,(g) + 3H_2O\,(l)$ is

$$\Delta H^o = \Delta H^o_{f(\text{products})} - \Delta H^o_{f(\text{reactants})}$$

$$= [2 \times \Delta H^o{}_f(CO_2) + 3 \times \Delta H^o{}_f(H_2O)] - [\Delta H^o{}_f(C_2H_5OH) + 3 \times \Delta H^o{}_f(O_2)]$$

$$= [2 \times (-393.5) + 3 \times (-285.8)] - [-277 + 0] = -787.0 - 857.4 + 277$$

$$= -1644.4 + 277 = \mathbf{-1367.4\ kJ.}$$

1.16.2. Heat (or enthalpy) of neutralization

The heat of neutralization of an acid by a base is defined as the heat change when one gram equivalent of the acid is neutralized by a base, the reaction being carried out in dilute aqueous solution.

The heat of neutralization of a base by an acid is defined in a similar manner.

For example, when one gram equivalent of HCl is neutralized by NaOH or one gram equivalent of NaOH is neutralized by HCl, in dilute and aqueous solution, 57.1 kJ of heat is produced.

$$NaOH + HCl \longrightarrow NaCl + H_2O, \quad \Delta H = -57.1\ kJ$$

Hence, the heat of neutralization of HCl with NaOH or NaOH with HCl is 57.1 kJ.

The heat of neutralization of any strong acid (HCl, HNO_3, H_2SO_4) with a strong base (NaOH, KOH) or vice versa, is always the same, *i.e.*, 57.1 kJ. This is because the strong acids, strong bases and the salts that they form are all completely ionized in dilute aqueous solution. Thus, the reaction between a strong acid and a strong base may be written as

$$Na^+ + OH^- + H^+ + Cl^- \longrightarrow Na^+ + Cl^- + H_2O, \quad \Delta H = -57.1\ kJ$$

Cancelling out common ions on the two sides, we have

$$H^+\,(aq) + OH^-\,(aq) \longrightarrow H_2O\,(l), \quad \Delta H = -57.1\ kJ$$

Thus, neutralization may be regarded as a reaction between the H^+ ions given by the acid with the OH^- ions given by the base. Since strong acids and strong bases ionize completely in dilute aqueous solution, the number of H^+ ions and OH^- ions produced by one gram equivalent of the strong acid and the strong base is always the same. Hence, the heat of neutralization between a strong acid and a strong base is always the same. *i.e.*, –57.1 kJ.

If the acid or the base is weak, the heat of neutralization is usually less than 57.1 kJ. Consider the neutralization of a weak acid like acetic acid with a strong base like NaOH. Acetic acid ionizes to a small extent whereas NaOH ionizes completely as

(*i*) $CH_3COOH \rightleftharpoons CH_3COO^- + H^+$ (*ii*) $NaOH \longrightarrow Na^+ + OH^-$

When H^+ ions given by the acid combine with the OH^- ions given by the base, the equilibrium (*i*) shifts to the right, *i.e.*, more of acetic acid dissociates. A part of the heat produced during the combination of H^+ ions and OH^- ions is used up for the complete dissociation of acetic acid. Hence, the net heat evolved in the above reaction is less than 57.1 kJ.

Compared to a strong acid, a weak acid produces a smaller number of H^+, hence, heat of neutralization is smaller.

Example 1. Determine the amount of heat released when :

(*a*) 0.75 mole of nitric acid is neutralised by 0.75 mole of sodium hydroxide.

(*b*) 0.75 mole of hydrochloric acid is neutralised by 0.5 mole of potassium hydroxide.

(*c*) 900 ml of 0.5 M hydrochloric acid solution is mixed with 500 ml of 0.4 M sodium hydroxide solution. Assuming the specific heat of water as 4.18 J K^{-1} g^{-1}, also calculate the rise in temperature.

(*d*) 200 ml of 0.2 M sulphuric acid is mixed with 300 ml of 0.1 M sodium hydroxide solution. Calculate the rise in temperature.

Solution. (*a*) The reaction is :

$$(0.75 \text{ moles}) \ HNO_3 + (0.75 \text{ moles}) \ NaOH \longrightarrow NaNO_3 + H_2O.$$

When one mole of HNO_3 is neutralised by 1 mole of NaOH, heat released is 57.1 kJ. Therefore, when 0.75 mole of HNO_3 is neutralised by 0.75 mole of NaOH, heat released will be

$$57.1 \times 0.75 = \mathbf{42.8 \ kJ.}$$

(*b*) Out of 0.75 mole of HCl, 0.5 mole will react with 0.5 mole of KOH and 0.25 mole of HCl will be left unreacted.

$$\text{Heat released} = 57.1 \times 0.5 = \mathbf{28.5 \ kJ.}$$

(*c*) $\quad 1000 \text{ ml of } 1 \text{ M HCl} = 1 \text{ mole}$

$$\therefore \quad 900 \text{ ml of } 0.5 \text{ M HCl} = \frac{900 \times 0.5}{1000} = 0.45 \text{ mole}$$

Similarly,

$$500 \text{ ml of } 0.4 \text{ M NaOH} = \frac{500 \times 0.4}{1000} = 0.2 \text{ mole}$$

Out of 0.45 mole of HCl, 0.2 mole of HCl reacts with 0.2 mole of NaOH

$$\therefore \quad \text{Heat evolved} = 57.1 \times 0.2 = 11.4 \text{ kJ.}$$

To calculate the rise in temperature, use the relation

Mass × Sp. heat × Rise in temp.= Heat evolved

Heat evolved is 11.4 kJ or 11400 J, mass of the mixture is 900 g + 500 g (assuming the densities of the solution as unity) and sp. heat is 4.18 J.

$\therefore$ Rise in temp. $= \dfrac{\text{Heat evolved}}{\text{Mass} \times \text{Sp. heat}} = \dfrac{11.4 \times 1000}{1400 \times 4.18} = 1.95°$

(*d*) 1 molecule of H_2SO_4 releases two H^+ ions

$$200 \text{ ml. of } 0.2\text{M } H_2SO_4 = \frac{2 \times 200 \times 0.2}{1000} \text{ moles of } H^+ = 0.08 \text{ moles of } H^+$$

$$300 \text{ ml of } 0.1 \text{ M NaOH} = 0.03 \text{ mole } OH^-$$

Out of 0.08 mole of H^+, 0.03 mole H^+ will react with 0.03 mole OH^-.

$\therefore$ Heat evolved $= 57.1 \times 0.03 = 1.7$ kJ.

$\therefore$ Rise in temperature $= \dfrac{1.7 \times 1000}{500 \times 4.18} = 0.8°$.

Example 2. Calculate the amount of heat evolved when :

(*i*) 600 cm^3 of 0.1 M hydrochloric acid is mixed with 200 cm^3 of 0.2 M sodium hydroxide solution.

(*ii*) 250 cm^3 of 0.2 M sulphuric acid is mixed with 400 cm^3 of 0.5 M potassium hydroxide solution.

Assuming that the specific heat of water is 4.18 J K^{-1} g^{-1} and ignoring the heat absorbed by the container, thermometer, stirrer, etc., what would be the rise in temperature in each of the above cases?

Solution. (*i*)

$$600 \text{ cm}^3 \text{ of } 0.1\text{M HCl} = \frac{0.1}{1000} \times 600 \text{ mole of HCl}$$

$$= 0.06 \text{ mole of HCl} = 0.06 \text{ mole of } H^+ \text{ ions}$$

$$200 \text{ cm}^3 \text{ of } 0.2\text{M NaOH} = \frac{0.2}{1000} \times 200 \text{ mole of NaOH}$$

$$= 0.04 \text{ mole of NaOH} = 0.04 \text{ mole of } OH^- \text{ ions}$$

Thus, 0.04 mole of H^+ ions will combine with 0.04 mole of OH^- ions to form 0.04 mole of H_2O and 0.02 mole of H^+ ions will remain unreacted.

Heat evolved when 1 mole of H^+ ions combines with 1 mole of OH^- ions = 57.1 kJ.

$\therefore$ Heat evolved when 0.04 mole of H^+ ions combines with 0.04 mole of OH^- ions.

$$= 57.1 \times 0.04 = \mathbf{2.284 \text{ kJ.}}$$

(*ii*) $$250 \text{ cm}^3 \text{ of } 0.2\text{M } H_2SO_4 = \frac{0.2}{1000} \times 250 \text{ mole of } H_2SO_4$$

$$= 0.05 \text{ mole of } H_2SO_4 = 0.10 \text{ mole of } H^+ \text{ ions}$$

$$400 \text{ cm}^3 \text{ of } 0.5\text{M KOH} = \frac{0.5}{1000} \times 400 \text{ mole of KOH}$$

$$= 0.2 \text{ mole of KOH} = 0.2 \text{ mole of } OH^- \text{ ions}$$

Thus, 0.10 mole of H^+ ions will neutralize 0.10 mole of OH^- ions, (out of 0.2 mole of OH^- ions) to form 0.10 mole of H_2O.

Hence, heat evolved $= 57.1 \times 0.1 =$ **5.71 kJ.**

In case (*i*), heat produced $= 2.284$ kJ $= 2284$ J

Total mass of the solution $= 600 + 200 = 800$ g

Specific heat $= 4.18\ JK^{-1}g^{-1}$

$$Q = m \times s \times \Delta t$$

$$\therefore \quad \Delta t = \frac{Q}{m \times s} = \frac{2284}{800 \times 4.18} = \mathbf{0.68^o}.$$

In case (*ii*), heat produced $= 5710$ J

Total mass of the solution $= 250 + 400 = 650$ g

$$\therefore \quad \Delta t = \frac{Q}{m \times s} = \frac{4568}{650 \times 4.18} = \mathbf{1.68^o}.$$

1.17. DIFFERENT KINDS OF HEAT (OR ENTHALPY) OF REACTION

1. Heat of Solution (or Enthalpy of solution) : *The heat of solution of a substance in a particular solvent is defined as the heat change (i.e., the amount of heat evolved or absorbed) when 1 mole of the substance is dissolved in such a large volume of the solvent that further addition of the solvent does not produce any more heat change.*

Thus, the thermochemical equations for the dissolution of KCl and $CuSO_4$ in water may represented as

$$KCl\,(s) + aq \longrightarrow KCl\,(aq), \quad \Delta H = +18.6 \text{ kJ}$$

$$CuSO_4\,(s) + aq \longrightarrow CuSO_4\,(aq), \quad \Delta H = -66.5 \text{ kJ}$$

Thus,

the heat of solution of KCl $= +\ 18.6$ kJ mol^{-1}

and heat of solution of $CuSO_4$ $= -\ 66.5$ kJ mol^{-1}

It is interesting to note that the salts like copper sulphate, calcium chloride, etc., when present in the hydrated state (*i.e.*, $CuSO_4.5H_2O$, $CaCl_2.6H_2O$, etc.) dissolve with the absorption of heat. For example,

$$CuSO_4.5H_2O + aq \longrightarrow CuSO_4\,(aq), \Delta H = +11.7 \text{ kJ}$$

Thus, it can be generalized that the process of dissolution is usually endothermic for :

(*i*) salts which do not form hydrates like NaCl, KCl, KNO_3, etc.

(*ii*) hydrated salts like $CuSO_4.5H_2O$, $CaCl_2.6H_2O$, etc.

2. Heat of Hydration. *The amount of heat change (i.e., the heat evolved or absorbed) when one mole of the anhydrous salt combines with the required number of moles of water so as to change into the hydrated salt, is called the heat of hydration.*

For example, the heat of hydration of copper sulphate is -78.2 kJ mol^{-1}. This is represented as follows:

$$CuSO_4\,(s) + 5H_2O \longrightarrow CuSO_4.5H_2O\,(s), \quad \Delta H = -78.2 \text{ kJ}$$

3. Heat of Fusion. *Heat of fusion is the heat change that takes place when one mole of a solid substance changes into its liquid state at its melting point.*

Heat of fusion (ΔH_{fus}) of ice (m.p. = 273 K) is 6.0 kJ mol^{-1}. It may be represented as follows :

$$\underset{\text{Ice}}{H_2O\,(s)} \longrightarrow \underset{\text{Water}}{H_2O\,(l)}, \quad \Delta H = +\,6.0 \text{ kJ}$$

4. Heat of Vaporisation. *It is the heat change that takes place when one mole of a liquid changes into its gaseous state at its boiling point.*

For example, the heat of vaporisation ($\Delta H_{vap.}$) of water into its gaseous state at the boiling point (373 K) is 40.7 kJ. It may be represented as follows :

$$\underset{\text{Water}}{H_2O\,(l)} \longrightarrow \underset{\text{Steam}}{H_2O\,(g)}, \quad \Delta H = +\,40.7 \text{ kJ}$$

5. Heat of Sublimation. *Heat of sublimation of a substance is the heat change that takes place when 1 mole of a solid changes directly into vapour phase at a given temperature below its melting point.*

Sublimation is a process in which a solid on heating changes directly into gaseous state below its melting point. For example, the heat of sublimation of iodine is 62.39 kJ mol^{-1}. It is represented as follows :

$$I_2\,(s) \longrightarrow I_2\,(g), \quad \Delta H = +\,62.39 \text{ kJ}$$

Most solids, that sublime readily are molecular solids, e.g., iodine and naphthalene, etc.

1.18. HESS'S LAW OF CONSTANT HEAT SUMMATION

G. H. Hess

G.H. Hess, a Russian chemist, in 1840, gave a law about the heats of reactions which states as follows :

The total amount of heat evolved or absorbed in a reaction is the same whether the reaction takes place in one step or in a number of steps. In other words, the total amount of heat change in a reaction depends only upon the nature of the initial reactants and the nature of the final products and not upon the path or the manner by which this change is brought about.

Hess's law may be illustrated with the help of following examples :

(1) Sulphur (Rhombic) burns to form SO_3 to evolve 395.4 kJ of heat.

$$S_R + \frac{3}{2}O_2\,(g) \longrightarrow SO_3\,(g), \quad \Delta H = -395.4 \text{ kJ}$$

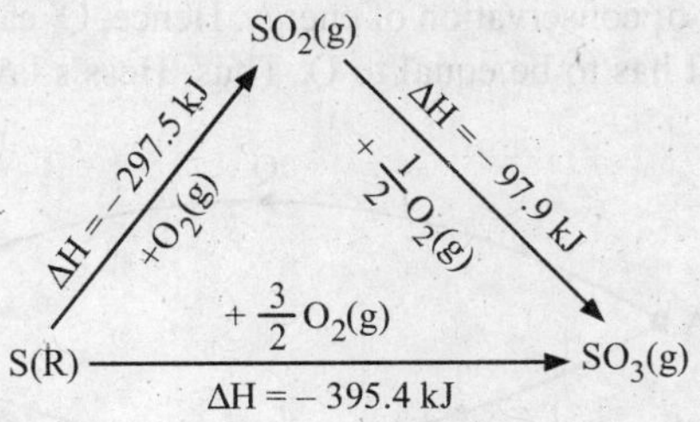

Sulphur may change to SO_3 in two steps as given below :

$$S_R + O_2\,(g) \longrightarrow SO_2\,(g), \quad \Delta H = -\,297.5 \text{ kJ}$$

$$SO_2(g) + \frac{1}{2}O_2\,(g) \longrightarrow SO_3\,(g), \quad \Delta H = -97.9 \text{ kJ}$$

(See accompanying figure)

Total heat evolved in the two steps is – 297.5 + (– 97.9) = – 395.4 kJ.

This is the same as for the direct reaction in one step.

(2) When carbon (graphite) burns to form carbon dioxide directly in one step, 393.5 kJ mol^{-1} of heat is evolved.

$$C\,(s) + C_2\,(g) \longrightarrow CO_2\,(g), \quad \Delta H = -393.5 \text{ kJ}$$

If carbon burns to form carbon monoxide first which then burns to form carbon dioxide, the heats evolved in the two steps are as follows (see accompanying figure).

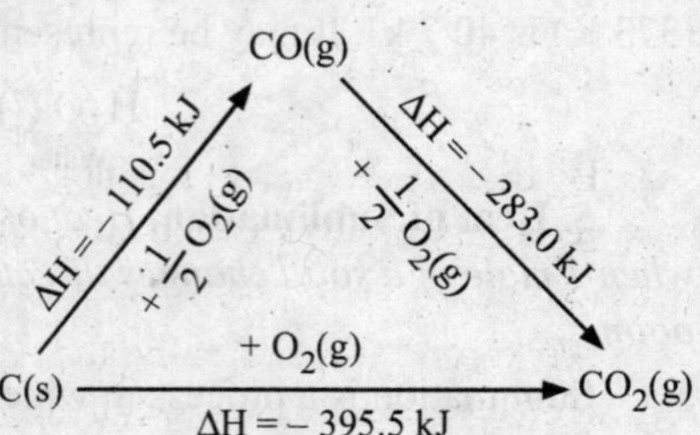

$$C\,(s) + \frac{1}{2}O_2\,(g) \longrightarrow CO\,(g), \quad \Delta H = -111.5 \text{ kJ}$$

$$CO\,(g) + \frac{1}{2}O_2\,(g) \longrightarrow CO_2\,(g), \quad \Delta H = -282.0 \text{ kJ}$$

Thus, the total heat evolved in the two steps comes out to be (–111.5) + (–282.0) = –393.5 kJ, which is the same when the reaction takes place directly in one step.

Proof of Hess's Law. Consider the general reaction : A $\longrightarrow$ D

Suppose the heat evolved in this reaction directly is Q joules.

Now suppose the same reaction takes place in three steps as follows :

$$A \xrightarrow{\text{Step 1}} B \xrightarrow{\text{Step 2}} C \xrightarrow{\text{Step 3}} D$$

Suppose the heats evolved in these three steps are q_1, q_2 and q_3 joules respectively (Fig. 1.11).

Thus, the total heat evolved from A to D = $q_1 + q_2 + q_3 = Q'$ joules (say)

According to Hess's law, we must have Q = Q′

Suppose Hess's law is not correct. In that case either Q′ < Q or Q′ > Q. Let us suppose Q′ > Q. We shall now complete the cycle, *i.e.,* we shall go from A to D through three steps and return to A in one step and we shall see what heat changes take place.

Heat change from D to A in one step is reverse of heat change from A to D in one step.

In the direct process from A to D, Q joules of heat are evolved. Therefore, in direct process from D to A, Q joules of heat are absorbed.

Now if we move from A to D in steps and return to A directly. Q′ – Q joules of heat is produced. Thus, by repeating the cyclic process a number of times, a large amount of heat can be created. This is, however, against the law of conservation of energy. Hence, Q′ cannot be greater than Q. Similarly. Q′ cannot be less than Q. It has to be equal to Q. Thus, Hess's law is proved.

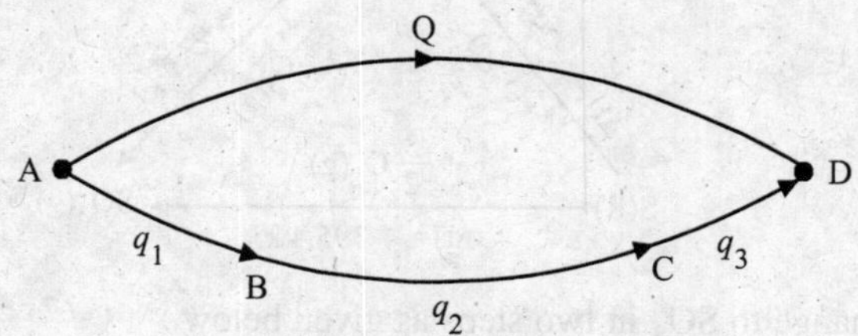

Fig. 1.11. Theoretical proof of Hess's law.

1.18.1. Applications of Hess's Law

Hess's law is very useful in determining the heats (enthalpies) of many reactions, which cannot otherwise be measured by direct methods. The thermochemical equations are added, subtracted, multiplied or divided just like other algebraic equations. Some important applications of Hess's law are :

***(i)* Determination of enthalpy of formation.** The heat of formation of some compounds can not be determined directly by experiment. This can be calculated by the application of Hess's law.

Suppose we want to determine the heat of formation of CO

$$C + \frac{1}{2}O_2 \longrightarrow CO; \quad \Delta H = ?$$

By direct experiment, it is not possible because in the reaction between carbon and oxygen, we may get a mixture of CO and CO_2. But we require only CO. Therefore, from the experiment it is not possible to determine the heat of formation.

We know experimentally, the heat of combustion of pure carbon and carbon monoxide. It is represented by the following equations :

(a) $$C + O_2 \longrightarrow CO_2; \quad \Delta H = -393.5 \text{ kJ}$$

(b) $$CO + \frac{1}{2}O_2 \longrightarrow CO_2; \quad \Delta H = -283.0 \text{ kJ}$$

Now, applying Hess's law on *(a)* and *(b)*, we get the required equation for the heat of formation of CO. This can be achieved by subtracting *(b)* from *(a)*, *i.e.*, by carrying out *(a – b)*.

$$C + \frac{1}{2}O_2 \longrightarrow CO; \quad \Delta H = -110.5 \text{ kJ}$$

Therefore, enthalpy of formation of CO is – 110.5 kJ.

***(ii)* Determination of heat (enthalpy) of transition.** Transition of an element from one allotropic form to another is a slow process and requires heating of one allotropic form of the element for a long time at constant temperature. Hence, enthalpies of transition of elements cannot be determined experimentally. They are, however, determined indirectly by means of Hess's law. Thus, Hess's law is useful in determining the enthalpies of transition of the allotropic forms such as rhombic sulphur to monoclinic sulphur, yellow phosphorus to red phosphorus, graphite to diamond, etc.

Suppose we want to determine the enthalpy of transition of sulphur (rhombic) into sulphur (monoclinic). Each variety is combustioned and heats of combustion are found out. This is represented by the following equations :

(a) $$S(R) + O_2 \longrightarrow SO_2; \quad \Delta H = -295.1 \text{ kJ}$$

(b) $$S(M) + O_2 \longrightarrow SO_2; \quad \Delta H = -296.4 \text{ kJ}$$

Subtracting *(b)* from *(a)*, we get

$$S(R) \longrightarrow S(M); \quad \Delta H = -295.1 - (-296.4) = +1.3 \text{ kJ}$$

Thus, transition of one gram atom of rhombic sulphur to monoclinic sulphur absorbs 1.3 kJ of heat.

***(iii)* Calculation of enthalpies from given data.** Thermochemical equations can be added, subtracted, multiplied or divided by any numerical factor like algebraic equations. Hess's law is, thus, used for the calculation of enthalpy changes for the reactions for which direct experimental measurements are not available.

Example 1. Calculate the heat of formation of acetic acid if its heat of combustion is 867 kJ mol^{-1}. The heats of formation of CO_2 (g) and H_2O *(l)* are –393.5 kJ mol^{-1} and –285.9 kJ mol^{-1} respectively.

Solution. The equations given are :

(*a*) $CH_3COOH + 2O_2 \longrightarrow 2CO_2 + 2H_2O; \quad \Delta H = -867 \text{ kJ}$

(*b*) $C + O_2 \longrightarrow CO_2; \quad \Delta H = -393.5 \text{ kJ}$

(*c*) $H_2 + \frac{1}{2} O_2 \longrightarrow H_2O; \quad \Delta H = -205.9 \text{ kJ}$

Our aim is to find out ΔH value for the reaction

$$2C + 2H_2 + O_2 \longrightarrow CH_3COOH; \quad \Delta H = ?$$

To obtain this equation, multiply equations (*b*) and (*c*) by 2 and add. We get equation (*d*)

$$2C + 2O_2 + 2H_2 + O_2 \longrightarrow 2CO_2 + 2H_2O$$

or

(*d*) $2C + 2H_2 + 3O_2 \longrightarrow 2CO_2 + 2H_2O; \Delta H = -1358.8 \text{ kJ}$

Subtracting (*a*) from (*d*), we get

$$2C + 2H_2 + 3O_2 \longrightarrow 2CO_2 + 2H_2O; \Delta H = -1358.8 \text{ kJ}$$

$$CH_3COOH + 2O_2 \longrightarrow 2CO_2 + 2H_2O; \Delta H = -867 \text{ kJ}$$

$$- \quad - \qquad - \quad -$$

$$2C + 2H_2 + O_2 \longrightarrow CH_3COOH \quad \Delta H = -491.8 \text{ kJ}$$

∴ Heat of formation of acetic acid = **– 491.8 kJ.**

Example 2. Find the heat of formation of SO_2 from the following thermochemical equations :

(*a*) $\mathbf{S\,(s) + \frac{3}{2} O_2\,(g) \longrightarrow SO_3\,(g) + X \text{ kJ}}$

(*b*) $\mathbf{SO_2\,(s) + O_2\,(g) \longrightarrow SO_3\,(g) + Y \text{ kJ}}$

Solution. Our aim is to find out ΔH value for the reaction

$$S + O_2 \longrightarrow SO_2 = \Delta H = ?$$

This can be obtained by subtracting (*b*) from (*a*), *i.e.*, by carrying out (*a* – *b*)

(*a*) $S\,(s) + \frac{3}{2} O_2\,(g) \longrightarrow SO_3\,(g); \quad \Delta H = -X \text{ kJ}$

(*b*) $SO_2\,(g) + \frac{1}{2} O_2\,(g) \longrightarrow SO_3\,(g); \quad \Delta H = -Y \text{ kJ}$

$$- \quad - \qquad -$$

$$S + O_2 \longrightarrow SO_2\,(g) \qquad \Delta H = -X - (-Y) = -X + Y$$

$$= -(X - Y) \text{ kJ}$$

Example 3. Calculate enthalpy of formation of anhydrous Al_2Cl_6 from the following data :

(*a*) $2Al(s) + 6HCl(aq) \longrightarrow Al_2Cl_6(aq) + 3H_2$; $\Delta H = 1004.4$ kJ

(*b*) $H_2(g) + Cl_2(g) \longrightarrow 2HCl(g)$; $\Delta H = 754.74$ kJ

(*c*) $HCl(g) + aq \longrightarrow HCl(aq)$; $\Delta H = -73.24$ kJ

(*d*) $Al_2Cl_6(s) + aq \longrightarrow Al_2Cl_6(aq)$; $\Delta H = -642.69$ kJ

Solution. Thermochemical equation for the formation of anhydrous Al_2Cl_6 is

$$2Al + 3Cl_2 \longrightarrow Al_2Cl_6(s); \quad \Delta H = ?$$

To obtain ΔH value for this equation from the given data we perform as

$$[a + (3 \times b) + (6 \times c)] - d$$

Rearranging the given equations, we get

$$2Al(s) + 6HCl(aq) + 3H_2 + 3Cl_2 \longrightarrow Al_2Cl_6(aq) + 3H_2(g) + 6HCl(g)$$
$$+ 6HCl(g) + aq - Al_2Cl_6(s) - aq \qquad + 6HCl(aq) - Al_2Cl_6(aq)$$

$$2Al(s) + 3Cl_2(g) \longrightarrow Al_2Cl_6(s), \Delta H = 1004.4 + 3 \times 754.74 + 6 \times (73.24) - (-642.69) = 3471.87 \text{ kJ}$$

Example 4. Calculate enthalpy of formation of methane. Given :

(*a*) $C(\text{graphite}) + O_2(g) \longrightarrow CO_2(g)$; $\Delta H = -393.5$ kJ

(*b*) $2H_2(g) + O_2(g) \longrightarrow 2H_2O(l)$; $\Delta H = -571.8$ kJ

(*c*) $CH_4(g) + 2O_2(g) \longrightarrow CO_2(g) + 2H_2O(l)$; $\Delta H = -890.3$ kJ

Solution. We want to find out the value of ΔH for the reaction

$$C(\text{graphite}) + 2H_2(g) \longrightarrow CH_4; \quad \Delta H = ?$$

This can be obtained by the operation $(a + b) - c$, we get

$$C(\text{graphite}) + O_2 + 2H_2 + O_2 \longrightarrow CO_2 + 2H_2O \qquad \text{.... (a + b)}$$

or $$C(\text{graphite}) + 2O_2 + 2H_2 \longrightarrow CO_2 + 2H_2O; \Delta H = -965.3 \text{ kJ}$$

$$CH_4 + 2O_2 \longrightarrow CO_2 + 2H_2O; \Delta H = -890.3 \text{ kJ}$$
$$- \quad - \quad - \quad -$$

$$C(\text{graphite}) + 2H_2 \longrightarrow CH_4 \qquad \Delta H = -75.0 \text{ kJ} \quad \text{... } (a + b) - c$$

Example 5. From the thermochemical equations :

(*a*) $N_2(g) + 3H_2(g) \longrightarrow 2NH_3(g)$; $\Delta H = -92.40$ kJ

(*b*) $$2H_2 (g) + O_2 (g) \longrightarrow 2H_2O; \quad \Delta H = -483.79 \text{ kJ}$$

Find ΔH_f for NH_3 (*g*) and ΔH_f for H_2O (*g*).

Solution. Heat of formation (ΔH_f) is for one mole

$$\therefore \quad \Delta H_f \text{ for } NH_3 = \frac{-92.4}{2} = -46.2 \text{ kJ}$$

$$\therefore \quad \Delta H_f \text{ for } H_2O = \frac{-483.79}{2} = \mathbf{-241.895 \text{ kJ.}}$$

Example 6. Calculate the heat of formation of carbon disulphide given that the heats of combustion of carbon disulphide, carbon and sulphur are –1109 kJ, –394.6 kJ and –298.7 kJ respectively.

Solution. The thermochemical equations for the combustion of CS_2, C and S are

(*a*) $$CS_2 + 3O_2 \longrightarrow CO_2 + 2SO_2; \quad \Delta H = -1109 \text{ kJ}$$

(*b*) $$C + O_2 \longrightarrow CO_2; \quad \Delta H = -394.6 \text{ kJ}$$

(*c*) $$S + O_2 \longrightarrow SO_2; \quad \Delta H = -298.7 \text{ kJ}$$

Our aim is to find out the value of ΔH for the following equation,

$$C + 2S \longrightarrow CS_2; \quad \Delta H = ?$$

Multiplying (*c*) by 2 and adding to (*b*), we get

$$2S + 2O_2 + C + O_2 \longrightarrow 2SO_2 + CO_2$$

or

(*d*) $$2S + C + 3O_2 \longrightarrow 2SO_2 + CO_2; \quad \Delta H = (-298.7) \times 2 + (-394.6) = -992.0 \text{ kJ}$$

Subtracting (*a*) from (*d*), we get

$$2S + C + 3O_2 \longrightarrow 2SO_2 + CO_2; \quad \Delta H = -992.0 \text{ kJ}$$

$$CS_2 + 3O_2 \longrightarrow CO_2 + 2SO_2; \quad \Delta H = -1109 \text{ kJ}$$

$$- \quad - \qquad - \quad -$$

$$2S + C \longrightarrow CS_2; \quad \Delta H = -992 - (-1109) = +117 \text{ kJ}$$

$$\therefore \quad \text{Heat of formation of } CS_2 = +117 \textbf{ kJ mol}^{-1}$$

Example. 7. The enthalpy of combustion of ethylene (C_2H_4) is – 1410 kJ. What will be calorific value of ethylene per gram?

Solution. The thermochemical equation for the combustion of C_2H_4 is

$$\underset{28 \text{ grams}}{C_2H_4} + O_2 \longrightarrow 2CO_2 + 2H_2O\,; \Delta H = -1410 \text{ kJ}$$

It is evident from the above equation that

$$28 \text{ grams of } C_2H_4 \text{ give heat} = -1410 \text{ kJ}$$

$$\therefore \quad 1 \text{ gram of } C_2H_4 \text{ will give heat} = \frac{-1410}{28} = -50.36 \text{ kJ}$$

$$\therefore \quad \text{Calorific value of ethylene} = -\ 50.36 \text{ kJ.}$$

Example 8. One gram of an organic liquid A (mol. wt. = 128) liberates 150 joules of heat on solidification. Calculate the enthalpy of fusion of A.

Solution. Molecular weight of liquid A=128 g

Heat evolved when 1 gram of liquid A solidifies = 150 J

∴ Heat evolved when 128 grams (1 mole) of liquid A solidify = 150 × 128 = 19200 J

$$\text{Liquid (A)} \longrightarrow \text{Solid (A)} ; \Delta H = -19200 \text{ J}$$

Rearranging this equation, we get

$$\text{Solid (A)} \longrightarrow \text{Liquid (A)} ; \Delta H = +\ 19200$$

∴ Enthalpy of fusion of **A = + 19200 joules.**

Example 9. Calculate the heat of formation of KCl from the following data :

$$\textbf{(i)} \quad KOH\ (aq) + HCl\ (aq) \longrightarrow KCl\ (aq) + H_2O\ (l); \quad \Delta H = -57.3 \text{ kJ}$$

$$\textbf{(ii)} \quad H_2\ (g) + \frac{1}{2}O_2\ (g) \longrightarrow H_2O\ (l); \quad \Delta H = -286.2 \text{ kJ}$$

$$\textbf{(iii)} \quad \frac{1}{2}H_2\ (g) + \frac{1}{2}\ Cl_2\ (g) + aq \longrightarrow HCl\ (aq); \quad \Delta H = -164.4 \text{ kJ}$$

$$\textbf{(iv)} \quad K\ (s) + \frac{1}{2}\ O_2\ (g) + \frac{1}{2}H_2\ (g) + aq \longrightarrow KOH\ (aq); \quad \Delta H = -487.4 \text{ kJ}$$

$$\textbf{(v)} \quad KCl\ (s) + aq \longrightarrow KCl\ (aq); \quad \Delta H = +\ 18.4 \text{ kJ}$$

Solution. Our aim is to find out ΔH for the following reaction.

$$\textit{(vi)} \quad K(s) + \frac{1}{2}\ Cl_2\ (g) \longrightarrow KCl\ (s), \Delta H = ?$$

In order to get this thermochemical equation, we follow the following two steps :

Step 1. Adding Eqns. (*iii*) and (*iv*) and subtracting Eqn. (*v*), we have

$$\textit{(vii)}\ K(s) + \frac{1}{2}\ Cl_2\ (g) + H_2(g) + \frac{1}{2}\ O_2\ (g) \longrightarrow KCl\ (s) + HCl\ (aq) + KOH\ (aq) - KCl\ (aq)$$

$$\Delta H = -487.4 + (-164.4) - (18.4) = \mathbf{-\ 670.2\ kJ.}$$

Step 2. To cancel out the terms of this equation which do not appear in the required equation (*vi*), add Eqn. (*i*) to Eqn. (*vii*) and subtract Eqn. (*ii*) from their sum. This gives

$$K\ (s) + \frac{1}{2}\ Cl_2\ (g) \longrightarrow KCl\ (s)\ ; \Delta H = -670.2 + 57.3 - (-286.2) = \mathbf{-441.3\ kJ.}$$

Example 10. Calculate the heat of hydration of anhydrous copper sulphate ($CuSO_4$) into hydrated copper sulphate ($CuSO_4.5H_2O$). Given that the heats of solutions of anhydrous copper sulphate and hydrated copper sulphate are – 66.5 and +11.7 kJ mol^{-1} respectively.

Solution. Data provided is :

(*i*) $CuSO_4\ (s) + aq \longrightarrow CuSO_4\ (aq);\quad \Delta H = -66.5$ kJ

(*ii*) $CuSO_4.5H_2O\ (s) + aq \longrightarrow CuSO_4\ (aq);\quad \Delta H = +11.7$ kJ

Our aim is to find out heat of the following reaction.

$CuSO_4\ (s) + 5H_2O\ (l) \longrightarrow CuSO_4.5H_2O\ (s);\quad \Delta H = ?$

Equation (*i*) can be written in two steps as :

(*iii*) $CuSO_4\ (s) + 5H_2O\ (l) \longrightarrow CuSO_4.5H_2O\ (s);\quad \Delta H = q_1$ kJ

(*iv*) $CuSO_4.5H_2O\ (s) + aq \longrightarrow CuSO_4\ (aq);\quad \Delta H = q_2$ kJ

According to Hess's law $q_1 + q_2 = -66.5$ kJ ...(*v*)

Further, equations (*ii*) and (*iv*) are same $\therefore\ q_2 = +11.7$ kJ

Putting this value in eq. (*v*) above, we get

$$q_1 + 11.7 = -66.5 \text{ or } q_1 = -66.5 - 11.7 \text{ kJ} = -78.2 \text{ kJ}$$

Thus, equation (*iii*) may be written as

$CuSO_4\ (s) + 5H_2O\ (l) \longrightarrow CuSO_4.5H_2O\ (s)\ ;\ \Delta H = -78.2$ kJ

This is what we aimed at. Hence, the required value of the heat of hydration is $\Delta H = -78.2$ kJ.

Example 11. Calculate the heat of formation of carbon monoxide (CO) from the following data :

(*i*) $\mathbf{C\ (s) + O_2\ (g) \longrightarrow CO_2\ (g)\ ;\quad \Delta H = -393\ kJ}$

(*ii*) $\mathbf{CO\ (g) + \frac{1}{2}\ O_2\ (g) \longrightarrow CO_2\ (g)\ ;\quad \Delta H = -282\ kJ}$

Solution. Our aim is to find out ΔH for the following reaction :

(*i*) $C\ (s) + \frac{1}{2} O_2\ (g) \longrightarrow CO\ (g)\ ;\ \Delta H = ?$

Subtracting equation (*ii*) from (*i*), we get

$C\ (s) + \frac{1}{2} O_2\ (g) - CO\ (g) \longrightarrow 0;\quad \Delta H = -393.3 - (-282.8) = -110.5$ kJ

or $C\ (s) + \frac{1}{2} O_2\ (g) \longrightarrow CO\ (g);\quad \Delta H = -110.5$ kJ

$\therefore$ Heat of formation of CO, ΔH_f = **110.5 kJ.**

Example 12. Calculate the heat change accompanying the transformation of C (graphite) to C (diamond). Given that the heats of combustion of graphite and diamond are 393.5 and 395.4 kJ mol^{-1} respectively.

Solution. Data provided is :

(*i*) C (graphite) + O_2 (*g*) $\longrightarrow$ CO_2 (*g*); $\Delta H = -393.5$ kJ

(*ii*) C (diamond) + O_2 (*g*) $\longrightarrow$ CO_2 (*g*); $\Delta H = -395.4$ kJ

Our aim is to find out ΔH for the following reaction

C (graphite) $\longrightarrow$ C (diamond); $\Delta H = ?$

Subtracting equation (*ii*) from (*i*), we get

C (graphite) – C (diamond) $\longrightarrow$ 0 ; $\Delta H = -393.5 - (-395.4) = +1.9$ kJ

or C (graphite) $\longrightarrow$ C (diamond) ; $\Delta H =$ **+ 1.9 kJ.**

Q. 13. The heat of combustion of ethyl alcohol (C_2H_5OH) is 1380.7 kJ mol^{-1}. If the heats of formation of CO_2 and H_2O are 394.5 and 286.6 kJ mol^{-1} respectively, calculate the heat of formation of ethyl alcohol.

Solution. Data provided is :

(*i*) $C_2H_5OH + 3O_2 \longrightarrow 2CO_2 + 3H_2O$; $\Delta H = -1380.7$ kJ

(*ii*) $C + O_2 \longrightarrow CO_2$; $\Delta H = -394.5$ kJ

(*iii*) $H_2 + \frac{1}{2}O_2 \longrightarrow H_2O$; $\Delta H = -286.6$ kJ

Our aim is to determine ΔH for the following reaction

$$2C + 3H_2 + \frac{1}{2}O_2 \longrightarrow C_2H_5OH$$

In order to get this thermochemical equation multiply Eqn. (*ii*) by 2 and Eqn. (*iii*) by 3 and substract Eqn. (*i*) from their sum, *i.e.,* operating : 2 × Eqn. (*ii*) + 3 × Eqn. (*iii*) – Eqn. (*i*), we get

$$2C + 3H_2 + \frac{1}{2}O_2 \longrightarrow C_2H_5OH; \quad \Delta H = 2(-394.5) + 3(-286.6) - (-1380.7) = -268.1 \text{ kJ}$$

Thus, the heat of formation of ethyl alcohol = ΔH_f = **–268.1 kJ mol^{-1}.**

Example 14. Calculate the standard enthalpy of formation of n-butane given that the standard enthalpies of combustion of n-butane (g), C (graphite) and H_2(g) are – 2878.5 kJ mol^{-1}, 393.5 kJ mol^{-1} and – 285.3 kJ mol^{-1} respectively.

Solution. We aim at the equation

$4C + 5H_2 \longrightarrow C_4H_{10}$; $\Delta H = ?$

The given data can be expressed in the form of the following equations.

(*i*) $C_4H_{10} + 6\frac{1}{2}O_2 \longrightarrow 4CO_2 + 5H_2O$; $\Delta H = -2878.5$ kJ

(*ii*) $4C + 4O_2 \longrightarrow 4CO_2$; $\Delta H = 4 \times (-393.5)$ kJ (*iii*) $5H_2 + \frac{5}{2}O_2 \longrightarrow 5H_2O$; $\Delta H = 5 \times (-285.3)$ kJ

Add (*ii*) and (*iii*) and subtract (*i*)

$$(-1574.0) + (-1426.5) - (-2878.5)$$
$$= -122.0 \text{ kJ}$$

Example 15. Heats of solution (ΔH) for $BaCl_2.2H_2O$ and $BaCl_2$ are 8.8 and –20.6 kJ respectively. Calculate the heat of hydration of $BaCl_2$ to $BaCl_2.2H_2O$.

Solution. We are given

$BaCl_2.2H_2O\ (s) + aq \longrightarrow BaCl_2\ (aq)$; $\Delta H = 8.8$ kJ ... (*i*)

$BaCl_2\ (s) + aq \longrightarrow BaCl_2\ (aq)$; $\Delta H = -20.6$ kJ ... (*ii*)

We aim at :

$BaCl_2\ (s) + 2H_2O \longrightarrow BaCl_2.2H_2O$ (s); ΔH = ? ... (*iii*)

Equation (*ii*) may be written in two steps as

$BaCl_2\ (s) + 2H_2O \longrightarrow BaCl_2.2H_2O\ (s)$; $\Delta H = \Delta H_1$ (say) ... (*iv*)

$BaCl_2.2H_2O\ (s)$ + aq $\longrightarrow BaCl_2\ (aq)$; $\Delta H = \Delta H_2$ (say) .... (*v*)

Then according to Hess's law

$$\Delta H_1 + \Delta H_2 = -20.6 \text{ kJ}$$

But $\Delta H_2 = 8.8$ kJ [∵ Equation (*i*) = Equation (*v*)]

∴ $\Delta H_1 = -20.6 - 8.8 = -29.4$ kJ

But Equation (*iii*) = Equation (*iv*)

Hence, the heat of hydration of $BaCl_2$ = **–29.4 kJ.**

Example 16. Given the following thermochemical equations :

(*i*) S (rhombic) + O_2 (*g*) $\longrightarrow$ SO_2 (*g*) ; ΔH = – 297.5 kJ

(*ii*) S (monoclinic) + O_2 (*g*) $\longrightarrow$ SO_2 (*g*) ; ΔH = –300.0 kJ

Calculate ΔH for the transformation of one gram atom of rhombic sulphur into monoclinic sulphur.

Solution. We aim at :

S (rhombic) $\longrightarrow$ S (monoclinic); ΔH = ?

Equation (*i*) – Equation (*ii*) gives

S (rhombic) – S (monoclinic) $\longrightarrow$ 0; ΔH = 297.5 – (–300.0) = 2.5 kJ

or S (rhombic) $\longrightarrow$ S (monoclinic); ΔH = **+2.5 kJ.**

Thus, for the transformation of one gram atom of rhombic sulphur into monoclinic sulphur, 2.5 kJ of heat is absorbed.

Example 17. Calculate the enthalpy of formation of N_2O_5(g) :

$$N_2(g) + \frac{5}{2}O_2(g) \longrightarrow N_2O_5(g)$$

The following information is given :

(i) $N_2(g) + O_2(g) \longrightarrow 2NO(g)$; $\Delta H° = +180$ kJ

(ii) $4NO_2(g) + O_2(g) \longrightarrow 2N_2O_5(g)$; $\Delta H° = -120.6$ kJ

(iii) $2NO(g) + O_2(g) \longrightarrow 2NO_2(g)$; $\Delta H° = -114.0$ kJ

Solution. Perform the operation (*i*) $+\frac{1}{2}$ (*iii*) to get the desired equation as

$$N_2(g) + \frac{5}{2}O_2(g) \longrightarrow N_2O_5(g)$$

Applying Hess's law, perform the same operation on enthalpies to obtain the enthalpy of formation of N_2O_5 (g)

$$\Delta H_f = [180 - \frac{1}{2}(120.6) - 114.0] \text{ kJ}$$

$$= 5.7 \text{ kJ}$$

PROBLEMS FOR PRACTICE

1. Calculate the enthalpy of formation of acetic acid from the following data:

(*i*) $C(s) + O_2(g) \longrightarrow CO_2(g)$, $\Delta H = -393.7$kJ

(*ii*) $H_2 + (g) + \frac{1}{2}O_2(g) \longrightarrow H_2O(l)$, $\Delta H = -285.8$ kJ

(*iii*) $CH_3COOH(l) + 2O_2(g) \longrightarrow 2CO_2(g) + 2H_2O(l)$, $\Delta H = -873.2$ kJ

[**Ans.** – 485.8 kJ mol $^{-1}$]

2. Calculate the standard enthalpy of formation of SO_3 at 298 K using the following reactions and enthalpies.

$S_8(s) + 8O_2(g) \longrightarrow 8SO_2(g)$, $\Delta H° = -2775$ kJ mol^{-1}

$2SO_2(g) + O_2(g) \longrightarrow 2SO_3(g)$, $\Delta H° = -198$ kJ mol^{-1}

[**Ans.** – 445.9 kJ mol^{-1}]

3. From the following thermochemical equations, calculate the standard enthalpy of formation of HCl (*g*).

(A) $H_2(g) \longrightarrow 2H(g)$, $\Delta H = +436.0$ kJ mol^{-1}

(B) $Cl_2(g) \longrightarrow 2Cl(g)$, $\Delta H = +242.7$ kJ mol^{-1}

(C) $HCl(g) \longrightarrow H(g) + Cl(g)$, $\Delta H = +431.8$ kJ mol^{-1}

[**Ans.** – 92.45 kJ mol^{-1}]

4. Calculate the enthalpy of formation of acetic acid if the enthalpy of combustion to CO_2 (g) and H_2O (*l*) is –867.0 kJ mol^{-1} and enthalpies of formation of CO_2 (*g*) and H_2O (*l*) are respectively –393.5 and –285.9 kJ mol^{-1} [**Ans.** –491.8 kJ mol^{-1}]

5. Calculate the enthalpy of formation of sucrose ($C_{12}H_{22}O_{11}$) from the following data:

(*i*) $C_{12}H_{22}O_{11} + 12O_2 \longrightarrow 12CO_2 + 11\ H_2O;\quad \Delta H = -5200.7\ \text{kJ mol}^{-1}$

(*ii*) $C + O_2 \longrightarrow CO_2,\quad \Delta H = -394.5\ \text{kJ mole}^{-1}$

(*iii*) $H_2 + \frac{1}{2}O_2 \longrightarrow H_2O,\quad \Delta H = -285.8\ \text{kJ mol}^{-1}$

[**Ans.** – 2677.1 KJ mol^{-1}]

6. Ethylene on combustion gives carbon dioxide and water. Its enthalpy of combustion is 1410.0 kJ mol^{-1}. If the enthalpy of formation of CO_2 and H_2O are 393.3 kJ and 286.2 kJ respectively. Calculate the enthalpy of formation of ethylene. [**Ans.** + 51.0 kJ mol^{-1}]

7. Calculate the enthalpy of formation of benzene, given that enthalpies of combustion of benzene, carbon and hydrogen are –3281.5 kJ, –394.9 kJ and –286.1 kJ/mol, respectively.

[**Ans.** +53.8 kJ mol^{-1}]

8. Calculate the enthalpy of formation of *n*-butane from the following data:

(*i*) $2C_4H_{10}\ (g) + 13O_2\ (g) \longrightarrow 8\ CO_2\ (g) + 10\ H_2O\ (l),\ \Delta H = -5757.2\ \text{kJ mol}^{-1}$

(*ii*) $C\ (s) + O_2\ (g), \longrightarrow CO_2\ (g),\quad \Delta H = -405.4\ \text{kJ mol}^{-1}$

(*iii*) $2H_2\ (g) + O_2\ (g) \longrightarrow 2H_2O\ (l),\quad \Delta H = -572.4\ \text{kJ mol}^{-1}$

On what law are your calculations based? [**Ans.** –174 kJ mol^{-1}]

9. Calculate the enthalpy of formation of anhydrous Al_2Cl_6 from the following data:

(*i*) $2Al\ (s) + 6HCl\ (aq) \longrightarrow Al_2Cl_6\ (aq) + 3\ H_2\ (g) + 1004.2\ \text{kJ mol}^{-1}$

(*ii*) $H_2\ (g) + Cl_2\ (g) \longrightarrow 2HCl\ (g) + 184.1\ \text{kJ mol}^{-1}$

(*iii*) $HCl\ (g) + aq \longrightarrow HCl\ (aq) + 73.2\ \text{kJ mol}^{-1}$

(*iv*) $2Al_2\ Cl_6(s) + aq \longrightarrow Al_2Cl_6\ (aq) + 643.1\ \text{kJ mol}^{-1}$

[**Ans.** –1352.6 kJ mol^{-1}]

10. Calculate the enthalpy of formation of carbon disulphide given that the enthalpy of combustion of carbon disulphide is 110.2 kJ mol^{-1} and those of sulphur and carbon are 297.4 kJ and 394.5 kJ/g atom respectively. [**Ans.** –879.1 kJ mol^{-1}]

1.19. BOND ENTHALPY OR BOND ENERGY

The bond energy of a particular bond is defined as the average amount of energy released when one mole of bonds are formed from isolated gaseous atoms or the amount of energy required when 1 mole of bonds are broken so as to get the separated gaseous atoms.

For diatomic molecules (like H_2, HCl, etc.), the bond energy is equal to the dissociation energy of the molecules but for a *polyatomic molecule* like CH_4, the bond dissociation energies of the four C — H bonds are different. Hence, an average value is taken. Bond energies of some common bonds are given in Table 1.3.

Table 1.3 Bond energies of some common bonds

Bond	Bond energy ($kJ\ mol^{-1}$)	Bond	Bond energy ($kJmol^{-1}$)
H — H	436	C — H	414
H — F	565	O — H	463
H — Cl	431	N — H	389
H — Br	364	C — C	347
H — I	297	C = C	619
F — F	155	C ≡ C	812
Cl — Cl	242	C — Cl	326
Br — Br	190	C — O	335
I — I	149	C = O	707
O = O	494	C — N	293
N ≡ N	941	C = N	616
		C ≡ N	879

Heat of reaction can be calculated from bond energy data. The following formula can be used directly in these calculations.

$$\Delta H_{reaction} = \Sigma \text{ Bond energies of reactants} - \Sigma \text{ Bond energies of products}$$

PROBLEMS ON CALCULATION OF BOND ENERGY

Example 1. Calculate the enthalpy change for the reaction

$$H_2(g) + Br_2(g) \longrightarrow 2HBr(g)$$

Given that the bond energies of H — H, Br — Br and H — Br are 435, 192 and 364 kJ mol^{-1} respectively.

Solution. First method. *By calculating the total energy absorbed and released.*

Energy absorbed in the dissociation of 1 mole of H – H bonds = 435 kJ

Energy absorbed in the dissociation of 1 mole of Br – Br bonds = 192 kJ

Total energy absorbed = 435 + 192 = 627 kJ

Energy released in the formation of 1 mole of H – Br bonds = 364 kJ

∴ Energy released in the formation of 2 moles of H – Br bonds = 2 × 364 = 728 kJ

Energy released is greater than energy absorbed.

Hence, net result is the release of energy.

Energy released = 728 kJ – 627 kJ = 101 kJ

Thus, for the given reaction,

$$\Delta H = \mathbf{-101\ kJ.}$$

Second method. *Using the formula directly.*

$$\begin{aligned}\Delta H_{reaction} &= \Sigma \text{ Bond energies of reactants} - \Sigma \text{ Bond energies of products}\\ &= [\text{Bond energy (H — H)} + \text{Bond energy (Br — Br)}]\\ &\quad -[2 \times \text{Bond energy (H— Br)}]\\ &= 435 + 192 - (2 \times 364) = 627 - 128 = \mathbf{-101\ kJ.}\end{aligned}$$

Hess's law can also be applied to solve the problem.

Example 2. Calculate the bond energy of C — H bond, given the heat of combustion of methane (CH_4), graphite and hydrogen are 891 kJ, 394 kJ and 286 kJ respectively while the heat of sublimation of graphite is 717 kJ and the heat of dissociation of hydrogen molecule is 436 kJ.

Solution. The data given is :

$$CH_4\,(g) + 2O_2\,(g) \longrightarrow CO_2\,(g) + 2H_2O\,(g);\quad \Delta H = -891 \text{ kJ} \quad ...(i)$$

$$C\,(s) + O_2\,(g) \longrightarrow CO_2\,(g);\quad \Delta H = -394 \text{ kJ} \quad ...(ii)$$

$$H_2\,(g) + \frac{1}{2}O_2\,(g) \longrightarrow H_2O\,(g);\quad \Delta H = -286 \text{ kJ} \quad ...(iii)$$

$$C\,(s) \longrightarrow C\,(g);\quad \Delta H = +717 \text{ kJ} \quad ...(iv)$$

$$H_2\,(g) \longrightarrow 2H\,(g);\quad \Delta H = +436 \text{ kJ} \quad ...(v)$$

We want ΔH for the following reaction

$$\frac{1}{4}\,[CH_4\,(g)] \longrightarrow C\,(g) + 4H\,(g)];\quad \Delta H = ? \quad ...(vi)$$

The problem is solved in two steps.

First we calculate ΔH for $CH_4\,(g) \longrightarrow C\,(g) + 4H\,(g)$

and then divide the result by 4.

Operating Eqn. (*i*) + Eqn. (*iv*) + 2 × Eqn. (*v*), we get

$$CH_4\,(g) + 2O_2\,(g) + C\,(s) + 2H_2\,(g) \longrightarrow CO_2\,(g) + 2H_2O\,(g) + C\,(g) + 2H\,(g) \quad ...(vii)$$

To cancel out the terms not needed now operate Eqn. (*vii*) – 2 × Eqn. (*iii*) – Eqn. (*ii*).

$$CH_4\,(g) \longrightarrow C\,(g) + 4H\,(g);$$

$$\Delta H = -891 + 717 + 2 \times 436 - 2 \times (-286) - (-394) = +\,1664 \text{ kJ}.$$

This is the energy required for the dissociation of 4 moles of C—H bonds.

$$\text{Bond dissociation energy for C—H bond} = \frac{1664}{4} \text{ kJ} = \mathbf{416\ kJ\ mol^{-1}}.$$

PROBLEMS FOR PRACTICE

1. ΔH for the reaction

$$H—C\equiv N\ (g) + 2\,H_2\,(g) \longrightarrow H—\underset{\underset{H}{|}}{\overset{\overset{H}{|}}{C}}—\overset{\overset{H}{|}}{N}—H\,(g)$$

is – 150 kJ. Calculate the bond energy of C ≡ N bond.

[Given bond energies of C — H = 414 kJ mol^{-1}; H — H = 435 kJ mol^{-1}; C — N = 293 kJ mol^{-1}, N — H = 396 kJ mol^{-1}] **[Ans.** 893 kJ mol^{-1}]

2. Calculate the enthalpy of hydrogenation of $C_2H_2\,(g)$ to $C_2H_4\,(g)$.

Given bond energies. C — H = 414.0 kJ mol^{-1}, C ≡ C = 827.6 kJ mol^{-1}, C = C = 606.0 kJ mol^{-1}, H — H = 430.5kJ mol^{-1}) **[Ans.** – 175.9 kJ mol^{-1}]

3. Calculate the C — C bond energy from the following data:

(*i*) $2C\ (\text{graphite}) + 3H_2\ (g) \longrightarrow C_2H_6\ (g), \Delta H = -84.67\ \text{kJ}$

(*ii*) $C\ (\text{graphite}) \longrightarrow C\ (g), \Delta H = 716.7\ \text{kJ}$

(*iii*) $H_2\ (g) \longrightarrow 2\ H\ (g), \Delta H = 435.9\ \text{kJ}$

Assume the C — H bond energy as 416 kJ. [**Ans.** – 964 kJ mol^{-1}]

4. Propane has the structure $H_3C — CH_2 — CH_3$. Calculate the change in enthalpy for the following reaction:

$$C_3H_8\ (g) + 5O_2\ (g) \longrightarrow 3\ CO_2\ (g) + 4\ H_2O\ (g)$$

Given that average bond enthalpies are :

C — C	C — H	C = O	O = O	O — H
347	414	741	498	464 kJ mol^{-1}

[**Ans.** – 1662 kJ mol^{-1}]

5. Calculate ΔH° for the reaction

$$CH_2 = CH_2 + 3\ O_2 \longrightarrow 2\ CO_2 + H_2O$$

Given that the average bond energies of the different bonds are

Bond	C — H	O = O	C = O	O — H	C = C
Bond energy (kJ mol^{-1})	414	499	724	460	619

[**Ans.** – 964 kJ mol^{-1}]

6. Calculate the enthalpy change for the following reaction

$$H_2\ (g) + Cl_2\ (g) \longrightarrow 2HCl\ (g)$$

Given that the bond dissociation energies of H — H, Cl — Cl and H — Cl are 437 kJ, 244 kJ and 433 kJ mol^{-1} respectively. [**Ans.** – 185 kJ]

MISCELLANEOUS EXAMPLES

Example 1. If a man takes a diet which gives him energy equal to 9500 kJ per day and he expends energy in all forms to a total of 12000 kJ per day, what is the change in internal energy per day? If the energy lost was stored as sucrose (1632 kJ per 100 g), how many days should it take to lose 1 kg? Ignore water loss.

Solution. Loss of energy per day = 12000 – 9500 = 2500 kJ

For a loss of 1632 kJ of energy, sucrose ($C_{12}H_{22}O_{11}$) lost = 100 g (given)

∴ For a loss of 2500 kJ of energy, sucrose lost $= \dfrac{100}{1632} \times 2500\ \text{g} = 153.2\ \text{g}$

Thus, loss of 153.2 g in weight takes place in 1 day.

∴ Time taken to lose 1 kg or 1000 g glucose $= \dfrac{1}{153.2} \times 1000 =$ **6.5 days.**

Example 2. From the thermochemical equation

$$C_6H_6\ (l) + 7\frac{1}{2}\ O_2\ (g) \longrightarrow 3H_2O\ (l) + 6CO_2\ (g)\ ; \qquad \Delta H = -3264.64 \text{ kJ mol}^{-1}$$

Calculate the energy evolved when 39 g of C_6H_6 are burnt in an open container.

Solution. 1 mole of benzene = 78 g

When 78 g of C_6H_6 burn, heat evolved = 3264.64 kJ

∴ When 39 g of C_6H_6 burn, heat evolved = $\frac{3264.64}{78} \times 39 =$ **1632.32 kJ.**

Example 3. The heat evolved in the combustion of methane is given by the following equation :

$$CH_4\ (g) + 2O_2\ (g) \longrightarrow CO_2\ (g) + 2H_2O\ (l); \qquad \Delta H = -890.3 \text{ kJ}$$

(*a*) How many grams of methane would be required to produce 445.15 kJ of heat on combustion?

(*b*) How many grams of carbon dioxide would be formed when 445.15 kJ of heat are evolved?

(*c*) What volume of oxygen at STP would be used in the combustion process (*a*) or (*b*)?

Solution. (*a*) From the given equation,

Heat produced from 1 mole of CH_4, *i.e.*, 16 g of CH_4 = 890.3 kJ

Heat produced from 8 g of CH_4 = **445.15 kJ.**

(*b*) From the given equation,

When 890.3 kJ of heat are evolved, CO_2 formed = 44 g

∴ When 445.15 kJ of heat are evolved, CO_2, formed = **22 g.**

(*c*) O_2 used in the production of 890.3 kJ of heat = 44.8 litres at STP.

Hence, O_2 used in the production of 445.15 kJ of heat = **22.4 litres at STP.**

Example 4. The heat evolved in the combustion of glucose is shown in the following equation :

$$C_6H_{12}O_6 + 6O_2\ (g) \longrightarrow 6CO_2\ (g) + 6H_2O\ (g); \ \Delta H = -2840 \text{ kJ}$$

What is the energy requirement for production of 0.36 g of glucose by the reverse reaction?

Solution. Data provided is

$$C_6H_{12}O_6 + 6O_2\ (g) \longrightarrow 6CO_2\ (g) + 6H_2O\ (g); \qquad \Delta H = -2840 \text{ kJ}$$

Writing the reaction in reverse direction

$$6CO_2\ (g) + 6H_2O\ (g) \longrightarrow C_6H_{12}O_6\ (s) + 6O_2\ (g); \qquad \Delta H = +2840 \text{ kJ}$$

For production of 1 mole of $C_6H_{12}O_6$ (180 g) heat required (absorbed) = 2840 kJ

∴ For production of 0.36 g of glucose, heat absorbed = $\frac{2840}{180} \times 0.36 =$ **5.68 kJ.**

Example 5. The thermochemical equation for solid and liquid rocket fuel are given below :

$$2Al\,(s) + 1\frac{1}{2}O_2\,(g) \longrightarrow Al_2O_3\,(s); \qquad \Delta H = -1667.8 \text{ kJ}$$

$$H_2\,(g) + \frac{1}{2}O_2\,(g) \longrightarrow H_2O\,(l); \qquad \Delta H = -285.9 \text{ kJ}$$

(*a*) If equal masses of aluminium and hydrogen are used, which is better rocket fuel?

(*b*) Determine ΔH for the reaction $Al_2O_3\,(s) \longrightarrow 2Al\,(s) + 1\frac{1}{2}O_2\,(g)$

Solution.

$$2 \text{ moles of Al} = 54 \text{ g}$$

$$54 \text{ g of Al on combustion give heat} = 1667.8 \text{ kJ}$$

$$\therefore \quad 1 \text{ g of Al on combustion gives heat} = \frac{1667.8}{54} = 30.9 \text{ kJ}$$

$$1 \text{ mole of } H_2 = 2 \text{ g}$$

$$2 \text{ g of } H_2 \text{ on combustion give} = 285.9 \text{ kJ of heat}$$

$$\therefore \quad 1 \text{ g of } H_2 \text{ on combustion will give} = \frac{285.9}{2} = 142.95 \text{ kJ of heat}$$

Thus, H_2 is a better rocket fuel.

(*b*) Writing the reverse of the first reaction, we have

$$Al_2O_3\,(s) \longrightarrow 2Al\,(s) + 1\frac{1}{2}O_2\,(g); \text{ H} = +\,1667.8 \text{ kJ}$$

Thus, for the reaction given in part (*b*) of the problem,

$$\Delta H = \mathbf{+\,1667.8 \text{ kJ.}}$$

1.20. VARIATION OF HEAT OF REACTION WITH TEMPERATURE KIRCHOFF'S EQUATION

Heat evolved or absorbed depends upon the temperature at which the reaction is carried out. The relationship showing the variation of the heat of reaction with temperature was given by Kirchoff in 1856.

Kirchoff's equation gives the relationship between variation in heat of reaction with temperature. It can be stated as :

The change in the heat of reaction at constant pressure for every degree change of temperature is equal to the change in the heat capacity at constant pressure, accompanying the reaction.

It can be derived as under.

Consider the general reaction

$$A \longrightarrow B$$

Kirchoffs

as shown in Fig.1.11

Suppose the heat of reaction (heat evolved) at temperature $T_1 = \Delta H_1$ and the heat of reaction (heat evolved) at temperature $T_2 = \Delta H_2$.

Starting from A at temperature T_1, we can reach B at temperature T_2 by two different paths :

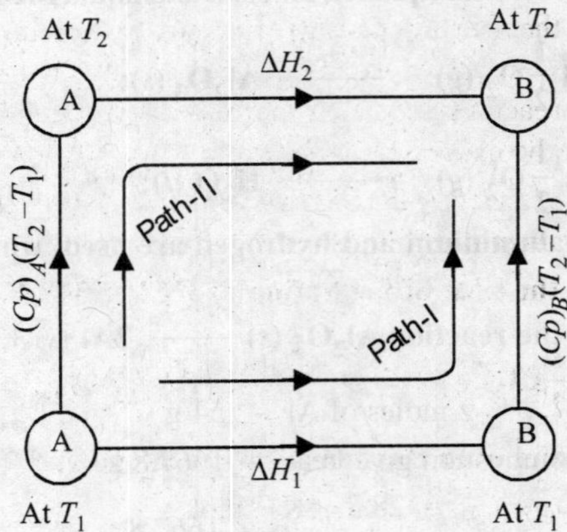

Fig 1.11 Changing A to T_1 to B at T_2 by two different paths.

Path I : We perform the reaction at temperature T_1 to obtain B at the same temperature and then heat this up to the temperature T_2. Then

Heat **evolved** in the first stage = ΔH_1

Heat **absorbed** in the second stage = $(C_P)_B\ (T_2 - T_1)$

where $(C_P)_B$ is the average molar heat capacity of the products.

$\therefore$ Net heat evolved = $\Delta H_1 - (C_p)_n(T_2 - T_1)$

Path II: We first heat up the reactants to the temperature T_2 and then perform the reaction to give the products at the same temperature. Then if $(C_P)_A$ is the average molar heat capacity of the reactant A,

$$\text{Heat absorbed in the first stage} = (C_p)_A\ (T_2 - T_1)$$

$$\text{Heat evolved in the second stage} = \Delta H_2$$

$$\therefore \quad \text{Net heat evolved} = \Delta H_2 - (C_p)_A\ (T_2 - T_1).$$

Since the initial and the final states are the same in both the above cases, therefore, by law of conservation of energy, we must have

Net heat evolved by path I = Net heat evolved by path II

$$i.e. \quad \Delta H_1 - (C_p)_B\ (T_2 - T_1) = \Delta H_2 - (C_p)_A\ (T_2 - T_1)$$

$$\text{or} \quad \frac{\Delta H_2 - \Delta H_1}{T_2 - T_1} = \Delta C_p$$

When the reaction is carried out at constant volume: We have the thermodynamic relation

$$\Delta H = \Delta U + P\Delta V$$

Under conditions of constant volume, $\Delta V = 0$

$$\therefore \quad \Delta H = \Delta U$$

Hence equation derived above takes the form

$$\frac{\Delta U_2 - \Delta U_1}{T_2 - T_1} = \Delta C_v$$

where $\Delta C_v = (C_v)_B - (C_v)_A$ is the change in the heat capacity at constant volume accompanying the reaction, $(C_v)_B$ and $(C_v)_B$ are the mean molar heat capacities at constant volume of the reactants and the products respectively, ΔU_1 and ΔU_2 are the internal energy changes (*i.e.*, heats of reaction at constant volume) when the reaction is carried out at temperatures T_1 and T_2 respectively. Thus Kirchoff's equation may also be expressed as follows :

The change in the heat of reaction at constant volume for every degree change of temperature is equal to the change in the heat capacity at constant volume, accompanying the reaction.

Example 1. Calculate the heat of formation of HCl at 348 K from the following data :

$$\frac{1}{2}\mathbf{H_2}(g) + \frac{1}{2}\mathbf{Cl_2}(g) \longrightarrow \mathbf{HCl}(g);\ \Delta H^0_{298} = -92300\text{J}$$

The mean heat capacities over this temperature range are

$$H_2\ (g);\ C_p = 28.53\ \text{JK}^{-1}\ \text{mol}^{-1}$$
$$Cl_2\ (g);\ C_p = 32.26\ \text{JK}^{-1}\ \text{mol}^{-1}$$
$$HCl\ (g);\ C_p = 28.49\ \text{JK}^{-1}\ \text{mol}^{-1}$$

Solution. The reaction under consideration is

$$\frac{1}{2}H_2(g) + \frac{1}{2}Cl_2(g) \longrightarrow HCl\ (g)$$

For this reaction,

$$\Delta C_p = (C_p)_{HCl} - \frac{1}{2}(C_p)_{H_2} - \frac{1}{2}(C_p)_{Cl_2}$$
$$= 28.49 - \frac{1}{2}(28.53) - \frac{1}{2}(32.26)$$
$$= 1.91\ \text{JK}^{-1}\ \text{mol}^{-1}$$

We are given that

$$\Delta H_1 = -92300\ \text{J}$$
$$T_1 = 298\ \text{K}$$
$$\Delta H_2 = ?$$

According to Kirchoff's equation

$$\frac{\Delta H_2 - \Delta H_1}{T_2 - T_1} = \Delta C_p$$

or

$$\Delta H_2 = \Delta H_1 + \Delta C_p\ (T_2 - T_1)$$
$$= -92300 + (-1{\cdot}91)\ (348 - 298)$$
$$= -923300 - 95{\cdot}5$$
$$= \mathbf{-92395.5\ J}$$

PROBLEMS FOR PRACTICE

1. Calculate the heat of formation of H_2O (liquid) at 383 K, if ΔH for the reaction

$$H_2\ (g) + \frac{1}{2}O_2\ (g) \longrightarrow H_2O\ (l)$$

at 298 K is –298.06 kJ. The average value of heat capacities between the two temperature for H_2 (*g*), O_2 (*g*) and H_2O (*l*) are 27.61, 29.50 and 75.31 $\text{JK}^{-1}\ \text{mol}^{-1}$ respectively.

[**Ans.** 283.92 kJ]

2. If the standard heat of formation of HCl gas at 25° C is –22.66 kcal mol^{-1}, calculate the value at 60° C. C_p for H_2 (g) = 6.82 cal mol^{-1} deg^{-1}

3. The heat of solution of ammonia at 25° C was found to be –11.04 k cal. Calculate the heat of solution at 348 K. Given that the mean heat capacities of N_2, H_2 and NH_3 are 6.80, 6.77 and 8.86 cal/degree/mol respectively.

[Hint. $\frac{1}{2} N_2 + \frac{3}{2} H_2 \longrightarrow NH_3$

$$\Delta C_p = (C_p)\,NH_3 - \frac{1}{2}(C_p)\,N_2 - \frac{3}{2}(C_p)H_2$$

$$T_1 = 25 + 273 = 298 \text{ K}, T_2 = 348 \text{ K}]$$

4. The molar heat capacities at constant pressure of H_2 (g), Cl_2 (g) and HCl (g) are respectively 29.3, 34.7 and 28.9 JK^{-1}. If the heat of formation of HCl (g) at constant pressure at 293 K is –91.2 kJ, what will be its heat of formation at 313 K?
[**Ans.** –91.27 kJ]

5. Consider the reaction

$$H_2O\,(g) \longrightarrow H_2 + \frac{1}{2} O_2\,;\ \Delta H° = +57780 \text{ cal at } 18° \text{ C.}$$

What would be ΔH° value at 25° C? The C_p values are

$$C_p\,(H_2O) = 8.02,\ C_p\,(H_2) = 6.89 \text{ and } C_p\,(O_2) = 7.96 \text{ cal deg}^{-1} \text{ mol}^{-1}$$

[**Ans.** 57796.45 kcal]

SOME SOLVED CONCEPTUAL PROBLEMS

Problem 1. Under what conditions, an extensive property may become intensive property? Give an example.

Solution. An extensive property might become intensive when unit amount of the substance is specified. For example, mass is extensive, but mass per unit volume (*i.e.* density) is intensive. Similarly, heat capacity is extensive but specific heat (which is heat capacity per unit mass) is intensive.

Problem 2. State the variables which are kept constant in the following processes

(*a*) isothermal (*b*) Isobaric (*c*) Isocaloric

Solution. (*i*) Temperature (*ii*) Pressure (*iii*) Volume

Problem 3. Which of the following expressions represent work done and under what conditions?

(*a*) PΔV (*b*) Δ (PV) (*c*) $P_{ext} \int_{V_1}^{V_2} dV$

Solution

(*i*) PΔV denotes work done against constant external pressure, P when the volume changes from V_1 to V_2 such that $\Delta V = V_2 - V_1$

(*ii*) It does not stand for work done

(*iii*) It represents the work done when there is reversible expansion of the gas from the volume V_1 to V_2 against extend pressure P_{ext}

Problem 4. What are the conditions under which q and w become state functions. Discuss briefly.

Solution.

(*i*) At constant volume, there is no work of expansion, hence $w = 0$. By first law of thermodynamics $\Delta U = q_v$. As ΔU is a state function, q_v is also a state function.

(*ii*) At constant external pressure $w = -P\Delta V$. Hence w becomes a state function.

(*iii*) $\Delta U = q_p + w$. As ΔU is a state function, w under constant pressure is also a state function [see (*ii*) above]. Therefore q_p is also a state function.

Problem 5. What are standard state conditions? Which form of carbon is assigned a value of zero for molar enthalpy in the standard state and why?

Solution. A substance is considered to be in standard state if it is at a temperature of 298.15 K and a pressure of one atmosphere. Graphite is assigned a value of zero for enthalpy in the standard state because it is the stable form.

Problem 6. In the case of polyatomic molecules, why is the bond energy of a particular type of bond, the average value?

Solution. Bond energy for the dissociation of a particular bond is not the same in different molecules. It is not the same even in the same molecule like CH_4. Bond energies of four C — H are not equal because after the dissociation of every C — H bond, the electronic environments change. Hence the average value is taken.

Problem 7. What is the Thermodynamic basis of Hess's law?

Solution. Enthalpy is a state function. That means enthalpy change during a reaction does not depend upon the reaction path in which the change is brought about. It depends only upon the nature of the reactants and the final products.

EXERCISES

[Based on Question Papers of Different Universities]

Multiple Choice Questions (Choose the correct option)

1. First law of thermodynamics is the law of
   (*a*) entropy (*b*) free energy
   (*c*) conservation of energy (*d*) none of these
2. The mathematical form of first law of thermodynamics is
   (*a*) dq = dU + dw (*b*) dq = dU – dw
   (*c*) dU = dq + dw (*d*) none of these
3. Standard state refers to
   (*a*) One atmosphere pressure and 25° C temperature
   (*b*) One atmosphere pressure and 100° C temperature
   (*c*) One atmosphere pressure and 0° C temperature
   (*d*) none of these
4. Endothermic reaction is one in which
   (*a*) heat is converted into electricity (*b*) heat is absorbed
   (*c*) heat is given out (*d*) heat is converted into mechanical work

5. Thermochemistry is the study of
(*a*) heat changes accompanying chemical reaction
(*b*) net entropy change in a reaction
(*c*) net free energy change in a chemical reaction
(*d*) none of these

6. A process during which there is no heat change is called
(*a*) an adiabatic process (*b*) reversible process
(*c*) irreversible process (*d*) none of these

7. A process in which a system comes to its initial position after a series of operations is called
(*a*) reversible (*b*) cyclic
(*c*) adiabatic (*d*) none of these

8. A reaction whose heat of reaction shows the bond energy of HCl is
(*a*) $HCl(g) \longrightarrow H(g) + Cl(g)$ (*b*) $2HCl(g) \longrightarrow H_2(g) + Cl_2(g)$
(*c*) $HCl(g) \longrightarrow \frac{1}{2}H_2(g) + \frac{1}{2}Cl_2(g)$ (*d*) $HCl(g) \longrightarrow H^+(g) + Cl^-(g)$

9. A process which proceeds infinitesimally slowly is called
(*a*) irreversible (*b*) reversible
(*c*) isothermal (*d*) adiabatic

10. The equations showing the effect of temperature change on the heat of reaction is known as
(*a*) Arrhenius equation (*b*) Kirchoff's equation
(*c*) Ostwald's equation (*d*) none of these

11. The heat of neutralisation is maximum for the reaction
(*a*) NaOH and HCl (*b*) NaOH and CH_3COOH
(*c*) HCl and NH_4OH (*d*) NH_4OH and CH_3COOH

12. "The resultant heat change in a chemical reaction is the same whether it takes place in one or several stages". This statement is called
(*a*) Le-Chatelier's principle (*b*) Hess's law
(*c*) Lavoisier-Laplace law (*d*) Joule Thomson effect

13. Standard heat of formation of an element is
(*a*) infinite (*b*) zero
(*c*) 100 kJ (*d*) 200 kJ

14. The apparatus used for measuring heat of reaction at constant volume is called
(*a*) calorimeter (*b*) colourimeter
(*c*) Pyrometer (*d*) pyknometer

15. The enthalpies of elements in their standard state are taken as zero. Thus, the enthalpy of formation of a compound
(*a*) will always be positive (*b*) will always be negative
(*c*) may be positive or negative (*d*) will always be zero

16. The heat constant of a system is called

(*a*) internal energy (*b*) enthalpy

(*c*) free energy (*d*) entropy

17. For the reaction $N_2 + 3H_2 \rightleftharpoons 2NH_3$, q_p is equal to

(*a*) $q_v - 2RT$ (*b*) $q_v + 2RT$

(*c*) $q_v + RT$ (*d*) $q_v - RT$

18. The least random state of water system is

(*a*) ice (*b*) liquid water

(*c*) steam (*d*) equal randomness in all

19. Identify the intensive quantities from the following

(*a*) Enthalpy (*b*) Temperature

(*c*) Volume (*d*) Refractive Index

20. The heat of neutralisation of HCl and NaOH is 13.7 kcal and that of NaOH and CH_3COOH is 13.2 kcal. Heat of ionisation of CH_3COOH is

(*a*) 13.3 kcal (*b*) 13.7 kcal

(*c*) 0.5 kcal (*d*) none of these

Answers to Multiple Choice Questions

1. (*c*)	**2.** (*c*)	**3.** (*a*)	**4.** (*b*)	**5.** (*a*)	**6.** (*a*)
7. (*b*)	**8.** (*a*)	**9.** (*b*)	**10.** (*b*)	**11.** (*a*)	**12.** (*b*)
13. (*b*)	**14.** (*a*)	**15.** (*d*)	**16.** (*b*)	**17.** (*a*)	**18.** (*a*)
19. (*d*), (*a*)	**20.** (*c*)				

Short Answer Questions

1. Point out the main difference between

(*a*) Open, closed and isolated system

(*b*) Reversible and irreversible processes

2. What do you understand by the state of a system? What are state variables? Why should it be sufficient to specify the values of a few variables to define the state of a system instead of specifying the values of all the variables?

3. Derive expression for the work done when a gas expands from volume V_1 to V_2 against external pressure, P.

4. Define Thermodynamics. What are its main objectives?

5. Distinguish between

(*i*) State function and path function.

(*ii*) Isothermal and adiabatic process.

(*iii*) Dependent and independent variables.

6. Explain that the work done during reversible expansion is the maximum work for a given increase of volume and minimum to compress the gas for a given decreases of volume.

7. What are extensive and intensive properties? Give three examples of each of them.

8. Give in brief the objectives and limitations of thermodynamics.

9. What is the relationship between Heat and Mechanical work? Define Joule's mechanical equivalent of heat. What is its value?

10. State First law of thermodynamics in two different ways. Derive its mathematical formulation.

11. Briefly explain the terms Enthalpy and Enthalpy change. Is it intensive or extensive property?

12. Prove that the heat capacity at constant volume (C_v) for an ideal gas having translational kinetic energy only is constant and equal to 3 cal mol^{-1} and is independent of temperature.

13. State and explain Joule's law about the variation of internal energy of an ideal gas with volume.

14. What do you understand by Inversion temperature ?

15. Define Zeroth law of thermodynamics. What is its most important application?

16. Explain the fact that internal energy is a state function but work and heat are path functions.

17. Derive expression for molar heat capacities C_v and C_P in terms of internal energy change and enthalpy change and hence show that $C_p - C_p = R$ for one mole of an ideal gas.

18. Prove that C_p for an ideal gas is constant and nearly equal to 5 cal degree^{-1} mol^{-1}.

19. Write a short note on Joule-Thomson Effect.

20. Derive expression for Joule Thomson coefficient and inversion temperature in terms of van der Waals' constants.

21. What generalization do you draw about the relationship between the atomicity of the gases and the ratio (γ) of their specific heats?

General Questions

1. State and explain Joule's law.
2. State First Law of Thermodynamics in at least three different ways. Derive the mathematical expression for it. Explain the fact that internal energy is a state function but heat and work are not.
3. State and explain the following terms with suitable examples:
   (*i*) Open, closed and isolated systems
   (*ii*) Thermodynamic equilibrium
   (*iii*) Thermodynamics Reversible and Irreversible processes
   (*iv*) Extensive and Intensive properties
   (*v*) State function and path function
4. What is Joule-Thomson effect? Justify that during this process, enthalpy of the system remains constant. Also explain what you understand by Joule-Thomson coefficient and Inversion temperature. Derive that Joule-Thomson coefficient is given by

$$\mu = -\frac{1}{C_p}\left(\frac{\partial H}{\partial P}\right)_T$$

5. Derive expressions for w, q, ΔU and ΔH when an ideal gas undergoes (*i*) isothermal reversible expansion (*ii*) isothermal irreversible expansion.
6. Derive expressions for w, ΔU, q and ΔH for isothermal reversible expansion of a real gas.
7. Define 'Bond energy'. How is bond energy of a bond calculated for a polyatomic molecule? How is it used to calculate the enthalpy change of a reaction?
8. Derive that for reversible adiabatic expansion of an ideal gas:
   (*i*) PV^γ = constant (*ii*) work done, $w = C_v\,(T_2 - T_1)$.

9. Define standard enthalpy of reaction, standard enthalpy of formation and standard enthalpy of combustion. What is Hess's law? What is the thermodynamic basis of Hess's law?

10. Briefly explain the terms internal energy, work and heat. Derive the expression for the work of expansion. Explain that work done by a system in a reversible process is the maximum work for a given increases of volume and for a given decreases of volume, it is the minimum work to reverse the process.

11. Explain the terms (*i*) 'Enthalpy' and 'Enthalpy change' (*ii*) Heat capacity. Derive expression for heat capacity at constant volume and that at constant pressure. Derive the relationship between them.

12. Prove thermodynamically that Joule-Thomson coefficient for an ideal gas is zero. Also deduce expression for Joule-Thomson coefficient for real gases in terms of van der Waals constants '*a*' and '*b*'

13. What do you understand by Heat of reaction at constant volume and that at constant pressure. Derive the relationship between them. Under what conditions, the two are equal?

14. Prove that for a monatomic gas (having transtational kinetic energy only), $C_v = \frac{3}{2}R.$

15. Derive expressions for ΔU and ΔH for adiabatic reversible expansion of real gas.

16. Briefly explain the terms enthalpy of neutralisation. Comment on their values if both are strong or one of them is weak or both are weak?

17. Derive thermodynamically Kirchoff's equation giving the variation of heat of reaction with temperature.

18. Derive thermodynamically Kirchoff's equation.

Or

Show that the temperature dependence of heat of reaction is given by the relation

$$\left(\frac{\partial \Delta H}{\partial T}\right)_P = \Delta C_p.$$

Name the equation.

19. (*a*) Distinguish between reversible isothermal expansion and reversible adiabatic expansion.

(*b*) Show that for adiabatic expansion of an ideal gas,

$$PV^{\gamma} = \text{constant}$$

20. Derive the following expression for the reversible adiabatic expansion of a real gas.

$$T_2 = T_1\left(\frac{V_1 - nb}{V_2 - nb}\right)^{R/C_p}$$

21. Define 'Bond energy' for a diatomic molecule and for a polyatomic molecule. How bond energy data helps to calculate the enthalpy change of a reaction?

22. How can the reversible isothermal expansion of an ideal gas be brought about? Derive an expression for the work of expansion of such a process.

23. What are the values/expressions for *q*, *w*, ΔU and ΔH for

(*i*) irreversible isothermal free expansion of an ideal gas?

(*ii*) irreversible isothermal intermediate expansion of an ideal gas?

Out of reversible and irreversible expansion which one will absorb less heat and why?

24. Derive the expression for the work done in an adiabatic expansion. Will the expression be different for adiabatic expansion along a reversible and an irreversible path ?

25. The work done in a reversible isothermal expansion of an ideal gas is greater than the work done in reversible adiabatic expansion. Explain.

26. What are the values/ expressions, for q, w, ΔU and ΔH for

(*i*) irreversible adiabatic free expansion of an ideal gas?

(*ii*) irreversible adiabatic intermediate expansion of an ideal gas?

27. Define 'Standard enthalpy of formation'. Taking a suitable example, prove that the standard enthalpy of a compound is equal to its standard enthalpy of formation.

28. How is the final temperature of irreversible adiabatic expansion of an idea gas calculated?

29. Define standard enthalpy of combustion. Explain with a suitable example.

30. Compare reversible isothermal expansion with reversible adiabatic expansion of an ideal gas.

31. Derive relationship between heat of reaction at constant pressure and that at constant volume.

32. Show the maximum work is produced in a reversible isothermal expansion of a gas.

33. How maximum amount of the work during isothermal expansion of a gas can be obtained? Derive an expression for the maximum work obtainable from the isothermal expansion of n moles of a perfect gas.

Time's arrow

Disorder is more probable than order

2

Thermodynamics-II

(Second and Third Laws of Thermodynamics)

2.1. INTRODUCTION – NEED FOR THE SECOND LAW

A major **limitation of the first law of thermodynamics** is that it provides no information about the feasibility of a process. For example, it does not tell us whether water can run uphill itself, whether a gas diffuse from a region of low pressure to a region of high pressure, whether heat can flow from a cold bold to a hot body etc. It does tell that there is an exact equivalence between various forms of energy and that heat gained is equal to heat lost. But that does not serve the whole purpose. Second law of thermodynamics answers the questions that remain unanswered by first-law.

The first law states that when heat is converted into work, the work performed is equivalent to heat absorbed. However, this is not precisely correct. It has been seen from our day-to-day exprience that heat absorbed cannot be completely converted into work without leaving some change in the system or surroundings.

In order to sort out these issues, second law of thermodynamics was proposed. It has a number of statements. Different statements of second law of thermodynamics supply answers to various questions.

2.1.1. Different statements of second law of thermodynamics

1. *All spontaneous processes like the flow of heat from hot end to cold end, diffusion of gas from high pressure to low pressure or the flow of water down a hill etc. are thermodynamically irreversible.*

Spontaneous process means a process which can take place without the help of any external agency. All *natural processes,* some of which are mentioned in the definition above, are spontaneous processes.

Further, the first law states that when heat is converted into work, the work obtained is equivalent to the heat absorbed. However it has been seen from experience that the heat absorbed cannot be completely converted into work without leaving some change in the system or the surroundings. Hence the second law is also stated as follows:

2. *The complete conversion of heat into work is impossible without leaving some effects elsewhere.*

3. *It is impossible to construct a machine, functioning in cycles, which can convert heat completely into the equivalent amount of work without producing changes elsewhere.*

4. *Without the use of an external agency, heat cannot by itself pass from a colder to a hotter body.*

2.1.2. Efficiency of a machine

The fraction of the heat absorbed by a machine that it can transform into work is called the **efficiency** *of the machine.* Thus if Q is the heat absorbed and W is the work done, then the efficiency of the machine is given by

$$\eta = \frac{W}{Q}$$

The machine used for the conversion of heat into work is called **Heat Engine.** In order to bring about this conversion, the engine absorbs heat from a heat reservoir at a higher temperature, called the **source**, converts a part of it into work and returns the remainder to the heat reservoir at a lower temperature, called the **sink**.

2.2. CARNOT CYCLE

A Carnot cycle is a process in which a system returns to its original state after a number of successive changes. A process conducted in this manner is called a cyclic process. The Carnot cycle consists of four different operations which can be shown on pressure-volume diagram as shown in Fig. 2.1.

A Carnot engine consists of a cylinder containing 1 mole of an ideal gas as the working substance and fitted with weightless, frictionless piston so shat all the operations in the cycle are carried out reversibly. For this reason, it is also called *reversible heat engine.* The cylinder is supposed to be insulated on all sides except at the bottom so that heat can flow to or from the system only through the bottom. Further, it is supposed that there are two heat reservoirs, one at a higher temperature T_2 (called the *source*) and the other at a lower temperature T_1 (called the *sink*). If some operation is carried out by placing the cylinder in the source or the sink, it can exchange heat with it and hence the temperature remains constant so that the process is *isothermal.* On the other hand, if the cylinder is placed on an insulating material, no heat exchange can take place between the system and the surroundings and hence the process is *adiabatic.*

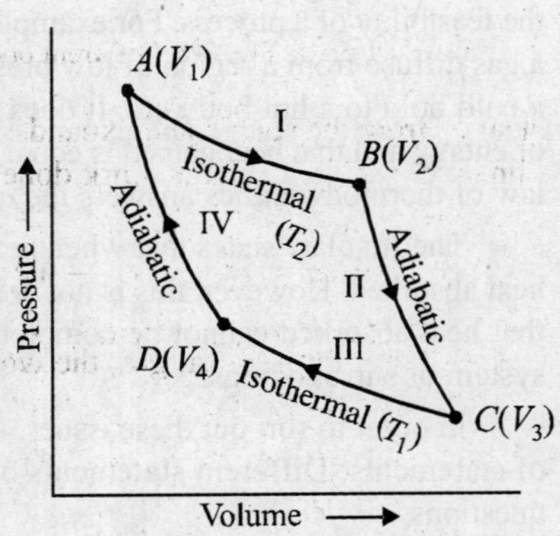

Fig. 2.1. Carnot cycle.

Four different operations are carried out as shown in Fig. 2.2 and are described below:

I. Isothermal expansion: The cylinder containing one mole of the ideal gas, occupying a volume V_1, is placed in contact with the source (*i.e.* heat reservoir at temperature T_2). The gas absorbs

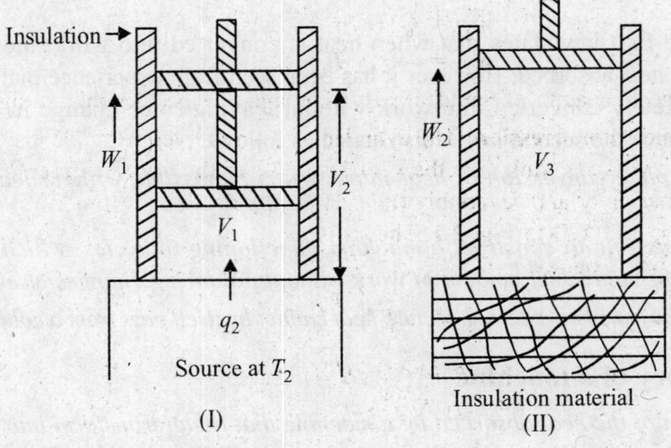

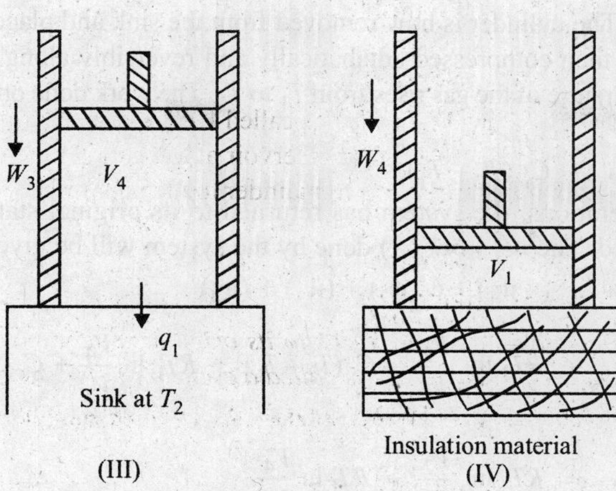

Sadi Carnot

Fig. 2.2. Different operations of the Carnot cycle.

heat q_2 from the source and expands *isothermally* and *reversibly* till its volume has increased to V_2. Since the gas is ideal, the work done w_1 by one mole of the ideal gas will be given by

$$- w_1 = RT_2 \ln \frac{V_2}{V_1} \qquad ...(i)$$

Further for an ideal gas, the work done is equal to the heat absorbed, q_2 so that

$$- w_1 = q_2 = RT_2 \ln \frac{V_2}{V_1} \qquad ...(ii)$$

Here, w_1 work done by the system has been taken as negative and q_2 as positive as per latest SI conventions.

II. Adiabatic expansion: The cylinder is now removed from the source and placed in perfectly insulating material so that the gas now expands adiabatically and reversibly. Work is done in the expansion but since no heat enters or leaves the system, the temperature must fall. The reversible adiabatic expansion is continued till the temperature has fallen to T_1 which is the temperature of the sink. Suppose the corresponding volume becomes V_3. The path is shown by the adiabatic curve *BC* in Fig. 2.1. The work done w_2 is be given by

$$- w_2 = C_v (T_1 - T_2)$$

or

$$w_2 = - C_v (T_2 - T_1) \qquad ...(iii)$$

where C_v is the heat capacity of the ideal gas. Here, again w_2 *is negative* by sign conventions.

III. Isothermal compression: The cylinder is now removed from the insulating material and placed in contact with the sink (*i.e.* the heat reservoir at the lower temperature T_1). The gas is compressed isothermally and reversibly till the volume decreases from V_3 to V_4. The process is represented by the path *CD* in Fig. 2.1. The work done on the system is taken as *positive* by new conventions and will be given by

$$w_3 = RT_1 \ln \frac{V_4}{V_3} \qquad ...(iv)$$

During the operation, an amount of heat q_1, exactly equal to w_3, will be returned to the sink at T_1. According to sign conventions, q_1 will be negative and w_3 will be positive.

$$- q_1 = w_3 = RT_1 \ln \frac{V_4}{V_3} \qquad ...(v)$$

IV. Adiabatic compression: The cylinder is now removed from the sink and placed again on the insulating material. The gas is then compressed adiabatically and reversibly along DA till the initial state A is regained. The temperature of the gas rises from T_1 to T_2. The work done on the system is *positive* and will be given by

$$w = C_v (T_2 - T_1) \qquad ...(vi)$$

As a result of these four operations, the system has returned to its original state so that a reversible cycle has been completed. The *net work* (w) done by the system will be given by

$$w = (-w_1) + (-w_2) + (w_3) + (w_4) \qquad ...(vii)$$

$$= RT_2 \ln \frac{V_2}{V_1} - C_v(T_2 - T_1) + RT_1 \ln \frac{V_4}{V_3} + C_v(T_2 - T_1)$$

$$= RT_2 \ln \frac{V_2}{V_1} + RT_1 \ln \frac{V_4}{V_3} \qquad ...(viii)$$

V_1 and V_4 lie on one adiabatic curve and V_3 and V_2 lie on another, applying volume-temperature relationships, we have

$$\left(\frac{V_4}{V_1}\right)^{\gamma-1} = \frac{T_2}{T_1} \qquad ...(ix)$$

and

$$\left(\frac{V_3}{V_2}\right)^{\gamma-1} = \frac{T_2}{T_1} \qquad ...(x)$$

Comparing equations (*ix*) and (*x*), we get

$$\left(\frac{V_4}{V_1}\right)^{\gamma-1} = \left(\frac{V_3}{V_2}\right)^{\gamma-1}$$

or

$$\frac{V_4}{V_1} = \frac{V_3}{V_2}$$

or

$$\frac{V_4}{V_3} = \frac{V_1}{V_2} \qquad ...(xi)$$

Substituting this value in equation (*viii*), we get

$$w = RT_2 \ln \frac{V_2}{V_1} + RT_1 \ln \frac{V_1}{V_2}$$

$$= RT_2 \ln \frac{V_2}{V_1} - RT_1 \ln \frac{V_2}{V_1}$$

$$= R(T_2 - T_1) \ln \frac{V_2}{V_1} \qquad ...(xii)$$

Dividing equation (*xii*) by equation (*ii*), we have

$$\frac{w}{q_2} = \frac{T_2 - T_1}{T_2} \qquad \text{...(xiii)}$$

This gives the *efficiency* of the Carnot cycle or engine.

From equation (*xiii*), it is clear that the efficiency of the reversible heat engine depends only upon the temperatures of the source and the sink and is independent of the nature of the working substance.

Since the quantity $\frac{T_2 - T_1}{T_2}$, which represents efficiency is always less than unit, the efficiency of the heat engine is thus always less than unity.

2.2.1. Carnot Theorem

The result that follows from Carnot cycle is called Carnot theorem.

Carnot theorem is stated as follows:

The efficiency of a reversible heat engine depends only upon the temperatures of the source and the sink and is independent of the nature of the working substance. In other words, all reversible heat engines working between the same two temperatures have the same efficiency.

Combining equations (*ii*), (*v*) and (*viii*) in the previous section, we get

$$w = q_2 - q_1$$

i.e. net work done by the system is equal to net heat absorbed by the system. Putting the value of w in equation (*xiii*) above, we have

$$\frac{q_2 - q_1}{q_2} = \frac{T_2 - T_1}{T_2}$$

Hence the efficiency of a heat engine may be given by any one of the following expressions:

$$\eta = \frac{w}{q_2} = \frac{q_2 - q_1}{q_2} = \frac{T_2 - T_1}{T_2}$$

Obviously, the efficiency of the engine can be increased by widening the difference in the temperatures of the source and sink.

SOLVED EXAMPLES ON CARNOT CYCLE

Example 1. Calculate the amount of heat supplied to Carnot's cycle working between 368 K and 288 K if the maximum work obtained is 895 joules.

Solution. It is given that

$$T_1 = 368 \text{ K}$$
$$T_2 = 288 \text{ K}$$
$$w = 895 \text{ Joules}$$

Substituting these values in the formula,

$$\eta = \frac{w}{q_2} = \frac{T_2 - T_1}{T_2}$$

we have

$$\frac{895}{q_2} = \frac{368 - 288}{368}$$

$$q_2 = \frac{895 \times 368}{80} = \mathbf{4117 Joules}$$

Example 2. Calculate the maximum efficiency of a steam engine operating between 110° and 25°C. What would be the efficiency of the engine if the boiler temperature is raised to 140°C, the temperature of the sink remaining the same.

Solution. First case. Given that

$$T_2 = 110°C = 110 + 273K = 383 \text{ K}$$
$$T_1 = 25°C = 25 + 273 \text{ K} = 298 \text{ K}$$

$$\therefore \quad \eta = \frac{T_2 - T_1}{T_2} = \frac{383 - 298}{383} = 0.222 = \mathbf{22.2\%}$$

Second case. Given that

$$T_2 = 140°C = 140 + 273 \text{ K} = 413 \text{ K}$$
$$T_1 = 25°C = 25 + 273 \text{ K} = 298 \text{ K}$$

$$\therefore \quad \eta = \frac{T_2 - T_1}{T_2} = \frac{413 - 298}{413} = 0.278 = \mathbf{27.8\%}$$

Example 3. Compare the thermodynamic efficiencies to be expected:

(*a*) when an engine is allowed to operate between 1000 K and 300 K;

(*b*) when an engine is allowed to operate between 1000 K and 600 K and then the waste heat is passed on to another engine which operates between 600 K and 300 K.

Solution. Case (*a*)

$$\eta = \frac{1000 - 300}{10000} = 0.70 = \mathbf{70\%}$$

Case (*b*)

(*i*) $$\eta = \frac{1000 - 600}{1000} = 0.40 = \mathbf{40\%}$$

(*ii*) $$\eta = \frac{600 - 300}{600} = 0.50 = 50\%$$ of the rejected 60% of (*i*) = 30%

$\therefore$ Total $\eta = 40 + 30$
$= 70$

2.3. THERMODYNAMIC SCALE OF TEMPERATURE

An ideal gas or a perfect gas ceases to exist at a temperature of – 273.15°C. This temperature is called zero degree absolute. The scale developed on this basis is called perfect gas temperature scale. It is found to be independent of the ideal gas taken.

Lord Kelvin

Lord Kelvin was the first to develop a scale of temperature that was independent of the nature of the gas. This was based on the efficiency of a reversible heat engine. This scale of temperature is called thermodynamic scale of temperature or Kelvin scale of temperature. It is found to be similar to the ideal gas temperature scale. Thermodynamic scale of temperature is developed as under.

Assume two heat reservoirs, one acting as a source and the other acting as a sink. Imagine a reversible heat engine operating between them. We further assume that the temperature of source on the new scale is proportional to the quantity of heat absorbed from it and the temperature of the sink is proportion to the heat lost to it.

Let the

$$\text{Heat absorbed from the source} = Q_2$$
$$\text{Heat lost to the sink} = Q_1$$
$$\text{Temperature of source on the thermodynamic scale} = \theta_2$$
$$\text{Temperature of sink on the thermodynamic scale} = \theta_1$$

According to the above discussion,

$$\frac{Q_2}{Q_1} = \frac{\theta_2}{\theta_1}$$

Taking reciprocal of both sides and subtracting the result from 1, we get

$$1 - \frac{Q_1}{Q_2} = 1 - \frac{\theta_1}{\theta_2} \quad \text{or} \quad \frac{Q_2 - Q_1}{Q_2} = \frac{\theta_2 - \theta_1}{\theta_2}.$$

Putting $\theta_1 = 0$ *i.e.* zero of the new scale,

$$\frac{Q_2 - Q_1}{Q_2} = 1$$

This signifies that zero on the Kelvin scale is the temperature of the sink for a reversible heat engine whose efficiency is 1 *i.e.* complete conversion of heat into work. But this is possible only at absolute zero on the perfect gas scale of temperature. We conclude from here that gas scale and Kelvin scale are similar, provided the gas taken is ideal (perfect).

Thermodynamic scale of temperature derived from second law of thermodynamics is more fundamental than that derived from ideal gas because we can use any fluid in the former.

2.4. CONCEPT OF ENTROPY

For a reversible Carnot cycle working between temperatures T_2 and T_1,

$$\frac{q_2 - q_1}{q_2} = \frac{T_2 - T_1}{T_2} \qquad \text{... }(i)$$

where q_2 is the heat absorbed isothermally and reversibly at temperature T_2 and q_1 is the heat lost isothermally and reversibly at temperature T_1.

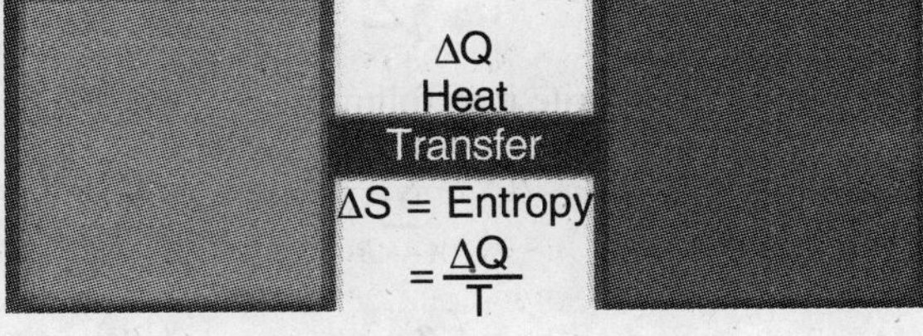

The above equation may be written as

$$1 - \frac{q_1}{q_2} = 1 - \frac{T_1}{T_2}$$

$$\text{or} \quad \frac{q_2}{T_2} = \frac{q_1}{T_1} \qquad \text{... }(ii)$$

$$\text{or} \quad \frac{q}{T} = \text{constant}$$

Thus heat absorbed or lost isothermally and reversibly divided by the temperature at which the heat is absorbed or lost is a constant quantity for a particular system.

In equation (*ii*), q_2 is the heat absorbed at temperature T_2 and q_1 is the heat lost at temperature T_1. If q_1 were the heat absorbed at temperature T_1, equation (*ii*) would be rewritten as

$$\frac{q_2}{T_2} = -\frac{q_1}{T_1}$$

or
$$\frac{q_2}{T_2} + \frac{q_1}{T_1} = 0$$

i.e.
$$\sum \frac{q}{T} = 0 \qquad \text{... }(iii)$$

Consider any reversible cycle *ABA*. It may be regarded as being made up of a number of Carnot cycles drawn across the diagram as shown in Fig. 2.3. Starting from A and going through all the cycles successively from *A* to *B* it can be shown that all paths inside the closed curve *ABA* cancel each other leaving only the zig-zag outer path. This zig-zag path is almost the same as the path *ABA* of the reversible cycle. By making each Carnot cycle smaller and increasing their number, it can be made to match more closely to *ABA*. The reversible cycle can be regarded as being constituted of an infinite number of Carnot cycles. For each of these cycles, we have

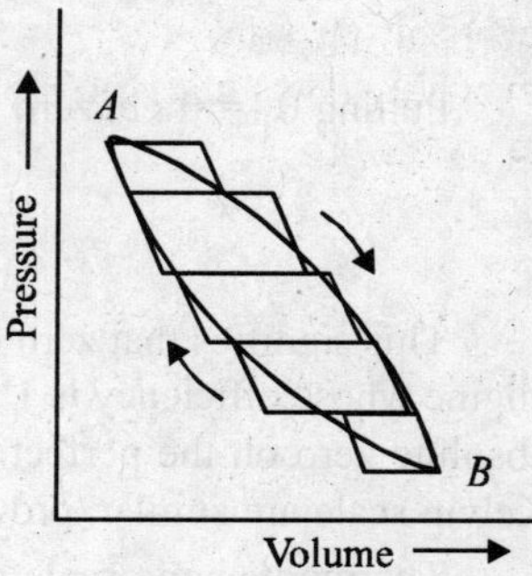

$$\sum \frac{q}{T} = 0.$$

Hence for an infinite number of Carnot cycles,

$$\sum \frac{\delta q}{T} = 0. \qquad \text{... }(iv)$$

Fig. 2.3. Carnot cycles drawn across the reversible cycle.

where δq is a very small quantity of the heat absorbed at temperature *T* during the small isothermal change of the reversible process.

The above term may be divided into two parts, one for the path *A* to *B* and the other for the back path *B* to *A*. Thus

$$\sum \frac{\delta q}{T} = \sum_{A \to B} \frac{\delta q}{T} + \sum_{B \to A} \frac{\delta q}{T} = 0$$

or we may write it as follows:

$$\sum_{A \to B} \frac{\delta q}{T} = -\sum_{B \to A} \frac{\delta q}{T}$$

or
$$\left(\frac{q}{T}\right)_{A \to B} = -\left(\frac{q}{T}\right)_{B \to A} \qquad \text{...}(v)$$

This means that the total value of the quantity $\frac{q}{T}$ for the path *A to B* is equal to the total value of the quantity $\frac{q}{T}$ for the reverse path *B* to *A*.

Thus *quantity* $\frac{q}{T}$ *is a state function, i.e. its value depends only upon the initial state* (*A*) *and the final state* (*B*) *and is independent of the path.* Obviously $\frac{q}{T}$ is the change in the value of a quantity or a function having some definite values at *A* and *B* depending upon the values of pressure, volume and temperature. This quantity or the function is called **entropy** and is usually represented by the symbol *S*. Thus if S_A is the value of the entropy at *A* and S_B is the value at *B*, then we must have

$$\frac{q}{T} = S_B - S_A = \Delta S \qquad \text{...}(vi)$$

where ΔS represents the total change in entropy in going from the initial state *A* to the final state *B*.

Hence **entropy** is a state function, the change in the value of which from the initial state to the final state is equal to the quantity *q/T i.e.* the total heat absorbed reversibly and isothermally in going from the initial state to the final state divided by the absolute temperature at which the heat is absorbed. From equation (*vi*), the **entropy change** *may be defined directly as the quantity of heat absorbed isothermally and reversibly divided by the absolute temperature* (*T*) *at which the heat is absorbed.*

Since the derivation of the above result is based upon Carnot cycle in which the heat is absorbed reversibly, therefore the expression (*vi*) for the entropy change is valid only when the heat is absorbed reversibly. Thus it is better to use the symbol q_{rev} in place of *q*. The expression (*vi*) may, therefore, be written as

$$\Delta S = \frac{q_{rev}}{T} \qquad \text{...}(vii)$$

Further, as entropy is a state function, a *small change* in its value can be represented by *dS*. Hence we may write

$$dS = \frac{q_{rev}}{T}$$

2.4.1. Entropy change in a reversible process

If q_{rev} is the heat absorbed by the system reversibly, then the heat lost by the surroundings will also be q_{rev}. If the process takes place isothermally at the absolute temperature *T*, then

(*i*) *Entropy change of the system* is given by

$$\Delta S_{\text{system}} = \frac{q_{rev}}{T} \qquad \text{...}(i)$$

(*ii*) *Entropy change of the surroundings* is given by

$$\Delta S_{\text{surroundings}} = -\frac{q_{rev}}{T} \qquad \text{...}(ii)$$

Thus the *total entropy change* for the combined system and the surroundings will be

$$\Delta S_{\text{system}} + \Delta S_{\text{surroundings}} = \frac{q_{rev}}{T} - \frac{q_{rev}}{T} = 0 \qquad \text{...}(iii)$$

We may conclude that *in a reversible process, the net entropy change for the combined system and the surroundings is zero.*

2.4.2. Entropy change in an irreversible process

Suppose that the total heat lost by the surroundings is q_{irrev}. This heat is absorbed by the system but the *entropy change* of the system *does not depend upon the heat actually absorbed but it depends upon the heat absorbed reversibly i.e.*, on the quantity q_{rev}. Thus if the heat is absorbed isothermally by the system at the absolute temperature *T*, the *entropy change of the system* is given by

$$\Delta S_{system} = \frac{q_{rev}}{T} \qquad \text{... (i)}$$

Further suppose that the loss of heat (q_{irrev}) by the surroundings takes place infinitesimally slowly because the surroundings are much bigger in size and magnitude compared to the system (*i.e.* reversibly) and isothermally at the temperature *T.* Then the *entropy change of the surroundings* is given by

$$\Delta S_{surroundings} = -\frac{q_{irrev}}{T} \qquad \text{... (ii)}$$

The *total entropy change for the combined system and the surroundings will, become*

$$\Delta S_{system} + \Delta S_{surroundings} = \frac{q_{rev}}{T} - \frac{q_{irrev}}{T} \qquad \text{... (iii)}$$

We know that the work done in a reversible process is the maximum work *i.e.*

$$w_{rev} > w_{irrev} \qquad \text{... (iv)}$$

Further, as the internal energy (*U*) is state function, the value of ΔU is same whether the process is carried out reversibly of irreversibly. Therefore

$$\Delta U = q_{rev} - w_{rev} = q_{irrev} - w_{irrev} \qquad \text{... (v)}$$

Combining results (*iv*) and (*v*), we come to the conclusion

$$q_{rev} > q_{irrev} \qquad \text{... (vi)}$$

$$\therefore \quad \frac{q_{rev}}{T} > \frac{q_{irrev}}{T} \qquad \text{... (vii)}$$

or

$$\frac{q_{rev}}{T} - \frac{q_{irrev}}{T} > 0 \qquad \text{... (viii)}$$

From (*iii*) and (*viii*), we have

$$\Delta S_{system} + \Delta S_{surroundings} > 0 \qquad \text{...(ix)}$$

Thus it may be concluded that *in an irreversible process, the entropy change for the combined system and the surroundings is greater than zero i.e., an irreversible process is accompanied by a net increase of entropy.*

Since all spontaneous processes are thermodynamically irreversible it may be stated that

All spontaneous processes are accompanied by a net increase of entropy.

2.4.4. Clausius Inequality

For an **irreversible** process, we can write

$$\Delta S_{system} + \Delta S_{surroundings} > 0 \qquad \text{... (i)}$$

or $\quad \Delta S_{system} > -\Delta S_{surroundings} \quad$ ... (*ii*)

If q_{irrev} is the heat change in surroundings at the temperature *T*, then

$$\Delta S_{surroundings} = -\frac{q_{irrev}}{T} \qquad \text{... (iii)}$$

Substituting this value in (*ii*) above, we get

$$\Delta S_{system} > \frac{q_{irrev}}{T} \qquad \text{... (iv)}$$

If the change in entropy is small, eq. (*iv*) can be written as

$$dS_{\text{system}} > \frac{q_{\text{irrev}}}{T} \quad \text{... (v)}$$

Expressions (*iv*) and (*v*) are known as Clausius inequality. Clausius assumed that the universe could be taken as isolated system, in which all naturally occurring processes are irreversible.

2.4.5. Entropy of the universe is increasing

All processes occurring in nature are thermodynamically irreversible and these are accompanied by increase of entropy. Hence it may be concluded that

The entropy of the universe is continuously increasing.

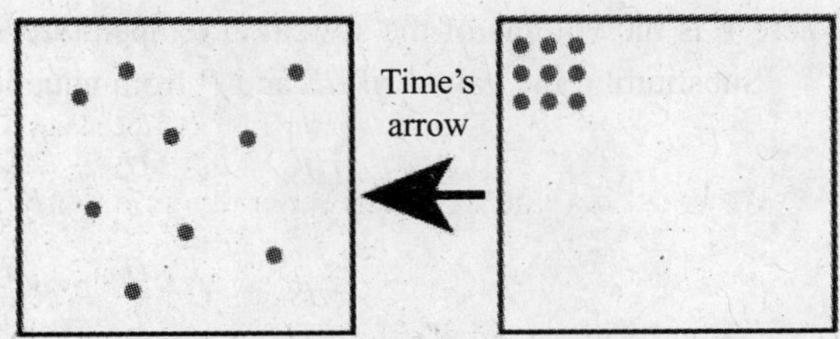

Thus the main ideas of the *first and the second law of thermodynamics* may be summed up as follows:

The energy of the universe is constant whereas the entropy of the universe is continuously increasing and tends to a maximum value.

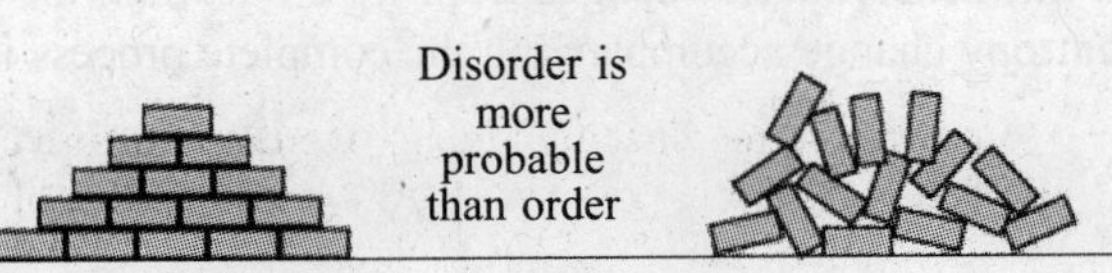

2.5. ENTROPY CHANGE FOR AN IDEAL GAS UNDER DIFFERENT CONDITIONS

Let us consider one mole of an ideal gas enclosed in a cylinder fitted with a frictionless piston. If a small amount of heat δq_{rev} is supplied to the system reversibly and isothermally at the temperature *T*, then the entropy change accompanying the process is given by

$$dS = \frac{\delta q_{rev}}{T} \quad \text{... (i)}$$

According to first law of thermodynamics, for an infinitesimal process,

$$\delta q = dE + \delta w \quad \text{... (ii)}$$

If the process is carried out *reversibly,* equation (*ii*) may be written as

$$\delta q_{rev} = dE + \delta w \quad \text{... (iii)}$$

Also, if the *work is restricted to the work of expansion,*

$$\delta w = PdV \quad \text{... (iv)}$$

where dV is the small increase in volume and P is the pressure of the system.

Putting the value of dw from equation (*iv*) in equation (*iii*), we get

$$\delta q_{re} = dE + PdV \quad \text{... (v)}$$

Substituting this value in equation (*i*), we have

$$dS = \frac{dE + PdV}{T}$$

or it may be written as

$$TdS = dE + PdV \quad \text{... (vi)}$$

For 1 mole of an ideal gas, we know that

$$C_v = \frac{dE}{dT}$$

i.e,

$$dE = C_v\, dT \quad \text{... (vii)}$$

where C_v is the molar heat capacity at constant volume and

$$PV = RT$$

i.e.

$$P = \frac{RT}{V} \quad \text{... (viii)}$$

where V is the volume of the system at temperature T and pressure P and R is the gas constant.

Substituting the values of dE and P from equation (*vii*) and (*viii*) in equation (*vi*), we get

$$TdS = C_v\, dT + \frac{RT}{V} dV$$

or

$$dS = C_v \frac{dT}{T} + R\frac{dV}{V} \quad \text{... (ix)}$$

If the volume changes from V_1 to V_2 when the temperature changes from T_1 to T_2, then the entropy change accompanying the complete process is given by the equation

$$\int_{S_1}^{S_2} dS = \int_{T_1}^{T_2} C_v \frac{dT}{T} + \int_{V_1}^{V_2} R\frac{dV}{V} \quad \text{... (x)}$$

Assuming that C_v remains constant in the temperature range T_1 to T_2, equation (*x*) may be put as

$$\int_{S_1}^{S_2} dS - = C_v \int_{T_1}^{T_2} \frac{dT}{T} + R\int_{V_1}^{V_2} \frac{dV}{V}$$

$$\boldsymbol{\Delta S = C_V \ln \frac{T_2}{T_1} + R \ln \frac{V_2}{V_1}} \quad \text{... (xi)}$$

This is an expression for the calculation of entropy change of 1 mole of an ideal gas accompanying a process when temperature changes from T_1 to T_2 and the volume changes from V_1 to V_2.

Further for 1 mole of an ideal gas, we may write

$$P_1V_1 = RT_1 \text{ for the initial state} \quad \text{... (xii)}$$

and

$$P_2V_2 = RT_2 \text{ for the final state} \quad \text{... (xiii)}$$

Dividing equation (*xiii*) by equation (*xii*), we have

$$\frac{P_2V_2}{P_1V_1} = \frac{T_2}{T_1}$$

or

$$\frac{V_2}{V_1} = \frac{T_2P_1}{T_1P_2} \quad \text{... (xiv)}$$

Substituting this value in equation (*xi*), we get

$$\Delta S = C_v \ln \frac{T_2}{T_1} + R \ln \frac{T_2P_1}{T_1P_2} \quad \text{... (xv)}$$

Also, we know that

$$C_P - C_v = R \quad \text{or} \quad C_v = C_P - R$$

Putting this value in equation (*xv*), we get

$$\Delta S = (C_P - R)\ln\frac{T_2}{T_1} + R\ln\frac{T_2 P_1}{T_1 P_2}$$

$$= C_P \ln\frac{T_2}{T_1} - R\ln\frac{T_2}{T_1} + R\ln\frac{T_2}{T_1} + R\ln\frac{P_1}{P_2}$$

or

$$\boldsymbol{\Delta S = C_P \ln\frac{T_2}{T_1} + R\ln\frac{P_1}{P_2}} \qquad \text{... (xvi)}$$

This equation gives the entropy change of 1 mole of an ideal gas accompanying a process when temperature changes from T_1 to T_2 and the pressure changes from P_1 to P_2.

(*i*) *If temperature is kept constant* (**isothermal process**)

$$T_1 = T_2$$

Equations (*xi*) and (*xvi*) are reduced to

$$\boldsymbol{\Delta S = R\ln\frac{V_2}{V_1} = R\ln\frac{P_1}{P_2}} \qquad \text{... (xvii)}$$

(*ii*) *If pressure is kept constant* (**isobaric process**)

$$P_1 = P_2$$

Equation (*xvi*) is reduced to

$$\boldsymbol{\Delta S = C_P \ln\frac{T_2}{T_1}} \qquad \text{... (xviii)}$$

(*iii*) *If volume is kept constant* (isochoric process)

$$V_1 = V_2$$

Equation (*xi*) converts to

$$\boldsymbol{\Delta S = C_v \ln\frac{T_2}{T_1}} \qquad \text{... (xix)}$$

2.5.1. Entropy change during phase change

A phase change involves the following:

(*i*) Melting of solid into liquid & freezing of liquid into solid.

(*ii*) Boiling of liquid into vapour and condensation of vapour into liquid

Entropy of melting

$$\boldsymbol{\Delta S_m = \frac{\Delta H_m}{T_m}}$$

where ΔH_m is the latent heat of melting and T_m is the melting temperature.

Entropy of vaporisation

$$\boldsymbol{\Delta S_v = \frac{\Delta H_v}{T_b}}$$

where ΔH_v is the latent heat of vaporisation and T_b is the boiling temperature.

Processes like melting and evaporation involve *absorption of heat* which is taken as *positive* by conventions of sign, therefore ΔS for these processes is positive *i.e.* these processes are accompanied by *increase of entropy.* On the other hand, the reverse processes like freezing and condensation involve *evolution of heat which* is taken as *negative,* therefore for these processes ΔS is negative *i.e.* these processes are accompanied by *decrease of entropy.*

SOLVED PROBLEMS ON ENTROPY CHANGE

Example 1. Calculate the increase in entropy in the evaporation of one mole of water at 3.73 K. Latent heat of vaporisation of water is 2.26 kJ/g (molecular mass of water = 18)

Solution. Given that

$$T_1 = 373 \text{ K}$$

$$\Delta H_v = 2260 \text{ Jg}^{-1}$$

$$= 2260 \times 18 \text{ Jmol}^{-1}$$

$$= 40680 \text{ J mol}^{-1}$$

$$\therefore \quad \Delta S = \frac{\Delta H_v}{T_b} - \frac{40680}{373} - 109 \text{ JK}^{-1}\text{mol}^{-1}$$

Example 2. Calculate the entropy change for the fusion of 1 mole of a solid which melts at 300 K. The latent heat of fusion is 2·51 kJ mol^{-1}.

Solution. Given that

$$\Delta H_m = 2510 \text{ J mol}^{-1}$$

$$T_m = 300 \text{ K}$$

$$\therefore \quad \Delta S = \frac{\Delta H_m}{T_m}$$

$$= \frac{2510}{300} = 8.37 \text{JK}^{-1}\text{mol}^{-1}$$

Example 3. Calculate the temperature at which the reaction $Ag_2O(s) \longrightarrow 2Ag\,(s) + \frac{1}{2}O_2\,(g)$ at 1 atmospheric pressure will be in equilibrium. The values of ΔH and ΔS for the reaction are 30.58 kJ and 66.11 JK^{-1} respectively and these values do not change much with temperature.

Solution. For a reaction in equilibrium

$$\Delta S = \frac{\Delta H}{T}$$

Given that

$$\Delta H = 30.58 \text{ kJ} = 30580 \text{ J}$$

$$\Delta S = 66.11 \text{ JK}^{-1}$$

Substituting the values in the above equation, we have

$$66.11 = \frac{30580}{T}$$

or

$$T = 462.6 \text{ K}$$

Example 4. Calculate the entropy change involved in the isothermal reversible expansion of 5 moles of an ideal gas from a volume of 10 litres to a volume of 100 litres at 300 K.

Solution. We have the following expression for ΔS in terms of temperature and volume

$$\Delta S = C_v \ln \frac{T_2}{T_1} + R \ln \frac{V_2}{V_1} \quad \text{for 1 mole of an ideal gas}$$

For isothermal process $T_1 = T_2$.

$$\therefore \quad \Delta S = R \ln \frac{V_2}{V_1} \quad \text{for 1 mole of an ideal gas}$$

or

$$\Delta S = nR \ln \frac{V_2}{V_1} \quad \text{for } n \text{ mole of an ideal gas}$$

or

$$\Delta S = 2.303\, nR \log \frac{V_2}{V_1}$$

Given that

$$n = 5$$
$$V_1 = 10 \text{ litres}$$
$$V_2 = 100 \text{ litres}$$
$$R = 8.314 \text{ JK}^{-1} \text{ mol}^{-1}$$

$$\therefore \quad \Delta S = 2.303 \times 5 \times 8.314 \times \log \frac{100}{10}$$
$$= \mathbf{95.75\ JK^{-1}\ mol^{-1}}$$

Example 5. 100 g of ice at 0°C was added to an insulated beaker containing 150 g of water at 100°C. Calculate

***(i)* The final temperature of the system and**

***(ii)* The entropy change of the process.**

Heat of fusion of ice is 80 cal/g and heat capacity of water is 1.0 cal/g/degree (Ignore the heat capacity of the beaker).

Solution. (*i*) Let the final temperature after mixing be = t°C

$$\text{Heat lost by water} = 150 \times (100 - t) \text{ cal}$$
$$\text{Heat gained by ice} = 100 \times 80 + 100 \times (t - 0)$$
$$= 8000 + 100\, t \text{ cal}$$
$$\text{Heat gained} = \text{Heat lost}$$
$$8000 + 100t = 150\,(100 - t)$$

or $t = \mathbf{28°C}$

(*ii*) the process involves the following steps :

(*a*) 100 g of ice melts at 0°C to give 100 g of liquid water.

(*b*) 100 g of liquid water gets heated up from 0°C to 28°C.

(*c*) 150 g of water gets cooled from 100°C to 28°C.

When temperature remains constant (*i.e.* for isothermal process), $\Delta S = \frac{q_{rev}}{T}$. When temperature varies (but volume or pressure remains constant), $\Delta S = C \ln \frac{T_2}{T_1}$,where C is C_v or C_p.

Entropy change in step (*a*)

$$= \frac{(100\text{g}) \times (80\text{cal/g})}{273} = \frac{8000}{273} = 29.30\,\text{cal deg}^{-1}$$

Entropy change in step (*b*)

$$= (100\text{ g})\ (1\text{ cal/g}) \times 2.303 \times \log \frac{28+273}{0+273}$$

$$= 230.3 \log \frac{301}{273} = 230.3 \times 0.0424$$

$$= 9.765\text{ cal deg}^{-1}$$

Entropy change in step (*c*)

$$= (150\text{ g}) \times (1\text{ cal/g}) \times 2.303 \log \frac{28+273}{100+273}$$

$$= 345.45\ \log \frac{301}{373} = 345.45 \times (-\,0.0931)$$

$$= -\,32.16\text{ cal deg}^{-1}$$

∴ Net entropy change in the process is given by

$$= 29.30 + 9.765 - 32.16$$

$$= \mathbf{6.905\ cal\ deg^{-1}}$$

PROBLEMS FOR PRACTICE

1. Calculate the change in entropy when 40 g of water at 50°C are mixed with 80 g of water at – 20°C. **[Ans.** 0.12 cal deg^{-1}]

2. For the reaction

$$H_2O\ (l) \longrightarrow H_2O\ (g)$$

$\Delta H = 9710$ cal mol^{-1}, $P = 1$ atm; $\Delta S = 26$ cal deg^{-1} mol^{-1}

Find out the temperature at which liquid water and water vapour at 1 atmospheric pressure are in equilibrium. **[Ans.** 100.46°C]

3. One mole of an ideal gas expands from a volume of 5 dm^3 to a volume of 60 dm^3 in an isothermal reversible process at 300 K. Calculate the entropy change during the process (R = 8.314 J K^{-1} mol^{-1}). **[Ans.** 20.67 JK^{-1} mol^{-1}]

4. Calculate the entropy change involved in the conversion of 1 mole of ice at 273 K to water at the same temperature.

Latent heat of fusion of ice is 335 joules per gram. **[Ans.** 22.1 J K^{-1}]

5. A heated copper block at 130°C loses 340 J of heat to the surroundings which are at room temperature of 32°C. Calculate

(*i*) entropy change of the copper block

(*ii*) entropy change of the surroundings

(*iii*) total entropy change of the universe due to this process.

Assume that the temperature of the block and the surroundings remains constant.

[Ans. (*i*) – 0.84 J K^{-1} (*ii*) + 1.11 J K^{-1} (*iii*) + 0.27 J K^{-1}]

6. Calculate the entropy change when 1 mole of solid SO_2 melts at –76°C. The latent heat of fusion is 1769 cal mol^{-1}. **[Ans.** 8.98 cal K^{-1} mol^{-1}]

7. 1 g of ice at 0°C is added to 10 g of water at the boiling point. What will be the final temperature and what is the entropy change accompanying this process ? Assume that the heat of fusion of ice is 80 cal/g and the specific heat of water is 1 cal/degree.

[**Ans.** $t = 83.6°C$, $\Delta S = 0.11$ cal deg^{-1}].

8. Calculate the entropy change when 2 moles of an ideal gas are allowed to expand from a volume of 1.0 litre to a volume of 10.0 litres at 27°C. [**Ans.** 38.29 J K^{-1} mol^{-1}]

9. For a certain ideal gas, $C_p = \frac{5}{2}$ R joules mol^{-1} deg^{-1}. Calculate the change in entropy suffered by 3 moles of the gas on being heated from 300 K to 600 K at (*a*) constant pressure (*b*) constant volume. [**Ans.** (*a*) 43.2 J K^{-1} (*b*) 25.9 J K^{-1}]

2.5.2. Physical significance of entropy

Various processes which are accompanied by a net increase of entropy are associated with an increased randomness of distribution. For example :

(*i*) **The melting of a solid** is accompanied by a net increase of entropy. We know that the molecules, atoms or ions in a solid have fixed positions and they become free to move about in the molten state. This is expressed by saying that there is no *disorder* among the molecules, atoms or ions in the solid but the disorder sets in when the solid changes into liquid. In other words, we can say that the *randomness* has increased.

(*ii*) **The vaporization of a liquid** is accompanied by a net increase of entropy. The disorder or the randomness also increases because the molecules of vapour are more random than the molecules in the liquid.

Thus

Entropy is a measure of randomness or disorder of the system.

This concept may be further understood with the help of the following analogy.

In a college when all classes are being held, all the students are sitting in their respective class rooms seriously and the disorder is minimum. As soon as the bell goes, the students of different classes move out to go to other rooms and disorder sets in. In other words, the disorder or the randomness increases.

SOME SOLVED CONCEPTUAL PROBLEMS ON SECOND LAW OF THERMODYNAMICS

Example 1. Is it possible to construct a reversible heat engine of unit efficiency? Give detailed reason for your answer.

Solution. $\eta = \frac{T_2 - T_1}{T_2}$. For η to be equal to unity, T_1 should be 0 K, *i.e.*, temperature of the sink should be zero degree absolute which is impossible. Hence, **it is not** possible to construct a reversible heat engine of unit efficiency.

Example 2. Write the expression for the efficiency of Carnot cycle in terms of work done, heat absorbed and temperatures of the source and sink.

Solution. $$\eta = \frac{w}{q_2} = \frac{q_2 - q_1}{q_2} = \frac{T_2 - T_1}{T_2}$$

where w = Net work done

q_2 = Total heat absorbed

$q_2 - q_1$ = Net heat absorbed $(\because w = q_2 - q_1)$

T_2 = Temperature of source

T_1 = Temperature of sinks

Example 3. How can the efficiency of Carnot engine be increased?

Solution. $\eta = \frac{T_2 - T_1}{T_2}$. Hence, to increase the efficiency of the heat engine, the temperature of the sink (T_1) should be minimum possible for a particular temperature of the source (T_2).

Example 4. Can the efficiency of a heat engine be greater than unity? If not why?

Solution. $\eta = \frac{T_2 - T_1}{T_2}$. Efficiency ($\eta$) cannot be greater than unity because temperature of sink (T_1) has to be less than the temperature of the source (T_2).

Example 5. What is the basis of thermodynamic scale of temperature? Why is it more basic than that based on perfect or ideal gas?

Solution. Thermodynamic scale of temperature as defined on the basis of second law of thermodynamics is more basic than that based on the ideal gas. This is because it is independent of the working substance.

Example 6. Which of the following have the higher value of S ?

(*i*) CO_2 at 15°C, 1 atm or dry ice at 1 atm.

(*ii*) a coiled spring or a spring relaxed.

(*iii*) 1 g of liquid water at 25°C, or 1 g of water vapour at 25°C.

(*iv*) silica glass or quartz (crystalline silica).

Solution. (*i*) CO_2 at 15° because gas is more random than solid.

(*ii*) A spring relaxed is more random.

(*iii*) Water vapour have greater randomness than liquid water.

(*iv*) Silica is more random than quartz because in the former SiO_4 tetrahedra are random while in the latter, they are arranged in a perfect order.

Example 7. Tell which will have greater entropy in each of the following cases. Justify your answer.

(*i*) Magnetised iron piece or non-magnetised piece of iron.

(*ii*) A mole of gas contained in a vessel of one litre capacity or that contained in a vessel of two-litre capacity.

(*iii*) A mole of $CaCO_3$ (*s*) or a system containing one mole of each of CaO (*s*) and CO_2 (*g*).

Solution. (*i*) Non-magnetised piece of iron.

(*ii*) A mole of gas contained in a vessel of 2 litre capacity has greater randomness and hence greater entropy.

(*iii*) One of mole of CaO + CO_2 will have greater entropy because gas has greater randomness.

Example 8. Write expressions for the entropy change in the following cases :

(*i*) Temperature changes from T_1 to T_2 and volume changes from V_1 to V_2 for 1 mole of an ideal gas.

***(ii)* Temperature changes from T_1 to T_2 and pressure changes from V_1 to V_2 for 1 mole of an ideal gas.**

***(iii)* 1 mole of a solid melts at the melting point.**

***(iv)* 1 mole of a liquid vaporises at the boiling point.**

Solution. *(i)* $\Delta S = C_v \ln \dfrac{T_2}{T_1} + R \ln \dfrac{V_2}{V_1}$ *(iii)* $\Delta_{fus} S = \dfrac{\Delta_{fus} H}{T_f}$

(ii) $\Delta S = C_v \ln \dfrac{T_2}{T_1} + R \ln \dfrac{P_1}{P_2}$ *(iv)* $\Delta_{vap} S = \dfrac{\Delta_{vap} H}{T_b}$

Example 9. Justify the 'Second Law of Thermodynamics' which states that "The net entropy of the universe tends to increase."

Solution. All spontaneous processes occurring in nature are thermodynamically irreversible and every irreversible process is accompanied by increase in entropy. Hence entropy of the universe is continuously increasing.

Example 10. In the following processes, state whether the entropy of the system increases, decreases or remains constant :

***(i)* Vaporization of a mole of water into steam at its boiling point.**

***(ii)* Solidification of a mole of liquid at its freezing point.**

***(iii)* Isothermal compression of a liquid.**

***(iv)* Separation of a mixture into its two constituents.**

Solution. *(i)* Entropy increases

(ii) Entropy decreases

(iii) Entropy remains almost constant

(iv) Entropy decreases.

Example 11. Select the most random system of the following :

***(i)* sugar cubes in a box**

***(ii)* sugar cubes scattered on a table**

***(iii)* sugar disssolved in water**

***(iv)* finely powdered sugar.**

Solution. Sugar dissolved in water is the most random system because the molecules are free to move about in the solution.

2.6. ENTROPY CHANGE ON MIXING OF IDEAL GASES

Suppose at constant temperature, n_1 moles of an ideal gas 1 at the initial pressure $P_1{}^o$ are mixed with n_2 moles of another ideal gas 2 at initial pressure $P_2{}^o$. After mixing let their partial pressures in the mixture be P_1 and P_2 respectively. We know, at constant temperature, the entropy change of an ideal gas when its pressure changes from initial pressure P_i to final pressure P_f is given by

$$\Delta S = R \ln \frac{P_i}{P_f} \text{ mol}^{-1}$$

∴ Entropy change of the first gas when the pressure of n_1 moles of the gas changes from P_1^0 to P_1 is given by

$$\Delta S_1 = n_1 R \ln \frac{P_i^0}{P_1} \quad \text{... } (i)$$

Similarly, entropy change of the second gas when pressure of n_2 moles of the gas changes from P_2^0. to P_2 is given by

$$\Delta S_2 = n_2 R \ln \frac{P_2^o}{P_2} \qquad \text{... (ii)}$$

Total entropy change on mixing the two gases will be the sum of the above two changes.

$$\Delta S_{mixing} = n_1 R \ln \frac{P_1^o}{P_1} + n_2 R \ln \frac{P_2^o}{P_2} \qquad \text{... (iii)}$$

Let P be the total pressure of the mixture. Then, $P = P_1 + P_2$. Further let x_1 and x_2 be the mole fractions of gases 1 and 2 in the mixture. By Dalton's law of partial pressures,

$$P_1 = x_1 P \quad \text{and} \quad P_2 = x_2 P$$

Substituting these values in eqn. (*iii*), we get

$$\Delta S_{mixing} = n_1 R \ln \frac{P_1^o}{x_1 P} + n_2 R \ln \frac{P_2^o}{x_2 P} \qquad \text{... (iv)}$$

Taking the simplest case in which each gas is taken at the same initial pressure. Under these conditions, after mixing, volume of the mixture will be the sum of their initial volumes, *i.e.*, $V = V_1 + V_2$. And final pressure of the mixture will be nearly the same as initial pressure of each gas, *i.e.*, $P_1^o = P_2^o = P$ as shown in the diagram below. Equation (*iv*) then is simplified to

$$\Delta S_{mixing} = -n_1 R \ln x_1 - n_2 R \ln x_2$$

or

$$\Delta S_{mixing} = -R\,(n_1 \ln x_1 + n_2 \ln x_2) \qquad \text{... (v)}$$

Gas 1		Gas 2		
Pressure = P Temp = T Volume = V_1 No. of moles = n_1	+	Pressure = P Temp = T Volume = V_2 No. of moles = n_2	On ⟶ mixing	Pressure = P Temp. = T Volume = $V_1 + V_2$ No. of moles = $n_1 + n_2$

For a mixture of a number of gases, eqn. (*v*) can be written in the general form as under :

$$\boxed{\Delta S_{mixing} = -R\,\Sigma\, n_i \ln x_i} \qquad \text{... (vi)}$$

Entropy change for *one mole* of the mixing is obtained by dividing eqn. (*v*) by $(n_1 + n_2)$

$$\Delta S_{mixing} = -R\left(\frac{n_1}{n_1+n_2}\ln x_1 + \frac{n_2}{n_1+n_2}\ln x_2\right)$$

or

$$\Delta S_{mixing} = -R\,(x_1 \ln x_1 + x_2 \ln x_2) \qquad \text{... (vii)}$$

which can be generalized to the form

$$\boxed{\Delta S_{mixing} = -R\,\Sigma\, x_i \ln x_i} \qquad \text{... (viii)}$$

We make the following conclusions from the above equations.

1. ΔS_{mixing} is independent of temperature.

2. As $x_i < 1$, ΔS_{mixing} will always be positive *i.e.* mixing of gases is accompanied by increase in entropy.

PROBLEM ON CALCULATION OF ENTROPY OF MIXING OF IDEAL GASES

Example 1. Calculate the molar entropy change of mixing 0.2 mole of oxygen and 0.6 mole of nitrogen at 25°C assuming that they are ideal gases.

Solution. $n_1 = 0.2$ mole, $n_2 = 0.6$ mole

$$\therefore \quad x_1 = \frac{0.2}{0.8} = 0.25, \quad x_2 = \frac{0.6}{0.8} = 0.75$$

$$\therefore \quad \Delta S_{mixing} = -R\,(x_1 \ln x_1 + x_2 \ln x_2)$$
$$= -2.303\,R\,(x_1 \log x_1 + x_2 \log x_2)$$
$$= -2.303 \times 8.314\,(0.25 \log 0.25 + 0.75 \log 0.75)$$
$$= +2.303 \times 8.314 \times 0.2443 \text{ J K}^{-1} \text{ mol}^{-1}$$
$$= \mathbf{4.678 \text{ J K}^{-1} \text{ mol}^{-1}}$$

2.7. THIRD LAW OF THERMODYNAMICS WORK FUNCTION (HELMHOLTZ FREE ENERGY)

Another thermodynamic quantity which is used in the study of thermodynamics is called work function or Helmholtz free energy

This is denoted by A and is defined by the equation

$$A = U - TS \qquad \text{... }(i)$$

where U is the internal energy of the system, T is the temperature and S is the entropy.

Since, U, T and S are the functions of the state of the system only and do not depend upon its previous history, therefore A also must be a state function.

In order to understand the *physical significance* of the work function, consider an isothermal change taking place at temperature T. then

$$A_1 = U_1 - TS_1 \textit{ for the initial state} \qquad \text{... }(ii)$$

and

$$A_2 = U_2 - TS_2 \textit{ for the final state} \qquad \text{... }(iii)$$

where A_1, U_1 and S_1 are respectively the values of work function, internal energy and entropy of the system in the initial state and A_2, U_2 and S_2 are the corresponding values in the final state.

$\therefore$ Change in the function A accompanying the process is given by

$$A_2 - A_1 = (U_2 - TS_2) - (U_1 - TS_1)$$
$$= (U_2 - U_1) - T\,(S_2 - S_1)$$

or

$$\mathbf{\Delta A = \Delta U - T\Delta S} \qquad \text{... }(iv)$$

According to the definition of entropy,

$$\Delta S = \frac{q_{rev}}{T} \qquad \text{... }(v)$$

According to the first law of thermodynamics as applied to an isothermal reversible process, we have

$$\Delta U = q_{rev} - w_{max} \qquad \text{... }(vi)$$

Substituting the values of ΔS and ΔU from equation (v) and (vi) in equation (iv), we get

$$\Delta A = (q_{rev} - w_{max}) - T.\frac{q_{rev}}{T}$$

or

$$\mathbf{-\Delta A = w_{max}} \qquad \text{... }(vii)$$

Thus *for a process occurring at constant temperature, the decrease in the work function A is equal to the maximum work done by the system.* It is for this reason that this thermodynamic quantity has been termed as work function.

2.8. GIBB'S FREE ENERGY

This is another thermodynamic quantity.

This is usually denoted by G and is defined the equation

$$\mathbf{G = H - TS} \quad \text{... (i)}$$

where H, T and S are the heat content, temperature and entropy of the system respectively.

Since H, T and S are the function of the state only, therefore G is also function of the state of the system only. Hence for the isothermal process occurring at temperature T, we can write

$$G_1 = H_1 - TS_1 \quad \text{for the initial state} \quad \text{... (ii)}$$

$$G_2 = H_2 - TS_2 \quad \text{for the final state} \quad \text{... (iii)}$$

or

$$G_2 - G_1 = (H_2 - TS_2) - (H_1 - TS_1)$$

$$= (H_2 - H_1) - T(S_2 - S_1)$$

or

$$\mathbf{\Delta G = \Delta H - T\,\Delta S} \quad \text{... (iv)}$$

where ΔG = Change in Gibb's free energy of the system

ΔH = Enthalpy change of the system

and ΔS = Entropy change of the system

The physical significance of Gibb's free energy may be understood as follows:

At constant temperature (T),

$$\Delta S = \frac{q_{rev}}{T}$$

i.e.

$$T\Delta S = q_{\text{rev}} \quad \text{... (v)}$$

At constant pressure (P),

$$\Delta H = \Delta U + P\,\Delta V \quad \text{...(vi)}$$

Substituting the values of $T\Delta S$ and ΔH from equation (v) and (vi) in equation (iv), we get

$$\Delta G = (\Delta U + P\Delta V) - q_{\text{rev}}$$

$$= (\Delta U - q_{\text{rev}}) + P\,\Delta V \quad \text{... (vii)}$$

Now, according to first law of thermodynamics

$$\Delta U = q_{rev} - w_{max}$$

(work done by a system carries negative sign)

or

$$\Delta U - q_{\text{rev}} = -\,w_{\text{max}}$$

Substituting this value in equation (vii), we get

$$\Delta G = -\,w_{\text{max}} + P\Delta V$$

$$\Delta G = w_{\text{max}} - P\Delta V$$

But $P\Delta V$ is the *work of expansion* done by the system corresponding to the increase in volume ΔV. Hence ($w_{\text{max}} - P\Delta V$) gives the maximum work other than the work of expansion. This is called the *maximum useful work* available from the process. We conclude that *for a process occurring at constant temperature and constant pressure, the decrease in Gibb's free energy is equal to the* **maximum useful work** *obtainable from the process i.e. the total work minus the pressure volume work* (*or work of expansion*).

2.8.1. Variation of work function with temperature and volume

The work function is given by the equation

$$A = U - TS \quad \text{... } (i)$$

Complete differentiation of this equation gives

$$dA = dU - TdS - SdT \quad \text{...} (ii)$$

Entropy change is given by

$$dS = \frac{\delta q_{\text{rev}}}{T} \quad \text{... } (iii)$$

From first law of thermodynamics,

$$\delta q_{\text{rev}} = dU - \delta w \quad \text{... } (iv)$$

and if the work is restricted to the work of expansion only,

$$-\delta w = PdV \quad \text{... } (v)$$

Putting this value in equation (*iv*), we have

$$\delta q_{\text{rev}} = dU + PdV \quad \text{... } (vi)$$

Putting this value in equation (*iii*), we get

$$dS = \frac{dU + PdV}{T}$$

or

$$TdS = dU + PdV \quad \text{... } (vii)$$

Putting this value in equation (*ii*), we obtain

$$dA = dU - dU - PdV - SdT$$

$$= -PdV - SdT \quad \text{... } (viii)$$

(*a*) *If temperature is kept constant, $dT = 0$. Equation* (*viii*) takes the form

$$(dA)_T = -(PdV)_T \quad \text{... } (ix)$$

or

$$\left(\frac{\partial A}{\partial V}\right)_T = -P \quad \text{... } (x)$$

(*b*) *If volume is kept constant, $dV = 0$. Equation* (*viii*) becomes

$$(dA)_V = -(SdT)_V$$

$$\left(\frac{\partial A}{\partial T}\right)_V = -S \quad \text{... } (xi)$$

Thus variation of work function with temperature and volume is given by eqs. (*x*) and (*xi*).

2.8.2. Variation of Gibb's free energy with temperature and pressure

The free energy is given by the equation

$$G = H - TS \quad \text{... } (i)$$

But,

$$H = U + PV \quad \text{... } (ii)$$

Substituting this value in equation (*i*), we get

$$G = U + PV - TS \quad \text{... } (iii)$$

Complete differentiation of this equation gives

$$dG = dU + PdV + VdP - TdS - SdT \quad \text{... } (iv)$$

But $$dS = \frac{\delta q_{rev}}{T} = \frac{dU + PdV}{T}$$

or $$TdS = dU + PdV \quad \text{... (v)}$$

Substituting this value in equation (*iv*), we get

$$dG = dU + PdV + VdP - (dU + PdV) - SdT = VdP - SdT$$

i.e. $$\mathbf{dG = VdP - SdT} \quad \text{... (vi)}$$

This is expression gives the change in free energy with change in pressure and change in temperature in a reversible process. Equation (*vi*) is called **total differential equation.**

(*a*) *If temperature is kept constant, $dT = 0$.* Equation (*vi*) takes the form

$$(dG)_T = (VdP)_T \quad \text{... (vii)}$$

or $$\left(\frac{\partial G}{\partial P}\right)_T = V \quad \text{... (viii)}$$

(*b*) *If pressure is kept constant, $dP = 0$.* Equation (*vi*) becomes

$$(dG)_p = -(SdT)_p$$

or $$\left(\frac{\partial G}{\partial T}\right)_P = -S \quad \text{... (ix)}$$

Thus variation of free energy with pressure and temperature is given equations (*viii*) and (*ix*) respectively.

2.8.3. Change in free energy for a process under isothermal condition

The change in free energy for a complete process under isothermal conditions can be calculated as under :

For an infinitesimal change, under isothermal conditions, we have the equation

$$(dG)_T = VdP \quad \text{... (i)}$$

For a complete process under isothermal conditions, when the pressure changes from P_1 to P_2, we have

$$(\Delta G)_T = \int_{P_1}^{P_2} VdP \quad \text{... (ii)}$$

For n moles of an ideal gas $PV = nRT$, *i.e.*, $V = \frac{nRT}{P}$

Substituting for V in eq. (*ii*), we have

$$(\Delta G)_T = \int_{P_1}^{P_2} \frac{nRT}{P} dP$$

$$= nRT \int_{P_1}^{P_2} \frac{dP}{P}$$

$$= nRT \ln \frac{P_2}{P_1} \quad \text{... (iii)}$$

Further for an isothermal process,

$$P_1V_1 = P_2V_2 \text{ or } \frac{P_2}{P_1} = \frac{V_1}{V_2}$$

Substituting this value in equation (*iii*), we get

$$(\Delta G)_T = nRT \ln \frac{P_2}{P_1} = nRT \ln \frac{V_1}{V_2} \quad \text{... (iv)}$$

SOLVED EXAMPLES ON FREE ENERGY CHANGE

Example 1. Calculate the free energy change accompanying the compression of 1 mole of CO_2 at 57°C from 5 atm to 50 atm. Assume that CO_2 behaves like an ideal gas.

Solution. Given that

$$n = 1 \text{ mole}$$
$$T = 57 + 273 \text{ K} = 330 \text{ K}$$
$$P_1 = 5 \text{ atm}$$
$$P_2 = 50 \text{ atm}$$

Taking $R = 1.987$ cal deg^{-1} mol^{-1}, we have

$$(\Delta G)_T = 2.303\, nRT \log \frac{P_2}{P_1}$$
$$= 2.303 \times 1 \times 1.987 \times 303 \times \log \frac{50}{5}$$
$$= \mathbf{1510.1 \text{ cal}}$$

Example 2. Calculate the free energy change which occurs when one mole of an ideal gas expands reversibly and isothermally at 300 K from the initial volume of 5 litres to 50 litres.

Solution. Given that

$$n = 1 \text{ mole}$$
$$T = 300 \text{ K}$$
$$V_1 = 5 \text{ litres}$$
$$V_2 = 50 \text{ litres}$$

Taking $R = 8.314$ JK^{-1} mol^{-1}

Substituting the values in the expression for $(\Delta G)_T$, we have

$$(\Delta G)_T = 2.303\, nRT \log \frac{V_1}{V_2}$$
$$= 2.303 \times 1 \times 8.314 \times 300 \times \log \frac{5}{50}$$
$$= \mathbf{-5744.1 \text{ joules}}$$

PROBLEMS FOR PRACTICE

1. During an isothermal reversible compression of one mole of an ideal gas, its volume decreases from 10 litres to 1 litre. The process was carried out at 27°C. Calculate the change in entropy and change in free energy of the gas. (R = 2 cal/degree/mol).

[**Ans.** $\Delta S = -4.606$ cal deg^{-1}, $\Delta G = +1381.8$ cal]

2. 1 mole of an ideal gas is allowed to expand at 25°C till its pressure falls to one-fifth of its original pressure. Find out the free energy change accompanying the process.

[**Ans.** – 953.2 cal]

3. The pressure of one mole of an ideal gas at 25°C falls from 5.0 bar to 0.2 bar. Evaluate the change in free energy. (R = 8.314 J K^{-1} mol^{-1}) [**Ans.** 7976.2 J]

4. Four moles of an ideal gas expand isothermally from 1 litre to 10 litres at 300 K. Calculate the change in free energy of the gas (R = 8.314 J K^{-1} mol^{-1}) [**Ans.** – 22.98 kJ]

2.9. CRITERIA FOR FEASIBILITY OR SPONTANEITY OF A PROCESS

In the earlier sections in this chapter, the following generalisations were derived.

(*i*) If $\Delta S_{system} + \Delta S_{surroundings} > 0$; the process is irreversible, *i.e.*, it is spontaneous or feasible.

(*ii*) If $\Delta S_{system} + \Delta S_{surroundings} = 0$; the process is reversible *i.e.*, the system is in equilibrium.

The above two results may be combined and represented as

$$\Delta S_{system} + \Delta S_{surroundings} \geq 0 \qquad \text{... } (i)$$

where the sign *greater than* refers to an irreversible process (*i.e.* spontaneous process) and the sign *equal to* for a reversible process.

Eq. (*i*) requires the knowledge of change in the entropy of the surroundings which is not so convenient to determine. Hence, we would search other criteria which involves change in entropy of the system only, in order to arrive at the criteria for feasibility of a process.

For an infinitesimal change, the above criterion may be written as

$$dS_{system} + dS_{surroundings} \geq 0 \qquad \text{... } (ii)$$

where the sign, > stands for irreversible process and the sign = for the reversible process.

If the surroundings lose heat δq reversibly and isothermally at the temperature T (which may be absorbed by the system reversibly or irreversibly), then

$$dS_{surrounds} = -\frac{\delta q_{rev}}{T} \qquad \text{... } (iii)$$

According to first law of thermodynamics,

$$\delta q_{rev} = dU + pdV \qquad \text{... } (iv)$$

Substituting this value in equation (*iii*), we get

$$dS_{surrounding} = -\frac{dU + PdV}{T} \qquad \text{... } (v)$$

Substituting this value in equation (*ii*), we get

$$dS_{system} - \frac{dU + PdV}{T} \geq 0$$

or

$$TdS - dU - PdV \geq 0$$

or

$$\boldsymbol{TdS \geq dU + PdV} \qquad \text{... } (vi)$$

where dS stands for the entropy change of the system without the subscript *system*.

Equation (*vi*) is the basic equation which leads to a number of criteria for predicting the feasibility of a process. In this equation, the sign '=' stands for the reversible process and the sign '>' for the irreversible process. From equation (*vi*), the various criteria may be deduced as under.

(*i*) *In terms of entropy change of the system.* If the internal energy and the volume of the system are kept constant,

$$dU = 0 \quad \text{and} \quad dV = 0$$

Then from equation (*vi*), we have

$$(TdS)_{U,V} \geq 0$$

or

$$\boldsymbol{(dS)_{U,V} \geq 0} \qquad \text{... } (vii)$$

where the sign '=' is for the reversible process and the sign '>' for the irreversible process. The subscripts U and V indicate that these properties remain constant.

(*ii*) *In terms of internal energy change of the system.* If the entropy and volume of the system are kept constant,

$$dS = 0, dV = 0$$

The equation (*vi*) becomes

$$0 \geq (dU)_{S,\,V}$$

or $$\mathbf{(dU)_{S,V} \leq 0} \qquad \dots (viii)$$

where the sign '=' stands for the reversible process and the sign '<' for the irreversible process. This implies that *if a process under constant entropy and volume is accompanied by a decrease of internal energy, the process is irreversible; if no change of internal energy takes place, the process is reversible.*

(*iii*) *In terms of enthalpy change of the system.*

$$H = U + PV$$

Complete differentiation of this equation gives

$$dH = dU + PdV + VdP$$

or $$dU + PdV = dH - VdP$$

Putting this value in equation (*vi*), we get

$$TdS \geq dH - VdP$$

If entropy and pressure of the system are kept constant,

$$dS = 0 \text{ and } dP = 0, \text{ so that}$$

$$0 \geq (dH)_{S,P}$$

or $$\mathbf{(dH)_{S,P} \leq 0} \qquad \dots (ix)$$

where the sign '=' refers to the reversible process and the sign '<' refers to the irreversible process.

(*iv*) *In terms of change in work function of the system.*

$$A = U - TS$$

Complete differentiation of this equation gives

$$dA = dU - TdS - SdT$$

or $$TdS = dU - SdT - dA$$

Putting this value in equation (*vi*), we get

$$dU - SdT - dA \geq dU + PdV$$

or $$-SdT - dA \geq PdV$$

or $$SdT + dA \leq -PdV$$

If temperature and volume of the system are kept constant, $dT = 0$, $dV = 0$, so that we have

$$(dA)_{T,V} \leq \mathbf{0} \qquad \dots (x)$$

whereas usual sign '=' is for the reversible process and the sign '<' for the irreversible process.

(*v*) *In terms of free energy change of the system.*

$$G = H - TS$$

Further $$H = U + PV$$

$\therefore$ $$G = U + PV - TS$$

Complete differentiation of this equation gives

$$dG = dU + PdV + VdP - TdS - SdT$$

or $$dU + PdV = dG - VdP + TdS + SdT$$

Substituting this value in equation (*vi*), we get

$$TdS \geq dG - VdP + TdS + SdT$$

or $$0 \geq dG - VdP + SdT$$

or $$dG - VdP + SdT \leq 0$$

If pressure and temperature of the system are kept constant, $dP = 0$, $dT = 0$, so that we obtain

$$(dG)_{P,T} \leq 0 \qquad \text{... (xi)}$$

where, as before, the sign '=' refers to the reversile process and the sign '<' refers to the irreversible process.

The criterion in terms of free energy change is most important because most of the processes take place at constant temperature and pressure.

(*i*) If $(dG)_{P,T} < 0$, the process is irreversible, *i.e.*, it is feasible.

(*ii*) If $(dG)_{P,T} = 0$, the process is reversible, *i.e.*, the system is in equilibrium.

(*iii*) If $(dG)_{P,T} > 0$, the process does not occur, *i.e.*, it is not feasible.

2.10 GIBB'S–HELMHOLTZ EQUATION

The Gibbs free energy (G) is given by the equation,

$$G = H - TS \qquad \text{... (i)}$$

Hence for an *isothermal* process, we can write

$$G_1 = H_1 - TS_1 \quad \text{for the initial state} \qquad \text{... (ii)}$$

and $$G_2 = H_2 - TS_2 \quad \text{for the final state} \qquad \text{... (iii)}$$

Subtracting equation (*ii*) from equation (*iii*), we get

$$G_2 - G_1 = (H_2 - TS_2) - (H_1 - TS_1)$$

$$= (H_2 - H_1) - T(S_2 - S_1)$$

or $$\Delta G = \Delta H - T\Delta S \qquad \text{... (iv)}$$

H.V. Helmholtz

where ΔG is the change in free energy of the system

ΔH is the change in enthalpy of the system

and ΔS is the entropy change of the system.

Putting $H = U + PV$, in eq. (*i*) we get

$$G = U + PV - TS \qquad \text{... (v)}$$

Differentiating this equation completely, we get

$$dG = dU + PdV + VdP - TdS - SdT \qquad \text{... (vi)}$$

But $$dS = \frac{\delta q_{\text{rev}}}{T} = \frac{dU + PdV}{T} \qquad \text{(From first law of thermodynamics)}$$

or $$TdS = dU + PdV$$

Substituting this value in equation (*vi*), we get

$$dG = VdP - SdT$$

If pressure is kept constant, $dP = 0$, so that the above equation reduces to

$$(dG)_P = -(SdT)_P$$

or $$\left(\frac{\partial G_1}{\partial T}\right)_P = -S \quad \text{... (vii)}$$

$\therefore$ For the initial state, $$\left(\frac{\partial G_1}{\partial T}\right)_P = -S_1 \quad \text{... (viii)}$$

For the final state, $$\left(\frac{\partial G_2}{\partial T}\right)_P = -S_2 \quad \text{... (ix)}$$

Subtracting equation (*viii*) from (*ix*),we get

$$\left(\frac{\partial G_2}{\partial T}\right)_P - \left(\frac{\partial G_1}{\partial T}\right)_P = -S_2 - (-S_1)$$

$$= -S_2 + S_1 = -(S_2 - S_1)$$

$$= -\Delta S$$

or it may be written as

$$\left[\frac{\partial (G_2 - G_1)}{\partial T}\right]_P = -\Delta S$$

$$\left[\frac{\partial (\Delta G)}{\partial T}\right]_P = -\Delta S \quad \text{... (x)}$$

Substituting this value of $-\Delta S$ in equation (*iv*), we get

$$\boldsymbol{\Delta G = \Delta H + T\left[\frac{\partial (\Delta G)}{\partial T}\right]_P} \quad \text{... (xi)}$$

This is the most common form of Gibbs-Helmholtz equation.

An analogous equation involving work function A can be obtained using the equation

$$A = U - TS$$

and proceeding in a similar manner as above :

$$\Delta A = \Delta U + \boldsymbol{T}\left[\frac{\boldsymbol{\partial (\Delta A)}}{\boldsymbol{\partial T}}\right]_V \quad \text{... (xii)}$$

Gibbs Helmholtz equation can be obtained in another form as under :

Writing the Gibb's-Helmholtz equation in its common form

$$\Delta G = \Delta H + T\left[\frac{\partial (\Delta G)}{\partial T}\right]_P$$

Dividing the above equation T^2

$$\frac{\Delta G}{T^2} = \frac{\Delta H}{T^2} + \frac{1}{T}\left[\frac{\partial (\Delta G)}{\partial T}\right]_P$$

or $$\frac{1}{T}\left[\frac{\partial (\Delta G)}{\partial T}\right]_P = \frac{\Delta G}{T^2} - \frac{\Delta H}{T^2}$$

or $$\left[\frac{\partial (\Delta G/T)}{\partial T}\right]_P = -\frac{\Delta H}{T^2}$$

2.10.1. Applications of Gibbs-Helmholtz equation

Gibbs-Helmholtz equation has the following applications.

1. Knowing the free energy change at any two different temperatures. the enthalpy change accompanying the process can be calculated.

2. When a redox reaction (*e.g.* $Zn + CuSO_4 \longrightarrow ZnSO_4 + Cu$) is allowed to take place in a single beaker, heat is evolved. It was believed earlier that when the same reaction is allowed to take place in an electrochemical cell, the electrical energy produced was equivalent to the heat evolved. However, it was revealed later that in a number of cases, the electrical energy produced is not equivalent to the heat evolved. This was explained by Gibbs and Helmholtz by stating that the *electrical energy produced was equal to the decrease in the free energy* ($-\Delta G$) *accompanying the cell reaction.* They pointed out that if n is the number of electrons given out by one of the electrodes (*e.g.* for $Zn \longrightarrow Zn^{+2} + 2e$, $n = 2$), or taken up by the other electrode (*e.g.* for $Cu^{+2} + 2e \longrightarrow Cu$, $n = 2$), then the quantity of electricity produced = nF faradays, where F = 1 Faraday = 96,500 coloumbs. If E is the E.M.F. of the cell, then

Electrical energy produced by the cell = nFE

$\therefore$ Decrease in free energy, $-\Delta G = nFE$

Substituting this value in the Gibbs-Helmholtz equation

$$\Delta G = \Delta H + T\left[\frac{\partial(\Delta G)}{\delta T}\right]_P$$

we have

$$-nFE = \Delta H + T\left[\frac{\partial(-nFE)}{\partial T}\right]_P$$

or

$$\boldsymbol{nFE = -\Delta H + TnF\left(\frac{\partial F}{\partial T}\right)_P}$$

Thus knowing the temperature coefficient of the E.M.F. of the cell *i.e.* $(\partial E/\partial T)_p$, ΔH can be calculated if E is known or *vice versa.* The electrical energy produced (nFE) is equal to the heat evolved ($-\Delta H$) only when $(\partial E/\partial T)_p = 0$.

SOLVED EXAMPLES ON GIBB'S HELMHOLTZ EQUATION

Example 1. The free energy change involved in a process is –1235 J at 300 K and – 1200 J at 310 K. Calculate the change in enthalpy of the process at 305 K.

Solution. Using Gibb's Helmholtz equation

$$\Delta G = \Delta H + T\left[\frac{\partial(\Delta G)}{\partial T}\right]_P$$

where

$$\left[\frac{\partial(\Delta G)}{\partial T}\right]_P = \frac{\Delta G_2 - \Delta G_1}{T_1 - T_1}$$. Substituting values, we have

$$\left[\frac{\partial(\Delta G)}{\partial T}\right]_P = \frac{-1200-(-1235)}{310-300} = \frac{35}{10} = 3.5\ JK^{-1}$$

The free energy change at 306 K can be taken as the average values of free energy change at 300 K and 310 K.

Thus, $\Delta G_{(\text{at } 305\text{ K})} = \frac{1}{2}[-1200 + (-1235)] = -$ **1217.5 J**

Gibbs Helmholtz equation can be written as

$$\Delta H = \Delta G - T\left[\frac{\partial(\Delta G)}{\partial T}\right]_P$$

Substituting values, we get

$$\Delta H = -1217.5 - 305 \times 3.5 = -2285 \text{ J}$$

Example 2. The free energy change of a reaction at 298 K is found to be 18.0 kJ. Calculate the free energy at 350 K. Given that the enthalpy change for the reaction is + 80 kJ. It is assumed that ΔH is independent of temperature.

Solution. Writing Gibbs Helmholtz equation in the form

$$\frac{\partial\left(\frac{\Delta G}{T}\right)}{\partial T} = -\frac{\Delta H}{T^2}$$

It can also be expressed as

$$d\left(\frac{\partial G}{T}\right) = -\frac{\Delta H}{T^2}dT$$

Integrating both sides of above equation, we get

$$\frac{\Delta G_2}{T_2} - \frac{\Delta G_1}{T_1} = \Delta H\left(\frac{1}{T_2} - \frac{1}{T_1}\right)$$

Given that

$$\Delta G_1 = 18.0 \text{ kJ}, \Delta H = 80.00 \text{ kJ}$$
$$T_1 = 298 \text{ K}, T_2 = 350 \text{ K}$$

Substituting the values, we get

$$\frac{\Delta G_2}{398} - \frac{18.0}{298} = 80\left(\frac{1}{350} - \frac{1}{298}\right)$$

$$\frac{\Delta G_2}{398} = -0.03988 + 0.060402$$

$$\Delta G_2 = 0.020522 \times 398$$

or $\Delta G_2 =$ **8.168 kJ.**

PROBLEM FOR PRACTICE

1. The free energy change accompanying a given process is – 85.8 kJ at 25°C and – 82.6 kJ at 35°C. Calculate the enthalpy change of the process at 25°C. **[Ans.** – 82.56 kJ]

2. Assuming that, for a certain reaction ΔH remains constant in the temperature range 427° to 627°C. Calculate ΔG for this reaction at 627°C from the following data : $\Delta H = -25$ kJ, ΔG (427°C) = – 40 kJ] **[Ans.** – 44.26 kJ]

2.11. NERNST HEAT THEOREM

This theorem gives the variation of enthalpy change (ΔH) and free energy change (ΔG) of a system with decrease of temperature.

According to Gibbs-Helmholtz equation,

$$\Delta G = \Delta H + T\left(\frac{\partial(\Delta G)}{\partial T}\right)_P$$

From this equation, it is evident that at the absolute zero *i.e.* when $T = 0$, $\Delta G = \Delta H$. Nernst observed that as the temperature is lowered towards absolute zero, the value of $\partial(\Delta G)/\partial T$ decreases and then approaches zero asymptotically. This means that ΔG and ΔH *are not only equal at absolute zero but the values approach each other asymptotically near this temperature.* This result is known as Nernst heat theorem. Mathematically, it may be expressed as

$$\lim_{T \to 0} \frac{d(\Delta G)}{dT} = \lim_{T \to 0} \frac{d(\Delta H)}{dT} = 0 \quad \text{... } (i)$$

Graphically, the result may be represented by Fig. 2.4.

In Fig. 2.4. ΔG has been shown as greater than ΔH at temperature away from absolute zero. However the reverse is also possible because $\partial(\Delta G)/\partial T$ can be both positive or negative.

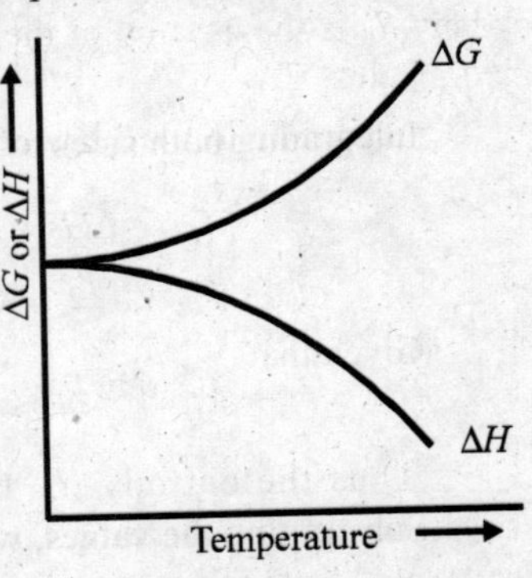

Fig. 2.4. Nernst heat theorem

Further $\left[\frac{\partial(\Delta G)}{\partial T}\right]_P = -\Delta S$ ... (*ii*)

and $\left[\frac{\partial(\Delta H)}{\partial T}\right]_P = \Delta C_p$ ... (*iii*)

(Kirchoff's equation)

From equations (*i*), (*ii*) and (*iii*), we can say, as $T \to 0$

$$\Delta S = 0 \text{ and } \Delta C_p = 0$$

i.e. as the temperature is lowered to absolute zero, the enthalpy change of the reaction and difference in the heat capacities of products and reactants also tends to be zero.

Since gases do not exist at the absolute zero, this means that the heat theorem is not applicable to gases. Similarly, it has been found to be inapplicable to liquids also. Nernst heat theorem is applicable to solids only.

2.12. THIRD LAW OF THERMODYNAMICS

Third law of thermodynamics states:

The entropy of all perfectly crystalline solids may be taken as zero at the absolute zero temperature.

The law is also stated as follows :

Every substance has finite positive entropy but at the absolute zero of temperature, the entropy may become zero and in fact it does become zero in case of perfectly crystalline solids.

Importance of the Third Law

Calculation of absolute entropies: It helps in the calculation of the absolute entropies of chemical compounds at any desired temperature. This is explained as under :

The infinitesimal entropy change is given by

$$dS = \frac{dq}{T} \quad \text{... (i)}$$

But $\qquad C_p = \frac{dq}{dT}$ so that $dq = C_p\, dT$

Substituting this value in equation (*i*), we have

$$dS = \frac{C_p dT}{T}$$

Entropy change of a substance when its temperature changes from absolute zero to the temperature (*T*) can be calculated using the equation

$$\int_{S=S_0}^{S=S} dS = \int_{T=0}^{T=T} C_p \frac{dT}{T}$$

or

$$S - S_0 = \int_0^T C_p \frac{dT}{T} \quad \text{... (ii)}$$

where S_0 is the entropy of the substance at absolute zero and *S* is the entropy of the substance at temperature *T*.

According to third law of thermodynamics, $S_0 = 0$.

$$\therefore \quad S_T = \int_0^T C_p \frac{dT}{T}$$

$$= \int_0^T C_p d \ln T \quad \text{... (iii)}$$

Thus the entropy (*S*) of the substance at temperature *T* can be calculated from the measurements of heat capacities (C_p), at a number of temperatures between 0 K to *T* K. The integral in equation (*iii*) can be evaluated by plotting C_p *vs* ln *T i.e.* 2·303 log *T* and then measuring the area under the curve between *T* = 0 to *T* = *T* as shown in Fig. 2.5.

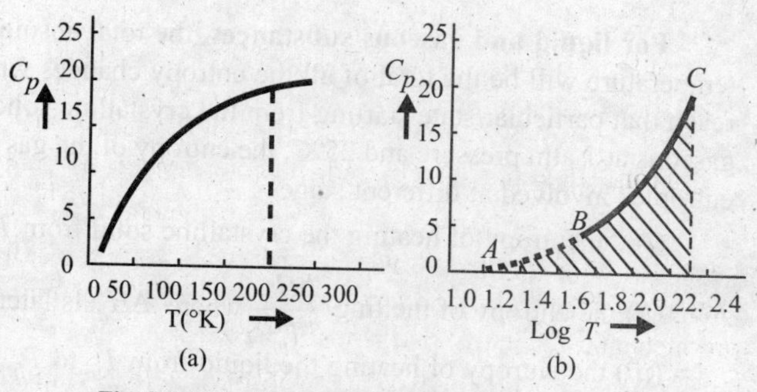

Fig. 2.5. Plot of heat capacities (C_p) *vs* in the first case and *vs* log T in the second case.

If C_p is supposed to remain almost constant in the temperature range 0 to *T*K, equation (*iii*) can be simplified to

$$S = C_p \ln T = 2{\cdot}303\, C_p \log T \quad \text{... (iv)}$$

This equation helps to calculate the value of *S* directly from the values of C_p and *T*.

However, heat capacities cannot be measured with accuracy below a temperature of 15 K. Thus part *CB* of the curve is experimental curve using values of C_p. The curve is extrapolated to get the value at 0K. Between the range 0–15 K, heat capacity is obtained by Debye *T*-cubed law

$$C_p = a\,T^3 \quad \text{... (v)}$$

where *a* is a constant. It is determined from the value of C_p at some low temperature.

Let T_1 be the temperature above which heat capacity can be measured. This corresponds to the point *B* in Fig. 2.5 (*b*). *C* corresponds to the temperature *T* at which the entropy of the solid is to be determined. In the Fig. 2.5 point *A* corresponds to 0K. The integral in eq. (*iii*) can be split as under

$$S_T = \int_0^{T_1} C_p \frac{dT}{T} + \int_{T_1}^{T} C_p \frac{dT}{T} \qquad \text{... (vi)}$$

The first integral in eq. (*vi*) can be calculated using eq. (*v*)

$$\int_0^{T_1} C_p \frac{dT}{T} = \int_0^{T_1} aT^3 \frac{dT}{T} = \int_0^{T_1} aT^2 dT = \frac{1}{3} aT_1^3$$

or $$\Delta S_1 = \frac{1}{3} C_p(at\ T_1) \qquad \text{...[See eq. (v)]}$$

The second integral in eq. (*vi*) can be evaluated by graphical method. C_p is plotted versus ln T. The area under the curve between the limits T_1 and T gives the value of the integral. Therefore, entropy of the solid between 0 K and T K is given by

$$S_T = \frac{1}{3} C_p(at\ T_1) + \int_{T_1}^{T} C_p\, d\ln T \qquad \text{... (vii)}$$

If there is some allotropic change between the temperature range 0 K – T K, then the entropy of transition $\Delta H_t / T_t$ is added to eq. (*vii*)

Thus, we have

$$S_T = \frac{1}{3} C_p(at\ T_1) + \int_{T_1}^{T} C_p\, d\ln T + \frac{\Delta H_t}{T_t}$$

For liquid and gaseous substances, the total absolute entropy of the substance at the given temperature will be the total of all the entropy changes which the substance undergoes in order to reach that particular state starting from the crystalline solid at absolute zero. Thus if a substance is gaseous at 1 atm pressure and 25°C, the entropy of the gas at 25°C will be the sum of the following entropies involved at different stages :

(*i*) the entropy of heating the crystalline solid from $T = 0$ to $T = T_m$ the melting point;

(*ii*) the entropy of melting $\frac{\Delta H_m}{T_m}$ where ΔH_m is latent heat of melting;

(*iii*) the entropy of heating the liquid from T_m to T_b, the boiling point;

(*iv*) the entropy of vaporization, $\frac{\Delta H_v}{T_b}$ where ΔH_v is latent heat of vaporisation;

(*v*) the entropy of heating the gas from T_b to 25°C (*i.e.* 298 K).

The complete expression for the calculation of entropy at temperature T may be expressed as

$$S = \underbrace{\int_0^{T_m} C_{p(s)} d\ln T}_{\text{I}} + \underbrace{\frac{\Delta H_m}{T_m}}_{\text{II}} + \underbrace{\int_{T_m}^{T_b} C_p(l)\, d\ln T}_{\text{III}} + \underbrace{\frac{\Delta H_v}{T_b}}_{\text{IV}} + \underbrace{\int_{T_b}^{T} C_{p(g)} d\ln T}_{\text{V}}$$

2.12.1. Testing validity of third law of thermodynamics

Validity of the third law can be tested as under :

(1) By calculation of equilibrium constants: The equilibrium constant (K) is related to standard free energy change ($\Delta G°$) according to the equation

$$-\Delta G° = RT \ln K \qquad \text{... (i)}$$

Also, the standard free energy change is related to the standard enthalpy change (ΔH°) and standard entropy change (ΔS°) according to the equation

$$\Delta G^\circ = \Delta H^\circ - T\,\Delta S^\circ \quad \text{... } (ii)$$

Substituting the experimental value of ΔH° and calculated value of ΔS°, using the absolute values of entropies obtained from the third law of thermodynamics, and applying the relation

$$\Delta S^\circ = \sum S^\circ_{\text{products}} - \sum S^\circ_{\text{reactants}}$$

ΔG° can be calculated for a reaction. Putting this value of ΔG° in Eq. (*i*), the equilibrium constant *K* can be calculated. Value of equilibrium constant as calculated above comes out to be in close agreement with the experimental value. That proves the validity of the law.

(2) By comparing with entropies calculated from statistical mechanics: The absolute entropies calculated on the basis of the third law are found to be in complete agreement with those calculated using statistical mechanics. This again proves the validity of third law.

2.13. RESIDUAL ENTROPY

Entropy calculated by using third law of thermodynamics as given in the previous section is called thermal entropy. We can also calculate the entropy of a substance by using the following relation of statistical mechanics :

$$S = k \ln W$$

where S = statistical entropy

W = Number of possible arrangement of atoms, molecules or ions in a crystal (also called thermodynamic probability)

k = Boltzmann constant

We observe that, in some cases, the experimental values of thermal entropy (obtained from third law of thermodynamics) are not in perfect agreement with those calculated from statistical methods. Thermal entropies are found to be slightly lower than statistical entropies. The lower values of third law entropies are due to the fact that even at $T = 0$, there is some disorder in the solid and it possesses a finite value of entropy.

The entropy which the crystal of a substance possesses at $T = 0$ is called residual entropy.

Some of the crystals, which do not show zero entropy at 0K are CO, NO, H_2 crystals and crystalline ice.

Calculation of residual entropy

Residual entropy can be calculated using the relation

$$S = k \ln W \quad \text{... } (i)$$

If *W* at is 0K known, we can calculate residual entropy.

Residual entropy of CO

Suppose, a sample of *CO* contains *N* molecules. Each molecule of *CO* can have two orientations.

CO or *OC*

Let there be *N* molecules in one mole of the substance.

Therefore *N* molecules can have 2^N orientations *i.e.* $W = 2^N$

Applying eq. (*i*) above, the entropy can be calculated as

$$S = k \ln 2^N \quad \text{or} \quad S = k N \ln 2 \quad \text{... } (ii)$$

$k\,N$ can be written as equal to *R*.

Thus, for one mole of the substance, the expression would be

$$S = R \ln 2 = 2.303 \times 8.314 \times 0.3010 \text{ JK}^{-1} \text{ mol}^{-1}.$$
$$= 5.85 \text{ JK}^{-1} \text{ mol}^{-1}.$$

The number of orientations for a molecule of the type ABO_3 (A and B are two different elements, O is oxygen) is 4. Hence $W = 4$ in this case.

In general, if there are x possible orientations of a molecule with nearly the same energy, the residual molar entropy will be

$$S = R \ln x$$

Example. $ClBrO_3$ molecule can have four possible arrangements with nearly the same energy. Calculate the residual entropy of $ClBrO_3$ molecule

Solution. Residual entropy $S = R \ln W = 2.303\, R \log W$

In this case $W = 4, \quad R = 8.314$

Substituting the values in the above equation, we have

$$S = 2.303 \times 8.314 \times 0.6021$$
$$= 11.5 \text{ Jk}^{-1} \text{ mol}^{-1}.$$

SOLVED EXAMPLES ON THIRD LAW OF THERMODYNAMICS

Example 1. Write expressions for the following :

(*i*) Variation of Helmholtz function with volume at constant temperature.

(*ii*) Variation of Helmholtz function with temperature at constant volume.

(*iii*) Total change in Helmholtz function at constant temperature when volume changes from V_1 to V_2.

Solution. (*i*) $\left(\dfrac{\partial A}{\partial V}\right)_T = -P$ (*ii*) $\left(\dfrac{\partial A}{\partial T}\right)_V = -S$ (*iii*) $(\Delta A)_T = -nRT \ln \dfrac{V_2}{V_1}$.

Example 2. Write expression for total change in Gibbs function when at constant temperature, volume changes from V_1 to V_2 or pressure changes from P_1 to P_2.

Solution. $$(\Delta G)_T = nRT \ln \frac{P_2}{P_1} = nRT \ln \frac{V_1}{V_2}.$$

Example 3. What is the difficulty in determining absolute entropy of a substance ? How has the problem been solved by Debye ?

Solution. In the determination of absolute entropy of a substance, heat capacity values are required at different temperatures, starting from 0K. But experimentally, heat capacities at low temperatures (< 15 K) cannot be determined. The problem was solved by Debye who gave third power law for heat capacities of solids according to which $C_p = a\,T^3$.

Example 4. What is the difference between Helmholtz function and Gibbs function? Under what condition, ΔG becomes equal to ΔA?

Solution. Helmholtz function $A = U - TS$

Gibbs function $G = H - TS$

$$-\Delta A = w_{max} \text{ and } -\Delta G = w_{max} - P\,\Delta V$$

Thus, if during the process there is no change in volume, $P\,\Delta V = 0$ so that $-\Delta G = -\Delta A$.

Example 5. What is total differential equation ? How does Gibbs function vary with pressure at constant temperature and with temperature at constant pressure?

Solution. Total differential equation is $dG = V\,dP - S\,dT$

$$\left(\frac{\partial G}{\partial P}\right)_T = V, \quad \left(\frac{\partial G}{\partial T}\right)_P = -S.$$

Example 6. Under what condition electrical energy produced in a cell reaction is equal to the heat evolved in the cell reaction?

Solution. $$nFE = -\Delta H + TnF\left(\frac{\partial E}{\partial T}\right)_P$$

Thus, $n\,FE = -\Delta H$ when $\left(\frac{\partial E}{\partial T}\right)_P = 0$, *i.e.*, when the temperature coefficient of the E.M.F. of the cell = zero.

Example 7. Write criteria for spontaneity and equilibrium conditions in terms of S, U, H, A and G.

Solution. $(dS)_{U,V} \geq 0$, $(dU)_{S,V} \leq 0$, $(dH)_{S,P} \leq 0$, $(dA)_{T,V} \leq 0$, $(dG)_{P,T} \leq 0$

Sign '=' refers to reversible process (*i.e.*, equilibrium) and > or < refers to irreversible process (spontaneity).

EXERCISES
(Based on Question Papers of Different Universities)

Objective Questions (Multiple Choice Questions)

1. For an isothermal process the total change in work function is given by the expression :

(*a*) $-(\Delta A)_T = -nRT\ln\frac{V_1}{V_2}$ (*b*) $(\Delta A)_T = -nRT\ln\frac{P_2}{P_1}$

(*c*) $(\Delta A)_T = -nRT\ln\frac{V_2}{V_1}$ (*d*) $(\Delta A)_T = -nRT\ln\frac{V_1}{V_2}$

2. Which of the following is not the criterion for spontaneity of a process :

(*a*) $(dG)_{T,P} \geq 0$ (*b*) $(dA)_{T,V} \leq 0$

(*c*) $(dG)_{T,P} \leq 0$ (*d*) $(dS)_{U,V} \geq 0$.

3. Nernst heat theorem can be mathematically stated as :

(*a*) $\lim\limits_{T\to 0}\frac{d(\Delta G)}{dT} = \lim\limits_{T\to 0}\frac{d(\Delta H)}{dT} = 0$ (*b*) $\lim\limits_{T\to 0}\frac{d(\Delta H)}{dT} = \lim\limits_{T\to 0}\frac{d(\Delta S)}{dT} = 0$

(*c*) $\lim\limits_{T\to 0}\frac{d(\Delta U)}{dT} = \lim\limits_{T\to 0}\frac{d(\Delta A)}{dT} = 0$ (*d*) $\lim\limits_{T\to 0}\frac{d(\Delta A)}{dT} = \lim\limits_{T\to 0}\frac{d(\Delta H)}{dT}$

4. Under what conditions change in work function is equal to change in Gibbs free energy *i.e.* $\Delta A = \Delta G$:

(*a*) When there is no change in volume, *i.e.* $\Delta V = 0$.

(*b*) When there is no change in enthalpy, *i.e.* $\Delta H = 0$.

(*c*) When there is no change in temperature.

(*d*) When there is no change in pressure.

5. The relationship between decrease in Gibbs free energy and electrical energy is :

(*a*) $-\Delta G = n\text{FE}_{\text{cell}}$ (*b*) $\Delta G = n\text{RT} \ln \text{E}_{\text{cell}}$

(*c*) $-\Delta G = -n\text{F E}_{\text{cell}}$ (*d*) $\Delta G = -n\text{RT} \ln \text{E}_{\text{cell}}$

6. Point out the correct relationship :

(*a*) $\left(\frac{\partial A}{\partial V}\right)_P = -T$ (*b*) $\left(\frac{\partial A}{\partial T}\right)_P = -P$

(*c*) $\left(\frac{\partial G}{\partial P}\right)_T = V$ (*d*) $\left(\frac{\partial G}{\partial T}\right)_P = -H$

7. Gibbs Helmholtz equation is :

(*a*) $\Delta G = \Delta H - T\Delta S$ (*b*) $\Delta G = \Delta H - T\left[\frac{(\partial G)}{\partial T}\right]_P$

(*c*) $\Delta G = \Delta U + T\left[\frac{\partial(\Delta G)}{\partial T}\right]_P$ (*d*) $\Delta G = \Delta H + T\left[\frac{\partial(\Delta G)}{\partial T}\right]_P$

8. At very low temperatures near the range of 0 K, the C_P of a solid can be calculated using the equation :

(*a*) $C_P \approx \alpha T^3$ (*b*) $C_P = \alpha T^2$

(*c*) $C_p = \frac{1}{3}\alpha T^3$ (*d*) $C_P = \frac{1}{3}\alpha T$

9. Residual entropy is :

(*a*) The entropy possessed by the in crystalline substance at $T = 0$.

(*b*) The entropy arising out of defects in crystalline substance

(*c*) The entropy which is in excess over the normal value

(*d*) The remaining entropy of a substance.

10. Which of the following is not a state function :

(*a*) H (*b*) S

(*c*) U (*d*) W

11. Residual molar entropy is given by expression :

(*a*) $S = R \ln W$ (*b*) $S = R \ln W^n$

(*c*) $S = \frac{R}{N} \ln W$ (*d*) $S = nR \ln W$.

12. The Boltzmann entropy equation is :

(*a*) $S = \frac{K}{\ln W}$ (*b*) $S = K \ln \frac{P_2}{P_1}$

(*c*) $S = K \ln \frac{V_1}{V_2}$ (*d*) $S = K \ln W$

13. When the limit $T \rightarrow 0$, the entropy of a crystalline solid is approximately equal to :

(*a*) $\frac{1}{2} C_P$ (*b*) $\frac{2}{3} C_P$

(*c*) $\frac{1}{3} C_P$ (*d*) $\frac{1}{5} C_P$

14. Choose the incorrect relationship :

(*a*) $(\Delta G)_T = nRT \ln \frac{P_2}{P_1}$ (*b*) $(\Delta A)_T = nRT \ln \frac{V_2}{V_1}$

(*c*) $(\Delta G)_P = nRT \ln \frac{V_1}{V_2}$ (*d*) $(\Delta A)_T = -nRT \ln \frac{P_2}{P_1}$

15. The most preferred criterion for spontaneity of a process is :

(*a*) $(dU)_{S,V} \leq 0$ (*b*) $(dH)_{S,P} \leq 0$

(*c*) $(\delta G)_{T,P} \leq 0$ (*d*) $(dS)_{U,V} \leq 0$

16. From third law, the entropy of a solid at temperature T is given as :

(*a*) $S_T = \frac{1}{3} C_P \text{ (at } T_1) + \int_{T_1}^{T} \frac{C_P}{T} dT$ (*b*) $S = C_P \text{ (at } T_1) + \int_{T_1}^{T} C_P \ln dT$

(*c*) $S = C_P \ln T + \int_{T_1}^{T} \frac{C_P}{T} dT$ (*d*) None of these

17. Point out the incorrect statement :

(*a*) $\Delta A = \Delta U - T\Delta S$ (*b*) $\Delta G = \Delta H - T\Delta S$

(*c*) $\Delta S = R \ln W$ (*d*) $-\Delta G° = RT \ln K$

18. Third law of thermodynamics may be stated as :

(*a*) At absolute zero the entropy of all perfectly crystalline solids tends to decrease.

(*b*) The entropy of a perfectly crystalline solid may be taken as zero at absolute zero.

(*c*) The entropy of every substance is zero at 0 K.

(*d*) The entropy of a substance is related to its heat capacity.

ANSWERS TO MULTIPLE CHOICE QUESTIONS

1. (*c*)	**2.** (*a*)	**3.** (*a*)	**4.** (*b*)	**5.** (*a*)	**6.** (*c*)
7. (*d*)	**8.** (*a*)	**9.** (*a*)	**10.** (*d*)	**11.** (*a*)	**12.** (*d*)
13. (*c*)	**14.** (*a*)	**15.** (*c*)	**16.** (*a*)	**17.** (*c*)	**18.** (*b*)

Second Law of Thermodynamics

Short Answer Questions

1. What is Carnot cycle ? How does it lead to the definition of entropy? Justify that

$$\Sigma \frac{q}{T} = 0 \text{ for Carnot cycle}$$

2. Explain the term 'Entropy'. Show that it is a state function.
   Write down the units of 'Entropy'.
3. Explain that in a reversible process, there is no net entropy change.
4. Explain that the entropy of the system and the surroundings increases in an irreversible process but remains constant in a reversible process.
5. Why there was a need for introduction of second law of thermodynamics ? Illustrate with a suitable example.
6. Show that the efficiency of a reversible Carnot cycle is $\frac{T_2 - T_1}{T_2}$ where T_2 and T_1 are the temperatures of the source and the sink.
7. State and explain 'Carnot's theorem'.
8. Calculate the total entropy changes (ΔS_{total}) during an irreversible process when heat flows from a body at higher temperature T_2 to a body at lower temperature T_1.

   Or

   Prove that for an irreversible process when heat flows from a body at temperature T_2 to a body at temperature T_1.

$$\Delta S_{total} = q_{rev}\left(\frac{1}{T_1} - \frac{1}{T_2}\right)$$

9. Explain how thermodynamic scale of temperature was developed. Why is it considered to be more basic than ideal or perfect gas scale of temperature ?
10. Explain giving reasons—the efficiency of a steam engine can be increased by using super-heated steam.
11. Show that for an irreversible process $\Delta S_{sys} + \Delta S_{surr} > 0$.

General Questions

1. Derive an expression for the entropy change on mixing of ideal gases. From this expression, what do you conclude about
   (*i*) effect of temperature on entropy change of mixing ?
   (*ii*) decrease or increase of entropy on mixing ?
2. Derive an expression for entropy change when n_1 mole of an ideal gas 1 are mixed with n_2 moles of an ideal gas 2.
3. Starting from basic principles, show that for the isothermal expansion of an ideal gas, $\Delta S = R \ln \frac{V_2}{V_1}$.
4. Derive an expression for the molar entropy change of mixing of two ideal gases.
5. Derive that for an ideal gas

$$\Delta S = C_v \ln \frac{T_2}{T_1} + R \ln \frac{V_2}{V_1}$$

6. Derive the required expression to show that entropy change on mixing of ideal gases is independent of temperature.
7. Derive an expression for the calculation of the entropy change of an ideal gas when the temperature changes from T_1 to T_2 and the pressure changes from P_1 to P_2.

8. Explain how entropy changes take place in a reverisble process and in an irreversible process. What do you conclude about the entropy change of the universe ?
9. What are the limitations of First law of thermodynamics ? How have these been overcome by the Second law of thermodynamics. Define it in as many ways as you can.
10. State and explain Carnot's theorem. Explain how on the basis of efficiency of a reversible heat engine, thermodynamic scale of temperature was developed by Lord Kelvin.
11. What is Clausius inequality ? Explain that in spontaneous expansion as well as spontaneous cooling, dS_{total} is positive.
12. What is Clausius inequality in thermodynamics ? Show that for irreversible adiabatic expansion, $dS_{total} > 0$.
13. Explain giving reasons that the entropy of the universe tends to increase.
14. Derive that for an ideal gas

$$\Delta S = C_p \ln \frac{T_2}{T_1} + R \ln \frac{P_2}{P_1}$$

15. Explain the term 'entropy'. Briefly explain the physical significance of the term 'entropy'.
16. Describe Carnot's cycle. Derive an expression for the efficiency of a reversible heat engine working between temperatures T_1 and T_2 $(T_2 > T_1)$.
17. On the basis of reversible Carnot cycle, how the concept of entropy was developed ? Define entropy and state its units.
18. Derive an expression for the calculation of the entropy change of an ideal gas when the temperature changes from T_1 to T_2 and volume changes from V_1 to V_2.
19. What is Clausius inequality ? Show that for spontaneous cooling, $dS_{total} > 0$.
20. Most of the processes taking place in nature are spontaneous in character and according to the second law of thermodynamics, all spontaneous processes are accompanied by a net increase of entropy. If entropy is regarded as a measure of randomness in a system, it would then mean that the universe is heading towards chaos. Is it true ? Explain clearly.
21. Derive the required expression to show that mixing of gases is always accompanied by increasing in entropy.

Third Law of Thermodynamics

Short Answer Questions

1. Differentiate between Gibbs free energy and Helmholtz free energy.
   How decrease in free energy signifies the chemical affinity of a reaction ? Explain with suitable examples. What do you understand by the terms 'free energy' and 'work function'? Derive a relationship between the two. Under what conditions ΔG becomes equal to ΔA?
2. Explain the meaning of each of the following expressions:
   (i) $(\partial G)_{T,P} < 0$ *(ii)* $(S)_{U,V} < 0$ *(iii)* $(\partial U)_{S,V} < 0$
3. Show that for an irreversible process, $dG_{P,T} < 0$.
4. Show that $\left(\frac{\partial G}{\partial P}\right)_T = V$
5. Derive thermodynamically the relationships

$$\Delta G = RT \ln \frac{P_2}{P_1} = RT \ln \frac{V_1}{V_2}$$

6. Derive the following thermodynamic expressions :

(i) $\left(\frac{\partial A}{\partial V}\right)_T = -P$ (ii) $\left(\frac{\partial A}{\partial T}\right)_V = -S$

where A is work function and other symbols have their usual meaning.

7. Show that the change in free energy is equal to net work done.

8. Explain the terms 'free energy' and 'Maximum work function'. Derive a relationship between the two.

9. "For a system in equilibrium at a given temperature and pressure, the free energy is minimum." Explain.

10. Find the change in free energy when a system undergoes a reversible change of pressure as well as temperature.

11. Show that $\left(\frac{\partial G}{\partial T}\right)_P = -S$

12. Show that $\Delta A \leq 0$ is criterion for spontaneity and equilibrium.

13. Show that $\Delta G \leq 0$ is a criterion of spontaneity and equilibrium.

14. Explain the term 'free energy'. What information does the free energy of a reaction give about the spontaneity of the reaction to occur ?

General Questions

1. What is Nernst heat theorem ? What result follows from it regarding entropy change and heat capacity change ? State 'third law of thermodynamics'. Briefly desribe its importance.

2. What is Residual entropy ? What is its origin and how can it be calculated ? Explain taking suitable examples.

3. Show that $\Delta G = \Delta H + T\left[\frac{\partial(\Delta G)}{\partial T}\right]_P$

4. First law of thermodynamics leads to the concept of energy content and second law leads to the concept of entropy. Does third law lead to any new concept ? If not, then why is it called a law ?

5. How can you test the validity of the third law of thermodynamics ?

6. What is Residual entropy ? How the concept of residual entropy originated ? How is it calculated ?

7. Explain the term 'Gibbs function'. Deduce that decrease in its value is equal to the maximum net work obtainable from the system. Also explain how does it vary with temperature and pressure ?

8. Bring out clearly the criteria for reversibility and irreversibility in terms of S, U, H, A and G. What do you mean by a reversible and an irreversible process ?

9. Derive an expression relating the changes in free energy and enthalpy of a closed thermodynamic system at constant pressure.

10. Derive 'total differential equation' and hence derive the following expression for an isothermal process when the volume changes from V_1 to V_2 or pressure changes from P_1 to P_2

$$(\Delta G)_T = nRT \ln\frac{P_2}{P_1} = nRT \ln\frac{V_1}{V_2}$$

11. What are the applications of Gibbs-Helmholtz equation ?
12. How the results of Nernst theorem led to the enunciation of third law of thermodynamics ?
13. Prove that the entropy of any substance at very low temperature is one third of its molar heat capacity at that temperature.
14. Explain the term 'Helmholtz function'. How can you deduce that for a process occurring at constant temperature, the decrease in Helmholtz function 'A' is equal to the maximum work done by the system. Also explain how it varies with temperature and volume.
15. Derive Gibbs-Helmholtz equation in terms of free energy and enthalpy change at constant pressure. Bring out its importance.
16. State and explain. 'Nernst heat theorem'.
17. Discuss Gibbs-Helmholtz equation with regard to its application to Galvanic cell.
18. State 'third law of thermodynamics'. How does this law help in the determination of absolute entropies of chemical compounds at any desired temperature ?
19. Write complete expression for the calculation of entropy of any substance in the gaseous state at temperature T starting from 0 K and assuming that a solid undergoes a polymorphic transition at temperature T_{tr}. Represent the various changes graphically (entropy versus temperature).
20. Show that $\Delta G = \Delta H + T\left[\frac{\partial(\Delta G)}{\partial T}\right]_P$

3

Chemical Equilibrium

3.1. INTRODUCTION

Most of the chemical reactions, when carried out in closed vessels do not go to completion. A process starts of its own or on initiation, continues for sometime with falling rate and ultimately appears to have stopped. Although, reactants are still there, but they don't appear to react anymore. The process or the reaction is said to be in a state of *equilibrium*. The composition of the reaction mixture at equilibrium at a given temperature is the same irrespective of whether we start from the reactant side or product side. This may be illustrated with the help of the following examples.

1. Decomposition of calcium carbonate: Solid calcium carbonate on heating in a closed vessel starts decomposing into solid calcium oxide and gaseous carbon dioxide. Pressure built up due to carbon dioxide can be recorded on a manometer attached to the vessel. The pressure goes on increasing and then becomes constant and remains so after that. Calcium carbonate is still present in the reaction mixture but does not appear to decompose further. A state of equilibrium, as represented by the following equation, is attained.

$$CaCO_3\ (s) \rightleftharpoons CaO\ (s) + CO_2\ (g)$$

2. Evaporation of water in a closed vessel: If a small amount of water is placed in a closed vessel at room temperature, evaporation starts. A pressure due to water vapours is built up in the vessel, which can be recorded by the attached manometer. After some time, we find that the pressure of water vapours remains constant, if the temperature does not change. Sufficient water is still present, but it does not appear to evaporate further. We say that a state of equilibrium between evaporation of water and condensation of vapours has been established. This can be represented as

$$H_2O\ (l) \rightleftharpoons H_2O\ (g)$$

3.1.1. Characteristics of chemical equilibrium

1. Chemical equilibrium at a given temperature, is characterised by constant values of certain observable properties such as pressure, concentration or colour. It is well known that hydrogen gas and iodine vapours react at higher temperature to give hydrogen iodide gas as per the following equation

$$H_2\ (g) + I_2\ (g) \rightleftharpoons 2HI\ (g)$$

It is also known that hydrogen iodide gas is not very stable and it dissociates into hydrogen and iodine as follows:

$$2HI\ (g) \rightleftharpoons H_2(g) + I_2(g)$$

The composition of the reaction mixture at equilibrium, is *1.56 moles of* HI, *0.22 mole of* H_2 *and 0.22 mole of* I_2 whether we carry out the reaction by taking one mole each of H_2 and Cl_2 as the reactants or two moles of HI as the reactant at 448°C.

This shows that *when equilibrium is attained at a given temperature, each reactant and each product has a fixed concentration and this is independent of the fact whether we start the reaction with the reactants or with the products. Thus, equilibrium can be attained from either side.*

2. *A catalyst can hasten the approach of equilibrium but does not alter the state of equilibrium. In other words, the relative concentrations of the products and the reactants remain the same irrespective of the presence or absence of a catalyst.*

3. *Chemical equilibrium is* **dynamic** *in nature.* Equilibrium involves two reactions proceeding in opposite directions. One of these reactions proceeds *from the reactants towards the products* and is known as the *forward reaction.* The other proceeds *from the products towards the reactants* and is known as the *reverse reaction.* When equilibrium is attained, there is no further change in the concentrations of the products or the reactants. This gives the impression that the reaction has come to a stand-still. But this is *not* the case. Actually, the two opposing reactions, the forward reaction and the reverse reaction, are proceeding simultaneously at *equal rates.*

3.1.2. Experimental proof of dynamic nature of chemical equilibrium

Consider the following reaction

$$H_2(g) + I_2(g) \rightleftharpoons 2HI(g)$$

When the equilibrium is attained at 448°C, a small amount of radioactive isotope of iodine is introduced into the reaction mixture. It is well known that radioactive isotopes have the same *chemical* properties as their non-radioactive counterparts. After sometime if the mixture is examined, it will be seen that HI *contains radioactive iodine.* There is, however, no change in the relative amounts of HI, H_2 and I_2. This shows that although there has been no change in the relative amounts of the reactants and the products, *chemical reaction has been taking place from left to right.* But since there has been no increase whatsoever in the amount of HI, it is evident that *the reaction has been taking place from the right to the left as well, at the same rate.* Thus, the two opposing reactions have been proceeding. This proves the dynamic nature of equilibrium.

3.2. THE STATE OF CHEMICAL EQUILIBRIUM

In a general reversible reaction

$$A + B \rightleftharpoons C + D$$

the reactants A and B react to form products. With the passage of time the concentrations of reactants decrease. Therefore the rate of forward reaction decreases. On the other hand, the concentration of C and D increases. Therefore rate of backward reaction increases. A stage is reached when the rate of backward reaction becomes equal to rate of forward reaction and the system attains a state of equilibrium. The variation of rate of forward and backward reaction is shown in Fig. 3.1.

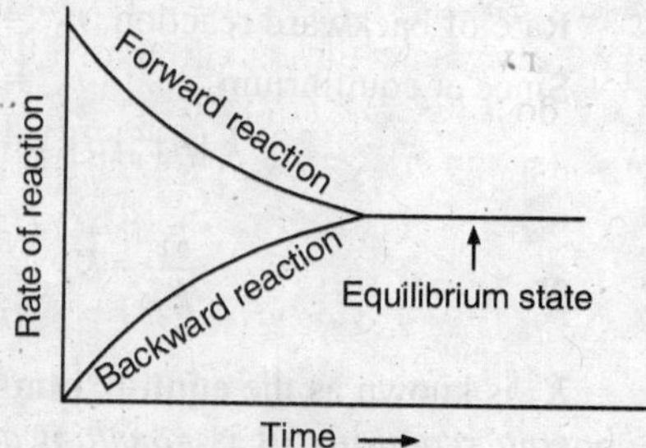

Fig. 3.1. Variation of rate of forward and backward reactions with time.

Thus, *the state of equilibrium is a state in which the measurable properties of the system do not undergo any noticeable change under a given set of conditions.* That, means the properties like concentration, pressure etc. of the system become constant. But this does not mean that the reaction has stopped. In fact, both the forward and backward reactions carry on even after attainment of equilibrium. However, the rates of forward and backward reactions become equal as shown in Fig. 3.1. Thus, *the equilibrium is dynamic in nature.*

Chemical Equilibrium

3.3. LAW OF MASS ACTION

According to the law of mass action, *the rate at which a substance reacts is proportional to its 'active mass' and the rate of a chemical reaction is proportional to the product of the active masses of the reacting substances.*

The term *active mass* used in the above statement implies *activity* which, for the sake of simplicity, may be taken as equal to *molar concentration.*

Consider the following reversible reaction, taking place at a constant temperature:

$$A + B \rightleftharpoons C + D$$

According to the law of mass action, the rate (r_1) at which A and B react is given by the equation

$$r_1 = k_1 [A] [B] \quad ...(i)$$

where k_1 is the proportionality constant, known as the **rate constant** of the reaction and the square brackets indicate molar concentrations. As the reaction proceeds further, the molar concentrations of A and B continue to decrease. Consequently, the velocity of the forward reaction goes on decreasing with time. The rate of the back reaction (r_2) is given by the equation

$$r_2 = k_2 [C] [D] \quad ...(ii)$$

where k_2 is the rate constant of the reverse reaction.

Thus, ultimately, a dynamic equilibrium is attained when the rate of the forward reaction becomes equal to that of the reverse reaction, *i.e.*, $r_1 = r_2$.

3.3.1. Law of chemical equilibrium

Consider the reaction

$$a\text{A} + b\text{B} \rightleftharpoons c\text{C} + d\text{D}$$

Rate of forward reaction, $r_1 = k_1 [A]^a [B]^b$

Rate of backward reaction, $r_2 = k_2 [C]^c [D]^d$

Since at equilibrium, $r_1 = r_2$, it follows that

$$k [A]^a [B]^b = k_2 [C]^c [D]^d$$

or

$$\frac{k_1}{k_2} = K_c = \frac{[\text{C}]^c [\text{D}]^d}{[\text{A}]^a [\text{B}]^b} \quad ... (i)$$

K_c is known as the **equilibrium constant.** It is simply the *ratio of the rate constants of two opposing reactions.* It is *constant at a given temperature.* It has the same value at a given temperature *irrespective of the direction* from which the equilibrium is reached. It has also the same value *irrespective* of the *pressure* on the system or the *concentrations* of the reactants and products.

In a general reaction represented by the equation

$$aA + bB + cC + \cdots \rightleftharpoons lL + mM + nN + \cdots,$$

the equilibrium constant is given by

$$K_c = \frac{[L]^l [M]^m [N]^n ...}{[A]^a [B]^b [C]^c ...} \quad ...(ii)$$

For gaseous reactions, it is more convenient to use *partial pressures* instead of concentrations. The equilibrium constant in that case is represented by K_p. Thus, for the general gaseous reaction

$$aA + bB + \cdots \rightleftharpoons lL + mM + \cdots,$$

$$K_P = \frac{[p_L]^l \times [p_M]^m ....}{[p_A]^a \times [p_B]^b ...} \quad ...(iii)$$

where p_L and p_M, etc stand for partial pressures of the products and p_A, p_B, etc. for the partial pressures of the reactants.

3.3.2. Types of equilibrium constants and their units

In eq. (*i*) above, the concentrations of reactants and products are generally expressed in terms of moles/litre.

1. The equilibrium constant is then written as K_c where the subscript *c* refer to the equilibrium constant expressed in terms of molar concentrations.

2. When the molar concentrations are replaced by activities of reactants and products, then; the equilibrium constant is written in *terms of activities* and is called K_a.

i.e.,
$$K_a = \frac{a_C^c \cdot a_D^d}{a_A^a \cdot a_B^b}, \; K_c = \frac{[C]^c \, [D]^d}{[A]^a \, [B]^b}$$

As activity is dimensionless, K_a has no units.

3. In gaseous reactions, the concentration terms in eq. (*i*) are replaced by the partial pressure. Then, the equilibrium constant may be expressed as

$$K_p = \frac{p_C^c \cdot p_D^d}{p_A^a \cdot p_B^b}$$

where p_A, p_B, p_C and p_D are the partial pressures of gases A, B, C and D respectively. K_p may not always be equal to K_c. Units of K_p are $(\text{Atmospheres})^{\Delta n}$.

As concentrations are expressed in moles-litre, according to eq. (*i*), dimensions of K_c are given by

$$\left(\frac{Moles}{litre}\right)^{\Delta n}$$

where Δn is the difference is the number of moles of products and the reactants [say $\Delta n = (c + d) - (a + b)$]

when $\Delta n = 0$, K_c and K_p both will become dimensionless *i.e.* will have no units, simply numbers.

4. When equilibrium constant is expressed in terms of mole fractions of reactants and products, it is expressed as K_x.

$$K_x = \frac{(x_C)^c \, (x_D)^d}{(x_A)^a \, (x_B)^b}$$

where x_A, x_B, x_C and x_D represent the mole fractions of A, B, C and D.

3.3.3. Relationship between K_p and K_c

For a general reaction,

$$aA + bB \rightleftharpoons cC + dD$$

The equilibrium constant in terms of concentrations (moles/litre) is

$$K_c = \frac{[C]^c [D]^d}{[A]^a [B]^b} \qquad \text{... (i)}$$

If the reactants and products are gaseous, then the concentration terms may be replaced by partial pressure. The equilibrium constant K_p is written as

$$K_p = \frac{p_C^c \, p_D^d}{p_A^a \, p_B^b} \qquad \text{... (ii)}$$

For an ideal gas, $pV = nRT$ or $p = \frac{n}{V} RT = CRT$,

where C is the molar concentration. For different gases A, B, C, D, we may write

$p_A = C_A RT, p_B = C_B RT, p_C = C_C RT, p_D = C_D RT$

Putting these values in equation (*ii*), we get

$$K_p = \frac{(C_C RT)^c \, (C_D RT)^d}{(C_A RT)^a \, (C_B RT)^b}$$

or
$$K_p = \frac{C_C^c \, C_D^d}{C_A^a \, C_B^b} (RT)^{(c+d)-(a+b)}$$

or
$$\boxed{K_p = K_c (RT)^{\Delta n}} \qquad \text{... (iii)}$$

where
$$\Delta n = (c+d) - (a+b)$$

$$= \left[\text{Number of moles of products}\right] - \left[\text{Number of moles of reactants}\right]$$

3.3.4. Relationship between K_x, K_p and K_c

The partial pressure p of a gas in an ideal gaseous mixture is related to total pressure P as

$$p = xP$$

where x is the mole fraction of the gas in the mixture

Substituting for the partial pressures of A, B, C and D in equation (*ii*), we get

$$K_p = \frac{p_C^c \, p_D^d}{p_A^a \, p_B^b} = \left[\frac{x_C^c \, x_D^d}{x_A^a \, x_B^b}\right] P^{(c+d)-(a+b)}$$

or
$$K_p = K_x (P)^{\Delta n} \qquad \text{... (iv)}$$

From eq. (*iii*) and (*iv*), we have

$$K_c (RT)^{\Delta n} = K_x P^{\Delta n}$$

or
$$K_c = K_x \left(\frac{P}{RT}\right)^{\Delta n}$$

or
$$\boxed{K_c = K_x V^{\Delta n}}$$

Where V is the volume of system containing 1 mole of ideal gas.

When $\Delta n = 0$, K_p, K_c and K_x become equal.

or $$K_p = K_x = K_c$$

Relationship between K_a and K_p or K_c

Activity and molar concentration are related as

$$a = \gamma C \qquad \text{... } (i)$$

where y is activity coefficient

$$K_a = \frac{a_C^c\, a_D^d}{a_A^a\, a_B^b} \qquad \text{... } (ii)$$

Substituting for activities of A, B, C and D in eq. (*ii*), we have

$$K_a = \left(\frac{\gamma_C^c\, \gamma_D^d}{\gamma_A^a\, \gamma_B^b}\right)\left(\frac{C_C^c\, C_D^d}{C_A^a\, C_B^b}\right) = K_\gamma \cdot K_c$$

or $$\boxed{K_a = K_\gamma K_c} \qquad \text{... } (iii)$$

For very dilute solutions as $C \to 0$, $\gamma \to 1$

Then $$K_a \to K_c$$

The activity of any gaseous component is related to its partial pressure as

$$a = \gamma_p \qquad \text{... } (iv)$$

Substituting (*iv*) in eq. (*ii*), we get

$$K_a = K_\gamma K_p$$

For gases at very low pressure $\gamma = 1$

$\therefore$ $K_\gamma = 1$, hence $K_a = K_p$

3.4. PARTIAL MOLAR QUANTITIES

The thermodynamic properties such as free energy, enthalpy, entropy, free energy etc. are extensive properties. Their values change with change in mass of the system. Consider an open system containing n_1, n_2, n_3 ...... moles of various components. Let the property X be a function of temperature., pressure and the number of moles of various constituents *i.e.*

$$X = f(T, P, n_1, n_2, n_3, \ldots n_i) \qquad \text{... } (i)$$

where $n_1 + n_2 + n_3 + \ldots n_i = N$

For a small change in temperature, pressure and the number of moles of constituents, the change dX in property X will be given by the following equation:

$$dX = \left(\frac{\partial X}{\partial T}\right)_{P, n_1, n_2 \ldots\ldots} dT + \left(\frac{\partial X}{\partial P}\right)_{T, n_1, n_2 \ldots\ldots} dP$$

$$+ \left(\frac{\partial X}{\partial n_1}\right)_{T, P, n_2, n_3 \ldots\ldots} dn_1 + \left(\frac{\partial X}{\partial n_2}\right)_{T, P, n_1, n_3 \ldots\ldots} dn_2 + \ldots\ldots \qquad \text{... } (ii)$$

At constant temp. and pressure of the system, $dT = 0$ and $dP = 0$.

Equation (*ii*), then becomes

$$(dX)_{T,P} = \left[\frac{\partial X}{\partial n_1}\right]_{T, P, n_2, n_3 \ldots\ldots} dn_1 + \left[\frac{\partial X}{\partial n_2}\right]_{T, P, n_1, n_3 \ldots\ldots} dn_2 + \ldots\ldots \qquad \text{... } (iii)$$

Each term on the right hand side of equation (*iii*) is called partial molar property of that particular component. It is represented by putting a bar over the concerned property *i.e.* $\bar{X}_1$, $\bar{X}_2$ for the first and second component respectively.

Thus,
$$\left[\frac{\partial X}{\partial n_1}\right]_{T, P, n_2, n_3 .....} = \bar{X}_1$$

$$\left[\frac{\partial X}{\partial n_2}\right]_{T, P, n_1, n_3 .....} = \bar{X}_2$$

In general, for an *i*th constituent, we can write

$$\left[\frac{\partial X}{\partial n_i}\right]_{T, P, n_1, n_2 .....} = \bar{X}_i$$

3.4.1. Partial molar free energy: Concept of chemical potential

The most important partial molar property is partial molar free energy. It is also called Chemical potential and represented by symbol μ. For an *i*th component, it is written as

$$\mu = \bar{G}_i = \left[\frac{\partial G}{\partial n_i}\right]_{T, P, n_1, n_2 ......} \quad ... (i)$$

If dn_i is taken as 1 mole,

Then
$$\mu = (dG)_{T, P, n_1, n_2 ......} \quad ... (ii)$$

The chemical potential of a given substance may be defined as the change in free energy of the system which results on the addition of ***one mole*** *of that particular substance at a constant temperature and pressure, to such a large quantity of the system that there is no appreciable change in the overall composition of the system.*

Taking free energy as the extensive property, change in property *i.e.* dG may be written as

$$(dG)_{T,P} = \left[\frac{\partial G}{\partial n_1}\right]_{T, P, n_2, n_3 ......} dn_1 + \left[\frac{\partial G}{\partial n_2}\right]_{T, P, n_1, n_3 ......} dn_2 + ...$$

or
$$(dG)_{T, P} = \mu_1\, dn_1 + \mu_2\, dn_2 + ... \quad ... (iii)$$

where μ_1, μ_2, ... are chemical potentials of constituents 1, 2, ...

For a system of definite composition having n_1, n_2, ... moles of components, equation (*iii*) on integration gives

$$G_{T, P, N} = n_1\mu_1 + n_2\mu_2 + ...$$

The subscript N stands for total number of moles.

Here μ_1, μ_2 etc. give the contribution per mole of components 1 and 2 to the total free energy of the system under conditions of constant temperature and pressure.

Thus, *the chemical potential of a constituent in a mixture is its contribution per mole to the total free energy of the system of a constant composition at constant temperature and pressure.*

It is observed that for one mole of pure component $G = \mu$ *i.e.* free energy is the same as chemical potential.

3.4.2. Gibbs-Duhem equation

Free energy is an extensive property. Besides temperature and pressure, it also depends on the composition of the system. For a system having n_1, n_2, ... moles of components 1, 2, ... we can write.

$$G = f(T, P, n_1, n_2, ...) \qquad ... (i)$$

Where $n_1 + n_2 + n_3 + ... = N$.

A small change in free energy with change in temp., pressure and no. of moles of various components can be obtained by partial differentiation of equation (*i*)

$$dG = \left(\frac{\partial G}{\partial T}\right)_{P, n_1, n_2..} dT + \left(\frac{\partial G}{\partial P}\right)_{T, n_1, n_2..} dP + \left(\frac{\partial G}{\partial n_1}\right)_{T, P, n_2, n_3..} dn_1 + \left(\frac{\partial G}{\partial n_2}\right)_{T, P, n_1, n_3..} dn_2 + .. \qquad ... (ii)$$

At constant temperature and pressure

$$dT = 0 \quad \text{and} \quad dP = 0$$

Equation (*ii*) becomes

$$(dG)_{T,P} = \left(\frac{\partial G}{\partial n_1}\right)_{T, P, n_2, n_3.....} dn_1 + \left(\frac{\partial G}{\partial n_2}\right)_{T, P, n_1, n_2....} dn_2 + ... \qquad ... (iii)$$

We can write

$$\left(\frac{\partial G}{\partial n_1}\right)_{T, P, n_2, n_3....} = \mu_1$$

$$\left(\frac{\partial G}{\partial n_2}\right)_{T, P, n_1, n_3....} = \mu_2$$

Thus, $$(dG)_{T,P} = \mu_1 dn_1 + \mu_2 dn_2 + ... \qquad ... (iv)$$

For a system having definite composition, the equation (*iv*) on integration gives

$$G_{T,P,N} = n_1\mu_1 + n_2\mu_2 + ... \qquad ... (v)$$

The total differential of G in equation (*v*) gives

$$dG = \mu_1 dn_1 + n_1\, d\mu_1 + \mu_2\, dn_2 + n_2\, d\mu_2 + ...$$
$$= (\mu_1\, dn_1 + \mu_2\, dn_2 + ...) + (n_1\, d\mu_1 + n_2\, d\mu_2 + ...)$$

From equations (*iv*) and (*vi*), we have

$$n_1 d\mu_1 + n_2\, d\mu_2 + ... = 0 \qquad ... (vi)$$

or $$\boxed{\Sigma n_i\, d\mu_i = 0}$$

The relationship is called as **Gibbs-Duhem equation.**

3.4.3. Variation of chemical potential with temperature and pressure

To study the variation of chemical potential with temperature and pressure, we need to derive some results for the variation of free energy with temperature and pressure.

Free energy is an extensive property. It is a function of temperature, pressure and composition of a system. Therefore we can write

$$G = f(T, P, n_1, n_2, n_3 ......)$$

$$\therefore \quad dG = \left(\frac{\partial G}{\partial T}\right)_{P,N} dT + \left(\frac{\partial G}{\partial T}\right)_{T,N} dP + \left[\frac{\partial G}{\partial n_1}\right]_{T,P,n_2,n_3...} dn_1 + \left(\frac{\partial G}{\partial n_2}\right)_{T,P,n_1,n_3...} dn_2 + ... (i)$$

In a closed system, there is no change in the number of moles of various components in the system *i.e.*

$$dn_1,\ dn_2,\ dn_3\ .... = 0$$

Equation (*i*) takes the form

$$dG = \left(\frac{\partial G}{\partial T}\right)_{P,N} dT + \left(\frac{\partial G}{\partial P}\right)_{T,N} dP \quad ... (ii)$$

For a closed system, we know

$$dG = VdP - SdT$$

or

$$dG = -SdT + VdP \quad ... (iii)$$

Equating coefficients of dT and dP in equations (*ii*) and (*iii*)

$$\left(\frac{\partial G}{\partial T}\right)_{P,N} = -S \quad ... (iv)$$

$$\left(\frac{\partial G}{\partial P}\right)_{T,N} = V \quad ... (v)$$

Variation of free energy with temperature and pressure respectively for a closed system is thus provided by equations (*iv*) and (*v*)

3.4.4. Variation of chemical potential with temperature

The chemical potential of a component *i* in a system is given by

$$\mu_i = \left(\frac{\partial G}{\partial n_1}\right)_{T,P,n_1,n_2.....} \quad ... (i)$$

Differentiate this equation with respect to temperature,

$$\left(\frac{\partial \mu_i}{\partial T}\right)_{P,N} = \frac{\partial^2 G}{\partial n_i\, \partial T} \quad ... (ii)$$

Differentiate equation (*ii*) with respect to n_i

$$\frac{\partial^2 G}{\partial T\, \partial n_i} = -\left[\frac{\partial S}{\partial n_i}\right]_{T,P,n_1,n_2....} \quad ... (iii)$$

$$= -\overline{S}_i$$

The term $\left(\frac{\partial S}{\partial n_i}\right)_{T,P,n_1,n_2.....} = \overline{S}_i$ is the partial entropy of the *i*th component.

From equations (*ii*) and (*iii*), we have

$$\left(\frac{\partial \mu_i}{\partial T}\right)_{P,N} = -\bar{S}_i \quad \text{... (iv)}$$

Thus, *the variation of chemical potential of a component of a system with temperature at constant pressure and composition is equal to the partial molar entropy of that component.*

3.4.5. Variation of chemical potential with pressure

The expression for chemical potential is

$$\mu_i = \left(\frac{\partial G}{\partial n_i}\right)_{T,P,n_1,n_2....} \quad \text{... (i)}$$

Differentiate μ_i with respect to pressure,

$$\left(\frac{\partial \mu_i}{\partial P}\right)_{T,N} = \frac{\partial^2 G}{\partial n_i\,\partial P} \quad \text{... (ii)}$$

The variation of free energy with pressure at constant T and V is expressed as

$$\left[\frac{\partial G}{\partial P}\right]_{T,N} = V \quad \text{... (iii)}$$

Differentiate equation (*iii*) with respect to n_i

$$\frac{\partial^2 G}{\partial P\,\partial n_i} = \left[\frac{\partial V}{\partial n_i}\right]_{T,P,n_1,n_2....} \quad \text{... (iv)}$$

But $$\left[\frac{\partial V}{\partial n_i}\right]_{T,P,n_1,n_2....} = \bar{V}_i \quad \text{... (v)}$$

$\bar{V}_i$ is the partial molar volume of the *i*th component.

From equations (*ii*) and (*iv*), we get

$$\left(\frac{\partial \mu_i}{\partial P}\right)_{T,N} = \bar{V}_i \quad \text{... (vi)}$$

Thus, *the variation of chemical potential of a component of a system, with pressure at constant temperature and composition is equal to the partial molar volume of that component.*

3.4.6. Chemical potential in a system of ideal gases

For an ideal gas,

$$PV = nRT \quad \text{... (i)}$$

Consider a system consisting of a number of ideal gases 1, 2, 3,.... etc.

Let n_1, n_2 .... be the number of moles of various constituents present in the mixture and n is the total number of moles.

Then $$n = n_1 + n_2 + n_3 \ldots\ldots + n_i$$

Substitute this value of n is equation (*i*)

$$PV = (n_1 + n_2 + \ldots.)\, RT$$

or $$V = (n_1 + n_2 + \ldots.)\frac{RT}{P} \qquad \ldots (ii)$$

Differentiate with respect to n_i at constant temperature and pressure,

$$\left(\frac{\partial V}{\partial n_i}\right)_{T,P,n_1,n_2 \ldots.} = \bar{V}_i = \frac{RT}{P} \qquad \ldots (iii)$$

Substituting the values of $\bar{V}_i$ in equation (*vi*) (Art. 3.4.5), we have

$$\left[\frac{\partial \mu_i}{\partial P}\right]_{T,N} = \frac{RT}{P} \qquad \ldots (iv)$$

Equation (*iv*) may be expressed in the form

$$d\mu_i = \left(\frac{RT}{P}\right) dP = RT\, d \ln P \qquad \ldots (v)$$

In a mixture of ideal gases, let p_i be the partial pressure of *i*th component in the mixture therefore,

$$p_i V = n_i RT \qquad \ldots (vi)$$

For a mixture of ideal gases having n moles, we have the relation

$$PV = nRT \qquad \ldots (vii)$$

From equation (*vi*) and (*vii*), we have

$$\frac{p_i}{P} = \frac{n_i}{n}$$

or $$p_i = \frac{n_i}{n} P$$

$$p_i = K.P \qquad \ldots (viii)$$

where $K = \frac{n_i}{n}$ is a constant as both n_i and n are constants for the given system

Taking logarithms of both sides of equation (*viii*), we have

$$\ln p_i = \ln K + \ln P$$

On differentiating

$$d \ln p_i = d \ln P \qquad (\because \; d \ln K = 0) \quad \ldots (ix)$$

From eq. (*v*) and (*ix*), we get

$$d\mu_i = RT\, d \ln p_i \qquad \ldots (x)$$

Integrating equation (*x*), we get

$$\boxed{\mu_i = \mu_i^0 + RT \ln p_i} \qquad \ldots (xi)$$

where μ_i^0 is the integration constant, the value of which depends upon the nature of gas and the temperature.

It is evident from equation (*xi*) that the chemical potential of a component in a mixture of ideal gases is determined by the partial pressure of the component in the mixture.

If the partial pressure is unity *i.e.* $p_i = 1$,

then $$\mu_i = \mu_i^0$$

Thus μ_i^0 gives the chemical potential of a gaseous component i when the partial pressure of the component is unity. It is called the chemical potential of the ith component in the standard state.

3.4.7. The free energy change of a reaction

For a general reaction

$$aA + bB \rightleftharpoons cC + dD$$

the free energy change is given by

ΔG = Free energy of products – Free energy of reactants

In terms of chemical potentials of reactants and products, the free energy change is given by the following equation

$$\Delta G = (c\,\mu_C + d\,\mu_D) - (a\,\mu_A + b\,\mu_B)$$

where μ_A, μ_B, μ_C and μ_D are the chemical potentials of A, B, C and D respectively.

Standard free energy change of a reaction

When all the reactants and products of a chemical reaction are taken in their standard state, the free energy change accompanying the reaction is called standard free energy change. It is represented by $\Delta G°$. Thus,

$$\Delta G° = G^0_{\text{Products}} - G^0_{\text{Reactants}}$$

In terms of chemical potentials, we can write it in the form

$$\Delta G° = \left(c\,\mu_C^o + d\,\mu_D^o\right) - \left(a\,\mu_A^o + b\,\mu_B^o\right)$$

where $\mu_A^o, \mu_B^o, \mu_C^o$ and μ_D^o are the standard chemical potentials of A, B, C and D respectively.

3.5. FREE ENERGY CHANGE AS A CRITERION OF SPONTANEITY

It has been discussed earlier (Chapter 2) that change in free energy can be used as a criterion of feasibility of a process. If

(*i*) $\Delta G < 0$, The process is spontaneous in forward direction.

(*ii*) $\Delta G > 0$, The process is non-spontaneous in forward direction. In fact it will proceed in backward direction *i.e.* products wil get converted into reactants.

(*iii*) $\Delta G = 0$, The reaction will be at equilibrium.

Consider a reversible reaction : Reactants $\rightleftharpoons$ Products

Free energy changes that take place when the reaction is in progress are depicted in Fig. 3.2.

In the begiing the concentration of reactants is maximum, therefore, the free energy associated with them is also maximum while the free energy associated with products is minimum. As the reaction proceeds, the concentration of reactants decreases and concentration of products increases. Hence the free energy associated with products increases.

Equilibrium is reached when the concentration of reactants and products becomes constant and the free energy of reactants is equal to the free energy of products. Thus, at equilibrium, $\Delta G = 0$.

In a reversible reaction, both the forward and backward reactions take place. The condition of spontaneity is that ΔG should be negative. Thus ΔG is negative for the forward as well as backward reaction. This is explained with the help of the following example.

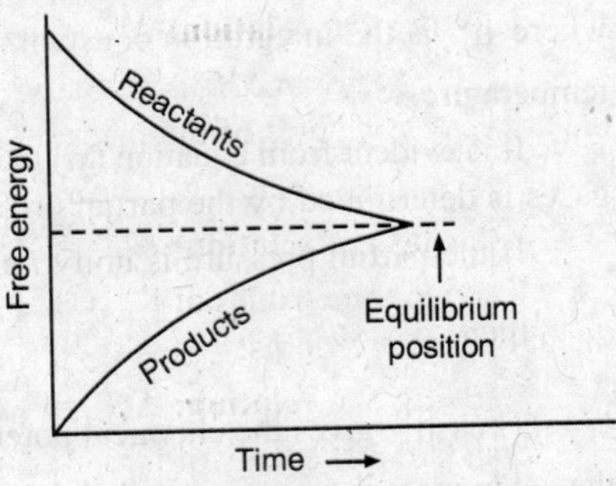

Fig. 3.2. Free energy change when reactants are converted into products.

Consider the following homogeneous gaseous reaction,

$$N_2O_4\ (g) \rightleftharpoons 2NO_2\ (g)$$

The variation of free energy of N_2O_4 and NO_2 versus the extent of reaction at constant temperature and pressure is shown in Fig. 3.3. At point A (where the extent of reaction is zero) only the reactant N_2O_4 is present. Point A represents the standard free energy of the reactant.

At point *B*, only products *i.e.* 2 moles of NO_2 are present. Point *B* therefore represents standard free energy of products. Point *C* is the equilibrium state where 16.6% of N_2O_4 and 83.40% NO_2 exist together. At this point the $\Delta G = 0$. The difference between the standard free energy of products (point *B*) and the standard free energy of reactants (point *A*) is $\Delta G° = +5.40$ kJ. The free energy of equilibrium mixture (point *C*) is 0.84 kJ lower than point *A*. That implies that when one mole of N_2O_4 (*g*) changes into equilibrium mixture, the value of $\Delta G° = -$ 0.84 kJ. Similarly when 2 moles of NO_2 change into equilibrium mixture, the value of $\Delta G°$.

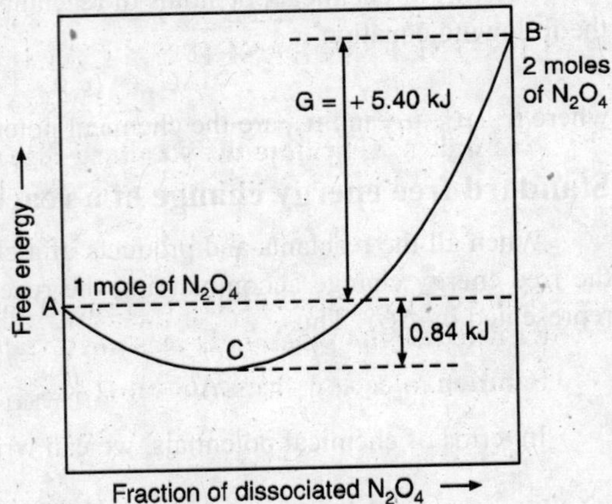

Fig. 3.3. Graph of free energy of system showing spontaneity of forward as well as backward reaction.

$$= -5.40 + (-0.84) = -6.24 \text{ kJ}$$

Thus, we find that in both the cases the value of $\Delta G°$ is negative. Therefore both the processes are spontaneous.

3.5.1. Calculation of standard free energy change in a reaction

The standard free energy change in a reaction may be calculated by one of the following methods

1. Using the relation: $\Delta G^\circ = \Delta H^\circ - T\Delta S^\circ$

Heat of reaction ΔH° is determined experimentally and entropy change ΔS° is calculated from the absolute entropies of the reactants and products, using the relation

$$\Delta S^\circ = \Sigma S^\circ \text{ (Products)} - \Sigma S^\circ \text{ (Reactants)}$$

2. Using the relation: $\Delta G^\circ = -RT \ln K_p$ or $\Delta G^\circ = -RT \ln K_c$.

Knowing the values of K_p or K_c, it is possible to calculate standard free energy change using the above relationships.

3. Using the relation: $\Delta G^\circ = \Sigma\, \Delta_f G^\circ$ (Products) $- \Sigma\, \Delta_f G^\circ$ (Reactants)

Standard free energy of formation is defined as the free energy change accompanying a reaction in which one mole of the substance in its standard state is formed from its elements in their standard states. Free energy of formation of an element in its standard state is taken as zero.

4. Free energy change ΔG of a gaseous reaction, at constant temperature can also be obtained, using the following relation

$$\Delta G = n\, RT \ln \frac{P_2}{P_1}$$

where P_1 and P_2 are initial and final pressures respectively.

PROBLEMS BASED ON CALCULATION OF STANDARD FREE ENERGY CHANGE

Example 1. Calculate the standard free energy change for the reaction

$$CO + \frac{1}{2} O_2\ (g) \longrightarrow CO_2\ (g) \quad , \quad \Delta H^\circ = 270 \text{ kJ}$$

Standard entropies of CO_2, CO and O_2 are 205, 190 and 200 J/degree/mole respectively. Predict whether the reaction is feasible or not.

Solution. Standard change in entropy ΔS° is given by

$$\Delta S^\circ = \Sigma S^\circ \text{ (Products)} - \Sigma S^\circ \text{ (Reactants)}$$

or

$$\Delta S^\circ = S^\circ(CO_2) - [S^\circ(CO) + \frac{1}{2} S^\circ(O_2)]$$

Substituting the values, we have

$$\Delta S^\circ = 205 - [190 + \frac{1}{2} \times 200] = -85 \text{ J/degree/mole}$$

Apply the relation

$$\Delta G^\circ = \Delta H^\circ - T\,\Delta S^\circ$$

Here

$$\Delta H^\circ = -270 \text{ kJ} = -270000 \text{ J}$$

$$T = 25 + 273 = 298 \text{ K}$$

$$\Delta S^\circ = -85 \text{ J/degree/mole}$$

$\therefore$

$$\Delta G^\circ = -270000 - 298\,(-85) \text{ J}$$

$$= -244670 \text{ J} = -244.67 \text{ kJ}$$

Since ΔG° is negative, hence the reaction is feasible.

Example 2. Calculate the standard free energy change of the reaction

$$NO\ (g) + \frac{1}{2} O_2\ (g) \longrightarrow NO_2\ (g)$$

Standard free energies of formation of NO(*g*) and NO_2(*g*) are 87 kJ and 52 kJ mol^{-1}.

Solution.

$$\Delta_f G^\circ = \Sigma\Delta_f G^\circ \text{ (Products)} - \Sigma\Delta_f G^\circ \text{ (Reactants)}$$

or

$$\Delta_f G^\circ = \Delta_f G^\circ (NO_2) - [\Delta_f G^\circ(NO) + \frac{1}{2}\,\Delta_f G^\circ(O_2)]$$

Substituting the values, we have

$$\Delta_f G^\circ = 52 - [87 + \frac{1}{2} \times (0)] \qquad \because \Delta_f G^\circ(O_2) = 0$$

or

$$\Delta_f G^\circ = -35 \text{ kJ}$$

PROBLEMS FOR PRACTICE

1. Calculate the standard free energy change at 25°C for the reaction

$$N_2(g) + 3H_2\,(g) \longrightarrow 2NH_3\,(g), \quad \Delta H^\circ = -22.08 \text{ kcal}$$

Given that ΔS° for the reaction is – 47.4 cal/degree

Predict the feasibility of the above reaction at the given state.

[**Ans.** $\Delta G^\circ = -7954.8$ kcal, Feasible]

2. Calculate the standard free energy change for the reaction

$$4NH_3\,(g) + 5O_2\,(g) \longrightarrow 4NO\,(g) + 6H_2O\,(l)$$

Given that the standard free energies of formation ($\Delta_f G^\circ$) for NH_3 (g), NO (g) and H_2O (*l*) are –16.74, 86.61 and –237.23 kJ mol^{-1} respectively.

Predict the feasibility (or spontaneity) of the above reaction at the standard state.

[**Ans.** $\Delta G^\circ = -1009.98$ kJ, Feasible]

3. Calculate the standard free energy for the reaction

$$Fe_2O_3\,(s) + 3H_2\,(g) \longrightarrow 2Fe\,(s) + 3H_2O\,(l)$$

Given that the standard free energies of formation of Fe_2O_3 and H_2O are –741.0 kJ and –237.2 kJ per mole respectively. Can the above reaction occur at the standard state?

[**Ans.** $\Delta G^\circ = +29.4$ kJ, No]

3.6. THERMODYNAMIC DERIVATION OF THE LAW OF CHEMICAL EQUILIBRIUM

Consider a general reversible reaction,

$$aA + bB \rightleftharpoons cC + dD$$

where the reactants and products are taken to be ideal gases. If μ_A, μ_B, μ_C, μ_D are the chemical potentials of A, B, C and D respectively at constant temperature and pressure, then

Total free energy of reactants = $G_{Reactants} = a\mu_A + b\mu_B$

Total free energy of products = $G_{Products} = c\mu_C + d\mu_D$

At constant temperature and pressure, the free energy change of the reaction is given by

$$\Delta G = G_{Products} - G_{Reactants}$$

or

$$\Delta G = (c\mu_C + d\mu_D) - (a\mu_A + b\mu_B) \qquad \text{... } (i)$$

When the system is in equilibrium state at constant temperature and pressure, $\Delta G = 0$, therefore equation (1) can be written as

$$(c\mu_C + d\mu_D) - (a\mu_A + b\mu_B) = 0$$

or
$$c\mu_C + d\mu_D = a\mu_A + b\mu_B \quad \text{... }(ii)$$

Chemical potential of any component in the ideal gas mixture is given by the relation

$$\mu = \mu^\circ + RT \ln p \quad \text{... }(iii)$$

where p is the partial pressure of the component and μ° is the chemical potential in the standard state (*i.e.* at 1 atm pressue). μ° depends only on temperature.

Substituting the value of μ from equation (*iii*) into equation (*ii*) we get

$$c(\mu_C^o + RT \ln p_C) + d(\mu_D^o + RT \ln p_D) = a(\mu_A^o + RT \ln p_A) + b(\mu_B^o + RT \ln p_B)$$

$$RT\,(c \ln p_C + d \ln p_D) + (c\,\mu_C^o + d\,\mu_D^o) = RT\,(a \ln p_A + b \ln p_B) + (a\mu_A^o + b\,\mu_B^o)$$

$$RT\,[(\ln P_C^c + \ln P_D^d)] - (\ln p_A^a + \ln p_B^b) = (a\,\mu_A^o + b\,\mu_B^o) - (c\,\mu_C^o + d\,\mu_D^o)$$

or
$$RT \ln \frac{p_C^c \cdot p_D^d}{p_A^a \cdot p_B^b} = G^o_{\text{Reactants}} - G^o_{\text{Products}}$$

$$= -\left(G^o_{\text{Products}} - G^o_{\text{Reactants}}\right) \quad \text{... }(iv)$$

$$= -\Delta G^\circ \quad \text{... }(v)$$

or
$$\ln \frac{p_C^c \cdot p_D^d}{p_A^a \cdot p_B^b} = \frac{-\Delta G^\circ}{RT}$$

or
$$\frac{p_C^c \cdot p_D^d}{p_A^a \cdot p_B^b} = e^{-\Delta G^\circ/RT} \quad \text{... }(vi)$$

The standard free energy change of a reaction ΔG° depends only on temperature, hence at constant temperature, the term $e^{-\Delta G^\circ/RT}$ is constant. Thus equation (*vi*) can be written as

$$\frac{p_C^c \cdot p_D^d}{p_A^a \cdot p_B^b} = \text{Constant } (K_p) \quad \text{... }(vii)$$

This is thermodynamic derivation of law of chemical equilibrium in terms of partial pressures. From equations (*vi*) and (*vii*), we arrive at the result

$$\boxed{\Delta G^\circ = -RT \ln K_p} \quad \text{... }(viii)$$

If molar concentrations are taken in place of partial pressures, the chemical potential of a component may be expressed as

$$\mu = \mu^\circ + RT \ln C$$

where C is the molar concentration of the given component in the reaction mixture and μ° is the chemical potential of the substance in the standard state (when the concentration is unity). Repeating the same steps as described above, we can show that

$$\frac{C_C^c \cdot C_D^d}{C_A^a \cdot C_B^b} = \text{Constant } (K_c)$$

or $$\frac{[C]^c\,[D]^d}{[A]^a \cdot [B]^b} = K_c$$

Similarly, we can obtain a relation between ΔG° and K_c as under

$$\boxed{\Delta G^\circ = -RT \ln K_c} \quad \text{....}(ix)$$

To derive the law of equilibrium in terms of activities, we write the chemical potential of each component as

$$\mu = \mu_x^o + RT \ln a$$

where a is the activity of the component and μ_x^o depends both on temperature and pressure.

Substituting the value of μ in eq. (*ii*), we get

$$\frac{a_C^c \cdot a_D^d}{a_A^a \cdot a_B^b} = \text{Constant } (K_a)$$

Also

$$\boxed{\Delta G^\circ = -RT \ln K_a} \quad \text{...}(x)$$

By knowing, ΔG° of the reaction we can calculate K_p, K_a or K_x and vice versa. $K_p = K_c$ only when the number of moles of reactants and products are equal. Therefore, the value of ΔG° obtained in equations (*vii*) and (*ix*) are different.

SOLVED PROBLEM ON STANDARD FREE ENERGY CHANGE AND EQUILIBRIUM CONSTANT

Example 1. At 2000 K, the standard state free energy change (ΔG°) for the reaction $N_2 + O_2 \rightleftharpoons 2NO$ is given by $\Delta G^\circ = 92048 - 10.46$ T J. Calculate K_p for the reaction at 2000 K ($R = 8.314$ J K^{-1} mol^{-1}).

Solution.

$$\Delta G^\circ = -RT \ln K_p = -2.303\, RT \log K_p$$

$$\log K_p = -\frac{\Delta G^\circ}{2.303\, RT}$$

$$= \frac{(-92048) + 10.46 \times 2000 \text{ Jmol}^{-1}}{(2.303)\,(8.314 \text{ J K}^{-1}\text{mol}^{-1})\,(2000 \text{ K})} = 1.857$$

$$K_p = 1.39 \times 10^{-2} \text{ atm}$$

Example 2. The value of K_p, for the water gas reaction

$$CO + H_2O \rightleftharpoons CO_2 + H_2$$

is 1.06×10^5 at 25°C. Calculate the standard state free energy change (ΔG°) of the reaction at 25°C. (R = 8.314 J K–1 mol^{-1})

Solution.

$$\Delta G^\circ = -\text{RT} \ln K_p = -2.303\, RT \log K_p$$

$$= -(2.303)\,(8.314 \text{ J K}^{-1})\,(298 \text{ K}) \log (1.06 \times 10^{-5})$$

$$= -28673.6 \text{ J} = -28.674 \text{ kJ}$$

Example 3. At 2000 K, the standard free energy change (ΔG°) for the reaction

$$N_2 + O_2 \rightleftharpoons 2NO$$

is given by $\Delta G^\circ = 22000 - 2.0$ T cal.

Calculate K_p for the reaction at 2000 K (R = 1.98 calorie)

Solution. Given

$$\Delta G^\circ = 22000 - 2.0\ T \text{ and } T = 2000\ K$$

$$\therefore \quad \Delta G^\circ = 220000 - 2.0 \times 2000 = 18000 \text{ cal}$$

Using the relationship between **ΔG°** and K_p

$$\Delta G^\circ = -TR \ln K_p = -2.303\ RT \log K_p,$$

Substituting the value, we have

$$18000 = -2.303 \times 1.98 \times 2000 \times \log K_p$$

or $$K_p = -1.9737 \text{ or } \log K_p = \bar{2}.0263$$

or $$K_p = \text{antilog}\ (\bar{2}.0263) = \mathbf{1.063 \times 10^{-2}}$$

Example 4. Calculate the standard state free energy change (ΔG°) of the reaction

$$2H_2O\ (g) \rightleftharpoons 2H_2\ (g) + O_2\ (g)$$

given that at 1000 K, water vapour at 1 atmosphere pressure has a degree of dissociation = 3×10^{-5} per cent (R = 8.314 J K^{-1} mol^{-1}).

Solution. The partial pressures of the various reactants in the equilibrium mixture will be evidently as below :

$$pH_2 = 3 \times 10^{-7} \text{ atm}, pO_2 = \frac{3}{2} \times 10^{-7} \text{ atm}$$

$$K_p = \frac{p_{H_2}^2 \times pO_2}{p_{H_2}O} = \frac{9 \times 10^{14} \times \frac{3}{2} \times 10^{-7}}{1}$$

$$= 13.5 \times 10^{-21} \text{ atm}$$

$$\Delta G^\circ = -RT \ln K_p = -2.303\ RT \log K_p$$

$$= -2.303\ (8.314\ J\ K^{-1}\ mol^{-1})\ (1000\ K) \log (13.5 \times 10^{-21})$$

$$= \mathbf{380448\ J = 380.45\ kJ}$$

PROBLEMS FOR PRACTICE

1. At 27°C, the standard free energy change for the reaction

$$CO\ (g) + H_2O\ (g) \rightleftharpoons CO_2\ (g) + H_2\ (g)$$

is – 6909 cal. Calculate K_p for the equilibrium. [**Ans.** 1.078×10^5]

2. The standard state free energy of a reaction at 298.15 K is –16.65 kJ. Calculate the value of equilibrium constant of the reaction at 298.15 K. [**Ans.** 8.252×10^2]

3. For the reaction

$$CCl_4\ (l) + H_2\ (g) \longrightarrow HCl(g) + CHCl_3\ (l)$$

at 25°C, $\Delta H^\circ = -21.83$ kcal/mol and $\Delta S^\circ = 9.92$ kcal K^{-1} mol^{-1}. Determine whether the reaction is favourable at 25° under standard conditions.

Calculate the value of K_p at 25°C.

[**Ans.** $\Delta G^\circ = -24.79$ kcal, Reaction is favourable, $K_p = 1.5 \times 10^{18}$]

4. Calculate K_p for the following reaction at 25°C

$$2SO_2\ (g) + O_2\ (g) \rightleftharpoons 2SO_3\ (g)$$

$$\Delta_f G^\circ\ (SO_2) = -71.79 \text{ kcal/mol}$$

$$\Delta_f G^\circ\ (SO_3) = -88.52 \text{ kcal/mol}$$

[**Ans.** 1.2×10^{22}]

5. The values of standard state free energy for a gaseous reaction

$$CO\ (g) + H_2O\ (g) \rightleftharpoons CO_2\ (g) + H_2\ (g)$$

at 298 K is – 28.53 kJ.

Calculate the value of the equilibrium constant K_p ($R = 8.314$ JK^{-1} mol^{-1}) [**Ans.** 1.0×10^5]

3.7. VAN'T HOFF REACTION ISOTHERM

Van't Hoff

Consider the following reaction

$$aA + bB + \cdots \rightleftharpoons lL + mM + \cdots$$

taking place *under any conditions* of temperature, pressure and composition. The free energy change of the reaction is given by the expression

$$\Delta G = (G)_{products} - (G)_{reactants} \quad \ldots(i)$$

$$= (l\mu_L + m\mu_M + \cdots) - (a\mu_A + b\mu_B + \cdots)$$

where μ_A, μ_B, μ_L, μ_M, etc., represent chemical potentials of the species concerned.

Let us suppose that various reactants and products are in gaseous state. Chemical potential of a gaseous substance in *any state* is given by the equation

$$\mu_{i(p)} = \mu^\circ_{i(p)} + RT \ln p_i \quad \ldots(ii)$$

where $\mu^\circ_{i(p)}$ is the chemical potential of a gaseous component *i*, in the *standard state* (partial pressure = 1 atmosphere) and $\mu_{i(p)}$ is the chemical potential of the gaseous component *i* in any state of partial pressure p_i.

Substituting the values of chemical potentials of various species at their partial pressures from eq. (*ii*) in (*i*)

$$\Delta G = \{l(\mu^\circ_{L(p)} + RT \ln p_L) + m(\mu^\circ_{M(p)} + RT \ln p_M) + \cdots\}$$
$$- \{a(\mu^\circ_{A(p)} + RT \ln p_A) + b(\mu^\circ_{B(p)} + RT \ln p_B) + \cdots\} \quad \ldots(iii)$$

Re-arranging, we have

$$\Delta G = \{l\mu^\circ_{L(p)} + m\mu^\circ_{M(p)} + \cdots) - (a\mu^\circ_{A(p)} + b\mu^\circ_{B(p)} + \cdots)\}$$
$$+ RT \ln \frac{(p_L)^l (p_M)^m \ldots}{(p_A)^a (p_B)^b \ldots} \quad \ldots(iv)$$

The first expression on the right hand side, evidently, is the free energy change of the reaction when the products and the reactants are all in their respective standard states. This expression may be substituted by ΔG°. Hence,

$$\mathbf{\Delta G = \Delta G^\circ + RT \ln Q_p} \quad \ldots(v)$$

where Q_p stands for the reaction quotient of partial pressures of the products and the reactants, *viz.*

$$\frac{(P_L)^l (p_M)^m \ldots}{(p_A)^a (p_B)^b \ldots}$$

Eq. (*v*) is known as **van't Hoff reaction isotherm.** This gives the *free energy change of a reaction* at any given temperature, pressure and composition of the reacting system.

Q_p is different from K_p. Here the values $p_L, p_M, p_A, p_B \ldots$ etc. are not the partial pressures at equilibrium, but for any state. If the partial pressure values correspond to equilibrium state, then Q_p will become equal to K_p. From eq. (*viii*) in section 3.6, we have

$$\Delta G^\circ = -RT \ln K_p$$

Substituting for ΔG° in eq. (*v*), we have

$$\Delta G = -RT \ln K_p + RT \ln Q_p \quad \ldots(vi)$$

Eq. (*v*) and (*vi*) are two forms of van't Hoff reaction isotherm.

Example. At 25°C, the ΔG° for the reaction

$$\mathbf{H_2(g) + I_2(g) \longrightarrow 2HI(g),}$$

is –3.1 kcal. Calculate ΔG when hydrogen gas at 0.1 atm and iodine vapour at 0.1 atm react to form hydrogen iodide at 10 atm, and 25°C.

Solution. For the given reaction,

$$K = \frac{p^2(\text{HI})}{p(\text{H}_2)\,p(\text{I}_2)}$$

where p terms represent partial pressures of the various species in the reaction mixture.

$$\Delta G = \Delta G° + RT \ln K = \Delta G° + 2.303\, RT \log \frac{p^2(\text{HI})}{p(\text{H}_2)\,p(\text{I}_2)}$$

$$= -3100 \text{ cal} + 2.303\,(1.987 \text{ cal K}^{-1} \text{ mol}^{-1})\,(298 \text{ K}) \log \frac{(10)^2}{(0.1)(0.1)}$$

$$= -3100 \text{ cal} + 1363.7 \log (10^4) \text{ cal} = -3100 \text{ cal} + 4\,(1363.7) \text{ cal}$$

$$= +2354.8 \text{ cal} = \mathbf{2.355\ kcal}$$

3.8. RELATION BETWEEN K_P AND K_C

The value of K_p for the reaction

$$aA + bB + cC + \cdots \rightleftharpoons lL + mM + \cdots$$

is given by the expression

$$K_p = \left[\frac{(p_L)^l\,(p_M)^m \ldots}{(p_A)^a\,(p_B)^b \ldots}\right]_{eq} \qquad \ldots(i)$$

The value of K_c for the above reaction would be given by the expression

$$K_c = \left(\frac{(c_L)^l\,(c_M)^m \ldots}{(c_A)^a\,(c_B)^b \ldots}\right)_{eq} \qquad \ldots(ii)$$

where p and c stand for partial pressure and concentration, respectively. Now, for an ideal gas,

$$pV = nRT.$$

or $$p = (n/V)\,RT = cRT$$

or $$c = p/RT \qquad \ldots(iii)$$

where c is the concentration. Thus, for any component i whose partial pressure is p_i,

$$c_i = p_i / RT \qquad \ldots(iv)$$

Substituting the concentrations of various species in terms of their partial pressures in Eq. (*ii*) we have

$$K_c = \left\{\frac{\left(\frac{p_L}{RT}\right)^l \left(\frac{p_M}{RT}\right)^m \ldots}{\left(\frac{p_A}{RT}\right)^a \left(\frac{p_B}{RT}\right)^b \ldots}\right\}_{eq}$$

$$= \left(\frac{(p_L)^l\,(p_M)^m \ldots}{(p_A)^a\,(p_B)^b \ldots}\right)_{eq} \times \left(\frac{1}{RT}\right)^{\{(l+m+\cdots)-(a+b+\cdots)\}}$$

$$= K_p \left(\frac{1}{RT}\right)^{\Delta n} \qquad ...(v)$$

where Δn= Number of moles of products — Number of moles of reactants

or $$K_p = K_c (RT)^{\Delta n} \qquad ...(vi)$$

Example. The value of K_p for the equilibrium

$$2H_2O\,(g) + 2Cl_2(g) \rightleftharpoons 4HCl\,(g) + O_2(g)$$

is 0.035 atm at 400°C when the partial pressures are expressed in atmospheres. Calculate the value of K_c for the same reaction.

Solution. $\Delta n = 4 + 1 - 2 - 2 = 1$

$\therefore$ We shall use the relation

$$K_p = K_c (RT)^{\Delta n}$$

or $$K_c = \frac{K_p}{(RT)^{\Delta n}}$$

Substitute the values

$$K_c = \frac{0.035 \text{ atm}}{(0.0820\, l \text{ atm K}^{-1} \text{mol}^{-1} \times 673 \text{K})}$$

$$= \mathbf{6.342 \times 10^{-4} \text{ mol } l^{-1}}$$

3.9. VAN'T HOFF EQUATION FOR THE TEMPERATURE DEPENDENCE OF EQUILIBRIUM CONSTANT (VAN'T HOFF REACTION ISOCHORE)

The equation for reaction isotherm when the reactants as well as the products are *gaseous* and are also in their *standard states,* is represented as

$$\Delta G° = - RT \ln K_p$$

Differentiating with respect to temperature at constant pressure, we have

$$\left\{\frac{\partial(\Delta G°)}{\partial T}\right\}_p = -R \ln K_p - RT \frac{d(\ln K_p)}{dT}$$

Multiplying throughout by *T,* we get

$$T\left\{\frac{\partial(\Delta G°)}{\partial T}\right\}_p = -RT \ln K_p - RT^2 \frac{d(\ln K_p)}{dT}$$

Substituting $\Delta G°$ for $-RT \ln K_p$, we have

$$T\left\{\frac{\partial(\Delta G°)}{\partial T}\right\}_p = \Delta G° - RT^2 \frac{d(\ln K_p)}{dT} \qquad ...(iii)$$

The well-known Gibbs-Helmholtz equation for substances in *standard states* may be written as

$$\Delta G^\circ = \Delta H^\circ + T\left\{\frac{\partial(\Delta G^\circ)}{\partial T}\right\}_p \qquad \text{... } (iv)$$

Comparing Eqs. (*iii*) and (*iv*), we get

$$RT^2 \frac{d(\ln K_p)}{dT} = \Delta H^\circ$$

or

$$\frac{d(\ln K_p)}{dT} = \frac{\Delta H^\circ}{RT^2} \qquad \text{...}(vi)$$

Eq.(*vi*) is known as the **van't Hoff's equation.** ΔH° is the enthalpy change for the reaction at constant pressure when the reactants as well as the products are in their standard states.

It is known from experiments that the enthalpy change, ΔH, accompanying a chemical reaction does not vary appreciably with change in partial pressures of the reactants or products. Therefore, *we may take* ΔH° *as equal to* ΔH, *where* ΔH is the enthalpy change of the reaction *whatever may be the partial pressures of the reactants or products.* Hence, the van't Hoff equation may be written as

$$\frac{d(\ln K_p)}{dT} = \frac{\Delta H}{RT^2} \qquad \text{...}(vii)$$

Integrating Eq. (*vii*) between temperatures T_1 and T_2 at which the equilibrium constants are K_p' and K_p'', respectively and assuming that ΔH remains constant over this range of temperature, we get

$$\int_{K_p'}^{K_p''} d(\ln K_p) = \frac{\Delta H}{R}\int_{T_1}^{T_2}\frac{dT}{T^2}$$

or

$$\ln K_p'' - \ln K_p' = \frac{-\Delta H}{R}\left[\frac{1}{T_1} - \frac{1}{T_2}\right]$$

$$= \frac{\Delta H}{R}\left[\frac{T_2 - T_1}{T_1\,T_2}\right] \qquad \text{... } (viii)$$

or

$$\log K_p'' - \log K_p' = \log\frac{K_p''}{K_p'} = \frac{\Delta H}{2.303}\left[\frac{T_2 - T_1}{T_1\,T_2}\right] \qquad \text{...}(ix)$$

Knowing the equilibrium constant at a temperature it is possible to calculate the equilibrium constant at another temperature provided the heat of reaction (ΔH) is known.

Alternatively, knowing the equilibrium constants of a reaction at two temperatures, the heat of reaction (ΔH) can be calculated.

SOLVED EXAMPLES ON VAN'T HOFF EQUATION

Example 1. The equilibrium constant K_p, for the reaction $H_2(g) + S(g) \rightleftharpoons H_2S(g)$ is 20.2 atm at 945°C and 9.21 atm at 1065°C. Calculate the heat of reaction.

Solution. Applying van't Hoff equation

$$\log K_p'' - \log K_p' = \frac{\Delta H}{2.303R}\left[\frac{T_2 - T_1}{T_1\, T_2}\right]$$

$$T_1 = 945 + 273 = 1218 \text{ K}$$
$$T_2 = 1065 + 273 = 1338 \text{ K}$$
$$K_p' = 20.2 \text{ atm}$$
$$K_p'' = 9.21 \text{ atm}$$

Substituting the values, we have

$$\log 9.21 - \log 20.2 = \frac{\Delta H}{(2.303)(8.134 \text{ JK}^{-1}\text{mol}^{-1})}\left[\frac{1338 - 1218}{(1218\text{K})(1338\text{K})}\right]$$

$$\mathbf{\Delta H = -\ 88126.3\ J = -\ 88.826\ kJ}$$

Example 2. The equilibrium constant K_p for the reaction $N_2 + 3H_2 \rightleftharpoons 2NH_3$ is 1.64×10^{-4} atm at 400°C. What will be the equilibrium constant at 500°C if the heat of reaction in this temperature range is – 105185.8 J ?

Solution. $\log K_p'' - \log K_p' = \dfrac{\Delta H}{2.303R}\left(\dfrac{T_2 - T_1}{T_1\, T_2}\right)$

$$T_1 = 400 + 273 = 673 \text{ K}$$
$$T_2 = 500 + 273 = 773 \text{ K}$$
$$\Delta H = -\ 105185.8 \text{ J}$$

$$\log K_p'' = \log(1.64 \times 10^{-4}) + \frac{-105185.8 \text{ J mol}^{-1}}{(2.303)(8.314 \text{J K}^{-1}\text{mol}^{-1})}\left(\frac{773\text{K} - 673\text{K}}{673\text{K} \times 773\text{K}}\right)$$

$\therefore$ $\mathbf{K_p' = 0.144 \times 10^{-4}\ atm}$

Example 3. The equilibrium constant K_p for the reaction

$$N_2 + 3H_2 \rightleftharpoons 2NH_3$$

is 1.64×10^{-4} at 400°C. What will be the equilibrium constant at 500°C if the heat of reaction in this temperature range is – 25140 calories?

Solution. Here, we have

$$T_1 = 400 + 273 \text{ K} = 673 \text{ K}$$
$$K_p' = 1.64 \times 10^{-4}$$
$$T_2 = 500 + 273 \text{ K} = 773 \text{ K}$$
$$K_p'' = \text{To be calculated}$$
$$\Delta H = -\ 25{,}140 \text{ cal.}$$

Substituting the values in the equation,

$$\log \frac{K_p''}{K_p'} = \frac{\Delta H}{2.303\, R}\left(\frac{T_2 - T_1}{T_1\, T_2}\right)$$

we have $\log \dfrac{K''_p}{1.64\times10^{-4}} = \dfrac{-25140}{2.303\times1.987}\left(\dfrac{773-673}{673\times773}\right) = -1.0560$

or $\log K''_p - \log(1.64\times10^{-4}) = -1.0560$

or $\log K''_p - \log 1.64 + 4 = -1.056$

or $\log K''_p = -4.8412 = \bar{5}.1588$

$\therefore \quad K''_p = 1.442\times10^{-5}$

PROBLEMS FOR PRACTICE

1. Calculate the ethalpy change for the reaction

$$N_2 + O_2 \rightleftharpoons 2NO$$

Given that the equilibrium constant for this reaction is 4.08×10^{-4} at 2000 K and 3.60×10^{-3} at 2500 K. [**Ans.** 181.1 kJ]

2. For a certain reaction $K_p = 1.21\times10^{-4}$ at 1800 K and 2.31×10^{-4} at 1900 K. Calculate K_p at 2000 K. [**Ans.** 4.13×10^{-4}]

3. The equilibrium constant K_p for a reaction is 3.0 at 400°C and 4.0 at 500°C. Calculate the value of ΔH° for the reaction ($R = 8.314$ J K^{-1} mol^{-1}). [**Ans.** 12.44 kJ]

4. For the reaction $N_2O_4 \rightleftharpoons 2NO_2$ $\Delta H = 14.6$ kcal.
if $K_p = 0.141$ at 298 K, calculate K_p at 338 K.[**Ans.** 2.607]

3.9.1. Van't Hoff's equation for temperature dependence of equilibrium constant in terms of K_c

K_p and K_c are related by the expression

$$K_p = K_c\,(RT)^{\Delta n} \qquad ...(i)$$

Taking logarithms of the above equation, we have

$$\ln K = \ln K_c + \Delta n \ln RT \qquad ...(ii)$$

Differentiating Eq. (*ii*) with respect to temperature, we get

$$\frac{d(\ln K_p)}{dT} = \frac{d(\ln K_c)}{dT} + \frac{\Delta n}{T} \qquad ...(iii)$$

or

$$\frac{d(\ln K_c)}{dT} = \frac{d(\ln K_p)}{dT} - \frac{\Delta n}{T}$$

$$= \frac{\Delta H^\circ}{RT^2} - \frac{n}{T} \qquad \text{Refer eq. (vii), Sec. 3.9}$$

$$= \frac{\Delta H^\circ}{RT^2} - \frac{nRT}{TRT}$$

$$= \frac{\Delta H^\circ - nRT}{RT^2}$$

But, $\Delta H^\circ - \Delta nRT = \Delta U^\circ$ (from thermodynamics)

$$\therefore \qquad \left(\frac{d \ln K_c}{dT}\right) = \frac{\Delta U^\circ}{RT^2} \qquad \ldots (iv)$$

Example. Explain the terms

(*i*) Homogeneous equilibrium

(*ii*) Heterogeneous equilibrium.

Solution. (*i*) An equilibrium which has all the reactants and products in the same phase is called homogeneous equilibrium. For example

$$H_2(g) + I_2(g) \rightleftharpoons 2HI\,(g)$$

Here H_2, I_2 and HI all exist in the same gaseous phase.

similarly $\quad CH_3COOH\,(l) + C_2H_5OH\,(l) \rightleftharpoons CH_3COOC_2H_5(l) + H_2O\,(l)$

is also an example of homogeneous equilibria as all the reactants and products are in the liquid phase.

(*ii*) An equilibrium which has the reactants and products in two or more phases is called heterogeneous equilibrium. For example

$$CaCO_3(s) \rightleftharpoons CaO\,(s) + CO_2(g)$$

3.10. LE-CHATELIER'S PRINCIPLE

Le-Chatelier

Le Chatelier, a noted French chemist, studied the effect of *concentration, temperature and pressure* on a large number of chemical equilibria. He summed up his observations in the form of a generalisation known as **Le-Chatelier's principle,** which states as follows:

If an equilibrium is subjected to a stress, the equilibrium shifts in such a way as to reduce the stress.

According to this principle, *if a system at equilibrium is subjected to a change of concentration, pressure or temperature, the equilibrium shifts in the direction that tends to* **undo** *the effect of the change.*

Effect of change of concentration

Consider the following equilibrium:

$$\underset{\text{Pale yellow}}{Fe^{3+}\,(aq)} + \underset{\text{Colourless}}{(SCN)^-\,(aq)} \rightleftharpoons \underset{\text{Dark brown}}{[Fe(SCN)]^{2+}\,(aq)}$$

Suppose, some ferric salt is added to this equilibrium. The colour of the solution will darken immediately showing that there is increase in the concentration of the coloured complex ion-$[Fe(SCN)]^{2+}$. This change is in accordance with Le-Chatelier's principle. Addition of more of Fe^{3+} ions has resulted in increasing the concentration of the complex ferrisulphocyanide.

Now, suppose a small amount of potassium ferrisulphocyanide capable of giving the complex ion, $[Fe(SCN)]^{2+}$ is added to the equilibrium. The solution will be seen to become less dark showing that the dark coloured $[(Fe(SCN)]^{2+}$ ion has been changed into Fe^{3+} and $(SCN)^-$ ions. Thus, increasing the concentration of the *product* shifts the equilibrium in favour of the *reactants*.

Effect of change of temperature

Consider the equilibrium

$$N_2O_4(g) \rightleftharpoons 2NO_2(g); \ \Delta H = +59.0 \text{ kJ}$$

In this equilibrium, the reaction favouring the product (NO_2) is seen to be endothermic. Therefore, the opposing reaction favouring the reactant (N_2O_4) must be exothermic. Now suppose the system is *heated* and its temperature is allowed to rise. According to Le Chatelier's principle, the equilibrium will shift in the direction which tends to undo the effect of heat, *i.e.*, which tends to produce *cooling.* Therefore, the equilibrium will shift in favour of NO_2, *i.e.*, the dissociation of N_2O_4 into NO_2 will increase. If the system is cooled, on the other hand, the equilibrium will shift in the direction which tends to produce *heat.* Therefore, the equilibrium will shift in the reverse direction, *i.e.*, in favour of N_2O_4. The dissociation of N_2O_4 will decrease.

The combination of nitrogen and hydrogen to give ammonia is represented by the following thermochemical equation :

$$N_2(g) + 3H_2(g) \rightleftharpoons 2NH_3(g) \ ; \Delta H = -92.38 \text{ kJ}$$

The forward reaction is *exothermic.* Therefore, the back reaction must be endothermic. By Le Chatelier's principle, the increase of temperature will favour the back reaction, *i.e.*, the dissociation of ammonia. Therefore, in order to get a better yield of ammonia, the temperature at equilibrium should be low. But since the reaction rate varies with the temperature, the time taken to reach the equilibrium state becomes very long if the temperature is kept low. Therefore, a temperature close to 500°C, which is neither too low nor too high, is maintained.

Effect of change of pressure

Consider another gaseous reaction involving *combination of nitrogen and hydrogen to give ammonia:*

$$\underbrace{\underset{1 \text{ mole}}{N_2\,(g)} + \underset{3 \text{ moles}}{3H_2\,(g)}}_{4 \text{ moles}} \rightleftharpoons \underset{2 \text{ moles}}{2NH_3}$$

As is evident from the above equation, the forward reaction is accompanied by a decrease in the number of moles. If the pressure is *increased,* the volume will decrease and, therefore, *the number of moles per unit volume will increase.* According to Le Chatelier's principle, therefore, the equilibrium will shift in the direction in which there is *decrease* in the number of moles, *i.e.*, in favour of the formation of ammonia. Thus, *the higher the pressure, the greater would be the yield of ammonia.* It is on account of this consideration that a pressure of about 200 atmospheres is maintained in the Haber's process for the manufacture of ammonia.

The combination of sulphur dioxide and oxygen to give sulphur trioxide:

$$2SO_2\,(g) + O_2\,(g) \rightleftharpoons 2SO_3(g)$$

is also accompanied by a *decrease* in the number of moles. According to Le Chatelier's principle, therefore, a high pressure would give a better yield of sulphur trioxide.

The dissociation of calcium carbonate

$$CaCO_3(s) \rightleftharpoons CaO\,(s) + CO_2(g)$$

is accompanied by an *increase* in the number of gaseous moles. The dissociation of calcium carbonate would, therefore, be *suppressed* by an *increase* of pressure.

Effect of addition of inert gas

The effect of addition of inert gas can be studied under two different conditions:

(*i*) *Addition of an inert gas at constant volume.* When an inert gas is added to the equilibrium state at constant volume, then the total pressure will increase. But the partial pressure of each component will remain unchanged. Under these conditions, there will be no effect on the equilibrium on addition of the inert gas.

(*ii*) *Addition of an inert gas at constant pressure.* When an inert gas is added to the system at constant pressure there will be an increase in the volume. As a result, the number of moles per unit volume of different components will decrease. The equilibrium will shift to the side where the number of moles are increased. For example consider the following equilibrium.

$$2SO_2\,(g) + O_2\,(g) \rightleftharpoons 2SO_3\,(g) \text{ at constant pressure}$$

The addition of inert gas at constant pressure will shift the equilibrium to backward direction. For the equilibrium

$$PCl_5\,(g) \rightleftharpoons PCl_3\,(g) + Cl_2\,(g) \text{ at constant pressure}$$

the addition of inert gas at constant pressure will shift the equilibrium to the forward direction.

However, addition of an inert gas to the following equilibrium

$$N_2\,(g) + O_2\,(g) \rightleftharpoons 2NO\,(g) \text{ at constant pressure}$$

will have no effect because the number of moles of reactants and products are same.

Effect of Catalyst

There is no effect of addition of a catalyst on the equilibrium state. This is because catalyst increases the rate of forward as well as backward reaction to the same extent. It simply helps to achieve the equilibrium quickly. It may be further noted that catalyst has no effect on the equilibrium concentration of a reaction mixture.

3.10.1. Applications of Le-Chatelier's principle to physical equilibrium

The following examples illustrate the application of Le-Chatelier's principle to physical equilibrium.

1. Liquid-vapour equilibrium. Consider water-vapour equilibrium

$$\underset{\text{(Less volume)}}{H_2O\,(g)} \rightleftharpoons \underset{\text{(More volume)}}{H_2O\,(\text{vapour})} \quad \text{(Endothermic process)}$$

The conversion of water to water vapour is an endothermic process.

Also as liquid water changes to vapour phase, it is accompanied by increase in volume. According to Le-Chatelier's principle, the increase in temperature shifts the equilibrium towards the right i.e. more of liquid water wil evaporate. This wil absorb the heat supplied.

On increasing pressure, the equilibrium shifts in a direction so as to lower the pressure. Thus increase of pressure favours the condensation of water vapour into liquid water. On the other hand decreasing the pressure will favour the vaporisation of water into water vapour.

2. Effect of pressure on boiling point of a liquid. Consider the equilibrium

$$\text{Liquid} \rightleftharpoons \text{Vapour}$$

The vaporisation of liquid is accompanied by increase of pressure. Thus, if pressure is increased, the equilibrium shifts towards condensation of vapour into liquid state at a given temperature. The vapour pressure will decrease. Higher temperature is thus needed to make the liquid boil. This is the reason that the boiling point of a liquid increases with increase in external pressure.

3. Effect of pressure on the freezing point of a liquid or melting point of a solid. Consider the equilibrium

$$\underset{\text{(More volume)}}{\text{Ice}} \rightleftharpoons \underset{\text{(less volume)}}{\text{Water}} \quad \text{(Endothermic process)}$$

When a solid melts, there is a change in volume. For example, when ice melts, there is decrease in volume.

Increase of pressure on ice $\rightleftharpoons$ water equilibrium will cause the equilibrium to shift towards right. Thus if at constant temperature, the pressure is increased, more of ice will melt. In order to retain ice in equilibrium with water, it is necessary to lower the temperature. Hence, the increase of pressure will lower the melting point of ice. As the process is endothermic, increase of temperature will shift the equilibrium to RHS. Thus more ice will melt.

Most of the solids show increase in volume on melting, e.g., sulphur (solid) $\rightleftharpoons$ sulphur (liquid) equilibrium. Therefore we can say that if the pressure on this equilibrium is increased, the melting point of sulphur is raised.

4. Effect of temperature on solubility. We can explain the effect of temperature on solubility of solids with the help of Le-Chatelier's principle. During the dissolution process if heat is evolved, the solubility decreases with increase in temperature. $CaCl_2$, NaOH, etc. are the examples of such solutes. The solubility of such substances decreases on increase of temperature. On the other hand dissolution of substances like NH_4Cl, $NaNO_3$ etc. is accompanied by absorption of heat. Their solubility increases with increase in temperature as per Le-Chatelier's principle.

5. Effect of pressure on solubility of gases in liquids. Consider the equilibrium involving dissolution of CO_2 in water.

$$CO_2\ (g) \rightleftharpoons CO_2\ (aq)$$

On increasing the pressure of CO_2 gas the equilibrium shifts in the direction which causes a decrease in the pressure of CO_2. The pressure of CO_2 will be lowered only if it dissolves more in water to form CO_2 (*aq*). Thus, according to Le-Chatelier's principle, the solubility of a gas in a liquid will increase with increase in pressure of the gas in equilibrium with its solution.

Example. State Le-Chatelier's principle and apply it for the following equilibrium :

$$\mathbf{PCl_5\ (g) \rightleftharpoons PCl_3\ (g) + Cl_2\ (g) - 17\ k\ cal}$$

Solution. For the statement of Le-Chatelier's principle, see relevant section

$$PCl_5(g) \rightleftharpoons PCl_3\ (g) + Cl_2\ (g) - 17\ \text{k cal}$$

The above reaction is endothermic and takes place with increase of pressure (number of gaseous product molecules is greater than the number of gaseous reactant molecules) in the forward direction.

Effect of pressure. On increasing the pressure, the reaction will shift in a direction where the effect can be neutralized i.e. in a direction where the pressure is decreased. Thus, the reaction will move in the backward direction. On decreasing the pressure, the reverse effect will be observed.

Effect of temperature. On increasing the temperature, the reaction will shift in a direction where it is endothermic i.e. heat can be absorbed. Thus, the reaction will shift in the forward direction. On decreasing the temperature, however, the reverse action will take place.

3.11. CLAUSIUS-CLAPEYRON EQUATION

When a system is in equilibrium, the free energy change accompanying the process is zero *i.e.*

$$\Delta G = 0, \text{ for a system in equilibrium} \qquad \text{... } (i)$$

The variation of free energy with temperature and pressure is given by the total differential equation, viz.

$$dG = VdP - SdT \qquad \text{... } (ii)$$

Rudolf Clausius

Clapeyron

Consider the system consisting of two phase I and II in equilibrium at temperature T and pressure P. Under these conditions, suppose that free energy of phase I is G_1 and that of phase II is G_2. If the temperature of the system is changed from T to $T + dT$ and pressure from P to $P + dP$, suppose the corresponding change in free energy of phase I is dG_1 and that of phase II is dG_2 (Fig. 3.4). Then according to equation (*ii*)

$$dG_1 = V_1 dP - S_1 dT \qquad \text{...}(iii)$$

and

$$dG_2 = V_2 dP - S_2 dT \qquad \text{...}(iv)$$

where V_1 and S_1 are the molar volume and entropy of phase I and V_2 and S_2 are the corresponding values for phase II.

But when phase I is in equilibrium with phase II,

$$\Delta G = 0$$

i.e.

$$G_2 - G_1 = 0$$

where G_1 *is* the molar free energy of phase I and G_2 that of phase II at temperature T and pressure P.

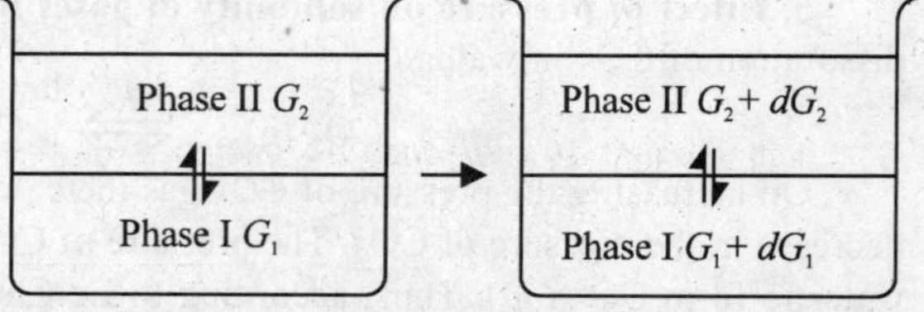

Fig. 3.4. Phase equilibria at T and P and at $T + dT$ and $P + dP$

Similarly, for the equilibrium at temperature $T + dT$ and pressure $P + dP$, we must have

$$(G_2 + dG_2) - (G_1 + dG_1) = 0$$

But

$$G_2 - G_1 = 0 \text{ (as already stated above)}$$

$\therefore$

$$dG_2 - dG_1 = 0$$

or

$$dG_2 = dG_1 \qquad \text{...}(v)$$

Substituting the values of dG_1 and dG_2 from equations (*iii*) and (*iv*), we get

$$V_2 dP - S_2 dT = V_1 dP - S_1 dT$$

or

$$(V_2 - V_1)\, dP = (S_2 - S_1)\, dT$$

or

$$\Delta V . dP = \Delta S . dT$$

or

$$\frac{dP}{dT} = \frac{\Delta S}{\Delta V} \qquad \text{...}(vi)$$

where $\Delta V = V_2 - V_1$ is the molar volume change

and $\Delta S = S_2 - S_1$ is the molar entropy change

where phase I changes into phase II.

If this change occurs at the temperature T (*i.e.* melting point, boiling point etc.) and ΔH is latent heat of transformation, then

$$\Delta s = \frac{\Delta H}{T}$$

Putting this value in equation (*vi*), we get

$$\boxed{\frac{dP}{dT} = \frac{\Delta H}{T\Delta V}} \quad \text{... (vii)}$$

This equation was first derived by Clapeyron (1934) and is called **Clapeyron equation**. It can be applied to various physical equilibria like melting, vaporisation and sublimation.

For the melting equilibrium *i.e.*

$$\text{Solid} \rightleftharpoons \text{liquid}$$

$\Delta H = \Delta H_m$, latent heat of melting and $T = T_m$, the melting point

Equation (*vii*) may then be written as

$$\frac{dP}{dT} = \frac{\Delta H_m}{T_m \Delta V} \quad \text{... (viii)}$$

Also $\Delta V = V_l - V_s$

where V_s = molar volume of the solid

and V_l = molar volume of the liquid (melt)

Hence equation (*viii*) may be written in the form

$$\frac{dP}{dT} = \frac{\Delta H_m}{T_m(V_l - V_s)}$$

or

$$\frac{dT}{dP} = \frac{T_m(V_l - V_s)}{\Delta H_m} \quad \text{...(ix)}$$

This equation gives the variation of melting point with pressure.

For the vaporisation equilibrium *i.e.*

$$\text{Liquid} \rightleftharpoons \text{Vapour}$$

The Clapeyron equation, *i.e.* equation (*viii*) takes the form

$$\frac{dP}{dT} = \frac{\Delta H_v}{T_b(V_v - V_l)} \quad \text{...(x)}$$

where ΔH_v = latent heat of vaporisation

T_b = boiling point of the liquid

V_l = molar volume of the liquid

and V_v = molar volume of the vapour.

Volume of the liquid (V_l) is very small as compared to the volume of the vapour (V_v), thus V_l can be neglected in comparison to V_v. Hence equation (*x*) may be written as (omitting the subscript in T_b).

$$\frac{dP}{dT} = \frac{\Delta H_v}{TV_v} \quad ...(xi)$$

Further assuming that the vapour behave like an ideal gas, we will have

$$PV_v = RT$$

or
$$V_v = \frac{RT}{P} \quad ...(xii)$$

Substituting this value in equation (*xii*), we get

$$\frac{dP}{dT} = \frac{\Delta H_v}{RT^2}.P$$

or
$$\frac{1}{P}.\frac{dP}{dT} = \frac{\Delta H_v}{RT^2}$$

or
$$\frac{\mathbf{d \ln P}}{\mathbf{dT}} = \frac{\mathbf{\Delta H_v}}{\mathbf{RT^2}} \quad ...(xiii)$$

This equation is called **Clausius-Clapeyron equation** as it was derived by Clausius during a detailed investigation of Clapeyron equation.

The equation (*xiii*) can also be written in another form called **integrated form of Clausius-Clapeyron equation** as explained below.

Equation (*xiii*) may be rewritten as

$$d \ln P = \frac{\Delta H_v}{RT^2} dT \quad ...(xiv)$$

If the temperature changes from T_1 to T_2 when the vapour pressure changes from P_1 to P_2 then integrating equation (*xiv*) between the appropriate limits, we get

$$\int_{P_1}^{P_2} d \ln P = \int_{T_1}^{T_2} \frac{\Delta H_v}{RT^2} dT$$

or
$$\ln \frac{P_2}{P_1} = \frac{\Delta H_v}{R} \int_{T_1}^{T_2} \frac{1}{T^2} dT$$

$$= \frac{\Delta H_v}{R}\left(\frac{1}{T_1} - \frac{1}{T_2}\right)$$

Changing from natural logarithm to the logarithm to base 10, we get

$$2.303 \log \frac{P_2}{P_1} = \frac{\Delta H_v}{R}\left(\frac{T_2 - T_1}{T_1 T_2}\right)$$

or
$$\mathbf{\log \frac{P_2}{P_1} = \frac{\Delta H_v}{2.303 R}\left(\frac{T_2 - T_1}{T_1 T_2}\right)} \quad ...(xv)$$

This is called the *integrated form of Clausius-Clapeyron equation.* It has a number of applications.

(*i*) If the vapour pressure of a liquid is known at one particular temperature, the value at any other temperature can be calculated, if the latent heat of vaporization of the liquid is known.

(*ii*) If the vapour pressures of a liquid at two different temperatures are known, the latent heat of vaporization of the liquid can be calculated from the equation.

(*iii*) If the boiling point of a liquid is known at some particular pressure, the boiling point at any other pressure can be obtained, provided the latent heat of vaporization of the liquid is known.

SOLVED EXAMPLES ON CLAUSIUS CLAYPEYRON EQUATION

Example 1. The normal boiling point of water is 100°C. Its vapour pressure at 80°C is 0.4672 atmosphere. Calculate the enthalpy of vaporisation per mole of water.

Solution. Here, we have

$$T_1 = 100°C = 100 + 273 = 373\ K$$
$$P_1 = 1\ atm$$
$$T_2 = 80°C = 80 + 273 = 353\ K$$
$$P_2 = 0.4672\ atm$$
$$\Delta H_v = \text{to be calculated}$$

Substituting the values in the equation,

$$\log \frac{P_2}{P_1} = \frac{\Delta H_v}{2.303R}\left(\frac{T_2 - T_1}{T_1 T_2}\right)$$

we get

$$\log \frac{0.4672}{1} = \frac{\Delta H_v}{2.303 \times' 8.314}\left(\frac{353 - -373}{353' \times 373}\right)$$

or

$$-0.3305 = \frac{\Delta H_v}{2.303 \times 8.314} \times \frac{-20}{353 \times 373}$$

or

$$\Delta H_v = \mathbf{41.66\ kJ\ mol^{-1}}$$

Example 2. Water boils at 373 K at one atmospheric pressure. At what temperature will it boil when atmospheric pressure becomes 528 mm of Hg at some space station.

Latent heat of H_2O = 2.28 kJ g^{-1}

$$\mathbf{R = 8.314\ JK^{-1}\ mol^{-1}}$$

Solution. We are given that

$$T_1 = 373\ K$$
$$P_1 = 1\ atm = 760\ mm$$
$$P_2 = 528\ mm,\ T_2 = \text{To be calculated}$$
$$\Delta H_v = 2.28\ kJ/g$$
$$= 2.28 \times 18\ kJ\ mol^{-1}$$
$$= 41.04\ kJ\ mol^{-1}$$
$$= 41040\ J\ mol^{-1}$$
$$R = 8.314\ JK^{-1}\ mo1^{-1}$$

Substituting the values in the, integrated form of clausius Clapeyron equation, we have

$$\log \frac{P_2}{P_1} = \frac{\Delta H_v}{2.303R}\left(\frac{T_2 - T_1}{T_2 T_1}\right)$$

we get $\log\dfrac{521}{760} = \dfrac{41040}{2.303\times 8.314}\times\left(\dfrac{T_2-373}{373T_2}\right)$

or $-0.1582 = 2143.4\times\left(\dfrac{T_2-373}{373T_2}\right)$

or $-59.0\,T_2 = 2143\,T_2 - 799488.2$

or $2202.4\,T_2 = 799488.2$

$\therefore$ $\boldsymbol{T_2 = 363\ \text{K}}$

Example 3. Water boils at 100°C at a pressure of 1 atm. Calculate the vapour pressure of water at 90°C. The heat of vaporisation of water is 9.80 kcal mol⁻¹.

Solution. Given that

$$T_1 = 100 + 273 = 373\ \text{K}$$
$$P_1 = 1\ \text{atm}$$
$$T_2 = 90 + 273 = 363\ \text{K}$$
$$P_2 = \text{To be calculated}$$
$$\Delta H_v = 9.80\ \text{kcal mol}^{-1} = 9800\ \text{cal mol}^{-1}$$
$$R = 1.987\ \text{cal deg}^{-1}\ \text{mol}^{-1}$$

Substituting these values in the Clausius-Clapeyron equation

$$\log\frac{P_2}{P_1} = \frac{\Delta H_v}{2.303R}\left(\frac{T_2-T_1}{T_1\,T_2}\right)$$

we get $\log\dfrac{P_2}{P_1} = \dfrac{9800}{2.303\times 1.987}\times\dfrac{363-373}{363\times 373}$

or $\log P_2 = -0{\cdot}1582 = \bar{1}.8418$

$\therefore$ $\boldsymbol{P_2 = 0{\cdot}6947\ \text{atm}}$

PROBLEMS FOR PRACTICE

1. At 373.6 K and 372.6 K the vapour pressures of water are 1.018 and 0.982 atm respectively. Calculate the heat of vaporisation of water. [**Ans.** 41.676 kJ mol^{-1}]

2. If the vapour pressure of toluene is 60 mm at 40.3°C and 20 mm at 18.4°C, calculate (*a*) the vapour pressure at 25°C and (*b*) the heat of vaporization. [**Ans.** (*a*) 28.4 mm (*b*) 38.08 kJ mol^{-1}]

3. The normal boiling point of cyclohexane is 80.7°C and the heat of vaporization is 30.08 kJ mol^{-1}. Calculate (*a*) the boiling point at 650 mm and (*b*) the vapour pressure at 25°C. [**Ans.** (*a*) 348.4 K (*b*) 0.1479 atm or 112.4 mm]

4. The vapour pressure of an organic liquid at 27°C and 37°C are 90 mm and 110 mm of mercury respectively. Calculate the molar enthalpy of vaporization (R = 1.987 cal degree^{-1} mol^{-1}). [**Ans.** 3711 cal mol^{-1}]

5. Using Clausius-Clapeyron equation, calculate the enthalpy of vaporisation of liquid X.

P (*torr*)	40	60
T (*K*)	294	302

[**Ans.** 37.422 kJ]

6. The vapour pressure of solid benzene is 2.24 mm at 243 K and 24.5 mm at 373 K. Calculate from these data the heat of fusion of benzene. [**Ans.** 43.99 kJ/mol]

SOME SOLVED CONCEPTUAL QUESTIONS

Example 1. Under what condition K_p, K_c, K_a and K_x are all equal?

Solution. It is when $\Delta n = 0$, *i.e.*, $n_p = n_r$ (gaseous).

Example 2. How thermodynamic equilibrium constant in terms of activities (K_a) is related to K_p or K_c? Under what condition K_a becomes equal to K_p or K_c?

Solution. For the reaction, $aA + bB + \ldots\ldots \rightleftharpoons mM + nN + \ldots\ldots,$

$$K_a = \frac{a_M^m\ a_N^n\ldots\ldots}{a_A^a\ a_B^b\ldots\ldots}$$

Activity = activity coefficient × composition variable (conc. or pressure)

i.e. $a = \gamma X$

Substituting the value of a in the above equation.

$$K_a = \left(\frac{\gamma_M^m\ \gamma_N^n\ldots\ldots}{\gamma_A^a\ \gamma_B^b\ldots\ldots}\right)\left(\frac{X_M^m\ X_N^n\ldots\ldots}{X_A^a\ X_B^b\ldots\ldots}\right) = K_\gamma\, K_X$$

$$\therefore \quad K_a = K_\gamma\, K_c \quad \text{or} \quad K_a = K_\gamma\, K_p$$

For ideal gases or real gases at very low pressures, $\gamma = 1$ for each gas.

Then $K_\gamma = 1$ and hence $K_a = K_p$

Similarly, for very dilute solutions, as $C \to 0$, $\gamma \to 1$ and $K_a \to K_c$.

Example 3. Can the equilibrium $CaCO_3$ (s) $\rightleftharpoons$ CaO (s) + CO_2 (g) be attained in an open vessel? Why or why not?

Solution. No, this equilibrium cannot be attained in open vessel because one of products, viz., CO_2 is gaseous which escapes out.

Example 4. How is that a reversible reaction is spontaneous in the forward as well as backward direction?

Solution. This is because the free energy of the equilibrium mixture is lower than that of reactants as well as that of products. Hence, ΔG is negative both for the forward direction and for the backward direction.

Example 5. Give Clausius-Clapeyron equation for liquid $\rightleftharpoons$ vapour equilibrium and write its applications.

Solution. The Clausius-Clapeyron equation for liquid $\rightleftharpoons$ vapour equilibrium is

$$\ln \frac{P_2}{P_1} = \frac{\Delta H_v}{R}\left[\frac{T_2 - T_1}{T_1 T_2}\right]$$

This equation can be used to calculate the molar heat of vapourisation, ΔH_v of a liquid if the vapour pressures at two temperatures are known. Further if the molar heat of vaporisation (enthalpy of vaporisation) is known and vapour pressure at one temperature is known, then vapour pressure at another temperature can be calculated from the equation.

If the boiling points of a liquid at one pressure is known that at any other pressure can also be calculated.

Example 6. How is the free energy of a gas determined (*i*) by studying its isothermal expansion (*ii*) from its equilibrium constant K_p (*iii*) from enthalpy change and entropy data ?

Solution. (*i*) $\Delta G = n\,RT \ln \dfrac{P_2}{P_1}$ (*ii*) $\Delta G° = -RT \ln K_p$

(*iii*) $\Delta G° = \Delta H° - T\,\Delta S°$ and $\Delta S° = \Sigma S°$ (Products) – $\Sigma S°$ (Reactants).

Example 7. On the basis of Le-Chatelier's principle discuss the favourable reaction conditions for the following reaction

$$N_2\,(g) + O_2\,(g) \rightleftharpoons 2NO(g)\ (\Delta H = 180.5\ kJ)$$

Solution. The given reaction is endothermic and therefore, it will be favoured by high temperature. Δn for the reaction is zero and therefore increase of pressure has no effect on equilibrium.

Example 8. Write Clausius-Clapeyron equation for solid $\rightleftharpoons$ vapour equilibria. What are applications of this equation for solid $\rightleftharpoons$ vapour equilibria?

Solution. Clausius-Clapeyron equation for solid $\rightleftharpoons$ vapour equilibrium is written as

$$\ln \frac{P_2}{P_1} = \frac{\Delta H_s}{R}\left[\frac{T_2 - T_1}{T_1 T_2}\right]$$

where ΔH_s is the molar enthalpy of sublimation of the solid.

This equation can be used to calculate the effect of pressure on the melting point of a solid. Further, if vapour pressures at two temperatures are known then ΔH_s can be calculated.

Example 9. Why $\Delta G°$ obtained from K_p and K_c has different values?

Solution. $\Delta G° = -RT \ln K_p$, Also $\Delta G° = -RT \ln K_c$. As $K_p \neq K_c$ unless $\Delta n = 0$, therefore, $\Delta G°$ values are different in the two cases.

Example 10. Which of the properties remain constant when equilibrium is attained?

Solution. Temperature, pressure and chemical potential of a particular substance remain constant in all the phases.

Example 11. Write van't Hoff equation. Explain that it leads to the same effect of temperature on equilibrium constant as predicted by Le-Chatelier.

Solution. van't Hoff equation is $\dfrac{d \ln K_p}{dT} = \dfrac{\Delta H°}{RT^2}$

For endothermic reactions, $\Delta H°$ is +ve. Hence, $d \ln K_p/dT$ is +ve. This means that $\ln K_p$ and hence K_p increases with increase of temperature, *i.e.*, equilibrium shifts towards products.

For exothermic reactions, $\Delta H°$ is –ve. Hence, $d \ln K_p/dT$ is –ve. This means that $\ln K_p$ and hence K_p decreases with increase of temperature, *i.e.*, equilibrium shifts towards reactants. This is what is stated by Le-Chatelier's principle.

Example 12. Write Clausius-Clapeyron equation both in the differential form and in the integrated form.

Solution. Differential form is $\dfrac{d \ln R}{dT} = \dfrac{\Delta_{vap} H}{RT^2}$

Integrated form is $\log \dfrac{P_2}{P_1} = \dfrac{\Delta_{vap} H}{2.303\,R}\left(\dfrac{1}{T_1} - \dfrac{1}{T_2}\right)$

Example 13. How is free energy change of a reaction in a given state related to its reaction quotient (pressure quotient) Q_p in that state and the equilibrium constant K_p?

Solution. $\Delta G = -RT \ln K_p + RT \ln Q_p$.

Example 14. What is van't Hoff reaction isotherm? Why is it so called?

Solution. van't Hoff reaction isotherm is $\Delta G = \Delta G° + RT \ln Q_p$

It gives free energy change of the reaction in terms of standard free energy change and the value of the reaction quotient Q_p under the given state at constant temperature. That is why it is called reaction isotherm.

Example 15. Write van't Hoff equation giving the variation of equilibrium constant with temperature both in the differential form and in the integrated form?

Solution. Differential form is $\dfrac{d \ln Kp}{dT} = \dfrac{\Delta H°}{RT^2}$

Integrated form is $\log \dfrac{(K_p)_2}{(K_p)_1} = \dfrac{\Delta H}{2.303R}\left(\dfrac{1}{T_1} - \dfrac{1}{T_2}\right)$

EXERCISES
(Based on Question Papers of Different Universities)

Objective Questions (Multiple Choice Questions)

1. According to law of Mass Action, for the reaction.

2A + B $\longrightarrow$ Products, which equation holds good?

(*a*) Rate = k $[A]^2$ [B] (*b*) Rate = *k* [A] $[B]^2$

(*c*) Rate = k [A] [B] (*d*) Rate = *k* [A] $[B]^{1/2}$

2. In which of the following reactions, K_p is less than K_c?

(*a*) N_2 (g) + $3H_2$ (g) $\rightleftharpoons$ $2NH_3$ (g)

(*b*) PCl_5 (g) $\rightleftharpoons$ PCl_3 (g) + Cl_2 (g)

(*c*) H_2 (g) + I_2 (g) $\rightleftharpoons$ 2HI (g)

(*d*) $2SO_3$ (g) $\rightleftharpoons$ $2SO_2$ (g) + O_2 (g)

3. What will be the equilibrium constant at 717 K for the reaction :

2HI (g) $\rightleftharpoons$ $\frac{1}{2}H_2$ (g) + $\frac{1}{2}I_2$ (g)

if its value for the reaction

H_2 (g) + I_2 (g) $\rightleftharpoons$ 2HI (g) at 717 K is 64?

(*a*) 8 (*b*) 64

(*c*) $\frac{1}{64}$ (*d*) $\frac{1}{8}$.

4. With increase in temperature, equilibrium constant of a reaction:

(*a*) always decreases (*b*) always increases

(*c*) may increase or decrease depending upon whether $n_p < n_r$ or $n_p > n_r$.

5. The relation between K_p and K_c for the reaction

$$2NO\,(g) + Cl_2\,(g) \rightleftharpoons 2NOCl\,(g) \text{ is}$$

(*a*) $K_p = K_c / RT$ (*b*) $K_p = K_c\,(RT)$

(*c*) $K_p = K_c / (RT)^2$ (*d*) $K_p = K_c$

6. K_1 and K_2 are equilibrium constants for reactions (1) and (2)

$$N_2\,(g) + O_2\,(g) \rightleftharpoons 2NO\,(g) \qquad \text{... (1)}$$

$$NO\,(g) \rightleftharpoons \frac{1}{2}N_2\,(g) + \frac{1}{2}O_2\,(g) \qquad \text{... (2)}$$

Then,

(*a*) $K_1 = K_2^2$ (*b*) $K_1 = \dfrac{1}{K_2}$

(*c*) $K_1 = \left(\dfrac{1}{K_2}\right)^2$ (*d*) $K_1 = (K_2)^0$.

7. Which of the following equilibria will shift to right side on increasing the temperature?

(*a*) $2SO_2\,(g) + O_2\,(g) \rightleftharpoons 2SO_3\,(g)$

(*b*) $H_2O\,(g) \rightleftharpoons H_2\,(g) + \frac{1}{2}O_2\,(g)$

(*c*) $4HCl\,(g) + O_2\,(g) \rightleftharpoons 2H_2O\,(g) + 2Cl_2\,(g)$

(*d*) $CO\,(g) + H_2O\,(g) \rightleftharpoons CO_2\,(g) + H_2O\,(g)$

8. At 490° C, the equilibrium constant for the synthesis of HI is 50, the value of K for the dissociation of HI will be:

(*a*) 2.0 (*b*) 20.0

(*c*) 0.002 (*d*) 0.02.

9. Consider the following reaction equilibrium.

$$2SO_2\,(g) + O_2\,(g) \rightleftharpoons 2SO_3\,(g); \quad \Delta H° = -198 \text{ kJ}.$$

On the basis of Le-Chatelier's principle, the condition favourable for the forward reaction is:

(*a*) decreasing the temperature and increasing the pressure.

(*b*) increasing temperature as well as pressure.

(*c*) lowering of temperature as well as pressure.

(*d*) any value of temperature and pressure.

10. In which of the following cases, does the reaction go farthest to completion?

(*a*) K = 10 (*b*) K = 1

(*c*) $K = 10^3$ (*d*) $K = 10^{-2}$.

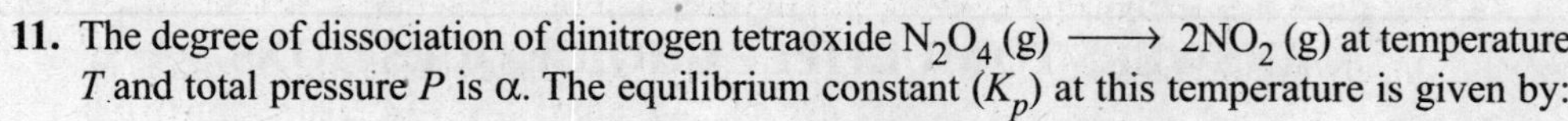

11. The degree of dissociation of dinitrogen tetraoxide N_2O_4 (g) $\longrightarrow$ $2NO_2$ (g) at temperature T and total pressure P is α. The equilibrium constant (K_p) at this temperature is given by:

(*a*) $\dfrac{2\alpha}{(1-\alpha^2)}$ (*b*) $\dfrac{\alpha^2 P}{(1-\alpha)}$

(*c*) $\dfrac{4\alpha^2}{(1-\alpha^2)}$ (*d*) $\dfrac{4\alpha^2 P}{(1-\alpha^2)}$.

12. The rate of forward reaction is two times that of the backward reaction at a given temperature and identical concentration.

$K_{equilibrium}$ is:

(*a*) 1.5 (*b*) 0.5

(*c*) 2.5 (*d*) 2.0.

13. In a reaction

$A + 2B \rightleftharpoons 2C$,

2.0 mole of 'A', 3.0 mole of 'B' and 2.0 mole of 'C' are placed in a 2.0 L flask and the equilibrium concentration of 'C' is 0.5 mole / L. The equilibrium constant (K) for the reaction is :

(*a*) 0.147 (*b*) 0.073

(*c*) 0.05 (*d*) 0.026.

14. For the reaction :

$$PCl_3\ (g) + Cl_2\ (g) \rightleftharpoons PCl_5\ (g)$$

the value of K_c is 26 at 250°C. The value of K_p at this temperature is:

(*a*) 0.83 (*b*) 0.57

(*c*) 0.46 (*d*) 0.61.

15. In the reaction $PCl_5\ (g) \rightleftharpoons PCl_3\ (g) + Cl_2\ (g)$, the equilibrium concentrations of PCl_5 and PCl_3 are 0.4 and 0.2 mole/litre respectively. If the value of K_c is 0.5, what is the concentration of Cl_2 in moles / litre?

(*a*) 1.0 (*b*) 1.5

(*c*) 0.5 (*d*) 2.0.

16. Which one of the following oxides of nitrogen will be the most stable one?

(*a*) $2NO_2\ (g) \rightleftharpoons N_2\ (g) + 2O_2\ (g)$ $K = 6.7 \times 10^{16}$ mol L^{-1}

(*b*) $2N_2O_5\ (g) \rightleftharpoons 2N_2\ (g) + 5O_2\ (g)$ $K = 1.2 \times 10^{24}$ $mol^5\ L^{-5}$

(*c*) $2N_2O\ (g) \rightleftharpoons 2N_2\ (g) + O_2\ (g)$ $K = 3.5 \times 10^{33}$ mol L^{-1}

(*d*) $2NO\ (g) \rightleftharpoons N_2\ (g) + O_2\ (g)$ $K = 2.2 \times 10^{30}$ mol L^{-1}.

17. At a given temperature, the equilibrium constant for the reactions:

$$NO\ (g) + \frac{1}{2}O_2\ (g) \rightleftharpoons NO_2\ (g)$$

and $2NO_2\ (g) \rightleftharpoons 2NO\ (g) + O_2\ (g)$ are K_1 and K_2 respectively. If $K_1 = 4 \times 10^{-3}$, then K_2 will be:

(*a*) 8×10^{-3} (*b*) 16×10^{-3}

(*c*) 6.25×10^{4} (*d*) 6.25×10^{6}.

ANSWERS TO MULTIPLE CHOICE QUESTIONS

1. (*a*)	**2.** (*a*)	**3.** (*d*)	**4.** (*d*)	**5.** (*b*)	**6.** (*c*)
7. (*b*)	**8.** (*d*)	**9.** (*a*)	**10.** (*c*)	**11.** (*d*)	**12.** (*d*)
13. (*c*)	**14.** (*d*)	**15.** (*a*)	**16.** (*a*)	**17.** (*c*)	

Short Answer Questions

1. Define the law of chemical equilibrium.

2. State the units in which K_p and K_c are expressed.

3. What do you understand by a reversible reaction? State with an example.

4. Explain the meaning of the term equilibrium constant.

5. What are irreversible reactions? Give an example.

6. State the law of mass action.

7. How is the standard free energy change of a chemical reaction related to its equilibrium constant?

8. Derive the relationship between:

(*i*) K_p and K_c.

(*ii*) K_a and K_p or K_c.

(*iii*) K_x with K_p and K_c.

9. What do you understand by equilibrium state?

10. Give two examples of each of the following:

(*a*) Reversible homogeneous reactions.

(*b*) Reversible heterogeneous reactions

(*c*) Irreversible reactions

Represent graphically how equilibrium is attained.

11. Write expressions for equilibrium constant for the following reactions:

(*i*) $N_2\ (g) + 3\ H_2\ (g) \rightleftharpoons 2\ NH_3\ (g)$

(*ii*) $CH_3COOH\ (l) + C_2H_5OH\ (l) \rightleftharpoons CH_3COOC_2H_5\ (l) + H_2O\ (l)$

(*iii*) $4\ Fe\ (s) + 4\ H_2O\ (g) \rightleftharpoons Fe_3O_4\ (s) + 4\ H_2\ (g)$

(*iv*) $CaCO_3\ (s) \rightleftharpoons CaO\ (s) + CO_2\ (g)$

12. Discuss any two applications of Le-Chatelier's principle as applied to physical equilibria.

13. Derive van't Hoff equation,

$$d\left(\frac{\ln K_p}{dT}\right) = \frac{\Delta H}{RT^2}$$

where the symbols have their usual meanings.

14. Derive the relationship between K_p and K_c.

15. Give the integrated form of the Clausius-Clapeyron equation and also list any two applications.

16. Prove that K_c is independent of pressure at constant temperature.

17. Derive the Clausius-Clapeyron equation.

General Questions

1. If K_1 and K_2 are the equilibrium constants of reaction at temperatures T_1 and T_2 respectively, derive thermodynamically the relation

$$\log \frac{K_2}{K_1} = \frac{\Delta H^\circ}{2.303\, R}\left(\frac{1}{T_1} - \frac{1}{T_2}\right)$$

2. With the help of Le-Chatelier—Braun principle, explain the following:
   (*i*) Effect of temperature on the solubility of sugar in water, given that the dissolution of sugar in water is an endothermic process.
   (*ii*) Effect of pressure on the melting of ice.
   (*iii*) Effect of temperature an pressure on the solubility of gases in liquids.
   (*iv*) Effect of pressure on the boiling point of a liquid.
   (*v*) Effect of pressure on the freezing point of a liquid.
3. State and explain Le-Chatelier's principle.
4. Derive van't Hoff reaction isotherm. How the free energy change of a reaction in any state other than state of equilibrium related to reaction quotient Q_p and equilibrium constant K_p? Derive this relationship.
5. Derive van't Hoff equation

$$\frac{d\,(\ln K_p)}{dT} = \frac{\Delta H^\circ}{RT^2}$$

for temperature dependence of equilibrium constant (K_p). What is the difference between ΔH and ΔH°?
6. Derive Clausius-Clapeyron equation for liquid-vapour equilibrium. Show how the equation can be expressed in the integrated form. What are its important applications?
7. Derive van't Hoff equation (reaction isochore) giving the effect of temperature on equilibrium constant.
8. Derive van't Hoff equation giving the variation of equilibrium constant with temperature.
9. Derive the criteria for thermodynamic equilibrium in terms of temperature, pressure and chemical potential.
10. Discuss 'Chemical Equilibrium' in terms of Free Energy Change.
11. For a gaseous reaction in equilibrium, derive the reaction $\Delta G^\circ = -RT \ln K_p$.
12. Derive van't Hoff equation in the integrated form.
13. Starting from Clapeyron equation, how Clausius-Clapeyron equation is obtained? Express it in the integrated form.
14. Starting from basic principles, derive the relationship $\Delta G^\circ = -RT \ln K_p$.
15. Write Clausius-Clapeyron equation in the integrated form. What are its important applications?
16. Define law of chemical equilibrium. How can it be derived thermodynamically?
17. Write expression for equilibrium constant
   (*a*) in terms of pressures (K_p)
   (*b*) in terms of concentrations (K_c)
   (*c*) in terms of activities (K_a)

(*d*) in terms of mole fractions (K_x)

Derive the relationship between

(*i*) K_p and K_c (*ii*) K_a and K_p or K_c (*iii*) K_c and K_x.

18. Briefly describe at least three different ways by means of which the standard free energy change of a reaction can be calculated.

19. For a gaseous reaction, derive the following relationship:

$$\Delta G = -RT \ln K_p + RT \ln Q_p.$$

20. Derive van't Hoff equation in terms of equilibrium constant K_c and standard internal energy change, $\Delta U°$.

21. What do you understand by reversible and irreversible reactions? What do you mean by saying that equilibrium is dynamic in nature? State the law of chemical equilibrium. Name at least three different types of equilibrium constants. Give the relationship between them. What are their units?

22. A reversible reaction is spontaneous in the forward direction as well in the backward direction (*i.e.*, ΔG is negative for both). Explain with a suitable example.

23. Derive the following for the gaseous reaction $aA + bB + ... \rightleftharpoons mM + nN + ...$

$$\frac{p_M^n \, p_N^n .....}{p_A^a \, p_B^b .....} = e^{-\Delta G°/RT}$$

24. For a gaseous reaction (not in equilibrium) derive the relationship $\Delta G = \Delta G° + RT \ln Q_p$.

25. How is free energy change and standard free energy change of a reaction calculated from the chemical potentials of the reactants and products involved? Taking suitable example, explain how free energy change of forward and backward reaction of a reversible reaction takes place.

26. Apply Le-Chatelier principle to predict suitable conditions for getting maximum yield of the product in each of the following cases :

(*i*) manufacture of ammonia by Haber's process.

(*ii*) manufacture of nitric oxide in Birkland–Eyde process for manufacture of nitric acid.

(*iii*) manufacture of hydrogen by Bosch process.

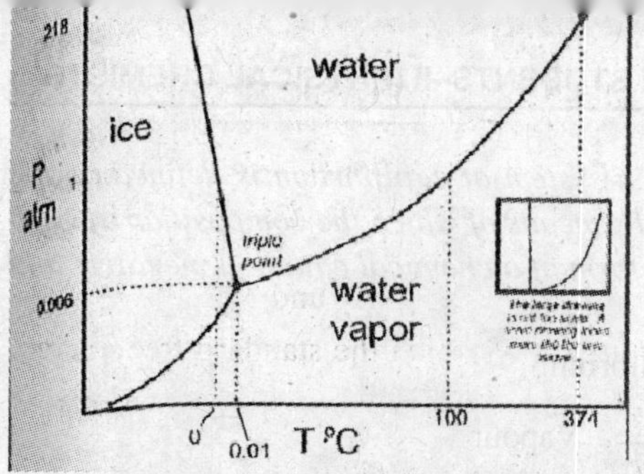

Phase Rule

4.1. INTRODUCTION

Phase equilibrium is relevant to heterogeneous systems. A heterogeneous system is defined as a system consisting of a number of phases in equilibrium. The number of phases that can exist together at equilibrium depends upon the conditions of temperature, pressure and concentrations of different phases. A relationship governing all heterogeneous equilibria was first given by Willard Gibbs. It is known as phase rule.

Willard Gibbs

Gibb's phase rule cannot be expressed in words. It is described by a mathematical equation as under:

For a heterogeneous system in equilibrium

$$F = C - P + 2$$

where F = No. of degrees of freedom
C = No. of components
P = No. of phases

4.2. EXPLANATION OF THE TERMS

(1) Phase: A phase may be defined as any part of a system which is

(*a*) *homogeneous and separated from other parts of the system by a distinct boundary,*

(*b*) *physically and chemically different from other parts of the same system,*

(*c*) *mechanically separable from other parts of the system.*

Examples illustrating phases are given below:

System	*Phases*
(*i*) Ice and water.	2 phases: 1 solid and 1 liquid.
(*ii*) Water and water vapour.	2 phases: 1 liquid and 1 gaseous.
(*iii*) Ice, water and water vapour.	3 phases: 1 solid, 1 liquid and 1 gaseous.
(*iv*) Two immiscible liquids, e.g., CS_2/H_2O, CCl_4/H_2O, etc.	2 phases: both liquids.
(*v*) Two miscible liquids, *i.e.*, water/alcohol	1 phase: only liquid.
(*vi*) $CaCO_3$, CaO and CO_2	3 phases: 2 solids and 1 gaseous.
(*vii*) Monoclinic and rhombic sulphur.	2 phases: both solids.

Miscible gases and miscible liquids constitute a *single* phase.

Solid + liquid, solid + gas, liquid + gas, two solids and two immiscible liquids constitute *two* phases.

Different solids are regarded as different phases, however intimately they may be mixed.

(2) Components: *The number of components of a system at equilibrium is defined as the smallest number of independently variable constituents by means of which the composition of each phase present can be expressed either directly or in the form of a chemical equation, negative and zero signs being permitted.*

Examples : (*i*) Water system consists of the equilibrium

$$\underset{\text{(Solid)}}{\text{Ice}} \rightleftharpoons \underset{\text{(Liquid)}}{\text{Water}} \rightleftharpoons \underset{\text{(Gaseous)}}{\text{Vapour}}$$

This system consists of one component only, *i.e., it is a one component system* because the composition of each of the three phases present can be directly expressed as H_2O.

(*ii*) **Sulphur system** consisting of four phases—Monoclinic (*s*), Rhombic (*s*), Liquid (*l*) and Vapours (*g*) is also *one component system* because the composition of each phases can be expressed in terms of one constituent—sulphur.

(*iii*) Decomposition of $CaCO_3$ by heat can be expressed according to the equilibrium

$$\underset{\text{(Solid)}}{CaCO_3} \rightleftharpoons \underset{\text{(Solid)}}{CaO} + \underset{\text{(Gas)}}{CO_2}$$

There are three different constituents forming three different phases. But the composition of each phase can be expressed in terms of any two of the constituents. For example,

(*a*) If the constituents chosen are CaO and CO_2, then

Phase		*Composition*
$CaCO_3$	$\longrightarrow$	$CaO + CO_2$
CaO	$\longrightarrow$	$CaO + 0\,CO_2$
CO_2	$\longrightarrow$	$0\,CaO + CO_2$

(*b*) If the constituents chosen are $CaCO_3$ and CO_2, then

Phase		*Composition*
$CaCO_3$	$\longrightarrow$	$CaCO_3 + 0\,CO_2$
CaO	$\longrightarrow$	$CaCO_3 - CO_2$
CO_2	$\longrightarrow$	$0\,CaCO_3 + CO_2$

Likewise composition of the three phases can be expressed in terms of $CaCO_3$ and CaO. Thus the *dissociation of $CaCO_3$ by heat* is a *two component system.*

(*iv*) **Sodium sulphate–water system.** This system at equilibrium contains a number of phases like $Na_2SO_4 \cdot 10H_2O$, $Na_2SO_4 \cdot 7H_2O$, Na_2SO_4 solution, ice, water and vapour, but the composition of each phase can be expressed, by chemical equations, in terms of only anhydrous sodium sulphate (Na_2SO_4) and water (H_2O) as follows:

Phase		*Composition*
$Na_2SO_4 \cdot 10H_2O$	$\longrightarrow$	$Na_2SO_4 + 10H_2O$
$Na_2SO_4 \cdot 7H_2O$	$\longrightarrow$	$Na_2SO_4 + 7H_2O$
Na_2SO_4	$\longrightarrow$	$Na_2SO_4 + 0H_2O$
Na_2SO_4 Solution	$\longrightarrow$	$x\,Na_2SO_4 + y\,H_2O$

$$\text{Ice} \longrightarrow 0\,Na_2SO_4 + H_2O$$
$$\text{Water} \longrightarrow 0Na_2SO_4 + H_2O$$
$$\text{Vapour} \longrightarrow 0Na_2SO_4 + H_2O$$

Thus the *smallest number* of independently variable constituents by which the composition of the phases present at equilibrium can be expressed is two —Na_2SO_4 and H_2O. Hence it is also a *two component system.*

In a chemically reactive system, the number of components is calculated as follows:

Number of components (C) = Total number of constituents or species, S (already present as well as produced due to reaction)

– Number of equations (E) representing equilibria between the constituents

– Number of restrictions (R) for electrical neutrality (if ions are considered)

Mathematically this may be written as $C = S - E - R$

The following examples will illustrate

1. Na_2SO_4 – H_2O

(*a*) *Assuming no dissociation*

S = 2 (Na_2SO_4 and H_2O)

E = 0 (No chemical reaction/equilibrium)

R = 0 (No restriction of electrical neutrality or material balance)

∴ C = S – E – R = 2 – 0 – 0 = 2 Thus no. of components = 2

(*b*) *Assuming complete dissociation of the salt.*

S = 3 (Na^+, SO_4^{2-} and H_2O)

E = 0

R = 1 (For electrical neutrality, 2 Na^+ = SO_4^{2-})

∴ C = S – E – R = 3 – 0 – 1 = 2 Thus no. of components = 2

(*c*) *Assuming dissociation of H_2O also.*

S = 6 (Na_2SO_4, Na^+, SO_4^{2-}, H_2O, H^+, OH^-)

E = 2 ($Na_2SO_4 \rightleftharpoons 2Na^+ + SO_4^{2-}$ and $H_2O \rightleftharpoons H^+ + OH^-$)

R = 2 ($2Na^+ = SO_4^{2-}$ *i.e*, number of moles of Na^+ = 2 × No. of moles of SO_4^{2-} and $H^+ = OH^-$)

C = S – E – R = 6 – 2 – 2 = 2. Thus no. of components = 2

2. KCl – NaBr – H_2O system

(*a*) *Assuming no dissociation*

S = 5 [KCl, NaBr, H_2O (already present) NaCl, KBr (produced due to reaction)]

E = 1 ($KCl + NaBr \rightleftharpoons KBr + NaCl$)

R = 0 (No ions have been considered)

∴ C = S – E – R = 5 – 1 – 0 = 4 or No. of components = 4

(*b*) *Assuming complete dissociation*

$S = 11$ (KCl, K^+, Cl^-, $NaBr$, Na^+, Br^-, KBr, $NaCl$, H_2O, H^+, OH^-)

$E = 5$ ($KCl \rightleftharpoons K^+ + Cl^-$, $NaBr \rightleftharpoons Na^+ + Br^-$, $NaCl \rightleftharpoons Na^+ + Cl^-$,

$KBr \rightleftharpoons K^+ + Br^-$, $H_2O \rightleftharpoons H^+ + OH^-$)

$R = 2$ (Electrical neutrality and $H^+ = OH^-$)

$\therefore \quad C = S - E - R = 11 - 5 - 2 = 4$ or No. of components = 4

If the salts KCl and NaBr are taken in equal amounts,

$R = 3$ (*viz.*, $Na^+ = K^+$, $Cl^- = Br^-$ and $H^+ = OH^-$)

In that case, $C = S - E - R = 11 - 5 - 3 = 3$ or No. of components = 3

Thus, $KCl - NaBr - H_2O$ is a four component system if salts are present in unequal amounts but it becomes a three-component system if the salts are taken in equal amounts.

3. $KCl - NaCl - H_2O$ system

(*a*) *Assuming no dissociation*

$S = 3$ (KCl, $NaCl$, H_2O)

$E = 0$

$R = 0$

$\therefore \quad C = S - E - R = 3 - 0 - 0 = 3$ (*three components*)

(*b*) *Assuming complete dissociation*

$S = 8$ (KCl, K^+, Cl^-, $NaCl$, Na^+, H_2O, H^+, OH^-)

$E = 3$ ($KCl \rightleftharpoons K^+ + Cl^-$, $NaCl \rightleftharpoons Na^+ + Cl^-$, $H_2O \rightleftharpoons H^+ + OH^-$)

$R = 2$ (Electrical neutrality and $H^+ = OH^-$)

$\therefore \quad C = S - E - R = 8 - 3 - 2 = 3$ (*three components*)

If the salts KCl and NaCl are taken in equal amounts, then there will be one more restriction ($Na^+ = K^+$) so that $R = 3$. Hence,

$C = S - E - R = 8 - 3 - 3 = 2$ or number of components = 2

From examples 2 and 3, we conclude that *if the salts are taken in unequal amounts, $KCl - NaBr - H_2O$ is a 4-component system and $KCl - NaCl - H_2O$ is a 3-component system and if they are taken in equal amounts in each case, the former is a 3-component system and the latter is a 2-component system.*

(3) Degrees of freedom or Variance. *Degrees of freedom or variance of a system is the smallest number of variable factors, i.e., temperature, pressure and concentration of the components which must be arbitrarily fixed in order that the conditions of the system may be completely defined.*

Examples : (*i*) Ice - water - vapour system. In this system, there are three phases of one component (H_2O). These three phases can co-exist in equilibrium only at one particular temperature and under one particular pressure. Any variation in these factors will make one or more of the phases disappear and, therefore, will destroy the system. Hence this system has no degree of freedom or in other words, it is *invariant.* The same result follows from Phase Rule *i.e.*,

$$F = C - P + 2 = 1 - 3 + 2 = 0$$

(*ii*) Water - vapour system. This system consists of two phases of one component. As the vapour pressure of water is definite at a definite temperature, independent of the amount of water present, it follows that if the temperature is fixed, the vapour pressure is also fixed and conversely, fixing the pressure fixes the temperature. Thus the system has only one degree of freedom or in other words, it is *univariant.*

***(iii)* For a gas.** The temperature, pressure and concentration (volume) are related according to the equation $PV=RT$. Thus in order to completely define the system any two factors out of P, V and T must be fixed. Hence it follows that a gas has two degrees of freedom or it is *blvariant.*

Example 1. What the number of components, phases and degrees of freedom in the following equilibrium system :

$$CaCO_3\,(s) \rightleftharpoons CaO\,(s) + CO_2\,(g)$$

Solution. No. of components $= 2$

No. of phases $= 3$ (2 solid + 1 gas)

Degree of freedom $F = C - P + 2$ or $F = 2 - 3 + 2$

$F = 1$

4.3. EQUILIBRIUM

A system is said to be in equilibrium if the properties like temperature, composition etc. of the various phases do not undergo any change with time. This is of the following two types:

(*i*) True Equilibrium *A system is said to be in a state of true equilibrium if the same state can be achieved by approach from either direction.* For example, the equilibrium between ice and water at 1 atm. pressure and 0°C is a true equilibrium, because it can be attained by melting of ice or by freezing of water.

(*ii*) Metastable equilibrium *If the state of a system can be attained only from one direction and that too by very careful change of conditions, it is called a state of* **metastable equilibrium.** For example, water at –2°C can be obtained only by very careful cooling of liquid water but not by melting of ice. Hence water at –2°C is said to be in a state of metastable equilibrium.

4.4. CRITERIA FOR PHASE EQUILIBRIUM

The following three equilibria must exist in phase equilibria in a multi-component system.

(*i*) **Thermal Equilibrium.** The temperature in every part of the system must be same as otherwise heat may flow from one part of the system to another. We arrive at this condition as under:

Suppose a system consists of two phases A and B and their temperatures are T_A and T_B respectively. Further, suppose that at equilibrium, a small amount of heat, δq, flows from phase A to phase B. Then

$$\text{Entropy change of phase A } (dS_A) = -\frac{\delta q}{T_A}$$

$$\text{Entropy change of phase B } (dS_B) = +\frac{\delta q}{T_B}$$

$$\therefore \quad \text{Total entropy change of the system } (ds) = -\frac{\delta q}{T_A} + \frac{\delta q}{T_B}$$

The condition for equilibrium in terms of entropy change is

$$(dS)_{E,V} = 0$$

$$\therefore \quad -\frac{\delta q}{T_A} + \frac{\delta q}{T_B} = 0 \text{ or } \frac{\delta q}{T_A} + \frac{\delta q}{T_B} \text{ or } T_A = T_B$$

***(ii)* Mechanical Equilibrium.** The pressure in every part of the system must be same as otherwise matter may flow one part of the system to another. We arrive at this condition as under:

Consider an isolated system consisting of two phases in equilibrium at constant volume and temperature. Suppose at equilibrium, a small amount of matter flows from phase *A* to phase *B* and as a result, the volume of phase *A* decreases by a small amount dV while that of phase *B* increases by the same amount dV. Let the pressure of the two phases now be P_A and P_B respectively. Then

Change in Helmoltz free energy of phase *A* $(dA_A) = -P_A dV$

Change in Helmholtz free energy of phase *B* $(dA_B) = +P_B dV$

But in terms of Helmholtz free energy change, the condition for equilibrium is $(dA)_{V,T} = 0$

$$\therefore \quad dA = dA_A + dA_B = -P_A dV + P_B dV = 0 \text{ or } P_A = P_B$$

***(iii)* Chemical Equilibrium.** At equilibrium, the chemical potential of any component in all the phases must be same. This is called **thermodynamic criteria for phase equilibrium**.

For a solution (or a phase) containing a number of constituents, the ***chemical potential*** *of any component is the increase or decrease of free energy that takes place, at constant temperature and pressure, for the addition or removal of one mole of that component, keeping the amount of all other components constant. It is usually represented by 'μ'*

As a system consisting of a number of phases in equilibrium constitutes a closed system, the total free energy of the whole system is constant at constant temperature and pressure. For any change in the composition of any component among the various phases, there is no change in the free energy of the whole system as long as temperature and pressure remain constant, i.e., $(\Delta G)_{T,P} = 0$.

Consider a simple case consisting of only two phase *A* and *B*. Suppose a number of components $(C_1, C_2 ... C_c)$ are distributed between them. Further, suppose that at constant temperature and pressure, a small amount δn_1 of component 1 is transferred from phase *A* to phase *B*. If $(\mu_1)_A$ and $(\mu_1)_B$ represent the chemical potentials of component 1 in phases *A* and *B* respectively, we have

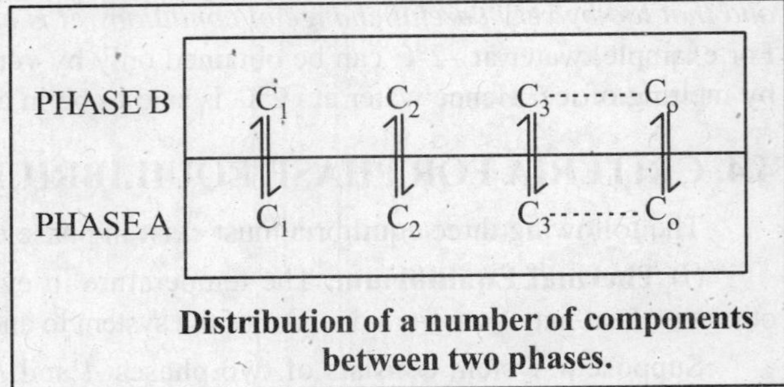

Distribution of a number of components between two phases.

Decrease in the free energy of component 1 in phase $A = (\mu_1)_A \, \delta n_1$

Increase in the free energy of component 1 in phase $B = (\mu_1)_B \, \delta n_1$

As the system is a closed one, $(\Delta G)_{T,P} = 0$

$$-(\mu_1)_A \, \delta n_1 + (\mu_1)_B \delta n_1 = 0 \qquad \text{(minus sign to represent the decrease)}$$

or $$(\mu_1)_A = (\mu_1)_B$$

Thus, for a system consisting of two phases in equilibrium, the chemical potential of any given component is same in both the phases.

In general, for a system consisting of a number of phases P_1, P_2, P_3 etc. in equilibrium (*i.e.*, a multiphase equilibria), for any component *i*,

$$(\mu_i)_{P_1} = (\mu_i)_{P_2} = (\mu_i)_{P_3} = ...$$

To conclude

For a multiphase equilibria containing a number of components distributed among them, the chemical potential of any component is the same in all the phases.

4.4.1. Criteria for two-phase equilibria for one-component system

If a system consists of one substance only which may exist in two different phases in equilibrium, e.g., liquid water $\rightleftharpoons$ water vapour or ice $\rightleftharpoons$ liquid water or sulphur (monoclinic) $\rightleftharpoons$ sulphur (rhombic)etc., the criterion is obtained in a simple manner in terms of *free energy* instead of chemical potential. Evidently, in such a case, if a certain amount of the substance is transferred from one phase to the other, the molar free energy of one phase decreases while that of the other phase increases by an equal amount. Hence the net result is that there is no change in free energy, *i.e.*,

$$\Delta G = 0$$

This implies that if at equilibrium, G_1 is the molar free energy of phase I and G_2 that of the phase II, we have

$$G_2 - G_1 = 0$$

$$G_1 = G_2$$

4.5. THERMODYNAMIC DERIVATION OF PHASE RULE

Let us consider a heterogeneous system consisting of C components (C_1, C_2, C_3 ...... C_c) distributed among P phases (P_1, P_2, P_3 ..... P_p) as shown in Fig.4.1. Degree of freedom of a system is equal to the number of independent variables which must be fixed arbitrarily to define the system completely. Clearly, the number of these variables is equal to the total number of variables of the system minus the number of variables which are defined by the system itself because of its being in equilibrium. Thus

Degrees of freedom = Total number of variables
– Number of variables defined by the system because of its being in equilibrium

The total number of variables can be calculated as follows:

Suppose a phase contains two components. If the molar concentration of one component is known, that of the other can be obtained by difference. Thus the phase containing two components

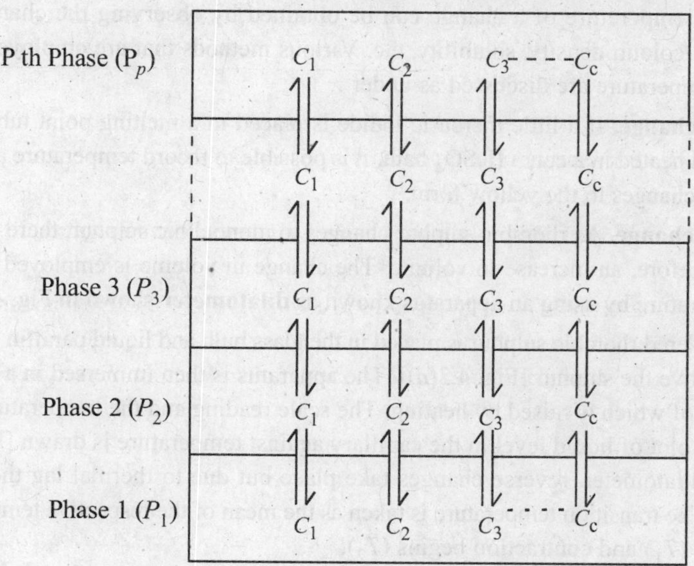

Fig. 4.1. Distribution of C components among P phases.

can be defined completely by one concentration term. Similarly, if a phase contains three components, two concentration terms should be known to define it (because then the concentration of the third can be found by difference). In general, if a phase contains C components it can be defined completely by $(C-1)$ concentration terms or variables. To define the system completely, *i.e.*, all the P phases, the total number of concentration terms or variables required will be $P(C-1)$. Besides these concentration variables, the temperature and pressure of the system, which are the same for all the phases, are also variables. Therefore

$$\text{Total number of variables} = P(C-1)+2.$$

The number of variables defined by the system itself, because of its being in equilibrium, can be calculated as follows:

For a system in equilibrium the *chemical potential* of a component (μ) is the same in all the P phases. Thus if a system consists of three phases, say A, B and C, then for any components i

$$(\mu_i)_A = (\mu_i)_B = (\mu_i)_C$$

These are in fact two equations *viz.*,

$$(i)\ (\mu_i)_A = (\mu_i)_B \text{ and } (ii)\ (\mu_i)_B = (\mu_i)_C$$

Thus for a system of three phases, two equations are known for each component, similarly for a system of four phases, three equations are known for each component. In general, for a system of P phases, $(P-1)$ equations are known for each component. Hence for a system of P phases and C components, the number of equations known will be $C(P-1)$. This is clearly the number of variables defined by the system itself. Thus, we have

Number of variables defined by the system itself = $C(P-1)$

Hence *the degree of freedom* F of the system is given by:

$$F = [P(C-1)+2]-[C(P-1)]$$
$$= C - P + 2$$

4.6. DIFFERENT TECHNIQUES FOR THE DETERMINATION OF TRANSITION POINT

Transition temperature of a change can be obtained by observing the change in physical property such as colour, density, solubility, etc. Various methods that are employed to determine the transition temperature are discussed as under :

1. Colour change. If a little mercuric iodide is placed in a melting point tube attached to a thermometer and heated in a cone. H_2SO_4 bath, it is possible to record temperature at which the red mercuric iodide changes to the yellow form.

2. Density change. As rhombic sulphur changes to monoclinic sulphur, there is a decrease in density and, therefore, an increase in volume. The change in volume is employed to measure the transition temperature by using an apparatus known as **dilatometer** shown in Fig. 4.2.

Some powdered rhombic sulphur is placed in the glass bulb and liquid paraffin (an inert liquid) is introduced above the sulphur [Fig. 4.2(*a*)]. The apparatus is then immersed in a hot water-bath, the temperature of which is raised by heating. The scale reading and the temperature are recorded every minute. A plot of liquid level in the capillary against temperature is drawn [Fig. 4.2(*b*)]. On cooling of the dilatometer, reverse changes take place but due to thermal lag the curve takes a different route. The transition temperature is taken as the mean of the respective temperatures where expansion starts (T_1) and contraction begins (T_2).

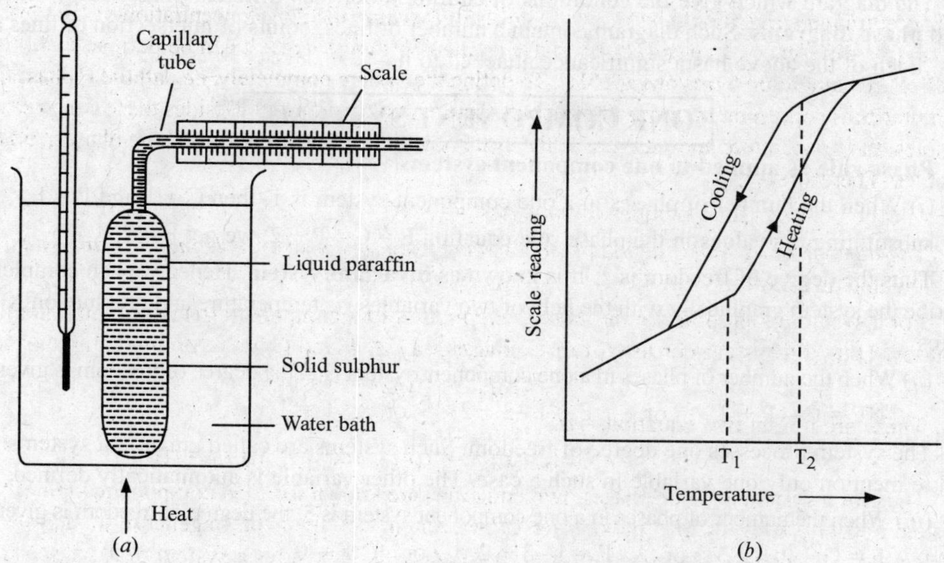

Fig. 4.2. (*a*) A dilatometer; (*b*) Plot of scale reading (volume) against temperature recorded with a dilatometer.

3. Solubility change. Two forms of the same substance may possess different solubilities but at the transition point they have identical solubility. Thus, if solubility-temperature graph is plotted for the two forms, it is found to consist of two parts with a sharp break. While one part represents the solubility curve for one form, the second part represents that for the other. At the meeting point of the two curves, the solubility of the two forms is the same. This is how the transition temperature is located.

For example, for the system sodium sulphate–water, the solubility curves of Na_2SO_4 (rhombic) and $Na_2SO_4.10H_2O$ meet at 32.2°C. Thus, 32.2°C is the transition temperature where $Na_2SO_4.10H_2O$ changes to Na_2SO_4.

(4) Cooling curve method. There is often an evolution or absorption of heat when one form changes into the other. Suppose form *A* is converted into form *B* on heating. Now let *B* be allowed to cool and a curve obtained by plotting the temperature against the time. **The otherwise steady curve has a distinct break at a temperature which corresponds to the transition point.**

This method is found to be suitable for determination of transition point for different hydrates of a salt or for a hydrate and anhydrous salt. The method can also be employed for different allotropes of a metal.

4.7. PHASE DIAGRAM

The number of phases that exist together when the system is in equilibrium are studied under the following variable conditions:

(*i*) Temperature and pressure

(*ii*) Temperature and composition (pressure being constant)

The variables temperature and pressure (or composition) governing the existence of various phases are obtained from the experiment. Then the temperature-pressure or temperature-composition diagrams are constructed for the different values of the variables.

The diagram which give the conditions of equilibria between different phases of system are called **phase diagrams**. Such diagram contain a number of lines, points of intersection of lines and areas. Each of the above has a significance attached to it.

ONE COMPONENT SYSTEMS

Phase rule as applied to one component systems.

(*i*) When the number of phases in a one component system is 1, then C = 1 and P = 1 substituting the values in the phase rule equation F = C – P + 2, we get F = 2.

Thus the degree of freedom is 2. It is known as **bivariant system**. Hence we can completely describe the system graphically with the help of two variables *viz* temperature and pressure on X and Y axes.

(*ii*) When the number of phases in a one component system is 2, the degree of freedom is given by

$$F = C - P + 2 \quad \text{or} \quad F = 1 - 2 + 2 \quad \text{or} \quad F = 1$$

The system processes one degree of freedom. Such systems are called univariant systems. We need to mention only one variable in such a case. The other variable is automatically defined.

(*iii*) When the number of phases in a one component system is 3, the degree of freedom is given by

$$F = C - P + 2 \quad \text{or} \quad F = 1 - 3 + 2 \quad \text{or} \quad F = 0$$

Such a system having no degree of freedom is called **non-variant** system. Univariant system are represented by lines, bivariant systems are represented by areas and non-variant system are represented by points in a one-component system.

4.8. PHASE DIAGRAM OF WATER SYSTEM

The system consists of three phases.

$$\underset{(s)}{\text{Ice}} \rightleftharpoons \underset{(l)}{\text{Water}} + \underset{(g)}{\text{Vapour}}$$

These three single phases may occur in four possible combinations in equilibrium as

(*i*) Liquid $\rightleftharpoons$ Vapour

(*ii*) Solid $\rightleftharpoons$ Vapour

(*iii*) Solid $\rightleftharpoons$ Liquid,

and (*iv*) Solid $\rightleftharpoons$ Liquid $\rightleftharpoons$ Vapour.

The number of phases that can exist in equilibrium at any point depends upon the conditions of temperature and pressure. These conditions have been determined experimentally and a phase diagram is constructed by plotting pressure against temperature (Fig.4.3). The phase diagram consists of :

(*i*) *Three stable curves OA, OB, OC and one metastable curve OA'*

(*ii*) *Three areas*

and (*iii*) *Definite point, O.*

These are separately discussed as under:

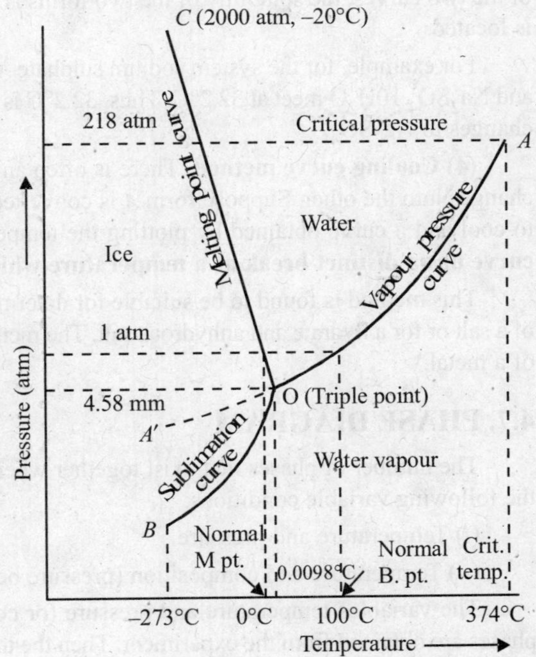

Fig. 4.3. The water system (pressure and temperature indicated are not to scale).

I. Curves: (*i*) ***OA.*** this is the *vapour pressure* curve of water. It represents the equilibrium between liquid water and vapour at different temperatures. At any point on the curve, the two phases are present. The curve starts from the point *O,* the freezing point of water and ends at *A,* the critical temperature (374°C) beyond which the two phases merge into each other. Consider any point on the curve. For a fixed value of pressure, temperature has also a fixed value. If by keeping the pressure constant, an attempt is made to raise the temperature then the liquid phase will disappear and if the temperature is lowered then the vapour phase will disappear. Thus, to define the system completely at any time, it is necessary to mention either temperature or pressure. In other words, pressure being fixed, it is not necessary to mention temperature; it gets automatically fixed. Therefore water-vapour system is univariant, or it has one degree of freedom. It also follows from the phase rule equation,

$$F = C - P + 2$$

$$F = 1 - 2 + 2 = 1$$

(*ii*) ***OA′.*** The dotted curve *OA′,* a continuation of curve *OA* represents the *vapour pressure curve of super-cooled water.* It is a metastable curve. On being disturbed, the supercooled phase at once changes to solid ice and curve merges into *OB.*

(*iii*) ***OB.*** It is the *sublimation curve* or the vapour pressure curve of ice. It gives values of temperature and pressure at which ice and vapours can exist together. It is obtained by studying the effect of pressure on the freezing point of water. The curve starts from the point *O,* the freezing point of water and ends at *B* (–273°C). As temperature decreases, the vapour pressure of ice tends to become negligible. We find that for a particular temperature, there can be one and only one pressure value and *vice versa.* In other words, ice-vapour system is univariant, or has one degree of freedom.

(*iv*) ***OC.*** It is *melting point curve* and represents the equilibrium between ice and liquid water at various pressures. In other words, it shows the effect of pressure on the melting point of ice. It should be noted that the line *OC* is inclined towards the pressure axis. The slope of the curve *OC* clearly indicates that melting point of ice is lowered by increase of pressure. There is, however, no limit for the curve *OC.* It goes up to a point corresponding to 2000 atmospheres and –20°C. At this point, one-type of ice changes into another solid modification but the solid-liquid equilibrium still remains.

At any point on the curve *OC,* two phases (ice and liquid water) are present in equilibrium. Hence the system is univariant, or degree of freedom is one.

II. Areas: The areas, *i.e.*, the regions bounded by two lines give the conditions of temperature and pressure under which the single phase *i.e.*, ice or liquid water or vapour is capable of stable existence. Area below *AOB* has water vapour, area *BOC* has ice and area *COA* has liquid water. These areas are bivariant because to locate any point in the area, temperature as well as pressure must be fixed, thus indicating two degrees of freedom. This also follows from phase rule equation.,

$$F = C - P + 2$$

$$F = 1 - 1 + 2 = 2$$

III. Triple point: It is a point where the three curves *OA, OB, OC* meet. At this point the three phases – ice, water and vapour are in equilibrium. It should be noted that it is only a point indicating that there is only one set of conditions under which it is possible for the three phases to exist together. The point *O* is a self-defined point corresponding to definite temperature of 0.0098°C and a definite pressure of 4.58 mm. If either pressure or temperature is changed, then one of the three phases will disappear. Since on changing either pressure or temperature one of the phases disappear, therefore, the system at the point *O* has no degree of freedom, that is, it is invariant. The same result follows from phase rule equation.

$$F = C - P + 2$$

$$\therefore \quad F = 1 - 3 + 2 = 0$$

It may be mentioned that triple point is different from normal melting point of ice i.e. 0°C. At the normal melting point of ice at 0°C and 1 atmospheric pressure, only ice and water (no vapours) exist together.

Example. Explain why the fusion curve of ice has a negative slope whereas the sublimation curve has a positive slope in the phase diagram of water.

Solution. This is because the freezing point of water is depressed by the increase of pressure. Therefore, the fusion curve is inclined towards the pressure axis *i.e.*, it has a negative slope.

But the sublimation curve has a positive slope *i.e.*, it is inclined away from the pressure axis. This is because the vapour pressure of water increases with the rise of temperature.

4.9. PHASE DIAGRAM OF CARBON DIOXIDE SYSTEM

Phase diagram of carbon dioxide system resembles that of water system with some differences.

The phase diagram for carbon dioxide system is shown in Fig. 4.4. It has three distinct areas in which carbon dioxide can exist either as *solid, liquid* or *gas.*

Areas

Area on the R.H.S. of the curve ABD consists of vapour phase. Area above the curve CBD consists of the liquid phase while the area on the L.H.S. of the curve ABC consists of solid phase.

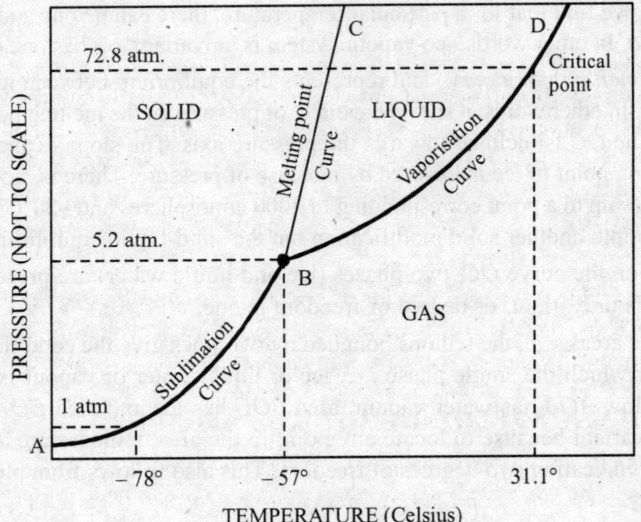

Fig. 4.4. The phase diagram for the carbon dioxide system.

Curves

AB is the *sublimation curve* along which solid carbon dioxide is in equilibrium with the gas, *BD* is the *vaporisation curve* along which liquid carbon dioxide is in equilibrium with the gas. This curve ends at 31.1°C, the critical temperature of CO_2 and corresponding to a pressure of 72.8 atmospheres. The curve *BC* is the *fusion curve* along which solid and liquid carbon dioxide are in equilibrium with each other.

Point

B is the *triple point* at which all the three phases of carbon dioxide co-exist in equilibrium with one another. The temperature of the system at this point is –57°C while the pressure is 5.2 atm. A slight variation in temperature or pressure at this point may result in the disappearance of one of the two phases. For example, a slight *increase* in temperature will result in the disappearance of the solid phase and the equilibrium will shift along the curve *BD* while a slight *decrease* in temperature will result in the disappearance of the liquid phase and the equilibrium will shift along the curve *BA*. Keeping the temperature constant, if the pressure is increased, the gaseous phase will disappear and the equilibrium will shift along the curve *BC*.

The phase diagram of carbon dioxide resembles that of water in showing three distinct areas for solid, liquid and gaseous phases. But it differs from the latter in several respects. In the first place, the melting point curve *slopes away from the pressure axis*. This indicates that increase of pressure raises the melting point of solid carbon dioxide.

The second difference is that solid carbon dioxide can exist in equilibrium with its liquid only at a very high pressure equal to 5·2 atm. This is unlike the water system in which ice and water can exist in equilibrium even at a very low pressure equal to 4·58 mm of Hg (Refer to phase diagram of water).

The third difference is that the vapour pressure of solid carbon dioxide even at extremely low temperature is very high and many times higher than that of ice.

A significant feature of carbon dioxide system is that even at atmospheric pressure, carbon dioxide gas can be directly solidified without the appearance of liquid phase merely on cooling to – 78°C (Fig. 4.4). It is for this reason that solid carbon dioxide is commonly known as **dry ice.**

Polymorphism

The existence of a given substance in more than one crystalline form with different physical properties is known as **polymorphism.** This phenomenon occurs in a number of elements and compounds. When it occurs in elements, it is referred to as **allotropy.**

Each polymorphic form constitutes a separate phase. The temperature at which one form changes into another, at a given pressure, is known as the **transition temperature.** For example, rhombic sulphur when heated under a pressure of one atm. changes into monoclinic sulphur at 95.6°C.

Similarly, monoclinic sulphur on cooling at a pressure of one atmosphere changes into rhombic sulphur at 95.6°C. Thus, 95.6°C is the transition temperature at which the two forms of sulphur (rhombic and monoclinic) change reversibly into each other.

4.10. PHASE DIAGRAM OF SULPHUR SYSTEM

It is a one component system comprising **four phases,** *viz.,* (*i*) *Rhombic sulphur,* (*ii*) *Monoclinic sulphur,* (*iii*) *Sulphur liquid* and (*iv*) *Sulphur vapour.* This a typical example of one component system exhibiting solid-solid transformation.

Sulphur exists in two well known crystalline forms, namely rhombic sulphur and monoclinic sulphur. The two forms are in equilibrium with each other at 95.5°C, which is called its **transition temperature.** It is defined as the *"temperature at which one stable form of a solid is transformed into another stable solid form of it, without any change in temperature, pressure on the system being one atmosphere."* Rhombic sulphur is stable below 95.5°C, whereas monoclinic is stable above this temperature.

According to phase rule equation, all the four phases cannot co-exist in equilibrium. This is because on substituting $P = 4$ and $C = 1$ in the phase rule equation, we get

$$F = C - P + 2$$
$$= 1 - 4 + 2 = -1$$

Thus, we get $F = -1$, which is meaningless. Hence for a system consisting of single component, it is not possible to have four phases together in equilibrium. However, three or fewer number of phases can exist together.

The phase diagram of the sulphur system as obtained experimentally by plotting pressure vs. temperature is shown in Fig. 4.5. The diagram consists of *six stable curves BO, OA, AE, OC, AC* and *CF* and *four metastable curves OD, OG, AD* and *CD*. The curves divide the diagram into *four areas, i.e.,* to the left of *BOCF,* below *BOAE,* to the right of *FCAE* and the one enclosed by *OCA*.

I. Areas : The areas are bivariant, that means to locate any point in the areas, both temperature and pressure are to be specified. Each area represents the conditions for the stable existence of a single phase. In areas thus there are two degrees of freedom, which also follows from the phase rule equation $F = C - P + 2$; putting the values $C = P = 1$, we get $F = 1 - 1 + 2 = 2$.

In the area to the left of *BOCF only* S_R (rhombic sulphur) exists while in the area below *BOAE,* only S_V (sulphur vapour) exists; to the right of *FCAE,* S_L (liquid sulphur) is present and in the region enclosed by *OCA,* S_M, (monoclinic sulphur) exists.

II. Curves: *(i)* ***OB***. It is *sublimation curve of rhombic sulphur.* The phases present in equilibrium along this curve are S_R and S_V. The curve starts from the point *O,* the transition temperature of S_R to S_M, and ends at *B* (50°C) below which vapour-pressure of S_R cannot be measured. If heating is done rapidly, then it is possible to heat rhombic sulphur above the transition point along *OD. OD is the metastable curve.* On slight disturbance it merges into curve *OA,* which is the vapour pressure curve of monoclinic sulphur.

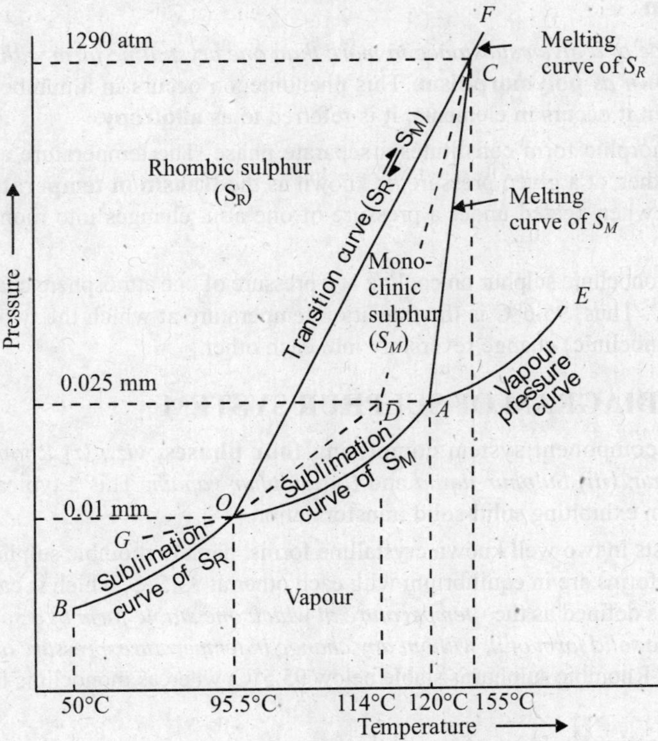

Fig. 4.5. Phase diagram sulphur system.

(ii) ***OA***. It is the *sublimation curve of monoclinic sulphur.* The two phases present in equilibrium along this curve are S_M, and S_V. The curve starts from the point *A,* melting point of S_M (120°) and ends at *O,* the transition temperature of S_R (95.5°C).

Except at the point O, where three phases S_R, S_M and S_V are in equilibrium, at any other point all along the curve only two phases S_V and S_M are in equilibrium. For a fixed value of temperature, pressure is automatically fixed. The degree of freedom, therefore, is one.

If the temperature of monoclinic sulphur is lowered, then it would change into rhombic sulphur at 95.5°C, but it is possible to prevent this conversion by careful cooling. Then we will move along the curve OG which is a *metastable vapour pressure curve of monoclinic sulphur.*

(iii) ***AE.*** *It is the vapour pressure curve of liquid sulphur.* The two phases present in equilibrium along this curve are liquid sulphur, S_L, and vapour sulphur, S_V. The curve starts at A, the melting point of S_M, and ends at E, the critical point beyond which out of the two phases S_L and S_V only S_V (vapour sulphur phase) can exist. For any point on the curve, for a fixed value of pressure, the temperature has a fixed value. Therefore the system also has one degree of freedom. The extended curve AD represents the metastable vaporisation curve of liquid sulphur.

(iv) ***OC.*** *It is the transition curve* and represents the equilibrium between rhombic and monoclinic sulphur and is obtained from the effect of pressure on the transition temperature of rhombic sulphur. The slope of the curve OC is away from the vertical pressure axis, indicating that the transition point is raised by the rise of pressure. The curve ends at the point C, after which monoclinic sulphur phase disappears.

(v) ***AC.*** *It is the melting point curve of monoclinic sulphur.* It represents the equilibrium between monoclinic sulphur and liquid sulphur and results from the effect of pressure on the melting point of monoclinic sulphur. The curve is inclined away from the pressure axis showing that the melting point is rises with the rise of pressure. The curve ends at C beyond which monoclinic sulphur does not exist in a stable condition. The system has degree of freedom equal to one.

(vi) ***CD.*** this curve represents the equilibrium between rhombic and liquid sulphur and is the *metastable melting point curve of rhombic sulphur.* The curves OC and AC meet at C, at 155°C and 1290 atmospheres. It follows therefore that if liquid sulphur were cooled at a pressure higher than 1290 atmospheres, then rhombic sulphur would crystallise at once, no monoclinic sulphur being formed. The curve passes through the point C and beyond this point it becomes a stable melting point curve CF of rhombic sulphur.

III. Triple points: There are three stable triple points *viz.,* O, A and C in the phase diagram. At the point O, three phases in equilibrium are sulphur rhombic, monoclinic and vapour. If one of the variables is changed, one of the phases disappears. Similarly A is also a triple point between monoclinic, vapour and liquid sulphur. This point is again invariant. Further, lines OC and AC intersect at C to yield a three phase equilibrium between rhombic, monoclinic and liquid sulphur. Since on changing one of the variables, either pressure or temperature, one of the phases disappears, therefore, the system at the triple point has no degree of freedom. We get the same result, from phase rule equation:

$$F = C - P + 2$$

$$F = 1 - 3 + 2 = 0$$

D is a metastable triple point; the phases in equilibrium being S_R, S_L and S_V.

TWO-COMPONENT SYSTEMS

4.11. APPLYING PHASE RULE TO TWO-COMPONENT SYSTEMS

A two-component system can be defined as one in which it is possible to express the composition of all the phases in terms of two substances. A two-component system may have as many as three degrees of freedom. This is because when $P = 1$, then

$$F = C - P + 2 = 2 - 1 + 2 = 3.$$

Thus the system will have, besides temperature and pressure variables, *one concentration variable for each phase present at the time of equilibrium.* A three dimensional model is required in which pressure, temperature and concentration would be plotted along the three axes at right angles to one another (Fig. 4.6). However, in actual practice, a simple plain diagram with two variables only is considered, the third variable is considered as constant. In this way, use is made of the following three types of diagrams (Fig. 4.7).

(*a*) Pressure-temperature diagrams

(*b*) Composition-pressure diagrams

(*c*) Temperature-composition diagram

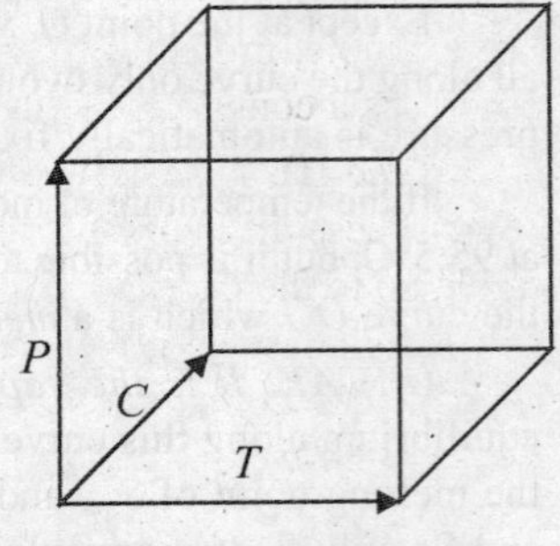

Fig. 4.6. Space model for Two-component System.

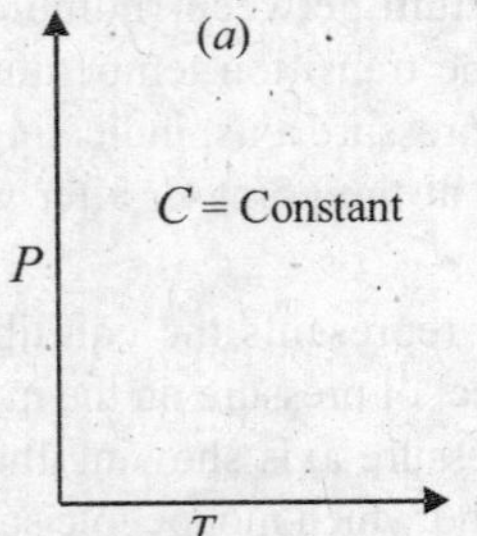

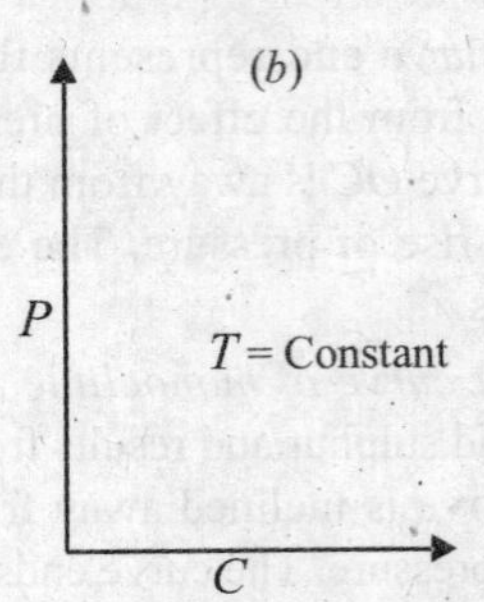

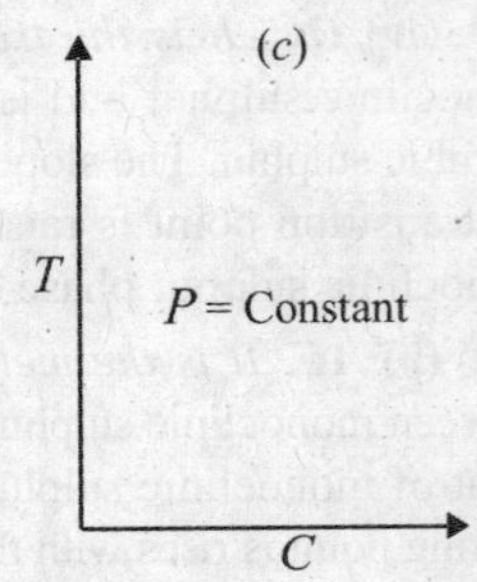

Fig. 4.7. Two dimensional diagrams.

However, it is usually convenient to keep pressure constant at the atmospheric value and thus a $T - C$ diagram is often used. A diagram drawn for a constant value of temperature is called *isothermal,* while one obtained by keeping composition constant, is called *isoplethal* diagram.

4.11.1. Reduced phase rule equation

In case of systems consisting of only solid and liquid phases, small changes of pressures have very little effect on the systems. Such systems are called *condensed systems*. As the degree of freedom in such a case is reduced by one as pressure is taken as fixed at one atmosphere pressure, therefore, we can write a *'reduced' phase rule equation* as

$$F' = C - P + 1$$

F' in the above equation represents the number of degrees of freedom which the system can possess in addition to the pressure.

4.12. TYPES OF TWO-COMPONENT SYSTEMS INVOLVING SOLID-LIQUID EQUILIBRIA

Broadly, there are three types of two-component system involving solid-liquid equilibria:

Type I. Those in which the two components do not react with each other but simply mix into each other in the molten state or in the solution. Examples of such systems are lead-silver system and $KI - H_2O$ system.

Type II. *Those in which the components react to form a compound with congruent melting point.* These are those systems in which the two pure components react to form a compound which is stable up to its melting point. At the melting point, it melts to give a liquid of the same composition as that of the solid compound. The compound formed is then said to have a *congruent melting point.*

For example, in $FeC1_3 - H_2O$ system, each hydrate formed is stable up to the melting point and possesses a congruent melting point.

Type III. *Those in which the components react to form a compound with incongruent melting point.* These are those two-component systems in which the compound formed does not remain stable up to the melting point. On heating, it starts decomposing before the melting point to give a new solid phase and a solution with composition different from that of the solid phase. When this happens, the compound is said to undergo a *transition* or *peritectic reaction* and is said to have *incongruent melting point.* In general, a peritectic reaction may be represented by the equation

$$C_2 \rightleftharpoons C_1 + \text{solution (or melt)},$$

C_2 represents is the compound formed as a result of reaction between the two pure components and C_1 is the new compound (or a pure constituent) formed as a result of decomposition of compound C_2 below its melting point. An example of this type of system is $Na_2SO_4 - H_2O$ system in which the hydrates formed undergo peritectic reaction.

4.12.1. General features of phase diagram of two components system (solids) miscible in liquid state

A number of mixtures of two components *A* and *B* of different composition are prepared and their freezing points determined. We carry out the process in an open vessel in each case so that the pressure remains equal to the atmospheric pressure. These freezing points are then plotted against the compositions of the liquid mixtures. A phase diagram of the type shown in Fig. 4.8 is obtained.

The diagram consists of the curves *AC, BC,* the areas enclosed by *ADC, BCE,* areas above *ACB,* below *DCE* and the point *C*. Each of these is described below.

I. Curves *AC* and *BC*. The point *A* and *B* represent the freezing points of pure *A* and pure *B* respectively. The addition of *B* to *A* lowers the freezing point along the curve *AC*. Similarly the addition of *A* to *B* lowers the freezing point along the curve *BC*. Thus, *AC* represents the freezing point curve of *A* to which different amounts of *B* are being added. Similarly, *BC* represents the freezing point curve of *B* to which different amounts of *A* are being added. It is evident that if liquid mixture rich in *A* (*i.e.*, for any composition between *B* and *C*) is cooled, solid *B* will separate out. Thus the curves *AC* and *BC* represent the conditions of temperature at which the liquid mixture is in equilibrium with solid *A* and solid *B* respectively.

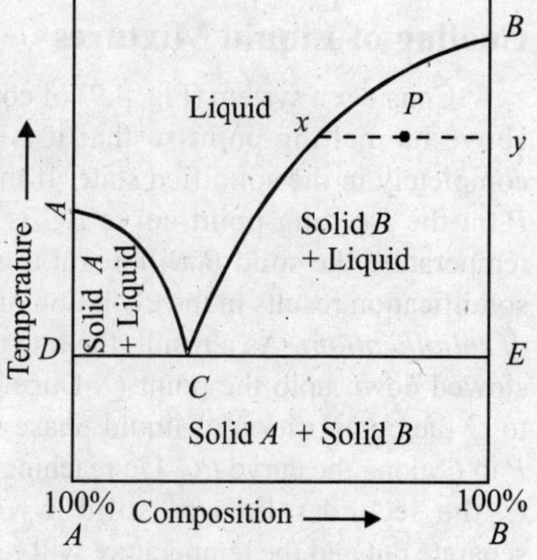

Fig. 4.8. General shape of the phase diagram of two-component system miscible in the liquid phase.

Ignoring the vapour phase, all along the curve *AC* or *BC* there are two phases in equilibrium, (solid *A* + liquid along the curve *AC* and solid *B* + liquid along the curve *BC*), *i.e.*, $P = 2$.

As $C = 2 \therefore \quad F = C - P + 1$ (reduced equation)

$$= 2 - 2 + 1 = 1$$

Thus the condensed phase has one degree of freedom. This is otherwise also evident because either the temperature or the composition of the liquid phase alone is sufficient to define the system completely.

II. Point *C*. *C* is the meeting point for curves *AC* and *BC*. At this point, three phases are in equilibrium *viz.*, solid *A*, solid *B* and the liquid mixture, *i.e.*, $P = 3$.

Hence $F = C - P - 1 = 2 - 3 + 1 = 0$,

i.e., condensed system becomes invariant at the point C. Thus, there is only one temperature, at the atmospheric pressure, where the liquid phase can be in equilibrium with both the solids. The point *C* corresponds to a temperature below which any liquid mixture of *A* and *B* will freeze as a whole. The point *C* is, therefore, called **eutectic point.** The composition corresponding to *C* is called *eutectic composition.*

III. Areas: *Area above the curve ACB.* In this area only liquid phase is present as the temperature is above the melting point of any mixture, *i.e.*, $P = 1$. Hence

$$F' = C - P + 1 \qquad \text{(reduced phase rule equation)}$$
$$= 2 - 1 + 1 = 2$$

Thus, there are two variables, *i.e.*, the areas are bivariant. This means, to define the system completely, both temperature and composition need to be specified. Evidently to define any point within the area, both the co-ordinates are required.

Area below DCE. As *DCE* is passing through the eutectic point *C*, so in the area below *DCE*, only the two solid phases *viz.*, pure *A* and pure *B* are present, *i.e.*, $P = 2$. Hence $F = C - P + 1 = 2 - 2 + 1 = 1$, *i.e.*, the condensed system is univariant. This is so because the composition of each phase is fixed (as each of them is a pure component) *i.e.*, composition is not variable. The only variable is the temperature.

Areas enclosed by ADC and BEC. Each of the these areas has two phases, being solid *A* + liquid in the area *ADC* and solid *B* + liquid in the area *BEC.* Thus for each of these areas, degree of freedom is one.

$$F' = C - P + 1 = 2 - 2 + 1 = 1$$

i.e., the condensed system is univariant.

The composition of the phases in equilibrium at any point (say *P*) within these areas can be obtained by drawing a horizontal line passing through that particular point to find where it cuts *AD* or *BE,* to know the nature of the solid and *AC* or *BC,* to know the composition of the liquid. Any horizontal line like *xy* through the point *P* in these areas connects the compositions of the solid and the liquid phases in equilibrium, and is called a *tie line.*

Cooling of Liquid Mixtures

Consider a system (Fig. 4.9) of composition represented by the line *LS, L* represents the system above its melting point so that it is the liquid phase, whereas *S* represents the same system completely in the solidified state. If the liquid *L* is cooled, no solid will separate out till the point *P* on the freezing point curve *BC* is reached. At this temperature the solid *B* will begin to separate out. The solidification results in the evolution of heat (called *heat of solidification).* As a result, the rate of cooling will be slowed down upto the point *Q*. During cooling from *P* to *Q* the composition of liquid phase will change from *P* to *C* along the curve *PC.* On reaching the eutectic point *C*, the second solid *viz.*, solid *A* will also begin to separate out and the temperature will remain constant till all the liquid has been solidified. According to phase rule, there is only one solid phase in a condensed system. Thus eutectic is point indicated by complete break in the rate of cooling. The temperature will further fall say from *Q* to *S* only after the whole of the system has been solidified.

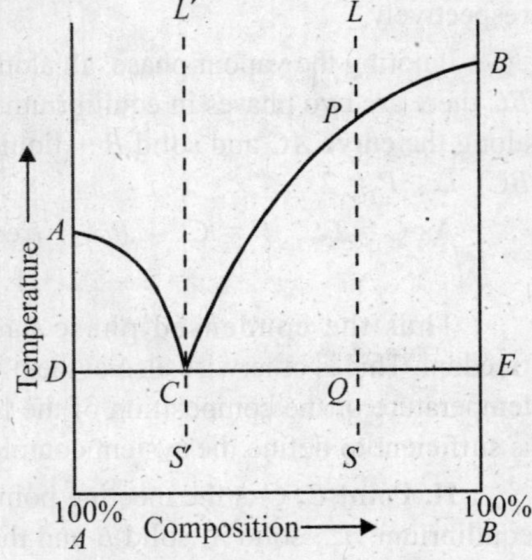

Fig. 4.9. Cooling of liquid mixtures.

Starting from a liquid mixture (molten mixture) corresponding to the temperature and composition represented by the point *L,* if cooling is allowed to take place and temperature is plotted versus time, a curve of the type shown in Fig. 4.10 is obtained.

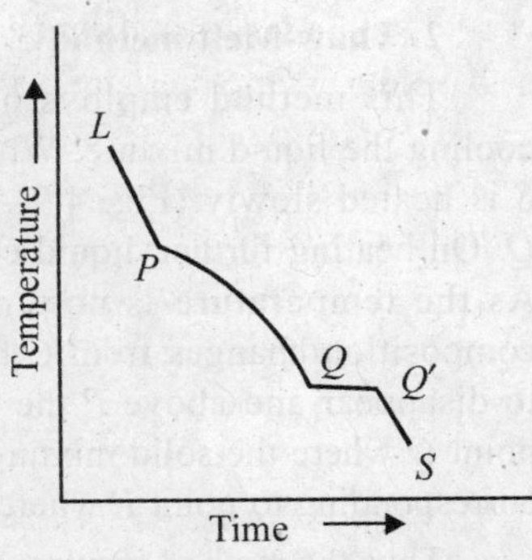

Fig. 4.10. Cooling curve.

From *L* to *P,* the liquid cools rapidly. At *P* solid begins to separate out, heat is evolved and the rate of cooling becomes slow. Solid *B* keeps on separating out and the composition of the *liquid phase keeps changing along the curve PQ.* After reaching the point *Q* which corresponds to the eutectic point, solid *A* also starts separating out and the temperature remains constant till whole of the liquid solidifies at the point *Q′*. Now further cooling results in the fall of temperature from *Q′* to *S*.

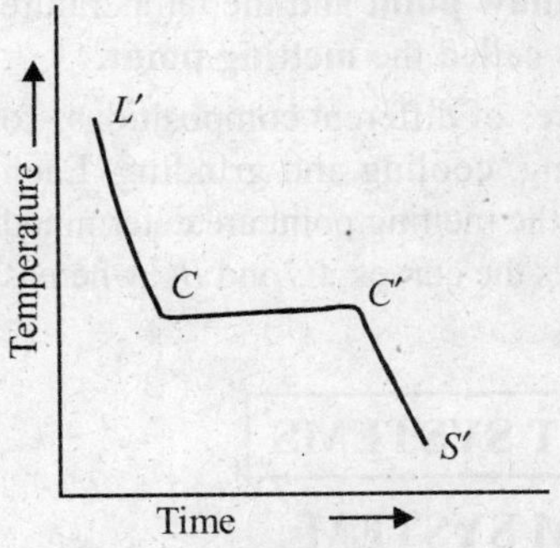

Fig. 4.11. Cooling curve of eutectic mixture.

If the composition of the liquid mixture is exactly same as that of the eutectic, as shown by point *L′* in Fig. 4.11, then on cooling, the solid will separate only when the eutectic point is reached at *C*. The system will now solidify completely. The cooling curve will show only one break as shown in Fig. 4.11. Solid separates at the point *C* and is complete at the point *C′*. Further cooling results in the fall of temperature of the solid along *C′S′*.

4.13. EXPERIMENTAL DETERMINATION OF PHASE DIAGRAMS OF TWO COMPONENT SYSTEMS

1. Cooling Curve method (Thermal analysis method)

This method is based on the principle illustrated in Fig. 4.9. A number of mixtures of the two components of different compositions are made. Each mixture is melted to form a homogeneous liquid and then allowed to cool. In this way, a cooling curve of the type shown in Fig. 4.10 is obtained for each of the mixtures. The point *P* corresponds to the freezing point and point *Q* corresponds to the eutectic point for the composition taken. In this way, the freezing point and the eutectic points for mixtures of different compositions are obtained. Freezing points are plotted versus composition to obtain the curves *AC* and *BC*, whereas plotting the eutectic temperatures versus composition, the line *DCE* is obtained. To complete the diagram the freezing points of pure *A* and pure *B* are also determined by the cooling curve method as a pure liquid solidifies at a constant temperature.

Limitations of the Cooling Curve Method

(*i*) In case of some liquid mixtures, the separation of the solid does not occur readily and supercooling takes place. Thus no break is observed at *P.* Instead, cooling continues beyond *P* along *PP′* and then the temperature rises suddenly and then follows the path *RQ*. The point *P* is then located by extrapolation of the curve *QR* as shown in Fig. 4.12.

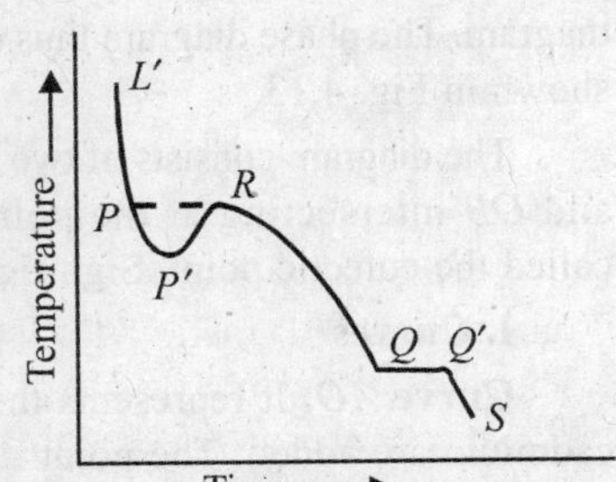

Fig. 4.12. Cooling curve with super cooling.

(*ii*) Even if there is no supercooling, the temperature of arrest at the eutectic point is sometimes difficult to observe correctly as the mixture contains a large amount of the solid and the stirring becomes difficult.

2. Thaw-Melt method

This method employs heating the solid mixture of a known composition instead of cooling the liquid mixture. When a mixture of solids *A* and *B* corresponding to the composition *S* is heated slowly (Fig. 4.9), no change in composition takes place upto the eutectic point *Q*. On heating further, liquid corresponding to the eutectic composition *C* begins to be formed. As the temperature is now raised from *Q* to *P*, the quantity of liquid increases and the composition changes from *C* to *P* (Refer to Fig. 4.9). At *P* the last trace of the solid is about to disappear and above *P* the system is completely liquid. The temperature corresponding to point *Q* where the solid mixture just *starts* melting is called the **thaw point** and the temperature corresponding to point *P* where the last trace of the solid melts is called the **melting point.**

Thus the method requires preparing a number of solid mixtures of different compositions. To ensure mixing of the solids, the mixtures are prepared by melting, cooling and grinding. Each mixture is packed into a wide capillary tube and the thaw point and the melting point are determined by usual method. The plot of melting point versus compositions gives the curves *AC* and *BC* whereas the plots of thaw point versus composition gives the line *DCE*.

STUDY OF SOME TWO-COMPONENT SYSTEMS

4.14. PHASE DIAGRAM OF BISMUTH-CADMIUM SYSTEM

This is an example of a two component system which forms a simple eutectic. The various phases that may be present are (*i*) solid bismuth (*ii*) solid cadmium (*iii*) solution of bismuth in cadmium or vice versa in the molten state (*iv*) vapour.

As small changes of pressure have negligible effect on this equilibrium we can apply reduced phase rule equation, $F' = C - P + 1$. When $F' = 0$, $P = 3$ (as $C = 2$). Thus the maximum number of phases (other than vapour) that can co-exist would be three.

Experimentaly a number of mixtures of the two metals, starting from 100% bismuth to 100% cadmium with intervals of 10% by mass are prepared in inert crucibles by heating them in an inert atmosphere in an electric furnace and then studying their cooling curves. For each mixture, the temperature at which solidification starts and the temperature at which solidification is complete are recorded by inserting a thermocouple in the melt. These are then plotted on the temperature-composition diagram. The phase diagram thus obtained is shown in Fig. 4.13.

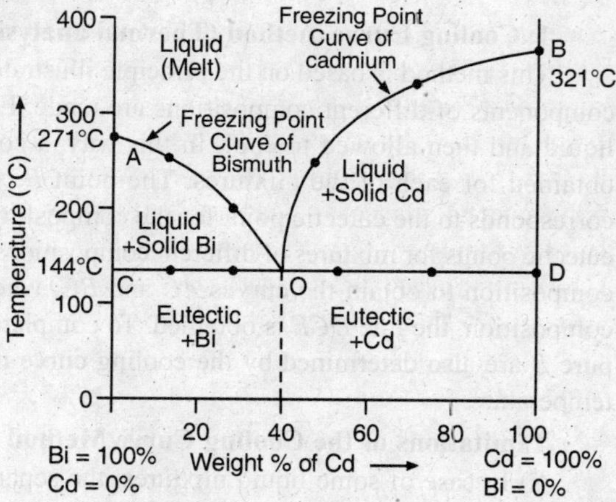

Fig. 4.13. Phase diagram of Bi-Cd system

The diagram consists of two curves *OA* and *OB* intersecting at the point *O*. *O* is called the eutectic point. Significance of the curves and point *O* is explained below:

I. Curves

Curve *AO*. It represents the *freezing point curve of bismuth* to which different amounts of cadmium are added. The point *A* represents the melting point or freezing point of pure bismuth (271°C). We find that the melting point (or freezing point) of bismuth is lowered gradually by addition of cadmium into it. The added cadmium goes into the solution and the separation of bismuth occurs till the point O is reached. At this point, no more cadmium goes into the solution as the solution becomes saturated with cadmium and hence the melting point of bismuth does not fall further. Thus the point O represents the composition with lowest melting point (144°C). Point O represents the eutectic point. Along the curve AO, two phases in equilibrium are solid bismuth and liquid melt (neglecting the vapour phase). Therefore the system is univariant as shown below:

$$F' = C - P + 1 = 2 - 2 + 1 = 1.$$

Curve *BO*. This is the *freezing point curve of cadmium* to which different amounts of bismuth are added. The point B represent the melting point or freezing point of pure cadmium (321°C). We find that melting point of cadmium decreases as more and more of bismuth is added into it. The added bismuth goes into the solution and the separation of solid cadmium takes place. This continues till the point O is reached. At this point no more bismuth goes into the solution and the solution is said to have become saturated with bismuth. Hence, the melting point does not fall any longer.

Along the curve, two phases, solid cadmium and liquid melt are in equilibrium. Therefore,

$$F' = C - P + 1 = 2 - 2 + 1 = 1 \text{ i.e. the system is univariant.}$$

It may be noted that the points on the curves AO and BO represent the intial freezing points while the points on the line CD represent the final freezing points. It may also be noted that the curve AO is not only freezing point curve of bismuth but may also be considered as *solubility curve of bismuth* because the points on this curve represent the solubilities of bismuth in the molten cadmium at various temperatures. Similarly, the curve BO may also be considered as the *solubility curve of cadmium.*

Eutectic point, *O*. This is the point where the two curves AO and BO meet. At this point, solid bismuth and solid cadmium are in equilibrium with the melt, *i.e.*, three phases are present. Degree of freedom at the point O is zero.

$$F' = C - P + 1 = 2 - 3 + 1 = 0$$

The point O represents the lowest possible temperature (144°C) below which a liquid phase cannot exist and above which the liquid phase cannot be enriched in either component by freezing out the other component. Such a liquid mixture of two components which has the lowest freezing point compared to all other liquid mixtures is called **eutectic mixture**. This mixture freezes out completely at constant temperature and on freezing gives the two components as solid phases in the same proportion in which they were present in the liquid phase. The temperature corresponding to the eutectic point O is called the **eutectic temperature**. This is always lower than the melting points of the either component. In this system, eutectic mixture (containing about 40% Cd and 60% Bi by mass) melts at 144°C, which is lower than the melting point of bismuth (271°C) and that of cadmium (321°C).

Note: *Eutectic* means melting readily (Greek word)

4.15. PHASE DIAGRAM OF LEAD-SILVER SYSTEM

The various phases possible are (*i*) solid lead (*ii*) solid silver (*iii*) solution of lead and silver in the molten state and (*iv*) vapour.

Since small changes of pressure have negligible effect on the equilibrium existing between lead and silver in the solid state or in the molten state, we can use the *reduced phase rule equation,* $F' = C - P + 1$. According to this equation, when $F' = 0$, P will be 3. Thus, the maximum number of phase (other than vapour) that can co-exist would be three.

The phase diagram of lead-silver system is shown in the Fig. 4.14. The diagram is drawn taking temperature and composition along the axes. The equilibrium diagram consists of two curves OA and OB, intersecting at point O.

Various curves and points are described as under:

I. Curves: (*i*) Curve *OA*. This is the *freezing point curve of silver* to which successive small quantities of lead are added. The point *A* indicates the melting point of pure silver (961°C). It is found that the melting point of silver is lowered gradually by the addition of lead into it. All along this curve the added lead goes into the solution and the separation of silver occurs when the point *O* is reached. At this point no more lead goes into the solution, as the solution becomes saturated with respect to lead and hence the melting point of silver does not fall any longer. Thus the point *O* represents the lowest possible temperature (303°C) in the system that corresponds to fixed composition, *viz.,* 2.4% Ag and 97.6% Pb. The point *O* is called the **Eutectic point.**

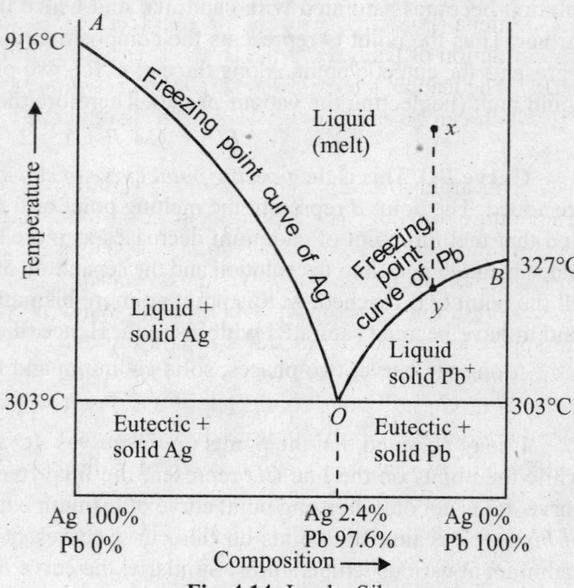

Fig. 4.14. Lead – Silver system.

All along the curve *AO,* two phases—solid silver and liquid (melt) are present in equilibrium neglecting the vapour phase. Therefore, the system is univariant as shown by the following equation.

$$F' = C - P + 1 = 2 - 2 + 1 = 1$$

(*ii*) Curve *BO*. This is the *freezing point curve of lead* to which successive small amounts of silver are added. The point *B* represents the melting point of pure lead (327°C). It is seen that the addition of silver to lead lowers the melting point of lead. All along this curve, the silver which is added goes into the solution while the separation of lead takes place. This state of affairs continues till the point *O* is reached. At this point, no more silver goes into the solution and the solution becomes saturated with respect to silver. Hence the melting point of lead does not fall any longer.

All along this curve, two phases — solid lead and liquid (melt) are present in equilibrium. Therefore, as before $F' = C - P + 1 = 2 - 2 + 1 = 1$, *i.e.*, the system is univariant.

II. Eutectic point, *O*. This is the point where two curves *AO* and *BO* meet. At this point solid lead and solid silver are in equilibrium with the fused mass, *i.e.,* three phases are present. Therefore, there are no degrees of freedom, as according to *reduced phase rule equation,*

$$F' = C - P + 1$$
$$= 2 - 3 + 1 = 0$$

The point *O* represents the lowest possible temperature (303°C) below which a liquid phase cannot exist and above which the liquid phase cannot be enriched in either component by freezing out the other component. *Such a liquid mixture of two components—Ag and Pb, which has the lowest freezing point compared to all other liquid mixtures is called* **eutectic mixture** (*eutectic* means melting easily). This mixture freezes out completely at a constant temperature and on freezing gives the two components as solid phases in the same proportion in which they are present in the liquid phase. The temperature corresponding to the eutectic point *O* is called the **eutectic temperature.** This is always lower than the melting points of the components constituting the eutectic mixture. For example, in the above system, we see that the alloy containing 2.6% silver and 97.4% lead melts at 303°C, a temperature lower than the melting point of Ag (961°C) and that of Pb (327°C).

4.15.1. Pattinson's process for desilverisation of lead

Galena (PbS, an ore of lead) is usually associated with silver. When galena is processed for the extraction of lead, some traces of silver pass into it, because silver is soluble in lead to some extent. The lead thus obtained is known as *argentiferous lead.* The desilverisation of lead is carried out which can be easily understood from the phase diagram of lead-silver system, shown in Fig. 4.14.

The argentiferous lead containing small percentage of silver is first of all heated to a temperature well above its melting point so that the system consists only of liquid phase represented by the point *x* in Fig. 4.14. It is then allowed to cool gradually along the line *xy* without any change in concentration till the point *y* is reached. When this point is reached, lead starts separating and solution will contain progressively increasing amounts of silver. Further cooling shifts the system along the line *y O*. Lead continues to separate out (which can be removed with the help of ladles/perforated spoons) and the melt continues to be richer and richer in silver, till the point *O* is reached where an alloy containing about 2.4% of silver is obtained.

The above principle is employed in Pattinson's process which is used for increasing the relative proportions of silver in the alloy. The relative increase in the amount of solid lead that separates out and in the proportion of silver in the melt on cooling is shown in Fig. 4.15.

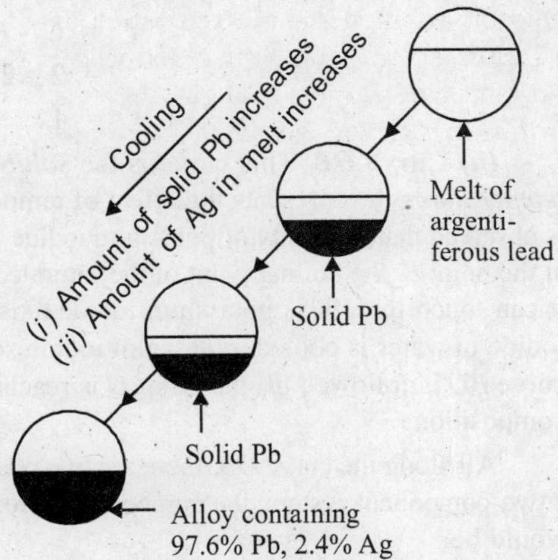

Fig. 4.15 Cooling the melt of argentiferous lead.

4.16. PHASE DIAGRAM OF KI–H_2O SYSTEM

The phase diagram of KI – H_2O system is shown in Fig. 4.16. Since KI does not form hydrates, the only phases that are possible are (*i*) solid KI, (*ii*) ice, (*iii*) solution of KI in water and (*iv*) vapour.

The diagram (Fig. 4.16) consists of the curves *AO, BO,* the point *O* and the various areas above and below the curves *AO* and *BO,* as described below:

I. Curves: (*i*) Curve *AO*. This curve represents *the freezing point curve of water to which successive quantities of potassium iodide are added* or in other words, the curve *AO* shows the effect of the addition of potassium iodide on the freezing point of water. The point *A* represents the freezing point of water (or m. pt. of ice) under normal conditions. At this point three phases—ice, water liquid and vapours are in equilibrium. It is observed that by gradual addition of

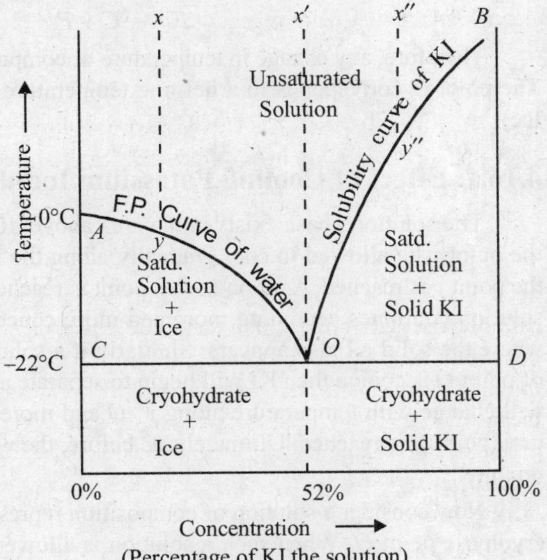

Fig. 4.16. KI – H_2O system.

potassium iodide to it, the freezing point of water is lowered and curve *AO* bends downwards. This, however, does not go indefinitely. As soon as the solution is saturated with respect to potassium iodide, the lowest temperature is reached at the point *O;* then further addition of KI will not produce further lowering in the freezing point of water.

The point *O* corresponding to a concentration of 52% potassium iodide represents the lowest temperature that can be attained in this system. This eutectic temperature in this case is called **cryohydric point** and the eutectic itself is called a **cryohydrate.**

All along the curve *OA,* two phases namely solution and ice are in equilibrium (neglecting the vapour phase). Therefore, the system is univariant as shown by reduced phase rule equation.

$$F = C - P + 1$$
$$= 2 - 2 + 1$$
$$= 1$$

***(ii)* Curve *OB*.** This curve is the *solubility curve of potassium iodide in water at different temperatures.* It represents the effect of temperature on the solubility of solid potassium iodide. It is observed that solubility of potassium iodide increases with the rise of temperature and is maximum at the point *B,* the boiling point of the saturated solution. However, the curve comes to an end before it can touch the 100% potassium iodide axis. Conversely, if the saturated solution of potassium iodide in water is cooled, potassium iodide separates out and the solution becomes dilute and the curve *BO* is followed till the point *O* is reached. Here the solution freezes as a whole with a fixed composition.

All along the curve *OB,* there are two phases — solid KI and solution in equilibrium, and being a two-component system, the number of degree of freedom according to reduced phase rule equation would be

$$F' = C - P + 1 = 2 - 2 + 1 = 1$$

II. Cryohydric point, *O*. It is the point where the two curves *AO* and *BO* intersect. Here, three phases—ice, solid salt and solution are in equilibrium. If the vapour phase is not neglected, then the fourth phase is also present. Point *O* is, therefore, called a *quadruple point.* If we apply the original phase rule, we find that the system is invariant, *i.e.*,

$$F' = C - P + 2 = 2 - 4 + 2 = 0$$

Therefore, any change in temperature or composition will cause one of the phases to disappear. The point *O* corresponds to a definite temperature (– 22°C) and composition (52% KI and 48% Ice).

4.16.1. Effect of Cooling Potassium Iodide Solution

The solution phase exists in the area above *AOB*. (Fig. 4.16) Let the solution represented by the point *x* be allowed to cool gradually along the line *xy* without any change in composition till the point *y* is reached. As soon as this point is reached, ice starts separating, and on further cooling, solution continues becoming more and more concentrated till the cryohydric point *O* is reached where the solid KI also appears. Similarly if a solution of composition *x″* lying on the right hand of point *O* is cooled then KI will begin to separate as soon as point *y″* is reached. The composition will change with temperature along *y″ O* and more and more of KI will continue to separate out until point *O* is reached. Ultimately, as before, the whole of the mixture will freeze to give eutectic mixture.

Now consider a solution of composition represented by the point *x′* lying vertically above the cryohyric point *O*. When such a solution is allowed to cool gradually along the line *x′O* without any change in composition, the solution will solidify as a whole, *i.e.*, on reaching the point *O,* both ice and KI start separating out simultaneously.

Hence we conclude that all solutions on cooling show no further change in temperature at the eutectic point *O*. Moreover, on cooling a solution of eutectic composition, the solution freezes at the eutectic temperature without any change in composition.

4.17. FREEZING MIXTURES

A good freezing mixture is one which satisfies the following conditions:

(*i*) *It should have a low cryohydric temperature*. This condition is fulfilled by having a salt which is highly soluble.

(*ii*) *The heat of solution of the salt should be high*. This condition is satisfied by salts whose, solubility rapidly increases on increasing the temperature.

(*iii*) *The components should be such that an intimate mixture can be obtained from them.*

(*iv*) *The material should be cheap.*

In actual practice, the common freezing mixture is made from ice and common salt (sodium chloride). Although the common salt is very cheap, it is not good salt to use because the heat of solution of this salt is very low and the heat absorbed is almost due to the heat of fusion of ice. Calcium chloride hexa-hydrate and ice form a good freezing mixture because this has a low cryohydric point and high heat of solution.

For attaining very low temperatures, (–50 to –70°C), **acetone-dry ice mixture** is quite often used. Acetone is taken in a Dewar or insulated flask and dry ice is added to it till the lowest temperature is attained.

4.18. PHASE DIAGRAM OF MAGNESIUM-ZINC SYSTEM

This is a two-component system in which the compound formation takes place with a congruent melting point. This system involves the formation of an intermetallic compound $MgZn_2$. It has four phases : solid magnesium, solid zinc, solid $MgZn_2$, and the liquid solution of Mg and Zn.

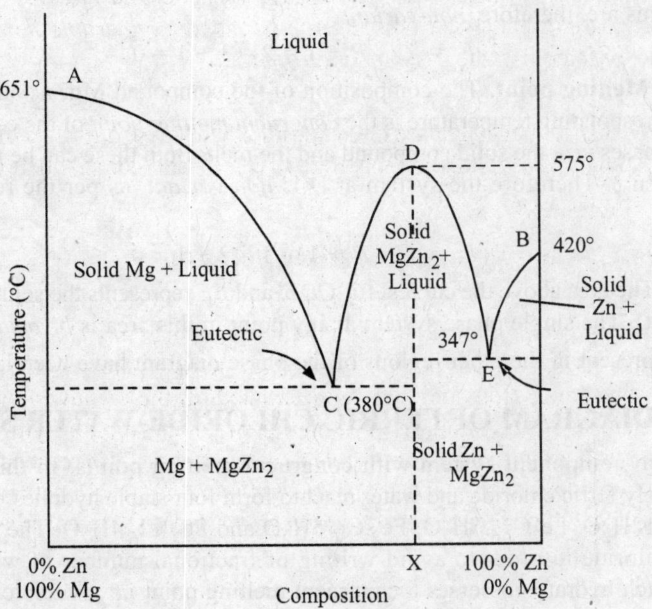

Fig. 4.17. Phase diagram of magnesium-zinc system.

The complete phase diagram of the system magnesium–zinc is shown in Fig. 4.17. It appears to be made of two simple eutectic diagrams. The one toward the left represents the eutectic system Mg–$MgZn_2$, while the one to the right is the system Zn–$MgZn_2$.

The curves AC, *CDE*, and *BE*. *AC* is the freezing point curve of magnesium; *BE* is the freezing point curve of zinc; and *CDE* is that of the compound $MgZn_2$.

The curve *AC* shows that the melting point of magnesium is (651°C). It is lowered on the addition of zinc. This continues until the point *C* is reached. Here a new phase, solid $MgZn_2$ appears.

The curve *CD* shows the increase of concentration of zinc in the melt with the rise of temperature. At the maximum point *D*, the composition of the melt and the solid compound become the same *i.e.,* $MgZn_2$. The point *D*, therefore, represents the melting point of $MgZn_2$ (575°C). The curve *DE* now shows the lowering of the melting point with the addition of zinc until the lowest point is attained. Here solid zinc appears.

The curve *BE* exhibits that the melting point of zinc (420°C) falls with the addition of magnesium until the point *E* is reached.

Along the freezing point curves *AC, CDE* and *BE*, there are two phases in equilibrium *viz.,* one solid phase (Mg, Zn, or $MgZn_2$) and the other liquid phase. Applying the reduced phase rule equation, and substituting values, we get the value of *F* as 1.

$$F = C - P + 1 = 2 - 2 + 1 = 1$$

This shows that the systems Mg/liquid, Zn/liquid, and $MgZn_2$/liquid are all *monovariant* having degree of freedom equal to one.

Eutectic points *C* and *E*. There are two eutectic points in the phase diagram. The systems at the points *C* and *E* have two components and three phases in equilibrium.

Point	Phases Present
C	Solid Mg, solid $MgZn_2$, Liquid
E	Solid Zn, solid $MgZn_2$, Liquid

These systems are, therefore, *non-variant.*

$$F = C - P + 1 = 2 - 3 + 1 = 0$$

Congruent Melting point. The composition of the compound $MgZn_2$ and the melt at *D* is identical. The corresponding temperature is the *congruent melting point* of the compound. Here the system has two phases *viz.*, the solid compound and the melt. Both these can be represented by one component ($MgZn_2$). Therefore the system at D is *non-variant,* as per the reduced phase rule equation.

$$F = C - P + 1 = 1 - 2 + 1 = 0$$

The Areas. The area above the curves *AC, CDE,* and *BE* represents the solution of magnesium and zinc (the melt). The single phase system at any point in this area is *bivariant.*

The phases present in the other regions of the phase diagram have been labelled.

4.19. PHASE DIAGRAM OF FERRIC CHLORIDE-WATER SYSTEM

This is a two-component system with congruent melting points. In this system the two components, namely ferric chloride and water react to form four stable hydrates, represented by the formulae $Fe_2Cl_6.12H_2O$, $Fe_2C1_6.7H_2O$, $Fe_2.Cl_6{\cdot}5H_2O$ and $Fe_2\ Cl_6{\cdot}4H_2O$. The double formula is used for ferric chloride in order to avoid writing of fractional number of water molecules of crystallization. Each hydrate possesses a congruent melting point *i.e.*, is stable up to the melting point and melts at the melting point to give liquid of the same composition as that of the solid compound or in other words, at this temperature, the saturated solution of the hydrate has the same composition as the solid phase. Besides the above four solid phases, there are two more solid phases, namely anhydrous Fe_2Cl_6 and ice, one liquid phase *viz.,* solution and one gaseous phase *viz.,* vapour. Thus in all there are *eight* possible phases. The different phases existing in equilibrium under different conditions of temperature and composition are shown in the phase diagram given in Fig. 4.18. Corresponding to the congruent melting point of each hydrate, a maximum is obtained in curve, as shown in the figure.

The various curves and points in the above diagram are explained as follows:

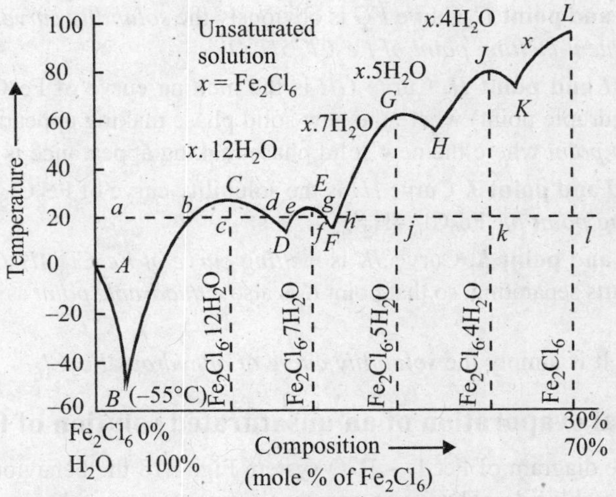

Fig. 4.18. $FeCl_3 - H_2O$ system.

Note: The points corresponding to capital letters *A, B, C,* ... lie on the curve while the points corresponding to small letters lie on the dotted horizontal line.

***(i)* Curve *AB* and Point *B*.** Point *A* represents the freezing point of water. Curve *AB* represents the *freezing curve of water to which increasing amounts of $FeCl_3$ have been added.*

When point *B* is reached, the solution is saturated with the hydrate $Fe_2Cl_6.12H_2O$ which then starts separating alongwith the ice. Thus *B* is a *cryohydric point* where the phases in equilibrium are ice, solid $Fe_2Cl_6.12H_2O$, saturated solution and vapour *i.e.,* there are 4 phases in equilibrium. Hence point *B* is *quadruple point.* The degree of freedom can be calculated as: $F = C - P + 2 = 2 - 4 + 2 = 0$ *i.e.,* the system is *invariant.* Point *B* is a *eutectic point.*

***(ii)* Curve *BC* and point *C*.** On adding more of $FeCl_3$ the ice phase disappears and now there are three phases in equilibrium *viz.,* solid $Fe_2Cl_6.12H_2O$, saturated solution and vapour.

Hence $F = C - P + 2 = 2 - 3 + 2 = 1$ i.e., the system becomes *univariant.* The equilibrium is represented by curve *BC.* Thus the curve *BC* represents the *solubility curve of $Fe_2Cl_2.12H_2O$.* At the point *C* (corresponding to 37°C) the saturated solution has the same composition as that of the solid $Fe_2Cl_6.12H_2O$. The point *C* thus represents a *congruent melting point.*

***(iii)* Curve *CD* and point *D*.** On further addition of $FeCl_3$, depresses the melting point of $Fe_2Cl_6.12H_2O$ and the curve *CD* is followed. Thus the curve *CD* represents the *melting curve of* $Fe_2Cl_6.12H_2O$. At the point *D,* a new solid phase *viz.,* $Fe_2Cl_6.7H_2O$ starts separating out. Hence at the point *D,* again there are four phases in equilibrium namely $Fe_2Cl_6.12H_2O$, $Fe_2Cl_6.7H_2O$, saturated solution and vapour *i.e.,* point *D* is a *quadruple point* and is *invariant.* Point *D* is in fact a *eutectic point* or a *transition point* representing the transition.

$$Fe_2Cl_6.12H_2O \rightleftharpoons Fe_2Cl_6.7H_2O$$

(*iv*) **Curve *DE* and point *E*.** Like the curve *BC*, the curve *DE* is the solubility curve of $Fe_2Cl_6.7H_2O$ and point *E* is the congruent melting point of $Fe_2Cl_6.7H_2O$.

(*v*) **Curve *EF* and point *F*.** Like the curve *CD*, the curve *EF* is the melting curve of $Fe_2Cl_6.7H_2O$. At the point *F*, a new solid *viz.*, $Fe_2Cl_6.5H_2O$ appears. The point *F* is thus a *quadruple point*, the four phases in equilibrium being $Fe_2Cl_6.7H_2O$, $FeCl_6.5H_2O$, saturated solution and vapour. Point *F* is another *transition point*, representing the change.

$$Fe_2Cl_6.7H_2O \rightleftharpoons Fe_2Cl_6.5H_2O.$$

(*vi*) **Curve *FG* and point *G*.** Curve *FG* is obviously the *solubility curve of $Fe_2Cl_6.5H_2O$ and point G is the congruent melting point of $Fe_2Cl_6.5H_2O$.*

(*vii*) **Curve *GH* and point *H*.** Curve *GH* is the melting curve of $Fe_2Cl_6.5H_2O$ and *H* is a *transition point* (quadruple point) where the new solid phase making appearance is $Fe_2Cl_6.4H_2O$ and *H* is a *transition point* where the new solid phase making appearance is $Fe_2Cl_6.4H_2O$.

(*viii*) **Curve *HJ* and point *J*.** Curve *HJ* is the solubility curve of $Fe_2Cl_6.4H_2O$ and point *J* is the *congruent melting point of* $Fe_2Cl_6.4H_2O$

(*ix*) **Curve *JK* and point *K*.** Curve *JK* is *melting curve of $Fe_2Cl_6.4H_2O$* and at point *K*, the anhydrous Fe_2Cl_6 starts separating, so that point *K* is also a *quadruple point* as well as the *transition point*.

(*x*) **Curve *KL*.** It is simply the *solubility curve of anhydrous Fe_2Cl_6.*

4.19.1. Isothermal evaporation of an unsaturated solution of ferric chloride

Using the phase diagram of $FeCl_3 - H_2O$ system Fig. 4.18 the behaviour on evaporation of an unsaturated ferric chloride solution at constant temperature can be explained. Suppose an unsaturated solution, at temperature between the temperatures of points *E* and *F* (say at 31°C) is evaporated at constant temperature. Suppose such a solution is represented initially by the point *a*. As the solution is heated, water evaporates off and the horizontal path *al* is followed. The various changes occurring on the way are as follows:

(*i*) The concentration of the solution increases till at the point *b*, the solid $Fe_2Cl_6.12H_2O$ begins to separate out.

(*ii*) Heating beyond the point *b*, causes more and more of water to evaporate off and more and more of the hydrate $Fe_2Cl_6.12H_2O$ separates out till the point *c* where there is only solid $Fe_2Cl_6.12H_2O$ and no water at all.

(*iii*) Heating beyond the point *c*, causes the water to come out of the hydrate $Fe_2Cl_6.7H_2O$. The latter keeps on dissolving into the former till a solution of composition corresponding to point *d* is obtained.

(*iv*) Continued evaporation leads to the point *e* where solid $Fe_2Cl_6.7H_2O$ starts separating out, then to the point *f* where there is only solid $Fe_2Cl_6.7H_2O$ which then begins to lose water and form solution till a solution having composition corresponding to point *g* is obtained.

(*v*) At *h*, solid $Fe_2Cl_6.5H_2O$ starts separating out till at the point *j*, there is only solid $Fe_2Cl_6.5H_2O$.

(*vi*) Beyond the point *j*, no solution is obtained because the point *j* lies at a temperature below that of the unsaturated solution. Evaporation beyond the point *j* results into the formation of next lower hydrate *viz.*, $Fe_2Cl_6.4H_2O$. Thus all along the path *jk* there are $Fe_2Cl_6.5H_2O$ and $Fe_2Cl_6.4H_2O$ in equilibrium. At the point *k*, there is only $Fe_2Cl_6.4H_2O$.

(*vii*) Heating beyond the point *k* leads to the gradual formation of anhydrous salt which will ultimately be the only residue.

4.20. PHASE DIAGRAM OF $Na_2SO_4 - H_2O$ SYSTEM

This is an example of a two-component system with incongruent melting point. Various phases present in this system are (*i*) solid $Na_2SO_4.10H_2O$ (*ii*) solid $Na_2SO_4.7H_2O$ (*iii*) solid Na_2SO_4 (rhombic) (*iv*) solid Na_2SO_4 (monoclinic) (*v*) ice (*vi*) solution in water and (*vii*) water vapour. The equilibrium diagram obtained between composition and temperature is shown in Fig. 4.19.

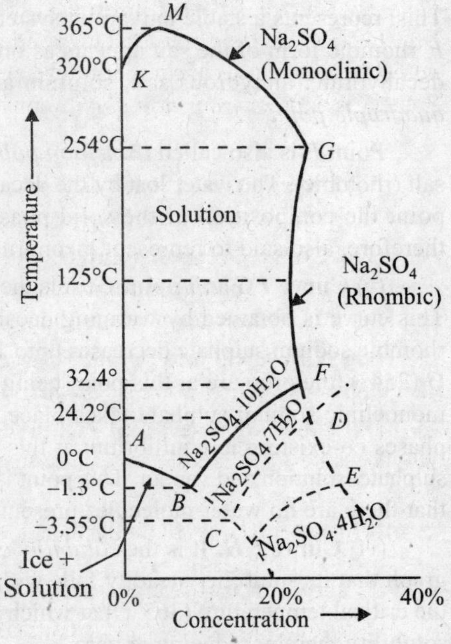

Fig. 4.19. $Na_2SO_4 - H_2O$ system.

The diagram consists of the curves *AB, BC, BF, CD, FD, FG* and *GH;* triple point *A* and quadruple points *B, C, D, F* and *G.* However, in the discussion of these curves and points given below, the vapour phase has been taken into consideration.

(*i*) **Curve *AB* and point *B*.** It is the *freezing point curve* or the curve showing the composition of solutions in equilibrium with ice at different temperatures. At the point *A,* ice is in equilibrium with water and vapour at 0°C. *A* is, therefore, a triple point. On adding anhydrous sodium sulphate, the temperature falls as shown by the freezing point curve *AB*. All along the curve *AB,* we have three phases namely ice, solution and vapour in equilibrium. Hence curve *AB* represents a univariant system.

$$F = C - P + 2$$
$$= 2 - 3 + 2 = 1.$$

At the point *B* (– 1.3°C) the fourth phase, namely sodium sulphate decahydrate, also appears, so that the system is now invariant,

$$F = C - P + 2 = 2 - 4 + 2 = 0.$$

The point *B* is, therefore, a stable quadruple point.

(*ii*) **Curve *BC* and point *C*.** Under certain careful conditions, it is possible to travel along the curve *BC* in which case decahydrate fails to appear at *B*. At the point *C* sodium sulphate heptahydrate separates out. The curve *BC* is, therefore, a *metastable curve* and the point *C* is a *metastable quadruple point ;* the four phases co-existing in metastable equilibrium at this point *C* are ice, heptahydrate, solution and vapour.

(*iii*) **Curve *CD* and point *D*.** If the temperature is raised and more of anhydrous sodium sulphate is added, ice phase disappears as we travel along the curve *CD.* The curve *CD* represents the *metastable solubility curve* of sodium sulphate heptahydrate. All along this curve, three phases *viz.,* heptahydrate, solution and vapour are present in equilibrium. Hence the curve *CD* represents a univariant system, as per phase rule equation.

$$F = C - P + 2$$
$$= 2 - 3 + 2 = 1.$$

At the *point D,* the fourth phase (anhydrous sodium sulphate) makes its appearance. *Point D,* therefore, represents a *second metastable quadruple point.* By lowering the temperature, it is theoretically possible to reach point *E,* at which sodium sulphate tetrahydrate would separate out. Practically, this has not been possible.

(*iv*) **Curve *BF* and point *F*.** It is the *stable solubility curve of sodium sulphate decahydrate.* All along this curve three phases namely sodium sulphate decahydrate, solution and vapour co-exist. This, represents a stable univariant system, however, only upto the point *F* (32.4°C). At the point *F*, rhombic form of the salt appears as one of the solid phases. At this point all the four phases—decahydrate, anhydrous salt, solution and vapour are in equilibrium. Point *F* is thus a *stable quadruple point.*

Point *F* is also called *transition point* because decahydrate changes into the solid anhydrous salt (rhombic). The water lost by the decahydrate is taken up by the solution and as a result, at this point the composition of the solid phase and liquid phase becomes different. The point *F* is, therefore, also said to represent incongruent melting point.

(*v*) **Curve *FG* and point *G*.** It is the *solubility curve for anhydrous rhombic sodium sulphate.* This curve is obtained by warming decahydrate. Shape of the curve shows that the solubility of rhombic sodium sulphate decreases upto 125°C, after which it again rises upto the *quadruple point G* (254°), the pressure at this point being 27.5 atmosphere. At this point, transition of rhombic to monoclinic sodium sulphate takes place. So point *G* is another *stable quadruple point,* the four phases co-existing in equilibrium at this point are rhombic sodium sulphate, monoclinic sodium sulphate, solution and vapour. This point is different from the previous quadruple points in the sense that there are no water molecules present in the two solid phases. It is also a transition point.

(*vi*) **Curve *GH*.** It is the *solubility curve of monoclinic sodium sulphate.* It is clear from the graph that its solubility steadily falls with the rise of temperature and finally the curve ends at *H,* the critical temperature (365°C) at which the solution and vapour phase merge into each other, the solubility thereby reducing to zero.

(*vii*) **Curve *HK*.** It shows the *solubility of anhydrous sodium sulphate in water vapour,* the two curves *GH* and *HK* being continuous.

4.21. SODIUM CHLORIDE-WATER SYSTEM

This is a system with incongruent melting point. Sodium chloride reacts with water to form the dihydrate, $NaCl.2H_2O$ which is not stable upto the melting and starts decomposing above 0.15°C according to the equation

$NaCl.2H_2O \rightleftharpoons NaCl$ + Saturated solution

Hence, this temperature (*viz.* 0.15°C) is called the incongruent melting point and represents the transition point for the above decomposition. (See Fig. 4.20)

The different possible phases which exist in this system are NaCl (*s*), $NaCl.2H_2O$ (*s*), Ice (*s*) and solution. The phase diagram showing the phases that exist under different conditions are shown in Fig. 4.20. Vapour phase has not been taken into consideration here.

The different curves, points and areas of this phase diagram are briefly explained below:

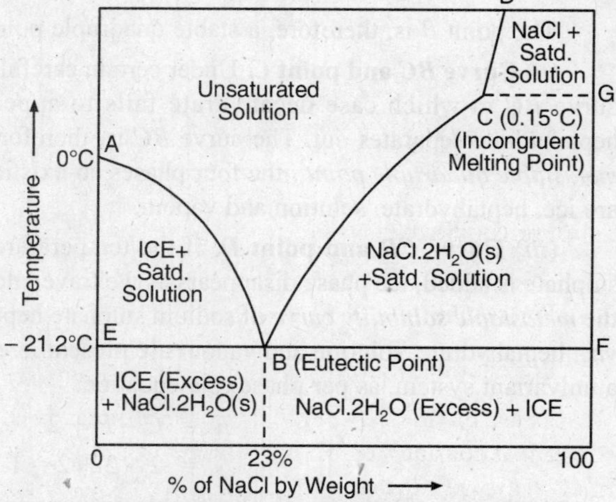

Fig. 4.20. Phase diagram of $NaCl–H_2O$ system

(*i*) **Curve *AB* and the point *B*.** The point *A* represents the freezing point of water *i.e.*, 0°C at one atmospheric pressure. The curve *AB* represents the *freezing point curve of water* to which increasing amounts of NaCl have been added. On adding more and more of NaCl to water, freezing point decreases along the curve *AB*. The NaCl added goes into the solution and ice keeps on separating til the point *B* is reached. Along the curve *AB*, the two phases in equilibrium are ice and solution. Applying reduced phase rule equation, we get $F' = C - P + 1 = 2 - 2 + 1 = 1$, *i.e.*, it is univariant.

The point *B* corresponds to a temperature of –21.2°C and has 23% NaCl by weight. Here the solution becomes saturated and NaCl. $2H_2O$ starts separating out as a solid phase. Below this point ice and NaCl.$2H_2O$ (*s*) separate out together as a whole. Hence, point *B* is the eutectic point. The three phases in equilibrium at this point are ice, NaCl.$2H_2O$ (*s*) and solution. Hence, applying reduced phase rule equation

$$F' = C - P + 1 = 2 - 3 + 1 = 0$$

(*ii*) Curve *BC* and the point *C*. The curve *BC* represents the *solubility curve of NaCl.2H$_2$O (s)*. From the curve, it is obvious that the solubility of NaCl.$2H_2O$ (*s*) increases with the increase of temperature. As the curve is steep, it shows that the solubility increases gradually with increase of temperature. Along the curve *BC*, the two phases in equilibrium are NaCl.$2H_2O$ (*s*) and solution. Hence, it is univariant.

The point *C* corresponds to a temperature of 0.15°C. Here NaCl.$2H_2O$ (*s*) starts decomposing to give anhydrous NaCl and saturated solution. This point *C* is transition temperature and also incongruent melting point.

(*iii*) Curve *CD*. It represents the *solubility curve of anhydrous NaCl*. It is very steep showing that the increase in the solubility of anhydrous NaCl is very very slow with increase of temperature. All along the curve *CD*, the two phases in equilibrium are anhydrous NaCl and solution.

4.22. PHASE DIAGRAM OF COPPER SULPHATE-WATER SYSTEM

This is a two-component system with incongruent melting points Copper sulphate and water combine to form hydrates such as $CuSO_4.5H_2O$, $CuSO_4.3H_2O$ and $CuSO_4.H_2O$. These salts start dissocating to form the lower hydrate and the anhydrous salt below their melting points which are also the transition temperatures. Starting with $CuSO_4.5H_2O$, the following equilibria are set up:

$$CuSO_4.5H_2O\ (s) \rightleftharpoons CuSO_4.3H_2O\ (s) + 2H_2O\ (g) \quad \ldots (i)$$

$$CuSO_4.3H_2O\ (s) \rightleftharpoons CuSO_4.H_2O\ (s) + 2H_2O\ (g) \quad \ldots (ii)$$

$$CuSO_4.H_2O\ (s) \rightleftharpoons CuSO_4\ (s) + H_2O\ (g) \quad \ldots (iii)$$

Applying the law of chemical equilibrium to each of these heterogeneous equilibria taking $p_{solid} = 1$, we get

For equilibrium (*i*), $(K_p)_1 = p^2_{H_2O}$ or $p_{H_2O} = \sqrt{(K_p)_1}$ = constant

For equilibrium (*ii*), $(K_p)_2 = p^2_{H_2O}$ or $p_{H_2O} = \sqrt{(K_p)_2}$ = constant

For equilibrium (*iii*), $(K_p)_3 = p_{H_2O}$ or $p_{H_2O} = (K_p)_3$ = constant

Thus, for each of the above equilibria, vapour pressure of water vapours would become constant at constant temperature. Study of the above equilibria is based on the above observation. **In contrast with the study of other salt hydrates, temperature is kept constant and the pressure over the salt is continuously decreased** till the dehydration of the salt starts and the dissociation equilibrium is set up. The experiment may be carried out by taking the crystals of $CuSO_4.5H_2O$ in a flask connected to a vacuum pump and a manometer and placing the flask in the thermostat at a fixed temperature (say 25°C) as shown in Fig. 4.21. The dissociation pressure is noted from time to time from the manometer till a constant value is obtained.

It is observed that as the suction is applied (*i.e.*, water vapour are sucked out), the pressure inside the flask decreases till at 7.8 mm of mercury, where it becomes constant at 25°C. It remains constant at this level till whole of $CuSO_4.5H_2O$ changes into $CuSO_4.3H_2O$. On applying the suction further, the pressure falls abruptly till at 5.6 mm it becomes constant. It remains constant till whole of $CuSO_4.3H_2O$ is converted into $CuSO_4.H_2O$. On continuing suction further, an abrupt decrease in pressure is observed till again at 0.8 mm, it becomes constant. It remains constant till the whole of $CuSO_4.H_2O$ is converted into the anhydrous $CuSO_4$ and the vapour pressure of water falls to zero.

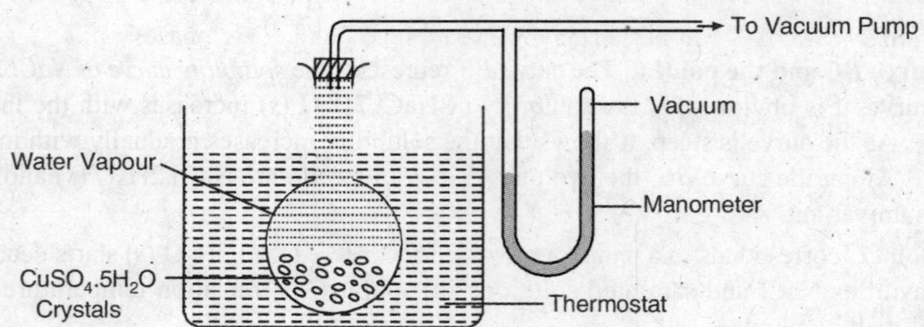

Fig. 4.21. A set up for the study of dissociation equilibrium of $CuSO_4.5H_2O$

Thus at 25°C.

For equilibria (*i*), $p_{H_2O} = 7.8$ mm

For equilibria (*ii*), $p_{H_2O} = 5.6$ mm

For equilibria (*iii*), $p_{H_2O} = 0.8$ mm

The various changes taking place are represented graphically as shown in Fig. 4.22.

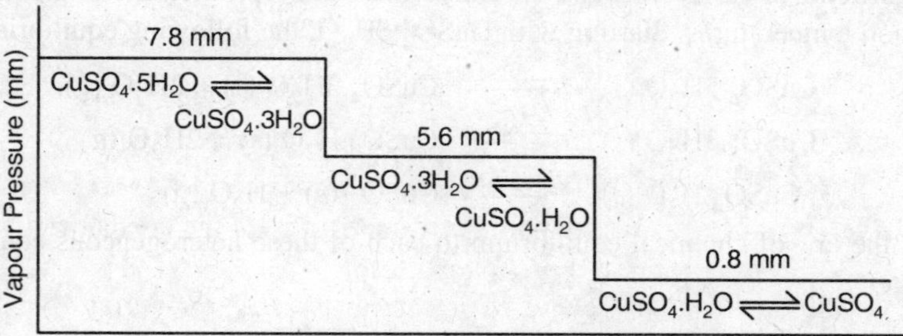

Fig. 4.22. Dehydration of $CuSO_4.5H_2O$ at 25°C

Each of the three equilibria occurring here involves two solids and one gas, *i.e.*, number of phases = 3. Hence, applying phase rule equation,

$$F = C - P + 2 = 2 - 3 + 2 = 1$$

The following points about equilibria involving salt hydrates may be noted:

(*i*) Higher the temperature, higher is the vapour pressure. For example, if dehydration of $CuSO_4.5H_2O$ is carried out at 50°C, the vapour pressures for equilibria (*i*) to (*iii*) are found to be 45.4, 30.9 and 4.5 mm respectively.

(*ii*) Above 102°C, pentahydrate is not stable. Hence, $CuSO_4.3H_2O$ is the first stable hydrate and only two horizontal portions would be found in the vapour pressure-dehydration diagram.

(*iii*) For a definite vapour pressure to exist, there must be two solid phases in equilibrium in addition to the water vapour. Hence, it is wrong to speak of "vapour pressure of a salt hydrate" because a hydrated salt has no vapour pressure of its own.

SOLVED CONCEPTUAL PROBLEMS

Problem 1. What will be the number of degrees of freedom for a system consisting of saturated aqueous solution of sodium chloride in contact with the solid salt and vapour?

Solution. $P = 3$ (solid, liquid and vapour), $C = 2$ (NaCl and H_2O)

$F = C - P + 2 = 2 - 3 + 2 = 1.$

Problem 2. List the three criteria for phase equilibrium of a multi-component system.

Solution. (*i*) Thermal equilibrium (Temperature is same in every phase)

(*ii*) Mechanical equilibrium (Pressure is same in every phase)

(*iii*) Chemical equilibrium (Chemical potential of a component is same in every phase).

Problem 3. What is meant by triple point of water? Why is it different from the normal melting point of ice?

Solution. Triple point of water represents a point in the phase diagram at which all the three phases *i.e.* solid ice, liquid water and water vapour exist together (at 0.0098 °C and 4.58 mm pressure). At the normal melting point of ice (0°C, 1 atmospheric pressure) only ice and water exists together.

Problem 4. Calculate the degrees of freedom of sulphur at the transition point.

Solution. At the transition point ($S_L \rightleftharpoons S_M$), $P = 2$. As $C = 1$, hence $F = C - P + 2 = 1 - 2 + 2 = 1$.

Problem 5. Suggest the possible maximum number of phases that can co-exist in the following system:

(*i*) Sulphur system **(*ii*) Lead and silver alloy system.**

(iii) Potassium iodide-water system.

Solution. (*i*) As $C = 1$ and minimum $F = 0$. Hence maximum $P = C - F + 2 = 3$ (Triple point)

(*ii*) $C = 2, F = 0 \therefore P = C - F + 2 = 2 - 0 + 2 = 4$ (Quadruple Point)

(*iii*) $C = 2, F = 0 \therefore P = C - F + 2 = 2 - 0 + 2 = 4$ (Quadruple Point)

Problem 6. What is the criteria for two phase equilibria for one-component system?

Solution. At a given temperature and pressure, free energy of the component is same in each phase

Problem 7. The melting point curve of water system has a negative slope whereas the melting point curve of sulpur has a positive slope. Explain.

Solution. Negative slope of melting point curve of ice means that it slopes towards the pressure axis. In this case as the pressure is increased, melting point of ice decreases. Positive slope of melting point curve of sulphur (monoclinic) means that it slopes away from the pressure axis. In this case, the melting point increases with increase of pressure.

Problem 8. Calculate the degrees of freedom of Pb – Ag system at the eutectic point.

Solution. At the eutectic point, $P = 3$ (solid Pb, solid Ag melts,) As $C = 2$, applying condensed phase rule equation, *viz.*, $F = C - P + 1$, we get $F = 2 - 3 + 1 = 0$. We neglect vapour phase here.

Problem 9. What types of systems are represented by curves (lines), point and areas in one- component systems?

Solution. Areas represent bivariant system, lines represent invariant system and points represent non-variant system.

Problem 10. Write down the number of components, the number of phases and the number of degrees of freedom (in addition to those already fixed) in the following examples:

***(i)* Solid carbon in equilibrium with gaseous CO, CO_2 and O_2 at 100°C.**

***(ii)* A 2 molar aqueous solution of sulphuric acid (partly dissociated into hydrogen, bisulphate and sulphate ions) at 1 atm.**

Solution. $C = S - E - R$ (In a chemically reactive system)

$S = 4(C, CO, CO_2, O_2)$ $E = 1, R = O$ so that $C = 3$

Degrees of freedom, $F = C - P + 2 = 3 - 2 + 2 = 3.$

(ii) $2\ H_2SO_4\ (aq) \rightleftharpoons 3H^+ + HSO_4^- + SO_4^{2-}$

Phases. Only one liquid phase (solution) *i.e.* $P = 1$

components $C = 1$

Degrees of freedom, $F = C - P + 2 = 1 - 1 + 2 = 2.$

Problem 11. How many number of phases and components are present in the following system?

***(i)* A mixture of molten Lead, Tin and Bismuth.**

***(ii)* Two ice cubes floating on water in a closed container in the presence of water vapour:**

Solution. *(i)* $P = 1$ (liquid phase only), $C = 3$ (Pb, Sn, *Bi*)

(ii) Ice $\rightleftharpoons$ Water $\rightleftharpoons$ vapour, $P = 3$, $C = 1$ (H_2O only).

Problem 12. Calculate the number of phase and degrees of freedom for the following systems:

***(i)* A gas in equilibrium with its solution in a liquid.**

***(ii)* A solution of one solid in one liquid in equilibrium with solvent vapour.**

***(iii)* Two partially miscible liquids in the absence of vapour.**

Solution.

(i) $P = 2$ (one gaseous, one liquid) $C = 2$ (gas and solvent).

(ii) $P = 3$ (one solid, one liquid, one gaseous) $C = 2$ (one solid component, one liquid component)

(iii) P = 2 (both liquid phases) $C = 2$ (both liquid component)

Problem 13. Find out the number of components present in the following system:

***(i)* Silver (*s*) $\rightleftharpoons$ Solution of silver and lead (*l*) + lead (*s*) + vapours of silver and lead (g).**

***(ii)* Water (*l*) $\rightleftharpoons$ Water vapours (*g*)**

***(iii)* NH_4Cl (*S*) $\rightleftharpoons$ NH_3 (*g*) + HCl (*g*)**

***(iv)* $CaCO_3$ (*s*) $\rightleftharpoons$ CaO (*s*) + CO_2 (*g*)**

Solution. *(i)* $C = 2$ (Pb and Ag)

(ii) $C = 1$ (H_2O only)

(iii) $C = 1$ (NH_4Cl only)

(iv) $C = 2$ (any two out of $CaCO_3$, CaO and CO_2 can describe the composition of any phase)

Problem 14. Is it possible to have a quadruple point in the phase diagram for a one-component system?

Solution. Quadruple point is a point where four different phases meet. Thus $P = 4$, substituting the values in the phase rule equation

$$F = C - P + 2$$

or $$F = 1 - 4 + 2$$

or $$F = -1$$

This is absurd. Therefore, a one-component system cannot have a quadruple point in the phase diagram.

Problem 15. Calculate the degree of freedom for the following:

(*i*) A mixture of nitrogen and oxygen gases contained in a vessel,

(*ii*) Rhombic sulphur in equilibrium with monoclinic sulphur.

Solution. (*i*) Mixture of nitrogen and oxygen

No. of phases = 1 (Gases form a single phase)

No. of component = 2

Degree of freedom is given by

$$F = C - P + 2$$
$$= 2 - 1 + 2$$
$$= 3$$

(*ii*) Rhombic sulphur in equilibrium with monoclinic sulphur.

No. of phases = 2

No. of components = 1

Degree of freedom is obtained as follows:

$$F = C - P + 2$$
$$= 1 - 2 + 2$$
$$= 1$$

Problem 16. Calculate the number of phases, components and degree of freedom in the following systems:

(*i*) N_2O_4 (g) $\rightleftharpoons$ $2NO_2$ (g)

(*ii*) Solid carbon in equilibrium with gaseous CO, CO_2 and O_2 at 100°C.

Solution. (*i*) N_2O_4 (g) $\rightleftharpoons$ $2NO_2$ (g)

No. of phases = 1 (gases mix uniformly to give one phase)

No. of components = 1

Degree of freedom $F = C - P + 2$

$$= 1 - 1 + 2$$

or $$F = 2$$

(*ii*) Solid carbon with gaseous CO, CO_2 and O_2

No. of phases = 2 (1 for solid carbon and 1 for gases)

No. of components = 2

Degree of freedom $F = C - P + 2$

$$= 2 - 2 + 2$$
$$= 2$$

EXERCISES

(Based On Question Papers of Different Universities)

Objective Questions (Multiple Choice Questions)

1. In the phase diagram of a one-component system, which of the following statements is correct?
   (*a*) The melting point of the solid increases with increase of pressure.
   (*b*) The solid can be in equilibrium with the vapour only at a temperature on or below the triple point.
   (*c*) The solid, if heated at 1 atm, will sublime.
   (*d*) The three lines meet at the critical point.

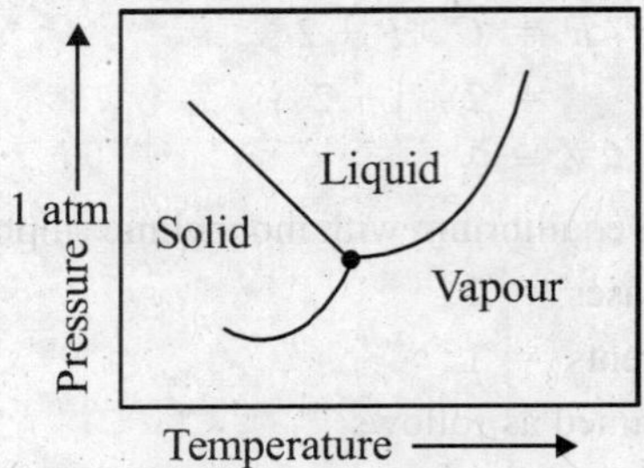

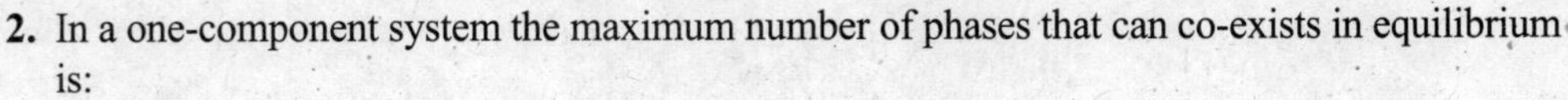

2. In a one-component system the maximum number of phases that can co-exists in equilibrium is:
   (*a*) 1 (*b*) 2
   (*c*) 3 (*d*) 4.
3. In terms of number of phases (P), components (C) and degrees of freedom (F), the phase rule is expressed as:
   (*a*) P + C = F + 2 (*b*) F = P + C – 2
   (*c*) P + F = C + 2 (*d*) P – F = C + 2.
4. The number of phases of a mixture of four gases enclosed in a container is:
   (*a*) 4 (*b*) 3
   (*c*) 1 (*d*) Zero.
5. The number of degrees of freedom for the system:
   $NH_4Cl\ (s) \rightleftharpoons NH_4Cl\ (g) \rightleftharpoons NH_3\ (g) + HCl\ (g)$ is:
   (*a*) 1 (*b*) 3
   (*c*) Zero (*d*) 2.
6. In the phase diagram of the benzene-naphthalene system, the point *C* is called the:
   (*a*) eutectic point (*b*) congruent point
   (*c*) triple point (*d*) critical point.
7. The temperature at which a compound melts into a liquid of the same composition as the solid is called the :
   (*a*) congruent melting point (*b*) incongruent melting point
   (*c*) peritectic temperature (*d*) metastable point.

8. The phase diagram for $FeCl_3$ – water system is shown in fig. below.

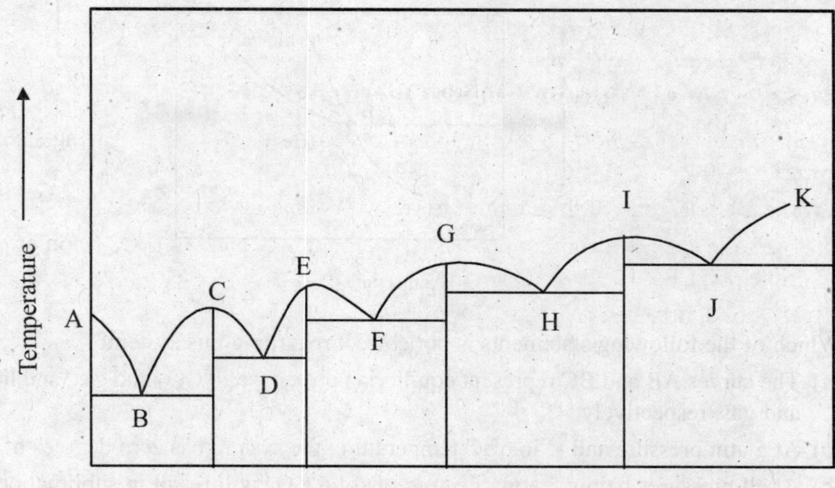

The diagram shows the formation of

(*a*) three hydrates (*b*) four hydrates

(*c*) six hydrates (*d*) five hydrates.

9. The phase diagram for the KI- water system is given below. The cryohydric point in the curve is:

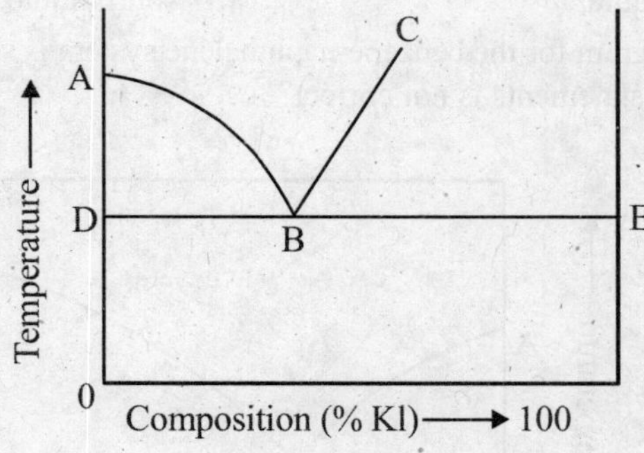

(*a*) A (*b*) B

(*c*) C (*d*) D and E.

10. In the phase diagram of KI – water system (given above), the solubility curve of KI is:

(*a*) AB (*b*) DB

(*c*) BC (*d*) BE

11. Lowest temperature is reached by using

(*a*) $CaCl_2.2H_2O$ (*b*) Acetone + dry ice

(*c*) NH_4Cl (*d*) Ether + dry ice

12. The phase diagram for CO_2 is shown below:

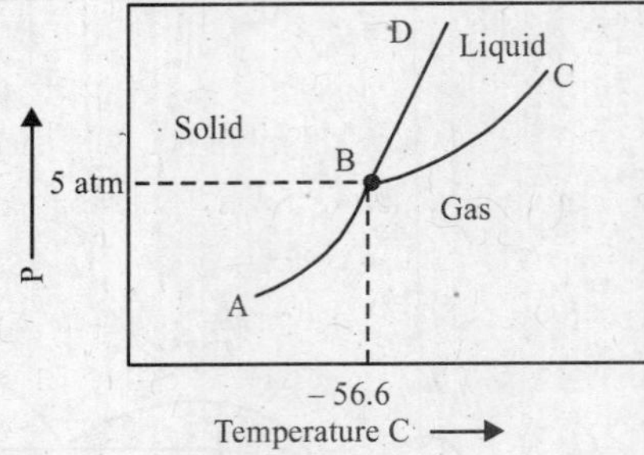

Which of the following statements is not correct regarding this system?

(*a*) The curves AB and BC represent equilibria between solid CO_2 and gas, and liquid CO_2 and gas, respectively.

(*b*) At 5 atm pressure and – 56.6° C temperature, the system has zero degrees of freedom.

(c) At all pressures below 5 atm., heating of solid CO_2 will result in sublimation.

(*d*) The melting point of CO_2 decreases with increases in pressure.

13. In the phase diagram of CO_2 (See above diagram) the system is bivariant:

(a) at point B
(*b*) at the critical point
(*c*) in the region ABD
(*d*) along BD.

14. In the phase diagram of CO_2 (See above diagram) the curve BD represents equilibrium between:

(*a*) solid and gas
(*b*) solid and liquid
(*c*) gas and liquid
(*d*) solid, liquid and gas.

15. The phase diagram for the benzene-naphthalene system is shown in figure below. Which of the following statements is not correct?

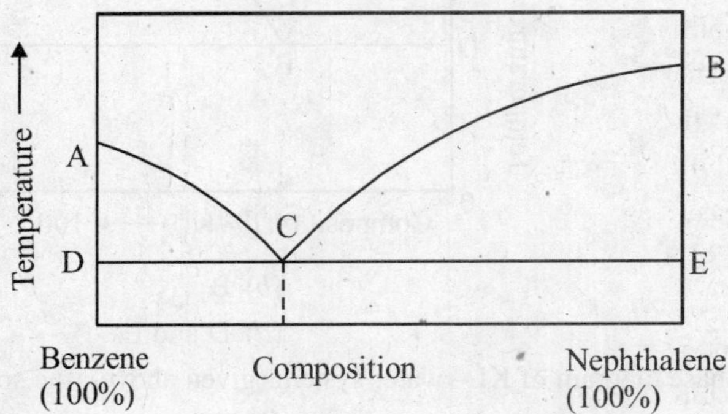

(*a*) Along curve AC, solid benzene is in equilibrium with liquid solution.

(*b*) Above curves AC and BC, the two components are present only as a homogenous liquid solution.

(*c*) The intensive properties of the system at C are entirely fixed under all conditions of pressure.

(*d*) At a given pressure, the system is monovariant along BC.

16. The number of eutectics in the phase diagram of the $Fe_2 Cl_6$ – water system is:

(*a*) 6 (*b*) 5

(*c*) 4 (*d*) 3.

17. In the phase diagram of Na_2SO_4 – water (See diagram in Question 18) the point D is the:

(*a*) Transition temperature at which rhombic Na_2SO_4 is converted into monoclinic Na_2SO_4.

(*b*) Transition temperature at which Glauber's salt changes into monoclinic sodium sulphur.

(*c*) Point at which Na_2SO_4. $10H_2O$ changes into $Na_2O_4.7H_2O$

(*d*) Temperature at which Na_2SO_4 becomes completely soluble in water.

18. In the phase diagram of Na_2SO_4 – water, the curve EF represents the

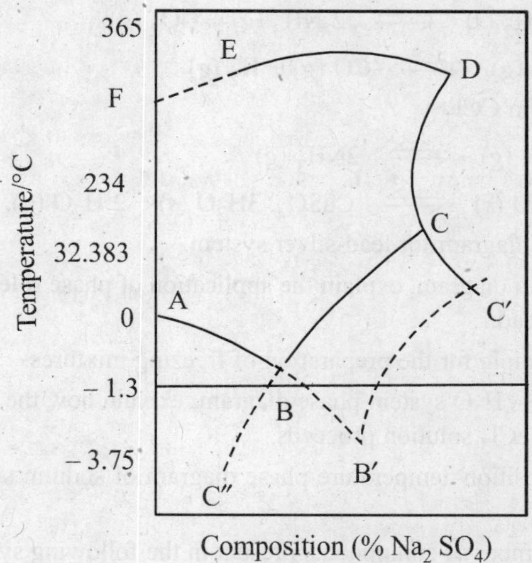

(*a*) Solubility curve of Na_2SO_4. $7H_2O$.

(*b*) Metastable solubility curve of monoclinic Na_2SO_4.

(*c*) Transition curve of monoclinic Na_2SO_4 to rhombic Na_2SO_4

(*d*) Solubility curve of monoclinic Na_2SO_4.

19. In the phase diagram of the Fe_2Cl_6 – water system, the point G corresponds to the compound:

(*a*) Fe_2Cl_6. 12 H_2O (*b*) $Fe_2Cl_6.7H_2O$

(*c*) $Fe_2 Cl_6$. $6H_2O$ (*d*) $Fe_2 Cl_6$. $5H_2O$.

20. A compound with an incongruent melting point decomposes on heating into:

(*a*) A liquid of the same composition as the solid.

(*b*) A new solid phase and a solution with a composition different from that of the solid phase.

(*c*) A new solid phase and a solution with the same composition as that of the solid phase.

(*d*) A solution of fixed composition.

ANSWERS TO MULTIPLE CHOICE QUESTIONS

1. (*b*)	**2.** (*c*)	**3.** (*c*)	**4.** (*c*)	**5.** (*a*)	**6.** (*a*)
7. (*a*)	**8.** (*b*)	**9.** (*b*)	**10.** (*c*)	**11.** (*d*)	**12.** (*d*)
13. (*c*)	**14.** (*b*)	**15.** (*c*)	**16.** (*b*)	**17.** (*c*)	**18.** (*b*)
19. (*d*)	**20.** (*b*)				

SUBJECTIVE QUESTIONS

Short Answer Questions

1. Calculate the number of phases in the following systems:

(a) $NH_4\,CO_2\,NH_2\,(s) \rightleftharpoons 2\,NH_3\,(g) + CO_2\,(g)$

(*b*) $C\,(s) + H_2O\,(g) \rightleftharpoons CO\,(g) + H_2\,(g)$

(*c*) I_2 dissolved in CCl_4

(*d*) $N_2\,(g) + 3H_2\,(g) \rightleftharpoons 2NH_3\,(g)$

(*e*) $CuSO_4 . 5H_2O\,(s) \rightleftharpoons CuSO_4 . 3H_2O\,(s) + 2\,H_2O\,(g)$.

2. Draw a labelled diagram for lead-silver system.

3. With the help of a diagram, explain the application of phase rule in the extraction of silver from an ore of lead.

4. Discuss the principle for the preparation of freezing mixtures.

5. Using the $FeCl_3 - H_2O$ system phase diagram, explain how the isothermal evaporation of an unsaturated $FeCl_3$ solution proceeds.

6. Draw the composition–temperature phase diagram of sodium sulphate–water system and label it properly.

7. Calculate the number of components present in the following systems.

(a) A solution of common salt

(*b*) $NH_3\,(g)$

(*c*) $NH_4Cl\,(s) \rightleftharpoons NH_3\,(g) + HCl\,(g)$

(*d*) $Mg\,CO_3\,(s) \rightleftharpoons MgO(s) + CO_2\,(g)$

(*e*) $H_2O\,(l) \rightleftharpoons H_2O\,(l)$.

8. Sketch the cooling curves for the following alloys of Pb-Ag system and explain what happens as their melts are cooled:

(*i*) An alloy of eutectic composition.

(*ii*) An alloy of any other composition.

9. Draw well labelled phase diagram of KI— H_2O system. Discuss how the cooling will proceed if solution contains (*i*) more than 52% KI (*ii*) less than 52% KI (iii) 52% KI.

10. Draw labelled phase diagram of Mg – Zn system. Indicate clearly the eutectic points and congruent melting point on the diagram.

11. Explain how NaCl – H_2O system is considered to be a system with incongruent melting point? Draw the labelled diagram for this system.

12. How does phase diagram of KI – H_2O system differ from that of Pb – Ag system and why? What do you understand by 'cryohydric point'?

13. Draw labelled diagram of $FeCl_3 - H_2O$ system. Where do the congruent melting points and eutectic points (transition points) lie ? Write the equilibria existing at different transition points.
14. How is the $CuSO_4 - H_2O$ system studied? Explain briefly.
15. Give two examples each of two-component system in which
    (*i*) the components do not react with each other.
    (*ii*) the components react to form a compound with congruent melting point.
    (*iii*) the components react to form a compound with incongruent melting point.
16. Draw a well labelled phase diagram of water system. What do you interpret from the slope of the melting point curve?
17. How are the number of components of a chemically reactive system calculated? Explain with two examples.
18. Calculate the total number of variables of a system of C components and P phases.
19. Draw a well labelled phase diagram of sulphur system. What do you interpret from the slopes of transition curve of S_R and melting curve of S_M?
20. What is a condensed system? Write the reduced phase rule equation.
21. Derive the criteria for phase equilibria for a multi-component and multi-phase system.
22. Calculate the number of variables of a system of C components and P phases, because of its being in equilibrium.
24. What do you understand by phase diagram? How are bivariant, univariant and non-variant systems represented in a phase diagram?
23. Explain the term ' degrees of freedom' with suitable examples as used in Phase Rule.
25. Derive the criteria for two phase equilibria for one component system.
26. Draw a well labelled phase diagram of water system. Discuss the effect of change of temperature at constant pressure on this system.
27. What are the main points of difference in the phase diagrams of CO_2 system and H_2O system?
28. Why three dimensional diagram are required for studying two-component system? How is the problem simplified?

Long Answer Questions

1. Define 'Gibbs Phase Rule'. How can you derive it thermodynamically?
2. Draw a phase diagram of water system. Label it and discuss the importance of various points, lines and areas at equilibrium. Discuss the effect of change of temperature at constant pressure.
3. Draw a labelled phase diagram of sulphur system and discuss its salient features.
4. Build the complete phase diagram for sulphur considering two solid phases, one liquid phase and one gas phase. Can all the four phases co-exist?
5. Briefly describe cooling curve method and thaw-melt method for construction of phase diagram of a two-component system miscible in the liquid phase (molten state).
6. State phase rule. Explain the terms used. Apply this rule to lead-silver system.
7. (a) Lead (m.p. 327°C) and silver (m.p. 961°C) form a eutectic mixture at 303°C with 2.6% by weight of silver.
    (*i*) Draw the phase diagram (temperature-composition) for the system.
    (*ii*) Label each region and discuss it fully.

(*iii*) Sketch and explain the type of cooling curve expected for 80% by weight of silver, and for 2.6% by weight of silver.

(*b*) Discuss the desilverization of lead on the basis of this phase diagram.

8. Draw and explain a phase diagram for one component system comparing more than one solid phase.

9. Draw labelled phase diagram of water system and discuss its salient features.

10. Draw the complete phase diagram of sulphur system and prove that the conclusions in regard to the degrees of freedom as derived from this diagram are the same as the deduction from phase rule.

11. How is phase diagram of Bi – Cd system constructed? Discuss the main features of this diagram.

12. Draw a well labelled diagram of carbon dioxide system. Discuss its main features. How does it differ from phase diagram of water system?

13. Draw a schematic phase diagram of sulphur system. Explain clearly the main points of information about the equilibria represented in various areas, along different lines and at various points of the diagram.

14. Draw a labelled temperature-pressure phase diagram of sulphur system. Describe two ways in which the triple point of the system in metastable equilibrium can be obtained.

15. How is phase diagram of a two-component system comprising of two solid components miscible in the liquid phase constructed? What are its main features?

16. Define condensed system from phase rule point of view and say how such a system can be dealt with the help of simple phase diagram. Draw the lead-silver phase diagram and explain how argentiferous lead containing a very low percentage of silver can be concentrated by cooling. State the maximum limit of concentration thus obtained.

17. Discuss the salient features of the phase diagram of lead–silver system.

18. Draw a schematic diagram of KI – H_2O system explaining the main points of information about the equilibria represented in various area along different lines and at various points in the diagram.

19. (*a*) Draw the phase diagram of Ferric chloride-water system. Briefly describe its important features.

(*b*) Briefly explain how the isothermal evaporation of an unsaturated solution of ferric chloride proceeds.

20. Draw a schematic phase diagram for water–sodium sulphate system. Clearly explain the main points of information about the equilibria represented in various areas, along different curves and at various points of the diagram.

21. Explain, with suitable example, the following terms used in the phase rule study of heterogeneous equilibria:

(*i*) Triple point | (*ii*) Transition point
(*iii*) Invariant system | (*vi*) Eutectic point
(*v*) Cryohydric point. | (*vi*) Metastable equilibrium.

22. Briefly describe a two-component system which involves compound formation with congruent melting point.

23. State phase rule. Discuss the application of phase rule to the system consisting of potassium iodide and water.

24. Draw the lead–silver phase diagram and illustrate the principle of 'Pattinson's Process' for enrichment of silver

25. Discuss how the equilibria involving salt hydrates are studied taking the example of $CuSO_4$ – H_2O system.

26. Describe the main features of NaCl – H_2O system after drawing a well labelled diagram.

27. Draw well labelled phase diagram of Mg – Zn system. Discuss the main features of this diagram.

28. Draw a well-labelled phase diagram of KI – H_2O system.
Using the above diagram, brief explain how the phase rule is applied in the preparation of 'Freezing Mixtures'.

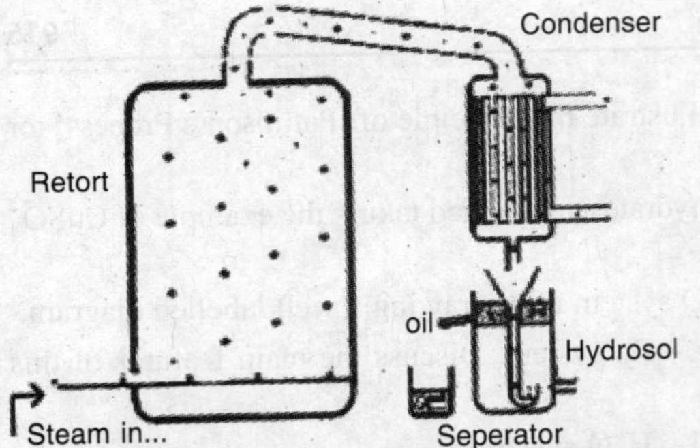

LIQUID–LIQUID MIXTURES

5.1. INTRODUCTION

Mixtures of liquids can be classified into three types:

1. Miscible liquids. If both the components of a mixture are either polar or non-polar, we get a liquid mixture of this type. Examples include methanol + water, benzene + toluene, *n*-hexane + *n*-heptane.

2. Immiscible liquids. If one of the liquids is polar and the other is non-polar, we get this type of mixtures. Examples are benzene + water, aniline + water, *n*-hexane + water, etc.

3. Partially miscible liquids. If one of the liquid is slightly polar and the second is polar or non-polar, we get this type of mixtures. Examples include phenol + water, triethylamine + water, nicotine + water.

5.2. RAOULT'S LAW

Raoult's law states :

In a solution the vapour pressure of a component at a given temperature is equal to the mole fraction of that component in the solution multiplied by the vapour pressure of that component in the pure state. Mathematically,

$$P_i = x_i \times x P^\circ$$

In a *binary solution*

Vapour pressure of the solvent in the solution

= Mole fraction of the solvent in solution ×

Vapour pressure of the pure solvent ...(*i*)

If the solute is non-volatile, it will not contribute to the total vapour pressure of the solution. Thus, the vapour pressure of the solution will be the vapour pressure due to solvent in the solution only.

Raoult

For such solutions, vapour pressure of the solution

= Mole fraction of the solvent in solution × Vapour pressure of the pure solvent

In terms of symbols, we can write

$$p_s = x_1 \times p^o \qquad ...(ii)$$

This can be re-written in the form

$$\frac{p_s}{p^o} = x_1 \qquad ...(iii)$$

If the solution contains n_2 moles of the solute dissolved in n_1 moles of the solvent, we have

Mole fraction of the solvent in solution (x_1)

$$= \frac{n_1}{n_1 + n_2}$$

Substituting this value in equation (*iii*), we get

$$\frac{p_s}{p^o} = \frac{n_1}{n_1 + n_2}$$

Subtracting each side from 1, we get

$$1 - \frac{p_s}{p^o} = 1 - \frac{n_1}{n_1 + n_2}$$

or
$$\boxed{\frac{p^o - p_s}{p^o} = \frac{n_2}{n_1 + n_2}} \quad \text{... (iv)}$$

In this expression, $p^o - p_s$ expresses the lowering of vapour pressure, $\frac{p^o - p_s}{p^o}$ is called relative lowering of vapour pressure, $\frac{n_2}{n_1 + n_2}$ represents the mole fraction of the solute in the solution. Mathematical equation (*iv*) may be stated as:

The relative lowering of vapour pressure of a solution containing a non-volatile solute is equal to the mole fraction of the solute in the solution.

This is another definition of Raoult's law.

NUMERICAL PROBLEMS ON RAOULT'S LAW

Example 1. Liquid *A* (molecular mass 46) and liquid *B* (molecular mass 18) form an ideal solution. At 293 K the vapour pressures of pure *A* and *B* are 44.5 and 17.5 mm of Hg respectively. Calculate (*a*) the vapour pressure of a solution of *A* in *B* containing 0.2 mole fraction of *A*, and (*b*) the composition of the vapour phase.

Solution. (*a*)
$$P_A^o = 44.5 \text{ mm}$$
$$X_A = 0.2$$
$$P_B^o = 17{\cdot}5 \text{ mm}$$
$$X_B = (1 - 0.2) = 0.8$$

According to Raoult's law

$$P_A = P_A^o \cdot X_A$$

Partial pressure of *A*, $P_A = P_A^o X_A = 44.5 \times 0.2 = 8.9$ mm.

Partial pressure of *B*, $P_B = P_B^o X_B$,
$$= 175 \times 0.8$$
$$= 14.0 \text{ mm.}$$

Total pressure $P = P_A + P_B$
$$= 8.9 + 14 = 22.9 \text{ mm.}$$

(*b*) In vapour phase

$$\text{Mole fraction of } A = \frac{\text{Partial pressure of A}}{\text{Total pressure}}$$

or $$X_A = \frac{P_A}{P_A + P_B} = \frac{8.9}{8.9+14} = \frac{8.9}{22.9}$$

$$= 0.39$$

$$X_B = 1 - 0.39 = 0.61.$$

Example 2. The vapour pressure of two pure liquids *A* and *B* are 15000 and 30000 Nm^{-2} at 298 K. Calculate the mole fraction of *A* and *B* in the vapour phase when an equimolar solution of the liquids is made.

Solution. $$P_A^o = 15000 \text{ Nm}^{-2}$$

$$P_B^o = 30000 \text{ Nm}^{-2}$$

In equimolar solution, mole fractions of *A* and *B, i.e.,* X_A and X_B are equal. Let

$$X_A = X_B = 0.5$$

$$X_A = 0.5$$

$$X_B = 1 - 0.5 = 0.5$$

Applying Raoult's law of ideal solution.

$$P_A = P_A^o X_A$$

$$= 15000 \times 0.5 = 7500 \text{ Nm}^{-2}$$

and $$P_B = P_B^o X_B$$

$$= 30000 \times 0.5 = 15000 \text{ Nm}^{-2}$$

Total pressure $$P = P_A + P_B$$

$$= 7500 + 15000 = 22500 \text{ Nm}^{-2}$$

In the vapour phase

$$\text{Mole fraction of } A = \frac{\text{Partial pressure A}}{\text{Total pressure}}$$

$$= \frac{7500}{22500} = 0.3333$$

$$\text{Mole fraction of } B = 1 - 0.3333 = 0.6667$$

Example 3. Benzene C_6H_6 (b.p. 353.1 K) and toluene C_7H_8 (b.p. 383.6 K) are two hydrocarbons that form a very nearly ideal solution. At 313 K, the vapour pressure of pure liquids are 160 mm Hg and 60 mm Hg respectively. Assuming an ideal solution behaviour, calculate the partial pressures of benzene and toluene and the total pressure over the following solutions:

(*i*) One made by combining equal number of toluene and benzene molecules.

(*ii*) One made by combining 4 mol of toluene and 1 mol of benzene.

(*iii*) One made by combining equal masses of toluene and benzene.

Solution. (*i*) When the number of molecules of toluene and benzene are equal, that means the number of moles of the two liquids are also equal :

Thus mole fraction of benzene $$X_A = \frac{1}{1+1} = 0.5$$

and mole fraction of toluene $X_B = 1 - 0.5 = 0.5$

According to Raoult's law

$$P = P^o . X$$

Partial pressure of benzene $P_A = P_A{}^o \cdot X_A$

$= 160 \times 0.5 = 80$ mm

Partial pressure of toluene $P_B = P_B{}^o X_B = 60 \times 0.5 = 30$ mm

Total vapour pressure $P_A + P_B = 80 + 30 = 110$ mm

(*ii*) Mole fraction of benzene $X_A = \dfrac{1}{1+4} = \dfrac{1}{5} = 0.2$

Mole fraction of toluene $X_B = 1 - 0.2 = 0.8$

Partial pressure of benzene $P_A = P^0{}_A X_A = 160 \times 0.2 = 32$ mm

Partial pressure of toluene $P_B = 60 \times 0.8 = 48$ mm

Total vapour pressure $P_A + P_B = 32 + 48 = 80$ mm

(*iii*) Here the masses of the two liquids are the same. Let the amount of each be *m* g

Then mole fraction of benzene $X_A = \dfrac{m/78}{m/78 + m/92} = 0.541$

Mole fraction of toluene $X_B = 1 - 0.541 = 0.459$

Partial pressure of benzene $P_A = P_A{}^o X_A = 160 \times 0.541$

$= 86.56$ mm of Hg

Partial pressure of toluene $P_B = P_B{}^o X_B$

$= 60 \times 0.459 = 27.54$ mm of Hg

Total vapour pressure $= 86.56 + 27.54 = 114.1$ mm

Example 4. The vapour pressures of pure components *A* and *B* are 120 mm and 96 mm Hg. What will be the partial pressures of the components and the total pressure when the solution contains 1 mole of component *A* and 4 mole of component *B* and the solution is ideal? What will be the composition in the vapour phase ?

Solution. $P_A{}^o = 120$ mm

Here $P_B{}^o = 96$ mm

Mole fraction of *A*, $X_A = \dfrac{\text{No. of moles of } A}{\text{No. of moles of } A + \text{no. of moles of } B}$

$= \dfrac{1}{1+4} = \dfrac{1}{5} = 0.2$

Mole fraction of *B*, $X_B = 1 - 0.2 = 0.8$

Partial pressure of component *A*, $P_A = P_A{}^o \times X_A$

$= 120 \times 0.2 = 24.0$ mm Hg

Partial pressure of *B*, $P_B = P_B{}^o \times X_B$

$= 96 \times 0.8 = 76.8$ mm

Total pressure $= P_A + P_B = 24 + 76.8$ mm $= 100.8$ mm

Mole fraction of *A* in the vapour phase $= \dfrac{P_A}{Total\ pressure} = \dfrac{24}{100.8} = 0.238$

Mole fraction of *B* in the vapour phase $= 1 - 0.238 = 0.762$

5.3. HENRY'S LAW

Henry's law states as :

The mass of a gas dissolved in a given volume of the liquid at constant temperature is directly proportional to the pressure of the gas present in equilibrium with the liquid.

Mathematically, $m \propto p$

or $m = kp$

where m = mass of the gas dissolved in a unit volume of the solvent

p = pressure of the gas in equilibrium with the solvent

k = constant called *Henry's law constant.* It is characteristic of the nature of the gas, the nature of the solvent and the temperature.

This is further illustrated with the help of Fig. 5.1

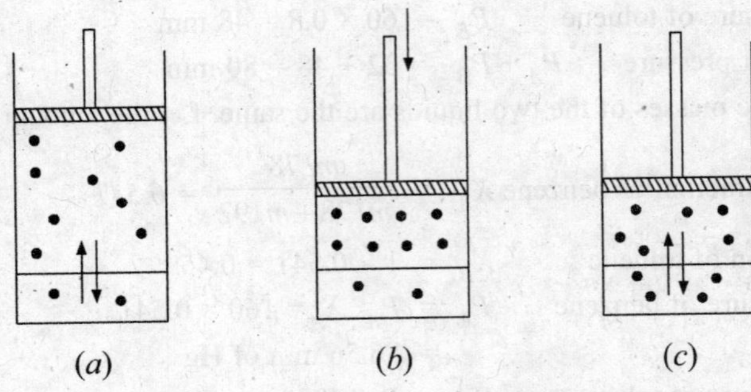

Fig. 5.1. (*a*) Equilibrium exists between the gas molecules and those in solution. (*b*) When pressure is applied, gas volume decreases and concentration in gas phase increases; (*c*) More gas molecules pass into solution to re-establish equilibrium whereby the pressure above the gas is decreased.

Figure 5.1 explains how the concentration of the gas in the solvent increases with rise of pressure.

Another statement of Henry's law is :

The volume of the gas dissolved, measured at the pressure used, is independent of the pressure.

The second statement of Henry's law is just a corollary of the first. This can be proved as under:

Suppose volume of the liquid taken = V ml

Pressure of the gas = p mm

Mass of the gas dissolved = m grams

Obviously, volume of the gas dissolved = V ml

Now, suppose the pressure is doubled *i.e.,* it becomes = $2p$

∴ The mass of the gas dissolved will be = $2\,m$

But volume of m grams of the gas at pressure p = V ml

∴ Volume of $2m$ grams of the gas at pressure p = $2\,V$ ml

or volume of $2m$ grams of the gas at pressure $2p$ = V ml

(by Boyle's law, volume and pressure are inversely related)

Thus the volume of the gas dissolved at pressure $2p$ is again the same *viz.* V ml

Third statement of Henry's law is as follows :

For a gas in equilibrium with a liquid, the ratio of the concentration of the gas in the solution phase to that in the gaseous phase is constant at constant temperature provided the gas remains in the same molecular state in both the phases.

Solubility of a gas can be defined in two different ways:

(*i*) *The solubility of a gas in a liquid (called the solubility co-efficient) is the volume of the gas in ml which will dissolve in 1 ml of the liquid to form a saturated solution, the volume of the gas being measured at the temperature and pressure at which the measurement of the solubility is made.*

(*ii*) *The solubility co-efficient of a gas in a liquid is also expressed as absorption co-efficient which is the volume of the gas in ml which will dissolve in 1 ml of the liquid at the temperature of the experiment under a partial pressure of the gas equal to one atmosphere to form the saturated solution, the volume of the gas being measured at N.T.P.* It is usually represented by α. Thus if v_0 is the volume of the gas dissolved, reduced to S.T.P., V is the volume of the solvent, p is the partial pressure of the gas in atmospheres, then the absorption co-efficient α will be given by

$$\alpha = \frac{v_0}{V_{Vp}}$$

Factors affecting solubility of a gas

The following factors affect the solubility:

(*i*) **Nature of the gas and the solvent:** Gases like hydrogen, oxygen, nitrogen etc. dissolve in water only to a small extent whereas gases like CO_2, HCl, NH_3 etc. are highly soluble. The greater solubility of the latter gases is due to their reaction with the solvent.

Again oxygen, nitrogen and carbon dioxide are much more soluble in ethyl alcohol than in water at the same temperature and pressure while H_2S and NH_3 are more soluble in water than in ethyl alcohol. Evidently, the greater solubility of gas in a solvent is again due to the chemical similarity between the gas and the solvent.

(*ii*) **Effect of temperature:** *The solubility of a gas decreases with increase in temperature.* This is expected because on heating the solution of a gas, some gas is usually expelled out of the solution. The same result also follows in an alternate manner as under :

The dissolution of a gas in a liquid is an exothermic process *i.e.*, it is accompanied by evolution of heat, *i.e.*

$$\text{Gas + Solvent} \rightleftharpoons \text{Solution + Heat}$$

Applying Le Chatelier's principle, it is evident that increase of temperature would shift the equilibrium in the backward direction *i.e.*, the solubility would decrease.

It may be pointed out that though generally the solubility of a gas in a liquid decreases with increase of temperature, there are some exceptions. For example, the solubility of some sparingly soluble gases, such as hydrogen and inert gases, increases slightly with increase of temperature especially in the non-aqueous solvents such as hydrocarbons, alcohols and acetone.

(*iii*) **Effect of pressure:** This is the most important factor influencing the solubility of a gas in a liquid at a particular temperature. It is governed by **Henry's law** according to which the solubility increases with the increase of pressure.

Limitations of Henry's law

Henry's law is applicable only if the following conditions are satisfied :

(*i*) *The pressure should be low and the temperature should be high i.e., the gas should behave like an ideal gas.*

(ii) The gas should not be involved in compound formation with the solvent. It should not undergo association or dissociation in the solvent.

For example, the law is not applicable in case of dissolution of ammonia in water because it forms a compound and then there is dissociation.

$$NH_3(g) + H_2O \longrightarrow NH_4OH\ (aq)$$

$$NH_4OH\ (aq) \rightleftharpoons NH_4^+\ (aq) + OH^-\ (aq)$$

Similarly, the law is not applicable to the dissolution of HCl gas in water because it undergoes dissociation after dissolution

$$HCl(g) + aq \longrightarrow HCl\ (aq)$$

$$HCl(aq) \longrightarrow H^+\ (aq) + Cl^-\ (aq)$$

NUMERICAL PROBLEM ON HENRY'S LAW

Example . The solubility of pure oxygen in water at 20°C and 1.00 atmosphere pressure is 1.38×10^{-3} mole/litre. Calculate the concentration of oxygen at 20°C and partial pressure of 0.20 atmosphere.

Solution. Calculation of *k* :

$$m = kP$$

or $$k = \frac{m}{P}$$

Substituting values for pure oxygen,

$$k = \frac{1.38 \times 10^{-3}\ \text{mole / litre}}{1.00\ \text{atm}} = 1.38 \times 10^{-3}\ \frac{\text{mol / litre}}{\text{atm}}$$

Calculation of concentration of O_2 at given pressure

$$m = kP$$

$$\text{Concentration of } O_2 = 1.38 \times 10^{-3} \frac{\text{mole / litre}}{\text{atm}} \times 0.20\ \text{atm}$$

$$= 2.76 \times 10^{-4}\ \text{mole / litre}$$

5.4. DUHEM-MARGULES EQUATION

Let us consider a liquid mixture of two components in equilibrium with their vapour. At *constant temperature and pressure,* the condition for an infinitesimal change of composition is given by Gibbs-Duhem equation in the form

$$n_1 d\mu_1 + n_2 d\mu_2 = 0 \qquad \text{... } (i)$$

where n_1 and n_2 are the number of moles of the constituents 1 and 2 respectively in the solution and μ_1 and μ_2 are their chemical potentials.

Dividing eqn. (*i*) by $n_1 + n_2$, we get

$$\frac{n_1}{n_1 + n_2} d\mu_1 + \frac{n_2}{n_1 + n_2} d\mu_2 = 0$$

or $$x_1 d\mu_1 + x_2 d\mu_2 = 0 \qquad \text{... } (ii)$$

where x_1 and x_2 represent the mole fractions of the components 1 and 2 in the mixture (solution).

The chemical potential of any component in the mixture depends upon temperature, pressure and composition of the mixture. If the temperature and pressure are kept constant, the chemical potential depends only upon the composition. Then for an infinitesimal change of composition, we may write

$$d\mu_1 = \left(\frac{\partial \mu_1}{\partial x_1}\right)_{T,P} dx_1 \qquad \text{... (iii)}$$

and

$$d\mu_2 = \left(\frac{\partial \mu_2}{\partial x_2}\right)_{T,P} dx_2 \qquad \text{... (iv)}$$

Substituting these values in eqn. (*ii*), we get

$$x_1\left(\frac{\partial \mu_1}{\partial x_1}\right)_{T,P} dx_1 + x_2\left(\frac{\partial \mu_2}{\partial x_2}\right)_{T,P} dx_2 = 0 \qquad \text{... (v)}$$

We know that the sum of mole fractions of components in a mixture is unity

or $\qquad x_1 + x_2 = 0$

Hence $\qquad dx_1 + dx_2 = 0$

or $\qquad dx_1 = -dx_2 \qquad$ ...(*vi*)

Substituting this value in eqn. (*v*) and cancelling the common factor dx_2 we get

$$x_1\left(\frac{\partial \mu_1}{\partial x_1}\right) - x_2\left(\frac{\partial \mu_2}{\partial x_2}\right)_{T,P} = 0$$

or

$$x_1\left(\frac{\partial \mu_1}{\partial x_1}\right)_{T,P} = x_2\left(\frac{\partial \mu_2}{\partial x_2}\right)_{T,P} \qquad \text{... (vii)}$$

The chemical potential of any constituent of a liquid mixture is given by

$$\mu = \mu_o + RT \ln p \qquad \text{...(viii)}$$

where p is the partial pressure of that constituent

Differentiating eqn. (*viii*) *w.r.t.* x at constant temperature, we get

$$\frac{\partial \mu}{\partial x} = RT\frac{d \ln p}{dx}$$

For the constituents 1 and 2, we can write two separate equations, given below

$$\frac{\partial \mu_1}{\partial x_1} = RT\frac{d \ln p_1}{dx_1} \qquad \text{... (ix)}$$

and

$$\frac{\partial \mu_2}{\partial x_2} = RT\frac{d \ln p_2}{dx_2} \qquad \text{... (x)}$$

Substituting these values in eqn. (*vii*) and simplifying, we get

$$\boldsymbol{x_1\frac{d \ln p_1}{dx_1} = x_2\frac{d \ln p_2}{dx_1}} \qquad \text{... (xi)}$$

It can also be written in the form

$$\frac{d \ln p_1}{d \ln x_1} = \frac{d \ln p_2}{d \ln x_2} \quad \text{... (xii)}$$

Equations (*xi*) and (*xii*) are two different forms of Duhem-Margules equation.

COMPLETELY SOLUBLE LIQUIDS

5.5. IDEAL AND NON-IDEAL SOLUTIONS

Ideal solutions: *A solution of two components is said to be ideal if each component of the solution obeys Raoult's law at all temperatures and concentrations. That is, the vapour pressure of each component in the solution at a given temperature is equal to the mole fraction of that component in the solution multiplied by the vapour pressure of that component in the pure state at the same temperature.* Thus if two liquids *A* and *B* are mixed together, the solution formed by them will be ideal if the following relations hold.

$$P_A = x_A \times P_A^o$$

and

$$P_B = x_B \times P_B^o$$

where p_A = vapour pressure of component *A* in the solution

x_A = mole fraction of the component *A* in the solution

p_A^o = vapour pressure of the liquid *A* in the pure state

P_B, x_B, P_B^o are the corresponding terms for the component *B*.

It is observed that *the solution formed by two liquids A and B is ideal if the interactions* (*i.e., the intermolecular forces of attraction*) *among the molecules of A and B in the solution are just the same as among the molecules of pure A or B.*

Further, it is observed that *the solution formed on mixing the two liquids is ideal if*

(*i*) *there is no volume change on mixing i.e.,* $\Delta V_{mix} = 0$

(*ii*) *there is no enthalpy change on mixing i.e.,* $\Delta H_{mix} = 0$

For example, if 50 ml of *n*-hexane are mixed with 50 ml of *n*-heptane, the total volume of the solution is found to be 100 ml and the temperature of the solution is found to be same as that of pure *n*-hexane and *n*-heptane before mixing. It shows that no heat is evolved or absorbed on mixing. Such a solution is an ideal solution.

Non-ideal solution: On the other hand, a solution formed by mixing two liquids is said to be non-ideal if it does not obey Raoult's law or the interactions of *A* and *B* molecules in the solution are not similar to those of pure *A* and pure *B* or $\Delta V_{mix} \neq 0$, and $\Delta H_{mix} \neq 0$.

5.6. VAPOUR PRESSURE-COMPOSITION DIAGRAM FOR IDEAL SOLUTIONS

In case of ideal solutions, the partial pressure of each component can be calculated using Raoult's law *i.e.*,

$$p_A = x_A \times p_A^o$$

and

$$p_B = x_B \times p_B^o$$

where p_A, p_A^o, p_B and p_B^o are all measured at the same temperature.

Thus

$$P_{sol} = x_A\, p_A^o + x_B\, p_B^o$$

At any particular temperature, we can determine the values of $p_{sol.}$ experimentally for different mole fractions. The values p_A^o and p_B^o for the pure liquids can be measured at the same temperature and hence p_A and p_B can be calculated for any particular mole fraction. For an ideal solution, the straight line plots are obtained for vapour pressure *vs.* composition as shown in Fig. 5.2. The dotted lines represent the plots of the partial pressures and the solid line that of the solution.

In these plots, for any mole fraction, the total vapour pressure of the solution is the sum of the partial pressures of its components A and B.

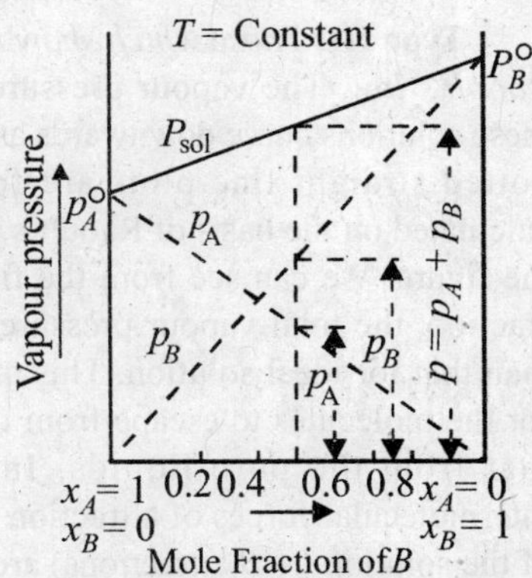

Fig. 5.2. Vapour pressure-composition diagram for an ideal solution.

5.7. VAPOUR PRESSURE-COMPOSITION CURVES FOR NON-IDEAL SOLUTIONS

(Deviation from ideal behaviour)

In case of non-ideal solutions, the plots of vapour pressure *vs.* composition (mole fraction) are curved lines instead of straight lines. Non-ideal solutions are divided into the following three types :

Type I: *Those which show small positive deviation from Raoult's law.* The vapour pressure-composition graphs of these solutions curve are slightly upwards as shown in Fig. 5.3. The dotted straight line plots are for ideal solution, calculated by using Raoult's law.

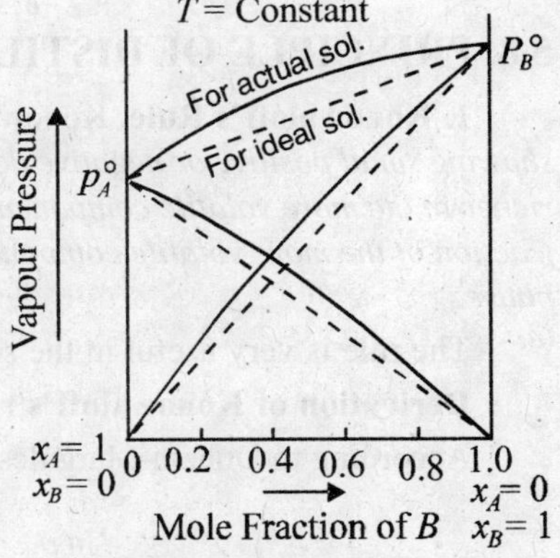

Fig. 5.3. Vapour pressure-composition plots for non-ideal solution showing small positive deviation.

Examples of solutions of this type are benzene + toluene, water + methyl alcohol, carbon tetrachloride + cyclohexane etc.

Type II: *Those which show large positive deviation from Raoult's law.* The vapour pressure-composition plots of these solutions curve upwards considerably as shown in Fig. 5.4. The dotted straight line plots, included in the figure are for ideal solution (calculated by means of Raoult's law).

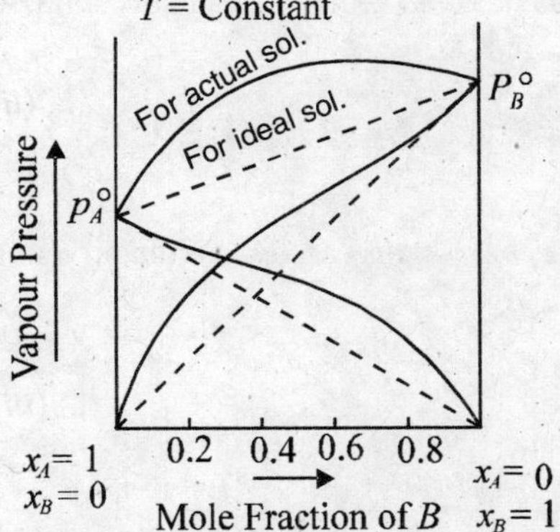

Fig. 5.4. Vapour pressure-composition plots for non-ideal solution showing large positive deviation

We can see from Figs. 5.3 and 5.4 that for any mole fraction, the total vapour pressure of the solution is more than that for ideal solution. This means that the tendency for the molecules to escape from the solution is more than from the pure liquids. In such solutions, the intermolecular forces of attraction between the molecules of the solution (*A-B* attractions) are weaker than those of either of the pure components (*A-A* attractions or *B-B* attractions). Such a behaviour is associated with an increase in volume and absorption of heat on mixing.

Some examples of liquid pairs showing large positive deviation include water + ethanol, ethanol + chloroform.

Type III: *Those which show negative deviation from Raoult's law.* The vapour pressure-composition plots of these solutions curve downwards as shown in Fig. 5.5. The dotted straight line plots are for the ideal solution, calculated on the basis of Raoult's law and are included in the figure. We can see from the figure that for any mole fraction, the total vapour pressure of the solution is less than that for ideal solution. This means that the tendency for the molecules to escape from the solution is less than that from the pure liquids. In such solutions, the intermolecular forces of attraction between the molecules of the solution (*A-B* attractions) are stronger than those of either of the pure components (*A-A* attractions or *B-B* attractions). It causes contraction in volume and evolution of heat in making the solution.

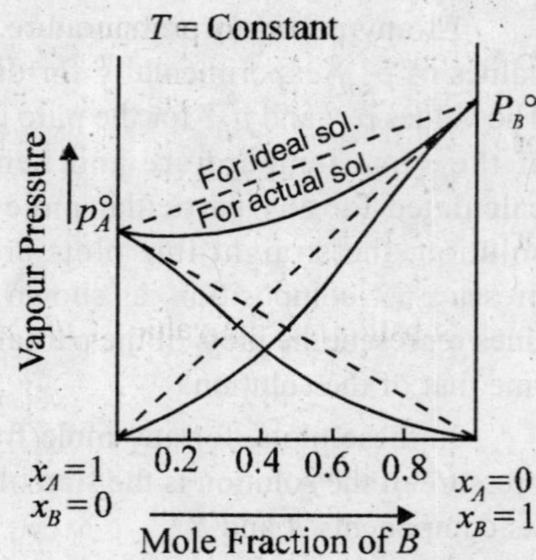

Fig. 5.5. Vapour pressure-composition plots for non-ideal solution showing negative positive deviation.

Some examples of the liquid pairs showing negative deviation are water + nitric acid, water + sulphuric acid, acetone + chloroform etc.

5.8. PRINCIPLE OF DISTILLATION OF BINARY MISCIBLE LIQUIDS

1. Konowaloff's Rule. Konowaloff's rule states that *in case of ideal solutions or solutions showing small positive or negative deviations, at any fixed temperature, the vapour phase is always richer in the more volatile component as compared to the solution phase.* In other words, *the mole fraction of the more volatile component is always greater in the vapour phase than in the solution phase.*

The rule is very useful in the separation of binary liquid solutions by distillation.

Derivation of Konowaloff's rule from the Duhem-Margules equation.

According to Duhem-Margules equation,

$$x_1 \frac{d \ln p_1}{dx_1} = x_2 \frac{d \ln p_2}{dx_2} \quad \text{... } (i)$$

It works out to

$$\frac{x_1}{dx_1} \frac{dp_1}{p_1} = \frac{x_2}{dx_2} \frac{dp_2}{p_2} \qquad \left(\because d \ln p = \frac{dp}{p} \right)$$

or

$$\frac{x_1}{p_1} \frac{dp_1}{dx_1} = \frac{x_2}{p_2} \frac{dp_2}{dx_2} \quad \text{... } (ii)$$

But $x_1 + x_2 = 1$ (sum of mole fractions = 1)

$\therefore \quad dx_1 + dx_2 = 0$ or $dx_1 = -dx_2$

$\therefore$ Equation (*ii*) takes the form

$$\frac{x_1}{p_1} \frac{dp_1}{dx_1} = -\frac{x_2}{p_2} \frac{dp_2}{dx_1} \quad \text{... } (iii)$$

or

$$\frac{dp_1}{dx_1} = -\frac{x_2 p_1}{x_1 p_2} \frac{dp_2}{dx_1} \quad \text{... } (iv)$$

The total vapour pressure p is the sum of partial pressures of the components.

$$P = p_1 + p_2$$

or

$$dP = dp_1 + dp_2$$

or

$$\frac{dP}{dx_1} = \frac{dp_1}{dx_1} + \frac{dp_2}{dx_1} \quad \text{... (v)}$$

Substituting the value of dp_1/dx_1 from equation (*iv*) in eq. (*v*), we get

$$\frac{dP}{dx_1} = -\frac{x_2 p_1}{x_1 p_2}\frac{dp_2}{dx_1} + \frac{dp_2}{dx_1}$$

$$= \frac{dp_2}{dx_1}\left(1 - \frac{x_2 p_1}{x_1 p_2}\right) \quad \text{... (vi)}$$

Also, we have $\quad dx_1 = -dx_2$

or

$$\frac{dp_2}{dx_1} = -\frac{dp_2}{dx_2}$$

Thus $\frac{dp_2}{dx_1}$ is a negative quantity. Hence according to equation (*vi*), $\frac{dP}{dx_1}$ can be positive only when

$$\frac{x_2 p_1}{x_1 p_2} > 1$$

or

$$\frac{p_1}{p_2} > \frac{x_1}{x_2}$$

This means that the vapour is richer in component 1 (or *A)* than is the liquid from which it vaporises. This is what is implied by Konowaloff's law.

2. Vapour pressure-liquid composition and vapour pressure-vapour composition curves for liquid mixtures

At any particular temperature, the composition of the vapour phase is not the same as that of the liquid phase. Hence we have different *vapour pressure-vapour composition curves.* At any particular temperature, these can be obtained by measuring the vapour pressure as well as the composition of the vapour phase corresponding to each composition of the liquid mixture.

For the real solutions, there are following three types of vapour pressure-composition diagrams:

Type I: *Those in which total vapour pressure is intermediate between those of the pure components and do not show any maximum or minimum.*

Type II: *Those which show a maximum in the total vapour pressure curve.*

Type III: *Those which show a minimum in the total vapour pressure curve.*

The vapour pressure-liquid composition curves as well as the vapour pressure-vapour composition curves of all the three types are shown in Fig. 5.6, in which liquid B is more volatile than liquid A

It may be observed that in all the three types, the vapour composition curve lies below the liquid composition curve. This is explained for each type as under :

(*i*) *In type I*, as B is more volatile than A, corresponding to composition a of the liquid phase, vapour phase will be richer in B *i.e.,* corresponding to composition a'.

(*ii*) *In type II,* upto the point C, behaviour is same as in Type I, at C, the liquid phase and the vapour phase have the same composition and after point C, the vapour phase is richer in A (or less rich in B) as seen from points c and c'.

(*iii*) *In type III,* upto the point D, the vapour phase is richer in A (or less rich in B) than the liquid phase, at the point D both have the same composition and after point D, the behaviour is similar to Type I. This can be seen from points d, d', e and e'.

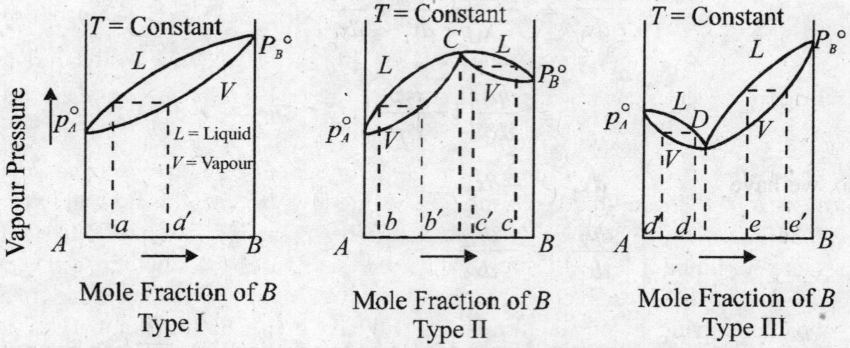

Fig. 5.6. Vapour pressure composition diagrams of binary miscible real solution at constant temperature.

3. Boiling point-composition diagrams for liquid mixtures

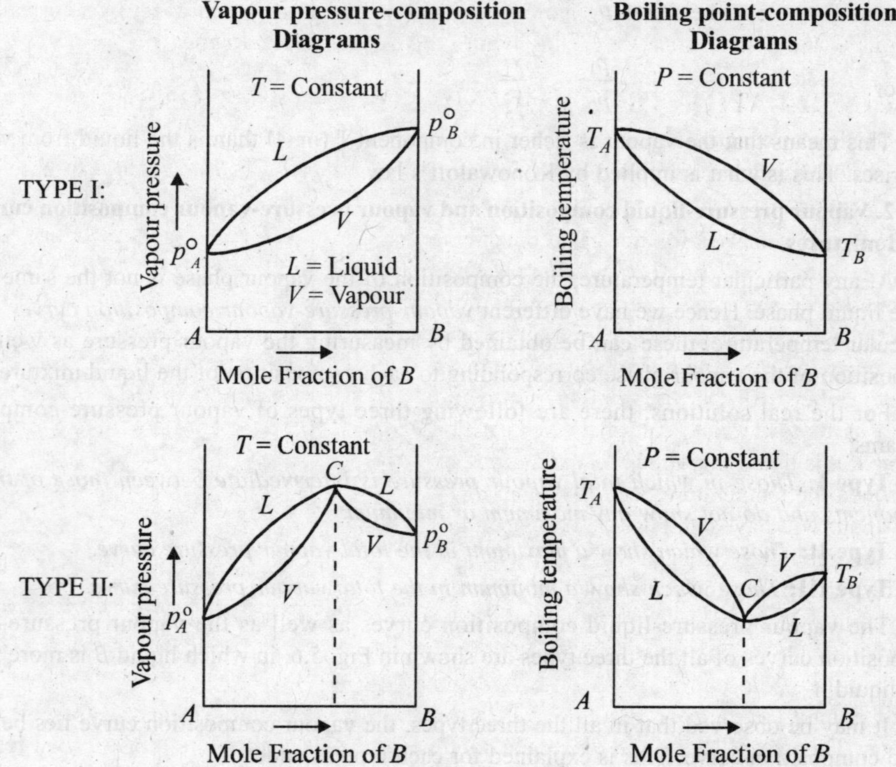

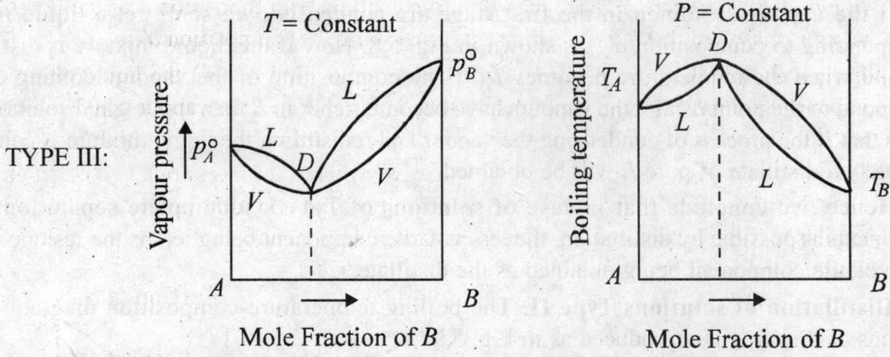

Fig. 5.7. Boiling point-composition diagrams constructed from the vapour pressure-composition diagrams of binary miscible solutions.

Boiling point is the temperature at which the total vapour pressure becomes equal to the atmospheric pressure. Thus a liquid or a solution having low vapour pressure will have a higher boiling point because it has to be heated more to make the vapour pressure equal to the atmospheric pressure. Thus pure liquid *A* which has less vapour pressure will have a higher boiling point and the pure liquid *B* which has higher vapour pressure will have lower boiling point. Similarly a solution of any composition having lower vapour pressure will have higher boiling point. These facts help us to construct the boiling point-composition diagrams from the vapour pressure-composition diagrams as shown in Fig. 5.7.

The point *C* in Type II which corresponds to maximum vapour pressure represents the lowest boiling point. Similarly, the point *D* in Type III which represents the lowest vapour pressure corresponds to the highest boiling point. The liquid mixtures having compositions corresponding to points *C* and *D* have constant boiling points and are called **azeotropes**.

5.9. DISTILLATION OF BINARY MISCIBLE SOLUTIONS

Distillation of solutions of Type I-Fractional distillation: The boiling temperature composition diagram for such a solution involving the liquids *A* and *B* in which *B* is more volatile than *A* is explained with the help of Fig. 5.8.

Suppose we have a solution corresponding to composition *a*. No boiling will start until temperature T_a is reached. The composition of the vapour phase at this stage will be *a′ i.e.*, it is richer in the component *B*. The residue, therefore, must become richer in *A i.e.*, the composition of the residue shifts towards *A*, say it becomes equal to *b*. Now if this liquid mixture is heated, it will boil only when the temperature becomes equal to T_b. The vapour will have composition corresponding to *b′ i.e.*, again it is richer in *B* and consequently the composition of the residue will be further enriched in *A*. If the process of heating the residue is continued, the boiling point of the solution will rise from the initial boiling point T_a towards the boiling point T_A of the pure liquid *A*. Every time the residue becomes richer in *A* than the original solution. This means that if the process is continued for sufficiently long time, a final residue of pure *A* can be obtained.

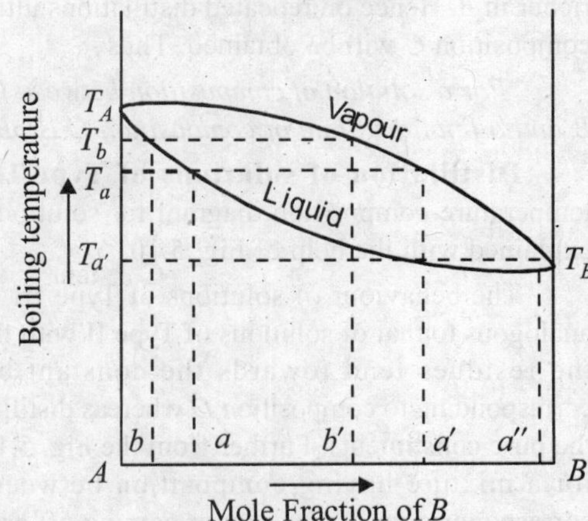

Fig. 5.8. Distillation behaviour of solutions of Type I.

If the vapours obtained in the first stage are condensed, we shall get a liquid mixture corresponding to composition a' , as shown in Fig. 5.8. Now if this liquid mixture is distilled, it will boil when the temperature becomes T_a' . The composition of the vapours coming off will correspond to the point a'' *i.e.,* the vapours have become richer in B than the original solution. This means that if the process of condensing the vapour and redistilling the liquid mixture is continued, ultimately a distillate of pure B will be obtained.

Hence, we conclude that in case of solutions of Type I, a complete separation of the components is possible by distillation, the less volatile component being left as the residue and the more volatile component being obtained as the distillate.

Distillation of solutions Type II: The boiling temperature-composition diagram for the solutions of this type is reproduced as in Fig. 5.9.

An example of this type is water (A)-ethyl alcohol (B) system. If a solution of composition between A and C, such as a, is distilled, the vapour coming off will have the composition a' and will be richer in B and the residue will, therefore, become richer in A *i.e.,* the composition of the residue shifts towards A till ultimately the *residue of pure A* is obtained. The liquid obtained on condensing the vapours (corresponding to composition a'), if distilled will give vapour richer in B. If the condensation of vapours and the distillation of the liquid is continued, ultimately the vapours of composition C will be obtained. If these vapours are condensed and the solution distilled, the distillate obtained will have the same composition as the solution. Hence no further separation is possible by distillation.

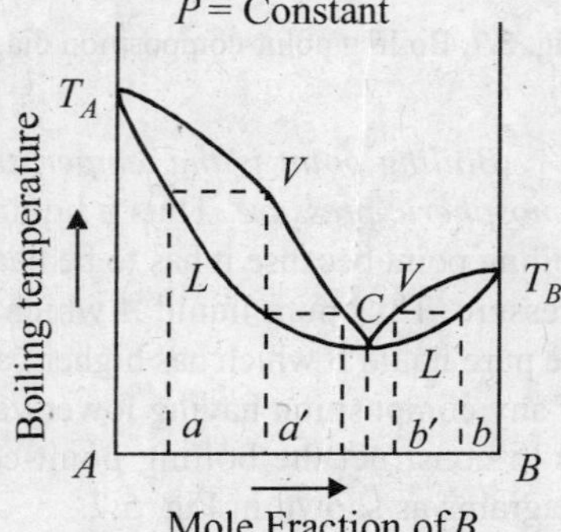

Fig. 5.9. Distillation of solutions of Type II.

Thus *"In case of solutions of Type II, a solution of composition between A and C on fractional distillation gives residue of pure A and a final distillate of composition C. No pure B can be recovered."*

Now consider the distillation of a solution having composition between C and B, say corresponding to the point b. The vapour coming off will be richer in A and so the residue will be richer in B. Hence on repeated distillations ultimately a residue of pure B and the final distillate of composition C will be obtained. Thus

"For a solution of composition between C and B, on fractional distillation, a residue of pure B and a final distillate of composition C is obtained. No pure A can be recovered.

Distillation of solutions of Type III: The boiling temperature-composition diagram for solutions of this type, is explained with the help of Fig. 5.10.

The behaviour of solutions of Type III on distillation is analogous to that of solutions of Type II with the exception that the residues tend towards the constant boiling mixture corresponding to composition D whereas distillates tend towards the pure constituents. Further from the Fig. 5.10, it is clear that for a mixture having composition between A and D (say corresponding to a), the vapour coming off are richer in A and hence the residue is richer in B. In other words, the composition of the residue shifts towards D and ultimately becomes equal to that of D, the composition of the vapours shifts towards A and

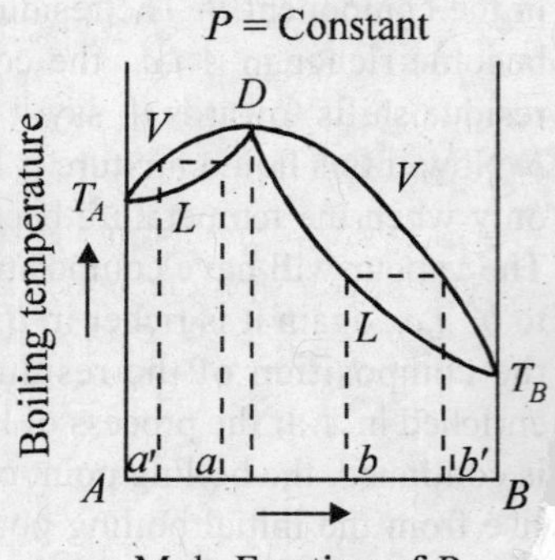

Fig. 5.10. Distillation of solutions of Type III.

finally a distillate of pure A is obtained. Thus *for a mixture having composition between A and D, distillate of pure A is obtained and the residue has composition corresponding to point D* (*i.e.*, a constant boiling mixture with maximum boiling point). In a similar manner, it can be seen that *for a mixture having composition between D and B, ultimately distillate of pure B and residue of composition corresponding to D is obtained.*

5.10. AZEOTROPES

Mixtures of liquids which boil at constant temperature like a pure liquid such that the distillate has the same composition as that of the liquid mixture are called constant boiling mixtures or azeotropic mixtures or simply azeotropes. Obviously, the components of an azeotrope cannot be separated by fractional distillation. These azeotropes are of two types.

(*a*) **Minimum boiling (point) azeotropes:** These azeotropes are formed by liquid pairs which show positive deviations from ideal behaviour. Such an azeotrope corresponds to an intermediate composition for which the total vapour pressure is the highest and hence the boiling point is the lowest. Such azeotropes have boiling points lower than either of the pure components. Ethyl alcohol-water is an example of this type of azeotrope.

(*c*) **Maximum boiling (point) azeotropes:** These azeotropes are formed by those liquid pairs which show negative deviations from ideal behaviour. Such an azeotrope corresponds to an intermediate composition for which the total vapour pressure is minimum and hence the boiling point is maximum. Such azeotropes have boiling points higher than either of the pure components. Water—HCl is an example of this type of azeotrope.

5.11. IMMISCIBLE LIQUIDS

Immiscible liquids mean those liquids which do not mix with each other at all. Thus, if such a mixture of two liquids is heated, each will behave as if the other were not present. Hence, the total vapour pressure above the mixture will be same as the sum of the vapour pressures of the two pure constituents *i.e.*,

$$P = p_A^o + p_B^o$$

where P is the total vapour pressure and p_A^o, p_B^o stand for the vapour pressures of the two pure liquids A and B at the same temperature.

Normal boiling point is the temperature at which the total vapour pressure becomes equal to the atmospheric pressure. In an immiscible liquid pair, since each liquid exerts vapour pressure, the total vapour pressure will become equal to atmospheric pressure at a temperature lower than the boiling point of either of the two liquids. Thus the liquid mixture boils at a temperature lower than the boiling point of either of the two liquids. As soon as one of the liquids is boiled away, the boiling temperature will rise from that of the mixture to either T_A or T_B depending upon which liquid has already boiled off.

At the boiling point of the mixture, the vapour pressure exerted by each liquid is proportional to its number of molecules and hence to the number of moles present in the vapour phase.

Therefore, we can write

$$\frac{p_A^o}{p_B^o} = \frac{n_A}{n_B}$$

where n_A and n_B are the number of moles of A and B respectively present in the vapour phase at the boiling point of the mixture.

Changing moles into weights, we have

$$n_A = \frac{W_A}{M_A}, n_B = \frac{W_B}{M_B}$$

where W_A is the weight of the vapours of A
M_A is the molecular weight of A
W_B is the weight of the vapours of B
M_B is the molecular weight of B

Thus
$$\frac{p_A^o}{p_B^o} = \frac{W_A / M_A}{W_B / M_B}$$

$$= \frac{W_A \cdot M_B}{W_B \cdot M_A}$$

or
$$\boxed{\frac{W_A}{W_B} = \frac{p_A^o \times M_A}{p_B^o \times M_B}}$$

If the vapours are condensed to give the distillate, the ratio of the weights of A and B in the distillate will thus be given by the above expression.

5.12. STEAM DISTILLATION

This method of distillation is based upon the principle of boiling of two immiscible liquids and is used in the laboratory or in the industry for the purification of organic liquids.

Applicability : The process is generally applied only under the conditions:

(*i*) The organic liquid to be purified should be immiscible with water

(*ii*) It should have high molecular weight

(*iii*) It should have vapour pressure at about 100°C

(*iv*) Impurities present should be non-volatile

Experimental details : The apparatus used as shown in Fig. 5.11.

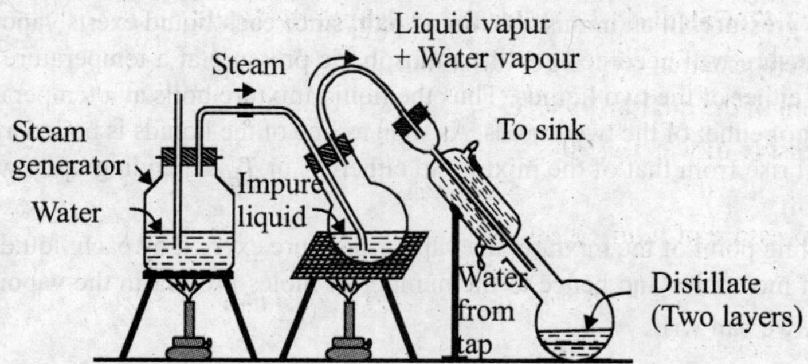

Fig. 5.11. Experimental set-up for steam distillation.

The steam is produced in the steam generator and is passed into the flask containing the impure liquid and kept in a slanting position. This flask is heated from below to prevent the excessive condensation of steam into it. Thus, in the space above the liquid in the flask, water vapours as well as the vapours of the liquid are present. The liquid water mixture present in the flask will therefore boil at a temperature lower than the boiling point of either of the two. The vapours on passing through the condenser will give distillate containing the liquid and the water in the ratio calculated using the formula given below :

$$\frac{W_l}{W_w} = \frac{p_l^o \cdot M_l}{p_w^o \cdot M_w}$$

The subscripts l and w stand for liquid and water respectively.

NUMERICAL PROBLEMS ON STEAM DISTILLATION

Example 1. A mixture of water and aniline boils at a temperature of 98.5°C at pressure of 760 mm. The vapour pressure of water at this temperature is 717 mm. Find the composition of the distillate.

Solution. Here, $p_w^0 = 717$ mm

$p_l^o = 760 - 717 = 43$ mm

Further $M_w = 18, M_l = 93$

$$\therefore \quad \frac{W_l}{W_w} = \frac{p_l^o \cdot M_l}{p_w^o \cdot M_w}$$

$$= \frac{43 \times 93}{717 \times 18} = 0.31$$

i.e. in the distillate, weight of aniline : weight of water = 0.31 : 1

Percentage by weight of aniline = $\frac{0.31}{1+0.31} \times 100 = \frac{0.31}{1.31} \times 100 = \mathbf{23.66}$

Percentage by weight of water = 100 – 23.66 = **76.34**

Example 2. An organic liquid was subjected to steam distillation. The liquid in the flask was boiled at 90°C. The external pressure (atmospheric pressure) was found to be 734.4 mm. The vapour pressure of water at 90°C is 526.0 mm. The ratio of the weights of liquid to water in the distillate was found to be 2.47. Calculate the molecular weight of the liquid.

Solution. Here, external pressure = 734.4 mm

Boiling point of the mixture = 90°C

Vapour pressure of water at 90°

i.e., $p_w = 526.0$ mm

∴ Vapour pressure of liquid at 90°

i.e., $p_l^o = 734.4 - 526.0 = 208.4$ mm

Ratio of the weights

$W_l / W_w = 2.47$

Molecular weight of water

$M_w = 18$

Substituting the values in the formula

$$\frac{W_l}{W_w} = \frac{p_l^o \times M_l}{p_w^o \times M_w}$$

we get
$$2.47 = \frac{208.4 \times M_l}{526.0 \times 18}$$

or
$$M_l = \frac{2.47 \times 526.0 \times 17}{208.4} = \mathbf{112.2}$$

5.13. PARTIALLY MISCIBLE LIQUIDS

Between the two extremes of completely miscible and completely immiscible liquid pairs, there is a category of liquids called partially miscible liquids. If a little phenol is added to water at ordinary temperature, we get a clear solution, on continued addition of phenol into water, a stage is reached when no more of phenol can be dissolved in water. We get two layers. One of these layers forms a saturated solution of water in phenol and the other layer is a saturated solution of phenol in water. The two solutions (layers) which are in equilibrium are called **conjugate solutions**.

Critical solution temperature

The temperature at which the two liquids, which are otherwise partially miscible at ordinary temperature, become completely miscible is called **critical solution temperature** (*CST*). There can be three types of partially miscible systems. Each type is briefly discussed below by taking one example.

Type I. System with an upper critical solution temperature (upper consolute temperature) Phenol-water system. If phenol and water are mixed in equal proportions, two layers are produced, one is a solution of phenol in water and the other is a solution of water in phenol. At any given temperature, the composition of these layers is fixed and can be determined by analytical methods. If the system is warmed, the amount of water in phenol layer increases as well as the amount of phenol in the water layer increases.

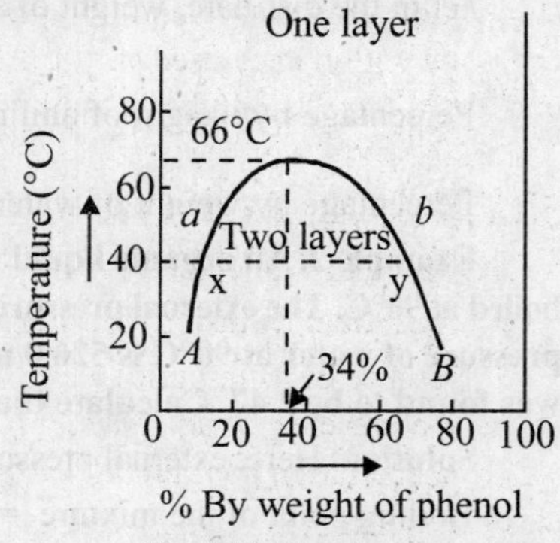

Fig. 5.12. Phenol-water system.

Ultimately, a temperature is reached where, the composition of both the layers becomes the same and thus the two become completely miscible. Experiments have shown that the minimum temperature at which phenol and water become completely miscible is 66°C and the composition is 34% of phenol by weight. Thus phenol-water system has an upper critical solution temperature of 66°C. The results on being plotted give a parabolic curve as shown in Fig. 5.12.

Alternatively, the same curve is obtained if solutions of phenol of different suitable compositions are prepared and the temperature at which the two liquids become completely miscible is determined in each case (from the disappearance of turbidity) and then plotted against the composition.

We observe that at temperatures above 66°C, phenol and water are completely miscible irrespective of composition. For a temperature below 66°C, with composition lying within the parabola, there are two layers in equilibrium. The points X and Y give the composition of the two layers.

Substituting the values in eqn. (*i*), we get

$$1 = k\, p_2^o \qquad \text{or} \qquad k = \frac{1}{p_2^o}$$

Example 6. When ethanol and cyclohexane are mixed, the vapour pressure is more than that expected from Raoult's law. Why?

Solution. Ethanol exhibits hydrogen bonding. On adding cyclohexane, its molecules get in between the molecules of ethanol thereby decreasing the hydrogen bonding, *i.e.*, ethanol-ethanol attractions are weakened. Hence, more molecules can go into the vapour phase.

EXERCISES
(Based On Question Papers of Different Universities)

Multiple Choice Questions (Choose the Correct Option)

1. In a solution, the vapour pressure of a component at a given temperature is equal to the mole fraction of that component in the solution multiplied by the vapour pressure of that component in the pure state. This is the statement of

(*a*) Henry's law (*b*) Raoult's law

(*c*) Konowaloff's law (*d*) None of the above.

2. Relative lowering of vapour pressure of a solution containing a non-volatile solute is equal to the mole fraction of

(*a*) solute (*b*) solvent

(*c*) solvent minus solute (*d*) solvent plus solute

3. Mathematical form of Henry's law is

(*a*) $m = kp$ (*b*) $p = km$

(*c*) $p = \frac{k}{m}$ (*d*) $m = \frac{1}{k}p$

4. A solution formed on mixing of two liquids is ideal if

(*a*) $\Delta V_{mix} = 0, \Delta H_{mix} \neq 0$ (*b*) $\Delta V_{mix} \neq 0, \Delta H_{mix} = 0$

(*c*) $\Delta V_{mix} = 0, \Delta H_{mix} = 0$ (*d*) $\Delta V_{mix} \neq 0, \Delta H_{mix} \neq 0$

5. Benzene + Toluene mixture shows

(*a*) negative deviation (*b*) no deviation

(*c*) large positive deviation (*d*) small positive deviation

6. The vapour phase is relatively richer in the component whose addition to the liquid mixture results in an increase of total vapour pressure. This is the statement of

(*a*) Duhem-Margules law (*b*) Raoult's law

(*c*) Konowaloff's rule (*d*) None of the above

7. Look at the figure below

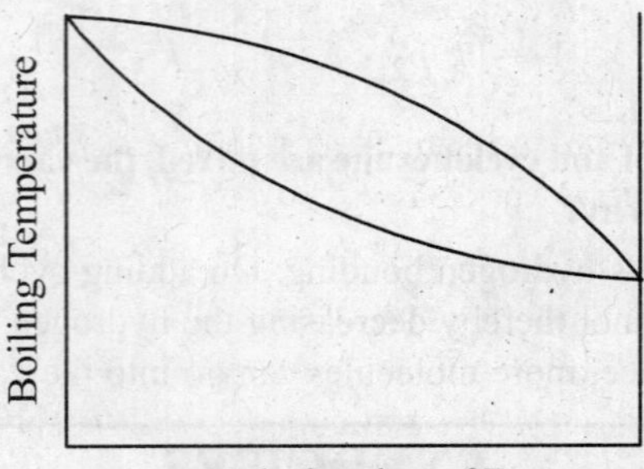

This figure illustrates the distillation behaviour of solutions of

(*a*) type I (*b*) type II
(*c*) type III (*d*) Both type I and III

8. Which of the following is not true about the applicability of steam distillation?

(*a*) The organic liquid to be purified should be immiscible with water.
(*b*) It should have high molecular weight.
(*c*) It should have high vapour pressure.
(*d*) The impurities present should be volatile

9. Addition of KCl to phenol-water system.

(*a*) raises the CST (*b*) lowers the CST
(*c*) has no effect on CST (*d*) none of the above

10. Triethylamine-water system is an example of

(*a*) upper CST (*b*) lower CST
(*c*) upper as well as lower CST (*d*) none of the above

ANSWERS

1. (*b*)	**2.** (*a*)	**3.** (*a*)	**4.** (*c*)	**5.** (*d*)	**6.** (*c*)
7. (*a*)	**8.** (*d*)	**9.** (*a*)	**10.** (*b*)		

Short Answer Questions

1. Differentiate between Ideal and Non-ideal solutions.

2. State and explain Raoult's law for volatile solutes as well as for non-volatile solutes.

3. State 'Henry's law'. What are the conditions under which it holds good? Show that the volume of a gas that dissolves in a fixed volume of a liquid at a given temperature is independent of pressure.

4. Describe Konowaloff's rule. Bring out its importance in the study of boiling point-composition curves of completely miscible ideal binary solutions.

5. Using boiling point-composition diagram, explain how the distillation of a binary miscible solution showing positive deviation takes place.

6. Explain the principle (or theory) underlying the process of steam distillation.

7. Briefly explain the effect of impurities on consolute temperature.

8. Using boiling point-composition diagram, explain how the distillation of a binary miscible solution showing negative deviation takes place.

9. State Henry's law in as many forms as you can. Give its limitations.

10. Derive expression for the total vapour pressure in terms of mole fractions of the components in the vapour phase for a binary miscible liquid mixture.

11. Prove that Raoult's law is a special case of Henry's law.

12. What do you understand by critical solution temperature'? Explain one system which has upper CST and one with lower CST.

13. Write a short note on 'Azeotropes'.

14. What are 'conjugate solution' and 'critical solution temperature? Explain a system which has both upper CST as well as lower CST. What is the effect of pressure on such a system?

15. Explain the vapour pressure-composition diagrams of ideal solutions.

16. Explain the process of 'steam distillation'.

General Questions

1. Discuss vapour pressure-composition curves of system containing binary mixtures of liquids miscible in all proportions. From these, construct boiling point-composition diagrams and describe briefly how their distillation will take place.

2. Briefly explain the principle (or theory) and the method of steam distillation. When is the method used?

3. Briefly explain the following:

(*i*) Effect of impurities on consolute temperature.

(*ii*) Effect of pressure on critical solution temperature.

(*iii*) Isotopic effect on critical solution temperature.

4. Derive Duhem-Margules equation.

5. State and explain Raoult's law and Henry's law. Prove that Raoult's law follows as a special case of Henry's law.

6. Derive the equation $\dfrac{d \ln p_1}{d \ln x_1} = \dfrac{d \ln p_2}{d \ln x_2}$

where x_1 and x_2 are the mole fractions of the solvent and the solute in the solution and p_1 and p_2 are their corresponding partial pressures of the vapours.

7. How are non-ideal solutions classified into different types? Briefly explain the behaviour of each type graphically giving reason with a suitable example in each case.

8. Explain the two main types of non-ideal solutions? How do they lead to the concept of a azeotropic mixtures. Explain briefly the types of azeotropic mixtures.

9. What are ideal and non-ideal solutions? How does the vapour pressure of the components and the total vapour pressure vary with mole fraction of any one of the components in case of an ideal solution?

10. Using boiling point-composition diagram, explain the principle of fractional distillation.

11. Derive the following relationships:

(*i*) Total vapour pressure in terms of mole fractions of the components in the vapour phase.

(*ii*) Between mole fractions of the components in the liquid phase and in the vapour phase.

12. Briefly explain giving one example each of the following types:

(*i*) System with upper CST

(*ii*) System with lower CST

(*iii*) System with upper as well as lower CST.

6

Distribution Law

6.1. NERNST DISTRIBUTION LAW

Walther Hermann Nernst

It has long been observed that when a substance is added to a mixture of two immiscible liquids, the substance distributes itself between the two liquids in such a manner that the concentration ratio of the substance in the two liquids is constant at constant temperature.

Nernst distribution law states as follows:

A solute distributes itself between two immiscible solvents in such a way that the ratio of the concentrations of the solute in the two solvents is constant at constant temperature, independent of any other molecular species present, provided that the solute has the same molecular state in both the solvents.

	Solvent B
X ⇅	Conc. = C_2
	Solvent A
X	Conc. = C_1

Fig. 6.1. Distribution of solute X between immiscible solvents A and B.

If a solute X distributes itself between the two immiscible solvents A and B (Fig. 6.1), then we have

$$\frac{\text{Concentration of } X \text{ in solvent } A}{\text{Concentration of } X \text{ in solvent } B} = \frac{C_1}{C_2} = K \text{ (constant)}$$

provided X remains in the same molecular species in both the solvents. The constant K is called *Distribution co-efficient* or *Partition co-efficient.*

Limitations of distribution law

The distribution law is valid only if the following conditions are fulfilled:

(*i*) The temperature remains constant throughout the experiment.

(*ii*) The two liquids (solvents) should be mutually immiscible.

(*iii*) The amount of the solute added should be small *i.e.*, the solution should be dilute. It is observed that in some cases, the solute added affects the mutual solubility of the solvents.

(*iv*) The solute must have the same molecular state in both the solvents. That is, it should not undergo association or dissociation in any of the solvents. Thus the law is not valid for the distribution of benzoic acid between water and benzene, because in benzene, benzoic acid undergoes association to form dimers *i.e.*, it exists as $(C_6H_5COOH)_2$. Similarly, the law is also not valid for the distribution of an electrolyte between water and a non-polar solvent, because the electrolytes undergo dissociation in water.

NUMERICAL PROBLEMS ON NERNST DISTRIBUTION LAW

Example 1. Succinic acid was shaken with a mixture of water and ether. The concentrations of the acid in the two layers are as follows (per 10 cm^3 of solution). Find out the partition coefficient.

In water layer :	**0.0244**	**0.071**	**0.121**
In ether layer :	**0.0046**	**0.013**	**0.022**

If succinic acid has normal molecular weight of 118 in water, find its mol. wt. in ether.

Solution. Partition co-efficient (distribution co-efficient) of succinic acid between water and ether is given by the ratio

$$K = \frac{\text{conc. in water } (C)}{\text{conc. in ether } (C)}$$

From first set : $$K = \frac{0.0244}{0.0046} = 5.30$$

From second set : $$K = \frac{0.071}{0.013} = 5.47$$

From third set : $$K = \frac{0.121}{0.022} = 5.50$$

$\therefore$ Mean value of Partition Co-efficient, $K = \dfrac{5.30 + 5.47 + 5.50}{3} = 5.42$

Since the value of $K = C_1/C_2$ is almost constant, succinic acid must have the same molecular weight in both the solvents.

As mol. wt. of succinic acid in water is equal to 118, therefore, mol. wt. of succinic acid in ether is also equal to 118

Example 2. When varying amounts of iodine were shaken with CCl_4 – water mixture, the following concentrations of iodine (in g/100 cm^3) were obtained:

CCl_4 layer	**5.1**	**10.2**	**15.2**	**20.3**
Water layer	**0.06**	**0.119**	**0.178**	**0.236**

Show that these results illustrate the distribution law.

Solution.

Conc. in CCl_4 layer (C_1)	*Conc. in H_2O layer* (C_2)	$C_1/C_2 = K$
5.1	0.06	5.1/0.06 = 85.0
10.2	0.119	10.2/0.119 = 85.7
15.2	0.178	15.2/0.178 = 85.4
20.3	0.236	20.3/0.236 = 86.0

Since, the value of K comes out to be almost constant, the given data illustrate the distribution law.

Example 3. An aqueous solution of succinic acid at 15°C containing 0.07 g in 10 cm^3 is in equilibrium with an ethereal solution containing 0.013 g in 10 cm^3. Succinic acid has its normal molecular weight in both the solvents. What will be the concentration of an ethereal solution which is in equilibrium with an aqueous solution containing 0.024 g in 10 cm^3 ?

Solution. First experiment

Conc. of succinic acid in water

$$(C_1) = 0.07 \text{ g in } 10 \text{ cm}^3$$

$$= \frac{0.07}{10} \text{ g / cm}^3$$

Conc. of succinic acid in ether

$$(C_2) = 0.013 \text{ g in } 10 \text{ cm}^3$$

$$= \frac{0.013}{10} \text{ g / cm}^3$$

Succinic acid has the same molecular weight in both the solvents, therefore, distribution law is applicable.

$$K = \frac{C_1}{C_2}$$

$$= \frac{0.07/10}{0.013/10} = 5.38$$

Second experiment

Conc. of succinic acid in water

$$(C_1) = 0.024 \text{ g in } 10 \text{ cm}^3$$

$$= \frac{0.024}{10} \text{ g / cm}^3$$

We want to calculate conc. of succinic acid in ether, *i.e.*, C_2.

According to the distribution law

$$\frac{C_1}{C_2} = K$$

Substituting the various values, we get

$$\frac{0.024/10}{C_2} = 5.35 \text{ (}K\text{ will have the same value as in the first experiment)}$$

or $$C_2 = \frac{0.024}{10} \times \frac{1}{5.38} = 0.00045$$

Thus, conc. of succinic acid in ether layer

$$= \mathbf{0.00045 \text{ g / cm}^3}$$

Example 4. If the distribution co-efficient of benzoic acid between water and benzene is 0.304 at 20°C, calculate the number of moles of benzoic acid which may be extracted from 100 cm^3 of 0.2 molar aqueous solution by 10 cm^3 of benzene.

Solution. 1000 cm^3 of aqueous solution contain benzoic acid = 0.2 mole

∴ 100 cm^3 of aqueous solution contain benzoic acid = 0.02 mole

Suppose the amount of benzoic acid extracted into benzene = x mole

∴ Amount of benzoic acid left in aqueous solution = $(0.02 - x)$ mole

∴ Conc. of benzoic acid in aqueous layer = $\frac{(0.02-x)}{100}$ mole /cm^3

and conc. of benzoic acid in benzene layer = $\frac{x}{10}$ mole/cm^3

Hence $\frac{(0.02-x)/100}{x/10} = 0.34$ which gives $x = 0.005$.

PROBLEMS FOR PRACTICE

1. The following data show the concentrations of solutions of iodine in water and CS_2 in equilibrium at 18°C :

Grams of I_2/10 cc of CS_2 :	1.64	1.29	0.63	0.40
Grams of I_2/10 cc of H_2O :	0.0041	0.0032	0.0016	0.0010

Show that these results are in agreement with the Distribution Law.

2. Calculate how much compound would be extracted from 100 ml of an aqueous solution containing 5 g of the compound if extracted with 50 ml of ether. The partition co-efficient of the compound between ether and water is 4. **[Ans.** 3.33 g]

3. In the distribution of succinic acid between ether and water at 20°C, 25 cc of the ethereal layer contained 0.08 g of the acid. Find the weight of the acid present in 50 cc of the aqueous solution in equilibrium with it if the value of K for succinic acid between water and ether is 5.3. **[Ans.** 0.848 g]

4. Succinic acid has the same molecular weight in water and ether. When varying amounts of the acid were shaken with ether-water mixture, the following results were obtained :

Conc. in water layer (C_1):	25.4	33.2	42.6
Conc. in ether layer (C_2):	4.2	5.5	7.1

Calculate the partition co-efficient and show that these figures illustrate the distribution law. **[Ans.** 6.03]

5. The distribution co-efficient of an organic acid between water and ether is 0.30 at 25°C. Calculate the number of moles of the acid which may be extracted from 100 ml of 0.025 molar aqueous solution by 10 ml of ether, provided the acid has normal molecular weight in both the solvents. **[Ans.** 0.000625 mole]

6.2. THERMODYNAMIC DERIVATION OF DISTRIBUTION LAW

The derivation is based upon the principle that *if there are two phases in equilibrium (in this case two immiscible solvents containing the same solute dissolved in them), the chemical potential of a substance must be same in both the phases.*

From thermodynamics, we know that the chemical potential (μ) of a substance is a solution given by

$$\mu = \mu° + RT \ln a$$

where μ° is the standard chemical potential and *'a'* is activity of the substance (solute) in the solution.

For the solute in liquid *A*, we have

$$\mu_A = \mu_A° + RT \ln a_A \qquad ...(i)$$

For the solute in liquid B we have

$$\mu_B = \mu_B{}^\circ + RT \ln a_B \quad \text{.... (ii)}$$

Since the liquids A and B are in equilibrium,

$$\mu_A = \mu_B \quad \text{.... (iii)}$$

or $$\mu_A{}^\circ + RT \ln a_A = \mu_B{}^\circ + RT \ln a_B$$

or $$RT \ln a_A - RT \ln a_B = \mu_B{}^\circ - \mu_A{}^\circ$$

or $$\ln \frac{a_A}{a_B} = \frac{\mu_B^0 - \mu_A^0}{RT} \quad \text{... (iv)}$$

At a given temperature, $\mu_A{}^\circ$ and $\mu_B{}^\circ$ are constant for given substance in the particular solvents. Hence at constant temperature, we have

$$\ln \frac{a_A}{a_B} = \text{constant} \quad \text{(from equation (iv))} \quad \text{... (v)}$$

or $$\frac{a_A}{a_B} = \text{constant} \quad \text{... (vi)}$$

This is the exact expression for the distribution law. However, if the solutions are dilute, the activities may be replaced by concentrations so that the expression (*vi*) is modified to

$$\frac{C_A}{C_B} = \text{constant}$$

which is the original form of the distribution law.

6.3. MODIFICATION IN DISTRIBUTION LAW IN CASE OF CHANGE IN MOLECULAR STATE

Case 1. *The solute undergoes association in one of the solvents*

Suppose the solute, X remains in the normal molecular state in the phase I but n molecules of it associate to form complex molecules X_n in phase II. Then the following two equilibria will exist simultaneously, as shown in the Fig. 6.2.

X	Phase I (No Change) Conc. = C_1
$nX = X_n$	Phase II (Association) Conc. = C_2

Fig. 6.2. Association in Phase II.

(*i*) $\underset{\text{Phase I}}{X} \rightleftharpoons \underset{\text{Phase II}}{X}$

(*ii*) $\underset{C_2(1-\alpha)}{nX} \rightleftharpoons \underset{\frac{1}{n}C_2\alpha}{X_n}$ in the same phase II.

Suppose the total conc. of the solute X in phase $I = C_1$

and the total conc. of solute X in phase $II = C_2$

Suppose α is the degree of association of the solute in phase II.

Then conc. of the associated molecules in phase II, *i.e.* $[X_n] = \frac{1}{n} C_2 \alpha$

and conc. of the unassociated (normal) molecules in phase $II = (1 - \alpha)\, C_2$.

Applying the distribution law to equilibrium (*i*), we get

$$\frac{\text{Conc. of solute } X \text{ in phase I}}{\text{Conc. of normal moles of } X \text{ in phase II}} = \text{constant}$$

We have to take into account concentration of identical molecular species in the two phases.

i.e.,
$$\frac{C_1}{(1-\alpha)C_2} = \text{constant} \quad \text{... } (i)$$

Now applying the law of chemical equilibrium to the equilibrium (*ii*), we get

$$\frac{\text{Conc. of associated molecules in phase II}}{(\text{Conc. of normal molecules in phase II})^n} = \text{constant}$$

i.e.,
$$\frac{\frac{1}{n}C_2\alpha}{[(1-\alpha)C_2]^n} = \text{constant}$$

or
$$\frac{C_2\alpha}{[(1-\alpha)C_2]^n} = \text{constant} \quad (\because n \text{ is constant})$$

Taking the nth root, we get

$$\frac{\sqrt[n]{C_2\alpha}}{(1-\alpha)C_2} = \text{constant} \quad \text{... } (ii)$$

Dividing eqn. (*i*) by eqn. (*ii*), we get

$$\frac{C_1}{\sqrt[n]{C_2.\alpha}} = \text{constant .... (Exact formula)}$$

When the molecules in the phase II are almost completely associated, $\alpha = 1$ and, therefore,

$$\frac{C_1}{\sqrt[n]{C_2}} = \text{constant .....(Approx formula)}$$

Thus, when the solute undergoes association in one of the solvents, the constant value for the partition co-efficient is obtained not for the ratio C_1/C_2 but for the ratio $C_1/\sqrt[n]{C_2}$.

The formula, $C_1/\sqrt[n]{C_2} = K$, is commonly used for finding the value of n, *i.e., the extent of association* of the solute molecules in one of the solvents. This is done by finding out the concentrations C_1 and C_2 of the solute in the two solvents in a number of experiments. The value of n is then calculated.

Case 2. *The solute undergoes dissociation in one of the solvents.*

Suppose the solute X dissolves in the phase I without any change in the molecular state but undergoes dissociation into Y, Z etc., in the phase II. Then the following two equilibria will be set up as shown in Fig. 6.3.

(*i*) $\underset{\text{Phase I}}{X} \rightleftharpoons \underset{\text{Phase II}}{X}$

(*ii*) $X \rightleftharpoons Y + Z$ in the same phase II

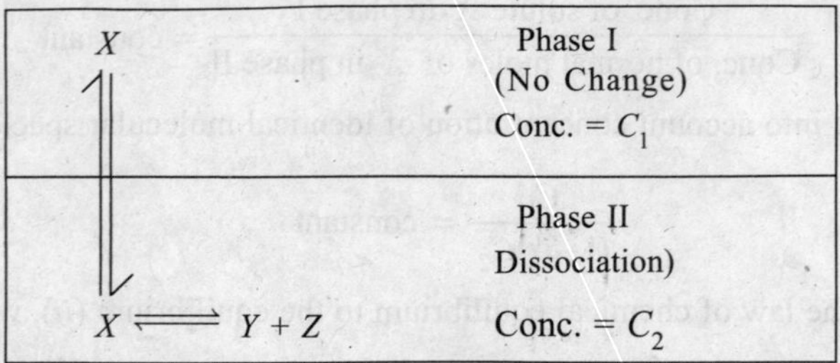

Fig. 6.3. Dissociation in Phase II.

Let us suppose, total conc.of solute X in phase I = C_1

and total conc. of solute X in phase II = C_2.

Also, suppose the degree of dissociation of the solute in phase II is α.

Then, conc. of undissociated (simple) molecules in phase II

$$= (1 - \alpha)\, C_2$$

Applying the distribution law to equilibrium (*i*), we get

$$\frac{C_1}{(1-\alpha)C_2} = K \text{ (Partition coefficient)}$$

It is understandable that if the solute undergoes dissociation in phase I (with degree of dissociation = α) but remains normal in phase II, then the formula obtained will be

$$\frac{C_1(1-\alpha)}{C_2} = K$$

From the above expression, the degree of dissociation, α, can be calculated.

Case 3. *When the solute combines with one of the solvents.*

Suppose the solute X remains in the normal molecular state in phase I but one molecule of it combines with n molecules of the solvent S to form the compound molecules, $X.nS$ in phase II. Then, the following two equilibria will be established as shown in Fig. 6.4.

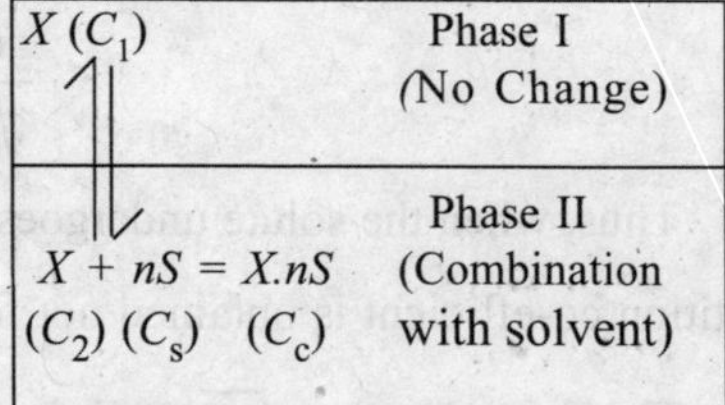

Fig. 6.4. Combination with solvent in Phase II.

(*i*) $\underset{\text{Phase I}}{X} \rightleftharpoons \underset{\text{Phase II}}{X}$

(*ii*) $X + nS \rightleftharpoons X.nS$ in the same phase II.

Let us suppose, conc. of solute X (normal molecules) in phase = C_1

conc. of the normal molecules in phase II = C_2

conc. of the solvent molecules in phaseII = C_s

and conc. of the compound molecules in phase II = C_c

Applying the distribution law to equilibrium (*i*), we get

$$\frac{C_1}{C_2} = K, \text{ the distribution co-efficient} \quad \text{... } (i)$$

Now, applying the law of chemical equilibrium to equilibrium (*ii*), we get

$$\frac{C_c}{C_2 \times (C_s)^n} = K_1 \text{ equilibrium constant} \qquad \text{... } (ii)$$

Since the solvent in phase II is present in excess, its concentration (C_s) remains almost constant despite of combination with the solute *X*, particularly if the solute is added in small quantity, *i.e.*, the solution is dilute. Thus $(C_s)^n$ is constant.

Eq. (*ii*) thus becomes $$\frac{C_c}{C_2} = K_1(C_s)^n = K_2, \text{ another constant}$$

Adding 1 to both the sides, we get

$$1 + \frac{C_c}{C_2} = 1 + K_2$$

or $$\frac{C_2 + C_c}{C_2} = 1 + K_2 \qquad \text{... } (iii)$$

Dividing eqn. (*i*) by eqn. (*iii*), we get

$$\frac{C_1}{C_2} \times \frac{C_2}{C_2 + C_c} = \frac{K}{1 + K_2}$$

or $$\frac{C_1}{C_2 + C_c} = \frac{K}{1 + K_2} = K_3, \text{ another constant.}$$

But $C_2 + C_c$ is the total conc. of the normal as well as the solvated molecules of the solute *X* in phase II. Thus the above equation, signifies that

$$\frac{\text{Total conc. of the solute } X \text{ in phase I}}{\text{Total conc. of the solute } X \text{ in phase II}} = \text{constant}$$

Thus we conclude that *when the solute enters into chemical combination with one of the solvents, there is no modification of the distribution law equation except that the numerical value of the partition constant is changed.*

6.4. APPLICATIONS OF DISTRIBUTION LAW

1. *Determination of solubility of a solute in a solvent*

When two immiscible solvents are in contact with each other and the solute is added in such a large excess that saturated solutions are formed, then the concentrations of the solutions at this stage will be equal to their solubilities. Assuming that the distribution law is applicable for such concentration, we have

$$\frac{C_1}{C_2} = \frac{S_1}{S_2} = K$$

where conc. $C_1 = S_1$ is the solubility of the solute in solvent 1.
and conc. $C_2 = S_2$ is the solubility of the solute in solvent 2.

Knowing the distribution co-efficient *K* and the solubility of the solute in one of the solvents, that in the other can be calculated.

SOLVED PROBLEMS ON SOLUBILITY OF SOLUTES

Example 1. At 15°C an aqueous solution of oxalic acid containing 5.0 g of oxalic acid per 100 cm^3 of water in equilibrium with an ethereal solution containing 0.50 g per 100 cm^3. The solubility of oxalic acid in water at 15°C is 10 g per 100 cm^3. Calculate the solubility in water.

Solution. Conc. of oxalic acid in water (C_1) = 5.0 g in 100 cm^3
= 50.0 g/litre

Conc. of oxalic acid in ether (C_2) = 0.50 g in 100 cm^3
= 5.0 g/litre

∴ Distribution co-efficient of oxalic acid between water and ether is given by

$$K = \frac{C_1}{C_2} = \frac{50}{5.0} = 10.0$$

In saturated solutions, the ratio of concentrations is equal to the ratio of solubilities

i.e.,

$$K = \frac{\text{Solubility in water } (S_1)}{\text{Solubility in ether } (S_2)}$$

But $S_1 = 10$ g/100 cm^3 (given)

∴

$$10 = \frac{10}{S_2}$$

or

$$S_2 = 1.0 \text{ g}$$

i.e. Solubility of oxalic acid in ether = **1.0 g/100 cm^3.**

PROBLEMS FOR PRACTICE

1. Find the solubility of iodine in CCl_4 at a temperature when water becomes saturated with 0.350 g of iodine per litre and the distribution co-efficient is 88 in favour of CCl_4.
   [**Ans.** 30.80 g/litre]
2. The solubility of iodine in carbon tetrachloride at 25°C is 29.1 g/litre. If a solution of iodine in water containing 0.103 g/litre is in equilibrium with a solution of carbon tetrachloride containing 8.2 g of iodine/litre, calculate the solubility of iodine in water at 25°C.
   [**Ans.** 0.366 g/litre]
3. At 25°C, an aqueous solution of iodine containing 0.0576 g per litre is in equilibrium with CCl_4 solution containing 4.412 g of iodine per litre. If the solubility of iodine in water is 0.340 g/litre, what will be its solubility in CCl_4 ? [**Ans.** 26.04 g/litre]

2. To determine the extent of association or dissociation of a solute in a solvent using distribution law

When a solute undergoes association in one of the solvents ($nX \rightleftharpoons X_n$) in which its total concentration is C_2 and has the normal molecular state in the second solvent in which its concentration is C_1, then distribution law formula is modified to

$$\frac{C_1}{\sqrt[n]{C_2}} = K, \text{ the distribution co-efficient.}$$

From this formula, the value of n can be calculated by either of the two methods given below.

(1) Hit and Trial method. The value of C_1 and C_2 obtained from different experiments are put in the equation

$$\frac{C_1}{\sqrt[n]{C_2}} = K.$$

Then putting $n = 2, 3, 4$ etc., one by one, the values of K are calculated in each case. The number which gives constant value for K is the correct value for n.

(2) Logarithmic method. Writing again the distribution law equation when association in one liquid takes place

$$\frac{C_1}{\sqrt[n]{C_2}} = K.$$

or $$\frac{C_1}{C_2^{1/n}} = K$$

or $$C_1 = (C_2)^{1/n}.K$$

Taking logarithm of both the sides, we get

$$\log C_1 = \frac{1}{n}\log C_2 + \log K$$

For two sets of values: C_1, C_2 and C_1', C_2', we can write two equations

$$\log C_1 = \frac{1}{n}\log C_2 + \log K \quad \text{... }(i)$$

and $$\log C_1' = \frac{1}{n}\log C_2' + \log K \quad \text{... }(ii)$$

Subtracting eqn. (*ii*) from eqn. (*i*), we get

$$\log C_1 - \log C_1' = \frac{1}{n}(\log C_2 - \log C_2')$$

or $$n = \frac{\log C_2 - \log C_2'}{\log C_1 - \log C_1'}$$

Hence n can be calculated.

Similarly, when a solute undergoes dissociation in one of the solvents, ($X \rightleftharpoons Y + Z$) in which its total concentration is C_2 and has the same normal molecular state in the other solvent, in which its concentration is C_1, then we have

$$\frac{C_1}{C_2(1-a)} = K,\ \text{the distribution co-efficient.}$$

SOLVED PROBLEMS ON ASSOCIATION OR DISSOCIATION OF SOLUTES

Example 1. In the distribution of benzoic acid between water and benzene, the following results were obtained :

C_1 (in water)	**1.50**	**1.95**	**2.97**
C_2 (in benzene)	**24.20**	**41.2**	**97.0**

Assuming that benzoic acid exists as single molecules in water, show that it exists as double molecules in benzene.

Solution. If benzoic acid exists as normal (single) molecules in water and as dimer (double molecules) in benzene, then the ratio $C_1/\sqrt{C_2}$ should be constant. This can be seen as follows :

C_1 (in water)	C_2 (in benzene)	$\dfrac{C_1}{\sqrt{C_2}} = K$
1. 5	24.20	$\dfrac{1.5}{\sqrt{24.20}} = 0.305$
1.95	41.2	$\dfrac{1.95}{\sqrt{41.2}} = 0.304$
2.97	97.0	$\dfrac{2.97}{\sqrt{97.0}} = 0.302$

As the ratio $C_1/\sqrt{C_2}$ comes out to be almost constant, this proves that benzoic acid exists as double molecules in benzene.

Example 2. In the distribution of a solute between water (C_1) and chloroform (C_2), the following results were obtained :

C_1	C_2
0.163	**0.761**
0.436	**5.43**

What information do you gather regarding the molecular state of the solute in chloroform?

Solution. If the solute exists as simple (normal) molecules in water as well as in chloroform then the ratio $C_1/C_2 = K$ should be constant.

For 1st set : $$\frac{C_1}{C_2} = \frac{0.163}{0.761} = 0.212$$

For 2nd set : $$\frac{C_1}{C_2} = \frac{0.436}{5.43} = 0.080$$

We observe that the ratio C_1/C_2 is not constant and hence the distribution law in its simple from is not applicable. In other words, the molecular state of the solute is not the same in the two solvents.

Suppose the solute is associated as double molecules in chloroform, then the ratio $C_1/\sqrt{C_2}$ should be constant.

For 1st set : $$K = \frac{C_1}{\sqrt{C_2}} = \frac{0.163}{\sqrt{0.761}} = 0.1869$$

For 2nd set : $$K = \frac{C_1}{\sqrt{C_2}} = \frac{0.436}{\sqrt{5.43}} = 0.1871$$

As the ratio $C_1 / \sqrt{C_2}$ comes out to be constant, this proves that the solute exists as *double* molecules in chloroform.

Example 3. Succinic acid shows normal molecular weight in water and is associated in benzene. The following data was obtained on the distribution of succinic acid between water and benzene.

C_1 in water (g / litre)	**1.50**	**1.95**	**2.89**
C_2 in benzene (g / litre)	**24.20**	**41.20**	**96.50**

Calculate the molecular weight of succinic acid in benzene.

Solution. *By hit and trial method.* Suppose succinic acid exists as double molecules (dimer) in benzene, then the ratio $C_1 / \sqrt{C_2}$ should be constant. Let us find it out

C_1 (in water)	C_2 (in benzene)	$\frac{C_1}{\sqrt{C_2}} = K$
1.50	24.20	$\frac{1.50}{\sqrt{24.20}} = 0.305$
1.95	41.20	$\frac{1.95}{\sqrt{41.20}} = 0.304$
2.89	96.50	$\frac{2.89}{\sqrt{96.50}} = 0.303$

As $C_1 / \sqrt{C_2} = K$ comes out to be almost constant, this proves that $n = 2$.

Molecular weight of succinic acid in benzene = 2 × 11.8 = 236

(ii) Logarithmic method. According to this method, the value of *n* is given by

$$n = \frac{\log C_2 - \log C_2'}{\log C_1 - \log C_1'}$$

The given data is tabulated as under :

S. No.	C_1	$\log C_1$	C_2	$\log C_2$
(1)	1.50	0.176	24.2	1.384
(2)	1.95	0.290	41.2	1.616
(3)	2.89	0.461	96.3	1.985

From (1) and (2),

$$n = \frac{1.616 - 1.384}{0.290 - 0.176} = 2.04$$

From (2) and (3),

$$n = \frac{1.985 - 1.616}{0.461 - 0.290} = 2.16$$

From (3) and (1),

$$n = \frac{1.985 - 1.384}{0.461 - 0.176} = 2.11$$

Thus the whole number value of $n = 2$

i.e. succinic acid exists as $\begin{bmatrix} CH_2COOH \\ | \\ CH_2COOH \end{bmatrix}_2$ in benzene. Hence mol. wt. of succinic acid in benzene $= 2 \times 118 = 236$.

PROBLEMS FOR PRACTICE

1. Experiments on the distribution of phenol between water and $CHCl_3$ gave the following results :

Conc. in aqueous soln. (C_1) :	0.094	0.163	0.274	0.436
Conc. in $CHCl_3$ soln. (C_2) :	0.254	0.761	1.85	5.53

What conclusion can be drawn from these results concerning the molecular condition of phenol in chloroform solution ? **[Ans.** Exists as double molecules]

2. Acetic acid was shaken with water and CCl_4 and following concentrations in moles per litre were found in the two layers :

Aqueous layer	5.02	7.8	10.70
CCl_4 layer	0.292	0.725	1.410

Assuming that acetic acid has single molecules in water, show that it has double molecules in CCl_4 solution.

3. In the distribution of a solute between water (C_1) and chloroform (C_2), the following results were obtained (in which C_1 and C_2 have been expressed in the same units) :

C_1	0.0160	0.0237
C_2	0.338	0.753

What information do you get regarding the molecular state of the solute in chloroform ? **[Ans.** Double molecules]

6.5. STUDYING CHEMICAL EQUILIBRIUM INVOLVING FORMATION OF COMPLEX COMPOUNDS

Some typical reactions involving complex ions are :

$$KI + I_2 \rightleftharpoons KI_3$$

$$CuSO_4 + 4NH_3 \rightleftharpoons [Cu(NH_3)_4]SO_4$$

Consider the following reaction

$$KI + I_2 \rightleftharpoons KI_3 \quad \text{or} \quad I^- + I_2 \rightleftharpoons I_3^-$$

Distribution law helps us to determine the equilibrium constant for such equilibrium reactions and to establish the formula of complex ion on the basis of the constancy in the value of the equilibrium constant. The study of this equilibrium involves the following steps:

(*i*) A solution of iodine in CS_2 is shaken with water to find out the distribution co-efficient of iodine between water and CS_2.

(*ii*) A solution of iodine in CS_2, of known concentration, is shaken with KI solution, also of known concentration (See Fig. 6.5)

(*iii*) After the distribution of iodine is complete between the two layers, the concentration of iodine left in the CS_2 layer is determined by titration against standard sodium thiosulphate solution.

Calculations

Suppose

Initial conc. of iodine in CS_2 = a moles/litre

Initial conc. of KI solution = b moles/litre

Conc. of I_2 left in CS_2 layer after distribution = c moles/litre

Distribution co-efficient of I_2 between water and CS_2 = K

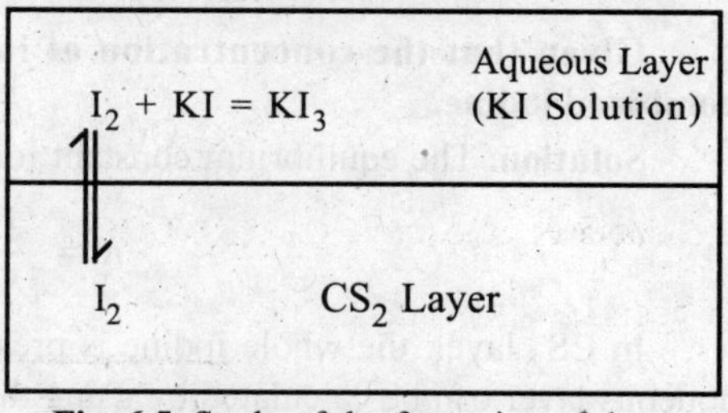

Aquevous Layer (KI Solution)

$I_2 + KI = KI_3$

I_2 CS_2 Layer

Fig. 6.5. Study of the formation of the complex KI_3.

Since iodine reacts with KI, the total concentration of iodine (free + combined) in the aqueous layer is much higher (than if no KI were present).

Thus, total conc. of I_2 (free + combined) in aqueous layer

= Initial conc. of I_2 in CS_2 – conc. of I_2 left in CS_2 layer

= $a - c$ moles/litre

Conc. of free I_2 in aqueous layer can be obtained from distribution law equation.

$$\frac{\text{Conc. of free } I_2 \text{ in aqueous layer}}{\text{Conc. of } I_2 \text{ in } CS_2 \text{ layer}} = K, \textit{i.e.}, \text{ distribution co-efficient.}$$

But conc. of I_2 in CS_2 layer = c moles/litre

∴ Conc. of *free* I_2 in aqueous layer = Kc mole/litre

∴ Amount of I_2 that *combined* with KI = $(a - c) - Kc$ moles/litre

= d moles/litre (say)

According to the reaction

$$KI + I_2 \rightleftharpoons KI_3$$

No. of moles of I_2 combined = No. of moles of KI combined

= No. of moles of KI_3 formed

∴ Conc. of KI_3 in the aqueous layer = d moles/litre

Conc. of KI in the aqueous layer = $(b - d)$ moles/litre

Conc. of I_2 in the aqueous layer = Kc moles/litre

Substituting these values in the equilibrium law expression *viz.*,

$$K' = \frac{[KI_3]}{[KI][I_2]}$$

the equilibrium constant K' can be calculated.

Starting with different initial concentrations of iodine and KI solutions, the value of K' is determined in each case. If we obtain nearly constant values of K' that confirms the formula of the complex (KI_3).

SOLVED PROBLEM ON REACTIONS INVOLVING COMPLEX IONS

Example 1. The partition co-efficient of iodine between CS_2 and H_2O is 410. A solution of KI containing 8 g of the salt per litre was taken with CS_2 until equilibrium was attained. It was found that at equilibrium, the aqueous layer contained 2.15 g of iodine per litre and the CS_2 layer 35.42 g iodine per litre. Calculate the equilibrium constant for the reaction.

$$KI + I_2 \rightleftharpoons KI_3$$

Given that the concentration of iodine in the aqueous layer is the sum of free and combined iodine.

Solution. The equilibrium constant for the reaction is given by

$$K = \frac{[KI_3]}{[KI][I_2]} \quad ...(i)$$

In CS_2 layer, the whole iodine is present as free iodine. Concentration of *free iodine* in the aqueous layer can be calculated by using the equation for partition co-efficient.

$$\frac{\text{Conc. of free iodine in } CS_2 \text{ layer } (C_1)}{\text{Conc. of iodine in aqueous layer } (C_2)} = K \text{ (Partition co-efficient)}$$

Substituting the values, we have

$$\therefore \quad \frac{35.42}{C_2} = 410$$

$$\text{or} \quad C_2 = \frac{35.42}{410} \text{ g/litre}$$

$$= \frac{35.42}{410 \times 254} \text{ moles/litre}$$

$$= 0.0003401 \text{ moles/litre}$$

$$\text{or} \quad [I_2] = 0.0003401 \text{ M}$$

$$\text{Total iodine in the aqueous layer} = 2.15 \text{ g/litre}$$

$$= \frac{2.15}{254} \text{ moles/litre}$$

$$= 0.008465 \text{ moles/litre}$$

$$\therefore \quad \text{Combined iodine in the aqueous layer} = 0.008465 - 0.0003401$$

$$= 0.008125 \text{ moles/litre}$$

$$\text{But} \quad [KI_3] = [\text{combined } I_2]$$

$$[KI_3] = 0.008125 \text{ M}$$

$$\text{Molar conc. of} \quad KI = \frac{8}{166} \text{ moles/litre}$$

$$= 0.04819 \text{ moles/litre}$$

$$\text{KI used in forming } KI_3 = 0.008125 \text{ moles/litre}$$

$$\therefore \text{ Conc. of free KI } i.e., \; [KI] = 0.04819 - 0.008125$$

$$= 0.04007 \text{ moles/litre}$$

Substituting the values in eqn. (*i*), we get

$$K = \frac{[KI_3]}{[KI][I_2]} = \frac{0.008125}{0.04007 \times 0.0003401}$$

$$= \mathbf{596.2}$$

6.6. UTILITY OF DISTRIBUTION LAW IN THE PROCESS OF EXTRACTION

Distribution law is of great practical utility in extraction and purification of substances. However the following conditions must be satisfied to achieve efficient extraction.

1. The value of distribution co-efficient of the substance between water and the extracting solvent should be greater in favour of the extracting substance. Distribution co-efficient of a substance in water can be varied. If it is desired to lower the distribution co-efficient in water *i.e.*, to make it less soluble in water, this is very much possible. Let us say the substance under question is an organic acid. Its solubility in water can be decreased by adding a small quantity of a mineral acid, which will release H^+ ions. The solubility of the organic acid will be depressed because of common ion effect.

2. The extracting liquid should not be used in one lot but in a number of smaller lots. The following example will establish the truth of this requirement.

Suppose X g of the substance are present in 1 litre of the aqueous solution. Also suppose 1 litre of ether is available as the extracting solvent and that the distribution co-efficient of the substance between ether and water is 2. If whole of the ether is used in one lot and x g be the amount of the substance extracted into ether, then we will have

$$\frac{\text{Conc. in ether}}{\text{Conc. in } H_2O} = \frac{x/1000}{(X-x)/1000} = 2$$

or
$$x = \frac{2}{3}X = 0.67\ X$$

Thus only 67% of the total amount will be extracted in this case.

Instead, suppose the whole of the ether is used in two instalments, each of 500 cm^3. After shaking with first instalment of 500 cm^3, suppose x_1 g of the substance passes into the ether layer. Then clearly,

$$\frac{x_1/500}{(X-x_1)/1000} = 2$$

or
$$x_1 = \frac{1}{2}X = 0.5\ X$$

Thus 50% of the total amount is extracted out in this operation. Therefore, the amount of the substance left unextracted $= X - \frac{1}{2}X = \frac{1}{2}X$. Now one litre of the aqueous solution containing $\frac{1}{2}$ X g of the substance is shaken with the remaining 500 cm^3 of ether. Suppose the amount extracted into ether by this operation is x_2 g. It is calculated as under

$$\frac{x_2/500}{\left(\frac{X}{2}-x_2\right)/1000} = 2$$

or
$$x_2 = \frac{1}{4}X = 0.25\ X$$

i.e. 25% of the total substance passes into the ether layer in this operation. Therefore, total amount of the substance extracted = 50 + 25 = 75%.

Thus when 1 litre of ether is used in two equal instalments, the percentage extraction increases from 67 to 75. Similarly, it can be seen that if 1 litre of ether is used in four equal instalments, the extraction can be increased to 93.75%. Thus keeping the total quantity of the extracting solvent as constant, we can extract more by using smaller lots of the solvent at a time.

6.7. GENERAL FORMULA FOR THE AMOUNT OF SUBSTANCE LEFT UNEXTRACTED AND THE VOLUME OF EXTRACTING LIQUID USED EACH TIME

Suppose *W* g of the substance is present in *V* ml of the solution made in solvent *A*. Also suppose that *v* ml of the extracting solvent *B* is used each time and *K* is the distribution co-efficient of the substance between the solvents *A* and *B*.

If w_1 g is the amount of the substance left *unextracted* in solvent *A* after the first operation, then we have

$$\frac{\text{Conc. in } A}{\text{Conc. in } B} = K, \text{ i.e., } \frac{w_1/V}{(W-w_1)v} = K$$

or $$K = \frac{(W-w_1)}{v} = \frac{w_1}{V}$$

or $$KV(W-w_1) = w_1 v$$

or $$w_1(v+KV) = KVW$$

or $$w_1 = \frac{KV}{(v+KV)}W \quad \text{... (i)}$$

After the second operation, if w_2 g be the amount of the substance left unextracted in solvent *A* so that the amount extracted is $(w_1 - w_2)$ g, then we have

$$\frac{w_2/V}{(w_1-w_2)v} = K$$

or $$w_2 = \left(\frac{KV}{v+KV}\right)w_1 \quad \text{... (ii)}$$

Substituting the value of w_1 from eqn. (*i*), we get

$$w_2 = \left(\frac{KV}{v+KV}\right)\left(\frac{KV}{v+KV}\right)W$$

$$= \left(\frac{KV}{v+KV}\right)^2 W$$

In general, after n^{th} operation, the amount of the substance that will be left unextracted is given by :

$$\boxed{w_n = \left(\frac{KV}{v+KV}\right)^n W} \quad \text{... (iii)}$$

In order that the extraction may be most efficient, clearly w_n should have the smallest possible value *i.e.* smaller lots of the solvent. This can be so only if the factor $\left(\frac{KV}{v+KV}\right)^n$ is as small as possible. In this factor, the denominator, $v + KV$ is greater than the numerator, KV. Therefore, $KV/(v + KV)$ is less than 1. Hence the above factor will be smallest possible when n has the largest possible value. It proves that the *multi-step extraction is more economical than the single-step extraction.*

Note for the Students. For solving numerical problems, it may be mentioned here that K is the distribution co-efficient of the substance between the solvent A already present and the extracting solvent B. If K is given for the distribution of the substance between solvents B and A, the reciprocal of this value should be used in equation (*iii*).

Parke's process for desilverisation of lead

Silver is more soluble in zinc than in lead. Zinc is immiscible with molten lead. Thus when zinc is added to argentiferous lead, a large quantity of silver goes into the molten zinc layer leaving little silver in the molten lead layer, because the distribution co-efficient of silver between zinc and lead is 300 at 880°C. Zinc containing silver forms the upper layer. Thus by repeating the process, almost total quantity of silver can be extracted from lead.

SOLVED PROBLEMS ON EXTRACTION OF A SUBSTANCE

Example 1. 0.83 g succinic acid was shaken up with 100 ml each of water and ether. The water layer was found to contain 0.70 g of succinic acid, the rest having passed into the ether layer. Calculate the quantity of succinic acid which can be extracted from 1000 ml of ether solution containing 1 g of the acid, using 100 ml of water in one lot and in two equal fractions.

Solution. Total amount of succinic acid taken = 0.83 g

Amount present in 100 ml of water layer = 0.70 g

∴ Amount gone into 100 ml ether layer = 0.83 – 0.70 = 0.13 g

Water is the extracting solvent, therefore, we shall use the relation

$$K = \frac{\text{Conc. in ether}}{\text{Conc. in water}} = \frac{0.13/100}{0.70/100} = 0.19.$$

(*a*) *When 100 ml of water is used in one lot :*

$$v = 100 \text{ ml}, n = 1$$

Also $\qquad W = 1 \text{ g}, V = 1000 \text{ ml}, K = 0.19$

Using the general equation for extraction,

Amount left unextracted will be

or $$w_n = \left(\frac{KV}{v+KV}\right)^n \times W = \left(\frac{0.19 \times 1000}{100 + 0.19 \times 1000}\right)^1 \times 1 = \frac{190}{290} = 0.655 \text{ g}$$

∴ Amount extracted 1 – 0.655 = **0.345 g.**

(*b*) *When 100 ml of water is used in two equal fractions:*

$$v = 50 \text{ ml}, n = 2$$

Also $\qquad W = 1 \text{ g}, V = 1000 \text{ ml}, K = 0.19$

Amount left unextracted will be given by

$$w_n = \left(\frac{KV}{v+KV}\right)^2 \times W = \left(\frac{0.19 \times 1000}{50 + 0.19 \times 1000}\right)^2 \times 1 = \left(\frac{190}{240}\right)^2 = 0.627 \text{ g}$$

∴ $\qquad$ Amount extracted = 1 – 0.627 = **0.373 g.**

Example 2. The distribution coefficient of a substance between chloroform and water is 5. Show that the amount of substance extracted from 100 ml of an aqueous solution is more by using 50ml chloroform at a time in two instalments rather than using 100 ml of in one lot.

Solution. Given that :

$$K = \frac{\text{Conc. in Chloroform } (C_1)}{\text{Conc. in Water } (C_2)} = \frac{5}{1}$$

Suppose the total amount of the substance to be extracted is W g, present in 100 ml of aqueous solution.

(*a*) *When 100 ml of chloroform is used in one lot:*

Suppose the amount of substance extracted into chloroform = w g

∴ Amount of the substance left unextracted in water = $(W - w)$ g

i.e., conc. of the substance in chloroform = $w/100$ g per ml and Conc. of the substance in water = $(W - w)/100$ g per ml.

∴ By distribution law

$$\frac{w/100}{(W - w)/100} = 5$$

or $$w = 5\,(W - w)$$

or $$w = \frac{5}{6}\,W = 0.833\,W \text{ g}$$

i.e., the amount extracted will be 0.833 W g

(*b*) *When 50 ml of chloroform is used at a time:*

(*i*) *In the first extraction,*

suppose the amount of the substance extracted into chloroform = w_1 g

∴ Amount of the substance left unextracted in water = $(W - w_1)$ g

i.e., conc. of the substance in chloroform layer = $\frac{w_1}{50}$ g /ml

and conc. of the substance in water layer = $\frac{(W - w_1)}{100}$ g/ml

∴ By distribution law

$$\frac{w_1/50}{(W - w_1)/100} = 5$$

or $$\frac{w_1}{50} \times \frac{100}{W - w_1} = 5$$

or $$2w_1 = 5\,(W - w_1)$$

or $$w_1 = \frac{5}{7}\,W = 0.714\,W \text{ g}$$

∴ Amount of the substance left unextracted in 100 ml of the aqueous solution

$$= W - 0.714\,W = 0.286\,W\text{g}$$

(*ii*) *In the second extraction,*

suppose the amount of the substance extracted into chloroform = w_2 g.

Then amount of substance left unextracted in water = $(0.286\ W - w_2)$ g

i.e., conc. of substance in chloroform = $w_2/50$ g per ml and conc. of substance in water

$$= (0.286\ W - w_2)/100 \text{ g per ml}$$

Applying distribution law, we get

$$\frac{w_2/50}{(0.286W - w_2)/100} = 5$$

or $$\frac{w_2}{50} \times \frac{100}{0.286W - w_2} = 5$$

or $$2w_2 = 5\ (0.286\ W - w_2)$$

or $$w_2 = 0.204\ W \text{ g}$$

∴ Total amount of the substance extracted

$$= 0.714\ W + 0.204\ W = 0.918\ W$$

Since 0.918 W g is more than 0.833 W g, the given statement is correct.

Example 3. At 25°C the distribution ratio of H_2S between water (C_1) and benzene (C_2) is 0.17.

(*i*) What is the minimum volume of C_6H_6 necessary at 25°C to extract in one step 90% of H_2S present in one litre of 0.1 molar aqueous solution?

(*ii*) What total volume of benzene will be required if the 90% extraction were to be completed in three steps of equal volumes of benzene?

(*iii*) Using 100 ml of benzene every time, how many extractions will be required to complete 90% extraction?

Solution. Suppose volume of benzene required = v litre

As extraction is 90%

∴ Amount left unextracted = 10%

$$\frac{w_n}{W} = \frac{10}{100} = 0.1$$

Now $$n = 1$$
$$V = 1 \text{ litre}$$
$$K = 0.17$$

Substituting these values in the general formula for extraction

$$\frac{w_n}{W} = \left(\frac{KV}{v + KV}\right)^n$$

or $$0.1 = \left(\frac{0.17 \times 1}{v + 0.17 \times 1}\right)^1$$

or $$v = 1.53 \text{ litres}$$

(*ii*) Suppose the total volume of benzene required = $3v$ litre

Then clearly $v = v$ litre, $n = 3$

Further, as the extraction = 90%

$\therefore$ Amount left unextracted = 10%

i.e. $$\frac{w_n}{W} = \frac{10}{100} = 0.1$$

Also $V = 1$ litre and $K = 0.17$

Putting these values in the formula

$$\frac{w_n}{W} = \left(\frac{KV}{v+KV}\right)^n, \text{ we get}$$

$$0.1 = \left(\frac{0.17\times1}{v+0.17\times1}\right)^3$$

or $$(v + 0.17)^3 = \frac{(0.17)^3}{0.1}$$

or $$(v + 0.17)^3 = 10 \times (0.17)^3$$

Taking cube roots of both the sides, we get

$$v + 0.17 = (10)^{1/3} \times 0.17$$

Taking logarithm of both the sides, we get

$$\log(v + 0.17) = \frac{1}{3}\log 10 + \log 0.17$$

$$= \frac{1}{3} + \bar{1}.2304$$

$$= 0.3333 + \bar{1}.2304$$

$$= \bar{1}.5637$$

$$v + 0.17 = \text{antilog}\,(\bar{1}.5637) = 0.3662$$

$\therefore$ $$v = 0.3662 - 0.17 = 0.1962 \text{ litres}$$

Total volume required $= 3v = 3 \times 0.1962$ litres

$= $ **0.5886 litres.**

(*iii*) Here, $\frac{w_n}{W} = \frac{1}{10} = 0.1$

$v = 100$ ml$= 0.1$ litre, $V = 1$ litre, $K = 0.17$, $n = ?$

Substituting these values in the formula

$$\frac{w_n}{W} = \left(\frac{KV}{v+KV}\right)^n, \text{ we get}$$

$$0.1 = \left(\frac{0.17\times1}{0.1+0.17\times1}\right)^n$$

$$0.1 = \left(\frac{0.17}{0.27}\right)^n$$

Taking logarithm of both the sides, we get

$$n \log \frac{0.17}{0.27} = \log 0.1$$

or $\quad n (\log 17 - \log 27) = \log 1/10$

or $\quad n (0.2304 - 0.4314) = -1$

or $\quad n (-0.2010) = -1$

or $\quad n = \frac{1}{0.201} = 5$

Thus 5 extractions are needed to accomplish the work.

PROBLEMS FOR PRACTICE

1. An aqueous solution contains 10 g of solute per litre. When 1 litre of the solution is treated with 100 cc of ether, 6 g of the solute are extracted. How much of the solute would be extracted from the aqueous solution by a further 100 cc ether ? Assume that the molecular state of solute is the same in ether and water **[Ans.** 2.4 g]

2. 1000 cc of an aqueous solution containing 30 g of an organic acid is extracted with 300 cc of ether using 100 cc in each operation. Calculate the amount of the organic acid left in aqueous solution. The partition co-efficient of organic acid between ether and water is 3.

[Ans. 13.65 g]

3. 100 ml of an aqueous solution of iodine containing 0.008 g iodine was extracted with 20 ml of carbon tetrachloride. If iodine is 15 times more soluble in carbon tetrachloride than in water, find out the amount of iodine remaining in the aqueous layer. If iodine is extracted a second time with 20 ml of carbon tetrachloride, how much of it will remain unextracted in the aqueous layer ? **[Ans.** 0.002 g, 0.0005 g]

4. When one litre of an aqueous solution containing 5 g of a solute is shaken with 50 cc of ether, it is found that 0.85 g of the solute passes into the ether layer. How much of the solute will be left unextracted when the same aqueous solution is shaken with a second instalment of 50 cc of ether ? What would be the difference if there were one extraction only with 100 cc of ether ? The solute may be supposed to have the same molecular weight in both water and ether.

[Ans. (*i*) 3.46 g (*ii*) 0.101 g more will be left unextracted]

6.8. DETERMINATION OF DEGREE OF HYDROLYSIS FROM DISTRIBUTION LAW

Let us consider the case of aniline hydrochloride. Determination of degree of hydrolysis of this salt involves the following steps :

(*i*) **Determination of the hydrolysis constant*.** Aniline hydrochloride is hydrolysed as

$$C_6H_5NH_3Cl + H_2O \rightleftharpoons C_6H_5\overset{+}{N}H_3\overset{-}{O}H + HCl$$
$$\downarrow$$
$$C_6H_5NH_2 + H_2O$$

The hydrolysis constant is given by

$$K_h = \frac{[\text{Free aniline}][\text{Free HCl acid}]}{[\text{Unhydrolysed salt}]}$$

To get the value of K_h, we need to find the molar concentrations of free aniline, free HCl acid and the unhydrolysed salt in an aqueous solution of aniline hydrochloride. This is done by shaking aqueous solution of aniline hydrochloride of known concentration with benzene. As a result, the free aniline distributes itself between the two solvents *i.e.*, water and benzene, whereas HCl acid and the unhydrolysed salt remain in the aqueous layer only as shown in Fig. 6.6.

(*a*) *Calculation of concentration of free aniline in the aqueous layer.*

The concentration of aniline present in the benzene layer is found by taking a known volume of this layer, evaporating off the benzene and finding weight of aniline left. Knowing the concentration of aniline in the benzene layer, that in the aqueous layer can be calculated if the distribution co-efficient of aniline between water and benzene is known.

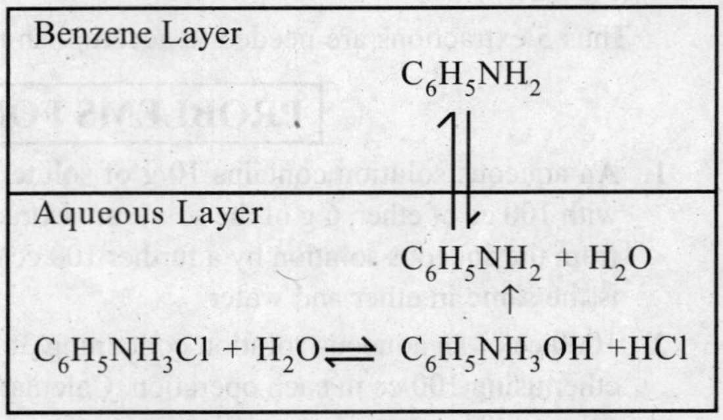

Benzene Layer

$C_6H_5NH_2$

⇅

Aqueous Layer

$C_6H_5NH_2 + H_2O$

↑

$C_6H_5NH_3Cl + H_2O \rightleftharpoons C_6H_5NH_3OH + HCl$

Fig. 6.6. Shaking an aqueous solution of aniline hydrochloride with benzene.

(*b*) *Calculation of concentration of free HCl produced*

From the hydrolysis equation,

Number of moles of HCl produced = Total number of moles of aniline produced.

But total amount of aniline produced = Amount of aniline present in the benzene layer + Amount of aniline present in the aqueous layer.

These quantities have been calculated in step (*a*) above. Hence, the total number of moles of aniline produced and the free HCl produced can be calculated. Knowing the volume of the aqueous layer the concentration of free HCl in the aqueous layer can be calculated.

(*c*) *Calculation of the concentration of the unhydrolysed salt.*

From the hydrolysis equation, we can say that the number of moles of the salt hydrolysed is equal to total number of moles of the aniline produced or free HCl produced, which has already been calculated in step (*b*) above. Knowing the initial number of moles of the salt taken, the number of moles of the salt left unhydrolysed can be calculated. Knowing the volume of the aqueous layer, the concentration of the unhydrolysed salt can be calculated.

(*ii*) **Calculation of the degree of hydrolysis*.** Knowing the hydrolysis constant K_h, the degree of hydrolysis of a solution of aniline hydrochloride having concentration *c* moles/litre can be calculated directly from the expression

$$h = \sqrt{\frac{K_h}{c}}$$

This was a case of salt of weak base and strong acid. When other salts are being considered *viz.,* salts of strong base and weak acid or weak base and weak acid, the relevant equation between hydrolysis constant and degree of hydrolysis may be used.

Relation between K_h and C

For salts of weak acid and strong base : $h = \sqrt{\frac{K_h}{C}}$

*To understand terms like hydrolysis constant and degree of hydrolysis, refer to chapter 8 in this book (Electrochemistry II)

For salts of weak acid and weak base : $h = \sqrt{K_h}$

assuming h as very small compared to 1.

SOLVED PROBLEM ON SALT HYDROLYSIS

Example 1. In an experiment to determine the hydrolysis constant of aniline hydrochloride at 20°C, 15 g of the salt were shaken with 1000 cm^3 of water and 100 cm^3 of benzene until equilibrium was established. 50 cm^3 of benzene layer was then found to contain 0.2360 g of aniline. The partition co-efficient of aniline between benzene and water at 20°C is found to be 11.3. Calculate the hydrolysis constant and the degree of hydrolysis of aniline hydrochloride.

Solution. Aniline hydrochloride is hydrolysed as

$$C_6H_5NH_3Cl + H_2O \rightleftharpoons C_6H_5\,NH_3\,OH + HCl$$
$$\downarrow$$
$$C_6H_5NH_2 + H_2O$$

Hence $$K_h = \frac{\text{[Free aniline][Free HCl acid]}}{\text{[Unhydrolysed salt]}}$$

Thus to find K_h, we need to find the molar concentrations of free aniline, free HCl and hydrolysed salt in the aqueous solution.

(i) To calculate the concentration of free aniline in the aqueous layer.

It is given that 50 cm^3 of benzene layer contains 0.2360 g of aniline *i.e.*, concentration of aniline present in the benzene layer

$$= 0.2360 \text{ g of aniline in 50 cm}^3 \text{ of benzene}$$

$$= \frac{0.2360}{93} \times \frac{1000}{50} \text{ moles/litre}$$

($\because$ Mol mass of aniline, $C_6H_5NH_2 = 93$)

$$= 0.05075 \text{ moles/litre}$$

As partition co-efficient of aniline between benzene and water is given to be 11.3, this means

$$\frac{\text{Conc. in the benzene layer}}{\text{Conc. in the aqueous layer}} = 11.3$$

But Conc. in benzene layer $= 0.05075$ moles/litre

$$\therefore \text{ Conc. in the aqueous layer } = \frac{0.05075}{11.3} \text{ moles/litre}$$

$$= 0.00449 \text{ moles/litre}$$

$\therefore$ Amount of aniline present in 1000 cm^3 of the aqueous layer

$$= 0.00449 \text{ moles}$$

As the volume of the aqueous layer is 1000 cm^3, therefore, concentration of free aniline is given by

$$\text{[Free aniline]} = 0.00449 \text{ moles/litre.}$$

(ii) To calculate the concentration of free HCl acid produced

We understand from the hydrolysis equation that

No. of moles of HCl produced = No. of moles of aniline produced

But total amount of aniline produced

= Amount of aniline present in the benzene layer

+ Amount of aniline present in the aqueous layer

Conc. of aniline in the benzene layer = 0.05075 moles/litre

∴ Amount of aniline present in 100 cm^3 of benzene layer = 0.005075 moles.

Also Conc. of aniline in aqueous layer = 0.00449 moles/litre

∴ Amount of aniline present in 1000 cm^3 of aqueous layer = 0.00449 moles

∴ Total amount of aniline produced = 0.005075 + 0.00449 = 0.009565 moles

Thus no. of moles of HCl produced = 0.009565 moles

As volume of the aqueous layer is 1000 cm^3.

[Free HCl] = 0.009565 moles/litre

(iii) To calculate the concentration of the unhydrolysed salt

From the hydrolysis equation, we can say

No. of moles of aniline hydrochloride hydrolysed

= No. of moles of aniline or HCl produced

= 0.009565 moles

As aniline hydrochloride taken was 15 g and molecular mass of aniline hydrochloride ($C_6H_5NH_3Cl$) is 129.5, Therefore,

No. of moles of aniline hydrochloride taken

$$= \frac{15}{129.5} \text{ moles} = 0.11583 \text{ moles}$$

∴ No. of moles of salt left unhydrolysed

= 0.11583 – 0.009565 = 0.10626 moles

As the volume of the aqueous layer is 1000 cm^3.

[Unhydrolysed salt] = 0.10626 moles/litre

Hence
$$K_h = \frac{\text{[Free aniline] [Free HCl acid]}}{\text{[Unhydrolysed salt]}}$$

$$= \frac{0.0049 \times 0.009565}{0.10626}$$

$$= \mathbf{4.04 \times 10^{-4}}$$

As aniline is a salt of weak base and strong acid, the degree of hydrolysis will be

$$h = \sqrt{\frac{K_h}{c}}$$

where c is the original concentration of the salt.

But c = 0.11583 moles/litre (calculated above)

$$\therefore \quad h = \sqrt{\frac{4.04 \times 10^{-4}}{0.11583}} = \mathbf{5.9 \times 10^{-2}}$$

Summary of applications of distribution law

(*i*) To determine the solubility of a solute in a solvent.

(*ii*) To determine the extent of association or dissociation of a solute in a solvent.

(*iii*) To study chemical equilibrium of reactions involving complex ions.

(*iv*) In the process of extraction.

(*v*) In the Parke's process of desilverisation of lead.

(*vi*) To determine hydrolysis constant and degree of hydrolysis.

6.9. SEPARATION OF A MIXTURE OF SMALL AMOUNTS OF ORGANIC SUBSTANCES USING LIQUID-LIQUID CHROMATOGRAPHY

This is a versatile technique for separation of a mixture of small amounts of organic substances. The separation is carried in a glass tube packed with an inert solid material (silica) soaked in solvent 1. The mixture solution is dropped at the column top. Then another immiscible solvent 2 is allowed to flow down the column (Fig. 6.7). A film of solvent 1 held by the silica forms the stationary liquid phase, while the running solvent 2 the mobile liquid phase. Each component of the mixture distributes itself between the stationary liquid phase and the mobile liquid phase.

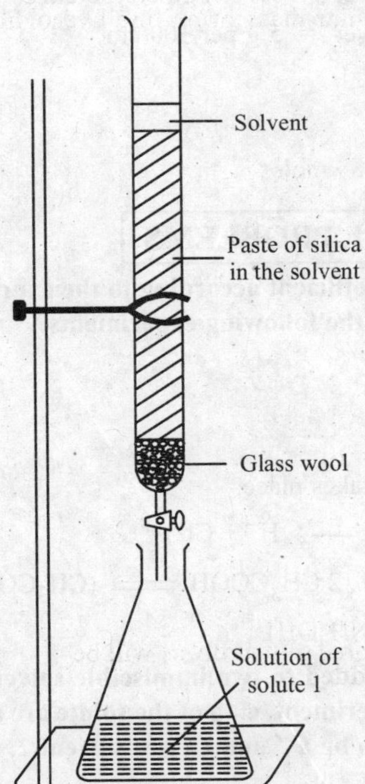

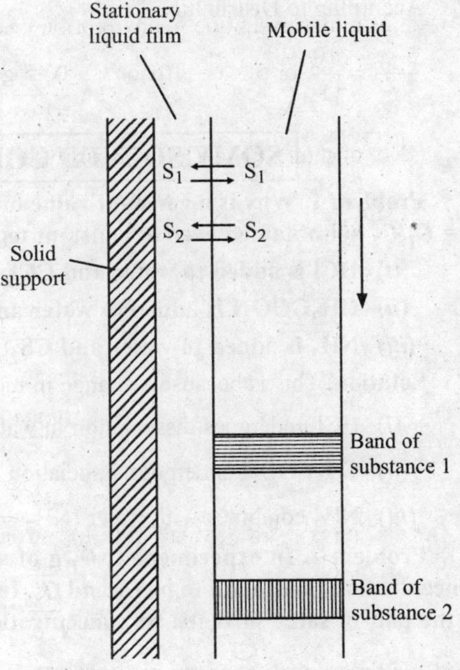

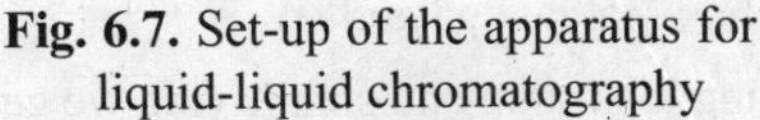

Fig. 6.7. Set-up of the apparatus for liquid-liquid chromatography

Fig. 6.8. Separation of substances 1 and 2 (S_1 and S_2) in the form of bands

The liquid film on the solid support acts as the **stationary** state while the second liquid which is immiscible with the first liquid forms the mobile liquid phase. The solute which is more soluble in solvent 2 comes down in the column first (Distribution co-efficient of substance 2 in the liquids 2 and 1 is more than unity).

Thus,
$$\frac{C_s}{C_m} = K_D$$

where C_s = concentration in stationary liquid phase and C_m = concentration in the mobile liquid phase. K_D = distribution co-efficient.

Different components in the mixture have different distribution co-efficients. They pass into the mobile phase in order of their distribution co-efficients. **The component (say S_2) with higher distribution co-efficient is extracted first and is the first to move down the column. The location of the components is easy if these are coloured.** Otherwise they are visualised with the help of a locating agent (Fig. 6.8). The portion of the solvent (or band) containing each component is received in a separate vessel and the solute recovered from it by distillation.

Example . A solute distributes in the ratio of 10 : 1 between ether and water. If 5 g of the solute is added to a liquid mixture containing 100 cm^3 each of ether and water, what is the mass of solute dissolved in aqueous layer?

Solution. Let the mass of the solute in aqueous layer be $= x$ g

Then, the mass of the solute in ether layer $= (5 - x)$ g

conc. of solute in ether layer $= (5 - x)$ g per 100 cm^3

conc. of solute in aqueous layer $=$ 5 g per 100 cm^3

According to Distribution Law

$$\frac{5-x}{x} = \frac{10}{1} \text{ or } 5 - x = 10x \text{ or } x = 0.45 \text{ g}$$

SOME SOLVED CONCEPTUAL PROBLEMS

Problem 1. Why is a constant value of partition co-efficient according to the expression $K = C_1/C_2$ not obtained even at constant temperature in the following experiments?

(*i*) HCl is added to water and CCl_4

(*ii*) CH_3COOH is added to water and benzene

(*iii*) NH_3 is added to water and CS_2

Solution. This is because a change in molecular state takes place.

(*i*) HCl undergoes dissociation in water ($HCl + aq \longrightarrow H^+ + Cl^-$).

(*ii*) CH_3COOH undergoes association in benzene, *i.e.*, $2\,CH_3COOH \rightleftharpoons (CH_3COOH)_2$.

(*iii*) NH_3 combines with water ($NH_3 + H_2O \rightleftharpoons NH_4OH$).

Problem 2. In experiment 1, w_1 g of a solute are added to two immiscible solvent. Its concentrations are found to be C_1 and C_2. In another experiment, w_2 g of the solute are added to the pair of same solvents. Its concentration is found to be C_1' and C_2'. In solvent 2, solute undergoes association as $nA \rightleftharpoons A_n$. Formulate an equation to calculate n.

Solution. $\frac{C_1}{\sqrt[n]{C_2}} = K$, *i.e.*, $\frac{C_1}{(C_2)^{1/n}} = K$. On taking logarithm and simplifying we get

$$n = \frac{\log C_2 - \log C_2'}{\log C_1 - \log C_1'}.$$

Problem 3. Why zinc is used in Parke's process for desilverisation of lead ?

Solution. This is because silver is more soluble in zinc than in lead. Also zinc is immiscible with molten lead.

Problem 4. In the expression for distribution law, is the value of partition co-efficient affected on changing the units of concentrations ?

Solution. No. As $K = C_1/C_2$, there is no effect on changing the units of concentrations C_1 and C_2 as long as they are in the same units.

Problem 5. Why 100% extraction is not possible howsoever large number of instalments of the extracting solvent may be used ?

Solution. Amount left unextracted after n instalments each of volume v cc from a solution containing W g of the solute in V cc of the solution is given by the equation

$$w_n = \left[\frac{Kv}{V + KV}\right]^n W$$

where K = Partition co-efficient of the solute between the original solvent and the extracting solvent. As $KV/(v + KV) < 1$, therefore, w_n will become smaller and smaller if n is larger. It can be seen from the equation that w_n can never be equal to zero however large value n we may have.

Problem 6. Can we find the distribution co-efficient of iodine between water and ethyl alcohol ?

Solution. No because ethyl alcohol and water are miscible. The two layers cannot be separated.

Problem 7. On shaking a solute with two immiscible liquids, it is observed that 10% of the solute in solvent 2 undergoes dimerization. Derive an expression for constant value of *K*.

Solution. $$K = \frac{C_1}{n\sqrt{C_2\,\alpha}}$$

$\alpha = 10\% = 0.10$ and $n = 2$, substituting the values of α and n, we get

$$K = \frac{C_1}{\sqrt{0.1 C_2}}$$

EXERCISES
(Based on Question Papers of Different Universities)

Multiple Choice Questions (Choose the correct option)

1. The Nernst distribution law equation $C_1/C_2 = K$ applies when
   (*a*) The molecular state of the solute is the same in both the solvents
   (*b*) The molecular state of the solute is different in both the solute
   (*c*) The molecular state of the solute may be the same or different in both the solute
   (*d*) None of the above
2. When a bottle of soda-water is opened, the partial pressure of CO_2 ________.
   (*a*) decreases (*b*) increases (*c*) remains the same (*d*) none of these
3. In the first extraction, the amount of the substance left unextracted is given by the formula (K is distribution co-efficient, V ml of the aqueous solution contain A gram of an organic substance)

(*a*) $x_1 = A\dfrac{KV}{KV+v}$ (*b*) $x_1 = A\dfrac{KV}{KV+v^2}$ (*c*) $x_1 = A^2\dfrac{KV}{KV+v}$ (*d*) $x_1 = A\dfrac{K^2V}{KV+v}$

4. The greater the distribution ratio in favour of the organic solvent, the ______ will be the amount extracted in any one operation

(*a*) greater (*b*) lesser (*c*) equal (*d*) none of these

5. Benzoic acid when shaken with mixtures of benzene and water undergoes dismerisation in benzene. The distribution law applicable here is

(*a*) $\dfrac{C_1}{C_2} = K_D$ (*b*) $\dfrac{C_1}{3\sqrt{C_2}} = K_D$ (*c*) $\dfrac{C_1}{2\sqrt{C_2}} = K_D$ (*d*) none of these

6. The ______ extraction is more efficient

(*a*) single (*b*) double (*c*) multiple (*d*) none of these

7. The Nernst distribution law $K_D = C_1/C_2$ is not applicable if the solute undergoes

(*a*) Association in one of the solvents

(*b*) Dissociation in one of the solvent

(*c*) Association or dissociation in one of the solvents

(*d*) None of the above

8. The distribution law is applied in the

(*a*) Haber's process for the manufacture of NH_3

(*b*) Parke's process for the extraction of Ag

(*c*) Contact process for the manufacture of H_2SO_4

(*d*) None of the above

9. When succinic acid or oxalic acid is shaken with ether and water it ______ in water.

(*a*) associates to form dimer (*b*) dissociates into ions

(*c*) remains the same (*d*) none of these

10. With the help of distribution law we can

(*a*) determine the association or dissociation of the solute in one of the solvents

(*b*) determine the solubility of solute in various solvents

(*c*) deduce the formula of complex ions

(*d*) all of the above

11. Multiple extraction is ______ efficient as compared to single extraction

(*a*) more (*b*) less (*c*) the same (*d*) none of these

ANSWERS

1. (*a*)	**2.** (*a*)	**3.** (*a*)	**4.** (*a*)	**5.** (*c*)	**6.** (*c*)
7. (*c*)	**8.** (*b*)	**9.** (*b*)	**10.** (*d*)	**11.** (*a*)	

Short Answer Questions

1. Define Distribution Law. Derive the expression showing how the law is modified if the solute undergoes association in one of the solvents with degree of association equal to α.

2. Explain the application of distribution law in the determination of degree of hydrolysis of aniline hydrochloride.

3. State Nernst Distribution Law. Derive the modified expression if the solute undergoes dissociation in one of the solvents with degree of dissociation equal to α.
4. How can distribution law be used to study the formation of complex ion I_3^- ?
5. A solute has normal molecular state in solvent 1 but one mole of it combines with n moles of the solvent 2 to form a compound. Derive the modified expression for distribution law and comment on the result obtained.
6. How do we use distribution law in desilverisation of lead.
7. Define Distribution Law. How is the law modified if the solute forms dimers in one of the solvents ? Derive it.
8. A solute undergoes association in a solvent according to the equation $nX \rightarrow X_n$. How can you determine the value of n by applying distribution law ?
9. Define 'chemical potential'. Applying the concept of 'chemical potential', how Nernst distribution law can be derived ?
10. Define Nernst Distribution Law. What are its limitations ?

General Questions

1. Derive the expression for the amount of solute left unextracted after n instalments of v *cc* each from a solution containing W *g* of the solute in V *cc* of the solution.
2. Define 'Hydrolysis constant' and 'degree of hydrolysis'. Applying distribution law, how the degree of hydrolysis of aniline hydrochloride at a particular concentration can be determined?
3. Define Distribution Law. How is the law modified if the solute undergoes.

   (*a*) association in one of the solvents (*b*) dissociation in one of the solvents ?
4. Define Distribution Law. Discuss its application to establish the formation of complex ion I_3^- according to the following equilibrium :

$$I^- + I_2 \rightleftharpoons I_3^- \text{ (or KI + I}_2 \rightleftharpoons \text{KI}_3)$$

5. Discuss the statement : "Multi-step extraction is more economical than a single step extraction."
6. State and explain Nernst Distribution Law. What are the conditions under which the law is applicable ? Derive the law thermodynamically.
7. State and explain Nernst Distribution Law. How can it be applied to determine

   (*a*) the solubility of a solute in a particular solvent ?

   (*b*) the extent of association or dissociation of the solute in a solvent ?
8. State Nernst Distribution Law. Prove that the law does not undergo modification if the solute combines with one of the solvents except that the numerical value of the partition co-efficient changes.
9. In order to extract a solute A from its solution in a solvent Y by means of a solvent X when both X and Y are immiscible, explain as to why it is always advised to use small lots of the solvent X at a time. Derive the necessary relation.

7

Electrochemistry–I

7.1. INTRODUCTION

There are broadly two types of substances as far as conduction of electricity through them is concerned. These are *conductors* and *insulators*.

Substances which allow the passage of electricity through them are called conductors while those which do not allow electricity to pass through them are called insulators.

Conductors can be classified into two types as stated below.

Electronic conductors. These substances conduct electricity without undergoing any decomposition. Electricity is conducted by the flow of electrons in this type of conductors which are mostly metals with the exception of graphite and certain minerals.

Electrolytic conductors (electrolytes). These substances undergo decomposition when electricity is passed through them. Acids, bases and salt solutions in water and fused salts come under this category. Electricity is conducted by ions in this case.

Electrolytes can be classified into two types :

Strong electrolytes. Substances like strong acids (H_2SO_4, HNO_3, HCl, etc), strong bases (NaOH, KOH, etc) and salts in solution or molten state are completely ionised and conduct electricity to a large extent. Such electrolytes are called strong electrolytes.

Weak electrolytes. Substances such as weak acids (like CH_3COOH), weak bases (like NH_4OH), etc are not completely ionised in solution and hence do not conduct electricity to a great extent and are called weak electrolytes.

Non-ionic substances like sugar, urea, etc do not conduct electricity at all and are called non-electrolytes.

7.1.1. Factors affecting electrolytic conduction

Factors affecting electrical conductivity of solutions are described below :

1. Interionic attraction. Interactions between the ions of the solute determine the conductivity. Greater the attractive forces between the ions of the solute, smaller will be the conductivity. Such interactions are called solute-solute interactions.

2. Solvation of ions. These are the interactions between the ions of the solute and the solvent, known as solute-solvent interaction. Greater the solute-solvent interactions (solvation of ions), smaller the electrical conductance.

3. Viscosity of the solvent. This factor involves solvent-solvent interactions. Greater the viscosity of the solvent, greater will be the hindrance to the movement of the solute ions and hence smaller will be the conductance.

4. Temperature. The two types of conductors show different behaviour with increase of temperature. While the conductance of electrolytic conductors increases, that of metallic conductors decreases with the rise of temperature. In the case of electrolytic conductors, the inter-ionic attractions and viscosity decrease with the increase of temperature. This results into increase of conductance. However, we observe a decrease in conductance of metallic conductors with temperature due to increase in the vibrations of kernls of metal atoms, which impede the flow of electrons.

7.1.2. Main points of difference between metallic and electrolytic conductors

The main points of difference between the two types of conductors are given in Table 7.1.

Table 7.1. Metallic and Electrolytic Conductors

Metallic Conductor	Electrolytic Conductor
1. In metallic conductors, the electricity is conducted by the flow of electrons from a negative potential to a positive (or less negative) potential.	**1.** Here, the electric current is carried by the charged ions moving towards the oppositely charged electrodes.
2. No chemical change takes place when a metallic (or electronic) conductor is conducting electricity. No new products are formed.	**2.** Conduction of electricity by electrolytic conductors is accompanied by some definite chemical process with the evolution of some gases or deposition of some metals at the electrodes.
3. In metallic conduction, temperature has the effect of decreasing the conductance, *i.e.*, conductance of a metallic conductor decreases with the increase of temperature. Metallic conductance becomes pronounced at low temperature. This is because at low temperature, the kernels of the atoms have the least vibrations and offer the least resistance to the flow of electrons.	**3.** In electrolytic conductors, the conductance increases with the rise of temperature. This is due to increase in the mobility of the ions and increase in the degree of ionisation of the electrolytes.

7.2. ELECTRICAL RESISTANCE AND ELECTRICAL CONDUCTANCE

Resistance. The law that gives exact relationship between the resistance, the current and voltage is called *Ohm's Law.* It states :

If to the ends of a conductor a voltage 'E' is applied and a current 'I' flows through it, then the resistance 'R' of the conductor is given by E / I.

Current is generally measured in amperes, whereas voltage is measured in volts. If one ampere current flows through a conductor when a voltage of one volt is applied to it, the resistance of the conductor is taken as 1 ohm (written Ω).

Mathematically, the law can be put as

$$R = \frac{E}{I}$$

All electrolytic solutions like metallic conductors obey Ohm's law.

Conductance. The reciprocal of the electrical resistance is called the conductance. It is usually represented by *C*. Thus,

$$C = \frac{1}{R}$$

It is a more common practice to speak of the conductance of a solution rather than its resistance.

Units. The units of conductance are siemen (*S*) or ohm^{-1} or mho ($1\ S = 1\Omega^{-1}$)

7.2.1. Specific conductance

It is found that resistance (R) of a conductor is

(*i*) directly proportional to its length (l) and

(*ii*) inversely proportional to its area of cross-section (a)

i.e., $$R \propto \frac{l}{a}$$

or $$R = \rho \frac{l}{a} \quad \text{... (i)}$$

where ρ is a constant of proportionality, called *Specific Resistance* or *Resistivity*. Its value depends upon the material of the conductor.

"The reciprocal of resistivity is known as *specific conductivity* or simply *conductivity.*" It is denoted by K. Thus, if K is the specific conductivity and C is the conductance of the solution, then

$$R = \frac{1}{C} \text{ and } \rho = \frac{1}{K}$$

Substituting the values of R and ρ in eq. (*i*) above

$$\frac{1}{C} = \frac{1}{K} \times \frac{l}{a}$$

$$K = C \times \frac{l}{a}$$

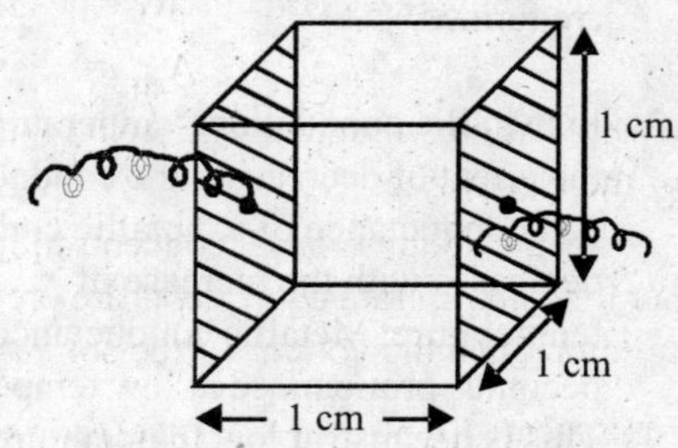

Fig. 7.1. Specific conductivity.

Now if l = 1 cm and a = 1 sq cm, then

$$K = C$$

Hence, specific conductivity of a solution is defined as the *conductance of a solution of 1 cm length and having 1 sq cm as the area of cross-section.* Alternatively, it may be defined as *conductance of one centimetre cube of the solution of the electrolyte.* This is illustrated in Fig. 7.1.

If the volume of the solution is V cm^3, the specific conductivity of such a solution at this dilution V is written as K_V.

Units. Resistivity or specific resistance (ρ)

$$= R\frac{a}{l} = \text{ohms } \frac{(\text{cm})^2}{\text{cm}} = \text{ohm cm or ohm m in SI units}$$

Specific conductivity $$(K) = \frac{1}{\rho} = \frac{1}{\text{ohm cm}}$$

$$= \text{ohm}^{-1}\ \text{cm}^{-1} \text{ or } \Omega^{-1}\ \text{cm}^{-1} \text{ or S cm}^{-1}$$

$$\text{or S m}^{-1} \text{ in SI units}$$

7.2.2. Equivalent conductance (conductivity) and molar conductance (conductivity)

Equivalent conductivity of a solution at a dilution V is defined as the *conductance of all the ions produced from one gram equivalent of the electrolyte dissolved in Vcm^3 of the solution when the distance between the electrodes is one cm and the area of the electrodes is so large that whole of the solution is contained between them.* It is represented by λ_{eq} (lambda).

Relationship between equivalent conductivity and specific conductivity. The equivalent conductivity of a solution is calculated from the specific conductivity. The relationship between equivalent conductivity and specific conductivity may be obtained as follows :

Consider a rectangular vessel with its two opposite walls one cm apart and made of some metal sheets so that they act as the electrodes.

Case I. Let us suppose 1cm^3 of the solution containing 1 g eq. of the electrolyte is taken in the vessel (Fig. 7.2). The conductance of this solution will be its specific conductivity. Further 1 cm^3 of the solution taken contains one gram equivalent of the electrolyte, the conductance of the solution will be its equivalent conductivity (by definition). Thus, when 1cm^3 of the solution containing one gram equivalent of the electrolyte is considered, the equivalent conductivity becomes equal to its specific conductivity.

Case II. Let us suppose 4 cm^3 of the solution containing one gram equivalent of the electrolyte is taken in the vessel. The conductance of the solution will be still equal to its equivalent conductivity at this dilution but now there will be four cubes each of volume 1cm^3 as shown in Fig. 7.2. The conductance of each 1cm^3 of the solution is equal to its specific conductivity so that the total conductance of the solution, *i.e.*, equivalent conductivity is four times the specific conductivity.

$$\text{Equivalent conductivity} = \text{Specific conductivity} \times V$$

$$\Lambda_{eq} = K_V \times V$$

where V is the volume in cm^3 containing one gram equivalent of the electrolyte.

If the solution has a concentration of c gram equivalent per litre, *i.e.*, c gram equivalents are present in 1000 cm^3 of the solution, then the volume of the solution containing one gram equivalent will be $\frac{1000}{c}$. Hence, the above expression may be written as

$$\Lambda_{eq} = K \times \frac{1000}{c_{eq}} = K \times \frac{1000}{\text{Normality}}$$

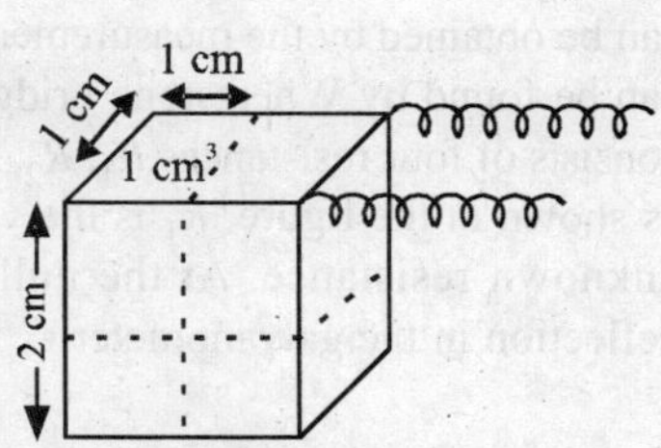

Fig. 7.2. Relationship between equivalent conductivity and specific conductivity.

Units of equivalent conductivity

$$\Lambda_{eq} = K \times \text{V}$$

$$= \text{ohms}^{-1}\ \text{cm}^{-1} \times \frac{\text{cm}^3}{\text{Gram Eq.}}$$

$$= \text{ohm}^{-1}\ \text{cm}^2\ (\text{g eq})^{-1} \text{ or } S\ \text{cm}^2\ \text{eq}^{-1} \text{ or } \Omega^{-1}\ \text{cm}^2\ \text{eq}^{-1}.$$

S.I. units

The units of Λ_{eq} are $S\ m^2\ eq^{-1}$

$$1\ S\ m^2\ eq^{-1} = 10^4 \times S\ \text{cm}^2\ eq^{-1}$$

Molar conductivity. The molar conductivity of a solution is defined as under :

The molar conductivity of a solution at a dilution V is the conductance of all the ions produced from one mole of the electrolyte dissolved in V cm^3 of the solution when the electrodes are one cm apart and the area of the electrodes is so large that the whole of the solution is contained between them. It is usually represented by Λ_m.

Relationship between molar conductivity and specific conductivity. Molar conductivity is related to the specific conductivity as follows.

Molar conductivity = Specific conductivity × Volume in cm^3 containing one mole of the electrolyte.

or
$$\Lambda_m = K \times V$$
$$= K \times \frac{1000}{C} = K \times \frac{1000}{\text{Molarity}}$$

where K is the specific conductivity and V is the volume of the solution containing one mole of the electrolyte and c is the molar concentration.

Units of Λ_m will be ohm^{-1} cm^2 mol^{-1} (or S cm^2 mol^{-1}) or Ω^{-1} cm^2 mol^{-1}.

S.I. units

Units of Λ_m are S m^2 mol^{-1}

$$1\,S\text{ m}^2\text{ mol}^{-1} = 10^4 \times S\text{ cm}^2\text{ mol}^{-1}$$

7.3. MEASUREMENT OF ELECTROLYTIC CONDUCTANCE, SPECIFIC CONDUCTIVITY, EQUIVALENT CONDUCTIVITY AND MOLAR CONDUCTIVITY (CONDUCTANCE)

(*i*) *Electrolyte conductance.* Conductance is the reciprocal of resistance. Hence, the conductance can be obtained by the measurement of the resistance and the latter can be found by Wheatstone bridge method shown in Fig. 7.3. It consists of four resistances R_1, R_2, R_3 and R_4 arranged in a manner as shown in the figure. R_1 is the variable resistance and R_2 is the unknown resistance. At the null point, *i.e.*, when there is no deflection in the galvanometer G, we have,

$$\frac{R_1}{R_2} = \frac{R_3}{R_4}$$

Knowing the values of R_1, R_3 and R_4, R_2 can be calculated.

To find the resistance of the solution of an electrolyte, a special type of cell has to be used such that the solution is present between the two electrodes. The cell thus used is called *conductivity cell.* It is made up of pyrex glass and two platinum electrodes at a fixed distance apart.

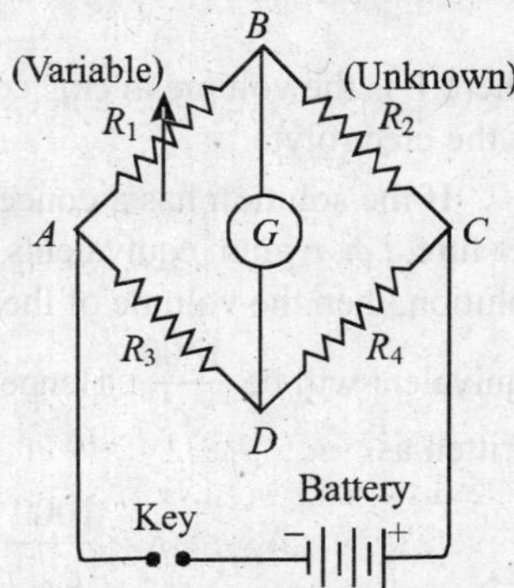

Fig. 7.3. Principle of Wheatstone bridge.

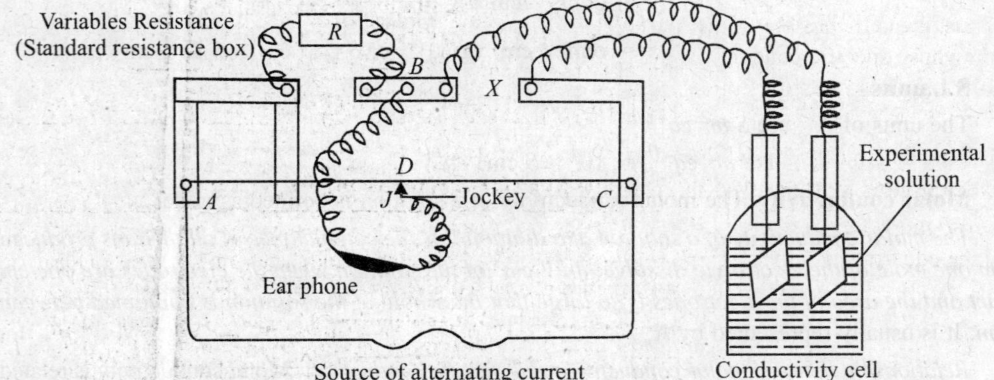

Fig. 7.4. Apparatus for the measurement of electrolytic conductance

Various designs of conductivity cells are available. If direct current is used, it causes electrolysis of the solution and this results in the change in the resistance of the solution. These effects are called *polarization effects.* To overcome this problem, an alternating current of frequency 50-200 cycles per second is used. Also we have to use a headphone instead of a galvanometer to detect the null point.

The complete assembly for the measurement of the electrolytic conductance is shown in Fig. 7.4.

A resistance R is introduced from the standard resistance box such that when the sliding contact, *i.e.*, the jockey J is moved along the stretched wire, the sound in the earphone is reduced to minimum at a point somewhere in the middle of the wire AC, say at the point D. This represents the null point. If X is resistance of the electrolytic solution, by Wheatstone bridge principle,

$$\frac{\text{Resistance } R}{\text{Resistance } X} = \frac{\text{Resistance of wire } AD}{\text{Resistance of wire } CD} = \frac{\text{Length } AD}{\text{Length } CD}$$

$$\text{Resistance } X = \text{Resistance } R \times \frac{\text{Length } CD}{\text{Length } AD}$$

Knowing the resistance R and the balance point D, the resistance X of the electrolytic solution can be calculated.

$$\text{Conductance, } C = \frac{1}{X}$$

(*ii*) *Specific conductivity.* Specific conductivity (K) is related to the conductance (C) by the expression

$$K = C \times \frac{l}{a}$$

Thus, the specific conductivity of a solution can be determined by measuring its conductance and the distance (l) between the electrodes and the area of cross-section (a) of each of the electrodes. However, for a particular cell, l/a is constant and this constant is called *cell constant.* The value of cell constant is generally written on the cell. It can also be determined experimentally.

$$\frac{l}{a} = \text{cell constant}$$

$$\therefore \quad \text{Specific conductivity } (K) = \text{Conductance} \times \text{cell constant.}$$

Thus, the cell constant of any particular cell can be found by measuring the conductance of a solution whose specific conductivity is known. The solution usually taken for this purpose is $N/50$ KCl solution whose specific conductivity at 298 K has been determined accurately to be 0.002765 $ohm^{-1}\ cm^{-1}$.

Having determined the value of the cell constant, the specific conductivity of the given solution can be determined by measuring its conductance and multiplying the value with the cell constant.

(*iii*) *Equivalent conductivity.* Equivalent conductivity (Λ_{eq}) is related to the specific conductivity (K) by the equation.

$$\Lambda_{eq} = K \times \frac{1000}{c_{eq}}$$

where c_{eq} is the concentration of the solution in gram equivalents per litre (*i.e.*, the normality of the solution). Thus, knowing the normality of the solution and the specific conductivity K, Λ_{eq} can be calculated.

(*iv*) *Molar conductivity.* The molar conductivity is related to the specific conductivity according to the relation

$$\Lambda_m = K \times \frac{1000}{c}$$

where c is the molar concentration. Thus, knowing molar concentration c and the specific conductivity K, Λ_m can be calculated.

NUMERICAL PROBLEMS ON SPECIFIC, EQUIVALENT AND MOLAR CONDUCTIVITIES

Formulas to be used :

(*i*) $$R = \rho \frac{l}{a} \quad \text{or} \quad \rho = R\frac{a}{l}$$

(*ii*) $$K = \frac{1}{\rho} = \frac{1}{R} \times \frac{l}{a} = C \times \frac{l}{a}$$

= observed conductance × cell constant.

(*iii*) $$\text{Cell constant} = \frac{l}{a} = \frac{\text{Specific conductivity}}{\text{Observed conductance}}$$

$$= \text{Specific conductivity} \times \text{Observed resistance}$$

(*iv*) $$\Lambda_{eq} = K \times \frac{1000}{c}$$

where c = concentration of the solution in gram equivalent/litre, *i.e.,* normality of the solution.

(*v*) $$\Lambda_m = K \times \frac{1000}{c}$$

where c = Molar concentration of the solution. Sometimes the term conductivity is used in place of specific conductivity. It can be understood from the units attached to it. If the units attached to conductivity are $ohm^{-1}\ cm^{-1}$, it is obvious that it is specific conductivity.

SOLVED EXAMPLES ON SPECIFIC, EQUIVALENT AND MOLAR CONDUCTIVITY

Example 1. The resistance of 0.01 N NaCl solution at 298 K is 200 ohm. Cell constant of the conductivity cell is unity. Calculate the equivalent conductance.

Solution. *Calculation of specific conductance*

Specific conductance, $$K = \frac{1}{R} \times \text{cell constant}$$

$$= \frac{1}{200\,\text{ohm}} \times 1\,\text{cm}^{-1}$$

$$= 5 \times 10^{-3}\ \text{ohm}^{-1}\ \text{cm}^{-1}$$

Calculation of equivalent conductance

$$\Lambda_m = \frac{K \times 1000}{\text{Normality}}$$

$$= \frac{5 \times 10^{-3} \times 1000}{0.01}$$

$$= 5 \times 10^2 \text{ ohm}^{-1} \text{ cm}^2 \text{ eq}^{-1}$$

Example 2. The conductivity of a solution containing 1 gram of anhydrous $BaCl_2$ in 200 cm^3 of water has been found to be 0.0058 mho cm^{-1}. What are the molar conductivity and equivalent conductivity of the solution (At. wt. of Ba = 137 and Cl = 35.5)?

Solution. Conductivity (K) = 0.0058 mho cm^{-1}

Mol. wt of $BaCl_2$ = 137 + 2 × 35.5 = 208

As 1 gram of $BaCl_2$ is present in 200 cm^3 of the solution,

Molar concentration, $$(c) = \frac{1}{208} \times \frac{1}{200} \times 1000 \text{ mol L}^{-1}$$

$$= 0.0240 \text{ mol L}^{-1}$$

Molar conductivity, $$\Lambda_m = K \times \frac{1000}{c} = 0.0058 \times \frac{1000}{0.0240}$$

$$= 241.67 \text{ ohm}^{-1} \text{ cm}^2 \text{ mol}^{-1}$$

In case of $BaCl_2$ equivalent weight $\frac{\text{Mol. wt.}}{2} = \frac{208}{2} = 104$

∴ Concentration of the solution in gram equivalent per litre

$$c_{eq} = \frac{1}{104} \times \frac{1}{200} \times 1000 = 0.0480$$

∴ Equivalent conductivity, $$\Lambda_{eq} = K \times \frac{1000}{c_{eq}}$$

$$= 0.0058 \times \frac{1000}{0.480}$$

$$= 120.83 \text{ ohm}^{-1} \text{ cm}^2 \text{ eq}^{-1}$$

Example 3. The resistance of a conductivity cell when filled with 0.02M KCl solution is 164 ohm at 298 K. However, when filled with 0.05 M $AgNO_3$ solution, its resistance is found to be 78.5 ohm. If specific conductivity of 0.02 M KCl is 2.768 × 10^{-3} ohm^{-1} cm^{-1}, calculate :

(*i*) The conductivity of 0.05 M $AgNO_3$.

(*ii*) The molar conductivity of $AgNO_3$ solution.

Solution. *Calculation of cell constant*

Specific conductivity of 0.02 M KCl = 2.768 × 10^{-3} ohm^{-1} cm^{-1}

Resistance = 164 ohm

$$\text{Specific conductivity} = \frac{1}{R} \times \text{cell constant.}$$

$$\text{Cell constant} = R \times \text{specific conductivity}$$
$$= 164 \times 2.768 \times 10^{-3}$$
$$= 0.4539 \text{ cm}^{-1}$$

$$\text{Conductivity} = \frac{1}{R} \times \text{cell constant}$$

$$= \frac{1}{78.5} \times 0.4539 = 0.00578 \text{ ohm}^{-1}\text{cm}^{-1}$$

Calculation of Λ_m of 0.05M $AgNO_3$

$$\Lambda_m = \frac{1000 \times K}{\text{Molarity}}$$

$$\Lambda_m = \frac{1000 \times 0.0058}{0.05}$$

$$= 115.6 \text{ ohm}^{-1} \text{ cm}^2 \text{ mol}^{-1}$$

Example 4. 0.5 normal solution of a salt placed between two platinum electrodes 3.0 cm apart and of area of cross-section 6.0 sq cm has a resistance of 25 ohms. Calculate the equivalent conductivity of solution.

Solution. *1st step — Calculation of specific conductivity.*

Here $l = 2.0$ cm, $a = 4.0$ sq cm, $R = 25$ ohms.

$$\text{Conductance, } C = \frac{1}{R} = \frac{1}{25} \text{ ohm}^{-1}$$

$$\text{Cell constant} = \frac{l}{a} = \frac{3}{6} = \frac{1}{2} \text{ cm}^{-1}$$

$$\text{Specific conductivity } (K) = \text{Observed conductance} \times \text{Cell constant}$$

$$= \frac{1}{25} \times \frac{1}{2} = 0.02 \text{ ohm}^{-1}\text{cm}^{-1}$$

2nd step — Calculation of equivalent conductivity.

$$\Lambda_{eq} = K \times \frac{1000}{c_{eq}}$$

$$c = 0.5 \text{ N}$$
$$K = 0.02 \text{ ohm}^{-1} \text{ cm}^{-1} \text{ (calculated above)}$$

$$\Lambda_{eq} = \frac{1000 \times 0.02}{0.5} = 40 \text{ ohm}^{-1} \text{cm}^2 \text{eq}^{-1}$$

Example 5. The resistance of a decinormal solution of a salt occupying a volume between two platinum electrodes 1·8 cm apart and 5·4 sq. cm in area was found to be 32 ohm. Calculate the equivalent conductance of the solution.

Solution. $R = \frac{1}{K} \cdot \frac{l}{a}$ or $K = \frac{1}{R} \cdot \frac{l}{a}$

Substituting the values

$$\text{Specific conductance } K = \frac{1}{32} \times \frac{1.8}{5.4} = 0.0104 \text{ ohm}^{-1} \text{ cm}^{-1}$$

Eq. conductance = Sp. conductance × Vol in ml containing 1gm eq.of electrolyte

$= 0.0104 \times 10000$

$= \mathbf{104 \text{ ohm}^{-1} \text{ cm}^2 \text{ eq}^{-1}}$

7.4. EFFECT OF DILUTION ON SPECIFIC CONDUCTANCE AND EQUIVALENT CONDUCTANCE

The specific conductance decreases with dilution. This is due to the fact that the number of current carrying particles *i.e.* ions per ml of the solution become less. Of course on dilution, the degree of dissociation also increases but the no. of ions per ml is decreased to such an extent that even complete dissociation does not produce as many ions per ml as were present before dilution.

The equivalent conductance and molecular conductance on the other hand increase with dilution and finally attains a maximum value at infinite dilution. This is due to the fact that this value is product of specific conductance and the volume V of the solution containing 1 gm equivalent (or 1 mole) of the electrolyte. Decreasing value of specific conductance is more than compensated by increasing value of V and therefore the value of $\wedge_V$ increases with dilution.

PROBLEMS FOR PRACTICE

1. The specific conductivity of 0.01 N KCl solution is 0.0014106 mho cm^{-1} at 25°C. When a conductivity cell was filled up with the same solution, it offered a resistance of 484 ohms at 25°C. The same cell was then filled with 0.001 N solution of NaCl at same temperature which gave a resistance of 5194 ohms. Calculate the value of equivalent conductivity of 0.001 N solution of NaCl. [**Ans.** 131.4 ohm^{-1} cm^2 eq^{-1}]
2. Calculate the equivalent conductivity of 1 M H_2SO_4 solution, if its conductivity is 26×10^{-2} ohm^{-1} cm^{-1} (Atomic weight of sulphur = 32). [**Ans.** 130 ohm^{-1} cm^2 eq^{-1}]
3. The conductivity of an aqueous solution of sodium chloride in a cell is 92×10^{-4} ohm^{-1} cm^{-2}. The resistance offered by this cell is 247.8 ohm. Calculate the cell constant for this cell. [**Ans.** 2.28 cm^{-1}]
4. Specific conductivity of a 0.12 normal solution of an electrolyte is 0.024 ohm^{-1} cm^{-1}. Determine its equivalent conductivity. [**Ans.** 200 ohm^{-1} cm^2 eq^{-1}]
5. The resistance of 0.01 N NaCl solution at 25°C is 200 Ω. Cell constant of the conductivity cell is unity. Calculate the equivalent conductance of the solution. [**Ans.** 500 Ω^{-1} cm^2]
6. The resistance of a decinormal solution of an electrolyte in a conductivity cell was found to be 245 ohms. Calculate the equivalent conductivity of the solution if the electrodes in the cell were 2 cm apart and cach has an area of 3.5 sq. cm. [**Ans.** 23.32 ohm^{-1} cm^2 eq^{-1}]
7. A cell with N/50 KCl solution showed a resistance of 550 ohms at 25°C. The specific conductivity of N/50 KCl at 25°C is 0.002768 ohm^{-1} cm^{-1}. The cell filled with N/10 $ZnSO_4$ solution at 25°C shows a resistance of 72.18 ohms. Find the cell constant and molar conductivity of $ZnSO_4$ solution. [**Ans.** 1.5224 cm^{-1}, 421.8 ohm^{-1} cm^2 mol^{-1}]
8. The specific conductivity of N/50 solution of KCl at 298 K is 0.002765 mhos cm^{-1}. If the resistance of the same solution placed in the cell is 2000 ohms, what is cell constant ? [**Ans.** 5.53 cm^{-1}]
9. In a particular cell, 0.01 M solution of potassium chloride gave a resistance of 150 ohms at 298 K while 0.01 M solution of hydrochloric acid gave a resistance of 51.40 ohms at the

same temperature. At 298 K the specific conductivity of 0.01 M potassium chloride solution is 0.0014088 ohm^{-1} cm^{-1}. Calculate equivalent conductivity of the given hydrochloric acid solution. [**Ans.** 411.12 ohm^{-1} cm^2 eq^{-1}]

10. The specific conductivity of N/50 KCl solution at 25°C is 2.768×10^{-3} mhos per cm. The resistance of this solution at 25°C when measured with a particular cell is 250.2 ohms. The resistance of M/100 $CuSO_4$ solution at 25°C measured with the same cell was 8331 ohms. Calculate the molar conductivity of copper sulphate solution.[**Ans.** 8.3 ohm^{-1} cm^2 mol^{-1}]

7.5. ARRHENIUS THEORY OF IONISATION

Postulates of Arrhenius theory of ionisation

The main points of Arrhenius theory are given below :

(i) When dissolved in water, an electrolyte splits or dissociates into positively and negatively charged particles called *ions.* This process is called electrolytic *dissociation or ionisation.*

$$NaCl \longrightarrow Na^+ + Cl^-$$

$$Mg\,Cl \longrightarrow Mg^{2+} + 2\,Cl^-$$

$$Al_2(SO_4)_3 \longrightarrow 2\,Al^{3+} + 3SO_4^{2-}$$

(ii) On passing electricity, the positive ions migrate towards the cathode and are therefore called *cations.* Similarly, negative ions called the *anions* move towards the positive electrode *viz.* anode.

(iii) The solution of the electrolyte is electrically neutral as a whole. That means total charge carried by the positive ions is equal to the charge carried by negative ions.

(iv) The ions obtained by dissociation of the electrolyte combine back to give the undissociated molecules. Ultimately an equilibrium is reached between the ions and the undissociated molecules.

$$Na\,Cl \rightleftharpoons Na^+ + Cl^-$$

$$Mg\,Cl_2 \rightleftharpoons Mg^{2+} + 2\,Cl^-$$

The fraction of the total number of molecules of the electrolyte that dissociates into ions is called the **degree of ionization or dissociation.** It is usually denoted by α, *i.e.*

$$\text{Degree of dissociation}\,(\alpha) = \frac{\text{No. of molecules dissociated}}{\text{Total No. of molecules taken}}$$

(v) The degree of dissociation increases with dilution and becomes maximum at the infinite dilution.

(vi) The conductance of an electrolytic solution is due to the presence of the ions, produced by the dissociation of the electrolyte. Greater the dissociation, the higher is the conductance.

(vii) The properties of an electrolyte are the properties of its ions. For example, the properties of an acid are due to the H^+ ions furnished by it in the aqueous solution. Similarly, the properties of a base are due to the OH^- ions produced in the aqueous solution.

(viii) The ions produced by the dissociation of an electrolyte behave like molecules towards colligative properties. For example, each NaCl molecule dissociates to give two ions *viz.* Na^+ and Cl^-. Hence the observed value of any colligative property is double the expected value.

7.5.1. Factors that govern degree of dissociation of an electrolyte

The degree of dissociation of an electrolyte in the solution depends upon the following factors :

(i) **Nature of the solute:** Some electrolytes like mineral acids, alkalis and most of the salts are almost completely ionized in water. These are called *strong electrolytes.* On the other hand, some

electrolytes like organic acids and bases and some inorganic acids and bases ionize only to a small extent. These are called *weak electrolytes.*

(*ii*) **Nature of the solvent:** Every solvent has some definite value of the *dielectric constant* which is defined as its power to weaken the forces of attraction present between the ions of the electrolyte. Water has the highest dielectric constant (≈ 80 compared with vacuum taken as unity). Thus the dissociation of an electrolyte is maximum in water.

(*iii*) **Concentration:** Degree of dissociation of an electrolyte increases with dilution and becomes maximum at infinite dilution *i.e.*, zero concentration.

(*iv*) **Temperature:** As the temperature is increased, more of the molecules split into ions *i.e.*, degree of dissociation increases.

(*v*) **Presence of common ions:** The degree of dissociation of an electrolyte decreases when another electrolyte is present which furnishes a common ion. This effect is called *common ion effect.* For example, dissociation of NH_4OH decreases in the presence of NH_4Cl.

7.5.2. Evidence in support of Arrhenius theory of ionisation

The following evidence lends support to the theory.

(*i*) **Electrolytic conductance:** Arrhenius theory says that the conductance of the solution of an electrolyte is due to the presence of the ions, produced by the electrolyte on dissociation. Strong electrolytes possess high degree of dissociation and hence show greater conductance whereas weak electrolytes dissociate to a small extent and hence show low conductance.

(*ii*) **Variation of conductance with dilution:** It is observed that the specific conductance decreases whereas conductance as well as equivalent conductance increases with dilution. The explanation is evidently based on postulates of the Arrhenius theory. The equivalent conductance and the dissociation of the electrolyte vary in such a similar manner with dilution that the degree of dissociation (α) of an electrolyte at any concentration can be expressed in terms of equivalent conductances as follows :

$$\alpha = \wedge_c / \wedge_0$$

where $\wedge_c$ is the equivalent conductance at the given concentration and $\wedge_0$ is the equivalent conductance at infinite dilution.

(*iii*) **Electrolysis:** According to Arrhenius theory, an electrolyte when dissolved in water dissociates into ions. Thus the function of the electricity is only to direct the ions towards the electrodes.

(*iv*) **Colligative properties:** Anomalous results are obtained while measuring colligative properties like osmotic pressure, elevation in boiling points etc. of the strong electrolytes. This is because every ion produced from the electrolyte acts as an independent particle as far as colligative properties are concerned.

(*v*) **Ionic nature of reactions:** According to Arrhenius theory, the properties of an electrolyte are the properties of its ions. Thus the reaction between NaCl solution and $AgNO_3$ solution forming a precipitate of AgCl may be explained as follows :

NaCl in aqueous solution ionizes as

$$NaCl \longrightarrow Na^+ + Cl^-$$

$AgNO_3$ also ionises in aqueous solution as

$$AgNO_3 \longrightarrow Ag^+ + NO_3^-$$

When the above two solutions are mixed, we have

$$Na^+ + Cl^- + Ag^+ + NO_3^- \longrightarrow AgCl + Na^+ + NO_3^-$$

The reaction between any two ionic compounds in aqueous solution takes place in a similar manner.

(*vi*) **Heat of neutralization:** *The heat of neutralization is defined as the amount of heat evolved when one gram equivalent of the acid is neutralized by the base or one gram equivalent of the base is neutralized by the acid.* The heat of neutralization of a strong acid with a strong base is always the same *viz.* 13.7 kcal (or 57.3 kJ). This is explained by the Arrhenius theory as follows :

If the strong acid is HCl and the strong base is NaOH, both are completely dissociated in the aqueous solution and the reaction between them may be written as

$$Na^+ + OH^- + H^+ + Cl^- \longrightarrow Na^+ Cl^- + H_2O$$

Cancelling out the common ions on both sides, the above reaction reduces to

$$H^+ + OH^- \longrightarrow H_2O$$

Thus neutralization may be defined simply as the process in which H^+ *ions given by acid combine with the* OH^- *ions given by the base to form unionized water molecules.*

One gram equivalent of any strong acid in aqueous solution gives the same number of H^+ ions as the number of OH^- ions given by one gram equivalent of any strong base. The neutralization involves the combination between the same number of H^+ ions and OH^- ions. Hence the heat of neutralization of a strong acid with a strong base is always the same.

(*vii*) **Salt hydrolysis:** *It is defined as the process in which the salt reacts with water to give back the acid and base.*

$$\text{Salt} + \text{Water} \longrightarrow \text{Acid} + \text{Base}$$

Depending upon the relative strengths of the acid and the base produced, the resulting solution is acidic, basic or neutral. These observations are satisfactorily explained on the basis of Arrhenius theory.

7.5.3. Limitations of Arrhenius theory

Arrhenius theory suffers from some limitations. These are :

(*i*) Arrhenius suggested that *ionization* of the electrolyte takes place when dissolved in water. However, X-ray studies have shown that many such electrolytes are already in the ionized state (*e.g.* Na^+Cl^-).

(*ii*) As an electrolyte can conduct electricity even in the molten state, it shows that the dissociation of the electrolyte takes place even in the molten state. As such the function of water could not be explained by Arrhenius theory.

(*iii*) The most serious objection against the Arrhenius theory *was its failure when applied to strong electrolytes:*

(*a*) Ostwald's dilution law which is derived on the basis of Arrhenius theory, using the concept of ionic equilibrium, fails for the strong electrolytes, *i.e.*, a constant value of the dissociation constant *K* is not obtained.

(*b*) The degree of dissociation (α) of strong electrolytes calculated from conductance measurements ($\alpha = \wedge_c / \wedge_0$) was found to be different from that calculated from colligative properties.

(*c*) Strong electrolytes are almost completely dissociated even at moderate concentrations, *i.e.*, $\alpha = 1$ or 100% at all concentrations. Hence Arrhenius' view that conductance increases with dilution because the dissociation increases is not valid in case of strong electrolytes. The increase in conductance with dilution is, therefore, due to some other factors.

7.6 OSTWALD DILUTION LAW

According to Arrhenius theory of ionization, when an electrolyte is dissolved in water, there is an equilibrium between the ions and the undissociated molecules. For example, for a binary electrolyte AB, we have

$$AB \rightleftharpoons A^+ B^-$$

Ostwald pointed out the law of chemical equilibrium is applicable to this *ionic equilibrium.*

Suppose the initial concentration of the binary electrolyte AB in the solution is *C* moles/litre and α is the degree of dissociation, then we will have

	AB $\rightleftharpoons$	A^+ +	B^-
Initial concentration :	*C* moles/litre	0	0
Concentrations at equilibrium :	$= C(1-\alpha)$	$C\alpha$	$C\alpha$ moles/litre

Applying the law of chemical equilibrium to the above process, we get

Wilhelm Ostwald

$$K = \frac{[A^+][B^-]}{[AB]} = \frac{C\alpha . C\alpha}{C(1-\alpha)}$$

or $$\boxed{K = \frac{C\alpha^2}{1-\alpha}} \quad (\textit{Exact Formula}) \qquad ... (i)$$

This is Ostwald dilution law.

The equilibrium constant *K* for the equilibrium $AB \rightleftharpoons A^+ + B^-$ is called the dissociation constant or ionization constant of the electrolyte AB and may be represented by K_d (or K_i). Like any other equilibrium constant, K_d is constant only at constant temperature. The units of K_d will be same as those of the concentration *C*.

If the degree of dissociation α is very small, as in the case of *weak electrolytes at ordinary concentrations,* α can be neglected in comparison to 1 in the denominator of expression (*i*). Then the approximate formula can be written as

$$K_d = C\alpha^2 \quad \text{or} \quad \alpha^2 = \frac{K_d}{C}$$

or $$\boxed{\alpha = \sqrt{\frac{K_d}{C}}} \qquad (\textit{Approx. Formula}) \qquad ... (ii)$$

Equations (*i*) and (*ii*) are the two expressions of Ostwald's dilution law. Expression (*ii*) can be stated as follows :

For a weak binary electrolyte, with a small degree of dissociation, the degree of dissociation is inversely proportional to the square root of the initial molar concentration.

7.6.1. Validity of Ostwald dilution law

To verify the law experimentally, the degree of dissociation α is determined from conductance measurements (using the formula $\alpha = \wedge_c / \wedge_0$) at different concentrations for a number of electrolytes. For each electrolyte, the data are substituted in equation (*i*) to see whether the value of K_d comes out to be constant or not. It is observed that :

(*i*) The value of K_d comes out to be constant only for weak electrolytes and that too when the concentration of the solution is not too high *i.e.*, for dilute solutions only.

(*ii*) The value of K_d does not come out to be constant for strong electrolytes.

It may be concluded that *Ostwald's dilution law is valid only for weak electrolytes in dilute solutions* (that is why it is called dilution law) *and it does not hold good for strong electrolytes.*

7.6.2. Importance of Ostwald's dilution law

Ostwald's dilution law is applicable to weak electrolytes only. It is very useful in the calculation of the dissociation constants (K_d) of the weak acids and weak bases, by determining the degree of dissociation (α) from conductance measurements ($\wedge_c / \wedge_0$) at any concentration *C*. Having determined the value of K_d, which is constant for a particular weak acid or weak base, at a particular temperature, the degree of dissociation (α) of that weak acid or weak base can be obtained at any concentration (*C*).

SOLVED PROBLEMS ON OSTWALD DILUTION LAW

Example 1. A base has a dissociation constant equal to 1.8 × 10^{-5} at 298 K. Calculate its degree of dissociation at a concentration of 0.1 N at the same temperature.

Solution. Suppose the base is BOH and α is its degree of dissociation. Then we have

	BOH ⟶	B^+ +	OH^-
Initial concentration	0.1 M	0	0
Concentration at equilibrium	0.1 (1 – α)	0.1α	0.1α

$$\therefore \quad K_d = \frac{[B^+][OH^-]}{[BOH]}$$

$$\text{or} \quad 1.8 \times 10^{-5} = \frac{0.1\alpha \times 0.1\alpha}{0.1(1-\alpha)}$$

$$= \frac{0.1\alpha^2}{1-\alpha} = 0.1\,\alpha^2 \text{ [Neglecting } \alpha \text{ compared to 1 in the denominator]}$$

$$\text{or} \quad \alpha^2 = \frac{1.8\times10^{-5}}{0.1} = 1.8\times10^{-4}$$

$$\therefore \quad \alpha = \sqrt{1.8\times10^{-4}} = 1.34\times10^{-2}$$

$$= \mathbf{0.0134}$$

Example 2. A solution of ammonia gives the following values of equivalent conductivities at different dilutions :

V (in litres)	**8**	**16**	**32**
$\wedge_v$	**3.20**	**4.45**	**6.32 ohm^{-1} cm^2**

If equivalent conductivity at infinite dilution of NH_4OH is 237 ohm^{-1} cm^2, show that the above given data establish the truth of Ostwald's dilution law.

Solution. The ionisation constant K_d for weak electrolytes is given by the following relation:

$$K_d = \frac{C\alpha^2}{1-\alpha} \quad \text{or} \quad K_d = \frac{\alpha^2}{(1-\alpha)V}$$

where C is the concentration of the solute in mol/litre and V is the volume of the solution in litre containing 1 mole of the solute.

Putting the values of α and V in the above expression, we have

S. No.	*V (litres)*	$\alpha = \dfrac{\wedge V}{\wedge_\infty}$	$K = \dfrac{\alpha^2}{(1-\alpha)V}$
1	8	$\dfrac{3.2}{237} = 0.0135$	$\dfrac{(0.0135)^2}{(1-0.0135)8} = 2.31\times10^{-5}$
2	16	$\dfrac{4.45}{237} = 0.0188$	$\dfrac{(0.0188)^2}{(1-0.0188)16} = 2.26\times10^{-5}$
3	32	$\dfrac{6.32}{237} = 0.0265$	$\dfrac{(0.0265)^2}{(1-0.0265)32} = 2.25\times10^{-5}$

We find that the value of K comes out to be almost constant. Hence the truth of the law is established.

PROBLEMS FOR PRACTICE

1. The weak acid HA in its 0.1 M aqueous solution is 50% dissociated. What will be the value of its dissociation constant ? [**Ans.** 0.05]

[**Hint:** Here α = 0.5 which is not small; hence exact formula, *viz.*, $K_d = \dfrac{C\alpha^2}{1-\alpha}$ should be used]

2. The dissociation constant of ammonium hydroxide is 2.5×10^{-5}. Calculate the hydroxyl ion concentration in an $\dfrac{N}{100}$ solution of the base. [**Ans.** 5×10^{-4} ion/litre]

3. A 0.04 N solution of acetic acid has specific conductance of $4.23 \times 10^{-4}\ \Omega^{-1}\ cm^{-1}$ while its equivalent conductivity at infinite dilution is 390 Ω^{-1} cm^2 equi^{-1}. Calculate the concentration of hydrogen ions in the solution. [**Ans.** α = 0.0271, $[H^+] = 1.08 \times 10^{-3}$]

4. A 0.10 M solution of ammonium hydroxide is 1.3 per cent ionised at 25°C. Calculate the dissociation constant of ammonium hydroxide. [**Ans.** 1.71×10^{-5}]

5. Calculate the H^+ ion concentration of 1.0 N solution of HCN if the dissociation constant of latter is 4.0×10^{-10}. [**Ans.** 2×10^{-5} g ion/litre]

6. A 0.01 N solution of a weak monobasic acid dissociates to the extent of 10%. Calculate the dissociation constant of the acid. [**Ans.** 1.11×10^{-4} moles/litre]

7. Calculate the hydrogen ion concentration of a solution of formic acid containing 0.092 g of acid per litre. The dissociation constant at 25°C is 2.14×10^{-4}.

[**Ans.** 6.54×10^{-4} g ion/litre]

8. Calculate the concentration of hydrogen ions in a solution containing 0.04 mole of acetic acid in 250 ml of solution. ($K = 1.8 \times 10^{-5}$ moles/litre). [**Ans.** 1.7×10^{-3} g ion/litre]

9. At 18°C the equivalent conductivities of 0.01 N solutions of HCl, NaCl, CH_3COONa and CH_3COOH are 369.3, 101.7, 71.7 and 14.3 ohm^{-1} cm^2 respectively. Calculate the hydrogen ion concentration in 0.01 N acetic acid at 18°C. [**Ans.** 4.2×10^{-4} g ion/litre]

7.7. VARIATION OF CONDUCTIVITY WITH CONCENTRATION OF STRONG ELECTROLYTES (DEBYE-HUCKEL-ONSAGAR EQUATION)

The following equation governs the change of conductance of a strong electrolyte with concentration.

$$\Lambda_m^c = \Lambda_m^0 - b\sqrt{C}$$

where, Λ_m^c = molar conductance of the solution at the concentration C.

Λ_m^0 = molar concentration at infinite dilution.

C = concentration of the solution. b is a constant for a particular solvent at a particular temperature.

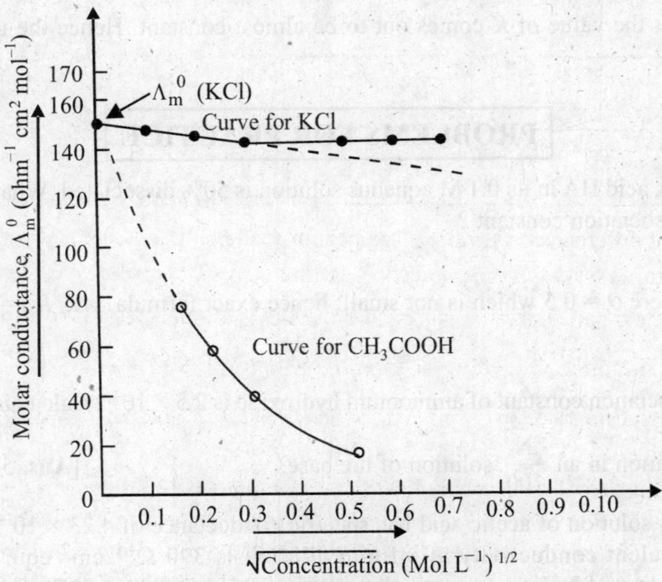

Fig. 7.5. Variation of molar conductivity of a strong electrolyte (KCl).

This equation is known as Debye-Huckel-Onsagar equation. A plot of Λ_m^c against $\sqrt{C}$ is a straight line as shown in Fig. 7.5. Taking the case of KC1, which is a strong electrolyte, we find that there is not much variation in conductance with dilution. This is because of the fact that a strong electrolyte is 100% ionised even in a concentrated solution. In a concentrated solution, there are interionic attractions which decrease the mobility of the ions. With dilution, these attractive forces are weakened resulting in greater mobility of the ions. Thus, there is slight increase in the conductance of the solution of a strong electrolyte on dilution.

In order to obtain the value of Λ_m^0 for a strong electrolyte, the plot which becomes linear at low concentration can be extrapolated to *y*-axis. (zero concentration) to obtain the value of Λ_m^0. However this is not possible for a weak electrolyte like CH_3COOH. Experimentally, it is impossible to determine Λ_m^0 for a weak electrolyte from the plot because it does not approach a limiting value. This problem was solved by Kohlrausch, as discussed in the next section.

Friedrich Kohlrausch

7.8. KOHLRAUSCH LAW

Kohlrausch studied the molar conductivities at infinite dilution, Λ_m^0 for a number of pairs of strong electrolytes each having a common cation or anion. Then, he calculated the differences of these Λ_m^0 values of each pair. A few of these results are given in Table 7.2.

Table 7.2. Λ_m^0 values of some pairs of electrolytes (at 298 K) having common ions (in ohm^{-1}cm^2 mol^{-1})

Electrolyte	Λ_m^0 *at 298 K*	*Difference*	*Electrolyte*	Λ_m^0 *at 298 K*	*Difference*
KCl NaCl	149.86 126.45	23.41	KBr KCl	151.92 149.86	2.06
KBr NaBr	151.92 128.51	23.41	NaBr NaCl	128.51 126.45	2.06
KNO_3 $NaNO_3$	144.96 121.55	23.41	LiBr LiCl	117.09 115.03	2.06

We find that difference between Λ_m^0 values for each pair of sodium and potassium salts having a common anion is same, irrespective of the anion. Similarly, the difference in the Λ_m^0 values for each pair of salts having the same different anions and a common cation is same, irrespective of the cation. In fact, column 3 in the Table gives the Λ_m^0 values for K^+ ion minus the Λ_m^0 values for Na^+ ion. Similarly, the last column gives Λ_m^0 value for Br^- ion minus Λ_m^0 value for Cl^- ion.

Based on this, Kohlrausch concluded that *each ion makes a definite contribution to the total molar conductivity of an electrolyte at infinite dilution, irrespective of the nature of the other ion of the electrolyte. This individual contribution of an ion towards the total molar conductivity is called molar ionic conductivity.* Kohlrausch in 1876 put forward a generalization, known as '*Kohlrausch Law*'. It states as under :

The molar conductivity of an electrolyte at infinite dilution is the sum of the ionic conductivities of the cations and the anions each multiplied with the number of ions present in one formula of the electrolyte. Mathematically,

$$\Lambda_m^0 \text{ for } A_xB_y = x\lambda^0_{A^+} + y\lambda^0_{B^-}$$

where Λ_m^0 is the molar conductivity of the electrolyte at infinite dilution, $\lambda^0_{A^+}$ and $\lambda^0_{B^-}$ are the molar conductivities of the cation and the anion respectively at infinite dilution.

Thus,

$$\Lambda_m^0 \text{ for NaCl} = \lambda^0_{Na^+} + \lambda^0_{Cl^-}$$

$$\Lambda_m^0 \text{ for BaCl}_2 = \lambda^0_{Ba^{2+}} + 2\lambda^0_{Cl^-}$$

$$\Lambda_m^0 \text{ for Al}_2(\text{SO}_4)_3 = 2\lambda^0_{Al^{3+}} + 3\lambda^0_{SO_4^{2-}}$$

In terms of equivalent conductivities, Kohlrausch's Law is defined as follows:

The equivalent conductivity of an electrolyte at infinite dilution is the sum of two values, one depending upon the cation and the other upon the anion, i.e.,

$$\Lambda^0_{eq} = \lambda^0_c + \lambda^0_a$$

where λ^0_c and λ^0_a are called the *ionic conductivities* at infinite dilution of the cation and the anion respectively.

Molar conductivity Λ_m is related to equivalent conductivity Λ_{eq} according to the equation

$$\Lambda_{eq} = \frac{\Lambda_m}{n_+z_+} \quad \text{or} \quad \Lambda_{eq} = \frac{\Lambda_m}{n_-z_-}$$

where n_+ and n_- are the number of cations and anions present per formula unit of the electrolyte and z_+ and z_- are the charges on the cation and the anion respectively.

7.8.1. Applications of Kohlrausch Law

Applications of Kohlrausch Law are discussed as under :

1. *Calculation of molar conductivity at infinite dilution* (Λ^0) *for weak electrolytes.* Molar conductivity of a weak electrolyte at infinite dilution cannot be determined experimentally. However this can be calculated using Kohlrausch's Law.

Consider the example of acetic acid [CH_3COOH] as the weak electrolyte. By Kohlrausch's Law.

$$\Lambda^0(\text{CH}_3\text{COOH}) = \lambda^0_{\text{CH}_3\text{COO}^-} + \lambda^0_{\text{H}^+}$$

This equation can be arrived at by knowing the molar conductivities at infinite dilution for the strong electrolytes KC1, CH_3COOK and HCl. As per Kohlrausch's Law,

$$\Lambda^0(\text{KCl}) = \lambda^0_{\text{K}} + \lambda^0_{\text{Cl}^-}$$

$$\Lambda^0(\text{CH}_3\text{COOK}) = \lambda^0_{\text{CH}_3\text{COO}^-} + \lambda^0_{\text{K}^+}$$

$$\Lambda^0(\text{HCl}) = \lambda^0_{\text{H}^+} + \lambda^0_{\text{Cl}^-}$$

Hence, we have

$$\lambda^0_{\text{CH}_3\text{COO}^-} + \lambda^0_{\text{H}^+} = (\lambda^0_{\text{CH}_3\text{COO}^-} + \lambda^0_{\text{K}^+}) + (\lambda^0_{\text{H}^+} + \lambda^0_{\text{Cl}^-}) - (\lambda^0_{\text{K}^+} + \lambda^0_{\text{Cl}^-})$$

i.e., $\Lambda^0(\text{CH}_3\text{COOH}) = \Lambda^0(\text{CH}_3\text{COOK}) + \Lambda^0(\text{HC1}) - \Lambda^0(\text{KCl})$

2. *Calculation of the degree of dissociation.* For a weak electrolyte, if Λ_m^0 is the molar conductivity of a solution at any concentration *c* and Λ_m^0 the molar conductivity at infinite dilution (*i.e.*, zero concentration); then we have

$$\text{Degree of dissociation,} \quad (\alpha) = \frac{\Lambda_m^c}{\Lambda_m^0}$$

The value of Λ_m^0 for the weak electrolytes can be calculated, using Kohlrausch's Law.

3. *Calculation of solubility of a sparingly soluble salt.* Salts such as AgC1, $BaSO_4$, $PbSO_4$ etc., which dissolve to a very small extent in water, are called sparingly soluble salts. As they dissolve very little, their solutions are considered as infinitely dilute. Further as their solutions are saturated, their concentration is equal to their solubility. Thus, by determining the specific conductivity (*K*) and the molar conductivity (Λ_m) of such a solution, we can obtain solubility as follows :

$$\Lambda_m^0 = K \times \frac{1000}{\text{Molarity}} = K \times \frac{1000}{\text{Solubility}}$$

or

$$\text{Solubility} = \frac{K \times 1000}{\Lambda_m^0}$$

Λ_m^0 can be obtained by using Kohlrausch Law.

SOLVED EXAMPLES ON KOHLRAUSCH'S LAW

Example 1. If the molar conductivities at infinite dilution of NaCl, HCl and CH_3COONa are 126.4, 426.1 and 91.0 ohm^{-1}cm^2 mol^{-1} respectively, what will be that of acetic acid ?

Solution. According to Kohlrausch Law.

$$\Lambda^0 \text{ for } CH_3COOH = \lambda^0_{CH_3COO^-} + \lambda^0_{H^+}$$

$$\Lambda^0 \text{ for NaCl} = \lambda^0_{Na^+} + \lambda^0_{Cl^-} = 126.4 \text{ ohm}^{-1}\text{cm}^2 \quad \text{... } (i)$$

$$\Lambda^0 \text{ for HCl} = \lambda^0_{H^+} + \lambda^0_{Cl^-} = 426.1 \text{ ohm}^{-1}\text{cm}^2 \quad \text{... } (ii)$$

$$\Lambda^0 \text{ for } CH_3COONa = \lambda^0_{CH_3COO^-} + \lambda^0_{Na^+} = 91.0 \text{ ohm}^{-1}\text{cm}^2 \quad \text{... } (iii)$$

Adding equations (*ii*) and (*iii*) and subtracting (*i*), we get,

$$\lambda^0_{H^+} + \lambda^0_{Cl^-} = \lambda^0_{CH_3COO^-} + \lambda^0_{Na^+} - \lambda^0_{Na^+} - \lambda^0_{Cl^-} = 426.1 + 91.0 - 126.4 \text{ ohm}^{-1}\text{ cm}^2$$

$$\lambda^0_{CH_3COO^-} + \lambda^0_{H^+} = 390.7 \text{ ohm}^{-1}\text{ cm}^2\text{ mol}^{-1}$$

$$\Lambda^0 \text{ for } CH_3COOH = 390.7 \text{ ohm}^{-1}\text{ cm}^2\text{ mol}^{-1}$$

Example 2. At 291 K, the molar conductivities, at infinite dilution of NH_4Cl, NaOH and NaCl are 129.8, 217.4 and 108.9 ohm^{-1} cm^2 respectively. If the molar conductivity of a centinormal solution of NH_4OH is 9.33 ohm^{-1} cm^2, what is the percentage dissociation of NH_4OH at this dilution ?

Solution. We are given that

$$\Lambda^0 \text{ for } NH_4C1 = 129.8 \text{ ohm}^{-1}\text{ cm}^2$$

$$\Lambda^0 \text{ for NaOH} = 217.4 \text{ ohm}^{-1}\text{ cm}^2$$

$$\Lambda^0 \text{ for NaCl} = 108.9 \text{ ohm}^{-1} \text{ cm}^2$$

By Kohlrausch law

$$\Lambda^0 \text{ for } NH_4OH = \lambda^0_{NH_4^+} + \lambda^0_{OH^-}$$

$$= \Lambda^0 (NH_4Cl) + \Lambda^0 (NaOH) - \Lambda^0 (NaCl)$$

Substituting the values

$$\Lambda^0(NH_4OH) = 129.8 + 217.4 - 108.9$$

$$= 238.3 \text{ ohm}^{-1} \text{ cm}^2$$

$$\Lambda_c = 9.33 \text{ ohm}^{-1} \text{ cm}^2 \text{ (Given)}$$

$$\text{Degreee of dissociation} \quad (\alpha) = \frac{\Lambda_c}{\Lambda^0} = \frac{9.33}{238.3} = 0.0392$$

Percentage dissociation = 0.0392 × 100 = 3.92

Example 3. From the following molar conductivities at infinite dilution:

$$\Lambda^0_m \text{ for } Ba(OH)_2 = 457.6 \text{ ohm}^{-1} \text{ cm}^2 \text{ mol}^{-1}$$

$$\Lambda^0_m \text{ for } BaCl_2 = 240.6 \text{ ohm}^{-1} \text{ cm}^2 \text{ mol}^{-1}$$

$$\Lambda^0_m \text{ for } NH_4Cl = 129.8 \text{ ohm}^{-1} \text{ cm}^2 \text{ mol}^{-1}$$

calculate Λ^0_m for NH_4OH.

Solution. Λ^0_m for $[Ba(OH)_2] = \lambda^0_{Ba^{2+}} + 2\lambda^0_{OH^-} = 457.6 \text{ ohm}^{-1} \text{ cm}^2 \text{ mol}^{-1}$ ...(*i*)

$$\Lambda^0_m \text{ for } (BaCl_2) = \lambda^0_{Ba^{2+}} + 2\lambda^0_{Cl^-} = 240.6 \text{ ohm}^{-1} \text{ cm}^2 \text{ mol}^{-1} \quad ...(ii)$$

$$\Lambda^0_m \text{ for } NH_4Cl = \lambda^0_{NH_4^+} + \lambda^0_{Cl^-} = 129.8 \text{ ohm}^{-1} \text{ cm}^2 \text{ mol}^{-1} \quad ...(iii)$$

In order to find $\Lambda^0_m (NH_4Cl)$, we need to find the value of $\lambda^0_{NH_4^+} + \lambda^0_{OH^-}$

Multiplying (*iii*) by 2, add to (*i*) and subtract eq. (*ii*) from the sum

$$2\lambda_{NH_4^+} + 2\lambda^0_{Cl^-} + \lambda^0_{Ba^{2+}} + 2\lambda^0_{OH^-} - \lambda^0_{Ba^{2+}} - 2\lambda^0_{Cl^-}$$

$$= (2 \times 129.8) + 457.6 - 240.6$$

$$2\left(\lambda^0_{NH_4^+} + 2\lambda^0_{OH^-}\right) = 476.6$$

$$\lambda^0_{NH_4^+} + \lambda^0_{OH^-} = 238.6 \text{ ohm}^{-1} \text{ cm}^2 \text{ mol}^{-1}$$

PROBLEMS FOR PRACTICE

1. From the following equivalent conductances at infinite dilution,

$$\Lambda_0 \text{ for } Ba(OH)_2 = 288.8 \text{ ohm}^{-1} \text{ cm}^2 \text{ equivalent}^{-1}$$

$$\Lambda_0 \text{ for } BaCl_2 = 120.3 \text{ ohm}^{-1} \text{ cm}^2 \text{ equivalent}^{-1}$$

$$\Lambda_0 \text{ for } NH_4Cl = 129.8 \text{ ohm}^{-1} \text{ cm}^2 \text{ equivalent}^{-1}$$

calculate Λ_0 for NH_4OH. [**Ans.** 238.3 $\text{ohm}^{-1} \text{ cm}^2 \text{ eq}^{-1}$]

2. The equivalent conductance of NH_4Cl at a certain dilution is 149.7 Ω^{-1} cm^2 and the ionic conductance of OH^- and Cl^- ions are 198 and 76.3 Ω^{-1} cm^2 respectively. Calculate the equivalent conductance of NH_4OH at this dilution. [**Ans.** 271.1 Ω^{-1} cm^2 eq^{-1}]

3. The equivalent conductivity at infinite dilution of HCl and CH_3COONa are 396 and 78 ohm^{-1} cm^2 equiv^{-1} respectively. If the equivalent conductivity of NaCl at infinite dilution is 109 ohm^{-1} cm^2 equiv^{-1}, calculate the equivalent conductivity of acetic acid at infinite dilution. [**Ans.** 365 ohm^{-1} cm^2 eq^{-1}]

MISCELLANEOUS SOLVED EXAMPLES

Example 1. The specific conductance of a 0.01 N solution of acetic acid at 25°C is 1.63 × 10^{-4} mhos cm^{-1} and the ionic conductance at infinite dilution but at the same temperature for H^+ and CH_3COO^- ions are 349.8 and 40.9 mhos cm^2 equiv^{-1} respectively. Calculate the dissociation constant of acetic acid at 25°C.

Also calculate the H^+ ion concentration in the solution.

Solution. Here we are given that

$$K = 1.63 \times 10^{-4} \text{ mhos cm}^{-1}$$

$$\lambda^0_{H^+} = 349.8 \text{ mhos cm}^2 \text{ equiv}^{-1}$$

$$\lambda^0_{CH_3COO^-} = 40.9 \text{ mhos cm}^2 \text{ equiv}^{-1}$$

Calculation of $\wedge_0$ for CH_3COOH

$$\wedge_0 \text{ for } CH_3COOH = 349.8 + 40.9 \qquad \text{(By Kohlrausch's law)}$$

$$= 390.7 \text{ mhos cm}^2 \text{ equiv}^{-1}$$

Calculation of $\wedge_c$ for CH_3COOH

As the solution is 0.01 N, volume of solution in cm^3 containing one gram equivalent of acetic acid $= \dfrac{1000}{0.01} = 10^5$ cm^3

$\therefore$ $\wedge_c$ = Sp. conductance × Vol. in cm^3 containing one gm equiv. of electrolyte.

$$\wedge_c = 1.63 \times 10^{-4} \times 10^5 = 16.3 \text{ mhos cm}^2 \text{ equiv}^{-1}$$

Calculation of α

$$\therefore \quad \alpha = \frac{\wedge_c}{\wedge_0} = \frac{16.3}{390.7} = 0.0417$$

Calculation of K_d

	CH_3COOH	$\rightleftharpoons$	CH_3COO^-	+	H^+
Initial concentration	0.01		0		0
Concentration at equilibrium	0.01(1– α)		0.01 α		0.01 α

$$K_d = \frac{[CH_3COO^-][H^+]}{[CH_3COOH]} = \frac{0.01\alpha \times 0.01\alpha}{0.01(1-\alpha)}$$

$$= \frac{0.01\alpha^2}{1-\alpha}$$

$$= 0.01\,\alpha^2 \text{ [Taking } 1-\alpha = 1\text{, as } \alpha \text{ is negligible compared to 1]}$$

$$= 0.01 \times (0.0417)^2$$

$$= \mathbf{1.74 \times 10^{-5}}$$

Calculation of H^+

$$[H^+] = 0.01\,\alpha = 0.01 \times 0.0417$$

$$= \mathbf{4.17 \times 10^{-4}} \textbf{ g ions/litre}$$

Example 2. At 18°C equivalent conductivities of 0.01 N solutions of hydrochloric acid, NaCl, CH_3COONa and CH_3COOH are 369.3, 101.7, 71.7 and 14.3 ohm^{-1} cm^2 respectively. Calculate the hydrogen ion concentration in 0.01 N acetic acid at 18°C.

Solution. Equivalent conductance at infinite dilution $\wedge_0$ of CH_3COOH can be obtained using Kohlrausch law

$$\wedge_0 (CH_3COOH) = 369.3 + 71.7 - 101.7$$

$$= 339.3$$

$$\text{Degree of dissociation of } CH_3COOH = \frac{\wedge_C}{\wedge_0}$$

$$= \frac{14.3}{339.3} = 0.042$$

$\therefore$ Hydrogen ion concentration in $CH_3COOH = 0.042 \times 0.01$

$$= \mathbf{4.2 \times 10^{-4}} \textbf{ g ion/litre}$$

Example 3. Calculate the hydrogen ion concentration of a solution of formic acid containing 0.092 g of acid per litre. The dissociation constant at 25°C is 2.14 × 10^{-4}.

Solution. Concentration of formic acid (HCOOH) in mole/litre $= \dfrac{0.092}{46} = 0.002$

Using Ostwald dilution law,

$$K = C\alpha^2$$

or $$2.14 \times 10^{-4} = 0.002\,\alpha^2$$

or $$\alpha^2 = \frac{2.14 \times 10^{-4}}{2 \times 10^{-3}}$$

or $$\alpha^2 = 0.107$$

or $$\alpha = 0.327$$

Hence $$[H^+] = 0.002 \times 0.327 \text{ g ion/litre}$$

$$= 2 \times 10^{-3} \times 3.27 \times 10^{-1}$$

$$= \mathbf{6.54 \times 10^{-4}}$$

7.9. ANOMALOUS BEHAVIOUR OF STRONG ELECTROLYTES DEBYE-HUCKEL THEORY

The increase of equivalent conductance with dilution in case of weak electrolytes can be explained on the basis of Arrhenius theory which suggests that the conductance increases because the dissociation of the weak electrolyte increases with dilution. However, in case of strong electrolytes, as they are almost completely dissociated even at moderate concentration, the increase in equivalent conductance with dilution must be due to some additional factors.

Debye-Huckel theory explains the increase in conductance of strong electrolytes on dilution based upon the following two effects:

1. **Relaxation effect or Asymmetry effect:** Each ion in a solution is surrounded by an **ionic atmosphere** of opposite charge. As long as no electric field is applied, the ionic atmosphere remains symmetrical around the central ion [Fig. 7.6 (*a*)]. However, when a current is passed through the solution, the central ion moves towards the oppositely charged electrode. As it is moving out of the ionic atmosphere, it has to rebuild an ionic atmosphere of opposite charge around it and the old ionic atmosphere is destroyed. However, the destruction of the old ionic atmosphere and the formation of the new ionic atmosphere do not take place at the same time. There is some time lag called *time of relaxation* between the destruction of the old and the formation of new ionic atmosphere. During this time, the old ionic atmosphere pulls the moving ion backward and hence retards its motion Fig. 7.6 (*b*)]. Hence this effect is called *relaxation effect.* Alternatively, it may be argued that as the central

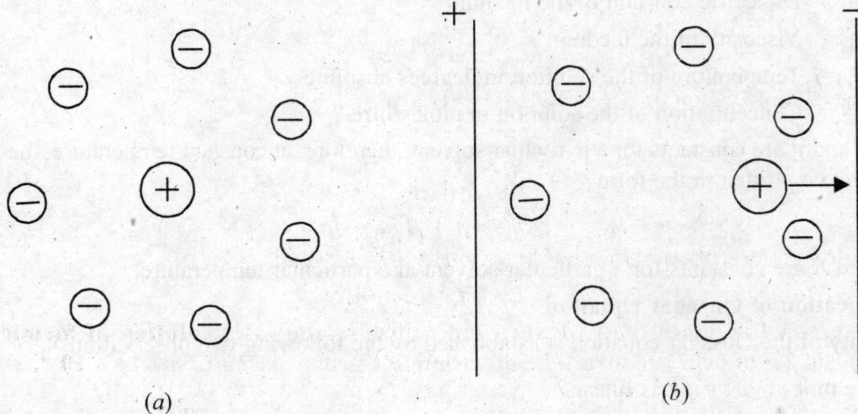

Fig. 7.6. Relaxation or asymmetry effect.

ion moves, the symmetry of the ionic atmosphere is lost; more ions of the ionic atmosphere are left behind than are present in the front [Fig. 7.6 (*b*)]. The excess ions of the ionic atmosphere present behind the moving ion drag the ion backward and retard its motion. Thus the effect arises because of the asymmetry of the ionic atmosphere of the moving ion and hence is also called *asymmetry effect.*

2. **Electrophoretic effect:** On applying EMF, the central ion moves in one direction and the oppositely charged ionic atmosphere moves in the opposite direction. As this ionic atmosphere moves, the solvent molecules associated with it also move. Thus the flow of the ionic atmosphere and that of the solvent molecules attached to it takes place in a direction opposite to that of the movement of the central ion. In other words, the central ion is moving against the stream. Thus motion of the ions is retarded. This effect is called *electrophoretic effect* (Fig. 7.7).

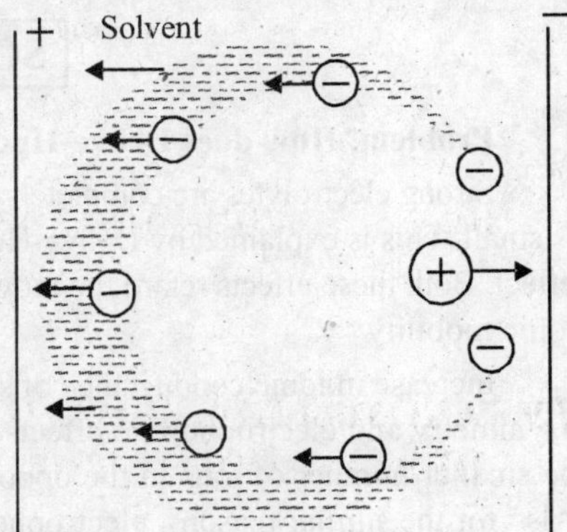

Fig. 7.7. Movement of a central ion in a direction opposite to that of the ionic atmosphere and the associated solvent molecules.

Besides these two effects, the third retarding force is the normal frictional resistance offered by the medium which depends on the viscosity of the medium, its dielectric constant etc.

Based upon the above ideas, Debye and Huckel (1923) derived a mathematical expression for the variation of equivalent conductance with concentration. This equation was further improved by Onsagar (1926-27) and now the equation is known as **Debye-Huckel-Onsagar equation** or simply **Onsagar equation.** For a uni-univalent electrolyte, it is written in the form

$$\wedge_c = \wedge_0 - \left[\frac{82.4}{(DT)^{1/2}\eta} + \frac{8.20\times10^5}{(DT)^{3/2}}\wedge_0\right]\sqrt{c} \quad \text{... } (i)$$

where $\wedge_c$ = Equivalent conductance at concentration c

$\wedge_0$ = Equivalent conductance at infinite dilution

D = Dielectric constant of the medium

η = Viscosity of the medium

T = Temperature of the solution in degrees absolute

c = Concentration of the solution in moles/litre.

As D and η are constants for a particular solvent, therefore, at constant temperature, the above equation can be written in the form

$$\wedge_c = \wedge_0 - (A + B\wedge_0)\sqrt{c} \quad \text{... } (ii)$$

where A and B are constants for a particular solvent at a particular temperature.

Verification of Onsagar equation

Validity of the Onsagar equation is established by the following two observations :

1. The plot of $\wedge c$ vs $\sqrt{c}$ is linear

2. The slope of the line obtained is equal to $A + B\wedge_0$ after putting the values of A and B.

Significance of Debye-Huckel Theory

The theory explains satisfactorily why the equivalent conductance of a strong electrolyte at any concentration $\wedge_c$ is less than the limiting value of $\wedge_0$ at infinite dilution, inspite of complete ionization of the strong electrolyte in terms of relaxation effect and asymmetry effect.

SOLVED PROBLEM

Problem. How does Debye-Huckel theory explain the behaviour of strong electrolytes?

Strong electrolytes are completely ionised even in concentrated solution. But their ionic mobility is small. This is explained by Debye-Huckel theory in terms of asymmetry effect and electrophoretic effect. Both these effects retard the movement of ions of the strong electrolyte thereby bringing down ionic mobility.

Increase in ionic conductance of strong electrolytes on dilution can be explained by saying that asymmetry and electrophoretic effects will be diluted when the solution is made dilute. There will be smaller number of ions in the oppositely charged ionic sphere per unit volume of the solution. Also for the similar reasons, electrophoretic effect will also be weakened. This will result in higher conductance on dilution.

7.10. MIGRATION OF IONS AND TRANSPORT NUMBER

On passing electricity through an electrolyte, the ions move towards the oppositely charged electrodes and are deposited there. This movement of ions towards oppositely charged electrodes on passing electricity is called migration of ions.

Take a U-tube. Fill its lower portion by $Cu_2Cr_2O_7$ jelly obtained by mixing a 5% solution of agar-agar and a mixture of $CuSO_4$ and $K_2Cr_2O_7$; and allow it to set (Fig. 7.8). The jelly is coloured green. To mark the position of jelly, some charcoal is scattered over the green jelly. Over charcoal in both the limbs, is added KNO_3 solution in agar-agar, *i.e.*, KNO_3 jelly and then aqueous solution of KNO_3. On passing current, blue coloured Cu^{2+} ions are seen moving towards negative electrode and yellow coloured $Cr_2O_7{}^{2-}$ ions are seen moving towards positive electrode. After some time the two types of ions are seen moving with well-defined boundaries. This experiment clearly shows the migration of ions.

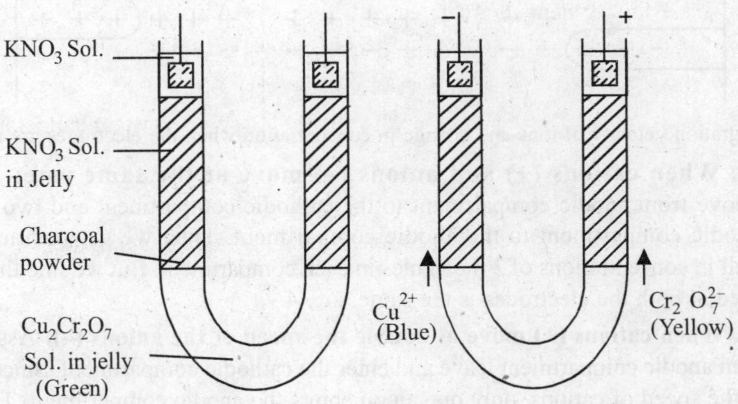

Fig. 7.8. Demonstration of the migration of ions.

7.11. HITTORF'S THEORETICAL DEVICE–CHANGE IN CONCENTRATION

Consider a cell containing the solution and provided with the anode *A* and the cathode *C* (Fig. 7.9). Let the solution lying between the electrodes *A* and *C* be divided into three imginary compartments (the anodic, the middle and the cathodic) with the help of porous partitions *a* and *c* shown by vertical dotted lines. Before electrolysis, suppose there are 13 pairs of ions, 4 being present in each of the anodic and the cathodic compartments and 5 in the middle (as shown in position I). The following cases are considered when the electrodes are not attacked and when they are attacked.

WHEN THE ELECTRODES ARE NOT ATTACKED

Case I. When only anions (–) move: Let two anions leave cathodic compartment and enter anodic compartment. We get the position II. Although the cations have not moved, still we see that the number of ions liberated is the same (*viz.* 2) at both the electrodes. But we see that concentration in anodic chamber has not altered while that in cathodic chamber has fallen by two molecules.

Similarly, if cations are capable of movement and two of them enter cathodic compartment from anodic compartment then there will be a fall in concentration of two molecules in the anodic chamber. Thus we see that the fall of concentration of chamber is proportional to the speed of the ion leaving that chamber.

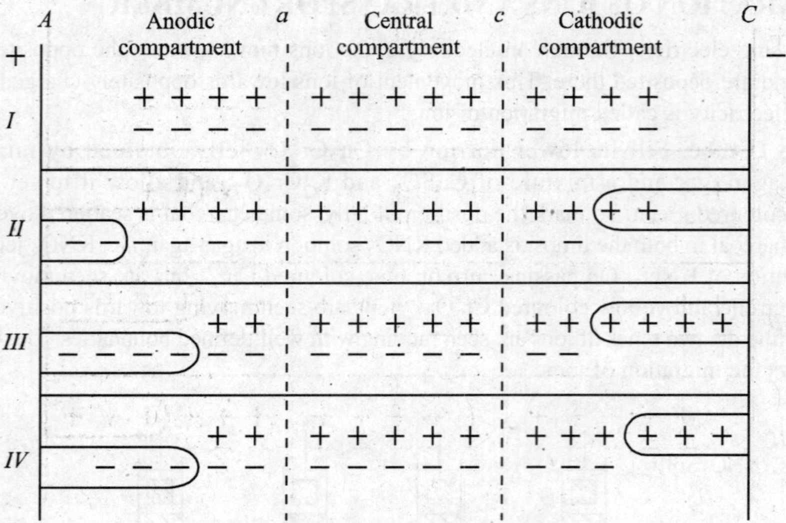

Fig. 7.9. Migration velocity of ions and change in concentration when the electrodes are not attacked

Case II. When cations (+) and anions (–) move at the same rate: Assume that two cations move from anodic compartment to the cathodic compartment and two anions move from the cathodic compartment to the anodic compartment. Then we get position III. In this case there is fall in concentrations of 2 molecules in each compartment. But we find that the number of ions liberated at both the electrodes is the same, *i.e.,* 4.

Case III. When cations (+) move at double the speed of the anions (–): Assume that two cations (+) from anodic compartment leave and enter the cathodic compartment. Since the speed of anions is half the speed of cations, only one anion enters the anodic compartment. The position is shown in IV. We also observe that the falls in concentration in anodic and cathodic compartments are 2 and 1 molecules respectively. The number of the ions discharged is again the same, *viz.,* 3, on each electrode.

From the above observation we draw the following conclusions :

(*i*) *Fall in concentration around any electrode* ∝ *the speed of the ion moving away from it.* It means: Fall in concentration around cathode ∝ speed of the anion (u_a) and fall in concentration around anode ∝ speed of the cation (u_c).

i.e.,
$$\frac{\text{Fall in conc. around anode}}{\text{Fall in conc. around cathode}} = \frac{\text{Speed of cation}}{\text{Speed of anion}}$$

(*ii*) *Irrespective of the speeds of the ions, the number of ions liberated on both the electrodes is equal.*

WHEN THE ELECTRODES ARE ATTACKABLE

In a silver voltameter (containing $AgNO_3$ solution and having silver electrodes) each NO_3^- ion that reaches the anodic compartment reacts with Ag of the electrode forming $AgNO_3$ and instead of decrease, there is an increase of concentration in the anodic compartment. In a silver voltameter, the above three cases are reproduced in Fig. 7.10.

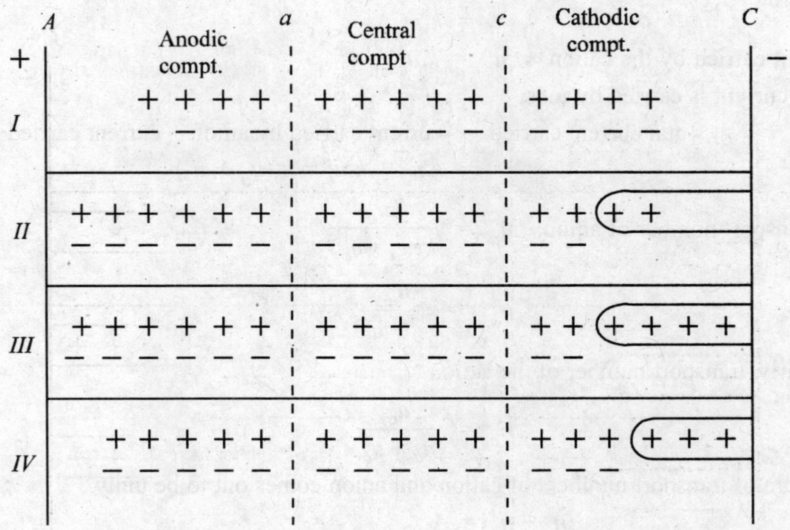

Fig. 7.10. Migration velocity of ions and change in concentration when the electrodes are attacked.

It is seen that in every case there is always an increase in concentration in the anodic compartment.

This increase in concentration takes place even when the Ag^+ ions migrate from the anodic compartment to the cathodic compartment (cases III and IV). In order to find the increase in concentration that would have taken place, if no Ag^+ ions had migrated from the anodic compartment to the cathodic compartment, a silver or a copper voltameter or coulometer is connected in series. Let the number of gram equivalents of Ag or Cu deposited in this voltameter is x. Obviously, x gram equivalents would have been the increase in the concentration of anodic compartment if no Ag^+ ions had migrated from this compartment. However the actual increase is less due to migration of Ag^+ ions. Suppose the actual increase in the concentration of Ag in the anodic compartment is y gram equivalents. Then

$$\left.\begin{array}{l}\textit{Fall in conc. in the anodic compartment} \\ \textit{due to migration of } Ag^+ \textit{ ions}\end{array}\right\} = (x-y) \textit{ gram equivalents.}$$

From Fig. 7.10, it follows that

Fall in conc. around cathode = Increase in conc. around anode = y gram equivalents

Thus the speed ratio in this case will be given by

$$\frac{\text{Speed of } Ag^+ \text{ ions}}{\text{Speed of } NO_3^- \text{ ions}} = \frac{x-y}{y}$$

7.12. TRANSPORT NUMBER

The fraction of the total current carried by an ion is called its **transport number** or **transference number** or **Hittorf's number**.

Greater the speed of the ion, greater is the amount of current it will carry with it and hence greater will be its transport number.

If u_a and u_c are the speeds of the anion and the cation respectively then

current carried by the anion $\propto u_a$

$= k.u_a$

Current carried by the cation $\propto u_c = k.u_c$

Since current is carried by ions,

$\therefore$ Total current carried = current carried by anion + current carried by cation

$$= k.u_a + k.u_c = k\,(u_a + u_c)$$

$\therefore$ Transport number of anion, $$n_a = \frac{ku_a}{k(u_a + u_c)}$$

or $$n_a = \frac{u_a}{u_a + u_c}$$

Similarly, transport number of the cation

$$n_c = \frac{u_c}{u_a + u_c}$$

The sum of transport numbers of cation and anion comes out to be unity.

or $$n_a + n_c = 1$$

DETERMINATION OF TRANSPORT NUMBER

7.13. HITTORF'S METHOD

The method is based upon the principle that the fall in concentration around an electrode is proportional to the speed of the ion moving away from it,

Wilhelm Hittorf

i.e., $$\frac{\text{Fall in conc. around cathode}}{\text{Fall in conc. around anode}} = \frac{\text{Speed of anion}}{\text{Speed of cation}} = \frac{u_a}{u_c}$$

Adding 1 to both sides, we get

$$1 + \frac{\text{Fall in conc. around cathode}}{\text{Fall in conc. around anode}} = \frac{u_a}{u_c} + 1$$

or

$$\frac{\text{Fall in conc. around anode + Fall in conc. around cathode}}{\text{Fall in conc. around anode}} = \frac{u_a + u_c}{u_c}$$

or $$\frac{\text{Total fall in conc. around both the electrodes}}{\text{Fall in conc. around anode}} = \frac{u_a + u_c}{u_c}$$

or $$\frac{\text{Fall in conc. around anode}}{\text{Total fall in conc. around both the electrodes}} = \frac{u_c}{u_a + u_c} = n_c$$

If the concentrations are expressed in gram equivalents, the above expression may be written as

$$n_c = \frac{\text{Number of gram equivalents lost from the anodic compartment}}{\text{Number of gram equivalents lost from both the compartments}}$$

It may be seen that the number of gram equivalents lost from both the compartments is equal to the number of gram equivalents discharged at each electrode. This can be found by connecting a silver or copper voltameter in series. According to Faraday's second law, when the same quantity of electricity flows through the silver voltameter, the same number of gram equivalents of Ag or Cu will be deposited. The above formula reduces to

$$n_c = \frac{\text{Number of gram equivalents lost from the anodic compartment}}{\text{Number of gram equivalents deposited in the voltameter}}$$

Hence, the determination of the transport number of the cation (n_c) involves the determination of the number of gram equivalents lost from the anodic compartment and the number of gram equivalents of Ag or Cu deposited in the voltameter placed in series with the apparatus.

To determine the transport number of the anion (n_a), there is no need to find the loss in concentration of the cathodic compartment because it can be calculated directly from the equation

$$n_a = 1 - n_c$$

Apparatus: It consists of a U-tube joined on its two sides to two vertical bulbed glass tubes (Fig. 7.11). The U-tube and the two vertical tubes have got the stop-cocks at their bottom. The U-tube is also provided with two stop-cocks, one on either side at the top so as to cut off the various compartments from each other when required. In the two vertical tubes are placed the anode and the cathode separately. These are made of platinum, if it is desired that these should not be attacked by the ions. Alternatively, the anode and the cathode of silver or copper are taken depending upon whether transport number of Ag^+ ions or Cu^{2+} ions is to be determined, in which case the anode is attacked by the anions. The apparatus is connected to a silver or copper voltameter in series with the apparatus along with a battery, a rheostat, a key and a milli-ammeter, as shown in Fig. 7.11.

Procedure: (*i*) The apparatus is filled with the solution of the electrolyte, containing ions whose transport number is to be determined.

(*ii*) A silver or copper voltameter is placed in series with the apparatus and the various connections are made as shown in Fig. 7.11.

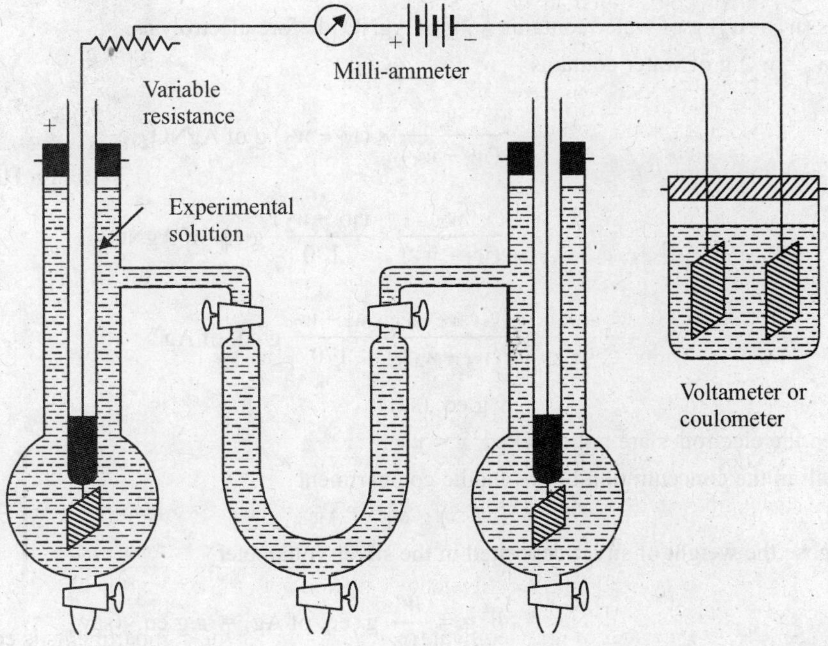

Fig. 7.11. Apparatus for the determination of transport number by Hittorf 's method

(*iii*) Before electrolysis, the concentration of the electrolyte is found by titrating a known weight of the solution against a suitable reagent.

(*iv*) A current of 0.01 to 0.02 ampere is passed for 2-3 hours. A low current is used so that the change in concentration around the electrodes is not too fast as otherwise the diffusion may take place.

(*v*) After electrolysis, the solution from the anodic compartment is taken out, weighed and then titrated as before to know the amount of the electrolyte present.

(*vi*) The weight of the silver deposited in the silver voltameter (or that of copper in case copper voltameter is used) is also determined.

Presumably there is no change in conc. of the middle compartment during the experiment.

Calculations: The method of calculation is illustrated by taking the example of $AgNO_3$ as the electrolyte and using silver voltameter in series.

When the electrodes are not attacked (*i.e., when platinum electrodes are used*).

After electrolysis :

Let the weight of the anodic solution taken out = w_1g

Weight of $AgNO_3$ found to be present (by titration) = w_2g

$\therefore$ Weight of water present = $(w_1 - w_2)$ g

Thus $(w_1 - w_2)$ g of water contains w_2 g of $AgNO_3$ after electrolysis.

$$w_2 \text{ g of } AgNO_3 = \frac{w_2}{170} \text{ g eq. of } AgNO_3 = \frac{w_2}{170} \text{ g of Ag} = x \text{ g eq. (say)}$$

Thus, we can say that after electrolysis $(w_1 - w_2)$ g of water contains *x g* eq. of Ag.

Before electrolysis :

Suppose w_3 g of solution contains w_4 g of $AgNO_3$

$\therefore$ Weight of water present $= (w_3 - w_4)$ g

Thus $(w_3 - w_4)$ g of water contains w_4 g of $AgNO_3$ before electrolysis.

$\therefore$ $(w_1 - w_2)$ g of water contains

$$= \frac{w_4}{(w_3 - w_4)} \times (w_1 - w_2) \text{ g of } AgNO_3$$

$$= \frac{w_4}{(w_3 - w_4)} \times \frac{(w_1 - w_2)}{170} \text{ g eq. of } AgNO_3$$

$$= \frac{w_4}{(w_3 - w_4)} \times \frac{w_1 - w_2}{170} \text{ g eq. of Ag}$$

$$= y \text{ g eq. (say)}$$

When the electrodes are not attacked, $x < y$

$\therefore$ Fall in the concentration of the anodic compartment

$$= (y - x) \text{ g eq. of Ag}$$

Suppose the weight of silver deposited in the silver voltameter

$$= W \text{ g} = \frac{W}{108} \text{ g eq. of Ag} = z \text{ g eq. (say)}$$

$\therefore$ Transport number of Ag^+ ion (n_{Ag^+})

$$= \frac{\text{Fall in concentration around anode in g eq.}}{\text{Amount of Ag deposited in voltameter in g eq.}}$$

$$= \frac{y-x}{z}$$

Transport number of NO_3^- ions $(n_{NO_3^-}) = 1 - n_{Ag^+}$

When the electrodes are attacked (*i.e. when silver electrodes are used*).

In this case, NO_3^- ions reaching the anode attack the silver anode. Therefore there is increase in the concentration of Ag^+ ions in the anodic solution (instead of decrease) *i.e.*, in this case we will have

$$x > y$$

$\therefore$ Increase in the concentration of the anodic solution

$$= (x - y) \text{ g eq. of Ag}$$

This increase in the concentration takes place in spite of the migration of Ag^+ ions from the anodic compartment to the cathodic compartment. If no Ag^+ ions had migrated from the anodic compartment, the increase in the concentration of Ag^+ ions in the anodic compartment would have been higher and equal to z g eq. of Ag (*i.e.*, the same as the amount deposited in the silver voltameter). Hence, we can say

Fall in concentration in the anodic solution due to migration of Ag^+ ions $= z - (x - y)$ g eq. of Ag.

$$\therefore \quad \text{Transport number of } Ag^+ \text{ ions} = \frac{z-(x-y)}{z}$$

$$\text{Transport number of } NO_3^- \text{ ions} = 1 - n_{Ag^+}$$

In case a copper voltameter is used, the number of gram equivalents of copper deposited in the voltameter is obtained by dividing the weight of copper deposited (Wg) by its equivalent weight *i.e.* 31.75.

SOLVED EXAMPLES ON HITTORF'S METHOD

Example 1. During the electrolysis of a solution of potassium chloride between platinum electrodes, 0.0137 g of the chloride was lost from the anodic compartment and 0.0857 g of silver was deposited in a silver coulometer connected in series with the cell. Determine the transport number of K^+ and Cl^- ions.

Solution. No. of g eq. of the Cl^- ions lost from the anodic compartment

$$= \frac{0.0137}{35.5} = 0.0003859. \qquad \{\because \text{Eq. wt. of Cl} = 35.5\}$$

No of g eq. of Ag deposited in the coulometer

$$= \frac{0.0857}{108} = 0.0007935 \qquad \{\because \text{Eq. wt. of Ag} = 108\}$$

$\because$ Transport number of the cation (*i.e.* K^+) is given by

$$n_{K^+} = \frac{\text{No. of g eq. lost from the anodic compartment}}{\text{No. of g eq. deposited in the coulometer}}$$

$$= \frac{0.0003859}{0.0007935} = \mathbf{0.486}$$

$$\therefore \quad n_{Cl^-} = 1 - n_{K^+}$$

$$= 1 - 0.486 = \mathbf{0.514}$$

Example 2. A solution containing 10.09 g of $CuSO_4$ in 189.9 g of water was electrolysed. After electrolysis, 275.4 g of the solution around the anode was found to contain 15.4 g of $CuSO_4$ and 3.096 g of silver was deposited in a coulometer placed in series. Calculate the Hittorf's numbers of Cu^{2+} and SO_4^{2-} ions.

Solution. The data for 'after electrolysis' and 'before electrolysis' is compiled as under:

After electrolysis:

Weight of the anode solution = 275.4 g

Weight of $CuSO_4$ present = 15.4 g

Weight of water present = 275.4 – 15.4 g = 260 g

Before electrolysis

Weight of water = 189.9 g

Weight of $CuSO_4$ = 10.09 g

i.e. 189.9 g of water are associated with = 10.09 g $CuSO_4$

∴ 260 g of water is associated with

$$= \frac{10.09}{189.9} \times 260 \text{ g } CuSO_4$$

$$= 13.815 \text{ g}$$

Increase in the concentration of $CuSO_4$ around anode

$$= 15.4 - 13.815 = 1.585 \text{ g}$$

$$\frac{1.585}{79.75} \text{ g eq.} \quad \left\{ \because \text{Eq. wt. of } CuSO_4 = \frac{63.5+32+64}{2} = 79.75 \right\}$$

$$= 0.01987 \text{ g eq.}$$

Since we find an increase in concentration around anode, it is a case of attack of electrode.

Silver deposited in coulometer = 3.096 g

$$= \frac{3.06}{108} \text{ g eq.} \qquad \{\because \text{ Eq. wt. of Ag} = 108\}$$

$$= 0.02867 \text{ g eq}$$

If no Cu^{2+} ions had migrated from the anodic compartment to the cathodic compartment, the increase around anode would have been = 0.02867 g eq. Actual increase =0.01987 g eq. Hence *fall in conc. around anode* due to migration of Cu^{2+} ions is given by

$$= 0.02867 - 0.01987 = 0.0088 \text{ g eq.}$$

$$\therefore \quad n_{Cu^{2+}} = \frac{\text{No. of g eq. of ion lost from anodic compartment}}{\text{No. of g eq. deposited in the coulometer}}$$

$$= \frac{0.0088}{0.02867} = \mathbf{0.307}$$

$$\therefore \quad n_{SO_4^{2-}} = 1 - 0.307$$

$$= \mathbf{0.693}$$

Example 3. A solution of silver nitrate was electrolysed between silver electrodes. Before electrolysis 10 g of the solution contained 0.01788 g of silver nitrate. After the experiment 20.09 g of the anodic solution contained 0.06277 g of silver nitrate and 0.00947 g of copper was deposited in the copper coulometer placed in series. Calculate the transport numbers of silver and nitrate ions. The atomic weights of Ag and Cu are 108 and 63.6 respectively.

Solution. *After electrolysis*:

Weight of the anodic solution = 20.09 g

Weight of $AgNO_3$ present = 0.06277 g

∴ Weight of water present = 20.09 – 0.06277 = 20.02723 g

Before electrolysis :

Weight of solution = 10 g

Weight of $AgNO_3$ present = 0.01788 g

∴ Weight of water present = 10 – 0.1788 = 9.98212 g

9.98212 g of water is associated with = 0.01788 g $AgNO_3$

∴ 20.02723 g of water is associated with

$$= \frac{0.01788}{9.98212} \times 20.02723 = 0.03587 \text{ g } AgNO_3$$

Increase in conc. of $AgNO_3$ around anode = 0.06277 – 0.03587

$$= 0.0269 \text{ g}$$

$$= \frac{0.0269}{170} \text{ g eq.} \qquad \{\because \text{ Eq. wt of } AgNO_3 = 170\}$$

$$= 0.0001582 \text{ g eq.}$$

Copper deposited in the coulometer = 0.00947 g

$$= \frac{0.00947}{31.8} \text{ g eq.} \qquad \{\because \text{ Eq. wt. of Cu} = \frac{63.6}{2} = 31.8$$

$$= 0.0002977 \text{ g eq.}$$

If no Ag^+ ions had migrated from the anodic compartment to the cathodic compartment, the increase in the conc. of the anodic compartment would have been = 0.0002977 g eq.

Actual increase in conc. = 0.0001582g eq. Hence fall in concentration around anode due to migration of Ag^+ ions from the anodic compartment

$$= 0.0002977 - 0.0001582 \text{ g eq.}$$

$$= 0.0001395 \text{ g eq.}$$

$$\therefore \quad n_{Ag^+} = \frac{\text{No. of g eq. lost from the anodic compartment}}{\text{No. of eq. deposited in the coulometer}}$$

$$= \frac{0.0001395}{0.0002977} = \mathbf{0.469}$$

$$\therefore \quad n_{NO_3^-} = 1 - n_{Ag^+} = 1 - 0.469 = \mathbf{0.531}$$

Example 4. A solution of $AgNO_3$ containing 0.5066 g of the salt in 60 g of solution was electrolysed between silver electrodes. After electrolysis 50 g of the anode liquid was found to contain 0.5186 g of $AgNO_3$. A current of one ampere was passed for 1 minute and 58 sec. What are the transport numbers of Ag^+ and NO_3^- ions ?

Solution. Here quantity of electricity passed = Current × time
(amp.) (sec)

$$= 1 \times 118 = 118 \text{ Coulomb}$$

96500 Coulomb of electricity deposit 1 gm eq. of Ag

$$\therefore \quad 118 \text{ Coulomb of electricity deposit Ag} = \frac{118}{96500} \text{ g eq.}$$

After electrolysis

Weight of anode solution = 50 g
Weight of $AgNO_3$ = 0.5186 g
Weight of water = 49.4814 g

Before electrolysis

Weight of anode solution = 60 g
Weight of $AgNO_3$ = 0.5066 g
Weight of water = 59.4934 g
59.4934 g water contain = 0.5066 g $AgNO_3$ (Before electrolysis)

$$\therefore \quad 49.4814 \text{ g water contain} = \frac{0.5066}{59.4934} \times 49.4814 \text{ g } AgNO_3$$

$$= 0.4213 \text{ g } AgNO_3$$

Increase in conc. of $AgNO_3$ = 0.0973 g $AgNO_3$

$$= \frac{0.0973}{170} \text{ g equivalent}$$

$$= 0.00057 \text{ g equivalent}$$

$$\text{Silver deposited in the coulometer} = \frac{118}{96500} \text{ g eq.}$$

$$= 0.00122$$

Hence decrease in the conc. of $AgNO_3$ due to migration of cations from the anodic compartment

$$= 0.00122 - 0.00057$$

$$= 0.00065 \text{ g eq.}$$

$$\therefore \quad \text{Transport number of } Ag^+ = \frac{0.00065}{0.00122} = \mathbf{0.533}$$

$$\text{Transport no. of } NO_3^- = 1 - 0.533 = \mathbf{0.467}$$

PROBLEMS FOR PRACTICE

1. The original strength of a solution of caustic soda was 0.06315 g per 100 cc. After electrolysis between platinum electrodes 100 cc of anode solution contained 0.059 g NaOH and the same current deposited 0.05216 g silver in silver voltameter. Find the transport number of Na^+ and OH^- ions.

As the solution is dilute, weights of solvent and solution may be taken as equal

[**Ans.** $n_{Na^+} = 0.215$, $n_{OH^-} = 0.785$]

2. During the electrolysis of silver nitrate solution between platinum electrodes, the fall in concentration of silver ions in the anode chamber is 1.06 g per unit volume while the corresponding fall in the cathode chamber is 0.1 g. What are the transport numbers of the silver ion and the nitrate ion ? [**Ans.** $n_{Ag^+} = 0.567$, $n_{NO_3^-} = 0.433$]

3. A solution of $AgNO_3$ containing 1.06 mg of silver per g of the solution was electrolysed between silver electrodes and the anode liquid after electrolysis contained 42.94 mg of silver in 25 g of the solution. In a silver voltameter in series 32.10 mg of Ag^+ were deposited. Calculate the transport number of Ag^+ ions. [**Ans.** 0.488]

4. A current which deposited 0.047 g of copper in a voltameter containing $CuSO_4$ solution was passed through a solution of $AgNO_3$ using silver electrodes. The liquid from the cathode contained before the experiment 1.315 g of silver and after the experiment 1.231 g of silver. Calculate the transport number of Ag^+ and NO_3^- ions. [**Ans.** $NO_3^- = 0.525$, $Ag^+ = 0.475$]

5. A solution containing 10.850 g silver nitrate per 1000 g water was electrolysed between silver electrodes. After electrolysis, the whole of the anode solution was removed and found to weigh 54.900 g. On analysis, it was found to contain silver equivalent to 0.7217 g silver nitrate. In a coulometer in series, 0.04672 g copper was deposited. Calculate the transport number of silver ions. The equivalent weight of copper is 31.77 and of $AgNO_3$ is 170.0.

[**Ans.** $Ag^+ = 0.464$; $NO_3^- = 0.536$]

6. A solution of $AgNO_3$ having the composition 28.435 g H_2O, 0.2099 g $AgNO_3$ was electrolysed using silver electrodes. The current passed deposited 0.000792 g eq. of silver. At the close of the electrolysis, the anode solution was found to have the composition : 28.435 g H_2O and 0.2874 g $AgNO_3$. Calculate the transport number of Ag^+ and NO_3^- ions.

[**Ans.** $Ag^+ = 0.424$, $NO_3^- = 0.576$]

7. A current, which deposited 30.0 mg of silver in a silver voltameter, was passed through a solution of $AgNO_3$ using silver electrodes. The increase in concentration of silver around anode corresponded to 15.9 mg of silver. Calculate the transport numbers of Ag^+ and NO_3^- ions. [**Ans.** $Ag^+ = 0.47$, $NO_3^- = 0.53$]

8. A solution of $AgNO_3$ containing 1.139 mg of silver per unit g of solution was electrolysed between silver electrodes and the anode liquid after electrolysis contained 39.66 mg silver in 20.09 g of solution. In a silver voltameter in series with the electrolytic cell, 32.10 mg of silver was deposited. Calculate the transport number of Ag^+ and NO_3^- ions.

[**Ans.** $Ag^+ = 0.477$, $NO_3^- = 0.523$]

9. In determining the transport number of silver and nitrate ions, the following results were obtained : Before experiment 1.0 g anode solution contained 0.001788 g of $AgNO_3$. After experiment, 20.0 g of anode solution contained 0.065227 g of $AgNO_3$. In a voltameter placed in circuit 0.0322 g of silver was deposited. Calculate the transport number of Ag^+ and NO_3^- ions. [**Ans.** $Ag^+ = 0.477$, $NO_3^- = 0.523$]

10. A solution of $AgNO_3$ containing 0.510 g of the salt in 60.40 g of the solution was electrolysed using silver electrodes. After electrolysis, 54 g of the anode solution was found to contain 0.56 g of $AgNO_3$. A current of 0.04 ampere was passed for 2950 seconds. Calculate the transport number of Ag^+ and NO_3^- ions. [**Ans.** $Ag^+ = 0.498$, $NO_3^- = 0.502$]

7.14 MOVING BOUNDARY METHOD FOR THE DETERMINATION OF TRANSPORT NUMBERS

This method is based upon the generalisation that the transport numbers of the cation and the anion are given by

$$n_c = \frac{u_c}{u_c + u_a}$$

and

$$n_a = \frac{u_a}{u_c + u_a}$$

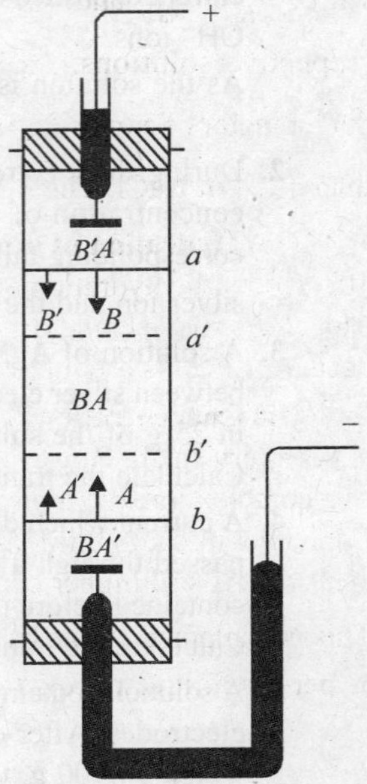

Fig. 7.12. Moving boundary method for determination of transport number.

Thus by determining the speeds of the cation and the anion (u_c and u_a), the transport numbers can be calculated.

The apparatus used for transport number determination is shown in Fig. 7.12. It consists of a long tube mounted vertically and fitted with the platinum electrodes on the two sides. The salt BA under investigation is dissolved in water and then placed between the solutions of two other salts $B'A$ and BA' one having a common cation and the other having a common anion. For example, if BA is KCl, $B'A$ may be LiCl and BA' may be CH_3COOK. The three solutions are placed in order of increasing density downwards to avoid mixing solutions. The original boundaries between the solutions are a and b. Even if the solutions are colourless, the boundaries can be observed because of the differences in the refractive indices of the solution. In order that the boundaries may remain sharp throughout the experiment, B' and A' must have smaller velocities than B and A respectively so that B' does not overtake B and A' does not overtake A. On passing the current, the boundaries move to a' and b'.

Speed of the cation $(B^+), u_c \propto$ length aa'

Speed of anion $(A^-), u_a \propto$ length bb'

Hence
$$n_c = \frac{aa'}{aa' + bb'}$$

and
$$n_a = \frac{bb'}{aa' + bb'}$$

7.15. FACTORS THAT INFLUENCE TRANSPORT NUMBER OF IONS

Following factors govern the transport number:

(*i*) **Nature of the ion:** Smaller the size of the ion, greater is its speed and hence higher is its transport number. For example, the transport number of H^+ ion in 0.1 N HCl is greater than that of Na^+ ion in 0.1 N NaCl or the K^+ ion in 0.1 N KCl etc.

(*ii*) **Nature of the other ion present:** The transport number of Cl^- ion (n_{Cl^-}) in HCl is different from that of Cl^- ion in NaCl. This may be explained as follows :

$$n_{Cl^-} \text{ in HCl} = \frac{u_{Cl^-}}{u_{H^+} + u_{Cl^-}} \qquad \text{... } (i)$$

$$n_{Cl^-} \text{ in NaCl} = \frac{u_{Cl^-}}{u_{Na^+} + u_{Cl^-}} \quad \text{... (ii)}$$

where u_{H^+}, u_{Na^+} and u_{Cl^-} represent the speeds of H^+ ion, Na^+ and Cl^- ions respectively in their respective solutions. As Na^+ ion is larger in size than H^+ ion $u_{H^+} > u_{Na^+}$. As a result, the denominator of expression (*i*) is greater than that of expression (*ii*) and hence n_{Cl^-} in HCl is smaller than n_{Cl^-} in NaCl. The same applies to other ions also.

(*iii*) **Hydration of the ions:** The speed of an ion and hence its transport number are very much affected by the hydration of the ions. For example, Li^+ ions being smaller in size than Na^+ ions are expected to have higher transport number than Na^+ ions but the actual position is otherwise. This is because Li^+ ions are more hydrated than Na^+ ions.

(*iv*) **Concentration:** The transport number of an ion changes with concentration but the change is very very small. However, if there is an increase in concentration and it results into the formation of a complex ion, the transport number changes significantly. For example, the transport number of Cd^{2+} ions in cadmium iodide solution falls from 0.444 in 0.005 M solution to 0.003 in 0.025 M solution. At still higher concentrations, the transport number appears to be zero or even negative. This is explained on the basis of the formation of the complex CdI_4^{2-} ions at higher concentrations as per the following equations :

$$CdI_2 \longrightarrow Cd^{2+}(aq) + 2I^-(aq)$$

$$Cd^{2+}(aq) + 4\,I^-(aq) \longrightarrow CdI_4^{2-}(aq)$$

As the concentration is increased, more and more of CdI_4^{2-} ions are formed, which migrate to the anodic compartment. A stage comes when as many CdI_4^{2-} ions migrate into the anodic compartment as the number of Cd^{2+} ions migrating from it to the cathodic compartment. Thus there is no change in the concentration of Cd^{2+} ions in the anodic compartment and so the Cd^{2+} ions appear to have zero transport number.

(*v*) **Temperature:** With increase of temperature, the ions tend to approach a transport number value of 0.5, *i.e.*, equal fraction of the current is carried by the cations and the anions. In case of ions having transport number less than 0.5 at ordinary temperature, the transport number increases with increase of temperature. In case of ions having transport number greater than 0·5 at ordinary temperature, the transport number decreases with increase of temperature.

7.15.1. Relationship between the ionic conductance and transport number

Ionic conductance (λ) $\propto$ ionic mobility (u)

i.e. $\lambda_c \propto u_c$ or $\lambda_c = ku_c$ ... (*i*)

and $\lambda_a \propto u_a$ or $\lambda_a = ku_a$ ... (*ii*)

where λ_c and λ_a are the ionic conductances and u_c and u_a are the ionic mobilities of the cation and the anion respectively and k is a constant.

Dividing equation (*i*) by equation (*ii*), we have

$$\frac{\lambda_c}{\lambda_a} = \frac{u_c}{u_a}$$

Adding 1 to both sides, we have

$$\frac{\lambda_c}{\lambda_a} + 1 = \frac{u_c}{u_a} + 1 \quad \text{or} \quad \frac{\lambda_c + \lambda_a}{\lambda_a} = \frac{u_c + u_a}{u_a}$$

Taking reciprocal of both sides, we have

$$\frac{\lambda_a}{\lambda_c + \lambda_a} = \frac{u_a}{u_c + u_a}$$

But $\frac{u_a}{u_c + u_a} = n_a$, transport number of the anion

and $\lambda_c + \lambda_a = \wedge_0$, eq. conductance at infinite dilution. *(By Kohlrausch's law)*

$\therefore$ $\frac{\lambda_a}{\wedge_0} = n_a$ or $\lambda_a = n_a \times \wedge_0$

Similarly $\lambda_c = n_c \times \wedge_0$

7.15.2. Ionic mobility and ionic conductance

Ionic mobility is the absolute velocity of an ion. It is the distance travelled by an ion in cm/sec under a potential gradient of 1 V/cm. Contribution by an ion towards total equivalent conductance of an electrolyte at infinite dilution is called ionic conductance. The relationship between the two is as follows :

$$\text{Ionic mobility} = \frac{\text{Ionic conductance}}{96500}$$

SOLVED EXAMPLES ON IONIC CONDUCTANCE AND IONIC MOBILITY

Example 1. The transport number of silver ion is 0.466 and that of nitrate ion is 0.534 at 18°C. The molar conductivity of a very dilute solution of $AgNO_3$ is 115.8 ohm^{-1} cm^2 at the same temperature. Calculate (*a*) the ionic conductance of silver and nitrate ions (*b*) the ionic mobilities of the two ions.

Solution. (*a*) Ionic conductance = Transport number of the ion × $\wedge_0$ for strong electrolyte containing that ion

i.e. $\lambda_a = n_a \times \wedge_0$

or $\lambda_{NO_3^-} = 0.533 \times 115.8 = 61.837$ ohm^{-1} cm^2

and $\lambda_c = n_c \times \wedge_0$

or $\lambda_{Ag+} = 0.466 \times 115.8$

$= 53.962$ ohm^{-1} cm^2

(*b*) $\text{Ionic mobility } (u) = \frac{\text{Ionic conductance}(\lambda)}{96,500}$

$\therefore$ $\text{Ionic mobility of Ag}^+ = \frac{53.962}{96,500} = \mathbf{0.000559\,cm/sec}$

and Ionic mobility of NO_3^- $= \dfrac{61.837}{96,500} = \mathbf{0.000641\,cm/sec}$

Example 2. The equivalent conductance of a very dilute solution of $NaNO_3$ at 18°C is 105.2 ohm^{-1} cm^2. If the ionic conductance of NO_3^- ion in the solution is 61.7 ohm^{-1} cm^2, calculate the transport number of Na^+ ions in the solution.

Solution. Given that

$$\wedge_0 \text{ for } NaNO_3 = 105.2 \text{ ohm}^{-1} \text{ cm}^2$$

$$\lambda_{NO_3^-} = 61.7 \text{ ohm}^{-1} \text{ cm}^2$$

Substituting the values in the equation

$$\lambda^0_{NO_3^-} = n_{NO_3^-} \times \wedge^0_{NaNO_3}$$

$$61.7 = n_{NO3-} \times 105.2$$

$$n_{NO_3^-} = \frac{61.7}{105.2} = 0.586$$

$\therefore$ Transport No. of $Na^+ = 1 - 0.516 =$ **0.414**

Example 3. At 18°C, the equivalent conductivities at infinite dilution of NH_4Cl, NaOH and NaCl are 129.8, 217.4 and 108.9 ohm^{-1} cm^2 respectively. If the equivalent conductivity of centi-normal solution of NH_4OH is 9.33 ohm^{-1} cm^2, what is the percentage dissociation of NH_4OH at this dilution ?

Solution. Given that

$$\wedge_0 \text{ for } NH_4Cl = 129.8 \text{ ohm}^{-1} \text{ cm}^2$$

$$\wedge_0 \text{ for NaOH} = 217.4 \text{ ohm}^{-1} \text{ cm}^2$$

$$\wedge_0 \text{ for NaCl} = 108.9 \text{ ohm}^{-1} \text{ cm}^2$$

By Kohlrausch's law

$$\begin{aligned}\wedge_0 \text{ for } NH_4OH &= \lambda_{NH_4^+} + \lambda_{OH^-} \\ &= \wedge_0 (NH_4Cl) + \wedge_0 (NaOH) - \wedge_0 (NaCl) \\ &= 129.8 + 217.4 - 108.9 \\ &= 238.3 \text{ ohm}^{-1} \text{ cm}^2\end{aligned}$$

Eq. conductance of centinormal NH_4OH ($\wedge$) = 9.33 ohm^{-1} cm^2 (Given)

$\therefore$ Degree of dissociation (α) $= \dfrac{\wedge}{\wedge_0} = \dfrac{9.33}{238.3} = 0.0392$

or % age dissociation $= 0.0392 \times 100 =$ **3.92%**

Example 3. If ionic conductance for H^+ and OH^- ions are 394.8 and 198.5 ohm^{-1} cm^2 respectively at 25°C and if the specific conductance of water at this temperature is 5.54×10^{-8} ohm^{-1} cm^{-1}, calculate the ionic product of water.

Solution. Given that, $\lambda_{H^+} = 394.8 \text{ ohm}^{-1} \text{ cm}^2$

$$\lambda_{OH^-} = 198.5 \text{ ohm}^{-1} \text{ cm}^2$$

$\therefore$ $\wedge_0 (H_2O) = \lambda_{H^+} + \lambda_{OH^-}$ *(By Kohlrausch's law)*

$$= 349.8 + 198.5 \text{ ohm}^{-1} \text{ cm}^2$$
$$= 548.3 \text{ ohm}^{-1} \text{ cm}^2.$$

Specific conductance of pure water at 25°C is found to be

$$K = 5.54 \times 10^{-8} \text{ ohm}^{-1} \text{ cm}^{-1}.$$

As water dissociates to a very small extent to give H^+ ions and OH^- ions, water may be considered to be a very dilute solution of H^+ ions and OH^- ions in water so that we have

Eq. conductance of water = Specific conductance of water × Vol. in cm^3 containing 1g eq. of ions

∴ Volume in cm^3 containing 1 g eq. of ions

$$= \frac{\text{Eq.conductance of water}}{\text{Sp.conductance of water}} = \frac{548.3}{5.54 \times 10^{-8}} = 9.897 \times 10^9$$

9.897×10^9 cm^3 of water contains 1g eq. of H^+ ions or OH^- ions.

∴ 1 cm^3 of water contains H^+ ions or OH^- ions

$$= \frac{1}{9.897 \times 10^9} = 1.01 \times 10^{-10} \text{ g eq.}$$

or 1000 cm^3 of water contains H^+ ions or OH^- ions

$$= 1.01 \times 10^{-10} \times 1000 = 1.01 \times 10^{-7} \text{ g eq.}$$

i.e.
$$[H^+] \text{ or } [OH^-] = 1.01 \times 10^{-7} \text{ g eq. per litre}$$
$$= 1.01 \times 10^{-7} \text{ moles/litre.}$$

Pure water dissociates to give equal number of H^+ ions and OH^- ions so that we have

$$[H^+] = [OH^-] = 1.01 \times 10^{-7} \text{ M at } 25°\text{C}$$

Hence
$$K_w = [H^+][OH^-]$$
$$= (1.01 \times 10^{-7}) \times (1.01 \times 10^{-7})$$
$$= 1.02 \times 10^{-14} \text{ at } 25° \text{ C.}$$

7.16. APPLICATIONS OF CONDUCTANCE MEASUREMENTS

Conductance measurements have the following applications :

1. Determination of equivalent conductance at infinite dilution ($\wedge_0$) for weak electrolytes. Equivalent conductance of a weak electrolyte at infinite dilution cannot be determined directly. It is obtained using Kohlrausch's law. This has been explained along with solved examples in previous sections in this chapter.

2. Determination of ionic conductance and absolute velocity of ions i.e. ionic mobilities. The relation between ionic mobility and ionic conductance is :

$$\text{Ionic mobility} = \frac{\text{Ionic conductance}}{96500}$$

This has also been explained along with solved examples in a previous section of the chapter.

3. Determination of degree of dissociation of a weak electrolyte. Degree of dissociation of a weak electrolyte (α) is given by

$$\alpha = \frac{\wedge_c}{\wedge_0}$$

where $\wedge_c$ = Equivalent conductance at the concentration c

$\wedge_0$ = Equivalent concentration at zero concentration (or infinite dilution)

Knowing the values of $\wedge_c$ and $\wedge_0$, α can be calculated.

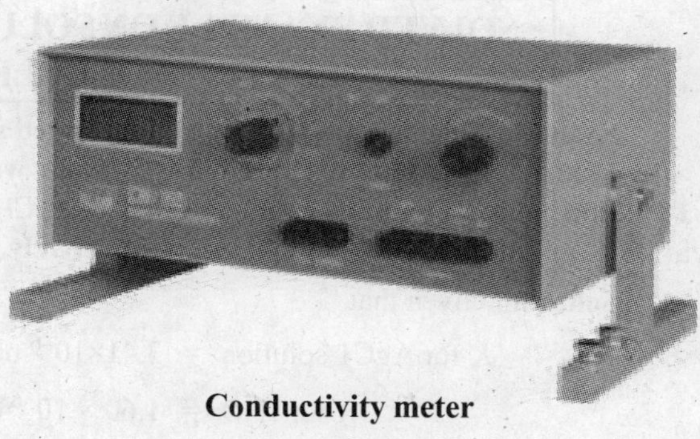

Conductivity meter

4. Determination of dissociation constant of weak acids (K_a) or (K_d). Relation between dissociation constant K_a and degree of dissociation α is given by

$$K_a = C\frac{\alpha^2}{1-\alpha}$$

Knowing the degree of dissociation (α), dissociation constant can be calculated.

α can be obtained from the relation : $\alpha = \wedge_c / \wedge_0$. Equivalent conductance at concentration c ($\wedge_c$) can be obtained using the relation $\wedge_c = K \times \frac{1000}{C}$ where K is specific conductance.

5. Determination of the solubility of a sparingly soluble salt..Salts such as $AgCl$, $BaSO_4$, $PbSO_4$ etc. dissolve in water to a very small extent. That is why these salts are called *sparingly soluble salts.* The solubility of such salts can be calculated by the application of Kohlrausch's law and conductance measurements as follows :

Specific conductance of the solution is determined. Suppose it comes out to be K. If S is the solubility of the salt in *gram equivalents per litre* then the volume of the solution in cm^3 containing 1 g eq will be 1000/S.

$$\text{Eq. conductance of the solution} = \text{Specific conductance} \times \frac{1000}{S}$$

i.e.,

$$\wedge = K \times \frac{1000}{S}$$

Since the salt dissolves very little, the solution is very dilute so that ∧ may be considered as equal to $\wedge_0$, *i.e.*, equivalent conductance at infinite dilution. Therefore,

$$\wedge_0 = K \times \frac{1000}{S}$$

So that

$$\boxed{S = \frac{1000 \times K}{\wedge_0}}$$

The value of $\wedge_0$ can be found by the application of Kohlrausch's law *e.g.* for $AgCl$, $\wedge_0 = \lambda_{Ag^+} + \lambda_{Cl^-}$

This will give the solubility in gram-equivalent per litre.

The solubility in grams/litre can be obtained by multiplying S with the equivalent weight of the substance.

SOLVED EXAMPLE ON SOLUBILITY OF SPARINGLY SOLUBLE SALT

Example. Specific conductivity of a saturated solution of AgCl at 25°C was found to be 3.41 × 10^{-5} ohm^{-2} cm^{-1}. The specific conductivity for water used to make up the solution was 1.60 × 10^{-6} ohm^{-1} cm^{-1}. Determine the solubility of AgCl in water. Ionic conductances of Ag^+ and Cl^- at 25°C are 60.3 and 78.0 ohm^{-1} cm^2 respectively.

Solution. Given that

$$K \text{ for AgC1 solution} = 3.41\times10^{-6}\ ohm^{-1}\ cm^{-1}$$

$$K \text{ for } H_2O = 1.60 \times 10^{-6}\ ohm^{-1}\ cm^{-1}$$

$$\lambda_{Ag^+} = 60.3\ ohm^{-1}\ cm^2$$

$$\lambda_{Cl^-} = 78.0\ ohm^{-1}\ cm^2.$$

By Kohlrausch's law, we have

$$\wedge_0 \text{ for AgCl} = \lambda_{Ag^+} + \lambda_{Cl^-}$$

$$= 60.3 + 78.0 = 138.3\ ohm^{-1}\ cm^2$$

$$K \text{ for AgCl} = K \text{ for solution} - K \text{ for } H_2O$$

$$= (3.41 - 1.60) \times 10^{-6} = 1.81 \times 10^{-6}\ ohm^{-1}\ cm^{-1}$$

Substituting the values in the formula

$$S = \frac{1000\times K}{\wedge_0}, \text{ we get}$$

$$S = \frac{1000\times1.81\times10^{-6}}{138.3}\ g\,eq/litre$$

$$= \frac{1000\times1.81\times10^{-6}}{138.3}\times 143.5\,g\ /litre$$

$$= \mathbf{0.001878\ g/litre}$$

7.17. CONDUCTOMETRIC TITRATIONS

Conductometric methods help us to reach accurate end points in titrations. The basic principle in conductometric titrations is that when one solution from the burette is added to the other solution in the titration flask, the conductance value changes. This helps us to arrive at the end point. The following examples will illustrate.

1. Conductometric titration of a strong acid with a strong base. Suppose a solution of HCl is to be titrated against NaOH solution. The acid solution is taken in a beaker and the NaOH solution in the burette. The conductance of the acid solution is noted initially as well as after successive additions of small amount of NaOH solution. Evidently, the conductance of the acid solution in the beginning is very high as it contains highly mobile H^+ ions (HC1→ $H^+ + Cl^-$). On adding NaOH solution to the HC1 solution, the H^+ ions are replaced by the slow moving Na^+ ions and hence the conductance of the solution keeps on falling till the end point is reached (*i.e.* all the H^+ ions are replaced by Na^+ ions).

$$H^+ + Cl^- + Na^+ + OH^- \longrightarrow Na^+ + Cl^- + H_2O$$

Beyond the end point, further addition of NaOH solution brings in fast moving OH^- ions and hence the conductance of the solution again starts increasing. If conductance values are plotted against the volume of the alkali added, a curve of type *ABC* is obtained as shown in Fig. 7.13. Point of intersection (*i.e.*, point *B*) corresponds to the end point.

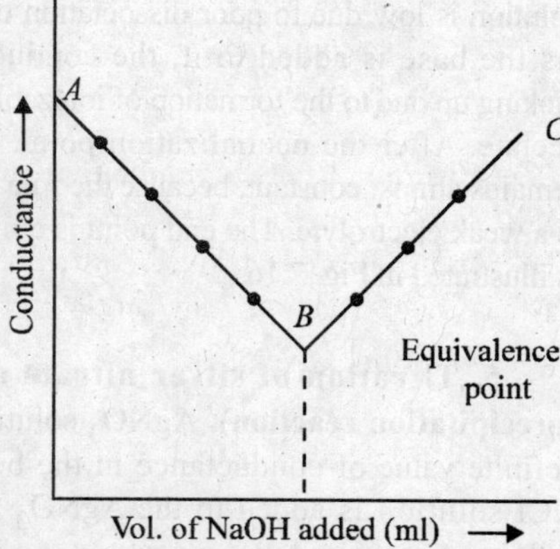

Fig. 7.13. Conductometric titration of a strong acid with a strong base.

2. Titration of a weak acid with a strong base. (Say CH_3COOH vs. NaOH)

Let us say a weak acid like acetic acid is titrated against a strong base like NaOH. The conductance of the acid initially is very low because of low ionization of the acid. However on adding NaOH solution to the acid solution, the salt produced *viz.* CH_3COONa is highly ionized and hence the conductance keeps on increasing along the line *AB* as shown in Fig. 7.14. When the whole of acetic acid has been neutralized, further addition of NaOH solution causes the conductance to increase sharply along the line *BC* because the NaOH added introduces the fast moving OH^- ions. Intersection of the two lines *AB* and *BC* at the point *B* gives the equivalent point.

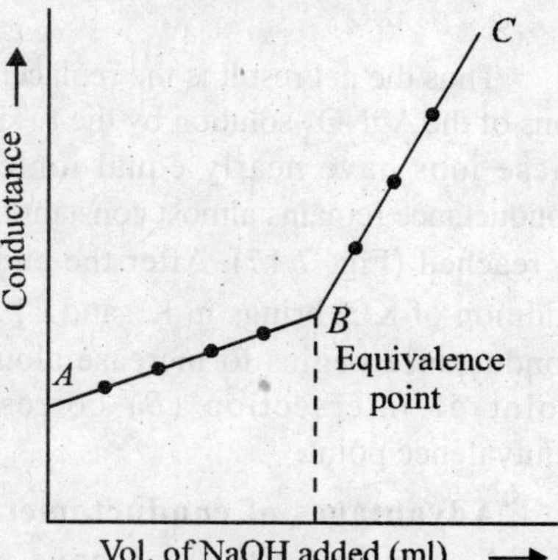

Fig. 7.14. Conductometric titration of a weak acid with a strong base.

3. Titration of a strong acid with a weak base. Let us consider the titration of HCl against NH_4OH solution. The conductance of HCl solution is initially high because of the presence of the fast moving H^+ ions. As NH_4OH solution is added, the fast moving H^+ ions are replaced by the slower NH_4^+ ions and hence the conductance falls along the line *AB*.

$$H^+ + Cl^- + NH_4OH \longrightarrow NH_4^+ + Cl^- + H_2O$$

When the end point is reached, further addition of NH_4OH does not cause much change in the conductance because NH_4OH is a weakly ionized substance. The line *BC* is thus almost horizontal (Fig. 7.15).

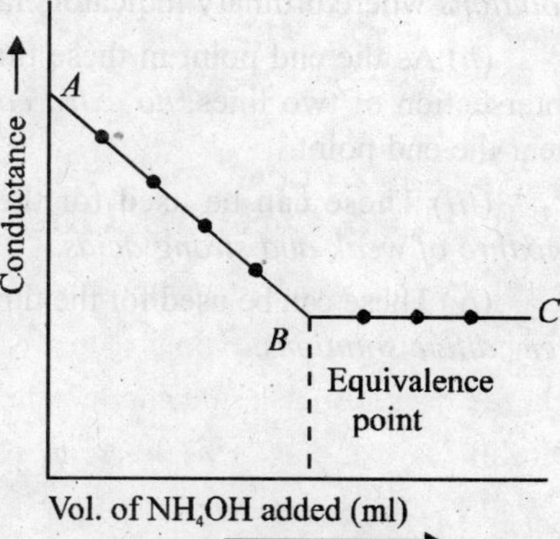

Fig. 7.15. Conductometric titration of a strong acid against a weak base.

4. Titration of a weak acid against a weak base. Conductometric titrations are particularly valuable because we do not get sharp end points with indicators in normal titrations. Consider the titration of acetic acid (weak acid) with ammonium hydroxide (weak base). Initially, the conductance of the solution is low due to poor dissociation of acetic acid. As the base is added to it, the conductance starts picking up due to the formation of ionizable ammonium acetate. After the neutialization point, conductance remains almost constant, because the free base NH_4OH is a weak electrolyte. The end point is quite sharp. This is illustrated in Fig. 7.16.

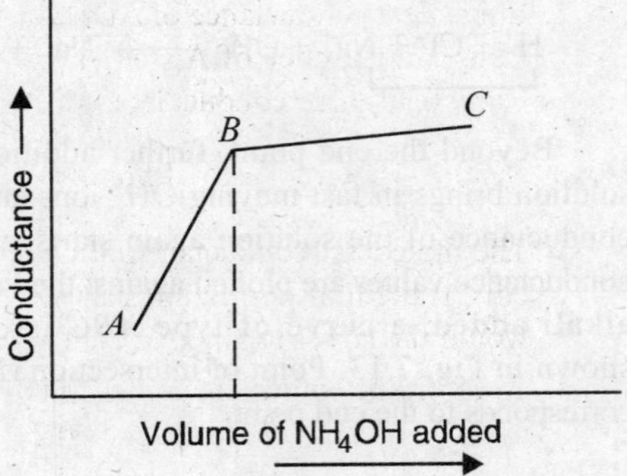

Fig. 7.16. Conductometric titration of a weak acid with a weak base

5. Titration of silver nitrate against KCl (precipitation reaction). $AgNO_3$ solution has some definite value of conductance in the beginning. As KCl solution is added to the $AgNO_3$ solution, the following reaction takes place :

$$Ag^+ + NO_3^- + K^+ + Cl^- \longrightarrow K^+ + NO_3^- + \underset{(ppt)}{AgCl}$$

Thus the net result is the replacement of the Ag^+ ions of the $AgNO_3$ solution by the K^+ ions. Since both these ions have nearly equal ionic mobility, the conductance remains almost constant till the end point is reached (Fig. 7.17). After the end point, further addition of KCl brings in K^+ and Cl^- ions. Thus the conductance begins to increase along the line *BC*. Point of intersection (*B*) corresponds to the equivalence point.

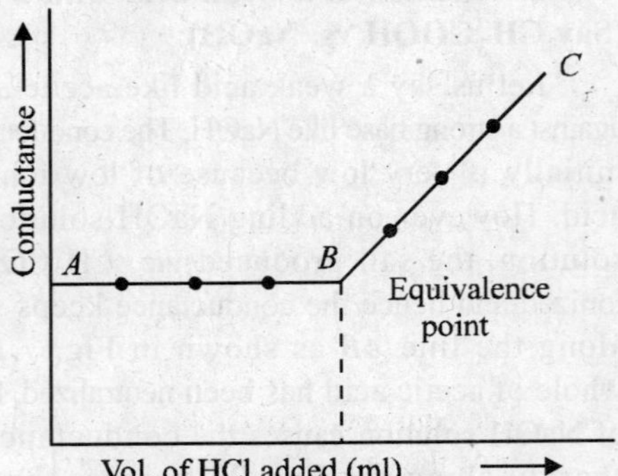

Fig. 7.17. Conductometric titration of $AgNO_3$ sol. against KCl sol.

Advantages of conductometric titrations. Conductometric titrations have a number of advantages over ordinary titrations as listed below.

(*i*) These titrations can be used for *coloured solutions* where ordinary indicators fail.

(*ii*) As the end point in these titrations is the intersection of two lines, *no extra care* is needed near the end point.

(*iii*) These can be used for the titration of *mixture of weak and strong acids.*

(*iv*) These can be used for the titration of even *very dilute solutions.*

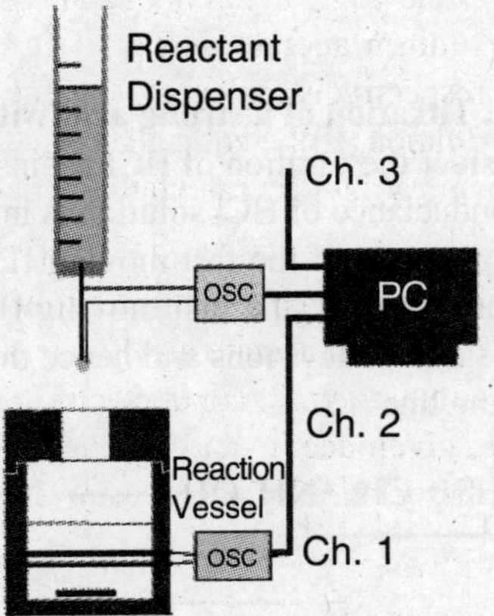

Conductometric titration

PROBLEMS FOR PRACTICE

1. The equivalent conductivity of a very dilute solution of KCl is 130.1 ohm^{-1} cm^2. The transport number of K^+ ion in such a solution is 0.496. Calculate the absolute velocities of the ions under unit fall of potential.

[Ans. K^+ = 0.000669 cm/sec. Cl^- = 0.000679 cm/sec]

2. Equivalent conductance of $AgNO_3$ at infinite dilution is 133.3 ohm^{-1} cm^2 equivalent^{-1}. The transport number of Ag^+ ions in very dilute $AgNO_3$ solution is 0.454. Calculate the equivalent ionic conductances of Ag^+ and NO^{3-} ions at infinite dilution.

[**Ans.** $\lambda^0_{Ag^+}$ = 60.158 ohm^{-1} cm^2, $\lambda^0_{NO_3^-}$ = 72.782 ohm^{-1} cm^2]

3. The molecular conductance of an aqueous solution of silver nitrate at 20°C is 120 ohm^{-1} cm^2. If the transport number of nitrate ions is 0.55 at this temperature, how much distance would the silver ions move under a unit potential gradient ? [**Ans.** 0.00056 cm/sec]

4. At 18°C, the equivalent conductivity of HCl acid solution is 379.6 ohm^{-1} cm^2. The transport number of chlorine in HCl is 0.16. Calculate the ionic conductances of hydrogen and chlorine. [**Ans.** $\lambda^0_{H^+}$ = 318.87 ohm^{-1} cm^2 and $\lambda^0_{Cl^-}$ = 60.73 ohm^{-1} cm^2]

5. The equivalent conductivity at infinite dilution of NH_4Cl was found to be 150.4 ohm^{-1} cm^2 at 25°C. The transport number of NH_4^+ ion was found to be 0.491. Calculate the ionic conductances of NH_4^+ and Cl^- ions.

[**Ans.** NH_4^+ = 73.84 ohm^{-1} cm^2, Cl^- = 76.54 ohm^{-1} cm^2]

6. A dilute solution of $AgNO_3$ has an equivalent conductivity of 115.2 ohm^{-1} cm^2 eq^{-1} and the transport number of silver ions is 0.47. Calculate the ionic conductances and ionic mobilities of silver and nitrate ions.

[**Ans.** $\lambda^0(Ag^+)$ = 54.144 ohm^{-1} cm^2, λ^0 (NO_3^-) = 61.056 ohm^{-1} cm^2
u (Ag^+) = 0.000561 cm/sec, u (NO_3^-) = 0.000633 cm/sec]

7. The equivalent conductivity of acetic acid at infinite dilution is 387 Ω^{-1} cm^2 (g equivalent)$^{-1}$. At the same temperature, but at a concentration of 1 g equivalent weight in 1000 litres, it is 55 Ω^{-1} cm^2 (g equivalent)$^{-1}$. What is the % dissociation of 0.001 N acetic acid ?

[**Ans.** 14.21%]

8. If the equivalent conductance at infinite dilution at 293 K for aqueous hydrochloric acid, sodium acetate and sodium chloride solution are 383.5, 78.4 and 102.0 ohm^{-1} cm^2 respectively, calculate the equivalent conductance of acetic acid at this temperature and dilution. If the equivalent conductance of acetic acid at some other dilution is 100.0 ohm^{-1} cm^2 at 293 K, calculate the degree of ionization of acetic acid at that dilution.

[**Ans.** Λ_0 = 359.9 ohm^{-1} cm^2, α = 0.278]

9. The molar conductance at infinite dilution of potassium chloride, hydrochloric acid and potassium acetate are 130.1, 379.4 and 95.6 ohm^{-1} cm^2 mol^{-1} respectively. Calculate the value of molar conductance at infinite dilution for acetic acid. If the molar conductance of a given acetic acid solution is 48.5 ohm^{-1} cm^2 mol^{-1} at 298 K, calculate the degree of dissociation of acetic acid at this temperature.

[**Ans.** Λ_0 = 344.9 ohm^{-1} cm^2 mol^{-1}; α = 0.141]

SOME SOLVED CONCEPTUAL PROBLEMS

Q.1. Molar conductivity of both strong and weak electrolytes increases with dilution. What is the difference in the cause of increase in the two cases ?

Ans. A strong electrolyte is completely ionized even at high concentration. But ions are not completely free to move. On dilution, the interionic attractions decrease and ions become more free. Hence, the conductance increases. A weak electrolyte is very little ionized at high concentration. On dilution, dissociation increases, *i.e.*, number of ions increases and hence the conductance increases.

Q.2. Why Ostwald's dilution law is not applicable to strong electrolytes ?

Ans. Strong electrolytes dissociate completely. Hence there is no equilibrium between ions and undissociated molecules, *i.e.*, degree of dissociation $\alpha = 1$ at all concentrations. Thus, the dilution law is not applicable.

Q.3. The transference number of Cl^- ion in an aqueous solution of HCl is 0.16 whereas its value is 0.62 in an aqueous solution of NaCl. Explain the difference.

Ans. The equationd for the transference numbers of Cl^- in the two electrolytes are given below :

$$n_{Cl^-} \text{ in HCl} = \frac{Cl^-}{u_{H^+} + u_{Cl^-}}, \qquad n_{Cl^-} \text{ in NaCl} = \frac{u_{Cl^-}}{u_{Na^+} + u_{Cl^-}}.$$

As H^+ ions are smaller in size than Na^+ ions, speed of H^+ ions (u_{H^+}) is much larger than u_{Na^+}. Hence n_{Cl^-} in HCl will be smaller than n_{Cl^-} in NaCl.

Q.4. In a conductometric titration, the solution to be added from above should be much stronger than the solution taken in the conductivity cell. Why ?

Ans. The solution taken in the burette is usually about ten times stronger than that taken in the titration flask so that the volume change of the latter solution is minimum on the addition of the former solution. This is required because conductance changes with dilution.

Q.5. Write Debye-Huckel-Onsagar equation in terms of dielectric constant (D) and coefficient of viscosity (η) of the medium at temperature T. Give two tests to ascertain the accuracy of this equation.

Ans. Debye-Huckel-Onsagar equation in terms of dielectric constant (D) and coefficient of viscosity (η) is given below.

$$\wedge_m = \wedge_0 - \left[\frac{82.4}{(DT)^{1/2}\eta} + \frac{8.20\times 10^5}{(DT)^{3/2}}\wedge_0\right]\sqrt{c} = \wedge_0 - (A + B\wedge_0)\sqrt{c}$$

1. Plot of $\wedge_m$ versus $\sqrt{c}$ should be linear with slope = A + B $\wedge_0$
2. Calculate the value of this quantity using the known values of different quantities involved and compare with the value of slope obtained from the graph, the two values should come out be nearly same.

Q.6. Li^+ ion is smaller in size than Na^+ ion, yet transport number of Li^+ in LiCl solution is less than that of Na^+ ion in NaCl solution. Why ?

Ans. This is because Li^+ ions are more hydrated than Na^+ ions. The effective size of Li^+ becomes much bigger than Na^+ ions.

Q.7. Why is it not possible to measure the limiting molar conductivity of a weak electrolyte experimentally ?

Ans. The dissociation of a weak electrolyte is not complete even at large dilutions. Hence, $\wedge_m^0$ cannot be determined. Moreover, the plot of $\wedge_m$ against $\sqrt{c}$ at large dilutions becomes parallel to $\wedge_m$-axis. Hence, it cannot be extrapolated to zero concentration.

Q.8. Specific conductivity decreases with dilution whereas molar conductivity increases with dilution. Explain why ?

Ans. Specific conductivity is the conductance of ions present in 1 cm^3 of the solution. On dilution, number of ions per cm^3 decreases. Hence specific conductivity decreases. Molar conductivity $(\wedge_m)$ is the product of Sp. conductivity (*K*) and volume (*V*) of the solution in cm^3 containing 1 mole of the electrolyte. While *K* decreases on dilution but *V* increases much more. Hence their product increases with dilution.

EXERCISES
(Based on Question Papers of Different Universities)

Multiple choice Questions (Choose the correct option)

1. On passing electrical current through an electrolyte solution, the cations
(*a*) move towards cathode with speed equal to that of anions towards anode
(*b*) move with faster speed than that of anions
(*c*) move with different speed as compared to that of anions
(*d*) move with slower speed than that of anions

2. The Hittorf's rule can be represented by the expression

(*a*) $\dfrac{\text{Fall around anode}}{\text{Rise around cathode}} = \dfrac{\text{speed of cation}}{\text{speed of anion}}$

(*b*) $\dfrac{\text{Fall around anode}}{\text{Fall around cathode}} = \dfrac{\text{speed of anion}}{\text{speed of cation}}$

(*c*) $\dfrac{\text{Fall around anode}}{\text{Fall around cathode}} = \dfrac{\text{speed of cation}}{\text{speed of anion}}$

(*d*) $\dfrac{\text{Rise around cathode}}{\text{Fall around anode}} = \dfrac{\text{speed of anion}}{\text{speed of cation}}$

3. The Hittorf's rule states that
(*a*) the loss of concentration around any electrode is proportional to the speed of the ions moving towards it
(*b*) the loss of concentration around any electrode is proportional to the speed of the ions moving away from it
(*c*) the loss of concentrations around both the electrodes is proportional to the sum of speed of cations and anions
(*d*) none of the above

4. The fraction of the total current carried by the cation or anion is termed as
(*a*) fractional number (*b*) speed number
(*c*) carrier number (*d*) transport number

5. The statement of Kohlrausch's law is
(*a*) the equivalent conductance of an electrolyte at infinite dilution is equal to the product of equivalent conductances of the component ions
(*b*) the equivalent conductance of an electrolyte at infinite dilution is equal to the difference of equivalent conductance of the component ions
(*c*) the equivalent conductance of an electrolyte at infinite dilution is equal to the sum of equivalent conductances of the component ions
(*d*) none of the above

6. Kohlrausch's law can be used to determine
(*a*) λ_α for weak electrolytes (*b*) absolute ionic mobilities
(*c*) solubility of sparingly soluble salts (*d*) all of these

7. When a strong acid is titrated against a strong base the end point is the point of

(*a*) zero conductance (*b*) maximum conductance

(*c*) minimum conductance (*d*) none of these

8. If the equivalent conductance of a certain solution of acetic acid is 39.07 ohm^{-1} cm^2 $eqvt.^{-1}$. If λ_α of CH_3COOH is 390.7, the degree of dissociation of acetic acid is

(*a*) 0.1 (*b*) 0.2 (*c*) 0.5 (*d*) 0.75

9. Kohlarausch's law can be expressed as

(*a*) $\lambda_\infty = \lambda_a - \lambda_c$ (*b*) $\lambda_\infty = \lambda_c - \lambda_a$ (*c*) $\lambda_\infty = \lambda_a + \lambda_c$ (*d*) $\lambda_\infty = \lambda_a \times \lambda_c$

10. If λ_α is the equivalent conductance at infinite dilution and λ_v the equivalent conductance of the electrolyte at the dilution v, the degree of dissociation is given by

(*a*) $\alpha = \dfrac{\lambda_\infty}{\lambda_v}$ (*b*) $\alpha = \dfrac{\lambda_v}{\lambda_\infty}$ (*c*) $\alpha = \lambda_\infty - \lambda_v$ (*d*) $\alpha = \lambda_v - \lambda_\infty$

11. During titration of a weak acid against a weak base, there is a sharp increase in ______ at the end point

(*a*) conductivity (*b*) equivalent conductance

(*c*) specific conductance (*d*) none of these

12. The equivalent conductance at infinite dilution of NaCl, HCl and CH_3COONa at 25°C are 126.0, 426.0 and 91.0 ohm^{-1} cm^2 respectively. The equivalent conductance of acetic acid at infinite dilution at 25°C will be

(*a*) 643.0 (*b*) 517.0 (*c*) 217.0 (*d*) 391.0

13. If λ_∞ and λ_v are the equivalent conductances at infinite dilution and at *V* dilution, the degree of dissociation, α, is given by

(*a*) $\alpha = \dfrac{\lambda_\infty}{\lambda_v}$ (*b*) $\alpha = \dfrac{\lambda_\infty}{\lambda_v^2}$ (*c*) $\alpha = \dfrac{\lambda_v}{\lambda_\infty}$ (*d*) none of these

14. The failure of Ostwald's dilution law in case of strong electrolytes is due to

(*a*) strong electrolytes are almost completely dissociated at all dilutions and λ_v/λ_α does not give accurate value of α

(*b*) the law of mass action in its simple form can not be applied when the concentration of ions is very high

(*c*) the ions get hydrolysed and affect the concentration terms

(*d*) all of the above

15. The effect that tends to retard the mobilities of ions in solution is

(*a*) asymmetry effect (*b*) relaxation effect

(*c*) electrophoretetic effect (*d*) all of these

16. Which of the following statement is true?

(*a*) Ostwald's dilution law holds good only for strong electrolytes and fails completely when applied to weak electrolytes

(*b*) Ostwald's dilution law holds good only for weak electrolytes and fails completely when applied to strong electrolytes

(*c*) Ostwald's dilution law holds good for both weak and strong electrolytes

(*d*) Ostwald's dilution law does not hold good for both weak and strong electrolytes

17. Which of the following postulates of Debye-Huckel thory is/are true?

(*a*) the strong electrolyte is completely ionised at all dilutions

(*b*) the oppositely charged ions are completely distributed in the solution but the cations tend to be found in the vicinity of anions and vice-versa

(*c*) decrease in equivalent conductance with increase in concentration is due to fall in mobilities of ions due to inter-ionic effect

(*d*) all of the above

ANSWERS

1. (*c*)	**2.** (*c*)	**3.** (*b*)	**4.** (*d*)	**5.** (*c*)	**6.** (*d*)
7. (*c*)	**8.** (*a*)	**9.** (*c*)	**10.** (*b*)	**11.** (*a*)	**12.** (*d*)
13. (*c*)	**14.** (*d*)	**15.** (*d*)	**16.** (*b*)	**17.** (*d*)	

Short Answer Questions

1. How are conductivity, equivalent conductivity and molar conductivity of an electrolytic solution related?

2. How do specific conductivity and equivalent conductivity vary with dilution and why ?

3. Give three postulates of Arrhenius theory of ionization.

4. State and explain Ostwald dilution law.

5. Explain Relaxation effect as used in Debye-Huckel theory.

6. What is transport number ? How is it related to velocity of the ions ?

7. Write Debye-Huckel-Onsagar equation in complete form explaining the meaning of different quantities.

8. What are the limitations of Arrhenius theory of ionization ?

9. Describe an experiment to show migration of ions towards the electodes on passing electricity.

10. Derive the relationship between equivalent conductance and specific conductance.

11. Give three points of difference between metallic conductor and electrolytic conductor.

12. Show graphically how the molar conductivity of a strong electrolyte and a weak electrolyte vary with concentration.

13. Draw a diagram to show electrophoretic effect as applied to Debye-Huckel theory.

14. Draw the titration curves obtained when

(*i*) HCl is titrated against NaOH

(*ii*) CH_3COOH is titrated against NH_4OH

15. What is meant by degree of dissociation of a weak acid ? How can we obtain it from conductance measurement ?

General Questions

1. What are the postulates of Arrhenius theory of ionization ? How does this theory explain the following ?

(*i*) Anomalous results about colligative properties of electrolytic solutions.

(*ii*) Constancy in the value of heat of neutralization of any strong acid by any strong base.

(*iii*) Acidic, basic or neutral nature of salt solutions.

2. What are the limitations of Arrhenius theory ? How does Debye-Huckel theory explain the anomalous behaviour of strong electrolytes ?

3. What do you mean by 'Transport Number' ? Briefly describe Hittorf's method for the determination of transport number of Ag^+ and NO_3^- in $AgNO_3$ solution when

(*a*) electrodes of platinum used (*b*) electrodes of silver are used.

4. State Kohlrausch's law. How can it be used to determine

(*i*) ionic product of water ? (*ii*) solubility of a sparingly soluble salt ?

5. Define 'Transport number'. Briefly describe 'moving boundary method' for the determination of transport number. Briefly explain the factors on which transport number depends.

6. (*a*) What is the principle underlying conductometric titrations ?

(*b*) Discuss the titration curve obtained in the conductometric titration of :

(*i*) an aqueous solution of HCl with an aqueous solution of NaOH.

(*ii*) NaOH against CH_3COOH.

(*iii*) CH_3COOH with NH_4OH solution.

(*iv*) HCl solution with NH_4OH solution.

(*v*) a mixture of HCl and CH_3COOH with NaOH.

(*vi*) oxalic acid solution with NaOH solution.

(*vii*) $AgNO_3$ solution with KCl solution.

(*viii*) Sodium acetate solution with hydrochloric acid.

7. Briefly explain the terms specific conductance, equivalent conductance and molar conductance. How are they inter-related ? How are they measured experimentally ?

8. State and explain Kohlrausch's law. Why is it called law of independent migration of ions ? How it helps in the calculation of degree of dissociation of a weak electrolyte ?

9. State and explain 'Ostwald's dilution law'. Briefly describe its limitations and importance.

10. Why limiting molar conductivity of weak electrolytes cannot be found experimentally. How has the problem been solved ? State and explain the law.

11. With the help of Hittorf's theoretical device, how can you justify that fall in concentration around any electrode is proportional to the speed of the ions moving away from it ? What happens if the electrodes are attackable ?

12. (*a*) What do you understand by 'ionic mobility' ? How is it related to ionic conductance ? Derive the relationship.

(*b*) What is the effect of the following on ionic conductance ?

(*i*) Viscosity of the solvent (*ii*) Temperature (*iii*) Pressure.

13. How do specific conductance, equivalent conductance and molar conductance vary with dilution ? Discuss the variation of molar conductivity of strong and weak electrolytes with concentration.

14. Explain the type of titration curves obtained when

(*i*) $AgNO_3$ solution is titrated against KCl solution

(*ii*) CH_3COONa solution is titrated against HCl solution.

15. Explain the method of calculation of transport number by Hittorf's method if the electrodes are attackable.

16. How conductance measurements can be used to find the degree of dissociation of a weak acid ?

17. What type of titration curve is obtained when

(*i*) Acetic acid is titrated against NaOH solution ?

(*ii*) HCl is titrated against NaOH solution ?

Explain why such curves are obtained.

18. Briefly explain the factors on which transport number of an ion depends.

19. List the important advantages of conductometric titrations over ordinary titrations.

20. Taking a suitable example, explain how conductometric titrations can be used for the titration of a mixture of a weak acid and a strong acid against NaOH solution.

21. Define Kohlrausch's law. How can it be applied to determine the ionic product of water ?

22. Briefly explain 'moving boundary method' for determination of transport numbers of K^+ and Cl^- in KCl solution.

23. Derive the relationship between ionic conductance and ionic mobility.

24. Briefly explain the principle of conductometric titrations taking the example of titration of HCl versus NaOH.

25. Derive the relationship between ionic conductance and transport number of an ion.

26. Define the terms ionic conductance, ionic mobility and transport number. How are they related to each other?

27. How conductance measurements can be used to find solubility of a sparingly soluble salt?

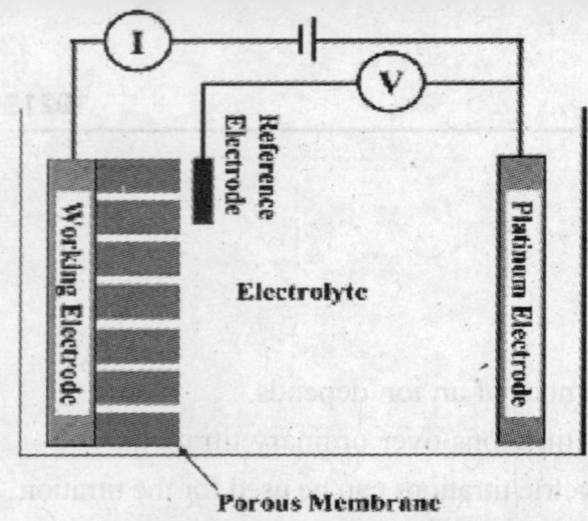

8

Electrochemistry-II

8.1. ELECTROCHEMICAL CELL

A device used to convert chemical energy of a reaction into electrical energy is called an electrochemical cell or Galvanic cell.

The following example explains the working of an electrochemical cell.

Lurgi Galvani Inventor of galvanic cell

Reaction between Zn and $CuSO_4$

The reaction that takes place between the two is

$$Zn + CuSO_4 \longrightarrow ZnSO_4 + Cu\downarrow$$

It may also be written as

$$Zn + Cu^{2+} \longrightarrow Zn^{2+} + Cu$$

The above reduction oxidation (redox) reaction can be split into two half-reactions as under

$$Zn \longrightarrow Zn^{2+} + 2e^-$$

(*Oxidation half reaction*)

$$Cu^{2+} + 2e^- \longrightarrow Cu \quad (\textit{Reduction half reaction})$$

The first reaction is called *oxidation half reaction* whereas the second reaction is called *reduction half reaction.* The addition of the two half reactions gives the *overall reaction.*

If a redox reaction is allowed to take place in such a way that oxidation half reaction takes place in one beaker and the reduction half reaction in the other breaker, the electrons given out by the former will be taken up by the latter and a current will flow.

A zinc rod is placed in zinc sulphate solution in a beaker. A copper rod is placed in copper sulphate solution in another beaker. The two rods are connected by a wire and the two solutions are connected by a *salt-bridge* (Fig. 8.1). A salt-bridge is a U-shaped tube containing concentrated solution of an electrolyte like KCl or K_2SO_4 . The function of the salt-bridge is to allow the movement of ions from one solution to the other, without mixing of the two solutions. Whereas electrons flow in the outer circuit in the wire, the inner circuit is completed by the flow of ions from one solution to the other, through the salt-bridge.

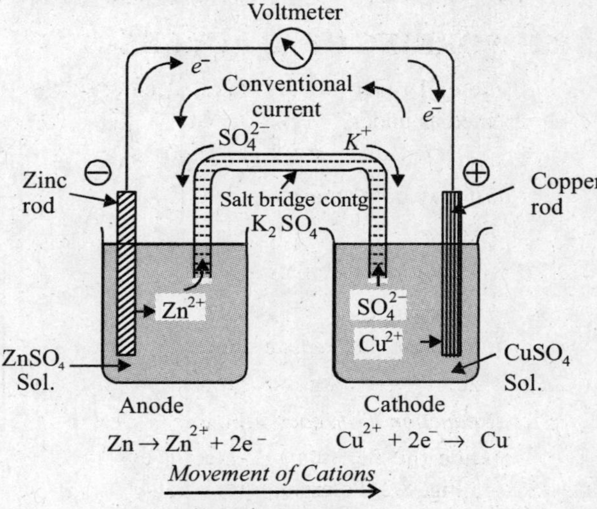

Fig. 8.1. Electrochemical cell based on the redox reaction $Zn + CuSO_4 \longrightarrow ZnSO_4 + Cu$

Some important features of the above cell are as follows:

(*i*) The zinc electrode at which *oxidation* takes place is called the *anode.* The copper electrode at which the *reduction takes place* is called the *cathode.*

(*ii*) Since electrons are produced at the zinc electrode, it is named as the *negative pole.* The other electrode *i.e.*, the copper electrode acts as the *positive pole.*

(*iii*) The electrons flow from the negative pole to the positive pole in the external circuit. Conventionally the current is said to flow in the opposite direction.

(*iv*) The oxidation of zinc into zinc ions produces excess of Zn^{2+} ions in the left beaker. Similarly, the reduction of copper ions to copper atoms leaves the excess SO_4^{2-} ions in the solution in the right beaker. To maintain electrical neutrality of the solutions in the two beakers, the cations and anions move through the salt bridge. This also completes the inner circuit.

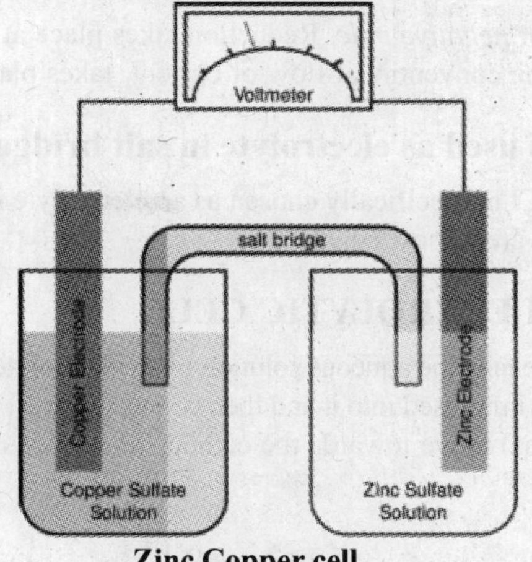

Zinc Copper cell

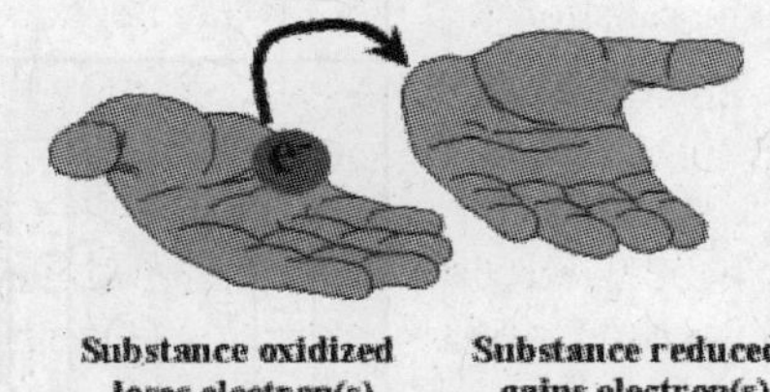

Oxidation and reduction, go hand in hand

(*v*) As copper from copper sulphate solution is deposited on the copper electrode and sulphate ions migrate to the other side, the concentration of copper sulphate solution decreases as the cell operates. Consequently the current falls with the passage of time.

(*vi*) The weight of copper rod will increase while that of zinc rod will decrease as the cell works.

8.2. WORKING OF Ag – Cu CELL

The cell reactions that take place are represented as under.

$$Cu + 2AgNO_3 \longrightarrow Cu(NO_3)_2 + 2Ag$$

or it may be written as

$$Cu + 2Ag^+ \longrightarrow Cu^{2+} + 2Ag$$

The two half reactions are

$$Cu \longrightarrow Cu^{2+} + 2e^-$$

(*Oxidation half reaction*)

$$2Ag^+ + 2e^- \longrightarrow Ag$$

(*Reduction half reaction*)

Hence the cell may be set up as shown in Fig. 8.2. Potassium nitrate may be taken as electrolyte in the salt bridge.

Here oxidation takes place at the copper electrode, therefore it acts as

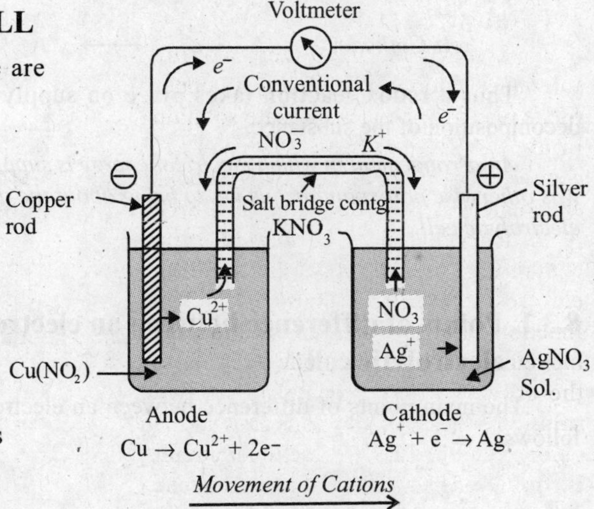

Fig. 8.2. Electrochemical cell based on the reaction $Cu + 2\,Ag\,NO_3 \longrightarrow Cu(NO_3)_2 + 2Ag$

anode or *negative pole.* Reduction takes place at the silver electrode, it acts as *cathode* or *positive pole.* The conventional flow of current takes place from silver to copper.

KCl is used as electrolyte in salt bridge

KCl is specifically chosen as an electrolyte in salt bridge because the transport numbers of K^+ and Cl^- are almost equal.

8.3. ELECTROLYTIC CELL

We take the aqueous solution of an electrolyte or the molten electrolyte in a vessel and two metal rods are immersed into it and then connected to a battery as shown in Fig. 8.3. We observe that cations (+ve ions) move towards the cathode and anions (–ve ions) towards the anode.

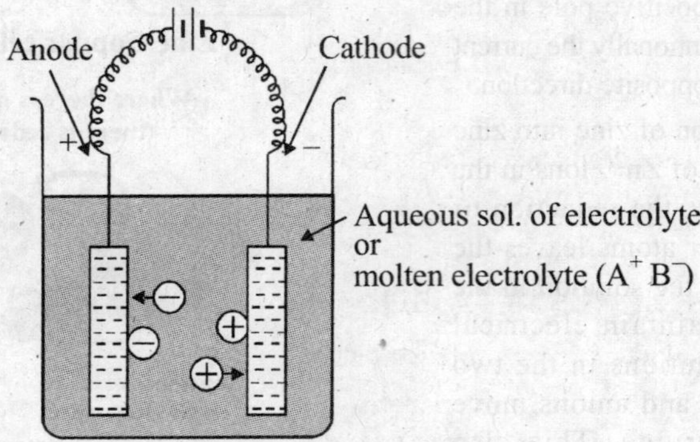

Fig. 8.3. Electrolytic cell.

The anions on reaching the anode, lose electrons. The cations on reaching the cathode gain electrons. Thus

At Anode : $B^- \longrightarrow B + e^-$ *(Oxidation)*

At Cathode: $A^+ + e^- \longrightarrow A$ *(Reduction)*

Thus a redox reaction takes place on supplying electrical energy resulting into the decomposition of the substance.

An arrangement in which electrical energy is supplied to bring about the redox reaction which was otherwise non-spontaneous i.e. to bring about the decomposition of the electrolyte is called an electrolytic cell.

8.3.1. Points of difference between an electrochemical cell and an electrolytic cell

The main points of difference between an electrochemical cell and electrolytic cell are as follows.

Electrochemical cell	Electrolytic cell
1. This device converts chemical energy into electrical energy *i.e.* electrical energy is produced as a result of the redox reaction taking place.	1. This device converts electrical energy into chemical energy *i.e.* electrical energy is supplied to the electrolytic solution to bring about the redox reaction.
2. It is based upon redox reaction which is spontaneous.	2. The redox reaction is non-spontaneous in this case
3. The two electrodes may be used in solutions of the same or different electrolytes.	3. The two electrodes are used in the same electrolytic solution.
4. The electrode on which oxidation takes place is called the anode and the electrode on which reduction takes place is called the cathode.	4. The electrode which is connected to the negative terminal of the battery is called cathode, the cations migrate to it which gain electrons and hence reduction takes place here. The other electrode is called the anode.
5. A salt bridge or a porous pot is required in this cell	5. No salt bridge is required in this case.

8.4. SCHEMATIC REPRESENTATION OF AN ELECTROCHEMICAL CELL

An electrochemical cell is represented as under for the Daniel cell

$$Zn \mid Zn^{2+} (C_1) \parallel Cu^{2+} (C_2) \mid Cu$$

Conventionally, the electrode on which oxidation takes place is written on the left hand side and the other electrode on which reduction takes place is written on the right hand side. The electrode on the left hand side is written by writing the symbol of the metal (or the gas) first followed by the symbol of the ion with its concentration in brackets. The electrode on the right hand side is written by first writing the ion alongwith its concentration in brackets followed by the symbol of the metal (or the gas). Single vertical lines show the phase boundaries of the electrodes and the double line represents the salt bridge. C_1 and C_2 represent the concentration of Zn^{2+} ions and Cu^{2+} ions respectively.

In a similar manner, Cu–$AgNO_3$ cell may be represented as

$Cu \mid Cu^{2+} (C_1) \parallel Ag^{+} (C_2) \mid Ag$

8.5. ELECTRODE POTENTIAL

The tendency of an electrode to lose or gain electrons when it is in contact with its own ions in the solution is called its **electrode potential**.

The electrode potential is further termed as **oxidation potential** if oxidation takes place at the electrode (with respect to normal hydrogen electrode) and is called **reduction potential** if reduction takes place at the electrode (with respect to normal hydrogen electrode). If in the half cell, the metal rod is suspended in a solution of unit activity and the temperature is kept at 298 K the electrode potential is called **standard electrode potential**. It is represented usually by $E°$. If a gas is involved, the standard conditions chosen are 1 atmosphere pressure.

8.6. EMF OF A CELL

The force which causes the flow of electrons from one electrode to the other and thus results in the flow of current is called electromotive force, or EMF, and measured in volts. This force arises because of the different potentials possessed by the two electrodes. Thus *electromotive force may also be defined as the difference of potential which causes the current to flow from an electrode at higher potential to an electrode at a lower potential.* If all the reactants and products are present in their standard state at 25°C, the EMF is called **standard EMF.**

Measurement

Potentiometric measurement method. This method is based upon *compensation principle and* is used for the accurate measurement of EMF. In this method, the unknown EMF is opposed by another known EMF until the two are equal. The general principle involved may be explained with the help of Fig. 8.4 given below:

AB is a stretched wire of uniform thickness, *S* is a standard cell whose EMF is known. It is connected to the end *A* and *B*. *X* is a cell whose EMF is to be determined. One end of the cell *X* is connected to the end *A* and the other to the sliding contact *J* through the galvanometer *G* is such a manner that it sends EMF in a direction opposite to that of the cell *S*. The sliding contact is moved along the wire *AB*. At a particular point, say *C*, no deflection is observed in the galvanometer. If E_s represents the EMF of the standard cell *S* and E_x that of the experimental cell *X*, then

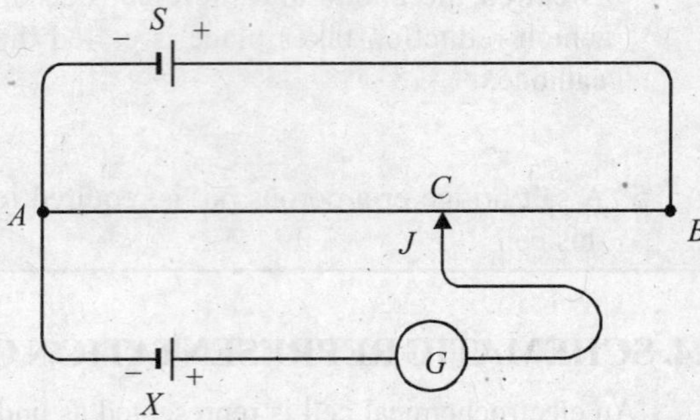

Fig. 8.4. Compensation method for measuring EMF.

$$\frac{E_x}{E_s} = \frac{\text{length } AC}{\text{length } AB}$$

or

$$E_x = \frac{\text{length } AC}{\text{length } AB} \times E_s$$

Thus knowing E_s, E_x can be calculated.

8.7. WESTON STANDARD CELL

It consists of an H-shaped glass vessel (Fig. 8.5). One of the limbs contains mercury, covered with a paste of mercurous sulphate and mercury, crystals of solid $CdSO_4.8/3H_2O$ and saturated solution of cadmium sulphate. This limb acts as the anode. The other limb which acts as the cathode contains cadmium amalgam (containing 12-14% of Cd by weight), covered with crystals of solid $CdSO_4.8/3H_2O$ and then saturated $CdSO_4$ solution. Platinum wires are sealed into the bottoms of the limbs to make contact with the active materials. The limbs are finally closed with corks and sealing wax. The purpose of placing solid crystals of $CdSO_4.8/3H_2O$ is to keep the $CdSO_4$ solution saturated at all temperatures.

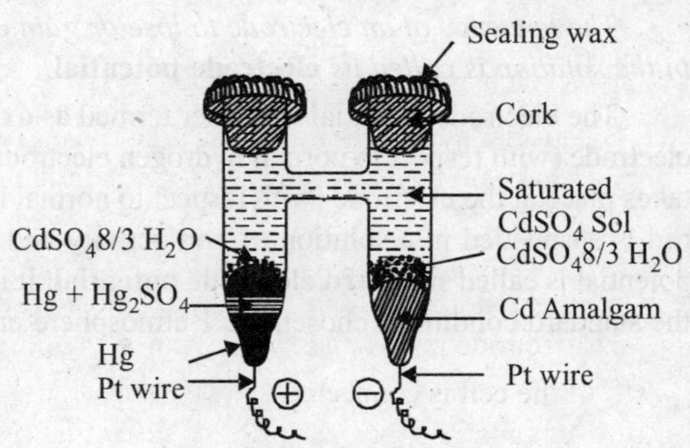

Fig. 8.5. Weston standard cell.

This cell has a constant reproducible value of e.m.f. which does not change with time. Also the temperature coefficient of e.m.f. is very low. that is variation in e.m.f per degree change of temperature is very small.

The e.m.f of the above cell is 1.0185 volt at 15°C and 1.0181 volt at 25°C.

The reaction taking place in the cell is as follows:

$$Cd + Hg_2SO_4 + aq \longrightarrow CdSO_4.8/3\ H_2O + 2Hg$$

8.8. REVERSIBLE CELL

A cell is said to be reversible if it satisfies the following three conditions:

(*i*) If an opposing EMF equal to that of the cell is applied to the cell, no current should flow in the circuit and hence no chemical reaction should take place.

(*ii*) If an opposing EMF slightly (infinitesimally) smaller than that of the cell is applied, an infinitesimally small current should flow from the cell and an infinitesimally small reaction should take place.

(*iii*) If an opposing EMF infinitesimally greater than that of the cell is applied, an infinitesimally small current should flow in the opposite direction and an infinitesimally small reaction in the reverse direction should occur.

This is illustrated in Fig. 8.6.

A well known example of a reversible cell is that of Daniell cell, consisting of a zinc electrode dipped in $ZnSO_4$ solution and a copper electrode dipped in $CuSO_4$ solution. When it is producing current, the reaction taking place in the cell is

$$Zn + Cu^{2+} \longrightarrow Zn^{2+} + Cu$$

The EMF of a reversible cell can be measured with the help of a potentiometer.

As no current flows at the time of measurement, the conditions of reversibility are satisfied in the potentiometric method for the determination of EMF.

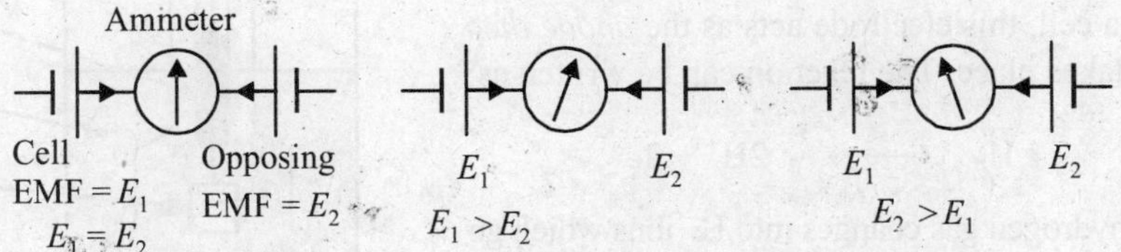

Fig. 8.6. Concept of reversible cell.

8.9. IRREVERSIBLE CELL

Consider a cell consisting of a zinc rod and a silver rod dipped in dilute sulphuric acid. When the two rods are connected with a wire, the following reaction takes place :

$$Zn + H_2SO_4 \longrightarrow ZnSO_4 + H_2$$

i.e., zinc is oxidized to Zn^{2+} ions and H^+ ions are reduced to H_2 gas. In other words, zinc dissolves at one electrode and H_2 gas is evolved at the other.

If the cell is connected to an external source of EMF whose EMF is slightly greater than that of the cell and is opposing the cell, silver dissolves at one electrode and H_2 gas is evolved at the other

$$2Ag + H_2SO_4 \longrightarrow Ag_2SO_4 + H_2$$

i.e. now oxidation takes place at the silver electrode and reduction at the other. Thus condition of reversibility is not satisfied or the chemical change is not reversed (*i.e.* $Zn + 2H^+ \longrightarrow Zn^{2+} + H_2$ does not become $Zn^{2+} + H_2 \longrightarrow Zn + 2H^+$). Instead, a new chemical reaction takes place. Hence the above cell is not categorised as reversible but is irreversible.

8.10. TYPES OF ELECTRODES

Different types of electrodes are described as under:

(*i*) **Metal–metal ion electrodes:** Such an electrode consists of a metal dipped in the solution of its ions *e.g.* zinc rod dipped in zinc sulphate solution, copper rod dipped in copper sulphate solution.

Taking the example of copper–copper ion electrode, when *oxidation* takes place at the electrode, the reaction may be represented as

$$Cu \longrightarrow Cu^{2+} + 2e^-$$

On the other hand, when *reduction* takes place at the electrode, the reaction would be represented as

$$Cu^{2+} + 2e^- \longrightarrow Cu$$

We can say that the electrode is reversible with respect to copper ions.

(*ii*) **Hydrogen electrode:** The design of the hydrogen electrode is shown in Fig. 8.7. In the standard or normal hydrogen electrode, hydrogen gas at 1 atmospheric pressure is passed into 1 M HCl acid in which the foil of platinized platinum remains immersed.

If in a cell, this electrode acts as the *anode then oxidation* takes place. The reaction can be written as:

$$H_2 \longrightarrow 2H^+ + 2e^-$$

i.e., some hydrogen gas changes into H^+ ions which go into the solution.

If this electrode acts as *cathode,* then, *reduction* takes place. The reaction can be writen as under:

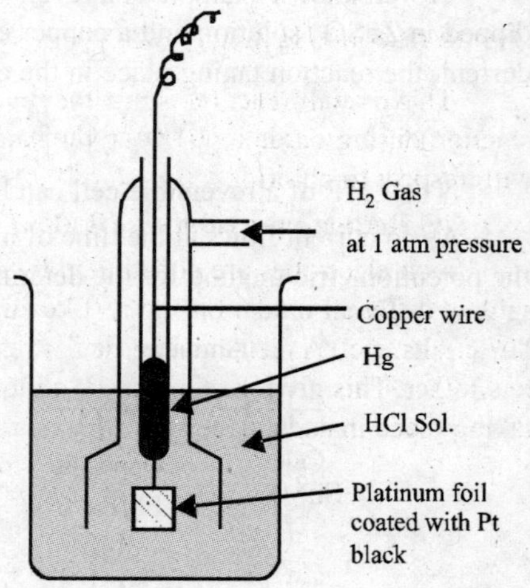

Fig. 8.7. Hydrogen electrode.

$$2H^+ + 2e^- \longrightarrow H_2$$

i.e. some H^+ ions from the solution change into H_2 gas. Thus the electrode is reversible with respect of H^+ ions. This electrode is usually represented as

$$Pt, H_2(g) \mid H^+ \text{ (conc. } = c)$$

(*iii*) **Hg – Hg_2Cl_2 – KCl electrode (Calomel electrode):** It is prepared as follows:

Pure mercury is placed at the bottom of a tube *T* (Fig. 8.8). It is covered with a paste of mercurous chloride (calomel) and then filled with either *decinormal* or *normal* or *saturated* solution of KCl from the side tube *A* such that the side tube *B* is also filled with the KCl solution. A tube *C* having a piece of platinum wire sealed at its end is fixed into the tube *T* such that the platinum wire remains dipped into the mercury. A little mercury is placed in the tube *C* and a copper wire put into it to make an electrical contact.

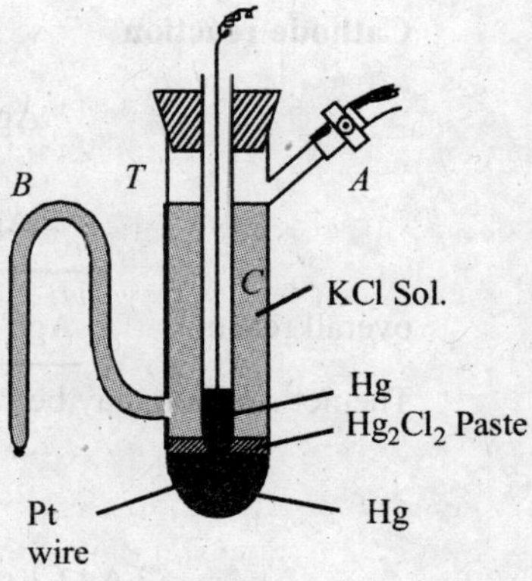

A more convenient and practical assembly of the calomel electrode used in the laboratory is shown in Fig. 8.9.

Oxidation occurs at this electrode through the following reactions:

$$2Hg(l) \longrightarrow Hg_2^{2+} + 2e^-$$

$$Hg_2^{2+} + 2Cl^- \longrightarrow Hg_2Cl_2(s)$$

Overall reaction $2Hg(l) + 2Cl^- \longrightarrow Hg_2Cl_2(s) + 2e^-$

Reduction occurs at this electrode as follows:

$$Hg_2Cl_2(s) \longrightarrow Hg_2^{2+} + 2Cl^-$$

$$Hg_2^{2} + 2e^- \longrightarrow 2Hg(l)$$

Overall reaction $Hg_2Cl_2(s) + 2e^- \longrightarrow 2Hg(l) + 2Cl^-$

This overall reaction is just the reverse of the overall reaction during oxidation. Hence the calomel is reversible with respect to chloride ions.

B T A C KCl Sol. Hg Hg_2Cl_2 Paste Pt wire Hg

Fig. 8.8. Calomel electrode.

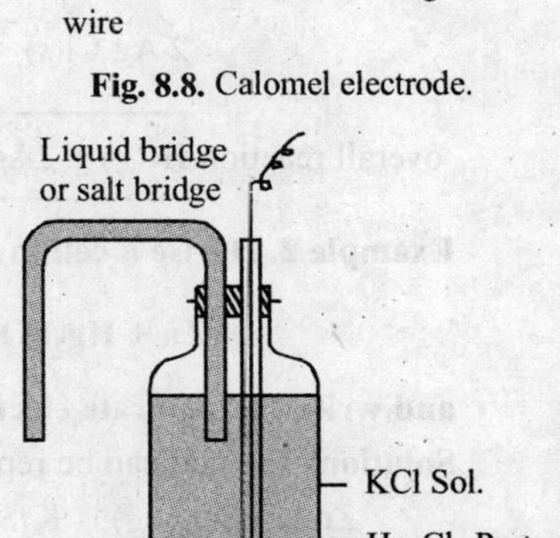

Fig. 8.9. Another form of calomel electrode.

(*iv*) **Reduction-oxidation (Redox) electrodes**

Such electrodes are constituted by dipping a platinum wire in a mixture of solution of metal salts in different oxidation states (like cuprous and cupric salts, aurous and auric salts, ferrous and ferric salts, etc.) A metal in a particular reaction has a tendency to transform from one oxidation state to another. This gives rise to emf. Platinum serves the purpose of electrical conductor. The reaction taking place in such electrodes may be representd as:

$$Sn^{2+} \rightleftharpoons Sn^{4+} + 2e^-$$

$$Cu^+ \rightleftharpoons Cu^{2+} + e$$

Example 1. Write down the reactions taking place separately at the two electrodes and the complete cell reaction

Pt, $H_2(g)$ | HCl (*aq*) || KCl (*aq*), AgCl(*s*) | Ag

Solution. According to the conventions, anode is written on the L.H.S. and cathode is written on the R.H.S. Hence in this case, hydrogen electrode is anode *i.e.* oxidation takes place here.

Anode reaction (Hydrogen electrode)

$$H_2(g) \longrightarrow 2H^+(aq) + 2e^-$$

The other electrode (R.H.S is metal–metal insoluble salt–salt anion electrode. Reduction takes place here.

Cathode reaction

$$AgCl(s) \longrightarrow Ag^+ + Cl^-$$

$$Ag^+ + e \longrightarrow Ag(s)$$

overall reaction $AgCl(s) + e^- \longrightarrow Ag(s) + Cl^-$

The net reaction may be written as under

$$H_2(g) \longrightarrow 2H^+ + 2e$$

$$2\,AgCl(s) + 2e \longrightarrow 2Ag(s) + 2Cl^-$$

overall reaction $H_2(g) + 2AgCl(s) \longrightarrow 2H^+ + 2Ag\,(s) + 2Cl^-$

Example 2. Devise a cell in which the overall reaction is

$$\mathbf{Zn + Hg_2SO_4(s) \longrightarrow ZnSO_4 + 2Hg(l)}$$

and write the separate electrode reactions.

Solution. The cell can be represented as under

$$Zn \mid ZnSO_4(ag) \parallel K_2SO_4(aq), Hg_2SO_4(s) \mid Hg$$

Thus the anode (L.H.S.) consists of zinc electrode and the cathode (R.H.S.) consists of calomel electrode. The electrode reactions are as under

Anode $Zn(s) \longrightarrow Zn^{2+}(aq) + 2e$ ...(*i*)

Cathode $Hg_2SO_4(s) \longrightarrow Hg_2^{2+} + SO_4^{2-}$

$Hg_2^{2+} + 2e \longrightarrow 2Hg\,(l)$

$Hg_2SO_4(s) + 2e \longrightarrow 2Hg(l) + SO_4^{-2}$ ...(*ii*)

The overall cell reaction can be obtained by adding the two electrode reactions (*i*) and (*ii*).

overall reaction : $Zn(s) + HgSO_4(s) \longrightarrow Zn^{2+}(aq) + 2Hg(l) + SO_4^{2-}(aq)$

8.11. RELATIONSHIP BETWEEN ELECTRICAL ENERGY AND CHEMICAL ENERGY

The electrical energy produced in a cell is defined as the product of the quantity of electricity that passes through the cell and the EMF of the cell i.e.

$$\text{Electrical energy} = \text{Quantity of electricity} \times \text{EMF}$$

For every one mole of electrons transferred in the cell reaction (*i.e.*, for every one gram equivalent of the ions deposited), the quantity of electricity that flows through the cell is one Faraday (1F = 96500 Coulombs). Thus if n moles of electrons are transferred in any cell reaction (*e.g.* for the reaction $Zn + Cu^{2+} \longrightarrow Zn^{2+} + Cu$, $n = 2$), the quantity of electricity passing through the cell will be nF Faradays. If E is the EMF of the cell, then

$$\text{Electrical energy} = nFE$$

For a large number of reactions it was noticed that the electrical energy produced was not equal to the chemical energy. It was suggested by Gibbs (1875) and Helmholtz (1882) that the electrical energy produced in a reversible cell is equal to the decrease in free energy accompanying the cell reaction. Thus if ΔG represents the change in free energy, we must have

$$-\Delta G = nFE \qquad \text{... }(i)$$

According to the Gibbs-Helmholtz equation, we have

$$\Delta G = \Delta H + T\left(\frac{\partial(\Delta G)}{\partial T}\right)_P \qquad \text{... }(ii)$$

Substituting the value of ΔG from equation (*i*) in equation (*ii*), we get

$$-nEF = \Delta H + T\left(\frac{\partial(-nFE)}{\partial T}\right)$$

$$nEF = -\Delta H + TnF\left(\frac{\partial E}{\partial T}\right)_P \qquad \text{... }(iii)$$

or

$$E = -\frac{\Delta H}{nF} + T\left(\frac{\partial E}{\partial T}\right)_P \qquad \text{... }(iv)$$

From equation (*iii*), it follows that whether electrical energy (*nFE*) produced is equal to, less than or greater than the enthalpy change (ΔH), it depends upon the sign of $\partial E/\partial T$ *i.e.*, the *temperature coefficient* of the EMF. Thus

(*i*) If $\partial E/\partial T$ is extremely small (almost equal to zero), as is the case in Daniell cell, the electrical energy produced is equal to the enthalpy change.

(*ii*) If $\partial E/\partial T$ is positive, the electrical energy produced is greater than the heat of reaction at constant pressure, the additional energy being absorbed from the surroundings.

(*iii*) If $\partial E/\partial T$ is negative, the electrical energy produced is less than the heat of reaction at constant pressure, the additional energy being absorbed by the surroundings.

8.12. EXPRESSIONS FOR ΔG, ΔH AND ΔS IN TERMS OF TEMPERATURE COEFFICIENT OF THE EMF

ΔG is related to EMF according to the expression

$$\Delta G = -nFE \qquad \text{... }(i)$$

Thus knowing the EMF, *E*, ΔG *can be calculated.*

Equation (*iii*) of Art. 8.11. can be rearranged as

$$\Delta H = nF\left[T\left(\frac{\partial E}{\partial T}\right)_P - E\right] \qquad \text{... }(ii)$$

Thus knowing the EMF, *E* and the temperature coefficient of the EMF, $\left(\frac{\partial E}{\partial T}\right)_P$, the heat of reaction, ΔH can be calculated.

Having known the values of ΔG and ΔH, the entropy change (ΔS) accompanying the reaction can be calculated using the relationship

$$\Delta G = \Delta H - T\Delta S \qquad \text{... }(iii)$$

Example. The EMF of the cell

$$\mathbf{Cd / CdCl_2 . 2\frac{1}{2} H_2O \text{ (saturated)} \,||\, AgCl(s) \,|\, Ag}$$

is found to be 0.6753 volt at 25°C and 0.6915 volt at 0°C. Calculate ΔG, ΔH and ΔS of the cell reaction at 25°C.

Solution. The cell reaction may be written as:

$$\text{Cd} + 2\text{AgCl} + aq \longrightarrow \text{CdCl}_2.2\frac{1}{2}\text{H}_2\text{O (sat.)} + 2\text{Ag}$$

or $$\text{Cd} + 2\text{Ag}^+ \longrightarrow \text{Cd}^{2+} + 2\text{Ag}$$

$\therefore$ $$n = 2$$

(i) Calculation of free energy change (ΔG)

$$\Delta G = -nFE$$
$$= -2 \times 96500 \times 0.6753$$
$$= -130333 \text{ joules}$$
$$= -130.33 \text{ kJ mol}^{-1}$$

(ii) Calculation of enthalpy change (ΔH)

At atmospheric pressure, the temperature coefficient of EMF in the range 0° – 25°C will be

$$\left(\frac{\partial E}{\partial T}\right)_P = \frac{E_2 - E_1}{T_2 - T_1} = \frac{0.6915 - 0.6753}{0 - 25}$$

$$= -\frac{0.0162}{25} = -0.00065 \text{ volt per degree}$$

Using the following relationship to calculate ΔH

$$E = -\frac{\Delta H}{nF} + T\left(\frac{\partial E}{\partial T}\right)_P$$

$$\Delta H = -nFE + TnF\left(\frac{\partial E}{\partial T}\right)_P$$

Using the data at 25°C, we get

$$\Delta H = -2 \times 96500 \times 0.6753 + 298 \times 2 \times 96500 \times (-0.00065)$$
$$= -167717 \text{ joules}$$
$$= \mathbf{-167.72 \text{ kJ mol}^{-1}}$$

(iii) Calculation of entropy change (ΔS)

Entropy change is related to enthalpy change as under:

$$\Delta G = \Delta H - T\Delta S$$

$\therefore$ $$\Delta S = \frac{\Delta H - \Delta G}{T}$$

$$= \frac{\mathbf{-167717 - (-130333)}}{\mathbf{298}} = \mathbf{-125.4 \text{ JK}^{-1} \text{ mol}^{-1}}$$

The negative signs with ΔG, ΔH and ΔS indicate the *decrease* in these quantities.

8.13. MEASUREMENT OF ELECTRODE POTENTIAL OF (*I*) ZINC ELECTRODE (*II*) COPPER ELECTRODE

(*i*) **Determination of standard electrode potential of Zn/Zn^{2+} electrode:** It is not possible to determine the absolute value of potential of a single electrode. It can only be measured relative to a reference electrode. One of the reference electrodes that can be used to measure the potential of an electrode is standard hydrogen electrode. This reference electrode is shown in Fig. 8.10 (R.H.S portion, H_2 gas at 1 atm pressure is bubbled through an acid solution of unit molarity (1 M) with platinum as electrical contact. The potential of this electrode is taken as zero)

Anode : $Zn \longrightarrow Zn^{2+} + 2e^-$ Cathode: $H_2 + 2e^- \longrightarrow 2H^+$

8.10 Measurement of standard electrode potential of Zn/Zn^{2+} electrode, using standard hydrogen electrode as the reference electrode.

A cell comprising of zinc electrode immersed in 1M $ZnSO_4$ solution and a standard hydrogen electrode is set up as shown in Fig. 8.10.

Standard electrode potential of hydrogen electrode is zero.

EMF of the above cell comes out to be 0.76 volt. Hence standard electrode potential (reduction) of Zn/Zn^{2+} electrode = – 0.76 volt. Further the direction of flow of current indicates that oxidation takes place at the zinc electrode.

(*ii*) **Determination of standard electrode potential of Cu/Cu^{2+} electrode:** A cell comprising of copper electrode immersed in 1 M $CuSO_4$ solution and a standard hydrogen electrode is set up as shown in Fig. 8.11.

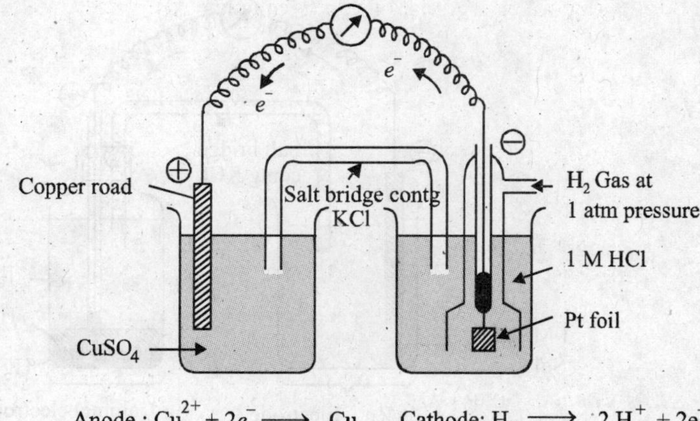

Anode : $Cu^{2+} + 2e^- \longrightarrow Cu$ Cathode: $H_2 \longrightarrow 2H^+ + 2e^-$

8.11. Measurement of standard electrode potential of Cu/Cu^{2+} electrode, using hydrogen electrode as the reference electrode.

Standard electrode potential of hydrogen electrode is zero.

EMF of the cell comes out to be 0.34 volt. The direction of flow of current indicates that reduction takes place at the copper electrode. Hence 0.34 volt is the **standard reduction potential** of Cu/Cu^{2+} electrode.

8.14. REFERENCE ELECTRODES

The potential of an unknown electrode can be obtained only by combining it with an electrode of known potential to make a cell. The EMF of this cell is determined and then the potential of the unknown electrode is calculated. *The electrode with a known potential used to determine the potential of unknown electrodes is called reference electrode.* One of the electrodes used as reference electrode is hydrogen electrode. Its electrode potential has been taken as zero.

But use of hydrogen electrode as the reference electrode is not convenient obviously because it involves the use of H_2 gas. Hence a number of other electrodes have been used as reference electrodes. The most common out of these is the **calomel electrode**. The electrode potentials of the calomel electrode at 25°C using 0.1 N KCl, 1.0 N KCl and saturated KCl solution have been determined accurately. These values are given below as *reduction potentials*

Conc. of KCl solution	Reduction Potential
0.1 N	+ 0.3338 volt
1.0 N	+ 0.2800 volt
Saturated	+ 0.2415 volt

To determine the electrode potential of any electrode, a cell is constituted using this as one of the electrodes and calomel electrode as the second electrode. The EMF of the cell is measured and then the following expression is used to calculate the required electrode potential.

EMF of the cell = Reduction potential of cathode – Reduction potential of anode.

Calomel electrode. This is another reference electrode as shown, Fig. 8.12. Standard electrode potentials of Zn/Zn^{2+} and Cu/Cu^{2+} are determined using Calomel electrode as under:

***(i)* Determination of the standard electrode potential of Zn/Zn^{2+} electrode:** A cell comprising of zinc electrode immersed in 1 M $ZnSO_4$ solution and the calomel electrode (say, using 1N KCl solution) is set up as shown in Fig. 8.12.

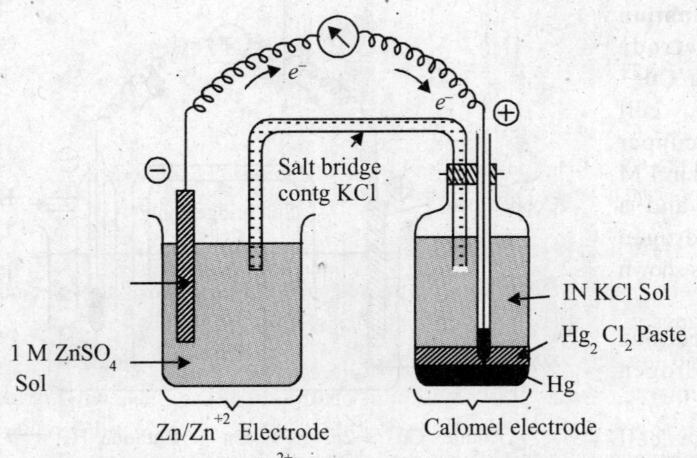

Anode : Zn $\longrightarrow$ Zn^{2+} + 2e^- Cathode: Hg_2Cl_2 + 2e $\longrightarrow$ 2Hg + Cl

Fig. 8.12. Measurement of standard electrode potential of Zn/Zn^{2+} electrode using calomel electrode as the reference electrode.

EMF of the above cell at 25°C comes out to be 1.083 volt. The direction of flow of current shows that oxidation takes place at the zinc electrode. Thus zinc electrode is the anode and calomel is cathode.

EMF of the cell = Reduction potential of cathode (calomel) – Reduction potential of anode

1.083 = 0.280 – Reduction potential of anode (Zn electrode)

or Reduction potential of anode (Zn electrode)

= 0.280 – 1.083

= – 0.803 volt

(*ii*) **Determination of the standard electrode potential of Cu/Cu^{2+} electrode:** A cell comprising of copper electrode immersed in 1 M $CuSO_4$ solution and the calomel electrode (say using 1N KCl solution) is set up as shown in Fig. 8.13.

EMF of the above cell comes out to be 0.018 volt. The direction of flow of current shows that reduction takes place at the copper electrode. Thus copper electrode is the cathode and calomel is the anode.

EMF of the cell = Reduction potential of cathode (copper) – Reduction potential of anode (calomel)

or 0.018 = Reduction potential of copper cathode – 0.280

or Reduction potential of copper cathode = 0.018 + 0.280

= 0.298 volt

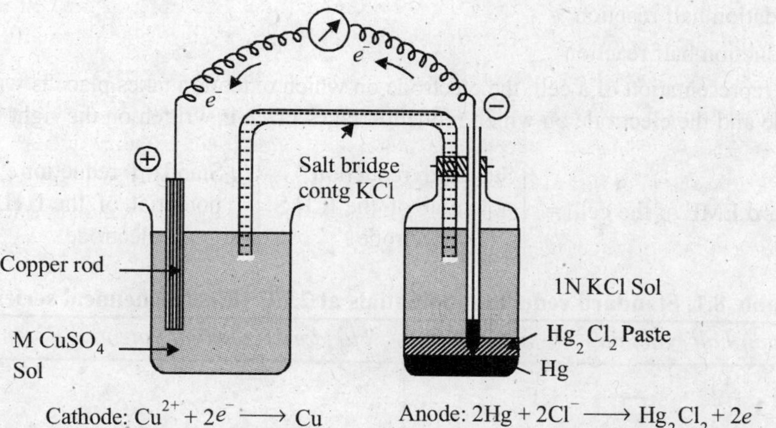

Fig. 8.13. Measurement of standard electrode potential of Cu/Cu^{2+} electrode using calomel electrode as the reference electrode.

8.15. ELECTROCHEMICAL SERIES

Various electrodes have been arranged in order of their decreasing values of standard reduction potentials. This arrangement is called electrochemical series (Table 8.1).

Applications of the Electrochemical Series

1. To predict the relative ease of oxidation or reduction of various elements: By convention positive sign is used to represent the reduction potential. This means greater the reduction potential, more easily is the substance (element or ion) reduced or in other words stronger oxidizing agent it is. F_2 has the highest reduction potential and Li^+ ion has the lowest reduction potential, this means that F_2 is reduced most easily whereas Li^+ ions are reduced with greatest difficulty. In other words, F_2 is the strongest oxidizing agent whereas Li^+ ions are the weakest oxidizing agent. Conversely, writing the reactions in the reverse order, it can be postulated that lithium (Li) will be oxidized most easily and hence is best reducing agent while fluoride ions (F^-) are oxidized with greatest difficulty and hence are weakest reducing agent.

2. To predict whether a metal reacts with acid to give hydrogen gas: In order that a metal (M) may react with an acid to give H_2 gas, the following reaction should take place:

$$M + H^+ \longrightarrow M^+ + 1/2\ H_2$$

This reaction can be split into two half reactions as

$$M \longrightarrow M^+ + e^- \qquad \textit{(Oxidation half reaction)}$$

$$H^+ + e^- \longrightarrow 1/2\ H_2 \qquad \textit{(Reduction half reaction)}$$

i.e. the metal should have the tendency to lose electrons *i.e.* it undergoes oxidation with respect to hydrogen. In other words, the metal would have a negative reduction potential. Thus all metals lying below hydrogen (having negative reduction potential) in the electrochemical series (Table 8.1) react with the acid to give hydrogen gas. Further evidently, lower the reduction potential, higher is the reactivity.

3. To calculate the standard EMF of any electrochemical cell: Electrochemical cell is based on a redox reaction which can be split into two half-reactions *viz.*,

(*i*) Oxidation half reaction

(*ii*) Reduction half reaction

In the representation of a cell, the electrode on which oxidation takes place is written on the left hand side and the electrode on which reduction takes place is written on the right hand side

$$\text{Standard EMF of the cell} = \begin{bmatrix}\text{Standard reduction}\\ \text{potential of the R.H.S.}\\ \text{electrode}\end{bmatrix} - \begin{bmatrix}\text{Standard reduction}\\ \text{potential of the L.H.S.}\\ \text{electrode}\end{bmatrix}$$

Table 8.1. Standard reduction potentials at 25°C (Electrochemical series)

Reduction half-reaction	*Standard reduction potential E° (in volts)*
$F_2 + 2e^- \longrightarrow 2F^-$	2.87
$Co^{3+} + e^- \longrightarrow Co^{2+}$	1.82
$Au^{3+} + 3e^- \longrightarrow Au$	1.50
$Cl_2 + 2e^- \longrightarrow 2Cl^-$	1.36
$Br_2 + 2e^- \longrightarrow 2Br^-$	1.09
$Ag^+ + e^- \longrightarrow Ag$	0.80
$Fe^{3+} + e^- \longrightarrow Fe^{2+}$	0.77
$I_2 + 2e^- \longrightarrow 2I^-$	0.53
$Cu^{2+} + 2e^- \longrightarrow Cu$	0.34

$AgCl + e^- \longrightarrow$	$Ag + Cl^-$	0.22
$Cu^{2+} + e^- \longrightarrow$	Cu^+	0.15
$Sn^{4+} + 2e^- \longrightarrow$	Sn^{2+}	0.13
$2H^+ + 2e^- \longrightarrow$	H_2	0.00
$Pb^{2+} + 2e^- \longrightarrow$	Pb	– 0.13
$Sn^{2+} + 2e^- \longrightarrow$	Sn	– 0.14
$Ni^{2+} + 2e^- \longrightarrow$	Ni	– 0.25
$Co^{2+} + 2e^- \longrightarrow$	Co	– 0.28
$PbSO_4 + 2e^- \longrightarrow$	$Pb + SO_4^{2-}$	– 0.31
$Cd^{2+} + 2e^- \longrightarrow$	Cd	– 0.40
$Fe^{2+} + 2e^- \longrightarrow$	Fe	– 0.44
$Cr^{3+} + 3e^- \longrightarrow$	Cr	– 0.74
$Zn^{2+} + 2e^- \longrightarrow$	Zn	– 0.76
$Mn^{2+} + 2e^- \longrightarrow$	Mn	– 1.18
$Al^{3+} + 3e^- \longrightarrow$	Al	– 1.66
$Mg^{2+} + 2e^- \longrightarrow$	Mg	– 2.37
$Na^+ + e^- \longrightarrow$	Na	– 2.71
$Ca^{2+} + 2e^- \longrightarrow$	Ca	– 2.87
$Ba^{2+} + 2e^- \longrightarrow$	Ba	– 2.90
$K^+ + 2e^- \longrightarrow$	K	– 2.93
$Li^+ + e^- \longrightarrow$	Li	– 3.05

4. To predict spontaneity of a reaction: To find out whether a reaction is feasible or not, EMF of the cell based upon the given redox reaction is calculated. If the EMF of the cell comes out to be positive, the reaction is feasible. If it comes out to be negative, the reaction is not feasible.

NUMERICAL PROBLEMS ON STANDARD EMF AND STANDARD ELECTRODE POTENTIALS

Example 1. A cell is prepared by dipping a copper rod in 1 M $CuSO_4$ solution and a nickel rod in 1 M $NiSO_4$ solution. The standard reduction potentials of copper electrode and nickel electrode are 0.34 volt and – 0.25 volt respectively.

(*a*) What will be cell reaction?

(*b*) What will be the standard EMF of the cell?

(*c*) Which electrode will be positive?

(*d*) How will the cell be represented?

(*a*) The cell reaction can be

either $Ni + CuSO_4 \longrightarrow NiSO_4 + Cu$

i.e. $Ni + Cu^{2+} \longrightarrow Ni^{2+} + Cu$

or $Cu + NiSO_4 \longrightarrow CuSO_4 + Ni$

i.e. $Cu + Ni^{2+} \longrightarrow Cu^{2+} + Ni$

The correct cell reaction will be the one which gives positive EMF.

If the first reaction is the cell reaction,

$$Ni \longrightarrow Ni^{2+} + 2e^-$$

Standard reduction potential = – 0.25 volt.

$$Cu^{2+} + 2e^- \longrightarrow Cu$$

Standard potential = + 0.34 volt

Overall reaction

$$Ni + Cu^{2+} \longrightarrow Ni^{2+} + Cu$$

Std. EMF = 0.34 – (– 0.25) volt = 0.59 volt

Thus EMF is positive for this reaction. Hence this is the correct cell reaction. (If the other reaction is checked, EMF will come out to be – 0.59 volts)

(*b*) Standard EMF of the cell as calculated above = 0.59 volt.

(*c*) Since oxidation takes place at the *nickel electrode,* therefore it acts as *anode* or since electrons are produced at this electrode it acts as a *negative pole.* Obviously, copper electrode will act as *cathode* or a *positive pole.*

(*d*) By convention, the cell will be represented as

$$Ni \mid NiSO_4\ (1M) \parallel CuSO_4(1M) \mid Cu$$

Example 2. Calculate the standard EMF of a cell which involves the following cell reaction

$$\mathbf{Zn + 2Ag^+ \longrightarrow Zn^{2+} + 2Ag}$$

Given that E^0_{Zn, Zn^+} = 0.76 volt and E^0_{Ag, Ag^+} = – 0.80 volt.

Solution. As oxidation takes place at the zinc electrode and reduction at the silver electrode, by convention, the cell may be represented as

$$Zn \mid Zn^{2+} \parallel Ag^+ \mid Ag$$

$$\text{Standard EMF of the cell } (E^0_{cell}) = \begin{bmatrix}\text{Standard reduction} \\ \text{potential of the R.H.S.} \\ \text{electrode}\end{bmatrix} - \begin{bmatrix}\text{Standard reduction} \\ \text{potential of the L.H.S.} \\ \text{electrode}\end{bmatrix}$$

As standard oxidation potentials of zinc electrode and silver electrode are given to be 0.76 volt and – 0.80 volt, their standard reduction potentials will be – 0.76 volt and + 0.80 volt respectively. Hence

$$E^0_{cell} = 0.80 - (-0.76) = 1.56$$

Example 3. Can a solution of 1 M copper sulphate be stored in a vessel made of nickel metal? Given that $E^0_{Ni, Ni^{2+}} = +0.25$ volt and $E^0_{Cu, Cu^{2+}} = -0.34$ volt.

Solution. In this problem, we want to see whether the following reaction takes place or not

$$Ni + CuSO_4 \longrightarrow NiSO_4 + Cu$$

i.e. $$Ni + Cu^{2+} \longrightarrow Ni^{2+} + Cu$$

The cell may be represented as

$$Ni \mid Ni^{2+} \parallel Cu^{2+} \mid Cu$$

Oxidation potentials are given as

$$E^0_{Ni, Ni^{2+}} = 0.25 \text{ volt and } E^0_{Cu,Cu^{2+}} = -0.34 \text{ volt}$$

Hence the reduction potentials will be

$$E^0_{Ni^{2+}, Ni} = -E^0_{Ni, Ni^{2+}} = -0.25 \text{ volt}$$

and $$E^0_{Cu^{2+},Cu} = -E^0_{Cu, Cu^{2+}} = +0.34 \text{ volt}$$

$\therefore$ $$E^0_{Cell} = E^0_{R.H.S.electrode} - E^0_{L.H.S.electrode}$$

$$= +0.34 - (-0.25) = 0.59 \text{ volt}$$

Thus EMF comes out to be positive. This implies that $CuSO_4$ reacts with nickel. Hence $CuSO_4$ cannot be stored in nickel vessel.

Example 4. Predict whether zinc and silver react with 1 N sulphuric acid to give out hydrogen gas or not. Given that the standard reduction potentials of zinc and silver are – 0.76 volt and 0.80 volt respectively.

Solution. (*a*) *To predict the reaction of zinc with sulphuric acid:* If it reacts, the following reaction should take place

$$Zn + H_2SO_4 \longrightarrow ZnSO_4 + H_2$$

i.e. $$Zn + 2H^+ \longrightarrow Zn^{2+} + H_2$$

The cell will be represented as

$$Zn \mid Zn^{2+} \parallel H^+ \mid H_2$$

$\therefore$ Standard EMF of the cell (E^0_{Cell})

$$\left[\begin{array}{l}\text{Standard reduction potential}\\ \text{of the R.H.S. electrode}\end{array}\right] - \left[\begin{array}{l}\text{Standard reduction potential}\\ \text{of the L.H.S. electrode}\end{array}\right]$$

$$= 0 - [-0.76] = +0.76 \text{ volt}$$

Thus the EMF of the cell comes out to be positive. Hence the reaction takes place.

(*b*) *To predict the reaction of silver with sulphuric acid:* Let us suppose that silver reacts with sulphuric acid to give the following reaction

$$2Ag + H_2SO_4 \longrightarrow Ag_2SO_4 + H_2$$

i.e. $$2Ag + 2H^+ \longrightarrow 2Ag^+ + H_2$$

The cell may be represented as

$$Ag \mid Ag^+ \parallel H^+ \mid H_2$$

$$E^0_{cell} = E^0_{H^+, H_2} - E^0_{Ag^+, Ag}$$

$$= 0 - 0.80 \text{ volt} = -0.80 \text{ volt.}$$

Thus EMF of the supposed cell comes out to be negative. Hence this reaction does not take place.

PROBLEMS FOR PRACTICE

1. Calculate the standard EMF of the following cell:

   $Cd \mid Cd^{2+} \parallel Ag^+ \mid Ag$

   Write the cell reaction, given that

   $E^o_{Cd^{2+}, Cd} = -0.40$ volt, $E^o_{Ag^+, Ag} = 0.80$ volt.

   [**Ans.** 1.20 volt. $Cd + 2Ag^+ \longrightarrow Cd^{2+} + Ag$]

2. Predict the reaction of 1N sulphuric acid with the follwing metals:

   (*a*) Copper (*b*) Lead (*c*) Iron

   Given: $E^\circ_{Cu^{2+},Cu} = 0.34$ volt, $E^\circ_{Pb^{2+},Pb} = -0.13$ volt and $E^\circ_{Fe^{2+},Fe} = -0.44$ volt.

   [**Ans.** (*i*) No (*ii*) Yes (*iii*) Yes]

3. The standard EMF of the cell

   $Ni \mid Ni^{2+} \parallel Cu^{2+} \mid Cu$

   is 0.59 volt. The standard electrode potential (reduction potential) of copper electrode is 0.34 volt. Calculate the standard electrode potential of nickel electrode.

   [**Ans.** $E^\circ_{Ni^{2+}, Ni} = -0.25$ volt]

4. Can a solution of 1 M $ZnSO_4$ be stored in a vessel made of copper?

   Given: $E^\circ_{Zn, Zn^{2+}} = 0.76$ volt, $E^\circ_{Cu^{2+},Cu} = 0.34$ volt. [**Ans.** Yes]

5. The EMFs of the cells

   Pt, H_2 (1 atm) | H^+ (1M) || KCl (saturated), Hg_2Cl_2 (s) | Hg, Pt and

   Zn | Zn^{++} (0.1 M) || KCl (saturated), Hg_2Cl_2 (*s*)| Hg, Pt

   are 0.242 volt and 0.548 volt respectively. What is the value of oxidation potential of Zn | Zn^{++} (0.1 M) electrode? The activity coefficient of H^+ ions in the first cell may be taken as unity. [**Ans.** 0.306 **volt**]

6. Can we stir copper sulphate solution with a silver spoon? Given standard electrode potentials of copper and silver electrodes are 0.34 volt and 0.80 volt respectively. [**Ans.** No]

$n = 2$. The overall reaction is

$$Zn + Cu^{2+} \longrightarrow Zn^{2+} + Cu$$

Hence we will have

$$E = E^0 - \frac{RT}{2F} \ln \frac{a_{Zn^{2+}} . a_{Cu}}{a_{Zn} . a_{Cu^{2+}}} \quad \text{... (vi)}$$

Putting the activities of solids equal to unity *i.e.*, $a_{Cu} = 1$ and $a_{Zn} = 1$, equation (*vi*) reduces to

$$E = E^0 - \frac{RT}{2F} \ln \frac{a_{Zn^{2+}}}{a_{Cu^{2+}}} \quad \text{... (vii)}$$

Writing concentrations in place of activities, equation (*vii*) becomes

$$E = E^0 - \frac{RT}{2F} \ln \frac{[Zn^{2+}]}{[Cu^{2+}]} \quad \text{... (viii)}$$

Substituting the values of $R = 8.314$ joule deg^{-1} mol^{-1} and $F = 96500$ coulombs and converting ln into log at 25°C, eq. (*viii*) changes to

$$E = E^0 - 0.0295 \log \frac{[Zn^{2+}]}{[Cu^{2+}]} \quad \text{... (ix)}$$

Important: If the potential is given for the reaction

$$M \longrightarrow M^{n+} + ne$$

it shows that it is oxidation potential. If the value of potential is given as $E_{M/M^{n+}}$ it again expresses oxidation potential. Reduction potential may be obtained by changing the sign.

NUMERICAL PROBLEMS ON NERNST EQUATION

Example 1. Calculate the EMF of a zinc-silver cell at 30°C when activity of Zn^{2+} ions is 0.5 and the activity of Ag^+ ions is 10.

Standard reduction potentials at 30°C are :

(*i*) Ag^+, Ag electrode = + 0.799 volt

(*ii*) Zn^{2+}, Zn electrode = – 0.760 volt

Solution. Since reduction potential of silver (0.799 volt) is greater than that of zinc (– 0.760 volt), this means that in the cell reaction, reduction takes place at the silver electrode and oxidation at the zinc electrode. Hence the cell reaction will be

$$Zn + 2Ag^+ \longrightarrow Zn^{2+} + 2Ag$$

Here $n = 2$.

The cell may be represented, as

$$Zn \mid Zn^{2+} (a = 0.5) \parallel Ag^+ (a = 10) \mid Ag$$

$\therefore$ Standard EMF of the cell, $E^0 = E^0_{Ag^+, Ag} - E^0_{Zn^{2+}, Zn}$

$$= + 0.799 - (- 0.760) \text{ volt} = 1.559 \text{ volt}$$

By Nernst equation, the EMF of the given cell is given by

$$E = E^0 - \frac{RT}{F} \ln \frac{a_{Zn^{2+}}}{a^2_{Ag^+}}$$

$$= 1.599 - \frac{8.314(30+273)2.303}{2 \times 96500} \log \frac{0.5}{(10)^2}$$

$$= 1.559 - 0.03006 \log (5 \times 10^{-3})$$

$$= 1.559 + 0.069 = \mathbf{1.628\ volt}$$

Example 2. Represent schematically the cell made up of the following half cell reactions:

$$\mathbf{Mg \longrightarrow Mg^{2+}\ (0.01\ \mathit{M}) + 2e^-,\ \mathit{E}^0 = +\ 2.34\ volts}$$

$$\mathbf{Sn^{2+}\ (0.1\ \mathit{M}) + 2e^- \longrightarrow Sn,\ \mathit{E}^0 = -\ 0.136\ volt}$$

Calculate the EMF of the above cell at 25°C.

Solution. Oxidation potential of Mg-electrode is +2.34 volt. Its reduction potential will therefore be – 2.34 volt. The potential of tin electrode given is reduction potential.

The cell will be represented as

$$Mg \mid Mg^{2+}\ (0.01\ M) \parallel Sn^{2+}\ (0.1\ M) \mid Sn$$

Standard EMF of the cell (E^0)

$$= -0.136 - (-2.34)$$

$$= 2.204 \text{ volts.}$$

The overall reaction is

$$Mg + Sn^{2+}\ (0.1\ M) \longrightarrow Mg^{2+}\ (0.01\ M) + Sn$$

Here $n = 2$

By Nernst equation, the EMF of the cell at 25°C is given by the following equation

$$E = E^0 - \frac{0.0591}{2} \log \frac{[Mg^{2+}]}{[Sn^{2+}]} \quad \text{[as [Mg] = [Sn] = 1]}$$

$$= 2.204 - \frac{0.0591}{2} \log \frac{0.01}{0.1}$$

$$= 2.204 - 0.02955 \times (-1)$$

$$= 2.23355 \text{ volt} \simeq \mathbf{2.234\ volt}$$

PROBLEMS FOR PRACTICE

1. The ΔE for the reaction

$$Fe\ (s) + Zn^{2+} \longrightarrow Zn\ (s) + Fe^{2+}$$

is – 0.0032 volt. What is the concentration of Fe^{++} reached when a piece of iron is placed in a 1 M Zn^{2+} solution?

The value of $\frac{2.303\,RT}{F}$ at 25°C is 0.059 [**Ans.** 1.283 M]

2. What will be the EMF of the cell
Zn, Zn^{2+} (a = 0.1) || Fe^{2+} (a = 0.005), Fe at 25°C?

Given : $E° = 0.323$ volts, $\frac{2.303RT}{F} = 0.059$ **[Ans.** 0.314]

3. Calculate the EMF of the cell
Pb | Pb^{2+} (a = 0.200) || Sn^{2+} (a = 0.800) | Sn

Given that $E°$ for Sn, Sn^{2+}, ($Sn \rightleftharpoons Sn^{2+} + 2e$) = + 0.140 volt and

$E°$ for Pb, Pb^{2+} ($Pb \rightleftharpoons Pb^{2+} + 2e^-$) = + 0.126 volt and log 4 = 0.60.

[Ans. 0.00373 volt]

and the reaction of the cell will be $Pb + Sn^{2+} \longrightarrow Pb^{2+} + Sn$.]

8.18. NERNST EQUATION FOR SINGLE ELECTRODE POTENTIAL

(*a*) **Metal-Metal Ion Electrode.** Take the example of zinc electrode in which zinc metal is in equilibrium with zinc ions. The electrode reaction, written as reduction reaction will be

$$Zn^{2+} + 2e^- \longrightarrow Zn \ (n = 2)$$

Applying Nernst equation, the electrode potential of the zinc electrode is given by

$$E_{Zn^{2+},Zn} = E^0_{Zn^{2+},Zn} - \frac{RT}{2F} \ln \frac{a_{Zn}}{a_{Zn^{2+}}} \quad \dots (i)$$

$$= E^0_{Zn^{2+},Zn} + \frac{RT}{2F} \ln \frac{a_{Zn^{2+}}}{a_{Zn}}$$

where $E^0_{Zn^{2+},Zn}$ represents the standard electrode potential of the zinc electrode. In the symbols, $E_{zn^{2+}}$, Zn and $E^0_{Zn^{2+},Zn}$ the ions have been written first because these represent *reduction* potentials.

Putting $a_{zn} = 1$ (because the activity of a pure solid is unity), equation (*i*) becomes

$$E_{Zn^{2+},Zn} = E^0_{Zn^{2+},Zn} + \frac{RT}{2F} \ln a_{Zn^{2+}}$$

Assuming that the solution is dilute, the activity of Zn^{+2} ions may be replaced by molar concentration.

$$E_{Zn^{2+},Zn} = E^0_{Zn^{2+},Zn} + \frac{RT}{2F} \ln [Zn^{2+}]$$

Similarly, the expression for the silver electrode may be derived. The electrode reaction, written as reduction reaction, is

$$Ag^+ + e^- \longrightarrow Ag \ (n = 1)$$

Applying Nernst equation and simplifying the result obtained, we have

$$E_{Ag^+,Ag} = E^0_{Ag^+,Ag} + \frac{RT}{F} \ln [Ag^+]$$

In general, for any metal electrode, consisting of the metal (M) in equilibrium with its ions, (M^{n+}), the electrode reaction, written as reduction reaction, is

$$M^{n+} + ne^- \longrightarrow M$$

The electrode potential will then be given by the expression

$$E_{M^{n+},M} = E^0_{M^{n+},\,M} + \frac{RT}{nF}\ln[M^{n+}]$$

Putting $R = 8{\cdot}314$ joules, $F = 96500$ coulombs, $\ln[M^{n+}] = 2.303 \log[M^{n+}]$ and $T = 25 + 273$K (assuming that the measurements are carried out at 25°C), the above equation takes the form

$$E_{M^{n+},M} = E^0_{M^{n+},M} + \frac{0.0591}{n}\log[M^{n+}]$$

This is **Nernst equation** for the calculation of electrode potential of metal–metal ion electrode.

(*b*) Hydrogen Electrode. Hydrogen electrode, consists of H_2 gas in equilibrium with H^+ ions. The electrode reaction, written as reduction reaction, is

$$H^+ + e^- \longrightarrow \frac{1}{2}H_2\ (n = 1)$$

Applying Nernst equation, the electrode potential of the hydrogen electrode is given by

$$E_{H^+,H_2} = E^0_{H^+,H_2} - \frac{RT}{F}\ln\frac{(a_{H_2})^{1/2}}{a_{H^+}} \quad \text{... } (i)$$

If H_2 gas is at 1 atmospheric pressure, $a_{H_2} = 1$

Equation (*i*) above takes the form

$$\therefore \quad E_{H^+,H_2} = E^0_{H^+,H_2} - \frac{RT}{F}\ln\frac{1}{a_{H^+}}$$

$$E_{H^+,H_2} = E^0_{H^+,H_2} + \frac{RT}{F}\ln a_{H^+}$$

Activity of H^+ ions may be replaced by its molar concentration, Thus, we have

$$E_{H^+,H_2} = E^0_{H^+,H_2} + \frac{RT}{F}\ln[H^+]$$

Since the standard electrode potential of hydrogen electrode is taken as zero

$$\therefore \quad E^0_{H^+,H_2} = 0$$

Hence $$E_{H^+,H_2} = \frac{RT}{F}\ln[H^+] = 0.0591\log[H^+] \text{ at } 25°C$$

(*c*) Calomel electrode. The electrode reaction, written as reduction reaction, is

$$Hg_2Cl_2(s) + 2e^- \longrightarrow 2Hg(l) + 2Cl^-\ (n = 2)$$

By Nernst equation, the electrode potential of this electrode is given by

$$E_{cal} = E^0_{cal} - \frac{RT}{2F}\ln\frac{a^2_{Cl^-} \times a^2_{Hg}}{a_{Hg_2Cl_2}} \quad \text{... } (i)$$

As Hg is a pure liquid and Hg_2Cl_2 is a pure solid, therefore $a_{Hg} = 1$ and $a_{Hg_2Cl_2} = 1$. Hence eq. (*i*) takes the form

$$E_{cal} = E^o_{cal} - \frac{RT}{2F} \ln a^2_{Cl^-}$$

$$= E^o_{cal} - \frac{RT}{F} \ln a_{Cl^-}$$

Taking molar concentration in place of activity, we have

$$E_{cal} = E^o_{cal} - \frac{RT}{F} \ln [Cl^-]$$

8.19. ELECTRODE POTENTIAL OF REDUCTION-OXIDATION ELECTRODE

Consider the electrode prepared by putting a platinum wire in a solution of a mixture of a ferrous salt and a ferric salt. The electrode reaction, expressed as reduction reaction, is

$$Fe^{3+} + e^- \longrightarrow Fe^{2+} \qquad (n = 1)$$

Applying Nernst equation, the electrode potential of this electrode is given by

$$E_{Fe^{3+}, Fe^{2+}} = E_{Fe^{3+}, Fe^{2+}} - \frac{RT}{2F} \ln \frac{a_{Fe^{2+}}}{a_{Fe^{3+}}}$$

Using molar concentrations in place of activities, we have

$$E_{Fe^{3+}, Fe^{2+}} = E^o_{Fe^{3+}, Fe^{2+}} - \frac{RT}{F} \ln \frac{[Fe^{2+}]}{[Fe^{3+}]}$$

or $$\boldsymbol{E_{Fe^{3+}, Fe^{2+}} = E^o_{Fe^{3+}, Fe^{2+}} - \frac{RT}{F} \ln \frac{[Fe^{3+}]}{[Fe^{2+}]}}$$

General expression for any reversible electrode

The electrode reaction can be written in a general form as

Oxidized state + n electrons = Reduced state

Applying Nernst equation, the expression for the electrode potential will be

$$E = E^0 - \frac{RT}{nF} \ln \frac{[\text{Reduced state}]}{[\text{Oxidized state}]}$$

or $$\boldsymbol{E = E^0 + \frac{RT}{nF} \ln \frac{[\text{Oxidized state}]}{[\text{Reduced state}]}}$$

Here E represents the electrode potential at the given concentrations of the ions and $E°$ represents the standard electrode potential.

NUMERICAL PROBLEMS ON NERNST EQUATION FOR ELECTRODE POTENTIAL

Example 1. Calculate the electrode potentials (reduction potentials) of each of the following single electrodes at 25°C.

(*i*) Sn | Sn^{++} (a = 0.01) $\qquad E^0_{Sn^{++},Sn} = -0.14V$

(*ii*) Ag | AgCl(s), Cl$^-$ (a = 0.0001) $\qquad E^0_{AgCl,Cl^-} = 0.22V$

Solution. (*i*) The electrode reaction expressed as reduction reaction is

$$Sn^{++} + 2e^- \longrightarrow Sn \ (n = 2)$$

$$\therefore \quad E_{Sn^{2+},Sn} = E^0_{Sn^{2+},Sn} + \frac{RT}{2F} \ln a_{Sn^{2+}}$$

Substituting the values, we have

$$E_{Sn^{2+},Sn} = -0.14 + \frac{0.0591}{2} \log 0.01$$

$$= -0.14 - 0.0591$$

$$= -0.1991 \text{ V}$$

(*ii*) The electrode reaction is

$$AgCl\,(s) + e^- \longrightarrow Ag(s) + Cl^- \ (n = 1)$$

$$\therefore \quad E_{AgCl,Cl^-} = E^0_{AgCl,Cl^-} - \frac{RT}{F} \ln a_{Cl^-}$$

Substituting the values

$$E_{AgCl,\,Cl^-} = 0.22 - 0.0591 \log 0.0001$$

$$= 0.22 + 0.2364$$

$$= 0.4564 \text{ V}$$

Example 2. Calculate the electrode potential of a copper wire dipped in 0.1 molar copper sulphate solution at 25°C. At this temperature, the standard electrode potential of copper is 0.34 volt. (F = 96500 coulombs, R = 8.314 joules/degree/mol). Assume $CuSO_4$ to be completely ionized and take the activity of copper ions equal to the molar concentration

Solution. The electrode reaction as reduction reaction may be written as

$$Cu^{2+} + 2e^- \longrightarrow Cu$$

Here $\qquad n = 2$

By Nernst equation, the electrode potential is given by

$$E = E° + \frac{RT}{nF} \ln \frac{[Cu^{2+}]}{[Cu]}$$

The concentration of solid is taken as unity, thus [Cu] = 1

Substituting for [Cu^{2+}] and [Cu], we have

$$E = E^0 + \frac{2.303RT}{nF} \log [Cu^{2+}]$$

or $$E = 0.34 + \frac{2.303 \times 8.314 \times 298}{2 \times 96500} \log [0.1]$$

or $$E = 0.34 + (-0.0296) = \mathbf{0.3104\ volt}$$

Example 3. A zinc electrode is placed in 0.1 M solution of zinc sulphate at 25°C. If the degree of dissociation of salt at this concentration is found to be 0.95, calculate the electrode potential of the electrode at 25°C. Given that $E^0_{Zn^2, Zn} = -0.76$ volt.

Solution. As $\alpha = 0.95$, the salt is 95% dissociated. Concentration of Zn^{2+} ions is given by

$$[Zn^{+2}] = \frac{95}{100} \times 0.1 = 0.095$$

$$E = E^0 + \frac{0.0591}{2} \log 0.095$$

$$= -0.76 + \frac{0.0591}{2} \times (-1.022)$$

$$= -0.76 - 0.03 = -0.79 \text{ volt}$$

PROBLEMS FOR PRACTICE

1. A zinc electrode is placed in 0.1 M solution of zinc sulphate at 25°C. If the degree of dissociation of the salt at this concentration is found to be 0.95, calculate the electrode potential of the electrode at 25°C. Given that $E^o_{Zn^{2+}, Zn} = -0.76$ volt.

   [**Hint** As $\alpha = 0.95$ i.e., the salt is 95% dissociated, $[Zn^{2+}] = \frac{95}{100} \times 0.1 = 0.095\,M$]

   [**Ans.** – 0.79 volt]

2. Calculate the electrode potential of the following electrode at 25° C.

   $$Zn \mid Zn^{2+} (a = 0.1)$$

   Given that $E^o_{Zn, Zn^{2+}} = 0.7618$ volt.

   [**Hint.** $E^o_{Zn^{2+}, Zn} = -0.7618$ volt] [**Ans.** $E_{Zn^{2+}, Zn} = -0.7914$ volt]

8.20. CALCULATION OF EQUILIBRIUM CONSTANT (K) FROM NERNST EQUATION

Nernst equation for a reversible cell in a state of equilibrium can be represented as:

$$E_{cell} = E^o_{cell} - \frac{2.303\, RT}{nF} \log K \qquad \text{... } (i)$$

K denotes the equilibrium constant. Consider the Zn–Cu cell as represented below:

$$Zn(s) \mid ZnSO_4\,(aq) \parallel CuSO_4\,(aq) \mid Cu(s)$$

Zinc acts as anode and copper acts as cathode in the above cell. The cell reaction can be written as :

$$Zn\ (s) \longrightarrow Zn^{2+} + 2e^- \quad \text{... }(ii)$$

$$Cu^{2+} + 2e \longrightarrow Cu\ (s) \quad \text{... }(iii)$$

As the reaction takes place, the concentration of $ZnSO_4$ increases while that of $CuSO_4$ decreases with time. Thus the potential of zinc and copper electrodes keep changing in accordance with Nernst equation. At the point of equilibrium, the two electrode potentials become equal. At this stage, the emf of the cell becomes zero (Fig 8.14). We observe no flow of electrons from the anode to the cathode in the outer circuit. The cell reaction in a state of equilibrium can be written by adding equations (*ii*) and (*iii*)

$$Zn\ (s) + Cu^{2+} \rightleftharpoons Zn^{2+}\ (aq) + Cu\ (s)$$

The equilibrium constant for the above reaction can be written as :

$$K = \frac{[Zn^{2+}]\,[Cu\,(s)]}{[Cu^{2+}]\,[Zn\,(s)]}$$

Taking the values of Cu (s) and Zn (*s*) as unity each, we have

$$K = \frac{[Zn^{2+}]}{[Cu^{2+}]}$$

Substituting the value of K in (*i*) above

$$E_{cell} = E^{\circ}_{cell} - \frac{2.303\,RT}{nF}\log\frac{[Zn^{2+}]}{[Cu^{2+}]} \quad \text{... }(iv)$$

At equilibrium $E_{cell} = 0$, Eq, (*iv*) becomes

$$E^{o}_{cell} - \frac{2.303\,RT}{nF}\log\frac{[Zn^{2+}]}{[Cu^{2+}]} = 0$$

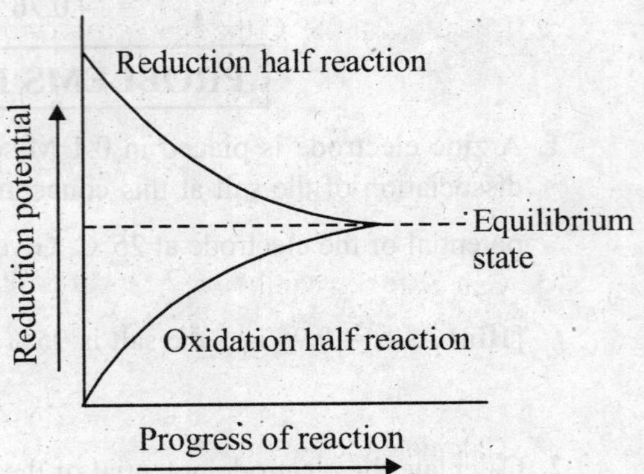

Fig. 8.14. The cell stops working after sometime when the equlbrium is reached

or
$$E^{o}_{cell} = \frac{2.303\,RT}{nF}\log\frac{[Zn^{2+}]}{[Cu^{2+}]}$$

In general, we can write

$$E^{0}_{cell} = \frac{2.303\,RT}{nF}\log K$$

Knowing the value of E^{0}_{cell}, we can calculate the value of *K*.

NUMERICAL PROBLEM ON CALCULATION OF K FROM E^{0}_{cell}

Example. Calculate the value of equilibrium constant for the reaction Zn (*s*) + Cu^{2+} (*aq*) $\rightleftharpoons$ Cu(*s*) + Zn^{2+} (*aq*) at 25° C. Standard emf of the cell is 1.10 V.

Solution. Writing the Nernst equation in terms of equilibrium constant,

$$E^0_{cell} = \frac{0.059}{n} \log K$$

For the Zn – Cu Cell, $n = 2$ and E^0_{cell} is given as 1.10 V

Substituting the values in the equation above, we have

$$1.10 = \frac{0.059}{2} \log K$$

or $$\log K = \frac{1.10 \times 2}{0.059} = 37.2 \text{ or } K = 1.6 \times 10^{37}$$

PROBLEMS FOR PRACTICE

1. Calculate the equilibrium constant for the reaction at 298 K

$$4Br^- + O_2 + 4H^+ \longrightarrow 2Br_2 + 2H_2O$$

Given that $E^0_{cell} = 0.16$ V. [**Ans.** 6.747×10^{10}]

2. Calculate at 25° C the equilibrium constant for the reaction

$$2Fe^3 + Sn^{2+} \longrightarrow 2Fe^{2+} + Sn^{4+}.$$

Given that [$E^0_{Fe^{3+}/Fe^{2+}}$] Pt = 0.771 V and [$E^0_{Sn^{4+}/Sn^{2+}}$] Pt = 0.150 V.

[**Ans.** 1.035×10^{21}]

3. Calculate the equilibrium constant for the reaction at 298 K :

$$2Fe^{3+} + 2I^- \longrightarrow 2Fe^{2+} + I_2 \text{ if } E^0_{cell} = 0.235\text{ V}.$$

[**Ans.** 8.967×10^7]

4. Calculate the equilibrium constant for the reaction

$$Cu\ (s) + 2\ Ag^+\ (aq) \rightleftharpoons Cu^{2+}\ (aq) + 2Ag\ (s),$$

Given that $E^0_{Ag^+/Ag} = 0.80$ V and $E^0_{Cu^{2+}/Cu} = 0.34$ V. [**Ans.** 3.688×10^{15}]

8.21. IRREVERSIBLE ELECTRODE PROCESSES

So far, we have studied reversible electrodes in which emf measurements are made under the conditions of "no flow of electricity" that we expect from the conditions of a reversible process. Practical applications of cells involve appreciable flow of current. Under these conditions, we come across irreversible electrode processes. Two phenomena that are associated with irreversible electrode processes in which there is a flow of electric current are discussed below.

1. Polarisation

In a reversible Galvanic process, the rate of formation and rate of deposition of ions are equal when the equilibrium has been established. There is a flow of only infinitesimally small current. In a situation when there is a flow of appreciable quantity of current, the rate of formation and rate of deposition of ions are no more equal. The disturbance of equilibrium associated with flow of current is called **polarisation.** We say that the electrode has been polarised. Polarisation causes change of concentration of ions around the electrode.

Consider an electrode M^{n+}/M. The electrode possesses certain potential depending upon the concentration of the ions, which can be obtained from Nernst equation, other factors remaining constant. We consider two situations.

(*i*) Suppose the metal is made the cathode and the potential applied is just greater than the reversible potential of the electrode. The metal ions will start getting discharged according to the reaction

$$M^{n+} + ne^- \longrightarrow M$$

This will result in decreases in the concentration of metal ions M^{n+}. The electrode potential will decreases according to the Nernst equation

$$E_{M^{n+}/M} = E^0_{M^{n+}/M} + \frac{RT}{nF}\ln[M^{n+}]$$

Likewise, if metal is made the anode and the potential applied is slightly greater than the reversible electrode potential, the metal will start changing into M^{n+} according to the equation:

$$M \longrightarrow M^{n+} + ne^-$$

The concentration of metal ions around the electrode will increase. Hence, the electrode potential will also increase.

Thus we define polarisation (or concentration polarisation) as the increase or decrease in electrode potential caused by increase or decrease in the concentration of metal ions.

2. Overpotential or Overvoltage

During the process of electrolysis, the potential at which metal ion deposits on the cathode is very near to reversible electrode potential for the metal. But when evolution of a gas like hydrogen or oxygen takes place at the anode, a potential is required which is much higher than reversible potential of hydrogen electrode or oxygen electrode

The difference between the potential of the electrode when gas evolution is actually observed and the reversible value for the same evolution is called overvoltage.

Consider the following electrochemical cell

Pt | H_2 (1 atm) | HCl (a = 1) | Pt | O_2 (1 atm)

Using the electrode potential data, its emf comes out to be 1.12 V.

Thus if an emf slightly greater than 1.12 V is applied in the opposing direction, it should reverse the reaction leading to the evolution of H_2 and O_2 gases. Actually, a voltage of 1.17 V is needed to evolve these gases.

The difference between the actual emf and the theoretical value (1.17 – 1.12 = 0.58 V) is called overvoltage or overpotential of the cell, Sometimes the term *bubble overvoltage* is also used for this phenomenon, because it refers to the stage at which the bubbles of the gas are evolved at the electrodes.

8.22. CONCENTRATION CELLS

Cells in which the EMF produced is only due to the difference in the concentrations of the electrodes or that of the solutions of the electrolytes with which they are in contact are called **concentration cells.** When the difference is in the concentration of the electrodes, the cells are called **electrode concentration cells.** When the difference is in the concentration of the electrolytes, the cells are called **electrolyte concentration cells.**

Example of electrode concentration cell:

$$H_2\left(P_{H_2} = P_1\right) | H^+ | H_2\left(P_{H_2} = P_2\right)$$

In the above cell both electrodes are hydrogen electrodes immersed in the same solution of H^+ ions but the H_2 gas passed in the two electrodes is at different pressures (Fig. 8.15)

In the above cell, there is no transference of electrolyte and hence this is called **concentration cell without transference.**

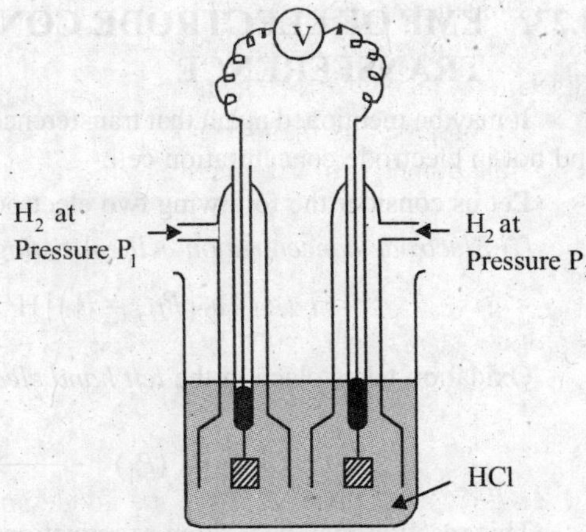

Fig. 8.15. Electrode concentration cell consisting of hydrogen electrodes.

Examples of electrolyte concentration cells :

Type A. *Those cells in which the two electrodes are same, having different concentrations of the electrolytes and connected by a salt bridge.*

For example, (*i*) H_2 (1 atm.) | H^+ (a_1) || H^+ (a_2) | H_2 | (1 atm). Both the electrodes are hydrogen electrodes and in each electrode, the H_2 gas at one atmospheric pressure is passed but the activity of the H^+ ions is different in the two electrodes. The two electrodes are connected by a salt bridge.

(*ii*) Zn | Zn^{2+} (a_1) || Zn^{2+} (a_2) | Zn

Here, both the electrodes are zinc electrodes but in one electrode, the zinc metal is in contact with Zn^{+2} ions having activity a_1, and in the other electrode, the zinc metal is in contact with Zn^{2+} ions having activity a_2. Again the two electrodes are connected by a salt bridge (Fig. 8.16).

Type A cells are concentration cells without transference.

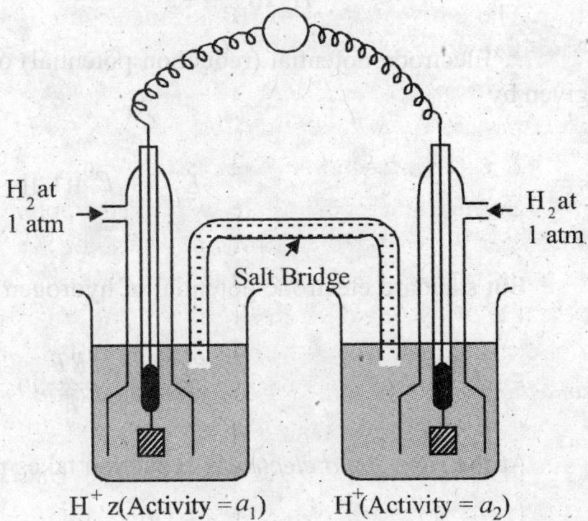

Fig. 8.16. Electrolyte concentration cells of hydrogen electrode connected by a salt bridge.

Type B. *In which the two electrodes are same having different concentrations of the electrolytes but the two solutions are directly in contact with each other.* This can be done by putting the two solutions in a tube very carefully so that there is no mixing of the two solutions. (Fig. 8.17)

In such a case, the cells are represented as

(*i*) H_2 (1 atm.) 1 HCl (a_1) ⋮ HCl(a_2) | H_2 (1 atm)

(*ii*) Zn | $ZnSO_4$ (a_1) ⋮ $ZnSO_4$ (a_2) | Zn *i.e.*, the double line is replaced by a dotted line. The junction at which the two solutions meet is called the *liquid junction.*

In this type of concentration cell, there is a direct transference of electrolyte from one solution to the other through the liquid junction. Hence these cells are called **concentration cell with transference.**

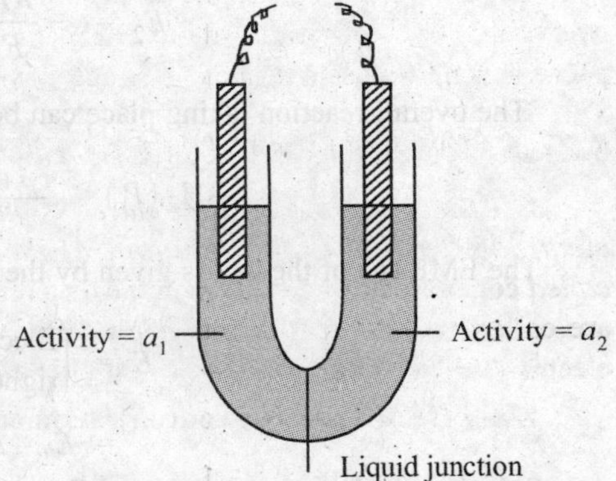

Fig. 8.17. Electrolyte concentration with a liquid junction.

8.23. EMF OF ELECTRODE CONCENTRATION CELL WITHOUT TRANSFERENCE

It may be mentioned again that transference can take place only in electrolyte concentration cell and not in electrode concentration cell.

Let us consider the following two electrode concentration cells.

(*i*) *Electrode concentration cell consisting of gas electrodes e.g.,*

$$H_2\left(P_{H_2}=P_1\right) | H^+\left(a_{H^+}\right) | H_2\left(P_{H_2}=P_2\right)$$

Oxidation takes place in the *left hand electrode i.e.,* the reaction in the left hand electrode is

$$\frac{1}{2}H_2\,(P_1) \longrightarrow H^+\,(a_{H^+})+e^- \quad \text{... (i)}$$

This reaction may be written as reduction reaction as

$$H^+(a_{H^+})+e^- \longrightarrow \frac{1}{2}H_2(P_1)(n=1) \quad \text{... (ii)}$$

∴ Electrode potential (reduction potential) of this electrode according to Nernst equation is given by

$$E_1 = E^0{}_{H^+,H_2} - \frac{RT}{F}\ln\frac{P_1^{1/2}}{a_{H^+}}$$

But standard electrode potential of hydrogen electrode, $E^0{}_{H^+,H_2}=0$

$$\therefore \qquad E_1 = -\frac{RT}{F}\ln\frac{P_1^{1/2}}{a_{H^+}} \quad \text{... (iii)}$$

At the *right hand electrode,* reduction takes place *i.e* the reaction is

$$H^+(a_{H^+})+e^- \longrightarrow \frac{1}{2}H_2\,(P_2)\,(n=1) \quad \text{... (iv)}$$

Applying Nernst equation, the electrode potential of this electrode, is given by

$$E_2 = -\frac{RT}{F}\ln\frac{P_2^{1/2}}{a_{H^+}} \quad \text{... (v)}$$

The overall reaction taking place can be obtained by adding equations (*i*) and (*iv*).

$$\frac{1}{2}H_2(P_1) \longrightarrow \frac{1}{2}H_2(P_2) \quad \text{... (vi)}$$

The EMF (E) of the cell is given by the relation

$$E = \begin{pmatrix}\text{Electrode pot. of the}\\ \text{right hand electrode}\end{pmatrix} - \begin{pmatrix}\text{Electrode pot. of the}\\ \text{left hand electrode}\end{pmatrix}$$

$$= E_2 - E_1$$

Substituting the values of E_1 and E_2 we have,

$$E = -\frac{RT}{F}\ln\frac{P_2^{1/2}}{a_{H^+}} + \frac{RT}{F}\ln\frac{P_1^{1/2}}{a_{H^+}}$$

$$= -\frac{RT}{F}\left(\ln P_2^{1/2} - \ln a_{H^+}\right) + \frac{RT}{F}\left(\ln P_1^{1/2} - \ln a_{H^+}\right)$$

$$= -\frac{RT}{F}\ln P_2^{1/2} + \frac{RT}{F}\ln P_1^{1/2}$$

$$= -\frac{RT}{2F}\ln P_2 + \frac{RT}{2F}\ln P_1$$

$$= \frac{RT}{2F}[\ln P_1 - \ln P_2]$$

$$\text{or} \quad E = \frac{RT}{2F}\ln\frac{P_1}{P_2} \quad \text{... (vii)}$$

From equation (*vii*), we find that there is no net chemical reaction. There is no transference of electrolyte. Also equation (*vii*) shows that the EMF of the cell depends only upon the pressures P_1 and P_2 and is independent of the activity of the H^+ ions.

(ii) Electrode concentration cell consisting of amalgam electrodes

$$Zn\ (Hg)(a_{Zn} = a_1)|\ Zn^{2+}\ (a_{Zn^{2+}})\ |\ Zn\ (Hg)(a_{Zn} = a_2)$$

For the left hand electrode, the reaction is

$$Zn\ (Hg)\ (a_1) \longrightarrow Zn^{2+}\ (a_{Zn^{2+}}) + 2e^- \quad \text{... (ix)}$$

Writing as reduction half reaction, we have

$$Zn^{2+}\ (a_{Zn^{2+}}) + 2e^- \longrightarrow Zn\ (Hg)(a_1)\ (n = 2).$$

By Nernst equation, electrode potential is given by

$$E_1 = E^0_{Zn^{2+},Zn} - \frac{RT}{2F}\ln\frac{a_1}{a_{Zn^{2+}}} \quad \text{... (x)}$$

For the right hand electrode, the reaction is

$$Zn^{2+}\left(a_{Zn^{2+}}\right) + 2e^- \longrightarrow Zn(Hg)(a_2) \quad \text{... (xi)}$$

By Nernst equation, electrode potential is given by

$$E_2 = E^0_{Zn^2,Zn} - \frac{RT}{2F}\ln\frac{a_2}{a_{Zn^2}} \quad \text{... (xii)}$$

Overall reaction is obtained by adding equations (*ix*) and (*xi*).

$$Zn(Hg)(a_1) \longrightarrow Zn\ (Hg)\ (a_2) \quad \text{... (xiii)}$$

EMF of the cell is obtained by the expression

$$E = E_2 - E_1$$

Substituting the values of E_1 and E_2

$$E = -\frac{RT}{2F}\ln\frac{a_2}{a_{Zn^{2+}}} + \frac{RT}{2F}\ln\frac{a_1}{a_{Zn^{2+}}}$$

or $$E = \frac{RT}{2F} \ln \frac{a_1}{a_2} \quad \text{... (xiv)}$$

From equation (*xiii*), it follows that the EMF of the cell is obtained due to the transfer of zinc from amalgam where its activity is a_1 to the amalgam where its activity is a_2. There is no transference of the electrolyte (namely Zn^{2+} ions). Also equation (*xiv*) reveals that the EMF of the cell depends only upon the activities of zinc in the two amalgams and not at all on the activity of the Zn^{2+} ions in the solution. There is one more observation, in the final expression for EMF of the cell, E^0 does not appear. This is true for all the concentration cells. We may, therefore, conclude that $E^0 = 0$ for concentration cells. Hence for a concentration cell in which each electrode reaction involves transference of n electrons, the EMF will be given by the general expression

$$E_{cell} = \frac{RT}{nF} \ln \frac{a_1}{a_2} \quad \text{... (xv)}$$

8.24. EMF OF ELECTROLYTE CONCENTRATION CELLS WITHOUT TRANSFERENCE

We shall explain it by taking three different concentration cells. Let us first take hydrogen concentration cell having different concentration of electrolytes.

(*i*) H_2 (1 atm) | H^+ (a_1) || H^+ (a_2) | H_2 (1 atm)

On the left hand electrode, by convention, oxidation takes place. Therefore, the electrode reaction is

$$\frac{1}{2} H_2 \text{ (1atm)} \longrightarrow H^+(a_1) + e^- \quad \text{... (i)}$$

This reaction may be written as reduction half reaction as

$$H^+ (a_1) + e^- \longrightarrow \frac{1}{2} H_2 \text{ (1 atm)} \; (n = 1)$$

∴ Electrode potential of the left hand electrode by Nernst equation will be

$$E_1 = E^0_{H^+,H_2} + \frac{RT}{F} \ln \frac{1}{a_1} \qquad [\because a_{H_2} = 1]$$

$$= E^0_{H^+,H_2} - \frac{RT}{F} \ln a_1$$

$$= -\frac{RT}{F} \ln a_1 \qquad \left[\because E^0_{H^+,H_2} = 0\right] \quad \text{... (ii)}$$

On the *right hand electrode,* by convention, reduction takes place. Therefore the electrode reaction is

$$H^+ (a_2) + e^- \longrightarrow \frac{1}{2} H_2 \text{ (1 atm)} \quad \text{... (iii)}$$

By Nernst equation, the EMF of this electrode is

$$E_2 = \frac{RT}{F} \ln a_2 \quad \text{... (iv)}$$

The *overall reaction* is obtained by adding the reactions (*i*) and (*iii*).

$$H^+ (a_2) \longrightarrow H^+ (a_1) \qquad \text{... }(v)$$

The EMF of the cell is given by the expression

$$E = E_2 - E_1$$

Substituting the values of E_1 and E_2, we get

$$E = \frac{RT}{F} \ln a_2 - \frac{RT}{F} \ln a_1$$

$$\text{or} \quad E = \frac{RT}{F} \ln \frac{a_2}{a_1} \qquad \text{.... }(vi)$$

From equation (*v*), it follows that for 1 faraday of electricity to be produced, there is a transfer of 1 mol of H^+ ions from solution having activity a_2 to the solution having activity a_1. The transfer takes place through the salt bridge. Secondly, from equation (*vi*), it is evident that for EMF to be positive, a_2 must be greater than a_1.

(*ii*) *For the zinc concentration cell*

$$Zn \mid Zn^{2+} (a_1) \parallel Zn^{2+} (a_2) \mid Zn$$

Following the same steps as in the hydrogen concentration cell, it can be seen that the overall reaction is

$$Zn^{2+} (a_2) \longrightarrow Zn^{2+} (a_1) \qquad \text{... }(vii)$$

and the EMF of the cell is

$$E = \frac{RT}{2F} \ln \frac{a_2}{a_1} \text{ (as } n = 2) \qquad \text{... }(viii)$$

In general, for any such cell, involving *n* faradays of electricity, the EMF is given by

$$E = \frac{RT}{nF} \ln \frac{a_2}{a_1} \qquad \text{... }(ix)$$

On replacing the activities of the ions by the products of their molalities and activity coefficients *i.e.*, putting $a_1 = m_1\gamma_1$ and $a_2 = m_2\gamma_2$, we get

$$E = \frac{RT}{nF} \ln \frac{m_2\gamma_2}{m_1\gamma_1} \qquad \text{... }(x)$$

(*iii*) *Cells obtained by combination of two chemical cells*

Consider the following chemical cell

$$H_2 \text{ (1 atm)} \mid HCl(a_1), AgCl(s) \mid Ag$$

i.e., the hydrogen electrode and the silver-silver chloride electrode are dipped in the same solution of HCl (of activity = a_1)

For this cell, the reaction may be expressed as

$$\frac{1}{2} H_2 \text{ (1 atm)} + AgCl(s) \longrightarrow Ag(s) + HCl (a_1) \qquad \text{... }(xi)$$

By Nernst equation, the EMF of this cell is given by

$$E_1 = E^0 - \frac{RT}{F} \ln a_1 \qquad \text{... }(xii)$$

$n = 1$ and the activities of solids Ag and AgCl and that of H_2 at 1 atm are taken as unity.

Now suppose we have the same cell but with a different activity of HCl acid (say a_2).

$$H_2 \text{ (1 atm.)} \mid HCl(a_2), AgCl(s) \mid Ag$$

The cell reaction will be

$$\frac{1}{2}H_2 \text{ (1atm.)} + AgCl(s) \longrightarrow Ag(s) + HCl(a_2) \qquad \text{... } (xiii)$$

By Nernst equation, the EMF of this cell is given by

$$E_2 = E^0 - \frac{RT}{F}\ln a_2 \qquad \text{... } (xiv)$$

If the above two cells are connected so as to oppose each other, we have the following cell

$$H_2 \text{ (1 atm.)} \; HCl\ (a_1),\ AgCl(s)\ Ag - Ag \mid AgCl(s),\ HCl\ (a_2) \mid H_2 \text{ (1 atm.)}$$

The *overall reaction* of this cell (a combination of two cells) will be the *difference* between equations (*xi*) and (*xiii*).

$$\frac{1}{2}H_2 \text{ (1 atm)} + AgCl(s) - \frac{1}{2}H_2 \text{ (1 atm)} - AgCl(s) = Ag\ (s) + HCl\ (a_1) - Ag(s) - HCl\ (a_2)$$

or $$HCl\ (a_2) = HCl\ (a_1) \qquad \text{... } (xv)$$

Similarly, the EMF (E) of combination will be the *difference* between equations (*xii*) and (*xiv*) *i.e.*,

$$E = E_1 - E_2 = \left(E^0 - \frac{RT}{F}\ln a_1\right) - \left(E^0 - \frac{RT}{F}\ln a_2\right)$$

$$= \frac{RT}{F}\ln\frac{a_2}{a_1} \qquad \text{... } (xvi)$$

From equation (*xv*), it is obvious that for one faraday of electricity passing through the cell, the overall change taking place is simply the transfer of 1 mol of HCl acid from the solution where its activity is a_2 to the solution where the activity is a_1. Hence it constitutes an *electrolyte concentration cell.* However, there is no direct transference of the electrolyte, HCl from one solution to the other. Hence strictly speaking, it is an *electrolyte concentration cell without transference.*

The cells of the above type are useful *in the determination of the activity coefficient of the electrolytes* involved. For the purpose, the concentration of the electrolyte in one of the cells is kept constant and the concentration of the electrolyte in the second cell is varied. For each concentration, the EMF of the cell is measured. Putting $a_1 = m_1^2\gamma_1^2$ and $a_2 = m_2^2\gamma_2^2$ in equation (*xvi*), we get

$$E = \frac{RT}{F}\ln\frac{m_2^2\gamma_2^2}{m_1^2\gamma_1^2} = \frac{2RT}{F}\ln\frac{m_2\gamma_2}{m_1\gamma_1}$$

$$= \frac{2RT}{F}\ln\frac{m_2}{m_1} + \frac{2RT}{F}\ln\frac{\gamma_2}{\gamma_1} \qquad \text{... } (xvii)$$

Knowing γ_2 corresponding to m_1 (which has been kept constant) the values of γ_2 can be calculated for different values of m_2.

8.25. EMF OF AN ELECTROLYTE CONCENTRATION CELL WITH TRANSFERENCE

Consider the cell (Fig. 8.18)

$$H_2 \text{ (1 atm)} \mid HCl(a_1) \;\vdots\; HCl\ (a_2) \mid H_2 \text{ (1 atm)}$$

Processes taking place at the electrodes: By convention, oxidation takes place on the left hand electrode. Therefore, the electrode reaction on the left hand electrode may be written as

$$\frac{1}{2}H_2 \text{ (1 atm)} \longrightarrow H^+ (a_1) + e^-$$

Reduction takes place on the right hand electrode. Therefore, the electrode reaction on the right hand electrode is

$$H^+ (a_2) + e^- \longrightarrow \frac{1}{2}H_2 \text{ (1 atm)}$$

The sum of the two electrode reactions gives

$$H^+ (a_2) \longrightarrow H^+ (a_1) \qquad \text{... } (i)$$

Processes taking place at the liquid junction: Electrons flow in the external circuit from left to right. As H^+ ions are produced in the left electrode and Cl^- ions are left in the right electrode, the inner circuit is completed by the transference of Cl^- ions from right to left and the transference of H^+ ions from left to right across the liquid junction.

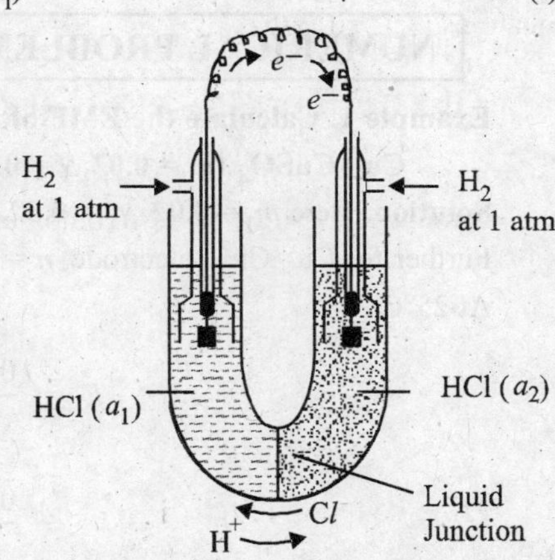

Fig. 8.18. Concentration cell with transference.

If the transport number of Cl^- ions is represented by t_-, then for one faraday of electricity passing through the cell, t_- gram equivalents of the Cl^- ions will be transferred from right to left (*i.e.* from solution of activity a_2 to the solution of activity a_1).

$$t_- Cl^- (a_2) \longrightarrow t_- Cl^- (a_1) \qquad \text{... } (ii)$$

Transport number of H^+ ions $= 1 - t_-$. Therefore during the same time, $(1 - t_-)$ gram equivalents of H^+ ions will be transferred from left to right, *i.e.*, (from solution of activity a_1 to the solution of activity a_2) *i.e.*

$$(1 - t_-)\, H^+ (a_1) \longrightarrow (1 - t_-)\, H^+ (a_2) \qquad \text{... } (iii)$$

Overall process: To calculate the *net transfer* of the material, equations (*i*), (*ii*) and (*iii*) may be added

$$H^+ (a_2) + t_-\, Cl^- (a_2) + (1 - t_-)\, H^+ (a_1) = H^+ (a_1) + t_-\, Cl^- (a_1) + (1 - t_-)\, H^+ (a_2)$$

or $$t_- H^+ (a_2) + t_- Cl^- (a_2) = t_- H^+ (a_1) + t_- Cl^- (a_1)$$

or $$t_- HCl(a_2) = t_- HCl (a_1) \qquad \text{... } (iv)$$

This equation tells that for one faraday of electricity passing through the cell, t_- gram equivalents of HCl from the solution of activity a_2 are transferred to the HCl solution of activity a_1. For the passage of one faraday of electricity through the cell, one *full gram equivalent* of HCl is transferred.

Applying Nernst equation to the process (*iv*), the EMF of the cell is given by

$$E = -\frac{RT}{F} \ln \frac{a_1^{t-}}{a_2^{t-}}$$

$$= \frac{t_- RT}{F} \ln \frac{a_2}{a_1} \qquad \text{... } (v)$$

For EMF to be positive, a_2 must be greater than a_1.

Replacing the activities of the electrolytes by the molalities and activity coefficients of the ions, we have

$$E = \frac{t_- RT}{F} \ln \frac{m_2^2 \gamma_2^2}{m_1^2 \gamma_1^2}$$

$$= \frac{2t_- RT}{F} \ln \frac{m_2 \gamma_2}{m_1 \gamma_1} \quad \text{... (vi)}$$

Thus the EMF of such a cell can be calculated if we know molalities, activity coefficients and transport number of the ions.

NUMERICAL PROBLEMS ON CONCENTRATION CELLS

Example 1. Calculate the EMF of the following concentration cell at 25°C

Cu | $CuSO_4$ (m = 0.02, γ = 0.32) || $CuSO_4$ (m = 0.2, γ = 0.11) | Cu

Solution. Here $m_1 = 0.02$, $\gamma_1 = 0.32$, $m_2 = 0.2$, $\gamma_2 = 0.11$

Further for Cu | Cu^{2+} electrode, $n = 2$.

At 25°C

$$E = \frac{0.0591}{n} \log \frac{a_2}{a_1} = \frac{0.0591}{n} \log \frac{m_2 \gamma_2}{m_1 \gamma_1}$$

$$= \frac{0.0591}{n} \log \frac{0.2 \times 0.11}{0.02 \times 0.32}$$

$$= 0.02955 \log 3.4375$$

$$= 0.02955 \times 0.5363 = \mathbf{0.0158 \ volt}$$

Example 2. What is the EMF of a concentration cell consisting of zinc electrodes, one immersed in a solution of 0.05 molar concentration and the other in a solution of 0.5 molar concentration of its ions at 25°C ? The two solutions are connected by a salt bridge.

Solution. Given that $C_2 = 0.5$ M, $C_1 = 0.05$ M.

As the cell involves Zn, Zn^{2+} ions, $n = 2$.

At 25°C,

$$E = \frac{0.0591}{n} \log \frac{C_2}{C_1}$$

$$= \frac{0.0591}{2} \log \frac{0.5}{0.05} = \mathbf{0.02955} \text{ volt}$$

PROBLEMS FOR PRACTICE

1. Calculate the emf of the following concentration cell at 25° C.

Cd | Cd SO_4 (m = 0.01, γ = 0.383) || Cd SO_4 (m =1.00, γ = 0.042) | Cd

[**Ans.** 0.0307 volt]

2. Calculate the voltage generated in a cell at 25° in which a half cell consists of a 1 M $AgNO_3$ solution and a silver wire electrode, in other half a silver wire dips into 0.01 M Ag NO_3 solution? [**Ans.** 0.1182 Volt]

3. Calculate the EMF of the concentration cell consisting of zinc electrodes, one immersed in a solution of 0.01 M concentration and other in a solution of 0.1 M concentration of its ions at 25° C. [**Ans.** 0.02955 volt]

8.26. LIQUID JUNCTION POTENTIAL

The potential set up at the junction of the two solutions because of the difference in the speeds of the ions moving across the boundary is called liquid junction potential.

The EMF of a cell is thus not only the sum of the oxidation potential (E_1) of the electrode where oxidation occurs and reduction potential (E_2) of the electrode where reduction occurs but also includes the junction potential (E_1). In other words, the EMF of the cell is

$$E = E_1 + E_2 + E_j$$

However, the type of junction employed most commonly is the salt bridge containing an electrolyte such as KC1. If we assume that the K^+ ions and Cl^- ions move almost with equal speed the junction potential is almost reduced to zero. In such a case, the EMF of the cell can be simple taken as $E_1 + E_2$.

8.26.1. Expression for liquid junction potential

Consider the following concentration cell with transference

H_2 (1 atm) | HCl (a_1) ⋮ HCl (a_2) | H_2 (1 atm)

Sum of the two electrode reactions gives

$$H^+ (a_2) \longrightarrow H^+ (a_1)$$

And the sum of the two electrode potentials *i.e.*, $E_1 + E_2$ is given by

$$E_1 + E_2 = -\frac{RT}{F} \ln \frac{(a_{H^+})_1}{(a_{H^+})_2}$$

$$= \frac{RT}{F} \ln \frac{(a_{H^+})_2}{(a_{H^+})_1}$$

$$= \frac{RT}{F} \ln \frac{(m_{H^+} \gamma_{H^+})_2}{(m_{H^+} \gamma_{H^+})_1} \quad \text{... } (i)$$

EMF of the cell considered above is given by

$$E = \frac{2t_- RT}{F} \ln \frac{m_2 \gamma_2}{m_1 \gamma_1} \quad \text{(Refer to eq. } (vi) \text{ Art. 8.25)} \quad \text{... } (ii)$$

Subtracting eqn. (*i*) from eqn. (*ii*), we get

$$E_j = E - (E_1 + E_2)$$

$$= \frac{2t_- RT}{F} \ln \frac{m_2 \gamma_2}{m_1 \gamma_1} - \frac{RT}{F} \ln \frac{(m_{H^+} \gamma_{H^+})_2}{(m_{H^+} \gamma_{H^+})_1} \quad \text{... } (iii)$$

But $(m_{H^+})_2 = m_2$ and $(m_{H^+})_1 = m_1$

If the activity coefficient of an ion is taken to be equal to the mean activity coefficient of the electrolyte, then

$(\gamma_{H^+})_2 = \gamma_2$ and $(\gamma_{H^+})_1 = \gamma_2$. Hence eqn. (*iii*) becomes

$$E_j = \frac{2t_- RT}{F} \ln \frac{m_2 \gamma_2}{m_1 \gamma_1} - \frac{RT}{F} \ln \frac{m_2 \gamma_2}{m_1 \gamma_1}$$

$$= (2t_- - 1) \frac{RT}{F} \ln \frac{m_2 \gamma_2}{m_2 \gamma_1} \quad ... (iv)$$

Also, we know $t_+ + t_- = 1$

$\therefore \quad 2t_- - 1 = t_- + (t_- - 1) = t_- + (-t_+)$

$= t_- - t_+$

Hence eqn. (*iv*) may be written in the form

$$\boldsymbol{E_j = (t_- - t_+) \frac{RT}{F} \ln \frac{m_2\gamma_2}{m_1\gamma_1}}$$

Thus we find that the liquid junction potential depends not only on the activities of the two solutions constituting the junction but also depends upon the difference between the two transport numbers of the electrolyte. When $t_- = t_+$, then $E_j = 0$. In case of an electrolyte for which $t_- > t_+$, E_j will be positive and thus E will be greater than $E_1 + E_2$ by E_j . In case of an electrolyte for which $t_- < t_+$, E_j will be negative and hence E will be less than $E_1 + E_2$ by E_j .

8.27. DETERMINATION OF ACTIVITY AND ACTIVITY COEFFICIENT FROM EMF MEASUREMENT

In order to know the activity of a solution corresponding to the given molality, we must measure the activity coefficient of the electrolyte in the given solution. To understand the method for this determination, consider the example of the *determination of activity coefficient of hydrochloric acid.* For this purpose, consider the following chemical cell without transference:

H_2 (1 atm) | HCl (activity $= a$), AgCl (s) | Ag

It consists of hydrogen and silver-silver chloride electrodes in hydrochloric acid as the electrolyte. As the two electrodes are reversible with respect to the ions of the electrolyte, they may be immersed directly in the acid to give a cell with no liquid junction. The cell reaction is given by

$$\frac{1}{2} H_2(1 atm) + AgCl(s) \longrightarrow Ag(s) + H^+ (a_{H^-}) + Cl^- (a_{Cl^-})$$

and the EMF of the cell is given by

$$E_{cell} = E^0_{Ag|AgCl} - \frac{RT}{F} \ln a_{H^+} a_{Cl^-}$$

$$= E^0_{Ag|AgCl} - \frac{RT}{F} \ln a \quad ... (i)$$

where a is the activity of HCl as a whole.

As the activity (a) of HCl at any molality m is related to the mean activity coefficiency $\gamma_\pm$ by the expression $a = m^2 \gamma_\pm^2$, substituting this value in eqn. (*i*), we get

$$E_{cell} = E^0_{Ag|AgCl} - - \frac{RT}{F} \ln m^2 \gamma_\pm^2$$

$$E_{cell} = E^0_{Ag\,|AgCl} \quad \frac{2RT}{F} \ln m \quad \frac{2RT}{F} \ln \gamma_\pm$$

$$\mathbf{E_{cell} + \frac{2RT}{F} \ln m = E^0_{Ag\,|AgCl} - \frac{2RT}{F} \ln \gamma_\pm} \quad \text{... }(ii)$$

All the quantities on the left hand side of this equation are determined experimentally. To calculate $\gamma_\pm$, the value of $E^0_{Ag\,|AgCl}$ is needed. To determine $E^0_{Ag\,|AgCl}$ the quantity $E_{cell} + \frac{2RT}{F}$ ln m is plotted against m and the result is extrapolated to $m = 0$. When $m = 0$, $\gamma_\pm = 1$, therefore from eqn. (*ii*) $E^0_{Ag\,|AgCl} = E_{cell} + \frac{2RT}{F} \ln m$ *i.e.* equal to the value of ordinate.

Problem. Calculate the mean ionic activity coefficient of 0.1 molar hydrochloric acid at 25°C given that the EMF of the cell

$$\mathbf{H_2\ (1\ atm)\ |\ HCl\ (\mathit{a}),\ AgCl(\mathit{s})\ |\ Ag}$$

is 0.3524 V at 25°C and that the standard electrode potential of Ag – AgCl is 0.2224 V at 25°C.

Solution. For the given cell

$$E_{cell} + \frac{2RT}{F} \ln m = E^0_{Ag\,|AgCl} - \frac{2RT}{F} \ln \gamma_2$$

$$\text{or } E_{cell} + \frac{2 \times 2.203RT}{F} \log m = E^0_{Ag\,|AgCl} - \frac{2 \times 2.203RT}{F} \log \gamma_\pm$$

Putting $T = 298$ K, $R = 8.314$ JK^{-1} mol^{-1} and $F = 96500$ C, this equation becomes

$$E_{cell} + 0.1183 \log m = E^0_{Ag\,|AgCl} - 0.1183 \log \gamma_\pm$$

Substituting $E_{cell} = 0.3524$ V, $E^0_{Ag\,|AgCl} = 0.2224$ V, and $m = 0.1$ mol kg^{-1}, we obtain

$$0.3524 + 0.1183 \log 0.1 = 0.2224 - 0.1183 \log \gamma_\pm$$

$$\text{or} \quad \log \gamma_\pm = \frac{-0.3524 + 0.2224 + 0.1183}{0.1183} = -0.0989$$

$$\text{or} \quad \boldsymbol{\gamma_\pm = 0.796}$$

8.28. APPLICATIONS OF EMF MEASUREMENTS

1. To calculate standard free energy change and equilibrium constant of a given reaction. Electrical energy produced in a cell is equal to the decrease in free energy accompanying the cell reaction *i.e.*

$$-\Delta G = nFE \quad \text{... }(i)$$

where ΔG is the change in free energy accompanying the reaction, n is the number of the moles of electrons transferred in the cell reaction, F is one faraday (= 96500 columns), E is the EMF of the cell.

If the reactants and products are taken in the standard state, we can write

$$-\Delta G^\circ = nFE^\circ \quad \text{.... }(ii)$$

Knowing the value of $E°$, the value of $\Delta G°$ can be calculated.

For a reversible reaction in equilibrium, ΔG° is related to the equilibrium constant K according to the equation

$$\Delta G^\circ = -RT \ln K = -2.303\, RT \log K \qquad \text{... (iii)}$$

Thus knowing ΔG°, K can be calculated at the given temperature T.

Problem. Calculate the free energy change of the following cell at 25°C.

$Sn|Sn^{2+}$ (a = 0.6) || Pb^{2+} (a = 0.3)| Pb

Standard EMF of the cell is 0.014 volt.

Solution. The cell reaction can be written as

$$Sn + Pb^{2+} \longrightarrow Sn^{2+} + Pb$$

Here $n = 2$

By Nernst equation, the EMF of this cell at 25°C is given by

$$E = E^0 - \frac{0.0591}{n} \ln \frac{a_{Sn^{2+}}}{a_{Pb^{2+}}}$$

$$= E^0 - \frac{0.0591}{n} \log \frac{a_{Sn^{2+}}}{a_{Pb^{2+}}}$$

$$= 0.014 - \frac{0.0591}{n} \log \frac{0.6}{0.3} = 0.0051 \text{volt}$$

ΔG can be obtained as follows:

$$\therefore \qquad \Delta G = -nFE = -2 \times 96500 \times 0.0051 \text{joules} = -\mathbf{984.3\ joules}$$

2. To determine the transport number of the ions:

For a cell with transference

H_2(1 atm) | HCl (a_1) : HCl (a_2) | H_2 (1 atm)

the EMF is given by

$$E = \frac{t\_RT}{F} \ln \frac{a_2}{a_1} \qquad \text{... (iv)}$$

By measuring the EMF of the cell and knowing the activities of the two solutions, the transport number of the anion ($t\_$) can be calculated.

3. To determine the valency of the ions:

EMF of a concentration cell without transference is given by

$$E = \frac{RT}{nF} \ln \frac{a_2}{a_1} \simeq \frac{RT}{nF} \ln \frac{c_2}{c_1} \qquad \text{... (v)}$$

In this expression n is the valency of the ion involved. Thus knowing all other quantities, n can be calculated.

For example, the valency of mercurous ions, which remained doubtful for a very long time, was determined by this method, by setting up the following cell :

$$\text{Hg}\left|\frac{N}{20}\ \begin{matrix}\text{Mercurous nitrate}\\ \text{solution}\end{matrix}\right|\left|\frac{N}{2}\ \begin{matrix}\text{Mercurous nitrate}\\ \text{solution}\end{matrix}\right|\text{Hg}$$

Here $$\frac{c_2}{c_1} = \frac{1/2}{1/20} = 10$$

Putting the values of the various constants in equation (*v*) at 25°C, we get

$$E = \frac{0.0591}{n} \log \frac{c_2}{c_1}$$

Putting $\frac{c_2}{c_1} = 10$ *i.e.*, $\log \frac{c_2}{c_1} = \log 10 = 1$, we get

$$E = \frac{0.0591}{n}$$

The EMF of the above cell was found to be 0.0295 volt.

Hence $$n = \frac{0.0591}{E} = \frac{0.0591}{0.0295} = 2$$

Thus the valency of mercurous ion is 2 and it is represented as Hg_2^{2+} and its salts such as nitrate by $Hg_2(NO_3)_2$.

4. To determine the solubility and solubility product of sparingly soluble salts.

The EMF of a concentration cell is given by

$$E = \frac{RT}{nF} \ln \frac{c_2}{c_1}$$

Thus knowing the EMF of the cell and the concentration of ions in one of the solutions, that in the other can be calculated. If the second solution is that of the sparingly soluble salt, its ionic concentration is equal to its solubility assuming that the substance is completely dissociated.

Suppose it is required to find the solubility of the sparingly soluble salt, AgC1. The following cell is set up and its EMF is measured :

$$\text{Ag} \left| \frac{\text{N}}{100} \begin{matrix} AgNO_3 \\ \text{solution} \end{matrix} \right\| \begin{matrix} \text{Saturated} \\ NH_4NO_3 \end{matrix} \left\| \frac{\text{N}}{10} \begin{matrix} \text{KCl solution saturated} \\ \text{with AgCl} \end{matrix} \right| \text{Ag}$$

This cell is prepared by dipping a silver electrode in N/100 $AgNO_3$ solution taken in one beaker and dipping another silver electrode in N/10 KCl solution to which 1 or 2 drops of $AgNO_3$ solution are added, forming AgC1 which is enough to make the solution saturated with AgCl. The two beakers are then connected by a salt bridge containing saturated NH_4NO_3 solution.

Calculations. The cell, as described above, is a concentration cell having higher concentration of Ag^+ ions in the left hand electrode and lower concentration of Ag^+ ions in the right hand electrode (produced by the dissociation of AgC1). By convention, we write the electrode with higher concentration of ions on the right hand side,

$$\text{Ag} \left| \frac{\text{N}}{10} \begin{matrix} \text{KCl solution saturated} \\ \text{with AgCl} \\ [Ag^+] = c\,(\text{say}) \end{matrix} \right\| \begin{matrix} \text{Saturated} \\ NH_4NO_3 \end{matrix} \left\| \frac{\text{N}}{100} \begin{matrix} AgNO_3 \text{ solution} \\ [Ag^+] = 0.01 \end{matrix} \right| \text{Ag}$$

EMF of this cell at 25°C is given by

$$E = 0.0591 \log \frac{0.01}{c}$$

From this equation, the value of c can be calculated.

Thus in the solution of AgCl in $\frac{N}{10}$ KC1,

$$[Ag^+] = c \text{ gram equivalents/litre.}$$

As AgC1 dissolves to a very small extent, the Cl^- ions produced form AgCl are negligible as compared to that produced from KCl. Therefore in the solution of AgCI in $\frac{N}{10}$ KCl,

$$[Cl^-] = [KCl] = 0.1$$

The solubility product (K_s) of AgC1 is, therefore, be given by the equation

$$K_s = [Ag^+][Cl^-]$$
$$= c \times 0.1$$

If the solubility of AgC1 is s moles/litre, then

$$[Ag^+] = [Cl^-] = s \text{ gram equivalent/litre}$$
$$K_s = [Ag^+][Cl^-] = s \times s = s^2$$

or
$$s = \sqrt{K_s}$$

Knowing the value of K_s, the value of s can be calculated.

Problem. Consider the cell :

Ag | 0.005 M $AgNO_3$ || KNO_3 (satd) || AgCl (satd) | Ag
salt bridge

If the EMF of this cell at 25°C is 0.1585 volt, calculate the solubility of AgCl at this temperature.

Solution. Solubility of AgCl in water is very small, therefore the concentration of Ag^+ ions in AgCl solution will be very small as compared to that of Ag^+ ion in 0.005 $AgNO_3$ solution. On writing the electrode with higher concentration of Ag^+ ions on the right hand side, we have

Ag | AgCl (satd)|| KNO_3 (satd) || 0.005M $AgNO_3$ | Ag

Let C be the concentration of Ag^+ ions in the left hand electrode of the above concentration cell. Substituting the values in the equation for EMF of concentration cell, we have

$$0.1585 = \frac{0.0591}{1}\log\frac{0.005}{C}$$

which gives
$$\log C = \log 0.0005 - 2.6819$$
$$= -3.301 - 2.6819$$
$$= -5.9829$$
$$= \bar{6}.0171$$

or
$$C = 1.040 \times 10^{-6} \text{ g eq/litre}$$
$$= 1.040 \times 10^{-6} \times 143.5 \text{ g /litre}$$
$$= \mathbf{1.49 \times 10^{-4} \text{ g/litre}}$$

8.29. DIFFERENT TYPES OF ELECTRODES IN PRACTICAL USE

(a) Dipping calomel electrode

Calomel electrode is now commercially available as a self-contained unit. It is used by simply dipping the tip of the electrode in the solution of another electrode whose emf is to be determined.

This type of electrode is often referred as the dipping calomel electrode (Fig. 8.19). This saves us from the cumbersome method of preparing the calomel.

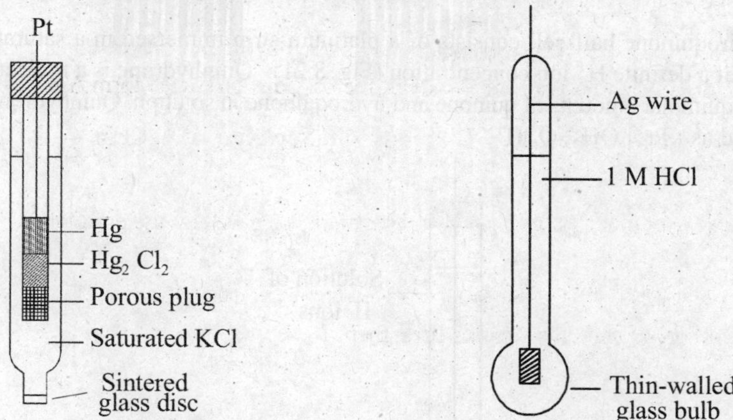

Fig. 8.19. A dipping calomel electrode **Fig. 8.20.** Glass-electrode

The mercury and calomel (Hg_2Cl_2) are held in a narrow central tube by a porous cotton wool plug. A platinum wire dips into the mercury. The KCl saturated solution contained in the electrode makes contact with the other electrode solution through a sintered glass disc. This disc functions as a 'salt bridge' between the two electrode solutions. Thus, a galvanic cell is set up and its emf is measured with the help of a voltmeter.

(*b*) Glass Electrode

A commonly used secondary standard electrode is glass electrode. Its emf is determined by coupling with a standard calomel electrode (SCE). The glass electrode provides one of the easiest methods for measuring the pH of a given solution.

A glass electrode (Fig. 8.20) consists of a glass tube having a thin-walled bulb at the lower end. The bulb contains a 1M HCl solution. A silver wire coated with silver chloride (or simply Pt were) is sealed into the glass tube at its lower end. The lower end of this silver wire dips into the hydrochloric acid, forming silver-silver chloride electrode. The glass electrode may be represented as

$$\text{Ag AgCl} \mid \text{1M HCl} \mid H^+ \quad \text{(Test Solution)}$$

When placed in a solution, the potential of the glass electrode depends on the H^+ ion concentration of the solution. The potential develops across the glass membrane as a result of a concentration difference of H^+ ions on the two sides of the membrane. This happens much in the same way as the emf of a concentration cell develops.

The potential of a glass electrode can be determined against a standard calomel electrode (SCE).

(c) Quinhydrone Electrode

It is a widely used secondary standard electrode. It involves the redox reaction between quinone (Q) and hydroquinone (QH_2),

$$\underset{\text{Quinone}}{O{=}C_6H_4{=}O} + 2H^+ + 2e^- \rightleftharpoons \underset{\text{Hydroquinone}}{HO{-}C_6H_4{-}OH}$$

or, simply

$$Q + 2H^+ + 2e^- \rightleftharpoons QH_2$$

The hydroquinone half-cell consists of a platinum strip immersed in a saturated solution of quinhydrone at a definite H^+ ion concentration (Fig. 8.21). Quinhydrone is a molecular compound which gives equimolar amounts of quinone and hydroquinone in solution. Quinhydrone half cell may be represented as : Pt | QH_2, Q, H^+

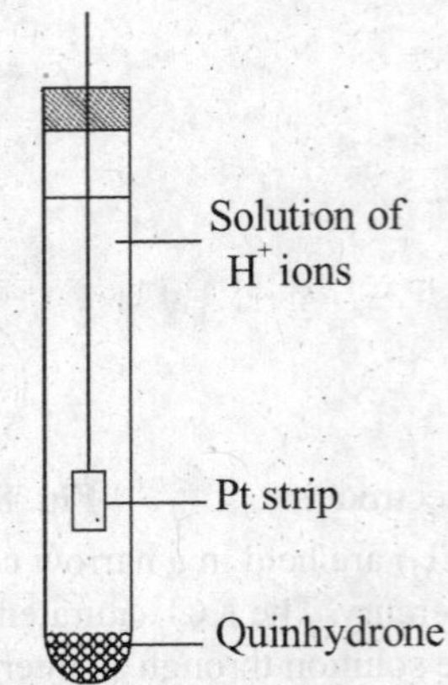

Fig. 8.21. The Quinhydrone Electrode

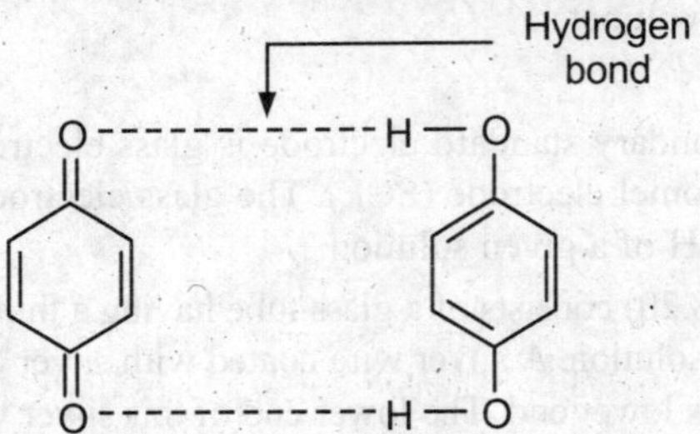

The potential developed is measured against a hydrogen electrode or calomel electrode. The emf with respect to a standard hydrogen electrode is 0.6996 V at 25°C.

8.30. DETERMINATION OF *p*H OR H^+ CONCENTRATION OF A SOLUTION BY EMF MEASUREMENT

The basic principle of the method is to set up a cell in which one of the electrodes is such that its potential depends upon the H^+ ion concentration in the solution under examination and the other electrode is a reference electrodes, either calomel electrode or standard hydrogen electrode. Knowing the emf of the constituted cell and the potential of reference electrode we can calculate the potential of the electrode containing the solution under examination and then the H^+ concentration using Nernst equation. The various electrodes whose potentials depend upon the H^+ ion concentration are *hydrogen electrode, quinhydrone electrode and glass electrode.* Using these

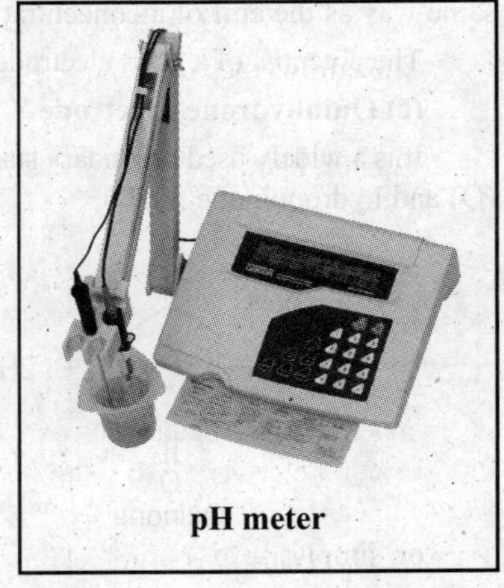

pH meter

electrodes against the reference calomel electrode, the various methods of determination of *p*H are briefly described below :

1. Using hydrogen electrode:

The electrode reaction, written as reduction reaction, is

$$H^+ + e^- \longrightarrow \frac{1}{2} H_2 \text{ (1 atm.)}$$

Potential of this electrode is given by

$$E_{H^+,H_2} = E^0_{H^+,H_2} + \frac{RT}{F} \ln[H^+] \text{ (Nernst equation)}$$

$$= E^0_{H^+,H_2} + 0.0591 \log [H^+] \text{ at } 25°C$$

$$= 0.0591 \log [H^+] \qquad [\text{as } E^0_{H^+,H_2} = 0]$$

$$= -0.0591\, pH \qquad [\text{as } -\log [H^+] = pH]$$

A standard hydrogen electrode is coupled with another hydrogen electrode which contains the solution of unknown pH (Fig. 8.22). In both half cells, hydrogen gas is used at 1 atm pressure and 25°C. The e.m.f. of the cell

$$Pt \mid H_2 \text{ (1 atm)} \mid H^+ \text{ (1M)} \mid\mid H^+ \text{ (unknown)} \mid H_2 \text{ (1 atm)} \mid Pt,$$

is found out from the experiment.

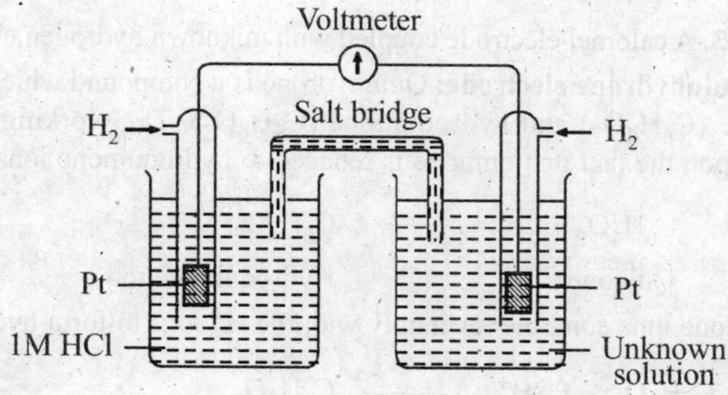

Fig. 8.22. Determination of pH with hydrogen electrode.

The emf of the cell, E_{cell} is given by

$$E_{cell} = E_{R.H.S} - E_{L.H.S}$$

The left hand electrode is standard hydrogen electrode and its e.m.f. is zero. Thus,

$$E_{cell} = E_{R.H.S} - E_{L.H.S} = -0.0591 \text{ pH}$$

or

$$pH = -\frac{E_{cell}}{0.0591}$$

2. Using Calomel Electrode

When the hydrogen electrode is coupled with the calomel electrode, oxidation takes place at the hydrogen electrode and reduction at the calomel electrode. Hence by convention, the cell may be written as

$$Pt, H_2 \text{ (1 atm)} \mid H^+ \text{ (unknown conc.)} \mid\mid KCl \text{ (saturated)} \mid Hg_2Cl_2(s) \mid Hg$$

EMF of this cell is given by the relation

$$E = \begin{bmatrix} \text{Electrode potential of} \\ \text{R.H.S. electrode} \end{bmatrix} - \begin{bmatrix} \text{Electrode potential of} \\ \text{the L.H.S. electrode} \end{bmatrix}$$

$$= 0.2415 - (-0.0591\,pH)$$

$$= 0.2415 + 0.0591\,pH$$

or $$\text{pH} = \frac{E - 0.2415}{0.0591}$$

Thus measuring the EMF (E) of the cell potentiometrically, the pH of the solution can be calculated (Fig. 8.23).

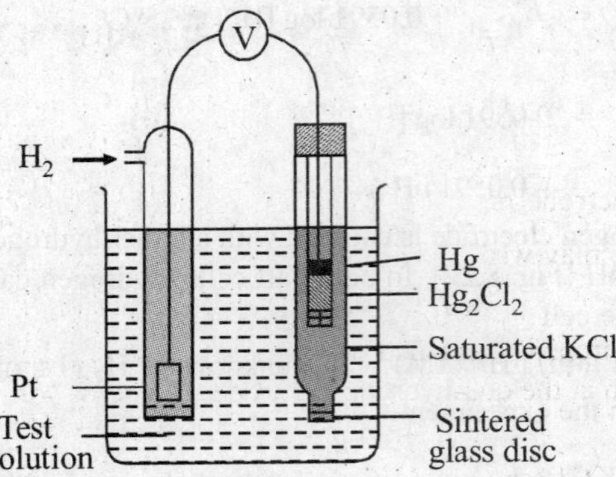

Fig. 8.23. A calomel electrode coupled with unknown hydrogen electrode.

3. Using the quinhydrone electrode: Quinhydrone is a compound which contains equimolar amounts of quinone ($C_6H_4O_2$) and hydroquinone ($C_6H_6O_2$). The working of the quinhydrone electrode is based upon the fact that quinone is reduced to hydroquinone ions

i.e., $$\underset{\text{Quinone}}{C_6H_4O_2} + 2e^- \longrightarrow \underset{\text{Hydroquinone ion}}{C_6H_4O_2^{\,2-}}$$

The hydroquinone ions combine reversibly with the H^+ ions to form hydroquinone

$$C_6H_4O_2^{\,2-} + 2H^+ \rightleftharpoons C_6H_6O_2$$

The overall reaction is

$$\underset{\text{Quinone}}{C_6H_4O_2} + 2H^+ + 2e^- \rightleftharpoons \underset{\text{Hydroquinone}}{C_6\,H_6\,O_2}$$

If quinone is represented by Q, then hydroquinone can be represented by QH_2. Therefore, the above reaction may be written as

$$Q + 2H^+ + 2e^- \rightleftharpoons QH_2$$

If a platinum wire is dipped into a solution containing H^+ ions and a pinch of a quinhydrone (*i.e.* $Q + QH_2$) is added, it acts as a reversible electrode (Fig 8.24). The reaction as written above is reduction reaction. Applying Nernst equation, the electrode potential of the above electrode is given by

$$E_Q = E_Q^0 - \frac{RT}{2F}\ln\frac{[QH_2]}{[Q][H^+]^2} \qquad [n = 2 \text{ for the above reaction}]$$

Quinhydrone is an equimolar mixture of Q and QH_2. Take [Q] = [QH_2] and put $\frac{[QH_2]}{[Q]} = 1$, in the above equation, we get

$$E_Q = E_Q^0 - \frac{RT}{2F} \ln \frac{1}{[H^+]^2}$$

$$= E_Q^0 + \frac{RT}{F} \ln[H^+]$$

$$= E_Q^0 + \frac{2.303RT}{F} \log[H^+]$$

$$= E_Q^0 + 0.0591 \log [H^+] \text{ at } 25° C$$

$$= E_Q^0 - 0.0591 \text{ pH}$$

The standard electrode potential (E_Q^0) of the quinhydrone electrode has been found to be 0.6996 volt. Hence we may write

$$E_Q = 0.6996 - 0.0591 \text{ pH}$$

When the quinhydrone is combined with calomel electrode, oxidation takes place at the calomel electrode and reduction at the quinhydrone electrode. Hence by convention, the cell may be written as

Hg | Hg_2Cl_2(s) | KC1 (saturated) || H^+ (unknown conc.) | Q| QH_2 | Pt

EMF of this cell is given by

$$E = \left[\begin{matrix}\text{Electrode potential of} \\ \text{R.H.S. electrode}\end{matrix}\right] - \left[\begin{matrix}\text{Electrode potential of} \\ \text{the L.H.S. electrode}\end{matrix}\right]$$

$$= (0.6996 - 0.0591 \text{ pH}) - 0.2415$$

$$= 0.4581 - 0.0591 \text{ pH}$$

or

$$\text{pH} = \frac{0.4581 - E}{0.0591}$$

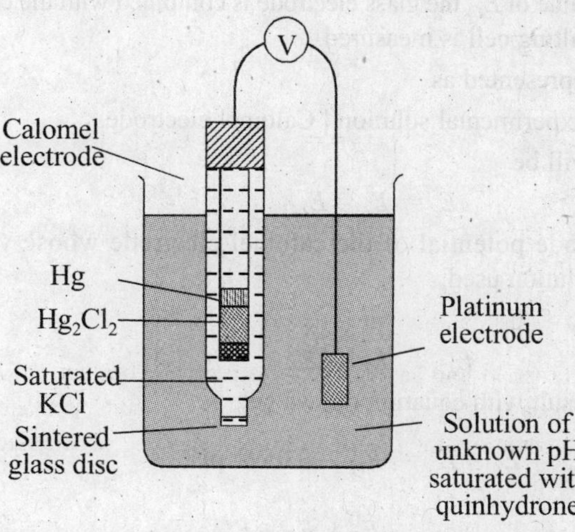

Fig. 8.24. Quinhydrone electrode coupled with standard calomel electrode.

EMF (E) of the cell is measured potentiometrically. from which the pH of the solution can be calculated.

4. Using glass electrode: If two solutions of different pH are present on the two sides of the glass surface, a potential is established across the glass membrane whose magnitude depends upon the difference in the pH of the two solutions. For a particular variety of a glass, if on one side of the glass surface, a solution of fixed pH value is placed, the potential developed across the glass membrane will depend only upon the pH of the other solution, placed in contact with the other side of the glass surface. The assembly thus obtained is called a glass electrode. (Fig. 8.25).

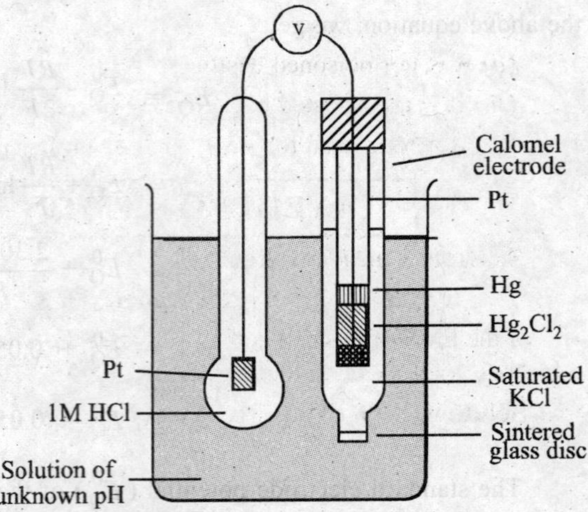

Fig. 8.25. A glass electrode coupled with standard calomel electrode.

A tube of a special variety of glass (usually soda glass) is taken which has low melting point and high electrical conductivity. At the end of this tube is blown a glass bulb having a very thin wall. This is filled with a solution of constant pH. To make the electrical contact with this solution, a platinum wire is dipped into this solution, the other end of which is attached to a terminal.

The electrode potential of the glass electrode, as prepared above, is given by the equation

$$E_G = E_G^0 + 0.0591 \log [H^+] \text{ at } 25°C$$

$$= E_G^0 - 0.0591\, p\text{H} \quad \text{... } (i)$$

where E_G^0 is a constant depending upon the nature of the glass and the pH of the solution taken inside the glass bulb. To determine the value of E_G^0 for a particular glass electrode, the value of E_G is determined for a solution of known pH value.

To determine the value of E_G the glass electrode is combined with the calomel electrode as usual and the EMF of the resulting cell is measured.

The cell may be represented as

Glass electrode | Experimental solution | Calomel electrode.

EMF of this cell will be

$$E = E_C - E_G \quad \text{... } (ii)$$

where E_C is the electrode potential of the calomel electrode whose value depends upon the concentration of KCl solution used.

From equation (ii),

$$E_G = E_C - E$$

Combining this result with equation (i), we get

$$E_C - E = E_G^0 - 0.0591 \text{ pH}$$

or

$$\text{pH} = \frac{E_G^0 - E_C + E}{0.0591} \quad \text{... } (iii)$$

Advantages of using glass electrodes

The advantages of the glass electrodes over the other electrodes are as mentioned below :

(*i*) It is easy to use.

(*ii*) It is not poisoned easily.

(*iii*) It is not affected by oxidizing or reducing agents.

(*iv*) It can be used for very small quantities of the solutions.

SOLVED PROBLEM ON CALCULATION OF pH

Example. Consider the cell

$$\textbf{Pt} \mid \textbf{H}_2\ \textbf{(g, 1 atm)} \mid \textbf{H}^+\ (x\ \textbf{molar}) \parallel \textbf{KCl (0.1 M)} \mid \textbf{Hg}_2\textbf{Cl}_2 \mid \textbf{Hg}$$

If the EMF of this cell is 0.50 volt at 25°C, what would be the pH of the *x* molar acid solution? (Electrode potential of the calomel electrode is 0.281 volt at 25°C)

Solution. EMF of the given cell is given by

$$E_{cell} = \begin{bmatrix}\text{Electrode potential} \\ \text{or R.H.S. electrode}\end{bmatrix} - \begin{bmatrix}\text{Electrode potential} \\ \text{or L.H.S. electrode}\end{bmatrix}$$

$$= E_{calomel} - E_{H^+,H_2}$$

$$\therefore \quad E_{H^+,H_2} = E_{calomel} - E_{cell}$$

But by Nernst equation, we have

$$E_{H^+,H_2} = 0.0591 - \log [H^+]$$

$$= -0.0591\ \text{pH} \qquad [\text{as pH} = -\log [H^+]]$$

$$\therefore \quad -0.0591\ \text{pH} = E_{calomel} - E_{cell}$$

or

$$\text{pH} = \frac{E_{Cell} - E_{calomel}}{0.0591}$$

$$= \frac{0.50 - 0.281}{0.0591} = \mathbf{3.71}$$

PROBLEMS FOR PRACTICE

1. If the EMF of the cell

Pt, H_2 (g) (1 atm) | H^+ (a = ?) || H^+ (a = 0.01) | H_2 (1 atm), Pt at 25°C is 0.861 volt, what is the pH of the solution around the left hand electrode? The value of 2.303 RT/F at 25°C is 0.0591. **[Ans.** 12.568]

2. EMF of cell

Pt, H_2 (1 atm) | solution of unknown pH || $Hg_2\ Cl_2$ (s) KCl (s) | Hg is 0.366 volt. Ignoring the junction potential, determine the pH of the given solution. Electrode potential of S.C.E. being +0.242 volt. **[Ans.** 2.098]

3. Determine the pH of the solution at 25° C from the following data:

Pt, H_2 (1atm) | H^+ ($a = x$) || Normal calomel electrode

$$E_{cell} = 0.6346\ \text{V}$$

Oxidation potential of normal calomel electrode = – 0.2800 V

$$\frac{2.300\,RT}{F} \text{ at } 25°\text{C} = 0.0591$$

[Ans. 6.0]

8.31. POTENTIOMETRIC TITRATIONS

In all types of titrations like acid-base titrations, redox titrations, precipitation titrations etc. one of the solutions is taken in the titration flask and the other solution is added to it from the burette. As a result, the concentration of some ion in the titration flask keeps on changing. If an electrode reversible with respect to this ion is set up in the titration flask, its electrode potential will keep on changing. If the electrode thus set up in the titration flask (or beaker) is coupled with a calomel electrode, the EMF of the cell will keep on changing and show a sudden change in value when the end point is reached or the reaction is complete.

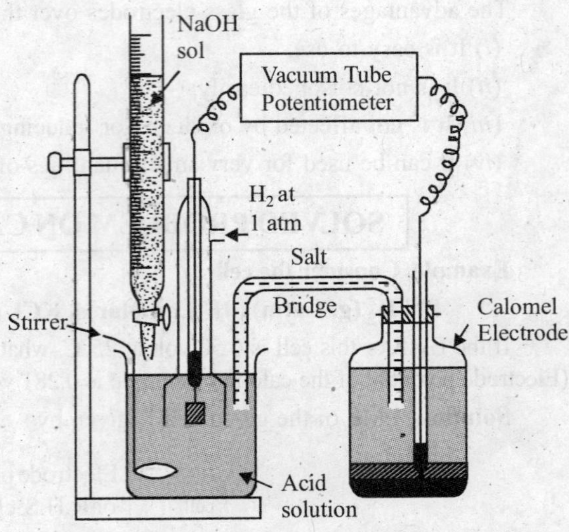

Fig. 8.26. Potentiometric titration of HCl *vs* NaOH.

Each type of titration is discussed hereunder :

(*i*) **Acid Base titrations:** Suppose a solution of HCl acid is to be titrated against NaOH solution. The acid solution is taken in a beaker and a hydrogen electrode (quinhydrone or glass or hydrogen gas) is set up into it. This electrode is then combined with the calomel electrode. A burette containing the NaOH solution is fitted over the beaker (Fig. 8.26). NaOH solution is added to the beaker first in the larger amounts (say 1 ml or 0.5 ml at a time) and then in very small amounts (say 0.1 ml or even less) near the end point. EMF of the cell is noted after each addition. The EMF is then plotted against the volume of NaOH added. A curve of the type shown in Fig. 8.27 is obtained. From this curve the volume of NaOH solution used corresponding to the equivalence point can be found. However, when the solutions are very dilute, or weak acids or bases are involved, the steepness of the curve is less marked and it is difficult to locate the end point. In such a case, we plot the slope of the curve $\Delta E / \Delta V$ against the volume of alkali used. The maximum of the curve indicates the end point (Fig. 8.28).

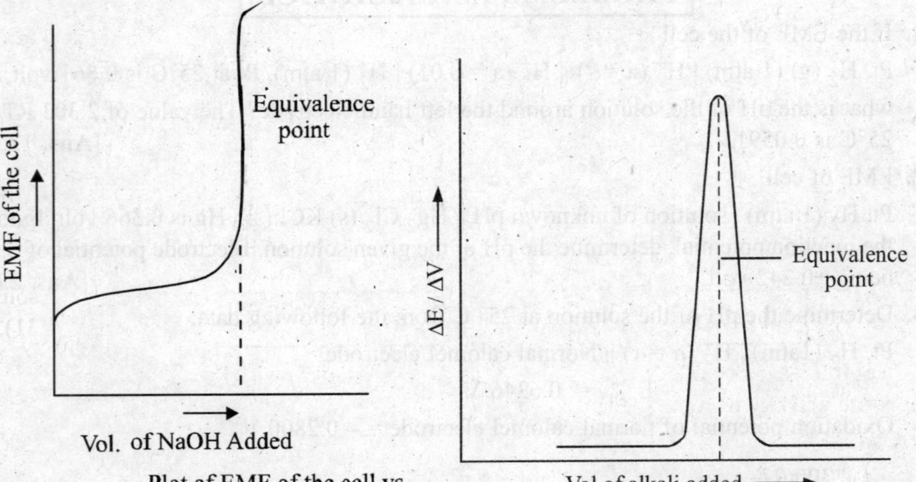

Fig. 8.27. Plot of EMF of the cell vs. vol. of the NaOH solution added.

Fig. 8.28. Titration curve of a weak acid and weak base

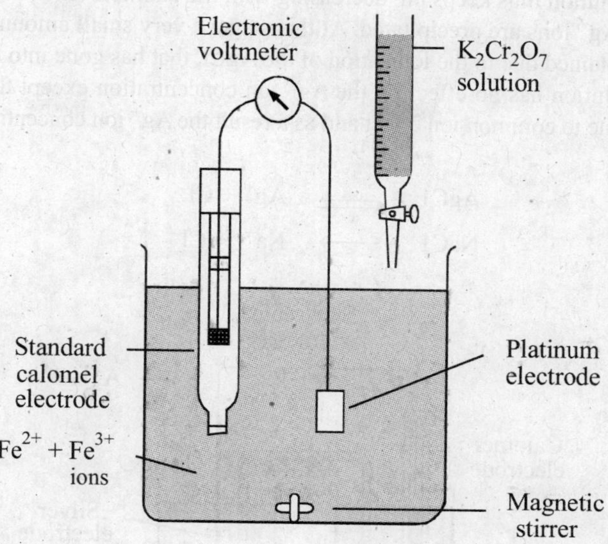

Fig 8.29 Apparatus for potentiometric titration of Fe^{2+} with $K_2Cr_2O_7$.

(*ii*) **Redox titrations:** Suppose a solution of ferrous sulphate is to be titrated against $K_2Cr_2O_7$ solution. The ferrous sulphate solution along with dilute H_2SO_4 is taken into a beaker and a platinum wire is inserted into it. The electrode thus formed is combined with the calomel electrode. (Fig. 8.29). A burette containing $K_2Cr_2O_7$ solution is fixed over the beaker and after every addition of $K_2Cr_2O_7$ solution, the EMF of the cell is noted. The end point is then determined in a manner similar to that of acid-base titrations.

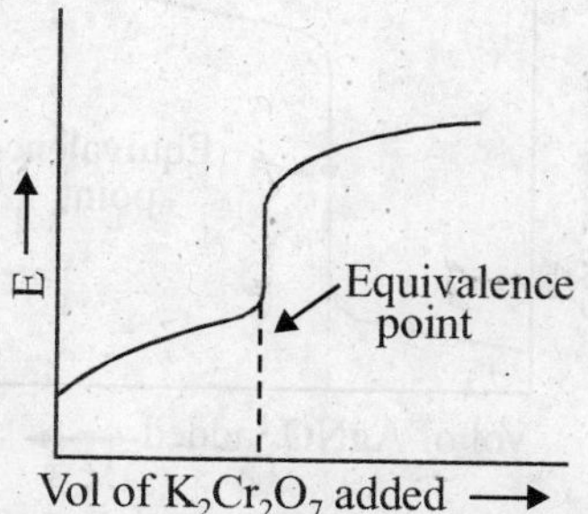

Fig. 8.30 Potentiometric titration curve of Fe^{2+} ions with $K_2Cr_2O_7$

Similarly Fe^{2+} can be estimated potentiometrically using a standard solution of $KMnO_4$. We shall take $KMnO_4$ solution in place of $K_2Cr_2O_7$ solution. Rest of the procedure remains the same. The equivalence point will be determined as in the case of Fe^{2+} — $K_2Cr_2O_7$ titration.

(*iii*) **Precipitation titrations:** Suppose a solution of $AgNO_3$ is to titrated against NaCl solution. The silver nitrate solution is taken in a beaker and a silver wire is inserted into it (Fig 8.31). The electrode thus formed is then combined with the calomel electrode. A burette containing the NaCl solution is fixed over the beaker and after every addition of NaCl solution, the EMF of the cell is noted. The end point is then determined as in the case of acid-base titrations (Fig 8.32).

We get a similar curve as in acid-base titration because the electrode potential of the silver electrode depends upon the concentration of Ag^+ ions in the solution. The addition of NaCl solution results into the precipitation of Ag^+ ions to form AgCl (which is sparingly soluble). The concentration of Ag^+ ions in the solution thus keeps on decreasing with the addition of NaC1 solution till the end point, when all the Ag^+ ions are precipitated. At this stage, a very small amount of Ag^+ ions is still present which are obtained due to the ionization of the AgCl, that has gone into the solution. Further addition of NaCl solution has no effect on the Ag^+ ion concentration except that it suppresses the ionization of AgCl due to common ion effect and as a result the Ag^+ ion concentration decreases very slightly.

$$AgC1 \rightleftharpoons Ag^+ + Cl^-$$

$$NaC1 \longrightarrow Na^+ + Cl^-$$

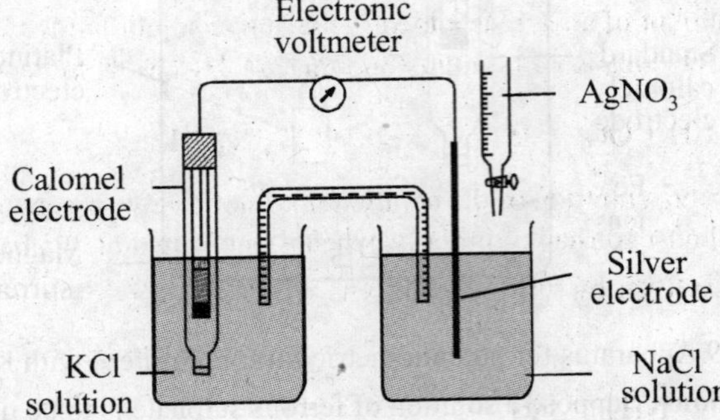

Fig. 8.31. Potentiometric titration of sodium chloride against silver nitrate solution.

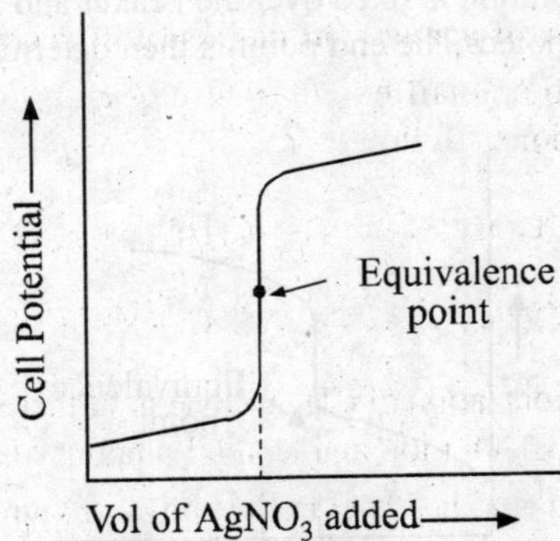

Fig. 8.32. Potentiometric titration curve for NaCl and $AgNO_3$.

Advantages of potentiometric titrations.

(*i*) These titration are very useful in case of coloured solutions where internal indicators fail.

(*ii*) For acid-base titration, no prior knowledge of the pH range in which the indicator changes colour is required.

8.32. BUFFER SOLUTION

A buffer solution is defined as a solution whose pH remains practically constant even when small amounts of the acid or the base are added to it.

Types of buffer solutions. There are two types of buffer solutions.

(*i*) Solutions of single substances: The solution of the salt of a weak acid and a weak base e.g. ammonium acetate (CH_3COONH_4) acts as a buffer.

(*ii*) Solutions of mixtures: These are further classified as under:

(*a*) *Acidic buffer* : It is the solution of a mixture of equimolar quantities of a weak acid and a salt of this weak acid with a strong base (e.g. $CH_3COOH + CH_3COONa$).

(*b*) *Basic buffer* : It is solution of a mixture of equimolar quantities of a weak base and a salt of this weak base with a strong acid (e.g. $NH_4OH + NH_4Cl$).

Buffer action of ammonium acetate

Ammonium acetate is almost completely dissociated in the aqueous solution as

$$CH_3COONH_4 \rightleftharpoons CH_3COO^- + NH_4^+$$

Thus in the solution, there are plenty CH_3COO^- ions and NH_4^+ ions.

When a small amount of an acid is added to the above solution, the H^+ ions given by the acid combine with CH_3COO^- ions to form almost unionised CH_3COOH molecules

$$CH_3COO^- + H^+ \longrightarrow CH_3COOH$$

Thus the H^+ ion concentration of the solution does not change practically and hence the pH of the solution remains almost constant. Similarly, when a small amount of a base is added to the above solution, the OH^- ions given by the base combine with the NH_4^+ ions to form almost unionised NH_4OH molecules.

$$NH_4^+ + OH^- \longrightarrow NH_4OH$$

Thus the OH^- ion concentration and hence the H^+ ion concentration or the pH of the solution remains almost constant

Buffer action of a mixture of acetic acid and sodium acetate (Acidic buffer).

Acetic acid dissociates to a small extent whereas sodium acetate is almost completely dissociated in the aqueous solution as follows:

$$CH_3COOH \rightleftharpoons CH_3COO^- + H^+ \quad \text{... (i)}$$

$$CH_3COONa \longrightarrow CH_3COO^- + Na^+ \quad \text{... (ii)}$$

By common ion effect, the ionization of CH_3 COOH is further suppressed. Thus in the solution, there are excess of acetate (CH_3COO^-) ion and a small amount of H^+ ions.

When a small amount of an acid is added to the above mixture solution, the H^+ ions given by the acid combine with the CH_3COO^- ions given by the acid to form CH_3COOH which is almost unionised.

$$CH_3COO^- + H^+ \longrightarrow CH_3COOH$$

Thus the H^+ ion concentration and hence the pH of the solution does not change much.

Similarly, when a small amount of a base is added, the OH^- ions given by the base combine with the H^+ ions already present to form almost unionised H_2O

$$\underset{\text{From [i]}}{H^+} + \underset{\text{[Given by the base]}}{OH^-} \longrightarrow H_2O$$

As the H^+ ions are consumed, the equilibrium (*i*) shifts towards right (according to Le chatelier's principle). Thus more of CH_3COOH dissociates to make up the loss of H^+ ions. Hence the H^+ ion concentration of the pH of the solution remains almost the same.

Buffer action of a mixture of ammonium hydroxide and ammonium chloride (Basic buffer).

NH_4OH dissociates to a small extent whereas NH_4Cl dissociates completely in the aqueous solution as follows:

$$NH_4OH \rightleftharpoons NH_4^+ + OH^- \quad \text{... }(i)$$

$$NH_4Cl \longrightarrow NH_4^+ + Cl^- \quad \text{... }(ii)$$

By common ion effect, the ionization of NH_4OH is further suppressed. Thus in the solution, there are excess of NH_4^+ ions and a small amount of OH^- ions.

When a small amount of a base is added, the OH^- ions given by it immediately combine to form the almost unionised NH_4OH.

$$NH_4^+ + OH^- \longrightarrow NH_4OH$$

Thus the H^+ ion concentration or the pH of the solution is not affected much.

Similarly, when a small amount of an acid is added, the H^+ ions given by it combine with the OH^- ions already produced by NH_4OH in equilibrium (*i*).

$$\underset{\text{[Given by the acid]}}{H^+} + \underset{\text{[Given by }(i)\text{]}}{OH^-} \longrightarrow H_2O$$

As the OH^- ions are consumed away the equilibrium (*i*) shifts in the forward direction. Thus more of NH_4OH dissociates to produce OH^- ions which make up the loss of OH^- ions. Hence the OH^- ion concentration and therefore the H^+ ion concentration or the pH of the solution does not change much.

Importance of buffer solutions

Importance of buffer solutions is discussed as under :

(1) In biological processes: The pH of our blood is maintained constant (at about 7.4) inspite of various acid and base producing reactions going on in our body. In the absence of its buffer nature, we can not eat variety of foods.

(2) In industrial processes: The use of buffers is an important part of many industrial processes, *e.g.*,

(*i*) in electroplating;

(*ii*) in the manufacture of leather, photographic materials and dyes ;

(*iii*) in bacteriological research, culture media are generally buffered to maintain the pH required for the growth of the bacteria being studied ;

(*iv*) in analytical chemistry.

8.33. HENDERSON EQUATION FOR THE pH OF A BUFFER MIXTURE

(*a*) **For acidic buffer mixture:** Suppose it is required to calculate the pH of an *acidic buffer mixture,* obtained by mixing a weak acid HA and its salt with a strong base, BA. The former being weak dissociates to a small extent whereas the latter being strong electrolyte dissociates almost completely in the solution *i.e.*, we have

$$HA \rightleftharpoons H^+ + A^-$$

$$BA \longrightarrow B^+ + A^-$$

If we apply the law of chemical equilibirum to the dissociation equilibrium of the weak acid, we get the following equation

$$K_a = \frac{[H^+][A^-]}{[HA]} \quad ... (i)$$

where K_a is the dissociation constant of the acid

$$\therefore \quad [H^+] = K_a \frac{[HA]}{[A^-]} \quad ... (ii)$$

Let the concentration of the acid and the salt taken originally in the buffer solution be c_1 and c_2 moles/litre respectively.

As HA dissociates to a small extent which is further suppressed by the common ion effect, the A^- ions in the mixture are mainly those produced from the salt BA. As the salt BA dissociates completely, we have

$$[A^-] = [BA] = c_2 \text{ moles/litre} \quad ... (iii)$$

As HA dissociates to a very small extent, the concentration of HA at equilibrium will be almost same as the original concentration

$$[HA] = c_1 \text{ moles/litre} \quad ... (iv)$$

Substituting the values of $[A^-]$ and [HA] from equations (*iii*) and (*iv*) in equation (*ii*), we get

$$\boxed{[H^+] = K_a \frac{c_1}{c_2}}$$

or

$$[H^+] = K_a \frac{[\text{Acid}]}{[\text{Salt}]} \quad ... (v)$$

Taking logarithm of both the sides, we get

$$\log[H^+] = \log K_a + \log \frac{[\text{Acid}]}{[\text{Salt}]}$$

or

$$-\log[H^+] = -\log K_a - \log \frac{[\text{Acid}]}{[\text{Salt}]}$$

or

$$\text{pH} = pK_a + \log \frac{[\text{Salt}]}{[\text{Acid}]} \quad ... (vi)$$

This equation is known as *Henderson equation.* Using this equation, we can calculate the pH of an acidic buffer mixture from the known concentrations of the acid and salt provided the dissociation constant of the acid is known.

***(b)* For basic buffer mixture:** Following the same steps as above, the expression for the *p*OH of a *basic buffer mixture* (BOH + BA) can be obtained. Henderson equation for basic buffer is

$$p\text{OH} = pK_b + \log \frac{[\text{Salt}]}{[\text{Base}]} \quad ... (vii)$$

where K_b is the dissociation constant of the base.

The pH of the basic buffer solution can be obtained from the following expression,

$$\text{pH} + \text{pOH} = 14$$

8.34. BUFFER CAPACITY AND BUFFER INDEX

Besides the general meaning of buffer capacity as given earlier, there are precise definitions of buffer capacity as given below:

Buffer capacity is defined as the number of moles of the acid or base which may be added to one liter or one dm^3 *of the buffer solution for changing its pH by unity.*

Alternatively

Buffer capacity is defined *as the change in the concentration of the buffer acid HA or the base buffer BOH required for changing its pH by unity, keeping the concentration of the acid + salt or base + salt, constant.*

Buffer Index

This term was introduced by Van slyke to express the buffer capacity of a buffer solution quantitatively. It is expressed by the letter β mathematically,

$$\text{Buffer index, } \beta = \frac{\text{No.of moles of the acid or base added to one litre of the buffer solution}}{\text{Change in pH value}}$$

PROBLEMS ON pH OF BUFFER SOLUTIONS

Example 1. Determine the pH of a solution obtained by mixing equal volumes of 0.10 N ammonium nitrate and 0.02 N ammonium hydroxide, K_b for NH_4OH being 1.8×10^{-5}.

Solution. *Before mixing,* the concentration of NH_4NO_3 adn NH_4OH are

$$[NH_4NO_3] = 0.10\text{ N} = 0.10\text{ M}$$

$$[NH_4OH] = 0.02\text{ N} = 0.02\text{ M}$$

After mixing, The concentrations change

As the volume of the solution becomes two-fold

$$[NH_4NO_3] = \frac{0.10}{2} = 0.05\text{M}$$

$$[NH_4OH] = \frac{0.02}{2} = 0.01\text{M}$$

Apply Henderson equation

$$\text{pOH} = pK_b + \log\frac{[\text{Salt}]}{[\text{Base}]}$$

$$= -\log(1.8 \times 10^{-5}) + \log\frac{0.05}{0.01}$$

$$= 4.7497 + 0.6990$$

$$= 5.4487$$

$$\therefore \quad \text{pH} = 14 - p\text{OH}$$

$$= 14 - 5.4487 = 8.5513 \simeq 8.55$$

Example 2. What will be the pH value of a solution obtained by mixing 5 g of acetic acid and 7.5 g of sodium acetate and making the volume to 500 ml (K_{CH_3COOH} =1.8 × 10^{-5} at 25°C).

Solution. Concentration of acetic acid is computed as under:

$$\text{Concentration} = 5 \text{ g in 500 ml}$$
$$= 10 \text{ g/litre}$$
$$= \frac{10}{60} \text{M} \qquad \{\because \text{Mol. wt. of } CH_3COOH = 60\}$$

i.e. $$[CH_3COOH] = \frac{1}{6} \text{ M}$$

Concentration of sodium acetate is computed as under:

$$\text{Concentration} = 7.5 \text{ g in 500 ml}$$
$$= 15 \text{ g/litre}$$

i.e., $$[CH_3OONa] = \frac{15}{82} \text{ M} \qquad \{\because \text{Mol. wt. of } CH_3COONa = 82\}$$

$$pK_a = -\log K_a$$
$$= -\log (1.8 \times 10^{-5})$$
$$= 4.7447$$

$$\therefore \qquad pH = pK_a + \log \frac{[\text{Salt}]}{[\text{Acid}]}$$
$$= 4.7447 + \log \frac{15/82}{1/6}$$
$$= 4.7447 + \log 1.097$$
$$= 4.7447 + 0.0402$$
$$= 4.7849 \simeq 4.78$$

Example 3. Calculate the *p*H of a mixture containing 0.01 M acetic acid and 0.03 M sodium acetate solutions (pK_a = 4.8).

Solution. *p*H of an acid-buffer is given by

$$pH = pK_a + \log \frac{[\text{Salt}]}{[\text{Acid}]}$$

Substituting the values

$$pH = 4.8 + \log \frac{0.03}{0.01}$$

or $$pH = 4.8 + \log 3$$

or $$pH = 4.8 + 0.4771$$
$$= 5.2771$$

PROBLEMS FOR PRACTICE

1. Calculate the pH at 18°C of a solution, prepared by the mixing of equimolar amounts of acetic acid and sodium acetate. The pK_a of acetic acid at 18°C is 4.74. [**Ans.** 4.74]

2. A buffer solution is prepared from an acid of dissociation constant 2×10^{-6} and its sodium salt. What will be the molar ratio of the salt to the acid per litre in a buffer solution of pH 6.8?

[**Ans.** $\frac{[\text{Salt}]}{[\text{Acid}]} = 12.62$]

3. Calculate the pH of a solution obtained by mixing 6.0 g of acetic acid and 12.3 g of sodium acetate and making the volume to 500 ml ($K_a = 1.9 \times 10^{-5}$) [**Ans.** 4.8973]

4. Calculate the hydrogen ion concentration of a solution obtained by mixing 100 ml of 0.3 N NH_4OH with 50 ml of 0.6 N NH_4Cl. Dissociation constant of NH_4OH is 2×10^{-5} and the ionic product of water 1×10^{-14} [**Ans.** 5×10^{-10} g ions/ lites]

[**Hints** First calculate pH]

5. Calculate the pH of a solution prepared by mixing 100 ml of 0.1 N acetic acid and 100 ml of 0.15 M sodium acetate. (Dissociation constant of acetic is 1.8×10^{-5}). [**Ans.** 4.9258]

6. What will be the hydrogen concentration of a solution obtained by mixing equal volumes of 0.05 M acetic acid and 0.5 M sodium acetate? Dissociation constant of acetic acid is 1.8×10^{-5}. [**Ans.** 1.779×10^{-6} g ions/litre]

[**Hint.** First calculate pH]

8.35. NEUTRALISATION PHENOMENON

Neutralisation of a strong acid with a strong base.

A reaction between an acid and a base is called neutralisation reaction. The acid could be strong or weak. Similarly, the base could be strong or weak. Correspondingly, we have four different types of neutralisation which are discussed separately.

(*a*) Neutralisation of a strong acid with a strong base.

The strong acids and strong bases are completely dissociated in the aqueous solutions to give H^+ and OH^- ions respectively. Thus the neutralization reaction taking place is simply

$$H^+ + OH^- \rightleftharpoons H_2O$$

The other ions like Na^+ or Cl^- (From say HCl and NaOH) do not participate in the neutralisation process.

Heat of neutralization: Obviously, the neutralization reaction, as written above, does not depend upon the nature of the acid and the base. If the solution containing equivalent amounts of the strong acid and the strong base are taken, they are dissociated completely and produce the same quantity of H^+ ions and OH^- ions respectively. On mixing the two solutions, the same amount of neutralization taken place and hence same amount of heat *i.e.*, 13.7 kcal/mole or 57.3 kJ mol^{-1} is produced always.

pH of the solution: In the present case, concentration of H^+ ions in the acid solution is equal to the concentration of OH^- ions present in the solution of the base. On mixing, the H^+ ions and OH^- ions combine with each other to form weakly ionised water.

The water formed will have equal concentration of H^+ and OH^- ions.

$$[H^+] = [OH^-] = 10^{-7}M$$

∴ pH of the resulting solution will be = 7 *i.e.* the resulting solution will be neutral.

(*b*) Neutralisation of a weak acid with a strong base.

Consider the reaction between CH_3COOH (weak acid) and NaOH (strong base). In this case, acetic acid is dissociated to a small extent while NaOH is dissociated completely in the aqueous solution as follows :

$$CH_3COOH \rightleftharpoons CH_3COO^- + H^+ \quad ...(i)$$

$$NaOH \longrightarrow Na^+ + OH^- \quad ...(ii)$$

The heat of neutralization and the pH of the resulting solution is explained below:

Heat of neutralization: The small amount of H^+ ions produced from the dissociation of acetic acid combine with the OH^- ions produced from NaOH. Thus as the H^+ ions are removed, the equilibrium (*i*) shifts towards right (according to Le Chatelier's principle). Thus, more of CH_3COOH dissociates producing more of H^+ ions which again combine with the OH^- ions. The process continues till all the H^+ ions produced combine with the OH^-. In this case, *some heat is used up for the dissociation of acetic acid.* Hence the net reaction may be represented as

$$CH_3COOH \rightleftharpoons CH_3COO^- + H^+ \qquad \Delta H = 0.5 \text{ kcal/mole } (2.1 \text{ kJ mol}^{-1})$$

$$H^+ + OH^- \longrightarrow H_2O \qquad \Delta H = -\ 13.7 \text{ kcal/mole } (57.3 \text{ kJ mol}^{-1})$$

$$CH_3COOH + OH^- \rightleftharpoons CH_3COO^- + H_2O, \quad \Delta H = -\ 13.2 \text{ kcal/mole } (55.2 \text{ kJ mol}^{-1})$$

Thus the net heat evolved is less than in the first case.

pH of the solution: The final reaction does not proceed to completion. Instead, it reaches a state of equilibrium. The backward reaction produces some acetic acid (CH_3COOH) molecules and OH^- ions. Thus the final solution contains some undissociated acetic acid molecules and an equal number of OH^- ions. Hence the solution obtained is alkaline *i.e.* has pH > 7.

(*c*) Neutralization of a strong acid with a weak base.

Consider the reaction between HCl (strong acid) and NH_4OH (weak base). NH_4OH is not completely dissociated, while HCl is completely dissociated. Thus we have

$$NH_4OH \rightleftharpoons NH_4^+ + OH^-$$

$$HCl \longrightarrow H^+ + Cl^-$$

Heat of neutralization: The reaction taking place may be represented as

$$NH_4OH \rightleftharpoons NH_4^+ + OH^- \qquad \Delta H = +\ 1.4 \text{ kcal/mole } (5.9 \text{ kJ mole}^{-1})$$

$$H^+ + OH^- \longrightarrow H_2O \qquad \Delta H = -\ 13.7 \text{ kcal/mole } (57.3 \text{ kJ mol}^{-1})$$

$$NH_4OH + H^+ \rightleftharpoons NH_4^+ + H_2O \qquad \Delta H = -\ 12.3 \text{ kcal/mole } (51.4 \text{ kJ mol}^{-1})$$

In this case, 1.4 kcal (5.9 kJ) of heat is used up for the dissociation of one mole of NH_4OH. Hence the net heat evolved is only 12.3 kcal/mole (51.4 kJ mol^{-1}).

pH of the solution : As represented by the final equilibrium equation above, the solution contains some undissociated NH_4OH molecules and an equal number of H^+ ions. Hence the final solution is acidic *i.e.* has pH < 7.

(*d*) Neutralization of a weak acid with a weak base.

Consider the reaction between CH_3COOH (weak acid) and NH_4OH (weak base). Both the substances dissociate to a small extent in the aqueous solution.

$$CH_3COOH \rightleftharpoons CH_3COO^- + H^+ \qquad \text{... } (i)$$

$$NH_4OH \rightleftharpoons NH_4^+ + OH^- \qquad \text{... } (ii)$$

Heat of neutralization: The small amount of H^+ ions and OH^- ions thus produced combine with each other as follows :

$$H^+ + OH^- \longrightarrow H_2O \qquad \text{... } (iii)$$

As the H^+ ions and the OH^- ions are being removed from equilibria (*i*) and (*ii*) respectively, these equilibria shift towards right producing more of H^+ ions and OH^- ions which again combine with one another and thus the process goes on. Here, some heat is required for the dissociation of CH_3COOH as well as NH_4OH. Hence the net heat evolved is much less.

pH of the solution: Although both the acid and the base involved dissociate to a small extent, yet depending upon their degrees of dissociation, one of the two may dissociate more than the other. Thus they may produce unequal number of H^+ ions and OH^- ions. Depending upon which of the two ions are in excess, the resulting solution may be alkaline or acidic.

8.36. SALT HYDROLYSIS

It is defined as the process in which a salt reacts with water to give back the acid and the base.

$$\text{Salt} + \text{Water} \longrightarrow \text{Acid} + \text{Base}$$

or

$$\underset{\text{Salt}}{BA} + \underset{\text{Water}}{H_2O} \longrightarrow \underset{\text{Acid}}{HA} + \underset{\text{Base}}{BOH}$$

All salts are strong electrolytes and thus ionize completely in the aqueous solution. If the acid (HA) produced is strong and the base (BOH) produced is weak, we can write the above equation as

$$B^+ + A^- + H_2O \longrightarrow H^+ + A^- + BOH$$

or

$$B^+ + H_2O \longrightarrow H^+ + BOH$$

Thus, in this case the cation reacts with water to give an acidic solution. This is called **cationic hydrolysis.**

Again, if the acid produced is weak and the base produced is strong, we can write

$$B^+ + A^- + H_2O \longrightarrow HA + B^+ + OH^-$$

or

$$A^- + H_2O \longrightarrow HA + OH$$

Here, the anion reacts with water to give the basic solution. This is called **anionic hydrolysis.**

Hence *salt hydrolysis may be defined as the reaction of the cation or the anion of the salt with water to produce acidic or basic solution.*

Salts can be divided into the following four categories

1. Salts of strong acids and strong bases. Examples of this category of salts are NaCl, KCl, K_2SO_4, Na_2SO_4, KNO_3 etc. Hydrolysis of NaCl can be represented as under

$$NaCl + aq \longrightarrow Na^+ + Cl^-$$

$$H_2O \rightleftharpoons H^+ + OH^-$$

or the overall process will be

$$NaCl + H_2O \rightleftharpoons Na^+ + Cl^- + H^+ + OH^-$$

i.e., there is only dissociation of sodium chloride and no hydrolysis.

2. Salts of weak acids and strong bases. Examples of this category are CH_3 COONa, Na_3PO_4, Na_2CO_3. Hyrolysis of CH_3COONa is given below

$$CH_3COONa + H_2O \rightleftharpoons CH_3COOH + NaOH$$

or $\quad CH_3COO^- + Na^+ + H_2O \rightleftharpoons CH_3OOH + Na^+ + OH^-$

or the overall reaction will be

$$CH_3COO^- + H_2O \rightleftharpoons CH_3COOH + OH^-$$

i.e., it is *anionic hydrolysis*

3. Salts of strong acids and weak bases. Examples of this category of salts are NH_4Cl, $CaCl_2$, NH_4NO_3. Hyrolysis of NH_4Cl is explained as under

$$NH_4Cl + H_2O \rightleftharpoons NH_4OH + HCl$$

or $\quad NH_4^+ + Cl^- + H_2O \rightleftharpoons NH_4OH + H^+ + Cl^-$

or the overall reaction will be

$$NH_4^+ + H_2O \rightleftharpoons NH_4OH + H^+$$

i.e., it is *cationic hydrolysis*

4. Salts of weak acids and weak bases. Examples of this category of salts are CH_3COONH_4, NH_4CN. Hydrolysis of CH_3COONH_4 is explained as under:

$$CH_3COONH_4 + H_2O \rightleftharpoons CH_3COOH + NH_4OH$$

or the overall reaction will be

$$CH_3COO^- + NH_4^+ + H_2O \rightleftharpoons CH_3COOH + NH_4OH$$

i.e., it involves both *anionic as well as cationic hydrolysis.*

8.37. HYDROLYSIS CONSTANT AND DEGREE OF HYDROLYSIS

Hydrolysis constant: Hydrolysis of a salt (BA) may be represented as

$$BA + H_2O \rightleftharpoons HA + BOH$$

Applying the law of chemical equilibrium, to the above reaction

$$\frac{[HA][BOH]}{[BA][H_2O]} = K, \text{ the equilibrium constant}$$

Since water is present in very large excess in the aqueous solution, its concentration $[H_2O]$ may be regarded as constant. Thus we have

$$\frac{[HA][BOH]}{[BA]} = K\,[H_2O] = K_h$$

where K_h is called the hydrolysis constant.

[HA] and [BOH] represent the molar concentrations of the free acid and the free base produced and [BA] represents the molar concentration of the unhydrolysed salt. The general expression for the hydrolysis constant may be written as :

$$K_h = \frac{[\text{Free Acid}][\text{Free Base}]}{[\text{Unhydrolysed salt}]}$$

(*ii*) **Degree of hydrolysis:** *The degree of hydrolysis of a salt is defined as the fraction (or percentage) of the total salt which is hydrolysed.* For example, if in a solution of aniline hydrochloride, 90% of the salt is hydrolysed into aniline and hydrochloric acid, the degree of hydrolysis is 0.90 or may be expressed as percentage *viz.* 90%. It is usually represented by the symbol *h*.

8.38. EXPRESSIONS FOR THE HYDROLYSIS CONSTANT, DEGREE OF HYDROLYSIS AND *p*H OF SOLUTION OF A SALT OF WEAK ACID AND STRONG BASE

(*a*) **Hydrolysis constant:** With the salt BA, the hydrolysis may be represented as follows :

$$\underset{\text{Salt}}{BA + H_2O} \rightleftharpoons \underset{\text{Strong}}{BOH + HA}$$

$$\downarrow \qquad\qquad \downarrow$$

or $$B^+ + A^- + H_2O \qquad B^+ + OH^- + HA$$

or $$A^- + H_2O \qquad OH^- + HA$$

It is a case of *anion hydrolysis.*

The hydrolysis constant K_h for the above reaction is given by

$$K_h = \frac{[OH^-][HA]}{[A^-]} \qquad \text{... } (i)$$

For the weak acid HA, the dissociation equilibrium is

$$HA \rightleftharpoons H^+ + A^-$$

∴ The dissociation constant K_a of the acid HA is given by

$$K_a = \frac{[H^+][A^-]}{[HA]} \qquad \text{... } (ii)$$

The ionic product of the water K_w is given by

$$K_w = [H^+][OH^-] \qquad \text{... } (iii)$$

Multiplying equation (*i*) with (*ii*) and dividing by (*iii*), we get

$$\frac{K_h K_a}{K_w} = \frac{[OH^-][HA]}{[A^-]} \times \frac{[H^+][A^-]}{[HA]} \times \frac{1}{[H^+][OH^-]} = 1$$

or $$K_h = \frac{K_w}{K_a} \qquad \text{... } (iv)$$

(*b*) **Degree of hydrolysis:** Let the original concentration of the salt in the solution be *c* moles litre and *h* be its degree of hydrolysis at this concentration. Then

	$A^- + H_2O$	$\rightleftharpoons$	OH^-	$+ HA$
Original molar conc.	c		0	0
Molar conc. at equilibrium	$c\,(1-h)$		ch	ch

The hydrolysis constant (K_h), is given by

$$K_h = \frac{[OH^-][HA]}{[A^-]}$$

$$= \frac{ch.ch}{c(1-h)} = \frac{ch^2}{1-h}$$

If h is very small as compared to 1, we can take $1 - h \simeq 1$ so that the above expression reduces to

$$ch^2 \simeq K_h$$

or $$h^2 \simeq \frac{K_h}{c}$$

or $$h \simeq \sqrt{\frac{K_h}{c}} \quad \text{... } (v)$$

i.e $$h \propto \sqrt{\frac{1}{c}}$$

Thus *the degree of hydrolysis of a salt of a weak acid and a strong base is inversely proportional to the square-root of its molar concentration.*

If V is the volume of the solution in litres containing 1 mole of the solute, we have $c = \frac{1}{V}$ so that the above expression can also be written as

$$h \simeq \sqrt{K_h.V}$$

Substituting the value of K_h from equation (*iv*) in eq. (*v*) we get

$$h \simeq \sqrt{\frac{K_w}{K_a.C}} \quad \text{... } (vi)$$

(*c*) **pH:** The hydrolysis equation is

	$A^- + H_2O$	$\rightleftharpoons$	OH^-	$+ HA$
Original concentration :	c		0	0
Conc. at equilibrium :	$c(1-h)$		ch	ch

i.e., $$[OH^-] = ch$$

$\therefore$ $$[H^+] = \frac{K_w}{[OH^-]} = \frac{K_w}{ch}$$

Substituting the value of h from equation (*vi*), we get

$$[H^+] = \frac{K_w}{c}\sqrt{\frac{K_a.C}{K_w}}$$

or $$[H^+] = \sqrt{\frac{K_w.K_a}{c}}$$

$\therefore$ $$pH = -\log[H^+] = -\log\sqrt{\frac{K_w\,K_a}{c}}$$

$$= -\log\left(\frac{K_w.K_a}{c}\right)^{1/2}$$

or $$\mathbf{pH = -\frac{1}{2}[\log K_w + \log K_a - \log c]}$$

8.39. EXPRESSIONS FOR THE HYDROLYSIS CONSTANT, DEGREE OF HYDROLYSIS AND *p*H OF SOLUTIONS OF SALTS OF STRONG ACID AND WEAK BASE

Hydrolysis constant: For the salt BA, the hydrolysis may be represented as

$$\underset{\substack{\text{Salt}\\ \downarrow}}{BA} + H_2O \rightleftharpoons \underset{\text{Weak}}{BOH} + \underset{\substack{\text{Strong}\\ \downarrow}}{HA}$$

or $$B^+ + A^- + H_2O \rightleftharpoons BOH + H^+ + A^-$$

or $$B^+ + H_2O \rightleftharpoons BOH + H^+$$

It is a case of *cation hydrolysis.*

The hydrolysis constant K_h is given by

$$K_h = \frac{[BOH][H^+]}{[B^+]} \quad ...(i)$$

For the weak base BOH, the dissociation equilibrium is

$$BOH \rightleftharpoons B^+ + OH^-$$

∴ The dissociation constant K_b of the weak base BOH is given by

$$K_b = \frac{[B^+][OH^-]}{[BOH]} \quad ...(ii)$$

The ionic product of water, K_w, is given by

$$K_w = [H^+][OH^-] \quad ...(iii)$$

Multiplying equation (*i*) with (*ii*) and dividing by (*iii*), we get

$$\frac{K_h k_b}{K_w} = 1$$

or $$K_h = \frac{K_w}{K_b} \quad ...(iv)$$

(*b*) **Degree of hydrolysis:** Proceeding as in the earlier case, we get the same relationship

$$h \simeq \sqrt{\frac{K_h}{c}}$$

Substituting the value of K_h from equation (*iv*), we get

$$h = \sqrt{\frac{K_w}{K_b.c}} \quad ...(v)$$

(*c*) **pH:** The hydrolysis equation is

	$B^+ + H_2O$	$\rightleftharpoons$	BOH	$+ H^+$
Original concentration:	c		0	0
Conc. at equilibrium:	$c(1-h)$		ch	ch

i.e. $$[H^+] = ch$$

Substituting the value of h from equation (v), we get

$$[H^+] = c.\sqrt{\frac{K_w}{K_b.c}} = \sqrt{\frac{K_w.c}{K_b}} = \left(\frac{K_w.c}{K_b}\right)^{1/2}$$

$$\therefore \quad pH = -\log[H^+] = -\log\left(\frac{K_w.c}{K_b}\right)^{1/2}$$

or $$pH = -\frac{1}{2}[\log K_w - \log K_b + \log c] \quad ...(vi)$$

8.40. EXPRESSIONS FOR HYDROLYSIS CONSTANT, DEGREE OF HYDROLYSIS AND *p*H OF SOLUTIONS OF SALTS OF WEAK ACID AND WEAK BASE

(*a*) **Hydrolysis constant:** Representing the salt by BA, we have

$$\underset{\text{Salt}}{BA} + H_2O \rightleftharpoons \underset{\text{Weak}}{BOH} + \underset{\text{Weal}}{HA}$$

$$\downarrow$$

$$B^+ + A^- + H_2O \rightleftharpoons BOH + HA$$

It involves both *anionic hydrolysis as well as cationic hydrolysis.* Writing the various equations, we have

$$K_h = \frac{[BOH][HA]}{[B^+][A^-]} \quad ...(i)$$

For the weak acid, $HA \rightleftharpoons H^+ + A^-$, so that

$$K_a = \frac{[H^+][A^-]}{[HA]} \quad ...(ii)$$

For the weak base, $BOH \rightleftharpoons B^+ + OH^-$, so that

$$K_b = \frac{[B^+][OH^-]}{[BOH]} \quad ...(iii)$$

Also $$K_w = [H^+][OH^-] \quad ...(iv)$$

Multiplying equation (*i*), (*ii*) and (*iii*) and dividing by equation (*iv*), we get]

$$\frac{K_h.K_a.K_b}{K_w} = 1$$

or $$Kh = \frac{K_w}{K_a K_b} \quad ...(v)$$

(*b*) **Degree of hydrolysis:** The hydrolysis equation is

	B^+	$+ A^-$	$+ H_2O \rightleftharpoons$	BOH	$+ HA$
Original concentration:	c	c		0	0
Conc. at equilibrium:	$c(1-h)$	$c(1-h)$		ch	ch

$$K_h = \frac{[BOH][HA]}{[B^+][A^-]} = \frac{ch.ch}{c(1-h).c(1-h)} = \frac{h^2}{(1-h)^2} \quad ... (vi)$$

Here, the relationship between K_h and h does not involve c. Thus the *degree of hydrolysis of such a salt is independent of the concentration of the solution.*

If h is very small in comparison to 1, we can take $1 - h \simeq 1$ so that equation (vi) becomes

$$K_h \simeq h^2$$

or

$$h \simeq \sqrt{K_h}$$

Substitute the value of K_h from eqn.(v)

$$h \simeq \sqrt{\frac{K_w}{K_a.K_b}} \quad ...(vii)$$

(c) **pH**: For the weak acid HA,

$$HA \rightleftharpoons H^+ + A^-$$

$\therefore$

$$K_a = \frac{[H^+][A^-]}{[HA]}$$

or

$$[H^+] = K_a \frac{[HA]}{[A^-]}$$

$$= K_a \frac{ch}{c(1-h)}$$

$$= K_a \frac{h}{1-h}$$

But from equation (vi), $\frac{h}{1-h} = \sqrt{K_h}$ so that we have

$$[H^+] = K_a \sqrt{K_h}$$

$$= K_a \sqrt{\frac{K_w}{K_a \cdot K_b}}$$

$$= \sqrt{\frac{K_a K_w}{K_b}}$$

$$pH = -\log H^+ = -\log \sqrt{\frac{K_a K_w}{K_b}}$$

$$= -\log \left(\frac{K_a K_w}{K_b} \right)^{1/2}$$

or

$$\mathbf{pH = -\frac{1}{2} [\log K_a + \log K_w - \log K_b]} \quad ... (viii)$$

NUMERICAL PROBLEMS ON HYROLYSIS CONSTANT, DEGREE OF HYDROLYSIS AND pH OF SALT SOLUTIONS

Example 1. What would be the pH of a 0.1 molar sodium acetate solution when the dissociation constant for acetic acid is 1.8×10^{-5} ?

Solution. The *p*H of the solution of a salt depends upon the products of hydrolysis. In the present case, sodium acetate is hydrolysed as follows :

$$CH_3COONa + H_2O \rightleftharpoons CH_3COOH + NaOH$$

$$\downarrow \qquad\qquad\qquad \downarrow$$

or $$CH_3COO^- + Na^+ + H_2O \rightleftharpoons CH_3COOH + Na^+ + OH^-$$

or $$CH_3COO^- + H_2O \rightleftharpoons CH_3COOH + OH^-$$

If c is the initial concentration of the salt and h is the degree of hydrolysis, we have

	CH_3COO^- + H_2O	$\rightleftharpoons$	CH_3COOH	+	OH^-
Before hydrolysis:	c		0		0
After hydrolysis :	$c(1-h)$		ch		ch

For a salt of weak acid and strong base

$$[OH^-] = ch = c\sqrt{\frac{K_h}{c}}$$

$$[OH^-] = \sqrt{K_h.c} = \sqrt{\frac{K_w}{K_a}.c}$$

Substituting values, we have

$$[OH^-] = \sqrt{\frac{10^{-14}}{1.8\times10^{-5}}\times 0.1} = \sqrt{55.55\times10^{-12}} = 7.45\times10^{-6}$$

$$[H^+] = \frac{K_w}{[OH^-]} = \frac{10^{-14}}{7.45\times10^{-6}}$$

$$= 1.342 \times 10^{-9}$$

$$\therefore \quad pH = -\log[H^+] = -\log(1.342 \times 10^{-9})$$

$$= \mathbf{8.8723}$$

Alternatively, We can obtain the value of pH by using the following relation directly.

$$pH = -\frac{1}{2}[\log K_w + \log K_a - \log c]$$

$$= -\frac{1}{2}[\log 10^{-14} + \log(1.8 \times 10^{-5}) - \log 10^{-1}]$$

$$= -\frac{1}{2}[-14 - 5 + 0.2553 + 1]$$

$$= \mathbf{8.8723}$$

Example 2. Ionization constant of acetic acid and ionic product of water at 25°C are 1.75×10^{-5} and 1×10^{-14} respectively. Calculate the hydrolysis constant of sodium acetate and its degree of hydrolysis in 0.1 molar solution at 25°C.

Solution. Sodium acetate is a salt of weak acid and strong base, therefore

$$K_h = \frac{K_w}{K_a}$$

Given that

$$K_w = 10^{-14} \text{ and } K_a = 1.75 \times 10^{-5}$$

$$\therefore \quad K_h = \frac{10^{-14}}{1.75\times10^{-5}} = \mathbf{5.7\times10^{-10}}$$

For a salt of weak acid and strong base,

$$h = \sqrt{\frac{K_h}{c}}$$

where c is the molar concentration *i.e.* number of moles/litre

$$\therefore \quad h = \sqrt{\frac{5.7\times10^{-10}}{0.1}} = 7.55\times10^{-5}$$

Example 3. At 25°C, the ionisation constant of anilinium hydroxide is 4.6×10^{-10}. Taking ionic product of water as 1×10^{-14}, calculate (*a*) the hydrolysis constant of anilinium chloride (*b*) the degree of hydrolysis and (*c*) *p*H value in 0.02 molar solution of the salt.

Solution. (*a*) *Calculation of hydrolysis constant.* As anilinium chloride $C_6H_5NH_3Cl$ is salt of weak base (aniline) and strong acid (HCl) hydrolysis constant is given by

$$K_h = \frac{K_w}{K_b}$$

Given that

$$K_w = 10^{-14} \text{ and } K_b = 4.6 \times 10^{-10}$$

$$\therefore \quad K_h = \frac{10^{-14}}{4.6\times10^{-10}} = \mathbf{2.17\times10^{-5}}$$

(*b*) *Calculation of degree of hydrolysis.* For such a salt

$$h = \sqrt{\frac{K_h}{c}} = \sqrt{\frac{2.17\times10^{-5}}{0.02}} = \mathbf{3.29\times10^{-2}}$$

(*c*) *Calculation of pH.* For a salt of weak base and strong acid

$$pH = -\frac{1}{2}[\log K_w - \log K_b + \log c]$$

$$= -\frac{1}{2}[\log10^{-14} - \log (4.6 \times 10^{-10}) + \log 2 \times 10^{-2}]$$

$$= -\frac{1}{2}[-14 - (0.6628 - 10) + (0.3010 - 2)]]$$

$$= \mathbf{3.18}$$

Example 4. If the dissociation constant of HCN be 7.2×10^{-10} and that of ammonia be 1.75×10^{-5}, find out the hydrolysis constant for ammonium cyanide and also its degree of hydrolysis.

Solution. For a salt of weak acid and weak base

$$K_h = \frac{K_w}{K_a.K_b}$$

In the present case $K_a = 7.2 \times 10^{-10}$

$K_b = 1.75 \times 10^{-5}$ and $K_w = 1 \times 10^{-14}$

Hence $$K_h = \frac{1\times10^{-14}}{7.2\times10^{-10}\times1.75\times10^{-5}}$$

or $$K_h = \frac{10}{7.2\times1.75} = 0.7937$$

Degree of hydrolysis $$h = \sqrt{\frac{K_w}{K_a \cdot K_b}}$$

or $$h = \sqrt{K_h}$$

or $$h = \sqrt{0.7937}$$
$$= 0.8909$$

PROBLEMS FOR PRACTICE

1. The degree of hydrolysis of decinormal solution of potassium cyanide is one per cent. What is the value of hydrolytic constant of the salt? [**Ans.** 1×10^{-5}]

2. 0.05 molar solution of the potassium salt of a weak monobasic acid was found to have a pH of 9.25 at 25°C. Calculate the degree of hydrolysis and the hydrolysis constant. Assuming the value of K_w to be 1×10^{-11} at 25° C, determine the dissociation constant of the weak acid.[**Ans.** $K_a = 1.585 \times 10^{-6}$; $K_h = 6.31 \times 10^{-9}$; $h = 3.55 \times 10^{-4}$]

3. Calculate the degree of hydrolysis of 0.01 M solution of NH_4Cl. The dissociation constant of NH_4OH is 1.7×10^{-5} and ionic product of water = 1×10^{-14}. Also calculate the pH of the solution. [**Ans.** $h = 2.42 \times 10^{-4}$, pH = 5.615]

4. If the dissociation constant of HCN be 7.2×10^{-10} and that of ammonia be 1.75×10^{-5}, find out the hydrolysis constant for ammonium cyanide and also its degree of hydrolysis. [**Ans.** $K_h = 0.794$; $h = 0.891$]

5. Calculate the degree of hydrolysis of 0.1 N solution of sodium acetate. (The ionic product of water is 1.0×10^{-14} and the dissociation constant of acetic acid is 1.8×10^{-5}). [**Ans.** 0.0002357]

6. The dissociation constant of aniline as a base at 25° C is 5.93×10^{-10}. The ionic product of water at 25°C is 1.02×10^{-14}. Calculate the percentage hydrolysis of aniline hydrochloride in 1.0 N solution at 25°. [**Ans.** 0.415%]

7. At 18° C the degree of hydrolysis of 0.01 N aniline acetate is 54%. Calculate the dissociation constant of aniline, given that K_a for acetic acid is 1.8×10^{-5} and $K_w = 1 \times 10^{-14}$. [**Ans.** 1.91×10^{-9}]

8. Calculate the percentage hydrolysis of 0.01 N solution of sodium salt of organic acid. The ionisation constant of water is 10^{-14} and dissociation constant of organic acid is 10^{-8}. [**Ans.** 10%]

9. Calculate the pH of a decinormal solution of NH_4CN at 25°C. Given K_a for acetic acid = 7.2 × 10^{-10}, K_b for NH_4OH = 1.80 × 10^{-5} and $K_w = 10^{-14}$. **[Ans.** pH = 9.2]

10. The hydrolysis constant of a solution of potassium formate is 5 × 10^{-11}. The dissociation constant for formic acid is 2 × 10^{-4}. Calculate the value of the ionic product (K_w) for water. **[Ans.** 1 × 10^{-14}]

11. In a decinormal solution of KCN at 25° C, the salt is hydrolysed to 1.18%. Calcualte (*a*) the hydrolysis consant (*b*) the degree of hydrolysis of N/20 solution of the salt at the same temperature. **[Ans.** (*a*) 1.392 × 10^{-5} (*b*) 0.0167]

8.41. CORROSION

The process of slow eating away of the metal due to the attack of atmospheric gases on the surface of the metal resulting into the formation of compounds such as oxides, sulphides and carbonates, etc. is called corrosion

Rusting of iron is the most common example of rusting. Rust is hydrated ferric oxide, $Fe_2O_3 \cdot x\,H_2O$. Other examples include tarnishing of silver, green coating on copper, bronze etc.

TYPES OF CORROSION

1. Galvanic Corrosion

This is the most common form of corrosion and can be dry or wet, chemical or electrochemical. When two different metals (*e.g.*, zinc and copper) are electrically connected and exposed to an electrolyte, the metal higher in electrochemical series undergoes corrosion. Such type of corrosion is called ***galvanic corrosion.*** In this corrosion, a uniform decrease in the volume of a metal takes place as a result of chemical action and soluble corrosion products are formed. (Fig 8.33)

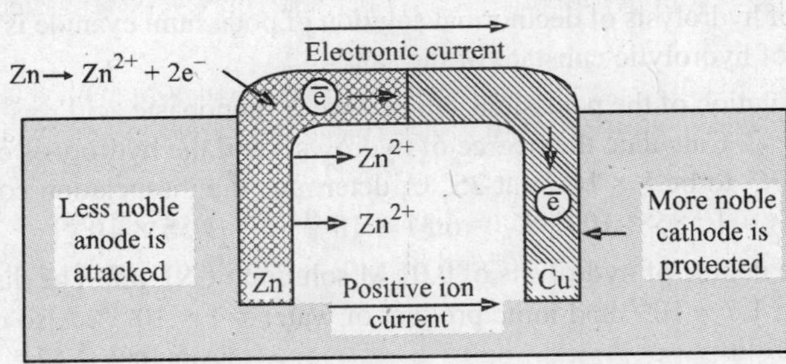

Fig. 8.33. Volatile oxide layer.

2. Crevice Corrosion

It is a local corrosion and is usually caused by dirt deposits, corrosion products, crack in paint coatings etc. It is usually attributed to changes in acidity in the crevice and lack of O_2 in the species in the crevice.

3. Erosion Corrosion

It is caused by the combined effect of turbulent flow of gases, vapour and liquids and the mechanical rubbing action of solids over a metal surface. The major cause for its corrosion is the removal of protective surface film.

4. Pitting Corrosion

Pitting corrosion is, usually due to the breakdown or cracking of the protective film on a metal at specific points. It is a localized accelerated attack, resulting in the formation of cavities around which the metal is relatively unattached and this corrosion results in the formation of pinholes, pits and cavities in the metal. Pitting corrosion gives rise to the formation of small anodic and large cathodic areas.

5. Intergranular Corrosion

Intergranular corrosion occurs along grain boundaries and only where the material, especially sensitive to corrosive attack exists, and corrosive liquid possesses a selective character of attacking only at the grain boundaries, but leaving the grain interiors untouched or only slightly attacked. This type of corrosion is due to the fact that the grain boundaries contain material, which shows electrode potential more anodic than that of the grain centre in the particular corroding medium. This may be due to the precipitation of certain compounds at the grain boundaries.

6. Stress-Corrosion Cracking

Stress corrosion is due to the combined effect of static tensile stresses and the corrosive environmental stresses on a metal that cause cracking result from residual cold working and quenching, welding, thermal treatment or due to applied loads during service. In such cases, the metal under stress becomes more anodic and tends to increase the rate of corrosion. For stress corrosion to occur presence of tensile stress and a specific corrosive environment are necessary. The corrosive agents are highly specific and selective.

7. Exfoliation and Selective Leaching

Exfoliation and selective leaching is a sub-surface corrosion which starts on a clean surface and spreads below it. It differs from pitting in that the attack has a rather laminated appearance and whole layers of material are corroded. The attack is usually recognized from a flaky or blistered surface. This type of corrosion is known to take place in aluminium alloys and can be avoided by heat treatment and proper alloying.

Selective leaching (parting) is the removal of one of the elements in an alloy. De-zincification from copper-zinc alloys is the most common example. This type of corrosion is highly undesirable as it yields a porous metal with poor mechanical properties. This can be remedied by the use of non-susceptible alloys.

This type of corrosion involves in a localized electrochemical corrosion, occurring along narrow paths, forming anodic areas with respect to the more cathodic areas at the metal surface.

8. Soil Corrosion

Corrosion by soils is very important in case of water mains, electric cables and other underground structures, which are embedded in the soil. The various factors that are responsible for soil corrosion include the following :

(*i*) Acidity of the soil
(*ii*) Moisture content
(*iii*) Content of electrolytes
(*iv*) Micro-organisms present
(*v*) Content of organic matter
(*vi*) Physical properties of the soil.

Soil corrosion is purely electrochemical in character. The texture of a soil is determined by the percentage of particles of various sizes.

9. Waterline Corrosion

Waterline corrosion results from differential aeration leading to the formation of oxygen concentration cells. It has generally been observed that maximum corrosion takes place in a steel tank containing water along a line just beneath the level of water, because access of oxygen is much less there (Fig. 8.34). The area above the waterline is highly oxygenated and hence acts as cathodic area.

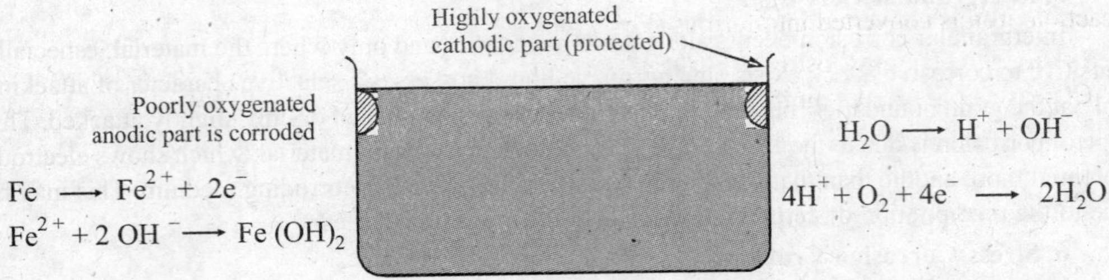

Fig. 8.34. Waterline corrosion occurs just underneath the meniscus and the water level.

Consequently, it is not corroded. However, little corrosion takes place when the water is relatively free from acidity.

Waterline corrosion is also caused in marine ships and is accelerated by marine plants which are attached to the sides of the ships.

10. Micro-Biological Corrosion

Corrosion caused by the metabolic activity of various micro-organisms, is called *microbiological corrosion.* The micro-organisms can develop in an environment with or without oxygen and are classed as aerobic or anaerobic.

Bacteria, fungi, algae, diatom etc. are capable of forming micro-biological film on the surface of iron and generally known as film-forming micro-organism. Such film maintains concentration gradients of dissolved salts, gases, acids etc. on the surface of iron and as a result, local biological concentration cells are developed on the surface leading to micro-biological corrosion.

8.42. ELECTROCHEMICAL THEORY OF CORROSION

The theory of corrosion can be explained by taking the example of rusting of iron. This theory explains the formation of rust on the basis of the formation of electrochemical cells on the surface of the metal.

The following steps are involved in the formation of rust :

(*i*) The water vapours on the surface of the metal dissolve CO_2 and O_2 from the air. Thus, the surface of metal is covered with the solution of CO_2 in water, *i.e.*, carbonic acid (H_2CO_3)

$$H_2O + CO_2 \longrightarrow H_2CO_3$$

This acts as an electrolytic solution of the cell. The carbonic acid and water ionise to a small extent as :

$$H_2CO_3 \rightleftharpoons 2H^+ + CO_3^{2-}$$

$$H_2O \rightleftharpoons H^+ + OH^-$$

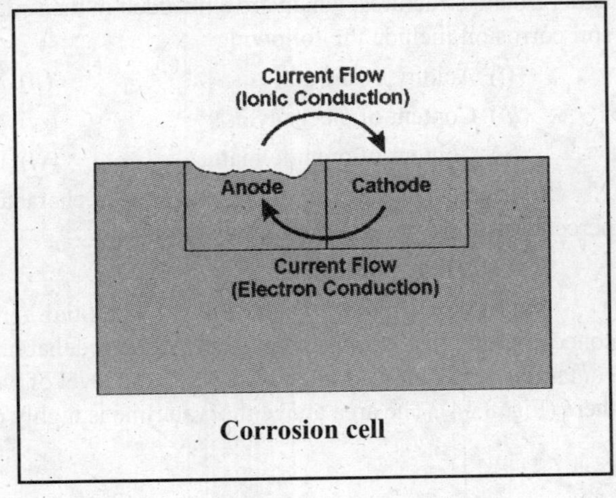

Corrosion cell

(*ii*) In contact with the dissolved CO_2 and oxygen iron undergoes oxidation as follows :

$$Fe \longrightarrow Fe^{2+} + 2e^- \qquad (E^0_{oxid} = 0.44\ V) \qquad ...(i)$$

Thus, the sites where the above reaction takes place act as *anodes*. As a result of the above reaction, iron is converted into ferrous (Fe^{2+}) ions.

(*iii*) The electrons lost by iron are taken up by the H^+ ions produced by the dissociation of H_2CO_3 and H_2O present on the surface of the metal. Thus, H^+ ions are converted into H-atoms.

$$H^+ + e^- \longrightarrow H \qquad \text{... }(ii)$$

These H-atoms react with the oxygen to form water.

$$4H + O_2 \longrightarrow 2H_2O \qquad \text{... }(iii)$$

Multiplying eqn. (*ii*) by 4 and adding to eqn. (*iii*), the complete reduction reaction may be written as

$$O_2 + 4H^+ + 4e^- \longrightarrow 2H_2O\ (E^0_{red} = 1.23\ V) \qquad \text{... }(iv)$$

The dissolved oxygen may take up electrons directly to form OH^- ions as follows :

$$O_2 + 2H_2O + 4e^- \longrightarrow 4OH^-$$

The sites where the above reactions take place act as *cathodes.*

Adding equations (i) and (iv), the overall reaction of the miniature cell will be

$$2\ Fe + O_2 + 4\ H^+ \longrightarrow 2\ Fe^{2+} + 2\ H_2O \quad E^0_{cell} = 1.67\ V$$

It may be mentioned here that if water contains some dissolved salts, it helps in the flow of current in the miniature cell and, hence, enhance the process of corrosion. This explains why there is a faster and greater corrosion by the sea-side.

(*iv*) The ferrous ions formed react with the dissolved oxygen or oxygen from the air to form ferric oxide

$$4Fe^{2+} + O_2 + 4H_2O \longrightarrow 2Fe_2O_3 + 8H^+(aq)$$

Ferric oxide then undergoes hydration to form rust.

$$Fe_2O_3 + xH_2O \longrightarrow \underset{\text{Hydrated ferric oxide (Rust)}}{Fe_2O_3.xH_2O}$$

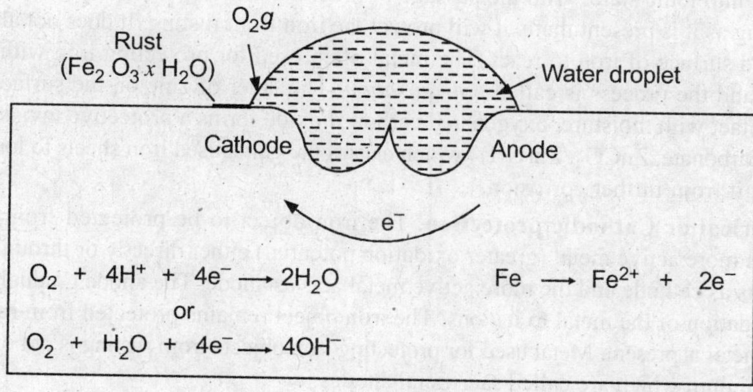

Fig. 8.35. Corrosion of iron.

It may be noted that the rust does not stick to the surface. It peels off exposing fresh iron surface for further rusting. It, thus, reduces the life of an article.

The rusting phenomenon may be represented diagrammatically as shown in Fig. 8.35

Factors responsible for corrosion.

(*i*) *Reactivity of the metal.* More active metals having high oxidation potential values are readily corroded.

(*ii*) *Presence of air and moisture.* Air and moisture accelerate corrosion phenomenon. Presence of gases like SO_2 and CO_2 in air enhance the process of corrosion. Iron does not rust when placed in vacuum.

(*iii*) *Presence of impurities.* Presence of impurities in metals enhances the chances of corrosion. Pure metals do not corrode.

(*iv*) *Presence of electrolytes.* Electrolytes increase the rate of corrosion. For example, iron rusts faster in saline water than in pure water.

(*v*) *Strains in metals.* Corrosion takes place rapidly at bends, scratches and cuts in the metal.

8.43. METHODS OF PREVENTION OF CORROSION

Some of the methods for prevention of corrosion are explained below :

By external coating.

(*i*) By applying a film of oil or greases on the surface of the iron tools and machinery, the rusting of iron can be prevented. It keeps the iron surface protected from moisture, oxygen and carbon dioxide.

(*ii*) The metal surface is coated with paint which prevents the action of air, moisture, etc., till the paint layer develops cracks.

(*iii*) The iron surface is coated with phosphate to give a tough adherent insoluble film which does not allow air and moisture to react with iron surface.

(*iv*) The iron surface is coated with non-corroding metals such as nickel, chromium, aluminium, etc., by electroplating with tin, zinc, etc., or by dipping the iron article in the molten metal. This again cuts off the supply of oxygen and water to iron surface.

2. Sacrificial Protection. Sacrificial protection implies covering the iron surface with a layer of metal which is *more active* (electropositive) than iron. It, thus, prevents the iron from losing electrons and getting oxidised. The more active metal loses electrons in preference to iron and converts itself into ionic state. With the passage of time, the more active metal gets consumed into ions but so long as it is present there, it will protect the iron from rusting. It does not allow even the nearly exposed surface of iron to react. The metal often used for protecting iron with more active metal is zinc and the process is called *galvanisation.* The layer of zinc on the surface of iron, on coming in contact with moisture, oxygen and carbon-dioxide forms a protective invisible thin layer of basic zinc carbonate, $ZnCO_3 . Zn(OH)_2$ which causes the galvanised iron sheets to lose their lustre and to protect it from further corrosion.

3. Electrical or Cathodic protection. The iron object to be protected from corrosion is connected to a more active metal (greater oxidation potential) either directly or through a wire. The iron object acts as cathode and the more active metal acts as anode. The anode gradually disappears due to the oxidation of the metal to its ions. The iron object remains protected from rusting as long as the active metal is present. Metal used for protecting iron objects from rusting is out of magnesium, zinc and aluminium which are called sacrificial anodes.

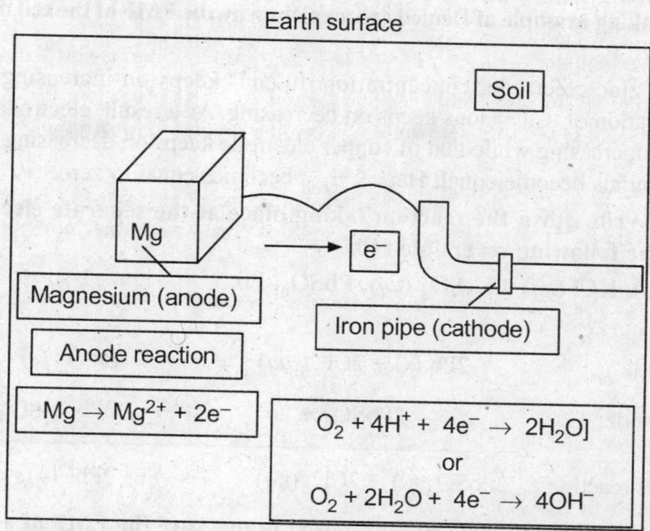

Fig. 8.36. Electrical protection of underground iron pipes.

Cathode protection is often employed in the protection of iron pipes buried in the moist soil, canals, storage tanks, etc. Pieces of magnesium are buried along the pipeline and connected to it by the wire as illustrated in Fig. 8.36. Electrode reactions are given alongside.

4. Using anti-rust solutions. These are alkaline phosphate and alkaline chromate solutions. The alkaline medium prevents availability of hydrogen ions. Phosphate deposits an insoluble protective film or iron phosphate on the iron. These solutions are used to prevent rusting of iron parts of the automobile engines and radiators.

SOME SOLVED CONCEPTUAL PROBLEMS

Example 1. Why calomel electrode is preferred over hydrogen electrode as reference electrode? Give the reactions occurring on the calomel electrode.

Solution. Use of hydrogen electrode is not convenient because it involves the use of hydrogen gas. Reactions occurring in calomel electrode are:

(*i*) When oxidation occurs : $2\text{ Hg }(l) + 2\text{Cl}^- \longrightarrow \text{Hg}_2\text{ Cl}_2\ (s) + 2e^-$

(*ii*) When reduction occurs : $\text{Hg}_2\text{Cl}_2\ (s)\ 2e^- \longrightarrow 2\text{ Hg}(l) + 2\text{Cl}^-\ (aq)$

Example 2. Write down the reactions taking place at the separate electrodes and in the complete reversible cell given below:

Pt, H_2 (*g*) | HCl (*aq*) || KCl (*aq*), Ag Cl (*s*) Ag.

Solution. At anode : $\quad H_2\ (g) \longrightarrow 2H^+\ (aq) + 2e^-$

At cathode: $\quad 2\text{ AgCl }(s) + 2e^- \longrightarrow \text{Ag }(s) + 2\text{Cl}^-\ (aq)$

overall reaction: $\quad H_2\ (g) + 2\text{AgCl }(s) \longrightarrow 2\text{ Ag }(s) + 2\text{ HCl }(aq)$

Example 3. Explain the meaning of an electrode potential being positive or negative with respect to hydrogen electrode.

Solution. Positive electrode potential with respect to hydrogen electrode means that reduction takes place on this electrode and oxidation on the hydrogen electrode. Negative electrode potential with respect to hydrogen electrode means just the opposite.

Example 4. Taking example of Deniell cell, explain why the EMF of the cell drops to zero after some time.

Solution. At zinc electrode, concentration of Zn^{2+} keeps on increasing while at copper electrode, concentration of Cu^{2+} ions keeps on decreasing. As a result, electrode potential of zinc electrode keeps on increasing while that of copper electrode keeps on decreasing till ultimately the two electrode potentials become equal. Hence, E_{cell} becomes equal to zero.

Example 5. Write down the reaction taking place at the separate electrode and in the complete cell in the following reversible cell:

Pb | $PbCl_2$ (*s*), KCl (*aq*) || K_2SO_4 (*aq*), $PbSO_4$| Pb.

Solution.

At anode $\quad 2Pb\ (s) + 2Cl^-\ (aq) \longrightarrow PbCl_2\ (s) + 2e^-$

A Cathode: $\quad 2PbSO_4 + 2e^- \longrightarrow 2Pb + SO_4^{2-}$

Overall reaction $\quad 2PbSO_4\ (s) + 2Cl^-\ (aq) \longrightarrow 2PbCl_2\ (s) + SO_4^{2-}$

Example 6. Why a voltmeter cannot be used to measure the EMF of a cell?

Solution. The voltmeter draws current from the cell thereby changing the EMF of the cell.

Example 7. Iron does not rust even if the zinc coating is broken in a galvanised iron pipe but rusting occurs much faster if tin coating over iron is broken. Explain.

Solution. This is because standard electrode of zinc is more negative than that of iron but that of tin is less negative than that or iron ($E^0_{Zn^{2+},Zn} = -0.76\,V, E^0_{Fe^{2+},\,Fe} = -0.44\,V, E^\circ_{Sn^{2+},Sn} = -0.14\ V$)

Example 8. Write down the cell reaction and expression for e.m.f of the following reversible cell:

(*i*) Cd (*s*) |Cd^{++} || H^+ | H_2 (*g*) | pt.

(*ii*) Zn (*s*) | Zn^{++} || Fe^{+++}, Fe^{++} | Pt.

Solution.

(*i*) At anode : $\quad Cd \longrightarrow Cd^{2+} + 2e^-$

At cathode: $\quad 2H^+ + 2e^- \longrightarrow H_2$

Overall reaction: $\quad Cd + 2H^+ \longrightarrow Cd^{2+} + H_2$

$$E^\circ_{cell} = E^\circ_{cathode} - E^0_{anode} = E^\circ_{H^+\frac{1}{2}H_2} - E^\circ_{Cd^{2+},Cd}$$

(*ii*) At anode : $\quad Zn \longrightarrow Zn^{2+} + 2e^-$

At cathode: $\quad 2Fe^{3+} + 2e^- \longrightarrow 2Fe^{2+}$

Overall reaction: $\quad Zn + 2\,Fe^{3+} \longrightarrow Zn^{2+} + 2Fe^{2+}$

$$E^\circ_{cell} = E^\circ_{Fe^{3+},\ Fe^{2+}} - E^\circ_{Zn^{2+},\ Zn}$$

Example 9. We can use aluminium in place of zinc for cathodic protection of rusting. Comment.

Solution. No. this is because aluminium forms an oxide layer which adheres to the metal surface and forms an impermeable layer. As a result, surface is sealed. The phenomenon is called *passivation* (through standard electrode potential of Al of highly negative being – 1.66 V).

EXERCISES
(Based on Question Papers of Different Universities)

Multiple choice question (Choose the correct option)

1. For the overall reaction

Cu^{2+} (C_1 aq) + Zn (s) $\longrightarrow$ Zn^{2+} (C_2 aq) + Cu (*s*)

of an electrochemical cell, the change in free energy, ΔG, at a given temp. is function of

(*a*) ln C_1
(*b*) ln (C_2/ C_1)
(*c*) ln C_2
(*d*) ln (C_1 + C_2).

2. Best way to prevent rusting of iron is by

(*a*) Making iron cathode
(*b*) Putting it in saline water
(*c*) both of these
(*d*) none of these.

3. The hydrogen electrode is dipped in a solution of pH = 3 at 25° C. The potential of the cell would be :

(*a*) 0.177 V
(*b*) 0.087V
(*c*) – 0.177 V
(*d*) 0.059 V.

4. In an electrochemical cell

(*a*) Potential Energy decreases
(*b*) Kinetic Energy decreases
(*c*) Potential Energy changes into electrical energy
(*d*) Chemical energy changes into electrical energy.

5. In which cell, the free energy of a chemical reaction is directly converted into electricity?

(*a*) Leclanche cell
(*b*) Concentration cell
(*c*) Fuel cell
(*d*) Lead storage battery.

6. If the standard electrode potential of Cu^{2+}/Cu electrode is 0.34 V, what is the electrode potential of 0.01 M concentration of Cu^{2+}, (T = 298 K)

(*a*) 0.399 V
(*b*) 0.281 V
(*c*) 0.222 V
(*d*) 0.176 V.

7. For a cell reaction involving $2e^-$ change, the standard e.m.f. of the cell is found to be 0.295 V at 25°C. The equilibrium constant. of the reaction at 25° C will be–

(*a*) 2.95×10^2
(*b*) 10
(*c*) 1×10^{10}
(*d*) 1×10^{-10}.

8. Three elements A, B and C have reduction potentials – 1.5, – 0.05 and + 1.50. Then the correct order of their reducing power is

(*a*) A > B > C
(*b*) B > A > C
(*c*) C > B > A
(*d*) B > C > A.

9. Saturated solution of KNO_3 is used to make 'salt bridge' because –

(*a*) velocity of K^+ is greater than that of NO_3^-
(*b*) velocity of NO_3^- is greater than that of K^+
(*c*) Velocities of K^+ and NO_3^- both are nearly the same.
(*d*) None of the above

10. The oxidation potential of Mg and Al are + 2.37 and + 1.66 volt respectively. The Mg in chemical reaction

(*a*) will be replaced by Al

(*b*) will replace Al

(*c*) will not be able to replace Al

(*d*) None of the above.

11. In the electrochemical cell, flow of electrons is from

(*a*) cathode to anode in the solution

(*b*) cathode to anode through external supply

(*c*) cathode to anode through internal supply

(*d*) anode to cathode through internal supply.

12. Cell reaction is spontaneous when

(*a*) E°_{red} is + ve (*b*) ΔG° is – ve

(*c*) ΔG° is + ve (*d*) E°_{red} is + ve

ANSWERS

1. (*b*)	**2.** (*a*)	**3.** (*c*)	**4.** (*d*)	**5.** (*b*)	**6.** (*b*)
7. (*c*)	**8.** (*a*)	**9.** (*c*)	**10.** (*b*)	**11.** (*c*)	**12.** (*a*)

Short Answer Questions

1. What is an electrolytic cell?
2. Explain the terms polarisation.
3. Which are the different types of Reversible Electrodes?
4. What are concentration cells? What are their different types?
5. Discuss the role of calomel electrode as a reference electrode for determination of electrode potential of any electrode.
6. What do you understand by liquid function potential?
7. Define Galvanic cell or Electrochemical cell. Give one example.
8. Define Buffer capacity and Buffer index.
9. Define EMF of a cell. How do you measure it experimentally?
10. Write a short note on corrosion.
11. What is a salt bridge? Give it's importance.
12. How do you represent and Electrochemical cell?
13. Define Reversible and Irreversible Cells giving example.
14. Derive a relationship between EMF of a cell and ΔG, ΔH and ΔS.
15. What is the important condition for a redox reaction to be spontaneous?
16. Define Electrode Potential.
17. What is an Electrochemical series. Discuss it's important uses.

General Questions

1. Briefly explain the terms (*i*) Polarisation, (*ii*) Decomposition potential and (*iii*) Discharge potential.
2. How is the standard electrode potential of an electrode measured using (a) standard hydrogen electrode (b) calomel electrode, as reference electrode? Explain with suitable examples.

3. What do you understand by the EMF of a cell? How can it be measured experimentally?
4. Derive relationship between electrical energy and chemical energy produced in a redox reaction. Explain the effect of temperature coefficient of EMF on the enthalpy change.
5. Define an electrochemical cell. Taking a typical electrochemical cell, explain its working.
6. What are 'Reversible electrodes'? Explain the different types of reversible electrodes with suitable example.
7. Derive Nernst equation for measuring EMF of cell.
8. What is 'Electrochemical series'? Describe its important applications.
9. What are reversible and irreversible cells? Explain each of them with a suitable examples.
10. What are different types of reversible electrodes? Taking a suitable example of each of them, derive Nernst equation for each case.
11. (*a*) Derive Henderson-Hasselbalch equation for the calculation of pH of an acidic buffer mixture.

    (*b*) How is the pH of a basic buffer mixture calculated?

    (*c*) What do you understand by 'buffer capacity' or 'buffer indeed'? When is the buffer capacity maximum?
12. What is corrosion? Give theory of rusting of iron. Briefly explain different methods of preventing corrosion.
13. What do you understand by overvoltage or overpotential? What is hydrogen overvoltage? Briefly explain the factors on which it depends. Give some of its important applications.
14. Taking the example of a suitable concentration cell with transference, derive expression for the EMF of the cell.
15. Describe how pH of a solution can be determined by using (*i*) Quinhydrone electrode (*ii*) Glass electrode.
16. What is the principle of potentiometric titrations? Briefly explain how the following titrations are carried out potentiometrically?

    (*i*) Acid-Base titrations (*ii*) Redox titrations (*iii*) Precipitation titrations.
17. What is 'Liquid Junction Potential'? Derive expression for the liquid junction potential. How can it be minimized or eliminated?
18. Explain the terms-activity and activity-coefficient. How are these determined using EMF measurement?
19. What is 'neutralization'? Comment on the heat of neutralisation and pH of the solution at end point for the titration of a strong acid with a strong base and also when either the acid or the base is weak.
20. What is meant by salt hydrolysis, hydrolysis constant and degree of hydrolysis? Derive expression for hydrolysis constant, degree of hydrolysis and pH of the solution if either the acid or the base is weak or both are weak.
21. What are concentration cells? How are they classified into different types? Give one example of each. Taking suitable example of a concentration cell, derive expression for the EMF of concentration cell.
22. What are Buffer solutions? What are different types of buffer solutions? Explain the buffer actions of each of them.
23. Describe the important applications of e.m.f. measurements.
24. Write detailed notes on the following

(*a*) Sacrificial protection

(*b*) Electrical or cathodic protection

APPENDIX 1

GREEK APLPHABET TABLE

Capital	*Small*	*Pronunication*	*English equivalent*
Α	α	alpha	a
Β	β	beta	b
G	γ	gamma	g, n
Δ	δ	delta	d
Ε	ε	eosilon	e
Ζ	ζ	zeta	z
Η	η	eta	ē
Θ	θ	theta	th
Ι	ι	iota	i
Κ	κ	kappa	k
Λ	λ	lamba	l
Μ	μ	mu	m
Ν	ν	nu	n
Ξ	ξ	xi	x
Ο	ο	omricon	o
Π	π	pi	p
Ρ	ρ	rho	r, rh
Σ	σ	sigma	s
Τ	τ	tau	t
Υ	υ	upsilon	y, u
Φ	ϕ	phi	ph
Χ	χ	chi	ch
Ψ	ψ	psi	ps
Ω	ω	omega	ō

APPENDIX 2

Table 1. Some useful conversion factors

$1\,\text{Å} = 10^{-10}\,\text{m}$	$1\,\text{atm} = 101325\,\text{pa or Nm}^{-2}$
	$= 1.013 \times 10^{6}\,\text{dynes / cm}^{2}$
$1\,\text{a.m.u} = 1.66053 \times 10^{-27}\,\text{kg}$	$1\,\text{mm or torr} = 133.32\,\text{Pa or Nm}^{-2}$
$\text{T} = (\text{t} + 273.16\,\text{K}) = (\text{t} + 273)\,\text{K}$	$1\,\text{calorie} = 4.184\,\text{J}$
$1\,\text{litre} = 10^{-3}\,\text{m}^{3} = 1\,\text{dm}^{3}$	$1\,\text{erg} = 10^{-7}\,\text{J}$
$1\,\text{dyne} = 10^{-5}\,\text{N}$	$1\,\text{electron volt}(\text{eV}) = 1.6022 \times 10^{-19}\,\text{J}$

Table 2. Some common prefixes used for S.I. units

Prefix	*Symbol*	*Multiplying factor*	*Example*
deci –	d	10^{-1}	$1\,\text{decimetre}(1\,\text{dm}) = 10^{-1}\,\text{m} = 0.1\,\text{m}$
centi –	c	10^{-2}	$1\,\text{centimetre}(1\,\text{cm}) = 10^{-2}\,\text{m} = 0.01\,\text{m}$
milli –	m	10^{-3}	$1\,\text{mil lim etre}(1\,\text{mm}) = 10^{-3}\,\text{m} = 0.001\,\text{m}$
micro –	μ	10^{-6}	$1\,\text{micrometre}(\mu\text{m}) = 10^{-6}\,\text{m}$
nano –	n	10^{-9}	$1\,\text{nanometre}(\text{nm}) = 10^{-9}\,\text{m}$
pico –	p	10^{-12}	$1\,\text{picometre}(\text{pm}) = 10^{-12}\,\text{m}$
femto –	f	10^{-15}	$1\,\text{femtometre}(\text{fm}) = 10^{-15}\,\text{m}$
deka –	da	10^{1}	$1\,\text{dekametre}(\text{dam}) = 10^{1}\,\text{m}$
hecto –	h	10^{2}	$1\,\text{hectometre}(\text{hm}) = 100\,\text{m}$
kilo –	k	10^{3}	$1\,\text{kilometre}(\text{km}) = 10^{3}\,\text{m} = 1000\,\text{m}$
mega –	M	10^{6}	$1\,\text{megametre}(\text{Mm}) = 10^{6}\,\text{m}$
giga –	G	10^{9}	$1\,\text{gigametre}(\text{Gm}) = 10^{9}\,\text{m}$
tera –	T	10^{12}	$1\,\text{terametre} = 10^{12}\,\text{m}$
peta –	P	10^{15}	$1\,\text{petametre} = 10^{15}\,\text{m}$